Hoefnagels Biology Concepts and Investigations

ON TRACK . . . Various types of interactives and quizzing keep you motivated and on track in mastering the key concepts.

- **Practice Tests** include a pre- and post-test for each chapter that allow you to quickly gauge your understanding of the concepts. The tests include feedback for incorrect answers to guide your learning.

BioTutorials

- **BioTutorials Animation Quizzes** present concepts through dynamic visuals. Watch biological processes and concepts come alive, and then take a quiz to determine your level of understanding.

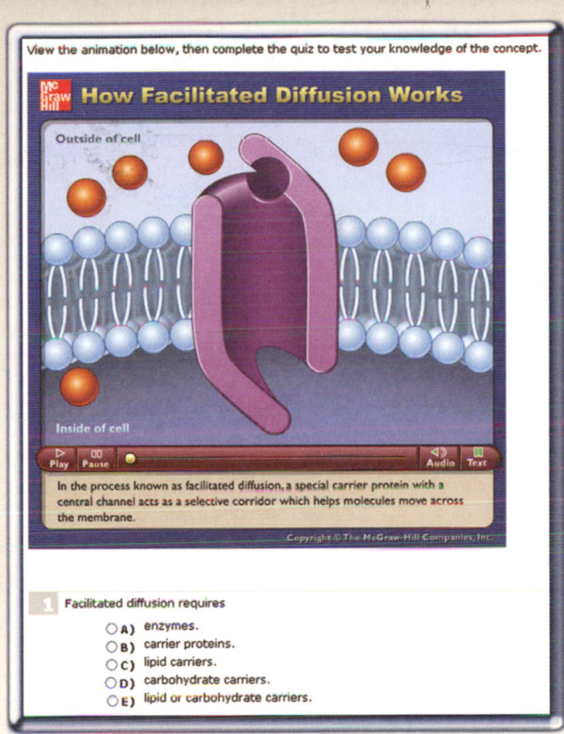
View the animation below, then complete the quiz to test your knowledge of the concept.
How Facilitated Diffusion Works
Outside of cell
Inside of cell
Play
Pause
Audio
Text
In the process known as facilitated diffusion, a special carrier protein with a central channel acts as a selective corridor which helps molecules move across the membrane.
Copyright © The McGraw-Hill Companies, Inc.
1 Facilitated diffusion requires
A) enzymes.
B) carrier proteins.
C) lipid carriers.
D) carbohydrate carriers.
E) lipid or carbohydrate carriers.

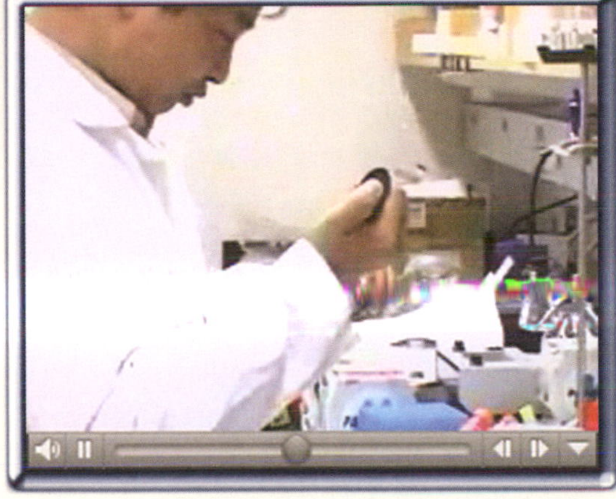

ScienCentral, Inc.

- **ScienCentral Video Quizzes** add relevancy to your study of biology. Learn how a photosynthesizing plant may one day keep your laptop and cell phone running longer!

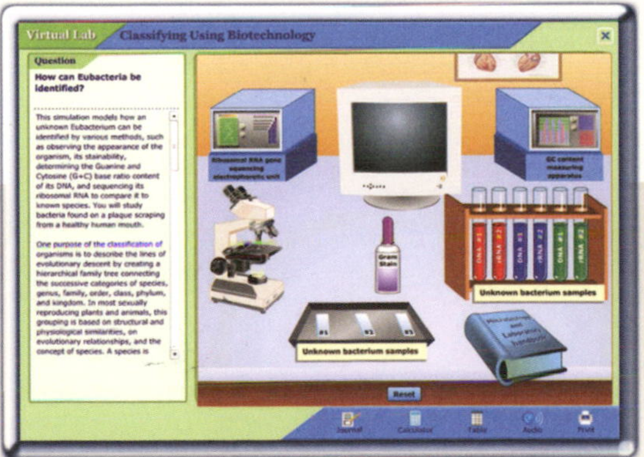
Virtual Lab
Classifying Using Biotechnology

- **Virtual Labs and Case Studies** provide unique online experiences. Step into the virtual lab to identify an unknown Eubacterium taken from a human mouth. Read and discuss case studies about medical, ecological, and social issues, including a case study about the effects of caffeine on the brain.

Mc Graw Hill
McGraw-Hill Higher Education

BIOLOGY

Concepts and Investigations

Mariëlle Hoefnagels
The University of Oklahoma

Boston Burr Ridge, IL Dubuque, IA New York San Francisco St. Louis
Bangkok Bogotá Caracas Kuala Lumpur Lisbon London Madrid Mexico City
Milan Montreal New Delhi Santiago Seoul Singapore Sydney Taipei Toronto

The McGraw·Hill Companies

BIOLOGY: CONCEPTS AND INVESTIGATIONS

Published by McGraw-Hill, a business unit of The McGraw-Hill Companies, Inc., 1221 Avenue of the Americas, New York, NY 10020.

 This book is printed on recycled, acid-free paper containing 10% postconsumer waste.

2 3 4 5 6 7 8 9 0 DOW/DOW 0 9 8

ISBN 978-0-07-291690-4
MHID 0-07-291690-7

Publisher: *Janice Roerig-Blong*
Executive Editor: *Michael S. Hackett*
Vice-President New Product Launches: *Michael Lange*
Senior Developmental Editor: *Lisa A. Bruflodt*
Marketing Manager: *Tamara Maury*
Senior Project Manager: *Sheila M. Frank*
Senior Production Supervisor: *Kara Kudronowicz*
Lead Media Project Manager: *Judi David*
Senior Freelance Design Coordinator: *Michelle D. Whitaker*
Cover/Interior Designer: *Elise Lansdon*
(USE) Cover Image: *© Tim Flach / Getty Images*
Senior Photo Research Coordinator: *John C. Leland*
Photo Research: *Emily Tietz / Editorial Images, LLC*
Project Coordinator: *Melissa M. Leick*
Art Studio and Compositor: *Electronic Publishing Services Inc., NYC*
Typeface: *10/12 Giovanni*
Printer: *R. R. Donnelley Willard, OH*

The credits section for this book begins on page C-1 and is considered an extension of the copyright page.

Library of Congress Cataloging-in-Publication Data

Hoefnagels, Marielle.
Biology : concepts and investigations / Marielle Hoefnagels. — 1st ed.
p. cm.
ISBN 978-0-07-291690-4 — ISBN 0-07-291690-7 (hard copy : alk. paper) 1. Biology. I. Title.
QH307.2.H64 2009
570--dc22

2007042447

www.mhhe.com

About the Author

Mariëlle Hoefnagels

Mariëlle Hoefnagels is an associate professor in the departments of Botany/Microbiology and Zoology at The University of Oklahoma where she teaches both traditional and online courses in introductory biology. She has received the University of Oklahoma General Education Teaching Award and the Longmire Prize, the Teaching Scholars Award from the College of Arts and Sciences. She has also been awarded honorary memberships in several student honor societies.

Dr. Hoefnagels received her B.S. in environmental science from the University of California at Riverside, her M.S. in soil science from North Carolina State University and her Ph.D. in plant pathology from Oregon State University in 1997. Her dissertation work focused on the use of bacterial biological control agents to reduce the spread of fungal pathogens on seeds. Her recent publications have focused on the creation of investigative teaching laboratories and methods for teaching experimental design in beginning and advanced biology classes. She frequently gives presentations on study skills and related topics to student groups across campus.

She has also served as Managing Editor and Chair of the Website Committee for the Association of Biology Laboratory Education and is a featured speaker at OU's annual freshman seminar for Women in Science and Engineering. Marielle is also a member of the National Association of Biology Teachers and the Mycological Society of America. Her hobbies include reading, traveling, gardening, and playing volleyball.

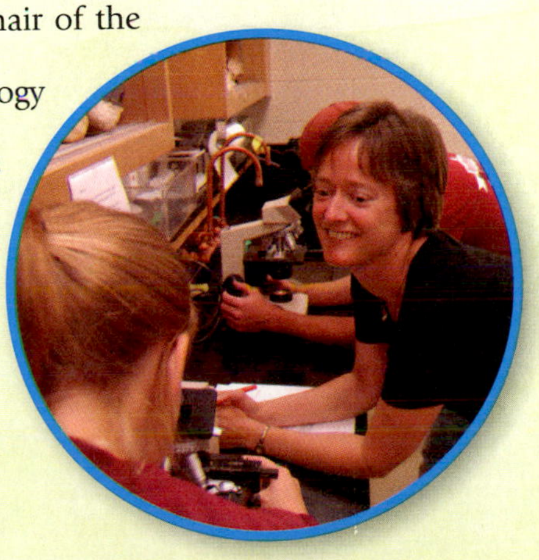

Dedication

To my students

MARIËLLE HOEFNAGELS

Brief Contents

Preface: Thinking for Life

Investigating Life

Biology—the science of life—is central to our lives and our planet. On a typical morning, you use a toothbrush to scrub the bacteria off your teeth, decide what to wear based (in part) on your ability to regulate your body temperature, choose clothes made partly of natural fibers such as cotton or wool, and eat a breakfast composed of foods produced by other organisms. And that's all before leaving the house! True, you could have done these things even without opening this book. But learning about biology should help you to understand much more about your world.

Nutrition, cancer, HIV/AIDS, global climate change, water quality, endangered species, stem cells, the spread of drug-resistant bacteria, and countless other matters have their foundation in biology. This book offers concrete medical applications such as understanding how to stop the spread of disease, why you need to take the whole prescription of antibiotics, and why there are few drugs without side effects. At the other end of the spectrum, it's important to evaluate the arguments and issues surrounding global climate change.

Connecting these two ideas is the relationship between environmental quality and the virulence of disease-causing organisms. I hope that after reading this book you will be better able to understand and evaluate items in the news, make these types of connections yourself, become a more thoughtful voter, and, most importantly, develop a greater appreciation for the amazing, ever-changing world around you. I designed this book to convey the general concepts of biology and to connect them to your life.

Biological Concepts are the Result of Scientific Inquiry

Every biology textbook explores the process of science as a way of learning about the natural world, but this book is unique in that each chapter reinforces the importance of scientific inquiry with a section titled "Investigating Life." These capstone concepts each explain one study that sheds light on an evolutionary topic related to the chapter's content. In each case, the focus is on how scientists developed and tested a specific hypothesis. You will see that the scientific community consists of a global team of clever and creative professionals.

Often, the experiment profiled in an Investigating Life section reinforces the connections between multiple fields of biology. Genetics and natural selection weave together in a discussion of speciation in monkeyflowers, for example, and DNA sequence analysis is critical to a study of the evolution of the human brain.

Model Organisms: A related feature of this textbook reinforces these connections and the process of science. "Focus on Model Organisms" boxes appear in each chapter of Unit 4, The Diversity of Life. Each box highlights one or two species that have made extraordinary contributions to biology. For example, Chapter 18 has a box on the bacterium *Escherichia coli*, Chapter 21 profiles *Arabidopsis thaliana*, and Chapter 22 has boxes for the nematode *Caenorhabditis elegans* and the fruit fly, *Drosophila melanogaster*.

The Process of Evolution Unifies the Field of Biology

On a road trip across the Midwestern United States, I watched swallows swoop over the Mississippi River with split-second precision. Fireflies flashed above the grass soon after sundown, and woodpeckers expertly hammered tree bark with their stout bills. A male bullfrog's guttural croaking signaled his availability to female frogs in a small pond. How do these animals know exactly what to do, and where, and how, and when to do it? Oak, poplar, sassafras, and hickory trees thrive in the forests of the Midwest, with raspberries growing in the shaded understory. Purple coneflowers populate the meadows. Why do these particular plants occur in the Midwest, but redwood trees and banana trees are absent?

All around us, we can see that life seems perfectly suited to its habitat. Centuries of scientific research—from observations and detailed note taking to, now, probing the base pairs of the DNA double helix—tell the compelling story of how this came to be. When Charles Darwin wrote *On the Origin of Species* in the mid-1860s, he set into motion the science of evolutionary biology. But it didn't stop there. Generations of scientists have built on that foundation, and we now have a richly detailed understanding of the evolutionary processes that have brought life to this point.

A famous journal article is titled, "Nothing in biology makes sense except in light of evolution," a profound statement that is not to be taken lightly. Evolution permeates this book because it permeates our understanding of biology. Evolution is in an animal's selection of mates, in a farmer's choice to cultivate the crops that grow well in a particular region of the world, and in our tendency to eat sweets and fats (even though we know we shouldn't). In addition to the evolution and diversity units in this textbook, the Investigating Life sections and many other concepts will allow you to discover the evolutionary forces behind biology at every scale, from chemistry to ecology.

Thinking as a Scientist

This book is full of features that will help you learn to think scientifically. Each chapter begins with an attention-grabbing essay and a learning outline that previews the main concepts that you will encounter. Each main section finishes with a summary and a set of questions designed to help you assess your understanding of the concept before moving on. Moreover, the Investigating Life section that concludes each chapter includes data and a critical thinking question. The end-of-chapter Multiple Choice, Testing Your Knowledge, and Thinking As a Scientist questions reinforce basic content and conceptual understanding. Illustrated tables and strategically placed mini-glossaries will help you organize the information and understand the connections between details.

Scattered throughout the book are "Can *You* Relate?" boxes, brief readings that explain the biology behind phenomena that you may have noticed for yourself. For example, I discuss why some (but not all) artificial sweeteners are calorie-free, why leaves change color in the fall, and why purebred dogs often suffer from health problems.

"Burning Question" boxes are based on questions that my own students have asked me over more than 10 years of teaching introductory biology at the University of Oklahoma. On the first day of class, I always ask my students to write on an index card a burning question that they would like to have answered during the semester. Their questions range from the quirky ("Is it true that you can lick toads to get high?") to the medical ("My grandmother just died of cancer, so will I get cancer too?") to the environmental ("How does human-made pollution harm other life forms, and how does this hurt ecosystems as a whole?").

I answer each of the questions at some time during the semester. My students enjoy seeing their questions projected on the big screen, and I find it exciting to catch a glimpse of what they have on their minds. This textbook contains a selection of my students' Burning Questions, which represent an immediate connection between my own classes at The University of Oklahoma and other introductory biology students who use this book.

Biology is a Visual Science

Biology is difficult to explain with words alone. This book features a new art program developed by a talented team of professional illustrators who considered my suggestions and improved on many of them. The illustrations are bright and colorful, often combining art and photos or micrographs into appealing and informative combinations. Repetition aids in learning, so the illustrators were careful to use consistent colors for membranes, DNA, proteins, cell organelles, molecules, atoms, and other structures that occur throughout the book. Numbered steps help you work through complex processes, and figure legends add additional explanation.

My Commitment to a Community of Educators and Students

Being a textbook author has made me a better teacher. In researching the topics in this book, I have developed a richer understanding of biological processes, and I have more examples to share with my students. At the same time, being an instructor makes me a better author. My students ask me questions that reveal which topics are the most difficult and confusing to them. I have devoted extra effort to developing new illustrations that offer extra help with these subjects. Sometimes, the solution is as simple as a small black-and-white context drawing that reminds students where a process occurs. In other cases, whole new illustrations were required. Throughout the book, I have tried to make sure that all the visual and textual information is accurate, complete, up-to-date, and explained at a level that a beginning student can understand.

Many of us who chose teaching as a profession are passionate about our subject, and we want our students to share our enthusiasm. I hope the text, art, and photos in this book will help bring biology to life for faculty and students alike. Therefore, this is a work in progress, and I welcome your suggestions on how to serve you better. I encourage you to write me at marielle_hoefnagels@mcgraw-hill.com with suggestions on how to improve this book and what you'd like to see in future editions. Perhaps one of your Burning Questions will become part of this story of life.

A Student's Guide to Using This Textbook

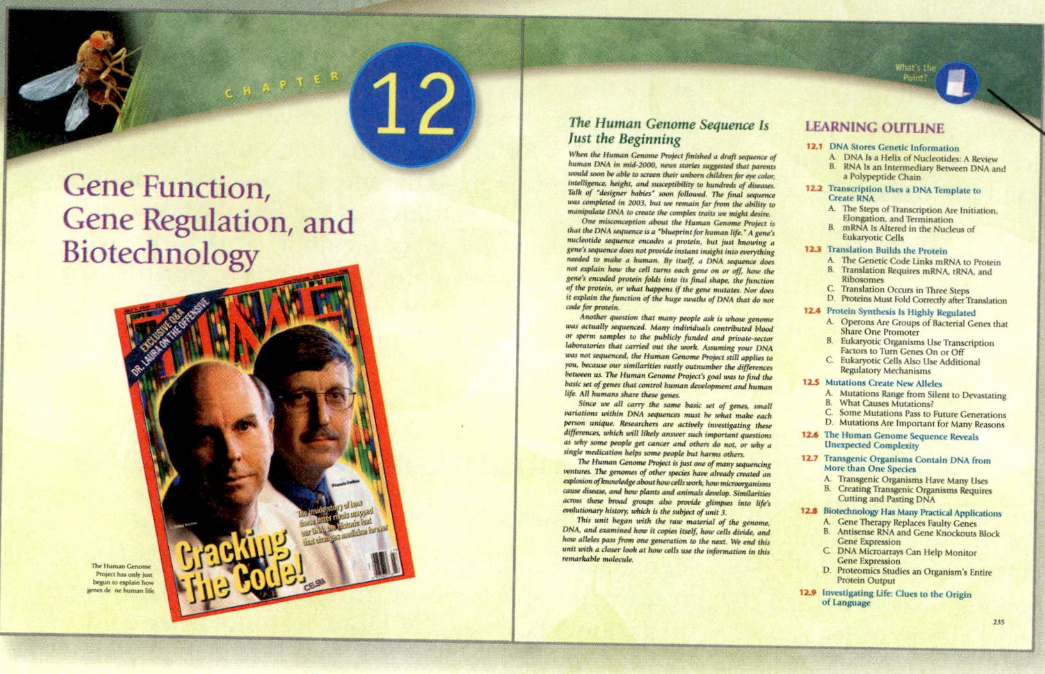

Ask yourself, "What's the Point?"

Try the digital learning aids available on the text's ARIS website.

Listen to the author briefly describe the key points of each chapter.

Check out the chapter Road Map.

The numbered concepts and detailed outline point the way.

The opening essay offers insights into why this chapter matters.

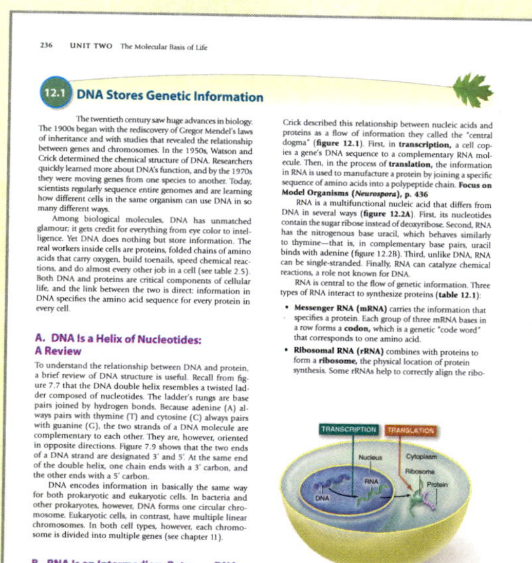

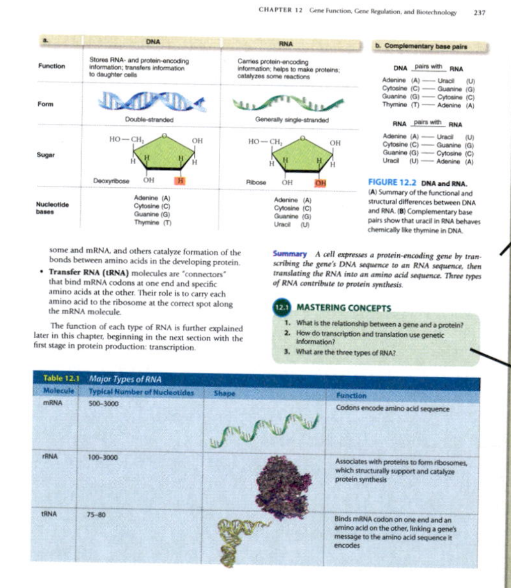

Build your understanding one concept at a time.

Focus on each concept's brief, summational statement.

Note key terms that appear in bold-faced, definitional sentences.

Take time to test your understanding after each concept.

Connect this material to your own life.

Burning Questions came from the author's own students.

Burning Questions

How do biologists use only DNA to clone mammals?

Unlike many other organisms, mammals do not naturally clone themselves. In 1996, however, researcher Ian Wilmut and his colleagues in Scotland used a new procedure to create Dolly the sheep, the first clone of an adult mammal. They removed the diploid nucleus of a cell taken from a donor sheep's mammary gland. They then transferred this nucleus to a sheep's egg cell whose own haploid nucleus had been removed (**figure 8.B**).

The cell divided mitotically to form an embryo. Inside a surrogate mother's uterus, the embryo developed into Dolly.

Dolly appeared normal, and she gave birth to four healthy lambs (via sexual reproduction). But she had arthritis in her hind legs, and she died of a lung infection in 2003. Although there is no evidence that Dolly's relatively early death was related to her being a clone, her short life has fueled speculation that clones may have hidden genetic abnormalities. Indeed, most clones die early in development, presumably because the gene regulation mechanisms in an adult cell's nucleus are somehow incompatible with those in the egg cell. Even the tiny percentage of clones that make it to birth often have abnormalities. These difficulties emphasize the significant ethical issues surrounding human cloning.

Have a Burning Question of your own? Submit it to marielle_hoefnagels@mcgraw-hill.com for possible inclusion in future editions of this book!

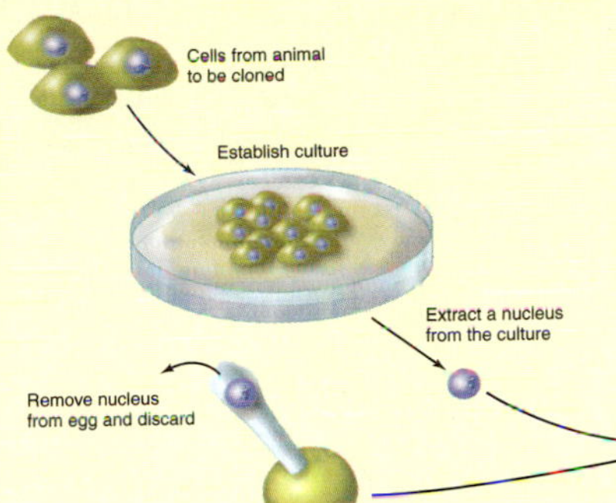

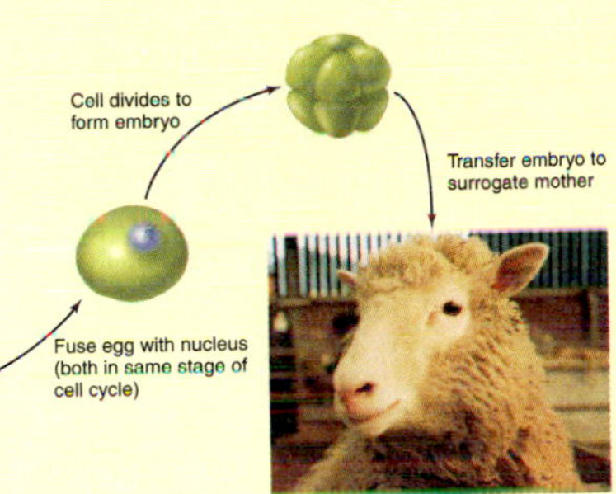

FIGURE 8.B **Creating Dolly.** Biologists cloned an adult sheep by obtaining a nucleus from a cell of a ewe's udder. They also removed the nucleus from an egg cell. Placing the adult cell's nucleus into the egg yielded a new cell genetically identical to the ewe. After being implanted into a surrogate mother sheep, the resulting embryo developed into Dolly.

Burning Questions

Is male baldness really from the female side of the family?

Male pattern baldness is the distinctive hair loss that many men (and some women) experience as they enter their 20s, 30s, and 40s. The baldness spreads outward from the temples and crown of the head in a characteristic pattern (**figure 11.A**).

Two conditions are required for male pattern baldness to develop. First, hormones called androgens must be present in normal amounts. Testosterone and dihydrotestosterone (DHT) are androgens; they bind to and enter hair follicle cells, interacting with the DNA

to stop growth of
have a genetic pred
Many people b
ness from his moth
ling this trait reside
Therefore, either pa
Why don't wo
The answer is that
condition, in which
alleles, yet express
testosterone is the
of testosterone, the stronger the influence
Men typically have more of this sex hormo
the name, *male* pattern baldness.

Have a Burning Question of your
marielle_hoefnagels@mcgraw-hill
inclusion in future editions o

FIGURE 11.A **Male Pattern Baldness.** Genes and hormones interact in this distinctive form of hair loss.

Can You Relate to the biology behind news items and your own experiences?

7.6 DNA Profiling Has Many Applications

DNA profiling detects genetic differences between individuals, with applications ranging from settling paternity claims to exonerating wrongly convicted criminals to identifying crime victims (see Can *You* Relate? Identifying Victims of the September 11, 2001, Attacks below).

Rather than sequencing and comparing entire genomes, DNA profiling considers just the most variable parts. In one approach, researchers use single-base sites that tend to vary in a population. DNA samples are treated with bacterial enzymes (called restriction enzymes) that cut DNA at specific sequences. If two samples differ at a cutting site, the restriction enzyme produces DNA fragments of different sizes and numbers. **Figure 7.17** illustrates how researchers use gel electrophoresis to reveal these underlying sequence differences.

Can *You* Relate?

Identifying Victims of the September 11, 2001, Attacks

Until September 11, 2001, the most challenging application of DNA profiling had been identifying plane-crash victims, a grim task eased by having lists of passengers. The terrorist attacks on the World Trade Center provided a staggeringly more
several reasons: the high number of ca-
of the remains, and the lack of a list of
e buildings. Overall, the disaster yielded
NA samples
ving September 11, somber researchers
ic., in Salt Lake City, who usually ana-
ancer genes, received frozen DNA from
om the disaster site. The laboratory also
gs from relatives of the missing, and tis-
toothbrushes, razors, and hairbrushes.
to determine the numbers of copies of
f DNA, called short tandem repeats, or
n the genome. The chance that any two
ame 13 markers by chance is one in 250
tern of a sample from the crime scene
m a victim's toothbrush, identification
A extracted from tooth and bone bits
nomics Corporation in Rockville, Mary-
nces were analyzed from mitochondria.
neration.
NA to identify about 850 of the more
orted missing. It was a very distressing
nicians and researchers, whose jobs had
detecting breast cancer and sequencing
recovery.

1 A restriction enzyme is added to three DNA samples, cutting each sample into a unique pattern of different-sized pieces.

Restriction enzyme

DNA sample 1 DNA sample 2 DNA sample 3

2 The three samples are loaded into an electrophoresis gel.

Longer fragments

Shorter fragments

Power source

3 Labeled DNA probes highlight certain fragments of interest by binding to specific fragments of sampled DNA. This way, DNA from different individuals may have different banding patterns.

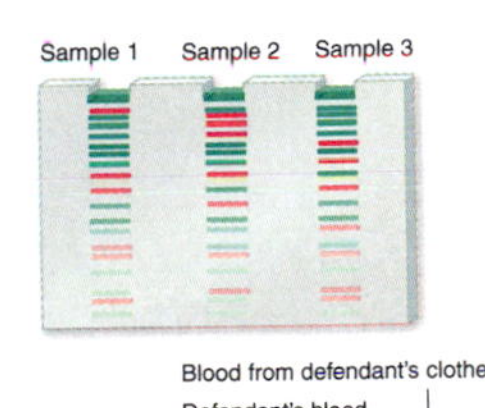

a.

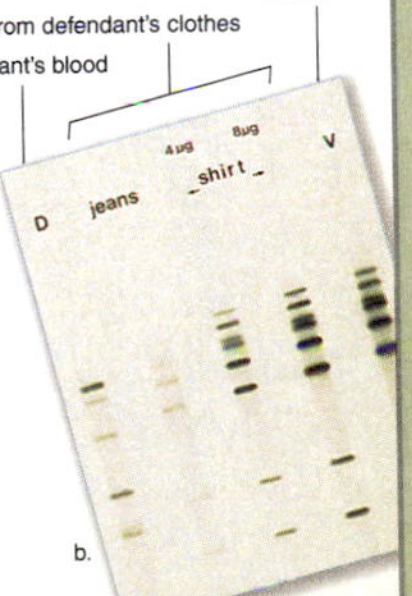

b.

FIGURE 7.17 **DNA Profiles from a Murder Case.** **(A)** DNA samples from different sources produce different patterns of fragments when cut with the same restriction enzyme. **(B)** DNA from bloodstains on the defendant's clothes matches the DNA profile of the victim but differs from the DNA profile of the defendant. This is evidence that the blood on the defendant's clothes came from the victim, not the defendant.

Can *You* Relate?

Dogs Are Products of Artificial Selection

The pampered poodle and graceful greyhound may win in the show ring, but they are poor specimens in terms of genetics and evolution. Behind carefully bred traits lurk small gene pools and extensive inbreeding, all of which may harm the health of purebred show animals (**table 13.A**).

The runny, sad eyes of the basset hound can be quite painful (**figure 13.A**). Short legs make the dog prone to arthritis, the long abdomen promotes back injuries, and the characteristic floppy ears often hide ear infections. The eyeballs of the Pekingese protrude so much that a mild bump can pop them out of their sockets. The tiny jaws and massive teeth of pugs and bulldogs cause dental and breathing problems, sinusitis, and their notorious "dog breath." Folds of skin on their abdomens easily become infected. Larger breeds, such as the Saint Bernard, have bone problems and short life spans. A Great Dane may suddenly die at a young age, its heart overworked from years of supporting a large body.

All of these examples make one truth clear: We may be able to breed desired characteristics into a dog, but we can't always breed other traits out.

Table 13.A *Purebred Plights*

Breed	Health Problems
Cocker spaniel	Nervousness, ear infections, hernias, kidney problems
Collie	Blindness, bald spots, seizures
Dalmatian	Deafness
German shepherd	Hip dysplasia
Golden retriever	Lymphatic cancer, muscular dystrophy, skin allergies, hip dysplasia, absence of one testicle
Great Dane	Heart failure, bone cancer
Labrador retriever	Dwarfism, blindness
Shar-pei	Skin disorders

Bulldog

Basset hound

FIGURE 13.A **Dog Traits Come from Genetic Variations Selected Through Breeding.** The mournful expression of the basset hound accompanies its heightened sense of smell, and the bulldog was selected for its flattened face and fierce demeanor. These traits originally occurred as natural genetic variation.

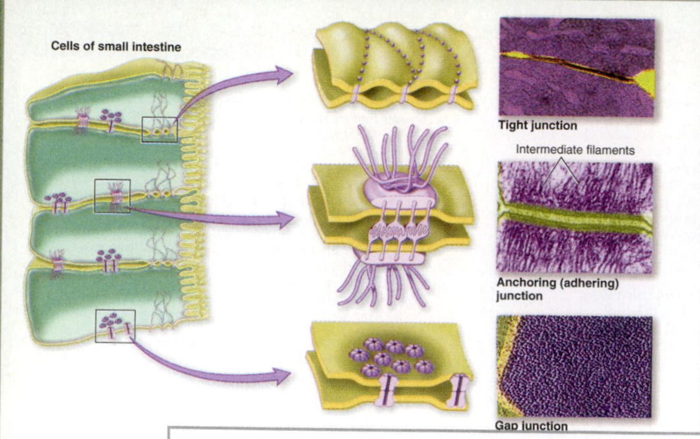
Cells of small intestine
Tight junction
Intermediate filaments
Anchoring (adhering) junction
Gap junction

Become an art expert—use your visual learning skills.

Combination Figures link art and photos to provide both perspectives.

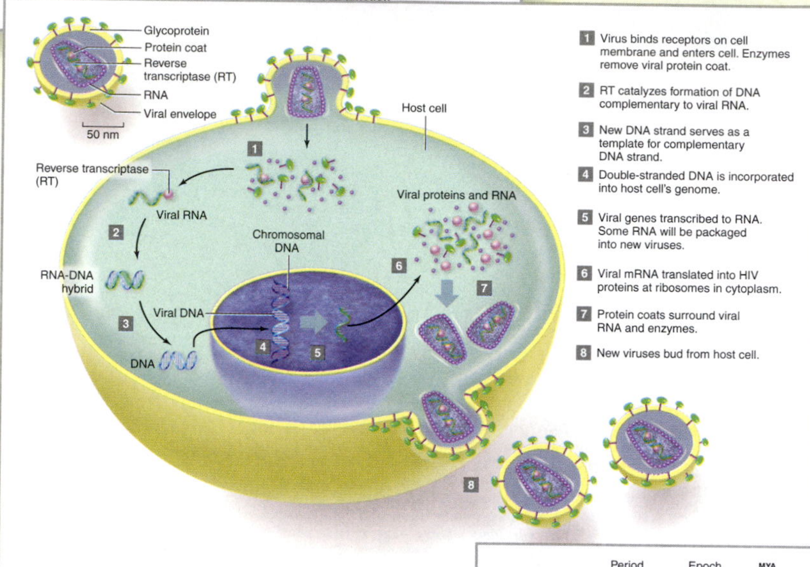
Glycoprotein
Protein coat
Reverse transcriptase (RT)
RNA
Viral envelope
50 nm
Host cell
Reverse transcriptase (RT)
Viral RNA
RNA-DNA hybrid
DNA
Chromosomal DNA
Viral DNA
Viral proteins and RNA
1 Virus binds receptors on cell membrane and enters cell. Enzymes remove viral protein coat.
2 RT catalyzes formation of DNA complementary to viral RNA.
3 New DNA strand serves as a template for complementary DNA strand.
4 Double-stranded DNA is incorporated into host cell's genome.
5 Viral genes transcribed to RNA. Some RNA will be packaged into new viruses.
6 Viral mRNA translated into HIV proteins at ribosomes in cytoplasm.
7 Protein coats surround viral RNA and enzymes.
8 New viruses bud from host cell.

Step-by-Step Figures present concepts in easy-to-follow steps.

Illustrated Tables help organize information and connect details.

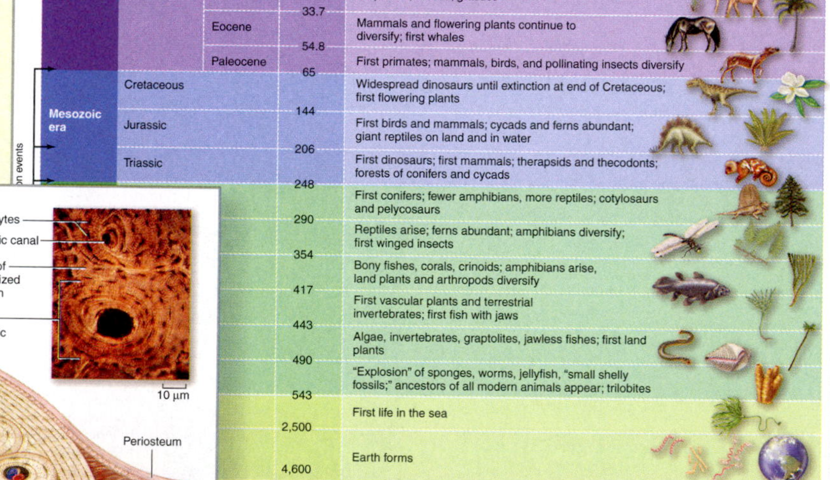

	Period	Epoch	MYA	Important events
Cenozoic era	Quaternary	Recent		Human civilization
			0.01	
		Pleistocene		*Homo sapiens*, large mammals; ice ages
			1.8	
	Tertiary	Pliocene		*Australopithecus*, modern whales
			5.3	
		Miocene		Hominoids; mammals continue to diversify; modern birds; expansion of grasslands
			23.8	
		Oligocene		Elephants, horses; grasses
			33.7	
		Eocene		Mammals and flowering plants continue to diversify; first whales
			54.8	
		Paleocene		First primates; mammals, birds, and pollinating insects diversify
			65	
Mesozoic era	Cretaceous			Widespread dinosaurs until extinction at end of Cretaceous; first flowering plants
			144	
	Jurassic			First birds and mammals; cycads and ferns abundant; giant reptiles on land and in water
			206	
	Triassic			First dinosaurs; first mammals; therapsids and thecodonts; forests of conifers and cycads
			248	
				First conifers; fewer amphibians, more reptiles; cotylosaurs and pelycosaurs
			290	
				Reptiles arise; ferns abundant; amphibians diversify; first winged insects
			354	
				Bony fishes, corals, crinoids; amphibians arise, land plants and arthropods diversify
			417	
				First vascular plants and terrestrial invertebrates; first fish with jaws
			443	
				Algae, invertebrates, graptolites, jawless fishes; first land plants
			490	
				"Explosion" of sponges, worms, jellyfish, "small shelly fossils;" ancestors of all modern animals appear; trilobites
			543	
				First life in the sea
			2,500	
				Earth forms
			4,600	

Cartilage
Spongy bone
Compact bone
Blood vessel
Nerve
Marrow Cavity
a.
Compact bone
Osteon
Spongy bone
Osteonic canal
Osteocytes
Osteonic canal
Matrix of mineralized collagen
10 µm
Periosteum
Nerve
Blood vessel
Osteonic canal
Nerve
Communicating canal
b.

Macro-to-Micro Figures provide context and perspective.

Become an investigator!

Study how science works in each chapter and in the Model Organisms feature.

FOCUS ON MODEL ORGANISMS: *Mus musculus*, the mouse

The history of the mouse, *Mus musculus*, as the stereotypical lab animal dates back to the early 1900s. Researchers discovered that its small size made it an excellent research animal. Also, mice are famous for their prolific breeding. They reach sexual maturity at the age of about 4 weeks, are sexually receptive every few days, and give birth to litters of 1 to 10 pups after a gestation of only about 3 weeks. Over a life span of 1.5 to 3 years, a single pair of mice can produce hundreds of offspring.

Researchers have benefited from biotechnology in their studies of mice. Transgenic mice have been available since the 1980s (see chapter 12). These mice are modified in countless ways, including altered susceptibility to human diseases. Mice were cloned in 1998, making possible the production of genetically identical animals ideal for testing new disease treatments. The mouse genome … 2002, revealing about 30,000 genes … mes. Not surprisingly, more than 99% … parts in the human genome. This sim… ne of the following ways in which *Mus* … biological research:

… ne 1930s, the discovery that mice … ll but their very close relatives led … ajor histocompatibility complex … biologists have discovered an array … ne function (see chapter 36). … ave been used to study human dis… en researchers discovered that mice … er. Vaccines were subsequently tested … of transgenic mice has opened new possibilities for research on the cause and treatment of human disease, including the protein deposits (beta-amyloid plaques) that accumulate in the brains of patients with Alzheimer disease; obesity; Parkinson disease; the role of mutated tumor suppressor genes such as *p53* in human cancer; and the effects of viruses such as HIV on the human immune system. **cancer, p. 166; HIV, p. 362**

- **X chromosome inactivation:** In the 1960s, biologist Mary Lyon proposed that in female mammals, one of the two X chromosomes is inactivated early in embryonic development. This phenomenon, often illustrated using calico cats, was first proved in mice with mottled coats. **X inactivation, p. 226**
- **Stem cells:** These undifferentiated cells, which can be derived from embryos or adults, can specialize into many other cell types. Mouse stem cell research has shown great promise in treating spinal cord injuries and many other ailments. **stem cells, p. 158**

13.5 INVESTIGATING LIFE
Size Matters in Fishing Frenzy

Studying the mechanisms of evolution helps us to understand life's history, but it also has practical consequences. A good example of natural selection is unfolding in fisheries worldwide. The selective force stems from a surprisingly mundane source—fishing regulations—but it affects everything from restaurant menus, to coastal economies, to the future of the ocean ecosystem.

The past several decades have seen devastating declines in the numbers of large predatory fishes such as swordfish, marlin, and sharks, as well as smaller animals including tuna, cod, and flounder. From a biological point of view, the reason for the fisheries decline is simple: the animals' death rate exceeds their reproductive rate. Industrial-scale fishing is the culprit. Since the 1950s, fishing fleets have employed larger ships and improved technologies in pursuit of their prey.

Regulations meant to protect fisheries allow the harvest of only those fish that exceed some minimum size. This measure is logical, because the smallest fish are most likely to be juveniles. Protecting the youngsters should permit the population to recover from the harvest of adult fish. Yet these regulations also have predictable evolutionary side effects. If fishing fleets selectively harvest the largest individuals, fish that are small at maturity are the most likely to survive long enough to reproduce. Large fish may become more scarce over many generations. The same policy should also select for slow-growing fish, since they would be last to exceed the minimum allowed size.

Fish ecologists David Conover and Stephan Munch of the State University of New York tested these predictions in a coastal fish called the Atlantic silverside (*Menidia menidia*; **figure 13.18**). These small, shiny fish live for about 2 years, eating small invertebrates such as shrimp and marine worms along the Atlantic coast of Canada and the United States. The researchers chose this species in part because large populations can be maintained in captivity. Also, Atlantic silversides reproduce rapidly, making multigeneration experiments practical.

Conover and Munch set up their experiment by randomly dividing a large, captive population of Atlantic silversides into six tanks, each containing about 1100 juvenile fish. This ensured that all treatment groups started with similar gene pools. After about 6 months, the researchers removed the largest 90% of the fish from two of the tanks, termed "large-harvested" tanks. This treatment simulated fishing policies that protect all fish below a certain size. In two "small-harvested" tanks, they removed the smallest 90% of the silversides. Two control tanks were "random-harvested," in which 90% of the fish were removed without size bias.

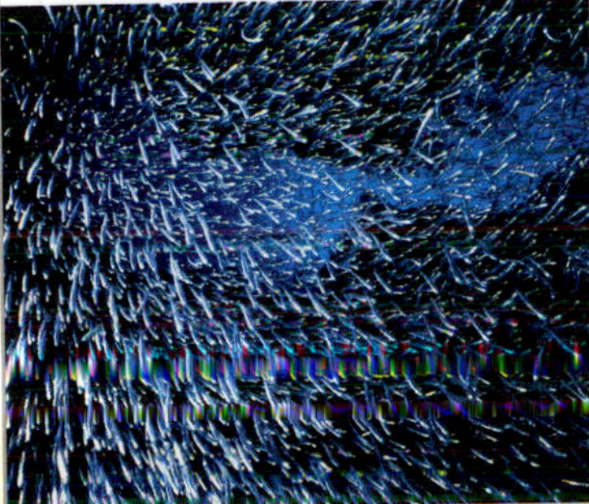

FIGURE 13.18 Atlantic Silverside. This small, silvery fish averages 9 cm in length, with a maximum of 15 cm.

Treatment Name	Removed …	Leaving to reproduce …
Large-harvested	Largest 90% of fish	Smallest 10% of fish
Small-harvested	Smallest 90% of fish	Largest 10% of fish
Random-harvested	Random 90% of fish	Random 10% of fish

After the harvests, about 100 survivors remained in each tank. These reproduced, and their descendants were reared in identical conditions until it was again time to harvest 90% of each population. The researchers repeated the large-harvest, small-harvest, and random-harvest treatments over four generations.

Predictably, both the total harvest weight and the weight of the average caught fish were initially highest for the large-harvested fish. Over four generations of size-biased fish removal, however, the average weight of the small-harvested … , while that of the large-harvested fish … 9A). Furthermore, by the end of the ex… rvested fish clearly grew the fastest (fig… all-harvested population, therefore, se… ize and rapid growth; the opposite was … sted population. The researchers con… treatments imposed different selective … e genetic structure of the populations. … populations of silversides do not ex… ne selective forces as do wild popula… rger fish. Nevertheless, the researchers … suggesting that the same evolutionary … in their study are probably also occur… ecies.

… unch's experiment is more than a … onstration of natural selection in ac… onomic and ecological applications. … , new fishing regulations that impose … maximum size limits, protecting fish … ertain size. Such a policy would con… nportant fish stocks, because it would … hat are critical to a species' future re… d also help fishing communities by … wing fish. Imposing a maximum size … e ecological benefits, by restoring the … other "ecosystem services" of the larg… nge of implications beautifully illustrates the powerful ideas that spring from understanding one fundamental idea: natural selection.

Conover, David O. and Stephan B. Munch. 2002. Sustaining fisheries yields over evolutionary time scales. *Science*, vol. 297, pages 94–96.

FIGURE 13.19 Size Matters in Fish Harvests. (A) Over four generations, the average per-fish weight in the large-harvested population was much less than the average for the small-harvested population. (B) Small harvested fish grow more rapidly than large-harvested fish as well. **Question:** A trait's heritability is the proportion of variability that can be explained by genes. Heritability of 1.0 means a trait is 100% under genetic control; 0 means a trait is entirely under environmental control. Conover and Munch estimated that, in Atlantic silversides, heritability of body size is about 0.2. How would the results of this experiment be different if heritability of body size were higher than 0.2? What if it approached 0?

Read and analyze the data in each chapter's capstone Investigating Life section.

CHAPTER SUMMARY

13.1 Evolutionary Thought Has Evolved for Centuries

- Biological **evolution** is change in allele frequencies in populations. Evolution has occurred in the past and is constant and ongoing.

A. Many Explanations Have Been Proposed for Life's Diversity

- Early attempts to explain life's diversity relied on belief in a creator. Geology laid the groundwork for evolutionary thought. Some people explained the distribution of rock strata with the idea of **catastrophism** (a series of floods). The more gradual **uniformitarianism** (continual remolding of Earth's surface) became widely accepted.
- The **principle of superposition** states that lower rock strata are older than those above, suggesting a time frame for fossils within them.
- Lamarck was the first to propose a mechanism of evolution, but it was erroneously based on **inheritance of acquired characteristics.**

B. Charles Darwin's Voyage Provided a Wealth of Evidence

- During the voyage of the HMS *Beagle*, Darwin observed the distribution of organisms in diverse habitats and their relationships to geological formations. He noted that similar adaptations could lead to **convergent evolution.** After much thought, and considering input from other scientists, he synthesized his theory of the origin of species by means of **natural selection.**

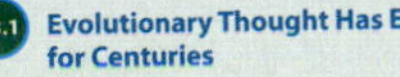

Use chapter study tools to check your understanding and prepare for the test.

Acknowledgments

360° Development

McGraw-Hill's 360° Development Process is an ongoing, never ending, market-oriented approach to building accurate and innovative print and digital products. It is dedicated to continual large scale and incremental improvement driven by multiple customer feedback loops and checkpoints. This is initiated during the early planning stages of our new products, and intensifies during the development and production stages, then begins again upon publication, in anticipation of the next edition.

This process is designed to provide a broad, comprehensive spectrum of feedback for refinement and innovation of our learning tools, for both student and instructor. The 360° Development Process includes market research, content reviews, faculty and student focus groups, course- and product-specific symposia, accuracy checks, and art reviews, all guided by a carefully selected Board of Advisors.

Contributors

Murray P. "Pat" Pendarvis, *Southeastern Louisiana University*—Protista chapter, Investigating Life essays, art development
Jan Jenner—Human evolution section, animal behavior chapter
Merri Lynn Casem—end-of-chapter questions
Jody Larson—chapter analysis

Board of Advisors

Don French, *Oklahoma State University*
Tom Jacobs, *University of Illinois–Urbana/Champaign*
A. Daniel Johnson, *Wake Forest University*
Pat Pendarvis, *Southeastern Louisiana University*
Brian Shmaefsky, *Kingwood College*
Cara Shillington, *Eastern Michigan University*
Jennifer Warner, *University of North Carolina–Charlotte*

Ancillary Authors

Instructor's Manual:

Brian Shmaefsky, *Kingwood College*

ARIS Quizzes:

Cara Shillington, *Eastern Michigan University*

Test Bank:

Richard Hanke, *Rose State College*
Scott Cooper, *University of Wisconsin–Lacrosse*
Richard Haro, *University of Wisconsin–Lacrosse*

Lecture Outlines:

Brenda Leady, *University of Toledo*

Reviewers

Charles Lee Biles, *East Central University*
Peggy Brickman, *University of Georgia*
Sharon Bullock, *Virginia Commonwealth University*
Beth Campbell, *Itawamba Community College*
Garry Davies, *University of Alaska*
Jessica L. DeGraff, *Gloucester County College*
Jean DeSaix, *University of North Carolina*
Dwight Dimaculangan, *Winthrop University*
Cathy A. Donald-Whitney, *Collin County Community College*
John A. Ewing, III, *Itawamba Community College*
Jerry L. Faulkner, *Chattanooga State Technical Community College*
Steven E. Fields, *Winthrop University*
Teresa G. Fischer, *Indian River Community College*
Jo-Elle Mogerman, *Harold Washington College*
Bob Harms, *St. Louis Community College*
Lee Kavaljian, *California State University*
Jerome A. Krueger, *South Dakota State University*
Thomas G. Lammers, *University of Wisconsin–Oshkosh*
John C. Landolt, *Shepherd University*
Nikki LoCascio, *Marshall University*
Michael P. Mahan, *Kean University*
John E. Marshall, *Pulaski Technical College*
Michael Masson, *Santa Barbara City College*
Caroline H. McNutt, *Schoolcraft College*
Judith M. Megaw, *Indian River Community College*
Dalivan Melendrez, *Harold Washington College*
David H. Mirman, *Mt. San Antonio College*
Kathy M. Monroe, *University of North Carolina*
Hao Nguyen, *California State University*
Nathan Opolot Okia, *Auburn University*
Frank H. Osborne, *Kean University*

John C. Osterman, *University of Nebraska*
Kathleen Pelkki, *Saginaw Valley State University*
Nirmala V. Prabhu, *Edison College*
Christopher L. Pritchett, *Northeastern State University*
Kirsten Raines, *San Jacinto Colleget*
Bruce Reid, *Kean University*
Jill D. Reid, *Virginia Commonwealth University*
Lynn J. Rivers, *Henry Ford Community College*
Steve J. Rothenberger, *University of Nebraska*
John Richard Schrock, *Emporia State University*
Shreekumar R. Pillai, *Alabama State University*
Larry J. Szymczak, *Chicago State University*
Christopher Tabit, *University of West Georgia*
Jack Waber, *Westchester University*
D. Alexander Wait, *Southwest Missouri State University*
Jennifer M. Warner, *University of North Carolina*
Randall Warwick, *Coastline Community College*
Edwin M. Wong, *Western Connecticut State University*
Kenneth Wunch, *Sam Houston State University*
Mark Wygoda, *McNeese State University*
Melissa Zwick, *Longwood University*

General Biology Symposia

Every year McGraw-Hill conducts several General Biology Symposia, which are attended by instructors from across the country. These events are an opportunity for editors from McGraw-Hill to gather information about the needs and challenges of instructors teaching non-majors level biology courses. It also offers a forum for the attendees to exchange ideas and experiences with colleagues they might not have otherwise met. The feedback we have received has been invaluable, and has contributed to the development of ***Biology: Concepts and Investigations***, and its supplements.

Norris Armstrong, *University of Georgia*
David Bachoon, *Georgia College and State University*
Sarah Bales, *Moraine Valley Community College*
Lisa Bellows, *North Central Texas College*
Joressia Beyer, *John Tyler Community College*
James Bidlack, *University of Central Oklahoma*
Mark Bloom, *Texas Christian University*
Paul Bologna, *Montclair University*
Linda Brandt, *Henry Ford Community College*
Marguerite Brickman, *University of Georgia*
Bradford Boyer, *Suffolk County Community College*
Art Buikema, *Virginia Polytechnic Institute*
Sharon Bullock, *Virginia Commonwealth University*
Raymond Burton, *Germanna Community College*
Nancy Butler, *Kutztown University of Pennsylvania*
Jane Caldwell, *West Virginia University*
Carol Carr, *John Tyler Community College*
Kelly Cartwright, *College of Lake County*
Rex Cates, *Brigham Young University*
Sandra Caudle, *Calhoun Community College*
Genevieve Chung, *Broward Community College*
Jan Coles, *Kansas State University*
Marian Wilson Comer, *Chicago State University*
Lewis Deaton, *University of Louisiana at Lafayette*
Jody DeCamilo, *St. Louis Community College*
Jean DeSaix, *University of North Carolina at Chapel Hill*
JodyLee Estrada-Duek, *Pima Community College–Desert Vista*
Laurie Faber-Foster, *Grand Rapids Community College*
Theresa Fischer, *Indian River Community College*
Theresa Fulcher, *Pellissippi State Technical College*
Dennis Fulbright, *Michigan State University*
Steven Gabrey, *Northwestern State University*
Cheryl Garett, *Henry Ford Community College*
Farooka Gauhari, *University of Nebraska–Omaha*
John Geiser, *Western Michigan University*
Cindy Ghent, *Towson University*
William Glider, *University of Nebraska–Lincoln*
Carla Guthridge, *Cameron University*
Bob Harms, *St. Louis Community College–Meramec*
Wendy Hartman, *Palm Beach Community College*
Tina Hartney, *California State Polytechnic University*
Kelly Hogan, *University of North Carolina–Chapel Hill*
Eva Horne, *Kansas State University*
David Huffman, *Texas State University–San Marcos*
Shelley Jansky, *University of Wisconsin–Stevens Point*
Tina Jones, *Shelton State Community College*
Arnold Karpoff, *University of Louisville*
Jeff Kaufmann, *Irvine Valley College*
Michael Koban, *Morgan State University*
Todd Kostman, *University of Wisconsin–Oshkosh*
Steven Kudravi, *Georgia State University*
Nicki Locascio, *Marshall University*
Dave Loring, *Johnson County Community College*
Janice Lynn, *Alabama State University*
Phil Mathis, *Middle Tennessee State University*
Daryl Miller, *Broward Community College–South Campus*
Marjorie Miller, *Greenville Technical College*
Meredith Norris, *University of North Carolina at Charlotte*
Mured Odeh, *South Texas College*
Nathan Olia, *Auburn University–Montgomery*
Rodney Olsen, *Fresno City College*
Alexander Olvido, *Virginia State University*
Clark Ovrebo, *University of Central Oklahoma*
Forrest Payne, *University of Arkansas at Little Rock*
Nancy Pencoe, *University of West Georgia*
Pat Pendarvis, *Southern Louisiana University*
Jennie Plunkett, *San Jacinto College*
Scott Porteous, *Fresno City College*
David Pylant, *Wallace State Community College*

Fiona Qualls, *Jones County Junior College*
Eric Rabitoy, *Citrus College*
Karen Raines, *Colorado State University*
Kirsten Raines, *San Jacinto College*
Jill Reid, *Virginia Commonwealth University*
Darryl Ritter, *Okaloosa–Walton College*
Robin Robison, *Northwest Mississippi Community College*
Bill Rogers, *Ball State University*
Vicki Rosen, *Utah State University*
Kim Sadler, *Middle Tennessee State University*
Cara Shillington, *Eastern Michigan University*
Greg Sievert, *Emporia State University*
Jimmie Sorrels, *Itawamba Community College*
Judy Stewart, *Community College of Southern Nevada*
Julie Sutherland, *College of DuPage*
Bill Trayler, *California State University–Fresno*
Linda Tyson, *Santa Fe Community College*
Eileen Underwood, *Bowling Green State University*
Heather Vance-Chalcraft, *East Carolina University*
Marty Vaughan, *IUPUI–Indianapolis*
Paul Verrell, *Washington State University*
Thomas Vogel, *Western Illinois University*
Brian Wainscott, *Community College of Southern Nevada*
Jennifer Warner, *University of North Carolina–Charlotte*
Scott Wells, *Missouri Southern State University*
Lan Xu, *South Dakota State University*
Robin Whitekiller, *University of Central Arkansas*
Allison Wiedemeier, *University of Illinois–Columbia*
Michael Windelspecht, *Appalachian State University*
Tom Worcester, *Mount Hood Community College*
Frank Zhang, *Kean University*
Michelle Zjhra, *Georgia Southern University*

I owe a huge debt of gratitude to Ricki Lewis, Bruce Parker, and Doug Gaffin for providing the manuscripts that are the foundation of this book. I also greatly appreciate the contributions of Pat Pendarvis and Jan Jenner. Doug is my husband, and he has done more than anyone to boost my confidence and encourage me when meeting my commitments seemed impossible.

Thanks to Jane Peden and JoAnn Mohr for administrative assistance and to Mari Anne Hartmann and Nicole Schlutt for assistance with reviews. Thanks also to my amazing book team, including Sheila Frank, Kara Kudronowicz, Michelle Whitaker, and John Leland. My developmental editors were Lisa Bruflodt, Anne Winch, Rose Koos, and Margaret Horn; all of whom made valuable contributions to keeping this project on track. Tom Lyon got this project started, and Michael Lange helped me finish it. I could not have come this far without Michael's intellectual and editorial input. Patrick Reidy, Suzanne Guinn, and Tamara Maury are among my many other friends at McGraw-Hill; I am happy to be able to talk things over with these "insiders."

I also appreciate The University of Oklahoma faculty who answered assorted technical questions for me: Bradley Stevenson, Tyrrell Conway, Ingo Schlupp, Rich Broughton, Scott Russell, Wayne Elisens, Ben Holt, and Laura Gibbs. Thank you also to Gordon Uno for helping me believe this project is a valuable use of my time as a faculty member. In addition, OU student Jessica Staley helped me track down information I needed to complete units 1, 6, and 7.

My family and friends—Art, Cees, Clarke, Cynthia, Dave, Doug, Helen, Karen, Lucelle, Marika, Robin, Scoops and Sidecar—have stayed interested in this project even when it seemed I would never finish it. I appreciate their love, pride, companionship, and support.

Supplements

Student Tools

Designed to help students maximize their learning experience in biology—we offer the following tools for students:

ARIS (aris.mhhe.com)

(*Assessment, Review, and Instruction System*) is an electronic study system that offers students a digital portal of knowledge. Students can readily access a variety of **digital learning objects** which include:

- Learning outcomes
- Chapter level quizzing with pre-test
- Answers to Mastering Concepts and end-of-chapter questions
- What's the Point? audio/video clips
- Tutorial Animations with quizzing
- ScienCentral Videos with quizzing
- Readings and References

Electronic Books

If you or your students are ready for an alternative version of the traditional textbook, McGraw-Hill and VitalSource have partnered to bring you innovative and inexpensive electronic textbooks. By purchasing E-books from McGraw-Hill & VitalSource, students can save as much as 50% on selected titles delivered on the most advanced E-book platform available, VitalSource Bookshelf.

E-books from McGraw-Hill & VitalSource are smart, interactive, searchable, and portable. VitalSource Bookshelf comes with a powerful suite of built-in tools that allow detailed searching, highlighting, note taking, and student-to-student or instructor-to-student note sharing. In addition, the media-rich E-book for *Biology: Concepts and Investigations* integrates relevant animations and videos into the textbook content for a true multimedia learning experience. E-books from McGraw-Hill & VitalSource will help students study smarter and quickly find the information they need. And they will save money. Contact your McGraw-Hill sales representative to discuss E-book packaging options.

How to Study Science

ISBN (13) 978-0-07-234693-0
ISBN (10) 0-07-234693-0

This workbook offers students helpful suggestions for meeting the considerable challenges of a science course. It gives practical advice on such topics as how to take notes, how to get the most out of laboratories, and how to overcome science anxiety.

Photo Atlas for General Biology

ISBN (13) 978-0-07-284610-2
ISBN (10) 0-07-284610-0

This atlas was developed to support our numerous general biology titles. It can be used as a supplement for a general biology lecture of laboratory course.

Supplements

Instructor Tools

Dedicated to providing high quality and effective supplements for instructors, the following Instructor supplements were developed for *Biology: Concepts and Investigations*.

Course Management

ARIS (*Assessment, Review, and Instruction System*) aris.mhhe.com is an electronic homework and course management system designed for greater flexibility, power, and ease of use than any other system. Whether you are looking for a preplanned course or one you can customize to fit your course needs, ARIS is your solution. In addition to having access to all student digital learning objects, ARIS offers the following resources for instructors:

- Instructor's Manual
- Virtual Labs
- Real People Doing Real Science
- Forensic Science
- Fostering Active Learning

ARIS also allows instructors to build assignments, track student progress, and share course materials with colleagues. The fully integrated grade book can also be downloaded to Excel, WebCT, or Blackboard.

Presentation Tools

Presentation Center The Presentation Center is an online digital library containing assets such as artwork, photos, animations, PowerPoints, and other media types that can be used to create customized lectures, visually enhanced tests and quizzes, compelling course websites, or attractive printed support materials. The dynamic search engine allows you to explore by discipline, course, textbook, chapter, asset type, or keyword. Simply browse, select, and download the files you need to build engaging course materials.

ScienCentral Videos McGraw-Hill has teamed up with ScienCentral, Inc. to provide brief biology new videos for use in lecture or for student study and assessment purposes. A complete set of ScienCentral videos are located within this text's ARIS Presentation Center site where each video includes a learning objective and quiz questions. These active learning tools enhance a biology course by engaging students in real life issues and applications such as developing new cancer treatments and understanding how methamphetamine damages the brain. ScienCentral, Inc., funded in part by grants from the National Science Foundation, produces science and technology content for television, video, and the web.

McGraw-Hill: Biology Digitized Video Clips

ISBN (13) 978-0-312155-0
ISBN (10) 0-07-312155-X

McGraw-Hill is pleased to offer adopting instructors an outstanding presentation tool—digitized biology video clips on DVD! License from some of the highest-quality science video producers in the world, these brief segments range from about 5 seconds to just under 3 minutes in length and cover all areas of general biology from cells to ecosystems. Engaging and informative, McGraw-Hill's digitized videos will help capture students' interest while illustrating key biological concepts and processes such as mitosis, how cilia and flagella work, and how some plants have evolved into carnivores.

Transparencies A set of acetate transparencies can be customized for your course. Please contact your McGraw-Hill sales representative for details.

Assessment

Test Bank A digital test bank that uses EZ Test software to quickly create customized exams is available for this text. This user-friendly program allows instructors to search for questions by topic or format, edit existing questions or add new ones; and scramble questions for multiple versions of the same test. Word files of the test bank questions are provided for those instructors who prefer to work outside the test-generator software.

Student Response System Wireless technology brings interactivity into the classroom or lecture hall. Instructors and students receive immediate feedback through wireless response pads that are easy to use and engage students. This system can be used to instructors to take attendance, administer quizzes and tests, create a lecture with intermittent questions, manage lectures and student comprehension through the use of the gradebook, and integrate interactivity into their PowerPoint presentations.

Contents

UNIT 1

The Cellular Basis of Life

CHAPTER 6 How Cells Release Energy 118

UNIT 2 *The Molecular Basis of Life*

CHAPTER 7 DNA Structure and Replication 136

CHAPTER 8 The Cell Cycle 154

CHAPTER 9 Sexual Reproduction and Meiosis 174

CHAPTER 10 Patterns of Inheritance 194

CHAPTER 11 Chromosomes and Human Inheritance Patterns 214

CHAPTER 12 Gene Function, Gene Regulation, and Biotechnology 234

UNIT 3 The Evolution of Life

CHAPTER 15 Evidence of Evolution 306

CHAPTER 16 The Origin and History of Life 330

UNIT 4

The Diversity of Life

CHAPTER 17 Viruses 354

CHAPTER 22 Invertebrate Animals 446

CHAPTER 23 Vertebrate Animals 476

UNIT 5 Plant Life

CHAPTER 24 Plant Form and Function 498

UNIT 6 *Animal Life*

UNIT 7 The Ecology of Life

BIOLOGY

UNIT 1

CHAPTER 1

The Scientific Study of Life

DNA technology has revolutionized biology.

What's the Point?

Biology Is Everywhere

Welcome to biology, the scientific study of life. Biology is everywhere—you are alive, and so are your friends, your pets, and the plants in your home and yard. Innumerable other organisms thrive on and in your body. The food you have eaten today was (until recently, anyway) alive. And the news is full of biology-related stories about newly discovered fossils, weight loss, cancer prevention, genetics, global climate change, and the environment.

Stories such as these enjoy frequent media coverage because this is an exciting time to study biology. Not only is the field changing rapidly, but its new discoveries and applications might change your life. DNA technology has brought us genetically engineered bacteria that can manufacture life-saving drugs—and genetically engineered plants that produce their own pesticides. The same technology may one day enable physicians to routinely cure hemophilia, cystic fibrosis, and other genetic diseases by replacing faulty DNA with a functional "patch."

The ability to sequence DNA has led to a wealth of new information. In a field of biology called genomics, scientists compare the DNA of many species, from bacteria to pine trees to humans. Genomics is yielding unprecedented insight into everything from the history of life to the function of individual disease-causing genes. And the biotechnology revolution doesn't stop with genomics. Each of your cells has the same DNA sequence, but cells don't all use that genetic information in the same way. DNA is organized into genes, each of which acts as a recipe for a specific protein. Thanks to different combinations of genes that are turned "on" and "off," each cell type has a specialized function. The study of the proteins that a cell produces has blossomed into yet another new field of biology: proteomics. Proteomics may someday save your life. Once we discover which proteins make breast cancer cells different from their healthy neighbors, perhaps we can find a better treatment or even a cure. Likewise, improved vaccines or antibiotics may follow from studies of the proteins that disease-causing microbes produce.

DNA technology, genomics, and proteomics are but a small part of modern biology. This book will bring you a taste of what we know about life and help you make sense of the science-related news you see every day. Chapter 1 begins your journey by introducing the scope of biology and explaining how science teaches us what we know about life.

LEARNING OUTLINE

1.1 What Is Life?
- A. Life Is Organized
- B. Life Requires Energy
- C. Life Maintains Internal Constancy
- D. Life Reproduces Itself, Grows, and Develops
- E. Life Evolves

1.2 A Taxonomic Hierarchy Describes Life's Diversity
- A. The Classification System Is Based on Shared Features
- B. Domains and Kingdoms Are the Most Inclusive Levels

1.3 Scientists Study the Natural World
- A. The Scientific Method Has Multiple Interrelated Steps
- B. An Experimental Design Is a Careful Plan
- C. Theories Are Comprehensive Explanations
- D. Scientific Inquiry Has Limitations

1.4 Investigating Life: Digital Organisms Mimic Evolution

1.1 What Is Life?

Biology is the scientific study of life. The second half of this chapter explores the meaning of the term *scientific,* but first we will consider the question, "What is life?" We all have an intuitive sense of what life is. If we see a rabbit on a rock, we know that the rabbit is alive and the rock is not. But it is difficult to state just what makes the rabbit alive. Likewise, in the instant after an individual dies, we may wonder what invisible essence has transformed the living into the dead.

The definition of life has stumped thinkers in many fields for centuries. Eighteenth-century French physician Marie François Xavier Bichat poetically but imprecisely defined life as "the ensemble of functions that resist death." Others less eloquent and still no more precise have suggested that life is a kind of "black box" that endows a group of associated biochemicals with the qualities of life. (Biochemicals are the chemicals that make up organisms; some, such as water, are also widespread in the nonliving world.) But "black box" thinking is not scientific; science seeks concrete explanations. Some have tried to break organisms down into the smallest parts that still exhibit the characteristics of life and then identify what makes those units different from their nonliving components. Researchers probe the genetic instructions—the genomes—of the simplest organisms to identify the minimal requirements for life.

One way to define life, then, is to list its basic components. The **cell** is the basic unit of life; every organism consists of one or more cells. Every cell has an outer membrane that separates it from its surroundings. Inside the membrane of every cell are water and other chemicals that carry out the cell's functions. One of those biochemicals, deoxyribonucleic acid (DNA), is the informational molecule of life (**figure 1.1**). Cells use genetic instructions—as encoded in DNA—to produce proteins, which enable them to specialize and to function in tissues, organs, and organ systems. A list of life's biochemicals, however, provides an unsatisfying definition of life. After all, placing DNA, water, proteins, and a membrane in a test tube does not create artificial life. And a crushed insect still contains all of the biochemicals it had immediately before it died.

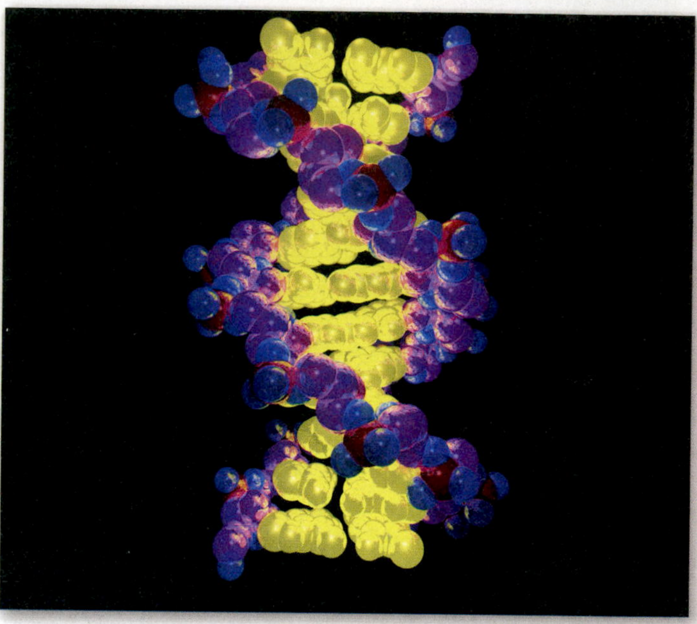

FIGURE 1.1 DNA: Informational Molecule of Life. All cells contain deoxyribonucleic acid, DNA, a series of "recipes" for proteins that each cell can make.

A concise definition continues to elude us, but scientists have settled on five qualities that, in combination, constitute life (**table 1.1**). An organism is a collection of structures that function together and exhibit these qualities.

A. Life Is Organized

Living matter consists of parts organized in a particular three-dimensional relationship. This organization often follows a hierarchical pattern of structures within structures within structures (**figure 1.2**). Bichat observed this organization

Table 1.1 *Characteristics of Life*

Characteristic	Example
Organization	Atoms make up molecules, which make up organelles, which occur inside cells, which make up tissues, and so on
Energy use	A kitten uses the energy from its mother's milk to fuel its own growth
Maintenance of internal constancy	Your kidneys regulate your body's water balance by increasing or decreasing the concentration of your urine
Reproduction, growth, and development	An acorn germinates, develops into an oak seedling, and, at maturity, reproduces sexually to produce its own acorns
Evolution	Increasing numbers of bacteria survive treatment with antibiotic drugs

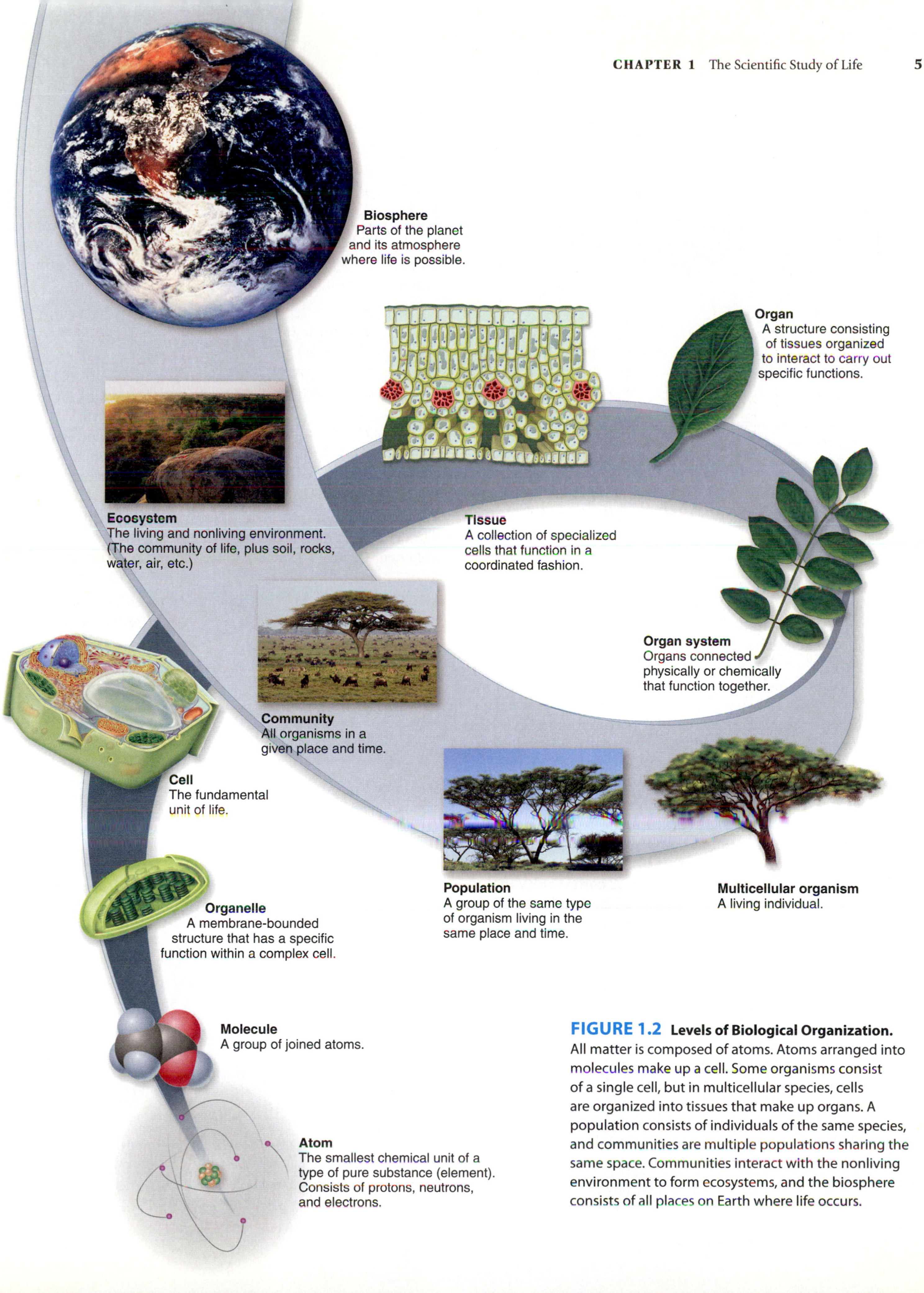

Biosphere
Parts of the planet and its atmosphere where life is possible.

Organ
A structure consisting of tissues organized to interact to carry out specific functions.

Ecosystem
The living and nonliving environment. (The community of life, plus soil, rocks, water, air, etc.)

Tissue
A collection of specialized cells that function in a coordinated fashion.

Organ system
Organs connected physically or chemically that function together.

Community
All organisms in a given place and time.

Cell
The fundamental unit of life.

Population
A group of the same type of organism living in the same place and time.

Multicellular organism
A living individual.

Organelle
A membrane-bounded structure that has a specific function within a complex cell.

Molecule
A group of joined atoms.

Atom
The smallest chemical unit of a type of pure substance (element). Consists of protons, neutrons, and electrons.

FIGURE 1.2 Levels of Biological Organization. All matter is composed of atoms. Atoms arranged into molecules make up a cell. Some organisms consist of a single cell, but in multicellular species, cells are organized into tissues that make up organs. A population consists of individuals of the same species, and communities are multiple populations sharing the same space. Communities interact with the nonliving environment to form ecosystems, and the biosphere consists of all places on Earth where life occurs.

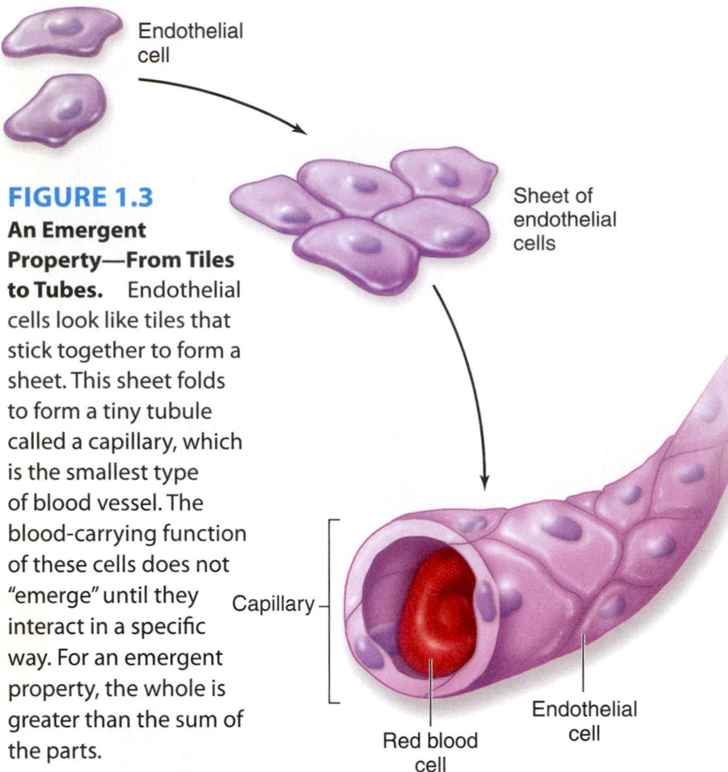

FIGURE 1.3 An Emergent Property—From Tiles to Tubes. Endothelial cells look like tiles that stick together to form a sheet. This sheet folds to form a tiny tubule called a capillary, which is the smallest type of blood vessel. The blood-carrying function of these cells does not "emerge" until they interact in a specific way. For an emergent property, the whole is greater than the sum of the parts.

in the human body. During the bloody French Revolution, as he performed autopsies, Bichat noticed that the body's heart, lungs, and other specialized **organs** were sometimes linked to form collections, or **organ systems,** but were also composed of simpler parts, which he named **tissues** (from the French word for "very thin"). Bichat may also have seen, with the aid of a microscope, that tissues themselves are composed of even smaller units, the cells. Some cells have compartments called **organelles** that carry out specialized functions. Ultimately, all living structures are composed of **molecules**, which are themselves made of particles called **atoms**.

We can also view organization in the living world beyond the level of the individual. A **population** includes two or more members of the same type of organism (species) living in the same place at the same time. A **community** includes the populations of different species in a region, and an **ecosystem** includes both the living and nonliving components of an area. Finally, the **biosphere** refers to all parts of the planet that can support life (see figure 1.2).

Biological organization is apparent in all life. Humans, eels, and evergreens, although outwardly very different, are all organized into specialized cells, tissues, organs, and organ systems. Bacteria, although less complex than animal or plant cells, are nevertheless highly organized.

An organism, however, is more than a collection of successively smaller parts. When those components interact, they create new, complex functions called **emergent properties** (**figure 1.3**). These characteristics are not magical. They arise from physical and chemical interactions among a system's components, much like flour, sugar, butter, and chocolate can become brownies—something not evident from the parts themselves.

Emergent properties explain why at all levels, and in all organisms, structural organization is closely tied to function. Disrupt a structure, and function ceases. Shaking a fertilized hen's egg, for instance, disturbs critical interactions and stops the embryo within from developing. Likewise, if a function is interrupted, the corresponding structure eventually breaks down. Unused muscles, for example, begin to atrophy (waste away). Biological function and form are interdependent.

B. Life Requires Energy

A constant stream of energy is needed to maintain the organized array of biochemicals that sustain life. Every organism must acquire and use energy to build new structures, repair old ones, and reproduce. This energy comes from the environment.

Biologists divide organisms into broad categories, based on their source of energy (**figure 1.4**). **Producers,** also called autotrophs, extract energy from the nonliving environment. The most familiar examples are the plants and microbes that capture light energy from the sun, but some bacteria can derive chemical energy from rocks. **Consumers** (also called heterotrophs),

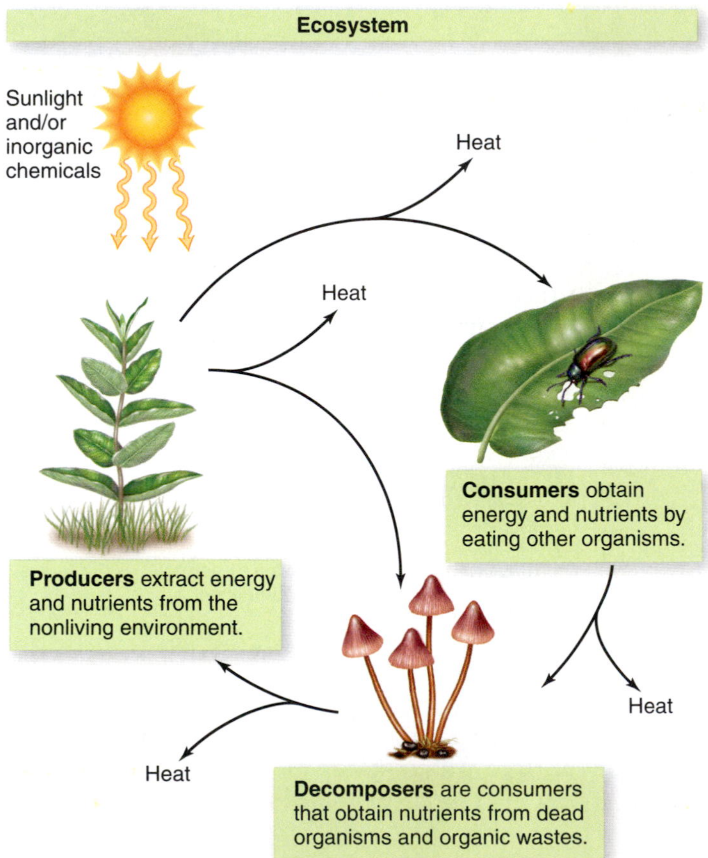

FIGURE 1.4 Life Is Connected. Organisms extract energy and nutrients from the nonliving environment or from other organisms. Decomposers recycle nutrients back to the nonliving environment. At every stage along the way, heat is lost to the system.

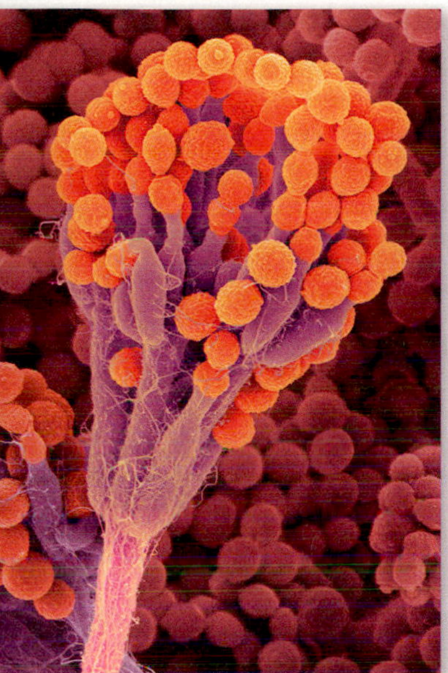

a. b. c.

FIGURE 1.5 Asexual and Sexual Reproduction. (**A**) This fungus, *Pencillium*, asexually produces identical cells on brushlike structures. Both a tree seedling (**B**) and a newborn mammal (**C**) are products of sexual reproduction.

in contrast, obtain energy by eating nutrients that make up other organisms. You are a consumer, using energy you extract from food to build your own molecules, move your muscles, send nerve signals, and maintain your body temperature. **Decomposers** are consumers that obtain energy from wastes or dead organisms. Fungi and many bacteria are decomposers.

Energy requirements link life into chains and webs of "who eats whom," beginning with producers and continuing through several levels of consumers and decomposers. But energy transfers are never 100% efficient; some energy is always lost in the form of heat. Because no organism can use it as an energy source, heat represents a permanent loss from the cycle of life. All ecosystems therefore depend on the continuous replacement of lost energy by an outside source—usually the sun.

C. Life Maintains Internal Constancy

An important characteristic of life is the ability to sense and react to stimuli. The conditions inside cells must remain within a constant range, even in the face of drastic changes in the surrounding environment. For example, to stay alive, a cell must maintain a certain water balance—not too much, and not too little. It must also take in nutrients, excrete wastes, and regulate its many chemical reactions to prevent deficiencies or excesses of essential substances. **Homeostasis** is this state of internal constancy, or equilibrium.

The most familiar instances of homeostasis are at the whole-organism level. Your body, for example, has several mechanisms that maintain your internal temperature at about 37 °C (98.6 °F). When you go outside on a cold day, you may begin to shiver; heat from these involuntary muscle movements warms the body. In severe cold, your lips and fingertips may turn blue as your circulatory system diverts blood away from your body's surface. Conversely, on a hot day, the evaporation of sweat from your skin helps cool your body.

D. Life Reproduces Itself, Grows, and Develops

Organisms reproduce, making other individuals like themselves. Reproduction transmits DNA from generation to generation, defining the characteristics of the offspring. The new individuals then grow and develop to maturity, when they can reproduce themselves.

Reproduction occurs in two basic ways: asexually and sexually. In **asexual reproduction,** an organism produces new individuals that are virtually identical to it. Unicellular (single-celled) organisms such as bacteria reproduce asexually by doubling and then dividing the contents of the cell. Some multicellular (many-celled) organisms also reproduce asexually. For example, a potato growing on an underground stem can sprout leaves and roots that form a new plant identical to the parent. The green, white, or black "powder" on moldy bread or cheese is made of the countless asexual spores of fungi (**figure 1.5A**). Even some animals, including sponges, reproduce asexually when a fragment of the parent animal detaches and develops into a new individual.

In **sexual reproduction,** genetic material from two individuals unites to form a third individual, which has a new combination of inherited traits. By mixing genetic traits at each generation, sexual reproduction results in tremendous diversity in a population. Sexual reproduction is a very successful strategy, especially in a changeable environment, and it is widespread among plants and animals (figure 1.5B, C). Many organisms can reproduce either asexually or sexually, depending on the conditions.

E. Life Evolves

One of the most intriguing ideas in biology is how organisms can seem so "perfectly" suited to their environments. Some organisms have color patterns that enable them to literally fade into the background (**figure 1.6**). Microorganisms that live in deep-sea hydrothermal vents have "extremozymes," proteins that permit life at very high temperatures. Tubular flowers have exactly the right shapes for the beaks of their hummingbird pollinators. These examples illustrate adaptations. An **adaptation** is an inherited characteristic or behavior that enables an organism to survive and reproduce successfully in a given environment. Where do these adaptive traits come from?

a.

b.

FIGURE 1.6 Blending In. (**A**) The superb camouflage of the adder snake, *Bitis peringueyi,* makes it virtually undetectable buried in the sand in the Namib Desert, Namibia. (**B**) It is little wonder that the sand lizard, *Aporosaura anchietae,* soon became the meal of the snake.

The answer lies in **natural selection**. The simplest way to think of natural selection is to consider two facts. First, because resources such as food and habitat are limited, populations produce many more offspring than will survive to reproduce. A single mature oak tree may produce thousands of acorns in one season, but only a few of those are likely to germinate, develop, and reproduce. The rest die. Second, no organism is exactly the same as any other. Sexual reproduction generates much more genetic variability among offspring than does asexual reproduction, but genetic mutations—changes in an organism's DNA sequence—occur in all organisms, even asexual ones.

Of all the offspring in a population, which will survive long enough to reproduce? The answer is those with the best adaptations to the current environment. Conversely, those that are not as well-suited are more likely to die before reproducing. A good definition of **natural selection**, then, is the enhanced reproductive success of certain individuals from a population based on inherited characteristics (**figure 1.7**). Over time, individuals with the best combinations of genes survive and reproduce, while those with less suitable characteristics fail to do so. As individuals that have inherited particularly adaptive traits contribute more offspring, they come to make up more of the population.

But the environment is constantly being remodeled. Continents shift, sea levels rise and fall, climates warm and cool. What happens to a population when the selective forces that drive natural selection change? Only some organisms survive—those with the "best" traits in the *new* environment. Features that may once have been rare become more common as the reproductive success of individuals with those traits improves. Notice, however, that this outcome depends on variability within the population. If no individual can reproduce in the new environment, the species may go extinct.

Natural selection is one mechanism of **evolution**, which is a change in the genetic makeup of a population. Although evolution also can occur in other ways, natural selection is the mechanism that selects for adaptations in a population. Charles Darwin achieved widespread fame in the 1860s after the publication of his book, *On the Origin of Species by Means of Natural Selection*, which outlined the theory of evolution by natural selection.

Evolution is the single most powerful idea in biology. As unit 3 describes in detail, evolution has been operating since life began, and it explains the current diversity of life. In fact, the chemical and cellular similarities among existing organisms strongly suggest that all species descend from a common ancestral type of organism. Evolution has molded the life that has populated the planet since the first cells formed almost 4 billion years ago, and it continues to act today.

As you review the characteristics of life, keep in mind that each of these traits may also occur in inanimate objects. A rock crystal is highly organized, but it is not alive. A fork placed in a pot of boiling water absorbs heat energy and passes it to the hand that grabs it, but this does not make the fork alive. A computer responds to stimuli, as does a self-flushing

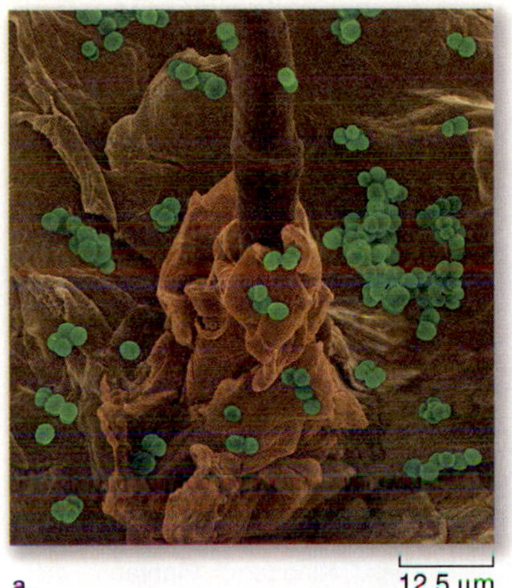

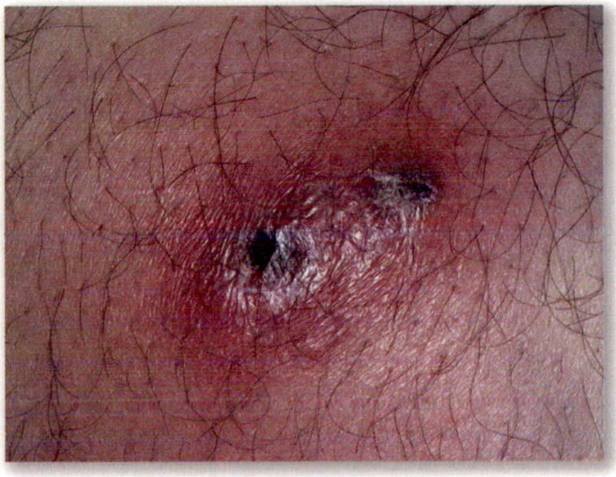

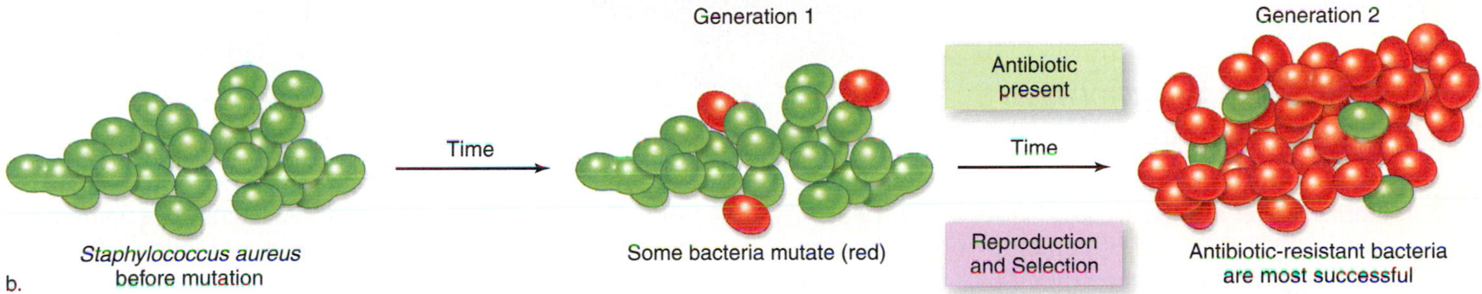

FIGURE 1.7 Natural Selection. (**A**) *Staphylococcus aureus* is a bacterium that causes skin infections. (**B**) By chance, some *S. aureus* bacteria are resistant to the antibiotic methicillin. The presence of the antibiotic increases the reproductive success of the resistant cells, which pass this trait to the next generation.

toilet, but neither is alive. A fire can "reproduce" very rapidly, but it lacks most of the other characteristics of life.

Summary *A distinctive set of characteristics distinguishes the living from the nonliving. All organisms are composed of one or more highly organized cells. Life also acquires energy, maintains internal constancy, grows and reproduces, and evolves. Life is always changing.*

1.1 MASTERING CONCEPTS

1. Why is it difficult to define life?
2. What are some of the ways people have tried to define life?
3. What combination of characteristics distinguishes the living from the nonliving?
4. How do natural selection and mutation guide evolution?

1.2 A Taxonomic Hierarchy Describes Life's Diversity

We humans love to classify things. We organize large stores by departments, sort our laundry according to color, and assign elementary-school students to grades and classrooms based on their ages, abilities, and interests. Similarly, the biological science of **taxonomy** classifies life.

A. The Classification System Is Based on Shared Features

In a taxonomic hierarchy, organisms are classified into progressively smaller groups based on common features. Just as a student is assigned to a school district, school, grade, and classroom, each organism is assigned to a **domain, kingdom, phylum, class, order, family, genus,** and **species** (**figure 1.8**). The most restrictive group, or taxon, is the species; it designates distinctive "types" of organisms, often defined by the ability to breed only among themselves and to produce fertile offspring. Each species has a unique name that consists of two descriptive words: its genus name and its species name. A human, for example, is *Homo sapiens.*

The more features two organisms share, the more taxonomic levels they share. A human, squid, and fly are all members of the animal kingdom, but they are clearly very different from one other, and they do not share lower taxonomic levels (phylum to species). A human, rat, and pig are more closely related—all belong to the same kingdom, phylum, and class (Mammalia). A

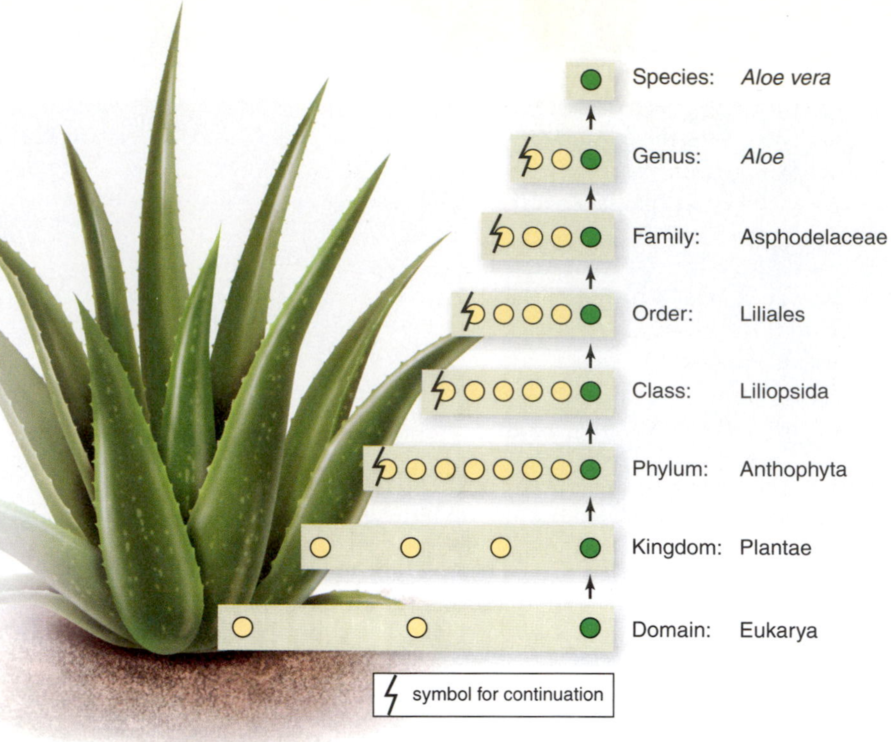

Species: *Aloe vera*
Genus: *Aloe*
Family: Asphodelaceae
Order: Liliales
Class: Liliopsida
Phylum: Anthophyta
Kingdom: Plantae
Domain: Eukarya
symbol for continuation

FIGURE 1.8 Taxonomic Hierarchy. Every species is classified based on a hierarchy in which similar species are grouped into genera, genera are grouped into families, and so on to the most inclusive category, domain. This diagram shows the complete classification for the plant *Aloe vera*.

human, orangutan, and chimpanzee are even more closely related, sharing the same kingdom, phylum, class, order, and family (Order Primates, Family Hominidae). As humans, our full classification is Eukarya-Animalia-Chordata-Mammalia-Primates-Hominidae-*Homo-Homo sapiens.*

B. Domains and Kingdoms Are the Most Inclusive Levels

The largest, or most inclusive, taxonomic categories are the three domains: Bacteria, Archaea, and Eukarya (**figure 1.9**). All eukaryotic cells have nuclei (membrane-bounded organelles that house DNA), and many members of Eukarya are multicellular. Members of Bacteria and Archaea are superficially similar to each other; all are single-celled organisms whose DNA is free in the cell, not confined to a nucleus. Archaeans, however, differ from members of both other domains in the chemistry of their cell walls, membranes, and some other structures. Nevertheless, they are somewhat similar to eukaryotes in the details of how they make proteins.

Domain Eukarya is further subdivided into four kingdoms: Protista, Plantae, Fungi, and Animalia (see figure 1.9). Except for Protista, organisms within a kingdom share the same general strategies for acquiring energy. For example, plants are autotrophs. Fungi and animals are consumers, although they differ from each other in the details of how they obtain food. Unlike the other three eukaryotic kingdoms, however, Protista contains a huge variety of unrelated species. Protista is a convenient but artificial "none of the above" category for the large number of eukaryotic organisms that are not plants, fungi, or animals.

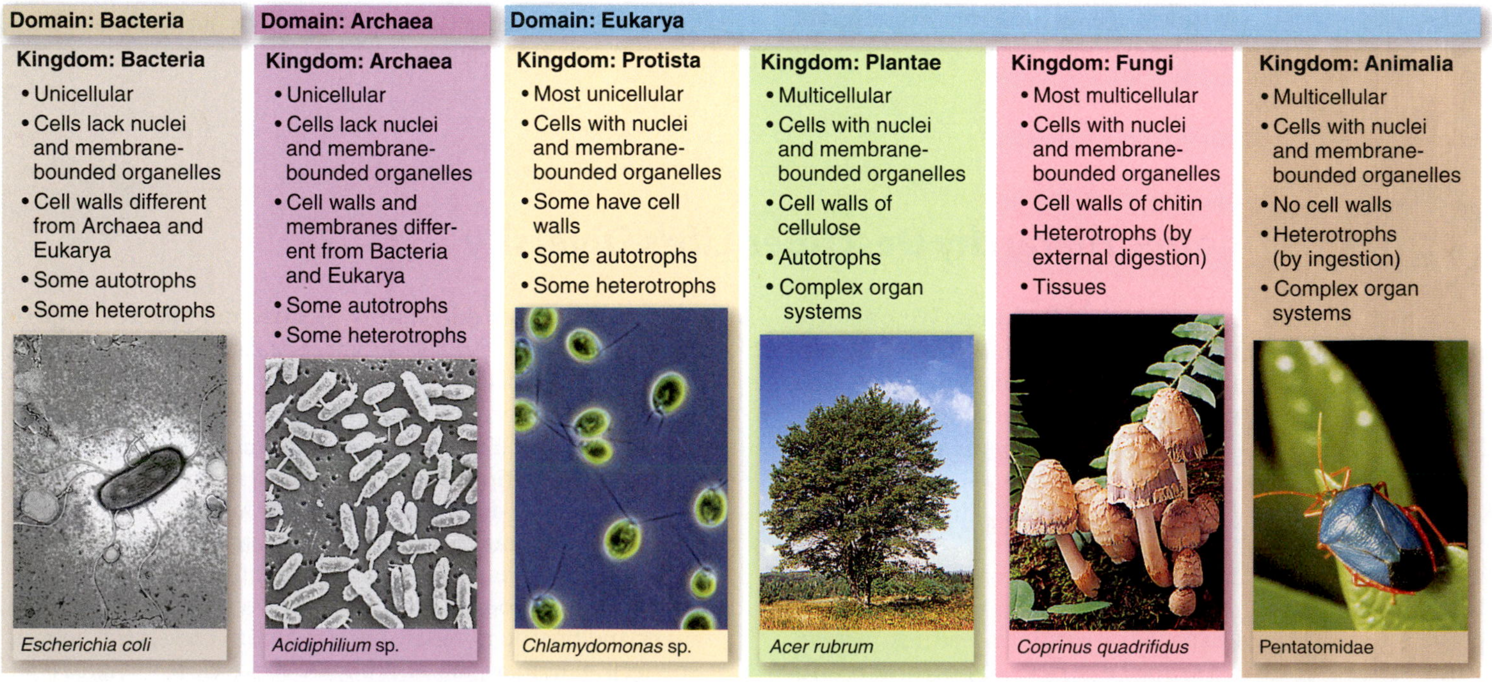

Domain: Bacteria	Domain: Archaea	Domain: Eukarya			
Kingdom: Bacteria	**Kingdom: Archaea**	**Kingdom: Protista**	**Kingdom: Plantae**	**Kingdom: Fungi**	**Kingdom: Animalia**
• Unicellular • Cells lack nuclei and membrane-bounded organelles • Cell walls different from Archaea and Eukarya • Some autotrophs • Some heterotrophs	• Unicellular • Cells lack nuclei and membrane-bounded organelles • Cell walls and membranes different from Bacteria and Eukarya • Some autotrophs • Some heterotrophs	• Most unicellular • Cells with nuclei and membrane-bounded organelles • Some have cell walls • Some autotrophs • Some heterotrophs	• Multicellular • Cells with nuclei and membrane-bounded organelles • Cell walls of cellulose • Autotrophs • Complex organ systems	• Most multicellular • Cells with nuclei and membrane-bounded organelles • Cell walls of chitin • Heterotrophs (by external digestion) • Tissues	• Multicellular • Cells with nuclei and membrane-bounded organelles • No cell walls • Heterotrophs (by ingestion) • Complex organ systems
Escherichia coli	*Acidiphilium* sp.	*Chlamydomonas* sp.	*Acer rubrum*	*Coprinus quadrifidus*	Pentatomidae

FIGURE 1.9 Three Domains, Six Kingdoms. These broad groupings are based on fundamental differences in cell structure, cell chemistry, and mode of nutrition.

Kingdom Protista notwithstanding, taxonomists strive to classify organisms according to what we know about the evolutionary relationships of organisms—that is, how recently one type of organism shared a common ancestor with another type of organism. The more recently they diverged from a shared ancestor, the more closely related we presume the two types of organisms to be. Researchers infer these relationships by comparing anatomical, behavioral, cellular, genetic, and biochemical characteristics to identify similarities. Because scientists are continuously collecting new data, classifications are always under review. For example, in the 1970s, taxonomists added the broadest category, the domain, following the recognition of Archaea. The advent of DNA and other molecular analyses since that time have led to the ongoing reclassification of many other types of organisms as well.

Summary ***Biologists use a taxonomic hierarchy from domain to species to classify life's diversity. Prokaryotic organisms occupy two domains, whereas the third domain contains all eukaryotes—protists, plants, fungi, and animals.***

1.2 MASTERING CONCEPTS

1. How are domains related to kingdoms?
2. What is the underlying basis of taxonomic groupings of organisms?

1.3 Scientists Study the Natural World

The idea of biology as a "rapidly changing field" may seem strange if you think of science as a collection of facts. After all, the parts of a frog are the same now as they were 50 or 100 years ago, right? But memorizing frog anatomy is not the same as thinking scientifically. Scientists use observations to ask questions about the natural world. For example, if you compare a frog to, say, a snake, can you determine how those animals are related? How can the frog live in both water and on land, and how does the snake survive in the desert? Knowledge of anatomy simply gives you the vocabulary you need to ask these and other interesting questions about life.

Biology is changing rapidly because new technology developed during the past few decades has expanded our ability to make observations. New microscopes allow us to spy on the inner workings of living cells, DNA sequencing machines are faster than ever, and powerful computers allow us to process huge amounts of data. Scientists can now answer questions about the natural world that previous generations could never have imagined.

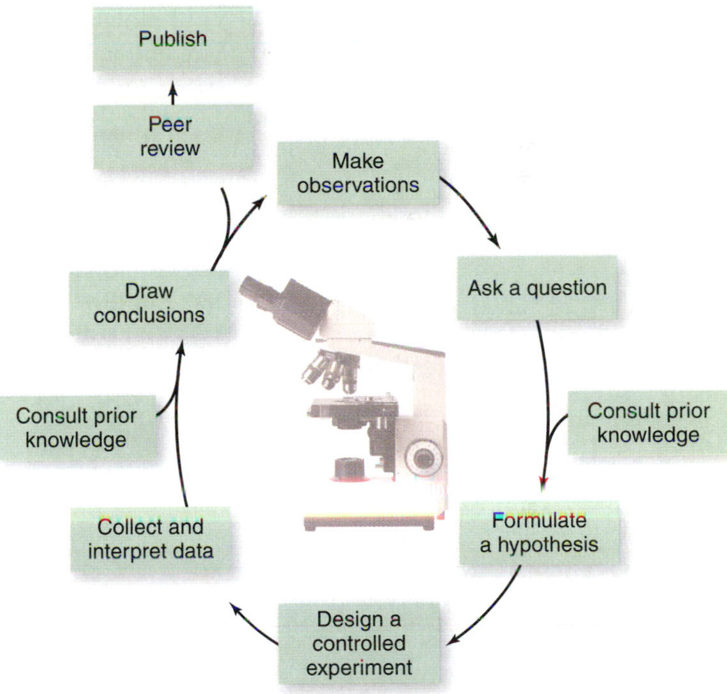

FIGURE 1.10 The Scientific Method. The power of the scientific method is its combined emphasis on logic and creativity. This figure is simplified because scientists often work on many of the "steps" simultaneously.

A. The Scientific Method Has Multiple Interrelated Steps

Scientific knowledge arises from application of the **scientific method,** which is a general way of organizing an investigation (**figure 1.10**). The scientific method is a framework in which to consider ideas and evidence in a repeatable way. Scientific inquiry consists of everyday activities—observing, questioning, reasoning, predicting, testing, interpreting, and concluding. It includes thinking, detective work, and noticing connections between seemingly unrelated events.

Observations

The scientific method begins with observations. These observations may be historical incidents or accidents, or they may be based on existing knowledge and experimental results. Often, a great leap in science happens when one person makes mental connections among previously unrelated observations. The idea of natural selection, for example, originated when Charles Darwin combined the study of geology with his detailed observations of different organisms. His understanding of Earth's long history and the variation he saw in life led him to the insight that organisms change over long periods.

Hypothesis

Observations lead to a **hypothesis**, a tentative explanation based on previous knowledge. The hypothesis is the essential "unit" of scientific inquiry. To be useful, the hypothesis must be testable—there must be a way to collect data that can support the hypothesis or prove it false. Interestingly, we cannot know something with absolute certainty. A scientist would not use the phrase "scientific proof," although advertisers commonly do. No hypothesis can be proven true with 100% certainty, because there is always the possibility of discovering additional information.

Experimentation and Data Collection

Many scientific investigations are based on discovery. A microbiologist finds a never-before-seen microorganism in a hot spring; an ecologist catalogs the species that survived severe weather; a cell biologist discovers a molecule that signals a cell to divide. Often, however, a scientist devises an **experiment** to test a hypothesis under controlled conditions.

Discovery and experimentation work hand in hand. For example, deciphering the sequence of DNA building blocks in the human genome—a multiyear project completed in 2000—is discovery. But demonstrating what individual genes do requires experiments. As another example, the discovery of the link between harmful chemicals and cancer began with noting what people have in common—that smokers are more likely than nonsmokers to develop lung cancer, for instance. Experiments on nonhuman animals and cells or tissues growing in culture then fill in the details of how the environmentally induced cancer develops.

Analysis and Peer Review

After collecting the data, the investigator reevaluates the hypothesis in light of the new information. Often, the most interesting results are those that are unexpected, because they force the scientist to rethink the hypothesis. The diagram in figure 1.10 shows this "feedback loop." Science advances as new information arises and explanations continue to improve.

Once a scientist has enough information to support or refute a hypothesis, he or she may write a paper and submit it for publication in a scientific journal. The journal's editors then send the paper to anonymous reviewers knowledgeable about the research. In a process called **peer review**, these scientists independently evaluate the validity of the methods, data, and conclusions. Peer review is not perfect. Some published papers are recalled or amended as unnoticed mistakes are later discovered. Overall, however, peer review ensures that published studies are high quality.

B. An Experimental Design Is a Careful Plan

One of the most important decisions that an investigator makes in designing an experiment is **sample size,** which is the number of individuals that he or she will study. In general, the larger the sample size, the more meaningful the results. But obtaining a large sample is not always practical or possible. When medical researchers study very rare disorders, only a few patients may be available. So they conduct small-scale experiments instead, which are valuable because they may indicate whether continuing research is likely to yield valid, meaningful results.

Variables

A systematic consideration of variables is important as well (**table 1.2**). A **variable** is a changeable element of an experiment, and there are several types. The investigator manipulates the levels of the **independent variable** to determine whether it influences some other phenomenon. For example, in an experiment designed to test whether a new headache remedy is effective, the independent variable would be the dose of the remedy. The simplest experiments test only one independent variable at a time. The **dependent variable** is the response that the investigator measures. In our test of the new remedy, the dependent variable would be the severity of headaches.

A **standardized variable** is anything that the investigator holds constant for all subjects in the experiment. For example, both dehydration and alcohol consumption can cause headaches. The best test of our new headache remedy would require that all subjects maintain similar hydration and abstain from alcohol. Holding standardized variables constant ensures the best chance of detecting the effect of the independent variable.

Controls

Distinguishing this effect, however, requires one more experimental component: a basis for comparison. Well-designed experiments compare a group of "normal" individuals or components to a group undergoing treatment. Ideally, the only difference between the normal group and the experimental group is the one factor being tested. The normal group is called an experimental **control** and provides a basis of comparison. Designing experiments to include controls helps ensure that a single independent variable causes the observed effect.

Table 1.2 *Types of Variables in an Experiment*

Type of Variable	Definition
Independent variable	What the investigator manipulates to determine whether it influences the phenomenon of interest
Dependent variable	What the investigator measures to determine whether the independent variable influenced the phenomenon of interest
Standardized variable	Any variable intentionally held constant for all subjects in an experiment, including the control group

Experimental controls may take several forms. Sometimes, the control group simply receives a "zero" value for the independent variable. If a gardener wants to test whether a new fertilizer improves her tomato yield, she may set up three treatment groups. Some plants would receive a high dose, some a low dose, and still others—the control plants—no fertilizer at all (**figure 1.11**). In other experiments, the control group might receive a **placebo**, an inert substance that resembles the treatment given to the experimental group. In medical research, a placebo is often a stand-in for a drug being tested—a sugar pill or a treatment already known to be effective.

Another safeguard used in medical experiments is a **double-blind** design, in which neither the researchers nor the participants know who received the substance being evaluated and who received the placebo. The researchers break the "code" of who received which treatment only after the data are tabulated or if one group does so well that it would be unethical to withhold treatment from the placebo group.

Statistical Analysis

Once the experiment is complete, the investigator compiles the data and decides whether the independent variable affected the dependent variable. Suppose the results of our tomato experiment were:

Fertilizer Application Level	Average Yield (tomatoes per plant)
High	20
Low	17
None (control)	15

Did fertilizer level influence tomato yield? Maybe. The only way to know for sure is to apply a statistical analysis, using a set of mathematical tools that help the researcher interpret the data. Researchers use many different statistical tests, all of which measure variation in the data. The high-fertilizer average of 20 tomatoes per plant, for example, could result from some of the plants producing only one tomato and others yielding 39—or all plants could have produced between 19 and 21 tomatoes. The less variation, the more likely that the fertilizer really is responsible for the difference between the treatments. The analysis considers both variation and sample size to yield a measure of **statistical significance**, which is the probability that the results arose purely by chance.

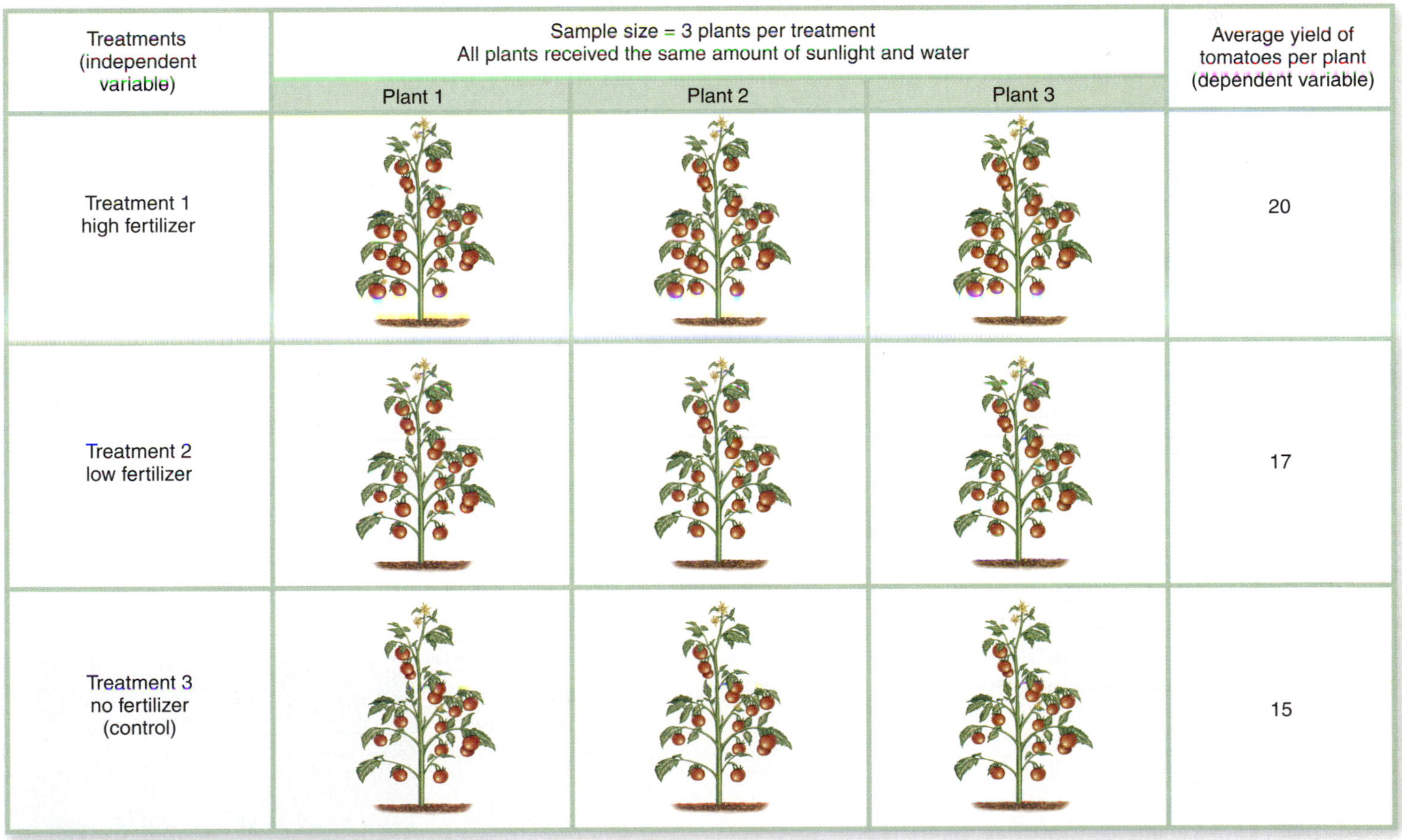

Treatments (independent variable)	Sample size = 3 plants per treatment All plants received the same amount of sunlight and water: Plant 1	Plant 2	Plant 3	Average yield of tomatoes per plant (dependent variable)
Treatment 1 high fertilizer				20
Treatment 2 low fertilizer				17
Treatment 3 no fertilizer (control)				15

FIGURE 1.11 Elements of an Experiment. A gardener testing whether fertilizer affects tomato yield might set up an experiment such as this. The amount of fertilizer is the independent variable, with control plants receiving none. Tomato yield is the dependent variable. Standardized variables might include the amount of sunlight and water, since the gardener knows these might also affect tomato yield.

Creativity and Logic in Experimental Design

Good experimental design requires both creativity and logic. Sometimes, the best experiments are very simple, prompting colleagues to wonder how they could have overlooked what seems obvious in retrospect. The creative inspiration may come at any stage of design: selecting the subjects, deciding on treatment levels and controls, or applying the treatments. Throughout the design process, sound logic ensures that the experiment is a convincing test of the hypothesis.

C. Theories Are Comprehensive Explanations

When asked to comment on the idea of biological evolution, former U.S. president Ronald Reagan dismissed its validity, saying, "It is a scientific theory only." In an informal sense, a theory may very well be little more than an opinion or a hunch. For instance, immediately after an airplane crash, experts often offer their "theories" about the disaster's cause. These tentative explanations are really hypotheses; in science, the word *theory* has a distinct meaning.

Like a hypothesis, a **theory** is an explanation for a natural phenomenon, but it is different in some ways. First, a theory is typically broader in scope than a hypothesis. For example, the germ theory—the idea that some microorganisms cause human disease—is the foundation for medical microbiology. Individual hypotheses relating to the germ theory are much narrower, such as the suggestion that human papilloma viruses cause cervical cancer. Not all theories are as "large" as the germ theory, but they generally encompass multiple hypotheses. Note also that the germ theory does not imply that *all* microbes cause disease, or that all diseases have microbial causes. But it does explain many types of human diseases.

A second difference between a hypothesis and a theory is acceptance. A hypothesis is tentative, whereas theories reflect broader agreement. This is not to imply that theories are not testable; in fact, the opposite is true. Every scientific theory is falsifiable, meaning that there must be a way to prove it wrong. The germ theory remains widely accepted because many observations support it, and no reliable tests have disproved it.

At some point, a theory is so widely accepted that people regard it as a fact. The line between theory and fact is fuzzy, but the late paleontologist Stephen Jay Gould articulated a useful difference: "In science 'fact' can only mean 'confirmed to such a degree that it would be perverse to withhold provisional consent.'" Although a theory can never be proven 100% true, some theories are so well-supported that no educated person questions its validity. Gravity, for example, is a fact.

Biologists also consider biological evolution to be a fact. Yet the phrase "theory of evolution" persists, because evolution is both a fact *and* a theory. Both terms apply equally well. The evidence for genetic change over time is so persuasive, and comes from so many different fields of study, that to deny its existence is unrealistic. Nevertheless, biologists do not understand everything about how evolution works. Many questions about life's history remain, but the debates swirl around *how*, not *whether*, evolution occurred.

One last quality of a scientific theory is its predictive power. A good theory not only ties together many existing observations, it also suggests predictions about phenomena that have yet to be observed. Charles Darwin, for example, used his theory of evolution by natural selection to predict the existence of a then-unknown pollinator for a plant whose flowers had a distinctive shape (**figure 1.12**). Only recently have scientists discovered the pollinator, a moth. A theory weakens if subsequent observations do not support its predictions.

Science is just one of many ways to know about the world, but its strength is its openness to new information. Theories change to accommodate new knowledge. The history of science is full of long-established ideas changing as we learned more about nature, often thanks to new technology. People thought that Earth was flat and at the center of the universe before inventions and data analysis revealed otherwise. Similarly, biologists thought all life was plant or animal until microscopes unveiled a world of organisms invisible to our eyes.

D. Scientific Inquiry Has Limitations

The generalized scientific method is neither foolproof nor always easy to implement. One problem is that experimental evidence may lead to multiple interpretations, and even the most carefully designed experiment can fail to provide a definitive answer. Consider the observation that animals fed large doses of vitamin E live longer than similar animals that do not ingest the vitamin. So, does vitamin E slow aging? Possibly, but excess vitamin E causes weight loss, and other experiments associate weight loss with longevity. Does vitamin E extend life, or does the weight loss? The experiment of feeding animals large doses of vitamin E does not distinguish between these possibilities. Can you think of further experiments to clarify whether vitamin E or weight loss extends life?

Another limitation is that researchers may misinterpret observations or experimental results. For example, centuries ago, scientists heated broth in a bottle, corked it shut, and observed bacteria in the broth a few days later. They concluded that life arose directly from the broth. The correct explanation, however, was that the cork did not keep airborne bacteria out. Science is self-correcting, in the sense that scientific thought is open to new data and new interpretations. But it is also fallible, especially in the short term. **spontaneous generation, p. 332**

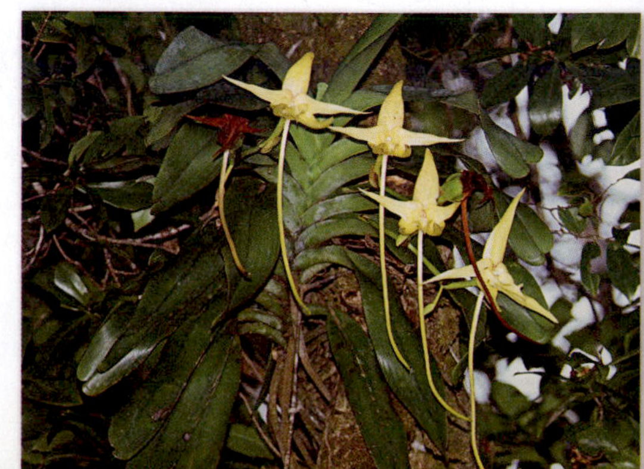

FIGURE 1.12
Prediction Confirmed. When Charles Darwin saw this orchid, he predicted that its pollinator would have mouthparts long and thin enough to reach the nectar deep within the flower.

Can *You* Relate?

Will This Hurt Me, or Won't It?

If you pay attention to health news, you probably hear reports now and then that some food previously thought to be healthy is actually bad for you—or vice versa. Eggs are good and caffeine is bad! No, they're both bad! No, they're both good! You may become impatient at these contradictory reports, thinking that scientific studies are not valid or that scientists are just out for publicity.

The reality is a bit more complex. Take, for example, the artificial sweetener saccharin (see Can You Relate: Sugar Substitutes and Fake Fats in chapter 2). Saccharin was discovered in 1879, and its popularity as a low-calorie sugar substitute peaked in the 1960s. But in 1977, the Food and Drug Administrated (FDA) proposed a ban on saccharin, based on a handful of studies suggesting that the sweetener caused bladder cancer in rats. Because few alternative artificial sweeteners were available at the time, however, Congress immediately placed a moratorium on the ban and instead required warning labels on products containing saccharin. In 1991, the FDA withdrew its proposed ban, and in 1998, the International Agency for Research on Cancer rated saccharin as "not classifiable as to its carcinogenicity to humans." Two years later, legislation removed the warning label requirement.

This tangled legislative history raises an important issue: Why can't science reply "yes" or "no" to the seemingly simple question of whether saccharin is bad for you? To understand the answer, consider one of the studies that prompted the FDA to propose the ban on saccharin in the first place. Researchers divided 200 rats into two groups. The control animals ate standard rodent chow, whereas the experimental group got the same food supplemented with saccharin. At reproductive maturity the animals were bred, and the researchers fed their offspring the same dose of saccharin throughout their lives as well. They measured the incidence of cancer in both generations of rats for 24 months or until the rats died, whichever came first. Here are the results:

	Rats with Tumors/Rats Examined (% with tumors)	
	First (parental) generation	Second (offspring) generation
Male rats		
Controls	1/36 (3%)	0/42 (0%)
Saccharin-fed	7/38 (19%)	12/45 (27%)
Female rats		
Controls	0/38 (0%)	0/47 (0%)
Saccharin-fed	0/40 (0%)	2/49 (4%)

(Data adapted from Office of Technology Assessment Report, October 1977, *Cancer Testing Technology and Saccharin*, page 52.)

At first glance, the conclusion seems inescapable: saccharin causes cancer in male lab rats. But closer study reveals several hidden complexities that make the data hard to interpret. First, the dose of saccharin was huge—5% of the rats' diets, for life. The equivalent dose in humans would require drinking hundreds of cans of saccharin-sweetened soda every day. In addition, the experimental rats weighed up to 20% less than the control rats by the end of the study, suggesting that such high doses of the sweetener are toxic. Rather than causing cancer directly, the saccharin may have simply weakened the animals and made them more susceptible to disease.

The researchers could have tested for that possibility by adding additional treatments with lower, less toxic saccharin concentrations. Then, if the sweetener really did cause cancer, they could have looked for a predicted "dose-response" relationship—low doses should yield just a few cases, and high doses should produce more. But the experiment was not designed to test for such a relationship. Furthermore, studies using mice, hamsters, and monkeys were inconclusive. The researchers who used these other animals created different designs for each experiment, so it is difficult to compare the results. The rat study, for example, followed two generations of animals; those with other animals used just one generation.

Given that animal studies did not yield uniform results, perhaps the scientists should have studied the saccharin-cancer connection in humans instead. Such research, however, is extremely difficult. It is obviously unethical to keep humans in captivity, control every facet of their environment and breeding, intentionally expose them to potentially harmful chemicals, and then kill and dissect them to check for tumors. The only way to approach the question in humans, therefore, would be to measure the incidence of cancer in saccharin users versus that in nonusers. But with so many other possible causes of cancer—smoking, poor diet, exposure to job-related chemicals, genetic predisposition—it is difficult to separate out just the effects of saccharin.

So what are we to make of these news reports on eggs, caffeine, chocolate, wine, and soy? It is hard to say, but one thing is certain: No matter what the newspaper headlines say, one study, especially a small one, cannot reveal the whole story. Good, bad, or neutral? The complexities of real-world science mean that in most cases, the jury is still out.

Burning Questions

Why am I here?

The Burning Questions featured in each chapter of this book came from my own students. On the first day of class, I always ask students to turn in a "Burning Question"—anything they have always wondered about biology. I save the cards and answer most of the questions as each relevant topic comes up during the semester.

Why not answer *all* of the questions? It is because at least one student often asks something like "Why am I here?" or "What is the meaning of life?" Such puzzles have fascinated humans throughout the ages, but they are among the many types of questions that we cannot approach scientifically. Biology can explain the mechanics of how you came to be—how you developed after a sperm from your father fertilized an egg cell from your mother. But no one can develop a testable hypothesis about life's meaning or the purpose of human existence. Science—and biology teachers—must therefore remain silent on such questions.

Other ways of knowing must therefore satisfy our curiosity about "why." Philosophers, for example, can help us see how others have considered these questions. Religion may also provide the meaning that many people seek. Part of the value of higher education (and your biology class) is to help you acquire the tools you need to find your own life's purpose.

Have a Burning Question of your own? Submit it to marielle_hoefnagels@mcgraw-hill.com for possible inclusion in future editions of this book!

A related problem is that the scientific community may be slow to accept new evidence that suggests unexpected conclusions. Every investigator should try to keep an open mind about observations, not allowing biases or expectations to cloud interpretation of the results. But it is human nature to be cautious in accepting an observation that does not fit what we think we know. The careful demonstration that life does not arise from broth surprised many people who believed that mice sprung from mud, flies from rotted beef, and beetles from cow dung. More recently, it took many years to dispel the common belief that stress causes ulcers. Today, we know that a bacterium (*Helicobacter pylori*) causes most ulcers.

Although science is a powerful tool for answering questions about the natural world, it cannot answer questions of beauty, morality, ethics, or religion. Nor can we directly study natural phenomena that occurred only long ago. For example, many experiments have attempted to re-create the sequence of chemical reactions that might have formed the chemicals that led to life on early Earth. Although the experiments produce interesting results and reveal ways that these early events may have occurred, we cannot know if they accurately re-create conditions at the beginning of life. **origin of life, p. 332**

Scientific research seeks to understand nature. Because humans are part of nature, we sometimes tend to view scientific research, and particularly biological research, as aimed at improving the human condition. But knowledge without any immediate application or payoff is valuable in and of itself—because we can never know when information will be useful.

Summary *The scientific method is a thinking process that guides our evaluation of what we observe. Curiosity, questioning, and systematic hypothesis testing enable scientists, and others, to learn how nature works.*

1.3 MASTERING CONCEPTS

1. What are the components of the scientific method?
2. How is scientific inquiry a continuous process?
3. What factors determine whether an experiment will provide meaningful information?
4. What is a scientific theory?
5. What are some limitations of scientific inquiry and experimentation?

1.4 INVESTIGATING LIFE
Digital Organisms Mimic Evolution

Each chapter of this book ends with a section that examines how biologists use systematic, scientific observations to solve a different evolutionary puzzle from life's long history. This first installment of "Investigating Life" fast-forwards to a time and a technology that Charles Darwin could not have imagined when he published *On the Origin of Species* in 1859. He knew of the countless tangible clues—including fossils of ancient life forms—that reveal evolution. Recently, however, computer scientists have jumped into the fray, demonstrating that the principles behind the evolution of organisms operate even in a virtual world.

Avida is one type of software useful in the study of artificial evolution. Avida is a sort of "digital world" inhabited by simple computer programs. These programs lack DNA, but like real organisms, they compete for energy, mutate randomly, and duplicate in their virtual world. Avida users employ the software to answer evolutionary questions. For example, many people wonder how evolution can produce complexity. If life started small and simple, how could mutations—which are usually but not always harmful—lead to improved function and increasing complexity? How could the descendants of bacteria have evolved into pine trees or parrots? How could the human eye have evolved from the simple light-detecting organs in simpler animals?

Michigan State University biologist Richard Lenski and his colleagues used Avida to tackle the origin of complexity. They began with 3600 identical copies of a simple "ancestral" computer program that was self-replicating, like a computer virus. Of the 50 instructions in the original program, 15 were required for self-replication. The other 35 were identical and performed no function when executed. Every time a program copied itself, it had a set probability of making a mistake. Instructions could also be inserted or deleted at random. All of the programs then competed for the "energy" (in the form of digital "permission slips") needed to execute the instructions. Avida awarded the most energy to the programs with the most instructions; additional energy went to programs that could perform increasingly complex logic operations. The more energy a program acquired, the more copies of itself would appear in the next generation (**figure 1.13**).

The ancestral programs could not perform any logic functions. As the programs replicated, only those descendants that evolved complex functions were rewarded with extra energy. After thousands of generations, Lenski and colleagues found that random changes in the programs' instructions, coupled with competition for computational "energy," eventually produced some programs that could perform very complex operations. Furthermore, not every program took the same path to complexity. This was an interesting parallel to biological evolution, which often produces similar adaptations in distantly related organisms—such as the wings of birds, butterflies, and bats.

A computer program is not life. It is not made of cells, and it does not maintain homeostasis. Although Avida is an extremely simplified simulation of life's evolutionary processes, the digital "organisms" did use energy, replicate, mutate, and undergo natural selection. Avida's digital landscape is a model that enables scientists to test assumptions and treatments over many generations in a very short time. The results yield helpful clues that suggest further studies of "real life" evolution. As computer processing power continues to increase, Avida and similar computer programs may reflect biology with even greater accuracy.

Lenski, Richard E., Charles Ofria, Robert T. Pennock, and Christoph Adami. May 8, 2003. The evolutionary origin of complex features. *Nature*, vol. 423, pages 139–144.

FIGURE 1.13 Avida. In this image generated by Avida, each color represents a different digital "organism." **Question:** How is Avida's digital evolution similar to and different from biological evolution?

CHAPTER SUMMARY

1.1 What Is life?

- A living organism is distinguished from an inanimate object by a combination of characteristics.

A. Life Is Organized

- An organism is organized as structures of increasing size and complexity, from **atoms** to **molecules**, to **organelles**, to **cells,** to **tissues, organs** and **organ systems,** to individuals, **populations, communities, ecosystems,** and the **biosphere.**
- **Emergent properties** arise as the level of organization of life increases. They result from interactions between smaller parts.

B. Life Requires Energy

- Life requires energy to maintain its organization and functions. **Producers** extract energy from the nonliving environment; **consumers** eat organic energy sources. **Decomposers** recycle nutrients to the nonliving environment.
- Because of heat losses, all ecosystems require constant energy input from an outside source.

C. Life Maintains Internal Constancy

- Organisms must maintain **homeostasis**, an internal state of constancy in changing environmental conditions.

D. Life Reproduces Itself, Grows, and Develops

- Organisms reproduce asexually, sexually, or both. **Asexual reproduction** yields virtually identical copies, whereas **sexual reproduction** generates genetic diversity.

E. Life Evolves

- **Natural selection** eliminates inherited traits that decrease the chance of survival and reproduction. The result of natural selection is **adaptations**, features that enhance reproductive success.
- **Evolution** through natural selection explains why organisms are alike yet diverse and how common ancestry unites all species.

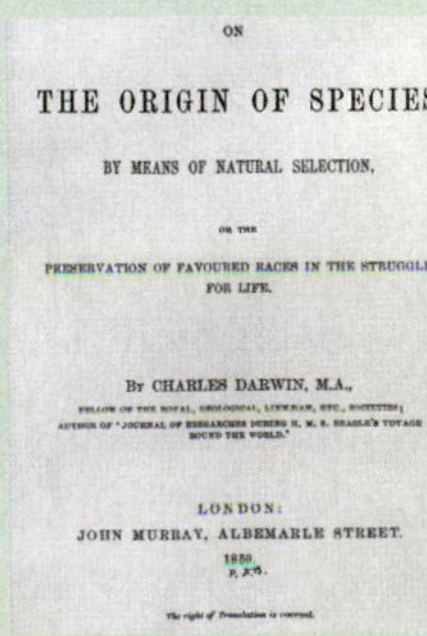
ON

THE ORIGIN OF SPECIES

BY MEANS OF NATURAL SELECTION,

OR THE

PRESERVATION OF FAVOURED RACES IN THE STRUGGLE FOR LIFE.

By CHARLES DARWIN, M.A.,

LONDON:

JOHN MURRAY, ALBEMARLE STREET.

1859.

1.2 A Taxonomic Hierarchy Describes Life's Diversity

A. The Classification System Is Based on Shared Features

- **Taxonomy** is the science of classification. Biologists classify organisms according to probable evolutionary relationships.
- The taxonomic levels are **domain**, **kingdom**, **phylum**, **class**, **order**, **family**, **genus**, and **species**.

B. Domains and Kingdoms Are the Most Inclusive Levels

- The three domains of life are distinguished by cell structure and organization. Mode of nutrition and other features distinguish members of kingdoms.

1.3 Scientists Study the Natural World

A. The Scientific Method Has Multiple Interrelated Steps

- Scientific inquiry, which uses the **scientific method,** is a way of thinking that involves observing, questioning, reasoning, predicting, testing, interpreting, concluding, and posing further questions.
- Scientific inquiry begins when a scientist makes an observation, raises questions about it, and uses reason to construct an explanation, or **hypothesis**.
- **Experiments** test the validity of the hypothesis, and conclusions are based on data analysis. **Peer review** ensures that published studies meet high-quality standards.

B. An Experimental Design Is a Careful Plan

- The larger the **sample size**, the more credible the results of an experiment.
- Experimental **controls** ensure that results reflect the effect of the **independent variable** in an experiment, and not other factors. The **dependent variable** in an experiment is the measurement that the investigator makes. **Standardized variables** are held constant for all subjects in an experiment.
- **Placebo**-controlled, **double-blind** experiments minimize bias.
- Experimental results are **statistically significant** if they are unlikely to be due to chance.

C. Theories Are Comprehensive Explanations

- A **theory** is more widely accepted and broader in scope than a hypothesis.

D. Scientific Inquiry Has Limitations

- The scientific method does not always yield a complete explanation, or it may produce ambiguous results. Science cannot answer all possible questions, only those about the observable natural world.

1.4 Investigating Life: Digital Organisms Mimic Life

- Computer programs that mimic natural selection help reveal how complex features can arise from simple ancestors.

MULTIPLE CHOICE QUESTIONS

1. All of the following are characteristics of life EXCEPT
   a. evolution.
   b. reproduction.
   c. homeostasis.
   d. movement.
2. Paleobotanists study fossilized plant material. Which property of life can a paleobotanist directly observe in a single plant fossil?
   a. Homeostasis
   b. Organization
   c. Energy use
   d. Growth
3. Which of the following correctly lists the levels of biological organization from smallest to largest?
   a. Cell < Tissue < Organelle < Individual < Community
   b. Community < Population < Ecosystem < Biosphere
   c. Organelle < Cell < Organ < Individual < Population
   d. Individual < Ecosystem < Community < Biosphere
4. The closest evolutionary relationship exists between individuals of the same
   a. genus.
   b. family.
   c. order.
   d. kingdom.
5. Evolution through natural selection will occur most rapidly for populations of organisms that
   a. are already well adapted to the environment.
   b. live in an unchanging environment.
   c. are in the same genus.
   d. reproduce sexually and live in an unstable environment.
6. Which of the following correctly lists taxonomic levels from largest to smallest?
   a. Kingdom > Order > Phylum > Genus
   b. Domain > Order > Class > Family
   c. Domain > Kingdom > Phylum > Genus
   d. Kingdom > Class > Phylum > Species
7. In designing an experiment, the variable that is measured by the scientist is the
   a. standardized variable.
   b. independent variable.
   c. dependent variable.
   d. control variable.
8. What is the role of a hypothesis in the scientific method?
   a. It is the answer to a scientific question.
   b. It is the control variable in the experimental design.
   c. It is a question that can be tested experimentally.
   d. It is a possible explanation that can be tested experimentally.

9. Can a theory be proven wrong?
   a. No, theories are the same as facts.
   b. No, because there is no good way to test a theory.
   c. Yes, a new observation or interpretation of data could disprove a theory.
   d. Yes, theories are the same as hypotheses.

10. Which of the following questions cannot be answered using the scientific method?
    a. What was the first living organism on Earth?
    b. How many genes control the aging process in humans?
    c. Why do monarch butterflies migrate?
    d. What effect does coastal development have on wetlands biodiversity?

TESTING YOUR KNOWLEDGE

1. List the five characteristics of life.
2. Name the levels of biological organization from smallest to largest, beginning with atoms.
3. Are humans producers or consumers?
4. What is homeostasis? Give an example.
5. Cite two ways that asexual and sexual reproduction differ.
6. How does natural selection cause evolution to occur?
7. Why do biologists assign taxonomic names to organisms, and what do those names reflect?
8. Name the levels of the taxonomic hierarchy, starting with domain.
9. What domain and kingdom are humans in?
10. What are the components of the scientific method?
11. What variables must you consider in designing an experiment?
12. In science, how is a theory more than an opinion and a hypothesis more than a guess?
13. Give two examples of questions that you cannot answer using the scientific method.

THINKING AS A SCIENTIST

1. In an episode of the television series *Star Trek,* three members of the crew of the starship USS *Enterprise* meet an enemy who dehydrates each one into a small box. He then crushes one of the boxes. He later attempts to restore the people to life but only the uncrushed ones reappear. How does this action illustrate emergent properties?
2. Is dissecting and labeling the parts of an earthworm science? Give an example of a testable hypothesis that could result from a dissection.
3. For the following examples, state whether each of the following faults occurred: (a) experimental evidence does not support conclusions; (b) inadequate controls; (c) biased sampling; (d) inappropriate extrapolation from the experimental group to the general population; (e) sample size too small.
   a) "I ran 4 miles every morning when I was pregnant with my first child," the woman told her physician, "and Jamie weighed only half as much as a normal baby. This time, I didn't exercise at all, and Jamie's sister had normal birth weight. Therefore, running during pregnancy must cause low birth weight."
   b) Eating foods high in cholesterol was found to be dangerous for a large sample of individuals with hypercholesterolemia, a disorder of the heart and blood vessels. It was concluded from this study that all persons should limit dietary cholesterol intake.
   c) Osteogenesis imperfecta (OI) is an inherited condition that causes easily fractured bones. In a clinical study, 30 children with OI were given a new drug for 3 years. The children all showed increased (improved) bone density and a lowered incidence of fractures compared with before treatment began, as well as less fatigue. The conclusion: The drug is effective in treating OI.
   d) Researchers studied HIV in blood and semen from 11 HIV-infected men. In eight of the men, the virus was resistant to several medications. In two men, virus from the blood was resistant to the class of drugs called protease inhibitors, but virus from semen was not resistant. The researchers concluded that protease inhibitors do not reach the male reproductive organs.
4. A 2004 study found that drug company-funded research is more favorable to new drugs than is publicly funded research. How can scientists avoid such systematic biases?
5. Design an experiment to test the following hypothesis: "Eating chocolate causes zits." Include sample size, independent variable, dependent variable, the most important variables to standardize, and an experimental control.

Visit www.mhhe.com/hoefnagels for practice quizzes, animations, videos, and activities designed to help you master the material in this chapter.

The Chemistry of Life

A tiny messenger molecule called nitric oxide can help relieve chest pain.

Just say NO

Among biological molecules, nitric oxide—abbreviated NO—has some odd traits. First, it is a gas. Second, it is tiny. NO consists of just two atoms, one of the element nitrogen, the other of oxygen, that come together in a way that makes NO highly reactive. Third, NO apparently can slip freely into and out of cells. Most other molecules must either enter the cell through special channels or bind to receptors on receiving cells.

Before 1980, NO was known mostly for its harmful presence in smog, cigarette smoke, and acid rain. But by the late 1980s, research had identified NO's function in animals, plants, fungi, and bacteria. A few roles of NO include:

- ***Blood Pressure.*** *NO causes blood vessels to widen by relaxing muscles in the vessel walls. On a whole-body scale, this activity lowers blood pressure by giving the blood more room to move. This effect has important medical applications. For example, angina is chest pain that occurs when the heart muscle does not get enough oxygen. The drug nitroglycerine increases NO production, relieving angina by widening the arteries that supply oxygen-rich blood to the heart.*
- ***Erectile Function.*** *A mammal's penis consists of three chambers of spongy tissue that surround blood vessels. Upon sexual stimulation, nerves and other cells lining the interiors of the blood vessels release NO. The NO then activates other chemicals that widen blood vessels, allowing blood to fill the spongy tissue. As a result, the penis becomes erect. In men with erectile dysfunction (impotence), an enzyme interferes with the chemical cascade that produces an erection. Sildenafil (Viagra), tadalafil (Cialis), and vardenafil (Levitra) treat this condition by blocking the action of this enzyme.*
- ***Plant Defenses.*** *Many bacteria, fungi, and viruses that cause disease in plants survive by parasitizing live plant cells. When a plant detects an attacking pathogen, NO triggers a cascade of reactions that kill the tissues in a zone surrounding the initial site of infection. This early reaction, called the hypersensitive response, stops the invader before it spreads.*

Life is made of chemicals, most of which are much larger than NO. Understanding biology is therefore impossible without an introduction to chemistry. This chapter describes how nitrogen, oxygen, and other atoms that compose all matter come together to form the molecules of life.

LEARNING OUTLINE

2.1 Atoms Are the Stuff of Life

If you have ever touched a plant in a restaurant to see if it's fake, you know that we all have an intuitive sense of what life is made of. A living leaf feels moist and pliable; a fake one is dry and stiff. But what does chemistry tell us about the composition of life?

Your desk, your book, your body, your sandwich, a plastic plant—indeed, all objects in the universe, including life on Earth—are composed of matter and energy. **Matter** is any material that takes up space, such as organisms, rocks, the oceans, and gases in the atmosphere. This chapter and the next concentrate on the composition of living matter. Physicists define energy, on the other hand, as the ability to do work. In this context, "work" means moving matter. Heat, light, and chemical bonds are all forms of energy; chapter 4 discusses the energy of life in detail.

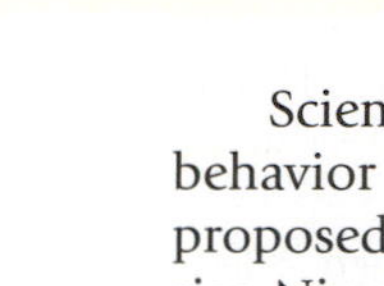

A. Elements Are Fundamental Types of Matter

The matter that makes up every object in the universe consists of one or more elements. A chemical **element** is a pure substance that cannot be broken down by chemical means into other substances. Examples include pure oxygen (O), carbon (C), sodium (Na), and hydrogen (H). Chemists recognize 92 elements that occur naturally on Earth and about 25 synthetic ones created in the laboratory.

Scientists had already noticed patterns in the chemical behavior of the elements by the mid-1800s, and several had proposed schemes for organizing the elements into categories. Nineteenth-century Russian chemist Dmitry Mendeleyev invented the **periodic table**, the chart that we still use today. The chart is "periodic" because the chemical properties of the elements repeat in each column of the table. **Figure 2.1** illustrates an abbreviated periodic table, emphasizing the elements that make up organisms. (Endsheet 1 contains a complete periodic table.)

Only about 25 of Earth's 91 naturally occurring elements are essential to life. **Bulk elements** such as carbon, hydrogen, oxygen, nitrogen (N), sulfur (S), and phosphorus (P) make up the vast majority of every living cell and are therefore required in large amounts. Those required in small amounts are called **trace elements**.

Many trace elements are important in ensuring that vital chemical reactions occur fast enough to sustain life. A person whose diet is deficient in these trace elements can become ill or die. The thyroid gland, for example, requires the element iodine (I). If the diet does not supply enough iodine, the thyroid may become enlarged, forming a growth called a goiter. Similarly, blood requires iron (Fe) to carry oxygen to the body's tissues. An iron-poor diet can cause anemia. **essential elements for plants, p. 522; animal nutrition, p. 702**

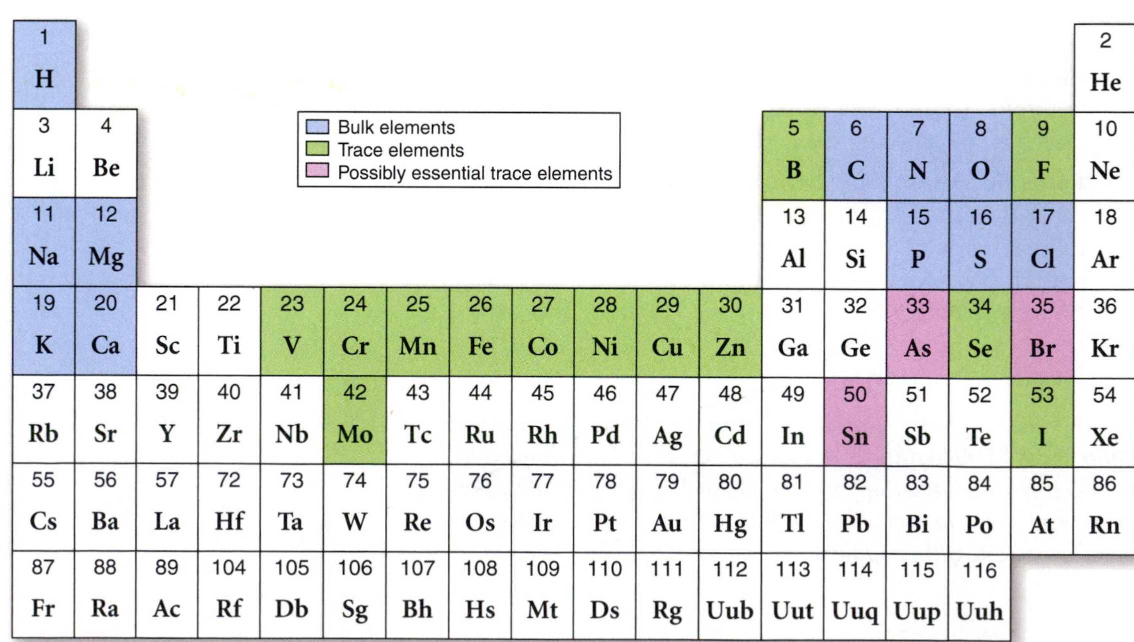

FIGURE 2.1 The Periodic Table of Elements. Each element has a symbol, which can come from the element's English name (He for helium, for example) or from a name in another language (Na for sodium, which is *natrium* in Latin). Elements 58 through 71 and 90 through 103 are omitted for clarity.

B. Atoms Are Particles of Elements

An **atom** is the smallest possible "piece" of an element that retains the characteristics of the element. An atom is composed of three major types of subatomic particles (**figure 2.2** and **table 2.1**). **Protons,** which carry a positive charge, and **neutrons,** which are uncharged, together form a central **nucleus.** Negatively charged **electrons** surround the nucleus. An electron is vanishingly small compared with a proton or a neutron.

For simplicity, most illustrations of atoms show the electrons closely hugging the nucleus. In reality, however, if the nucleus of a hydrogen atom were the size of a meatball, the electron belonging to that atom could be about 1 kilometer (0.62 miles) away from it! Thus, most of an atom's mass is concentrated in the nucleus, while the electron cloud occupies virtually all of its volume.

Each element has a different **atomic number**, the number of protons in the nucleus. Hydrogen, the simplest type of atom, has an atomic number of 1. In contrast, an atom of uranium has 92 protons. Elements are arranged sequentially in the periodic table by atomic number, which appears above each element's symbol (see figure 2.1).

When the number of protons equals the number of electrons, the atom is electrically neutral—that is, it has no net charge. An **ion** is an atom (or group of atoms) that has gained or lost electrons and therefore has a net negative or positive charge. Hydrogen (H^+), sodium (Na^+), potassium (K^+), and chloride (Cl^-) ions participate in many biological processes, including the transmission of messages in the nervous system. They also form ionic bonds, discussed in section 2.2. **action potential, p. 580**

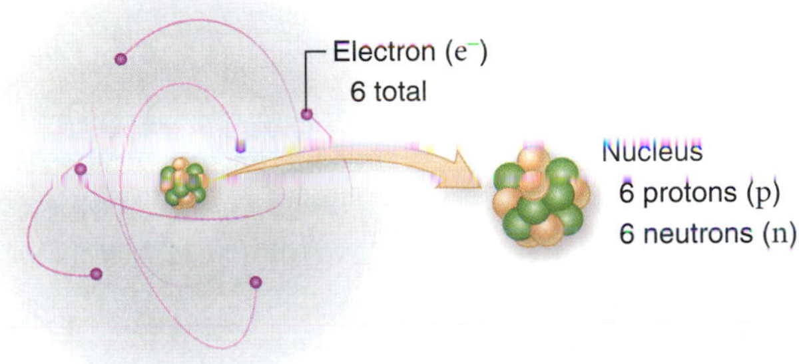

FIGURE 2.2 Atom Anatomy. An atom is composed of a nucleus made of protons and neutrons, surrounded by a cloud of electrons. This is a carbon atom.

Table 2.1 *Subatomic Particles*			
Particle	**Charge**	**Mass**	**Location**
Electron	–	0	Surrounding nucleus
Neutron	None	1	Nucleus
Proton	+	1	Nucleus

C. Isotopes Have Different Numbers of Neutrons

An atom's **mass number** is the total number of protons and neutrons in its nucleus. We can approximate the mass of a proton (and that of a neutron) to be equal to one. By comparison, the contribution of an electron to an atom's mass is negligible. Subtracting the atomic number (the number of protons) from the mass number (the number of protons and neutrons) yields the number of neutrons.

All atoms of a given element have the same atomic number, but not necessarily the same number of neutrons. An **isotope** is any of these different forms of a single element. All isotopes of an element have the same chemical properties, but different mass numbers because of the different numbers of neutrons.

Often one isotope of an element is very abundant, and others are rare. For example, 99% of carbon isotopes have six neutrons, and only 1% have seven or eight neutrons. An element's **atomic mass** (also called atomic weight) is the average mass of all isotopes. Because the vast majority of carbon atoms contains six neutrons, its atomic mass is very close to 12 in the periodic table.

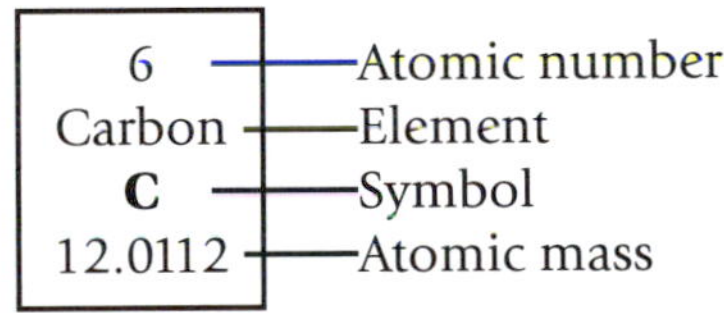

Many of the known isotopes are unstable and **radioactive**, which means they emit energy as rays or particles when they break down into more stable forms. Every radioactive isotope has a characteristic half-life, which is the time it takes for half of the atoms in a sample to emit radiation, or "decay" to a different, more stable form. The radiation energy is relatively easy to detect, even from very small quantities of isotopes.

Because radioactive isotopes have the same chemical properties as stable isotopes, they have a variety of uses, some of which are listed in **table 2.2**. All in all, at least 35 different isotopes are used in medicine to diagnose and treat disease. But the same properties that make radioactive isotopes useful can also make them dangerous. Exposure to excessive radiation can lead to radiation sickness, and radiation-induced mutations of a cell's DNA can cause cancer (see chapter 8). The lead-containing "bib" that a dentist lays over your chest during mouth X-rays protects that area from radiation.

Table 2.3 reviews the terminology of matter.

Table 2.2 *Selected Uses of Radioactive Isotopes*

Use	Sample Applications
Kill disease-causing organisms	Use of radiation from cobalt-60 to sterilize hospital equipment and kill microorganisms on food surfaces.
Tracers	Certain isotopes, used as "tracers," allow biologists to understand life processes. The tracer emits radiation or particles, which the researcher can track. • A plant biologist might expose a leaf to radioactive carbon-14, then track the isotope's progress during and after the chemical reactions of photosynthesis. Likewise, a biologist might learn how plants transport and use nutrients by supplying radioactive phosphorus-32 to roots. • A physician might give a patient a radioactive tracer (orally, by inhalation, or by injection), then track how the isotope moves in the body to search for tumors or examine a physiological process. (This is the basis of a PET scan, which uses fluorine-18 as a tracer).
Radiometric dating	Archaeologists and paleontologists use the known half-lives of radioactive isotopes to determine the ages of artifacts and fossils (see chapter 15). Isotopes with relatively short half-lives date younger materials, whereas those with long half-lives date more ancient items.
Cancer therapy	Directing radiation at a tumor kills cancer cells (see chapter 8). Cobalt-60 targets cancer cells deep in the body, whereas phosphorus-32, which emits lower energy radiation, treats skin cancers. A patient being treated for Graves disease (overactive thyroid gland) or thyroid cancer might ingest radioactive iodine-131, which concentrates in the thyroid and kills some of its cells.

Summary *Like all matter, organisms consist of atoms, although they use only a subset of the elements found in nature. Protons, neutrons, and electrons make up atoms. Although every atom of a given element has the same number of protons, the number of neutrons can vary.*

2.1 MASTERING CONCEPTS

1. Which chemical elements do organisms require in large amounts?
2. Where in an atom are protons, neutrons, and electrons located?
3. What does an element's atomic number indicate?
4. What is the relationship between an atom's mass number and an element's atomic mass?
5. How are different isotopes of the same element different from one another?

Table 2.3 *A Mini-Glossary of Matter*

Designation	Definition
Element	A fundamental type of substance
Atom	The smallest unit of an element that retains the characteristics of that element
Atomic number	The number of protons in an atom's nucleus
Mass number	The number of protons plus neutrons in an atom's nucleus
Isotope	Any of the different forms of the same element, distinguished from each other by the number of neutrons in the nucleus
Atomic mass	The average mass of all isotopes of an element

2.2 Chemical Bonds Link Atoms

Like all organisms, you are composed mostly of carbon, hydrogen, oxygen, and nitrogen atoms. But the arrangement of these atoms is not random. Instead, your atoms are organized into molecules (see figure 1.2). A **molecule** is two or more chemically joined atoms.

Some molecules, such as the gases hydrogen (H_2), oxygen (O_2), or nitrogen (N_2), are "diatomic," meaning that they consist of two atoms of the same element. More often, however, the elements in a molecule are different. A **compound** is a molecule composed of two or more different elements. Nitric oxide, described in the chapter opening essay, is a compound consisting of one nitrogen and one oxygen atom. Likewise, water (H_2O) is made of two atoms of hydrogen and one of oxygen. Many large biological compounds, including DNA and proteins, consist of thousands of atoms.

A compound's characteristics can differ strikingly from those of its constituent elements. Consider table salt, sodium chloride. Sodium (Na) is a silvery, highly reactive solid metal, whereas chlorine (Cl) is a yellow, corrosive gas. But when equal numbers of these two atoms combine, the resulting compound forms the familiar white salt crystals that we sprinkle on food—an excellent example of an emergent property (chapter 1). The same is true for methane, the main component of natural gas. Its components are carbon, a black sooty solid, and hydrogen, a light, combustible gas.

Scientists describe molecules by writing the symbols of their constituent elements and indicating the numbers of atoms of each element in one molecule as subscripts. For example, methane is written CH_4, which denotes one carbon atom attached to four hydrogen atoms. This representation of the atoms in a molecule is termed a molecular formula. Table salt's formula is NaCl, that of water is H_2O, and that of the gas carbon dioxide is CO_2.

What forces hold together the atoms that make up each of these molecules? To understand the answer, we must first learn more about how electrons are arranged around the nucleus.

A. Electrons Determine Bonding

Electrons occupy distinct energetic regions around the nucleus. They are constantly in motion, so it is impossible to determine the exact location of any single electron at any instant in time. Instead, chemists use the term **orbitals** to describe the most likely location for an electron relative to its nucleus. Each orbital can hold up to two electrons. Consequently, the more electrons in an atom, the more orbitals they occupy.

Electron orbitals exist in several energy levels; an **energy shell** is a group of orbitals that share the same level. The number of orbitals in each shell determines the number of electrons the shell can hold. The lowest energy shell, for example, contains just one orbital and thus holds up to two electrons. The two next shells each contain four orbitals (eight electrons total).

Electrons occupy the lowest energy level available to them. As each energy shell fills, any additional electrons must reside in higher energy shells. For example, hydrogen has only one electron in the lowest energy orbital, and helium has two. Lithium has three electrons; two occupy the lowest energy orbital, and the third is in the next energy shell. Oxygen, with eight electrons total, has two electrons in the lowest energy orbital and six in three orbitals at the next higher energy level.

We can thus envision any atom's electrons as occupying a series of concentric energy shells, each having a higher energy level than the one inside it. In accordance with this view, electrons often are illustrated as dots moving in two-dimensional circles around a nucleus (**figure 2.3**). These depictions, called Bohr models, are useful for visualizing the interactions between atoms to form bonds. However, most orbitals are not spherical (**figure 2.4**). Bohr models therefore do not accurately portray the three-dimensional structure of atoms.

An atom's **valence shell** is its outermost occupied energy shell. Atoms are the most stable—that is, least likely to combine with other atoms—when their valence shells are full. The gases helium (He) and neon (Ne), for example, are inert. Because their outermost shells are full, they exist in nature without combining with other atoms.

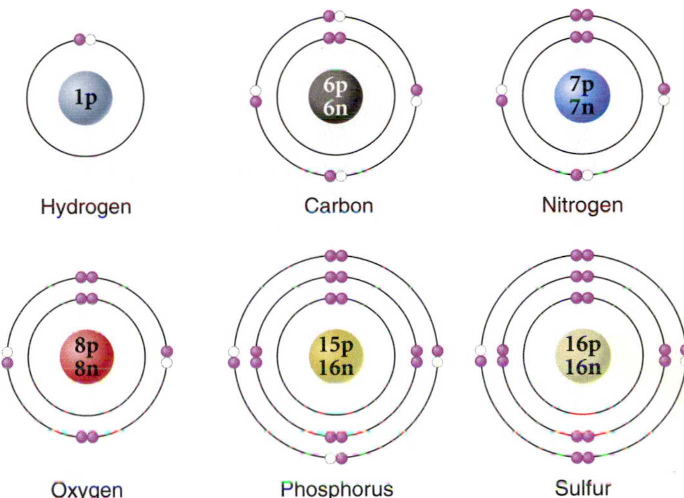

FIGURE 2.3 Energy Shells. Shown here are Bohr models of the six most common atoms in organisms.

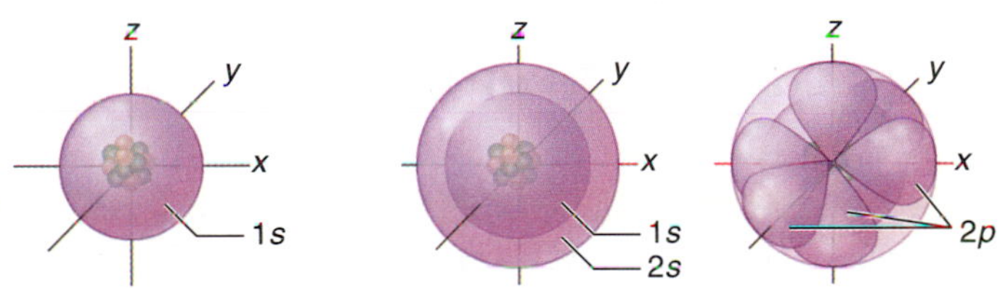

a. The first shell contains up to 2 electrons in one spherical orbit (1*s*).

b. The second shell contains up to 8 electrons. Two electrons occupy one spherical orbit (2*s*) and each of 3 perpendicular dumbbell-shaped orbits (2*p*).

FIGURE 2.4 Electron Orbitals. An electron is constantly in motion, so it is impossible to pinpoint its exact position. Orbitals depict the probability that an electron is in any given location. (**A**) The first (lowest) energy level consists of one spherical (1*s*) orbital containing up to two electrons. (**B**) The second energy level has four orbitals (one spherical [2*s*] and three dumbbell-shaped [2*p*]), each containing up to two electrons. The nucleus of the atom is at the center, where the axes intersect.

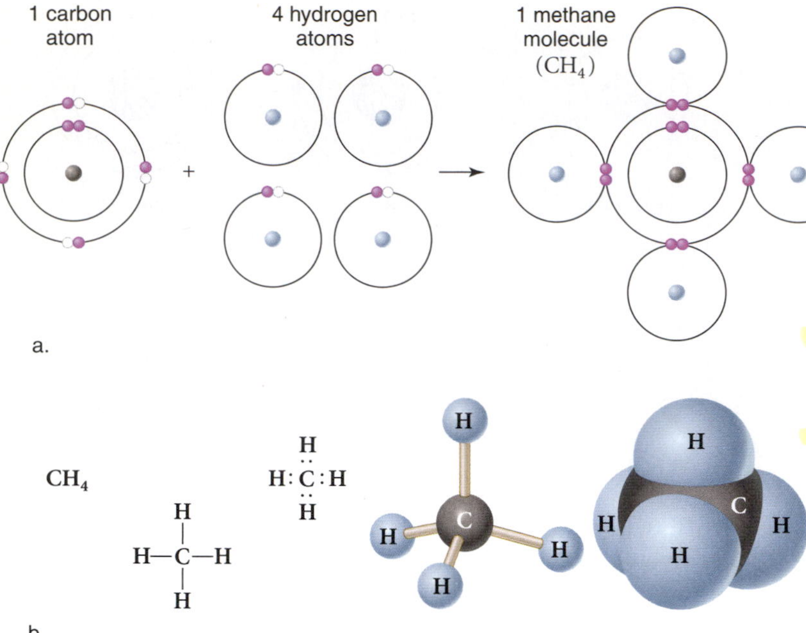

FIGURE 2.5 Atoms Share Electrons in Covalent Bonds. (**A**) Methane (CH_4) contains four covalent bonds, formed when one carbon and four hydrogen atoms complete their outermost shells by sharing electrons. (**B**) Different types of diagrams can show the same molecule in different ways.

For most atoms, however, the valence shell is only partially filled. Such an atom will become most stable if its valence-shell "vacancies" fill. To arrive at exactly the right number, atoms share, steal, or donate electrons. The result is a **chemical bond**: an attractive force that holds atoms together in a molecule. The remainder of this section describes three types of chemical bonds that are important in biology.

B. In a Covalent Bond, Atoms Share Electrons

A **covalent bond** forms when two atoms share electrons. The shared electrons travel around both nuclei, strongly connecting the atoms together.

Methane provides an excellent example of how atoms share electrons to fill their valence shells. A carbon atom has six electrons, two of which occupy its innermost shell. That leaves four electrons in its valence shell, which has a capacity of eight. Carbon therefore requires four more electrons to fill its outermost shell. A carbon atom can attain the stable eight-electron configuration by sharing electrons with four hydrogen atoms, each of which has one electron in its only shell (**figure 2.5A**). The resulting molecule is methane, CH_4. Figure 2.5B shows several ways to represent its chemical structure.

Covalent bonds are usually depicted as lines between the interacting atoms, with each line representing one bond. Each single bond contains two electrons: one from each atom. Atoms can also share two or three electron pairs, forming double and triple covalent bonds, respectively (**figure 2.6**). The diatomic molecule O_2, for example, has one double bond; a strong triple bond holds together the two atoms in N_2.

Covalent bonding means "sharing," but the partnership is not necessarily equal. **Electronegativity** is a measure of an atom's ability to attract electrons (**figure 2.7**). Oxygen, for example, strongly attracts electrons. Carbon and hydrogen have low electronegativity relative to oxygen. A **nonpolar covalent bond** is a "bipartisan" union in which both atoms exert approximately equal pull on their shared electrons. Carbon and hydrogen atoms, for example, have similar electronegativity. A carbon–hydrogen bond is therefore nonpolar, as is a bond between two atoms of the same element; after all, a bond between two identical atoms must be electrically balanced. H_2, N_2, and O_2, for example, are all nonpolar molecules.

A **polar covalent bond**, in contrast, is a lopsided union in which one nucleus exerts a stronger pull on the shared electrons than does the other nucleus. Polar bonds form whenever a highly electronegative atom such as oxygen shares electrons—albeit unequally—with a less electronegative partner like carbon or hydrogen. Like a battery, a polar covalent bond has a positive end and a negative end.

Polar covalent bonds are critical to biology. They are responsible for the shape of DNA and proteins, as well as for the unique role of water in life (see section 2.3).

C. In an Ionic Bond, One Atom Takes Electrons from Another Atom

So far, we have seen covalent bonds in which atoms share electrons either equally (nonpolar) or unequally (polar). Is it possible for two atoms to have such different electronegativities that one actually takes control of one or more of its partner's electrons?

The answer is yes. Recall that an atom is most stable if its outermost shell is full. Not surprisingly, the most electronegative atoms, such as chlorine (Cl), are usually those whose valence shells have only one "vacancy." Likewise, sodium (Na) and other weakly electronegative atoms have only one electron in their outermost shells. Neither chlorine nor sodium would benefit from sharing. Instead, sodium is most stable if it simply releases its extra electron to chlorine, which needs this "scrap" electron to complete its own valence shell.

Recall that an ion is an atom that has lost or gained electrons. The atom that has lost electrons carries a positive charge, whereas the one that has gained electrons acquires a negative charge. An **ionic bond** results from the electrical attraction between two ions with opposite charges. In gen-

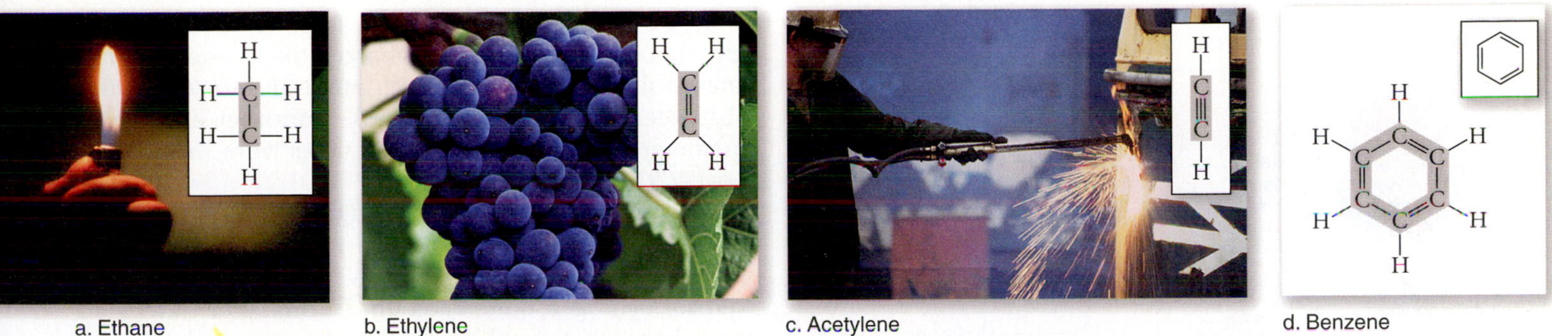

FIGURE 2.6 Carbon Atoms Form Four Covalent Bonds. Two carbon atoms can bond, forming single, double, or triple bonds. Note that as the number of bonds between carbon atoms increases, the number of bonded hydrogens decreases. (**A**) Ethane, a component of natural gas, is a hydrocarbon built around two singly bonded carbon atoms. Breaking the atoms apart releases enough energy to heat a home or light a fire. (**B**) Ethylene consists of two carbon atoms linked by a double bond. As described in chapter 26, ethylene is a plant hormone, triggering flowers to drop and fruit to ripen. (**C**) Acetylene, consisting of two carbons held by a triple bond, is a flammable gas used in metal-cutting torches. Tremendous heat energy is released when a triple bond is broken. (**D**) Benzene illustrates a ring structure. The inset shows an abbreviation of the ring that assumes carbon atoms are at the corners, each with an implied hydrogen atom attached.

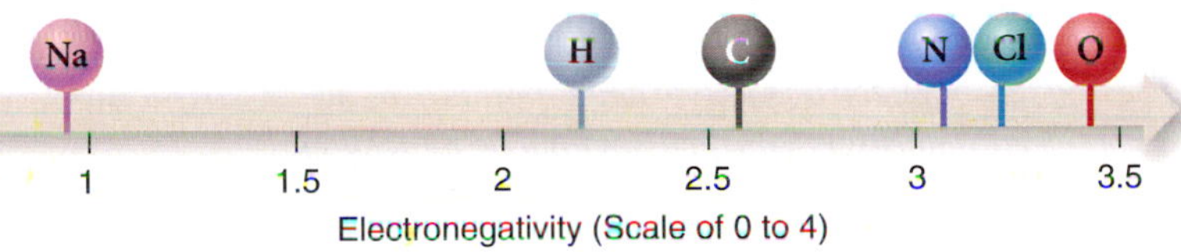

FIGURE 2.7 Unequal Attraction. Atoms vary widely in their electronegativity, the ability to attract electrons.

eral, such bonds form between an atom whose outermost shell is almost empty and one whose valence shell is nearly full. Thus, when sodium donates its electron to an atom of chlorine, the two atoms bond ionically to form NaCl (**figure 2.8**). In NaCl, the oppositely charged ions Na^+ and Cl^- attract each other in such an ordered manner that a three-dimensional crystal results.

Ionic bonds in crystals are strong, as demonstrated by the stability of the salt in your shaker. Those same crystals, however, dissolve when you stir them into water. As described in section 2.3, water molecules pull ionic bonds apart.

Nonpolar covalent bonds, polar covalent bonds, and ionic bonds represent points along a continuum. Two atoms of similar electronegativity share electrons equally in nonpolar covalent bonds. If one atom tugs at the shared electrons much more than the other, the covalent bond is polar. And if one atom is so electronegative that it rips electrons from another atom's valence shell, an ionic bond forms. Notice that the bond type depends on the *difference* in electronegativity, so the same element can participate in different types of bonds. Oxygen, for example, forms nonpolar bonds with itself (as in O_2) and polar bonds with hydrogen (as in H_2O).

D. Partial Charges on Polar Molecules Create Hydrogen Bonds

When a covalent bond is polar, the negatively charged electrons spend more time around the nucleus of the more electronegative atom than around its partner. The "electron-hogging" atom therefore has a partial negative charge, and

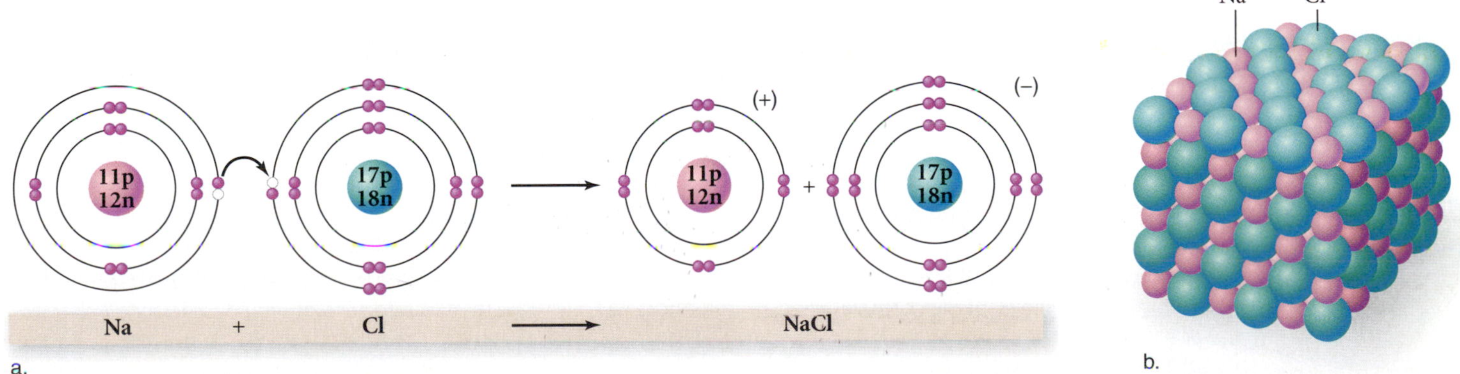

FIGURE 2.8 Table Salt, an Ionically Bonded Molecule. (**A**) A sodium atom (Na) can donate the one electron in its valence shell to a chlorine atom (Cl), which has seven electrons in its outermost shell. Notice that the valence shells of both atoms are now full. The resulting ions (Na^+ and Cl^-) bond to form the compound sodium chloride (NaCl). (**B**) The ions that constitute NaCl occur in a repeating pattern that produces crystals.

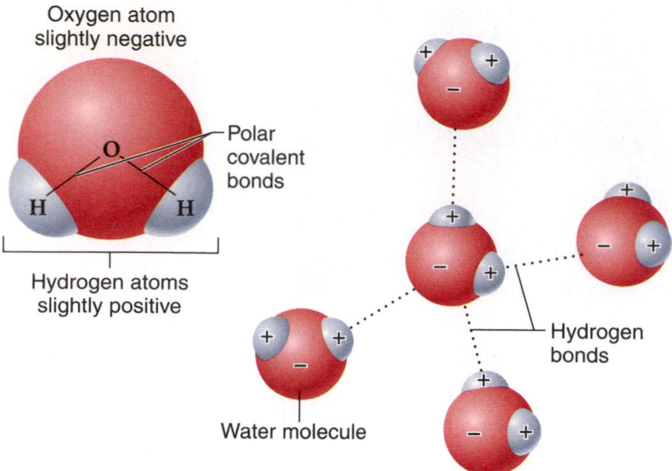

FIGURE 2.9 Water Molecules Form Hydrogen Bonds. Polar covalent bonds hold together the two hydrogen atoms and one oxygen atom of water (H_2O). Because the oxygen attracts the negatively charged electrons more strongly than the hydrogen nuclei do, the oxygen atom bears a partial negative charge, and the hydrogens carry a partial positive charge. These partial charges attract one molecule to another; this attraction is the hydrogen bond.

the less-electronegative partner has an electron "deficit" and a partial positive charge.

In a **hydrogen bond**, opposite partial charges on *adjacent molecules*—or within a single large molecule—attract each other. The name comes from the fact that the atom with a partial positive charge is usually hydrogen.

Water provides the simplest illustration of hydrogen bonds (**figure 2.9**). Each water molecule has a "boomerang" shape, owing to the oxygen atom's two pairs of unshared valence electrons. Moreover, the two O—H bonds in water are polar, with the nucleus of each oxygen atom attracting the shared electrons more strongly than do the hydrogen nuclei. Each hydrogen atom therefore has a partial positive charge, which attracts the partial negative charge of the oxygen atom on an adjacent molecule. The bent shape, plus the partial charges on O and H, cause water molecules to stick to each other and to some other substances. (This slight stickiness is another example of an emergent property, because it arises from interactions between O and H.)

Hydrogen bonds are relatively weak. In one second, the hydrogen bonds between a single water molecule and its nearest neighbors form and re-form some 500 billion times. Even so, hydrogen bonds account for many of water's unusual characteristics—the subject of the next section. In addition, multiple hydrogen bonds help stabilize some large molecules, including proteins and DNA.

Table 2.4 summarizes the chemical bonds important in life.

Summary *Atoms bond together to form molecules. An atom's electron arrangement determines the type and number of bonds it will form with other atoms. Covalent bonds link atoms to one another by sharing electrons; ionic bonds result from the transfer of electrons. Hydrogen bonds are weak attractions between oppositely charged regions of molecules.*

2.2 MASTERING CONCEPTS

1. How are atoms, molecules, and compounds related?
2. What is the difference between an orbital and an energy shell?
3. How does the number of valence electrons determine an atom's tendency to form bonds?
4. Explain how electronegativity differences between atoms result in nonpolar covalent bonds, polar covalent bonds, and ionic bonds.
5. What is the relationship between polar covalent bonds and hydrogen bonds?

Table 2.4 *Chemical Bonds*

Type	Chemical Basis	Strength	Example
Covalent bond	Atoms share electron pairs	Strong	O—H bond within water molecule
Ionic bond	One atom donates one or more electrons to another atom, forming oppositely charged ions that attract each other	Strong, but break easily in water	Sodium chloride (NaCl)
Hydrogen bond	Atom with partial negative charge attracts atom with partial positive charge; form between adjacent molecules or between different parts of a large molecule	Weak	Attracts adjacent water molecules to each other

2.3 Water Is Essential to Life

Although water may seem to be a rather ordinary fluid, it is anything but. The tiny, three-atom water molecule has extraordinary properties that make it essential to all organisms. Indeed, life on Earth could not have begun without it, which is why the search for life on other planets begins with the search for water. This section explains some of the properties that make water central to biology.

FIGURE 2.11 Defying Gravity. Both cohesion and adhesion allow water to rise in this narrow glass tube.

A. Water Is Cohesive and Adhesive

Hydrogen bonds contribute to a property of water called **cohesion**—the tendency of water molecules to stick together. Without cohesion, water would evaporate instantly in most locations on Earth's surface. Cohesion also contributes to the observation that you can sometimes fill a glass so full that water is above the rim, yet it doesn't flow over the side unless disturbed.

This tendency of a liquid to hold together at its surface is called surface tension, and not all liquids exhibit it. Water has high surface tension because it is cohesive. At the boundary between water and air, the water molecules form hydrogen bonds with neighbors to their sides and below them in the liquid. These bonds tend to hold the surface molecules together, creating a thin "skin" that is strong enough to support small insects without breaking through (**figure 2.10**).

A related property of water is **adhesion**, the tendency to form hydrogen bonds with other substances. Both cohesion and adhesion are at work when water seemingly defies gravity as it rises in a small-diameter tube (**figure 2.11**), soaks into a paper towel, or moves from a plant's roots to its highest leaves. As described in chapter 25, this movement depends upon cohesion of water within the plant's conducting tubes. Water entering roots is drawn up through these tubes as water evaporates from leaf cells. In addition to cohesion, adhesion to the walls of the conducting tubes also helps lift water to the topmost leaves of trees.

FIGURE 2.10 Running on Water. A lightweight body and water-repellent legs allow this water strider to "skate" across a pond without breaking the water's surface tension.

B. Polar Substances Dissolve in Water

Another reason that water is vital to life is that it can dissolve a wide variety of chemicals. To illustrate this process, picture the slow disappearance of table salt as it dissolves in water. Although the salt crystals appear to vanish, the sodium and chloride ions remain. Water molecules surround each ion individually, separating them from each other (**figure 2.12**).

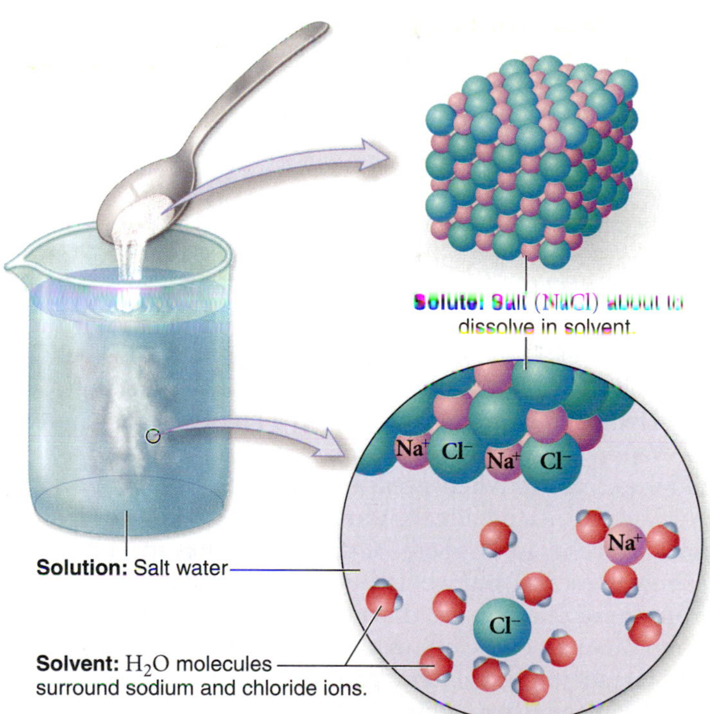

FIGURE 2.12 Solutions Are Mixtures of Molecules. As salt crystals dissolve, water molecules surround the individual sodium and chloride ions.

In this example, water is a **solvent:** a chemical in which other substances, called **solutes**, dissolve. A **solution** consists of one or more solutes dissolved in a liquid solvent. In a so-called aqueous solution, water is the solvent. But not all solutions are aqueous. According to the rule "Like dissolves like," polar solvents such as water dissolve polar molecules; similarly, nonpolar solvents dissolve nonpolar substances.

Scientists divide chemicals into two categories, based on their affinity for water. **Hydrophilic** substances readily dissolve in water (the term literally means "water-loving"). Examples include sucrose (table sugar), salt, and ions. Electrolytes are ions in the body's fluids, and the salty taste of sweat is a testament to water's ability to dissolve them. Sports drinks replace not only water, but also sodium, potassium, magnesium, and calcium ions, which are lost in perspiration during vigorous exercise. Electrolytes are essential to many processes, including heart and nerve function.

Not every substance, however, is water-soluble. Nonpolar molecules made mostly of carbon and hydrogen, such as fats and oils, are called **hydrophobic** ("water-fearing") because they do not dissolve in water. This is why water alone will not remove grease from hands, dishes, or clothes. Dry cleaning companies use nonpolar solvents to remove oily spots from fabric. Detergents contain molecules that attract both water and fats, so they can dislodge greasy substances and carry the mess down the drain with the wastewater.

C. Water Regulates Temperature

Another unusual property of water is its ability to resist temperature changes. When molecules absorb energy, they move faster. Water's hydrogen bonds tend to counter this molecular movement; as a result, more heat is needed to raise water's temperature than is required for most other liquids, including alcohols. Because an organism's fluids are aqueous solutions, the same effect holds: an organism may encounter considerable heat before its body temperature becomes dangerously high. Likewise, the body cools slowly in cold temperatures.

Hydrogen bonds also mean that a lot of heat is required to evaporate water. **Evaporation** is the conversion of a liquid into a vapor. When sweat evaporates from skin, individual water molecules break away from the liquid droplet and float into the atmosphere. Surface molecules must absorb energy to escape, and when they do, heat energy is removed from those that remain, drawing heat out of the body—an important part of the homeostatic mechanism that regulates body temperature. **homeostasis, p. 567**

Water's unusual tendency to expand upon freezing also affects life. In ice, water's solid form, the molecules are farther apart than they are in the liquid form. Therefore, ice is less dense and floats on the surface of a frozen pond (**figure 2.13**). This characteristic benefits aquatic organisms. When the air temperature drops sufficiently, a small amount of water freezes at the air–water interface, forming a solid cap of ice that retains heat in the water below. Many organisms can then survive in the depths of such shielded lakes and ponds. If ice were to become denser upon freezing, it would sink to the bottom. The lake would then gradually turn to ice from the bottom up, entrapping the organisms that live there.

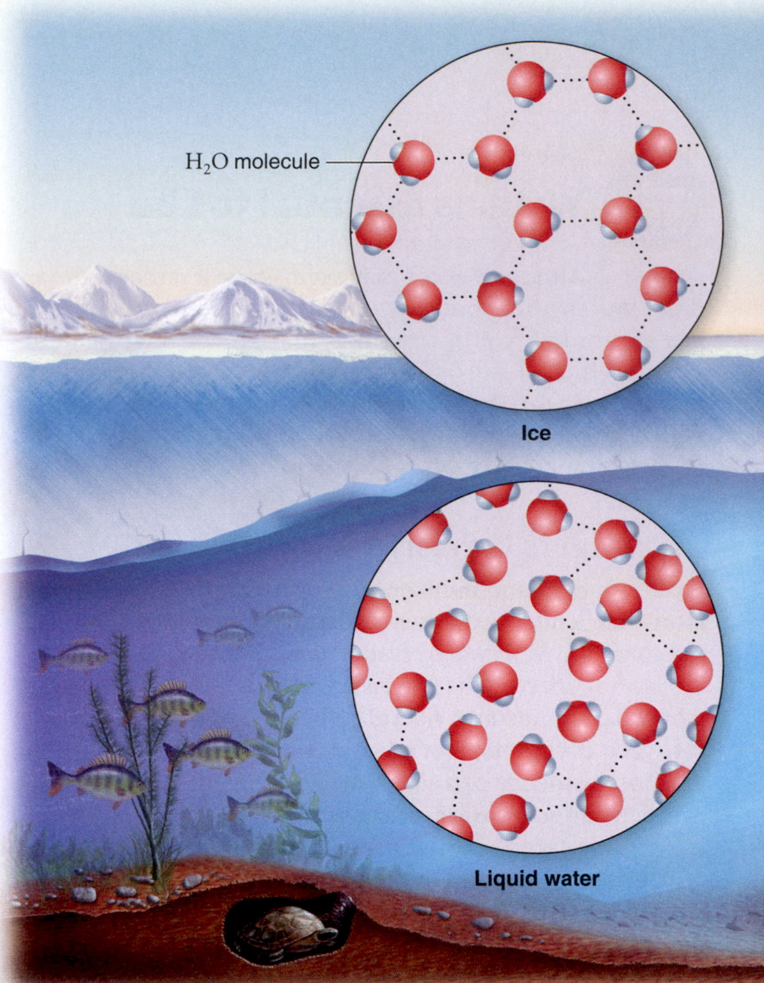

FIGURE 2.13 Ice Floats. Thanks to hydrogen bonds, ice crystals are less dense than liquid water. Ice therefore floats to the top of a freezing lake.

Some organisms do freeze and die, however, because water inside cells expands when it freezes and breaks apart the cells. Many organisms have adaptations that enable them to survive freezing, such as producing biochemicals that act as antifreeze. Another strategy is to dehydrate temporarily, as do small aquatic animals called tardigrades. These organisms can survive drying for 10 or more years, resuming activity when water again becomes available. The larvae of flies called midges exhibit yet another adaptation to freezing—ice forms between their cells, not inside them.

D. Water Participates in Life's Chemical Reactions

Life exists because of thousands of simultaneous chemical reactions. In a **chemical reaction**, two or more molecules "swap" their atoms to yield different molecules; that is, some chemical bonds break, and new ones form. Chemists depict reactions as equations with the **reactants**, or starting materials, on the left; the **products**, or results of the reaction, are listed to the right of the arrow. Consider what happens when the methane in natural gas burns—say, inside a heater or gas oven:

$$CH_4 + 2O_2 \longrightarrow CO_2 + 2H_2O$$
$$\text{methane} + \text{oxygen} \longrightarrow \text{carbon dioxide} + \text{water}$$

In words, this says that one methane molecule combines with two oxygen molecules to produce a carbon dioxide molecule and two molecules of water. The bonds of the methane and oxygen molecules have broken, and new bonds formed to make the products.

Note that the total number of atoms of each element must always be the same on either side of the equation: that is, one carbon, four hydrogens, and four oxygens. In chemical reactions, atoms are neither created nor destroyed.

Nearly all of life's chemical reactions occur in the watery solution that fills and bathes cells. Moreover, water participates directly in many of these reactions, as either a reactant or a product. In photosynthesis, for example, plants use the sun's energy to assemble food out of just two reactants: carbon dioxide and water (see chapter 5). Section 2.5 describes two other water-related reactions, hydrolysis and dehydration synthesis, that are vital to life.

Summary *Water has distinctive properties that make it essential for life: cohesion, adhesion, the ability to dissolve hydrophilic substances, and resistance to temperature change. Water also participates in the chemical reactions that sustain life.*

2.3 MASTERING CONCEPTS

1. How is cohesion different from adhesion?
2. What is the difference between hydrophilic and hydrophobic molecules?
3. Which properties of water enable it to regulate body temperature?
4. How does the fact that water expands upon freezing affect life?
5. What happens in a chemical reaction?
6. How does water participate in the chemistry of life?

2.4 Organisms Balance Acids and Bases

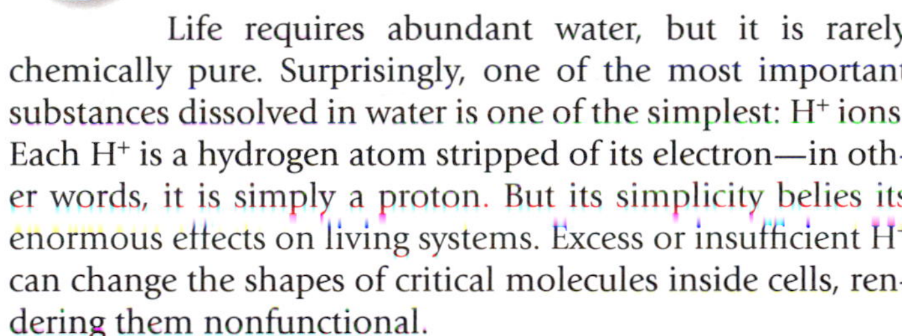

Life requires abundant water, but it is rarely chemically pure. Surprisingly, one of the most important substances dissolved in water is one of the simplest: H^+ ions. Each H^+ is a hydrogen atom stripped of its electron—in other words, it is simply a proton. But its simplicity belies its enormous effects on living systems. Excess or insufficient H^+ can change the shapes of critical molecules inside cells, rendering them nonfunctional.

One source of H^+ is pure water. At any time, perhaps one in a million water molecules spontaneously breaks into two pieces. When this happens, the highly electronegative oxygen atom keeps the electron from the breakaway hydrogen atom. The result is one hydrogen ion (H^+) and one hydroxide ion (OH^-):

$$H_2O \longrightarrow H^+ + OH^-$$

In pure water, the number of hydrogen ions must exactly equal the number of hydroxide ions. A **neutral** solution likewise has exactly the same amount of H^+ as OH^-.

Some substances, however, alter this balance. An **acid** is a chemical that adds H^+ to a solution, making the concentration of H^+ ions exceed the concentration of OH^- ions. Examples include hydrochloric acid (HCl), sulfuric acid (H_2SO_4), and many common, sour-tasting household products such as vinegar and lemon juice. Adding sulfuric acid to pure water releases H^+ ions into the solution:

$$H_2SO_4 \longrightarrow 2H^+ + SO_4^{-2}$$

Because no OH^- ions were added at the same time, the balance of H^+ to OH^- skews toward extra H^+.

A **base** is the opposite of an acid—it makes the concentration of OH^- ions exceed the concentration of H^+ ions. Bases work in one of two ways. They come apart to directly add OH^- ions to the solution, or they absorb H^+ ions. Either way, the result is the same: the balance between H^+ and OH^- shifts toward OH^-. A common base is sodium hydroxide (NaOH), an ingredient in oven and drain cleaners. When it dissolves in water, it releases OH^- into solution:

$$NaOH \longrightarrow Na^+ + OH^-$$

What happens if a person mixes an acid with a base? The acid releases protons, while the base either absorbs the H^+ or releases OH^-. Acids and bases therefore neutralize each other.

Both acids and bases are important in life. As you will see in section 2.5, two of life's building blocks are amino acids and nucleic acids. Your stomach produces hydrochloric acid that helps you digest food. Antacids contain bases that neutralize excess acid, relieving an upset stomach. In the environment, some pollutants react with water in the atmosphere and return to Earth as acid precipitation, which weakens or kills plants and aquatic life as well as damaging buildings and outdoor sculptures. **acid deposition, p. 847**

A. The pH Scale Expresses Acidity or Alkalinity

Scientists use a system of measurement called the **pH scale** to gauge how acidic or basic a solution is. The pH scale ranges from 0 to 14, with 7 representing a neutral solution such as pure water (**figure 2.14**). An acidic solution has a pH lower than 7, whereas an **alkaline**, or basic, solution has a pH greater than 7. Note that it is a "reverse scale," in that the higher the H^+ concentration of a solution, the lower its pH. Thus, 0 represents a strongly acidic solution and 14 represents an extremely basic one (low H^+ concentration).

Each unit on the pH scale represents a 10-fold change in H^+ concentration. A solution with a pH of 4, therefore, is 10 times more acidic than one with a pH of 5, and 100 times more acidic than one with a pH of 6.

All species have optimal pH requirements. Some organisms, such as the bacteria that cause ulcers in human stomachs, are adapted to low-pH environments. In contrast, the normal pH of human blood is 7.35 to 7.45. Illness or death can occur when the blood's pH strays too far from normal. Extremely shallow breathing or kidney failure can cause the blood's pH to drop below 7, a condition called acidosis. Vomiting, hyperventilating, or taking some types of alkaloid drugs, on the other hand, can raise the blood's pH above 7.8.

B. Buffer Systems Regulate pH in Organisms

Maintaining the correct pH of body fluids is critical to an organism's survival. But in the course of life, organisms frequently encounter conditions that could alter their internal pH. How do they maintain homeostasis? The answer lies in **buffer systems**, pairs of weak acids and bases that resist pH changes.

Hydrochloric acid is a strong acid because it releases all of its H^+ when dissolved in water. As you can see in figure 2.14, the pH of pure HCl is 0. A weak acid, in contrast, does not release all of its H^+ into solution. An example is carbonic acid, H_2CO_3, which forms one part of the human body's buffer system:

$$\underset{\text{carbonic acid}}{H_2CO_3} \longleftrightarrow H^+ + \underset{\text{bicarbonate}}{HCO_3^-}$$

The dual arrow indicates that the reaction can proceed in either direction, depending on the pH of the fluid. If a base removes H^+ from the solution, the reaction moves to the right to produce more H^+. Alternatively, if an acid contributes H^+ to the solution, the reaction proceeds to the left and consumes

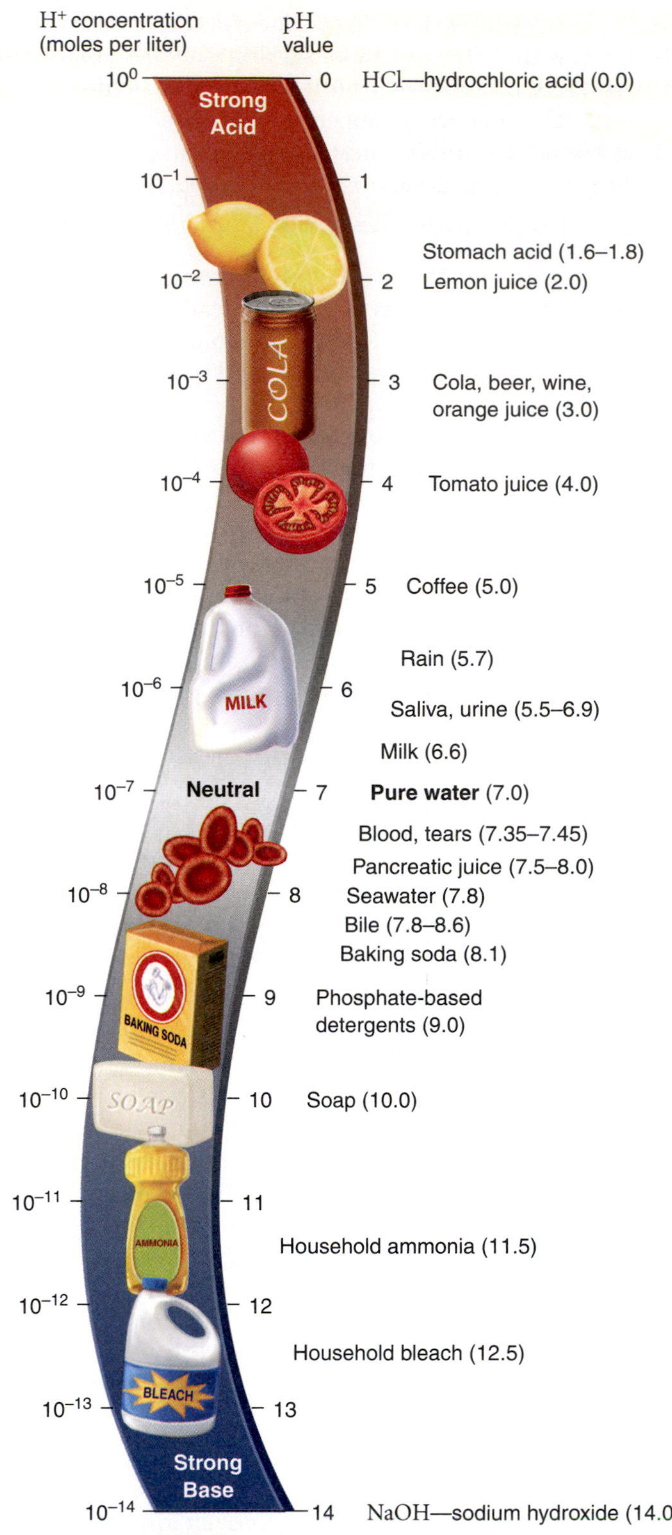

FIGURE 2.14 The pH Scale. The pH scale indicates the concentration of hydrogen ions (H^+). The lower the pH, the higher the concentration of free H^+ becomes and the more acidic the solution. Conversely, the higher the pH, the more free hydroxide (OH^-) ions there are and the more alkaline (basic) the solution.

the H^+. This action keeps the pH of the solution relatively constant. Carbonic acid is just one of several buffers that maintain the pH of body fluids within an optimum range.

Summary *pH is a measure of the concentration of hydrogen ions in a solution. Acids lower pH; bases raise it. The correct pH is critical to life's functions. Buffers help cells maintain proper pH.*

2.4 MASTERING CONCEPTS

1. How do acids and bases affect a solution's H^+ concentration?
2. How does a pH value of 3 compare with a pH of 9?
3. How do buffer systems regulate the pH of a cell?

2.5 Organic Molecules Generate Life's Form and Function

Organisms are composed mostly of water and **organic molecules**, chemical compounds that contain both carbon and hydrogen. (This chapter's Burning Questions box discusses two different uses of the term organic.) As you will learn later in this unit, most plants can produce all the organic molecules they require, whereas animals—including humans—must obtain them from food.

Life uses a tremendous variety of organic molecules. Organic molecules consisting almost entirely of carbon and hydrogen are called hydrocarbons; methane (CH_4) is the simplest example. Because a carbon atom forms four covalent bonds, however, this element can assemble into much more complex molecules, including long chains, intricate branches, and rings. Many organic molecules also include other elements, such as oxygen, nitrogen, phosphorus, or sulfur.

The four most abundant types of organic molecules in organisms are carbohydrates, lipids, proteins, and nucleic acids. Vitamins are also biologically important organic molecules, but they are required in smaller amounts. Vitamin deficiencies can cause illnesses such as scurvy (vitamin C), beriberi (vitamin B_1), and pellagra (vitamin B_3). Chapter 4 describes the function of vitamins in more detail.

Proteins, nucleic acids, and some carbohydrates all share a property in common with each other—they are chains of small molecular subunits called **monomers**. Linked together, these monomers form **polymers**, just as a train is made of individual railcars.

How does your body produce new muscle proteins and other polymers? Your cells use a chemical reaction called dehydration synthesis, also called a condensation reaction, to link the monomers together. In a **dehydration synthesis** reaction, a protein called an enzyme removes an —OH (hydroxyl group) from one molecule and a hydrogen atom from another, forming H_2O and a new bond between the two smaller components. (The term *dehydration* means that water is lost).

Burning Questions

What does it mean when food is "organic" or "natural?"

The word *organic* has multiple meanings. To a chemist, the term describes compounds containing carbon atoms. All food is organic in this sense. To a farmer or consumer, however, the word refers to foods produced according to a defined set of standards.

The U.S. Department of Agriculture (USDA) certifies crops as organically grown if the farmer did not use pesticides (with few exceptions), petroleum-based fertilizers, or sewage sludge. Organically raised cows, pigs, and chickens cannot receive growth hormones or antibiotics, and they must have access to the outdoors and receive organic feed. In addition, no food labeled "organic" may be genetically engineered or treated with ionizing radiation.

Nutritionally, organically grown plants and animals are equivalent to conventional foods, but the groups differ in their effects on the environment. Pesticides and fertilizers can kill nontarget organisms, pollute waterways, and leave residues on crops. Additionally, conventional farms feed animals antibiotics, which may contribute to the spread of antibiotic-resistant bacteria. Finally, organic methods emphasize soil and water conservation and the use of renewable resources.

A term that consumers frequently confuse with organic is natural. A *natural* food may or may not be organic, because the term *natural* refers to the way in which foods are processed, not how they are grown. Standards for what constitutes a natural food are fuzzy. The USDA specifies that meat and poultry labeled as natural cannot contain artificial ingredients or added color, but no such standards exist for other foods.

Have a Burning Question of your own? Submit it to marielle_hoefnagels@mcgraw-hill.com for possible inclusion in future editions of this book!

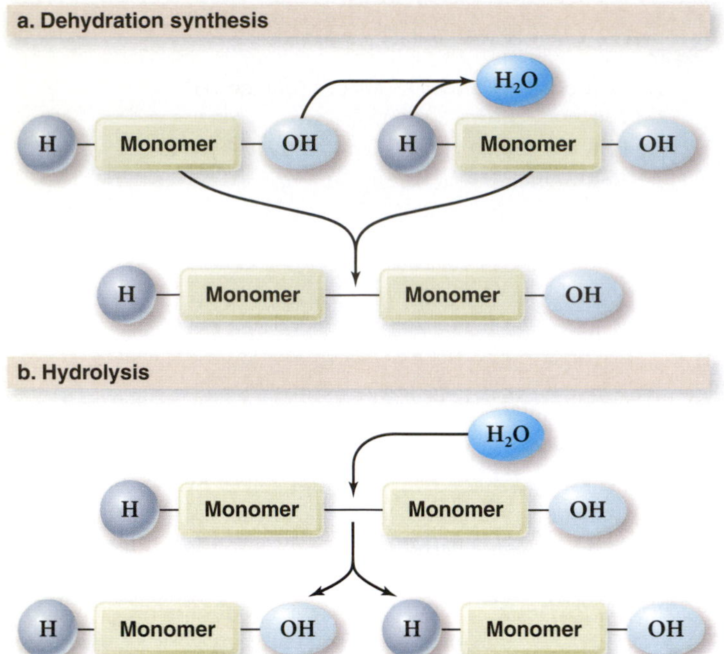

FIGURE 2.15 Opposite Reactions. (**A**) In dehydration synthesis, water is removed and a new covalent bond forms between two monomers. (**B**) In hydrolysis, water is added when the bond between monomers is broken.

Figure 2.15A shows an example of dehydration synthesis. By repeating this reaction many times, cells can build extremely large polymers consisting of thousands of monomers.

The reverse reaction also occurs, breaking the covalent bonds that link monomers (figure 2.15B). In **hydrolysis**, enzymes use atoms from water to add a hydroxyl group to one molecule and a hydrogen atom to another (*hydrolysis* means "breaking with water.") Hydrolysis happens in your body when digestive enzymes in your stomach and intestines break down the proteins and other polymers in food.

The rest of this section takes a closer look at each of the four main types of organic molecules.

A. Carbohydrates Include Simple Sugars and Polysaccharides

Anyone following a "low-carb" diet can recite a list of the foods to avoid: potatoes, pasta, bread, cereal, sugary fruits, and sweets. All are rich in **carbohydrates**, organic molecules that consist of carbon, hydrogen, and oxygen, often in the proportion 1:2:1.

Carbohydrates are the simplest of the four main types of organic molecules, mainly because just a few monomers account for the most common types in cells. The two main groups of carbohydrates are simple sugars and complex carbohydrates.

Simple Sugars

Monosaccharides, the smallest carbohydrates, usually contain five or six carbon atoms (**figure 2.16A**). Monosaccharides with the same number of carbon atoms can differ from one another by how their atoms are bonded. For example, glucose (blood sugar), galactose, and fructose (fruit sugar) are all six-carbon monosaccharides with the molecular formula $C_6H_{12}O_6$, but their chemical structures differ.

A **disaccharide** ("two sugars") is two monosaccharides joined by dehydration synthesis. Figure 2.16B shows how sucrose (table sugar) forms when a molecule of glucose bonds to a molecule of fructose. Lactose, or milk sugar, is a disaccharide formed from glucose and galactose.

Together, monosaccharides and disaccharides are called sugars, or simple carbohydrates. Their function in cells is to provide a ready source of energy, which is released when their bonds are broken (see chapter 6). Sugarcane sap and sugar beet roots contain abundant sucrose, which the plants use to fuel growth. Maltose, a disaccharide formed from two glucose molecules, provides energy in sprouting seeds; beer brewers also use it to promote fermentation.

Oligosaccharides are carbohydrates of intermediate length, consisting of 3 to 100 monomers. They attach to proteins on cell membranes, as discussed in the next chapter. A person's blood type—A, B, or O—refers to the combination of such glycoproteins ("sugar proteins") on blood cells. Among other functions, glycoproteins on cell surfaces are important in immunity.

Complex Carbohydrates

Polysaccharides ("many sugars"), also called complex carbohydrates, are huge molecules consisting of hundreds of monosaccharide monomers. The most common polysaccharides are cellulose, chitin, starch, and glycogen. They are all long chains of glucose, but they differ from one another by the orientation of the bonds that link the monomers.

Cellulose (figure 2.16C) forms part of plant cell walls. Although it is the most common organic compound in nature, humans cannot digest it. Yet cellulose is an important component of the human diet, making up much of what nutrition labels refer to as "fiber." Cotton, wood, and paper consist largely of cellulose. **from wood to paper, p. 513**

Chitin also supports cells, forming the flexible exoskeletons of insects, spiders, and crustaceans and part of the cell wall in fungi. Chitin is the second most common polysaccharide in nature. It resembles a glucose polymer, but a group that contains nitrogen replaces one —OH group in each subunit. Because it is tough, flexible, and biodegradable, chitin is used in the manufacture of surgical thread.

Starch and glycogen store energy (see figure 2.16C). Most plants store energy as starch, which humans exploit as food sources—potato tubers, rice, and wheat are all high-energy, starchy staples in the human diet. Animal and fungal cells store energy in a slightly different form, as glycogen. In humans, muscle cells and the liver store glycogen, which readily breaks down into glucose subunits when cells need energy.

a. Monosaccharides: simple sugars composed of carbon, hydrogen, and oxygen in the proportions 1:2:1.

Glyceraldehyde $C_3H_6O_3$

Ribose $C_5H_{10}O_5$

Glucose $C_6H_{12}O_6$

Fructose $C_6H_{12}O_6$

Galactose $C_6H_{12}O_6$

b. Disaccharides: molecules composed of two monosaccharides joined by dehydration synthesis. Hydrolysis converts disaccharides into their component monosaccharides. (The structures of the molecules are simplified to emphasize the joining process.)

Glucose $C_6H_{12}O_6$ + Fructose $C_6H_{12}O_6$ ⇄ Sucrose $C_{12}H_{22}O_{11}$

Dehydration (H_2O)

Hydrolysis (H_2O)

c. Polysaccharides: complex carbohydrates composed of long chains of simple sugars, usually glucose. Their chemical characteristics are determined by the orientation and location of the bonds between the monomers.

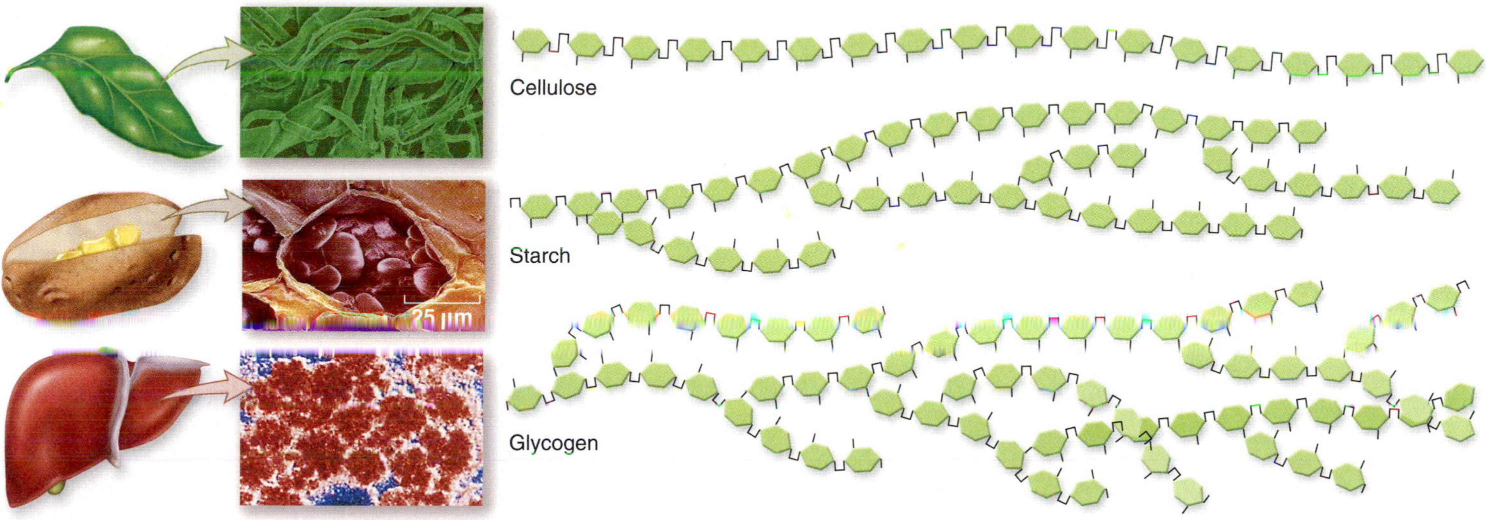

FIGURE 2.16 Carbohydrates—Simple and Complex. (**A**) Monosaccharides are composed of single molecules, such as glucose or fructose. (**B**) Disaccharides form by dehydration synthesis, which binds two monosaccharides and removes water. For instance, glucose and fructose bond to form sucrose. (**C**) Polysaccharides are long chains of monosaccharides such as glucose. Different orientations of these bonds produce different characteristics in the molecules.

B. Lipids Are Hydrophobic and Energy-Rich

Lipids are organic molecules with one property in common: they do not dissolve in water. They are hydrophobic because they contain large areas dominated by nonpolar carbon–carbon and carbon–hydrogen bonds. Unlike the other three groups of organic molecules, lipids are not polymers consisting of long chains of monomers. Instead, they have extremely diverse chemical structures.

This section discusses several groups of lipids: triglycerides, sterols, and waxes. Another important group, phospholipids, forms the majority of cell membranes; chapter 3 describes them.

Triglycerides

A **triglyceride** consists of three long hydrocarbon chains called **fatty acids** bonded to **glycerol**, a three-carbon molecule that forms the triglyceride's backbone. Although triglycerides do not consist of long strings of similar monomers, cells nevertheless use dehydration synthesis to produce them (**figure 2.17**). Each fatty acid has a **carboxyl group**, a carbon atom double-bonded to one oxygen and single-bonded to another oxygen carrying a hydrogen atom (abbreviated as —COOH). Enzymes link the —OH from one fatty acid to each of glycerol's three carbon atoms, yielding three water molecules per triglyceride.

Many dieters try to avoid triglycerides, commonly known as fats. Butter, margarine, oil, cream, cheese, lard, fried foods, and chocolate are all examples of high-fat foods. Nutrition labels divide these fats into two groups: saturated and unsaturated. The degree of saturation is a measure of a fatty acid's hydrogen content. A **saturated** fatty acid contains all the hydrogens it possibly can—that is, single bonds connect all the carbons, and each carbon has two hydrogens (see the straight chains in figure 2.17). Animal fats are saturated and tend to be solid, like bacon fat and butter. Most nutritionists recommend a diet low in saturated fats, citing their tendency to clog arteries.

A fatty acid is **unsaturated** if it has at least one double bond between carbon atoms. These double bonds cause kinks to form in the fatty acid "tails," producing an oily (liquid) consistency at room temperature. Olive oil, for example, is an unsaturated fat, as are most plant-derived lipids. Such fats are healthier than are their saturated counterparts.

Food chemists have discovered how to turn oils into solid fats. A technique called partial hydrogenation, used to produce some brands of margarine and shortening, adds hydrogen to a vegetable oil to solidify it—in essence, partially saturating a formerly unsaturated fat. An unfortunate byproduct of this process is a high proportion of trans fats, which are unsaturated fats whose fatty acid tails are straight, not kinked. Trans fats are common in fast foods, fried foods, and many snack products, and they raise the risk of heart disease even more than saturated fats. Nutritionists therefore recommend a diet as low as possible in trans fats. New York City's Board of Health voted to ban the use of trans fats in restaurants in 2006.

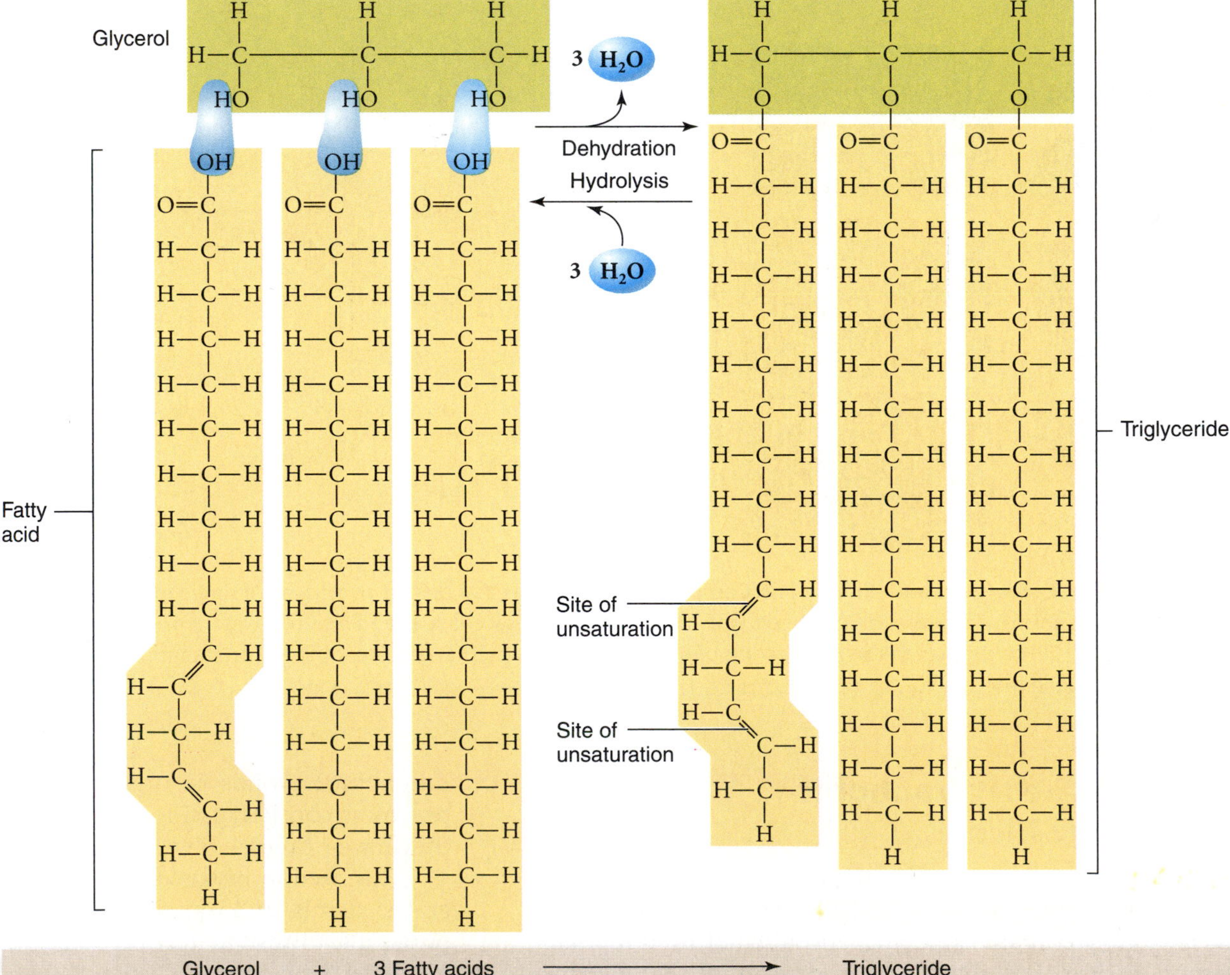

FIGURE 2.17
A Fat Molecule Forms. A triglyceride forms by bonding fatty acids to glycerol. The number of carbon atoms in each fatty acid "tail" ranges from 4 to about 24. Double bonds that bend the fatty acid tails make the lipid more fluid at room or body temperature.

Despite their unhealthful reputation, fats and oils are vital to life in many ways. Fat is an excellent energy source in all organisms, providing more than twice as much energy as equal weights of carbohydrate or protein. In animals, dietary fat is necessary for growth—human milk is rich in lipids, partly to suit the rapid growth of the brain in the first 2 years of life. Fats also slow digestion, thereby delaying hunger, and they are required for the use of some vitamins and minerals.

Fat cells aggregate as adipose tissue in animals. White adipose tissue forms most of the fat in human adults, cushioning organs and helping to retain body heat as insulation. Another type, brown adipose tissue, releases energy as heat and keeps organisms warm, particularly mammals that hibernate.

Sterols

Sterols are lipids that have four interconnected carbon rings. Vitamin D and cortisone are examples of sterols, as is cholesterol (**figure 2.18**). Cholesterol is a key part of animal cell membranes. In addition, animal cells use it as a starting material to synthesize other lipids, including the sex hormones testosterone and estrogen.

Although cholesterol is essential, an unhealthy diet can easily contribute to cholesterol levels that are too high. Excess cholesterol sometimes adheres to the inner linings of blood vessels and impedes blood flow (see Can *You* Relate? " 'Bad' and 'Good' Cholesterol"). Because the liver essentially converts saturated fat into cholesterol, it is important to limit dietary intake of saturated fats as well as cholesterol.

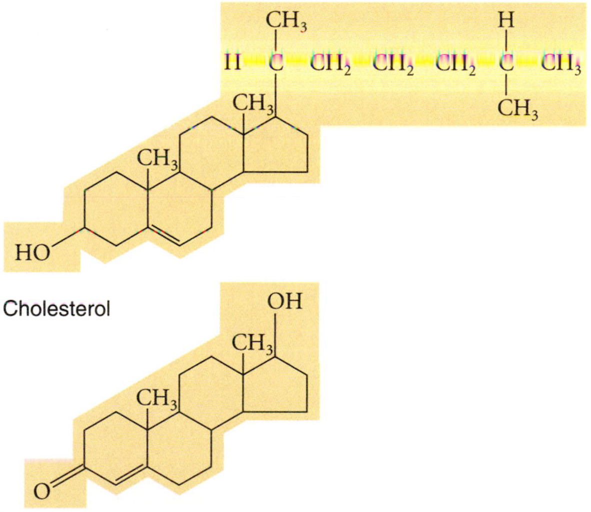

Cholesterol

Testosterone

FIGURE 2.18 **Cholesterol.** In animal cells, cholesterol adds to membrane fluidity and provides raw material for production of steroid hormones such as testosterone.

FIGURE 2.19 **Covered with Wax.** Waxes waterproof the coat of the otter and the cuticles of the grasses growing in the background.

Waxes

Waxes are fatty acids combined with either alcohols or other hydrocarbons, usually forming a stiff, water-repellent covering. These lipids help waterproof fur, feathers, leaves, fruits, and some stems (**figure 2.19**). Jojoba oil, used in cosmetics and shampoos, is unusual in that it is a liquid wax.

C. Proteins Are Complex and Highly Versatile

As varied as the functions of lipids are, they do not approach the diversity of proteins, which do more jobs in the cell than any other type of biological molecule. **Table 2.5** lists a few of the thousands of kinds of proteins in the human body. Proteins literally control all the activities of life, so much so that illness or death can result if even one is missing or faulty. To name just one example, the protein insulin controls the amount of sugar in the blood. The failure to produce insulin leads to one form of diabetes, an illness that can be deadly.

Can *You* Relate?

"Bad" and "Good" Cholesterol

Cholesterol is not water-soluble, so it travels in the bloodstream encased in proteins. The resulting packets of cholesterol (a lipid) and protein are called lipoproteins, and they occur in two varieties.

Low-density lipoprotein (LDL) particles carry cholesterol to the arteries. Excess LDL cholesterol that does not enter cells may accumulate on the inner linings of blood vessels, eventually impeding blood flow, so elevated levels of LDL cholesterol (commonly called "bad cholesterol") increase the risk of heart disease.

High-density lipoproteins (HDL), in contrast, carry cholesterol to the liver, which removes it from the bloodstream. High levels of HDL cholesterol ("good cholesterol") promote heart health.

Table 2.5 *Protein Diversity in the Human Body*

Proteins	Function	Proteins	Function
Actin, myosin, dystrophin	Muscle contraction	Fibrin, thrombin	Blood clotting
Antibodies, cytokines	Immunity	Growth factors	Promote cell division
Carbohydrases, lipases, proteases, nucleases	Digestive enzymes*	Hemoglobin, myoglobin	Transport and storage of oxygen
		Insulin, glucagon	Control of blood glucose level
Casein	Milk protein	Keratin	Structure of hair, fingernails
Collagen, elastin	Connective tissue	Transferrin	Iron transport in blood
Colony-stimulating factors	Blood cell formation	Tubulin, actin	Cell movements
DNA and RNA polymerase	Enzymes* required for DNA replication, gene expression	Tumor suppressors	Block cell division

*Enzymes, discussed further in chapter 4, are proteins that speed chemical reactions. Without enzymes, most of the cell's reactions would proceed much too slowly to sustain life.

Amino Acid Structure and Bonding

A **protein** is a chain of monomers called **amino acids**. Each amino acid has a central carbon atom bonded to four other atoms or groups of atoms(**figure 2.20A**):

- a hydrogen atom;
- a carboxyl group (acid);
- an **amino group**, which is a nitrogen atom single-bonded to two hydrogen atoms (—NH_2);
- a side chain, or **R groups**, which can be any of 20 chemical groups.

Organisms use 20 types of amino acids, even though many others exist. Figure 2.20B shows three of them (see appendix C for a complete set of amino acid structures). The chemical composition of the R group distinguishes the amino acids from one another. An R group may be as simple as the lone hydrogen atom in glycine or as complex as the two organic rings of tryptophan. Some R groups contain nitrogen, and two contain sulfur. Humans can synthesize many of the amino acids, but eight are essential and must come from protein-rich foods such as meat, fish, dairy products, beans, and tofu.

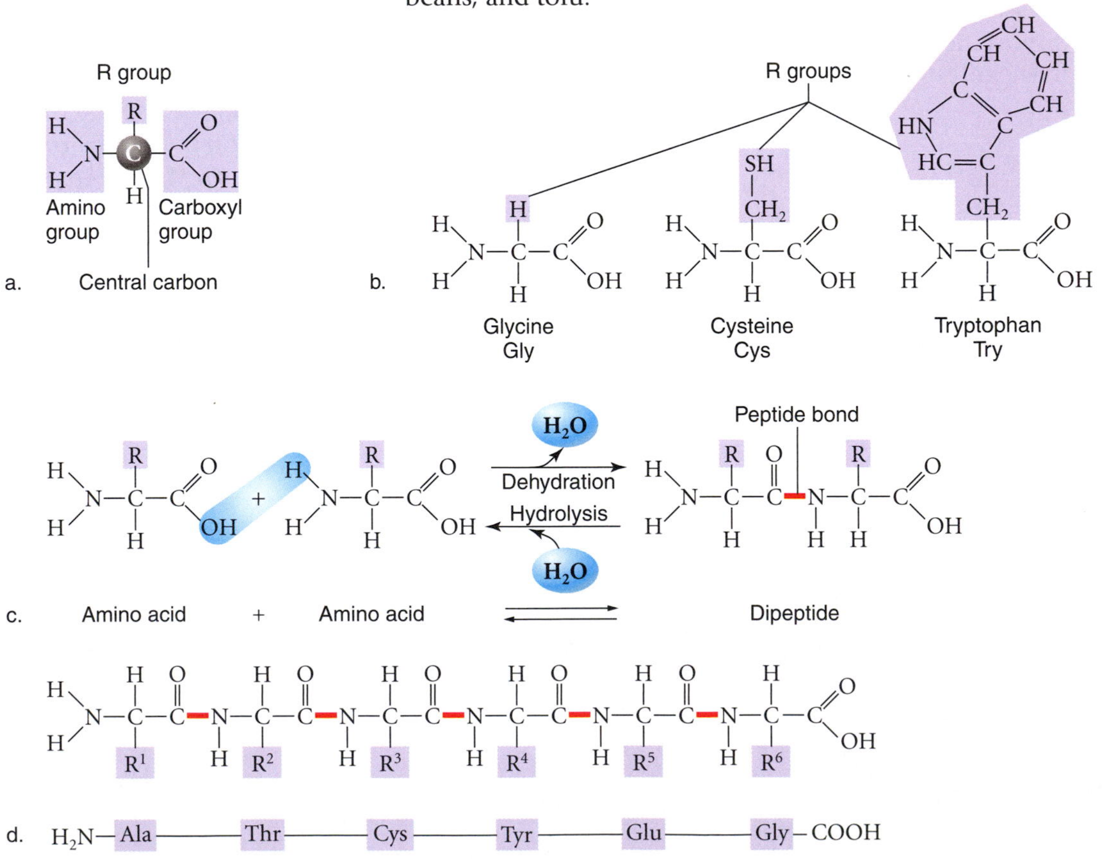

FIGURE 2.20 Amino Acids Join to Form Peptides. Amino acids are the monomer subunits of proteins. (**A**) An amino acid is composed of an amino group, an acid (carboxyl) group, and one of 20 R groups attached to a central carbon atom. (**B**) The composition of the R groups contributes different functions to the final protein. (**C**) A peptide bond forms when an —OH from a carboxyl group of one amino acid combines with a hydrogen from the amino group of another amino acid, creating a water molecule (dehydration synthesis) and linking the carboxyl carbon of the first amino acid to the nitrogen of the other. (**D**) Short chains of amino acids are peptides.

Just as the 26 letters in our alphabet combine to form a nearly infinite number of words in many languages, the 20 different amino acids make possible a nearly infinite number of unique proteins. This variety means that proteins have a seemingly limitless array of structures and functions.

Dehydration synthesis links amino acids together. The **peptide bond** is the covalent bond that links the carbon from the carboxylic acid group of one amino acid to the nitrogen of the other (figure 2.20C). Two linked amino acids form a dipeptide; three form a tripeptide. Chains with fewer than 100 amino acids are peptides (figure 2.20D), and finally, those with 100 or more amino acids are **polypeptides**. A polypeptide is called a protein once it folds into its functional shape; a protein may consist of one or more polypeptide chains. Hydrolysis breaks peptide bonds, releasing the constituent amino acids.

Protein Folding

Unlike polysaccharides, most proteins do not exist as long chains inside cells. Instead, as it is synthesized in a cell, the peptide chain folds into a unique three-dimensional structure determined by the order and kinds of amino acids. Because of this structural complexity, biologists describe the conformation of a protein at four levels (**figure 2.21**):

- **Primary (1°) Structure:** The amino acid sequence of a polypeptide chain, held together by covalent bonds.

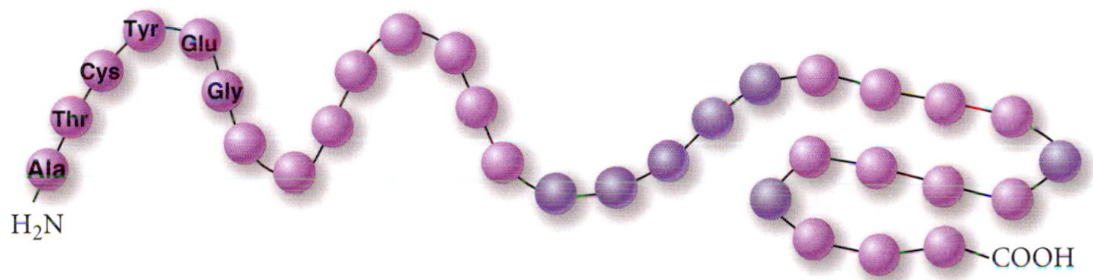

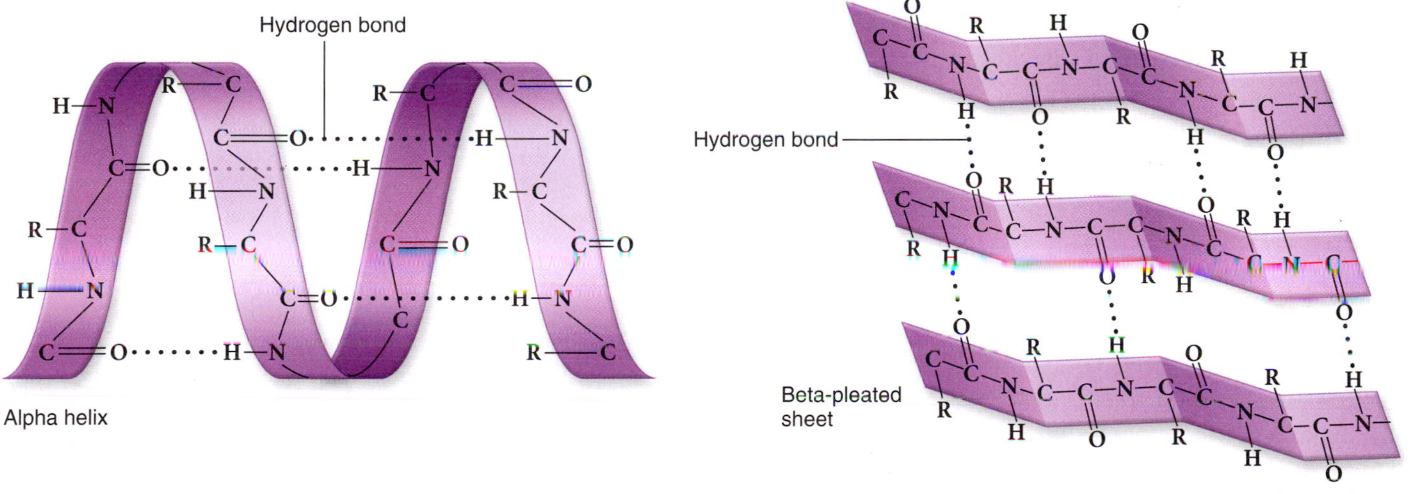

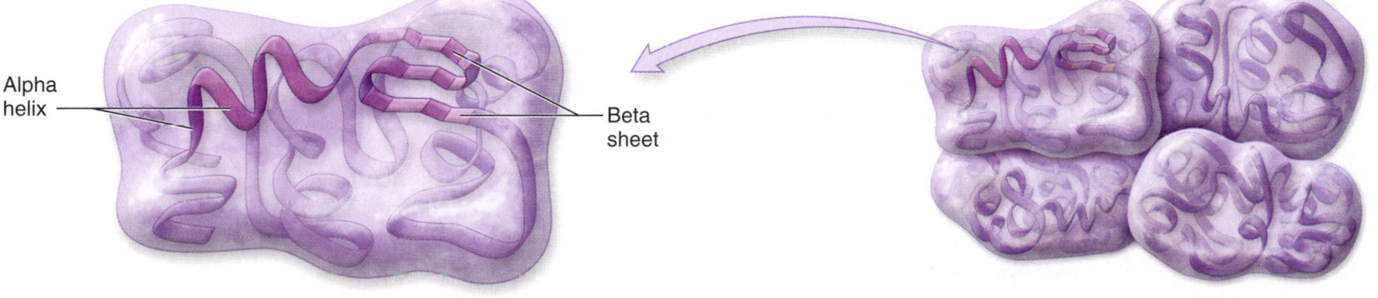

FIGURE 2.21 Four Levels of Protein Structure. (**A**) The amino acid sequence of a polypeptide forms the primary structure, while (**B**) hydrogen bonds between non-R groups create secondary structures such as helices and sheets. The tertiary structure (**C**) is the overall three-dimensional shape of a protein. (**D**) The interaction of multiple polypeptides forms the quaternary structure of a protein.

As detailed in chapter 12, an organism's genetic code specifies the primary structure of each of the organism's proteins. A genetic mutation can cause misfolded, nonfunctional proteins by changing the primary structure at one or more critical locations.

- **Secondary (2°) Structure:** A "substructure" with a defined shape, resulting from hydrogen bonds between parts of the peptide backbone. These interactions fold the polypeptide into coils, sheets, loops, and combinations of these shapes. Figure 2.21B shows two of the more common shapes—alpha helices (singular: helix) and beta pleated sheets. One protein can have many areas of secondary structure.
- **Tertiary (3°) Structure:** The overall shape of a polypeptide, arising primarily through interactions between R groups and water. Inside a cell, thousands of water molecules surround a polypeptide, contorting it as its hydrophobic R groups move away from water toward the protein's interior. Bonds that form within the protein also contribute to this folding process. Ionic and hydrogen bonds form between the peptide backbone and some R groups, and covalent bonds between sulfur atoms in some R groups further stabilize the structure. Called disulfide bonds, these links are abundant in structural proteins such as keratin, which forms hair, scales, beaks, feathers, wool, and hooves (**figure 2.22**).
- **Quaternary (4°) Structure:** The shape arising from interactions between multiple polypeptide subunits of the same protein. For example, the oxygen-toting blood protein hemoglobin is composed of four polypeptide chains, forming a quartet of globular, or ball-shaped, subunits.

Many biologists devote their careers to deducing protein structures, in part because the research has so many practical applications. Misfolded infectious proteins called prions, for instance, cause mad cow disease (see chapter 17). Knowledge of protein structure can also aid in the treatment of infectious disease. If scientists can determine the shape of a protein unique to the organism that causes malaria, for example, they may be able to use that information to create effective new drugs with few side effects. Some consumer products also exploit protein shape. "Permanent wave" solutions and hair straighteners break disulfide bonds in keratin. These bonds then re-form once the hair is in the desired conformation.

Denaturation: Loss of Function

A protein's overall shape makes possible its function. A digestive enzyme holds a large nutrient molecule in just the right way to break the nutrient apart. An antibody binds to a very specific part of a molecule on the surface of a bacterium. Muscle proteins form long, aligned fibers that slide past one another, shortening their length to create muscle contractions.

FIGURE 2.22 A Gallery of Keratin-Based Structures. Keratin forms (**A**) the beak and feathers of a bird, (**B**) the scales of a snake, and (**C**) the horns and fur of a ram.

Proteins are therefore vulnerable to conditions that alter their shapes. High temperatures, excessive salt, or the wrong pH can disrupt the hydrogen bonds that maintain the protein's secondary and tertiary structures. The protein is **denatured** if its structure is modified enough to destroy its function. Cooking an egg, for example, denatures its proteins—the proteins unfold in the heat, then clump and refold randomly as the once-clear egg protein turns solid white. Similarly, fish turns from translucent to opaque as it cooks.

Most denatured proteins will not renature; there is no way to uncook an egg. Gentle denaturation, however, is sometimes reversible. Edible gelatin, for example, is a protein derived from pig and cow collagen. Short chains of amino acids in powdered gelatin wrap around each other, forming minuscule "ropes." When a cook dissolves the powder in hot water, the ropes unwind. As the gelatin cools, some—but not all—of the ropes re-form. Pockets of liquid trapped within the tangled strands create a jellylike texture in the finished product.

As different as carbohydrates, lipids, and proteins are, food chemists have discovered ways to use all three substances to make artificial sweeteners and fat substitutes, as Can *You* Relate? "Sugar Substitutes and Fake Fats" describes.

D. Nucleic Acids Store and Transmit Genetic Information

Making a protein is a more complex task than synthesizing a carbohydrate or fat because of the great variability of amino acid sequences. How does an organism "know" which amino acids to string together to form a particular protein? The answer is that each protein's primary structure is encoded in the sequence of a **nucleic acid**, a polymer consisting of monomers called nucleotides. Cells contain two types of nucleic acids, deoxyribonucleic acid (DNA) and ribonucleic acid (RNA).

Can *You* Relate?

Sugar Substitutes and Fake Fats

Many weight-conscious people turn to artificial sweeteners and fat substitutes to cut calories without sacrificing favorite foods. Chemically, how do these sugar and fat replacements compare with the originals?

Artificial Sweeteners

Table sugar delivers about 4 Calories per gram (as described in chapter 4, a nutritional Calorie—with a capital C—is a measure of energy that represents 1000 calories). The use of artificial sweeteners reduces calorie intake, but not always because they are truly calorie-free. Instead, most are hundreds of times sweeter-tasting than sugar, so a tiny amount of artificial sweetener achieves the same effect as a teaspoon of sugar. A few popular artificial sweeteners include:

- **Saccharin** (sold as Sweet-n-low and Sugar Twin): This sweetener, which has only 1/32 of a Calorie per gram, was originally derived from coal tar in 1879. It consists of a double ring structure that includes nitrogen and sulfur. (Saccharin's eventful history as a food additive is the topic of the Can You Relate? box, "Will This Hurt Me or Won't It?" in chapter 1).
- **Aspartame** (sold as NutraSweet and Equal): Surprisingly, aspartame's chemical structure does not resemble sugar: It consists of two amino acids, phenylalanine and aspartic acid. Like sugar, it delivers about 4 Calories per gram, but it is about 200 times sweeter than sugar, so less is needed. Another dipeptide that may appear in U.S. markets is alitame (alanine and aspartic acid), which tastes 2000 times as sweet as sugar.
- **Sucralose** (sold as Splenda): This sweetener is a close relative of sucrose, except that three chlorine (Cl) atoms replace three of sucrose's —OH groups. Sucralose is about 600 times sweeter than sugar, but the body digests little if any of it, so it is virtually calorie-free.
- **Acesulfame-K** (Sweet One, Sunett): Structurally similar to saccharin, acesulfame-K is about 200 times sweeter than sugar. However, it is essentially calorie-free because the body cannot absorb or use it as an energy source. (The "K" in the name stands for potassium, since acesulfame is sold as its potassium salt.)

Fat Substitutes

Because fat is so calorie-dense (about 9 Calories per gram), cutting fat is a quick way to trim calories from the diet. Excess dietary fat can be harmful, leading to weight gain and increasing the risk of heart disease and cancer. It is important to remember, however, that some dietary fat is essential for good health. Fat aids in the absorption of some vitamins and provides fatty acids that human bodies cannot produce. Fats also lend foods taste and consistency.

Fat substitutes are chemically diverse. The most common ones are based on carbohydrates, proteins, or even fats, and a careful reading of nutrition labels will reveal their presence in many processed foods.

- **Carbohydrate-based fat substitutes:** Modified food starches, dextrins, guar gum, pectin, and cellulose gels are all based on polysaccharides, and they all mimic fat's "mouth feel" by absorbing water to form a gel. Depending on whether they are digestible (starches) or indigestible (cellulose), these fat substitutes deliver 0 to 4 Calories per gram. They cannot be used to fry foods.
- **Protein-based fat substitutes:** These food additives are derived from egg whites or whey (the watery part of milk). When ground into "microparticles," these proteins mimic fat's texture as they slide by each other in the mouth. Protein-based fat substitutes deliver about 4 Calories per gram, and they cannot be used in frying.
- **Fat-based fat substitute:** Olestra (marketed as Olean) is a hybrid molecule that combines a central sucrose molecule with six to eight fatty acids. Its chief advantage is that it tastes and behaves like fat—even for frying. Olestra is currently approved only for savory snacks such as chips. It is indigestible and calorie-free, but some people have expressed concern that olestra removes fat-soluble vitamins as it passes through the digestive tract. Others have publicized its reputed mineral oil-like laxative properties. Most people, however, do not experience problems after eating small quantities of olestra.

Sugar and fat substitutes can be useful for people who cannot eat much of the real thing—or do not wish to. But nutritionists warn that they should not take the place of a healthy diet and moderate eating habits.

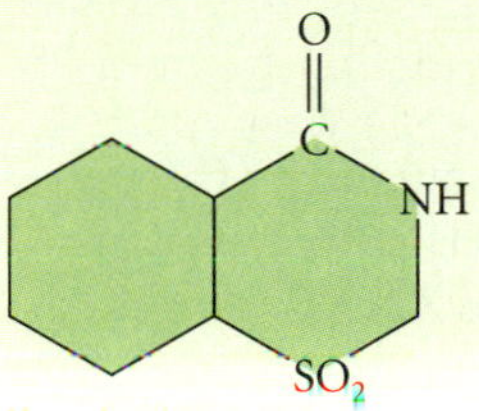

Saccharin

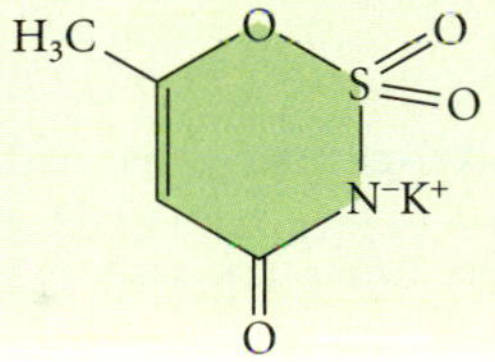

Acesulfame-K (potassium salt)

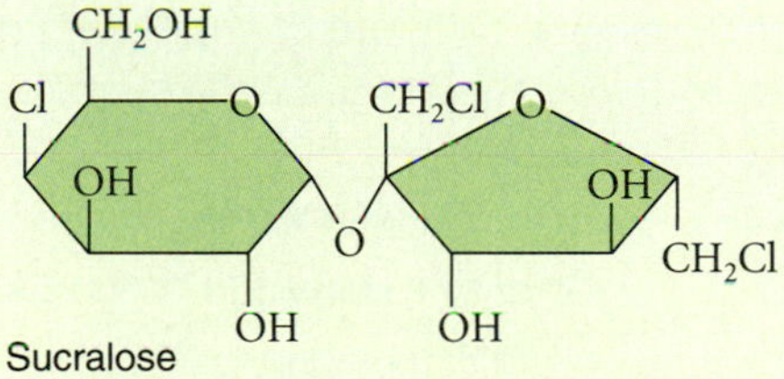

Sucralose

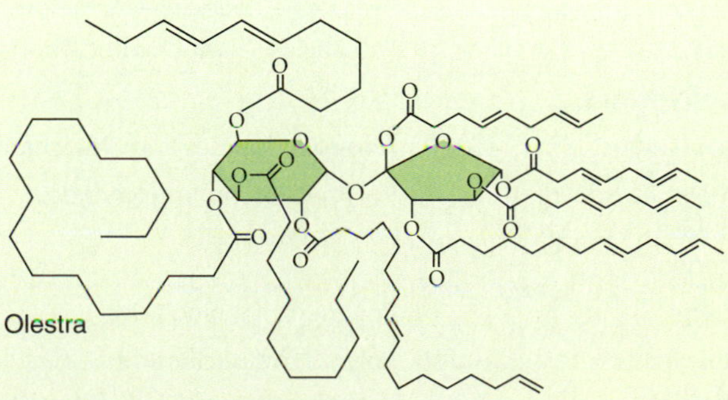

Olestra

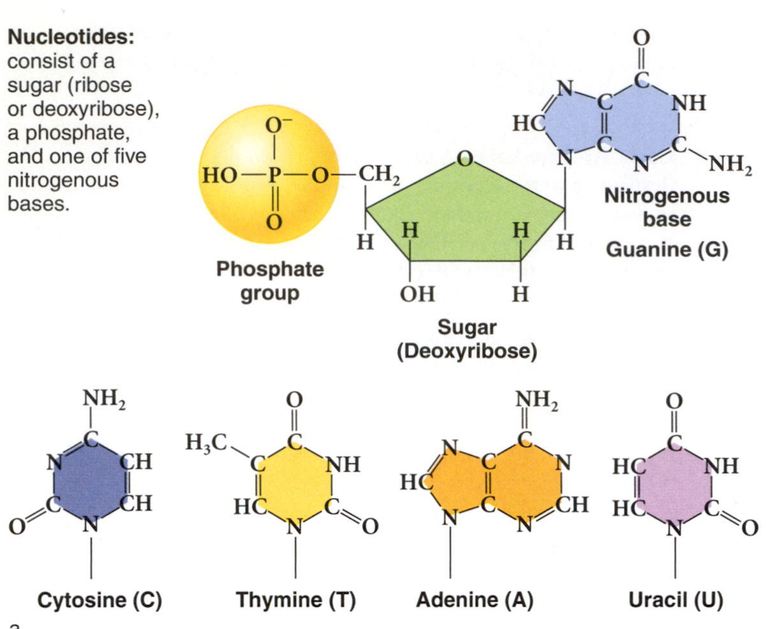

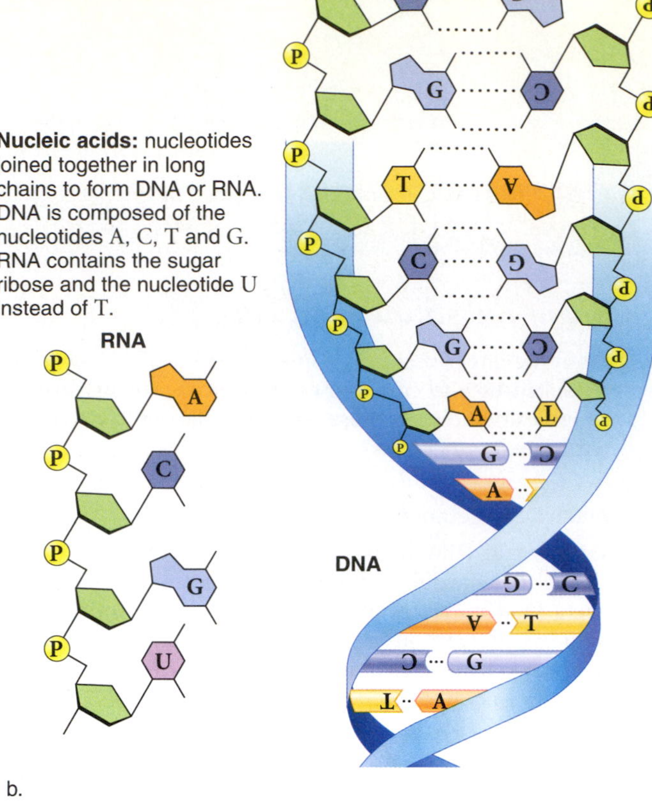

Nucleic acids: nucleotides joined together in long chains to form DNA or RNA. DNA is composed of the nucleotides A, C, T and G. RNA contains the sugar ribose and the nucleotide U instead of T.
RNA
DNA
b.

FIGURE 2.23 DNA and RNA Are Nucleic Acids. (**A**) Nucleic acids are composed of monomers called nucleotides. (**B**) A DNA molecule consists of two strands entwined to form a double-helix shape. Unlike DNA, RNA is usually single-stranded and contains a different sugar (ribose) and a different base (uracil instead of thymine).

Each **nucleotide** monomer consists of three components (**figure 2.23A**):

- a five-carbon sugar—ribose in RNA, and deoxyribose in DNA
- a phosphate group (PO_4)
- a **nitrogenous base:** adenine (A), guanine (G), thymine (T), cytosine (C), or uracil (U). DNA contains A, C, G, and T, whereas RNA contains A, C, G, and U.

The DNA polymer is a double helix and resembles a spiral staircase in which alternating sugars and phosphates form the rails, and nitrogenous bases form the rungs (figure 2.23B). Hydrogen bonds between the bases hold the two strands of nucleotides together: A with T, C with G. If one strand contains an A, the corresponding base on the complementary strand must be a T. The two strands are therefore complementary, or "opposites," of each other. Because of complementary base pairing, one strand of DNA contains

Table 2.6 *The Macromolecules of Life*

Type of Molecule	Chemical structure	Function(s)
Lipids		
Triglycerides (fats, oils)	Glycerol + 3 fatty acids	Store energy
Phospholipids	Glycerol + 2 fatty acids + phosphate group (see chapter 3)	Form major part of biological membranes
Sterols	Four fused rings, mostly of C and H	Stabilize animal membranes; sex hormones
Waxes	Fatty acids + other hydrocarbons or alcohols	Provide water-proofing
Carbohydrates		
Simple sugars	Monosaccharides and disaccharides	Provide quick energy
Complex carbohydrates (cellulose, chitin, starch, glycogen)	Polymers of monosaccharides	Support cells and organisms (cellulose, chitin); store energy (starch, glycogen)
Proteins	Polymers of amino acids	Carry out nearly all the work of the cell (see table 2.5)
Nucleic acids (DNA, RNA)	Polymers of nucleotides	Store and use genetic information and transmit it to the next generation

the information for the other, providing a mechanism for the molecule to replicate. **DNA replication, p. 143**

How does DNA specify the primary structure of a protein? When a cell needs to produce a protein, enzymes copy part of the base sequence of one DNA strand to RNA molecules. As described in chapter 12, the cell uses the RNA to guide the assembly of amino acids into polypeptide chains. In a correspondence called the genetic code, each group of three adjacent DNA bases specifies one amino acid. A complete sequence of DNA bases that specifies one polypeptide is a gene.

DNA, therefore, is the material that stores genetic information. Every organism inherits DNA from its parents (or parent, in the case of asexual reproduction). Slight changes in DNA from generation to generation, coupled with natural selection, account for many of the evolutionary changes that have occurred throughout life's history.

Even though DNA is the genetic material, RNA is, in some ways, even more important. RNA has several functions because its single-stranded structure enables it to assume different shapes. In its various guises, RNA enables the information in DNA to be expressed (see chapter 12). RNA can also function as an enzyme. In addition, one RNA nucleotide, adenosine triphosphate (ATP), carries the energy that cells use in nearly all biological functions. Because of its eclectic roles, RNA, or a molecule similar to it, may have been a bridge between complex groups of chemicals and the first organisms. **ATP, p. 85; RNA world, p. 334**

Table 2.6 reviews the characteristics of the major types of organic molecules in life.

Summary *The organic molecules that make up organisms—carbohydrates, lipids, proteins, and nucleic acids—provide support, energy, protection, and an information system to carry out life's functions. Dehydration synthesis reactions link small molecules together to form larger ones; hydrolysis reactions do the opposite.*

2.5 MASTERING CONCEPTS

1. Draw an example of how monomers link together to form polymers.
2. List examples of carbohydrates, lipids, proteins, and nucleic acids, and name the function of each.
3. Describe the monomers that form polysaccharides, proteins, and nucleic acids.
4. What are the components of a triglyceride?
5. What is the significance of a protein's shape, and how can that shape be destroyed?
6. What are some of the differences between RNA and DNA?

2.6 INVESTIGATING LIFE
A Left Hand from Mars?

The unity of life is reflected in its chemistry: the same elements form the same types of molecules in all organisms. This observation is consistent with evolution by common descent. After all, if one or a few ancient organisms gave rise to all of life's diversity, then contemporary life should reflect the chemistry of that ancestor.

The chemical unity of life extends to the symmetry of its molecules. Some organic molecules can exist in two forms that are mirror images of each other but are not superimposable—like a right and left hand (**figure 2.24**). This relationship, called chirality, occurs when a central carbon bonds to four different chemical groups (see figure 2.20A). Each configuration of a chiral molecule is either "right-handed" or "left-handed," depending on whether light passing through it rotates in a clockwise or counterclockwise direction. (The word *chiral* comes from the Greek word for hand.)

You can demonstrate chirality for yourself by sniffing spearmint leaves and caraway seeds. Both contain an organic molecule called carvone. Spearmint, however, contains only the left-handed version, while carvone in caraway seeds is

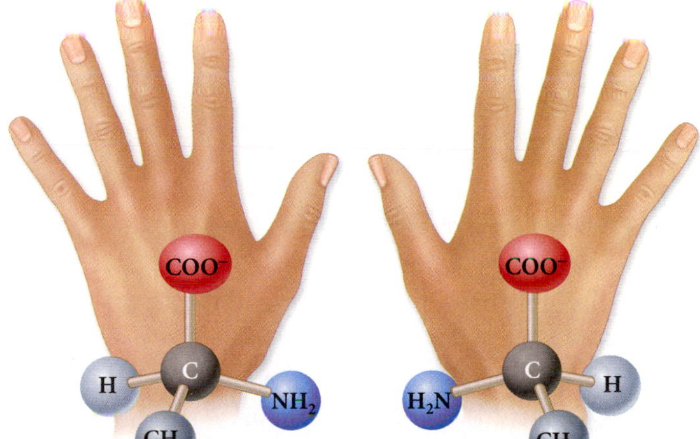

FIGURE 2.24 Living Systems Recognize "Handedness." Molecules that are composed of identical atoms can be mirror images of each other, like human hands. **Question**: Besides your hands, what other everyday mirror-image objects can you think of?

right-handed. The chemical formulas of the two forms are identical, but the sensory proteins of your nose can tell the difference.

When a chemist synthesizes chiral molecules in the laboratory, half of each batch comes out in one configuration, half in the other. Oddly, however, the proportions are not equal in organisms. Nineteen of the 20 amino acids of life are left-handed (the exception, glycine, is not chiral at all). Life's sugars, including the ribose and deoxyribose found in RNA and DNA, are always right-handed.

Biologists argue about what this asymmetry means and how and when it came about. One hypothesis is that meteorites or comets carrying just one form of chiral molecules seeded Earth with the precursors of life. Evidence for this explanation comes from a meteorite that fell to Earth near an Australian town called Murchison in 1969. Researchers discovered that the meteorite had predominantly the left-handed form of three amino acids, including alanine. But did these molecules come from Earth life that contaminated the meteorite after it fell? Or did the amino acids come from space?

Michael Engel, from the University of Oklahoma, and the University of Virginia's Stephen Macko tested the hypothesis that the Murchison meteorite's amino acids are extraterrestrial. They chemically extracted amino acids from the interior of a meteorite stone. Next, they looked for an isotope of nitrogen, ^{15}N, in the amino acids. Other researchers had already discovered that extraterrestrial organic molecules have more ^{15}N than do Earthly organic molecules. Engel and Macko therefore predicted that the Murchison amino acids should be enriched in ^{15}N. That is exactly what they found, for each of the 12 amino acids they studied. Therefore, their study supported the hypothesis that the Murchison meteorite carried the left-handed amino acids to Earth.

Does this mean that life (or its key molecules) originally came from outer space? Not necessarily. Another explanation may be that chirality was part of the origin of life. Perhaps, when the first self-replicating molecules formed on early Earth, either a left-handed or a right-handed version of a key constituent had an advantage. If one form prevailed over the other form early on, it may have persisted without extraterrestrial intervention.

The exact events leading to the single-handedness of life's amino acids and sugars may forever remain a mystery. But chirality's importance extends to the practical, everyday realm of pharmaceutical drugs, many of which are chiral molecules. Often, one of the configurations is effective, and the other is either useless or toxic. One example is the drug thalidomide, which was used to treat morning sickness in pregnant women in the 1940s and 1950s. Sadly, its right-handed version caused severe birth defects in the children of the women who took the drug during pregnancy. **birth defects, p. 765**

To get around this "evil twin" phenomenon, many contemporary drugs include only the right- or left-handed conformation—whichever works best and has the fewest side effects. But even this strategy would not have prevented the birth defects plaguing "thalidomide babies." Each version of thalidomide converts easily to the other in the body, meaning that even a pure preparation of the left-handed conformation would not have been safe.

Evolutionary biologists and drug companies are not alone in their interest in chirality. Food chemists also use molecular "handedness" to develop new flavorings and sweeteners; the caraway–spearmint example is just one of many instances in which a flavor molecule's configuration matters. The wide range of reasons to understand chirality reinforces the importance of chemistry to the study of biology.

Engel, M. H. and S.A. Macko. 1997. Isotopic evidence for extraterrestrial nonracemic amino acids in the Murchison meteorite. *Nature*, vol. 389, pages 265–268.

CHAPTER SUMMARY

2.1 Atoms Are the Stuff of Life

- Any substance that occupies space is called **matter**. Matter can be broken down into pure substances called **elements**.

A. Elements Are Fundamental Types of Matter

- **Bulk elements**, those essential to life in large quantities, include carbon, hydrogen, oxygen, nitrogen, sulfur, and phosphorus. **Trace elements** are required in smaller amounts.

B. Atoms Are Particles of Elements

- An **atom** is the smallest unit of an element. Positively charged **protons** and neutral **neutrons** form the **nucleus**, and the negatively charged, much smaller **electrons** circle the nucleus.
- Elements are organized in the **periodic table** according to **atomic number** (the number of protons).
- An **ion** is an atom that gains or loses electrons.

C. Isotopes Have Different Numbers of Neutrons

- The **mass number** of an atom is the combined number of protons and neutrons. **Isotopes** of an element differ by the number of neutrons. A **radioactive** isotope is unstable.
- An element's **atomic mass** reflects the average mass number of all isotopes, weighted by the proportions in which they naturally occur.

2.2 Chemical Bonds Link Atoms

- A **molecule** is two or more atoms joined together; if they are of different elements, the molecule is called a **compound**. A compound's characteristics differ from those of its constituent elements.

A. Electrons Determine Bonding

- Electrons move constantly; they are most likely to occur in volumes of space called **orbitals**. Orbitals are grouped into **energy shells**.
- An atom's tendency to fill its outermost or **valence shell** with electrons drives it to form **chemical bonds** with other atoms.

B. In a Covalent Bond, Atoms Share Electrons

- **Covalent bonds** form between atoms that can fill their valence shells by sharing one or more pairs of electrons.
- Atoms in a **nonpolar covalent bond** share electrons equally. **Electronegative** atoms in covalent bonds attract electrons away from less electronegative atoms, forming **polar covalent bonds**.

C. In an Ionic Bond, One Atom Takes Electrons from Another Atom

- An **ionic bond** is an attraction between two oppositely charged ions, which form when one atom donates one or more electrons to another atom.

D. Partial Charges on Polar Molecules Create Hydrogen Bonds

- Polar covalent bonds produce opposite partial charges on different parts of a molecule. **Hydrogen bonds** result from the attraction between opposite partial charges on adjacent molecules or between oppositely charged parts of a large molecule. Each hydrogen bond is relatively weak.

2.3 Water Is Essential to Life

A. Water Is Cohesive and Adhesive

- Water is **cohesive** and **adhesive**, sticking to itself and other materials.

B. Polar Substances Dissolve in Water

- A **solution** consists of a **solute** dissolved in a **solvent**.
- Water dissolves **hydrophilic** (polar and charged) substances but not **hydrophobic** (nonpolar) substances.

C. Water Regulates Temperature

- Water helps regulate temperature in organisms because it resists temperature change and **evaporation**. Ice is less dense than liquid water, a property that protects aquatic life in freezing weather.

D. Water Participates in Life's Chemical Reactions

- In a **chemical reaction**, the **products** are different from the **reactants.**
- Some compounds break down and others form, but the total number of atoms of each element remains the same. Most biochemical reactions occur in a watery solution.

2.4 Organisms Balance Acids and Bases

- In pure water, the numbers of H^+ and OH^- in water are equal, and the solution is **neutral**. An **acid** adds H^+ to a solution, and a **base** adds OH^- or removes H^+.

A. The pH Scale Expresses Acidity or Alkalinity

- The **pH scale** measures H^+ concentration. Pure water has a pH of 7, acidic solutions have a pH below 7, and an **alkaline** solution has a pH between 7 and 14.

B. Buffer Systems Regulate pH in Organisms

- **Buffers** consist of weak acid–base pairs that maintain the pH ranges of body fluids.

2.5 Organic Molecules Generate Life's Form and Function

- Many large **organic molecules** are composed of small subunit molecules called **monomers**, which possess characteristics distinct from the resulting **polymers**. Monomers link to form polymers by **dehydration synthesis** or are released from polymers by **hydrolysis**.

A. Carbohydrates Include Simple Sugars and Polysaccharides

- **Carbohydrates** consist of carbon, hydrogen, and oxygen in the proportions 1:2:1.
- **Monosaccharides** are single-molecule sugars such as glucose. Two bonded monosaccharides form a **disaccharide**. These simple sugars provide quick energy.
- **Oligosaccharides** are short chains of monosaccharides.
- **Polysaccharides** are complex carbohydrates consisting of hundreds of monosaccharides. They provide support and store energy.

B. Lipids Are Hydrophobic and Energy-Rich

- **Lipids** are diverse hydrophobic compounds consisting mainly of carbon and hydrogen.
- **Triglycerides** (fats and oils) consist of **glycerol** and three **fatty acids**, which may be **saturated** (no double bonds) or **unsaturated** (at least one double bond). They store energy, slow digestion, cushion organs, and preserve body heat.
- **Sterols**, including cholesterol and sex hormones, are lipids containing four carbon rings.
- **Waxes** are hard, waterproof coverings consisting of fatty acids combined with other molecules.

C. Proteins Are Complex and Highly Versatile

- **Proteins** consist of **amino acids**, each of which has a central carbon atom bonded to a hydrogen atom, an **amino group**, a carboxyl group, and one of 20 variable **R groups**. Amino acids join into **polypeptides** by forming **peptide bonds** through dehydration synthesis.

- A protein's three-dimensional shape is vital to its function and is determined by the amino acid sequence (**primary structure**), hydrogen bonds (**secondary structure**), hydrophobic interactions, and both ionic and covalent bonds (**tertiary** and **quaternary structure**). A **denatured** protein has a ruined shape.
- Proteins have a great variety of functions, participating in all the work of the cell.

D. Nucleic Acids Store and Transmit Genetic Information

- **Nucleic acids**, including DNA and RNA, are polymers consisting of **nucleotides**. DNA's nucleotides include deoxyribose and the **nitrogenous bases** adenine, cytosine, guanine, and thymine. RNA contains ribose and has uracil instead of thymine.
- DNA carries genetic information and transmits it from generation to generation. RNA copies the information, enabling the cell to synthesize proteins.

2.6 Investigating Life: A Left Hand from Mars?

- Most amino acids, along with ribose and deoxyribose, are chiral—they can occur in "left-handed" or "right-handed" forms. Biologists puzzle over the reasons that life always uses one form or the other of these chiral molecules.

MULTIPLE CHOICE QUESTIONS

1. The *atomic mass* of an element represents the total number of
   a. electrons.
   b. protons.
   c. neutrons.
   d. protons + neutrons.
2. The atomic number of the element neon (Ne) is 10. How many electrons does a neutral atom of neon contain?
   a. 5
   b. 10
   c. 20
   d. It can't be determined from this information.
3. A *covalent bond* is formed when
   a. electrons are present in a valence shell.
   b. a valence electron is removed from one atom and added to another.
   c. a valence electron is shared between two atoms.
   d. the electronegativity of one atom is greater than that of another atom.
4. The atomic number of silicon (Si) is 14. Use the idea of energy shells to predict the number of covalent bonds that Si could form.
   a. 2
   b. 3
   c. 4
   d. 8
5. An *ionic bond* is formed when
   a. an electrical attraction occurs between two atoms of different charge.
   b. a nonpolar attraction is formed between two atoms.
   c. a valence electron is shared between two atoms.
   d. two atoms have similar electronegativities.
6. A hydrophilic substance is one that can
   a. form covalent bonds with hydrogen.
   b. dissolve in water.
   c. buffer a solution.
   d. interact with nonpolar solvents.
7. What type of chemical bond is being broken when methane is burned?
   a. Ionic
   b. Hydrogen
   c. Polar covalent
   d. Nonpolar covalent
8. What type of chemical bond is formed during a dehydration synthesis reaction?
   a. Covalent
   b. Ionic
   c. Hydrogen
   d. Polymer
9. A sugar is an example of a __________ molecule while a molecule of DNA is a __________ .
   a. protein; nucleic acid
   b. nucleic acid; lipid
   c. lipid; protein
   d. carbohydrate; nucleic acid
10. The shape of a protein is determined by
    a. the sequence of amino acids.
    b. chemical bonds between amino acids.
    c. temperature and pH.
    d. all of the above.

TESTING YOUR KNOWLEDGE

1. Define the following terms: *atom, element, molecule, compound, isotope*, and *ion*.
2. The vitamin biotin contains 10 atoms of carbon, 16 of hydrogen, 3 of oxygen, 2 of nitrogen, and 1 of sulfur. What is its molecular formula?
3. Distinguish between nonpolar covalent bonds, polar covalent bonds, and ionic bonds.
4. If oxygen is highly electronegative, why is a covalent bond between two oxygen atoms considered nonpolar?
5. Why does carbon usually form exactly four covalent bonds?
6. Can nonpolar molecules such as CH_4 participate in hydrogen bonds? Why or why not?
7. Define *solute, solvent*, and *solution*.
8. Explain why each of the following properties of water is essential to life: cohesion, adhesion, ability to dissolve solutes, resistance to temperature change.
9. What is the physical basis of pH?
10. Is a shampoo labeled "nonalkaline" more likely to have a pH of 3, 6, or 12?
11. Why are buffer systems important in organisms?
12. Compare and contrast the chemical structures and functions of carbohydrates, lipids, proteins, and nucleic acids.
13. How is an amino acid's R group analogous to a nucleotide's nitrogenous base?

THINKING AS A SCIENTIST

1. Consider the following atomic numbers: nitrogen (N) = 7; oxygen (O) = 8; fluorine (F) = 9; neon (Ne) = 10; magnesium (Mg) = 12. Build a Bohr model of each atom, and then predict how many covalent (or ionic) bonds each atom should form.
2. On a scale of 0 to 4, potassium (K) has electronegativity of 0.82, and chlorine's (Cl) is 3.16 (see figure 2.7). Would a bond between K and Cl be nonpolar covalent, polar covalent, or ionic? Explain your answer.
3. Using your knowledge of the properties of water, explain the quote "Hydrogen bonds sank the *Titanic*."
4. Pickles and several other foods are preserved in acids such as vinegar. Why is an acid a good preservative? [*Hint*: Consider the effect of acids on protein shape].
5. A topping for ice cream contains fructose, hydrogenated soybean oil, salt, and cellulose. What types of chemicals are in it?
6. Why are proteins extremely varied in organisms, but carbohydrates and lipids are not?
7. Amyotrophic lateral sclerosis, also known as ALS, or Lou Gehrig's disease, paralyzes muscles. An inherited form of the illness is caused by a gene (sequence of DNA) that encodes an abnormal enzyme that contains zinc and copper. The abnormal enzyme fails to rid the body of a toxic form of oxygen. Which of the molecules mentioned in this description is a:

   (a) protein? (c) bulk element?
   (b) nucleic acid? (d) trace element?
8. A man on a very low-fat diet proclaims to his friend, "I'm going to get my cholesterol down to zero!" Why is achieving this goal impossible (and undesirable)?
9. Using information in the Can *You* Relate? box "Sugar Substitutes and Fake Fats" on page 41 and the amino acid structures in Appendix C, draw the dipeptide called aspartame (NutraSweet).
10. Three very different proteins are silk, hair, and collagen (used to plump lips in plastic surgery). Chemically, how are they similar, and how are they different?
11. Name three examples of emergent properties (see chapter 1) in chemistry.

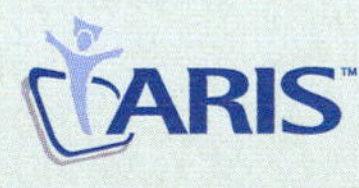

Visit www.mhhe.com/hoefnagels for practice quizzes, animations, videos, and activities designed to help you master the material in this chapter.

CHAPTER 3

Cells

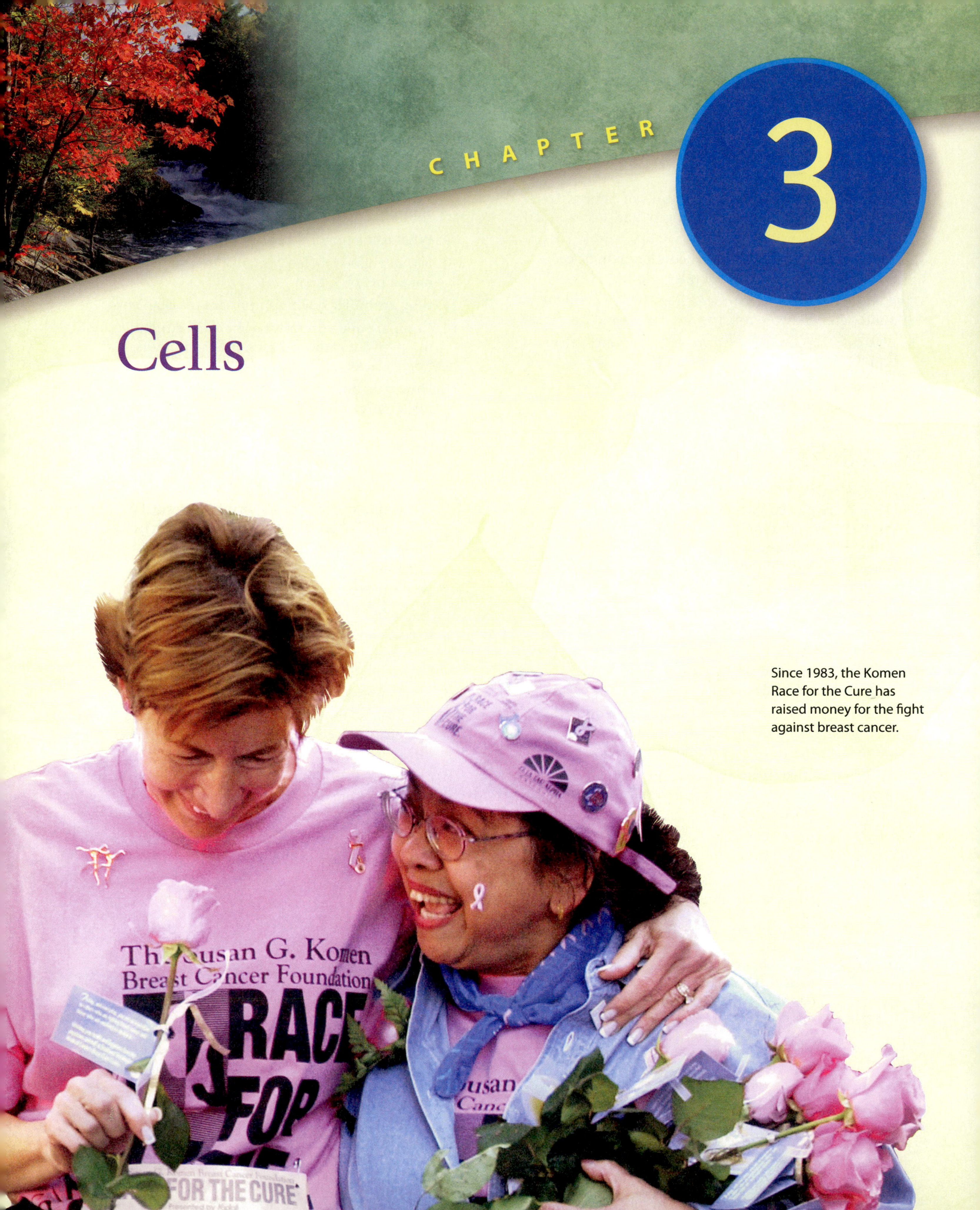

Since 1983, the Komen Race for the Cure has raised money for the fight against breast cancer.

Cancer Cells: A Tale of Two Drugs

Preventing and conquering cancer are compelling reasons to study cell biology. In cancer, a person's own cells multiply out of control, invading nearby tissues or spreading to other parts of the body (see chapter 8). Research revealing how cancer cells differ from normal cells has yielded spectacular new treatments that target these differences.

A healthy cell has a characteristic shape, with a membrane boundary that allows entry to some substances, yet blocks others. Not so the misshapen cancer cell, with its fluid surface and less discriminating boundaries. The cancer cell squeezes into spaces where other cells cannot, secreting biochemicals that blast pathways through healthy tissue, even creating its own blood supply. The renegade cell's genetic controls change, and it transmits these mutations when it divides. Cancer cells disregard the "rules" of normal cell division and multiply unchecked, forming tumors or uncontrolled populations of cells.

Research unraveling how deranged signals prompt the out-of-control cell division of cancer has produced new drugs that inhibit only abnormal cells; trastuzumab (Herceptin) is one. This drug, which was created to treat some forms of breast cancer, got its name from its target: HER2, a receptor protein on the surface of breast cells. The HER2 receptor binds to a molecule that stimulates the cell to divide. Normal breast cells have thousands of HER2 receptors, but cells of one form of breast cancer have many millions. Cells with too many HER2 receptors divide and spread rapidly. Herceptin prevents this by binding specifically to HER2 receptors. The drug also stimulates the immune system to kill the cancer cell.

Another successful drug is imatinib (Gleevec), which treats some forms of leukemia and gastrointestinal cancers. Leukemia is a disease in which the body produces many abnormal white blood cells. In a type of leukemia called chronic myeloid leukemia, a DNA change causes cells to produce an abnormal protein. This protein prompts the body to produce cancerous cells. Gleevec blocks the protein, selectively slowing division of the abnormal cells without harming normal cells.

Both Herceptin and Gleevec took decades to develop. Each interferes with an aspect of cell biology unique to the cancer cells, producing fewer side effects than older treatments that destroy healthy cells, too. These drugs owe their success to generations of cell biologists who painstakingly documented the structures in and on cells—the subject of this chapter.

LEARNING OUTLINE

3.1 Cells Are the Units of Life
- A. Discovering the Cellular Basis of Life
- B. The Cell Theory Emerges
- C. Microscopes Reveal Cell Structure
- D. Features Common to All Cells

3.2 A Membrane Separates Each Cell from Its Surroundings
- A. Lipids and Proteins Form the Cell Membrane
- B. Signal Transduction Transmits Messages to a Cell's Interior

3.3 Different Cell Types Characterize Life's Three Domains
- A. Domain Bacteria
- B. Domain Archaea
- C. Domain Eukarya

3.4 Eukaryotic Organelles Divide Labor
- A. Organelles Interact to Secrete Substances
- B. Lysosomes, Vacuoles, and Peroxisomes Are Cellular Digestion Centers
- C. Chloroplasts Are Glucose Factories
- D. Mitochondria Extract Energy from Nutrients

3.5 The Cytoskeleton Supports Eukaryotic Cells
- A. Microtubules Participate in Cell Division and Cell Movement
- B. Microfilaments Move Other Parts Inside a Cell
- C. Intermediate Filaments Maintain Shape and Connect Cells

3.6 Cells Stick Together and Communicate with One Another
- A. Cell Walls Are Strong, Flexible, and Porous
- B. Animal Cell Junctions Occur in Several Forms

3.7 Investigating Life: Did the Cytoskeleton Begin in Bacteria?

3.1 Cells Are the Units of Life

A human, a hyacinth, a mushroom, and a bacterium appear to have little in common other than being alive. However, on a microscopic level, these organisms share many similarities. All organisms consist of microscopic structures called **cells,** the smallest unit of life that can function independently. Within cells, highly coordinated biochemical activities carry on the basic functions of life. This chapter introduces the cell, and the chapters that follow delve into specific cellular events.

A. Discovering the Cellular Basis of Life

The study of cells—cell biology—began in 1660, when English physicist Robert Hooke melted strands of spun glass to create lenses that he focused on bee stingers, fish scales, fly legs, feathers, and any type of insect he could hold still. When he looked at cork, which is bark from a type of oak tree, it appeared to be divided into little boxes, left by cells that were once alive. Hooke called these units "cells" because they looked like the cubicles (Latin, *cellae*) where monks studied and prayed. Although Hooke did not realize the significance of his observation, he was the first person to see the outlines of cells.

In 1673, Antony van Leeuwenhoek of Holland improved lenses further (**figure 3.1A**). He used only a single lens, but it was more effective at magnifying and produced a clearer image than most two-lens microscopes then available. One of his first objects of study was tartar scraped from his own teeth, and his words best describe what he saw there:

> *To my great surprise, I found that it contained many very small animalcules, the motions of which were very pleasing to behold. The motion of these little creatures, one among another, may be likened to that of a great number of gnats or flies disporting in the air.*

Leeuwenhoek opened up a vast new world to the human eye and mind (figure 3.1B). He viewed bacteria and protista that people hadn't known existed. However, he failed to see the single-celled "animalcules" reproduce, and therefore he perpetuated the idea that life arises from nonliving matter or from nothing. Nevertheless, he described with remarkable accuracy microorganisms and microscopic parts of larger organisms, including human red blood cells and sperm.

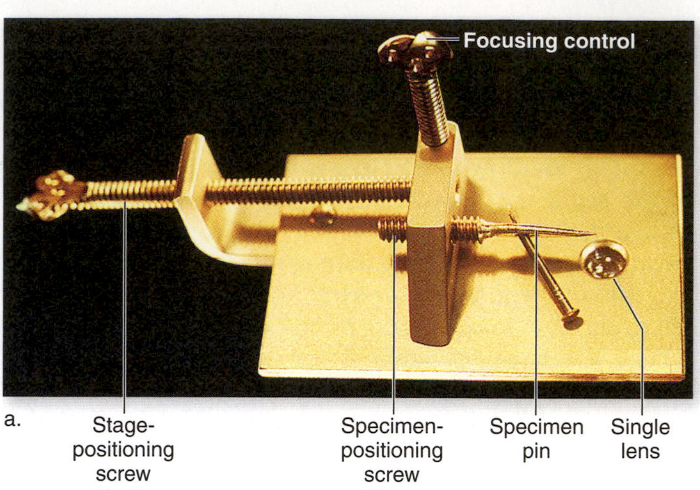

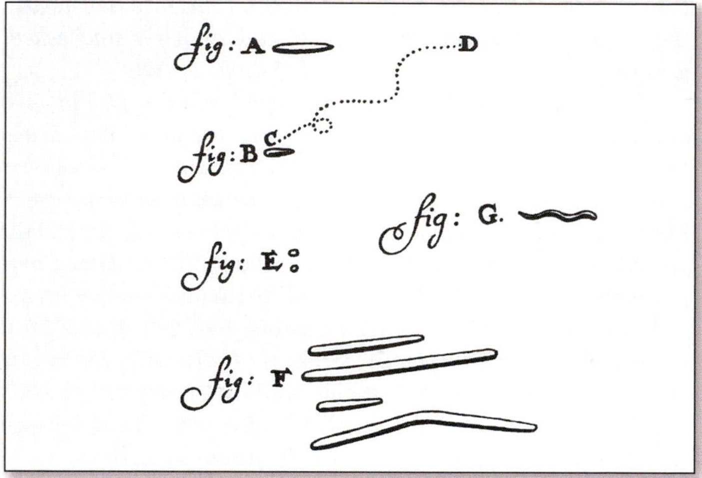

FIGURE 3.1 A First Microscope. (**A**) Antony van Leeuwenhoek made many simple microscopes such as this example, which opened a previously unseen world to view. (**B**) He drew what he saw and, in doing so, made the first record of microorganisms.

B. The Cell Theory Emerges

In the nineteenth century, more powerful microscopes, with better magnification and illumination, revealed details of life at the subcellular level. In the early 1830s, Scottish surgeon Robert Brown noted a roughly circular structure in cells from orchid plants. He saw the structure in every cell, then identified it in cells of a variety of organisms. He named it the "nucleus," a term that stuck. Today, we know that in complex cells the **nucleus** houses DNA. Soon microscopists distinguished the translucent, moving material that made up the rest of the cell, calling it the cytoplasm.

In 1839, German biologists Mathias J. Schleiden and Theodor Schwann proposed cell theory based on many observations made with microscopes. Schleiden first noted that cells were the basic units of plants, and then Schwann compared animal cells to plant cells. After observing similarities in many different plant and animal cells, they concluded that cells were "elementary particles of organisms, the unit of structure and function."

Schleiden and Schwann articulated two of the main tenets of **cell theory**. First, all organisms are made of one or more cells (Can *You* Relate? One Cell, Two Cells … a Trillion Cells … and More! suggests just how many cells make up a human). Second, the cell is the fundamental unit of all life. Many cell biologists extended Schleiden and Schwann's observations and ideas. German physiologist Rudolf Virchow added a third component in 1855: all cells come from preexisting cells. This contradicted spontaneous generation, the then-widespread idea that life can arise from inanimate matter or from nothingness. The French chemist and microbiologist, Louis Pasteur, finally disproved spontaneous generation in 1859, thereby providing important evidence in support of cell theory.

The existence of cells is an undisputed fact, yet cell theory is still evolving. For 150 years after its formulation, research focused on documenting the parts of a cell—the topic of this chapter—and the mechanics of cell division. Since the discovery of DNA's structure and function in the 1950s, however, cell theory has focused on the role of genetic information in dictating what happens inside cells. Modern cell theory therefore adds the ideas that all cells have the same basic chemical composition (see chapter 2), use energy (see chapters 4, 5, and 6), and contain DNA that is duplicated and passed on as each cell divides (see chapters 7, 8, and 9).

Like any scientific theory, cell theory is *potentially* falsifiable—yet many lines of evidence support each of its components, making it one of the most powerful ideas in biology.

Can *You* Relate?

One Cell, Two Cells . . . a Trillion Cells . . . and More!

How many cells are in the human body? For adults, estimates range widely from about 10 trillion to 100 trillion, indicating that this question is harder to answer than it appears. First, the number changes throughout life. A child's growth comes from cell division that adds new cells, not from the expansion of existing ones. Second, no one has found a good way to count them all. Cells come in so many different shapes that it is hard to extrapolate from a small sample to the whole body. Also, new cells arise as old cells die, so a "true" count is a moving target.

Surprisingly, nonhuman cells vastly outnumber the body's own cells. Microbiologists estimate that the number of bacteria living in and on a typical human is *10 times* the number of human cells! Although some of these bacteria can cause disease, most exist harmlessly on the skin and in the mouth and gastrointestinal tract. These inconspicuous guests, which so vastly outnumber your own cells, also can help extract nutrients from food and prevent disease.

C. Microscopes Reveal Cell Structure

Studying life at the cellular and molecular levels requires magnification, because most cells are too small for the unaided human eye to see. Cell biologists use a variety of microscopes to magnify different types of images of cell contents. Following is a survey of several types; **figure 3.2** provides a sense of the size of objects that they can image.

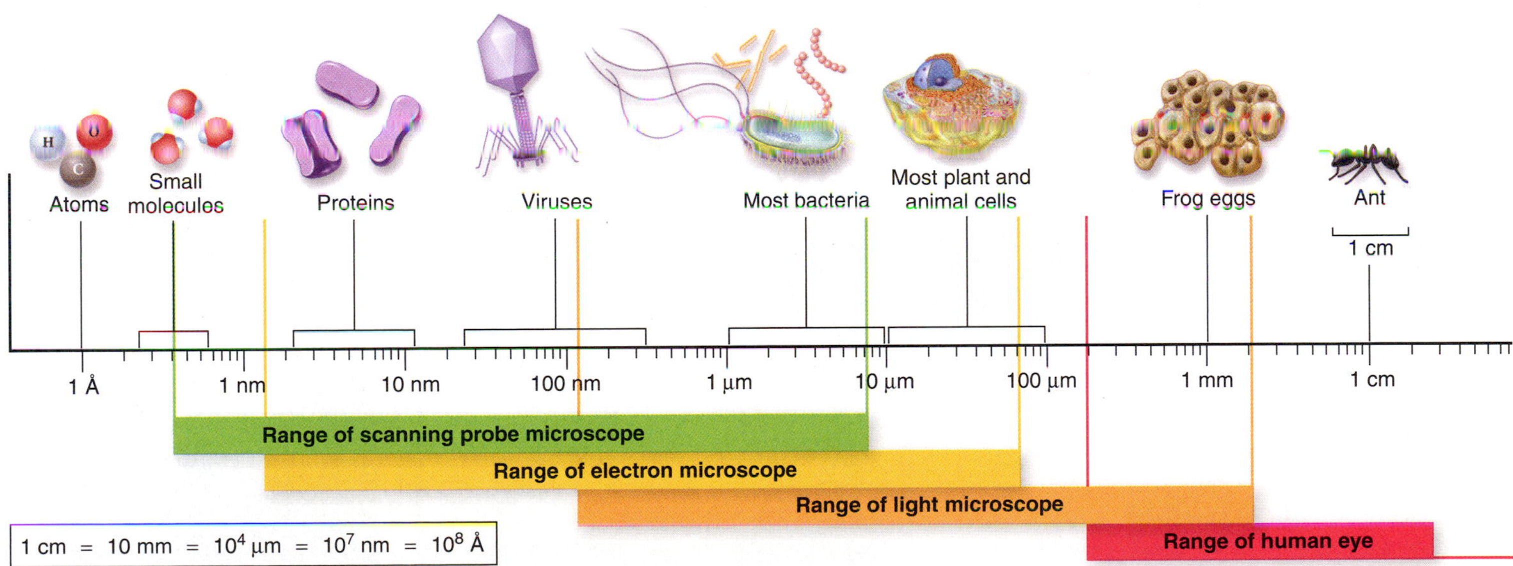

FIGURE 3.2 Ranges of the Light, Electron, and Scanning Probe Microscopes. Biologists use the metric system to measure size (see appendix B). The basic unit of length is the meter (m), which is slightly longer than a yard. Smaller metric units measure many chemical and biological structures. A centimeter (cm) is 0.01 meter; a millimeter (mm) is 0.001 meter; a micrometer (μm) is 0.000001 meter; a nanometer is 0.000000001 meter; and an angstrom unit (Å) is 1/10 of a nanometer. In the scale shown, each segment represents only 1/10 of the length of the segment to its right.

Light Microscopes

The compound light microscope (**figure 3.3A**) uses glass lenses to focus visible light through a specimen. Different regions of the object scatter the light differently, producing an image. This type of microscope can resolve objects that are 0.2 μm (8 millionths of an inch) apart. A limitation of light microscopy is that it focuses on only one two-dimensional plane at a time, diminishing the sense of depth.

A confocal microscope is a type of light microscope that enhances resolution by focusing white or laser light through a lens to the object. The image then passes through a pinhole, thereby eliminating the possible blurring caused by light reflecting from regions of the specimen near the object of interest. The result is a scan of highly focused light on one tiny part of the specimen. Computers can integrate many confocal images of specimens exposed to fluorescent dyes to produce spectacular three-dimensional peeks at living structures (figure 3.3B).

Transmission and Scanning Electron Microscopes

Electron microscopes provide greater magnification and better resolution than light microscopes. Both transmission and scanning electron microscopes require that a specimen be killed, chemically fixed, and placed in a vacuum—treatment that can distort natural structures.

Instead of focusing light, the transmission electron microscope (TEM) sends a beam of electrons through a very thin slice of a specimen, using a magnetic field rather than a glass lens to focus the beam. Different parts of the specimen interact with electrons differently. When electrons transmitted through the specimen hit a fluorescent screen coated with a chemical, light rays are given off, translating the contrasts in electron transmission into a high-resolution, two-dimensional image (figure 3.3C).

The scanning electron microscope (SEM) produces lower resolution images than the TEM, but it has much greater depth of field. It scans a beam of electrons over the surface of a metal-coated, three-dimensional specimen, generating an image that highlights crevices and textures (figure 3.3D).

Scanning Probe Microscopes

Scanning probe microscopes work on a different principle than light or electron microscopes. They move a probe over a surface and translate the distances into an image—a little like moving your hands over someone's face to get an idea of his or her appearance. The amount of detail is exquisite—down to the nanometer (nm) scale. One type of scanning probe microscope, the atomic force microscope, uses a diamond-tipped probe that presses a molecule's surface with a very gentle force. As the force is kept constant, the probe moves, generating an image of the molecule (figure 3.3E).

D. Features Common to All Cells

Microscopes and other tools clearly reveal that although cells can appear very different, they all have some seemingly indispensable traits—not surprising, given their common evolutionary history. All cells have the following structures and mol-

FIGURE 3.3 Different Microscopes Reveal Different Details. (**A**) A conventional light microscope focuses visible light on small objects. (**B**) The confocal microscope produces amazingly detailed images of inner cell workings. (**C**) A transmission electron microscope delivers high-magnification and high-resolution images of a slice through a cell, whereas a scanning electron microscope (**D**) helps biologists visualize surfaces. (**E**) The scanning probe microscope can reveal detail of individual molecules.

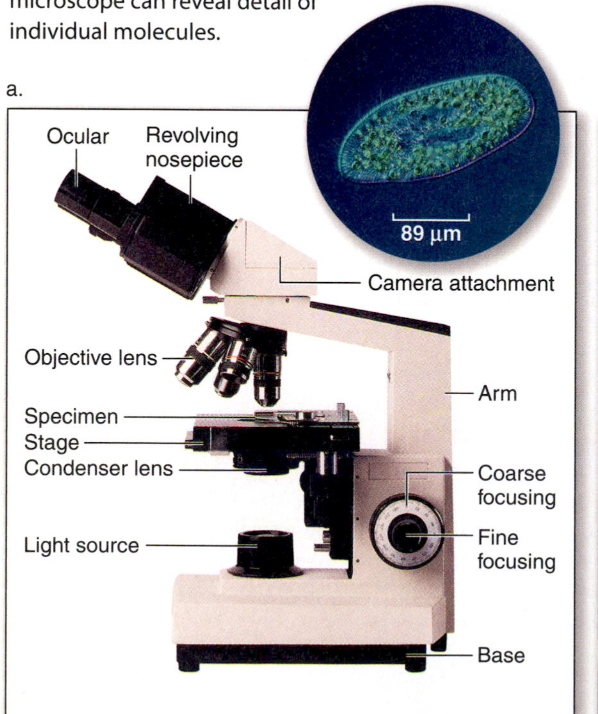

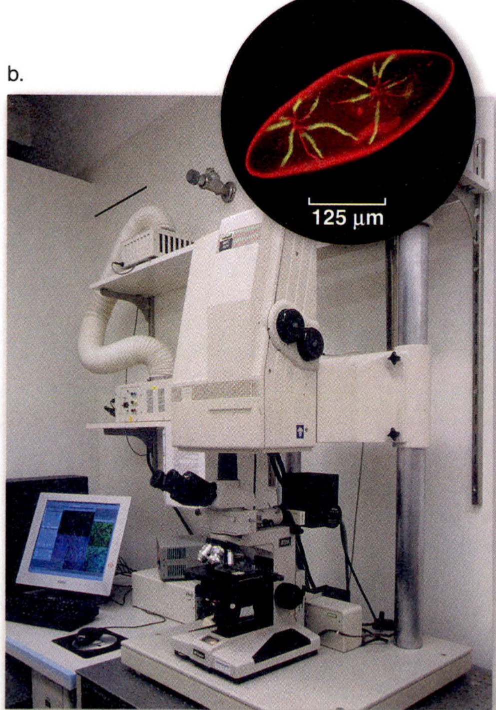

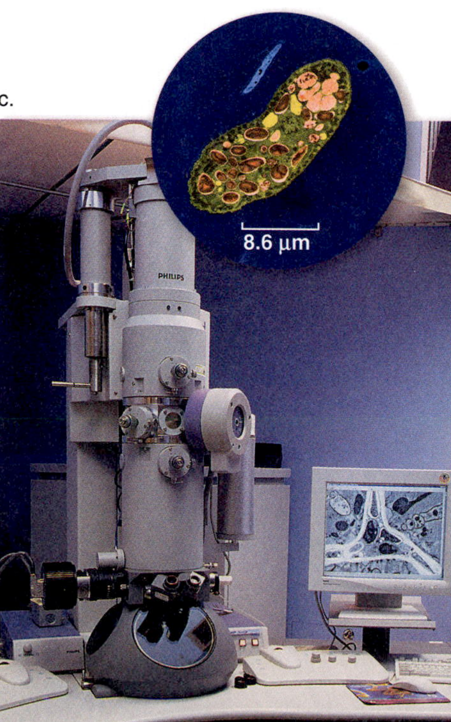

ecules in common that allow them to reproduce, grow, respond to stimuli, and obtain energy and convert it to a usable form:

- genetic information, that is, DNA
- proteins that carry out all of the cell's work, from orchestrating reproduction to processing energy to regulating what enters and leaves the cell (see section 2.5)
- RNA, which participates in the production of the cell's proteins (see chapter 12)
- **ribosomes,** structures that use RNA to manufacture proteins
- **cytoplasm,** the watery soup of salts, organic molecules, and other substances inside the cell
- a lipid-rich **cell membrane** (sometimes called the plasma membrane) that forms a boundary between living matter and the environment (see section 3.2).

In addition to these universal cell parts, the cytoplasm of complex cells such as those of animals, fungi, and plants is divided into **organelles**, compartments that carry out specialized functions. Section 3.4 describes these cellular partitions in detail.

One other feature common to nearly all cells is small size, typically less than 0.1 mm in diameter. Cells are small because nutrients, water, oxygen, carbon dioxide, and waste products enter or leave a cell through its surface. All cells therefore require relatively large surface areas through which they interact with the environment. As a cell grows, its volume increases at a faster rate than its surface area (**figure 3.4**).

Cells avoid surface area limitations in several ways. Nerve cells are long (up to a meter or so) and extremely thin, so the ratio of surface area to volume remains high. The flattened shape of a red blood cell and the many fingerlike extensions of an intestinal cell achieve the same effect. An efficient transportation system that quickly circulates materials throughout the cell also helps. A mature plant cell also illustrates another tactic: a large, water-filled compartment called a vacuole occupies most of its volume. Because the water in the vacuole does not demand nutrients or produce wastes, it lowers the effective volume of the cell.

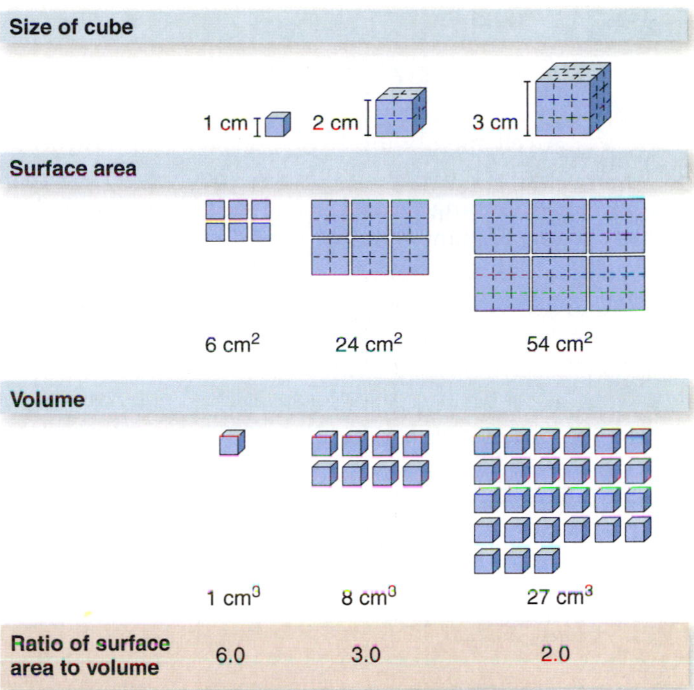

FIGURE 3.4 The Relationship Between Surface Area and Volume. This simple model shows that larger objects have less surface area *relative to their volume* than do smaller objects with the same overall shape.

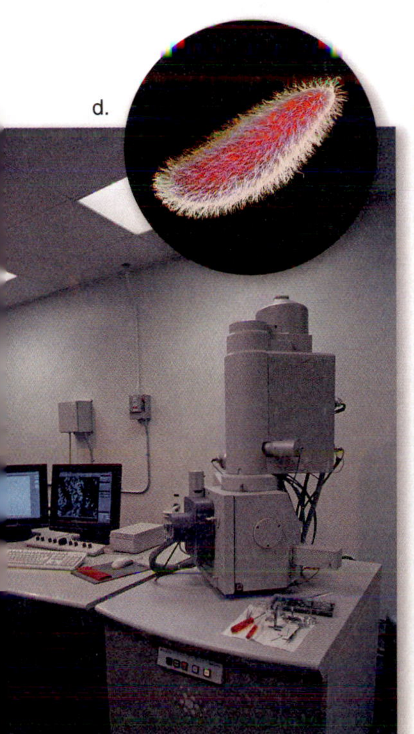

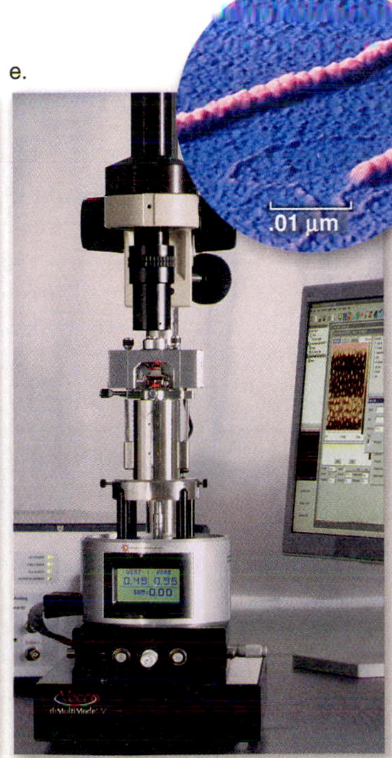

One other strategy for avoiding a surface area limitation is to improve efficiency. A cell can reduce waste production and nutrient needs by dividing its work among specialized organelles. In part because of this division of labor, most plant and animal cells are considerably larger than bacterial cells (see figure 3.2).

Summary *Microscopes permitted the formulation of cell theory, and they remain essential to the contemporary study of the cell. All cells have several features in common. Cells are typically small because the surface area must be large relative to the cell's volume.*

3.1 MASTERING CONCEPTS

1. What is a cell?
2. How did Hooke and Leeuwenhoek contribute to the study of cell biology?
3. What are the main components of cell theory?
4. Describe the five contemporary types of microscopes.
5. Which molecules and structures occur in all cells?
6. Describe adaptations that increase the ratio of surface area to volume in cells.

3.2 A Membrane Separates Each Cell from Its Surroundings

A cell membrane is one feature common to all cells. The membrane separates the cytoplasm from the environment surrounding the cell, whether it is a one-celled organism or one of many cells inside a multicellular organism. The cell's surface also transports substances into and out of the cell (see chapter 4), and it receives and responds to external stimuli. Inside a eukaryotic cell, membranes form a network that delimits organelles.

A. Lipids and Proteins Form the Cell Membrane

The cell membrane is composed of phospholipids, organic molecules that resemble triglycerides (see section 2.5). Like a fat, each phospholipid includes a three-carbon glycerol molecule. But in a **phospholipid**, glycerol bonds to only two fatty acids, whereas the third carbon atom binds to

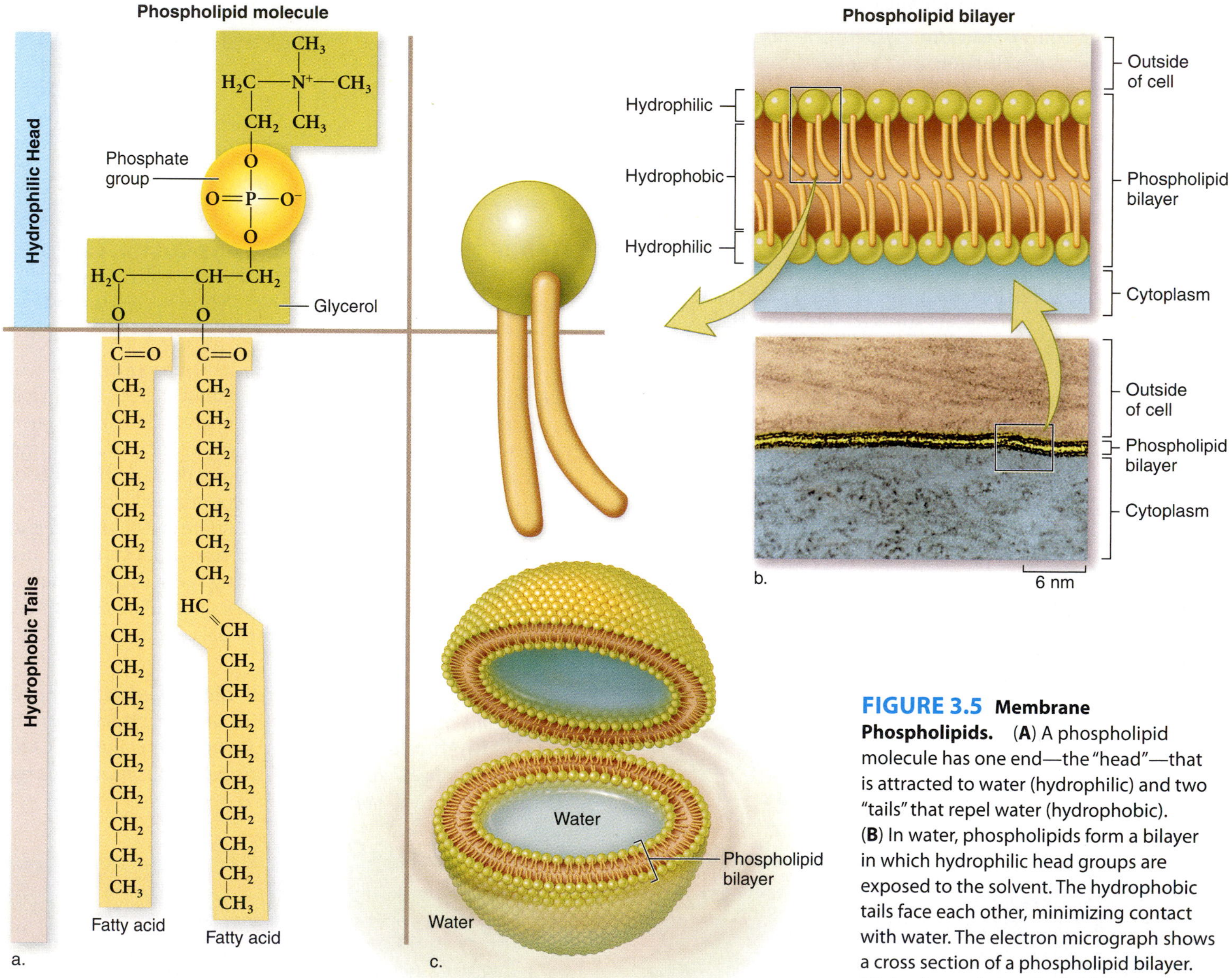

Phospholipid molecule

Hydrophilic Head

Hydrophobic Tails

Phosphate group

Glycerol

Fatty acid

Fatty acid

a.

Phospholipid bilayer

Hydrophilic

Hydrophobic

Hydrophilic

Outside of cell

Phospholipid bilayer

Cytoplasm

Outside of cell

Phospholipid bilayer

Cytoplasm

b.

6 nm

Water

Phospholipid bilayer

Water

c.

FIGURE 3.5 Membrane Phospholipids. (**A**) A phospholipid molecule has one end—the "head"—that is attracted to water (hydrophilic) and two "tails" that repel water (hydrophobic). (**B**) In water, phospholipids form a bilayer in which hydrophilic head groups are exposed to the solvent. The hydrophobic tails face each other, minimizing contact with water. The electron micrograph shows a cross section of a phospholipid bilayer. (**C**) A sphere of phospholipids forms the basis for the cell membrane.

a phosphate group (**figure 3.5A**). This chemical structure gives phospholipids unusual properties in water. The phosphate "head" end, with its polar covalent bonds, is attracted to water; that is, it is hydrophilic. The other end, consisting of two fatty acid "tails," repels water—it is hydrophobic. **lipids, p. 35**

Because of these opposing preferences, phospholipid molecules in water spontaneously arrange into the most energy-efficient organization, a two-layered, sandwich-like structure. In this **phospholipid bilayer**, the hydrophilic surfaces form the "bread" layers of the sandwich, exposed to the watery medium outside and inside the cell. The hydrophobic tails face each other on the inside of the sandwich, away from water (figure 3.5B).

Cell membranes consist of phospholipid bilayers and associated sterols, proteins, and other molecules (**figure 3.6**). Sterols, including cholesterol in animal cell membranes, increase membrane fluidity. Membrane proteins may lie completely within the phospholipid bilayer or traverse the membrane to extend out of one or both sides. The cell membrane is often called a **fluid mosaic** because both proteins and phospholipids are free to move laterally within the bilayer.

The proteins in the membrane have diverse functions:

- **Transport proteins**: Transport proteins embedded in the phospholipid bilayer create passageways through which water-soluble molecules and ions pass into or out of the cell.
- **Enzymes:** These proteins facilitate chemical reactions that otherwise would not proceed quickly enough to sustain life. In some membranes, different enzymes are physically laid out in the order in which they participate in chains of chemical reaction.
- **Recognition proteins**: Carbohydrates attached to cell surface proteins serve as "name tags" that help the body recognize its own cells. The immune system attacks cells with unfamiliar surface molecules, which is why transplant recipients often reject donated organs. Surface structures also distinctively mark cells of different tissues in an individual, so a bone cell's surface is different from that of a nerve cell or a muscle cell.
- **Adhesion proteins**: These membrane proteins enable cells to stick to one another (see section 3.6).
- **Receptor proteins**: Some membrane proteins exposed on the outer face of the membrane are receptors, binding molecules outside the cell and triggering a reaction inside the cell. HER2, described in the chapter opening essay, is a receptor. The next section describes the role of receptor proteins in signal transduction.

Researchers estimate that about one-third of every organism's genome encodes membrane proteins, and studying

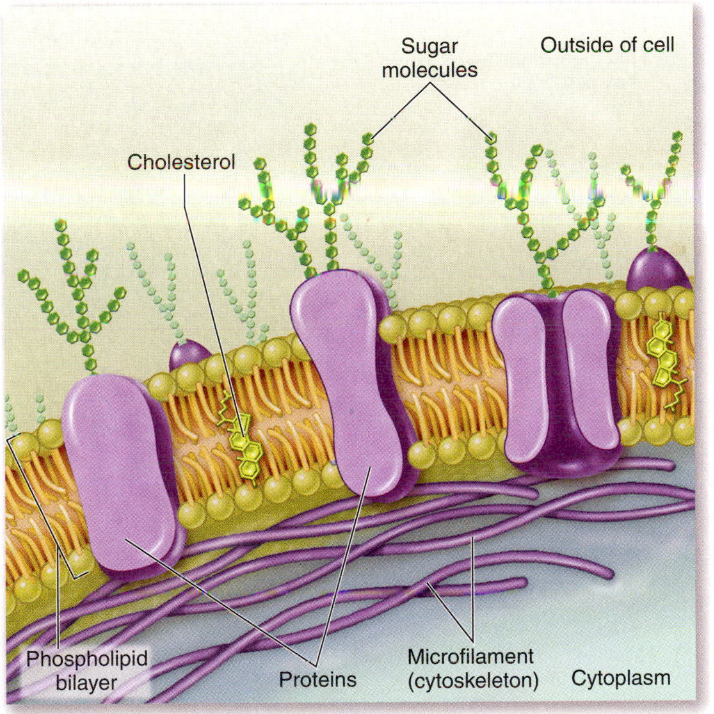

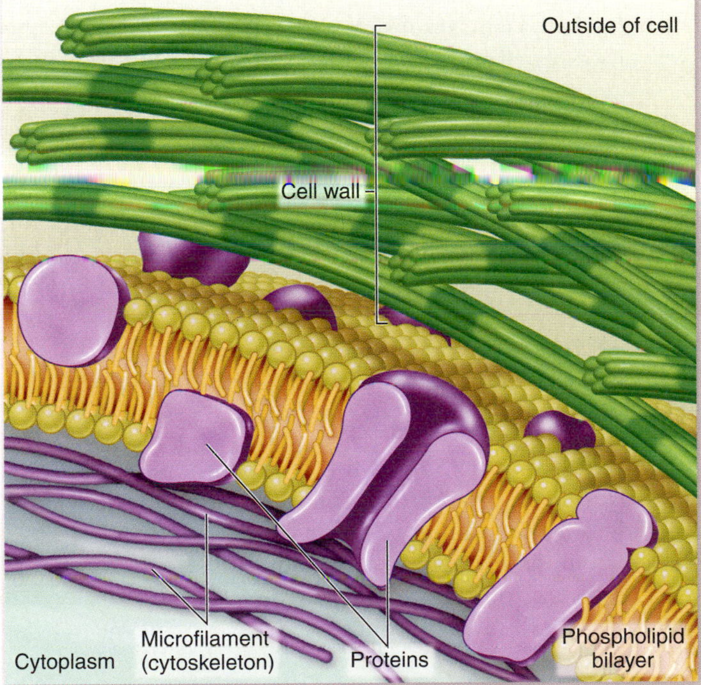

FIGURE 3.6 Anatomy of a Cell Membrane. Mobile proteins embedded in the phospholipid bilayer produce a somewhat fluid structure. An underlying mesh of protein fibers supports the cell membrane. Jutting from the animal cell membrane's outer face are carbohydrate molecules linked to proteins. A network of cellulose fibers—the cell wall—surrounds a plant cell. Notice that the plant cell membrane lacks cholesterol.

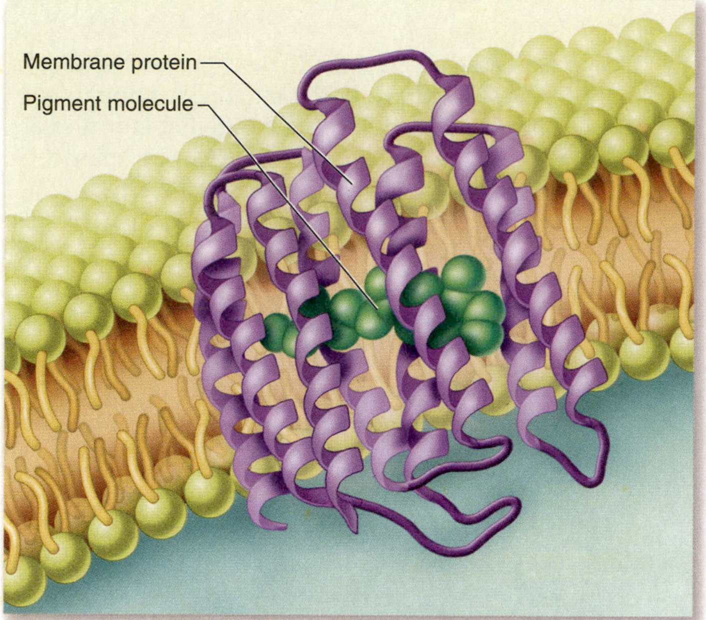

FIGURE 3.7 Pigment-Protein Complex. Membrane-bound proteins and pigments work together in photosynthesis, which uses the sun's energy to make food. Lipid-soluble regions of the molecules remain within the hydrophobic portion of the membrane. Water-soluble regions extend into the cytoplasm. These pigment-protein complexes occupy the membanes of photosynthetic organelles called chloroplasts in plants and algae.

them has led to fantastic insights into photosynthesis and other cell processes (**figure 3.7**). Understanding these proteins is also a vital part of human medicine, in part because at least half of all drugs bind to them. One example is omeprazole (Prilosec). This drug relieves heartburn and gastric reflux by blocking some of the transport proteins that pump acid into the stomach. Another is the antidepression drug fluoxetine (Prozac), which prevents receptor proteins on brain cells from binding to a mood-altering biochemical called serotonin.

B. Signal Transduction Transmits Messages to a Cell's Interior

One important function of the cell membrane is to detect and respond to the environment. A cell's environment can include outside influences such as light, sounds, odors, touch, temperature, and water. But it also includes each cell's surroundings *within* a multicellular organism. The fluid that bathes nearly every cell in your body contains water, oxygen and other dissolved gases, nutrients, and hormones—all part of the environment of each individual cell.

In a process called **signal transduction,** a cell receives an external "message" and converts it into an internal signal (**figure 3.8**). Membrane proteins are crucial to signal transduction. The process begins when a receptor protein binds to a stimulus molecule, the first messenger. The responding receptor then contacts a nearby protein, which triggers a chemical reaction inside the cell. The product of this chemical reaction is called the second messenger, and it lies at the crux of the entire process. Approximately 1 second after the stimulus arrives, the second messenger provokes the cell's response, typically by activating particular genes or enzymes.

Some classes of hormones act by binding to cell receptors. Examples include epinephrine (the "fight-or-flight" hormone) and glucagon, which stimulates the breakdown of stored glycogen into glucose. In addition, sildenafil (Viagra), discussed at the beginning of chapter 2, works by altering signal transduction. The drug binds to a second messenger molecule, which prevents an enzyme from breaking the molecule down. The second messenger stays around longer, and so does its effect of relaxing muscles in the blood vessels of the penis.

FIGURE 3.8 Signal Transduction. A small external molecule, the first messenger, binds to a receptor on the outer surface of the cell. This event triggers the production or release of second messengers inside the cell. The resulting cascade of chemical reactions stimulates the cell's response.

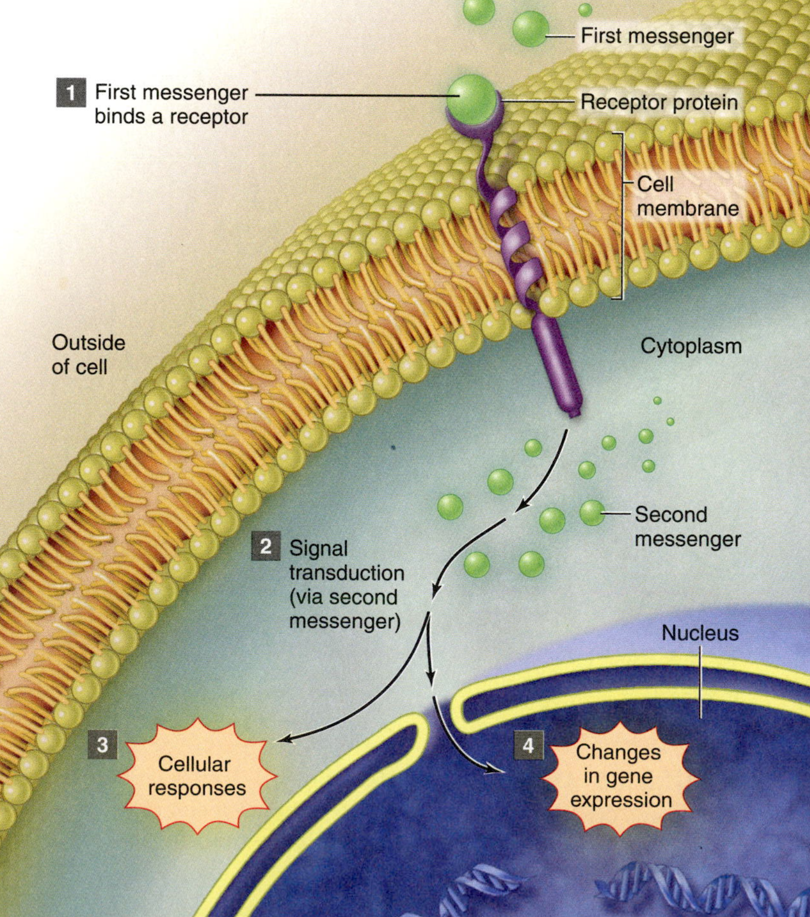

Faulty signal transduction is at the heart of cancer and other diseases. As chapter 8 describes, healthy (noncancerous) cells stop dividing when they receive inhibitory signals from neighboring cells. But if a cell fails to receive or act on the message, it may continue to divide, forming a tumor. New drugs that alter signal transduction may target these abnormal cells.

Summary ***A biological membrane is a phospholipid bilayer containing proteins with various functions, including receiving and transmitting stimuli from outside the cell.***

3.2 MASTERING CONCEPTS

1. How does the chemical structure of phospholipids enable them to form a bilayer in water?
2. What are the main components of the cell membrane?
3. What are some functions of membrane proteins?
4. How do cells detect and respond to external signals?

3.3 Different Cell Types Characterize Life's Three Domains

Until recently, biologists recognized just two types of organisms. **Prokaryotes**, the simplest and most ancient forms of life, are organisms whose cells lack organelles (see chapter 18). At least 1.5 billion years ago, prokaryotes gave rise to **eukaryotes,** whose cells contain organelles. But in 1977, physicist-turned-microbiologist, Carl Woese, studied key molecules and detected differences that were great enough to suggest that some prokaryotes represented a completely different form of life. Biologists subsequently divided life into three domains, based on the characteristics of their cells: Bacteria, Archaea, and Eukarya (**figure 3.9**).

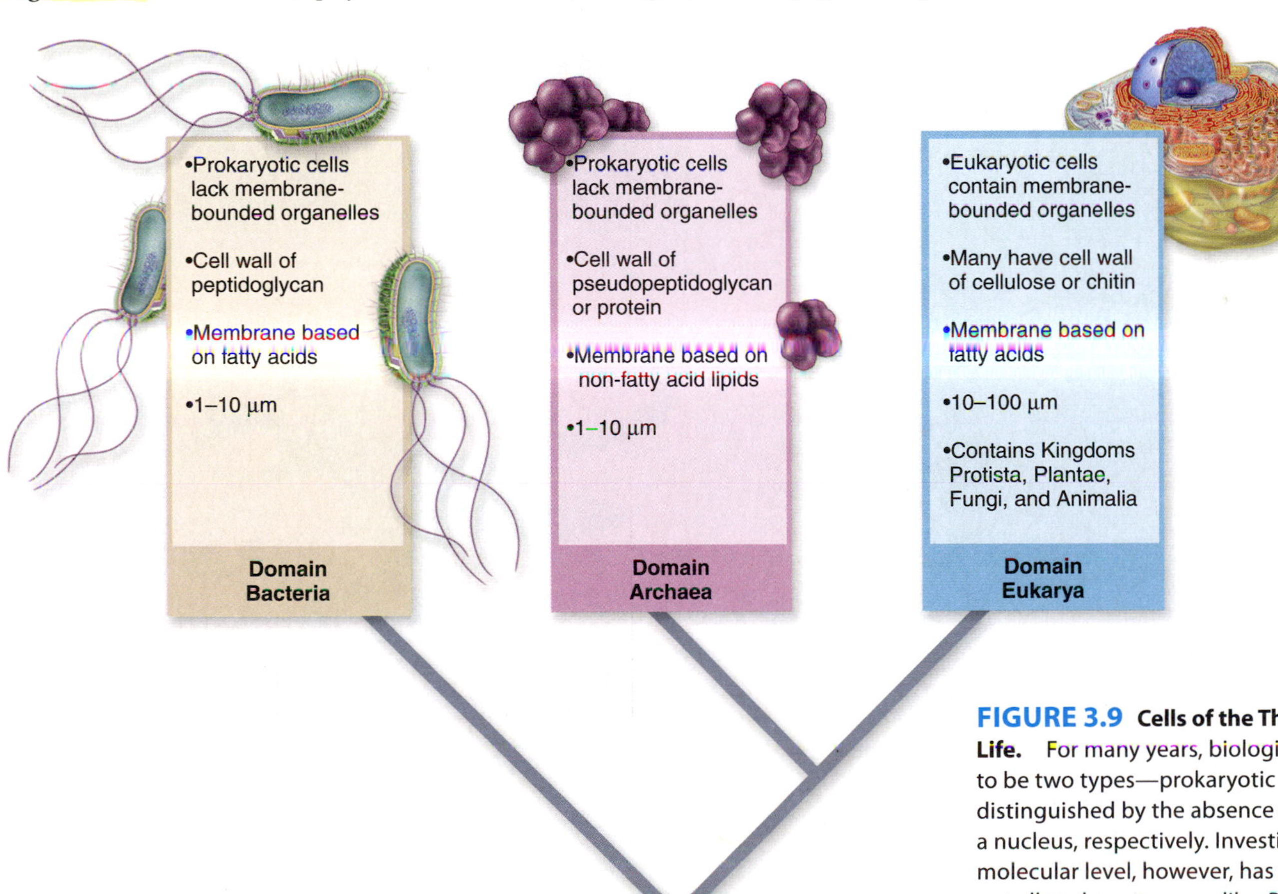

•Prokaryotic cells lack membrane-bounded organelles
•Cell wall of peptidoglycan
•Membrane based on fatty acids
•1–10 µm
Domain Bacteria
•Prokaryotic cells lack membrane-bounded organelles
•Cell wall of pseudopeptidoglycan or protein
•Membrane based on non-fatty acid lipids
•1–10 µm
Domain Archaea
•Eukaryotic cells contain membrane-bounded organelles
•Many have cell wall of cellulose or chitin
•Membrane based on fatty acids
•10–100 µm
•Contains Kingdoms Protista, Plantae, Fungi, and Animalia
Domain Eukarya
Common Ancestor

FIGURE 3.9 Cells of the Three Domains of Life. For many years, biologists considered cells to be two types—prokaryotic or eukaryotic—distinguished by the absence or presence of a nucleus, respectively. Investigation at the molecular level, however, has revealed that not all prokaryotes are alike. Biologists now recognize three types of cells: bacteria, archaea, and eukaryotes.

A. Domain Bacteria

Bacteria lack membrane-bounded nuclei (**figure 3.10**). Each cell does have an area, the **nucleoid,** where its genetic material—one circular DNA molecule—congregates. Nearby are RNA molecules and ribosomes. In part because the DNA, RNA, and ribosomes in prokaryotic cells are in close contact, protein synthesis is rapid compared with the process in more complex cells, whose cellular components are separated. (Chapter 12 describes protein synthesis in detail.)

A rigid **cell wall** surrounds the cell membrane of most bacterial cells. The cell wall protects the cell, prevents it from bursting, and gives it a distinctive shape: round, rod-shaped, spiral, comma-shaped, or spindle-shaped. In some bacteria, polysaccharides on the cell wall form a capsule that protects the cell or enables it to attach to surfaces. Many antibiotic drugs, including penicillin, halt bacterial infection by interfering with the microorganism's ability to construct its cell wall. Anchored in the cell wall and underlying cell membrane of some bacteria are one or more **flagella,** tail-like appendages that enable the cell to move.

Members of domain Bacteria are extremely abundant and diverse (**figure 3.11**), and they include the smallest known cells. (To learn just how small, see this chapter's Burning Question: What is the smallest living organism?).

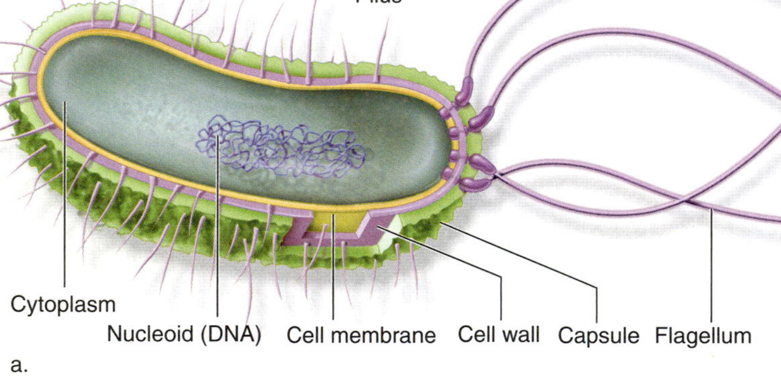

a.

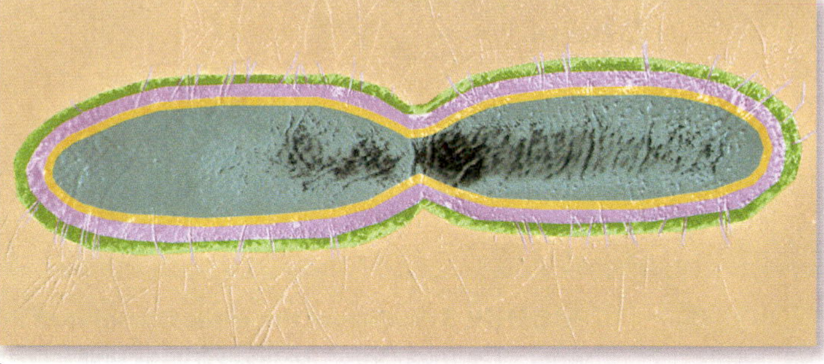

b.

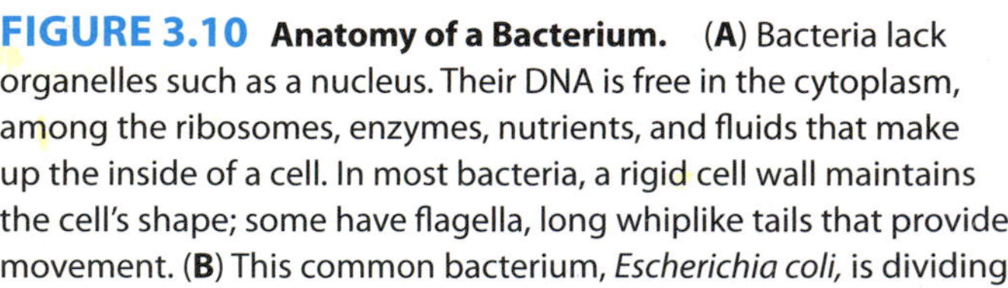

FIGURE 3.10 Anatomy of a Bacterium. (**A**) Bacteria lack organelles such as a nucleus. Their DNA is free in the cytoplasm, among the ribosomes, enzymes, nutrients, and fluids that make up the inside of a cell. In most bacteria, a rigid cell wall maintains the cell's shape; some have flagella, long whiplike tails that provide movement. (**B**) This common bacterium, *Escherichia coli,* is dividing.

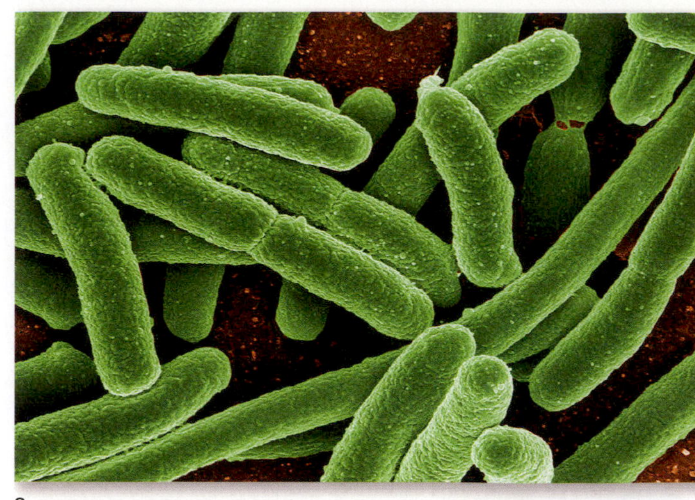

a.

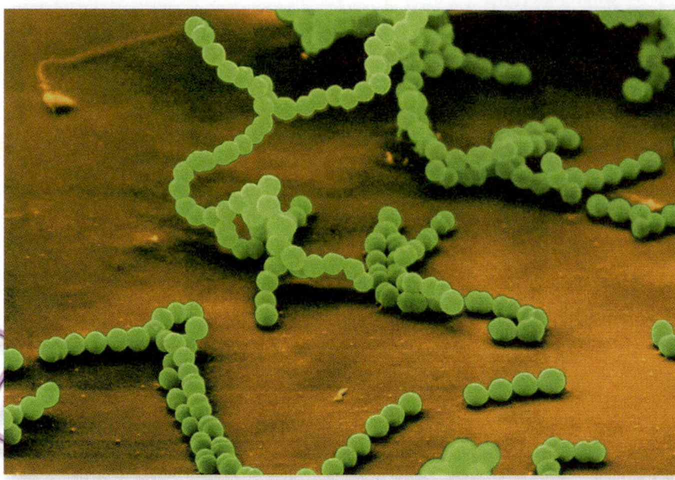

b.

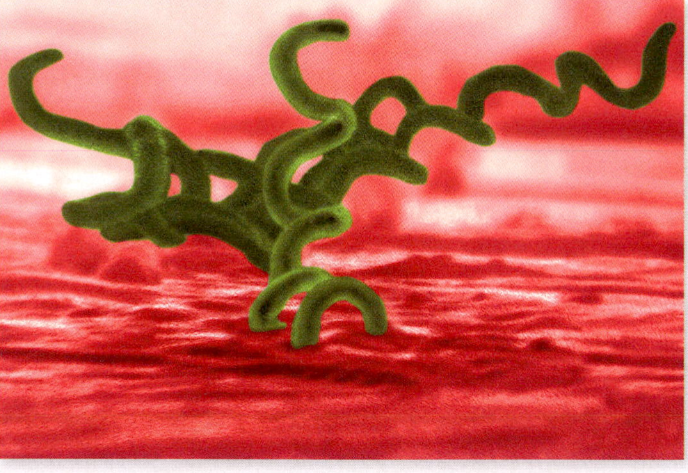

c.

FIGURE 3.11 A Trio of Bacteria. (**A**) *E. coli* inhabits the intestines of some animal species, including humans. (**B**) *Streptococcus pyogenes* infects humans. (**C**) The sexually transmitted disease syphilis is caused by the corkscrew-shaped *Treponema pallidum*.

Some species, such as *Streptococcus* and *Escherichia coli*, cause illnesses, but others living on your skin and inside your intestinal tract are essential for good health. Bacteria are also very valuable in research, food and beverage processing, and pharmaceutical production. In ecosystems, bacteria play critical roles as decomposers and producers.

B. Domain Archaea

Archaean cells resemble bacterial cells in many ways (**figure 3.12**). They are smaller than most eukaryotic cells, and they lack a membrane-bounded nucleus and other organelles. Most have cell walls, and flagella are also common. Because of these similarities, Woese first named his newly recognized group Archaebacteria. The name later changed to Archaea when it became apparent that the resemblance to bacteria was only superficial.

Burning Questions

What is the smallest living organism?

Since the advent of microscopes, investigators have wondered just how small an organism can be and still sustain life. This seemingly simple question is hard to answer; "life" is hard to define. Some people consider viruses alive because they share some, but not all, characteristics with cells (see chapter 17). Viruses are indeed miniscule: The smallest are less than 20 nm in diameter (see figure 3.2). Yet most biologists do not consider them alive, in part because viruses do not consist of cells or reproduce on their own.

Some scientists consider "nanobes" to be the world's smallest microorganisms, but others are skeptical. At about 20 to 150 nm long, these structures are hard to analyze for hallmarks of life such as DNA, RNA, ribosomes, and protein (**figure 3.A**). Their status remains controversial.

For now, the smallest certifiable living organisms are bacteria called mycoplasmas. Besides their small size (150 nm and larger), these microorganisms are unusual among bacteria because they lack cell walls. Biologists have studied mycoplasmas in detail for two reasons. First, some cause human disease such as urinary tract infections and pneumonia. Second, mycoplasmas have the smallest amount of genetic material of any known free-living cell. They may therefore give insight into which genes are minimally required to sustain life.

Have a Burning Question of your own? Submit it to marielle_hoefnagels@mcgraw-hill.com for possible inclusion in future editions of this book!

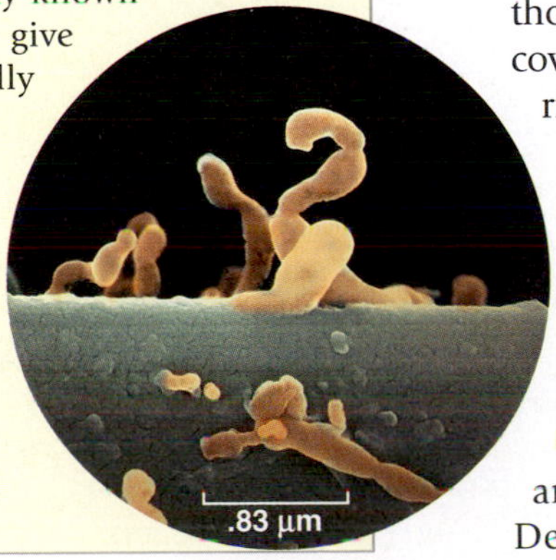

.83 µm

FIGURE 3.A Nanobes. Alive or not?

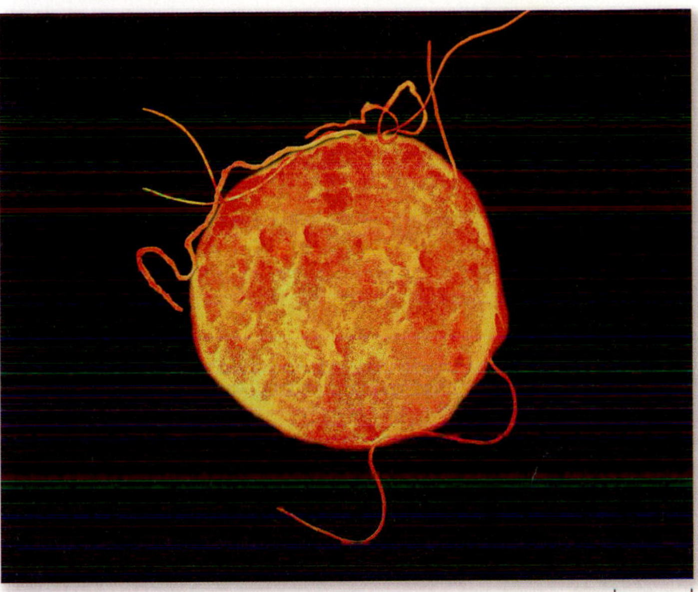

FIGURE 3.12 An Archaeon. *Sulfolobus acidocaldarius* thrives in hot springs, at a temperature of 80°C and a pH of 2.0. Note the prominent flagella.

The first members of Archaea to be described were microorganisms that use carbon dioxide and hydrogen from the environment to produce methane—and hence are called methanogens. When researchers deciphered all of the genes of a methanogenic archaeon in 1996, they found that more than half of the genes had no counterpart among bacteria or eukaryotes. The fact that nearly half of the genes *do* correspond indicates that the three forms of life branched from a shared ancestor long ago.

Archaea have their own domain because they build their structural components out of biochemicals that are slightly different from those that either bacteria or eukaryotes use. Their phospholipids, cell walls, and flagella are all chemically unique. Their ribosomes, however, are more similar to those of eukaryotes than to those of bacteria. Archaea may therefore be the closest relatives of eukaryotes.

Archaea are famous as "extremophiles" because they are common in environments that have extremes of temperature, pressure, pH, or salinity. This characterization is somewhat misleading, however, because some bacteria also live in those environments. In addition, researchers have since discovered archaea in a variety of habitats, including swamps, rice paddies, and throughout the oceans.

C. Domain Eukarya

An astonishing diversity of other organisms, including humans, belong to domain Eukarya. Our fellow animals are eukaryotes, as are yeasts, mushrooms, and other fungi. Plants are also eukaryotes, and so are one-celled protists such as *Amoeba* and *Paramecium*. Despite their great differences in external appearance, all eukaryotic organisms share many features on a cellular level. **Figures 3.13** and **3.14** depict generalized animal and plant cells.

One obvious feature that sets eukaryotic cells apart is their large size, typically 10 to 100 times greater than prokaryotic cells. The other main difference is that eukaryotic cells have elaborate systems of internal membranes, which create compartments—organelles—where specialized biochemical reactions can occur. In general, organelles keep related biochemicals and structures sufficiently close together to make them function efficiently, without altering or harming other cellular contents (see section 3.4). Compartmentalization also means that the cell must maintain high concentrations of each biochemical only in certain organelles, not throughout the entire cell.

If all life shares a common ancestor, how did eukaryotes first acquire organelles more than 1.5 billion years ago? According to the endosymbiosis theory, some ancient organism (or organisms) engulfed other cells and, rather than digesting them, kept them on as partners. The simi-

FIGURE 3.13 An Animal Cell. The large, generalized view shows the relative sizes and locations of the cell components. The inset shows a human white blood cell with a prominent nucleus and many mitochondria.

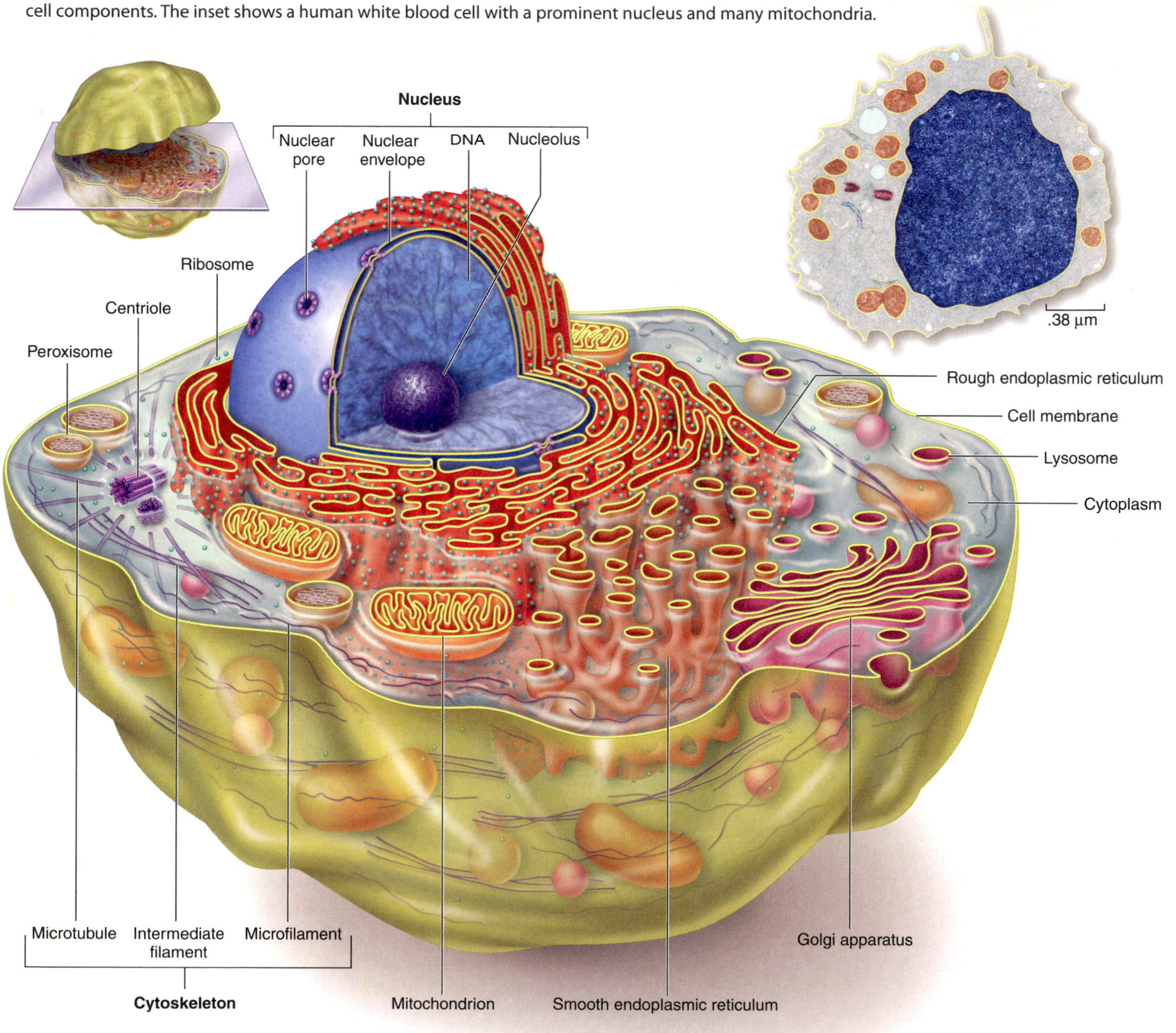

larities between bacteria and some eukaryotic organelles, including mitochondria and chloroplasts, lend powerful support to this theory. **endosymbiosis, p. 338**

The rest of this chapter describes the structure of the eukaryotic cell in greater detail.

Summary *Prokaryotic cells, which are small and lack nuclei and other organelles, are in domains Archaea and Bacteria. The archaea have unique characteristics but also share features with bacteria and the third domain, Eukarya. Eukaryotic cells are larger than prokaryotic cells, and they contain specialized organelles.*

3.3 MASTERING CONCEPTS

1. How do prokaryotic cells differ from eukaryotic cells?
2. How are bacteria and archaea similar to and different from each other?
3. How do organelles contribute to efficiency in eukaryotic cells?

FIGURE 3.14 A Plant Cell. The large, generalized view illustrates key features of the plant cell. The inset shows a leaf cell; note the prominent nucleus and chloroplasts.

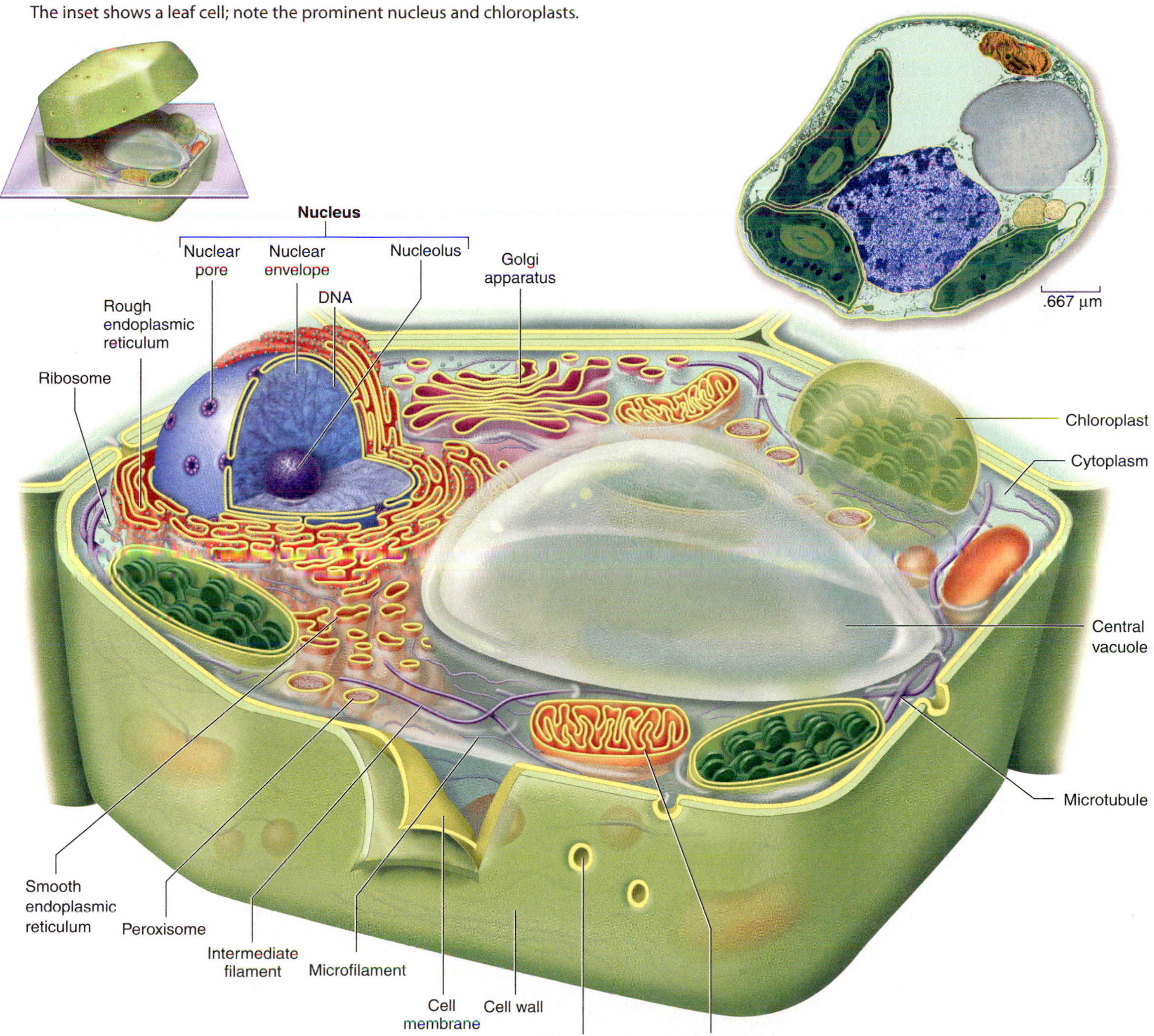

3.4 Eukaryotic Organelles Divide Labor

Organelles were discovered shortly after biologists began using microscopes to examine cells. In eukaryotic cells, organelles have specialized functions that carry out the work of the cell. One or more membranes surround each of these compartments. Many of these membranes are studded with enzymes that catalyze chemical reactions on their surfaces. In many organelles, intricate folds in the membrane provide tremendous surface area.

About half of the volume of an animal cell is organelles; in contrast, some plant cells contain up to 90% water, much of it within a large organelle called a vacuole. This section will describe the structures and functions of the most important organelles.

A. Organelles Interact to Secrete Substances

Coordinated interactions between organelles enable cells to produce, package, and release complex mixtures of biochemicals, such as milk (**figure 3.15**). Special cells in the mammary glands of female mammals produce milk, which contains proteins, fats, carbohydrates, and water in a proportion ideal for development of a newborn. Human milk is rich in lipids, which the rapidly growing baby's nervous system requires. (Cows' milk contains a higher proportion of protein, better suited to a calf's rapid muscle growth.) Dormant most of the time, these special cells of the mammary glands increase their activity during pregnancy and then undergo a burst of productivity shortly after the fe-

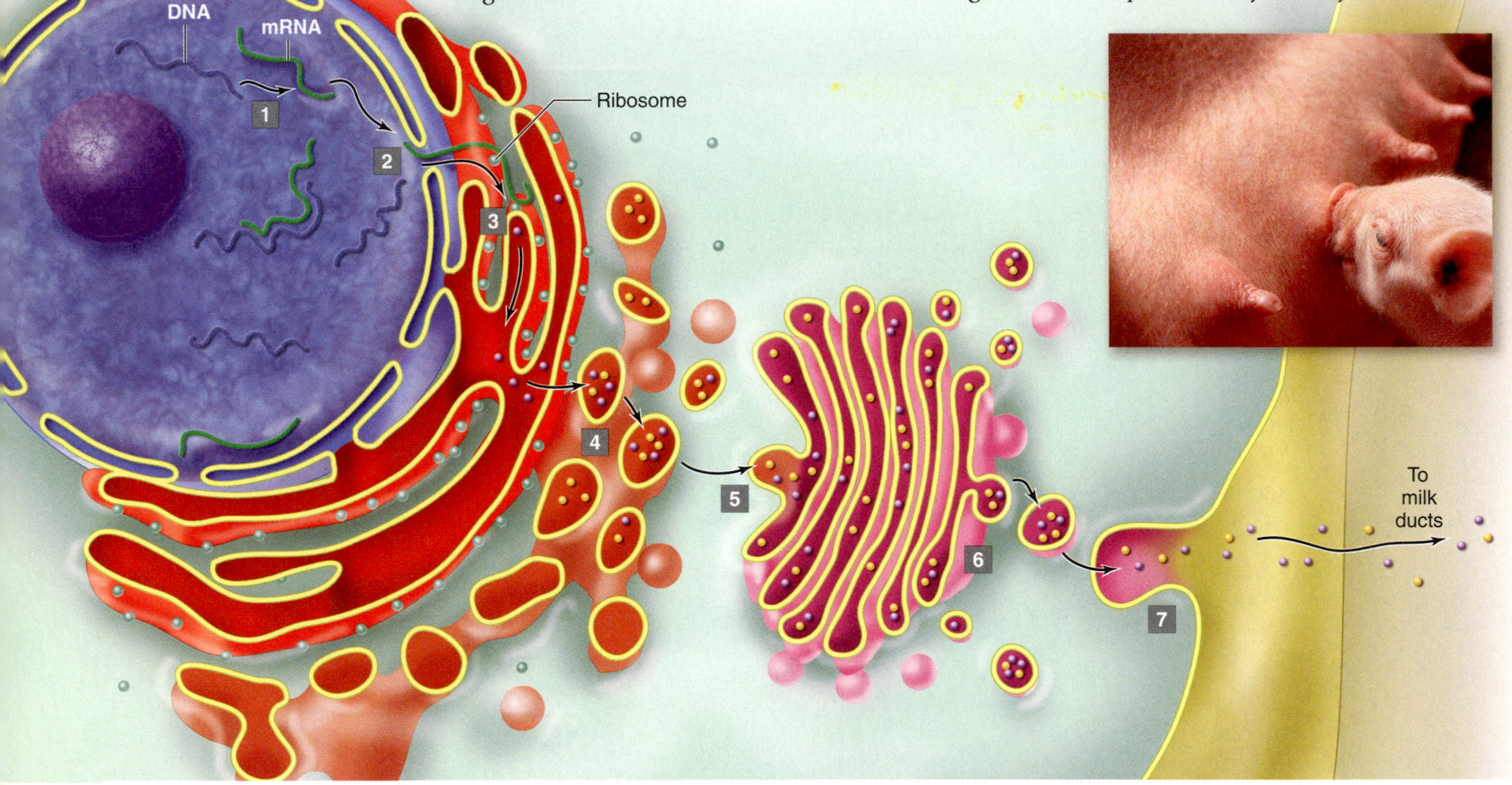

1 Milk protein genes transcribed to mRNA

2 mRNA exits through nuclear pore

3 At ribosomes on surface of rough ER, information in mRNA is used to produce milk protein

4 Enzymes in smooth ER manufacture lipids

5 Milk proteins and lipids are packaged into vesicles from both rough and smooth ER for transport to Golgi

6 Final processing of proteins in Golgi and packaging for export out of cell

7 Proteins and lipids released from cell when vesicles fuse with cell membrane

FIGURE 3.15 Making Milk. Milk production and secretion illustrate organelle functions and interactions in a cell from a mammary gland; (*1*) through (*7*) indicate the order in which organelles participate in this process. The inset shows a piglet suckling from a sow.

male gives birth. How do the organelles work together to manufacture milk? **mammals, p. 491**

The Nucleus

Milk secretion begins in the nucleus (see figure 3.15, step 1), the most prominent organelle in most eukaryotic cells. The nucleus contains DNA, an informational molecule that specifies the "recipe" for every protein a cell can make (such as milk protein and enzymes required to synthesize carbohydrates and lipids). A milk-producing cell copies the genes for these proteins into another nucleic acid, messenger RNA (mRNA). The mRNA molecules exit the nucleus through **nuclear pores**, which are holes in the two-layered **nuclear envelope** that separates the nucleus from the cytoplasm (figure 3.15, step 2 and **figure 3.16**). Nuclear pores are not merely perforations but highly specialized channels composed of more than a hundred types of proteins. Traffic through the nuclear pores is busy, with millions of proteins and mRNA molecules passing in or out each minute.

Also inside the nucleus is the **nucleolus,** a dense spot that assembles the components of ribosomes.

The Endoplasmic Reticulum and Golgi Apparatus

The remainder of the cell, between the nucleus and cell membrane, is the cytoplasm. In all cells, the cytoplasm is a watery soup of ions, enzymes, RNA, and other dissolved substances. In eukaryotes, the cytoplasm also includes organelles and arrays of protein rods and tubules called the cytoskeleton (see section 3.5).

Once in the cytoplasm, mRNA coming from the nucleus binds to a ribosome, which manufactures proteins (see figure 3.15, step 3). The ribosome is composed of multiple proteins and ribosomal RNA (rRNA). Once attached to mRNA, ribosomes begin synthesizing new proteins. If the proteins are destined for secretion (in milk, for example), the entire complex of ribosome, mRNA, and partially made protein anchors to the surface of the **endoplasmic reticulum**, a network of sacs and tubules composed of membranes. (*Endoplasmic* means "within the cytoplasm," and *reticulum* means "network.")

The endoplasmic reticulum (ER) originates at the nuclear membrane and winds throughout the cell. Close to the nucleus, the membrane surface is studded with ribosomes making proteins that enter the inner compartment of the ER. This section of the network is called the **rough ER** because

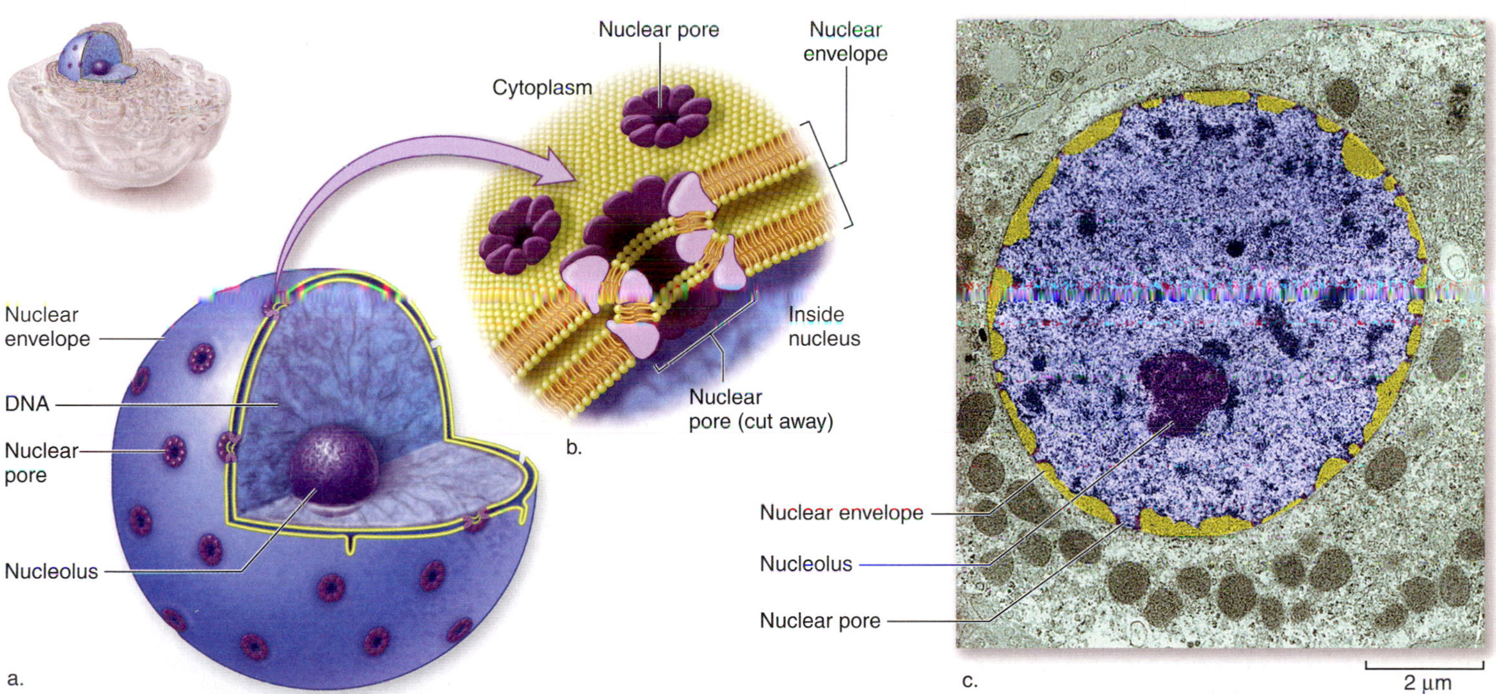

FIGURE 3.16 The Nucleus. (**A**) The nucleus is surrounded by two membrane layers, which make up the nuclear envelope (**B**). Pores through the envelope allow some molecules to move in and out of the nucleus. (**C**) The darkly staining nucleolus is the site of ribosome manufacture and assembly.

the ribosomes give these membranes a roughened appearance (**figure 3.17**). Enzymes inside this compartment fold and modify the proteins that enter; if they fail to form properly, the proteins are removed and destroyed.

Adjacent to the rough ER, a section of the network called **smooth ER** synthesizes lipids—such as those that will end up in the milk—and other membrane components (see figure 3.15, step 4 and figure 3.17). The smooth ER also houses enzymes that detoxify drugs and poisons.

The lipids and proteins made by the ER exit the organelle in **vesicles,** which are small, membranous spheres that transport materials inside the cell. A loaded vesicle pinches off of the tubular endings of the ER membrane (see figure 3.15, step 5) and takes its contents to the next stop in the production line, the **Golgi apparatus** (**figure 3.18**). This organelle is a stack of flat, membrane-enclosed sacs that functions as a processing center. Proteins from the ER pass through the series of Golgi sacs, where they complete their intricate folding and become functional (see figure 3.15, step 6). Enzymes in the Golgi apparatus also manufacture and attach carbohydrates to proteins or lipids, forming glycoproteins or glycolipids.

The Golgi apparatus sorts and packages materials into vesicles, which move toward the cell membrane. Some of the proteins it receives from the ER will become membrane surface proteins; other substances (such as milk protein and fat) are packaged for secretion from the cell. In the production of milk, these vesicles fuse with the cell membrane and release the proteins outside the cell (see figure 3.15, step 7). Fat droplets retain a layer of surrounding membrane when they leave the cell.

This entire process happens simultaneously in countless specialized cells lining the milk ducts of the breast, beginning shortly after a baby's birth. When the infant suckles, hormones released in the mother's system stimulate muscles surrounding balls of these cells to contract, squeezing milk into the ducts that lead to the nipple. In addition to proteins, fat, and carbohydrates, the milk also contains calcium, potassium, and antibodies that help jumpstart the baby's immunity to disease.

All eukaryotic cells use secretory pathways in one way or another. It isn't surprising that cells that secrete copiously have a large ER and numerous Golgi apparatuses.

B. Lysosomes, Vacuoles, and Peroxisomes Are Cellular Digestion Centers

Besides producing molecules for export, the cells of eukaryotes also break down molecules in specialized compartments. All of these "digestion center" organelles are sacs surrounded by a single membrane.

Lysosomes

Lysosomes are organelles containing enzymes that dismantle captured bacteria, worn-out organelles, and debris

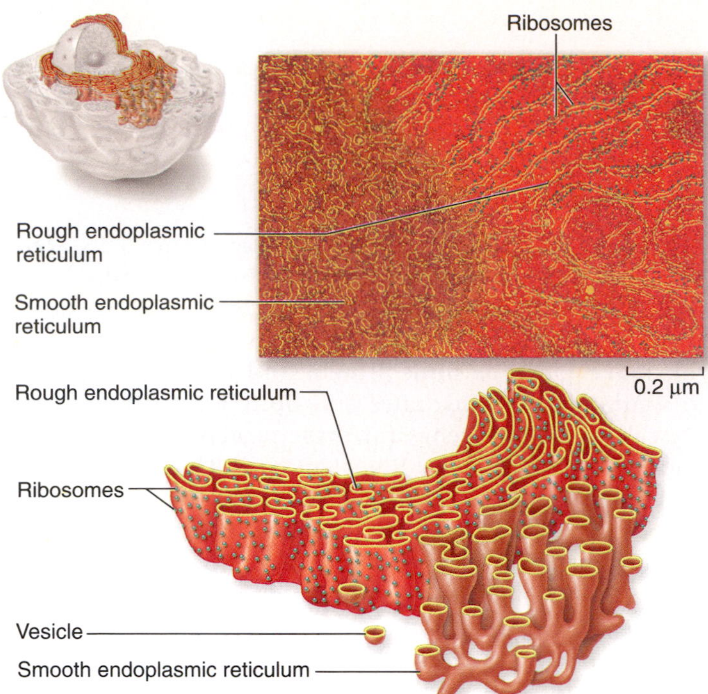

FIGURE 3.17 The Endoplasmic Reticulum. The rough ER is an extension of the outer membrane of the nuclear envelope and is the site for manufacturing secreted proteins. Ribosomes dot the surface of the ER membrane, giving it a "rough" appearance. The smooth ER is a series of interconnecting tubules and is the site for lipid production and other metabolic processes.

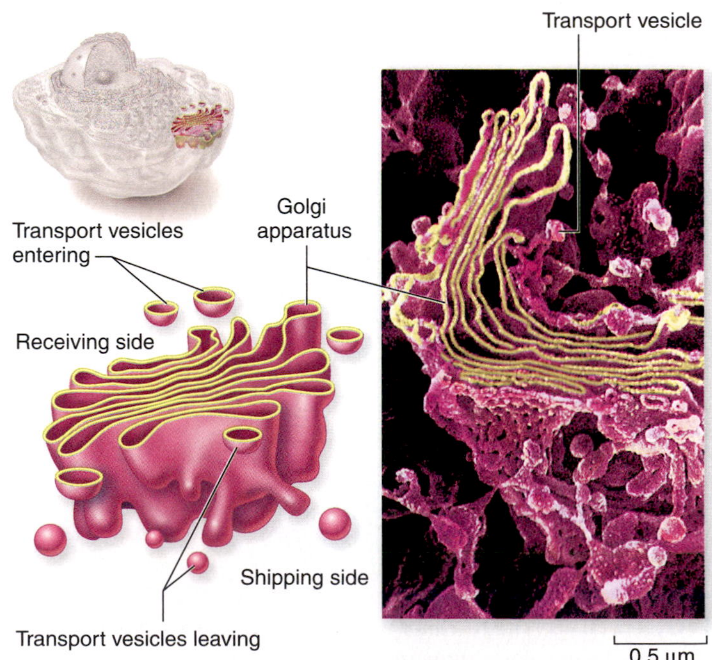

FIGURE 3.18 The Golgi Apparatus. The Golgi apparatus is composed of a series of membrane vesicles and flattened sacs. Proteins are sorted and processed as they move through the Golgi apparatus on their way to the cell surface or lysosomes.

(**figure 3.19**). They are so named because their enzymes lyse, or cut apart, their substrates.

The rough ER manufactures the enzymes inside lysosomes. The Golgi apparatus detects these enzymes by recognizing a sugar attached to them, then packages them into vesicles that eventually become lysosomes. Lysosomes fuse with vesicles carrying debris from outside or from within the cell. The lysosomal enzymes then break down the carbohydrates, lipids, proteins, and nucleic acids into smaller forms that the cell can use, releasing them into the cytoplasm.

What keeps a lysosome from digesting the entire cell? The surrounding membrane maintains the pH of the organelle's interior at about 4.8, much more acidic than the neutral pH of the rest of the cytoplasm. If a lysosome were to burst, the liberated enzymes would no longer be at their optimum pH, so they could not digest the other cellular constituents.

Different types of cells have varying numbers of lysosomes. White blood cells, for example, have many lysosomes because these cells engulf and dispose of debris and bacteria. Liver cells require many lysosomes to process cholesterol.

Malfunctioning lysosomes can cause illness. In Tay-Sachs disease, for example, one defective lysosomal enzyme allows a lipid to accumulate to toxic levels in nerve cells of the brain. The nervous system deteriorates, and an affected person eventually becomes unable to see, hear, or move. In the most severe forms of the illness, death usually results by age 5.

Vacuoles

Most plant cells lack lysosomes, but they do have an organelle that serves a similar function. In mature plant cells, the large central **vacuole** contains a watery solution of enzymes that degrade and recycle molecules and organelles (see figure 3.14).

The vacuole also has other roles. Most of the growth of a plant cell comes from an increase in the volume of its vacuole. As the vacuole acquires water, it exerts pressure (called turgor pressure) against the cell membrane. This pressure helps plants stay rigid.

Besides water and enzymes, the vacuole also contains a variety of salts, sugars, and weak acids. Therefore, the pH of the vacuole's solution is usually somewhat acidic. In citrus fruits, the solution is very acidic, producing the tart taste of lemons and oranges. Water-soluble pigments also reside in the vacuole, producing blue, purple, and magenta colors in leaves, flowers, and fruits.

Some protists have vacuoles, although their function is different from that in plants. The contractile vacuole in *Paramecium*, for example, pumps excess water out of the cell. In *Amoeba*, a food vacuole digests nutrients that the cell has engulfed.

Peroxisomes

All eukaryotic cells contain **peroxisomes,** organelles that contain several types of enzymes that dispose of toxic substances

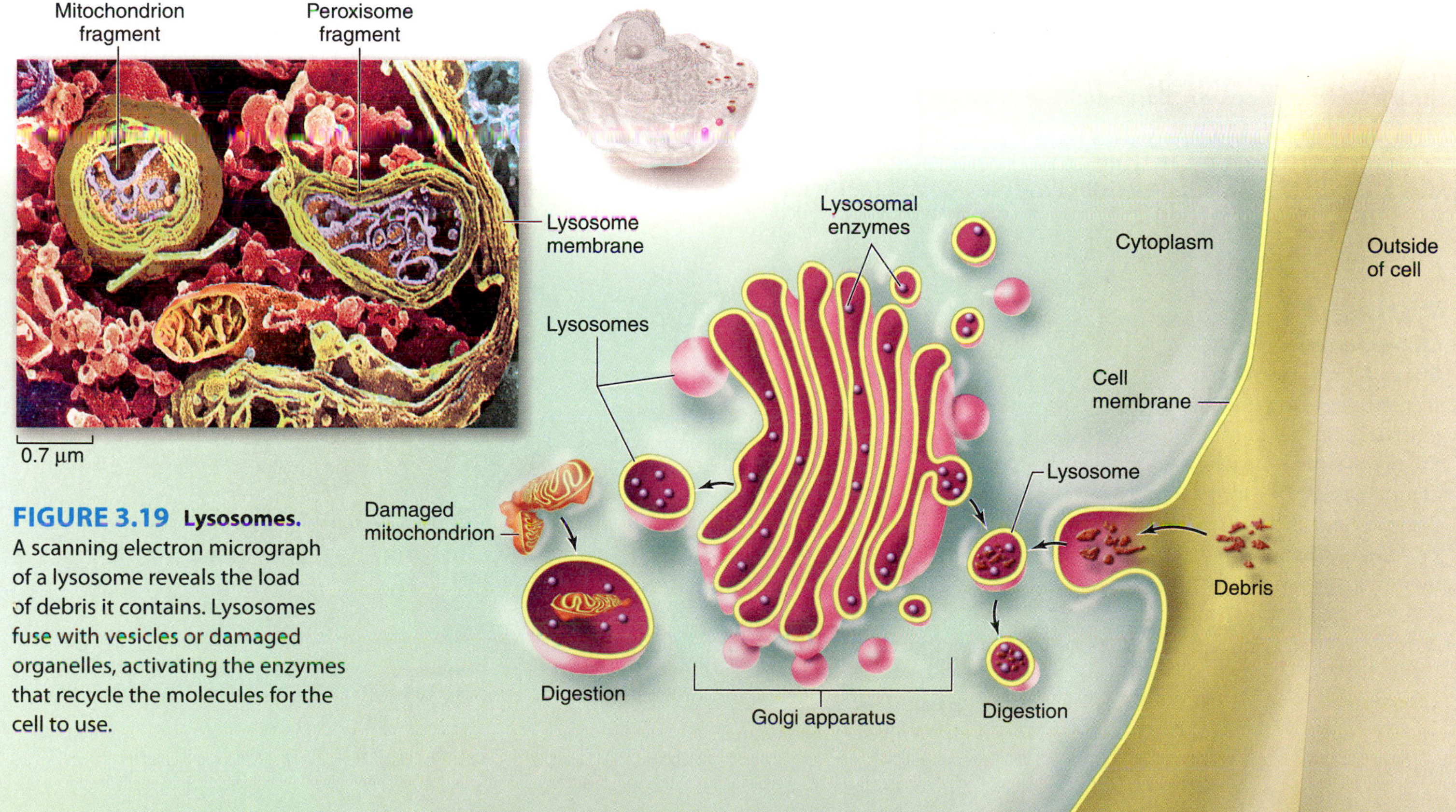

FIGURE 3.19 Lysosomes. A scanning electron micrograph of a lysosome reveals the load of debris it contains. Lysosomes fuse with vesicles or damaged organelles, activating the enzymes that recycle the molecules for the cell to use.

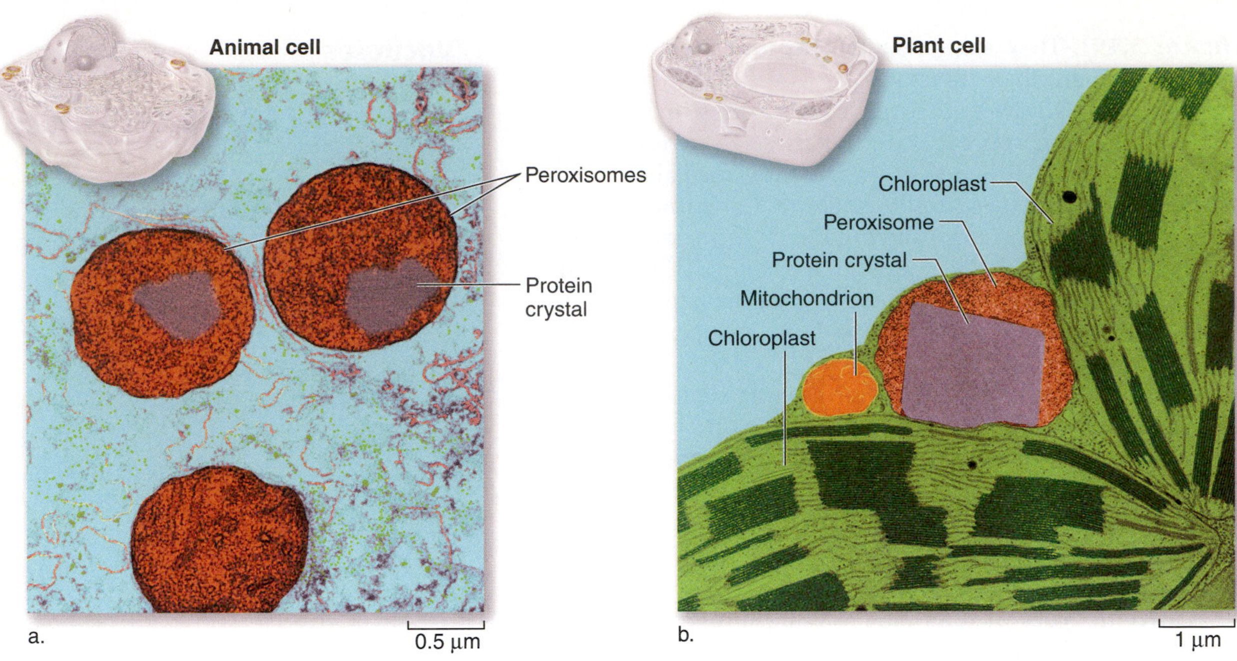

FIGURE 3.20 **Peroxisomes.** (**A**) The high concentration of enzymes inside peroxisomes results in crystallization of the proteins, giving these organelles a characteristic appearance. (**B**) In plants, peroxisomes have enzymes that catalyze many reactions that assist in photosynthesis and defense.

(**figure 3.20**). Peroxisomes also break down fatty acids and produce cholesterol and some other lipids.

Some of the reactions in the peroxisome produce hydrogen peroxide (H_2O_2). This compound can produce oxygen free radicals. Because the oxygen atoms in free radicals have unpaired electrons, they can damage the cell. To counteract the free-radical buildup, peroxisomes contain an enzyme that removes an oxygen atom from hydrogen peroxide and combines it with hydrogen atoms, producing harmless water molecules.

Liver and kidney cells contain many peroxisomes, which help dismantle toxins from the blood. In some plant cells, the concentration of enzymes reaches such high levels that the protein condenses into easily recognized crystalline arrays. Peroxisomes in plants help to break down and process organic molecules.

Peroxisomal abnormalities also can cause illness. In a disease called X-linked adrenoleukodystrophy, a faulty enzyme causes fatty acids to accumulate to toxic levels in the brain. The film *Lorenzo's Oil* depicted a boy with this disease and his parents' struggle to find treatment options.

C. Chloroplasts Are Glucose Factories

Plants and many protists carry out photosynthesis, a process that uses energy from sunlight to produce glucose and other food molecules. These nutrients sustain not only the producers but also the animals (including humans) and fungi that eat them.

The **chloroplast** (**figure 3.21**) is the site of photosynthesis in eukaryotes. Each chloroplast contains multiple membrane layers. Two outer membrane layers enclose a space known as the stroma. Within the stroma is a third membrane system folded into flattened sacs called thylakoids. The thylakoids are stacked and interconnected

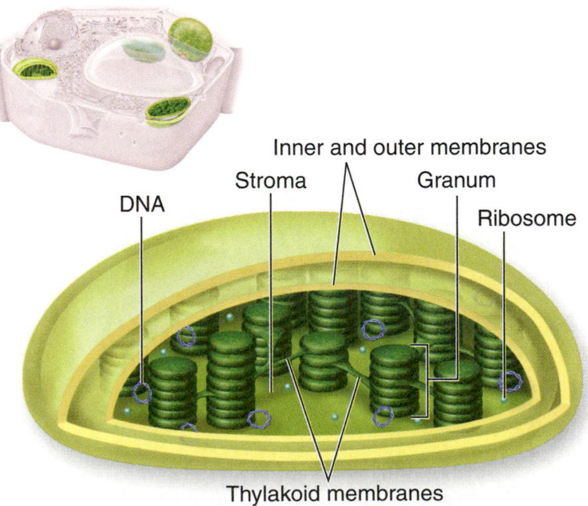

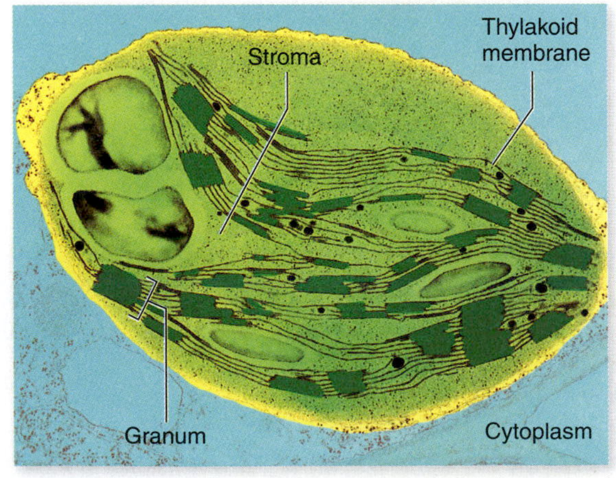

FIGURE 3.21 **Chloroplasts Are the Sites of Photosynthesis.** A chloroplast contains stacks of thylakoids that form the grana within the inner compartment, the stroma. Enzymes and light-harvesting proteins embedded in the membranes of the thylakoids convert sunlight to chemical energy.

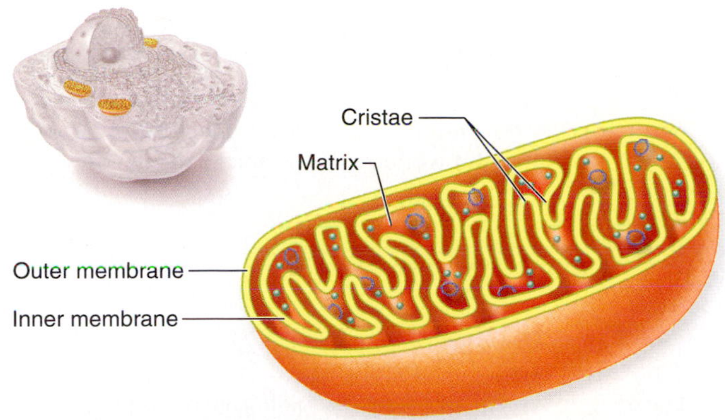

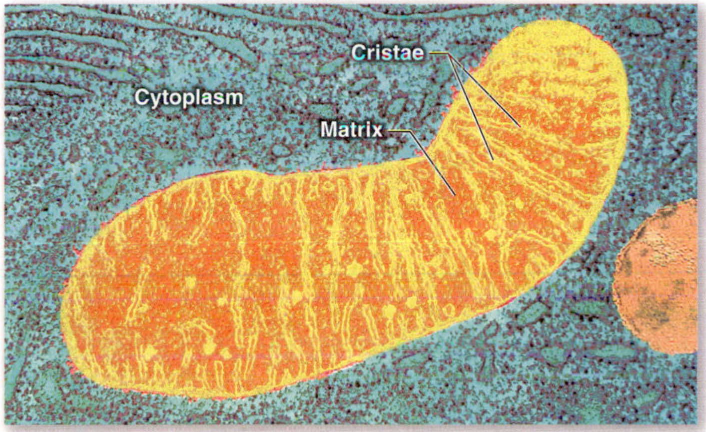

FIGURE 3.22 Mitochondria Extract Energy from Food. This transmission electron micrograph of a mitochondrion clearly shows the cristae—infoldings of the inner membrane that increase the available surface area for the reactions of cellular respiration.

in structures called grana. Photosynthesis, the subject of chapter 5, occurs in these thylakoids.

Chloroplasts also contain DNA, a feature that supports the endosymbiosis theory and helps biologists understand the evolution of eukaryotic cells (see chapter 16). This DNA encodes proteins unique to chloroplast structure and function, including some of the enzymes required for photosynthesis.

Plant cells also contain other types of plastids—organelles closely related to chloroplasts—that store food and other molecules. Some synthesize lipid-soluble red, orange, and yellow carotenoid pigments, such as those found in carrots and ripe tomatoes. Plastids that assemble starch molecules not only store food but also help plants detect the direction of gravity. Interestingly, any plastid can convert into any other type. **gravitropism, p. 552**

D. Mitochondria Extract Energy from Nutrients

Protein production, secretion, and the many chemical reactions in the cytoplasm all require a steady supply of energy. Organelles called **mitochondria** (singular: mitochondrion) use a process called cellular respiration to extract this needed energy from food (see chapter 6). Nearly all eukaryotic cells have mitochondria.

A mitochondrion has two membrane layers: an outer membrane and an intricately folded inner membrane (**figure 3.22**). The folds of the inner membrane, called **cristae,** contain enzymes that catalyze the biochemical reactions of cellular respiration

This organelle also resembles a chloroplast in that it contains its own DNA. In most mammals, mitochondria are inherited from the female parent only. Mitochondria occur in the middle regions of sperm cells, but not in the head region, which is the portion that enters the egg at fertilization. In humans, inherited diseases whose symptoms result from abnormal mitochondria always pass from mother to offspring. These mitochondrial illnesses usually produce extreme muscle weakness because muscle is a highly active tissue dependent on the functioning of many thousands of mitochondria in every cell.

Organelles divide a cell's work, just as departments in a large store group related items together. Some specialty stores, however, sell only shoes or women's clothing; likewise, cells can also have specialized functions (**figure 3.23**). For example, an active muscle cell contains many

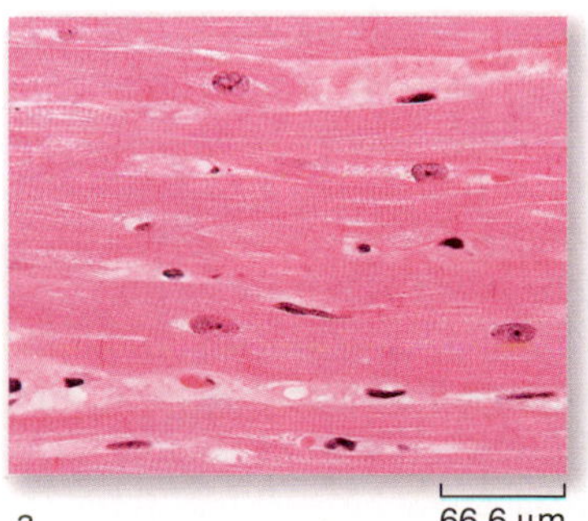

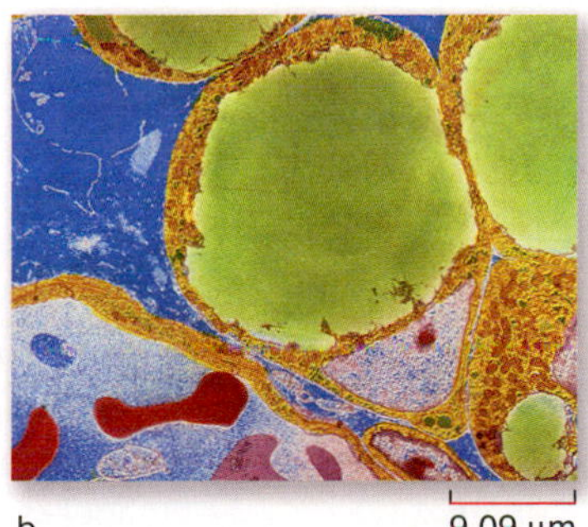

FIGURE 3.23 Specialized Cells. (**A**) Muscle cells in the heart contract in unison, thanks to electrical signals that spread rapidly from cell to cell. (**B**) The cytoplasm of this fat cell contains a giant lipid droplet. (**C**) The cholorplasts in these leaf cells carry out photosynthesis. (**D**) Root hairs have enormous surface area through which plants absorb water and minerals from soil.

Table 3.1 *Structures and Functions of Eukaryotic Organelles*

Organelle	Structure	Function
Nucleus	Perforated sac containing DNA, proteins, and RNA	Separates DNA from rest of cell; site of transcription (first step in protein synthesis); nucleolus produces ribosomes
Ribosome	Two associated globular subunits of RNA and protein	Scaffold for protein synthesis
Endoplasmic reticulum (ER)	Membrane network; rough ER has ribosomes, smooth ER does not	Site of some protein synthesis and folding; lipid synthesis
Golgi apparatus	Stacks of membrane-enclosed sacs	Links sugars to form starches or joins them to lipids or proteins; completes protein folding; stores secretions
Lysosome	Sac containing digestive enzymes	Degrades debris; recycles cell contents
Central vacuole	Sac containing watery solution of enzymes, acids, water-soluble pigments, and other solutes	Produces turgor pressure; recycles cell contents; contains pigments
Peroxisome	Sac containing enzymes	Disposes of toxins; breaks down fatty acids; eliminates hydrogen peroxide
Chloroplast	Two membranes enclose inner stacks of flattened membrane sacs called thylakoids, which contain pigments	Carries out photosynthesis to produce food (glucose)
Mitochondrion	Two membranes; inner one folded into enzyme-studded cristae	Releases energy from food

more mitochondria than does an adipose cell, which is little more than a blob of stored fat. A mesophyll cell in a leaf of a flowering plant is packed with chloroplasts, organelles that capture the sun's energy in photosynthesis. A root cell has few, if any, chloroplasts; instead, it specializes in absorption.

Table 3.1 summarizes the structures and functions of the eukaryotic organelles.

Summary ***Organelles compartmentalize a cell's activities, improving efficiency and protecting cell contents from harsh chemicals. Specialized organelles store and secrete substances, derive energy from nutrients, and degrade debris.***

3.4 MASTERING CONCEPTS

1. Which organelles interact to produce and secrete a complex substance such as milk?
2. What is the function of the nucleus and its contents?
3. Which organelles are the cell's "recycling centers?"
4. What are some functions of plastids?
5. Which organelle houses the reactions that extract chemical energy from nutrient molecules?
6. Which three organelles contain DNA?

3.5 The Cytoskeleton Supports Eukaryotic Cells

Once considered simply a watery soup, biologists have discovered that the cytoplasm of a eukaryotic cell contains a **cytoskeleton**, an intricate network of internal protein "tracks" and tubules. The cytoskeleton is a structural framework with numerous functions. It is a transportation system, and it provides the structural support necessary to maintain the cell's characteristic three-dimensional shape (**figure 3.24**). The cytoskeleton also enables cells—or parts of a cell—to move. It aids in cell division and helps connect cells to one another.

The protein girders of the cytoskeleton include three major components: microtubules, microfilaments, and intermediate filaments (**figure 3.25**). They are distinguished by protein type, diameter, and how they aggregate into larger structures. Other proteins connect these components to one another, creating an intricate meshwork.

A. Microtubules Participate in Cell Division and Cell Movement

A **microtubule** is composed of a protein, called tubulin, that is assembled into a hollow tube 23 nm in diameter (see figure 3.25). Long microtubules provide many cellular movements. The cell can change the length of the tubule rapidly by adding or removing tubulin molecules. For example, microtubules pull a dividing cell's duplicated chromosomes apart. Some anticancer drugs either prevent tubulin from assembling into microtubules or prevent microtubules from breaking down into free tubulin molecules. In each case, cell division stops. **mitosis, p. 160**

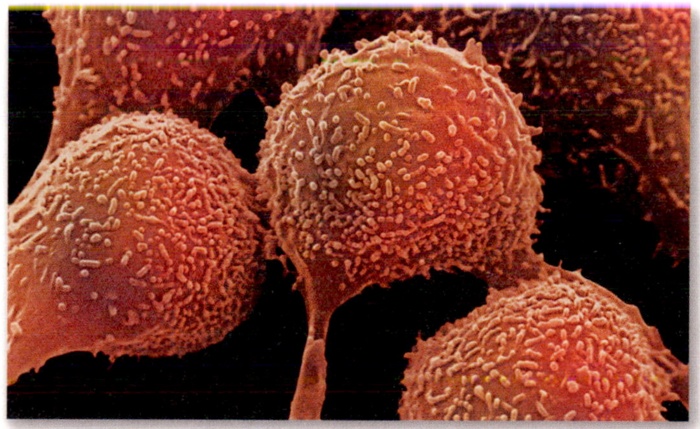

FIGURE 3.24 Cellular Architecture. A white blood cell's inner skeleton and surface features enable it to move in the body and to recognize "foreign" cell surfaces, such as those of transplanted tissue.

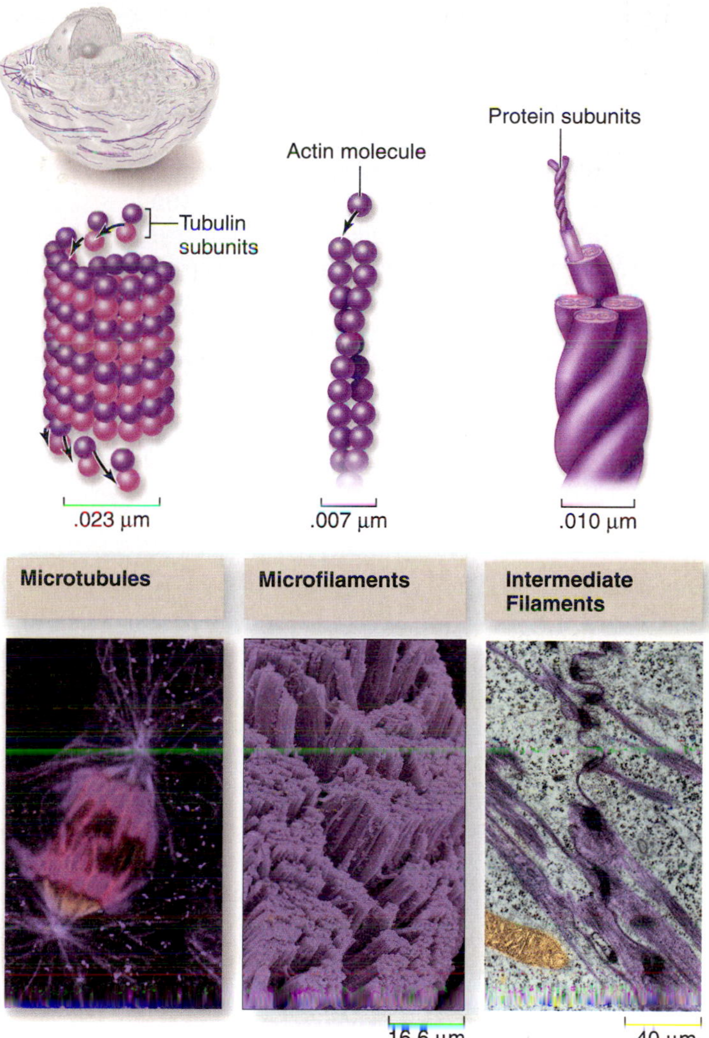

FIGURE 3.25 The Cytoskeleton Is Made of Protein Rods and Tubules. The three major components of the cytoskeleton are microtubules, microfilaments, and intermediate filaments. (**A**) Microfilaments are purple and chromosomes are pink in this confocal micrograph of a cell preparing to divide. (**B**) This scanning electron micrograph shows tightly packed actin filaments in a cross section of skeletal muscle. (**C**) Intermediate filaments stretch diagonally across this transmission electron micrograph. Along the center of the image, filaments join at specialized junctions between cells, an arrangement that gives high tensile strength to skin.

Microtubules also serve as a type of "trackway" within a cell to move organelles and proteins rapidly from the center to the periphery and back. Some organisms, such as the squid, can change colors rapidly by using this process to rearrange pigment molecules in their skin cells.

Some cells have cilia or flagella that enable them to move. Both types of appendages contain microtubules (**figure 3.26**). **Cilia** are short and numerous, like a fringe. Coordinated movement of cilia sets up a wave that propels particles up and out of respiratory tubules or moves an egg cell through the female reproductive tract. Some single-celled organisms may have thousands of individual cilia, enabling them to "swim" in water. In contrast, a flagellum is much longer than a cilium. Flagella are more like tails, and their whiplike movement propels cells. Sperm cells in many species have prominent flagella. **ciliates, p. 400**

An individual cilium or flagellum is constructed of nine microtubule pairs that surround a central, separated pair and form a pattern described as "9 + 2" (see figure 3.26). A type of motor protein called dynein connects the outer microtubule pairs and also links them to the central pair, a little like a wheel. Dynein molecules shift in a way that slides adjacent microtubules against each other. This movement bends the cilium or flagellum. (The flagellum in prokaryotic cells does not have this structure.)

B. Microfilaments Move Other Parts Inside a Cell

A second component of the cytoskeleton is the **microfilament,** a long, thin rod composed of the protein actin. In contrast to microtubules, microfilaments are not hollow and are only about 7 nm in diameter (see figure 3.25). Actin microfilament networks are part of nearly all eukaryotic cells, providing the machinery to move should an appropriate signal arrive. Muscle contraction, for example, relies on actin filaments and another protein, myosin. **sliding filaments, p. 644**

Microfilaments also provide strength for cells to survive the stretching and compression that often occurs in multicellular organisms. These tiny rods also help to anchor one cell to another (see section 3.6).

C. Intermediate Filaments Maintain Shape and Connect Cells

Intermediate filaments are so named because their 10-nm diameters are intermediate between those of microtubules and microfilaments (see figure 3.25). Unlike microtubules and microfilaments, which consist of a single protein type, intermediate filaments are made of different proteins in different specialized cell types. They form an internal scaffold in the cytoplasm and resist mechanical stress, both functions that maintain a cell's shape. Intermediate filaments also help bind some cells together (see section 3.6).

Given the cytoskeleton's many functions, it is not surprising that defects can cause disease. For example, people with Duchenne muscular dystrophy lack a protein called dystrophin, part of the cytoskeleton in muscle cells. Without dystrophin, muscles—including those in the heart—

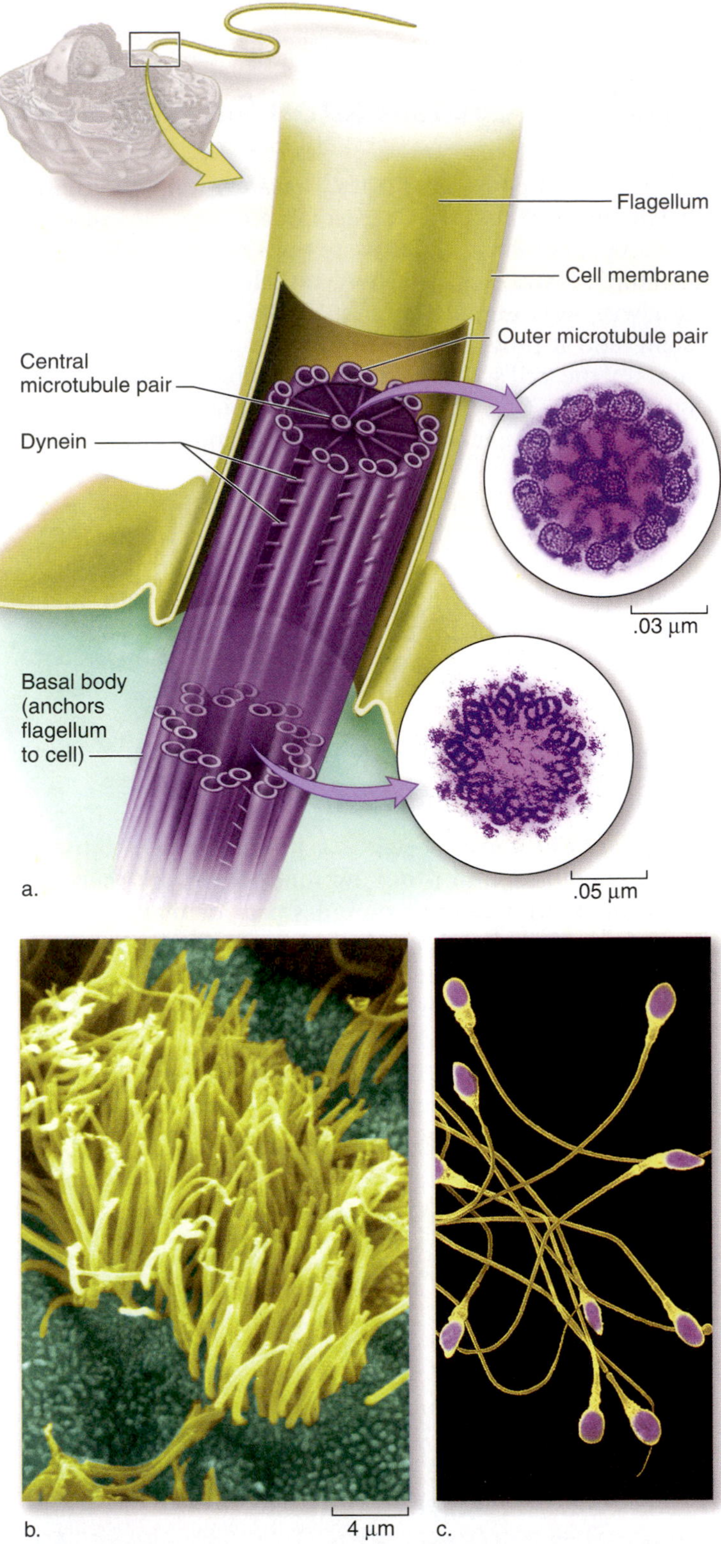

FIGURE 3.26 Microtubules Move Cells. (**A**) The microtubules that form cilia and the cytoskeletons of eukaryotic flagella have a characteristic "9 + 2" organization. Dynein joins the outer microtubule doublets to each other and to the central pair of microtubules. (**B**) These cilia line the human respiratory tract, where their coordinated movements propel dust particles upward so the person can expel them. (**C**) The flagella on human sperm cells enable them to swim.

degenerate. Patients rarely survive past the age of 30. Another faulty cytoskeleton protein, ankyrin, causes a genetic disease of blood. In healthy red blood cells, the cytoskeleton maintains a concave disk shape. When ankyrin is missing, red blood cells are small, fragile, and misshapen, greatly reducing their ability to carry oxygen.

Summary *Proteins form a vast cytoskeleton, a network of tubules and filaments inside the eukaryotic cell.*

3.5 MASTERING CONCEPTS

1. What are some functions of the cytoskeleton?
2. What are the major components of the cytoskeleton?
3. How are cilia and flagella similar, and how are they different?

3.6 Cells Stick Together and Communicate with One Another

So far, this chapter has described individual cells. But multicellular organisms, including plants and animals, are made of many cells that work together. How do these cells adhere to one another so that your body—or that of a plant—doesn't disintegrate in a heavy rain?

In addition, how do cells communicate with one another to coordinate development and respond to the environment? A large part of the answer is hormones, described in chapter 26 (for plants) and chapter 30 (for animals). Most hormones, however, act over long distances: one tissue releases signal molecules, which exert their effects at another location. How do cells in direct contact with one another communicate?

This section describes how the cells of plants and animals stick together and how neighboring cells share signals.

A. Cell Walls Are Strong, Flexible, and Porous

Cell walls surround the cell membranes of nearly all bacteria, archaea, fungi, algae, and plants. But *cell wall* is a misleading term—it is not just a barrier that serves only to outline the cell. Cell walls impart shape, regulate cell volume, prevent bursting when a cell takes in too much water, and interact with other molecules to help determine how a cell in a complex organism specializes. In plants, for example, whether a cell specializes to become a root, shoot, or leaf depends on which cell walls it touches.

Cell walls are built of different components. Bacterial cell walls are composed of peptidoglycan, and those of fungi contain chitin. Much of the plant cell wall consists of cellulose (see figure 2.16C). Cellulose molecules align into fibrils, which in turn aggregate and twist to form larger fibrils. This fibrous organization imparts great strength. Other molecules, including the polysaccharides hemicellulose and pectin, impart additional strength, add flexibility, and glue adjacent cells together (**figure 3.27**). Plant cell walls also contain glycoproteins, enzymes, and many other proteins.

A plant cell secretes many of the components of its wall, so the older layer of a cell wall is on the exterior of the cell, and the newer layers hug the cell membrane. Some cells have

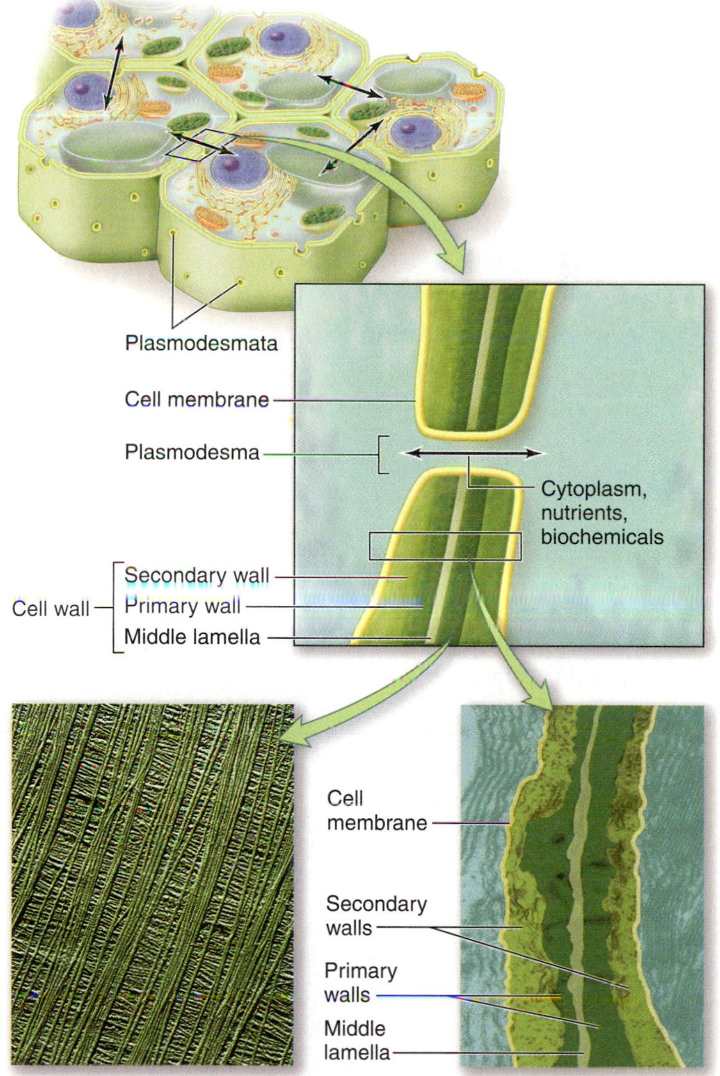

FIGURE 3.27 Plant Cell Connections. The walls of adjoining cells are quite complex. They are composed of layers that each cell lays down, joined by a layer called the middle lamella. Plasmodesmata connect the cytoplasms of adjacent cells. The inset shows cellulose fibrils that make up the cell wall.

rigid secondary cell walls beneath the initial, more flexible primary one. The rigidity of the secondary cell wall comes from lignin, a polymer so complex that only a few types of organisms can decay it. Wood's toughness comes from secondary cell walls. **wood, p. 512**

How do plant cells communicate with their neighbors through the wall? **Plasmodesmata** are channels that connect adjacent cells. They are essentially "tunnels" in the cell wall, through which the cytoplasm and some of the organelles of one plant cell can interact with those of another (see figure 3.27). They therefore facilitate cell-to-cell communication and coordination of function. Plasmodesmata are particularly plentiful in parts of plants that conduct water or nutrients and in cells that secrete oils and nectars.

B. Animal Cell Junctions Occur in Several Forms

Unlike plants and fungi, animal cells lack cell walls. Instead, many animal cells secrete a complex extracellular matrix that holds them together and coordinates many aspects of cellular life. The protein collagen, for example, is a major component of the extracellular matrix in mammalian skin, ligaments, cartilage, and bone (see chapter 27).

In other tissues, however, the plasma membranes of adjacent cells directly connect to one another via several types of intercellular junctions (**figure 3.28**):

- A **tight junction** fuses cells together, forming an impermeable barrier between them. Proteins anchored in membranes connect to actin in the cytoskeleton and join cells into sheets, such as those lining the inside of the human digestive tract and the tubules of the kidneys. These connections allow the body to control where biochemicals move, since fluids cannot leak between the joined cells. Extensive tight junctions create the blood–brain barrier, which protects the brain from chemical fluctuations. However, this barrier readily admits lipid-soluble drugs such as heroin, nicotine, alcohol, and cocaine across its cell membranes, accounting for these drugs' rapid action (see Can *You* Relate? Nicotine Addiction). Water-soluble molecules must take other routes across the barrier.
- An **anchoring (or adhering) junction** connects adjacent cells by linking their intermediate filaments in single spot. Somewhat like rivets, these junctions hold skin cells in place by anchoring them to the extracellular matrix.
- A **gap junction** is a protein channel that links the cytoplasm of adjacent cells, allowing exchange of ions, nutrients, and other small molecules. It is therefore analogous to the plasmodesmata in plants. Gap junctions join heart muscle cells as well as muscle cells that line the digestive tract, allowing groups of cells to act together. Because cells "talk" to each other to signal when it is time to stop dividing, problems with gap junction communication can lead to uncontrolled cell division—cancer.

Table 3.2 summarizes cell–cell connections for plants and animals, and **table 3.3** compares prokaryotic, plant, and animal cells.

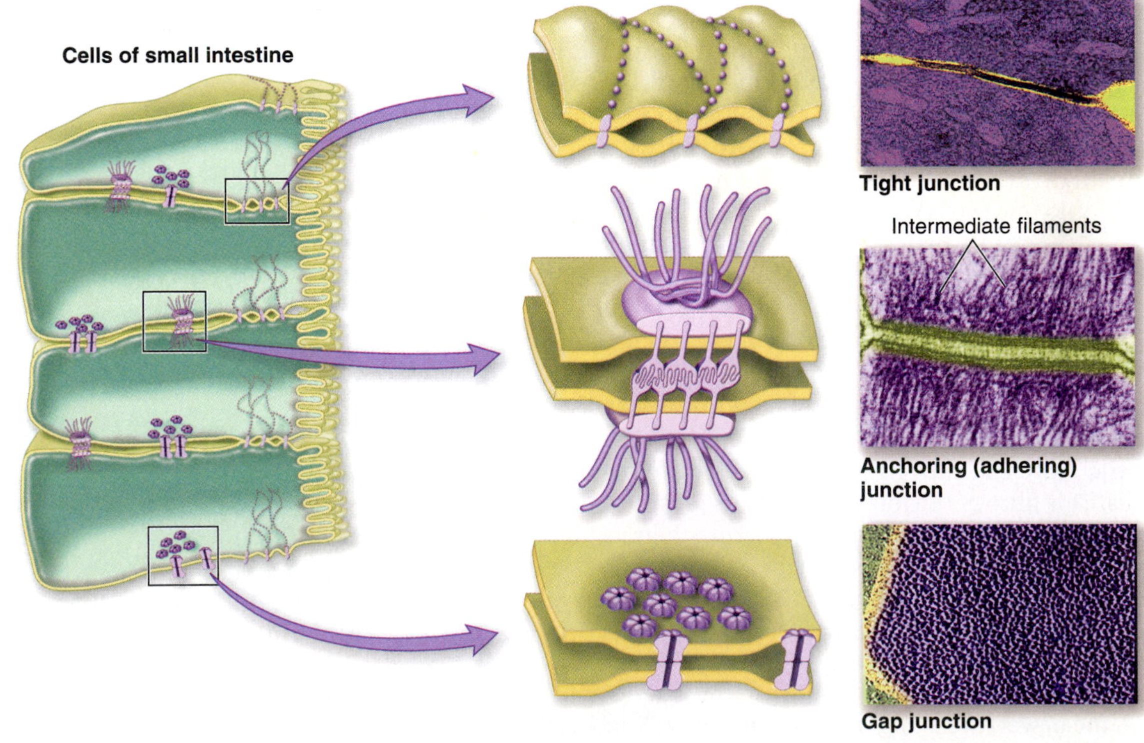

FIGURE 3.28
Animal Cell Connections. These cells illustrate all three types of animal cell junctions. Tight junctions fuse neighboring cell membranes, anchoring junctions form "spot welds," and gap junctions allow small molecules to move between the cytoplasms of adjacent cells.

Can *You* Relate?

Nicotine Addiction

In the United States, nearly one-fourth of adults smoke cigarettes. Yet the effects of smoking on health, such as greatly increased risks of developing heart disease, cancer, stroke, and lung disease, are well known. Why, then, do so many people smoke?

The answer is that they are addicted to nicotine, a naturally occurring chemical in the tobacco leaves in cigarettes (see the opening essay for chapter 24). Cell biology can explain how a pleasant experience can become a physical dependency.

Nicotine reaches the brain within seconds of the first inhalation, binding to proteins that are parts of the cell membranes of some nerve cells. These receptors normally bind a brain chemical, acetylcholine, which is produced in response to nerve impulses. When sufficient nicotine binds to the receptors instead, a channel within the receptor opens and admits positively charged ions into the nerve cell. In response, the cell releases another chemical, dopamine, from its other end. Dopamine provides the pleasurable feelings associated with smoking. Addiction stems from two sources: seeking the dopamine release and avoiding painful withdrawal symptoms.

When a person smokes more frequently, the number of nicotine receptors on brain cells increases. Nicotine binding impairs the recycling of receptor proteins, so that new receptors form faster than they are taken apart. But after a period of steady nicotine exposure, many receptors malfunction and no longer trigger nerve transmission. This may be why as time goes on it takes more nicotine to produce the same physical effects.

Summary ***Cells must permanently attach to one another to build most tissues. Strands of cytoplasm connect adjacent plant cells through rigid cell walls. Animal cells lack walls, but several types of specialized junctions keep the cells connected and communicating.***

3.6 MASTERING CONCEPTS

1. What functions do cell walls provide?
2. What is the chemical composition of a plant cell wall?
3. What are plasmodesmata?
4. What are the three types of junctions that link cells in animals?

Table 3.2 *Intercellular Junctions*

Type	Function	Location
Plasmodesmata	Allow substances to move between plant cells	Plant cell walls
Tight junctions	Close spaces between animal cells by fusing cell membranes	Inside lining of small intestine
Anchoring (adhering) junctions	Spot weld adjacent animal cell membranes	Outer skin layer
Gap junctions	Form channels between animal cells, allowing exchange of substances	Muscle cells in heart and digestive tract

Table 3.3 *Prokaryotic, Plant, and Animal Cells Compared*

Cell Feature	Prokaryotic Cells (Bacteria and Archaea)	Plant Cells	Animal Cells
Nucleus	No	Yes	Yes
Ribosome	Yes	Yes	Yes
Endoplasmic reticulum	No	Yes	Yes
Golgi apparatus	No	Yes	Yes
Lysosome	No	Rarely	Yes
Central vacuole	No	Yes	No
Peroxisome	No	Yes	Yes
Chloroplast	No	Yes	No
Mitochondrion	No	Yes	Yes
Cell wall	Usually	Yes	No

3.7 INVESTIGATING LIFE
Did the Cytoskeleton Begin in Bacteria?

Biologists constantly ask questions about life, and some of the most intriguing evolutionary puzzles arise when one group of organisms has a feature that its ancestors do not. The lure of an angler fish, the stinger on a scorpion's tail, and the spines of a cactus prompt the question, "Where did that come from?" The origins of the cell's internal structures are equally perplexing.

The presence of a cytoskeleton is one feature that distinguishes eukaryotic from prokaryotic cells. As described in section 3.5, this supportive framework has three major components: microtubules composed of tubulin, microfilaments made of actin, and intermediate filaments consisting of multiple types of proteins. Because all eukaryotic cells have these cytoskeletal elements, they must have been present in their last common (shared) ancestor. But if they are not present in prokaryotic cells, which were the first life forms, where did these elements come from?

Researchers are beginning to solve this mystery by questioning the assumption that prokaryotic cells lack a cytoskeleton. In the late 1990s, for example, researchers discovered a protein in the bacterium *E. coli* that has a structure similar to that of tubulin. Further study revealed that the protein is essential for cell division in both bacteria and archaea.

More recently, researchers have found two bacterial proteins similar to actin. Laura Jones, Rut Carballido-López, and Jeffery Errington at the University of Oxford discovered that the proteins form filaments that lie just inside the cell surface, helping the bacterium keep its shape. Their use of a variety of tools, including altered genes, microscopy, and comparisons of molecules from several species, illustrates how researchers use multiple lines of evidence to arrive at a conclusion.

The scientists studied the proteins in a bacterium called *Bacillus subtilis*. This soil-dwelling organism ordinarily forms rod-shaped cells, but the researchers noticed that when they experimentally "turned off" either of the two genes, the cells had abnormal shapes (**figure 3.29A**). Recall from chapter 2 that a gene is a DNA sequence that specifies the amino acid sequence of one protein. When a gene is switched off, a cell cannot make the corresponding protein. When the scientists turned off the gene encoding the protein called MreB, the cells were the correct length but appeared abnormally inflated or rounded. When they turned off the gene encoding the protein Mbl instead, the cells were bent and twisted.

Another way to study protein function is to mutate the gene that encodes it. Mutating a gene can have several results: the encoded protein may remain unchanged, the protein's function may change slightly, or the protein may be completely ruined. The researchers mutated both genes and found, again, that many of the mutant bacteria had abnormal shapes.

Next, Jones and her colleagues used microscopy to see where in the *B. subtilis* cell the two proteins are located. Because an individual protein is far too small for even the most powerful microscope to resolve, the researchers used glowing fluorescent "tags" to reveal exactly where MreB and Mbl occurred. This method exploits the ability of antibodies to recognize specific molecules. Antibodies are a critical part of the human immune system, binding to cells or molecules that the body does not recognize as its own. Each antibody binds to just one type of molecule. The scientists created antibodies that had two critical abilities: they could bind to MreB or Mbl, and they could glow under ultraviolet light. The results were remarkable: each protein, marked with its fluorescent tag, clearly formed a helix just inside the surface of the cell (figure 3.29B).

These experiments showed that MreB and Mbl together help determine cell shape in *B. subtilis*. Could the same proteins be at work in other prokaryotes as well? The researchers searched the DNA sequences of dozens of other bacteria and archaea, looking for genes similar to the one encoding MreB. They found that most of the species with such a gene had cells shaped like rods, filaments, or corkscrews. Most species without it had spherical cells.

Finally, the scientists searched computerized databases of amino acid sequences for proteins of similar structure from other species. Both MreB and Mbl were similar to actin from both yeast and human cells. Although the sequences were not identical, they were similar at critical locations, suggesting related functions.

This research has brought biologists closer to answering the question of "Where did that come from?" for the cytoskeleton. The function, location, and amino acid sequences of MreB and Mbl in *B. subtilis* strongly suggest that the actin in eukaryotic cells has a prokaryotic counterpart with a function that persists to this day. At least some elements of the cytoskeleton apparently evolved before the two cell types diverged some 2 billion years ago.

FIGURE 3.29 Cytoskeleton Precursor? (**A**) Mutated genes produced misshapen *Bacillus subtilis* cells. (**B**) Mbl protein tagged with a fluorescent label clearly accumulates in a helix just inside the cell wall. **Question:** How would these cells look different if Mbl occurred throughout the cytoplasm?

Jones, Laura J. F., Rut Carballido-López, and Jeffery Errington. 2001. Control of cell shape in bacteria: helical, actin-like filaments in *Bacillus subtilis*. *Cell*, vol. 104, pages 913–922.

CHAPTER SUMMARY

3.1 Cells Are the Units of Life

- **Cells** are the microscopic components of all organisms.

A. Discovering the Cellular Basis of Life

- The first person to see cells was Robert Hooke, who viewed cork with a crude lens in the late seventeenth century. Antony van Leeuwenhoek viewed many cells under the light microscope.

B. The Cell Theory Emerges

- Schleiden, Schwann, and Virchow's formulation of **cell theory** states that all life is composed of cells, that cells are the functional units of life, and that all cells come from preexisting cells.
- Contemporary cell biology focuses on the processes that occur inside cells.

C. Microscopes Reveal Cell Structure

- Light microscopes, electron microscopes, and scanning probe microscopes are essential tools in cell biology.

D. Features Common to All Cells

- All cells have DNA, RNA, **ribosomes** that build proteins, **cytoplasm**, and a **cell membrane** that is the interface between the cell and the outside environment. Complex cells also have specialized compartments called **organelles**.
- The volume of a cell must be small relative to its surface area.

3.2 A Membrane Separates Each Cell from Its Surroundings

A. Lipids and Proteins Form the Cell Membrane

- A biological membrane consists of a **phospholipid bilayer** embedded with movable proteins and sterols, forming a **fluid mosaic**.
- Membrane proteins carry out a variety of functions.

B. Signal Transduction Transmits Messages to a Cell's Interior

- In **signal transduction,** receptors in the cell membrane receive input and transmit the messages to the interior of the cell. Eventually this signaling stimulates the cell to carry out a specific function.

3.3 Different Cell Types Characterize Life's Three Domains

- Cells are **prokaryotic** (lacking a **nucleus** and other organelles) or **eukaryotic** (having a nucleus and other organelles). Prokaryotic cells include bacteria and archaea.

A. Domain Bacteria

- Bacterial cells are structurally simple, but they are abundant and diverse. Most have a **cell wall** and one or more **flagella**. DNA occurs in an area called the **nucleoid**.

B. Domain Archaea

- Archaea share some characteristics with bacteria and eukaryotes but also have unique structures and biochemistry.

C. Domain Eukarya

- Eukaryotic cells include those of protista, fungi, animals, and plants. Most eukaryotic cells are larger than prokaryotic cells.

3.4 Eukaryotic Organelles Divide Labor

A. Organelles Interact to Secrete Substances

- A eukaryotic cell houses DNA in a nucleus. **Nuclear pores** permeate the **nuclear envelope**; ribosome assembly occurs in the **nucleolus**.
- The **smooth** and **rough endoplasmic reticulum** and the **Golgi apparatus** work together to synthesize, store, transport, and release molecules. **Vesicles** transport materials within cells.

B. Lysosomes, Vacuoles, and Peroxisomes Are Cellular Digestion Centers

- A eukaryotic cell degrades wastes and digests nutrients in **lysosomes.**
- In plants, a watery **vacuole** degrades wastes, exerts turgor pressure, and stores acids and pigments.
- **Peroxisomes** process toxins and oxygen.

C. Chloroplasts Are Glucose Factories

- Cells of plants and algae have **chloroplasts**, organelles that use solar energy to make food.

D. Mitochondria Extract Energy from Nutrients

- Nearly all eukaryotic cells have **mitochondria**. The **cristae** (folds) of the inner mitochondrial membrane house the reactions of cellular respiration.

3.5 The Cytoskeleton Supports Eukaryotic Cells

- The **cytoskeleton** is a network of rods and tubules that provides cells with form, support, and the ability to move.

A. Microtubules Participate in Cell Division and Cell Movement

- **Microtubules** self-assemble from hollow tubulin subunits to form **cilia,** flagella, and the fibers that separate chromosomes during cell division.

B. Microfilaments Move Other Parts Inside a Cell

- **Microfilaments** are solid and smaller than microtubules. They are composed of the protein actin and provide contractile motion when they interact with myosin.

C. Intermediate Filaments Maintain Shape and Connect Cells

- **Intermediate filaments** are intermediate in diameter between microtubules and microfilaments. They consist of various proteins, and they strengthen the cytoskeleton.

3.6 Cells Stick Together and Communicate with One Another

A. Cell Walls Are Strong, Flexible, and Porous

- Most organisms other than animals have cell walls, which provide shape and mediate signals. Plant cell walls consist of cellulose fibrils connected by hemicellulose, plus pectin and various proteins.
- **Plasmodesmata** are continuations of cell membranes between cells through thinned parts of the cell wall.

B. Animal Cell Junctions Occur in Several Forms

- Junctions connecting animal cells include **tight junctions, anchoring junctions,** and **gap junctions.** Tight junctions create a seal between adjacent cells. Anchoring junctions are "spot welds" that secure cells in place. Gap junctions allow adjacent cells to exchange cytoplasmic material.

3.7 Investigating Life: Did the Cytoskeleton Begin in Bacteria?

- Although the cytoskeleton occurs only in eukaryotic cells, bacteria do have actin-like proteins that help control cell shape.

MULTIPLE CHOICE QUESTIONS

1. Why are cells considered to be the smallest unit of life?
   a. Because you need a microscope to see them.
   b. Because a cell is the smallest thing that carries out all the functions of life.
   c. Because they have a structure.
   d. Because all cells have a nucleus with DNA.

2. Which of the following is NOT a feature found in all cells?
   a. Proteins
   b. Ribosomes
   c. Cell wall
   d. Cell membrane

3. A cell membrane is said to be a *fluid mosaic* because
   a. there is water in the membrane
   b. the membrane is made of lipids and proteins that move.
   c. it forms a bilayer.
   d. transport proteins allow for the movement of water-soluble molecules.

4. One property that distinguishes cells in Domain Bacteria from those in Domain Eukarya is
   a. the presence of a cell wall.
   b. the presence of DNA.
   c. the presence of flagella.
   d. the presence of membrane-bounded organelles.

5. Which of the following organelles is/are associated with the job of cellular digestion?
   a. Lysosomes and peroxisomes
   b. Golgi apparatus and vesicles
   c. Nucleus and nucleolus
   d. Smooth endoplasmic reticulum

6. What type of protein would you expect to find on the membrane of a vacuole?
   a. Enzyme
   b. Receptor protein
   c. Transport protein
   d. Cytoskeleton

7. What type of microscope would you use if you wished to study how the cytoskeleton of a cell changed during its growth?
   a. Confocal microscope
   b. Light microscope
   c. Transmission electron microscope
   d. Scanning electron microscope

8. Which of the following organelles does not contain its own DNA?
   a. Nucleus
   b. Chloroplast
   c. Rough endoplasmic reticulum
   d. Mitochondrion

9. What cellular process leads to the production of milk-specific mRNA molecules?
   a. Protein synthesis
   b. Signal transduction
   c. Lipid synthesis
   d. Transport

10. Anchoring junctions are stabilized by the intermediate filament cytoskeleton. What else is required for a cell to form an anchoring junction?
    a. A cell wall
    b. Extracellular matrix
    c. A receptor protein
    d. A cell adhesion protein

TESTING YOUR KNOWLEDGE

1. What features do all cells share?
2. Why are large organisms made of numerous small cells instead of a few large ones?
3. What types of chemicals make up cell membranes?
4. What is signal transduction? What is the cell membrane's role in this process?
5. Until recently, biologists thought that there were only two types of cells. How has that view changed?
6. Name three structures or activities found in eukaryotic cells but not in bacteria or archaea.
7. Which three organelles in a eukaryotic cell are surrounded by a double membrane?
8. List the components and functions of the cytoskeleton.
9. How do plant cells interact with their neighbors through the rigid cell wall?
10. Describe how animal cells use junctions in different ways.

THINKING AS A SCIENTIST

1. How has improved technology enabled the expansion of cell theory from Schleiden and Schwann's original formulation?
2. The simplest viruses consist only of a protein coat surrounding DNA or RNA. What does such a virus have in common with cells? What does it lack that all cells have?
3. A liver cell has a volume of 5000 μm^3. Its total membrane area, including the inner membranes lining organelles as well as the cell membrane, is 110,000 μm^2. A cell in the pancreas that manufactures digestive enzymes has a volume of 1,000 μm^3 and a total membrane area of 13,000 μm^2. Which cell is probably more efficient in carrying out activities that require extensive membrane surfaces, and why?
4. What advantages does compartmentalization confer on a large cell?
5. Choose one of the functions of membrane proteins mentioned in this chapter. If a person was born with a faulty version of a protein with that function, what symptoms might you predict?
6. In what ways is a prokaryotic cell like a baseball stadium, but a eukaryotic cell is more like an office building?
7. Why does a muscle cell contain many mitochondria and a white blood cell (an immune cell that engulfs bacteria) contain many lysosomes?
8. Your friend claims that an ostrich egg is the largest single cell, but you are skeptical that one cell can be that large. If you had access to an ostrich egg and a high-quality light microscope, what features would you look for to resolve the argument?
9. Would a substance that destroys the blood–brain barrier be dangerous? Why or why not?

Visit www.mhhe.com/hoefnagels for practice quizzes, animations, videos, and activities designed to help you master the material in this chapter.

The Energy of Life

Comedy duo Stan Laurel and Oliver Hardy spend energy on a performance in 1927.

What's the Point?

Whole-Body Metabolism: Energy on an Organismal Level

"I wish I had your metabolism!" Perhaps you have overheard a calorie-counting friend make a similar comment to someone who stays slim on a diet of fattening foods.

In that context, the word metabolism *means how fast a person burns food. But biochemists define metabolism as the chemical reactions that build and break down molecules within any cell. How are these two meanings related? Each cell's interlocking networks of metabolic reactions supply the energy it needs to stay alive. In humans, teams of metabolizing cells perform specialized functions such as digestion, muscle movement, hormone production, and countless other activities. It all takes a reliable energy supply—food, which we "burn" at an ever-changing rate.*

Minimally, a body needs energy to maintain heartbeat, temperature, breathing, brain activity, and other basic life requirements. For an adult human male, the average energy use is 1750 Calories in 24 hours; for a female, 1450 Calories. These measurements do not include the energy required for physical activity or digestion, so the number of Calories needed to get through a day generally exceeds the basal requirement.

These averages mask the fact that people have different metabolic rates. Age, sex, weight, and regular activity level all influence metabolic rates, as does body fat composition. All other things being equal, a person with the most lean tissue (muscle, nerve, liver, and kidney) will have the highest metabolic rate, because lean tissue consumes more energy than relatively inactive fat tissue. The amount of a thyroid hormone, thyroxine, also affects energy expenditure.

So why do some people gain weight? The deceptively simple answer is that if energy input exceeds energy loss, you gain weight; in the reverse situation, you slim down. This is why reducing caloric intake and exercising are two basic weight-loss recommendations. Unfortunately, metabolic differences make it difficult to translate this simple energy balance into a one-size-fits-all plan.

This chapter brings together the fundamental principles of metabolism, including how cells organize, regulate, and fuel the chemical reactions that sustain life.

LEARNING OUTLINE

4.1 All Cells Capture and Use Energy

You're running late. You overslept, you have no time for breakfast, and you have a full morning of classes. You rummage through your cupboard and find something called an "energy bar"—just what you need to get through the morning. But what is energy?

A. Energy Allows Cells to Do Life's Work

Physicists define **energy** as the ability to do work—that is, to move matter. This idea, abstract as it sounds, is fundamental to biology. Life depends on rearranging atoms and trafficking substances across membranes in precise ways. These intricate movements represent work, and they require energy.

The total amount of energy in any object is the sum of energy's two forms: potential and kinetic (**figure 4.1** and **table 4.1**). **Potential energy** is stored energy available to do work. A snake poised to strike at its prey illustrates potential energy, as does a baseball player about to throw a ball. Likewise, unburned gasoline—and that energy bar you grabbed—contains potential energy stored in the chemical bonds of its constituent molecules. A chemical gradient is another form of potential energy (see section 4.5).

Kinetic energy is energy being used to do work; any moving object possesses kinetic energy. The snake striking and the soaring baseball both demonstrate kinetic energy. A chameleon shooting out its sticky tongue to capture a butterfly illustrates kinetic energy, as does a Venus's flytrap closing its leaves around an insect (see chapter 25's opening essay—Carnivorous Plants). Moving pistons and a rolling bus have kinetic energy. Heat, light, and sound are also types of kinetic energy.

Calories are units used to measure energy. One **calorie** (cal) is the amount of energy required to raise the temperature of 1 gram (g) of water from 14.5°C to 15.5°C. The most common unit for measuring the energy content of food and the heat output of organisms, however, is the **kilocalorie** (kcal), which is the energy required to raise the temperature of a kilogram (kg) of water by 1°C. The kilocalorie equals 1000 calories. (In nutrition, one food Calorie—with a capital "C"—is actually a kilocalorie.)

Table 4.1 *Examples of Energy in Biology*

Potential Energy	Kinetic Energy
Chemical energy (stored in bonds)	Light
Concentration gradient across a membrane	Heat
	Sound
	Muscle movement
	Random molecular movement

B. The Laws of Thermodynamics Describe Energy Transfer

Thermodynamics is the study of energy transformations. The laws of thermodynamics regulate the energy conversions vital for life, as well as those that occur in the nonliving world. They apply to all energy transformations—gasoline combustion in a car's engine, a burning chunk of wood, or a cell breaking down glucose.

Energy Can Change Form

The **first law of thermodynamics** is the law of energy conservation. It states that energy cannot be created or destroyed but only converted to other forms. This means that the total amount of energy in the universe is constant.

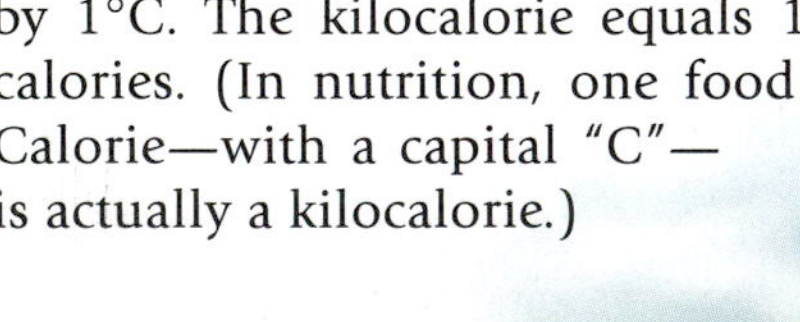

FIGURE 4.1 Potential Energy and Kinetic Energy. (**A**) Potential energy in the chemical bonds of food is converted to kinetic energy as muscles push the cyclist to the top of the hill. The potential energy of gravity provides a free ride by conversion to kinetic energy on the other side. (**B**) The ball this pitcher is about to throw has potential energy. (**C**) A chameleon's tongue whips out to ensnare a butterfly, illustrating kinetic energy.

Every aspect of life centers on converting energy from one form to another. Plants and some microorganisms convert energy in sunlight to potential energy stored in the chemical bonds of carbohydrates. They translate that potential energy into the kinetic energy of molecular motion (although not at 100% efficiency), and use that burst of kinetic energy to do work.

Although it may seem strange to think of a "working" plant, all organisms do tremendous amounts of work on a cellular and molecular scale. For example, a plant cell assembles glucose molecules into enormous cellulose fibers, moves ions across its membranes, and performs thousands of other tasks simultaneously. Likewise, a gazelle grazes on a plant's tissues to acquire potential energy that will enable it to do its own cellular work. A crocodile eats that gazelle for the same reason.

The energy transformations that sustain life are similar in all organisms. The most important are photosynthesis and cellular respiration, and they are intimately related (**figure 4.2**). In photosynthesis, plants and some other organisms use kinetic energy in light to assemble carbon dioxide and water into carbohydrates, which contain potential energy in their chemical bonds. During cellular respiration, the energy-rich carbohydrate molecules change back to carbon dioxide and water, liberating the energy necessary to power life.

In a practical sense, the first law of thermodynamics explains why we can't get something for nothing. That is, the amount of energy an organism uses cannot exceed the amount it takes in. The energy released when a baseball hurtles toward the outfield doesn't appear out of nowhere. It comes from a batter's muscles, which converted the potential energy in food into the kinetic energy of muscle movement. Likewise, green plants do not manufacture glucose from nothingness; they trap kinetic energy in sunlight and store it as potential energy in chemical bonds. The amount of chemical energy that a plant's leaves produce cannot exceed the amount of energy in the light it has absorbed.

Most organisms obtain energy from the sun, either directly through photosynthesis or indirectly by consuming other organisms. Even the potential energy in fossil fuels originated as solar energy. However, a few species of microorganisms can extract potential energy from the chemical bonds of inorganic chemicals, then synthesize organic compounds that are nutrients for them and the organisms that consume them (**figure 4.3**).

Some Energy Is Always Lost as Heat

The **second law of thermodynamics** states that all energy transformations are inefficient because every reaction loses some energy to the surroundings as heat. If you eat your energy bar on the way to your first class, your cells can use the potential energy in its chemical bonds to make proteins, divide, or do other forms of work. According to

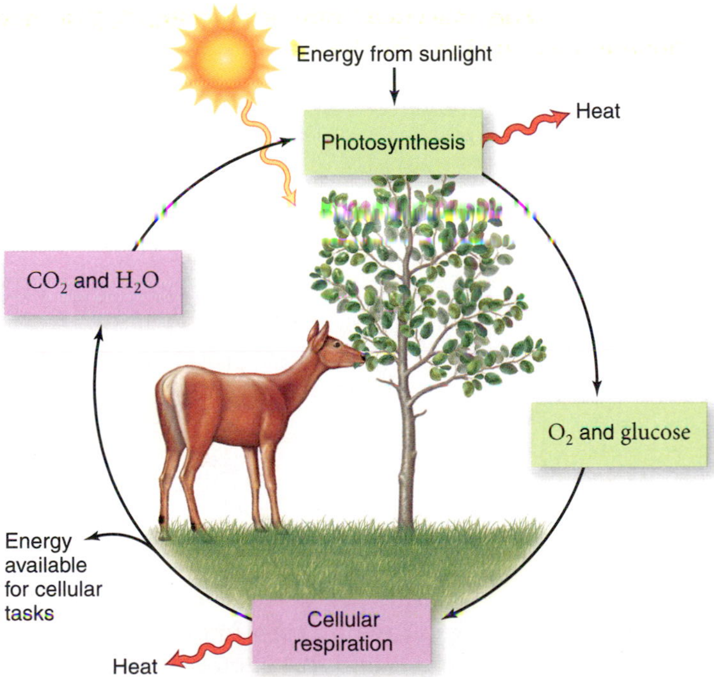

FIGURE 4.2 Energy Can Take Many Forms. Plants capture kinetic energy in sunlight to form energy-rich covalent bonds in glucose. Those chemicals, in turn, provide energy for the plants themselves and the organisms that eat plants. In this way, energy is continually flowing through one organism to another.

FIGURE 4.3 Chemical Energy Supports Life. Some bacteria derive energy from inorganic chemicals, such as hydrogen sulfide, that come from Earth's interior at deep-sea hydrothermal vents. The bacteria then synthesize organic compounds that sustain them and the organisms that consume them, such as the tube worms and crab shown here.

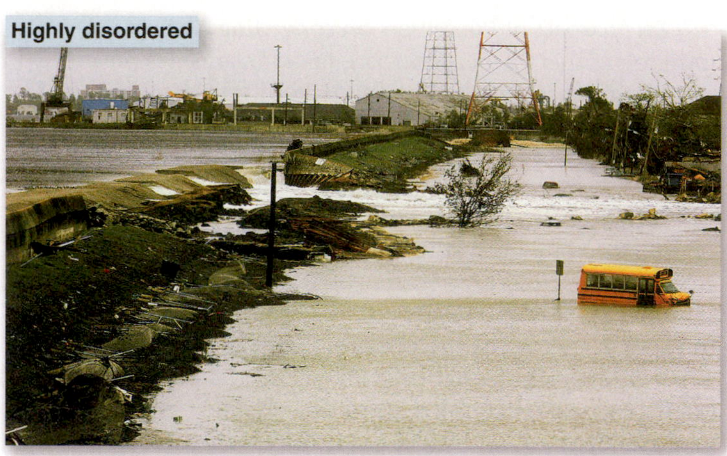

FIGURE 4.4 Entropy Represents Disorder. The destruction of a violent storm symbolizes entropy—extreme disorder. It takes great energy to rebuild because specific objects must be placed in specific places.

the second law of thermodynamics, however, you will lose some energy as heat with every chemical reaction.

This heat is forever lost to the system. Unlike other forms of energy, heat energy results from random molecule movements. Because all energy eventually becomes heat, and heat is disordered, all energy transformations head toward increasing disorder. **Entropy** is another word for the tendency toward randomness. In general, the more disordered a system is, the higher its entropy (**figure 4.4**).

Because organisms are highly organized, they may seem to defy the second law of thermodynamics—but only when considered alone as closed systems. Organisms can remain organized only because they are *not* closed systems. They use incoming energy and matter from sources such as sunlight and food to maintain their organization and stay alive. The second law of thermodynamics implies that organisms can increase in complexity *as long as something else decreases in complexity by a greater amount*. Ultimately, life remains ordered and complex because the sun is constantly decreasing in complexity and releasing energy. The entropy of the universe as a whole is always increasing.

The ideas in this chapter, and the two that follow, describe how organisms acquire and use the energy they need to sustain life.

Summary *All life processes require energy, the ability to move matter. Potential energy is stored, whereas kinetic energy is motion. Energy cannot be created or destroyed, but it can change form. As it does so, some energy is lost in the form of heat, irreversibly increasing the disorder (entropy) in the universe.*

4.1 MASTERING CONCEPTS

1. If energy is the capacity to do work, what are some examples of the "work" of a cell?
2. Give an example of how your body has both potential and kinetic energy.
3. What are the first and second laws of thermodynamics?
4. Why does the amount of entropy in the universe always increase?

4.2 Networks of Chemical Reactions Sustain Life

The number of chemical reactions occurring in even the simplest cell is staggering. Thousands of reactants and substrates form interlocking pathways that resemble complicated roadmaps.

The word **metabolism** encompasses all of these chemical reactions in cells, including those that build new molecules and those that break down existing ones. Each reaction rearranges atoms into new compounds, and each reaction either absorbs or releases energy. Digesting your morning energy bar and using its carbohydrates to fuel muscle movement are part of your metabolism. Photosynthesis and respiration are part of the metabolism of the grass under your feet as you walk to class.

These chemical reactions do not occur randomly; instead, cells organize them into chains and cycles. In step-by-step sequences called **metabolic pathways,** the product of one reac-

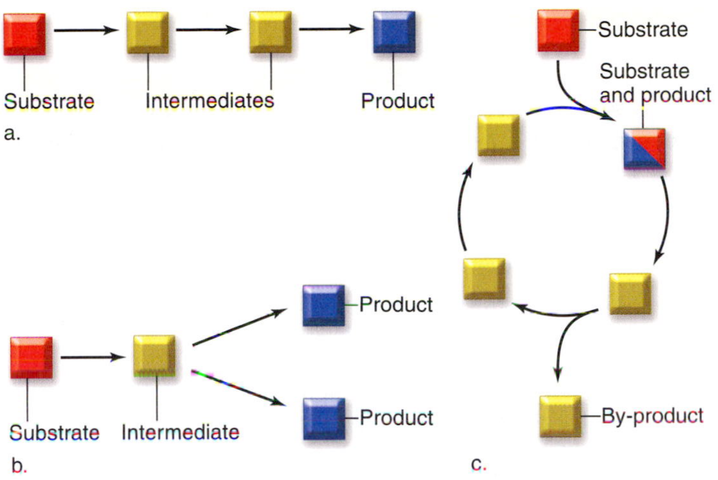

FIGURE 4.5 Metabolic Pathways. The chemical reactions of life form (**A**) straight chains, (**B**) branched chains, and (**C**) cycles of metabolic pathways. Note that in a cycle, the product of the last reaction is also the starting material of the first. Cycles release energy and by-products that are important in other biochemical pathways.

tion becomes the starting point, or substrate, of another (**figure 4.5**). Pathways may branch or form cycles. Proteins called enzymes enable metabolic reactions to proceed fast enough to sustain life, a point we return to later in the chapter.

A. Chemical Reactions Absorb or Release Energy

Biologists group metabolic reactions into two categories based on energy requirements: endergonic and exergonic (**figure 4.6**). An **endergonic reaction** ("energy inward") requires an input of energy to proceed; that is, the products contain more energy than the reactants. Typically, endergonic reactions build complex molecules from simpler components.

An example of an endergonic reaction is the assembly of sucrose, a disaccharide, from the two monosaccharides glucose and fructose (see figure 2.16). The product—the sucrose molecule—is more ordered and contains more energy in its bonds than the individual monosaccharides. For the products to gain energy, the reactants must have absorbed energy from the surroundings. Photosynthesis is another

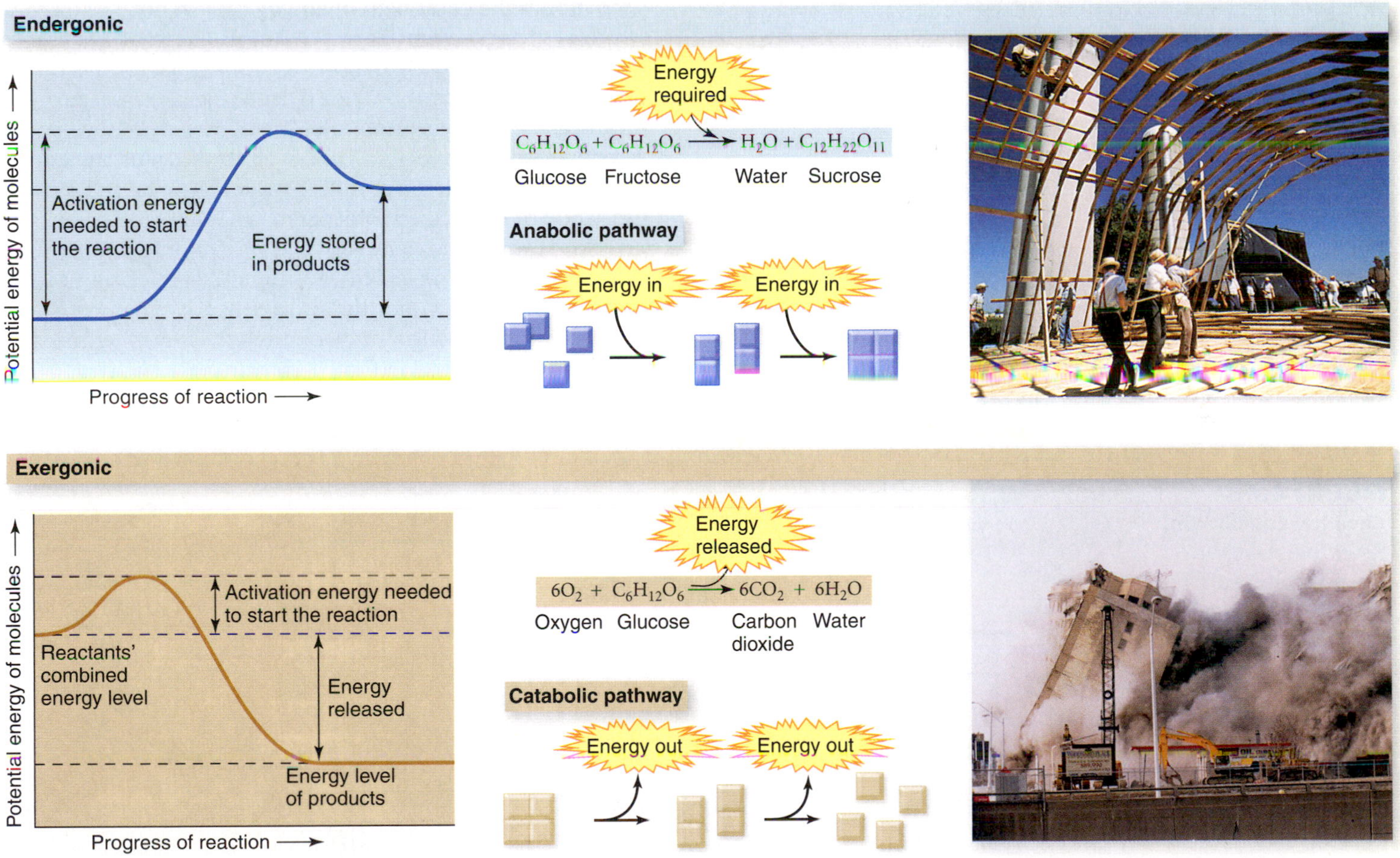

FIGURE 4.6 Endergonic and Exergonic Reactions. Endergonic reactions store energy by building complex molecules from small components, like building a barn from bricks and boards. Exergonic reactions release energy by dismantling complex molecules. Both types of reactions require a certain amount of energy—the activation energy—to get started.

example of an endergonic reaction. Glucose ($C_6H_{12}O_6$), the product of photosynthesis, contains more potential energy than carbon dioxide (CO_2) and water (H_2O), the reactants. The energy source that powers this reaction is sunlight.

In contrast, an **exergonic reaction** ("energy outward") releases energy; the products contain less energy than the reactants. Such reactions break large, complex molecules into their smaller, simpler components. Cellular respiration, the breakdown of glucose to carbon dioxide and water, is an example. The carbon dioxide and water products contain less energy than glucose.

Every compound stores potential energy in its chemical bonds. According to the first law of thermodynamics, energy is not created or destroyed. It therefore takes the same amount of energy to break a bond as it does to form that same bond. Some chemical bonds are stronger (store more potential energy) than others. When a stronger bond breaks, it releases more energy. Conversely, forming a stronger bond requires a greater input of energy.

What happens to the energy released in an exergonic reaction? According to the second law of thermodynamics, some is lost as heat; entropy always increases. But some of the energy released can be used to do work—for example, to form other bonds or to power other types of endergonic reactions. As we shall see, life's biochemistry is full of endergonic reactions that proceed at the expense of exergonic ones.

B. At Chemical Equilibrium, Reaction Rates Are in Balance

Most chemical reactions can proceed in both directions; that is, when enough product forms, some of it converts back to reactants. Arrows between reactants and products going in both directions indicate reversible reactions (**figure 4.7**). If reactants accumulate, the reaction is more likely to go forward; an excess of products, on the other hand, increases the chance that the reaction will proceed in reverse. At **chemical equilibrium,** the reaction goes in both directions at the same rate. Equilibrium does not necessarily mean that the *amounts* of products and reactants are equal—rather, their *rate of formation* equalizes.

Cells must remain far from chemical equilibrium for their metabolic processes to occur. They do this by continually preventing the accumulation of products in metabolic pathways. For example, in metabolic pathways, one enzyme might quickly use the products of a previous enzyme's reaction. This product "disappearance" allows the cell to stave off equilibrium. In this way, reactions keep moving in the direction that the cell requires.

C. Linked Oxidation and Reduction Reactions Form Electron Transport Chains

Electrons can carry energy. Most energy transformations in organisms occur in **oxidation–reduction ("redox") reactions,** which transfer energized electrons from one molecule to another.

Oxidation means the loss of electrons from a molecule, atom, or ion. The name comes from the observation that many reactions in which molecules lose electrons involve oxygen. Oxidation is the equivalent of adding oxygen because oxygen, which is strongly electronegative, attracts electrons away from their original atom. Oxidation reactions, such as the breakdown of glucose to carbon dioxide and water, release energy as they degrade complex molecules into simpler products. Conversely, **reduction** means a gain of electrons (plus any energy contained in the electrons). Reduction reactions, such as the formation of lipids, therefore require a net input of energy.

Oxidations and reductions occur simultaneously because electrons removed from one molecule during oxidation join another molecule and reduce it. That is, if one molecule is reduced (gains electrons), then another must be oxidized (loses electrons).

Many energy transformations in living systems involve carbon oxidations and reductions. Reduced carbon contains

FIGURE 4.7
Chemical Equilibrium. Reversible chemical reactions proceed toward a "middle point" known as equilibrium. At equilibrium, a reaction is equally likely to proceed in either direction. By manipulating the concentrations of reactants and products, living cells can "drive" a reaction in one direction.

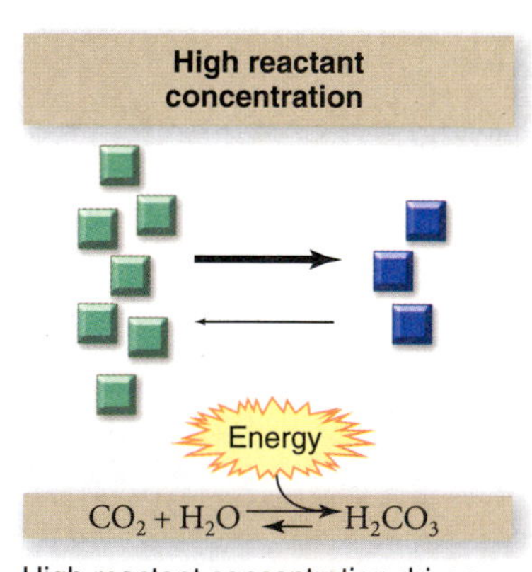

High reactant concentration drives reaction forward.

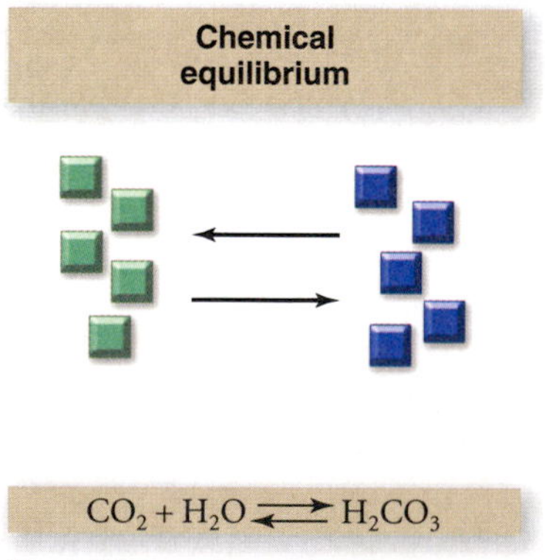

At equilibrium, forward and reverse reactions are occurring at the same rate.

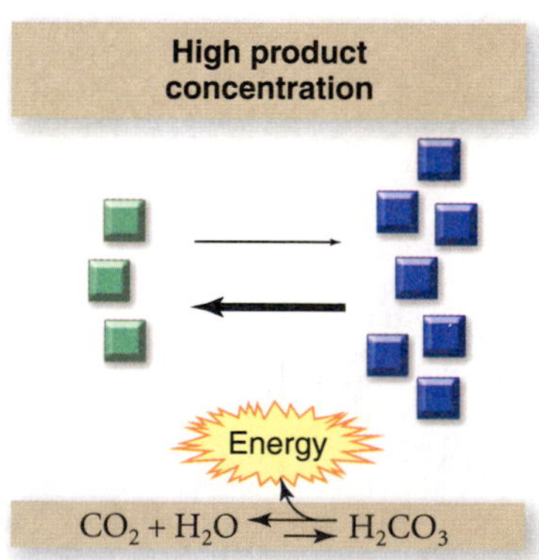

High product concentration drives reaction in reverse.

more energy than oxidized carbon. This is why reduced molecules such as methane (CH_4) are explosive, whereas oxidized molecules such as carbon dioxide (CO_2) are not. Similarly, saturated fats are highly reduced, and they contain more than twice as many Calories by weight as proteins or carbohydrates. Thus, a fatty meal of a bacon double cheeseburger with fries and a shake may contain the same amount of energy as a bathtub full of lettuce.

Some molecules are electron-shuttling "specialists." Cytochromes, for example, are iron-containing proteins that transfer electrons in metabolic pathways. Groups of cytochromes align in membranes with other proteins to form **electron transport chains**, with each protein accepting an electron from the molecule before it and passing an electron to the next. Small amounts of energy are released at each step of an electron transport chain, and the cell uses this energy in other reactions.

Cytochromes take part in many energy transformations in life, including photosynthesis and respiration in both eukaryotes and prokaryotes, suggesting that this cellular strategy for energy transformation is ancient. Given cytochromes' vital role in metabolism, it is not surprising that abnormal cytochromes can cause a wide variety of illnesses, from muscle weakness to mental retardation and blindness.

Summary *The life of a cell is a complex web of interacting biochemical reactions that build new molecules and dismantle existing ones. Synthesizing the new requires energy; breaking down the old releases energy. A cell avoids chemical equilibrium by removing reaction products. Redox reactions transfer electrons and energy from one compound to another.*

4.2 MASTERING CONCEPTS

1. What does metabolism mean in a cellular sense?
2. Distinguish between endergonic and exergonic reactions.
3. What distinguishes a reaction that has reached chemical equilibrium?
4. What are oxidation and reduction, and why are they always linked?
5. What role do cytochromes play in metabolism?

4.3 ATP Is Cellular Energy Currency

All cells contain a maze of interlocking chemical reactions—some releasing energy, and others absorbing it. The covalent bonds of **adenosine triphosphate,** a compound more commonly known as **ATP,** temporarily store much of the released energy of life. ATP is a "go-between" molecule that holds the energy released in exergonic reactions, and then applies it to endergonic ones. All cells depend on ATP to power metabolism and many other cellular activities.

ATP is structurally similar to a nucleotide (**figure 4.8**). It contains the nitrogen-containing base adenine and the

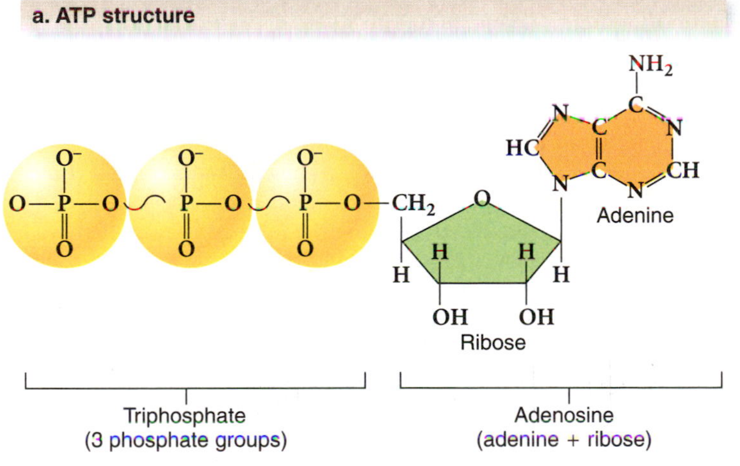

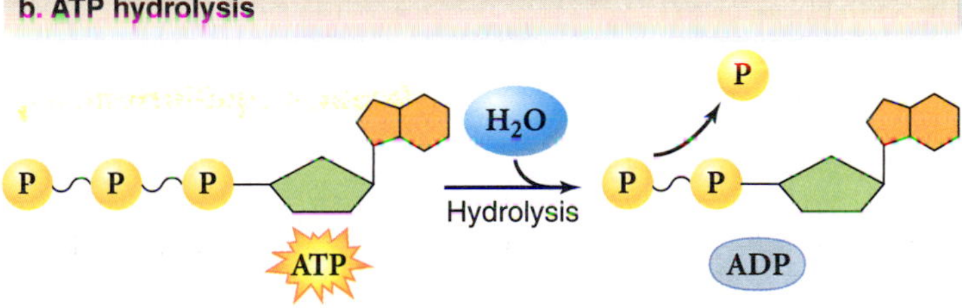

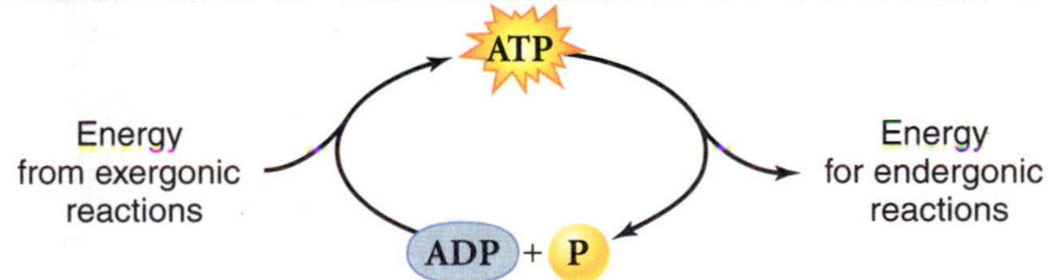

FIGURE 4.8 ATP Hydrolysis and Formation Link Biochemical Reactions. The phosphates bound to ATP (**A**) represent a significant source of potential energy. Removing the endmost phosphate group (**B**) lowers the energy level of ATP (now ADP), and the cell can use the released energy to do work. A cell can also store energy by replacing the phosphate to re-form ATP. By adding and removing phosphates (**C**), cells can use an exergonic reaction to fuel an endergonic reaction.

five-carbon sugar ribose, but where a nucleotide has just one phosphate group (PO_4), ATP has three. These three phosphate groups place three negative charges very close to one another. This arrangement destabilizes the molecule, which releases energy when the covalent bonds between the phosphates break.

A. Coupled Reactions Release and Store Energy in ATP

When a cell requires energy for an endergonic reaction, it "spends" ATP by removing the endmost phosphate group. The products of this exergonic hydrolysis reaction are adenosine *di*phosphate (ADP, since only two phosphate groups remain attached to ribose), the liberated phosphate group, and a burst of energy:

$$\text{ATP} + H_2O \longrightarrow \text{ADP} + \textcircled{\text{P}} + \text{energy}$$

In the reverse situation, energy can be temporarily stored by adding a phosphate to ADP, forming ATP:

$$\text{ADP} + \textcircled{\text{P}} + \text{energy} \longrightarrow \text{ATP} + H_2O$$

The energy for this endergonic reaction comes from molecules broken down in other reactions, such as those in cellular respiration (see chapter 6).

These reactions are fundamental to biology because ATP is the "go-between" that links endergonic to exergonic reactions. **Coupled reactions**, as their name implies, are simultaneous reactions in which one provides the energy that drives the other. Cells couple the hydrolysis of ATP to endergonic reactions that occur at the same time. The hydrolysis reaction drives the endergonic one, which does work or synthesizes new molecules.

As an example, consider again the formation of sucrose and water from glucose and fructose. This reaction is endergonic—it requires a net input of energy. When ATP provides additional energy, the reactants now have more energy available than the products, and the reaction proceeds. ATP hydrolysis, an exergonic reaction, is therefore coupled to sucrose synthesis.

B. Transfer of Phosphate Completes the Energy Transaction

How does this coupling work? A cell uses ATP as an energy source by **phosphorylating** (transferring its phosphate group to) another molecule. This transfer may have either of two effects. The presence of the phosphate may energize the target molecule, making it more likely to bond with other molecules. In this way, ATP fuels endergonic reactions. The other possible consequence of phosphorylation is a change in the shape of the target molecule. For example, adding phosphate to a protein can force that protein into a different shape, and removing phosphate allows that molecule to return to its original shape. The cell uses these changes to move substances throughout the cell. Muscle contraction is the large-scale effect of millions of small molecules changing shape in a coordinated way. ATP provides the energy.

ATP thus provides an energy "currency" for the cell. Just as you can use money to purchase a great variety of

Can *You* Relate?

Summer Light Show

Bioluminescence converts chemical energy into light energy. Many marine organisms, including fishes, squid, and jellyfish, generate their own light by harboring pockets of glowing bacteria or protists in special light organs.

Although bioluminescence is much less common on land than in the sea, a familiar example in many parts of the United States is the summertime glow of fireflies. Members of each of the more than 1900 species of fireflies attract mates with a distinctive repertoire of light signals. Typically, flying males emit a pattern of flashes. Wingless females, called glowworms, usually are on leaves, where they emit light in response to the male (see section 39.5).

To generate the glow, a molecule called *luciferin* reacts with ATP, yielding an intermediate compound (**figure 4.A**). The enzyme *luciferase* then catalyzes a reaction of this intermediate with molecular oxygen (O_2) to yield *oxyluciferin* and a flash of light. Oxyluciferin is then reduced to luciferin, and the cycle starts over.

Although we understand the biochemistry of the firefly's glow, the ways that animals use their bioluminescence are still very much a mystery. This is particularly true for the bioluminescent synchrony seen in fireflies in the same trees. When night falls, first one firefly, then another, then more, begin flashing from the tree. Soon the tree twinkles like a Christmas tree. But then, order slowly descends. In small parts of the tree, the lights begin to blink on and off together. The synchrony spreads. A half hour later, the entire tree seems to blink on and off every second. Biologists studying animal behavior are trying to determine just what the fireflies are doing—or saying—when they synchronize their glow.

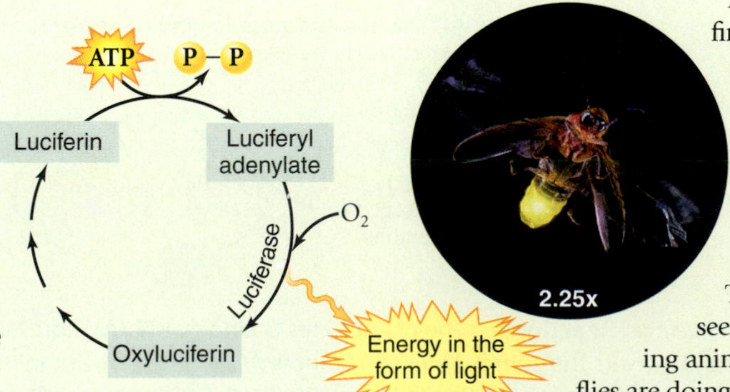

FIGURE 4.A Fireflies Exhibit a Unique Use of Energy. ATP hydrolysis creates flashes of light as energy is transferred to a specialized molecule called luciferin.

different products, all cells use ATP in many chemical reactions to do different kinds of work. For this reason, organisms require huge amounts of ATP; a typical adult uses the equivalent of 2 billion ATP molecules a minute just to stay alive. Organisms recycle ATP at a furious pace, adding phosphate groups to ADP to reconstitute ATP, using the ATP to drive reactions, and turning over the entire supply every minute or so. If you ran out of ATP, you would die instantly.

This chapter's Can *You* Relate?—Summer Light Show—addresses the role of ATP in bioluminescence, the reaction that gives fireflies their "glow."

Summary *ATP's high-energy phosphate bonds temporarily store energy that a cell uses for a wide variety of activities. Energy released in exergonic reactions fuels ATP formation; conversely, ATP hydrolysis powers the cell's endergonic reactions.*

4.3 MASTERING CONCEPTS

1. What is the molecular structure of ATP?
2. How does ATP hydrolysis supply energy for cellular functions?

4.4 Enzymes Speed Biochemical Reactions

Among the most important of all biological molecules are enzymes. An **enzyme** is a protein that catalyzes (speeds) a chemical reaction without being consumed. Most enzymes catalyze reactions that either dismantle or build other molecules. Enzymes copy DNA, build proteins, digest food, and recycle a cell's worn-out parts. Without enzymes, these biochemical reactions would proceed far too slowly to support life; untreated waste products would build to toxic levels, and the cell would die.

A. Enzymes Are Molecular "Matchmakers"

Enzymes speed reactions by lowering the **energy of activation,** the amount of energy required to start a reaction (**figure 4.9**). Even exergonic reactions, which ultimately release energy, require an initial energy "kick" to get started. The enzyme brings reactants into contact with one another, so that less energy is required for the reaction to proceed. By reducing the energy of activation, some enzymes increase reaction rates a billion times.

Most enzymes can catalyze only one or a few chemical reactions. An enzyme that dismantles a fatty acid, for example, cannot break down starch. The key to this specificity lies in the shape of the enzyme's **active site,** the region to which the reactants (also called substrates) bind. The substrate fits into the active site, but not as precisely as a key fits a lock. Rather, the active site contorts slightly, as if it is hugging the substrate (**figure 4.10**). A short-lived enzyme–substrate complex forms, and

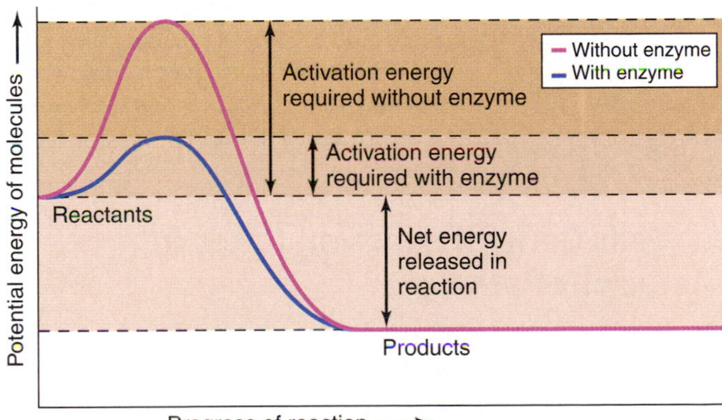

FIGURE 4.9 Enzymes Lower Activation Energy. Enzymes speed chemical reactions by increasing the chance that reactants will come together. This decreases the activation energy required to start the reaction.

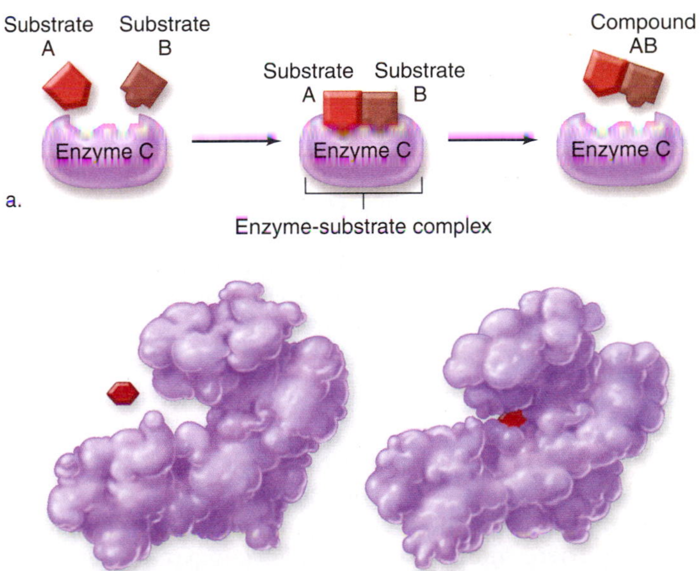

FIGURE 4.10 Enzymes Are Specific. (**A**) The shape of an enzyme creates an active site that binds to one or more substrates. (**B**) When binding to its substrate, an enzyme changes shape very slightly, as illustrated here with the binding of glucose to the enzyme hexokinase.

then the complex releases the products of the reaction. Note that the reaction does not consume or alter the enzyme. After the protein releases the products, its active site is empty and ready to pick up more substrate.

Enzymes are so critical to life that many human diseases stem from just one faulty or missing enzyme. Lactose intolerance is one example. People whose intestinal cells do not secrete an enzyme called lactase cannot digest milk sugar—lactose. Fortunately, a product called LactAid can supply the missing enzyme. Phenylketonuria (PKU) is a much more serious disease. A PKU sufferer lacks an enzyme required to break down an amino acid called phenylalanine. When this amino acid accumulates in the bloodstream, it causes brain damage. People with PKU must avoid foods containing phenylalanine, including the artificial sweetener aspartame (NutraSweet). **artificial sweeteners, p. 41**

Enzymes also have many everyday applications. Many detergents contain enzymes that break down food stains on clothing. Raw pineapple contains an enzyme that breaks down protein, which explains why you cannot put this fruit in gelatin. The pineapple's enzymes will destroy the gelatin, which will not solidify. Some meat tenderizers contain the same enzyme, which breaks down muscle tissue and makes the meat easier to chew. Infant formula manufacturers use an enzyme called trypsin to process whey protein, breaking the molecules into smaller pieces so they are easier for an infant to digest.

Preventing the production of an enzyme may be profitable, too. To meet market demand, for example, plant breeders are trying to create coffee plants that lack the enzyme required for the synthesis of caffeine. Coffee beans that grow caffeine-free would save processors the expense and flavor loss associated with chemical caffeine extraction.

B. Enzymes Have Partners

Nonprotein "helpers" called **cofactors** are substances that must be present (in addition to water and substrates) for an enzyme to catalyze a chemical reaction. Cofactors are often oxidized or reduced during the reaction, but, like enzymes, they are not consumed—they return to their original state when the reaction is complete.

Some cofactors are ionic forms of metals such as zinc, iron, and copper. Magnesium ions (Mg^{2+}), for example, help to stabilize many important enzymes. Organic cofactors are called coenzymes. The cell uses many water-soluble vitamins, including B_1, B_2, B_6, B_{12}, niacin, and folic acid, to produce coenzymes; vitamin C is a coenzyme itself. Diets lacking in vitamins can therefore lead to reduced enzyme function, producing illnesses such as scurvy (vitamin C deficiency), pellagra (niacin deficiency), and beriberi (vitamin B_1 deficiency). NAD^+, $NADP^+$, and FAD are vitamin-derived coenzymes; we return to these vital molecules in chapters 5 and 6.

a. Negative Feedback

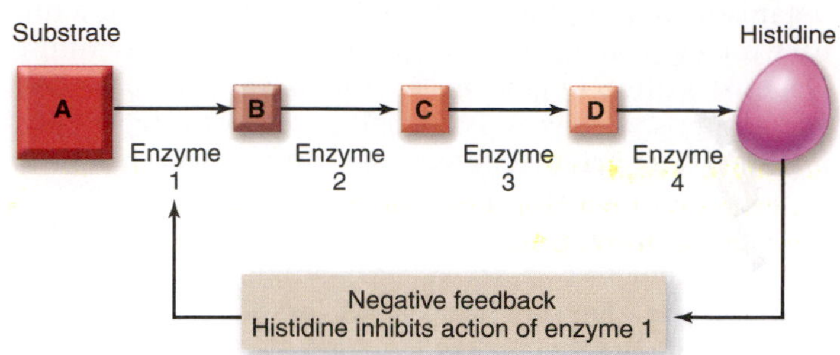

b. Noncompetitive inhibition

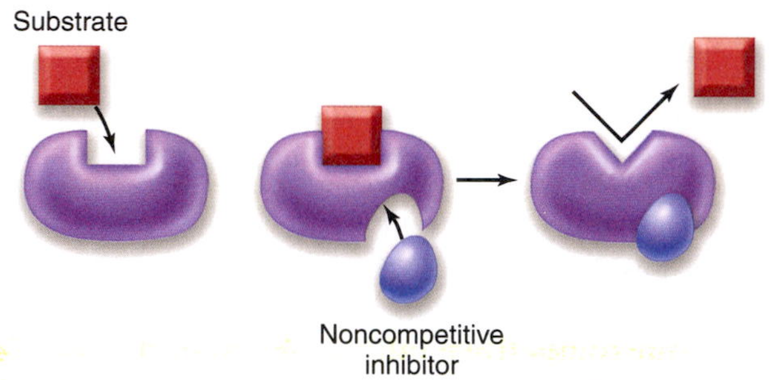

c. Competitive Inhibition

Substrate

Competitive inhibitor

FIGURE 4.11 Negative Feedback. (**A**) This diagram illustrates the enzyme-catalyzed, sequential reactions leading to histidine production. When sufficient histidine accumulates, it inhibits the activity of the first enzyme in the pathway, temporarily halting further production. (**B**) In noncompetitive inhibition, a product binds to the enzyme in a way that alters the shape of the active site. (**C**) Competitive inhibition is a negative feedback mechanism in which a product blocks the active site of an enzyme.

C. Cells Control Reaction Rates in Metabolic Pathways

The intricate network of metabolic pathways may seem chaotic, but in reality it is just the opposite. Cells precisely control the rates of their chemical reactions. If they did not, some vital compounds would always be in short supply, and others might accumulate to toxic levels. Several regulatory mechanisms that govern enzymes preserve this delicate balance.

Some enzymes control entire biochemical pathways by functioning as pacesetters. The enzyme with the slowest reaction rate sets the pace for the pathway's productivity, just as the slowest runner on a relay team limits the overall pace for the whole team. This is because each subsequent reaction in the metabolic pathway (like each subsequent relay runner) requires the product of the preceding reaction.

Another way to regulate a metabolic pathway is by **negative feedback** (or feedback inhibition), in which an excess of a reaction's product inhibits the enzyme that controls its formation. Once the product's level falls, the pathway resumes its activity. A thermostat works by negative feedback—when the temperature in a room reaches a preset level, the thermostat shuts the heat off for a while. A falling temperature cues the thermostat to turn on the furnace again (see figure 27.9).

Figure 4.11A illustrates negative feedback. It shows the pathway that some bacteria use to synthesize the amino acid histidine. When excess histidine accumulates, it binds to an enzyme that regulates the pathway. This temporarily destabilizes the enzyme and, for a time, histidine synthesis ceases. When the level of histidine in the bacterium falls, the block on the enzyme lifts, and the cell can once again synthesize the amino acid.

Negative feedback works in two general ways to prevent too much of a substance from accumulating (figure 4.11 B and C). In **noncompetitive inhibition**, product molecules bind to the enzyme at a site other than the active site, but in a way that alters the shape of the enzyme so that it can no longer bind substrate. The inhibitor does not directly compete to occupy the active site. Alternatively, in **competitive inhibition**, the product of the reaction binds to the enzyme's active site, preventing it from binding substrate. It is "competitive" because the product competes with the substrate to occupy the active site.

In **positive feedback,** a product activates the pathway leading to its own production. Blood clotting, for example, begins when a biochemical pathway synthesizes fibrin, a threadlike protein. The products of the later reactions in the clotting pathway stimulate the enzymes that activate earlier reactions. Because of positive feedback, fibrin accumulates faster and faster, until there is enough to stem the blood flow. When the clot forms and the blood flow stops, however, the clotting pathway shuts down—an example of negative feedback. Positive feedback is much rarer than negative feedback in organisms.

D. Environmental Conditions Affect Enzyme Activity

Because enzymes are proteins, they are very sensitive to conditions in the cell. Enzymes work faster as the temperature increases, but only to a limit. If it gets too hot, if the pH changes, or if the salt concentration becomes too high or too low, an enzyme can denature (become misshapen) and stop working. Very high temperatures destroy most enzymes, which makes cooking an excellent food preservation method. Organisms adapted to hot springs, however, have special heat-tolerant enzymes. **protein folding, p. 39**

Substances from outside the body, such as drugs and poisons, can also inhibit enzyme function. Many antibiotics, including sulfanilamide and triclosan (an ingredient in antibacterial soap), kill microorganisms—but not people—by inhibiting enzymes not present in our own cells. Aspirin relieves pain by binding to an enzyme that cells use to produce pain-related molecules called prostaglandins. Likewise, a chemical called glyphosate (the active ingredient in an herbicide called Roundup) competitively inhibits an enzyme found in plant cells but not in animals. And some nerve gases bind to an enzyme that normally removes a nerve cell molecule after it sends its message. The nerve gas prevents the enzyme from binding its substrate, causing an overload of the messenger. As a result, neural transmission ceases.

Summary *Enzymes are proteins that lower the activation energy required to start reactions. Some enzymes require the help of cofactors. The cell uses negative feedback and other mechanisms to regulate enzyme action. Unfavorable environmental conditions can denature an enzyme, and some drugs and poisons kill by binding to critical enzymes.*

4.4 MASTERING CONCEPTS

1. How does an enzyme lower a reaction's activation energy?
2. What is the difference between an enzyme and a coenzyme?
3. In what two ways can the product of a biochemical pathway turn off its own synthesis?
4. List three examples of conditions that can influence an enzyme's activity.

4.5 Membrane Transport May Release Energy or Cost Energy

A cell without a membrane is no cell at all; once the difference between the inside of a cell and its environment is erased, the cell no longer functions. Cells spend tremendous amounts of energy maintaining this all-important difference between themselves and the outside world.

The fluid inside a cell—or inside an organelle—is an aqueous solution (recall from chapter 2 that an aqueous solution consists of one or more solutes dissolved in water). The solutes can be ions or molecules, large or small. Concentrations of some solutes are higher inside the cell than outside, and others are lower. Likewise, the inside of each organelle in a eukaryotic cell may be chemically quite different from the solution in the rest of the cell.

The term *gradient* describes any such difference between two neighboring regions. For example, differences in pH, electrical charge, and pressure all create gradients. In a **concentration gradient**, a solute is more concentrated in one region than in a neighboring region. When you first place a tea bag in a cup of hot water, for example, you can immediately see a concentration gradient: near the tea bag, there are many more brown tea molecules than elsewhere in the cup.

Over time, the brownish color spreads to create a uniform brew. This occurs because a concentration gradient dissipates *unless energy is expended to maintain it*. Random molecular motion always increases the amount of disorder (entropy). It costs energy to counter this tendency toward disorder. For the same reason, however, an existing concentration gradient represents a form of potential energy. As the molecules disperse, the amount of potential energy in the gradient declines.

All forms of transport across membranes involve gradients. As this section explains, in the simplest types of transport, a cell simply allows a gradient to dissipate across the membrane. A substance moving from an area where it is more concentrated to an area where it is less concentrated is said to be "moving down" or "following" its concentration gradient. In other situations, a cell spends energy to maintain a concentration difference.

The cell membrane oversees the cell's concentration gradients. As described in chapter 3, a biological membrane is a phospholipid bilayer studded with proteins. Membranes are "choosy," or selectively permeable—they freely admit some substances (either between the molecules of the phospholipid bilayer or through protein-lined channels) but not others. The following section describes three basic forms of traffic across the membrane.

A. Passive Transport Does Not Require Energy Input

In **passive transport**, a substance moves across a membrane without the direct expenditure of cellular energy. All forms of passive transport involve **diffusion**, the spontaneous movement of a substance from a region where it is more concentrated to a region where it is less concentrated (**figure 4.12**). Because diffusion represents the dissipation of a chemical gradient—and the loss of potential energy—it does not require energy input.

How does any solute "know" which way to diffuse? The answer is, of course, that the solute knows nothing. Diffusion occurs because atoms and molecules have kinetic energy; that is, they are in constant, random motion. To simplify the tea example, suppose each molecule can move randomly along one of 10 possible paths (in reality, the number of possible directions is infinite). Assume further that only one path leads back to the tea bag. Since 9 of the 10 possibilities point away from the tea bag, the tea molecules tend to spread out; that is, they move down their concentration gradient.

If simple diffusion continues long enough, the gradient disappears, and the concentration of the substance is the same on both sides of the membrane. Diffusion *appears* to stop, but the molecules do not stop moving. Instead, they continue to travel randomly back and forth across the membrane at the same rate, so at equilibrium the concentration remains equal on both sides.

If gradients dissipate without energy input, how can a cell use diffusion to acquire essential solutes? The cell maintains the gradients, either by continually consuming the substances as they diffuse in, or by producing more of the substances that diffuse out. For example, cellular respiration consumes O_2 shortly after the gas diffuses into the cell, maintaining the O_2 gradient that drives its diffusion. Respiration also produces CO_2, which diffuses out because its concentration remains higher in the cell than outside.

Simple Diffusion: No Proteins Required

Simple diffusion is a form of passive transport in which a solute moves down its concentration gradient without the use of a carrier molecule. Lipids and small, nonpolar molecules such as oxygen (O_2) and carbon dioxide (CO_2), for example, diffuse easily across the hydrophobic portion of biological membranes.

Gas exchange in the lungs illustrates simple diffusion. Deep inside your lungs, O_2 from your most recent breath diffuses across an air sac membrane and into the blood, which has a lower concentration of O_2 than does air. At the same time, the blood carries waste CO_2 from the body's tissues. This gas diffuses freely out of the blood, across the air sac membrane, and into the lungs. Exhaling pushes the CO_2 out of the body.

FIGURE 4.12 Diffusion Results from Random Movement. Solute particles from the tea bag can move in any direction. Only a few paths lead back to the source. Eventually the solutes are distributed uniformly throughout the cup.

Osmosis: Simple Diffusion of Water Across a Semipermeable Membrane

Simple diffusion applies to solutes that freely cross a membrane. Suppose, however, that two solutions of different solute concentrations are separated by a membrane that allows water, but not the solute, to pass through (**figure 4.13**). In this case, water diffuses across the membrane in the direction that dilutes the solute on the side where it is more concentrated. **Osmosis** is the simple diffusion of water across a biological membrane.

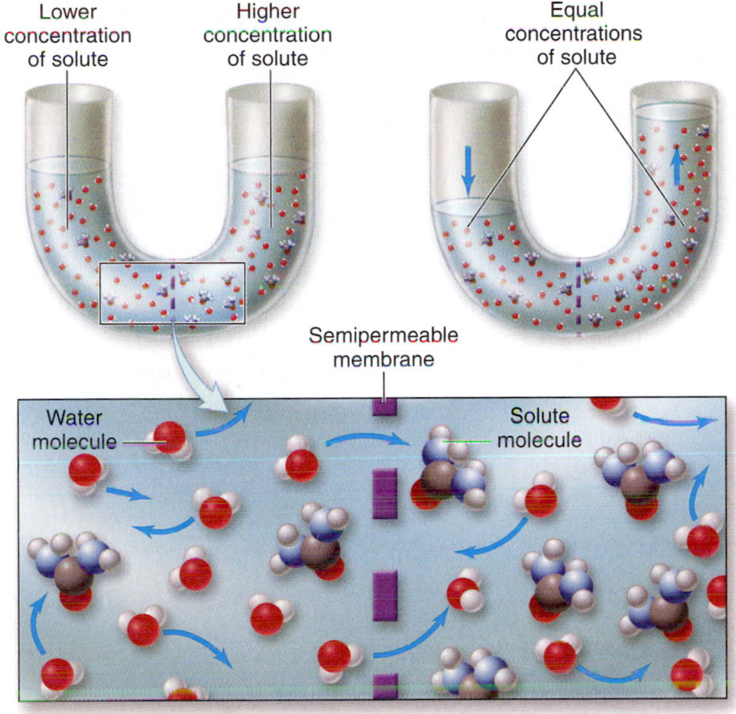

FIGURE 4.13 Osmosis. An artificial membrane dividing a beaker permits water to pass from one chamber to another, but prevents solutes from doing the same. Water will flow from an area of low salt (solute) concentration toward an area of high salt concentration. Eventually, the volume on each side of the membrane will be different, but the final concentrations (amount of solute per unit of volume) will be the same. At equilibrium, water flow is equal in both directions.

A human red blood cell, which is normally suspended in a solution called plasma, demonstrates the effects of osmosis (**figure 4.14**). The cell's interior is **isotonic** to the surrounding fluid, which means that the plasma's solute concentration is the same as the inside of the red blood cell (*iso-* means "equal", and *tonicity* is the ability of a solution to cause water movement). Water does not flow into or out of the cell, which retains its normal doughnut-shaped form. A 0.9% salt solution is also isotonic to human red blood cells.

A cell can also be in a **hypotonic** environment, in which the concentration of solutes is lower than inside the cell (*hypo-* means "under," as in hypodermic). Water will enter a red blood cell immersed in pure water; the cell may even burst. In the opposite situation (for example, placing a cell in a solution of 2% salt), water leaves the cell. This occurs because **hypertonic** surroundings have a higher concentration of solutes than the cell's cytoplasm (*hyper-* means "over," as in hyperactive). The cell shrivels and may die for lack of water.

Hypotonic and *hypertonic* are relative terms that can refer to the surrounding solution or to the solution inside the cell. The same solution might be hypertonic to one cell but hypotonic to another, depending on the solute concentrations inside the cells.

Because shrinking and swelling cells may not function normally (and may even die), unicellular organisms maintain their shapes by regulating osmosis. The cell of a pond-dwelling microorganism such as *Paramecium* has a higher solute concentration than its freshwater habitat, so water flows into the organism. A special organelle called a

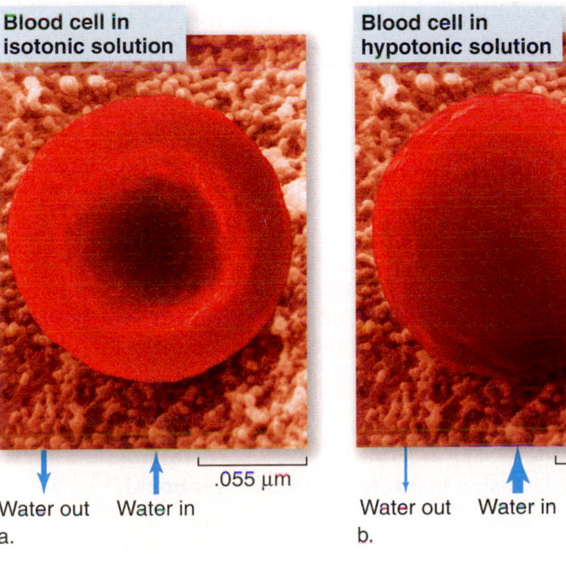

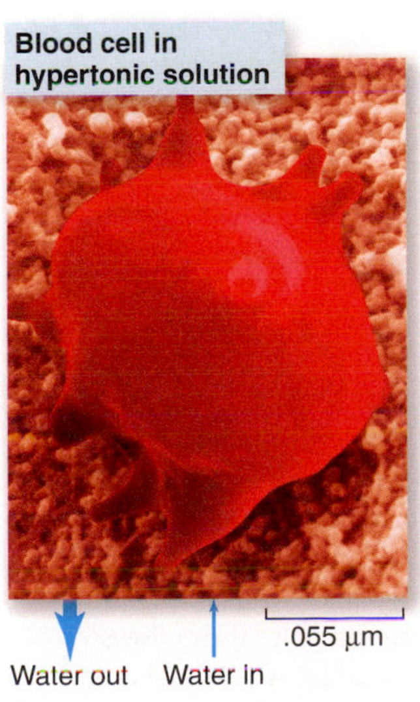

FIGURE 4.14 Osmosis Affects Cell Shape. A red blood cell changes shape in response to changing plasma solute concentrations. (**A**) A human red blood cell is normally isotonic to the surrounding plasma. When water enters and leaves the cell at the same rate, the cell maintains its shape. (**B**) When the salt concentration of the plasma decreases relative to the salt concentration inside the cell, water flows into the cell faster than it leaves. The cell swells and may even burst. (**C**) When the salt concentration of the plasma increases, water leaves the cells to dilute the outside solute faster than water enters the cell. The cell shrinks.

contractile vacuole pumps the extra water out—a process that requires the expenditure of energy.

Likewise, a plant's roots are often hypertonic to the soil, particularly after a heavy rain. Instead of expelling the extra water that rushes in, as *Paramecium* does, central vacuoles of the plant cells expand until the cell walls restrain their growth. **Turgor pressure** is the resulting force of water against the cell wall (**figure 4.15**). A limp, wilted piece of lettuce demonstrates the effect of lost turgor pressure. But the leaf becomes crisp again if placed in water, as individual cells expand like inflated balloons. Turgor pressure helps keep plants erect.

Water balance is just as important in the human body. As described in chapter 35, kidneys use osmosis to regulate water balance. This chapter's Burning Question (What Causes Headaches?) describes the role of dehydration and osmosis in some types of headaches.

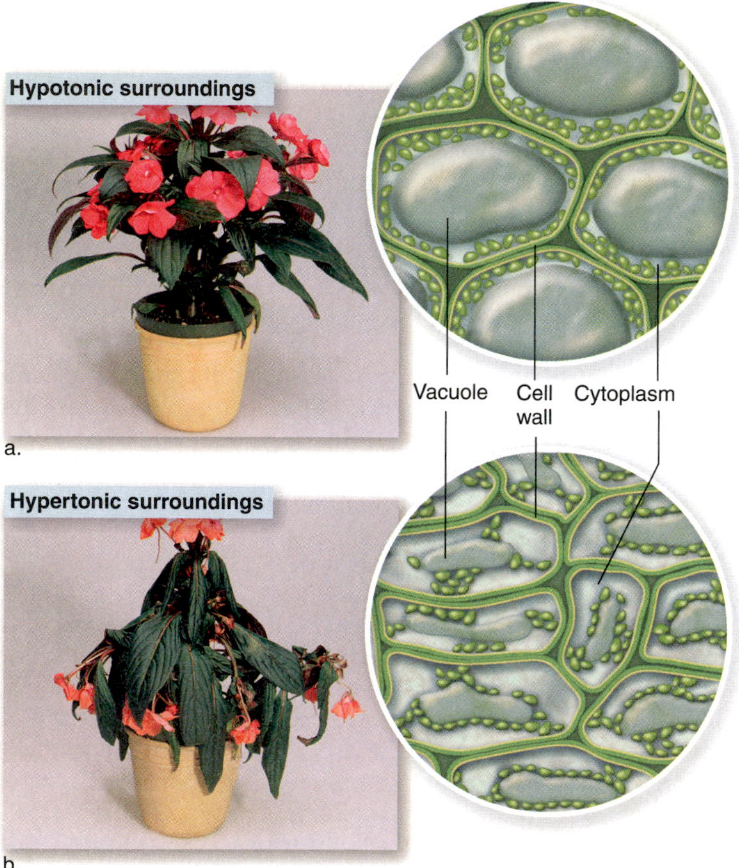

FIGURE 4.15 Plant Cells Keep Their Shapes by Regulating Diffusion. Plant cells usually contain more concentrated solutes than their surroundings, drawing water into the cell. (**A**) In a hypotonic environment, water enters the cell and collects in vacuoles. The cell swells against its rigid cell wall, generating turgor pressure. (**B**) When a plant cell is placed in a hypertonic environment (so that solutes are more concentrated outside the cells), water flows out of the vacuoles, and the cell shrinks. Turgor pressure is low, and the plant wilts.

Facilitated Diffusion: Proteins Required

The hydrophobic part of the phospholipid bilayer is a barrier to ions and polar molecules, but membrane-spanning transport proteins help these solutes cross. **Facilitated diffusion** is a form of passive transport in which a membrane protein assists the movement of a polar solute along its concentration gradient. Facilitated diffusion does not require energy expenditure because the solute moves from where it is more concentrated to where it is less concentrated.

Glucose is an example of a molecule that moves into red blood cells via facilitated diffusion. It is too hydrophilic to pass freely across the membrane, but glucose transporter proteins form channels that allow it in. Respiration inside the red blood cells consumes the glucose and maintains the concentration gradient.

Membrane proteins can even enhance osmosis. Although membranes are somewhat permeable to water, osmosis can be slow. The cells of many organisms, including bacteria, plants, and animals, use water-specific membrane proteins called aquaporins to increase the rate of water flow. Kidney cells control the amount of water that enters urine by changing the number of aquaporins in kidney cell membranes.

Burning Questions

What causes headaches?

Headaches have diverse causes, including muscle tension, nighttime teeth-grinding, stress and anxiety, bright lights, and some food ingredients, to name just a few. One other major cause is dehydration. When a person does not replenish fluids lost to urine, sweat, and breathing, the body becomes dehydrated. Among the many effects of dehydration is constriction of the blood vessels, resulting in less blood and oxygen traveling to the brain.

The kidneys regulate the blood's solute concentration by adjusting the amount of water removed from the blood and lost as urine. When the brain detects that body fluids are too concentrated, it releases a hormone that acts on the kidneys to conserve water. Instead of leaving the body in urine, the water returns to the bloodstream. **kidney function, p. 716**

The infamous "hangover headache" associated with a night of heavy drinking is likely a result of dehydration. Alcohol consumption contributes to dehydration because alcohol acts as a diuretic. By interfering with the hormone that triggers water conservation in the kidneys, alcohol increases urine output. To prevent or cure this type of headache, experts recommend drinking more water, both with the alcohol and after the merriment ends.

Have a Burning Question of your own? Submit it to marielle_hoefnagels@mcgraw-hill.com for possible inclusion in future editions of this book!

Venom from animals such as spiders, scorpions, snakes, and jellyfish point to the importance of membrane transport proteins. Not all venoms work in the same way, but many block channels that admit calcium, sodium, and potassium ions into nerve cells. By interfering with the body's communication network, the most potent of these toxins can cause paralysis and even death.

B. Active Transport Requires Energy Input

Both simple diffusion and facilitated diffusion dissipate an existing concentration gradient. Often, however, a cell needs to do the opposite: create and maintain a concentration gradient. A plant's root cell, for example, may need to absorb nutrients from soil water that is much more dilute than the cell's interior. In **active transport,** a cell uses a transport protein to move a solute *against* its concentration gradient—from where it is less concentrated to where it is more concentrated (**figure 4.16**). Because a gradient represents a form of potential energy, the cell must expend energy to create it. Energy for active transport often comes from ATP.

One active transport system in the cell membranes of most animal cells is an enzyme called the **sodium–potassium pump** (**figure 4.17**), which uses ATP as an energy source to expel three sodium ions (Na^+) for every two potassium ions (K^+) it admits. Cells must contain high concentrations of K^+ and low concentrations of Na^+ to perform many functions. Maintaining these gradients is costly: the million or more sodium–potassium pumps embedded in a cell's membrane use some 25% of the cell's ATP. But it is also necessary. In animals, sodium and potassium ion gradients are essential for nerve and muscle function.

A drug called digitalis treats congestive heart failure by inhibiting the sodium–potassium pump in cells of the heart.

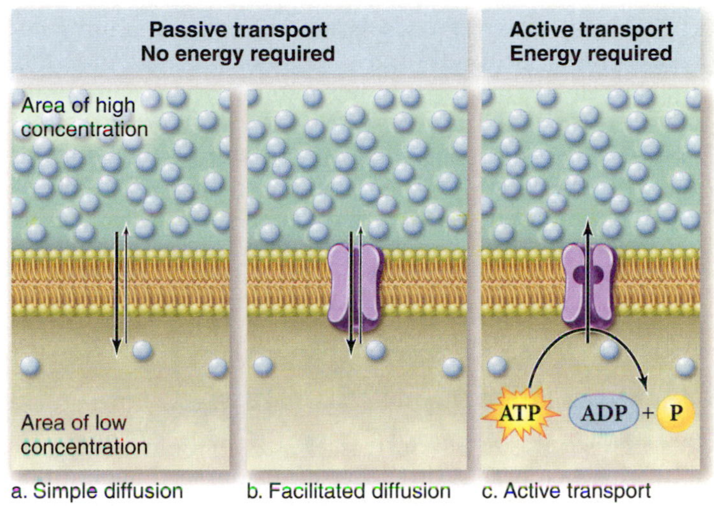

FIGURE 4.16 Passive and Active Transport Compared. Simple diffusion, facilitated diffusion, and active transport move ions and molecules across cell membranes. (**A**) In simple diffusion, solutes move down their concentration gradients by squeezing between the membrane lipids. (**B**) In facilitated diffusion, solutes use a membrane protein to move down the concentration gradient without direct energy input. (**C**) In active transport, a molecule or ion crosses a membrane against its concentration gradient, using energy and carrier proteins.

This action indirectly increases the concentration of calcium ions inside the heart's muscle cells. Because muscle contraction requires a high concentration of calcium, the net effect of digitalis is to increase the force with which the heart muscle contracts.

Concentration gradients are an important source of potential energy that cells can use to do work. A photosynthesizing cell uses energy from the sun to establish a concentration gradient

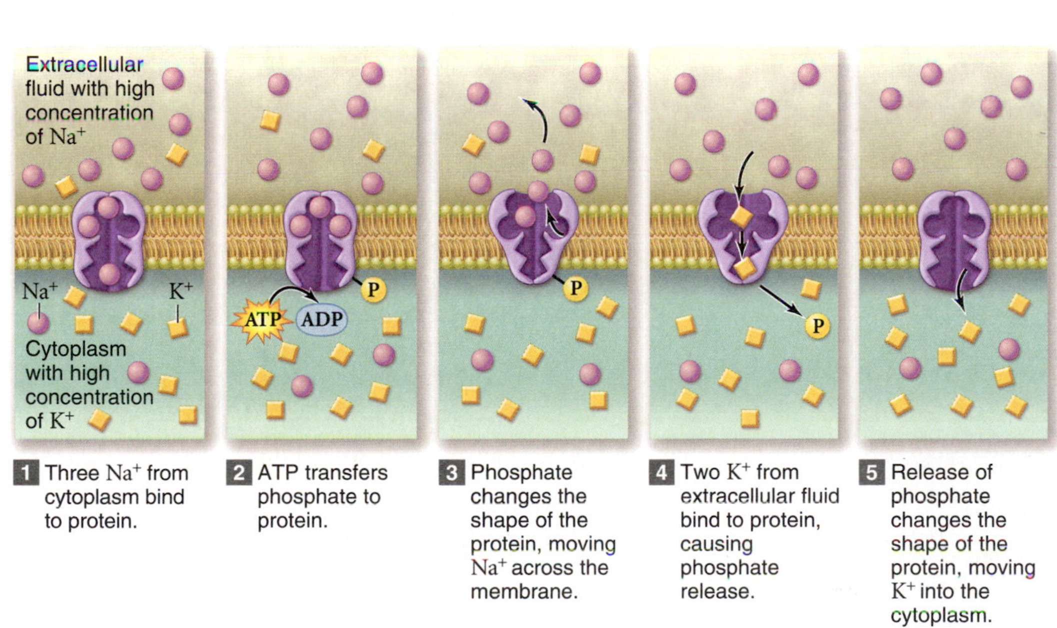

Extracellular fluid with high concentration of Na+
Na+
K+
Cytoplasm with high concentration of K+
ATP
ADP
P
1 Three Na+ from cytoplasm bind to protein.
2 ATP transfers phosphate to protein.
3 Phosphate changes the shape of the protein, moving Na+ across the membrane.
4 Two K+ from extracellular fluid bind to protein, causing phosphate release.
5 Release of phosphate changes the shape of the protein, moving K+ into the cytoplasm.

FIGURE 4.17 The Sodium–Potassium Pump. This "pump," actually an enzyme embedded in the cell membrane, uses energy (ATP) to move potassium ions (K^+) into the cell and sodium ions (Na^+) out of the cell. The pump first binds Na^+ on the inside face of the membrane (*1*). ATP is split to ADP, and a phosphate group is transferred to the carrier or pump. (*2*) This binding alters the conformation of the pump and causes it to release the Na^+ to the outside. The altered pump can now take up K^+ from outside the cell (*3*). Next, the pump releases the bound phosphate (*4*), which again alters the conformation of the carrier protein. This change in shape releases K^+ to the cell's interior (*5*). The pump is back in the proper shape to bind intracellular Na^+.

of hydrogen ions (H^+) across the chloroplast membrane. Likewise, a mitochondrion uses potential energy in food to establish an H^+ gradient during respiration. By controlling how and when H^+ diffuses back across the membrane, the mitochondrion can convert the potential energy stored in the gradient into another form of potential energy—chemical energy in the bonds of ATP. An enzyme called ATP synthase catalyzes this essential reaction in all cells. (Chapter 6 describes ATP production in more detail).

C. Endocytosis and Exocytosis Use Vesicles to Transport Substances

Most molecules dissolved in water are small, and they can cross cell membranes by simple diffusion, facilitated diffusion, or active transport. Large molecules (and even bacteria) can also enter and leave cells, with the help of vesicles that form directly from cell membranes.

Endocytosis

Endocytosis allows a cell to engulf fluids and large molecules and bring them into the cell. A small indentation forms in the cell membrane and closes in on itself, forming an enclosed vesicle that traps whatever is outside the membrane (**figure 4.18A**).

The two main forms of endocytosis are pinocytosis and phagocytosis. In pinocytosis, the cell engulfs small amounts of fluids and dissolved substances. In **phagocytosis,** the cell captures and engulfs large particles, such as debris or even another cell (figure 4.18B). The vesicle then fuses with a lysosome, activating the enzymes within. Once digested, the contents of the vesicle enter the cytoplasm for the cell to use. Whereas many cells use pinocytosis to acquire food and molecules, only specialized cells (including the immune system's white blood cells) use phagocytosis.

When biologists first viewed endocytosis in white blood cells in the 1930s, they thought a cell would gulp in anything at its surface. We now recognize a more specific form of the process called receptor-mediated endocytosis. A receptor protein on a cell's surface binds a biochemical; the cell membrane then indents, embracing the substance and drawing it into the cell (**figure 4.19**).

Receptor-mediated endocytosis enables liver cells to absorb cholesterol from the bloodstream. Particles called lipoproteins carry cholesterol in blood. One type, a low-density lipoprotein (LDL), binds to receptors clustered in protein-lined pits in the surfaces of liver cells. The cell membranes envelop the LDL particles, forming loaded vesicles and bringing them into the cell, where they move toward lysosomes. Within lysosomes, enzymes liberate the cholesterol from the LDL carriers. The receptors are recycled to the cell surface, where they can bind more cholesterol-laden LDL particles. **lysosomes, p. 68**

Exocytosis

Exocytosis, the opposite of endocytosis, uses vesicles to transport fluids and large particles out of cells. Inside a cell, the Golgi apparatus produces vesicles filled with substances to be secreted. The vesicle moves to the cell membrane and joins with it, releasing the substance outside the membrane (**figure 4.20**). For example, exocytosis in the front tip of a sperm cell releases enzymes that enable the tiny cell to penetrate the much larger egg cell. The secretion of milk into milk ducts, depicted in figure 3.15, is another example.

Endocytosis

1 A small portion of the cell membrane buds inward, entrapping particles.

Cytoplasm

2 A vesicle forms, which brings particles into the cell.

3 Vesicle surrounds the imported particles.

a. Endocytosis

b. Phagocytosis

FIGURE 4.18 Endocytosis (A) Endocytosis brings large particles into a cell. A small portion of the cell membrane buds inward (*1*), entrapping the particles (*2*), and a vesicle forms, which brings the substances into the cell (*3*). **(B)** The white blood cell (*in yellow*) is engulfing a yeast cell by phagocytosis, a form of endocytosis.

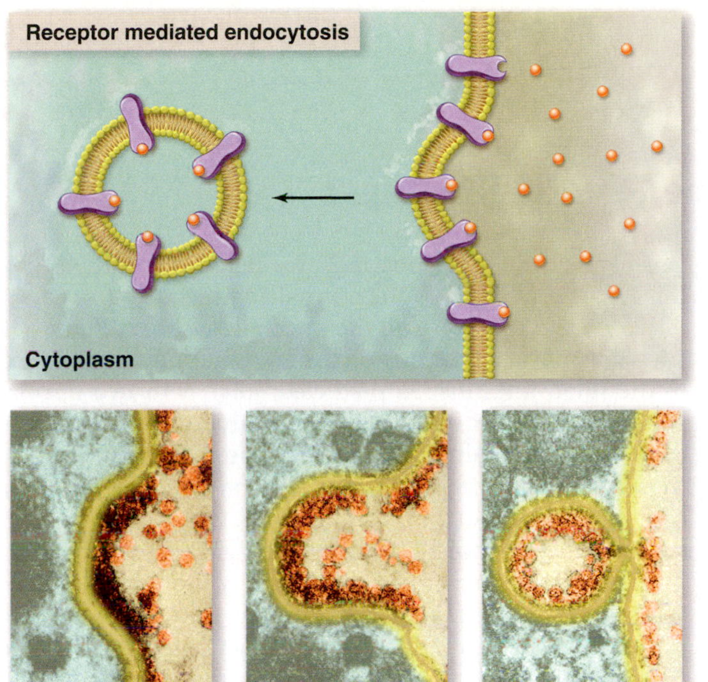

FIGURE 4.19 Receptor-Mediated Endocytosis. The binding of a molecule to a receptor protein triggers receptor-mediated endocytosis.

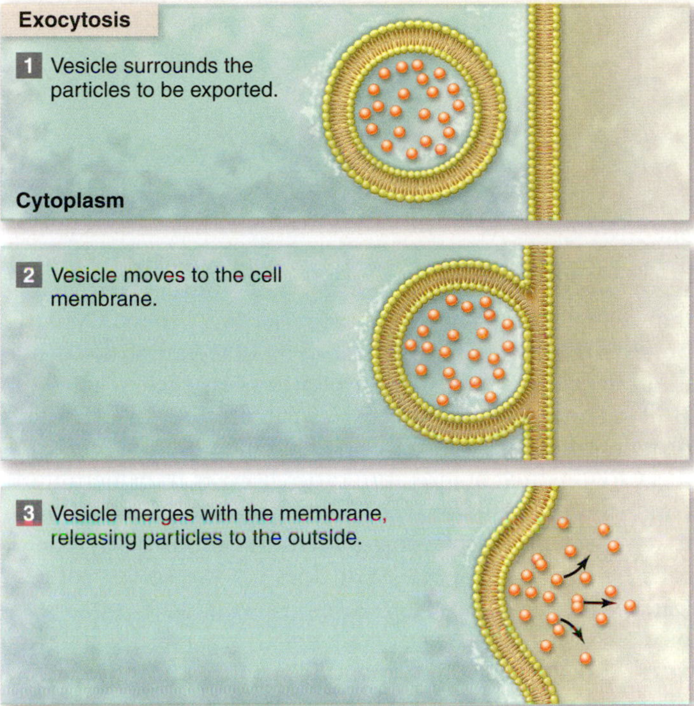

Exocytosis
1 Vesicle surrounds the particles to be exported.
Cytoplasm
2 Vesicle moves to the cell membrane.
3 Vesicle merges with the membrane, releasing particles to the outside.

FIGURE 4.20 Exocytosis. Biochemicals and particles exit cells by exocytosis. Vesicles surround the structures to be exported (*1*), then move to the cell membrane (*2*) and merge with it, releasing the particles to the outside (*3*).

Table 4.2 summarizes the transport mechanisms that move substances across membranes.

Summary *Cells must regulate what enters and leaves them. A few types of small molecules, including water, freely cross membranes by simple diffusion. Osmosis is the simple diffusion of water across a membrane that does not admit solutes. Transport proteins move solutes across membranes, either down the concentration gradient (facilitated diffusion) or against it (active transport). Endocytosis and exocytosis move liquids and large particles across membranes.*

4.5 MASTERING CONCEPTS

1. How do differing concentrations of solutes in neighboring solutions drive the osmotic movement of water molecules?
2. If a solution is hypertonic to a cell's interior, in which direction will water move?
3. Why does it cost energy for a cell to maintain a concentration gradient?
4. How do facilitated diffusion and active transport differ from each other and from simple diffusion?
5. How do exocytosis and endocytosis use vesicles to transport materials across cell membranes?

Table 4.2 *Movement Across Membranes*

Mechanism	Characteristics	Example
Passive Transport	Net movement is down concentration gradient. Does not require energy input.	
Simple diffusion	Solutes to which the membrane is permeable move across membrane without assistance of transport proteins.	Oxygen gas diffuses from lung into blood vessel
Osmosis	Simple diffusion of water (note that protein channels called aquaporins enhance osmotic rate in some cells)	Water reabsorbed from kidney tubules
Facilitated diffusion	Dissolved substances to which the membrane is not permeable move across membrane with assistance of transport proteins.	Glucose diffuses into red blood cells
Active transport	Transport protein moves substances against concentration gradient. Requires energy input, often from ATP.	Salts reabsorbed into blood from kidney tubules
Endocytosis (pinocytosis and phagocytosis)	Membrane engulfs substance and draws it into cell in membrane-bounded vesicle	White blood cells ingest bacteria
Exocytosis	Membrane-bounded vesicle fuses with cell membrane, releasing substances outside of cell	Sperm secretes enzymes that help it penetrate egg

INVESTIGATING LIFE
Does Natural Selection Maintain Some Genetic Illnesses?

An individual enzyme or membrane protein may seem too small to be very important—until you consider that a single faulty one can cause serious illness. Cystic fibrosis, a disease that affects about 30,000 Americans, is one example. One in every 2500 babies born each year in the United States has cystic fibrosis. Each affected person lacks a protein called CFTR in his or her cell membranes.

CFTR, which stands for "*c*ystic *f*ibrosis *t*ransmembrane conductance *r*egulator," is a membrane transport protein that moves negatively charged chloride ions (Cl^-) out of cells by active transport. As it does so, the solute concentration outside the cell increases, drawing water out by osmosis. Because CFTR indirectly helps cells expel water, it is not surprising that this protein occurs in tissues that secrete mucus, perspiration, and other fluids.

One of the many locations where CFTR does its job is in cells lining the airspaces of the lung. Water moving out of these cells thins the lung's mucus, which beating cilia clear away. Patients with cystic fibrosis, however, lack a working CFTR protein. The mucus in the lungs remains thick, making breathing difficult and creating an ideal breeding ground for bacteria. The patient eventually succumbs to chronic infections, often before age 30.

Cystic fibrosis may render patients too sick to have children or even take their lives before they are old enough to reproduce. Why hasn't natural selection eliminated this disease from the human population? A possible answer to this evolutionary mystery may lie not in the lungs but in another place where CFTR proteins occur: the cells lining the digestive tract.

Many disease-causing bacteria affect the digestive tract, and some of these exploit CFTR. For example, bacteria that cause cholera (*Vibrio cholerae*) produce a toxin that overstimulates CFTR, triggering Cl^- and water to pour from the lining of the small intestine. The water and ions leave the body in watery diarrhea; the resulting dehydration can kill if left untreated.

Researcher Sherif Gabriel and his colleagues at the University of North Carolina hypothesized that people with normal CFTR may be the most vulnerable to cholera. Because cholera is deadly, it would be unethical to test this hypothesis by infecting healthy people and cystic fibrosis sufferers with *Vibrio cholera*. Instead, Gabriel's team bred mice with different numbers of copies of the gene for the CFTR protein.

Mice, like humans, have two versions (alleles) of every gene, one inherited from each parent. A person suffers from cystic fibrosis only if he or she receives a defective copy of the CFTR-encoding gene from both parents. Inheriting just one normal CFTR gene prevents cystic fibrosis from developing. To test their hypothesis, the researchers therefore bred three groups of mice. One line had two normal (functioning) CFTR gene copies. Another line had two defective copies, and a third group of mice had one normal and one defective copy.

The team then gave all the mice *Vibrio cholera* toxin (via a feeding tube) and measured the amount of fluid produced in the small intestine. As predicted, mice with two normal copies of the CFTR-encoding gene produced the most fluid. Mice with two faulty genes resisted the toxin's effects, and those with two different copies lost intermediate amounts of fluid. The amount of normal CFTR was correlated with susceptibility to cholera (**figure 4.21**).

Another disease linked to CFTR is typhoid fever, whose symptoms include high fever, weakness, and headache. The cause is a bacterium called *Salmonella typhi*, which enters the digestive tract in food or water contaminated by human feces. Unlike most disease-causing bacteria, *S. typhi* can live inside human cells. Researcher Gerald Pier and his colleagues at the Harvard Medical School and the Universities of Bristol and Cambridge wondered whether *S. typhi* uses CFTR to enter the cells lining the small intestine. That is, could *S. typhi* have a "key" to fit into the CFTR "lock"?

Pier's team used several strategies to answer this question. First, they measured how many *S. typhi* cells entered human cells in culture. Cells that expressed CFTR took in the most bacteria, while cells without the protein took in the fewest. Furthermore, when they added chemicals that specifically blocked CFTR, the bacteria could no longer enter the cells. Experiments

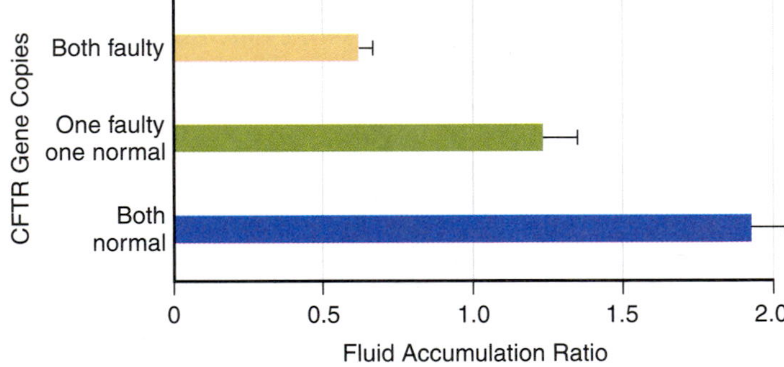

FIGURE 4.21 Cholera Toxin and CFTR. Mice with two normal cystic fibrosis transmembrane conductance regulator (CFTR) genes are more susceptible to cholera toxin than those with one or two faulty genes. The fluid accumulation ratio represents the mass of the fluid in the mouse small intestine 6 h after administering cholera toxin, divided by the mass of the gut itself. **Question:** How do you think the results would change if, before adding cholera toxin, the researcher added a chemical that blocked the site at which the toxin binds to CFTR? [Data from Gabriel, S. E., K. N. Brigman, B.H. Koller, et al., 1994, Cystic fibrosis heterozygote resistance to cholera toxin in the cystic fibrosis mouse model. *Science*, vol. 266, pages 107–109]

using specially bred mice similar to those in Sherif Gabriel's studies confirmed the findings: the more normal CFTR on cell membranes, the greater the susceptibility to *S. typhi*.

These studies help explain how natural selection might maintain apparently harmful alleles in populations. On average, cystic fibrosis sufferers leave fewer offspring than healthy people do. But a person without cystic fibrosis may still be a "carrier"—that is, he or she may have one copy of the disease-causing gene without knowing it. (Chapter 10 explains that such a person is called a heterozygote). Evolutionary biologists suggest that, in some areas of the world, symptomless carriers of the faulty CFTR gene have a reproductive advantage over people with two copies of the normal gene.

This so-called heterozygote advantage would occur wherever diseases such as cholera and typhoid fever threaten human populations. The evolutionary tradeoff of improved resistance to infectious disease apparently offsets losing some children to cystic fibrosis. A similar phenomenon occurs in human populations with a high frequency of the sickle cell trait. In that case, carriers have some resistance to another infectious disease, malaria.

Membrane proteins are among the most important components of cells. It is not surprising that microbes exploit these vital molecules to their own advantage. But it is also valuable to learn of their connection to genetic diseases such as cystic fibrosis, because this knowledge may someday lead to a cure. The Human Genome Project, along with the relatively new field of proteomics, will undoubtedly teach us much more about protein structure than biologists a generation ago thought possible. This rich harvest of information may eventually help medical researchers relieve the suffering of cystic fibrosis patients.

Gabriel, Sherif E, K. N. Brigman, B. H. Koller, et al. Oct. 7, 1994. Cystic fibrosis heterozygote resistance to cholera toxin in the cystic fibrosis mouse model. *Science*, vol. 266, pages 107–109.

Pier, Gerald B., M. Grout, T. Zaidi, et al. May 7, 1998. *Salmonella typhi* uses CFTR to enter intestinal epithelial cells. *Nature*, vol. 393, pages 79–82.

CHAPTER SUMMARY

4.1 All Cells Capture and Use Energy

A. Energy Allows Cells to Do Life's Work

- **Energy** is the ability to do work. **Potential energy** is stored energy, and **kinetic energy** is action.
- Energy is measured in units called **calories**. One food Calorie is 1000 calories, or one **kilocalorie**.

B. The Laws of Thermodynamics Describe Energy Transfer

- The **first law of thermodynamics** states that energy cannot be created or destroyed but only converted to other forms.
- The **second law of thermodynamics** states that all energy transformations are inefficient because every reaction results in increased **entropy** (disorder) and the loss of usable energy as heat.

4.2 Networks of Chemical Reactions Sustain Life

- **Metabolism** is the sum of the energy and matter conversions in a cell. It consists of chemical reactions organized into interconnected **metabolic pathways**.

A. Chemical Reactions Absorb or Release Energy

- In **endergonic reactions,** products have more energy than reactants. These reactions require energy input because they synthesize large, organized molecules from small components.
- In **exergonic reactions,** products have less energy than reactants, and energy is released. These reactions break down molecules into small pieces.

B. At Chemical Equilibrium, Reaction Rates Are in Balance

- At **chemical equilibrium,** a reaction proceeds in both directions at the same rate. Cells avoid chemical equilibrium by consuming reaction products, driving reactions forward.

C. Linked Oxidation and Reduction Reactions Form Electron Transport Chains

- Many energy transformations in organisms occur via **oxidation-reduction (redox) reactions**. **Oxidation** is the loss of electrons; **reduction** is the gain of electrons. Oxidation and reduction reactions occur simultaneously.
- In both photosynthesis and respiration, cytochromes shuttle electrons along **electron transport chains.**

4.3 ATP Is Cellular Energy Currency

A. Coupled Reactions Release and Store Energy in ATP

- **ATP** stores energy in its high-energy phosphate bonds.
- Many energy transformations involve **coupled reactions,** in which the cell uses the energy released by ATP to drive another reaction.

B. Transfer of Phosphate Completes the Energy Transaction

- **Phosphorylation** is the transfer of a phosphate group from ATP to another molecule, causing the recipient to become energized or to change shape.

4.4 Enzymes Speed Biochemical Reactions

A. Enzymes Are Molecular "Matchmakers"

- **Enzymes** are proteins that speed biochemical reactions by lowering the **energy of activation.** Substrate molecules fit into the enzyme's **active site**.

B. Enzymes Have Partners

- **Cofactors** are inorganic or organic substances that enzymes require to catalyze reactions. Like enzymes, cofactors are not consumed in the reaction.

C. Cells Control Reaction Rates in Metabolic Pathways

- In **negative feedback,** a reaction product temporarily shuts down its own synthesis whenever its levels rise. Negative feedback may occur by **competitive** or **noncompetitive inhibition**.
- In **positive feedback,** a product stimulates its own further production.

D. Environmental Conditions Affect Enzyme Activity

- Enzymes have narrow ranges of conditions in which they function. Under extreme conditions, enzymes denature and become unable to function. Some poisons and drugs work by binding to essential enzymes.

4.5 Membrane Transport May Release Energy or Cost Energy

- A **concentration gradient** is a difference in solute concentration between two neighboring regions, such as across a membrane. Gradients dissipate without energy input.

A. Passive Transport Does Not Require Energy Input

- All forms of **passive transport** involve **diffusion**, the dissipation of a chemical gradient by random molecular motion.
- In **simple diffusion**, a substance passes through a membrane along its concentration gradient without the aid of a transport protein.
- **Osmosis** is the simple diffusion of water across a semipermeable membrane. Terms describing tonicity (**isotonic, hypotonic,** and **hypertonic**) predict whether cells will swell or shrink when the surroundings change. When plant cells lose too much water, the resulting loss of **turgor pressure** causes the plant to wilt.
- In **facilitated diffusion,** a membrane protein admits a substance along its concentration gradient without expending energy.

B. Active Transport Requires Energy Input

- In **active transport,** a carrier protein uses energy (ATP) to move the substance against its concentration gradient. In animals cells, the **sodium-potassium pump** uses active transport to exchange sodium ions for potassium ions.

C. Endocytosis and Exocytosis Use Vesicles to Transport Substances

- In **endocytosis,** a cell engulfs liquids or large particles. Pinocytosis brings in fluids; **phagocytosis** brings in solid particles.
- In **exocytosis,** vesicles inside the cell carry substances to the cell membrane, where they fuse with the membrane and release the cargo outside.

4.6 Investigating Life: Does Natural Selection Maintain Some Genetic Illnesses?

- The faulty membrane protein that causes cystic fibrosis may persist in the human population by helping to protect against cholera.

MULTIPLE CHOICE QUESTIONS

1. Which of the following is the best example of potential energy in a cell?
   a. Cell division
   b. A molecule of glucose
   c. Movement of a flagellum
   d. Assembly of a cellulose fiber
2. An *endergonic* reaction is one that is characterized
   a. by a rapid release of energy.
   b. as needing an input of energy.
   c. by phosphorylation.
   d. as occurring spontaneously.
3. How does a cytochrome protein contribute to the function of an electron transport chain?
   a. It becomes oxidized and reduced.
   b. It undergoes osmosis.
   c. It is involved in the active transport of electrons.
   d. It catalyzes the transport of electrons.
4. Where in a molecule of ATP is the stored energy that is used by the cell?
   a. Within the nitrogenous base, adenine
   b. Within the five-carbon ribose sugar
   c. In the covalent bonds between the phosphate groups
   d. In the bond between adenine and ribose

5. What is the role of an enzyme in a cell?
   a. To speed up chemical reactions
   b. To become consumed during a chemical reaction
   c. To increase the energy required to make a reaction occur
   d. To provide energy to the cell
6. Which of the following is true regarding noncompetitive inhibition?
   a. The cofactors of an enzyme are altered.
   b. Excess product blocks the active site.
   c. The active site is unable to bind substrate.
   d. Excess product enhances enzyme activity.
7. The movement of water molecules during osmosis is due to
   a. simple diffusion.
   b. active transport.
   c. pinocytosis.
   d. endocytosis.
8. What would happen to a cell that was placed into a *hypertonic* environment?
   a. There would be no change.
   b. It would swell and burst.
   c. It would exhibit turgor pressure.
   d. It would shrink.
9. What type of transport is responsible for the movement of a polar molecule from a region of high concentration to a region of low concentration?
   a. Simple diffusion
   b. Facilitated diffusion
   c. Active transport
   d. Phagocytosis
10. A concentration gradient is an example of
    a. oxidation-reduction.
    b. potential energy.
    c. entropy.
    d. equilibrium.

TESTING YOUR KNOWLEDGE

1. Which units are used to measure energy?
2. What are some sources of energy?
3. Give one example of potential energy and one of kinetic energy.
4. Cite everyday illustrations of the first and second laws of thermodynamics. How do the laws of thermodynamics underlie every organism's ability to function?
5. Give an example of entropy.
6. Why isn't heat usable energy?
7. State the differences between endergonic and exergonic reactions.
8. What is chemical equilibrium?
9. Why are oxidation and reduction reactions linked?
10. Why is ATP called the cell's "energy currency?"
11. How does an enzyme speed a chemical reaction?
12. Explain how the product of a biochemical pathway can control its own rate of synthesis.
13. Explain the differences among diffusion, facilitated diffusion, active transport, and endocytosis.

THINKING AS A SCIENTIST

1. Some people claim that life's high degree of organization defies the second law of thermodynamics. What makes this statement false?
2. Cells contain many different cytochromes. One is cytochrome *c*, an electron carrier that is nearly identical in all species. No known disorders affecting cytochrome *c* exist. What does this suggest about the importance of this molecule?
3. When a person eats a fatty diet and excess cholesterol accumulates in the bloodstream, cells temporarily turn off their synthesis of cholesterol. What phenomenon described in the chapter does this control of cholesterol illustrate?
4. A restriction enzyme is a protein that cuts DNA at a specific, unique sequence. Since DNA is made of only four chemicals (A, C, G, and T), how does the restriction enzyme "know" when it finds its target sequence?
5. Liver cells are packed with glucose. If the concentration of glucose in a liver cell is higher than in the surrounding fluid, what mechanism could the cell use to transport even more glucose into a liver cell? Why would only this mode of transport work?
6. A drop of a 5% salt (NaCl) solution is added to a leaf of the aquatic plant *Elodea*. When the leaf is viewed under a microscope, colorless regions appear at the edges of each cell as the cell membranes shrink from the cell walls. What is happening to these cells?
7. Why would a cell's fat-digesting enzymes not be able to digest an artificial fat like Olestra (see chapter 2)?
8. Why does poking a hole in a cell's membrane kill the cell?
9. Your fingers get wrinkly after you take a long swim in a lake, because water enters some of your skin cells and swells them. Do you think your fingers would get wrinkly after an equally long swim in the ocean?
10. Diffusion is an efficient means of transport only over small distances. How does this relate to a cell's need for a high surface-area-to-volume ratio (see chapter 3)?

Visit www.mhhe.com/hoefnagels for practice quizzes, animations, videos, and activities designed to help you master the material in this chapter.

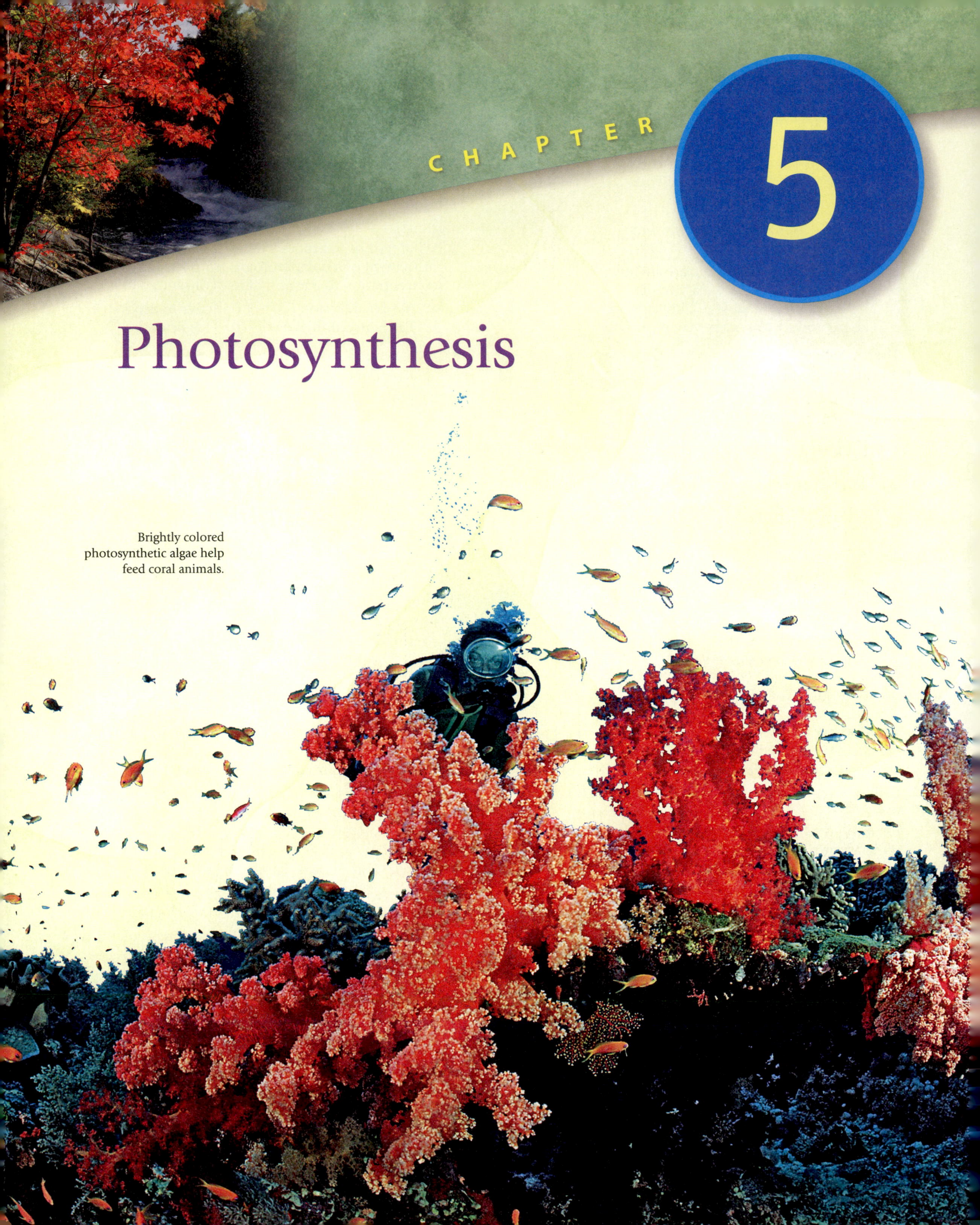

Photosynthesis

Brightly colored photosynthetic algae help feed coral animals.

What's the Point?

Is It Easier Being Green?

If you could have a photosynthetic child, would you? Imagine the benefits. Like a houseplant, he could make his own food, free of charge, simply by sitting outside. Of course, your child would look odd: his skin would be green, for starters, and he might move a little slowly, especially at night. He might even have skin flaps that capture extra sunlight.

Maybe photosynthetic cows, pigs and chickens—or pets like dogs and cats—would be a better idea. Feed-free animals would be a commercial and environmental triumph, costing less to own and generating less waste than the animals we raise now.

Fortunately or unfortunately, scientists will probably never be able to create photosynthetic people, chickens, or pooches. Mammals and birds move, breathe, pump blood, and maintain high body temperatures. All of this activity would likely require energy beyond what photosynthesis alone could supply.

Some animals, however, have adopted the "green" lifestyle by harboring live-in photosynthetic partners. The closest to a true plant–animal hybrid is probably the sea slug Elysia chlorotica, *a solar-powered mollusk with chloroplasts (photosynthetic organelles) in the cells lining its digestive tract. The chloroplasts come from algae in the slug's diet. As it grazes, the animal punctures the algal cells and discards everything but the chloroplasts, which migrate into the animal's cells. Light passes through the slug's skin and strikes the food-producing chloroplasts. Once its "solar panels" are in place, the animal may not eat again for months!*

Perhaps the most famous animals to "farm" photosynthetic partners are corals. Inside the cells lining the coral animal's digestive tract are tiny one-celled eukaryotes called dinoflagellates. These protists lend vivid colors to the reef and use the sun's energy to feed the coral animals. In exchange, the animals provide a home.

Sometimes, however, the partners break up. Corals under stress sometimes expel their dinoflagellates, or the protists may leave on their own. The reef then turns white. The coral animals eventually die, endangering the entire reef ecosystem. Pollution, disease, shading, excessively warm water, and ultraviolet radiation all trigger coral bleaching. Biologists predict that global warming will only make this problem worse.

Corals and sea slugs are not the only animals whose lives depend on photosynthesis. Yours does, too, as you will learn in the next two chapters.

LEARNING OUTLINE

5.1 Life Depends on Photosynthesis

It is spring. A seed germinates, its tender roots and pale yellow stem extending rapidly in a race against time. For now, the seedling's sole energy source is food stored along with the embryonic plant in the seed itself. If its shoot does not reach light before its reserves run out, the seedling will die. But if it makes it, the shoot will quickly turn green and unfurl leaves that spread and catch the light. The seedling begins to feed itself, and an independent new life begins.

Self-feeding organisms underlie every ecosystem on Earth. It is not surprising, therefore, that if asked to designate the most important metabolic pathway, most biologists would not hesitate to cite **photosynthesis:** the process by which plants, algae, and some microorganisms harness solar energy and convert it into chemical energy (**figure 5.1**). With the exception of deep-ocean hydrothermal vent communities, all life on this planet ultimately depends on photosynthesis.

Earth without photosynthesis would not long be a living world. If a disaster such as a nuclear holocaust, eruption of a huge volcano, or a massive meteor impact were to blacken the sky, light reaching Earth's surface would have about a tenth of its normal intensity. Plants would die as they depleted their energy reserves faster than they could manufacture more food. Animals that normally ate these producers would go hungry, as would the animals that ate them. A year or even two might pass before enough life-giving light could penetrate the hazy atmosphere, but by then, it would be too late. The lethal chain reaction that began with the cessation of photosynthesis would already be well into motion, destroying food webs at their bases.

A. Photosynthesis Builds Glucose out of Carbon Dioxide and Water

Most plants are easy to grow (compared with animals, anyway) because their needs are simple. Give a plant the right amount of water, essential elements in soil, carbon dioxide, and light, and it will produce food and oxygen not only for itself but also for a host of consumers. How can they do so much with such simple raw materials?

In photosynthesis, specialized pigment molecules in plant cells capture energy from the sun. Enzymes in those cells then use the energy to build the carbohydrate glucose ($C_6H_{12}O_6$) from carbon dioxide (CO_2) molecules. The plant uses water in the process and releases oxygen gas (O_2) as a byproduct. The reactions of photosynthesis can be summarized as follows:

$$6CO_2 + 6H_2O \xrightarrow{\text{light energy}} C_6H_{12}O_6 + 6O_2$$

Photosynthesis is an oxidation–reduction (redox) process. As you will see in sections 5.3 and 5.4, electrons stripped from water reduce CO_2. Because oxygen atoms attract electrons more strongly than do carbon atoms (as explained in chapter 2), oxygen is highly electronegative. Moving electrons from water to CO_2 therefore requires energy input. The energy source for this endergonic reaction is, of course, sunlight. **redox reactions, p. 84**

A variety of fates awaits the glucose produced in photosynthesis. A plant's cells use about half of the glucose as fuel for their own cellular respiration (the subject of chapter 6), generating ATP to power biochemical reactions. Roots, flowers, fruits, seeds, and other nonphotosynthetic plant parts could not grow without sugar shipments from green leaves and stems. Plants also combine glucose with other substances to manufacture a variety of other compounds, including amino acids and a host of economically important products from rubber to medicines and spices.

If a plant produces more glucose than it immediately needs, it may store the excess as starch. Carbohydrate-rich tubers and grains, such as potatoes, rice, and wheat, are all energy-storing plant organs. Some plants, including sugar cane and sugar beets store energy as sucrose instead. Table sugar comes from these crops, just as sweet maple syrup comes from the sucrose-rich sap of a sugar maple tree. **maple syrup, p. 531**

Plants also use glucose as a raw material to build a cellulose wall for each of its cells. Wood is the remains of dead cells (see chapter 24), and it is mostly made of cellulose. The timber in the world's forests therefore stores enormous amounts of carbon. So do vast deposits of coal and other fossil fuels, which are the remains of plants and other organisms that lived

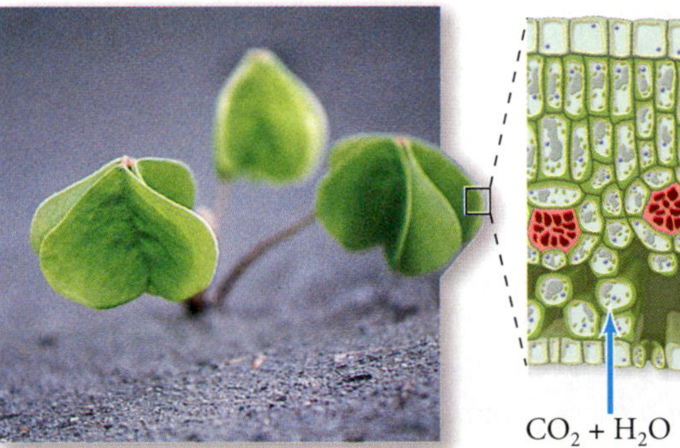

FIGURE 5.1 Life Depends on Photosynthesis. Plants, algae, and many bacteria can use light energy to produce organic molecules such as glucose from inorganic starting materials. The products of photosynthesis feed not only the producer but also most consumers on Earth.

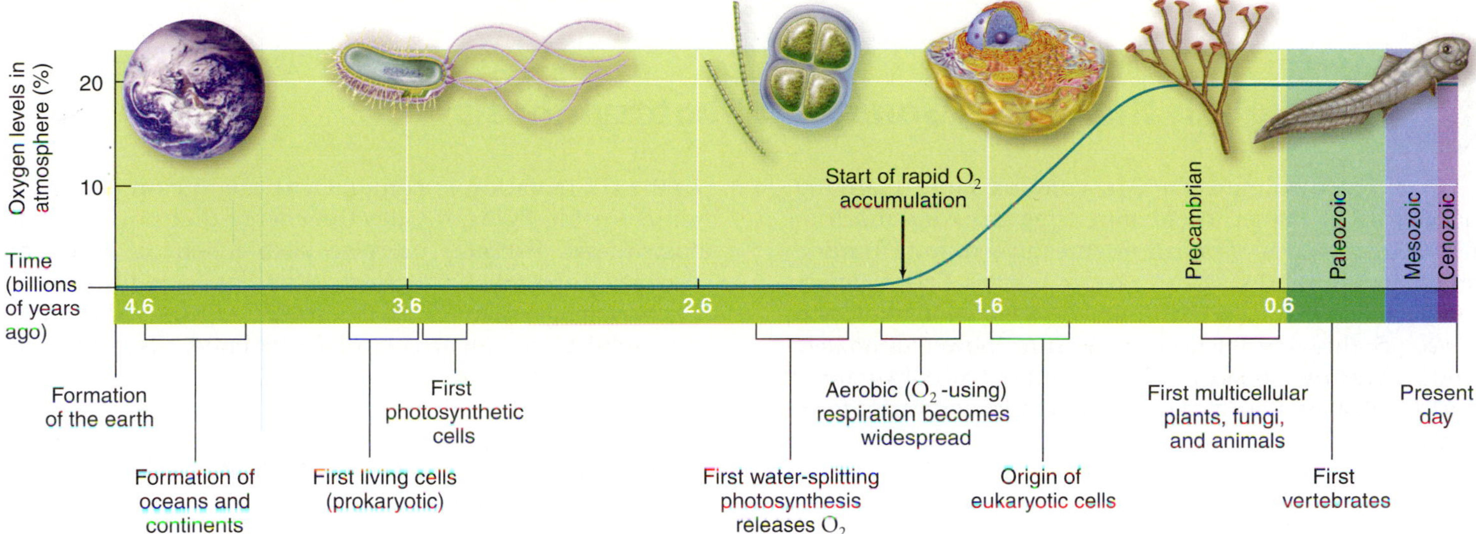

FIGURE 5.2 Oxygen Changed the World. Earth's atmosphere was not always rich in oxygen. Photosynthesis, which evolved about 3.5 billion years ago, pumped oxygen into the atmosphere and profoundly altered life's diversity. *Source*: Data from *Teaching About Evolution and the Nature of Science*, 1998, National Academy of Sciences.

long ago. Burning wood or fossil fuels releases this stored carbon into the atmosphere as CO_2. As the amount of CO_2 in the atmosphere has increased, Earth's average temperature has risen. Living forests help reduce climate change by keeping carbon locked in wood. **global climate change, p. 849**

Animals, fungi, and other consumers eat the leaves, stems, roots, flowers, nectar, fruits, and seeds of the world's producers. Even the waste product of photosynthesis, O_2, is essential to much life on Earth. Some scientists consider tropical rainforests to be the world's "lungs," because they produce much of the oxygen that we breathe. Photosynthesis therefore provides energy, raw materials, and oxygen for most animal life, including ours. No wonder biologists consider it the most important metabolic process on Earth.

B. The Evolution of Photosynthesis Changed Planet Earth

Before photosynthesis evolved, all organisms were **heterotrophs**, meaning that they obtained carbon by consuming preexisting organic molecules. As these early heterotrophs oxidized the carbon compounds from their surroundings, they released CO_2 into the environment. Because these organisms could not use the carbon in CO_2, they faced extinction as soon as they depleted the organic compounds in their habitats.

The evolution of photosynthesis some 3.5 billion years ago gave life a new energy source (**figure 5.2**). Photosynthetic organisms began to use the energy in sunlight. These were the first photosynthetic **autotrophs**, organisms that make their own organic compounds from inorganic substances such as water and CO_2. The autotrophs' ability to convert light energy into chemical energy soon supported most other forms of life.

The rise of photosynthetic organisms radically altered Earth. It decreased the concentration of CO_2 in the atmosphere and therefore lowered global temperature, which added to the polar ice caps and lowered sea level. Photosynthesis also filled the atmosphere with a new waste product: oxygen gas. The proportion of O_2 in the atmosphere gradually rose from a tiny fraction of a percent to 20% today (see figure 5.2). In this way, oxygen allowed the evolution of aerobic respiration. Our evolutionary look at photosynthesis continues at the end of chapter 6.

Because humans live on land, we are most familiar with the contribution that plants make to Earth's terrestrial ecosystems. In fact, however, more than half of the world's photosynthesis occurs in the ocean, courtesy of countless algae and bacteria. Several groups of bacteria are photosynthetic, some using pigments and metabolic pathways that are completely different from those in plants—some of these organisms do not use water as an electron source or generate oxygen gas. For simplicity, this chapter focuses on photosynthesis as it occurs in plants and algae.

Summary *Photosynthesis enables plants, algae, and many bacteria to harness light energy and convert it to chemical energy. Light therefore provides the energy that powers nearly all life. The evolution of photosynthesis changed the history of life.*

5.1 MASTERING CONCEPTS

1. Why is photosynthesis essential to life on Earth?
2. What is photosynthesis? Describe the reactants and products in words and in chemical symbols.
3. What happens to the glucose that plants produce in photosynthesis?
4. How did the origin of photosynthesis alter Earth's atmosphere and the evolution of life?

5.2 Sunlight Is the Energy Source for Photosynthesis

Each minute, the sun converts more than 100 metric tons of matter to radiant energy, releasing much of it outward as waves of radiation. After an 8-minute journey, about two-billionths of this energy reaches Earth's upper atmosphere. Of this, only about 1% is used for photosynthesis, yet this tiny fraction of the sun's power ultimately produces nearly 2 quadrillion kilograms of carbohydrates a year! Insubstantial as it may seem, light is a powerful force on Earth.

A. What Is Light?

Visible light is a small sliver of a much larger **electromagnetic spectrum**, the range of possible frequencies of radiation (**figure 5.3**). All electromagnetic radiation, including light, consists of **photons**, discrete packets of kinetic energy. A photon's **wavelength** is the distance it moves during a complete vibration. The shorter a photon's wavelength, the more energy it contains.

Sunlight consists of three main components: ultraviolet (UV) radiation, visible light, and infrared (IR) radiation. Each has different characteristics and different effects on organisms. UV radiation has wavelengths that are shorter than those of visible light. Its high-energy photons damage DNA and cause sunburn and skin cancer. The ozone (O_3) in the upper atmosphere absorbs some of the UV radiation in sunlight, making life on land possible. Visible light provides the energy that powers photosynthesis. Humans perceive visible light of different wavelengths as distinct colors. IR radiation, with its longer wavelengths, contains too little energy per photon to be useful to organisms. Most of its energy is converted immediately to heat.

B. Pigment Molecules in Chloroplasts Capture Light Energy

Although all eukaryotic cells share many similarities (see chapter 3), photosynthetic cells have something that the others lack: **chloroplasts**, the organelles of photosynthesis in plants and algae. Most photosynthetic cells contain 40 to 200 chloroplasts, which add up to about 500,000 per square millimeter of leaf—an impressive array of solar energy collectors.

Each chloroplast contains folded membranes that provide tremendous surface area for the reactions of photosynthesis. Two outside membranes enclose the **stroma**, a gelatinous matrix containing ribosomes, DNA, and enzymes (**figure 5.4**). Suspended in the stroma of each chloroplast are between 10 and 100 **grana** (singular: granum), each composed of a stack of 10 to 20 disk-shaped thylakoids. Each **thylakoid**, in turn, consists of a membrane studded with

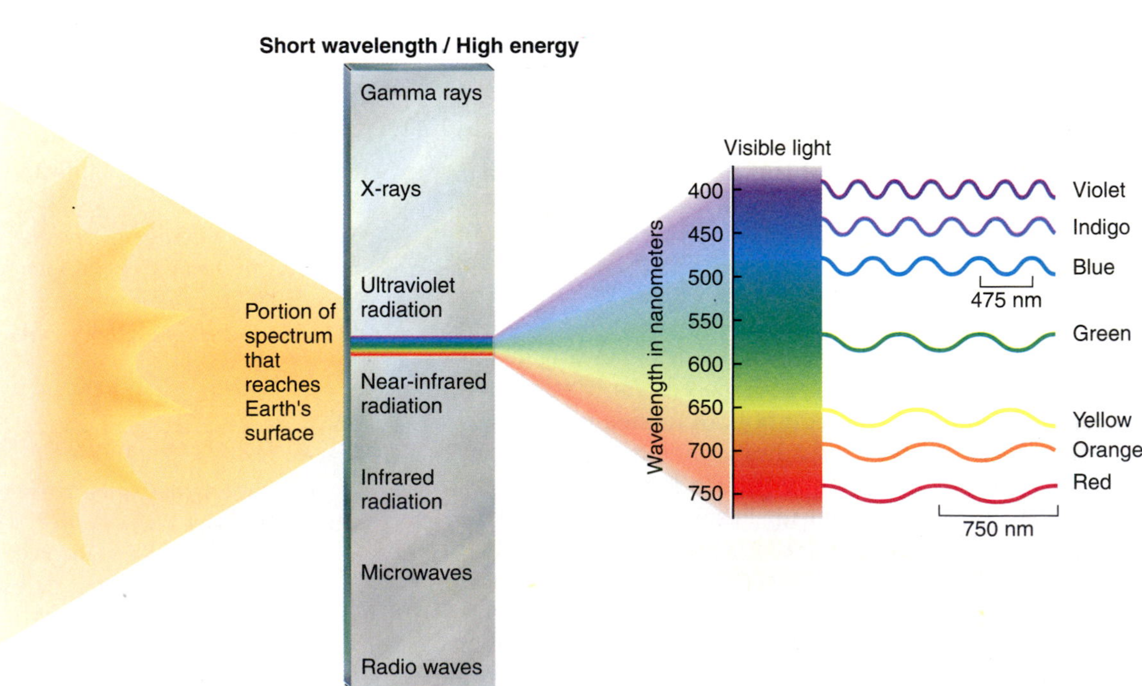

FIGURE 5.3 The Electromagnetic Spectrum. Sunlight reaching Earth consists of about 4% ultraviolet (UV) radiation, 44% visible light, and 52% infrared (IR) radiation, all of which is just a small part of a continuous spectrum of electromagnetic radiation. Wavelength is measured in nanometers (nm), or billionths of a meter. The shorter the wavelength, the more energy associated with the radiation.

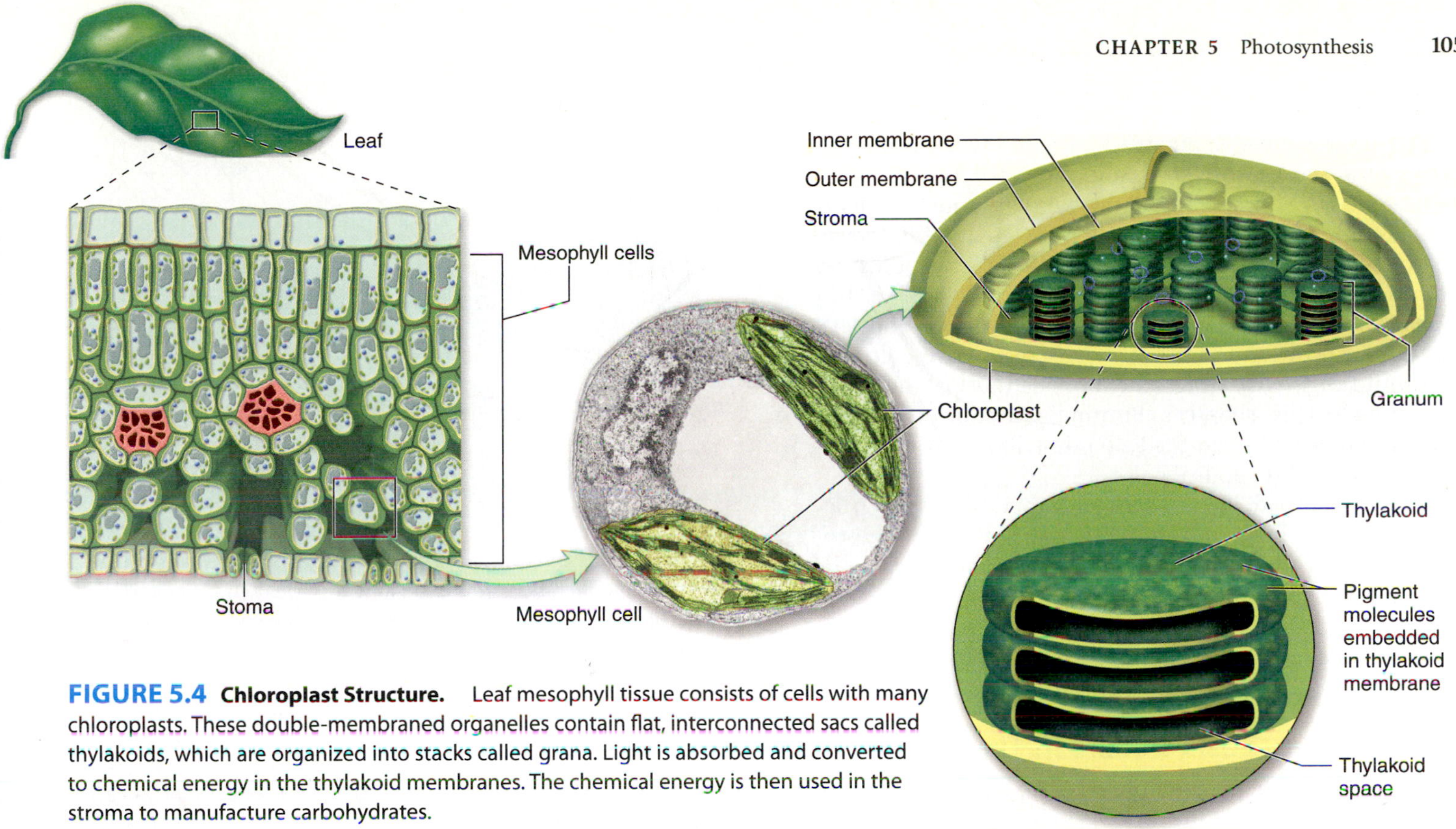

FIGURE 5.4 Chloroplast Structure. Leaf mesophyll tissue consists of cells with many chloroplasts. These double-membraned organelles contain flat, interconnected sacs called thylakoids, which are organized into stacks called grana. Light is absorbed and converted to chemical energy in the thylakoid membranes. The chemical energy is then used in the stroma to manufacture carbohydrates.

photosynthetic pigments and enclosing a volume called the **thylakoid space**.

The thylakoids contain several pigment molecules that capture light energy. The most abundant is **chlorophyll *a***, a green photosynthetic pigment in plants, algae, and cyanobacteria. Chlorophyll *a* is a huge molecule (**figure 5.5**). It includes a flat ringlike end that consists of a central magnesium (Mg) atom surrounded by four nitrogen (N) atoms and several organic rings. Transfer of energy occurs in this part of the molecule. A long, hydrophobic hydrocarbon tail anchors the pigment in lipids within the thylakoid membrane.

Photosynthetic organisms usually also have several types of **accessory pigments**, which are energy-capturing pigment molecules other than chlorophyll *a* (**table 5.1**). Chlorophyll *b* and carotenoids are accessory pigments in plants.

Table 5.1 *Pigments of Photosynthesis*		
Pigment	**Color**	**Organisms**
Major pigment		
Chlorophyll *a*	Blue-green	Plants, algae, cyanobacteria
Accessory pigments		
Chlorophyll *b*	Yellow-green	Plants, green algae
Carotenoids (carotenes and xanthophylls)	Red, orange, yellow	Plants, algae, bacteria, archaea

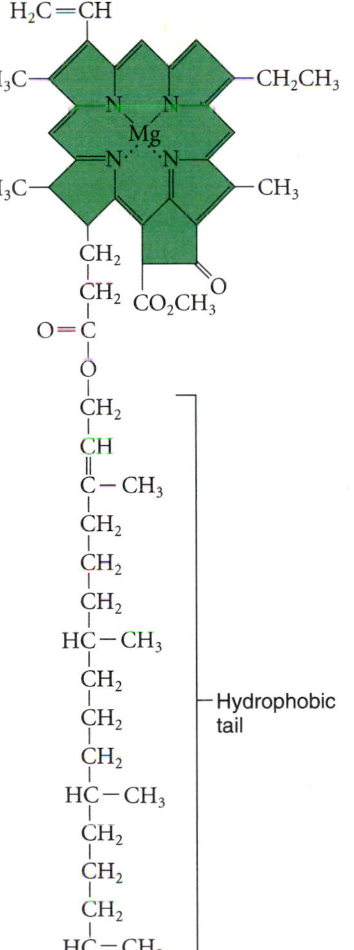

FIGURE 5.5 Chlorophyll *a*. Chlorophyll *a* is the dominant pigment in the photosynthetic cells of plants. A long hydrophobic tail anchors the molecule in the chloroplast's thylakoid membranes.

The photosynthetic pigments each have distinctive colors because they absorb only some wavelengths of visible light, while transmitting or reflecting others. Thus, pigments such as chlorophylls *a* and *b* that absorb red and blue wavelengths of light appear green (**figure 5.6**). Carotenoids, on the other hand, reflect longer wavelengths of light, so they appear red, orange, or yellow. (Carrots, tomatoes, lobsters, and the flesh of salmon all owe their distinctive colors to carotenoid pigments, which the animals must obtain from their diets). This chapter's Burning Question (Why Do Leaves Change Colors in the Fall?) describes how accessory pigments contribute to leaf color.

Only absorbed light is photosynthetically active. Because accessory pigments absorb wavelengths that chlorophyll *a* cannot, they extend the range of light wavelengths that a cell can harness. This is a little like the members of the same team on a quiz show, each contributing answers from a different area of expertise.

C. Photosynthesis Occurs in Two Stages

Inside a chloroplast, photosynthesis occurs in two stages: the light reactions and the carbon reactions (**figure 5.7**). The **light reactions**, which occur in the thylakoid membranes, convert solar energy to chemical energy. Pigment molecules capture energy from photons and pass it to proteins that use this energy to synthesize molecules of ATP and NADPH. Recall from chapter 4 that ATP stores potential energy in the covalent bonds between its phosphate groups. **NADPH** is a coenzyme that carries energized electrons.

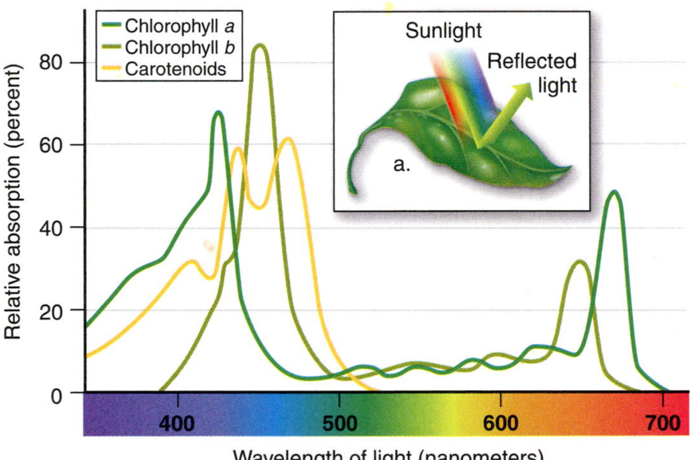

FIGURE 5.6 Everything *but* Green. (**A**) Leaves appear green because chlorophyll molecules in leaf cells reflect green and yellow wavelengths of light and absorb the other wavelengths. (**B**) Although less abundant than chlorophyll *a*, accessory pigments absorb some wavelengths of light that chlorophyll *a* cannot.

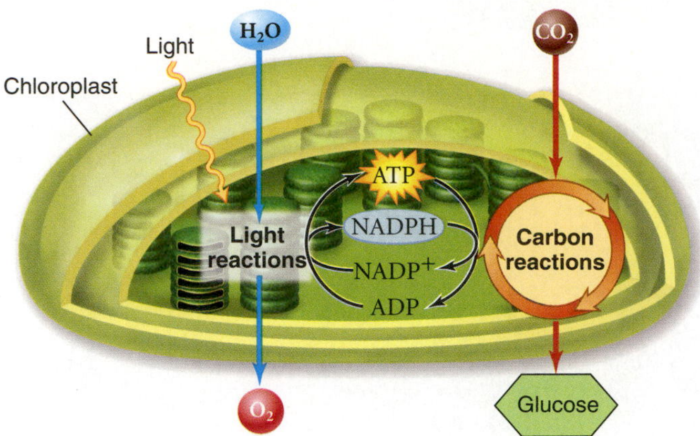

FIGURE 5.7 Overview of Photosynthesis. Molecules in the thylakoid membranes capture sunlight energy and transfer it to molecules of ATP and NADPH. In the stroma, the enzymes of the carbon reactions use this energy to capture carbon dioxide and build molecules of glucose.

These two resources (energy and "loaded" electron carriers) set the stage for the second stage of photosynthesis. The **carbon reactions** use the ATP and NADPH produced in the light reactions to reduce CO_2 to glucose. These reactions occur in the chloroplast's stroma.

Because the carbon reactions do not require light, they are sometimes called the "dark reactions" of photosynthesis. This term is misleading, however, because the carbon reactions can occur at any time of day or night, as long as ATP and NADPH are available. A more accurate alternative would be the "light-independent reactions."

The next two sections of this chapter explain the stages of photosynthesis in more detail.

Summary *Light is a form of kinetic energy. Chlorophyll and other pigment molecules in chloroplasts absorb light energy. The chloroplast channels this light energy into biochemical pathways in two main stages: the light reactions and the carbon reactions.*

5.2 MASTERING CONCEPTS

1. What are the three main components of sunlight?
2. Describe the relationship among the chloroplast, stroma, grana, and thylakoids.
3. How does it benefit a photosynthetic organism to have more than one type of pigment?
4. Where in the chloroplast do the light reactions and the carbon reactions of photosynthesis occur?

Burning Questions

Why do leaves change colors in the fall?

Most leaves are green throughout a plant's growing season, although there are exceptions—some ornamental plants, for example, have yellow or purple leaves. The near-ubiquitous green color comes from chlorophyll *a*, the most abundant pigment in photosynthetic plant parts.

But the leaf also has other photosynthetic pigments, including carotenoids and xanthophylls, with brilliant yellow, orange, and red colors. Purple pigments, such as anthocyanins, are not photosynthetically active, but they do protect leaves from damage by ultraviolet radiation.

Since these other pigments are less abundant than chlorophyll, they usually remain invisible to the naked eye during the growing season. As winter approaches, however, some plants prepare to shed their leaves. The chlorophyll degrades, and the now "unmasked" accessory pigments reveal their colors for a short time as a spectacular autumn display.

Trees store glucose they produced during the growing season in the form of starch. When spring returns, they use this reserve to fuel the regrowth of fresh, green, photosynthetically active leaves. The food these organs produce throughout the spring and summer allows the tree to grow—both aboveground and below—and to invest in fruits and seeds.

Have a Burning Question of your own? Submit it to marielle_hoefnagels@mcgraw-hill.com for possible inclusion in future editions of this book!

5.3 The Light Reactions Begin Photosynthesis

A plant placed in a dark closet literally starves. Without light, the plant cannot generate ATP or NADPH. And without these two critical energy and electron carriers, the plant cannot feed itself. Once its stored reserves are gone, the plant dies.

The plant's life thus depends on the light reactions of photosynthesis, which require the coordination of many different types of pigments and proteins. This section begins by describing how these molecules are arranged in the chloroplast's thylakoid membranes; it then explains how the various components work together to produce ATP and NADPH.

A. The Light Reactions Require Photosystems and Electron Transport Chains

The pigments and proteins that participate in the light reactions require a specific layout in the thylakoid membrane. A **photosystem** is a unit consisting of chlorophyll *a* aggregated with other pigment molecules and proteins that anchor the entire complex in the membrane (**figure 5.8**). Within each photosystem are some 300 chlorophyll molecules and 50 accessory pigments.

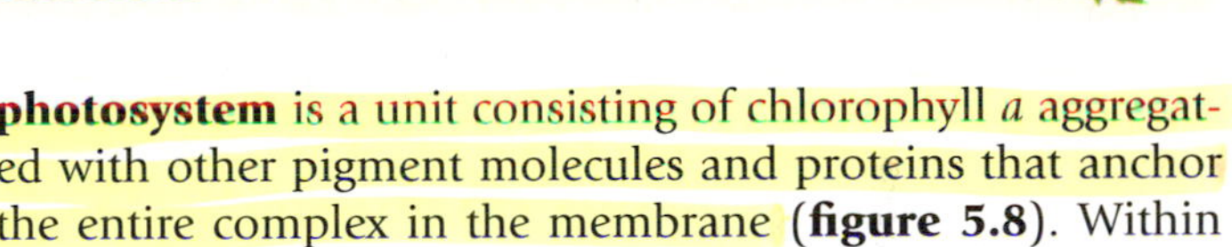

Although all of the pigment molecules absorb light energy, only one chlorophyll *a* molecule per photosystem actually uses the energy in photosynthetic reactions. The photosystem's **reaction center** is this chlorophyll molecule and its associated proteins. All other pigments in the photosystem are called **antenna pigments** because they capture photon energy and funnel it to the reaction center. If the different pigments are like a quiz show team, then the reaction center is analogous to the one member who announces the team's answer to the show's moderator.

Why does only one chlorophyll molecule out of few hundred actually participate in photosynthetic reactions? A single chlorophyll *a* molecule can absorb only a small

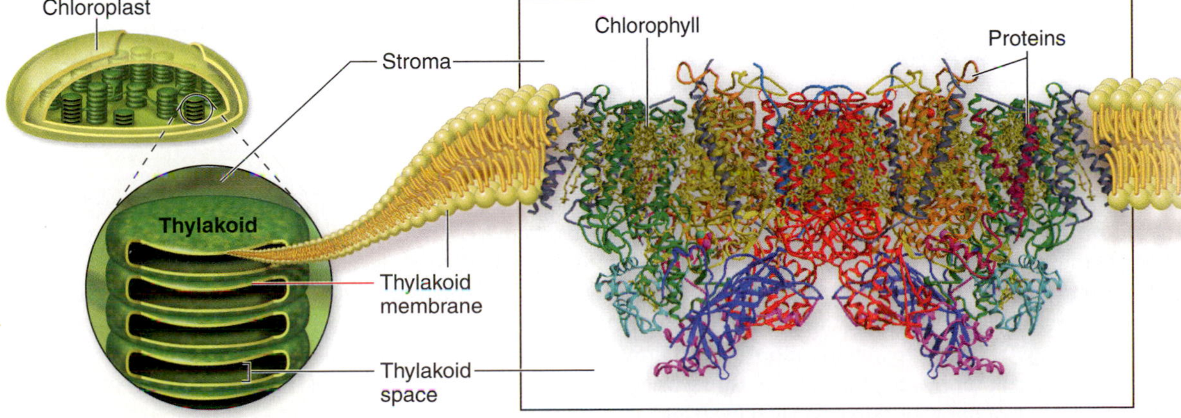

FIGURE 5.8 Photosystem II. This diagram of photosystem II shows a complex aggregation of proteins (depicted as cylinders and ribbon-like structures), chlorophyll (small green "checkerboards"), and other pigments embedded in the thylakoid membrane.

amount of light energy. Several pigment molecules near each other capture much more energy because they can pass the energy on to the reaction center, freeing them to absorb other photons as they strike. Thus, the photosystem's organization greatly enhances the efficiency of photosynthesis.

The thylakoid membranes of algae and higher plants contain two types of photosystems, I and II, connected by an electron transport chain (**figure 5.9**). Recall from chapter 4 that an **electron transport chain** is a group of aligned cytochromes and other proteins that shuttle electrons, releasing energy with each step. The electron transport chain that links photosystems I and II provides energy for ATP synthesis. A second electron transport chain ends in the production of NADPH.

B. Photosystem II Produces ATP

Photosynthesis begins in the cluster of pigment molecules of photosystem II. This may seem illogical, but the two photosystems were named as they were discovered; photosystem II was discovered after photosystem I, but it functions first in the overall process.

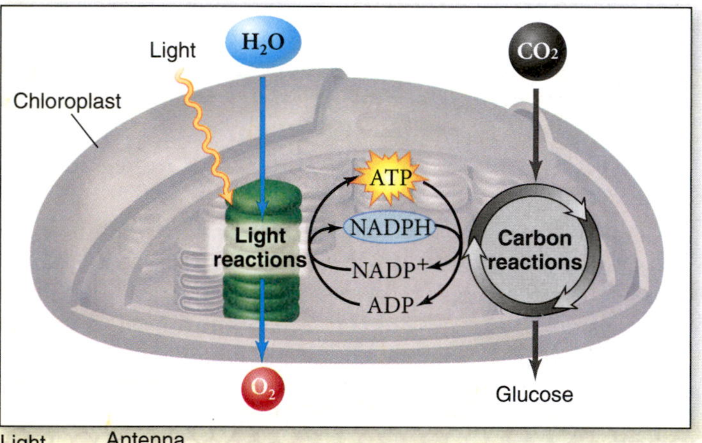

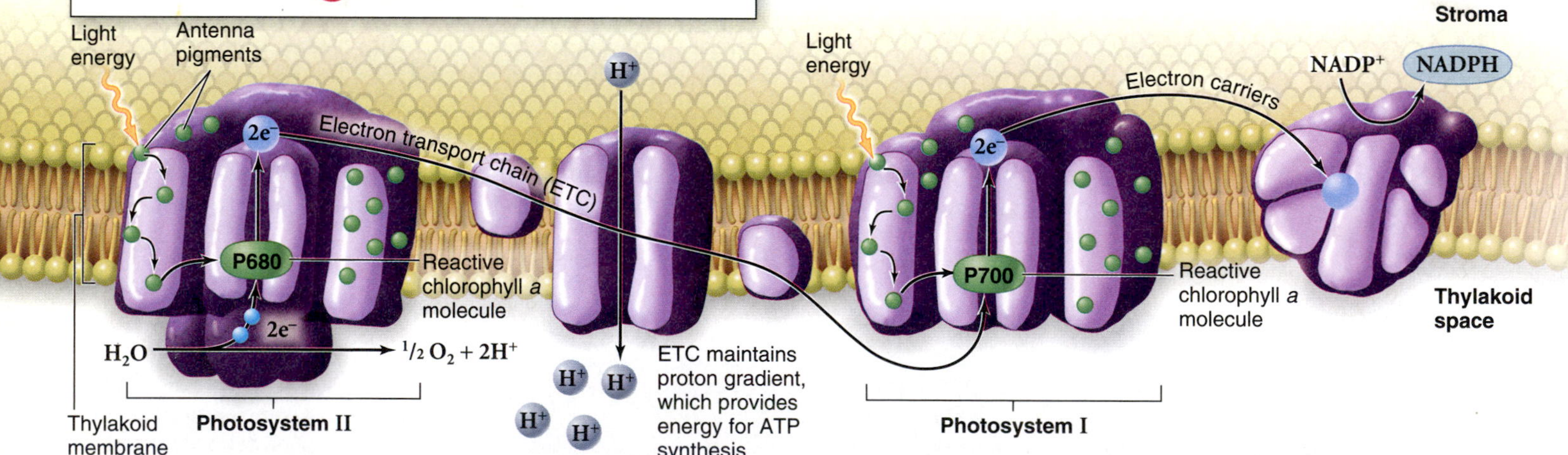

FIGURE 5.9 The Light Reactions of Photosynthesis. Chlorophyll molecules in photosystem II capture the sun's energy and transfer it to electrons ripped from water molecules (figure 5.8 shows a more realistic depiction of photosystem II). Oxygen is released as a byproduct. The energy-rich electrons pass to photosystem I via a series of carriers that make up an electron transport chain (ETC). At each transfer, a small amount of energy is removed and used to pump protons (hydrogen ions taken from water) into the thylakoid space. As the thylakoids release the hydrogen gradient, phosphate is added to ADP, forming ATP. In photosystem I, more solar energy is added to the electrons, which are passed to $NADP^+$, creating the energy-rich NADPH. The reactive chlorophyll of photosystem I absorbs light energy mostly at 700 nm and is therefore called P700 (P stands for pigment). The reactive chlorophyll of photosystem II is called P680 and absorbs energy of 680 nm.

Pigment molecules in photosystem II absorb light energy that is then transferred from one pigment molecule to another until it reaches a chlorophyll *a* reaction center, where it excites a pair of electrons. The excited electrons, now packed with potential energy, are ejected from this chlorophyll *a* molecule and grabbed by the first protein in the electron transport chain that links the two photosystems (see figure 5.9).

The reactive chlorophyll *a* molecule has now lost two electrons, which it must replace. It gets them when water (H_2O) is split into oxygen gas (½ O_2) and two protons (H^+), releasing two electrons. The O_2 is a waste product that the plant releases to the environment.

As electrons pass along the electron transport chain, the energy they lose drives the active transport of protons from the stroma into the thylakoid space (**figure 5.10**). The protons cannot easily leak back out because the membrane is relatively impermeable to them. The proton concentration therefore increases in the thylakoid space. This gradient between the inside of the thylakoid and the stroma of the chloroplast is a form of potential energy. **active transport, p. 93**

The chloroplast is now ready to capture the gradient's potential energy as chemical energy. A membrane-bound enzyme complex called **ATP synthase** uses the gradient to produce ATP (see figure 5.10). ATP synthase forms a channel for the protons trapped inside the thylakoid space. As the protons move through the channel, they alter the enzyme's shape in a way that adds phosphate to ADP. This mechanism is similar to using a dam to generate electricity. As water accumulates, tremendous pressure (a form of potential energy) builds on the face of the dam. That pressure is released by diverting water through a large pipe at the base of the dam, turning massive blades that spin an electric generator.

The coupling of ATP formation to the release of energy from a proton gradient is called **chemiosmotic**

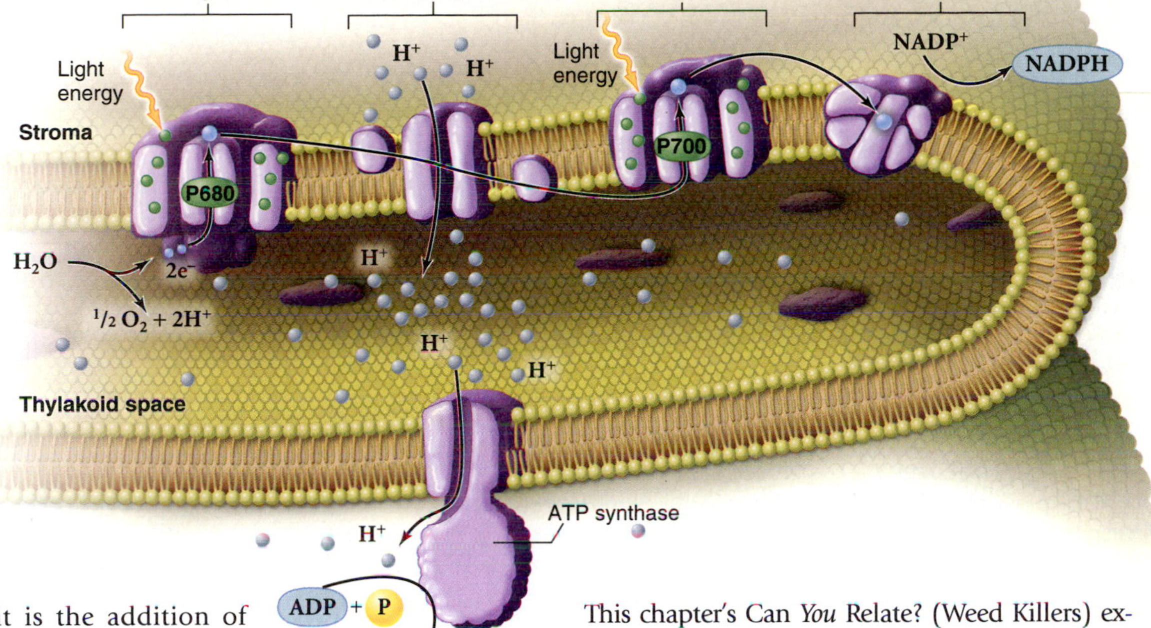

FIGURE 5.10 Chemiosmotic Phosphorylation. As energized electrons pass along electron transport chains, they release energy, some of which is used to pump hydrogen ions into the thylakoid space. ATP synthase is a channel through which the protons can escape. As they do, ADP is phosphorylated to ATP.

phosphorylation because it is the addition of a phosphate to ADP (phosphorylation) using energy from the movement of protons across a membrane (chemiosmosis). As described in chapter 6, the same process also occurs in cellular respiration.

C. Photosystem I Produces NADPH

Photosystem I functions much as photosystem II does, although the product is different. Photon energy strikes energy-absorbing molecules of chlorophyll *a*, which pass the energy to the reaction center. The reactive chlorophyll molecules eject electrons to an electron carrier molecule in a second electron transport chain. The boosted electrons in photosystem I are then replaced with electrons passing down the first electron transport chain from photosystem II.

Unlike in photosystem II, however, the second electron transport chain does not generate ATP, nor does it pass its electrons to yet another photosystem. Instead, the electrons reduce a molecule of $NADP^+$ to NADPH. This NADPH is the electron carrier that will reduce carbon dioxide in the carbon reactions, while the ATP that photosystem II generates will provide the energy.

This chapter's Can *You* Relate? (Weed Killers) explains how some herbicides kill plants by damaging their photosystems.

Summary *A photosystem consists of chlorophyll, other pigments, and proteins. In plants, two linked photosystems capture light energy and store it in the chemical bonds of ATP and NADPH. Water provides the necessary electrons, yielding* O_2 *as a waste product.*

5.3 MASTERING CONCEPTS

1. What is the relationship between the reaction center chlorophyll and the antenna pigments in a photosystem?
2. Describe the events in photosystem II, beginning with light striking a photosynthetic pigment, and ending with the production of ATP.
3. How do electrons pass from photosystem II to photosystem I?
4. How are the boosted electrons from photosystem II replaced?
5. What happens in photosystem I?

Can *You* Relate?

Weed Killers

A plant literally starves to death if it cannot photosynthesize. One low-tech way to kill a plant is to deprive it of light. Gardeners who want to convert a lawn into a garden, for example, might kill the grass by covering it with layers of newspaper for several weeks. The light reactions of photosynthesis do not occur in the dark, and the plants die.

Many herbicides also stop the light reactions. For example, DCMU (short for 3-(3,4-dichlorophenyl)-1,1-dimethylurea and known by the name diuron) blocks electron flow in photosystem II. Paraquat, noted for its use in destroying marijuana plants, diverts electrons from photosystem I.

Other herbicides take a different approach. Accessory pigments called carotenoids protect plants from damage caused by free radicals. Triazole herbicides kill plants by blocking carotenoid synthesis. No longer protected from free-radical damage, the cell's organelles are destroyed.

Still other weed killers exploit pathways not directly related to photosynthesis. For instance, chapter 4 mentions glyphosate (Roundup), which inhibits an enzyme that plants require for amino acid synthesis. Another herbicide, 2,4-D (short for 2,4-Dichlorophenoxyacetic acid), mimics a plant hormone called auxin, as described in chapter 26.

5.4 The Carbon Reactions Produce Glucose

The carbon reactions use ATP and NADPH from the light reactions, along with gaseous CO_2, to produce glucose. Where does the CO_2 come from? Earth's atmosphere consists of about 0.03% CO_2, which also dissolves in water. Algae, bacteria, and some aquatic plants absorb CO_2 directly from the water that surrounds them. Air enters land plants through **stomata** (singular: stoma), tiny openings in the epidermis of a leaf or stem (see figure 5.4). Once inside the leaf, CO_2 diffuses into a cell and across the chloroplast membrane into the stroma, where the carbon reactions occur.

A. The Calvin Cycle Produces 3-Carbon Molecules from CO_2

The **Calvin cycle,** named after its discoverer (American biochemist Melvin Calvin), is the metabolic pathway that assembles CO_2 molecules into glucose. The first step of the Calvin cycle is **carbon fixation**—the initial incorporation of carbon from CO_2 into an organic compound. Specifically, CO_2 combines with **ribulose bisphosphate (RuBP)**, a five-carbon sugar with two phosphate groups (**figure 5.11**). The enzyme that catalyzes this reaction is RuBP carboxylase/oxygenase, also known as **rubisco**. As an essential component of every plant, rubisco is one of the most abundant and important proteins on Earth.

The six-carbon product of the rubisco-catalyzed reaction immediately breaks down into two 3-carbon molecules called phosphoglyceric acid (PGA). Further steps in the cycle convert PGA to phosphoglyceraldehyde (PGAL), which is the carbohydrate product that leaves the Calvin cycle. The cell can use PGAL to synthesize more complex nutrient molecules, such as glucose and sucrose. Some of the PGAL, however, is rearranged to form RuBP again. This regeneration of the five-carbon starting material is important because it perpetuates the Calvin cycle.

ATP and NADPH, the products of the light reactions, provide the potential energy and electrons necessary to reduce CO_2. As long as ATP and NADPH are plentiful, the Calvin cycle continuously "fixes" the carbon from CO_2 into small organic molecules, in both darkness and light.

FIGURE 5.11 Calvin Cycle. ATP and NADPH from the light reactions power the Calvin cycle, simplified here. Rubisco catalyzes the first carbon fixation step, the reaction between RuBP and CO_2. The overall cycle of the carbon reactions generates a three-carbon molecule, PGAL, which is used to build glucose and other organic molecules and also regenerates RuBP.

B. C_3 Plants Use Only the Calvin Cycle to Fix Carbon

The Calvin cycle is also known as the **C_3 pathway** because the three-carbon molecule, PGA, is the first stable compound in the pathway. Although all plants use the Calvin cycle to generate glucose, C_3 plants use *only* the Calvin cycle to fix carbon from CO_2. About 95% of plant species are C_3, and they include cereals, peanuts, tobacco, spinach, sugar beets, soybeans, most trees, and lawn grasses. C_3 photosynthesis is obviously a successful adaptation, but it does have a weakness in hot, dry habitats—as the next section describes.

Summary *The reactions of the Calvin cycle reduce carbon dioxide to carbohydrates, using ATP for energy and NADPH as an electron source. These reactions can occur both day and night.*

5.4 MASTERING CONCEPTS

1. How do plants acquire the CO_2 that enters the carbon reactions?
2. What are the roles of rubisco and ribulose bisphosphate in the Calvin cycle?
3. Why is the Calvin cycle also called the C_3 cycle?
4. What is the relationship between the light reactions and the carbon reactions?

5.5 The C_4 and CAM Pathways Save Carbon and Water

According to the second law of thermodynamics, energy acquisition is always inefficient because some is always lost to heat. Even if each photosystem absorbs the maximum possible number of photons, theoretical calculations yield an efficiency rate of only 30%. In reality, however, photosynthesis falls far short of that. On cloudy days, field measurements show that individual plants average from 0.1% to 3% photosynthetic efficiency. The plant with the greatest efficiency (8%) is *Oenothera claviformis*, the annual winter evening primrose grown in Death Valley, California. Sugarcane follows at 7%.

How do plants waste so much solar energy? One contributing factor is a process that counters photosynthesis. In a metabolic pathway called **photorespiration**, the rubisco enzyme uses O_2 as a substrate instead of CO_2, starting a process that removes already-fixed carbon from the carbon reactions (**figure 5.12**). This phenomenon has no known benefit to plants; perhaps it is a holdover from ancient times, when the atmosphere contained less O_2.

Since CO_2 and O_2 compete for rubisco's active site, the greatest losses to photorespiration occur when CO_2 is scarce or O_2 is abundant. Therefore, as plants use CO_2 and release O_2 during photosynthesis, photorespiration becomes more likely. Hot, dry conditions increase photorespiration because plants close their stomata to conserve water, causing waste O_2 from the light reactions to build up inside the leaves. Therefore, plants living in hot climates face a tradeoff. If they open their stomata in dry, hot climates to acquire more CO_2 and minimize photorespiration, they risk losing water at a dangerously high rate.

A. C_4 Plants Fix Carbon Twice, in Separate Cells

Plants may lose as much as 30% of their fixed carbon to photorespiration, and avoiding this loss can give a significant competitive advantage in hot climates. Some plants do this by adding a preliminary carbon fixation reaction to the C_3 cycle. In this add-on reaction, called the **C_4 pathway**, CO_2 first combines with a three-carbon "ferry" molecule to form oxaloacetate, a four-carbon compound (hence the name C_4;

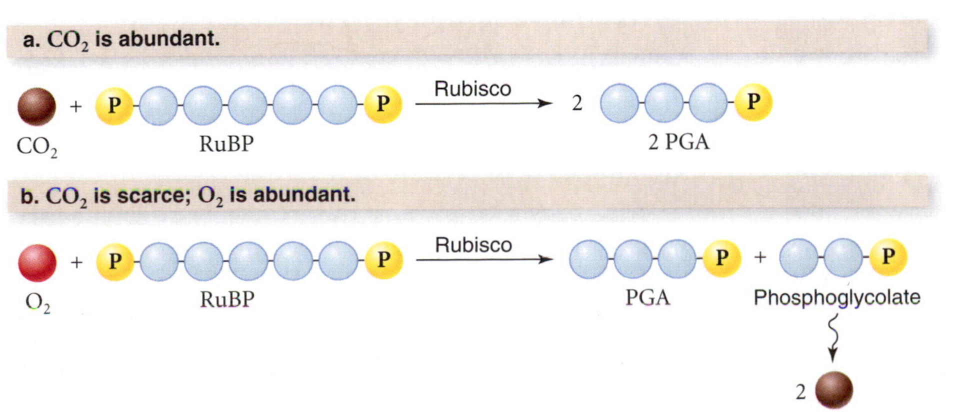

FIGURE 5.12 Photorespiration Wastes Carbon and Energy. (**A**) When CO_2 is abundant, rubisco pairs it with the five-carbon molecule RuBP to form two 3-carbon molecules (PGA). This is carbon fixation, and it begins the Calvin cycle. (**B**) When O_2 is abundant, rubisco joins it with RuBP, forming one PGA molecule plus a two-carbon molecule called phosphoglycolate. Cells may scavenge some phosphoglycolate and return it to the Calvin cycle, but much of it moves to the mitochondria, where the carbon atoms are liberated as CO_2.

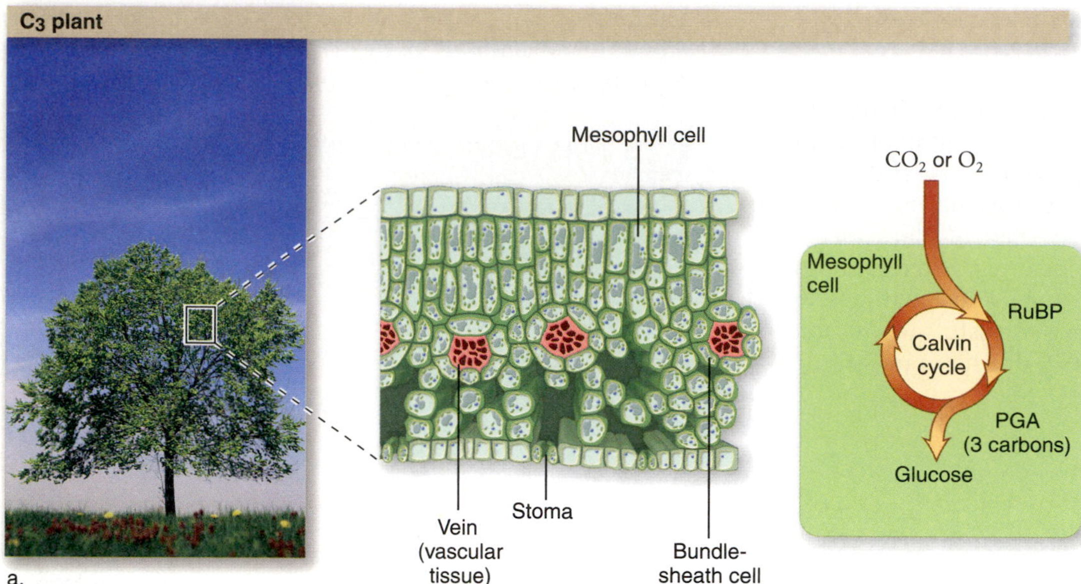

FIGURE 5.13 A Comparison of C_3 and C_4 Pathways. (**A**) In C_3 plants, the Calvin cycle occurs in leaf mesophyll cells. (**B**) The leaves of C_4 plants such as corn have a characteristic cellular organization that physically separates the light reactions and the carbon reactions. A mesophyll cell captures CO_2, combining it with a three-carbon molecule to yield a four-carbon compound—oxaloacetate. This molecule moves to adjacent bundle-sheath cells. Here, the carbon reactions make glucose from the released CO_2. By concentrating CO_2 in the bundle-sheath cells, C_4 plants reduce the effects of photorespiration.

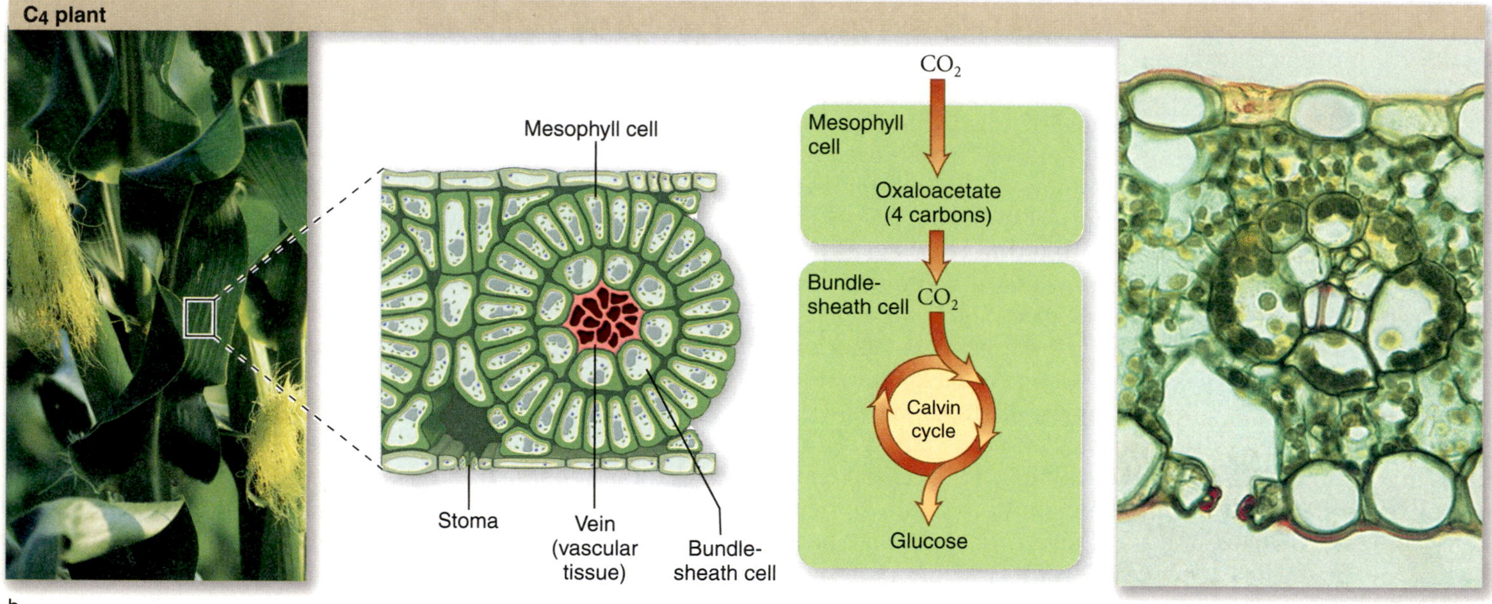

figure 5.13). The oxaloacetate is usually reduced to malate, another four-carbon molecule. These reactions occur inside mesophyll cells, which make up the majority of a leaf's interior. (The light reactions of photosynthesis also occur in these cells).

Malate then moves into adjacent **bundle-sheath cells** that surround the leaf veins. The CO_2 is liberated inside these cells, and the Calvin cycle (the second carbon fixation) occurs there. Meanwhile, at the cost of two ATP molecules, the three-carbon "ferry" returns to the mesophyll to pick up another CO_2. Figure 5.13B shows how the leaf anatomy of a C_4 plant accommodates the extra biochemical pathway.

C_4 plants lose less carbon to photorespiration than do C_3 plants in hot, dry, sunny weather. They owe their increased efficiency to the arrangement of cells in their leaves. Unlike mesophyll cells, bundle-sheath cells are not exposed directly to the air inside the leaf. They therefore can accumulate high concentrations of CO_2 while avoiding exposure to atmospheric O_2. Rubisco is much more likely to bind CO_2 instead of O_2, thereby reducing photorespiration.

As a bonus, C_4 plants require only about half as much water as C_3 plants. The enzyme that first fixes CO_2 in the mesophyll cells has a high affinity for CO_2, and O_2 does not compete for its active site. C_4 plants can therefore have fewer, smaller stomata than C_3 plants and still acquire the CO_2 they need. Since water loss occurs primarily through stomata, the water savings are significant.

Only about 1% of plants use the C_4 pathway, but they include important crop plants such as sugarcane and corn. All are flowering plants growing in hot, open environments. They represent diverse species, suggesting that C_4 photosynthesis may have evolved independently several times. In cooler, moister habitats, however, C_4 plants are not as abundant. In those environments, the ATP cost of ferrying each CO_2 from a mesophyll cell to a bundle-sheath cell apparently exceeds the benefits of reduced photorespiration.

B. CAM Plants Acquire CO_2 at Night

Another energy- and water-saving strategy was discovered in desert plants of family Crassulaceae, and so it is called crassulacean acid metabolism (CAM). Plants that use the **CAM pathway** add a new twist to the C_4 cycle: they open their stomata to fix CO_2 only at night, then fix it again in the Calvin cycle during the day. Unlike in C_4 plants, however, both fixation reactions occur in the same cell.

At night, when temperature drops and humidity rises, a CAM plant's stomata open, and CO_2 diffuses into cells. The cells incorporate the CO_2 into malate, which they store in large vacuoles (**figure 5.14**). When daylight returns, the stomata close to conserve water, but the plant already has CO_2 from its nighttime activity. The stored malate moves from the vacuole to a chloroplast in the same cell, and releases its CO_2. The chloroplast then fixes the CO_2 in the Calvin cycle.

About 3% to 4% of plant species, including pineapple and cacti, use the CAM pathway. All CAM plants are adapted to dry environments. This pathway saves water, and it reduces photorespiration by generating high CO_2 concentrations inside chloroplasts. But CAM plants cannot compete with C_3 plants in cool habitats because their stomata are only open at night. CAM plants therefore have much less carbon available to their cells for growth and reproduction.

Table 5.2 summarizes the differences between C_3, C_4, and CAM plants.

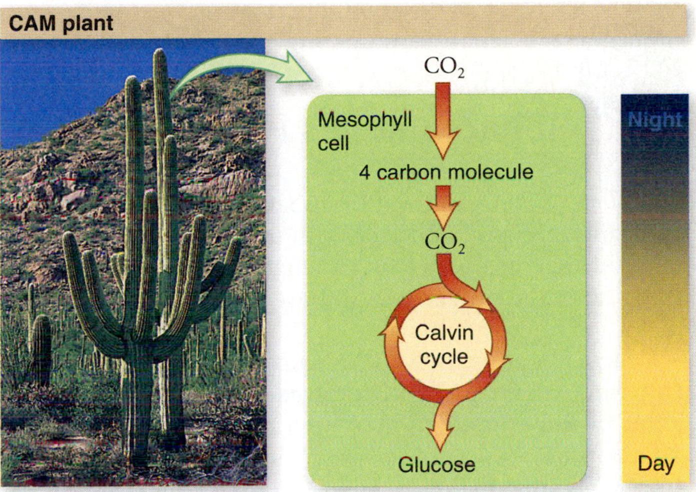

FIGURE 5.14 CAM Plants Acquire CO_2 at Night. CAM plants open their stomata at night to acquire CO_2 without losing too much water. The plant incorporates the CO_2 into a four-carbon molecule, which it stores in a vacuole. The next day, the cell releases the stored CO_2 to a chloroplast.

Summary *Photosynthesis is not highly efficient, largely due to a process called photorespiration. C_4 and CAM plants, however, have carbon fixation pathways that minimize photorespiration.*

5.5 MASTERING CONCEPTS

1. How does photorespiration counter photosynthesis?
2. Under what conditions is photorespiration most likely to occur?
3. Describe how the anatomy and biochemical reactions of a C_4 plant minimize photorespiration.
4. How is the CAM pathway like C_4 metabolism, and how is it different?

Table 5.2 *C_3, C_4, and CAM Plants Compared*

	C_3	C_4	CAM
Number of C atoms in first product of CO_2 fixation	3 (PGA)	4 (oxaloacetate)	4 (oxaloacetate)
How plant avoids photorespiration	—	Light reactions and carbon reactions occur in separate cells.	Light reactions occur during the day, and carbon reactions occur at night
Limitation of strategy	Carbon and energy losses to photorespiration	ATP cost to transport malate from mesophyll to bundle-sheath cell	Reduced carbon availability (stomata open only at night)
Habitat	Cool, moist	Hot, dry	Hot, dry
Species	95%	1%	3–4%

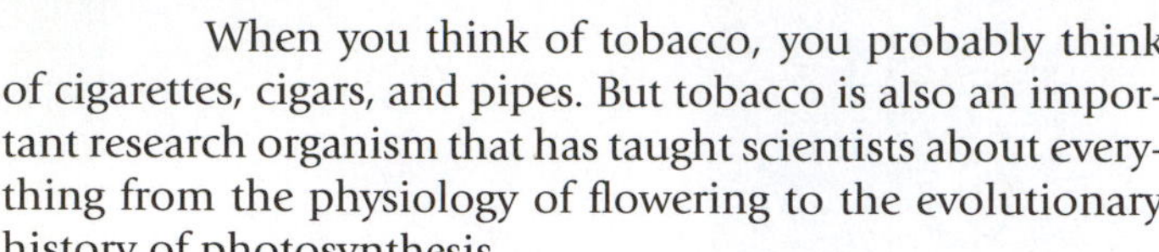

5.6 INVESTIGATING LIFE

Tobacco Stems Hold Clues to C_4 Pathway's Origin

When you think of tobacco, you probably think of cigarettes, cigars, and pipes. But tobacco is also an important research organism that has taught scientists about everything from the physiology of flowering to the evolutionary history of photosynthesis.

Most plants use only the C_3 pathway to fix carbon, but the C_4 pathway has apparently evolved independently in many plant families. In addition to their distinctive biochemistry, C_4 plants also have unique arrangements of leaf cells. Since C_3 plants came first, they must have had the anatomical and biochemical precursors for the C_4 plants' specializations. But what were they?

A study by Julian Hibberd and W. Paul Quick at the Universities of Cambridge and Sheffield in the United Kingdom may yield part of the answer. They studied tobacco, a C_3 plant, using a variety of techniques. First, they used a microscope to study the anatomy of tobacco, concentrating on stems and leaf stalks (petioles). These organs have thick strands of specialized transport tissue, called vascular bundles, that branch into veins as they enter the leaf (see chapter 25). Hibberd and Quick demonstrated that cells containing chloroplasts were present within the vascular bundles of both stems and petioles.

This arrangement is a bit surprising, however, because a photosynthetic cell buried within a stem's vascular bundle is likely to be far from the nearest open stoma or air space. Where would its chloroplasts acquire CO_2 for photosynthesis? Could water moving in the "pipes" of the vascular bundles carry dissolved CO_2 from roots to these "buried" photosynthetic cells?

The researchers hypothesized that tobacco roots could take up carbon in two forms: sodium bicarbonate ($NaHCO_3$) and glucose. $NaHCO_3$ is an inorganic molecule that breaks down to CO_2, whereas glucose is an organic source of carbon that roots could use in cellular respiration (see chapter 6) to liberate CO_2. They tested their hypothesis by supplying roots with $NaHCO_3$ and glucose that was marked with a traceable "tag." The marker was a radioactive isotope of carbon, ^{14}C (see chapter 2). In each case, 30 minutes after exposure to the radioactive carbon, cells near the vascular bundles had stored the ^{14}C in starch. This finding was consistent with the idea that chloroplast-containing cells in the vascular tissue can use CO_2 obtained from sources other than the atmosphere (in this case, from roots).

Does this result mean that tobacco is really a "closet" C_4 plant? Remember, the Calvin cycle in C_4 plants does not directly use atmospheric CO_2. Instead, the leaf mesophyll cells first fix carbon into the four-carbon compound oxaloacetate, which is converted to malate. The malate moves into the bundle-sheath cells, where a CO_2 molecule is stripped off for use in the Calvin cycle. Hibberd and Quick wondered whether they could find the origin of this biochemical "trick" in tobacco plants.

They therefore supplied tobacco plants with radioactive ^{14}C in the form of malate. They injected it directly into the vascular bundle and found that, again, nearby cells incorporated the ^{14}C into starch (**figure 5.15**). The tobacco cells had removed a single carbon atom from the malate and used it in photosynthesis. Tobacco, a C_3 plant, has at least some of the reactions that were previously thought to be exclusively associated with the C_4 pathway.

Taken together, these results reveal both anatomical and biochemical parallels between tobacco and C_4 plants. In both groups, photosynthetic cells that can use malate as a CO_2 source surround vascular tissue. But there are also some differences. The photosynthetic tobacco cells were near the vascular bundles of stems and petioles, not the leaf veins as in C_4 plants. The source of the malate also differs. In C_4 plants, malate comes from adjacent mesophyll cells, but in tobacco, it arrives from distant cells via the vascular tissue.

Does this work suggest that tobacco is the ancestor of C_4 plants? Absolutely not. After all, the C_4 pathway arose independently many times in the evolutionary history of plants. Hibberd and Quick's simple, elegant experiments revealed that the biochemical machinery required for the evolution of C_4 plants are already present in C_3 plants. This study shows, as do so many others, that the roots of an evolutionary innovation may be visible, if we only take the time to look.

Hibberd, Julian M., and W. Paul Quick. January 24, 2002. Characteristics of C_4 photosynthesis in stems and petioles of C_3 flowering plants. *Nature*, vol. 415, pages 451–454.

FIGURE 5.15 Carbon Fixation Near Vascular Tissues. Carbon that has been "fixed" from ^{14}C-labeled malate ends up in cells surrounding the petiole and central vein of tobacco leaves (dark areas of leaves). **Question:** In these leaves, could the incorporation of $^{14}CO_2$ into glucose occur in the dark, or would light have to strike the leaf to drive this reaction?

CHAPTER SUMMARY

5.1 Life Depends on Photosynthesis

- **Photosynthesis** converts kinetic energy in light to potential energy in the covalent bonds of glucose. Plants, algae, and some bacteria are photosynthetic.

A. Photosynthesis Builds Glucose out of Carbon Dioxide and Water

- Photosynthesis is a redox reaction in which water is oxidized and CO_2 is reduced to glucose.
- Plants use glucose to generate ATP, grow, nourish nonphotosynthetic plant parts, and produce cellulose and many other biochemicals. Most store excess glucose as starch.
- Most life ultimately depends on photosynthesis.

B. The Evolution of Photosynthesis Changed Planet Earth

- Before photosynthesis evolved, organisms were **heterotrophs** that relied on organic molecules as a carbon source. The first **autotrophs** developed the ability to produce their own organic molecules from atmospheric CO_2.
- Over billions of years, oxygen produced in photosynthesis changed Earth's climate and the history of life.

5.2 Sunlight Is the Energy Source for Photosynthesis

A. What Is Light?

- Visible light is a small part of the **electromagnetic spectrum**.
- **Photons** move in waves. The longer the **wavelength**, the less kinetic energy per photon. Visible light occurs in a spectrum of colors representing different wavelengths.

B. Pigment Molecules in Chloroplasts Capture Light Energy

- A **chloroplast** consists of a gelatinous matrix called the **stroma**, which contains stacks of **thylakoid** membranes called **grana.** Photosynthetic pigments are embedded in the thylakoid membranes, which enclose the **thylakoid space.**
- **Chlorophyll *a*** is the primary photosynthetic pigment in plants. **Accessory pigments** absorb wavelengths of light that chlorophyll *a* cannot absorb, extending the range of wavelengths useful for photosynthesis.

C. Photosynthesis Occurs in Two Stages

- The **light reaction**s of photosynthesis produce ATP and **NADPH;** these molecules provide energy and electrons for the glucose-producing **carbon reactions**.

5.3 The Light Reactions Begin Photosynthesis

A. The Light Reactions Require Photosystems and Electron Transport Chains

- A **photosystem** consists of **antenna pigments** and a **reaction center**. An **electron transport chain** joins photosystem II to photosystem I.

B. Photosystem II Produces ATP

- Photosystem II captures light energy and sends electrons from reactive chlorophyll *a* to an electron transport chain, replacing them with electrons from water. O_2 is the waste product.
- The energy released in the electron tranport chain drives the active transport of protons into the thylakoid space. The protons diffuse out through channels in **ATP synthase.** This movement powers the phosphorylation of ADP to ATP.
- The coupling of the proton gradient and ATP formation is called **chemiosmotic phosphorylation.**

C. Photosystem I Produces NADPH

- Photosystem I receives electrons from the electron transport chain and uses them to reduce $NADP^+$, producing NADPH. Light provides the energy.

5.4 The Carbon Reactions Produce Glucose

- The carbon reactions use energy from ATP and electrons from NADPH to **fix** carbon into organic compounds.
- CO_2 enters plants through pores called **stomata.**

A. The Calvin Cycle Produces 3-Carbon Molecules from CO_2

- In the **Calvin cycle**, **rubisco** catalyzes the reaction of CO_2 with **ribulose bisphosphate** (RuBP) to yield two molecules of PGA. These are converted to PGAL, the immediate carbohydrate product of photosynthesis. PGAL later becomes glucose.

B. C_3 Plants Use Only the Calvin Cycle to Fix Carbon

- The Calvin cycle is also called the **C_3 pathway**. Most plants species are C_3 plants, which use only this pathway to fix carbon.

5.5 The C_4 and CAM Pathways Save Carbon and Water

- Relative to the energy available in light, photosynthesis produces very little carbohydrate. **Photorespiration** contributes to this inefficiency by fixing oxygen instead of carbon from CO_2. Photorespiration wastes both CO_2 and energy.

A. C_4 Plants Fix Carbon Twice, in Separate Cells

- The **C_4 pathway** resists photorespiration by separating the light and carbon reactions into different cells. C_4 plants fix CO_2 twice. In mesophyll cells, CO_2 is fixed as a four-carbon molecule, which moves to a **bundle-sheath cell** and liberates CO_2 to be fixed again in the Calvin cycle.

B. CAM Plants Acquire CO_2 at Night

- In the **CAM pathway**, desert plants such as cacti open their stomata and take in CO_2 at night, storing the fixed carbon in vacuoles. During the day, they split off CO_2 and fix it in chloroplasts in the same cells.

5.6 Investigating Life: Tobacco Stems Hold Clues to C_4 Pathway's Origin

- Although tobacco is a C_3 plant, its photosynthetic cells contain some parts of the C_4 carbon fixation pathway.

MULTIPLE CHOICE QUESTIONS

1. Where does the energy come from to drive photosynthesis?
   a. A chloroplast
   b. ATP
   c. The sun
   d. Glucose
2. Photosynthesis is an example of an ________________ chemical reaction because
   a. exergonic; energy is released by the reaction center pigment.
   b. endergonic; light energy is used to build chemical bonds.
   c. exergonic; light energy is captured by pigment molecules.
   d. endergonic; the reactions occur inside a cell.
3. The evolution of photosynthesis resulted in
   a. an increase in the amount of O_2 in the atmosphere.
   b. the initial appearance of heterotrophs.
   c. global warming.
   d. an increase in the amount of CO_2 in the atmosphere.
4. A plant appears green because
   a. it contains chloroplasts.
   b. chlorophyll *a* absorbs red and blue light.
   c. chlorophyll *a* absorbs ultraviolet light.
   d. both a and c.
5. Only high-energy light can penetrate the ocean and reach photosynthetic organisms in coral reefs. What color light would you predict these organisms use?
   a. Red
   b. Yellow
   c. Blue
   d. Orange
6. Which part of the chloroplast is associated with the production of glucose?
   a. The thylakoid
   b. The grana
   c. The thylakoid space
   d. The stroma
7. The ATP that is produced as a result of the light reactions is used by the cell to
   a. reproduce and grow.
   b. build a glucose molecule.
   c. move electrons through the electron transport chain.
   d. split water into H^+ and O_2.
8. Can carbon fixation occur at night?
   a. Yes, because CO_2 can always enter a leaf.
   b. No, because a plant cell is not active at night.
   c. Yes, if there is a source of ATP and NADPH.
   d. No, photorespiration occurs at night.
9. What happens to the enzyme rubisco during photorespiration?
   a. The enzyme speeds up the formation of glucose.
   b. The enzyme's active site can't distinguish between O_2 and CO_2.
   c. It becomes denatured.
   d. The enzyme catalyzes the breakdown of glucose.
10. A plant that only opens its stomata at night to allow diffusion of CO_2 is a
    a. C_2 plant.
    b. C_3 plant.
    c. C_4 plant.
    d. CAM plant.

TESTING YOUR KNOWLEDGE

1. Explain whether each of the following molecules is involved in the light reactions, the carbon reactions, or both: O_2, CO_2, carbohydrate, photons, chlorophyll *a*, NADPH, ATP, H_2O.
2. Define these terms and arrange them from smallest to largest:
   a. thylakoid membrane
   b. chloroplast
   c. reaction center
   d. photosystem
   e. electron transport chain

3. How does photorespiration counter photosynthesis?
4. How is CAM photosynthesis adaptive in a desert environment?
5. Explain how C_4 photosynthesis is based on a spatial arrangement of structures, whereas CAM photosynthesis is temporally based.
6. Explain why each of the following misconceptions about photosynthesis is false:
   a. Only plants are autotrophs.
   b. Plants do not need cellular respiration because they are photosynthetic.
   c. Chlorophyll is the only photosynthetic pigment.

THINKING AS A SCIENTIST

1. Photosynthesis takes place in plants, algae, and some microbes. How does it affect a meat-eating animal?
2. What color would plants be if they absorbed all wavelengths of visible light? Why?
3. When vegetables and flowers are grown in greenhouses in the winter, their growth rate greatly increases if the CO_2 level is raised to two or three times the level in the natural environment. What is the biological basis for the increased rate of growth?
4. Over the past decades, the CO_2 concentration in the atmosphere has increased.
   a. Predict the effect of increasing carbon dioxide concentrations on photorespiration.
   b. Scientists suggest that increasing CO_2 concentrations are leading to higher average global temperatures. If temperatures are increasing, does this change your answer to part (a)?
5. One of the first investigators to explore photosynthesis was Flemish physician and alchemist Jan van Helmont. In the early 1600s, he grew willow trees in weighed amounts of soil, applied known amounts of water, and noted that in 5 years the trees gained more than 45 kg (100 pounds), but the soil lost only a few ounces. Because he had applied large amounts of water, van Helmont concluded (incorrectly) that plants grew solely by absorbing water. What is the actual source of the added biomass?
6. One of the classic experiments in photosynthesis occurred in 1771, when Joseph Priestley found that if he placed a mouse in an enclosed container with a lit candle, the mouse would die. But if he also added a plant to the container, the mouse could live. Priestley concluded that plants "purify" air, allowing animals to breathe. What is the biological basis for this observation?
7. In 1941, biologists exposed photosynthesizing cells to water containing a heavy oxygen isotope, designated ^{18}O. The "labeled" isotope appears in the O_2 gas released in photosynthesis, showing that the oxygen came from the water. Where would the ^{18}O have ended up if the researchers had used ^{18}O-labeled CO_2 instead of H_2O?
8. Of the many groups of photosynthetic bacteria, only cyanobacteria use chlorophyll *a*. How does this observation support the idea that cyanobacteria gave rise to the chloroplasts of today's plants and algae?

Visit www.mhhe.com/hoefnagels for practice quizzes, animations, videos, and activities designed to help you master the material in this chapter.

CHAPTER 6

How Cells Release Energy

Swallowing and digesting a meal almost as big as oneself requires a huge energy investment. This African rock python is consuming a Thomson's gazelle.

What's the Point?

Eating for Life

The African rock python lay in wait for the lone gazelle. When the gazelle came close, the snake moved suddenly, positioning the victim's head and holding it in place while it swiftly entwined its 9-meter (30-foot) long body snugly around the mammal. Each time the gazelle exhaled, the snake squeezed, shutting down the victim's heart and lungs in less than a minute.

Thanks to the adaptations of its digestive system, the snake can swallow and digest a meal over half its own size. The reptile begins by opening its jaws at an angle of 130 degrees (compared with 30 degrees for the most gluttonous human) and places its mouth over the gazelle's head, using strong muscles to gradually envelop and push along the carcass. Saliva coats the prey, easing its journey to the snake's stomach. After several hours, the huge meal arrives at the stomach, and the remainder of the digestive tract readies itself for several weeks of dismantling the gazelle. Hydrochloric acid (HCl) builds up in its stomach, lowering the pH sufficiently for the digestive enzymes to function, and the output of digestive enzymes in the intestines increases 60-fold.

The snake pays dearly for its meal. Although most organisms that eat frequently invest 10% to 23% of a meal's energy in digesting it and assimilating its nutrients, a snake invests 32% in energy acquisition. First, it must use considerable muscle power to capture, subdue, and swallow its prey. The reptile also expends energy in the rapid buildup of HCl and enzymes in its digestive tract.

As the gazelle passes through the snake's digestive system, it breaks into clumps of cells. These cells disintegrate, releasing proteins, carbohydrates, and lipids. After the snake digests these macromolecules into their component amino acids, monosaccharides, and fatty acids, they are small enough to enter the blood and move to the body's tissues. When these smaller nutrient molecules enter the animal's cells, they are broken down further. Then, in cellular respiration, energy in the bonds of glucose and other nutrients is transferred to the high-energy phosphate bonds of ATP. Afterward, only a few chunks of hair and bone will remain to be eliminated.

In humans, snakes, and every other organism, all activities depend on ATP. Yet nothing eats ATP directly. This chapter describes how cells convert what we do eat—glucose and other food molecules—into those little ATP molecules that nothing can live without.

LEARNING OUTLINE

6.1 Cells Use Energy in Food to Make ATP

No cell can survive without **ATP**—adenosine triphosphate. Without this energy carrier, you could not have developed from a fertilized egg into an adult. You could not breathe, chew, talk on the phone, circulate your blood, blink your eyes, walk, or listen to music. Without ATP, a plant could not take up soil nutrients, grow, or produce flowers, fruits, and seeds. A fungus could not absorb food or produce mushrooms. Like a car without gasoline, a cell without ATP simply dies.

ATP powers every activity that requires energy input in the cell: synthesis of DNA, RNA, proteins, carbohydrates, and lipids; active transport across the membranes surrounding cells and organelles; unwinding of the DNA double helix and movement of chromosomes during cell division; movement of cilia and flagella; muscle contraction; and many others. This constant need for ATP explains the need for a steady food supply—all organisms use the potential energy in food to make ATP.

Although all cells need ATP, they don't all produce it in the same way. The ATP-generating pathways fall into three categories, each beginning with glucose. In **aerobic cellular respiration**, the main subject of this chapter, a cell uses oxygen and the potential energy in the bonds of glucose to generate ATP. Plants, animals, and many microbes, especially those in oxygen-rich environments, use aerobic respiration. Anaerobic respiration generates ATP in much the same way as aerobic respiration, but in the absence of oxygen. The third pathway, fermentation, does not use oxygen either; it generates little energy compared with either type of respiration. Both anaerobic respiration and fermentation are most common in microorganisms.

The overall equation for aerobic respiration is:

$$\text{glucose} + \text{oxygen} \longrightarrow \text{carbon dioxide} + \text{water} + \text{ATP}$$

$$C_6H_{12}O_6 + 6O_2 \longrightarrow 6CO_2 + 6H_2O + 30\text{ATP}$$

This means that the potential energy in glucose is transferred to ATP, consuming oxygen and yielding carbon dioxide and water as byproducts.

To most people, the word *respiration* means "breathing," that is, inhaling and exhaling. Aerobic cellular respiration and breathing are very much related (**figure 6.1**). Both processes involve gas exchange, taking in oxygen (O_2) and releasing carbon dioxide (CO_2). In humans and many other animals, the respiratory and circulatory systems work together to carry inhaled oxygen gas to cells, where gas exchange can occur. Carbon dioxide leaves through the bloodstream and is exhaled. Aerobic cellular respiration, then, explains why our lungs obtain O_2 and expel CO_2. **gas exchange, p. 676**

Many people mistakenly believe that plants do not respire because they are photosynthetic. In fact, plants respire about half of the glucose they produce. Why do plants have a reputation for producing O_2, if they also consume it? Because they store so much carbon as cellulose, starch, and other organic molecules, they produce much more O_2 in photosynthesis than they consume in respiration, and they absorb more CO_2 than they release. They are net O_2-producers and CO_2-consumers.

The rest of this chapter describes how cells use food to generate ATP. Like photosynthesis, the journey entails several overlapping metabolic pathways and many different chemicals. But if we consider energy release in major stages, the logic emerges.

Summary *All organisms require ATP, which stores energy in a form that the cell can use. Three ATP-generating pathways are aerobic respiration, anaerobic respiration, and fermentation. Aerobic respiration occurs in animals, plants, and many microbes, and it requires the exchange of O_2 and CO_2.*

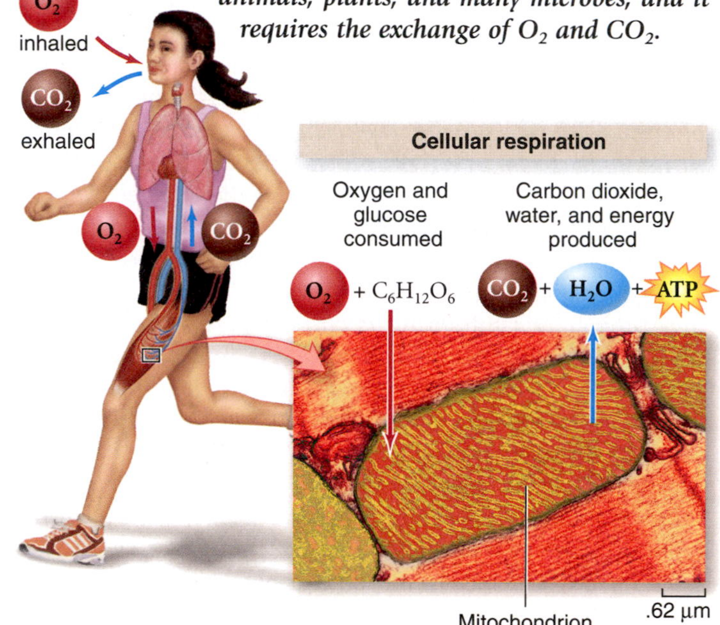

FIGURE 6.1 Breathing and Cellular Respiration Are Linked. The athlete breathes in oxygen (O_2) and exhales carbon dioxide (CO_2), a metabolic waste. Oxygen enters the bloodstream in the lungs and is distributed to all cells. There, in mitochondria, oxygen enables the reactions of cellular respiration to occur, generating ATP from potential energy in food. ATP powers the contractions of the athlete's muscles.

6.1 MASTERING CONCEPTS

1. Why do all organisms need ATP?
2. What is the overall equation that describes cellular respiration?
3. What is the relationship between cellular respiration and breathing?
4. How can plants release more O_2 in photosynthesis than they consume in respiration?

6.2 Cellular Respiration Includes Three Main Processes

The chemical reaction that generates ATP is straightforward: an enzyme tacks a phosphate group onto ADP, yielding ATP. As described in chapter 4, however, ATP synthesis requires an input of energy. It comes from the metabolic pathways of respiration, which harvest potential energy from food molecules and transfer it to ATP. This section briefly describes these pathways; later sections explain the reactions of these respiratory pathways in more detail.

Like photosynthesis, respiration is an oxidation–reduction (redox) reaction. Oxygen is highly electronegative; that is, it attracts electrons strongly compared with carbon. The pathways of aerobic respiration oxidize (remove electrons from) glucose and use the electrons to reduce oxygen. Because of oxygen's electronegativity, this reaction is "easy," like pushing a bike downhill. It releases energy, which the cell traps in the bonds of ATP.

This reaction does not happen all at once. If a cell released all the potential energy in glucose's chemical bonds in one uncontrolled step, the sudden release of heat would destroy the cell—in effect, it would act like a tiny bomb. Rather, the chemical bonds and atoms in glucose are rearranged one step at a time, releasing a tiny bit of energy with each transformation. According to the second law of thermodynamics, some of this energy is lost as heat. But much of it is stored in the chemical bonds of ATP. Biologists organize these intricate biochemical pathways into three main groups: glycolysis, the Krebs cycle, and electron transport (**figure 6.2**).

In **glycolysis** (literally, "breaking sugar"), glucose splits into two 3-carbon molecules of **pyruvate**. This process harvests energy in two forms. First, some of the electrons from glucose are transferred to an electron carrier called nicotine adenine dinucleotide (**NADH**). Second, glycolysis generates two molecules of ATP. Glycolysis occurs in virtually all living cells.

An additional set of reactions, including the **Krebs cycle**, release all of the carbon in pyruvate as CO_2. Enzymes rearrange atoms and bonds in ways that transfer the potential energy in pyruvate to ATP, NADH, and another electron carrier—flavin adenine dinucleotide ($FADH_2$).

By the time the Krebs cycle is complete, the carbon atoms that made up the glucose are gone—liberated as CO_2. The cell has generated a few molecules of ATP, but most of the potential energy from glucose now lingers in the high-energy electron carriers (NADH and $FADH_2$) produced in glycolysis and the Krebs cycle. The cell uses them to generate more ATP.

The **electron transport chain** transfers energy-rich electrons from NADH and $FADH_2$ through a series of cytochromes and other membrane proteins. As they pass from carrier to carrier, the energy is used to create a gradient of hydrogen ions. Chemiosmotic phosphorylation uses the potential energy stored in this proton gradient to generate ATP. A complex enzyme called **ATP synthase** forms a channel in the membrane, releasing the protons and using their potential energy to phosphorylate ADP. In the meantime, the "spent" electrons are transferred to oxygen or to another electron acceptor, depending on whether respiration is aerobic or anaerobic.

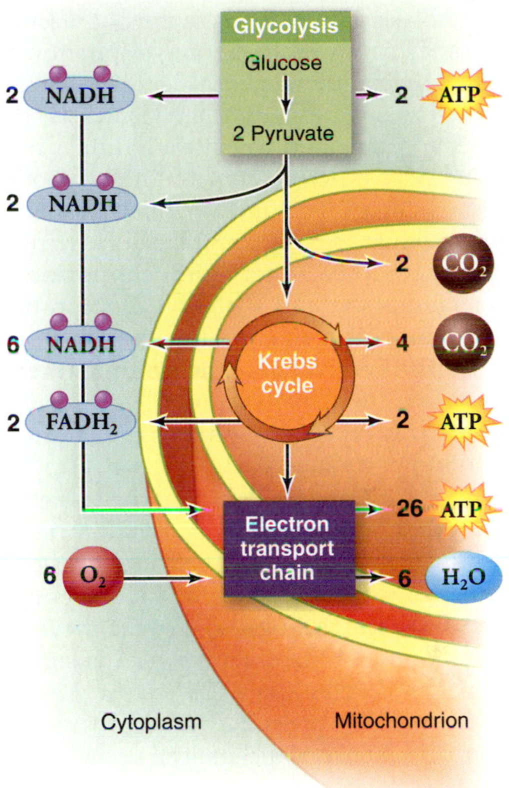

FIGURE 6.2 Overview of Aerobic Cellular Respiration. Glucose is broken down to carbon dioxide through a series of enzyme-catalyzed reactions. The energy released phosphorylates ADP to ATP. As the chapter progresses, each of the biochemical pathways will become more detailed. Look to the insets that repeat this diagram with different sections highlighted to follow the part of the overall pathway under discussion.

Summary *Glycolysis splits glucose in half, then the Krebs cycle and the electron transport chain extract additional energy from the products.*

6.2 MASTERING CONCEPTS

1. Why do the reactions of respiration occur step-by-step instead of all at once?
2. What are the roles of the electron transport chain and ATP synthase in chemiosmotic phosphorylation?

6.3 In Eukaryotic Cells, Respiration Occurs in Mitochondria

Glycolysis always occurs in the cytoplasm, but the location of other pathways in aerobic respiration depends on the cell type. Bacteria and archaea lack specialized organelles. In those prokaryotic cells, the enzymes for the Krebs cycle and electron transport are in the cytoplasm and embedded in the cell's outer membrane. In eukaryotic organisms (protista, plants, fungi, and animals) organelles called **mitochondria** house the reactions of cellular respiration.

Recall from chapter 3 that a mitochondrion consists of an outer membrane and a highly folded inner membrane, an organization that creates two compartments (**figure 6.3**). The **intermembrane compartment** is the area between the two membranes. The mitochondrial **matrix** is the area enclosed by the highly folded inner mitochondrial membrane.

In a eukaryotic cell, the two pyruvate molecules produced in glycolysis cross both mitochondrial membranes and move into the matrix. Here, enzymes cleave pyruvate and carry out the Krebs cycle. Then, $FADH_2$ and NADH (both from glycolysis and the Krebs cycle) move to the highly folded inner mitochondrial membrane, which is studded with electron carrier molecules and ATP synthase. The extensive folding greatly increases the surface area on which the reactions of the electron transport chain can occur.

ATP synthase occurs not only in all mitochondria but also in the thylakoid membranes of chloroplasts, which use it to generate ATP in the light reactions of photosynthesis (see chapter 5). Enzymes with essentially the same structure and function also operate in the cell membranes of respiring bacteria and archaea, making ATP synthase one of the most highly conserved proteins over evolutionary time.

Given the role of mitochondria in the Krebs cycle and electron transport chain, it is not surprising that a person with abnormal mitochondria may be very ill or even die. Faulty versions of ATP synthase, electron transport proteins, and enzymes required for glycolysis and the Krebs cycle cause at least 40 diseases, many of which strike in early childhood.

Summary *In eukaryotic cells, glycolysis occurs in the cytoplasm, while the other reactions of respiration happen in the mitochondria.*

6.3 MASTERING CONCEPTS

1. What are the parts of a mitochondrion?
2. Which respiratory reactions occur in each part of the mitochondrion?

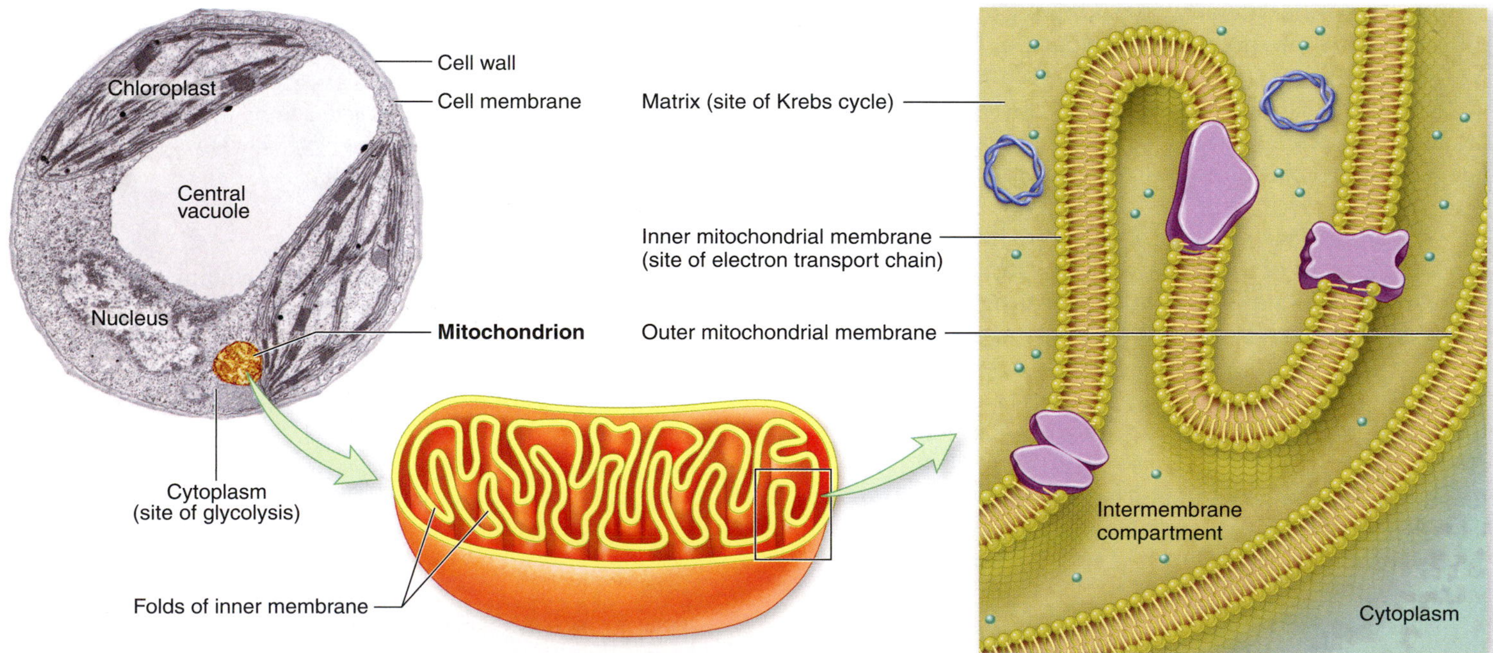

FIGURE 6.3 Cellular Respiration Occurs in Mitochondria. In eukaryotic cells (such as the leaf cell pictured), mitochondria provide most of the ATP for cellular functions. Enzymes in the matrix and membrane oxidize pyruvate (from glycolysis) to carbon dioxide and transfer the released energy to ATP. An electron transport chain in the inner mitochondrial membrane forms the proton gradient for chemiosmotic phosphorylation, similar to events at the thylakoid membrane in photosynthesis (see figure 5.10).

6.4 Glycolysis Breaks Down Glucose to Pyruvate

Glycolysis is a more-or-less universal metabolic pathway that splits glucose into two 3-carbon pyruvate molecules. The entire process requires 10 steps, all of which occur in the cytoplasm (**figure 6.4**). The first five steps use ATP to "activate" glucose, redistributing energy in the molecule. The rest of the pathway then extracts some of this energy, regaining the ATP molecules invested earlier plus two more. None of the steps requires oxygen, so cells can use glycolysis in both oxygen-rich and anaerobic environments.

The first step of glycolysis uses one molecule of ATP to add a phosphate group to glucose. This activates glucose, enabling the appropriate enzyme to catalyze the next step. Next, the atoms are rearranged; in step 3, a second ATP transfers a phosphate to this new molecule. An enzyme splits this compound into two 3-carbon compounds in steps 4 and 5, and each of the three-carbon products has one phosphate. The formation of these two molecules (phosphoglyceraldehyde, or PGAL) marks the halfway point of glycolysis (steps 1 through 5 in figure 6.4). So far, energy in the form of ATP has been invested, but no ATP has been produced.

The first energy-obtaining step of the glycolysis pathway occurs when enzymes reduce NAD^+ to NADH using electrons stripped from PGAL (step 6). This reaction releases enough energy to add a second phosphate group to PGAL. One of the phosphates is then transferred to ADP, producing

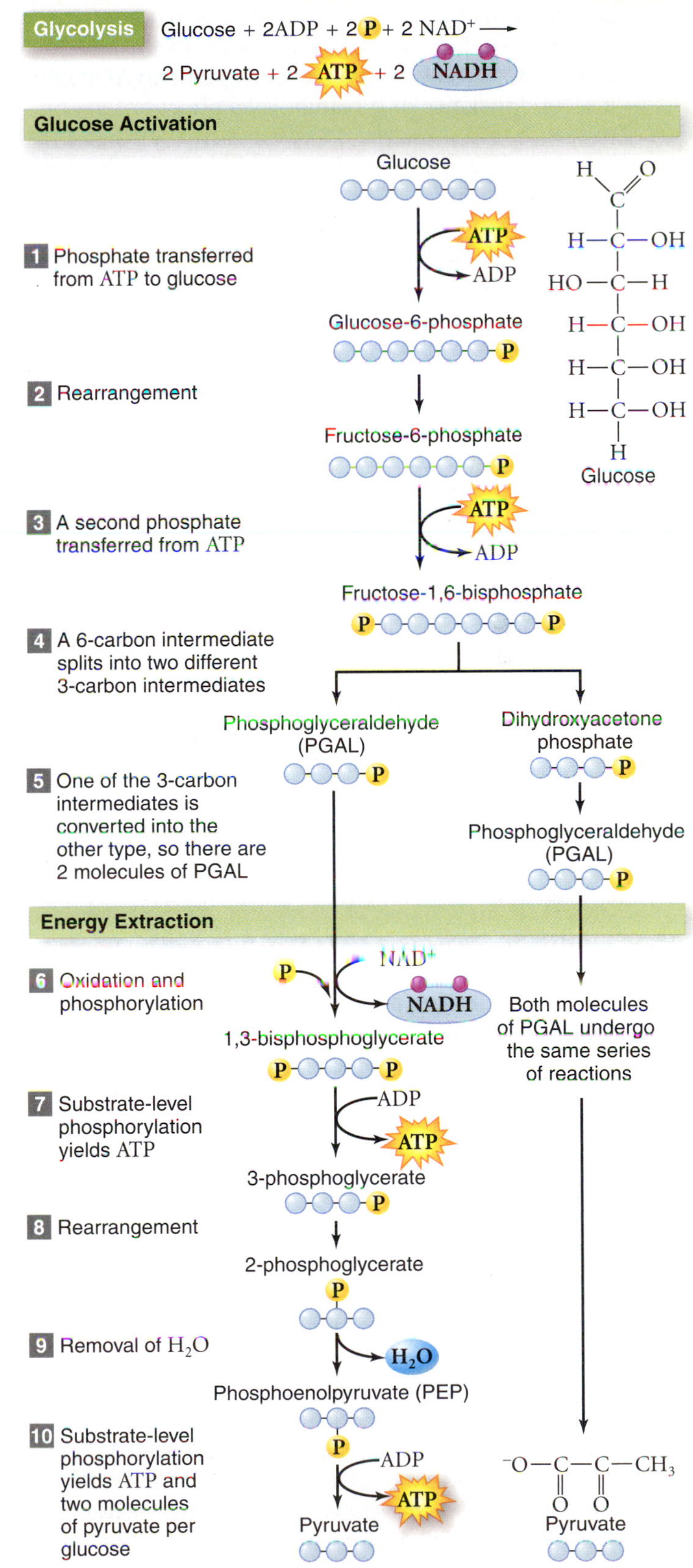

FIGURE 6.4 Glycolysis. In the glycolysis reactions, glucose splits into two molecules of pyruvate. Along the way, the pathway produces four ATPs and two NADHs. Because activating glucose consumes two ATPs, however, the net yield is two ATP molecules per molecule of glucose. (Each light blue sphere represents a carbon atom.)

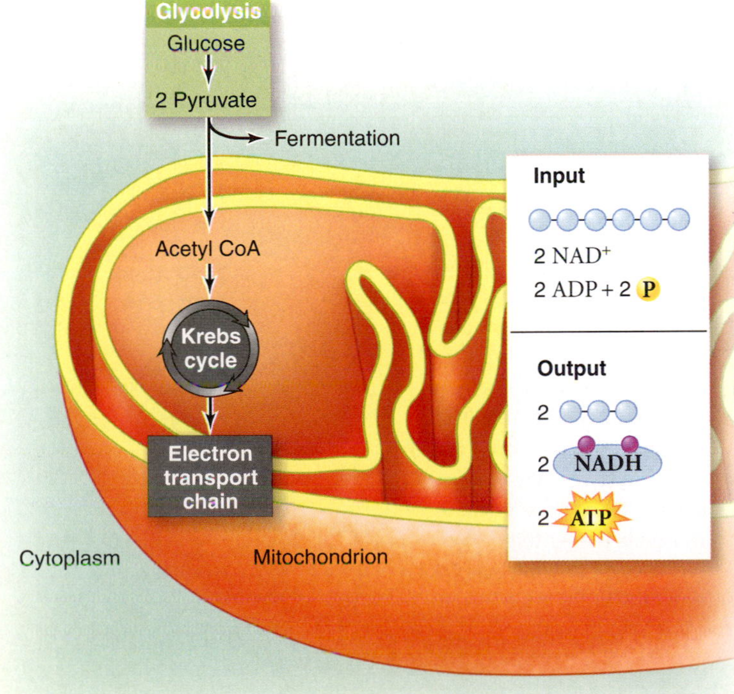

one molecule of ATP (step 7). This way of producing ATP is called **substrate-level phosphorylation**, which means that a high-energy "donor" molecule (in this case, PGAL) physically transfers a phosphate group to ADP (**figure 6.5**).

The remaining three-carbon molecule is then rearranged (step 8). This compound then loses water (step 9) and becomes pyruvate when it donates its phosphate to ADP (step 10)—a second instance of substrate-level phosphorylation.

Thus, each PGAL from step 5 eventually yields two ATPs, one NADH, and one pyruvate. Because each molecule of glucose becomes two molecules of PGAL, glycolysis produces four ATPs and two pyruvates per glucose. However, because the early steps of glycolysis require two ATPs, the net gain is two ATPs per molecule of glucose.

Summary *The reactions of glycolysis start the energy-releasing process by splitting one molecule of glucose into two molecules of pyruvate.*

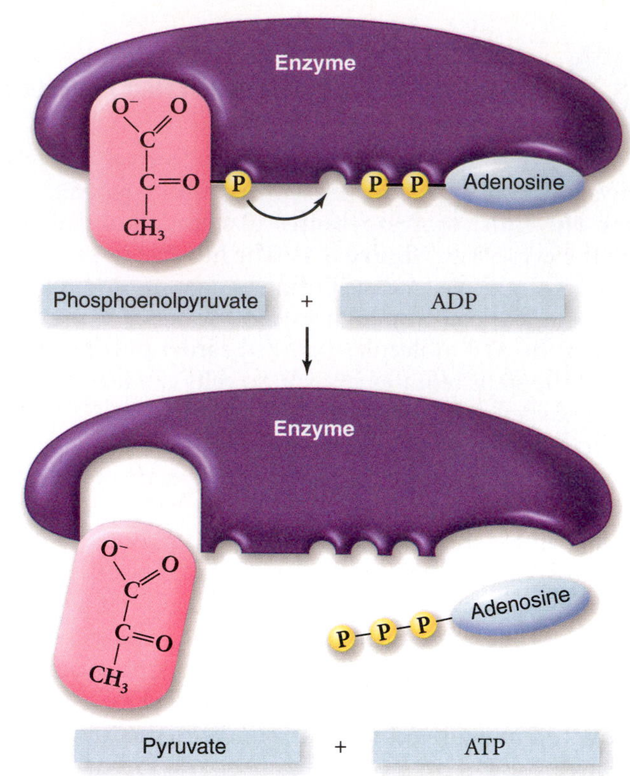

FIGURE 6.5 Substrate-Level Phosphorylation. In substrate-level phosphorylation, one molecule directly transfers a phosphate group to ADP, forming ATP.

6.4 MASTERING CONCEPTS

1. What are the starting materials of glycolysis?
2. How is substrate-level phosphorylation different from chemiosmotic phosphorylation?
3. What is the net gain of ATP and NADH for each glucose molecule undergoing glycolysis?

6.5 Aerobic Respiration Yields Much More ATP Than Glycolysis Alone

Glucose contains considerable bond energy, but cells recover only a small portion of it as ATP and NADH during glycolysis; most remains in the bonds of pyruvate. For cells that use fermentation, such as yeasts that produce wine and beer, glycolysis is the sole source of ATP (see section 6.6). Respiring organisms, however, tap much more of this potential energy in the Krebs cycle and electron transport. This section explains how.

A. The Krebs Cycle Produces ATP and Electron Carriers

Pyruvate, the final product of glycolysis, moves into the mitochondrial matrix, but it is not directly used in the Krebs cycle. First, a molecule of carbon dioxide is removed when NAD^+ is reduced to NADH. The remaining molecule, called an acetyl group, is transferred to a coenzyme to form acetyl coenzyme A, abbreviated acetyl CoA (**figure 6.6**). **Acetyl CoA** is the compound that enters the Krebs cycle.

In the first step of the Krebs cycle, the two-carbon acetyl group is transferred to a four-carbon molecule called

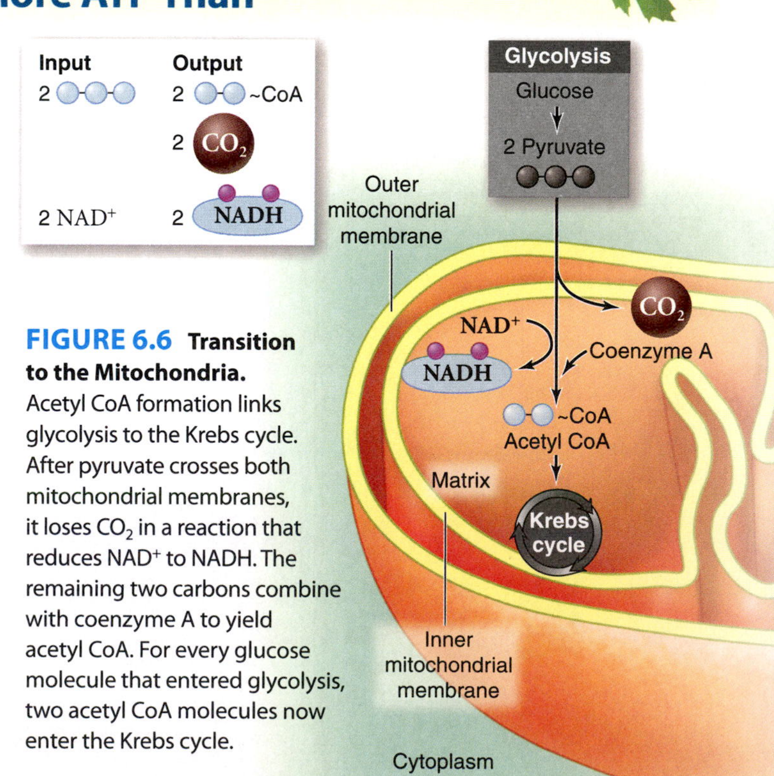

FIGURE 6.6 Transition to the Mitochondria. Acetyl CoA formation links glycolysis to the Krebs cycle. After pyruvate crosses both mitochondrial membranes, it loses CO_2 in a reaction that reduces NAD^+ to NADH. The remaining two carbons combine with coenzyme A to yield acetyl CoA. For every glucose molecule that entered glycolysis, two acetyl CoA molecules now enter the Krebs cycle.

oxaloacetate (**figure 6.7**). The new six-carbon molecule is citrate (this is why the Krebs cycle is also called the citric acid cycle). The remaining steps in the Krebs cycle rearrange and oxidize citrate through several intermediates. Some of these transformations transfer electrons to the carriers NADH and $FADH_2$; others produce ATP by substrate-level phosphorylation. As the molecules rearrange, the two carbons that entered the cycle are released as CO_2. The molecules in the Krebs cycle are eventually altered to re-create the original acceptor molecule, oxaloacetate. The cycle can now repeat.

Since glycolysis splits each glucose into two molecules of pyruvate, the Krebs cycle turns twice per glucose molecule. Thus, the combined net output to this point (glycolysis, acetyl CoA formation, and the Krebs cycle) is 4 ATP molecules, 10 NADH molecules, 2 $FADH_2$ molecules, and 6 molecules of CO_2. Of course, this process does not capture all of the potential energy in glucose. According to the second law of thermodynamics, some is always lost as heat.

Besides continuing the breakdown of glucose, the Krebs cycle also has another function, not directly related to respiration. The cell uses intermediate compounds formed in the Krebs cycle to manufacture other organic molecules, such as amino acids or fats. Later in this section, you will see that the reverse process also occurs; amino acids and fats can enter the Krebs cycle to generate energy from nonglucose food sources.

B. The Electron Transport Chain Drives ATP Formation

Of the products generated in glycolysis and the Krebs cycle, the cell ejects CO_2 as waste and uses ATP to fuel essential processes. But what becomes of NADH and $FADH_2$? The cell requires ATP as an energy source, not these two electron carriers. As it turns out, the cell transfers their potential energy to ATP as well. An electron transport chain, embedded in the inner mitochondrial membrane, accomplishes this feat.

The electron transport chain extracts most of the potential energy in NADH and $FADH_2$ by removing the energy from their electrons in incremental steps. The first carrier molecule in the chain accepts electrons from NADH, extracts some of their energy, uses that energy to perform

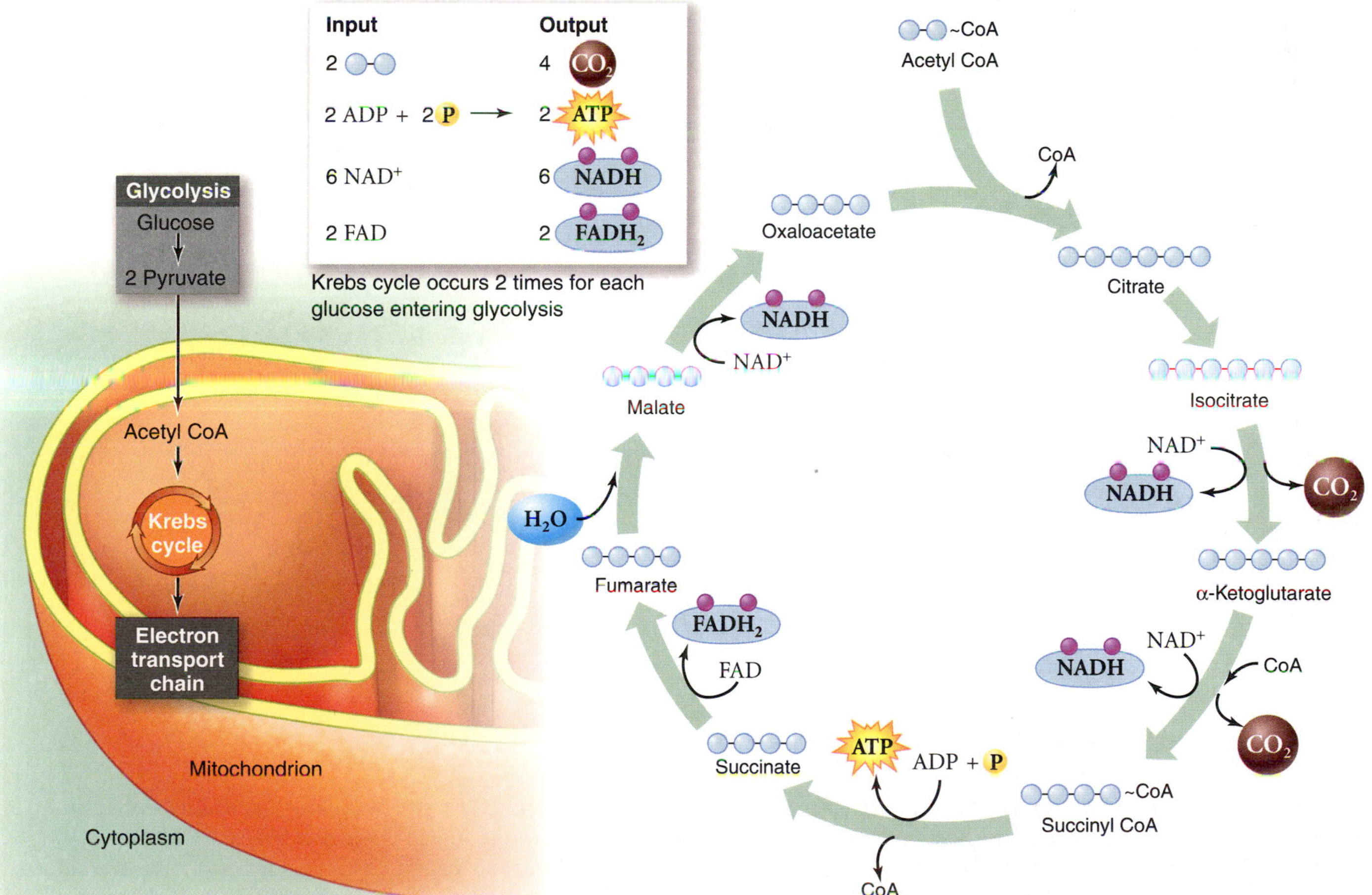

FIGURE 6.7 Krebs Cycle. Acetyl CoA enters the Krebs cycle by combining with oxaloacetate to form six-carbon citrate. A progression of reactions in the Krebs cycle generates one molecule of ATP, three molecules of NADH, one molecule of $FADH_2$, and two molecules of CO_2. Because one glucose molecule yields two molecules of acetyl CoA, each glucose molecule is associated with two turns of the Krebs cycle.

work, and then transfers the electrons to the next set of molecules (**figure 6.8**). The process repeats over a series of carriers. In aerobic respiration, the final acceptor of the moving electrons is oxygen, which combines with hydrogen ions to form water. Breathing provides the oxygen for this final electron acceptor.

What "work" do the proteins of the electron transport chain do as they pass the electrons along? The answer is that they pump hydrogen ions (protons, or H^+) from the matrix of the mitochondrion into the intermembrane compartment. The membrane is not very permeable to protons, so they do not readily leak back across it. The resulting proton gradient is a form of potential energy. Then, in **chemiosmotic phosphorylation,** protons move down their gradient through ATP synthase channels back into the matrix, and ADP is phosphorylated to ATP. Thus, the interaction between electron transport and ATP synthesis is indirect: the electron transport chain produces a proton gradient, and that gradient drives ATP synthesis.

Just as herbicide manufacturers can kill plants by inhibiting photosynthesis (see chapter 5), it is easy to see how disrupting electron transport could deprive a respiring organism of ATP. This chapter's Can *You* Relate? (How Poisons Kill) describes a few poisons that do just that.

C. How Many ATPs Can One Glucose Molecule Yield?

In tracing the energy pathways, it is easy to lose track of the overall function of the process—converting the potential energy in a molecule of glucose into ATP, a form the cell can easily use. How productive is cellular respiration? That is,

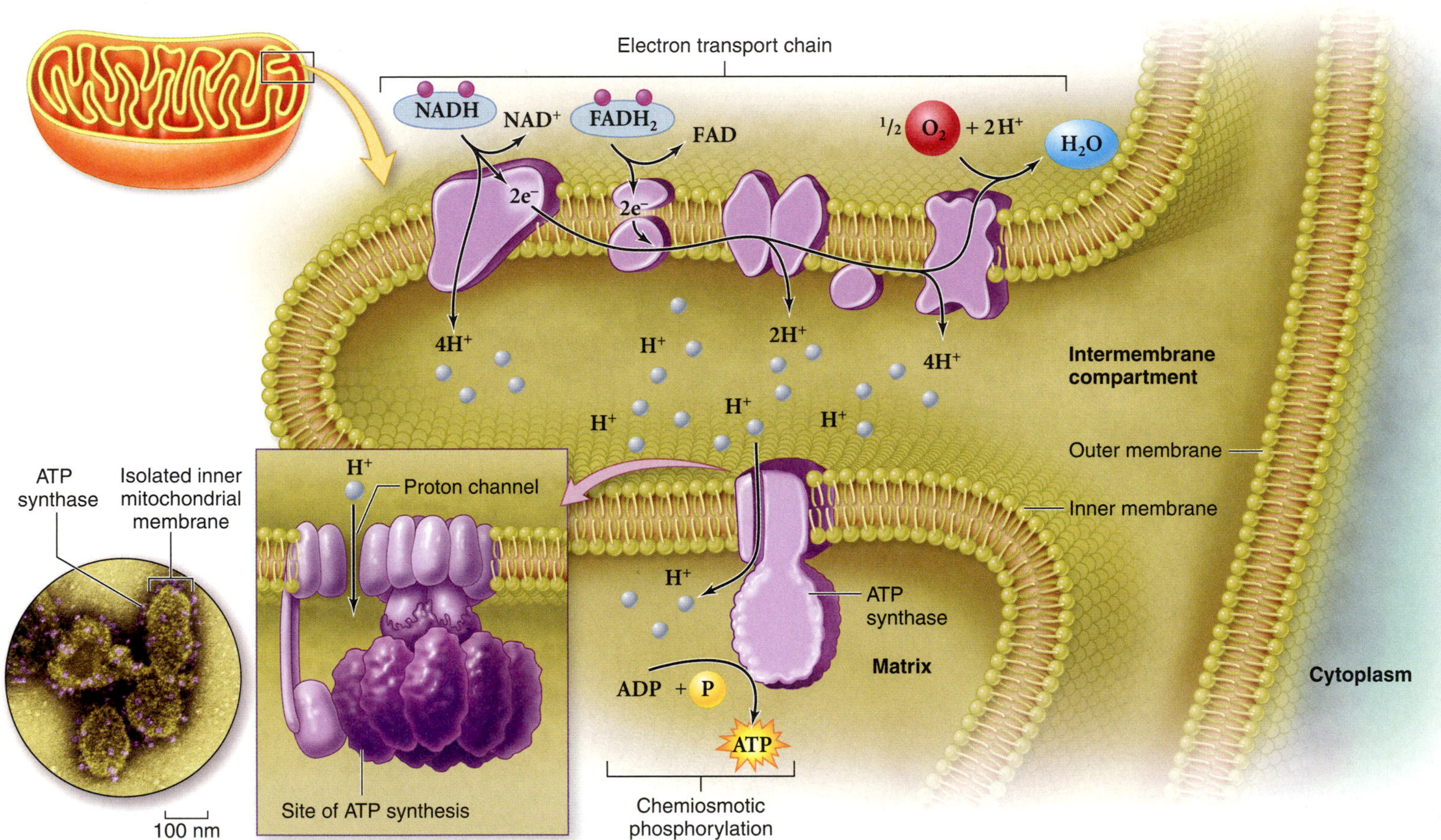

FIGURE 6.8 The Electron Transport Chain. Energy-rich electrons removed from NADH and $FADH_2$ slowly release their energy as they are transferred along the inner membrane of the mitochondrion. Membrane-bound enzymes use the energy to pump protons (H^+) from the matrix side of the inner mitochondrial membrane to the intermembrane compartment, establishing a gradient of charge, pH, and atoms across the membrane. As the protons pass through a channel in ATP synthase, ADP is phosphorylated to form ATP. The electron micrograph shows ATP synthase molecules, resembling lollipops, protruding from the surface of mitochondrial membrane fragments.

Can *You* Relate?

How Poisons Kill

Many toxic chemicals kill by blocking one or more reactions in respiration. Poisons are therefore the tools of murderous villains—and of biochemists. Since most individual molecules are too small to observe directly, biologists must take indirect routes to test hypotheses about the reactions of respiration. The judicious use of poisons in isolated cells (or even isolated mitochondria) can reveal much about the chemistry of the Krebs cycle and electron transport. The following lists a few examples of chemicals that kill by inhibiting respiration:

Krebs cycle inhibitor:

- Arsenic interferes with several essential chemical reactions. For example, arsenic binds to part of a biochemical needed for the formation of acetyl CoA. It therefore blocks the Krebs cycle.

Electron transport inhibitors:

- Rotenone stops an oxidation–reduction reaction early in the electron transport chain. Gardeners and reservoir managers use this pesticide to kill insects and unwanted fish.
- Mercury compounds are toxic for several reasons. For example, they inhibit electron transport near the same point as rotenone. Mercury is used in some thermometers, some types of lights, and many industrial applications.
- Cyanide blocks the final transfer of the electrons to O_2. This highly poisonous compound has no household uses, but it is used in mining and some other industries.
- Carbon monoxide (CO) blocks electron transport at the same point as cyanide. A byproduct of incomplete fuel combustion, toxic levels of CO in homes come from unvented heaters, stoves, and fireplaces. Car exhaust and cigarette smoke are other sources.

Chemiosmosis inhibitors:

- The insecticide 2,4-dinitrophenol (DNP) kills by making the inner mitochondrial membrane permeable to protons, blocking formation of the proton gradient necessary to drive ATP synthesis.
- Oligomycin blocks the phosphorylation of ADP by inhibiting the part of the ATP synthase enzyme that lets the protons through. Oligomycin is mostly used in laboratory studies of respiration.

how many ATPs can one molecule of glucose generate? To estimate the yield of ATPs, we can add the presumed maximum net number of ATPs generated from glycolysis, the Krebs cycle, and chemiosmotic phosphorylation. **Figure 6.9** summarizes this theoretical calculation.

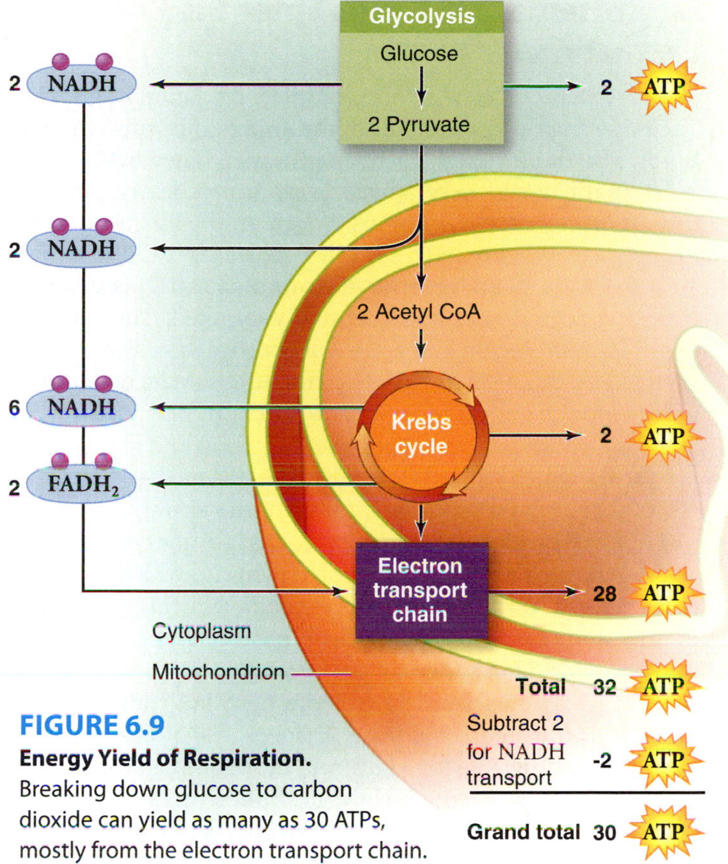

FIGURE 6.9
Energy Yield of Respiration. Breaking down glucose to carbon dioxide can yield as many as 30 ATPs, mostly from the electron transport chain.

Substrate-level phosphorylation yields two ATPs from glycolysis and two ATPs from the Krebs cycle (one ATP each from two turns of the cycle). These are the only steps that produce ATP directly. In addition, each glucose yields two NADH molecules from glycolysis, two NADHs from acetyl CoA production, and 6 NADHs and 2 $FADH_2$s from two turns of the Krebs cycle.

As you have just seen, most of the ATP generated from cellular respiration comes from chemiosmotic phosphorylation. The average net ATP yield from electron transport is about 2.5 ATPs per NADH and 1.5 ATPs per $FADH_2$. The 10 NADHs from glycolysis and the Krebs cycle therefore yield 25 ATPs; the two $FADH_2$s yield 3 more. Add the 4 ATPs from substrate-level phosphorylation and the total is 32 ATPs. However, NADH from glycolysis must be moved into the mitochondrion by active transport, usually at a cost of one ATP for each NADH. This reduces the net production of ATPs from a molecule of glucose to 30.

How efficient is aerobic respiration? The number of kilocalories stored in 30 ATPs is about 32% of the total kilocalories stored in the glucose bonds. The rest of the potential energy in glucose is lost as heat. This may seem wasteful, but for a biological process, it is reasonably efficient. To put this energy yield into perspective, an automobile uses only about 20% to 25% of the energy contained in gasoline's chemical bonds—the rest is lost as heat.

D. Proteins and Lipids Enter the Energy-Extracting Pathways

So far, we have focused on the complete oxidation of glucose. The other major components of the animal diet, proteins and lipids, also fit into these energy pathways (**figure 6.10**).

Usually, a cell uses amino acids from dietary proteins to manufacture more proteins. When an organism exhausts immediate carbohydrate supplies, however, cells may use amino acid as an energy source. Ammonia (NH_3) is stripped from the amino acid and eventually excreted. The remainder of each molecule enters the energy pathways as pyruvate, acetyl CoA, or an intermediate of the Krebs cycle, depending on the type of amino acid. **amino acids, p. 38**

Meanwhile, enzymes digest the fat in food into glycerol and fatty acids, which enter the bloodstream. Plants digest lipids as well. To fuel activities such as germination, for example, enzymes inside a seed break down oily triglycerides into glycerol and fatty acids. Enzymes convert the glycerol to pyruvate, which then proceeds through the rest of cellular respiration as though it came directly from glucose. The fatty acids enter cells and move into mitochondria, where they are cut into many two-carbon pieces that are released as acetyl CoA. From here, the pathways continue as they would for glucose. The reason that fats can store so much energy is that long fatty acid molecules can yield many acetyl CoA groups for the Krebs cycle. **lipids, p. 35**

This chapter's Burning Question (What does ephedra do to a person's metabolism?) considers how weight loss remedies such as diet pills influence the body's fat-burning pathways.

Summary *In eukaryotes, pyruvate from glycolysis enters a mitochondrion and is used to form acetyl CoA. This molecule enters the Krebs cycle. NADH and $FADH_2$ from glycolysis and the Krebs cycle donate electrons to the electron transport chain. All together, glycolysis, the Krebs cycle, and electron transport yield 30 ATP molecules per glucose. Cells can also use amino acids and fatty acids to generate ATP.*

6.5 MASTERING CONCEPTS

1. Pyruvate contains three carbon atoms; an acetyl group has only two. What happens to the other carbon atom?
2. How does the Krebs cycle generate CO_2, ATP, NADH, and $FADH_2$?
3. How do the electrons captured in NADH and $FADH_2$ power ATP formation?
4. What is the role of oxygen in the electron transport chain of aerobic respiration?
5. Explain how to arrive at the estimate that each glucose molecule yields 30 ATPs.
6. At which points in the energy pathways do digested proteins and fats enter?

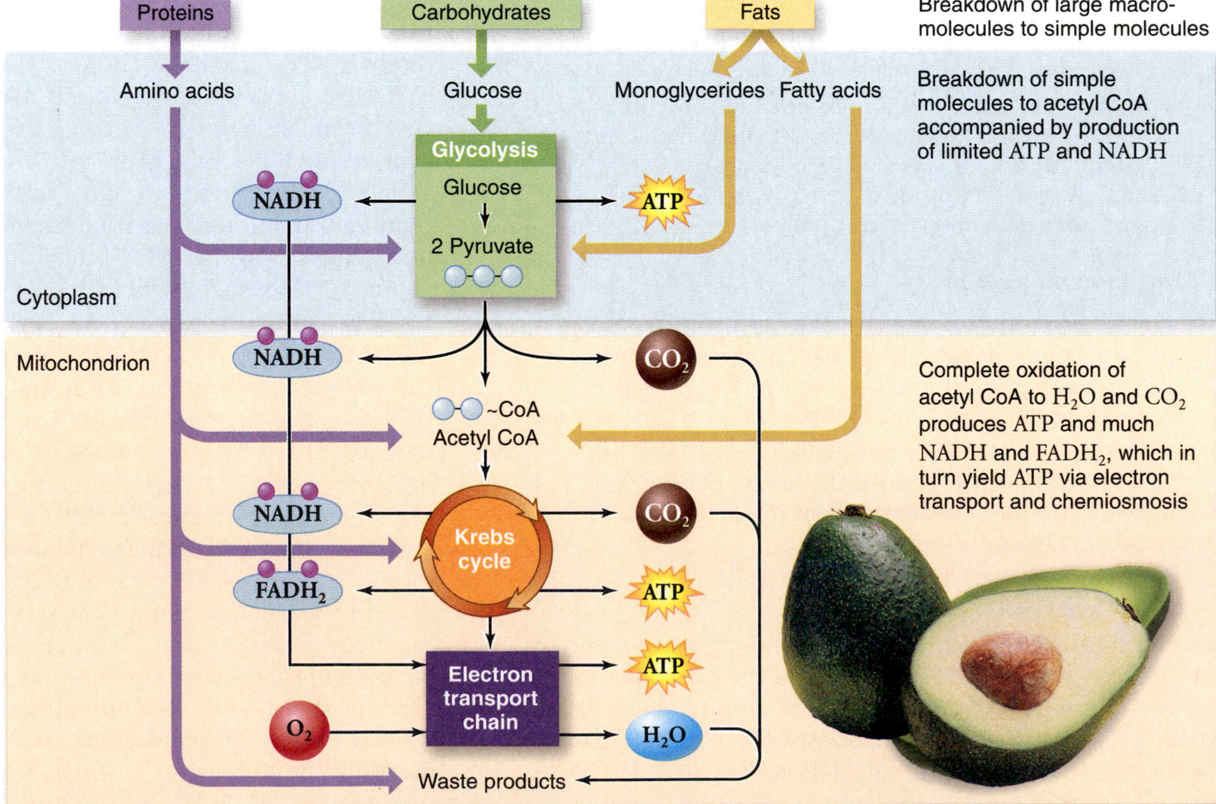

FIGURE 6.10
How Proteins and Fats Enter the Energy Pathways. Most cells use carbohydrates as a primary source of energy, but amino acids can enter the energy pathways after conversion to pyruvate, acetyl CoA, or intermediates of the Krebs cycle. Glycerol from fats is converted to pyruvate, and fatty acids are snipped into two-carbon fragments that become acetyl CoA.

Burning Questions

What does ephedra do to a person's metabolism?

Most over-the-counter weight loss remedies work in one of two ways: they either decrease appetite or increase metabolic rate. One of the most commonly used ingredients, phenylpropanolamine, suppresses appetite by an unknown mechanism. (It also treats stuffy noses.) Another common ingredient is ephedra, an herbal supplement derived from a plant native to Asia (*Ephedra sinica*, also called *ma huang*). **gymnosperms, p. 417**

Ephedra's reputation for promoting weight loss comes partly from its "thermogenic" (heat-releasing) properties. Thermogenesis is any form of metabolic heat production; shivering, for example, is an involuntary thermogenic response to cold. In experiments, subjects who ingested ephedra used more O_2 than those who did not, indicating an increased metabolic rate. Other studies showed that ephedra and caffeine together increased fat metabolism. (These studies, however, were limited to clinically obese subjects, and therefore did not include healthy, athletic people wanting to lose weight for cosmetic reasons).

Ephedra is also marketed as an "energy booster" because it stimulates the "fight-or-flight" part of the central nervous system, which controls heart rate and blood vessel dilation. Amphetamines, commonly found in some other weight loss aids, have the same effect.

These products, and their promises of effortless weight loss, may seem to be a dream come true for many people. Unfortunately, they have serious, and sometimes fatal, side effects. Phenylpropanolamine is associated with an increased risk of bleeding in the brain. Studies have linked ephedra to sometimes fatal seizures, strokes, and heart attacks. Therefore, in April 2004 the Food and Drug Administration banned the sale of both phenylpropanolamine and ephedra in the United States.

Have a Burning Question of your own? Submit it to marielle_hoefnagels@mcgraw-hill.com for possible inclusion in future editions of this book!

6.6 Anaerobic Respiration and Fermentation Do Not Require Oxygen

Most of the known organisms on Earth, including humans, are aerobes. We are so accustomed to our absolute need for oxygen that it is easy to forget that many habitats lack it. Waterlogged soils, deep puncture wounds, sewage treatment plants, and your own digestive tract are just a few of the places where life thrives without O_2. In the absence of oxygen, the microbes in these habitats generate ATP using anaerobic metabolic pathways. Two examples of such pathways are anaerobic respiration and fermentation (**figure 6.11**). Although many organisms use these pathways, neither extracts as much potential energy from glucose as does aerobic respiration.

A. Anaerobic Respiration Uses an Alternative Electron Acceptor

Anaerobic respiration is essentially the same as aerobic respiration, except that an inorganic molecule other than O_2 is the electron acceptor at the end of the electron transport chain. The number of ATPs generated per molecule of glucose depends on the electron acceptor, but it is always lower than the ATP yield for aerobic respiration.

Examples of alternative electron acceptors include NO_3^- (nitrate), SO_4^{2-} (sulfate), and even CO_2. Many prokaryotic organisms generate ATP in this way, and they play an important role in global nutrient cycles. For example, in waterlogged, oxygen-poor soils, bacteria that use NO_3^- as an electron acceptor begin a chain reaction that ends with the production of nitrogen gas (N_2). This gas drifts into the atmosphere, leaving the soil less fertile for plant growth. Bacteria that live in wetlands may use SO_4^{2-}, producing smelly hydrogen sulfide (H_2S) as a byproduct. And archaea living inside the intestinal tracts of cattle use CO_2 as an electron acceptor, generating methane gas (CH_4) as

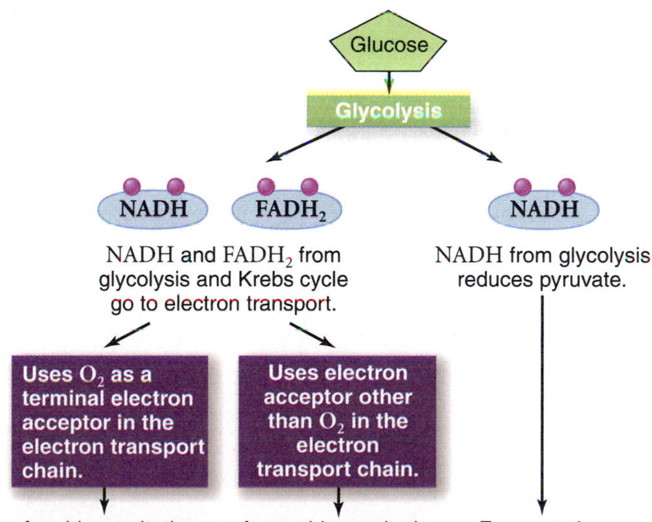

FIGURE 6.11 Alternative Metabolic Pathways. If oxygen is available, most organisms generate ATP in aerobic respiration. In the absence of O_2, however, some organisms use anaerobic respiration. Others use fermentation, which generates ATP only from glycolysis. Because fermentation is much less efficient than respiration, most fermenters occupy environments rich in carbohydrates.

a byproduct. The methane, which the cattle emit as belches and flatulence, is an important greenhouse gas. **carbon cycle, p. 816; global climate change, p. 849**

B. Fermenters Acquire ATP Only from Glycolysis

Some microorganisms, including many inhabitants of your digestive tract, are fermenters. In these organisms, glycolysis yields two ATPs, two NADHs, and two molecules of pyruvate per molecule of glucose, just as it does in every other cell. But the NADH does not donate its electrons to an electron transport chain. Instead, in **fermentation,** electrons from NADH reduce pyruvate. This process regenerates NAD^+ so that glycolysis can continue, but it also "wastes" the potential energy in both pyruvate and NADH, and it generates no additional ATP. Fermentation is therefore far less efficient than respiration.

Some microorganisms make their entire living by fermentation. An example is *Entamoeba histolytica,* a protist that causes a form of dysentery in humans. Others, including the gut-dwelling bacterium *E. coli,* use O_2 when it is available, but switch to fermentation when it is not. Most multicellular organisms, however, require too much energy to rely on fermentation exclusively.

Many fermentation pathways exist, often with industrial applications. Two of the most common products of fermentation are ethanol (an alcohol), CO_2, and lactic acid (**figure 6.12**).

In **alcoholic fermentation,** NADH reduces pyruvate to produce NAD^+, ethanol, and CO_2. Manufacturers of baked goods and alcoholic beverages owe their jobs to fermentation by yeast cells. Depending on the substance being fermented, the variety of yeast used, and whether carbon dioxide is allowed to escape during the process, fermentation may produce the airy texture of breads; wine or champagne from grapes; the syrupy drink called mead from honey; or cider from apples. Fermenting grain—barley, rice, or corn—produces beer, sake, whisky, and other spirits.

In **lactic acid fermentation,** a cell uses NADH to reduce pyruvate, but in this case, the products are NAD^+ and the three-carbon compound lactic acid. The bacterium *Lactobacillus,* for example, ferments the lactose in milk, producing the lactic acid that gives yogurt its sour taste.

Lactic acid fermentation also occurs in human muscle cells that are working so strenuously that they consume their available oxygen supply. In this "oxygen debt" condition (e.g., during the second half of a 100-m dash) the muscle cells can acquire ATP only from glycolysis. The cells use lactic acid fermentation to generate NAD^+ so that glycolysis can continue. If too much lactic acid accumulates, however, the muscle fatigues

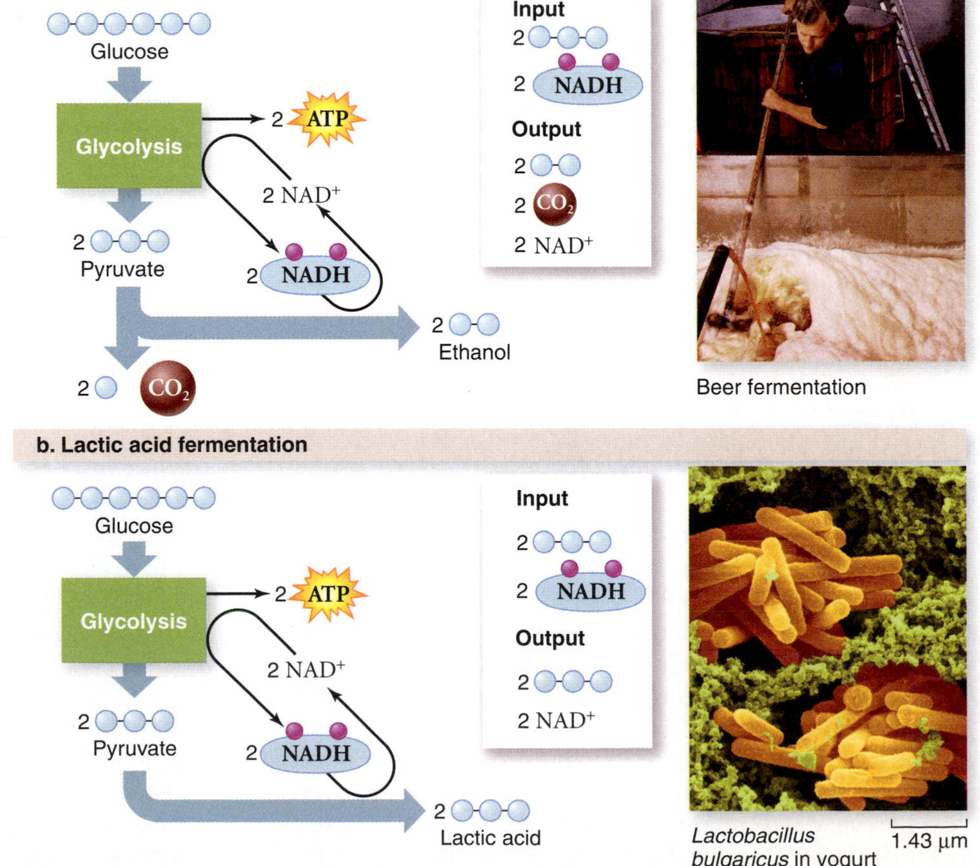

FIGURE 6.12 Fermentation. In fermentation, ATP comes only from glycolysis. Fermentative cells use NADH from glycolysis to reduce pyruvate and regenerate NAD^+, which allows glycolysis to continue. (**A**) The beer and wine industry relies on yeasts to produce ethanol and carbon dioxide. The man in the photograph is stirring a large vat of fermenting beer. (**B**) Lactic acid fermentation occurs in some bacteria and, occasionally, in mammalian muscle cells. The photograph shows *Lactobacillus bulgaricus* bacteria in yogurt.

and cramps. After the race, when the circulatory system catches up with the muscles' demand and oxygen is once again present, liver cells convert lactic acid back to pyruvate. Mitochondria then process the pyruvate as usual, by aerobic respiration.

Summary ***Anaerobic respiration uses an electron acceptor other than oxygen in the electron transport chain. Fermentation uses NADH from glycolysis to reduce pyruvate, yielding alcohol or lactic acid.***

6.6 MASTERING CONCEPTS

1. What are some examples of alternative electron acceptors used in anaerobic respiration?
2. How many ATP molecules per glucose does a fermenting organism produce?

6.7 Photosynthesis and Respiration Are Ancient Pathways

Photosynthesis, glycolysis, and cellular respiration are intimately related (**figure 6.13**). The organic product of photosynthesis—glucose—is the starting material for glycolysis. The oxygen released in photosynthesis becomes the final electron acceptor in aerobic respiration. CO_2 generated in the Krebs cycle enters the carbon reactions in chloroplasts. Finally, photosynthesis splits water produced by aerobic respiration, releasing electrons that replace those boosted out of chlorophyll molecules. Together, these biological energy reactions sustain life. How might they have arisen?

Glycolysis is probably the most ancient of the energy pathways because it is common to virtually all cells. Glycolysis evolved when the atmosphere lacked or had very little oxygen. These reactions enabled the earliest organisms to extract energy from simple organic compounds in the nonliving environment. Photosynthesis may have evolved from glycolysis; some of the reactions of the Calvin cycle are the reverse of some of those of glycolysis. Broadly speaking, glycolysis breaks down a 6-carbon compound into two 3-carbon compounds; some of the reactions of the Calvin cycle do the opposite. **Calvin cycle, p. 110**

The origins of photosynthesis remain unknown, but fossil evidence of cyanobacteria show that it arose at least 3.5 billion years ago. Once it started, photosynthesis altered life on Earth forever. First, organisms no longer depended on organic compounds in their surroundings for energy; they now had a way to produce nutrients constantly from sunlight. Second, photosynthesis released oxygen into the primitive atmosphere, paving the way for an explosion of new species capable of using this new atmospheric component. Third, O_2 from photosynthesis reacted with free oxygen atoms to produce ozone (O_3). As ozone accumulated high in the atmosphere, it blocked harmful ultraviolet radiation from reaching the planet's surface, which prevented some genetic damage and allowed new varieties of life to arise.

The first photosynthetic organisms could not have been plants, because such complex organisms were not present on the early Earth. Rather, photosynthesis may have debuted in an anaerobic cell that used hydrogen sulfide (H_2S) instead of water as an electron donor. These first photosynthetic microorganisms would have released sulfur, rather than oxygen, into the environment. Eventually, changes in pigment molecules enabled some of these organisms to use water instead of H_2S. Then, if a large cell engulfed one of those ancient microbes, it may have become a eukaryotic-like cell, complete with chloroplasts. Perhaps this was the ancestor of modern plants. Mitochondria might have evolved in a similar way, when larger cells engulfed bacteria capable of using oxygen. **endosymbiosis, p. 338**

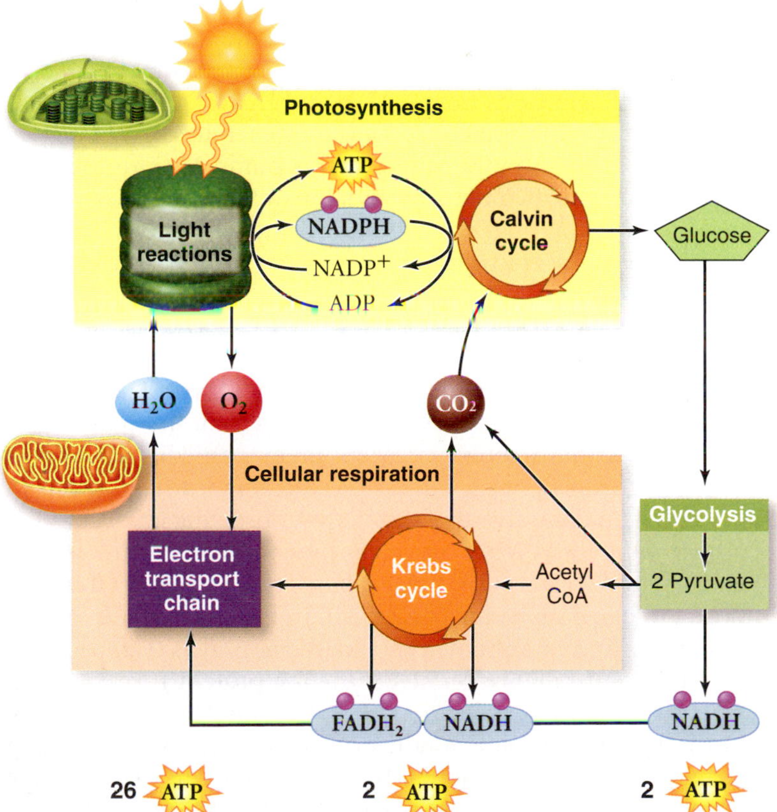

FIGURE 6.13 The Energy Pathways and Cycles Connect Life. An overview of energy metabolism illustrates how biological energy reactions are interrelated.

Afterward, different types of complex cells probably diverged, leading to the evolution of a great variety of eukaryotic organisms. Today, the interrelationships of the biological reactions of photosynthesis, glycolysis, and aerobic respiration and the great similarities of these reactions in diverse species demonstrate a unifying theme of biology: All types of organisms are related at the biochemical level.

Summary *Interactions and similarities among the reactions of photosynthesis and the energy-releasing pathways suggest a sequence in which they might have originated and evolved.*

6.7 MASTERING CONCEPTS

1. Which energy pathway is probably the most ancient? What is the evidence?
2. Why must the first metabolic pathways have been anaerobic?
3. What is the evidence that photosynthesis may have evolved from glycolysis?
4. How did photosynthesis forever alter life on Earth?

6.8 INVESTIGATING LIFE

Plants' "Alternative" Lifestyles Yield Hot Sex

Think of an organism that feels warm. Did you think of yourself? A puppy? Your cat? Chances are you thought of a mammal, or perhaps a bird, but certainly not a plant. Yet some plants, including *Philodendron*, do warm themselves (or at least their reproductive parts) to several degrees above ambient temperature (**figure 6.14**). How do they do it, and more importantly, what do they get out of it?

Philodendron flowers generate heat with a metabolic pathway involving the electron transport chain. As described in section 6.5, electrons from NADH and $FADH_2$ pass along a series of proteins embedded in the inner mitochondrial membrane. Along the way, the proteins pump H^+ into the space between the two mitochondrial membranes; ATP synthase uses the resulting proton gradient to generate ATP. The last protein in the electron transport chain dumps the electrons on oxygen, yielding water as a waste product.

Plants and a few other types of organisms have another pathway, unimaginatively dubbed "alternative oxidase," that diverts electrons from the electron transport chain. NADH and $FADH_2$ still donate electrons to a protein in the chain, but that electron acceptor transfers them immediately to O_2 instead of to the next carrier. (Researchers learned how alternative oxidase diverts electrons by observing that the pathway functions even if cyanide is present. As the Can You Relate? [How Poisons Kill] box explains, cyanide blocks the transfer of electrons to the last protein in the electron transport chain. Since the alternative oxidase pathway is cyanide-resistant, it must branch off before the last protein.)

The alternative oxidase pathway generates heat, but it does not help the mitochondrion produce ATP. Although this pathway reduces energy efficiency, it benefits the plant in several other ways. First, it regenerates NAD^+ and FAD even if the electron transport chain is restricted, ensuring that glycolysis and the Krebs cycle can continue. Second, it reduces production of harmful oxygen free radicals. Third, leaf mesophyll cells in CAM plants apparently use the pathway early in the morning. Stripping CO_2 from malate generates NADH, which the CAM plant metabolizes via the alternative oxidase pathway. **CAM plants, p. 113**

But none of those reasons explains what *Philodendron* gains by warming its flowers. One clue, however, comes from the observation that the plant heats *just* its flowers, and not its leaves, stem, or roots. Since flowers are reproductive parts, could the hot blooms somehow improve the plant's reproductive success?

Many plants give away free meals of sweet nectar to lure pollinators such as insects, birds, mammals. As the animal collects the offering, it brushes against the pollen-producing (male) flower parts. It then deposits the pollen on the female part of the next flower it visits. Australian researcher Roger Seymour and his colleagues wondered whether heat from *Philodendron solimoesense*'s flowers serve as a different kind of reward for its beetle pollinator (*Cyclocephala colasi*). They did a simple set of experiments to find out. First, they measured the temperature of *Philodendron* flowers over 24 hours. The central spike of the flower peaked at 40°C (104°F), about 15° above ambient temperature, while the surrounding floral chamber was consistently at least a few degrees warmer than the air around it.

Next, they used a device called a respirometer to measure the amount of CO_2 generated by active and resting beetles at a range of temperatures from 20°C to 35°C (68°F to 95°F). Since respiration generates CO_2 as a waste product, the respirometer indirectly measures how much

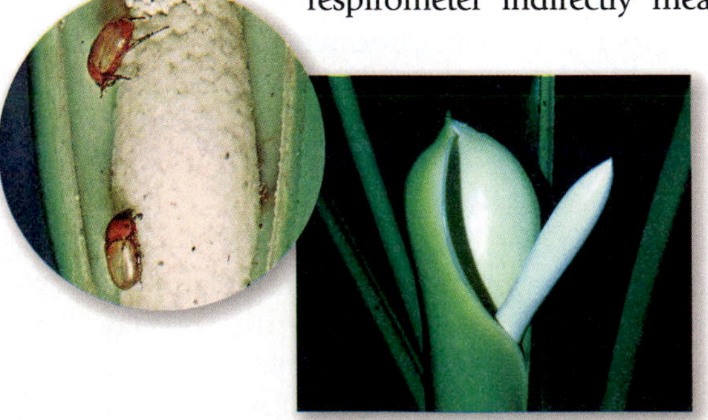

FIGURE 6.14 **Hot Bloom.** The central part of this *Philodendron solimoesense* flower lends much-needed warmth to its beetle pollinators.

energy an organism uses. Although resting beetles emitted approximately the same amount of CO_2 at all temperatures, active ones (such as those that would visit flowers) needed at least 10 times more energy at 20°C than at 30°C (**figure 6.15**).

Finally, the researchers used their data to calculate the "energy-saving factor" attributed to floral heat. They concluded that the beetles used 2.0 to 4.8 times more energy at ambient temperature than at the temperature of the warmed flower, depending on time of night. The beetles therefore save energy simply by loitering on or near the flowers, energy that they can use to find food or lure mates even as they pollinate the plant. The hot flowers—courtesy of the seemingly wasteful alternative oxidase pathway—therefore enhance the reproductive success of both *Philodendron* and the beetles.

Seymour, Roger S., Craig R. White, and Marc Gibernau. Nov. 20, 2003. Heat reward for insect pollinators. *Nature*, vol. 426, pages 243–244.

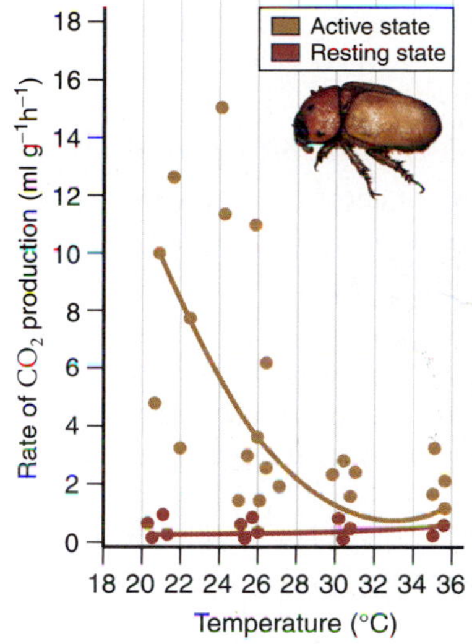

FIGURE 6.15 Energy Saver. Resting beetles respired at the same rate no matter what the temperature, but active beetles saved energy in warmer surroundings. **Question:** Suppose you hold one group of active beetles at 20°C, and another group at 30°C. After several hours, you place each beetle in a device that measures how far the animal can fly in a 20°C environment. Which group of beetles do you predict will be able to fly farther?

CHAPTER SUMMARY

6.1 Cells Use Energy in Food to Make ATP

- Every cell requires **ATP** to power reactions that require energy input.
- **Aerobic cellular respiration** is a biochemical pathway that extracts energy from glucose in the presence of oxygen.
- The overall reaction for cellular respiration is

$$C_6H_{12}O_6 + 6O_2 \longrightarrow 6CO_2 + 6H_2O + 30ATP$$

- In humans and many other animals, the respiratory system provides the oxygen that aerobic cellular respiration requires.
- Photosynthetic organisms such as plants use aerobic respiration to generate ATP.

6.2 Cellular Respiration Includes Three Main Processes

- Respiration gradually strips electrons from carbon atoms in glucose, eventually using them to reduce O_2.
- All cells begin energy release from nutrients with **glycolysis** in the cytoplasm. The **Krebs cycle** and an **electron transport chain** follow. In eukaryotes, these steps occur in **mitochondria.**
- In the electron transport chain, ATP synthesis occurs by **chemiosmotic phosphorylation,** in which a proton gradient powers phosphorylation of ADP to ATP by the enzyme **ATP synthase.**

6.3 In Eukaryotic Cells, Respiration Occurs in Mitochondria

- The two membranes of a mitochondrion enclose the **matrix,** where the Krebs cycle occurs.
- The electron transport chain establishes a proton gradient in the **intermembrane compartment,** and ATP synthase spans the inner membrane.

6.4 Glycolysis Breaks Down Glucose to Pyruvate

- In the first half of glycolysis, glucose is broken down into two molecules of the three-carbon compound PGAL.
- In the second half of glycolysis, the PGALs are oxidized as NAD^+ is reduced to **NADH.** They also contribute phosphate groups to form two ATPs by **substrate-level phosphorylation** (phosphate transfer between organic compounds) and are rearranged to form two molecules of **pyruvate**.

6.5 Aerobic Respiration Yields Much More ATP than Glycolysis Alone

A. The Krebs Cycle Produces ATP and Electron Carriers

- In the mitochondria, pyruvate is broken down into **acetyl CoA** in a coupled reaction that produces CO_2 and reduces NAD^+ to NADH.
- Acetyl CoA enters the Krebs cycle, a series of oxidation–reduction reactions that produces ATP, NADH, $FADH_2$, and CO_2. Substrate-level phosphorylation produces ATP in the Krebs cycle.

B. The Electron Transport Chain Drives ATP Formation

- Energy-rich electrons from NADH and $FADH_2$ fuel an electron transport chain. Electrons move through a series of carriers that release energy at each step. The terminal electron acceptor, oxygen, is reduced to form water.
- Electron transport energy establishes a proton gradient that pumps protons from the mitochondrial matrix into the intermembrane compartment. As protons

diffuse back into the matrix through channels in ATP synthase, their potential energy drives chemiosmotic phosphorylation of ADP to ATP.

C. How Many ATPs Can One Glucose Molecule Yield?

- For the combined pathways of aerobic respiration, each glucose molecule yields about 30 ATP molecules.

D. Proteins and Lipids Enter the Energy-Extracting Pathways

- Amino acids enter the energy pathways as pyruvate, acetyl CoA, or an intermediate of the Krebs cycle. Fatty acids enter as acetyl CoA, and glycerol enters as pyruvate.

6.6 Anaerobic Respiration and Fermentation Do Not Require Oxygen

A. Anaerobic Respiration Uses an Alternative Electron Acceptor

- In the absence of oxygen, some organisms can use an alternative electron acceptor such as nitrate or sulfate.

B. Fermenters Acquire ATP Only from Glycolysis

- **Fermentation** pathways oxidize NADH to NAD^+, which is recycled to glycolysis, but these pathways do not produce additional ATP. **Alcoholic fermentation** reduces pyruvate to ethanol and loses carbon dioxide. **Lactic acid fermentation** reduces pyruvate to lactic acid.

6.7 Photosynthesis and Respiration Are Ancient Pathways

- The energy pathways are interrelated, with common intermediates and some reactions that mirror the reactions of other pathways.
- Glycolysis may be the oldest energy pathway because it is most widespread; the other pathways are more specialized.
- The advent of photosynthesis forever changed the history of life on Earth.
- Eukaryotes may have arisen when larger cells engulfed prokaryotes that were forerunners to mitochondria and chloroplasts.

6.8 Investigating Life: Plants' "Alternative" Lifestyles Yield Hot Sex

- *Philodendron* plants use a modified respiratory pathway to create a "heat reward" for their insect pollinators.

MULTIPLE CHOICE QUESTIONS

1. Which of the following best describes *anaerobic* respiration?
   a. The production of ATP energy from glucose in the presence of oxygen
   b. The production of very little energy in the absence of oxygen
   c. The production of ATP energy from the sun in the presence of oxygen
   d. The production of ATP energy from glucose in the absence of oxygen

2. Which stage in cellular respiration directly requires the presence of oxygen?
   a. Glycolysis
   b. The Krebs cycle
   c. Electron Transport
   d. Both a and b

3. What is the role of ATP synthase?
   a. It uses ATP to make glucose
   b. It uses a hydrogen ion gradient to make ATP
   c. It uses ATP to make a hydrogen ion gradient
   d. It synthesizes ATP directly from glucose

4. How many ATP are made as a result of glycolysis?
   a. Two ATP are made
   b. Four ATP are made
   c. Two ATP are made, but two are consumed for a net gain of 0
   d. Four ATP are made, but the net gain is two

5. Which of the following molecules has the greatest amount of potential energy?
   a. Pyruvate
   b. Acetyl CoA
   c. Glucose
   d. CO_2

6. Which of the following molecules can be used to generate ATP energy?
   a. Carbohydrates
   b. Amino acids
   c. Lipids and fats
   d. All of the above

7. The difference between anaerobic and aerobic respiration is
   a. the amount of NADH that is produced.
   b. the electron carriers.
   c. the electron acceptors.
   d. the presence of $FADH_2$ instead of NADH.

8. Why is it important to regenerate NAD^+ during fermentation?
   a. It helps maintain the reactions of glycolysis
   b. So it can transfer an electron to the electron transport chain
   c. In order to maintain the alcohol levels of pyruvate in a cell
   d. In order to produce alcohol or lactic acid for the cell

9. Why is glycolysis considered to be the oldest metabolic reaction?
   a. Because glucose is a simple molecule
   b. Because it occurs in the absence of oxygen
   c. Because it is common to all cells
   d. Because it requires sunlight

10. What is *endosymbiosis*?
   a. A type of fermentation
   b. The transport of pyruvate into the matrix of the mitochondria
   c. A possible explanation for the origin of mitochondria
   d. The movement of electrons along the electron transport chain

TESTING YOUR KNOWLEDGE

1. How are breathing and cellular respiration similar? How are they different?
2. How do chemiosmotic phosphorylation and substrate-level phosphorylation each generate ATP? In which pathways does each occur?
3. Why does aerobic respiration yield more ATP from each glucose molecule than does glycolysis alone?
4. Cite a reaction or pathway that occurs in each of the following locations:
   a. cytoplasm
   b. mitochondrial matrix
   c. inner mitochondrial membrane
   d. intermembrane compartment
5. At what point does oxygen (O_2) enter the energy pathways of aerobic respiration? What is its role?
6. What energy pathways are available for cells living in the absence of O_2?
7. How are photosynthesis, glycolysis, and cellular respiration interrelated?

THINKING AS A SCIENTIST

1. Health-food stores sell a product called "pyruvate plus," which supposedly boosts energy. Why is this product unnecessary? What would be a much less expensive substitute that would accomplish the same thing?
2. A student runs 5 km each afternoon at a slow, leisurely pace. One day, she runs 2 km as fast as she can. Afterward she is winded and feels pain in her chest and leg muscles. She thought she was in great shape! What, in terms of energy metabolism, has she experienced?
3. In a properly functioning mitochondrion, is the pH in the matrix lower than, higher than, or the same as the pH in the intermembrane compartment? If you add one or more poisons described in this chapter's Can *You* Relate? How Poisons Kill, does your answer change?
4. Explain the fact that species as diverse as humans and yeasts use the same biochemical pathways to extract energy from nutrient molecules.
5. A chemical works as a disinfectant by poking holes in bacterial cell membranes. Why would this stop the cells from making ATP? Why would the inability to make ATP kill a cell?
6. Which of the following processes stops in the absence of oxygen: glycolysis, Krebs cycle, electron transport chain, fermentation. What assumptions are you making in your answer?
7. A seed is a plant embryo packaged with a food supply. Soaking a seed in water prompts the embryo to begin respiring, metabolizing its food supply to fuel its growth. Suppose that Anna has 50 soaked seeds. She boils half of them, killing their embryos, and lets them return to room temperature. She then places the dead seeds in one container and live seeds in another. If she later measures the temperature in the two containers, will they be different? Explain your answer.
8. Birds and mammals are endotherms; they maintain a constant internal body temperature no matter whether the environment is cold or hot (within limits, of course). An endotherm that gets too cold will increase its metabolic rate to generate heat. An ectothermic animal such as a reptile, on the other hand, allows its body temperature to fluctuate with the environment. If you own a pet rat and a pet snake of equal weight, which will require more food, and why?

Visit www.mhhe.com/hoefnagels for practice quizzes, animations, videos, and activities designed to help you master the material in this chapter.

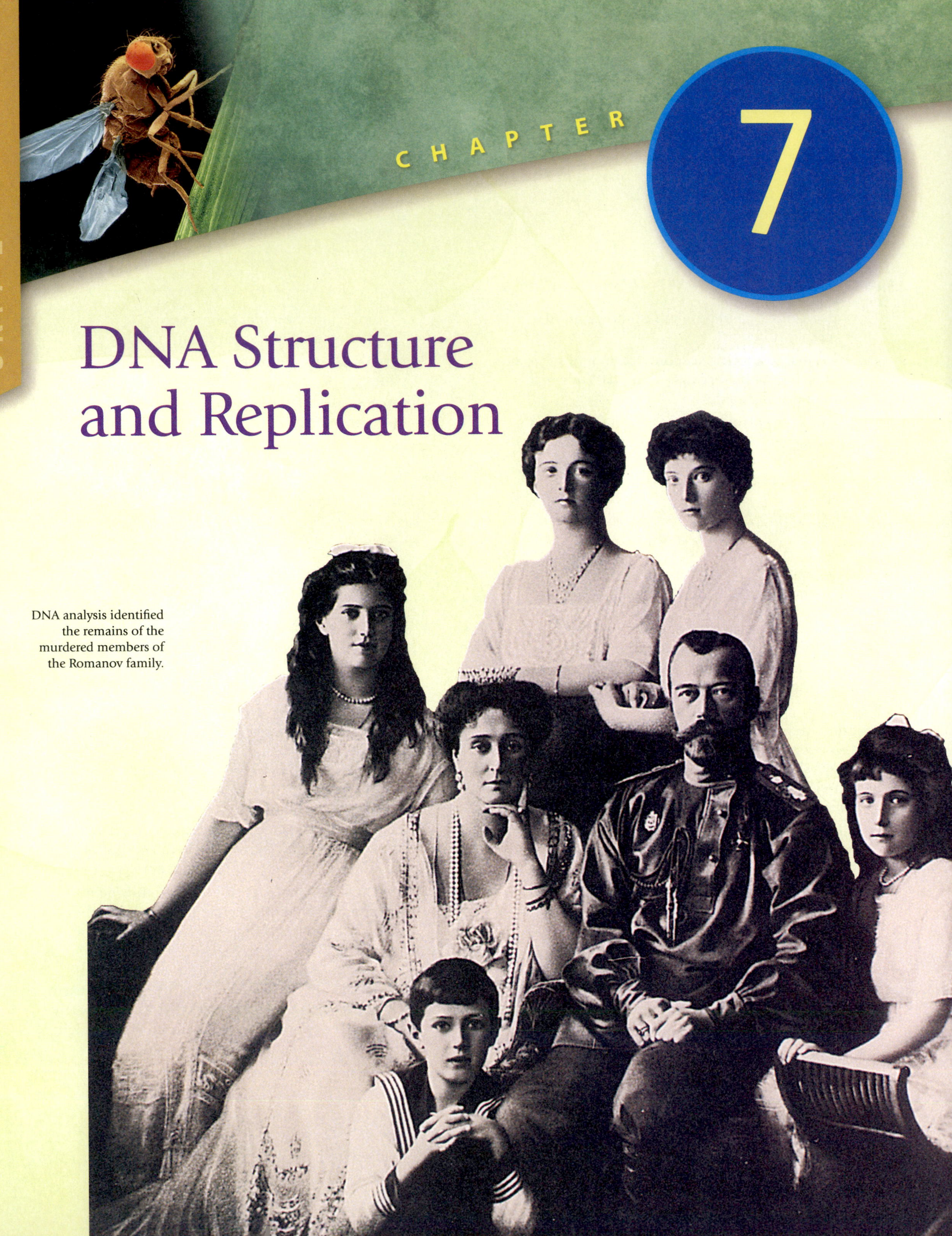

CHAPTER 7

DNA Structure and Replication

DNA analysis identified the remains of the murdered members of the Romanov family.

What's the Point?

DNA Analysis Solves a Royal Mystery

One night in July 1918, Tsar Nicholas II of Russia and his family met gruesome deaths at the hands of Bolsheviks in Siberia. Captors led the tsar and tsarina, their four daughters and one son, the family physician, and three servants to a cellar and shot and bayoneted them. The executioners then stripped the bodies, placed them in a shallow grave, and poured sulfuric acid over their faces so they could not be identified.

In July 1991, two amateur historians found the grave and alerted the government that they might have unearthed the long-sought bodies of the Romanov family. Researchers soon determined that the 1000 pieces of bone at the scene came from nine individuals. The sizes of the skeletons indicated that three were children. The porcelain, platinum, and gold in some of the teeth suggested that some of the people were royalty. The remains were so badly damaged, however, that some conventional forensic tests were not possible. Yet one very valuable type of evidence survived—DNA.

The researchers first used DNA specific to the Y chromosome, which occurs only in males, to distinguish male from female skeletons. Then, they used DNA from mitochondria—organelles inherited only from the mother—to identify Tsarina Alexandra and her children. Moreover, the tsarina's DNA had some key sequences that matched those of England's Prince Philip, a modern member of the same royal family.

The next step was to prove that a particular male skeleton belonged to the tsar. The researchers knew that the tsar had a brother, Grand Duke of Russia Georgij Romanov, who had died in 1899 of tuberculosis. His body was exhumed in July 1994, and researchers sequenced DNA from mitochondria in bone cells from his leg. They also extracted mitochondrial DNA from the male skeleton. When they compared the sequences, the researchers found a match! They calculated the probability that the remains are those of the tsar, rather than resembling Georgij by chance, as 130 million to 1. The mystery of the Russian royal family's demise was solved.

Much more than a tool for solving century-old puzzles, DNA holds in its sequence the recipe for life. This chapter explains how cells copy this intriguing molecule, enabling them to pass those critical instructions to the next generation.

LEARNING OUTLINE

7.1 Experiments Identified the Genetic Material

A. Griffith Discovered that Bacteria Can Transfer Genetic Information
B. Avery, MacLeod, and McCarty Showed that Genetic Information Is DNA
C. Hershey and Chase Confirmed the Genetic Role of DNA

7.2 DNA Is a Double Helix That Encodes "Recipes" for Proteins

A. Biochemists and Physicists Discovered DNA's Structure
B. DNA Contains the Information Needed for Life's Functions

7.3 DNA Replication Maintains Genetic Information

A. Replication Requires Many Enzymes
B. Mutations May Occur During Replication

7.4 PCR Replicates DNA in a Test Tube

7.5 DNA Sequencing Reveals the Order of Bases

A. Sanger Used Fragments to Determine a DNA Sequence
B. DNA Microarrays Speed Sequencing

7.6 DNA Profiling Has Many Applications

7.7 Investigating Life: Genetic Messages from the Dead Tell Tales of Ancient Ecosystems

7.1 Experiments Identified the Genetic Material

The nucleic acid **DNA (deoxyribonucleic acid)** is one of the most familiar molecules, the subject matter of movies and headlines (**figure 7.1**). Fictional scientists reconstruct dinosaurs from DNA preserved in an ancient mosquito's gut. Criminal trials hinge on DNA evidence; the idea of cloning animals raises questions about the role of DNA in determining who we are; and DNA-based discoveries are yielding new diagnostic tests, treatments, and vaccines.

More important than DNA's role in society today is its role in life itself. Of all the characteristics that distinguish the living from the nonliving, the one most important to the continuance of life is the self-replicating cell. At the molecular level, reproduction depends on DNA, a biochemical that has dual abilities. First, it directs the activities of the cell by controlling protein synthesis. Second, it manufactures an exact replica of itself, copying those instructions for the next generation of cells.

The recognition of DNA's vital role in life was a long time in coming. By the early 1900s, researchers had recognized the connection between inheritance and protein. For example, the English physician, Archibald Garrod, noted that people with inherited "inborn errors of metabolism" lacked certain enzymes. Other researchers added supporting evidence: they linked abnormal or missing enzymes to unusual eye color in fruit flies and nutritional deficiencies in bread mold. But how were enzyme deficiencies and inheritance linked? Experiments in bacteria would answer the question.

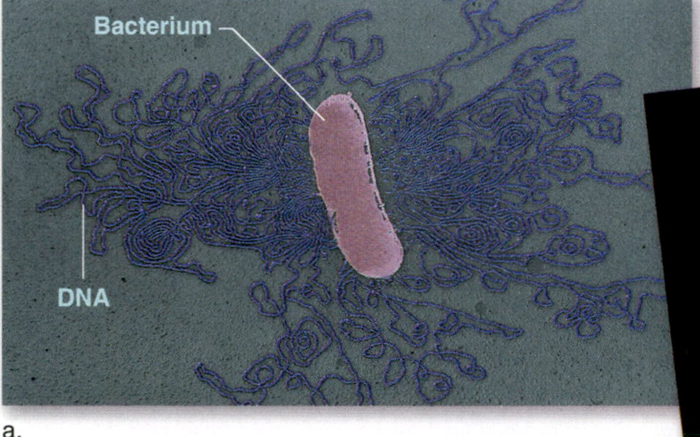

a.

b.

FIGURE 7.1 DNA—The Molecule in the Media. (**A**) DNA bursts forth from this treated bacterial cell, illustrating just how much DNA is tightly wound into a single cell. (**B**) *Jurassic Park* was a 1993 blockbuster movie in which fictional scientists recreated dinosaurs. The dinosaur DNA came from blood found in ancient mosquitoes entombed in amber.

A. Griffith Discovered that Bacteria Can Transfer Genetic Information

In 1928, English microbiologist Frederick Griffith contributed the first step in identifying DNA as the genetic material. Griffith studied mice with pneumonia caused by a bacterium, *Streptococcus pneumoniae.* He identified two types of bacteria: type R and type S (**figure 7.2**). Type R bacteria form rough-shaped colonies, and when injected into mice, they do not cause pneumonia. Type S bacteria form smooth colonies because they are encased in a polysaccharide capsule. When injected into mice, type S bacteria cause pneumonia. Therefore, the smooth polysaccharide coat seemed to be necessary for infection.

When Griffith heated type S bacteria ("heat-killing" them) and injected them into mice, they no longer caused pneumonia. However, when he injected mice with a mixture of type R bacteria plus heat-killed type S bacteria, neither of which was able to cause pneumonia alone, the mice died of pneumonia. Their bodies contained live type S bacteria encased in polysaccharide. How had the previously harmless bacteria acquired the ability to cause disease? In the 1940s, U.S. physicians Oswald Avery, Colin MacLeod, and Maclyn McCarty offered an explanation.

B. Avery, MacLeod, and McCarty Showed that Genetic Information Is DNA

Avery, MacLeod, and McCarty hypothesized that something in the heat-killed type S bacteria entered and "transformed" the normally harmless type R strain into a killer. Was this "transforming principle" a protein? Treating the solution from the type S strain with a protein-destroying enzyme (a protease) failed to keep the type R strain from being transformed into a killer (**figure 7.3**). Therefore, a protein was not responsible for transmitting the killing trait. Treating the solution from the heat-killed

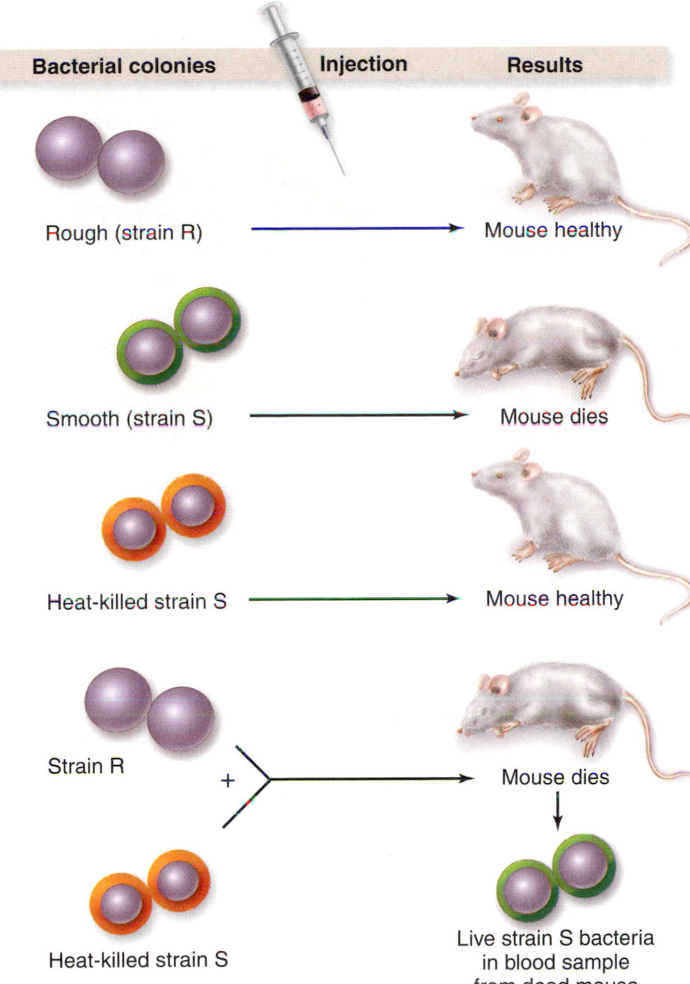

FIGURE 7.2 A Tale of Two Microbes. Griffith's experiments showed that a molecule in a lethal strain of bacteria (type S) could transform nonkilling bacteria (type R) into killers.

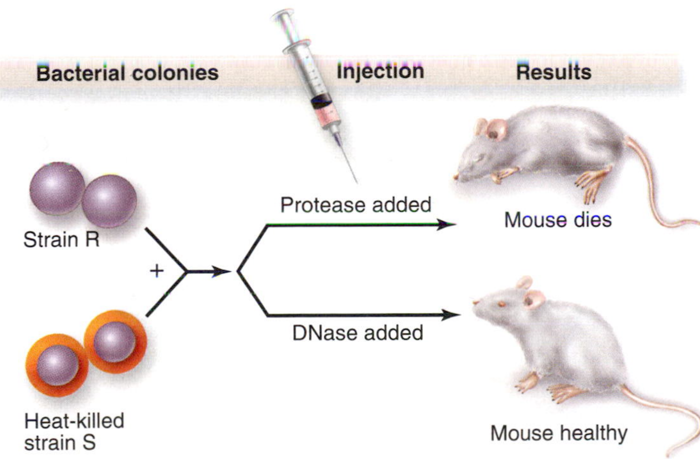

FIGURE 7.3 DNA Is the "Transforming Principle." Avery, MacLeod, and McCarty identified Griffith's transforming principle as DNA. By adding enzymes that destroy either proteins (protease) or DNA (DNase) to the types of mixtures that Griffith used in his experiments, they demonstrated that DNA, and not protein, transforms bacteria.

S bacteria with a DNA-destroying enzyme (DNase) first, however, prevented the killing ability.

Avery, MacLeod, and McCarty confirmed that DNA transformed the bacteria by isolating DNA from heat-killed type S bacteria and injecting it along with type R bacteria into mice. The mice died, and their bodies contained active type S bacteria. The conclusion: Type S DNA altered the type R bacteria, enabling them to manufacture the smooth coat necessary to cause infection.

At first, biologists hesitated to accept DNA as the biochemical of heredity. They knew more about proteins than about nucleic acids. They also thought that protein, with its 20 building blocks, was able to encode many more traits than DNA, which includes just four types of building blocks. In 1950, however, U.S. microbiologists Alfred Hershey and Martha Chase conclusively showed that DNA—not protein—is the genetic material.

C. Hershey and Chase Confirmed the Genetic Role of DNA

Hershey and Chase used a very simple system; they infected the bacterium *Escherichia coli* with a bacteriophage called T4. A bacteriophage is a virus that infects only bacteria; most consist of only a protein coat and a nucleic acid core (DNA in this case). We now know that when the virus infects the bacterial cell, it injects its DNA, and the protein coat remains loosely attached to the bacterium (**figure 7.4**). The

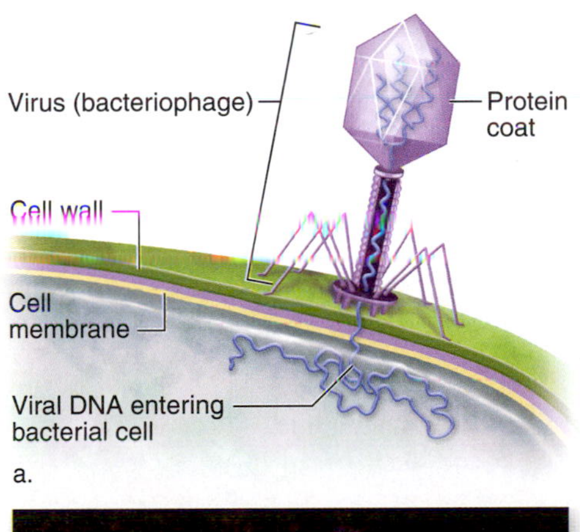

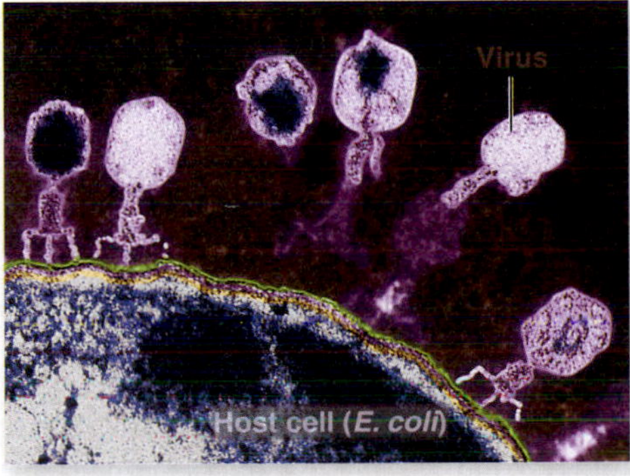

FIGURE 7.4 A Virus (Bacteriophage) Infects a Bacterium. (**A**) A bacteriophage is a virus that infects only bacteria. It consists of a nucleic acid in a protein coat. The virus uses the protein coat to attach to and inject its DNA into a bacterial cell. (**B**) This photo shows several bacteriophages infecting a bacterium.

viral DNA uses the bacterial cell's energy and raw materials to manufacture more of itself. New virus particles then burst from the cell. Much of this information was not available in 1950. **bacteriophages, p. 356; Focus on Model Organisms (*E. coli*), p. 378**

Hershey and Chase wanted to know which part of the virus controls its replication: the DNA or the protein coat. The researchers "labeled" two batches of viruses, one with radioactive sulfur that marked protein, and the other with radioactive phosphorus that marked DNA. They used each type of labeled virus to infect a separate batch of bacteria and allowed several minutes for the virus particles to bind to the bacteria and inject their DNA into them (**figure 7.5**). Then they agitated each mixture in a blender, which removed the unattached viruses and empty protein coats from the surfaces of the bacteria. They poured the mixtures into test tubes and spun them at high speed. This settled the infected bacteria at the bottom of each test tube because they were heavier than the liberated viral protein coats.

Hershey and Chase examined the contents of the bacteria that had settled to the bottom of each tube. In the test tube containing sulfur-labeled virus, the virus-infected bacteria were not radioactive, but the fluid portion of the material in the tube was. In the other tube, where the virus contained radioactive phosphorus, the infected bacteria were radioactive, but the fluid was not. The "blender experiments" therefore showed that the part of the virus that could enter the bacteria and direct them to mass produce more viruses was the part with the phosphorus label—namely, the DNA. The genetic material, therefore, was DNA and not protein.

Summary *A series of experiments revealed DNA, and not protein, to be the genetic material. Further experiments provided clues to its structure.*

7.1 MASTERING CONCEPTS

1. How did Griffith's research, coupled with the work of Avery and his colleagues, demonstrate that DNA, not protein, is the genetic material?
2. How did the Hershey-Chase "blender experiments" confirm Griffith's results?

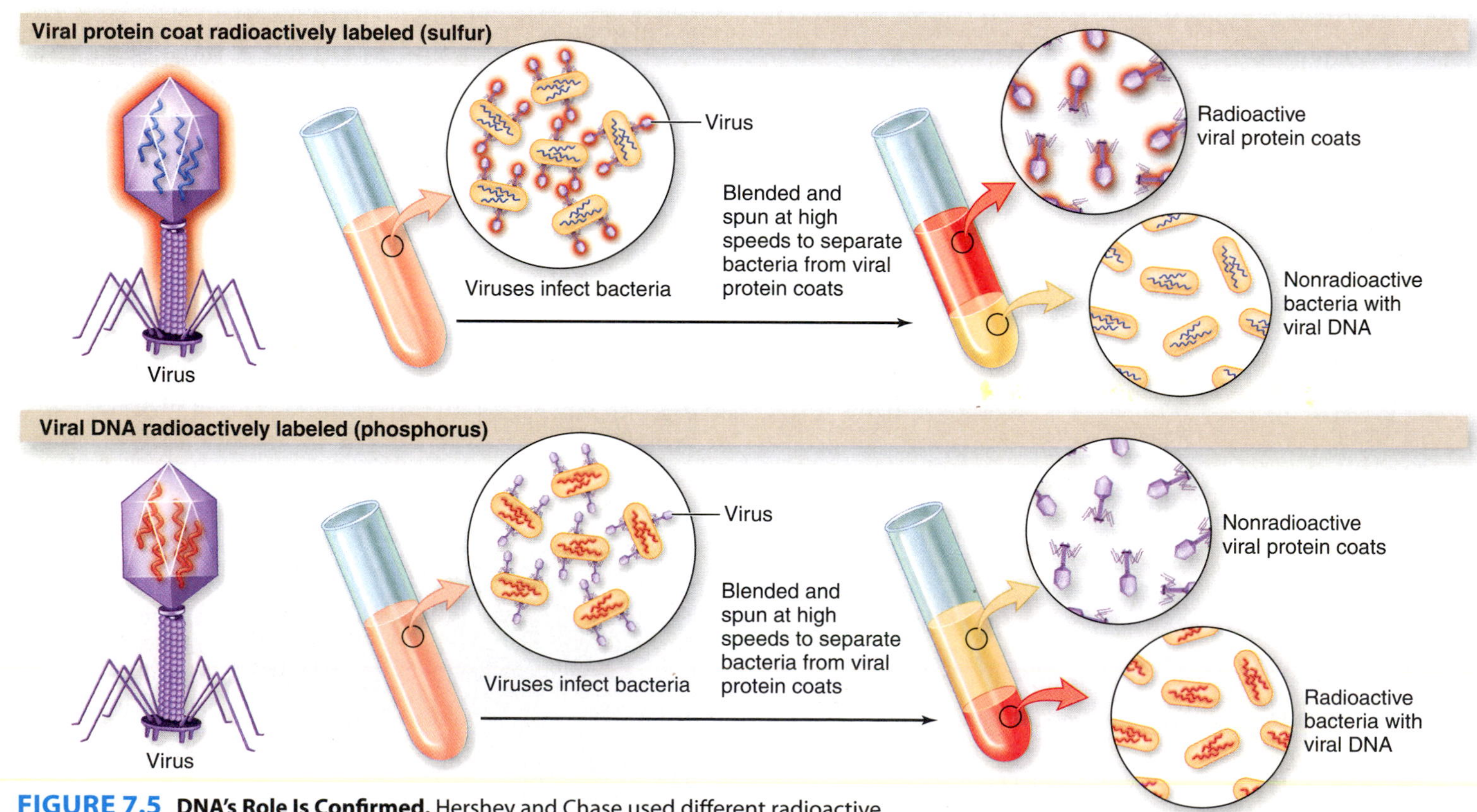

FIGURE 7.5 DNA's Role Is Confirmed. Hershey and Chase used different radioactive isotopes to distinguish the viral protein coat from the DNA. Their experiments showed that the virus transfers DNA (not protein) to the bacterium, and that viral DNA alone could cause bacterial cells to produce viruses.

7.2 DNA Is a Double Helix That Encodes "Recipes" for Proteins

The early twentieth century also saw corresponding advances in the study of the structure of DNA. By 1929, biochemists had discovered the distinction between ribonucleic acid (RNA) and DNA, the two types of nucleic acid. Later, they determined that **nucleotides**, the building blocks of nucleic acids, included sugars, nitrogen-containing groups, and phosphorus-containing components. Another important clue was the observation that DNA and RNA nucleotides always contain the same sugars and phosphates, but they may contain any one of four different nitrogen-containing bases.

A. Biochemists and Physicists Discovered DNA's Structure

In the early 1950s, two lines of evidence together revealed DNA's chemical structure. Austrian-American biochemist Erwin Chargaff showed that DNA contains equal amounts of the bases adenine (A) and thymine (T) and equal amounts of the bases guanine (G) and cytosine (C). English physicist Maurice Wilkins and chemist Rosalind Franklin bombarded DNA with X-rays, using a technique called X-ray diffraction to determine the three-dimensional shape of the molecule. The X-ray diffraction pattern revealed a regularly repeating structure of building blocks (**figure 7.6**).

Watson and Crick's Model Fits the Data

In 1953, U.S. biochemist James Watson and English physicist Francis Crick, working at the Cavendish laboratory in Cambridge in the United Kingdom, used these clues to build a ball-and-stick model of the DNA molecule. The now familiar double helix included equal amounts of G and C and of A and T, and it had the sleek symmetry revealed in the X-ray diffraction pattern.

The DNA double helix resembles a twisted ladder (**figure 7.7**). The twin rails of the ladder, also called the

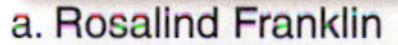

a. Rosalind Franklin

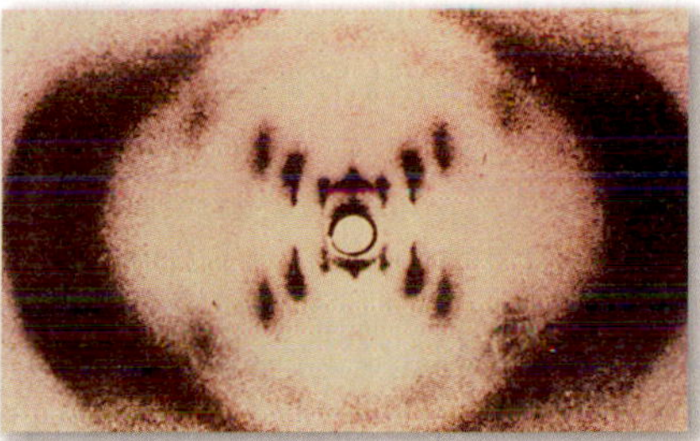

b. X-ray diffraction

c.

FIGURE 7.6 Discovery of DNA's Structure. (**A**) Rosalind Franklin produced high-quality X-ray images of DNA (**B**) that were crucial in the discovery of DNA's structure. (**C**) Maurice Wilkins, Francis Crick, and James Watson (first, third, and fifth from the left) shared the Nobel Prize in physiology or medicine for their now-famous discovery. Franklin had died in 1958, and by the rules of the award, she could not be included.

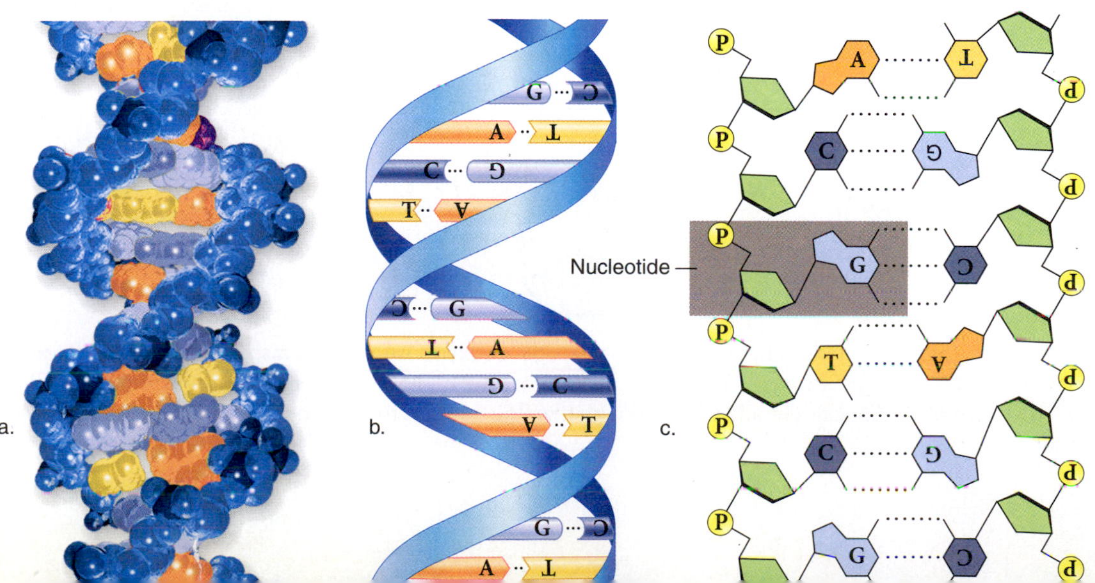

FIGURE 7.7 Three Ways to Represent DNA. (**A**) This photo is a space-filling model of a DNA molecule that shows the three-dimensional relationships of the component atoms. (**B**) This representation shows the helical structure of DNA. The sugar–phosphate rails are identical in all DNA molecules and run in opposite directions. (**C**) The helix is unwound to show the base pairs and the sugar–phosphate backbone in more detail.

sugar-phosphate "backbones," are alternating units of deoxyribose and phosphate joined with covalent bonds. The ladder's rungs are A–T and G–C base pairs joined by hydrogen bonds.

DNA Has Paired Bases in Complementary Strands

The A–T and G–C pairs arise from their chemical structures (**figure 7.8**). Adenine and guanine are purines, bases with a double ring structure. Cytosine and thymine are pyrimidines, which have a single ring. Each A–T pair is the same width as a C–G pair because each includes a purine and a pyrimidine.

The two strands of a DNA molecule are **complementary** to each other because the sequence of each strand defines the sequence of the other; that is, an A on one strand means a T on the opposite strand, and a G on one strand means a C on the other. Complementary base pairing is also the basis of gene function, as chapter 12 describes.

Although they are parallel to each other, the two chains of the DNA double helix are oriented in opposite directions, like the northbound and southbound lanes of a highway (**figure 7.9**). This head-to-tail arrangement is apparent when the carbon atoms in deoxyribose are numbered. Opposite ends of each strand are designated "3 prime" (3′) and "5 prime" (5′). At the same end of the double helix, one chain ends in the 3′ carbon, while the opposite chain ends in the 5′ carbon.

B. DNA Contains the Information Needed for Life's Functions

The amount of DNA in any cell is immense; in humans, each nucleus contains some 6.4 billion base pairs. An organism's **genome** is all of the genetic material in its cells. The genome of a prokaryotic cell consists of one circular DNA molecule (see figure 7.1A). In a eukaryotic cell, the genome is divided into multiple **chromosomes**, long DNA molecules that associate closely with proteins. The mitochondria and chloroplasts of eukaryotic cells also contain DNA (see chapter 3).

What does all of that DNA do? Much of it has no known function, but some of it encodes the cell's RNA and proteins. A **gene** is a sequence of DNA nucleotides that codes for a specific protein or RNA molecule; the human genome includes 20,000 to 25,000 genes scattered on its 23 pairs of chromosomes.

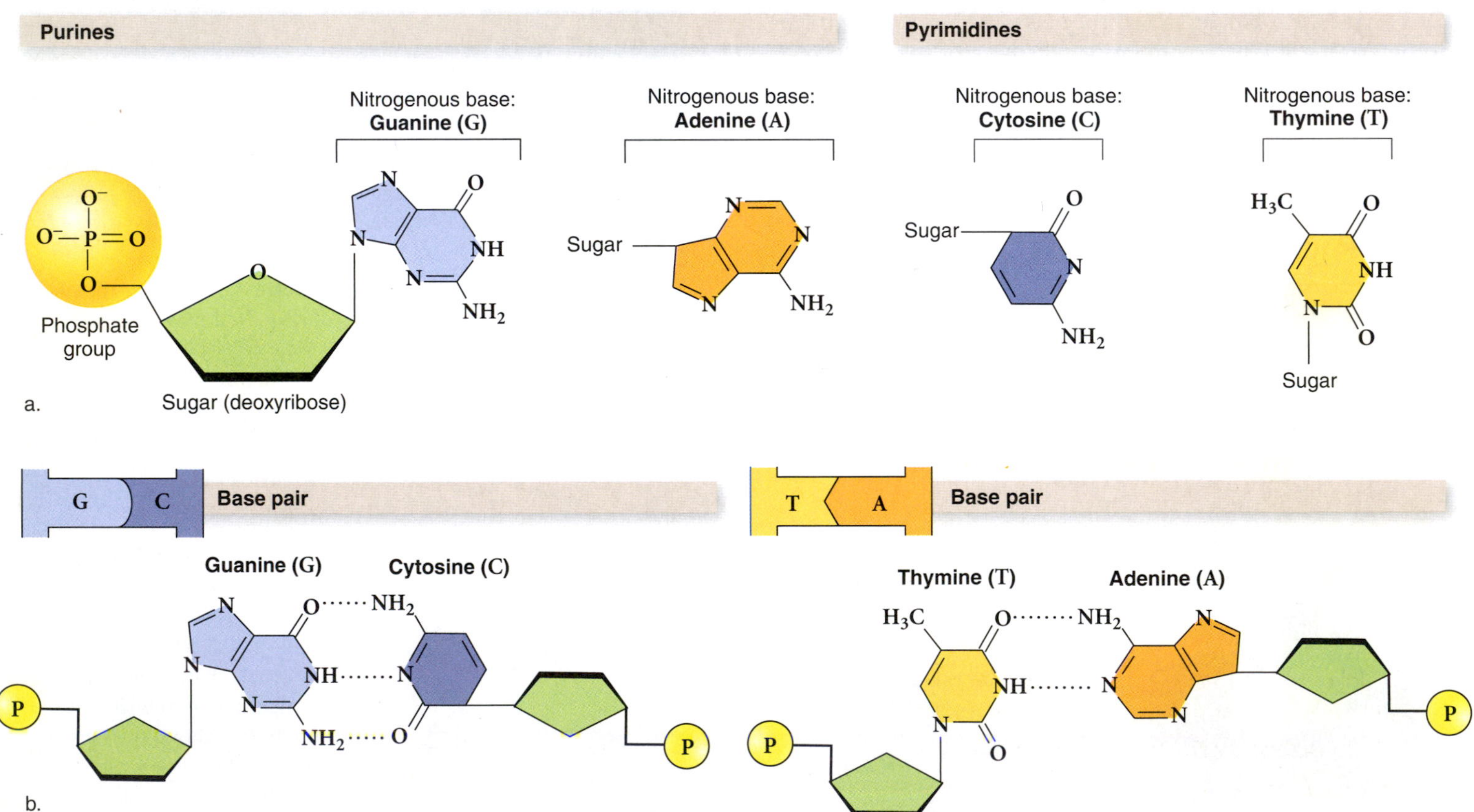

FIGURE 7.8 **DNA Base Pairing.** (**A**) In a nucleotide, a nitrogenous base is joined to the sugar deoxyribose, which is also bonded to a phosphate. Guanine and adenine are purines, each composed of two organic rings. Cytosine and thymine are pyrimidines, each built of one six-membered ring. (**B**) In the DNA double helix, purines pair with pyrimidines. Specifically, adenine pairs with thymine through two hydrogen bonds, and cytosine pairs with guanine through three hydrogen bonds.

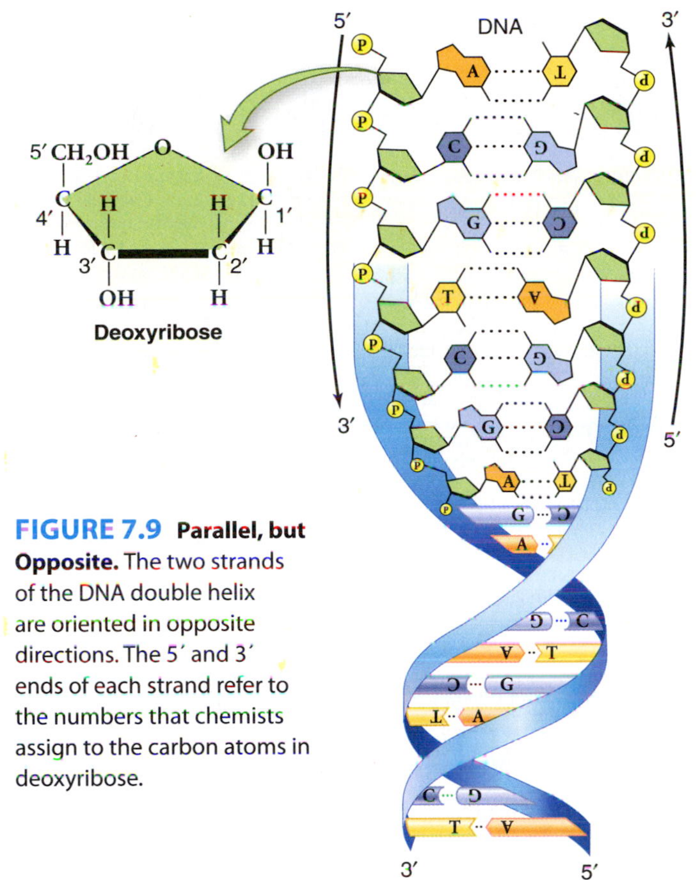

FIGURE 7.9 Parallel, but Opposite. The two strands of the DNA double helix are oriented in opposite directions. The 5′ and 3′ ends of each strand refer to the numbers that chemists assign to the carbon atoms in deoxyribose.

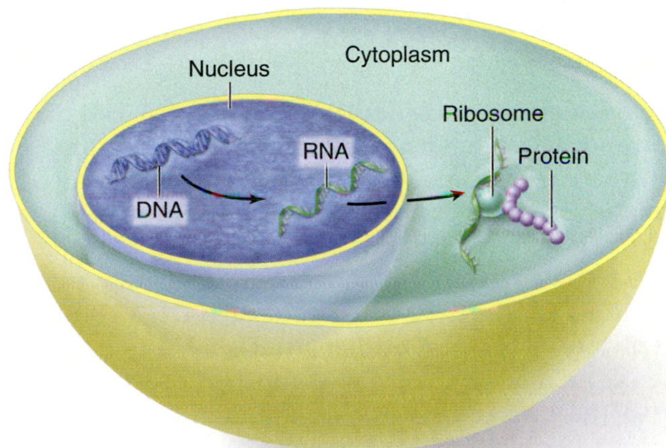

FIGURE 7.10 Genes Code for Proteins. This simplified figure illustrates DNA's function. In the nucleus, enzymes copy a gene to a molecule of RNA. The RNA leaves the nucleus and binds to a ribosome, where the cell uses the sequence of nucleotides in RNA to create a particular sequence of amino acids—a protein.

To illustrate DNA's function, suppose a cell in a female mammal's breast is producing milk to feed an infant (see figure 3.15). One of the many proteins in milk is albumin. The steps below summarize the production of albumin, starting with its genetic "recipe" (**figure 7.10**):

1. Inside the nucleus, an enzyme first copies the albumin gene's DNA sequence to a complementary sequence of RNA.
2. After some modification, the RNA emerges from the nucleus and binds to a ribosome in the cytoplasm.
3. At the ribosome, amino acids are assembled in a specific order to produce the albumin protein, dictated by the sequence of nucleotides in the RNA molecule.

In this case, the result is albumin, but the sequence of nucleotides in DNA specifies the sequence of amino acids in every other protein in the cell as well. The details of this process are the subject of chapter 12.

Summary *The DNA molecule is a double helix with sugar–phosphate rails and pyrimidine–purine pairs as rungs. The two strands of a DNA molecule are oriented in opposite directions. Genes are portions of DNA that code for specific proteins or RNA molecules.*

7.2 MASTERING CONCEPTS

1. What are the components of DNA and its three-dimensional structure?
2. What evidence enabled Watson and Crick to decipher the structure of DNA?
3. How does a gene encode a protein?

7.3 DNA Replication Maintains Genetic Information

Before a cell divides, its DNA must replicate so that each daughter cell receives the same set of genetic instructions. Clues to the self-replication mechanism came from Watson and Crick's report on DNA's chemical structure. The paper ends with the tantalizing statement, "It has not escaped our notice that the specific pairing we have postulated immediately suggests a possible copying mechanism for the genetic material." They envisioned DNA unwinding, exposing unpaired bases that would attract their complements, and neatly knitting two double helices from one. This route to replication (which turned out to be essentially correct) is called semiconservative because each DNA double helix conserves half of the original molecule (**figure 7.11**).

A. Replication Requires Many Enzymes

An army of enzymes copies DNA just before a cell divides (**figure 7.12**). Enzymes called helicases unwind and hold apart replicating DNA so that other enzymes can guide the assembly of new DNA strands. Another enzyme breaks the hydrogen bonds that connect a base pair. A primase enzyme builds a short complementary piece of RNA, an RNA primer, at the start of each DNA segment to be replicated. The RNA primer attracts **DNA polymerase**, the enzyme that adds new DNA nucleotides complementary to the bases on the exposed strand. The primer is necessary because DNA polymerase can only add nucleotides to an existing strand. As the new DNA strand grows hydrogen bonds form between the complementary bases. **cell cycle, p. 160**

DNA polymerase "proofreads" as it goes, discarding mismatched nucleotides and inserting correct ones. At the same time, another enzyme removes each RNA primer and replaces it with the correct DNA nucleotides. Enzymes called **ligases** form covalent bonds between the resulting DNA segments.

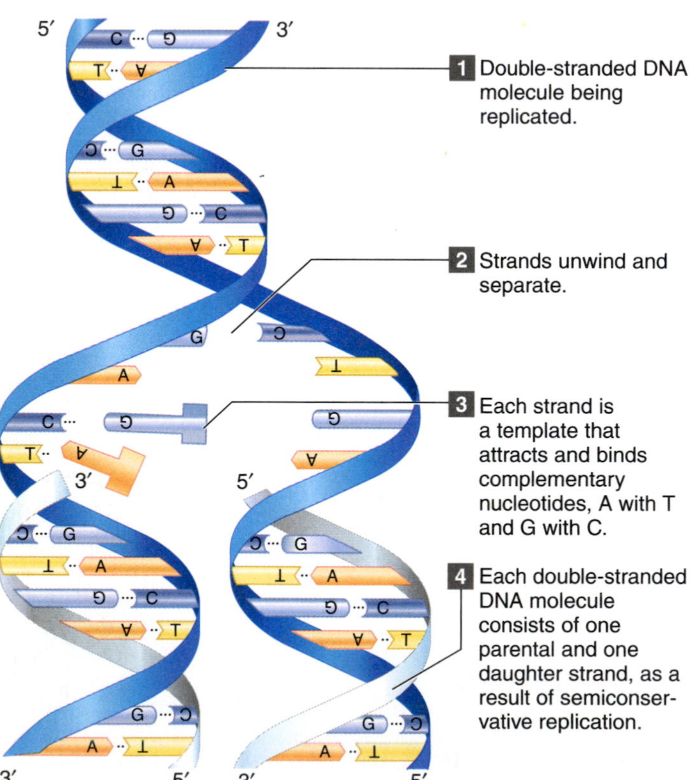
5′
3′
1 Double-stranded DNA molecule being replicated.
2 Strands unwind and separate.
3 Each strand is a template that attracts and binds complementary nucleotides, A with T and G with C.
4 Each double-stranded DNA molecule consists of one parental and one daughter strand, as a result of semiconservative replication.
3′
5′
3′
5′

FIGURE 7.11 DNA Replication Is Semiconservative. In this simplified view of DNA replication, DNA strands separate. New nucleotides form complementary base pairs with each exposed strand. DNA replication requires RNA primers and many enzymes, which are not shown here.

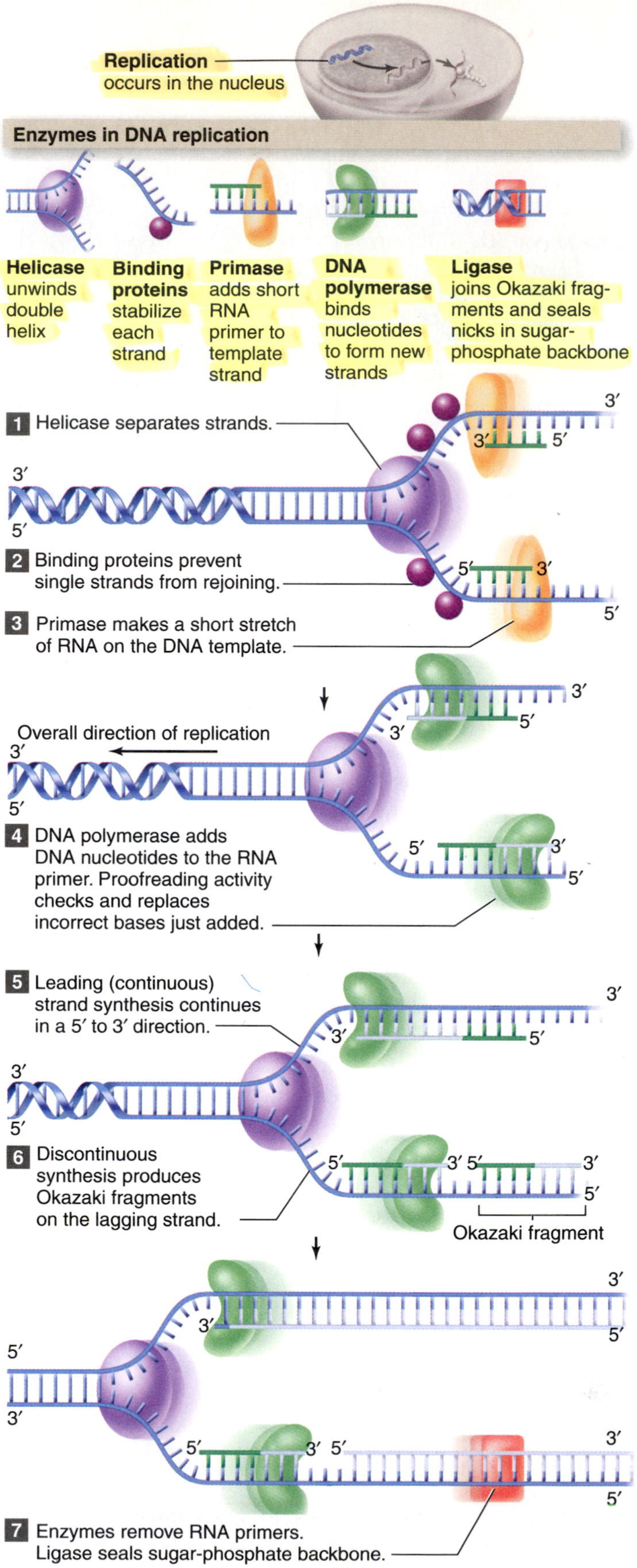

Replication occurs in the nucleus

Enzymes in DNA replication

Helicase unwinds double helix

Binding proteins stabilize each strand

Primase adds short RNA primer to template strand

DNA polymerase binds nucleotides to form new strands

Ligase joins Okazaki fragments and seals nicks in sugar-phosphate backbone

1 Helicase separates strands.

2 Binding proteins prevent single strands from rejoining.

3 Primase makes a short stretch of RNA on the DNA template.

Overall direction of replication

4 DNA polymerase adds DNA nucleotides to the RNA primer. Proofreading activity checks and replaces incorrect bases just added.

5 Leading (continuous) strand synthesis continues in a 5′ to 3′ direction.

6 Discontinuous synthesis produces Okazaki fragments on the lagging strand.

Okazaki fragment

7 Enzymes remove RNA primers. Ligase seals sugar-phosphate backbone.

FIGURE 7.12 DNA Replication Requires Many Enzymes. This process occurs simultaneously at many points along a cell's DNA. Although these enzymes are depicted separately for clarity, in reality they form a single cluster.

Enzymes copy DNA simultaneously at hundreds of points, called origins of replication, on a long DNA molecule. Replication proceeds in both directions at once from each origin of replication. DNA polymerase, however, can add new nucleotides only to the exposed 3′ end—never the 5′ end—of a growing strand. DNA replication therefore proceeds continuously from each point of origin on one strand, but discontinuously (in short 5′ to 3′ pieces) on the other strand. The short pieces on the discontinuous strand are called Okazaki fragments.

Of course, none of this occurs for free. DNA replication requires a great deal of energy because a large, organized molecule contains much more potential energy than do many individual nucleotides. Energy is required to synthesize nucleotides and to create the covalent bonds that join them together in the new strands of DNA. Many of the enzymes that participate in DNA replication, including helicase and ligase, also require energy in the form of ATP to catalyze their reactions. **ATP, p. 85**

DNA replication is incredibly accurate; after proofreading, DNA polymerase incorrectly incorporates only about 1 in a billion nucleotides. Other repair enzymes help ensure the accuracy of DNA replication by cutting out and replacing incorrect nucleotides. Nevertheless, mistakes occasionally remain. The result is a **mutation,** which is any change in a cell's DNA sequence.

B. Mutations May Occur During Replication

A mutation in a gene sometimes changes the structure of its encoded protein so much that the protein can no longer do its job. Inherited diseases, including cystic fibrosis and sickle cell anemia, stem from such DNA sequence changes. In other cases, mutations either do not affect the encoded proteins at all, or may even improve their function.

Mutations are extremely important. They are the raw material for evolution because they create new **alleles**, or variants of genes. Except for identical twins, everyone has a different combination of alleles for the 25,000 or so genes in the human genome. The same is true for any genetically variable group of organisms. In any population, individuals with the "best" combination of alleles—that is, those alleles that improve their chances of successful reproduction—are most likely to pass those alleles to the next generation.

Errors in DNA replication can cause mutations, as can exposure to radiation or harmful chemicals. In addition, chapter 9 describes how chromosomes can sometimes break, causing mutations. If repair enzymes do not fix the error, a dividing cell can pass the error to its descendants.

Some of the most harmful mutations affect the genes encoding the proteins that repair DNA. Additional mutations then accumulate in the cell's DNA, which can kill the cell or lead to cancer. For example, DNA repair defects cause skin cancer and a disease called xeroderma pigmentosum (see this chapter's Burning Question: What causes skin cancer?). Such disorders reveal how essential precise DNA replication is to cell survival. **cancer, p. 166**

Burning Questions

What causes skin cancer?

Cancer is a family of disorders in which cells divide out of control (see chapter 8). Cancer has many forms, some inherited and others caused by exposure to radiation or harmful chemicals. Exposure to ultraviolet radiation from the sun or from tanning beds, for example, increases the risk of skin cancer. Several types of skin cancer exist; melanoma is the most dangerous because it can quickly spread to other parts of the body.

Sometimes, cells fail to detect or repair the damaged DNA. The result is a genetic mutation (discussed further in chapter 12). If the mutation occurs in genes coding for proteins that control the pace of cell division, cells may begin dividing out of control, forming a tumor. Problems with the enzymes that repair DNA therefore predispose people to cancer.

A rare disorder called xeroderma pigmentosum (XP) illustrates the vital role of these enzymes in protecting against skin cancer. For people with XP, exposure to sunlight is extremely dangerous. They lack the proteins that would otherwise repair damage caused by UV radiation in sunlight. People with XP therefore have a 2000-fold higher risk of developing skin cancer than the population at large.

Have a Burning Question of your own?
Submit it to marielle_hoefnagels@mcgraw-hill.com
for possible inclusion in future editions of this book!

Summary *When DNA replicates, the two strands separate and each is a template for a new, complementary strand. A helicase begins the process, RNA polymerase and then DNA polymerase fill in the nucleotides of the new strand, and ligases join the sugar–phosphate backbone. Enzymes repair mutations that arise during DNA replication.*

7.3 MASTERING CONCEPTS

1. Why does DNA replicate?
2. What is semiconservative replication?
3. What are the steps of DNA replication?
4. What is a mutation, and why are mutations important?
5. Why is DNA repair necessary?

7.4 PCR Replicates DNA in a Test Tube

An extremely powerful and useful tool, the **polymerase chain reaction (PCR)**, taps into the cell's DNA copying machinery to rapidly produce millions of copies of a DNA sequence of interest (**figure 7.13**). PCR is useful whenever a small amount of DNA would provide information if it was mass-produced. Thanks to this tool, a single hair or a few skin cells left at a crime scene can yield enough genetic material for DNA profiling.

PCR rapidly replicates a selected sequence of DNA in a test tube. The requirements include the following:

- A target DNA sequence to be replicated.
- *Taq* polymerase, a heat-tolerant DNA polymerase produced by *Thermus aquaticus*, a bacterium that inhabits hot springs. (Other heat-tolerant polymerases can also be used.)
- Two types of short, laboratory-made primers that are complementary to opposite ends of the target sequence. The primers are necessary because DNA polymerase can only attach nucleotides to an existing strand.
- A supply of the four types of DNA nucleotides.

PCR occurs in an automated device called a thermal cycler that controls key temperature changes. In the first step of PCR, heat separates the two strands of the target DNA. Next, the temperature is lowered, and the short primers attach to the separated target strands by complementary base pairing. *Taq* DNA polymerase adds nucleotides to the primers and builds sequences complementary to the target sequence. The newly synthesized strands then act as templates in the next round of replication, which is initiated immediately by raising the temperature to separate the strands once more. The number of pieces of DNA doubles with every round of PCR, so that after just 20 cycles, a million-fold increase occurs in the number of copies of the target sequence.

PCR finds a wide variety of applications. Its greatest strength is that it works on crude samples, such as a bit of brain tissue on the bumper of a car, which in one criminal case led to identification of a missing person. In forensics, PCR is used routinely to establish genetic relationships, identify remains, convict criminals, and exonerate the falsely accused. When used to amplify the nucleic acids of microorganisms, viruses, and other parasites, PCR is important in agriculture, veterinary medicine, environmental science, and human health care. In genetics, PCR is both a crucial basic research tool and a way to identify disease-causing genes.

PCR's greatest weakness, ironically, is its exquisite sensitivity. A blood sample contaminated by leftover DNA from a previous run or by a stray eyelash dropped from the person running the reaction can yield a false result.

Summary *Polymerase chain reaction (PCR) uses primers, nucleotides, and heat-tolerant DNA polymerase to generate copies of DNA.*

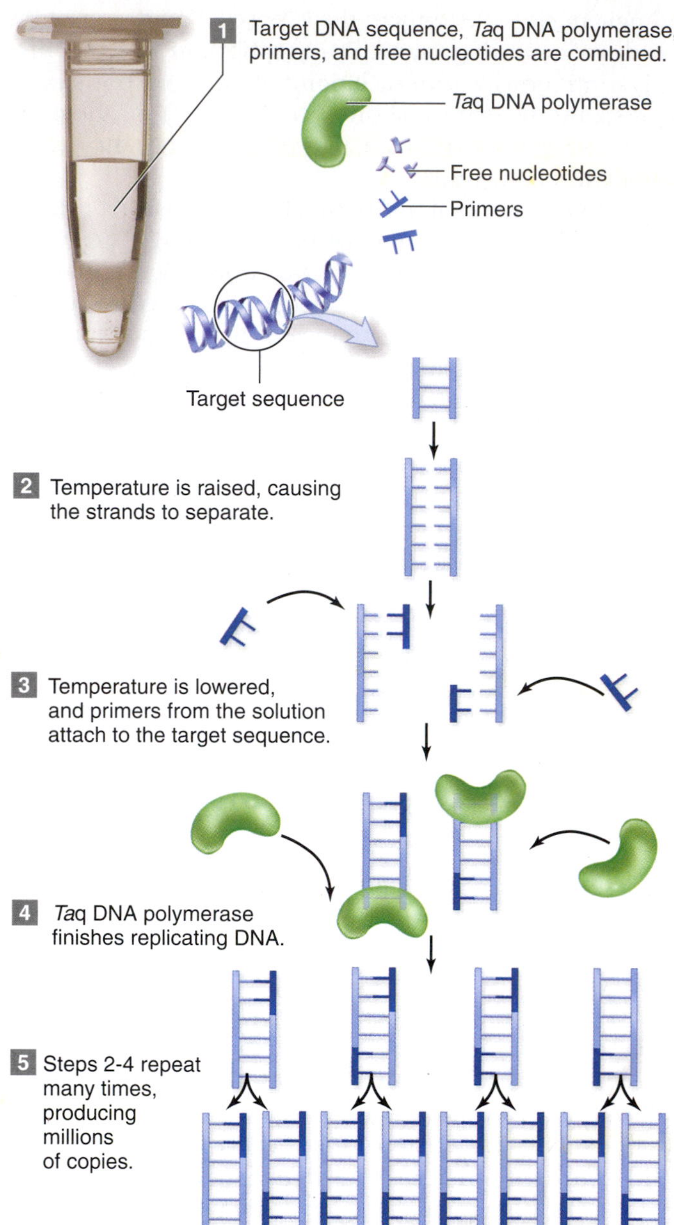
1 Target DNA sequence, Taq DNA polymerase, primers, and free nucleotides are combined.
Taq DNA polymerase
Free nucleotides
Primers
Target sequence
2 Temperature is raised, causing the strands to separate.
3 Temperature is lowered, and primers from the solution attach to the target sequence.
4 Taq DNA polymerase finishes replicating DNA.
5 Steps 2-4 repeat many times, producing millions of copies.

FIGURE 7.13 PCR Amplifies a Specific DNA Sequence. In the polymerase chain reaction, primers bracket a DNA sequence of interest. A heat-stable *Taq* DNA polymerase uses these primers, and plenty of nucleotides, to build up millions of copies of the target sequence.

7.4 MASTERING CONCEPTS

1. How do target DNA, primers, nucleotides, and *Taq* DNA polymerase interact in PCR?
2. Why is PCR useful?

7.5 DNA Sequencing Reveals the Order of Bases

The uses of DNA sequences, from individual genes to entire genomes, seem endless. Researchers can apply sequence information to everything from identifying corpses, to predicting protein sequences, to determining evolutionary relationships. How do investigators get the DNA sequence information they need?

A. Sanger Used Fragments to Determine a DNA Sequence

Modern DNA sequencing instruments use a highly automated version of a basic technique Frederick Sanger developed in 1977. The overall goal is to generate a series of DNA fragments that are complementary to the DNA being sequenced. These fragments differ in length from one another by one end base. Once a collection of such pieces is generated, a technique called electrophoresis can be used to separate the fragments by size. Reading the end bases in order by size reveals the sequence of the complement, and deriving the original DNA sequence is then easy.

Sanger invented a way to generate the DNA pieces (**figure 7.14**). He separated a DNA fragment into single strands and added a primer. He then divided the DNA into four test tubes, each of which also contained DNA polymerase and all four types of nucleotides. Sanger's innovation was to supplement the "regular" A, C, T, and G with a small amount of chemically modified nucleotides. DNA synthesis halts when DNA polymerase encounters a modified nucleotide by chance, leaving the strand only partially replicated. Since the modified nucleotides made up only about 1% of the total nucleotide content in each tube, DNA replication generated a series of strands of different lengths in each test tube.

After the replication reactions were complete, Sanger ran the contents of the tubes in four separate lanes of an electrophoresis gel. Sanger used radioactive labels to visualize the fragments, but researchers today use fluorescent labels, one for each of the four base types. The data appear as a sequential readout of the wavelengths of the fluorescence from the labels (**figure 7.15**).

B. DNA Microarrays Speed Sequencing

DNA chips offer a second way to sequence DNA (**figure 7.16**). Short DNA fragments of known sequence are immobilized on a small glass square called a DNA microchip or a DNA microarray. In one version, the 4096 possible six-base combinations of DNA are placed onto a 1-cm × 1-cm microchip. Copies of an unknown DNA segment incorporating a

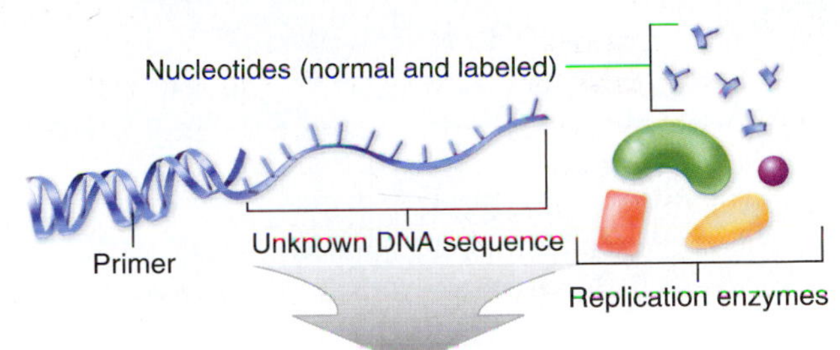

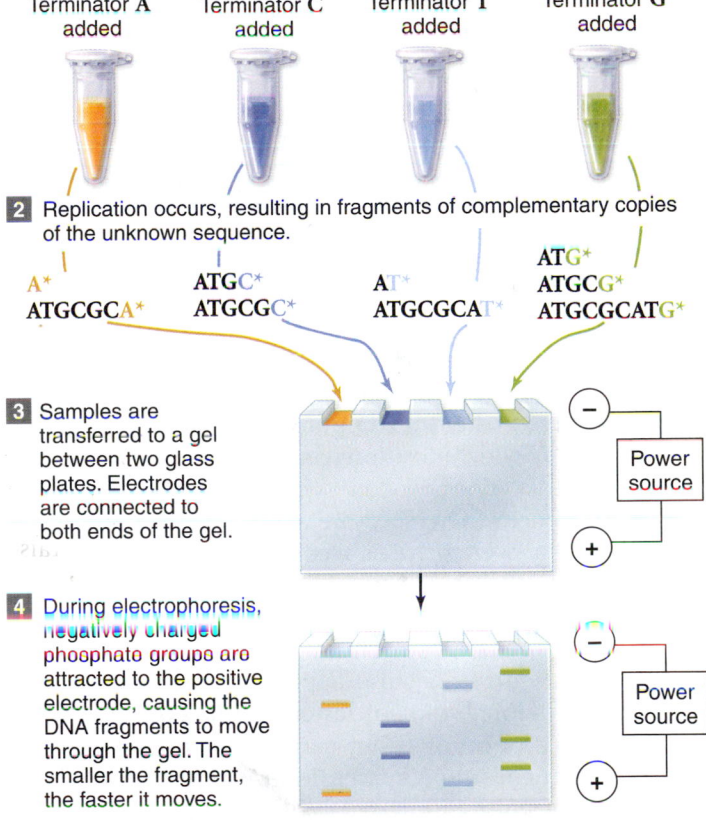

1 Four solutions contain unknown DNA sequence, primers, normal nucleotides (A,C,G, and T), labeled nucleotides, replication enzymes, and a small amount of "terminator" nucleotide.

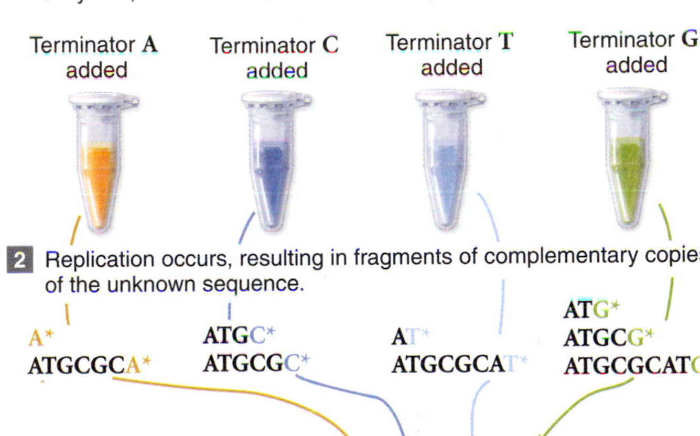

2 Replication occurs, resulting in fragments of complementary copies of the unknown sequence.

A*
ATGCGCA*
ATGC*
ATGCGC*
AT*
ATGCGCAT*
ATG*
ATGCG*
ATGCGCATG*

3 Samples are transferred to a gel between two glass plates. Electrodes are connected to both ends of the gel.

4 During electrophoresis, negatively charged phosphate groups are attracted to the positive electrode, causing the DNA fragments to move through the gel. The smaller the fragment, the faster it moves.

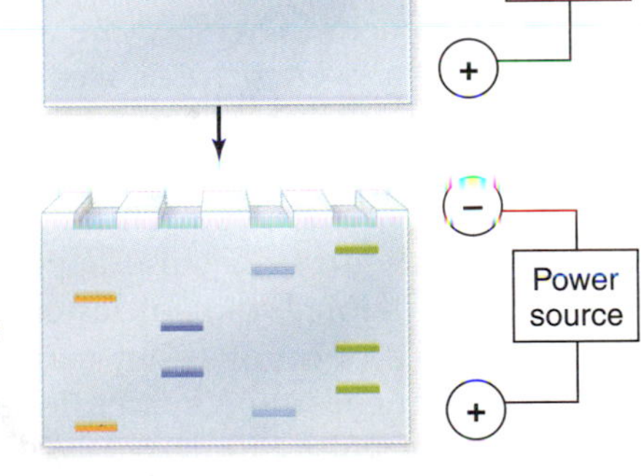

5 The fragments are read off by size, and the original sequence can be deduced.

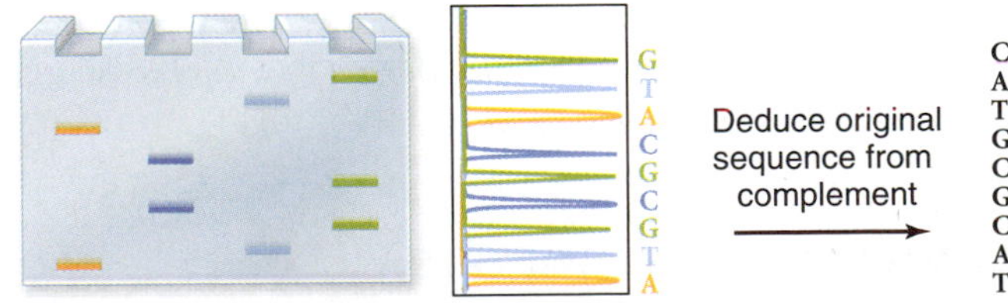

FIGURE 7.14 Determining the Sequence of DNA. In the Sanger method of DNA sequencing, complementary copies of an unknown DNA sequence are terminated early because of the addition of chemically modified "terminator" nucleotides. Placing the fragments in order by size reveals the sequence.

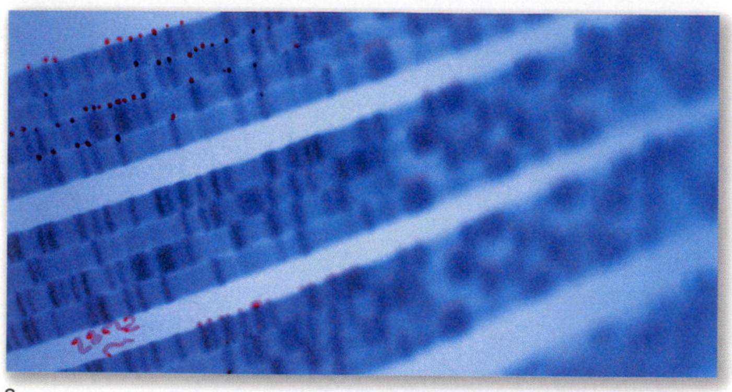

a.

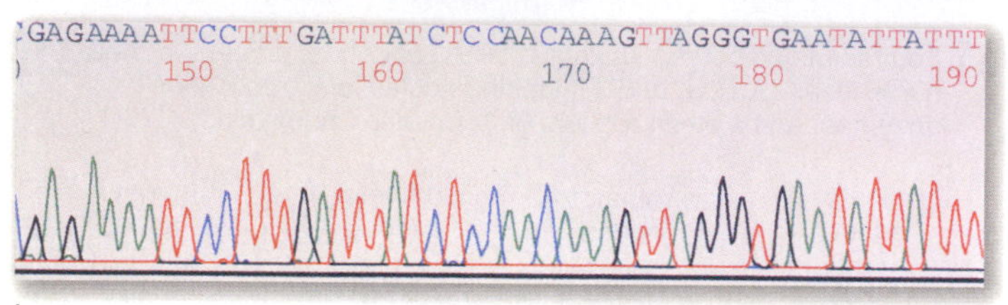
GAGAAAATTCCTTTGATTTATCTCCAACAAAGTTAGGGTGAATATTATTT
150
160
170
180
190

b.

FIGURE 7.15 Two Ways to Read DNA. (**A**) A DNA sequencing gel contains four lanes, one each for A, C, G, and T. Radioactive labels reveal the bands in this gel. (**B**) A computerized readout of a DNA sequence, made possible by fluorescent labels.

fluorescent label are then also placed on the microchip. The copies stick to any six-base strands on the chip whose sequences are complementary to the unknown DNA segment's sequences. Under laser light, the bound sequences fluoresce. Because the researcher (or computer program) knows which strands occupy which positions on the microchip, a scan of the chip reveals which six-base sequences make up the unknown sequence. Then, software aligns the identified fragments by their overlaps. This reconstructs the complement of the entire unknown sequence.

Microarrays are increasingly common in research because they offer a fast, inexpensive way to generate sequence data. One day, they may also appear in doctors' offices. If medical professionals can quickly determine a patient's DNA sequence at key locations, they may be able to pinpoint the cause of a patient's cancer or customize drug treatments to minimize side effects.

Summary ***The Sanger method uses modified nucleotides to reveal the DNA sequence. Alternatively, DNA fragments of unknown sequence can stick to short DNA pieces of known sequence arranged on microchips.***

7.5 MASTERING CONCEPTS

1. How does the Sanger method use modified nucleotides to produce truncated DNA fragments, and how do researchers use these fragments to deduce a DNA sequence?
2. What is a DNA microchip?

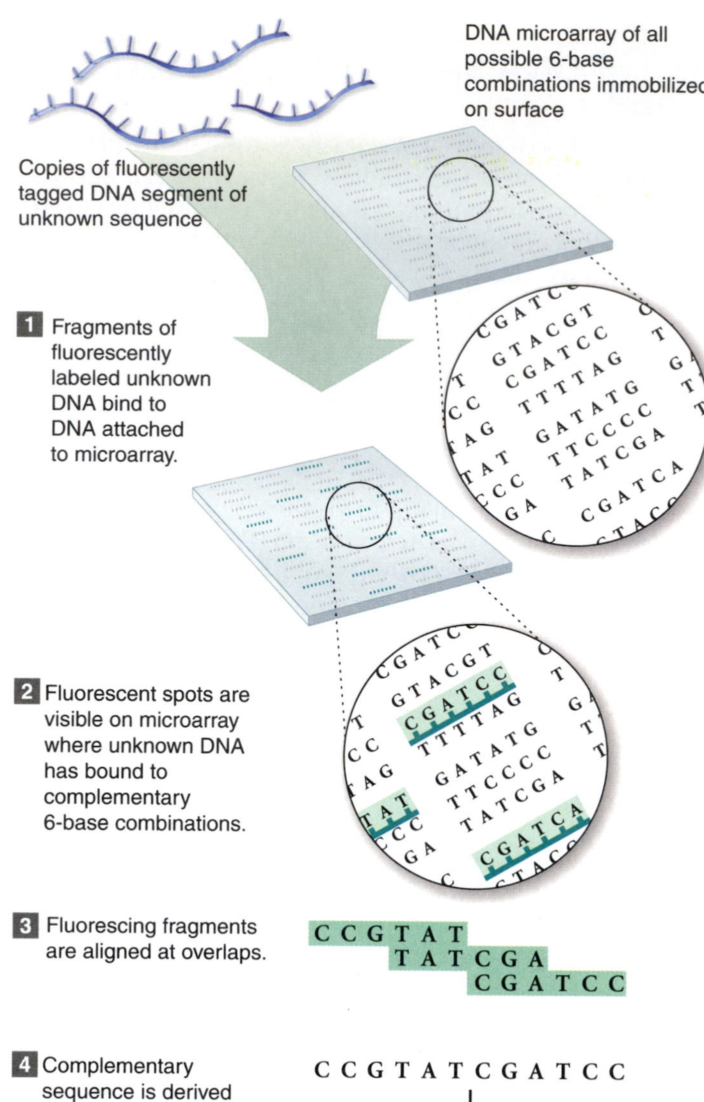
DNA microarray of all possible 6-base combinations immobilized on surface
Copies of fluorescently tagged DNA segment of unknown sequence
1 Fragments of fluorescently labeled unknown DNA bind to DNA attached to microarray.
2 Fluorescent spots are visible on microarray where unknown DNA has bound to complementary 6-base combinations.
3 Fluorescing fragments are aligned at overlaps.
C C G T A T
T A T C G A
C G A T C C
4 Complementary sequence is derived from overlaps, revealing the original unknown sequence.
C C G T A T C G A T C C
G G C A T A G C T A G G

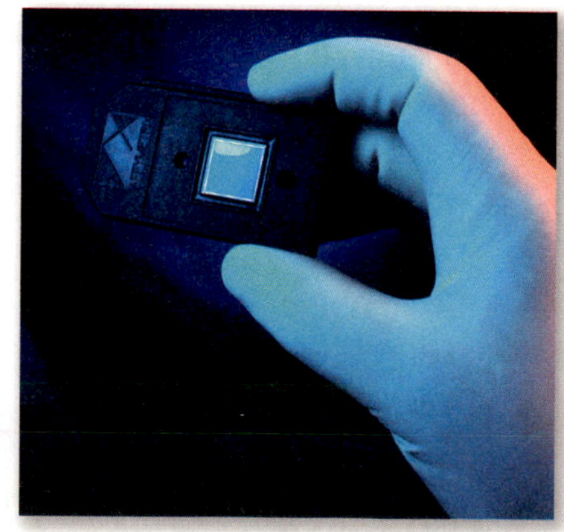

FIGURE 7.16 Sequencing on a Chip. A labeled DNA segment of unknown sequence binds to short, known DNA sequences immobilized on a small glass microchip. Identifying areas of overlap among the bound sequences reveals the unknown DNA sequence.

7.6 DNA Profiling Has Many Applications

DNA profiling detects genetic differences between individuals, with applications ranging from settling paternity claims to exonerating wrongly convicted criminals to identifying crime victims (see Can *You* Relate? Identifying Victims of the September 11, 2001, Attacks below).

Rather than sequencing and comparing entire genomes, DNA profiling considers just the most variable parts. In one approach, researchers use single-base sites that tend to vary in a population. DNA samples are treated with bacterial enzymes (called restriction enzymes) that cut DNA at specific sequences. If two samples differ at a cutting site, the restriction enzyme produces DNA fragments of different sizes and numbers. **Figure 7.17** illustrates how researchers use gel electrophoresis to reveal these underlying sequence differences.

1 A restriction enzyme is added to three DNA samples, cutting each sample into a unique pattern of different-sized pieces.

Restriction enzyme

DNA sample 1 DNA sample 2 DNA sample 3

2 The three samples are loaded into an electrophoresis gel.

Longer fragments

Shorter fragments

Power source

−

+

3 Labeled DNA probes highlight certain fragments of interest by binding to specific fragments of sampled DNA. This way, DNA from different individuals may have different banding patterns.

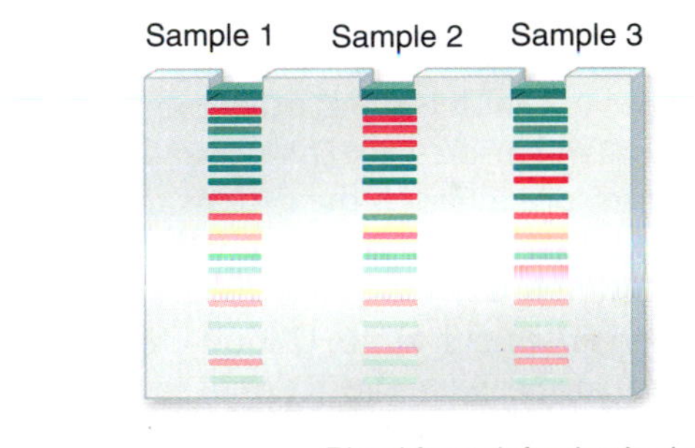

a.

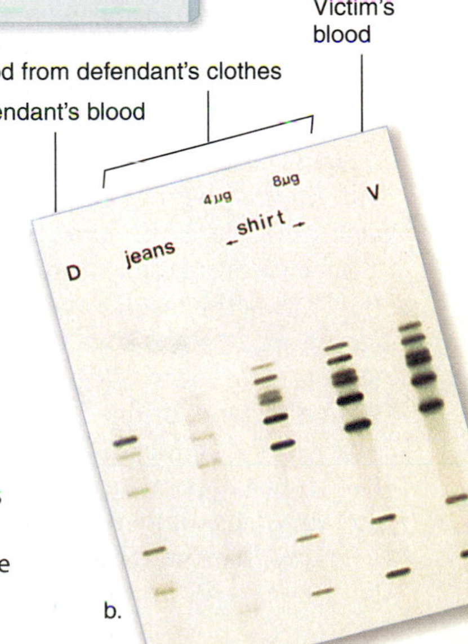

FIGURE 7.17 DNA Profiles from a Murder Case. (**A**) DNA samples from different sources produce different patterns of fragments when cut with the same restriction enzyme. (**B**) DNA from bloodstains on the defendant's clothes matches the DNA profile of the victim but differs from the DNA profile of the defendant. This is evidence that the blood on the defendant's clothes came from the victim, not the defendant.

Can *You* Relate?

Identifying Victims of the September 11, 2001, Attacks

Until September 11, 2001, the most challenging application of DNA profiling had been identifying plane-crash victims, a grim task eased by having lists of passengers. The terrorist attacks on the World Trade Center provided a staggeringly more complex situation, for several reasons: the high number of casualties, the condition of the remains, and the lack of a list of who was actually in the buildings. Overall, the disaster yielded more than a million DNA samples

In the days following September 11, somber researchers at Myriad Genetics, Inc., in Salt Lake City, who usually analyze DNA for breast cancer genes, received frozen DNA from soft tissue recovered from the disaster site. The laboratory also received cheek scrapings from relatives of the missing, and tissue from the victims' toothbrushes, razors, and hairbrushes. The workers used PCR to determine the numbers of copies of four-base sequences of DNA, called short tandem repeats, or STRs, at 13 locations in the genome. The chance that any two individuals have the same 13 markers by chance is one in 250 trillion. If the STR pattern of a sample from the crime scene matched a sample from a victim's toothbrush, identification was fairly certain. DNA extracted from tooth and bone bits was sent to Celera Genomics Corporation in Rockville, Maryland. Here, DNA sequences were analyzed from mitochondria, which can survive incineration.

The labs used DNA to identify about 850 of the more than 2700 people reported missing. It was a very distressing experience for the technicians and researchers, whose jobs had suddenly shifted from detecting breast cancer and sequencing genomes to helping in recovery.

Suppose that DNA extracted from a hair found on a murder victim's body matches DNA from a suspect's white blood cells at five restriction sites. What is the probability that the two matching DNA patterns come from the same person—the suspect—rather than from two individuals who resemble each other by chance? To find out, investigators consult databases that compile the frequency of each allele in the population. If each of the five alleles occurs at a rate of 10% in the population, the chance of the match occurring by chance is $0.1 \times 0.1 \times 0.1 \times 0.1 \times 0.1$, or 1 in 100,000. There is therefore a 99.999% chance that the samples came from the same person.

Analysis of mitochondrial DNA is also sometimes useful. Mitochondrial DNA is typically only about 16,500 base pairs long, far shorter than nuclear DNA. But because each cell contains multiple mitochondria, each of which contains many DNA molecules, mitochondria can therefore often yield useful information even when nuclear DNA is badly degraded. Investigators extract mitochondrial DNA from hair, bones, and teeth, then use PCR to amplify the variable regions for sequencing.

Because everyone inherits mitochondria only from his or her mother, this technique cannot discriminate between siblings or many other close family members. It is very useful, however, for verifying the relationship between woman and child. For example, children who were kidnapped during infancy can be matched to their biological mothers or grandmothers. The study of human evolution also has benefited from mitochondrial DNA. And, as you learned at this chapter's beginning, mitochondrial DNA helped identify members of the Russian royal family. **human evolution, p. 344**

Summary *DNA profiling reveals genetic differences between individuals by analyzing variable portions of the genome.*

7.6 MASTERING CONCEPTS

1. How are variable DNA sequences used in DNA profiling?
2. Why does DNA profiling require an understanding of probability?
3. Why does mitochondrial DNA provide different information from nuclear DNA?

7.7 INVESTIGATING LIFE

Genetic Messages from the Dead Tell Tales of Ancient Ecosystems

Psychics claim to be able to communicate with the spirits of the dead. Although scientists do not assert the same spiritual connection, they can bring back the genetic remnants of species that lived and died long ago.

When an organism dies, its DNA usually degrades rapidly. But in some special cases, the genetic material remains intact indefinitely. Millions of years ago, for example, sticky tree resin smothered and preserved insects, protecting the animals' DNA in an amber tomb. (Blood-engorged mosquitoes preserved in amber were the fictional source of dinosaur DNA in *Jurassic Park*). Mineral crystals in bones and teeth can safeguard DNA for thousands of years too.

Freezing also preserves DNA. An ideal source of diverse ancient DNA would therefore be a landscape that once teemed with life but that has since become permanently frozen. One such example is the land bridge, called Beringia, that once connected present-day northeastern Siberia to Alaska.

During the last ice age, which ended about 10,000 years ago, much of Earth's water was locked in polar ice caps and glaciers. As a result, global sea levels were low, exposing the Beringian land bridge. Although glaciers scoured most of North America during the ice age, Beringia's dry climate kept it ice-free. Giant mammals such as mammoth, bear, bison, and large cats roamed the grassy Beringian landscape.

Now that the ice age has ended, the land bridge is under the sea, and much of the rest of Beringia is permanently frozen land in Siberia, Alaska, and the Yukon. Could DNA trapped in frozen Siberian soil reveal an ancient ecosystem's organisms? Danish researchers Eske Willerslev, Anders Hansen, and their colleagues decided to find out.

To study the ancient remains, the scientists drove metal cylinders deep into the permafrost and removed long, thin rods (called "cores") of ice, soil, and organic material (**figure 7.18**). Because new sediments accumulate over old ones, the deepest holes yield the oldest deposits. Radiometric dating (a technique that uses radioactive decay of isotopes as a type of "clock"), pollen analysis, and other techniques helped them estimate the age of each layer of material in the cores. **radiometric dating, p. 312**

Willerslev and company tried to extract DNA from eight small sediment samples ranging in age from modern to 1.5 to 2 million years old. Wherever DNA was present, they used PCR primers to amplify two sets of genes. One target sequence was part of the gene encoding the protein rubisco, which is essential for photosynthesis (see chapter 5). Because this gene occurs in chloroplasts, its presence indicates

FIGURE 7.18 **Core Sample.** Cores of frozen sediment taken from permafrost may contain ancient DNA.

plant material in a sample. The other three target DNA sequences were fragments of genes found in mitochondria of vertebrate animals. (Vertebrates are animals with bony backbones, such as fishes, reptiles, and mammals).

The PCR results showed that sediments from 300,000 to 400,000 years old contained plant DNA. When the scientists sequenced the DNA and compared it to genes cataloged in GenBank, a public database of nucleotide sequences, the scientists found that the ancient sediments harbored tremendous diversity: 11 classes, 23 orders, and 28 families representing mosses, herbs, shrubs, and trees.

The oldest mitochondrial DNA was much younger, about 20,000 to 30,000 years old. Some came from grazers that still exist today, such as horse, lemming, hare, musk ox, and reindeer, but the researchers also found genes from extinct mammoth and bison. (Previous researchers studying fossils of these huge herbivores had deposited the gene sequences in GenBank.) The presence of genes from animals that vanished long ago suggests that the sediment DNA was authentic and not simply a modern contaminant.

These gene fragments are important, but not for *Jurassic Park*-style reasons. After all, it is not possible to bring back the mammoth from just a tiny piece of its mitochondrial DNA. Rather, the study is significant because the long-buried sediments can help scientists reconstruct ancient ecosystems. The chloroplast DNA, for example, reveals that herbs (grasses and other nonwoody green plants) dominated the Beringian landscape 300,000 or so years ago but lost ground to shrubs over time (**figure 7.19**). The most dramatic decline of grasses occurred in the past 10,000 years, a time that coincided with the extinction of the mammoth and bison. Did one event cause the other? Or did the end of the ice age cause both? What role did increasing human populations play? These questions remain unanswered for now.

Willerslev's team hopes their work will inspire others to extract DNA from ancient sediments around the world. The resulting patchwork of gene fragments, pieced together, will reveal valuable information about the changes that shaped ancient ecosystems—without the help of a psychic.

Willerslev, E., A. J. Hansen, J. Binladen, et al. May 2, 2003. Diverse plant and animal genetic records from Holocene and Pleistocene sediments. *Science*, vol. 300, pages 791–795.

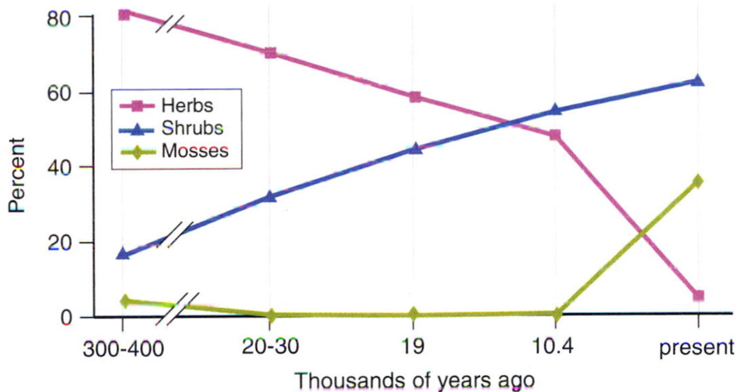

FIGURE 7.19 Communities Change. Chloroplast DNA isolated from sediment samples shows that grasses and other herbs have become much less common, while shrubs have become more common, over the past 400,000 or so years. **Question:** What are some alternative hypotheses for why the researchers failed to recover any DNA from sediments that were more than 300,000 to 400,000 years old? How would you test your hypotheses?

CHAPTER SUMMARY

7.1 Experiments Identified the Genetic Material

- DNA encodes the information necessary for a cell's survival and specialization; it must be able to replicate for a cell to divide. Many experiments described DNA and showed it to be the genetic material.

A. Griffith Discovered that Bacteria Can Transfer Genetic Information

- Griffith determined that an unknown substance transmits a disease-causing trait between two types of bacteria.

B. Avery, MacLeod, and McCarty Showed that Genetic Information Is DNA

- With the help of protein- and DNA-destroying enzymes, scientists subsequently showed that Griffith's "transforming principle" is DNA.

C. Hershey and Chase Confirmed the Genetic Role of DNA

- Using viruses that infect bacteria, Hershey and Chase confirmed that the genetic material is DNA and not protein.

7.2 DNA Is a Double Helix That Encodes "Recipes" for Proteins

A. Biochemists and Physicists Discovered DNA's Structure

- Chargaff discovered that A and T, and G and C, occur in equal proportions in DNA. Wilkins and Franklin provided X-ray diffraction data. Watson and Crick combined these clues to propose the double-helix structure of DNA.
- DNA consists of building blocks called **nucleotides.** The rungs of the DNA double helix consist of hydrogen-bonded **complementary** base pairs (A with T, and C with G). The rails are antiparallel chains of alternating deoxyribose and phosphate.

B. DNA Contains the Information Needed for Life's Functions

- An organism's **genome** is all of the DNA in its cells. In eukaryotic cells, the genome is divided into **chromosomes.**
- **Genes** are sequences of DNA that encode a cell's proteins.

7.3 DNA Replication Maintains Genetic Information

- DNA replication is semiconservative because each double-stranded molecule contains one strand from the original DNA.

A. Replication Requires Many Enzymes

- To replicate, DNA unwinds, and the hydrogen bonds between strands break. **DNA polymerase** adds DNA nucleotides to a short RNA primer. **Ligase** seals the sugar–phosphate backbone after the RNA primer is replaced with DNA.
- Replication proceeds only in a 5′ to 3′ direction, so the process is discontinuous in short stretches on one strand.
- Enzymes repair mistakes in replication or damaged DNA.

B. Mutations May Occur During Replication

- **Mutations** are changes in a cell's DNA sequence; they give rise to new **alleles** (gene variants).
- DNA repair disorders raise cancer risk.

7.4 PCR Replicates DNA in a Test Tube

- In the **polymerase chain reaction (PCR),** DNA separates into two strands, then a primer provides DNA polymerase with a starting point for replication. Repeated cycles of heating and cooling allow for rapid amplification of the target DNA sequence.
- PCR finds many applications in research, forensics, medicine, agriculture, and other fields.

7.5 DNA Sequencing Reveals the Order of Bases

A. Sanger Used Fragments to Determine a DNA Sequence

- The Sanger method uses modified nucleotides to generate DNA fragments of various lengths. Sorting the fragments by size reveals the DNA sequence.

B. DNA Microarrays Speed Sequencing

- DNA chips contain all possible combinations of short nucleotide sequences. DNA of unknown sequence sticks to some of the chip-bound DNA, then computers reconstruct the original sequence.

7.6 DNA Profiling Has Many Applications

- Individuals vary genetically in single bases and short repeated DNA sequences. **DNA profiling** detects these differences.
- Investigators can use known frequencies of alleles in the population to calculate the probability that two DNA samples match.
- Analysis of mitochondrial DNA can verify maternal relationships.

7.7 Investigating Life: Genetic Messages from the Dead Tell Tales of Ancient Ecosystems

- Researchers have extracted DNA from plants and animals buried long ago in frozen sediments. The DNA evidence reveals changes in the landscape over hundreds of thousands of years.

MULTIPLE CHOICE QUESTIONS

1. A molecule of DNA is composed of all of the following EXCEPT
   a. a sugar.
   b. a nitrogen-containing group.
   c. a sulfur-containing group.
   d. a phosphorus-containing group.
2. If one strand of DNA has the sequence ATTGTCC, then the sequence of the complementary strand would be
   a. TAACAGG.
   b. CGGAGTT.
   c. ACCTCGG.
   d. CCTGTTA.
3. Which of the following correctly describes the relationship between the terms?
   a. DNA nucleotide > genome > chromosome > gene
   b. Genome > chromosome > gene > DNA nucleotide
   c. Chromosome > genome > gene > DNA nucleotide
   d. Gene > genome > DNA nucleotide > chromosome
4. The job of a ligase enzyme is to
   a. unwind the DNA.
   b. build a short sequence of RNA using the DNA as a template.
   c. form covalent bonds between DNA nucleotides.
   d. add DNA nucleotides to an existing strand.
5. What would happen to DNA replication if the helicase enzyme did not function?
   a. Replication could occur, but there would be errors.
   b. The new DNA strand would not be held together by covalent bonds.
   c. Replication would occur in a single direction.
   d. Replication would not occur at all.
6. Are mutations bad?
   a. Yes, because the DNA is damaged.
   b. No, because changes in the DNA result in better alleles.
   c. Yes, because mutated proteins don't function.
   d. It depends on how the mutation affects the protein's function.
7. Why is PCR useful?
   a. Because it replicates all the DNA in a cell
   b. Because it can create large amounts of DNA from small amounts
   c. Because it uses a heat-tolerant, *Taq* polymerase
   d. Because it occurs in an automated device

8. Why did the DNA from the heat-killed S-type cells transform the R-type cells?
   a. Because it was mutated
   b. Because DNA determines what proteins are made in a cell
   c. Because the R-type cells did not have DNA
   d. Because DNA is less complex than proteins

9. What is a DNA microarray used for?
   a. Determining the number of base pairs in a person's chromosome
   b. Determining the sequence of a segment of DNA
   c. Creating genetic mutations
   d. Identifying a DNA profile

10. In 1920, a woman claiming to be a surviving member of the royal family of Tsar Nicholas II of Russia appeared in Europe. What modern method of DNA profiling would you use to determine if this person was a member of the Romanov family?
    a. Analysis of variable single-base sites
    b. Analysis of the sequence of the woman's entire genome
    c. Analysis of mitochondrial DNA sequences
    d. Comparison of allele frequencies to population databases

TESTING YOUR KNOWLEDGE

1. Describe the three-dimensional structure of DNA.
2. What is the function of DNA?
3. Write the complementary DNA sequence of each of the following base sequences:
   a. T C G A G A A T C T C G A T T
   b. C C G T A T A G C C G G T A C
   c. A T C G G A T C G C T A C T G
4. If a cell contains all the genetic material it needs to synthesize protein, why must the DNA also replicate?
5. State the functions of the following participants in DNA replication: primer, DNA polymerase, ligase, and helicase.
6. Explain how the ingredients of a PCR reaction tube replicate DNA.
7. How does the Sanger method help researchers determine the sequence of a fragment of DNA?
8. Explain how DNA from two genetically different individuals will produce different-sized fragments when cut with a restriction enzyme.

THINKING AS A SCIENTIST

1. Choose an experiment mentioned in the chapter and analyze how it follows the scientific method.
2. Give an example from the chapter of different types of experiments used to address the same hypothesis. Why might this be necessary?
3. The experiments that revealed DNA structure and function used a variety of organisms. How can such diverse organisms demonstrate the same genetic principles?
4. How would the results of the Hershey-Chase experiment have differed if protein was the genetic material?
5. A person with deficient DNA repair may have an increased cancer risk and chromosomes that cannot heal breaks. The person is, nevertheless, alive. How long would an individual lacking DNA polymerase be likely to survive?
6. To diagnose encephalitis (brain inflammation) caused by West Nile virus infection, a researcher needs a million copies of a viral gene. She decides to use PCR on a sample of cerebrospinal fluid, which bathes the person's infected brain. If one cycle of PCR takes 2 minutes, how long will it take the researcher to obtain her million-fold amplification if she starts with a single copy of the viral gene?
7. As the climate warms, permafrost in Siberia and other northern regions could melt. How might global warming affect future DNA-based studies of ancient ecosystems whose remains are trapped in the frozen soil?

Visit www.mhhe.com/hoefnagels for practice quizzes, animations, videos, and activities designed to help you master the material in this chapter.

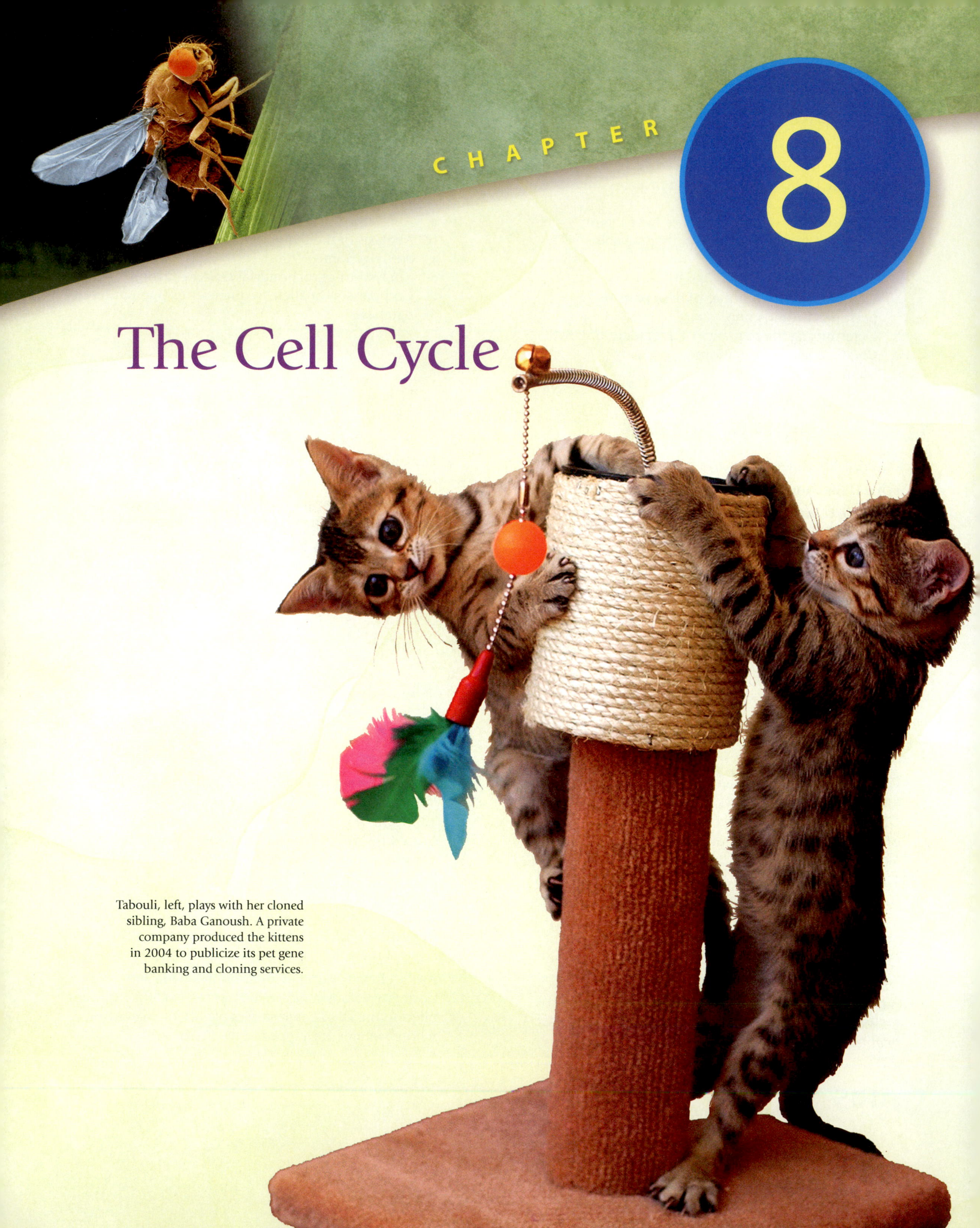

The Cell Cycle

Tabouli, left, plays with her cloned sibling, Baba Ganoush. A private company produced the kittens in 2004 to publicize its pet gene banking and cloning services.

What's the Point?

The Clones Are Here

Imagine being able to grow a new individual, genetically identical to yourself, from a bit of skin or the root of a hair. Although humans cannot reproduce in this way, many organisms do the equivalent. They develop parts of themselves into genetically identical individuals—clones—that then detach and live independently.

Cloning, or asexual reproduction, has been a part of life since the first cell arose billions of years ago. Long before sex evolved, each individual simply reproduced by itself, without a partner to contribute half the offspring's genetic information.

In its simplest form, asexual reproduction consists of the division of a single cell. In prokaryotes and single-celled eukaryotes such as Amoeba, *the cell's DNA replicates, and then the cell splits into two identical, individual organisms. Although the details of cell division differ between prokaryotes and eukaryotes, the result is the same: one individual becomes two.*

Most multicellular eukaryotes (plants, fungi, animals, and some types of protists) reproduce sexually, but at least some organisms in each kingdom also use asexual reproduction. This strategy is especially common in plants and fungi. Hobbyists and commercial plant growers clone everything from fruit trees to African violets by cultivating cuttings from a parent plant's stems, leaves, and roots. Many fungi also are phenomenal breeders, producing countless microscopic spores on bread, cheese, and every other imaginable food supply (see chapter 21).

Asexual reproduction is much less common in animals. Sponges, coral animals, hydra, and jellyfishes "bud" new individuals that break away from the parent. Mammals, however, normally reproduce only sexually. That is why biologists attracted widespread public attention in the 1990s by creating a lamb called Dolly, the first clone of an adult mammal (see the Burning Question later in this chapter).

Since Dolly was born, many have wondered whether humans can and should be cloned, either to help infertile couples or to permit the artificial growth of new organs for transplant. Despite its potential benefits, however, human cloning also carries unresolved ethical questions.

All eukaryotes rely on a process called mitosis, which enables cells to copy themselves faithfully. Without mitosis, no eukaryote could copy itself, grow, or repair injuries. This chapter describes how and when cells divide—and explains how they die.

LEARNING OUTLINE

8.1 Cells Divide, and Cells Die

"In a sense we contain ourselves, wrapped up within ourselves, trillions of times repeated."

So wrote noted geneticist Herman J. Müller in 1947, referring to the astonishing precision with which each of a human's trillions of cells retains the genetic information that was present in the fertilized egg. Every cell in the body results from countless rounds of cell division, each time forming two genetically identical cells from one. **cell theory, p. 50**

The development of a multicellular organism, however, requires more than just cell division. Cells also die in predictable ways, carving distinctive structures. In an animal's body, **apoptosis** (from the Greek word for "falling off") is cell death that is a normal part of development. Like cell division, it is a precise, tightly regulated sequence of events (see section 8.5). Apoptosis is therefore also called "programmed cell death."

Throughout an animal's life, cell division and cell death are in balance, so tissue neither overgrows nor shrinks (**figure 8.1**). During early development, both cell division and apoptosis shape new structures. A foot starts out as a webbed triangle of tissue, for example, from which toes form as cells between the digits die. Likewise, cells in the tail of a tadpole die as the young frog develops into an adult. Cell division also compensates for the death of skin and blood cells, a little like adding new snow (cell division) to a snowman that is melting (apoptosis). Both cell division and apoptosis also help protect the organism. For example, cells divide to heal a scraped knee; apoptosis peels away sunburnt skin cells that might otherwise become cancerous.

This chapter explores the opposing but coordinated forces of cell division and cell death and considers the consequences of either process gone awry.

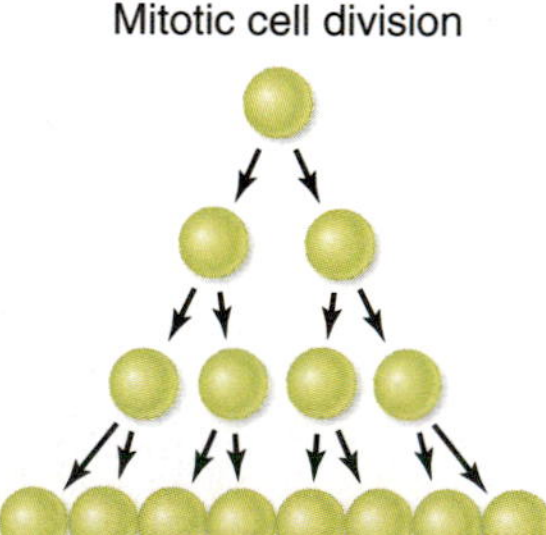

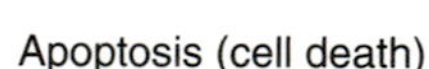

FIGURE 8.1 Cell Division and Apoptosis Regulate Development. Cells divide during development, producing new tissues. Meanwhile, apoptosis eliminates cells, maintaining the shapes and sizes of body parts. The balance between these opposing processes forms organs in the embryo and maintains their shapes during the rapid growth of the fetus.

A. Cell Division Requires Chromosome Duplication

When a cell divides, it must first duplicate its entire **genome,** which consists of all of its genetic material. This DNA contains the instructions for the proteins that sustain the cell's life. As the cell splits in two, it must therefore ensure that each new "daughter" cell receives a full set of DNA; if they do not, the new cells may die.

In prokaryotic cells, the genome consists of a single circular DNA molecule, so cell division is relatively simple (see figure 18.8). In a eukaryotic cell, however, distributing the DNA into daughter cells is a bit more complicated because the genetic information is divided among multiple chromosomes housed inside the cell's nucleus. A **chromosome** is a discrete package of DNA and associated proteins. The human genome consists of 46 chromosomes, whereas chickens have 78 and rice plants have just 24.

With so many chromosomes, every eukaryotic cell must balance two needs. On the one hand, to produce the proteins it requires, the cell must have access to the information in its DNA. On the other hand, if the cell is to divide, it must package its DNA into a portable form that can easily move into the two daughter cells. To understand how cells maintain this balance, we must take a closer look at the structure of the eukaryotic chromosome.

Eukaryotic chromosomes consist of **chromatin,** which is a collective term for DNA and its associated proteins in the nucleus. Among other functions, the proteins help to pack the DNA efficiently inside the cell. Stretched end to end, the DNA in one human cell would form a thread some 2 meters long. If the DNA bases of all 46 human chromosomes were typed as A, C, T, and G, the several billion letters would fill 4000 books of 500 pages each! How can a cell only 100 microns in diameter contain so much material?

The explanation is that chromatin is organized into units called **nucleosomes,** each consisting of a stretch of DNA wrapped around eight proteins (histones). A continuous thread of DNA connects nucleosomes like beads on a string. When the cell is not dividing, chromatin is barely visible because the nucleosomes are loosely packed together. The information in the DNA is therefore accessible for the cell to produce the enzymes and other proteins that it needs for all of its metabolic activities. DNA replication in preparation for cell division also requires that the cell's DNA be unwound (see chapter 7).

After DNA replication, but shortly before cell division occurs, the nucleosomes begin to fold into progressively larger structures, eventually forming discrete chromosomes that become visible in the microscope (**figure 8.2**). DNA

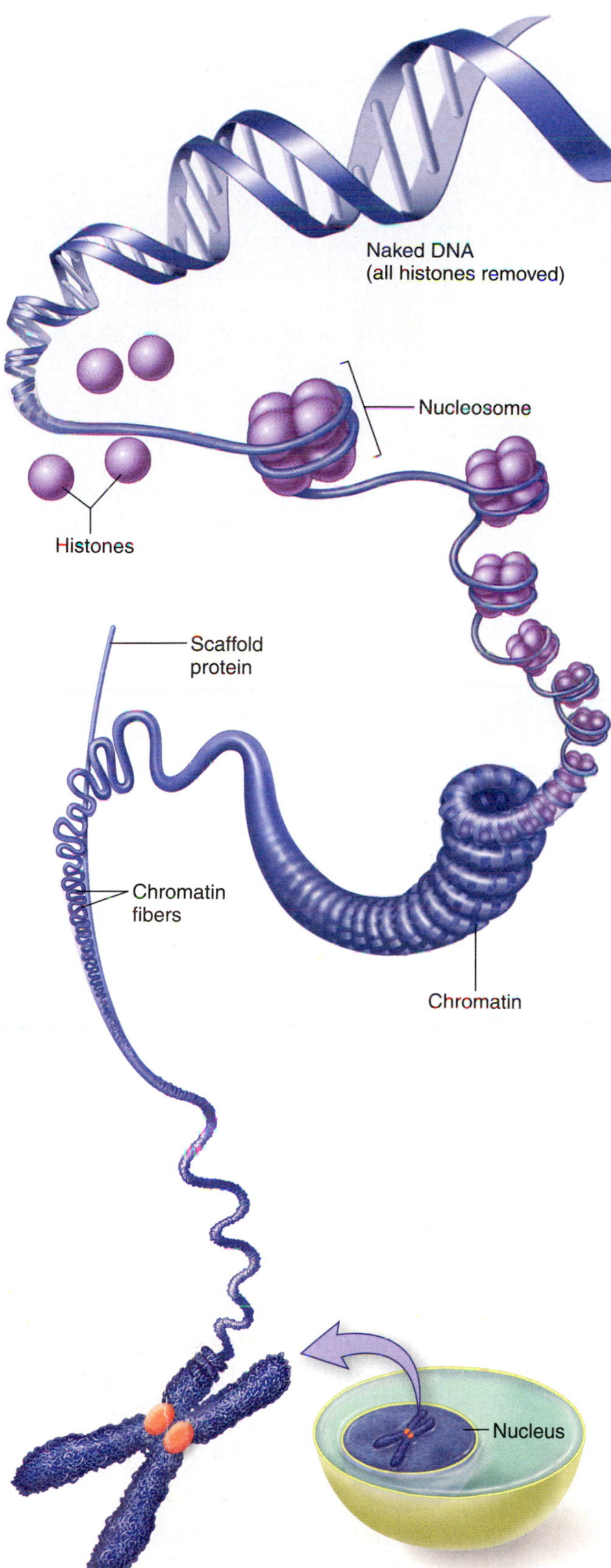

FIGURE 8.2 Chromatin Condenses to Form Visible Chromosomes in Eukaryotic Cells. Different degrees of chromatin packing produce a visible chromosome that consists of a very tightly wound molecule of DNA, plus associated proteins.

packing is somewhat similar to winding a very long length of thread around a wooden spool. Just as spooled thread occupies less space and is easier to move than a disorganized wad, so is condensed DNA much easier for the cell to manage than is unwound chromatin.

Once condensed, a chromosome has readily identifiable parts (**table 8.1** and **figure 8.3**). A replicated chromosome consists of two **chromatids,** each with a DNA sequence identical to the other. The two chromatids of a replicated chromosome are called "sister chromatids." The **centromere** is the small section of DNA and associated proteins that attaches the two sister chromatids to each other. It often appears as a constriction in a replicated chromosome. As a cell's nucleus divides, the centromere splits, and the sister chromatids separate from each other to become individual chromosomes, each with its own centromere.

Table 8.1 *Miniglossary of Chromosome Terms*

Term	Definition
Chromatin	Nucleic acids and associated proteins in the nucleus
Chromosome	A discrete, continuous molecule of DNA wrapped around protein. Eukaryotic cells contain multiple linear chromosomes, whereas prokaryotic cells each contain one circular chromosome.
Chromatid	One of two identical attached copies of a replicated chromosome
Centromere	A small part of a chromosome that attaches sister chromatids to each other

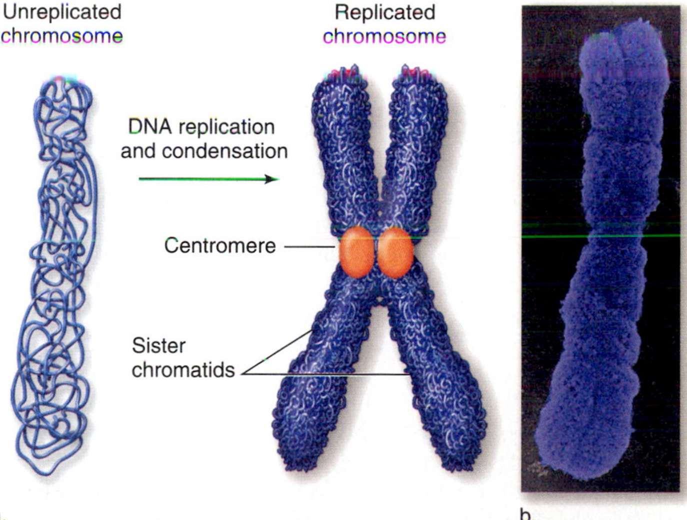

FIGURE 8.3 Parts of a Chromosome. (**A**) DNA replicates just before a cell divides. The two genetically identical chromatids of a replicated chromosome attach at the centromere. (**B**) This human chromosome is in the midst of forming sister chromatids. A longitudinal furrow extends from the chromosome tips inward.

FIGURE 8.4 Human Chromosomes. A chart called a karyotype displays the 46 chromosomes in a cell from a human male, arranged by size. Note the X and Y sex chromosomes. A cell from a female has two X chromosomes in addition to pairs 1 through 22.

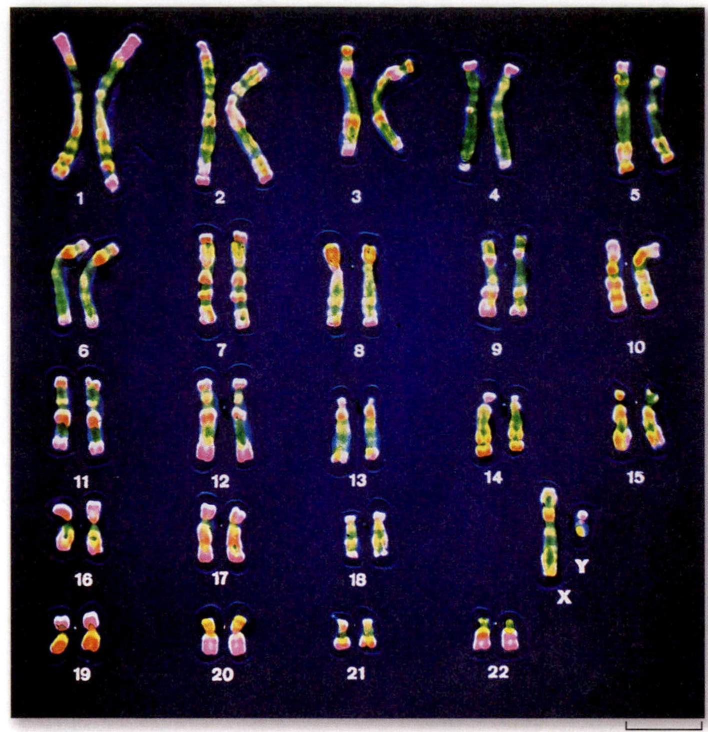

B. Two Parents, Two Sets of Chromosomes

Most human cells have 46 chromosomes, 23 from each parent. Thus, a human cell with 46 chromosomes is **diploid;** that is, it contains two full sets of genetic information, one set from each parent. In most animals and higher plants, the vast majority of cells are diploid.

Figure 8.4 shows that with just one exception, the two members of each chromosome pair look very much alike. (They are not, however, genetically identical to each other). The exception is the unmatched pair of sex chromosomes near the lower-right corner, in which the chromosome labeled "X" is clearly larger than the one labeled "Y." By itself, this observation identifies the individual as a male. A female, on the other hand, has two matching sex chromosomes (both X). In humans and many other sexually repro-

Can *You* Relate?

Why researchers study stem cells

The public debate over embryonic stem cells combines science, philosophy, religion, and politics in ways that few other modern issues do. What is the biology behind the issue?

A human develops from a single fertilized egg into an embryo and then a fetus—and eventually into an infant, child, and adult—thanks to mitotic cell division. As development continues, more and more cells become permanently specialized into muscle, skin, liver, brain, and other cell types. All contain the same DNA, but some genes become irreversibly "turned off" in specialized cells. Once committed to a fate, a mature cell rarely reverts to another type.

Stem cells are immature cells that retain the potential to develop into multiple cell types. When a stem cell divides to yield two daughter cells, one remains a stem cell, able to divide again. The other specializes. Many people believe that stem cells have great potential to treat Parkinson disease, spinal cord injuries, diabetes, heart disease, and many other illnesses that are currently incurable. With the correct combination of chemical signals, researchers should theoretically be able to coax stem cells to divide in the laboratory and produce any cell type.

The practical benefits would extend beyond treating illness. Currently, pharmaceutical companies test new drugs primarily on whole organisms, such as mice and rats. The ability to test on just kidney or brain cells, for example, would allow researchers to predict with much more precision the likely side effects of a new drug. It might also reduce the need for laboratory animals.

Where do stem cells come from? In fertility clinics, technicians fertilize eggs *in vitro*, but only a few of the resulting embryos are ever implanted into a woman's uterus. Some of the "spare" embryos are destroyed at 4 to 5 days old to harvest their stem cells. (The other embryos are either stored for possible later implantation, or they are destroyed without using the stem cells.) But many people consider it unethical to destroy human embryos for use in medical research. Researchers are therefore investigating adult stem cells in the brain, skin, small intestinal lining, and bone marrow (**figure 8.A**).

Both adult and embryonic stem cells have advantages and disadvantages for medical use. A patient's immune system would not reject tissues derived from his or her own adult stem cells. Adult stem cells are less abundant, however, and they usually give rise to only some cell types. In contrast, embryonic stem cells are more abundant and more versatile. However, a patient's immune system would probably reject tissues derived from an embryo's stem cells.

One possible solution to this last limitation might be to combine cloning technology (see this chapter's Burning Question: How do biologists use only DNA to clone mammals?) with stem cells. Transferring the nucleus of a patient's cell into an embryonic stem cell would create "customized" stem cells that the patient's immune system would not reject. This procedure, however, also raises ethical concerns.

ducing animals, the sex chromosomes alone determine an individual's sex.

If most human cells have 46 chromosomes, and if a baby arises from the union of a man's sperm and a woman's egg, then why does a human baby not have 92 chromosomes per cell (46 from each parent)? The answer is that the special cells required for sexual reproduction, sperm cells and egg cells, are not diploid. Rather, they are **haploid;** that is, they contain only one set of genetic information. These haploid cells, called **gametes,** are sex cells that combine to form the first (diploid) cell of the offspring. When a haploid sperm cell fertilizes a haploid egg cell, the new generation begins.

Only some cells can produce gametes. In humans and other animals, these specialized cells, called **germ cells,** occur only in the ovaries and testes. Plants don't have the same reproductive organs as animals, but they do have specialized gamete-producing cells in flowers and other reproductive parts. The rest of the body's cells, called **somatic cells,** do not produce sperm or eggs. Leaf cells, root cells, skin cells, and neurons are examples of somatic cells.

Thus, the life of a sexually reproducing, multicellular organism requires two ways to package DNA into daughter cells. **Meiosis,** described in chapter 9, forms genetically variable gametes that each contain half the number of chromosomes as the organism's somatic cells. **Mitosis,** by contrast, divides a eukaryotic cell's chromosomes into two identical daughter cells. This process happens over and over as an organism grows, develops, repairs damage, and (in many organisms) reproduces asexually. The next section of this chapter describes the events of mitotic cell division.

Summary *Proper development reflects a balance between cell division and cell death. DNA duplicates and coils tightly before a cell divides. Diploid cells contain two full sets of chromosomes, whereas haploid cells contain only one set.*

8.1 MASTERING CONCEPTS

1. Why are both cell division and apoptosis necessary for the development of an organism?
2. How do chromatin and histones interact?
3. What are the main parts of a chromosome?
4. Which cells in the body are diploid, and which are haploid?
5. How are somatic cells different from germ cells?

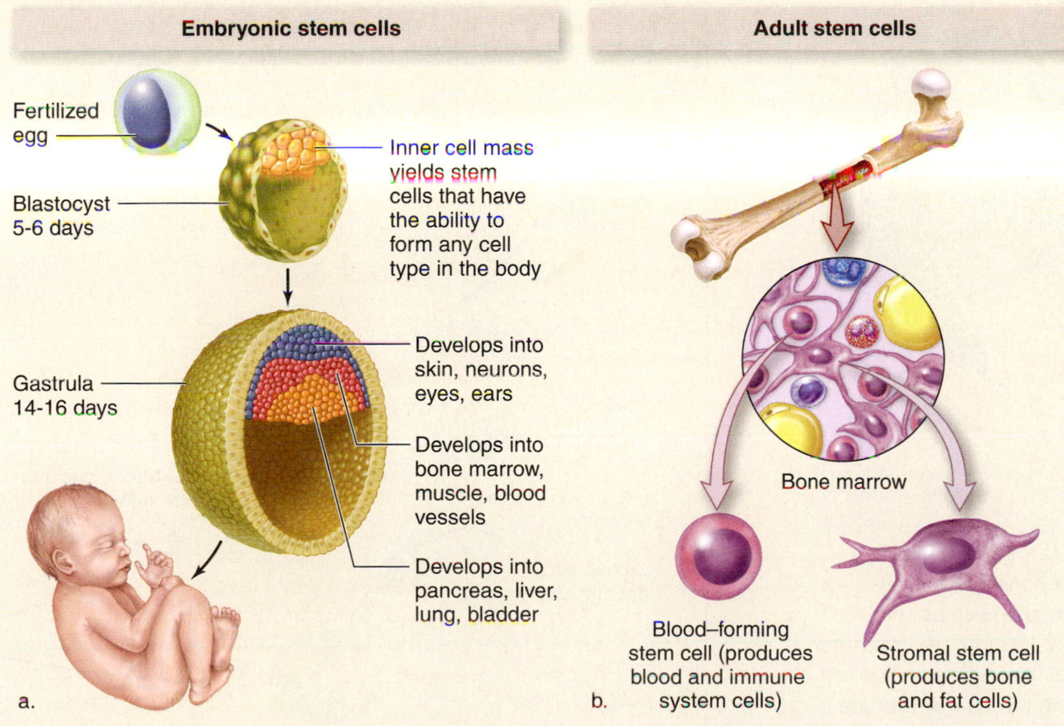

FIGURE 8.A Stem Cells. (**A**) Human embryonic stem cells are derived from the inner cell mass of a blastocyst, a stage in development that occurs several days after fertilization. Shortly thereafter, cells begin to specialize. (**B**) The adult body also contains stem cells, but they do not have the potential to develop into as many different cell types as do embryonic stem cells.

8.2 DNA Replicates, the Nucleus Divides, and the Cell Splits in Two

Suppose you scrape your leg while sliding into second base during a softball game. At first, the wound bleeds, but the blood soon clots and forms a scab. Underneath the dried crust, cells of the immune system clear away trapped dirt and dead cells. At the same time, undamaged skin cells bordering the wound begin to divide repeatedly, producing fresh, new daughter cells that eventually fill the damaged area.

Those actively dividing skin cells beautifully illustrate the **cell cycle,** which describes the events that occur between one cell division and the next. Biologists divide the cell cycle into two major stages (**figure 8.5**). During **interphase,** the cell is not dividing, although DNA replicates and many other events also occur. Immediately following interphase is the other major stage, cell division, which consists of mitosis and cytokinesis. Mitosis is the division of the nucleus, and **cytokinesis** is the splitting of the cell itself. The products of cell division are two daughter cells, each receiving complete, identical genetic instructions, as well as the molecules and organelles they need to maintain metabolism.

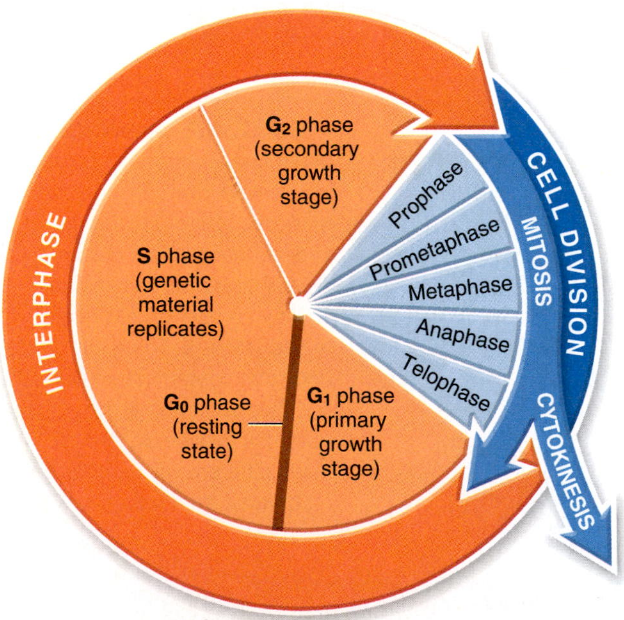

FIGURE 8.5 The Cell Cycle. The cell cycle is divided into two main stages. During interphase, cellular components replicate. During cell division (mitosis and cytokinesis), the nucleus and cytoplasm split in two, distributing their contents into two daughter cells. Interphase includes two gap phases (G_1 and G_2), when the cell grows and some organelles duplicate. During the synthesis phase (S) of interphase, DNA replicates. Mitosis consists of overlapping stages: prophase, prometaphase, metaphase, anaphase, and telophase.

A. Interphase Is a Time of Great Activity

Biologists once mistakenly described interphase as a time when the cell is at rest. The cell appears inactive because the chromatin is unwound and therefore barely visible. However, interphase is actually a very active time in the life of a cell. All of the cell's basic biochemical functions continue, and

Burning Questions

How do biologists use only DNA to clone mammals?

Unlike many other organisms, mammals do not naturally clone themselves. In 1996, however, researcher Ian Wilmut and his colleagues in Scotland used a new procedure to create Dolly the sheep, the first clone of an adult mammal. They removed the diploid nucleus of a cell taken from a donor sheep's mammary gland. They then transferred this nucleus to a sheep's egg cell whose own haploid nucleus had been removed (**figure 8.B**).

FIGURE 8.B Creating Dolly. Biologists cloned an adult sheep by obtaining a nucleus from a cell of a ewe's udder. They also removed the nucleus from an egg cell. Placing the adult cell's nucleus into the egg yielded a new cell genetically identical to the ewe. After being implanted into a surrogate mother sheep, the resulting embryo developed into Dolly.

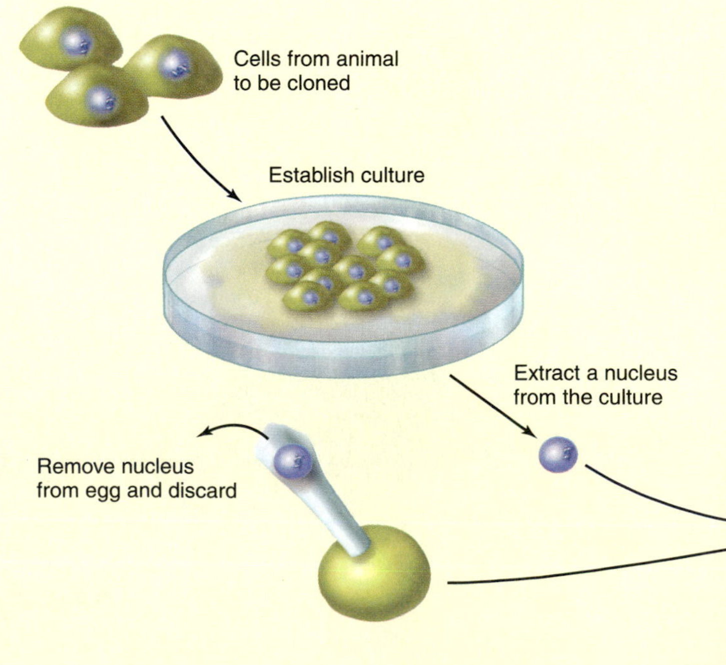

the cell replicates its genetic material, a process discussed in detail in chapter 7.

Interphase is divided into two "gap" phases (designated G_1 and G_2), separated by a "synthesis" (S) phase. During **G_1,** the cell grows, carries out basic functions, and produces molecules needed to build new organelles and other components it will require if it divides. A cell in G_1 is exquisitely sensitive to extracellular signals that "tell" it whether it should divide, stop to repair damaged DNA, die, or enter a quiescent stage called **G_0.** A cell in G_0 can continue to function, but it does not replicate its DNA or divide.

The cell divided mitotically to form an embryo. Inside a surrogate mother's uterus, the embryo developed into Dolly.

Dolly appeared normal, and she gave birth to four healthy lambs (via sexual reproduction). But she had arthritis in her hind legs, and she died of a lung infection in 2003. Although there is no evidence that Dolly's relatively early death was related to her being a clone, her short life has fueled speculation that clones may have hidden genetic abnormalities. Indeed, most clones die early in development, presumably because the gene regulation mechanisms in an adult cell's nucleus are somehow incompatible with those in the egg cell. Even the tiny percentage of clones that make it to birth often have abnormalities. These difficulties emphasize the significant ethical issues surrounding human cloning.

Have a Burning Question of your own? Submit it to marielle_hoefnagels@mcgraw-hill.com for possible inclusion in future editions of this book!

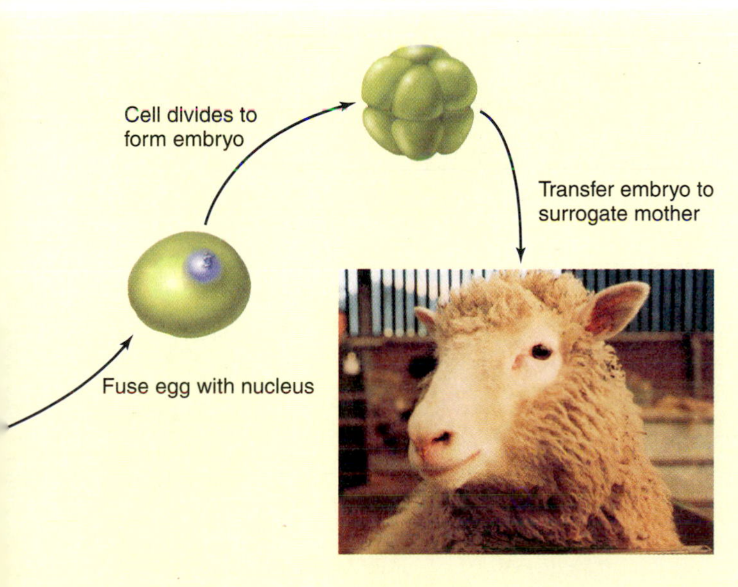

At any given time, most cells in the human body are in G_0, which may or may not be reversible. Nerve cells in the brain, for example, are permanently in G_0, which is why brain damage is often irreparable. Understanding how G_0 fits into the cell cycle was crucial in the first cloning of a mammal from the nucleus of a cell from an adult(see the Burning Question: How do biologists use only DNA to clone mammals?)

During **S phase,** the cell replicates the genetic material. In most human cells, assembling billions of DNA nucleotides takes 8 to 10 hours. By the end of S phase, each chromosome consists of two attached sister chromatids, although they are not yet visible with a light microscope. Enzymes also repair damaged DNA during S phase.

In **G_2 phase,** the cell prepares to divide, producing the proteins that will coordinate the movements of the chromosomes during mitosis. The DNA winds more tightly around its associated proteins, and this start of chromosome condensation signals impending mitosis. Interphase has ended.

B. Mitosis Distributes Chromosomes and Divides the Nucleus

Overall, mitosis separates the genetic material that replicated during S phase. For replicated chromosomes to be evenly distributed into two cells, they must line up in a way that enables them to be split equally into two sets that are then pulled to opposite poles of the cell. The **mitotic spindle** is a portion of the cytoskeleton that accomplishes this task (**figure 8.6**). Two structures called **centrosomes** organize the protein subunits of the mitotic spindle. In many animal cells, each centrosome includes a pair of barrel-shaped centrioles. (Plant cells and many cells of animal embryos lack centrioles.) Proteins called **kinetochores** attach the chromosomes to the spindle.

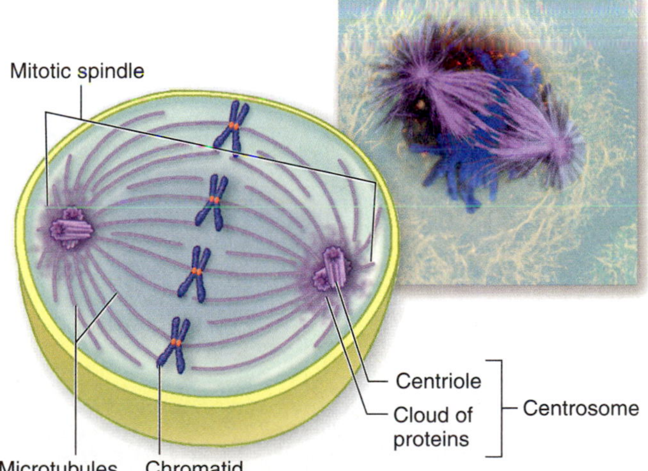

FIGURE 8.6 The Spindle Aligns Chromosomes. The mitotic spindle consists of microtubules that form the fibers that grow outward from two centrosomes. The spindle fibers push and pull to align the chromosomes. The inset shows an amphibian cell during metaphase.

Mitosis is a continuous process, but biologists divide it into stages for ease of understanding (**figure 8.7**).

During **prophase,** DNA coils very tightly around its histone "spools," shortening and thickening the chromosomes. As they condense, the chromosomes become visible when stained and viewed under a microscope. The nucleolus (the darkened area in the nucleus) disappears. The two centrosomes migrate toward oposite poles of the cell, and the mitotic spindle begins to form.

Prometaphase occurs immediately after the formation of the spindle. Complexes of kinetochores begin to grow on each chromosome's centromere, and the nuclear envelope breaks into small pieces.

As **metaphase** begins, the mitotic spindle aligns the chromosomes down the center, or equator, of the cell. Because of this alignment, when cell division is complete, each resulting cell will contain one chromatid from each duplicated chromosome.

In **anaphase,** the centromeres split as the mitotic spindle pulls one chromatid from each pair to opposite poles of the cell. As the chromatids separate, some microtubules in the spindle shorten and others lengthen in a way that moves the poles farther apart, stretching the dividing cell.

In **telophase,** the final stage of mitosis, the mitotic spindle disassembles and chromosomes begin to unwind. In addition, nucleoli and nuclear envelopes re-form at each end of the stretched-out cell.

After telophase, division of the genetic material is complete, and the cell contains two nuclei—but not for long.

C. Cytokinesis Distributes and Divides the Cytoplasm

In cytokinesis, organelles and macromolecules are distributed into the two forming daughter cells, which then physically separate. After cytokinesis is complete, the daughter cells enter interphase, and the cell cycle begins anew.

In an animal cell, the first sign of cytokinesis is the **cleavage furrow,** a slight indentation around the cell at the equator (**figure 8.8A**). This indentation results from a contractile ring of actin and myosin that forms beneath the cell membrane. It contracts like a drawstring, separating the daughter cells.

Plant cells must construct a new cell wall that separates the two daughter cells (figure 8.8B). In plants, the first sign of cell wall construction is a line, called a cell plate, separating the forming cells. Vesicles deliver structural materials

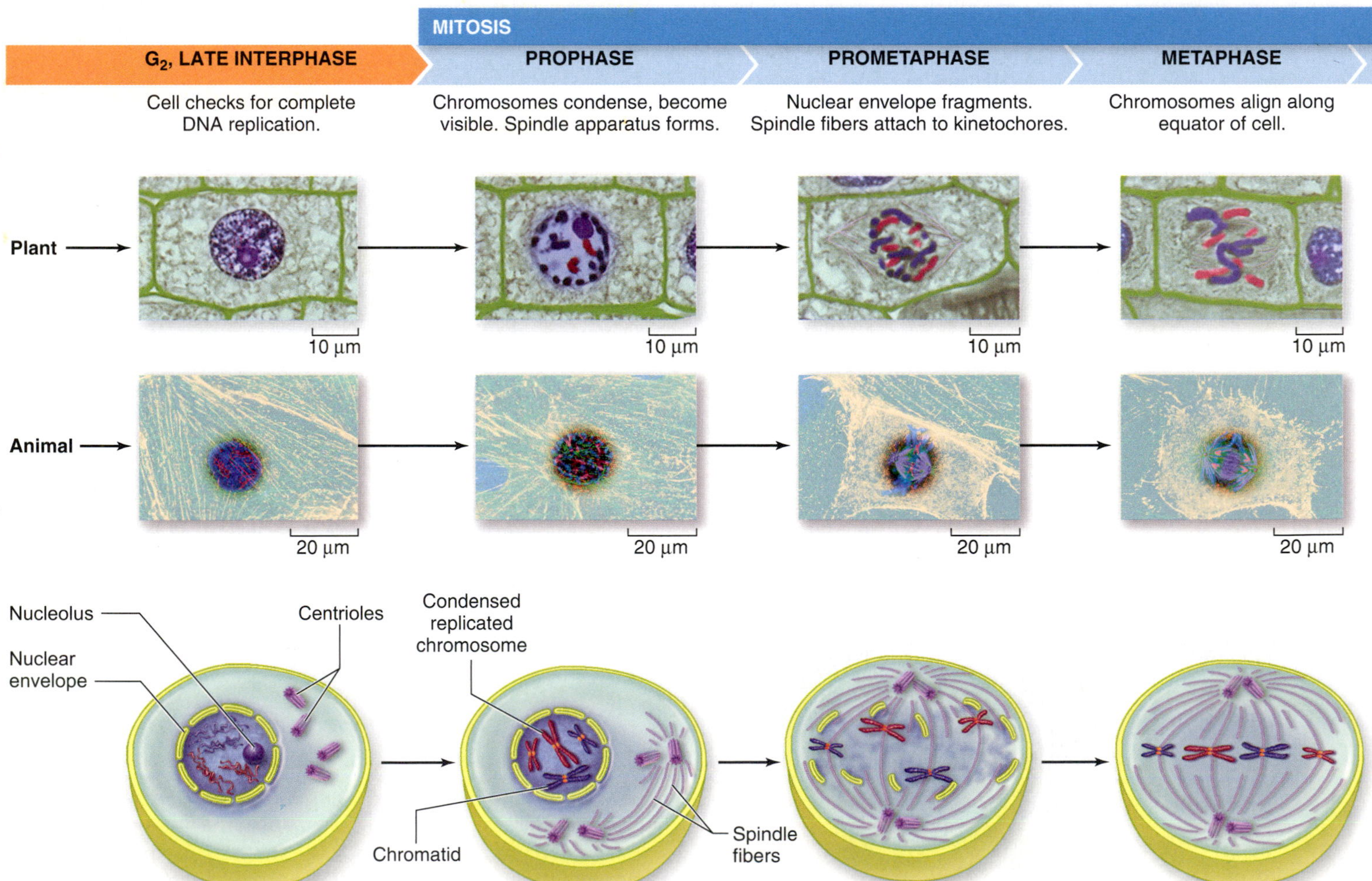

FIGURE 8.7 **Steps of Mitosis.** Mitotic cell division includes similar stages in all eukaryotes, including plants and animals.

Contractile ring
Nucleus
Nucleus
Cleavage furrow
Nucleus
Nucleus
Daughter cells
Spindle fibers
Chromosome
Chromosome
a.
Contractile ring

Vesicle
Primary cell wall
Cell plate
Nucleus
Two primary cell walls
Cell plate
Nucleus
b.
10 µm

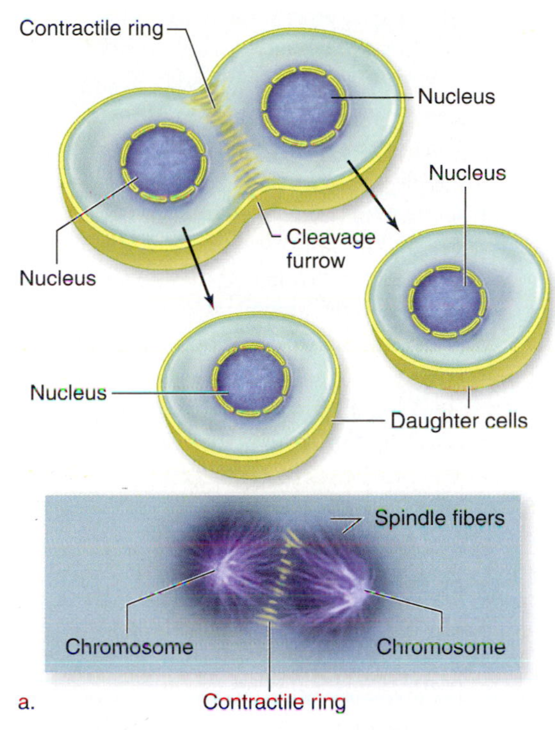

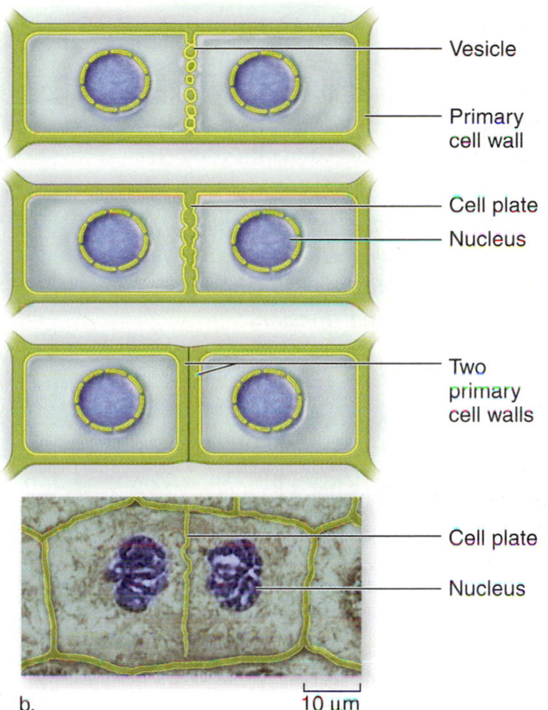

FIGURE 8.8
Cytokinesis Begins as the Chromosomes Separate. (**A**) In an animal cell, the first sign of cytokinesis is the formation of an indentation called the cleavage furrow. A contractile ring consisting of actin and myosin proteins forms the cleavage furrow. (**B**) A cell plate is an early sign of cell wall formation during cytokinesis in a plant cell.

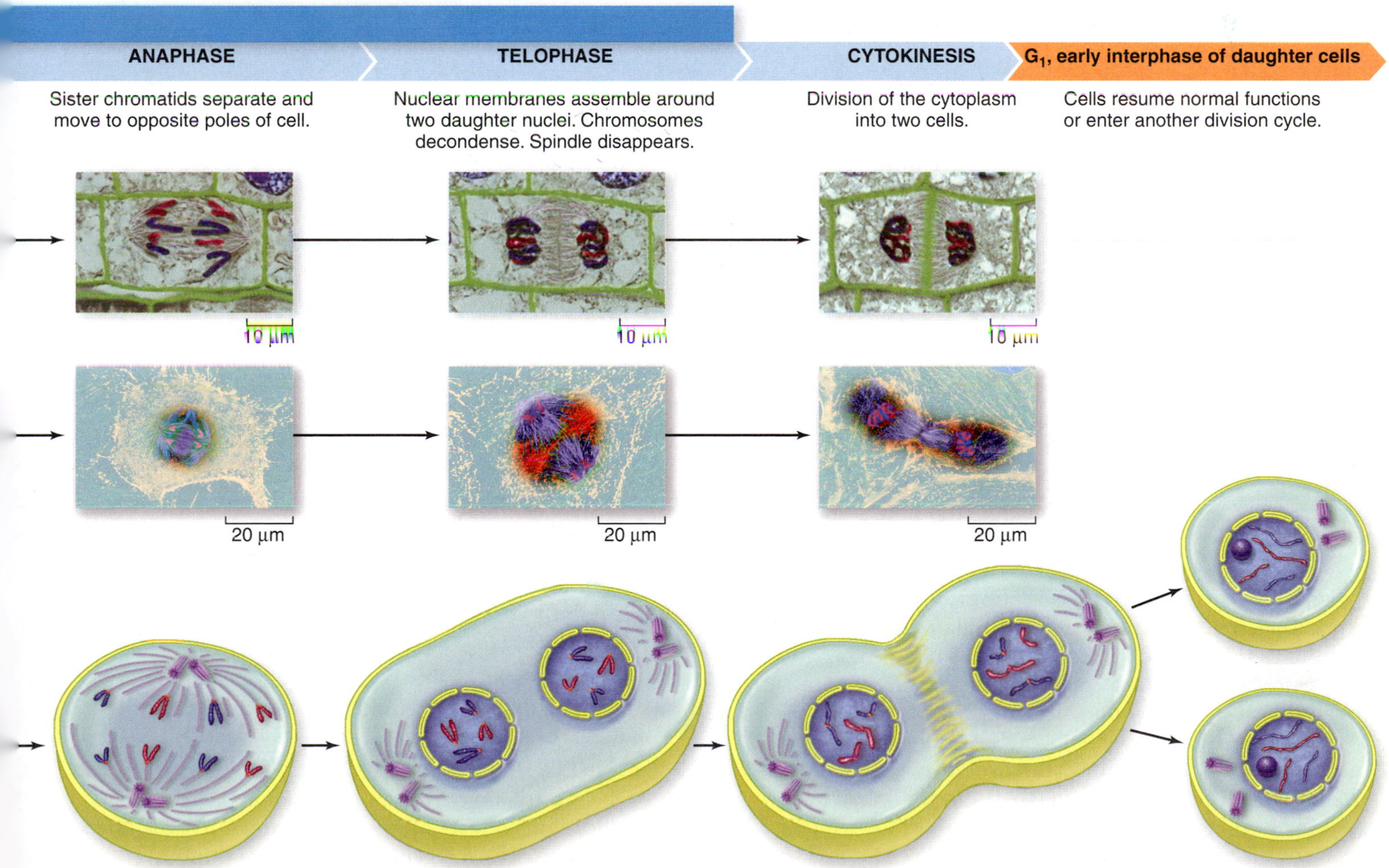

ANAPHASE
Sister chromatids separate and move to opposite poles of cell.
TELOPHASE
Nuclear membranes assemble around two daughter nuclei. Chromosomes decondense. Spindle disappears.
CYTOKINESIS
Division of the cytoplasm into two cells.
G1, early interphase of daughter cells
Cells resume normal functions or enter another division cycle.
10 µm
10 µm
10 µm
20 µm
20 µm
20 µm

such as cellulose fibers, other polysaccharides, and proteins. The layer of cellulose fibers embedded in surrounding material makes a strong, rigid wall that gives plant cells their rectangular shapes. **cell wall, p. 71**

Table 8.2 reviews vocabulary related to the cell cycle.

Summary *The cell cycle (interphase, mitosis, and cell division) describes the events occurring in an actively dividing cell. The number of chromosomes in a cell before mitotic cell division is the same as in each of its daughter cells after division.*

8.2 MASTERING CONCEPTS

1. What happens during interphase?
2. How does the mitotic spindle form, and what is its function?
3. What happens during each stage of mitosis?
4. Distinguish between mitosis and cytokinesis.

Table 8.2 *Miniglossary of Cell Division Terms*

Term	Definition
Diploid	Containing two sets of chromosomes, one from each parent; somatic cells are diploid
Haploid	Containing one set of chromosomes; gametes are haploid
Meiosis	Division of a diploid nucleus into four genetically different haploid nuclei
Mitosis	Division of a nucleus into two identical nuclei
Cytokinesis	Distribution of cytoplasm to daughter cells following division of a cell's nucleus
Mitotic spindle	Part of the cytoskeleton that moves chromosomes during mitosis
Kinetochore	Part of the centromere to which the mitotic spindle attaches
Centrosome	Structure that organizes the mitotic spindle; it consists of tubulin and other proteins, and sometimes centrioles

8.3 Cells Tightly Regulate the Cell Cycle

Some cells divide more or less constantly. The cells at the tips of a plant's roots, for example, may continue to divide throughout the growing season, exploring the soil for water and nutrients. Likewise, stem cells in your bone marrow constantly produce new blood cells. On the other hand, the skin cells bordering a wound quit dividing once healing is complete; brain cells simply do not divide at all once they are mature. How do any of these cells "know" what to do?

Regulation of mitosis is a complex task involving several mechanisms that chemically "tell" a somatic cell to divide or to cease dividing. Precise timing is essential. Too little cell division, and an injury may go unrepaired; too much, and an abnormal growth forms. An understanding of these signals may help reveal how diseases such as cancer arise.

Signals to divide usually come from outside the cell. **Growth factors** are proteins that stimulate cell division. They bind to receptors on a receiving cell's membrane, and then a cascade of chemical reactions inside the cell initiates cell division. At a wound site, for example, epidermal growth factor stimulates cells to divide and produce new skin underneath a scab. **signal transduction, p. 56**

Some growth factors remove internal blocks to cell division. Other signals suppress the cell cycle by preventing the removal of the same blocks. The remainder of this section describes two of these internal cell cycle control mechanisms.

A. Checkpoints Keep the Cell Cycle on Track

Several biochemical "checkpoints" control the cell cycle, ensuring that a cell does not enter the next stage of the cell cycle until the previous stage is complete. These checkpoints are somewhat like the guards that check passports and other documents at border crossings. **Figure 8.9** illustrates a few:

- The G_1 checkpoint screens for DNA damage. If the DNA is damaged beyond repair, however, a protein called p53 triggers apoptosis, and the cell dies.
- Several S phase checkpoints ensure that DNA replication occurs properly. If the cell does not have enough nucleotides to complete replication, or if a DNA molecule breaks, the cell cycle may pause or stop at this point.
- The G_2 checkpoint is the last one before the cell begins mitosis. Does the cell contain two full sets of identical DNA? Can damaged DNA be repaired? Is the spindle-making machinery in place? If not, the cell cycle may be delayed, or the p53 protein triggers apoptosis.
- At the metaphase checkpoint, the cell makes sure that all chromosomes are aligned and that the spindle fibers attach correctly to the chromosomes. If everything checks out, the cell proceeds to anaphase.

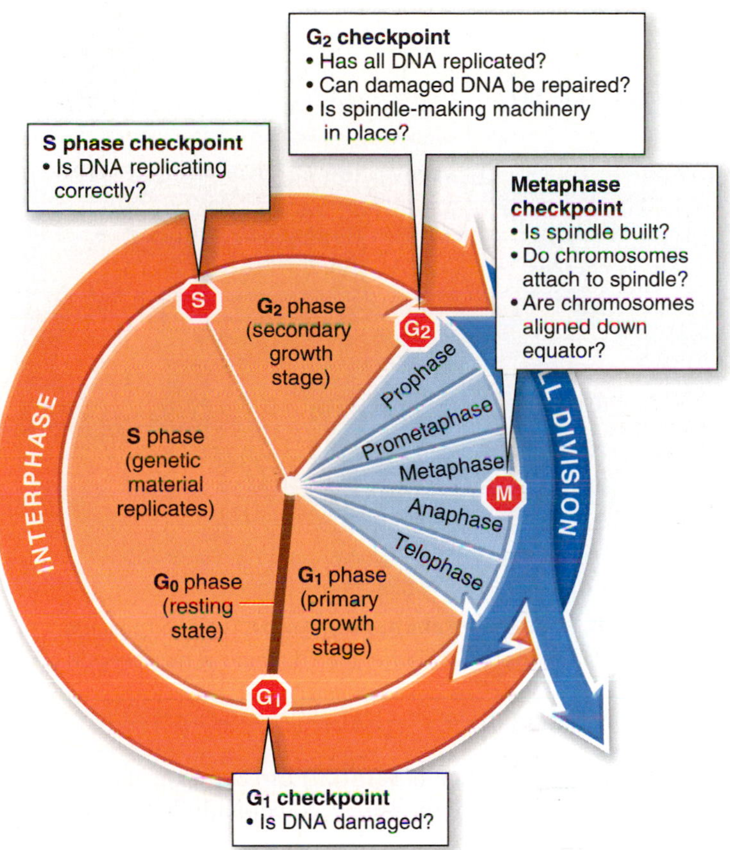

FIGURE 8.9 Cell Cycle Control. Cell cycle checkpoints ensure that the cell completes each stage correctly before proceeding to the next. Many types of cancer result from missing or faulty proteins that cause damaged checkpoints.

B. Telomeres Provide a Built-in Limit to Cell Division

Many cells grown in the laboratory obey an internal "clock" that allows them to divide a maximum number of times. It is as if cells "know" how many times they have divided and how many more divisions they can undergo. In the early 1990s, cell biologists discovered that the cellular clock resides in **telomeres,** the tips of eukaryotic chromosomes.

Telomeres consist of hundreds to thousands of repeats of a specific DNA sequence. At each cell division, the telomeres in a cell lose nucleotides from their ends, and the chromosomes gradually become shorter. After about 50 divisions, the cumulative loss of telomere DNA signals cell division to cease.

Cells with shrinking telomeres lack an enzyme called **telomerase,** which can continually add DNA to chromosome tips (**figure 8.10**). When cells make telomerase, their telomeres stay long, which enables them to divide beyond the 50-or-so division limit. In humans, telomerase levels are high in sperm-forming cells, blood cells, and the cells that line the intestine. In cancer cells, which divide many more times than normal cells, the level of telomerase rises as the disease progresses. Inactivating telomerase in cancer cells, or boosting its activity in aging cells, could have tremendous medical potential.

Plant cells produce telomerase, and many divide beyond 50 divisions. Lack of a telomere clock makes sense for the way that plants develop. Many plant parts have abundant meristem tissue, which has an unlimited ability to divide. If plants could not keep their telomeres long, the chromosomes of meristem cells would wear down and the cells would cease to divide. The plant's growth would stop. **meristems, p. 503**

Summary *Signals from outside the cell trigger cell cycle control mechanisms, including biochemical checkpoints. A "clock" provided by shortening chromosome tips limits the number of times a cell can divide.*

8.3 MASTERING CONCEPTS

1. What are growth factors?
2. What happens at cell cycle checkpoints?
3. How does telomere length serve as a cell division clock in some organisms?

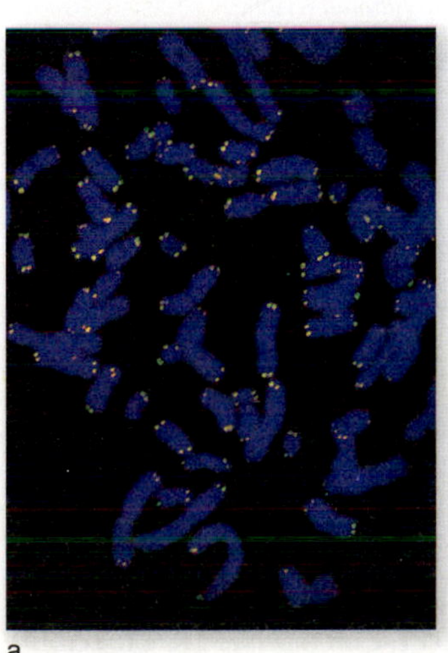
a.

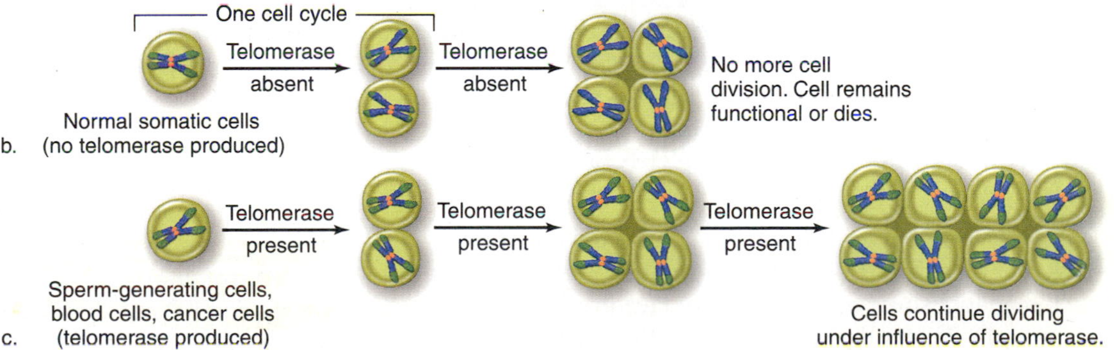

FIGURE 8.10 Telomeres. (**A**) Fluorescent tags indicate the telomeres on these stained human chromosomes. (**B**) In normal somatic cells of vertebrates, telomeres shorten with each cell division because the cells do not produce telomerase. When the telomeres shrink too much, the cell no longer divides. (**C**) Sperm-generating cells, blood cells, and cancer cells produce telomerase and continually extend their telomeres, resetting the cell division clock.

8.4 Cancer Arises When Cells Divide out of Control

What happens when the body loses control over the balance between cell division and cell death? Sometimes, a **tumor**—an abnormal mass of tissue—forms. Biologists classify tumors into two main groups. Benign tumors are usually slow-growing and harmless, unless they become large enough to disrupt nearby tissues or organs. They do not invade nearby tissues or spread to other parts of the body. In contrast, a **malignant** tumor invades adjacent tissue. It also is likely to **metastasize,** meaning that some of its cells break away from the original mass and travel in the bloodstream or lymphatic system to colonize other areas of the body. **Cancer** is a class of diseases characterized by malignant tumors.

Cancer begins with a single cell that breaks through its death and division controls. The cell continues to divide, and it grows into a malignant tumor. Each cancerous cell passes its loss of cell cycle control to its daughter cells.

Any malignant tumor's growth rate is slow at first because fewer cells are dividing. However, not all tumors continue to grow at the same rate. The smallest detectable fast-growing tumor is about 0.5 cm in diameter and can contain hundreds of millions of cells, dividing at a rate that produces a million or so new cells an hour. Other cancers are very slow to develop. Lung cancer, for example, may take three to four decades to develop.

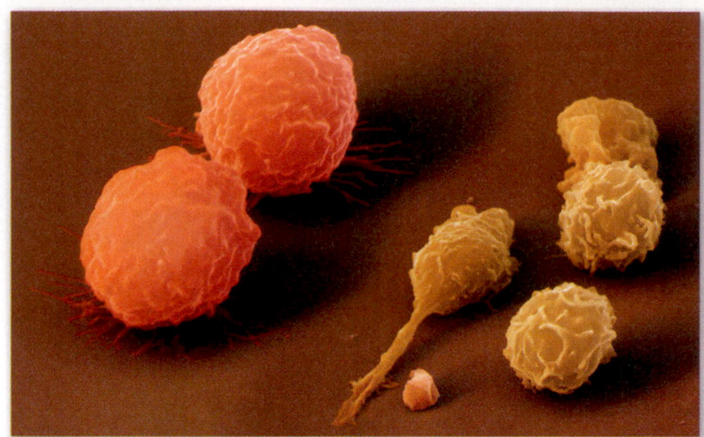

FIGURE 8.11 Cancer Cells Are Abnormal. The two cancerous leukemia cells on the left are larger than the normal marrow cells on the right.

A. Cancer Cells Differ from Normal Cells in Many Ways

Given sufficient nutrients and space, cancer cells can divide uncontrollably and eternally. The cervical cancer cells of a woman named Henrietta Lacks vividly illustrate these characteristics. Shortly before she died in 1951, researchers removed some of her cancer cells and began to propagate them in a laboratory at Johns Hopkins University. Lacks's cells grew so well, dividing so often, that they quickly became a favorite of cell biologists seeking cells to culture that would divide beyond the normal 50-division limit. Still used today, "HeLa" (for *He*nrietta *La*cks) cells replicate so vigorously that if just a few of them contaminate a culture of different cells, within days they completely take over.

Cancer cells differ from normal cells in many ways.

- A cancer cell looks different from a normal cell (**figure 8.11**). It is rounder, its cell membrane is more fluid, and it may lose some of the specialized features of its parent cells. These differences allow pathologists to detect cancerous cells by examining tissue under a microscope.
- Unlike normal cells, many cancer cells are essentially immortal. Cancer cells ignore the "clock" that limits normal cells to 50 or so divisions.
- Whereas normal cells respond only to external growth factors, many cancer cells produce their own signals to divide.
- Normal cells growing in culture exhibit **contact inhibition,** meaning that they stop dividing when they touch one another in a one-cell-thick layer. Cancer cells lack contact inhibition, so they tend to pile up in culture.
- A normal cell dies (undergoes apoptosis) when badly damaged, but many cancer cells do not.
- Cancer cells send signals that stimulate the development of new blood vessels. In this way, a tumor builds its own blood supply that delivers nutrients and removes wastes.

B. Inheritance and Environment Both Can Cause Cancer

Proteins control both the cell cycle and apoptosis. Genes encode proteins, so genetic mutations (changes in genes) play a key role in causing cancer. So far, researchers know of hundreds of genes that contribute to cancer. Two classes of cancer-related genes are oncogenes and tumor suppressor genes.

Oncogenes are abnormal variants of genes that normally control cell division (**figure 8.12**). They can encode many types of proteins, from the receptors that normally bind growth factors outside the cell, to any of the participants in the series of reactions that trigger cell division. If expressed at the wrong time or in the wrong tissues, oncogenes may accelerate the cell cycle and cause cancer. Oncogenes are especially

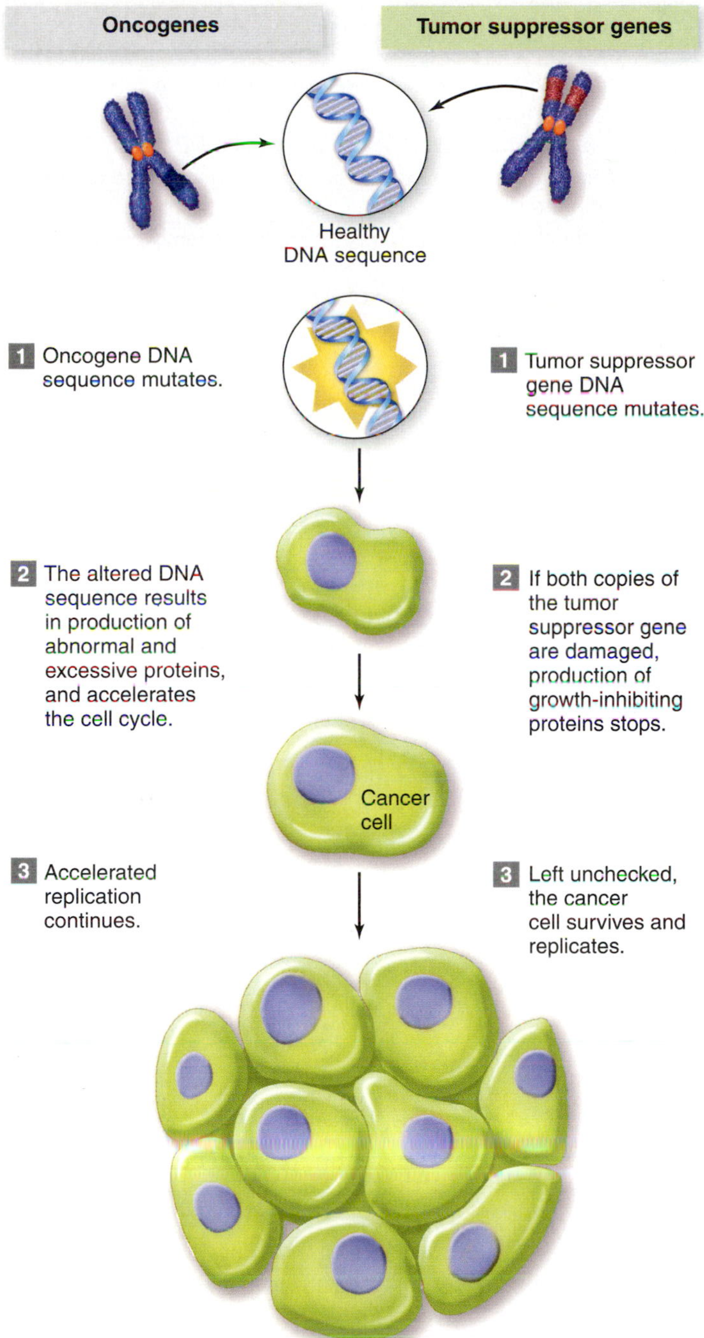

Oncogenes
Tumor suppressor genes
Healthy DNA sequence
1 Oncogene DNA sequence mutates.
1 Tumor suppressor gene DNA sequence mutates.
2 The altered DNA sequence results in production of abnormal and excessive proteins, and accelerates the cell cycle.
2 If both copies of the tumor suppressor gene are damaged, production of growth-inhibiting proteins stops.
Cancer cell
3 Accelerated replication continues.
3 Left unchecked, the cancer cell survives and replicates.

FIGURE 8.12 Two Types of Cancer Related Genes. Oncogenes and tumor suppressor genes both influence the cell cylcle. Oncogenes are abnormal genes that accelerate cell division. Tumor suppressor genes normally inhibit cell divison, but when these genes are mutated, cancer can develop.

dangerous because only one of a cell's two copies needs to be damaged for cancer to develop. Mutated oncogenes cause some cancers of the cervix, bladder, and liver.

Tumor suppressor genes encode proteins that normally block cancer development; that is, they promote normal cell death or prevent cell division. *BRCA1*, a gene associated with some types of breast cancer, is one example. Inactivation, deletion, or mutation of a tumor suppressor gene can cause cancer by eliminating these crucial limits on cell division. Unlike oncogenes, usually both copies of a tumor suppressor gene in a cell must be damaged for cancer to develop.

Where do the cancer-causing mutations in oncogenes and tumor suppressor genes come from? Sometimes, a person inherits mutant versions of the genes from one or both parents. The parent may have had cancer, or the mutations may have arisen spontaneously in his or her sperm- or egg-producing cells. Often, however, people develop cancer after exposure to harmful chemicals, radiation, and viruses, all of which may alter their genes. Poor diet and exercise habits, sun exposure, and cigarette smoking can also raise cancer risks. Chapter 12 explores mutations that predispose people to cancer in more detail.

C. Cancer Treatments Remove or Kill Abnormal Cells

Traditional cancer treatments, often used in combination, include surgical tumor removal, drugs (chemotherapy), and radiation. Chemotherapy drugs, usually delivered intravenously, are intended to stop cancer cells anywhere in the body from dividing. Radiation therapy uses directed streams of energy from radioactive isotopes to kill tumor cells in limited areas. **isotopes, p. 23**

Chemotherapy and radiation are relatively "blunt tools" that target rapidly dividing cells, both cancerous and healthy. The death of healthy cells in the bone marrow, digestive tract, and hair follicles causes the most common side effects from cancer treatments, including fatigue, nausea, and hair loss. Fortunately, the healthy cells usually return after the treatment concludes. Some patients, especially those who receive high doses of chemotherapy or radiation, receive bone marrow transplants to speed the replacement of healthy blood cells.

The future has never been brighter for new treatments, many of which come from basic research into the cell cycle. New, more targeted drugs home in on the receptors for growth factors that are overabundant on some cancer cells (see the opening essay for chapter 3). Such new drugs are very successful in treating some forms of breast cancer and leukemia. In addition, drugs called angiogenesis inhibitors block a tumor's ability to recruit blood vessels, starving the cancer cells of their support system. Cancer patients may one day receive gene therapy treatments that replace faulty genes with functional copies.

The success of any cancer treatment depends on many factors, including the type of cancer and the stage in which it is detected. Surgery can cure cancers that have not spread, or that have spread only to local lymph nodes. Once cancer metastasizes, however, it becomes more difficult to treat. The reason is that as cancer cells spread, their DNA often mutates. Treatment that shrank the original tumor may have no effect on this new, changed growth. Also, a treatment that kills 99.9% of a tumor's cells can still leave millions of cells to divide and regrow (see section 8.6).

Summary *Cancer reflects abnormal cell cycle control. A cancer cell has characteristic features, some of which are visible in a microscope. Mutations in oncogenes, tumor suppressor genes, and other genes eliminate a cancer cell's mitotic "brake." These mutations may be inherited or triggered by environmental factors. Cancer treatments include surgery, drugs, and radiation.*

8.4 MASTERING CONCEPTS

1. What is the difference between a benign and a malignant tumor?
2. How do cancer cells differ from normal cells?
3. What is the difference between an oncogene and a tumor suppressor gene?
4. Distinguish between surgery, radiation, chemotherapy, and gene therapy.

8.5 Cell Death Is a Part of Life

Development relies on a balance between cell division and cell death. So far, this chapter has described cell division, its control, and the breakdown of control that leads to cancer. The other side of the equation is apoptosis, or programmed cell death.

A. Why Cells Die

Apoptosis has two main functions. First, if cell division continued unchecked and some cells didn't die, all organisms would be massive, shapeless lumps. Instead, apoptosis eliminates excess cells to carve out functional structures such as fingers, toes, nostrils, and ears. The disappearance of a tadpole's tail as it transforms into an adult frog illustrates apoptosis, as does the webbing that vanishes as a chicken's foot forms **(figure 8.13)**.

Second, apoptosis weeds out cells that otherwise might harm the organism. In fact, apoptosis appears to be a "default" option in the cell cycle, which must be overcome at a checkpoint for mitosis to occur (see figure 8.9). A good example of the protective function of apoptosis is the skin peeling that follows a sunburn. Sunlight contains UV radiation that damages DNA and may cause cancer. If the damage is very severe, the G_1 checkpoint protein p53 sends the cell along a pathway toward apoptosis. The killed skin cells peel away.

Abnormal apoptosis harms health. In the fetus, apoptosis fine-tunes the immune system by killing the white blood cells that react with the body's own molecules (see chapter 36). If this were not the case, the immune cells would attack the body, resulting in an autoimmune disorder such as lupus or rheumatoid arthritis. On the other hand, excessive apoptosis can also upset cell number balance. In the brain, a stroke triggers apoptosis in the cells surrounding the directly affected area, extending the damage as cells die.

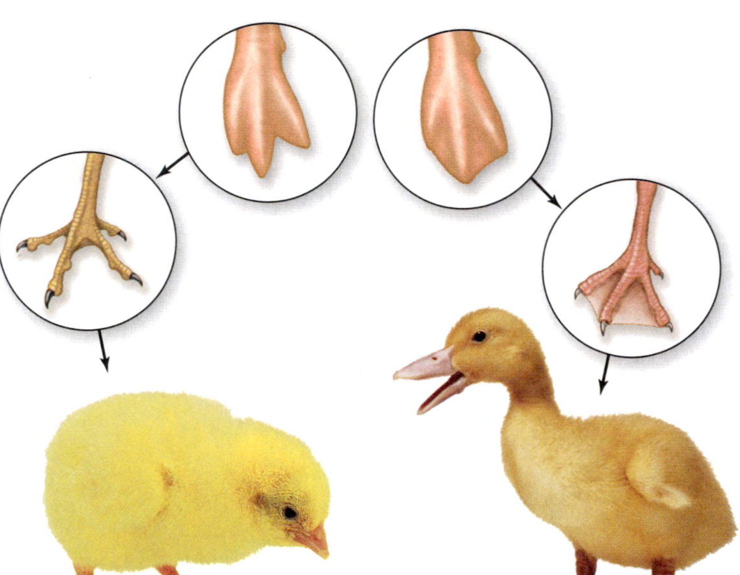

FIGURE 8.13 Apoptosis Carves Toes. A developing chicken foot undergoes extensive apoptosis (**A**), but a developing duck's foot doesn't (**B**), retaining the webbing.

B. Killer Enzymes Dismantle the Cell

Overall, apoptosis dismantles a cell into pieces that immune system cells can engulf and destroy. Apoptosis begins when a "death receptor" protein on a doomed cell's membrane receives

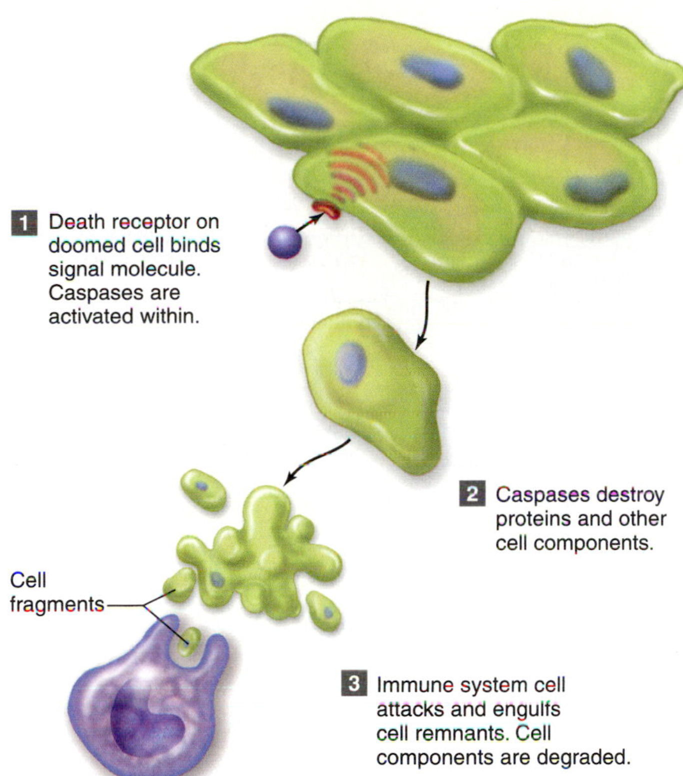
1 Death receptor on doomed cell binds signal molecule. Caspases are activated within.
2 Caspases destroy proteins and other cell components.
Cell fragments
3 Immune system cell attacks and engulfs cell remnants. Cell components are degraded.

FIGURE 8.14 Death of a Cell. A cell undergoing apoptosis loses its characteristic shape, bulges out, and finally falls apart. Caspases destroy the cell's insides. Immune system cells digest the remains.

a signal to die. Within seconds, a series of chemical reactions activates apoptosis-specific enzymes called **caspases,** which cut apart the cell's proteins and destroy the cell **(figure 8.14)**.

Within an hour of when the death receptor first received the bad news, the event is over. Immune system cells descend, and the cell is soon gone.

Apoptosis is clearly part of animal life, but many of the genes encoding proteins that carry out apoptosis are also present in such diverse organisms as fungi and slime molds. Plant cells die too, but not in precisely the way that animal cells meet their programmed fate. Instead, plant cells are digested by enzymes in their own vacuoles. Plants also use a form of cell death to kill cells infected by fungi or bacteria, limiting the spread of the pathogen.

Summary *Apoptosis is programmed death that occurs as a normal part of development or that weeds out damaged cells. Cells that perish go through a characteristic sequence of events.*

8.5 MASTERING CONCEPTS

1. How can cell death be a normal part of development?
2. Give an example of a protective function of apoptosis.
3. What events happen in a cell undergoing apoptosis?
4. How can too much or too little apoptosis affect health?

8.6 INVESTIGATING LIFE
Cutting Off a Tumor's Supply Lines in the War on Cancer

When Charles Darwin proposed natural selection as a mechanism of evolutionary change, he envisioned selective forces operating on tortoises, flowering plants, and other whole organisms. But the power of natural selection extends to a much, much smaller scale, including the individual cells that make up a tumor. The advance, retreat, resurgence, and death of these renegade cells command dramatic headlines in the war on cancer.

Our weapons against cancer include powerful chemotherapy drugs, but the evolution of drug-resistant tumor cells is a significant barrier to successful treatment. To illustrate why, consider a woman with ovarian cancer. She receives the drug cisplatin, and her tumor shrinks. But if just a few tumor cells can detoxify this drug, they will continue to divide. The tumor will regrow, and subsequent treatment with cisplatin will be ineffective. The physician will then need to find a new drug to treat the woman's cancer.

Unfortunately, rapidly dividing tumor cells frequently develop resistance to drugs because frequent cell division produces abundant opportunities for mutations. An alternative cancer-fighting strategy, therefore, might be to launch an indirect attack on the tumor's slower growing support tissues instead.

Any tumor larger than 1 or 2 mm^3 needs a blood supply to carry nutrients, oxygen, and wastes. How can a growing tumor acquire the blood supply it needs? Blood travels throughout the body in vessels lined with endothelial tissue. For a blood vessel to grow, its endothelial cells must divide, which happens rarely in adults. Cancer cells secrete molecules that change the behavior of the blood vessels. The tumor's chemical signals bind to receptor proteins on endothelial cell membranes, triggering a series of reactions. The stimulated endothelial cells secrete enzymes that break down the material surrounding the cells, making room for new endothelial cells to migrate toward the source of the signal—the tumor. The cells divide as they migrate, forming a hollow tube that will eventually become a new blood vessel. This sprouting of new "supply lines" is called angiogenesis.

This discovery suggested a new cancer-fighting strategy, in the form of drugs called angiogenesis inhibitors. These drugs stop blood vessel growth in one of several ways. Some block the chemical signals that trigger cell division, while others prevent endothelial cells from making room for the new cells. Still others, including one called endostatin, inhibit endothelial cells directly.

Endostatin is a 184-amino-acid protein that keeps endothelial cells from dividing but does not affect resting endothelial cells or other cells in the body. It should therefore choke off a tumor's supply lines without toxic side effects. The drug became the focus of intense media attention in 1997 when cancer researchers Thomas Boehm, Judah Folkman, and their colleagues at the Dana Farber Cancer Center and Harvard Medical School reported that endostatin suppressed tumor development in mice—and that the cancer cells did not develop resistance to it.

To test endostatin, the researchers first induced cancer in 6-week-old male mice by injecting each animal with one of three types of cancer cells. After tumors developed, the researchers injected some of the mice with endostatin, while control mice received a placebo (inert injections of saline solution). Injections continued for several days, until the tumors in endostatin-treated mice were barely detectable. Whenever the tumors regrew, the researchers repeated the injections.

The results were astounding. For each of the three cancer types, the tumors never developed resistance **(figure 8.15A)**. Each time the tumors grew back and the researchers administered endostatin, the tumors again shriveled. Standard chemotherapy drugs might temporarily shrink a tumor or delay its development, but resistant cells soon cause the tumor to bounce back (figure 8.15B). Moreover, after two to six treatments with endostatin, the tumors never grew back. For all three cancer types, the endostatin-treated mice remained healthy and gained weight normally.

The results of subsequent clinical trials with human cancer patients, however, were mixed. Endostatin shrank tumors, without side effects, in a handful of people. But the drug was ineffective in most patients, and its U.S. manufacturer eventually stopped making it, citing high production costs. (No one knows why it worked so much better in mice than in people, but such disparities are common in cancer research.) In the meantime, however, Chinese researchers reported in 2005 that a 27-amino-acid fragment of endostatin has similar cancer-fighting properties and costs much less to produce. Time will tell whether this development will rekindle initial excitement about endostatin.

What does endostatin have to do with evolution? The logic behind its use as an anticancer drug relies on basic concepts of natural selection. Because DNA may mutate every time it replicates, rapidly dividing cancer cells are genetically slightly different from one another. A conventional chemotherapy drug may kill most cancer cells in a tumor, but a few have a mutation that lets them survive, divide, and give rise to a new, much more resistant tumor. Endostatin is different. It does not target the tumor itself; instead, it affects a blood vessel's endothelial cells. These cells rarely divide and therefore accumulate mutations very slowly, so the chance that they will become resistant to endostatin is small.

This may seem comforting, but evolution will not stand still for our convenience. Tumor cells may still defeat endostatin with new mutations that will enable them to either inactivate or break down the drug. Understanding natural selection helps researchers know what to look for—and perhaps even launch new offensives in the war on cancer.

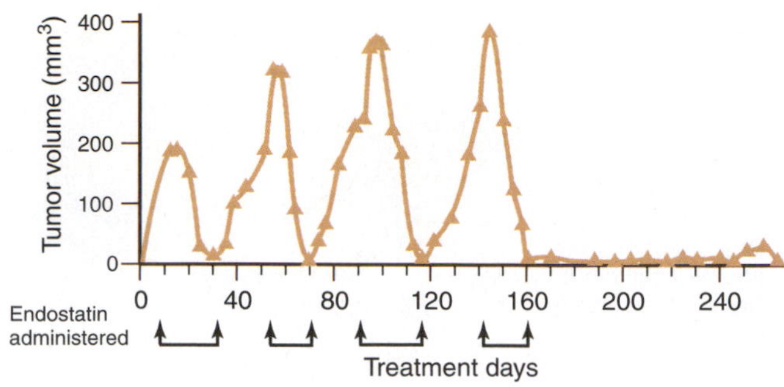

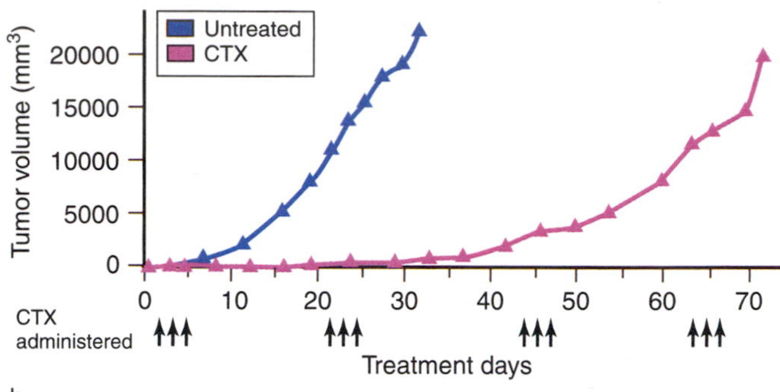

FIGURE 8.15 No Resistance. **(A)** Endostatin repeatedly shrank tumors in mice, and the tumors never developed resistance to the drug. **(B)** A traditional chemotherapy drug, cyclophosphamide (CTX), delayed but did not prevent tumor growth in mice. **Question:** Suppose a friend shows you the results of a study in which ginger slowed tumor growth in mice for 30 days. What additional questions would you ask before deciding whether to recommend that your cancer-stricken relative eat more ginger?

Boehm, Thomas, Judah Folkman, Timothy Browder, Michael S. O'Reilly. November 27, 1997. Antiangiogenic therapy of experimental cancer does not induce acquired drug resistance. *Nature*, vol. 390, pages 404–407.

CHAPTER SUMMARY

8.1 Cells Divide, and Cells Die

- Cell division produces identical copies of eukaryotic cells, while **apoptosis** is programmed cell death. Both occur during the normal development of an organism.

A. Cell Division Requires Chromosome Duplication

- A dividing cell must first duplicate its **genome**, which may consist of one or more **chromosomes**. Most human cells contain 46 chromosomes.
- A chromosome consists of **chromatin** (DNA plus protein). In eukaryotic cells, chromatin is organized into **nucleosomes,** which enable the cell to pack a lot of DNA into a small space.
- Once replicated, a chromosome consists of two identical **chromatids** attached at a section of DNA called a **centromere**.

B. Two Parents, Two Sets of Chromosomes

- **Diploid** cells contain two full sets of chromosomes, one from each parent. **Haploid** cells contain only one set.
- **Germ cells** produce haploid **gametes** by **meiosis;** all other cells in the body are **somatic** and divide **mitotically**.

8.2 DNA Replicates, the Nucleus Divides, and the Cell Splits in Two

- The **cell cycle** is a sequence of events in which a cell is preparing to divide (interphase), dividing its genetic material (mitosis), or dividing its cytoplasm (cytokinesis).

A. Interphase Is a Time of Great Activity

- **Interphase** includes two gap periods, $\mathbf{G_1}$ and $\mathbf{G_2}$, when the cell makes proteins, carbohydrates, and lipids, and a synthesis period (**S**), when it replicates genetic material.

B. Mitosis Distributes Chromosomes and Divides the Nucleus

- Microtubules assemble to build the **mitotic spindle,** which arises from paired **centrosomes.**
- **Mitosis** consists of five stages. In **prophase,** the chromosomes condense, the nucleolus disassembles, and the mitotic spindle forms. In **prometaphase,** the nuclear envelope breaks up and moves out of the way, and spindle fibers attach to **kinetochores.** In **metaphase,** spindle fibers align replicated chromosomes down the cell's equator. In **anaphase,** the chromatids of each replicated chromosome separate, sending a complete set of genetic instructions to each end of the cell. In **telophase,** the spindle breaks down, and nuclear envelopes form.

C. Cytokinesis Distributes and Divides the Cytoplasm

- **Cytokinesis** is the physical separation of the two daughter cells, and it usually begins during anaphase or telophase. When an animal cell undergoes cytokinesis, a **cleavage furrow** forms, and a contractile band draws the two cells apart. In a plant cell, a cell plate separates the daughter cells, marking the site where a new cell wall will form.

8.3 Cells Tightly Regulate the Cell Cycle

- **Growth factors** are external molecular signals that stimulate cell division; other chemical signals inhibit it.

A. Checkpoints Keep the Cell Cycle on Track

- Cell cycle control checkpoints allow the cell to ensure that each stage of the cell cycle is complete before the next begins. The cell may pause briefly to repair errors. Alternatively, if the damage is too great to repair, the checkpoints may trigger apoptosis.

B. Telomeres Provide a Built-in Limit to Cell Division

- **Telomeres** are DNA sequences at the ends of chromosomes that track the number of divisions a cell has undergone. When telomeres become very short, division ceases. An enzyme called **telomerase** adds DNA to telomeres in some cells. Cancer cells and some rapidly dividing cells retain long telomeres and divide continually.

8.4 Cancer Arises When Cells Divide out of Control

- **Tumors** can result from excess cell division or deficient apoptosis. A **malignant** tumor infiltrates nearby tissues and **metastasizes** if it reaches the bloodstream.

A. Cancer Cells Differ from Normal Cells in Many Ways

- A **cancer** cell has an altered surface, loses specialization, lacks **contact inhibition**, and divides uncontrollably to yield other cancer cells.

B. Inheritance and Environment Both Can Cause Cancer

- Cancer can result from an overexpressed **oncogene** or a mutated **tumor suppressor gene.** Mutations in cancer-related genes may be inherited or occur in response to environmental triggers.

C. Cancer Treatments Remove or Kill Abnormal Cells

- Surgery, chemotherapy, and radiation are the most common cancer treatments. Newer treatments target cell membrane proteins. Success depends on the type of cancer and whether it has spread. Gene therapy may provide new treatment options in the future.

8.5 Cell Death Is a Part of Life

A. Why Cells Die

- Apoptosis shapes structures and protects by killing cells that could become cancerous.

B. Killer Enzymes Dismantle the Cell

- **Caspases** destroy an apoptotic cell, then immune system cells mop up the remains.

8.6 Investigating Life: Cutting off a Tumor's Supply Lines in the War on Cancer

- Natural selection occurs inside tumors. As chemotherapy drugs eliminate susceptible cells, resistant ones survive and divide to regrow the tumor.
- Endostatin starves tumors by stopping the growth of blood vessels. Because endothelial cells in blood vessels divide much more slowly than tumor cells, they are much less likely to become resistant to endostatin.

MULTIPLE CHOICE QUESTIONS

1. A chromosome is made of
   a. DNA.
   b. histones.
   c. chromatin.
   d. all of the above.
2. Why is a chromosome composed of *two* chromatids?
   a. Because the nucleosomes are folded.
   b. Because the chromosome contains the entire genome of a cell.
   c. Because the DNA has replicated.
   d. Because the cell is diploid.
3. A ______ cell is ______ and a gamete is ______
   a. sperm; diploid; haploid.
   b. somatic; diploid; haploid.
   c. germ; haploid; diploid.
   d. somatic; haploid; diploid.
4. How is G_0 different from G_1?
   a. In G_0 the cell is replicating its DNA.
   b. A G_1 cell is getting ready to divide, but a G_0 cell has already divided.
   c. G_0 occurs at the end of interphase.
   d. A G_1 cell can continue to divide, but a G_0 cell does not.
5. Which is the correct order of phases in mitosis?
   a. prometaphase, prophase, metaphase, anaphase, telophase
   b. prometaphase, metaphase, telophase, anaphase, prophase
   c. prophase, prometaphase, metaphase, anaphase, telophase
   d. telophase, anaphase, metaphase, prometaphase, prophase
6. What would happen to a dividing animal cell if there was no cytokinesis?
   a. The number of nuclei in the cell would increase over time.
   b. The amount of DNA in the cell would decrease over time.
   c. The cell would enter G_0.
   d. The cell would not form a new cell wall.
7. Why is the metaphase checkpoint so important?
   a. Because it determines how quickly a cell goes through mitosis.
   b. Because it ensures that the chromosomes will be properly separated.
   c. Because it ensures that cytokinesis will proceed properly.
   d. Because it can trigger apoptosis.
8. Predict how excess telomerase activity would affect a cell.
   a. It would cause the telomeres of the chromosomes to rapidly shrink.
   b. It would reduce the number of chromosomes in the cell.
   c. It would increase the number of times the cell could divide.
   d. It would inhibit growth of the organism.
9. What is an *oncogene*?
   a. A gene that normally regulates the cell cycle
   b. A gene that regulates cell death
   c. An abnormal allele of a gene that can influence the cell cycle
   d. A gene that responds to the signals associated with contact inhibition
10. Why are the caspases so important to apoptosis?
    a. They kill the cell by destroying its DNA and proteins.
    b. They function as the "death receptor" on the surface of the cell.
    c. They are part of the immune response that eliminates the cells.
    d. They cause the cell to swell and burst.

TESTING YOUR KNOWLEDGE

1. Explain the relationship among chromatin, chromosome, chromatid, and centromere.
2. What is the difference between haploid and diploid cells? Are your skin cells haploid or diploid? What about gametes?
3. Biologists once thought interphase is a time of cellular rest, but it is not. What happens during interphase?
4. Why is G_1 a crucial time in the life of a cell?
5. Describe the events that take place during mitosis.
6. How do biochemicals from inside and outside the cell control the cell cycle?
7. List four ways that cancer cells differ from normal cells.
8. Why do cancer treatments have unpleasant side effects?
9. What is the role of apoptosis in the development of a human from fertilized egg to adult? What role does apoptosis play in an adult's body?

THINKING AS A SCIENTIST

1. Can mitosis occur in both haploid and diploid cells?
2. Describe what will happen to a cell if interphase happens, but mitosis does not.
3. Which contains the most DNA, a cell in G_1 or a cell in G_2 phase?
4. Cytochalasin B is a drug that blocks cytokinesis by disrupting the microfilaments in the contractile ring. What effect would this drug have on cell division?
5. A cell from a newborn human divides 19 times in culture and is then frozen for 10 years. After thawing, how many times is the cell likely to divide?
6. How might the observation that more advanced cancer cells have higher telomerase activity be developed into a test that could help physicians treat cancer patients?
7. A protein called p53 regulates the cell cycle in multicellular organisms. Explain the observation that many cancer cells have mutations in the *p53* gene.
8. A bacterium called *Agrobacterium tumefaciens* infects plant cells. It causes a plant disease called crown gall by inserting a piece of DNA that causes the plant cell to divide rapidly and produce molecules that the bacteria eat. In what ways is the gall similar to a malignant tumor? In what ways is it different?
9. A researcher removes a tumor from a mouse and breaks it up into individual cells. He injects each cell into a different mouse. Although all of the mice in the experiment are genetically identical and were raised in the same environment, they develop cancers that spread at different rates. Some mice die quickly, some linger, and others recover. What do these results indicate about the cells that made up the original tumor?
10. Why can combining a traditional cancer treatment with an angiogenesis inhibitor be more effective than either treatment alone?
11. How would your body look if cells divided, but apoptosis never occurred?

Visit www.mhhe.com/hoefnagels for practice quizzes, animations, videos, and activities designed to help you master the material in this chapter.

Sexual Reproduction and Meiosis

This salamander, *Ambystoma tremblayi*, reproduces by parthenogenesis.

What's the Point?

Fatherless Salamanders, Mice, and Blackberries

Are males optional? This question may sound strange, because humans cannot reproduce unless a sperm cell fertilizes an egg cell. The females of some species, however, can produce armies of identical daughters in the absence of a male. The general term for this type of asexual reproduction is parthenogenesis, *Greek for "virgin birth."*

Parthenogenesis is unusual in animals, yet it occurs in a surprising variety of species, including some snakes, lizards, salamanders, fish, insects, roundworms, and flatworms. Some species of salamanders, for example, consist entirely of females that reproduce only by parthenogenesis. They do "mate" with males of a closely related species. The sperm cells activate the females' oocytes, but the male cells usually die before they can contribute their DNA. Thus, the offspring are genetically identical to their mother.

Natural parthenogenesis in mammals is unknown, but in 2004, Japanese researchers artificially stimulated mouse egg cells to undergo parthenogenesis. The result: two fatherless mice. This type of research suggests scientists may one day be able to coax unfertilized human egg cells to divide in the laboratory and yield stem cells for research and medical applications (see chapter 8's Can You Relate? Why Researchers Study Stem Cells). If it works, this technique could reduce the need to harvest stem cells from discarded human embryos.

About 400 species of plants, including blackberries and dandelions, produce seeds by apomixis, which is the botanical equivalent of parthenogenesis. Normally, sexual reproduction in plants requires a cell inside a flower's ovary to develop into an egg cell. Pollination then brings a sperm cell to the egg cell, and the fertilized egg cell develops into an embryo. In apomixis, the ovarian cell does not produce an egg cell. It instead divides to become an embryo that is genetically identical to its parent.

Logically, asexual reproduction should be more common than sex, because it saves the energy cost of attracting a mate. Indeed, asexual reproduction works for organisms that are well adapted to their environment. Yet it is also risky. The organisms are genetically alike, and all may perish if conditions change. Parthenogenetic organisms may therefore not survive long-term competition with their sexually reproducing, genetically variable counterparts.

Chapter 8 described how cells make virtually identical copies of themselves. This chapter explains another type of cell division, meiosis, which generates the genetically variable sperm and egg cells (gametes) that lie at the heart of sexual reproduction.

LEARNING OUTLINE

9.1 Why Sex?

Organisms must reproduce—generate other individuals like themselves—for a species to persist. The most straightforward and ancient way for a single-celled organism to reproduce is **asexually,** by replicating its genetic material and splitting the contents of one cell into two. Except for the occasional mutation, asexual reproduction generates genetically identical offspring. Bacteria and archaea reproduce in this way (see figure 18.8), as do single-celled eukaryotes such as the amoeba in **figure 9.1A.** Many multicellular organisms also reproduce asexually, as described in the opening essays to chapter 8 and this chapter. **genetic mutations, p. 248**

Sexual reproduction, in contrast, is the production of offspring whose genetic makeup comes from two parents. A female parent contributes an egg, and the male produces sperm. The fusion of these sex cells signals the start of the next generation. Because sexual reproduction mixes up and recombines traits, the offspring are genetically different from each other (figure 9.1B).

Learning how diverse organisms reproduce and exchange genetic material can provide clues to how sexual reproduction may have evolved. The earliest process that combines genes from two individuals appeared about 3.5 billion years ago, in a form of bacterial gene transfer that is still prevalent today. In **conjugation**, one bacterial cell uses an outgrowth called a sex pilus to transfer genetic material to another bacterium (**figure 9.2**). The unicellular eukaryote *Paramecium* uses a variation on this theme, exchanging entire nuclei via a bridge of cytoplasm.

Both bacteria and *Paramecium* reproduce asexually, but they can still acquire new genetic information from their neighbors. Unicellular green algae of the genus *Chlamydomonas*, however, exhibit a simple form of true sexual reproduction, in which two genetically different cells fuse to form a new individual. The earliest sexual reproduction, which may have begun about 1.5 billion years ago, may have been similar to that of *Chlamydomonas*.

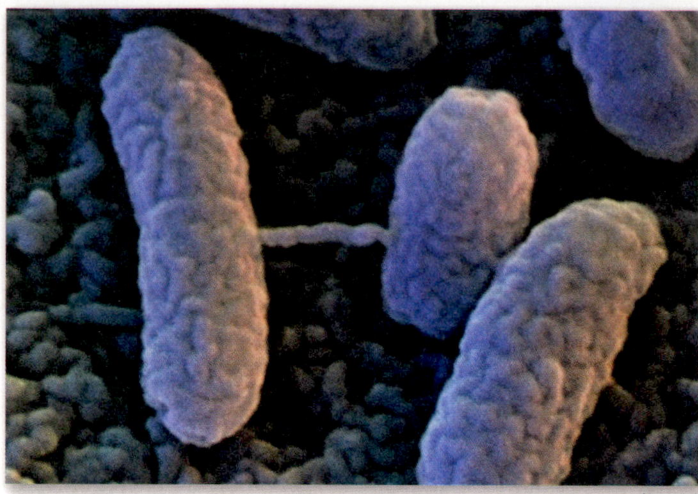

FIGURE 9.2 Conjugation. The bacterium *Escherichia coli* reproduces asexually, but *E. coli* cells can also exchange genetic material via an appendage called a sex pilus. DNA transfer between bacterial cells is called conjugation.

Attracting mates takes a lot of energy, as does producing and dispersing sperm and egg cells. Yet the persistence of sexual reproduction over billions of years and in many diverse species attests to its success. Why does such a costly method of reproducing persist, and why is asexual reproduction comparatively rare?

The mass production of identical offspring makes sense in an unchanging environment, but environmental condi-

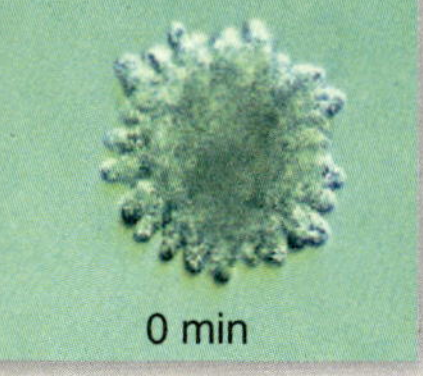

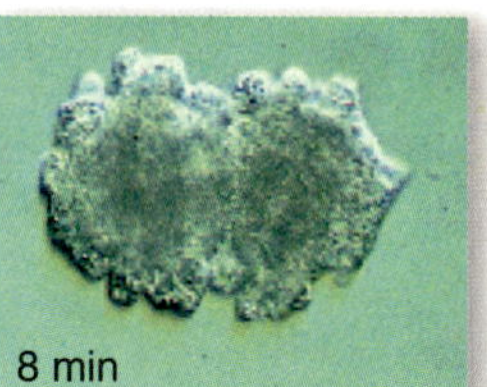

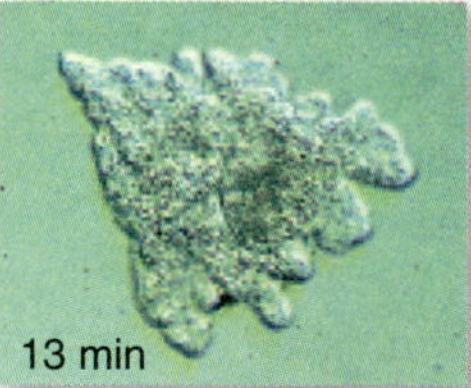

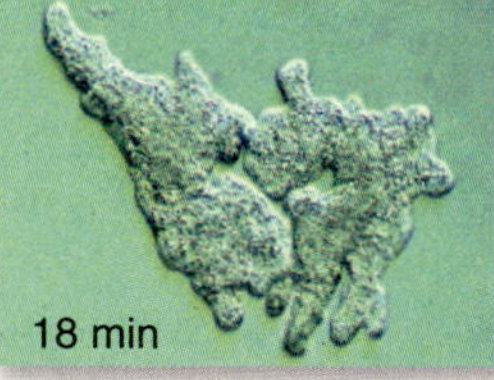

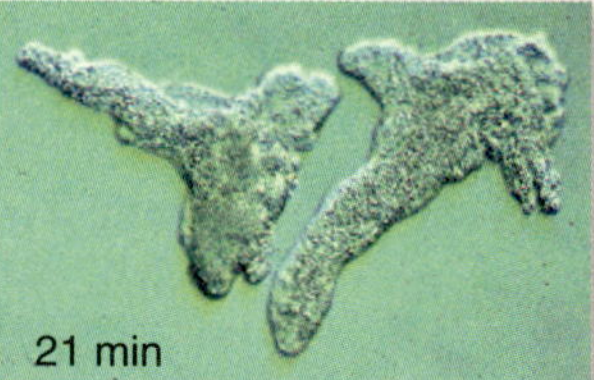

a.

b.

FIGURE 9.1 Asexual and Sexual Reproduction. (**A**) The single-celled *Amoeba proteus* reproduces asexually by splitting in two. (**B**) These kittens differ from each other because they were conceived sexually, so each received different combinations of the parents' genes.

tions rarely remain constant in the real world. Faced with a drastic environmental change, poorly suited individuals will die, but others might have a combination of traits that allows them to survive and reproduce (**figure 9.3**). Asexual reproduction cannot create or maintain this diversity, but sexual reproduction can.

Summary *Reproduction may yield offspring identical to a parent or offspring that combine the traits of two parents. Both asexual and sexual reproduction persist because they are successful in different conditions.*

9.1 MASTERING CONCEPTS

1. How do asexual and sexual reproduction differ?
2. How can asexually reproducing organisms acquire new genetic information?
3. What circumstances select for asexual reproduction? Sexual reproduction?

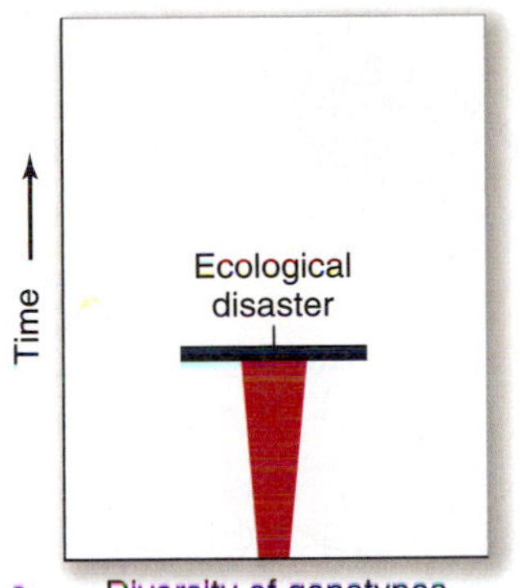

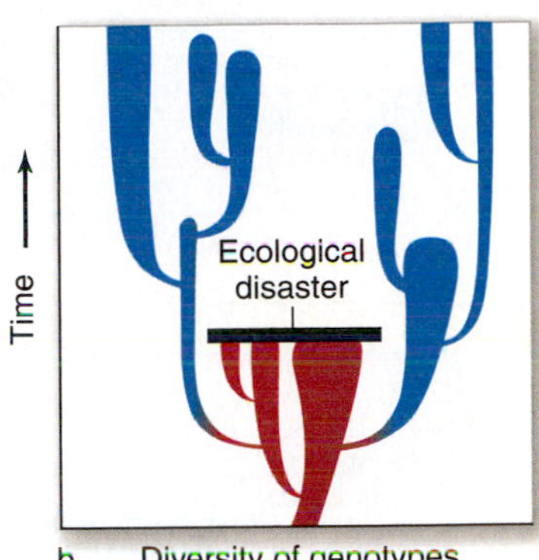

FIGURE 9.3 **Why Sex?** (**A**) In asexually reproducing organisms, the members of a population are usually very similar to one another; a single change in the environment can spell disaster. (**B**) Sexual reproduction generates genetic variability, which increases the chance that at least some members of the population will survive in a changing environment.

9.2 Meiosis Is Essential in Sexual Reproduction

Sexual reproduction produces genetic diversity, but it also poses a practical problem: maintaining the correct chromosome number. Recall from chapter 8 that most cells in a sexually reproducing organism are **diploid** (abbreviated $2n$), which means they have two full sets of chromosomes, one from each parent. Of the 46 chromosomes in each of your diploid cells, for example, 23 came from your mother and 23 came from your father.

Now consider what would happen if human sperm and egg cells were diploid. A sperm carrying 46 chromosomes would fertilize an egg cell with the same number, and the baby would inherit 92. Worse still, if that offspring later reproduced, cells in the next generation would have 184 chromosomes! Clearly, sexual reproduction must include a built-in mechanism to avoid this problem, since the normal chromosome number does not change with every generation.

A. Meiosis Halves the Chromosome Number

Most of the body's diploid cells are **somatic**, meaning they do not participate directly in reproduction. Examples of somatic cells are nerve, muscle, and skin cells. One way to maintain a constant chromosome number is to set aside a few diploid **germ cells** that specialize in producing **gametes**—sperm and egg cells. (In humans, the germ cells are located in the testes and ovaries.) Unlike somatic cells, gametes are **haploid** ($1n$), which means that they contain only one set of chromosomes. Such cells are an essential part of sexual reproduction.

Every sexual life cycle, no matter the species, includes three main events: meiosis, gamete formation, and fertilization.

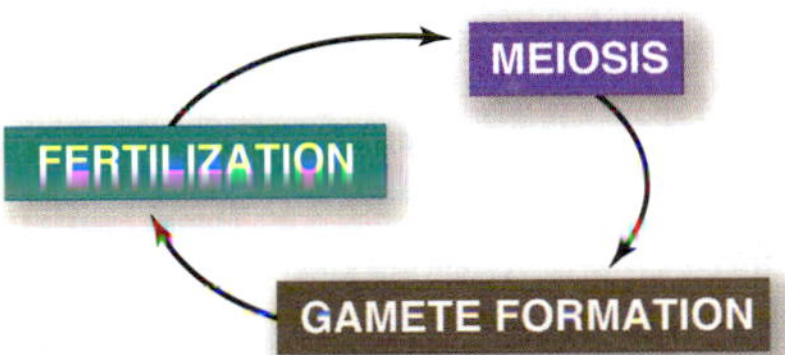

In **meiosis,** a diploid germ cell reduces its chromosome number by half to generate four haploid nuclei. As it does so, it scrambles its genetic information in ways that generate astonishing diversity (see section 9.4). The organism then packages these genetically different haploid nuclei into individual sperm or egg cells. **Fertilization** merges the gametes from two parents, creating a new cell: the diploid **zygote,** which is the first cell of the new organism. The zygote has two full sets of chromosomes, one set from each parent.

To make sense of this, consider your own life. It began when a small, swimming sperm cell carrying 23 chromosomes from your father wriggled toward your mother's comparatively enormous egg cell, also containing 23 chromosomes. You were conceived when the sperm fertilized the egg cell. At that moment, you were a one-celled zygote, with 46 chromosomes. That one cell then began dividing mitotically,

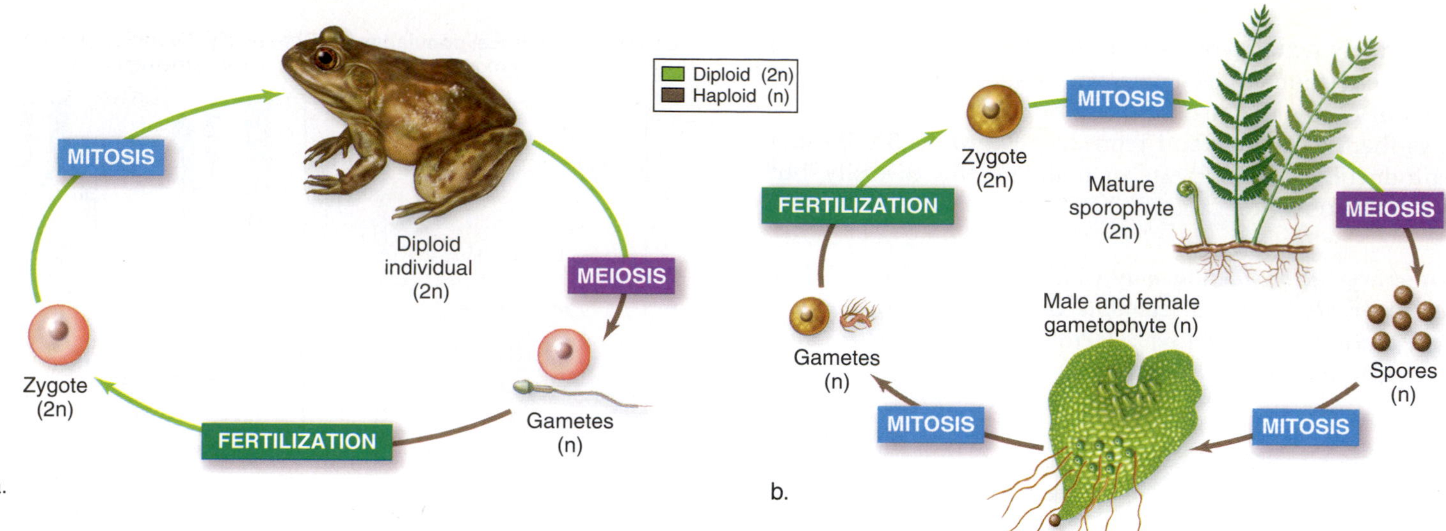

FIGURE 9.4 Sexual Reproduction. All sexual life cycles include meiosis, gamete formation, and fertilization. The difference between the life cycles of animals (**A**) and plants (**B**) is that plants have a multicellular haploid stage, called the gametophyte generation, that animals almost never have.

generating identical copies of itself to form an embryo, then a fetus, infant, child, and eventually an adult. Once you reached reproductive maturity, germ cells in your testes or ovaries produced haploid gametes of your own, perpetuating the cycle.

B. Many Organisms Have a Multicellular Haploid Stage

The human life cycle is of course most familiar to us, and many animals reproduce in essentially the same way (**figure 9.4A**). Gametes are the only haploid cells in this life cycle; all other cells are diploid. Sexual reproduction, however, takes many other forms as well.

In some organisms, both the haploid and the diploid stages are multicellular. The life cycle of a sexually reproducing plant, for example, includes an **alternation of generations** (figure 9.4B). Meiosis occurs in the diploid (sporophyte) generation, yielding haploid spores. These spores divide mitotically to form the haploid (gametophyte) generation. Gametophytes produce haploid gametes by mitosis, not meiosis. Sperm fertilizes egg to form a diploid zygote, which divides mitotically and develops into a sporophyte. The cycle begins anew.

These examples remind us that mitosis and meiosis interact to produce the full sexual life cycle. Only mitosis yields identical cells needed for growth, development, and tissue repair. Meiosis halves the chromosome number and leads to the formation of genetically different cells.

This chapter examines two of the main events in sexual reproduction: meiosis and the formation of gametes. The next two sections explain how meiosis simultaneously halves the chromosome number and produces genetically variable nuclei. We then turn to problems that can occur in meiosis and describe gamete formation in humans and plants.

Summary *Sexually reproducing organisms produce gametes with half the chromosome number of somatic cells. Sexual life cycles are diverse, but all include meiosis, gamete formation, and fertilization.*

9.2 MASTERING CONCEPTS

1. What is the difference between somatic and germ cells?
2. How do haploid and diploid nuclei differ?
3. What are the roles of meiosis, gamete formation, and fertilization in sexual life cycles?
4. What is a zygote?
5. How is the sexual life cycle of humans different from that of plants?

9.3 In Meiosis, DNA Replicates Once, But the Nucleus Divides Twice

The mitotic cell cycle described in chapter 8 includes three main parts: interphase, mitosis, and cytokinesis. During interphase, the cell grows, synthesizes molecules, carries out its functions, and replicates its DNA (see chapter 7). During mitosis and cytokinesis, the nucleus and cytoplasm split into two. Mitotic cell division creates identical copies by replicating a cell's DNA once and then dividing once.

Meiosis is closely related to mitosis. Interphase also occurs just before meiosis, and the names of the meiotic phases are similar to those in mitosis. One major difference is that meiosis includes two divisions, not just one (**figure 9.5**). The first division is meiosis I, and it reduces the number of chromosomes by half: in humans, from 46 to 23. The second division, meiosis II, produces four haploid nuclei from the two formed in meiosis I. The other major difference is that meiosis I shuffles genetic information (see section 9.4), setting the stage for each haploid nucleus to receive a unique mixture of genes.

A. In Meiosis I, Homologous Chromosomes Separate

Before embarking on a tour of meiosis, a quick look at a human cell's chromosomes is in order. Of the 23 chromosome pairs in a human cell, 22 pairs are **autosomes;** they do not determine whether an individual is male or female. The other chromosome pair, the **sex chromosomes,** carries genes that determine an individual's sex. A person with two X chromosomes is female, whereas a male has one X and one Y chromosome.

The two members of a **homologous pair** of chromosomes look alike and carry the same sequence of genes for the same traits. (The word *homologous* means "having the same basic structure.") Chromosome 21, for example, is among the smallest chromosomes. It includes 367 genes, always in the same order. Humans normally inherit one version of chromosome 21 from their father and a different version from their mother; the same is true for the other autosomes.

Unlike the autosome pairs, X and Y chromosomes are not homologous to each other. X is much larger than Y, and the genes are completely different. Nevertheless, in males, the sex chromosomes behave as homologous chromosomes during meiosis.

For meiosis to work, a human germ cell must first duplicate all 46 chromosomes, then make sure that each haploid nucleus gets exactly 23. As you shall see, this feat is possible because of the unique arrangement of the cell's chromosomes during meiosis I. Before the first meiotic division, every chromosome duplicates and then pairs up with its counterpart: chromosome 1 with chromosome 1, X chromosome with Y, and so forth. (Mules are sterile because their germ cells cannot complete this stage, as described in this chapter's Burning Question: If mules are sterile, how are they produced?) The homologous pairs split up during meiosis I, then meiosis II partitions one chromosome into each haploid nucleus.

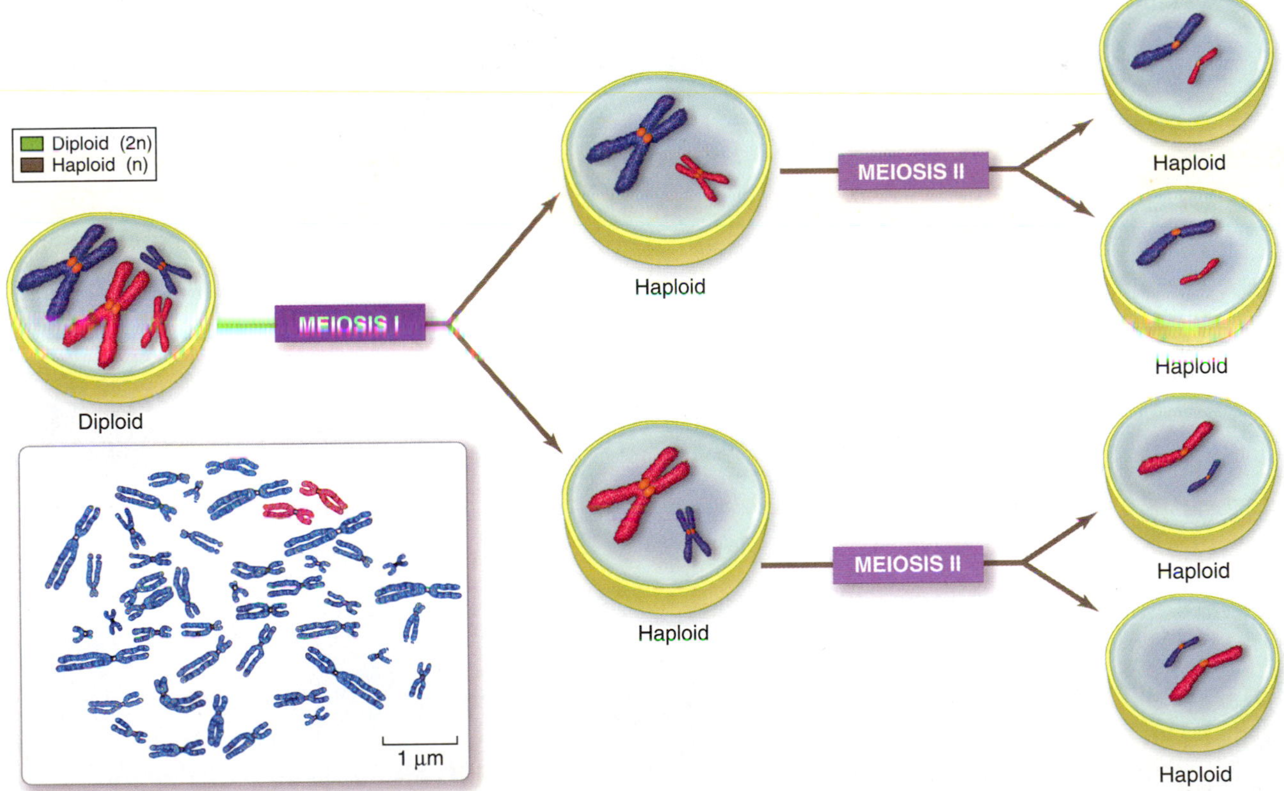

FIGURE 9.5 **Summary of Meiosis.** In meiosis, a diploid nucleus gives rise to four haploid nuclei. The first meiotic division reduces the chromosome number by half. In the second meiotic division, the cells essentially undergo mitosis. In the cells shown here, different colors distinguish the parental origins of members of homologous pairs, and different pairs are distinguished by size. The figure is simplified in the sense that the diploid cell contains only two pairs of homologous chromosomes. In reality, eukaryotic cells typically contain more chromosomes. For example, a diploid human cell (inset) contains 23 pairs of homologous chromosomes. The figure also omits the effects of crossing over (see figure 9.7).

The key to understanding meiosis, therefore, is paying careful attention to the movements of the chromosome pairs. **Figure 9.6** diagrams the process.

Interphase

The interphase that comes before meiosis is similar to interphase in the mitotic cell cycle. The cell grows during G_1 phase and synthesizes the molecules necessary for division. DNA replicates during S phase, and the cell produces proteins and other enzymes necessary to divide the cell. Afterward, each of the cell's chromosomes consists of two identical sister chromatids attached at a centromere (review table 8.1 for a reminder of the specialized terminology associated with chromosome structure). Finally, in G_2, chromatin begins to condense, and the cell produces the spindle proteins that will move the chromosomes.

Prophase I

During prophase I (that is, the prophase of meiosis I), replicated chromosomes condense. The homologous chromosomes, or homologs, line up next to one another, gene by gene. Section 9.4 describes how this arrangement allows for a gene-shuffling mechanism called crossing over. A spindle begins to form from microtubules, spindle attachment points called kinetochores grow on each centromere, and the nuclear membrane breaks up.

Metaphase I

In metaphase I, the paired homologs align down the center of the cell. Each member of a homologous pair attaches to a spindle fiber stretching to one pole. The stage is set for the homologous pairs to be separated, reducing the number of chromosomes in the daughter cells.

Anaphase I, Telophase I, and Cytokinesis

Homologs separate in anaphase I, and the chromosomes complete their movement to opposite poles in telophase I. In most species, cytokinesis occurs after telophase I to produce two haploid cells.

B. In Meiosis II, Haploid Cells Divide

A second interphase precedes meiosis II in many species. During this time, the chromosomes unfold into very thin threads. The cell manufactures proteins, but the genetic material does not replicate a second time.

Prophase II marks the start of the second meiotic division (see figure 9.6). The chromosomes again condense and become visible. In metaphase II, the chromosomes align down the center of the cell. In anaphase II, the centromeres part, and the separated sister chromatids move to opposite poles. In telophase II, nuclear envelopes form around the

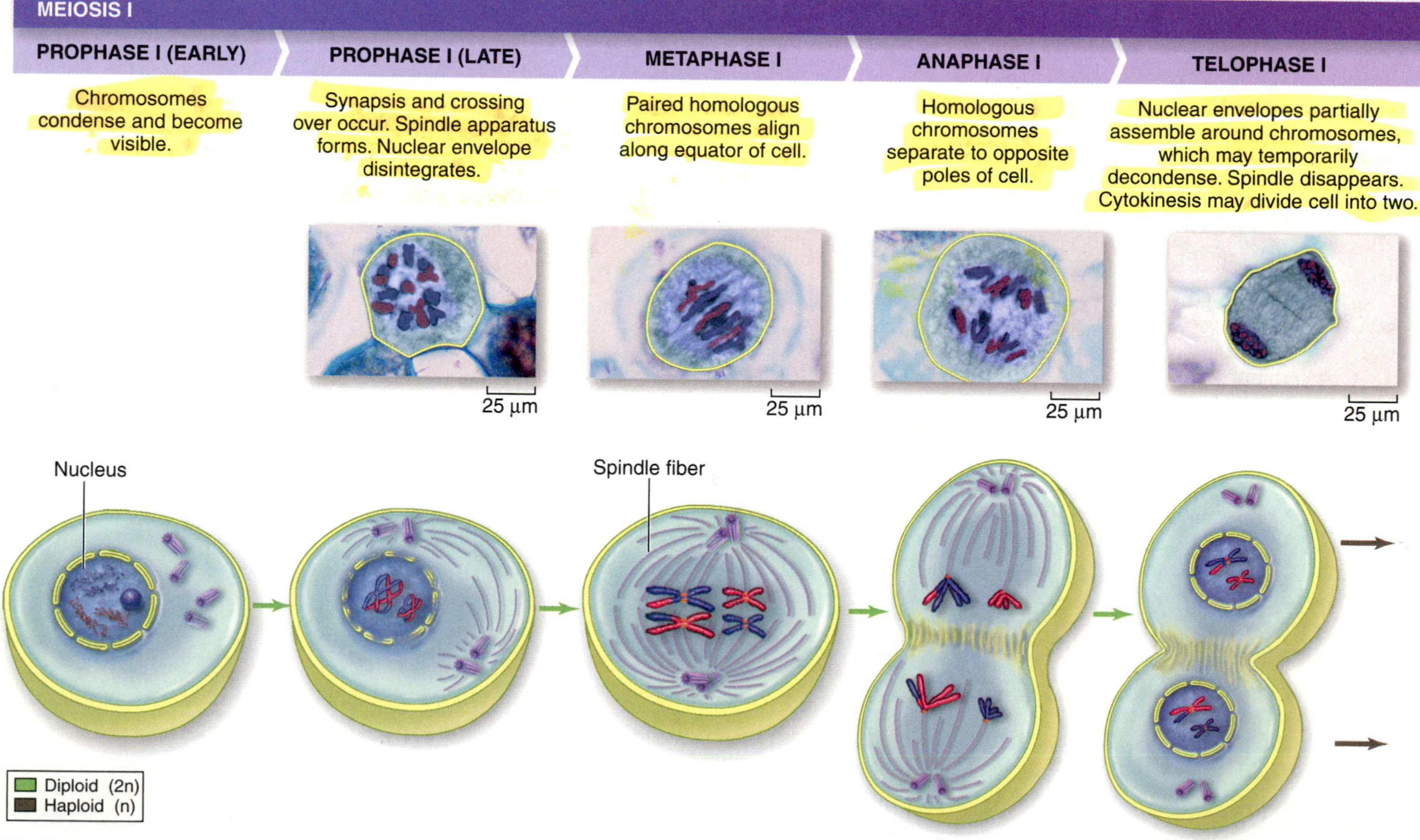
MEIOSIS I
PROPHASE I (EARLY)
PROPHASE I (LATE)
METAPHASE I
ANAPHASE I
TELOPHASE I
Chromosomes condense and become visible.
Synapsis and crossing over occur. Spindle apparatus forms. Nuclear envelope disintegrates.
Paired homologous chromosomes align along equator of cell.
Homologous chromosomes separate to opposite poles of cell.
Nuclear envelopes partially assemble around chromosomes, which may temporarily decondense. Spindle disappears. Cytokinesis may divide cell into two.
25 µm
25 µm
25 µm
25 µm
Nucleus
Spindle fiber
Diploid (2n)
Haploid (n)

FIGURE 9.6 **The Stages of Meiosis.**

separated sets of chromosomes. Cytokinesis then separates the nuclei into individual cells.

Summary *A diploid cell undergoing meiosis replicates its DNA once but divides twice. Meiosis therefore halves the chromosome number, producing four haploid cells. Because meiosis also shuffles genes, the result of meiosis is four genetically different haploid nuclei.*

9.3 MASTERING CONCEPTS

1. How are autosomes different from sex chromosomes?
2. What is a homologous pair of chromosomes?
3. What happens during interphase and each of the stages of meiosis I and meiosis II?

Burning Questions

If mules are sterile, then how are they produced?

A mule is the hybrid offspring of a mating between a male donkey and a female horse. The opposite cross (female donkey with male horse) yields a hybrid called a hinny.

Mules and hinnies may be male or female, but they are usually sterile. Why?

A peek at the parents' chromosomes reveals the answer. Donkeys have 31 pairs of chromosomes, whereas horses have 32 pairs. When gametes from horse and donkey unite, the resulting hybrid zygote has 63 chromosomes (31 + 32). The zygote divides mitotically to yield the cells that make up the mule or hinny.

These hybrid cells cannot undergo meiosis for two reasons. First, they have an odd number of chromosomes, which disrupts meiosis because at least one chromosome lacks a homologous partner. Second, donkeys and horses have slightly different chromosome structures, so the hybrid's parental chromosomes cannot align properly during prophase I. The result: an inability to produce sperm and egg cells. The only way to produce more mules and hinnies is to again mate horses with donkeys.

Have a Burning Question of your own? Submit it to marielle_hoefnagels@mcgraw-hill.com for possible inclusion in future editions of this book!

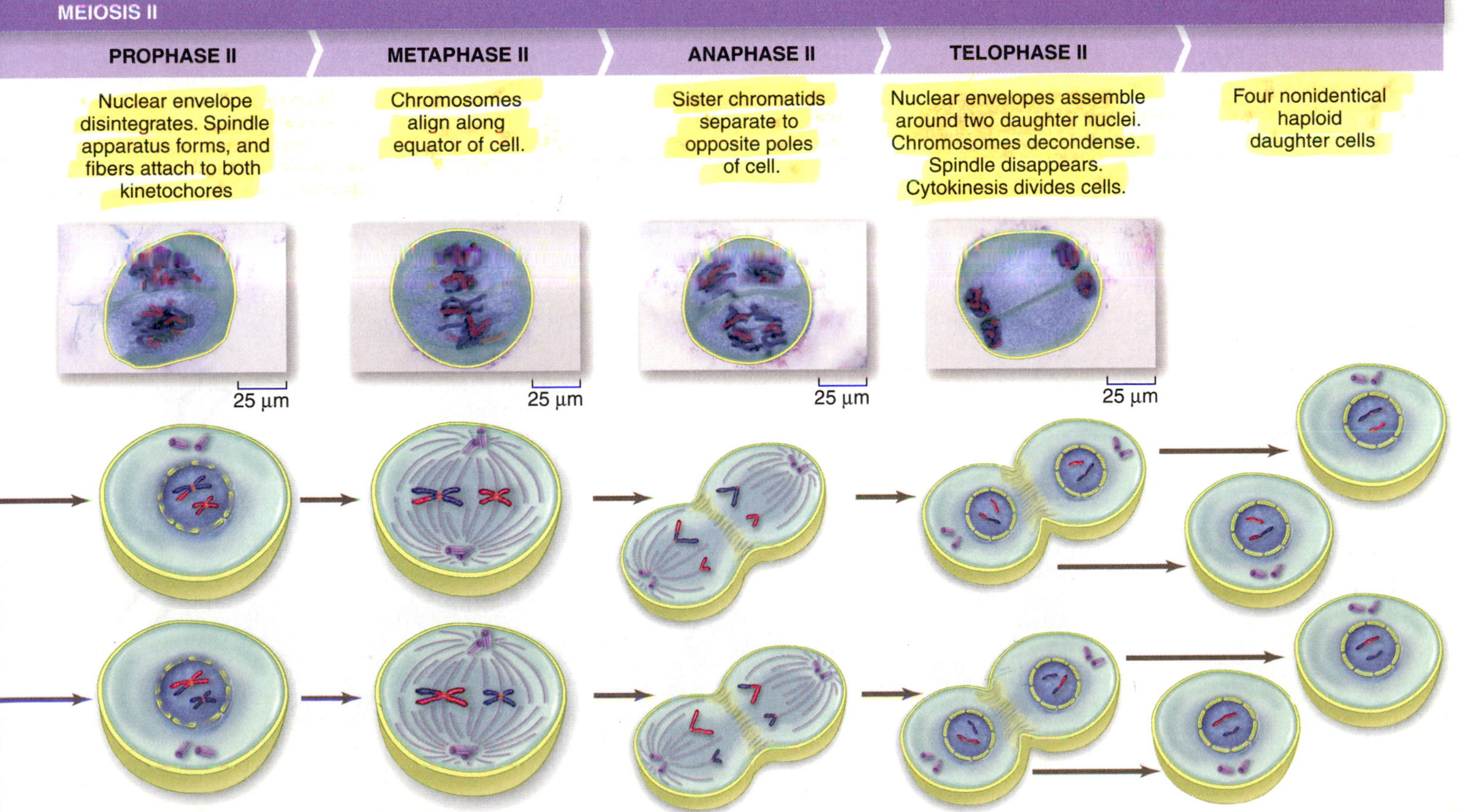
MEIOSIS II
PROPHASE II
Nuclear envelope disintegrates. Spindle apparatus forms, and fibers attach to both kinetochores
METAPHASE II
Chromosomes align along equator of cell.
ANAPHASE II
Sister chromatids separate to opposite poles of cell.
TELOPHASE II
Nuclear envelopes assemble around two daughter nuclei. Chromosomes decondense. Spindle disappears. Cytokinesis divides cells.
Four nonidentical haploid daughter cells
25 µm
25 µm
25 µm
25 µm

9.4 Meiosis Generates Enormous Variability

Meiosis generates astounding genetic variety among the offspring from just two parents. Three mechanisms account for this diversity: crossing over, independent assortment, and random fertilization.

A. Crossing over Shuffles Genes

Crossing over is a process in which two homologous chromosomes exchange genetic material (**figure 9.7**). During prophase I, the homologs align themselves precisely, gene by gene, in a process called synapsis. The chromosomes are attached at a few points along their lengths, called chiasmata, where the homologs exchange chromosomal material.

Because each homolog comes from a different parent, crossing over produces chromosomes that have some genes from the mother and some from the father. New gene combinations arise when the parental chromosomes carry different **alleles,** or alternative forms of a single gene.

Consider a simplified example of how crossing over mixes trait combinations. Suppose that one chromosome carries the genes for hair color, eye color, and finger length. One of the homologs (perhaps the one that came from the father) has alleles that specify blond hair, blue eyes, and short fingers. The homolog from the mother may have different alleles for the same genes, perhaps dictating black hair, brown eyes, and long fingers. Now, suppose that crossing over occurs between the homologous chromosomes. Afterward, one chromatid might bear alleles for blond hair, brown eyes, and long fingers; the other would have the alleles for black hair, blue eyes, and short fingers.

The two chromatids that did not form chiasmata, however, remain unchanged. The result: four genetically different chromatids in place of two pairs of identical chromatids. As meiosis continues after crossing over, each chromatid will end up in a separate haploid cell. Thus, crossing over ensures that each haploid cell will be genetically different from the others.

B. Chromosome Pairs Align Randomly During Metaphase I

A look at figure 9.6 reveals a second way that meiosis creates genetic variability. At metaphase I, the paired chromosomes line up at the cell's center, each red chromosome from one parent attached (for the moment) to its blue homolog from the other parent. Examine the orientation of these chromosomes. Notice that the blue chromosome is "on top" in the pair on the left, whereas the red chromosome occupies that position in the pair on the right. In anaphase I, the chromosomes separate, and the resulting nuclei have a mixture of paternal and maternal genetic material.

The next time a germ cell in the same individual undergoes meiosis, the orientation of the chromosomes may be the same, or it may not be. The alignment of chromosomes at metaphase I is a random process, and all possible combinations are equally probable (**figure 9.8**). The number of possible arrangements is related to the number of chromosomes. For two pairs of homologs, four (2^2) different metaphase I configurations are possible. For three pairs of homologs, as in figure 9.8, eight (2^3) configurations can occur.

Extending this formula to humans, with 23 chromosome pairs, there are 8,388,608 (2^{23}) possible combinations of chromosomes during metaphase I—all equally likely. The number of genetically different gametes that one person can produce rises even higher when the potential genetic diversity from crossing over is also considered.

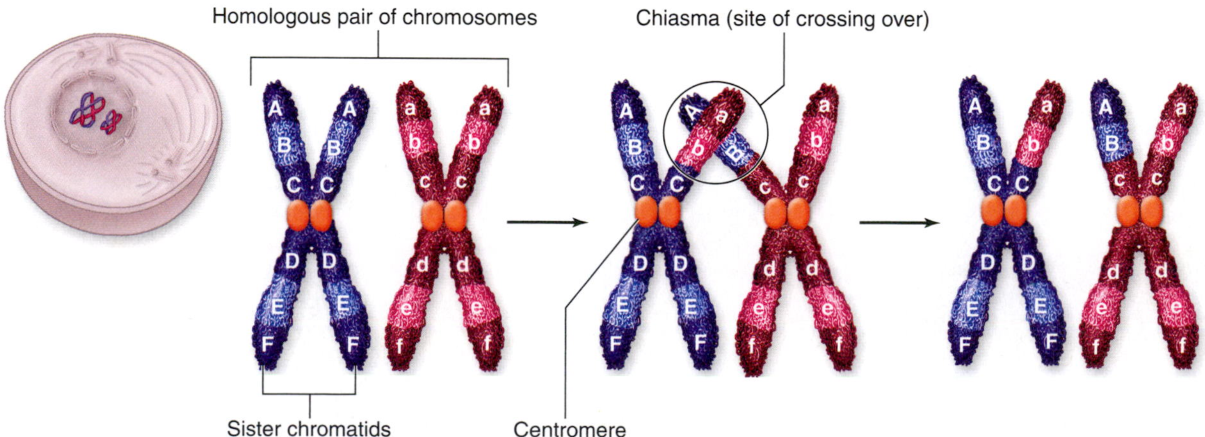

FIGURE 9.7 Crossing Over Recombines Genes. Crossing over helps to generate genetic diversity by mixing up parental traits. The capital and lowercase forms of the same letter represent different versions (alleles) of the same gene.

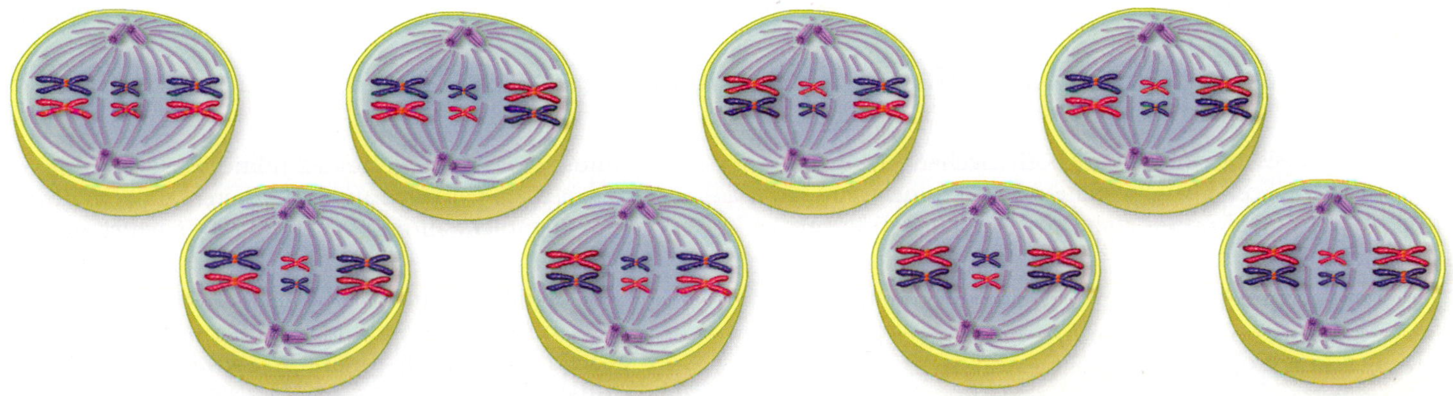

FIGURE 9.8 Many Possibilities. The arrangement of homologous pairs of chromosomes during metaphase I determines the combination of chromosomes in the daughter cells. The potential variability generated by meiosis skyrockets when one considers all 23 chromosome pairs (in humans), plus the effects of crossing over.

C. Random Fertilization Multiplies the Diversity

Every human germ cell undergoing meiosis is likely to produce haploid nuclei with different combinations of chromosomes. Furthermore, it takes two to reproduce. In one mating, any of a woman's 8,388,608 possible egg cells can combine with any of the 8,388,608 possible sperm cells of a partner. One couple could therefore theoretically create more than 70 trillion ($8,388,608^2$) genetically unique individuals! Crossing over contributes even more genetic variability to gametes.

With so much potential variability, the chance of two parents producing genetically identical individuals seems exceedingly small. How do the parents of identical twins defy the odds? The answer is that identical twins result from just one fertilization event. The resulting zygote or embryo "clones" itself, resulting in two separate, identical babies (**figure 9.9**). (When the embryos fail to separate completely, the twins remain conjoined, or physically attached to one another). Identical twins are called "monozygotic" because they derive from one zygote. They are natural clones.

In contrast, nonidentical (fraternal) twins occur when two sperm cells simultaneously fertilize two separate egg cells. The twins are therefore called "dizygotic." Triplets and higher order multiple births occur when three or more babies develop at the same time (see this chapter's Can *You* Relate? Multiple Births).

Summary *In crossing over, parts of homologous chromosomes switch places. Additional variety arises when paired chromosomes align at random during metaphase I. Random fertilization ensures that two individuals can produce huge numbers of genetically different offspring.*

9.4 MASTERING CONCEPTS

1. How does crossing over shuffle genes?
2. Explain how different chromosome alignments during metaphase I can result in over 8 million genetically different gametes in a human.
3. How are identical twins different from fraternal twins?

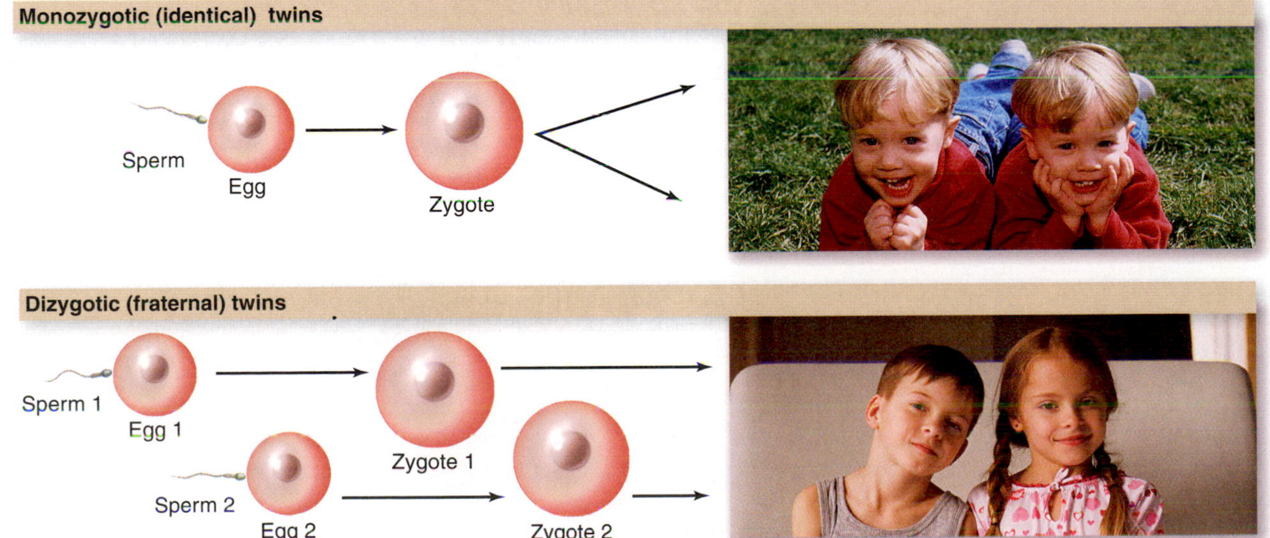

FIGURE 9.9 Two Origins for Twins. Monozygotic twins are genetically identical because they come from the same zygote. Dizygotic, or fraternal, twins, are no more alike than nontwin siblings because they start as two different zygotes.

9.5 Mitosis and Meiosis Have Different Functions: A Summary

Mitosis and meiosis are both mechanisms that divide a eukaryotic cell's genetic material, and some events are common to both. However, there are also many differences (**figure 9.10**):

- Mitosis occurs in somatic cells throughout the life cycle, whereas meiosis occurs only in germ cells, and only at some stages of life (see section 9.7).
- Only one cell division follows mitosis, yielding two daughter cells. Two divisions occur in meiosis, so one cell yields four daughter cells.
- After mitosis, the chromosome number in the daughter cells is the same as in a parent cell. In contrast, only diploid cells divide meiotically, producing four haploid daughter cells.
- Following mitosis, cytokinesis occurs once for every DNA replication event. In meiosis, cytokinesis occurs twice, but the DNA has replicated only once.
- Mitosis does not require that homologous chromosomes align with one another, whereas meiosis does. This alignment allows for crossing over, which occurs only in meiosis.
- Mitosis yields identical daughter cells for growth, repair, and asexual reproduction. Meiotic division generates genetically variable daughter cells used in sexual reproduction.

Summary ***Mitosis and meiosis both divide replicated DNA in eukaryotic cells, but the two processes occur in different types of cells and have different outcomes.***

9.5 MASTERING CONCEPTS

1. In what ways are mitosis and meiosis similar?
2. In what ways are mitosis and meiosis different?

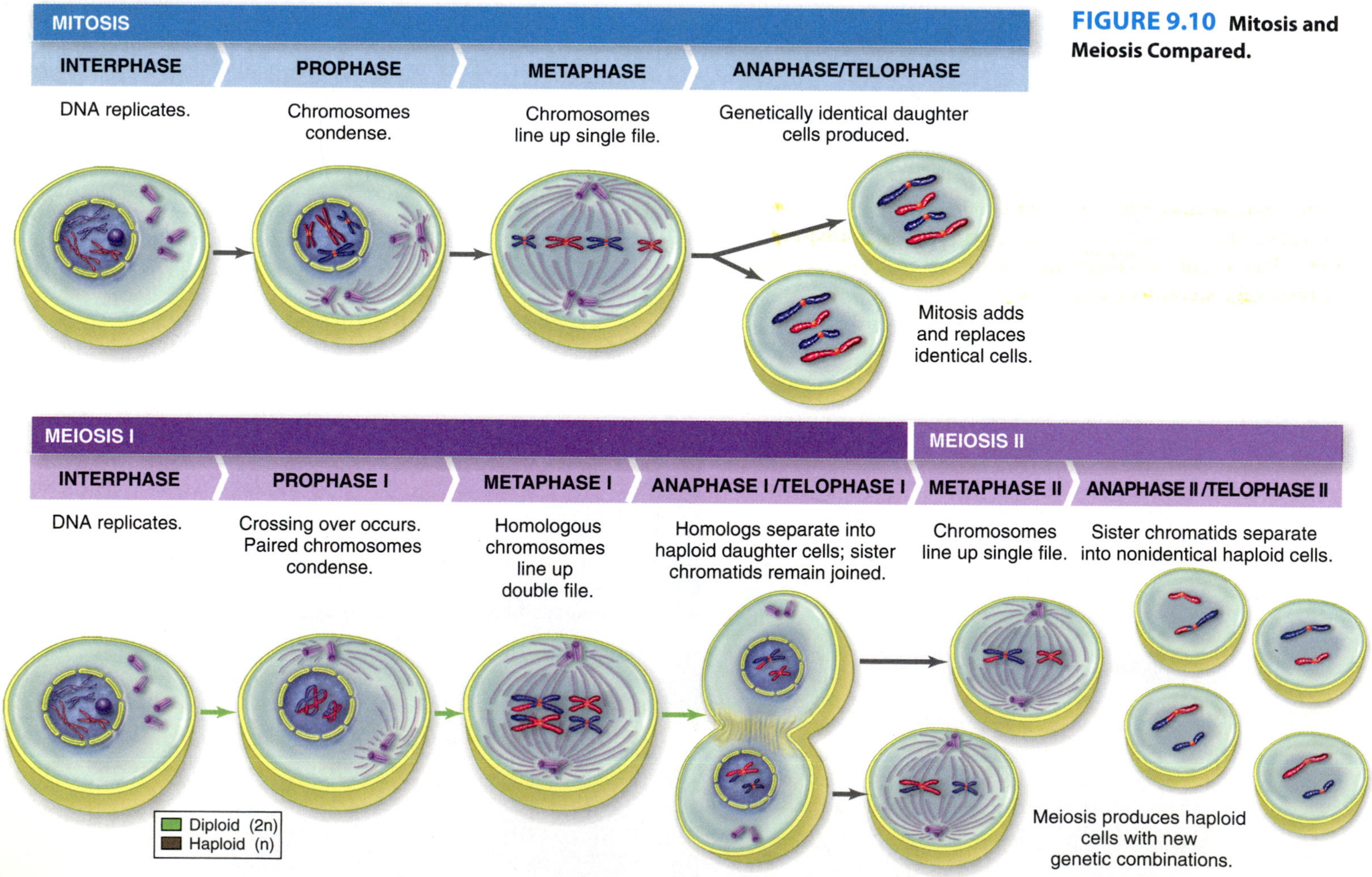

FIGURE 9.10 Mitosis and Meiosis Compared.

MITOSIS
INTERPHASE
PROPHASE
METAPHASE
ANAPHASE/TELOPHASE
DNA replicates.
Chromosomes condense.
Chromosomes line up single file.
Genetically identical daughter cells produced.
Mitosis adds and replaces identical cells.

MEIOSIS I
MEIOSIS II
INTERPHASE
PROPHASE I
METAPHASE I
ANAPHASE I /TELOPHASE I
METAPHASE II
ANAPHASE II /TELOPHASE II
DNA replicates.
Crossing over occurs. Paired chromosomes condense.
Homologous chromosomes line up double file.
Homologs separate into haploid daughter cells; sister chromatids remain joined.
Chromosomes line up single file.
Sister chromatids separate into nonidentical haploid cells.
Meiosis produces haploid cells with new genetic combinations.
Diploid (2n)
Haploid (n)

9.6 Errors Sometimes Occur in Meiosis

Considering the number of separate events that take place in meiosis, it is not surprising that things occasionally take a wrong turn. The result can be gametes with extra or missing chromosomes. Even small chromosomal abnormalities can have devastating effects on health.

A. Polyploidy Means Extra Chromosome Sets

An error in meiosis, such as the failure of the spindle to form properly, can produce a **polyploid** gamete with one or more complete sets of extra chromosomes (*polyploid* means "many sets"). For example, if a sperm with the normal 23 chromosomes fertilizes an abnormal egg cell with two full sets (46), the resulting zygote will have three copies of each chromosome (69 total), a type of polyploidy called triploidy. Most human polyploids cease developing as embryos or fetuses.

In contrast to humans, about 30% of flowering plant species tolerate polyploidy well, and many crop plants are polyploids. The duram wheat in pasta is tetraploid (it has four sets of seven chromosomes), and the wheat species in bread is a hexaploid, with six sets of seven chromosomes. Polyploidy is an important force in plant evolution, as section 14.7 describes.

B. Nondisjunction Results in Extra or Missing Chromosomes

Some gametes have just one extra or missing chromosome. The cause of the abnormality is an error called **nondisjunction**, which occurs when chromosomes fail to separate at either the first or the second meiotic division (**figure 9.11**). The result is a sperm or egg cell with two copies of a particular chromosome or none at all, rather than the normal one copy. When such a gamete fuses with another at fertilization, the resulting zygote has either 45 or 47 chromosomes instead of the normal 46.

Most embryos with incorrect chromosome numbers cease developing before birth; they account for about 50% of all spontaneous abortions (miscarriages). Extra genetic material, however, causes fewer problems than missing material. This is why most children with the wrong number of chromosomes have an extra one—a trisomy—rather than a missing one. **when a pregnancy ends, p. 758**

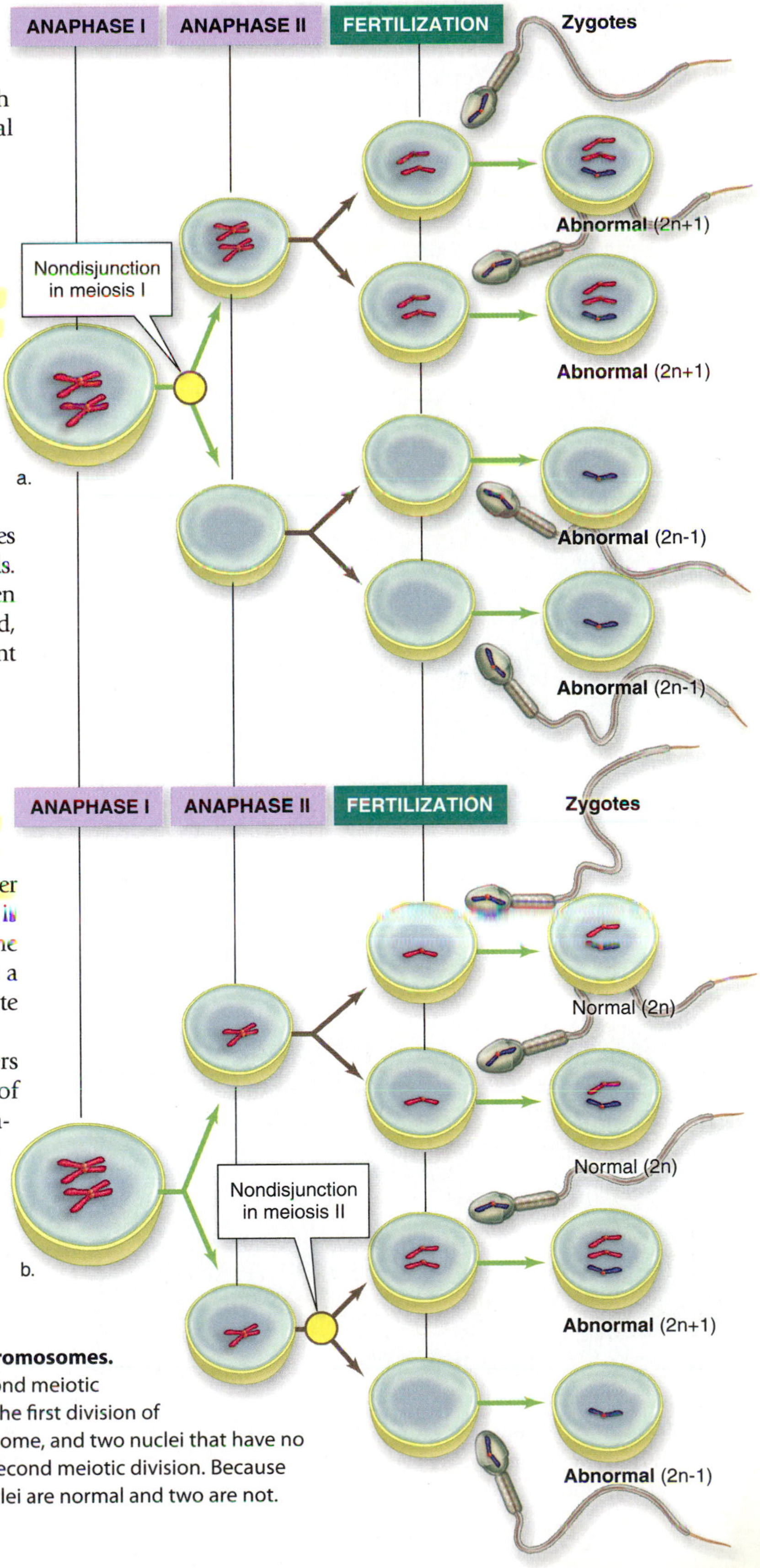

FIGURE 9.11 Nondisjunction Results in Extra or Missing Chromosomes. Unequal division of chromosome pairs can occur at the first or second meiotic division. (**A**) A single pair of chromosomes fails to separate during the first division of meiosis. The result: two nuclei that have two copies of the chromosome, and two nuclei that have no copies of that chromosome. (**B**) This nondisjunction occurs at the second meiotic division. Because the two products of the first division are unaffected, two of the nuclei are normal and two are not.

Following is a look at some syndromes in humans resulting from too many or too few chromosomes.

Extra Autosomes: Trisomy 21, 18, or 13

A person with trisomy 21, the most common cause of Down syndrome, has three copies of chromosome 21 (**figure 9.12**). An affected individual has a flat face, slanted eyes, a protruding tongue, and thick lips. He or she also has an abnormal pattern of hand creases, loose joints, and poor reflex and muscle tone. Intelligence varies greatly; some of these children have profound mental impairment, whereas others learn well. Nearly 50% of affected children die before their first birthdays, often because of congenital heart defects. People with Down syndrome also have an above average risk for leukemia and Alzheimer disease.

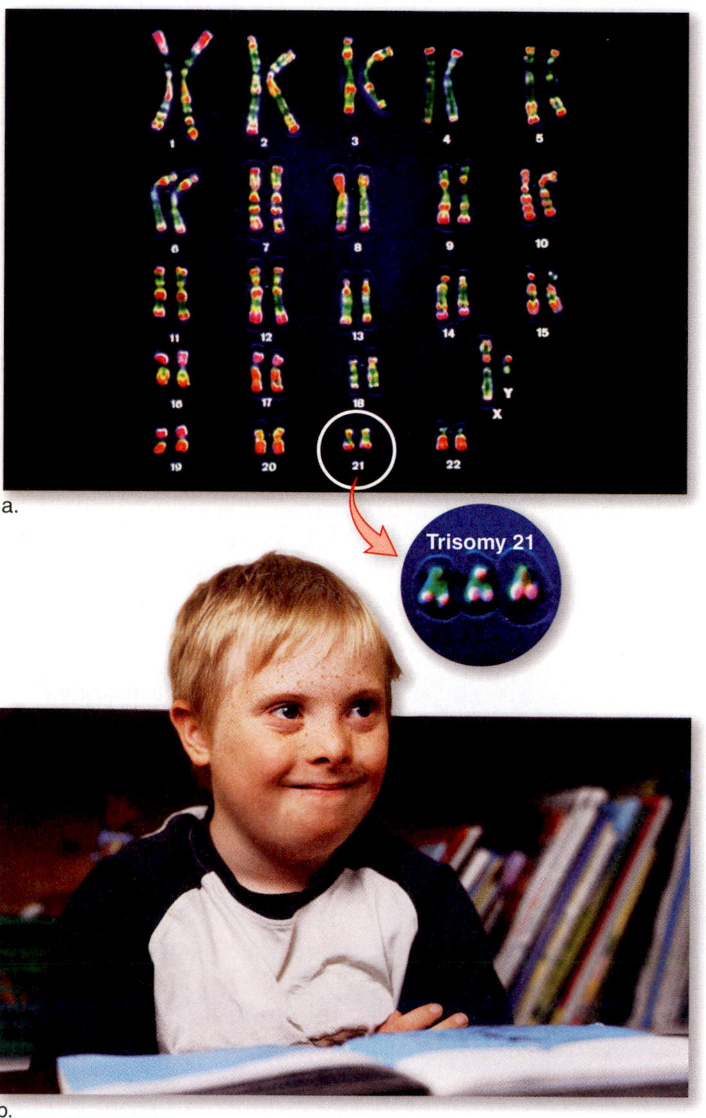

FIGURE 9.12 **Trisomy 21 Produces Down Syndrome.** (**A**) A normal human karyotype reveals 46 chromosomes, in 23 pairs. (**B**) A child born with three copies of chromosome 21 has Down syndrome.

The likelihood of giving birth to a child with trisomy 21 increases dramatically as a woman ages. For women younger than 30, the chances of conceiving a child with the syndrome are 1 in 3000. For a woman of 48, the incidence jumps to 1 in 9. An increased likelihood of nondisjunction may account for this age association. However, about 40% of trisomy 21 cases result from sperm with an extra chromosome.

Trisomy 21 is the most common autosomal trisomy, but only because it is the least likely to kill a fetus before birth. Trisomies 18 and 13 are the next most common, but few infants with these genetic abnormalities survive infancy. Trisomies undoubtedly occur with other chromosomes, but the embryos fail to develop.

Extra or Missing Sex Chromosomes: XXX, XXY, XYY, and XO

About 1 in every 1000 to 2000 females has an extra X chromosome in each cell, a condition called triplo-X. Symptoms are tallness, menstrual irregularities, and a normal-range IQ that is slightly lower than that of other family members. A woman with triplo-X may produce some egg cells bearing two X chromosomes, which increases her risk of giving birth to triplo-X daughters or XXY sons.

Males with an extra X chromosome have Klinefelter, or XXY, syndrome. The syndrome varies greatly, and some affected individuals do not realize anything is wrong until well into adulthood. Often, however, they are sexually underdeveloped, with rudimentary testes and prostate glands and no pubic or facial hair. They also have very long limbs and large hands and feet, and they may develop breast tissue. Individuals with XXY syndrome may be slow to learn, but they are usually not mentally retarded unless they have more than two X chromosomes, which is rare. The syndrome occurs in 1 out of every 500 to 2000 male births.

One male in 1000 has an extra Y chromosome, a condition called Jacobs, or XYY, syndrome. The vast majority of XYY males are apparently normal, although they may be very tall. They may also have acne and problems with speech and reading.

One sex chromosome can also be missing. If one gamete contains an X chromosome and the other gamete lacks a sex chromosome, the resulting zygote is XO. More than 99% of XO fetuses do not survive to birth, and the other 1% account for the 1 in 2000 newborn girls with XO (also called Turner) syndrome. Young women with the missing chromosome are short and sexually undeveloped. Many do not know they have a chromosomal abnormality until they lag in sexual development. They are usually of normal intelligence and, if treated with hormone supplements, lead fairly normal lives, but they are infertile.

Medical researchers have never reported a person with one Y and no X chromosome. When a zygote lacks an X chromosome, so much genetic material is missing that it probably cannot sustain more than a few cell divisions.

C. Smaller-Scale Chromosome Abnormalities Include Deletions, Duplications, Inversions, and Translocations

Because chromosomes consist of hundreds or thousands of genes, even small changes in a chromosome's structure can affect an organism. Parts of a chromosome may be deleted, duplicated, inverted, or moved to a new location (**figure 9.13**).

A chromosomal **deletion** results in the loss of one or more genes. Cri du chat syndrome (French for "cat's cry"), for example, is associated with deletion of several genes on chromosome 5. An affected child has an odd cry, similar to the mewing of a cat, and pinched facial features, severe mental retardation, and developmental delay.

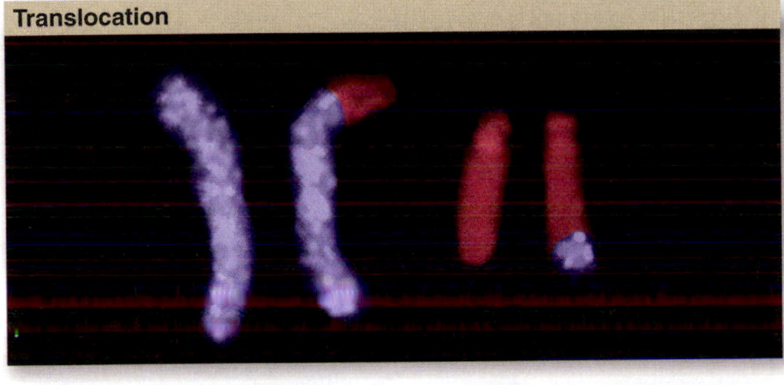

FIGURE 9.13 Chromosomal Abnormalities. (**A**) Portions of a chromosome can be deleted, duplicated, or inverted. (**B**) In translocation, two nonhomologous chromosomes exchange parts. The micrograph shows a portion of chromosome 5 (larger pair) that has switched places with part of chromosome 14 (smaller pair). In the diagram, genes C, D, and E on the blue chromosome are exchanging positions with genes M and N on the red chromosome.

In the opposite situation, a **duplication** produces multiple copies of one or more genes. Like deletions, duplications are more likely to cause symptoms if they affect large amounts of genetic material. Gene duplication probably plays an important role in evolution. While one copy of the original gene continues to do its old job, the "spare" copy is free to mutate. These mutations often ruin the gene, but occasionally they lead to new functions (see section 27.6).

In an **inversion,** part of a chromosome flips and reinserts, changing the gene sequence. Unless they break genes, inversions are usually less harmful than deletions, because all the genes are still present. If an adult has an inversion in one chromosome but its homolog is normal, however, he or she will probably have fertility problems. During prophase I, the inverted chromosome and its noninverted partner may twist around each other in a way that generates chromosomes with deletions or duplications. Because the gametes will have extra or missing genes, the result may be a miscarriage or birth defects.

In a **translocation,** nonhomologous chromosomes exchange parts (figure 9.13B). If no genes are broken, then the person has the normal amount of genetic material—it is simply rearranged. Such a person is healthy but may have fertility problems. Some sperm or egg cells will receive one of the translocated chromosomes but not the other, causing a genetic imbalance—some genes are duplicated, and others are deleted. The consequences depend on which genes the rearrangement disrupts.

Translocations often break genes, and sometimes the result is leukemia or other cancers. In about 95% of people with chronic myelogenous leukemia (CML), for example, a translocation has occurred between chromosomes 9 and 22. As a result, part of one gene from chromosome 9 fuses with another on chromosome 22. The combined gene encodes a protein that speeds cell division and suppresses apoptosis, causing leukemia (a form of cancer in which blood cells divide out of control). The modified version of chromosome 22 is called the "Philadelphia chromosome," named after the city where it was discovered.

Summary *Spindle problems and nondisjunction can lead to errors in which gametes have the wrong number of chromosomes; the consequences to the offspring can be fatal. Chromosome breakage can result in mutated genes and problems in meiosis.*

9.6 MASTERING CONCEPTS

1. What is polyploidy?
2. How can nondisjunction during meiosis lead to gametes with extra or missing chromosomes?
3. How can deletions, duplications, inversions, and translocations cause illness?
4. How do inversions and translocations cause fertility problems?

9.7 Haploid Nuclei Are Packaged into Gametes

The events of meiosis explain how a diploid germ cell produces four genetically different haploid nuclei. The same process occurs in both sexes, yet sperm and egg cells typically look very different from each other. Usually, a sperm can move and is lightweight; an egg cell is huge by comparison and packed with nutrients and organelles (**figure 9.14**). How do males and females package those haploid nuclei into such different-looking gametes?

FIGURE 9.14 A Human Sperm Contacts an Egg Cell. Note the size difference between the gametes.

A. In Humans, Gametes Form in Testes and Ovaries

Formation and specialization of sperm cells is called spermatogenesis (**figure 9.15**). Inside the testes, spermatogonia are diploid stem cells that divide mitotically to produce more spermatogonia and also specialized germ cells called primary spermatocytes—the cells that actually undergo meiosis. During interphase, primary spermatocytes accumulate cytoplasm and replicate their DNA. The first meiotic division yields two equal-sized haploid cells called secondary spermatocytes.

After a short second interphase, each secondary spermatocyte completes its second meiotic division. The products are four equal-sized spermatids, each of which specializes into a mature, tadpole-shaped sperm cell. The entire process, from spermatogonium to sperm, takes about 74 days.

In comparison to a sperm cell, an egg cell is massive. The female produces these large cells by unequally packaging the cytoplasm from the two meiotic divisions. The egg cell gets most of the cytoplasm, and the other products of meiosis are tiny.

The formation of egg cells is called oogenesis. It occurs in the ovaries and begins with a diploid stem cell, an oogonium. This cell can divide mitotically to produce more oogonia or a germ cell called a primary oocyte. In meiosis I, the primary oocyte divides into a small haploid cell with very little cytoplasm, called a polar body, and a much larger haploid cell called a secondary oocyte (**figure 9.16**). In meiosis II, the secondary oocyte divides unequally to produce another polar body and the mature egg cell, or ovum, which contains a large amount of cy-

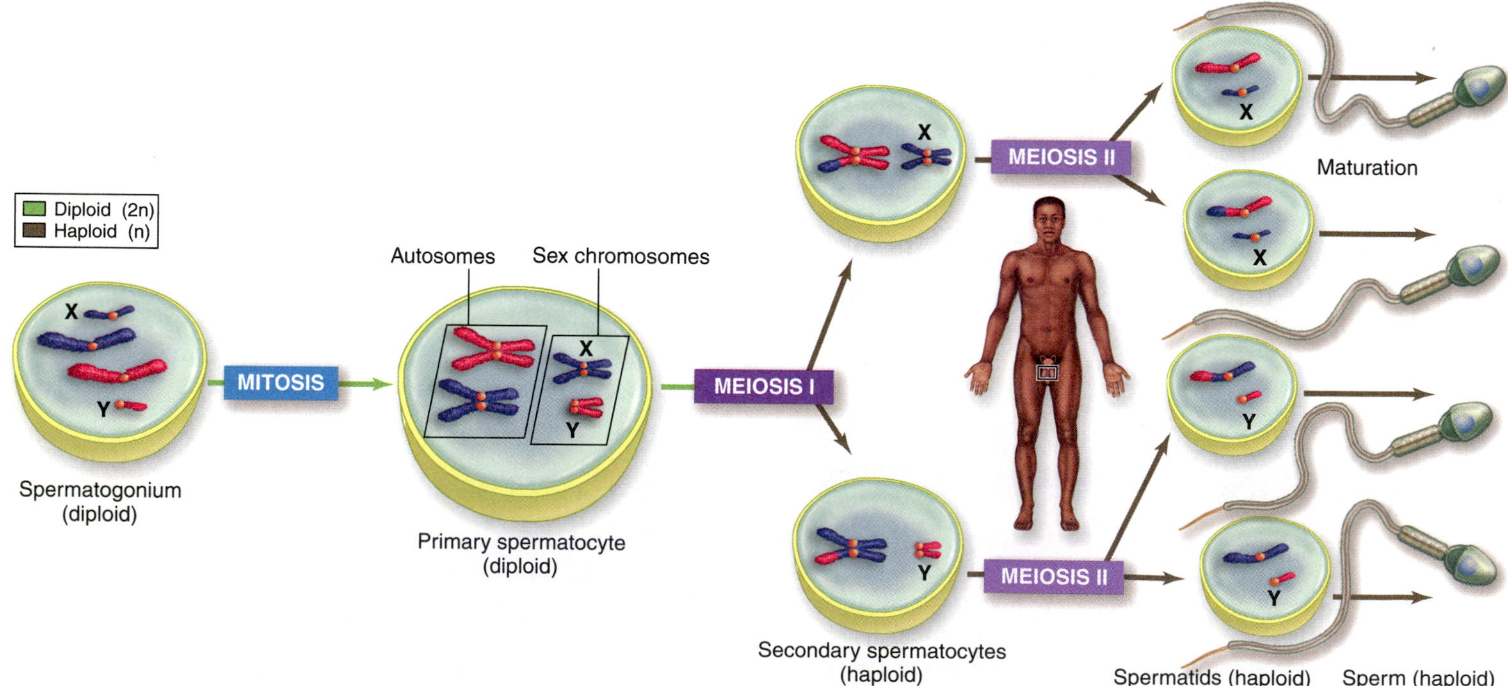

FIGURE 9.15 Sperm Formation (Spermatogenesis). In humans, diploid primary spermatocytes undergo meiosis, eventually yielding four equal-sized, haploid sperm. Only two of the normal 23 pairs of chromosomes are shown.

toplasm. The tiny polar bodies normally play no further role in reproduction.

Chapter 37 explores human reproduction and development further.

B. In Plants, Gametophytes Produce Gametes

Plant life cycles alternate between multicellular haploid and diploid generations (see figure 9.4). Meiosis occurs in the diploid plant, or sporophyte, to produce haploid cells called spores. The spores germinate, dividing mitotically to produce a multicellular haploid generation. The haploid gametophyte then produces sperm or egg cells by mitotic cell division.

In mosses, the gametophytes are easily visible with the unaided eye (**figure 9.17**). In flowering plants, however, the gametophyte is microscopic and relies on the sporophyte for nutrition. The egg-producing female gametophyte, for example, is buried deep within a flower.

Some plants produce sperm cells with flagella. In mosses and ferns, the male gametes use flagella to swim in a film of water to the stationary egg cell. The sperm of conifers and flowering plants, however, do not swim. Instead, these plants produce pollen grains—male gametophytes—that travel in wind or on animals to reach female plant parts. Pollen germination delivers sperm cells directly to the stationary egg cell. Chapters 20 and 26 further describe plant reproduction.

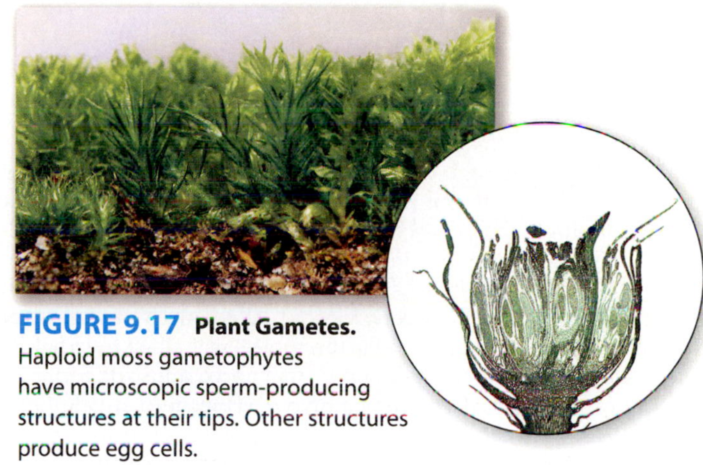

FIGURE 9.17 Plant Gametes. Haploid moss gametophytes have microscopic sperm-producing structures at their tips. Other structures produce egg cells.

Summary *Haploid nuclei are packaged into gametes. In animals, spermatogenesis produces lightweight, mobile sperm cells, whereas oogenesis yields a comparatively huge egg cell. In plants, the haploid gametophyte generation produces gametes by mitotic cell division.*

9.7 MASTERING CONCEPTS

1. What are the stages of sperm development in humans?
2. What are the stages of development of an egg cell in humans?
3. How does gamete production in plants differ from that in animals?

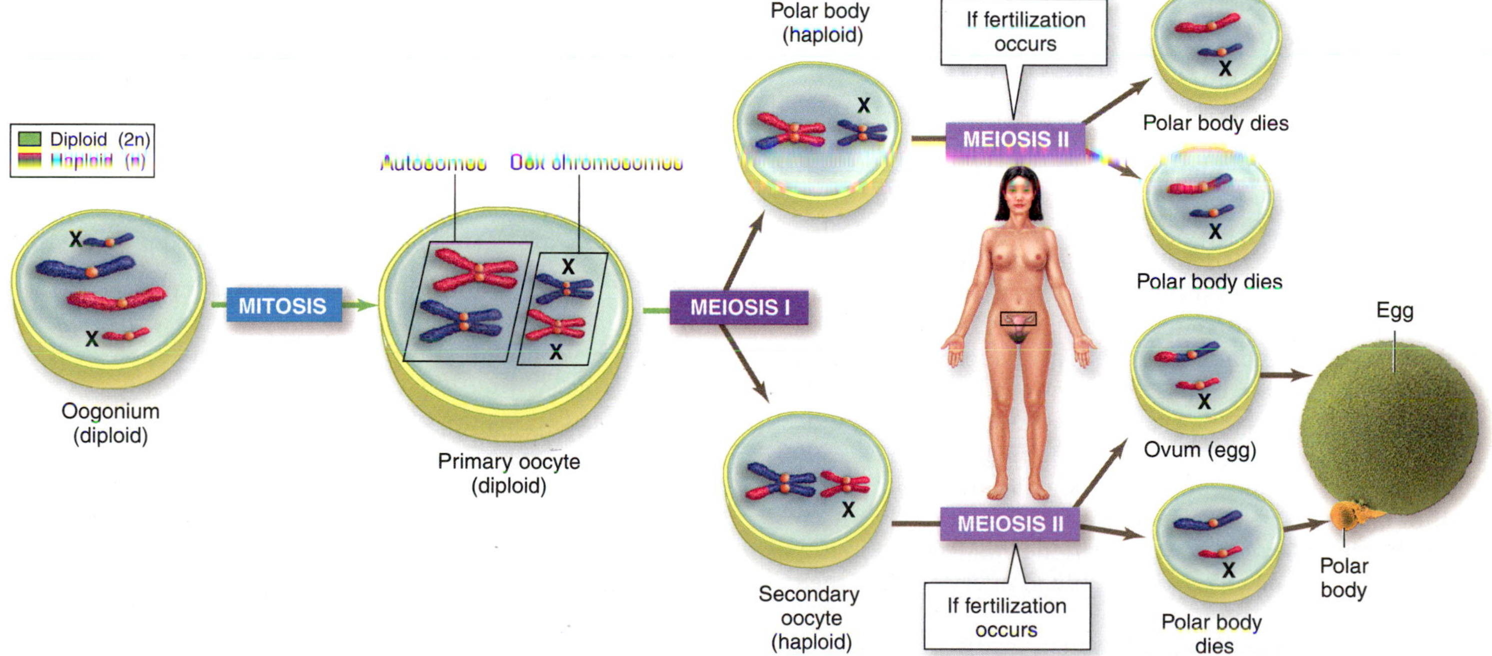

FIGURE 9.16 Ovum Formation (Oogenesis). In humans, diploid primary oocytes undergo meiosis. Meiotic division in females is uneven, concentrating most of the cytoplasm into one large egg cell. The body discards the other products of meiosis, called polar bodies, which contain the other sets of chromosomes. As in figure 9.15, only two of 23 chromosome pairs are shown here.

Can *You* Relate?

Multiple Births

Twins can be fraternal (no more alike genetically than siblings born singly) or identical. But how do triplets, quadruplets, and higher order multiple births arise?

Triplets come about in several ways. The least common route is for a single embryo to split and develop into three genetically identical babies (monozygotic triplets). Alternatively, if three sperm fertilize three separate egg cells, the triplets will all be fraternal (trizygotic). Most commonly, however, an embryo splits and forms two identical babies, and another embryo develops into an additional, nonidentical baby.

Identical quadruplets are exceedingly rare, occurring perhaps once in 11 million deliveries. Monozygotic quintuplets are even more unusual, with only one set ever known to have been born. Just as for triplets, higher order multiples usually form combinations of identical and fraternal siblings.

Multiple births have become more common since the 1980s, for two reasons. First, older women are more likely to have multiple births, and childbearing among these women has become more common. Second, treatment for infertility has increased. Some fertility drugs stimulate a woman to release one or more egg cells. If sperm fertilize all of them, a multiple birth could result. Another infertility therapy is *in vitro* fertilization, in which sperm fertilize egg cells harvested from a woman's ovaries in the lab. One or more embryos judged most likely to result in a live birth are then implanted into the woman's uterus. Multiple births often result.

9.8 INVESTIGATING LIFE
An Arms Race at a Snail's Pace

Sexual reproduction is a hassle. Why spend energy to attract mates and build costly reproductive structures when you could just make identical copies of yourself and ensure that your genome makes it to the next generation?

Some scientists suggest that parasites such as bacteria and worms explain the evolution and persistence of sexual reproduction. These parasites often reproduce much faster than their hosts, so mutations frequently produce new variants. Perhaps host species that can generate new gene combinations in each generation have the best chance to survive in an environment full of constantly changing parasites.

Several testable predictions follow from this hypothesis. If genetic diversity does help organisms avoid parasites, then a host population should change continuously. As parasites kill off susceptible hosts, resistant individuals should become more abundant. In turn, that change should select for parasites that can colonize these now-common hosts. By the time the number of resistant hosts has become large, the more rapidly reproducing parasite population has already evolved a variant that can infect them. Therefore, the most common host varieties should be the most heavily infected at any given time. Likewise, varieties that were recently common should be more susceptible to infection than their rarer neighbors.

The predicted endless evolutionary battle between host and parasite gives this proposition its common name: the Red Queen hypothesis. In Lewis Carroll's book *Through the Looking Glass*, the Red Queen remarks to Alice, "It takes all the running you can do, to keep in the same place." Perhaps sexual reproduction gives a species just enough variation to hold its own against rapidly evolving parasites. But does any evidence support this idea?

Indiana University researchers Mark Dybdahl and Curtis Lively tested the predictions of the Red Queen hypothesis by studying small freshwater snails (*Potamopyrgus antipodarum*) in a New Zealand lake (**figure 9.18A**). It may seem odd to choose a seemingly obscure species to test such an important hypothesis, but the snail offered numerous advantages:

- Most snail populations consist of multiple clones, each made up of a female and its genetically identical offspring.
- Each clone has a unique set of genetic "markers" that the researchers can use to identify the asexually reproducing subpopulation to which any individual snail belongs.
- A wormlike parasite called *Microphallus piriformis* colonizes the snails (see figure 9.18A). The parasites eat the reproductive organs of their hosts, causing sterility. Some clones are more susceptible to infection than others.

Once a year for 5 years, Dybdahl and Lively gathered snails along the shores of New Zealand's Lake Poerua. They collected a random sample of the entire snail population and another sample consisting only of infected snails. Of the 112 clones that they found, only 4 were common (a clone was considered "common" if it made up 15% or more of the sample). All other clones were considered "rare."

Each year, the researchers tracked the frequency of the four common clones in the random sample. They also checked whether any clone was overrepresented among the infected snails. They found that each clone's abundance did indeed change from year to year, as did its infection with *Microphallus* (figure 9.18B). Three of the four clones became less abundant in the year after making up a large proportion of the infected snails, suggesting just the type of trend that the Red Queen hypothesis predicts.

Because the snails reproduce asexually, this study does not help us understand the origin of sexual reproduction itself. But it does reveal a potential benefit of genetic diversity: without it, a population might fall victim to parasites and pathogens. In the evolutionary "arms race" between organisms and their parasites, sexual reproduction produces new combinations of genes that improve the chance that host organisms will evade extinction in a rapidly changing pool of parasites. Sex costs time and energy, but the alternative—easy reproduction of a doomed gene combination—may be even costlier.

Dybdahl, Mark F. and Curtis M. Lively. August 1998. Host-parasite coevolution: evidence for rare advantage and time-lagged selection in a natural population. *Evolution*, vol. 52, pages 1057–1066.

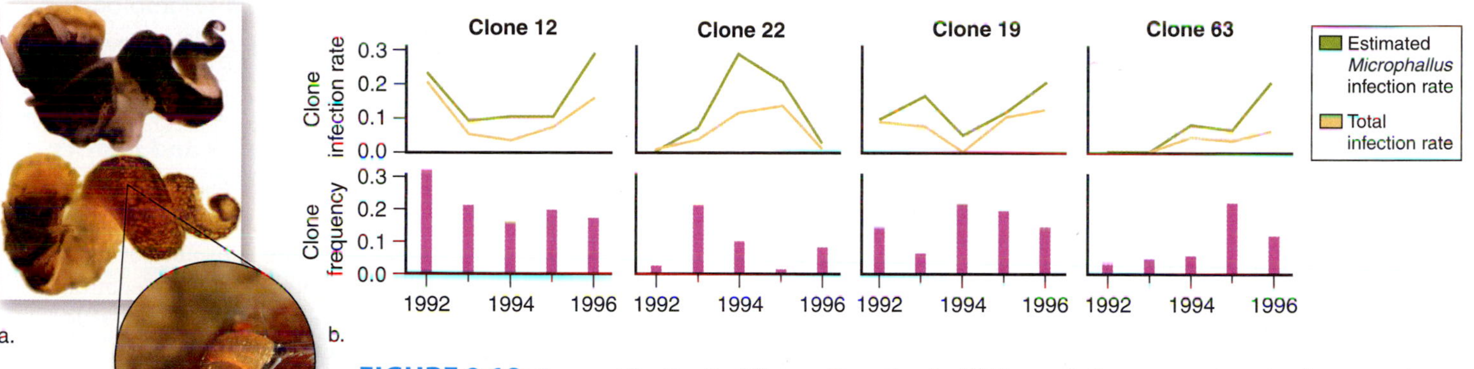

FIGURE 9.18 **Support for the Red Queen Hypothesis.** (**A**) The snail, *Potamopyrgus antipodarum,* and its parasite, *Microphallus piriformis.* The snails have been removed from their shells. (**B**) In Lake Poerua, the abundance of different snail clones fluctuates from year to year, as does the infection rate. **Question:** Using the Red Queen hypothesis, which clones do you predict were less abundant in 1997 than they were in 1996?

CHAPTER SUMMARY

9.1 Why Sex?

- **Asexual reproduction** is reproduction without sex. **Sexual reproduction** mixes traits from two parents as it increases the number of organisms.
- **Conjugation,** a form of gene transfer in some microorganisms, is sexual because one individual transfers genetic material to another, but it is not reproduction because no additional individual forms. *Chlamydomonas* undergoes cell fusion, which may have been a forerunner of sexual reproduction.
- Asexual reproduction can be successful in a stable environment, but a changing environment selects for sexual reproduction.

9.2 Meiosis Is Essential in Sexual Reproduction

A. Meiosis Halves the Chromosome Number

- **Diploid** cells have two full sets of chromosomes, one from each parent. **Somatic** cells do not participate in reproduction, but diploid **germ cells** produce **haploid gametes**.
- Three key events in sexual life cycles are **meiosis,** which halves the genetic material; gamete formation; and **fertilization,** which occurs when gametes fuse and form the diploid **zygote**.

B. Many Organisms Have a Multicellular Haploid Stage

- In most animals, gametes are the only haploid cells. In plants, however, sexual reproduction involves an **alternation of generations** with multicellular haploid and diploid phases.

9.3 In Meiosis, DNA Replicates Once, But the Nucleus Divides Twice

- In humans, the **sex chromosomes** (X and Y) determine whether an individual is male or female. The 22 **homologous pairs** of **autosomes** do not determine sex.

A. In Meiosis I, Homologous Chromosomes Separate

- Interphase happens before meiosis. Homologous pairs of chromosomes align during prophase I, then split apart during anaphase I.

B. In Meiosis II, Haploid Cells Divide

- In meiosis II, the two products of meiosis I undergo essentially a mitotic division to yield four haploid cells.

9.4 Meiosis Generates Enormous Variability

A. Crossing over Shuffles Genes

- **Crossing over,** which occurs in prophase I, produces variability as portions of homologous chromosomes switch places. After crossing over, the chromatids carry new combinations of parental **alleles**.

B. Chromosome Pairs Align Randomly During Metaphase I

- Every possible orientation of homologous pairs of chromosomes at metaphase I is equally likely. In humans, there are over 8 million possible arrangements of paternal and maternal chromosomes at metaphase I.

C. Random Fertilization Multiplies the Diversity

- Because any sperm can fertilize any egg cell, a human couple can produce over 70 trillion genetically different offspring.
- Identical (monozygotic) twins arise when a zygote splits into two embryos.

9.5 Mitosis and Meiosis Have Different Functions: A Summary

- Mitotic division produces identical copies of a cell and occurs throughout life.
- Meiosis produces genetically different haploid cells. It occurs only in specialized cells and only during some parts of the life cycle.

9.6 Errors Sometimes Occur In Meiosis

A. Polyploidy Means Extra Chromosome Sets

- **Polyploid** cells have one or more extra sets of chromosomes.

B. Nondisjunction Results in Extra or Missing Chromosomes

- **Nondisjunction** is the failure of chromosomes to separate in meiosis, and it causes gametes to have incorrect chromosome numbers. An incorrect number of sex chromosomes is less severe than an abnormal number of autosomes.

C. Smaller-Scale Chromosome Abnormalities Include Deletions, Duplications, Inversions, and Translocations

- Chromosomal rearrangements can **delete** or **duplicate** genes. An **inversion** flips gene order, possibly disrupting vital genes. In **translocations**, two nonhomologs exchange parts. Some translocations cause cancer.

9.7 Haploid Nuclei Are Packaged into Gametes

A. In Humans, Gametes Form in Testes and Ovaries

- Spermatogenesis begins in the testes with spermatogonia, which divide and become primary spermatocytes. After meiosis I, the cells are haploid secondary spermatocytes. In meiosis II, the secondary spermatocytes divide, each yielding two spermatids, which differentiate along the male reproductive tract, becoming sperm.
- In oogenesis, oogonia become primary oocytes. In meiosis I, the primary oocyte divides, distributing cytoplasm to one large secondary oocyte and a much smaller polar body. In meiosis II, the secondary oocyte divides, yielding the large ovum and another small polar body. Oogenesis occurs in the ovaries.

B. In Plants, Gametophytes Produce Gametes

- In plants, meiosis occurs in the sporophyte to yield haploid spores, which develop into the haploid gametophyte generation. The gametophyte produces gametes by mitotic cell division.

9.8 Investigating Life: An Arms Race at a Snail's Pace

- Studies of snails in a New Zealand lake suggest that over multiple generations, sexual reproduction may allow populations to change just fast enough to evade parasites.

MULTIPLE CHOICE QUESTIONS

1. The unique feature of sex is
   a. the ability of a cell to divide.
   b. the production of offspring.
   c. the ability to generate new genetic combinations.
   d. all of the above.

2. How is a somatic cell different from a germ cell?
   a. Somatic cells are diploid and germ cells are haploid.
   b. Somatic cells are haploid and germ cells are diploid.
   c. Somatic cells do not divide.
   d. Germ cells produce gametes.

3. Fertilization results in the formation of a
   a. diploid zygote.
   b. haploid gamete.
   c. diploid somatic cell.
   d. haploid zygote.
4. Why can a gametophyte produce gametes by mitosis?
   a. Because a plant's gametes are diploid.
   b. Because the gametophyte cells are already haploid.
   c. Because the gametes will go through meiosis later.
   d. Because spores function like germ cells.
5. What is the relationship between homologous chromosomes?
   a. They are exact copies.
   b. They carry the same genes, but in different order.
   c. They came from a single parent.
   d. They carry different versions of the same genes.
6. Crossing over occurs during which phase of meiosis?
   a. Prophase I
   b. Metaphase I
   c. Metaphase II
   d. Anaphase II
7. Which of the following is *not* a mechanism that contributes to diversity?
   a. Random fertilization
   b. Crossing over
   c. Cytokinesis
   d. Independent assortment
8. Nondisjunctions are most likely due to an error at which stage of meiosis?
   a. Prophase
   b. Metaphase
   c. Anaphase
   d. Telophase
9. A translocation occurs when
   a. a part of the chromosome flips and re-inserts into the same chromosome.
   b. nonhomologous chromosomes exchange DNA.
   c. a section of a chromosome is lost.
   d. multiple copies of a gene become incorporated into a chromosome.
10. Down syndrome results from which of the following examples of polyploidy?
    a. An extra chromosome 21
    b. An extra X chromosome
    c. The absence of a Y chromosome
    d. The absence of chromosome 18

TESTING YOUR KNOWLEDGE

1. Distinguish between asexual reproduction and sexual reproduction.
2. What is the evidence that sexual reproduction has been successful over evolutionary time?
3. What are the three main events of a sexual life cycle?
4. Describe how a plant life cycle may include a multicellular haploid and a diploid phase.
5. Define the following terms:
   a. crossing over
   b. gamete
   c. haploid
   d. homologous pair
   e. synapsis
   f. diploid
   g. autosome
6. How are mitosis and meiosis different?
7. Draw all possible metaphase I chromosomal arrangements for a cell with a diploid number of 8.
8. What is the difference between monozygotic and dizygotic twins?
9. What are some examples of chromosomal abnormalities, and how does each relate to an error in meiosis?
10. How does spermatogenesis differ from oogenesis, and how are the processes similar?

THINKING AS A SCIENTIST

1. A dog has 39 pairs of chromosomes. Considering only the orientation of homologous chromosomes during metaphase I, how many genetically different puppies are possible from the mating of two dogs? Is this number an underestimate or an overestimate? Why?
2. Many male veterans of the Vietnam War claim that their children born years later have birth defects caused by a contaminant in the herbicide Agent Orange used as a defoliant in the conflict. What types of cells would the chemical have to have affected in these men to cause birth defects years later? Explain your answer.
3. Is it possible for a boy–girl pair of twins to be genetically identical? Why or why not?
4. Some fungi reproduce asexually while nutrients are abundant, but switch to sexual reproduction when conditions are not as good. Explain this observation.

Visit www.mhhe.com/hoefnagels for practice quizzes, animations, videos, and activities designed to help you master the material in this chapter.

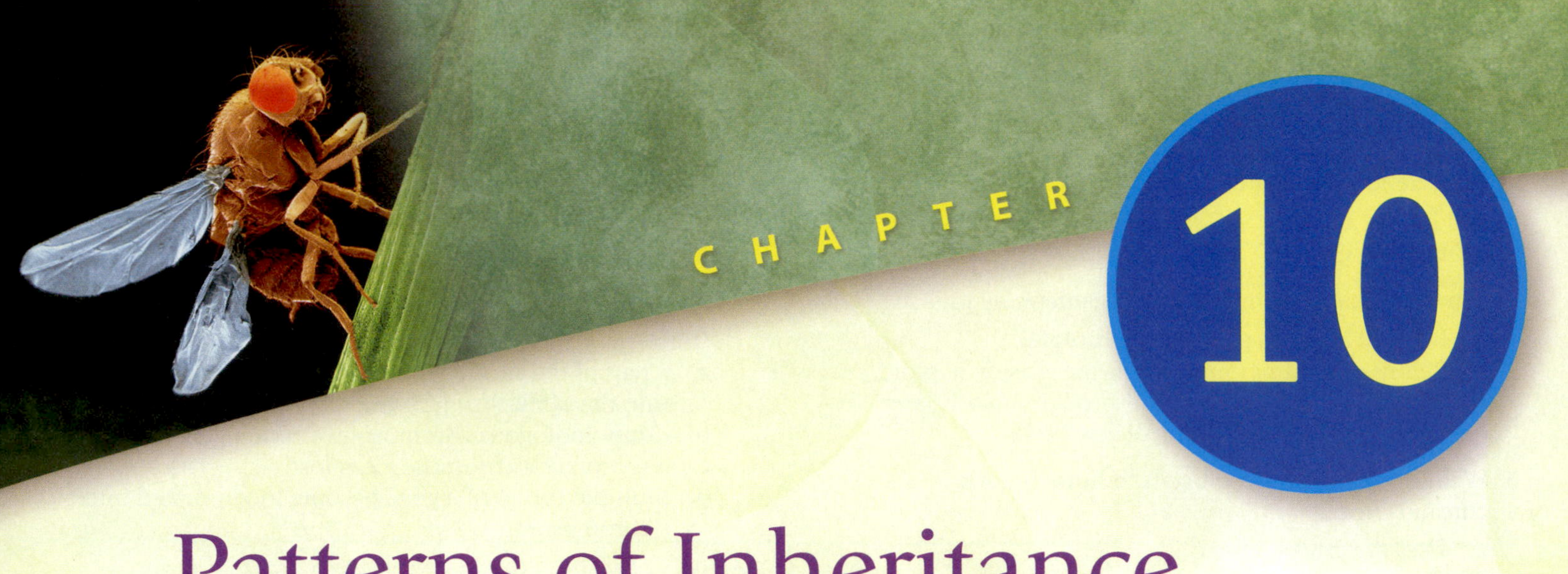

Patterns of Inheritance

Gregor Mendel (1822–1884) used experiments with pea plants to discover the basic principles of inheritance.

What's the Point?

From Mendel to Medical Genetics

Interest in heredity is probably as old as humankind itself. People throughout time have wondered at their similarities. Of all the people who have studied inheritance, one nineteenth-century investigator, Gregor Mendel, made the most lasting impression on what would become the science of genetics.

Mendel was born in 1822 and spent his early childhood in a small village in what is now the Czech Republic, where he learned early how to tend fruit trees. After finishing school ahead of schedule, Mendel became a priest at a monastery where he could teach and do research in natural science.

The young man eagerly learned how to artificially pollinate crop plants to control their breeding. The monastery sent him to earn a college degree at the University of Vienna, where courses in the sciences and statistics fueled his interest in plant breeding. Mendel began to think about experiments to address a compelling question that had confounded other plant breeders: Why did some traits disappear, only to reappear a generation later?

From 1857 to 1863, Mendel crossed and cataloged some 24,034 plants through several generations. He observed consistent ratios of traits in the offspring and deduced that the plants transmitted distinct units, or "elementen" (now called genes). Mendel described his work to the Brno Medical Society in 1865 and published it in the organization's journal the next year.

Biologists at the time did not appreciate the significance of Mendel's findings. In 1900, however, three botanists working independently rediscovered the principles of inheritance. They eventually found Mendel's paper. Over the next 10 years, researchers discovered the molecular basis of Mendel's traits, for the first time appreciating the role of chromosomes in inheritance. Thomas Hunt Morgan and other geneticists began mapping genes to their locations on chromosomes and explained the role of crossing over in meiosis, a topic we pick up in chapter 11.

Today, genetics and DNA are familiar to nearly everyone, and the entire set of genetic instructions to build a person—the human genome—has been deciphered. Even so, when a family meets with a genetic counselor or physician to learn about an inherited illness, they encounter the same principles of heredity that Mendel derived in experiments with peas. Our look at genetics begins the traditional way, with Gregor Mendel, but we can now appreciate his genius in light of what we know about DNA.

LEARNING OUTLINE

10.1 Mendel's Experiments Uncovered Basic Laws of Inheritance

A. Why Peas?
B. Alleles May Be Dominant or Recessive
C. For Each Gene, a Cell's Two Alleles May Be Identical or Different

10.2 The Two Alleles of Each Gene End Up in Different Gametes

A. Monohybrid Crosses Track the Inheritance of One Gene
B. Meiosis Explains Mendel's Law of Segregation

10.3 Two Genes on Different Chromosomes Are Inherited Independently

A. Dihybrid Crosses Track the Inheritance of Two Genes at Once
B. Alleles of Different Genes Move Independently into Gametes

10.4 Gene Expression Can Appear to Alter Mendelian Ratios

A. Incomplete Dominance and Codominance Add Phenotype Classes
B. One Genotype Can Yield Multiple Phenotypes
C. Multiple Genes Can Yield Identical Phenotypes
D. Gene Products Can Interact

10.5 Most Traits Are Influenced by the Environment and Multiple Genes

A. The Environment Can Alter the Phenotype
B. Polygenic Traits Depend on More Than One Gene

10.6 Investigating Life: Heredity and the Hungry Hordes

10.1 Mendel's Experiments Uncovered Basic Laws of Inheritance

A healthy young couple, both with family histories of cystic fibrosis, visits a genetic counselor before deciding whether to have children. The counselor suggests genetic tests, which reveal that both the man and the woman are symptomless carriers of cystic fibrosis. The counselor tells the couple that each of their future children has a 25% chance of inheriting this serious illness. How does the counselor arrive at that one-in-four chance? This chapter will explain the answer.

First, however, it may be useful to review some concepts from chapter 7. The nucleus of a eukaryotic cell contains chromosomes, which are long strands of DNA associated with proteins. A gene is a portion of DNA that encodes a specific protein; each gene can have many **alleles,** or alternative forms.

Each person inherits two alleles—one from each parent—for each of the 25,000 or so genes in the human genome. Except for identical twins, everyone has a different combination of alleles. For example, you may have black hair and brown eyes, whereas your best friend has brown hair and green eyes. You and your friend do not look alike because you have different alleles for the hair and eye color genes.

Recall also from chapter 9 that a sexual life cycle has three main events: meiosis, gamete formation, and fertilization. Each gamete arises from a diploid cell containing two sets of homologous chromosomes, with one set coming from each parent (**figure 10.1**). The two members of a homologous pair may or may not carry the same alleles for each trait. Each gamete contains only one set of chromosomes, but because of crossing over and the random alignment of chromosome pairs in metaphase I of meiosis, all the gametes may carry a different combination of alleles.

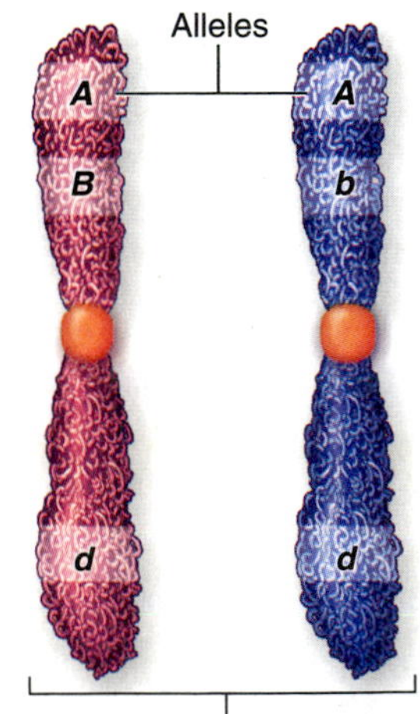

FIGURE 10.1
Homologous Chromosomes. A diploid cell in an organism has two alleles for every gene, located on a homologous pair of chromosomes. One allele came from each of the organism's parents. On these chromosomes, the alleles for genes *A* and *D* are the same, but the alleles for gene *B* are different.

No one can examine a single gamete and say for sure which allele it carries for every gene. As we shall see in this chapter, however, for some traits, we can use knowledge of a person's characteristics and family history to say that a gamete has a 100% chance, 50% chance, or 0% chance of carrying a specific allele. With this information for both parents, it is simple to calculate the probability that a child will inherit the allele. Although Gregor Mendel knew nothing of chromosomes or meiosis, he nevertheless discovered how to calculate these probabilities, at least for some traits.

A. Why Peas?

As Mendel discovered, the pea plant (*Pisum sativum*) is a good choice for studying heredity. It is easy to grow, develops quickly, produces many offspring, and has many traits that appear in two easily distinguishable forms. For example, seeds may be round or wrinkled, yellow or green. Pods may be smooth or may conform to the shape of the peas inside. Stems may be tall or short.

Pea plants also have another advantage for studies of inheritance: it is easy to control which plants mate with which (**figure 10.2**). An investigator can take pollen from the male flower parts of one plant and apply it to the female part of another plant, then allow the offspring (seeds) to develop. After planting the seeds, the investigator can observe the traits that each one inherited.

B. Alleles May Be Dominant or Recessive

Mendel's first experiments with peas dealt with single traits that have two expressions, such as short and tall. He set up all possible combinations of crosses: tall with tall, short with short, and tall with short (**figure 10.3**). Mendel noted that short plants crossed with other short plants were "true-breeding"; that is, they always produced short plants that looked like the parents. The crosses of tall plants were more variable. Sometimes, tall plants were true-breeding, but in other cases, the offspring of two tall plants were a mix of short and tall. Sometimes the short trait vanished in one generation, only to reappear in the next.

Mendel noticed a similar mode of inheritance when he studied other pea plant characteristics: one trait seemed to obscure the other. Mendel called the masking trait *dominant;* the trait being masked is called *recessive*. The tall trait, for example, is dominant to the recessive trait.

Although Mendel referred to *traits* as dominant or recessive, modern biologists reserve these terms for *alleles*. A **dominant** allele is one that exerts its effects whenever it is

1 Stamens (male parts) removed from flowers of short pea plant to prevent self-pollination.
2 Pollen from tall pea plant flower transferred to female part of short pea plant flower.
3 Pod from cross-pollinated plant contains seeds, each representing an independent offspring.
4 Mature plants developed from seeds can reveal inheritance pattern for gene controlling plant height.

FIGURE 10.2 Mendel's Experimental Approach for Breeding Peas. One of the advantages of working with pea plants is that an investigator can easily control which plants breed with each other. Gregor Mendel used this technique to set up crosses of pea plants, so he could observe the appearance of traits in the next generation.

present; a **recessive** allele is one whose effect is masked if a dominant allele is also present. When a gene has two alleles, it is common to symbolize the dominant allele with a capital letter (such as *T* for tall) and the recessive with the corresponding lowercase letter (*t* for short).

The "dominance" of an allele may seem to imply that it "dominates" in the population as a whole. The most common allele, however, is not always the dominant one. In humans, the allele that causes a form of dwarfism called achondroplasia is dominant, but it is very rare—as is the dominant allele that causes Huntington disease. In contrast, blue eyes are the norm in people of northern European origin, but the alleles that produce this eye color are recessive.

C. For Each Gene, a Cell's Two Alleles May Be Identical or Different

Mendel chose traits encoded by genes with only two possible alleles, but some genes have hundreds of forms. Regardless of the number of possibilities, however, a diploid cell can have only two alleles for each gene. After all, each diploid individual has inherited one set of chromosomes from each parent, and each chromosome carries only one allele per gene.

The two alleles in a diploid cell may be identical or different. An individual that is **homozygous** for a particular gene has two identical alleles, meaning that both

FIGURE 10.3 Mendel Crossed Short and Tall Pea Plants. (A) When Mendel crossed short pea plants with short pea plants, all of the offspring were short. (B) Sometimes, tall plants crossed with tall plants yielded only tall plants. Other times, the short trait appeared among the offspring. (C) Some tall plants crossed with short plants produced only tall plants. Other tall plants crossed with short plants produced some tall plants and some short plants.

parents contributed the same gene version. If both alleles are dominant, the individual is homozygous dominant; if both are recessive, the individual is homozygous recessive. A **heterozygous** individual has two different alleles for the gene; that is, the two parents contributed different genetic information. (Burning Question: What does "recessive" really mean? explains why recessive alleles seem to "vanish" in heterozygotes.)

Burning Questions

What does "recessive" really mean?

How can a recessive allele seem to hide when a dominant allele is present, then emerge from its hiding place if the dominant allele is absent? How does it "know" what to do?

In fact, a recessive allele does not hide, emerge, or know anything. It remains a part of the diploid cell's DNA, regardless of the presence of a dominant allele. It only seems to hide because it codes for a nonfunctional protein.

For example, consider the disorder phenylketonuria. The dominant allele codes for an enzyme that converts the amino acid phenylalanine into another amino acid. The recessive allele codes for an abnormal, nonfunctional enzyme. People who have just one copy of the recessive, disease-causing allele appear normal because the cell has enough of the normal protein, thanks to the dominant allele. Because heterozygotes have a normal phenotype, the dominant allele appears to mask the recessive allele, just as in Mendel's pea plants.

Individuals who are homozygous recessive, however, do not have a normal allele, and they therefore do not produce the normal enzyme. The disease phenotype appears in homozygous recessive people as phenylalanine accumulates to toxic levels, causing mental retardation and other problems. Foods containing the artificial sweetener aspartame carry a warning to people with phenylketonuria because aspartame contains phenylalanine (**figure 10.A**).

Have a Burning Question of your own? Submit it to marielle_hoefnagels@mcgraw-hill.com for possible inclusion in future editions of this book!

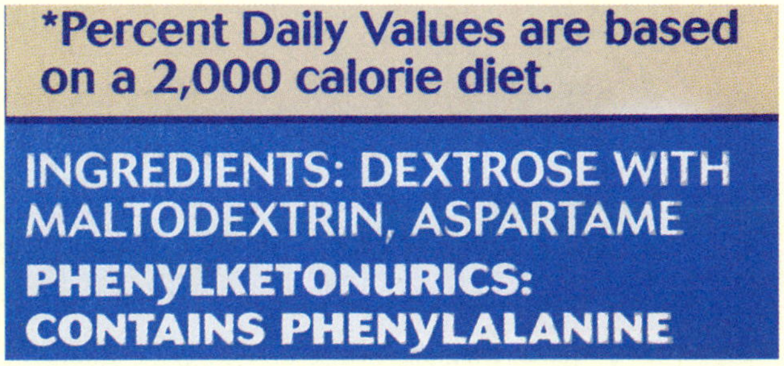

*Percent Daily Values are based on a 2,000 calorie diet.

INGREDIENTS: DEXTROSE WITH MALTODEXTRIN, ASPARTAME

PHENYLKETONURICS: CONTAINS PHENYLALANINE

FIGURE 10.A Warning Label. Products containing the artificial sweetener aspartame warn people with phenylketonuria of the presence of phenylalanine.

Table 10.1 *Miniglossary of Genetic Terms*

Term	Definition
Generations	
P	The parental generation
F_1	The first filial generation; offspring of P generation
F_2	The second filial generation; offspring of F_1 generation
Chromosomes and Genes	
Chromosome	A dark-staining body that consists of a continuous double helix of DNA plus associated proteins; each chromosome carries many genes.
Gene	A sequence of DNA that encodes a protein
Allele	An alternative form of a gene
Dominant and Recessive	
Dominant allele	An allele that masks the expression of another allele
Recessive allele	An allele whose expression is masked by another allele
Identical or Different Alleles	
Homozygous	Possessing identical alleles of one gene
Heterozygous	Possessing different alleles of one gene
Genotypes and Phenotypes	
Genotype	The allele combination in an individual
Phenotype	The observable expression of an allele combination
Wild type	The most common phenotype or allele for a gene in a population
Mutant	A phenotype or allele resulting from a change (mutation) in a gene

Homozygous dominant, homozygous recessive, and heterozygous are examples of **genotypes,** which express the genetic makeup of an individual. The organism's genotype is distinct from its **phenotype,** or outward expression of an allele combination. Your own phenotype includes your height, eye color, shoe size, number of fingers and toes, blood type, skin color, hair texture, and every other observable characteristic.

A pea plant with a short phenotype always has a homozygous recessive genotype, written *tt* for the two recessive alleles in each diploid cell. Because *T* is dominant, however, a tall pea plant can be either homozygous dominant (*TT*) or heterozygous (*Tt*) for the stem length gene.

Mendel's observation that only some tall pea plants were true-breeding (see figure 10.3) arises from the two possible genotypes for the tall phenotype. All homozygous plants are true-breeding because they always pass on the same allele. Heterozygous plants, however, are not true-breeding because a gamete from a heterozygous plant may contain either allele.

Today, biologists use additional terms to describe organisms. A **wild-type** allele, genotype, or phenotype is the most common form or expression of a gene in a population. Wild-type fruit flies, for example, have two antennae and one pair of wings. A **mutant** allele or phenotype is a variant that arises when a gene undergoes a **mutation,** or sequence change. Mutant phenotypes for fruit flies, for example, would include having legs instead of antenna growing out of the head, or having multiple pairs of wings (see figure 12.13). Both wild-type and mutant phenotypes can reflect the expression of either dominant or recessive alleles.

Table 10.1 summarizes the important terms encountered so far. The remainder of the chapter uses this basic vocabulary to integrate Mendel's findings with what biologists now know about chromosomes and reproduction.

Summary *Mendel used peas to demonstrate basic principles of inheritance: genes can have multiple alleles; every diploid individual has two alleles for every gene; an individual's genotype may be homozygous or heterozygous; and if one allele is dominant to a recessive one, a homozygous individual has the same phenotype as a heterozygote.*

10.1 MASTERING CONCEPTS

1. Why did Gregor Mendel choose pea plants as his experimental organism?
2. Distinguish between dominant and recessive; heterozygous and homozygous; phenotype and genotype; wild type and mutant.

10.2 The Two Alleles of Each Gene End Up in Different Gametes

Part of Mendel's genius was that he kept careful tallies of the offspring from countless crosses. This required a systematic accounting of multiple generations of plants. Standardized names for these generations help biologists keep track of inheritance patterns. The purebred **P generation** (for "parental") is the first set of individuals being mated; the **F_1 generation,** or first filial generation, is the offspring from the P generation (*filial* derives from the Latin word for "child"). The **F_2 generation** is the offspring of the F_1 plants, and so on. (Although these terms are applicable only to lab crosses, they are analogous to human family relationships. If you consider your grandparents the P generation, your parents are the F_1 generation, and you and your siblings are the F_2 generation.)

A. Monohybrid Crosses Track the Inheritance of One Gene

Mendel used a systematic series of crosses to deduce the rules of inheritance. He began with a P generation consisting of true-breeding short plants (*tt*) and true-breeding tall plants (*TT*). When he crossed these plants, the resulting seeds grew into tall F_1 offspring (genotype *Tt*). That is, the short trait seemed to disappear in the F_1 generation.

Next, he used the F_1 plants to set up a **monohybrid cross:** a mating between two individuals that are both heterozygous for one gene. The resulting F_2 generation had both tall and short phenotypes, in a ratio of 3:1; that is, for every three tall plants, Mendel observed one short plant.

A diagram called a **Punnett square,** which uses the genotypes of the parents to reveal which alleles the offspring may inherit, shows how the short phenotype reappeared in the F_2 generation (**figure 10.4**). In a monohybrid cross, both parents are heterozygous (*Tt*) for the stem length gene. Each therefore produces some gametes carrying the *T* allele and some gametes carrying *t*. All three possible genotypes may therefore appear in the F_2 generation, in the ratio 1 *TT*: 2 *Tt*: 1 *tt*. The corresponding phenotypic ratio is three tall plants

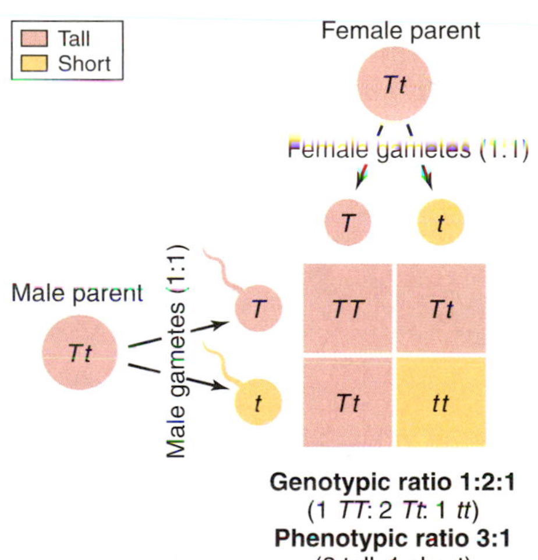

FIGURE 10.4 Punnett Square. A diagram of gametes and how they can combine is helpful in following the inheritance of traits. The possible types of female gametes are listed along the top of the square; possible male gametes are listed on the left-hand side. Each compartment within the square contains the genotype that results when the corresponding gametes join. The Punnett square here describes Mendel's monohybrid cross of two tall, heterozygous (*Tt*) pea plants.

Table 10.2 *Mendel's Law of Segregation: Crossing Heterozygotes Produces a 3:1 Phenotypic Ratio*

Experiment	Total	Plants Expressing Dominant Allele	Plants Expressing Recessive Allele	Ratio*
1. Seed form	7324	5474 round (*R*)	1850 wrinkled (*r*)	2.96:1
2. Seed color	8023	6022 yellow (*Y*)	2001 green (*y*)	3.01:1
3. Pod form	1181	882 inflated (*V*)	299 restricted (*v*)	2.95:1
4. Pod color	580	428 green (*G*)	152 yellow (*g*)	2.82:1
5. Flower position	858	651 axial (*F*)	207 terminal (*f*)	3.14:1
6. Seed coat color	929	705 gray (*A*)	224 white (*a*)	3.15:1
7. Stem length	1064	787 tall (*L*)	277 short (*l*)	2.84:1
				Average = 2.98:1

* Each ratio deviates slightly from the expected 3:1 because inheritance reflects the rules of probability. Repeating each experiment would likely yield slightly different ratios, each very close to 3:1.

to one short plant, or 3:1. Mendel saw similar results for all seven traits that he studied (**table 10.2**).

Mendel could tally the number of plants with each phenotype, but he had no way to tell by looking at a tall plant whether it was *TT* or *Tt*. He knew that short plants were always true-breeding because they are always homozygous recessive (*tt*). To distinguish between *TT* and *Tt* plants, Mendel set up a **testcross**—a mating between a homozygous recessive individual and an individual of unknown genotype (**figure 10.5**). If a tall plant crossed with a *tt* plant produced both tall and short offspring, Mendel knew it was genotype *Tt*; if the cross produced only tall plants, he knew it must be *TT*.

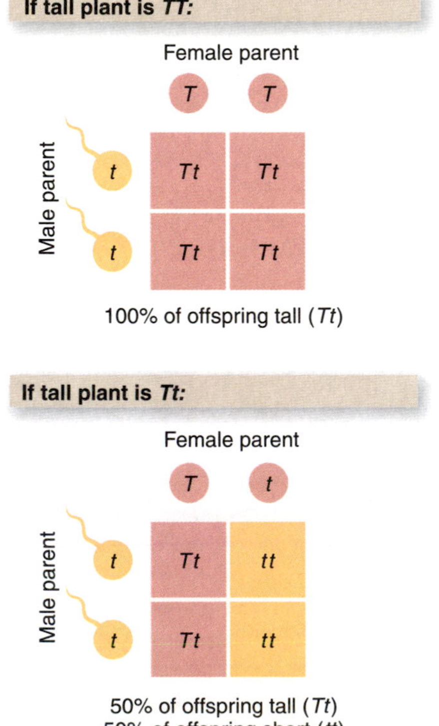

FIGURE 10.5
Testcross. A testcross reveals the genotype of a tall pea plant, which may be *TT* or *Tt*. The unknown plant is mated with a homozygous recessive (*tt*) plant. If the unknown plant is *TT*, all offspring of the testcross are tall; if the unknown plant is *Tt*, about half the offspring are likely to be short.

B. Meiosis Explains Mendel's Law of Segregation

All of Mendel's breeding experiments and calculations added up to a brilliant description of basic genetic principles. Without any knowledge of chromosomes or genes, Mendel used his data to conclude that genes occur in alternative versions (what we now call alleles). He further determined that each individual inherits two alleles for each gene, and that these alleles may be the same or different. Finally, he deduced his **law of segregation,** which states that the two alleles of each gene are packaged into separate gametes; that is, they "segregate," or move apart from each other, during gamete formation. (In science, a *law* is a statement about a phenomenon that is invariable, at least as far as anyone knows. Unlike a theory, a law does not necessarily explain the phenomenon).

Mendel's law of segregation makes perfect sense in light of what we now know about reproduction. When a diploid cell undergoes meiosis, the homologous chromosomes separate, carrying with them two alleles of each gene. In a plant of genotype *Tt*, for example, gametes carrying either *T* or *t* form in equal numbers (**figure 10.6**). When gametes meet to start the next generation, they combine at random. That is, both a *t*-bearing sperm and a *T*-bearing sperm have an equal chance at fertilizing a *t*- or *T*-bearing egg.

This basic principle of inheritance applies to all diploid species, including humans. Return for a moment to the couple and their genetic counselor introduced in section 10.1. Cystic fibrosis arises when a person has two recessive alleles for a particular gene on chromosome 7 (see section 4.6). Genetic testing revealed that the man and the woman are both symptomless carriers. In genetic terms, this means that although neither has the disease, both are heterozygous for the gene that causes cystic fibrosis. Just as in Mendel's monohybrid crosses, each of their children has a 25% chance of inheriting two recessive alleles (**figure 10.7**). Each child also has a 50% chance of being heterozygous and a 25% chance of inheriting two dominant alleles.

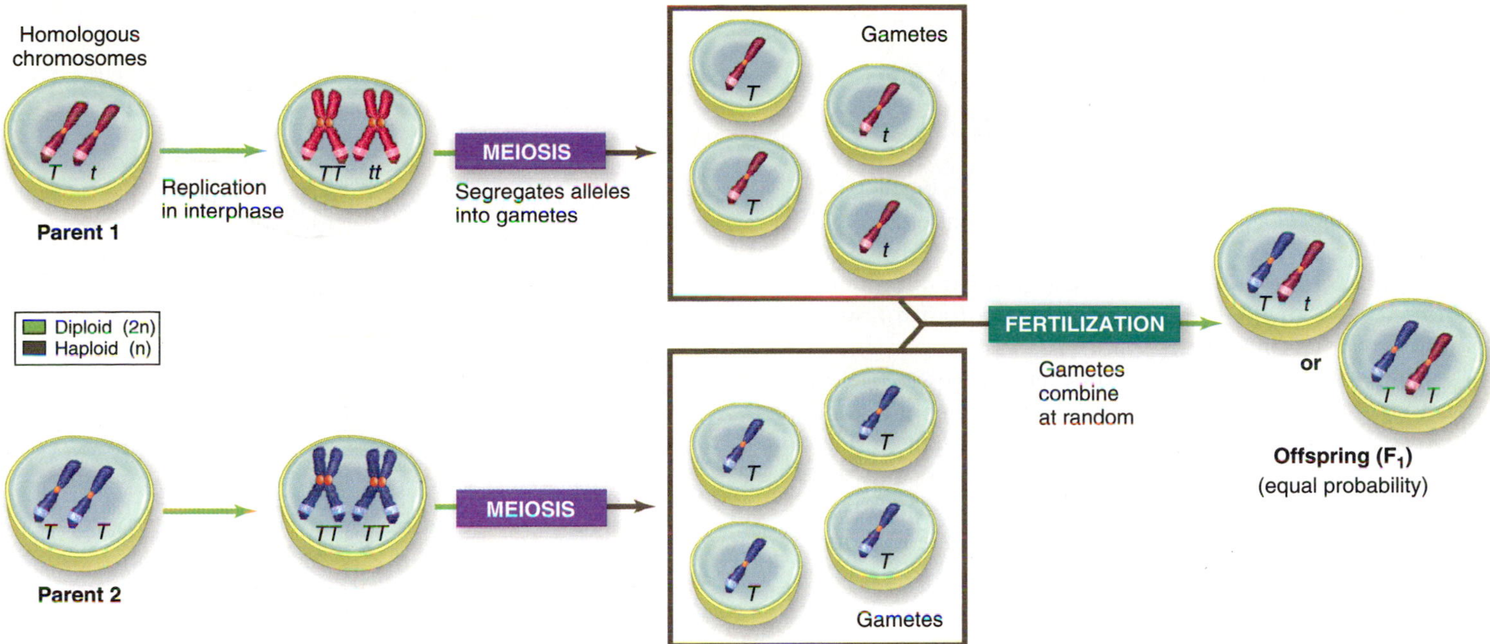

FIGURE 10.6 Mendel's Law of Segregation. During meiosis, homologous pairs of chromosomes (and the genes they carry) separate from one another and are packaged into separate gametes. At fertilization, gametes combine at random to form the next generation. Red and blue denote different parental origins of the chromosomes. Mendel did not know about meiosis, but observed its results in his studies of pea plants.

Note that all Punnett squares, including the one in figure 10.7, show the *probabilities* that apply to each offspring. That is, if the couple has four children, there will not necessarily be exactly one with genotype *CC*, two with *Cc*, and one with *cc*. Similarly, the chance of tossing a fair coin and seeing "heads" is 50%, but two tosses will not necessarily yield one head and one tail. If you toss the coin 1000 times, however, you will likely approach the expected 1:1 ratio of heads to tails. As Mendel discovered, pea plants are ideal for genetics studies in part because they produce many offspring in each generation.

Summary *Mendel's experiments with single-gene, two-allele traits in peas led to the law of segregation, which states that alleles of the same gene separate as they are packaged into gametes. Each offspring therefore has a known probability of inheriting a given allele from each parent.*

10.2 MASTERING CONCEPTS

1. What is a monohybrid cross, and what are the genotypic and phenotypic ratios expected in the offspring of the cross?
2. How are Punnett squares helpful in following inheritance of single genes?
3. What is a testcross, and why is it useful?
4. How does the law of segregation reflect the events of meiosis?

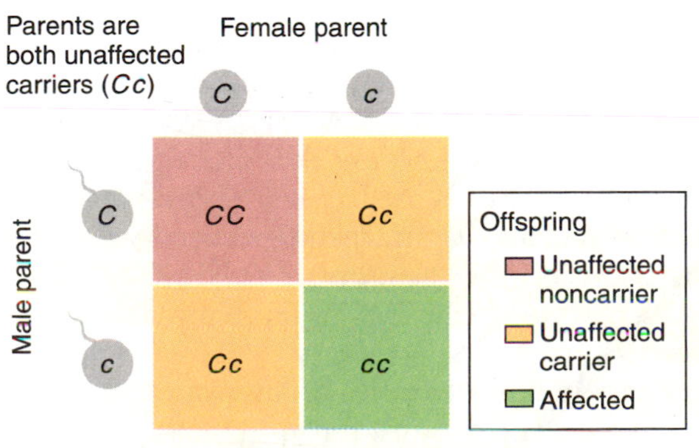

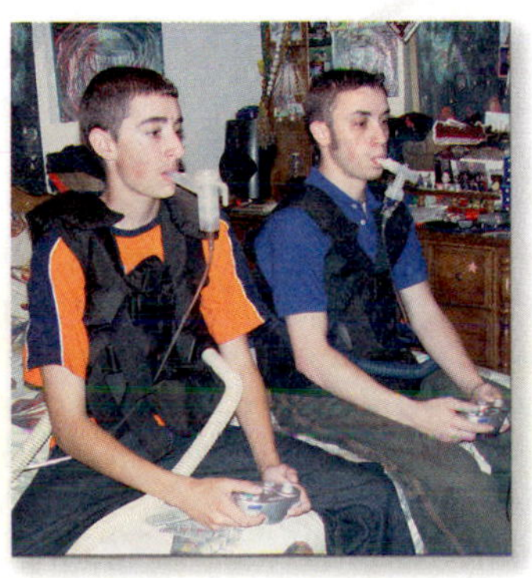

FIGURE 10.7 Mendel's Law Applied to Humans. In cystic fibrosis, sticky mucus accumulates in the respiratory tract and pancreas, triggering frequent lung infections and great difficulty in breathing and digesting (among other symptoms). When both parents are carriers, each child faces a 25% risk of inheriting the condition, a 50% chance of being a carrier like the parents, and a 25% chance of inheriting two dominant alleles.

10.3 Two Genes on Different Chromosomes Are Inherited Independently

Mendel's law of segregation arose from his studies of the inheritance of single traits. He next asked himself whether the same law would apply if he followed two different characters at the same time. Would one trait influence the inheritance of the other, or would each trait follow its own independent inheritance pattern?

Mendel therefore began another set of breeding experiments in which he simultaneously examined the inheritance of two characteristics of peas: shape and color. A pea's shape may be round or wrinkled (determined by the *R* gene, with the dominant allele specifying round shape). At the same time, its color may be yellow or green (determined by the *Y* gene, with the dominant allele specifying yellow).

A. Dihybrid Crosses Track the Inheritance of Two Genes at Once

As he did before, Mendel began with a P generation consisting of true-breeding parents. He crossed plants that had round, yellow seeds (homozygous dominant for genes *R* and *Y*, denoted *RRYY*) with plants that produced wrinkled, green seeds (homozygous recessive for both genes, *rryy*). All F_1 plants were heterozygous for both genes (*RrYy*) and had round, yellow seeds (**figure 10.8A**).

Next, Mendel crossed F_1 plants with each other. This is a **dihybrid cross,** a mating between two individuals that are each heterozygous for two genes. Four phenotypes appeared in the F_2 generation (figure 10.8B): round, yellow (315 plants); round, green (108 plants); wrinkled, yellow (101 plants); and wrinkled, green (32 plants). This is an approximate phenotypic ratio of 9:3:3:1.

The large Punnett square in figure 10.8 explains the origin of this 9:3:3:1 ratio. Each *RrYy* individual in the F_1 generation produced equal numbers of gametes of four different types: *RY*, *Ry*, *rY*, and *ry*. The Punnett square predicts that the four phenotypes—round, yellow (*RRYY*, *RrYY*, *RRYy*, and *RrYy*); round, green (*RRyy*, *Rryy*); wrinkled, yellow (*rrYY*, *rrYy*); and wrinkled, green (*rryy*)—will occur in the ratio 9:3:3:1, just as Mendel found.

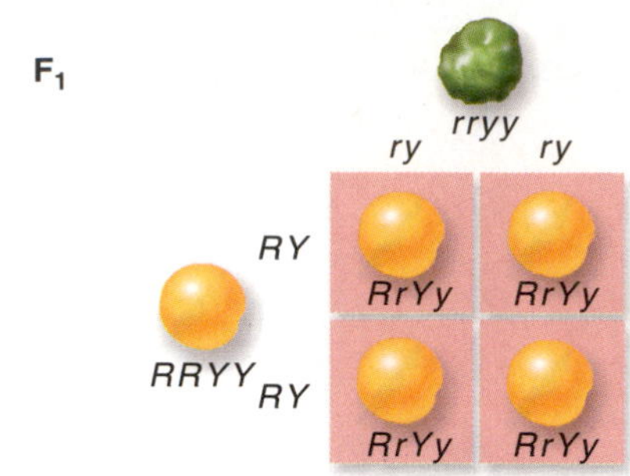

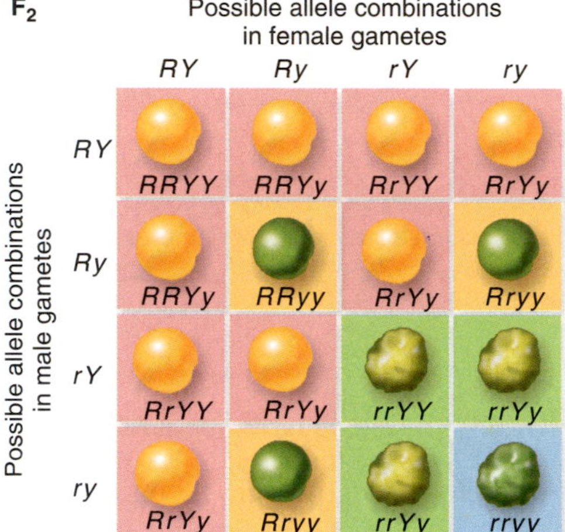

FIGURE 10.8 Plotting a Dihybrid Cross. (**A**) The parental generation consists of true-breeding plants. One parent is homozygous dominant for both genes; the other is homozygous recessive. The F_1 generation is heterozygous for both. (**B**) When plants from the F_1 generation are crossed with each other, phenotypes occur in an approximate ratio of 9:3:3:1. A Punnett square explains this phenotype ratio.

B. Alleles of Different Genes Move Independently into Gametes

Based on the results of the dihybrid cross, Mendel proposed what we now know as the **law of independent assortment.** It states that during gamete formation, the segregation of the alleles for one gene does not influence the segregation of the alleles for another gene. That is, alleles for two different genes are randomly packaged into gametes with respect to each other. With this second set of experiments, Mendel had again inferred a principle of inheritance based on meiosis (**figure 10.9**).

Punnett squares become cumbersome when analyzing more than two genes. A Punnett square for three genes has 64 boxes; for four genes, 256 boxes. An easier way to predict genotypes and phenotypes is to use the rules of probability on which Punnett squares are based. The **product rule** states that the chance that two independent events will both occur (for example, an offspring inheriting two particular alleles) equals the product of the individual chances that each event will occur.

The product rule can predict the chance of obtaining a wrinkled, green (*rryy*) plant from dihybrid (*RrYy*) parents.

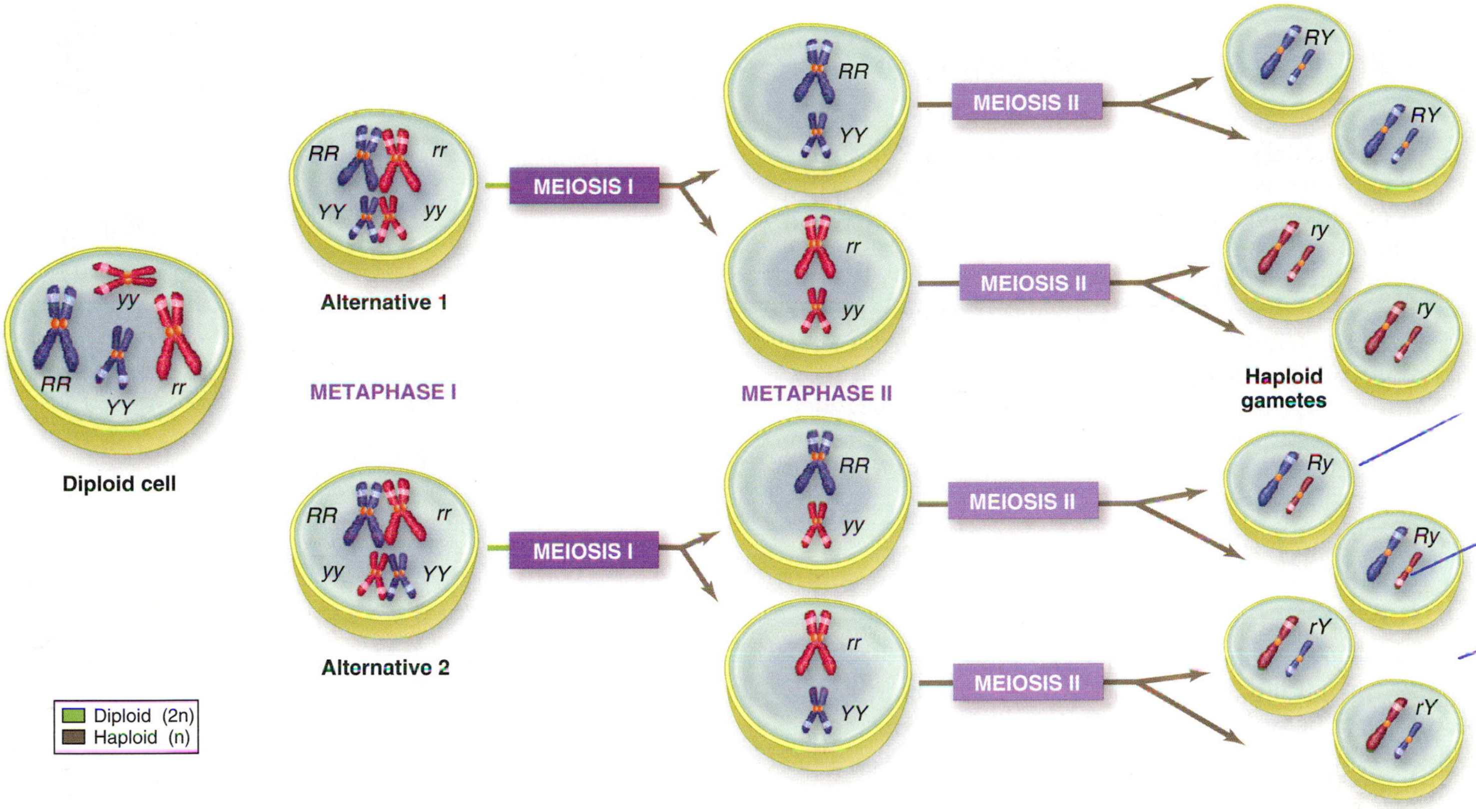

FIGURE 10.9 Mendel's Law of Independent Assortment. The independent assortment of genes carried on different chromosomes results from the random alignment of homologous chromosome pairs during metaphase I of meiosis. An individual of genotype *RrYy*, for example, produces four types of gametes: *RY*, *ry*, *Ry*, and *rY*. The exact allele combination in a gamete depends on which chromosomes happen to be packaged together—and this happens at random.

Consider the dihybrid individual one gene at a time. The probability that two *Rr* plants will produce *rr* offspring is 25%, or 1/4. Similarly, the chance of two *Yy* plants producing a *yy* individual is 1/4. According to the product rule, the chance of dihybrid parents (*RrYy*) producing homozygous recessive (*rryy*) offspring is 1/4 multiplied by 1/4, or 1/16. Now consult the 16-box Punnett square for Mendel's dihybrid cross (see figure 10.8). As expected, only one of the 16 boxes contains *rryy*. **Figure 10.10** applies the product rule to three traits.

Interestingly, Mendel found some trait combinations for which a dihybrid cross did not yield the expected 9:3:3:1 ratio. Mendel could not explain this result. No one could, until Thomas Hunt Morgan's work led to the chromosomal theory of inheritance. As you'll see in chapter 11, the law of independent assortment does not apply to genes that are close together on the same chromosome.

Summary *Mendel's dihybrid crosses provided evidence for his law of independent assortment: a gene on one chromosome does not influence inheritance of a gene on a different chromosome.*

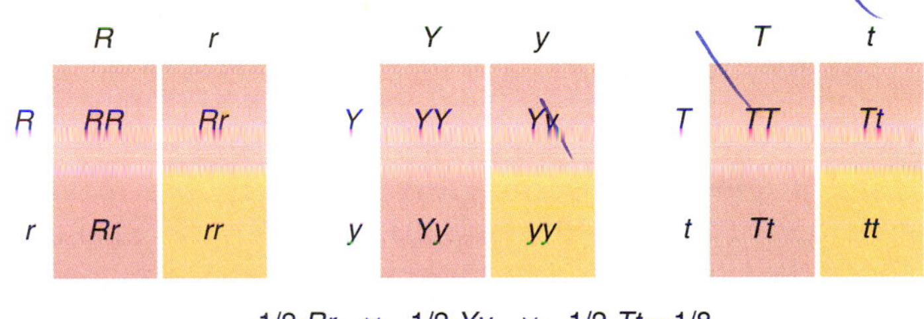

FIGURE 10.10 The Product Rule. Both parents are trihybrid, so they are heterozygous for three genes (*Rr Yy Tt*). Multiplying probabilities derived from Punnett squares for each individual gene reveals the chance that the offspring plant is also a trihybrid: one in eight.

10.3 MASTERING CONCEPTS

1. What is a dihybrid cross, and what is the phenotypic ratio expected in the offspring of the cross?
2. How does the law of independent assortment reflect the events of meiosis?
3. How can the product rule be used to predict the results of crosses in which multiple genes are studied simultaneously?

10.4 Gene Expression Can Appear to Alter Mendelian Ratios

Mendel's crosses yielded easily distinguishable offspring. A pea is either yellow or green, round or wrinkled; a plant is either tall or short. Often, however, offspring traits do not occur in the proportions that Punnett squares or probabilities predict. It may appear that Mendel's laws do not apply—but they do. The underlying genotypic ratios are there, but the nature of the phenotype, other genes, or the environment alter how traits appear.

This section describes some situations that may alter Mendelian ratios. Chapter 11 explains two other complications, linked genes and X-linked traits, in the context of human inheritance patterns.

A. Incomplete Dominance and Codominance Add Phenotype Classes

For the traits Mendel studied, one allele is completely dominant, and the other is completely recessive. The phenotype of a heterozygote is therefore identical to that of a homozygous dominant individual. In many cases, however, additional phenotype classes appear in the heterozygous offspring of a cross.

When a gene shows **incomplete dominance,** the heterozygous phenotype is intermediate between those of the two homozygotes. For example, a red-flowered snapdragon plant of genotype r_1r_1 crossed with a white-flowered r_2r_2 plant gives rise to a pink-flowered r_1r_2 (**figure 10.11**). The single copy of allele r_1 in the pink heterozygote directs less pigment production than the two copies in a red-flowered r_1r_1 plant.

In **codominance,** a heterozygote fully expresses two different alleles. For example, a person's ABO blood type is determined by the *I* gene, which has three possible alleles: I^A, I^B, and *i* (**figure 10.12**). The *I* gene encodes enzymes that insert either an "A" or a "B" molecule onto the surfaces of red blood cells (such cell surface molecules are termed *antigens*). Allele *i* is recessive, so a person with genotype *ii* produces neither antigen A nor antigen B and therefore has type O blood. A person who produces only antigen A (genotype I^AI^A or I^Ai) has type A blood; likewise, someone with only antigen B (genotype I^BI^B or I^Bi) has type B blood. Genotype I^AI^B yields type AB blood. The I^A and I^B alleles are codominant because both are equally expressed when both are present.

What is the difference between the recessive *i* allele and the codominant I^A and I^B alleles? Recall from Burning Question: What does "recessive" really mean? that a recessive allele encodes a nonfunctional protein. In codominance, however, neither allele is "silent." Both alleles I^A and I^B code for functional proteins, so people of blood type AB have both antigens A and B on the surfaces of their red blood cells.

The ABO blood type example also illustrates another condition that can alter phenotypic ratios: a gene with three or more possible alleles can yield many phenotypes. In the ABO blood type system, two codominant alleles (I^A and I^B) and one recessive allele (*i*) produce six possible genotypes and four phenotypes (blood types A, B, AB, and O).

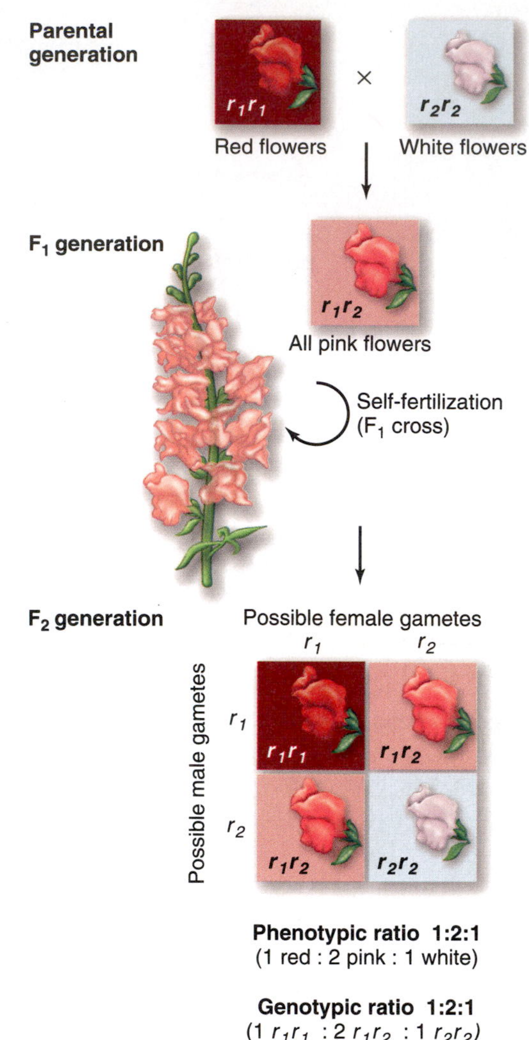

FIGURE 10.11 Incomplete Dominance in Snapdragon Flowers. A cross between a plant with red flowers (r_1r_1) and a plant with white flowers (r_2r_2) produces a heterozygous plant with pink flowers (r_1r_2). When these pink plants are crossed, one quarter of the offspring are red-flowered (r_1r_1), one half are pink-flowered (r_1r_2), and one quarter are white-flowered (r_2r_2). The phenotypic ratio of this monohybrid cross is 1:2:1 because the heterozygotes have a phenotype different from that of the homozygous plants.

B. One Genotype Can Yield Multiple Phenotypes

A **pleiotropic** gene has multiple phenotypic expressions. Pleiotropy arises when one protein is important in different biochemical pathways or affects more than one body part or

Genotypes	Phenotypes	
	Surface antigens	ABO blood type
I^AI^A I^Ai	Only A	Type A
I^BI^B I^Bi	Only B	Type B
I^AI^B	Both A and B	Type AB
ii	None	Type O

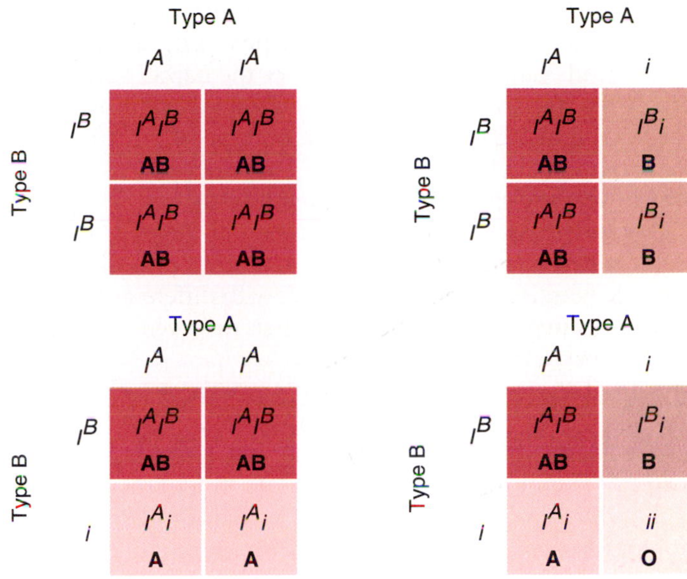

FIGURE 10.12 Codominance. The I^A and I^B alleles of the I gene are codominant, meaning that both are expressed in a heterozygote. Yet they still follow Mendel's law of segregation. These Punnett squares follow the genotypes and phenotypes that could result by crossing a person with type A blood with a person with type B blood.

process. For example, a single connective tissue protein abnormality causes Marfan syndrome, with symptoms including long limbs, spindly fingers, a caved-in chest, a weakened aorta, and lens dislocation. Volleyball player and Olympic silver medalist Flo Hyman had the disease; she died in 1986 during a game in Japan. Abraham Lincoln may also have had Marfan syndrome. (Like achondroplasia and Huntington disease, Marfan syndrome is an example of a rare disorder caused by a dominant allele.) **connective tissue, p. 562**

Another famous example is a disorder that is believed to have afflicted Britain's King George III in the late 1700s. Starting at age 50, he experienced recurrent bouts of a long list of symptoms (**figure 10.13**). In the twentieth century, researchers realized that one of George's symptoms, dark-

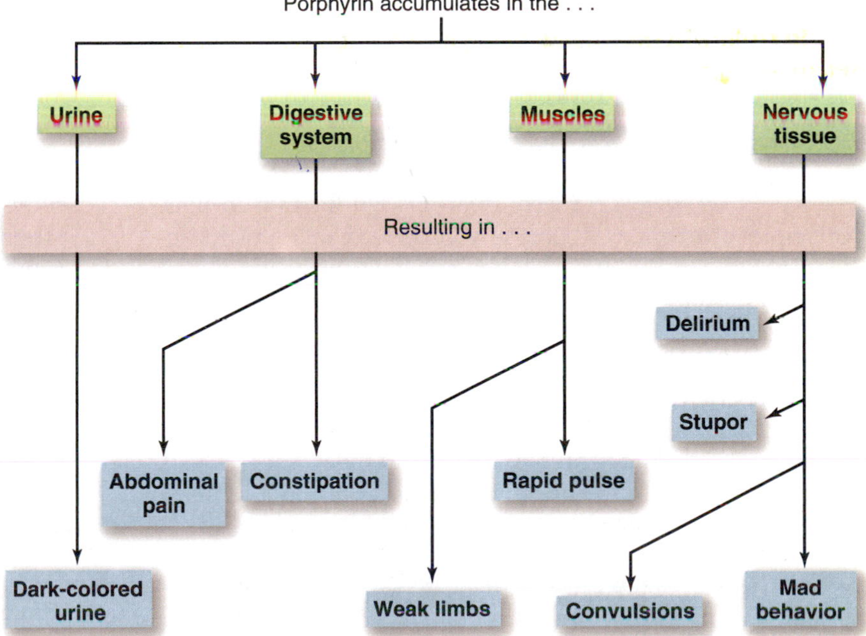

FIGURE 10.13 Pleiotropy. King George III is thought to have suffered from porphyria variegata, along with several other members of the Hanover royal family of which he was a member. In King George, symptoms appeared every few years, always in the same order. Because of pleiotropy, the family's varied symptoms and quirks appeared to be different, unrelated disorders. The king also suffered from periodic bouts of depression and paranoia, which led to the popular conclusion that he was mad. He was eventually deemed unfit to rule and was dethroned.

colored urine, may have indicated an illness known as porphyria variegata. Lack of a liver enzyme results in buildup of compounds called porphyrins (hence the name, porphyria). Biologists still do not understand how this condition causes the symptoms. Careful study of George's relatives revealed that some of them had different subsets of the symptoms, which were thought to be unrelated diseases.

Although Mendelian rules of inheritance apply to pleiotropic genes, the conditions they cause can be difficult to trace through families because individuals with different subsets of symptoms may appear to have different disorders.

C. Multiple Genotypes Can Yield Identical Phenotypes

Another situation that can complicate inheritance patterns is that different genes can produce similar or identical phenotypes. For example, blood clot formation requires eleven separate biochemical reactions. A different gene encodes each enzyme in the pathway, and clotting disorders may result from abnormalities in any of these genes. The phenotypes are the same (poor blood clotting), but the genotypes differ. **metabolic pathways, p. 82**

Porphyria likewise has several forms, one of which seems to have afflicted King George III. Eight enzymes participate in the production of porphyrins. Mutations in any of the genes encoding these enzymes can cause different forms of porphyria.

D. Gene Products Can Interact

In **epistasis,** one gene affects the expression of another, which may appear to disrupt Mendel's laws. As a simple example, male pattern baldness (a genetic condition) hides the effects of the allele for a "widow's peak" hairline. Epistasis is different from a dominant allele masking a recessive allele of the same gene. Instead, the masking happens between two different genes.

Epistasis is also responsible for some ABO blood type inconsistencies between parents and their children. Most people with alleles I^A or I^B have blood type A, B, or AB (see figure 10.12). However, a protein must physically link the A and B antigens to the cell surface. If a gene called *H* is mutant, part of that protein is missing, and antigens A and B cannot attach. As a result, a person with the extremely rare genotype *hh* has blood that always tests as type O, even though he or she may have a genotype indicating type A, B, or AB blood. Confusion over paternity might arise if a type O offspring is not possible, given the parents' ABO genotypes.

Summary *Mendelian ratios can be disrupted if one allele is not completely dominant to another, a single genotype yields multiple phenotypes, unrelated genes yield identical phenotypes, or the product of one gene alters the phenotype produced by another.*

10.4 MASTERING CONCEPTS

1. How do incomplete dominance and codominance increase the number of phenotypes?
2. What is pleiotropy?
3. How can the same phenotype stem from many different genotypes?
4. How can epistasis decrease the number of phenotypes?

10.5 Most Traits Are Influenced by the Environment and Multiple Genes

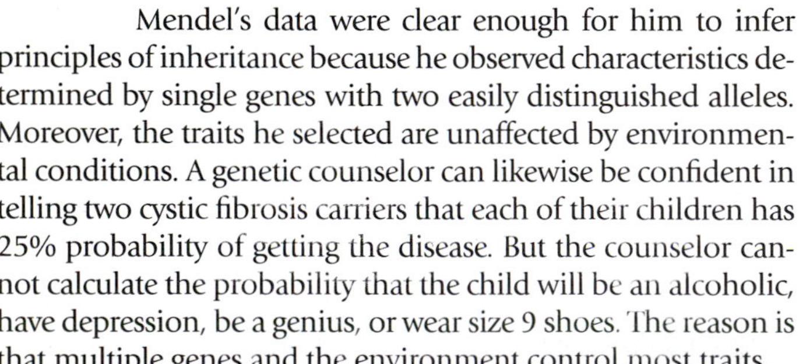

Mendel's data were clear enough for him to infer principles of inheritance because he observed characteristics determined by single genes with two easily distinguished alleles. Moreover, the traits he selected are unaffected by environmental conditions. A genetic counselor can likewise be confident in telling two cystic fibrosis carriers that each of their children has 25% probability of getting the disease. But the counselor cannot calculate the probability that the child will be an alcoholic, have depression, be a genius, or wear size 9 shoes. The reason is that multiple genes and the environment control most traits.

A. The Environment Can Alter the Phenotype

The environment often affects gene expression. As a simple example, temperature influences the quantity of pigment molecules in the fur of some animals. Siamese cats and Himalayan rabbits have dark ears, noses, feet, and tails because these parts are colder than the animals' abdomens (**figure 10.14**).

Likewise, a person's upbringing and personal circumstances certainly influence susceptibility to depression or alcoholism, although both diseases also have a genetic component. Adoption and twin studies offer other means to help sort out the relative contributions of "nature" and "nurture" (see section 38.3). This chapter's Can *You* Relate?: The Roots of Addiction describes how scientists use these tools in the study of addiction.

Even human diseases with simple, single-gene inheritance patterns can have an environmental component. Cystic fibrosis, for example, is a single-gene disorder. Because cystic fibrosis patients are very susceptible to infection, however, the course of the illness depends on which infectious agents a person encounters.

FIGURE 10.14 Environment Affects Phenotype. In Siamese cats, the gene encoding an enzyme required for pigment production is mutated. The heat-sensitive enzyme is active in cool areas, such as the paws, ears, snout, and tail. But the enzyme is inactive at body temperature, so the cat's fur remains light-colored where the skin is warmer.

B. Polygenic Traits Depend on More Than One Gene

Unlike cystic fibrosis, most inherited traits are **polygenic;** that is, the phenotype reflects the activities of more than one gene. Eye color is an example of a polygenic trait because multiple enzymes, encoded by multiple genes, influence the production and distribution of the pigment melanin in the eye's iris. Eye color is among the few traits unaffected by external conditions.

To complicate matters, the environment often profoundly affects the expression of both single-gene and polygenic traits. For example, in plants, polygenic traits include flower color, the density of leaf pores (stomata), and crop yield. But these traits do not remain static throughout a plant's life. Soil pH can affect flower color, CO_2 concentration can

Can *You* Relate?

The Roots of Addiction

Many people wonder what causes addictions to alcohol, heroin, gambling, and other stimuli. Is addiction rooted in character flaws? An illness? Do some people have a genetic predisposition to addiction? Or is culture more important? Perhaps addictions are different for everyone, reflecting a blend of all of these causes.

Clearly, many addictions run in families. For example, half of those diagnosed as alcoholic come from families with a history of heavy alcohol use. It is difficult, however, to measure the relative contributions of genetics and environment. After all, most family members are related genetically and share the same (or similar) surroundings.

Adopted children have helped geneticists tease apart the genetic and environmental components of addiction. A child adopted by nonrelatives shares environmental influences, but not alleles, with his or her adoptive family. On the other hand, adopted individuals share alleles, but not the exact same environment, with their biological parents. Information on both sets of parents can reveal to what degree heredity and the environment contribute to an addiction.

Twins have also helped geneticists distinguish between nature and nurture. If a trait occurs more frequently in both members of identical (monozygotic) twin pairs than in both members of fraternal (dizygotic) twin pairs, it is at least partly controlled by genes. In contrast, both identical and fraternal twins have an equal probability of having traits molded mostly by the environment. **twins, p. 183**

The best way to distinguish between the effects of nature and nurture is to study twins separated at birth. Genetics accounts for much of what they have in common, especially if their environments have been very different. Any differences between the twins reflect differences in their upbringing, not in genetics. The "twins-reared-apart" approach, however, also has limitations. Identical twins share an environment in the uterus that may affect later development. Furthermore, siblings do not always share identical home environments. Differences in age, sex, general health, school and peer experiences, temperament, and personality affect each individual's perception of such important environmental influences as parental affection and discipline.

Studies of adopted children are not perfect either. Adoption agencies do not assign children to homes at random; rather, they often search for families who have socioeconomic or religious backgrounds similar to those of the biological parents. Thus, even when different families adopt twins separated at birth, their environments may not be as different as they might be for two unrelated adopted people.

Despite these complications, adoption and twin studies, coupled with experiments in laboratory animals, clearly indicate that addiction involves a genetic vulnerability. However, the exact genes remain unknown, and environmental influences are substantial. Researchers hope that further study of all the factors contributing to addiction will improve prevention and treatment.

Separated at birth, the Mallifert twins meet accidentally.

a.

b.

FIGURE 10.15 Height Is Polygenic and Environmental. (**A**) These students from the University of Connecticut at Storrs lined up by height in 1920, demonstrating the characteristic bell-shaped distribution of polygenic traits. (**B**) A similar photo taken in 1997 reveals more tall students than lineups photographed early in the twentieth century, illustrating the influence of environment on height. In 1920, the tallest student was 6′2″; in 1997, the tallest was 6′5″.

change the number of stomata per square centimeter, and nutrient and water availability greatly influence crop production. **stomata, p. 508**

When the frequencies of all the phenotypes associated with a polygenic trait are plotted on a graph, they form a characteristic bell-shaped curve. **Figure 10.15** shows a bell curve for height, which is a product of genetics, childhood nutrition, and health care. **Figure 10.16** shows a continuum of gene expression for skin color, another trait affected by genes and the environment—in this case, exposure to sunlight. Body weight and intelligence are other traits that are both polygenic and influenced by the environment.

Summary *The environment often influences phenotype, and many traits are determined by more than one gene. The resulting inheritance patterns are complex.*

10.5 MASTERING CONCEPTS

1. How can the environment affect a phenotype?
2. What is a polygenic trait?

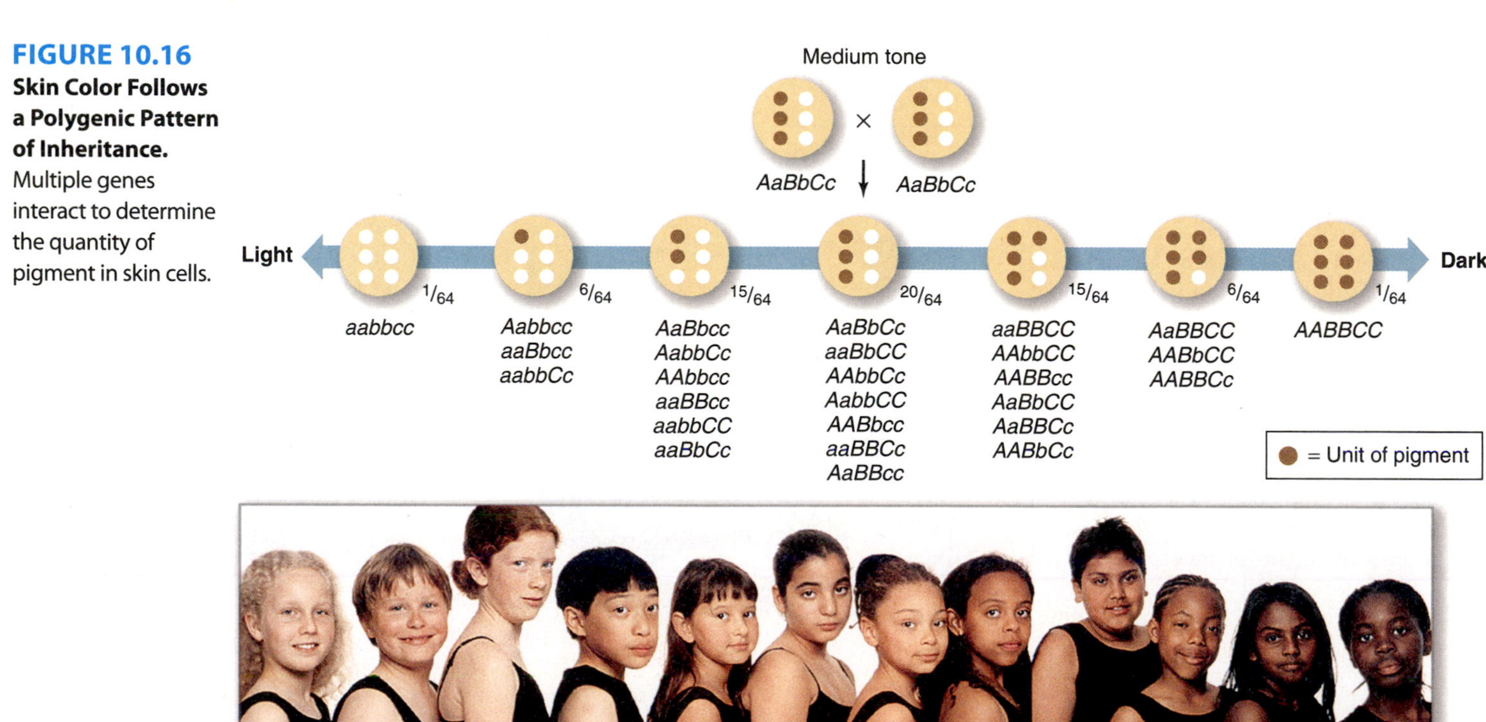

FIGURE 10.16 Skin Color Follows a Polygenic Pattern of Inheritance. Multiple genes interact to determine the quantity of pigment in skin cells.

10.6 INVESTIGATING LIFE

Heredity and the Hungry Hordes

Agriculture provides a steady food supply, but not just to humans. Hungry insects and other animals can devastate crops by eating the leaves, roots, seeds, and fruits of the plants we grow for food or fiber. Farmers continually seek new ways to kill these competitors, but each tactic selects for new adaptations in the insects. It is a long-standing, and seemingly unavoidable, evolutionary battle.

The larvae of butterflies and moths are particularly voracious. A good example is the pink bollworm (*Pectinophora gossypiella*; **figure 10.17**). The adults of this species are moths that lay eggs on cotton bolls. When the eggs hatch, the pink caterpillars tunnel into the boll and eat the seeds, damaging the cotton fibers. This insect is a major pest. Cotton producers in the United States lose more than $21 million a year in the fight against the pink bollworm. In developing countries, the caterpillars destroy as much as 20% of the cotton crop each year.

One tool that keeps bollworms and other caterpillars at bay is a soil bacterium called *Bacillus thuringiensis*, abbreviated Bt. This microbe produces a toxic protein that pokes holes in a caterpillar's intestinal tract, leaving the animal vulnerable to infection and unable to digest food. Bt does not affect humans, because we lack the specific molecule to which the toxin binds. Bt is among the few insecticides that organic farmers can spray on their plants.

Bt is not perfect. Some caterpillars (those that tunnel inside an ear of corn, for example) escape death because they never eat the Bt sprayed on a plant's surface. In the 1990s, however, biotechnology solved this problem when scientists inserted the bacterial gene for the Bt toxin into plant cells (see chapter 12). Every cell in these so-called Bt plants produces the toxin. Nibbling on any hidden or exposed plant part spells death to the caterpillar.

Genetically modified Bt corn and cotton are enormously popular with farmers in the United States, greatly increasing the selection pressure for Bt resistance in insect populations. When Bt was available only as a spray, insect populations encountered it infrequently. But genetically modified crops produce the toxin throughout their lives, so caterpillars are exposed from the moment they begin feeding. Large fields of Bt plants strongly select against susceptible individuals, leaving resistant caterpillars to produce the next generation.

FIGURE 10.17 Hungry Caterpillar. Pink bollworms cost cotton producers tens of millions of dollars each year.

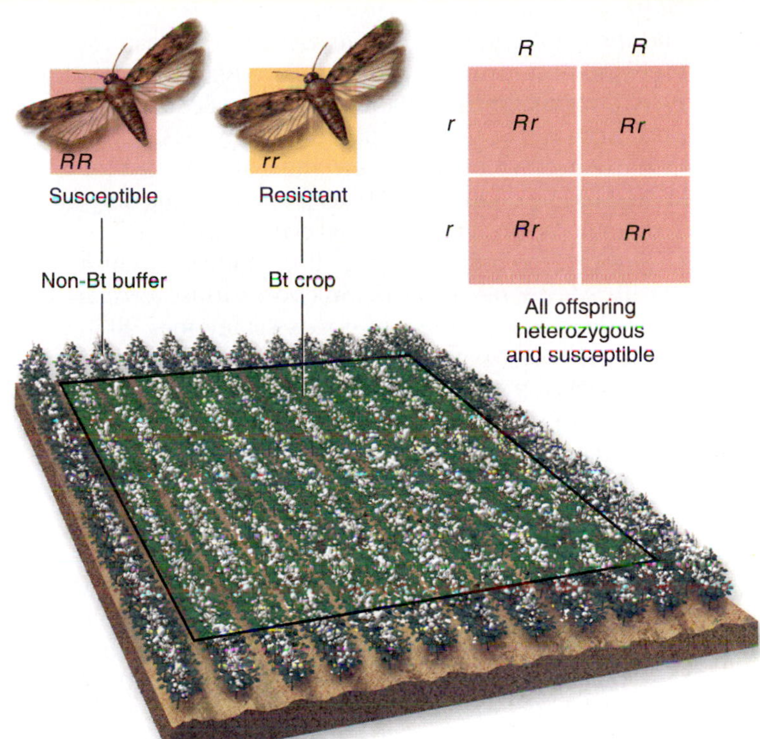

FIGURE 10.18 Bt Crops Require Buffers. This Punnett square shows the logic of using buffer strips around Bt crops to help keep the incidence of recessive alleles low.

To combat this selective pressure, farmers growing Bt crops must agree to surround each field with a buffer strip planted with a conventional (non-Bt) variety of the same crop—a refuge. A simple Punnett square illustrates the logic of this strategy (**figure 10.18**). If the strip is large enough, a homozygous recessive (resistant) moth has a good chance of encountering a susceptible mate from the buffer strip. All of their heterozygous offspring will die if they eat the Bt plants. According to mathematical models, a ready pool of susceptible mates from the refuge should therefore keep the recessive allele rare.

How can scientists test whether refuges really do keep recessive alleles rare in the real world? One simple approach would be to capture a sample of insects from a cotton field, feed them Bt crops, and count how many die. This method, however, does not distinguish between the two possible genotypes of a susceptible caterpillar: homozygous dominant or heterozygous. Because the resistance allele can "hide" in a susceptible animal, it is impossible to measure its incidence simply by classifying a caterpillar as resistant or susceptible.

A better solution would be to find a way to measure a bollworm's genotype directly. Biologists Shai Morin, Bruce Tabashnik, and their colleagues at the University of Arizona tackled this problem by studying Bt resistance in several populations of pink bollworms. One population, reared for decades in a lab without exposure to Bt, was susceptible. Other groups, originally collected in Arizona and Texas, were artificially selected for resistance by feeding them Bt-laced meals and allowing only the survivors to reproduce. **artificial selection, p. 267**

The researchers arranged breeding studies to confirm that the allele for Bt resistance in pink bollworms is recessive. When they bred susceptible moths to resistant ones, all individuals in the heterozygous F_1 generation were susceptible. Then they backcrossed F_1 offspring with the resistant strains. Half of the resulting insects were susceptible, and half were resistant (**figure 10.19**). The two sets of crosses confirmed that the allele that confers Bt resistance is recessive.

The next task was to count the number of different Bt resistance alleles in their population. From previous studies, the researchers knew that Bt resistance comes from changes in a protein called cadherin, which is the target molecule to which the toxin binds. They also knew the DNA sequence of the gene that encodes cadherin. They extracted DNA from resistant pink bollworms, determined the cadherin gene sequence for each, and compared it with the gene from susceptible insects. The results suggested that there are three unique resistance alleles, each encoding a different variation on the cadherin protein's shape. An insect with two such resistance alleles is immune to the toxin.

The unique DNA sequence associated with each allele is a useful "tag" for identifying Bt resistance in pink bollworms. Researchers can simply extract an insect's DNA and test for the presence of the recessive alleles. As a result, we can now spy on the pink bollworm's evolution as it happens. The stakes are high. If the resistance alleles become very common, Bt may become useless as a control measure. Organic farmers will have lost one of the few insecticides available to them, and other growers will switch back to broad-spectrum pesticides that kill many beneficial insects. Careful monitoring will be crucial as we continue to wage war against the insects that will forever compete for our crops.

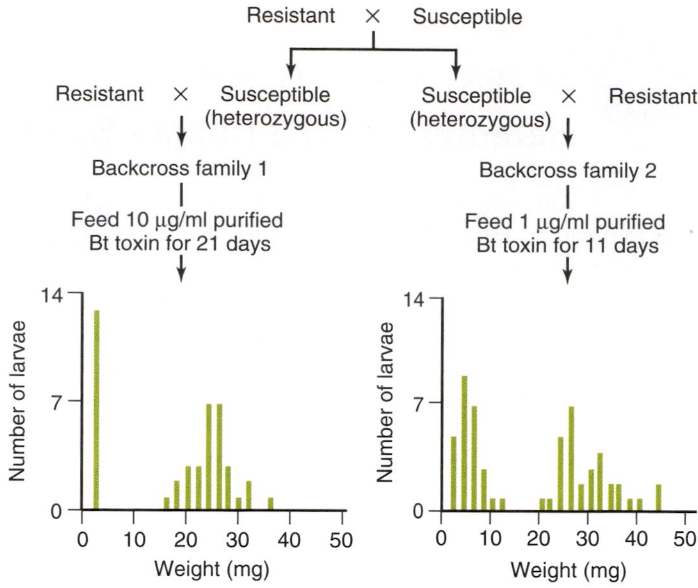

FIGURE 10.19 Bt Resistance Is Recessive. When a heterozygous susceptible insect is bred with a resistant mate, half the offspring should be resistant and half should be susceptible. Tests with Bt toxin show that this is indeed the case: half of the insects thrived in the presence of Bt toxin, while the susceptible ones died or were very small. **Question:** What do you predict will happen to the incidence of resistance alleles in pink bollworm populations if farmers choose not to plant the required refuge?

Morin, Shai, Robert W. Biggs, Mark S. Sisterson, and 10 other authors (including Bruce E. Tabashnik). April 29, 2003. Three cadherin alleles associated with resistance to *Bacillus thuringiensis* in pink bollworm. *Proceedings of the National Academy of Sciences*, vol. 100, pages 5004–5009.

CHAPTER SUMMARY

10.1 Mendel's Experiments Uncovered Basic Laws of Inheritance

A. Why Peas?

- Gregor Mendel studied inheritance patterns in pea plants because they are easy to grow, develop quickly, and produce abundant offspring. The genes for the traits he studied had only two forms, or **alleles.** It is also easy to control crosses between pea plants.

B. Alleles May Be Dominant or Recessive

- An allele whose expression masks another is **dominant;** an allele whose expression is masked by a dominant allele is **recessive.** Dominant does not always mean "most common."

C. For Each Gene, a Cell's Two Alleles May Be Identical or Different

- A **heterozygote** has two different alleles of a gene. A **homozygous** recessive individual has two recessive alleles. A homozygous dominant individual has two dominant alleles.
- The combination of alleles for a gene is the individual's **genotype,** and the observable expression

of a genotype is the organism's **phenotype.** A **wild-type** allele is the most common in a population. A change in a gene is a **mutation** and may result in a **mutant** phenotype.

10.2 The Two Alleles of Each Gene End Up in Different Gametes

- In genetic crosses, the purebred parental generation is designated **P**; the next generation is the first filial generation, or $\mathbf{F_1}$; and the next is the second filial generation, or $\mathbf{F_2}$.

A. Monohybrid Crosses Track the Inheritance of One Gene

- A **monohybrid cross** between two heterozygotes yields a genotypic ratio of 1:2:1 and a phenotypic ratio of 3:1.
- **Punnett squares** are useful for calculating the probability of each possible outcome in a genetic cross.
- A **testcross** reveals an unknown genotype by breeding the individual to a homozygous recessive individual.

B. Meiosis Explains Mendel's Law of Segregation

- Mendel's **law of segregation** states that the two alleles of the same gene separate into different gametes. Each individual receives one allele of each gene from each parent.

10.3 Two Genes on Different Chromosomes Are Inherited Independently

A. Dihybrid Crosses Track the Inheritance of Two Genes at Once

- A **dihybrid cross** between individuals heterozygous for two genes yields a 9:3:3:1 phenotypic ratio if the genes are on different chromosomes.

B. Alleles of Different Genes Move Independently into Gametes

- According to Mendel's **law of independent assortment,** the inheritance of one gene does not affect the inheritance of another gene on a different chromosome. This law reflects meiosis, in which homologous pairs of chromosomes (and the genes they carry) align randomly during metaphase I.
- The **product rule** is an alternative to Punnett squares for following inheritance of two or more traits at a time.

10.4 Gene Expression Can Appear to Alter Mendelian Ratios

- In some crosses, the ratio of offspring phenotypes does not seem to follow Mendel's principles.

A. Incomplete Dominance and Codominance Add Phenotype Classes

- Heterozygotes of **incompletely dominant** alleles have phenotypes intermediate between those of the two homozygotes. **Codominant** alleles are both expressed in a heterozygote.

B. One Genotype Can Yield Multiple Phenotypes

- A **pleiotropic** gene has multiple effects on the body and therefore produces many different phenotypes.

C. Multiple Genotypes Can Yield Identical Phenotypes

- Many biochemical reactions require multiple proteins. Mutated genes encoding any of the proteins can stop the pathway, producing the same phenotype.

D. Gene Products Can Interact

- In **epistasis,** one gene masks the effect of another.

10.5 Most Traits Are Influenced by the Environment and Multiple Genes

A. The Environment Can Alter the Phenotype

- Unlike traits that Mendel studied, many traits have environmental as well as genetic influences.

B. Polygenic Traits Depend on More Than One Gene

- A **polygenic trait** varies continuously in its expression, and the frequencies of the phenotypes form a bell-shaped curve.

10.6 Investigating Life: Heredity and the Hungry Hordes

- Researchers have developed a way to test for the presence of recessive alleles in insect pests of cotton. This genetic test allows researchers to monitor caterpillars for resistance to Bt, a toxin in genetically modified cotton.

MULTIPLE CHOICE QUESTIONS

1. According to Mendel, if an individual is *heterozygous* for a specific gene, the phenotype will be that of
   a. the recessive trait alone.
   b. the dominant trait alone.
   c. a blend of the dominant and recessive traits.
   d. a wild type trait.
2. If an individual is *homozygous* for a specific gene, then the genotype will contain
   a. only the recessive allele.
   b. only the dominant allele.
   c. both a dominant and a recessive allele.
   d. either a or b could be true.

3. What can you conclude if the offspring of a testcross all show the dominant phenotype?
   a. One parent was homozygous dominant.
   b. One parent was heterozygous.
   c. The offspring are all homozygous dominant.
   d. Both b and c are correct.
4. What cellular process is responsible for Mendel's law of segregation?
   a. Mitosis
   b. Mutation
   c. Meiosis
   d. Metaphase
5. Which of the following is a possible gamete for an individual with the genotype *PPrr*?
   a. *PP*
   b. *Pr*
   c. *pr*
   d. *rr*
6. Use the product rule to determine the chance of obtaining an offspring with the genotype *RrYy* from a dihybrid cross between parents with the genotype *RrYy*.
   a. 1/2
   b. 1/4
   c. 1/8
   d. 1/16
7. How does incomplete dominance affect the phenotype of a heterozygote?
   a. It results in a blend of the dominant and recessive phenotypes.
   b. It results in the expression of only the recessive phenotype.
   c. The dominant phenotype is still expressed, but only in patches.
   d. The trait is not observed in the individual.
8. One gene affecting the expression of another gene is an example of
   a. pleiotropy.
   b. mutation.
   c. codominance.
   d. epistasis.
9. What is a polygenic trait?
   a. A trait that reflects expression of both dominant and recessive alleles
   b. A trait that reflects the influence of the environment
   c. A trait that reflects the expression of many alleles of the same gene
   d. A trait that reflects the expression of many different genes
10. In the ABO blood type system, an individual with the blood type AB expresses both alleles I^A and I^B. This situation illustrates
    a. epistasis.
    b. independent assortment.
    c. codominance.
    d. pleiotropy.

TESTING YOUR KNOWLEDGE

1. What advantages do pea plants have for genetics studies? Why aren't humans equally suitable?
2. Given the relationship between genes, alleles, and proteins, how can a recessive allele appear to "hide" in a heterozygote?
3. Is the size of your ears an example of a genotype or a phenotype?
4. How did Mendel use evidence from monohybrid and dihybrid crosses to deduce his laws of segregation and independent assortment? How do these laws relate to meiosis?
5. In a dihybrid cross, the predicted phenotype ratio is 9:3:3:1; the "9" represents the proportion of plants expressing at least one dominant allele for both traits. How would you use testcrosses to determine whether these plants are homozygous dominant or heterozygous for one or both genes?
6. Explain how each of the following appears to disrupt Mendelian ratios: incomplete dominance, codominance, pleiotropy, epistasis.
7. Explain the following "equation":

   Genotype + Environment = Phenotype

THINKING AS A SCIENTIST

1. Many plants are polyploid (see chapter 9); that is, they have more than two sets of chromosomes. How would having four (rather than two) copies of a chromosome more effectively mask expression of a recessive allele?
2. In an attempt to breed winter barley that is resistant to barley mild mosaic virus, agricultural researchers cross a susceptible domesticated strain with a resistant wild strain. The F_1 plants are all susceptible, but when the F_1 plants are crossed with each other, some of the F_2 individuals are resistant. Is the resistance allele recessive or dominant? How do you know?
3. Springer spaniels often suffer from canine phosphofructokinase (PFK) deficiency. The dogs lack an enzyme that is crucial in extracting energy from glucose molecules. Affected pups have extremely weak muscles and die in weeks. A DNA test is available to identify male and female dogs that are carriers. Why would breeders wish to identify carriers if these dogs are not affected?

4. A white woman with fair skin, blond hair, and blue eyes and a black man with dark brown skin, hair, and eyes have fraternal twins. One twin has blond hair, brown eyes, and light skin, and the other has dark hair, brown eyes, and dark skin. What Mendelian principle does this real-life case illustrate?
5. In humans, more than 100 forms of deafness are inherited as recessive alleles on many different chromosomes. Suppose that a woman who is heterozygous for a deafness gene on one chromosome has a child with a man who is heterozygous for a deafness gene on a different chromosome. Does the child face the general population risk of inheriting either form of deafness, or the 25% chance that Mendelian ratios predict for a monohybrid cross? Explain your answer.
6. Suppose a single trait is controlled by a gene with four codominant alleles. A person can inherit any combination of two of the four alleles. How many phenotypes are possible for this trait?

GENETICS PROBLEMS

See answers in Appendix D.

1. Holstein cattle suffer from the condition citrullinemia, in which homozygous recessive calves die within a week of birth because they cannot break down ammonia that is produced when amino acids are metabolized. If a cow that is heterozygous for the citrullinemia gene is inseminated by a bull that is homozygous dominant, what is the probability that a calf inherits citrullinemia?
2. A man and a woman each have dark eyes, dark hair, and freckles. The genes for these traits are on separate chromosomes. The woman is heterozygous for each of these genes, but the man is homozygous. The dominance relationships of the alleles are as follows:

   B = dark eyes; *b* = blue eyes
   H = dark hair; *h* = blond hair
   F = freckles; *f* = no freckles

   a. What is the probability that their child will share the parents' phenotype?
   b. What is the probability that the child will share the same genotype as the mother? As the father?

   Use the product rule or a Punnett square to obtain your answers. Which method do you think is easier?
3. A woman with type AB blood has children with a man who has type O blood. What are the chances that a child they conceive will have type A blood? Type B? AB? O?
4. A male cat has short hair, a stubby tail, and extra toes. A female cat has long hair, a long tail, and extra toes. The genes and alleles are as follows:

   L = short hair; *l* = long hair
   M = stubby tail (Manx); *m* = long tail
   Pd = extra toes; *pd* = normal number of toes

   The two cats have kittens. One has long hair, a long tail, and no extra toes. Another has short hair, a stubby tail, and extra toes. The third kitten has short hair, a long tail, and no extra toes. What is the genotype of the father?

Visit www.mhhe.com/hoefnagels for practice quizzes, animations, videos, and activities designed to help you master the material in this chapter.

5. Wild type canaries are yellow. A dominant mutant allele of the color gene, designated *W*, causes white feathers. Inheriting two dominant alleles is lethal to the embryo. If a yellow canary is crossed to a white canary, what is the probability that an offspring will be yellow? What is the probability that it will be white?
6. One in 20 people has restless leg syndrome, in which the legs feel tingly or achy nearly all the time, causing great daytime fatigue from interrupted sleep. Another inherited condition causes extremely red hair and an inability to tan. The affected person has an unusual variant of the pigment molecule melanin, called "red" melanin. The alleles that cause both restless leg syndrome and red melanin are recessive, and their genes are on different chromosomes. Use the following information to answer questions about the inheritance of restless leg syndrome.
   - Suzanne and Michael met at the original Woodstock rock festival in the summer of 1969, drawn to each other because of their striking red hair and very pale skin. Neither has a family history of restless leg syndrome. A few years later, they married, and in 1975, Marvin and Gary, monozygotic (identical) red-headed twins who do not tan, were born.
   - Jackie and David were at Woodstock, too. They each have dark brown hair, and David suffers from restless leg syndrome. Their daughter, Eileen, has dark brown hair and legs that aren't restless.
   - Gary and Eileen get together and have a son, Todd. He has red hair, doesn't tan, and has restless leg syndrome

   a. Taking all of the preceding information into account, which individuals *must* be heterozygous for restless leg syndrome?
   b. Which individuals *might* be heterozygous for restless leg syndrome gene?
   c. Which individuals *must* be heterozygous for the red melanin gene?
   d. If Marvin marries a woman who has restless leg syndrome and red hair and cannot tan, what is the probability that a child of theirs will inherit both traits?

Chromosomes and Human Inheritance Patterns

Expectant Mother. Thanks to medical technology, a pregnant woman can see her baby's chromosomes long before the infant is born.

What's the Point?

Prenatal Diagnosis Highlights Ethical Dilemmas

Barbara is pregnant, and like many women, she periodically has her fetus examined by ultrasound. Her latest scan has revealed a possible abnormality, but her physician cannot be sure without ordering a test of the fetus's chromosomes.

Technicians can collect cells containing fetal chromosomes by extracting a small amount of the fluid or tissue surrounding the developing fetus. These cells can then be used to prepare a karyotype, a size-ordered chart of the fetus's chromosomes. The karyotype may reveal several types of chromosomal abnormalities. For example, problems during meiosis can yield gametes with extra or missing chromosomes. Often, the resulting embryo dies early in development. An extra copy of chromosome 21, however, causes Down syndrome (see figure 9.12); extra or missing sex chromosomes cause Klinefelter syndrome, Turner syndrome, and other disorders. Karyotypes can also reveal large-scale translocations, in which DNA has moved from one chromosome to a nonhomologous chromosome (see figure 9.13). **errors in meiosis, p. 185**

If the karyotype reveals that Barbara's fetus has a chromosomal abnormality, she may seek a counselor who can advise her on how best to prepare for the birth of her baby. If the baby's abnormality will be severe, Barbara may decide to seek an abortion, ending the pregnancy. But this choice raises many difficult issues.

Prenatal diagnosis illustrates the intersection between morality and science. Few people would argue against Barbara's use of prenatal diagnosis to learn more about a possible illness. But should parents have the right to expose a fetus to the small risks of prenatal screening simply to determine its sex? Should parents be allowed to abort a fetus of the "wrong" sex? What if an expectant mother lives in a country where having a second female child can bring economic ruin?

Furthermore, what constitutes a "severe" abnormality, and who decided on the definition? Many defects are not survivable, and the child will die shortly after birth (if not before). On the other hand, many Down syndrome people have happy lives. Or consider cri-du-chat syndrome, in which part of chromosome 5 is missing. Symptoms may range from mild to severe, and a karyotype cannot always predict the severity. And what if a mother or family lacks the resources to care for a special needs individual?

These are difficult questions without scientific answers. Science can, however, help us understand the role of chromosomes in inheritance—the subject of this chapter.

LEARNING OUTLINE

11.1 A Cell's DNA Is Divided Among Chromosomes

Although Gregor Mendel's research was unappreciated at the time, his careful observations of pea plants in the mid-1800s laid the foundation for modern genetics. Soon after the rediscovery of his work in 1900, other scientists demonstrated Mendel's ratios again and again in several species—at about the same time that advances in microscopy were allowing chromosomes to be observed and described for the first time.

A **chromosome** is a discrete package of DNA and associated proteins. Some of the proteins serve as scaffolds around which DNA entwines. Other proteins include the many enzymes that help replicate the DNA and transcribe it to a sequence of RNA (see chapters 7 and 12).

Each species has a characteristic number of chromosomes. A mosquito has six chromosomes; grasshoppers, rice plants, and pine trees all have 24; humans have 46; a dog has 78; a carp has 104. Each of these chromosome numbers is even because every diploid cell contains two homologous sets of chromosomes, one from each parent.

Of the 23 chromosome pairs in a diploid human cell, 22 are **autosomes**—chromosomes that are the same for both sexes. The remaining pair is made up of the two **sex chromosomes,** which determine whether an individual is male or female. Females have two X sex chromosomes, whereas males have an X and a Y chromosome.

When a cell is not dividing, its chromosomes are very loosely packed in the nucleus and are therefore not visible. Just before the cell's nucleus divides, however, the chromosomes replicate and then begin to wind tightly around their scaffold proteins. Only then do they take on their characteristic, familiar, compact shapes (**figure 11.1**). These shapes are easily visible in a **karyotype,** a size-ordered chart of all the chromosomes in any cell of an organism (**figure 11.2**).

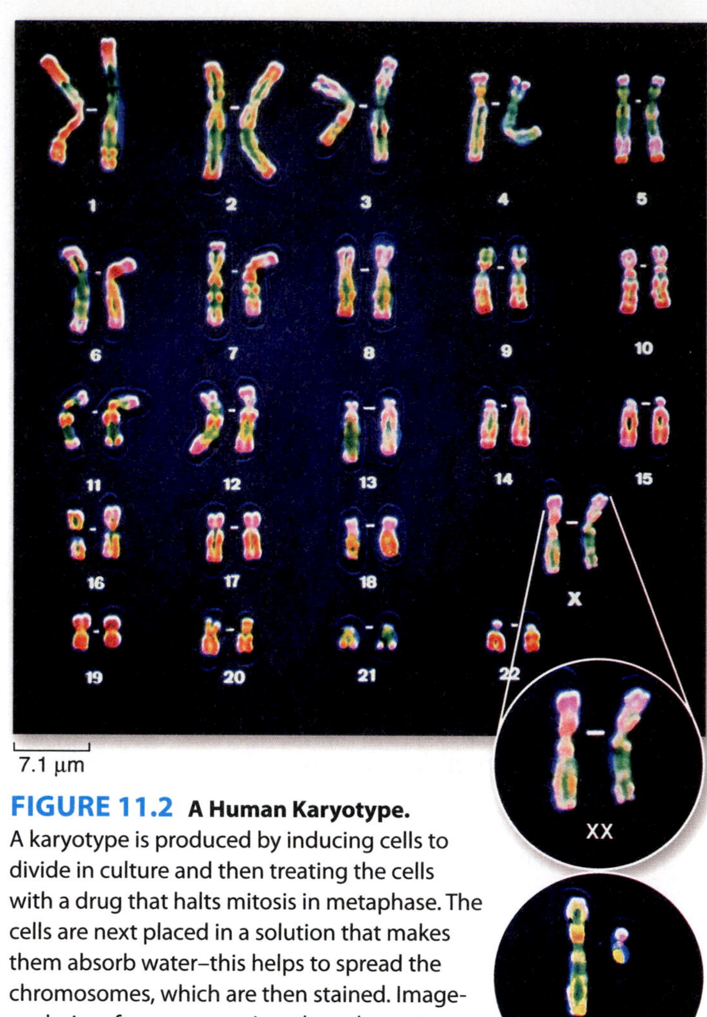

FIGURE 11.2 A Human Karyotype. A karyotype is produced by inducing cells to divide in culture and then treating the cells with a drug that halts mitosis in metaphase. The cells are next placed in a solution that makes them absorb water–this helps to spread the chromosomes, which are then stained. Image-analysis software recognizes the color pattern of each chromosome pair and size-orders them into a chart.

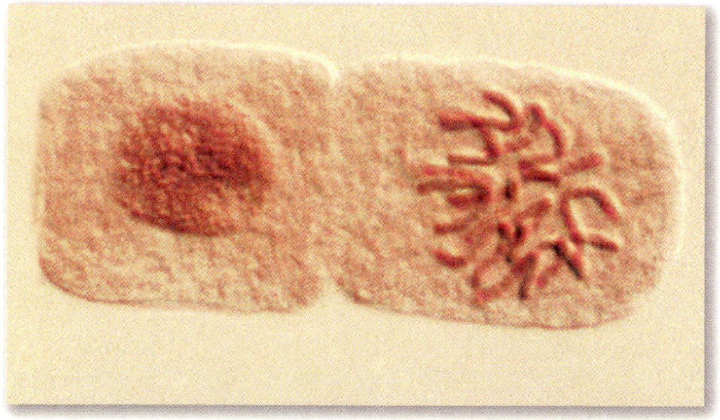

FIGURE 11.1 Two Views of DNA. In the cell on the left, DNA is loosely packed in the nucleus. DNA replication and protein synthesis occur at this stage of the cell cycle. Before a cell divides, however, the DNA winds into the compact, portable chromosomes visible in the cell on the right.

Karyotypes illustrate the major characteristics that uniquely identify each chromosome: size, banding pattern, and centromere position.

- **Size:** Autosomes are numbered based on size. Chromosome 1 is the largest, and chromosome 22 is the smallest. (The sex chromosomes are not numbered).

- **Banding pattern:** Stains applied to chromosomes highlight unique patterns of light- and dark-staining bands, which differ among the chromosome types (**figure 11.3**).
- **Centromere position:** A centromere is a characteristically located constriction in a chromosome (see figure 11.3). The centromere may be close to a chromosome tip, near the center, or somewhere in between.

These characteristics, taken together, uniquely identify each chromosome. Medical professionals use this knowledge to learn which genetic material is missing or extra in a fetus with a chromosomal abnormality.

Summary ***A chromosome is a long, continuous strand of DNA, plus several types of associated proteins. Each species has a characteristic chromosome number. Size, banding pattern, and centromere position distinguish each chromosome.***

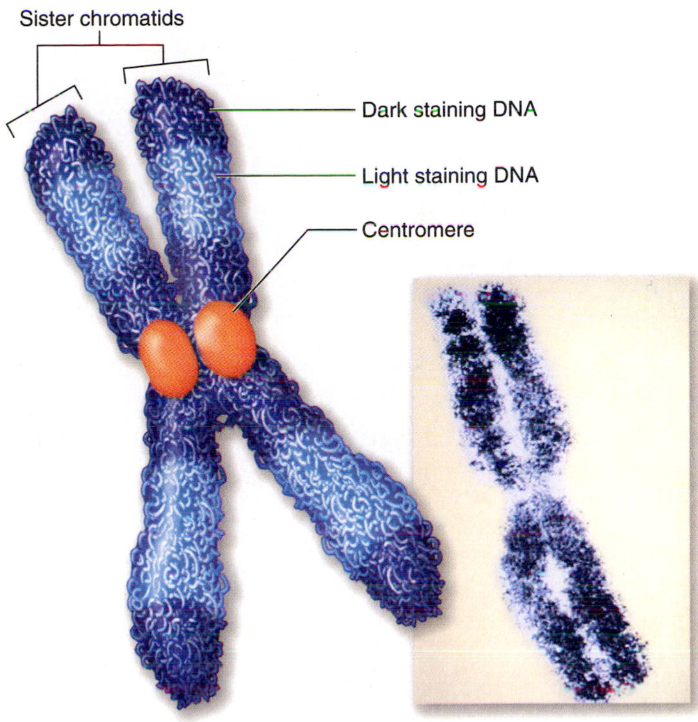

FIGURE 11.3 Anatomy of a Chromosome. Dark-staining chromosomal material alternates with lighter-staining material to create a distinctive series of bands. Before crossing over occurs during meiosis, the sister chromatids are identical.

11.1 MASTERING CONCEPTS

1. What is the difference between autosomes and sex chromosomes?
2. What is a karyotype?
3. What is the origin of the banding pattern on a chromosome?

11.2 Studies of Linked Genes Have Yielded Chromosome Maps

Shortly after chromosomes were described, it became apparent that what Mendel called "elementen" (later renamed "genes") and chromosomes had much in common. Both genes and chromosomes, for example, come in pairs. In addition, alleles of a gene are packaged into separate gametes, as are the members of a homologous pair of chromosomes. Finally, both genes and chromosomes are inherited in random combinations.

As biologists cataloged traits and the chromosomes that transmit them in several species, it soon became clear that the number of traits far exceeded the number of chromosomes. Fruit flies, for example, have four pairs of chromosomes, but dozens of different bristle patterns, body colors, eye colors, wing shapes, and other characteristics. How might a few chromosomes control so many traits? The answer: each chromosome carries many genes.

A. Genes on the Same Chromosome Are Linked

Linked genes are carried on the same chromosome; they are therefore inherited together. Unlike genes on different chromosomes, they do not assort independently during meiosis. The seven traits that Mendel followed in his pea plants were all transmitted on separate chromosomes. Had the same chromosome carried these genes, Mendel would have generated markedly different results in his dihybrid crosses. **dihybrid cross, p. 202**

The different inheritance pattern of linked genes was first noticed in the early 1900s, when William Bateson and R. C. Punnett observed offspring ratios in pea plants that were different from the ratios Mendel's laws predicted. They crossed true-breeding plants with crimson flowers and long pollen

a. *PpLl* self-cross, genes not linked

F_1

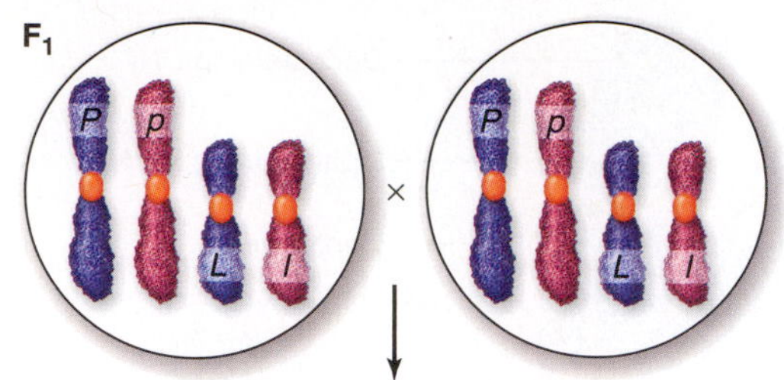

F_2

Possible female gametes (columns); Possible male gametes (rows)

	PL	*Pl*	*pL*	*pl*
PL	*PPLL*	*PPLl*	*PpLL*	*PpLl*
Pl	*PPLl*	*PPll*	*PpLl*	*Ppll*
pL	*PpLL*	*PpLl*	*ppLL*	*ppLl*
pl	*PpLl*	*Ppll*	*ppLl*	*ppll*

9:3:3:1 phenotypic ratio

b. *PpLl* self-cross, genes linked

F_1

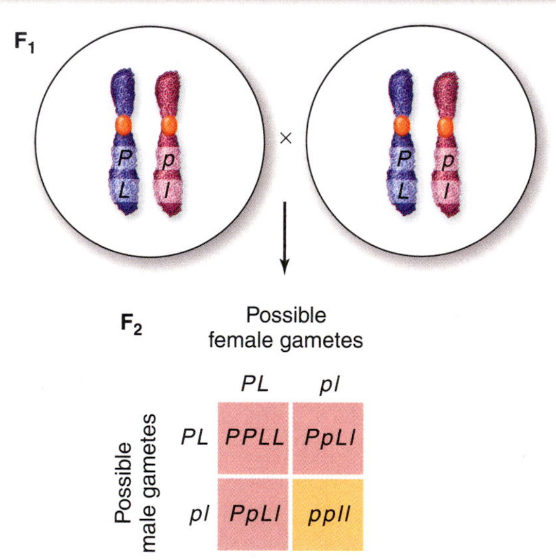

	PL	*pl*
PL	*PPLL*	*PpLl*
pl	*PpLl*	*ppll*

3:1 phenotypic ratio

FIGURE 11.4 Gene Linkage Changes the Results of a Dihybrid Cross. (**A**) When genes are not linked on the same chromosome, they assort independently. The gametes then represent all possible allele combinations, and the expected phenotypic ratio of the offspring from a dihybrid cross is 9:3:3:1. (**B**) If genes are linked on the same chromosome only two allele combinations are expected in the gametes. The expected phenotypic ratio would be 3:1.

grains (genotype *PPLL*) with true-breeding plants with red flowers and round pollen grains (genotype *ppll*). Then they crossed the heterozygous F_1 plants, of genotype *PpLl,* with each other. Surprisingly, the F_2 generation did not show the expected 9:3:3:1 phenotypic ratio for an independently assorting dihybrid cross (**figure 11.4**).

Two types of F_2 peas—those with the same phenotypes as the P generation—were more abundant than predicted. The other two classes of offspring, with a mix of phenotypes (genotypes *ppL_* and *P_ll*) were less common. Bateson and Punnett hypothesized that this pattern reflected two genes on the same chromosome.

If the two genes were on the same chromosome, why did the researchers see offspring with trait combinations not seen in either parent? These offspring classes arise because of another event, crossing over. Recall from chapter 9 that **crossing over** is an exchange of genetic material between homologous chromosomes during meiosis (see figure 9.7). After crossing over, no two chromosomes are identical (**figure 11.5**). **Recombinant chromosomes** have a mix of maternal and paternal alleles, whereas **parental chromosomes** retain the allele combinations from each parent.

While Bateson and Punnett were studying linkage in peas, Thomas Hunt Morgan at Columbia University was breeding the fruit fly *Drosophila melanogaster.* He chose flies in part because of their short life cycle (just 14 days), large numbers of offspring, and the ability to keep thousands of flies in a small amount of space. **Focus on Model Organisms (*Drosophila*), p. 464**

Morgan and his colleagues bred fruit flies and studied the inheritance of pairs of traits. The data began to indicate four **linkage groups,** collections of genes that tended to be inherited together. Within each linkage group, dihybrid crosses did not produce the proportions of offspring that Mendel's law of independent assortment predicts. Because the number of linkage groups was the same as the number of homologous pairs of chromosomes, scientists eventually realized that each linkage group was simply a set of genes transmitted together on the same chromosome.

B. Linkage Maps Derive from Crossover Frequencies

Morgan wondered why some crosses produced a higher proportion of recombinant offspring than others. Might the differences reflect the physical relationships between the genes on the chromosome? Alfred Sturtevant, Morgan's undergraduate assistant, explored this idea. In 1911, he developed a theory and technique that would profoundly affect the fledgling field of genetics in his day and the medical genetics of today. Sturtevant proposed that the farther apart two alleles are on the same chromosome, the more likely crossing over is to separate them—simply because more space separates the genes (**figure 11.6A**).

Sturtevant's idea became the basis for mapping genes on chromosomes. By determining the percentage of recombinant offspring, investigators can infer how far apart the

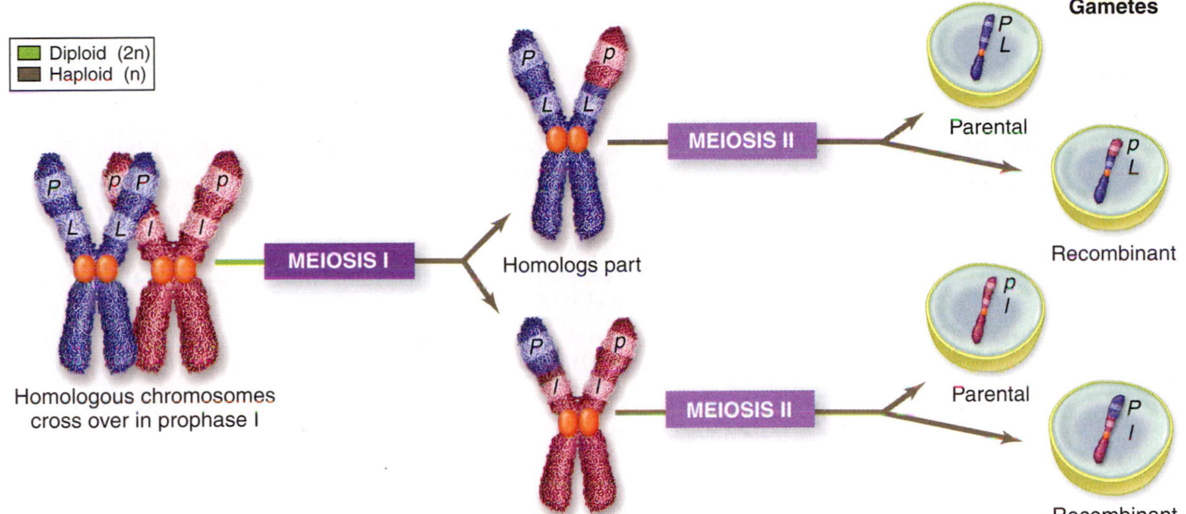

FIGURE 11.5 **Crossing Over.** Alleles linked closely on the same chromosome are usually inherited together. Linkage between two alleles can be interrupted if the chromosome they are located on crosses over with its homolog at a point between the two genes. This packages recombinant arrangements of the alleles into gametes.

genes are on one chromosome. Crossing over frequently separates alleles on opposite ends of the same chromosome, so recombinant offspring occur frequently. In contrast, a crossover would rarely separate alleles lying very close together on the chromosome, and the proportion of recombinant offspring would be small. Geneticists use this correlation between crossover frequency and the distance between genes to construct **linkage maps,** which are diagrams of gene order and spacing on chromosomes.

In 1913, Sturtevant published the first genetic linkage map, depicting the order of five genes on the X chromosome of the fruit fly (figure 11.6B). Researchers then rapidly mapped genes on all four fruit fly chromosomes. Linkage maps for the human chromosomes followed over the next half century.

At one time, phenotypes were the basis for linkage maps. Now, however, genetic marker technology can associate known, detectable DNA sequences with particular phenotypes. The marker does not have to be part of the gene that controls the phenotype; the two must simply be located close enough together that presence of the marker correlates strongly with the phenotype.

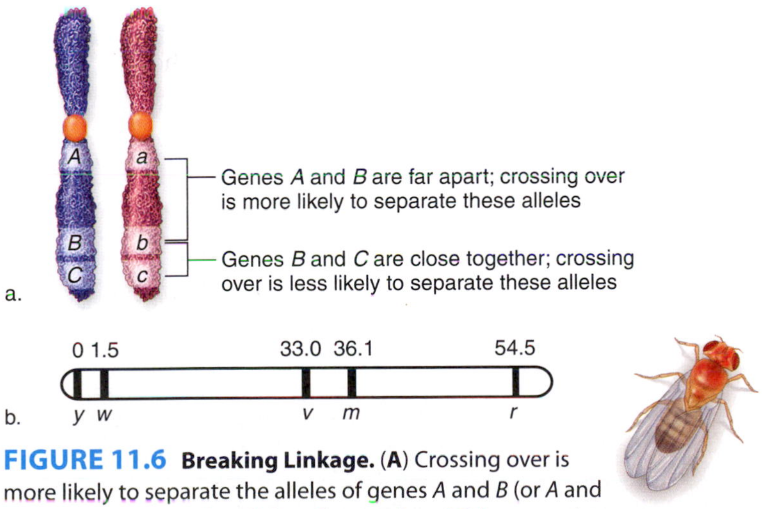

a. Genes *A* and *B* are far apart; crossing over is more likely to separate these alleles

Genes *B* and *C* are close together; crossing over is less likely to separate these alleles

b. 0 1.5 33.0 36.1 54.5 — *y w v m r*

FIGURE 11.6 **Breaking Linkage.** **(A)** Crossing over is more likely to separate the alleles of genes *A* and *B* (or *A* and *C*) than to separate the alleles of genes *B* and *C*, because there is more room for an exchange to occur. **(B)** A linkage map of a fruit fly chromosome, showing the locations of five genes. The numbers represent crossover frequencies relative to the leftmost gene, *y*.

Genetic markers are useful tools for predicting the chance that a person will develop a particular inherited illness. The first such use of genetic markers occurred in the 1980s. An extensive study of a small Venezuelan village with a high incidence of Huntington disease revealed that all family members with the disorder also carry a unique DNA sequence (the "marker") on chromosome 4. Healthy relatives do not have this sequence. The DNA sequence "marks" the presence of the disease-causing allele because the two are closely linked. The actual gene was not identified until the 1990s, but in the meantime, researchers could identify who was likely to carry the Huntington disease-causing allele long before the symptoms developed.

The linkage maps of the mid-twentieth century provided the rough drafts to which DNA sequence information was added later in the century. Today, entire genomes are routinely sequenced using powerful computers that assemble many short overlapping DNA sequences. **DNA sequencing, p. 147**

Summary *Linked genes (those carried on the same chromosome) do not follow Mendel's law of independent assortment. Linkage maps are constructed based on how often crossovers separate alleles for pairs of linked genes.*

11.2 MASTERING CONCEPTS

1. How do patterns of inheritance differ when pairs of genes are linked, versus when they are not linked?
2. What is the difference between recombinant and parental chromosomes, and how do they arise?
3. How do biologists use linkage and crossover frequency to map genes on chromosomes?

11.3 Pedigrees Show Modes of Inheritance

Although Gregor Mendel did not study human genetics, our species nevertheless has "Mendelian traits": those determined by single genes with alleles that are either dominant or recessive. Several thousand phenotypes fit these criteria, and most of the corresponding genes are on autosomes. Because both sexes have two copies of each autosome, genes on those chromosomes affect both sexes equally. Genes on the X and Y chromosomes, however, show different inheritance patterns, as described later in section 11.5.

Autosomal Mendelian traits exhibit two modes of inheritance: autosomal dominant and autosomal recessive (**table 11.1**). To inherit an **autosomal dominant** disorder, a person can receive the disease-causing allele from either parent. An affected individual's mother or father must have the disorder, unless the disease-causing allele arose by mutation. If a generation arises in which no individuals inherit the allele, transmission of the disorder stops in that family. This chapter's Burning Question: Is male baldness really from the female side of genetics? describes a common example of an autosomal dominant condition—male pattern baldness.

Inheriting an **autosomal recessive** disorder requires that a person receive the disease-causing allele from both parents. Each parent must therefore have at least one copy of the allele, either because they are homozygous recessive and have the disease, or because they are unaffected heterozygotes ("carriers"). If both parents are carriers, the disorder appears to skip generations.

Table 11.1 *Some Autosomal Dominant and Autosomal Recessive Disorders in Humans*

Disorder	Genetic Explanation	Characteristics
Autosomal recessive inheritance		
Albinism	Mutant allele of gene on chromosome 11 encodes faulty gene in biochemical pathway required for pigment production	Lack of pigmentation in skin, hair, and eyes
Cystic fibrosis	Mutant allele of gene on chromosome 7 encodes faulty chloride channel protein	Lung infections and congestion, poor fat digestion, infertility, poor weight gain, salty sweat
Phenylketonuria	Mutant allele of gene on chromosome 12 causes enzyme deficiency in biochemical pathway that breaks down the enzyme phenylalanine	Buildup of metabolic byproducts causes mental retardation
Sickle cell disease	Mutant allele of gene on chromosome 11 causes abnormally shaped hemoglobin protein	Joint pain, spleen damage, high risk of infection
Tay-Sachs disease	Mutant allele of gene on chromosome 15 causes deficiency of lysosome enzyme	Buildup of byproducts causes nervous system degeneration
Autosomal dominant inheritance		
Achondroplasia	Mutant allele of gene on chromosome 4 causes deficiency of receptor protein for growth factor	Dwarfism with short limbs, normal size head and trunk
Familial hypercholesterolemia	Mutant allele of gene on chromosome 2 encodes faulty cholesterol-binding protein	High cholesterol, heart disease
Huntington disease	Mutant allele of gene on chromosome 4 encodes protein with extra amino acids that cause it to misfold and form clumps in brain cells	Progressive uncontrollable movements and personality changes, beginning in middle age
Marfan syndrome	Mutant allele of gene on chromosome 15 causes connective tissue disorder	Long limbs, sunken chest, lens dislocation, spindly fingers, weakened aorta
Neurofibromatosis (type 1)	Mutant allele of gene on chromosome 17 encodes faulty cell signaling protein	Brown skin marks (café-au-lait spots), benign tumors beneath skin
Polydactyly	Multiple genes on multiple chromosomes; unknown mechanism	Extra fingers or toes or both

To determine a disorder's mode of inheritance, modern researchers track its incidence over multiple generations, much as Mendel did with his pea plants more than a century ago. **Pedigree** charts depicting family relationships and phenotypes are useful tools in this research (**figure 11.7**). In a pedigree chart, squares indicate males, and circles denote females. Colored shapes indicate individuals with the disorder, and half-filled shapes represent known carriers. Horizontal lines connect parents. Siblings connect to their parents by vertical lines and to each other by an elevated horizontal line.

The different modes of inheritance have characteristic pedigree patterns. Autosomal dominant disorders such as polydactyly (extra fingers or toes) typically appear in every generation (see figure 11.7A). Autosomal recessive conditions such as albinism, however, may seem to disappear in one generation, only to reappear in the next (figure 11.7B).

Summary *Pedigrees are charts that simultaneously depict family relationships and inheritance patterns for a phenotype.*

11.3 MASTERING CONCEPTS

1. What is the difference between autosomal dominant and autosomal recessive modes of inheritance?
2. How are pedigrees helpful in determining a disorder's mode of inheritance?

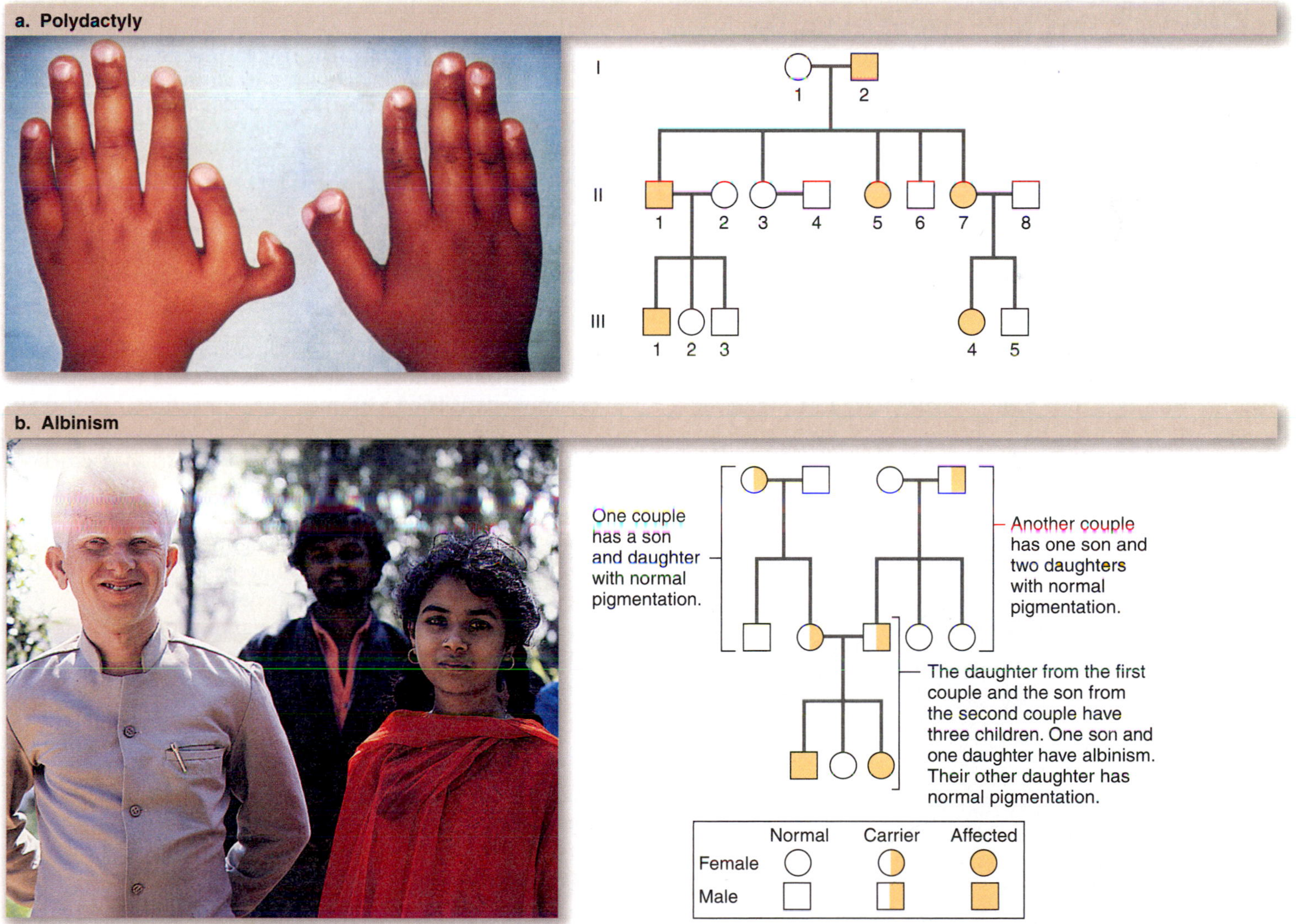

FIGURE 11.7 Pedigrees Reveal Mode of Inheritance. (**A**) A pedigree for a disorder with an autosomal dominant mode of inheritance such as polydactyly. (**B**) A pedigree for a disorder with an autosomal recessive mode of inheritance, albinism.

Burning Questions

Is male baldness really from the female side of the family?

Male pattern baldness is the distinctive hair loss that many men (and some women) experience as they enter their 20s, 30s, and 40s. The baldness spreads outward from the temples and crown of the head in a characteristic pattern (**figure 11.A**).

Two conditions are required for male pattern baldness to develop. First, hormones called androgens must be present in normal amounts. Testosterone and dihydrotestosterone (DHT) are androgens; they bind to and enter hair follicle cells, interacting with the DNA to stop growth of the hair follicle. Second, the individual must have a genetic predisposition for the condition.

Many people believe that a man inherits the tendency for baldness from his mother's side of the family, but the gene(s) controlling this trait reside on autosomes, not on the sex chromosomes. Therefore, either parent can pass the baldness allele(s) to a child.

Why don't women suffer from baldness as often as males? The answer is that pattern baldness is a so-called "sex-influenced" condition, in which males and females can carry the same pair of alleles, yet express them differently. In this case, the amount of testosterone is the deciding factor. The higher the concentration of testosterone, the stronger the influence of the "baldness allele." Men typically have more of this sex hormone than women—hence the name, *male* pattern baldness.

FIGURE 11.A Male Pattern Baldness. Genes and hormones interact in this distinctive form of hair loss.

Have a Burning Question of your own? Submit it to marielle_hoefnagels@mcgraw-hill.com for possible inclusion in future editions of this book!

11.4 X and Y Chromosomes Determine Sex in Humans

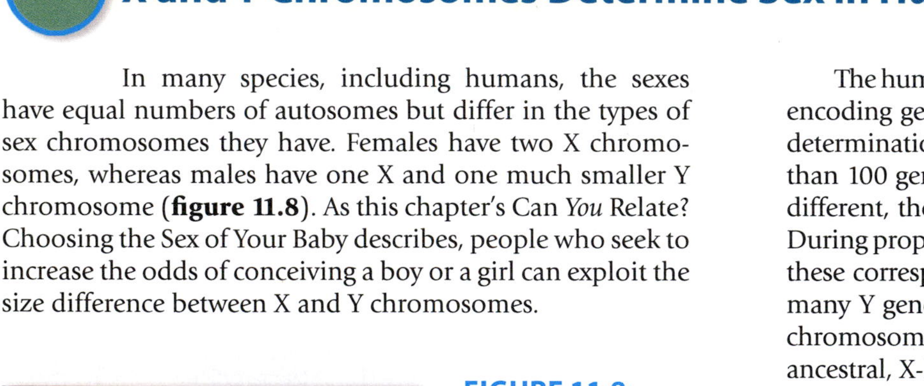

In many species, including humans, the sexes have equal numbers of autosomes but differ in the types of sex chromosomes they have. Females have two X chromosomes, whereas males have one X and one much smaller Y chromosome (**figure 11.8**). As this chapter's Can *You* Relate? Choosing the Sex of Your Baby describes, people who seek to increase the odds of conceiving a boy or a girl can exploit the size difference between X and Y chromosomes.

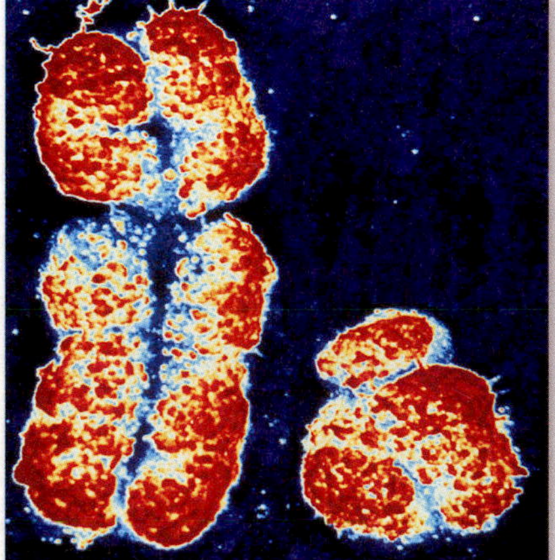

FIGURE 11.8
Sex Chromosomes. The human Y chromosome on the right is much smaller than the X chromosome on the left.

The human X chromosome carries more than 1000 protein-encoding genes, most of which have nothing to do with sex determination. The Y chromosome, in contrast, carries fewer than 100 genes. Although the two sex chromosomes are very different, they do have small regions of matching sequences. During prophase I of meiosis, crossing over can occur between these corresponding areas of the X and Y chromosomes. The many Y genes that are similar or identical to genes on the X chromosome may be evidence that the Y is a remnant of an ancestral, X-like chromosome (see section 11.6).

The Y chromosome plays the largest role in human sex determination. All human embryos start with rudimentary female structures. An embryo having a working copy of a Y-chromosome gene called *SRY* (for *s*ex-determining *r*egion of the *Y*) develops into a male (see figure 37.15). *SRY* encodes a protein that switches on other genes that direct the undeveloped testes to secrete the male sex hormone testosterone. Cascades of other gene activities follow, promoting the development of male sex organs. The SRY protein also turns on a gene that encodes a protein that dismantles embryonic female structures.

Other organisms have different sex determination mechanisms. Grasshoppers, for example, have only one type of sex chromosome, designated X. A female has two sex chromosomes (XX); a male has only one (XO). In bees, the queen determines

FIGURE 11.9 Sex Change. In breeding groups of clown anemonefish (*Amphiprion percula*), the largest fish is a female. The second-largest individual is her mate. If the female dies, the male changes into a female, and the next-largest fish takes over as the breeding male.

the sex of her offspring: the eggs that she allows to be fertilized become diploid females. Unfertilized eggs develop into haploid males. In still other species, the environment determines sex. For alligators and some turtles, the temperature of the ground where the eggs are laid determines sex; eggs incubated at high temperatures hatch females, while cooler eggs produce males.

Unlike most land animals, many marine fishes can change sex during their lives. Some are born female and later become male; others are born male and later become female. This transformation, which can occur within days, allows these fish to continue breeding even if the population's male–female balance suddenly changes (**figure 11.9**).

Sex determination in plants is complex. Most flowering plants combine male and female structures in the same individual; each offspring therefore receives the full set of chromosomes. Other flowering plants, however, have separate male and female individuals. In these plants, the females always produce separate male and female offspring. Sex determination mechanisms include X-to-autosome ratios, X and Y chromosomes, and autosomal genes that determine sex.

Summary *Sex chromosomes carry genes that direct development of reproductive structures of one sex, while suppressing development of structures of the other sex. In humans and other mammals, the X and Y chromosomes determine sex, but other organisms use different mechanisms.*

11.4 MASTERING CONCEPTS

1. What is sex determination?
2. What is the role of the Y chromosome's *SRY* gene in human sex determination?
3. What are some nonchromosomal mechanisms of sex determination?

Can *You* Relate?

Choosing the sex of your baby

Scientists have used the size difference between the X and Y chromosomes to develop technologies that may help people choose the sex of their babies. All of the methods rely on the fact that sperm cells with a Y chromosome weigh slightly less than those carrying an X chromosome.

Technologies for sex selection include:

- **MicroSort:** Chromosomes in a sperm sample are stained with a fluorescent dye. A sperm cell with an X chromosome absorbs more dye and therefore glows more brightly than one with a Y chromosome.
- **Ericsson method:** Sperm swim through a container holding a thick fluid. The Y-carrying sperm are lighter and should be faster swimmers than the heavier X-carrying sperm. Therefore, the Y-carrying sperm should reach the bottom of the container faster than their slower counterparts. Removing the sperm from the top or the bottom of the container should produce a sample enriched in X- or Y-carrying sperm cells, respectively.
- **Spin method:** A sperm sample is placed in test tube, which is placed into a centrifuge. The centrifuge spins the test tube so that the heavier (X-containing) sperm fall to the bottom, whereas the lighter sperm remain near the top.

After Y-carrying sperm are separated from X-carrying sperm, the woman is inseminated with the desired fraction of the sperm. Alternatively, if a woman is using *in vitro* fertilization, the egg is fertilized in the laboratory with the sperm most likely to give a baby of the desired sex. None of the methods, however, works all the time; at best, the sperm samples are enriched in sperm that favor one sex over the other.

11.5 Sex-Linked Genes Have Unique Inheritance Patterns

Huntington disease, cystic fibrosis, and other diseases caused by genes that are on autosomes affect both sexes equally. A few conditions, including red–green colorblindness and hemophilia, however, occur much more frequently in males than females. These phenotypes are **sex-linked;** that is, the alleles controlling them are on X or Y chromosomes.

Thomas Hunt Morgan was the first to unravel the unusual inheritance patterns associated with genes on the X chromosome (**figure 11.10**). The eyes of fruit flies are normally red, but one day Morgan discovered a male with white eyes. To study the inheritance of this odd phenotype, he created true-breeding lines of flies with each eye color. When he mated a parental generation of white-eyed males with red-eyed females, the F_1 flies were all red-eyed. The F_2 flies had a 3:1 ratio of red-eyed to white-eyed flies—but all the flies with white eyes were male. Morgan also did the reverse

a. Cross of true breeding white-eyed male with red-eyed female

(W) Dominant allele; encodes red eyes
(w) Recessive allele; encodes white eyes

P — Female with red eyes × Male with white eyes

	X^W	X^W
X^w	$X^W X^w$	$X^W X^w$
Y	$X^W Y$	$X^W Y$

All offspring have red eyes

F_1 — Female with red eyes × Male with red eyes

	X^W	X^w
X^W	$X^W X^W$	$X^W X^w$
Y	$X^W Y$	$X^w Y$

All females have red eyes
50% of males have red eyes
50% of males have white eyes

b. Cross of true breeding red-eyed male with white-eyed female

(W) Dominant allele; encodes red eyes
(w) Recessive allele; encodes white eyes

P — Female with white eyes × Male with red eyes

	X^w	X^w
X^W	$X^W X^w$	$X^W X^w$
Y	$X^w Y$	$X^w Y$

All females have red eyes
All males have white eyes

F_1 — Female with red eyes × Male with white eyes

	X^W	X^w
X^w	$X^W X^w$	$X^w X^w$
Y	$X^W Y$	$X^w Y$

50% of females have red eyes
50% of females have white eyes
50% of males have red eyes
50% of males have white eyes

FIGURE 11.10 Fly Eyes. Thomas Hunt Morgan first described the unusual inheritance patterns of X-linked traits. In the crosses shown, *W* represents the dominant allele, which encodes red eyes, and *w* is the recessive allele for white eyes. (**A**) A cross between a true-breeding white-eyed male with a red-eyed female. (**B**) A cross between a true-breeding red-eyed male and a white-eyed female. Had this eye color gene been on an autosome, the outcome of the cross shown in (**A**) would have been the same as that in(**B**).

cross, mating a P generation of red-eyed males with white-eyed females. That time, all the males of the F_1 generation had white eyes, and all the females had red eyes.

Female fruit flies pass X chromosomes to all of their offspring, whereas males produce sperm bearing either X or Y. Morgan reasoned that the recessive white-eye allele must be on the X chromosome, and that males with white eyes had no corresponding dominant allele on the Y chromosome. With two X chromosomes, a female will express the white-eye phenotype only if *both* of her eye color alleles are recessive.

A. X-Linked Recessive Disorders Affect More Males than Females

Since only males have a Y chromosome, it is easy to jump to the conclusion that the genes controlling sex-linked disorders must be on the Y chromosome. The human Y chromosome, however, has few genes. Scientists therefore know of very few Y-linked disorders; most involve defects in sperm production, not traits such as blood clotting or color vision. (Nevertheless, researchers use DNA sequences on the Y chromosome to trace father-to-son inheritance. For example, comparing Y chromosome sequences enabled researchers to identify Thomas Jefferson's descendants through his slave Sally Hemings.)

Because the X chromosome has many more genes than the Y, most human sex-linked traits are **X-linked;** that is, they are controlled by genes on the X chromosome. Recessive alleles cause most X-linked disorders, although a few are associated with dominant alleles (**table 11.2**).

Recessive X-linked traits have unusual inheritance patterns. Whereas a female inherits two X chromosomes (one from each parent), a male inherits his single X chromosome from his mother (**figure 11.11**). A male expresses every allele on his X chromosome (whether dominant or recessive) because he lacks a second allele that could mask the expression of recessive alleles. A female, in contrast, exhibits an X-linked recessive disorder only if she inherits recessive alleles from both parents.

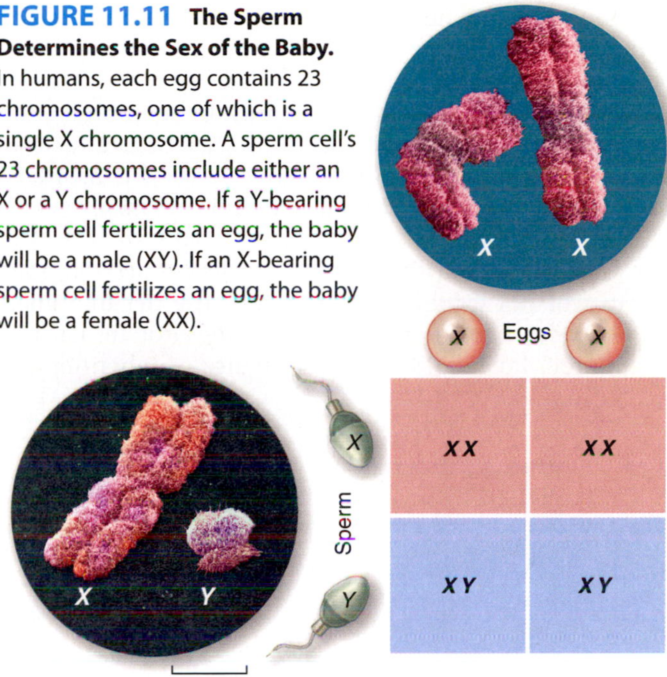

FIGURE 11.11 The Sperm Determines the Sex of the Baby. In humans, each egg contains 23 chromosomes, one of which is a single X chromosome. A sperm cell's 23 chromosomes include either an X or a Y chromosome. If a Y-bearing sperm cell fertilizes an egg, the baby will be a male (XY). If an X-bearing sperm cell fertilizes an egg, the baby will be a female (XX).

Table 11.2 Some X-Linked Disorders in Humans

Disorder	Genetic Explanation	Characteristics
X-linked recessive inheritance		
Duchenne muscular dystrophy	Mutant allele for gene encoding dystrophin	Rapid muscle degeneration early in life
Fragile X syndrome	Unstable region of X chromosome has unusually high number of CCG repeats	Most common form of inherited mental retardation
Hemophilia A	Mutant allele for gene encoding blood clotting protein (factor VIII)	Uncontrolled bleeding, easy bruising
Red–green color blindness	Mutant alleles for genes encoding receptors for red or green (or both) wavelengths of light	Reduced ability to distinguish between red and green
X-linked dominant inheritance		
Extra hairiness (congenital generalized hypertrichosis; some forms)	Mechanism unknown	Many more hair follicles than normal
Hypophosphatemic rickets (some forms)	Mutant allele for gene involved in phosphorus absorption	Low blood phosphorus level causes defective bones
Retinitis pigmentosa (some forms)	Mutant allele for cell-signaling protein; mechanism unknown	Defects in retina cause partial blindness

Punnett squares and pedigrees illustrate transmission of sex-linked traits. **Figure 11.12** shows the inheritance of hemophilia A, a disorder with an X-linked recessive mode of inheritance, in an extended royal family. In hemophilia, the absence or deficiency of a protein clotting factor greatly slows blood clotting. As is typical for X-linked recessive disorders, the only family members with hemophilia are males.

Females rarely have hemophilia, but many are heterozygous "carriers" for the disease-causing allele. A heterozygous woman will not exhibit symptoms because her dominant allele encodes a functional blood-clotting protein. Each of her children, however, has a 50% chance of inheriting the recessive allele from her (see figure 11.12). If her son receives the recessive allele, he will have hemophilia. Any daughter who inherits the recessive allele from the mother, however, will be a carrier—unless she also inherits the disease-causing allele from the father.

B. X Inactivation Prevents "Double Dosing" of Gene Products

Relative to males, female mammals have a "double dose" of every gene on the X chromosome. In 1961, geneticist Mary

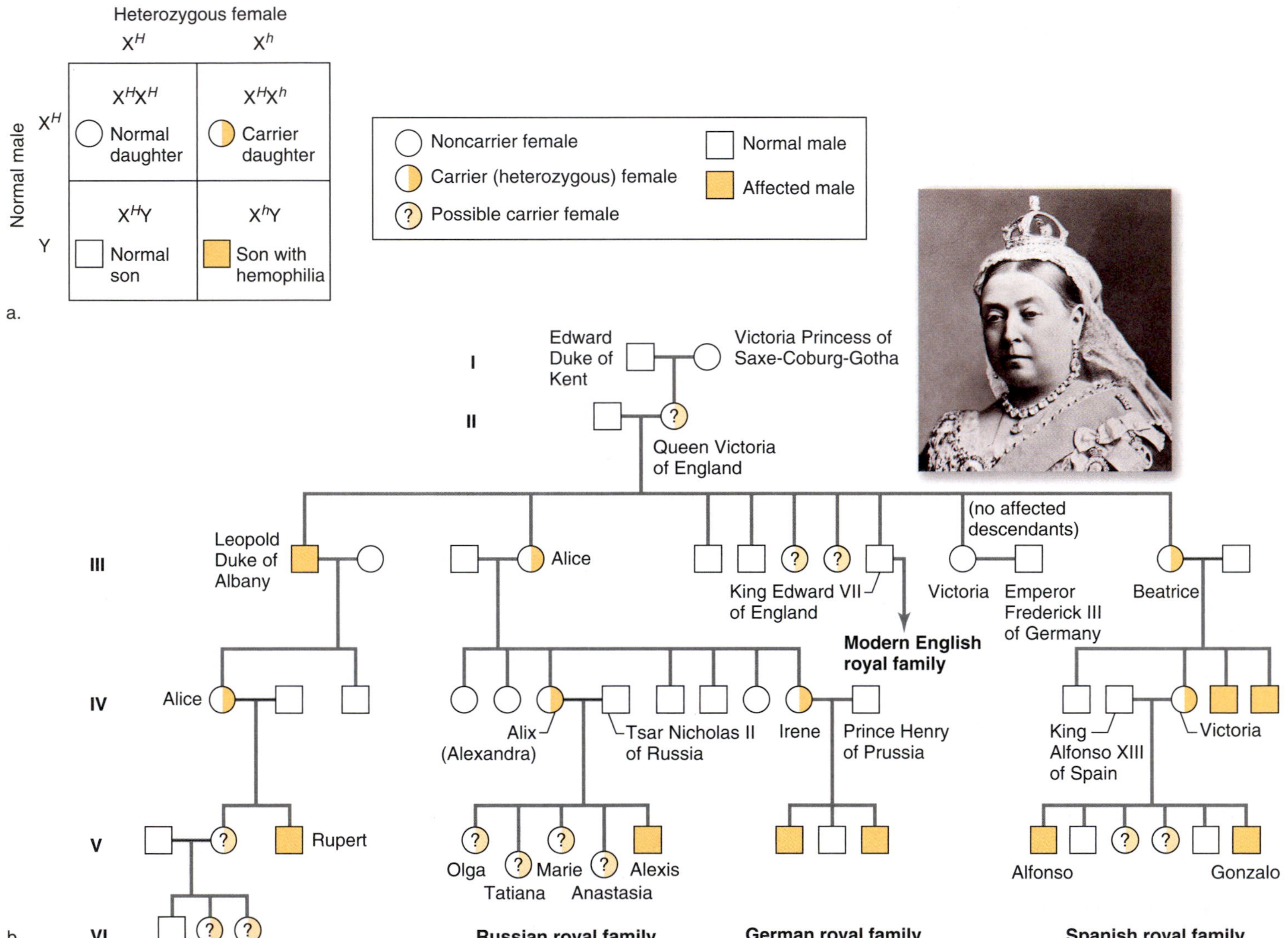

FIGURE 11.12 **Inheritance of Hemophilia A.** (**A**) This Punnett square depicts a cross between a heterozygous ("carrier") woman and a normal male, the most common way to transmit any X-linked recessive allele. (**B**) This figure depicts only part of the enormous pedigree that follows the inheritance of hemophilia A in the royal families of England, Germany, Spain, and Russia. The mutant allele apparently arose in Queen Victoria, who was a carrier. The modern royal family in England does not carry hemophilia.

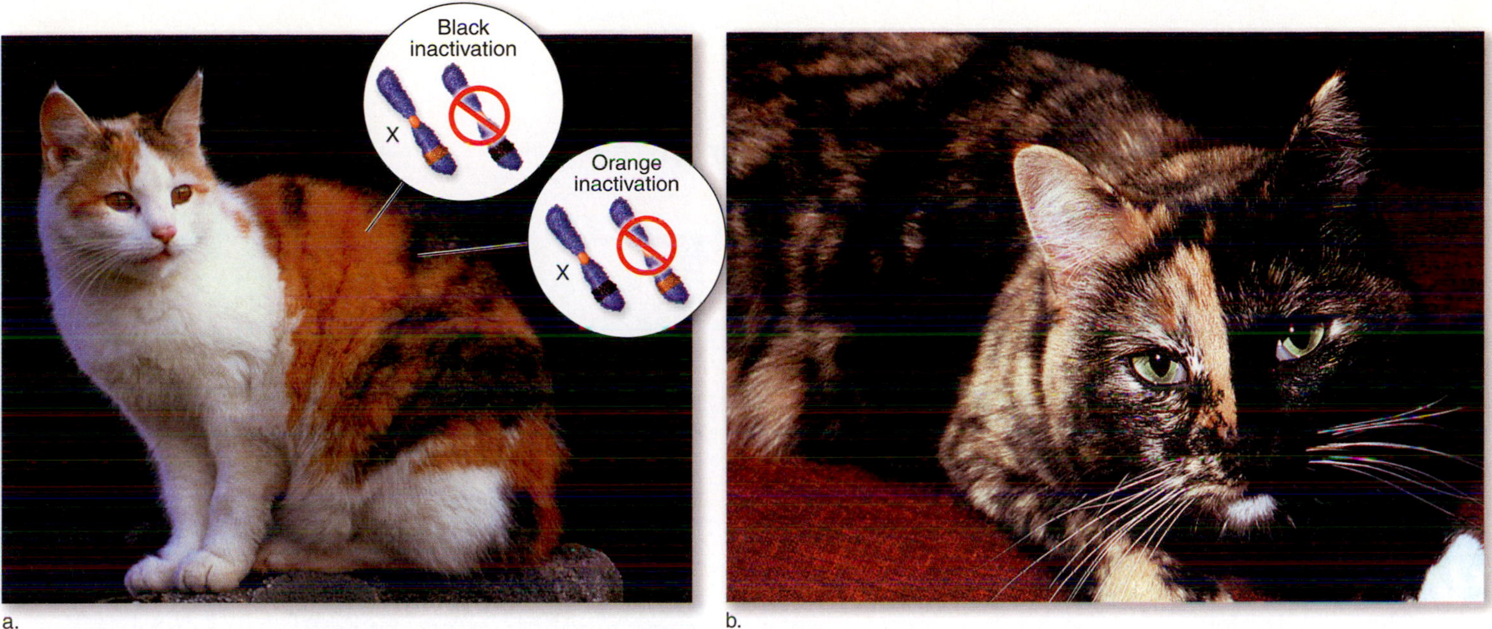

FIGURE 11.13 Calico and Tortoiseshell Cats Reveal Patterns of X Inactivation. The X chromosome carries a coat color gene with alleles for black or orange coloration. Calico and tortoiseshell cats are heterozygous for this gene. Each orange patch is composed of cells in which the X chromosome carrying the coat color allele for black was inactivated. Conversely, each black patch is made of cells in which the X chromosome carrying the orange allele was turned off. A different gene accounts for the white background. (**A**) X inactivation happened early in development, resulting in large patches. (**B**) X inactivation occurred later in development, producing smaller patches and a "tortoiseshell" coloration pattern. Can you see why calico cat clones do *not* share identical fur color patterns?

Frances Lyon proposed a mechanism for how cells balance this inequality. In **X inactivation,** a cell shuts off all but one X chromosome in each cell, a process that happens early in the embryonic development of a mammal.

X inactivation is directly observable in cells because a turned-off X chromosome absorbs a stain much more readily than an active X chromosome does; the inactivated X forms a Barr body. A normal male cell has no Barr bodies because the single X chromosome remains active. Likewise, patients with Turner syndrome have only one X chromosome (and no Y chromosome); their cells therefore also lack Barr bodies. Cells of a female with extra X chromosomes, on the other hand, have extra Barr bodies, as do males with Klinefelter syndrome (XXY). **sex chromosome disorders, p. 186**

Which X chromosome becomes inactivated—the one inherited from the father or the one from the mother—is a random event. As a result, a female expresses the paternal X chromosome alleles in some cells and the maternal alleles in others. Moreover, when a cell with an inactivated X chromosome divides mitotically, all of the daughter cells have the same X chromosome inactivated. Because the inactivation occurs early in development, females have patches of tissue that differ in their expression of X-linked alleles. **Figure 11.13** shows how X inactivation of either of two alleles of a coat color gene causes the distinctive appearance of calico and tortoiseshell cats, which are always female (except for rare XXY males). The earlier X inactivation occurs, the larger the patches.

X chromosome inactivation explains why females suffer less than males from some sex-linked disorders with a dominant mode of inheritance. Although some of the female's tissues express the mutant dominant allele, others do not. As a result, the female experiences less severe symptoms than an affected male, who expresses the dominant allele in every cell.

Summary *X-linked recessive disorders are more common in males than in females, because males have only one X chromosome to a female's two. In mammals, all but one X chromosome is inactivated in each cell early in development.*

11.5 MASTERING CONCEPTS

1. Why do males and females express recessive X-linked alleles differently?
2. How does X inactivation in mammals equalize the contributions of X-linked genes between the sexes?

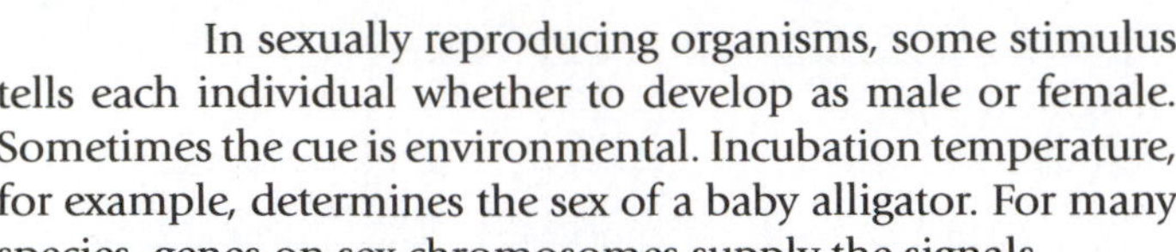

11.6 INVESTIGATING LIFE

Papaya Sex—Is It a Boy or a Girl? Or Both?

In sexually reproducing organisms, some stimulus tells each individual whether to develop as male or female. Sometimes the cue is environmental. Incubation temperature, for example, determines the sex of a baby alligator. For many species, genes on sex chromosomes supply the signals.

Most plants also reproduce sexually. In the largest and most recently evolved group of plants, the angiosperms, flowers produce sex cells. For most species, a single flower has both sperm- and egg-producing parts. About 6% of flowering plant species, however, produce male and female flowers on separate individuals. These plants have separate sexes, just as humans do. Sex chromosomes tell these plants what type of flower to develop.

Flowering plants arose hundreds of millions of years later than did the four-legged ancestors from which mammals evolved. Sex chromosomes therefore probably evolved independently in these groups (an example of convergent evolution). All sexually reproducing organisms have asexual ancestors. Where did sex chromosomes come from? Did they somehow arise from autosomes? **convergent evolution, p. 266**

Part of the answer to this question comes from a surprising source: papaya trees (*Carica papaya;* **figure 11.14**). These flowering plants have three sexes. Two of the three sexes are analogous to our own: a plant produces flowers that are exclusively male or exclusively female. Hermaphrodites are the third sex, and they produce flowers with both male and female parts.

Unlike in mammals, however, examining a male papaya plant's 18 chromosomes with a microscope does not reveal an obvious pair of different-sized sex chromosomes. Instead, sex is thought to be determined by just one gene with three alleles, two dominant (M and M^h) and one recessive (m). Papaya plants are diploid, so they inherit one allele from each parent. The following table shows the possible genotypes and phenotypes:

Genotype	Phenotype
mm (homozygous recessive)	Female
Mm (heterozygous)	Male
M^hm (heterozygous)	Hermaphrodite
MM, MM^h, or M^hM^h	(Lethal)

FIGURE 11.14 **The Papaya Plant, *Carica papaya*.**

If a sex chromosome arose from an autosome, then it should carry genes that control at least some traits unrelated to sex. In papayas, only one gene is thought to control the sex of a plant, and all other genes on the same chromosome have nonsex functions. In addition, biologists would expect sex chromosomes to have regions that do not cross over during meiosis. After all, X and Y could not remain different from each other if they were to recombine along their entire length—and they must remain different to specify sex.

To learn more about sex determination in papaya, biologist Zhiyong Liu, Ray Ming, and their colleagues at the Hawaii Agriculture Research Center mapped the papaya

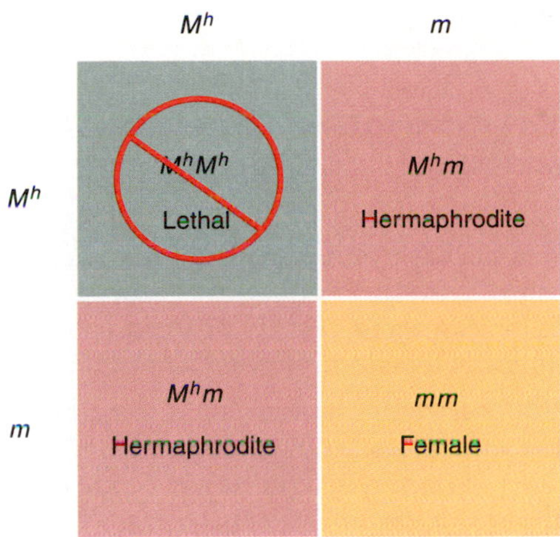

FIGURE 11.15 Breeding Papayas. Because of a lethal allele combination that omits one genotypic class, crossing two hermaphrodites together yields about 2/3 hermaphrodites and 1/3 female papaya trees. **Question:** What would be the outcome of a cross between a hermaphrodite and a female?

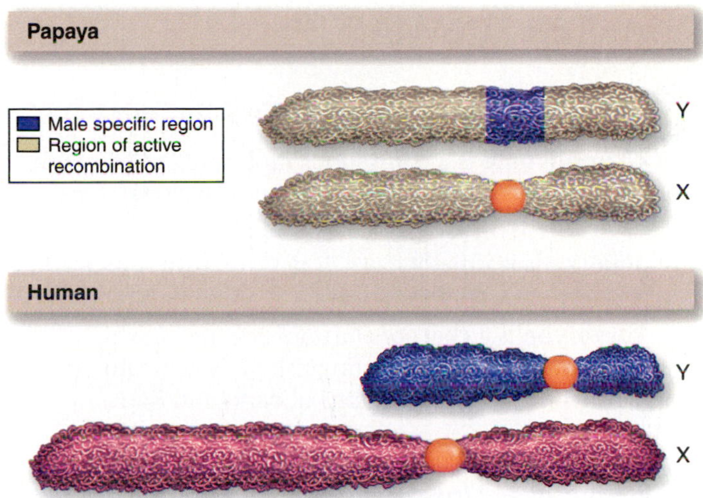

FIGURE 11.16 The Birth of a Sex Chromosome. Recombination occurs along most of the papaya sex chromosomes, except for a small area around the gene that actually determines the plant's sex. In contrast, the area of recombination between the human X and Y chromosomes is tiny, and multiple genes on the human Y contribute to sex determination. (The centromere location on the papaya Y chromosome is unknown).

chromosome that carries the sex-determining gene. After studying DNA from dozens of plants, they found an extremely high level of genetic variation in the chromosomal region around the critical *M* gene. This was consistent with the "no crossing over" prediction, because areas of a chromosome that do not recombine accumulate mutations rapidly.

Next, the researchers looked for direct evidence of crossing over between the "X-like" chromosome (the one with allele *m*) and the "Y like" chromosome (with allele *M* or M^h). They began by breeding hermaphroditic papaya plants with each other (**figure 11.15**). The researchers had already identified DNA markers that occurred only on the Y-like chromosome and that were physically near the sex-determining gene. They reasoned that if crossing over occurred, the markers should turn up on the X-like chromosome in some of the offspring of each cross. Of the 2190 plants they examined, however, the researchers did not find a single plant bearing a recombinant chromosome. They therefore concluded that crossing over does not occur in the chromosomal region near the sex-determining gene (**figure 11.16**).

This papaya study supplies direct evidence that sex chromosomes can arise from autosomes. The researchers called the *M*- or M^h-carrying chromosome a "primitive Y chromosome" because it shares some features with the mammalian Y. For example, part of the Y chromosome in both papaya and mammals is degenerating (i.e., mutating rapidly) because of a lack of recombination with X. But the plant and mammalian chromosomes are also different (see figure 11.16). Only 10% of the papaya Y-like chromosome has high genetic variability and suppressed recombination, compared with 95% of Y chromosome in humans. Also, the human X is much larger than the Y, whereas the papaya's sex chromosomes are equal in size.

Why would anyone study papaya sex chromosomes at all? Liu and his colleagues did not set out to test a hypothesis about the evolution of X and Y. Rather, they were hoping to help fruit producers. To a papaya farmer, hermaphrodites are the most desirable because they are most productive and yield the tastiest fruit. The original aim of the study was to find a way to genetically engineer true-breeding hermaphrodites. In the process, however, the researchers stumbled on an important clue to the evolutionary history of sex—a good example of science leading in new and unexpected directions.

Liu, Zhiyong, Paul H. Moore, Hao Ma, and 10 coauthors, including Ray Ming. 22 January, 2004. A primitive Y chromosome in papaya marks incipient sex chromosome evolution. *Nature*, vol. 427, pages 348–352.

CHAPTER SUMMARY

11.1 A Cell's DNA Is Divided Among Chromosomes

- A **chromosome** is a continuous molecule of DNA with associated proteins. Each species has a characteristic number of chromosomes.
- Human cells normally have 22 pairs of **autosomes** and one pair of **sex chromosomes**.
- A **karyotype** is a chart of chromosome pairs, which are ordered by size and distinguished by centromere position and banding patterns of dark- and light-staining regions.

11.2 Studies of Linked Genes Have Yielded Chromosome Maps

A. Genes on the Same Chromosome Are Linked

- Dihybrid crosses for pairs of **linked genes** produce more offspring with **parental** genotypes than with **recombinant** genotypes.
- **Linkage groups** are collections of genes that are often inherited together because they are on the same chromosome.

B. Linkage Maps Derive from Crossover Frequencies

- The farther apart two genes are on a chromosome, the more likely **crossing over** is to separate their alleles. Breeding studies reveal the crossover frequencies used to create **linkage maps**—diagrams that show the order of genes on a chromosome.

11.3 Pedigrees Show Modes of Inheritance

- An **autosomal dominant** disorder affects both sexes and is inherited from one affected parent. An **autosomal recessive** disorder can also appear in either sex, is passed from parents who are either carriers or are affected, and can skip generations.
- **Pedigrees** trace phenotypes in families and reveal mode of inheritance.

11.4 X and Y Chromosomes Determine Sex in Humans

- In humans, the male has X and Y sex chromosomes, and the female has two X chromosomes. The *SRY* gene on the Y chromosome controls other genes that stimulate development of male structures and suppress development of female structures.
- Sex determination mechanisms are diverse. Some species use sex chromosomes, whereas others use autosomes or environmental cues to determine sex. Some organisms can change sex, and some have both male and female reproductive parts.

11.5 Sex-Linked Genes Have Unique Inheritance Patterns

- Genes controlling **sex-linked** traits are located on the X or Y chromosomes.

A. X-Linked Recessive Disorders Affect More Males than Females

- An **X-linked** gene passes from mother to son because the male inherits his X chromosome from his mother and his Y chromosome from his father. Scientists know of many more X-linked disorders than Y-linked disorders.

B. X Inactivation Prevents "Double Dosing" of Gene Products

- **X inactivation** shuts off all but one X chromosome in the cells of female mammals, equalizing the number of active X-linked genes in each sex. A female is a "mosaic" for X-linked genes because the maternal or paternal X chromosome is inactivated at random in each cell.

11.6 Investigating Life: Papaya Sex—Is It a Boy or a Girl? Or Both?

- Genetic studies show that the sex chromosome of papaya plants originated as an autosome.

MULTIPLE CHOICE QUESTIONS

1. Which of the following is a difference between an autosome and a sex chromosome?
   a. An autosome has more DNA.
   b. A sex chromosome is only present in the germ cells.
   c. Only autosomes can be diploid.
   d. There are more autosomes than sex chromosomes in a cell.

2. How many chromosome pairs would you find in a karyotype from a human?
   a. 46
   b. 23
   c. 22
   d. 12

3. Why do homologous chromosomes have a similar appearance in a karyotype?
   a. Because they share the same banding pattern
   b. Because they share the same amount of DNA
   c. Because their centromeres are in the same location
   d. All of the above
4. When two genes are linked
   a. they are found on the same chromosome.
   b. they interact in a pleiotropic manner to affect phenotype.
   c. they are located on homologous chromosomes.
   d. the expression of one influences the expression of the other.
5. Recombination is most likely to occur between
   a. closely linked genes.
   b. genes on nonhomologous chromosomes.
   c. linked genes that are far apart.
   d. parental chromosomes.
6. What is a linkage map?
   a. A diagram of the DNA sequence of a gene
   b. A diagram of the banding patterns on a chromosome
   c. A diagram of the order and spacing of genes on a chromosome
   d. A diagram of the pattern of inheritance of a trait in a family
7. What would the genotype be for a symptomless carrier of an autosomal recessive disease?
   a. Homozygous recessive
   b. Heterozygous
   c. Homozygous dominant
   d. Both a and b
8. What is the role of the Y chromosome in sex determination in humans?
   a. It controls the expression of other genes.
   b. It carries the gene for the male sex hormone, testosterone.
   c. It encodes proteins that inactivate the X chromosome.
   d. It carries genes that control the formation of male body structures.
9. What are the chances of a daughter inheriting a recessive X-linked disease if her mother is a symptomless carrier and her father has the disease?
   a. 100%
   b. 50%
   c. 25%
   d. 0
10. How does X inactivation contribute to genetic diversity?
    a. It controls the number and kind of genes expressed on an X chromosome.
    b. It determines which X chromosome is expressed in an individual.
    c. It allows for expression of either the maternal or paternal X in different cells.
    d. It enhances expression of Y-linked genes in males.

TESTING YOUR KNOWLEDGE

1. What are two criteria that a scientist can use to tell two similarly sized chromosomes apart?
2. Why is *Drosophila* a good model organism to use in studies of inheritance?
3. How does gene linkage interfere with Mendel's law of independent assortment? Why doesn't the inheritance pattern of linked genes disprove Mendel's law?
4. How does crossing over "unlink" genes?
5. What is the difference between the sex chromosomes of human males and females?
6. How are X-linked genes inherited differently in male and female humans?
7. What does X inactivation accomplish?
8. A track runner has a karyotype constructed before an Olympic competition. Her cells have two Barr bodies. How is this unusual?
9. In the following pedigree, is the disorder's mode of inheritance autosomal dominant, autosomal recessive, or X-linked recessive?

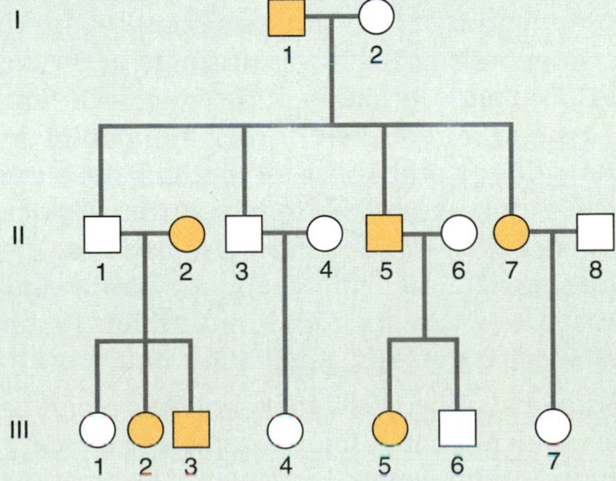

THINKING AS A SCIENTIST

1. The radish has nine groups of traits. Within each group, dihybrid crosses do not yield a 9:3:3:1 phenotypic ratio. Instead, such crosses yield an overabundance of phenotypes like those of the parents. What does this information reveal about the chromosomes of this plant?
2. Genes *J*, *K*, and *L* are on the same chromosome. The crossover frequency between *J* and *K* is 19%, the crossover frequency between *K* and *L* is 2%, and the crossover frequency between *J* and *L* is 21%. Use this information to create a linkage map for the chromosome.
3. Pedigree charts can sometimes be difficult to construct and interpret. People may refuse to supply information, and adoption or serial marriages can produce blended families. Artificial insemination may involve anonymous sperm donors. How does each of these factors complicate the use of pedigrees?
4. Do you agree with the statement that all alleles on the Y chromosome are dominant? Why or why not?
5. Will a fetus that has X and Y chromosomes but that lacks receptors for the *SRY* gene product develop as a male or as a female?
6. A family has an X-linked dominant form of congenital generalized hypertrichosis (excessive hairiness). Although the allele is dominant, males are more severely affected than females. Moreover, the women in the family often have asymmetrical, hairy patches on their bodies. How does X chromosome inactivation explain this observation?
7. Why are male calico cats rare?
8. Why don't genes in mitochondria and chloroplasts follow Mendelian laws of inheritance?
9. If two different but linked genes are located very far apart on a chromosome, how may the inheritance pattern create the appearance of independent assortment?

GENETICS PROBLEMS

See answers in Appendix D.

1. A normal-sighted woman with a color-blind father marries a color-blind man. What is the probability that their son will be color-blind? Their daughter?
2. New parents Gloria and Michael were startled when their son Will's diapers turned blue when he urinated. Fortunately, this occurred for the first time in the hospital, where tests determined that the newborn had inherited "blue diaper syndrome." Because of abnormal transport of the amino acid tryptophan across the small intestinal lining, bacteria act on urine precursors to produce a compound that turns blue on contact with the air. Gloria's sister Edith was pregnant at the time of Will's diagnosis and became concerned that her child might inherit the disorder. The family doctor assured Gloria and her sister that this wasn't possible because each parent had to be a carrier. However, Edith and Archie's son Aaron also had blue diaper syndrome. Draw a pedigree for this family and describe how this disorder is most likely inherited. How was the doctor's explanation incorrect?
3. In Scandinavia, people who do not have blond hair and blue eyes stand out. A study considered hair and eye color in 100 families in Copenhagen, Denmark. In each of 50 couples (the F_1 generation), one partner has the common blond hair and blue eyes, and the other has brown hair and brown eyes. The brown-haired and -eyed parents each have a parent who looks or looked like them, and the other parent has or had blond hair and blue eyes (the P generation). The 50 F_1 parents produce 260 children (the F_2 generation). The 260 children have the following phenotypes and frequencies:

Phenotype	Number of Children
Brown hair, blue eyes	4
Blond hair, brown eyes	6
Brown hair, brown eyes	120
Blond hair, blue eyes	130

The allele for brown hair (*H*) is dominant to the allele for blond hair (*h*), and the allele for brown eyes (*E*) is dominant to the allele for blue eyes (*e*).

a. What is the genotype or genotypes of the blond-haired, blue-eyed parents of the F_1 generation?
b. Are the genes for these two traits linked or unlinked?
c. Explain or diagram how the unusual appearing children in the F_2 generation arose.

4. In 1875, Charles Darwin described a family with anhidrotic ectodermal dysplasia:

Ten men, in the course of 4 generations, were furnished, in both jaws taken together, with only 4 small and weak incisor teeth and with 8 posterior molars. The men thus affected have very little hair on the body, and become bald early in life. . . . Although the daughters are never affected, they transmit the tendency to their sons; and no case has occurred of a son transmitting it to his sons.

In a group of affected families, some females lack hair and sweat glands in patches, with only some areas of skin affected.

a. Why is expression of this allele in females "patchy"?
b. What is the probability that a son of a woman who has patchy symptoms inherits anhidrotic ectodermal dysplasia?

5. Ron has mild hemophilia A that he can control by taking a clotting factor. He marries Lydia, whom he met at the hospital where he and Lydia's brother, Marvin, receive their treatment. Lydia and Marvin's mother and father, Emma and Clyde, do not have hemophilia. What is the probability that Ron and Lydia's son inherits the disorder?
6. How does the mode of inheritance differ for each of the following syndromes in humans?
   - In BPES syndrome (which stands for *b*lepharophrimosis, *p*tosis, *e*picanthus inversus *s*yndrome), a person has low-set ears; arched brows; and droopy, overgrown eyelids. In 26 two-generation pedigrees analyzed in one study, 86 females and 92 males were affected among the offspring, and in each pedigree, one parent was affected.
   - A person with lysinuric protein intolerance cannot digest 3 of the 20 types of dietary amino acids due to a defect in the small intestinal wall. Symptoms include poor weight gain, vomiting, diarrhea, enlarged liver and spleen, and coma. It is deadly in childhood. The condition is extremely rare, affects males and females, and skips generations.
7. Wayne and Marge have identical twin sons, Wally and Todd, whose huge toes make it difficult to find shoes. Wayne's mother, Cecile, recalled that she had similar problems fitting Wayne and his sister Colleen with shoes when they were children. Cecile's deceased husband, George, had very bizarre feet. Colleen and her husband, Jack, have a 4-year-old daughter, Leah, who has the family's large toes too. Colleen, a geneticist, thinks that Wally, George, Todd, Wayne, Leah, and she inherited acrocephalosyndactyly. She suggested that the relatives have their toes x-rayed. As she suspected, they all have double bones in each toe.
   a. If the "double bone" condition has an autosomal dominant mode of inheritance and Leah marries John, a ballet dancer with exquisite, slim toes, what is the probability that a child of theirs would inherit normal toes?
   b. Construct a pedigree of this family if the "big toe" condition's mode of inheritance is autosomal dominant.
   c. Construct a pedigree of this family if the "big toe" condition's mode of inheritance is autosomal recessive. Include individuals who must be carriers.
8. Marjorie and Joe Winthrop are healthy, but only two of their six children are well. Michael, the eldest, was born without arms and has an abnormal heart. Jacob, born next, is healthy. Marianne, the third eldest, has a severe allergy to cow's milk. She also has impaired blood clotting and an excess of a type of white blood cell. The next child, Adam, is healthy. The youngest two children, Mira and Pete, are twins. Pete is allergic to cow's milk and has a minor heart abnormality. Mira was born without arms but is otherwise healthy. When Pete's cardiologist asks for a family history and Joe mentions the children's symptoms, the doctor realizes that the family has thrombocytopenia and absent radius syndrome. Marjorie and Joe's parents do not have any of these symptoms.
   a. What is the most likely mode of inheritance for this condition, and how do you know this?
   b. If Pete marries a woman who is homozygous for the wild-type allele of the gene that causes this condition, what is the probability that a child of theirs would be a carrier?
9. Consider a woman whose brother has hemophilia A but whose parents are healthy. What is the chance that she has inherited the hemophilia allele? What is the chance that the woman will conceive a son with hemophilia?

Visit www.mhhe.com/hoefnagels for practice quizzes, animations, videos, and activities designed to help you master the material in this chapter.

CHAPTER 12

Gene Function, Gene Regulation, and Biotechnology

The Human Genome Project has only just begun to explain how genes define human life.

What's the Point?

The Human Genome Sequence Is Just the Beginning

When the Human Genome Project finished a draft sequence of human DNA in mid-2000, news stories suggested that parents would soon be able to screen their unborn children for eye color, intelligence, height, and susceptibility to hundreds of diseases. Talk of "designer babies" soon followed. The final sequence was completed in 2003, but we remain far from the ability to manipulate DNA to create the complex traits we might desire.

One misconception about the Human Genome Project is that the DNA sequence is a "blueprint for human life." A gene's nucleotide sequence encodes a protein, but just knowing a gene's sequence does not provide instant insight into everything needed to make a human. By itself, a DNA sequence does not explain how the cell turns each gene on or off, how the gene's encoded protein folds into its final shape, the function of the protein, or what happens if the gene mutates. Nor does it explain the function of the huge swaths of DNA that do not code for protein.

Another question that many people ask is whose genome was actually sequenced. Many individuals contributed blood or sperm samples to the publicly funded and private-sector laboratories that carried out the work. Assuming your DNA was not sequenced, the Human Genome Project still applies to you, because our similarities vastly outnumber the differences between us. The Human Genome Project's goal was to find the basic set of genes that control human development and human life. All humans share these genes.

Since we all carry the same basic set of genes, small variations within DNA sequences must be what make each person unique. Researchers are actively investigating these differences, which will likely answer such important questions as why some people get cancer and others do not, or why a single medication helps some people but harms others.

The Human Genome Project is just one of many sequencing ventures. The genomes of other species have already created an explosion of knowledge about how cells work, how microorganisms cause disease, and how plants and animals develop. Similarities across these broad groups also provide glimpses into life's evolutionary history, which is the subject of unit 3.

This unit began with the raw material of the genome, DNA, and examined how it copies itself, how cells divide, and how alleles pass from one generation to the next. We end this unit with a closer look at how cells use the information in this remarkable molecule.

LEARNING OUTLINE

12.1 DNA Stores Genetic Information
- A. DNA Is a Helix of Nucleotides: A Review
- B. RNA Is an Intermediary Between DNA and a Polypeptide Chain

12.2 Transcription Uses a DNA Template to Create RNA
- A. The Steps of Transcription Are Initiation, Elongation, and Termination
- B. mRNA Is Altered in the Nucleus of Eukaryotic Cells

12.3 Translation Builds the Protein
- A. The Genetic Code Links mRNA to Protein
- B. Translation Requires mRNA, tRNA, and Ribosomes
- C. Translation Occurs in Three Steps
- D. Proteins Must Fold Correctly after Translation

12.4 Protein Synthesis Is Highly Regulated
- A. Operons Are Groups of Bacterial Genes that Share One Promoter
- B. Eukaryotic Organisms Use Transcription Factors to Turn Genes On or Off
- C. Eukaryotic Cells Also Use Additional Regulatory Mechanisms

12.5 Mutations Create New Alleles
- A. Mutations Range from Silent to Devastating
- B. What Causes Mutations?
- C. Some Mutations Pass to Future Generations
- D. Mutations Are Important for Many Reasons

12.6 The Human Genome Sequence Reveals Unexpected Complexity

12.7 Transgenic Organisms Contain DNA from More than One Species
- A. Transgenic Organisms Have Many Uses
- B. Creating Transgenic Organisms Requires Cutting and Pasting DNA

12.8 Biotechnology Has Many Practical Applications
- A. Gene Therapy Replaces Faulty Genes
- B. Antisense RNA and Gene Knockouts Block Gene Expression
- C. DNA Microarrays Can Help Monitor Gene Expression
- D. Proteomics Studies an Organism's Entire Protein Output

12.9 Investigating Life: Clues to the Origin of Language

12.1 DNA Stores Genetic Information

The twentieth century saw huge advances in biology. The 1900s began with the rediscovery of Gregor Mendel's laws of inheritance and with studies that revealed the relationship between genes and chromosomes. In the 1950s, Watson and Crick determined the chemical structure of DNA. Researchers quickly learned more about DNA's function, and by the 1970s they were moving genes from one species to another. Today, scientists regularly sequence entire genomes and are learning how different cells in the same organism can use DNA in so many different ways.

Among biological molecules, DNA has unmatched glamour; it gets credit for everything from eye color to intelligence. Yet DNA does nothing but store information. The real workers inside cells are proteins, folded chains of amino acids that carry oxygen, build toenails, speed chemical reactions, and do almost every other job in a cell (see table 2.5). Both DNA and proteins are critical components of cellular life, and the link between the two is direct: information in DNA specifies the amino acid sequence for every protein in every cell.

A. DNA Is a Helix of Nucleotides: A Review

To understand the relationship between DNA and protein, a brief review of DNA structure is useful. Recall from figure 7.7 that the DNA double helix resembles a twisted ladder composed of nucleotides. The ladder's rungs are base pairs joined by hydrogen bonds. Because adenine (A) always pairs with thymine (T) and cytosine (C) always pairs with guanine (G), the two strands of a DNA molecule are complementary to each other. They are, however, oriented in opposite directions. Figure 7.9 shows that the two ends of a DNA strand are designated 3′ and 5′. At the same end of the double helix, one chain ends with a 3′ carbon, and the other ends with a 5′ carbon.

DNA encodes information in basically the same way for both prokaryotic and eukaryotic cells. In bacteria and other prokaryotes, however, DNA forms one circular chromosome. Eukaryotic cells, in contrast, have multiple linear chromosomes. In both cell types, however, each chromosome is divided into multiple genes (see chapter 11).

B. RNA Is an Intermediary Between DNA and a Polypeptide Chain

In the 1940s, biologists working with the fungus *Neurospora crassa* deduced that a single **gene** somehow controls the production of each protein. In the next decade, Watson and Crick described this relationship between nucleic acids and proteins as a flow of information they called the "central dogma" (**figure 12.1**). First, in **transcription,** a cell copies a gene's DNA sequence to a complementary RNA molecule. Then, in the process of **translation,** the information in RNA is used to manufacture a protein by joining a specific sequence of amino acids into a polypeptide chain. **Focus on Model Organisms (*Neurospora*), p. 436**

RNA is a multifunctional nucleic acid that differs from DNA in several ways (**figure 12.2A**). First, its nucleotides contain the sugar ribose instead of deoxyribose. Second, RNA has the nitrogenous base uracil, which behaves similarly to thymine—that is, in complementary base pairs, uracil binds with adenine (figure 12.2B). Third, unlike DNA, RNA can be single-stranded. Finally, RNA can catalyze chemical reactions, a role not known for DNA.

RNA is central to the flow of genetic information. Three types of RNA interact to synthesize proteins (**table 12.1**):

- **Messenger RNA (mRNA)** carries the information that specifies a protein. Each group of three mRNA bases in a row forms a **codon,** which is a genetic "code word" that corresponds to one amino acid.
- **Ribosomal RNA (rRNA)** combines with proteins to form a **ribosome,** the physical location of protein synthesis. Some rRNAs help to correctly align the ribo-

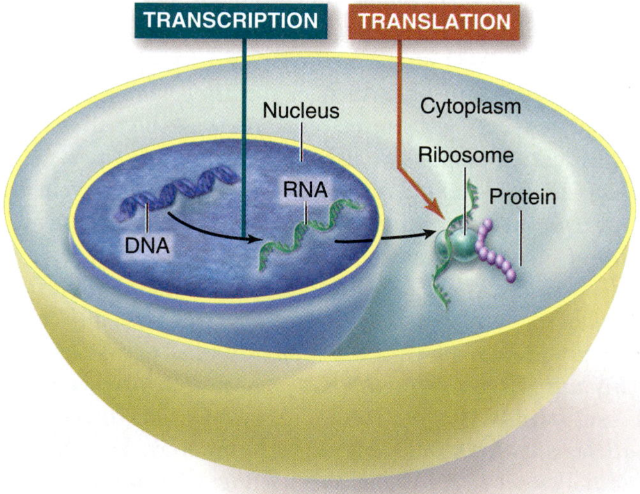

FIGURE 12.1 DNA to RNA to Protein. The central dogma of biology states that information stored in DNA is copied to RNA (transcription), which is used to assemble proteins (translation).

a.	DNA	RNA
Function	Stores RNA- and protein-encoding information; transfers information to daughter cells	Carries protein-encoding information; helps to make proteins; catalyzes some reactions
Form	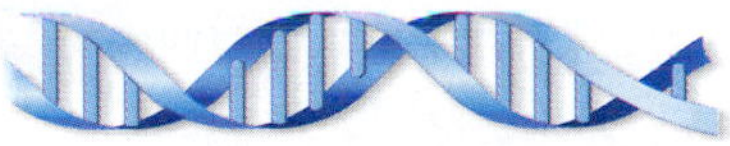 Double-stranded	 Generally single-stranded
Sugar	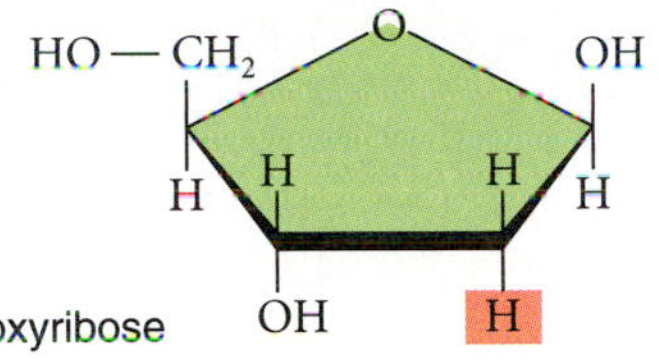 Deoxyribose (HO—CH_2, OH, H, H, H, H, OH, **H**)	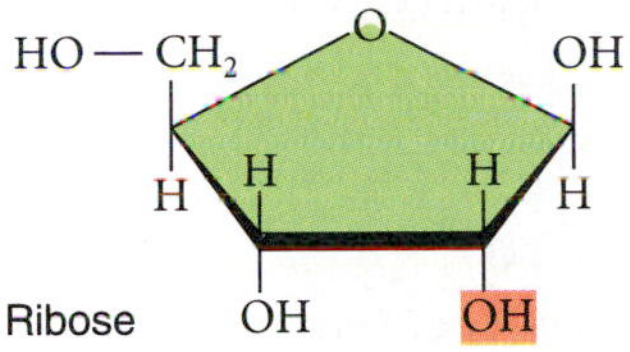 Ribose (HO—CH_2, OH, H, H, H, H, OH, **OH**)
Nucleotide bases	Adenine (A) Cytosine (C) Guanine (G) Thymine (T)	Adenine (A) Cytosine (C) Guanine (G) Uracil (U)

b. Complementary base pairs

DNA	pairs with	RNA
Adenine (A)	—	Uracil (U)
Cytosine (C)	—	Guanine (G)
Guanine (G)	—	Cytosine (C)
Thymine (T)	—	Adenine (A)

RNA	pairs with	RNA
Adenine (A)	—	Uracil (U)
Cytosine (C)	—	Guanine (G)
Guanine (G)	—	Cytosine (C)
Uracil (U)	—	Adenine (A)

FIGURE 12.2 DNA and RNA. (**A**) Summary of the functional and structural differences between DNA and RNA. (**B**) Complementary base pairs show that uracil in RNA behaves chemically like thymine in DNA.

some and mRNA, and others catalyze formation of the bonds between amino acids in the developing protein.

- **Transfer RNA (tRNA)** molecules are "connectors" that bind mRNA codons at one end and specific amino acids at the other. Their role is to carry each amino acid to the ribosome at the correct spot along the mRNA molecule.

The function of each type of RNA is further explained later in this chapter, beginning in the next section with the first stage in protein production: transcription.

Summary *A cell expresses a protein-encoding gene by transcribing the gene's DNA sequence to an RNA sequence, then translating the RNA into an amino acid sequence. Three types of RNA contribute to protein synthesis.*

12.1 MASTERING CONCEPTS

1. What is the relationship between a gene and a protein?
2. How do transcription and translation use genetic information?
3. What are the three types of RNA?

Table 12.1 *Major Types of RNA*

Molecule	Typical Number of Nucleotides	Shape	Function
mRNA	500–3000	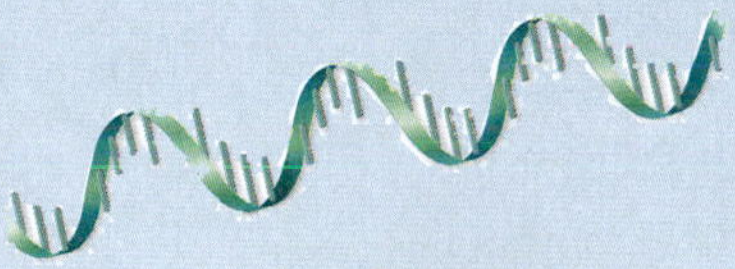	Codons encode amino acid sequence
rRNA	100–3000	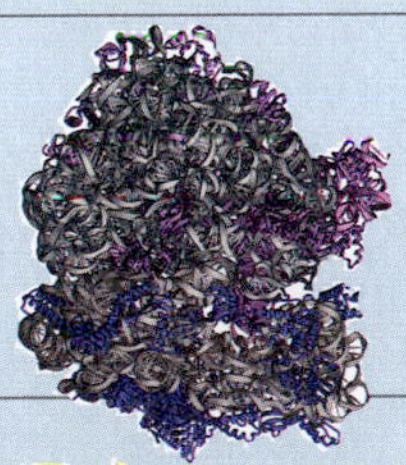	Associates with proteins to form ribosomes, which structurally support and catalyze protein synthesis
tRNA	75–80	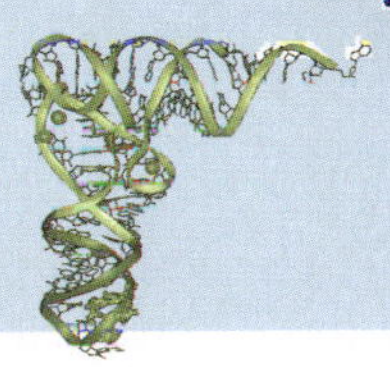	Binds mRNA codon on one end and an amino acid on the other, linking a gene's message to the amino acid sequence it encodes

12.2 Transcription Uses a DNA Template to Create RNA

Transcription produces an RNA copy of one gene. Like DNA replication, transcription requires that the cell's genetic material be unwound, not compacted into visible chromosomes. Not surprisingly, both transcription and DNA replication occur during interphase of the cell cycle, the most active time of a eukaryotic cell's life. **interphase, p. 160**

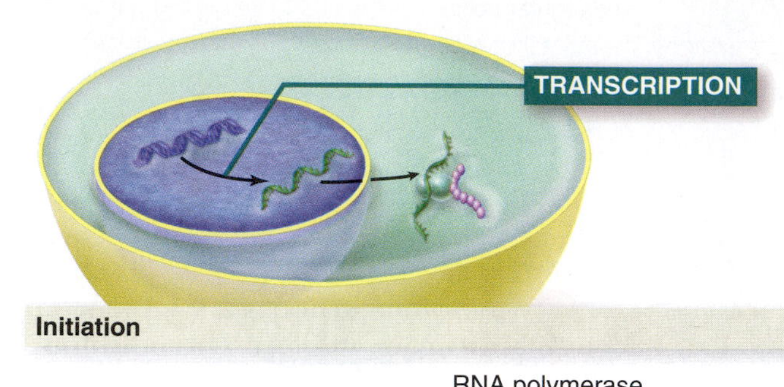

A. The Steps of Transcription Are Initiation, Elongation, and Termination

Complementary base pairing underlies transcription, just as it does DNA replication (see chapter 7). In fact, transcription resembles DNA replication, with two main differences: (1) the product of transcription is RNA, not DNA; (2) transcription copies just one gene from one DNA strand, rather than copying both strands of an entire chromosome.

In transcription, RNA nucleotide bases bond with exposed complementary bases on the DNA template strand (**figure 12.3**). The process occurs in three stages:

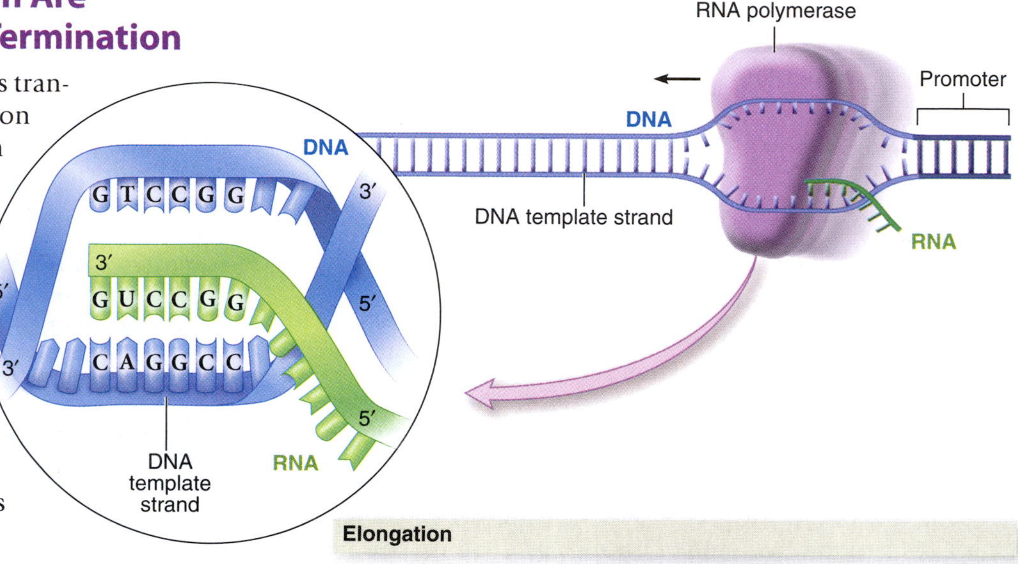

1. In the first stage of transcription, called *initiation*, enzymes unwind the DNA double helix, exposing the **template strand** that encodes the RNA molecule. **RNA polymerase** (the enzyme that builds an RNA chain) binds to the **promoter,** a DNA sequence that signals the gene's start.
2. During the *elongation* stage of transcription, RNA polymerase moves along the DNA strand in a 3′-to-5′ direction, adding nucleotides to the growing molecule in a 5′-to-3′ direction.
3. Transcription ends at the *termination* stage, when the RNA polymerase enzyme reaches a **terminator** sequence that signals the end of the gene. RNA, RNA polymerase, and the DNA template separate from each other, and the DNA molecule resumes its usual double helix shape.

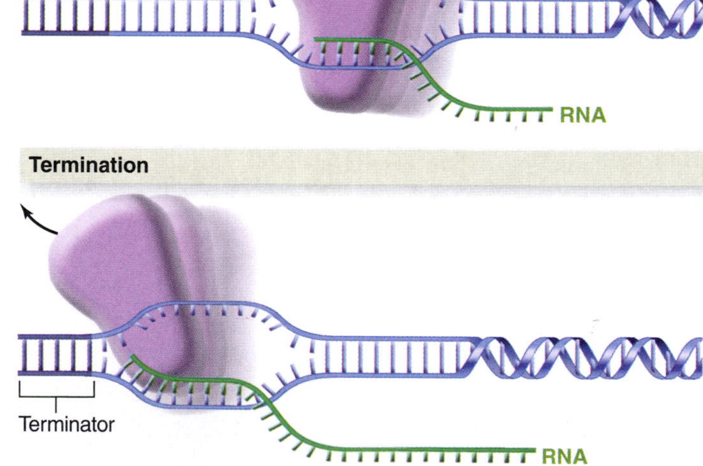

FIGURE 12.3 Transcription of RNA from DNA. Transcription occurs in three stages: initiation, elongation, and termination. Initiation is the control point that determines which genes are transcribed and when. RNA nucleotides are added during elongation, and a terminator sequence in the gene signals the end of transcription.

As the RNA molecule is synthesized, it curls into a three-dimensional shape dictated by complementary base pairing within the molecule. The final shape determines whether the RNA functions as mRNA, tRNA, or rRNA. In fact, the definition of gene includes any DNA sequence that is transcribed to any type of RNA. Because most genes are transcribed to mRNA, and because mRNA encodes protein, a common "shorthand" definition for gene is a DNA sequence that encodes a protein.

B. mRNA Is Altered in the Nucleus of Eukaryotic Cells

In bacteria and archaea, ribosomes begin translating mRNA to a protein as soon as transcription is complete. This can happen in eukaryotes too, but usually mRNA is altered before it leaves the nucleus (**figure 12.4**).

5′ Cap and Poly A Tail

After transcription, a short sequence of modified nucleotides, called a cap, is added to the 5′ end of the mRNA molecule. At the 3′ end, 100 to 200 adenines are added, forming a "poly A tail." Together, the cap and poly A tail enhance translation by helping ribosomes attach to the 5′ end of the mRNA molecule. The length of the poly A tail may also determine how long an mRNA lasts before being degraded.

Intron Removal

In eukaryotic cells, only part of an mRNA molecule is translated into an amino acid sequence. Small catalytic RNAs and proteins first remove sequences called **introns** (for *int*e*r*vening sequences) and splice together the remaining portions of the mRNA, called **exons**. The spliced together exons form the mature mRNA that leaves the nucleus to be translated. (A tip for remembering this is that *ex*ons *ex*it the nucleus.)

The amount of genetic material devoted to introns can be immense. The average exon is 100 to 300 nucleotides long, whereas the average intron is about 1000 nucleotides long. Some mature mRNA molecules consist of 70 or more spliced-together exons; the cell therefore simply discards much of the RNA created in transcription.

Summary *Transcription creates an RNA copy of a gene. In eukaryotic cells, mRNA is processed in the nucleus before it carries the information to a ribosome in the cytoplasm.*

12.2 MASTERING CONCEPTS

1. What are the steps of transcription?
2. What is the role of RNA polymerase in transcription?
3. What are the roles of the promoter and terminator sequences in transcription?
4. How is mRNA modified before it leaves the nucleus of a eukaryotic cell?

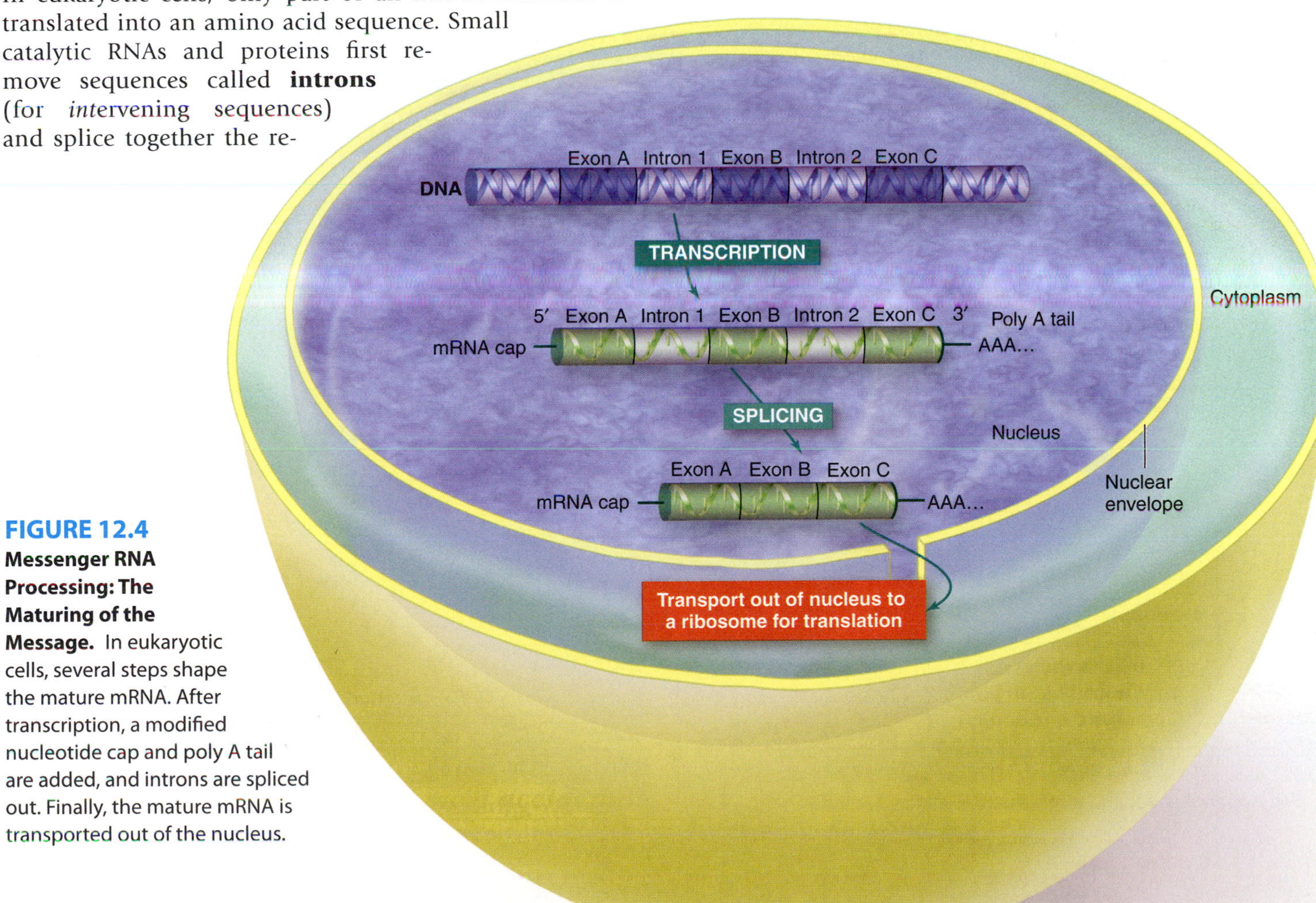

FIGURE 12.4 **Messenger RNA Processing: The Maturing of the Message.** In eukaryotic cells, several steps shape the mature mRNA. After transcription, a modified nucleotide cap and poly A tail are added, and introns are spliced out. Finally, the mature mRNA is transported out of the nucleus.

12.3 Translation Builds the Protein

Transcription copies the information encoded in a DNA base sequence into the complementary language of mRNA. Once transcription is complete and mRNA is processed, the cell is ready to translate the mRNA "message" into a sequence of amino acids.

A. The Genetic Code Links mRNA to Protein

In the 1960s, researchers knew the structure and function of DNA, but they did not yet understand exactly how the **genetic code** worked. One question was the number of RNA bases that specify each amino acid in a protein. Researchers reasoned that RNA contains only four different nucleotides, so a genetic code with a one-to-one correspondence of mRNA bases to amino acids could specify only four different amino acids—far fewer than the 20 amino acids that make up biological proteins. A code consisting of two bases per codon could specify only 16 different amino acids. A code with three bases per codon, however, yields 64 different combinations, more than enough to specify the 20 amino acids in life. Experiments later confirmed the triplet nature of the genetic code.

A second, and more difficult, problem was to determine which codons correspond to which amino acids. In

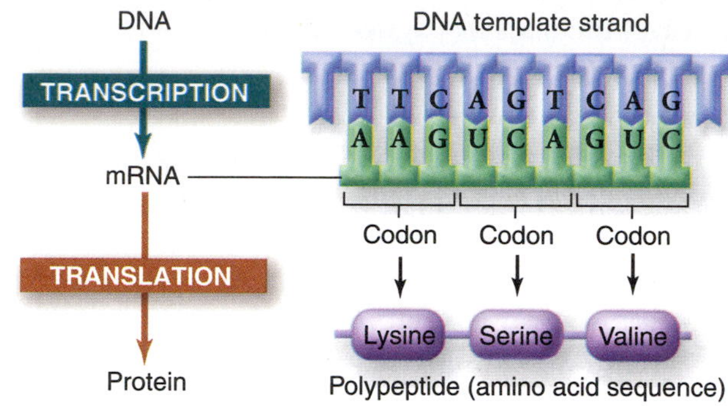

FIGURE 12.5 How Cells Use the Genetic Code. Messenger RNA is transcribed from DNA. Then, in translation, transfer RNA matches mRNA codons with amino acids as specified in the genetic code (see table 12.2).

Table 12.2 *The Genetic Code*

First Letter in Codon	Second Letter in Codon: U		C		A		G		Third Letter in Codon
U	UUU	Phenylalanine (Phe; F)	UCU	Serine (Ser; S)	UAU	Tyrosine (Tyr; Y)	UGU	Cysteine (Cys; C)	U
	UUC		UCC		UAC		UGC		C
	UUA	Leucine (Leu; L)	UCA		UAA	"Stop"	UGA	"Stop"	A
	UUG		UCG		UAG	"Stop"	UGG	Tryptophan (Trp; W)	G
C	CUU	Leucine (Leu; L)	CCU	Proline (Pro; P)	CAU	Histidine (His; H)	CGU	Arginine (Arg; R)	U
	CUC		CCC		CAC		CGC		C
	CUA		CCA		CAA	Glutamine (Gln; Q)	CGA		A
	CUG		CCG		CAG		CGG		G
A	AUU	Isoleucine (Ile; I)	ACU	Threonine (Thr; T)	AAU	Asparagine (Asn; N)	AGU	Serine (Ser; S)	U
	AUC		ACC		AAC		AGC		C
	AUA		ACA		AAA	Lysine (Lys; K)	AGA	Arginine (Arg; R)	A
	AUG	Methionine (Met; M) and **"start"**	ACG		AAG		AGG		G
G	GUU	Valine (Val; V)	GCU	Alanine (Ala; A)	GAU	Aspartic acid (Asp; D)	GGU	Glycine (Gly; G)	U
	GUC		GCC		GAC		GGC		C
	GUA		GCA		GAA	Glutamic acid (Glu; E)	GGA		A
	GUG		GCG		GAG		GGG		G

See Appendix C for chemical structures of amino acids.

the 1960s, researchers answered this question by synthesizing mRNA molecules in the laboratory. They added these synthetic mRNAs to test tubes containing all the ingredients needed for translation, extracted from *E. coli* cells. Analyzing the resulting polypeptides allowed scientists to finish deciphering the genetic code in less than a decade—a monumental task. Chemical analysis eventually showed that the genetic code also contains directions for starting and stopping translation. The codon AUG signals "start," and the codons UGA, UAA, and UAG each signify "stop." **Table 12.2** shows the complete genetic code.

Nearly all species use the same mRNA codons to specify the same amino acids. Mitochondria and a handful of species use alternative codes that differ only slightly from the code in table 12.2. The most logical explanation for this observation is that all life on Earth evolved from a common ancestor.

B. Translation Requires mRNA, tRNA, and Ribosomes

Translation—the actual construction of the protein—requires the following participants:

- **mRNA:** This product of transcription carries the genetic information that encodes a protein, with each three-base codon specifying one amino acid (**figure 12.5**).
- **tRNA molecules:** tRNA is a "bilingual" molecule that binds to both mRNA codons and amino acids (**figure 12.6**). The **anticodon** is a three-base loop that is complementary to one mRNA codon. The other end of the tRNA molecule forms a covalent bond to the amino acid corresponding to that codon. For example, a tRNA with the anticodon sequence AAG always picks up the amino acid phenylalanine.
- **Ribosome:** The ribosome, built of rRNA and proteins, anchors mRNA during translation. Each ribosome has two subunits that join at the initiation of protein synthesis **(figure 12.7)**.

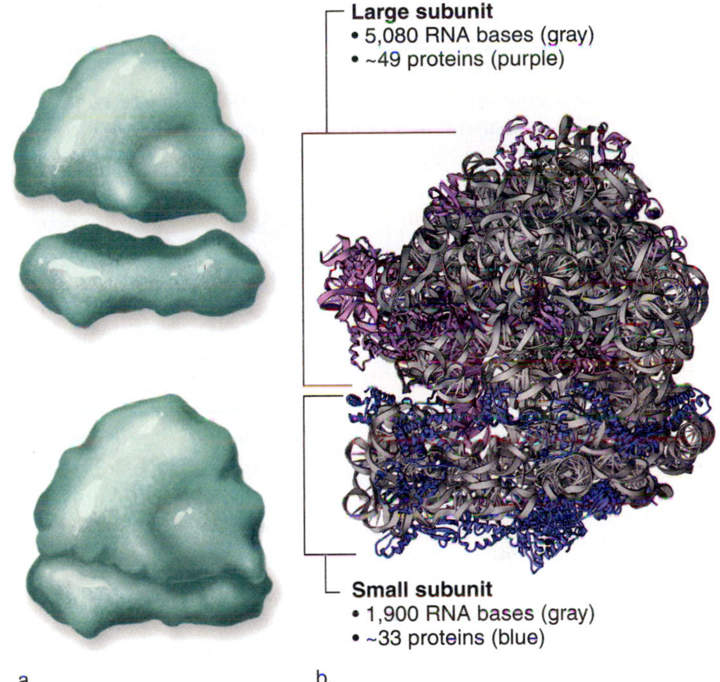

FIGURE 12.7 The Ribosome. A ribosome from a eukaryotic cell, shown here, has two subunits containing a total of 82 proteins and four rRNA molecules.

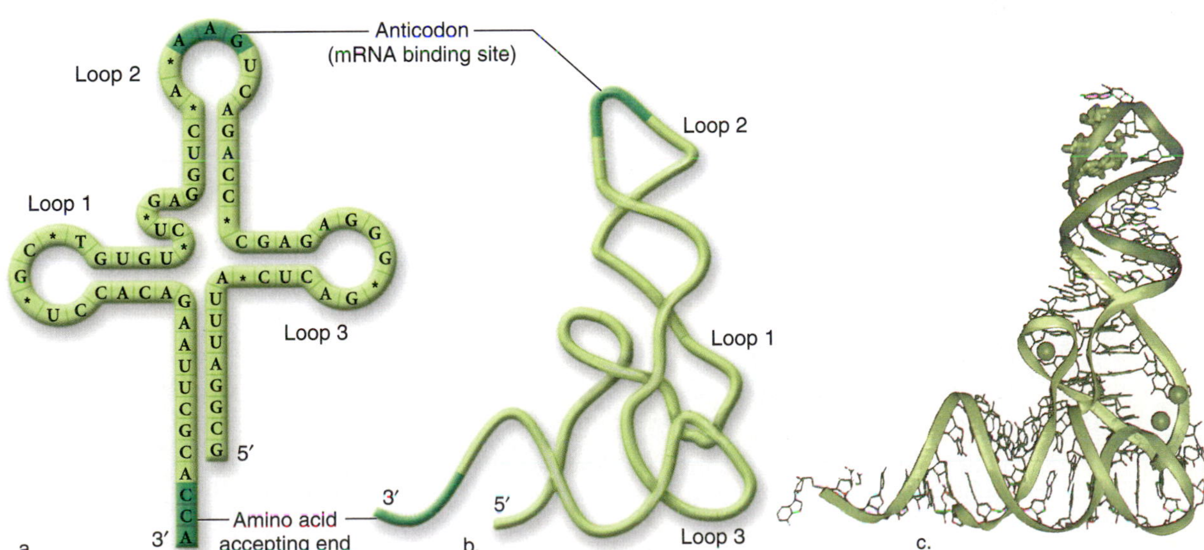

FIGURE 12.6 Transfer RNA. (**A**) Complementary base pairing within a tRNA molecule gives the molecule a "cloverleaf" shape. At one end, the anticodon is the sequence that binds a complementary mRNA codon. At its opposite end, each tRNA also binds the amino acid corresponding to that codon. (**B**) and (**C**) Three-dimensional views of tRNA depict the loops that interact with the ribosome during translation.

C. Translation Occurs in Three Steps

The process of translation can be divided into three stages, during which mRNA, tRNA molecules, and ribosomes come together, link amino acids into a chain, and then dissociate again.

1. The first step is *initiation* (**figure 12.8A** and **B**). The leader sequence of the mRNA molecule bonds with a small ribosomal subunit. The first mRNA codon to specify an amino acid is usually AUG, which attracts a tRNA that carries the amino acid methionine. This methionine signifies the start of a polypeptide.
2. To start the next stage, *elongation*, a large ribosomal subunit attaches to the small subunit. The second codon, which is GGA in figure 12.8C, then bonds to the anticodon of a tRNA molecule carrying the

FIGURE 12.8 Translation.
(**A**) Initiation of translation brings together a small ribosomal subunit, mRNA, and an initiator tRNA. (**B**) Once aligned in the proper orientation to begin translation, (**C**) a large ribosomal subunit joins the cluster. As elongation begins, a tRNA molecule bearing a second amino acid (glycine, in this example) forms hydrogen bonds between its anticodon and the mRNA's second codon. (**D**) The methionine brought in by the first tRNA forms a peptide bond with the amino acid brought in by the second tRNA. (**E**) A third tRNA has arrived, in this example carrying the amino acid cysteine, as a fourth approaches. (**F**) The fourth amino acid, lysine, links to the growing polypeptide chain as a fifth tRNA approaches. (**G**) A protein release factor binds to the stop codon, signaling termination. (**H**) The completed protein releases from the last tRNA, and all components of the translation machine are liberated.

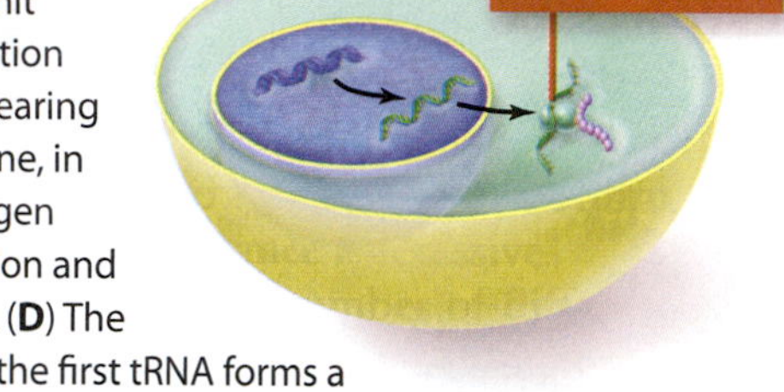

Initiation

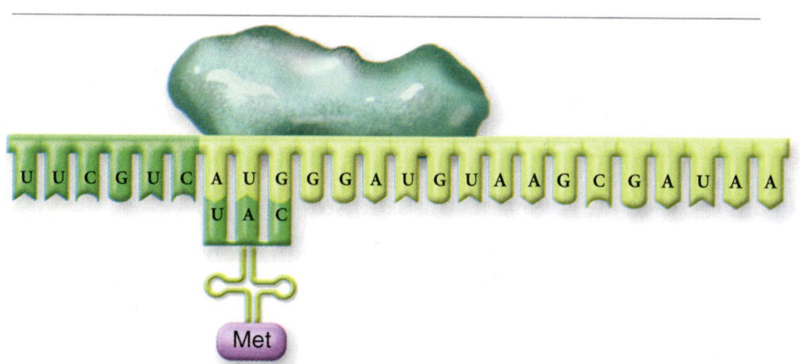

Elongation

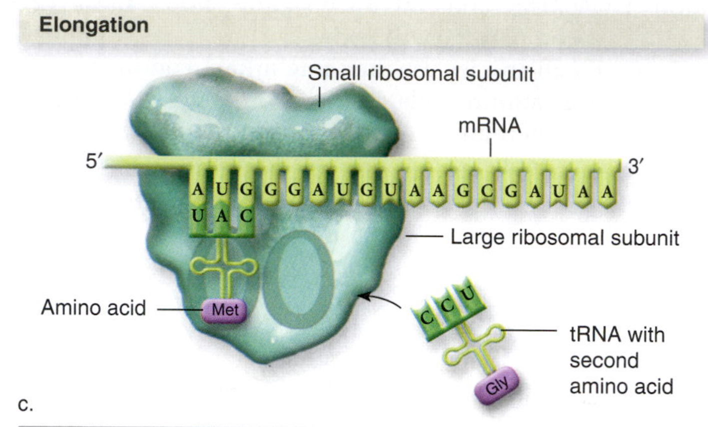

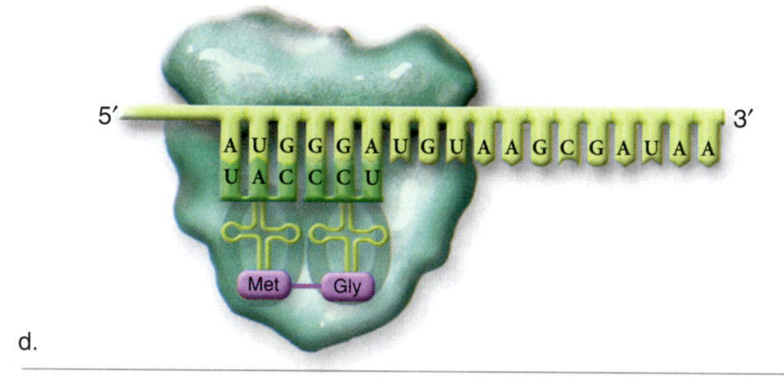

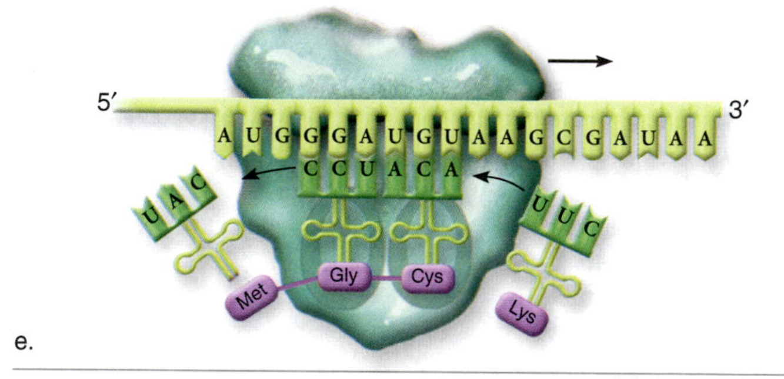

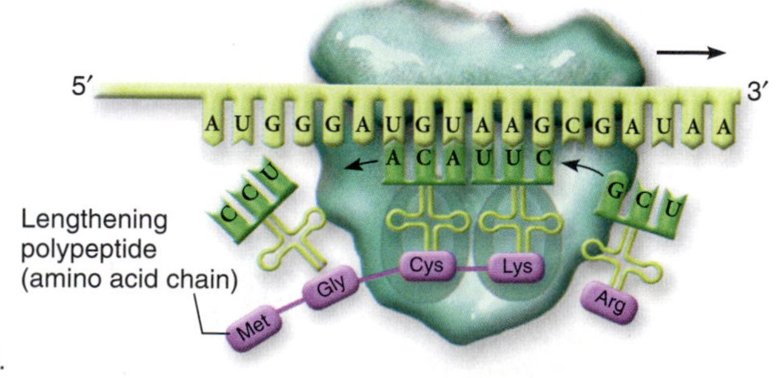

amino acid glycine. The two amino acids (methionine and glycine in the example), which are still attached to their tRNAs, align. A covalent bond forms between them, and the ribosome releases the first tRNA. This tRNA will pick up another methionine and be used again. The ribosome and its attached mRNA are now bound to a single tRNA, with two amino acids extending from it. This begins the polypeptide.

Next, the ribosome moves down the mRNA by one codon. A third tRNA enters, carrying its amino acid (cysteine in figure 12.8E). This third amino acid aligns with the other two and forms a covalent bond to the second amino acid in the growing chain. The tRNA attached to glycine is released and recycled. With the help of proteins called elongation factors, the polypeptide grows one amino acid at a time, as tRNAs continue to deliver their cargo (figure 12.8F).

Termination

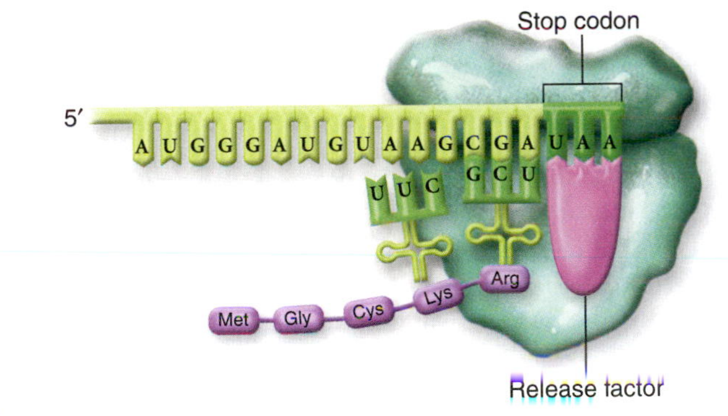

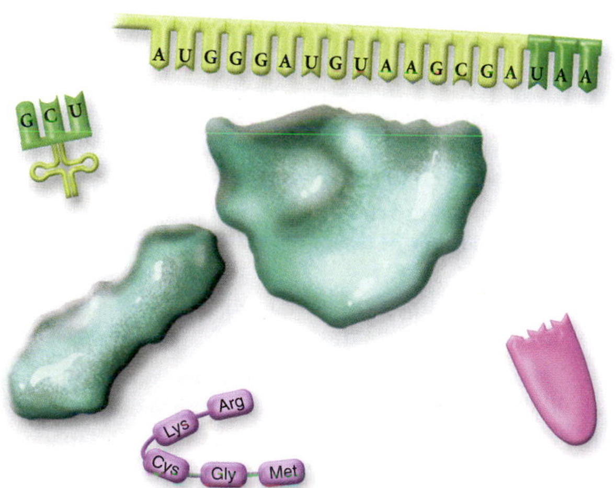

3. Elongation halts at a "stop" codon (UGA, UAG, or UAA). This is the *termination* stage. No tRNA molecules correspond to these stop codons. Instead, proteins called release factors bind to the stop codon (figure 12.8G), prompting the release of the last tRNA from the ribosome. The ribosomal subunits separate from each other and are recycled, and the new polypeptide is released (figure 12.8H).

D. Proteins Must Fold Correctly after Translation

The newly synthesized protein cannot do its job until it folds into its final shape. Some regions of the amino acid sequence attract or repel other parts, contorting the polypeptide's overall shape. Enzymes catalyze the formation of chemical bonds, and "chaperone" proteins stabilize partially folded regions. **protein folding, p. 39**

Errors in protein folding can cause illness. In some forms of cystic fibrosis, for example, a membrane protein that normally controls the flow of chloride ions does not fold correctly into its final form. Alzheimer disease is associated with a protein called amyloid that forms an abnormal mass in brain cells because of improper folding. Mad cow disease and similar conditions in sheep and humans are caused by abnormal clumps of proteins called prions in nervous system cells. **prions, p. 365**

In addition to folding, some proteins must be altered in other ways before they become functional. For example, insulin, which is 51 amino acids long, is initially translated as the 80-amino-acid polypeptide, proinsulin. Enzymes cut proinsulin to form insulin. A different type of modification occurs when polypeptides join to form larger protein molecules. The oxygen-carrying blood protein hemoglobin, for example, consists of four polypeptide chains (two alpha and two beta) encoded by separate genes.

Summary *The genetic code consists of three-base mRNA codons, each of which corresponds to a single amino acid or a "stop" signal. To synthesize a protein, tRNA molecules carrying amino acids form base pairs with mRNA molecules; ribosomes align the amino acids. A protein must fold into a specific shape to function.*

12.3 MASTERING CONCEPTS

1. How did researchers determine that the genetic code is a triplet and learn which codons specify which amino acids?
2. What is the significance of the near-universality of the genetic code?
3. What are the steps of translation?
4. How does a polypeptide fold into its finished shape?

12.4 Protein Synthesis Is Highly Regulated

Protein synthesis can be very speedy. A plasma cell in the human immune system can manufacture 2000 identical antibody proteins per second. How can protein synthesis occur fast enough to meet all of a cell's needs? First, it is efficient. Transcription produces multiple copies of each mRNA, and dozens of ribosomes may simultaneously bind to a single mRNA copy (**figure 12.9**). A cell can therefore make many copies of a protein from the same mRNA. Second, ribosomes zip along mRNA molecules, incorporating some 15 amino acids per second.

Producing proteins costs tremendous amounts of energy. For example, an *E. coli* cell spends 90% of its ATP on protein synthesis. Transcription and translation require energy, as does the synthesis of the nucelotides, tRNA, rRNA, enzymes, and other molecules that make protein synthesis possible. In eukaryotes, splicing out introns and other modifications of the mRNA require still more energy. The expenditure is essential to life, as described in this chapter's Can *You* Relate? Poisons. **ATP, p. 85**

Nevertheless, given the enormous cost of making protein, it makes sense that cells save energy by not producing unneeded proteins. Of course, genes encoding proteins that are essential to life must be expressed all the time, such as those for the enzymes involved in many energy pathways (see chapter 6). Cells transcribe other genes only under some conditions. This section describes some examples of the many mechanisms that regulate gene expression in cells.

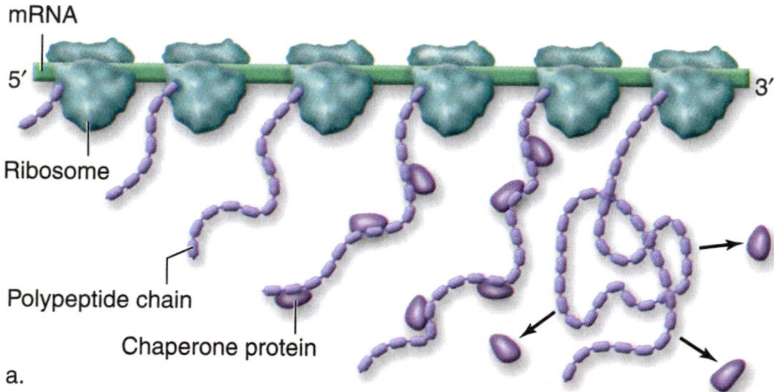

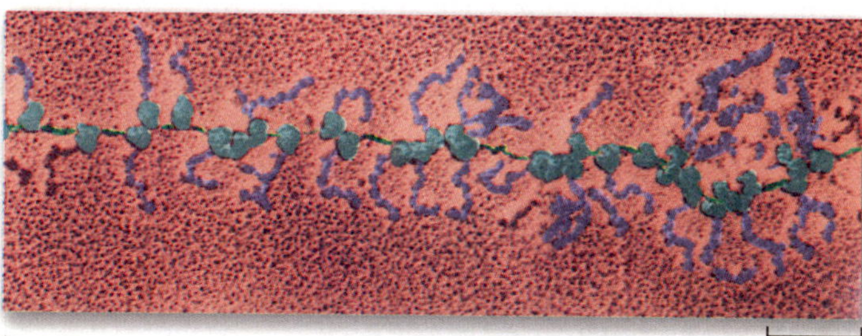

FIGURE 12.9 Efficient Translation. (**A**) Several ribosomes can simultaneously translate the same mRNA. The closer a ribosome is to the end of an mRNA, the longer its polypeptide. Chaperone proteins help fold the polypeptide into its characteristic shape. (**B**) In the micrograph, the ribosomes toward the left have just begun translation, and the polypeptides are just barely visible. Farther along in translation, the polypeptides are longer. The chaperones are not visible.

A. Operons Are Groups of Bacterial Genes that Share One Promoter

It wasn't long after the structure of DNA was published that researchers began to discover the controls of gene expres-

Can *You* Relate?

Poisons

We learned in chapter 6 that some poisons kill because they interfere with respiration. Here we list a few poisons that inhibit protein synthesis. A cell that cannot make proteins quickly dies.

- **Amanatin:** This toxin naturally occurs in the "death cap mushroom," *Amanita phalloides*. Amanatin inhibits RNA polymerase, making transcription impossible.
- **Diphtheria toxin:** Bacteria called *Corynebacterium diphtheriae* secrete a toxin that causes the respiratory illness diphtheria. The toxin inhibits an elongation factor, a protein that helps add amino acids to a polypeptide chain during translation.
- **Antibiotics:** Antibiotics that bind to prokaryotic ribosomes include clindamycin, chloramphenicol, tetracyclines, and gentamicin. When its ribosomes are disrupted, a bacterial cell cannot make proteins, and it dies.
- **Ricin:** Derived from seeds of the castor bean plant (*Ricinus communis*), ricin is a potent natural poison that consists of two parts. One part binds to a cell, and the other enters the cell and inhibits protein synthesis by an unknown mechanism. Interestingly, the part of the molecule that enters the cell is apparently more toxic to cancer cells than to normal cells, making ricin a potential cancer treatment.
- **Trichothecenes:** Fungi in genus *Fusarium* produce toxins called trichothecenes. During World War II, thousands of people died after eating bread made from moldy wheat, and many researchers believe trichothecenes were used as biological weapons during the Vietnam War. The mode of action is unclear, but the toxins seem to interfere somehow with ribosomes.

sion. In 1961, French biologists François Jacob and Jacques Monod described how *E. coli* bacteria produce the three enzymes that they need to degrade the sugar lactose, but only when lactose is present in the cell's surroundings. What signals "tell" a simple bacterial cell to transcribe all three genes at precisely the right time? **Focus on Model Organisms (*E. coli*), p. 378**

In *E. coli* and other bacteria, related genes are organized as operons. An **operon** is a group of genes plus a promoter and operator that control the transcription of the entire group at once (figure 12.10A). The promoter, as described earlier, is the site to which RNA polymerase can attach to begin transcription. The **operator** is a DNA sequence located between the promoter and the three protein-encoding regions. If a protein called a **repressor** binds to the operator, it prevents the transcription of the three genes. *E. coli*'s ***lac* operon** consists of the three lactose-degrading genes plus the promoter and operator that control their transcription.

To understand how the *lac* operon works, consider *E. coli* in an environment lacking lactose. Expressing the lactose-degrading genes would be a waste of energy. The repressor protein therefore binds to the operator, preventing RNA polymerase from transcribing the genes (figure 12.10B). The genes are effectively "off."

When lactose is present, however, a modified form of the sugar attaches to the repressor, changing its shape so that it detaches from the DNA. RNA polymerase is now free to transcribe the genes (figure 12.10C). After translation, the resulting enzymes enable the cell to absorb and degrade the sugar. Lactose, in a sense, causes its own dismantling.

Soon, geneticists discovered other groups of genes organized as operons. Some, like the *lac* operon, negatively control transcription by removing a block. Others produce factors that turn on transcription. As Jacob and Monod stated in 1961, "The genome contains not only a series of blueprints, but a coordinated program of protein synthesis and means of controlling its execution."

B. Eukaryotic Organisms Use Transcription Factors to Turn Genes On or Off

In multicellular eukaryotes, genetic control is more complex than in bacteria, because different cell types express different subsets of genes. A cell in an early animal embryo, for example, must express the proteins that dictate the formation of body parts in the correct places. A differentiated skin cell in an adult would not need those proteins, but would need to produce pigments that protect the body from the sun's ultraviolet radiation.

To manage such complexity, groups of regulatory proteins called **transcription factors** bind DNA at the promoter and regulate transcription. The transcription factors form a pocket for RNA polymerase. In eukaryotes (unlike in prokaryotes), RNA polymerase cannot bind to a promoter in the absence of transcription factors.

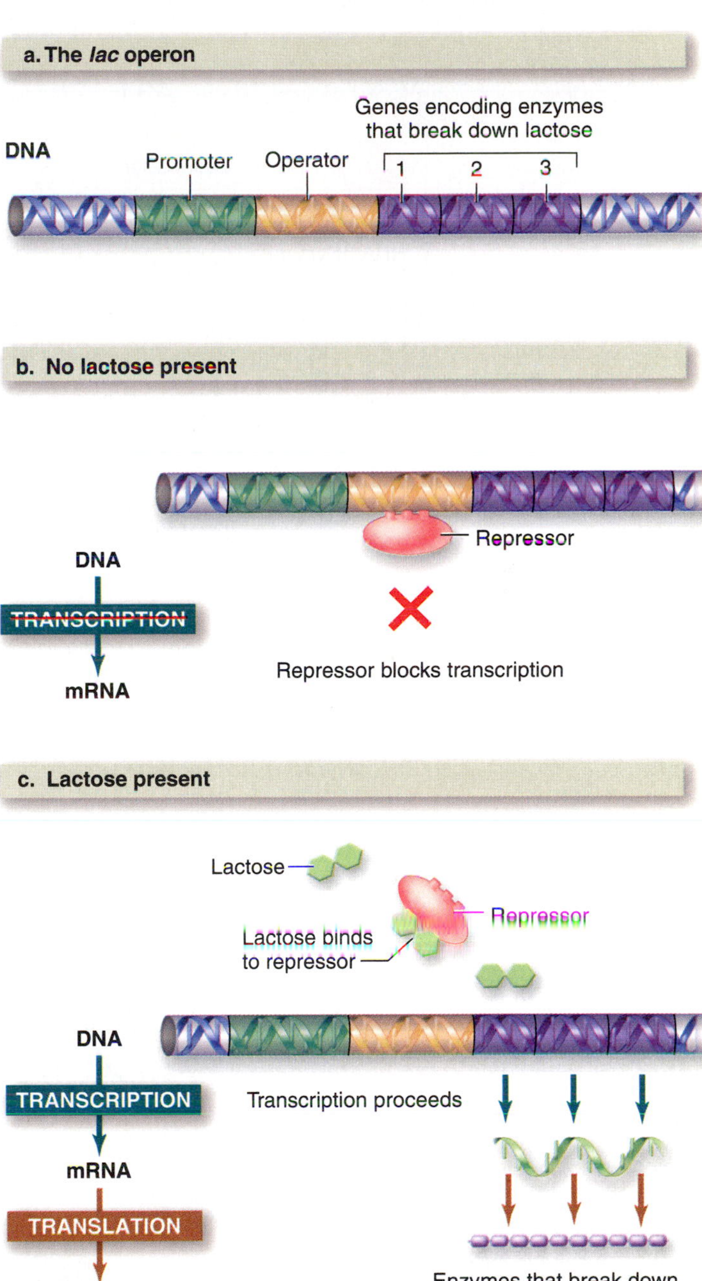

FIGURE 12.10 The *Lac* Operon. **(A)** In the *lac* operon, a single promoter controls expression of three genes whose encoded proteins metabolize the sugar lactose. **(B)** In the absence of lactose, a repressor protein binds to the operator region of the DNA, preventing transcription of the genes. **(C)** In the presence of lactose, the repressor binds only to lactose, and transcription of the three genes can proceed.

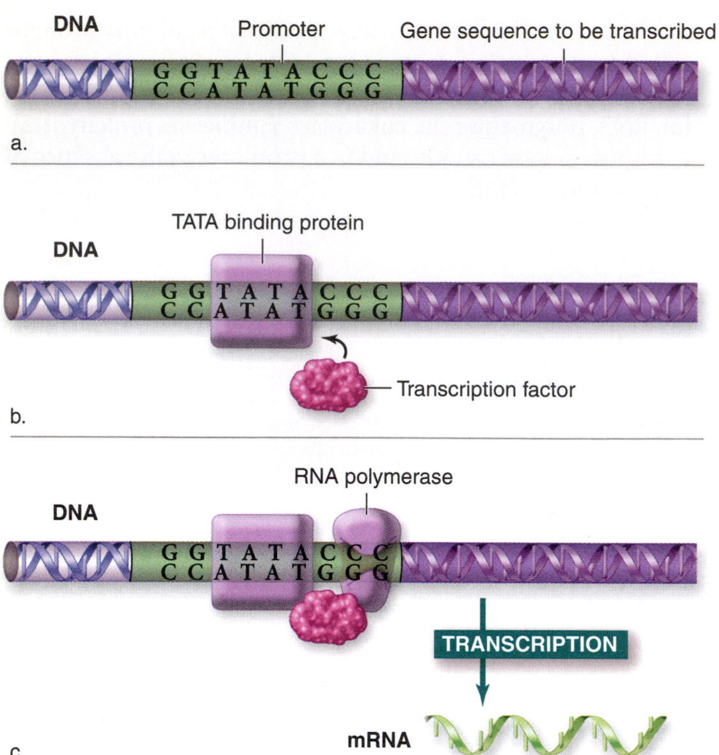

FIGURE 12.11 Transcription Factors Regulate Eukaryotic Gene Expression. (**A**) Proteins that initiate transcription recognize certain sequences in the promoter region of a gene. (**B**) A binding protein recognizes the TATA region and binds to the DNA. This allows other transcription factors to bind. (**C**) The presence of the necessary transcription factors allows RNA polymerase to bind and begin making RNA.

Figure 12.11 shows how transcription factors prepare a site to receive RNA polymerase. The first transcription factor to bind is attracted to a DNA sequence called a TATA box. This transcription factor attracts others. Finally RNA polymerase joins the complex, binding just in front of the start of the gene sequence.

Transcription factors make signal transduction possible. Recall from chapter 3 that signal transduction is the process by which a cell responds to external stimuli (see figure 3.8). Once a signaling molecule binds to the outside of a target cell, a series of chemical reactions occurs inside the cell. The last step in the series can be the activation or deactivation of a transcription factor. Because a single transcription factor may influence many genes, one stimulus may trigger many simultaneous changes in the cell. One example is the series of events that follows the fusion of egg and sperm (see chapter 37).

Hundreds of transcription factors are known, and in humans, defects in them underlie some diseases, including cancers. This makes sense, because the signals that trigger cell division are proteins. In addition, some drugs interfere with transcription factors. The "abortion pill" RU486, for example, indirectly blocks the action of transcription factors needed for the development of an embryo. **cancer, p. 166**

C. Eukaryotic Cells Also Use Additional Regulatory Mechanisms

In addition to transcription factors, eukaryotic cells also have several other ways to control gene expression:

- A cell can "tag" unneeded DNA with methyl groups ($—CH_3$). Proteins inside the cell bind to the tagged

FIGURE 12.12 Alternative Splicing. The same mRNA can trigger the development of a male or female fruit fly, depending on which introns are spliced out. Spliced the "female" way, the first mRNA induces the others to also splice in the "female" way. A different way of splicing the first mRNA produces a male.

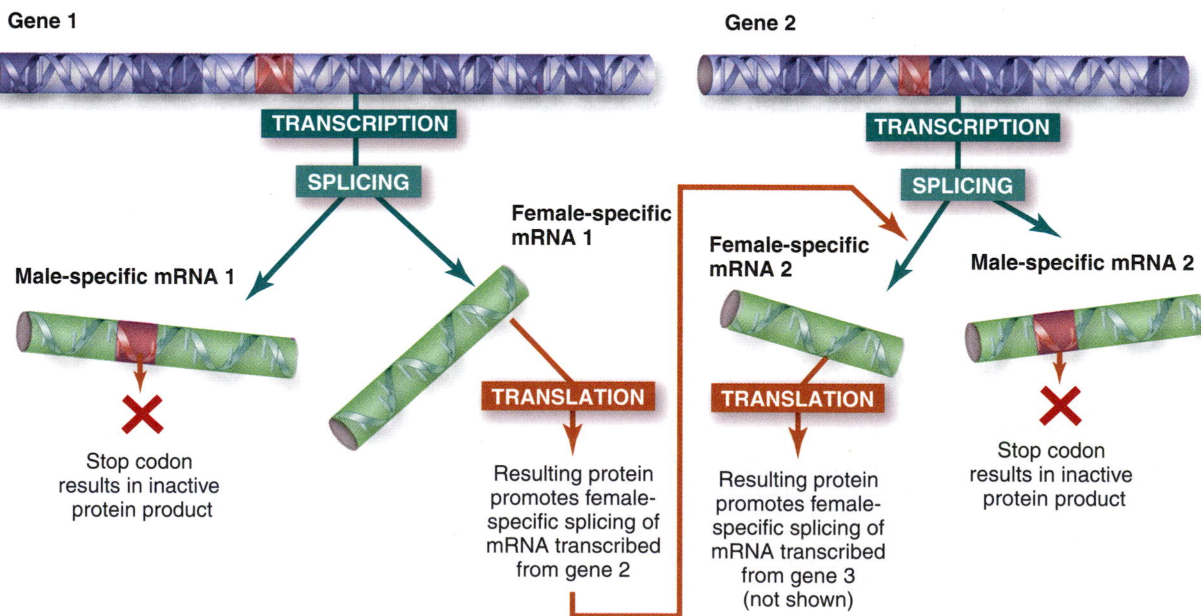

Burning Questions

Is there a gay gene?

Research linking human behavior to individual genes is extremely difficult. First, genes encode proteins, not behaviors, so the question of a "gay gene" is somewhat misleading. The relationship between any gene and any behavior is necessarily indirect. Second, to establish a clear link to DNA, the researcher must be able to define and measure the behavior. This in itself is difficult, because people disagree about what it means to be homosexual. Third, multiple genes are likely to be involved. Fourth, an individual that possesses an allele associated with a trait will not necessarily express the allele; many genes in each cell remain "off" at any given time. To complicate matters, the environment contributes mightily to gene expression.

Because of these complications, research into the genetic contribution to homosexuality has yielded ambiguous results. A male homosexual's identical twin is more likely to also be homosexual than is a nonidentical twin, indicating a significant genetic component. In addition, anatomical studies of cadavers have revealed differences in the size of a particular brain structure between heterosexual and homosexual men, but the relative contribution of genes and environment to this structure is unknown. One study linked homosexuality in males, but not in females, to part of the X chromosome; a subsequent study did not support that conclusion.

So is there a gay gene? The answer remains elusive. But we can say without a doubt that both the environment and genetics play important roles.

Have a Burning Question of your own? Submit it to marielle_hoefnagels@mcgraw-hill.com for possible inclusion in future editions of this book!

DNA, preventing gene expression and signaling the cell to fold that section of DNA more tightly. Transcription factors and RNA polymerase cannot access highly compacted DNA, effectively turning off the genes.

- One gene can encode multiple proteins if different introns are removed from the mRNA. For example, alternative splicing explains how fruit flies develop into males or females (**figure 12.12**).
- For a protein to be produced, mRNA must leave the nucleus and attach to a ribosome. If the mRNA is not allowed to leave, the gene is effectively silenced.
- Not all mRNA molecules are equally stable. Some are rapidly degraded, perhaps before they can be translated, whereas others are more stable. Similarly, some proteins are degraded shortly after they form, whereas others persist longer.
- Some proteins, such as insulin, must be altered before they become functional. If these modifications do not occur correctly, the protein cannot function.
- To do its job, a protein must move from the ribosome to where the cell needs it. For example, a protein secreted in milk must be escorted to the Golgi apparatus and be packaged for export (see figure 3.15). A gene is effectively silenced if its product never moves to the correct destination.

A human cell may express hundreds to thousands of genes at once. Unraveling the complexities of the many regulatory mechanisms that control the expression of each of these genes is an enormous challenge. As described in section 12.8, biologists now have the technology to begin navigating this regulatory maze. The payoff will be a much better understanding of cell biology, along with many new medical applications. The same research may also help scientists understand how external influences on gene expression contribute to complex traits, such as the one described in this chapter's Burning Question: Is there a gay gene?.

Summary *Protein synthesis represents a substantial, but vital, energy cost to the cell. Cells therefore continually express some genes, but turn on others only when needed. In bacterial cells, operons are groups of genes that switch on and off simultaneously. In eukaryotes, proteins called transcription factors bind to DNA and turn individual genes on or off. Regulatory mechanisms apply to other steps in protein synthesis as well.*

12.4 MASTERING CONCEPTS

1. Which steps in protein synthesis require energy?
2. Why do cells regulate which genes are expressed at any given time?
3. How do proteins determine whether a bacterial operon is expressed?
4. What is the role of transcription factors in signal transduction?
5. What are other mechanisms by which eukaryotic cells control gene expression?

12.5 Mutations Create New Alleles

A **mutation** is a change in a cell's DNA sequence. Many people think that all mutations are always harmful, perhaps because some cause such dramatic changes in phenotype (**figure 12.13**). Although some mutations do cause illness, they also provide the variation that makes life interesting (and makes evolution possible).

A. Mutations Range from Silent to Devastating

A mutation can be anything from a single-base change, to an insertion or deletion that shifts the reading frame, to the expansion of repeated sequences. Some may not be detectable except by DNA fingerprinting, while others may have lethal consequences for a developing embryo.

Point Mutations

A **point mutation** is a substitution of one DNA base for another. Such a mutation is "silent" if the mutated gene encodes the same protein as the original gene version. Silent mutations are possible because more than one codon encodes most amino acids.

Often, however, a point mutation changes a base triplet so that it specifies a different amino acid. This change is called a missense mutation. The substituted amino acid may drastically alter the protein's shape, changing its function. Sickle cell disease results from this type of mutation (**figure 12.14**).

In other cases, called nonsense mutations, a base triplet specifying an amino acid changes into one that encodes a "stop" codon. This shortens the protein product, which can profoundly influence the phenotype. At least one of the mutations that gives rise to cystic fibrosis, for example, is a nonsense mutation. Instead of the normal 1480 amino acids, the faulty protein has only 493 and therefore cannot function.

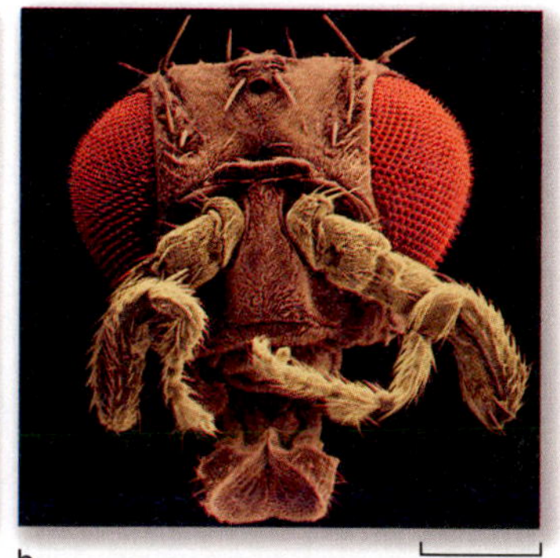

FIGURE 12.13 One Mutation Can Make a Big Difference. A homeotic gene affects development by influencing the fate of each body segment. Mutating a homeotic gene can cause parts to develop in the wrong places. The fruit fly on the right has legs growing where antennae should be.

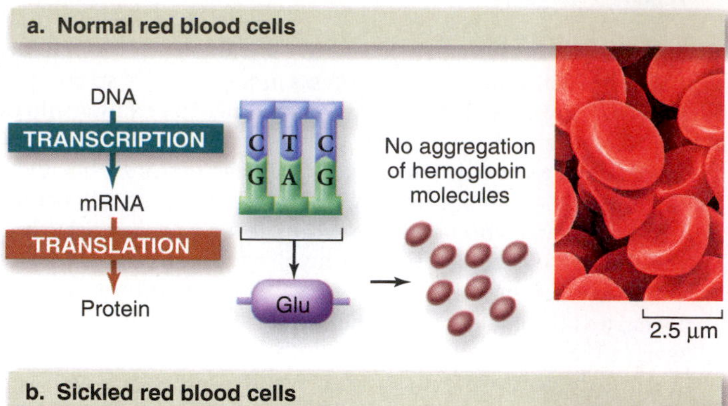

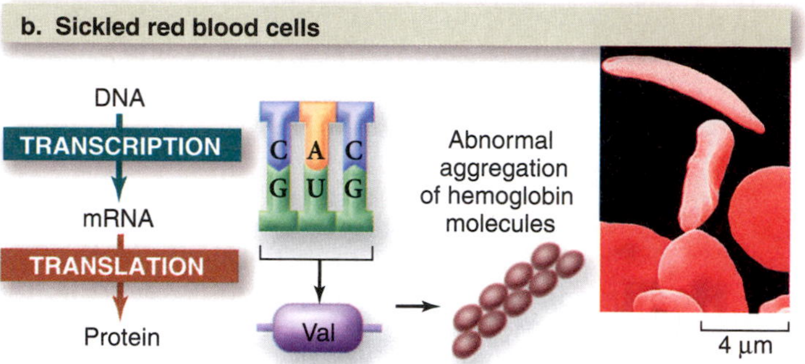

FIGURE 12.14 Sickle Cell Disease Results from a Single Base Change. Two genes encode different parts of the hemoglobin protein, which carries oxygen throughout the body. The most common form of sickle cell anemia results from a mutation in one of the two genes, on chromosome 11. (**A**) Normal hemoglobin molecules do not aggregate, enabling the cell to assume a rounded shape. (**B**) In sickle cell disease, a single substitution mutation replaces one amino acid (glutamic acid) with a different one (valine). This change causes hemoglobins to aggregate into long, curved rods that deform the red blood cell. The abnormal cells bend into sickle shapes that cause anemia, joint pain, and organ damage when the cells lodge in narrow blood vessels, cutting off local blood supplies.

Base Insertions and Deletions

One or more nucleotides can be added to, or deleted from, a gene. In a **frameshift mutation,** nucleotides are added or deleted by any number other than a multiple of three (**figure 12.15**). Because triplets of DNA bases specify amino acids, such an addition or deletion disrupts the reading frame. It therefore also disrupts the sequence of amino acids and usually devastates a protein's function. Some mutations that cause cystic fibrosis result from the addition or deletion of just one or two nucleotides in the *CFTR* gene.

Even if a small insertion or deletion does not shift the reading frame, the effect on the phenotype might still be severe if

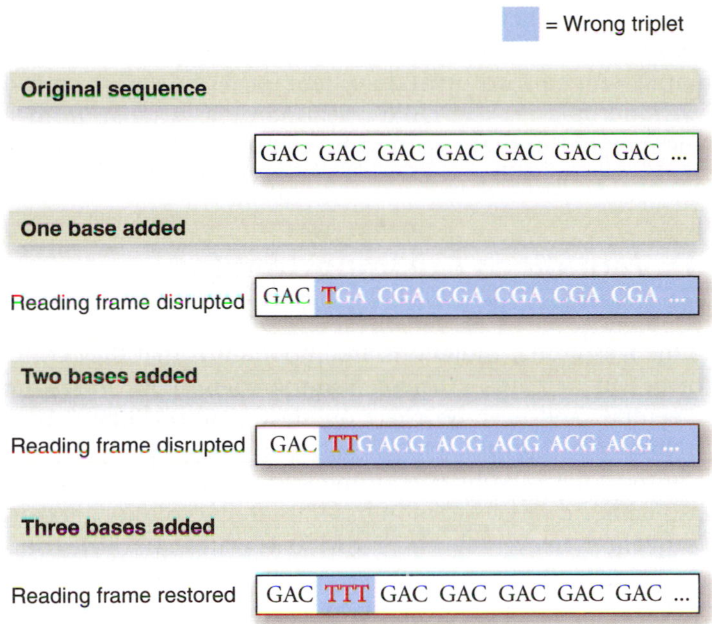

FIGURE 12.15 Frameshift Mutations. Insertions or deletions of one or two nucleotides dramatically alter a gene. However, adding or deleting three nucleotides does not disrupt the reading frame.

Table 12.3 *Types of Mutations. A sentence of three-letter words can serve as an analogy to demonstrate the effects of mutations on gene sequence.*

Wild type	THE ONE BIG FLY HAD ONE RED EYE
Missense	TH**Q** ONE BIG FLY HAD ONE RED EYE
Nonsense	THE ONE BIG
Frameshift	THE ONE **Q**BI GFL YHA DON ERE DEY
Deletion of three letters	THE ONE BIG HAD ONE RED EYE
Duplication	THE ONE BIG FLY **FLY** HAD ONE RED EYE
Insertion	THE ONE BIG **WET** FLY HAD ONE RED EYE
Expanding repeat mutation	P: THE ONE BIG FLY HAD ONE RED EYE
	F_1: THE ONE BIG FLY **FLY FLY** HAD ONE RED EYE
	F_2: THE ONE BIG FLY FLY FLY **FLY FLY FLY** HAD ONE RED EYE

the change drastically alters the protein's shape. The most common mutation that causes severe cystic fibrosis, for example, deletes only a single group of three nucleotides. The resulting protein lacks just one amino acid, but it cannot function.

Expanding Repeats

In an **expanding repeat mutation,** the number of copies of a three- or four-nucleotide sequence increases over several generations. Expanding genes underlie several inherited disorders, including fragile X syndrome and Huntington disease. In Huntington disease, expanded repeats of GTC cause extra glutamines (an amino acid) to be incorporated into the gene's protein product. The abnormal protein forms fibrous clumps in the nuclei of some brain cells, which causes the symptoms of uncontrollable movements and personality changes.

Table 12.3 summarizes the major types of mutations, all of which create new alleles. What determines whether the new allele will be dominant or recessive to the original gene version? In general, if the mutated allele loses its function entirely (as in the case of the *CFTR* mutations that cause cystic fibrosis) the allele is usually recessive. If the resulting protein gains a new function, such as in Huntington disease, the new allele is usually dominant.

B. What Causes Mutations?

Some mutations form spontaneously—that is, without outside causes. A spontaneous point mutation usually originates as a DNA replication error. Replication errors can also cause insertions and deletions, especially in genes with repeated base sequences, such as GCG CGC . . . . It is as if the molecules that guide and carry out replication become "confused" by short, repeated sequences, as a proofreader scanning a manuscript might miss the spelling errors in the words "happpiness" and "bananana." **DNA replication, p. 143**

The average rate of replication errors for most genes is about 1 in 100,000 bases, but it varies in different organisms and for different genes. The larger a gene, the more likely it is to mutate. In addition, the more frequently DNA replicates, the more it mutates. Bacteria accumulate mutations faster than cells of complex organisms simply because their DNA replicates more often.

Mutations may also occur during meiosis. Genes containing repeated sequences sometimes misalign during prophase I, producing chromosomes with more or fewer gene copies than normal. The genes encoding the alpha subunit of hemoglobin are especially likely to mutate because they are repeated in their entirety next to each other on each human chromosome 16 (**figure 12.16**). **crossing over, p. 182**

Chromosome inversions and translocations also cause mutations if they bring together gene segments that were not previously joined (see figure 9.13). When part of chromosome 9 breaks off and fuses with chromosome 22, for example, the resulting "Philadelphia chromosome" has a fused gene whose protein product causes a type of leukemia. Movable DNA sequences called transposons, which insert themselves randomly into chromosomes, are yet another source of spontaneous mutations (see section 12.6).

Exposure to chemicals or radiation may also damage DNA. A **mutagen** is any external agent that induces

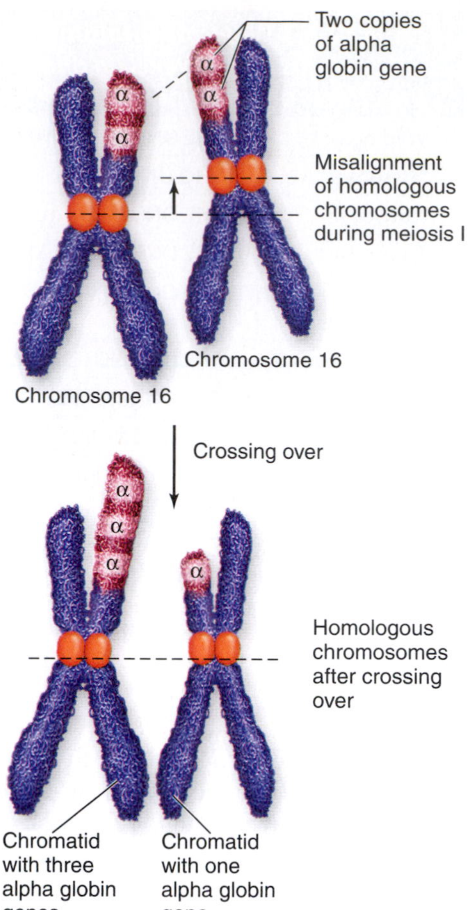

FIGURE 12.16 Gene Duplication and Deletion in Crossing Over. The repeated nature of the alpha globin genes makes them vulnerable to mutation by mispairing during meiosis. A person missing one alpha globin gene can develop anemia.

mutations, such as the ultraviolet radiation in sunlight, X rays, radioactive fallout from atomic bomb tests and nuclear accidents, chemical weapons such as mustard gas, and chemicals in tobacco. The more contact a person has with mutagens, the higher the risk for cancer. Coating skin with sunscreen, wearing a lead "bib" during dental X-rays, and stopping smoking all lower cancer risk by reducing exposure to mutagenic chemicals and radiation.

C. Some Mutations Pass to Future Generations

Not all mutations are passed from generation to generation. A **somatic mutation** occurs in nonsex cells, such as those that make up the skin or intestinal tract. All cells derived from the altered one will also carry the mutation, but the mutation does not pass to the organism's offspring. The children of a cigarette smoker with mutations that cause lung cancer, for example, do not inherit the parent's damaged genes.

A **germline mutation,** on the other hand, occurs in cells that give rise to gametes (see section 9.7). Germline mutations are heritable because the mutated allele will appear in gametes that the organism produces. As a result, every cell of the organism's affected offspring will carry the mutation as well. Such mutations may appear suddenly in a family. For example, two healthy people of normal height may have a child with an autosomal dominant form of dwarfism called achondroplasia. The child's achondroplasia arose from a new mutation that occurred by chance in the mother's or father's germ cell.

D. Mutations Are Important for Many Reasons

One reason that mutations are important is that they create new gene variants (alleles). Some of these new alleles are "neutral" and have no effect on an organism's fitness. Your reproductive success, for example, does not generally depend on the color of your eyes or the size of your feet. As unit 3 explains, however, variation has important evolutionary consequences. In every species, individuals with some allele combinations reproduce more successfully than others. Natural selection "edits out" the less favorable allele combinations.

Mutations in disease-causing bacteria and viruses have enormous medical importance. Antibiotic drugs kill bacteria by targeting prokaryotic membrane proteins, enzymes, and other structures. Random mutations in bacterial DNA encode new versions of these proteins, and the descendants of some of the mutated cells become new strains that are not susceptible to these antibiotics. Likewise, random mutations enable viruses to jump from other animals to humans. Evolving viruses have caused the global epidemics of HIV, influenza, and other diseases (see section 17.6).

Geneticists frequently induce mutations to learn how genes normally function. For example, biologists discovered how genes control flower formation by studying mutant *Arabidopsis* plants in which flower parts form in the wrong places (see figure 26.28).

Plant breeders also induce mutations to create new varieties of many crop species (**figure 12.17**). Some kinds of rice, grapefruit, oats, lettuce, begonias, and many other plants owe their existence to breeders treating cells with radiation and then selecting interesting new varieties from the mutated individuals.

a.

b.

c.

FIGURE 12.17 Beautiful Mutants. (**A**) Rio Red grapefruits and several varieties of (**B**) rice and (**C**) cotton are among the many plant varieties that have been created by using radiation to induce mutations.

Summary *A change in a gene's sequence—a mutation—can alter its function and affect the organism's phenotype. Mutations may occur spontaneously as an error in DNA replication, or they may be induced. Cells have some safeguards against mutations. Mutations have practical uses in basic scientific research and plant breeding, and they are the source of all new alleles in a population.*

12.5 MASTERING CONCEPTS

1. What are the types of mutations, and how does each alter the encoded protein?
2. What causes mutations?
3. What is the difference between a germline mutation and a somatic mutation?
4. How are mutations important tools in biological research?

12.6 The Human Genome Sequence Reveals Unexpected Complexity

The Human Genome Project revealed unexpected complexities and seeming contradictions in our 3.2 billion base pairs. Although our genome includes approximately 25,000 genes, our cells can produce some 400,000 different proteins. Furthermore, only about 1.5% of the human genome sequence encodes protein.

Two questions emerge from these observations. How can so few genes encode so many proteins? And what does the other 98.5% of the human genome do? One answer to the first question lies in introns (see section 12.4). By removing different combinations of introns from an mRNA molecule, a cell can produce several proteins from one gene—a departure from the old idea that each gene encodes exactly one protein. So far, no one understands exactly how a cell "decides" which introns to remove.

The second question—the function of 98.5% of the genome—has several answers. Some of this DNA encodes rRNA, tRNA, and the regulatory sequences that control gene expression. Human chromosomes also contain many **pseudogenes,** DNA sequences that are very similar to protein-encoding genes and that are transcribed, but the mRNA is not translated into protein. Pseudogenes may be remnants of old genes that once functioned in our nonhuman ancestors, but that mutated too far from the normal sequence to encode a working protein.

The human genome is also riddled with highly repetitive sequences that have no known function. The most abundant type of repeat is a **transposable element,** or transposon for short, a DNA sequence that can "jump" within the genome. Originally identified in corn by Barbara McClintock in the 1940s (**figure 12.18**), and then in bacteria in the 1960s, transposons make up about 45% of the human genome. Transposons can act as mutagens. A transposon that lands in a gene can disrupt the gene's function while it is inserted, and it can cause a gap when it leaves.

The genome also contains many tandem repeats (or "satellite DNAs"), sequences of one or more bases repeated many times without interruption, such as CACACA or ATTCGATTCG. Huge clusters of tandem repeats, up to 100 million base pairs long, occur at telomeres and centromeres and on the Y chromosome. Thousands of shorter tandem repeats (a few dozen to 30,000 base pairs long) also litter much of the rest of the genome. The exact number of repeats varies from person to person; DNA fingerprinting technology exploits these differences to match suspects with evidence left at the scene of a crime. Some expanding repeated sequences may cause illness such as Huntington disease. In other cases, the repeats appear to have no phenotypic effect. **DNA profiling, p. 149**

Summary *Only a small percentage of the 3.2 billion base pairs in the human genome form genes that encode protein, yet the number of proteins far exceeds the number of genes.*

12.6 MASTERING CONCEPTS

1. How does the human genome maximize its protein-encoding informational content?
2. What is the function of the 98.5% of the human genome that does not specify protein?

FIGURE 12.18 Jumping Genes. Transposons cause mutations in the genes encoding pigments in these corn kernels.

12.7 Transgenic Organisms Contain DNA from More than One Species

The fact that virtually all species use the same genetic code means that one type of organism can express a gene from another. Biologists take advantage of this fact by coaxing cells to take up **recombinant DNA,** which is genetic material that has been spliced together from multiple organisms. Once the recombinant DNA is taken up, the cells are grown in culture, mass-producing the gene or its protein product. Scientists first accomplished this feat of "genetic engineering" in *E. coli* in the 1970s, but many other organisms have since been genetically modified. **Focus on Model Organisms (*E. coli*), p. 378**

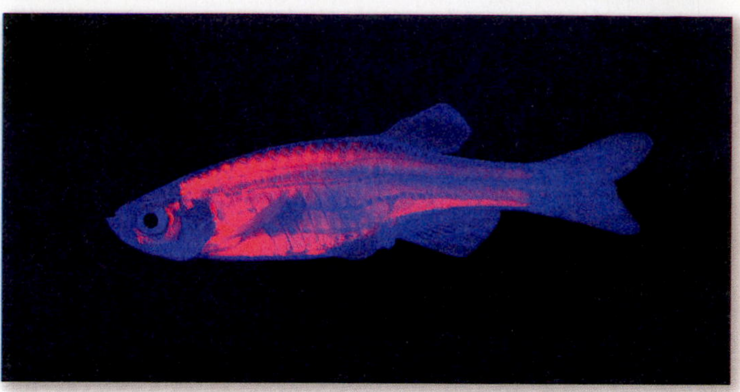

FIGURE 12.19 **Transgenic Animal.** Glow-in-the-dark zebra fish were the first transgenic animals marketed as pets.

A. Transgenic Organisms Have Many Uses

A **transgenic** organism is one that receives recombinant DNA. In the pharmaceutical industry, transgenic bacteria produce several dozen drugs, including human insulin to treat diabetes, blood clotting factors to treat hemophilia, immune system biochemicals, and fertility hormones. Other genetically modified bacteria produce the amino acid phenylalanine, which is part of the artificial sweetener aspartame, whereas still others degrade petroleum, pesticides, and other soil pollutants. Transgenic yeast cells produce a milk-curdling enzyme called chymosin used by many U.S. cheese producers. **artificial sweeteners, p. 41**

Transgenic crop plants may resist pests, survive harsh environmental conditions, or contain nutrients that they otherwise wouldn't. A large portion of the corn and soybeans grown in the United States is transgenic, containing genes that help the plants resist herbicide applications or fight off insect pests (see section 10.6). "Golden Rice" is a genetically engineered plant that contains genes from petunia and bacteria that enable it to produce beta-carotene (a vitamin A precursor) and extra iron, making the rice grains gold in color and more nutritious.

Transgenic animals also have diverse applications. A glow-in-the-dark zebra fish is the first genetically modified house pet (**figure 12.19**). On a more practical note, a transgenic mouse "model" for a human gene can reveal how a disease begins, enabling researchers to develop drugs that treat the early stages of diseases. Transgenic farm animals can secrete human proteins in their milk or semen, yielding abundant, pure supplies of otherwise rare substances that are useful as drugs.

B. Creating Transgenic Organisms Requires Cutting and Pasting DNA

The first step in the creation of a transgenic organism is to obtain DNA from a source cell—usually a bacterium, plant, or animal. The researcher may synthesize the DNA in the laboratory or extract it directly from the source cell.

Complementary DNA (cDNA)

Extracting the DNA directly poses a problem if the gene's source is a eukaryotic cell and the recipient is a bacterium.Bacterial cells cannot remove introns from mRNA, so the gene would encode a defective protein in bacteria. Researchers therefore first isolate a mature mRNA molecule from a eukaryotic cell, with the introns already removed. Then, they use an enzyme called **reverse transcriptase** to make a DNA copy of the mRNA. (Retroviruses such as HIV use this enzyme when they infect cells.) The resulting complementary DNA, or cDNA, still encodes the eukaryotic protein but avoids the intron problem.

The Cloning Vector

Next, the researcher inserts the source DNA into a **cloning vector,** a self-replicating genetic structure. (In molecular biology, "cloning" means to make many identical copies of a DNA sequence.) A common type of cloning vector is a **plasmid,** which is a small circle of double-stranded DNA separate from the cell's chromosome. Viruses are also used as vectors. They are altered so that they transport DNA but cannot cause disease.

How do researchers create a recombinant plasmid? They use **restriction enzymes,** proteins that cut double-stranded DNA at a specific base sequence. Some restriction enzymes generate single-stranded ends that "stick" to each other by complementary base pairing (**figure 12.20**). The natural function of restriction enzymes is to protect bacteria by cutting up DNA from infecting viruses.

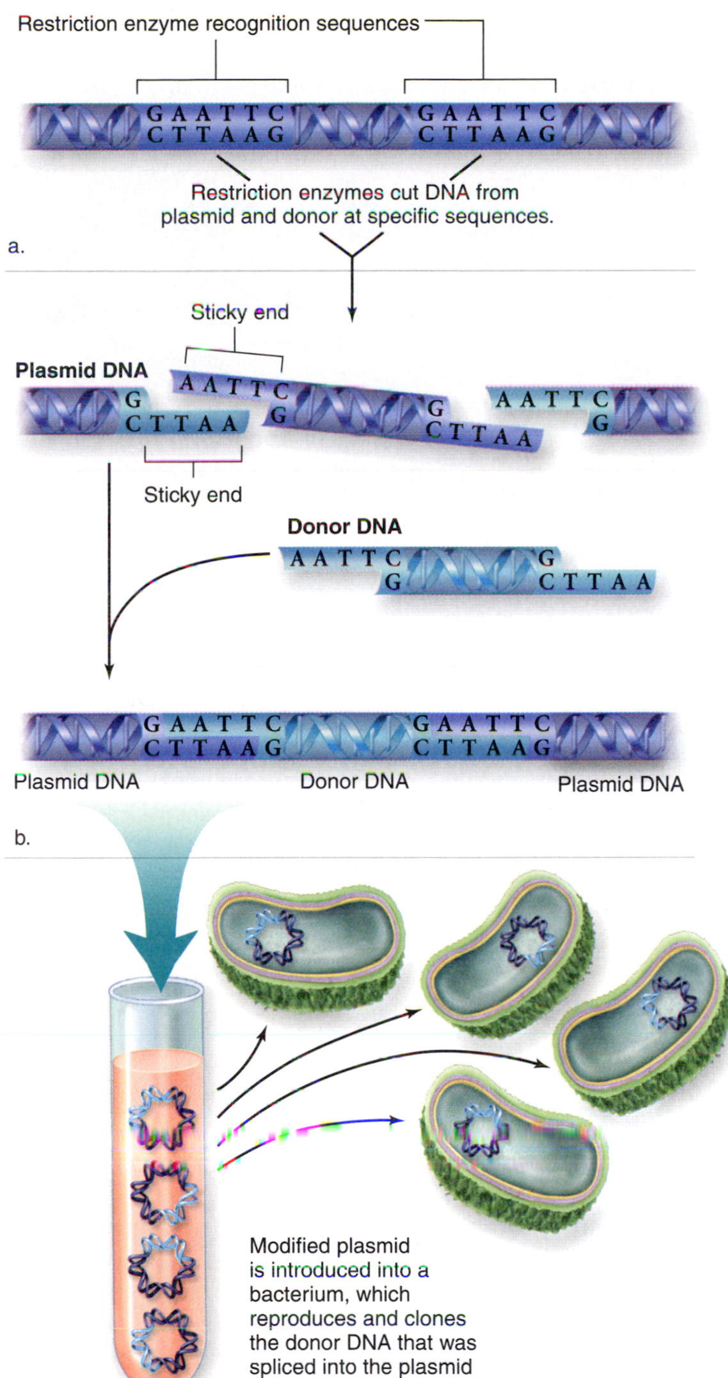

FIGURE 12.20 Recombining DNA. (**A**) A restriction enzyme cuts DNA at specific sequences. The enzyme *Eco*RI, for example, cuts the sequence GAATTC between the G and the A. This staggered cutting pattern produces "sticky ends" of single-stranded DNA. (**B**) Researchers use the same restriction enzyme to cut DNA from a donor cell and a plasmid. When the pieces are mixed, the ends attract through complementary base pairing and pieces join, forming recombinant DNA molecules. (**C**) After the plasmid is sent into a bacterium, it is mass-produced as the bacterium divides.

Biologists, however, use them to cut and paste segments of DNA from different sources (see figure 12.20). When plasmid and donor DNA is cut with the same restriction enzyme and the fragments are mixed, the single-stranded sticky ends of some plasmids form base pairs with those of the donor DNA. Another enzyme, DNA ligase, seals the segments together.

Insertion of the Vector

Next, the researchers move the cloning vector with its recombinant DNA into a recipient cell. Zapping a bacterial cell with electricity opens temporary holes that admit naked DNA. Alternatively, "gene guns" shoot DNA-coated pellets directly into cells. DNA can also be packaged inside a fatty bubble called a liposome that fuses with the recipient cell's membrane, or it can be hitched to a virus that subsequently infects the recipient cell.

One tool for introducing new genes into plant cells is a bacterium called *Agrobacterium tumefaciens*. In nature, these bacteria enter the plant at a wound and inject a plasmid into the host's cells. The plasmid normally encodes genes that stimulate the infected plant cells to divide rapidly, producing a tumorlike gall that produces food and habitat for the bacteria. (The name of the plasmid, T_i, stands for "tumor inducing.") Scientists can replace some of the plasmid's natural genes with other DNA, such as a gene encoding a protein that confers herbicide resistance. They allow the transgenic *Agrobacterium* to inject these modified plasmids into plant cells (**figure 12.21**). All plants derived from the infected cells should express the new herbicide-resistance gene.

Plants can grow from individual cells, but most animals cannot. Therefore, biologists create transgenic animals by using viruses to introduce genes into a gamete or fertilized egg. The organism that develops carries the foreign genes in every cell.

The New Transgenic Organism

Regardless of whether the recipient of recombinant DNA is a bacterium, plant, or animal, the result is the same: When cells containing the DNA divide, all of their daughter cells also harbor the new genes. These transgenic organisms express their new genes just as they do their own, producing the desired protein along with all of the others that they normally make.

Although transgenic organisms have many practical uses, some people question whether their benefits outweigh their potential dangers. Some fear that ecological disaster could result if genetically modified organisms displace other species in the wild. Others worry that unfamiliar protein combinations in genetically modified crops could trigger food allergies. Still others object to the

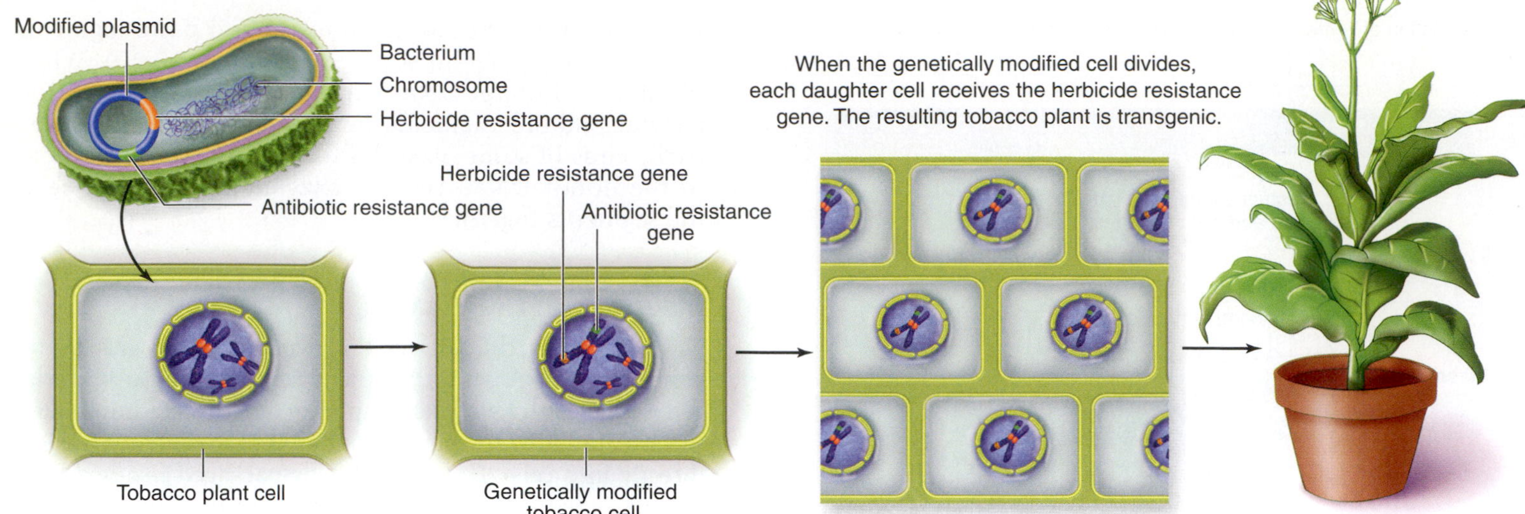

FIGURE 12.21 Creating a Transgenic Plant. *Agrobacterium* cells infect plant cells with a genetically modified Ti plasmid containing genes that confer a new trait, such as herbicide or insect resistance. Since bacteria do not infect 100% of cells, the modified Ti plasmid also contains a gene that confers antibiotic resistance. This helps the researcher screen out tobacco cells that do not contain the plasmid.

"unnatural" practice of combining genes from organisms that would never breed in nature.

Summary *Recombinant DNA comes from two or more different species. Transgenic organisms contain recombinant DNA and have a variety of practical uses.*

12.7 MASTERING CONCEPTS

1. What are some uses for transgenic organisms?
2. What are the steps in creating a recombinant plasmid?
3. How are bacteria, plant, and animal cells induced to take up recombinant DNA?
4. How does reverse transcriptase help scientists produce recombinant plasmids containing eukaryotic genes?

12.8 Biotechnology Has Many Practical Applications

Biotechnology holds out hope for an unprecedented ability to treat some genetic diseases. It is also an incredibly powerful tool for research. This section describes some applications.

A. Gene Therapy Replaces Faulty Genes

Thousands of diseases are caused by faulty genes: cancer, cystic fibrosis, sickle cell anemia, hemophilia, and Tay-Sachs disease are just a few examples. Most genetic illnesses currently have no cure, but **gene therapy** may someday provide one by replacing the faulty gene in a person's somatic cells.

Gene therapy is challenging for several reasons. The new, therapeutic gene must be delivered directly to the cell type that needs correction. Viruses may be ideal for carrying DNA into target cells, but for gene therapy to be safe, the viruses must not alert the immune system. In addition, the gene therapy patient must express the repaired genes long enough for his or her health to improve.

Gene therapy trials in humans have proceeded very slowly since 1999, when 18-year-old Jesse Gelsinger received a massive infusion of viruses carrying a gene to correct an inborn error of metabolism. He died in days from an overwhelming immune system reaction. Gelsinger's unfortunate death prompted a temporary halt to several gene therapy studies and stricter rules for conducting experiments. Nevertheless, gene therapy research and clinical trials continue.

B. Antisense RNA and Gene Knockouts Block Gene Expression

Sometimes it is useful to block the expression of a gene, perhaps to silence a harmful gene or to learn a gene's normal function. Antisense and knockout technologies do this; both have potential applications in agriculture and health care.

Antisense RNA

Like DNA, RNA can form a double-stranded molecule. Ribosomes cannot translate double-stranded RNA, however, and cells normally destroy RNA in this form. Artificially adding an RNA sequence complementary to a messenger RNA therefore blocks a gene's expression (this type of gene inactivation is called RNA interference, or RNAi). mRNA is sometimes called "sense" RNA; its complement is therefore called "antisense" RNA (**figure 12.22**). Theoretically, antisense RNA can squelch the activity of any gene, if the RNA can persist long enough and if it can be delivered to the appropriate tissue.

Gene Knockouts

Knockout technology blocks a gene's function by replacing a normal copy of the gene with a disabled version.

In mice, researchers perform the alteration on isolated cells of a very early embryo. These genetically altered cells are then transferred to embryos and implanted into the uterus of a female mouse. Breeding the resulting mice to each other then yields some individuals who have two copies of the knocked-out gene in every cell.

Researchers can compare knockout organisms with their normal counterparts to learn the deleted gene's function. Researchers routinely knock out mouse genes that have disease-causing counterparts in humans to better understand how the disease arises. Knockouts have also yielded surprises, when animals with genes thought to be vital do quite well without them. Such results suggest that gene functions can be redundant; that is, if one gene is disabled, another might substitute for it.

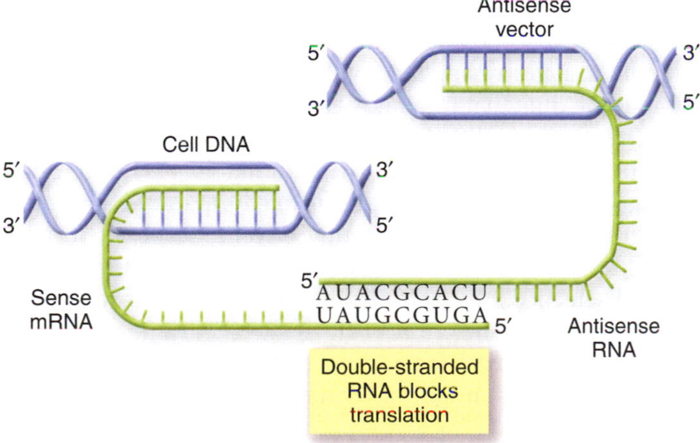

FIGURE 12.22 Silencing Gene Expression. Antisense RNA is a complementary nucleic acid sequence that bonds with mRNA, thereby preventing protein synthesis.

C. DNA Microarrays Can Help Monitor Gene Expression

A **DNA microarray,** also known as a "DNA chip," is a collection of short DNA fragments of known sequence placed in tens of thousands of defined spots on a small square of glass or other inert material. Chapter 7 described the use of DNA microarrays in gene sequencing. The same tool can monitor gene expression in different cell types or in cells under different conditions.

As a simple example, suppose that a researcher wants to know exactly how gene expression differs between kidney and liver cells. The first step would be to make fluorescently tagged cDNA copies of the mRNAs in both types of cells. The tagged DNA would then be applied to a chip containing sequences representing the human genome. Dots of light would appear wherever the sample DNA matches the chip's DNA. A computer then analyzes the pattern of colored dots on the square.

DNA microarrays promise to individualize medicine. A DNA chip for leukemia, for example, scans expressed genes in a patient's white blood cells. The pattern reveals the cancer subtype, whether the person's cells will admit a particular drug, whether the drug will be safe and effective, and how the person's immune system is likely to respond to both the cancer and the drug. Similar DNA chips identify which antibiotics are most likely to kill a strain of bacteria causing an infection, allowing targeted treatment.

D. Proteomics Studies an Organism's Entire Protein Output

Proteomics is an extension of genomics. An organism's **proteome** is all of the proteins that it expresses, either under defined conditions or throughout its entire life. Although an organism's genome changes little throughout its life, the combination of proteins that its cells produce depend on the cell type, the organism's stage of development, and the influence of other gene products and the environment.

Proteomic studies are important for basic research into cell biology, but they also promise medical applications. Understanding how breast cancer cells differ from normal cells, for example, may reveal new targets for anticancer drugs.

Summary *Gene therapy means replacing a faulty gene with a normally functioning copy. Antisense RNA and gene knockouts suppress the activity of specific genes. DNA microarrays help researchers monitor the activity of multiple genes simultaneously.*

12.8 MASTERING CONCEPTS

1. How does gene therapy work, and why is it difficult to accomplish?
2. How do antisense RNA and gene knockouts silence genes?
3. How are DNA microarrays useful?
4. What is the difference between a genome and a proteome?

12.9 INVESTIGATING LIFE

Clues to the Origin of Language

FIGURE 12.23
Can We Talk? DNA from humans, other primates, and mice contains small but critical differences.

As you chat with your friends and study for your classes, you may take language for granted. Although communication is not unique to humans, a complex spoken language does set us apart from other organisms. Every human society has language; without it, people could not transmit information from one generation to the next, so culture could not develop. Its importance to human evolutionary history is therefore incomparable. But how and when did such a crucial adaptation arise?

In the early 1990s, scientists described a family with a high incidence of an unusual language disorder. Affected family members had difficulty controlling the movements of their mouth and face, so they could not pronounce sounds properly. They also had lower intelligence compared with unaffected individuals, and they had trouble applying simple grammatical rules. Later studies confirmed the presence of brain abnormalities. Researchers traced the language disorder to one mutation in a single gene on chromosome 7. (Overblown media reports incorrectly dubbed this "the language gene" or "the grammar gene," even though many genes control language capabilities.) Inheritance patterns suggested that the mutated allele was dominant. Therefore, a person exhibiting the disorder had to inherit a mutated copy of the gene from only one affected parent.

Further research revealed that the gene belongs to the large *fo*rkhead bo*x* family of genes, abbreviated *FOX*. All members of the *FOX* family code for transcription factors, proteins that bind to DNA and control the expression of other genes. The "language gene" on chromosome 7, eventually named *FOXP2*, is not solely responsible for language acquisition. But the fact that it is a transcription factor explains how it can simultaneously affect both muscle control and brain structure.

To learn more about the evolution of language, scientists Wolfgang Enard, Svante Pääbo, and colleagues at Germany's Max Planck Institute and at the University of Oxford compared the amino acid sequence of the FOXP2 protein in humans, several other primates, and mice (**figure 12.23**). Out of 715 amino acids in the protein, the human version differs from the mouse's by only three amino acids (**figure 12.24**). Chimpanzees, gorillas, and the rhesus macaque monkey all have identical FOXP2 proteins; their version differs from the mouse's by only one amino acid. This result showed that in the 70 million or so years since the mouse and primate lineages split, the FOXP2 protein changed by only one amino acid. Yet in the 5 or 6 million years since humans split from the rest of the primates, the *FOXP2* gene changed twice.

Initially, the new, human-specific *FOXP2* version would have been rare, as are all mutations. Today, however, nearly everyone is homozygous for the same allele of *FOXP2*, meaning that the new allele somehow became "fixed" in the human population. This observation raised an intriguing question. Did the new, human-specific *FOXP2* allele confer such improved language skills that individuals with the allele consistently produced more offspring than those without it? That is, did natural selection "fix" the new, beneficial allele in the growing human population?

To detect natural selection that occurred in the past, scientists search for evidence of a so-called selective sweep. Recall from chapter 11 that the closer together two linked alleles are on the same chromosome, the less often crossing-over separates them. As natural selection quickly eliminated other, less beneficial alleles of *FOXP2* from the human population, a selective sweep should have simultaneously reduced genetic variability near *FOXP2*'s location on chromosome 7.

Enard and his colleagues tested this hypothesis by sequencing part of chromosome 7 from 20 humans from around the world. They found that the amount of variation around *FOXP2* was unusually low compared with the variation around hundreds of other genes. This result suggested that a selective sweep fixed the *FOXP2* allele, along with nearby "hitchhiking" regions of chromosome 7, in the human population. The research team used mathematical models to estimate that the original mutation happened within the past 200,000 years—around the time that humans became anatomically modern.

The study of *FOXP2* is important because it helps us understand a critical period in human history. The gene changed after humans diverged from chimpanzees, and then individuals with the new, highly advantageous allele produced more offspring than those with any other version. By natural selection, the new allele quickly became fixed in the human population. Without those events, human communication and culture (including everything you chat about with your friends) might never have happened.

Enard, Wolfgang, Molly *Przeworski*, Simon E. *Fisher*, and five coauthors, including Svante Pääbo. August 22, 2002. Molecular evolution of *FOXP2*, a gene involved in speech and language. *Nature*, volume 418, pages 869–872.

```
Human      MNQESATETI SNSSMNQNGM STLSSQLDAG SRDGRSSGDT SSEVSTVELL HLQQQALQA  ARQLLLQQQD SGLKSPKSSD KQRPLQVPVS
Chimp      .......... .......... .......... .......... .......... .......... .......... .......... ..........
Gorilla    .......... .......... .......... .......... .......... .......... .......... .......... ..........
Orangutan  .....V.... .......... .......... .......... .......... .......... .......... .......... ..........
Rhesus     .......... .......... .......... .......... .......... .......... .......... .......... ..........
Mouse      .......... .......... .......... .......... .......... .......... .......... .........E ..........

Human      VAMMTPQVZT PQQMQQILQQ QVLSPQQLQA LLQQQAVML  QQQQLQEFYK KQQEQLHLQL LQQQQQQQQ  QQQQQQQQQQ QQCQ-QQQQQ
Chimp      .......... .......... .......... .......... .......... .......... .......... .......... ....Q.....
Gorilla    .......... .......... .......... .......... .......... .......... ....-..... ...-...... ....-.....
Orangutan  .......... .......... .......... .......... .......... .......... ....-..... .......... ....-.....
Rhesus     .......... .......... .......... .......... .......... .......... ....-..... .......... ....-.....
Mouse      .......... .......... .......... .......... .......... .......... ....-..... .......... ....Q.....

Human      QQQQQQQCQQ QQHPGKQAKE QQQQQQQQQQ LAAQQLVFQQ QLLQMQQLQQ QQHLLSLQRQ GLISIPPGQA ALPVQSLPQA GLSPAEIQQL
Chimp      .......... .......... .......... .......... .......... .......... .......... .......... ..........
Gorilla    .......... .......... .......... .......... .......... .......... .......... .......... ..........
Orangutan  .......... .......... .......... .......... .......... .......... .......... .......... ..........
Rhesus     .......... .......... .......... .......... .......... .......... .......... .......... ..........
Mouse      .......... .......... .......... .......... .......... .......... .......... .......... ..........

Human      WKEVTGVESM EDNGIKHGGL DLTTNNSSST TSSNTSKASP PITHHSIVNG QSSVLSARRD
Chimp      .......... .......... .......... ...T...... .......... .....N....
Gorilla    .......... .......... .......... ...T...... .......... .....N....
Orangutan  .......... .......... .......... ...T...... .......... .....N....
Rhesus     .......... .......... .......... ...T...... .......... .....N....
Mouse      .......... .......... .......... ...T...... .......... .....N....
```

FIGURE 12.24 FOXP2 Protein Compared. This figure shows the first 330 amino acids of the primate and mouse versions of the 715 amino acid long FOXP2 protein. The one-letter amino acid abbreviations correspond to those in table 12.2. **Question**: Scientists use genetic mutations to discover the function of proteins in animal and plant development. What insights could scientists gain by intentionally disrupting the *FOXP2* gene in a developing human? Would such an experiment be ethical?

CHAPTER SUMMARY

12.1 DNA Stores Genetic Information

A. DNA Is a Helix of Nucleotides: A Review

- DNA is a double helix consisting of nucleotides. Hydrogen bonds between adenine and thymine, and between cytosine and guanine, hold the two strands together.

B. RNA Is an Intermediary Between DNA and a Polypeptide Chain

- A **gene** is a stretch of DNA that is transcribed to RNA. To produce a protein, a cell **transcribes** the gene's information to mRNA, which is **translated** into a sequence of amino acids.
- RNA is similar to DNA, but it contains uracil and ribose rather than thymine and deoxyribose.
- Three types of RNA (**mRNA, rRNA,** and **tRNA**) participate in gene expression.

12.2 Transcription Uses a DNA Template to Create RNA

A. The Steps of Transcription Are Initiation, Elongation, and Termination

- Transcription begins when **RNA polymerase** binds to a **promoter** on the DNA **template strand.** RNA polymerase then builds an RNA molecule. Transcription ends when RNA polymerase reaches a **terminator** sequence in the DNA.

B. mRNA Is Altered in the Nucleus of Eukaryotic Cells

- After transcription, the cell adds a cap and poly A tail to mRNA. **Introns** are cut out of RNA, and the remaining **exons** are spliced together.

12.3 Translation Builds the Protein

A. The Genetic Code Links mRNA to Protein

- The correspondence between codons and amino acids is the **genetic code.**
- Each group of three consecutive mRNA bases is a **codon** that specifies one amino acid (or a stop codon).
- Experiments with synthetic mRNA enabled scientists to match each codon with its corresponding amino acid.

B. Translation Requires mRNA, tRNA, and Ribosomes

- mRNA carries a protein-encoding gene's information. rRNA associates with proteins to form **ribosomes,** which support and help catalyze protein synthesis.
- On one end, tRNA has an **anticodon** sequence complementary to an mRNA codon; the corresponding amino acid binds to the other end.

C. Translation Occurs in Three Steps

- Translation begins when mRNA joins with a small ribosomal subunit and a tRNA, usually carrying methionine.
- In the elongation stage, a large ribosomal subunit joins the small one. A second tRNA binds to the next codon, and its amino acid bonds with the methionine that the first tRNA brought in. The ribosome moves down the mRNA as the chain grows.
- Upon reaching a "stop" codon, the ribosome is released, and the new polypeptide breaks free.

D. Proteins Must Fold Correctly after Translation

- Chaperone proteins help fold the polypeptide, which may be shortened or combined with others.

12.4 Protein Synthesis Is Highly Regulated

- Protein synthesis requires substantial energy input because large, ordered molecules are created from many small components.

A. Operons Are Groups of Bacterial Genes that Share One Promoter

- In bacteria, **operons** coordinate expression of grouped genes whose encoded proteins participate in the same metabolic pathway. *E. coli's* **lac operon** is a well-studied example. Transcription does not occur if a **repressor** protein binds to the **operator** sequence of the DNA.

B. Eukaryotic Organisms Use Transcription Factors to Turn Genes On or Off

- In eukaryotic organisms, proteins called **transcription factors** bind to DNA and regulate which genes a cell transcribes.

C. Eukaryotic Cells Also Use Additional Regulatory Mechanisms

- Other regulatory mechanisms include inactivating regions of a chromosome; alternative splicing; and control over mRNA stability, translation, and protein folding and movement.

12.5 Mutations Create New Alleles

- A **mutation** adds, deletes, alters, or moves nucleotides.

A. Mutations Range from Silent to Devastating

- A **point mutation** alters a single DNA base. The resulting mRNA may encode a different amino acid or substitute a "stop" codon for an amino acid–coding codon. Point mutations can also be "silent."
- Altering the number of nucleotides in a gene (a **frameshift mutation**) may disrupt the reading frame, altering the amino acid sequence of the encoded protein.
- **Expanding repeat mutations** cause some inherited illnesses.

B. What Causes Mutations?

- A gene can mutate spontaneously, particularly if it contains regions of repetitive DNA sequences. **Mutagens,** such as chemicals or radiation, induce mutations.
- Problems in meiosis can cause mutations if portions of chromosomes are deleted, inverted, or moved.

C. Some Mutations Pass to Future Generations

- A **germline mutation** originates in cells that give rise to gametes and therefore appears in every cell of the offspring. A **somatic mutation** occurs in nonsex cells and affects a subset of cells in the body, but not the offspring.

D. Mutations Are Important for Many Reasons

- Mutations create new alleles, which are the raw material for evolution.
- Induced mutations help scientists deduce gene function and help plant breeders produce new varieties of fruits and flowers.

12.6 The Human Genome Sequence Reveals Unexpected Complexity

- Only 1.5% of the 3.2 billion base pairs of the human genome encode protein, yet those 25,000 or so genes specify hundreds of thousands of distinct proteins.
- Alternative splicing of introns explains how a set number of genes can encode a larger number of proteins.
- The 98.5% of the human genome that does not encode protein encodes RNA, control sequences, **pseudogenes, transposable elements,** and other repeats.

12.7 Transgenic Organisms Contain DNA from More than One Species

A. Transgenic Organisms Have Many Uses

- **Transgenic** organisms are important in industry, research, and agriculture.

B. Creating Transgenic Organisms Requires Cutting and Pasting DNA

- **Restriction enzymes, cloning vectors** such as **plasmids,** and **reverse transcriptase** are tools that help researchers create **recombinant DNA** and introduce it to recipient cells.

- Several methods induce cells to take up recombinant DNA and become transgenic.

12.8 Biotechnology Has Many Practical Applications

A. Gene Therapy Replaces Faulty Genes

- **Gene therapy** requires placing a functional gene into cells expressing a faulty gene.

B. Antisense RNA and Gene Knockouts Block Gene Expression

- Silencing specific genes can treat illness or help researchers understand gene function.

C. DNA Microarrays Can Help Monitor Gene Expression

- **DNA microarrays** help researchers visualize the expression of many genes simultaneously.

D. Proteomics Studies an Organism's Entire Protein Output

- An organism's **proteome** is the set of proteins that it actually expresses.

12.9 Investigating Life: Clues to the Origin of Language

- A family with a language disorder led researchers to discover a gene that is apparently involved in the acquisition of language.
- Comparing the human version of the gene with that in other primates suggests that the gene apparently began evolving rapidly soon after modern humans arose. Eventually one allele became fixed in the human population.

MULTIPLE CHOICE QUESTIONS

1. Which of the following is NOT a type of RNA?
   a. Messenger RNA
   b. Transcription RNA
   c. Ribosomal RNA
   d. Transfer RNA

2. Choose the mRNA sequence that is complementary to the gene sequence GGACTTACG.
   a. CCTGAATGC
   b. AACUGGCUA
   c. GGTCAATCG
   d. CCUGAAUGC

3. The segments of a eukaryotic mRNA that are translated into protein are called
   a. promoters.
   b. introns.
   c. exons.
   d. caps.

4. What might happen if you changed one nucleotide in a codon?
   a. The protein would stop being made.
   b. The protein would have the wrong amino acid sequence.
   c. There would be no effect on the protein.
   d. All of the above are possible.

5. What is the job of the tRNA during translation?
   a. It carries amino acids to the mRNA.
   b. It triggers the formation of a covalent bond between amino acids.
   c. It binds to the small ribosomal subunit.
   d. It triggers the termination of the protein.

6. What could cause the *lac* operon to shut off after it has been activated?
   a. The binding of the sugar lactose to the promoter
   b. The inactivation of RNA polymerase by the addition of a modified sugar
   c. The re-binding of the repressor to the operator after all the lactose is degraded
   d. The binding of the repressor to the promoter

7. Which of the following types of mutations would alter the sequence of amino acids in a protein?
   a. Point mutation
   b. Frameshift mutation
   c. Expanding repeat mutation
   d. All of the above are possible.

8. What is a pseudogene?
   a. A gene that encodes an rRNA or a tRNA
   b. A sequence of DNA involved in the regulation of gene expression
   c. A gene that has been disrupted by a transposon
   d. A gene that is transcribed, but not translated

9. What type of mutation is caused by the insertion of a transposon?
   a. A point mutation
   b. A frameshift mutation
   c. A germline mutation
   d. An expanding repeat mutation

10. Which biotechnological method would you use to temporarily stop a gene from being expressed?
    a. Antisense RNA
    b. Gene knockout
    c. Cloning
    d. Transgenic

TESTING YOUR KNOWLEDGE

1. List the differences between RNA and DNA.
2. Define and distinguish between transcription and translation.
3. List the three major types of RNA and their functions.
4. During what stage of the eukaryotic cell cycle does protein synthesis occur?
5. Where in a eukaryotic cell do transcription and translation occur?
6. List the sequences of the mRNA molecules transcribed from the following template DNA sequences:
   a. T T A C A C T T G C T T G A G A G T T
   b. G G A A T A C G T C T A G C T A G C A
7. Given the following partial mRNA sequences, reconstruct the corresponding DNA template sequences:
   a. G U G G C G U A U U C U U U U C C G G G U A G G
   b. A G G A A A A C C C C U C U U A U U A U A G A U
8. Refer to the figure to answer these questions:
   a. Label the mRNA and the tRNA molecules, and draw in the ribosomes.
   b. What are the next three amino acids to be added to peptide *b*?
   c. Fill in the codons in the mRNA complementary to the template DNA strand.
   d. What is the sequence of the DNA complementary to the template strand (as much as can be determined from the figure)?
   e. Does this figure show the end of the peptide that this gene encodes? How can you tell?
   f. What might happen to peptide *b* after its release from the ribosome?
9. What are some ways that cells regulate gene expression?
10. A protein-encoding region of a gene has the following DNA sequence:

    G T A G C G T C A C A A A C A A A T C A G C T C

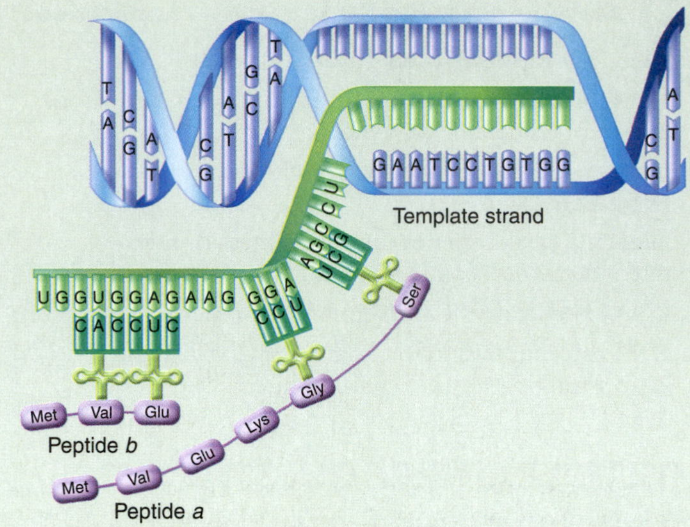

Determine how each of the following mutations alters the amino acid sequence:

a. substitution of a T for the C in the 10th position
b. substitution of a G for the C in the 19th position
c. insertion of a T between the 4th and 5th DNA bases
d. insertion of a GTA between the 12th and 13th DNA bases
e. deletion of the first DNA nucleotide

11. What are two ways that genomes are more complex than had been thought, based on knowledge from sequenced genomes?
12. How do researchers create recombinant DNA and transgenic organisms, and what are some applications of this technology?
13. Define gene therapy, antisense RNA, gene knockout, and DNA microarray.

THINKING AS A SCIENTIST

1. Put the following objects in order from smallest to largest: nucleotide, nitrogenous base, gene, nucleus, cell, codon, chromosome.
2. The pea plant phenotypes that Gregor Mendel studied trace their origins to mutations. One gene in pea plants, for example, enables the plant to produce a stem-elongating hormone. The wild-type allele encodes a tall plant. How could a mutation in that gene produce a short plant? Why is the mutant allele recessive?
3. Unneeded genes in an adult animal cell are permanently inactivated, making it impossible for most specialized cells to turn into any other cell type. How does this arrangement save energy? Why does the ability to clone an adult mammal depend on techniques for reactivating these "dormant" genes?
4. How can a mutation alter the sequence of DNA bases in a gene but not produce a noticeable change in the gene's polypeptide product? How can a mutation alter the amino acid sequence of a polypeptide yet not alter the phenotype?

5. Point mutations usually occur during interphase, but most chromosomal abnormalities arise during meiosis. Review how gamete formation differs in males and females (see chapter 37). Given these differences, why is it reasonable to predict that more point mutations occur during sperm production, and more chromosomal abnormalities appear in egg cells?
6. Which biotechnology might be able to accomplish the following goals? More than one answer may be possible.
   a. Shut off HIV genes integrated into the chromosomes of people with HIV infection (which leads to AIDS).
   b. Create bacteria that produce human growth hormone, used to treat extremely short stature.
7. Many patients waste precious time taking anticancer drugs that are ineffective or too toxic. How might DNA microarray technology refine the treatment of cancer?
8. In the early 1900s, scientists began to experiment with radiation as a cancer treatment. Many physicians who administered the treatment subsequently died of cancer. Why?

GENETICS PROBLEMS

See answers in Appendix D.

1. Titin is a muscle protein whose gene has the largest known coding sequence—80,781 DNA bases. How many amino acids long is titin?
2. On the television program *The X Files,* Agent Scully discovers an extraterrestrial life form that has a triplet genetic code, but with five different bases instead of the four of earthly inhabitants. How many different amino acids can this code specify?
3. A mutation in a gene that encodes PGR-interacting protein causes vision loss. The protein is 1259 amino acids long. What is the minimum size of this gene?
4. Parkinson disease causes rigidity, tremors, and other motor symptoms. Only 2% of cases are inherited, and these tend to have an early onset of symptoms. Some inherited cases result from mutations in a gene that encodes the protein parkin, which has 12 exons. Indicate whether each of the following mutations in the parkin gene would result in a smaller protein, a larger protein, or not change the size of the protein.
   a. deletion of exon 3
   b. deletion of six consecutive nucleotides in exon 1
   c. duplication of exon 5
   d. disruption of the splice site between exon 8 and intron 8
   e. deletion of intron 2

ARIS™ **Visit www.mhhe.com/hoefnagels for practice quizzes, animations, videos, and activities designed to help you master the material in this chapter.**

5. Genomics uses computer algorithms that search DNA sequences for indications of specialized functions. Explain the significance of detecting the following sequences:
   a. a promoter
   b. a sequence of 75 to 80 nucleotides that folds into a cloverleaf shape
   c. a gene with a sequence very similar to that of a known protein-encoding gene but that is not translated into protein
   d. RNAs with poly A tails
6. In gyrate atrophy, cells in the retina begin to degenerate in late adolescence, causing night blindness that progresses to blindness. The cause is a mutation in the gene that encodes an enzyme, ornithine aminotransferase (OAT). Researchers sequenced the *OAT* gene for five patients with the following results:
   - Patient A: A change in codon 209 of UAU to UAA
   - Patient B: A change in codon 299 of UAC to UAG
   - Patient C: A change in codon 426 of CGA to UGA
   - Patient D: A two-nucleotide deletion at codons 64 and 65 that results in a UGA codon at position 79
   - Patient E: Exon 6, including 1071 nucleotides, is entirely deleted.

   a. Which patient(s) have a frameshift mutation?
   b. How many amino acids is patient E missing?
7. Consult the genetic code to write codon changes that could account for the following changes in amino acid sequence.
   a. tryptophan to arginine
   b. glycine to valine
   c. tyrosine to histidine

CHAPTER 13

The Forces of Evolutionary Change

The Antibiotic Revolution. Penicillin, a fungus-produced antibiotic compound, revolutionized health care during World War II by enabling people to survive otherwise fatal bacterial infections. Penicillin destroys the cell walls of susceptible bacteria. Today, penicillin and many other antibiotics are losing their effectiveness because bacteria have become drug-resistant. The reason for antibiotic resistance lies in the basic concepts of evolution.

What's the Point?

The Rise of Antibiotic Resistance

Do you owe your life to antibiotics? If you have never had a serious infection, you may not think so. Chances are that one or more of your ancestors did, however, and that antibiotics saved their lives. Had it not been for these "wonder drugs," you may never have been born.

Many antibiotics are naturally occurring chemicals. Soil fungi and bacteria secrete these compounds into their surroundings, giving them an edge against microbial competitors. Biologists discovered antibiotics in the early 1900s, but it took decades for chemists to figure out how to mass-produce them. Once that occurred, antibiotics revolutionized medical care in the twentieth century and enabled people to survive many once deadly bacterial infections.

Unfortunately, the miracle of antibiotics is under threat. The overuse and misuse of these drugs in medicine is one culprit. Some physicians prescribe antibiotics for viral infections (often under pressure from patients), and many patients do not take the drugs as directed. The other half of the problem comes from agriculture. Producers of cattle, chickens, and other animals use antibiotics to treat and prevent disease, even adding small amounts to the animals' feed to promote growth. Antibiotics also prevent and treat infections in fruit- and vegetable-producing plants.

The widespread use of antibiotics for these purposes has profoundly affected the evolution of bacteria, many of which are now resistant to the drugs that once defeated them. The drugs create a situation in which only some bacteria flourish, while the rest die. If a particular antibiotic dismantles a molecule in a bacterial cell wall, for example, a resistant mutant might have a slightly different version of the molecule that the antibiotic does not alter. The drug kills the susceptible bacteria and leaves behind the resistant ones. The survivors multiply, producing a new generation of antibiotic-resistant bacteria. This is an example of natural selection. **antibiotics, p. 374**

Antibiotic-resistant bacteria appeared just 4 years after these drugs entered medical practice in the late 1940s, and researchers responded by discovering new drugs. But the microbes kept pace. Today, 40% of hospital Staphylococcus *infections resist all antibiotics but one, and already some laboratory strains are resistant to all antibiotics. Researchers fear that the discovery of new antibiotics will not keep up with the global spread of resistant strains.*

Natural selection is just one mechanism of evolution, a process that is ongoing in every species. This chapter explains how evolution occurs.

LEARNING OUTLINE

13.1 Evolutionary Thought Has Evolved for Centuries

Scientific reasoning has profoundly changed human thinking about our own origins. Just 250 years ago, people believed Earth was about 6000 years old. A century later, scientists accepted evidence that Earth is much older (millions of years old or more), but still believed that a Creator made all life on Earth in its present form. Contemporary scientists, using evidence from many fields of research, now accept evolution as the best explanation for life's diversity.

But what *is* evolution? **Evolution** occurs in a population when some alleles (versions of a gene) become more common, and others less common, from one generation to the next. In other words, evolution is genetic change in a population over time. (A population is a group of interbreeding members of a species that live in the same area; see figure 1.2.) As the chapters in this unit will repeatedly demonstrate, evolution is everywhere, and it is obvious in many ways. It serves as such a compelling conceptual framework for many observations about life that geneticist Theodosius Dobzhansky gave this title to a much quoted article he wrote: "Nothing in biology makes sense except in the light of evolution."

Evolution does not, however, answer one question that fascinates many people: how life began in the first place. Because little evidence remains from life's ancient origin, this question is difficult to answer scientifically; chapter 16 describes some of what we do know.

A. Many Explanations Have Been Proposed for Life's Diversity

People have tried to explain the diversity of life for a very long time (**figure 13.1**). In ancient Greece, Aristotle (384 BCE to 322 BCE) recognized that all organisms are related in a hierarchy of simple to complex forms, but he believed that all members of a species were created identical in form and capacity. This idea influenced scientific thinking for nearly 2000 years.

Several other ideas were also considered fundamental principles of science well into the 1800s. Among them was the concept of a "special creation," the sudden appearance of organisms on Earth. People believed that this creative event was planned and purposeful, that species were fixed and unchangeable, and that Earth was relatively young. The idea of a special creation also implied that there could be no extinctions.

What about Fossils?

Scientists struggled to reconcile these beliefs with compelling evidence that species could in fact change. Fossils, which had been discovered at least as early as 500 BCE, were at first thought to be oddly shaped crystals or faulty attempts at life that arose spontaneously in rocks. By the mid-1700s, the increasingly obvious connection between organisms and fossils argued against these ideas.

To explain how fossils came to be, yet not deny the role of a creator, scientists suggested that fossils represented organisms killed during the Biblical flood. Yet some of the fossils depicted organisms not seen before. Because people believed that species created by God could not become extinct, these fossils presented a paradox. The conflict between ideology and observation widened as geologists discovered that different rock layers revealed different groups of fossils, all now extinct. How could this be?

New Ideas from Geology

In 1749, French naturalist Georges-Louis Buffon (1707 to 1788) became one of the first to openly suggest that closely

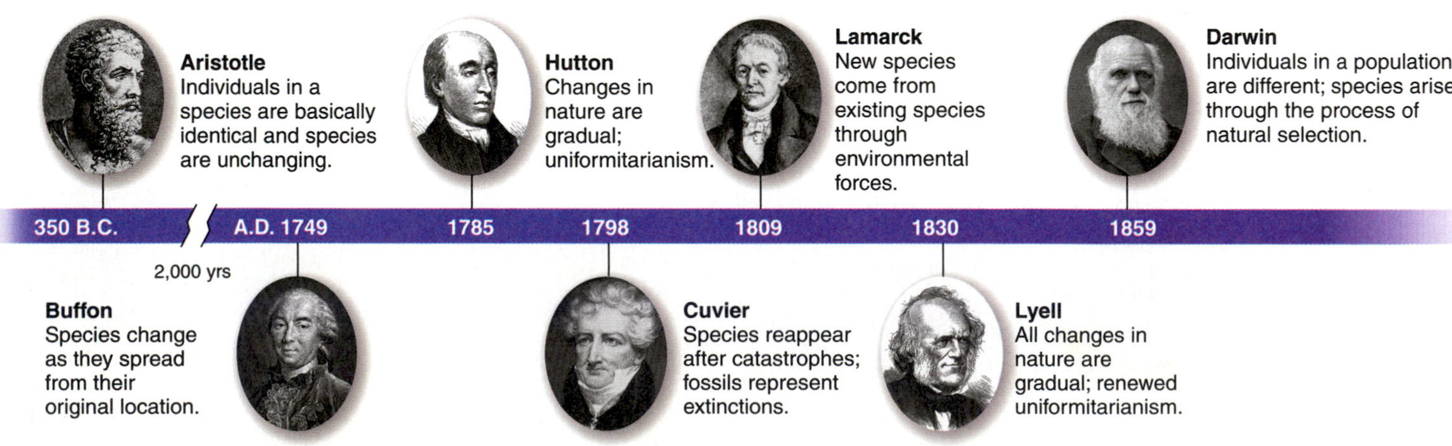

FIGURE 13.1 History of Evolutionary Theory Before Darwin. Many scientists made significant contributions, over many years, to develop the foundation Darwin used to describe natural selection as the mechanism for evolution.

FIGURE 13.2 Rock Layers Reveal Earth's History. Layers of sedimentary rock formed from sand, mud, and gravel that were deposited in ancient seas. Such sediment layers are visible along the Grand Canyon. The rock layers on the bottom are older than those on top. Rock strata sometimes contain fossil evidence of organisms that lived (and died) when the layer was formed, providing clues about when the organism lived.

related species arose from a common ancestor and were changing—a radical idea at the time. By moving the discussion into the public, he made possible a new consideration of evolution and its causes from a scientific point of view.

Meanwhile, in the 1700s and 1800s, much of the study of nature focused on geology. In 1785, physician James Hutton (1726 to 1797) proposed the theory of **uniformitarianism,** which suggested that the processes of erosion and sedimentation that act in modern times have also occurred in the past, producing profound changes in Earth over time.

On the other side, Georges Cuvier (1769 to 1832) was convinced of **catastrophism,** the theory that a series of brief upheavals such as floods, volcanic eruptions, and earthquakes were responsible for most geological formations.

Cuvier also used his knowledge of anatomy to identify fossils and to describe the similarities among organisms. He was the first to recognize the **principle of superposition**—the idea that lower layers of rock (and the fossils they contain) are older than those above them (**figure 13.2**). Although he had to accept that some species must have become extinct, he refused to believe that they were not originally formed through creation. He argued that catastrophes would destroy most of the organisms in an area, but then new life would arrive from surrounding areas.

Early Ideas about the Origin of Species

Once fossils were recognized as evidence of extinct life, it became clear that species could in fact change. Still, no one had proposed how this might happen. Then, in 1809, French taxonomist Jean Baptiste de Lamarck (1744 to 1829) proposed a radical new theory. He reasoned that organisms that used one part of their body repeatedly would increase their abilities, very much like weight lifters developing strong arms. He proposed that the resulting changes in individuals would give them the ability to get more food in a changing environment. To explain how those traits were passed to the next generation, Lamarck applied the (then) accepted theory of the **inheritance of acquired characteristics**—the idea that an organism can inherit the traits that its parent acquired during its lifetime. By this theory, for example, the offspring of wading birds have long legs because their parents stretched their own legs to keep their bodies above the water. This mechanism is absurd in light of what we know today about genetics, but Lamarck remains important because he was the first to suggest that animals could change or become extinct in response to interactions with their environment.

Geologist Charles Lyell (1797 to 1875) renewed the argument for uniformitarianism in 1830, suggesting that natural processes are slow and steady. One obvious conclusion from his contribution is that gradual changes in some organisms could be represented in successive fossil layers. Lyell was so persuasive that many scientists began to support the idea of gradual geologic change.

B. Charles Darwin's Voyage Provided a Wealth of Evidence

With these new theories and ideas, people were beginning to accept the concept of evolution but could not understand how it could result in the formation of new species. Ultimately, Charles Darwin (1809 to 1882) recognized their application to the changing diversity of life on Earth. Darwin was the son of a physician and grandson of noted physician and poet Erasmus Darwin. He attended Cambridge University in England and, at the urging of his family, completed studies to enter the clergy.

Meanwhile, he also followed his own interests. He joined several geological field trips and met several eminent geology professors. Eventually Darwin was offered a position aboard the HMS *Beagle.* Before the ship set sail for its 5-year voyage in 1831 (**figure 13.3**), the botany professor who had arranged Darwin's position gave the young man the first volume of Lyell's *Principles of Geology.* Darwin picked up the second and third volumes in South America. By the time he finished reading, Darwin was an avid proponent of uniformitarianism.

He recorded his observations as the ship journeyed around the coast of South America. He noted forces that uplifted new land, such as earthquakes and volcanoes, and the constant erosion that wore it down. He marveled at forest plant fossils interspersed with sea sediments, and at shell fossils in a mountain cave. Darwin tried to reconstruct the past from contemporary observations and wondered how each fossil had arrived where he found it.

He was particularly aware of similarities and differences among organisms. If there had been a single special creation, then why was one sort of animal or plant created to live on a mountaintop in one part of the world, yet another type on mountains elsewhere? Even more puzzling was the resemblance between organisms living in similar habitats in different parts of the world. We now know that such species have undergone **convergent evolution,** which means that they have similar characteristics because they evolved in similar environmental conditions, although they are not closely related (**figure 13.4**; also see figure 14.10).

In the fourth year of the voyage, the HMS *Beagle* spent a month in the Galápagos Islands, off the coast of Ecuador. The notes and samples he brought back would form the seed of Darwin's theory of evolution by natural selection.

C. The *Origin of Species* Proposed Natural Selection as an Evolutionary Mechanism

Toward the end of the voyage, Darwin began to assimilate all he had seen and recorded. Pondering the great variety of organisms in South America and their relationships to fossils and geology, he began to think that these were clues to how species originate.

Descent with Modification

Darwin returned to England in 1836, and by 1837 began assembling his notes in earnest. In March 1837, Darwin consulted ornithologist (bird expert) John Gould about the finches the *Beagle* brought back from the Galápagos Islands. Gould could tell from bill structures that some of the birds ate small seeds, whereas others ate large seeds, fruits, or insects. In all, he described 14 distinct types of finch, each different from the finches on the mainland, yet sharing some features.

Darwin thought that the different varieties of finch on the Galápagos had probably descended from a single ancestral type of finch that had flown to the islands and, finding a relatively unoccupied new habitat, flourished. Gradually, the finch population branched in several directions, with different groups eating insects, fruits, and seeds of different sizes, depending on the resources each island offered. Darwin noted similar changes in the length of the Galápagos tortoise's neck. He coined the phrase **descent with modification** to describe this gradual change from an ancestral type.

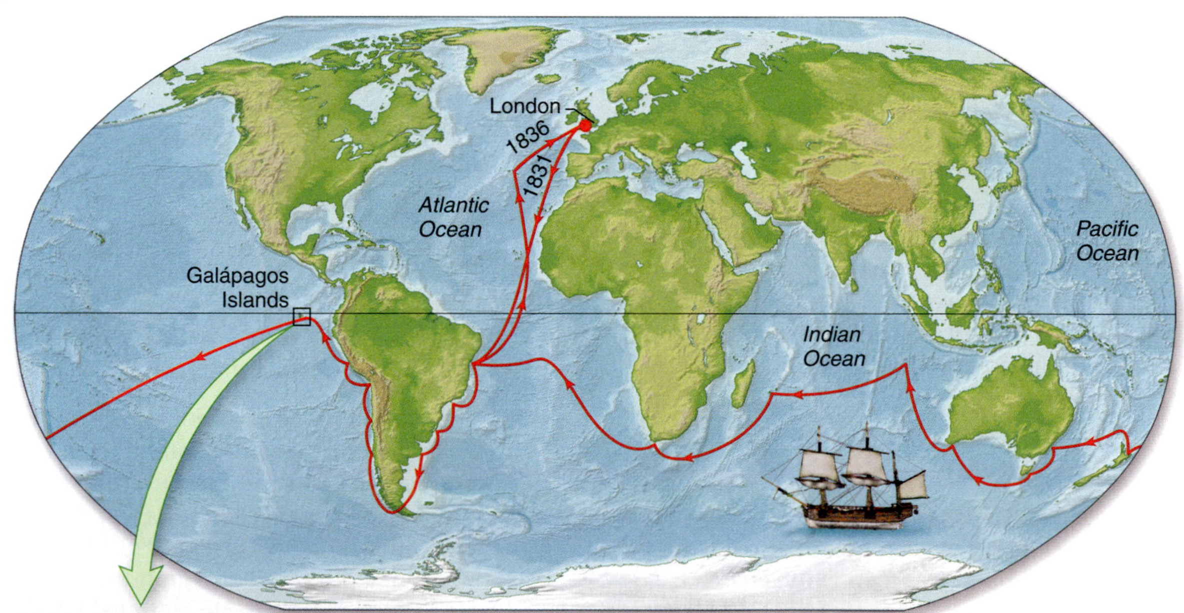

FIGURE 13.3 The Voyage of the *Beagle.* Darwin formulated his theories about organisms and evolution by observing life and geology throughout the world during the journey of the HMS *Beagle.* Many of his ideas had their origins in the observations Darwin made of the different species inhabiting the Galápagos Islands.

a.

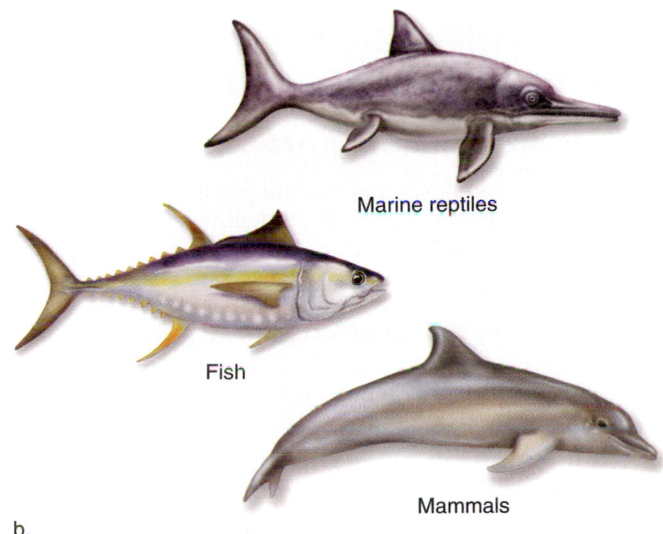

b.

FIGURE 13.4 Convergent Evolution. (**A**) Plants in different parts of the world have similar mechanisms of protection, as this ocotillo from California and allauidi from Madagascar illustrate. (**B**) Each of these three animals has a back fin, flippers, a tail with two lobes, and a streamlined body, all adaptations to aquatic habitats. Yet the animal on the top is an ancient reptile (an ichthyosaur), the middle animal is a fish, and the animal on the bottom is a dolphin, a mammal.

Malthus's Ideas on Populations

In September 1838, Darwin read a work that helped him understand the diversity of finches on the Galápagos Islands. Economist and theologian Thomas Malthus's *Essay on the Principle of Population,* written 40 years earlier, stated that food availability, disease, and war limit the size of a human population. Wouldn't populations of other organisms face similar limitations? If so, then individuals who could not obtain essential resources would die.

The insight Malthus provided was that individual members of a population were not all the same, as Aristotle had taught. Instead, individuals better able to obtain resources were more likely to survive and reproduce. This would explain the observation that more individuals are produced in a generation than survive; they do not all obtain enough vital resources to live. Over time, environmental challenges would "select" out the more poorly equipped variants, and gradually, the population would change.

The Concept of Natural Selection

Darwin used the term *natural selection* to describe "this preservation of favourable variations and the rejection of injurious variations." Biologists later modified the definition to add modern genetics terminology, defining **natural selection** as the differential reproductive success of individuals with particular genotypes. Darwin got the idea of natural selection from thinking about **artificial selection** (also called selective breeding). In artificial selection, a human chooses one or a few desired traits, such as milk production or seed size, and then allows breeding of only the individuals that best express those qualities. Artificial selection is responsible for many breeds of domestic dogs and cats (see Can *You Relate?* Dogs Are Products of Artificial Selection) and agriculturally important varieties of plants and animals (**figure 13.5**). Darwin himself raised pigeons and created several new breeds by artificial selection.

Natural selection explained the diversity of finches on the Galápagos. Originally, some finches flew from the

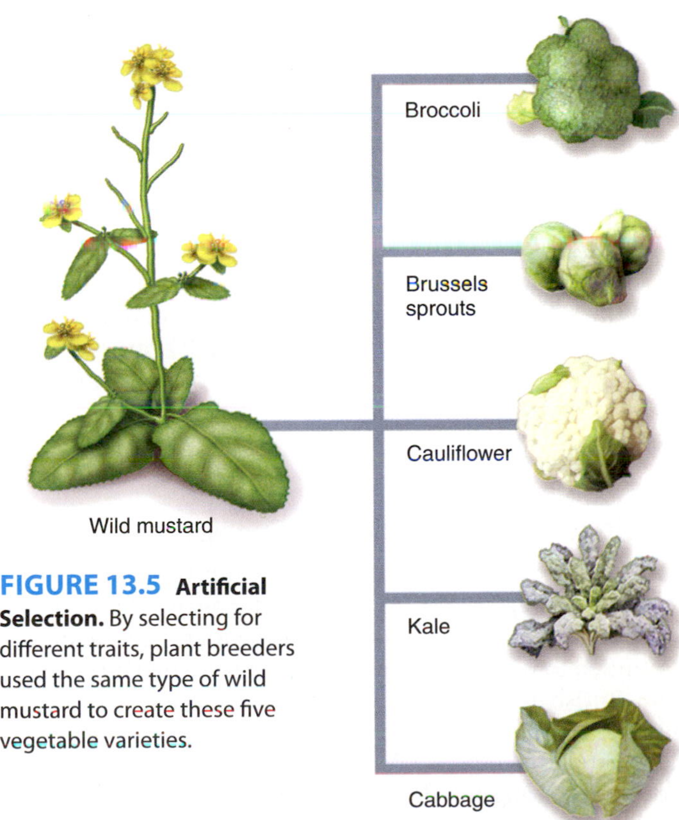

FIGURE 13.5 Artificial Selection. By selecting for different traits, plant breeders used the same type of wild mustard to create these five vegetable varieties.

Can *You* Relate?

Dogs Are Products of Artificial Selection

The pampered poodle and graceful greyhound may win in the show ring, but they are poor specimens in terms of genetics and evolution. Behind carefully bred traits lurk small gene pools and extensive inbreeding, all of which may harm the health of purebred show animals (**table 13.A**).

The runny, sad eyes of the basset hound can be quite painful (**figure 13.A**). Short legs make the dog prone to arthritis, the long abdomen promotes back injuries, and the characteristic floppy ears often hide ear infections. The eyeballs of the Pekingese protrude so much that a mild bump can pop them out of their sockets. The tiny jaws and massive teeth of pugs and bulldogs cause dental and breathing problems, sinusitis, and their notorious "dog breath." Folds of skin on their abdomens easily become infected. Larger breeds, such as the Saint Bernard, have bone problems and short life spans. A Great Dane may suddenly die at a young age, its heart overworked from years of supporting a large body.

All of these examples make one truth clear: We may be able to breed desired characteristics into a dog, but we can't always breed other traits out.

Table 13.A *Purebred Plights*

Breed	Health Problems
Cocker spaniel	Nervousness, ear infections, hernias, kidney problems
Collie	Blindness, bald spots, seizures
Dalmatian	Deafness
German shepherd	Hip dysplasia
Golden retriever	Lymphatic cancer, muscular dystrophy, skin allergies, hip dysplasia, absence of one testicle
Great Dane	Heart failure, bone cancer
Labrador retriever	Dwarfism, blindness
Shar-pei	Skin disorders

FIGURE 13.A Dog Traits Come from Genetic Variations Selected Through Breeding. The mournful expression of the basset hound accompanies its heightened sense of smell, and the bulldog was selected for its flattened face and fierce demeanor. These traits originally occurred as natural genetic variation.

mainland to one island. When that first island population grew too large for all individuals to obtain enough small seeds, those who could eat nothing else starved. Birds that could eat other things, perhaps because of an inherited quirk in bill structure, survived. Since the new food was plentiful, these once-unusual birds gradually came to make up more of the population. Because each of the islands had slightly different habitats, different varieties of finches predominated on each one.

Publication of *The Origin of Species*

Darwin described his theory of evolution by natural selection in a 35-page sketch in 1842, and 2 years after that as a 230-page analysis. He did not publish either account, but he continued to work on his ideas until 1858, when he received from British naturalist Alfred Russel Wallace a manuscript entitled *On the Tendency of Varieties to Depart Indefinitely from the Original Type.* Like Darwin, Wallace had observed evolution among the diverse species of South America and southeast Asia. Darwin submitted his own paper, along with Wallace's, to the Linnaean Society meeting later that year. In 1859, Darwin finally published the 490-page-long *On the Origin of Species by Means of Natural Selection, or Preservation of Favoured Races in the Struggle for Life.* It would form the underpinning of modern life science.

Although some members of the scientific community happily embraced Darwin's efforts, others were less appreciative. Some of his ideas were perceived to clash with religious beliefs that all life arose from separate special creations, that species did not change, and that nature is harmonious and purposeful. Perhaps most disturbing to many people was the idea that humans were just one more species competing for resources.

Despite these objections, overwhelming evidence supports the theory of evolution by natural selection. Contemporary biologists therefore accept evolution as the best explanation for the fact that diverse organisms use the same genetic code, the same chemical reactions to extract energy from nutrients, and many of the same (or very similar) enzymes and other proteins. Descent from a common ancestor explains both this great unity of life and the spectacular diversity of organisms today. Coupled with a wide variety of changing habitats and enormous amounts of time, the result of natural selection is a planet packed with millions of variations of the same underlying biochemical theme.

Summary *Various thinkers attempted to explain the appearances and disappearances of different types of organisms over vast amounts of time. In the nineteenth century, people began to notice how changes on Earth might have influenced life. Charles Darwin formulated his theory of the origin and diversification of species by natural selection based on his own observations and the ideas of others.*

13.1 MASTERING CONCEPTS

1. What is evolution?
2. What were some of the ways that people thought species arose and diversified before Charles Darwin published his theory of evolution by natural selection?
3. What did Darwin observe about the distribution of organisms and geology that led him to develop his theory of the origin of species by natural selection?
4. How is artificial selection different from natural selection?
5. How did Darwin's ideas challenge prevailing beliefs about life's diversity and the status of humans?

13.2 Natural Selection Molds Evolution

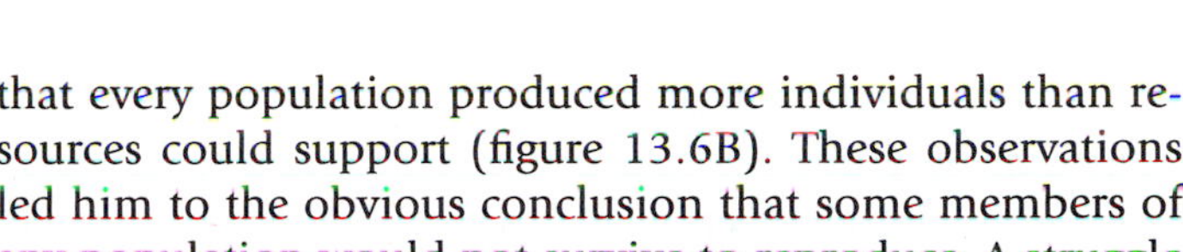

Biological evolution includes large-scale events such as the appearance of new species, the topic of chapter 14. **Microevolution,** the relatively short-term changes in allele frequencies within a population or species, occurs on a much smaller scale. Natural selection is the most famous, and often the most important, mechanism of microevolution; section 13.4 describes several others.

A. Selection Results from Differential Reproductive Success

Darwin envisioned natural selection as the mechanism whereby new species arise through modifications of existing species. He observed that organisms of the same species are different from one another (**figure 13.6A**), and that every population produced more individuals than resources could support (figure 13.6B). These observations led him to the obvious conclusion that some members of any population would not survive to reproduce. A struggle for existence was inevitable.

Darwin further observed that each population always had some individuals who were better than others at obtaining nutrients and water, avoiding predators, tolerating temperature changes, attracting mates, and reproducing. He recognized that these traits must be inherited from parents. Darwin concluded that those with the best **adaptations**—features that provide a selective advantage because they improve the organism's ability to survive and reproduce—would pass that advantage to their offspring. Over time, because of this "differential reproductive success," the population would change. The best adaptations

a.

b.

FIGURE 13.6 **Requirements for Natural Selection.** (**A**) This basketball player and referee illustrate genetic variation within the human population. (**B**) Dandelions produce many offspring, but few survive.

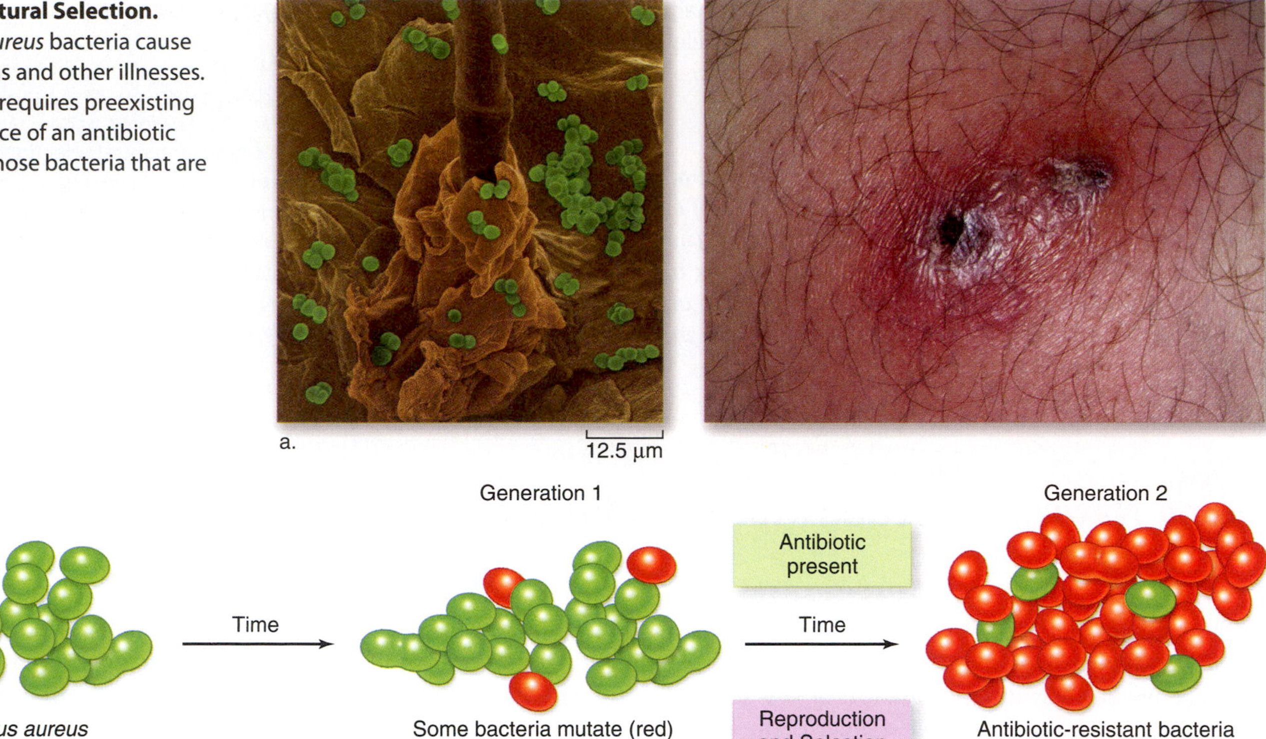

FIGURE 13.7 Natural Selection. (**A**) *Staphylococcus aureus* bacteria cause serious skin infections and other illnesses. (**B**) Natural selection requires preexisting variation. The presence of an antibiotic strongly selects for those bacteria that are immune to the drug.

to the existing environment would become more common (**figure 13.7**).

Darwin reasoned that changes in the environment, such as a drought or the availability of a new food source, put some individuals at a disadvantage and allowed others to survive or even flourish. A new species might arise when a population adapted to so many new conditions that its members could no longer breed with the original group (see chapter 14). Throughout the history of life, more and more species evolved as populations adapted to new and different resources. This was the genius of Darwin's theory. He had provided a simple, clear explanation for evolution that accounted for the observations, even though he knew nothing about genetics. **Table 13.1** summarizes Darwin's main ideas.

Natural selection requires genetic diversity, which arises largely by chance. Although Darwin could not know it at the time, the ultimate source of genetic variation is mutation (described in chapter 12 and section 13.4). Mutations (changes in DNA sequence) occur at random in all organisms, both asexual and sexual. In organisms that reproduce sexually, each generation reshuffles parental alleles to produce genetically different offspring (see chapter 9).

Although genetic variation arises at random, natural selection itself is not random. Instead, it selectively eliminates most of the individuals that are least able to compete for resources or cope with the prevailing environment.

B. Natural Selection Eliminates Certain Phenotypes

Although Mendel and Darwin were contemporaries, Mendel's work remained obscure until after Darwin's death. Sci-

Table 13.1 *Darwin's Main Ideas*

Observations of Nature
1. Organisms are varied, and some variations are inherited. Within a species, no two individuals (except identical siblings) are exactly alike.
2. More individuals are born than survive to reproduce.
3. Individuals compete with one another for the limited resources that enable them to survive.
Inferences from Observations
1. Within populations, the inherited characteristics of some individuals make them more likely to survive and produce fertile offspring.
2. Because of the environment's selection against nonadaptive traits, only individuals with adaptive traits live long enough to transmit their genes to the next generation. Over time, natural selection can change the characteristics of populations, even giving rise to new species.

entists have learned much more about genetic variation, its origin, and its spread since that time.

A population's **gene pool** is its entire collection of genes and their alleles. The proportion of different alleles for each gene determines the characteristics of that population. A Swedish population, for example, might include a large proportion of hair color alleles conferring blondness; a population of Asians would have very few, if any, such alleles but would have many alleles conferring darker hair. When allele frequencies change (e.g., if Swedes migrate to Asia and interbreed with the locals), evolution happens. Because an individual cannot change his or her alleles, *evolution occurs in populations, not individuals.*

By "weeding out" individuals with poorly adapted phenotypes, natural selection changes allele frequencies in the population. Recall from chapter 10 that an individual's phenotype is its observable properties, most of which arise from a combination of environmental influences and the action of multiple genes. (Only the genetic portion, however, is subject to natural selection.)

A phenotype that is adaptive in one set of circumstances may become a liability in another. Consider the finches on the tiny Galápagos Island of Daphne Major (**figure 13.8**). In a very dry season in the early 1980s, birds with large beaks were more likely to survive because they could eat the large, tough seeds that remained after the small seeds were depleted. In 1983, however, many small seeds accumulated following 8 months of extremely heavy rainfall. Over the next 2 years, small-beaked finches, which could easily eat the tiny seeds, came to predominate in the population. Constantly changing conditions mean that evolution never really stops.

Burning Questions

Why doesn't natural selection produce one superorganism?

Natural selection cannot produce one "perfect" organism that is supremely adapted to every possible habitat on Earth. The simple reason is that the adaptations that seem "perfect" in one habitat would be completely wrong in another. To take an extreme example, a trout's adaptations that work so well in a cold mountain stream are useless in the sands of the Sahara. The variety of habitats on Earth—from oceans, to freshwater, to the tundra, prairie, desert, and forests—is just too great for one species to be able to thrive everywhere.

Some people believe that evolution actually has produced a superorganism: humans. True, we have the intelligence, dexterity, and cultural background to live on every continent on Earth. But humans can only visit Earth's waters and the highest mountains, areas where other organisms thrive, for brief periods. And few organisms can live in the extreme heat of Earth's interior.

Nevertheless, organisms that have demonstrated a remarkable ability to thrive in a variety of habitats may threaten Earth's ecosystems. Weeds such as cheat grass and dandelions, and pests such as cockroaches and rats, crowd out native species and appear to be able to live anywhere. Yet even they can't live where the weather is too hot, too cold, too dry, or too wet. Because different habitats have such different conditions, it seems unlikely that any organism will ever evolve with the combination of traits that would enable it to live everywhere.

Have a Burning Question of your own? Submit it to marielle_hoefnagels@mcgraw-hill.com for possible inclusion in future editions of this book!

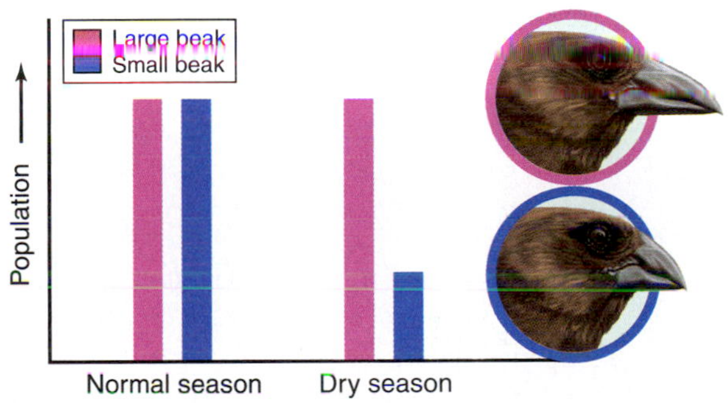

FIGURE 13.8 Finch Beak Shape Reflects Natural Selection. Since the early 1970s, Princeton University researchers Peter and Rosemary Grant have continued Darwin's observations of changes in the finch populations of the Galápagos. The Grants captured, banded, and recaptured medium ground finches (*Geospiza fortis*) and monitored their beak sizes. The average beak size in a population (an inherited trait) can change appreciably in as short a time as a year. Following a very dry season, when seeds were sparse, the small, easy-to-crack ones were eaten rapidly. Birds with large beaks survived because only they were strong enough to open the large, tough-to-crack seeds.

(Learn more about the finches on Daphne Major in chapter 15's Investigating Life: The Shrinking, Growing Beaks of Darwin's Finches Reveal Ongoing Evolution).

C. Natural Selection Does Not Have a Goal

Because species have become more complex over life's long evolutionary history, many people erroneously believe that natural selection leads to ever more "perfect" organisms, or that evolution works toward some long-term goal. (Some people also wonder whether evolution could produce one organism that could thrive in every habitat, a topic discussed in this chapter's Burning Question: Why doesn't natural selection produce one superorganism?.)

Evolution, however, does not have a goal. How could it? No known mechanism allows the environment to tell DNA how to yield the alleles needed to confront future conditions. Nor does natural selection strive for perfection; if it did, the vast majority of species in life's history would still exist. Instead, most are extinct.

FIGURE 13.9 Extinction. Sea scorpions once thrived worldwide. These animals became extinct some 250 million years ago, during one of Earth's several major mass extinction events.

Several factors combine to prevent natural selection from producing all of the traits that a species might find useful. First, every genome has limited potential, imposed by its evolutionary history. The structure of the human skeleton, for example, will not allow for the sudden appearance of wheels, no matter how useful they might be on paved roads. Second, no gene pool contains every allele needed to confront every possible change in the environment. If the right alleles aren't available at the right time, an environmental change may quickly wipe out a species (**figure 13.9**). Third, disasters like floods and volcanic eruptions can indiscriminately wipe out the best allele combinations, simply by chance. And finally, some harmful genetic traits are out of natural selection's reach, such as diseases that appear only after reproductive age.

D. What Does "Survival of the Fittest" Really Mean?

Natural selection is often called "the survival of the fittest," but this phrase is not entirely accurate or complete. In everyday language, the "fittest" individual is the one in the best physical shape: the strongest, fastest, or biggest. Physical fitness, however, is not the key to natural selection (although it may play a part). Rather, in an evolutionary sense, **fitness** refers to an organism's contribution to the next generation's gene pool. A large, quick, burly elk scores zero on the evolutionary fitness scale if poor eyesight makes it vulnerable to an early death in the jaws of a predator. On the other hand, a mayfly that dies in the act of producing thousands of offspring is highly fit.

These examples illustrate an important point: *by itself, survival is not enough*. Because successful reproduction is the only way for an organism to perpetuate its genes, fitness depends on the ability to survive just long enough to reproduce. Plants that germinate, grow, flower, produce seeds, and die within just a few weeks may have fitness equal to a redwood that lives for centuries (**figure 13.10**).

Many adaptations contribute to an organism's overall fitness. The ability to overcome poor weather conditions, combat parasites and pathogens (see chapter 9's Investigating Life: An Arms Race at a Snail's Pace), evade predators, and compete for resources all enhance an organism's chances of reaching reproductive age. At that point, the ability to attract mates (or pollinators in the case of many flowering plants) affects the number of offspring an organism produces. But fitness includes not only the total number of offspring produced, but also the proportion that reach reproductive age. Some organisms have few offspring, but invest large amounts of energy in each one. Others produce thousands of young, but invest minimally in each. Chapter 39 further describes this evolutionary tradeoff between "quality" and "quantity."

a.

b.

FIGURE 13.10 Fitness Is Reproductive Success. One key to fitness is living long enough to reproduce. (**A**) For a redwood, that may take centuries. (**B**) In an annual plant, it may take just a few weeks.

Successful reproduction is so important that an adaptation that gives a male a mating advantage will persist, even though that trait may virtually guarantee his death. For example, the male praying mantis does not resist if the female begins to eat his head during copulation. The male's passive behavior is adaptive because the extra food the female obtains in this way will enhance the chance of survival for their young. Although he does not survive, his alleles will.

E. There Are Three Modes of Natural Selection

Different modes of natural selection—directional, disruptive, and stabilizing—are distinguished by their effects on phenotypes (**figure 13.11**). Natural selection may favor the intermediate phenotype (stabilizing selection) or one or more extreme phenotypes (directional and disruptive selection).

Directional Selection

In **directional selection,** one extreme phenotype is fittest, and the environment selects against the others (see figure 13.11A). For example, populations of approximately 100 insect species have undergone color changes enabling them to blend into polluted backgrounds. The rise of antibiotic resistance among infection-causing bacteria also reflects directional selection, as does the increase in herbicide-resistant plants. In directional selection, the fittest phenotype may initially be rare. Its frequency increases in the population over multiple generations as the environment changes—for example, after exposure to the antibiotic or herbicide.

Disruptive Selection

In **disruptive selection** (sometimes called diversifying selection), two or more extreme phenotypes are fitter than the intermediate phenotype. The extreme phenotypes therefore come to predominate (see figure 13.11B). For example, in a population of marine snails that live among tan rocks encrusted with white barnacles, the animals near the barnacles are white and camouflaged, and those on the bare rock are tan and likewise blend in. The snails that are not white or tan, or that lie against the oppositely colored background, are more often seen and eaten by predatory shorebirds.

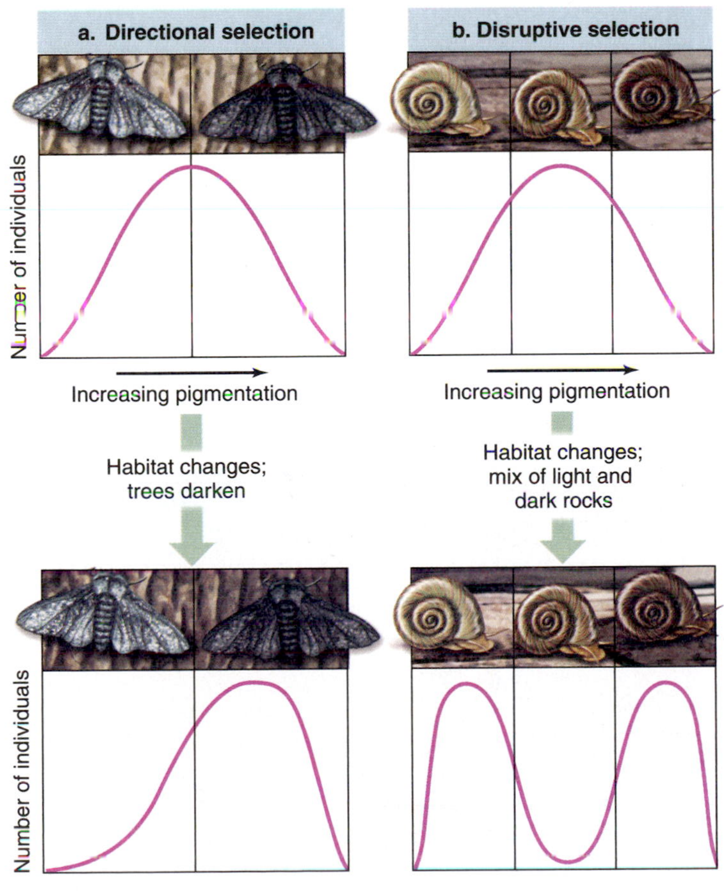

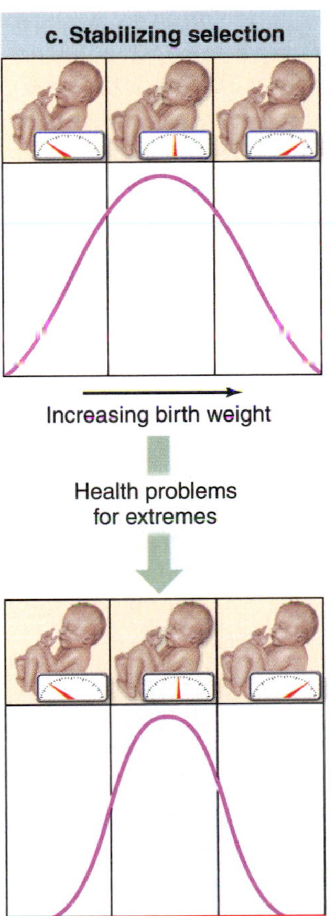

FIGURE 13.11 Types of Natural Selection. (**A**) Directional selection results from selection against one extreme phenotype. (**B**) In disruptive selection, two extreme phenotypes each have a selective advantage, and both persist. (**C**) Stabilizing selection maintains an intermediate expression of a trait by selecting against extreme variants.

Stabilizing Selection

In a third form of natural selection, called **stabilizing selection** (or normalizing selection), extreme phenotypes are less fit than the optimal intermediate phenotype (see figure 13.11C). Human birth weight illustrates this tendency to stabilize. Very small or very large newborns are less likely to survive than babies of intermediate weight. By eliminating all but the individuals with the optimal phenotype, stabilizing selection tends to reduce the variation in a population. It is therefore most common in stable, unchanging environments.

F. Balanced Polymorphism Maintains Multiple Alleles for One Gene

The three models of natural selection might seem to suggest that, for each trait, only one or a few alleles ought to persist in the population. The disadvantageous alleles should gradually become less common until they disappear, while the most beneficial one should become "fixed" in the population (see, for example, chapter 12's Investigating Life: Clues to the Origin of Language). Instead, however, natural selection often maintains a **balanced polymorphism,** in which multiple alleles of a gene persist indefinitely in the population. (*Polymorphism* means "multiple forms").

How does natural selection allow seemingly harmful alleles of a gene to remain in a population? One way this can occur is called a **heterozygote advantage,** in which a heterozygous individual (one with two different alleles for the gene) has greater fitness than homozygotes, whose two alleles are identical. A heterozygote maintains the presence of a disease-causing allele in a population, even though the illness usually reduces the fitness of affected individuals (**table 13.2**).

The best known example of a heterozygote advantage is sickle cell disease, a disorder that causes anemia, joint pain, a swollen spleen, and frequent, severe infections. In all the red blood cells of a person with this disease, and in about half of the cells of a carrier, abnormal hemoglobin chains stick together to form aggregates that bend the cell into the characteristic sickle shape (see figure 12.14). People who are homozygous for the sickle cell allele may not live long enough to reproduce. Why doesn't natural selection remove this apparently harmful allele from the population? It is because heterozygotes, despite their slight anemia, have a reproductive advantage over homozygotes: they are resistant to malaria.

Malaria is an infection by any of four species of the protistan genus *Plasmodium*. When a mosquito carrying *Plasmodium* feeds on a human with normal hemoglobin, the parasite enters the red blood cells. Eventually, infected blood cells burst, spreading the parasite throughout the body. Sickle cell disease changes the shape of a person's red blood cells, making them break down more rapidly and therefore resisting the spread of the parasite. **malaria, p. 400**

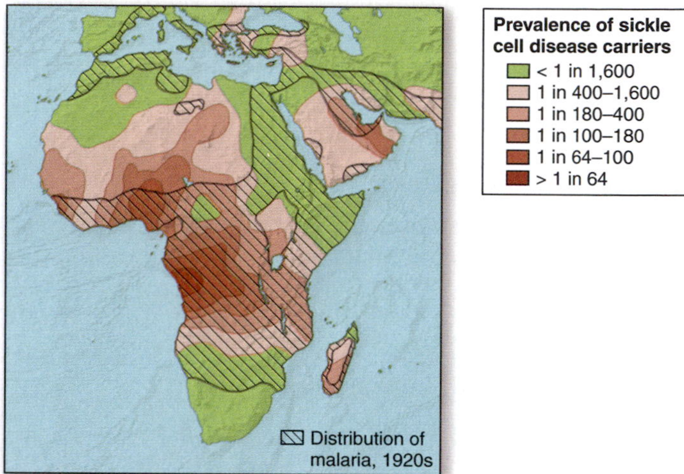

FIGURE 13.12 Heterozygote Advantage Produces Balanced Polymorphism. Being a carrier for an inherited illness can protect against another type of condition. A map that compares the frequency distributions of both disorders provides evidence of this phenomenon. The map of malaria prevalence, for example, overlaps closely with the distribution of people heterozygous for the gene associated with sickle cell trait, indicating that being a sickle cell carrier protects against malaria.

Table 13.2 *Balanced Polymorphism*

Person Who Has or Carries	Is Protected From	Possibly Because
Cystic fibrosis	Diarrheal disease	Carriers have too few functional chloride channels in intestinal cells, blocking toxin
G6PD deficiency	Malaria	Red blood cells inhospitable to malaria parasite
Phenylketonuria (PKU)	Miscarriage induced by ochratoxin A, a fungal toxin	Excess amino acid (phenylalanine) in carriers inactivates toxin
Sickle cell disease	Malaria	Red blood cells inhospitable to malaria parasite
Tay-Sachs disease	Tuberculosis	Unknown
Noninsulin-dependent diabetes mellitus	Starvation	Tendency to gain weight protects against starvation during famine

Anemia and malaria are two opposing selective forces that maintain balanced polymorphism for the sickle cell trait. In malaria-free areas, the sickle cell allele is rare, because anemia is the predominant selective force, and heterozygotes have no advantage. Where malaria rages, however, sickle cell carriers remain healthiest; they are resistant to malaria but not sick from sickle cell disease. In those areas, the frequency of the sickle cell allele is high, because carriers have more children than people homozygous for the "normal" hemoglobin allele (**figure 13.12**). Unfortunately, two carriers have a 25% chance of producing a child who suffers from sickle cell disease—a homozygote. These children pay the price for this genetic protection. (Cystic fibrosis, the topic of section 4.6, provides another example of a balanced polymorphism.)

Summary *Natural selection means differential reproductive success: some individuals contribute more alleles than others to the next generation. Natural selection is not goal-oriented; it simply eliminates some individuals from each generation, leaving those that are best adapted to the prevailing conditions. Natural selection may favor one phenotype, two extreme phenotypes, or an intermediate phenotype. Balanced polymorphism maintains multiple alleles in a population, as when a heterozygote has an advantage over homozygotes.*

13.2 MASTERING CONCEPTS

1. What is an adaptation, and how do adaptations become more common within a population?
2. How does genetic variation arise, and how does it contribute to differential reproductive success?
3. How can natural selection favor different phenotypes at different times?
4. Why doesn't natural selection produce perfectly adapted organisms?
5. What is evolutionary fitness?
6. Distinguish among directional, disruptive, and stabilizing selection.
7. How can being a carrier for an inherited disease be beneficial?

13.3 Evolution Is Inevitable in Real Populations

Shifting allele frequencies in populations are the small steps of change that collectively drive evolution. Given the large number of genes in any organism and the many factors that can alter allele frequencies (including but not limited to natural selection), evolution is not only possible but unavoidable.

Hardy–Weinberg equilibrium is the highly unlikely situation in which allele frequencies do not change from one generation to the next. It occurs only in populations that meet the following assumptions: (1) mutations do not occur (and thus no new alleles arise); (2) individuals mate at random; (3) individuals do not migrate into or out of the population; (4) the population is infinitely large, or at least large enough to eliminate random changes in allele frequencies (genetic drift); and (5) natural selection does not occur.

These conditions do not occur together in real populations. The concept of Hardy–Weinberg equilibrium is nevertheless important, because it serves as a basis of comparison to reveal when microevolution is occurring.

A. At Hardy–Weinberg Equilibrium, Allele Frequencies Do Not Change

Hardy–Weinberg equilibrium is named after mathematician Godfrey H. Hardy and physician Wilhelm Weinberg. In 1908, they independently proposed that the expression $p + q = 1$ could represent the frequency of both of the alleles $(p + q)$ for a gene in a population of diploid organisms, if only two alleles exist for that gene. For example, suppose that in a population of ferrets, the frequency for the dark fur allele (D) is 0.7; the frequency of the alternative allele d, which confers tan fur, is 0.3. The two frequencies add up to 1 because the two alleles represent all the possibilities in the population.

Knowing allele frequencies is important because the measure of evolution is a change in allele frequencies. We can also use allele frequencies to calculate genotype frequencies. If p represents the frequency of the allele D, then the proportion of the population with genotype DD equals p^2, which in the ferret example equals 0.7×0.7, or 0.49. Likewise, the proportion of population members with genotype dd equals 0.3×0.3 (q^2), or 0.09. To calculate the proportion of the population that is heterozygous, subtract the total proportion of homozygotes ($0.49 + 0.09 = 0.58$) from 1, which gives 0.42. That is, 42% of the ferrets in the population have genotype Dd and dark fur. This result is significant because we would be unable to determine the percentage of heterozygotes based on phenotype alone.

Hardy and Weinberg used an equation to represent the proportions of genotypes that make up a population:

$$p^2 + 2pq + q^2 = 1$$

In this equation, $2pq$ represents the frequency of the heterozygous class. Note that $2pq$ is equal to $2 \times 0.7 \times 0.3$,

p = frequency of D (dominant allele) = dark fur = 0.7
q = frequency of d (recessive allele) = tan fur = 0.3

Algebraic Expression	What It Means
$p + q = 1$	Frequency of all dominant alleles plus frequency of all recessive alleles for this gene.
$p^2 + 2pq + q^2 = 1$ ($DD + 2Dd + dd = 1$)	For a particular gene, the frequencies of all the homozygous dominant individuals (p^2) plus heterozygotes ($2pq$) plus all homozygous recessives (q^2) add up to all of the individuals in the population.

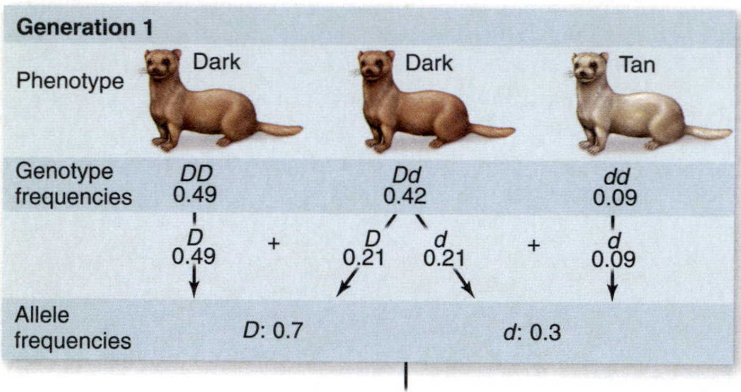

Population mates at random

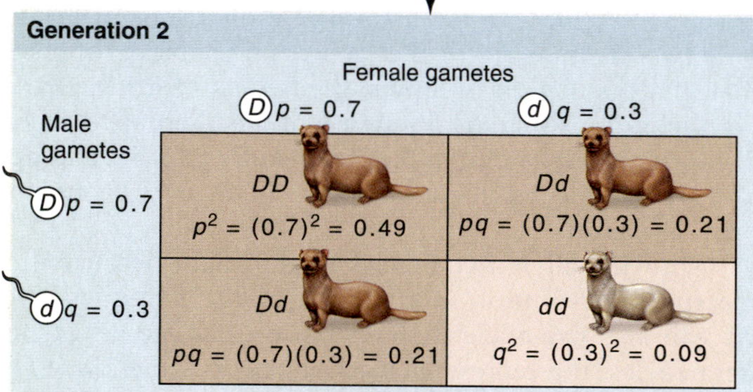

All possible crosses			Genotype frequency in offspring			Total
Male		Female	*DD*	*Dd*	*dd*	
0.49 *DD*	×	0.49 *DD*	0.2401			0.2401
0.49 *DD*	×	0.42 *Dd*	0.1029	0.1029		0.2058
0.49 *DD*	×	0.09 *dd*		0.0441		0.0441
0.42 *Dd*	×	0.49 *DD*	0.1029	0.1029		0.2058
0.42 *Dd*	×	0.42 *Dd*	0.0441	0.0882	0.0441	0.1764
0.42 *Dd*	×	0.09 *dd*		0.0189	0.0189	0.0378
0.09 *dd*	×	0.49 *DD*		0.0441		0.0441
0.09 *dd*	×	0.42 *Dd*		0.0189	0.0189	0.0378
0.09 *dd*	×	0.09 *dd*			0.0081	0.0081
			0.49	**0.42**	**0.09**	**1.0000**
			DD +	*Dd* +	*dd* = 1	

FIGURE 13.13 Hardy–Weinberg Equilibrium. At Hardy–Weinberg equilibrium, allele frequencies remain constant from one generation to the next; evolution does not occur. This figure depicts the random mating that underlies Hardy–Weinberg equilibrium. In the lower part of the figure, the "male" and "female" columns show every possible combination of gametes. The other columns show how genotype frequencies of the next generation are derived from the random matings.

or 0.42, the same result that we obtained by subtracting the frequencies of the two types of homozygotes (*DD* and *dd*) from 1.

If conditions of Hardy–Weinberg equilibrium are met, allele and genotype frequencies will not change in future generations (**figure 13.13**). When the ferrets produce gametes, the proportion of *D* alleles will equal that of the homozygous dominant (*DD*) class (0.49) plus one half of the gametes from the heterozygotes (*Dd*), which equals one half of 0.42, or 0.21. Therefore, the proportion of *D* alleles would be $p = 0.49 + 0.21 = 0.70$. If $p = 0.7$, then $q = 1 - p$, or 0.3. We are back at the beginning: in the next generation, the same proportion of ferrets will have dark fur, Hardy–Weinberg equilibrium persists, and evolution is not occurring.

B. In Reality, Allele Frequencies Always Change

A theoretical ferret population at Hardy–Weinberg equilibrium does not evolve. A real ferret population, however, violates some or all of the assumptions of Hardy-Weinberg equilibrium. Mutation, nonrandom mating, migration, genetic drift, and natural selection are common. In addition, no population is infinitely large. Natural populations never fulfill the conditions necessary for Hardy–Weinberg equilibrium. Allele frequencies always change over multiple generations, which is another way of saying that evolution is inevitable. The next section describes the other mechanisms of microevolution in more detail.

Summary *Evolution occurs at the population level as allele frequencies change. Algebra can be used to represent Hardy–Weinberg equilibrium, in which evolution does not occur. Natural populations are never at Hardy–Weinberg equilibrium.*

13.3 MASTERING CONCEPTS

1. What are the five conditions required for Hardy–Weinberg equilibrium?
2. Why is the concept of Hardy–Weinberg equilibrium important?
3. Explain the components and meaning of the equation $p^2 + 2pq + q^2 = 1$.
4. Why doesn't Hardy–Weinberg equilibrium occur in real populations?

13.4 Evolutionary Changes Occur in Several Ways

The effects of natural selection are easy to see in most populations, because they result in the adaptations that enhance survival and reproduction. This section describes four additional mechanisms of microevolution: mutation, nonrandom mating, gene flow, and genetic drift (**figure 13.14**). They all occur frequently, and each can, by

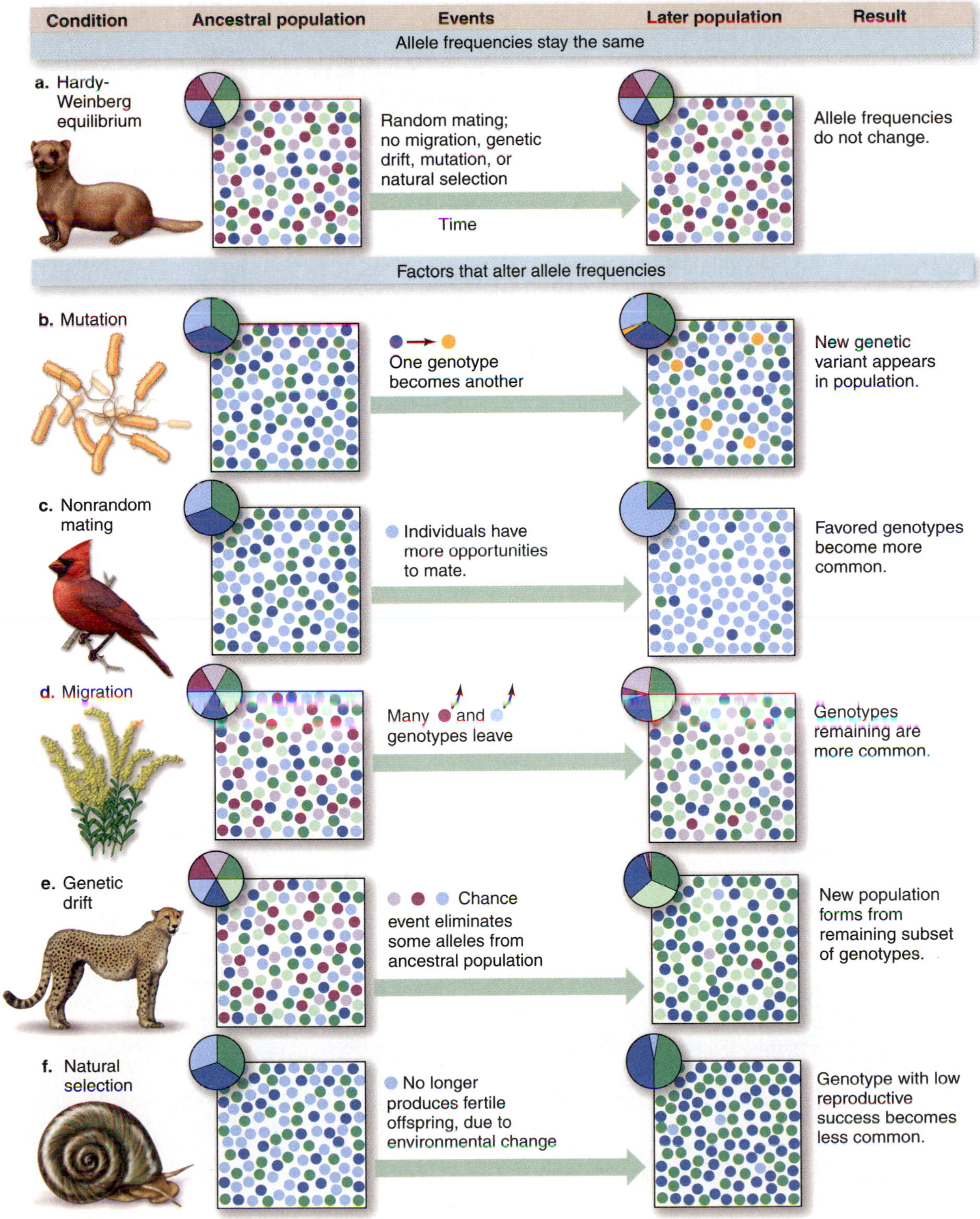

Condition	Ancestral population	Events	Later population	Result
Allele frequencies stay the same				
a. Hardy-Weinberg equilibrium		Random mating; no migration, genetic drift, mutation, or natural selection Time		Allele frequencies do not change.
Factors that alter allele frequencies				
b. Mutation		One genotype becomes another		New genetic variant appears in population.
c. Nonrandom mating		Individuals have more opportunities to mate.		Favored genotypes become more common.
d. Migration		Many and genotypes leave		Genotypes remaining are more common.
e. Genetic drift		Chance event eliminates some alleles from ancestral population		New population forms from remaining subset of genotypes.
f. Natural selection		No longer produces fertile offspring, due to environmental change		Genotype with low reproductive success becomes less common.

FIGURE 13.14
Factors That Alter Allele Frequencies and Thereby Contribute to Evolution. (**A**) At Hardy–Weinberg equilibrium, allele frequencies stay constant. (**B**) Mutation creates new alleles. (**C**) Nonrandom mating increases some allele frequencies and decreases others, because individuals with certain phenotypes are more attractive to the opposite sex. (**D**) Migration removes alleles from or adds alleles to populations. (**E**) Genetic drift randomly samples a portion of a population, altering allele frequencies. (**F**) Natural selection operates when environmental conditions prevent individuals with certain genotypes from reproducing successfully.

itself, disrupt Hardy–Weinberg equilibrium. The changes in allele frequencies that constitute microevolution therefore occur nearly all the time.

A. Mutation Fuels Evolution

A change in an organism's DNA sequence introduces a new allele to a population. Mutations are the raw material for evolution because genes contribute to phenotypes, and natural selection acts on phenotypes—for example, bacterial populations become resistant to antibiotics as described in the chapter opening essay. Mutations are also responsible for the emergence of new pathogens, such as HIV.

A common misconception is that a mutation produces a novel adaptation precisely when a population "needs" it to confront a new environmental challenge. For example, many people mistakenly believe that antibiotics *create* resistance; that is, that resistance arises in bacteria *in response* to exposure to the drugs. In reality, genes do not "know" when to mutate; the chance that a mutation will occur is independent of whether a new phenotype would benefit the organism. The only way antibiotic resistance arises is if some bacteria happen to have a mutation that confers antibiotic resistance *before* exposure to the drug. The drug creates a situation in which these variants can flourish. That trait will then become more common within the population by natural selection. If no bacteria start out resistant, the drug kills the entire population. Because bacterial populations are often enormous, however, it is likely that at least a few individuals carry such a mutation.

The rate at which mutations occur varies, both among different genes and within a gene. The average rate is around one DNA sequence change per 10^9 base pairs. At first, this number may seem too low to pose a significant force in evolution. Each genome, however, has an enormous number of base pairs, and a large number of cell divisions occur throughout life. In humans, for example, each genome is estimated to have at least 120 new mutations per generation.

A mutation affects evolution only if subsequent generations inherit it. In asexually reproducing organisms such as bacteria, each mutated cell gives rise to mutant offspring (if the mutation does not prevent reproduction). In a multicellular organism, however, a mutation can pass to the next generation only if it arises in a germ cell (i.e., one that will give rise to gametes; see chapter 12). For example, a cigarette smoker with lung cancer will not pass any smoking-induced mutations to his children, because his sperm cells will not contain the altered DNA.

B. Nonrandom Mating Concentrates Alleles Locally

To achieve and maintain Hardy–Weinberg equilibrium, a population must have completely random mating, in which each individual has an equal chance of mating with any other member of the population.

In reality, mating is rarely random. Many factors influence mate choice, including geographical restrictions, access to the opposite sex, and behavior. Most species also exhibit some form of preference in mate choice. **Sexual selection** is a type of natural selection resulting from variation in the ability to obtain mates (**figure 13.15**).

For example, the vivid feathers of male cardinals make the birds obvious to predators, which would seem to reduce their survival. But female cardinals prefer bright red males, so showy plumage directly increases a male's chances of reproducing. Because the brightest males get the most chances to reproduce, alleles that confer red plumage are common in the population.

Why do males usually show the greatest effects of sexual selection? In most (but not all) animal species, females

FIGURE 13.15 Sexual Selection. The individual who is most successful in attracting a mate and reproducing is the most fit. (**A**) The male bowerbird builds intricate towers of sticks and grass to attract a mate. (**B**) These Hercules beetles use horns to battle rivals for access to the hornless sex. (**C**) The male bird-of-paradise displays bright plumes and capes in his quest for sexual success. (**D**) A female long-horned, wood-boring beetle chooses her mate based on the size of his territory. A better territory means more food for offspring.

a.

b.

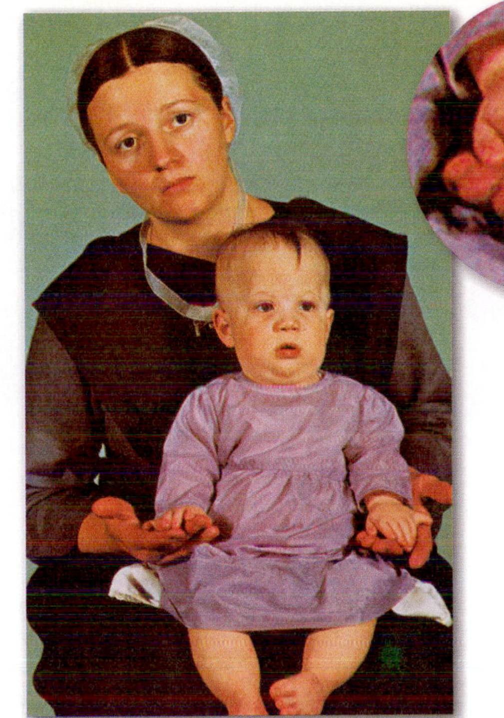

FIGURE 13.16 Nonrandom Mating Concentrates Alleles in Subpopulations. This Amish child from Lancaster County, Pennsylvania, has inherited Ellis-van Creveld syndrome. He has short-limbed dwarfism, extra fingers, heart disease, and fused wrist bones, and he had teeth at birth. Ellis-van Creveld is an autosomal recessive disorder that occurs in 7% of the people of this Amish community. The proportion is high because the Amish tend to marry among themselves.

spend more time and energy rearing each offspring than do males. Because of this high investment in reproduction, females tend to be selective about their mates. Males, which must compete for access to females, therefore usually show the greatest effects of sexual selection. The result of sexual selection is often **sexual dimorphism,** a difference in appearance between males and females. Common sexually dimorphic features include body size, coloration, and structures such as horns.

Humans also give great consideration to selecting mates; it is hardly a random process. Cultural factors are particularly important in choosing partners. The Amish people provide an extreme example of nonrandom mating. They tend to marry within their group, so some alleles occur more frequently among the Amish than in other human populations (**figure 13.16**).

The practice of artificial selection is another way to reduce random mating. Humans select those animals or plants that have a desired trait and then prevent them from mating with those lacking that trait. The result, as discussed in Can *You* Relate? Dogs Are Products of Artificial Selection, is a wide variety of subpopulations that humans maintain by selective breeding.

C. Gene Flow Moves Alleles Between Populations

Under Hardy–Weinberg equilibrium, no alleles ever leave or enter the population. In reality, **gene flow** moves alleles between populations. Migration is one common way that gene flow occurs. Departing members of a population take their alleles with them. Likewise, members entering a population potentially add new alleles. But gene flow does not require the movement of entire individuals. Wind can carry a plant's pollen for miles, for example, spreading one individual's alleles to a new population.

Large cities defy Hardy–Weinberg equilibrium by their very existence. Waves of immigration built the population of New York City, for example. The original Dutch settlers of the 1600s lacked many of the alleles present in today's metropolis; immigrants from other parts of Europe, Africa, Central and South America, and Asia introduced them. Within the city, pockets of ethnicity illustrate nonrandom mating as people have children with mates who are most like themselves.

Geographical barriers to migration can yield differences in allele frequencies between populations occupying

c.

d.

relatively close regions, such as on either side of a mountain range. For example, the allele frequencies for many genetic diseases vary among different European populations as a result of geographical barriers. As worldwide transportation removes barriers to migration, however, these regional differences should disappear.

D. Genetic Drift Occurs by Chance

Genetic drift is a change in allele frequencies that occurs purely by chance. To illustrate genetic drift, suppose a new allele appears in a sexually reproducing population. If the individual carrying the mutation fails to reproduce, the allele disappears again. Even if the individual does reproduce, however, the new allele still might not pass to the next generation—not because it affects fitness, but simply by chance. For the same reason, allele frequencies for any gene will fluctuate by chance from generation to generation. Hardy–Weinberg equilibrium requires populations to be very large (approaching infinity) to minimize the effects of these random fluctuations in allele frequencies.

The Founder Effect

Because genetic drift relies on the random loss of alleles, it is especially important in small populations. One cause of genetic drift is the **founder effect,** which occurs when small groups of individuals leave their home population and establish new settlements, mating only among themselves. If the new group has allele frequencies that are not representative of the original population, some traits that were rare in the original population may become more frequent in the new population. Likewise, other traits will become less frequent or even disappear.

A disease called porphyria (see figure 10.13) illustrates the founder effect in a human population in South Africa. In the 1680s, a Dutch immigrant to South Africa carried the allele that causes porphyria. Today, tens of thousands of South Africans have the allele, and all can trace its inheritance to the original immigrant from centuries ago.

The Bottleneck Effect

Genetic drift also may result from a **population bottleneck,** which occurs when many members of a population die, causing the loss of much of the genetic diversity that was present in the larger ancestral population. Even if the few remaining individuals mate and eventually restore the population's numbers, the loss of genetic diversity is permanent.

Cheetahs are currently undergoing a population bottleneck (**figure 13.17**). Until 10,000 years ago, these cats were common in many areas. Today, just two isolated populations live in South and East Africa, numbering only a few thousand animals. The South African cheetahs are so genetically alike that even unrelated animals can accept skin grafts from each other. Researchers attribute the genetic uniformity of cheetahs to two bottlenecks: one that occurred at the end of the most recent ice age, when habitats changed drastically, and another when humans slaughtered many cheetahs during the nineteenth century. The loss of genetic diversity among the cheetahs presents a potential disaster: a single change in the environment might doom them all.

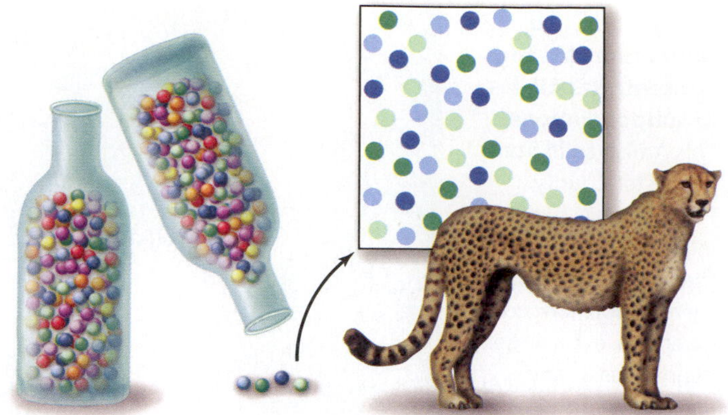
Original cheetah population with 25 different alleles of a particular gene.
Cheetah population drastically reduced.
Repopulation occurs. Only four different alleles remain.

FIGURE 13.17 The Bottleneck Effect. A population bottleneck occurs when the size of a genetically diverse population drastically falls, eliminating some alleles at random. Even if the population rebuilds, some genetic diversity is lost. The two dwindling cheetah populations in South and East Africa vividly illustrate bottlenecks. Because of the lack of genetic diversity, sperm quality is poor, and many newborns die. Cheetahs are therefore difficult to breed in captivity.

Summary *Several conditions, in addition to natural selection, prevent Hardy–Weinberg equilibrium. Mutations introduce new alleles, mate choice is usually nonrandom, alleles move between populations, and changes in gene pools occur purely by chance.*

13.4 MASTERING CONCEPTS

1. What are some ways that mutations affect an organism's phenotype?
2. Under what conditions does a mutation in one organism pass to subsequent generations?
3. How does sexual selection promote traits that would seem to decrease fitness?
4. What is gene flow, and how does it disrupt Hardy–Weinberg equilibrium?
5. What is the difference between the founder effect and a population bottleneck?

13.5 INVESTIGATING LIFE
Size Matters in Fishing Frenzy

Studying the mechanisms of evolution helps us to understand life's history, but it also has practical consequences. A good example of natural selection is unfolding in fisheries worldwide. The selective force stems from a surprisingly mundane source—fishing regulations—but it affects everything from restaurant menus, to coastal economies, to the future of the ocean ecosystem.

The past several decades have seen devastating declines in the numbers of large predatory fishes such as swordfish, marlin, and sharks, as well as smaller animals including tuna, cod, and flounder. From a biological point of view, the reason for the fisheries decline is simple: the animals' death rate exceeds their reproductive rate. Industrial-scale fishing is the culprit. Since the 1950s, fishing fleets have employed larger ships and improved technologies in pursuit of their prey.

Regulations meant to protect fisheries allow the harvest of only those fish that exceed some minimum size. This measure is logical, because the smallest fish are most likely to be juveniles. Protecting the youngsters should permit the population to recover from the harvest of adult fish. Yet these regulations also have predictable evolutionary side effects. If fishing fleets selectively harvest the largest individuals, fish that are small at maturity are the most likely to survive long enough to reproduce. Large fish may become more scarce over many generations. The same policy should also select for slow-growing fish, since they would be last to exceed the minimum allowed size.

FIGURE 13.18 Atlantic Silverside. This small, silvery fish averages 9 cm in length, with a maximum of 15 cm.

Fish ecologists David Conover and Stephan Munch of the State University of New York tested these predictions in a coastal fish called the Atlantic silverside (*Menidia menidia*; **figure 13.18**). These small, shiny fish live for about 2 years, eating small invertebrates such as shrimp and marine worms along the Atlantic coast of Canada and the United States. The researchers chose this species in part because large populations can be maintained in captivity. Also, Atlantic silversides reproduce rapidly, making multigeneration experiments practical.

Conover and Munch set up their experiment by randomly dividing a large, captive population of Atlantic silversides into six tanks, each containing about 1100 juvenile fish. This ensured that all treatment groups started with similar gene pools. After about 6 months, the researchers removed the largest 90% of the fish from two of the tanks, termed "large-harvested" tanks. This treatment simulated fishing policies that protect all fish below a certain size. In two "small-harvested" tanks, they removed the smallest 90% of the silversides. Two control tanks were "random-harvested," in which 90% of the fish were removed without size bias.

Treatment Name	Removed ...	Leaving to reproduce ...
Large-harvested	Largest 90% of fish	Smallest 10% of fish
Small-harvested	Smallest 90% of fish	Largest 10% of fish
Random-harvested	Random 90% of fish	Random 10% of fish

After the harvests, about 100 survivors remained in each tank. These reproduced, and their descendants were reared in identical conditions until it was again time to harvest 90% of each population. The researchers repeated the large-harvest, small-harvest, and random-harvest treatments over four generations.

Predictably, both the total harvest weight and the weight of the average caught fish were initially highest for the large-harvested fish. Over four generations of size-biased fish removal, however, the average weight of the small-harvested fish grew dramatically, while that of the large-harvested fish declined (**figure 13.19**A). Furthermore, by the end of the experiment, the small-harvested fish clearly grew the fastest (figure 13.19B). In the small-harvested population, therefore, selection favored large size and rapid growth; the opposite was true in the large-harvested population. The researchers concluded that the three treatments imposed different selective forces that changed the genetic structure of the populations.

Of course, captive populations of silversides do not experience all of the same selective forces as do wild populations, especially of larger fish. Nevertheless, the researchers point to other studies suggesting that the same evolutionary trends they observed in their study are probably also occurring in longer lived species.

Conover and Munch's experiment is more than a straightforward demonstration of natural selection in action; it also has economic and ecological applications. Imagine, for example, new fishing regulations that impose both minimum and maximum size limits, protecting fish below and above a certain size. Such a policy would conserve commercially important fish stocks, because it would retain the juveniles that are critical to a species' future reproduction. It would also help fishing communities by selecting for fast-growing fish. Imposing a maximum size limit would also have ecological benefits, by restoring the feeding patterns and other "ecosystem services" of the largest fish. This wide range of implications beautifully illustrates the powerful ideas that spring from understanding one fundamental idea: natural selection.

Conover, David O. and Stephan B. Munch. 2002. Sustaining fisheries yields over evolutionary time scales. *Science*, vol. 297, pages 94–96.

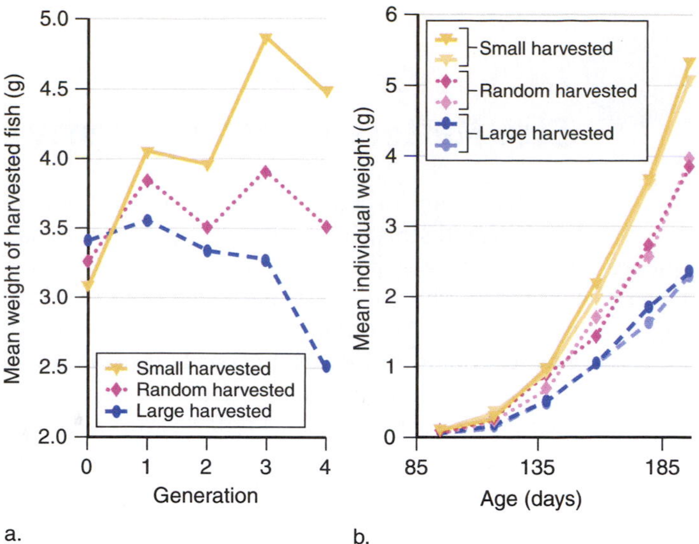

FIGURE 13.19 Size Matters in Fish Harvests. (**A**) Over four generations, the average per-fish weight in the large-harvested population was much less than the average for the small-harvested population. (**B**) Small-harvested fish grow more rapidly than large-harvested fish as well. **Question**: A trait's heritability is the proportion of variability that can be explained by genes. Heritability of 1.0 means a trait is 100% under genetic control; 0 means a trait is entirely under environmental control. Conover and Munch estimated that, in Atlantic silversides, heritability of body size is about 0.2. How would the results of this experiment be different if heritability of body size were higher than 0.2? What if it approached 0?

CHAPTER SUMMARY

13.1 Evolutionary Thought Has Evolved for Centuries

- Biological **evolution** is change in allele frequencies in populations. Evolution has occurred in the past and is constant and ongoing.

A. Many Explanations Have Been Proposed for Life's Diversity

- Early attempts to explain life's diversity relied on belief in a creator. Geology laid the groundwork for evolutionary thought. Some people explained the distribution of rock strata with the idea of **catastrophism** (a series of floods). The more gradual **uniformitarianism** (continual remolding of Earth's surface) became widely accepted.
- The **principle of superposition** states that lower rock strata are older than those above, suggesting a time frame for fossils within them.
- Lamarck was the first to propose a mechanism of evolution, but it was erroneously based on **inheritance of acquired characteristics.**

B. Charles Darwin's Voyage Provided a Wealth of Evidence

- During the voyage of the HMS *Beagle,* Darwin observed the distribution of organisms in diverse habitats and their relationships to geological formations. He noted that similar adaptations could lead to **convergent evolution.** After much thought, and considering input from other scientists, he synthesized his theory of the origin of species by means of **natural selection.**

C. *The Origin of Species* Proposed Natural Selection as an Evolutionary Mechanism

- Darwin's theory of evolution by natural selection was based on the observations that populations include individuals that vary for inherited traits; that many more offspring are born than survive; and that life is a struggle for use of limited resources. **Artificial selection** operates on similar principles, with a human breeder taking the place of the environment.
- *The Origin of Species* offered abundant evidence for the idea of **descent with modification**. However, people who believed that Earth is young and that humans are unique had difficulty accepting his ideas.

13.2 Natural Selection Molds Evolution

- Natural selection is one mechanism of **microevolution,** the small-scale genetic changes within a species,

A. Selection Results from Differential Reproductive Success

- Individuals with the best **adaptations** to the current environment are most likely to leave fertile offspring, and therefore their alleles become more common in the population over time.
- Natural selection requires variation, which arises ultimately from random mutations. Several events in sexual reproduction scramble alleles from generation to generation.

B. Natural Selection Eliminates Certain Phenotypes

- A **gene pool** includes all the alleles for all the genes in a population. Natural selection weeds out some phenotypes, causing changes in allele frequencies over multiple generations.

C. Natural Selection Does Not Have a Goal

- Natural selection does not work toward a goal, nor can it achieve perfectly adapted organisms.

D. What Does "Survival of the Fittest" Really Mean?

- Organisms with the highest evolutionary **fitness** are the ones that have the greatest reproductive success. Many traits contribute to an organism's fitness.

E. There Are Three Modes of Natural Selection

- In **directional selection,** an extreme phenotype becomes more prevalent in a population. In **disruptive selection,** extreme phenotypes survive at the expense of intermediate forms. In **stabilizing selection,** an intermediate phenotype has an advantage.

F. Balanced Polymorphism Maintains Multiple Alleles for One Gene

- In **balanced polymorphism,** natural selection indefinitely maintains more than one allele for a gene. Harmful recessive alleles may remain in a population because of a **heterozygote advantage,** in which carriers have a reproductive advantage over homozygotes.

13.3 Evolution Is Inevitable in Real Populations

A. At Hardy–Weinberg Equilibrium, Allele Frequencies Do Not Change

- At **Hardy–Weinberg equilibrium,** we can calculate the proportion of genotypes and phenotypes in a population by inserting known allele frequencies into an algebraic equation: $p^2 + 2pq + q^2$.
- If a population meets all assumptions of Hardy–Weinberg equilibrium, evolution does not occur because allele frequencies do not change from generation to generation.

B. In Reality, Allele Frequencies Always Change

- The conditions for Hardy–Weinberg equilibrium do not occur together in natural populations, suggesting that allele frequencies always change.

13.4 Evolutionary Changes Occur in Several Ways

A. Mutation Fuels Evolution

- Mutation alters allele frequencies by changing one allele into another, sometimes providing new phenotypes for natural selection to act on. Many mutations do not pass to the next generation.

B. Nonrandom Mating Concentrates Alleles Locally

- Nonrandom mating causes some alleles to concentrate in subpopulations. It sometimes results from **sexual selection,** a form of natural selection in which certain inherited traits—even those that seem nonadaptive—make an individual more likely to attract mates. The result may be **sexual dimorphisms** that differentiate the sexes.

C. Gene Flow Moves Alleles Between Populations

- Allele movement between populations, as by migration, is **gene flow.**

D. Genetic Drift Occurs by Chance

- In **genetic drift,** allele frequencies change purely by chance events, especially in small populations. The **founder effect** and **population bottlenecks** are forms of genetic drift.

13.5 Investigating Life: Size Matters in Fishing Frenzy

- Fishing regulations that spare only the smallest fish in a population select for small, slow-growing individuals. Studies of Atlantic silversides suggest that protecting the largest fish as well would increase fishery productivity in the long run.

MULTIPLE CHOICE QUESTIONS

1. The idea that geological processes have occurred in the past as they are occurring today is characterized as
   a. catastrophism.
   b. uniformitarianism.
   c. the principle of superposition.
   d. inheritance of acquired characteristics.
2. How are artificial selection and natural selection similar?
   a. They both rely on human intervention.
   b. They both work toward a specific goal.
   c. They both select for specific traits within a population.
   d. They are processes that only affect animals.
3. Microevolution applies to changes that occur
   a. only within small populations of organisms.
   b. in small regions of DNA.
   c. to the allele frequencies in a population or species.
   d. to small cells like bacteria.
4. How does natural selection influence a gene pool?
   a. It selects for alleles that will be useful in the future.
   b. It triggers mutations leading to increased diversity.
   c. It alters the frequency of alleles within an individual.
   d. It alters the frequency of alleles within a population.
5. Which of the following is the best definition of evolutionary fitness?
   a. The ability to increase the number of alleles in a gene pool
   b. The ability to survive for a long period of time
   c. The ability of an individual to adapt to a changing environment
   d. The ability to produce many offspring
6. Selection that favors two or more extreme phenotypes is called
   a. directional selection.
   b. stabilizing selection.
   c. disruptive selection.
   d. normalizing selection.
7. Huntington disease is caused by a lethal dominant allele that affects individuals late in life. Why is this disease-causing allele still around?
   a. The heterozygous individual might have some advantage.
   b. The effect of the disease allele occurs after the individual has reproduced.
   c. New mutations generate the allele.
   d. Both a and b.
8. Assume a population is 36% homozygous dominant and 16% homozygous recessive. What percent of the population is heterozygous based on the Hardy–Weinberg equilibrium?
   a. 64%
   b. 52%
   c. 48%
   d. 26%
9. Darwin observed that different types of organisms were found on either side of a geographic barrier. In this case the barrier was preventing
   a. gene flow.
   b. genetic drift.
   c. sexual selection.
   d. mutation.
10. The loss of genetic diversity is associated with
    a. a population bottleneck.
    b. a heterozygote advantage.
    c. the founder effect.
    d. balanced polymorphism.

TESTING YOUR KNOWLEDGE

1. How did James Hutton, Georges Cuvier, Georges-Louis Buffon, Jean Baptiste de Lamarck, Charles Lyell, and Thomas Malthus influence Charles Darwin's thinking?
2. How do artificial and natural selection lead to "descent with modification"?
3. How does variation arise in an asexually reproducing population? A sexually reproducing population?
4. What sorts of traits make up the portion of an organism's phenotype that is subject to natural selection?
5. People sometimes describe evolution as though it were goal-oriented (e.g., "The bacteria evolved antibiotic resistance in order to survive.") Is this an accurate way to describe how evolution works? Why or why not?
6. What sorts of adaptations contribute to an organism's evolutionary fitness?
7. Give examples of directional, disruptive, and stabilizing selection other than those mentioned in the chapter.
8. What is one reason that harmful recessive alleles persist in populations, even though they prevent individuals from reproducing when present in two copies?
9. Explain how the algebraic expression $p^2 + 2pq + q^2$ represents the proportions of different genotypes in a population.
10. The Fraggles are a population of mythical, mouselike creatures that live in underground tunnels and chambers beneath a large vegetable garden that supplies their food. Of the 100 Fraggles in this population, 84 have green fur, and 16 have gray fur. A dominant allele *F* confers green fur, and a recessive allele *f* confers gray fur. Assuming Hardy–Weinberg equilibrium is operating, answer the following questions:
    a. What is the frequency of the gray allele *f*?
    b. What is the frequency of the green allele *F*?
    c. How many Fraggles are heterozygotes (*Ff*)?
    d. How many Fraggles are homozygous recessive (*ff*)?
    e. How many Fraggles are homozygous dominant (*FF*)?

11. One spring, a dust storm blankets the usually green garden of the Fraggles in gray. Under these conditions, the green Fraggles become very visible to the Gorgs, who tend the gardens and try to kill the Fraggles to protect their crops. The gray Fraggles, however, blend into the dusty background and find that they can easily steal radishes from the garden. How might this event affect microevolution in this population of Fraggles? What mode of natural selection does this represent?
12. Define the following: gene flow, genetic drift, gene pool.
13. How does sexual selection maintain seemingly harmful traits in one sex?

THINKING AS A SCIENTIST

1. You have a pet cocker spaniel and believe that if you snip off the end of his tail and then breed him, his puppies will be born with snipped off tails. Is this idea consistent with Darwinian evolution? Why or why not?
2. The giant anteater lives in South America, the scaly anteater lives in tropical parts of Asia and Africa, and the spiny anteater lives in Australia. The three animals are not at all closely related, but resemble each other closely. They have long, sticky tongues that they use to eat ants, no teeth, large salivary glands, and long, bald snouts. Which phenomenon that Darwin observed do these animals illustrate?
3. Which theory best explains Darwinian evolution: catastrophism or uniformitarianism? Give a reason for your answer.
4. Many articles about the rise of antibiotic-resistant bacteria claim that overuse of antibiotics creates resistant strains. How is this statement incorrect?
5. A thin-shelled crab can more readily move to escape a predator than can a thick-shelled crab, but it is more vulnerable to predators that drill through the shell. Because of these opposing forces, shell thickness for many types of crabs has remained within a narrow range over a long time. What type of natural selection does crab shell thickness illustrate?
6. Can natural selection occur among genetically identical clones? Why or why not?
7. What happens to a population if the environment changes and no individuals have the allele combinations required to survive and produce offspring?
8. Give an example in which humans have countered natural selection by medically correcting phenotypes that would otherwise prevent some of us from surviving long enough to reproduce.
9. Explain the observation that asexual reproduction occurs most often in environments that experience little change.
10. Explain the following statement: "DNA's ability to mutate has provided, and continues to provide, the variation on which natural selection acts."

ARIS™ **Visit www.mhhe.com/hoefnagels for practice quizzes, animations, videos, and activities designed to help you master the material in this chapter.**

11. In a few countries, antibiotics are available without a prescription. How does this practice contribute to the spread of antibiotic-resistant bacteria?
12. How does the appearance of new infectious diseases or resurgence of old ones illustrate continuing evolution?
13. How do the following situations or practices disrupt Hardy–Weinberg equilibrium?
    a. Couples who find out that they are carriers for the same illness decide not to have children together.
    b. Several dozen young adults in a large Midwestern city discover they are half-siblings. Each was conceived by artificial insemination, with sperm from the same donor.
    c. Members of a very close-knit Amish community are forbidden to marry outside the community.
    d. A new viral illness kills only people who have a certain blood type.
14. Because of global travel, a new strain of antibiotic-resistant bacteria arising in Malaysia may soon appear in New York City. Would the introduction of the new resistance alleles into New York's bacteria be an example of gene flow, genetic drift, mutation, or natural selection?
15. Which factors contributing to evolution discussed in this chapter do the following science fiction film plots illustrate?
    a. In *When Worlds Collide,* Earth is about to be destroyed. One hundred people, chosen for their intelligence and fertility, leave to colonize a new planet.
    b. In *The Time Machine,* set in the distant future on Earth, one group of people is forced to live on the planet's surface and another group is forced to live in caves. After many years, they look and behave differently. The Morlocks, who live below ground, have dark skin, dark hair, and are very aggressive, whereas the Eloi, who live aboveground, are blond, fair-skinned, and meek.
    c. In *The War of the Worlds,* Martians cannot survive on Earth because they are vulnerable to infection by terrestrial microbes.

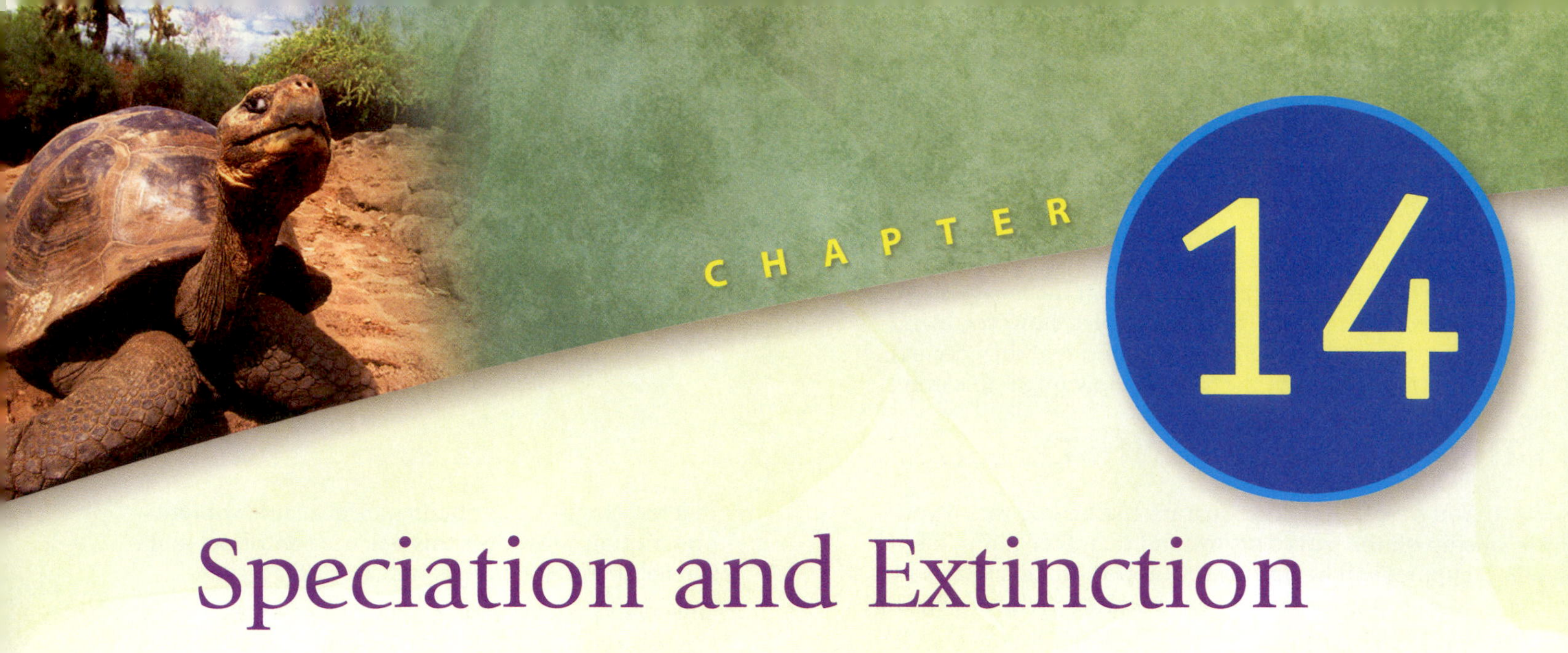

Speciation and Extinction

The 'I'iwi is a honeycreeper native to the Hawaiian islands. Like many island species, this scarlet bird is endangered by habitat loss, introduced predators, and disease.

What's the Point?

Islands Provide Windows on Speciation and Extinction

Over billions of years, Earth has seen many new species originate (a process called speciation). Yet most species that have ever lived are now extinct. The opposing, ongoing processes of speciation and extinction have defined life's history.

Islands provide ideal opportunities to study speciation. Their small land areas house populations that are relatively easy to monitor. In addition, few organisms can cross vast distances across oceans to reach isolated islands. The descendants of those that have managed to colonize the island diversified and exploited multiple habitats. Islands located far away from a mainland therefore often have groups of closely related species found nowhere else.

Many examples illustrate the spectacular diversity that islands may host. For example, about 800 species of Drosophila *flies and their close relatives inhabit the Hawaiian Islands. The 28 species of plants called silverswords thrive in every imaginable Hawaiian island habitat, yet all apparently descended from one ancestor that colonized Hawaii long ago. The same islands are also home to birds called honeycreepers, each with a bill adapted to a different food source.*

At the same time, island species are especially vulnerable to extinction. Many factors conspire against them. A single hurricane, fire, or flood may destroy a small island population, as can the extinction of a prey species or the introduction of a new predator. Even random fluctuations in birth and death rates can doom a small population.

A volcanic island called Mauritius, in the Indian Ocean, illustrates how humans increase the extinction risk that island populations face. Until the sixteenth century, Mauritius teemed with dense, tall forests, colorful birds, scurrying insects, and basking reptiles. One inhabitant was the large, flightless dodo bird, described by one scientist as "a magnificently overweight pigeon." When European sailors arrived in the 1500s, however, the men ate dodo meat, and their pet monkeys and pigs ate dodo eggs. Rats and mice swam ashore from ships and attacked native insects and reptiles. The sailors' Indian myna birds inhabited nests of the native echo parakeet, while imported plants crowded the seedlings of native trees. By the mid-1600s, only 11 of the original 33 species of birds remained. The dodo was exterminated by 1681, the first of many recorded extinctions caused by human activities.

Both speciation and extinction have likely been a part of evolution since life began. This chapter explains how these processes occur.

LEARNING OUTLINE

14.1 The Definition of "Species" Has Evolved over Time

Throughout the history of life, the types of organisms have changed: new ones have appeared, and existing ones have vanished. The term **macroevolution** describes these large, complex changes in life's panorama. Macroevolutionary events tend to span very long periods, whereas the microevolutionary processes described in chapter 13 happen so rapidly that we can sometimes observe them over just a few years. Nevertheless, the many small changes that accumulate in a population by microevolution eventually lead to large-scale, macroevolutionary events.

Macroevolution has produced an obvious diversity of life. A bacterium, for example, is clearly distinct from a tree or a bird (**figure 14.1A**). At the same time, some organisms are more closely related than others, as the cats in figure 14.1B illustrate. To make sense of these observations, biologists recognize the importance of grouping similar individuals into **species**—that is, distinct types of organisms. This task requires agreement on what the word *species* means. Perhaps surprisingly, the definition has changed considerably over time.

A. Linnaeus Devised the Binomial Naming System

Swedish botanist Carolus Linnaeus (1707 to 1778) was not the first to ponder this question, but his contributions last to this day. Linnaeus defined species as "all examples of creatures that were alike in minute detail of body structure." Importantly, he was the first investigator to give every species a two-word biological name. Each species' name combines the broader classification *genus* with the term *species*. The scientific name for humans, for example, is *Homo sapiens*. Linnaeus eventually published his species descriptions in 10 editions of huge volumes called *Systema Naturae*. **taxonomic hierarchy, p. 9**

Linnaeus also devised a hierarchical system of classification. He grouped similar genera into orders, classes, and kingdoms (scientists now use additional categories, including domains, phyla, and families). His classifications organized the great diversity of life and helped scientists communicate with one another, although he did not consider the role of evolutionary relationships. He thought that each species was created separately and that species could not change. Therefore, species could not appear or disappear, nor were they related to one another.

Charles Darwin finally connected species diversity to evolution, writing that "our classifications will come to be, as far as they can be so made, genealogies." As the theory of evolution by natural selection became widely accepted in the nineteenth and twentieth centuries, scientists no longer viewed classifications merely as ways to organize God's creation. Instead, they considered them to be hypotheses about the evolutionary history of life.

B. Ernst Mayr Developed the Biological Species Concept

In the 1940s, Harvard biologist Ernst Mayr amended the work of Linnaeus and Darwin by considering reproduction and genetics. Mayr defined a **biological species** as a population, or group of populations, whose members can interbreed and produce fertile offspring. **Speciation** occurs (that is, new species arise) when members of a population can no longer successfully interbreed. How might this happen? A new species could form if a population some-

a.

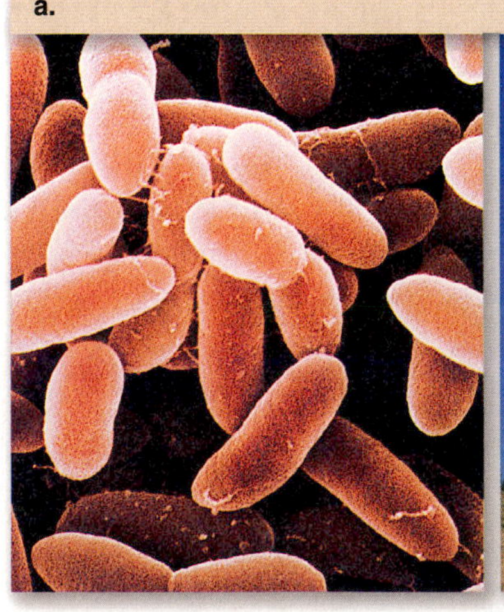

FIGURE 14.1 Species Are Distinctive Types of Organisms. The more recently different species diverged from a common ancestor, the more characteristics they share. (**A**) The bacteria, tree, and bird are about as dissimilar as three types of organisms can be; their last common ancestor must have lived before cells diverged into the three basic types (bacterial, archaean, and eukaryotic). (**B**) These three cats shared their last common ancestor much more recently. They have many characteristics in common.

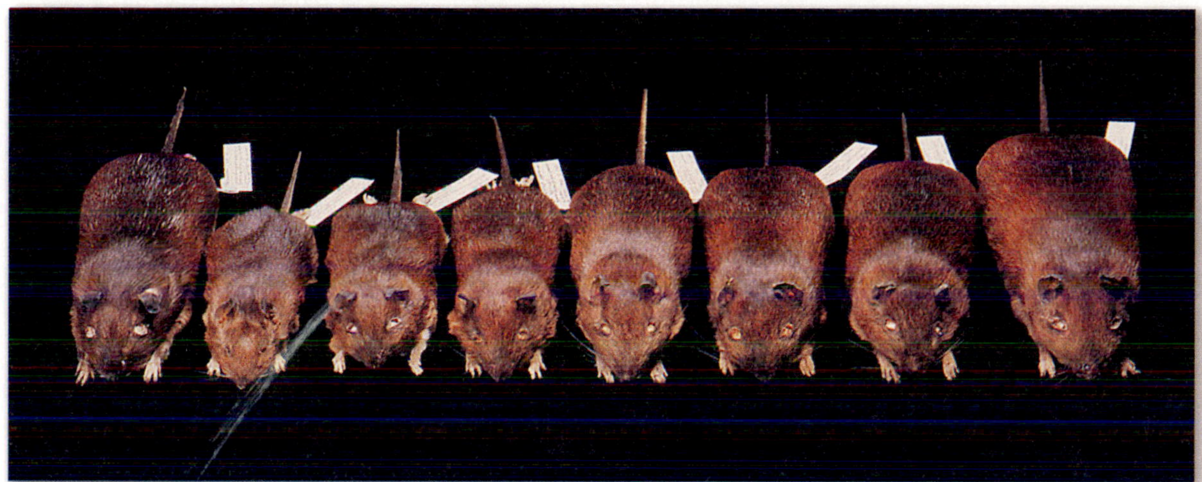

FIGURE 14.2
Which is Which? These spiny rats are so alike that researchers have difficulty telling them apart. The biological species concept gives an objective basis by which to assign them to species.

how became divided, and microevolutionary forces led to genetic divergence between the groups. With the accumulation of enough differences, the two groups could no longer produce fertile offspring even if they came into contact once again. In this way, microevolution would become macroevolution.

The biological species definition does not rely on physical appearance to assign species, so it is much less subjective than Linnaeus's observations. Under the system of Linnaeus, it would be impossible to determine whether two similar-looking rats belong to different species (**figure 14.2**). Using Mayr's definition, however, if the two groups can produce fertile offspring together, they belong to one species.

Today, scientists designate most species based on the ability to produce fertile offspring. Nevertheless, Mayr's species definition raises several difficulties. First, how can we designate species for organisms that have the *potential* to interbreed (e.g., in captivity) but do not do so in nature? Second, how are asexually reproducing organisms, such as bacteria and fungi, assigned to species? Third, it is not possible to apply the biological species definition to extinct organisms known only from fossils. Fourth, for some species, reproductive isolation is not absolute. Many closely related species of plants, for example, sometimes produce fertile offspring together. DNA sequence analysis has allowed scientists to fill in some of these gaps.

Summary *The concept of "species" has changed over time. According to the biological species concept, different species cannot produce fertile offspring with each other. Species designations identify types of organisms, and modern classifications depict their evolutionary relationships.*

14.1 MASTERING CONCEPTS

1. What is the relationship between macroevolution and microevolution?
2. How does today's definition of species differ from Linnaeus's definition?
3. What are some of the challenges in defining and describing species?

b.

14.2 Reproductive Barriers Separate Species from One Another

In keeping with Mayr's biological species concept, new species form when one portion of a population can no longer breed with the rest of the population. The precise point at which two populations become reproductively isolated varies, because successful reproduction requires many complex events: courtship, fertilization, embryo formation, and development to sexual maturity of the next generation. For ease of study, biologists divide mechanisms of reproductive isolation into two broad groups (**figure 14.3**): prezygotic or postzygotic. (Recall that a zygote is a fertilized egg.)

A. Prezygotic Barriers Prevent Fertilization

Mechanisms of **prezygotic reproductive isolation** affect the ability of two species to combine gametes. These reproductive barriers include:

- **Ecological (or habitat) isolation:** A difference in habitat preference can separate two populations in the same geographic area. For example, if a portion of an insect population begins to use a different food source—say, pears instead of apples—its members might be physically separated from the original population, and speciation would be possible.
- **Temporal isolation:** Two related species may never encounter each other, even though they live in the same habitat, simply because one is active during the day and the other at night. Temporal isolation is therefore based on differences in timing.
- **Behavioral isolation:** Even if two species share a habitat and are active at the same time, their different behaviors may prevent them from mating. An obvious example is the intricate mating dances of some insects and birds. Any variation in the ritual from one group to another could prevent them from even being attracted to one another.
- **Mechanical isolation:** In many animal species, male and female parts fit together almost like a key in a lock. Any change in the shape of the gamete-delivering structures may prevent individuals from mating with the original population. In plants, males and females do not copulate, but mechanical barriers may still apply. For example, some orchids deliver pollen in specialized ways to insect carriers. A slight change in the shape of the flower may lead to inefficient pollen transfer, and therefore present a reproductive barrier. **pollination, p. 539**
- **Gametic isolation:** If a sperm cannot penetrate an egg cell, then no reproduction will occur. For example, many marine organisms simply release gametes into the water. Each species' gametes display unique patterns of surface molecules that enable sperm to recognize eggs of the same species. In the absence of a "match," fertilization will not occur. As another example, in plants, the pollen tubes of one species may be too short to deliver sperm to the egg cell of a related species.

Prezygotic Reproductive Isolation

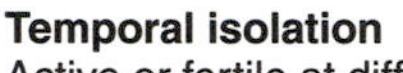

Ecological (habitat) isolation
Different environments
(e.g. desert fox and arctic fox)

Temporal isolation
Active or fertile at different times
(e.g. two species of field crickets reproduce in different seasons)

Behavioral isolation
Different activities
(e.g. different flash patterns in firefly species)

Mechanical isolation
Mating organs incompatible
(e.g. Great Dane and Chihuahua)

Gametic isolation
Gametes cannot unite
(e.g. mouse and rat)

Postzygotic Reproductive Isolation

Hybrid inviability
Gametes unite, but development cannot produce a viable embryo
(e.g. goat and sheep)

Hybrid infertility
Hybrids lack the ability to make or deliver viable gametes
(e.g. liger, the hybrid offspring of a lion and a tiger)

FIGURE 14.3 Separating Species. Reproductive isolating mechanisms act along the continuum of reproduction and development. Prezygotic barriers prevent fertilization; postzygotic barriers prevent a hybrid embryo from developing into a fertile adult.

B. Postzygotic Barriers Prevent Viable or Fertile Offspring

Individuals of two different species may produce a hybrid zygote. (A hybrid, in this case, is the offspring of individuals from two different species.) Even then, **postzygotic reproductive isolation** may keep the species separate. Reproductive barriers that prevent the formation of fertile offspring include:

- **Hybrid inviability** (also called **hybrid breakdown):** A hybrid embryo may die early in development because the genes of its parents are incompatible.
- **Hybrid infertility (sterility):** Some hybrids are infertile. For example, a mule is the hybrid offspring of a female horse and a male donkey. Mules are infertile because a horse's egg has one more chromosome than a donkey's sperm cell. The difference in chromosome number does not prevent mitotic cell division, so the animal can grow and develop. Meiosis is impossible in the mule's germ cells, however, because the chromosomes are not homologous. **meiosis, p. 178; mules, p. 181**

Two animal species may produce hybrids under unusual circumstances. For example, sometimes in zoos, a male lion will mate with a female tiger, producing a "liger." Likewise, when zebras and horses mate, they produce "zebroids" that look like a mixture of the two. Like mules, ligers and zebroids are infertile.

Although successful hybridization is rare in animals, it frequently occurs in plants (see section 14.7). One of the problems with introducing nonnative species of plants into a region is the potential production of hybrids that displace the native plants. On the other hand, many of our food crops are the result of hybridization. The tangelo, for example, is the hybrid offspring of a tangerine and a pomelo (grapefruit). **how can a fruit be seedless? p. 542**

Summary *Many events must occur for two individuals to reproduce successfully. Reproductive barriers occur at all points along the continuum from mating to the development of fertile offspring.*

14.2 MASTERING CONCEPTS

1. How can reproductive isolation occur in different ways?
2. Name five modes of prezygotic reproductive isolation.
3. What are two ways that postzygotic reproductive isolation may occur?

14.3 Spatial Arrangements Define Three Main Modes of Speciation

Reproductive barriers keep related species apart. How do those barriers arise in the first place? More specifically, how could two populations of the same species evolve along different pathways, eventually yielding two species?

The most obvious solution is to physically separate the populations so that they have no contact with each other. Over time, natural selection and genetic drift would likely yield substantial differences between the two populations. Yet speciation can also separate populations that have physical contact with each other. They may divide into species even though they inhabit neighboring regions or even share a habitat. Biologists recognize these different circumstances by dividing the geographic setting of speciation into three categories: allopatric, parapatric, and sympatric (**figure 14.4**).

A. Allopatric Speciation Reflects a Geographic Barrier

In **allopatric speciation,** a geologic event or structure physically separates a population into two groups that cannot

FIGURE 14.4 Speciation and Geography. Allopatric and sympatric speciation represent two extremes along a continuum of contact between populations. Parapatric speciation is intermediate: two populations mingle within a shared border area.

interbreed (*allo-* means "other," and *patria* means "fatherland"). The two groups therefore cannot contact each other. Agents of allopatric speciation include volcanoes, earthquakes, storms, tidal waves, glaciers, floods, and the formation or destruction of mountains or bodies of water.

After the event, the forces of microevolution act independently on the geographically separated groups. The result may be one or more reproductive barriers. If the geographic barrier is lifted and descendants of the original two populations cannot interbreed, one species has branched into two.

In the Amazon jungle, the huge rivers are the source of much of the area's striking biodiversity. A study of mitochondrial DNA sequences in tamarin monkeys supports the hypothesis that rivers divide populations, driving speciation. Where the Amazon is very wide, monkeys on one side of the river are brown; on the other side, they are white. Where the banks are close together, however, populations on either side of the river have both coat colors, presumably because the animals have easy access to each other (**figure 14.5A**).

The Devil's Hole pupfish, which inhabits a warm spring at the base of a mountain near Death Valley, California, illustrates one way that allopatric speciation occurs. The spring was isolated from other bodies of water about 50,000 years ago, preventing genetic exchange between the fish trapped in the spring and those in the original population. In that time, the gene pool has shifted sufficiently so that a Devil's Hole pupfish cannot mate with fish from another spring. It has become a distinct species (figure 14.5B).

Allopatric speciation has been considered the most common mechanism because the evidence for it is the most abundant and obvious. The diversification of species on island archipelagos, described in the chapter opening essay, provides many striking examples. So do the world's fishes: of the 29,000 or so known species of fishes, 36% live in freshwater habitats, although these places account for only 1% of Earth's surface. Compared with the vast oceans, the countless lakes, ponds, streams, and rivers provide diverse habitats and ample barriers to genetic exchange.

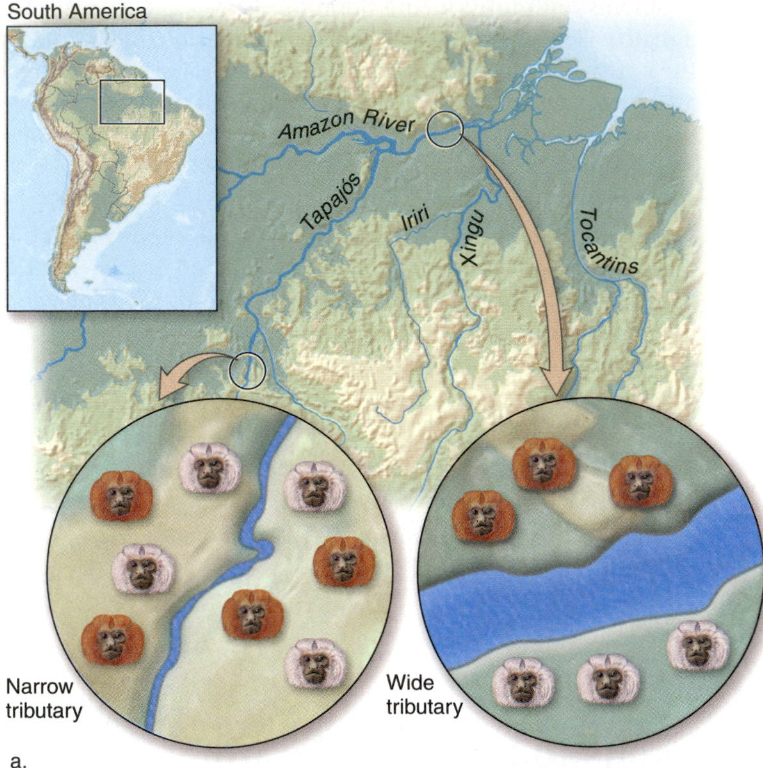

FIGURE 14.5 Allopatric Speciation. (A) Tamarin monkey populations on two widely separated sides of the Amazon River are diverging toward speciation. The populations can still mix where the river is narrow. **(B)** Devil's Hole is a pool in a limestone cavern east of Death Valley National Park. The pupfish species that lives there is a product of allopatric speciation.

B. Parapatric Speciation Occurs in Neighboring Regions

In **parapatric speciation,** part of a population enters a new habitat bordering the range of the parent species (*para-* means "alongside"). Most individuals mate within their own populations, but a few may venture into the shared border zone. The resulting genetic divergence between the group sharing the border and the two original populations can be an initial step toward speciation. Since the border group possesses a unique combination of traits, natural selection could eventually isolate the new group from the two original populations.

Consider the little greenbul (*Andropadus virens*), a small green bird that lives in the tropical rain forest of Cameroon, West Africa (**figure 14.6**). The birds also inhabit patches of forest in the transitional border areas (called ecotones) between rain forest and grassland. In the 1990s, researchers captured birds from six tropical rain forest sites and six ecotone sites. The birds in the ecotone patches had greater weight,

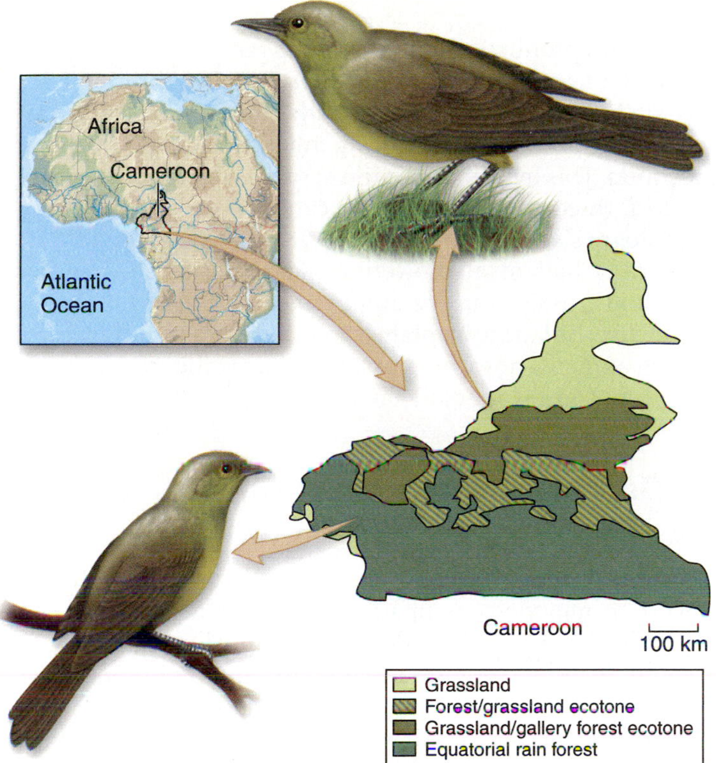

FIGURE 14.6 Parapatric Speciation. The little greenbul of Cameroon lives in the rain forest and in patches of "gallery forests" in the border zone with grasslands. Birds from the ecotone are larger overall and have longer bills, legs, and wings than their rain forest counterparts. These differences indicate that the two populations are diverging genetically and one day may become separate species.

deeper bills, and longer legs and wings than their rain forest counterparts. The little greenbuls from the ecotones can still mate with those from the rain forest, so speciation has not yet occurred. Nevertheless, the researchers concluded that the forces of natural selection are greater than the gene flow between the two populations, gradually taking the groups farther apart. We are likely seeing speciation in action.

C. Sympatric Speciation Occurs in a Shared Habitat

In **sympatric speciation,** populations diverge genetically while living in the same physical area (*sym-* means "together"). Among evolutionary biologists, the idea of sympatric speciation is often controversial. After all, how can a new species arise in the midst of an existing population?

Often sympatric speciation reflects the fact that a habitat that appears uniform (e.g., a body of water) actually consists of many microenvironments. Fishes called cichlids, for example, have diversified into many species within the same large lakes in Cameroon (**figure 14.7A**). Some populations feed exclusively on the lake bottom, whereas others prefer the regions near roots of aquatic plants, or closer to the surface. They are reproductively isolated: no genetic exchange occurs between members of each population. Many biologists would argue, however, that this case of cichlid speciation is the consequence of ecological isolation. It therefore reflects allopatric speciation on a small scale, not true sympatric speciation.

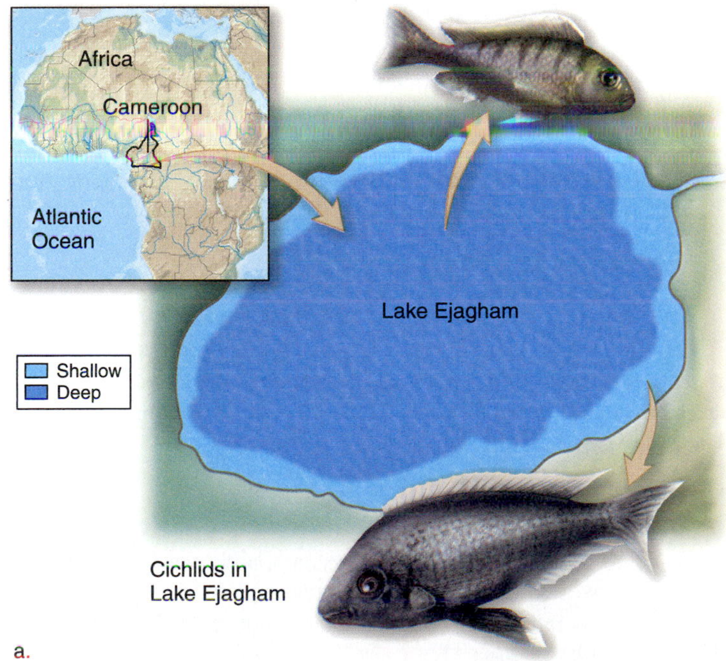

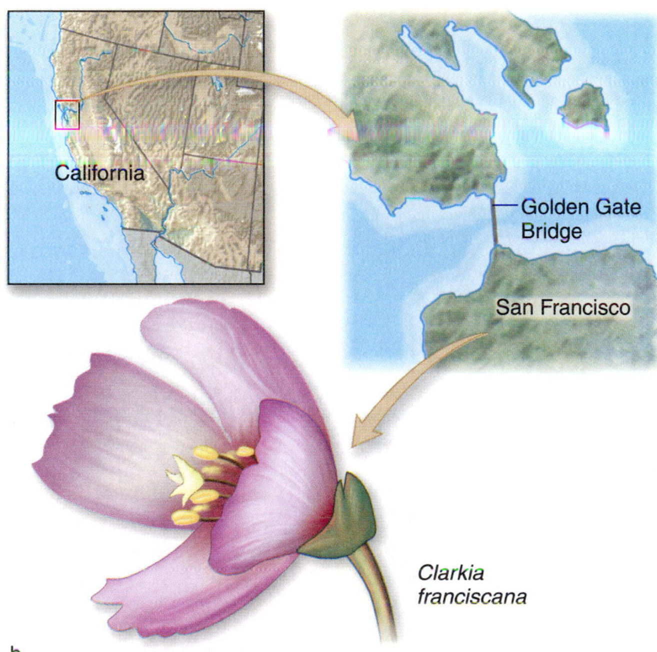

FIGURE 14.7 Sympatric Speciation. (A) Cichlids in the deep waters of Cameroon's Lake Ejagham have smaller bodies and larger eyes than shallow water fish, an indication of sympatric speciation in progress. **(B)** *Clarkia franciscana* is a new species that arose following a drought in California's San Francisco Bay area.

Sympatric Speciation after a Bottleneck Event

Sometimes a large-scale genetic change produces an undisputed case of sympatric speciation. This happened when a population of the plant *Clarkia rubicunda,* common along the coast of central California, underwent a bottleneck event. A severe drought in the Golden Gate Bridge region in San Francisco nearly wiped out the local population of *C. rubicunda.* The only survivors had several chromosomal abnormalities. These plants cross-fertilized among themselves and established a new population. When the drought ended, *C. rubicunda* plants encroached from the surrounding regions, but they could not reproduce with the Golden Gate group. The genetic material was organized too differently in the two groups, although both types of plants descended from the same ancestors. A new species, *C. franciscana,* had arisen (figure 14.7B).

Sympatric Speciation Due to Polyploidy

An even more drastic chromosomal change is **polyploidy,** which occurs when the number of sets of chromosomes increases. Polyploids sometimes form when gametes from two different species fuse (see section 14.7). For example, an "Old World" species of cotton has 26 large chromosomes, whereas one from Central and South America has 26 small chromosomes. The cotton commonly cultivated for cloth is a polyploid derived from both types. It has 52 chromosomes: 26 large and 26 small (**figure 14.8**).

Polyploidy can also occur when meiosis fails. A diploid individual may occasionally produce diploid sex cells. Self-fertilization will produce tetraploid offspring (with four sets of chromosomes) that represent an "instant species" because it is reproductively isolated from its diploid ancestors. When the tetraploid plant's gametes (each containing two sets of chromosomes) combine with haploid gametes from the parent species, each offspring inherits three sets of chromosomes. This hybrid is infertile because it cannot complete meiosis: three chromosome sets cannot form the required homologous pairs. **polyploidy, p. 185**

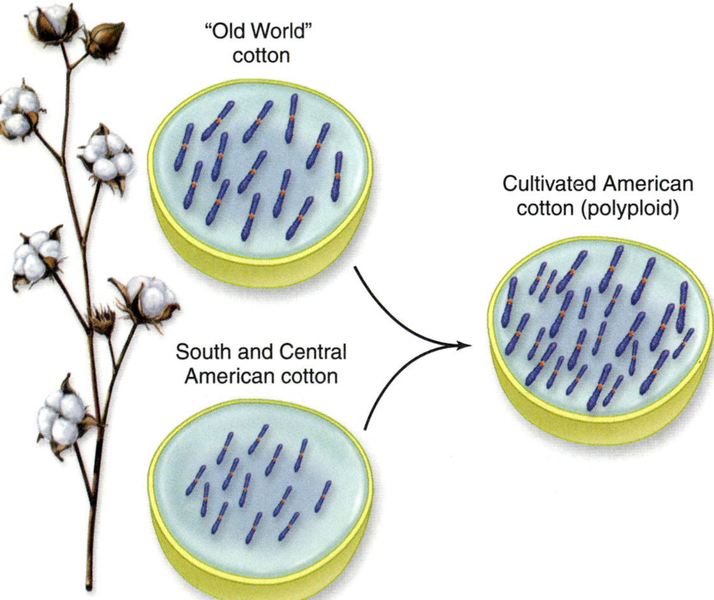

FIGURE 14.8 Useful Polyploid. Cultivated American cotton is a polyploid species derived from Old World and New World ancestors.

Nearly half of all flowering plant species are natural polyploids, indicating the importance of this form of reproductive isolation in plant evolution. Polyploidy is rare among animals, however, possibly because the extra "dose" of chromosomes is usually fatal.

D. Determining the Mode of Speciation May Be Difficult

Biologists sometimes debate whether a speciation event is allopatric, parapatric, or sympatric. One reason for the disagreements is that the definitions represent three points along a continuum, from complete isolation to continuous intermingling. Another difficulty is that we may not be able to detect the barriers that are important to other species. Is an apparently uniform patch of forest a potential setting for sympatric speciation, because it appears the same throughout? Our perspective may differ from that of a soil-dwelling insect, whose habitat differs greatly from the environment in the treetops.

The problem of perspective also leads to debate over the size of the geographic barrier needed to separate two populations, which depends on the distance over which a species can spread its gametes. A plant with windblown pollen, or a fungus producing lightweight spores, encounters few barriers to gene flow. Pollen and spores can travel thousands of miles in the upper atmosphere. On the other hand, a desert pupfish has few opportunities for gene flow beyond its aquatic habitat, so the simple isolation of its pool from other water bodies instantly creates an insurmountable geographic barrier. The same circumstance would probably not deter gene flow in species that can walk or fly between pools.

Summary *Species may arise when a geographical barrier separates a single population into two populations that then diverge genetically. Speciation may also occur when groups are neighbors and share a border but mate mostly among themselves. Reproductive isolation can even produce new species within a single area.*

14.3 MASTERING CONCEPTS

1. Distinguish among allopatric, parapatric, and sympatric speciation, and provide examples of each.
2. How can changes in chromosome number contribute to sympatric speciation?
3. What are some difficulties in considering the effects of geography on speciation?

14.4 Evolution May Be Gradual or Occur in Bursts

Speciation can happen quickly or gradually—or at any rate in between. On a global level, Earth has seen times of relatively little change in the living landscape, but also periods of rapid speciation and times of mass extinctions. Even within a short time, one species can change rapidly, while another hardly changes at all (this chapter's Burning Question: Why does evolution occur rapidly in some species but slowly for others? explains this disparity).

A. Gradualism and Punctuated Equilibrium Are Two Models of Speciation

Darwin envisioned evolution as one species gradually transforming into another through a series of intermediate stages. The pace as Darwin saw it was slow, although not necessarily constant. This idea, which became known as **gradualism,** held that evolution proceeds in small, incremental changes over many generations (**figure 14.9**).

If the gradualism model is correct, "slow and steady" evolutionary change should be evident in the fossil record. In fact, however, many steps in species formation did not leave fossil evidence, and so we do not know of many intermediate or transitional forms between species. Much of the fossil record instead suggests the opposite: that many species have appeared relatively suddenly, followed by long periods over which the species changed little.

What accounts for the "missing" transitional forms? One explanation is that the fossil record is incomplete, for many reasons: poor preservation of biological material, natural forces that destroyed fossils, and the simple fact that

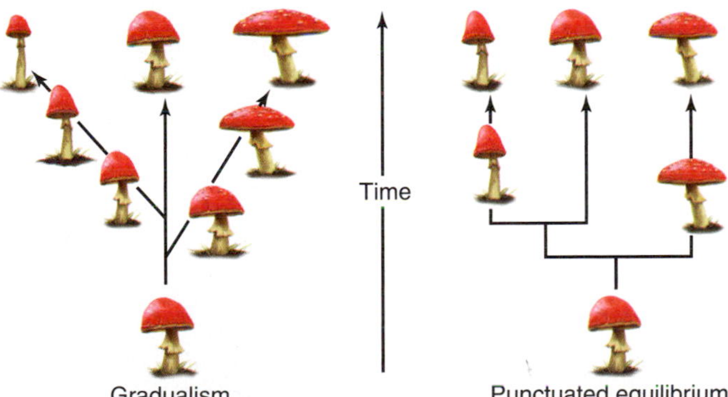

FIGURE 14.9 Evolution—Both Gradual and Dramatic. In gradualism, three fungus species arise from one ancestor by way of small, incremental steps. Punctuated equilibrium produces the same result, except that the new species arise in rapid bursts followed by periods of little change.

Burning Questions

Why does evolution occur rapidly in some species but slowly for others?

Even though the same basic forces of microevolution operate on all populations, different species evolve at different rates. This may seem like a paradox until you think of all the different ways that the genetic structure of a population can change.

The following factors speed evolution:

- genes with high mutation rates
- small populations (they are most susceptible to genetic drift and other microevolutionary forces)
- tendency toward chromosome breakage, polyploidy, or both
- a short generation time, such as in bacteria
- rapid environmental change
- rapid immigration or emigration

A mutation in a single gene that controls the timing of early developmental events can have an evolutionary influence by producing new phenotypes (see figure 12.13 and **figure 14.A**). It is not difficult to imagine how a single genetic "switch" that alters the timing of mitotic cell division could have produced the prolonged brain growth that characterizes our own species, or a multitude of other significant changes.

Have a Burning Question of your own? Submit it to marielle_hoefnagels@mcgraw-hill.com for possible inclusion in future editions of this book!

FIGURE 14.A A Single Genetic Switch Can Change Phenotype. The leg on the left is a normal, untreated embryonic chicken leg. The leg on the right has had some of its cell surface receptors blocked, so they cannot receive the message to die. As a result, the webbing that joins the digits remains intact. The blocked leg also developed feathers instead of scales. The experiment shows that a single genetic change can exert a profound effect on phenotype, one great enough to have influenced evolution.

we haven't discovered all there is to discover. Chapter 15 explores these reasons in more detail.

Another explanation for the absence of some predicted transitional forms is that such "missing links" may have been too rare to leave many fossils. Evolution may not always occur gradually, and speciation may sometimes occur in bursts. Periods of rapid biological changes would not leave much fossil evidence of transitional forms. In 1972, paleontologists Stephen Jay Gould and Niles Eldredge coined the term **punctuated equilibrium** to describe long periods of stasis alternating (*punctuated*) with relatively brief bursts of fast evolutionary change (figure 14.9). The fossil record, they argued, lacks some transitional forms because they never existed in a particular location, or because there were simply too few organisms to leave fossils.

The punctuated equilibrium model fits well with the concept of allopatric speciation. Imagine the isolated population of desert pupfish that, over time, became genetically distinct from its ancestral population. If the climate changed, and the pool containing the new species rejoined its "old" pool, the fossil record might show that a new fish species suddenly appeared with its ancestors. Unless paleontologists were lucky enough to recover fossils from the formerly isolated pool, there would be no apparent transitions between the two species. Afterward, unless the environment changed, no new selective forces would drive the formation of new species. Thus, a period of stability would ensue.

The fossil record reveals that both punctuated equilibrium and gradualism occur. Fossils of microscopic protists such as foraminiferans and diatoms, for example, reveal a gradualistic pattern of evolution (see chapter 19's Investigating Life: Glassy Fossils Reveal the Birth of a Species). These asexually reproducing organisms live in vast populations that span the oceans. Since isolated populations would rarely form under those circumstances, it is perhaps not surprising that speciation has occurred only gradually among those protists. On the other hand, the fossils of diverse animals such as bryozoans, mollusks, and mammals all reveal many examples of rapid speciation followed by periods of stability.

B. Bursts of Speciation Occur During Adaptive Radiation

Speciation can happen in rapid bursts during an **adaptive radiation,** in which a population faced with a diverse environment gives rise to multiple specialized forms in a relatively short time. Such an event might occur in several ways. For example, a few individuals might colonize a new, isolated habitat such as a mountaintop or island. The presence of multiple potential food sources (plants with different-sized seeds, for example) would simultaneously select for different phenotypes. Over time, multiple species would develop.

The examples noted in the chapter opening essay, and the finches and tortoises that Darwin observed in the Galápagos, illustrate the adaptive radiation that is common in island groups. Another stunning island example of adaptive radiation occurred in the Greater Antilles, which consist of Cuba, Jamaica, Puerto Rico, and Hispaniola. On each of the islands, small lizards called anoles have adapted in precisely the same ways to different parts of the trees on which they live (**figure 14.10**). Adaptive radiation occurred separately on each island.

FIGURE 14.10
Adaptive Radiation on Islands. Adaptive radiation of anoles has occurred on the islands of the Greater Antilles. On each island, different species of the lizards have adapted to living in different parts of trees, in strikingly similar ways.

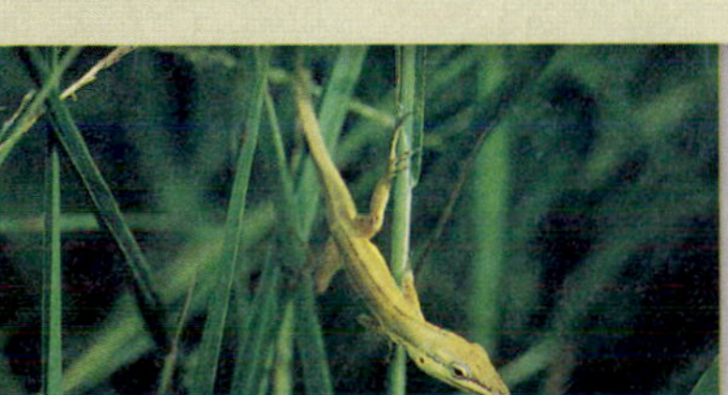

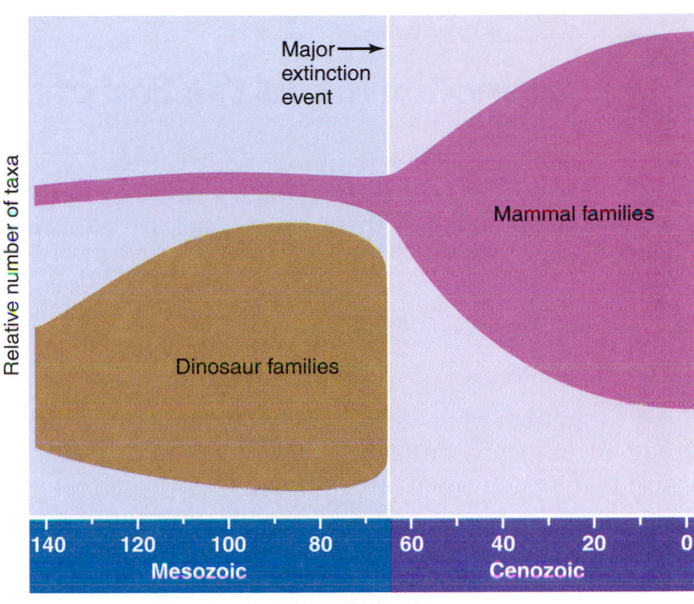

FIGURE 14.11 Adaptive Radiation of Mammals. With many ecological niches vacated when dinosaurs became extinct about 65 million years ago, mammals flourished.

As another path to adaptive radiation, some members of a population may inherit a structure or ability that gives them an advantage, such as a large beak that enables a bird to eat seeds too large for others of its kind. The birds might then diversify as they exploit the new food source. On a larger scale, the evolution of bird flight likewise opened an entirely new set of options for feeding and shelter, prompting rapid diversification. The evolution of photosynthesis billions of years ago likewise led to a burst of speciation.

A third type of adaptive radiation occurs when some members of a population have a combination of adaptations that enable them to survive a major environmental change. After the poorly suited organisms perished, the survivors would diversify as they exploited the new resources in the changed environment. Mammals, for example, underwent an enormous adaptive radiation when the extinctions of the dinosaurs opened up many new habitats (**figure 14.11**).

Summary *Speciation may be gradual or fast; the fossil record supports both gradualism and punctuated equilibrium theories. In adaptive radiation, speciation occurs in response to the availability of new resources.*

MASTERING CONCEPTS

1. Describe the theories of gradualism and punctuated equilibrium. How can they both occur?
2. What are three ways that adaptive radiation can occur?

14.5 Extinction Marks the End of the Line

A species goes **extinct** when all of its members have died. If speciation is the birth of a species, extinction represents its death.

A. Many Factors Can Combine to Put a Species at Risk

Many factors can cause extinction, but all amount to a failure to adapt to environmental change. A species eventually vanishes if its individuals are not adapted well enough to a changing environment to produce fertile offspring and sustain the population. The change that wipes out a species may be habitat loss, new predators, or new diseases. Extinction may also be a matter of bad luck: sometimes no individual of a species survives a volcanic eruption or asteroid impact.

When faced with a shifting environment, what is the chance that a species will become extinct? The answer depends on how fast the environment is changing relative to the rate at which the population evolves. Species such as elephants, which reach reproductive maturity at 30 years, are much less likely to survive a sudden change than mice, which produce three generations a year. In addition, the smaller the initial size of a population, the less likely it is to endure a major challenge. Smaller populations experience fewer genetic mutations, which are the ultimate source of new adaptations (see section 13.4). Low genetic diversity within a population poses two problems. A new disease-causing organism may leave so few survivors that the host population cannot reproduce sufficiently to maintain its size. Secondly, inbreeding tends to bring together lethal alleles, which can drastically weaken the organisms.

B. Extinction Rates Have Varied over Time

Biologists distinguish between two different types of extinction events. The **background extinction rate** results from the gradual loss of species as populations shrink in the face of new challenges. From consulting the fossil record, paleontologists calculate that most species exist from 1 to 10 million years before extinction. Thus, the rate of background extinctions is roughly 0.1 to 1.0 species per year per million species. Most extinctions overall occur as part of this background rate.

Earth has also witnessed several periods of **mass extinctions,** when a great number of species disappeared over relatively short expanses of time (**figure 14.12**). Mass extinctions have had a great influence on Earth's history because they have periodically opened vast new habitats for adaptive radiation to occur.

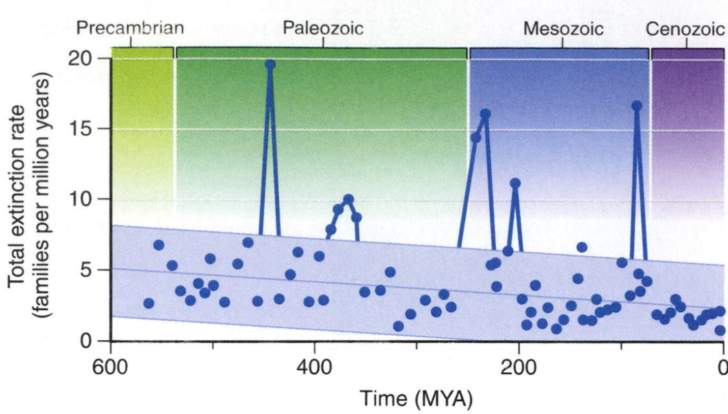

FIGURE 14.12 Catastrophes Affect Evolution. Earth has seen several mass extinctions that have wiped out whole families of species, including marine animals. This graph shows the extinction rate for marine animals over the past 600 million years. The shaded area near the bottom of the graph estimates the background extinction rate; peaks show five mass extinctions.

Paleontologists study clues in Earth's sediments to understand the catastrophic events that contribute to mass extinctions. Two theories have emerged to explain these events, although several processes have probably contributed to mass extinctions.

The first explanation, called the **impact theory,** suggests that a meteorite or comet crashed to Earth, sending dust, soot, and other debris into the sky, blocking sunlight and setting into motion a deadly chain reaction. Without sunlight, plants died. The animals that ate plants, and the animals that ate those animals, then perished. Evidence for the impact theory of the extinction at the end of the Cretaceous period includes centimeter-thin layers of earth that are rich in iridium, an element rare on Earth but common in meteorites (**figure 14.13**).

FIGURE 14.13 Impact Theory Evidence. This distinctive layer of rock marks the end of the Cretaceous period.

A second theory is that movements of Earth's crust may explain some mass extinctions. The crust, or uppermost layer of the planet's surface, is divided into many pieces, called tectonic plates. During Earth's history, movement of tectonic plates caused continents to drift apart, then come back together. Oceans mixed and separated. The result was dramatic environmental change that profoundly affected life. Organisms that had thrived in their habitats had new competitors for limited resources. Weather conditions changed; ice ages and droughts killed many. Shifting continents altered shorelines, diminishing shallow sea areas packed with life. **plate tectonics, p. 314**

The role of *Homo sapiens* in causing extinctions is evident today. For many types of organisms, ecologists are documenting an alarming increase in background extinction rates to 20 to 200 extinctions per million species per year. Habitat loss, introduced species, and overharvest combine to imperil many species (see chapter 42). This chapter's Can *You* Relate? Recent Species Extinctions lists a few of the many vertebrate species that have recently become extinct worldwide. Only time will tell if we are on the brink of another mass extinction or are just at a peak in the many ebbs and flows of biodiversity that have characterized life on Earth.

Summary *Extinctions occur continuously as populations lose genetic diversity or cannot adapt to changing environmental conditions. Historically, mass extinctions have reflected global disturbances.*

MASTERING CONCEPTS

1. What factors can cause or hasten extinction?
2. Distinguish between background extinction and mass extinctions.
3. How have humans influenced extinctions?

Can *You* Relate?

Recent Species Extinctions

Species extinctions have occurred throughout life's long history. They continue today, many times accelerated by human activities. Overharvesting contributes to species extinctions, as does habitat loss to agriculture or urbanization. Less obvious is the effect of introduced plants and animals, which often deplete native species by competing with or preying on them. Many biologists estimate that we are currently experiencing a sixth global mass extinction; global climate change may make the effects even worse (see chapter 42).

The following table lists a few selected species of vertebrate animals that have disappeared during the past few centuries. This list is far from complete; many more species of animals (both vertebrate and invertebrate) and plants (both vascular and nonvascular) have become extinct during the same time. Countless others are threatened or endangered, meaning that they are at risk for extinction.

Common Name	Scientific Name	Former Location
Fishes:		
Snake River sucker	*Chasmistes muriei*	North America
Las Vegas dace	*Rhinichthys deaconi*	North America
Amphibians:		
Palestinian painted frog	*Discoglossus nigriventer*	Israel
Southern day frog	*Taudactylus diurnus*	Australia
Reptiles:		
Yunnan box turtle	*Cuora yunnanensis*	China
Martinique lizard	*Leiocephalus herminieri*	Martinique
Birds:		
Dodo	*Raphus cucullatus*	Mauritius
Moa	*Megalapteryx diderius*	New Zealand
Laysan honeyeater	*Himatione sanguinea*	Hawaii
Black mamo	*Drepanis funerea*	Hawaii
Passenger pigeon	*Ectopistes migratorius*	North America
Great auk	*Alca impennis*	North Atlantic
Mammals:		
Quagga	*Equus quagga quagga*	South Africa
Steller's sea cow	*Hydrodamalis gigas*	Bering Sea
Bali tiger	*Panthera tigris balica*	Indonesia
Javan tiger	*Panthera tigris sondaica*	Indonesia
Caspian tiger	*Panthera tigris virgata*	Central Asia

14.6 Tree Diagrams Describe the Comings and Goings of Species

Darwin proposed that evolution occurs in a branched fashion, with each species giving rise to other species as populations occupy and adapt to new habitats. Ample evidence has since shown him to be correct.

Figure 14.14A shows an example of the current view of evolution as a series of branches that form a treelike diagram. Evolutionary trees, also called **phylogenies,** depict species' relationships based on descent from shared ancestors. A branching pattern of lines represents populations that diverge genetically, splitting off a new species or subspecies (a species-in-the-making). Discontinuation of a line indicates extinction. Evolutionary tree diagrams also distinguish between gradualism and punctuated equilibrium by the angle of the branch points. A small angle indicates gradualism, and a horizontal offshoot depicts rapid change.

Phylogenies were originally derived from fossil data. For many years, they depicted evolutionary relationships without pinpointing exactly when the branching occurred. Newer molecular data, however, have enabled researchers to place approximate timescales on evolutionary tree diagrams (figure 14.14B).

Much of the molecular data consists of comparing sequences of DNA found in mitochondria. This "mtDNA" represents only the mother's lineage, because sperm do not usually donate mitochondria to eggs during fertilization. Also, mtDNA mutates more rapidly than DNA from a cell's nucleus. Comparing the number of mtDNA nucleotide base differences between pairs of species provides a way to estimate how long ago the species diverged, because we can estimate the rate at which this DNA mutates. That is, the more

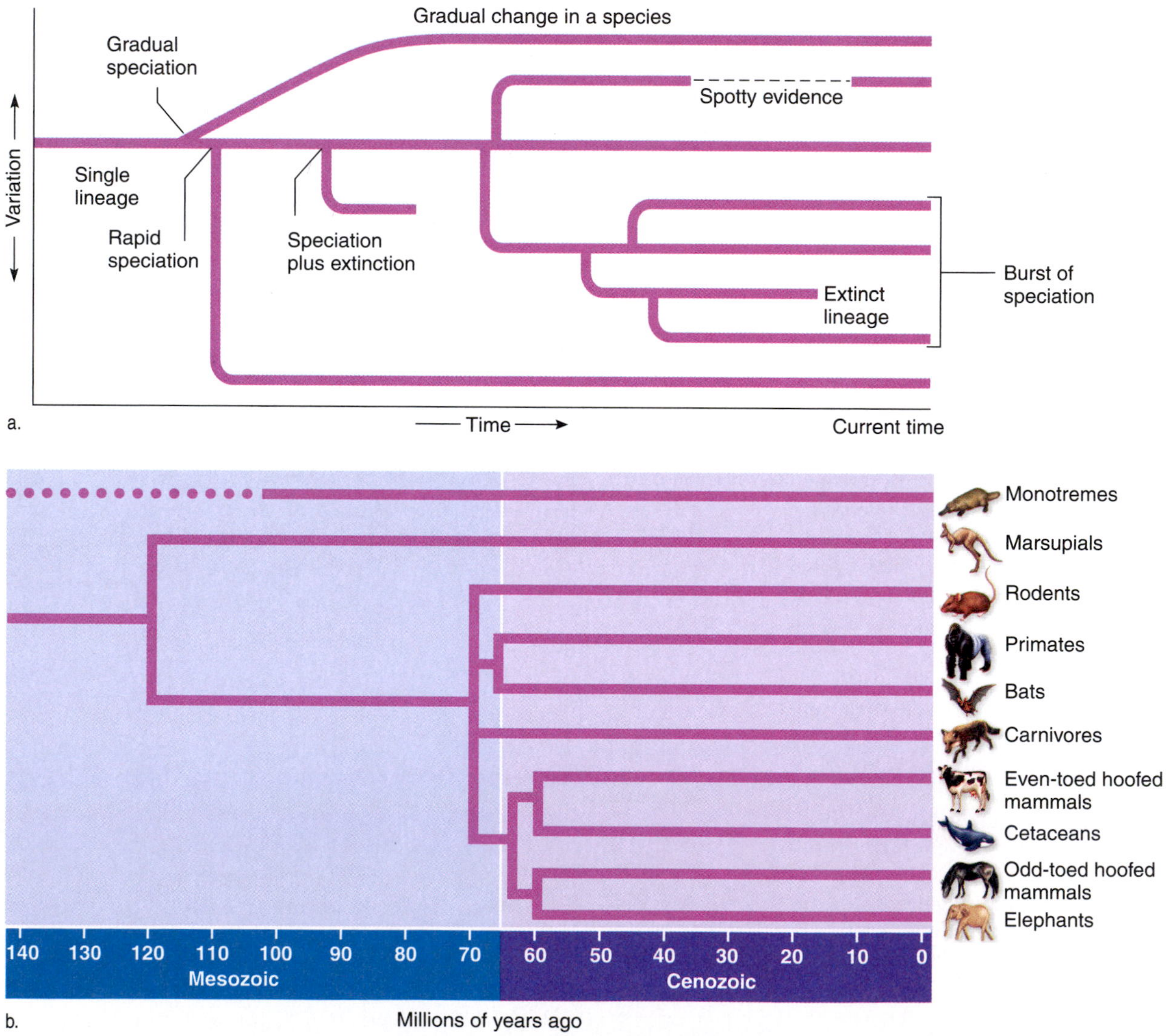

Gradual change in a species
Gradual speciation
Spotty evidence
Variation
Single lineage
Rapid speciation
Speciation plus extinction
Extinct lineage
Burst of speciation
a.
Time
Current time

Monotremes
Marsupials
Rodents
Primates
Bats
Carnivores
Even-toed hoofed mammals
Cetaceans
Odd-toed hoofed mammals
Elephants
140 130 120 110 100 90 80 70 60 50 40 30 20 10 0
Mesozoic
Cenozoic
b.
Millions of years ago

FIGURE 14.14
Plotting Evolution. (**A**) An evolutionary tree diagram can depict rates and times of speciation and extinction events. (**B**) This phylogenetic tree shows that only a few species of mammals existed before the extinction of dinosaurs. Afterward, however, the number of species increased dramatically.

different the mtDNA sequences, the longer ago the two species shared an ancestor. **molecular clocks, p. 320**

Chapter 15 explores in more detail how biologists use multiple lines of evidence to construct phylogenetic trees.

Summary ***Phylogenetic trees describe the evolutionary relationships among existing and extinct species.***

14.6 MASTERING CONCEPTS

1. How do tree diagrams represent evolutionary history, including speciation and extinction?
2. How do DNA sequences help biologists refine phylogenies?

14.7 INVESTIGATING LIFE

A New Species Is Born, but Who's the Daddy?

The title of Charles Darwin's book, *On the Origin of Species*, conveys a central question in biology: Where do new types of organisms come from? All species descend from a common ancestor, so each must have arisen from a preexisting one. But how? According to the biological definition, a new species forms when a reproductive barrier arises between two existing populations. Unfortunately, since most new species formed long ago, details about the exact events of speciation usually remain lost to history.

Sometimes, however, science catches a lucky break, as in the case of a flowering plant called goat's beard, or *Tragopogon*. This weedy plant is a type of dandelion. Europeans introduced three *Tragopogon* species to North America in the 1900s, cultivating the plants for their edible roots. The seeds ride the wind on a parachute-like crown of fluff, so the plants spread widely across the continent. The most common of the three introduced plants is *T. dubius*, whereas two rarer species are *T. pratensis* and *T. porrifolius*.

A brief lesson on plant reproduction will help clarify why *Tragopogon* is so important to biologists who study speciation. Plants in the three introduced species are diploid, and each individual has two parents, just as you do. Pollen carries haploid sperm cells to a female flower part containing an egg cell. After fertilization, the diploid zygote develops into an embryo, which is packaged along with a food supply into a seed.

Errors occasionally occur during gamete production, and a plant may produce sex cells with two sets of chromosomes (diploid) instead of just one (haploid). If a diploid sperm cell fertilizes a diploid egg cell, the resulting zygote has four sets of chromosomes, not two; in other words, it is tetraploid.

Biologists have known since the 1950s that this type of "mistake" sometimes occurs in *Tragopogon*. The tetraploid offspring formed two new species, called *T. miscellus* and *T. mirus* (**figure 14.15**), neither of which can mate with the "parental" diploid plants. By Mayr's biological definition, therefore, they are undoubtedly new species—the first reported example of speciation occurring in nature **(table 14.1)**.

Because *Tragopogon* did not exist in North America before the 1900s, the two tetraploid species must have arisen in just half a century. Researchers have therefore studied the plants' structure, genetics, and chemistry since the 1950s, hoping to learn some of speciation's intimate details. By the late 1980s, a new tool had emerged, one that Darwin could never have imagined: DNA analysis.

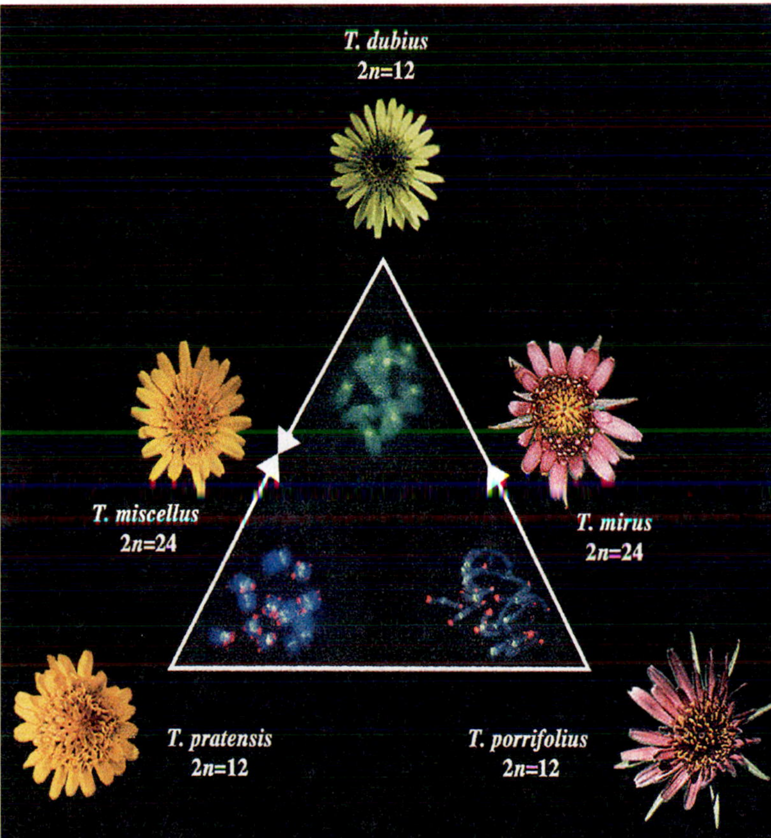

FIGURE 14.15 **The *Tragopogon* Triangle.** The three diploid species of *Tragopogon* have hybridized to form two tetraploid species. **Question:** *Tragopogon dubius* fathered all known populations of *T. mirus*, and most populations of *T. miscellus*. The Soltises suggest that because *T. dubius* is so much more common than the other two diploid species, its pollen is also the most abundant. How would you design an experiment to test the hypothesis that the most common plant is most likely to be the father of a tetraploid hybrid?

Table 14.1 *Tragopogon* Species Origins

Species	Diploid or Tetraploid?	Parents
T. dubius	Diploid	—
T. pratensis	Diploid	—
T. porrifolius	Diploid	—
T. mirus	Tetraploid	*T. dubius* (paternal) and *T. porrifolius* (maternal)
T. miscellus (most populations)	Tetraploid	*T. dubius* (paternal) and *T. pratensis* (maternal)
T. miscellus (two populations)	Tetraploid	*T. pratensis* (paternal) and *T. dubius* (maternal)

Washington State University biologists Douglas and Pamela Soltis studied DNA extracted from chloroplasts, the organelles of photosynthesis. Each plant inherits chloroplast DNA from just one parent—the female. The egg contributes cytoplasm, containing mitochondria and chloroplasts, to the zygote. The comparatively tiny sperm cell contains few organelles.

Chloroplast DNA can therefore answer a question about the short history of North American *Tragopogon*: Which diploid species was the father, and which was the mother, of the two tetraploid species? To find an answer, the Soltises collected seeds from 39 natural populations of all five *Tragopogon* species. After allowing the seeds to germinate in large trays of soil, they isolated the chloroplast DNA from each plant's leaves. The researchers then digested the DNA with 18 restriction enzymes, each of which cut the DNA at a different sequence. Electrophoresis separated the fragments, and stains made the bands of DNA visible. The Soltises knew that different patterns of DNA fragments reflected underlying differences in chloroplast DNA sequences. **DNA profiling, p. 149**

The gels revealed a unique fragment pattern for each diploid *Tragopogon* species. When Soltis and Soltis compared these genetic "fingerprints" to those of the tetraploid hybrids, it was clear that the female parent of *T. mirus* was *T. porrifolius*, whereas the male parent was *T. dubius*. The same species also fathered most populations of *T. miscellus*, with *T. pratensis* being the female parent (**figure 14.16**). Interestingly, however, two samples of *T. miscellus* had chloroplast DNA like that of *T. dubius*. *Tragopogon miscellus* has therefore arisen more than once.

The story of goat's beard is important because it shows that reproductive barriers do not always arise gradually; instead, a sudden genetic change can instantly separate a brand new species from its parents. The observation that this process has occurred more than once in 50 years, at least for *Tragopogon*, is tantalizing. How many times has it happened in life's history? We will probably never know. Nearly 150 years after *The Origin of Species*, however, our window on speciation is clearer than ever.

Soltis, Douglas E. and Pamela S. Soltis. 1989. Allopolyploid speciation in *Tragopogon*: insights from chloroplast DNA. *American Journal of Botany*, vol. 76, pages 1119–1124.

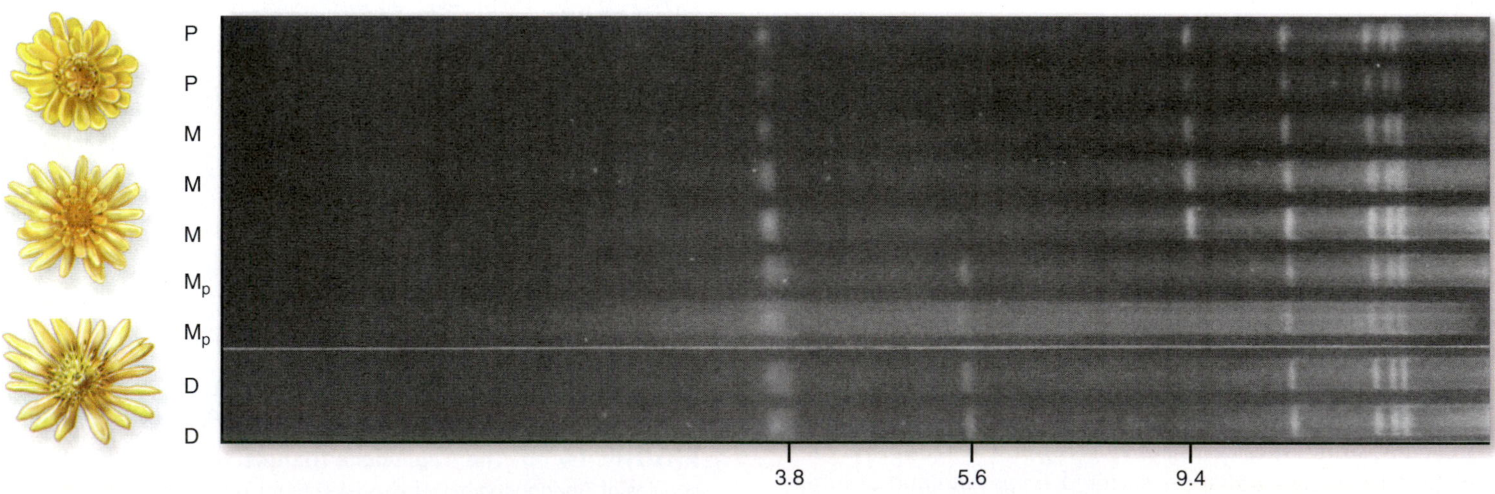

FIGURE 14.16 Chloroplast DNA: The Soltises purified DNA from chloroplasts of *T. pratensis* (P), multiple populations of *T. miscellus* (M and M_p), and *T. dubius* (D). They then cut the DNA into fragments with a restriction enzyme. DNA with different sequences should yield different-sized fragments, which migrate to different locations on a gel. Because chloroplast DNA comes from only one of a plant's parents, a match reveals the identity of the mother. *T. miscellus* may have either *T. pratensis* or *T. dubius* as a maternal parent, indicating that this hybrid species has arisen more than once.

CHAPTER SUMMARY

14.1 The Definition of "Species" Has Evolved over Time

- **Macroevolution** refers to large-scale changes in life's diversity, including the appearance of new **species** and higher taxonomic levels.

A. Linnaeus Devised the Binomial Naming System

- Linnaeus's species designations and classifications helped scientists communicate. Darwin added evolutionary meaning.

B. Ernst Mayr Developed the Biological Species Concept

- Mayr added the requirement for reproductive isolation to define **biological species.**
- **Speciation** is the formation of a new species. The appearance of a reproductive barrier provides objective evidence of speciation.

14.2 Reproductive Barriers Separate Species from One Another

A. Prezygotic Barriers Prevent Fertilization

- **Prezygotic reproductive isolation** occurs before or during fertilization. It includes obstacles to mating such as space, time, and behavior; mechanical mismatches between male and female; and molecular mismatches between gametes.

B. Postzygotic Barriers Prevent Viable or Fertile Offspring

- **Postzygotic reproductive isolation** results in offspring that die early in development or are infertile.

14.3 Spatial Arrangements Define Three Main Modes of Speciation

A. Allopatric Speciation Reflects a Geographic Barrier

- **Allopatric speciation** occurs when a geographic barrier separates a population. The two populations then diverge genetically to the point that their members can no longer produce fertile offspring together.

B. Parapatric Speciation Occurs in Neighboring Regions

- **Parapatric speciation** occurs when two populations live in neighboring areas but share a border zone. Genetic divergence between the two groups exceeds gene flow, driving speciation.

C. Sympatric Speciation Occurs in a Shared Habitat

- **Sympatric speciation** enables populations that occupy the same area to diverge, often via drastic genetic change such as **polyploidy.**

D. Determining the Mode of Speciation May Be Difficult

- The distinction between allopatric, parapatric, and sympatric speciation is not always straightforward, partly because it is difficult to define the size of a geographic barrier.

14.4 Evolution May Be Gradual or Occur in Bursts

A. Gradualism and Punctuated Equilibrium Are Two Models of Speciation

- Evolutionary change occurs at many rates, from slow and steady **gradualism,** to the periodic bursts that characterize **punctuated equilibrium.**

B. Bursts of Speciation Occur During Adaptive Radiation

- In **adaptive radiation,** an ancestral species rapidly branches into several new species, reflecting the sudden availability of new habitats or resources or a particularly beneficial adaptation.

14.5 Extinction Marks the End of the Line

- **Extinction** is the disappearance of a species.

A. Many Factors Can Combine to Put a Species at Risk

- A slow reproductive rate and low genetic diversity make species vulnerable to extinction during times of rapid environmental change.

B. Extinction Rates Have Varied over Time

- The **background extinction rate** reflects ongoing losses of species on a local scale.
- Historically, **mass extinctions** have resulted from global changes such as continental drift. The **impact theory** suggests that a meteorite or comet changed Earth's climate and caused a mass extinction at the end of the Cretaceous period.
- Human activities are increasing the extinction rate.

14.6 Tree Diagrams Describe the Comings and Goings of Species

- Diagrams of branching lines represent evolutionary relationships among species. Such evolutionary trees are called **phylogenies.**
- Sequence comparisons of mtDNA help clarify and supplement knowledge of species' relationships based on visible characteristics.

14.7 Investigating Life: A New Species is Born, but Who's the Daddy?

- Analysis of chloroplast DNA has revealed the parental origin of two species of *Tragopogon* plants that have arisen since the 1950s.

MULTIPLE CHOICE QUESTIONS

1. The term *macroevolution* refers to
   a. large-scale changes in the diversity of biological organisms.
   b. large-scale changes in the DNA of biological organisms.
   c. evolutionary changes that affect larger biological organisms.
   d. evolutionary changes that can be observed.
2. Why is the ability to produce fertile offspring a key part of the definition of a species?
   a. Because of the role of mutation in evolution.
   b. Because the individual offspring should be adapted to their environment.
   c. Because fertile offspring are more fit than unfertile ones.
   d. Because the offspring maintain the gene pool of the population.
3. Which of the following is NOT a form of prezygotic reproductive isolation?
   a. Temporal isolation
   b. Genetic isolation
   c. Mechanical isolation
   d. Ecological isolation
4. How does a difference in chromosome number lead to hybrid infertility?
   a. The difference affects mitotic cell division.
   b. The cells of the hybrid cannot grow, so the embryo dies.
   c. Meiosis is blocked, so gametes cannot form.
   d. Mitosis is altered, so the gametes are inviable.
5. Home construction and urban development often lead to habitat fragmentation. What type of speciation might occur as a result of this human activity?
   a. Sympatric speciation
   b. Allopatric speciation
   c. Parapatric speciation
   d. Both a and c
6. Polyploidy represents a ______ type of reproductive barrier.
   a. mechanical
   b. temporal
   c. postzygotic
   d. prezygotic
7. Adaptive radiation can best be defined as the ability of a population
   a. to evolve a phenotype that helps them adapt to a new environment.
   b. with a specific adaptation to find new environments.
   c. to migrate to a better environment.
   d. to undergo speciation in response to new or changing environments.
8. Why is a species with a small population more likely to undergo an extinction?
   a. Because they cannot produce enough offspring.
   b. Because there will be less genetic diversity.
   c. Because the individuals may be isolated from one another.
   d. Because the time required to produce offspring may be too long.
9. The difference between a mass extinction and the background extinction rate is
   a. the number of species affected.
   b. the time period involved.
   c. the role of humans.
   d. Both a and b.
10. How does mitochondrial DNA contribute to our understanding of phylogenies?
    a. It allows scientists to identify the maternal source for an organism.
    b. It allows for a calculation of the evolutionary time scale.
    c. It demonstrates whether gradualism or punctuated equilibrium occurred.
    d. It identifies extinctions.

TESTING YOUR KNOWLEDGE

1. What did Linnaeus, Darwin, and Mayr contribute to the meaning of the term *species*?
2. Whenever two particular types of animals mate, pregnancy occurs, but the embryos stop developing when they are just balls of cells. Is this prezygotic or postzygotic reproductive isolation?
3. Distinguish among allopatric, parapatric, and sympatric speciation and give an example of each.
4. Examine the lizards and their characteristics shown in Figure 14.10. Propose an advantage for each of the variations, given the environment. How could each of these species have arisen from a common ancestor?

5. How does natural selection predict a gradualistic mode of evolution? Does the presence of fossils that are consistent with punctuated equilibrium mean that natural selection does not occur?
6. Why do species become extinct?
7. Describe the relationships among the species represented by letters in these evolutionary tree diagrams:

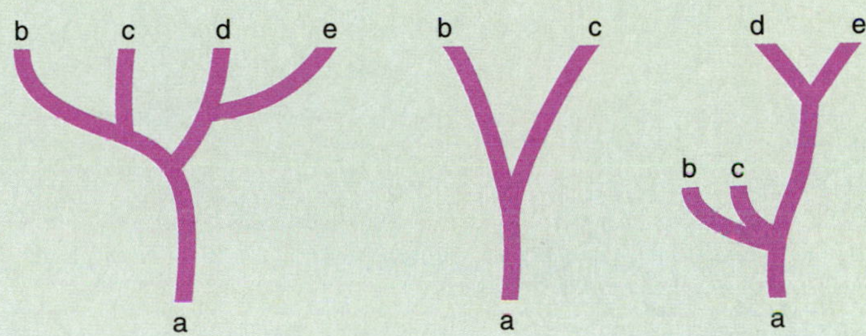

THINKING AS A SCIENTIST

1. A researcher sets up an experiment in which two genetically very different populations of fruit fly are placed in two enclosed areas connected by a narrow tube, through which the flies can fit. Suggest two scenarios that can occur in the entire setup, over time.
2. Each island of the Greater Antilles is home to a similar-appearing assortment of anoles (see figure 14.10). However, mtDNA analysis showed that the anoles from different islands are not alike genetically, even though they resemble one another. How is this possible?
3. Humans introduced apple trees to North America in the 1800s. Insects called hawthorn flies, which feed and mate on hawthorn plants, quickly discovered the new fruits. Some flies preferred the taste of apples to their native host plants. Because these flies mate where they eat, this difference in food preference quickly led to a reproductive barrier. Which type of reproductive isolation does this situation represent? If the apple-feeding flies form a different species from their hawthorn-feeding relatives, which mode of speciation (allopatric, parapatric, or sympatric) does this illustrate?
4. Investigation of fossilized eggshells from a huge bird (*Genyornis newtoni*) shows that it became extinct about 50,000 years ago in Australia, a time that coincided with human colonization. What types of evidence would support or refute the hypothesis that humans hunted the bird into extinction?
5. What information would you need to determine the background extinction rate hundreds of millions of years ago? What about to determine the current extinction rate?.

Visit www.mhhe.com/hoefnagels for practice quizzes, animations, videos, and activities designed to help you master the material in this chapter.

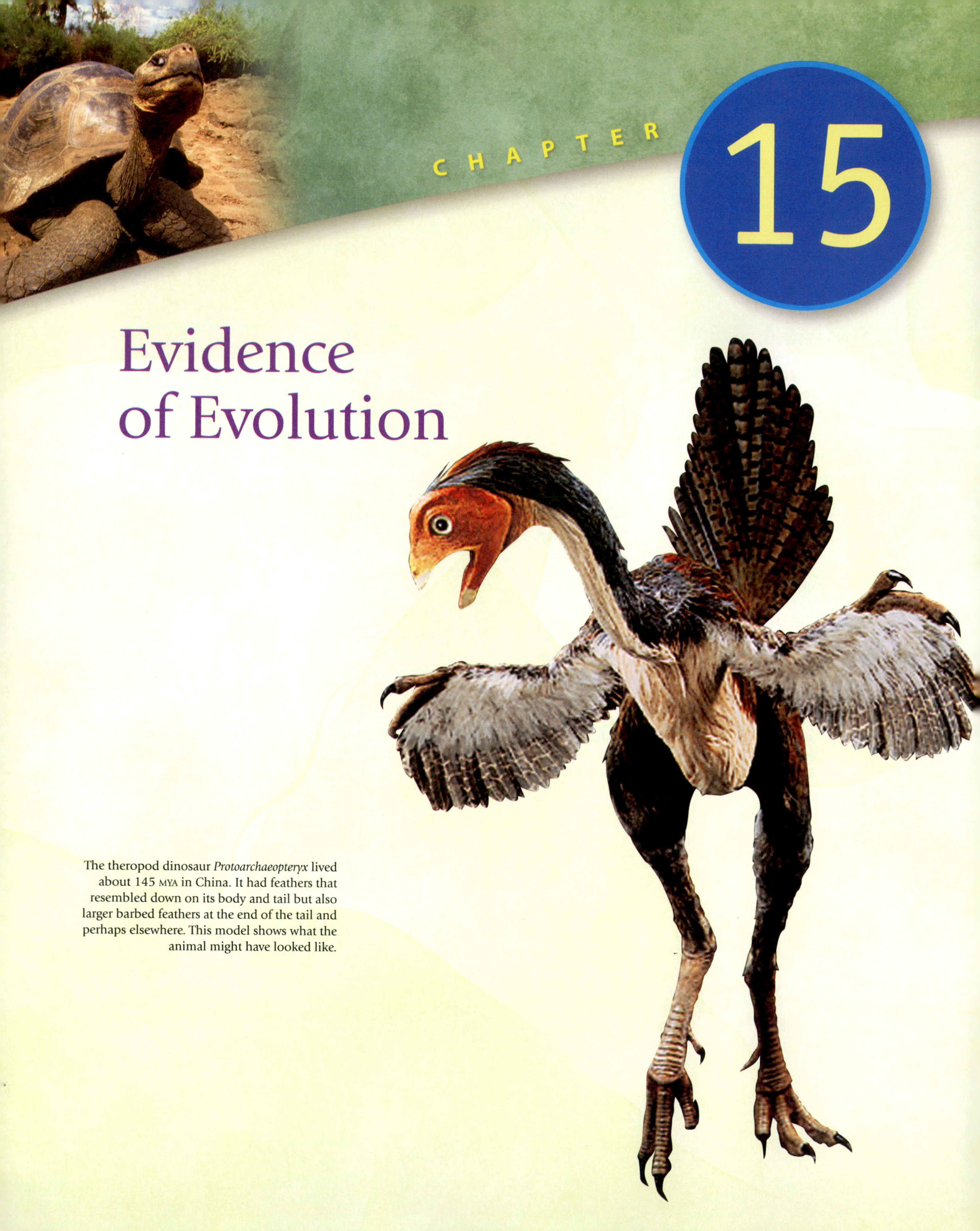

CHAPTER 15

Evidence of Evolution

The theropod dinosaur *Protoarchaeopteryx* lived about 145 MYA in China. It had feathers that resembled down on its body and tail but also larger barbed feathers at the end of the tail and perhaps elsewhere. This model shows what the animal might have looked like.

What's the Point?

Are Birds Dinosaurs?

Pictures and stories of life's history emerge as scientists collect and assemble bits of evidence over time. Like a puzzle with many pieces missing, the history of life has many possible interpretations. But as more information surfaces, the pieces fit together, and the story becomes clearer. Investigating the relationship between birds and dinosaurs is one such story of life.

The hypothesis that birds are closely related to small, bipedal (standing on two legs) meat-eating dinosaurs called theropods began with the discovery of a 150-million-year-old fossil of a theropod with feathers in Bavaria. It was 1860, the year after Darwin published On the Origin of Species. *The animal, named* Archaeopteryx lithographica, *was the size of a blue jay, with a mix of features seen in birds and reptiles today: wings and feathers, a toothed jaw, and a long, bony tail.*

In 1870, Thomas Henry Huxley reported that he had identified 35 features that only ostriches and theropod skeletons shared, which he interpreted as evidence of a close relationship. Others dismissed Huxley's ideas, largely because dinosaurs were not known to have flown. By 1916, the hypothesis seemed even less likely, when Gerhard Heilmann, a Danish physician and fossil collector, noted that theropods lack clavicles, which form a bird's wishbone. Fossils of theropods with clavicles had simply not yet been discovered.

In the 1960s, however, Yale University paleontologist John Ostrom strengthened the bird–dinosaur link with an exhaustive comparison of bones discovered in the 1920s, including the clavicles of theropods. Since then, the similarities between dinosaurs and birds have added up. Many of the structures that would make flight possible were present in dinosaurs that lived millions of years before birds. For example, some theropods had hollow bones, a bipedal stance in which they stood on their toes, a horizontal back, long arms, and a short tail. Rare finds of dinosaur nests reveal behavioral and reproductive similarities to birds.

For years, all that was missing was evidence of feathered dinosaurs. In 1996, however, researchers discovered fossils of Sinosauropteryx *in Liaoning, a region of China rich in dinosaur fossils. About the size of a turkey, this animal had a fringe of downlike structures along its neck, back, and flanks. In 1998, a bonanza fossil find introduced* Protoarchaeopteryx *and* Caudipteryx, *also from China, and both even more feathery than* Sinosauropteryx.

The surprising bird–dinosaur link illustrates the changeable nature of scientific knowledge and reinforces the importance of evidence. Every chapter of this textbook contains a section titled Investigating Life that explains how biologists test hypotheses about evolution. This chapter summarizes the main types of evidence they use.

LEARNING OUTLINE

15.1 Clues to Evolution Lie in the Earth, Body Structures, and Molecules

At about 4.6 billion years, Earth's history is almost unimaginably long. Scientists describe long-ago events in the context of the **geological time scale,** which divides history into a series of eras defined by major geological or biological events such as mass extinctions (**figure 15.1**). Within the eras, time is further divided into periods and epochs. **mass extinctions, p. 298**

The history of life is nearly as long as that of Earth. The millions of species alive today did not just pop into existence all at once; they are the result of continuing macroevolution-

Era	Period	Epoch	MYA	Important events
Cenozoic era	Quaternary	Recent	0.01	Human civilization
		Pleistocene	1.8	*Homo sapiens*, large mammals; ice ages
	Tertiary	Pliocene	5.3	*Australopithecus*, modern whales
		Miocene	23.8	Hominoids; mammals continue to diversify; modern birds; expansion of grasslands
		Oligocene	33.7	Elephants, horses; grasses
		Eocene	54.8	Mammals and flowering plants continue to diversify; first whales
		Paleocene	65	First primates; mammals, birds, and pollinating insects diversify
Mesozoic era	Cretaceous		144	Widespread dinosaurs until extinction at end of Cretaceous; first flowering plants
	Jurassic		206	First birds and mammals; cycads and ferns abundant; giant reptiles on land and in water
	Triassic		248	First dinosaurs; first mammals; therapsids and thecodonts; forests of conifers and cycads
Paleozoic era	Permian		290	First conifers; fewer amphibians, more reptiles; cotylosaurs and pelycosaurs
	Carboniferous		354	Reptiles arise; ferns abundant; amphibians diversify; first winged insects
	Devonian		417	Bony fishes, corals, crinoids; amphibians arise, land plants and arthropods diversify
	Silurian		443	First vascular plants and terrestrial invertebrates; first fish with jaws
	Ordovician		490	Algae, invertebrates, graptolites, jawless fishes; first land plants
	Cambrian		543	"Explosion" of sponges, worms, jellyfish, "small shelly fossils;" ancestors of all modern animals appear; trilobites
Precambrian: Proterozoic era			2,500	First life in the sea
Precambrian: Archean era and earlier			4,600	Earth forms

Major extinction events (arrows at the ends of the Paleocene/Cretaceous, Jurassic/Triassic, Triassic/Permian, Devonian/Carboniferous, and Ordovician/Silurian boundaries)

FIGURE 15.1 The Geological Time Scale. Scientists divide Earth's 4.6-billion-year-long history into five eras. Notice that the lengths of the eras and epochs are not to scale. That is, the Cenozoic era is but a flicker of time compared with all of Earth's existence. By contrast, the Archean and Proterozoic eras are by far the longest, yet we know the least about them.

ary change in organisms that lived billions of years in the past. Many types of clues enable us to hypothesize about how modern species evolved from extinct ancestors and to understand the relationships between organisms that live today. (See this chapter's Burning Question: Is evolution really testable?).

Traditionally, evidence for evolution came mostly from **paleontology,** the study of fossil remains or other clues to past life (**figure 15.2**). The discovery of many new types of fossils in the early 1800s created the climate that allowed Charles Darwin to make his tremendous breakthrough (see chapter 13). As people recognized that fossils must represent a history of life, scientists developed theories to explain that history. The geographical locations of fossils and modern species provided additional clues.

Scientists began extensively using molecular techniques, including protein and DNA sequencing, in the 1960s and 1970s. After the development of the polymerase chain reaction (PCR) in the 1980s made possible the analysis of tiny quantities of DNA, biologists have produced an explosion of molecular sequences that also provide clues to evolutionary relationships. **polymerase chain reaction, p. 146**

Molecular, fossil, and biogeographical studies each provide different types of evidence in support of evolution. Molecular evidence tells us how species are related to each other. Fossil and biogeographical evidence introduce the element of time, revealing when species most likely diverged from common ancestors against the backdrop of other events happening on Earth.

Chapter 13 explained how natural selection and other mechanisms account for microevolutionary changes within species, and chapter 14 described how new species arise. This chapter expands on these relatively small-scale evolutionary events. Here we will examine the different approaches to studying macroevolution in both living and extinct species. The chapter ends with a description of how scientists assemble the many clues into hypotheses of the relationship between species. Chapter 16 continues on this theme by offering a brief history of life on Earth.

Summary ***Earth's history spans some 4.6 billion years. Evolutionary biologists assemble clues from many sources, both ancient and modern, to paint portraits of past life.***

15.1 MASTERING CONCEPTS

1. What is the geological time scale?
2. What types of information provide the clues that scientists use in investigating evolutionary relationships?

FIGURE 15.2 Fossil Evidence Is Diverse. Plants and animals alike have left a rich fossil record spanning hundreds of millions of years. Clockwise from left: *Archaefructus*, an early flowering plant that lived in China 138 milllion years ago; petrified wood; a 190-million-year-old dinosaur embryo encased in its egg; 20,000-year-old fossilized feces of an extinct giant sloth; trilobites, arthropods that became extinct some 250 million years ago; a skull of *Triceratops*, a dinosaur that lived in North America until 65 million years ago; a *Ginkgo* leaf; an exceptionally well-preserved fish.

Burning Questions

Is evolution really testable?

Some people mistakenly believe that since evolution happened in the past, it cannot be scientifically tested. Although it is true that no experiment can recreate the past conditions that led to the evolution of today's organisms, it is not true that evolution is untestable.

Hypothesis testing does not rely exclusively on experimentation, but it does require investigators to deduce predictions. For example, an evolutionary biologist might predict that fishes should appear in the fossil record before reptiles or mammals. A fossil find that contradicts the prediction would then require investigators to form a new hypothesis. On the other hand, fossils that confirm the prediction lend additional weight to the theory's validity. Other types of evidence, such as molecular evidence, may also either refute or confirm the predictions.

Experiments on organisms that exist today support the case for evolution. As you learned earlier, the increased incidence of antibiotic resistance in bacteria illustrates microevolution (see chapter 13). Although experiments, by themselves, do not prove that microevolution occurred in the past, they do suggest a likely mechanism for evolutionary change. At the very least, they do not contradict the predictions of evolutionary theory.

Evaluating common descent is somewhat similar to solving a crime that no one witnessed. A criminal may leave footprints, tire tracks, fingerprints, and DNA at a crime scene. Detectives develop hypotheses about possible suspects by piecing together the physical and biological evidence. Innocent suspects can exonerate themselves by providing evidence, such as a strong alibi, that contradicts the hypothesis. The hypothesis that best explains the crime is the one that is consistent with all available evidence.

Similarly, extinct organisms have left traces of their existence, both as fossils and as the genetic legacy that all current organisms have inherited. No other scientifically testable hypothesis explains and unifies all of these observations as well as common descent.

Have a Burning Question of your own? Submit it to marielle_hoefnagels@mcgraw-hill.com for possible inclusion in future editions of this book!

15.2 Fossils Record Evolution

A **fossil** is any evidence of an organism from more than 10,000 years ago. Fossils come in all sizes, documenting the evolutionary history of everything from prokaryotes to dinosaurs to humans. These remains give us our only direct evidence of organisms that preceded human history. They occur all over the world and represent all major groups of organisms, revealing much about the geological past. For example, the abundant remains of extinct marine animals called ammonites in Oklahoma indicate that a vast, shallow ocean once submerged the central United States (**figure 15.3**).

FIGURE 15.3 Big Change. Ammonite fossils such as these are common in land-locked Oklahoma, indicating that what is now the central United States was once covered by an ocean.

A. Fossils Form in Many Ways

Evidence of past life comes in many forms (**figure 15.4**). Often, a fossil forms when an organism dies, becomes buried in sediments, and then is chemically altered. For example, coal, oil, and natural gas (also called fossil fuels) are the decomposed remains of plants and other organisms preserved by compression. Alternatively, minerals can replace the organic matter left by a decaying organism, which literally "turns to stone." Petrified wood forms in this way.

Impression fossils form when an organism presses against soft sediment, which then hardens after the organism decays, leaving an outline of the organism. An impression fossil may also be evidence of an animal's movements, such as footprints. If the imprint later fills with mud that hardens into rock, the resulting cast is a rocky replica of the ancient organism. Teeth, bones, arthropod exoskeletons, and tree trunks often leave casts.

Rarely, a whole organism is preserved intact. Fossil pollen in waterlogged lake sediments provides clues to long-ago climates. Sticky tree resin entraps plants and animals, hardening them in translucent amber tombs. The La Brea Tar Pits in Los Angeles have preserved more than 660 species of animals and plants that became stuck in the gooey tar thousands of years ago.

The most striking fossils formed when sudden catastrophes, such as mudslides and floods, rapidly buried organisms in an oxygen-poor environment. Decomposition and tissue damage were minimal in the absence of oxygen; scavengers could not reach the dead. In those conditions,

a. Compression

Leaf sinks.

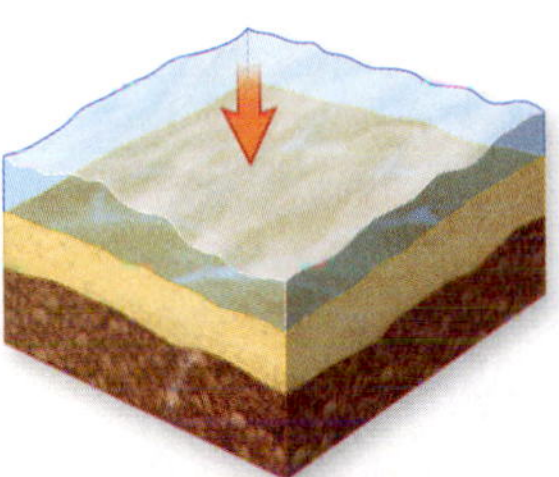
Fine sediment covers leaf.

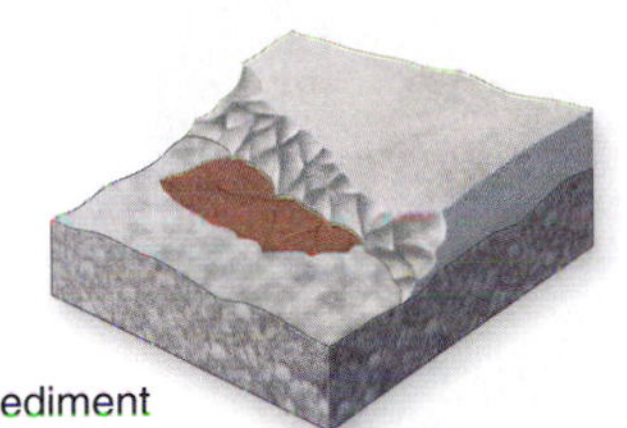
Sediment compresses, forming sedimentary rock.

b. Impression

Animal dies, making impression in mud.

Animal decays away.

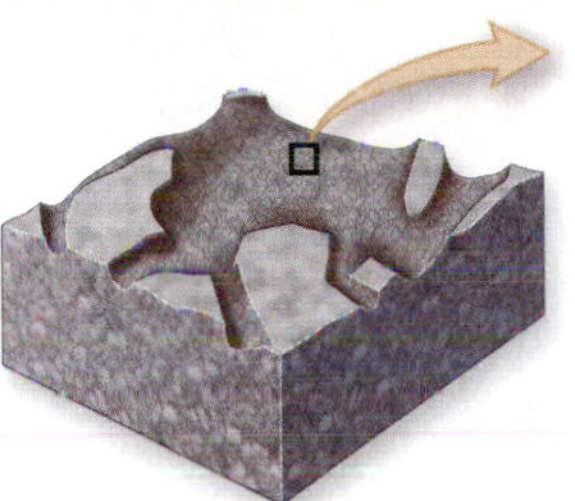
Mud hardens to rock.

c. Cast

Animal dies and sinks into soft sediment.

Animal decays away.

Imprint fills with mud.

Mud hardens to rock.

d. Petrifaction

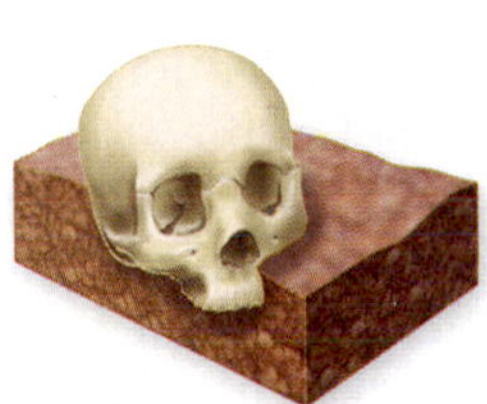
Animal dies, decays, and is buried.

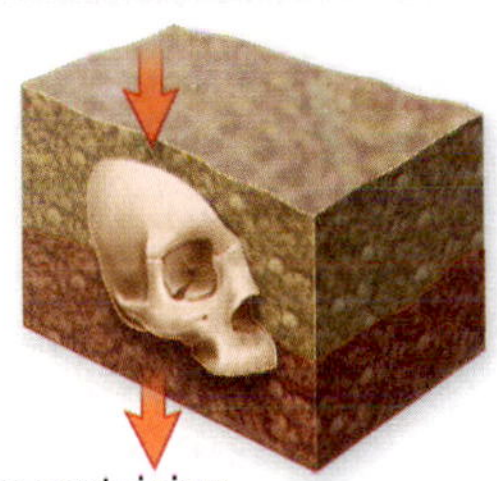
Water containing dissolved minerals seeps through.

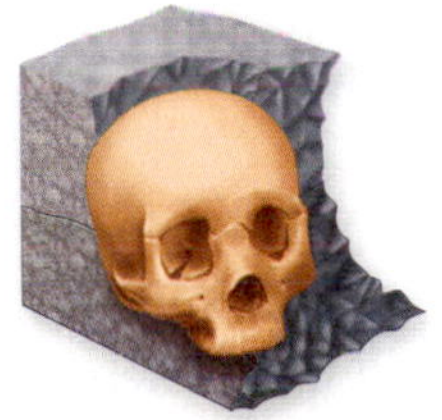
Organic matter replaced by minerals "turns to stone."

e. Intact preservation

Oozing sap traps an insect.

FIGURE 15.4 How Fossils Form. (**A**) A compression fossil of a leaf preserves part of the plant. (**B**) An impression fossil reveals anatomical details as an imprint. A dinosaur's scaly skin left this imprint fossil. (**C**) This horn coral is a cast. Once-living material dissolved and was replaced by mud that hardened into rock. (**D**) Most fossils of our immediate ancestors consist of mineralized bones and teeth, usually found in fragments. (**E**) Fossils can also be preserved intact in tree resin, which hardens to form amber.

a.

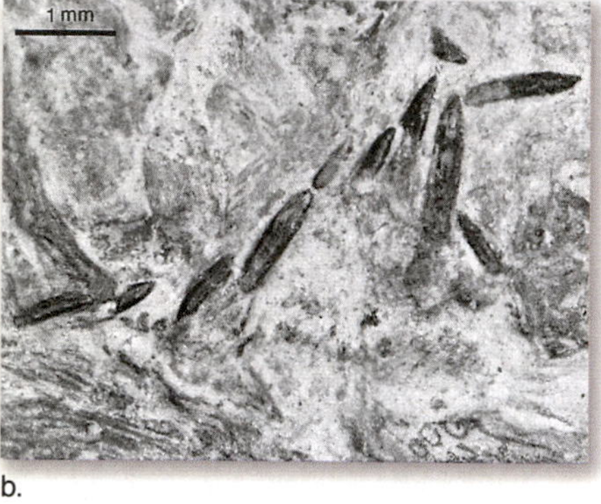

b.

c.

FIGURE 15.5 A Field of Dinosaur Embryos Trapped in Mud and Time. (**A**) Thousands of dinosaur eggs containing about-to-hatch babies were buried as streams overflowed this plain, which was a valley in present-day Argentina about 89 MYA. The newborns would have been only 40 cm (about 15 inches) long, and would attain an adult length of over 13 m (about 45 feet). (**B**) Preserved embryonic teeth were so tiny that a mouthful of 32 of them would fit into an "o" in this sentence. (**C**) The preserved skin from the embryos resembles that from a modern-day 13-week-old Nile crocodile; it does not yet have the hard parts that formed the characteristic armor of the adults.

even soft-bodied organisms have left detailed anatomical portraits (**figure 15.5**).

B. The Fossil Record Is Often Incomplete

Occasionally, researchers find fossils that document, step-by-step, the evolution of one species into another (see chapter 19's Investigating Life: Glassy Fossils Reveal the Birth of a Species). Usually, however, the fossil record is incomplete, meaning that some of the features marking the transition from one group to another are not recorded in fossils.

Several reasons account for this partial history. First, the vast majority of organisms never leave a fossil trace. Soft-bodied organisms, for example, are much less likely to be preserved than are those with teeth, bones, or shells. Organisms that decompose or are eaten after death, rather than being buried in sediments, are also unlikely to fossilize. Second, erosion or the movements of Earth's continental plates have destroyed many fossils that did form. Third, scientists are unlikely to ever discover the many fossils that must be either buried deep in the Earth or submerged under water.

C. The Age of a Fossil Can Be Estimated in Two Ways

Scientists use two general approaches to estimate when a fossilized organism lived: relative dating and absolute dating.

Relative Dating

Relative dating places a fossil into a sequence of events without assigning it a specific age. It is usually based on the principle of superposition, with lower rock strata presumed to be older than higher layers. The farther down a fossil is, therefore, the longer ago the organism it represents lived—a little like a memo at the bottom of a pile of papers on a desk being older than one nearer the top. Relative dating therefore places fossils in order from "oldest" to "most recent."

Absolute Dating and Radioactive Decay

Researchers use **absolute dating** to determine the age of a fossil in years. The dates usually are expressed in relation to the present. For example, scientists studying fossilized *Archaeopteryx* showed that this animal lived about 150 million years ago (MYA). (Although the term *absolute* dating seems to imply pinpoint accuracy, absolute dating techniques typically return a range of likely dates.)

Radiometric dating is a type of absolute dating that uses radioactive isotopes as a "clock." Recall from chapter 2 that different isotopes of the same element have different numbers of neutrons. Some isotopes are naturally unstable, which causes them to emit radiation as they "radioactively decay." Each radioactive isotope decays at a characteristic and unchangeable rate, called its half-life. The **half-life** is the time it takes for half of the atoms in a sample of a radioactive substance to decay. If an isotope's half-life is 1 year, for example, 50% of the radioactive atoms in a sample will have decayed during that time. In another year, half of the remaining radioactive atoms will decay, leaving 25%, and so on. If we measure the amount of a radioactive isotope in a fossil, and compare it with the amount typically found in living tissues, we can use the isotope's half-life to deduce when the fossil formed. **isotopes, p. 23**

One radioactive isotope often used to assign dates to fossils is carbon-14 (^{14}C; **figure 15.6**), which forms in the atmosphere when cosmic rays from space bombard nitrogen gas. Carbon-14 has a half-life of 5730 years; it decays to the more stable nitrogen-14 (^{14}N). Organisms accumulate ^{14}C during photosynthesis or by eating organic mat-

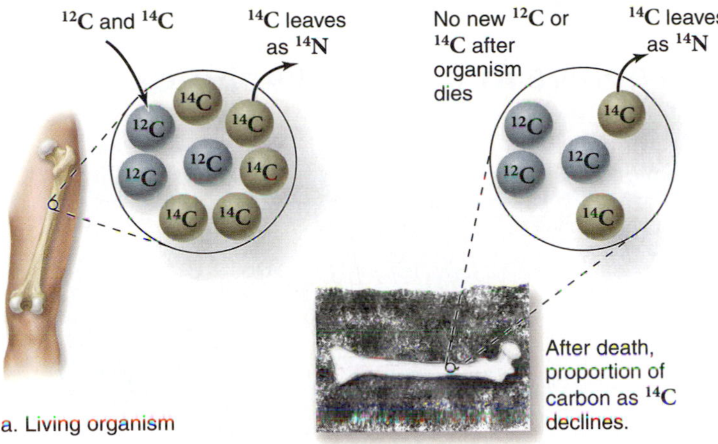

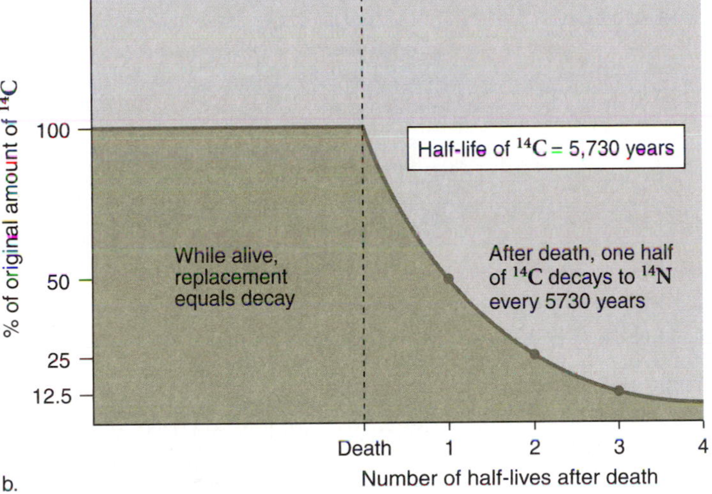

FIGURE 15.6 Carbon-14 Dating. (A) Living organisms accumulate radioactive carbon-14 (^{14}C) through photosynthesis or by eating other organisms. While the organism is alive, ^{14}C is decaying to nitrogen-14 (^{14}N), but the ^{14}C is continually replaced. After death, no new ^{14}C enters the body, **(B)** so the proportion of ^{14}C to ^{12}C in the organism begins to decline. **(C)** Measuring that proportion allows scientists to determine how long ago a fossilized organism—such as this woolly mammoth—died.

ter. One in every trillion carbon atoms present in living tissue is ^{14}C; most of the rest are ^{12}C, a nonradioactive isotope. When an organism dies, however, its intake of carbon, including ^{14}C, stops. As the ^{14}C decays without being replenished, the ratio of ^{12}C to ^{14}C increases. This ratio is then used to determine when death occurred, up to about 40,000 years ago.

For example, radioactive carbon dating determined the age of fossils of vultures that once lived in the Grand Canyon. The birds' remains have about one fourth the ^{14}C-to-^{12}C ratio of a living organism. Therefore, about two half-lives, or about 11,460 years, passed since the animals died. It took 5730 years for half of the ^{14}C to decay, and another 5730 years for half of what was left to decay to ^{14}N.

Another widely used radioactive isotope, potassium-40 (^{40}K), decays to argon-40 (^{40}Ar) with a half-life of 1.3 billion years, so it is valuable in dating very old rocks containing traces of both isotopes. Chemical analyses can detect ^{40}Ar in amounts small enough to correspond to fossils that are about 300,000 years old or older.

One limitation of ^{14}C and potassium-argon dating is that they leave a gap, resulting from the different half-lives of the radioactive isotopes. Several techniques, such as tree-ring comparisons and using other isotopes with intermediate half-lives, cover the missing years. **tree rings, p. 515**

Summary *Fossils—the remains of organisms living long ago—form in several ways. Although the fossil record is incomplete, it is important because it provides our only direct look at extinct life. The time that a fossilized organism lived can be estimated from its position in rock strata or by radiometric dating.*

15.2 MASTERING CONCEPTS

1. What are some of the ways that fossils form?
2. Why will the fossil record always be incomplete?
3. Distinguish between relative and absolute dating of fossils.
4. How does radiometric dating work?

15.3 Large-Scale Changes in Earth's Surface Have Influenced Evolution

Since geographical barriers greatly influence speciation, it is not surprising that the studies of geography and biology overlap. Many examples illustrate how small- and large-scale changes in Earth's features have influenced the rise and fall of species. **allopatric speciation, p. 291**

A. The Theory of Plate Tectonics Describes Earth's Shifting Continents

Despite the occasional hurricanes, tornados, and earthquakes, Earth's geological history might seem rather uneventful. Not so. Fossils tell the story of ancient seafloors rising all the way to Earth's "ceiling": the Himalayan Mountains. Littering the Kali Gandaki River in Nepal are countless fossilized ammonites, large mollusks similar to the chambered nautilus. How did fossils of marine animals end up more than 3600 meters above sea level?

According to the theory of **plate tectonics,** Earth's surface consists of several rigid layers, called tectonic plates, which move in response to forces acting deep within the planet (**figure 15.7**). In some areas where plates come together, mountain ranges form as the plates become wrinkled and distorted. Long ago, the Indo-Australian plate (which includes the subcontinent of India) moved slowly north and eventually collided with the Eurasian plate. The mighty Himalayas—once an ancient seafloor—rose at the boundary (see figure 16.25).

In other places, called subduction zones, one plate dives beneath the other. New plate material forms at areas where plates move apart and molten rock seeps to Earth's surface at the seam. As a result, wide oceans now separate continents that were once joined together.

It may seem hard to imagine that Earth's continents have not always been located where they are now. But a wealth of evidence, including the locations of the world's earthquake-prone and volcanic "hot spots," indicates that the continents have moved—and they continue to do so. This slow-motion dance of the continental plates has dramatically affected life's history as oceans shifted, land bridges formed and disappeared, and mountain ranges emerged.

B. Biogeography Considers Species' Geographical Locations

Biogeography is the study of the distribution pattern of species across the planet. Together, biogeographical studies and plate tectonics have shed light on past large-scale evolutionary events.

The Case of the Missing Marsupials

The evolutionary history of marsupials illustrates the importance of geography in dictating a species' fate (**figure 15.8**).

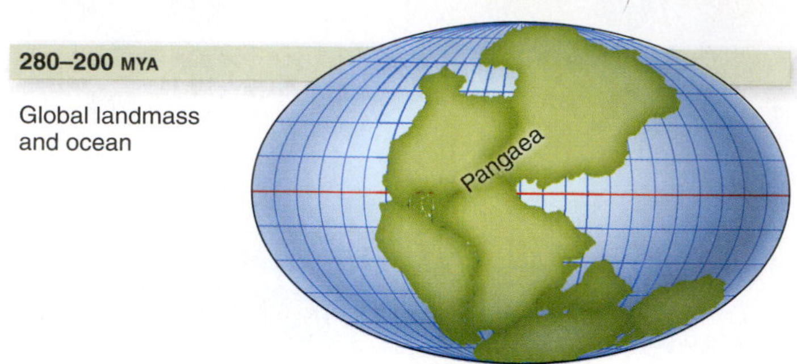

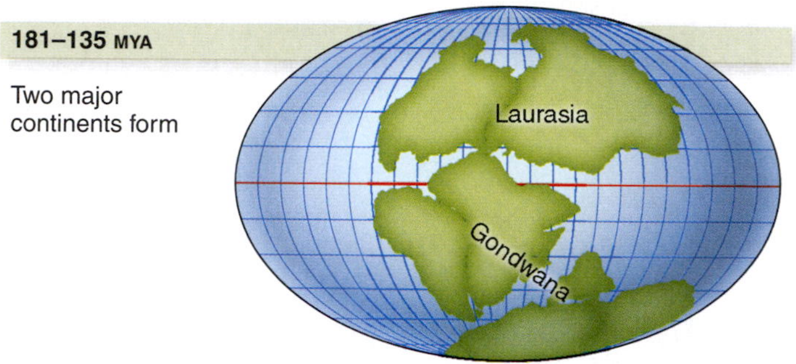

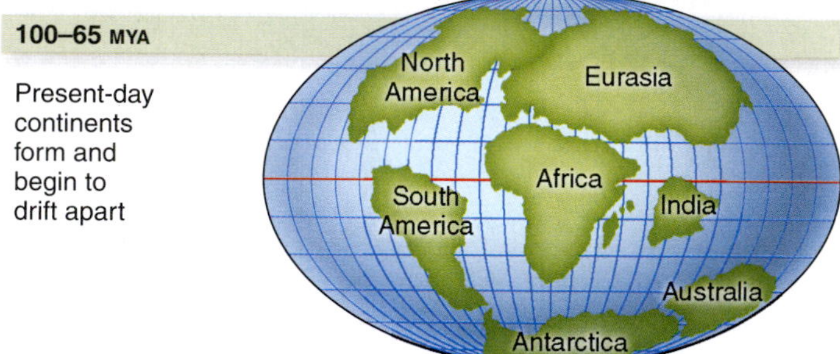

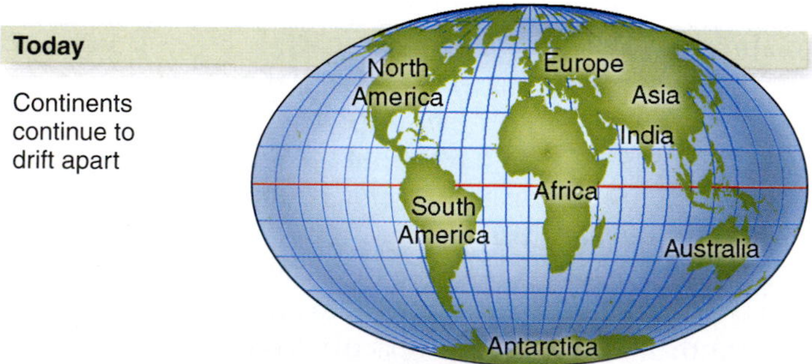

FIGURE 15.7 A Changing World. The distribution of landmasses on Earth has changed with time, due to shifting tectonic plates. About 255 MYA, Earth consisted of a large landmass, Pangaea, and a huge ocean, Panthalassa. Later, Pangaea broke into two continents: Laurasia and Gondwana. Around 100 MYA, Earth's landmasses moved to their contemporary arrangement.

Marsupials are mammals such as kangaroos, koalas, and sugar gliders that carry their young in a pouch. The young are born tiny, hairless, blind, and helpless. As soon as they are born, they crawl along the mother's fur to tiny, milk-secreting nipples inside the pouch. Australia is unique in that most of its native mammals are marsupials, a stark contrast to the rest of the world, where most native marsupials are extinct.

The explanation for this pattern lies in the slow shift of Earth's tectonic plates. Up until about 140 MYA, Antarctica and Australia were part of Gondwana. Then, about 60 or 70 MYA, Australia separated from Antarctica and began drifting toward Eurasia. Over the next 30 million years or so marsupials became widespread, abundant, and diverse on Australia. Fossil evidence suggests that marsupials were also diverse and abundant in South America until about 1 or 2 MYA. Yet most South American marsupials are now extinct. Why the difference between Australia and South America?

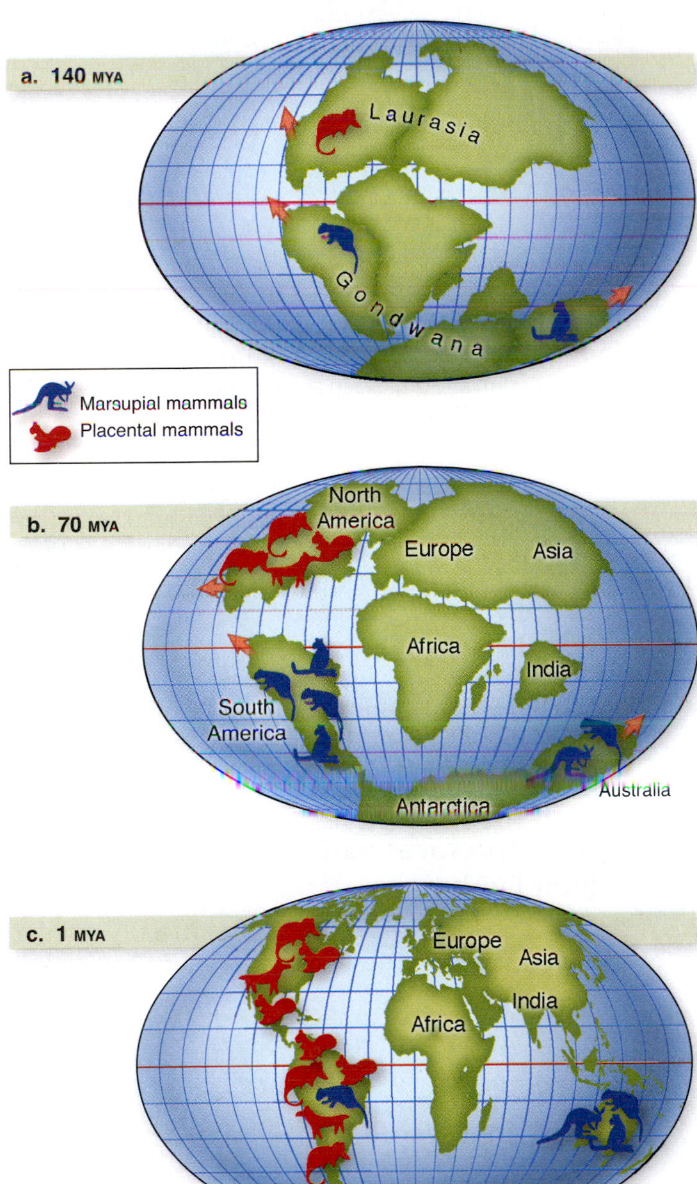

FIGURE 15.8 Where Did the Marsupials Go? Marsupials were once widespread, but the more recent placental mammals eventually outcompeted their pouched relatives. Diverse marsupials remained on Australia, however, because that continent had already broken away from the other land masses by the time the placentals evolved.

Until about 7 or 8 MYA, water separated South America from North America. When sea levels went down, a new land bridge emerged, permitting migration between North and South America. In North America at that time, placental mammals were dominant. The young of these animals develop within the female's body, nourished in the uterus by the placenta. Thus, baby placental mammals are born more fully developed than are marsupials, giving them a better chance of survival after birth. Because of this reproductive advantage, the placental mammals migrating to South America displaced the marsupials, which largely became extinct. Australia's marsupials remained isolated from competition from placental mammals for much longer.

Adaptive Radiation on Islands

Another, smaller scale application of biogeography is the study of adaptive radiation on island chains (see chapter 14's opening essay: Islands Provide Windows on Speciation and Extinction). Why should island groups such as Hawaii and the Galápagos house unique groups of species that appear closely related to those on the nearest mainland? Again, the answer relates to geography. The only organisms that can colonize a newly formed volcanic island are those that can swim, fly, or raft from an inhabited location. The exchange of organisms between an isolated island and the mainland is probably quite rare. Those organisms that do successfully colonize the island encounter conditions that are different from those on the mainland. With limited migration between the island and the ancestral populations, the relocated organisms evolve and diversify into new species. Just as for the marsupials, geographical barriers influence the course of evolution.

Summary *The continents have drifted over Earth's surface for hundreds of millions of years, influencing speciation. Biogeography studies the large-scale changes wrought by continental drift as well as smaller scale phenomena such as adaptive radiation on remote islands.*

15.3 MASTERING CONCEPTS

1. How have the positions of Earth's continents changed over the past 200 million years?
2. How has continental drift affected the distribution of marsupials on Earth?
3. How do biogeographical observations help biologists interpret the evolutionary history of species on isolated islands?

15.4 Homologous Structures Reflect Common Ancestry

Many clues to the past come from the present. As Unit 1 explains, all life is made of cells, and eukaryotic cells are very similar in the structure and function of their membranes and organelles. On a molecular scale, cells share many similarities in their enzymes, signaling proteins, and metabolic pathways. Unit 2 describes another set of common features: the mechanisms of inheritance and the relationship between DNA and proteins. On a whole-body scale, comparisons of anatomy and physiology also reveal many important commonalities among modern species.

Two structures are termed **homologous** if the similarities between them reflect common ancestry. (Recall from chapter 9 that two chromosomes are homologous if they have the same genes, though not necessarily the same alleles, arranged in the same order.) Homologous genes, chromosomes, anatomical structures, or other features are similar in their configuration, position, or developmental path (**figure 15.9**).

Homology is a powerful tool for discovering evolutionary relationships. For example, as described in section 15.5, a newly sequenced gene or genome can be compared with homologous genes from other species to infer how closely related any two species are. Likewise, fossilized structures are often compared with homologous parts in known species. Taken together, comparisons based on anatomy, embryonic development, and molecular sequences can add multiple lines of evidence that can support or refute a hypothesis about descent.

A. Comparing Anatomical Parts Can Reveal Evolutionary Relationships

Different modern species share many anatomical similarities. For example, all vertebrate skeletons support the body, are made of the same materials, and consist of many of the same parts (**figure 15.10**). The simplest explanation for this similarity is that modern vertebrates descended from a common ancestor that originated this skeletal organization. Fishes, amphibians, reptiles, birds, and mammals gradually modified the skeleton as they adapted to their environments.

FIGURE 15.10 Homologous Limbs. Although a human walks erect, a lion walks on all fours, a bird flies, and a seal swims, all have skeletons that are similarly organized and composed of the same type of tissue.

Although they share a common origin, homologous structures may not have the same function. The middle ear bones of mammals, for example, originated as bones that supported the jaws of primitive fishes, and they still exist as such in some vertebrates. These bones are homologous and reveal our shared ancestry with fishes.

Analogous Structures and Convergent Evolution

Some anatomical parts that appear superficially similar among different species are not homologous but rather

FIGURE 15.9 Homology Reflects Shared Ancestry. Mice, cats, humans, and other species sometimes have a similar combination of features: light eye color, hearing or other neurological impairment, and a fair forelock of hair. The genes that control these similar phenotypes may be homologous.

FIGURE 15.11 Homology and Analogy. A bat wing and a bird wing are homologous structures, composed of bone and presumably inherited from a recent common ancestor. The wings of birds and butterflies both function in flight, but they are analogous structures because they are not made of the same materials, nor are they organized in the same way. They are not inherited from a recent common ancestor.

analogous, meaning that the structures evolved independently. Recall from chapter 13 that **convergent evolution** produces similar adaptations in organisms that do not share the same evolutionary lineage. Flight, for example, evolved independently in birds and in insects. Both have wings, but the bird's wing is a modification of vertebrate limb bones, whereas the insect's wing is an outgrowth of the cuticle that covers its body (**figure 15.11**). The wings have the same function—flight—and enhance fitness in the face of similar environmental challenges. The differences in structure, however, indicate they do not have a common developmental pathway. They are analogous, not homologous.

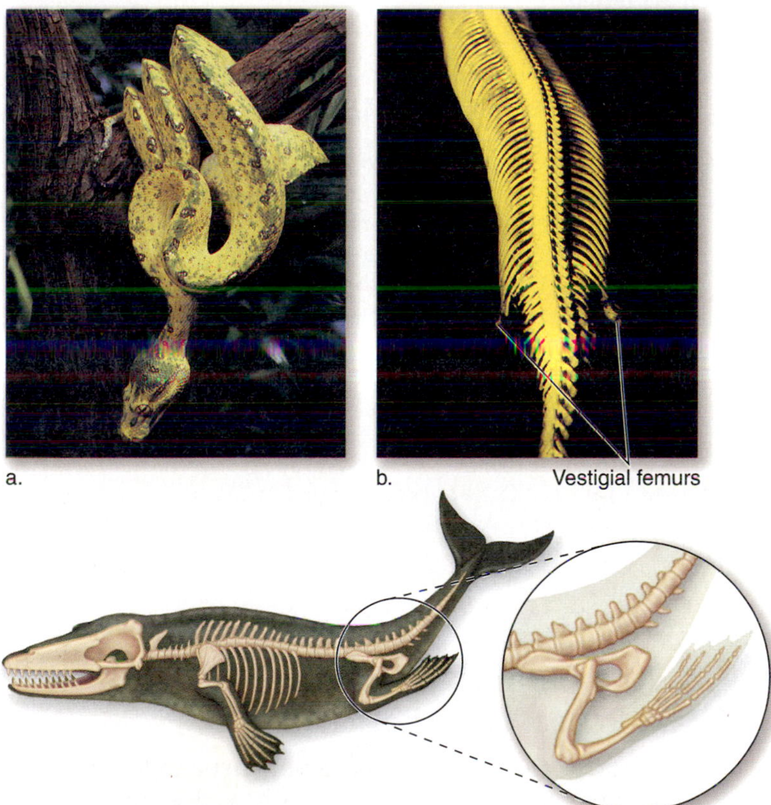

FIGURE 15.12 Vestigial Structures. (**A**) Some snakes have tiny femurs (leg bones) that are vestigial, but (**B**) detectable only in the skeleton. (**C**) Some whales likewise have a vestigial pelvis and hind limbs.

Vestigial Structures

Evolution is not a perfect process. As environmental changes select against some structures, others persist even if they are not used. A **vestigial structure has no apparent function** in one species, yet is homologous to a functional organ in another. (Darwin compared vestigial structures to silent letters in a word; they are not pronounced, but they offer clues to the word's origin.) In some whales and snakes, tiny leg bones are vestigial, retained from vertebrate ancestors that used their legs to walk on land (**figure 15.12**).

Humans have several vestigial organs. In the digestive system, the appendix serves no known function, but may once have helped process different types of foods. The tiny muscles that make hairs stand on end helped our furry ancestors conserve heat or show aggression; in us, they apparently serve only as the basis of goose bumps. Human embryos have tails, which usually disintegrate long before birth; in other vertebrates, tails persist into adulthood. Above our ears, a trio of muscles (that most of us can't use) helps other mammals move their ears in a way that improves hearing. Each vestigial structure links us to other animals that still use these features.

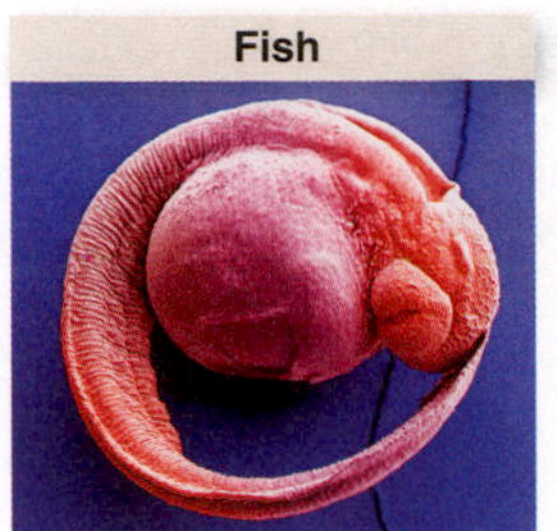

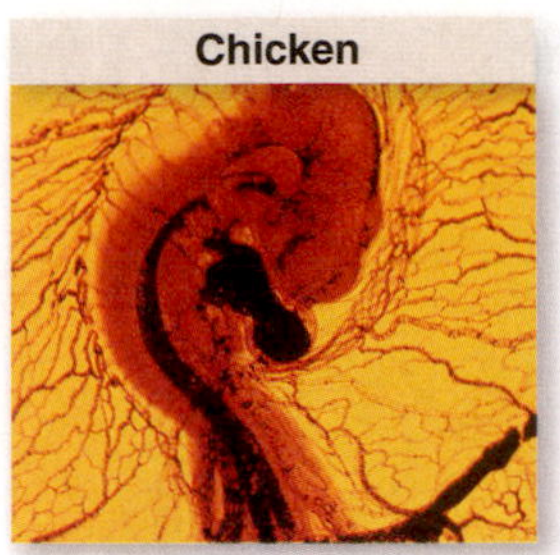

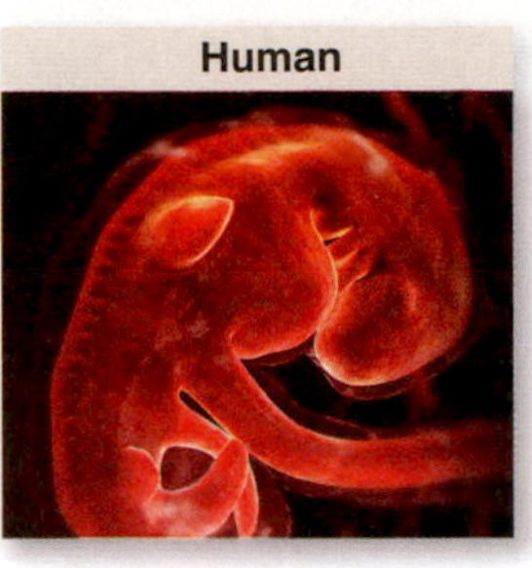

FIGURE 15.13 Embryo Resemblances. Vertebrate embryos appear alike early in development, reflecting similar basic processes as cells divide and specialize. As development continues, parts grow at different rates in different species, and the embryos do not look as similar. © Dr. Richard Kessel/Visuals Unlimited

B. Embryonic Development Patterns Provide Evolutionary Clues

Because related organisms share many physical traits, they must also share developmental processes that produce those traits. By comparing embryos at different stages, it should be possible to reconstruct some of the steps that have led to differences among species.

Historically, the field of comparative embryology has seen some controversy. In 1874, German naturalist Ernst Haeckel published nearly identical drawings of embryos from different vertebrate species, including fishes, reptiles, birds, and mammals. Shortly thereafter, Haeckel admitted that he altered some details to make the embryos look more alike. He also failed to represent scale. That is, although embryos might start out with the same basic parts, different growth rates (such as the very fast growth of the head in the primate fetus) distinguish species as development proceeds.

Haeckel's fudging fueled skepticism about the true degree of similarity among vertebrate embryos. Years later, however, developmental biologists photographed embryos and fetuses of a variety of vertebrate species throughout development (**figure 15.13**). The data show that there really are similarities in embryonic structures, supporting the concept of common ancestry. Contemporary work in evolutionary developmental biology ("evo-devo"), described in section 15.5, has added abundant additional evidence.

Summary *Homologous structures are similar in construction and reflect shared ancestry. Analogous structures reflect convergent evolution rather than shared ancestry. Comparing functional and vestigial anatomical parts and observing similarities among embryos also provide evidence of evolution.*

15.4 MASTERING CONCEPTS

1. What can homologies reveal about evolution?
2. How can distinguishing between homologous and analogous structures be difficult?
3. What is a vestigial structure? What are some examples of vestigial structures in humans and other animals?
4. How does the study of embryonic development reveal clues to a shared evolutionary history?

15.5 Molecules Reveal Relatedness

The techniques used to study molecular evolution are based on the comparison of nucleotide and amino acid sequences among species. One advantage of such comparisons is that they are less subjective than deciding whether two anatomical structures are homologous or analogous. As a result, scientists are modifying some evolutionary trees to reflect these more definitive molecular data (see this chapter's Can *You* Relate? Reevaluating the Mammalian Family Tree).

A. Comparing DNA and Protein Sequences May Reveal Close Relationships

The fact that all species use the same genetic code argues for a common ancestry to all life on Earth, as does the fact that all use the same 20 amino acids. In addition, many proteins have only minor sequence differences from one species to another. The keratin of sheep's wool, for example, is virtually identical to that of human hair.

It is highly unlikely that two unrelated species would evolve precisely the same DNA and protein sequences by chance. It is more likely that the similarities were inherited from a common ancestor, and that differences arose by mutation after the species diverged from the ancestral type. Thus, an underlying assumption of molecular evolution studies is that the greater the molecular similarities between two modern species, the closer their evolutionary relationship.

DNA differences can be assessed for a few bases, for a single gene, for families of genes with related structures or functions, and even for entire genomes. Likewise, comparisons of amino acid sequences often support fossil and anatomical evidence of evolutionary relationships. One study, for example, found 7 of 20 selected proteins to be identical in amino acid sequence in humans and chimps. One of these proteins is cytochrome *c*, which is part of the electron transport chain in mitochondria (see section 6.2). **Figure 15.14** shows that the more closely related two species are, the more alike is their cytochrome *c* amino acid sequence. **DNA sequencing, p. 147**

B. "Evo-devo" Bridges Evolution and Developmental Biology

Developmental biology, the study of how the adult body takes shape from its single-celled beginning, has yielded ample evidence for evolution (see section 15.4). More recently, the discovery of genes that contribute to development has spawned the field of evolutionary developmental biology (or "evo-devo" for short). Evo-devo studies how these genes change and give rise to new body forms (**figure 15.15**).

How can a new developmental pathway—and a new body plan—arise? One possibility is the evolution of genes with new functions (see section 27.6). Another possibility is the evolution of new gene regulation pathways. Early in development, gradients of several different proteins distinguish parts of the developing embryo, eventually molding them into limbs and other body parts (**figure 15.16**). New phenotypes might arise simply from a change in the timing or location of these signals, each of which is encoded by a gene.

Not surprisingly, mutations in the genes encoding proteins that regulate development can produce dramatic abnormalities. **Homeotic** is a general term describing any gene that, when mutated, leads to organisms with structures in abnormal or unusual places (see figure 12.13). Many homeotic genes encode proteins that regulate the transcription of other genes. Homeotic genes occur in all animal phyla studied to date, as well as in plants and fungi.

Homeotic genes provide clues about development in diverse species. The genes that prompt limb formation in mouse and chicken embryos, for example, do so by turning on in cells at specific points in the middle tissue layer of the embryo. In snake embryos, these genes turn on all along the length of the tissue layer, rather than at defined locations. In the absence of a developmental signal telling the embryo where to start sprouting limbs, the snake's legs remain mere buds (see figure 15.12).

Cytochrome *c* Evolution

Organism	Number of amino acid differences from humans
Chimpanzee	0
Rhesus monkey	1
Rabbit	9
Cow	10
Pigeon	12
Bullfrog	20
Fruit fly	24
Wheat germ	37
Yeast	42

FIGURE 15.14 Cytochrome *c* Similarities. Similarities in amino acid sequence for the respiratory protein cytochrome *c* in humans and other species parallel the degree of relatedness among them.

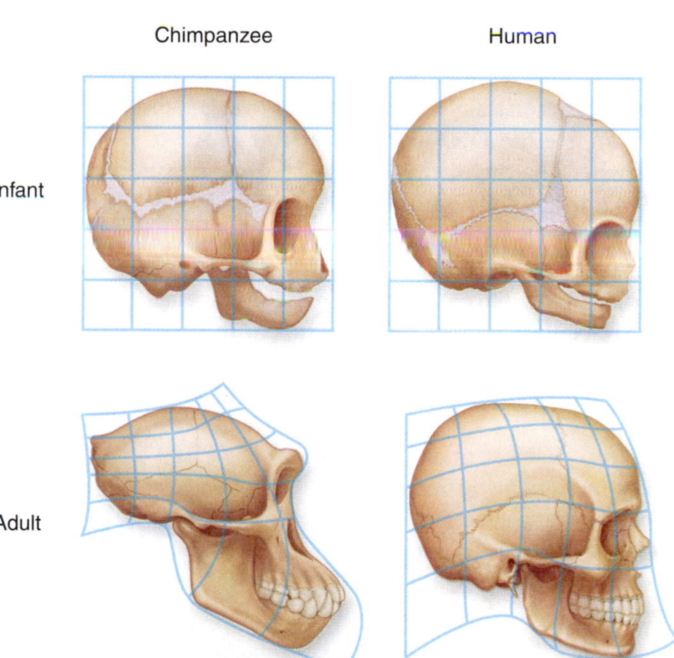

FIGURE 15.15 Same Parts, Different Proportions. Infant humans and chimpanzees have very similar skulls. As the organism ages, however, chemical signals specify slightly different developmental pathways. In humans (*right*), the brain becomes larger, and the jaw smaller, than in chimpanzees (*left*).

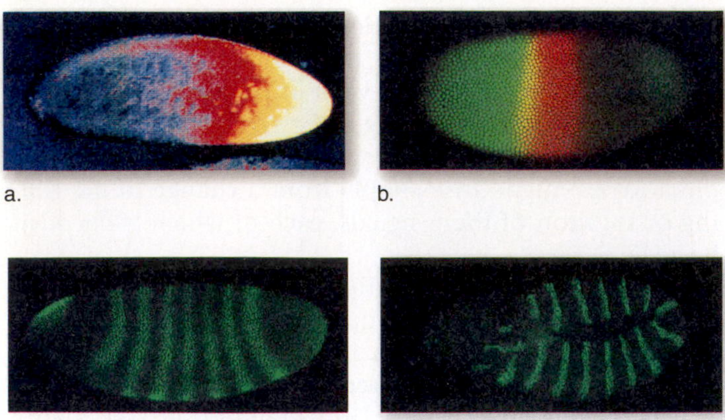

FIGURE 15.16 Early Development in Fruit Flies. Genes encode proteins, which are expressed unequally along the length of an embryo. Protein gradients set up distinct environments that profoundly influence differentiation. (**A**) Very early in the development of a fruit fly, a protein called bicoid distinguishes the embryo's head from its rear. Different colors represent different concentrations of bicoid. (**B**) Bicoid protein directs the synthesis of two other proteins, one shown in red and one in green. (**C**) A half hour later, another gene directs production of a protein that divides the embryo into seven stripes. (**D**) Yet another gene oversees dividing each existing section of the embryo in two.

C. Molecular Clocks Help Assign Dates to Evolutionary Events

A biological molecule such as DNA can act as a "clock." If biologists know the mutation rate for a gene, plus the number of differences in the DNA sequences for that gene in two species, they can use the DNA as a **molecular clock** to estimate the time when the organisms diverged from a common ancestor. For example, many human and chimpanzee genes differ in about 4% to 6% of their bases, and substitutions occur at an estimated rate of 1% per 1 million years. Therefore, about 4 to 6 million years have passed since the two species diverged.

Molecular clock studies are not quite as straightforward as glancing at a wristwatch. DNA replication errors occur in different regions of a chromosome at different rates, and some genes affect phenotypes more than others. Therefore, to construct molecular tree diagrams incorporating molecular clocks, researchers often consider DNA sequences that are not subject to natural selection, such as genes encoding ribosomal RNA. Carl Woese used ribosomal RNA to deduce the existence of the three domains (Archaea, Bacteria, and Eukarya).

A special type of molecular clock measures mutations in mitochondrial DNA. Mitochondria contain only about 16,500 DNA base pairs, but mitochondrial DNA (mtDNA) is particularly valuable in tracking recent evolutionary events because its molecular clock "ticks" 5 to 10 times faster than the nuclear DNA clock. Mitochondrial DNA presents unique possibilities because it is inherited from mothers only, in the egg. Biologists use it for everything from matching kidnapped children with their birth mothers to tracing human origins. For example, many studies show that people from Africa have the most diverse mtDNA sequences. This suggests that Africans have existed longer than other modern populations because it takes time for mutations to accumulate. The idea that early humans originated in Africa and then migrated to the other continents is called the single origin ("out of Africa") hypothesis (**figure 15.17**). **DNA profiling, p. 149; human evolution, p. 344**

Summary ***Molecular evidence for evolution includes similarities at the DNA and protein levels. Evolutionary developmental biologists observe both genes and development patterns to study how new developmental pathways (and thus new phenotypes) arise. Gene sequence differences among species can be placed in a time frame derived from mutation rates.***

15.5 MASTERING CONCEPTS

1. How does analysis of DNA and proteins support other evidence for evolution?
2. Why are evolutionary biologists interested in how homeotic genes influence development?
3. What is the basis of using a molecular clock to determine when two species diverged from a common ancestor?
4. What is an advantage of using mtDNA instead of nuclear DNA in tracing evolution?

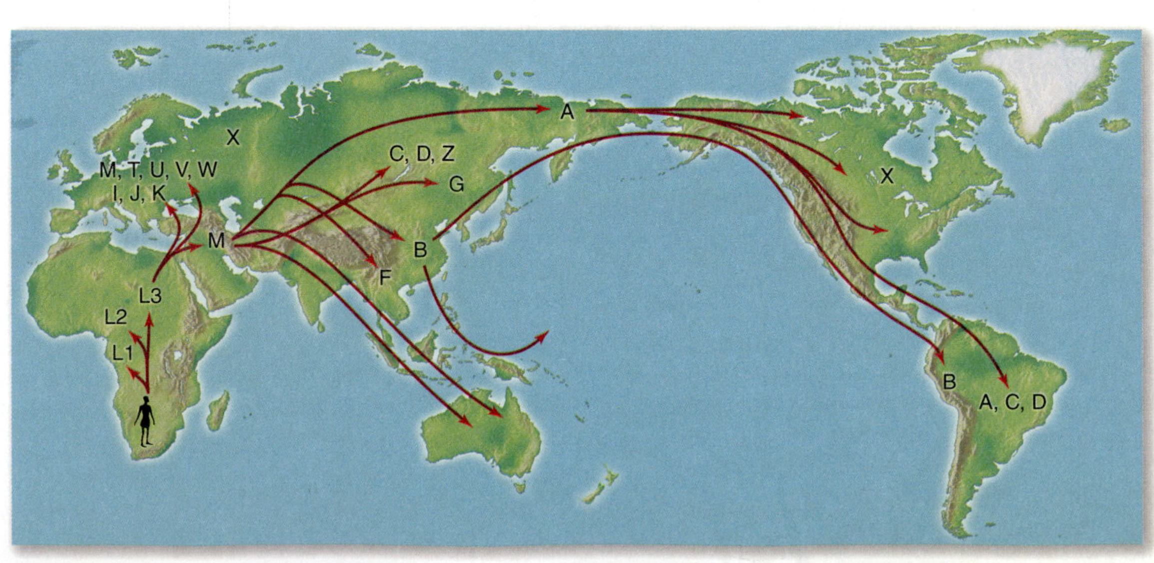

FIGURE 15.17 Out of Africa. This world map depicts different genetic "types" of human mitochondrial DNA. Analysis of mtDNA suggests that groups L1, L2, and L3, each originating in Africa, gave rise to all other groups worldwide.

Can *You* Relate?

Reevaluating the Mammalian Family Tree

Since the 1980s, a wealth of molecular data has led to the reclassification of many types of organisms. Two examples in the mammalian family tree are the duck-billed platypus and cetaceans, the group of marine mammals that includes whales, porpoises, and dolphins.

The Case of the Egg-Laying Duck-Billed Platypus

The duck-billed platypus (*Ornithorhynchus anatinus*) is an unusual mammal by any measure (**figure 15.A**). Other than the two species of echidna, the platypus is the only egg-laying mammal. A female platypus typically lays one to three eggs that hatch 10 days later. The young drink their mother's milk for about 4 months.

Echidnas and platypuses have traditionally been considered to make up a group of primitive mammals called monotremes, separate from marsupials (pouched mammals such as kangaroos) and placental mammals. In recent years, researchers have sequenced the entire mitochondrial genomes of many mammalian species, and this has led to the revision of the classification of the monotremes (**figure 15.B**). The monotremes are more like marsupials at the genetic level than had been suspected based on their unusual appearance. Instead of a tiny side branch with the echidnas, mtDNA sequencing places both of these egg-layers along the lineage of marsupials.

FIGURE 15.A **Duck or Mammal?** The platypus shares features with diverse animals, making it difficult to classify.

The Terrestrial Origin of Cetaceans

Classifying cetaceans has always been a problem. Fossil evidence suggests these animals went from land to water very rapidly, in just a few million years. About a century ago, biologists placed the cetaceans closest to the ruminants (hoofed, grazing mammals such as deer, cows, sheep, and giraffes), based on skeletal similarities. The ruminants are part of a larger group, the artiodactyls, which also includes hippos, pigs, peccaries, camels, and llamas. All of these mammals first appeared in modern form about 48 to 50 MYA.

The cetaceans have many adaptations to life in the water that complicate comparisons to terrestrial vertebrates. The most defining feature of the artiodactyls is a very mobile heel joint, something that is obviously not present in the legless cetaceans. The axis of symmetry that splits artiodactyl feet is also nonexistent in the footless whales, porpoises, and dolphins. A triple row of cusps (raised areas) on the back molars also characterizes the artiodactyls, a trait that may have disappeared in the cetaceans because of dietary differences from their land-dwelling relatives.

The traditional family tree, based largely on superficial resemblances, placed the cetaceans closest to the ruminants, and then most closely related to the hippos, which, in turn, were closest to pigs and peccaries (**figure 15.C**). In commonsense terms, a hippo looks more like a pig than it does a porpoise! But molecules told a different story. DNA sequences support a tree that places the cetaceans, ruminants, and hippos as a single evolutionary lineage. Definitive molecular evidence came in the form of repetitive DNA sequences that are so species-specific that it is very unlikely for them to appear in different species by chance alone. Nearly identical repeats occur in the genomes of cetaceans, ruminants, and hippos, but they are notably absent in the genomes of pigs, peccaries, camels, and llamas.

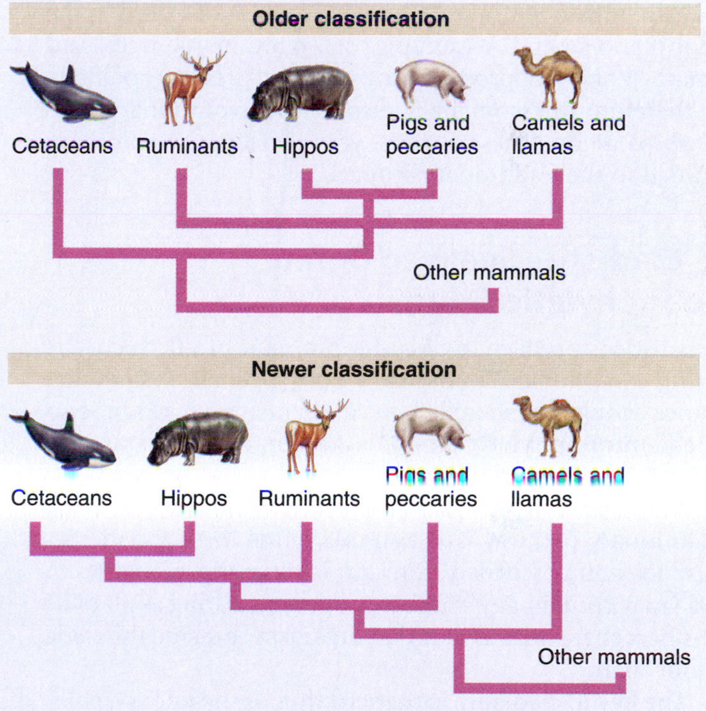

FIGURE 15.C **Cetacean Shift.** Molecular evidence led to reconsideration of the relationships among hippos, pigs, and whales.

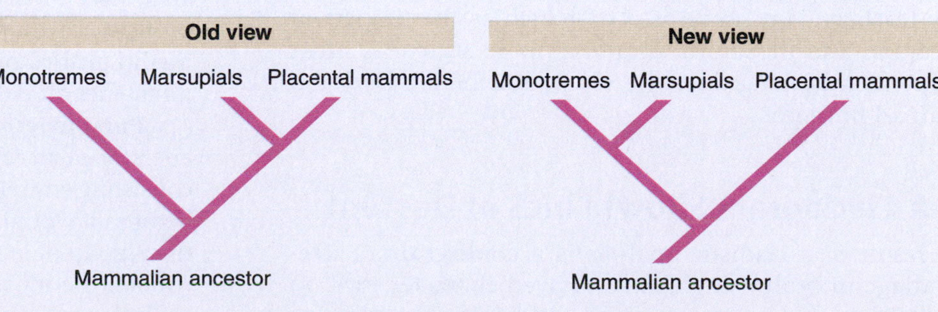

FIGURE 15.B **Placing the Platypus.** Mitochondrial DNA from 19 mammalian species revealed that the monotremes are a side branch of the marsupials, rather than a separate lineage. The three groups are distinguished by the way that offspring are born. Monotremes lay eggs, marsupials have pouches, and placental mammals nurture fetuses within the female.

15.6 Systematics Reconstructs the Stories of Life

How do biologists apply the evidence for evolution to the monumental task of organizing life's diversity into groups that reflect evolutionary history? The answer lies in the field of **systematics,** which includes both taxonomy (describing and naming species) and phylogenetics. **Phylogenetics,** in turn, attempts to explain the evolutionary relationships among species. Systematists have long constructed phylogenetic trees, which are hypotheses about evolutionary relationships. (Charles Darwin included an example in *The Origin of Species.*)

In the past, systematists constructed phylogenetic tree diagrams by comparing as many characteristics as possible among species. Those organisms with the most characteristics in common would be neighbors on the tree's branches. A problem with this approach is that **ancestral characters** (i.e., those already present in the ancestor of a group) count as much as **derived characters,** which are not found in the group's ancestors. For example, oak trees, maple trees, and tomato plants all produce flowers. The presence of flowers therefore does not help resolve the evolutionary relationships among these species, yet this feature would have counted in the traditional approach.

A. Cladistics Seeks to Define Monophyletic Groups

A cladistics approach avoids the problem of all characters having equal value. **Cladistics** is a phylogenetic system that defines groups by shared derived characters. A **clade,** also called a **monophyletic** group, is a group of organisms consisting of a common ancestor and all of its descendants; in other words, it is a group of species united by a single evolutionary pathway. For example, birds form a clade because they all descended from the same group of reptiles. A clade may contain any number of species, as long as all of its members share an ancestor that organisms outside the clade do not share.

The key to cladistics is the selection of the derived characters that define the clades. For example, the presence of feathers is suitable as the basis of a clade because only birds have feathers. Not useful is a trait such as the flippers of a penguin and those of a porpoise; these structures arose independently, long after the separation of the bird and mammal lineages.

B. A Cladogram Shows Lines of Descent

The result of a cladistic analysis is a **cladogram,** a tree-like diagram built using shared derived characteristics. To construct a cladogram, a researcher begins by selecting traits that probably reflect descent from a shared ancestor. The next step is to make a chart listing which species have which traits. In constructing the tree, species sharing the most derived characters occupy the branches farthest from the root. The nodes indicate where two new groups arise from a common ancestor. **Figure 15.18** demonstrates how to construct a cladogram, using the example from the chapter opening essay. It depicts key dinosaurs on the path to birds, using derived characters that are important in the acquisition of flight.

In contrast to a cladogram, a traditional tree is based on the overall similarities among organisms, whether or not traits are derived. **Figure 15.19** depicts the distinction between a traditional evolutionary tree diagram and a cladogram for vertebrates (animals with backbones). The traditional approach separates mammals and birds from reptiles. The cladogram, however, places reptiles and birds together.

A problem with cladistics is that the diagrams can become enormously complicated when many species and derived characters are included. Mathematically, several trees can accommodate any one data set. The most **parsimonious** tree, which is the one that requires the fewest steps to construct, is selected to be the closest to reality. Cladograms, however, are based on limited and sometimes ambiguous information. They are therefore not peeks into the past, but rather tools that researchers can use to construct hypotheses about the relationships of different types of organisms. These investigators can then add other approaches to test the hypotheses.

C. Many Traditional Groups Are Not Monophyletic

Despite the cladistics approach that contemporary systematists use, many familiar groups of species are not monophyletic. For example, kingdom Protista is **paraphyletic** because it contains a common ancestor and some, but not all, of its descendants. That is, plants, fungi, and animals are also descended from the same eukaryotic ancestor as the protists, yet they are assigned to different kingdoms. Domain Eukarya, on the other hand, is monophyletic because it includes all eukaryotes.

Polyphyletic groups exclude the most recent common ancestor of all members of the group. For example, a group consisting entirely of "flying animals" would be polyphyletic because it would include only birds, bats, insects, and an extinct group of reptiles called pterosaurs. This group does not contain the most recent common ancestor that all of these animals share.

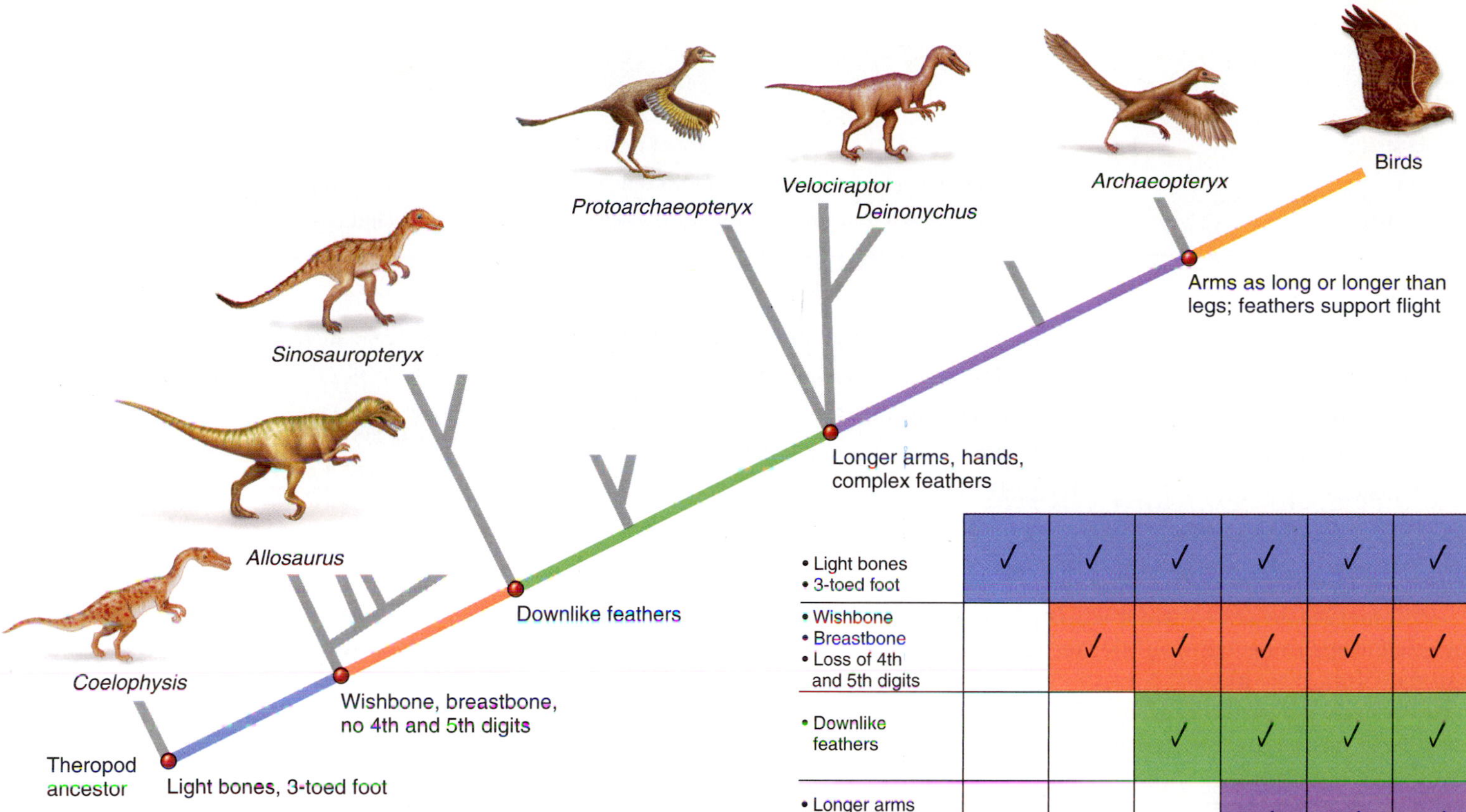

Birds

Archaeopteryx

Velociraptor

Deinonychus

Protoarchaeopteryx

Sinosauropteryx

Allosaurus

Coelophysis

Theropod ancestor

Light bones, 3-toed foot

Wishbone, breastbone, no 4th and 5th digits

Downlike feathers

Longer arms, hands, complex feathers

Arms as long or longer than legs; feathers support flight

	Coelophysis	Allosaurus	Sinosauropteryx	Protoarchaeopteryx	Archaeopteryx	Birds
• Light bones • 3-toed foot	✓	✓	✓	✓	✓	✓
• Wishbone • Breastbone • Loss of 4th and 5th digits		✓	✓	✓	✓	✓
• Downlike feathers			✓	✓	✓	✓
• Longer arms • Hands • Complex feathers				✓	✓	✓
• Arms as long or longer than legs • Feathers support flight					✓	✓

FIGURE 15.18 Constructing a Cladogram. To build a cladogram, tally up the derived characters that species share. Pairs that share the greatest number of derived characters are the closest on the cladogram. Each branch on the cladogram is a clade distinguished by characteristics that appear only in its members. *Source*: Cladogram data based on Kevin Padian, "When Is a Bird Not a Bird?" *Nature*, June 25, 1998, vol. 393, page 729.

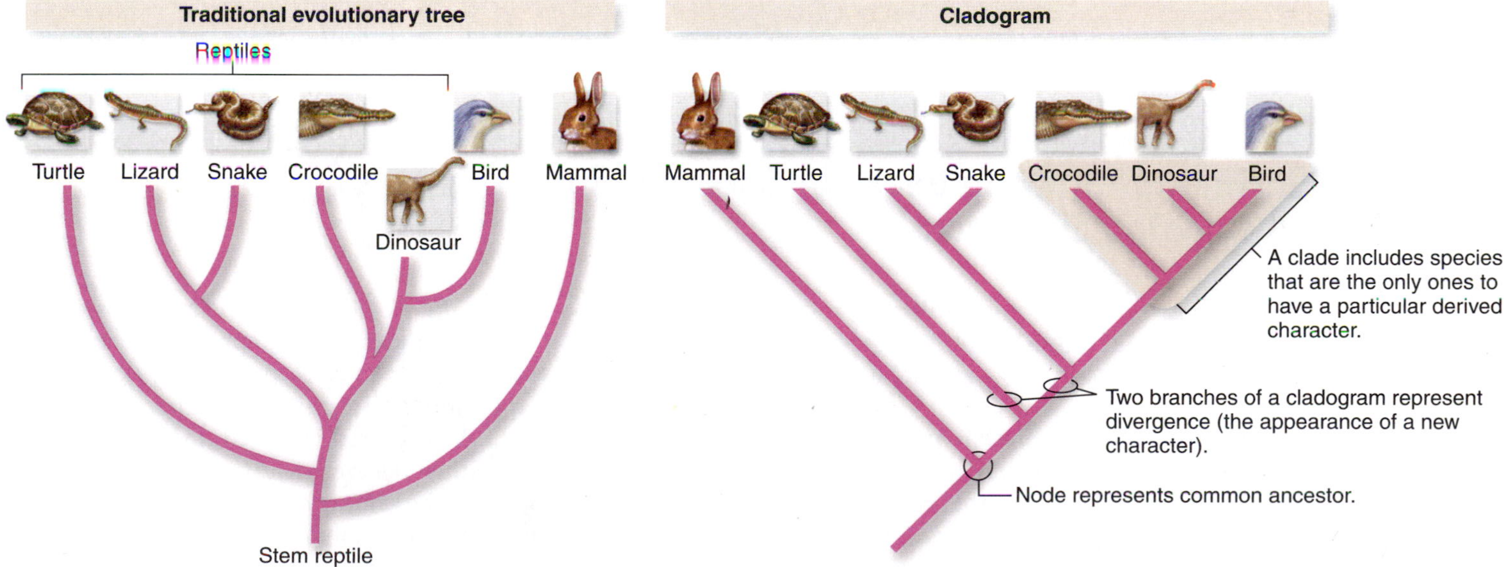

FIGURE 15.19 Two Types of Evolutionary Tree Diagrams. A traditional evolutionary tree excludes birds from the group labeled "reptiles," even though all of these animals descend from the same common ancestor. The cladogram uses derived characters to place these related animals on a continuum.

Table 15.1 *Miniglossary of Systematics Terms*

Term	Definition
Ancestral characters	Features present in the common ancestor of a clade
Clade	Group of organisms consisting of a common ancestor and all of its descendants; a monophyletic group
Cladistics	Phylogenetic system that groups organisms by characteristics that best indicate shared ancestry
Cladogram	Phylogenetic tree built on shared derived characteristics
Derived characters	Features not found in a clade's ancestors; also called apomorphies
Monophyletic group	Group of organisms consisting of a common ancestor and all of its descendants; a clade
Paraphyletic group	Group of organisms consisting of a common ancestor and some, but not all, of its descendants
Phylogenetic tree	Diagram depicting hypothesized evolutionary relationships
Polyphyletic group	Group of species that excludes the most recent common ancestor
Systematics	The combined study of taxonomy and evolutionary relationships among organisms

Systematists do not consider either paraphyletic or polyphyletic groups to be "natural" because they do not reflect a shared evolutionary history. Nevertheless, many such group names remain in everyday usage. Kingdom Protista is one example of a paraphyletic group, as is class Reptilia, which excludes birds even though they share a common ancestor with reptiles. Likewise, the term *algae* reflects a polyphyletic grouping of many unrelated species of aquatic organisms that carry out photosynthesis (see chapter 19).

Table 15.1 summarizes some of the language of systematics.

Summary *Systematists describe, name, and organize species based on their evolutionary (phylogenetic) relationships. Cladistics is a system of phylogeny that groups species based on characteristics that they have inherited from a recent shared ancestor.*

15.6 MASTERING CONCEPTS

1. What is the advantage of a cladistics approach over a more traditional approach to phylogeny?
2. Distinguish between ancestral and derived characters.
3. How is a cladogram constructed?
4. What is the difference between monophyletic, paraphyletic, and polyphyletic groups?

15.7 INVESTIGATING LIFE

The Shrinking, Growing Beaks of Darwin's Finches Reveal Ongoing Evolution

Fossils, anatomical structures, and molecules reveal past evolution, but Darwin's finches on the Galápagos Islands unmistakably show that we can also watch it happen in real time, in the real world.

Thirteen of the 14 known species of Darwin's finches inhabit the Galápagos Islands. All have similar size (10 to 20 cm) and coloration (brown to black), but they differ in their food sources, behaviors, and song melodies. They received their name not only because Charles Darwin collected them during *Beagle's* voyage, but also because subsequent research on the birds has taught scientists so much about natural selection and evolution.

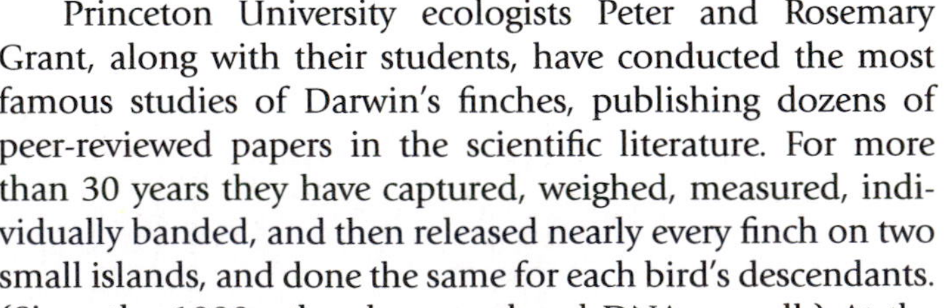

Princeton University ecologists Peter and Rosemary Grant, along with their students, have conducted the most famous studies of Darwin's finches, publishing dozens of peer-reviewed papers in the scientific literature. For more than 30 years they have captured, weighed, measured, individually banded, and then released nearly every finch on two small islands, and done the same for each bird's descendants. (Since the 1990s, they have analyzed DNA as well.) At the same time, they have also measured rainfall, plant cover, seed diversity, seed abundance, and seed size on the islands.

The main goal of their massive project has been to learn more about how the 14 bird species diverged from a common ancestor over the past 3 million years. But the long-term data on individual birds, coupled with ecological information, also enable the Grants to analyze the island's short-term "natural experiments." That is, whenever the climate changes dramatically (an unusually severe drought, for instance, or abnormally heavy rainfall), they can compare the birds before and after the event, documenting evolutionary changes over multiple generations.

This essay focuses on just one of the Grants' many studies on the uninhabited island of Daphne Major. Two permanent residents of Daphne Major are the medium ground finch (*Geospiza fortis*), and the slightly larger cactus finch (*Geospiza scandens*) (**figure 15.20**). The beaks of the two birds reflect their different diets. The medium ground finch uses its crushing bill to eat seeds of various sizes, whereas the cactus finch has a probing bill that it uses to dine on the pollen, nectar, and medium-sized seeds of the *Opuntia* cactus.

In late 1982 to mid-1983, a severe El Niño event triggered heavy rains for 8 months (**figure 15.21A**). The downpour changed the assortment of seeds available for finches to eat on the island. Large, hard seeds dominated before the rainy year, but smaller, softer ones became much more abundant in the years that followed (figure 15.21B).

The change in food source was a selective force on the medium ground finches' population for several years after 1983. During and immediately after the heavy rains, breeding conditions for the birds were excellent. A dry spell followed from 1984 to 1986, however, and finch populations declined. Finches with small bills, which easily handle small seeds, had more food available than larger beaked birds. Since they could eat the most, they could raise the most offspring. As a result, average beak width declined from 8.86 mm for medium ground finches alive in 1984 to 8.74 mm for birds born in 1987 (once they were fully grown).

Geospiza fortis *Geospiza scandens*

FIGURE 15.20 Two of Darwin's Finches.

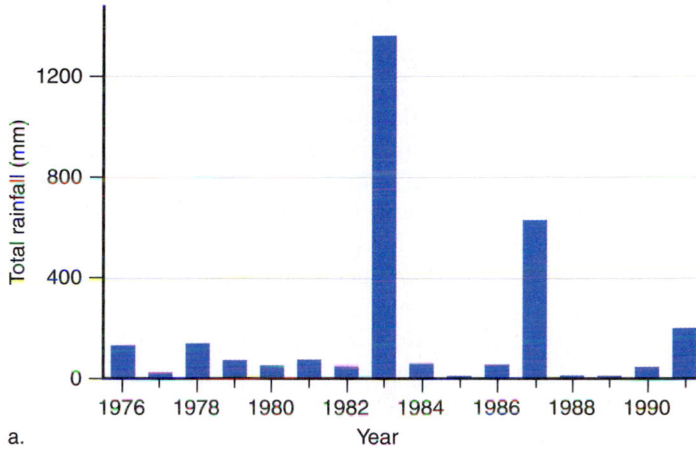

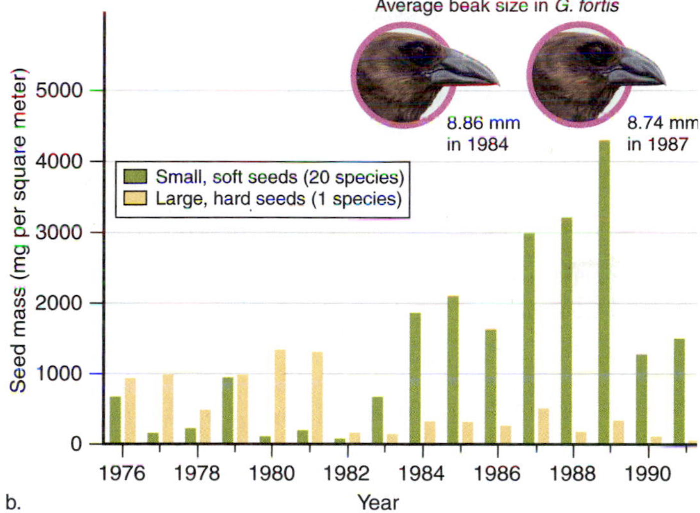

FIGURE 15.21 Evolution in Action. (**A**) Rainfall was extraordinarily heavy on the Galápagos island of Daphne Major in 1983. (**B**) As a result, small, soft seeds became much more abundant than large, hard seeds for several years. Average beak width in *G. fortis* declined during the same period. **Question:** Would careful examination of finch fossils be a practical way to address the questions of short-term evolution that the Grants are studying? Why or why not? What about studying DNA?

What about the cactus finches, *G. scandens*? They did not change, because their diets did not shift after the El Niño event; they simply kept eating medium-sized *Opuntia* seeds. Although their populations also declined when food became scarce from 1983 to 1987, directional selection did not alter the size of their beaks.

This study beautifully complements a famous earlier study documenting a severe drought in 1977 that wiped out most plants on the island. In the resulting food shortage, finches rapidly devoured the small, easy-to-eat seeds. Afterward, only birds with beaks that could crack the largest, toughest seeds survived long enough to reproduce. They passed the "large-beak" alleles to their offspring, and data from subsequent years clearly show a shift toward larger, stronger bills.

The biotechnology revolution has recently enabled the Grants and their colleagues to delve even deeper into the evolution of Darwin's finches. For example, in 2004, researchers reported that they had discovered a gene that influences the development of a finch chick's beak. Finch species with the widest beaks expressed the gene earlier, and at a higher level, than species with narrower beaks. To confirm the gene's role in beak development, the researchers increased its expression in a chicken embryo, and the experimental bird responded by growing a wide, deep, finchlike beak.

The data that the Grants have collected in the Galápagos Islands clearly reveal that heritable traits really do evolve measurably in response to short-term climatic changes, even in a natural setting. It is exciting to consider that technology now allows us to link the measurable features of generations of Darwin's finches to the genes that dictate their development. In the future, it may be possible to track the frequencies of specific alleles over time, taking us on yet another step along the journey that Darwin himself started some 150 years ago.

Grant, B. Rosemary and Peter R. Grant. February 22, 1993. Evolution of Darwin's finches caused by a rare climatic event. *Proceedings of the Royal Society of London* B., vol. 251, pages 111–117.

CHAPTER SUMMARY

15.1 Clues to Evolution Lie in the Earth, Body Structures, and Molecules

- The **geological time scale** divides life's history on Earth into eras defined by major events such as mass extinctions.
- Evidence for evolutionary relationships comes from **paleontology** (the study of past life) and comparing anatomical and biochemical characteristics of species.

15.2 Fossils Record Evolution

- **Fossils** are the remains of ancient organisms.

A. Fossils Form in Many Ways

- Fossils may form when mineral replaces tissue gradually, or they may be indirect evidence such as footprints or feces. Rarely, entire organisms are preserved whole.

B. The Fossil Record Is Often Incomplete

- Many organisms that lived in the past have not left fossil evidence.

C. The Age of a Fossil Can Be Estimated in Two Ways

- The position of a fossil in the context of others provides a **relative date.**
- The ratio of a radioactive isotope to its breakdown product gives an **absolute date,** which is a range of time when an organism lived. Radioactive isotopes with different **half-lives** are useful for **radiometric dating** of different-aged fossils.

15.3 Large-Scale Changes in Earth's Surface Have Influenced Evolution

A. The Theory of Plate Tectonics Describes Earth's Shifting Continents

- The **plate tectonics** theory indicates that forces deep inside Earth have moved the continents through much of life's history, creating and eliminating geographical barriers.

B. Biogeography Considers Species' Geographical Locations

- **Biogeography,** the study of where species occur on Earth, provides insight into both large- and small-scale evolutionary events.

15.4 Homologous Structures Reflect Common Ancestry

- **Homologous** anatomical structures and molecules have similarities that indicate they were inherited from a shared ancestor, although they may differ in function.

A. Comparing Anatomical Parts Can Reveal Evolutionary Relationships

- Skeletal remains of past and present organisms reveal many homologous structures.
- **Analogous** structures are similar in function but do not reflect shared ancestry. **Convergent evolution** can produce analogous structures, such as the wings of birds and insects.
- **Vestigial** structures have no function in an organism but are homologous to functioning structures in related species.

B. Embryonic Development Patterns Provide Evolutionary Clues

- Similar structures in embryos of different species reflect actions of genes retained from ancestors.

15.5 Molecules Reveal Relatedness

A. Comparing DNA and Protein Sequences May Reveal Close Relationships

- Molecular sequences contain so much information that it is unlikely that similarities happened by chance; descent from a shared ancestor is more likely.

- DNA sequence comparisons provide an indication of the relationship between species, as can the amino acid sequences of proteins.

B. "Evo-devo" Bridges Evolution and Developmental Biology

- Evolutionary developmental biology combines the study of development with the study of DNA sequences. Many genes, including **homeotic genes,** influence the development of new phenotypes.

C. Molecular Clocks Help Assign Dates to Evolutionary Events

- A **molecular clock** compares DNA sequences to estimate the time when two species diverged from a common ancestor. Molecular clocks based on mitochondrial DNA are used to date recent events because mtDNA mutates faster than nuclear DNA.

15.6 Systematics Reconstructs the Stories of Life

- The study of **systematics** includes naming and describing species (taxonomy) and the study of species relationships (**phylogeny**).

A. Cladistics Seeks to Define Monophyletic Groups

- **Cladistics** defines groups based on shared **derived characters.** A group of species related by recent common descent is a **clade,** or **monophyletic** group.

B. A Cladogram Shows Lines of Descent

- **Cladograms** have largely replaced evolutionary tree diagrams that consider both **ancestral** and derived characters.
- The most **parsimonious** cladogram is the simplest tree that fits the data.

C. Many Traditional Groups Are Not Monophyletic

- **Paraphyletic** and **polyphyletic** group names remain in common usage, but they do not reflect complete evolutionary relationships.

15.7 Investigating Life: The Shrinking, Growing Beaks of Darwin's Finches Reveal Ongoing Evolution

- Peter and Rosemary Grant's classic studies of Darwin's finches on the Galápagos Islands show how evolution unfolds in real time.
- A short-term change in climate on the island of Daphne Major triggered changes in food availability. As a result, the average beak size in a species of seed-eating birds declined.

MULTIPLE CHOICE QUESTIONS

1. You are living in which of the following geological eras?
   a. Archean
   b. Mesozoic
   c. Paleozoic
   d. Cenozoic

2. Which form of a fossil would be best at providing material for DNA analysis?
   a. Petrifaction
   b. Impression
   c. Intact preservation
   d. Cast

3. The distribution of species around the globe can be explained by
   a. plate tectonics.
   b. changes in ocean levels.
   c. adaptive radiation.
   d. All of the above

4. The wing of a bird and the wing of a bat are
   a. homologous structures.
   b. vestigial structures.
   c. analogous structures.
   d. convergent structures.

5. What term is used to describe an anatomical structure with no known function?
   a. homologous
   b. vestigial
   c. convergent
   d. analogous

6. What is the advantage to comparing *both* DNA and protein sequences in evaluating evolutionary relatedness?
   a. They both provide a measure of the rate of mutation.
   b. Small differences in DNA sequence can still produce proteins of the same sequence.
   c. DNA changes more quickly than proteins.
   d. Differences in protein sequence may exist even when the DNA sequences are similar.

7. How does the activity of a homeotic gene relate to evolution?
   a. Homeotic genes serve as a marker for convergent evolution
   b. Organisms with similar homeotic genes have similar body plans
   c. Homeotic genes are easily mutated
   d. Homeotic genes are retained between evolutionarily distinct organisms

8. How does a molecular clock work?
   a. It estimates when two species diverged using mutation rates
   b. It measures the loss of ^{14}C isotopes in a sample
   c. It estimates the time required for a population to undergo speciation
   d. All of the above

9. Kingdom Protista contains many lineages but does not include all descendants of a common ancestor. Kingdom Protista is therefore
   a. homeotic.
   b. paraphyletic.
   c. polyphyletic.
   d. a clade.

10. What assumption is made regarding all of the organisms within a clade?
    a. They all belong to the same species
    b. They share the same DNA and protein sequences with a common ancestor
    c. They share a characteristic that is derived through evolution from a common ancestor
    d. They share a common developmental pathway

TESTING YOUR KNOWLEDGE

1. What types of information are used to hypothesize how species are related to one another by descent from shared ancestors?
2. Describe six types of fossils and how they form. What present environmental conditions might preserve today's organisms to form tomorrow's fossils?
3. Why is the fossil record useful, even if it doesn't represent every type of organism that ever lived?
4. How does the type of data that molecular sequences provide differ from the type of information that relative and absolute dating provide?
5. How have geological events such as continental movements and the emergence of new volcanic islands influenced the history of life on Earth?
6. Why is it important for evolutionary biologists to be able to distinguish between homologous and analogous anatomical structures?
7. How do biologists use sequences of proteins and genes to infer evolutionary relationships?
8. How did the discovery of homeotic genes help launch the new field of "evo-devo?"
9. Why are molecular clocks useful, and what are their limitations?
10. The following cladogram shows the relationships among a tree species (*Hymenaea protera*) with leaves preserved in amber; three living relatives; and six other types of plants:
    a. Which organism is *H. protera*'s closest relative?
    b. Is *H. protera* more closely related to tobacco or to palm?
    c. Which organism depicted is ancestral to all the others?

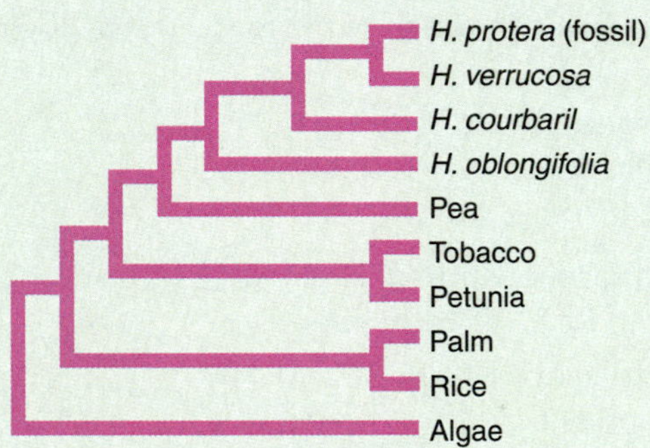

THINKING AS A SCIENTIST

1. Suggest a type of genetic change that could have a drastic effect on the evolution of a species.
2. Suppose biologists find a fossilized lizard bone containing one-eighth the amount of carbon-14 present in the atmosphere. How long ago did the lizard die?
3. Why can't scientists use potassium–argon dating on a preserved woolly mammoth that died 20,000 years ago?
4. Give an example of how molecular and anatomical evidence can support each other.
5. Provide examples of homologous and analogous structures other than those mentioned in this chapter.
6. What evidence was important in placing birds and dinosaurs in the same clade? What evidence could you use to test the hypothesis that humans, chimps, and gorillas belong in the same clade?
7. If scientists could extract DNA from dinosaur fossils, how could they use the sequence to learn more about the origin of birds?
8. Why do you think that molecular evidence often yields an earlier date for a speciation event than fossil evidence?
9. Some genes are more alike between human and chimp than other genes are from person to person. Does this mean that chimps are humans or that humans with different alleles are different species? What other explanation fits the facts?
10. Inheritance can be traced through females by tracking mtDNA sequences. Considering what you know about the chromosomal differences between males and females, suggest a way to follow male inheritance.
11. Why is the DNA sequence of one gene a less accurate indicator of the evolutionary relationship between two species than a comparison of large portions of the two genomes?

Visit www.mhhe.com/hoefnagels for practice quizzes, animations, videos, and activities designed to help you master the material in this chapter.

CHAPTER 16

The Origin and History of Life

Meteoroids enter Earth's atmosphere every day, as they have done for eons. Some scientists hypothesize that ancient objects from space carried the organic molecules needed for life to begin on Earth.

What's the Point?

Life from Space

In a 1908 book entitled Worlds in the Making, *Swedish physical chemist Svante Arrhenius suggested that life came to Earth from the cosmos. He later broadened the idea, calling it "panspermia," and proposing that life-carrying spores arrived on interstellar dust, comets, asteroids, and meteorites. Later versions of panspermia proposed that instead of spores, which are protected cells, only the organic chemicals required for life arrived from space.*

What is the evidence for panspermia? Modern proponents point to several intriguing clues:

- *Prokaryotic cells can survive under extreme conditions, which would be necessary to endure the high radiation and cold temperatures of space during a journey that could take millions of years. A bacterium called* Deinococcus radiodurans, *for example, tolerates a thousand times the radiation level that a person can; it even lives in nuclear reactors! Microbes also live in pockets of water within the ice of Antarctic lakes, surroundings not unlike the icy insides of a comet. In addition, researchers have revived some bacteria after a dormancy lasting millions of years.*
- *Meteorites that have fallen to Earth from space sometimes contain organic compounds such as polycyclic aromatic hydrocarbons (PAHs) or amino acids (see section 2.6).*
- *Some meteorites contain intriguing fossil-like shapes. A famous example is ALH84001, a meteorite that left Mars millions of years ago and fell to Antarctica some 13,000 years ago. In 1996, researchers studying ALH84001 discovered structures that resemble the remains of bacteria. After initial excitement, most scientists now believe the "fossils" have chemical—not biological—origins.*
- *In 1998, the orbiting* Mars Global Surveyor *detected on Mars a large stretch of hematite, a mineral that forms in the presence of warm water. Then in 2004, another orbiter, the* Mars Express, *found that high concentrations of water vapor and methane often occur together in the Martian atmosphere. This finding led to speculation that methane-producing microbes might live in liquid water beneath the ice on Mars' surface.*

The evidence for panspermia, though sparse, is enough to keep the debate alive. Yet even if the evidence mounts, it would not reveal how life began, *just how it might have gotten here. This chapter explains some of what scientists do know about life's origin and history, based on current evidence.*

LEARNING OUTLINE

16.1 Life's Origin Remains Mysterious

Reconstructing life's start is like reading all the chapters of a novel except the first. A reader can get some idea of the events and setting of the opening chapter from clues throughout the novel. Similarly, scattered clues from life through the ages reflect events that may have led to the origin of life.

Panspermia may explain how life started on Earth, but it is not widely accepted, in part because it sidesteps the question of life's ultimate origin. Instead, most scientists accept that life probably arose from simple chemical substances on Earth. This idea may seem to contradict cell theory, which says that cells come only from preexisting cells (see section 3.1 and this chapter's Can *You* Relate? Spontaneous Generation Debunked). An ancient chemical origin for life on Earth, however, is consistent with cell theory because the conditions under which life gradually formed billions of years ago no longer exist. All available evidence supports the validity of cell theory in today's world.

Can *You* Relate?

Spontaneous Generation Debunked

Modern knowledge of cell division and genetics clearly supports the idea that cells come only from preexisting cells. For centuries, however, proponents of an idea called spontaneous generation believed that frogs, cockroaches, mice, and other complex organisms arose suddenly from the nonliving. Why not? Centuries ago, people had no way to know that animals produce reproductive cells too small to see with the unaided eye. Although spontaneous generation was disproved long ago, its story remains instructive because it illustrates yet again that simple experiments can transform scientific understanding of the world. This transformation, however, did not occur overnight.

In 1668, Italian physician Francesco Redi conducted experiments designed to see if life arose from invisible eggs. Redi put meat into four containers open to the air and in four sealed containers. When maggots appeared only in the open flasks, Redi, thinking that flies had laid eggs on the meat, repeated the experiment covering the open flasks with gauze. As he had predicted, keeping out the flies kept out the maggots.

The idea of spontaneous generation arose anew with the invention of microscopes. Perhaps animals did not appear from nowhere, but what about bacteria? In 1768, Italian biologist Lazzaro Spallanzani boiled mutton broth in sealed glass vessels. Bacteria did not appear. Still, some people claimed that boiling killed a "vital principle" required for life, and that sealing the flasks kept it out. Finally, in 1859, Louis Pasteur boiled meat broth in flasks that had S-shaped, curved, open necks that allowed air (and the vital principle, whatever it was) in, but kept out dust particles that carry bacterial spores (see **figure 16.A**). No bacteria grew. People finally believed what they were seeing: life does not arise spontaneously.

FIGURE 16.A Pasteur's Swan-Neck Flask. This ingenious design allowed air, but not bacteria-laden dust, to enter the broth.

The study of the chemistry that led to life begins with astronomy and geology. Earth and the solar system's other planets formed about 4.6 BYA (billion years ago) as solid matter condensed out of a vast expanse of dust and gas swirling around the early Sun. The red-hot ball that became Earth cooled enough to form a crust by about 4.2 to 4.1 BYA, when the surface temperature ranged from 500°C to 1,000°C (932°F to 1832°F), and atmospheric pressure was 10 times what it is now. During the planet's first 500 to 600 million years, comets, meteors, and possibly asteroids bombarded the surface, repeatedly boiling off the seas and vaporizing rocks to carve the features of the fledgling world (**figure 16.1**).

The geological evidence paints a chaotic picture of volcanic eruptions, earthquakes, cosmic bombardment, and ultraviolet radiation. Still, there were probably protected pockets of the environment where organic molecules could aggregate and perhaps interact. The clues from geology and paleontology suggest that from 4.2 to 3.85 BYA, simple cells (or their precursors) arose. So harsh and unsettled was the early environment of Earth that organized groups of chemicals may have formed many times and at many places, only to be torn apart by heat, debris from space, or radiation. We can't know. Somehow, however, an entity arose that could survive, thrive, reproduce, and diversify. **organic molecules, p. 33**

The following sections discuss some of the major steps in the chemical evolution that eventually led to the first cell.

A. The First Organic Molecules May Have Formed in a Chemical "Soup"

Early Earth was not only different geologically from today's planet, but it was also different chemically. The atmosphere today is rich in carbon dioxide (CO_2), nitrogen (N_2), water (H_2O), and oxygen (O_2). What might it have been like about 4 BYA?

Soviet chemist Alex I. Oparin hypothesized in a 1938 book, *The Origin of Life*, that a hydrogen-rich, or reducing, atmosphere was necessary for organic molecules to form on Earth. Oparin thought that this long-ago atmosphere included methane (CH_4), ammonia (NH_3), water, and hydrogen (H_2), similar to the atmospheres of the outer planets today. Due to the extreme conditions on Earth immediately after cooling, little O_2 would have been available, because most of

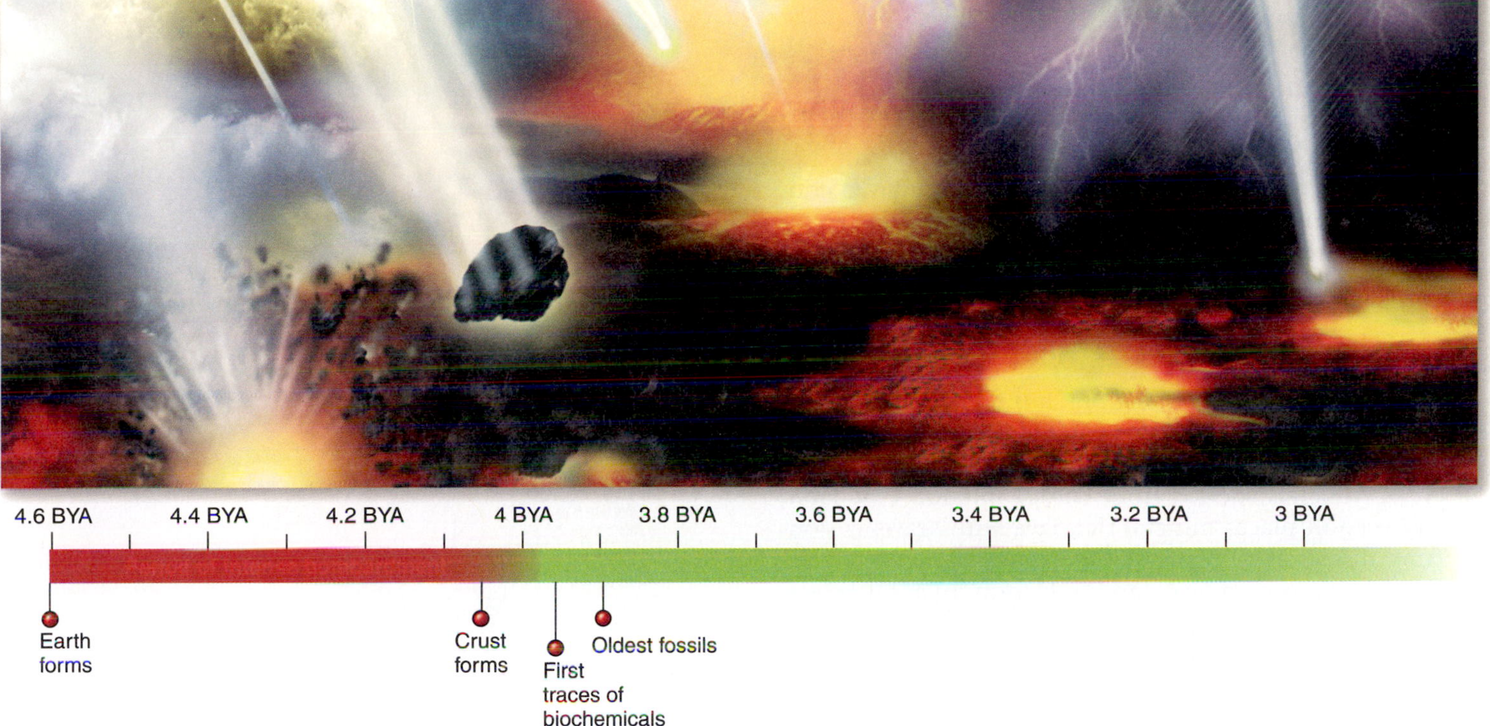

FIGURE 16.1 A Time Line of Early Earth. Originally a violent place, Earth's surface eventually supported life. (Billions of years ago = BYA)

the newly formed chemicals were highly reactive with oxygen. Without O_2, Oparin suggested, chemical reactions that form amino acids and nucleotides would have been possible.

Miller's Experiment

In 1953, Stanley Miller, a graduate student in chemistry at the University of Chicago, and his mentor, Harold Urey, decided to test whether Oparin's atmosphere could indeed give rise to organic molecules. Miller built a glass enclosure to contain Oparin's four gases, through which he passed electric discharges to simulate lightning (**figure 16.2**). He condensed the gases in a narrow tube and passed them over an electric heater, a laboratory version of a volcano. After a few failures and adjustments, Miller saw the condensed liquid turn yellowish. Chemical analysis showed that he had made glycine, the simplest amino acid in organisms. When he let the brew cook a full week, the solution turned varying shades of red, pink, and yellow-brown, which turned out to be several more amino acids, some found in life. A prestigious journal published the work, which Urey gallantly refused to put his name on. The 25-year-old Miller made headlines in newspapers and magazines reporting (incorrectly) that he'd created "life in a test tube."

Life is far more than just a few amino acids, but "the Miller experiment" would go down in history as the first **prebiotic simulation,** an attempt to re-create chemical conditions on Earth before life arose. Miller and many others later extended his results by altering conditions or using different starting materials. For example, methane and ammonia could form clouds of hydrogen cyanide (HCN), which produced amino acids in the presence of ultraviolet light and water. Prebiotic "soups" that included phosphates yielded nucleotides, including the biological energy molecule ATP. Other experiments produced carbohydrates and phospholipids similar to those

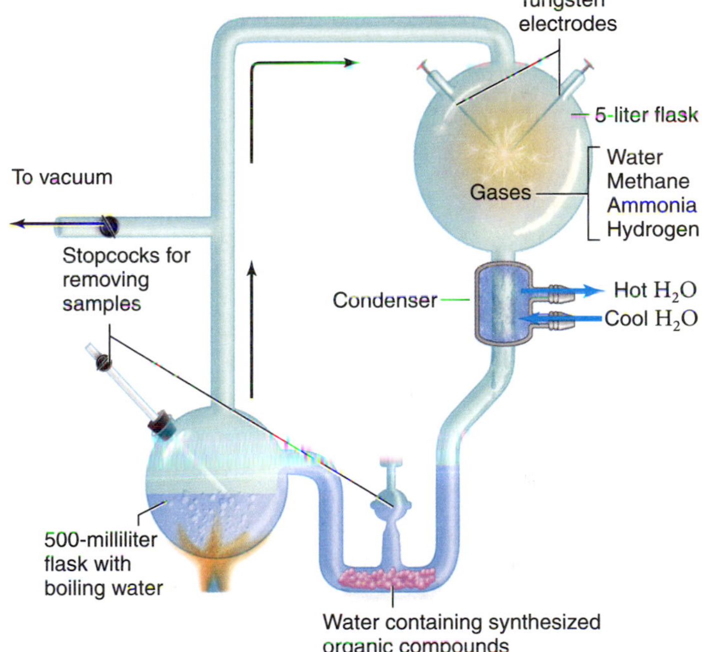

"You have to define 'simulate.' One has to reconstruct an historical event—how did it happen on the primitive earth? You don't even need to argue if you can construct exactly how it happened, but what you can do is to go through a plausible process, from an initial atmosphere, through something that is capable of self-replication. When you do a good prebiotic experiment, you see biological material—amino acids, purines, pyrimidines, and sugars—just fall out. That is telling us something."

—Stanley Miller

FIGURE 16.2 The First Prebiotic Simulation. When Stanley Miller passed an electrical spark through heated gases, he generated amino acids and other organic molecules that may have played a role in the origin of life.

in biological membranes. The experiment has survived criticisms that Earth's early atmosphere actually contained abundant CO_2, a gas not present in Miller's original setup. Organic molecules still form, even with an adjusted gas mixture.

Hydrothermal Vents as a Model

More recent prebiotic simulations mimic deep-sea hydrothermal vents. Here, in a zone where hot water meets cold water, chemical collections could have encountered a rich brew of minerals spewed from Earth's interior. One laboratory version places mineral-rich lava with seawater containing dissolved CO_2 under high temperature and pressure, and simple organic compounds form. In another vent model, nitrogen compounds and water mix with an iron-containing mineral under high temperature and pressure. The iron catalyzes reactions that produce ammonia—one of the components of the original Miller experiment.

The Possible Role of Clays

Once the organic building blocks of macromolecules were present, they had to have linked into chains (polymerized). This may have happened on hot clays or other minerals that provided ample, dry surfaces.

Clays may have played an important role in early organic chemistry for several reasons. They can form templates on which chemical building blocks could have linked to build larger molecules. Some types of clay also contain minerals (e.g., iron pyrite, or "fool's gold") that could release electrons, providing energy to form chemical bonds. These minerals may also have acted as catalysts to speed chemical reactions.

Prebiotic simulations demonstrate that the first RNA molecules could have formed on clay surfaces (**figure 16.3**). Not only do the positive charges on clay's surface attract and hold negatively charged RNA nucleotides, but clays also promote formation of the covalent bonds that link the nucleotides into chains. They even attract other nucleotides to form a complementary strand. About 4 BYA, clays might have been fringed with an ever-increasing variety of growing polymers. Some of these might have become the macromolecules that would eventually build cells.

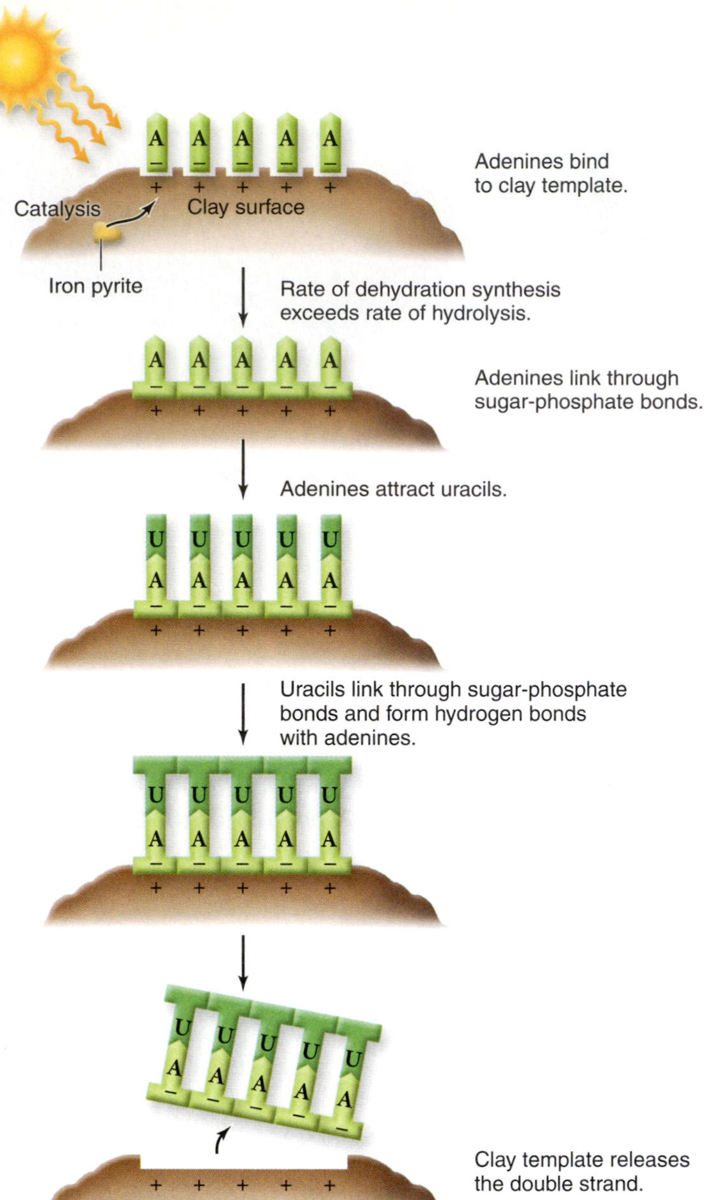
Adenines bind to clay template.
Catalysis
Clay surface
Iron pyrite
Rate of dehydration synthesis exceeds rate of hydrolysis.
Adenines link through sugar-phosphate bonds.
Adenines attract uracils.
Uracils link through sugar-phosphate bonds and form hydrogen bonds with adenines.
Clay template releases the double strand.

FIGURE 16.3 From "Soup" to "Crepes." Clays may have provided templates on which chains of nucleotides could form. In one hypothesized scenario, the dry environment enabled the rate of polymer formation (dehydration synthesis) to exceed that of polymer breakdown (hydrolysis). Energy came from the sun, and iron pyrite ("fool's gold") was the catalyst.

B. Some Investigators Suggest an "RNA World"

Life requires an informational molecule. That molecule may have been RNA, or something like it, because RNA is the most versatile molecule that we know of. It stores genetic information and uses it to manufacture proteins. RNA can also catalyze chemical reactions and duplicate on its own. The term **"RNA world"** has come to describe how self-replicating RNA may have been the first independent form of life on Earth.

As time passed, pieces of RNA would have continued to form and accumulate, growing longer, becoming more complex in sequence, and changing as replication errors led to mutations. Some members of this accumulating community of molecules would have been more stable than others. As Stanley Miller sums it up, "The origin of life is the origin of evolution, which requires replication, mutation, and selection. Replication is the hard part. Once a genetic material could replicate, life would have just taken off."

At some point, RNA might have begun encoding proteins, just short chains of amino acids at first. An RNA molecule may eventually have grown long enough to encode the enzyme reverse transcriptase, which copies RNA to DNA. With DNA, the chemical blueprints of life found a much more stable home. Protein enzymes eventually took over some of the functions of catalytic RNAs.

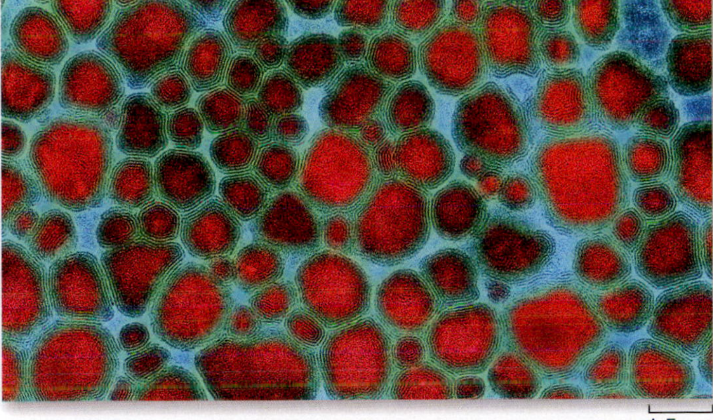
1.5 μm

FIGURE 16.4 Liposomes. The boundaries of these bubbles consist of a phospholipid bilayer similar to a cell membrane. Each liposome contains a watery solution.

C. Membranes Enclosed the Molecules

Meanwhile, lipids would have been entering the picture. Under the right temperature and pH conditions, and with the necessary precursors, phospholipids could have formed membrane-like structures, some of which left evidence in ancient sediments. Experiments using lipids show that pieces of membrane can indeed grow on structural supports and break free, forming a bubble called a liposome (**figure 16.4**). **phospholipids, p. 54**

Might an ancient liposome have enclosed a collection of nucleic acids and proteins to form a cell-like assemblage (**figure 16.5**)? Carl Woese, who described the domain Archaea, gave the term **progenotes** to these hypothetical, ancient aggregates of RNA, DNA, proteins, and lipids. Also called protocells or protobionts, these were precursors of cells, but not nearly as complex.

D. The Origin of Metabolism Would Have Involved Early Enzymes

The next step in the origin of life was the evolution of metabolism. As described in unit 1, metabolism is the ability to acquire and use energy to maintain the organization necessary for life. In a simple sense, this means converting one kind of molecule to another. The capacity of nucleic acids to mutate may have enabled progenotes to become increasingly self-sufficient, giving rise eventually to the reaction pathways of metabolism.

Imagine a progenote that fed on a molecule, nutrient A, that was abundant in its environment. As the progenote reproduced, it began to use up nutrient A. The ancient seas, however, probably held more than one genetic variety of the progenote. One type might have had an enzyme that could convert another nutrient, B, into the original nutrient A. The second progenote would have had a nutritional advantage because it could extract energy from two food sources: B and A.

CO_2
H_2
H_2O
CH_4
NH_3
Lipids
Carbohydrates
Nucleotides
C
A
G
U
Amino acids
Gly
Ala
Ala
U
A
G
C
RNA
A
T
C
G
Complementary DNA
Gly Pro Ala Gly Glu Val Pro
Polypeptides
Lipid spheres
DNA
RNA
Proteins
1 Prebiotic chemistry
Prebiotic chemicals react to form small organic molecules in a watery environment ("soup").
2 Pre-RNA world
Polymerization and dehydration ("crepes") form nucleic acids and polypeptides.
3 RNA world
Reverse transcriptase enzyme copies RNA into complementary DNA.
4 DNA/protein world
DNA replicates; RNA from DNA encodes proteins. Lipids form spheres.
5 Primordial cell
Self-replicating system enclosed in a permeable protective lipid sphere.

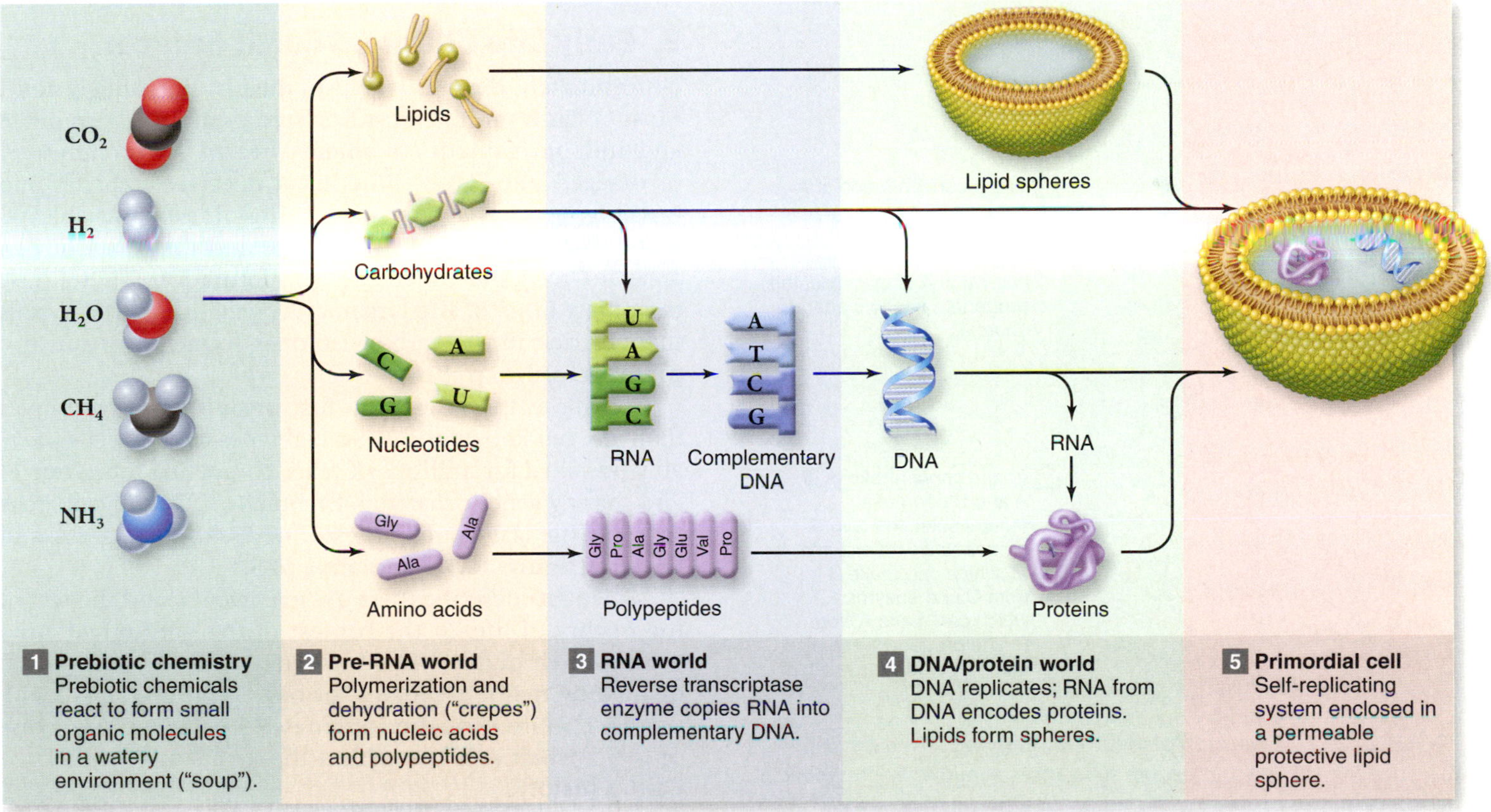

FIGURE 16.5 Pathway to a Cell. The steps leading to the origin of life on Earth may have started with the formation of organic molecules from simple precursors. However it originated, the first cell would have contained self-replicating molecules enclosed in a phospholipid bilayer membrane.

Soon the first type of progenote, totally dependent on nutrient A, would die out as its food vanished. For a while, the progenote that could convert nutrient B to A would flourish. In time, however, nutrient B would also become scarce. Meanwhile, perhaps a new type of progenote would have arisen with an additional enzyme that converted nutrient C to B (and then, using the first enzyme, B to A). In time, this progenote would flourish; then another, and yet another. Over time, an enzyme-catalyzed sequence of connected chemical reactions, a biochemical pathway linking D to C to B to A, would arise. More pathways would form. As intermediates of one pathway spawned others, metabolism evolved (**figure 16.6**).

No one knows what these first metabolic pathways were or how they eventually led to respiration, photosynthesis, and thousands of other chemical reactions that support life today. Despite intriguing similarities between glycolysis and some photosynthetic reactions (see section 6.7), these early stages of metabolism may not have left enough evidence for us ever to understand their origins.

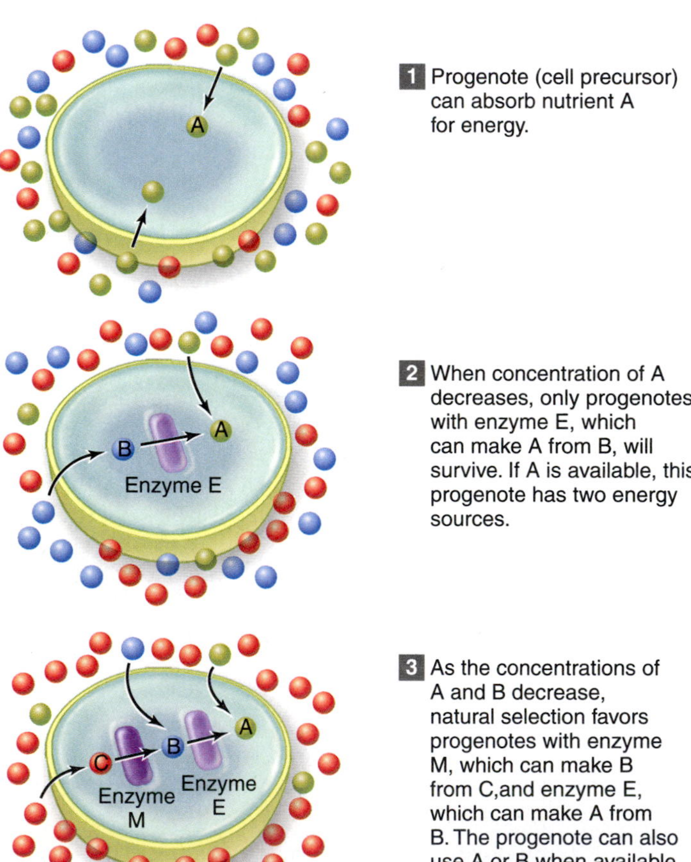
A
1 Progenote (cell precursor) can absorb nutrient A for energy.
B
A
Enzyme E
2 When concentration of A decreases, only progenotes with enzyme E, which can make A from B, will survive. If A is available, this progenote has two energy sources.
C
B
A
Enzyme M
Enzyme E
3 As the concentrations of A and B decrease, natural selection favors progenotes with enzyme M, which can make B from C,and enzyme E, which can make A from B. The progenote can also use A or B when available.

FIGURE 16.6 Evolution of Metabolic Pathways. As early cells developed the ability to use more substances as nutrients, metabolic pathways may have emerged. Natural selection would have favored progenotes that could use more diverse food sources.

Burning Questions

Does new life spring from inorganic molecules now, as it did in the past?

It is intriguing to think of the possibility that new life could be forming now, just as it did long ago in Earth's history. Unfortunately, such an event would be hard to detect, for several reasons. First, when Earth was young, no life existed, so newly evolving life encountered no competition. Now, however, life thrives nearly everywhere on Earth (see chapter 18's Burning Question: Are there areas on Earth where no life exists?). Perhaps new life *is* forming, but before it has a chance to become established, a hungry microbe gobbles it up.

Or maybe the conditions that were once conducive for the development of new life no longer exist. After all, conditions now are very different from what they were on the young Earth. The atmosphere contains abundant O_2, and the ozone layer protects life from the sun's ultraviolet radiation. In any event, scientists have never observed the formation of life from nonliving matter, but that does not mean that it did not happen in the past.

Have a Burning Question of your own? Submit it to marielle_hoefnagels@mcgraw-hill.com for possible inclusion in future editions of this book!

E. Early Life Changed Earth Forever

Unfortunately, direct evidence of the first life is likely gone. Most of Earth's initial crust has been destroyed: torn down and built up again into sediments, heated and compressed, or dragged into Earth's interior at deep-sea trenches and possibly recycled to the surface. The oldest rocks that remain today date to about 3.85 BYA, and they are from an area of Greenland called the Isua formation. They house the oldest hints of life, organic deposits in quartz crystals that are rich in the carbon isotopes found in organisms. **plate tectonics, p. 314**

Whatever they were, the first organisms were simpler than any cell known today. Several types of early cells probably prevailed for millions of years, competing for resources and sharing genetic material. Eventually, a type of cell arose that was the last shared ancestor of Archaea, Bacteria, and Eukarya (**figure 16.7**). **domains, p. 57**

Scientists describe these ancient events, and the rest of the history of life, in the context of the **geological time scale,** which divides time into eras, periods, and epochs defined by major geological or biological events (see figure 15.1). Life probably originated in the Archean, one of two ancient eras that reflect the first 2 billion years of Earth's history.

Some of the oldest fossils are from 3.7-billion-year-old rock in Warrawoona, Australia, and Swaziland, South Africa.

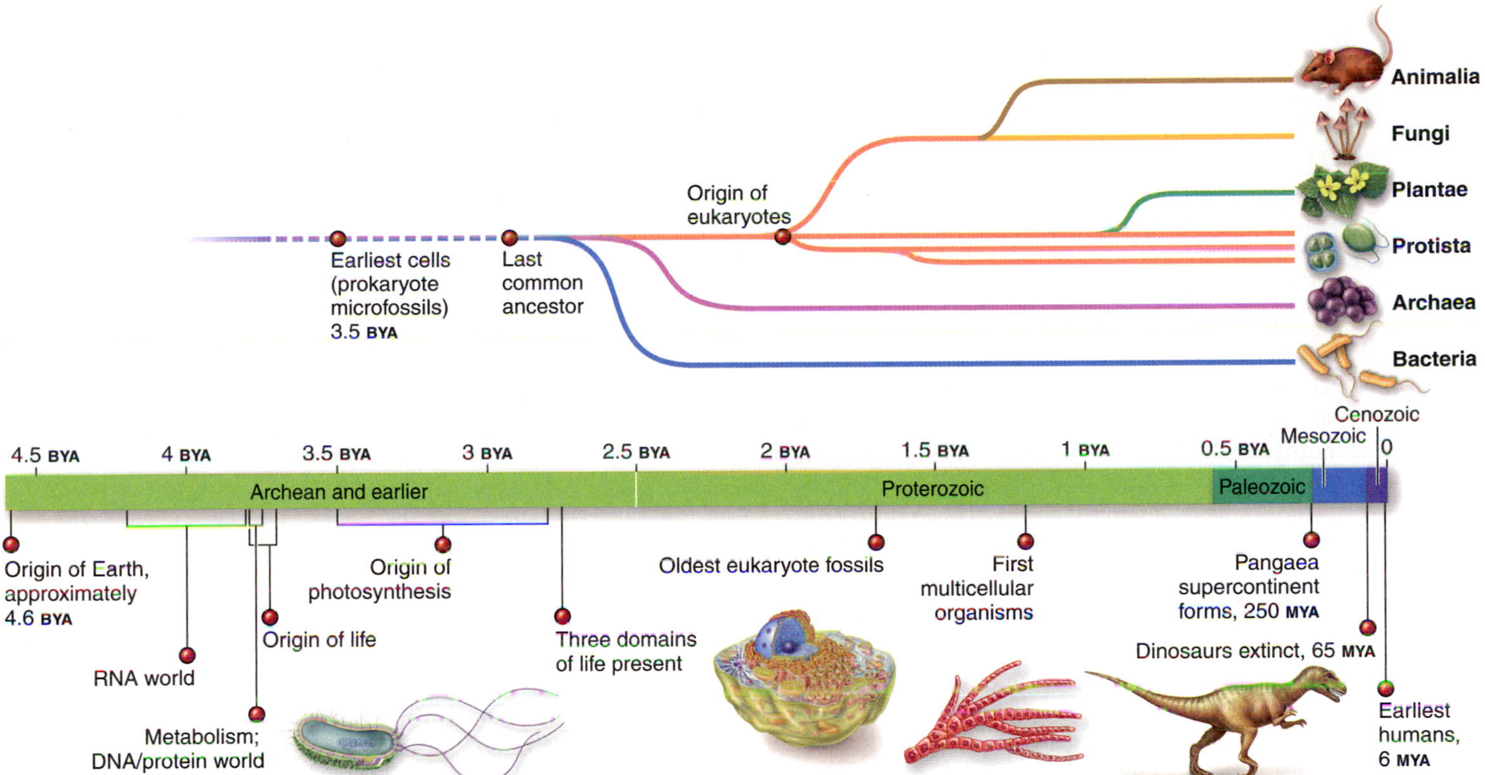

FIGURE 16.7 Timeline for the Emergence of Life's Diversity. We know little about the earliest types of organisms, but they were single-celled. Although life has increased in complexity, simpler organisms such as bacteria and archaea still flourish. The time when scientists think life on Earth has been abundant—the Paleozoic, Mesozoic, and Cenozoic eras—accounts for only one-sixth of the planet's history, and humans are very recent arrivals. (BYA = billion years ago, MYA = million years ago.)

They are large formations of cyanobacteria-like cells called stromatolites. By about 3 BYA, however, the geological turmoil had calmed somewhat, and unicellular organisms inhabited nearly every part of the planet.

The presence of cyanobacteria during the Archean era suggests photosynthetic organisms that generated O_2 as a byproduct. Photosynthesis probably originated, however, in an anaerobic cell that used hydrogen sulfide (H_2S) instead of water as an electron donor. These first photosynthetic microorganisms would have released sulfur, rather than oxygen, into the environment. Eventually, changes in pigment molecules enabled some of these organisms to use water instead of hydrogen sulfide as an electron donor (see chapter 5).

Once this new form of photosynthesis arose, it added O_2 to the atmosphere over millions of years during the Archean and Proterozoic eras, forever altering life on Earth. Large numbers of organisms could now use light as an energy source to produce their own food, instead of relying on organic compounds in their surroundings. In addition, natural selection began to favor aerobic organisms that could use oxygen in metabolism, while anaerobic species would persist in pockets of the environment away from oxygen. Ozone (O_3) also formed from O_2 high in the atmosphere, blocking the sun's damaging ultraviolet radiation. The overall result: an explosion of new life (see this chapter's Burning Question: Does new life spring from inorganic molecules now, as it did in the past?).

Summary *Geology, paleontology, and biochemistry provide evidence for the gradual formation of the first cells on long-ago Earth. The first step on the path to life probably included the formation of simple organic chemicals. RNA or a similar molecule encoded information, replicated, and mutated. Eventually membranes enclosed the molecules, and metabolic pathways developed. Early life changed Earth forever.*

16.1 MASTERING CONCEPTS

1. How were conditions on Earth before life began different from current conditions?
2. What types of information can prebiotic simulations provide?
3. Why is RNA the molecule most likely to have been pivotal in life's beginnings?
4. What is a progenote?
5. How might metabolic pathways have originated and evolved?
6. About when did the first cells probably originate?
7. How did early life change Earth?

16.2 Complex Cells and Multicellularity Arose over a Billion Years Ago

Until this point, we have considered the origin of prokaryotic cells. Fossil evidence shows that eukaryotic cells emerged during the Proterozoic era, at least 1.9 to 1.4 BYA (**figure 16.8**). Australian fossils consisting of organic residue 1.69 billion years old are chemically similar to eukaryotic membrane components and may have come from a very early unicellular eukaryote.

Recall from chapter 3 that prokaryotic cells are structurally simple compared with compartmentalized eukaryotic cells (see figures 3.10, 3.13, and 3.14). We may never know the origin of the nuclear envelope, endoplasmic reticulum, Golgi apparatus, and other membranes within the eukaryotic cell. The membranes of these organelles consist of phospholipids and proteins, as does the cell's outer membrane. Could the outer membrane of an ancient cell have folded in on itself, over and over, until a complex internal network formed (**figure 16.9**)? Unfortunately, that hypothesis is difficult or impossible to test, so we can only speculate about that aspect of eukaryotic cell evolution.

Some details, however, are becoming clear. Chapter 3's Investigating Life: Did the Cytoskeleton Begin in Bacteria? describes the discovery of cytoskeleton-like proteins in prokaryotic cells. In addition, the endosymbiont theory may explain the origin of mitochondria and chloroplasts.

A. Endosymbiosis Explains the Origin of Organelles

The **endosymbiont theory** proposes that mitochondria and chloroplasts originated as free-living bacteria that were engulfed by other prokaryotic cells (**figure 16.10**). (This process would be similar to endocytosis, pictured in figure 4.18.) The term *endosymbiont* derives from *endo-*, meaning "inside," and *symbiont*, meaning "to live together." Biologist Lynn Margulis at the University of Massachusetts at Amherst proposed this theory in the late 1960s.

1 cm

FIGURE 16.8 Fossil Evidence of Eukaryotic Cells. A very ancient eukaryote, *Grypania spiralis,* left this spiral of organic matter in a 1.4-billion-year-old rock in China. No one knows whether it consists of one cell or many.

Cell membrane
Nucleus
Cytoplasm

FIGURE 16.9 Membrane Infolding. This simplified diagram shows a possible step toward the origin of eukaryotic cells. A highly folded cell membrane may have formed internal structures such as the endoplasmic reticulum and the nuclear envelope.

The compelling evidence supporting the idea that mitochondria and chloroplasts in present-day eukaryotic cells originated as independent organisms includes:

- similarities in size, shape, and membrane structure between the organelles and some types of prokaryotes;
- the observation that mitochondria and chloroplasts reproduce by splitting in two, as do prokaryotic cells;
- the similarity between the photosynthetic pigments in chloroplasts and those in cyanobacteria;
- the observation that mitochondria and chloroplasts contain DNA, RNA, and ribosomes, which are similar to those in prokaryotic cells; and
- similarities in DNA sequences between mitochondrial and bacterial genomes.

Once ancient cells had acquired their endosymbiont organelles, genetic changes made the captured microorganisms unable to live on their own outside their hosts. Over time, they came to depend on one another for survival. The result of this biological interdependency, according to the endosymbiont theory, is the compartmentalized cells of modern eukaryotes.

B. Multicellularity May Also Have Its Origin in Cooperation

One critical step leading to the evolution of plants, fungi, and animals was the origin of multicellularity, which occurred about 1.2 BYA. The earliest fossils of multicellular life are from a red alga that lived about 1.25 BYA to 950 MYA in Canada

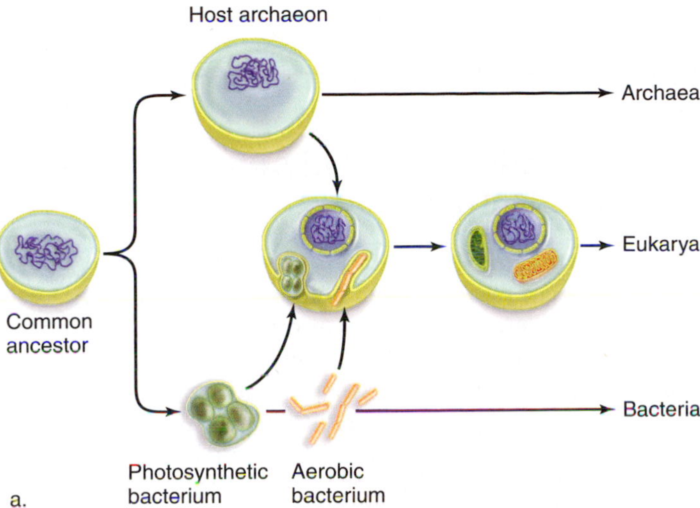

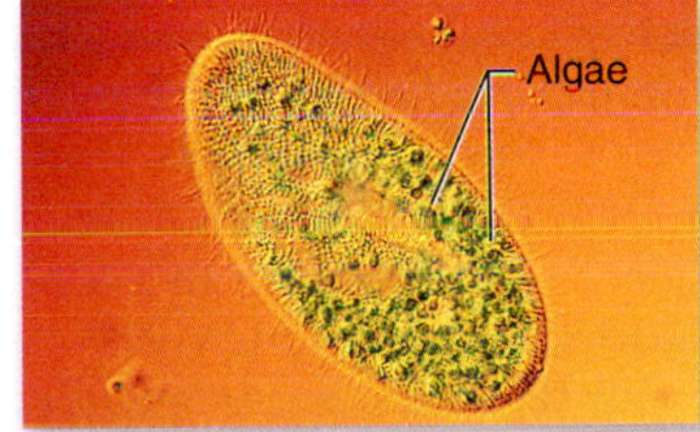

b. *Paramecium bursaria* 50 µm

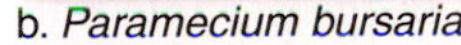

c. *Reclinomonas americana* 15 µm

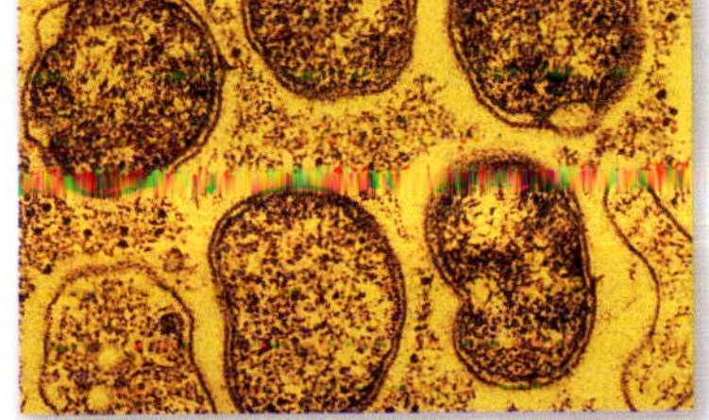

d. *Rickettsia prowazekii* 40 µm

FIGURE 16.10 The Endosymbiont Theory. (**A**) Mitochondria and chloroplasts may have originated from an ancient union of bacterial cells with archaean cells. (**B**) Modern examples of endosymbionts, such as the algae living inside this protist, support the theory. (**C**) *Reclinomonas americana* is a eukaryote whose mitochondrial genome is organized like that of a bacterium. (**D**) Compelling evidence for endosymbiosis also comes from genome sequences, which indicate that the bacterium *Rickettsia prowazekii* carries out energy reactions very much like a mitochondrion does.

(**figure 16.11**). Abundant fossil evidence of multicellular algae also comes from Siberia, dating from a billion years ago.

Exactly how life proceeded from the single-celled to the many-celled is a mystery. Perhaps many individual cells came together, joined, and took on specialized tasks to form a multicellular organism. Slime molds, described in section 19.3, are modern-day protists that may be a good model for how this may have occurred. Alternatively, a single-celled organism may have divided, and the daughter cells may have remained stuck together rather than separating. After many rounds of cell division, these cells may have begun expressing different subsets of their DNA. The result would be a multicellular organism with specialized cells—similar to the way in which modern animals and plants develop from a single fertilized egg cell.

FIGURE 16.11 Early Multicellularity. *Bangiomorpha pubescens*, known only from 1.2-billion-year-old fossils, represents one of the oldest known examples of a multicellular organism. It resembles *Bangia*, a contemporary red alga. Multicellularity made possible the development of specialized cells, such as a "holdfast." An upright orientation opened new habitats for other organisms and introduced new evolutionary possibilities for *Bangiomorpha*.

Whatever way it happened, the origin of multicellularity led to an explosion in the variety of body sizes and forms representing life. The remainder of this chapter describes the diversification and spread of these complex organisms.

Summary ***The endosymbiont theory proposes that eukaryotic cells arose from ancestral cells that acquired mitochondria and chloroplasts by engulfing bacteria; membrane infolding may also have played a role in the origin of eukaryotes. Some time later, multicellular life formed and diverged.***

16.2 MASTERING CONCEPTS

1. When did eukaryotic cells arise?
2. How might the endoplasmic reticulum, nuclear envelope, and other internal membranes have arisen in eukaryotic cells?
3. What is the evidence that mitochondria and chloroplasts descend from simpler cells engulfed long ago?
4. When did the first multicellular organisms appear in the fossil record?
5. What are two ways that multicellular organisms may have originated?

16.3 Life's Diversity Exploded in the Past Five Hundred Million Years

It would take many thousands of pages to capture all of the events that passed from the rise of the first cells to life today—if we even knew them. In this section, we highlight a few key events in the history of multicellular, eukaryotic life.

A. The Strange Ediacarans Flourished During the Precambrian

Paleontologists divide history informally into two parts: the Precambrian, and everything that came during and after the Cambrian (the first period of the Paleozoic era). The reason for this distinction is that fossils of all the major phyla of animals appeared within a few million years in the Cambrian (a phenomenon sometimes called the "Cambrian explosion"). It seems highly unlikely, however, that the late Precambrian was devoid of animals. Instead, whatever lived then was soft-bodied and did not readily fossilize.

The 4 billion years of the Precambrian were a very eventful time in life's history. During this time, life originated, reproduced, and diversified; photosynthesis evolved; O_2 accumulated in Earth's atmosphere; and eukaryotes arose, as did the first multicellular algae and animals.

Perhaps the most famous Precambrian residents were the mysterious Ediacaran organisms, which left no known modern descendants (**figure 16.12**). One example, *Dickinsonia,* was a meter in diameter but less than 3 mm thick. Biologists have interpreted these fossils as everything from worms to ferns to fungi. The Ediacarans vanished from the fossil record about 544 MYA. (In 2004, geologists named the last portion of the Precambrian, from 543 to about 600 MYA, the Ediacaran period in honor of these strange creatures.)

B. Paleozoic Plants and Animals Emerged onto Land

Cambrian Period (543 to 490 MYA)

The Cambrian seas exploded with a spectacular diversity of life very different from what came before. Remnants of the Ediacaran world coexisted with abundant red and green algae, sponges, jellyfish, and worms. Most notable were the earliest known organisms with hard parts, such as insect-like trilobites, nautiloids, scorpion-like eurypterids, and brachiopods, which resembled clams. Many of these invertebrates left remnants identified only as "small, shelly fossils." The early Cambrian seas were also home to diverse wormlike, armored animals, some of which would die out. Others gave rise to modern mollusks, worms, and arthropods.

The Burgess Shale from British Columbia preserves a glimpse of life from this time. A mid-Cambrian mudslide buried enormous numbers of organisms, including animals with skeletons and soft-bodied invertebrates not seen elsewhere. The Burgess Shale animals were abundant, diverse, and preserved in exquisite detail (**figure 16.13**).

Ordovician Period (490 to 443 MYA)

During the Ordovician period, the seas continued to support huge communities of algae and invertebrates such as sponges, corals, snails, clams, and cephalopods. The first vertebrates to leave fossil evidence, jawless fishes called ostracoderms, appeared at this time. Fossilized spores indicate that life had ventured onto land, in the form of primitive plants that may have resembled modern liverworts (**figure 16.14**).

FIGURE 16.12 Ediacarans Were... Different. (**A**) A "typical" Ediacaran organism, *Dickinsonia*. It had segments, two different ends, and internal detail that paleontologists have interpreted as remnants of a simple circulatory or digestive system. But just what it was remains unclear. (**B**) No one knows what type of organism *Spriggina* was either.

FIGURE 16.13 **Cambrian Life.** *Opabinia* is one of many strange animals whose fossils have been discovered in the Burgess Shale. Its body was segmented, and its head sported stalked eyes and a long, flexible snout.

Silurian Period (443 to 417 MYA)

The first vascular plants evolved during the Silurian (see chapter 20). The newly arrived plants would have provided food and shelter for animals. The first terrestrial animals to leave fossils resembled scorpions, which may have preyed upon other small animals exploring the land. Fungi likely colonized land at the same time. **vascular plants, p. 415**

Aquatic life also continued to change. Fish with jaws arose, as did the first freshwater fish, but jawless fishes were still widespread during the Silurian. The oceans also contained abundant corals, trilobites, and mollusks.

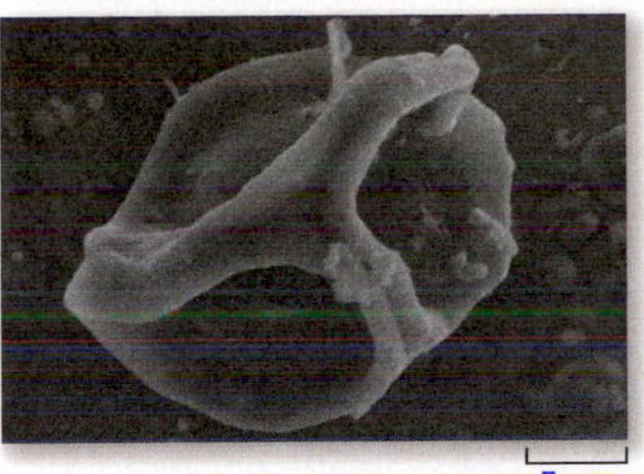

FIGURE 16.14 **Plants Settle the Land.** Plants likely evolved from freshwater algae and left the first traces of life on land in the late Ordovician. This spore from that time is from *Tetrahedraletes medinensis*, a plant related to the liverworts.

Devonian Period (417 to 354 MYA)

The Devonian period was the "Age of Fishes." The seas continued to support more life than did the land. The now prevalent invertebrates were joined by fishes with skeletons of cartilage or bone. Corals and animals called crinoids that resembled flowers were abundant.

The fresh waters of the Devonian were home to the lobe-finned fishes (**figure 16.15A**). These animals had fleshy, powerful fins and could obtain oxygen through both gills and primitive lunglike structures. Toward the end of the Devonian period, about 360 MYA, the first amphibians appeared (figures 16.15B and C). *Acanthostega* had a fin on its tail like a fish and used its powerful tail to move underwater, but it also had hips, legs, and toes. Preserved footprints indicate that the animal could venture briefly onto land. A contemporary of *Acanthostega*, called *Ichthyostega*, had more powerful legs and a rib cage strong enough to support the animal's weight on land, yet it had a skull shape and finned tail reminiscent of fish ancestors.

Many fossils indicate that by this time, plants were diversifying to ferns, horsetails, and seed plants. Scorpions, millipedes, and other invertebrates lived on the land.

a. Coelacanth

b. *Acanthostega*

c. *Ichthyostega*

FIGURE 16.15 **From an Aquatic to a Terrestrial Existence.** **(A)** Lobe-finned fishes lived during the Devonian period and had lungs, an adaptation that would permit their descendants to venture onto the land. **(B)** *Acanthostega* stayed mostly in the water but had adaptations that permitted it to spend short periods on land, including legs with eight toes. **(C)** *Ichthyostega* could spend longer periods on land because its rib cage was stronger. It retained the skull shape and finned tail of its ancestors.

Carboniferous Period (354 to 290 MYA)

The amphibians flourished from about 300 to 350 MYA, giving this period the name the "Age of Amphibians." These animals spent time on land, but they had to return to the water to wet their skins and to lay eggs. At the same time, some types arose that coated their eggs with a hard shell. These animals branched from the other amphibians, eventually giving rise to reptiles, birds, and mammals. The first animals capable of living totally on land, the primitive reptiles, appeared about 300 MYA.

The swamps of this time had fernlike plants and conifers, which towered to 40 m (**figure 16.16A**). The air was alive with the sounds of dragonflies, grasshoppers, and crickets, some of them giant versions of their familiar modern descendants. Land snails and other invertebrates flourished in the sediments. By the end of the period, many of the plants had died, buried beneath the swamps to form, over the coming millennia, coal beds. Today, a split piece of coal will sometimes reveal an impression left by an ancient plant (figure 16.16B).

In the oceans, the bony fishes and sharks were beginning to resemble modern forms, and protists called foraminiferans were abundant. Bryozoans and brachiopods were plentiful, but trilobites were becoming less common, and the armored fishes had become extinct.

Permian Period (290 to 248 MYA)

During the Permian, seed plants called gymnosperms became more prominent. Reptiles were also becoming more prevalent. The reptile introduced a new biological structure, the amniote egg, in which an embryo could develop completely on dry land. Amniote eggs persist today in reptiles, birds, and a few mammals. **amniote, p. 480**

The Permian period foreshadowed the dawn of the dinosaur age. Cotylosaurs were early Permian reptiles that gave rise to the dinosaurs, as well as to modern reptiles, birds, and mammals. They coexisted with their immediate descendants, the pelycosaurs, or sailed lizards.

The Permian Extinction

The Paleozoic era (and Permian period) ended dramatically with what paleontologists call "the mother of mass extinctions." It affected marine life the most, forcing more than 90% of species in the shallow seas into extinction. On the land, many types of insects, amphibians, and reptiles disappeared, paving the way for the age of dinosaurs. **mass extinctions, p. 298**

Paleontologists hypothesize that the Permian extinctions were the result of a drop in sea level, which dried out coastline communities. CO_2 accumulated in the atmosphere from oxidation of organic molecules, which raised global temperature and depleted the sea's dissolved O_2. A long series of volcanic eruptions beginning 255 MYA, and lasting a few million years, further altered global climate. Finally, sea level rose again, drowning coastline communities.

C. Reptiles and Flowering Plants Thrived During the Mesozoic Era

Triassic Period (248 to 206 MYA)

During the Triassic period, small animals called thecodonts flourished. Thecodonts shared the forest of cycads, ginkgos, and conifers with other animals called therapsids, which were the ancestors of mammals (**figure 16.17**).

At the close of the Triassic period, thecodonts and therapsids were becoming rarer, as much larger animals began to infiltrate a wide range of habitats. These new, well-adapted animals were the dinosaurs, and they would dominate for the next 120 million years.

FIGURE 16.16 The Carboniferous, or Coal Age. (**A**) About 300 MYA, lush forests dominated the landscape. The fernlike plants in the foreground are ancient seed-bearing plants. Other plants and trees, also extinct, gave rise to modern club mosses and ground pines. (**B**) These forests were eventually preserved in massive coal beds containing the remains of Carboniferous plants, such as this fossilized fern frond.

a.

b.

FIGURE 16.17 Life in the Mesozoic. The dawn of the Mesozoic era saw many small animals, called therapsids, that eventually gave rise to mammals. Mammals did not suddenly appear with the demise of the dinosaurs; they just became larger, more diverse, and more common.

Jurassic Period (206 to 144 MYA)

By the Jurassic period, giant reptiles were everywhere. Ichthyosaurs, plesiosaurs, and giant marine crocodiles swam in the seas alongside sharks and rays, feasting on fish, squid, and ammonites. Apatosaurs and stegosaurs roamed the land. Carnivores, such as allosaurs, preyed on the herbivores. Pterosaurs glided through the air, as did *Protoarchaeopteryx* and then *Archaeopteryx*—the first birds (see chapter 15's opening essay: Are Birds Dinosaurs?).

At the same time, the first flowering plants (angiosperms) appeared on land. The forests, however, still consisted largely of tall ferns and conifers, ginkgos, club mosses, and horsetails. The first frogs and the first true mammals, which were no larger than rats, appeared as well.

Cretaceous Period (144 to 65 MYA)

The Cretaceous period was a time of great biological change. Beginning about 100 MYA, flowering plants spread in spectacular diversity; many modern insects arose at about the same time. Marine reptiles hunted mollusks and fish, and birds and pterosaurs roamed the skies. Duck-billed maiasaurs traveled in herds of thousands in what is now Montana. Huge herds of apatosaurs migrated on the plains of Alberta to the Arctic, northern Europe, and Asia, which were joined as one continent at the time. By the end of the period, *Triceratops* was so widespread that some paleontologists call it the "cockroach of the Cretaceous."

The Cretaceous Extinction

The reign of the giant reptiles ended about 65 MYA, with the extinction of ichthyosaurs, plesiosaurs, mososaurs, pterosaurs, and nonavian dinosaurs. Ammonites vanished too, as did many types of foraminiferans, sea urchins, and bony fishes. In all, nearly 75% of species perished. The mass extinction opened up habitats for many species that survived, including flowering plants, mollusks, amphibians, some smaller reptiles, birds, and mammals. Many of these groups, including the mammals that gave rise to our own species, subsequently flourished.

We do not know what caused this famous mass extinction, but it coincides with an asteroid impact near the Yucatán peninsula (see figure 14.13). The asteroid, which was about 10 km (6.2 miles) in diameter, left a debris- and clay-filled crater offshore and a huge semicircle of sinkholes onshore. Biologists estimate that photosynthesis was almost nonexistent for 3 years, as debris that was thrown into the sky circulated in the atmosphere, blocking the sunlight. Plankton, which provide microscopic food for many larger marine dwellers, died as well, causing a devastating chain reaction.

D. Mammals Radiated During the Cenozoic Era

Tertiary Period (65 to 1.8 MYA)

The dawn of the Cenozoic era was a time of great adaptive radiation for mammals, according to the fossil record (see figure 14.11). At the start of the Cenozoic era, diverse hoofed mammals grazed the grassy Americas. Many may have been marsupials (pouched mammals) or egg-laying monotremes, ancestors of the platypus (see figure 15.A). Then placental mammals appeared, and fossil evidence indicates that they rapidly dominated the mammals. Within just 1.6 million years during the Tertiary period, 15 of the 18 modern orders of placental mammals arose. **adaptive radiation, p. 296**

Geology and the resulting climate changes molded the comings and goings of species throughout the Cenozoic. The era began with formation of new mountains and coastlines as tectonic plates shifted. The wet warmth of the Paleocene epoch, which opened up many habitats for mammals, continued into the Eocene, providing widespread forests and woodlands. Grasslands began to replace the forests by the end of the Eocene, when the temperature and humidity dropped. Extinctions of some mammals paralleled the changing plant populations as the forests diminished, but grazing mammals thrived throughout the Oligocene, Miocene, and Pliocene.

Quaternary Period (1.8 MYA to present)

The Pleistocene epoch accounts for all but the last 10,000 years of the Quaternary period. During several Pleistocene ice ages, huge glaciers covered about 30% of Earth's surface, and then withdrew again.

Many organisms of the time were similar to those that are familiar now, including flowering plants, insects, birds, and mammals. Some Pleistocene species, however, are now extinct. The woolly mammoths, mastodons, saber-toothed cats, and other large mammals that once roamed North America, Asia, and Europe are known only from their fossils and the occasional DNA fragment (see chapter 7's Investigating Life: Genetic Messages from the Dead Tell Tales of Ancient Ecosystems).

Our species, *Homo sapiens* ("the wise human"), probably first appeared about 200,000 years ago in Africa and had spread throughout most of the world by about 10,000 years ago. Other species of *Homo*, including Neanderthals, vanished during the Pleistocene, leaving *Homo sapiens* as the sole human species. But the history of our species began long before that time, some 60 MYA, when the order Primates arose. We pick up the story of human evolution in more detail in the next section.

Summary *Cells first appeared in the Precambrian era. Life diversified dramatically during the Paleozoic era as plants, fungi, and animals colonized land and became better adapted to their terrestrial habitat. During the Mesozoic era, reptiles ruled, flowering plants began to diversify, and mammals appeared. The Cenozoic era saw extensive adaptive radiation of mammals, including the appearance of our own species.*

16.3 MASTERING CONCEPTS

1. When did the Ediacarans live, and what were they like?
2. What types of organisms flourished in the Cambrian?
3. How did Paleozoic life diversify during the Ordovician, Silurian, Devonian, Carboniferous, and Permian periods?
4. How did the Paleozoic era end?
5. Which organisms came and went during the Mesozoic era?
6. How did the Mesozoic era end?
7. Which new organisms arose during the Cenozoic era?
8. When did *Homo sapiens* first appear?

16.4 Fossils and DNA Tell the Human Evolution Story

In many ways, humans are Earth's dominant species. We are not the most numerous—more microbes occupy one person's intestinal tract than there are people on Earth. In the short time of human existence, however, we have colonized most continents, altered Earth's surface, eliminated many species, and changed many others to fit our needs. Where did we come from?

A. Humans Are Primates

If you watch the monkeys or apes in a zoo for a few minutes, it is almost impossible to ignore how similar they seem to humans. Young ones scramble about and play. They sniff and handle food. Babies cling to their mothers. Adults gather in small groups or sit quietly, snoozing or staring into space.

Flat nails and grasping hands with opposable thumbs

Opposable thumb

Flat fingernails

Tarsier

Gorilla

Homo sapiens

Large brains

Human brain

Chimpanzee brain

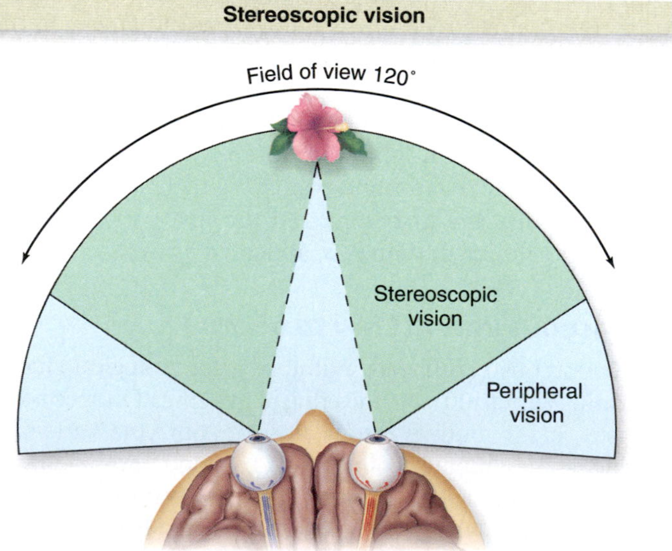

FIGURE 16.18 Primate Characteristics. Primates share several physical characteristics, including opposable thumbs, stereoscopic vision, and large brains.

It is no surprise that we see ourselves reflected in the behaviors of monkeys and apes. As **primates,** monkeys, apes, and humans share a suite of physical characteristics (**figure 16.18**). First, primates have grasping hands with opposable thumbs that can bend inward to touch the pads of the fingers. Some primates also have grasping feet with opposable big toes. Second, eyes set in the front of the skull give primates overlapping fields of sight that result in excellent depth perception. Third, the primate brain is large by comparison with body size. Finally, primates have flat nails instead of claws on their fingers and toes.

Compared with many other groups of mammals, primate anatomy is unusually versatile. For instance, bat wings are useful for flight but not much else; likewise, horse hooves are best for fast running. In contrast, primates have multipurpose fingers and toes that are useful not only for locomotion but also for grasping and manipulating small objects. Primate limbs are similarly versatile.

The Primate Lineage

The primate lineage contains three groups (**figure 16.19**). **Prosimian** is an informal umbrella term for lemurs, aye-ayes, lorises, tarsiers, and bush babies. **Simian** is the corresponding term for monkeys, both Old World (native to Africa and Asia) and New World (native to South and Central America). **Hominoids** include all apes, including humans.

Hominoids are further divided into two groups (**table 16.1**). One contains the gibbons, or "lesser apes." The other, the **hominids,** contains all of the "great apes": orangutans, gorillas, chimpanzees (including bonobos), and humans. Orangutans, however, are not as closely related to the other great apes. **Hominines** include only gorillas, chimpanzees, and humans.

Table 16.1 *Miniglossary of Primate Terminology*

Term	Animals Included in Group*
Primates	Prosimians, monkeys, apes
Hominoids	"Lesser apes" (gibbons) and great apes
Hominids	"Great apes" (orangutans, gorillas, chimpanzees, humans)
Hominines	Gorillas, chimpanzees, humans

*Each group also contains extinct representatives known only from fossils

Of the many sources of information indicating how these groups are related, paleontologists have used one the longest: examining physical characteristics, especially of skeletons (**figure 16.20**). Most human fossils consist of bones and teeth. Comparing these remains with existing primates reveals surprisingly detailed information about locomotion and diet. For this reason, knowledge of primate skeletal anatomy is essential to interpreting human fossils.

Primate Locomotion

Among the most important characteristics in hominoid skeletons are adaptations related to locomotion. Brachiation is swinging from one arm to the other while the body dangles below. Many hominoids move through the treetops in this way; in contrast, monkeys run on all fours along the tops of branches. Orangutans spend most of their lives in trees and move by brachiation when they are in treetops. Gibbons, the most superbly acrobatic hominoids, have long arms and hands. The size and opposability of the thumb are reduced, but their arms connect to the shoulders by ball-and-socket

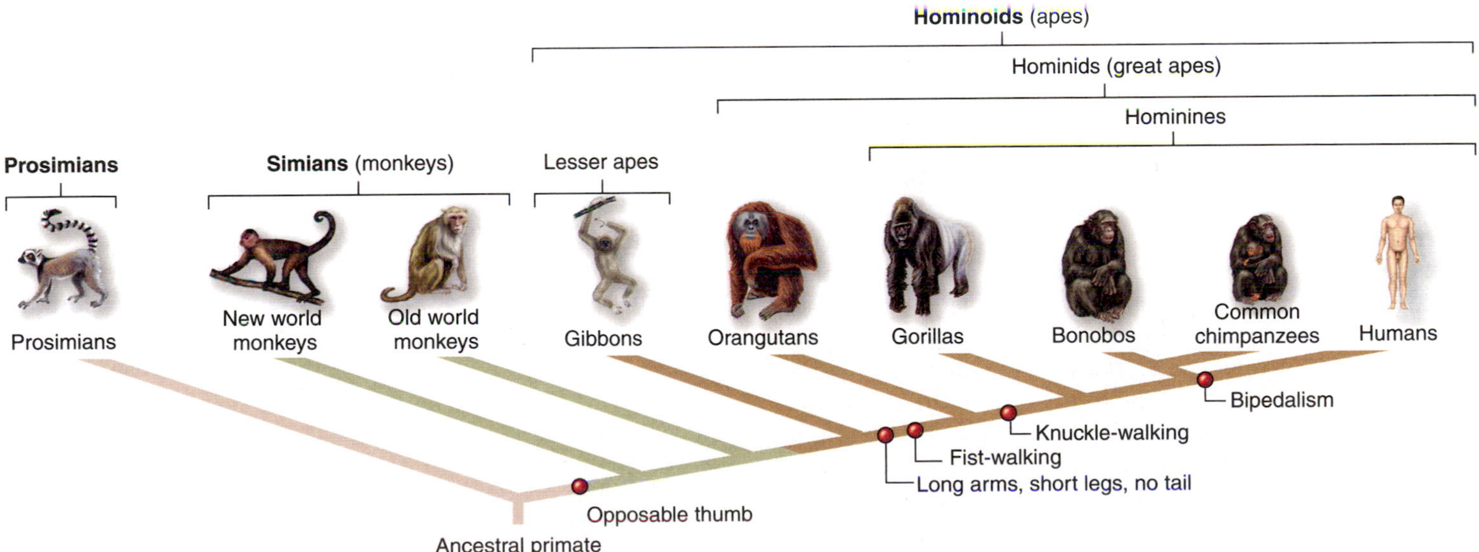

FIGURE 16.19 **Primate Lineages.** This cladogram shows that hand structure, locomotion, and other physical traits differentiate the three main groups of primates: prosimians, simians, and hominoids.

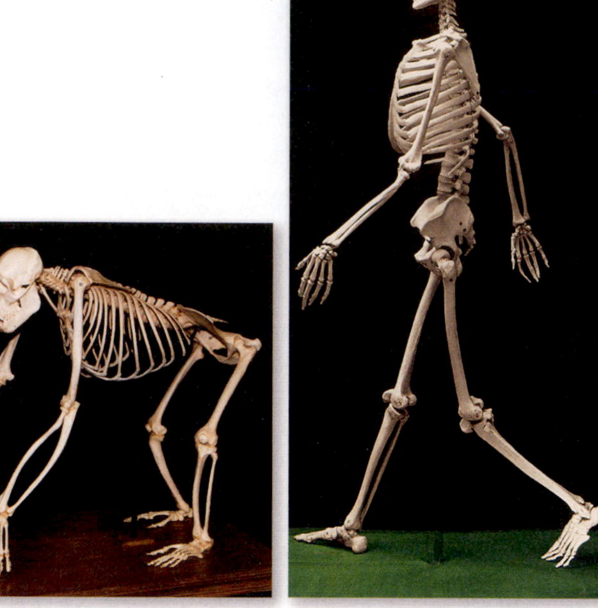

FIGURE 16.20 Clues Are in the Bones. The skeleton on the left, from a chimpanzee, shares many similarities with the human skeleton on the right.

joints that allow free movement of the arms in 360 degrees. In addition, a long collarbone acts as a brace and keeps the shoulder from collapsing toward the chest.

Heavier bodied chimpanzees and gorillas don't brachiate as much as gibbons and orangutans, but like humans, they can do so. With small, light-bodied children playing on schoolyard "monkey bars" as exceptions, humans seldom brachiate. Adult human arms are too weak to support their heavy legs.

Chimpanzees and gorillas move by knuckle-walking, a behavioral modification that allows an animal to run rapidly on the ground on all fours, with their weight resting on the knuckles. The proportionately longer arms of chimps and gorillas are an adaptation to knuckle-walking.

One important feature distinguishes humans from the other great apes: bipedalism, or the ability to walk upright on two legs. Adaptations to bipedalism include proportionally shorter arms and longer, stronger leg bones. Foot bones form firm supports for walking, with the big toe fixed in place and not opposable. The bowl-shaped pelvis supports most of the weight of the body, and lumbar vertebrae are robust enough to bear some body weight. In a modern human the foramen magnum, the large hole in the skull where the spinal cord leaves the brain, is tucked beneath the skull. In gorillas and chimps, the foramen magnum is located nearer the rear of the skull; in animals that run on all fours, like horses and dogs, the foramen magnum is at the rear of the skull.

Dietary Adaptations in Primates

Other skeletal characteristics, including the size and shape of teeth, are related to diet (**figure 16.21**). Teeth fossilize well because they are very durable. Upper and lower molar teeth have ridges that fit together, much as the teeth of gears intermesh. Food caught between these surfaces is ground, crushed, and mashed. The size of these teeth is an adaptation that reflects the toughness of the diet.

Closely related to teeth are bony ridges on the skull that serve as attachments for groups of muscles. Figure 16.21 shows the skulls and teeth of some existing hominids. As you examine the skulls, note differences in the sagittal crests. This bony ridge runs lengthwise along the top of the skull. It is an attachment point for muscles, and its presence signals particularly strong jaws. Other important features include the size of jaw bones, the prominence of the ridge of bone above the eye, the degree to which the jaw protrudes, and the shape of the curve of the tooth row. All of these characteristics allow paleoanthropologists, the scientists who study fossil humans, to identify different hominid species.

B. Molecular Evidence Documents Primate Relationships

Fossil evidence and anatomical similarities were once the only lines of evidence that paleoanthropologists could analyze in tracing the course of human evolution. Around 1960, however, scientists began to use molecular sequences to investigate relationships between species of primates. Studies of blood proteins and DNA presented a new picture of primate evolution that grouped humans as one species of great ape. One of the astounding findings of these molecular studies was that the genes of humans and chimpanzees are 99% identical.

The distinctions between humans and other great apes further eroded in the 1970s. Previously, humans had been placed in a separate group, supposedly characterized by upright walking, tool-making, and language. Chimpanzees and wild gorillas were observed to use tools, and captive great apes learned to use sign language to communicate with their trainers. The only characteristic that now remains unique to *Homo* is bipedal locomotion.

As scientists accepted the new molecular evidence, they began to view primate relationships differently. **Figure 16.22** shows relationships of living primates that take into account the new data. As you examine these trees, keep in mind two important concepts. First, note that *humans are not descended from other groups of modern apes*. Instead, all living humans and chimpanzees share a common ancestor and diverged from that ancestor perhaps 4.5 MYA. Second, note that gibbons, orangutans, gorillas, and chimpanzees are not "less evolved" than humans. All living species are on an equal evolutionary footing, although some may belong to older lineages.

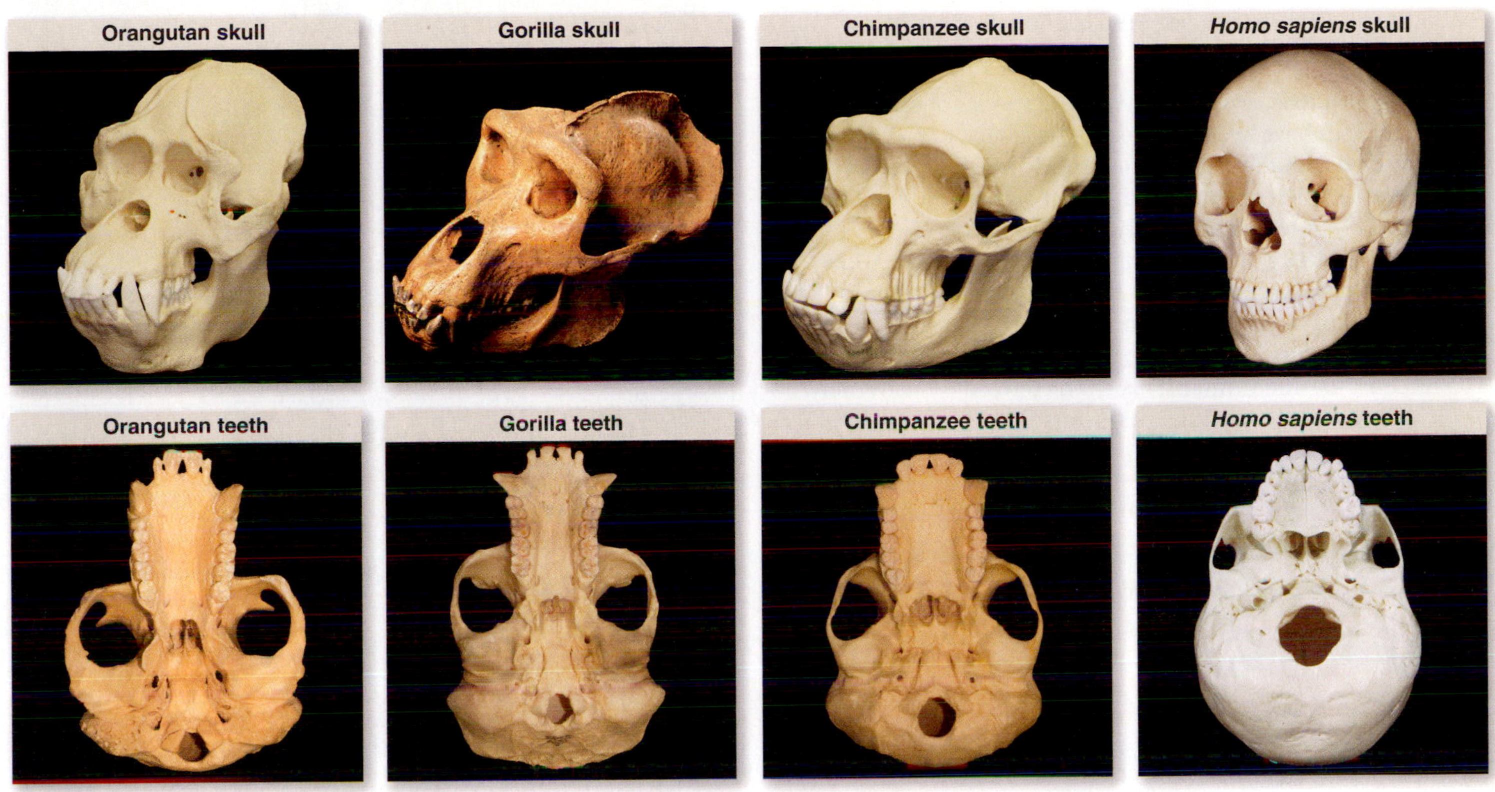

FIGURE 16.21 Skulls and Teeth. The skulls and teeth of an orangutan, gorilla, chimpanzee, and human reveal details about diet and jaw strength. Paleoanthropologists compare fossils of extinct species to bones of existing primates to learn how our ancestors lived.

C. Hominine Evolution Is Partially Recorded in Fossils

Even though DNA and proteins provide overwhelming evidence of the relationships between living primates, these molecules deteriorate with time. Scientists therefore cannot usually use molecular data to establish relationships of prehistoric hominines. For this, we must turn to studies of fossilized remains.

To interpret human fossils, paleoanthropologists compare details of skeletal features and try to reconstruct as much as they can of the organisms from the fossils that are

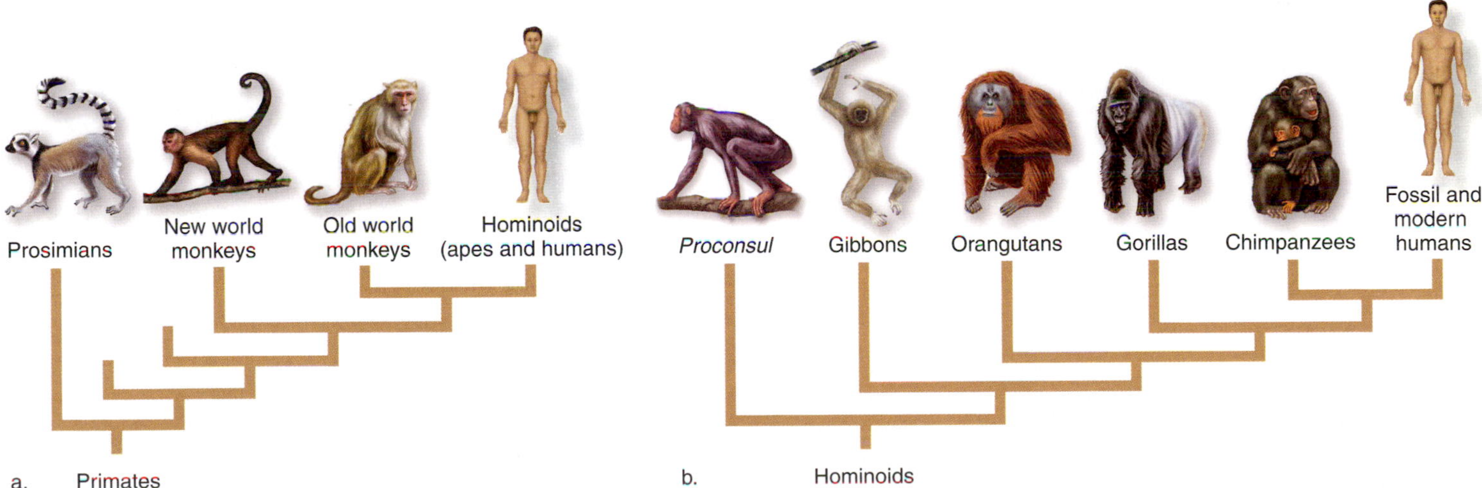

FIGURE 16.22 Primate Relationships. Molecular data have led scientists to the current view of evolutionary relationships among existing primates. (**A**) This phylogenetic tree shows the relationships among hominoids and other major groups of primates. (**B**) An expanded tree of the hominoids shows the place of humans among the great apes.

available. Fossil hominines in the human family tree include three groups (**figure 16.23**):

- **Australopiths.** Fossils of four or five species of extinct small apes have been assigned to the genus *Australopithecus*, meaning "Southern ape-man." The downward position of the foramen magnum indicates that these apes walked bipedally. A remarkable trail of fossilized footprints is additional evidence of their hominine heritage (**figure 16.24**). *Australopithecus afarensis* and *A. africanus* are members of this group, which dates from about 4 to 2.5 MYA.
- ***Paranthropus.*** This extinct group, whose name literally means "beside humans," is characterized by extremely large teeth, protruding jaws, and skulls that have a sagittal crest. All of these specializations probably relate to the large jaw muscles needed to crush tough plants or crack nuts. Most researchers hypothesize that *Paranthropus* descended from *Australopithecus*. *Paranthropus aethiopicus, P. boisei,* and *P. robustus* are members of this group, which dates from about 3 to 1.5 MYA. *Paranthropus* seems to be an evolutionary dead end that gave rise to no other group.
- ***Homo.*** Fossils in this group are associated with stones thought to have been tools. *Homo* species tend to have larger bodies and larger brains than do australopiths. All members of genus *Homo* are considered humans, and *Homo habilis, H. ergaster,* and *H. erectus* belong to the cluster of extinct species that are called "early *Homo*." These species lived from about 2.5 MYA to about 1 MYA and gave rise to "recent *Homo*." Recent species of *Homo* have smaller teeth, lighter jaws, larger braincases, less protruding jaws, and lighter brow ridges. Their fossils are associated with evidence of culture. *Homo heidelbergensis, H. neanderthalensis, H. floresiensis, and H. sapiens* are recent *Homo* species. The only human species alive today is *Homo sapiens*.

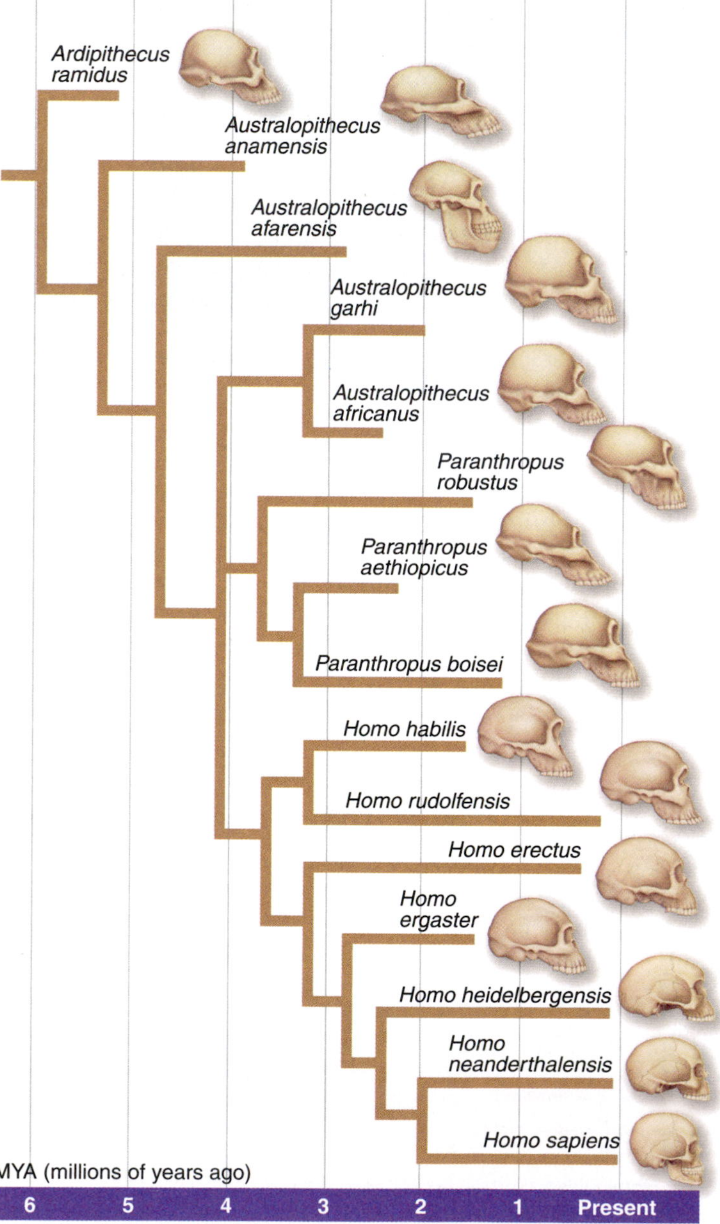

FIGURE 16.23 Fossil Hominines. Fossilized teeth and skeletons place hominines into three main groups: australopiths, *Paranthropus*, and *Homo*.

FIGURE 16.24 Fossil Footprints. The so-called Laetoli footprints date to about 3.7 MYA. They were discovered in Tanzania in 1978.

One interesting trend in hominine evolution has been an adaptive radiation of species followed by a reduction in the number of species. Fossil evidence shows that about 1.8 MYA, as many as five species of hominines lived together in Africa. Similarly, about 200,000 years ago, three species of recent *Homo* lived in Europe. Today, however, only *Homo sapiens* remains worldwide. Scientists had speculated that Neanderthals disappeared because *H. sapiens* interbred with them, but an analysis of Neanderthal DNA has contributed to the rejection of that hypothesis. What happened to the other *Homo* species? No one is certain, but many anthropologists speculate that *H. sapiens* may have contributed to their extinction.

D. Environmental Changes Have Spurred Hominine Evolution

What provoked the hominid ancestors of humans to abandon brachiation in favor of bipedal, upright walking? What allowed the large brains that are characteristic of recent *Homo* to develop? To find these answers, we have to consider a related question: Where did hominids evolve?

Charles Darwin was one of the first to speculate that humans evolved in Africa. About 12 MYA, tectonic movements caused a period of great mountain building. The continental plates beneath India and the Himalayan region collided and ground together, heaving up the Himalayas (**figure 16.25**). The resulting climatic shift had enormous ecological consequences. Cooler temperatures reduced the thick tropical forests that had covered much of Europe, India, the Middle East, and East Africa. Open plains appeared, bringing new opportunities for species that could live there. These included less competition for food in the treetops, a different assortment of foods, and a different group of predators. Experts speculate that one small ape, perhaps *Australopithecus* or an as-yet-unidentified hominine, moved out of the trees and began life on the African savannas.

Perhaps at first this species alternated between running on all fours and bipedal walking. On open plains, however, there are advantages to bipedal walking, especially the elevated vantage point for sensing danger and spotting food and friends. This environment would have selected for apes with the best skeletal adaptations to bipedalism, and the trait would have been preserved and honed in the plains. Bipedalism also freed hominine hands to carry objects and use the tools that are so characteristic of *Homo* species.

What about the large brains that characterize humans? There are several lines of speculation about what might have spurred the evolution of a large brain. Some experts relate the development of a large brain to tool use; others relate it to life in social groups and language.

Once social and communication skills improved, a byproduct of the large brain was **culture:** the knowledge, beliefs, and behaviors that humans transmit from generation to generation. Cave art from about 14,000 years ago indicates that our ancestors had developed fine hand coordination and could use symbols. By 10,000 years ago, people had migrated from the Middle East across Europe. Depending on the availability of native plants and animals that early farmers could domesticate, agriculture began to replace a hunter–gatherer lifestyle in many places. Agriculture meant increased food production, which enabled societies to change in profound ways. Freed from the necessity of producing their own food, specialized groups of political leaders, soldiers, weapon-makers, religious leaders, artists, and many other types of workers arose. Specialized workers meant improved technologies and the ability to explore the world for new lands and new resources. In a sense, many of today's civilizations owe their existence to the development of agriculture.

From a biological standpoint, humans are a special species because we can alter the environment much more than, for example, a slime mold or an earthworm, and we can alter natural selection. Despite our unique set of features, however, we are a species, descended from ancestors with which we share many characteristics. It is intriguing to think about where the human species is headed, which species will vanish, and how life will continue to diversify.

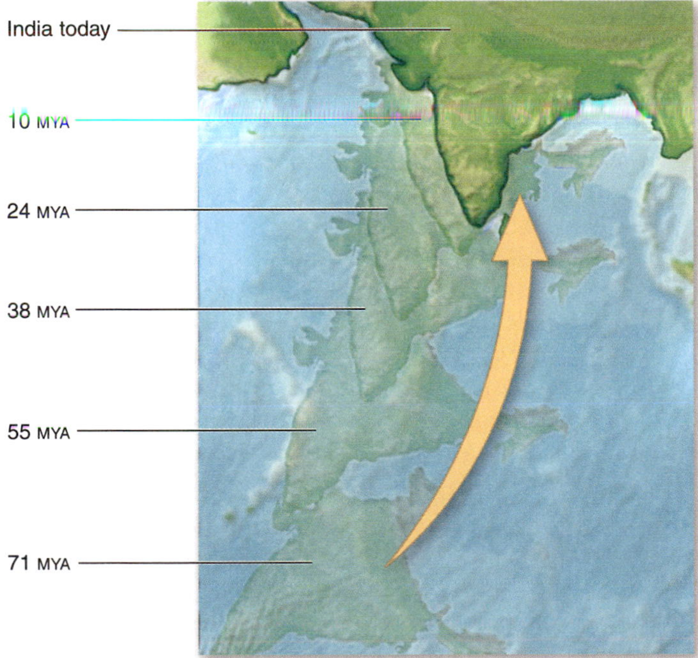

FIGURE 16.25 Changing Times. India was once an island, but over millions of years, it moved toward Eurasia. When the two continents collided, the Himalayas formed, changing the African climate.

Summary *Humans share several traits with other primates. Comparisons of fossilized skeletons and existing primates reveal how extinct species lived. DNA also helps tell the story of the human family tree, which includes many extinct lineages. Today, only* Homo sapiens *remains.*

16.4 MASTERING CONCEPTS

1. Name and describe the three groups of contemporary primates. To which group do humans belong?
2. What can an understanding of skeletal anatomy in existing primates tell us about the study of human evolution?
3. What are the three groups of hominines in the human family tree, and which still exist today?
4. Which conditions may have contributed to the evolution of humans?

16.5 INVESTIGATING LIFE
What Makes Us Human?

Perhaps no scientific issue is more tantalizing, and more entangled with philosophy, than the question of what makes us human. One way to look for answers is to study the similarities and differences between humans and chimpanzees, our closest living relatives on the evolutionary tree (**figure 16.26**). A team of scientists has sequenced the 3 billion or so DNA nucleotides that make up the chimpanzee genome and has begun comparing it with our own.

The story begins with the Chimpanzee Sequencing and Analysis Consortium, a group of 67 researchers in the United States, Europe, and Israel who collaborated to determine the genetic sequence of one chimpanzee (*Pan troglodytes*). The subject was a captive male chimp named Clint, who lived at a primate research center in Atlanta until he died of heart failure in 2004. (For comparison, the Human Genome Project, completed in 2003, used DNA from many anonymous donors.)

The scientists used the "whole-genome shotgun approach" to sequence the genome. They first isolated DNA from Clint's blood cells, then broke the genetic material into many small fragments. Next, they inserted each fragment into a separate plasmid, which is a small ring of DNA (see figure 18.3). Each plasmid that carried chimp DNA was placed into a different bacterial cell. The researchers allowed the bacteria to replicate on culture plates, producing countless copies of the plasmid. Then, when it was a fragment's "turn" to be processed, a technician simply retrieved a sample of the bacteria, extracted the plasmid, and determined the sequence of A, C, G, and T (see section 7.5). Finally, the scientists used powerful computers to assemble the sequences from tens of millions of fragments.

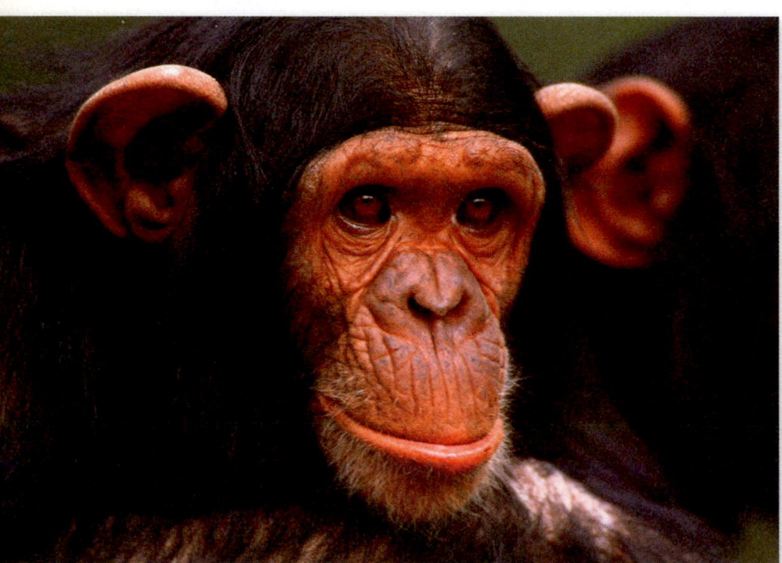

FIGURE 16.26 Close Relatives. Humans share 99% of our protein-encoding genes with chimpanzees. What are the differences that make us human?

Long before this project began, the startling genetic similarities between humans and chimps were well known (**figure 16.27**). The complete DNA sequences for both species, however, reveal exactly how much we have in common: The two genomes are 96% alike, with the differences concentrated in the noncoding regions. The coding regions (the sequences that specify proteins) are 99% alike.

Although sequencing the chimp and human genomes is a noteworthy accomplishment in itself, the work has just begun. Scientists must still scrutinize both genomes, not only to identify the 25,000 or so coding regions and their regulatory sequences, but also to determine which sequences correspond to previously discovered genes. They must also sequence the genomes of additional individuals to locate the variable regions, and they must determine which gene variants (alleles) contribute to which traits.

It all represents an enormous investment of time, money, and energy. Biologists consider the effort worthwhile, however, because it will reveal an unprecedented view of human biology and evolution. Here is a sampling of questions we may soon be able to answer:

- **Which genes define humans?** Previously, scientists could only compare the human genome with those of bacteria such as *E. coli*, plants such as *Arabidopsis*, and animals such as nematodes (*Caenorhabditis elegans*), fruit flies (*Drosophila melanogaster*), and mice (*Mus musculus*). Those comparisons gave important insights into the traits that are common to all animals, all eukaryotes, or all cells. With the chimp genome complete, we can finally determine precisely how human DNA *differs* from that of our closest relative. To answer this question, scientists will search the two genomes for regions that have been duplicated, inserted, deleted, and otherwise changed since humans and chimps last shared a common ancestor.
- **What accounts for bipedalism, large brains, complex language, and other uniquely human features?** Considering how genetically similar humans and chimps are, we have strikingly different phenotypes. At least two hypotheses attempt to explain this curious observation. Some scientists suggest that human and chimp proteins are essentially the same, but that we turn the genes encoding those proteins on and

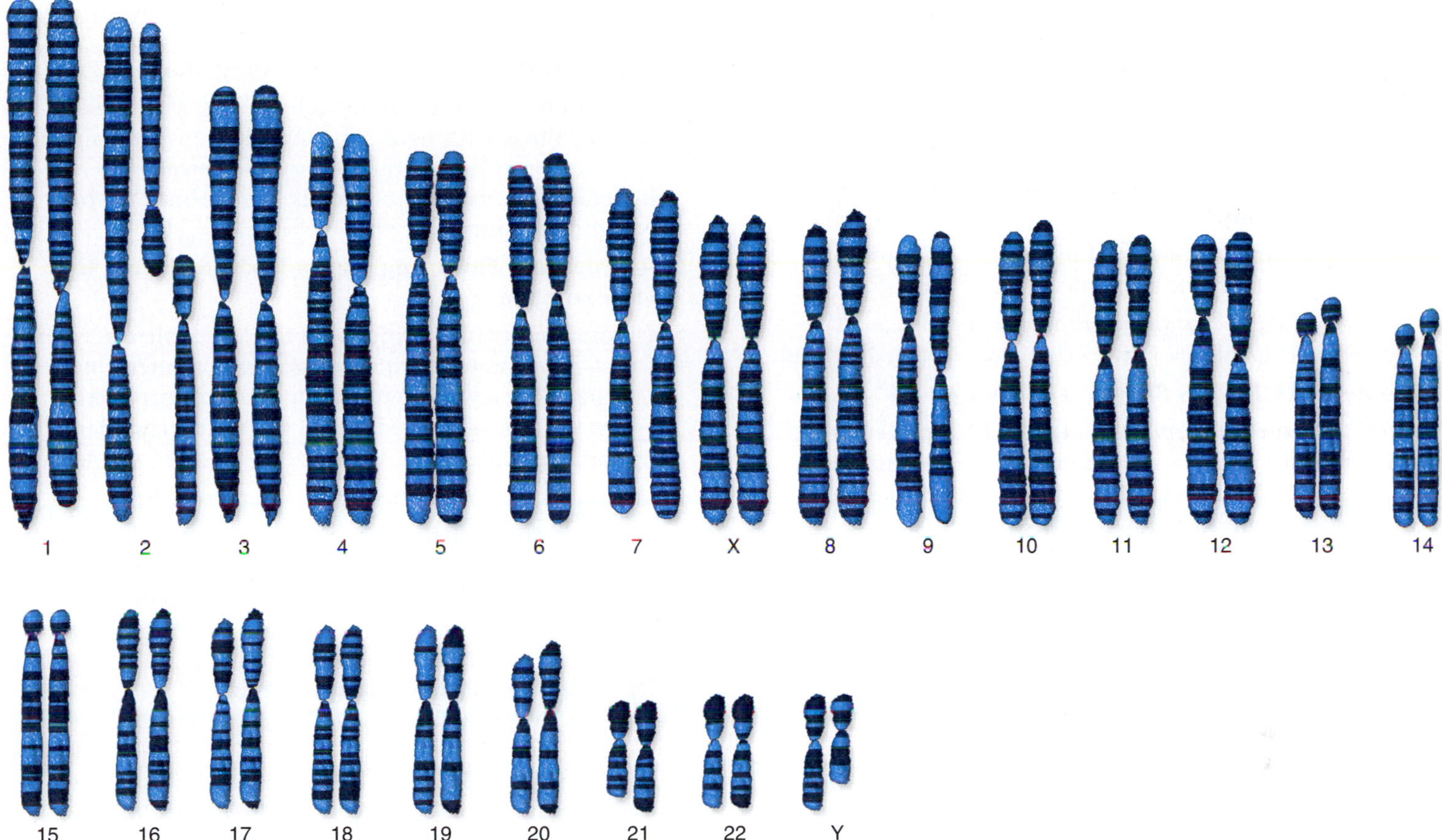

FIGURE 16.27 Chimpanzee and Human Chromosomes. The chromosomes of chimpanzees and humans look virtually identical, and molecular studies have indicated extremely high sequence similarity between us and chimpanzees. **Question**: A side-by-side comparison of the chimpanzee and human genomes does not reveal which species has the "ancestral" form of each gene (the version present in the last common ancestor of humans and chimpanzees). How will additional genome sequences from primates such as orangutans or gorillas help scientists determine which gene variants are ancestral?

off at different times. For example, chapter 12's Investigating Life: Clues to the Origin of Language illustrates how a small change in just one transcription factor can affect a person's ability to use language. A competing hypothesis is that after humans and chimps diverged, changes in the human lineage caused some previously functional genes to stop working. Such "degenerate" genes may make us less muscular, and less hairy, than chimpanzees. Comparing the chimp and human genotypes will help biologists test these hypotheses and yield insights into human evolution.

- **Why do humans and chimps have different diseases?** Because our two species are so closely related, it is reasonable to suppose that we suffer from the same illnesses. Yet the number of differences is surprising. For example, humans are susceptible to Alzheimer disease, heart attack, and carcinomas (cancers of the epithelium), but the great apes are not. HIV progresses to AIDS in humans, but not in the great apes. At least some of the differences will certainly lie in our genes, and their discovery may yield new disease cures.

Whatever the source of our humanity, it is revealed partly in our ability to make reasoned decisions and in our compassion for others. Chimpanzees are endangered in the wild, where they succumb to habitat loss, hunting, the pet trade, and biomedical research. The Chimpanzee Sequencing and Analysis Consortium's paper advocates the protection of chimpanzees in the wild, and it ends with this statement: "We hope that elaborating how few differences separate our species will broaden recognition of our duty to these extraordinary primates that stand as our siblings in the family of life."

The Chimpanzee Sequencing and Analysis Consortium. September 1, 2005. Initial sequence of the chimpanzee genome and comparison with the human genome. *Nature*, vol. 437, pages 69–87.

Additional reference: Olson, Maynard V. and Ajit Varki. January 2003. Sequencing the chimpanzee genome: insights into human evolution and disease. *Nature Reviews Genetics*, vol. 4, pages 20–28.

CHAPTER SUMMARY

16.1 Life's Origin Remains Mysterious

- The solar system formed about 4.6 BYA, and life first left evidence on Earth some 700 million years later.

A. The First Organic Molecules May Have Formed in a Chemical "Soup"

- **Prebiotic simulations** combine simple inorganic chemicals to form life's organic building blocks.
- Amino acids may have polymerized into peptides, and nucleotides into nucleic acids, on hot clay or mineral surfaces.

B. Some Investigators Suggest an "RNA World"

- The **RNA world** theory proposes that RNA preceded formation of the first cells. Proteins provided enzymes and structural features. Reverse transcriptase could have copied RNA's information into DNA.

C. Membranes Enclosed the Molecules

- Phospholipid sheets that formed bubbles around proteins and nucleic acids may have formed cell precursors, or **progenotes.**

D. The Origin of Metabolism Would Have Involved Early Enzymes

- Metabolic pathways may have originated when progenotes mutated in ways that enabled them to use alternative or additional nutrients.

E. Early Life Changed Earth Forever

- The earliest fossils of true cells formed about 3.7 BYA. Early organisms permanently changed conditions for the further evolution of life.

16.2 Complex Cells and Multicellularity Arose over a Billion Years Ago

- The internal membranes of eukaryotic cells may have formed when the outer membrane folded in on itself repeatedly.

A. Endosymbiosis Explains the Origin of Organelles

- The **endosymbiont theory** proposes that chloroplasts and mitochondria originated as free-living bacteria that were engulfed by larger archaea.

B. Multicellularity May Also Have Its Origin in Cooperation

- The evolution of multicellularity, which occurred about 1.2 BYA, is poorly understood.

16.3 Life's Diversity Exploded in the Past Five Hundred Million Years

A. The Strange Ediacarans Flourished During the Precambrian

- The Ediacarans were soft, flat organisms that were completely unlike modern species. They lived in the late Precambrian and early Cambrian periods.

B. Paleozoic Plants and Animals Emerged onto Land

- The Cambrian explosion introduced many species, notably those with hard parts. Amphibian-like animals ventured onto land about 360 MYA, followed by reptiles, birds, and mammals. Invertebrates, ferns, and forests flourished.

C. Reptiles and Flowering Plants Thrived During the Mesozoic Era

- Dinosaurs prevailed throughout the Mesozoic era, when forests were largely cycads, ginkgos, and conifers. In the middle of the era, flowering plants became prevalent. When the dinosaurs died out 65 MYA, resources opened up for mammals.

D. Mammals Radiated During the Cenozoic Era

- Mammals diversified during the Tertiary period. Humans arose during the Pleistocene epoch of the Quaternary period. Repeated ice ages, and the extinction of many large mammals, also occurred during the Pleistocene.

16.4 Fossils and DNA Tell the Human Evolution Story

A. Humans Are Primates

- **Primates** have grasping hands, opposable thumbs, binocular vision, large brains, and flat nails. The three groups are **prosimians, simians,** and **hominoids.**
- **Hominids** are the "great apes," whereas **hominines** are gorillas, chimpanzees, and humans.
- Fossil bones and teeth reveal how extinct species moved and what they ate.

B. Molecular Evidence Documents Primate Relationships

- Protein and DNA analysis has altered how scientists draw the human family tree.

C. Hominine Evolution Is Partially Recorded in Fossils

- Three groups of hominines are australopiths, *Paranthropus*, and *Homo*.

D. Environmental Changes Have Spurred Hominine Evolution

- Millions of years ago, new mountain ranges arose, causing climate shifts. Savannas replaced tropical forests, and apes—the ancestors of humans—moved from the trees to the savanna.
- Humans owe our success to language and **culture**.

16.5 Investigating Life: What Makes Us Human?

- Scientists completed the chimpanzee genome project in 2005. Comparing the chimp and human genomes will lead to unprecedented insight into the genes, phenotypes, and diseases that are unique to humans.

MULTIPLE CHOICE QUESTIONS

1. The first important step leading to life on Earth was most likely the formation of
   a. membrane enclosed structures.
   b. O_2 gas.
   c. organic molecules.
   d. CO_2 gas.
2. What must occur in order for natural selection to occur in an "RNA world"?
   a. RNA must encode proteins.
   b. RNA molecules must undergo mutations.
   c. RNA molecules must replicate.
   d Both b and c.
3. How did photosynthetic cells affect early Earth?
   a. It resulted in the addition of O_2 into the atmosphere.
   b. It resulted in an increase of hydrogen sulfide in the early oceans.
   c. It depleted the ozone layer.
   d. It altered the pH of the early oceans.
4. The *endosymbiont theory* attempts to explain the origin of
   a. cellular life.
   b. the nuclear membrane in eukaryotes.
   c. mitochondria and chloroplasts in eukaryotes.
   d. the plasma membrane of prokaryotes.
5. Which of the following provides the strongest support for the idea that mitochondria were once independent organisms?
   a. Their size is similar to that of prokaryotes.
   b. They are surrounded by a membrane.
   c. They are shaped like a prokaryote.
   d. They have their own DNA and ribosomes.
6. How have mass extinctions contributed to the evolution of life on Earth?
   a. Extinctions create a selective pressure.
   b. Extinctions eliminated only the abundant species.
   c. Extinctions allow for adaptive radiation.
   d. Both a and c.
7. Which group of the hominine family is the oldest?
   a. Austrolopiths
   b. *Paranthropus*
   c. *Homo*
   d. Simian

TESTING YOUR KNOWLEDGE

1. About how soon after Earth formed did life first appear, according to fossil evidence?
2. Review the structures of nucleic acids and proteins in chapter 2. What chemical elements had to have been in primordial "soup" to generate these organic molecules?
3. How might the first metabolic pathway have arisen?
4. List the major life events of the Precambrian, Paleozoic, Mesozoic, and Cenozoic.
5. Distinguish between the terms *primate, hominid, hominine,* and *Homo*.
6. What can scientists learn by comparing the fossilized skeletons of extinct primates with the bones of modern species?
7. In what ways has culture been an important factor in human evolution?

THINKING AS A SCIENTIST

1. Panspermia suggests that life on Earth developed from primitive "spores" that arrived from outer space. Suppose it did so on multiple planets, eventually giving rise to humans (on Earth) and *Star Trek*'s Vulcans (on planet Vulcan). In that case, would humans be more closely related to Vulcans or to Earth species such as crabgrass and bread mold? Given your answer, does it seem likely that *Star Trek*'s Mr. Spock could have one Vulcan and one human parent?
2. The amoeba *Pelomyxa palustris* is a single-celled eukaryote with no mitochondria, but it contains symbiotic bacteria that can live in the presence of O_2. How does this observation support or argue against the endosymbiont theory of the origin of eukaryotic cells?
3. At one time, several species of *Homo* existed at the same time. Propose at least two hypotheses that might explain why only *Homo sapiens* remains.
4. How do you predict a scientist would respond to a question about whether humans "evolved from monkeys?"

Visit www.mhhe.com/hoefnagels for practice quizzes, animations, videos, and activities designed to help you master the material in this chapter.

CHAPTER 17

Viruses

Guarding Against the Flu. The 1918 flu epidemic spread so quickly that cities took drastic steps. Policemen in London wore surgical masks to avoid infection.

What's the Point?

From the Birds: Influenza

Most viruses infect the cells of only a few, closely related species, but some—including those that cause influenza—jump back and forth between birds and humans.

Influenza, known as "flu" for short, probably began in China, where the influenza virus moved from wild to domesticated ducks thousands of years ago. In the seventeenth century, the Chinese brought domesticated ducks to live among rice paddies, where they were close to people, pigs, and chickens.

Pigs contribute to influenza epidemics because cells that line a pig's throat carry receptors that bind to both the avian (bird) and human versions of flu viruses. Long ago, an avian flu virus mutated in pig throats in a way that enabled the virus to infect humans. Avian and human flu viruses infecting pigs commonly exchange segments of their genomes, generating new strains that can evade the human immune system.

Sometimes a new viral variety causes a serious epidemic. The flu pandemic of 1918 killed more people in the United States than World Wars I and II, the Korean War, and the Vietnam War combined. Unlike modern flu outbreaks, which typically kill children and the elderly, most victims of the 1918 flu were 20 to 40 years old. Researchers are trying to determine why this flu was so deadly by studying viruses in lung tissue from Alaskan victims preserved in permafrost graves.

Flu pandemics also occurred in 1957, 1968, and 1977. In 1997, epidemiologists feared yet another pandemic when a new flu variant appeared to jump from birds directly to humans. A 3-year-old boy in Hong Kong died of a fierce flu caused by a virus never before seen in humans, but known in birds. When two other Hong Kong children fell ill with the chicken virus, panic set in. To avert an epidemic, the government killed every chicken in Hong Kong. Fortunately, the outbreak never went beyond 18 people, 6 of whom died. Another outbreak of avian flu in Asia occurred in 2003 to 2004, again triggering the widespread slaughter of poultry.

So far, the avian influenza virus has not acquired the genes necessary to spread directly from person to person. Viruses, however, are highly changeable, and most humans do not have immunity against viruses that normally infect other species. Epidemiologists therefore watch carefully for new variants that might trigger much more serious outbreaks.

Viruses infect every type of organism, not just humans, birds, and pigs. This chapter explains what viruses are, how they replicate and cause disease, and why they have proven so hard to defeat.

LEARNING OUTLINE

17.1 Viruses Are Infectious Particles of Genetic Information and Protein

Smallpox, influenza, the common cold, rabies, polio, chickenpox, warts, mononucleosis, AIDS—this diverse list includes illnesses that range from merely inconvenient to deadly. All have one thing in common: they are infectious diseases caused by viruses.

Many people mistakenly lump viruses and bacteria together as "germs." Viruses, however, are not bacteria. In fact, they are not even cells. A **virus** is a small, infectious agent that is simply genetic information (DNA or RNA) enclosed in a protein coat. The 2000 or so known species of viruses therefore straddle the boundary between the chemical and the biological.

Table 17.1 *Some Viruses That Infect Humans and the Diseases They Cause*

DNA—Single Stranded
B19 virus (erythema infectiosum)
DNA—Double Stranded
Variola major (smallpox)
Herpesviruses (oral and genital herpes; chickenpox)
Epstein–Barr virus (mononucleosis, Burkitt lymphoma)
Papillomaviruses (warts, cervical cancer)
Hepatitis B virus
RNA—Single Stranded
Human immunodeficiency virus (AIDS)
Coronaviruses (e.g., SARS)
Poliovirus
Influenza viruses
Measles virus
Mumps virus
Rabies virus
Ebola virus
Rhinovirus (common cold)
West Nile virus
Hepatitis A and C viruses
RNA—Double Stranded
Rotavirus (respiratory and gastrointestinal infections)

A. Viruses Are Smaller and Simpler than Cells

A virus is much smaller than a cell (see figure 3.2). At about 10 μm (microns) in diameter, an average human cell is perhaps one-tenth the diameter of a human hair. A bacterium is about one-tenth again as small, at about 1 μm (1000 nm) long. The average virus, with a diameter of about 80 nm, is more than 12 times smaller than a bacterium. Put another way, it would take millions of viral particles to form a dot the size of a printed period. (One exception is the mimivirus, a comparatively huge virus with a diameter of about 600 nm. It infects a type of amoeba.)

A virus does not have a nucleus, organelles, ribosomes, or even cytoplasm. Only a few types of viruses contain enzymes. All viruses share two features:

- **Genetic information.** All viruses contain genetic material (either DNA or RNA) that carries instructions to make their molecular components. The major criterion for classifying viruses is whether the genetic material is DNA or RNA. Either type of nucleic acid may be single- or double-stranded (**table 17.1**).
- **Protein coat.** The **capsid,** or protein coat, surrounds the genetic material; it is composed of repeated subunits called capsomeres. The capsid's shape determines a virus's overall form, which is another characteristic used in classification (**figure 17.1**). Many viruses are spherical or icosahedral (a 20-faced shape built of triangular sections). Others are rod-shaped, oval, or filamentous.

Some viruses have other features besides genetic material and a protein coat. For example, some have an **envelope**, a layer of membrane outside the protein coat. The envelope, which is derived from the host cell membrane, may include embedded proteins that help a virus invade a host cell. The human immunodeficiency virus (HIV), the virus that causes acquired immunodeficiency syndrome (AIDS), is an enveloped virus, as is the influenza virus. The presence or absence of an envelope is another criterion for virus classification.

Despite having relatively few components, a virus's overall structure can be quite intricate and complex. For example, some **bacteriophages,** which are viruses that infect bacteria, have parts that resemble tails, legs, and spikes. These viruses look like the spacecrafts once used to land on the moon (see figure 17.1B).

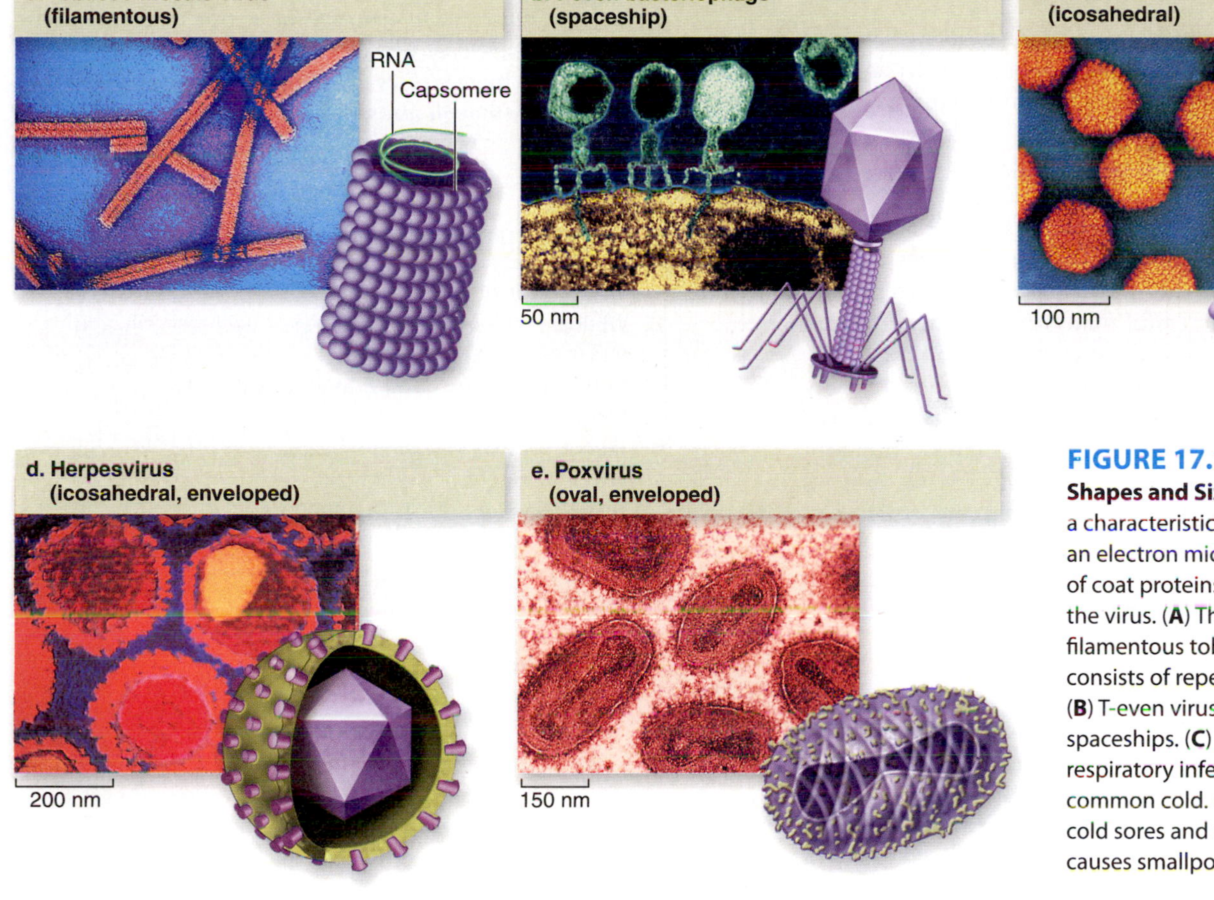

FIGURE 17.1 Viruses of Many Shapes and Sizes. Each type of virus has a characteristic structure, visible only with an electron microscope. The organization of coat proteins determines the shape of the virus. (**A**) The capsid of the filamentous tobacco mosaic virus (TMV) consists of repeated protein subunits. (**B**) T-even viruses look like tiny spaceships. (**C**) Adenoviruses cause respiratory infections similar to the common cold. (**D**) Herpesviruses cause cold sores and rashes. (**E**) A poxvirus causes smallpox.

B. A Virus's Host Range Consists of the Organisms It Infects

The **host range** of a virus is the kinds of organisms or cells that it can infect. Viruses can enter only cells that have a specific target attachment molecule, or receptor, on their surfaces. Animals, fungi, plants, protists, and bacteria all get viral infections, but a bacteriophage cannot attack human cells because our cells do not share enough surface molecules.

Some target molecules are on a very small subset of cells in an organism, whereas others occur in a group of related organisms. HIV, for example, infects only human helper T cells (a type of white blood cell). The rabies virus, on the other hand, can infect humans, skunks, raccoons, bats, and dogs because all of these mammals have common target molecules on their cells (**figure 17.2**).

Sometimes a virus can replicate within a host species without causing symptoms. Such carriers of a virus are called **reservoirs** because they act as a continual source of the virus to other host species. Examples of reservoirs for viruses that cause diseases in humans include mosquitoes (West Nile encephalitis), wild birds (influenza), rodents (hantavirus pulmonary syndrome), and raccoons (rabies).

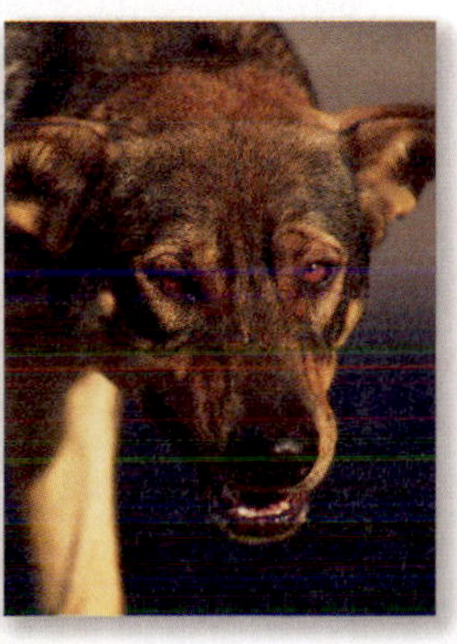

FIGURE 17.2 The Rabies Virus Has Many Hosts. All mammals, including raccoons, dogs, and skunks, are potential hosts for rabies. The bullet-shaped virus, which contains RNA as its genetic material, casuses the brain of an infected animal to become inflamed. Nerves and saliva carry the virus, which humans may acquire after being bitten by a rabid animal.

C. Are Viruses Alive?

Most biologists do not consider a virus alive because it does not metabolize, respond to stimuli, or reproduce on its own. A virus must enter a living host cell to manufacture more of itself. **what is life? p. 4**

Nevertheless, viruses do have some features in common with life, including genetic material. Both DNA and RNA can mutate, which means that viruses evolve just as life does. Each time a virus replicates inside a host cell, some new variants arise. These new viral types are subject to natural selection, in which some are more successful than others at infecting and replicating in host cells. The most successful variants spread.

Although viruses evolve, their extreme genetic diversity suggests that they do not share a single common ancestor. For this reason, viruses are not part of any domain or kingdom (see figure 1.9). Scientists group viruses into species, genera, and families based on the type of nucleic acid, the structure of the virus (including its shape and whether it has an envelope), how it replicates, and the type of disease it causes. Although some families have been assigned to orders, most have not.

Summary *A virus is much smaller and simpler than a cell. It consists of a nucleic acid wrapped in a protein capsid; some also have a fatty envelope. Viruses can infect only specific cells of particular species; some viruses persist, without causing harm, in reservoir species. Although viruses are not alive, they do evolve.*

17.1 MASTERING CONCEPTS

1. How are viruses similar to and different from bacteria?
2. What are the relative sizes of viruses, bacterial cells, and eukaryotic cells?
3. How does an enveloped virus acquire its outer membrane layer?
4. What is a virus's host range?
5. Why is it important to know the reservoir species for a virus that causes disease in humans?
6. How do viruses evolve?
7. Why don't viruses belong to a domain or kingdom of life?

17.2 Viral Replication Occurs in Five Stages

The production of new viruses is very different from cell division. When a cell divides, it doubles all of its components and splits in two. Viral production, on the other hand, resembles the assembly of cars in a factory. After a virus infects a cell, its host manufactures many copies of the viral proteins and nucleic acids, then assembles these components into new viruses. A cell infected with one virus may produce hundreds of new viral particles; each can go on to infect another cell.

Unlike bacteria, yeasts, and other microorganisms that are easy to grow in pure culture, viruses must replicate inside a living cell. To produce viruses in a laboratory, scientists must therefore inoculate host cells and wait for the cells to produce the desired number of viruses. The choice of host depends on the virus; bacteria, cultured animal cells, live animals and plants, and fertilized chicken eggs (**figure 17.3**) are all candidates.

Whatever the host species or cell type, the same basic processes occur during a viral infection (**figure 17.4**):

1. **Attachment:** A virus attaches to a host cell by adhering to a surface molecule that has some other function. Generally, the virus can attach only to a cell within which it can reproduce. HIV cannot infect human skin cells because the virus requires receptors found only on the surface of human helper T cells.

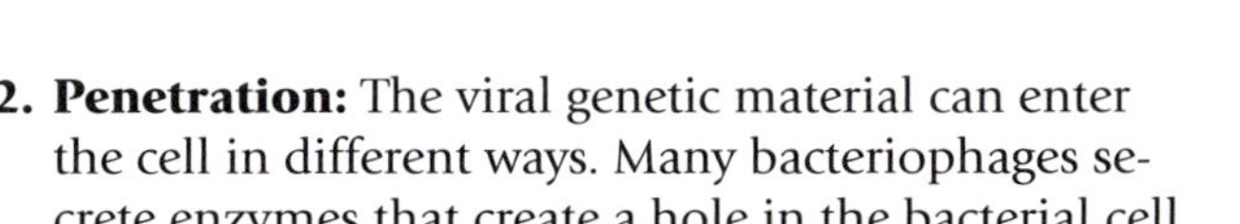

2. **Penetration:** The viral genetic material can enter the cell in different ways. Many bacteriophages secrete enzymes that create a hole in the bacterial cell wall through which they inject their genetic material,

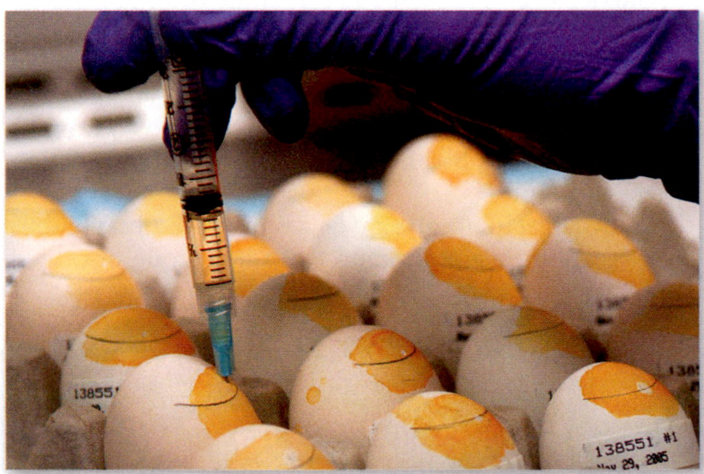

FIGURE 17.3 Viruses from Eggs. One way to produce huge quantities of influenza viruses for research or vaccine production is to allow them to replicate inside fertilized chicken eggs. People with egg allergies should therefore avoid certain vaccines.

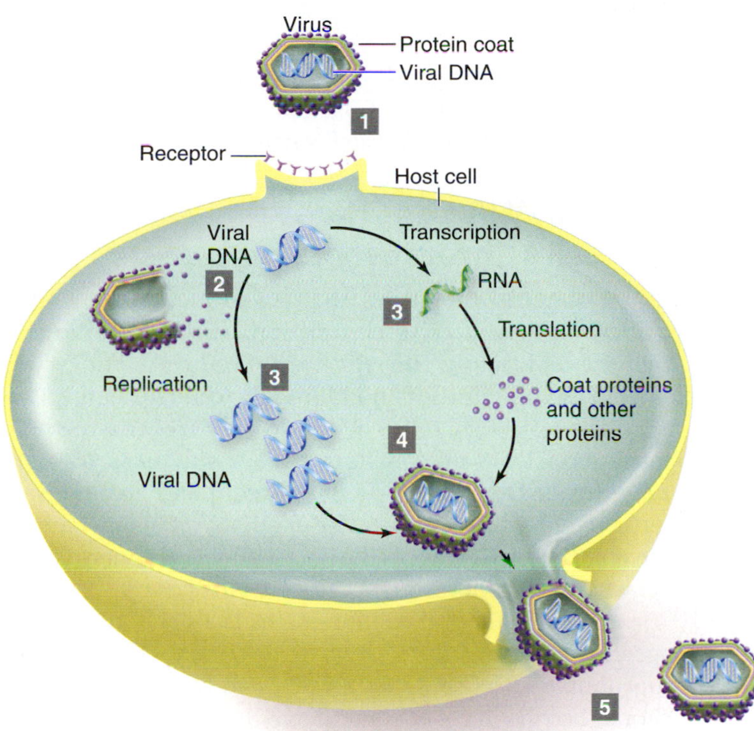

FIGURE 17.4 Generalized Viral Replication Cycle. (1) To enter host cells, viruses must first attach to receptor molecules. (2 and 3) Viral DNA or RNA enters the host cell and serves as a template to produce viral proteins and additional copies of the nucleic acid. (4) These components self-assemble to produce new viruses, which (5) leave the cell ready to infect a new host.

1 **Attachment:**
Virus binds cell surface receptor.

2 **Penetration:**
Viral nucleic acid is released inside host cell.

3 **Synthesis:**
Host cell manufactures viral nucleic acids and proteins.

4 **Assembly:**
New viruses are assembled from newly synthesized coat proteins, enzymes, and nucleic acids.

5 **Release:**
New viruses leave the host cell.

somewhat like a syringe. Viruses that infect plants often contaminate the mouthparts of insects such as aphids. When the animal chews on the plant's cell walls, viruses enter host cells. Animal cells, which lack cell walls, engulf virus particles and bring them into the cytoplasm via endocytosis. **endocytosis, p. 94**

3. **Synthesis:** The host cell provides all of the resources required for the production of new viruses: ATP, tRNA, ribosomes, nucleotides, amino acids, and enzymes. The cell produces multiple copies of the viral genome and transcribes viral DNA to mRNA, which is translated at the host cell's ribosomes (see chapter 12).
4. **Assembly:** The subunits of the capsid join, and then genetic information is packed into each protein coat. Enveloped viruses such as HIV are not complete until they bud from the host cell, acquiring their outer coverings from the host cell membrane.
5. **Release:** Once the virus particles are assembled, they are ready to leave the cell. Some bacteriophages induce production of an enzyme that breaks down the host's cell wall, killing the cell as it releases the new viruses. A virally infected eukaryotic cell may also lyse all at once as viruses burst out. Other viruses, such as HIV and herpesviruses, bud from the host cell by exocytosis. Budding often destroys the host cell, as enveloped viruses carry off segments of the cell membrane. **exocytosis, p. 94**

These events damage or destroy the host cell, which is how disease begins. The amount of time between initial infection and cell death varies. Bacteriophages need as little as a half hour to infect a cell and replicate, forming dozens to hundreds of new viruses in that time. At the other extreme, for some animal viruses, years may elapse between initial attachment and the final burst of viral particles. The next section describes how such a long delay may arise.

Summary *The steps in viral replication are attachment to the cell surface, penetration of viral genetic material, synthesis of viral components, assembly of new viruses, and release.*

17.2 MASTERING CONCEPTS

1. Why is it inaccurate to refer to the "growth" of viruses?
2. What factors determine whether a virus infects a cell?
3. How does a virus's genetic material enter a cell?
4. What is the source of energy and raw materials for the synthesis of viruses in a host cell?
5. How does the assembly and release of completed viruses occur?

17.3 Cell Death May Be Immediate or Delayed

Following attachment to the host cell and penetration of the viral genetic material, viruses follow two major replication strategies: lytic and lysogenic. Bacteriophages offer the simplest examples of these types of viral infections (**figure 17.5**).

A. Some Viral Infections Kill Cells Immediately

In a **lytic infection,** a virus enters a cell, immediately replicates, and causes the host cell to burst (lyse) as it releases a flood of new viruses. The newly released viruses infect other cells, repeating the process until all of the bacteria in a culture are dead (**figure 17.6**).

Some researchers have investigated the possibility of using lytic bacteriophages to treat bacterial infections in people. "Phage therapy" would have two main advantages over antibiotics. First, unlike drugs, viruses evolve along with their bacterial hosts, and they keep killing until all host cells are dead. Bacterial populations are therefore unlikely to acquire resistance to the treatment. Second, each bacteriophage targets only one or a few strains of bacteria, making the treatment tailored to the infection. Phage therapy's main weakness is that each bacteriophage attacks only one or a few strains of bacteria, so medical personnel must first identify the exact strain of bacteria causing infection before beginning treatment. This delay could be deadly.

B. Viral DNA Can "Hide" in a Cell's Chromosome

In a **lysogenic infection,** the genetic material of a virus "hides" in a host cell's chromosome until conditions are right for replication. The virus then reverts to a lytic cycle, releasing the new viruses and killing the cell (see figure 17.5).

After infecting a cell, a lysogenic virus uses enzymes to cut the host cell DNA and join its own DNA with the host's. A **prophage** is the DNA of a lysogenic bacteriophage that is inserted into the host chromosome. The viral DNA remains **latent,** or integrated into the DNA of the host without causing symptoms. When the cell divides, it replicates the viral genes too.

As long as the viral DNA remains latent, it does not damage the host cell. Only a few viral proteins are produced, most functioning as a "switch" that determines whether the virus should become lytic. At some signal, such as stress from DNA damage or cell starvation, these viral proteins cut the viral DNA out of the host genome. A lytic infection cycle then begins, killing the cell and releasing new viruses that infect other cells. The

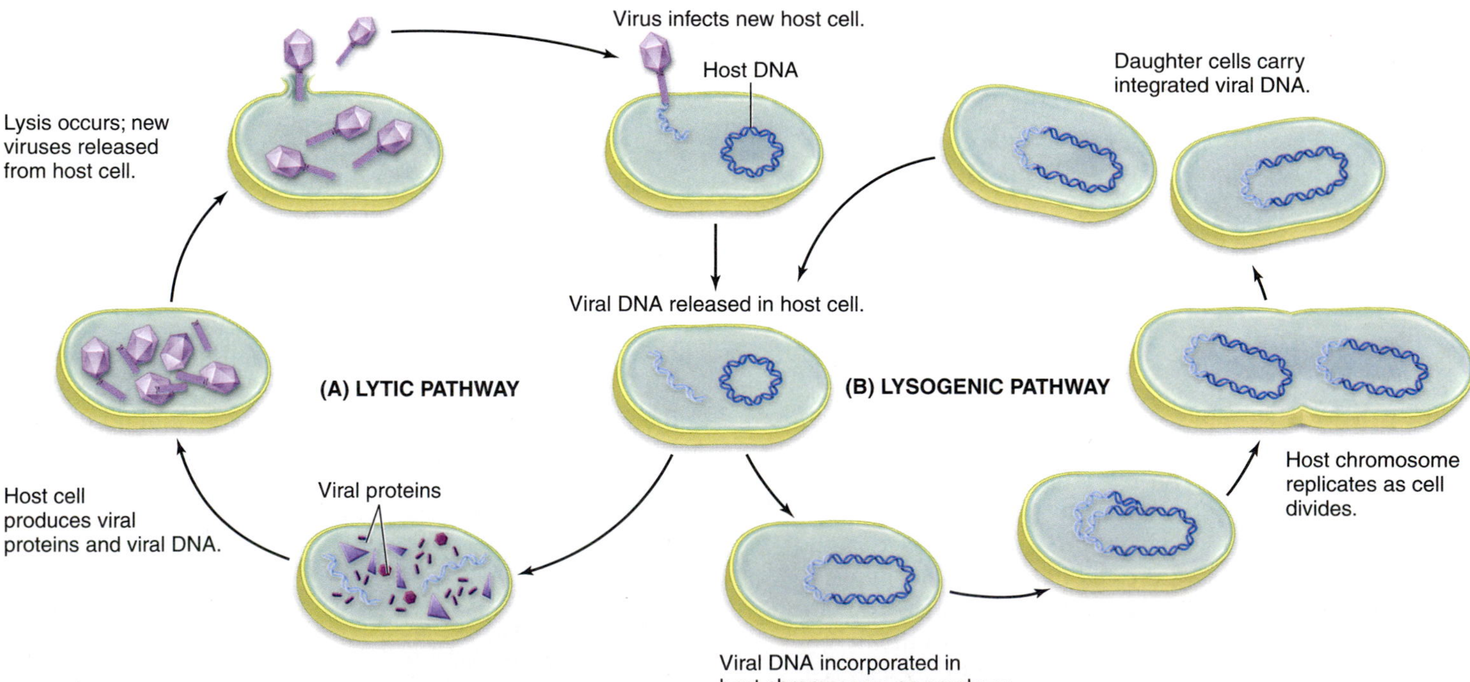

FIGURE 17.5 Lysis and Lysogeny. (**A**) Viral genetic information is used by a host cell's enzymes to produce many copies of the virus. Lysis occurs when the virus particles assemble within the cell, causing it to burst. (**B**) In lysogeny, a virus inserts its DNA into the host chromosome. It may remain integrated there, even through many rounds of cell division. An environmental change may trigger a lysogenic virus to become lytic.

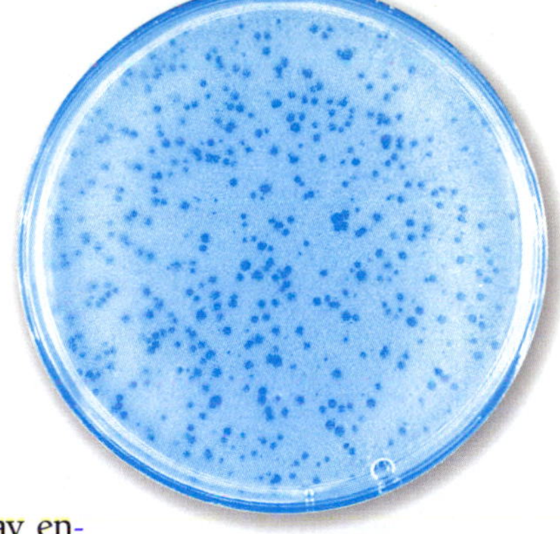

FIGURE 17.6 Sick Bacteria. Each spot in the culture of bacteria on this dish represents a zone of killed cells that started from one virus infecting one cell. This dish is about 15 cm in diameter.

next generation of viruses may enter a lytic or lysogenic replication cycle, depending on the condition of the cells.

Much of what biologists know about viral replication comes from research on the viruses that infect bacteria. The Focus on Model Organisms: Bacteriophage Lambda describes some of what scientists have learned from these studies.

C. Some Animal Viruses Linger for Years

Some animal viruses can remain latent within a cell until conditions make it possible, or necessary, to replicate. An example is the herpesvirus that causes cold sores. After initial infection, the viral DNA remains in the cells indefinitely. When the cell becomes stressed or damaged, new viruses are assembled and leave the cell to infect other cells. Cold sores often recur at the site of the original infection.

HIV is another virus that can remain latent inside a human cell (**figure 17.7**). HIV belongs to a family of viruses called retroviruses, all of which have an RNA genome. After HIV attaches to and penetrates a helper T cell, the viral reverse transcriptase enzyme transcribes this RNA to DNA. The DNA then incorporates into the host cell's DNA. Shortly after infection, many HIV viruses are produced and released by budding. Eventually, an infection persists as infected cells produce small numbers of virus particles. Infected individuals have almost no symptoms, yet HIV is present in their bloodstreams. This persistent latent phase can last for many years. Finally, in response to some (still unidentified) stimulus, the virus enters a fully productive stage. Many viruses are produced at that time, so that nearly all of the helper T cells in the body die. This destroys the immune system, leaving the body unable to defend itself from infections or cancer. AIDS is the result. (The Burning Question: How Do Anti-HIV Drugs Work? explains how some HIV-fighting treatments interfere with viral replication).

Because latent viruses persist by signaling their host cells to divide continuously, some cause cancer. A latent infection by some strains of human papillomavirus, which causes genital warts, can lead to cervical cancer. (A new vaccine prevents infection with the most dangerous strains.) Epstein–Barr virus is another example. More than 80% of the human population

FOCUS ON MODEL ORGANISMS: *Bacteriophage Lambda*

Although viruses are not organisms, they nevertheless have contributed enormously to the scientific understanding of life. This box focuses on a virus that has never made you sick, because it does not infect human cells. Rather, it kills *E. coli* bacteria that live in the intestines of humans and other mammals. As with all viruses, bacteriophage lambda (Greek letter λ) injects its genetic material into its host cell and turns the bacterium's cell components into a virus-making factory. In so doing, it has revealed many processes fundamental to all life.

Phage lambda's double-stranded DNA genome of 50,000 base pairs contains all the information it takes to make more phages, and its 60 or so genes must turn on and off in proper sequence for the new viruses to form properly. Biologists learn the functions of viral proteins by studying viral mutants—phages with missing proteins. Phage lambda's proteins fall into three general groups:

- **Capsid proteins:** Lambda phage's protein coat consists of a nearly spherical head (enclosing the viral DNA) and a tube-shaped tail with protein fibers that bind to *E. coli*'s surface proteins (**figure 17.A**). DNA enters the host cell through the tail.
- **Regulatory proteins:** Some of lambda phage's proteins bind to viral DNA and either promote or prevent transcription of particular genes. Other regulatory proteins prevent the virus from becoming latent or prevent the replication of other viruses that may have also infected the cell.

FIGURE 17.A Bacteriophage Lambda. This virus, deadly to the bacterium *E. coli*, is harmless to humans.

- **Enzymes:** A protein called integrase helps integrate lambda phage's DNA into the host cell's chromosome when it enters the lysogenic phase. Another enzyme cuts the viral DNA back out of the chromosome when the lytic phase begins. The balance between these two enzymes determines whether the virus enters the lytic or lysogenic phase of the life cycle.

Why study gene regulation in bacteriophages? Biologists can learn a lot about complex systems by studying the simplest models, since similar mechanisms of gene repression and activation also work in our own cells. When those mechanisms fail, cancer may result.

Phage lambda has taught scientists about not only gene regulation but also recombination and protein folding, both of which are fundamental life processes. In addition, this virus has become something of a laboratory workhorse. Phage lambda is unusual because it can complete its life cycle even if a large amount of foreign DNA is inserted into its genome. This discovery has made phage lambda an extremely important tool for ferrying recombinant DNA into *E. coli*.

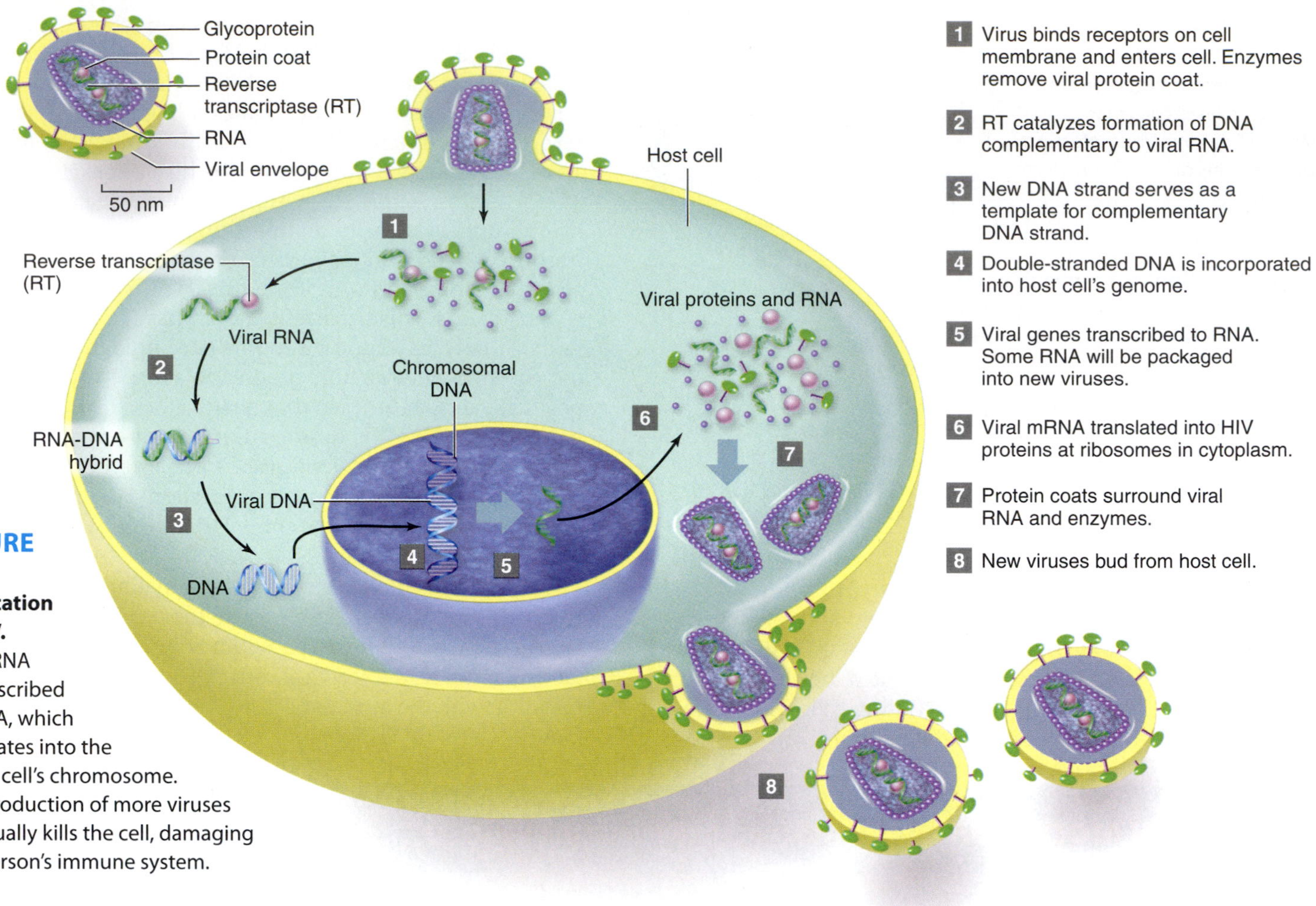

Glycoprotein
Protein coat
Reverse transcriptase (RT)
RNA
Viral envelope
50 nm
Host cell
Reverse transcriptase (RT)
Viral RNA
RNA-DNA hybrid
Viral DNA
DNA
Chromosomal DNA
Viral proteins and RNA

1 Virus binds receptors on cell membrane and enters cell. Enzymes remove viral protein coat.

2 RT catalyzes formation of DNA complementary to viral RNA.

3 New DNA strand serves as a template for complementary DNA strand.

4 Double-stranded DNA is incorporated into host cell's genome.

5 Viral genes transcribed to RNA. Some RNA will be packaged into new viruses.

6 Viral mRNA translated into HIV proteins at ribosomes in cytoplasm.

7 Protein coats surround viral RNA and enzymes.

8 New viruses bud from host cell.

FIGURE 17.7 Replication of HIV. HIV's RNA is transcribed to DNA, which integrates into the host T cell's chromosome. The production of more viruses eventually kills the cell, damaging the person's immune system.

carries this virus, which infects B cells of the human immune system. A person who is initially exposed to the virus may develop mononucleosis. The virus later maintains a latent infection in B cells. In a few people, especially those with weakened immune systems, the virus eventually causes a form of cancer called Burkitt lymphoma.

Summary *In a lytic viral infection, the virus sets into motion its own reproduction, eventually bursting the cell to release new viruses. In a lysogenic infection, viral genetic information inserts into the host's DNA and persists as the cell divides until an event triggers active viral replication. Animals can harbor latent viruses in their cells for long periods.*

17.3 MASTERING CONCEPTS

1. How is a lytic viral infection similar to, and different from, a lysogenic cycle?
2. What is a latent animal virus, and what are some examples?
3. Describe how HIV replicates in host cells.
4. How are some latent viral infections linked to cancer?

17.4 Effects of a Viral Infection May Be Mild or Severe

A person can acquire a viral infection by inhaling the respiratory droplets of an ill person or by ingesting food or water contaminated with viruses. Some viruses enter a person's bloodstream via a blood transfusion, sexual contact, or the use of contaminated needles. (Many of these pathways operate in the transmission of the various forms of hepatitis, as Can *You* Relate? Hepatitis explains.)

Once a viral infection is established, the death of infected cells produces a wide range of symptoms that reflect the types of host cells destroyed. If enough cells die, the disease may be severe or even fatal.

In addition to the direct effects of cell death, the immune system's response to a viral infection also causes symptoms. As described in chapter 36, animal immune systems attack

Burning Questions

How Do Anti-HIV Drugs Work?

Because viruses invade living cells, drugs that inhibit viruses without killing the host cell are hard to find. Nevertheless, anti-HIV medications provide excellent examples of drugs with well-understood mechanisms of action. Here are some examples of strategies drug developers use to prevent or slow symptoms in HIV-positive patients:

- **Keep viruses out of host cells.** Entry inhibitors prevent viruses from entering uninfected host cells. They block viral or host molecules that the virus requires to recognize and enter the host cell. The drug enfuvirtide (Fuzeon) is an example.
- **Inhibit replication of viral genetic information.** Azidothymidine (AZT) stops reverse transcriptase from making a DNA copy of HIV's RNA genome.
- **Inhibit viral DNA integration into host DNA.** Integrase inhibitors keep viral DNA from inserting into the host cell's chromosome. Researchers are in the early stages of investigating this new class of drugs.
- **Inhibit assembly of new viruses inside infected host cells.** Protease inhibitors prevent viral enzymes called proteases from cleaving the proteins that make up HIV's protein coat. If the protein coat cannot form properly, few mature viruses will leave infected cells, slowing the rate of infection.
- **Inhibit host cell reproduction.** Cellular inhibitors stop the replication of infected host cells. Hydroxyurea, for example, keeps T cells from replicating in AIDS patients. In doing so, it reduces the number of viruses produced. Not surprisingly, this drug should be used with caution in AIDS patients with low T-cell counts. **cancer, p. 166**

Because of the way viruses replicate, a patient must take antiviral drugs early in the course of a viral illness, before too many body cells become infected. Vaccination solves these problems by preventing viral infection in the first place. Unfortunately, vaccines against many common diseases, including AIDS, remain elusive.

Have a Burning Question of your own? Submit it to marielle_hoefnagels@mcgraw-hill.com for possible inclusion in future editions of this book!

viruses in several ways. The immune response makes the body inhospitable to the virus, raising body temperature, causing the aches and pains of inflammation, and increasing mucous secretions, among other actions. Unfortunately, these responses can also make the host rather miserable!

A. Drugs and Vaccines Help Fight Viral Infections

Antibiotics never work against viral infections because viruses lack the cell walls, ribosomes, and enzymes that these drugs target. Although antibiotics are useless against viruses, many patients nevertheless demand that physicians prescribe them for viral infections. This behavior contributes to the spread of antibiotic-resistant bacteria, an enormous and growing public health problem. **antibiotic resistance, p. 263**

Halting a viral infection is a challenge. The low number of antiviral drugs, along with the large number of incurable viral diseases, testify to the difficulty of developing drugs that stop the virus without harming infected host cells. The task is complicated by the genetic variability of many viruses. Consider the common cold. Many different cold viruses exist, and their genomes mutate rapidly. As a result, a different virus strain is responsible every time you get the sniffles. Developing drugs that work against all of these variations has so far proven impossible. Even if a drug inactivated 99.99% of cold-causing viruses, the remaining 0.01% would be resistant. Because viruses have high rates of both mutation and replication, natural selection would rapidly render the drug ineffective.

Scientists have developed some antiviral drugs that interfere with enzymes or other proteins that are unique to viruses, but vaccination remains our most potent weapon against many viral diseases. Vaccines "teach" the immune system to recognize one or more molecular components of a virus, without actually exposing the person to the disease. Smallpox has vanished from human populations, and polio is nearly defeated, thanks to successful global vaccination programs. Unfortunately, researchers have been unable to develop vaccines against many viruses, including HIV. **vaccines, p. 738**

B. Viruses Also Cause Diseases in Plants

Plants can have viral infections too (**figure 17.8**). To infect a plant cell, a virus must penetrate waxy coats and thick cell walls. Most viral infections spread when plant-eating insects such as leafhoppers and aphids move virus-infested fluid from plant to plant on their mouthparts. Once inside a plant, viruses spread through plasmodesmata (bridges of cytoplasm between plant cells) and in the vascular tissues that distribute sap. Symptoms may include blotchy, mottled leaves or abnormal growth. A few symptoms, such as the streaking of some tulip petals, appear beautiful to us.

Although plants do not have the same forms of immunity as do animals, they can also fight off viral infections. For example, plant cells use a process called "posttranscriptional gene silencing" that prevents the expression of viral genomes. Researchers are learning more about the role of

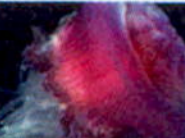

Can *You* Relate?

Hepatitis

Hepatitis, which literally means "inflammation of the liver," is a family of diseases whose causes range from viral infection to alcohol abuse. Symptoms include jaundice (yellowing of the skin and whites of the eyes), nausea, diarrhea, fever, headache, abdominal pain, and fatigue.

At least five viruses cause hepatitis. Each is from a different viral family and has a different transmission route. The three hepatitis viruses that are common in the United States are

- **Hepatitis A:** This virus, which has a single-stranded RNA genome, is acquired by ingesting food or water contaminated by the feces of an infected person. The immune system eliminates the disease soon after exposure, after which the patient is immune to hepatitis A for life. A vaccine is available to help prevent the disease.
- **Hepatitis B:** The virus that causes hepatitis B has a double-stranded DNA genome. It spreads by contact with blood and body fluids. Common routes of exposure include unprotected sexual contact and injecting drugs (or tattoo ink) with a contaminated needle. In about 5% of patients, the viral infection results in chronic liver disease. The hepatitis B vaccine helps prevent infection in people at risk for this disease.
- **Hepatitis C:** Like hepatitis A, the virus that causes hepatitis C has a single-stranded RNA genome. Like hepatitis B, this virus spreads by contact with blood. Unlike either virus, however, a person with hepatitis C usually does not experience symptoms immediately after infection. Instead, in most people, hepatitis C may become a long-term infection that can develop into liver cancer. No vaccine is available for this virus, which is the leading cause of liver transplants in the United States.

FIGURE 17.8 Sick Plant. Infection with tobacco mosaic virus causes a characteristic mottling (spotting) of the leaf.

posttranscriptional gene silencing in the defense against viruses in both plants and animals.

Summary *A person can acquire a viral infection in several ways. Once established, symptoms of a viral illness arise from the effects of cell death and from the immune system's response. Vaccines and antiviral drugs supplement our defenses against viral infection. Viruses also cause diseases in plants.*

17.4 MASTERING CONCEPTS

1. How can one person transmit a viral infection to another?
2. Why do different viral infections lead to different types of symptoms?
3. How does the immune response to a viral infection cause symptoms?
4. Why don't antibiotics cure viral infections?
5. How do viruses enter plant tissues?

17.5 Viroids and Prions Are Other Noncellular Infectious Agents

The idea that something as comparatively simple as a piece of DNA or RNA wrapped in protein—a virus—can cause devastating illness may seem amazing. Yet some infectious agents are even simpler than viruses.

A. A Viroid Is an Infectious RNA Molecule

A **viroid** is a highly wound circle of RNA that lacks a protein coat; it is simply naked RNA that can infect a cell. The RNA coils tightly, bonding with itself to form double-stranded RNA, preventing degradation by host cell enzymes.

Although viroid RNA does not encode protein, it can nevertheless cause severe disease in plants and ruin crops (**figure 17.9**). Transmission of the viroid usually occurs in seeds or pollen. Alternatively, viroids can spread in many of the same ways as viruses do: wind causes contaminated plants to rub against their uninfected neighbors, insects feed on multiple plants, and farm equipment spreads contaminated plant sap throughout a field.

FIGURE 17.9
Viroids Infect Plants. The plant on the left has a viroid-caused disease called "tomato bunchy top;" the one on the right is healthy.

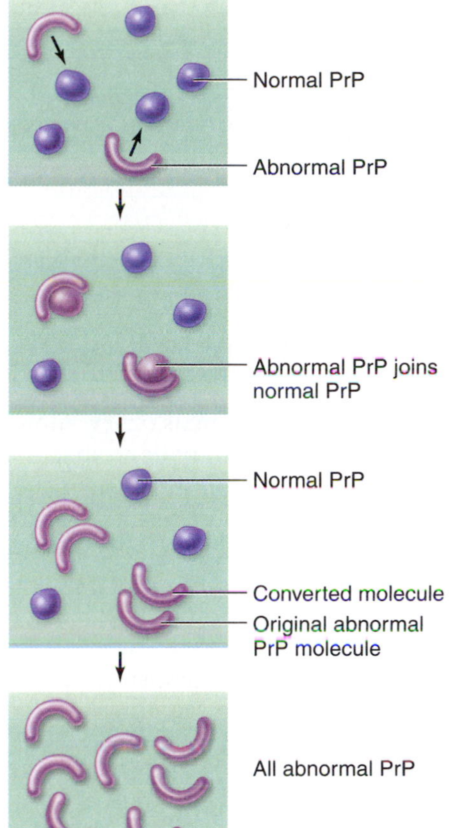

FIGURE 17.10
Prions Change Shape. A prion disease may begin when a single infectious prion protein (PrP) contacts a normal PrP and causes it to switch into the abnormal conformation. As the change spreads, disease occurs in susceptible species.

B. A Prion Is an Infectious Protein

Another type of infectious agent is a **prion,** which stands for "proteinaceous infectious particle." A prion protein (PrP for short) is a normal cellular protein, but it can cause disease because it can exist in several conformations, or three-dimensional shapes. Some of the shapes are "normal," but at least one is "infectious" because it converts one conformation into another (**figure 17.10**). **protein folding, p. 39**

A prion disease results from a chain reaction in which an abnormal form of PrP triggers normal PrP molecules to change. The resulting condition is a **transmissible spongiform encephalopathy,** a disease in which the brain becomes riddled with holes, like a sponge (**figure 17.11**).

Prions occur in more than 80 types of mammals. Two very rare human diseases caused by prions are kuru, which is associated with cannibalism, and Creutzfeldt–Jakob disease (CJD). The now-infamous "mad cow disease" is another example. Affected cattle grow fearful, then aggressive, and then have difficulty standing. The animals rapidly lose weight and die. Animals acquire these diseases by ingesting an infected animal or receiving a transplant of infected tissue. Because of mad cow disease, governments now ban the practice of feeding cattle the processed remains of other cattle.

Prions are especially scary because heat, radiation, and chemical treatments that destroy other infectious agents (including bacteria and viruses) have no effect on prions. Luckily, common-sense precautions that keep animal brains and

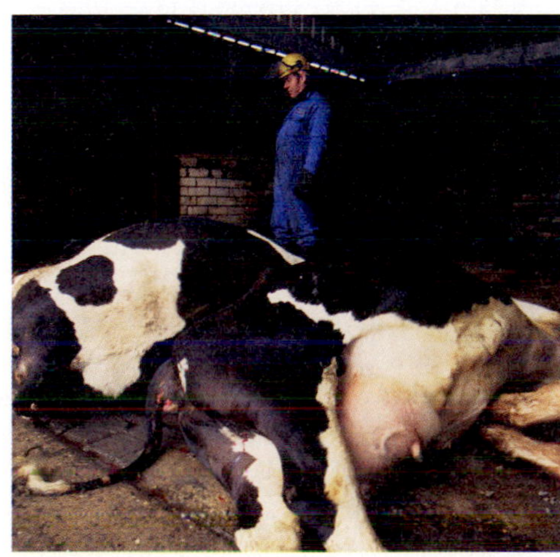

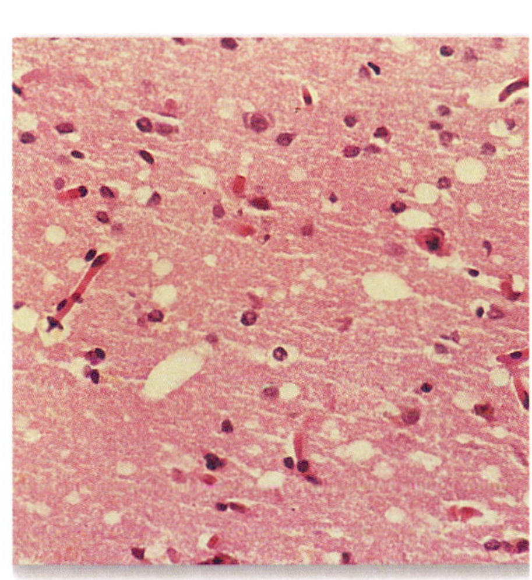

FIGURE 17.11 Transmissible Spongiform Encephalopathy. The holes and clumps of protein fibrils are evident in a brain from a cow that died of bovine spongiform encephalopathy (BSE), also known as "mad cow disease."

spinal cords out of the human food chain can prevent the spread of prion diseases.

Summary ***Viroids are infectious RNA molecules that harm plants, and prions are infectious proteins that cause nervous system diseases in mammals.***

17.5 MASTERING CONCEPTS

1. How is a viroid different from a virus?
2. How do prions cause disease?
3. What is the best way to avoid prion diseases?

17.6 INVESTIGATING LIFE
Scientific Detectives Follow HIV's Trail

Anyone who watches crime dramas on TV knows that detectives ask many questions when trying to match a suspect to a crime. Do footprints found at the scene match the suspect's shoes? Do skin cells found under the victim's fingernails contain DNA matching the suspect's? Was the suspect in the victim's neighborhood at the time of the crime? Do bank records suggest a financial motive?

Epidemiologists use similar tactics when they test hypotheses about the evolution of viruses. Scientists have suggested that a virus that causes symptomless infections in monkeys and apes, simian immunodeficiency virus (SIV), is the ancestor of HIV. They tested this hypothesis by asking five independent questions about SIV and HIV:

1. Do the genomes of the two viruses consist of the same type of genetic material, and do they have the same order, number, and types of genes?
2. Is there an evolutionary link between the viruses? That is, do they have similar nucleotide sequences in their genes, and do their proteins have similar amino acid sequences?
3. Is SIV common enough in its natural host that it has a chance of spreading to other hosts?
4. Do the two viruses occur in the same geographic region?
5. Is there a plausible transmission pathway for SIV to spread from its original host to humans?

HIV occurs in two major strains, HIV-1 and HIV-2, that apparently arose independently. HIV-2 is the milder of the two types, and it occurs primarily in West Africa. By the late 1980s and early 1990s, scientists had used the five criteria to identify a strain called SIVsm as the source of HIV-2. (The "sm" stands for sooty mangabey, a monkey that is SIVsm's primary host.) SIVsm and HIV-2 are genetically very similar (criteria 1 and 2). Furthermore, SIVsm is common in the sooty mangabey (criterion 3) in the part of Africa where the HIV-2 epidemic started (criterion 4). Finally, people keep these monkeys as pets and hunt them for meat. This intimate contact suggests a ready transmission route between the two species (criterion 5).

The origin of HIV-1, however, remained mysterious until 1999. Scientists suspected it arose from a different source, SIVcpz (for chimpanzee), because HIV-1 and SIVcpz share some genetic similarities. Yet the genomes of the two viruses also have regions with many differences. Furthermore, few chimpanzees test positive for SIVcpz infection, and they live in parts of Africa not known to be the epicenter of the HIV-1 epidemic.

University of Alabama medical researcher Beatrice Hahn, along with a multinational team of scientists, wanted to learn more about HIV-1's evolutionary history. They used frozen tissue taken at necropsy from a chimpanzee named Marilyn, who was born in Africa (no one knows which country) around 1960 and exported to the United States as an infant. Marilyn had never been used in AIDS research, yet a 1985 study showed that she was the only one out of 98 chimpanzees who tested positive for antibodies against HIV-1.

Using the polymerase chain reaction (PCR) to amplify viral DNA from Marilyn's tissue, Hahn and her colleagues found sequences that were similar to both HIV-1 and SIV, but not identical to either. They gave Marilyn's virus a new name: SIVcpzUS. This was only the fourth SIV strain ever found in chimpanzees. Two of the other three infected chimpanzees belonged to the same subspecies as Marilyn—the central chimpanzee. The third chimpanzee, named Noah, was a so-called eastern chimpanzee (**figure 17.12**). **polymerase chain reaction, p. 146**

The researchers found that the viral genome sequences reflected the animals' origins. All three SIV strains from central chimpanzees were more closely related to one another than to Noah's strain. This was evidence that SIV evolved differently in chimp subspecies from different parts of Africa. Furthermore, all known subgroups of HIV-1 (M, N, and O) were closely related to those same three SIV strains (**figure 17.13A**). Had the researchers found the origin of HIV-1?

To make sure, the researchers scrutinized the genome of the least common type of HIV-1, subgroup N, which occurs only in western Africa. They found that about half of its genome resembled HIV-1 subgroup M, and the other half looked like that of SIVcpz (figure 17.13B). The researchers concluded that multiple strains of SIV combined in a chimpanzee to yield a new type of virus (later named HIV-1 subgroup N) that somehow jumped to humans—just as influenza viruses from pigs have spread to humans after recombining in birds. Presumably, HIV-1 subgroups M and O independently arose in the same way as subgroup N.

The researchers thus used the same criteria to demonstrate the origin of HIV-1 as others had done for HIV-2. Clearly, SIVcpz and HIV-1 share genetic similarities (criteria 1 and 2). Moreover, the habitat of central chimpanzees overlaps with the region of Africa where HIV-1 occurs (criterion 4), and humans who hunt chimps for meat provide a likely transmission route (criterion 5). Only the third criterion remains questionable: If SIV is common in chimpanzees, why have scientists found it in only four animals out of more than 1500 tested? Hahn and her coauthors suggest that this low infection rate is a predictable consequence of studying chimps that were either wild-caught as infants or born in captivity. Compared with wild chimps, these captive animals would have fewer opportunities to acquire the virus. More intensive study of wild populations, which itself presents ethical problems, may reveal a much higher infection rate.

Even when television cops nab the criminal, some details of the crime remain obscure. Similarly, Beatrice Hahn and her colleagues paint a convincing portrait of HIV-1's origin from SIV, but many questions remain. How and when did the virus jump from chimpanzees to humans? Why do HIV-1 subgroups N and O remain rare, whereas subgroup M is causing the global AIDS epidemic? Can we use HIV-1 and HIV-2 to learn about the potential for other viruses to emerge as human pathogens? The scientific detectives continue to search for the answers.

Gao, Feng, Elizabeth Bailes, David L. Robertson, and 9 coauthors, including Beatrice Hahn. 1999. Origin of HIV-1 in the chimpanzee *Pan troglodytes troglodytes*. *Nature*, vol. 397, pages 436–441.

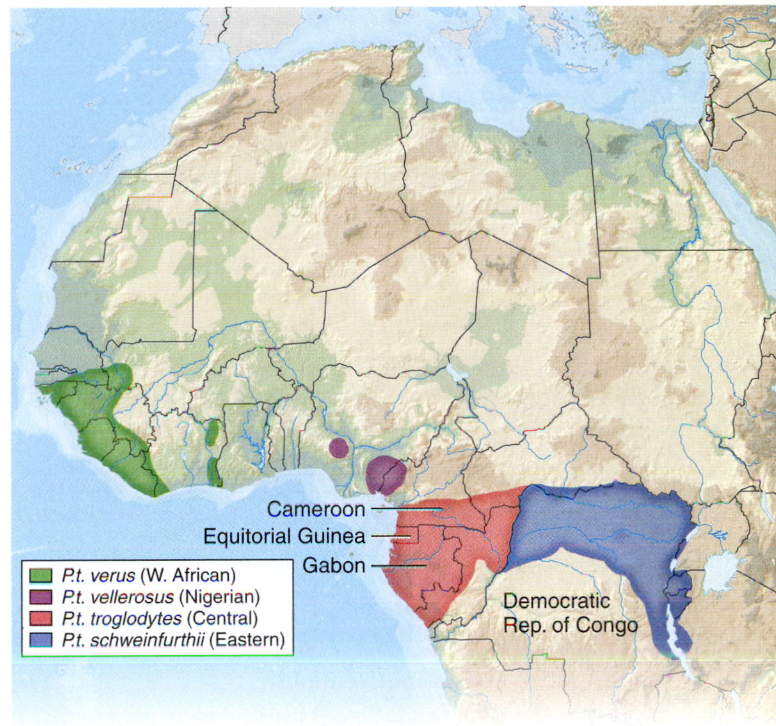

FIGURE 17.12 Chimpanzee Subspecies Ranges. Marilyn and two chimps from Gabon belonged to the "central" subspecies of chimpanzee (red area on map); Noah was born in the Democratic Republic of Congo and belonged to the "eastern" subspecies (blue area on map).

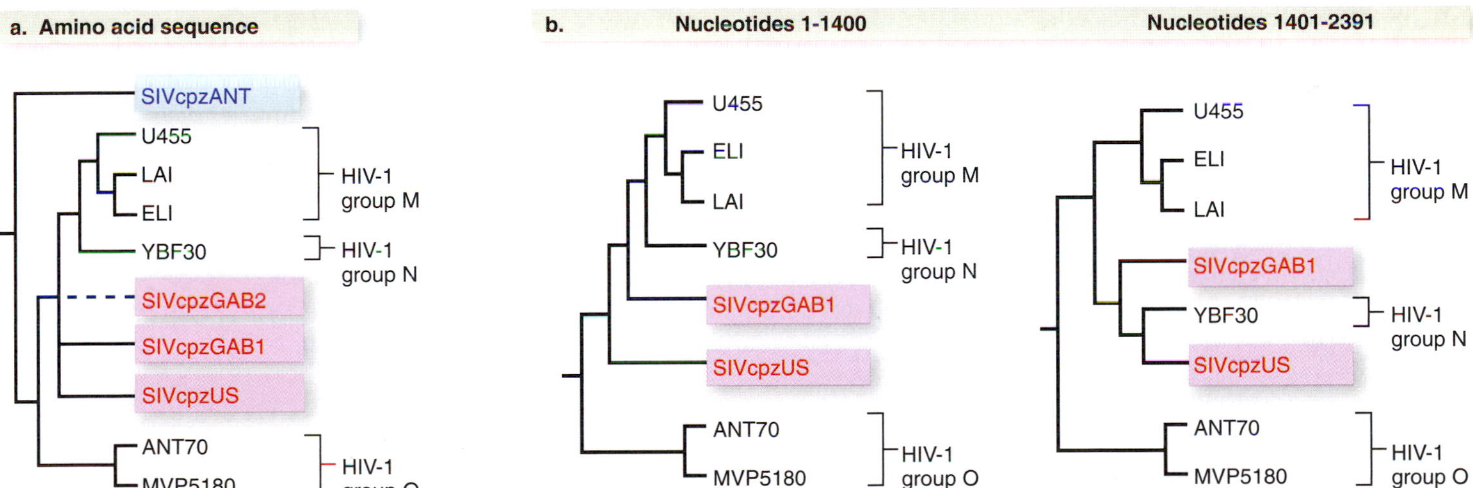

FIGURE 17.13 HIV-1 from SIV. (**A**) Amino acid sequences of one viral protein shows that Noah's strain of SIV (blue) is distinct from all other SIVcpz strains (red) and from all known HIV-1 subgroups. (**B**) The sequence of first 1400 nucleotides in HIV-1 subgroup N's genome (YBF30) is very similar to those in HIV-1 subgroup M (many representatives of subgroup M are omitted for simplicity). The sequence of the remaining 1000 or so nucleotides from YBF30 clusters with SIV from central chimpanzees. **Question:** How could researchers use a similar strategy to determine the origin of a new influenza virus?

CHAPTER SUMMARY

17.1 Viruses Are Infectious Particles of Genetic Information and Protein

A. Viruses Are Smaller and Simpler than Cells

- A **virus** is a nucleic acid (DNA or RNA) in a **capsid** and perhaps an **envelope.** It must infect a living cell to reproduce.
- Many viruses, including some **bacteriophages,** have relatively complex structures.

B. A Virus's Host Range Consists of the Organisms It Infects

- The types of species that a virus infects constitute its **host range.** A virus may be carried in a **reservoir** species, in which it does not cause symptoms.

C. Are Viruses Alive?

- Viruses are intracellular parasites that most biologists do not consider alive.
- Many viral genomes mutate rapidly.
- Scientists classify viruses based on the type of genetic material, the shape of the capsid, the presence or absence of an envelope, the replication strategy, and the type of disease.

17.2 Viral Replication Occurs in Five Stages

- The five stages of viral replication within a host cell are attachment, penetration, synthesis, assembly, and release.

17.3 Cell Death May Be Immediate or Delayed

A. Some Viral Infections Kill Cells Immediately

- In a **lytic infection,** new viruses are immediately manufactured, assembled, and released.

B. Viral DNA Can "Hide" in a Cell's Chromosome

- In a **lysogenic infection,** the virus' nucleic acid integrates as a **prophage** into the host DNA and remains hidden, or **latent.**

C. Some Animal Viruses Linger for Years

- Some viruses, including HIV, may remain latent inside animal cells.
- Some latent viruses are associated with cancer.

17.4 Effects of a Viral Infection May Be Mild or Severe

- The effects of a virus depend on the cell types it infects. Viruses cause disease by killing infected cells and by stimulating immune responses.

A. Drugs and Vaccines Help Fight Viral Infections

- Antibiotics that kill bacteria are ineffective against viruses.
- Antiviral drugs and vaccines combat some viral infections.

B. Viruses Also Cause Diseases in Plants

- Viruses infect plant cells then spread via plasmodesmata.

17.5 Viroids and Proteins Are Other Noncellular Infectious Agents

A. A Viroid Is an Infectious RNA Molecule

- **Viroids** are naked RNA molecules that infect plants.

B. A Prion Is an Infectious Protein

- A **prion** protein can assume different conformations, some of which can cause **transmissible spongiform encephalopathies.** Treatments that destroy other infectious agents have no effect on prions.

17.6 Investigating Life: Scientific Detectives Follow HIV's Trail

- Scientists have tested hypotheses about HIV's origin by collecting five types of information.
- HIV-2 apparently evolved from a simian immunodeficiency virus (SIV) in monkeys called sooty mangabeys. Different subgroups of HIV-1 evolved independently from SIV infecting chimpanzees.

MULTIPLE CHOICE QUESTIONS

1. Which of the following is NOT a feature associated with viruses?
   a. Cytoplasm
   b. Genetic information
   c. Protein coat
   d. Membrane

2. An organism that carries a virus without having symptoms is called
   a. a harbor.
   b. an agent.
   c. a host.
   d. a reservoir.

3. Why are viruses not considered to be alive?
   a. Because they depend on host cells to reproduce
   b. Because they do not evolve
   c. Because they do not contain genetic information
   d. Both a and c
4. At which stage in viral replication does the genetic information enter the host cell?
   a. Penetration
   b. Synthesis
   c. Assembly
   d. Release
5. A virus can jump from one host species to a different host species because
   a. viruses are small and numerous.
   b. mutations in the virus let it recognize a new host surface molecule.
   c. the host cell provides ATP, ribosomes, nucleotides and amino acids.
   d. viruses leave the host by exocytosis.
6. Which step in viral replication would be the best target to prevent the spread of a virus?
   a. Attachment
   b. Penetration
   c. Synthesis
   d. Release
7. What occurs during a lysogenic infection?
   a. Viral particles attach, but do not penetrate a host cell.
   b. Viral particles fill a host cell and cause it to burst.
   c. Viral genetic material inserts into the host DNA.
   d. The viral prophage DNA is packaged into a capsid.
8. The severity of the symptoms associated with a viral infection is related to
   a. the response of the immune system.
   b. the number of viruses released.
   c. the number and types of cells that become infected.
   d. Both a and c.
9. What is a prion?
   a. A highly wound circle of RNA
   b. An "escaped intron"
   c. A protein that can alter the shape of a second protein
   d. The protein associated with a latent virus
10. What property of a virus makes it useful for making transgenic organisms?
    a. The ability to replicate inside a host cell
    b. The insertion of viral genes into the host's DNA
    c. The ability of the host cell to release viruses
    d. The ability of viral genes to undergo rapid mutation

TESTING YOUR KNOWLEDGE

1. What are the four types and configurations of nucleic acids in viruses?
2. Describe how a virus's envelope can include lipid molecules that are also part of host cell membranes.
3. What events occur in each of the five stages of viral replication?
4. Rhinoviruses replicate in the mucus-producing cells in a person's nose, throat, and lungs, causing the common cold. Papillomaviruses, which infect skin cells, cause growths called warts. HIV infects T cells and causes AIDS. How do these three types of viruses "know" which human cells to infect?
5. Distinguish between lytic and lysogenic infections.
6. How is a virus similar to, and different from, a cell, a viroid, and a prion?

THINKING AS A SCIENTIST

1. Why do antibiotics such as penicillin kill bacteria but leave viruses unharmed?
2. Why are lytic viruses better suited as agents of "phage therapy" than are lysogenic viruses? How would you test whether such a treatment would be effective? Would you be willing to take a "viral antibiotic?"
3. The National Center for Biotechnology Information maintains a list of viruses for which genome sequence data are available. Choose one, and describe some discoveries that have come from research on this virus.
4. Your biology lab instructor gives you a petri dish of agar covered with visible colonies. Your lab partner says the colonies are viruses, but you disagree. How do you know the colonies are bacteria?
5. At about 600 nm, mimiviruses are enormous compared with other viruses. Their genomes consist of about 1.2 million base pairs and encode more than 1000 genes—more than some bacteria. If you encountered a mimivirus-like object in your research, what sorts of studies could you carry out to determine whether the object was a virus or a bacterium?

Visit www.mhhe.com/hoefnagels for practice quizzes, animations, videos, and activities designed to help you master the material in this chapter.

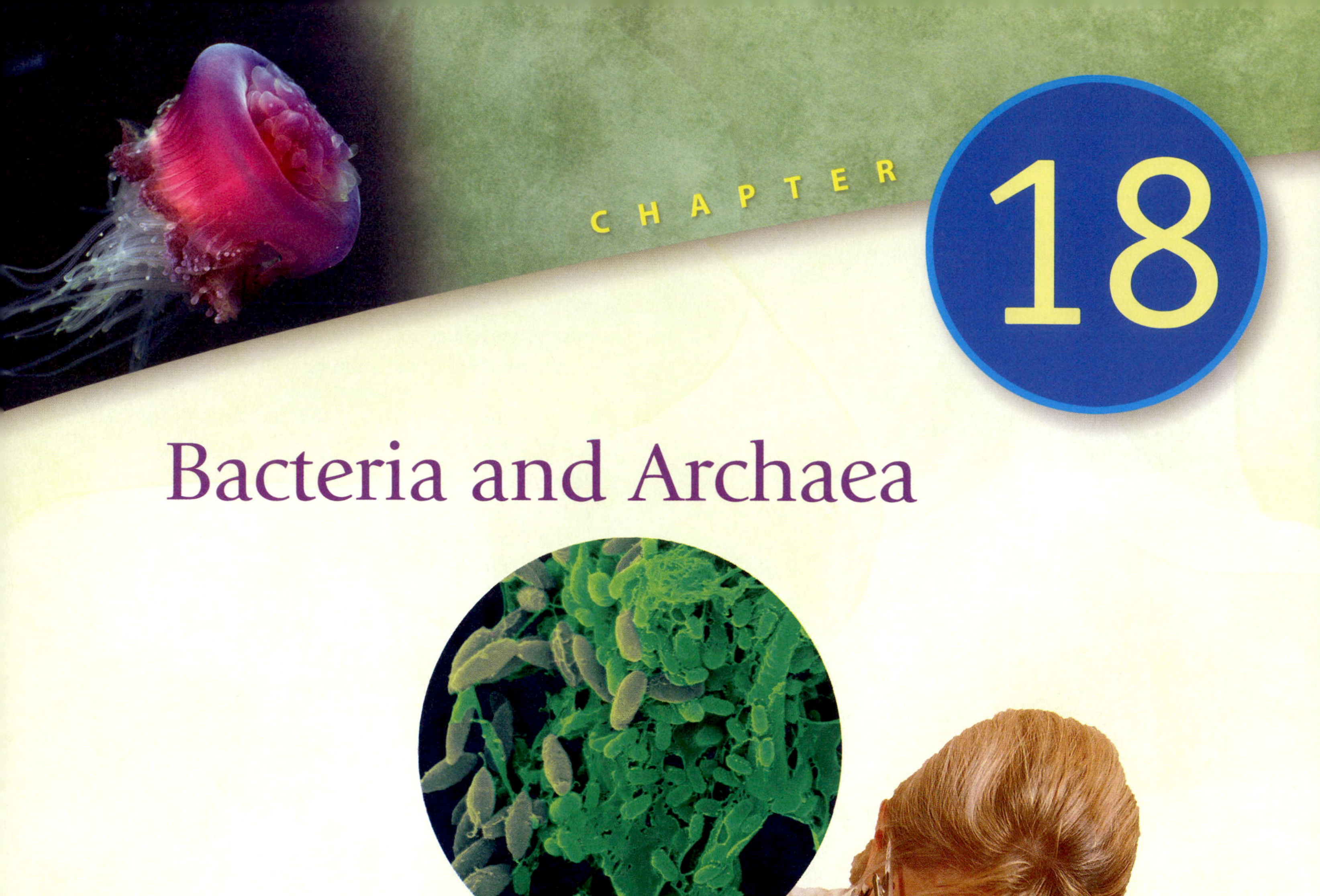

Bacteria and Archaea

The plaque that accumulates on teeth is a familiar example of a bacterial biofilm.

What's the Point?

Bacterial Biofilms: "Mob Mentality" on a Microscopic Scale

Bacteria are one-celled organisms, so it is tempting to think of them as rugged individualists. Instead, however, they often build complex communities in which cells of the same species communicate with one another, protect one another, and even form differentiated structures with specialized functions. These organized aggregations of bacterial cells are called biofilms.

A biofilm forms when bacteria settle and reproduce on a solid surface. Once the cells reach a critical density, they express genes that trigger the secretion of a sticky slime made of polysaccharides. As cells continue to divide, three-dimensional mushroom-shaped structures form. Occasionally, cells released from the biofilm colonize new habitats, starting the process anew.

Microbiologists are still learning about the cues that trigger biofilm formation. For example, bacteria use signaling molecules to detect the density of cells around them. These "quorum-sensing" signals allow the bacteria in a biofilm to coordinate their activities. Thus, within the first 6 hours of biofilm formation, the bacteria turn off genes encoding the proteins that form the flagellum—after all, success in a sedentary lifestyle does not require the ability to swim. In contrast, genes encoding proteins that build attachment structures called pili are activated.

Interest in bacterial biofilms extends far beyond idle curiosity about prokaryotic life. These microbial mats degrade sewage in the trickling filters of community wastewater treatment plants, corrode water pipes, help mine copper ore, and coat the surfaces of plants' roots. Much of the attention that biofilms receive, however, is medical. Persistent biofilms can form on medical implants such as heart valves and catheters. They can also form dental plaque, provoke gum disease, and colonize the thick mucus that accumulates in the lungs of cystic fibrosis patients. Bacteria in biofilms are particularly resistant to immune defenses and antibiotic treatment.

As microbiologists learn more about biofilm formation, they may be able to develop new treatments to combat medically important bacteria. For example, it may be possible to disrupt biofilm formation by silencing quorum-sensing signals. Conversely, learning how to trigger biofilm formation in beneficial soil bacteria may enhance efforts to clean up toxic wastes.

The more scientists learn about prokaryotes—bacteria and archaea—the more we can appreciate the amazing capabilities of Earth's simplest organisms. As this chapter explains, microbes are much more than just "germs."

LEARNING OUTLINE

18.1 Prokaryotes Are a Biological Success Story

A **prokaryote** is a single-celled organism that lacks a nucleus and membrane-bounded organelles. Microbiologists estimate that 100,000 to 10,000,000 species exist in the two prokaryotic domains: **Bacteria** and **Archaea** (**figure 18.1**).

Prokaryotic cells are simpler than those of eukaryotes, yet they have had a far greater influence on Earth's natural history. The first cells probably resembled existing prokaryotes more than any other type of organism. Along the road of evolution, prokaryotes were probably the precursors of the chloroplasts and mitochondria of eukaryotic cells. Ancient photosynthetic prokaryotes also contributed oxygen gas (O_2) to Earth's atmosphere, creating a protective ozone layer and paving the way for aerobic respiration. **aerobic respiration, p. 124; endosymbiosis, p. 338**

Virtually no place on Earth is prokaryote-free; their cells live within rocks and ice, high in the atmosphere, far below the ocean's surface, in thermal vents, nuclear reactors, hot springs, animal intestines, plant roots, and practically everywhere else. Many species prefer habitats that humans consider "extreme" (**table 18.1**). This chapter's Burning Question: Are there areas on Earth where no life exists? describes some of the few places where prokaryotes do not live.

Most of what we know about prokaryotes comes from the relatively few species that microbiologists can culture in the laboratory. However, culturable microorganisms do not represent the natural diversity of the prokaryotes. Laboratory culture does not accurately mimic the chemical and physical conditions of the world's countless natural habitats, so it cannot support the diversity of microbial species that occupy those habitats.

Table 18.1 *Identifying Prokaryotes by Habitat*

Designation	Environment	Sample Habitats
Acidophile	Low pH (1.0–5.4)	Hot springs
Halophile	Extreme salt (3.5–30%)	Ocean, Dead Sea, evaporation ponds
Thermophile	Extreme heat (50°C–110°C)	Compost heaps, boiling springs

Summary *Prokaryotes have life's simplest cells, yet they are the most abundant and diverse organisms on Earth. They were the first organisms, and they have profoundly influenced evolutionary history. They live practically everywhere. Nevertheless, they remain poorly understood in many ways.*

18.1 MASTERING CONCEPTS

1. What are two domains that contain prokaryotes?
2. List several ways that prokaryotes have influenced evolution.
3. In what habitats do prokaryotes live?

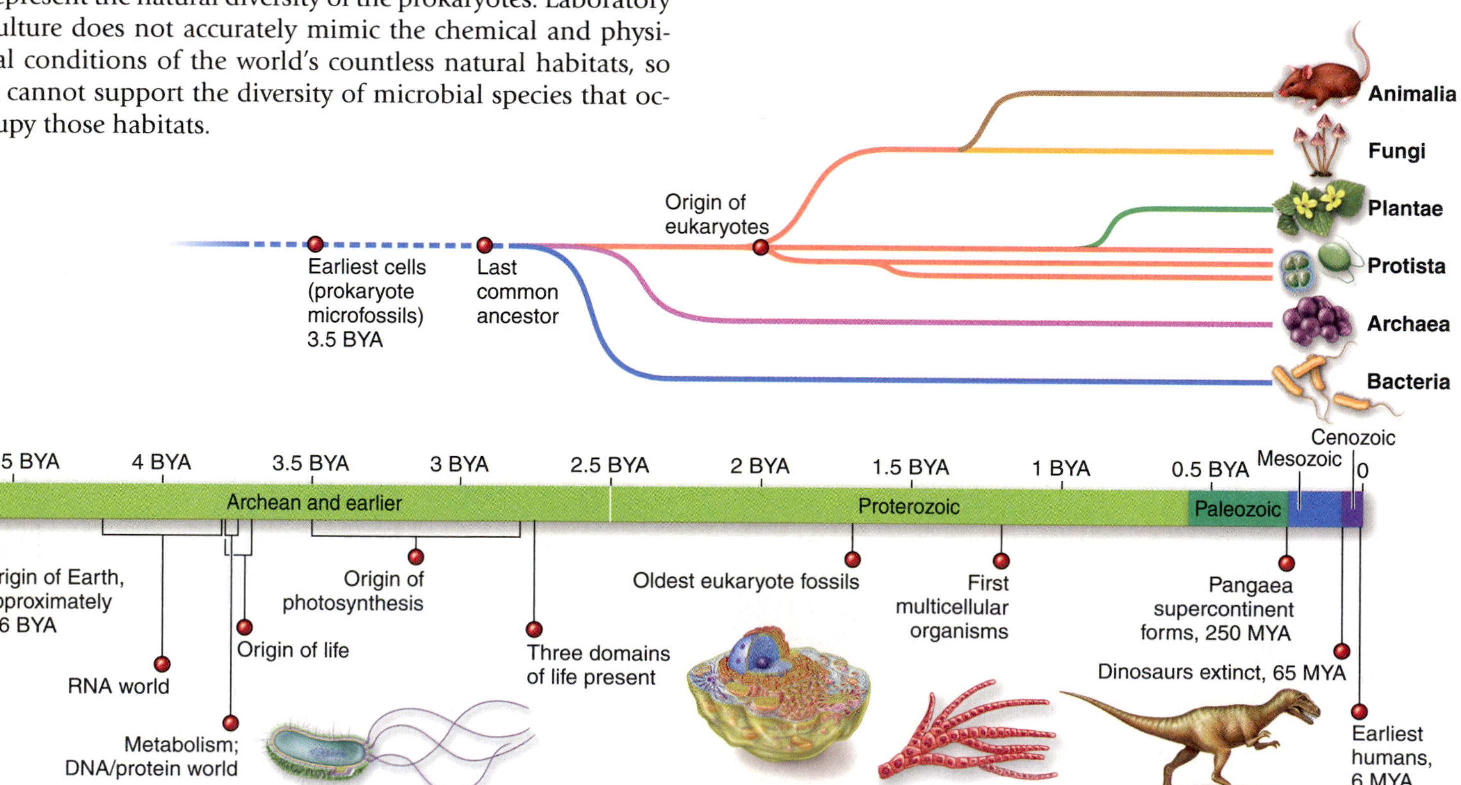

FIGURE 18.1 Prokaryotes at the Base of the Tree. Domains Bacteria and Archaea are two of the three main branches of life.

Burning Questions

Are there areas on Earth where no life exists?

Sand, bare rock, and polar ice may seem devoid of life, but they are not. Scientists using microscopes and molecular tools have discovered prokaryotes living in the hottest, coldest, wettest, driest, saltiest, highest, most radioactive, and most pressurized places on the planet. Prokaryotes live in places where no other organism can.

Although prokaryotes live nearly everywhere, many body fluids and tissues normally remain microbe-free. Examples include the sinuses, muscles, brain and spinal cord, ovaries and testes, blood, cerebrospinal fluid, urine in kidneys and the bladder, and semen before it enters the urethra.

Artificial sterilization of some items is an important tool in disease control. People in many professions regularly use autoclaves, radiation, and filters to sterilize everything from surgical tools, to medicines and bandages, to processed foods.

Have a Burning Question of your own? Submit it to marielle_hoefnagels@mcgraw-hill.com for possible inclusion in future editions of this book!

18.2 Prokaryotic Cells Are Structurally Simple

Like the cells of other organisms, all prokaryotic cells are bounded by a cell membrane that encloses cytoplasm, DNA, and ribosomes. At about 1 to 10 μm long, however, a typical prokaryotic cell is 10 to 100 times smaller than most eukaryotic cells (see figure 3.2). Prokaryotes also lack the membrane-bounded organelles that characterize eukaryotic cells (see table 3.3). **Figure 18.2** depicts bacterial cell structures; a given cell may have some or all of the structures pictured.

A. Internal Structures Include a Chromosome, Plasmids, and Ribosomes

A prokaryotic cell's DNA typically consists of one circular chromosome. The **nucleoid** of a prokaryotic cell is the region where this DNA is located, along with some RNA and a few proteins. Unlike the nucleus of a eukaryotic cell, a membranous envelope does not surround the nucleoid.

Many prokaryotic cells also contain one or more **plasmids,** circles of DNA apart from the chromosome (**figure 18.3**). Plasmids may include genes that make possible their own transfer and replication. Other plasmid-encoded genes may provide the ability to resist a drug or toxin, cause disease, or alter the cell's metabolism. Recombinant DNA technology uses plasmids to ferry genes from one kind of prokaryote to another. **transgenic organisms, p. 252**

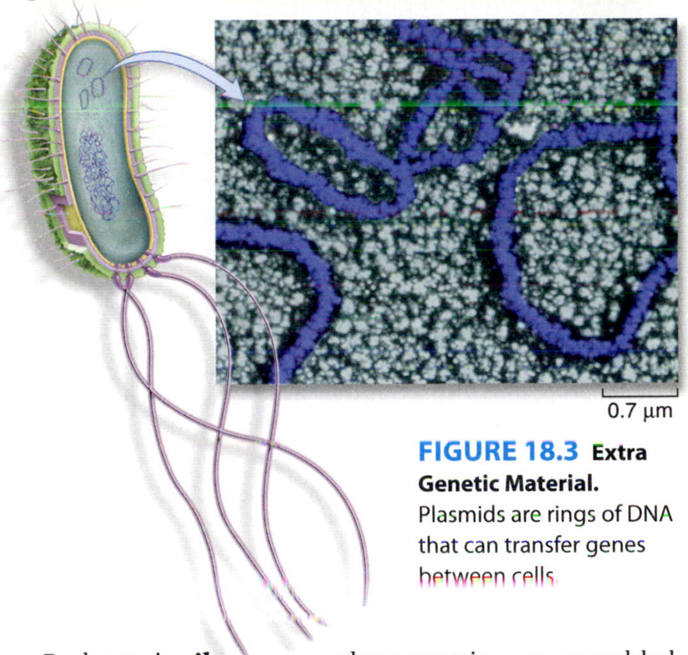

FIGURE 18.3 Extra Genetic Material. Plasmids are rings of DNA that can transfer genes between cells.

Prokaryotic **ribosomes,** where proteins are assembled, are structurally different from eukaryotic ribosomes. Some antibiotics, such as streptomycin, kill bacteria without harming eukaryotic host cells by exploiting this structural difference. Can *You* Relate? Antibiotics and Other Germ Killers describes more examples of how antibiotics work.

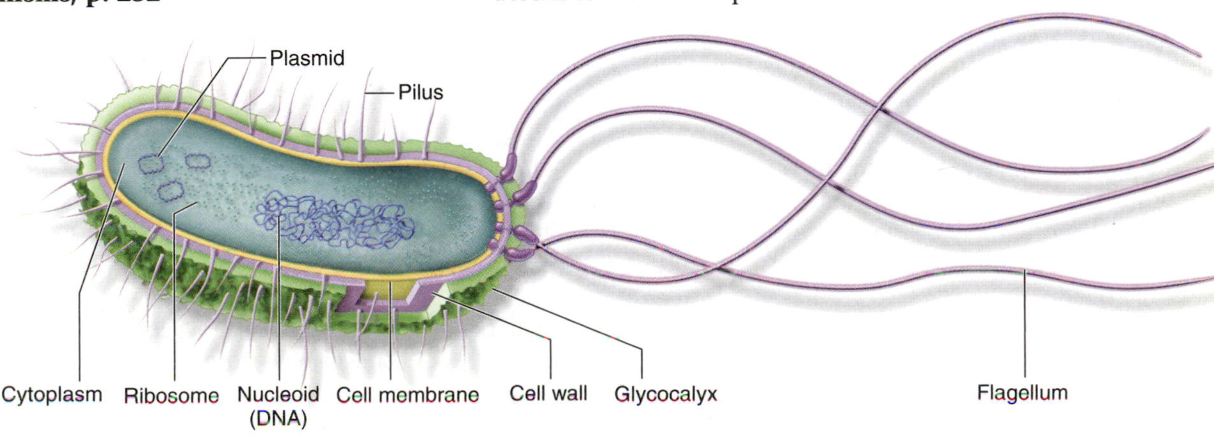

FIGURE 18.2 Bacterial Cell Structure. The bacterial chromosome occurs in the nucleoid region of the cell. Ribosomes are free in the cytoplasm. Some cells have flagella, pili, and a slimy glycocalyx around the cell wall. A plasmid is a circle of DNA apart from the cell's chromosome.

Can *You* Relate?

Antibiotics and Other Germ Killers

When a person develops a bacterial infection, a physician may prescribe antibiotics. These drugs typically exploit structures and functions in bacterial, but not host, cells. Some mechanisms of action include:

- **Inhibiting cell wall synthesis:** A bacterium that cannot make a rigid cell wall will burst. Penicillin is one antibiotic that interferes with cell wall formation.
- **Disrupting cell membranes:** All life depends on an intact membrane that regulates what enters and leaves a cell. Polymyxin antibiotics exploit differences between bacterial and eukaryotic cell membranes. **membranes, p. 54**
- **Inhibiting protein synthesis:** No organism can survive if it cannot make proteins. The antibiotics streptomycin, chloramphenicol, and erythromycin bind to different parts of bacterial ribosomes, but all have the same effect—they kill bacteria without killing us. **ribosomes, p. 241**
- **Inhibiting transcription:** Gene expression requires RNA synthesis. Rifamycin antibiotics prevent RNA synthesis in bacteria by binding to a bacterial form of RNA polymerase. **transcription, p. 238**
- **Inhibition of metabolic enzymes:** Theoretically, antibiotics could block any bacterial metabolic pathway that does not occur in host cells. Sulfanilamide, for example, mimics the substrate of a bacterial enzyme that participates in an essential chain of chemical reactions. **enzyme inhibition, p. 88**

B. External Structures Aid in Protection, Attachment, and Movement

A **cell wall** is a rigid barrier that surrounds most prokaryotic cells. Bacterial cell walls contain **peptidoglycan,** a complex polysaccharide that does not occur in the cell walls of archaea. The antibiotic penicillin inhibits the reproduction of bacteria (but not archaea) because it interferes with the final steps in peptidoglycan synthesis.

The **Gram stain** reaction distinguishes between two major groups of bacteria based on differences in cell wall architecture (**figure 18.4**). After the staining procedure is complete, gram-positive cells appear purple. Their walls are made primarily of a thick layer of peptidoglycan. In contrast, gram-negative cells have much thinner cell walls and stain pink. Medical technicians often use Gram staining as a first step in identifying bacteria that cause infections. The distinction is important because gram-positive and gram-negative bacteria are susceptible to different antibiotic drugs.

Besides the thin inner layer of peptidoglycan, the cell walls of gram-negative bacteria have an outer membrane covering of lipid, protein, and polysaccharide. Parts of this outer membrane are responsible for the toxic effects of many medically important gram-negative bacteria, including *Salmonella.*

A **glycocalyx** (also sometimes called a capsule or slime layer) is a sticky layer composed of proteins or polysaccharides that may surround the prokaryotic cell wall (**figure 18.5A**). The glycocalyx has many functions, including attachment, resistance to drying, and protection from immune system cells. It also plays a role in biofilm formation on teeth and other surfaces (see the chapter opening essay).

Some cells have **pili** (singular: pilus), which are short, hairlike projections made of protein. Attachment pili enable cells to adhere to objects (figure 18.5B). The bacterium that produces a toxin that causes cholera, for example, attaches to a human's intestinal wall using pili. Sex pili aid in the transfer of DNA from cell to cell (see section 18.4).

Another external structure is a **flagellum,** which is a whiplike extension that moves the cell (figure 18.5C); a cell may have none, one, or more. In a response called **taxis,** prokaryotes rotate their flagella to move toward or away from a stimulus such as food or a toxin. (Some eukaryotic cells also have flagella, but they are not homologous to those on prokaryotic cells.)

C. Endospores Help Some Bacteria Survive Heat and Drying

Some prokaryotes can form **endospores,** which are dormant, thick-walled structures that enable them to survive

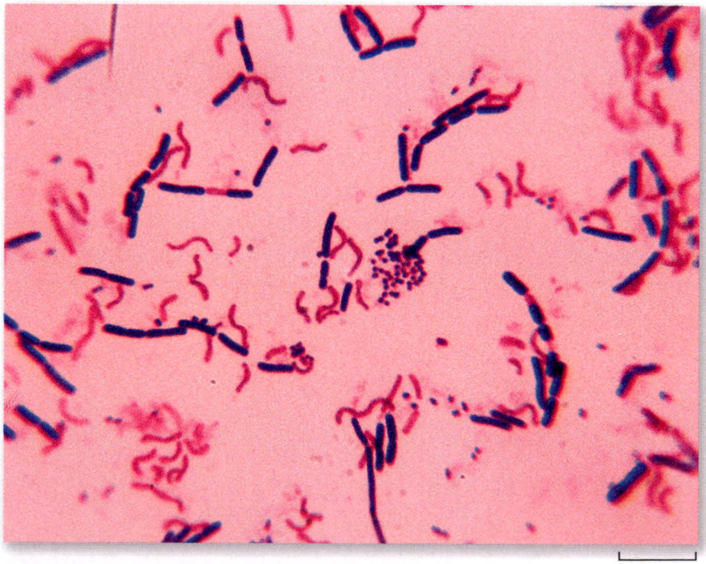

FIGURE 18.4 Gram Stain. The Gram stain distinguishes bacteria on the basis of cell wall structure. Gram-positive cells have a thick layer of peptidoglycan that retains the purple dye. The pink gram-negative cells have a thinner peptidoglycan layer, coated with an outer membrane. This stained smear contains both gram-positive and gram-negative bacteria.

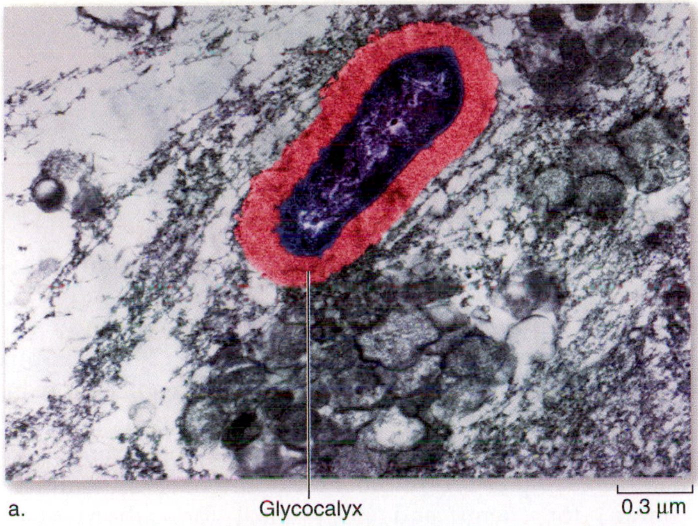

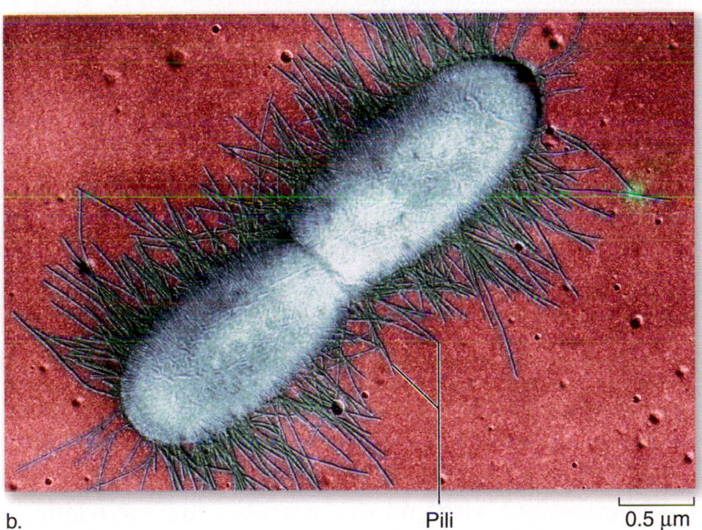

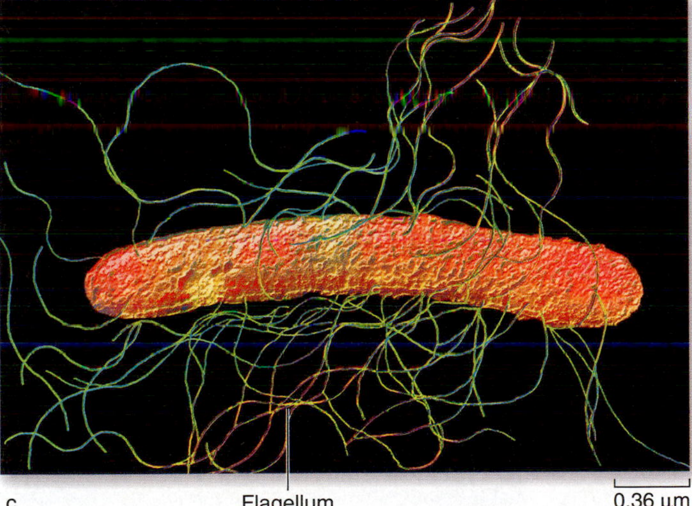

FIGURE 18.5 External Structures of Prokaryotic Cells. (**A**) A glycocalyx is a sticky layer surrounding the cell wall that enables cells such as this *Bacteroides* to adhere to surfaces. (**B**) Pili are extensions that enable cells to attach to objects, surfaces, and other cells. This is *Escherichia coli*. (**C**) The numerous flagella on this *Proteus mirabilis* enable it to move.

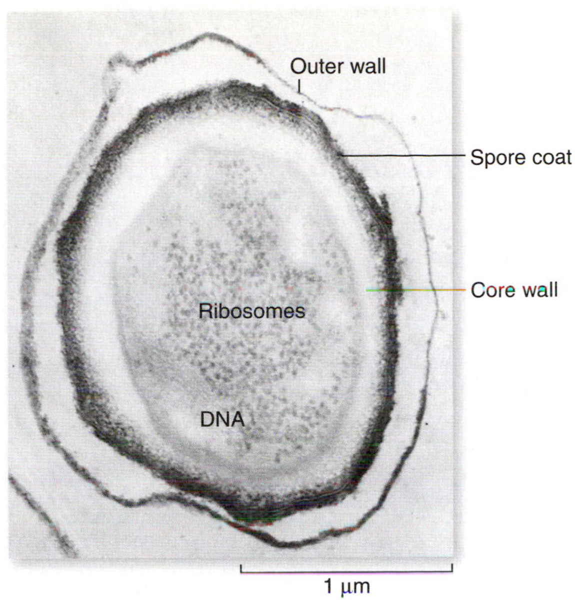

FIGURE 18.6 An Endospore. Bacteria that cause the disease anthrax, *Bacillus anthracis*, survive environmental extremes by forming thick-walled endospores.

harsh conditions, often for long periods (**figure 18.6**). An endospore can withstand boiling, drying, ultraviolet radiation, and disinfectants. The normal cellular form returns when environmental conditions improve. Food manufacturing processes typically include a 10- to 15-minute superheated steam treatment to destroy endospores of a bacterium called *Clostridium botulinum*. If any endospores survive, they can germinate, producing cells that thrive in the absence of oxygen. These cells produce a toxin that causes botulism, a severe (and sometimes deadly) form of food poisoning.

Another example of a spore-forming bacterium is *Bacillus anthracis*. This organism, ordinarily found in soil, can cause a deadly disease when inhaled. Cultures of *B. anthracis* can be dried to induce formation of endospores, then ground into a fine powder that remains infectious for decades. Anthrax received widespread attention in late 2001 when anthrax-tainted mail killed five people.

Summary ***Basic features of prokaryotes include chromosomal and sometimes plasmid DNA, ribosomes, and a cell membrane. Most prokaryotic cells have cell walls; some also have sticky outer layers, pili, or flagella. A few species form survival structures called endospores.***

18.2 MASTERING CONCEPTS

1. What is the difference between a nucleoid and the nucleus of a eukaryotic cell?
2. What are plasmids, and how are they important?
3. What is the basis of Gram staining?
4. What are the functions of a glycocalyx, pili, and flagella?
5. What is the advantage of endospores to species that make them?

18.3 Prokaryotes Include Two Domains with Enormous Diversity

The tendency to lump together all prokaryotic organisms hid much of the diversity in the microbial world for many decades. New molecular data have led scientists closer to a classification system that reflects evolutionary relationships among all organisms, including prokaryotes. This section begins by explaining how microbiologists have traditionally classified microorganisms, then briefly describes a small sampling of the extraordinary diversity of prokaryotic species.

A. Prokaryote Classification Traditionally Relies on Observable Features

For hundreds of years, microbiologists classified microbes based on characteristics they could easily observe. For example, prokaryotic cells can take a variety of forms that are easily distinguished with a microscope (**figure 18.7**). Three of the most common shapes are **cocci** (spherical), **bacilli** (rod-shaped), and **spirilla** (spiral). The arrangement of the cells in pairs, clusters (*staphylo-*), or chains (*strepto-*) also is sometimes important. *Staphylococcus* causes infections in humans; its spherical cells form grapelike clusters.

Scientists also use stains to divide prokaryotes into groups. For example, the Gram stain reaction places bacteria into major subgroups based on cell wall structure. Other stains can reveal the presence of flagella, endospores, or a glycocalyx.

The methods by which prokaryotes acquire carbon and energy form another basis for classification. **Autotrophs** acquire carbon from inorganic sources such as carbon dioxide (CO_2), and **heterotrophs** get carbon from organic molecules produced by other organisms. An autotroph or heterotroph may also be a phototroph or chemotroph, which refers to the organism's energy source. **Phototrophs** derive energy from the sun; **chemotrophs** oxidize inorganic or organic chemicals. Combining these terms describes how an organism fits into the environment. Cyanobacteria, for example, are photoautotrophs; they use sunlight (photo) for energy and CO_2 (auto) for carbon. Many disease-causing bacteria are chemoheterotrophs, because they use organic molecules from their hosts to acquire both carbon and energy.

Oxygen requirements may also be important in classification. **Obligate aerobes** require O_2 for generating ATP. For **obligate anaerobes,** O_2 is toxic, and they live in habitats that lack it. *Clostridium tetani* (the bacterium that causes tetanus) is one example. **Facultative anaerobes,** which include *E. coli* and *Salmonella*, can live either with or without O_2. (*E. coli* is the subject of this chapter's Focus on Model Organisms: *Escherichia coli*).

These traditional classification criteria remain in widespread use, even though they almost certainly group together organisms that are only distantly related to one another. They remain useful, however, because they are based on characteristics that are relatively easy to observe, unlike molecular sequences. Nevertheless, the importance of molecular data in microbial taxonomy is difficult to overstate. Once microbiologists began to compare ribosomal RNA (rRNA) sequences for different species in the late 1970s, they ended up realigning all organisms

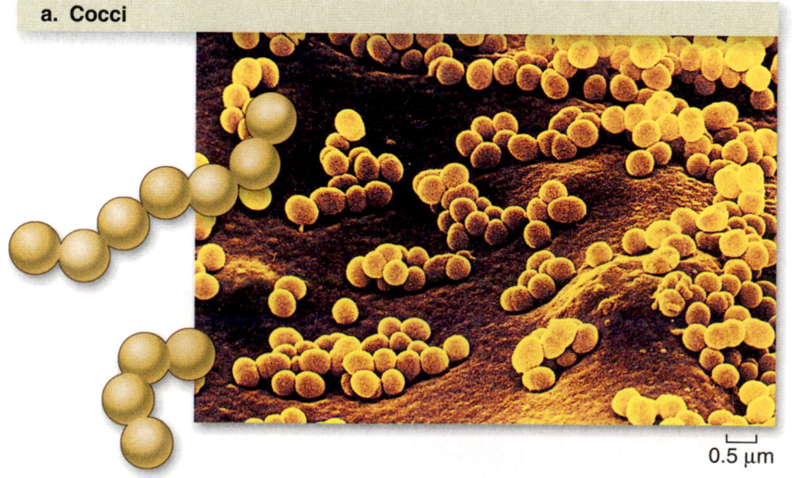

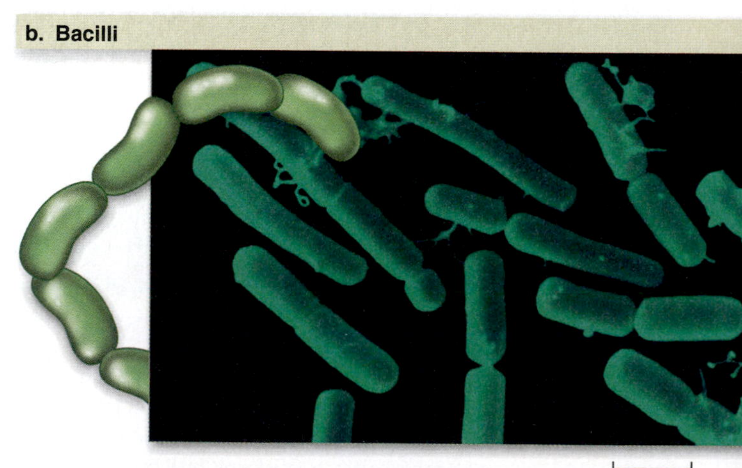

FIGURE 18.7 Cell Shapes. *Micrococcus* (**A**) are spherical (cocci). *Bacillus megaterium* (**B**) are rods (bacilli). *Rhodospirillum rubrum* (**C**) are spiral-shaped (spirilla).

Table 18.2 *Bacteria and Archaea Compared*

	Bacteria	Archaea
Similarities		
Predominantly unicellular	Yes	Yes
Cell size	1–10 μm	1–10 μm
Nucleus and other membrane-bounded organelles	No	No
Circular chromosome	Yes	Yes
Able to grow at temperatures above 80°C	Yes (some)	Yes (some)
Nitrogen fixation?	Yes (some)	Yes (some)
Differences		
Cell wall composition	Peptidoglycan	Pseudopeptidoglycan, protein
Membrane composition	Based on fatty acids	Based on nonfatty acid lipids (isoprenes)
Use chlorophyll in photosynthesis	Yes (some)	No
Generate methane as byproduct of metabolism	No	Yes (some)
Sensitive to streptomycin	Yes	No
Introns	No	Yes

into three domains. The most unexpected discovery was that two of the domains—Archaea and Bacteria—contain only prokaryotes. These two groups of prokaryotes are similar in some ways, but different in many others (**table 18.2**).

B. Domain Bacteria Includes Many Diverse Groups

Scientists have identified at least 23 phyla within domain Bacteria, a few of which are listed in **table 18.3** A phylum called the proteobacteria exemplifies the overall diversity within the domain. Some proteobacteria, including the purple sulfur bacteria, carry out photosynthesis. Others play important roles in nitrogen or sulfur cycling, whereas still others form a medically important group that includes enteric bacteria and vibrios. *Helicobacter*, the bacterium that causes ulcers in humans, is a proteobacterium, as are *E. coli* and *Salmonella*.

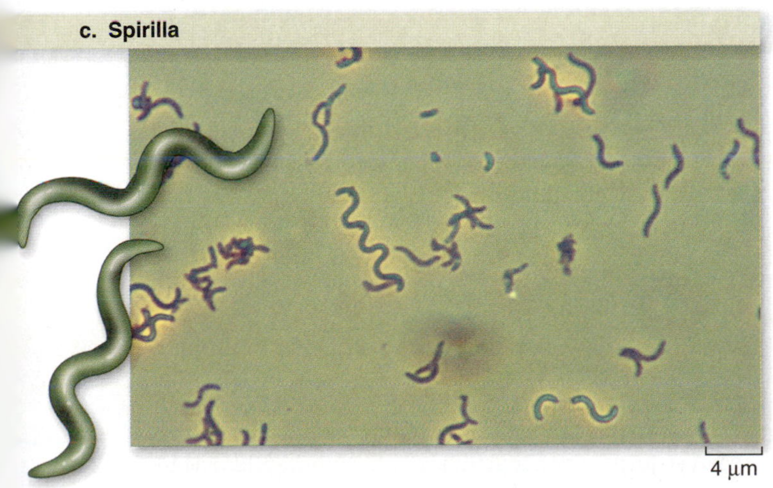

A second phylum consists of the spirochetes, some of which are medically important. These spiral-shaped organisms include *Borrelia burgdorferi* (the cause of Lyme disease) and *Treponema pallidum* (the cause of the sexually transmitted disease, syphilis).

The cyanobacteria form a third phylum. Billions of years ago, these autotrophs were the first to produce O_2 as a byproduct of photosynthesis. They remain important in ecosystems, forming the base of aquatic food chains and participating in symbiotic relationships with fungi and plants on land (see section 18.5).

An informal grouping, the gram-positive bacteria, includes endospore-forming microbes such as *Bacillus anthracis* (the cause of anthrax) and *Clostridium tetani* (the cause of tetanus). Unrelated to these spore-formers are the gram-positive actinobacteria. These filamentous, soil-dwelling bacteria are also medically important, but not because they cause disease. Rather, they are the source of infection-fighting antibiotics, including streptomycin.

C. Many, But Not All, Archaea Are "Extremophiles"

Archaea are often collectively described as "extremophiles" because scientists first found them in habitats that lacked oxygen or that were extremely hot, acidic, or salty (see table 18.1). The newly discovered organisms

Table 18.3 *Selected Groups in Domain Bacteria*

Group	Features	Example(s)
Proteobacteria		
Purple sulfur bacteria	Bacterial photosynthesis using H_2S (not H_2O) as electron donor	*Chromatium vinosum*
Enteric bacteria	Rod-shaped, facultative anaerobes in animal intestinal tracts	*Escherichia coli, Salmonella* species (cause gastrointestinal disease)
Vibrios	Comma-shaped, facultative anaerobes common in aquatic environments	*Vibrio cholerae* (causes cholera)
Spirochetes	Spiral-shaped; some pathogens of animals	*Borrelia burgdorferi* (causes Lyme disease) *Treponema pallidum* (causes syphilis)
Cyanobacteria	Photosynthesis releases O_2; some fix nitrogen; free-living or symbiotic with plants, fungi (lichens), or protists	*Nostoc, Anabaena*
Gram-positive bacteria		
Endospore-forming bacteria	Aerobic or anaerobic; rods or cocci	*Bacillus anthracis* (causes anthrax), *Clostridium tetani* (causes tetanus)
Actinobacteria	Filamentous	*Streptomyces*

were informally divided into three groups: methanogens, thermophiles, and halophiles. As more archaea are discovered in moderate environments such as soil or the open ocean, however, formal taxonomic descriptions become more important.

Microbiologists now tentatively divide domain Archaea into three phyla. One phylum contains archaea that live in stagnant waters and the anaerobic intestinal tracts of many animals, generating large quantities of methane gas. The same phylum also includes halophilic archaea that live in very salty habitats such as seawater, evaporating ponds, and salt flats. A second phylum not only includes thermophiles that thrive in hot springs or at hydrothermal vents on the ocean floor, but also contains a wide variety

FOCUS ON MODEL ORGANISMS: *Escherichia coli*

Of all the model organisms profiled in this unit, *Escherichia coli*, commonly called *E. coli*, is easily the best understood (**figure 18.A**). Dr. Theodor Escherich (1857 to 1911) discovered this normal resident of the human intestinal tract in 1885, and it quickly became a popular lab organism. Biologists have studied every aspect of this bacterium for over a century, and its contributions to biology are enormous. The following are a few highlights:

- **DNA is genetic material:** In 1950, Hershey and Chase used bacteriophage-infected *E. coli* cells to demonstrate that DNA, not protein, is the genetic material. **Hershey and Chase, p. 139**
- **Genetic exchange:** In the 1940s and 1950s, biologists studying *E. coli* discovered conjugation and transduction (see section 18.4). Both phenomena have important implications for bacterial evolution and transgenic technology.
- **DNA replication:** In the late 1950s, Meselson and Stahl used *E. coli* in their famous experiments that demonstrated how DNA copies itself. **DNA replication, p. 143**
- **Gene regulation:** The 1961 description of the *lac* operon in *E. coli* revealed how cells can turn genes on or off, depending on environmental conditions such as the type of sugar present. **operons, p. 244**
- **Transgenic technology:** In 1973, *E. coli* became the first organism to receive a gene from another species. The researchers used the newly discovered restriction enzyme *Eco*R1, derived from *E. coli*, to produce the recombinant plasmids. Since then, *E. coli* cells have received countless genes from many other species.
- **Gene function:** The *E. coli* genome sequence was completed in 1997, and many scientists are working to describe the functions of hundreds of *E. coli* genes. Many genes in other organisms, including humans, will no doubt have similar DNA sequences and functions. Such studies yield new insights into disease and cell function.

FIGURE 18.A ***Escherichia coli.***

of soil and water microorganisms with moderate temperature requirements. Other thermophiles are classified into a third phylum, known mostly from genes extracted from their habitats.

Summary ***Microbiologists classify bacteria and archaea based on molecular sequence data, cell shape, stains, nutrition, physiological characteristics, and habitat.***

18.3 MASTERING CONCEPTS

1. In what ways are bacteria and archaea similar and different?
2. How do microbiologists use cell shapes and different stains to classify prokaryotes?
3. What are the different ways that organisms acquire energy and carbon?
4. What is the difference between an obligate aerobe, an obligate anaerobe, and a facultative anaerobe?
5. What are some examples of phyla within domains Bacteria and Archaea?

18.4 Genetic Diversity in Prokaryotes Has Multiple Sources

Imagine a world in which humans could acquire useful genes from other humans, living or dead. Given our trillion-cell bodies, gene-swapping remains out of our grasp (gene therapy notwithstanding). Not so for prokaryotes. Like all organisms, these simplest of cells can transmit DNA from generation to generation. But prokaryotes can also transfer genes "horizontally"—that is, to cells that are not their offspring.

A. Prokaryotes Reproduce Asexually by Binary Fission

Prokaryotes reproduce by **binary fission,** an asexual process that replicates DNA and distributes it and other cell parts into two daughter cells. Binary fission therefore transfers genes from one generation to the next.

In prokaryotic cells, the chromosome is attached to the inner face of the cell membrane (**figure 18.8**). The DNA replicates as the cell prepares to divide. The cell membrane grows between the two DNA molecules and separates them. Then the cell membrane dips inward, pinching off two daughter cells from the original one. Formation of cell walls completes the binary fission process. In optimum conditions, some bacterial cells can produce a new generation every 20 minutes. This rapid reproduction explains how the oral microbes that survive your nightly tooth-brushing regimen produce countless descendants (and the notoriously unpleasant "morning breath") as you sleep.

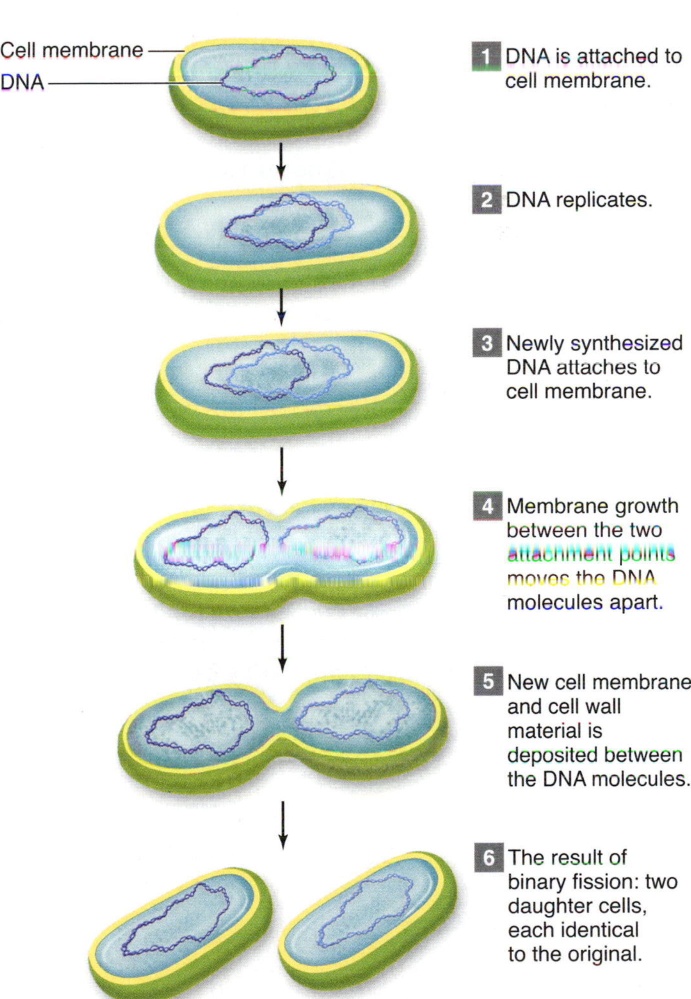
Cell membrane
DNA
1 DNA is attached to cell membrane.
2 DNA replicates.
3 Newly synthesized DNA attaches to cell membrane.
4 Membrane growth between the two attachment points moves the DNA molecules apart.
5 New cell membrane and cell wall material is deposited between the DNA molecules.
6 The result of binary fission: two daughter cells, each identical to the original.

FIGURE 18.8 Prokaryotic Cell Division. In binary fission, the cell grows and then indents as the DNA replicates, separating one cell into two.

B. Horizontal Gene Transfer Occurs in Three Ways

In many asexually reproducing organisms, genetic diversity arises only from random mutations in a cell's DNA, limiting a population's ability to adapt to changing conditions. Prokaryotic populations, however, have additional ways to generate diversity. In **horizontal gene transfer,** a cell receives DNA from another cell that is not its ancestor. It occurs in three ways:

- **Transformation.** In transformation, a prokaryote takes up naked DNA without cell-to-cell contact (**figure 18.9A**). It occurs when prokaryotes die and pieces of their DNA enter other cells.
- **Transduction.** Bacteriophages sometimes mistakenly package host cell DNA along with their own. In transduction, a phage transfers this combined DNA to another bacterial cell (figure 18.9B). Because phages have narrow host ranges, transduction moves DNA only between closely related cells.
- **Conjugation.** In conjugation, a prokaryote receives DNA via direct contact with another cell of a related species (figure 18.9C). In gram-negative bacteria such as *E. coli*, a **sex pilus** is a specialized appendage through which DNA passes from donor to recipient. Conjugation is sometimes called "bacterial sex," but it is not sexual reproduction. Neither the donor nor the recipient cell produces gametes, and no zygote forms.

Horizontal gene transfer has profound implications in fields as diverse as origin of life research, medicine, biodiversity, evolution, systematics, and biotechnology. Both transduction and conjugation, for example, spread antibiotic-resistance genes among bacteria, a serious and growing public health problem. Yet at the same time, biotechnologists take advantage of horizontal gene transfer when they send new genes into bacteria in recombinant DNA technology. **antibiotic resistance, p. 263**

Summary *Prokaryotes transfer genes from generation to generation in binary fission. They can also transfer DNA to cells other than their offspring by three mechanisms: transformation, transduction, and conjugation.*

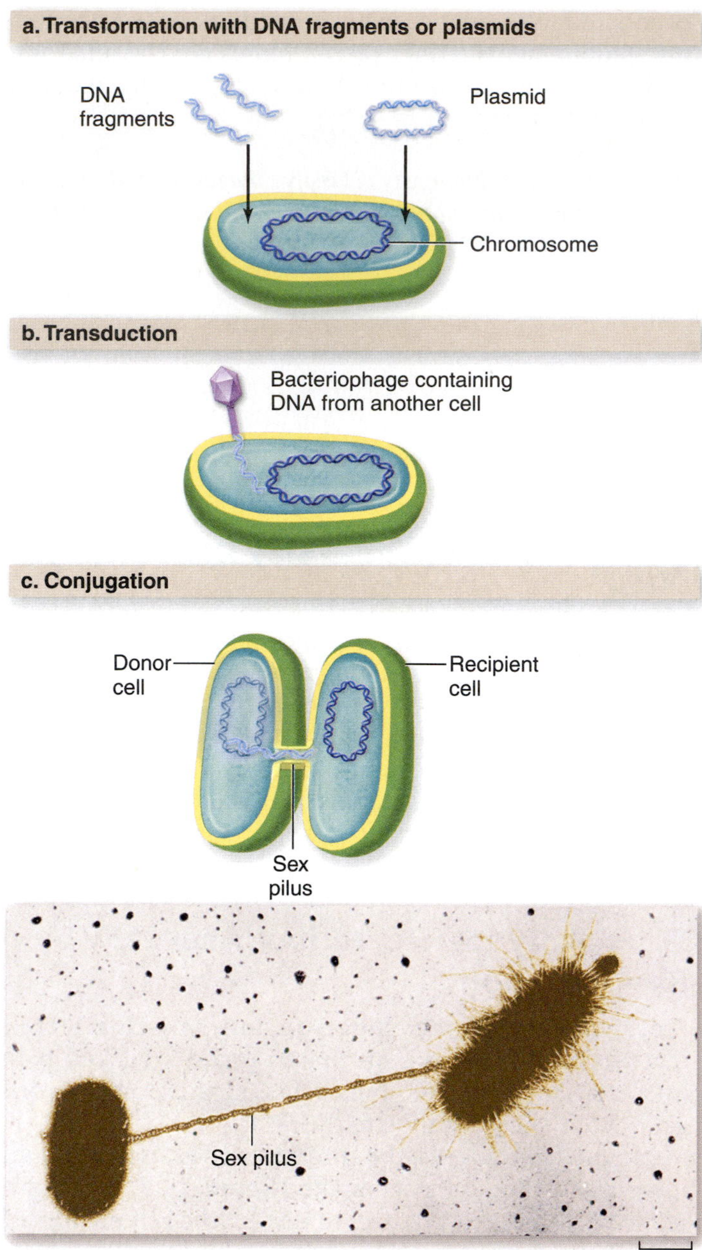

FIGURE 18.9 Horizontal Gene Transfer. (**A**) In transformation, a cell picks up DNA fragments or plasmids released from another cell. (**B**) In transduction, a bacteriophage (virus) picks up DNA from one cell and transfers it to another. (**C**) In conjugation, a copy of some of one cell's DNA moves into another via a cytoplasm bridge (sex pilus).

18.4 MASTERING CONCEPTS

1. What are the events of binary fission?
2. What are the differences among transformation, transduction, and conjugation?
3. How is horizontal gene transfer important?

18.5 Prokaryotes Are Essential to Human Life

Many people think of prokaryotes as organisms that cause disease in humans. However, the number of prokaryotes in all animals combined probably accounts for less than 1% of the total prokaryotes on Earth. Many of the others play vital roles in global nutrient cycles by decomposing organic matter in soil and water, carrying out photosynthesis, and feeding countless organisms in every imaginable habitat.

All other species would die without bacteria and archaea. For example, **nitrogen fixation** is a process in which prokaryotes convert atmospheric nitrogen gas (N_2) to ammonia (NH_3), which other organisms can absorb and incorporate into amino acids and other organic molecules. Some nitrogen-fixing bacteria live in soil or water. Others, such as those in the genus *Rhizobium*, induce the formation of nodules in the roots of clover and some other host plants (**figure 18.10**). Inside the nodules, *Rhizobium* cells share the nitrogen that they fix with their hosts; in exchange, the bacteria receive nutrients and protection from their hosts.

Although many prokaryote species are essential to life, this section focuses on the two groups of which we are most aware: the bacterial species that are pathogenic (cause disease) and the organisms that we harness for their products and processes.

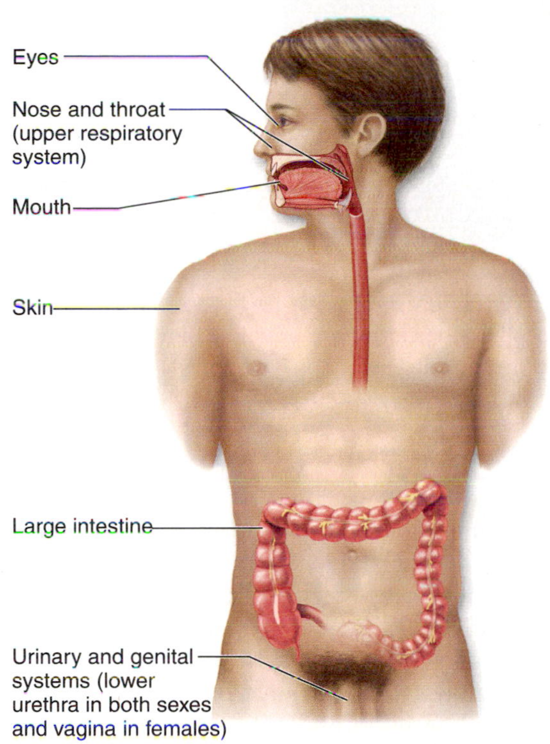

FIGURE 18.11 Your Microscopic Companions. Many parts of the body house thriving populations of microorganisms. The blood, nervous system, stomach, urinary bladder, and other areas of the body, however, normally remain sterile.

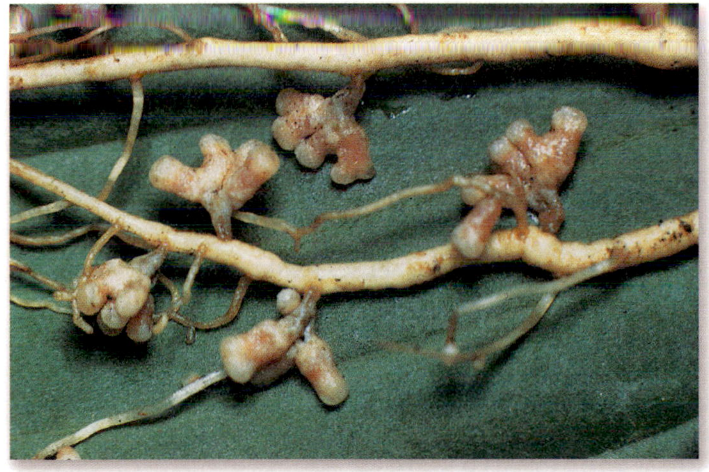

a.

b.

FIGURE 18.10 Nitrogen-Fixing Bacteria. (**A**) *Rhizobium* bacteria infect these sweet clover roots, producing root nodules where nitrogen fixation occurs. (**B**) Cross section of a root nodule, showing bacteria inside the plant's cells.

a.

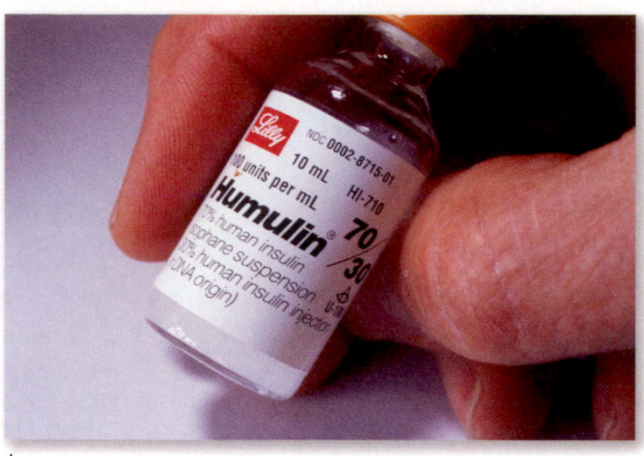

b.

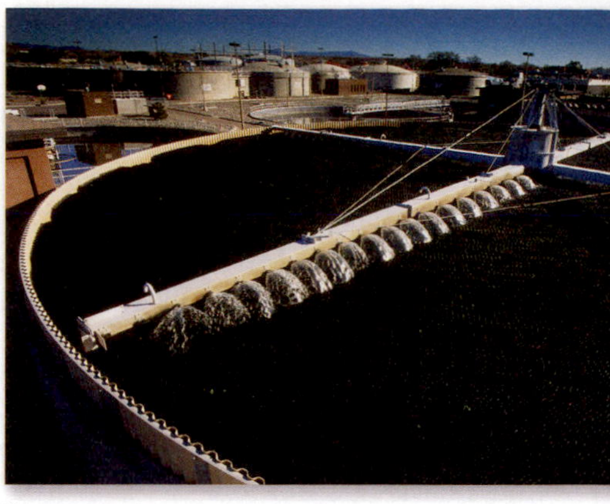

c.

FIGURE 18.12 Bacteria at Work. (**A**) Bacteria of genus *Lactococcus* are used to manufacture cheddar cheese from fermenting milk. (**B**) Genetically modified bacteria produce many drugs, including human insulin. (**C**) Raw sewage is sprayed on a trickling filter at a municipal waste water treatment plant. Biofilms on the rocks degrade the organic matter in the sewage.

A. Some Bacteria Cause Disease

No matter how hard you scrub, it is impossible to escape the fact that you are a habitat for bacteria. A menagerie of prokaryotes lives on your skin and in your mouth, large intestine, urogenital tract, and upper respiratory tract (**figure 18.11**). As long as your immune system remains healthy, you should welcome these companions because they help keep pathogens out. If you take antibiotics to fight an infection, you may also kill off some of these beneficial residents. As they die, harmful ones take their place. The resulting microbial imbalance causes unpleasant side effects like diarrhea or a vaginal yeast infection.

Although most prokaryotes inside the human body are harmless, some bacteria cause disease. (So far, no archaea are known to cause disease in humans.) To cause infection, bacteria must enter the body. Insect and tick bites, sexual activity, ingestion of food or water, and inhalation all can transmit bacteria; the cells can also enter through cuts or skin abrasions. Once inside the host, adaptations such as pili enable the pathogens to attach to host cells, then the bacteria produce enzymes that break down host tissues. As the host's immune system reacts to an infection, the characteristic signs and symptoms of a bacterial infection develop: fever, swollen lymph nodes, pain, and nausea, among others.

Some bacteria produce toxins that harm the host's circulatory, digestive, or nervous system. These poisonous molecules can act at or near the site of infection, as in the case of the harmful *E. coli* strain, O157:H7. Alternatively, a person may ingest the toxins in food. *Staphylococcus aureus* and *Bacillus cereus* are two bacteria that thrive in improperly handled food. Their toxins—not infection with the bacteria themselves—produce the vomiting and diarrhea associated with food poisoning.

One component of a pathogen's success is its ability to spread from host to host. Just as bacteria enter the body through multiple portals, they also exit in many ways: in respiratory droplets, feces, vaginal discharge, or semen, for example. Blood-feeding animals such as ticks and mosquitoes also carry bacteria between hosts.

B. Humans Put Many Prokaryotes to Work

Humans have used prokaryotes to help make food for centuries, and scientists continue to invent new applications. Bacteria are essential in the production of vinegar, sauerkraut, pickles, olives, yogurt, and cheese (**figure 18.12A**). They also help produce enormous quantities of vitamins and useful chemicals such as ethanol and the solvent acetone. Transgenic bacteria mass-produce human proteins such as insulin and blood-clotting factors (figure 18.12B).

In addition, heat-, acid-, and salt-tolerant enzymes isolated from microbes from extreme environments have many industrial applications. Dishwashing and laundry detergents contain enzymes from these sources. A heat-tolerant bacterial enzyme revolutionized modern biology by improving the efficiency of PCR, the polymerase chain reaction (see section 7.4).

Water and waste treatment also use prokaryotes. Sewage treatment plants in most communities, for example, harbor countless aerobic and anaerobic microbes that degrade organic wastes (figure 18.12C). Bioremediation uses microorganisms that can metabolize and detoxify pollutants such as polychlorinated biphenyls (PCBs) and heavy metals such as cadmium, lead, copper, zinc, silver, and mercury.

Summary *Because of their diverse ecological roles, prokaryotes are essential to eukaryotic life. Few species of prokaryotes cause disease; those that do can enter our bodies and invade our tissues in many ways. Many industries rely on the metabolic diversity of prokaryotes.*

18.5 MASTERING CONCEPTS

1. In what ways are prokaryotes essential to eukaryotic life?
2. How are the microbes that colonize your body beneficial?
3. What adaptations enable pathogenic bacteria to enter the body and cause disease?
4. What are some industrial uses of prokaryotes?

18.6 INVESTIGATING LIFE

A Romp Through the *Staphylococcus aureus* Genome Solves Two Mysteries

DNA has astonishing power as a tool for detecting evolutionary change. Genetic material from organisms that lived hundreds of thousands of years ago can provide clues about life in the distant past (see chapter 7's Investigating Life: Genetic Messages from the Dead Tell Tales of Ancient Ecosystems). The same molecule can also teach us about events that occurred within the past 30 years.

Consider, for example, a bacterium called *Staphylococcus aureus*. Most people harbor this common gram-positive organism without becoming ill. If the bacteria breach the skin's defenses or enter the bloodstream, however, they can cause everything from painful boils to blood infection and death. Toxic shock syndrome, scalded skin syndrome, and damage to the heart's inner lining are a few of the potentially life-threatening consequences of a "staph" infection.

Because *S. aureus* cells are invisible to the unaided eye, it is easy to forget to take precautions against their spread. And spread they do. Athletes in contact sports such as football and wrestling transmit the bacteria directly to each other. Even athletes who do not touch their opponents can spread *S. aureus* by sharing unwashed equipment. The bacteria also linger on the linens and possessions of prison inmates, and they are especially common in college dormitories and hospitals. Medical personnel and contaminated equipment transmit the germs between patients, who often have weakened immune systems and are therefore vulnerable to infection.

The many possible symptoms of a staph infection stem from *S. aureus*'s genetic diversity. Each strain has a different set of genes and therefore produces a unique set of toxins and other weapons. Researcher James Musser and his colleagues at the National Institute of Allergy and Infectious Diseases wanted to learn more about the relationship between genetic variation, pathogenicity, and evolution in *S. aureus*. His team studied 36 strains representing the range of known genetic variability in *S. aureus* isolated in the United States, Europe, Canada, and Japan.

The design of their study was simple. The researchers constructed identical microarrays loaded with short DNA sequences from a reference strain of *S. aureus* whose genome was sequenced in 2001 (see section 7.5). They washed the microarrays with DNA extracted from each of the test strains. If a bacterium's nucleotide sequence matched a spot on the microarray, its DNA stuck there; if not, it simply washed away. The more matches, the more genetically similar each test strain was to the reference. Moreover, by comparing regions on the microarray where the DNA matched a test train, the researchers could deduce how similar the strains were to one another.

Musser and his team were especially interested in the origin of *S. aureus* strains that resist treatment with antibiotics. One such group is called methicillin-resistant *S. aureus*, abbreviated MRSA. Medical professionals first noticed MRSA in 1961, shortly after the introduction of the antibiotic methicillin. The bacteria earned widespread attention in the 1990s when they began causing frequent infections in hospital patients (**figure 18.13**). These bacteria have gradually acquired resistance to additional antibiotics, such as penicillin, cephalosporins, and amoxicillin.

Scientists have offered two competing explanations for the origin of MRSA. One hypothesis is that MRSA evolved many times through horizontal transfer of the gene that confers antibiotic resistance. An alternative hypothesis is that the resistance gene arose by random mutation only once in a formerly susceptible cell. The descendants of that cell then acquired additional variability over many generations, eventually giving rise to multiple resistant strains. The researchers tested both hypotheses by assembling their microarray data into a treelike diagram showing clusters of the most closely related organisms (**figure 18.14**).

Musser's study of 35 *S. aureus* strains included 12 MRSA strains. If a single lineage (tree "branch") had contained all of the resistant strains, it would have supported the hypothesis that all MRSA originated from a single ancestor. Yet that

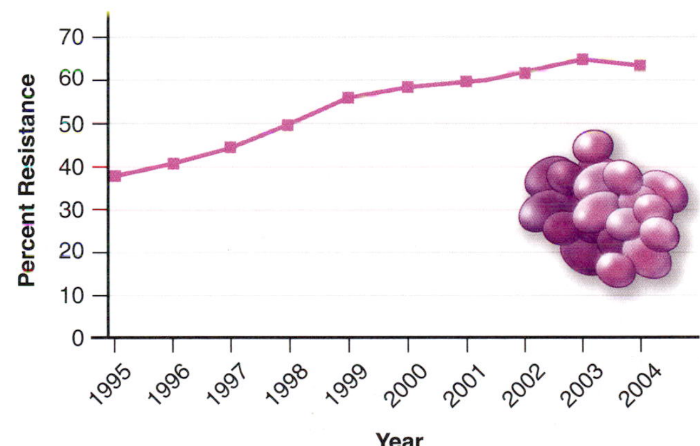

FIGURE 18.13 MRSA on the Rise. The percent of *S. aureus* (isolated from patients in intensive care) that were classified as MRSA strains increased from less than 40% to nearly 70% between 1995 and 2004. **Question:** If you worked for a school confronting an outbreak of *S. aureus*, how would you determine whether the strains were MRSA? What measures would you recommend to control the outbreak?

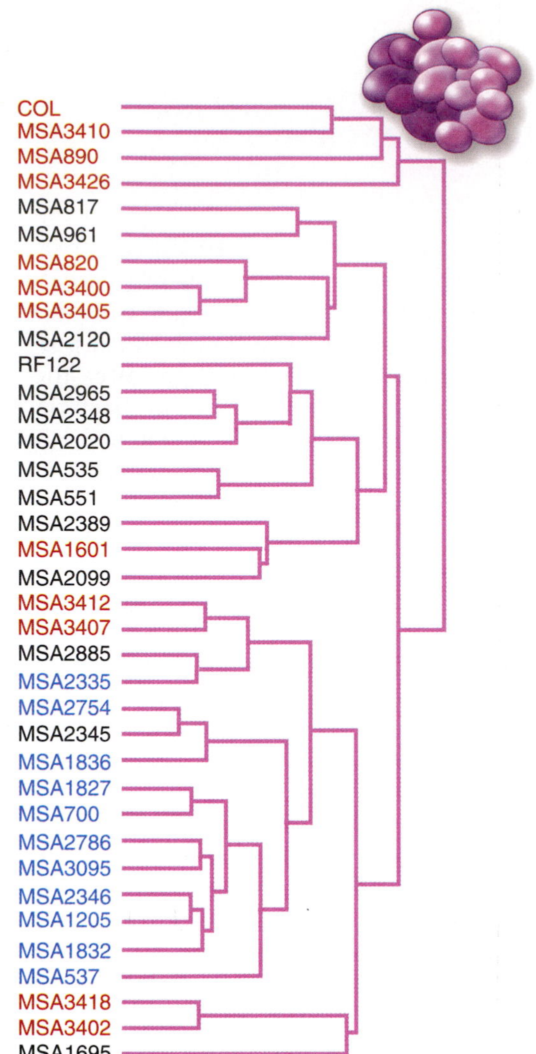

FIGURE 18.14 ***Staphylococcus aureus* Relationships.** This tree shows the genomic relationships among 35 strains of *S. aureus*. Strains in red are MRSA; strains in blue were isolated from patients with toxic shock syndrome.

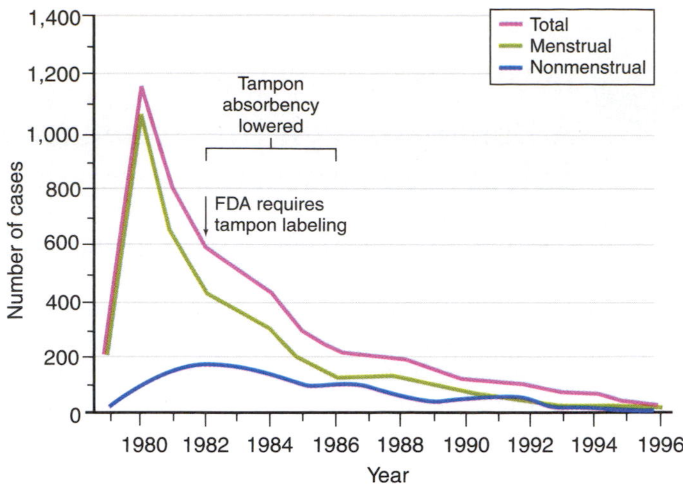

FIGURE 18.15 **Toxic Shock Syndrome Epidemic.** The use of ultra high-absorbency tampons was linked to a spike in the number of toxic shock syndrome cases in 1980. Better education and standardized tampon absorbency ratings soon caused the number of cases to decline.

is not what the researchers observed. Instead, the resistant strains were scattered throughout the tree diagram, supporting the multiple-origin hypothesis. Apparently, the antibiotic resistance gene has moved many times into previously susceptible cells, conferring resistance anew each time.

A related, hotly debated question is the origin of the toxic shock syndrome epidemic of 1980 (**figure 18.15**). Most cases were linked with the use of high-absorbency tampons among previously healthy young women. When technicians cultured bacteria from the vaginas of the ill women, they found *S. aureus* 90% of the time. Studies of the pathogens' proteins suggested that the bacteria were closely related.

Why the sudden epidemic in 1980? Epidemiologists have proposed two hypotheses. One possibility is that the increasing use of ultra high-absorbency tampons at that time created a new type of environment, and that many similar strains of *S. aureus* independently exploited this change. An alternative explanation is that just one cell mutated and became a highly successful "superpathogen," whose descendants swept across multiple continents.

The researchers' tree diagram again led to the answer (see figure 18.14). The 11 strains that came from toxic shock syndrome patients were not genetically identical, as would be expected if all were simply clones of the same superpathogen. The team therefore rejected the second hypothesis in favor of the first, concluding that multiple *S. aureus* strains together caused the toxic shock syndrome epidemic.

DNA evidence is perhaps most famous for its role in solving crimes, but it is equally valuable to the "detectives" who study infectious disease. Musser's team has revealed the origins of different pathogenic strains of *S. aureus*. Understanding how bacteria have evolved in the recent past gives medical professionals a head start in anticipating future epidemics. Given the enormous adaptability of bacterial populations, we probably will need all the help we can get.

Fitzgerald, J. Ross, Daniel E. Sturdevant, Stacy M. Mackie, et al. July 17, 2001. Evolutionary genomics of *Staphylococcus aureus*: Insights into the origin of methicillin-resistant strains and the toxic shock syndrome epidemic. *Proceedings of the National Academy of Sciences*, vol. 98, pages 8821–8826.

CHAPTER SUMMARY

18.1 Prokaryotes Are a Biological Success Story

- **Prokaryotes** (domains **Bacteria** and **Archaea**) are important in many ways. They were the first organisms, and they may have given rise to mitochondria and chloroplasts in eukaryotic cells more than a billion years ago.
- Prokaryotes are very abundant and diverse, and they occupy a great variety of habitats.

18.2 Prokaryotic Cells Are Structurally Simple

A. Internal Structures Include a Chromosome, Plasmids, and Ribosomes

- Like other organisms, prokaryotic cells contain DNA and **ribosomes,** and they are bounded by a cell membrane.
- The chromosome, along with some RNA and proteins, is located in an area called the **nucleoid. Plasmids** are circles of DNA in addition to the chromosome.

B. External Structures Aid in Protection, Attachment, and Movement

- A **cell wall** surrounds most prokaryotic cell membranes. **Gram staining** reveals differences in cell wall architecture. Gram-positive bacteria have a thick **peptidoglycan** layer; gram-negative bacteria have an outer membrane surrounding the thinner peptidoglycan cell wall.
- A **glycocalyx** outside the cell wall provides attachment to surfaces or protection from host immune system cells.
- **Pili** are projections that allow cells to adhere to surfaces or transfer DNA to other cells.
- **Flagella** spin to provide **taxis**, which is movement toward or away from a stimulus.

C. Endospores Help Some Bacteria Survive Heat and Drying

- Some bacteria survive harsh conditions by forming protective **endospores.**

18.3 Prokaryotes Include Two Domains with Enormous Diversity

A. Prokaryote Classification Traditionally Relies on Observable Features

- Prokaryotes are spherical **cocci**, rod-shaped **bacilli,** spiral-shaped **spirilla**, or variations of these. Stains can reveal the cell wall structure or the presence of other cell structures important in classification.
- **Autotrophs** acquire carbon from inorganic sources, and **heterotrophs** obtain carbon from other organisms. A **phototroph** derives energy from the sun, and a **chemotroph** acquires energy by oxidizing organic or inorganic chemicals.
- **Obligate aerobes** require oxygen, **facultative anaerobes** can live whether or not oxygen is present, and **obligate anaerobes** cannot function in the presence of oxygen.
- Microbiologists are using molecular data to reconsider traditional classification of prokaryotes.

B. Domain Bacteria Includes Many Diverse Groups

- Proteobacteria, spirochetes, cyanobacteria, and gram-positive bacteria are a few examples of diversity within domain Bacteria.

C. Many, But Not All, Archaea Are "Extremophiles"

- Domain Archaea contains methanogens, halophiles, and thermophiles, but also contains organisms that thrive in more moderate conditions.

18.4 Genetic Diversity in Prokaryotes Has Multiple Sources

A. Prokaryotes Reproduce Asexually by Binary Fission

- **Binary fission** is division of a prokaryotic cell, yielding two daughter cells. DNA is therefore transferred from one generation to the next.

B. Horizontal Gene Transfer Occurs in Three Ways

- Prokaryotes can also acquire new DNA directly from the environment (**transformation**), a virus (**transduction**), or from another cell via a **sex pilus** (**conjugation**). All three are routes of **horizontal gene transfer**.

18.5 Prokaryotes Are Essential to Human Life

- All life depends on the prokaryotes that contribute gases to the atmosphere, recycle organic matter, and **fix nitrogen.**

A. Some Bacteria Cause Disease

- Pathogenic bacteria adhere to host cells and colonize tissues. Disease symptoms often result from the immune system's reaction to the infection.
- Toxins produced by bacteria, plus the ability to spread to new hosts, also contribute to disease.

B. Humans Put Many Prokaryotes to Work

- Prokaryotes are used in the manufacture of many foods, drugs, and other chemicals, in sewage treatment, and for cleaning the environment.

18.6 Investigating Life: A Romp Through the *Staphylococcus aureus* Genome Solves Two Mysteries

- Molecular studies revealed that *Staphylococcus aureus* strains that resist multiple antibiotics, and those that caused the toxic shock syndrome epidemic of 1980, arose independently several times.

MULTIPLE CHOICE QUESTIONS

1. A *prokaryotic* cell is one that
   a. lacks DNA.
   b. has membrane-bounded organelles.
   c. lacks a nucleus.
   d. lacks a plasma membrane.
2. What is the cell wall of a Gram negative cell made of?
   a. Pectin
   b. Peptidoglycan
   c. Glycocalyx
   d. Plasmids
3. The external structure that attaches the prokaryote to its environment is a
   a. pilus.
   b. flagellum.
   c. plasmid.
   d. endospore.
4. Which of these is a distinguishing characteristic between the domains Bacteria and Archea?
   a. Their size
   b. Their use of chlorophyll for photosynthesis
   c. Their ability to grow at high temperatures
   d. The presence of membrane-bounded organelles
5. Which property of an archaean cell accounts for its resistance to the antibiotic streptomycin?
   a. The composition of its plasma membrane
   b. Its ability to live in extreme environments
   c. The structure of its ribosomes
   d. The presence of introns within its DNA
6. What name is used to describe prokaryotic cells that are rod-shaped?
   a. Spirilla
   b. Bacilli
   c. Cocci
   d. Both a and b
7. Which form of prokaryote would be the most challenging to isolate and culture?
   a. A facultative anaerobe
   b. A photoautotroph
   c. An obligate aerobe
   d. An obligate anaerobe
8. Which of the following is the best explanation of why binary fission can occur without a spindle structure like that found in mitotic cells?
   a. The cell is small, so there is less material to divide.
   b. There is only one chromosome, and it attaches to the membrane.
   c. The prokaryotic DNA does not need to replicate.
   d. The DNA is transferred through the sex pilus.
9. Which of the following is the bacterial equivalent to sex?
   a. Conjugation
   b. Transformation
   c. Transduction
   d. Transfection
10. What property of a prokaryote makes it useful for bioremediation?
    a. The ability to reproduce quickly
    b. The ability to live in extreme environments
    c. The diversity of metabolic pathways found in various prokaryotes
    d. All of the above

TESTING YOUR KNOWLEDGE

1. What structures do prokaryotic cells have that do not occur in eukaryotic cells?
2. Distinguish between the following pairs of terms: (a) a phototroph and a chemotroph; (b) bacteria and archaea; (c) a halophile and a methanogen; (d) a gram-negative and a gram-positive bacterium; (e) an autotroph and a heterotroph; (f) an obligate anaerobe and a facultative anaerobe; and (g) transformation and transduction.

3. What are the criteria that microbiologists use in classifying prokaryotes?
4. How do prokaryotes reproduce, and what are three ways they can acquire genes other than by inheritance?
5. Give five examples that illustrate how bacteria and archaea are important to other types of organisms.
6. What adaptations in pathogenic bacteria enable them to cause disease?
7. How have humans harnessed the metabolic diversity of prokaryotes for military and industrial purposes?

THINKING AS A SCIENTIST

1. How can the polymerase chain reaction (PCR; see chapter 7) be useful in studying prokaryotes that do not survive in laboratory culture?
2. A young child develops a very high fever and an extremely painful sore throat. Knowing that the child could have an infection with a strain of *Streptococcus* that could be deadly, the physician seeks a very specific diagnosis. What three approaches might the doctor (or a laboratory) use to tell whether this infection is viral or bacterial and, if the latter, identify the bacterium?
3. Scientists are studying bacteria discovered on the sea floor. In the absence of light, these bacteria can use organic molecules (including human waste) to generate electricity. NASA hopes they may one day be useful in producing some of the electricity on spacecraft. What mode of nutrition do these bacteria probably use?
4. Stomach ulcers, once thought to be entirely a product of spicy food or high stress, are now known to be caused by bacteria (*Helicobacter pylori*). How has ulcer treatment changed because of this new knowledge?
5. Researchers can raise mice in the complete absence of microorganisms. When they subsequently expose the animals to pathogenic microbes, they are more likely to develop disease than are mice that have been raised in a normal (nonsterile) environment. Explain this finding.
6. Coral animals that live off the coast of Florida have developed an infection that researchers are calling a "plague." What steps should they take to identify a causative microorganism?
7. Genome sequencing projects are complete or in progress for many prokaryotes other than *E. coli*. Choose one of these bacteria or archaea, and describe some new discoveries that have come from research on this organism.
8. "Botox" is a toxin produced by the bacterium *Clostridium botulinum*. When ingested with tainted food, it can kill by paralyzing muscles needed for breathing and heartbeat. Physicians inject small quantities of diluted botox into facial muscles to paralyze them and reduce the appearance of wrinkles. Some people have expressed concern about a trend in which people come together for "botox parties" at hair salons and other nonmedical settings. What are the risks of getting injections in such a setting?

Visit www.mhhe.com/hoefnagels for practice quizzes, animations, videos, and activities designed to help you master the material in this chapter.

Protista

***Pfiesteria* Fish Lesions.**
Toxins produced by the protist *Pfiesteria* cause gaping sores on fish.

What's the Point?

Science Lessons from a Killer Cell

Of the many thousands of species of protists, the vast majority are harmless to animals. One exception, however, is Pfiesteria piscicida*, which lives in the waters off the east coast of the United States.* Pfiesteria *belongs to a group of protists called dinoflagellates. Typically,* Pfiesteria *is nontoxic, feeding on algae and bacteria in coastal waters. But a few of the stages in its life cycle can produce extremely potent toxins that kill fish and accumulate in shellfish, making them poisonous to humans.*

Pfiesteria *produces at least two toxins. One is a powerful neurotoxin, and the other causes the skin of the fish to disintegrate. Although the toxin breaks down rapidly, the infected fish often develop open sores. Many fish die. In humans,* Pfiesteria *toxins can produce rashes, open sores, fatigue, erratic heartbeat, breathing difficulty, personality changes, and extreme memory loss.*

The on-again, off-again nature of the toxins has made them very difficult for scientists to study. Controversy erupted in 2002, when scientists reported that Pfiesteria shumwayae *did not produce them. In making their case, the researchers used several experimental strategies. They exposed fish to fluid in which* P. shumwayae *had grown (after first removing the cells). They attempted to extract toxic chemicals from water in which* P. shumwayae *grew. They searched the microbe's genome for genes known to encode toxins in other organisms. Finally, they grew fish and* P. shumwayae *cells in separate halves of two-part tanks, divided by a membrane through which toxins, but not cells, could pass. None of the experiments yielded evidence of toxins.*

Later in 2002, however, other researchers refuted these results. They argued that the first group raised the Pfiesteria *cells under conditions that suppressed toxin production (the cells must grow with live fish to produce toxins). This second group used the same form of* P. shumwayae *as the first group, but they also included two control forms: one known to make toxins, and the other known to be nontoxic. Their research showed that, under some conditions,* P. shumwayae *did indeed produce toxins that harmed fish.*

These conflicting results illustrate two important features of science. First, the conditions under which scientists conduct their experiments can greatly affect their conclusions. Second, communication is vital. When the first group published their results, they described their Pfiesteria *forms and experimental techniques in detail. The second group could then use this information to identify conditions that they knew would suppress toxin production. The careful reporting of methodology enables scientists to evaluate each other's work.*

LEARNING OUTLINE

19.1 Kingdom Protista Lies at the Crossroads Between Simple and Complex Organisms

What do green algae, kelp, slime molds, and *Paramecium* have in common? They all belong to kingdom **Protista,** an extremely diverse collection of organisms that share just one feature: they are **eukaryotes**, so their cells possess nuclei and other membrane-bounded organelles (see table 3.3).

The protists represent a landmark in the evolution of life on Earth. Eukaryotes evolved from prokaryotes some 2 BYA and eventually gave rise to many diverse descendants, including plants, fungi, and animals (**figure 19.1**). Unlike the other kingdoms, however, biologists do not define kingdom Protista by a unique set of unifying features. Rather, protists are defined by *exclusion:* they are eukaryotes that do not belong in any other kingdom. Not surprisingly, the nearly 100,000 named species of protists are the most metabolically diverse of the four eukaryotic kingdoms, displaying great variety in size, nutrition, locomotion, reproduction, and cell surfaces.

Protists are important ecologically, medically, and industrially. Algae photosynthesize, producing much of the O_2 in Earth's atmosphere. Free-living protists in oceans, lakes, rivers, and ponds form the basis of most aquatic food webs. Some algae live with fungi or even in the fur of sloths. Parasitic protists infect plant and animal hosts (including humans), costing billions of dollars, incalculable suffering, and millions of deaths annually. Protists also make paints reflective, point the way to petroleum reserves, and help make chocolate smooth and creamy. **food webs, p. 812**

Protists are especially interesting to evolutionary biologists. First, protists are ancestors of plants, fungi, and animals; studying protists therefore sheds light on the evolutionary his-

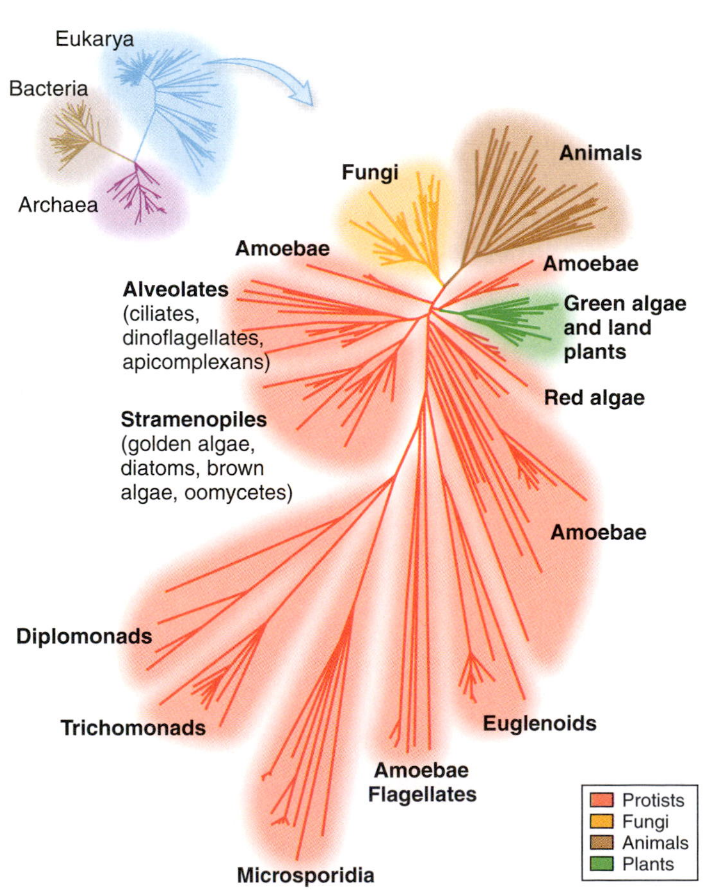

FIGURE 19.1 Kingdom Protista at the Crossroads. Protista is by far the most diverse of the four eukaryotic kingdoms. Plants, fungi, and animals trace their ancestry to protists, living or extinct.

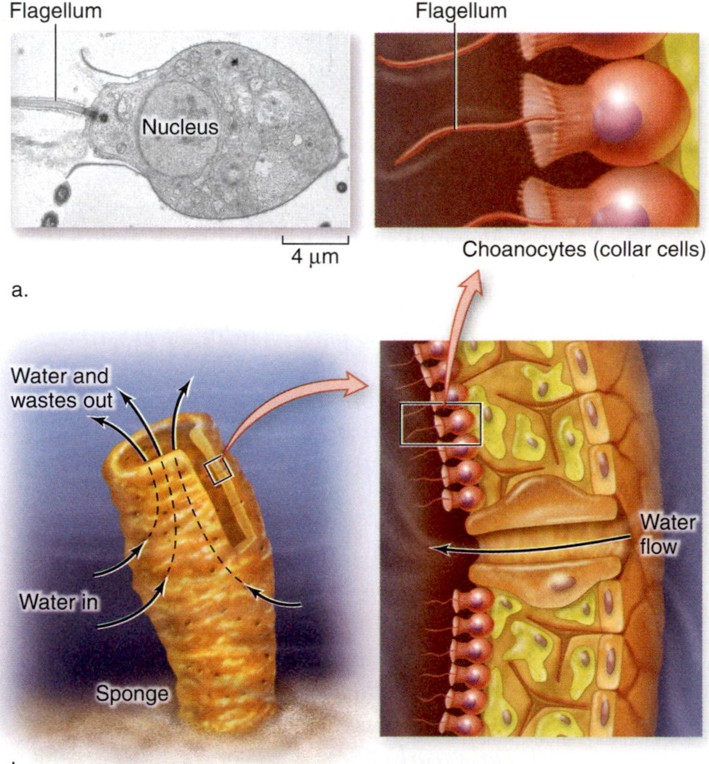

FIGURE 19.2 Are Choanoflagellates the Ancestors of Sponges? Choanoflagellates (**A**) are protists that greatly resemble choanocytes (collar cells) in sponges (**B**). The photo in (**A**) depicts a member of genus *Salpingoeca.* The organism's flagellum and tentacles from the collar are partly visible near the left side. Movement of the flagellum produces water currents that carry food particles into the cell.

tory of these important groups of organisms. Second, current protists may retain clues to important milestones in eukaryote history. Some protists lack mitochondria, whereas others have mitochondria with genomes that are very much like those of bacteria. These organisms may provide glimpses into what cells were like around the time that eukaryotes first acquired organelles by endosymbiosis (see figure 16.10).

At the other end of the evolutionary spectrum are the **choanoflagellates,** protists that may be similar to the forerunners of animals. These organisms bear an uncanny resemblance to "collar cells" in sponges, which are the simplest animals (**figure 19.2**). (Their name originates from the Greek word for collar, *choanos*.) Because some choanoflagellates form colonies in which individual organisms seem to interact to move and obtain food, they might represent a step toward multicellularity. **origin of multicellularity, p. 338; sponges, p. 454**

Textbooks and taxonomists have traditionally considered the protists in terms of the more familiar organisms that they resemble: the plantlike algae, funguslike slime molds, and animal-like protozoa. These designations, however, do not accurately depict evolutionary relationships (see figure 19.1). Biologists are currently reconsidering the classifications within kingdom Protista to reflect nucleic acid sequences, which provide the most objective measure of relatedness. Some classification systems propose splitting the protists into as many as 20 kingdoms.

Because the classification of protists is in transition and many of the new groupings are not universally accepted, this chapter uses the traditional approach to protist classification. The examples for each section represent just a small sampling of some of its most intriguing members. Section 19.5 describes modern trends in protist classification.

Summary ***Kingdom Protista includes the simplest of the eukaryotes. They form a diverse group of organisms that vary widely in size, metabolism, cell structures, and reproductive strategies. The classification of the protists remains in flux.***

19.1 MASTERING CONCEPTS

1. What features define kingdom Protista?
2. For what reasons are evolutionary biologists interested in protists?
3. Where do protists live?
4. Give examples of how protists are important, both economically and ecologically.

19.2 Many Protists Are Photosynthetic

Most people probably think of algae as pond scum, but **algae** is a general term that refers to any photosynthetic protist that lives in water. Algae range in size from microscopic to 30 meters in length, and their photosynthetic pigments are yellow, gold, brown, red, and green.

A. Dinoflagellates Are "Whirling Cells"

The marine protists known as **dinoflagellates** are characterized by two flagella of different lengths (**figure 19.3A, B**). One of the flagella beats in a way that propels the cell with a whirling motion (the Greek *dinein* means "to whirl"). In

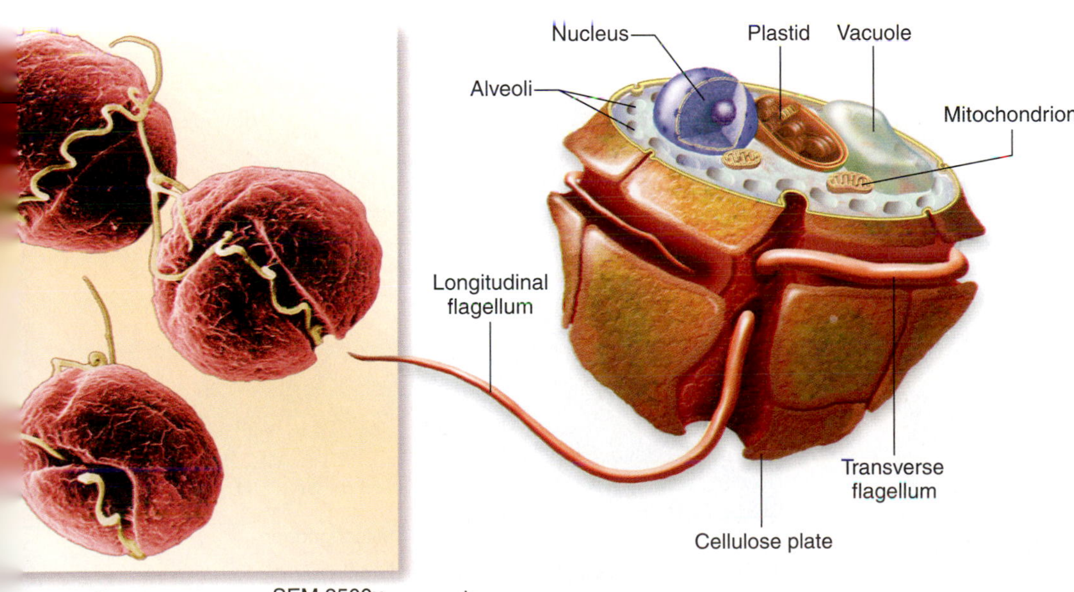

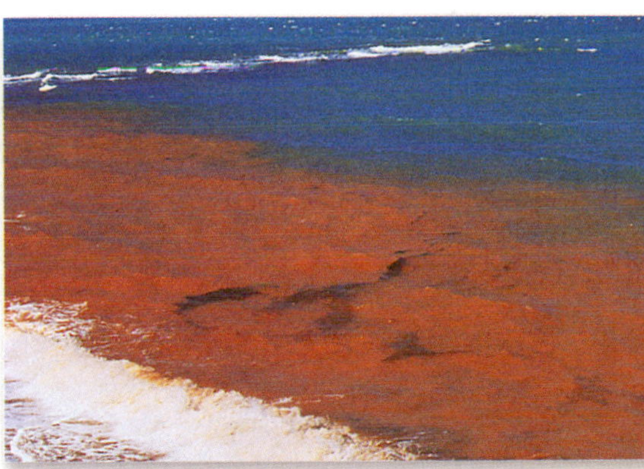

FIGURE 19.3 Dinoflagellate Features. (**A** and **B**) *Gymnodinium* sp. exhibits classic dinoflagellate structure. Note the characteristic transverse and longitudinal flagella. (**C**) Huge populations of dinoflagellates cause red tides.

addition, many dinoflagellates have cell walls that consist of overlapping cellulose plates.

These photosynthetic organisms are a major component of plankton, the microscopic food that supports the ocean's vast food webs. Several species live within the tissues of jellyfishes, corals, sea anemones, or giant clams, providing carbohydrates to their host animals. Other species of dinoflagellates are predators or parasites. Some are bioluminescent, producing flashing lights in tropical waters. **cnidarians, p. 455; mollusks, p. 458**

A red tide results from a sudden population explosion, or "bloom," of dinoflagellates that turn the water red, orange, or brown (figure 19.3C). Usually these blooms occur after the nutrient content of the water increases (see this chapter's Burning Question: Why and how do algae form?). Toxins from dinoflagellates may become concentrated in the tissues of clams, scallops, oysters, and mussels. If a person eats tainted shellfish, the toxins cause the numb mouth, lips, face, and limbs of paralytic shellfish poisoning.

Burning Questions

Why and how do algae form?

Algae are common aquatic organisms, but they are often inconspicuous. Sometimes, however, their populations grow so large that they seem to take over; ponds and poorly maintained swimming pools can turn bright green with algal overgrowth (**figure 19.A**). This population explosion, also called an algal bloom, occurs whenever nutrients and sunlight are abundant.

Algal blooms are normal in some ecosystems, such as in many ponds. A bloom where water is normally clear, however, usually indicates that nutrients from sewage, fertilizer, or animal waste are polluting the waterway. The use of lawn fertilizers is a common cause of algal blooms in ponds in residential settings. **eutrophication, p. 834**

Have a Burning Question of your own? Submit it to marielle_hoefnagels@mcgraw-hill.com for possible inclusion in future editions of this book!

FIGURE 19.A Algae Bloom.

B. Euglenoids Are Heterotrophs and Autotrophs

The **euglenoids** are unicellular flagellates with elongated cells that primarily inhabit fresh water. Most have a long whiplike flagellum used in locomotion and a short flagellum that does not extend from the cell. Supporting the cell membrane is a pellicle, a protective layer made of rigid or elastic protein strips.

About one third of the species are photosynthetic, and the rest feed on organic compounds suspended in the water. But these metabolic roles are not always fixed. Photosynthetic euglenoids such as *Euglena* (**figure 19.4**) may occasionally feed on organic matter, and those placed in darkness become entirely heterotrophic until light returns.

C. Golden Algae Contain Carotenoid Pigments

The **golden algae** are named for their color, which results from the presence of yellowish carotenoid accessory pigments. The cells usually have two flagella. Most are unicellular, but filamentous and colonial forms do exist. *Dinobryon* is an example of a freshwater genus of golden algae, in which individual vase-shaped cells stack end to end to produce branched or unbranched chains.

Golden algae are a significant source of food for zooplankton in freshwater and marine ecosystems. When light or nutrient supplies dwindle, however, many golden algae can consume bacteria or diatoms.

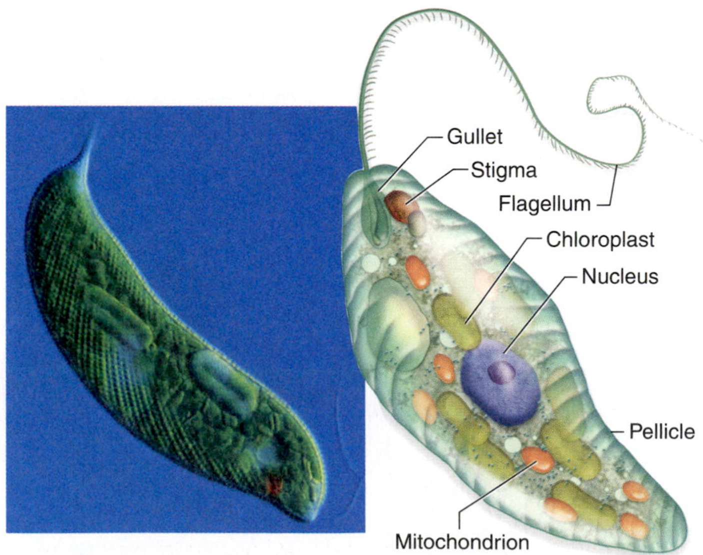

FIGURE 19.4 ***Euglena.*** The pond-dwelling *Euglena* has a flagellum and chloroplasts, but it can also ingest food particles. A typical *Euglena* cell is about 50 μm long.

FIGURE 19.5 Diatoms. The silica walls of these photosynthetic protists exhibit a dazzling variety of forms.

FIGURE 19.6 A Red Alga. *Bossiella* is a coralline red alga that secretes calcium carbonate and helps build coral reefs.

D. Diatoms Have Silica Cell Walls

Diatoms are primarily unicellular algae with two-part silica walls that confer a variety of ornate shapes (**figure 19.5**). Their cells contain photosynthetic pigments that give them a yellow or brown color.

Most diatoms live in oceans. Through the eons, the glassy shells of the diatoms have accumulated on the ocean floor, forming deposits over 200 meters thick in some regions. The abrasive shells make diatoms useful ingredients in swimming pool filters, polishes, toothpaste, and many other products. Diatoms also impart the distinct quality of paint used in reflective roadway signs and license plates.

Few microorganisms fossilize as well as diatoms, and biologists know of some 35,000 extinct species. Section 19.6 describes how silica shells have left a complete record of the formation of a new species in a Wyoming lake.

E. Red Algae Contain Unique Pigments

The mostly marine **red algae** are somewhat similar to green plants in that they store carbohydrates as a modified form of starch, have cell walls containing cellulose, and produce chlorophyll *a*. Red algae also have reddish and bluish photosynthetic pigments that allow the algae to live at depths that may exceed 200 meters, using wavelengths of light that chlorophyll cannot capture (**figure 19.6**).

Humans use red algae in many ways. One polysaccharide in the cell wall of some species is agar. This jellylike substance is used as a culture medium for microorganisms, an inert ingredient in medications, a gel in canned meats, and a thickener in ice cream and yogurt. Another useful product is carrageenan, a polysaccharide that emulsifies fats in chocolate bars and stabilizes paints, cosmetics, and creamy foods. A red alga called nori is used for wrapping sushi.

F. Brown Algae Are Large and Complex

The **brown algae** are the most complex and largest protists. They owe their distinctive olive green to brown color to an accessory pigment called fucoxanthin. Brown algae reproduce by means of swimming spores, each with two flagella.

These seaweeds live along marine shorelines all over the world. The **kelps,** which are the largest of the brown algae, can exceed 30 meters in length. They produce enormous underwater forests that provide food and habitat for many animals. The distinctive body form of a kelp consists of a holdfast organ, a stemlike region, a balloonlike area called a bladder, and a flattened leaflike blade that may extend from each bladder (**figure 19.7**). The Sargasso Sea in the northern Atlantic Ocean is named after floating masses of brown algae called *Sargassum*.

FIGURE 19.7 A Giant Kelp. A holdfast organ anchors brown algae to surfaces when tides and currents are strong. The bladder is a gas-filled float that enables the upper parts of the organism to rise above the holdfast.

Humans consume several species of kelp. *Laminaria digitata*, for example, is an ingredient in many Asian dishes. Algin, a chemical extracted from the cell walls of brown algae, is used as an emulsifying, thickening, and stabilizing agent in many products including ice cream, candies, chocolate, salad dressings, sauces, soft drinks, beer, cough syrup, toothpaste, cosmetics, polishes, latex paint, and paper.

G. Green Algae Are Ancestors of Plants

The **green algae** are the most plantlike of the protists. They use chlorophyll *a* and *b* as photosynthetic pigments, use starch as a storage carbohydrate, and have cell walls containing cellulose. Like plants, many green algae also have **alternation of generations,** with alternating multicellular haploid (gametophyte) and diploid (sporophyte) phases in their life cycles (**figure 19.8**).

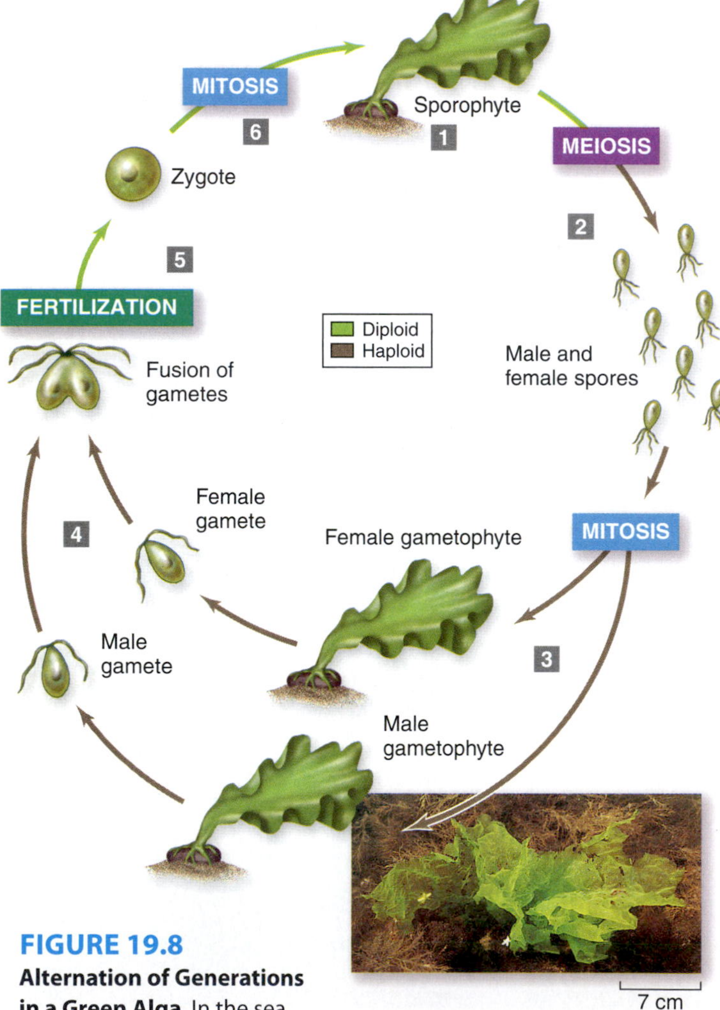

FIGURE 19.8 **Alternation of Generations in a Green Alga.** In the sea lettuce *Ulva*, (*1*) a diploid sporophyte undergoes meiosis (*2*) and yields haploid reproductive male and female spores. The spores divide mitotically, forming male and female gametophytes (*3*) that look like each other (and like the sporophyte). The gametophytes produce male and female gametes (*4*), which fuse (*5*), yielding a diploid zygote (*6*) that grows by mitotic cell division and matures to form a sporophyte. The cycle begins anew.

The habitats and body forms of green algae are diverse (**figure 19.9**). Most live in fresh water or in moist habitats on land, although some live in symbiotic relationships with fungi, forming lichens. Green algae range in size from the smallest eukaryote (*Micromonas*), only 1 μm in diameter, to sea lettuce (*Ulva*) exceeding 1 meter in length. Green algae may be

FIGURE 19.9 **Diversity of Green Algae.** Green algae have a variety of body forms, from solitary microscopic cells to complex multicellular forms. Clockwise from top: *Spirogyra, Volvox, Codium, Acetabularia, Chlamydomonas, Hydrodictyon.*

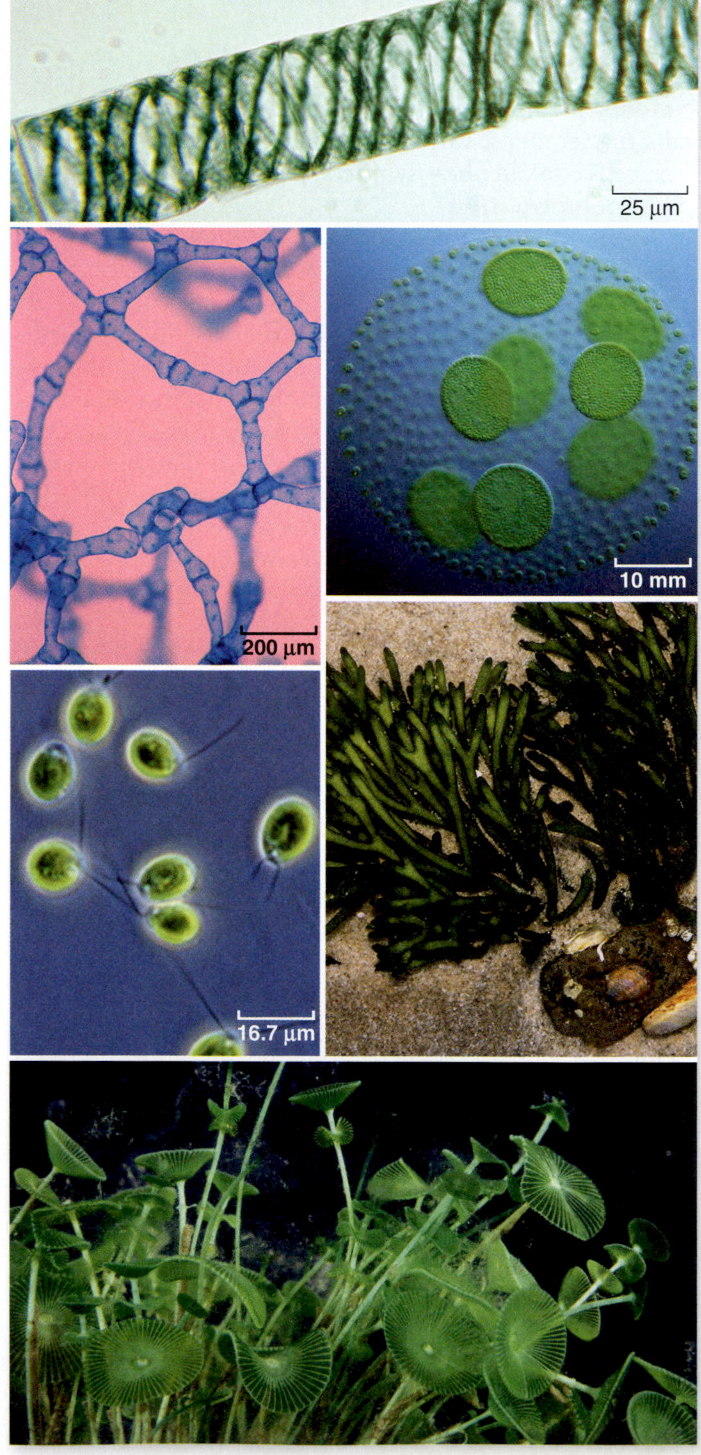

unicellular, filamentous, colonial, or multicellular. The multicellular species may have rootlike and stemlike structures, but they are far less specialized than plants. **lichens, p. 441**

One well-studied green alga is *Chlamydomonas,* a unicellular organism that reproduces asexually and sexually. Scientists study these algae to learn about the evolution of sex, how an individual's sex is determined, and how cells of opposite sexes recognize each other. A classroom favorite is the colonial green alga, *Volvox*. Hundreds to thousands of *Volvox* cells form hollow balls; the cells move their flagella in coordinated waves to move the sphere. New colonies remain within the parental ball of cells until they burst free.

Other green algae include the geometrically shaped desmids, striking water nets (*Hydrodictyon*), mermaid's wineglass (*Acetabularia*), the ribbonlike *Spirogyra*, and the tubular *Codium*. One species, *Chlorella,* is being considered as a food and oxygen source on prolonged space flights. An advantage of *Chlorella* over plants is that green algae can multiply very quickly, as evidenced by the rapid "greening" of a poorly maintained aquarium or swimming pool.

Summary ***Although diverse in form, most algae are unicellular, aquatic, photosynthetic eukaryotes. All use a form of chlorophyll in photosynthesis, but they also use a variety of accessory pigments.***

19.2 MASTERING CONCEPTS

1. Describe several criteria for classifying the algae.
2. List and describe the seven phyla of algae.
3. How do the cells of diatoms differ from those of other algae?
4. How does a red tide occur, and why are red tides important?
5. What are some useful products made from algae?

19.3 Some Heterotrophic Protists Were Once Classified as Fungi

Slime molds and water molds are protists that resemble fungi in some ways: they are not photosynthetic, and some produce filamentous feeding structures similar to those in fungi. They also commonly occur alongside fungi in many habitats. Nevertheless, molecular evidence clearly indicates that they are not fungi. In addition, their cell walls lack chitin, a characteristic of the cell walls of fungi.

A. Slime Molds Are Unicellular and Multicellular

The slime molds are informally divided into two groups whose relationship to each other remains unclear. Both types live in damp habitats such as forest floors. In addition, each type of organism exists as single, amoeboid cells and as large masses that behave as one multicellular organism. The major difference between the two types is reflected in their names: plasmodial and cellular slime molds.

Plasmodial Slime Molds

The feeding stage of a **plasmodial slime mold** consists of a **plasmodium,** a mass of thousands of diploid nuclei enclosed by a single cell membrane. (This structure gives these organisms their other common name, the "acellular" slime molds.) The plasmodium may be a conspicuous, slimy, bright yellow or orange mass up to 25 cm in diameter. It migrates along the forest floor, engulfing bacteria and other microorganisms on leaves, debris, and rotting logs.

In times of drought or food shortages, the plasmodium halts and forms stalks topped with reproductive structures (sporangia) that produce resistant haploid spores. When favorable conditions return, the spores germinate and form haploid cells. Two of these cells may fuse, forming a diploid zygote nucleus that divides repeatedly by mitosis, forming a new multinucleate plasmodium. Scientists use the plasmodial slime mold, *Physarum polycephalum*, to study mitosis and cytoplasmic streaming (**figure 19.10**). **mitosis, p. 161**

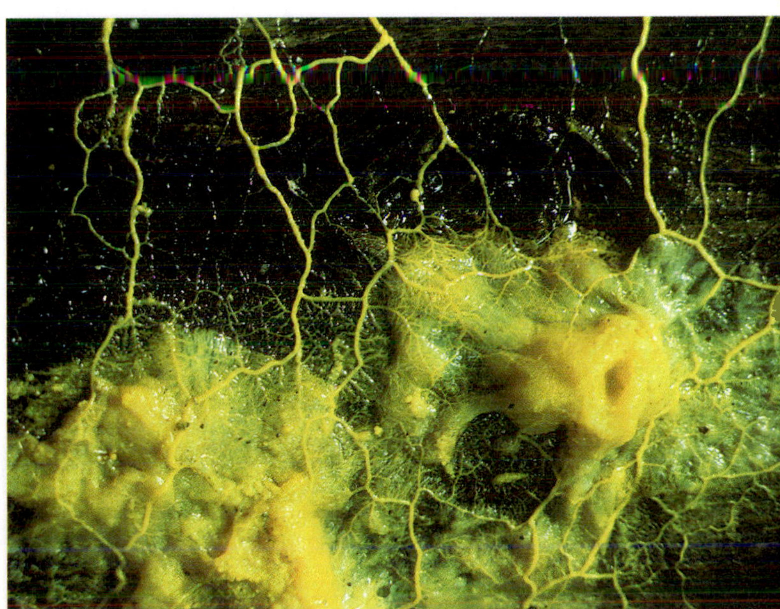

FIGURE 19.10 Plasmodial Slime Mold. Streaming masses of the plasmodial slime mold *Physarum* commonly move across dead logs and leaf litter.

Cellular Slime Molds

In contrast to the plasmodial slime molds, individual cells of a **cellular slime mold** retain their membranes throughout the life cycle. The cells exist as haploid amoebae, engulfing bacteria and other microorganisms in fresh water, moist soil, and decaying vegetation.

When food becomes scarce, the amoebae secrete chemical attractants, which stimulate the neighboring amoebae to aggregate into a sluglike structure (**figure 19.11**). The "slug" moves toward light, stops, and forms a stalk topped by a fruiting body that produces haploid spores. The cells of the stalk perish, but the spores survive; water, soil animals, or birds spread them to new habitats. The spores then germinate to form haploid amoebae, and the cycle begins anew. Sexual reproduction also occurs in some conditions. *Dictyostelium discoideum* is a cellular slime mold used in many scientific studies (see this chapter's Focus on Model Organisms: *Dictyostelium discoideum*).

B. Water Molds Are Decomposers and Parasites

The **water molds,** or oomycetes, are decomposers or parasites of plants and animals in moist environments. The filaments of water molds secrete digestive enzymes into their surroundings and absorb the nutrients. In addition, like fungi called chytrids, water molds produce swimming spores that aid their dispersal in water and wet soil. Despite the similarities, however, water molds are unlike fungi in many ways. The filaments of water molds are diploid, for example, whereas fungal filaments are haploid. Fungi have cell walls containing chitin, but water mold cell walls contain cellulose.

The best known water molds are those that ruin crops, causing such diseases as downy mildew of grapes and lettuce. In the 1870s, downy mildew of grapes, caused by *Plasmopara viticola,* nearly destroyed the French wine industry. The water mold *Phytophthora infestans,* which means "plant destroyer," causes late blight of potato (**figure 19.12A, B**). This disease

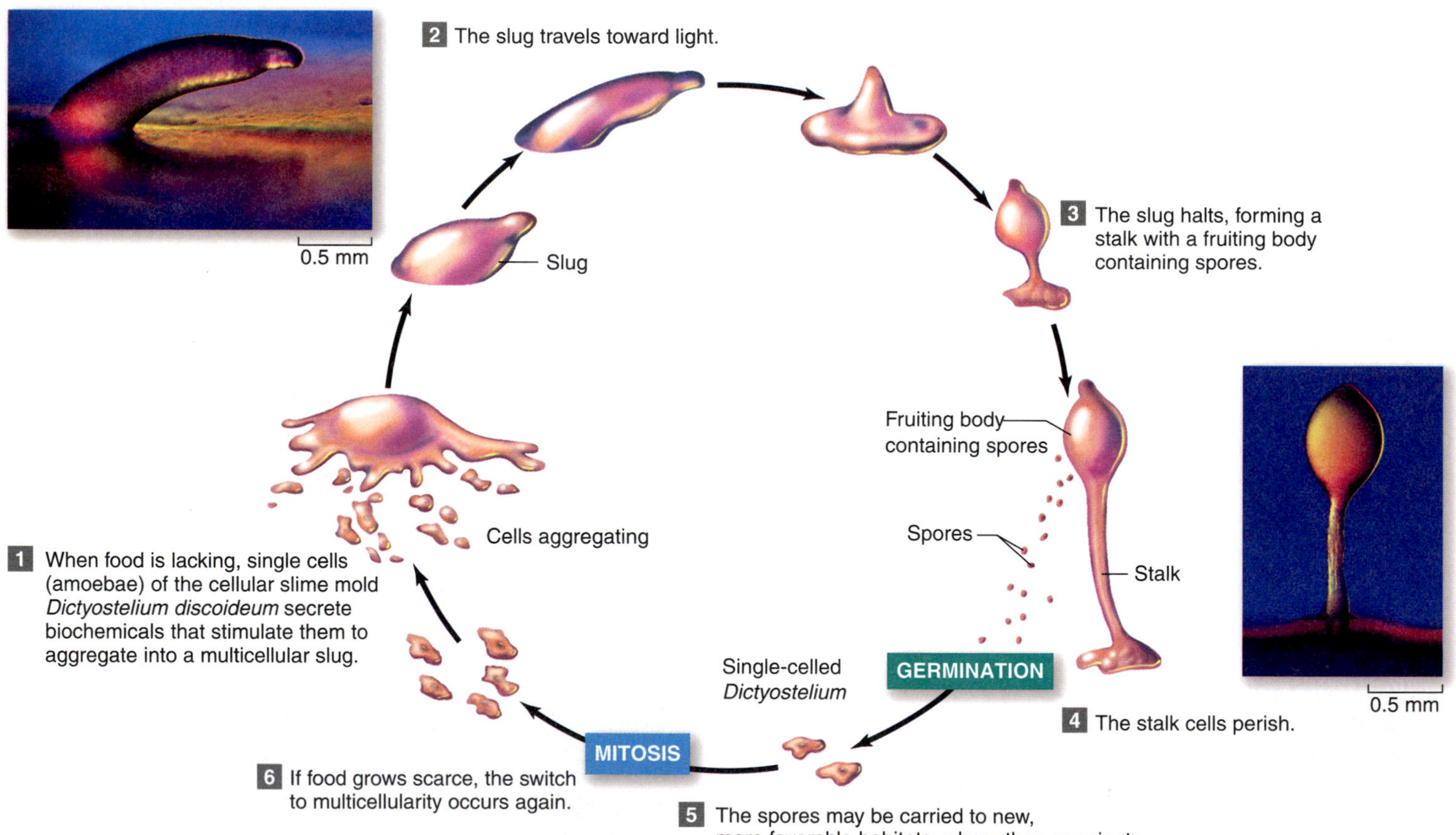

FIGURE 19.11 Life Cycle of the Cellular Slime Mold, *Dictyostelium discoideum*. The amoeboid cells of this species feed on bacteria. Starvation stimulates the cells to aggregate into a multicellular "slug," which crawls to a new habitat. Only the asexual portion of the life cycle is shown, although a sexual phase may also occur.

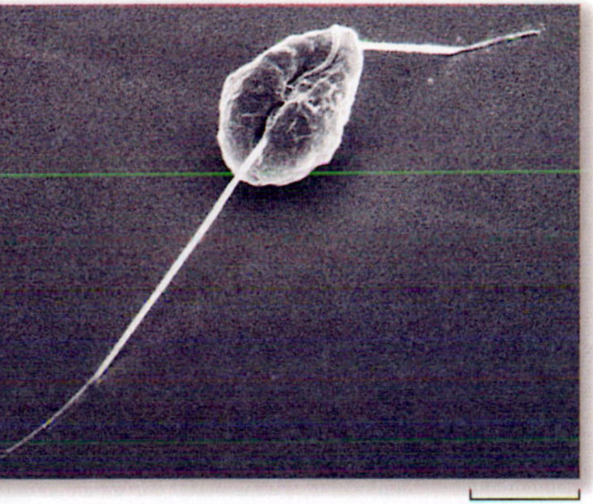

FIGURE 19.12 Oomycetes. (**A**) *Phytophthora infestans* is a water mold that has biflagellated zoospores. (**B**) It causes "late blight" of potatoes and was responsible for the Irish potato famine in the mid 1840s. (**C**) *Saprolegnia* infects a dead insect.

caused the devastating Irish potato famine from 1845 to 1847, during which more than a million people starved and millions more emigrated from Ireland. The Irish potato famine followed several rainy seasons, which fostered the rapid and devastating spread of the plant disease. A newly discovered relative of *P. infestans*, called *Phytophthora ramorum*, causes a tree disease called sudden oak death. Another well-known water mold is *Saprolegnia*, a common protist that forms cottony masses on fishes and other aquatic organisms (figure 19.12C).

Summary *Slime molds and water molds were once erroneously considered fungi, but they are not closely related. Slime molds engulf small cells, whereas water molds decay organic matter or parasitize plants and animals.*

19.3 MASTERING CONCEPTS

1. In what ways do slime molds and water molds resemble fungi?
2. How are the plasmodial and cellular slime molds different and similar?
3. What has been the role of water molds in the environment and history?

FOCUS ON MODEL ORGANISMS: *Dictyostelium discoideum*

Slime molds have such an unappealing name that it may seem hard to imagine why anyone would study them. But *Dictyostelium discoideum* is an unusual organism, one that straddles the boundary between the unicellular and the multicellular. Its feeding phase consists of individual amoeba-like cells that move independently, feeding on bacteria by phagocytosis. When the food runs out, cells begin to aggregate into a multicelled structure that migrates toward light. The cells differentiate into a base, stalk, and spores; only the spores survive to colonize a new habitat.

Dictyostelium discoideum is useful as a model because, like other model organisms, it is easy to grow in the laboratory and has a short generation time. In addition, its cells are readily accessible to microscopy and genetic studies. As a result, *D. discoideum* (affectionately called "Dicty" by its researchers) remains a useful and fascinating organism. Discoveries by researchers working on *D. discoideum* include:

- **Cell movement.** A *Dictyostelium* cell eats by producing extensions that engulf and absorb food particles by phagocytosis. Scientists have discovered that this movement is possible because proteins such as actin and myosin move rapidly within the cell. These same proteins produce muscle movement in animals. **muscle movement, p. 642**
- **Cytokinesis.** Researchers observing cell division in *D. discoideum* have discovered that myosin is also required for cytokinesis (the physical division of one cell into two). **cytokinesis, p. 162**
- **Chemotaxis.** Starving Dicty cells move toward each other and form a multicellular "slug." This movement toward a chemical stimulus, called chemotaxis, requires cell membrane proteins that detect the signal and transmit the information to the inside of the cell. Similar systems occur in many other organisms. **signal transduction, p. 56**
- **Cell differentiation.** When individual Dicty cells come together, chemical signals presumably determine which cells will become stalk cells (and die) and which will become spore cells (and survive). Researchers have discovered a sterol-like compound that induces the differentiation of stalk cells. Such research may help answer questions about the origin of multicellularity. **sterols, p. 37; multicellularity, p. 338**

19.4 Protozoa Are Diverse Heterotrophic Protists

Finding a list of characteristics that unite the diverse **protozoa** is difficult. Most are unicellular, and the vast majority are heterotrophs, but several autotrophic species exist. They move by flagella, cilia, or pseudopodia. Some are free-living, and others are obligate parasites. Most are asexual, but sexual reproduction occurs in many species.

This chapter describes four groups of unrelated protozoa that are defined by locomotion and morphology. New molecular techniques are redefining the protozoa, but until the newer system of classification is better defined and more widely accepted, these groups remain practical for general biology, education, and medicine.

A. Amoeboid Protozoa Produce Pseudopodia

The **amoeboid protozoa** produce cytoplasmic extensions known as pseudopodia (Latin, meaning "false feet"), which are important in locomotion and capturing food via phagocytosis. The most studied species is *Amoeba proteus,* a common freshwater microbe that eats by engulfing bacteria, algae, and other protists within its pseudopodia (**figure 19.13**). The human digestive tract may be invaded by another species, *Entamoeba histolytica,* which can cause amoebic dysentery. **phagocytosis, p. 94**

The **foraminiferans,** or forams, are an ancient group of mostly marine amoeboid protozoa with complex, brilliantly colored shells made primarily of calcium carbonate (**figure 19.14A**). Their populations are immense: about one third of the ocean floor is made of the shells of the marine foram, *Globigerina.* The White Cliffs of Dover, among other limestone and chalk deposits, are made largely of the tests (shells) of forams and other marine organisms. Paleontologists studying extinct forams have learned which species correlate with oil and gas deposits. The tests are also useful in dating rock strata.

The **radiolarians** are among the oldest protozoa. They are planktonic organisms with intricate tests made of silica (figure 19.14B); pseudopodia extend through holes in the shells. "Radiolarian ooze" is sediment consisting of large numbers of their tests. On the ocean floor, radiolarian ooze can be as thick as 4000 m.

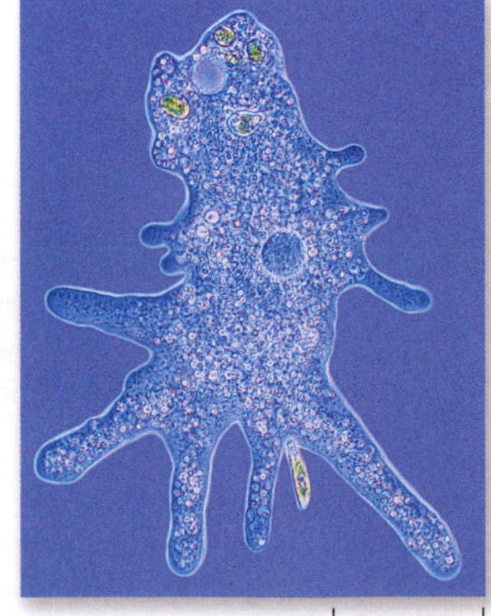

FIGURE 19.13 "False Feet." Pseudopodia are temporary projections from the cells of amoeboid organisms. A pseudopod enables the organism to move or take in food. This organism, *Amoeba proteus,* is consuming *Euglena,* another protist (small green cell at bottom center).

FIGURE 19.14 Foraminiferans and Radiolarians. (**A**) Foraminiferans such as *Globigerina* form vast amounts of ocean sediment. The White Cliffs of Dover are composed of mostly foraminiferan tests. (**B**) Radiolarians have intricate silicon-based shells.

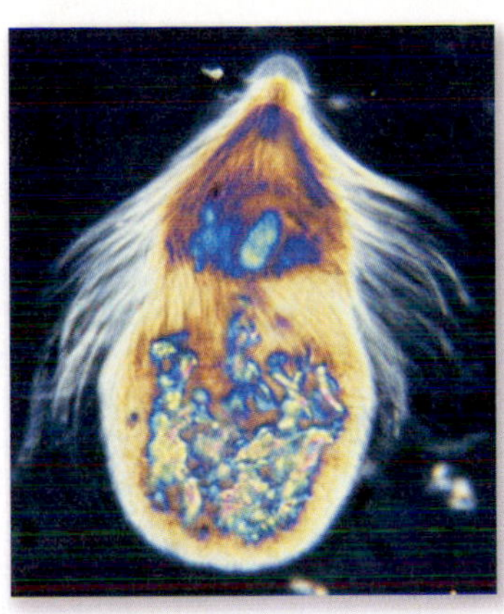

FIGURE 19.15 Life Inside a Termite. A termite cannot digest the cellulose in its wood meal without symbiotic protists, such as this *Trichonympha,* which, in turn, harbors symbiotic bacteria. Exposing termites to high oxygen or high temperature kills their protist symbionts. The insects soon die, with guts full of undigested wood.

B. Several Flagellated Protozoa Cause Disease

The **flagellated protozoa** include photosynthetic species such as the euglenoids and dinoflagellates described in section 19.2. We now turn to some of the heterotrophic species.

Although most flagellated protozoa are free-living in fresh water, the ocean, and soil, a few live inside animals. One example is *Trichonympha,* a large protist that lives in the intestines of termites (**figure 19.15**). The cells of *Trichonympha,* in turn, harbor bacteria that digest cellulose. This bacterium-within-protist living organization enables termites to "digest" wood.

Some parasitic flagellated protozoa cause disease. For example, *Trichomonas vaginalis* resides in the urogenital tracts of both men and women (**figure 19.16**). It is sexually transmitted and causes vaginitis in females. *Giardia intestinalis* (also known as *Giardia lamblia*) causes "hiker's diarrhea," or giardiasis (**figure 19.17**). People ingest the cysts of the organism in contaminated water. As *Giardia* cells divide in the small intestine, they impair the host's ability to absorb nutrients, resulting in diarrhea and cramping.

FIGURE 19.16 ***Trichomonas vaginalis.*** This organism causes the sexually transmitted disease trichomoniasis.

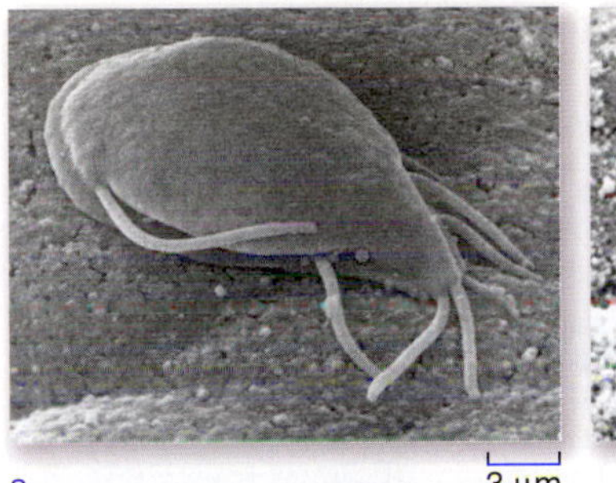

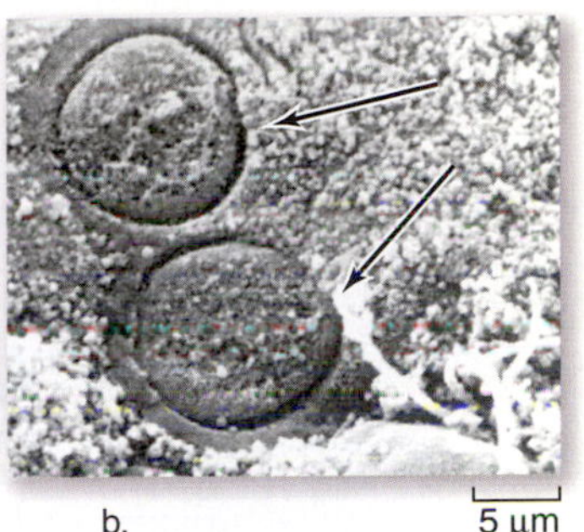

FIGURE 19.17 ***Giardia intestinalis.*** This flagellated protist can cause giardiasis, also called "hiker's diarrhea." (**A**) It attaches to the lining of a person's small intestine and (**B**) leaves an impression (*arrows*) when it detaches.

Another group of disease-causing flagellated protozoa are the **trypanosomes** (**figure 19.18**). Tsetse flies transmit African sleeping sickness, caused by *Trypanosoma brucei. Trypanosoma cruzi* causes Chagas disease in South and Central America, killing 45,000 people annually. Animals as diverse as rodents, armadillos, and dogs are reservoirs for *T. cruzi*, and the kissing bug is the vector. The sand fly transmits a related parasite, *Leishmania*.

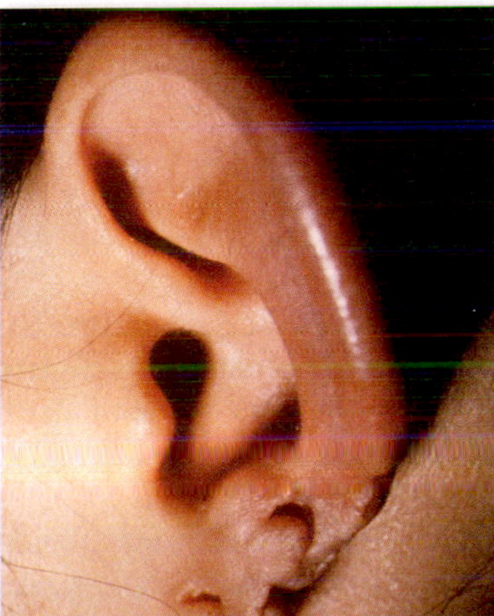

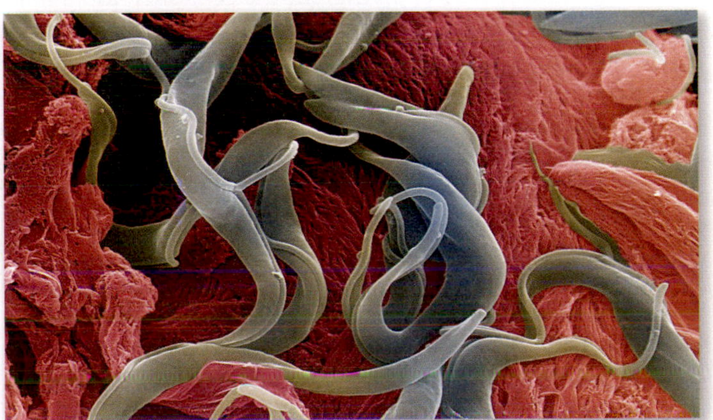

FIGURE 19.18 Trypanosomes. (**A**) Some species of *Leishmania* are responsible for mucocutaneous leishmaniasis, which can cause grotesque deformities. (**B**) *Trypanosoma brucei* causes African sleeping sickness.

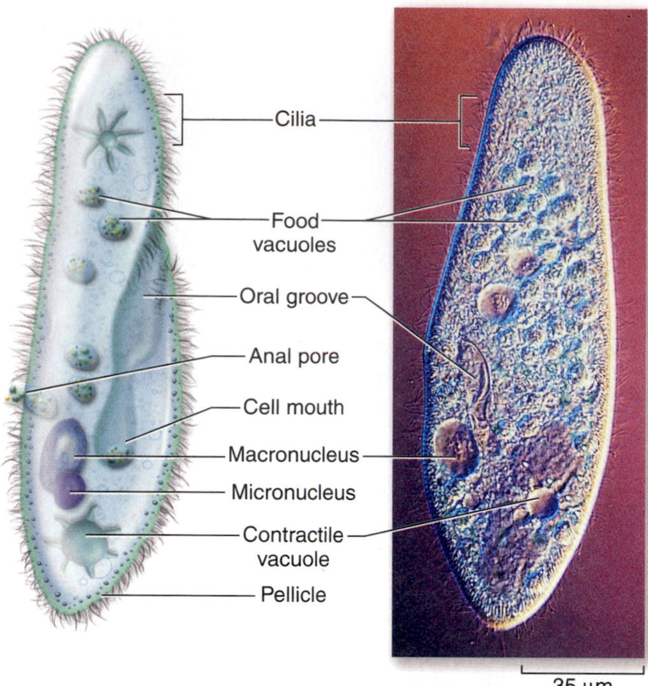

FIGURE 19.19 Anatomy of *Paramecium*. Structures in *Paramecium* include a micronucleus and a macronucleus; an oral groove with a "mouth" into which cilia wave food; food vacuoles for storage; contractile vacuoles that maintain solute concentrations; and an anal pore, which releases wastes.

C. Ciliates Are Common Protozoa with Complex Cells

The **ciliates** are complex, mostly unicellular protists that are characterized by abundant hairlike cilia (**figure 19.19**). When the cilia beat, they propel the organism through the water. Cilia also sweep food (bacteria, algae, and other ciliates) into the cell's gullet. A food vacuole then surrounds and transports the captured meal inside the cell. In some species, a permanent anal pore releases the wastes. Contractile vacuoles pump excess water out of the cell. **cilia, p. 70**

Most ciliates are free-living, motile cells such as *Paramecium* and its predator, *Didinium*. Several ciliate species, such as *Stentor*, are sessile or attached forms living on a variety of substrates. Nearly one third of ciliates are symbiotic, living in the bodies of crustaceans, mollusks, and vertebrates. Some inhabit the stomachs of cattle, where they house bacteria that break down the cellulose in grass. *Ichthyophthirius multifilis* causes a common epidermal disease in freshwater fish, called "ich" for short.

Ciliates can reproduce sexually and asexually. Many have two types of nuclei, a small haploid micronucleus and a larger, diploid macronucleus. The micronuclei are important during sexual reproduction, and the macronucleus is important in both metabolic and developmental functions.

Can *You* Relate?

Malaria

Four species of *Plasmodium* cause malaria in humans, the most common and deadly of which is *Plasmodium falciparum*. (*Plasmodium* is a genus of apicomplexans, not to be confused with the plasmodium produced by some slime molds.) From 300 to 500 million people a year suffer from malaria, with more than 90% of them in sub-Saharan Africa. Globally, 2 to 3 million people die each year from malaria. Among very ill children, the death rate is 50%. Even survivors may retain dormant forms of the parasite, becoming ill again months or years after the initial infection.

A cycle of malaria begins when an infected female mosquito of any of 60 *Anopheles* species feeds on human blood (see figure 19.21). The insect's saliva contains small haploid cells called sporozoites that enter the host as the mosquito feeds. The sporozoites enter the human host's liver cells, where they multiply rapidly, eventually emerging as merozoites. Some merozoites infect red blood cells, where they feed on hemoglobin as trophozoites. In the infected red blood cell another generation of merozoites forms. Every 48 to 72 hours they burst from the host cells and infect other red blood cells (**figure 19.B**). This release is often synchronized throughout the victim's body, causing the characteristic recurrent chills and fever of malaria.

Other merozoites become specialized as male and female gametocytes (sexual forms). When mosquitoes ingest the gametocytes from an infected person's blood, the cells unite in the insect's stomach. After several additional steps, sporozoites form. These move to the mosquito's salivary glands, ready to enter a new host when the insect seeks its next blood meal.

Not everyone is equally susceptible to malaria. In areas of the world where this killer disease is endemic, human populations have a relatively high incidence of the sickle cell allele. People with one copy of this recessive allele are much less likely to get malaria than are people with two dominant alleles. In malaria-free areas, the sickle cell allele is much rarer. This pattern illustrates the selective force that malaria exerts on the human population.

Despite decades of research, malaria continues to be the world's most significant infectious disease. *Plasmodium* continues to develop resistance to drugs that were effective in the past, just as the use of insecticides selects for resistance in *Anopheles* mosquitoes. Increased international travel has spread malaria worldwide, and nations troubled by poverty and civil unrest struggle to distribute drugs to prevent or treat malaria.

FIGURE 19.B Emerging Parasites. *Plasmodium* cells burst from this red blood cell in a laboratory culture dish.

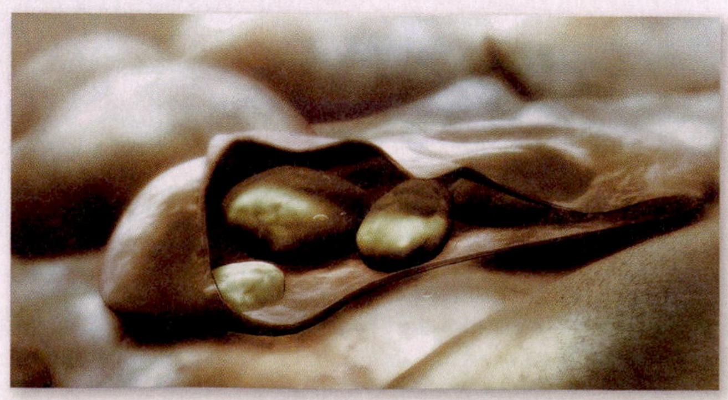

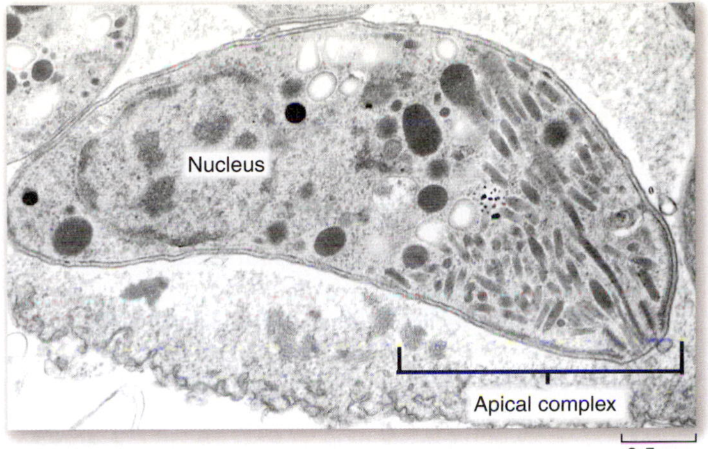
Nucleus
Apical complex
0.5 μm

FIGURE 19.20 Apical Complex. A longitudinal section through the invasive stage of an apicomplexan (*Toxoplasma gondii*) illustrates organelles that make up the characteristic apical complex. Supplied by D. J. P. Ferguson, 2000. Oxford University, with permission from *The Biologist* vol. 47 pages 234–238.

D. Apicomplexans Include Nonmotile Animal Parasites

The **apicomplexans** are nonmotile, spore-forming, internal parasites of animals. The name "apicomplexa" comes from the **apical complex,** a cluster of microtubules and organelles at one end of the cell (**figure 19.20**). This structure, visible only with an electron microscope, apparently helps the parasite attach to and invade host cells.

The apicomplexans have complicated life cycles, often involving many stages in multiple hosts, and usually including both asexual and sexual reproduction. **Figure 19.21** shows a simplified life cycle for *Plasmodium,* which causes mosquito-borne malaria. This killer disease is the subject of this chapter's Can *You* Relate? Malaria.

Another apicomplexan is *Toxoplasma gondii,* which causes toxoplasmosis. This organism can spread to humans who

SEXUAL CYCLE
ASEXUAL CYCLE
Mosquito infects human by injecting saliva containing sporozoites
Injected sporozoites migrate to liver
Sporozoites enter liver cells, multiply, and emerge as merozoites
Stages in liver cells
Merozoites released
Merozoites enter red blood cells
Merozoites released
Stages in red blood cells (repeats)
Some merozoites become gametocytes
Female mosquito ingests gametocytes with blood meal
Salivary glands
Sporozoites are released and migrate to salivary glands
Male gametocyte
Female gametocyte
FERTILIZATION
MEIOSIS
Diploid (2n)
Haploid (n)

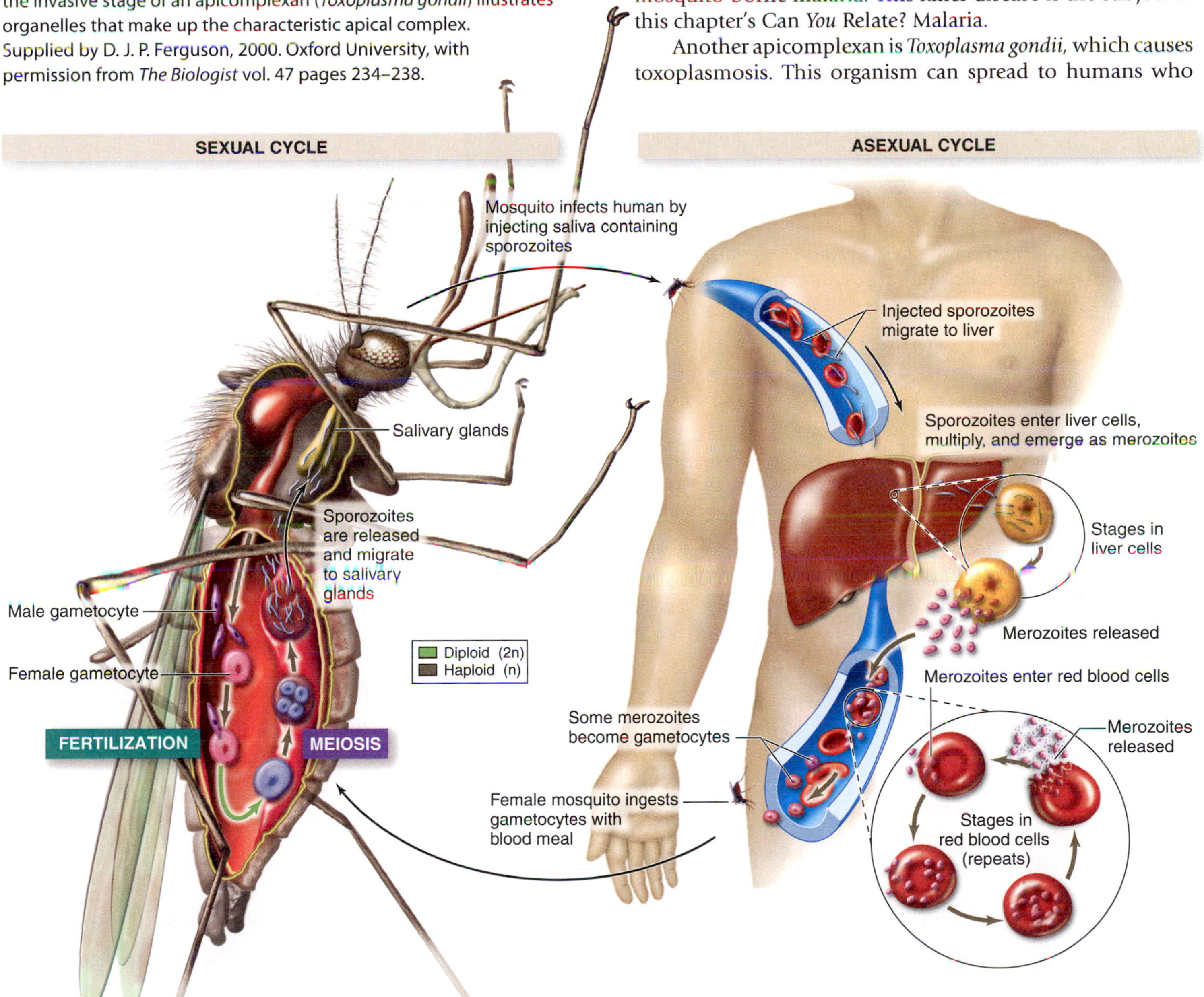

FIGURE 19.21 ***Plasmodium* Causes Malaria.** A mosquito ingests *Plasmodium* gametocytes (sexual forms) that join and form sporozoites, which the insect transmits while feeding. The sporozoites migrate to the liver, where they divide to form merozoites, which may continue to cycle in red blood cells for some time. At some point, gametocytes form from merozoites, and mosquitoes ingest them, starting the cycle anew.

handle feces from infected cats. Healthy pregnant women can unknowingly become infected and pass on the disease to the fetus. About 3500 infants are born with toxoplasmosis each year in the United States.

Summary ***Protozoa is an informal term for a diverse group of heterotrophic, mostly unicellular protists. Most can move via pseudopodia, flagella, or cilia. Protozoa are very important in food webs, industry, and medicine.***

19.4 MASTERING CONCEPTS

1. What are the characteristics of each of the major groups of protozoa?
2. Compare and contrast foraminiferans and radiolarians.
3. List three diseases caused by flagellated protozoa.
4. How do ciliates move and eat?
5. What are the distinguishing characteristics of apicomplexans?

19.5 Protist Classification Is Changing Rapidly

This chapter illustrates some of the difficulties in classifying the protists. For example, the euglenoids and dinoflagellates could easily fall into either of two groups: the algae (because they photosynthesize) and the flagellated protozoa (because they have flagella). Likewise, the water molds are traditionally grouped with slime molds because they share a habitat with fungi, but water molds are actually closely related to brown algae.

Clearly, the traditional scheme groups unrelated organisms. New research is helping to assign each species into a lineage with its closest relatives, but biologists have not yet firmly established either the number of kingdoms or their names. **Table 19.1** reviews and summarizes some of the widely recognized groups, and figure 19.1 shows one possible tree based on ribosomal RNA (rRNA).

One group that unites organisms once thought to be dissimilar is the **stramenopiles,** which include water molds, diatoms, brown algae, and golden algae. The word *stramenopile* means "flagellum-hair" (*stramen* = straw or flagellum, and *pilos* = hair). At some point in their life cycles, stramenopiles produce cells with two flagella, one of which is covered with tubular hairs. Another new grouping is the **alveolates,** which are eukaryotes that have a series of flattened sacs, or alveoli, just beneath the cell membrane. The alveolates include dinoflagellates, ciliates, apicomplexans, foraminiferans, and radiolarians.

The other protists form dozens of groups whose relationships to one another remain unclear. These other lineages include:

- **Parabasalia.** These flagellated protists lack mitochondria. They include *Trichomonas vaginalis* (in humans) and *Trichonympha* (in termites), both of which are exclusively anaerobic and live only in association with animals.
- **Diplomonads.** The diplomonads are also flagellates without mitochondria. The cells of diplomonads

Table 19.1 ***Summary of Proposed Protist Groups***

Protist Group	Some Distinguishing Features	Examples
Stramenopiles	Cells have two flagella, one of which has tubular hairs	Water molds, diatoms, brown algae, golden algae
Alveolates	Alveoli beneath cell membrane	Dinoflagellates, apicomplexans, ciliates
Parabasalia	Cells lack mitochondria; flagella are clustered at one end of cell	*Trichomonas*
Diplomonadida	Cells lack mitochondria and have two nuclei	*Giardia*
Euglenozoa (euglenoids and trypanosomes)	Cells have two flagella; euglenoids are photosynthetic and have pellicle made of protein; trypanosomes are parasitic	*Euglena*, trypanosomes
Red algae	Multicellular; reddish or bluish accessory pigments	*Bossiella*
Green algae	Alternation of generations; chlorophyll *a* and *b*; starch; cellulose	*Ulva, Chlamydomonas, Volvox*
Plasmodial slime molds	Feeding phase is multinucleate mass; forms spores upon starvation	*Physarum*
Cellular slime molds	Individual amoebae aggregate into movable mass during starvation	*Dictyostelium*

have two nuclei. They live in stagnant fresh water or in the intestines of animals. *Giardia intestinalis* is a diplomonad.

- **Euglenozoa.** This group includes the euglenoids and their close relatives, the trypanosomes.
- **Red algae.**
- **Green algae.** These photosynthetic protists share a clade with land plants (the subject of chapter 20).
- **Plasmodial slime molds.**
- **Cellular slime molds.**

The placement of many protists remains unresolved. Figure 19.1 clearly shows, for example, that unicellular "amoebae" occur in several totally unrelated lineages. Protistan classification will continue to evolve as research reveals new molecular sequences, but it will likely remain a "work in progress" for years to come.

Summary *Molecular sequences have triggered a revolution in protistan classification. New groupings unite several protists that were previously thought to be unrelated.*

19.5 MASTERING CONCEPTS

1. What are the alveolates and stramenopiles?
2. In what ways are parabasalids and diplomonads similar and different?
3. What are some of the other lineages of protists?

19.6 INVESTIGATING LIFE
Glassy Fossils Reveal the Birth of a Species

Gazing at a serene lake may not conjure images of evolution—nor do most people give a thought to the muddy sediments under the water's surface. But it might be a good idea to think again, especially at Yellowstone Lake in Wyoming (**figure 19.22**). Preserved in its muck is a fossil record that spans thousands of years and documents not only the pace of evolutionary change but also the transition of one species into another.

The story begins about 14,000 years ago, at the end of the last ice age. The glaciers that had covered what is now Wyoming retreated, exposing Yellowstone Lake. Although the lake still freezes each winter, the ice melts during late spring, and a thriving aquatic community exploits the warm summer weather. Meanwhile, soil particles and organic matter enter the lake from surrounding land, mixing with the wastes and remains of dead aquatic organisms. Since the ice age ended, these sediments have rained down on the lake bottom at an average rate of about 0.6 to 0.7 mm per year.

Most of the dead organisms that fall to the lake bottom simply decompose, but traces of others remain. Fossils of diatoms—single-celled protists with two-part silica walls—are especially likely to become preserved. Their glassy remains provide a continuous, 14,000-year record of evolution in Yellowstone Lake.

A team led by diatom experts Edward C. Theriot, from the University of Texas, and Sherilyn C. Fritz of the University of Nebraska studied both modern and fossilized diatoms in the lake. They were especially interested in two species in the genus *Stephanodiscus* (the name comes from the Greek words for "crown disk," after the pillbox-shaped diatoms' ring of spines). *S. yellowstonensis* was named after its home lake, where it was first discovered in 1984. Its close relative, *S. niagarae*, was described in 1845 at Niagara Falls, and it lives in many lakes worldwide. Fossils show that although *S. niagarae* once occupied Yellowstone Lake, it does not live there today.

Theriot and Fritz knew from previous research that Yellowstone Lake's diatom community had changed during the past 14,000 years, but they wanted to know more. How fast had diatoms evolved? And was their evolutionary rate related to changes in their environment?

To find out, they collected an 8.5-meter-long sediment core from the approximate center of the lake floor. They cut

FIGURE 19.22 Yellowstone Lake. Sediments at the bottom of Wyoming's Yellowstone Lake yielded a fossil record of diatoms that spans over 10,000 years.

the core into nine chunks and brought the pieces to a laboratory at the University of Minnesota. There, they took small sediment samples at 16-cm intervals, decreasing the interval to 4 or 8 cm where the rate of change was highest. After treating the muck with hydrogen peroxide to remove organic matter, they mounted the diatoms on slides and examined them with a compound microscope.

The researchers carefully examined any diatom resembling *S. yellowstonensis* or *S. niagarae.* They measured the diameter of the silica wall, the number of ridges radiating from the center of each wall, and the number of spines. They compared each with modern *S. niagarae* diatoms collected from nearby lakes and with two other *Stephanodiscus* species.

The researchers collected other data as well. They used carbon-14 (^{14}C) analysis of organic material taken from several locations in the core to determine the age of the sediments, which ranged from nearly 14,000 years old at the deepest part to less than 1000 years old at the shallowest. They also analyzed pollen from different parts of the core. Pollen grains that blow or wash into the lake become part of the sediments, recording changes in the nearby plant community over time. Because biologists know which plants thrive in different conditions, pollen analysis gives clues to climatic changes. **absolute dating, p. 312**

The diatom fossils, ^{14}C analysis, and pollen painted a detailed portrait of evolution and environmental change over thousands of years. Photographs and measurements of the diatoms show that *S. niagarae* occupied the lake until about 12,400 years ago. The species then began to change, with many "transitional" individuals whose appearance is between that of *S. niagarae* and *S. yellowstonensis*. By 10,000 years ago, the transition was complete: *S. niagarae* was gone, and *S. yellowstonensis* had taken its place. One species had evolved into another, leaving only fossils behind to tell the tale (**figure 19.23**).

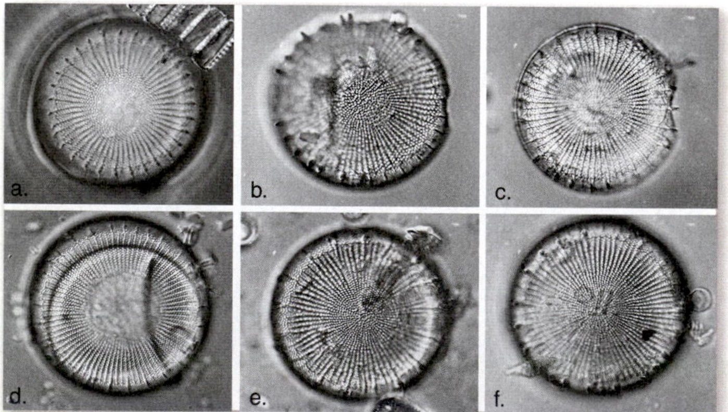

FIGURE 19.23 Diatoms in Transition. (**A**) Modern *S. niagarae.* (**B**)–(**D**) Transitional forms from 13,728 years ago, 11,730 years ago, and 11,215 years ago. (**E**) *S. yellowstonensis* from 4254 years ago. (**F**) Modern *S. yellowstonensis.* **Question**: Propose a possible selective force that could account for the loss of spines from *S. niagarae.* How might you test your hypothesis in a laboratory setting?

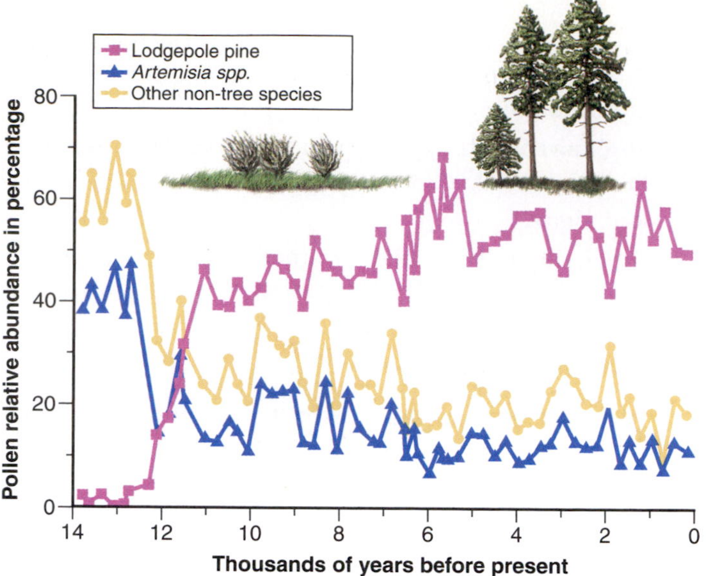

FIGURE 19.24 Changing Climate. As the climate warmed some 12,000 years ago, the plant community surrounding Yellowstone Lake also changed. Pine trees became more abundant as populations of *Artemisia* (an herb in the daisy family) and grasses declined.

The pollen record reveals a shift in the plant community over the same period (**figure 19.24**). Grasses and scattered trees surrounded the lake until about 12,600 years ago, but then the climate began to warm, and the forests expanded. Tree pollen became increasingly abundant from around 12,600 to 10,700 years ago, after which the plant community changed little.

The researchers concluded that, as the climate warmed some 12,000 years ago, the winter ice melted sooner, allowing more time for diatom growth in the spring and summer. For some reason, the environmental changes selected for different diatom traits, and *S. niagarae* evolved into the less-spiny *S. yellowstonensis.* No one knows why the loss of spines occurred during the period of rapid environmental change, or why it stopped after the climate stabilized.

Like the *Tragopogon* plants in Washington and Idaho (see chapter 14's Investigating Life: A New Species is Born, but Who's the Daddy?), the diatoms at Yellowstone Lake teach us that speciation does not always require millions of years of slow, steady change. The diatoms changed measurably in just 2000 to 3000 years. Besides confirming that rapid speciation is possible, these diatoms are also valuable for their unusually complete fossil record. Biologists can thank the decay-resistant silica shells of diatoms—and a continuous, gentle, rain of sediment spanning 10,000 years—for this extraordinary fossil collection.

Theriot, Edward C., Sherilyn C. Fritz, Cathy Whitlock, and Daniel J. Conley. 2006. Late Quaternary rapid morphological evolution of an endemic diatom in Yellowstone Lake, Wyoming. *Paleobiology*, vol. 32, pages 38–54.

CHAPTER SUMMARY

19.1 Kingdom Protista Lies at the Crossroads Between Simple and Complex Organisms

- All **protists** are **eukaryotes** that do not fit into any of the other three eukaryotic kingdoms.
- The protists are very diverse in anatomy, size, and ecological role. Some may resemble early eukaryotes; **choanoflagellates** may provide clues to the origin of multicellularity.
- The relationships among the many lineages of protists remain unclear. Classification of protists is changing as molecular sequence data are considered along with traditional traits.

19.2 Many Protists Are Photosynthetic

- The photosynthetic protists are known as **algae.** Algae range in size from microscopic to dozens of meters long.

A. Dinoflagellates Are "Whirling Cells"

- The **dinoflagellates** have two different-sized flagella at right angles that generate a whirling movement. Many have outer plates made of cellulose. Dinoflagellates cause red tides.

B. Euglenoids Are Heterotrophs and Autotrophs

- **Euglenoids** are unicellular, elongated flagellates that commonly inhabit fresh water.

C. Golden Algae Contain Carotenoid Pigments

- The **golden algae** are photosynthetic but can consume other microorganisms when light or nutrient supplies decline.

D. Diatoms Have Silica Cell Walls

- **Diatoms** are microscopic phytoplankton with intricate silica shells.

E. Red Algae Contain Unique Pigments

- **Red algae** contain unique reddish and bluish photosynthetic pigments that expand their photosynthetic range.

F. Brown Algae Are Large and Complex

- **Brown algae** are large, multicellular seaweeds such as **kelp.**

G. Green Algae Are Ancestors of Plants

- **Green algae** store carbohydrates as starch and use the same pigments as plants. Many have **alternation of generations.** Unlike plants, green algae lack true roots, stems, and leaves, and they have less specialized cells. The green algae have diverse body forms, ranging from microscopic, one-celled organisms to large, multicellular seaweeds.

19.3 Some Heterotrophic Protists Were Once Classified as Fungi

A. Slime Molds Are Unicellular and Multicellular

- **Plasmodial slime molds** form **plasmodia,** brightly colored masses containing thousands of diploid nuclei. A plasmodium feeds by engulfing other cells.
- In **cellular slime molds,** individual cells retain their separate cell membranes throughout the life cycle.

B. Water Molds Are Decomposers and Parasites

- **Water molds** are filamentous heterotrophs that live in moist or wet environments. Some are pathogens of plants, and some decompose organic matter.

19.4 Protozoa Are Diverse Heterotrophic Protists

- Most **protozoa** are heterotrophs, and most have motile cells.

A. Amoeboid Protozoa Produce Pseudopodia

- **Amoeboid protozoa** move by means of "false feet," or pseudopodia. This group includes amoebae, **foraminiferans,** and **radiolarians.**

B. Several Flagellated Protozoa Cause Disease

- Several species of **flagellated protozoa,** such as *Giardia*, *Trichomonas*, *Trypanosoma* (the **trypanosomes**), and *Leishmania*, cause disease in humans.

C. Ciliates Are Common Protozoa with Complex Cells

- **Ciliates** have complex cells with cilia, food vacuoles, contractile vacuoles, and two types of nuclei.

D. Apicomplexans Include Nonmotile Animal Parasites

- **Apicomplexans** are obligate parasites of animals. They have an **apical complex** of organelles that helps them attach to or penetrate host cells. Malaria and toxoplasmosis are diseases caused by apicomplexans.

19.5 Protist Classification Is Changing Rapidly

- The traditional means of classifying protists is giving way to a newer scheme based on molecular evidence.
- **Stramenopiles** have cells that produce two flagella, one smooth and one with tubular "hairs." This group includes water molds, golden algae, brown algae, and diatoms.
- **Alveolates** have flattened sacs beneath the cell membrane and include the dinoflagellates, ciliates, apicomplexans, foraminiferans, and radiolarians.
- Parabasalia and diplomonads are two lineages of anaerobic, flagellated eukaryotes that lack mitochondria.

- The relationships among most lineages of protists remain unclear.

19.6 Investigating Life: Glassy Fossils Reveal the Birth of a Species

- Diatoms in Wyoming's Yellowstone Lake have left a continuous fossil record that spans more than 10,000 years.
- Researchers studying the fossils have documented the formation of a new species, *S. yellowstonensis*, from its direct ancestor, *S. niagarae*. The period over which speciation happened coincided with a period of climatic change.

MULTIPLE CHOICE QUESTIONS

1. Which of the following is NOT a characteristic of the Kingdom Protista?
   a. All are unicellular
   b. Cells contain membrane-bounded organelles
   c. Cells contain a nucleus
   d. Most are aerobic
2. The toxic bloom associated with a red tide is the product of which type of protist?
   a. Euglenoids
   b. Diatoms
   c. Red algae
   d. Dinoflagellates
3. The different colors associated with the various phyla of photosynthetic protist result from
   a. variations in the cell walls of these organisms.
   b. the presence of different photosynthetic pigments.
   c. differences in the habitats occupied by these cells.
   d. the presence of silica or polysaccharides on the outside of the cell.
4. How is a plasmodial slime mold different from a cellular slime mold?
   a. The ability to form resistant spores
   b. The use of photosynthetic pigments
   c. The composition of their cell walls
   d. The ability to form a single cell with multiple nuclei
5. What property distinguishes an alga from a protozoan?
   a. The presence of a flagellum
   b. The ability to live in water
   c. The ability to use photosynthesis
   d. Both a and b
6. What form of motility is associated with amoebas?
   a. Flagella
   b. Cilia
   c. Pseudopods
   d. Kinetoplasts
7. Which of the following diseases is not caused by a flagellated protozoan?
   a. Vaginitis
   b. Hiker's diarrhea
   c. Malaria
   d. African sleeping sickness
8. Cilia can be used by protozoans to
   a. collect food.
   b. move through the environment.
   c. attach to a surface.
   d. both a and b.
9. How do the merozoites of an Apicomplexa differentiate into gametocytes?
   a. They undergo meiotic cell division.
   b. They undergo mitotic cell division.
   c. They undergo loss of the apical complex.
   d. They undergo fertilization.
10. Why is the classification of protists based on rRNA sequences considered useful?
    a. Because only protists have ribosomal RNAs
    b. Because rRNA evolves rapidly, the relationship will be more obvious
    c. Because some protists have prokaryotic-like ribosomes
    d. Because rRNA evolves slowly, making it easier to identify related phyla

TESTING YOUR KNOWLEDGE

1. Explain why the choanoflagellates and green algae are of evolutionary importance.
2. Name an organism that uses each form of locomotion:
   a. beating cilia
   b. amoeboid motion
   c. flagella movement
3. Name at least one type of organism that has each of the following structures:
   a. cyst
   b. plasmodium
   c. apical complex
   d. pseudopod
   e. silica cell wall

4. Label the parts of the *Euglena* cell in the figure.

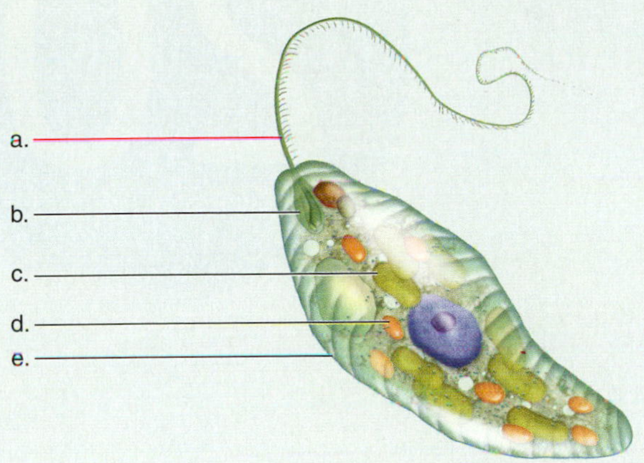

5. Which protist causes each of the following diseases?
   a. malaria
   b. grape downy mildew
   c. amoebic dysentery
   d. late blight of potato
   e. African sleeping sickness
   f. toxoplasmosis
   g. paralytic shellfish poisoning
   h. leishmaniasis
   i. hiker's diarrhea
   j. Chagas disease
6. Give three examples of protists for which the classifications have recently changed. In each case, what was the justification for the old category, and what is the justification for the change?

THINKING AS A SCIENTIST

1. A strange event occurred on a South African beach. Half a million spiny rock lobsters, which are a delicacy, became stranded on the sand, and police had to set up roadblocks to prevent people from plucking free dinners from the sea. The lobsters had fled the ocean due to a "red tide" that tinged the water red. What type of protist might have caused the lobster stranding? Are they safe to eat?
2. How is it adaptive for a red alga to have pigments other than chlorophyll?
3. How might *Volvox*'s colonial lifestyle be adaptive?
4. What adaptations enable cellular slime molds and dinoflagellates to survive a temporarily harsh environment?
5. How does rainy weather foster the spread of *Phytophthora*, the protist that causes late blight of potato?
6. The fossil record for diatoms is much more complete than that of other protists, such as amoebas and slime molds. Explain this observation.
7. Give one example for which molecular evidence unites organisms once thought to be dissimilar, and one example for which such evidence indicates that organisms once thought to be alike due to their appearance are actually not closely related.
8. Insecticides such as DDT kill mosquitoes that transmit the protist that causes malaria. The use of DDT, however, causes serious reproductive problems in eagles, ospreys, and other fish-eating birds. The insecticide is therefore banned for use in the United States. In countries where malaria still rages, what information would you need to help you decide whether to allow the use of DDT in the fight against malaria?
9. Explain why scientists are interested in the presence or absence of mitochondria in some lineages of protists.
10. The flagellated protist *Giardia* does not have true mitochondria, but in 2005, researchers reported that it does have tiny organelles called mitosomes. Chapter 6 describes the structure and function of the mitochondrion. What features would you look for to test the hypothesis that mitosomes are derived from mitochondria?

Visit www.mhhe.com/hoefnagels for practice quizzes, animations, videos, and activities designed to help you master the material in this chapter.

CHAPTER 20

Plants

Sphagnum, the source of peat moss, grows in bogs. This man is cutting peat for fuel.

What's the Point?

Peat Moss, Pot Scrubbers, Drugs, and More: The Many Uses of Plants

Most people know that plants are essential to animal life. Vegetation provides food and habitat, and photosynthesis produces oxygen. Yet plants serve us in many unexpected ways as well. Here are some interesting examples from the four main plant lineages.

- ***Peat moss:*** *Gardeners and houseplant lovers recognize peat moss as a major ingredient in potting mixes. Peat comes from partially decomposed sphagnum moss harvested from enormous bogs. The dried moss is unusually spongy, absorbing 20 times its weight in water. When mixed with soil, peat slowly releases water to plant roots.*
- ***Horsetail:*** Equisetum, *or the horsetail, is a seedless plant related to ferns. Some horsetails are called "scouring rushes" because their stems and leaves contain abrasive silica particles. Native Americans used horsetails to polish bows and arrows, and early colonists and pioneers used them to scrub pots and pans. Some people believe that horsetails, taken as a dietary supplement, can strengthen fingernails and prevent osteoporosis.*
- ***Pacific yew:*** Taxus brevifolia, *the Pacific yew, is a conifer that contains the compound paclitaxel in its bark. This compound has been found to have anti-cancer properties, particularly for treating breast cancer. The yew is one of the slowest growing trees, however, and harvesting them for their bark would mean their extinction. Fortunately, paclitaxel can be synthesized in the laboratory—it is marketed under the trade name Taxol.*
- ***Cotton:*** *You probably own many garments made of cotton, a cloth that comes from flowering plants in genus* Gossypium. *This plant's seeds develop in a dense web of cellulose fibers, which textile manufacturers spin into threads that make up T-shirts, blue jeans, underwear, towels, sheets, and many other cloth products. Cotton seeds also produce cooking oil. About three fourths of the U.S. cotton crop is genetically modified to produce its own insecticides (see chapter 10's Investigating Life: Heredity and the Hungry Hordes).*

Peat moss, horsetails, the Pacific yew, and cotton are just four of the many plants that humans use, and they represent only a tiny percentage of the diverse kingdom Plantae. This chapter highlights some of the history and diversity of these essential organisms.

LEARNING OUTLINE

20.1 Plants Have Changed the World

If you glance at your surroundings in almost any habitat, the first thing you see is plants: grasses, trees, shrubs, ferns, or mosses exist nearly everywhere, at least on land (**figure 20.1**). They are so familiar that it is difficult to imagine a time during which plants did not exist.

All plants, from mosses to maple trees, are multicellular organisms with eukaryotic cells. With the exception of a few parasitic species, plants are autotrophs: they use photosynthesis to produce the sugars that provide energy, build their bodies, and ultimately feed most other species. Nearly all plant species live on land, from moist bogs to parched deserts. Where did these essential organisms come from? **photosynthesis, p. 102**

A. Plants Descend from Green Algae

Less than a billion years ago, oceans and fresh water teemed with life, including many algae. About 480 to 470 MYA, or perhaps earlier, one group of green algae called the **charophytes** likely gave rise to land plants. Although the exact events remain uncertain, molecular and other evidence clearly suggest that green algae and plants descend from a common ancestor; that is, they form a clade. **green algae, p. 394; systematics, p. 322**

Like their green algae ancestors, a plant's photosynthetic cells have chloroplasts that contain chlorophyll *a* and other pigments (see this chapter's Burning Question: Why are plants green?). In addition, like green algae, plants have cellulose-rich cell walls and use starch as a nutrient reserve. **polysaccharides, p. 34**

Plants and some algae also have some reproductive similarities. In a life cycle called **alternation of generations,** a multicellular diploid stage alternates with a multicellular haploid stage (**figure 20.2**). The **sporophyte** (diploid) generation develops from a zygote that forms when gametes come together at fertilization. At maturity, some cells in the sporophyte undergo meiosis and produce haploid **spores,** which divide mitotically to form the gametophyte. The haploid **gametophyte** produces gametes by mitotic cell division; these fuse at fertilization, starting the cycle anew. **mitosis, p. 161; meiosis, p. 177**

A prominent evolutionary trend among land plants is a change in the relative sizes of the gametophyte and sporophyte generations. A moss, for example, is one of the simplest types of land plants. In these organisms, the green gametophyte stage is the most obvious; the small, brown sporophyte depends on the gametophyte for nutrition. In complex plants, by contrast, the sporophyte generation is green and much larger than the gametophytes. A fern frond, pine tree, and rose bush are all sporophytes. Ferns and pines have gametophyte generations that are barely visible to the unaided eye. In rose bushes and other flowering plants, the gametophyte is microscopic.

FIGURE 20.1 Plants Are Everywhere. Nearly every ecosystem on land is dominated by plants.

Burning Questions

Why are plants green?

Most plants appear green because of the way their pigments reflect light. Visible light from the sun contains wavelengths corresponding to all colors, from violet (the shortest wavelength of visible light) to red (the longest wavelength). Chlorophyll *a*, the main photosynthetic pigment in chloroplasts, absorbs red and blue wavelengths and reflects green light, which we perceive as the plant's color (see figure 5.6).

Leaves may be other colors besides green. Some leaves are purple, red, or yellow throughout their lives. Others change from green to yellow, orange, or red during autumn. These spectacular seasonal displays occur as deciduous plants degrade their photosynthetic pigments in preparation for winter. Chlorophyll disappears before the red, orange, and yellow carotenoid pigments, leaving them visible until the plant sheds its leaves.

Have a Burning Question of your own? Submit it to marielle_hoefnagels@mcgraw-hill.com for possible inclusion in future editions of this book!

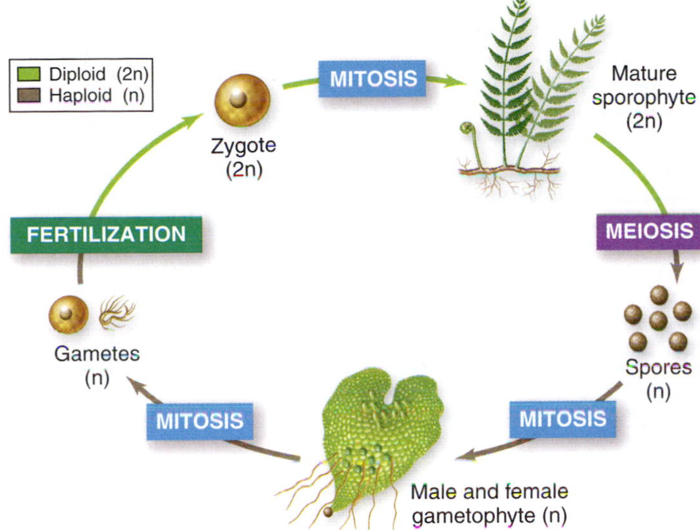

FIGURE 20.2 Alternation of Generations. Unlike most animals, plants have multicellular haploid and diploid generations. The haploid, or gametophyte, generation produces gametes by mitotic cell division. After fertilization, the diploid zygote develops into the sporophyte, which undergoes meiosis and produces haploid spores. The spores germinate and divide mitotically, producing gametophytes. The relative sizes of the sporophyte and gametophyte generations depend on the type of plant. In the simplest plants, the sporophyte is much smaller than the gametophyte, but in complex plants, the gametophyte consists only of a few cells.

Once plants had settled the land, they set into motion a complex series of changes that would profoundly affect both the living and nonliving worlds. The explosion of photosynthetic activity from plants altered the atmosphere, lowering carbon dioxide levels and raising oxygen content. Plants eventually formed the bases of intricate food webs, providing diverse and nutrient-packed habitats for many types of animals. Leaf litter accumulating on forest floors fed countless soil microorganisms, insects, and worms. When washed into streams and rivers, this litter fueled a spectacular diversification of fishes and other aquatic animals. In short, our planet would not be what it is today were it not for plants.

B. Plants Have Adapted to Life on Land

The terrestrial landscape presents selective forces that are far different from those in the aquatic habitat of the green algae. Water, minerals, and dissolved gases surround the whole body of a green alga, and the buoyancy of water provides physical support. Algae reproduce sexually by simply releasing gametes into the water. On land, soil stores water and minerals essential for plant growth, but the aboveground part of a plant has limited access to these resources. Air not only provides little physical support, but it also dries out the plant's aboveground tissues. Sexual reproduction becomes more complicated in the absence of free water.

Terrestrial habitats have selected for many adaptations that enable plants to hold their bodies erect, retain moisture, survive, and reproduce without being immersed in water (**figure 20.3**). In most plants, these adaptations include:

- leaves that capture solar energy
- root systems that anchor plants and absorb water and nutrients from soil
- a **cuticle,** which is a waxy coating that minimizes water loss from the aerial parts of a plant
- **stomata,** specialized pores in the leaves and stems. When open, stomata enable the plant to exchange gases (especially CO_2 and O_2) with the atmosphere. In dry weather, plants reduce water loss by closing the pores. **plant epidermis, p. 507**
- **lignin,** a complex polymer that strengthens and supports cell walls. Lignin enables plants to grow tall and upright and to form branches, an important adaptation in the intense competition for sunlight.
- **vascular tissue** that transports water and nutrients throughout the bodies of most plants. As described in chapters 24 and 25, xylem and phloem are the two types of vascular tissue. **Xylem** conducts water and dissolved minerals from the roots up to the leaves. The thick, lignin-rich walls of xylem cells also help keep a plant upright. **Phloem** transports sugars produced in photosynthesis to the nongreen parts of the plant, including the roots.

Not all plants have all of these adaptations, and many have additional ones. The most complex plants have specialized structures, including pollen and seeds, that enable them to reproduce in the absence of water. Not surprisingly, these plants

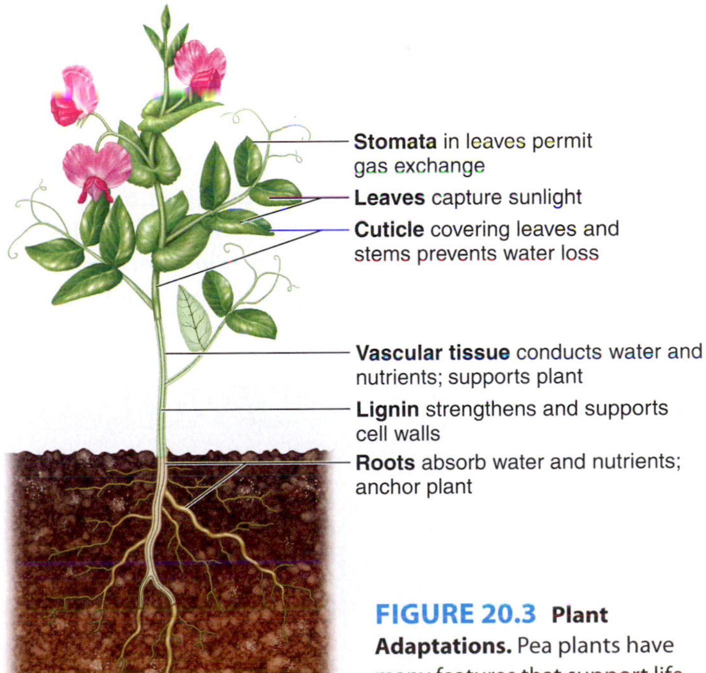

FIGURE 20.3 Plant Adaptations. Pea plants have many features that support life on dry land.

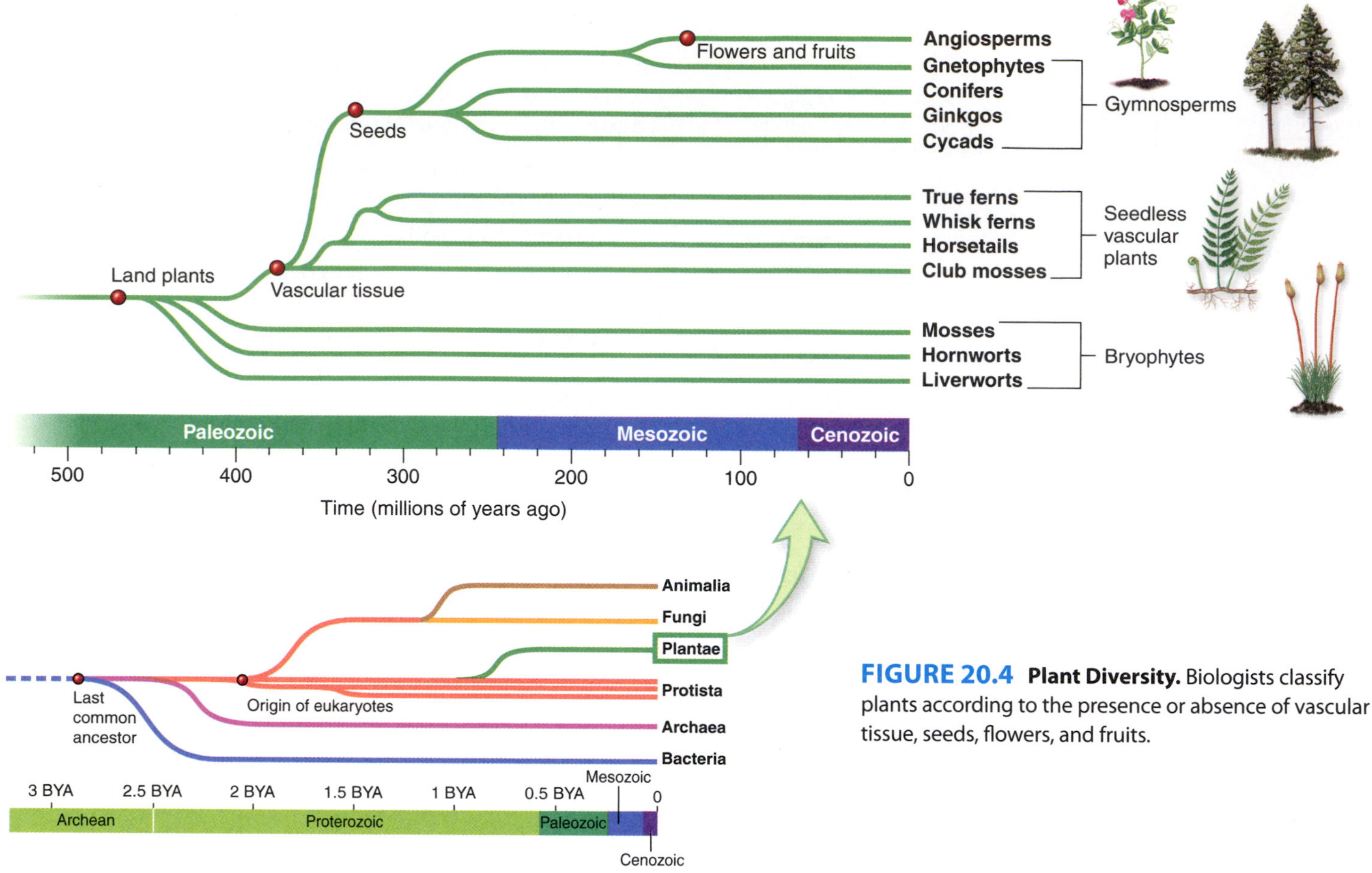

FIGURE 20.4 Plant Diversity. Biologists classify plants according to the presence or absence of vascular tissue, seeds, flowers, and fruits.

have colonized the driest habitats on Earth. The next four sections of this chapter describe the main groups of land plants: bryophytes, seedless vascular plants, gymnosperms, and angiosperms (**figure 20.4** and **table 20.1**).

Summary *Plants have changed the world since they colonized land nearly 500 MYA. They are multicellular eukaryotes that photosynthesize, have cellulose cell walls, store starch, and have life cycles with alternation of generations. Over hundreds of millions of years, land plants have become highly specialized and adapted to terrestrial life. Plants are classified based on the presence or absence of transport tissues, seeds, flowers, and fruits.*

20.1 MASTERING CONCEPTS

1. How have plants changed the landscape on Earth, and how are they still vital to life on Earth today?
2. What evidence suggests that plants evolved from green algae?
3. What is alternation of generations?
4. List some adaptations that enable plants to thrive on dry land.
5. What is vascular tissue, and how does it adapt plants to land?
6. What features differentiate the four major groups of plants?

Table 20.1 *Four Groups of Plants*

Group	Swimming Sperm?	Vascular Tissue?	Pollen?	Seeds?	Flowers?	Fruits?
Bryophytes	Yes	No	No	No	No	No
Seedless vascular plants	Yes	Yes	No	No	No	No
Gymnosperms	No	Yes	Yes	Yes	No	No
Angiosperms	No	Yes	Yes	Yes	Yes	Yes

20.2 Bryophytes Are the Simplest Plants

The earliest plants probably resembled modern bryophytes. Although all bryophytes are seedless, nonvascular plants, **figure 20.5** shows that they do not form a clade. This section describes them together, however, because they share some important features.

A. Bryophytes Are Small and Lack Vascular Tissue

Bryophytes are small, compact plants without vascular tissue. Water and nutrients move from cell to cell within the plant by diffusion and osmosis, not within specialized transport tissues. Lignin hardens bryophyte cell walls, but it does not provide enough support to allow these plants to become very large.

Although bryophytes lack true leaves and roots, many have structures that are similar to these organs. For example, photosynthesis occurs at flattened leaflike areas. In addition, **rhizoids** are hairlike extensions that cover a bryophyte's lower surface, anchoring the plant and absorbing water and minerals.

The 24,000 or so species of bryophytes are classified into three phyla:

- **Mosses** are the bryophytes that are most closely related to the vascular plants. The gametophytes resemble a short "stem" with many "leaves." The brown or green sporophyte does not resemble the gametophyte at all (**figure 20.6A**).
- **Liverworts** have flattened leaflike structures that usually lie close to the ground (figure 20.6B). The diverse liverworts may be the bryophytes most closely related to ancestral land plants.
- **Hornworts** are the smallest group of bryophytes, with only about 100 species. They are named for their sporophytes, which are shaped like tapered horns (figure 20.6C).

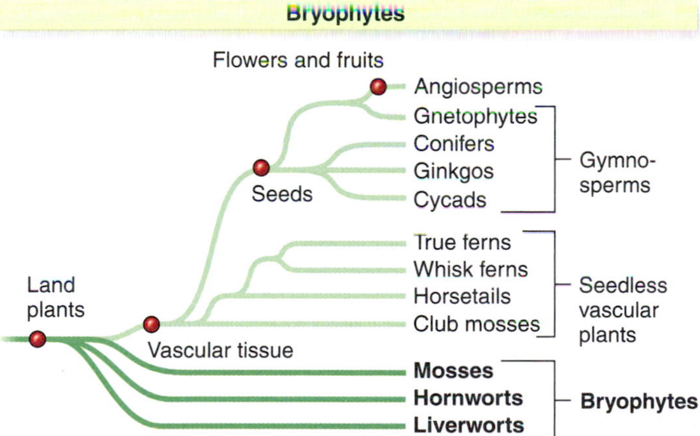

FIGURE 20.5 Bryophytes. The earliest land plants probably resembled existing bryophytes, a collective term for mosses, hornworts, and liverworts.

a.

b.

c.

FIGURE 20.6 Bryophyte Diversity (**A**) This moss, *Splachnum luteum,* looks like a flowering plant, but each "flower" is actually the broad base of a spore capsule. (**B**) The gametophyte of this liverwort resembles small leaves. The umbrella-shaped structures produce sperm and egg cells. (**C**) In the hornwort *Anthoceros,* the tapered hornlike structures are sporophytes, below which the flat gametophytes are visible.

B. Bryophytes Have a Conspicuous Gametophyte and Swimming Sperm

Most reproduction in bryophytes occurs asexually. For example, mosses and liverworts asexually produce **gemmae,** small pieces of tissue that detach and grow into new plants (**figure 20.7**).

Bryophytes also reproduce sexually (**figure 20.8**). These plants are unlike other land plants in that the most conspicuous generation is the haploid gametophyte. Gametes form by mitosis on the gametophyte, in separate sperm- and egg-producing structures. Sperm swim to the egg cell, and the sporophyte generation begins at fertilization. The diploid zygote divides mitotically, producing a stalk that remains attached to the gametophyte. At the tip of the stalk, specialized cells inside a **sporangium** undergo meiosis and produce haploid spores. After the spores are released, they germinate, dividing mitotically and giving rise to new haploid gametophytes.

This life cycle explains the observation that mosses usually occur in moist habitats. Because bryophytes have swimming sperm, a film of water must coat these plants for sexual reproduction to occur.

Summary *The bryophytes are simple, small, seedless plants that lack vascular tissue. The entire plant absorbs nutrients and water directly from its surroundings. Bryophytes reproduce both asexually and sexually; most live in moist habitats.*

FIGURE 20.7 Asexual Reproduction in a Liverwort. The gametophytes of liverworts such as this *Lunularia* can asexually produce fragments called gemmae. Raindrops splash the haploid gemmae from the cups. In its new habitat, each gemma can form a new plant.

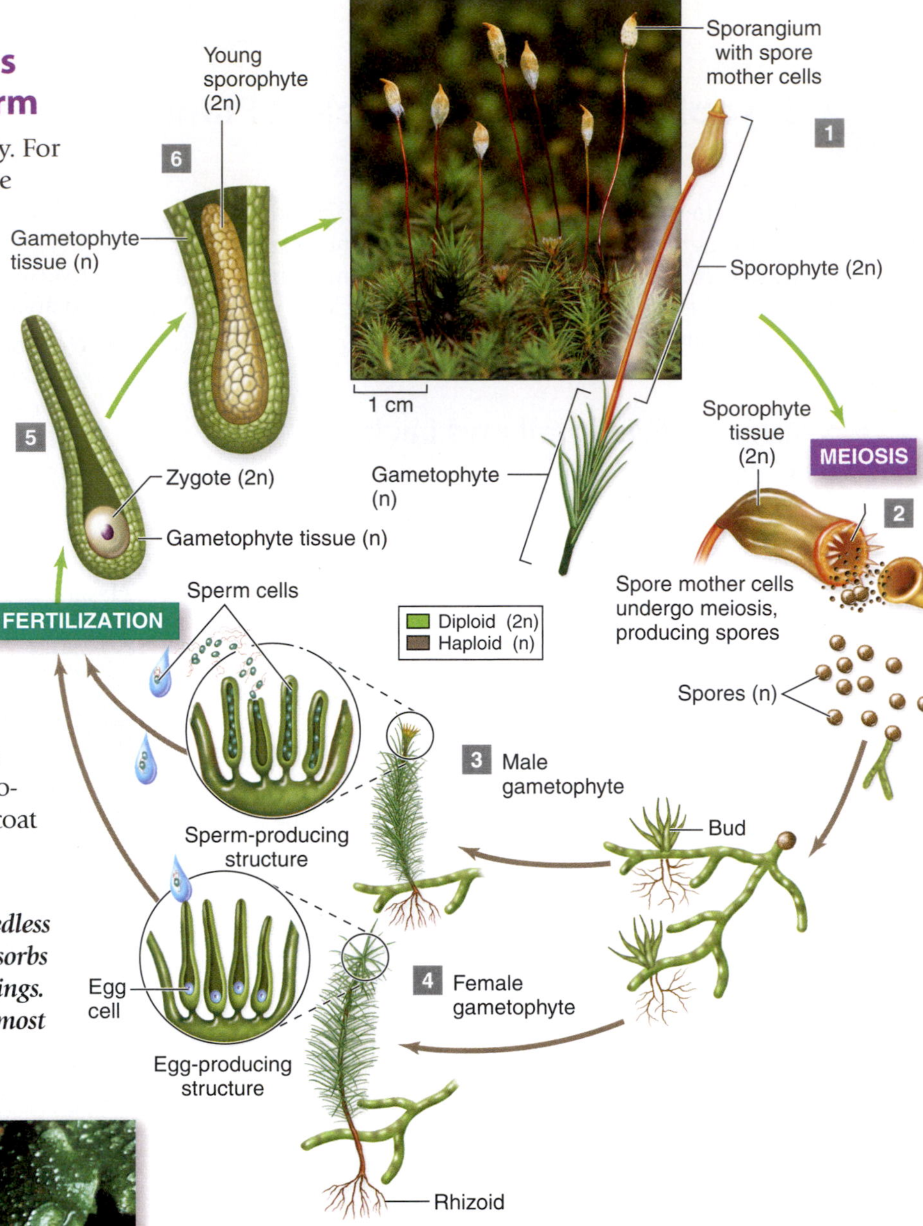

FIGURE 20.8 Sexual Reproduction in a Bryophyte. In mosses, the gametophyte is dominant, as is typical for bryophytes. The sporophyte generation, although distinct, is attached to and dependent on the gametophyte. (*1*) In the sporophyte, spore mother cells in sporangia undergo meiosis, which yields haploid spores (*2*) that generate the male and female gametophytes. (*3*) Male gametophytes produce swimming sperm, and (*4*) egg cells are produced at the female gametophyte tip. Gametes join and form a zygote (*5*), which develops into a new sporophyte (*6*).

20.2 MASTERING CONCEPTS

1. Describe the three types of bryophytes.
2. How do bryophytes reproduce asexually and sexually?
3. Explain the observation that mosses occur more frequently in moist, shady habitats than in hot, sunny places.

20.3 Seedless Vascular Plants Have Xylem and Phloem but No Seeds

Because they are small and typically limited to moist habitats, most bryophytes are easily overlooked. Not so for the vascular plants, which occur in drier locations and have much larger representatives than their bryophyte relatives. Xylem and phloem are transport tissues that allow vascular plants to develop true roots, stems, and leaves with specialized functions. Not only does vascular tissue permit life in drier habitats, but thick cell walls in xylem also physically support the growth of large plants. Vascular plants also have a well-developed cuticle and stomata that minimize water loss.

A. Seedless Vascular Plants Include Ferns and Their Allies

The **seedless vascular plants** are the 12,000 species that have xylem and phloem but do not produce seeds. **Figure 20.9** shows the evolutionary position of ferns and other seedless vascular plants. The earliest species are extinct, but many fossils remain from the Devonian Period (see section 24.6). Their successful adaptations to life on land persist in their descendants.

The seedless vascular plants include four lineages:

- **True ferns** make up the largest group of seedless vascular plants, with about 11,000 species. The **fronds,** or leaves, of ferns are their most obvious feature (**figure 20.10A,B**). Each year, new fronds grow from underground stems called rhizomes. Ferns were especially widespread and abundant during the Carboniferous period, when their huge fronds dominated warm, moist forests. Their remains form most coal deposits.
- **Club mosses,** or lycopods, are divided into two groups: the club mosses (*Lycopodium*) and the spike mosses (*Selaginella*), which take their names from their club- or spike-shaped reproductive structures (figure 20.10C). Because they have vascular tissue, they are different from the bryophyte mosses.
- **Horsetails** grow along streams or at the borders of forests. The only living genus of horsetails, *Equisetum*, includes plants with branched rhizomes that give rise to green aerial stems that bear spores at their tips (figure 20.10D).

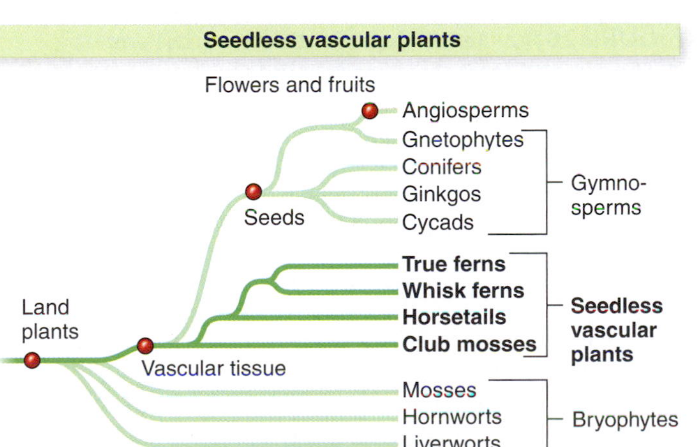

FIGURE 20.9 Seedless Vascular Plants. The seedless vascular plants include club mosses, horsetails, whisk ferns, and true ferns.

a.

b.

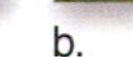

c.

d.

FIGURE 20.10 Seedless Vascular Plant Diversity (A) The narrow beech fern (*Phegopteris connectilis*) is a true fern. (**B**) A fiddlehead is a developing fern frond. (**C**) This is the club moss *Lycopodium selage.* (**D**) *Equisetum fluvatile* is a horsetail.

- **Whisk ferns** are simple plants that have rhizomes but not roots. Most species have no obvious leaves. Their name comes from the highly branched stems of *Psilotum,* which resemble whisk brooms (**figure 20.11**).

B. Seedless Vascular Plants Have a Conspicuous Sporophyte and Swimming Sperm

Like other seedless vascular plants, ferns reproduce sexually (**figure 20.12**). The sporophyte produces haploid spores by meiosis in collections of sporangia on the underside of each frond. Once shed, the spores germinate and develop into tiny, heart-shaped gametophytes that produce gametes by mitotic cell division. The swimming sperm require a film of free water to reach the egg cell. The gametes fuse, forming a zygote. This diploid cell divides mitotically and forms the sporophyte, which quickly dwarfs the gametophyte.

Like bryophytes, most seedless vascular plants live in shady, moist habitats. Both groups of plants produce swimming sperm and therefore cannot reproduce sexually in the absence of free water.

Summary *The seedless vascular plants have vascular tissue and true leaves, stems, and roots, but they do not produce seeds. They also have a waterproof cuticle and stomata. Their swimming sperm cells require a film of free water to reach the egg cell.*

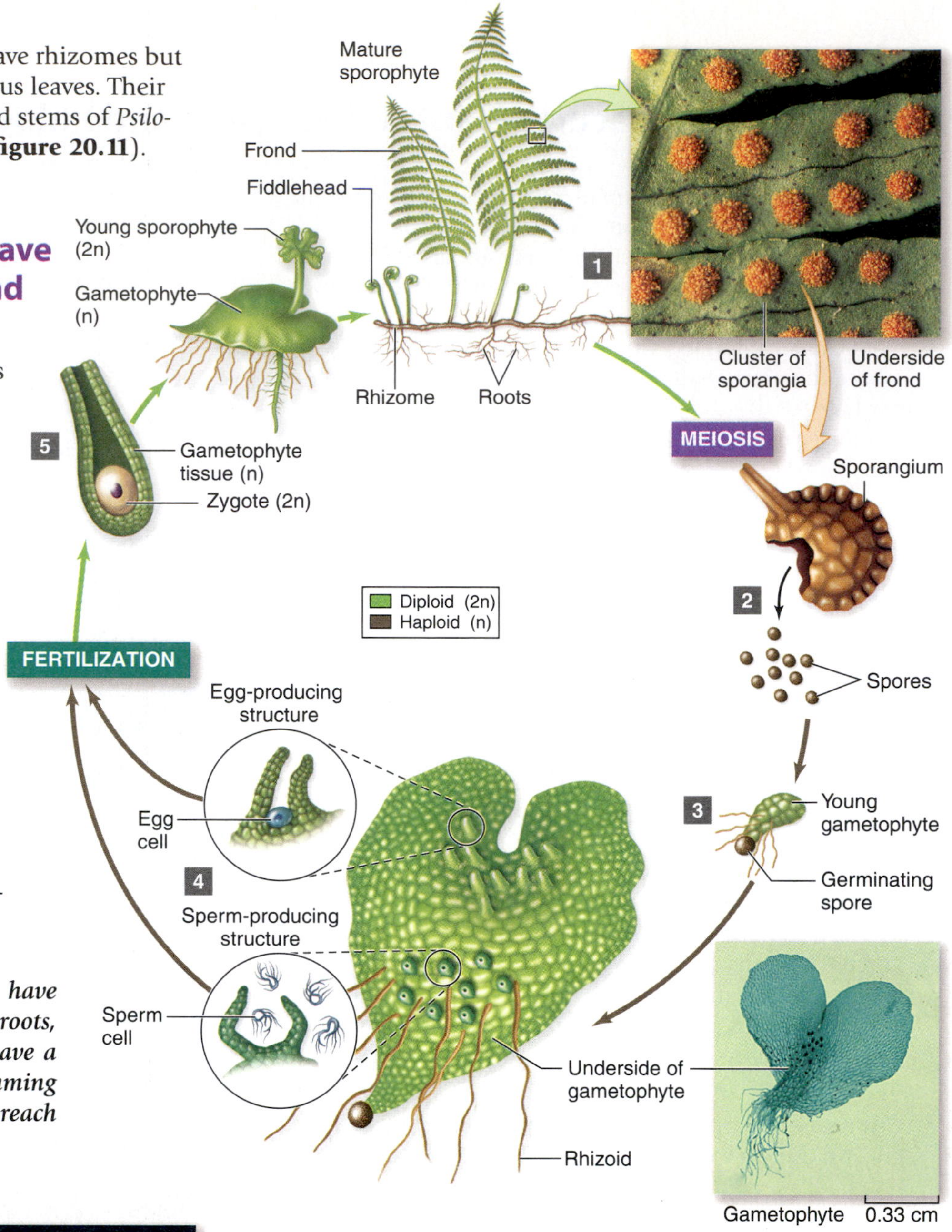

FIGURE 20.12 Sexual Reproduction in a True Fern. (*1*) Sporangia on sporophyte fronds house spore mother cells that produce spores through meiosis (*2*). The haploid spores develop into gametophytes (*3*). The gametophyte produces egg cells and swimming sperm (*4*). These gametes join and produce a zygote, which develops into the sporophyte (5). The large sporophyte quickly dwarfs the tiny gametophyte.

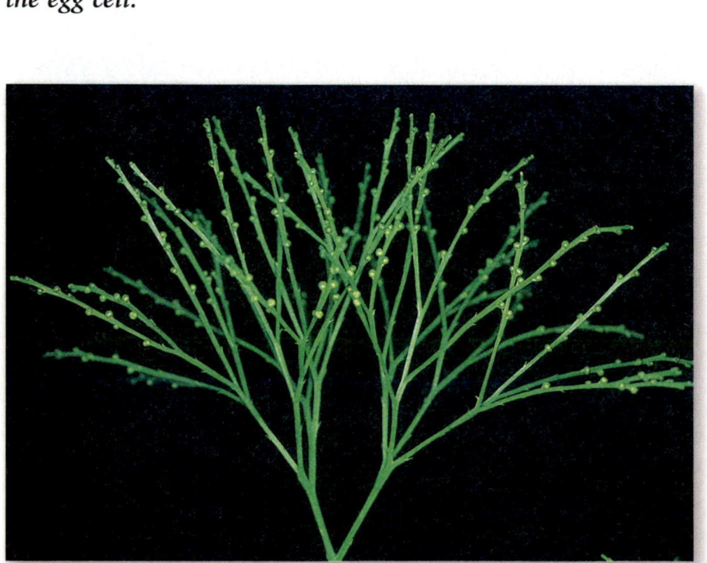

FIGURE 20.11 The Whisk Fern, *Psilotum nudum*.

20.3 MASTERING CONCEPTS

1. List two advantages of vascular tissue to plants that have it.
2. Describe the four groups of seedless vascular plants.
3. How do seedless vascular plants reproduce?
4. How are seedless vascular plants similar to and different from bryophytes?

20.4 Gymnosperms Are "Naked Seed" Plants

Seedless vascular plants dominated Earth's vegetation for more than 250 million years. Eventually, however, they were overtaken by the seed plants, which have two adaptations that enable them to live and reproduce in much drier habitats than either bryophytes or seedless vascular plants:

- **pollen grains** that produce sperm cells. Wind and animals deliver the pollen directly to female plant parts, freeing the sperm from the need for a film of water.
- the **seed**, which is a plant embryo (young sporophyte) packaged with a food supply inside a tough outer coat. Seeds can remain dormant for years, breaking dormancy when favorable conditions occur.

The origin of pollen and seeds was among the most significant events in the evolution of plants, because these adaptations allowed plants to colonize even the driest habitats on Earth's surface. Two groups of vascular plants produce pollen and seeds: gymnosperms (naked seed plants) and angiosperms (flowering plants, which produce seeds inside fruits; see section 20.5). **Figure 20.13** shows the evolutionary position of gymnosperms.

A. Gymnosperms Include Conifers and Three Related Groups

The term **gymnosperm** derives from the Greek words *gymnos,* meaning "naked," and *sperma,* meaning "seed." The seeds of these plants are termed "naked" because they are not enclosed in fruits.

Living gymnosperms are remarkably diverse in reproductive structures and leaf types. The sporophytes of most gymnosperms are woody trees or shrubs, although a few species are more vinelike. Leaf shapes range from tiny reduced scales to needles, flat blades, and large fernlike leaves. The 800 or so species of gymnosperms group into four phyla:

- **Conifers** such as pine trees are by far the most familiar gymnosperms. These plants often have needlelike or scalelike leaves, and they produce egg cells and pollen in cones (**figure 20.14A**). Conifers are commonly called "evergreens" because most do not shed all their leaves at once each year, as deciduous trees do. This term is somewhat misleading, however, because conifers do shed their needles. They just do it a few needles at a time, turning over their entire needle supply every few years.
- **Cycads** are trees that live primarily in tropical and subtropical regions. They have palmlike leaves, and they produce large cones (figure 20.14B). Many cycads are

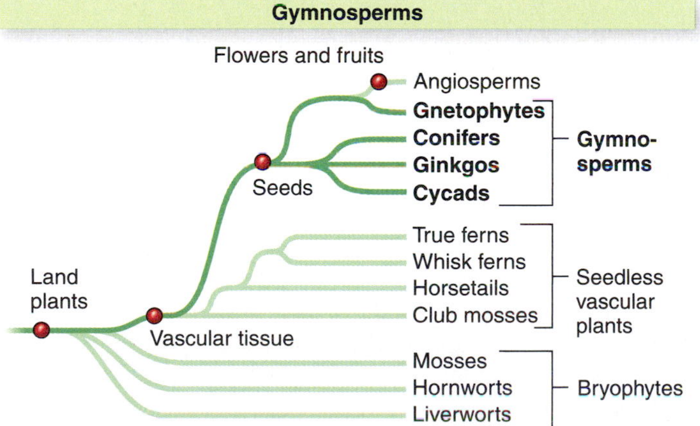

FIGURE 20.13 Gymnosperms. The gymnosperms, or "naked seed" plants, include cycads, ginkgos, conifers, and gnetophytes. Botanists continue to debate the position of the gnetophytes.

FIGURE 20.14 Pines and Cycads. (**A**) This pinyon pine is an example of a conifer. The seed cone has woody scales. (**B**) Cycad trees are ancient seed plants. The cones form in the center of a crown of large leaves. A seed cone is shown here.

planted as ornamentals, but only two species are native to the United States. Cycads dominated Mesozoic era landscapes; today, many species are near extinction because of their slow growth, low reproductive rates, and shrinking habitats.

- The **ginkgo,** also called the maidenhair tree, has distinctive, fan-shaped leaves that have remained virtually unchanged for 80 million years (**figure 20.15A**). Only one species exists: *Ginkgo biloba*. Although it no longer grows wild in nature, it is a popular cultivated tree. Ginkgos have male and female organs on separate plants; landscapers avoid planting female ginkgo trees because the fleshy seeds produce a foul odor. Although clinical trials have not conclusively supported its medical benefits, some people believe that extracts of *Ginkgo biloba* leaves may improve memory and concentration.
- **Gnetophytes** include some of the most distinctive (if not bizarre) of all seed plants. Botanists have struggled with the classification of these plants. Some details of their life history have led to speculation that gnetophytes are closely related to the flowering plants. Molecular evidence, however, places these puzzling plants with the conifers. *Ephedra* (figure 20.15B) is a gnetophyte, as is *Welwitschia*, a slow-growing plant that lives in African deserts. Mature *Welwitschia* plants have a single pair of large, strap-shaped leaves that persist throughout the life of the plant.

a.

b.

FIGURE 20.15 Ginkgo and Gnetophytes. (**A**) *Ginkgo biloba* is also called the maidenhair tree. Male trees are preferred for cultivation because seeds on female trees emit an odor of rot. (**B**) *Ephedra* is a gnetophyte with cones that resemble tiny flowers.

B. Conifers Produce Pollen and Seeds in Cones

Pines illustrate the gymnosperm life cycle (**figure 20.16**). Large female **cones** bear two sporangia, called **ovules,** on the upper surface of each scale. Through meiosis, each ovule produces four haploid structures called megaspores, only one of which develops into a female gametophyte. Over many months, the female gametophyte undergoes mitosis and gives rise to two to six egg cells. At the same time, small male cones bear sporangia on thin, delicate scales. Through meiosis, these sporangia produce microspores, which eventually become pollen grains (immature male gametophytes). Wind carries the pollen grains, sometimes over large distances. Pollination occurs when airborne pollen grains settle between the scales of female cones and adhere to drops of a sticky secretion.

After pollination occurs, the pollen grain germinates, giving rise to a **pollen tube** that grows through the ovule toward the egg cell. Inside the pollen tube, a specialized cell divides mitotically and gives rise to two haploid sperm cells, one of which fertilizes the haploid egg cell (the other sperm cell disintegrates). The resulting zygote is the first cell of the sporophyte generation. The whole process is so slow that fertilization occurs about 15 months after pollination.

Within the ovule, the haploid tissue of the female gametophyte nourishes the developing diploid embryo. Following a period of metabolic activity, the embryo becomes dormant, and the ovule develops a tough, protective seed coat. The seed may remain in the cone for another year. Eventually, however, the seed is shed and dispersed by wind or animals. If conditions are favorable, the seed germinates, giving rise to a new tree.

Summary ***The evolution of pollen and seeds continued the trend of adapting to drier habitats on land. The gymnosperms produce seeds that are not enclosed in fruits.***

20.4 MASTERING CONCEPTS

1. How do seeds and pollen enable plants to live and reproduce in dry climates?
2. What are the characteristics of gymnosperms?
3. What are the four groups of gymnosperms?
4. What is the role of cones in conifer reproduction?
5. What happens during and after pollination in gymnosperms?

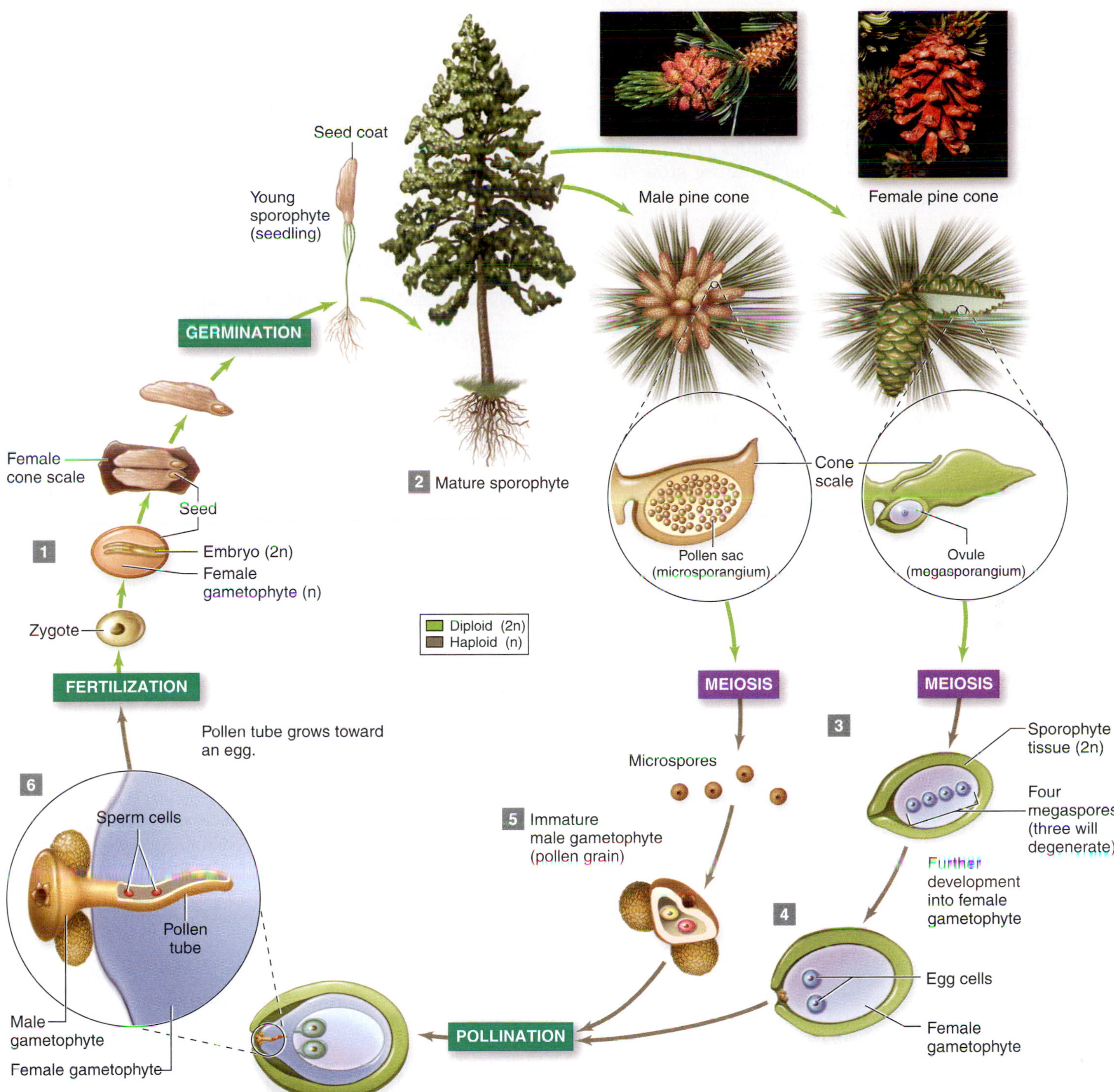

FIGURE 20.16 Sexual Reproduction in Pines. A pine tree's seed contains the embryo, or young diploid sporophyte (*1*). After the seed germinates, the sporophyte grows into a mature, cone-producing tree (*2*). Cells in the male and female cone scales undergo meiosis, producing spores (*3*), which develop into haploid gametophytes that consist of just a few cells each. On the larger female cones, each scale has two ovules (only one is visible in the figure), each of which yields an egg-producing gametophyte (*4*). The smaller male cones produce pollen, the male gametophytes (*5*). A mature pollen grain delivers a sperm cell (*6*) to an egg cell via a pollen tube. More than a year after pollination, each fertilized egg (zygote) completes its development into a seed.

20.5 Angiosperms Produce Seeds in Fruits

More than 95% of modern plants are angiosperms. Apple trees, corn, roses, petunias, lilies, grasses, and many other familiar plants—including those we grow for our own food—are angiosperms.

A. Angiosperms Are Flowering Plants

Like gymnosperms, **angiosperms** are seed plants (**figure 20.17**). Unlike gymnosperms, however, angiosperms have **flowers,** reproductive structures that produce pollen and egg cells. After fertilization, parts of the flower develop into a **fruit** that contains the plant's seeds. Flowers and fruits are adaptations that help angiosperms disperse both their pollen and their offspring.

Many people think of flowers only as decorations for the human habitat. Roses, tulips, petunias, and many other plants are the products of selective breeding for their spectacular blooms. Many flowers, however, are not showy, sweet-smelling beauties (**figure 20.18**). The flowers of wind-pollinated plants such as grasses are plain and easily overlooked.

Biologists group angiosperms into two classes:

- **Monocotyledons** have one cotyledon (the first leaf structure to arise in the embryo). Examples of the 70,000 species of monocots include orchids, lilies, grasses, bananas, and ginger. The grasses include not only lawn plants, but also sugar cane and important grains such as rice, wheat, barley, and corn (see this chapter's Can *You* Relate? Corn, Corn Everywhere). The monocots form a clade.
- **Dicotyledons** have two cotyledons. The diverse dicots include magnolias, water lilies, sunflowers, oaks, tomatoes, roses, beans, and many others. The 184,000 or so species of dicots are not monophyletic, but most belong to one clade, called the eudicots. *Arabidopsis thaliana*, the subject of this chapter's Focus on Model Organisms, is a eudicot.

Monocots and eudicots also differ by other characteristics, further described in chapter 24.

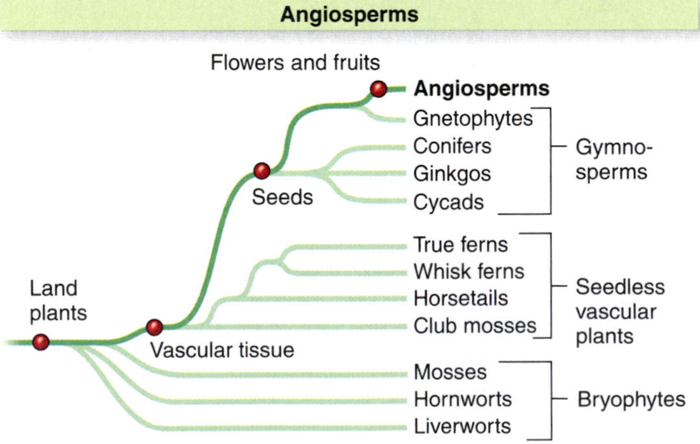

FIGURE 20.17 Angiosperms. The angiosperms are distinguished by flowers, which develop into fruits that enclose the seeds.

Can *You* Relate?

Corn, Corn, Everywhere

The contemporary corn plant (*Zea mays*) is a product of thousands of years of artificial selection. Its ancestor, a wild grass called teosinte, had small, loosely packed kernels, each contained in its own husk. Farmers cultivated this monocot in Mexico at least 7000 years ago. Native people settling North America brought it with them; somewhere along the line, farmers selected for the cob and kernels so familiar today.

Even if you don't eat the kernels off the cob, you likely have much more corn in your diet than you imagine. Besides fresh corn, frozen corn, canned corn, corn meal, corn oil, and corn starch, some of the many unexpected sources of corn include:

- baking powder and confectioners ("powdered") sugar, which often contain corn starch.
- corn syrup, a liquid sugar derived from corn starch, that sweetens soft drinks and other foods.
- vanilla extract, which is often made with corn syrup.
- the simple sugars dextrose (glucose) and fructose.
- dextrin and maltodextrins, polysaccharides that result from the partial hydrolysis of starch. They thicken syrups, provide texture to low-fat foods, and are an ingredient in postage stamp adhesive, among many other uses. **hydrolysis, p. 34**
- grain alcohol from corn, the traditional ethanol source for bourbon whiskey.

Corn appears in other surprising places as well. Corn is gaining popularity as a source of ethanol to fuel vehicles, and the corn plant has also played a major role in the history of biology. In the 1940s, Barbara McClintock used multicolored Indian corn (**figure 20.A**) to discover transposons, "jumping genes" that move within a chromosome from one location to another. She won the 1983 Nobel Prize for physiology or medicine for this work, which has led to new ways to mutate genes and, in turn, new insights into gene function. **transposable element, p. 251**

FIGURE 20.A Indian Corn. *Zea mays* is an angiosperm that is important not only as food but also in the history of biology.

a.

b.

c.

d.

FIGURE 20.18 Angiosperm Diversity. The angiosperms exhibit an astonishing variety of flowers and fruit. (**A**) In bananas (genus *Musa*), flowers occur in clusters. Yellow flowers and green developing fruits are visible in this photograph. (**B**) Red maple (*Acer rubrum*) produces small, bright red flowers that develop into winged fruits after fertilization. (**C**) Passion vines are tropical plants with very showy flowers. (**D**) Cattails (*Typha latifolia*) are familiar wetland plants. The brown cylindrical "tails" are actually spikes of tiny brown flowers.

FOCUS ON MODEL ORGANISMS: *Arabidopsis thaliana*

The choice of a "star" model organism for plants is simple: *Arabidopsis thaliana* easily beats out all others (**figure 20.B**). This tiny angiosperm, a mustard relative, has become a staple of plant biology laboratories because of its small size, easy cultivation, rapid and prolific reproduction, and small genome size (125 million bases). In addition, biologists can study gene function by using a bacterium called *Agrobacterium tumefaciens* to carry new genes into *Arabidopsis* cells.

The history of *Arabidopsis* as a model organism dates only to 1985, following years of research into its physiology, biochemistry, and development. Its genome sequence was completed in 2000. The amount of attention devoted to this plant may seem extravagant, considering its insignificance as a commercial plant. Because angiosperms are closely related to each other, however, discoveries in *Arabidopsis* will likely apply directly to economically important crop plants. Plant scientists expect improved understanding of *Arabidopsis* to contribute directly to better varieties of food staples such as rice, barley, wheat, and corn.

A few of the important discoveries resulting from work on *Arabidopsis* include:

- **Flowering:** Angiosperms delay flowering until they reach reproductive maturity. How do they "know" when the time comes, and how do they build flower parts in the right places? *Arabidopsis* research has revealed genes that control the timing of flowering, the differentiation of stem cells that give rise to flowers, the development of individual flower parts, and the development of ovules inside the flower. For example, some genetic mutations induce flowers to develop into shoots; others promote early flowering (see section 26.9).
- **Response to the environment:** Because a plant cannot avoid extremes of temperature, light availability, and salinity by moving to a better location, it must adjust. *Arabidopsis* has genes that control its response when the weather turns cold, a finding that could help crop plants survive freezing. The plant also has genes whose expression changes throughout each day. Together with light-detecting proteins, these genes help *Arabidopsis* determine whether the season is right for flowering.
- **Circadian rhythms:** Circadian processes occur in 24-hour cycles. In *Arabidopsis*, for example, proteins encoded by clock genes ensure that the expression of genes needed for photosynthesis peak at around noon. The same genes are repressed at night. Chapter 26 describes how pigments called phytochromes "reset" the clock each day.
- **Hormones:** Ethylene is a gas that helps control fruit ripening and senescence (aging) in plants. Mutant *Arabidopsis* plants that do not respond to ethylene have helped researchers find ethylene receptor proteins. Researchers have also discovered that ethylene response requires copper, which the plant transports using a protein similar to one that transports copper in humans. (When faulty in humans, this protein causes Menkes disease.) **plant hormones, p. 545**
- **Disease resistance:** Some plants construct a sort of "fire break" around the spot where a bacterium or fungus has entered, killing small areas of surrounding plant tissue. The resulting zone of dead cells prevents the invader from spreading throughout the plant. Study of this response in *Arabidopsis* has led to new insights into how genes regulate this form of programmed cell death. **apoptosis, p. 168**

FIGURE 20.B
Arabidopsis thaliana.

B. Flowers and Fruits Are Unique to Angiosperms

The angiosperm life cycle is similar to that of gymnosperms in some ways (**figure 20.19**). The sporophyte is the only conspicuous generation, and pollen grains produce sperm cells.

Yet the life cycles differ in other important ways. Most obviously, the reproductive organs in angiosperms are flowers, not cones. Many flowers attract animal pollinators with bright colors, alluring scents, and food rewards such as nectar. The relationship between angiosperms and their pollinators is sometimes so tight that one cannot reproduce or survive without the other. This situation can lead to **coevolution,** in which a genetic change in one species selects for subsequent change in another species. For example, an alteration in the shape or color of a flower can select for new adaptations in its pollinator. The reverse can also occur: a change in the curve of a hummingbird's beak can select for corresponding modifications in the flower that it pollinates.

Another difference between gymnosperms and flowering plants is that angiosperm seeds develop inside the flower's ovary following fertilization, and other floral parts develop into a fruit that houses the seeds. Thus, an angiosperm's seeds are not "naked." The fruit's main functions are seed protection and dispersal. Some fruits, such as those of dandelions, have "parachutes" that promote wind dispersal. Others have burrs that cling to animal fur. Still others are sweet and fleshy, attracting animals that eat the fruits and later discard the seeds in their feces. **seed dispersal, p. 543**

The angiosperm life cycle also includes other complexities, such as double fertilization, which yields an embryo plus endosperm. The endosperm supplies nutrients to the germinating seedling. Chapter 26 considers angiosperm reproduction in more detail.

Summary *Angiosperms have flowers that produce pollen and egg cells; they enclose their seeds in fruits that protect developing seeds and aid in their dispersal. Monocots and dicots are the two main groups of angiosperms.*

20.5 MASTERING CONCEPTS

1. In what ways are angiosperms similar to and different from gymnosperms?
2. What is the difference between monocots and dicots?
3. How do angiosperms and their pollinators illustrate coevolution?

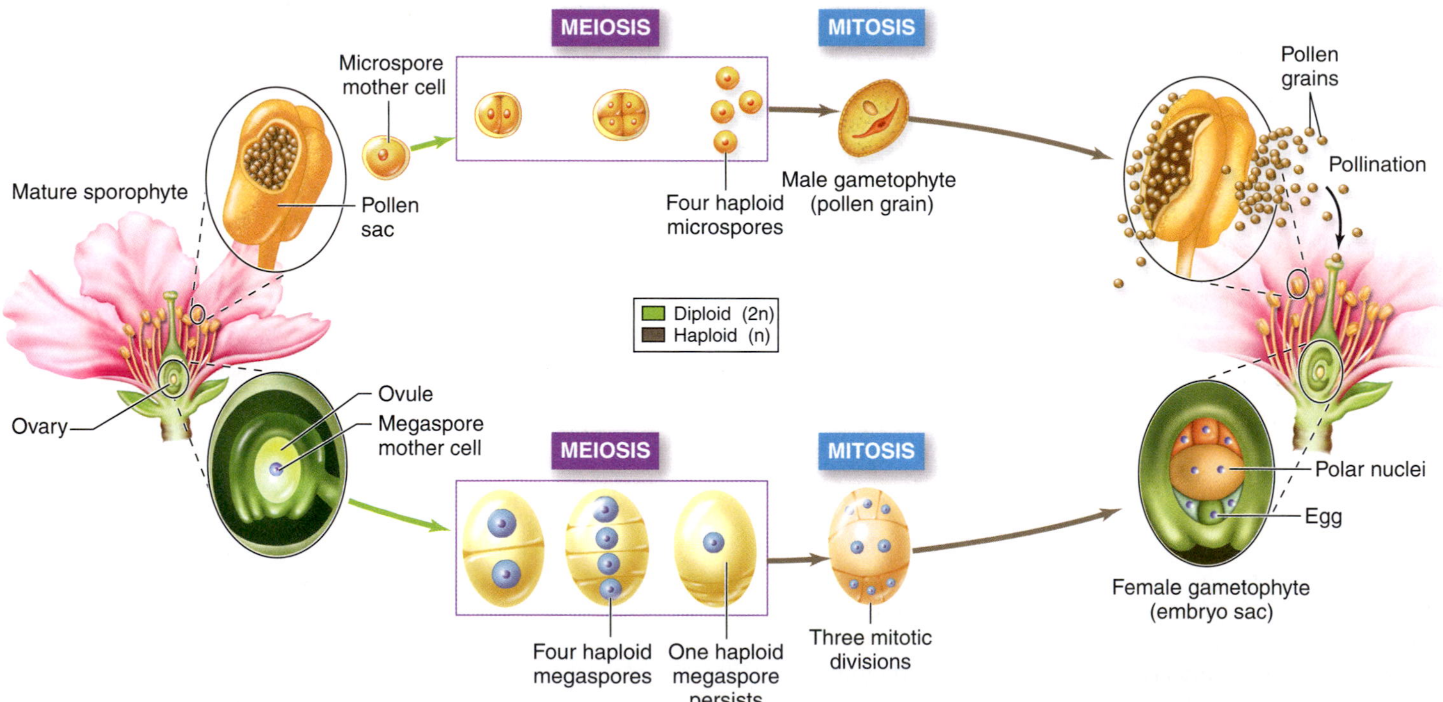

FIGURE 20.19 Sexual Reproduction in Angiosperms. Like gymnosperms, angiosperms produce pollen and seeds. Angiosperm life cycles also include flowers. These structures produce the gametes, house the developing seeds, and develop into fruits that aid in seed dispersal. Chapter 26 discusses angiosperm reproduction in more detail.

20.6 INVESTIGATING LIFE

Birds Do It, Bees Do It … for the Flowers

How do new species form? This question has long intrigued biologists, even before Charles Darwin published *On the Origin of Species*. The formation of a new reproductive barrier (anything that prevents successful interbreeding or the production of fertile offspring) signals the birth of a new species. It is easy to understand how mountains, glaciers, and other physical obstacles can prevent interbreeding. But how can a new reproductive barrier arise within a single population, creating two species from one?

Plants offer an ideal opportunity to answer this question. Many species of angiosperms depend on animals such as birds or bees as pollen carriers. Closely related angiosperm species may employ different pollinators, thus creating a reproductive barrier. Consider two species of wildflowers native to the western United States (**figure 20.20**): the purple monkeyflower (*Mimulus lewisii*) and the scarlet monkeyflower (*Mimulus cardinalis*). Bumblebees pollinate *M. lewisii*, whereas hummingbirds prefer the red-flowered *M. cardinalis*. If, by chance, a hummingbird carries *M. cardinalis* pollen to *M. lewisii*, the resulting hybrid offspring are viable. Such cross-breeding rarely happens in the wild, however, thanks to the more-or-less exclusive relationship between each plant and its pollinators.

Because the two monkeyflowers are so closely related, it is reasonable to suppose that, in the past, one ancestral *Mimulus* species gave rise to both types. The reproductive barrier that separates them may have arisen after the slow, constant buildup of mutations in many genes. Alternatively, a mutation in one or a few "major" genes (those that control flower form or color, for example) may have had the same effect. We know that a single mutation can dramatically affect an organism's appearance (see the animals with homeotic mutations in figure 12.13). Can a single mutation also create a reproductive barrier by attracting a new pollinator?

Biologist H. D. "Toby" Bradshaw, of the University of Washington, and Michigan State University's Douglas Schemske studied the two *Mimulus* species to find out. They considered a gene locus (area of a chromosome) that controls the concentration of yellowish orange pigments called carotenoids in the flower petals. The locus, which may consist of one or a few genes, is called *YUP* (for yellow upper). In the purple monkeyflower, the dominant allele prevents carotenoids from forming, leaving the pink anthocyanin pigments to provide the petal color. Scarlet monkeyflowers have two copies of the recessive *yup* allele. In these plants,

FIGURE 20.20 Wild-Type and Mutant *Mimulus*. The flowers of wild-type *Mimulus lewisii*, the purple monkeyflower, are pinkish purple; those of wild-type *M. cardinalis* are red. Mutant *M. lewisii* flowers resemble the wild-type ones, but they are orange; mutant *M. cardinalis* flowers are dark pink.

carotenoid pigments accumulate, turning the flowers red in the presence of anthocyanins. (**table 20.2**)

Bradshaw and Schemske used a breeding trick to create plants that had different-colored flowers from their parents. They began by crossing purple with scarlet monkeyflowers; the resulting hybrid offspring were heterozygous for the *YUP* locus. They then repeatedly backcrossed the hybrids with the wild-type parents, selecting each time for the desired allele. After four generations of backcrossing, the research-

Table 20.2 *Monkey Flower Genotypes*

Species	Flower Color	Genotype at Flower Color Locus	Pollinator
Purple monkeyflower, *Mimulus lewisii*	Pink-purple	*YUP/YUP* or *YUP/yup*	Bumblebee
Scarlet monkeyflower, *Mimulus cardinalis*	Red	*yup/yup*	Hummingbird

ers had produced two new plant lineages: *M. lewisii* that were homozygous for the recessive *yup* allele, and *M. cardinalis* plants that expressed the dominant *YUP* allele. Each was 97% identical to its parent strain, but with one obvious difference—they had different flower colors. The new variety of *Mimulus lewisii* had pale yellow-orange petals, while the new *M. cardinalis* plants had dark pink blooms (see figure 20.20). The different-colored plants mimicked the effect of a mutation in the *YUP* allele.

To test the hypothesis that a change in the *YUP* locus could prompt a pollinator shift in the two species, Bradshaw and Schemske planted all four *Mimulus* varieties (the two wild-type species and their mutant siblings) at a California location where both species normally occur. For about a week and a half, the researchers observed the plants from dawn until evening, recording the animal species that visited each flower. They found that the substituted *YUP* alleles did indeed alter pollinator preferences (**table 20.3**). The mutant yellow-orange *M. lewisii* flowers attracted fewer bees, but more hummingbirds, than did their wild-type purple counterparts. *M. cardinalis* drew far more bumblebee visits with its new dark pink petals than did the wild-type red plants, while hummingbird visits stayed about the same.

This experiment supports the hypothesis that a change in just one gene locus may have jump-started speciation in *Mimulus*. Additional mutations, and natural selection, eventually sculpted the flower shapes and petal positions that define the purple and scarlet monkeyflower species. It is tempting to extend this finding to other species. Can a single mutation modify an animal's mating ritual just enough to create a new reproductive barrier? Or cause a flower to open its petals at a different time of day? Or enable an animal to exploit a new food source, extending its habitat? This experiment cannot answer these questions. Every now and then, however, small mutations may fuel the birth of an entirely new species.

Table 20.3 *Pollinator Visits to Wild-Type and Mutant* Mimulus *Plants*

	Bumblebee Visits (10^{-3} visits/flower/hour)	Hummingbird Visits (10^{-3} visits/flower/hour)
M. lewisii		
Wild-type (pink-purple)	15.4	0.0212
Mutant (yellow-orange)	2.63	1.44
M. cardinalis		
Wild-type (red)	0.148	189
Mutant (dark pink)	10.9	168

Question: Rather than using plant breeding to swap alleles between the species, the researchers would like to mutate just the gene that controls carotenoid concentration. If they could do so, they would have pairs of plants that differed only in that one gene, improving on the 97% similarity they achieved by backcrossing hybrids with wild-type plants. Explain why it is so important to use plants that differ by only one gene in experiments such as the one described in this study.

Bradshaw, H. D., Jr. and Douglas W. Schemske. November 13, 2003. Allele substitution at a flower colour locus produces a pollinator shift in monkeyflowers. *Nature*, vol. 426, pages 176–178.

CHAPTER SUMMARY

20.1 Plants Have Changed the World

A. Plants Descend from Green Algae

- Current evidence suggests that plants evolved about 480 MYA from green algae called **charophytes.**
- Plants are multicellular eukaryotes that have cellulose cell walls and use starch as a carbohydrate reserve. Except for a few parasitic species, nearly all plants photosynthesize.
- Plant life cycles have an **alternation of generations,** with multicellular **sporophyte** (diploid) and **gametophyte** (haploid) phases. The sporophyte produces haploid **spores** by meiosis; the gametophyte produces haploid sperm and egg cells by mitosis. Fertilization restores the diploid number.
- In the simplest plants, the gametophyte generation is most prominent; in more complex plants, the sporophyte dominates.

B. Plants Have Adapted to Life on Land

- Adaptations that made life on dry land possible include roots, leaves, a waterproof **cuticle, stomata, lignin,** and **vascular tissue** (**xylem** and **phloem**).
- Plants are classified by presence or absence of vascular tissue, seeds, flowers, and fruits.

20.2 Bryophytes Are the Simplest Plants

A. Bryophytes Are Small and Lack Vascular Tissue

- **Bryophytes** are small green plants lacking vascular tissue, leaves, roots, and stems. Lignin hardens bryophyte cell walls; **rhizoids** anchor the plants to the ground.
- The three groups of bryophytes are **mosses, liverworts,** and **hornworts.**

B. Bryophytes Have a Conspicuous Gametophyte and Swimming Sperm

- In bryophytes, the gametophyte stage is dominant. Many bryophytes reproduce asexually by fragmentation of the gametophyte. In sexual reproduction, sperm require free water to swim to egg cells.

20.3 Seedless Vascular Plants Have Xylem and Phloem but No Seeds

A. Seedless Vascular Plants Include Ferns and Their Allies

- **Seedless vascular plants** have vascular tissue but lack seeds. This group includes **true ferns, club mosses, horsetails,** and **whisk ferns.**

B. Seedless Vascular Plants Have a Conspicuous Sporophyte and Swimming Sperm

- The diploid sporophyte generation is the most obvious stage of a fern life cycle, but the heart-shaped haploid gametophyte still forms a tiny separate plant.
- In the sexual life cycle of ferns, collections of sporangia appear on the undersides of **fronds.** Meiosis occurs in the sporangia and yields haploid spores, which germinate in soil and develop into gametophytes. The gametophytes produce egg cells and sperm.
- Like bryophytes, seedless vascular plants have swimming sperm. Unlike bryophytes, however, they have a dominant sporophyte generation.

20.4 Gymnosperms Are "Naked Seed" Plants

A. Gymnosperms Include Conifers and Three Related Groups

- **Gymnosperms** are vascular plants with **seeds** that are not enclosed in fruits. The four groups of gymnosperms are **conifers, cycads, ginkgos,** and **gnetophytes.**

B. Conifers Produce Pollen and Seeds in Cones

- In pines (a type of conifer), male **cones** produce **pollen grains,** and female cones produce egg cells inside **ovules.** Pollen germination yields a **pollen tube,** through which a sperm cell fertilizes the egg cell. The resulting embryo remains dormant in a seed until germination.
- In conifers, the sperm do not require free water to swim to the egg cell. Instead, most gymnosperms rely on wind to spread pollen grains.

20.5 Angiosperms Produce Seeds in Fruits

A. Angiosperms Are Flowering Plants

- **Angiosperms** are vascular plants that produce **flowers.** These reproductive structures, which may or may not be showy, produce pollen and egg cells. As seeds develop inside, floral parts develop into a **fruit.**
- The two major groups of angiosperms are **monocotyledons** and **dicotyledons.**

B. Flowers and Fruits Are Unique to Angiosperms

- Like gymnosperms, angiosperms use pollen to transport sperm to egg cells; both groups also produce seeds. Only angiosperms, however, have flowers that produce gametes. The fruits of angiosperms protect developing seeds and aid in dispersal.

- Many animal-pollinated angiosperms have **coevolved** with their pollinators.

20.6 Investigating Life: Birds Do It, Bees Do It ... for the Flowers

- Researchers have studied *Mimulus* plants to determine how the reproductive barrier that separates two closely related species may have arisen.
- A single mutation in the gene that controls petal color can induce flowers to attract new pollinators.

MULTIPLE CHOICE QUESTIONS

1. Which of the following is NOT a property common to both land plants and green algae?
   a. Photosynthesis using chlorophyll *a*
   b. Starch as a storage form of energy
   c. Cellulose cell walls
   d. The presence of a cuticle and stomata

2. In the alternation of generations in plants, the gametophyte is _______ and produces gametes by _______.
   a. haploid; mitosis
   b. diploid; mitosis
   c. haploid; meiosis
   d. diploid; meiosis

3. Which group of bryophytes is most like its green algae ancestor?
   a. Mosses
   b. Liverworts
   c. Hornworts
   d. Both a and c

4. When spores are released from a sporangium, they form
   a. a sperm cell.
   b. a gametophyte.
   c. a sporophyte.
   d. a gemma.

5. How does the presence of vascular tissue (xylem and phloem) affect a plant?
   a. It reduces the plant's dependence on a moist environment.
   b. It allows specialization of roots, leaves, and stems.
   c. It allows for the growth of larger plants.
   d. All of the above.

6. Why do ferns require a shady, moist environment?
   a. They lack vascular tissue.
   b. They use swimming sperm for sexual reproduction.
   c. They lack a water-retaining cuticle layer.
   d. The production of spores within the sporangium requires moisture.

7. Which type of plant produces a seed that is surrounded by a fruit?
   a. Bryophyte
   b. Gymnosperm
   c. Angiosperm
   d. True fern

8. Reproduction in a gymnosperm is associated with
   a. male and female cones.
   b. airborne pollen grains.
   c. the formation of pollen tubes.
   d. All of the above.

9. What is one key adaptation of the angiosperms?
   a. Dominant gametophyte generation
   b. Pollen grains that use pollen tubes to fertilize egg cells
   c. The use of flowers as reproductive structures
   d. Both a and b

10. How does the concept of co-evolution apply to the angiosperms?
    a. Angiosperms co-exist with simpler land plants like bryophytes.
    b. Evolutionary changes to the environment have altered the diversity of fruits.
    c. The shape of a flower selects for the type of pollinator that can visit it.
    d. The type of pollinator selects for the form of seed dispersal.

TESTING YOUR KNOWLEDGE

1. What characteristics do all land plants have in common?
2. Which organisms are considered the ancestors of land plants?
3. List the adaptations that enable plants to live on land.
4. What are the four major groups of plants and the characteristics that distinguish them?
5. What are the main groups of plants within the bryophytes, seedless vascular plants, gymnosperms, and angiosperms?
6. Describe how sexual reproduction in each of the four groups of plants illustrates alternation of generations.
7. What adaptations in gymnosperms and angiosperms enable them to live in drier habitats than bryophytes and seedless vascular plants?
8. How do angiosperms differ from gymnosperms?
9. What is the function of a flower? Of a fruit? Of a seed?

THINKING AS A SCIENTIST

1. List at least two reasons that bryophytes are much smaller than most vascular plants.
2. Which adaptations enabled seedless vascular plants to invade new habitats?
3. A fern plant can produce as many as 50 million spores a year. How are these spores similar to, and different from, seeds? In a fern population that is neither shrinking nor growing, what proportion of these spores is likely to survive long enough to reproduce? What factors might determine whether an individual spore successfully produces a new fern plant?
4. What advantage did the development of seeds offer in early plant evolution?
5. Some websites for gardeners warn against the "foul-smelling fruits" that female ginkgo trees produce. In what way is this statement incorrect?
6. Some viroids (infectious RNA molecules that cause diseases in plants) move to new plant hosts in pollen. Which of the four main groups of plants could "catch" viroids in this way?
7. The immature fruit of the opium poppy produces many chemicals that affect animal nervous systems. In what way might this benefit the plant?
8. Some scientists hypothesize that a geological catastrophe, such as an asteroid impact, killed the dinosaurs. If this is true, how might angiosperms have survived such a catastrophe?
9. Some angiosperm species have exclusive relationships with just one species of pollinator. How would this relationship benefit the plant? What are the risks to the plant?
10. Other than *Arabidopsis thaliana* and corn (*Zea mays*), what are some examples of plants that have contributed to scientists' knowledge of general biology and plant biology? What are some examples of plants and plant parts important in your everyday life?

Visit www.mhhe.com/hoefnagels for practice quizzes, animations, videos, and activities designed to help you master the material in this chapter.

CHAPTER 21

Fungi

Mushrooms Galore. This man is harvesting white button mushrooms in a "grow room."

The Mushroom Mystique

For many people, the word fungi conjures images of mushrooms, along with emotions that range from delight to revulsion. Enthusiasts know that mushrooms can be delectable, yet fairy tales associate toadstools with dank, lonely forests teeming with witches and evil spirits.

Mushroom imagery is common in everyday language. A phrase like "the mushrooming of private enterprise in China," for example, evokes the amazing growth rates of fungi. When warm weather follows a rain, mushrooms seem to pop up overnight on lawns and around trees. This extraordinarily rapid growth adds to their mystique.

How can mushrooms appear so suddenly? Like an apple on a tree, a mushroom is a reproductive structure attached to a much larger organism. Most of the fungal body, however, is underground. Fungi produce filaments that feed by penetrating their food, absorbing nutrients as they go. Sometimes, the two threads unite, and the combined organism prepares to reproduce. When the temperature and moisture are just right, the cells of tiny premushrooms absorb water, expanding rapidly. Mushrooms erupt out of the ground and shed reproductive cells called spores. Afterward, the mushrooms wilt, vanishing as quickly as they appeared.

Picking and eating wild mushrooms can be extremely dangerous. The deadliest mushrooms, such as the false morel (Gyromitra esculenta) *and the death angel* (Amanita virosa), *kill by preventing a person's cells from producing RNA or vital enzymes. The poisons can be deceptive. Initial symptoms of nausea and diarrhea are followed by a lag of about 24 hours, during which the toxins quietly destroy the liver, kidney, and other organs. By the time symptoms return, the damage is irreversible, and an emergency organ transplant is the only hope for survival. Unfortunately, there is no simple way to tell poisonous species from safe ones.*

Commercial mushroom growers cultivate many edible fungi. The familiar white "button mushroom" (Agaricus brunnescens) *grows on a compost of horse manure and straw. Shiitakes* (Lentinus edodes) *live on hardwood logs or compressed sawdust; oyster mushrooms* (Pleurotus *species*) *prefer wheat or rice straw. Hobbyists can purchase kits to grow these species at home.*

Mushrooms represent only a small subset of kingdom Fungi. Diverse fungi spoil food and cause Dutch elm disease, chestnut blight, athlete's foot, diaper rash, ringworm, and yeast infections. On the other hand, they are essential decomposers in ecosystems, and they help humans manufacture cheese, bread, alcoholic beverages, dyes, plastics, soaps, toothpaste, drugs, soy sauce, flavorings, fuel, and food coloring. Moreover, research on yeasts and other simple fungi has helped biologists learn about genetics and basic cell processes. This chapter opens the door to this fascinating group.

LEARNING OUTLINE

21.1 Fungi Are Essential Decomposers

Fungi live nearly everywhere—in soil, in and on plants and animals, in water, and in dung. Microscopic fungi infect living plant cells, and massive fungi extend enormous distances (**figure 21.1**). For example, a single underground fungus extends over 600 hectares (about 1500 acres) in Washington State, coating tree roots.

In addition to the many ways fungi benefit humans directly (see the chapter opening essay), fungi are vitally important in ecosystems. They secrete digestive enzymes that break down dead plants and animals in soil, releasing inorganic nutrients and recycling them to plants. These organisms are, in a sense, the garbage processors of the planet.

a.

b.

FIGURE 21.1 Fungi Range Greatly in Size. (**A**) Powdery mildew fungi live on leaf surfaces and produce microscopic reproductive structures (inset). (**B**) Other fungi are enormous. One individual of *Armillaria gallica* in Michigan extends underground for about 15 hectares (37 acres). Here, *A. gallica* mushrooms emerge from a parasitized white birch tree (*Betula papyrifera*).

A. Fungi Are Heterotrophs with Cell Walls of Chitin

Molecular evidence clearly places the fungi closer to animals than to plants. This finding may surprise those who notice the superficial similarities between plants and fungi. Unlike plants, however, fungi do not photosynthesize, and these two groups of organisms differ in many other respects.

Fungi share several characteristics:

- Like animals, fungi are heterotrophs. Their cells secrete enzymes that break down organic matter, then "eat" by absorbing the nutrients that the enzymes release.
- Fungal cell walls are composed primarily of the modified carbohydrate chitin, a tough, flexible molecule that also forms the exoskeletons of some animals. **carbohydrates, p. 34**
- The storage carbohydrate of fungi is glycogen, the same as for animals (see figure 2.16).
- In most fungi, the zygote is the only diploid cell. Meiosis occurs in the zygote and yields haploid nuclei, which subsequently divide mitotically as the organism grows. Some fungi remain haploid throughout most of their life cycles. In other fungi, two individuals fuse and form a **dikaryotic** stage in which each cell retains two separate nuclei. Dikaryotic cells are unique to the fungi. **mitosis, p. 161; meiosis, p. 177**

FIGURE 21.2 Yeast Cells. A yeast is a unicellular fungus. These cells are reproducing by budding.

B. Most Fungi Are Filamentous Spore-Formers

Although most fungi are multicellular, the **yeasts** are unicellular. The bread yeast *Saccharomyces cerevisiae,* for example, is a single-celled fungus that reproduces by budding (**figure 21.2**). The distinction between yeasts and multicellular fungi is not absolute; some species can switch between uni- and multicellular phases.

Hyphae (singular: hypha) are microscopic filaments that make up the bulk of a multicellular fungus (**figure 21.3**). A fungus may have an enormous number of hyphae that branch rapidly within a food source, growing and absorbing nutrients at their tips. A **mycelium** is a mass of aggregated hyphae that may form visible strands in soil or decaying wood.

While the feeding hyphae remain hidden in the fungus's food, the reproductive structures emerge. Most fungi produce **spores,** which are microscopic reproductive cells. When they land in a suitable habitat, the spores germinate and give rise to new feeding hyphae. In many species, such as those that form mold on food, individual hyphae produce asexual spores called conidia (**figure 21.4**; see the Burning Question: Why does bread get moldy?). In the most complex species of fungi, hyphae aggregate to form a **fruiting body,** a specialized sexual spore-producing organ such as a mushroom, puffball, or truffle.

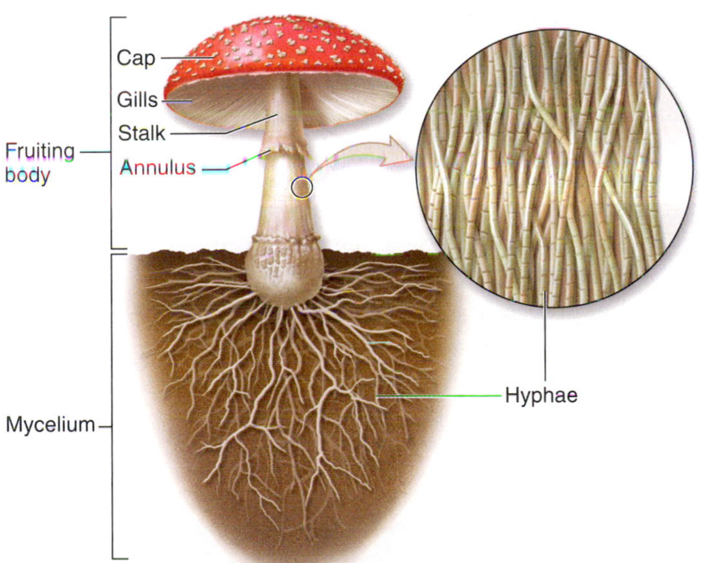

FIGURE 21.3 The Fungal Body. A mushroom is one type of fruiting body, or reproductive structure. It arises from the mycelium, which is a collective term for the extensive network of hyphae that penetrates the fungus's food source. Like the mycelium, the mushroom is made of hyphae, but they are tightly aligned to form a solid structure.

FIGURE 21.4 Conidia. This fungus is producing abundant asexual spores.

C. Fungal Classification Is Based on Reproductive Structures

Despite the widespread distribution and varying habitats of fungi, we know little of their diversity. Mycologists (biologists who study fungi) have identified about 80,000 species, but 1.5 million or so are thought to exist.

Burning Questions

Why does bread get moldy?

Fungal spores are everywhere. They germinate and grow into colonies only if they land on a surface that provides enough food and moisture. Fresh bread, cheese, and fruit are all perfect for fungal growth.

Perhaps a better question would be: why *don't* some foods get moldy? Humans have devised many ways to preserve food. Refrigeration dramatically slows the rate of fungal growth. Salt and sugar, in sufficiently high concentrations, also retard mold growth by limiting the fungus' ability to take up water by osmosis. Dried foods are preserved in the same way. Cooking and pickling foods prevent food spoilage by damaging microbial enzymes. **osmosis, p. 91**

One additional method is to add chemical preservatives to foods. Organic acids such as sodium benzoate are common food additives that inhibit mold growth by disrupting fungal cell membranes. Many processed foods are so laden with preservatives that their shelf lives extend for years, a remarkable accomplishment in a world full of hungry microbes! **membranes, p. 54**

Have a Burning Question of your own? Submit it to marielle_hoefnagels@mcgraw-hill.com for possible inclusion in future editions of this book!

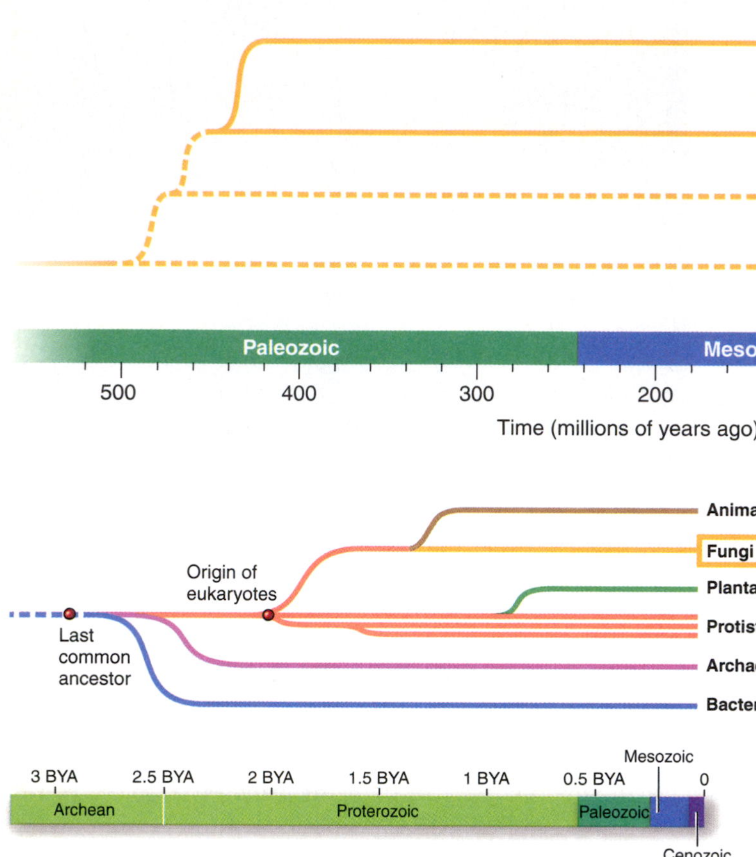

FIGURE 21.5 Fungal Phyla. The fungal kingdom contains four phyla, distinguished mainly on the basis of spore type. Broken lines indicate uncertain relationships among phyla. The time at which fungi originated remains uncertain, but they evidently colonized land around the same time as plants.

Mycologists classify fungi into four phyla based on sexual structures (**table 21.1** and **figure 21.5**). The **chytridiomycetes** (chytrids) are the simplest fungi; they produce sexual and asexual spores that have flagella. **Zygomycetes** produce thick-walled sexual zygospores. **Ascomycetes** produce sexual spores in characteristic sacs, and **basidiomycetes** release sexual spores from club-shaped structures. A fifth group, called the **deuteromycetes** (or "imperfect fungi"), contains fungi that have no known sexual phase. Most produce abundant asexual spores; many common molds fall into this category. Molecular data have now placed many asexual fungi, such as *Penicillium* and *Candida*, within the ascomycetes.

Biologists know little about the evolutionary relationships of the chytrids and zygomycetes within the fungi. The ascomycetes and basidiomycetes, however, are clearly sister groups, meaning that they share derived characters that no other groups have. They diverged most recently from a shared ancestor, as figure 21.5 indicates. These are the most complex fungi. Their fruiting bodies, which include morels, truffles, mushrooms, and puffballs, are usually visible with the unaided eye.

Summary *Fungi are heterotrophic eukaryotes that profoundly influence ecosystems. Their filamentous bodies grow within their food, secreting digestive enzymes and absorbing nutrients. They also produce reproductive cells called spores. Fungal classification is based largely on sexual structures.*

21.1 MASTERING CONCEPTS

1. In what ways are fungi important in ecosystems?
2. What characteristics define the fungi?
3. What evidence suggests that fungi are more closely related to animals than to plants?
4. Describe the major structures of a fungus.
5. What are the four main groups of fungi?

Table 21.1 *Four Groups of Fungi*

Group	Sexual Spore Type	Feeding Hyphae
Chytridiomycetes	Flagellated zoospore	Variable. Some species do not have true hyphae; others produce haploid or diploid hyphae.
Zygomycetes	Diploid zygospore formed from union of two cells	Haploid, without cross walls separating nuclei
Ascomycetes	Haploid ascospore formed in ascus after meiosis	Usually haploid, with cross walls separating nuclei
Basidiomycetes	Haploid basidiospore formed on basidium after meiosis	Usually dikaryotic, with cross walls separating pairs of nuclei

21.2 Chytrids Are Fungi that Produce Swimming Spores

The chytrids are sometimes called primitive fungi, and they may provide a glimpse of what the earliest members of this kingdom were like. Their body forms vary from single cells to slender hyphae, and they differ from other fungi in that they produce **zoospores**—motile spores—each with a single flagellum (**figure 21.6**). Biologists have yet to discover the details of most chytrid life cycles.

These microscopic fungi are powerful decomposers, secreting enzymes that degrade cellulose, chitin, and keratin. One ecosystem where resident chytrids are particularly valuable is a ruminant's digestive tract. There, anaerobic chytrids start digesting the cellulose in a cow's grassy meal, paving the way for bacteria to continue the process.

Chytrids also contribute to the ongoing worldwide decline in amphibian populations (see section 42.5). The chytrid *Batrachochytrium dendrobatidis* causes a lethal skin disease called cutaneous chytridiomycosis in frogs (**figure 21.7**). Once in contact with the frog, the fungus feeds on keratin, coating the host's legs and undersides and impairing the frog's ability to breathe through its skin. The fungus spreads to new hosts by releasing zoospores into the water.

amphibians, p. 486

Summary *Chytrids are microscopic and widespread, but their life cycles are variable and poorly understood. They are unique among fungi in producing flagellated spores.*

21.2 MASTERING CONCEPTS

1. What are zoospores, and how do they adapt chytrids to moist environments?
2. What do chytrids eat?

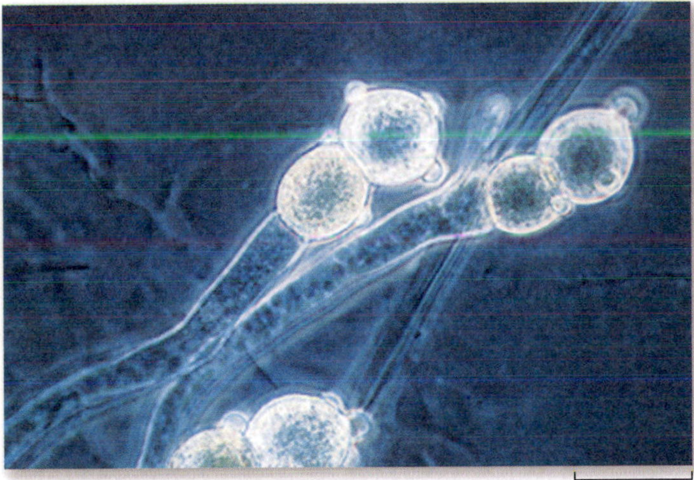

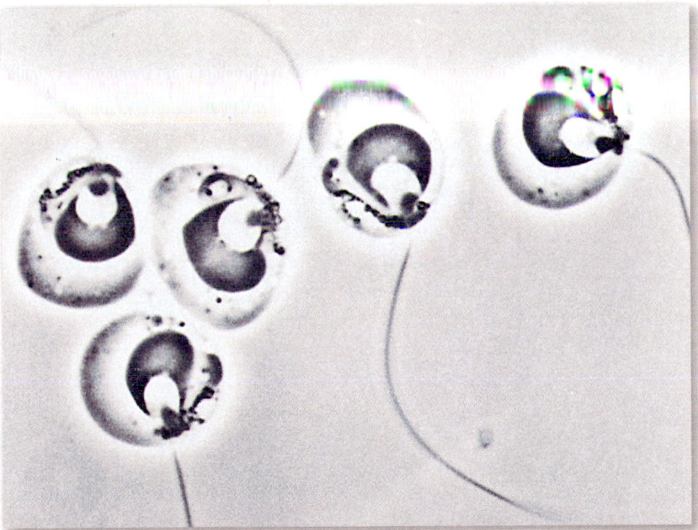

FIGURE 21.6 Chytrids. (**A**) Hyphae and sporangia of the chytrid *Allomyces*. Flagellated zoospores will emerge from the rounded sporangia at the tips of the hyphae. (**B**) Zoospores from the chytrid *Blastocladiella*.

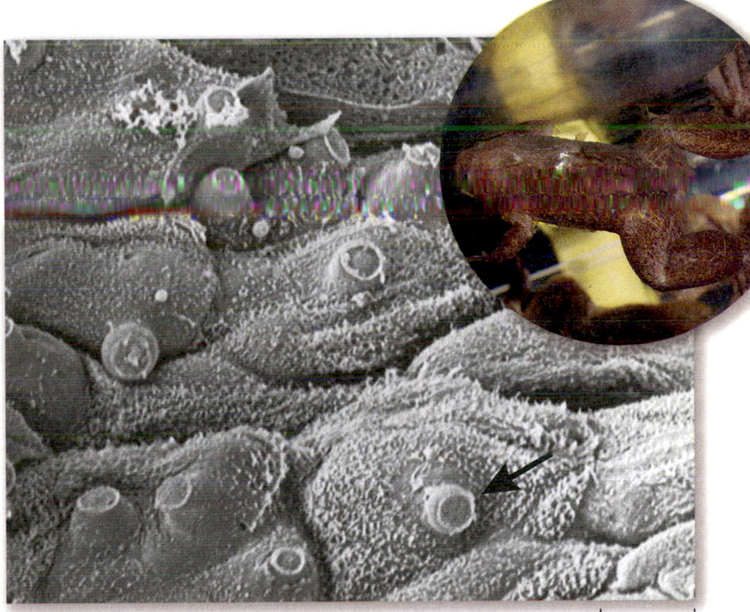

FIGURE 21.7 Amphibian Infection. A chytrid is partially responsible for the decline in amphibian populations. In these infected frog skin cells, the arrow shows a tube through which the chytrid's zoospores leave the host. The inset shows a frog infected with the chytrid.

21.3 Zygomycetes Are Fast Growing and Prolific

The 900 or so species of zygomycetes account for only about 1% of identified fungi, but they include some familiar organisms. The black mold, *Rhizopus stolonifer*, which grows on bread, fruits, and vegetables, is a zygomycete. In addition to forming black fuzz on refrigerated leftovers, zygomycetes occur on decaying plant and animal matter in soil. Some are parasites of insects, and others colonize dung (**figure 21.8**).

An extremely important group of fungi lives inside plant roots, forming a mutually beneficial relationship with many grasses, trees, shrubs, and other plants (see section 21.6). They are unusual in that they can live only in association with roots, they will not grow in pure culture, and they have no known sexual phase. These fungi were once classified as zygomycetes, but biologists are considering placing them into a fifth phylum, the Glomeromycota.

The zygomycetes are known for their spectacular growth rates, but they take their name from their mode of sexual reproduction: *zygon* is Greek for "yoke," implying the joining of two parts. This is what happens when two hyphae fuse [location (*1*) in **figure 21.9**]. Then their haploid nuclei merge into a new structure, a diploid **zygospore** with a distinctive spiny, dark wall (*2*). After the merger, the cells from which they arose appear as two vacated areas that hug the zygospore (see figure 21.9 inset). The diploid zygospore nucleus undergoes meiosis, and a haploid hypha emerges (*3*). The hypha immediately produces a spore sac, which breaks open and releases numerous haploid spores (*4*). Each spore then gives rise to a hypha that grows into a haploid mycelium (*5*).

In zygomycetes, asexual spores are much more common than zygospores. The abundant haploid conidia spread easily to new habitats, as the explosive spores of *Pilobolus* show (see figure 21.8). Spore germination produces a hypha, which quickly grows and branches as it digests its food. Within days, the new mycelium sprouts its own spore sacs, and the cycle begins anew.

Summary *The zygomycetes reproduce asexually from lightweight spores and reproduce sexually when haploid hyphae of different mating types merge.*

21.3 MASTERING CONCEPTS

1. Where do zygomycetes occur?
2. What is a zygospore, and how does it fit into the life cycle of a zygomycete?

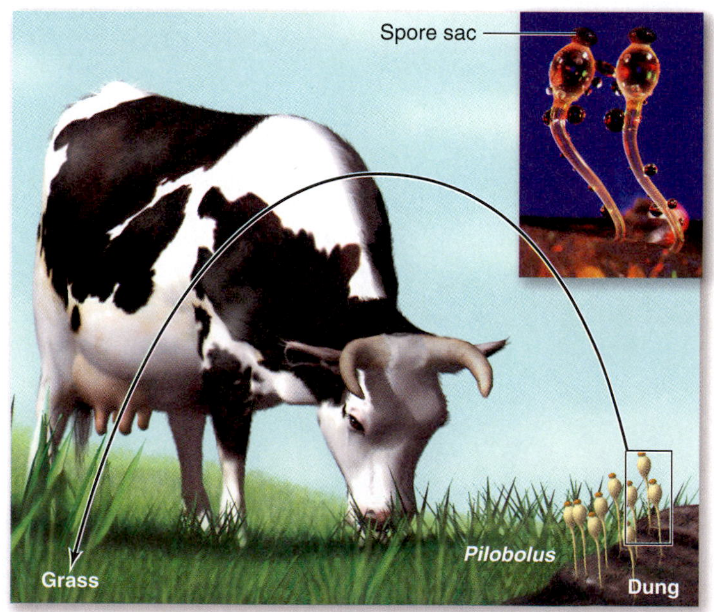

FIGURE 21.8 Diverse Zygomycetes. (**A**) Spores of the parasitic fungus *Entomophthora muscae* germinate on a fly's exoskeleton and penetrate the fly's armor, killing its host within days. A few hours later, the fungus emerges from between the host's body segments and produces a shower of spores (*inset*). More flies become infected when they explore the corpses. (**B**) *Pilobolus*, the "hat thrower," produces spores that must pass through a cow's digestive system to germinate. The fungus grows a stalk topped with a black case containing thousands of spores (*inset*). Exposure to sunlight causes water pressure to build until the stalk literally blows its top. The spore case, accompanied by a sticky substance, may fly as far as 2.5 m (about 8 feet). The spores land on grass, where cows will eat them and pass them, undigested, in their dung.

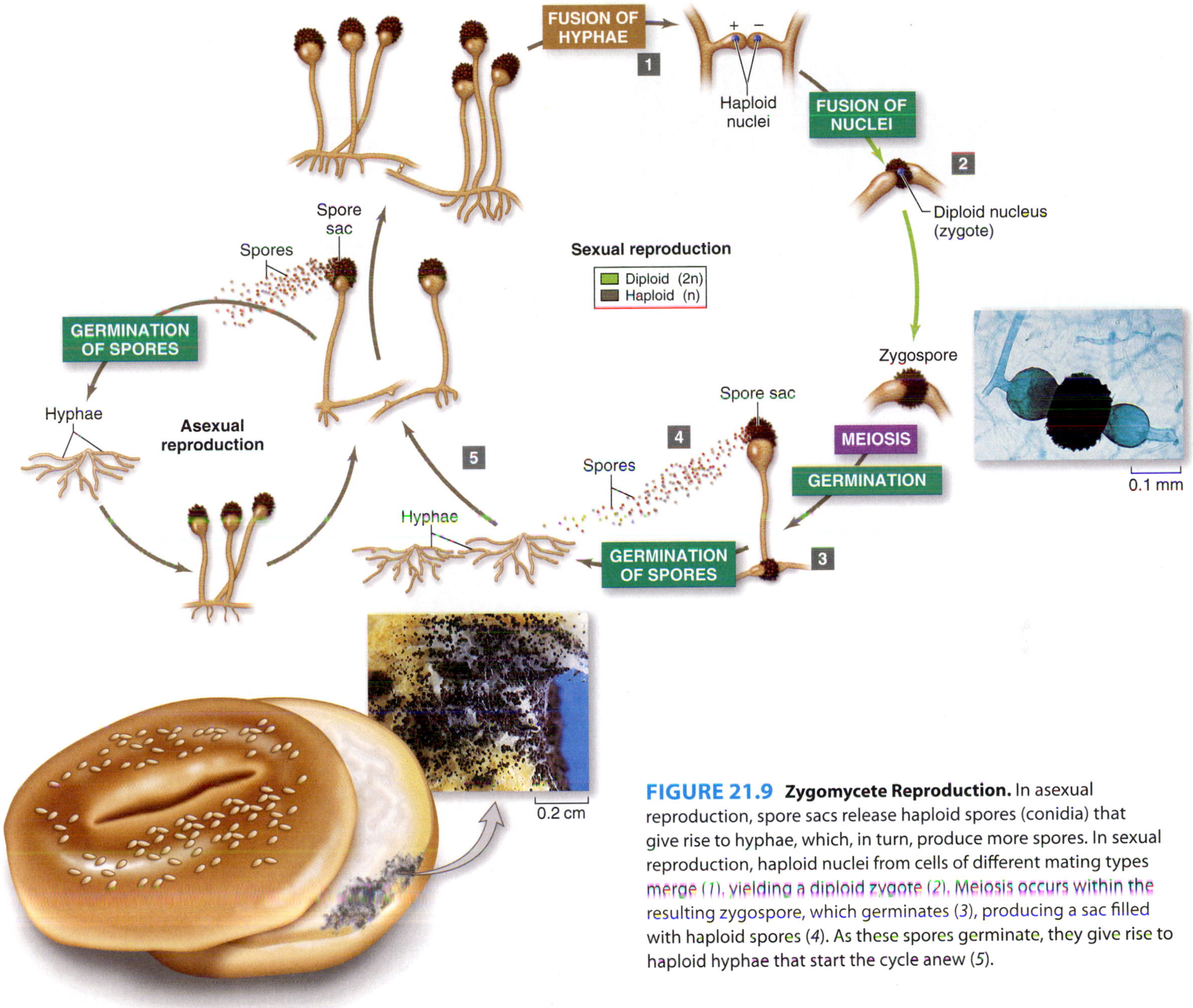

FIGURE 21.9 Zygomycete Reproduction. In asexual reproduction, spore sacs release haploid spores (conidia) that give rise to hyphae, which, in turn, produce more spores. In sexual reproduction, haploid nuclei from cells of different mating types merge (*1*), yielding a diploid zygote (*2*). Meiosis occurs within the resulting zygospore, which germinates (*3*), producing a sac filled with haploid spores (*4*). As these spores germinate, they give rise to haploid hyphae that start the cycle anew (*5*).

21.4 Ascomycetes Are the Sac Fungi

The more than 30,000 ascomycete species make up the largest group of known fungi. Some recycle leaves and wood in soil, and others form partnerships with photosynthetic organisms (see section 21.6). A few are carnivores (**figure 21.10**).

Many ascomycetes are important to humans. They cause most fungal diseases of plants, including Dutch elm disease and chestnut blight. A few species cause skin infections such as athlete's foot and other

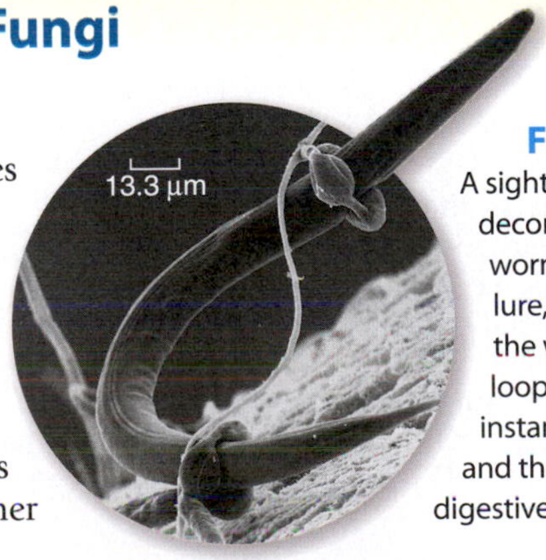

FIGURE 21.10 A Carnivorous Fungus. A sightless nematode worm burrowing through decomposing wood detects a sweet odor, and the worm approaches it. But the promise of food is a lure, set by the fungus *Arthrobotrys anchonia*. When the worm touches the sweet-smelling threadlike loops emanating from the fungus, it sticks and instantly becomes trapped. The loops constrict, and the fungus threads itself into its prey, releasing digestive enzymes and eating it from within.

a.

b.

FIGURE 21.11 Edible Ascomycetes. (**A**) Truffles are familiar ascomycetes. A typical truffle is about 2 to 4 cm in diameter and produces asci on internal folds of tissue. (**B**) Morels are also edible. This morel is about 10 cm tall. Each "pit" in the morel's cap produces thousands of asci, each containing eight ascospores.

human diseases. Nevertheless, their benefits to humans are immeasurable. The common mold *Penicillium* is famous for producing penicillin, the first antibiotic ever discovered; other *Penicillium* species lend their sharp flavors to Roquefort and other cheeses. Fermentation by yeasts such as *Saccharomyces* is essential in the baking, beer brewing, and winemaking industries. *Aspergillus oryzae* ferments soybean pulp in the production of soy sauce. Cyclosporine, a drug that suppresses the immune systems of organ transplant recipients, comes from an ascomycete. Truffles and morels are prized for their unique and delicious taste (**figure 21.11**).

Ascomycetes can produce enormous numbers of asexual spores. *Neurospora crassa*, for example, produces abundant orange conidia, giving this fungus its common name of "red bread mold." The Focus on Model Organisms box describes some of the lessons biologists have learned from this organism.

FOCUS ON MODEL ORGANISMS: *Neurospora crassa*

Why would biologists study *Neurospora crassa*, a filamentous ascomycete commonly known as red bread mold? First, since all fungi share a common ancestor, *N. crassa* can tell us about other ecologically, industrially, and medically important fungi. Second, *N. crassa* has the traits that all model organisms share: it is small and easy to grow, and it has a rapid reproductive cycle. Third, this fungus has special advantages for tracing inheritance patterns. Because its hyphae are haploid, *N. crassa* expresses every recessive or mutated allele in its DNA (in diploid organisms, dominant alleles can cause their recessive counterparts to appear to "hide").

A full accounting of *Neurospora*'s contributions to basic biology could literally fill a book. The following list, however, details some ways this fungus has enhanced our understanding of life:

- **Crossing over.** In the 1920s and 1930s, *Neurospora* provided the first clear evidence of crossing-over, which occurs during prophase I of meiosis (see chapter 9). *Neurospora*'s narrow, tube-shaped asci are especially well suited to such studies because they retain the products of meiosis in the original order in which the cells divided (see figure 21.12).
- **One-gene/one-enzyme hypothesis.** In the 1940s, George Beadle, Edward Tatum, and their colleagues isolated *Neurospora* mutants with unusual nutritional requirements. Their observation that each mutant had a single enzyme deficiency provided convincing evidence for the idea that a single gene directs the production of each protein in a cell (see chapters 4 and 12).

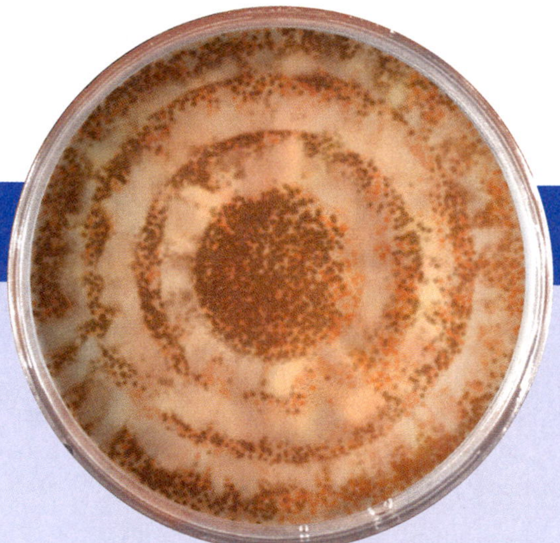

FIGURE 21.A Conidia of *Neurospora crassa*. *Neurospora* produces chains of orange asexual spores (conidia) in rhythmic cycles. Here, the fungus was initially inoculated onto the center of a petri dish filled with agar. The hyphae produced spores as they grew outward; thus, a ring of conidia marks each day's growth.

- **Circadian rhythms.** As shown in **figure 21.A,** *Neurospora* cultures produce conidia at roughly 24-hour intervals, a phenomenon called a circadian rhythm (*circa*, "about;" *dies*, "day;" see chapter 26). Mutants that produce conidia constantly or not at all have yielded insights into how cells of many organisms regulate so-called clock-controlled genes.
- **Gene regulation.** Recall from chapter 12 that some genes in a cell are "silent" at any given time. Work on *Neurospora* has helped reveal some of the ways cells regulate gene expression. For example, cells can "tag" DNA with chemical groups that prevent transcription, or they may destroy RNA after transcription.

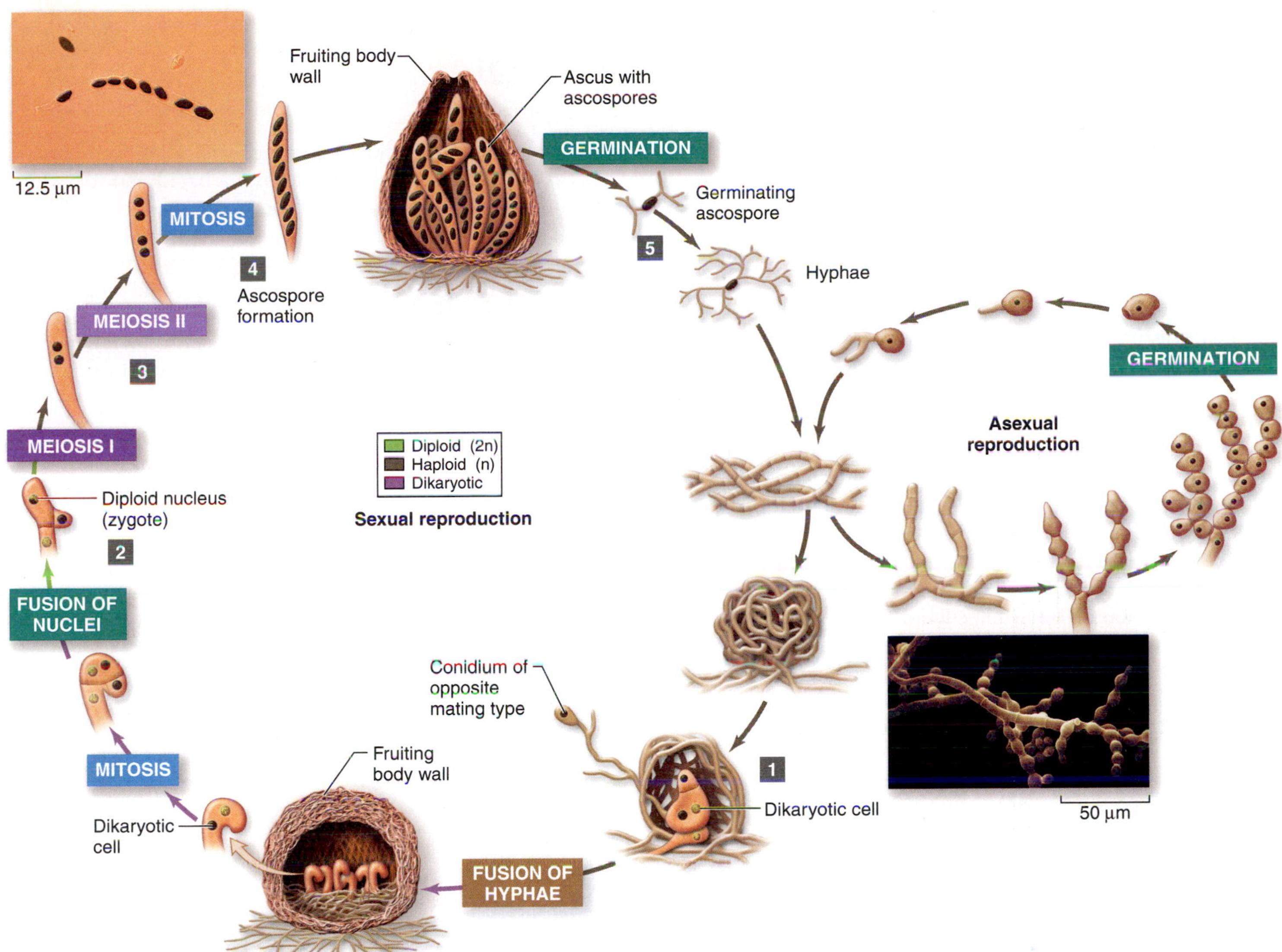

FIGURE 21.12 Ascomycete Reproduction. Many ascomycetes reproduce asexually by producing haploid conidia on haploid hyphae. In sexual reproduction, two compatible haploid hyphae fuse (*1*). For a short time, the nuclei remain separate as the cells continue to divide, producing dikaryotic cells. Eventually the nuclei fuse, producing a diploid zygote (*2*). This cell undergoes meiosis (*3*); a subsequent mitotic division (*4*) usually yields eight haploid spores inside an ascus. Once released, each ascospore germinates (*5*) and gives rise to new haploid hyphae, beginning the cycle again.

When ascomycetes reproduce sexually, the hyphae of compatible mating types fuse [location (*1*) in **figure 21.12**], but the individual nuclei from the two parents do not immediately merge. The resulting cell is dikaryotic. When the cell divides, the two nuclei undergo mitosis separately, each retaining its genetic identity. The fruiting body consists of tightly woven hyphae, often in a cuplike or bottlelike shape, that form a fertile layer of dikaryotic cells. Eventually, the two nuclei in each dikaryotic cell fuse, forming a diploid zygote (*2*). The zygote immediately undergoes meiosis, producing four haploid nuclei (*3*), each of which usually divides once by mitosis. The result is eight haploid **ascospores** (*4*), so named because they form in a saclike **ascus** (plural: asci). After dispersal by wind, water, or animals, ascospore germination yields a new haploid individual (*5*).

Summary *The sac fungi include many important organisms. Ascomycetes reproduce asexually by producing conidia, and sexually by producing ascospores inside a saclike ascus.*

21.4 MASTERING CONCEPTS

1. List examples of ascomycetes that are important to humans.
2. How do dikaryotic cells fit into the life cycle of an ascomycete?
3. How does an ascus come to contain eight ascospores?

21.5 Basidiomycetes Are the Familiar Club Fungi

The 30,000 or so species of basidiomycetes account for about a third of all described fungi. Familiar examples include mushrooms, toadstools, puffballs, stinkhorns, shelf fungi, and bird's nest fungi (**figure 21.13**). Some mushrooms are edible, some are deadly, and others are hallucinogenic (see the Can *You* Relate? Fungi That Harm Human Health). Basidiomycetes called smuts and rusts are plant pathogens, causing serious diseases of cereal crops such as corn and wheat. In forests, wood decay fungi play a vital role in Earth's carbon cycle. Unfortunately, their talent for degrading the cellulose and lignin in fallen logs also makes them serious pests in another context—they cause dry rot in wooden wall studs and other building materials.

Basidiomycetes can reproduce asexually, but the sexual portion of the life cycle is usually the most prominent. As **figure 21.14** shows, the fusion of two haploid hyphae creates a dikaryotic mycelium (*1*). This mycelium typically grows unseen within its food source; but when environmental conditions are favorable, it produces one or more mushrooms (*2*). Lining a mushroom's gills are numerous dikaryotic, club-shaped cells called **basidia** (singular: basidium) (*3*). Inside each basidium, the haploid nuclei fuse, giving rise to a diploid zygote (*4*). The zygote immediately undergoes meiosis, yielding four haploid nuclei (*5*). Each nucleus migrates into a **basidiospore,** which germinates after dispersal and produces a haploid hypha (*6*). The cycle begins anew.

Sometimes, a circle of mushrooms emerges from the ground all at once. The growth pattern of the underground fungal mycelium explains this phenomenon. The hyphae extend in all directions from their food source. Mushrooms poke up above these spreading mycelia, creating the "fairy rings" of folklore.

Summary ***The club fungi include many familiar species; they often produce large, easily recognized fruiting bodies. They have dikaryotic hyphae for most of the life cycle. Spore-bearing structures called basidia produce sexual spores.***

FIGURE 21.13 A Gallery of Club Fungi (Basidiomycetes). (**A**) Puffballs (genus *Lycoperdon*), (**B**) stinkhorns (*Phallus impudicus*), (**C**) turkey tail bracket fungi (*Trametes versicolor*), and (**D**) bird's nest fungi (order Nidulariales).

21.5 MASTERING CONCEPTS

1. What are some familiar basidiomycetes?
2. Which features distinguish the basidiomycetes from other fungi?
3. In what structure does the zygote occur in the basidiomycete life cycle?
4. How does a "fairy ring" form?

Can *You* Relate?

Fungi That Harm Human Health

Although most fungi are harmless (or even beneficial) to humans, some can threaten human health by causing mild to fatal infections, allergic reactions, or poisonings.

Infection: Fungi that degrade the protein keratin can infect our skin, hair, and nails, causing ringworm, athlete's foot, and other irritating diseases. Some fungi, such as *Candida albicans,* normally inhabit the body's mucous membranes. When populations of the body's other microbial companions are disrupted, such as when a person uses antibiotics to fight a bacterial infection, populations of these fungi can grow out of control. The result is a vaginal, intestinal, or oral yeast infection.

FIGURE 21.B *Coccidioides immitis* Causes Valley Fever. This microscopic, thick-walled structure, called a spherule, produces hundreds of fungal spores in infected tissues.

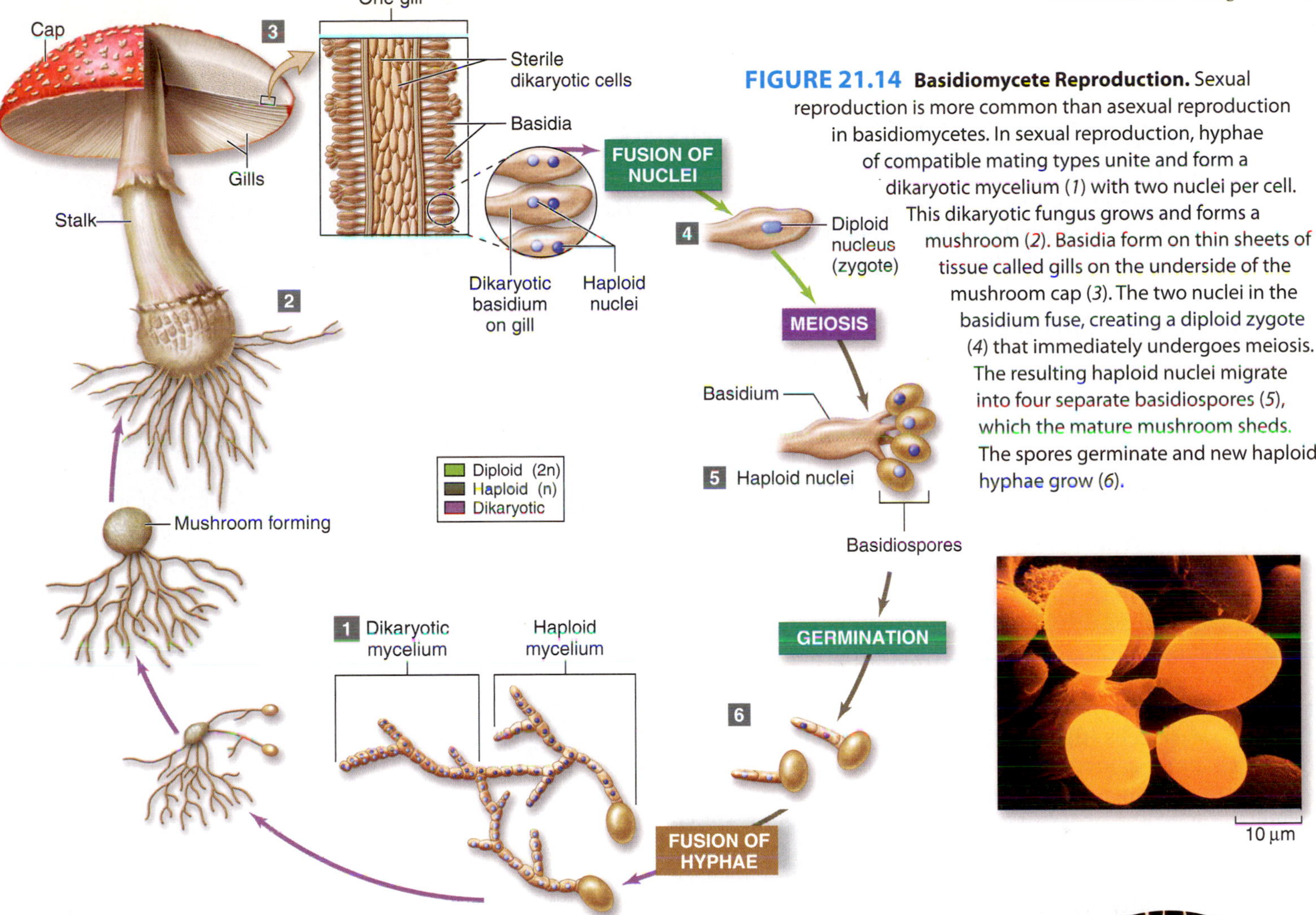

FIGURE 21.14 Basidiomycete Reproduction. Sexual reproduction is more common than asexual reproduction in basidiomycetes. In sexual reproduction, hyphae of compatible mating types unite and form a dikaryotic mycelium (*1*) with two nuclei per cell. This dikaryotic fungus grows and forms a mushroom (*2*). Basidia form on thin sheets of tissue called gills on the underside of the mushroom cap (*3*). The two nuclei in the basidium fuse, creating a diploid zygote (*4*) that immediately undergoes meiosis. The resulting haploid nuclei migrate into four separate basidiospores (*5*), which the mature mushroom sheds. The spores germinate and new haploid hyphae grow (*6*).

Other fungi enter body tissues via wounds or the lungs. People who inhale dust containing spores of a fungus called *Coccidioides immitis* may develop a potentially fatal disease called valley fever (**figure 21.B**). Similarly, *Histoplasma capsulatum* inhabits bird droppings; inhaling the spores may result in a typically mild infection called histoplasmosis.

Allergic reactions: People in many parts of the United States receive "mold spore counts" along with their weather reports, and residents with mold allergies avoid going outside when the counts are high. Common allergenic fungi include species of *Alternaria* and *Aspergillus,* both of which produce abundant asexual spores. They can produce enormous spore masses in mold-infested "sick buildings." **allergies, p. 740**

Toxicity: Some people report symptoms such as headaches and memory loss after living or working in moldy buildings. *Stachybotrys atra,* the toxic "black mold," is a commonly cited culprit, but proof that fungal toxins cause these symptoms is lacking.

Poisonous mushrooms such as those described in the chapter opening essay are the most famous toxic fungi, but others produce potent chemicals as well. *Claviceps purpurea* is an ascomycete that causes a plant disease called ergot (**figure 21.C**). People who eat bread made from ergot-contaminated grain can develop the convulsions, gangrene, and psychotic delusions associated with ergotism. Some of the women in Salem, Massachusetts, who were burned at the stake as witches in the late 1690s may have been suffering from ergotism; people thought their uncontrollable movements meant they were possessed by demons. The drug lysergic acid diethylamide (LSD) comes from lysergic acid, a chemical found in ergots.

FIGURE 21.C *Claviceps purpurea* Infects Grain. These black structures are ergots, masses of hyphae. If the ergots become mixed in with grain used to mill flour, people who eat the flour may become ill.

Another toxin-producing fungus is the ascomycete *Aspergillus flavus.* Hyphae of *A. flavus* feed on peanuts or other stored crops and release aflatoxin, which can be deadly at sufficiently high doses. At low doses, aflatoxin is one of the most carcinogenic compounds known.

21.6 Fungi Interact with Other Organisms

Fungi are important decomposers in ecosystems, but they also form mutually beneficial relationships with live organisms. Here we profile three interesting examples.

A. Mycorrhizal Fungi Live on or in Roots

Mycorrhizae (literally, "fungus-roots") are associations between fungal hyphae and plant roots. Some of the oldest plant fossils show evidence of mycorrhizae, indicating that plants and fungi moved onto land together hundreds of millions of years ago.

About 80% of all land plants have mycorrhizae. Some plants, such as orchids, cannot live without their fungal associates. The relationship benefits both partners: the plant obtains water and minerals that the hyphae absorb from the soil, and the fungus gains carbohydrates that the plant produces in photosynthesis.

Some types of mycorrhizal fungi wrap around roots and may extend between root cells, but the hyphae do not penetrate individual cells (**figure 21.15A, B**). More commonly, the fungi pierce the host plant's root cells and produce highly branched structures through which they absorb nutrients (figure 21.15C).

Many edible fungi resist commercial cultivation because they depend on their mycorrhizal relationships with live tree roots. The popularity and high price of these wild delicacies have lured many mushroom pickers into the woods. Scientists are currently debating whether the ever-growing harvest of wild mushrooms will harm the populations of fungi—and the trees that depend on them—in the long term.

B. Some Ants Cultivate Fungi with the Help of Bacteria

The leaf-cutter ants of Central and South America and the southern United States cultivate the basidiomycete *Lepiota* in special underground chambers. Into each chamber, the ants transport a paste made from saliva and the discs that

FIGURE 21.16 Ants at Work. Instead of eating the vegetation they harvest, leaf-cutter ants use it to farm fungi—their food source.

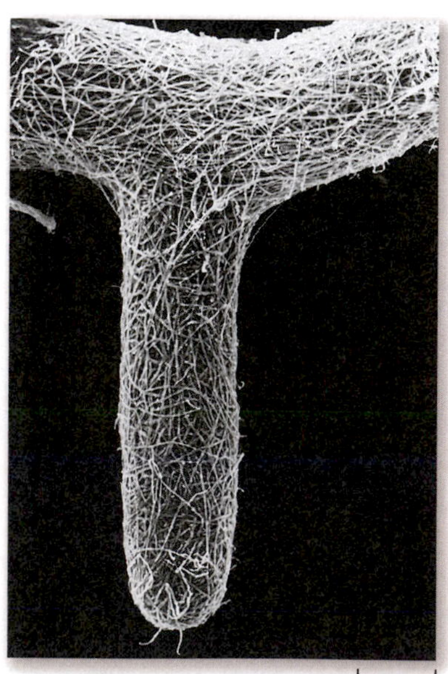

FIGURE 21.15 Mycorrhizae. (**A**) The roots of this lodgepole pine (*Pinus contorta*) seedling are colonized by an unidentified basidiomycete, which has produced a mushroom (*arrow*). The ectomycorrhizal root tips are recognizable by their pale creamy to light brown color and Y-shaped branching pattern. (**B**) Hyphae of an ectomycorrhizal fungus wrap around a host's root tip. (**C**) This structure, which is inside a root cell, is an arbuscule produced by a mycorrhizal fungus.

they cut from green leaves (**figure 21.16**). The fungi grow on the paste, and both adult and larval ants eat the hyphae so quickly that mushrooms never get a chance to form, although they will when grown in a laboratory.

The ants and their fungal gardens constitute a mutualistic partnership; the ants eat, and the fungi have a home. Both benefit. But how do the ants keep out competitors or pathogens? Biologists once thought that the ants simply ate all interlopers, but then they discovered a third partner. *Streptomyces* bacteria coat parts of the ants' cuticles and secrete a potent antibiotic that kills an ascomycete, *Escovopsis*, that attacks the cultivated fungus.

C. Lichens Are Distinctive Dual Organisms

A **lichen** is a dual organism that forms when a fungus, either an ascomycete or a basidiomycete, harbors green algae or cyanobacteria (**figure 21.17**). Lichens are typically flattened growths that are green, orange, yellow, or black, although many form upright structures that somewhat resemble mosses. The photosynthetic partner contributes food to the lichen, and the fungus absorbs essential minerals.

Lichens are important ecologically because they break rock down into soil, which can then support plant growth. Many also harbor nitrogen-fixing bacteria, which make usable nitrogen available to plants. Lichens grow on trees and rocks, in the driest deserts, and in the wettest tropical rain forests. They survive extreme environmental conditions by drying out; they revive when moisture returns. **nitrogen fixation, p. 381**

One type of habitat, however, is hostile to lichens—polluted areas. Lichens absorb toxins, but cannot excrete them. Toxin buildup hampers photosynthesis, and the lichen dies. Disappearance of native lichens is a sign that pollution is disturbing the environment; scientists therefore use lichens to monitor air quality.

Summary *Mycorrhizal fungi wrap around plant roots or penetrate root cells. Leaf-cutter ants cultivate fungi in underground chambers. A fungus in intimate association with a green alga or a cyanobacterium forms a lichen.*

21.6 MASTERING CONCEPTS

1. What do the plant and the fungus gain from a mycorrhizal relationship?
2. How do ants, plants, fungi, and bacteria interact in leaf-cutter ant colonies?
3. What is the physical arrangement between a fungus and its photosynthetic partner in a lichen?
4. How do scientists use lichens to monitor pollution?

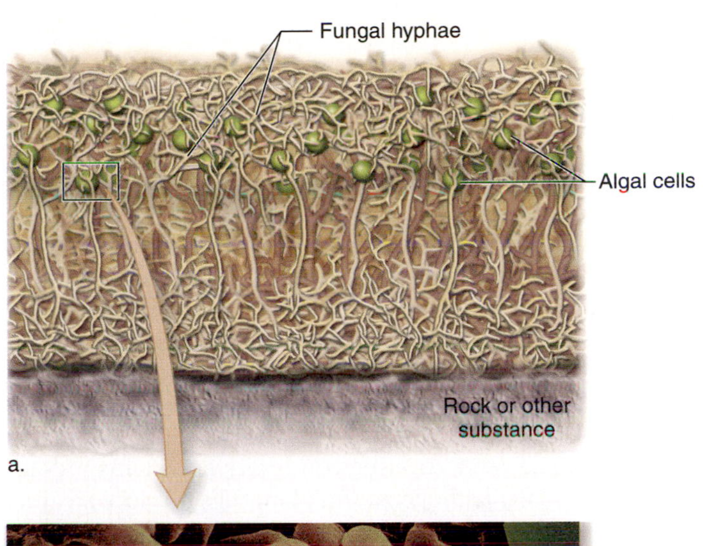

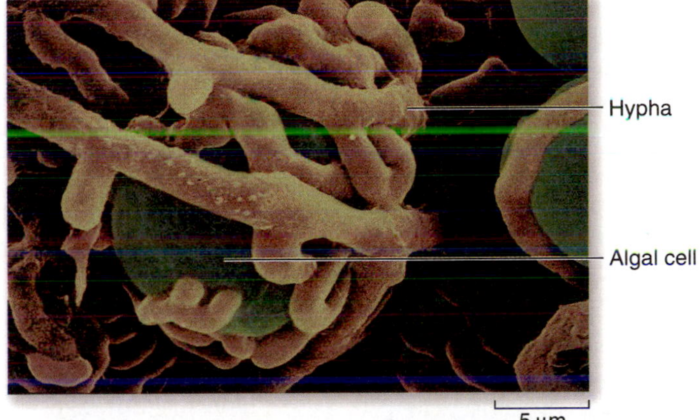

FIGURE 21.17 Anatomy of a Lichen. Fungi associate with green algae or cyanobacteria, forming a composite organism called a lichen. (**A**) Cross section of a lichen encrusting a rock. Note how the fungal hyphae wrap tightly around their photosynthetic "partner" cells. (**B**) A lichen looks like a single organism when viewed at a large scale.

21.7 INVESTIGATING LIFE
The Battle for Position in Cacao Tree Leaves

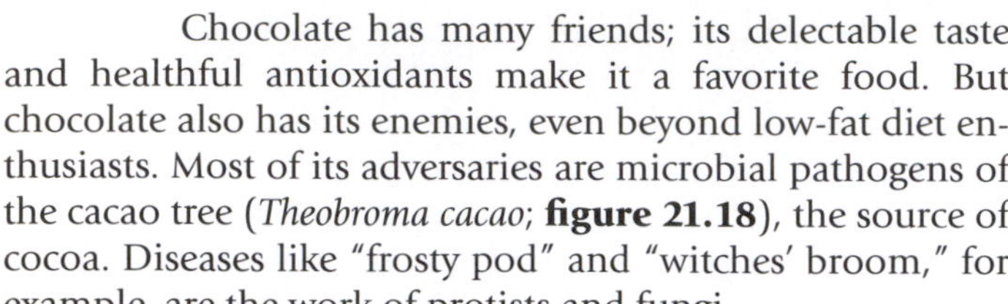

Chocolate has many friends; its delectable taste and healthful antioxidants make it a favorite food. But chocolate also has its enemies, even beyond low-fat diet enthusiasts. Most of its adversaries are microbial pathogens of the cacao tree (*Theobroma cacao*; **figure 21.18**), the source of cocoa. Diseases like "frosty pod" and "witches' broom," for example, are the work of protists and fungi.

All organisms, including plants, defend themselves against disease. Waxy coverings on leaves and stems prevent the entry of many pathogens, and plant cells produce an array of noxious chemicals that deter those microbes that manage to enter. Besides these innate protections, some plants also enlist allies in the war against predators and pathogens. Some fungi live inside plant tissues without triggering disease symptoms. Scientists have discovered that these fungi, which are called endophytes ("endo" means inside, and "phyte" means plant), are everywhere. Every known plant, from mosses to angiosperms, harbors them inside its stems and leaves (**figure 21.19**). The ubiquity of endophytes has led some researchers to comment that "all plants are part fungi."

Some plants pass endophytes to their offspring in seeds, but the cacao tree acquires its resident fungi from the environment. A brand-new leaf is endophyte-free, but fungal spores soon arrive in wind or rain and germinate on the young foliage. The hyphae penetrate the waxy leaf cuticle and set up shop within the plant's tissue, absorbing water and nutrients that leak out of the leaf veins. Oddly, the endophytes do not trigger the plant's defenses. This raises an interesting question: What does the tree gain from allowing plant-eating fungi to live inside its tissues?

University of Arizona biologist A. Elizabeth Arnold, along with a research team at the Smithsonian Tropical Research Institute in Panama, wondered whether the endophytes help protect cacao trees from disease. To find out, they collected seeds from cacao plants and soaked them briefly in dilute bleach to kill any fungal spores sticking to the seed coats. Because they needed endophyte-free plants to test their hypothesis, they planted the seeds in a greenhouse and protected them from fungal spores that might blow in from outside. Once the trees were 100 days old, the researchers sampled the leaves to ensure that the plants were endophyte-free.

FIGURE 21.18 The Cacao Tree, *Theobroma cacao*. The pods of this tree are the source of chocolate.

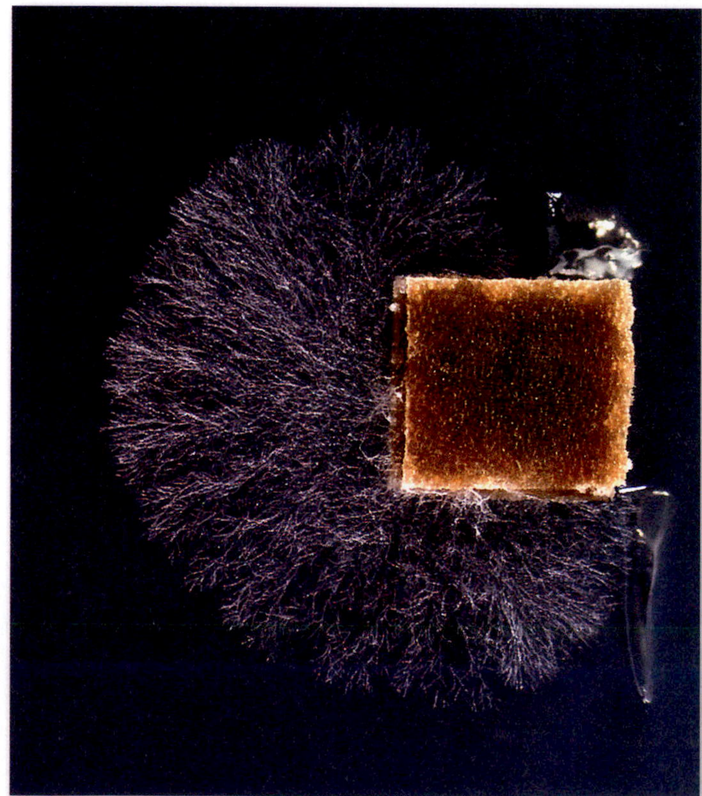

FIGURE 21.19 Endophytes. An endophytic fungus grows out of a tiny piece of leaf tissue.

Next, they sprayed a mixture of spores from several species of endophyte fungi on some of the leaves, so each tree had some treated leaves and some that remained free of the fungi. A couple of weeks later, once the fungi were thriving in the sprayed leaves, the researchers were ready to test their hypothesis. They inoculated both endophyte-treated and control leaves with spores of a parasite called *Phytophthora*, which causes black pod disease of cacao (a different *Phytophthora* species causes late blight of potatoes; see figure 19.12).

Fifteen days later, Arnold and her team counted the number of dead leaves on each tree and measured the *Phytophthora*-damaged area on surviving leaves. These combined measures showed that plant parts without endophytes lost twice as much leaf area as those with the resident fungi (**figure 21.20**).

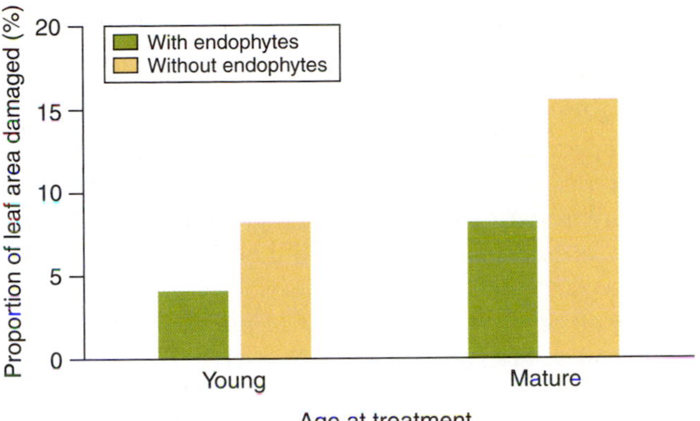

FIGURE 21.20 Helpful Partners. Endophytic fungi protected young and mature cacao leaves from damage by the pathogen, *Phytophthora*. **Question:** This study used a mix of endophyte species. How would you design an experiment to answer the question of whether one species of endophyte, or some combination of multiple species, is necessary to protect the leaves against pathogens?

No one is sure what the endophytes gain from the relationship, other than the nutrients they absorb from inside the leaf. One hypothesis is that their biggest gain comes after their home leaf ages, dies, and falls to the ground. Like devoted fans who camp overnight to buy prime tickets for a concert, perhaps the endophytes colonize a living leaf early to get "first dibs" on the dead tissue.

Win–win relationships, such as the one between the cacao tree and its resident fungi, are common in life. Most plants share their roots with mycorrhizal fungi; our own bodies teem with beneficial microbes that protect us from pathogens. Scientists are now trying to develop some of these cooperative organisms into biological control agents that might prevent disease in cacao plants without the use of harmful chemicals. Our endophyte allies are already living quietly in the stems and leaves of our crop plants. Perhaps one day farmers will enlist these friendly fungi in a more aggressive fight to preserve our chocolate fix.

Arnold, A. Elizabeth, Luis Carlos Mejia, Damond Kyllo, et al. December 23, 2003. Fungal endophytes limit pathogen damage in a tropical tree. *Proceedings of the National Academy of Sciences*, U.S.A., vol. 100, pages 15649–15654.

CHAPTER SUMMARY

21.1 Fungi Are Essential Decomposers

- Fungi are widespread and profoundly affect ecosystems by feeding on living and dead organic material.

A. Fungi Are Heterotrophs with Cell Walls of Chitin

- Fungi are more closely related to animals than to plants.
- Fungal characteristics include heterotrophy, chitin cell walls, and glycogen. Fungi have both asexual and sexual reproduction, and some have unique **dikaryotic** cells with two genetically different nuclei.

B. Most Fungi Are Filamentous Spore-Formers

- A fungal body includes a **mycelium** built of threads called **hyphae,** which may form a **fruiting body.** Some fungi occur both as filamentous hyphae and as unicellular **yeasts.**
- Fungi reproduce using asexual and sexual **spores**.

C. Fungal Classification Is Based on Reproductive Structures

- The four main groups of fungi are **chytridiomycetes, zygomycetes, ascomycetes,** and **basidiomycetes. Deuteromycetes** are fungi with no known sexual stage; molecular evidence places most with the ascomycetes.
- Molecular evidence indicates that ascomycetes and basidiomycetes are sister groups.

21.2 Chytrids Are Fungi that Produce Swimming Spores

- Chytrids are microscopic fungi that produce flagellated, motile **zoospores.**
- Chytrids decompose major biological carbohydrates such as cellulose, chitin, and keratin.

21.3 Zygomycetes Are Fast Growing and Prolific

- Zygomycetes reproduce rapidly via asexually produced, haploid, thin-walled spores.
- Sexual reproduction occurs when hyphae of different mating types fuse their nuclei into a **zygospore.** The zygospore undergoes meiosis, then generates a spore sac. Spore germination yields haploid hyphae, continuing the cycle.

21.4 Ascomycetes Are the Sac Fungi

- Ascomycetes include important plant pathogens, but they also have many uses in industry.
- Hyphae are haploid for most of the life cycle. Asexual reproduction occurs via spores (conidia) or by budding.
- In sexual reproduction, haploid hyphae join, producing a brief dikaryotic stage. After the nuclei fuse, meiosis yields haploid **ascospores** in saclike **asci**.

21.5 Basidiomycetes Are the Familiar Club Fungi

- Basidiomycetes include mushrooms and other familiar fungi. Wood decay fungi are important in ecosystems but also rot building materials. Smuts and rusts are important plant pathogens.
- Basidiomycetes are dikaryotic for most of the life cycle. Asexual reproduction occurs by budding, by fragmentation, or with asexual spores.
- Sexual reproduction occurs when haploid nuclei in **basidia** along the gills of mushrooms fuse. The resulting zygote then undergoes meiosis, producing haploid **basidiospores.** Shortly after spore germination, hyphae fuse, regenerating the dikaryotic state.

21.6 Fungi Interact with Other Organisms

A. Mycorrhizal Fungi Live on or in Roots

- **Mycorrhizae** are mutually beneficial associations between fungi and roots. Mycorrhizal plants have extra surface area for absorbing nutrients, and the fungus acquires carbohydrates from its plant host.

B. Some Ants Cultivate Fungi with the Help of Bacteria

- Leaf-cutter ants cultivate basidiomycetes on a nutritious paste made from leaf disks, and bacteria on the ants kill a parasitic ascomycete.

C. Lichens Are Distinctive Dual Organisms

- A **lichen** is a compound organism that consists of a fungus in intimate association with a cyanobacterium or a green alga.
- Lichens are useful air pollution indicators.

21.7 Investigating Life: The Battle for Position in Cacao Tree Leaves

- Endophytes are fungi that live inside plant tissues without triggering disease symptoms.
- Experimental evidence indicates that endophytes protect cacao tree leaves from *Phytophthora* pathogens.

MULTIPLE CHOICE QUESTIONS

1. How are fungi similar to animals?
   a. They have a cell wall made of chitin.
   b. They store energy in the form of glycogen.
   c. They are predominantly haploid organisms.
   d. Both a and b.
2. Fungi get nutrients through their
   a. fruiting body.
   b. hyphae.
   c. spores.
   d. all of the above.
3. Fungi are considered ______ because they get their energy from ______.
   a. autotrophs; photosynthesis
   b. autotrophs; organic matter
   c. heterotrophs; organic matter
   d. heterotrophs; photosynthesis
4. Which phylum of fungus is most commonly eaten by humans?
   a. Chytridiomycetes
   b. Deuteromycetes
   c. Basidiomycetes
   d. Zygomycetes
5. Why is the zygospore diploid?
   a. Because it was formed through meiosis
   b. Because it was formed when two haploid nuclei fused
   c. Because it was formed by mitosis of the compatible mating types
   d. Because it was formed from a multicellular stage in the life cycle
6. If a cell is *dikaryotic* it
   a. contains two nuclei.
   b. is diploid.
   c. functions as a spore.
   d. is a zygospore.

7. Are the eight ascospores of an ascomycete typically genetically identical?
   a. Yes, because they are produced by mitosis.
   b. Yes, because they are the product of a single dikaryotic cell.
   c. No, because they are the products of meiosis, then mitosis.
   d. No, because they undergo mutations.
8. The tissue that forms a mushroom is
   a. haploid.
   b. diploid.
   c. dikaryotic.
   d. all of the above.
9. What is a lichen?
   a. A type of photosynthetic fungus
   b. A combination of a fungus and an alga or cyanobacterium
   c. A combination of two phyla of fungi
   d. A combination of a fungus and a root
10. How does a mycorrhizal fungus differ from free-living basidiomycetes?
   a. It gains its nutrients from a living plant.
   b. It has no sexual stage in its life cycle.
   c. It is not edible.
   d. All of the above.

TESTING YOUR KNOWLEDGE

1. What evidence suggests that fungi are more closely related to animals than to plants?
2. Which characteristics do all types of fungi share?
3. List the characteristics that distinguish the four phyla of fungi.
4. What are two functions of spores?
5. Sketch each of the following structures: mycelium; ascus with ascospores; zoospore; zygospore; basidium with basidiospores.
6. Distinguish between the following pairs of terms: yeast and mycelium; ascomycete and basidiomycete; mycorrhiza and lichen; zoospore and zygospore; diploid and dikaryotic.
7. Other than shared DNA sequences, what characteristics place ascomycetes and basidiomycetes together as sister groups?
8. Describe the relationships that connect:
   a. roots and mycorrhizal fungi
   b. leaf-cutter ants, their fungal food, bacteria, and pathogenic fungi
   c. the fungal and photosynthetic partners in a lichen
9. Give examples of fungi that are important economically, ecologically, as human pathogens, as plant pathogens, and as food for humans.

THINKING AS A SCIENTIST

1. Penicillin, an antibiotic derived from a fungus, kills bacteria by destroying their cell walls. Why doesn't penicillin harm fungal cell walls?
2. Describe how experiments might show that:
   a. chytrids are killing amphibians
   b. fungi benefit more from a lichen relationship than do algae
   c. bacteria help leaf-cutting ants cultivate one fungus while killing another
   d. overharvesting of mycorrhizal basidiomycetes harms forest health
3. Review alternation of generations in plants (figure 20.2). Compare and contrast the life cycles of zygomycetes, ascomycetes, and basidiomycetes to the basic plant life cycle.
4. Genome sequencing projects are complete or in progress for many fungi other than *Neurospora crassa*. Choose one of these fungi, and describe some new discoveries that have come from research on this organism.

Visit www.mhhe.com/hoefnagels for practice quizzes, animations, videos, and activities designed to help you master the material in this chapter.

CHAPTER 22

Invertebrate Animals

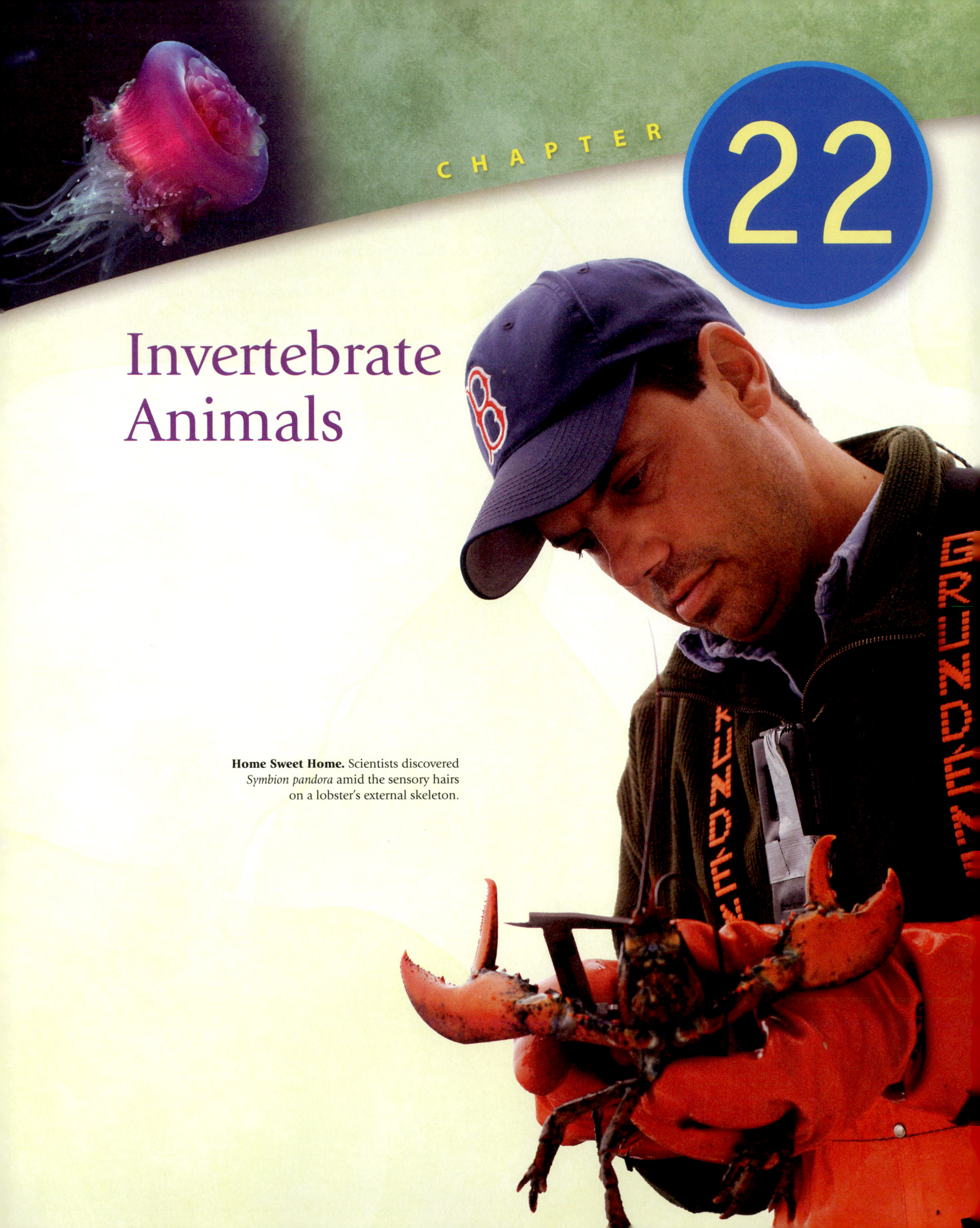

Home Sweet Home. Scientists discovered *Symbion pandora* amid the sensory hairs on a lobster's external skeleton.

What's the Point?

Life on a Lobster's Lips

Discovering and describing never-before-seen organisms is nothing new to biologists. Finding* Symbion pandora *was unusual, however, not only because of its habitat on the mouthparts of the Norwegian lobster,* Nephrops norvegicus, *but also because it did not fit into any known animal phylum. Danish researchers reported discovering the animal in late 1995.

The researcher who codiscovered S. pandora, *Reinhardt Kristensen, says that many animals await our detection and description. In 1983, Kristensen discovered animals in another phylum, the Loricifera, living between grains of sand. Amazingly, even as other scientists are exploring the remotest reaches of Earth looking for new life, both of these phyla were discovered "hiding in plain sight." The lobster that hosts* S. pandora *lives in a busy Scandinavian shipping lane.*

Among the smallest of animals, S. pandora *measures less than a millimeter. Its saclike body adheres to a lobster's mouthparts. This unusual animal was assigned to a new phylum, Cycliophora, which means "wheel bearer" in Greek. The "wheel" refers to the ciliated ring that forms the animal's mouth, which opens into a funnel-shaped structure atop the animal's body.*

Symbion pandora*'s life is full of competition and frequent relocation. Living on lobster mouthparts means ready access to nutrients—as the lobsters crush their food in their jaws,* S. pandora *sweeps stray particles into its ciliated mouth ring. Because food and oxygen availability vary among the bristles surrounding the lobster's mouth, competition for the prime dining spots is fierce.*

Victory at finding food doesn't last, however, because like all arthropods, lobsters molt. When they do, what was once prime real estate becomes worthless. Fortunately, S. pandora *is adapted to this regular upheaval. Its complex life cycle includes short swimming stages in addition to the lengthier immobile (sessile) stages. As soon as the lobster's new exoskeleton is complete, the animals resettle, renewing the competition for the best locations.*

What exactly is S. pandora? *DNA sequence comparisons suggest it may be closest to a group of animals called rotifers. Additional research is needed to determine whether these curious organisms will be absorbed into an existing phylum or remain in their own group. In the meantime, it's intriguing to think of how much we do not know about Animalia—the most familiar of biological kingdoms.*

LEARNING OUTLINE

22.1 Animals Live Nearly Everywhere

Think of any animal. There's a good chance that the example that popped into your head was a mammal such as a dog, cat, horse, or cow. Although it makes sense that we think first of our most familiar companions, the mammals represent only a tiny subset of organisms in kingdom Animalia.

Biologists have described about 1,200,000 animal species, and their diversity is astonishing (**table 22.1**). The vast majority are **invertebrates** (animals without backbones). The phylum Arthropoda alone, for example, includes more than 1 million identified species of invertebrates such as insects, crustaceans, and spiders. Only about 45,000 known animal species are **vertebrates** (animals with backbones). Of those, only about 5,500 are mammals.

Animals live in us, on us, and around us (see this chapter's Can *You* Relate? Your Tiny Companions on page 452). They are extremely diverse in size, habitat, body form, and intelligence. Whales are immense; roundworms are microscopic. Bighorn sheep scale mountaintops; crabs scuttle on the deep ocean floor. Earthworms are squishy; clams surround themselves in heavy armor. Sponges are witless; humans, chimps, and dolphins are clever. This chapter and the next explore some of this amazing variety.

Table 22.1 *Nine Phyla of Animals*

Phylum	Examples	Number of Existing Species
Porifera (sponges)	Sponges	5000
Cnidaria	*Hydra*, jellyfishes, corals, sea anemones	11,000
Platyhelminthes (flatworms)	Planaria, tapeworms, flukes	25,000
Mollusca	Bivalves, chitons, snails, slugs, squids, octopuses	112,000
Annelida	Earthworms, leeches, polychaetes	15,000
Nematoda (roundworms)	Pinworms, hookworms, *C. elegans*	20,000
Arthropoda	Horseshoe crabs, spiders, scorpions, crustaceans, insects	More than 1,000,000
Echinodermata	Sea stars, sea urchins, sand dollars	7,000
Chordata (see chapter 23)	Tunicates, lancelets, fishes, amphibians, reptiles, birds, mammals	50,000

A. The First Animals Likely Evolved from Protists

The animal you thought of a moment ago probably lives on land, since terrestrial animals are the ones we see most often. Nevertheless, only 10 of the 37 known phyla include species that live on land, and no phylum contains only terrestrial animals. All of today's animals clearly have their origins in aquatic ancestors.

The first animals, which arose about 570 MYA, may have been related to protists called choanoflagellates (**figure 22.1**), aquatic organisms that strongly resemble sponge cells. Although no one knows exactly what the first animal looked like, the Ediacaran organisms that thrived during the Precambrian left some of the oldest animal fossils ever found (see figure 16.12). Animal life diversified spectacularly during the Cambrian, which ended about 490 MYA. Most of today's major groups of animals, including sponges, jellyfishes, arthropods, mollusks, and many types of worms, originated in the Cambrian seas. Their fossils are exceptionally abundant in an area of British Columbia called the Burgess Shale (see figure 16.13).

Aquatic animals were already diverse by the time plants and fungi colonized the land about 475 MYA. Arthropods, vertebrates, and other animals soon followed, diversifying further as they adapted to new food sources and habitats. Chapter 23 picks up the story of vertebrate evolution at this point.

B. Animals Share Several Characteristics

All animals, from sponges to chimpanzees, share a combination of features. First, they are multicellular organisms with

FIGURE 22.1 **Animal Ancestor?** (**A**) An immediate animal ancestor may have resembled a choanoflagellate. (**B**) Whatever the ancestor was, it eventually gave rise to all modern animals, including monarch butterflies.

eukaryotic cells, as are plants, fungi, and some protists. Unlike the cells of plants and fungi, animal cells lack cell walls. **cell structure, p. 60**

Second, all animals are heterotrophs, obtaining both carbon and energy from organic compounds that other organisms have produced. Most animals ingest their food, break it down in a digestive tract, absorb the nutrients, and eject the indigestible wastes.

Third, animal development is unlike that of any other type of organism. After fertilization, the zygote (the first cell of the new organism) divides rapidly. The early animal embryo begins as a solid ball of cells that quickly hollows out to form a **blastula,** a sphere of cells surrounding a fluid-filled cavity. (Embryonic stem cells, the source of so much controversy, come from embryos at this stage of development.) No other organisms go through a blastula stage of development. **stem cells, p. 158**

Fourth, animal cells secrete and bind to a nonliving substance called the extracellular matrix (see chapter 27). This matrix enables some cells to move, others to assemble into sheets, and yet others to embed in supportive surroundings, such as bone or shell.

C. Biologists Classify Animals Based on Organization, Morphology, and Development

This book considers 9 of the 37 known animal phyla: 8 in this chapter, and 1 in the next. **Figure 22.2** compiles them into a phylogenetic tree. This section explains the features that biologists use to construct the tree; section 22.1D lists other features that are also sometimes important in describing animals.

Cell and Tissue Organization

The first major branching point of the tree in figure 22.2 separates animals into two clades. The simplest animals, the sponges, have several specialized cell types, but the cells do not interact to provide specific functions as they would in a true tissue. The other clade contains **eumetazoans,** animals with true tissues such as nerve and muscle tissues. In complex animals, multiple tissue types interact to form organs such as a heart, brain, or kidney. The organs, in turn, interact in systems that circulate and distribute blood, dispose of

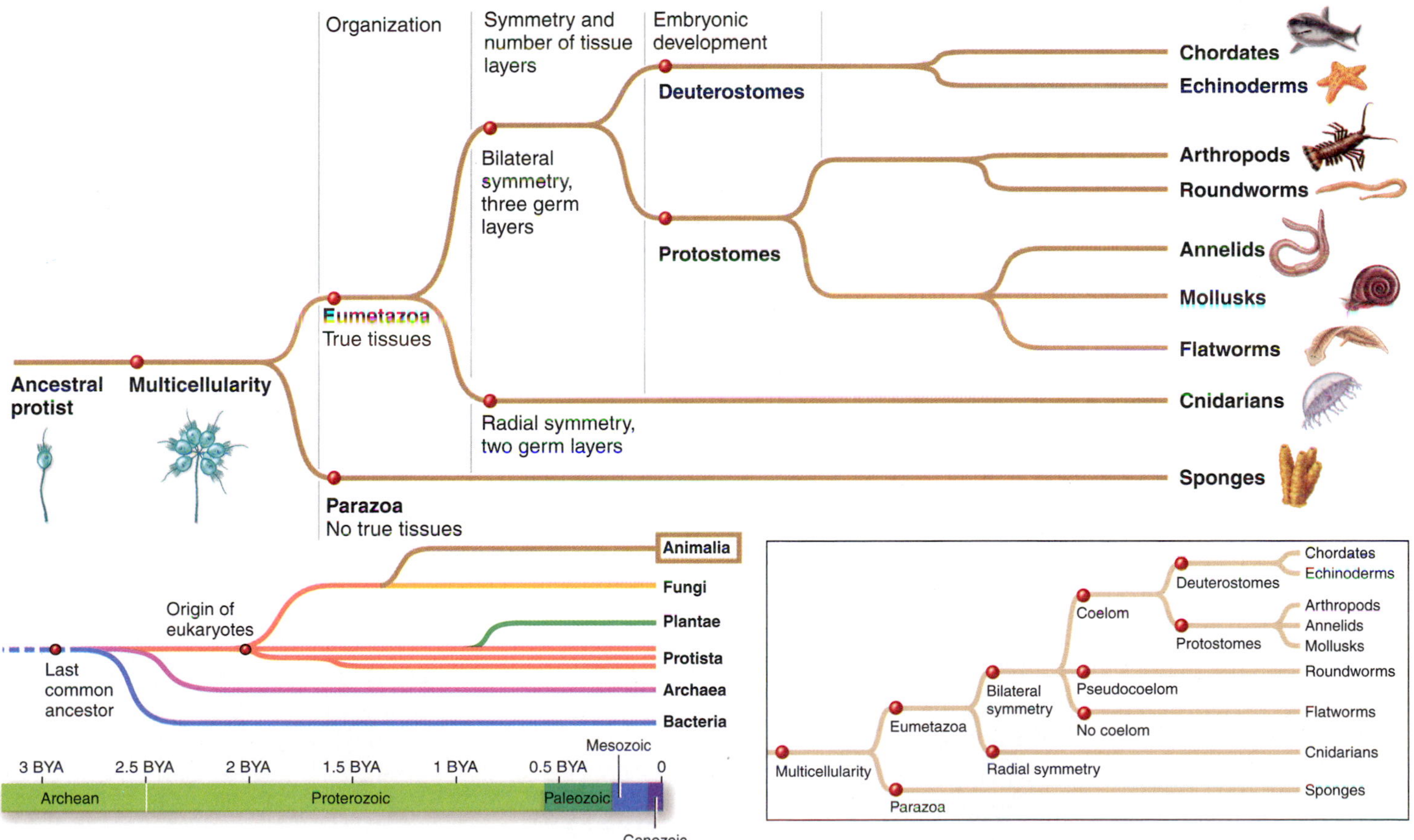

FIGURE 22.2 Animal Classification. The relationships among the animal phyla reflect complexity of body form, developmental characteristics, and DNA sequences. The inset shows a traditional scheme based on anatomical characteristics.

wastes, and carry out other functions. In general, the larger and more active an animal, the more complex and specialized its organ systems are.

Body Symmetry and Cephalization

Body symmetry is another major criterion used in animal classification (see figure 22.2). Most sponges are asymmetrical; that is, they lack symmetry (**figure 22.3**A). Hydra, jellyfishes, and their close relatives have **radial symmetry,** in which any plane passing through the body from the mouth to the opposite end divides the body into mirror images (figure 22.3B). Adult sea stars also have radial symmetry. All other animals have **bilateral symmetry,** in which only one plane divides the animal into mirror images (figure 22.3C). Bilaterally symmetrical animals such as crayfish and humans have head (anterior) and tail (posterior) ends, and they typically move "head-first" through their environment. This behavior is correlated with **cephalization,** the tendency to concentrate sensory organs and a brain at the head end of the animal.

Embryonic Development: Two or Three Germ Layers

Early embryos give other clues to evolutionary relationships (see figure 22.2). In eumetazoans, the blastula folds in on itself to generate the **gastrula,** which is composed of two or three tissue layers called primary germ layers (**figure 22.4**). The gastrulas of jellyfishes and their relatives have two germ layers: **ectoderm** to the outside, and **endoderm** to the inside. All other groups of animals (except sponges) have a third germ layer, **mesoderm,** that forms between the ectoderm and endoderm. These germ layers give rise to all of the body's tissues and organs. Ectoderm develops into the skin and nervous system, whereas endoderm becomes the digestive tract and the organs derived from it. Mesoderm gives rise to the muscles, reproductive system, and many other specialized structures.

Embryonic Development: Protostomes and Deuterostomes

Biologists divide bilaterally symmetrical animals into two main clades based on events that occur after the embryo has begun to fold into a gastrula. As development proceeds, the inner cell layer of the gastrula fuses with the opposite side of the embryo, forming a tube with two openings. This cylinder of endoderm will develop into the animal's digestive tract. In **protostomes,** the first indentation to form develops into the mouth, and the anus develops from the second opening. In **deuterostomes,** such as echinoderms and chordates, the first indentation becomes the anus, and the mouth develops from the second opening. (The words *protostome* and *deuterostome* derive from this difference in embryonic development: protostome means "mouth first," and deuterostome means "mouth second"). These two groups differ in other details of their early embryological development as well.

D. Biologists Also Consider Additional Characteristics

Other features that describe animals include the body cavity, the organization of the digestive tract, segmentation, and patterns of reproduction and development.

Body Cavity (Coelom)

Bilaterally symmetrical animals were once classified based on the presence or absence of a **coelom,** which is a fluid-filled body cavity that forms completely within the mesoderm. Flatworms lack a coelom (**figure 22.5**A), although evidence suggests their ancestors did have body cavities.

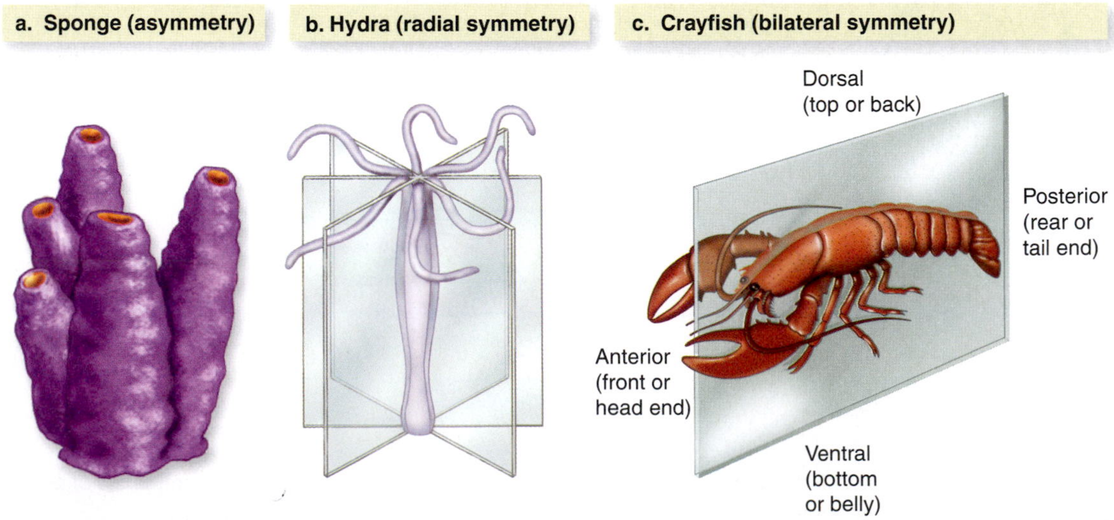

a. Sponge (asymmetry)

b. Hydra (radial symmetry)

c. Crayfish (bilateral symmetry)

Dorsal (top or back)

Posterior (rear or tail end)

Anterior (front or head end)

Ventral (bottom or belly)

FIGURE 22.3 Types of Symmetry. (**A**) Sponges are asymmetrical. (**B**) *Hydra* has radial symmetry: any plane passing from the mouth through the opposite end divides the animal into mirror images. (**C**) A crayfish has bilateral symmetry. Only one plane divides the animal into mirror images. Animals with bilateral symmetry have a front (anterior) and rear (posterior) end, and a dorsal (back or top) and ventral (bottom or belly) side.

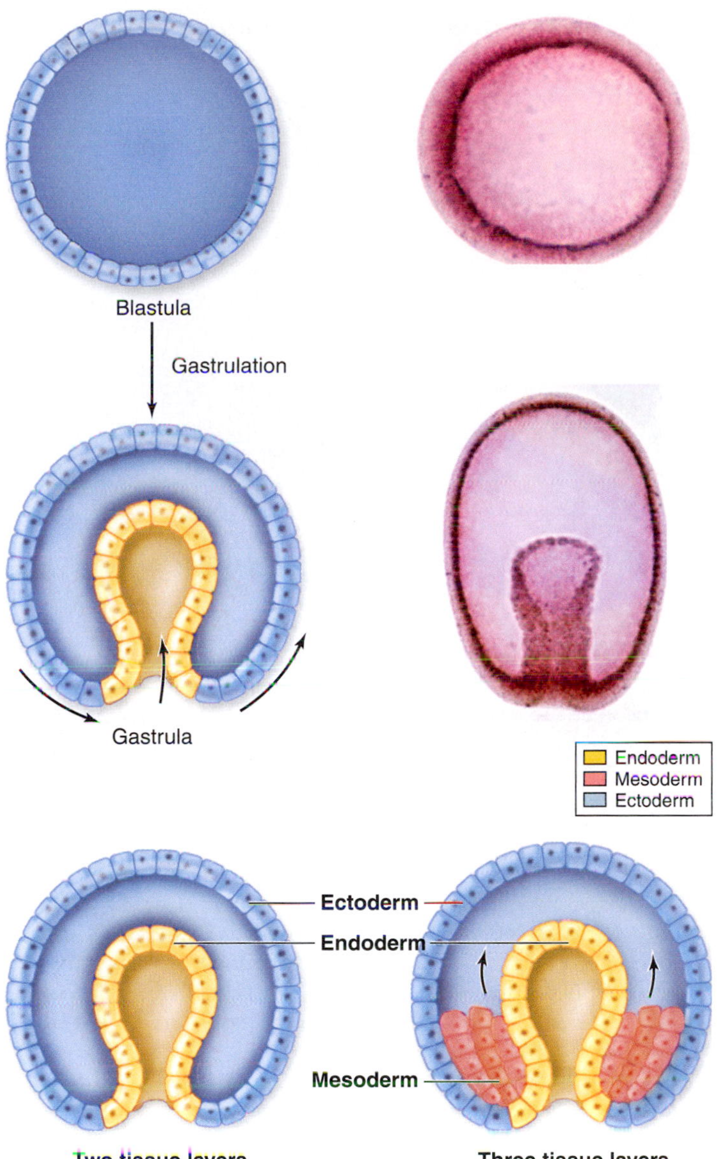

FIGURE 22.4 Two or Three Primary Germ Layers. During early development of eumetazoans, a fluid-filled ball of cells called a blastula folds in on itself and forms the gastrula. (The two photos show a sea star's blastula and gastrula.) Animals with two primary germ layers have ectoderm (outer, *blue*) and endoderm (inner, *yellow*). In other animals, mesoderm (*red*) forms between the ectoderm and endoderm.

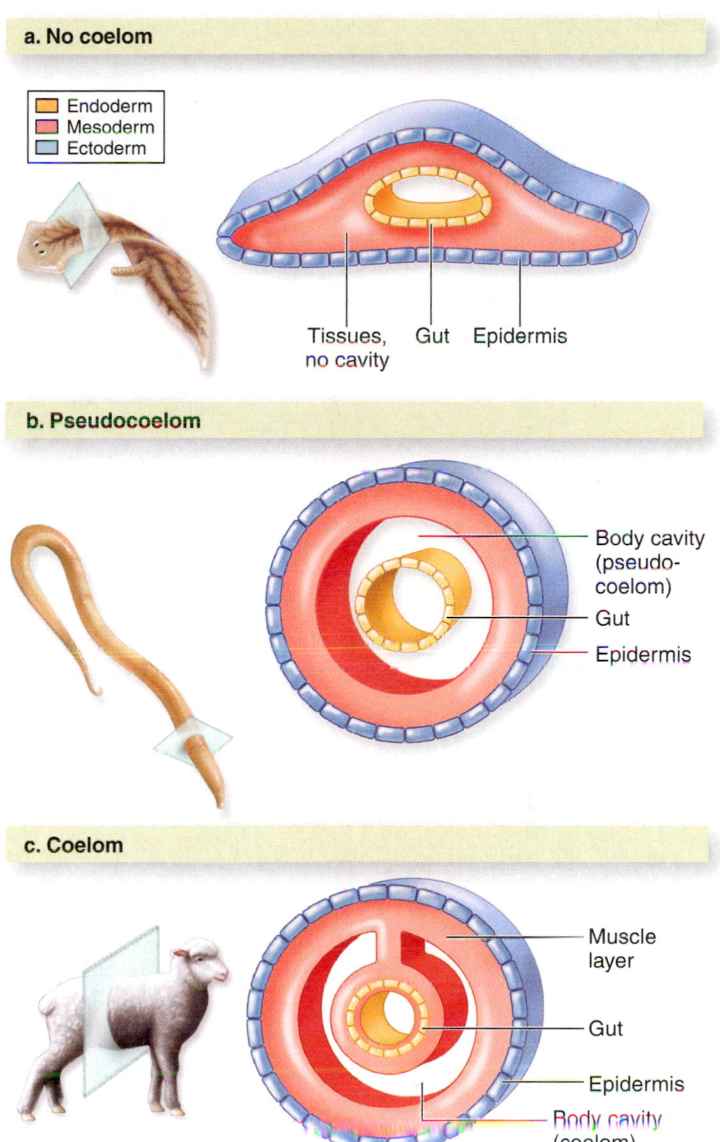

FIGURE 22.5 Body Cavities. (**A**) A flatworm lacks a coelom. (**B**) A roundworm has a pseudocoelom, a body cavity that is not completely lined with mesoderm. (**C**) Like many other animals, a sheep has a true coelom, completely lined with mesoderm. Note that these drawings are abstractions. In the sheep, for example, the internal organs grow into the coelom, greatly distorting its shape.

Roundworms have a body cavity called a **pseudocoelom** ("false coelom") that is lined partly with mesoderm and partly with endoderm (figure 22.5B). Animals that have a true coelom include earthworms, snails, insects, sea stars, and chordates (figure 22.5C).

The coelom's chief advantage is flexibility. As internal organs such as the heart, lungs, liver, and intestines develop, they push into the coelom. The fluid of the coelom cushions the organs, protects them, and enables them to shift as the animal bends and moves. Moreover, the coelom serves as a hydrostatic skeleton in many animals. In a **hydrostatic skeleton,** muscles near the body cavity push against the fluid inside, providing both support and movement. The earthworm, for example, burrows through soil by alternately contracting and relaxing muscles surrounding its fluid-filled coelom.

Digestive Tract

A sponge does not have a digestive tract, but rather a simple opening through which filtered water leaves the body. Cnidarians and flatworms have an **incomplete digestive tract,**

FIGURE 22.6 **Digestive Tracts.** (**A**) An animal with an incomplete digestive tract takes in food and ejects undigested wastes through its mouth. (**B**) A complete digestive tract has one-way flowt from mouth to anus.

a. Incomplete digestive tract

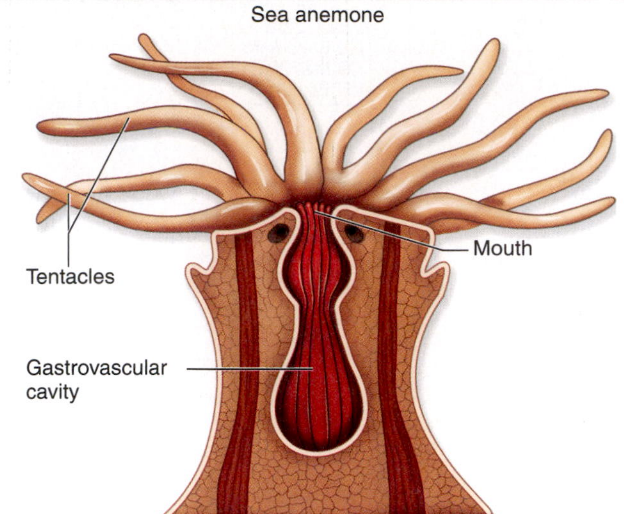

b. Complete digestive tract

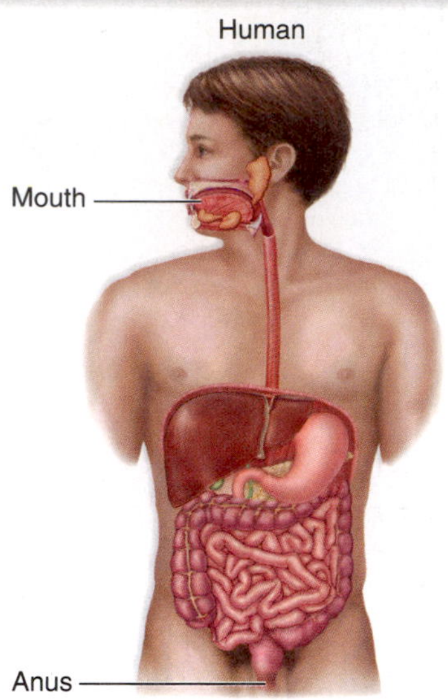

in which one opening (the mouth) both takes in food and ejects wastes (**figure 22.6**A). Digestion occurs in the **gastrovascular cavity,** which distributes nutrients to all parts of the animal's body. In animals with a **complete digestive tract,** food passes in one direction from mouth to anus (figure 22.6B). A complete digestive tract allows for increased specialization. Cells near the mouth can secrete digestive enzymes into the tract, "downstream" cells can absorb nutrients, and those near the anus can help eject wastes.

Segmentation

Centipedes, millipedes, and earthworms all illustrate **segmentation,** the division of an animal body into repeated parts (**figure 22.7**). Insects and vertebrates also have segmented bodies, although the subdivisions may be less obvious. Segmentation adds not only flexibility but also

FIGURE 22.7 **Segmentation.** This millipede illustrates segmentation—the division of the body into repeated parts.

Can *You* Relate?

Your Tiny Companions

Even when you think you are alone, you aren't; your body hosts a diverse assortment of invertebrates. Head lice and body lice are biting insects that cause skin irritation. Ticks latch onto the skin and suck your blood, sometimes transmitting the bacteria that cause Lyme disease. The tiny larvae of chigger mites produce saliva that digests small areas of skin tissue, causing intense itching.

Lice, ticks, and chiggers are hard to ignore, but you may never notice one inconspicuous companion: the follicle mite (genus *Demodex*). This arachnid, which is less than half a millimeter long, lives in hair follicles and nearby oil glands (**figure 22.A**), where it eats skin secretions and dead skin cells. *Demodex* mites are by no means rare. Nearly everyone has them, and each follicle may house up to 25 of the tiny animals. (If you would like to see your own tiny companions, carefully remove an eyebrow hair or eyelash and examine it with a compound microscope.) Luckily, the infestation is typically symptomless, although occasionally the mites may cause a rash.

FIGURE 22.A **Follicle Mites.** Tiny *Demodex* mites live in skin pores and hair follicles.

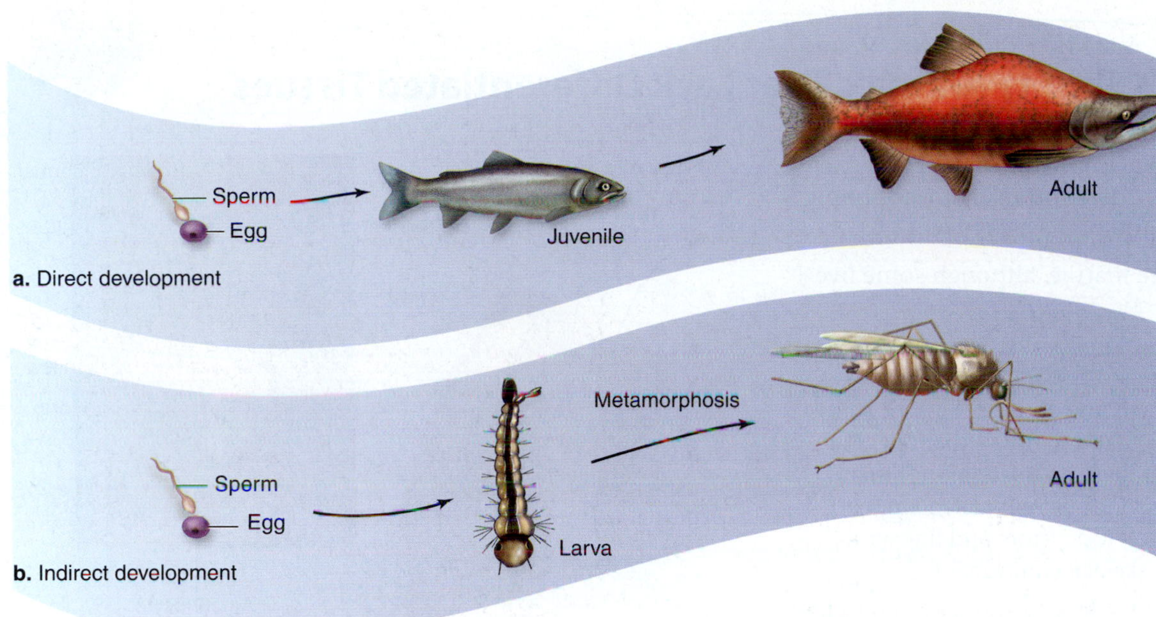

FIGURE 22.8 Animal Development. (**A**) The young of animals that undergo direct development resemble the adult. (**B**) In indirect development, metamorphosis transforms a mosquito larva into a very different-looking adult.

enormous potential for specialization. Activating different combinations of genes can produce body parts with very different functions. Antennae can form on a fly's head, for example, while wings or legs sprout from another segment.

Reproduction and Development

Most animals reproduce sexually, and the development of the resulting embryo follows either of two paths (**figure 22.8**). Animals that undergo **direct development** have no larval stage; at hatching or birth, they already resemble adults. A newborn elephant, for example, looks like a smaller version of an adult elephant. In contrast, an animal with **indirect development** may spend part of its life as a **larva,** which is an immature stage that does not resemble the adult. These larvae eventually undergo **metamorphosis,** in which they change greatly as they mature into adults. Tadpoles and caterpillars are familiar larvae. Chapter 37 discusses animal development in more detail.

Summary *The animal kingdom includes more than a million diverse species. Animals probably evolved from protists called choanoflagellates, and all animal phyla still contain aquatic representatives. All animals are multicellular, heterotrophic eukaryotes with cells that lack walls. Biologists classify animals based on cell and tissue organization, body symmetry, and patterns in embryonic development. Other important features include body cavities, digestive tracts, segmentation, reproduction, and development.*

22.1 MASTERING CONCEPTS

1. When and in what habitat did animals likely originate, and when did today's major groups of animals arise?
2. What is the difference between a blastula and a gastrula?
3. Which animals have true tissues, and which have organs?
4. What is the difference between radial and bilateral symmetry?
5. How do protostomes and deuterostomes differ?
6. What is the difference between a coelom and a pseudocoelom?
7. What are the two main types of digestive tracts?
8. What advantages does segmentation confer?
9. How does direct development differ from indirect development?

22.2 Sponges Are Simple Animals That Lack Differentiated Tissues

The **sponges** belong to phylum Porifera, which means "pore-bearers"—an apt description of these simple animals that lack true tissues and organs (**Figure 22.9**).

Habitat: Aquatic. Most are marine, although some live in fresh water.

Body Structure: The simplest sponges have asymmetrical, hollow, porous bodies (**figure 22.10**). The body wall has an outer layer of flattened cells and an inner layer of flagellated "collar cells" (the cells that resemble choanoflagellates). Sandwiched between these layers is a noncellular, jellylike matrix that contains mobile cells called amoebocytes. Amoebocytes can digest food, store and transport nutrients, divide, or secrete skeletal components.

Feeding: Sponges are filter feeders. Movement of flagella on the collar cells produces a current of water, which flows toward the sponge's central cavity. Cells trap and partially digest particles of organic matter. The food particles then pass to amoebocytes, which distribute the food to other cells. Water and wastes exit the sponge through a large hole at the top (see figure 22.10).

Support and Movement: Sponge skeletons consist of protein fibers and spicules, which are sharp slivers of silica or calcium carbonate (see figure 22.10). These animals lack muscles and are therefore generally sessile (immobile).

Reproduction and Development: Sponges are hermaphrodites; each animal releases sperm into water currents and retains eggs. After fertilization, the zygote develops into a blastula, which is released and may drift for some time before settling into a new habitat. Sponges also commonly reproduce asexually by budding or fragmentation.

Defense: Spicules and toxic chemicals can help sponges fight off predators.

Impact on Humans: Some people use natural sponges in bathing, but collecting sponges can harm ecosystems. Also, the chemicals that protect sponges from predators may yield useful anticancer and antimicrobial drugs.

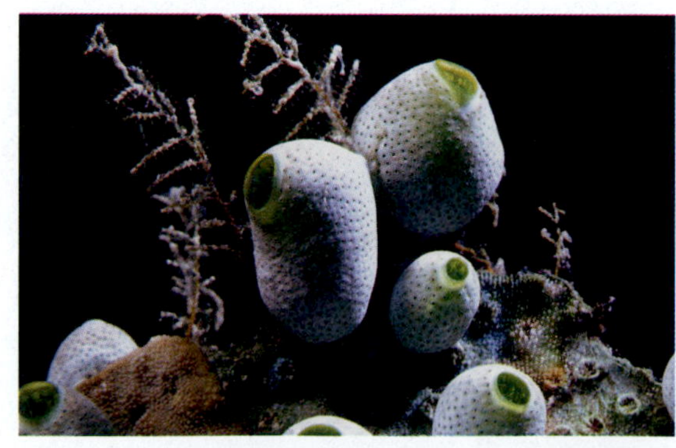

FIGURE 22.9 **Porous Animals.** Sponges are the simplest animals, lacking true tissues. Note the lack of symmetry in these sponges.

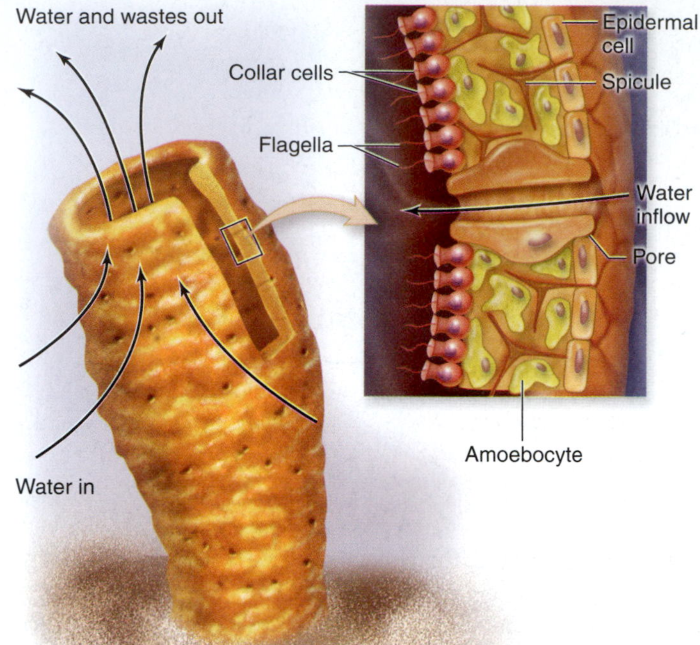

FIGURE 22.10 **Sponge Anatomy.** Flagellated collar cells capture suspended food particles from water that enters through pores in the body wall. Water and wastes exit through the large hole at the top. Multifunctional amoebocytes secrete spicules, which make up the sponge's skeleton.

Sponges	
Level of organization	Cellular
Symmetry	Asymmetrical or radial
Cephalization	Absent
Coelom	Absent
Type of digestive tract	Absent
Segmentation	Absent

Summary *Sponges are simple, asymmetrical aquatic animals that lack true tissues and organs. Their hollow bodies feed by straining food particles from water.*

22.2 MASTERING CONCEPTS

1. How is a sponge's body different from that of other animals?
2. What are two cell types in a sponge?
3. What is the function of spicules?
4. How do sponges reproduce sexually and asexually?
5. In what ways are sponges important?

22.3 Cnidarians Are Radially Symmetrical, Aquatic Animals

Phylum Cnidaria takes its name from the Greek word for "nettle," a stinging plant. **Cnidarians** include diverse aquatic animals (**figure 22.11**).

Habitat: Aquatic (mostly marine).

Body Structure: Cnidarian bodies typically take one of two radially symmetrical forms (**figure 22.12**). A **polyp** consists of a sessile stalk with tentacles on one end, whereas a **medusa** is a free-swimming, bell-shaped, "tentacles-down" body form typical of jellyfishes. In both body forms, two tissue layers sandwich a jellylike, noncellular substance called mesoglea. One opening, the mouth, leads to the dead-end gastrovascular cavity. Tentacles surrounding the mouth can sting by discharging microscopic capsules from cells called **cnidocytes.**

Diversity: Many cnidarians, such as *Hydra*, corals, and sea anemones, exist exclusively as polyps. In contrast, jellyfishes exist only as medusas. Still other cnidarians alternate between the two body forms.

Feeding: Cnidarians are carnivores that use their tentacles to grab and sting passing prey and stuff it into their gastrovascular cavities. Later, they eject indigestible matter through the same opening.

Support and Movement: In all cnidarians, the mesoglea forms a hydrostatic skeleton. Some cnidarians also secrete calcium carbonate exoskeletons. Over many generations, the exoskeletons secreted by countless coral animals build magnificent coral reefs. **coral reefs, p. 836**

Cnidarians can swim and move their tentacles courtesy of groups of linked neurons, called nerve nets, that coordinate their movements. The contraction of specialized cells forces water out of the bell of a jellyfish, propelling the animal through the water. The same mechanism enables a sea anemone to stuff food into its gastrovascular cavity.

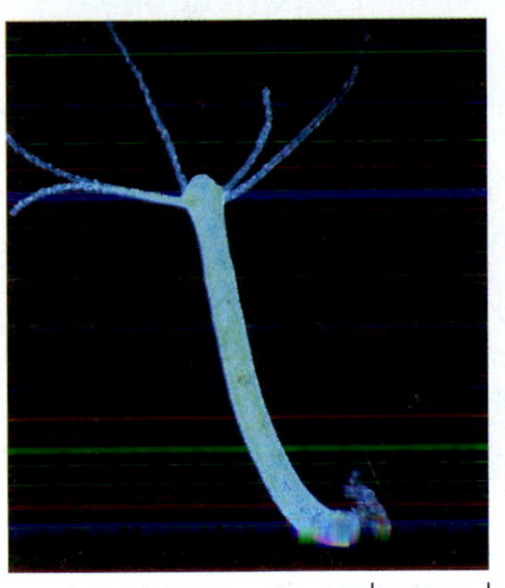

FIGURE 22.11 Jellyfishes and More. Cnidarians are radially symmetrical animals. Phylum Cnidaria includes (**A**) *Hydra*, (**B**) jellyfishes, and (**C**) corals.

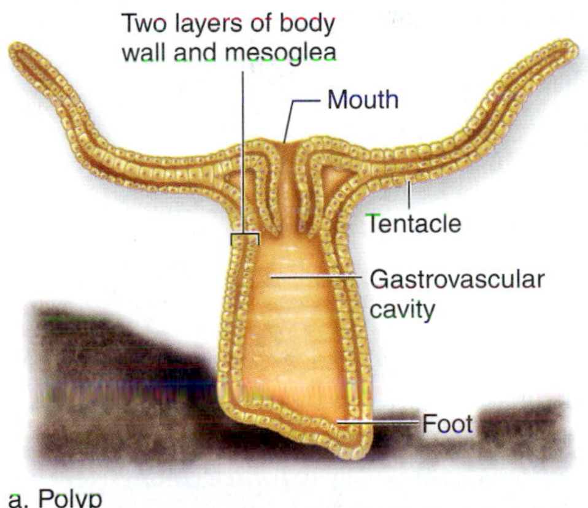

FIGURE 22.12 Cnidarian Anatomy. Cnidarians may have two body forms: (**A**) the polyp and (**B**) the medusa. Some species have only one form; others alternate between the two.

Reproduction and Development: Cnidarians reproduce sexually and asexually (**figure 22.13**). In the moon jelly *Aurelia*, for example, male and female medusae release gametes into the water. After fertilization, a cilia-fringed larva develops, attaches to a surface, and metamorphoses into the polyp form.

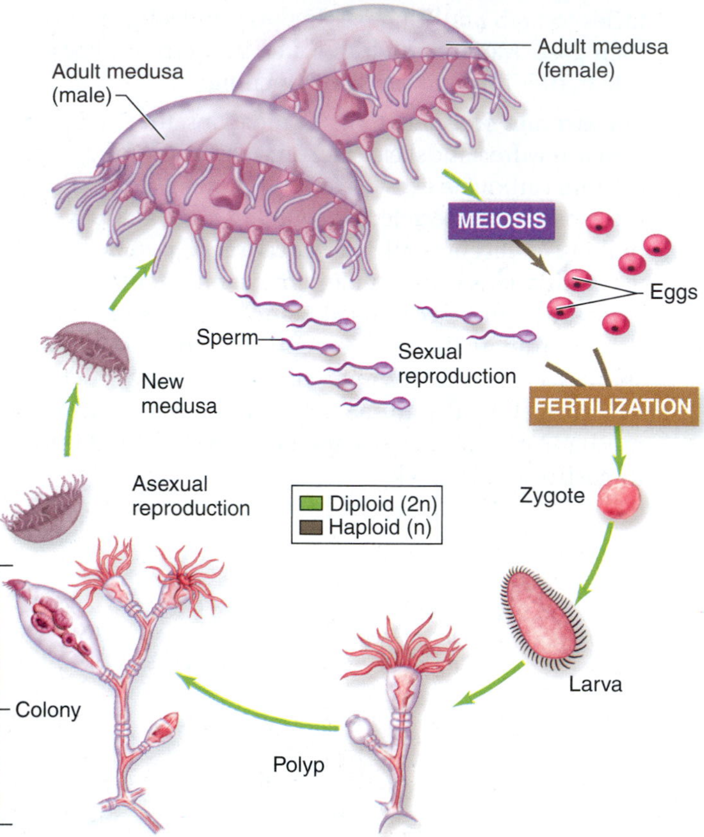

FIGURE 22.13 Life Cycle of the Moon Jelly *Aurelia*. In *Aurelia*, male and female medusae release sperm and eggs, and after fertilization, a larva develops. It attaches to a surface and becomes a polyp, which reproduces asexually to form a colony. New medusae form asexually from the polyp colony.

The polyp then buds asexually, generating a colony of additional polyps and, eventually, medusae. The cycle begins anew.

Defense: Stinging cnidocytes are the main defense against predators.

Impact on Humans: Jellyfish stings may cause skin irritation or cramps; some species have toxins that can be lethal. Coral reefs house many commercially important species of fishes and other animals, and they protect coastlines from erosion. As they build their calcium carbonate reefs, corals remove carbon from the atmosphere. A molecule originally isolated from corals (but now produced in the laboratory) is being developed into a sunscreen for human use.

Cnidarians	
Level of organization	Tissue
Symmetry	Radial
Cephalization	Absent
Coelom	Absent
Type of digestive tract	Incomplete
Segmentation	Absent

Summary *Cnidarians have radial symmetry, incomplete digestive tracts, and two body forms. Their tissues do not form organs. Cnidarian tentacles bear cnidocytes.*

22.3 MASTERING CONCEPTS

1. What are cnidocytes, and how do cnidarians use them?
2. What is the difference between a polyp and a medusa?
3. How do cnidarians feed, move, and reproduce?
4. In what ways are cnidarians important?

22.4 Flatworms Have Bilateral Symmetry and Incomplete Digestive Tracts

Phylum Platyhelminthes includes the **flatworms.** Because flatworms have incomplete digestive tracts and lack a coelom, biologists once assigned them to their own lineage (see figure 22.2, inset). Molecular data, however, have prompted the reclassification of protostomes. Flatworms now share a clade with annelids and mollusks.

Habitat: Free-living (usually in aquatic ecosystems) or parasitic on other animals.

Body Structure: Their unsegmented, flattened bodies enable their cells to exchange CO_2 and O_2 with their environment.

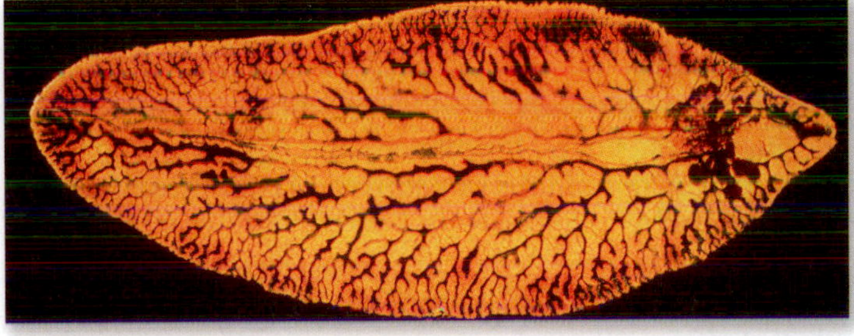

FIGURE 22.14 Flatworms. Flatworms (phylum Platyhelminthes) are the simplest protostomes. (**A**) A planarian is an example of a free-living flatworm. (**B**) This liver fluke, *Fasciola hepatica*, is a parasite that infects sheep, cattle, and humans. (**C**) Tapeworms in genus *Taenia* have many adaptations to life as internal parasites, including suckers that enable them to latch onto their host's intestines. © Dr. Richard Kissel & Dr. Gene Shih/Visuals Unlimited.

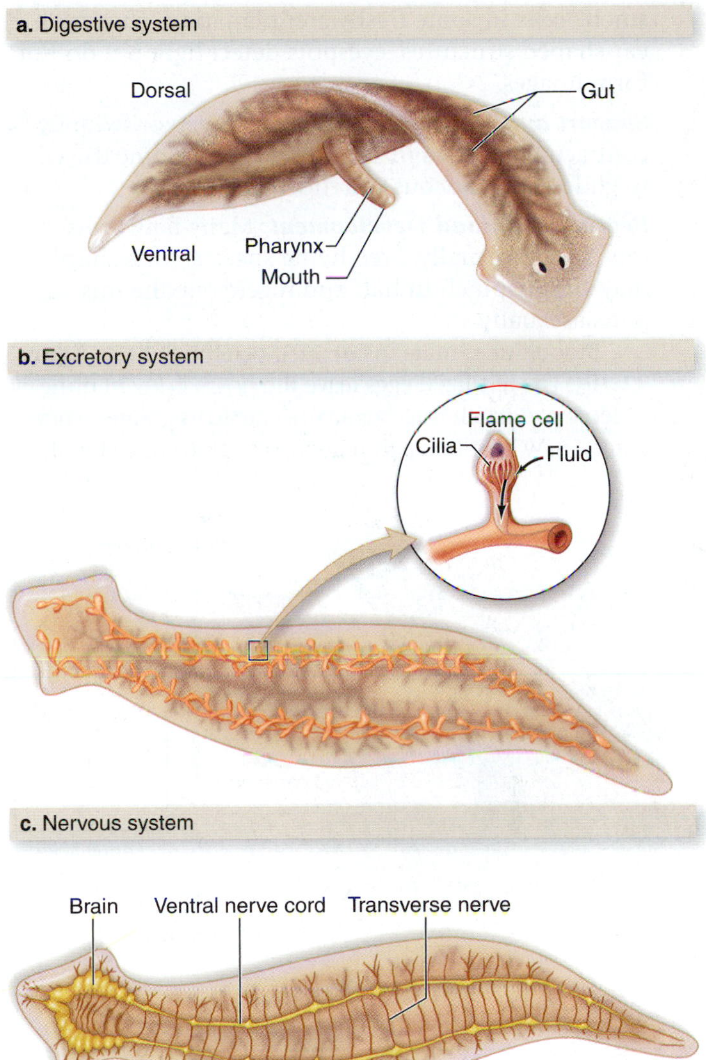

FIGURE 22.15 Simple Organ Systems. (**A**) In a planarian, the pharynx is a muscular tube that opens into the incomplete digestive tract. (**B**) Flame cells form the excretory system. (**C**) A planarian has a brain and ladderlike nerve cords.

Diversity: **Figure 22.14** shows the three groups of flatworms, including the **free-living flatworms** (such as a planarian) and the parasitic **flukes** and **tapeworms.**

Feeding: A fluke feeds on blood and other host tissues, using its muscular, tubelike pharynx to pull food into the gastrovascular cavity. A tapeworm lacks a mouth and digestive system; instead, it attaches to the host's intestine by a scolex (see figure 22.14C) and absorbs food through its body wall. Free-living flatworms usually are predators or scavengers. A muscular pharynx takes in food and ejects wastes from the highly branched gut (**figure 22.15**A).

Circulation and Respiration: Flatworms do not have specialized circulatory or respiratory systems. CO_2 and O_2 simply diffuse through the body wall.

Excretion: Protonephridia are structures that maintain internal water balance and excrete nitrogenous wastes. Each consists of numerous flame cells, so named because their shape and beating cilia resemble a flickering candle (figure 22.15B). The tubules of protonephridia open to pores on the body surface. **nitrogenous wastes, p. 715**

Nervous System: Some flatworms have nerve cords and concentrations of nerve cell bodies in their anterior ends, forming a simple brain (figure 22.15C). Others have a nerve net. Chemosensory (taste and smell) cells in some freshwater planarians cluster on ear-shaped structures; eyespots detect light but do not form images.

Support and Movement: Flatworms creep or swim by contracting muscles in a rolling motion. Some use cilia to glide along mucous secretions.

Reproduction and Development: Many flatworms reproduce asexually. Free-living species, for example, may simply pinch in half and regenerate the missing parts asexually.

Blood flukes mate inside their vertebrate host (**figure 22.16**). The fertilized eggs leave the host's body in urine or feces and hatch into larvae upon reaching water. They enter a new host through ingestion or skin contact with contaminated water. Tapeworms use a different strategy. The tapeworm's body consists of units called proglottids, each with a complete set of male and female reproductive structures. Proglottids with fertilized eggs break off of the worm and leave the host in feces. When a new host swallows proglottids in contaminated water, the eggs hatch. The resulting larvae can migrate to the host's muscles; this is how people acquire tapeworm infections by eating undercooked fish, beef, or pork.

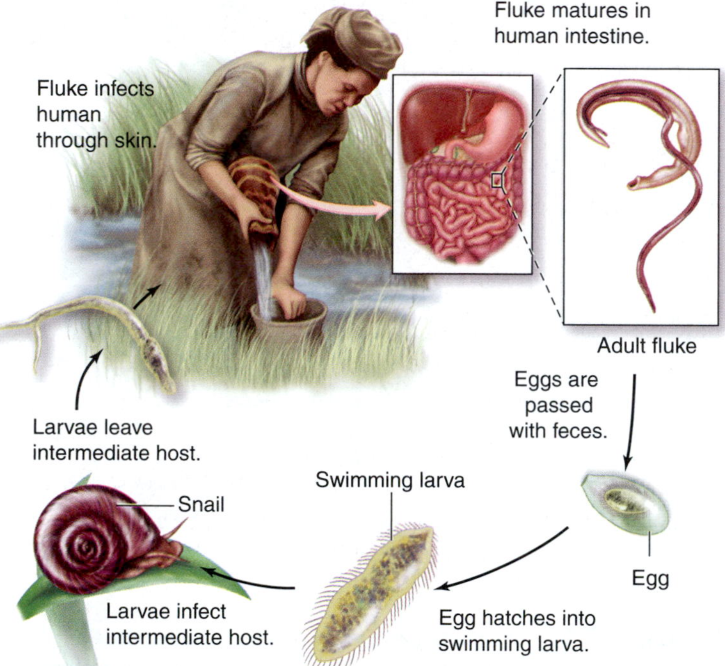
Fluke matures in human intestine.
Fluke infects human through skin.
Adult fluke
Larvae leave intermediate host.
Eggs are passed with feces.
Swimming larva
Snail
Egg
Larvae infect intermediate host.
Egg hatches into swimming larva.

FIGURE 22.16 Life Cycle of the Blood Fluke, *Schistosoma*. Humans acquire this parasite through openings in the skin. The worms live and reproduce inside the host's blood vessels, leaving the body in urine or feces. The eggs hatch in water and infect an intermediate host, usually a snail, which houses the parasite's larvae. Another organism may serve as a second intermediate before the fluke moves to its final vertebrate host.

Defense: When inside the host, parasitic flatworms are safe from predators; free-living forms secrete a protective mucus.

Impact on Humans: Worldwide, infections with blood flukes, lung flukes, liver flukes, and tapeworms affect hundreds of millions of people and countless domesticated and wild animals.

Flatworms	
Level of organization	Organ system
Symmetry	Bilateral
Cephalization	Present
Coelom	Absent
Type of digestive tract	Incomplete (when present)
Segmentation	Absent

Summary *Flukes, tapeworms, and free-living flatworms belong to phylum Platyhelminthes. These animals lack circulatory and respiratory systems, but they have excretory structures and concentrations of nerve cells.*

22.4 MASTERING CONCEPTS

1. How does the body shape of a flatworm enhance gas exchange with the environment?
2. Describe the three groups of flatworms.
3. How do free-living flatworms, tapeworms, and flukes eat?
4. What types of specialized cells and tissues do flatworms have in their excretory and nervous systems?
5. How do flatworms move and reproduce?
6. In what ways are flatworms important?

22.5 Mollusks Exhibit Soft, Unsegmented Bodies and Complex Organ Systems

Mollusks share a clade with annelids and flatworms. This fascinating phylum contains clams, snails, and the largest known invertebrate, the giant squid, which may be up to 20 m (65 feet) long. It also includes the octopus, the most intelligent invertebrate (**figure 22.17**).

Habitat: Terrestrial, marine, and freshwater.

Body Structure: Mollusks have several body structures in common (**figure 22.18**). One is the **mantle,** a dorsal fold of tissue that secretes a shell in most species. A mus-

FIGURE 22.17 Mollusk Diversity. Mollusks are unsegmented protostomes with complex organ systems. (**A**) The sea cradle chiton comes from the west coast of the United States. (**B**) The Pacific giant clam, a bivalve. (**C**) The land snail is a gastropod. (**D**) This octopus is a cephalopod, as are squids and chambered nautiluses.

a.

c.

b.

d.

cular **foot** provides movement, and an area called the **visceral mass** contains the digestive and reproductive organs. Many mollusks have a **radula,** a tonguelike strap with teeth made of chitin (a tough polysaccharide). They use the radula to scrape food into their mouths.

Diversity: Biologists classify mollusks into four groups. **Chitons** (see figure 22.17A) are marine animals with eight flat shells that overlap like shingles. **Bivalves,** such as oysters, clams, scallops, and mussels, have two-part, hinged shells (see figure 22.17B). **Gastropods** ("stomach-foot")

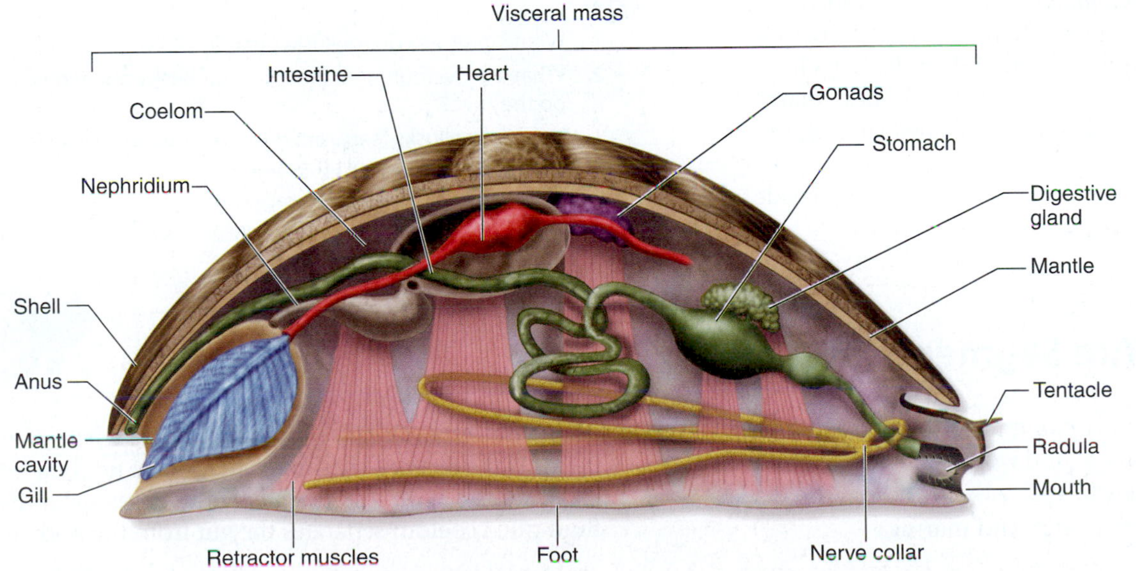

FIGURE 22.18 Molluscan Organs. The digestive, circulatory, excretory, and reproductive systems are all contained within the visceral mass. The muscular foot provides locomotion, and the mantle secretes a protective shell.

are snails, slugs, sea slugs (nudibranchs), and limpets. Their name comes from the broad, flat foot on which they crawl (see figure 22.17C). The **cephalopods** include marine animals such as octopuses, squids, and nautiluses. Cephalopod means "head-foot," a reference to the eight arms connecting directly to the head of the animal (see figure 22.17D).

Feeding: Bivalves filter organic particles and small organisms out of the water. Most slugs and snails are herbivores, whereas cephalopods are active predators of fast-moving prey such as fishes.

Circulatory System: Most mollusks have an open circulatory system, in which a heart pumps blood to tissues throughout the body cavity instead of within vessels. In contrast, cephalopods have a closed circulatory system, with blood confined to vessels. These animals are the speediest invertebrates, in part because their blood vessels efficiently exchange O_2 and CO_2 with respiring muscle cells.

Respiratory System: Aquatic mollusks have gills that exchange O_2 and CO_2 with the environment. Snails and slugs have a lung derived from a space called the mantle cavity.

Excretory System: An excretory organ called a nephridium ("little kidney") filters blood and produces urine.

Nervous System: The molluscan nervous system varies from simple and ladderlike to complex and cephalized. An octopus's nervous system includes a brain, a highly developed visual system, nerve cords that extend into each of its eight arms, and an excellent sense of touch.

Support and Movement: All mollusks have a hydrostatic skeleton, and most have an internal or external shell. The animal moves when muscles act on the constrained fluid of the coelom. Cephalopods can move by "jet propulsion," squirting water out of their siphons. Bivalves use their muscular foot to burrow into sediments. On land, snails and slugs glide on a trail of mucus.

Reproduction and Development: Many bivalves shed gametes into the water, where external fertilization occurs. Gastropods and cephalopods mate and fertilize eggs internally (see section 22.10). Following fertilization, many marine mollusks have a ciliated, pear-shaped larval stage, which settles to the bottom of the sea and develops into an adult. In cephalopods and snails, larvae develop inside the fertilized egg, and hatchlings resemble adults.

Defense: A hard shell protects against many predators, but mollusks also have other defenses. Squids and octopuses, for example, can change their color and shape to match their background with uncanny camouflage. When alarmed, cephalopods can squirt a melanin-pigmented "ink" that cloaks their escape.

Impact on Humans: Mollusks have diverse effects on human life, health, and environmental quality. We harvest pearls from oysters, and we eat clams, mussels, oysters, snails, squids, and octopuses. Bivalves can become poisonous if they accumulate pollutants or toxins produced by microorganisms called dinoflagellates. Snails and slugs are voracious consumers of garden and crop plants. Some aquatic snails host parasitic worms (see figure 22.16), and cone snails can kill humans with a venomous "harpoon." Zebra mussels have disrupted the ecology of the Great Lakes since their introduction in the 1980s. The cuttlebones that supply calcium to pet birds come from the thick internal shells of cephalopods called cuttlefish.
invasive species, p. 851; dinoflagellates, p. 391

Mollusks	
Level of organization	Organ system
Symmetry	Bilateral
Cephalization	Present
Coelom	Present
Type of digestive tract	Complete
Segmentation	Absent

Summary *Although mollusks appear quite different from one another, they all share basic body parts such as a mantle, muscular foot, and visceral mass. They have complete digestive tracts, open circulatory systems, and well-developed nervous systems.*

22.5 MASTERING CONCEPTS

1. What structures do mollusks share?
2. What are the four main groups of mollusks, and where do they live?
3. How do mollusks feed, move, excrete metabolic wastes, reproduce, and protect themselves?
4. In what ways are mollusks important?

22.6 Annelids Are Segmented Worms

Segmented worms belong to the phylum Annelida. Like mollusks, flatworms, round worms, and arthropods, the **annelids** are protostomes.

Habitat: Terrestrial, freshwater, and marine.

Body Structure: The most obvious characteristic of annelids is the repeated segments that make up their bodies. They also have a complete digestive system, and a fluid-filled coelom separates the gut from the body wall.

FIGURE 22.19 Segmented Worms. Annelids are segmented protostomes. (**A**) Oligochaetes, such as this earthworm (*Lumbricus terrestris*), have obvious segments. (**B**) Leeches have suckers and lack parapodia and setae. (**C**) *Nereis* is a polychaete; the fleshy paddles are parapodia.

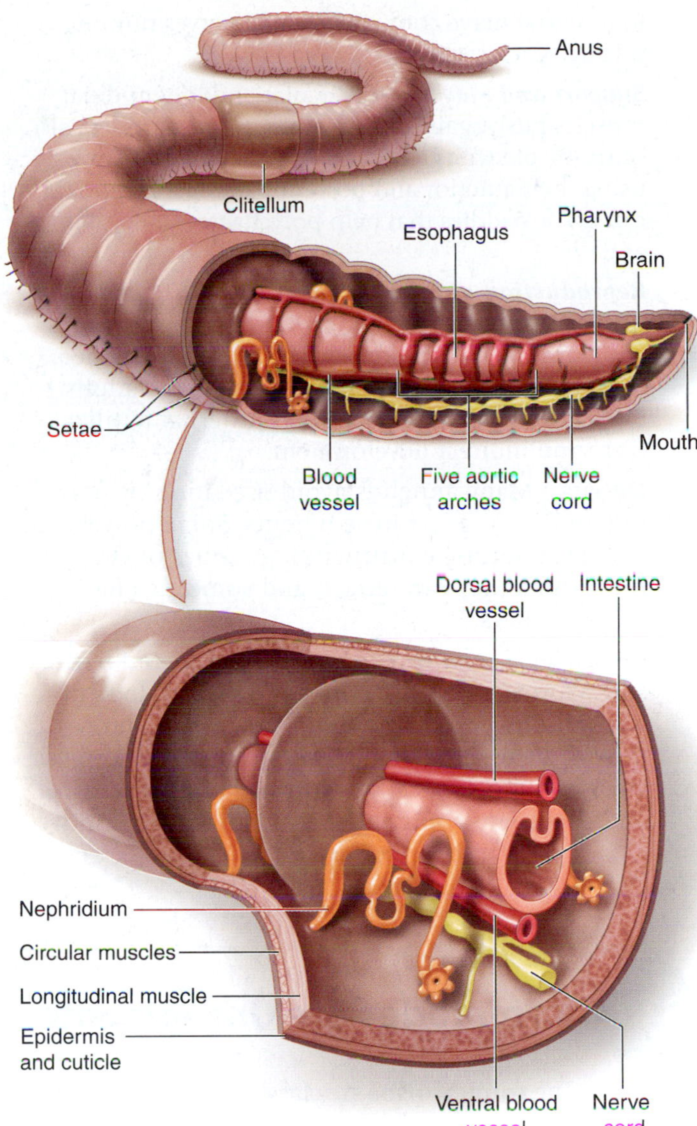

FIGURE 22.20 Earthworm Anatomy. Earthworms show the body segmentation that is typical of annelids. Setae are bristles that provide traction in locomotion. Groups of nerve cell bodies above the pharynx fuse to form a primitive brain that connects by nerve rings to the ventral nerve cord. The circulatory system is closed. The anterior end of the dorsal blood vessel contracts rhythmically, forcing blood through aortic arches that connect with the ventral blood vessel. Pairs of nephridia in each segment remove metabolic wastes from the coelom and blood.

Diversity: Biologists recognize three groups of annelids. **Oligochaetes,** such as earthworms, have a saddlelike thickening called a clitellum that is visible at all times (**figure 22.19A**). "Oligo-" means "few," a reference to the small number of bristles (setae) on the side of each segment. **Leeches** have a clitellum that is visible only during breeding season. Their segments lack bristles (figure 22.19B), but they have suckers and superficial rings called annuli within each segment. Most live in fresh water. **Polychaetes** are marine segmented worms. Most polychaetes have pairs of fleshy, paddlelike appendages called parapodia that they use in locomotion (figure 22.19C). Parapodia have many long bristles embedded in them.

Feeding: Earthworms are deposit feeders, ingesting soil and straining out organic material for food. Many leeches suck blood from vertebrates, but most eat small organisms such as arthropods, snails, or other annelids. Polychaetes are often filter feeders, although some are predators with formidable jaws.

Circulatory System: All annelids except leeches have a closed circulatory system. Earthworms have multiple hearts that pump blood throughout the body in vessels (**figure 22.20**)

Respiratory System: Some polychaetes have feathery gills, but oligochaetes and leeches exchange gases by diffusion through the body wall. They must therefore remain moist, which explains why earthworms that crawl onto sidewalks during a heavy rain die once the sun comes out.

Excretory System: The nephridium draws in fluid from the coelom, returns some ions and other substances to the blood, and discharges the waste-laden fluid outside the body through a pore (see figure 22.20).

Nervous System: The nervous system includes a simple "brain," a mass of nerve cells at the head end of the animal. These cells connect around the digestive tract to a ventral nerve cord, with lateral nerves running through each segment.

Support and Movement: Circular and longitudinal muscles push against the coelom as the worm crawls, burrows, or swims. Leeches crawl, inchworm-style, by using their anterior and posterior suckers. Parapodia can act as paddles that help polychaetes walk, swim, and dig.

Reproduction and Development: Leeches and oligochaetes are hermaphrodites. Two individuals copulate, each discharging sperm that its partner temporarily stores. Juveniles resemble the adults. Polychaetes have separate sexes, external fertilization, and indirect development.

Defense: Many annelids avoid predation by burrowing underground or in sediments. Some polychaetes (the tubeworms) construct tough tubes of chitin into which they can retract, and some also have powerful jaws.

Impact on Humans: Earthworms aerate and fertilize soil. Worm farms raise oligochaetes for sale as fishing bait or soil conditioners. In medicine, a blood-thinning chemical from leeches is used to stimulate circulation in surgically reattached digits and ears, and physicians sometimes apply leeches to remove excess blood that accumulates after damage to the nervous system.

Annelids	
Level of organization	Organ system
Symmetry	Bilateral
Cephalization	Present
Coelom	Present
Type of digestive tract	Complete
Segmentation	Present

Summary *Annelids are segmented worms with complex organ systems. They include earthworms, leeches, and marine worms called polychaetes.*

22.6 MASTERING CONCEPTS

1. What are the distinguishing features of the three groups of annelids, and where do they live?
2. How do annelids feed, exchange gases, excrete metabolic wastes, move, reproduce, and defend themselves?
3. In what ways are annelids important?

22.7 Nematodes Are Unsegmented, Cylindrical Worms

The **roundworms** are classified in Phylum Nematoda (**figure 22.21**). Most are barely visible to the unaided eye, but they are extremely abundant. A square meter of soil can yield millions of nematodes, with dozens of species coexisting in the same habitat.

Because nematodes have a pseudocoelom, biologists once classified them in their own lineage, separate from animals with true coeloms. They now share a clade with arthropods, partly based on molecular data and partly because both nematodes and arthropods periodically molt.

Habitat: Most nematodes are free-living in soil or in sediments of aquatic ecosystems, both freshwater and marine. Some roundworms parasitize plants or animals. Most parasitic species live in only one host, although mosquitoes can transmit *Wuchereria bancrofti*, the nematode that causes elephantiasis (**figure 22.22**).

Body Structure: Nematodes are unsegmented, cylindrical worms with tapered ends. An external layer of tissue

FIGURE 22.21 Roundworms. Nematodes are among the most abundant animals; they are unsegmented protostomes. Millions of humans harbor the intestinal roundworm *Ascaris lumbricoides*. The parasite is acquired through contact with human feces containing eggs, which hatch in the intestine. In this photograph, the female on the left dwarfs the male; a female can be up to 40 cm. long.

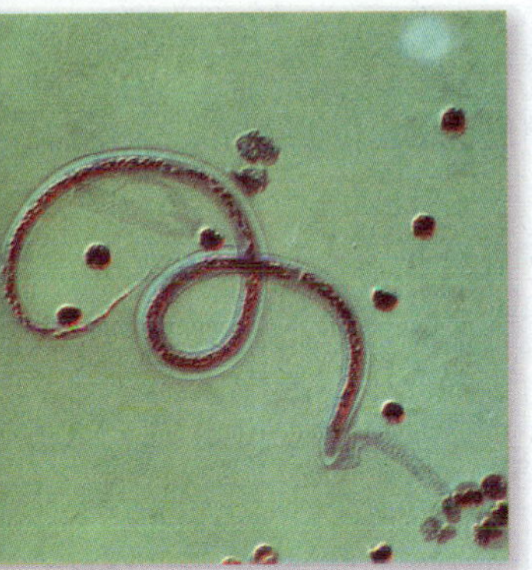
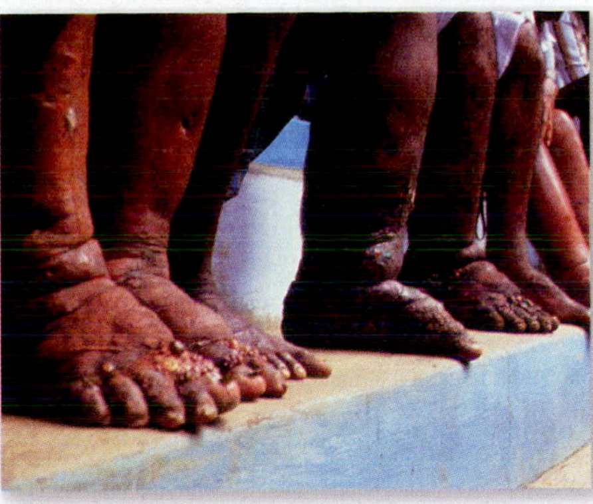

FIGURE 22.22 Elephantiasis. *Wuchereria bancrofti* causes elephantiasis. The worms inhabit lymph nodes and cause the accumulation of excess fluid in the legs and feet.

secretes a surrounding protein-rich cuticle. Like arthropods, nematodes molt; that is, they shed the cuticle and grow a new one several times during their lives.

Diversity: Nematodes have a few external features that are useful in classification; for example, plant-pathogenic nematodes are classified partly based on mouthpart shapes. Most nematode classification relies on genetic sequences.

Feeding: In soil, nematodes feed on fungi, bacteria, roots, or almost any other plant part. Some prey on insect larvae. Nematodes that parasitize mammals suck blood or eat digested food in the intestines, competing with the host for nutrients.

Circulation and Respiration: Nematodes do not have specialized circulatory or respiratory organs; the body cavity distributes nutrients, O_2, and CO_2 throughout the body (**figure 22.23**).

Excretory System: A series of interconnected cells maintain salt balance and remove nitrogenous wastes from the body cavity. The canals open to the outside via an excretory pore.

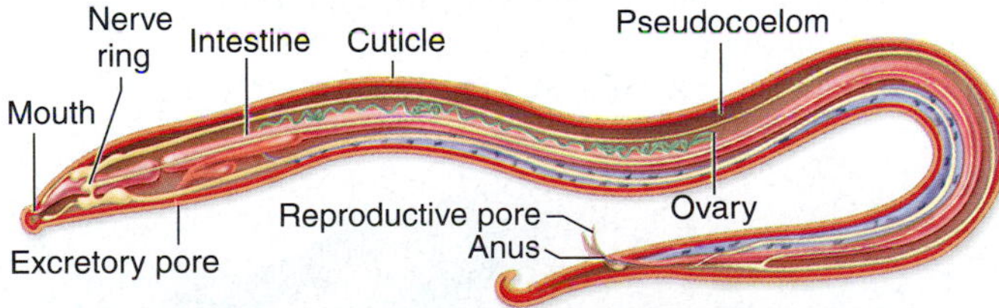

FIGURE 22.23 Nematode Organs. Although a free-living nematode is only about 1 mm long, its body nevertheless contains a complete digestive tract, a simple nervous system, and reproductive organs.

Nervous System: Nematodes have an anterior brain and two nerve cords along the length of the body. They can sense chemicals and touch.

Support and Movement: The fluid-filled pseudocoelom acts as a hydrostatic skeleton. Nematodes are limited to back-and-forth, thrashing motions because they have only longitudinal (lengthwise) muscles acting on the pseudocoelom.

Reproduction and Development: Most species have separate sexes. Females produce large numbers of tough eggs that survive drying and exposure to damaging chemicals. The juveniles undergo direct development.

Defense: The tough cuticle may protect nematodes. Also, parasitic roundworms are protected from predators while inside their hosts.

Impact on Humans: Biologists use the nematode *Caenorhabditis elegans* in scientific research (see Focus on Model Organisms: *Caenorhabditis elegans* and *Drosophila melanogaster*). Pinworms, hookworms, and *Ascaris* inhabit the intestines of humans and other animals, and the worms that cause elephantiasis live in lymph nodes. Still other parasitic nematodes inhabit the tissues beneath the skin and cause river blindness. Some cause diseases in important food crops, whereas others attack insect pests.

Roundworms	
Level of organization	Organ system
Symmetry	Bilateral
Cephalization	Present
Coelom	Pseudocoelom
Type of digestive tract	Complete
Segmentation	Absent

Summary *Nematodes are extremely abundant, unsegmented worms that lack respiratory and circulatory organs and have complete digestive tracts.*

22.7 MASTERING CONCEPTS

1. Compare and contrast the roundworm body structure with those of a flatworm and an annelid.
2. What evidence places roundworms in a clade with arthropods?
3. How do nematodes feed, excrete metabolic wastes, move, and reproduce?
4. In what ways are roundworms important?

FOCUS ON MODEL ORGANISMS: *Caenorhabditis elegans* and *Drosophila melanogaster*

Two invertebrates, a roundworm and a fruit fly, share the spotlight in this box. Both have the characteristics common to all model organisms: small size, easy cultivation in the lab, and rapid, prolific life cycles. Each has provided crucial insights into life's workings.

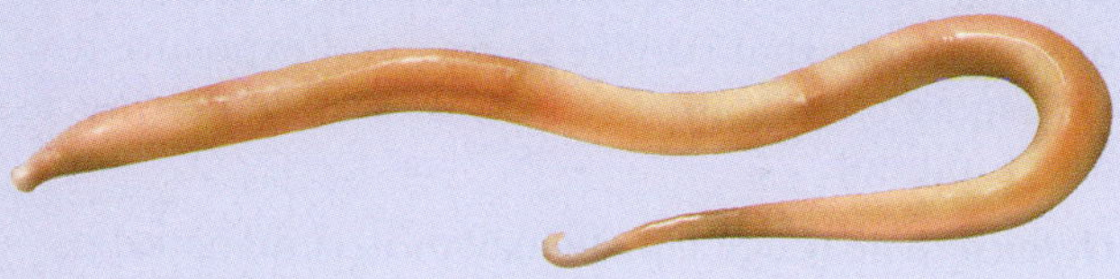

The nematode: *Caenorhabditis elegans*

This roundworm, a soil inhabitant, is arguably the best understood of all animals. This was the first animal to have its genome sequenced (in 1998), revealing about 18,000 genes. A small sampling of the contributions derived from research on *C. elegans* includes:

- **Animal development:** An adult *C. elegans* consists of only about 1000 cells. Because the animal is transparent, biologists have been able to trace the ancestry of each cell back to the zygote. Eventually, researchers hope to understand every gene's contribution to the development of this worm.
- **Apoptosis:** Programmed cell death, or apoptosis, is the planned "suicide" of cells as a normal part of development. Researchers observing *C. elegans* development know exactly which cells will die, and when. Learning about genes that promote apoptosis may lead researchers to a better understanding of cancer, a family of diseases in which cell division is unregulated. **cancer, p. 166; apoptosis, p. 168**
- **Muscle function:** The first *C. elegans* gene to be cloned, *unc-54*, revealed the amino acid sequence of one part of myosin, a protein required for muscle contraction. **myosin, p. 643**
- **Drug development:** Nematodes provide a good forum for preliminary testing of new pharmaceutical drugs. For example, researchers might identify a *C. elegans* mutant lacking a functional insulin gene, then test new diabetes drugs to see whether they replace the function of the missing gene. **diabetes, p. 615**
- **Aging:** Worms with mutations in some genes have life spans that are twice as long as normal. The selective destruction of neurons can also expand or reduce the life span, depending on which cells are destroyed. Insights on aging in *C. elegans* may eventually help increase the human life span.

The fruit fly: *Drosophila melanogaster*

Drosophila melanogaster is only about 3 mm long, but like *C. elegans*, it is a giant in the biology lab. The flies are easy to rear in plugged jars containing rotting fruit or a mix of water, yeast, sugar, cornmeal, and agar. Its genome sequence was completed in 1999; many of its 13,600 genes have counterparts in humans. These findings belie *Drosophila*'s century-long history as a model organism. Some of the most important research areas include:

- **Heredity:** In the early 1900s, Thomas Hunt Morgan and his colleagues used *Drosophila* to show that chromosomes carry the information of heredity. Studies on mutant flies with different-colored eyes led to the discovery of sex-linked traits. Morgan's group also demonstrated that genes located on the same chromosome are often inherited together. In the process, they discovered crossing over. **meiosis, p. 177; Mendelian genetics, p. 196; chromosomes, p. 216; gene linkage, p. 217; sex linkage, p. 224**
- **Human disease:** The similarity of some *Drosophila* genes to those in the human genome has led to important insights into muscular dystrophy, cancer, and many other diseases. For example, researchers have studied the fly version of the human *p53* gene, which induces damaged cells to commit suicide (apoptosis). When that gene is faulty, the cell may continue to divide uncontrollably. The result: cancer.
- **Animal development:** Homeotic genes are "master switch" genes that regulate the overall development of the body, including segmentation and wing placement. Researchers discovered these genes in mutant flies with dramatic abnormalities, such as legs growing in place of antennae on the fly's head (see figure 12.13). Later, researchers discovered comparable genes in many organisms, including mice, leading to new insights into mammalian development. **the mouse as model organism, p. 493**
- **Circadian rhythms:** The expression of some genes in bacteria, plants, fungi, and animals cycles throughout a 24-h day. How do the rhythmically expressed genes "know" what time it is? In *Drosophila*, clock genes called *period* and *timeless* encode proteins that turn off their own expression, much like a thermostat turns off a heater when the temperature is too high. This "master clock" controls the animal's other daily cycles of hormone secretion and behavior.

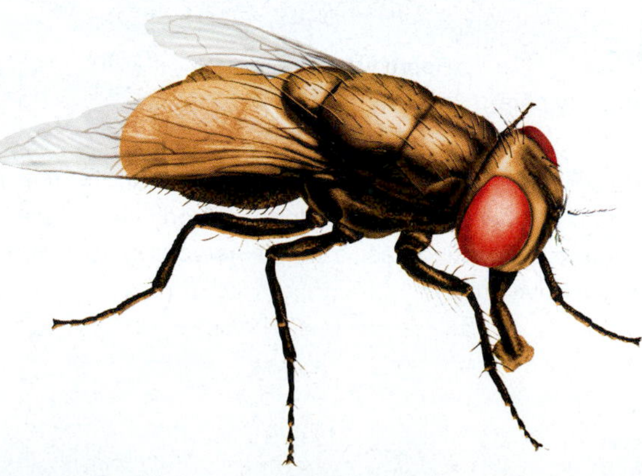

22.8 Arthropods Have Exoskeletons and Jointed Appendages

If diversity and sheer numbers are the measure of biological success, then the phylum Arthropoda certainly is the most successful group of animals (see figure 22.1). More than 1,000,000 species of **arthropods** have been recorded already, and biologists speculate that this number could double. Arthropods share a clade with nematodes, another extraordinarily successful group.

A. Arthropods Have Efficient Organ Systems

Habitat: Terrestrial, freshwater, and marine.

Body Structure: Arthropod bodies are segmented. In many arthropods, the segments group into three major body regions: head, thorax (chest), and abdomen (**figure 22.24**). The most distinctive feature of arthropods, however, is their jointed appendages. Although Arthropoda means "jointed foot," the appendages include legs, mouthparts, wings, antennae, copulatory organs, ornaments, and weapons.

The versatile, lightweight **exoskeleton** is made mostly of chitin, protein, and (sometimes) calcium salts. Thin, flexible areas create moveable joints between body segments and within appendages. An exoskeleton has a drawback, though—to grow, an animal must molt and secrete a bigger one.

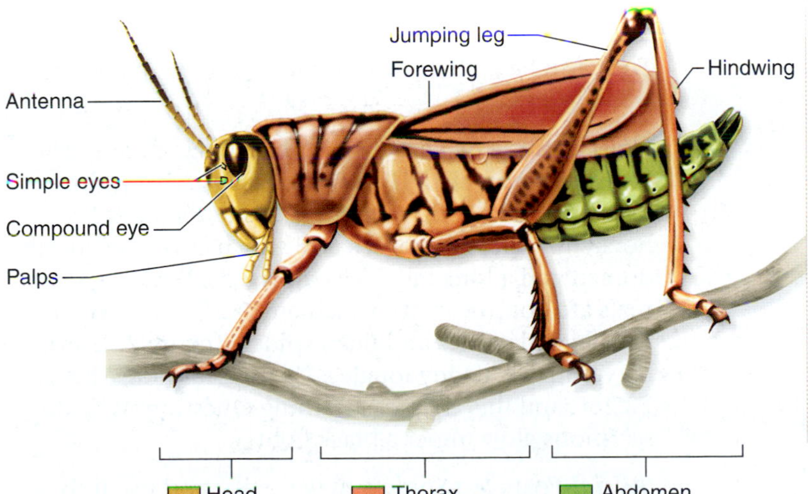

FIGURE 22.24 Arthropods. Arthropods are segmented protostomes with exoskeletons and jointed appendages. The grasshopper displays the external anatomy of an arthropod.

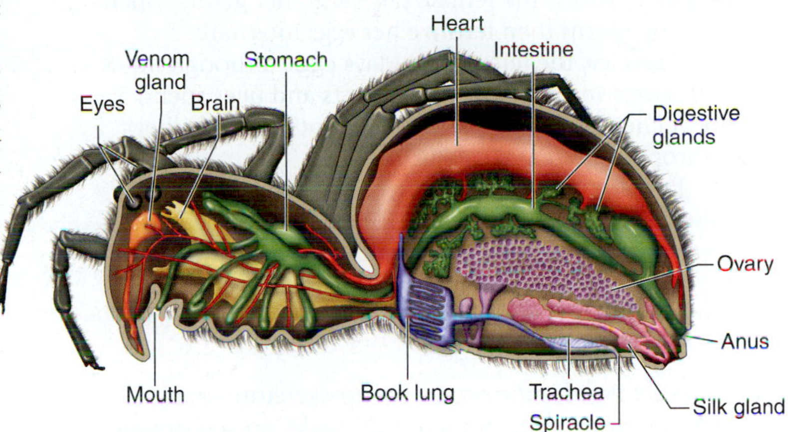

FIGURE 22.25 Arthropod Organs. The internal anatomy of a spider is surprisingly complex.

Feeding: Arthropods use chewing or sucking mouthparts to eat almost everything imaginable, including dead organic matter, plant parts, and other animals.

Circulatory System: Arthropods have open circulatory systems (see figure 32.2). A heart propels the blood, which circulates freely through the body cavity (**figure 22.25**).

Respiratory System: In most land arthropods, the body wall is perforated with holes (spiracles) that open into a series of branching tubes, transporting oxygen and carbon dioxide to and from tissues. Aquatic arthropods have gills, and spiders and scorpions have stacked plates called book lungs.

Excretory System: Insects, spiders, and other terrestrial arthropods have organs called Malpighian tubules that collect and remove nitrogenous wastes while reabsorbing water. (The "green gland" in crayfish and other aquatic arthropods has a similar function.) The tubules deposit dry, nitrogen-rich waste into the posterior end of the digestive tract. The animal ejects the waste, together with undigested food, through its anus.

Nervous System: Many arthropods are active, fast, and sensitive to their environment, thanks to a nervous system with a dorsal brain and a ventral nerve cord. Their sensory systems detect touch, air currents, and body position. Vibration, light, and chemicals signal potential meals or mates. **firefly bioluminescence, p. 799**

Support and Movement: The tough exoskeleton protects the animal and gives it its shape. Internal muscles span the joints between body segments and within appendages, creating hollow lever systems that can generate enormous force.

Reproduction and Development: Most arthropods have separate sexes. The male commonly produces a packet of sperm, which the female takes into her genital opening. The sperm then fertilize her eggs internally. In most species, the female then lays eggs, although mites and scorpions bear live young. Ants and bees spend a lot of energy tending their young, but in most other arthropods, parental care is minimal.

Development may be direct or indirect. Moths, butterflies, houseflies, and some others undergo indirect development, changing dramatically during a metamorphosis from larva to pupa (cocoon) to adult. Others, such as crickets, change only gradually from molt to molt.

Defense: Besides the protective exoskeleton, many arthropods can bite, sting, pinch, make noises, or emit foul odors or toxins that deter predators. Some have excellent camouflage that enables them to blend into their surroundings. They may jump, run, roll into a ball, dig into soil, or fly away when threatened. Some moths unfurl wings with dramatic eyespots that startle or confuse predators.

Impact on Humans: Arthropods intersect with human society in about every way imaginable. Mosquitoes, flies, fleas, and ticks spread infectious diseases as they consume human blood. Bees and scorpions sting, termites chew wood in our homes, and many insect larvae destroy crops. Yet, entire industries rely on arthropods—consider beeswax, honey, silk, and delicacies such as shrimp, crabs, and lobsters. Insects pollinate many plants, and spiders eat crop pests. The fruit fly, *Drosophila melanogaster*, is important in biological research. On a much smaller scale, dust mites eat flakes of skin that we shed as we move about our homes.

FIGURE 22.26 Trilobites. These extinct marine arthropods are called "three-lobed" because of the three distinct body regions running the length of the body: a long central lobe plus flanking right and left lobes.

FIGURE 22.27 Arthropod Mouthparts. Some arthropods, such as spiders, have chelicerae. Others, such as ants, have mandibles.

B. Arthropods Are the Most Diverse Animals

Of the four subphyla in Arthropoda, one contains the extinct **trilobites** (**figure 22.26**). The other subphyla are classified based on mouthpart shape (**figure 22.27**). Spiders, scorpions, and other **chelicerates** have clawlike mouthparts called chelicerae. **Mandibulates** have jawlike mouthparts termed mandibles.

Chelicerates: Spiders and Their Relatives

Habitat: Terrestrial and aquatic.

Body Structure: Most chelicerates have two major body regions: an abdomen and a fused head and thorax. They also have chelicerae and four or more pairs of walking legs.

Diversity: The two groups of chelicerates are horseshoe crabs and arachnids. **Horseshoe crabs** are primitive-looking animals whose name refers to their hard, horseshoe-shaped exoskeleton, which covers a wide abdomen and a long tailpiece (**figure 22.28**A). These animals are not true crabs, which are crustaceans. **Arachnids** include mites and ticks; spiders (figure 22.28B); harvestmen ("daddy longlegs"); and scorpions (figure 22.28C and this chapter's Burning Question: Why do scorpions glow under a black light?).

Special Features: Spiders make "silk" and use it to produce webs, tunnels, egg cases, and spiderling nurseries. Some spiderlings use silk to "balloon" to a new habitat.

a. b. c.

FIGURE 22.28 Chelicerate Arthropods. (**A**) Horseshoe crabs are chelicerates, as are (**B**) spiders and (**C**) scorpions. The scorpion is fluorescing green under ultraviolet light.

Mandibulates: Crustaceans

Habitat: Mostly aquatic. Isopods, commonly known as pill bugs or "roly polies," are the only terrestrial **crustaceans.**

Body Structure: **Crustaceans** have mandibles, branched appendages, two pairs of antennae, and two or three major body segments (**figure 22.29A**).

Diversity: This group contains many familiar animals, including crabs, shrimp, and lobsters. Other crustaceans include isopods, brine shrimp, water fleas (*Daphnia*), copepods and barnacles.

Mandibulates: Insects, Millipedes, and Centipedes

These arthropods colonized land shortly after plants, about 475 MYA, and they have diversified greatly ever since. This largest group of arthropods includes around 1 million species of **insects** and about 13,000 species of **millipedes** and **centipedes** (figure 22.29B).

Habitat: Millipedes and centipedes are terrestrial, as are most insects, but many insects live or reproduce in fresh water. The ocean, high altitudes, and extremely cold habitats are about the only places that are nearly devoid of insects.

Body Structure: All members of this group have mandibles and one pair of antennae. Insects have a head, thorax, and abdomen, six legs, and usually two pairs of wings. Centipede and millipede bodies include a head and trunk, which is further divided into repeating subunits, each with one or two pairs of appendages.

a.

b.

d.

c.

e.

FIGURE 22.29 Mandibulate Arthropods. (**A**) The lobster is a familiar crustacean. (**B**) The centipede has strong mandibles and venomous claws. (**C**) This molting cicada is an insect, as is (**D**) this beetle, which before metamorphosis was a grub (**E**).

Diversity: Insect diversity almost defies description (see figure 22.29C-E and **table 22.2**). They range in size from wingless soil-dwellers less than 1 mm long to fist-sized beetles, foot-long walking sticks, and flying insects with foot-wide wingspans. Some extinct dragonflies were even larger—one had a wingspan of about 75 cm (30 inches)!

Special Features: Insects were the first animals to fly, using their wings to disperse to new habitats, escape predators, court mates, and find food. Many of today's flowering plants evolved in conjunction with insect flight, trading nectar for pollination services. **pollination, p. 539**

Table 22.2 *Some Major Groups of Insects*

Name of Group	Name of Subgroup	Examples
Thysanura		Silverfish
Ephemeroptera		Mayflies
Odonata		Dragonflies and damselflies
Neoptera	Orthoptera	Roaches, crickets, grasshoppers
	Phthiraptera	Lice
	Hemiptera	Cicadas, aphids
	Coleoptera	Beetles
	Hymenoptera	Ants, wasps, bees
	Lepidoptera	Moths, butterflies
	Diptera	True flies
	Siphonaptera	Fleas

Arthropods	
Level of organization	Organ system
Symmetry	Bilateral
Cephalization	Present
Coelom	Present
Type of digestive tract	Complete
Segmentation	Present

Summary *Arthropods are segmented animals with jointed appendages and exoskeletons of chitin. They have complex organ systems, including sensitive sensory structures. Trilobites are extinct arthropods; living species are classified as chelicerates or mandibulates. Insects are extremely abundant and diverse mandibulates.*

22.8 MASTERING CONCEPTS

1. What are the main body regions of an arthropod?
2. How do arthropods use their jointed appendages?
3. Describe an arthropod's organ systems and how these animals feed, respire, excrete metabolic wastes, sense their environment, move, reproduce, and defend themselves.
4. What is the function of the exoskeleton?
5. In what ways are arthropods important?
6. How are chelicerates different from mandibulates?
7. Give an example of an animal in each of the four subphyla of arthropods.

Burning Questions

Why do scorpions glow under a black light?

Unlike most other arthropods, all scorpions emit an eerie greenish glow under ultraviolet light (see figure 22.28C). The glow comes from fluorescent compounds, called beta-carbolines, that scorpions produce in their exoskeletons. (Interestingly, the same substances also occur in human lenses that have developed cataracts.)

No one knows whether or how this fluorescence benefits the nocturnal scorpion, but hypotheses abound. One idea is that scorpions "capture" ultraviolet radiation from the night sky and re-emit it as green light to attract prey, including moths and other insects. Or perhaps the fluorescence somehow allows the animal to sense or measure ultraviolet radiation or helps the animal avoid harmful light levels.

Have a Burning Question of your own? Submit it to marielle_hoefnagels@mcgraw-hill.com for possible inclusion in future editions of this book!

22.9 Echinoderm Adults Have Five-Part, Radial Symmetry

The phylum Echinodermata contains some of the most colorful sea animals. Their name means "spiny skin." **Echinoderms** and chordates group together as deuterostomes because of similarities in early embryonic stages. Comparing DNA sequences has confirmed this close evolutionary tie.

Habitat: Marine.

Body Structure: Most adult sea stars and brittle stars have five arms (**figure 22.30A**). Although sea urchins, sea cucumbers, and sand dollars are armless, they retain the five-part symmetry (figure 22.30B-D).

In the echinoderm's unique **water vascular system,** seawater enters a series of enclosed canals that end in hollow, suction-cup-like tube feet (**figure 22.31**). Coordinated muscle contractions extend and retract each foot, bending it from side to side or creating a suction-cup effect when the foot is applied to a hard surface.

Diversity: The five most common and familiar groups of echinoderms are sea lilies; sea stars; brittle stars; sea urchins and sand dollars; and sea cucumbers.

Feeding: Some echinoderms are predators. A sea star, for example, attaches its tube feet to a bivalve's shell and steadily pulls until the prey's muscles tire and the shell opens. The sea star then everts its stomach through its mouth into the bivalve, secretes digestive enzymes, and absorbs the liquefied food. Sea cucumbers eat dead organic matter, and sea urchins scrape algae from rocks.

FIGURE 22.30 Spiny Skin. Echinoderms are deuterostomes that are more closely related to chordates than to any other animals. Echinoderms include (**A**) sea stars, (**B**) sea cucumbers, and (**C**) sea urchins. (**D**) The underside of this Indonesian sea urchin clearly reveals its five-part symmetry.

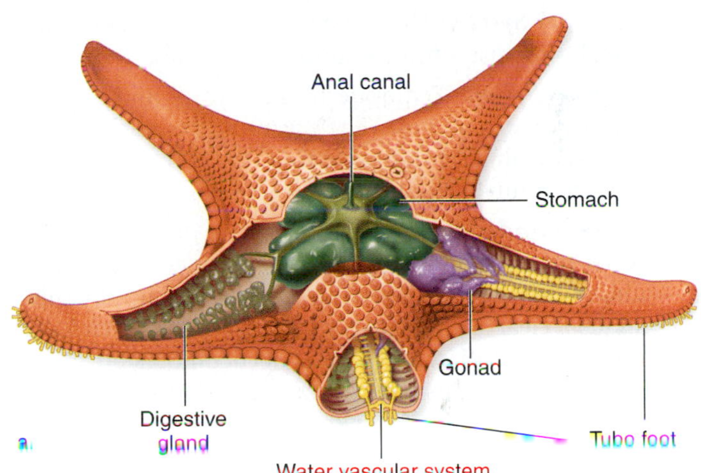

FIGURE 22.31 Water Vascular System. (**A**) The echinoderm water vascular system connects to tube feet. (**B**) The tube feet of this sea cucumber help the animal move along the ocean floor, acquire oxygen, and feel its surroundings.

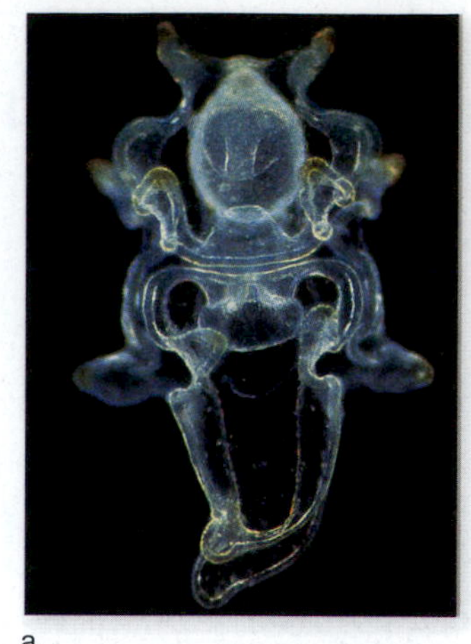

FIGURE 22.32 Echinoderm Metamorphosis.
(**A**) This sea star larva does not resemble the adult organism. These bilaterally symmetrical larvae reveal the evolutionary relationship between echinoderms and other animals.
(**B**) As development proceeds, the five-part symmetry becomes apparent. This sea star has recently cast off its larval body.

a. b.

Circulatory and Respiratory Systems: Tube feet function like gills, absorbing oxygen from the water. Respiration may also occur by means of tiny skin gills, as in sea stars, or by a network of passageways called a respiratory tree, as in sea cucumbers.

Excretory System: The water vascular system helps flush metabolic wastes from echinoderm bodies.

Nervous System: Echinoderms lack heads and brains, but nerves extend down their arms and tentacles and connect with a central nerve ring that surrounds the gut. Their tube feet can serve as sense organs.

Support and Movement: The wavelike pumping of water in and out of the tube feet allows echinoderms to glide slowly while maintaining a firm grip on the substrate, a clear advantage in a wave-pounded environment. Sand dollars and some sea cucumbers burrow into soft sediments, and some brittle stars can swim by using their appendages as oars.

Reproduction and Development: Echinoderms usually reproduce sexually, with male and female gametes from separate individuals combining in the sea. The larvae start out with bilateral symmetry (**figure 22.32**), but then a group of cells assembles into a five-sided disc that turns inside out and consumes the remainder of the larva. The animal is now a tiny juvenile, a replica of the adult form.

Defense: Echinoderms have spiny or spiky skin, sometimes equipped with small pinchers. Many echinoderms, such as the sea star, have internal skeletal plates that are strong but lightweight. In sea urchins and sand dollars, the plates fuse into a protective shell. Sea cucumbers have soft bodies, but they often produce poisonous chemicals. Many echinoderms can also regenerate severed body parts.

Impact on Humans: People harvest some species of sea urchins for their eggs, an ingredient in Japanese sushi. People also eat parts of some sea cucumbers. The crown-of-thorns sea star, *Acanthaster planci*, can cause painful wounds in humans who touch them. These animals eat corals and have wiped out large areas of Australia's Great Barrier Reef.

Echinoderms	
Level of organization	Organ system
Symmetry	Bilateral larvae; radial adults
Cephalization	Absent
Coelom	Present
Type of digestive tract	Complete
Segmentation	Absent

Summary *With their spiny skin and five-part radial symmetry, adult echinoderms look very different from all other animals. Their embryos and larvae reveal a close relationship with the chordates.*

22.9 MASTERING CONCEPTS

1. Where do echinoderms live?
2. What is a water vascular system?
3. What are some examples of echinoderms?
4. How do echinoderms eat, respire, excrete metabolic wastes, sense their environment, move, reproduce, and defend themselves?
5. In what ways are echinoderms important?

22.10 INVESTIGATING LIFE

The "Cross-Dressers" of the Reef

The scene on TV is familiar: Two bighorn sheep rear up on their hind legs and bash their heads together with bone-crunching force. Only one will win the prize—the exclusive "right" to mate with a female. But fighting is not the only form of competition that determines a male's reproductive success. If a female mates with many males in a short period, the competition shifts to a much smaller field: the reproductive tract of the individual female. Only one sperm will fertilize each egg cell. Which will it be?

Sometimes, numbers decide the winner. Males that produce the most sperm are most likely to fertilize an egg cell, just as the owner of several raffle tickets has a better chance of winning a prize than someone who buys only one. But behavior also often plays a role. In some species, for example, a male may try to block other males from approaching a female he has already mated with. Males also do the equivalent of destroying raffle tickets that others have purchased. That is, they remove other males' sperm before depositing their own.

Sperm competition has selected for other interesting behavioral adaptations in a squidlike cephalopod called the Australian giant cuttlefish *(Sepia apama)*. These animals, which grow to nearly a meter long, usually live alone. During the winter mating season, however, they congregate by the hundreds of thousands on the reefs of southern Australia.

The two cuttlefish sexes look different from each other (**figure 22.33**). A female has shorter arms than a male, and her skin has dark patches on a white background. The male's arms are longer and whiter, and he displays moving patterns of zebralike stripes on his skin during courtship rituals. Specialized pigment sacs called chromatophores create the ever-changing color palette. Muscles surround each chromatophore, which holds yellow, reddish orange, or brownish black pigment. Rapid expansion and contraction of different-colored chromatophores creates the appearance of shifting patterns.

When a female accepts a male's mating attempt, the two animals align head-to-head, and he inserts a sperm packet into a pouch near her mouth. He also tries to flush out the sperm packets that other males have deposited. After mating, she retreats to her den and removes eggs, one by one, from her mantle cavity. She passes the eggs through her sperm pouch, then hangs them from the roof of her den.

With a sex ratio of at least four males to every female, the competition to father offspring is fierce. The largest males mate with females and guard them afterward, fighting off rivals to prevent subsequent insemination of their mates. A study led by Marie-José Naud of Australia's Flinders University and Roger Hanlon of the Marine Biological Laboratory at Woods Hole, Massachusetts, showed that males who guarded a female for up to 40 minutes after mating fertilized more eggs than those who guarded for less than 20 minutes (**figure 22.34**).

Nevertheless, smaller males do get to mate, and they sometimes use surprising strategies to gain access to

FIGURE 22.33 ***Sepia apama*, the Australian Giant Cuttlefish.** The larger male is guarding the female as she lays eggs.

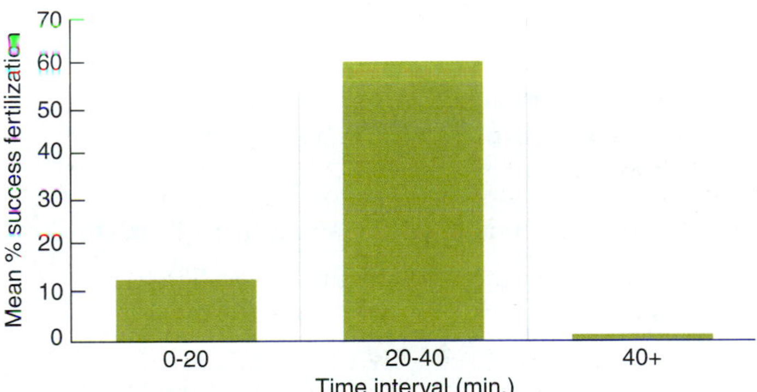

FIGURE 22.34 Mate Guarding Works. Researchers timed the mate-guarding activities of male cuttlefish and used DNA to determine the paternity of the resulting fertilized eggs. Males who guarded their mates from 20 to 40 min had the greatest reproductive success. **Question**: How would you test the hypothesis that the offspring of a mate-guarding male is more likely to mate-guard than the offspring of a "sneaker" male? What does this hypothesis assume about the inheritance of mate-guarding?

females. A team of Australian biologists, led by Mark Norman, noticed that some males are female impersonators. By hiding the arms that reveal his sex and manipulating his chromatophores to change his skin color, a small male can convincingly disguise himself as a female. (The ploy is so realistic that other males, including other female impersonators, often try to mate with the mimic.) Norman's team observed more than 20 examples of sexual mimicry, but they were unable to tell how often mimics slipped past a guard, approached a female, mated, and successfully fertilized eggs.

Hanlon, Naud, and their colleagues began to answer these questions. They videotaped sexual encounters on the reef, but they also used DNA analysis to determine which male fathered each female's next laid egg. They found that impersonators deceived a guarding male and approached a female about 30 times out of 62 attempts. Five of the 30 mimics tried to mate with the female. The guard male interrupted one attempt, and the female rejected another, but three were successful. DNA analysis showed that two of these three female impersonators fathered the first egg laid after mating. The third did not. Because females store sperm, however, the team could not rule out the possibility that sperm from the third mimic fertilized a subsequent egg.

Sexual mimicry illustrates an important point: in many species, there is more than one path to reproductive success. In polygamous species such as the cuttlefish, natural selection favors the male with the most successful sperm—and he may not be the largest or the strongest individual. True, the browniest males can fight off rivals and preserve their own access to a female, but the fierce competition and constant distractions often leave females with additional mating opportunities. A small male has little chance of winning a head-to-head competition with a larger rival, but he does have another tool—deception. Thanks to his changeable skin, even a small cuttlefish has an excellent shot at winning in the great raffle of life.

Hanlon, Roger T., Marie-José Naud, Paul W. Shaw, Jonathan N. Havenhand. 2005. Transient sexual mimicry leads to fertilization. *Nature*, vol. 433, page 212.

Naud, Marie-José, Roger T. Hanlon, Karina C. Hall, et al. 2004. Behavioural and genetic assessment of reproductive success in a spawning aggregation of the Australian giant cuttlefish, *Sepia apama*. *Animal Behaviour*, vol. 67, pages 1043–1050.

Norman, Mark D., Julian Finn, and Tom Tregenza. 1999. Female impersonation as an alternative reproductive strategy in giant cuttlefish. *Proceedings of the Royal Society of London B.*, vol. 266, pages 1347–1349.

CHAPTER SUMMARY

22.1 Animals Live Nearly Everywhere

- Of the 1,200,000 or so animal species described, most are **invertebrates;** only one phylum contains **vertebrates,** which have a segmented backbone.

A. The First Animals Likely Evolved from Protists

- The immediate ancestor of animals was likely a choanoflagellate.
- The earliest fossil evidence of animals is from about 570 MYA. These animals lived in water, and existing animal diversity strongly reflects this aquatic heritage.

B. Animals Share Several Characteristics

- Animals are multicellular, eukaryotic heterotrophs whose cells secrete extracellular matrix but do not have cell walls. Most digest their food internally.
- The **blastula** is a stage in embryonic development that is unique to animals.

C. Biologists Classify Animals Based on Organization, Morphology, and Development

- Animal bodies exhibit degrees of organization into tissues, organs, and organ systems. **Eumetazoans** are animals with true tissues.
- Body symmetry may be **radial** or **bilateral,** with a **cephalized** (head) end. Sponges are asymmetrical.
- An animal zygote divides mitotically to form a blastula and then usually a **gastrula.** In some animals, the gastrula has two tissue layers (**ectoderm** and **endoderm**). In others, a third layer (**mesoderm**) forms between the other two.
- The two major lineages of bilaterally symmetrical animals are protostomes and deuterostomes. In **protostomes,** the gastrula's first indentation forms into the mouth. In **deuterostomes,** the first indentation develops into the anus.

D. Biologists Also Consider Additional Characteristics

- Biologists also describe animals based on the presence or absence of a body cavity (**coelom** or **pseudocoelom**). The body cavity can act as a **hydrostatic skeleton.**

- Digestive tracts are **incomplete** or **complete.**
- **Segmentation** improves flexibility and increases the potential for specialized body parts.
- Most animals reproduce sexually. The resulting embryo may undergo **direct development** or **indirect development,** in which a **larva** undergoes **metamorphosis** to develop into an adult.

22.2 Sponges Are Simple Animals That Lack Differentiated Tissues

- **Sponges** are asymmetrical, aquatic animals. Their porous bodies filter small food particles out of water. Although they lack tissues, sponges have specialized cell types, including collar cells and amoebocytes.
- A sponge's skeleton consists of spicules or organic fibers (or both).
- Sponges reproduce sexually and may bud asexually.

22.3 Cnidarians Are Radially Symmetrical, Aquatic Animals

- **Cnidarians** are mostly marine animals with radial symmetry and incomplete digestive tracts. They capture prey with tentacles and stinging **cnidocytes** and digest their food in a gastrovascular cavity.
- A cnidarian body form is a **polyp** or a **medusa.**
- Examples of cnidarians include corals, *Hydra,* and jellyfishes.
- Cnidarians move by contracting muscle cells that act on a hydrostatic skeleton.
- In species that alternate between polyp and medusa forms, sexual reproduction produces a larva, which attaches and becomes a polyp. The polyp reproduces asexually to generate a colony that yields medusae.

22.4 Flatworms Have Bilateral Symmetry and Incomplete Digestive Tracts

- **Flatworms** are unsegmented animals that lack coeloms. Their flattened shape allows individual cells to exchange gases with their environment.
- Flatworms include **free-living** species such as a planarian; **flukes** and **tapeworms** are parasitic.
- They lack circulatory and respiratory systems, but protonephridia maintain water balance. They have simple nervous systems and hydrostatic skeletons.
- Flatworms reproduce asexually and sexually, and most are hermaphrodites.

22.5 Mollusks Exhibit Soft, Unsegmented Bodies and Complex Organ Systems

- **Mollusks** have bilateral symmetry and complete digestive tracts.
- The diverse mollusks (**chitons, bivalves, gastropods,** and **cephalopods**) have **mantles,** muscular **feet,** and **visceral masses.** Most have shells, and many have a tonguelike **radula.** They are filter feeders, herbivores, or predators.
- Cephalopods have complex sensory and nervous systems.
- Sexes are usually separate in the mollusks; fertilization may be internal or external.

22.6 Annelids Are Segmented Worms

- **Annelid** bodies consist of repeated segments. They include **oligochaetes, leeches,** and **polychaetes,** and they feed in diverse ways.
- The complex organ systems include a complete digestive tract, a closed circulatory system, and respiratory, excretory, and nervous systems. The coelom acts as a hydrostatic skeleton.
- Leeches and oligochaetes are hermaphrodites; polychaetes have separate sexes.

22.7 Nematodes Are Unsegmented, Cylindrical Worms

- **Roundworms** are unsegmented worms that molt periodically. They include parasitic and free-living species in soil and aquatic sediments.
- Nematodes have diverse diets and complete digestive tracts. The pseudocoelom is a hydrostatic skeleton. Most have separate sexes and undergo direct development.

22.8 Arthropods Have Exoskeletons and Jointed Appendages

A. Arthropods Have Efficient Organ Systems

- **Arthropods** are segmented animals with jointed appendages and a chitin-rich **exoskeleton.**
- Arthropods exhibit great diversity in feeding, respiratory systems, excretory systems, nervous systems, and reproduction. They have open circulatory systems.

B. Arthropods Are the Most Diverse Animals

- The extinct **trilobites** were arthropods, as are the **chelicerates** (**horseshoe crabs** and **arachnids**). **Crustaceans, insects, centipedes,** and **millipedes** are **mandibulate** arthropods.
- Insects are by far the most diverse arthropods.

22.9 Echinoderm Adults Have Five-Part, Radial Symmetry

- **Echinoderms** are spiny-skinned marine animals whose adults have radial symmetry. Molecular evidence and similarities in embryonic development classify echinoderms as deuterostomes with the chordates.

- Echinoderms move by using tube feet, which are part of the **water vascular system.** This network of canals also aids in circulation and gas exchange.
- Most echinoderms reproduce sexually. The bilaterally symmetrical larvae look very different from the adults.

22.10 Investigating Life: The "Cross-Dressers" of the Reef

- In many animal species, females store sperm from multiple males. Natural selection favors the males that produce the most sperm or that prevent other males from fertilizing the female's eggs.
- In cephalopods called Australian giant cuttlefish (*Sepia apama*), some males disguise themselves as females, increasing their opportunities to mate.

MULTIPLE CHOICE QUESTIONS

1. Which of the following characteristics is associated with echinoderms?
   a. Protostome
   b. Two germ layers (ectoderm and endoderm)
   c. Bilateral symmetry
   d. All of the above
2. If an organism lacks mesoderm then it cannot form which type of tissue?
   a. Skin
   b. Digestive tract
   c. Nervous tissue
   d. Muscle
3. When an earthworm is moving through the soil, its muscles are pushing against its
   a. digestive tract.
   b. coelom.
   c. bony skeleton.
   d. segmentation.
4. Sponges belong to which phylum?
   a. Cnidarian
   b. Platyhelminthes
   c. Porifera
   d. Nematoda
5. Which animal phylum has bilateral symmetry but an incomplete digestive tract?
   a. Porifera
   b. Platyhelminthes
   c. Nematoda
   d. Cnidaria
6. What is a radula?
   a. A stinging cell
   b. A tongue-like structure
   c. The sessile form of a cnidarian
   d. A stage in animal development
7. What is a key characteristic of all arthropods?
   a. Six legs
   b. Pseudocoelom
   c. Hydrostatic skeleton
   d. Exoskeleton
8. How is the body structure of an annelid different from an arthropod?
   a. Annelids lack jointed appendages.
   b. Annelids have a complete digestive tract.
   c. Annelids have cephalization.
   d. Annelids are bilateral.
9. Why is it necessary for an arthropod to molt?
   a. Because they undergo indirect development.
   b. Because the exoskeleton becomes damaged over time.
   c. Because the exoskeleton prevents the organism from growing.
   d. Because they have an open circulatory system.
10. What is a *water vascular system*?
    a. A network of canals that enables echinoderms to move
    b. The circulatory system of an echinoderm
    c. The excretory system of an echinoderm
    d. All of the above

TESTING YOUR KNOWLEDGE

1. Cite the criteria used to distinguish each of the following:
   a. animals from other organisms
   b. vertebrate from invertebrate
   c. protostomes from deuterostomes
   d. a tapeworm from a nematode and an earthworm
2. Distinguish between the following:
   a. radial and bilateral symmetry
   b. blastula from gastrula
   c. direct and indirect development
   d. complete and incomplete digestive tract
   e. coelom and pseudocoelom

3. Name an animal that has the following:
   a. cnidocytes
   b. flame cells
   c. mantle
   d. segmentation
   e. water vascular system
4. What are the three existing groups of arthropods, and what are the characteristics of each?
5. On an episode of the television series *The X-Files,* two FBI agents chase a bizarre creature that has the form of a human but with a scolex atop its head. From what animal phylum did the writer get inspiration for including a scolex?
6. Compare feeding in a sponge and a sea urchin.
7. A marine flatworm somewhat resembles a mollusk called a sea slug (nudibranch). What features would you look for to tell the two types of animal apart?

THINKING AS A SCIENTIST

1. Compare the major animal phyla in the order in which the chapter presents them, listing the new features for each group.
2. Give an example from the animal kingdom where molecular sequence information has either confirmed a classification based on physical resemblance or altered previous phylogenetic groupings.
3. The Chincharro people lived on the northern coast of Chile from 5500 to 500 BCE. Many of them died at very young ages due to a number of parasitic diseases. Their preserved feces (coprolites) and mummies contain eggs and other remains of tapeworms, roundworms, and flukes. How can a researcher tell these worms apart?
4. Use the internet to answer this question. Other than *Caenorhabditis elegans* and *Drosophila melanogaster,* what are some examples of invertebrate animals that have contributed to scientists' knowledge of general biology and animal biology? What other genomes of invertebrate animals have scientists sequenced, and what are some resulting discoveries?
5. Explain how a sessile or slow-moving lifestyle, such as that of sponges and sea cucumbers, might select for bright colors and an arsenal of toxic chemicals.
6. Give three examples of interactions between animals classified in different phyla.
7. To molt, an arthropod must crawl out of its old exoskeleton; an animal may die if it becomes trapped in its old exoskeleton. Molting behavior is under genetic control. Use natural selection to explain why defects in molting behavior are uncommon.
8. In 1902, paleontologist E. H. Sellards found a treasure trove of insect fossils in Dickinson County, Kansas. The animals lived during the Permian, about 300 to 250 MYA. If you examined one of the fossils, what features would you look for to verify that the animal was an insect and not another type of arthropod?
9. What is the evidence for the surprisingly close relationship between echinoderms and chordates?

Visit www.mhhe.com/hoefnagels for practice quizzes, animations, videos, and activities designed to help you master the material in this chapter.

Vertebrate Animals

Painted Frog. Many new species of amphibians and reptiles are being discovered in the forests and jungles of Vietnam. The Asian painted frog *Microhyla pulchra* blends in with its environment.

What's the Point?

Vietnam's Diverse Vertebrates

Vietnam is one of the few places on Earth where biologists can discover new species of large animals. Since the country's jungles opened for exploration in 1982, researchers have described dozens of species of insects, amphibians, reptiles, and mammals.

Sometimes an animal is new to science, but familiar to native people. This was the case for Sus bucculentus, *a long-snouted, warty-faced, short-legged pig that biologists thought had been extinct for more than a century. But in 1998, a researcher found one in the Annamite Range along the Laos–Vietnam border. The pig was a popular dinner dish among the natives! DNA sequence comparisons between someone's dinner and a museum specimen revealed that the old and new pigs were of the same species.*

In the Tam Dao reserve near Hanoi, biologists have cataloged more than 100 new species of snakes. Other reptiles and amphibians are abundant too, including blind snakes, legless lizards, snakes that eat lizards, green tree snakes, giant salamanders, and frogs that resemble moss. Their camouflage is astounding. Frogs blend in with bark, and a lizard's face has the uncanny mosaic green pattern of a leaf.

The reason for the incredible biodiversity at Tam Dao is that here, three very different types of ecosystems intersect: the eastern Himalayan alpine forest, the south China temperate forest, and the southeast Asian tropics. The many peaks, valleys, furrows, and ridges create microecosystems, small isolated areas where new species can easily arise.

Trees grow taller here than elsewhere in the world, and they are farther apart, not linked by vines as they are in other forests. In a spectacular example of convergent evolution, very different types of animals have strikingly similar adaptations to this forest topography. Lizards, snakes, frogs, and squirrels have extensive webbing between their toes and skin flaps on their abdomens that turn them into living parachutes, able to glide short distances from tree to tree.

The teams of biologists combing Vietnam's forests and jungles are cataloging species, taking population counts, and studying diets and reproductive strategies. The hope is that the attention of biologists will attract tourists interested in natural history, which will bring money for environmental conservation of these rare habitats and their residents.

LEARNING OUTLINE

23.1 Vertebrates (and a Few Invertebrates) Are Chordates

Many people find phylum Chordata to be the most interesting of all, at least in part because it contains humans and many of the animals that we eat, keep as pets, and enjoy observing in zoos and in the wild. The **chordates** are a diverse group of about 55,000 species (**table 23.1**). From the tiniest tadpole to fearsome sharks and lumbering elephants, chordates are dazzling in their variety of forms.

No one knows what the common ancestor of chordates was, but it was certainly an aquatic animal, and it apparently arose along with most other animal phyla during the Cambrian explosion. Like all other groups of animals, the chordates began as **invertebrates**—animals lacking a backbone. This chapter, however, focuses on the **vertebrates** (fishes, amphibians, reptiles, birds, and mammals). All of these familiar chordates have an internal skeleton (endoskeleton) that includes a segmented backbone. **Paleozoic life, p. 340**

A. Four Features Distinguish Chordates

Chapter 22 introduced the animals, described what all members of our kingdom have in common, and showed the evolutionary relationships among 9 of the 37 animal phyla. A glance back at figure 22.2 reveals that phylum Chordata includes animals with true tissues (eumetazoa), bilateral symmetry, a gastrula with three germ layers, and a true coelom. Like echinoderms, chordates are deuterostomes.

What sets Chordata apart from the other phyla? Every chordate has the following four features at some point during its life (**figure 23.1**):

1. **Notochord:** The notochord is a flexible rod that extends dorsally (along the back) down the length of a chordate's body.
2. **Dorsal, hollow nerve cord:** The dorsal, hollow nerve cord is parallel to the notochord. In many chordates, the nerve cord enlarges at the anterior end, forming a brain.

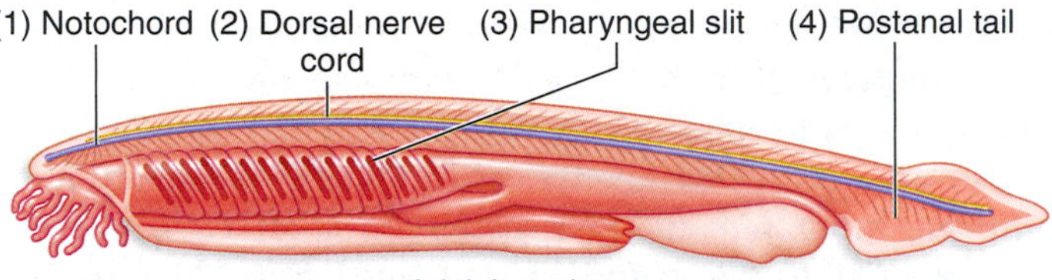

FIGURE 23.1 Defining Characteristics of a Chordate. At some time in its life, every chordate has each of these four features. (*1*) A notochord is a flexible rod that extends down the body dorsally. (*2*) The nerve cord in a chordate is hollow and dorsal to the digestive tract. (*3*) Pharyngeal slits are openings in the pharynx of a chordate. (*4*) A muscular tail extends past the anus.

Table 23.1 *Major Groups of Chordates*

Group	Examples	Features	Number of Existing Species
Tunicates	Sea squirts	Invertebrate filter feeders with protective tunic and two siphons	3,000
Lancelets	Amphioxus	Invertebrate filter feeders with all four chordate features in adults	25
Hagfishes	Slime hags	Wormlike bodies with cranium but not vertebrae	60
Fishes	Lampreys, sharks, salmon, lungfishes, coelacanth	Scale-covered bodies with fins and gills	29,000
Amphibians	Frogs, salamanders, caecilians	Naked (scaleless) tetrapods that breathe through their moist skin and spend time both on land and in water	6,000
Reptiles	Turtles, lizards, snakes, tuataras, crocodilians	Tetrapods with amniote eggs and dry body scales composed of keratin	8,000
Birds	Ostriches, penguins, hummingbirds, eagles	Two-legged tetrapods in which the forelimbs are modified as wings; produce amniote eggs; feathers composed of keratin cover the body	9,000 to 10,000
Mammals	Platypus, kangaroo, dog, whale, human	Milk-producing tetrapods with hair composed of keratin; amnion surrounds developing embryo	5,500

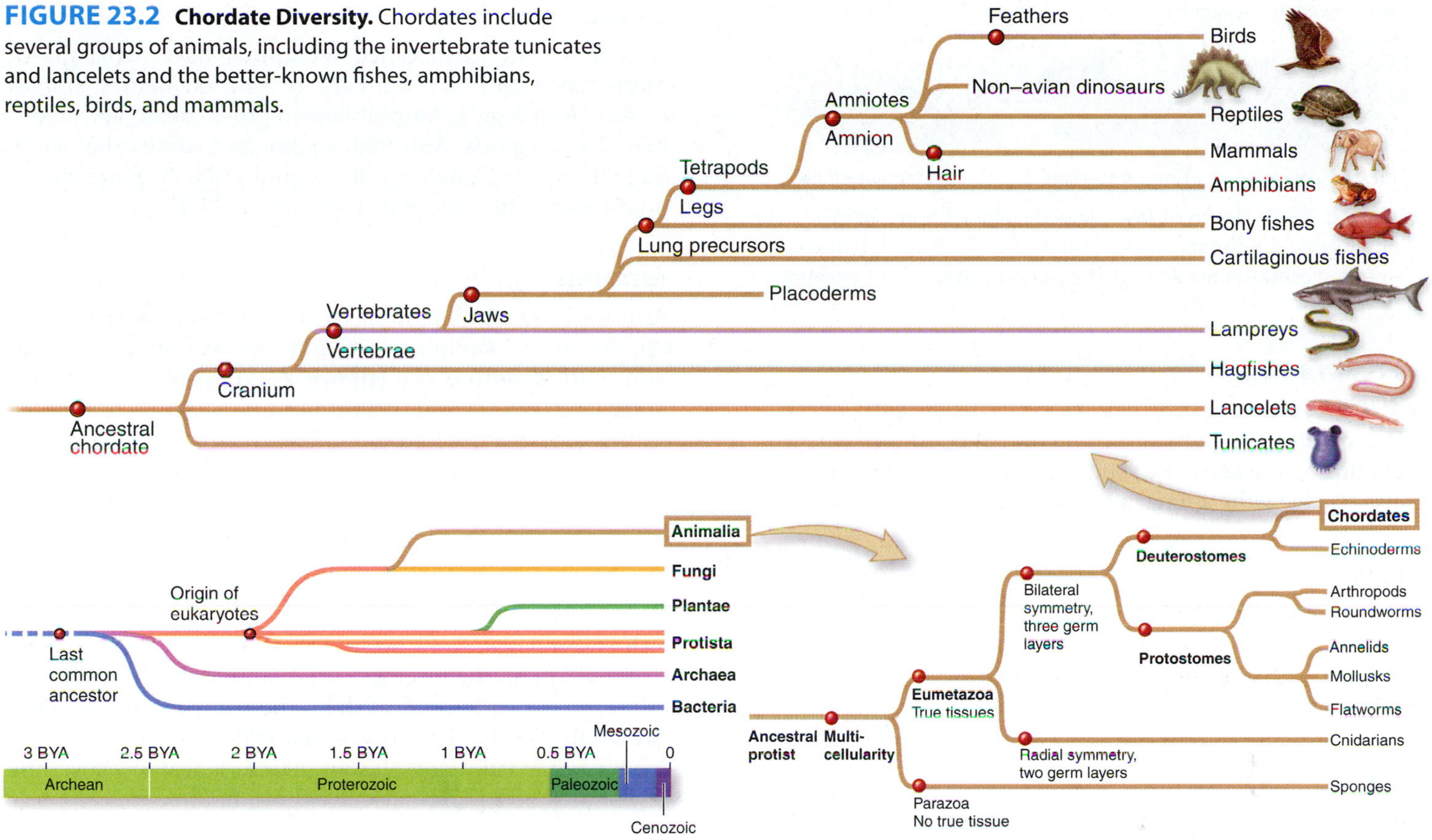

FIGURE 23.2 Chordate Diversity. Chordates include several groups of animals, including the invertebrate tunicates and lancelets and the better-known fishes, amphibians, reptiles, birds, and mammals.

Feathers
Birds
Non–avian dinosaurs
Amniotes
Amnion
Reptiles
Hair
Mammals
Tetrapods
Legs
Amphibians
Lung precursors
Bony fishes
Cartilaginous fishes
Jaws
Placoderms
Vertebrates
Vertebrae
Lampreys
Cranium
Hagfishes
Ancestral chordate
Lancelets
Tunicates

Animalia
Fungi
Origin of eukaryotes
Plantae
Protista
Last common ancestor
Archaea
Bacteria
3 BYA
2.5 BYA
2 BYA
1.5 BYA
1 BYA
0.5 BYA
0
Mesozoic
Archean
Proterozoic
Paleozoic
Cenozoic

Chordates
Deuterostomes
Echinoderms
Bilateral symmetry, three germ layers
Arthropods
Roundworms
Protostomes
Annelids
Mollusks
Eumetazoa
True tissues
Flatworms
Ancestral protist
Multi-cellularity
Cnidarians
Radial symmetry, two germ layers
Sponges
Parazoa
No true tissue

3. **Pharyngeal pouches or slits:** In most chordate embryos, pouches or slits form in the **pharynx,** the muscular tube that begins at the back of the mouth. Invertebrate chordates filter food particles from water that passes through the openings. In vertebrates, the pouches develop into gills, the middle ear cavity, or other structures.
4. **Postanal tail:** A muscular tail extends past the anus in all chordate embryos. In humans, the tail normally withers away before birth, leaving only the tailbone as a vestige. (Rarely, a human baby is born with a tail, which is removed in minor surgery.) In fishes, salamanders, lizards, cats, and many other species, adults retain the tail.

B. Biologists Use Many Features to Classify Chordates

Figure 23.2 depicts the evolutionary relationships within phylum Chordata. This section explains the features that biologists use to construct the tree, and sections 23.2 to 23.8 consider each chordate group separately.

Cranium

In most chordates, a bony or cartilage-rich **cranium** surrounds and protects the brain (**figure 23.3**). Hagfishes and vertebrates form two clades of **craniates,** animals that have a cranium.

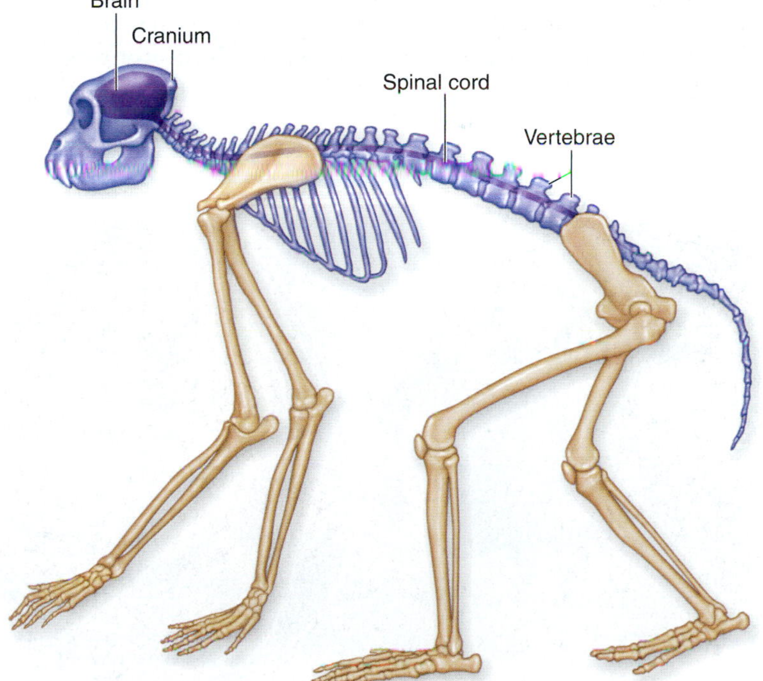

FIGURE 23.3 Skull and Backbone. A cranium and vertebrae are two of the characteristics that define the different groups of chordates.

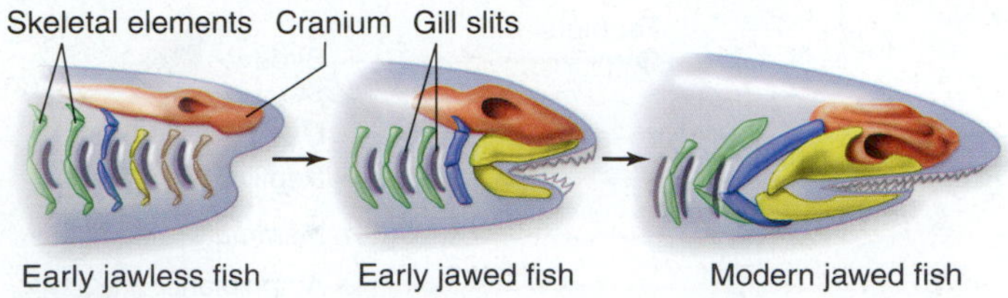

FIGURE 23.4 Origin of Jaws. Jaws developed from skeletal elements that supported gill slits near the mouth of the fish. The two elements closest to the mouth of the jawless fish were lost over time.

Vertebrae

Vertebrates are chordates that have a vertebral column, or backbone, composed of cartilage or bone (see figure 23.3). **Vertebrae** protect the nerve cord and provide attachment points for muscles, giving the animal a greater range of movement.

Jaws

Jaws are the bones that frame the entrance to the mouth. The development of hinged jaws from gill supports, shown in **figure 23.4**, greatly expanded the ways that animals could feed.

Lungs

Aquatic vertebrates, such as most fishes, have gills that exchange gases with the atmosphere (see figure 33.3). In contrast, air-breathing vertebrates have internal saclike **lungs.** Gills and lungs are not homologous to each other. Instead, lungs derived from the swim bladders that help fishes maintain their buoyancy in water. **homologous structures, p. 316**

Limbs

Terrestrial vertebrates called **tetrapods** have two pairs of limbs that enable the animals to walk on land (tetrapod means "four legs"). Amphibians, reptiles, birds, and mammals are tetrapods. Although snakes and some other vertebrates lack limbs, anatomical and molecular evidence clearly links them to the tetrapods (see section 23.9).

Amnion

Reptiles, birds, and mammals form a clade of **amniotes** that can live in arid environments, in part because of the evolution of the **amniote egg** (**figure 23.5**). Its leathery or hard outer layer surrounds a yolk that nourishes the developing embryo and enables it to survive away from water. The mammalian **amnion,** an internal membrane that protects the developing fetus, has a counterpart in the amniote egg (see figure 23.5). **human development, p. 758**

Body Covering

Fish scales derive from bone, and amphibians have naked, unscaly skin. In the three groups of amniotes, the body coverings all are composed of the same protein—keratin (**figure 23.6**). Reptiles have dry, tough scales all over their bodies. Birds have similar scales on their legs; feathers cover the rest of the body. Mammals have fur or hair.

Other Characteristics That Describe Chordates

The regulation of body temperature is an additional characteristic that is important in animal biology. The body temperature of an **ectotherm** tends to fluctuate with the environment; these animals have no internal mechanism keep-

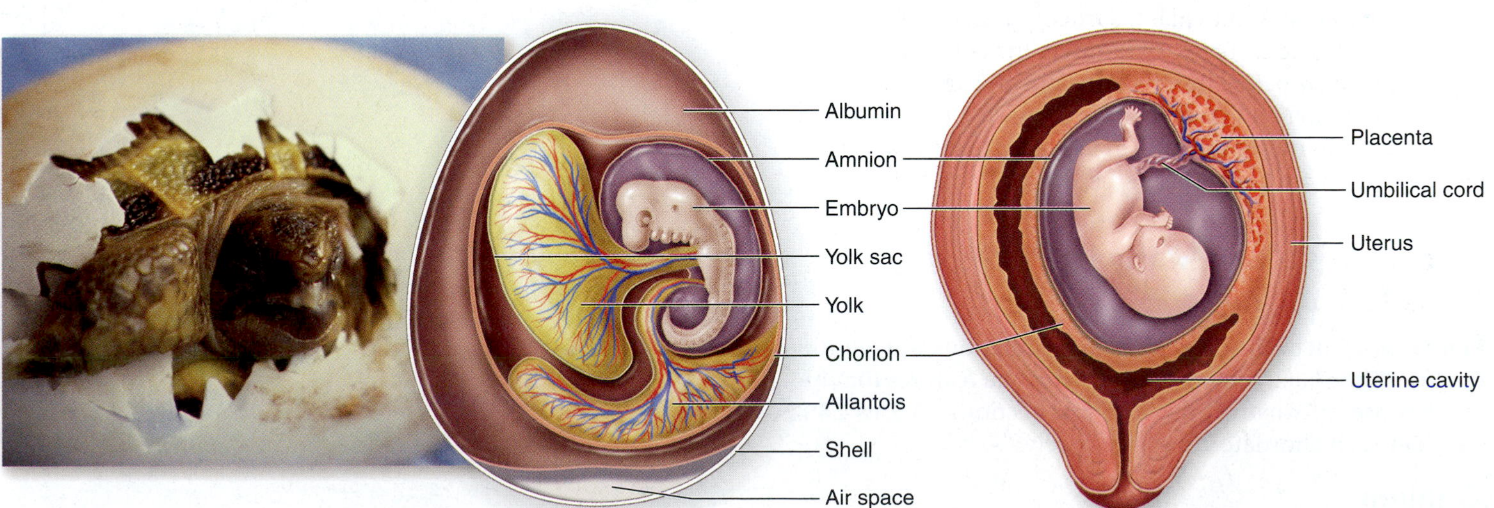

FIGURE 23.5 The Amnion. The amnion is a sac that encloses the developing embryo of a reptile, bird, or mammal. In an amniote egg, the embryo is encased in a hard, protective shell, and it is supported internally by three membranes—the amnion, allantois, and chorion. Placental mammals also enclose their embryos in an amnion.

FIGURE 23.6 Body Coverings. Vertebrate body coverings include (clockwise from upper left) the bony scales of a fish; the dry, keratin-rich scales of a reptile; the fur of a mammal; the feathers of a bird; and the naked, unscaly skin of an amphibian.

ing their temperature within a narrow range. Invertebrates, fishes, amphibians, and reptiles are ectotherms. Many behaviors, such as basking in the sun or burrowing into the ground during the hottest part of the day, help an ectotherm adjust its temperature. Birds and mammals are **endotherms,** which means they maintain their body temperature by using heat generated from their own metabolism. Endothermy requires an enormous amount of energy, which explains why birds and mammals must eat so much more food than ectotherms of the same size. Section 35.4 explores the origin of fur and feathers, which help endothermic animals conserve body heat.

The number of chambers in the heart is also important (see figure 32.3). Fishes have a two-chambered heart, with one atrium and one ventricle. In amphibians and most reptiles, the heart has two atria and one ventricle. Four chambers (two atria and two ventricles) make up the hearts of crocodiles, birds, and mammals. The more heart chambers, the better the separation of oxygen-rich blood from oxygen-poor blood, and the greater the efficiency with which blood delivers oxygen—a necessity in an energy-hungry endothermic animal.

Chapter 22 listed the characteristics of animals in eight phyla, including their habitat, body structure, organ systems, defense, and effects on humans. Much of the focus was on differences in the internal organ systems, which highlight the dramatic transitions between the simplest animals and the most complex. This chapter takes a slightly different approach. Most chordates have complex respiratory, digestive, excretory, and nervous systems. So we focus here on the evolutionary transitions between the main groups of chordates and on the diversity of animals within the phylum.

Chordates	
Level of organization	Organ system
Symmetry	Bilateral
Cephalization	Present
Coelom	Present
Type of digestive tract	Complete
Segmentation	Present (except in tunicates)

Summary *Many familiar animals belong to phylum Chordata. Except for tunicates, lancelets, and hagfishes, most chordates are vertebrates. The vertebrates are classified based on the presence or absence of a cranium, vertebrae, jaws, lungs, limbs, and amnion, and on the type of body covering. Body temperature regulation and the number of heart chambers are also important in describing vertebrates.*

23.1 MASTERING CONCEPTS

1. What are the four defining characteristics of chordates?
2. Which chordates are craniates, and which of those are also vertebrates?
3. How did the origin of jaws, lungs, limbs, and the amnion affect the course of vertebrate evolution?
4. How do the body coverings of fishes, amphibians, reptiles, birds, and mammals differ?
5. What is the difference between an ectotherm and an endotherm?
6. How does the number of heart chambers affect the efficiency of oxygen delivery to body tissues?

23.2 Tunicates and Lancelets Have Neither Cranium Nor Backbone

Tunicates and lancelets form two clades of invertebrates that probably would attract little attention if not for one fact: they are the modern organisms that most resemble the ancestral chordates. Biologists continue to debate which group is more closely related to vertebrates.

The **tunicates** take their name from the tunic, a protective, flexible body covering that the epidermis secretes. The best studied tunicates, the ascidians, are marine animals that resemble a bag with two siphons (**figure 23.7**). Cilia pull water in through one siphon. As the water moves across gill slits in the pharynx, oxygen diffuses into nearby blood vessels, and carbon dioxide diffuses out. Mucus covering the gill slits traps suspended food particles, and the water exits through the other siphon. These animals are also called sea squirts because they can forcibly eject water from their siphons if disturbed. Some tunicate species are a major nuisance in coastal areas, forming huge colonies that coat dock pilings and smother desirable shellfish species.

The free-swimming tunicate larva resembles a tadpole, and it has all four chordate characteristics. Once it settles headfirst onto a solid surface, however, the tail and notochord disappear, and the nerve cord shrinks to nearly nothing. Adults retain only the pharyngeal slits. Neither adult nor larva is segmented.

Lancelets (also called amphioxus) resemble small, eyeless fishes with translucent bodies (**figure 23.8**). They live in shallow seas, with their tails buried in sediment and their mouths extending into the water. Cilia on the gills, coupled with the mucus secreted by the pharynx, trap and move food particles into the digestive tract. The filtered water passes out through the gill slits, leaving the body through a separate pore. Blood distributes nutrients to body cells, but gas exchange occurs directly across the skin. To reproduce, males and females release gametes into the water; the larvae, like the adults, resemble tiny fish.

Lancelets clearly display all four major chordate characteristics, as well as inklings of the organ systems that appear in the vertebrates. Like vertebrates, lancelets have segmented blocks of muscles. Furthermore, the lancelet nervous system consists of a nerve cord with a slight swelling at the head end, plus sensory receptors on the body. The lancelet "brain" appears to share many of the same divisions as the more elaborate vertebrate brain. However, these animals lack the sophisticated sensory organs, brain, and mobility of vertebrates.

Tunicates and Lancelets	
Cranium	Absent
Vertebrae	Absent
Jaws	Absent
Skeletal composition	Does not apply
Lungs	Absent
Limbs	Absent
Amnion	Absent
Body temperature regulation	Ectotherm

Summary *Tunicates and lancelets are aquatic invertebrate chordates. Larval tunicates and adult lancelets clearly express the four defining chordate features.*

23.2 MASTERING CONCEPTS

1. How do tunicates use their siphons in feeding and gas exchange?
2. What is the relationship between lancelets and the vertebrate chordates?

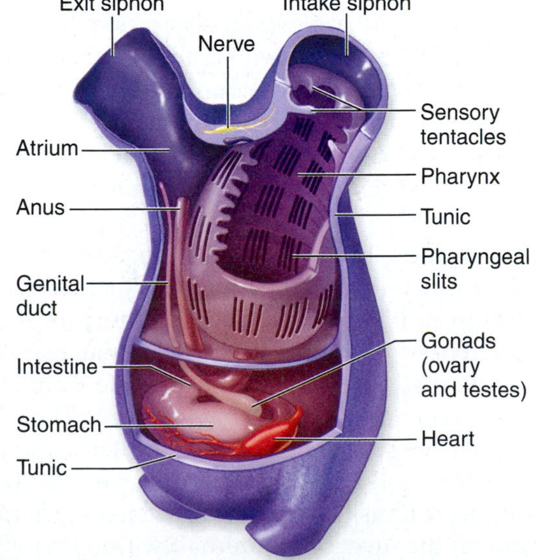

FIGURE 23.7 Tunicates. The sessile adult shows none of the chordate-defining characteristics that the free-swimming larva does. The outer covering (tunic) and two siphons are major features of the adult tunicate body.

a.

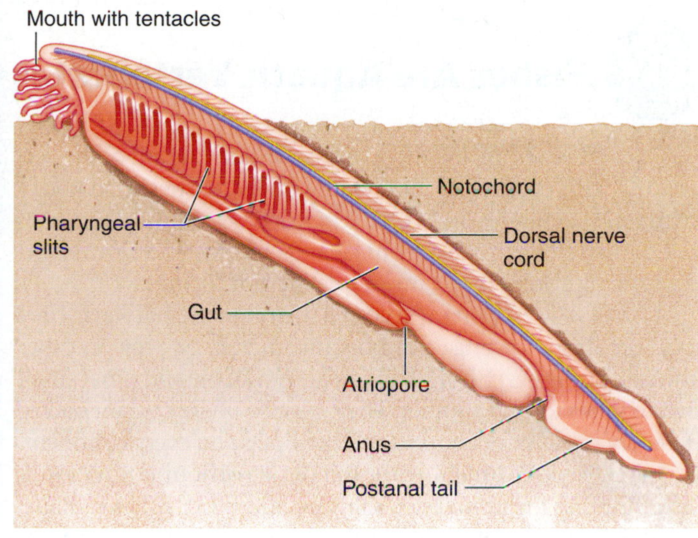

b.

FIGURE 23.8 **Lancelets.** (**A**) The faint stripes on these lancelets are formed by blocks of muscle. The largest lancelets are about 8 cm long. (**B**) This lancelet is in its suspension-feeding posture, with its head sticking up out of the substrate and its tail buried. The internal anatomy of the lancelet reveals the chordate characteristics.

23.3 Hagfishes Have a Cranium, but Lack a Backbone

The common ancestor of the craniates may have resembled a **hagfish.** The long, slender hagfish looks something like an eel (**figure 23.9**), and hagfish skin is even marketed as "eel skin" in boots and wallets. But hagfishes are not eels. In a hagfish, cartilage makes up the notochord and supports the tail, but because vertebrae do not surround the nerve cord, hagfishes are not vertebrates. An eel is a true fish—a vertebrate.

Hagfishes live in cold ocean waters, eating marine invertebrates such as shrimp and worms, or using their tongues to scavenge the soft tissues of dead or near-dead animals. Tentacles near the mouth help locate food. Scientists once classified hagfishes with other jawless fishes (see section 23.4), but molecular evidence, and the fact that hagfishes lack vertebrae, justify their place in a separate subphylum.

Hagfishes have some unusual abilities. They can slide their flexible bodies in and out of knots to pull on food, escape predation, or clean themselves. Hagfishes are also called "slime hags," in recognition of the slime glands that release a sticky white substance when the animal is disturbed.

Hagfishes	
Cranium	Present
Vertebrae	Absent
Jaws	Absent
Skeletal composition	Cartilage
Lungs	Absent
Limbs	Absent
Amnion	Absent
Body temperature regulation	Ectotherm

Summary *Hagfishes resemble eels and they have a cranium, but they lack vertebrae.*

23.3 MASTERING CONCEPTS

1. What features separate hagfishes from the jawless fishes?
2. How do hagfishes eat and defend themselves?

FIGURE 23.9 **Hagfishes.** These marine animals secrete a sticky slime and scavenge dead animals, including whales.

23.4 Fishes Are Aquatic Vertebrates with Gills and Fins

Fishes are the most diverse and abundant of the vertebrates, with more than 29,000 known living species that vary greatly in size, shape, and color. They occupy nearly all types of water, from fresh to salty, from clear to murky, and from frigid to warm, although they cannot tolerate hot springs.

Fishes play important roles in their aquatic habitats. They graze on algae, scavenge dead organic matter, or prey on other animals, eating everything from mosquito larvae and other small invertebrates to one another. They are also an important source of dietary protein for people (and their pets) on every continent. For many people, lunch is not the same without a tuna sandwich. Angling for trout, bass, salmon, and other fishes remains a popular sport. Fishes also inspire a wide range of emotions, from an intense fear of sharks to the peace and tranquility that come from watching tropical fish in a home aquarium.

A. Fishes Changed the Course of Vertebrate Evolution

Fishes originated some 500 MYA from an unknown ancestor. Several features arose in this group that would have profound effects on vertebrate evolution. A segmented backbone, with its multiple muscle attachment points, expanded the range of controlled, flexible motion. Jaws opened new feeding opportunities, which selected for a more complex brain that could develop a hunting strategy or plan an escape route.

Two adaptations that eventually enabled vertebrates to thrive on land originated in fishes: lungs and limbs. Lungs developed in a few species of fishes, and the air-breathing descendants of these animals eventually colonized the land. No fish has true limbs, but some fishes have fins with stronger bones and more flesh than the delicate, swimming fins of other fishes. These robust fins may have enabled tetrapod ancestors to move along the sediments of their shallow water homes. Whatever their original selective advantage, those fins eventually evolved into the limbs that define tetrapods.

B. Fishes Are Divided into Two Main Groups

Biologists recognize two clades of jawless fishes and three clades of fishes with jaws.

Jawless Fishes

Ostracoderms (Extinct). The earliest vertebrates to leave fossil evidence were jawless fishes called **ostracoderms.** These animals, now extinct, were filter feeders that lived on ocean bottoms. They had notochords and brains, and their bodies were encased in bony, armorlike plates. They lived during the Cambrian period, but they vanished when jawed fishes, which were better able to feed, diversified during the Silurian period.

Lampreys. The modern jawless fishes include **lampreys** (**figure 23.10**). Lampreys are the simplest organisms to have cartilage around the nerve cord, and they are most like the first true vertebrates. They spend most of their lives as larvae, straining food from the water column. The adults feed on small invertebrates, although some species use their suckers to consume the blood of fish. Over the past century, some lampreys have ventured beyond their natural Lake Ontario range into the other Great Lakes, where they have been largely responsible for the decline in populations of lake trout and whitefish. **invasive species, p. 851**

Jawed Fishes

Placoderms (Extinct). Around the time that ostracoderms lived, another group of fishes arose: the **placoderms.** Like the ostracoderms, the placoderms had a notochord and armor. They also had another feature: fearsome jaws.

Cartilaginous Fishes. The **cartilaginous fishes** include sharks, skates, rays, and chimaeras (**figure 23.11**). This group arose some 450 MYA and has persisted even as other groups became extinct. As the name implies, their skeletons are made of a soft material called cartilage. Although some sharks feed on plankton, the carnivorous species are notorious for their ability to detect blood in the water. They also have a **lateral line system,** a network of canals that extend along both sides of the fish and house vibration-detecting receptor organs.

Bony Fishes. This group includes 96% of existing fish species. They have skeletons of bony tissue reinforced with mineral deposits of calcium phosphate. Like sharks, **bony fishes** have a lateral line system. In addition, by altering the volume of gas

FIGURE 23.10 Jawless Fish. A lamprey has a distinctive sucker mouth.

a.

b.

FIGURE 23.11 Cartilaginous Fishes. This group includes (**A**) skates and rays and (**B**) sharks.

confined in the swim bladder, a bony fish can adjust its buoyancy. Bony fishes are divided into three groups (**figure 23.12**):

- The **ray-finned fishes** include nearly all familiar fishes: eels, minnows, catfish, trout, tuna, salmon, and many others. Their diversity and abundance are testament to their superb adaptations to a watery world.
- During droughts, the **lungfishes** burrow into the mud beneath stagnant water, gulping air and temporarily slowing their metabolism. The lungfishes are the bony fishes most closely related to the tetrapods.
- The **lobe-finned fishes** have only one surviving member, the coelacanth. This fish moves its fins as if they are limbs, and it has ears that are similar to those of tetrapods. Tissue samples from living specimens are providing clues that will help clarify the coelacanth's evolutionary status.

Fishes	
Cranium	Present
Vertebrae	Present
Jaws	Usually present
Skeletal composition	Cartilage or bone
Lungs	Usually absent
Limbs	Absent
Amnion	Absent
Body temperature regulation	Ectotherm
Number of heart chambers	Two

a.

b.

c.

FIGURE 23.12 Bony Fishes. Most fishes belong to this group, which includes (**A**) ray-finned fishes, (**B**) lungfishes, and (**C**) lobe-finned fishes.

Summary *Fishes account for half of the species of vertebrates. They are exquisitely adapted to an aquatic lifestyle. The fishes are classified in two main groups: the jawless and the jawed. Their skeletons are composed of cartilage or bone.*

23.4 MASTERING CONCEPTS

1. Which structures in fishes were the precursors to limbs?
2. What are two groups of jawless fishes? Which is extinct and which still exists?
3. What are three groups of jawed fishes?
4. How were placoderms different from ostracoderms?
5. What are the major types of cartilaginous and bony fishes?

23.5 Amphibians Lead a Double Life on Land and in Water

The word ***amphibian*** is Greek for "double-life," referring to the ability of these tetrapod vertebrates to live in fresh water and on land. Amphibians are probably the least familiar vertebrates because most people do not eat them or keep them as pets. Nevertheless, they are important in ecosystems, controlling both algae and populations of insects that spread human disease. Their chorus of mating calls, ranging from squeaks to grunts and croaks, adds incomparable ambience to springtime evenings in many areas. Scientists are also studying toxins in amphibian skin as possible painkilling drugs.

Many biologists are concerned about dramatic declines in amphibian populations worldwide. The destruction of wetlands devastates amphibians' breeding areas, and infection with a chytrid fungus has killed many of them outright (see section 42.5). Collecting animals for the pet trade also takes a toll (see this chapter's Can *You* Relate? Wild-Caught Pets). In addition, amphibians have porous skin that makes them especially vulnerable to pollution. They are therefore valuable as indicators of environmental quality. **chytrids, p. 433**

A. Amphibians Were the First Four-Legged Vertebrates

It is easy to envision how gulping lungfishes might have foreshadowed the lungs of amphibians, or how the fleshy fins of lobe-finned fishes might have become legs. New fossil finds continue to fill in the pieces of the fish–amphibian transition, which occurred about 375 MYA (see figure 16.15 and section 23.9).

Life on land provided amphibian ancestors with space, protection, food, and plentiful oxygen, but it also presented new challenges. The animals faced wider swings in temperature, and delicate gills collapsed without the buoyancy of water. The new habitat selected for new adaptations. Lungs improved, and circulatory systems (including a three-chambered heart) grew more complex and powerful. The skeleton became denser and better able to withstand the force of gravity. Natural selection also favored acute hearing and sight, with tear glands and eyelids keeping eyes moist.

Yet amphibians retain a strong link to the water. Amphibian eggs, which lack protective shells and membranes, will die if they dry out. Also, the larvae (tadpoles) respire through external gills, which require water. Although most adults have lungs, many supplement oxygen intake by gas exchange through their thin skin, which must therefore remain moist.

B. Amphibians Include Three Main Lineages

The amphibians are grouped into three main lineages: frogs and toads; salamanders and newts; and caecilians (**figure 23.13**).

Frogs and Toads

Most amphibian species are **frogs** or toads. Adults have large mouths, and they are "neckless"; that is, their heads are fused to their trunks. Their bodies lack both tails and scales. A female frog lays her eggs directly in the water as a male clasps her back and releases sperm. The fertilized eggs hatch into legless tadpoles that feed on algae. As they mature, the tadpoles undergo a dramatic change in body form—a metamorphosis. They develop legs and lungs, lose the tail, and acquire carnivorous tastes.

Frogs and toads have remarkable adaptations that help them avoid predation. Whereas some have convincing camouflage, others display vibrant colors that warn of toxins secreted from glands in the skin. (Some of these toxins kill bacteria, making them a potential source of antibiotic drugs.) Frogs and toads also escape predation by jumping away, playing dead, and inflating their mouths so that a snake cannot swallow them.

Salamanders and Newts

Salamanders and newts have tails and four legs, so they resemble lizards. Unlike lizards, however, their skin lacks scales, their digits lack claws, and they always live near water. Many salamanders lay their eggs in water, which hatch out free-swimming larvae with a finlike tail. Both adults and young are carnivores, eating arthropods, worms, snails,

a.

c.

b.

FIGURE 23.13 Amphibian Diversity. (**A**) This tree frog is native to tropical forests. (**B**) Salamanders are small amphibians with tails and four delicate legs. (**C**) Caecilians are legless amphibians that resemble earthworms or snakes.

fish, and other salamanders. In some groups of salamanders, the adults have larval features. The North American mudpuppy, for example, swims on pond bottoms and retains external gills.

Can *You* Relate?

Wild-Caught Pets

Wild animals may seem to make fascinating and unique pets, but owning one is not usually a good idea. Many arguments justify laws against owning animals caught in the wild:

- **Population pressures:** The harvest of exotic coral reef fishes depletes their populations and damages the entire reef ecosystem. Trading in wild-caught parrots and macaws likewise threatens their native populations. Huge numbers of North American box turtles are captured for the pet trade each year. These animals have low reproductive rates, so it is difficult for the wild populations to recover. In addition, long life spans make owning a box turtle a major commitment. Sadly, many wild-caught animals die before reaching pet owners.
- **Threats to native species:** Virtually every animal in the pet industry is an introduced species, including birds, ferrets, gerbils, sugar gliders, reptiles, amphibians, fishes, and invertebrates. If an exotic pet escapes, or if its owner releases it, the animal may prey on, compete with, or spread disease to native organisms. Goldfish, for example, have become a nuisance species in lakes of the Pacific Northwest.
- **Physical dangers:** Large animals such as crocodiles and large cats represent a physical threat to their owners, becoming more dangerous as they grow.
- **Disease:** Wild-caught animals can carry disease. In 2003, monkeypox spread among people who kept prairie dogs as pets. The animals caught the disease from African rodents imported into the United States. Wild birds can carry diseases that can infect native birds, poultry, and people. Reptiles and amphibians can carry *Salmonella*.

Caecilians

The limbless **caecilians** resemble giant earthworms. Most species burrow under the soil in tropical forests, but a few inhabit shallow freshwater ponds. They are the only amphibians that use internal fertilization, in which a male uses a sex organ to deliver sperm inside a female's body. Caecilians are carnivores, eating insects and worms.

Amphibians	
Cranium	Present
Vertebrae	Present
Jaws	Present
Skeletal composition	Bone
Lungs	Present
Limbs	Present
Amnion	Absent
Body temperature regulation	Ectotherm
Number of heart chambers	Three

Summary *Amphibians are adapted to life in water and on land. In the amphibians, lungs became the most important organs of gas exchange, circulatory systems grew more efficient, the skeleton became denser, and senses sharpened. Nevertheless, most amphibians live in moist habitats and must return to water to reproduce.*

23.5 MASTERING CONCEPTS

1. What role does water play in amphibian gas exchange and reproduction?
2. What features distinguish the three major groups of amphibians?
3. What are some reasons for the recent decline in amphibian populations worldwide?

23.6 Reptiles Were the First Vertebrates to Thrive on Dry Land

Along with amphibians, **reptiles** may be among the least familiar vertebrates. Movies depict snakes, alligators, and crocodiles as terrifying, deadly killers, but most reptiles are inconspicuous animals that cannot harm people. Yet they are an important link in ecosystems, controlling the populations of rodents and insects while providing food for owls and other predatory birds. In addition, some people keep snakes, lizards, or turtles as pets. In some parts of the world, reptiles have an even greater economic effect; the skins of farm-raised snakes and crocodiles are the raw material for boots, belts, and wallets, and some restaurants serve alligator meat.

A. Reptiles Descended from Amphibians

Reptiles evolved from amphibians during the Carboniferous period, between 363 and 290 MYA. They dominated animal life during the Mesozoic era, until their decline beginning 65 MYA. Although many reptile species survived to the present day, many others are known only from fossils. The extinct groups include the terrestrial dinosaurs, the marine ichthyosaurs and plesiosaurs, and the flying pterosaurs. Of these, the dinosaurs especially capture the imagination (see the Burning Question: What were dinosaurs?). As the next section describes, however, biologists consider birds one dinosaur lineage that did not become extinct 65 MYA. **Mesozoic life, p. 342**

Like fishes and amphibians, reptiles are ectothermic. Unlike their aquatic ancestors, however, most live on dry land. Several adaptations enable reptiles to thrive in arid habitats. Their excretory systems, including paired kidneys, conserve water, as do their tough scales. In addition, internal fertilization meant that reptiles no longer deposited sperm in water. Their amniote eggs are adapted to dry conditions (see figure 23.5). Finally, well-developed lungs and enhanced circulation increased their respiratory capacity beyond that of their aquatic ancestors.

Burning Questions

What were dinosaurs?

Dinosaurs were reptiles that lived during the Mesozoic era, between 245 and 65 MYA. They were extremely diverse, ranging from chicken-sized to truly gargantuan—the largest could have peeked into the sixth story of a modern-day building. Biologists divide the hundreds of dinosaur species into two main lineages. One clade includes the beaked plant-eaters like *Stegosaurus* and *Triceratops*. The other contains long-necked herbivores like *Apatosaurus* and the theropods, the only carnivorous dinosaurs. (*Tyrannosaurus rex* belonged to this group.) Theropods, which walked upright on two legs, are most closely related to modern birds.

Many people mistakenly believe that all reptiles (or even mammals) that lived during the Mesozoic era were dinosaurs. In reality, lizards, snakes, crocodiles, and other terrestrial reptiles lived alongside the dinosaurs. The Mesozoic also saw marine reptiles like plesiosaurs and flying reptiles like pterodactyls, but neither of these were dinosaurs, which were all terrestrial. Nor did all dinosaurs live at the same time. As some species appeared, others went extinct throughout the Mesozoic.

Despite movie plots suggesting the contrary, humans and dinosaurs have never coexisted. The last of the dinosaurs were gone by the time our own mammalian lineage—the primates—was just getting started. Tens of millions of years later, humans finally roamed the Earth and found the fossils that prove that these huge reptiles once existed.

Have a Burning Question of your own? Submit it to marielle_hoefnagels@mcgraw-hill.com for possible inclusion in future editions of this book!

B. Non-Avian Reptiles Include Three Main Groups

Other than birds, which are described in section 23.7, the major existing lineages of reptiles include turtles and tortoises; lizards, snakes, and tuataras; and crocodilians (**figure 23.14**).

Turtles and Tortoises

With a reputation for being "slow and steady," **turtles** and tortoises may not seem to have a recipe for success. Yet they have persisted in marine, freshwater, and terrestrial habitats since the Triassic period. The turtle's trademark feature is its shell, made of bony plates and a covering derived from the animal's epidermis. The shell's plates are fused to the animal's vertebrae and ribs, so it forms an integral part of the skeleton.

The largest member of this group is the giant leatherback sea turtle. Earth's magnetic field guides their migration over vast distances of open ocean between their breeding and feeding areas. These animals are endangered worldwide because they are hunted for food, become trapped in fishing nets,

FIGURE 23.14 Reptile Diversity. Reptiles include (**A**) turtles and tortoises, (**B**) lizards, (**C**) snakes, and (**D**) crocodiles and alligators. The extinct dinosaurs were also reptiles.

swallow garbage, and ingest toxic chemicals. The destruction of nesting areas also takes its toll. Habitat destruction likewise threatens many land turtles, as does the pet trade.

Lizards, Snakes, and Tuataras

Almost 95% of reptile species are **snakes** or **lizards.** These animals are similar to each other, except that lizards usually have legs, and snakes do not (see section 23.9). Also, snakes never have external ear openings or moveable eyelids, and only a few lizards have a forked, snakelike tongue. Several adaptations help lizards and snakes feed and avoid becoming food. A lizard might detach its tail as it escapes a predator. Camouflage and the ability to hold perfectly still enable snakes to surprise their prey, then subdue it by injecting venom or wrapping coils around the victim and squeezing the life out of it. Snakes then demonstrate a unique skill; they can unhinge their jaws, allowing them to swallow animals much larger than they are.

The **tuatara** is closely related to the lizards. The group containing tuataras once contained many other species, all of which are now extinct. Only the tuatara, a lizardlike animal that lives on a few islands near New Zealand, remains.

Crocodilians

The **crocodilians** (crocodiles, alligators, caimans, and gavials) all live in or near water. Their horizontally held heads have eyes on top and nostrils at the end of the elongated snout. Heavy scales cover their bodies; four legs project from the sides. All crocodilians are carnivores, swallowing their prey whole or in huge chunks. Unlike most other reptiles, they have a four-chambered heart.

These reptiles look somewhat primitive, and indeed they are ancient animals, dating back to 230 MYA. Yet they have acute senses and complex behaviors, comparable to those of birds and mammals. For example, like birds, crocodilians lay eggs in nests, which the adults guard. The adults also care for the hatchlings, which stay with their mothers while they are young and call to the adults when they are in danger. Adults roar, growl, and hiss, and they mark their territories by smacking their heads or jaws on the water surface. Like many birds and mammals, they even have dominance hierarchies.

The similarity between crocodilians and birds is unsurprising in light of the evolutionary history of reptiles. Birds, dinosaurs, and crocodilians all belonged to a group called archosaurs, of which only the birds and crocodilians survive today.

Reptiles	
Cranium	Present
Vertebrae	Present
Jaws	Present
Skeletal composition	Bone
Lungs	Present
Limbs	Present
Amnion	Present
Body temperature regulation	Ectotherm
Number of heart chambers	Three (four in crocodilians)

Summary *Reptiles are fully adapted to dry land, with scaly body coverings and amniote eggs. Although dinosaurs and many other reptiles became extinct about 65 MYA, many species remain.*

23.6 MASTERING CONCEPTS

1. How do dry scales and the amniote egg adapt reptiles to dry land?
2. What features characterize each of the three main groups of reptiles?

23.7 Birds Are Warm, Feathered Reptiles

Birds are a part of everyday human life. U.S. consumers eat hundreds of millions of chickens and turkeys every year, along with countless chicken eggs. We keep parrots, finches, canaries, and many other caged birds as pets, and we use bird feathers in everything from hats to down blankets. Songbirds enrich the lives of many birdwatchers, and hunters pursue wild turkeys, doves, and ducks. Birds are important in ecosystems as well. Some pollinate plants and disperse fruits and seeds, whereas others eat rodents, insects, and other vermin. But birds can also be pests. Starlings and pigeons are a nuisance in cities, fouling buildings and sidewalks with their droppings and speeding the rusting of bridges. Moreover, ducks and other domesticated birds transmit bird flu and other diseases to humans.

A. Birds Have Many Adaptations to Flight

Birds originated before the end of the Triassic, or about 146 MYA. They belong to the same clade as reptiles. As described in chapter 16, *Archaeopteryx* and fossils of feathered reptiles are important clues to the evolutionary history of birds, as are skeletal similarities between birds and reptiles. Also, birds retain scales and egg-laying from their reptilian ancestors. The bird's amniote egg, with its hard calcium-rich shell, protects the developing embryo inside. The adults care for, feed, and protect their hatchlings.

Birds are the only modern animals that have **feathers,** epidermal structures that provide insulation and enable birds to fly. They are also important in mating behavior, as anyone who has watched a peacock show off his plumage can attest. Like a reptile's scales, a feather is built of the protein keratin. Different-sized and different-shaped feathers serve distinct functions, such as flight or insulation.

Most birds have anatomical adaptations to flight (**figure 23.15**). Tapered bodies offer a streamlined profile, and hollow bones reduce weight. The powerful four-chambered heart and unique lungs supply the oxygen that supports the high metabolic demands of flight (see figure 33.5). Feather-covered wings provide lift in the air, while highly developed muscles power flight.

Like mammals, birds are endothermic, a major departure from all preceding vertebrates. Paleontologists debate whether dinosaurs, the ancestors of birds, were endotherms or ectotherms. The posture, body size, habitats, growth rate, bone structure, and other features of dinosaurs seem to argue for the endothermic point of view, but the question remains open.

FIGURE 23.16 Two Groups of Birds. (**A**) A kingfisher is a carinate, a bird with a keeled breastbone. (**B**) This running ostrich is an example of a ratite.

B. Birds Are Classified Based on Breastbone Shape

Biologists divide birds into two main groups (**figure 23.16**). Most living birds, from the tiny hovering nectar-feeding hummingbirds to soaring eagles, are **carinates**. The breastbone of these birds resembles the keel, or axis, of a sailboat and serves as the structural support to which flight muscles attach. Most carinates can fly, although this group also includes the flightless puffins, penguins, and rails. **Ratites** are flightless birds such as emus, ostriches, rheas, and kiwis. They have reduced wings and breastbones that are not keeled.

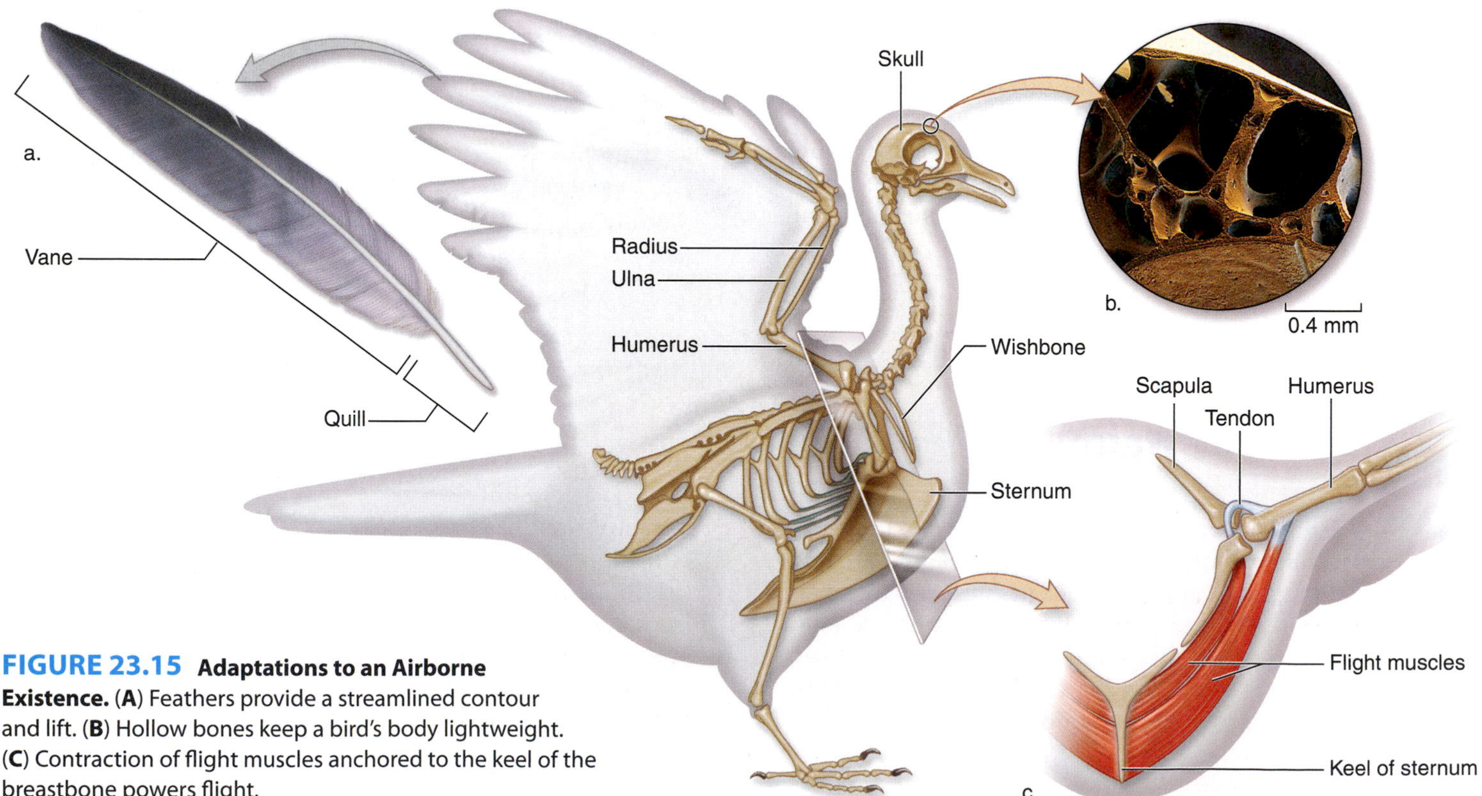

FIGURE 23.15 Adaptations to an Airborne Existence. (**A**) Feathers provide a streamlined contour and lift. (**B**) Hollow bones keep a bird's body lightweight. (**C**) Contraction of flight muscles anchored to the keel of the breastbone powers flight.

Birds	
Cranium	Present
Vertebrae	Present
Jaws	Present
Skeletal composition	Bone
Lungs	Present
Limbs	Present
Amnion	Present
Body temperature regulation	Endotherm
Number of heart chambers	Four

Summary *Birds belong to the same clade as reptiles, but they have a collection of adaptations to a life spent partially in flight.*

23.7 MASTERING CONCEPTS

1. What characteristics do birds retain from their reptilian ancestors?
2. What are the functions of feathers?
3. What adaptations enable birds to fly?
4. What are the two main groups of birds?

23.8 Mammals Are Warm, Furry Milk-Drinkers

Mammals are by far the most familiar vertebrates, not only because we *are* them, but also because we surround ourselves with them. We keep dogs, cats, rabbits, gerbils, hamsters, and ferrets as pets. Farms raise many types of mammals for their meat or milk, including cows, pigs, and goats. Lambs and adult sheep provide meat and wool, and leather from cows makes up everything from upholstery to shoes. Horses, oxen, mules, and dogs are important work animals. Many people enjoy hunting deer for food and sport, and trappers kill mink, fox, beaver, and other mammals for their fur.

Mammals are also important in ecosystems. Coyotes, wolves, and foxes keep populations of herbivorous deer, rodents, rabbits, and other mammals in check. Some bats eat countless insects, and others pollinate plants. At the same time, mammals can be pests. Rats, mice, and skunks thrive alongside human populations; some transmit diseases, including hantavirus and rabies.

A. Mammals Descended from Reptiles

Mammals arose before the end of the Triassic period (about 200 MYA). DNA sequence similarities and the existence of egg-laying mammals suggest that mammals arose from reptiles. Fossil evidence also provides clues to their origin. Mammals and their immediate ancestors are **synapsids;** that is, they have a single hole on each side of the skull, behind the eye orbits. (Birds and other reptiles have two such openings on each side.) One group of synapsids, called therapsids, had characteristics of both mammals and reptiles; mammals are the only remaining members of the therapsid clade. Most mammals were small until after the mass extinction that occurred 65 MYA. The loss of so many reptile species paved the way for the diversification of many larger species of mammals.

The word *mammal* derives from the Latin *mammae* for "breast," and refers to the milk-secreting **mammary glands** of the female. Infant mammals are nourished by their mother's milk. Mammals are also distinguished from other vertebrates by hair. Hair is composed of keratin and helps conserve body heat. Even whales and dolphins have hair at birth, but they lose it as they mature into their streamlined shapes.

Because fur and breasts do not leave fossils, mammalian remains are distinguished from reptilian fossils in other ways. A mammal has three middle ear bones compared with the reptilian one or two, and the lower jaw consists of one bone, compared with the reptile's several. Mammalian teeth are distinctive too, with four types: molars, premolars, canines, and incisors. Reptiles have only one type of teeth.

A few other features also characterize living mammals. Like birds, mammals have a four-chambered heart, which evolved independently in the two groups. The outer layer of the brain is very well developed, enabling mammals to learn, remember, and even think. As a result, many mammals have unparalleled ability to plan and purposefully respond to stimuli. In addition, only mammals have a dome-shaped muscular diaphragm that separates the chest cavity from the abdominal cavity. Contracting the muscles of the diaphragm draws air into the lungs. Reptiles and birds use other mechanisms to ventilate their lungs.

B. Mammals Include Three Main Groups

Biologists divide mammals into three groups (**figure 23.17**).

Monotremes

The **monotremes** include the duck-billed platypus and spiny anteaters of Australia. The amniotic eggs in this group form a direct link to the reptiles. When the young hatch from the eggs, they are helpless, tiny, hairless, and blind. A newborn monotreme crawls along its mother's fur until it reaches sweat glands modified as tiny nipples, which secrete milk.

FIGURE 23.17 Mammal Diversity. (**A**) Platypuses retain some characteristics thought to have been present in the earliest mammals. (**B**) A kangaroo is a marsupial. (**C**) This mammal, a human baby, is deriving nourishment from his mother's milk. (**D**) Dolphins are aquatic mammals, and (**E**) bats are flying ones.

a.

b.

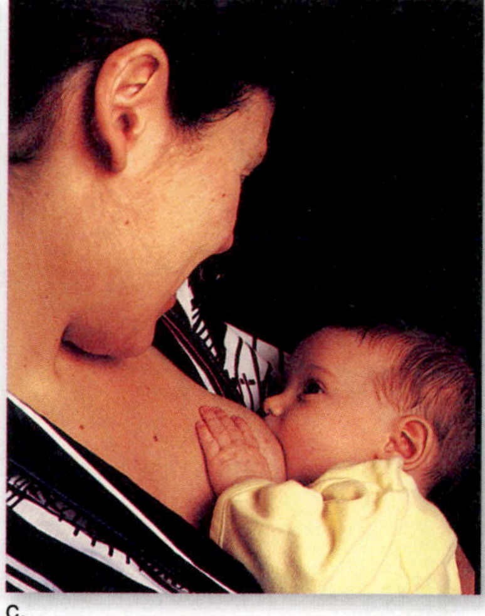
c.

d.

e.

Marsupials

Marsupials, such as kangaroos and opossums, give birth to tiny, immature young about 4 to 5 weeks after conception. The babies crawl from the mother's vagina to a marsupium, or pouch, where they nurse and continue developing.

Placental Mammals

The **placental mammals,** also called eutherians, are the most diverse of the three groups. This group includes bats, rodents, cetaceans (whales and dolphins), carnivores (dogs and cats), hoofed mammals, elephants, and many others. (The mouse, *Mus musculus,* is a placental mammal and the topic of this chapter's Focus on Model Organisms.) Female placental mammals carry their young inside the uterus, where a **placenta** connects the maternal and fetal circulatory systems. The placenta nourishes and removes wastes from the developing offspring. Placental mammals diversified and displaced most marsupials early in the Cenozoic era, which began 65 MYA.

Humans are placental mammals. We belong to a class, Primates, that arose some 60 MYA. Section 16.4 describes how this group gave rise to the characteristics and abilities that distinguish our own species.

Mammals	
Cranium	Present
Vertebrae	Present
Jaws	Present
Skeletal composition	Bone
Lungs	Present
Limbs	Present
Amnion	Present
Body temperature regulation	Endotherm
Number of heart chambers	Four

Summary *Like birds, mammals have reptilian ancestors. Hair and mammary glands distinguish mammals from other vertebrates.*

23.8 MASTERING CONCEPTS

1. Other than hair and mammary glands, which characteristics define a mammal?
2. What evidence suggests that mammals arose from reptiles?
3. How do monotremes, marsupials, and placental mammals differ in how they reproduce?

FOCUS ON MODEL ORGANISMS: *Mus musculus*, the mouse

The history of the mouse, *Mus musculus*, as the stereotypical lab animal dates back to the early 1900s. Researchers discovered that its small size made it an excellent research animal. Also, mice are famous for their prolific breeding. They reach sexual maturity at the age of about 4 weeks, are sexually receptive every few days, and give birth to litters of 1 to 10 pups after a gestation of only about 3 weeks. Over a life span of 1.5 to 3 years, a single pair of mice can produce hundreds of offspring.

Researchers have benefited from biotechnology in their studies of mice. Transgenic mice have been available since the 1980s (see chapter 12). These mice are modified in countless ways, including altered susceptibility to human diseases. Mice were cloned in 1998, making possible the production of genetically identical animals ideal for testing new disease treatments. The mouse genome sequence was completed in 2002, revealing about 30,000 genes divided among 20 chromosomes. Not surprisingly, more than 99% of mouse genes have counterparts in the human genome. This similarity has made possible some of the following ways in which *Mus musculus* has contributed to biological research:

- **Immune function:** In the 1930s, the discovery that mice reject transplants from all but their very close relatives led to the discovery of the major histocompatibility complex (MHC). Since that time, biologists have discovered an array of genes related to immune function (see chapter 36).
- **Human disease:** Mice have been used to study human disease since the 1930s, when researchers discovered that mice could contract yellow fever. Vaccines were subsequently tested in mice. The availability of transgenic mice has opened new possibilities for research on the cause and treatment of human disease, including the protein deposits (beta-amyloid plaques) that accumulate in the brains of patients with Alzheimer disease; obesity; Parkinson disease; the role of mutated tumor suppressor genes such as *p53* in human cancer; and the effects of viruses such as HIV on the human immune system. **cancer, p. 166; HIV, p. 362**
- **X chromosome inactivation:** In the 1960s, biologist Mary Lyon proposed that in female mammals, one of the two X chromosomes is inactivated early in embryonic development. This phenomenon, often illustrated using calico cats, was first proved in mice with mottled coats. **X inactivation, p. 226**
- **Stem cells:** These undifferentiated cells, which can be derived from embryos or adults, can specialize into many other cell types. Mouse stem cell research has shown great promise in treating spinal cord injuries and many other ailments. **stem cells, p. 158**

23.9 INVESTIGATING LIFE
Limbs Gained, and Limbs Lost

Every now and then, a spectacular fossil find grabs headlines worldwide. That is exactly what happened in April 2006, when fossils shed light on two important questions in vertebrate evolutionary biology: how did tetrapods gain their limbs, and how did snakes lose them?

The idea that early amphibians crawled onto land some 375 MYA is intriguing, especially since the descendants of those early colonists are today's amphibians, reptiles, birds, and mammals. Fossils discovered over the past half-century, including *Acanthostega* and *Ichthyostega*, have clarified the fish-amphibian transition. Still, some details of this fascinating event remain poorly understood.

Scientists Edward T. (Ted) Daeschler of Philadelphia's Academy of Natural Sciences, Neil Shubin of the University of Chicago, and Harvard University's Farish Jenkins added new insights when they published back-to-back papers in the journal *Nature* in April 2006. The articles described fossils of an extinct animal, *Tiktaalik roseae*, that the researchers had

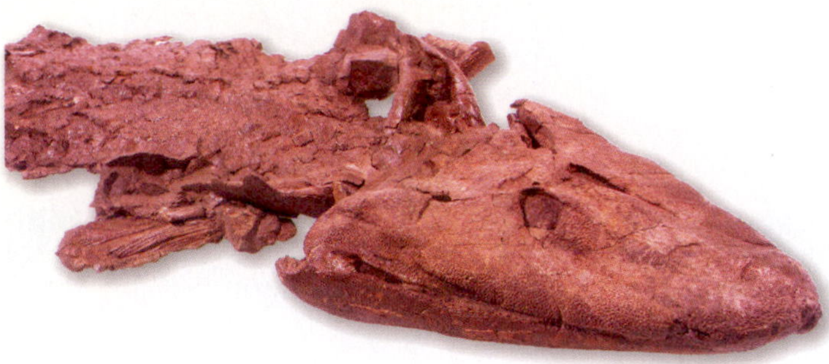

FIGURE 23.18 Fossil "Fishapod." (**A**) This photo shows a portion of the *Tiktaalik* fossil discovered in 2006. (**B**) *Tiktaalik's* limbs clearly contained bones, yet they were fringed with fins—one clue to the animal's aquatic heritage. **Question**: How might the ability to crawl on land for short periods have enhanced the reproductive fitness of *Tiktaalik*?

unearthed in Arctic Canada (**figure 23.18**A). *Tiktaalik* either crawled or paddled in shallow tropical streams about 380 MYA, during the late Devonian period, right in the middle of the fish–amphibian transition. (Although today's Canada is anything but tropical, the entire North American continent was near the equator during the Devonian.)

Scientists jokingly call the animal a "fishapod" because of its uncanny mix of fish and tetrapod characteristics. Like a fish, *Tiktaalik* had scales and gills. Like a tetrapod, it had lungs, and its ribs were robust enough to support its body. It could also move its head independently of its shoulders, something that fish cannot do. But the appendages got the most attention. *Tiktaalik* had moveable wrist bones that were sturdy enough to support the animal in shallow water or on short excursions to land. Although the bones were clearly limblike, the "limbs" were fringed with fins, not toes (figure 23.18B).

The *Tiktaalik* fossils caught the world's eye because they were extraordinarily complete and exquisitely preserved. Scientifically, *Tiktaalik* is important for two reasons. First, it adds to our knowledge about the fish–amphibian transition. Second, it highlights the predictive power of evolutionary biology. The researchers did not simply stumble on *Tiktaalik* by accident. Instead, they were looking for a fossil representing the fish–amphibian transition, based on previous knowledge of how and when vertebrates moved onto land hundreds of millions of years ago. Finding *Tiktaalik* confirmed the prediction.

Fossils have answered the question of how tetrapods got their limbs, but until recently the same was not true for another important issue in vertebrate evolution: how did snakes *lose* their limbs? Molecular and anatomical information, including vestigial legs in some snakes (see figure 15.12), clearly indicated that snakes evolved from lizards. Yet the precise four-legged ancestor has remained elusive.

Scientists proposed two competing hypotheses to explain the origin of snakes. Noting that snakes resemble two existing groups of burrowing lizards, some scientists suggested that

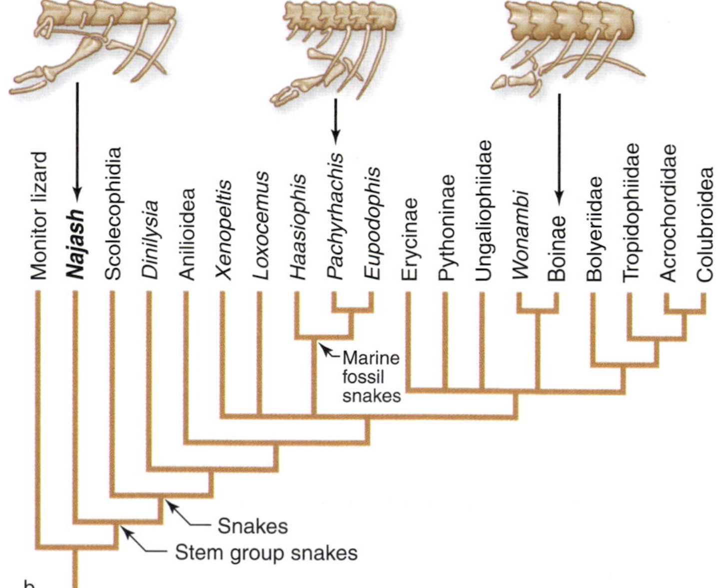

FIGURE 23.19 ***Najash,* the Fossil Snake** (**A**) *Najash* had hindlimbs, a pelvis, and a sacrum. (**B**) The phylogenetic tree based on skeletal similarities clearly places *Najash* as one of the most primitive snakes, whereas the three marine fossil snakes with legs (*Pachyrhachis, Haasiophis*, and *Eupodophis*) belong to a more "derived" group. **Question:** How might the loss of hind limbs enhance the reproductive fitness of a burrowing animal such as *Najash*?

snakes evolved on land. Opposing scientists contend that snakes descend from mosasaurs, marine reptiles that thrived during the Cretaceous period, based on skull and jaw similarities.

Argentinian paleontologist Sebastián Apesteguía, along with Brazilian colleague Hussam Zaher, added a critical clue to the debate over snake origins in April 2006, when they reported finding three fossilized snakes in the Patagonia region of Argentina (**figure 23.19**A). The snakes, which they named *Najash rionegrina*, lived during the Upper Cretaceous period, about 90 MYA.

Najash is different from other ancient snakes for at least two reasons. First, it is the first snake ever found to have not only functional legs and a pelvis but also a sacrum—a bone connecting the pelvis to the spine. (In other fossil snakes with limbs, the pelvis is "free floating" and not connected to the backbone.) The sacrum is important because lizards and other tetrapods, the ancestors of snakes, have the same bone. *Najash* is therefore more primitive than any snake ever found (figure 23.19B). Second, both the fossil's features and the rock where it was found suggest that it was terrestrial, not marine. Taken together, these two pieces of evidence seem to settle the question: snakes originated on land, not in water.

Spectacular fossils such as *Tiktaalik* and *Najash* thrust researchers into the limelight, at least for a short time. The flurry of excitement, however, ignores the countless hours of tedious, labor-intensive work and deep understanding of anatomy needed to interpret fossils. Researchers scrutinize the scales, limbs, skull, jaw, teeth, vertebrae, and other parts of each new find to glean every possible piece of information about what the animal was and how it lived. The analysis may confirm our previous understanding, as in the case of *Tiktaalik*, or, like *Najash*, it may help settle a long-standing argument. Either way, fossils contribute immeasurably to our understanding of life's long history.

Apesteguía, Sebastián, and Hussam Zaher. April 20, 2006. A Cretaceous terrestrial snake with robust hindlimbs and a sacrum. *Nature*, vol. 440, pages 1037–1040.

Daeschler, Edward B., Neil H. Shubin, and Farish A. Jenkins, Jr. April 6, 2006. A Devonian tetrapod-like fish and the evolution of the tetrapod body plan. *Nature*, vol. 440, pages 757–763.

Shubin, Neil H., Edward B. Daeschler, and Farish A. Jenkins, Jr. April 6, 2006. A pectoral fin of *Tiktaalik roseae* and the origin of the tetrapod limb. *Nature*, vol. 440, pages 764–771.

CHAPTER SUMMARY

23.1 Vertebrates (and a Few Invertebrates) Are Chordates

- Tunicates, lancelets, hagfishes, and vertebrates are all **chordates; vertebrates** include fishes, amphibians, reptiles, birds, and mammals.

A. Four Features Distinguish Chordates

- Chordates share four characteristics: a **notochord,** a **dorsal hollow nerve cord, pharyngeal pouches** or slits in the **pharynx,** and a **postanal tail.**

B. Biologists Use Many Features to Classify Chordates

- The absence of a **cranium** distinguishes tunicates and lancelets from the **craniates** (hagfishes and vertebrates).
- Other features that distinguish chordates from one another include **vertebrae, jaws, lungs,** and the presence of limbs in **tetrapods.**
- Reptiles, birds, and mammals are **amniotes;** in these animals, the **amniote egg** or **amnion** protects the developing embryo. The type of body covering is also important in classifying chordates.
- Only vertebrates have a backbone of cartilage or bone.
- **Ectotherms** use the environment to regulate body temperature, whereas **endotherms** use metabolic heat to maintain body temperature.
- Fishes have two heart chambers, amphibians and most reptiles have three, and birds and mammals have four.

23.2 Tunicates and Lancelets Have Neither Cranium Nor Backbone

- The tunicates and lancelets are **invertebrate** chordates.
- **Tunicates** obtain food and oxygen with a siphon system. Only the larva has all four chordate characteristics; adults retain only the pharyngeal slits.
- **Lancelets** resemble eyeless fishes; adults have all four chordate characteristics.

23.3 Hagfishes Have a Cranium, but Lack a Backbone

- **Hagfishes** have a cranium of cartilage, but they are not vertebrates. They secrete slime and lack jaws.

23.4 Fishes Are Aquatic Vertebrates with Gills and Fins

- **Fishes** are abundant and diverse vertebrates.

A. Fishes Changed the Course of Vertebrate Evolution

- Adaptations in fishes that allowed vertebrates to move onto land include lungs and fleshy, paired fins that were later modified as limbs. Jaws and a vertebral column also originated in fishes.

B. Fishes Are Divided into Two Main Groups

- Jawless fishes include the extinct **ostracoderms** and the modern **lampreys.**

- **Placoderms** are an extinct class of fishes with jaws.
- The **cartilaginous fishes** include skates, rays, and sharks. Sharks detect vibrations from prey with a **lateral line system.**
- The **bony fishes** account for 96% of existing fish species. Bony fishes have lateral line systems and swim bladders, which enable them to control their buoyancy. The three groups of bony fishes include **ray-finned fishes, lungfishes,** and **lobe-finned fishes.**

23.5 Amphibians Lead a Double Life on Land and in Water

A. Amphibians Were the First Four-Legged Vertebrates

- **Amphibians** retain some characteristics of fishes. They breed in water and must keep their skin moist to breathe. Adaptations to life on land include a sturdy skeleton, lungs, four limbs, and a three-chambered heart.

B. Amphibians Include Three Main Lineages

- Amphibians include **frogs** and toads, **salamanders** and newts, and **caecilians.**

23.6 Reptiles Were the First Vertebrates to Thrive on Dry Land

A. Reptiles Descended from Amphibians

- **Reptilian** adaptations to life on land include efficient excretory, respiratory, and circulatory systems. Internal fertilization and amniote eggs make reproduction on dry land possible.

B. Non-Avian Reptiles Include Three Main Groups

- Reptiles include **turtles** and tortoises; **lizards, snakes,** and **tuataras; crocodilians;** and dinosaurs and birds.

23.7 Birds Are Warm, Feathered Reptiles

A. Birds Have Many Adaptations to Flight

- **Birds** retain scales and egg-laying from reptilian ancestors. Hollow bones, streamlined bodies, and **feathers** are adaptations that enable flight.
- Like mammals but unlike reptiles, birds are endothermic.

B. Birds Are Classified Based on Breastbone Shape

- The two main groups of birds are the **carinates** and the **ratites.**

23.8 Mammals Are Warm, Furry Milk-Drinkers

A. Mammals Descended from Reptiles

- **Mammals** have fur, secrete milk from **mammary glands,** and have distinctive teeth and highly developed brains. They are **synapsids.**

B. Mammals Include Three Main Groups

- **Monotremes** are mammals that hatch from an egg. The young of **marsupial** mammals develop inside the mother's marsupium, whereas **placental mammals** are nourished by the **placenta** in the mother's uterus.

23.9 Investigating Life: Limbs Gained, and Limbs Lost

- *Tiktaalik roseae* is an extinct "fishapod." Its fossils clearly reflect the transition between fishes and amphibians, which occurred about 375 MYA.
- Another fossil, *Najash rionegrina,* has helped researchers understand the origin of snakes. The features of *Najash* suggest that snakes arose on land from burrowing ancestors.

MULTIPLE CHOICE QUESTIONS

1. What is a notochord?
   a. The spine of a chordate animal
   b. The segments of the chordate body plan
   c. A type of germ layer
   d. A fibrous rod that runs down the back of a chordate
2. How are birds, reptiles, and mammals similar?
   a. They all have an amnion.
   b. Their body coverings are made of keratin.
   c. They are craniates.
   d. All of the above.
3. Why is a four-chambered heart more efficient than a two-chambered heart?
   a. Because it separates oxygen-rich and oxygen-poor blood.
   b. Because it is more evolved.
   c. Because it can transport larger volumes of blood.
   d. Because it can generate a stronger contraction.
4. Since a tunicate is considered to be a chordate, it must have a
   a. cranium.
   b. notochord.
   c. amniote egg.
   d. lung.
5. Which of the following is an example of a cartilaginous fish?
   a. Eel
   b. Trout
   c. Shark
   d. Hagfish

6. Why is it important for amphibians to live in a moist habitat?
   a. To promote gas exchange across their skin
   b. To maintain their eggs
   c. To maintain their bouyancy
   d. Both a and b
7. A four-chambered heart is a characteristic of
   a. frogs and toads.
   b. turtles and tortoises.
   c. crocodilians.
   d. lizards and snakes.
8. What is the evolutionary link between birds and reptiles?
   a. Four-chambered hearts
   b. Egg laying
   c. Scales
   d. All of the above
9. Which group of mammals lays eggs?
   a. Monotremes
   b. Marsupials
   c. Placental mammals
   d. Both a and b
10. Since a whale is a mammal, it must
   a. have scales.
   b. have gills.
   c. produce milk.
   d. all of the above.

TESTING YOUR KNOWLEDGE

1. What are the four distinguishing characteristics of chordates?
2. How do tunicates and lancelets differ from fishes and tetrapods?
3. Trace the evolutionary history of vertebrates from fishes to amphibians to reptiles, birds, and mammals.
4. If you found an eel-like animal at the beach, what features would you look for in deciding whether you had a hagfish or an eel (a true fish)?
5. List five adaptations that enable
   a. fishes to live in water.
   b. amphibians to live on land.
   c. reptiles to live in the desert.
   d. birds to fly.
6. How is an amphibian's skin both an advantage and a liability?
7. How are fishes, amphibians, reptiles, birds, and mammals important to humans? How are they important in ecosystems?

THINKING AS A SCIENTIST

1. Tunicates and lancelets did not leave fossil evidence because their parts were not hard. The distinguishing characteristics of mammals (hair and breasts) also are not hard, yet mammals have a rich fossil record. Explain this difference.
2. How are a fish's and a bird's skeletons similar in structure and function?
3. Fishes are adapted to life in water, and tetrapods to life on land. Cite two criteria for assessing which group has been more successful.
4. Give two examples in which DNA sequence evidence contradicts traditional classification of vertebrates based on easy-to-observe characteristics.
5. What is the evidence for the idea that birds are reptiles?
6. Lungfishes and therapsids are thought to be transitional forms. Which types of animals does each resemble?
7. How does a group of organisms that has lost a specialized characteristic complicate biological classification? Cite two examples of this phenomenon. Suggest a way that biologists can overcome this confusion.
8. If you found a fossil and were not sure whether it was from a reptile or a mammal, how might you tell the difference?
9. Other than the mouse, *Mus musculus*, what are some examples of vertebrates that have contributed to scientists' knowledge of general biology and animal biology? What other genomes of vertebrate animals have scientists sequenced, and what are some resulting discoveries?
10. It is rare to discover a new species of large mammal, but scientists often report naming and describing new species of small vertebrates. What is the scientific value of cataloging the diversity of the world's species of animals?

Visit www.mhhe.com/hoefnagels for practice quizzes, animations, videos, and activities designed to help you master the material in this chapter.

CHAPTER 24

Plant Form and Function

Cash Crop. Women harvest coca leaves in Peru.

What's the Point?

Tea, Herbs, and Drugs: Plants Are Chemical Factories

Food, beverages, paper, lumber, textiles, oils, rope ... the list of useful products derived from plants is long and familiar. But another item on the list may be surprising: a vast array of potent chemicals. These compounds may come from the cells of roots, stems, leaves, flowers, or fruits, and they fall into several chemical classes.

If you enjoy peppermint tea or peppermint oil aromatherapy, for example, you are using the product of peppermint plant leaves. In plants of the mint family, leaf hairs produce and store chemicals called terpenes. The most abundant terpene in peppermint is menthol, which produces a cooling sensation when eaten or inhaled. Other aromatic herbs in the mint family, including spearmint, sage, basil, lavender, rosemary, thyme, and oregano, produce different terpenes.

Another leaf hair extract is hashish, which is purified resin from the leaves and flowers of Cannabis sativa. *A chemical called tetrahydrocannabinol (THC) produces the narcotic effect of hashish and marijuana. THC is part of a class of chemical compounds called phenolics.*

Alkaloids are another diverse class of powerful plant-derived compounds. Examples include caffeine (from coffee beans), nicotine (from tobacco leaves), morphine and codeine (from opium poppy flowers), and cocaine (from coca leaves). The antimalarial medicine quinine is an alkaloid from the bark of Cinchona trees. The alkaloid capsaicin makes chili peppers taste "hot," and vincristine is an antileukemia drug from leaves of Catharanthus roseus.

The terpenes, phenolics, alkaloids, and other potent chemicals that plants produce are called secondary metabolites, and the plant kingdom boasts many other examples. These chemicals do not participate directly in photosynthesis, respiration, reproduction, or other essential processes. So why do plant cells invest the energy required to make them?

The answer lies in natural selection. Many secondary metabolites are bad-tasting or toxic to disease-causing organisms, insects, and other animals. Quite simply, plants that defend themselves with these weapons have a reproductive advantage over those that don't. A chemical arsenal that repels or poisons hungry animals is just one defense against predation. As you'll see in this chapter, a plant's growth pattern and structural defenses play important roles as well.

LEARNING OUTLINE

24.1 Plant Parts Depend on One Another

Imagine a rose bush that produces a bounty of sweet-smelling flowers. Biologists would divide your plant into two sets of parts. The **vegetative,** or nonreproductive, parts are the roots, stems, and leaves. These organs, and the tissues that compose them, are the subject of this chapter. The gorgeous flowers, which will eventually give rise to inconspicuous fruits called rose hips, are the reproductive parts of the plant. Chapter 26 describes them in detail.

The vegetative organs work together in a living plant. The stem and leaves together constitute the **shoot,** or aboveground part of a plant. The **stem** supports the **leaves,** which produce carbohydrates such as sucrose by photosynthesis. A large portion of this sugar moves down the stem and nourishes the **roots,** which are usually below ground. Root cells depend completely on the shoots to provide energy for their metabolism. At the same time, however, roots anchor the plant and absorb water and minerals that move via the stem to the leaves.

In all of the chapters in unit 5, the emphasis is on flowering plants (angiosperms), the most diverse and abundant plants. Biologists divide angiosperms into two classes: monocotyledons (monocots) and dicotyledons (dicots). The dicots are not monophyletic, but most do belong to one clade, called the eudicots. Although monocots and eudicots differ in many ways (summarized in **figure 24.1**), the same general structures occur in both.

A. Vegetative Plant Parts Include Stems, Leaves, and Roots

A close look at a stem reveals that it consists of repeating units consisting of alternating nodes and internodes (**figure 24.2**). A **node** is a point at which one or more leaves attach to the stem. Along with at least one leaf, each node also has an **axillary bud,** an undeveloped shoot that forms in the angle between the stem and leaf stalk. Although a bud has the potential to elongate to form a branch or flower, many remain small and dormant. **Internodes** are the stem areas between the nodes.

Most leaves have two main parts: the flattened **blade** and the supporting, stalklike **petiole.** The broad, flat blade maximizes surface area for the plant to capture solar energy. For example, a large maple tree has approximately 100,000 leaves, with a total surface area that would cover six basketball courts (about 2500 m^2).

Biologists categorize leaves according to their basic forms (**figure 24.3**). A **simple leaf** has an undivided blade. **Compound leaves** are divided into leaflets, typically either paired along a central line or all attached to one point at the top of the petiole, like fingers on a hand. How can you tell the difference between a simple leaf and one leaflet of a compound leaf? A leaf has an axillary bud, whereas an individual leaflet does not.

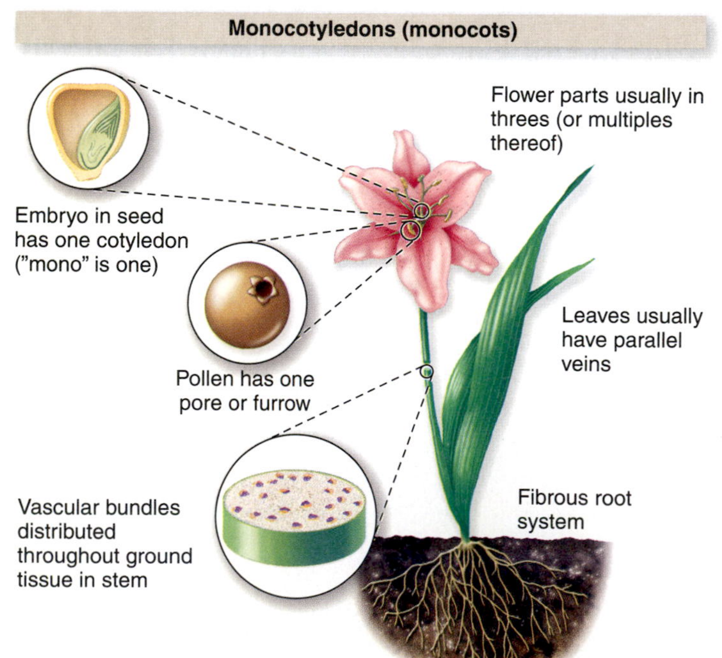

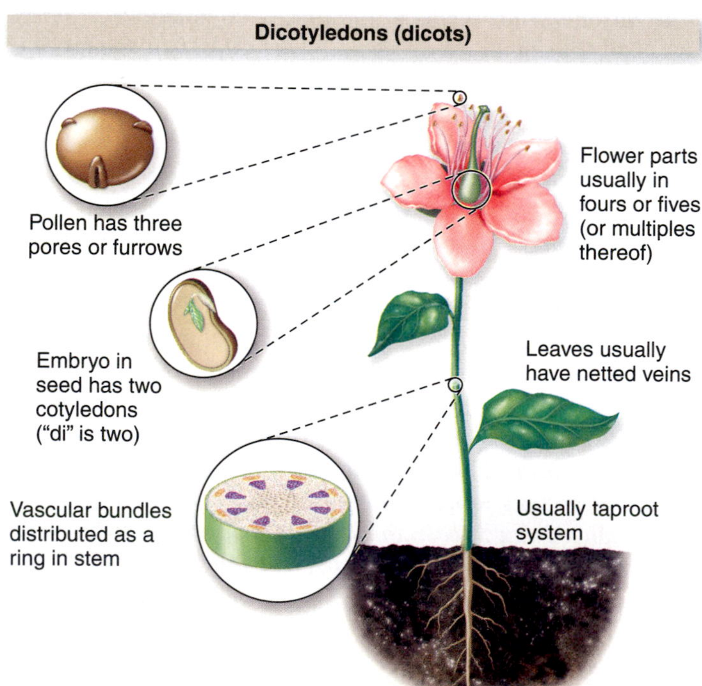

FIGURE 24.1 Monocots and Eudicots Compared. Although many variations occur, most monocots and eudicots differ from each other in flower and leaf morphology, stem anatomy, pollen structure, root system architecture, and seed structure.

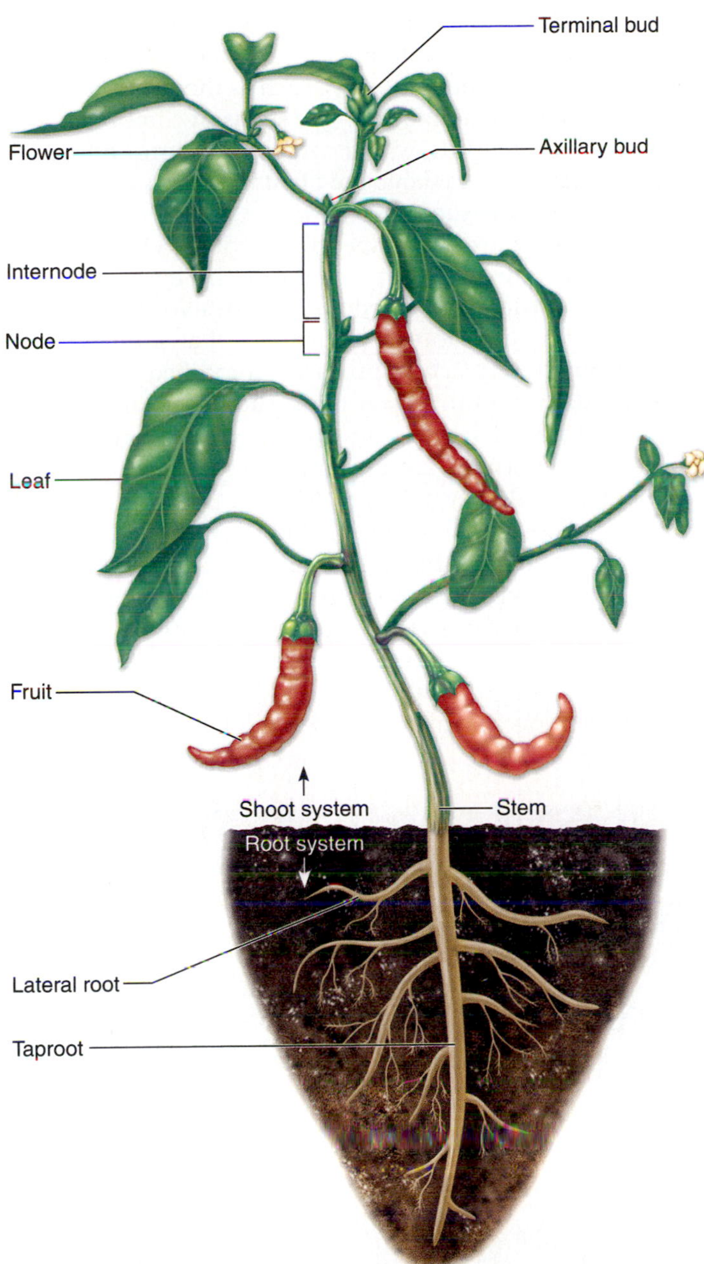

FIGURE 24.2 Parts of a Flowering Plant. A plant consists of a root system and a shoot system. Roots, stems, and leaves are vegetative organs; flowers and fruits are reproductive structures, the subject of the chapter 26.

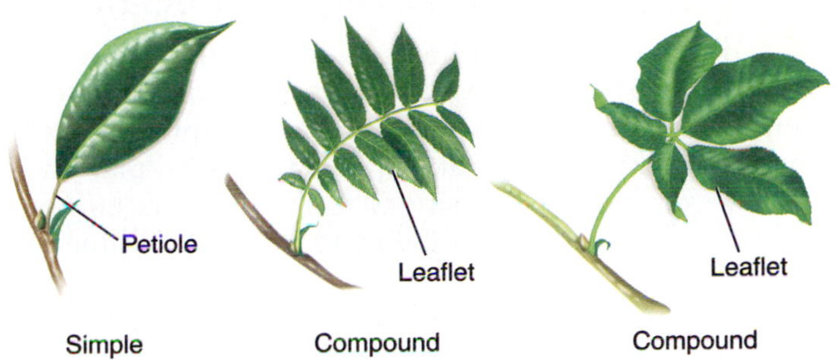

FIGURE 24.3 Leaf Forms. A simple leaf has an undivided blade, whereas compound leaves consist of multiple leaflets. An axillary bud defines the base of each leaf.

FIGURE 24.4 Two Main Patterns of Root Organization. (A) The taproot system of a dandelion, a eudicot. **(B)** The fibrous root system of barley, a monocot.

Roots are also critical to the plant's life. Two categories of root systems are based on the fate of the primary root (the first root to develop after a seed germinates). In a **taproot system,** the primary root enlarges to form a major root that persists throughout the life of the plant (**figure 24.4**A). Branches emerge from this main root. Taproots grow fast and deep, maximizing support and enabling a plant to use minerals and water deep in the soil. Most eudicots develop taproot systems. In a **fibrous root system,** branching roots arise from the base of the stem and replace the short-lived primary root (figure 24.4B). The roots in a fibrous system are examples of **adventitious roots,** a general term for roots that arise from nonroot plant parts such as stems or leaves. (If you have ever rooted a houseplant by placing a stem cutting in water, you have seen adventitious roots.) Grasses and other monocots usually have fibrous root systems. Because they are relatively shallow, fibrous root systems rapidly absorb minerals and water near the soil surface and prevent erosion.

B. Plant Parts Can Be Highly Specialized

In addition to hungry animals, plants contend with everything from extreme drought to continuous flooding to frozen winters. These selective forces have sculpted the stems, leaves, and roots that make up the vegetative plant body into a tremendous diversity of forms. Farmers have also cultivated many plants for their wide variety of edible parts (**table 24.1**).

Stems support leaves, but they may also help the plant reproduce, store nutrients, protect itself, and compete for light (**figure 24.5**). Stolons grow along the soil surface, asexually forming new plants at their nodes. Rhizomes are underground stems that produce roots and new shoots. Tubers, such as potatoes, are swollen regions of underground stems that store starch. Succulent, fleshy stems of plants such as cacti stockpile large volumes of water. Still other stems are protective. Some types of thorns, such as those on hawthorn plants, are modified branches. The stems of climbing plants may also form tendrils, which coil around surrounding objects for support and enable the plant to maximize exposure of the leaves to the sun.

Leaves may also have special functions. Onion bulbs, for example, are collections of the fleshy bases of leaves that store nutrients. Cotyledons are embryonic leaves; in some species, they store carbohydrates that supply energy for germination. Cactus spines are modified leaves that deter predators. Some flower parts, such as sepals and petals, are modified leaves. And in a few carnivorous plant species, leaves attract, capture, and digest prey, as described in chapter 25. **flower parts, p. 538**

Specialized root functions include storage, gas exchange, and support (**figure 24.6**). Beet and carrot roots stockpile starch, and desert plant roots may store water. In oxygen-poor habitats such as swamps, specialized roots form underground and grow up into the air, allowing oxygen to diffuse in. Thick buttress roots at the base of a tree provide support, as do prop roots that arise from a corn plant's stem.

Table 24.1 *Many Parts Are Edible*

	Plant Part*	Edible Example(s)
Vegetative	Axillary (lateral) bud	Brussels sprout
	Terminal bud	Cabbage
	Bark	Cinnamon
	Bulb	Onion, garlic
	Leaf blade	Basil, parsley, spinach, chives, kale
	Leaf petiole	Celery, rhubarb
	Rhizome	Ginger
	Root	Carrot, parsnip, beet, sweet potato
	Seedling	Alfalfa sprout, bean sprout
	Stem	Asparagus, kohlrabi, sugar cane
	Tuber	Potato
Reproductive	Flower bud	Broccoli, cauliflower, artichoke
	Style and stigma (flower parts)	Saffron
	Seeds and fruits	Apple, raspberry, tomato, string bean, olive, cucumber, pumpkin, corn kernel, bell pepper, black peppercorn, rice, wheat, walnut, peanut, pea, nutmeg

*Why no wood? Because it's mostly empty cell walls made of cellulose and lignin, neither of which we can digest.

FIGURE 24.5 Modified Stems. (**A**) Tendrils may be stems modified to coil around objects, supporting and anchoring plants. (**B**) The rhizome of an iris is an underground stem. (**C**) The stem of the fishhook barrel cactus is the primary organ of photosynthesis, and it is highly modified to store water. Its thorns are modified leaves, not stems. (**D**) The thorns that protect this honey locust are outgrowths of the stem.

a.

b.

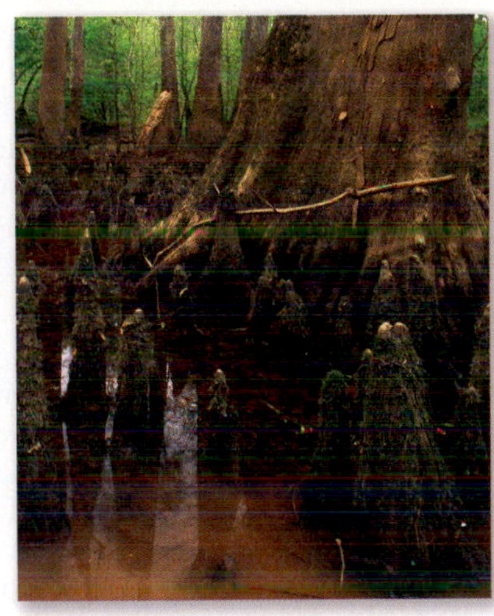
c.

FIGURE 24.6
Root Modifications. (**A**) Prop roots on corn arise from the stem and support the plant. (**B**) Aerial roots of orchids grow aboveground. (**C**) The roots of bald cypress trees form pneumatophores, commonly called "knees."

Roots often form beneficial relationships with other organisms that increase their capacity to obtain nutrients. In a mycorrhizal association, for example, fungal hyphae extend into soil from inside roots, greatly increasing the surface area for absorption (see chapter 21). Some plant species, such as peas and beans, form root nodules in association with nitrogen-fixing bacteria (see figure 25.4). These bacteria function as built-in fertilizer. **nitrogen cycle, p. 817**

Summary *The vegetative plant body consists of three types of organs: stems, roots, and leaves. Modification of these basic parts yields a tremendous variety of plant forms.*

24.1 MASTERING CONCEPTS

1. How do stems, leaves, and roots support one another?
2. What is the difference between the node and the internode of a stem?
3. List the main types of leaf form.
4. Compare and contrast the development of taproot and fibrous root systems.
5. Give examples of stems, leaves, and roots that specialize in storage, protection, asexual reproduction, or absorption.

24.2 Plants Have Flexible Growth Patterns, Thanks to Meristems

Consider the plight of a green plant. Rooted in place, it seems vulnerable and defenseless against drought, flood, wind, fire, and hungry herbivores. Yet plants dominate nearly every habitat on land. How do they do it? Some plant defenses come from a diverse chemical arsenal, as described at the start of this chapter. But in many ways, plants owe their success to modular growth.

A. Plants Grow by Adding New Modules

To understand modular growth, imagine a landowner who wants to build a motel. Money is tight at first, so she starts with just a few units. As her business grows, however, she adds more units to the basic plan. Plant growth is similar. Shoots become larger by adding units ("modules") consisting of repeated nodes and internodes (see figure 24.2).

One result of modular growth is extreme flexibility in form. Suppose a shrub is growing in a dense forest, shaded from all sides. When a neighboring tall tree dies and topples, the resulting hole in the canopy allows more sunlight to reach one side of the shrub. The shrub can add new segments to the sunny side, while the shaded half remains unchanged. In this way, the loss of a branch does not affect a tree as much as, say, the loss of a leg affects a cat. Neighboring branches can add modules to compensate for the lost tree limb, but the cat's body cannot regenerate a leg. Modular growth is one feature that distinguishes plants from animals.

Some plants stop growing after they reach their mature size, a pattern called **determinate growth.** Many types of plants, however, can grow indefinitely by adding module after module. Such **indeterminate growth** can persist as long as environmental conditions allow it. Determinate growth is most common in **herbaceous plants,** so named for their green, soft stems. Examples include daisies, dandelions, radishes, and other plants with little or no woody tissue. Indeterminate growth is most common in **woody plants** such as elm and cedar trees, whose stems are made of tough wood covered with bark.

B. Plant Growth Occurs at Meristems

Regardless of whether growth is determinate or indeterminate, the source of a plant's new cells is the same: **meristems,** which are regions of the plant that undergo active mitotic cell division. Like stem cells in animals, meristems are patches of "immortality" within a plant. **stem cells, p. 158; mitosis, p. 161**

Plants have three main types of meristems: apical, lateral, and intercalary (**table 24.2**). **Apical meristems** are small patches of actively dividing cells near the tips of roots and shoots in all plants, both herbaceous and woody (**figure 24.7** in a shoot, and figure 24.16 in a root). When cells in the apical meristem divide, they give rise to cells that become the protoderm, procambium, and ground meristem. These three primary meristems, in turn, produce all tissue types (see section 24.3). **Primary growth** lengthens the root or shoot tip by adding cells produced by the primary meristems.

Lateral meristems (also called cambia) produce cells that thicken a stem or root. A lateral meristem is usually an internal cylinder of cells extending along most of the length of the plant. When the cells divide, they typically produce tissues to both the inside and the outside of the meristem. The resulting increase in girth is called **secondary growth,** and it occurs in woody plants (see section 24.5).

In some plants, **intercalary meristems** occur between the nodes of a mature stem, usually at the base of an internode. (*Intercalary* means "inserted between," referring to the position of these meristems). Grasses, for example, tolerate grazing and mowing because they have intercalary meristems whose cells divide to regrow a leaf from its base when its tip is munched off.

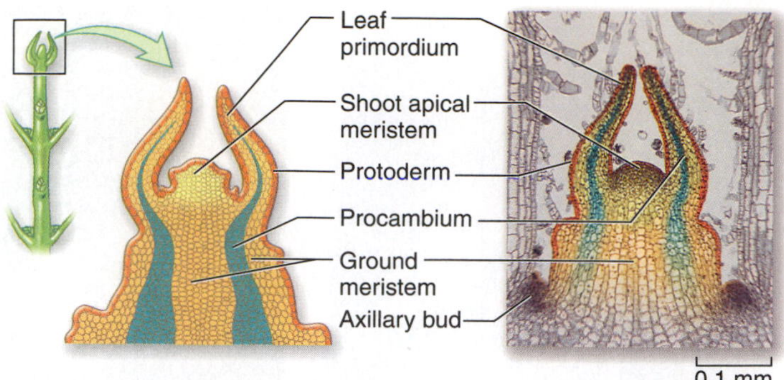

FIGURE 24.7 Primary Growth of a Eudicot Shoot. The tissue layers of the growing shoot tip reveal how apical meristems give rise to the three primary meristems. The protoderm, procambium, and ground meristem, in turn, form mature, specialized tissues visible in older parts of the shoot.

Meristems are vitally important to plant life. Although plants cannot run away from danger, they can tap the power of indeterminate growth to replace damaged parts. Meristems also help plants adjust to the environment. For example, a houseplant may produce the most leaves on its "sunny side," or its roots may grow the most in pockets of soil with the richest nutrients. The plants owe this flexibility to their numerous meristems.

Summary *Plant stems display modular growth, with alternating nodes and internodes. Plants grow and adjust to environmental challenges by adding modules. The added length and girth comes from meristems, regions of cells that retain the ability to divide throughout the life of the plant.*

24.2 MASTERING CONCEPTS

1. What is the difference between determinate and indeterminate growth?
2. What are the locations and functions of the three types of meristems in plants?

Table 24.2 *Meristem Types*

Type	Locations	Function
Apical	Terminal and lateral buds of shoots; root tips	Gives rise to primary meristems (protoderm, procambium, ground meristem) that produce epidermis, vascular tissue, and ground tissue; the resulting primary growth lengthens roots and shoots
Lateral (vascular cambium and cork cambium)	Internal cylinder along length of some roots and stems	Thickens roots and stems in woody plants
Intercalary	Between nodes of mature stems in grasses and other monocots	Allows regrowth of tissue at base of mature stem or leaf if stem tip is removed

24.3 Plant Cells Build Tissues

A cactus, an elm tree, and a dandelion have distinctly different stems, leaves, roots, and growth patterns. They may seem to have little in common, but a closer look reveals that all consist of the same types of cells and tissues. This section turns to these building blocks.

A. Plants Have Several Cell Types

Plants consist of several cell types, all arising from meristems. This section lists the most common ones.

Parenchyma

Parenchyma cells are the most abundant cells in the primary plant body. Parenchyma cells are alive at maturity, and they retain the ability to divide, which enables them to differentiate in response to injury or a changing environment. It is parenchyma cells, for example, that divide to produce the adventitious roots in a houseplant cutting. Although structurally unspecialized, parenchyma cells have vital functions, including respiration, photosynthesis, and storage (**figure 24.8**A).

Collenchyma

Collenchyma cells are elongated living cells with unevenly thickened primary walls that can stretch as the cells grow (figure 24.8B). These cells provide support without interfering with the growth of young stems or expanding leaves. Collenchyma strands often form near vascular tissue or at the angular ridges on the stems of some plants, such as those in the mint family. Collenchyma is perhaps most familiar, however, as the tough, flexible "strings" in celery stalks.

Sclerenchyma

Sclerenchyma cells also provide support, with thick, rigid secondary cell walls that occupy most of the cell's volume. These cells are dead at maturity. Two types of sclerenchyma are fibers and sclereids; both support plant parts that are no longer growing. **cell walls, p. 71**

- **Fibers** are elongated cells that usually occur in strands (figure 24.8C), often associated with vascular bundles. Linen, for example, comes from soft, cellulose-rich fibers from the stems of *Linum usitatissimum*, or flax. In contrast, a tough molecule called lignin strengthens the cell walls of "hard fibers." Sisal, which comes from the leaves of *Agave sisalana*, is a hard fiber used in coarse fabrics and rope.
- **Sclereids** have many shapes, although they are generally shorter than fibers. Small groups of sclereids create

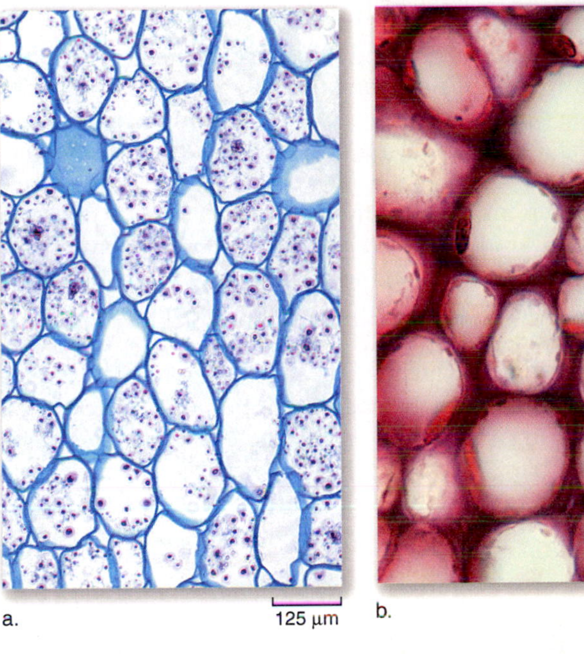

FIGURE 24.8 Plant Cells. (**A**) Parenchyma cells are usually structurally unspecialized and store carbohydrates and other important biochemicals. The photo shows parenchyma cells in the root of a buttercup plant. Starch grains are visible inside many of the cells. (**B**) The unevenly thickened, elastic walls of collenchyma cells support growing shoots. The collenchyma pictured is from the stem of a sunflower plant. (**C**) Sclerenchyma supports plant parts that are no longer growing. Fibers are one type of sclerenchyma. These fibers are from a spruce tree. (**D**) Sclereids, which provide the gritty texture of pear flesh, are also sclerenchyma cells.

a pear's gritty texture (figure 24.8D). Sclereids also form hard layers in nutshells, apple cores, and the pits of cherries and plums.

Water-Conducting Cells: Xylem

Xylem transports water and dissolved minerals from the roots to all parts of the plant (see chapter 25). The water-conducting cells of xylem are elongated and have thick secondary walls. They are dead at maturity, which means that no cytoplasm blocks water flow. The lignin-rich cell walls also help keep the plant upright.

The two kinds of water-conducting cells in xylem are tracheids and vessel elements (**figure 24.9**). **Tracheids** are long, narrow cells that overlap at their tapered ends. Water moves from tracheid to tracheid through pits, areas where secondary cell walls are absent and primary cell walls are thin. Water moves slowly in tracheids because of their small diameters and overlapping end walls. **Vessel elements** are short, wide, barrel-shaped conducting cells that stack end to end. Their side walls have pits, but their end walls either are perforated or disintegrate completely. Because of their greater diameter, and the fact that water can pass easily through each vessel element's end wall, water moves much faster in vessels than in tracheids.

Sucrose-Conducting Cells: Phloem

Phloem transports dissolved organic compounds, primarily sugars produced in photosynthesis (see chapter 25). Phloem sap also contains hormones, ions, and sometimes viruses. The conducting cells of phloem are alive at maturity. Their cell walls have thin areas perforated by many pores. Strands of cytoplasm pass through the openings, allowing carbohydrates to pass from cell to cell.

The main phloem conducting cells are **sieve tube elements,** which align end to end to make a single functional unit called a sieve tube. Most sieve tube elements have **sieve plates,** which are areas of large pores grouped at the ends of the cells (**figure 24.10**). Sieve tube elements carry on metabolism but lack nuclei. Adjacent to each sieve tube element is at least one **companion cell,** a specialized parenchyma cell that retains its nucleus and helps transfer carbohydrates into and out of the sieve tube elements.

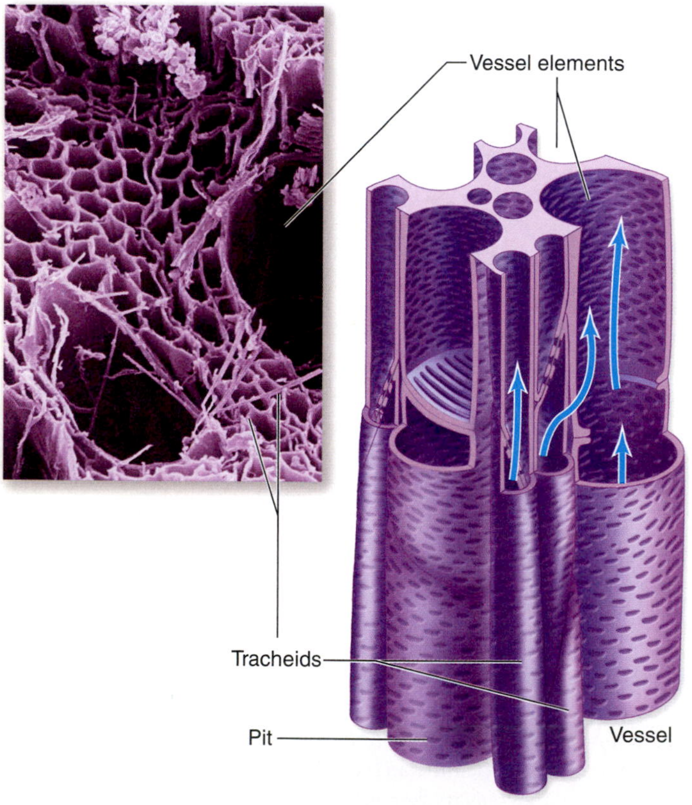

FIGURE 24.9 Two Types of Xylem Cells. Xylem consists of dead, hollow cells that transport water and dissolved minerals from roots to shoots. Tracheids are long, narrow, and relatively unspecialized. Water travels through pits in tracheid walls. Vessel elements are more specialized than tracheids. They are barrel-shaped and have fully or partially disintegrated end walls, easing water flow.

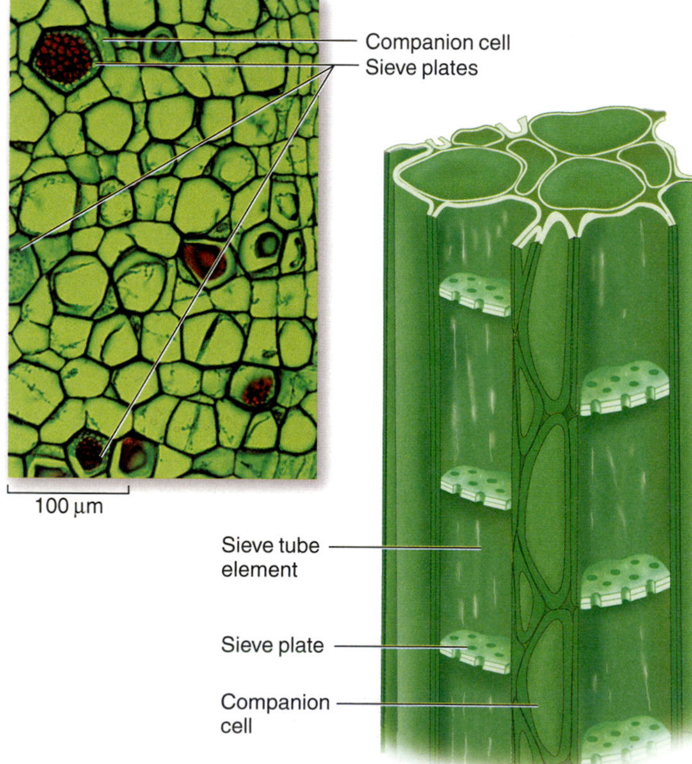

FIGURE 24.10 Phloem Cells. Phloem consists of living cells and transports a watery solution of carbohydrates, hormones, ions, and various other substances. In flowering plants, sieve tube elements with perforated end walls form sieve tubes that carry phloem sap.

B. Plant Cells Form Three Main Tissue Types

The cells that make up a plant form three main tissue types: ground, dermal, and vascular (**figure 24.11** and **table 24.3**). For comparison, animals have four types of tissues (see chapter 27). Each tissue type derives its properties from a unique combination of specialized cells. Together, these cells carry out all of the plant's functions.

Ground Tissue

Ground tissue makes up the majority of the primary body of a plant, and it consists mostly of parenchyma cells derived from the ground meristem. Ground tissue often fills the spaces between more specialized cell types inside roots, stems, leaves, fruits, and seeds. Although most ground tissue cells are structurally unspecialized, they are important sites of photosynthesis, respiration, and storage.

Dermal Tissue

Dermal tissue covers the plant. In the primary plant body, dermal tissue consists of the epidermis, which is derived from the protoderm. But in plants with secondary growth, lateral meristems produce tissues that replace the epidermis in stems and roots (see section 24.5).

The **epidermis,** usually only one cell layer thick, covers the leaf, stem, and roots of the primary plant body. Epidermal cells are flat, transparent, and tightly packed. On aboveground parts of the plant, epidermal cells secrete a waxy layer called the **cuticle** (**figure 24.12**A). The cuticle

FIGURE 24.11 Three Tissue Types Build Plant Organs. Apical meristems ultimately give rise to all cells that make up the primary plant body, including dermal, ground, and vascular tissues.

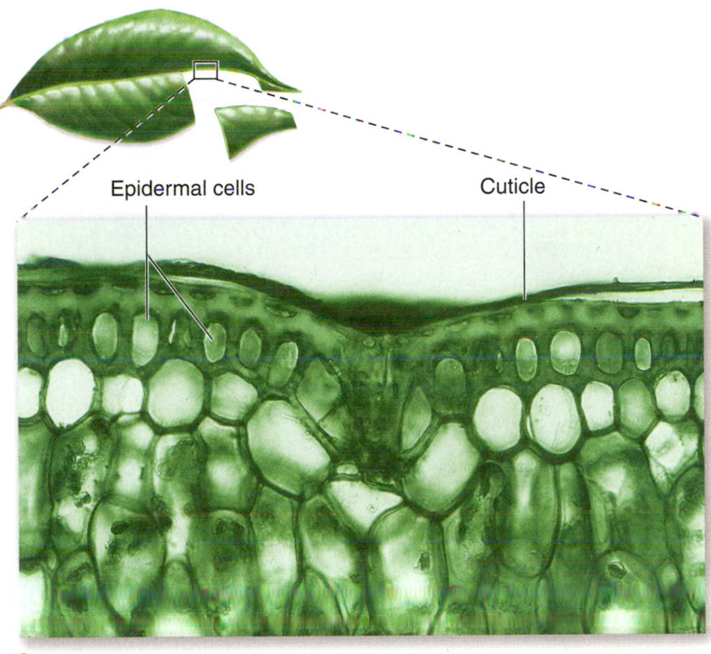

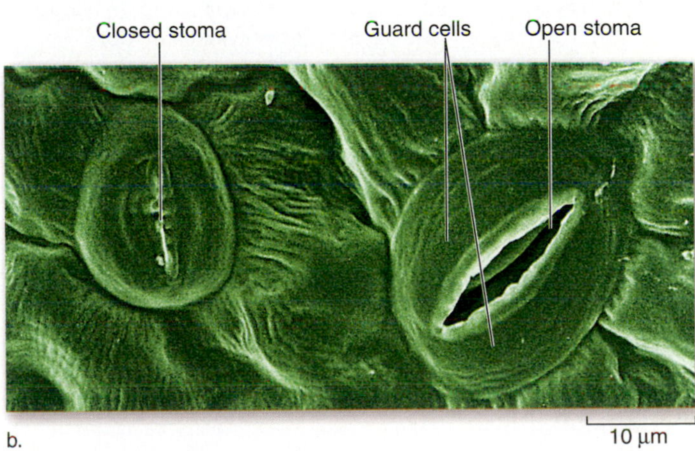

FIGURE 24.12 Epidermal Specializations. (**A**) A waterproof cuticle protects the epidermis. (**B**) Open stomata admit carbon dioxide, which is used in photosynthesis.

conserves water and protects the plant from predators and infectious agents. In addition, **stomata** (singular: stoma) are pores through which leaves and stems exchange gases with the atmosphere (figure 24.12B). A pair of specialized **guard cells** surrounds each stoma and control its opening and closing. Open stomata let carbon dioxide diffuse into a leaf for photosynthesis but also allow water to diffuse out. Plants constantly balance the need for carbon against the need to conserve water in dry conditions. **transpiration, p. 525**

Vascular Tissue

Vascular tissues are specialized conducting tissues that transport water, minerals, carbohydrates, and other dissolved compounds throughout the plant. Xylem and phloem cells derived from the procambium form a continuous distribution system embedded in the ground tissue of shoots and roots.

In stems and leaves, a **vascular bundle** consists of xylem and phloem together with parenchyma and sclerenchyma. Lignified fibers often associate with xylem in vascular bundles. Section 24.4 describes the organization of the cells that make up vascular tissue in roots, stems, and leaves, and chapter 25 further explores their function.

Summary *Plant bodies consist of parenchyma, collenchyma, sclerenchyma, and the water- and sucrose-conducting cells of xylem and phloem. These cells form the three basic tissue types: ground tissue, dermal tissue, and vascular tissue.*

24.3 MASTERING CONCEPTS

1. Which cell types function primarily in support?
2. What are some functions of parenchyma?
3. Where in the plant is ground tissue found?
4. How does the structure of dermal tissue contribute to its functions?
5. What are the functions of vascular tissue in plants?

Table 24.3 *Plant Tissue and Cell Types*

Tissue Type	Location	Cell Types	Function
Ground	Forms bulk of interior of stems, leaves, and roots, including cortex, pith, and leaf mesophyll	Parenchyma	Photosynthesis, storage, respiration
		Collenchyma	Elastic support
		Sclerenchyma (fibers and sclereids)	Nonelastic support
Dermal	Surface of roots and stems	Epidermis	Protects primary plant body; controls gas exchange in stems and leaves; absorbs water and minerals in roots
		Periderm	Protects woody roots and stems
Vascular: xylem	Vascular bundles in stems; veins in leaves; vascular bundles or cylinder in roots	Tracheids, vessel elements	Conduct water and minerals in stems, leaves, and roots
		Parenchyma	Storage
		Sclerenchyma (fibers and sclereids)	Support
Vascular: phloem	Vascular bundles in stems; veins in leaves; vascular bundles or cylinder in roots	Sieve tube elements	Transport of organic molecules
		Companion cells (specialized parenchyma)	Transfer carbohydrates to/from sieve tube elements
		Sclerenchyma (fibers and sclereids)	Support

24.4 Tissues Build Stems, Leaves, and Roots

The tissue types described in section 24.3 make up the stems, leaves, and roots of vascular plants. We now return to these organs to examine their structures more closely.

A. Stems Support Leaves

Stems grow and differentiate at their tips, with new cells originating at the apical meristem in the terminal bud (see figure 24.7). The daughter cells eventually give rise to ground tissue, the epidermis, and the primary xylem and phloem.

The stem elongates as the vacuoles of these new cells absorb water, pushing the apical meristem upward. Remnants of the apical meristem remain, however, in the axillary buds that form at stem nodes. These buds may either remain dormant or "awaken" to form side branches. When a shoot loses its terminal bud, hormones described in chapter 26 signal meristems in dormant axillary buds to begin dividing. Gardeners often intentionally pinch off the tips of basil or tomato plants to promote bushy growth. **apical dominance, p. 547**

Inside the stem of a herbaceous plant, loosely packed ground tissue allows for gas exchange and stores water and starch. Vascular tissue is organized into bundles, usually with phloem to the outside and xylem toward the inside (**figure 24.13**). Thick-walled sclerenchyma fibers often strengthen the vascular tissue and prevent animals from tapping the rich sugar supply in the phloem. Most eudicots have a single ring of vascular bundles in the stem. Ground tissue consisting primarily of parenchyma cells occupies most of the rest of the stem: the **cortex** fills the area between the epidermis and vascular tissue, and **pith** occupies the center. Parenchyma cells also occupy the space between the vascular bundles. In most monocots, vascular bundles are scattered throughout the ground tissue of the stem.

B. Leaves Are the Primary Organs of Photosynthesis

Leaves originate as bumps called "leaf primordia" on the flanks of the apical meristem (see figure 24.7). Each leaf primordium develops into a small, dormant leaf bud at a stem node, where vascular bundles from the stem pass into the leaf's petiole. When it is time for the leaf to expand, meristem cells begin dividing. The new cells eventually absorb water, elongate, and specialize to become the cells of the mature leaf.

The leaf epidermis may contain many stomata—more than 11 million in a cabbage leaf, for example. Several adaptations minimize water loss from these pores. In many species, stomata are most abundant on the shaded undersides of horizontal leaves. In pine needles, they are recessed in grooves.

The ground tissue inside the leaves is called **mesophyll,** and it is composed mostly of parenchyma cells

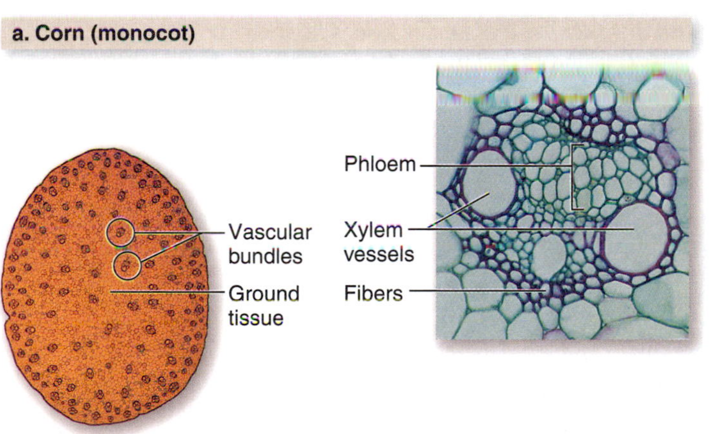

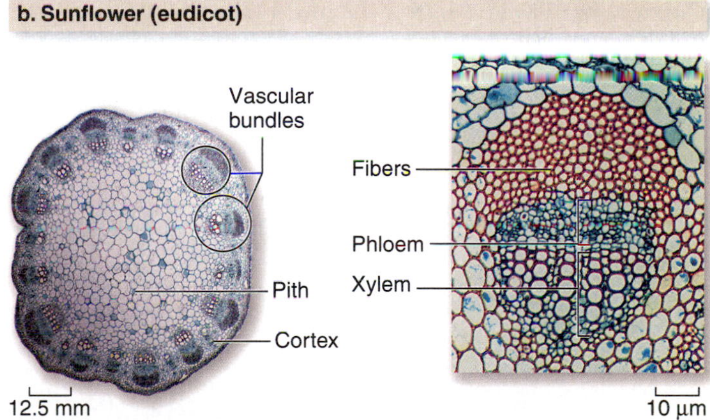

FIGURE 24.13 Anatomy of Primary Stems. Cross sections of (**A**) a monocot stem (corn) and (**B**) a eudicot stem (sunflower). Notice the scattered vascular bundles in the monocot stem and the ring of vascular bundles in the eudicot stem.

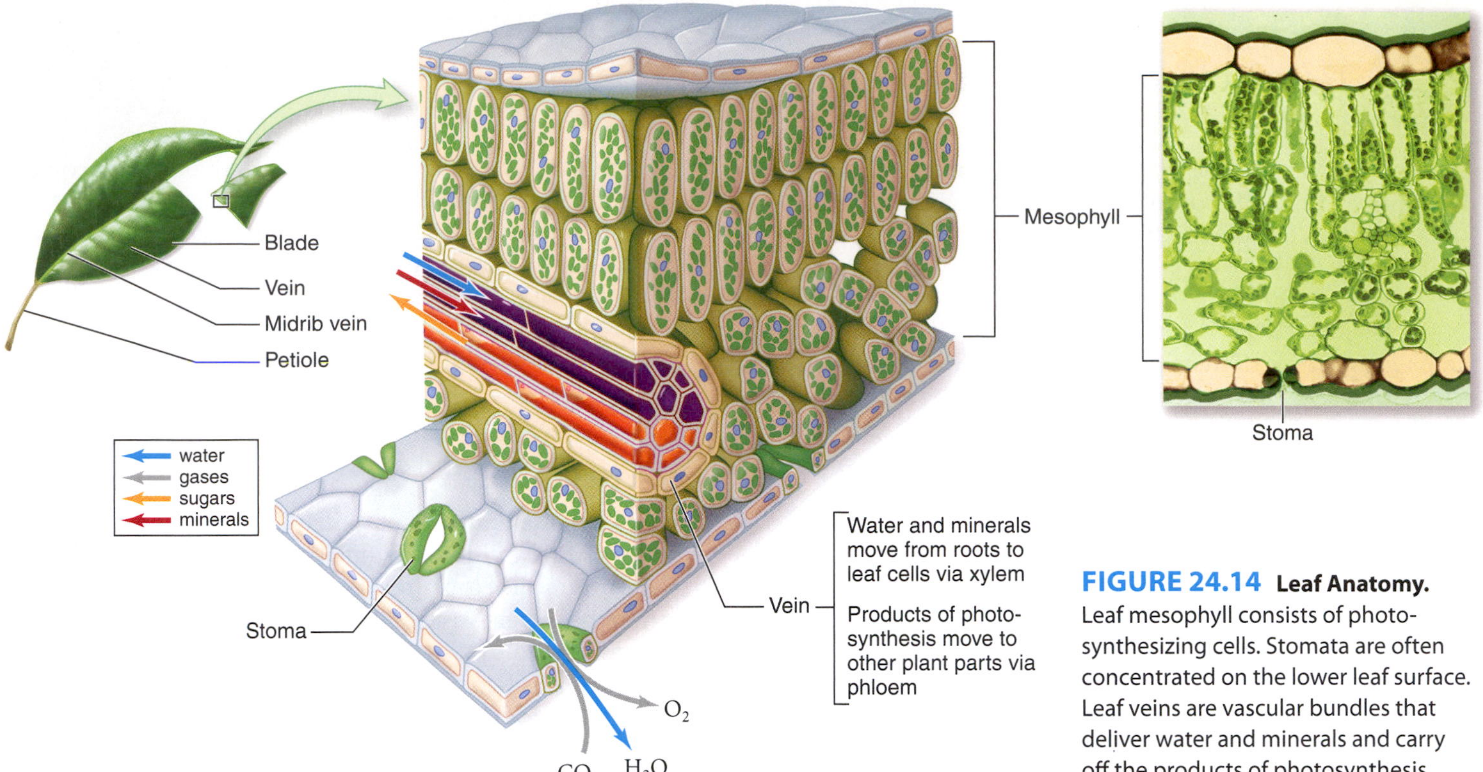

FIGURE 24.14 Leaf Anatomy. Leaf mesophyll consists of photosynthesizing cells. Stomata are often concentrated on the lower leaf surface. Leaf veins are vascular bundles that deliver water and minerals and carry off the products of photosynthesis.

(**figure 24.14**). Most mesophyll cells have abundant chloroplasts and produce sugars by photosynthesis. The long, column-shaped mesophyll cells along the upper side of a leaf maximize light absorption. Below this layer are irregularly shaped mesophyll cells separated by large air spaces in direct contact with the atmosphere (via open stomata).

Embedded in the mesophyll is a prominent network of **veins,** which are vascular bundles inside leaves. A layer of parenchyma cells called a bundle sheath usually surrounds each vein; these cells lend celery "strings" their green color. In both eudicots and monocots, a leaf vein connects to the stem's vascular tissue at a node. The vein emerges into the leaf via the petiole, then branches into intricate networks that occur in two main patterns: netted and parallel (**figure 24.15**). Most eudicots have netted veins, with minor veins branching in all directions from larger, prominent midveins. Many monocots have parallel veins, with several major longitudinal veins connected by smaller minor veins.

Xylem at the ends of the tiniest leaf veins delivers water and ions collected at the roots to nearby mesophyll cells. Meanwhile, sugars produced at mesophyll cells move in the opposite direction, via the phloem, for later delivery to nonphotosynthetic plant parts.

a.

b.

FIGURE 24.15 Leaf Veins in Eudicots and Monocots. (**A**) Leaves of eudicots, such as this pumpkin plant, have netlike venation. (**B**) Leaves of many monocots, such as this lily, have prominent parallel veins interconnected by many tiny veins.

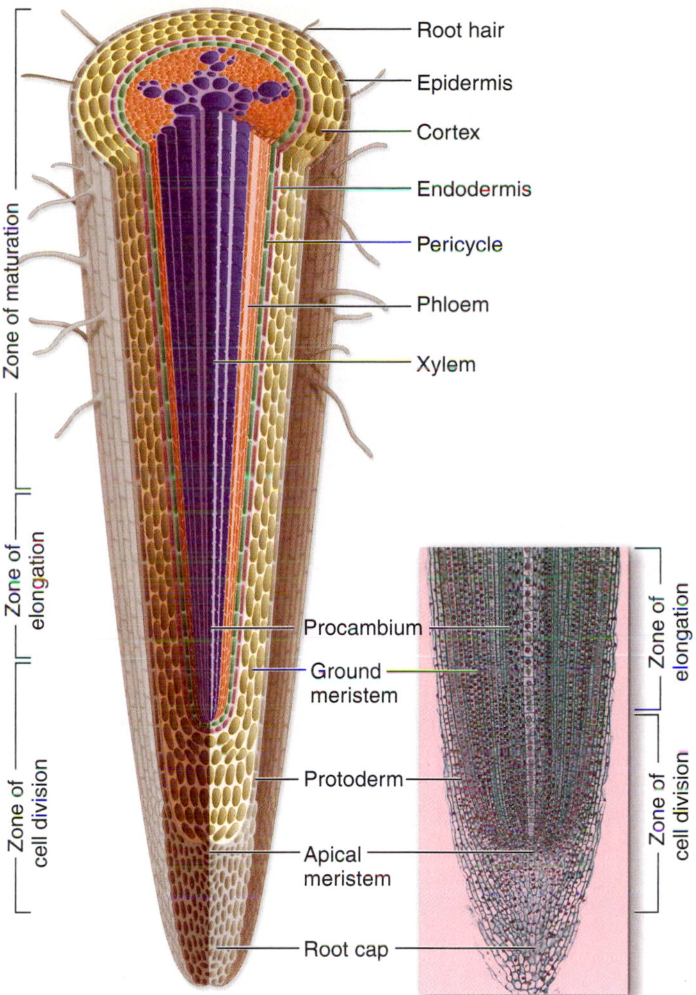

FIGURE 24.16 The Root Apical Meristem. The root apical meristem produces root cap cells toward the root tip and the three primary meristems just above the root cap. The primary meristems produce cells that mature into ground tissue, vascular tissue, and the root epidermis.

C. Roots Absorb Water and Minerals and Anchor the Plant

A root's apical meristem is located just behind the root tip (**figure 24.16**). Toward the tip, cells produced at the apical meristem differentiate into the **root cap,** which protects the meristem from abrasions. Root cap cells, which slough off and are continually replaced as the root grows through the soil, secrete a slimy substance that lubricates the root as it grows. The root cap also plays a role in sensing gravity. **gravitropism, p. 552**

Just above the root cap, the apical meristem produces cells that differentiate into ground meristem, protoderm, and procambium. The root grows as these new cells elongate by absorbing water into their vacuoles, pushing the root farther into the soil. Beyond this zone of elongation is a zone of maturation, in which cells complete their differentiation into ground tissue, the epidermis, and primary xylem and phloem.

The epidermis surrounds the entire primary root except the root cap, and its adaptations maximize the absorption of water and minerals from soil. Root epidermal cells have a very thin cuticle or none at all. In addition, root hairs in the zone of maturation provide extensive surface area for absorption (**figure 24.17**).

Just internal to the epidermis is the cortex, which makes up most of the primary root's bulk. It consists of loosely packed, interconnected parenchyma cells that may store starch or other materials. The spaces between the cells allow for aeration and water movement. The **endodermis** is the innermost cell layer of the cortex. It includes a single layer of tightly packed cells with walls that contain a ribbon of waxy, waterproof material. The waxy deposits form a barrier, called the **Casparian strip,** which ensures that all materials entering the vascular cylinder pass through the cytoplasm of endodermal cells first.

Internal to the endodermis is the **pericycle,** the outermost layer of the root's vascular cylinder. Cells in the pericycle divide to produce lateral branch roots. As each branch root develops, it reserves at its tip a small patch of dividing cells that become its apical meristem. The branch root grows

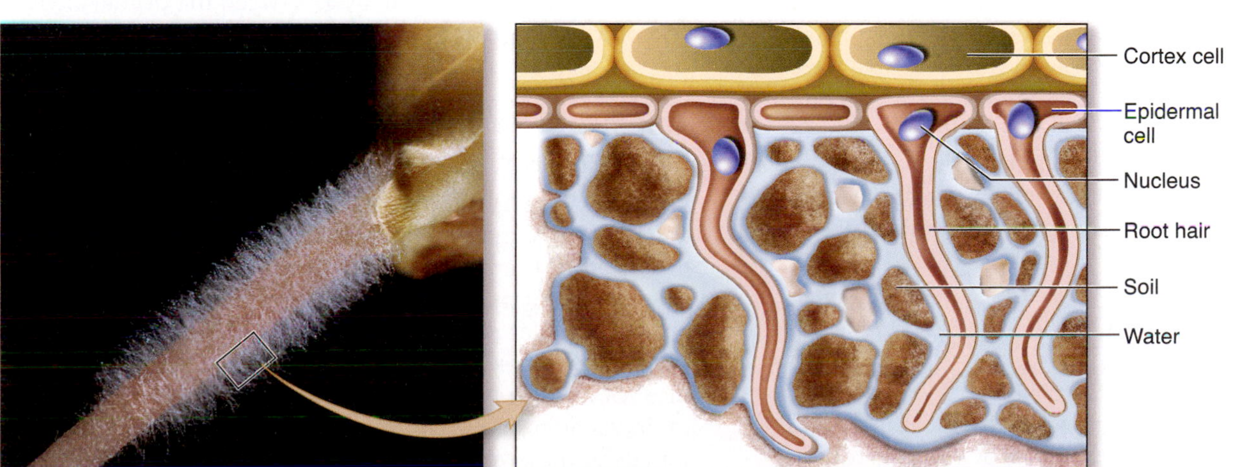

FIGURE 24.17 Root Hairs. These epidermal cell outgrowths extend through the soil, greatly increasing the absorptive surface area of this corn seedling's root. Root hairs give the zone of cell maturation a fuzzy appearance.

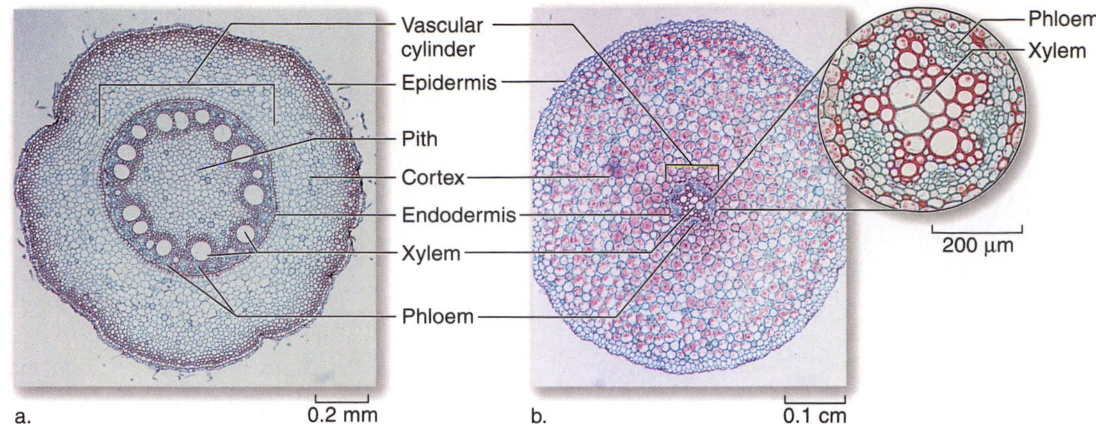

FIGURE 24.18
Anatomy of Primary Roots. Cross sections of (**A**) a monocot root (corn) and (**B**) a eudicot root (buttercup). The inset shows a close-up of the eudicot's vascular cylinder.

as these cells divide, pushing through the cortex and epidermis and into the soil.

The vascular cylinders of eudicot and monocot roots have different arrangements (**figure 24.18**). In most eudicots, the vascular cylinder consists of a solid core of xylem, with ridges that project to the pericycle. Phloem strands are generally located between the "arms" of the xylem core. In most monocot roots, a ring of vascular tissue surrounds a central core (pith) of parenchyma cells.

Summary ***Stem apical meristems lengthen stems and give rise to leaves, just as root apical meristems lengthen roots. Monocots and eudicots differ in the external and internal structures of their stems, leaves, and roots.***

24.4 MASTERING CONCEPTS

1. List the locations of parenchyma cells in stems, leaves, and roots.
2. Name the cell layers that occur in the stem of a eudicot and a monocot, moving from the epidermis to the innermost tissues.
3. How does the arrangement of tissues of a leaf maximize photosynthesis?
4. What are the regions and structures of a root?
5. Compare and contrast the formation of branches in stems and roots.
6. How do the stems, leaves, and roots of monocots differ from those of eudicots?

24.5 Lateral Meristems Produce Wood and Bark

In many habitats, plants compete for sunlight. The tallest plants reach the most light, so selection for height has been a powerful force in the evolutionary history of plants. But primary tissue is not strong enough to support a tall plant. As plants grew taller, therefore, a new selective pressure arose for tissues that could increase the girth of stems and roots—that is, for secondary growth.

This last section turns to wood and bark, tissues that arise from secondary growth. Wood and bark come from two types of lateral meristems: vascular cambium and cork cambium.

A. The Vascular Cambium Produces Xylem and Phloem in Woody Plants

The **vascular cambium** is an internal cylinder of meristem tissue that produces most of the diameter of a woody root or stem. It originates from procambium in woody eudicots and gymnosperms. In the roots and stems of these plants, the vascular cambium is a thin layer between the primary xylem and phloem (**figure 24.19**).

Xylem and Phloem Production in the Vascular Cambium

When a cell in the vascular cambium divides, it produces two daughter cells, one of which remains a meristem cell. Cells produced to the inside of the cambium become secondary xylem; secondary phloem forms on the outer side. Overall, the vascular cambium produces much more secondary xylem than secondary phloem, so the stem or root acquires most of its girth from growth internal to the cambium. Secondary xylem, more commonly known as wood, can accumulate to massive proportions. For example, a giant sequoia

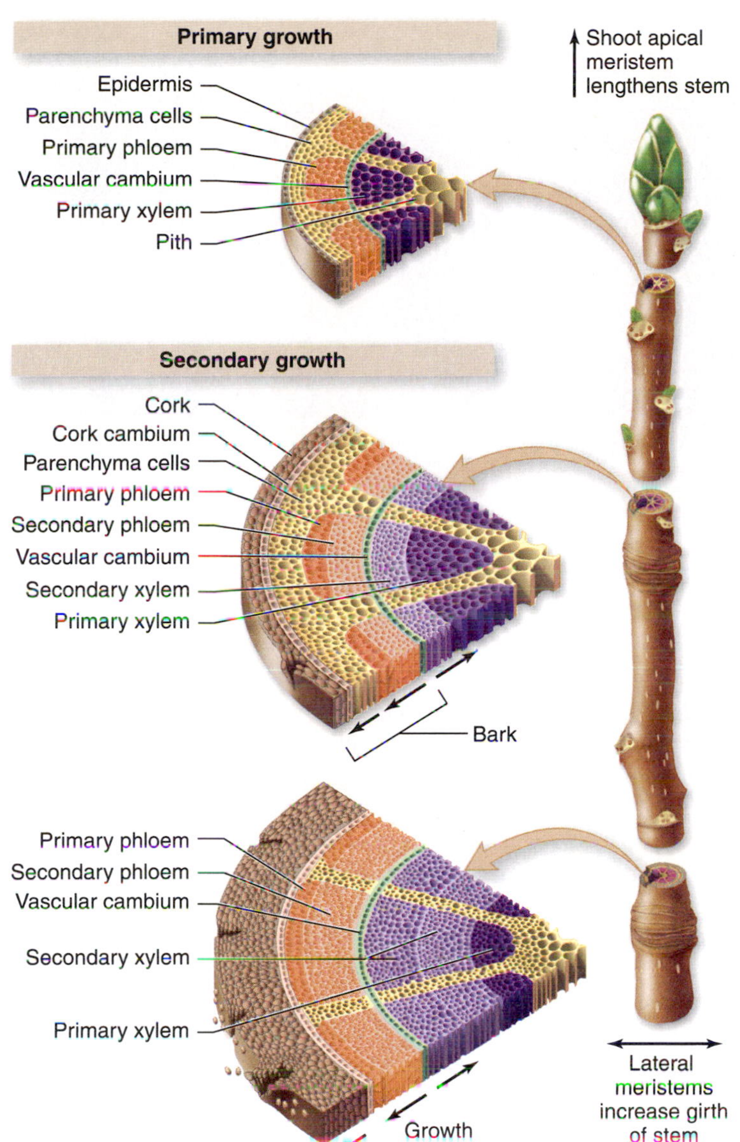

FIGURE 24.19 Secondary Growth Produces Wood. In the top cross section, the microscopic vascular cambium has not yet started producing secondary vascular tissue. The two lower diagrams show the vascular cambium producing secondary xylem toward the inside of the stem and secondary phloem toward the outside. These tissues eventually destroy the primary vascular tissue, cortex, and epidermis. Cork cambium, meanwhile, produces cork cells to the outside and parenchyma to the inside.

(Sequoiadendron giganteum) in California is 100 m (328 feet) tall and more than 7 m (23 feet) in diameter.

The vascular cambium also produces **rays,** bands of parenchyma cells that extend from the center of the stem or root (like spokes on a bicycle wheel). Rays transport water and nutrients laterally within secondary xylem and phloem.

It does not take long before secondary growth destroys the stem or root's primary phloem, cortex, and epidermis.

Can *You* Relate?

From Wood to Paper

Along with food, paper is probably the most prevalent plant product in our lives. We use it to package food, as currency, to communicate, to decorate walls, and to absorb fluids. Currently, people in the United States use almost 181 million kg (200,000 tons) of paper a day.

Manufacturing paper on a commercial scale demands a constant supply of wood pulp from millions of trees—mostly softwoods such as pine, spruce, and hemlock, but also hardwoods such as birch. To make paper, wood is cooked in water, washed, and sometimes bleached. The slurry may be mixed with additives such as dyes, then spread on a fine mesh. As the water drains, the cell walls entangle, adjacent cellulose molecules bond, and paper forms. The material is compressed into a sheet, then dried.

Paper recycling programs keep over a third of discarded newspapers, magazines, catalogs, office paper, and junk mail out of landfills. At paper mills, this "postconsumer" waste paper is shredded and placed into huge tanks. Solvents and detergents are added to remove inks, and paper clips and other unwanted objects are removed. The material is then reassembled into new paper. Using recycled paper requires 60% to 70% less energy and about half as much water as using virgin wood pulp. It also produces less water pollution, because recycled pulp requires less bleaching than does wood pulp.

Unfortunately, recycling cannot supply all of the world's paper needs. Pulp deteriorates during recycling, limiting the number of possible uses and re-uses. De-inked pulp from newsprint, for example, is too degraded for use in high-quality printer paper. Also, contamination problems restrict the use of recycled fibers in certain types of food packaging.

But secondary phloem produced to the outside of the vascular cambium takes over responsibility for phloem sap transport. Secondary tissue forms the live, innermost layer of **bark,** a collective term for all tissues to the outside of the vascular cambium (see figure 24.19). The outer layer of bark includes the periderm, the subject of the next section.

Durable, Useful Wood

Few plant products are as versatile or economically important as wood. Lumber forms the internal frame that supports many buildings. Firewood provides heat and cooking fuel. Most paper comes from wood, as this chapter's Can *You* Relate? From Wood to Paper describes. Throughout history, humans have fashioned wood from various tree species into furniture, pencils, cabinets, boats, baseball bats, serving bowls, roofing shingles, jewelry, picture frames, and countless other items.

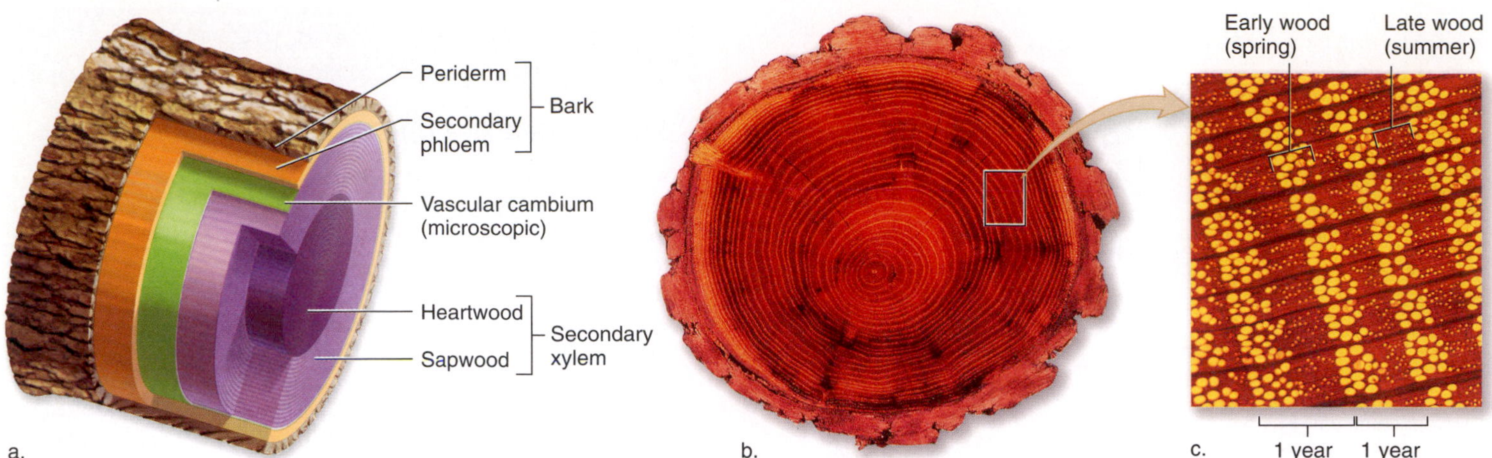

FIGURE 24.20 Anatomy of a Secondary Stem. (**A**) Wood is secondary xylem, and bark is all the tissue outside the vascular cambium. At the center of the stem, the darker colored heartwood is nonfunctioning secondary xylem. (**B**) Differences in available soil moisture when wood forms in spring and summer result in different-sized cells, which are visible as growth rings (**C**).

Wood varies in hardness. Angiosperms such as oak, maple, and ash are often called hardwood trees, whereas gymnosperms such as pine, spruce, and fir are called softwood trees. Sclerenchyma fibers account for the difference. Angiosperm wood contains fibers in addition to tracheids and vessels, whereas wood from softwood trees consists mainly of tracheids. As a result, eudicot wood is usually stronger and denser than wood from gymnosperms. The soft, light wood from the balsa tree, an angiosperm, is a notable exception.

A quick glance at a cross section of a tree trunk reveals that the innermost wood is darker than the outer portion. This color difference arises as the tree ages. As the years pass, the oldest secondary xylem gradually becomes unable to conduct water. This darker colored, nonfunctioning region is called **heartwood** (**figure 24.20**A). The lighter colored **sapwood,** located nearest the vascular cambium, transports water and dissolved minerals.

Another feature of a trunk's cross section is growth rings. In temperate climates, cells in the vascular cambium are dormant in winter, but they divide to produce wood during the spring and summer. Seasonal differences in xylem cell size yield annual growth rings (figure 24.20B,C). This chapter's Burning Question: What causes tree rings? explains how these rings form and how biologists use them to deduce tree age and past climate.

B. The Cork Cambium Produces the Outer Layer of a Woody Stem or Root

Periderm is the protective layer of dermal tissue that covers a woody stem or root. The periderm replaces the epidermis, which is destroyed as the plant's girth increases. It originates from the **cork cambium,** a lateral meristem that gives rise to cork to the outside and parenchyma to the inside (see figure 24.19). Periderm thus consists of three layers: a living parenchyma layer, the cork cambium, and the nonliving cork.

Cork consists of layers of densely packed, waxy cells on the surfaces of mature stems and roots. The cells are dead at maturity and form waterproof, insulating layers that protect plants. In trees and shrubs with smooth bark, the cork is more or less continuous. Breaks in the cork cambium produce scaly or rough bark.

The cork used to stopper wine bottles comes from cork oak trees that grow in the Mediterranean region. Every 10 years, harvesters remove much of the cork cambium and cork, which grows back. Cork is also important in the history of biology; in 1665, Robert Hooke became the first person to see cells when he used a primitive microscope to gaze at cork.

Summary *Wood supports the stems, branches, and roots of many plants, and bark protects their outer surfaces. Wood is secondary xylem, and it transports water and minerals within the plant. Bark consists of secondary phloem and the periderm. These secondary tissues arise from two lateral meristems, the vascular and cork cambia.*

24.5 MASTERING CONCEPTS

1. What type of tissue occupies most of the volume of a woody stem or root?
2. How is the vascular cambium similar to the cork cambium, and how are these meristems different?
3. How does the structure of bark contribute to its function?

Burning Questions

What causes tree rings?

Tree rings, also called growth rings, result from seasonal differences in the size of cells produced in wood. During the moist days of spring, the vascular cambium produces wood made of large, water-conducting cells. During the drier days of summer, new wood has smaller cells. The contrast between the summer wood of one year and the spring wood of the next highlights each annual growth ring (see figure 24.20). By counting these rings, a forester can estimate a tree's age.

Growth rings also provide historical clues. A fire leaves behind a charred "burn scar." A thick growth ring indicates plentiful rainfall and good growing conditions. Stress from herbivory or disease, or fierce competition for light or water, can all lead to narrow growth rings. These patterns provide clues about altitude, temperature, and length of the growing season in times past.

Researchers also use tree ring data to look into ancient history. They can align distinctive patterns in tree rings from living trees that are thousands of years old with growth rings in dead trees or timbers and in charcoal recovered from archaeological sites. For example, the oldest known living tree is a bristlecone pine (*Pinus longaeva*) growing in the White Mountains of California. It is more than 4760 years old. Combining its tree ring information with information from older, dead trees in the area has provided rainfall data going back 8200 years!

Have a Burning Question of your own? Submit it to marielle_hoefnagels@mcgraw-hill.com for possible inclusion in future editions of this book!

24.6 INVESTIGATING LIFE
Ancient Plants Tell Tales of How They Came to Be

Hundreds of millions of years before dinosaurs tromped on Earth's surface, there were plants. How do we know? First, all animals eat, and plants must have formed the base of the food chain for the many land animals that preceded dinosaurs. Second, an amazing collection of fossils tells us so.

Plant fossils have formed worldwide, but paleontologists (fossil experts) have found some of the most spectacular fossils in a formation called the Rhynie Chert in northeast Scotland (**figure 24.21A**). Dense stands of now-extinct plants lived during the Devonian period, some 400 MYA, when Scotland was not in its current northerly position. Instead, it was

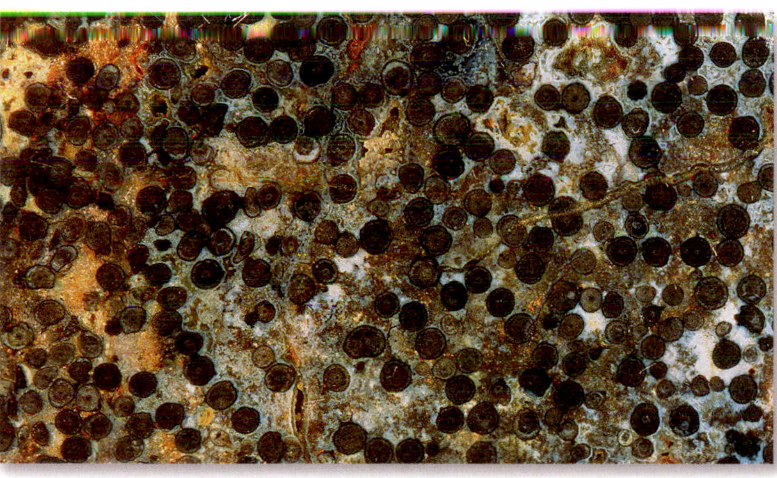
a.

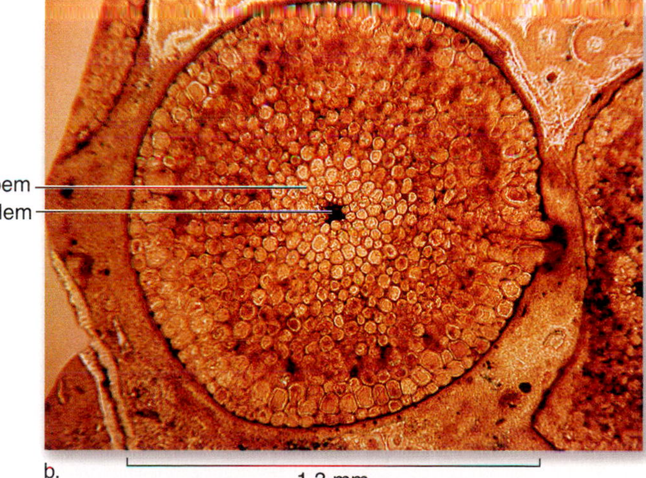

b.

FIGURE 24.21 Fossil Plants from the Rhynie Chert. (**A**) Dense patches of *Rhynia* grew hundreds of millions of years ago in what is now Scotland. This piece of chert shows dozens of fossilized stems in cross section. (**B**) A magnified cross section of one stem clearly shows the resemblance between *Rhynia* and modern plants. **Question:** What type of evidence would you look for to test the hypothesis that the *Rhynia* fossils were buried where the plants lived (as opposed to being transported from another location and then buried)?

about 28 degrees south of the equator and joined with other landmasses as part of a much larger continent.

At that time, near what is now the town of Rhynie, geysers spurted hot, silica-rich water that periodically bathed nearby plants, animals, and microorganisms. When the water cooled and evaporated on a plant, silica surrounded and filled individual cells but left cell walls intact. Over millions of years, additional silica buried the preserved fossils, and the surrounding sediments turned to a type of rock called chert.

A geologist named William Mackie first discovered the Rhynie Chert in 1912. When he saw that the rocks contained fossil plant stems, he alerted paleontologists. Fossil collection began in earnest later that year and continues even today. So far, paleontologists have described seven species of extinct land plants from Rhynie Chert.

The fossils at Rhynie are so amazingly well preserved that biologists can see individual cells of the ancient plants. Microscopic examination reveals cuticles, guard cells, air spaces between parenchyma cells, tracheids, sporangia (reproductive structures in early vascular plants), spores, and even mycorrhizal fungi.

The plant most commonly found at Rhynie is appropriately called *Rhynia*. Its branched stems were up to 3 mm in diameter and 20 cm tall. Cross sections clearly show a central core of vascular tissue surrounded by a cortex, epidermis, and cuticle (figure 24.21B). Another fossil plant is *Aglaophyton*. At about 15 cm tall, it was a bit shorter than *Rhynia*, but its stems were about twice as thick. In some fossils of *Aglaophyton*, beautifully preserved strands of mycorrhizal fungi snake between the plant cells. In others, kidney-shaped guard cells surround a stoma.

The fossils at Rhynie are extraordinary not only for their anatomical detail; they also showcase ancient life cycles. Vast numbers of plants are "caught" in every reproductive stage from spore to gametophyte to sporophyte (see figure 20.2). The Rhynie Chert also preserves entire communities of interacting plants, microbes, and arthropods, revealing intimate details of an ecosystem that thrived hundreds of millions of years ago. **alternation of generations, p. 411**

Plants from the Rhynie Chert clearly had xylem. But what about phloem? Phloem cells are often not well preserved in fossil plants, because their cell walls are more delicate than are those of xylem. In the 1970s, biologists Donna Satterthwait and J. William Schopf of the University of California, Los Angeles, used a light microscope to examine very thin sections of *Rhynia*. They focused on the "phloem zone," where phloem should occur, based on its position in a modern plant. Satterthwait and Schopf identified two interesting cell types: parenchyma cells of unknown function, and elongated cells with thin, pitted walls. Based on the shape and position of the cells, the researchers concluded that the tissue was indeed phloem. Both phloem and xylem had therefore evolved by the Devonian.

Many other questions remain. How did vascular tissue evolve? What about seeds, or the apical meristem, or modular growth? How were the first land plants related to the plants with which we now share Earth? The Rhynie fossils and many others from around the world provide important clues.

Satterthwait, Donna F., and J. William Schopf. 1972. Structurally preserved phloem zone tissue in *Rhynia*. *American Journal of Botany*, vol. 59, pp. 373–376.

CHAPTER SUMMARY

24.1 Plant Parts Depend on One Another

A. Vegetative Plant Parts Include Stems, Leaves, and Roots

- The **vegetative** plant body consists of a **shoot** and **roots** that depend on each other.
- The **stem** is the central axis of a shoot and consists of **nodes,** where **leaves** attach, and **internodes** between leaves, where the stem elongates. An **axillary bud** is located at each node.
- Leaves are the main sites of photosynthesis. A stalklike **petiole** supports each leaf **blade.** A **simple leaf** has one undivided blade, and a **compound leaf** forms leaflets.
- Roots absorb water and dissolved minerals. **Taproot systems** have a large, persistent major root, whereas **fibrous root systems** are shallow, branched, and shorter lived. **Adventitious roots** arise from stems or leaves.

B. Plant Parts Can Be Highly Specialized

- Stem modifications include stolons, rhizomes, tubers, succulent stems, thorns, and tendrils.
- Leaf modifications include tendrils, storage leaves, cotyledons, spines, and insect-trapping leaves.
- Some specialized roots store starch or water, absorb oxygen from air, or provide support. Mycorrhizal fungi

help roots take up soil nutrients; nitrogen-fixing bacteria in root nodules provide some plants with nitrogen.

24.2 Plants Have Flexible Growth Patterns, Thanks to Meristems

A. Plants Grow By Adding New Modules

- Plants with **determinate growth** stop growing when they reach a certain size. **Herbaceous plants** often have determinate growth. Plants with **indeterminate growth** grow indefinitely, a pattern that is most common in **woody plants.**

B. Plant Growth Occurs at Meristems

- **Meristems** are localized collections of cells that retain the ability to divide throughout the life of the plant. **Apical meristems** at the plant's root and shoot tips provide **primary growth. Lateral meristems** are cylinders of cells at the periphery of a stem or root that add girth, or **secondary growth.** In some plants, **intercalary meristems** occur between nodes.

24.3 Plant Cells Build Tissues

A. Plants Have Several Cell Types

- **Parenchyma** cells are alive at maturity. They are relatively unspecialized and have thin cell walls. They often function in metabolism or storage.
- **Collenchyma** cells are also alive. Their thick primary cell walls provide elastic support to growing shoots.
- **Sclerenchyma** cells, including long **fibers** and shorter **sclereids,** are dead at maturity. Their thick secondary cell walls support plant parts that are no longer growing.
- Water-conducting cells in **xylem** include long, narrow, less specialized **tracheids** and more specialized, barrel-shaped **vessel elements.** Both cell types have thick walls and are dead when functioning. Water moves through pits in tracheids but through the end walls of vessel elements.
- Sucrose-conducting cells in **phloem** include **sieve tube elements.** Pores cluster at **sieve plates,** allowing nutrient transport between adjacent cells via cytoplasmic strands. **Companion cells** help transfer carbohydrates.

B. Plant Cells Form Three Main Tissue Types

- The tissues of a flowering plant are ground tissue, dermal tissue, and vascular tissue (phloem and xylem). Most of the primary plant body consists of **ground tissue,** relatively unspecialized parenchyma cells that fill the space between dermal and vascular tissues.
- **Dermal tissue** includes the **epidermis,** a single cell layer covering the plant. The epidermis secretes a waxy **cuticle** that coats aerial plant parts. Gas and water exchange in the shoot occur through **stomata** bounded by **guard cells.**
- **Vascular tissue** is conducting tissue. Xylem transports water and dissolved minerals from roots upward. Phloem is a living tissue that transports dissolved carbohydrates and other substances throughout a plant. Xylem and phloem occur together with other tissues to form **vascular bundles.**

24.4 Tissues Build Stems, Leaves, and Roots

A. Stems Support Leaves

- The apical meristem at the shoot apex yields protoderm, ground meristem, and procambium (the primary meristems), which produce cells that become dermal tissue, vascular tissue, and ground tissue in stems and leaves.
- Vascular bundles are scattered in the ground tissue of monocot stems but form a ring of bundles in eudicot stems. Between a eudicot stem's epidermis and vascular tissue lies the **cortex,** made of ground tissue. **Pith** is ground tissue in the center of a stem.

B. Leaves Are the Primary Organs of Photosynthesis

- Leaves arise at the flanks of the apical meristem. Leaf epidermis is tightly packed, transparent, and mostly nonphotosynthetic. Leaf ground tissue includes **mesophyll** cells that carry out photosynthesis. Stomata enable gas exchange. **Veins** are vascular bundles in leaves; they may be in either netted or parallel formation.

C. Roots Absorb Water and Minerals and Anchor the Plant

- The root apical meristem produces cells that differentiate into epidermis, cortex, and vascular tissues. A **root cap** protects the tip of a growing root. The root cortex consists of storage parenchyma and **endodermis.** The waxy **Casparian strip** ensures that the solution entering the root passes through the cells of the endodermis. The **pericycle** gives rise to branch roots. In monocots, pith fills the center of the root.

24.5 Lateral Meristems Produce Wood and Bark

A. The Vascular Cambium Produces Xylem and Phloem in Woody Plants

- The **vascular cambium** increases the girth of the stem or root. It produces secondary xylem (wood), secondary phloem, and **rays.**

- **Heartwood** is the central, nonfunctioning wood in a tree. The light-colored **sapwood** transports water and minerals within a tree.

B. The Cork Cambium Produces the Outer Layer of a Woody Stem or Root

- The **cork cambium** produces the **periderm,** which makes up the majority of a woody plant's **bark.**

24.6 Investigating Life: Ancient Plants Tell Tales of How They Came to Be

- Plant fossils in Scotland's Rhynie Chert show that plants already had stomata, guard cells, parenchyma, vascular tissue, gametophytes, and sporophytes 400 MYA.

MULTIPLE CHOICE QUESTIONS

1. Which of the following is NOT a vegetative organ in a plant?
   a. The stems
   b. The leaves
   c. The flowers
   d. The roots

2. A carrot is an example of a(n)
   a. fibrous root.
   b. taproot.
   c. adventitious root.
   d. root nodule.

3. The ability of a plant like a sunflower to grow very tall is most likely due to its
   a. apical meristems.
   b. lateral meristems.
   c. intercalary meristems.
   d. both a and c.

4. Which of the following is a living cell type that physically supports the growing plant?
   a. Parenchyma
   b. Collenchyma
   c. Sclerenchyma
   d. Both b and c

5. The sugars that make up sap travel through ______ while water and dissolved minerals travel through ______.
   a. the cytoplasm of xylem cells; sieve cells.
   b. tracheids and vessels; companion cells.
   c. the cytoplasm of the phloem sieve tube system; companion cells.
   d. the cytoplasm of the phloem sieve tube system; tracheids and vessels.

6. Root hairs are an example of what type of plant tissue?
   a. Ground tissue
   b. Dermal tissue
   c. Vascular tissue
   d. All of the above

7. The apical meristem of a plant is capable of producing
   a. epidermal cells.
   b. mesophyll cells.
   c. xylem cells.
   d. all the cells of the plant.

8. What is the function of the Casparian strip?
   a. It allows the cells of the endodermis to function as a filter for the plant.
   b. It is made up of meristematic cells that produce the vascular tissue in a plant.
   c. It blocks the movement of water out of the vascular bundle.
   d. It is a layer of cells that minimizes the loss of water from the leaves of a plant.

9. What type of tissue occupies most of the volume of a woody stem or root?
   a. Secondary xylem
   b. Secondary phloem
   c. Bark
   d. Vascular cambium

10. Where does gas exchange occur along the surface of a woody stem?
    a. At the stomates
    b. At the nodes
    c. At the lenticels
    d. At the periderm

TESTING YOUR KNOWLEDGE

1. List the main vegetative organs of a plant, and explain how each relies on the others.
2. Describe how leaves and roots maximize surface area.
3. How could you determine whether a plant's growth is determinate or indeterminate?
4. Which parts of a plant provide the potential for continuous growth?
5. Biologists often say that "form follows function;" that is, the form of a biological structure facilitates its function. List a function of each organ, tissue, and cell type described in this chapter, and then list at least one feature that facilitates that function.
6. How are tracheids and vessel elements similar to sieve elements, and how are they different?
7. Which plant tissue is the most abundant in an herbaceous plant? In a large woody plant?
8. Cite a stem specialization and a leaf specialization that provide protection.
9. Which plant structures are adaptations that conserve water? Which plant structures are adaptations that transport water?
10. Corn is a monocot and sunflower is a eudicot. How do these plants differ in stem structure, leaf venation, and root organization?
11. Suppose you drive a metal spike from the outermost bark layer to the center of a tree's trunk. Which tissues does your spike encounter as it moves through the stem, and what type of meristem produced each type?

THINKING AS A SCIENTIST

1. Wood pulp used to make paper may include fibers, tracheids, and vessel elements. Does a slurry that contains tracheids and just a few fibers come from a hardwood or softwood tree?
2. Which plant tissues have cells that are dead at maturity? Why is it advantageous to the plant for these cells to die?
3. Many biology labs use preserved slides of root tips to demonstrate the stages of mitosis. Why is this a better choice than using a slide of a mature leaf?
4. Girdling is cutting away or severing the living bark in a ring around a tree's trunk. Which part of a girdled tree do you expect to die first, the roots or the shoot?
5. When you brush past the leaves or stems of a stinging nettle (*Urtica* species), you break off the tips of hollow leaf hairs that act as miniature hypodermic needles, injecting a painful cocktail of toxins and irritants into the skin. What benefit does the plant derive from its investment in these chemicals?
6. What problems might you encounter in trying to develop a foliar herbicide intended to be absorbed through leaves?
7. Thorns, spines, and tendrils are so highly modified that it can be difficult to tell whether they derive from leaves or stems. How could a biologist use knowledge of internal plant anatomy to determine the origin of these structures?

Visit www.mhhe.com/hoefnagels for practice quizzes, animations, videos, and activities designed to help you master the material in this chapter.

CHAPTER 25

Plant Nutrition and Transport

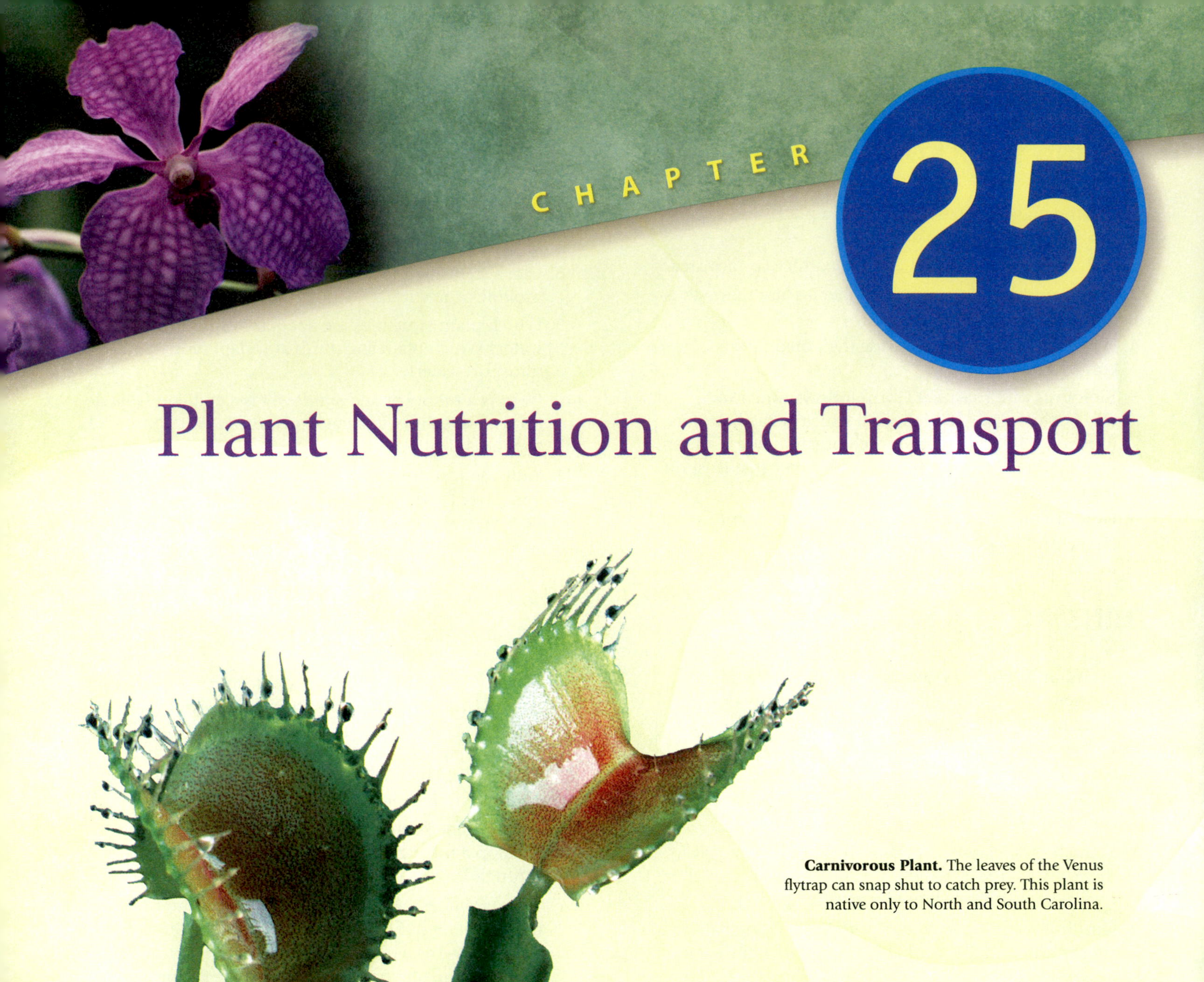

Carnivorous Plant. The leaves of the Venus flytrap can snap shut to catch prey. This plant is native only to North and South Carolina.

What's the Point?

Carnivorous Plants

Little Shop of Horrors *tells the story of a small, innocent-looking plant that grows to enormous size and develops a taste for humans. Fortunately, such bloodthirsty plants exist only in science fiction ... right?*

Although it is true that no plants eat people, some do consume small animals, especially insects and other arthropods. The following examples illustrate just a few of the world's 450 or so species of carnivorous plants:

- *The Venus flytrap (***Dionaea muscipula***) has two-sided leaves, each of which has highly sensitive trigger hairs. When an insect wanders onto a leaf and bends the trigger hairs, the two halves snap shut. The leaf's epidermis secretes enzymes that digest the captured animal. These plants, which are native only to North and South Carolina, are endangered in the wild because of habitat loss. Collecting them is illegal. (Venus flytraps sold commercially are propagated asexually in greenhouses.)*
- *The sundew plant (genus* **Drosera***) has paddle-shaped leaves with tiny nectar-covered leaf hairs that lure insects. Once the insect alights on the plant and starts dining on its sweet, sticky meal, the surrounding leaf hairs begin to move. Gradually they fold inward, entrapping the helpless visitor and forcing it down toward the leaf's center. Here, powerful digestive enzymes dismantle the insect's body and release its component nutrients. Afterward, all that remains of the guest are a few indigestible bits.*
- *The pitcher plant (genus* **Sarracenia***) entices insects to explore a large, slippery, tube-shaped leaf that collects rainwater. When a hapless insect falls in, digestive enzymes go to work. By summer's end, pitcher plants contain many leftover insect parts.*

It may seem strange that a plant would consume insects. Plants are, after all, photosynthetic, so they can make their own food. The answer to this puzzle is that carnivorous plants use insect protein as a source of nitrogen, not energy. Carnivorous plants are most abundant in boggy, acidic habitats, and their leafy traps are adaptations that allow them to extract scarce nitrogen from their prey instead of from the soil. As will become clear in this chapter, most plants have less exotic ways of acquiring nutrients.

LEARNING OUTLINE

25.1 Soil and Air Provide Water and Nutrients

To stay healthy, a person needs water and the right dietary mix of fat, protein, carbohydrates, vitamins, and minerals (see section 34.4). Plants have similar needs, but they do not acquire these raw materials by eating and drinking. Instead, they are autotrophs that assemble their own organic molecules from water and elements that they absorb from their environment.

This section describes the sources of the elements that plants require. The next section focuses on how plants absorb and transport water and minerals within their tissues. Finally, section 25.3 turns to the transportation of organic chemicals, notably sucrose, within plants.

A. Plants Require 16 Essential Elements

Like every organism, a plant requires certain **essential nutrients,** which are chemicals that are vital for metabolism, growth, and reproduction. Biologists have identified at least 16 elements essential to all plants (**figure 25.1**). Nine of these are **macronutrients,** meaning that they are needed in fairly large amounts. The macronutrients are carbon (C), hydrogen (H), oxygen (O), phosphorus (P), potassium (K), nitrogen (N), sulfur (S), calcium (Ca), and magnesium (Mg). The others are **micronutrients,** which are required in much smaller amounts.

Among the essential elements, C, H, and O are by far the most abundant, together accounting for about 96% of the dry weight of a plant. The six other macronutrients account for another 3.5%. Of these, N, P, and K are the most common ingredients in commercial fertilizers (see the Can *You* Relate? Fertilizers Boost Plant Growth). Gardeners and farmers use fertilizers to prevent or treat nutrient deficiencies such as those shown in **figure 25.2.**

B. Soils Have Distinct Layers

Roots extract water and mineral nutrients from **soil,** which consists of small particles of rocks and clay minerals mixed with decaying organisms and organic molecules. Air and water occupy the spaces between soil particles. Besides plant roots, soil is also home to bacteria, fungi, protists, and animals.

Soil formation begins when rocks disintegrate into soil particles on the earth's surface. The texture of a soil depends on the size of these particles. Sandy soils have coarse particles; silty soils have finer ones. Clay particles are the smallest. The finer the particles, the more surface area per unit of soil volume, and the higher the soil's water-holding capacity. Soil scientists use the relative amounts of sand, silt, and clay when classifying soils.

Carbon, oxygen, and hydrogen (96% of dry weight)
Other macronutrients (~3.5%)
Micronutrients (~0.5%)

Macronutrients	Form taken up by plants	Percent dry weight	Selected functions
Carbon (C)	CO_2	45	Part of organic compounds
Oxygen (O)	H_2O, O_2	45	Part of organic compounds
Hydrogen (H)	H_2O	6	Part of organic compounds
Nitrogen (N)	NO_3^-, NH_4^+	1.5	Part of nucleic acids, amino acids, coenzymes, chlorophyll, ATP
Potassium (K)	K^+	1.0	Controls opening and closing of stomata, activates enzymes
Calcium (Ca)	Ca^{2+}	0.5	Cell wall component, activates enzymes, second messenger in signal transduction, maintains membranes
Magnesium (Mg)	Mg^{2+}	0.2	Part of chlorophyll, activates enzymes, participates in protein synthesis
Phosphorus (P)	$H_2PO_4^-$, HPO_4^{2-}	0.2	Part of nucleic acids, sugar phosphates, ATP, coenzymes, phospholipids
Sulfur (S)	SO_4^{2-}	0.1	Part of cysteine and methionine (amino acids), coenzyme A
Micronutrients	**Form taken up by plants**	**Percent dry weight**	**Selected functions**
Chlorine (Cl)	Cl^-	0.01	Water balance
Iron (Fe)	Fe^{3+}, Fe^{2+}	0.01	Chlorophyll synthesis, cofactor for enzymes, part of electron carriers
Boron (B)	BO_3^-, $B_4O_7^{2-}$	0.002	Growth of pollen tubes, sugar transport, regulates certain enzymes
Zinc (Zn)	Zn^{2+}	0.002	Hormone synthesis, activates enzymes, stabilizes ribosomes
Manganese (Mn)	Mn^{2+}	0.005	Activates enzymes, electron transfer, photosynthesis
Copper (Cu)	Cu^{2+}	0.0006	Part of plastid pigments, lignin synthesis, activates enzymes
Molybdenum (Mo)	MoO_4^{2-}	0.00001	Nitrate reduction

FIGURE 25.1 Essential Nutrients for Plants. The nine most abundant elements in plants are called macronutrients. The micronutrients occur in much lower concentrations but are also essential for plant survival. The table lists the nutrients present in all plants; some plants require additional micronutrients not listed here.

FIGURE 25.2 Nutrient Deficiencies Produce Distinctive Symptoms. (**A**) Phosphorus deficiency causes dark green or purple leaves in seedlings. (**B**) Iron deficiency causes yellowed leaves, but the veins remain green. A deficiency of manganese produces similar symptoms.

a.

b.

As soils form, they develop distinct layers (**figure 25.3**). A layer of dead, decomposing leaves and stems, called litter, lies on the soil's surface. Microorganisms release most of the carbon in decaying litter to the atmosphere as carbon dioxide (CO_2). Some carbon, however, remains in the upper layer of soil in the form of **humus,** a chemically complex, hard-to-digest, spongy organic substance. Most humus is in the upper soil layer, the **topsoil,** also called the A horizon. Below topsoil is the B horizon, which has less organic matter, although roots extend to this depth. Still lower is the C horizon, which consists almost entirely of partially weathered pieces of rocks and minerals. It is an interface between the bedrock below and the soil above.

Topsoil is a major source of water and nutrients for most plants growing in the soil; without it, little plant growth is possible. Plant roots normally stabilize the topsoil and resist **erosion,** the "wearing away" of soil by water and wind. Many human activities, including construction, agriculture, logging, and road-building, remove plants and promote erosion. The soil washes away, choking streams and lakes with sediment. The loss of the nutrient-rich topsoil makes it difficult for plants to regrow.

FIGURE 25.3 Soil Horizons. Soil consists of layers called horizons, designated A through C. Above the rich topsoil is a litter layer. Water passing through the topsoil deposits clay and minerals in the next layer, and below this is a layer of partially weathered rock.

C. Leaves and Roots Absorb Essential Elements

Plants obtain their three most abundant elements (C, H, and O) from water and the atmosphere. Water (H_2O) enters the plant through the roots, as described in the next section. Carbon and oxygen atoms come from the atmosphere in the form of CO_2 gas, which enters the leaf or stem by diffusing in through pores called stomata. The plant's cells use CO_2 and H_2O to produce glucose by photosynthesis. **diffusion, p. 90**

Roots take up all other elements from soil. Many nutrients, including P, Ca, and Mg, become available to plants because the minerals in rocks dissolve in water as they weather in the soil. Others are released when microbes decompose dead organisms.

The source of nitrogen deserves special mention. Plants require large amounts of nitrogen to manufacture amino acids, proteins, nucleic acids, and chlorophyll. Even though nitrogen gas (N_2) makes up 78% of the atmosphere, N availability often limits plant growth. A triple covalent bond holds together the two nitrogen atoms, and this bond requires more energy to break than a plant can muster. N_2 is therefore unavailable to plants. Instead, roots take up nitrogen from soil in the form of nitrate (NO_3^-) or ammonium ions (NH_4^+).

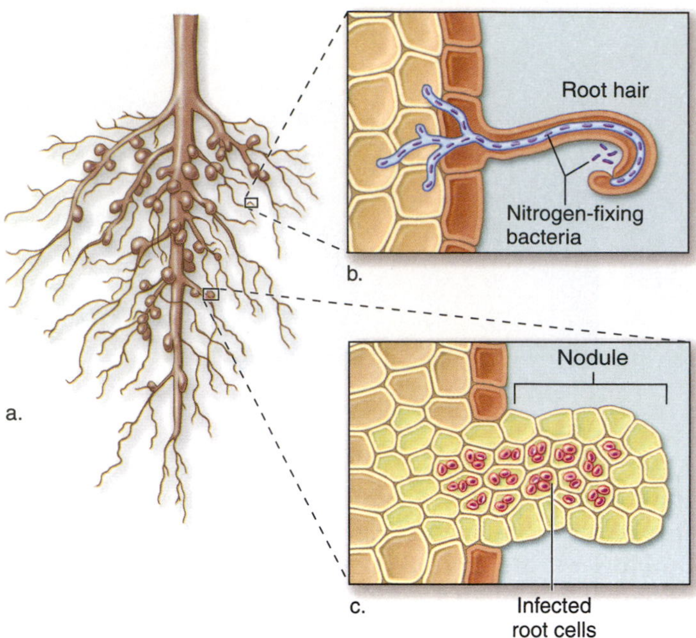

FIGURE 25.4 Bacteria Provide Usable Nitrogen to Plants. (**A**) In response to signals from a bean plant or other legume, *Rhizobium* triggers the development of root nodules. (**B**) The bacteria enter through root hairs and (**C**) live symbiotically within the plant's cells. The plant provides the energy the bacteria need to break the triple covalent bond in N_2.

Fortunately for plants (and ultimately animals), several types of bacteria use **nitrogen-fixing** enzymes to convert N_2 to more accessible forms. Some nitrogen-fixing bacteria live in growths called **nodules** on the roots of some types of plants (**figures 25.4** and 18.10). Most famously, bacteria in genus *Rhizobium* stimulate nodule formation on the roots of plants called legumes (beans, peas, peanuts, soybeans, and alfalfa). Powered by sugar that the plant produces by photosynthesis, *Rhizobium* reduces atmospheric N_2 to NH_4^+.

The plant, in turn, incorporates the nitrogen atoms from NH_4^+ into its own tissues. When the plant's tissues die, decomposers make the nitrogen available to other organisms. Biological nitrogen fixation therefore jump-starts the nitrogen cycle, bringing otherwise inaccessible nitrogen to other organisms (see figure 40.15 for a detailed look at the nitrogen cycle).

A few plants acquire nitrogen from more unusual sources. For example, carnivorous plants, described at the beginning of the chapter, often live in waterlogged, acidic soils. They obtain nitrogen from the insects that they consume.

Summary ***All plants require at least 16 essential elements. Plants obtain carbon and oxygen atoms from the atmosphere; other elements, including additional oxygen, come from water or soil. Bacteria convert atmospheric nitrogen to forms that plants can take up and use.***

25.1 MASTERING CONCEPTS

1. What are 16 essential macronutrients and micronutrients for plants?
2. Which of the layers in a typical soil horizon are richest in organic matter?
3. How do nitrogen-fixing bacteria form a critical link in the nitrogen cycle?

Can *You* Relate?

Fertilizers Boost Plant Growth

To boost soil productivity, farmers and gardeners may apply nutrient-rich organic matter, such as manure or compost, or a commercial synthetic fertilizer. Plant growth may surge if the soil amendment provides a nutrient that was previously scarce.

Commercial fertilizer labels prominently display three numbers that indicate the content of nitrogen, phosphorus, and potassium, the three nutrients most commonly deficient in soils (**figure 25.A**). For example, a product marked 10-20-10 contains 10% elemental nitrogen (N), 20% phosphate (P_2O_5), and 10% potash (K_2O) by weight; the rest is filler. The label also lists other macro- and micronutrients in the fertilizer.

Nitrogen promotes leaf growth, and potassium and phosphorus encourage flowering and fruit development. A typical synthetic lawn fertilizer might therefore have an N-P-K ratio of 29-3-4. A fertilizer formulated for vegetables, on the other hand, would be more balanced at 10-20-20. By comparison, household compost is much less concentrated at about 0.5-0.5-0.5.

Chemically, nutrients from inorganic fertilizer are equivalent to those from manure or compost. So why should a gardener bother with organic matter when commercial fertilizers are so much more potent? The answer is that organic matter not only improves fertility, but it also aerates the soil, increases the soil's water-holding capacity, and provides food for beneficial soil-dwelling microbes and animals. Organic matter also releases its nutrients more slowly than chemical fertilizers, resulting in less pollution from fertilizer runoff into storm drains, streams, lakes, and other waterways.

6-12-6 LIQUID PLANT FOOD PLUS
Contains Plant Food Supplements PLUS Humate Humic Acid and Medina Soil Activator
GUARANTEED ANALYSIS
Total Nitrogen (N) 6.00%
Available Phosphoric Acid (P2O5) 12.00%
Soluble Potash (K2O) 6.00%
Copper (Cu) .... .02%
.02% Chelated Copper
Iron (Fe) .... .05%
.05% Chelated Iron
Manganese (Mn) .... .05%
.05% Chelated Manganese
Molybdenum (Mo) .... .0005%
Zinc (Zn) .... .05%
.05% Chelated Zinc
Humic Acid 1.00%
Derived from Urea, Phosphoric Acid, Potassium Hydroxide, Aqua Ammonia, EDTA Copper, EDTA Iron, EDTA Manganese, Sodium Molybdate, EDTA Zinc and Leonardite Ore.
ONE QUART (.95L) NET WEIGHT 2.5 LBS. (1.13Kg)

FIGURE 25.A The Label Says It All. Commercial fertilizer labels show which nutrients the product contains, and in what quantities.

25.2 Water and Dissolved Minerals Are Pulled Up to Leaves

The amount of water that passes through plants each day is staggering. The corn plants on 0.4 hectares (1 acre) of land, for example, might use 15,000 L (4000 gallons) in one summer day, and a typical tree might consume 265 L (70 gallons) daily. Throughout a growing season, a plant might use 200 to 1000 L of water to produce just 1 kilogram of tissue.

Why so much water? A plant's cells are mostly water, which is a medium for most of its metabolic reactions. Water participates in some of those reactions, including hydrolysis and photosynthesis. Furthermore, turgor pressure that the watery cytoplasm exerts on the cell wall enables plant cells to elongate and helps plants stay upright. Finally, the surfaces of leaf mesophyll cells must remain moist for CO_2 to diffuse inside. These uses, however, add up to only a small fraction of the water that a plant's roots pull in. As you will see in this section, the rest simply evaporates. **hydrolysis, p. 34; photosynthesis, p. 106**

As you learned in chapter 24, xylem and phloem together form a continuous system of microscopic pipes that extend from the tips of roots to the ends of the tiniest veins in leaves, connecting the plant's roots, stems, leaves, flowers, and fruits (**figure 25.5**). Phloem distributes the products of photosynthesis, as described in section 25.3. Xylem transports **xylem sap,** a dilute solution consisting of water and dissolved minerals absorbed from soil.

A. Water Vapor Is Lost from Leaves Through Transpiration

The easiest way to visualize how plants acquire and transport water is to begin at the end. Plants lose water through **transpiration,** the evaporation of water from a leaf. Heat from the sun and from the plant's metabolism causes water

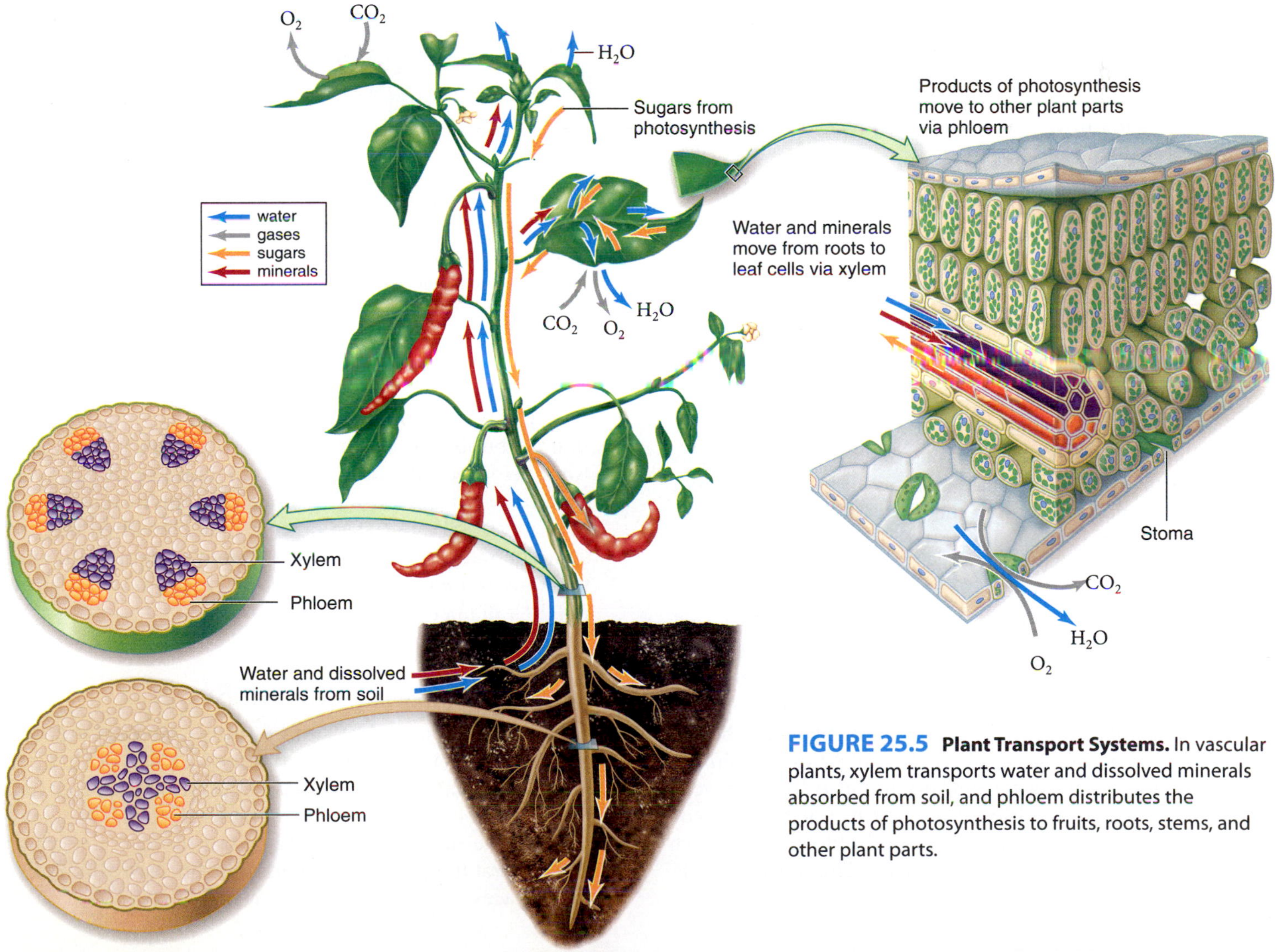

FIGURE 25.5 Plant Transport Systems. In vascular plants, xylem transports water and dissolved minerals absorbed from soil, and phloem distributes the products of photosynthesis to fruits, roots, stems, and other plant parts.

in the cell walls to evaporate into the spaces between the leaf's cells. This evaporation helps cool the leaf, but it also establishes a gradient. That is, the concentration of water molecules in the intercellular spaces is higher than in the air surrounding the leaf. Water vapor diffuses down the gradient, from inside the leaf to the outside air. Most transpiration occurs through a leaf's open stomata, although a small amount escapes through the cuticle.

Any environmental factor that increases the evaporation rate of water will also speed transpiration. Low humidity, for example, increases transpiration rates by increasing the gradient between the leaf's interior and the surrounding air. Wind and high temperatures do the same, although stomata close when the weather becomes too hot or the air too dry.

If sufficient moisture is available, water molecules that evaporate from the leaf are immediately replaced by water molecules that are drawn up through the stem. If water lost in transpiration is not replaced, however, the cells will quickly lose turgor, and the leaf will wilt. This explains why cut flowers begin to droop if they are not kept in water.

B. Xylem Transport Relies on Cohesion and Adhesion

The functional cells of the xylem, tracheids and vessel elements, are only the remaining walls of cells that died when they reached maturity (see figure 24.9). Because these cells are dead, the metabolic activity of the plant is *not* what drives the movement of water. How then, does water in the xylem get from roots to the mesophyll in the leaves?

The **cohesion–tension theory** explains how water moves within a plant. As its name implies, the cohesion–tension theory hinges on the cohesive properties of water—the tendency for water molecules to "cling" together. As water molecules diffuse out of a leaf vein and into the mesophyll, they attract water molecules adjacent to them in the xylem, pulling them under tension (negative pressure) toward the leaf. Each water molecule tugs on the one behind it, eventually pulling water in the roots up through the xylem. In this way, the water in xylem forms a continuous hydraulic system that rises in columns through the plant body (**figure 25.6**). Notice that this mechanism of water movement exploits the physical properties of water, so the plant does not spend energy hauling water from soil to the tips of its leaves.

How does a plant counter the powerful gravitational forces that resist the upward movement of water? The answer lies in another property of water, adhesion. Recall from chapter 2 that adhesion is a molecule's attraction to another type of substance. In this case, water forms hydrogen bonds with the walls of the xylem tubes. This attraction, combined with cohesion, draws water upward and also helps prevent the formation of air bubbles in the xylem. Together, the cohesive and adhesive properties of water hold a narrow water column up against the force of gravity. This movement is called capillary action.

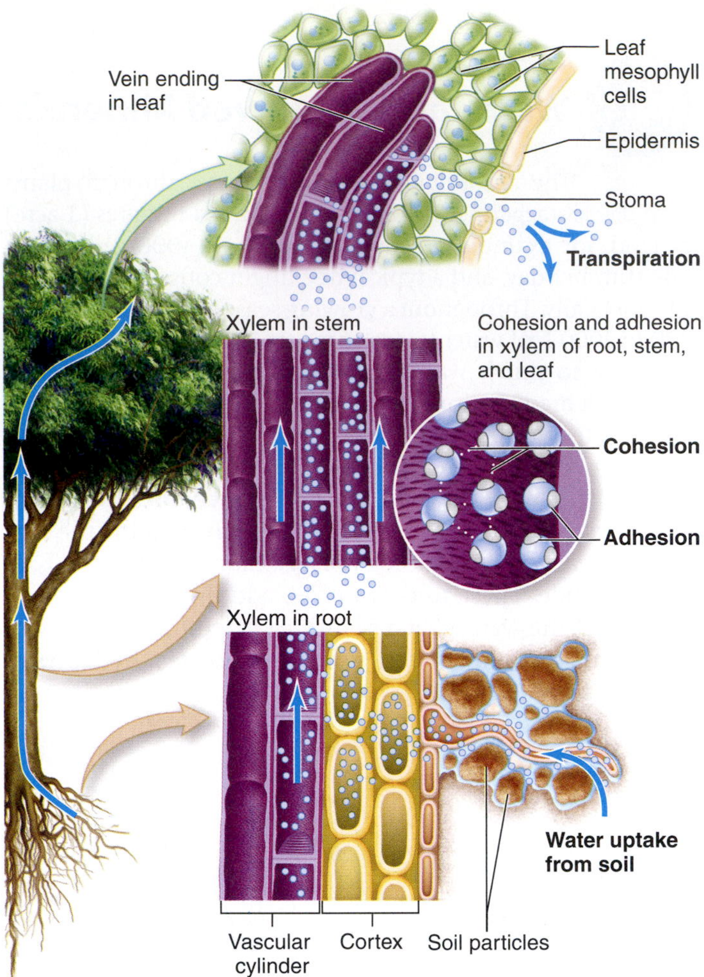

FIGURE 25.6 Water and Dissolved Minerals Are Pulled to Treetops. Transpiration of water from leaves creates a force that pulls water up a plant's body from the roots. The cohesiveness and adhesiveness of water make this possible.

Experimental evidence supports the cohesion–tension theory of water movement in plants. For example, we know that air bubbles can interrupt upward water flow. If water were being pushed from below, air bubbles in the xylem stream would not pose a problem. In addition, scientists have measured a decrease in stem diameter when the transpiration rate is highest. Water adheres to xylem walls; when the water is under tension, it pulls the walls inward, narrowing their diameter.

C. Water Enters Plants in Roots

As evaporation from leaf surfaces pulls water up the stem, additional water enters roots from the soil. Just behind each growing root tip, the epidermis is fringed with many root hairs (see figure 24.17). The plant's millions of root hairs greatly increase the surface area for water and mineral absorption. Mycorrhizal fungi associated with roots increase this absorptive surface even more (see figure 21.15).

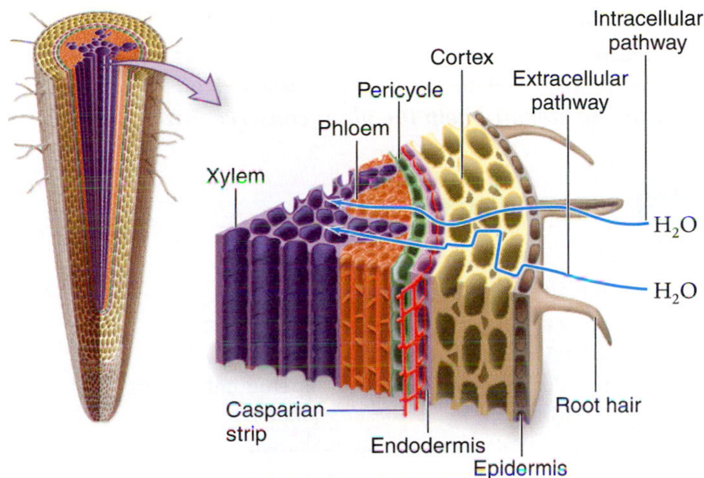

FIGURE 25.7 Two Routes into Roots. Water and dissolved minerals travel through a root's epidermis and cortex in intercellular spaces, along cell walls, or through living cytoplasm. The waxy Casparian strip ensures that all of this incoming material passes through the living cells of the endodermis to reach the xylem.

The concentration of solutes is generally lower in the soil than in the root cells, so water enters roots by osmosis. Water and the dissolved minerals it carries can move through the root's epidermis and cortex in two ways (**figure 25.7**). In the extracellular pathway, water and its solutes move in the spaces between cells or along the cell walls. In the intracellular pathway, the solution moves from cell to cell via plasmodesmata (see figure 3.27). **osmosis, p. 91**

Eventually, the incoming water and minerals contact the endodermis, the innermost layer of the cortex. The endodermis's impermeable **Casparian strip** forces the water that had gone around cells to now enter them. Transport proteins that form channels in the membranes of the endodermal cells admit only certain ions. The water and dissolved minerals that cross the endodermis continue into the xylem, enter the transpiration stream, and move up the plant. The water eventually returns to the atmosphere through the open stomata in the leaves and stem. Figure 25.6 summarizes the process.

D. The Cuticle and Stomata Help Conserve Water

The cuticle is an important water-saving adaptation in land plants. This waxy layer covers the epidermis of a plant's leaves and primary stem (see figure 24.12). Impermeable to water and gases, the cuticle prevents the plant's tissues from drying out.

If the cuticle formed a continuous, impermeable barrier, however, CO_2 could not enter leaves. Instead, pores called stomata permeate the cuticle and permit the leaf to exchange gases with the atmosphere. As long as the soil contains enough moisture to replace water moving out of the leaf through the stomata, the plant will flourish. However, if the soil dries out, the plant will wilt and die unless it closes its stomata and temporarily shuts down the transpiration stream.

The guard cells bordering each stoma determine whether a stoma is open or closed (**figure 25.8**). When water is plentiful, membrane proteins in the guard cells pump potassium ions (K^+) from adjacent cells in the epidermis. The concentration of K^+ therefore becomes greater inside than outside the guard cells, and water enters the guard cells by osmosis. The incoming water swells the guard cells, which opens the stoma. During drought stress, a plant hormone called abscisic acid binds to the guard cell membranes, activating pathways that ultimately trigger the loss of K^+ from the guard cells. Water follows, the guard cells lose turgor and collapse against one another, and the pore between them closes. The transpiration stream stops. **abscisic acid, p. 547**

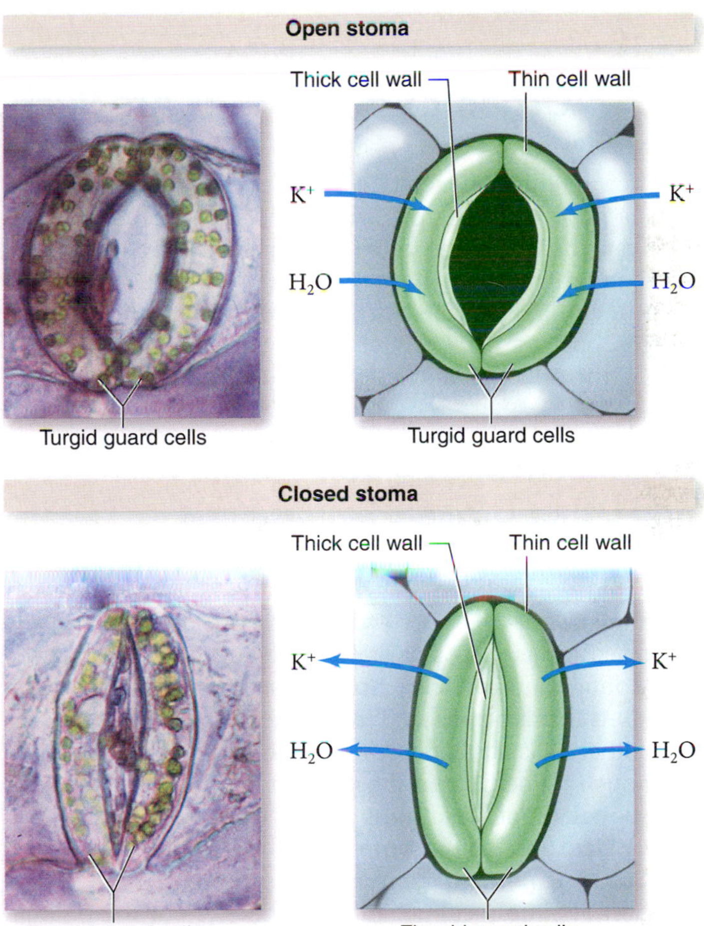

FIGURE 25.8 Opening and Closing of Stomata. Relative concentrations of potassium ions (K^+) in guard cells and adjacent cells determine whether the stoma is open or closed. When water is abundant, guard cells use energy to import K^+ from adjacent cells. Water follows by osmosis, and the guard cells swell, opening the stoma. When water is scarce, hormonal signals stimulate K^+ to leave the guard cells. Water follows, and the stoma closes as the flaccid guard cells collapse.

When a plant closes its stomata, it conserves water. But it also eliminates its source of CO_2, which enters leaves only through open stomata. Most plants prevent nighttime transpiration by closing their stomata after dark, when photosynthesis cannot occur. Cacti and some other drought-tolerant plants, however, close their stomata during the day and open them during the cooler, more humid night. In a photosynthetic pathway called CAM, CO_2 enters the plant at night, and the plant uses a special metabolic pathway to store it until light returns. During the day, the stomata close, and the plant uses its stored CO_2 in photosynthesis. **CAM plants, p. 113**

Summary *Water travels from the roots, up the stems, and to the leaves of plants. The cohesion–tension theory says that water molecules used in metabolism or exiting through leaf stomata in transpiration pull up water from farther down in the plant. This system continually lifts water and dissolved nutrients absorbed in the roots to the top of the plant. The cuticle and stomata help the plant conserve water.*

25.2 MASTERING CONCEPTS

1. Which tissues make up plant transport systems?
2. Describe how transpiration occurs.
3. How do the properties of water and xylem cells enable xylem to pull up water from roots?
4. How do water and dissolved minerals enter roots and move through the endodermis to the xylem?
5. How do roots maximize surface area?
6. How do the cuticle and stomata help plants conserve water?

25.3 Organic Compounds Are Pushed to Nonphotosynthetic Cells

With sufficient light, water, and essential nutrients, a photosynthetic cell will produce a surplus of sugars that can then be transported in phloem to the nonphotosynthetic cells of the plant. The major transport structures of phloem, as described in the preceding chapter, are the microscopic sieve tubes (see figure 24.10). Unlike cells of xylem, the cells that make up sieve tubes are alive.

A. Phloem Sap Contains Sugars and Other Organic Compounds

The organic compounds carried in phloem are dissolved in the **phloem sap,** a solution that also includes water and minerals from the xylem. The carbohydrates in phloem sap are mostly dissolved sugars such as sucrose (see figure 2.16). Phloem sap also contains amino acids, hormones, enzymes, and messenger RNAs that are translated into proteins in apical meristems and other cells distant from the site of their transcription. (Although phloem sap is the most common vehicle for sugar transport, it is not the only one, as this chapter's Burning Question: Where does maple syrup come from? explains.)

Studying phloem composition is difficult. An investigator cannot simply cut a stem and squeeze out the phloem sap because the plant quickly plugs the wound. Biologists receive help, however, from a surprising source: aphids. These insects feed on phloem sap without triggering the wound response, and they excrete a sticky, sweet material called "honeydew" (**figure 25.9**). Scientists found that if they amputated the feeding tube of an aphid while it was dining on a plant, sugary drops continued to flow from the cut end of the tube. In this way, phloem sap could be harvested and analyzed.

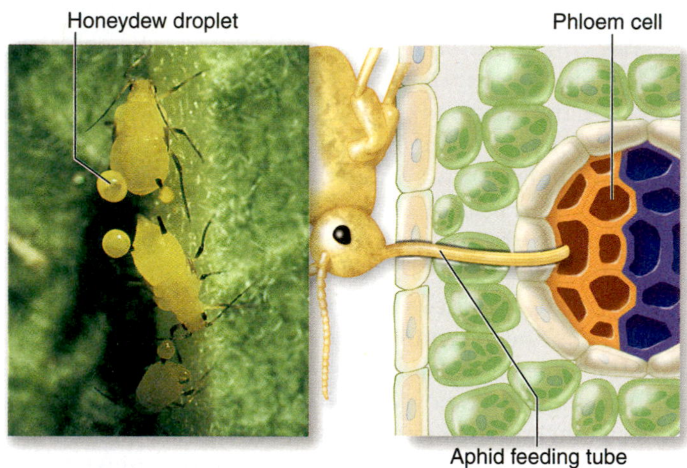

FIGURE 25.9 Aphids Reveal Sieve Tube Pressure. (**A**) Biologists learned about the composition of phloem sap with the help of aphids. These insects excrete drops of "honeydew," which is derived from phloem sap. (**B**) The insects' feeding tubes penetrate phloem cells in the host plant.

B. The Pressure Flow Theory Explains Phloem Function

The explanation of phloem transport is called the **pressure flow theory.** This theory suggests that phloem sap moves under positive pressure from "sources" to "sinks." A **source** is any plant part that produces or releases sugars; a **sink** is any plant part that does not photosynthesize. Examples of sinks include flowers, fruits, shoot apical meristems, roots, and storage organs. If these cells do not receive enough sugar to generate the ATP they require, the plant may die or fail to reproduce. **meristems, p. 503**

Inside a leaf or other sugar source, companion cells load sucrose into sieve tube elements by active transport. Because sucrose becomes so much more concentrated in the sieve tubes than in the adjacent xylem, water moves by osmosis out of the xylem and into the phloem sap. The resulting increased turgor pressure drives phloem sap through the sieve tubes (**figure 25.10**). **active transport, p. 93**

At a root, flower, fruit, or other sink, cells take up the sucrose (and other compounds in the phloem sap) through facilitated diffusion or active transport. As the sucrose is unloaded from the sieve tubes, the concentration of solutes in the phloem sap becomes more and more dilute. Water therefore moves by osmosis from the sieve tube to the surrounding tissue (often xylem). Movement of water out of the sieve tube relieves the pressure, and the pressurized phloem sap in the sieve tube continues to flow toward the sink.

A given organ may act as either a sink or a source. For example, a developing potato tuber is a sink, storing the plant's sugars in the form of starch. Later, when the plant uses those stored reserves to fuel the growth of new tissues, that same tuber becomes a source. The starch in the tuber breaks down into simple sugars, which are loaded into phloem sap for transport to other plant parts. A similar mobilization occurs each spring when a deciduous tree produces new stems and leaves, using carbohydrates stored in roots. Later in the growing season, leaves approach their mature size and produce sugar of their own. The leaves are then sources, and the roots are again sinks.

Evidence for the pressure flow theory includes the observation that phloem sap moves through aphids under positive pressure, a little like toothpaste being squeezed from a tube. Microscopes that produce images of living tissue "in action" have also confirmed some details of phloem function.

Summary *Phloem sap contains sugars produced in photosynthesis, along with other organic substances. Incoming fluid from the xylem pushes phloem sap from photosynthetic sources to sinks throughout the plant.*

25.3 MASTERING CONCEPTS

1. What are some examples of sources and sinks in a plant?
2. How do researchers know that phloem sap is under pressure?
3. What are the roles of osmosis and active transport in phloem sap movement from source to sink?

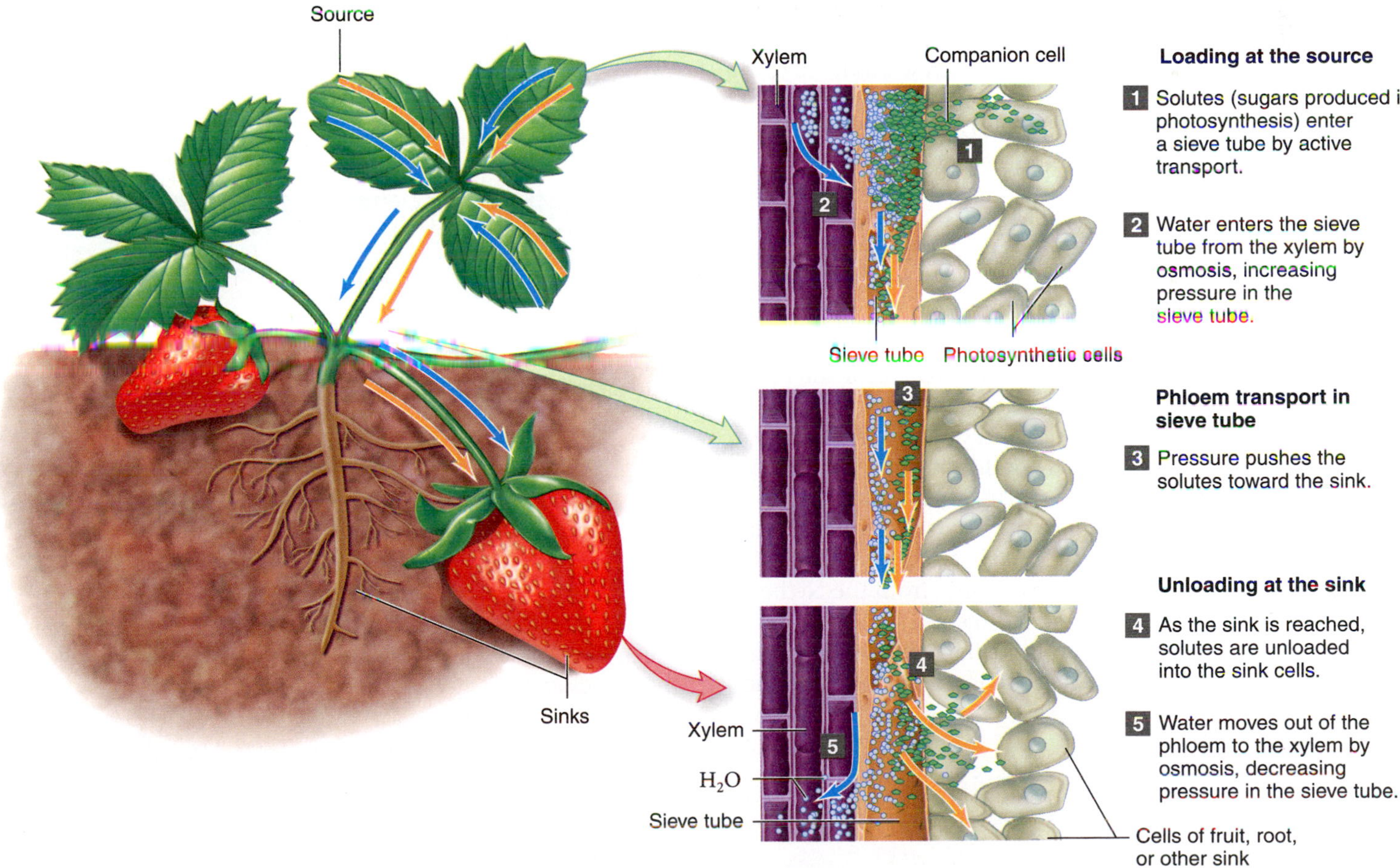
Source
Sinks
Xylem
Companion cell
Sieve tube
Photosynthetic cells
Loading at the source
1 Solutes (sugars produced in photosynthesis) enter a sieve tube by active transport.
2 Water enters the sieve tube from the xylem by osmosis, increasing pressure in the sieve tube.
Phloem transport in sieve tube
3 Pressure pushes the solutes toward the sink.
Unloading at the sink
4 As the sink is reached, solutes are unloaded into the sink cells.
5 Water moves out of the phloem to the xylem by osmosis, decreasing pressure in the sieve tube.
Xylem
H_2O
Sieve tube
Cells of fruit, root, or other sink

FIGURE 25.10 Organic Compounds Move Under Pressure from a Source to a Sink. Sugar produced in green "source" organs such as leaves moves via phloem to roots, fruits, and other nonphotosynthetic "sinks." According to the pressure flow theory, phloem sap is pushed from sources to sinks.

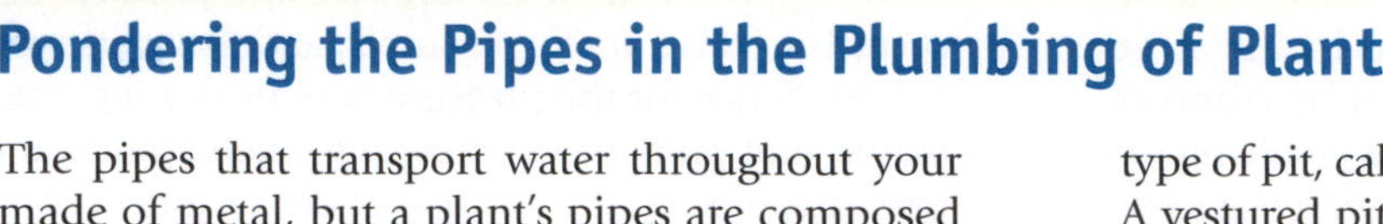

25.4 INVESTIGATING LIFE

Pondering the Pipes in the Plumbing of Plants

The pipes that transport water throughout your home are made of metal, but a plant's pipes are composed of organic materials such as cellulose and lignin. In woody plants, this "plumbing" can accumulate to massive proportions, supplying water to the enormous leaf canopy of the largest trees. As you learned in chapter 24, wood from angiosperms (flowering plants) is typically more durable than wood from conifers. The difference originates in their microscopic components, such as tracheids, vessel elements, and fibers. But how did these anatomical differences come about? Evolutionary biologists would like to know.

Some aspects of xylem evolution seem clear. Fossil evidence shows that plants already had vascular tissue at least 400 MYA. *Rhynia* and the other plants collected from the Rhynie Chert (see section 24.6) clearly had xylem. The first xylem cells were early tracheids (see figure 24.9). These tapered, narrow cells with overlapping end walls originally had two important traits: thick walls made them strong, and pits and hollow centers enabled water to travel from cell to cell. But as plants grew larger over evolutionary time, some tracheids developed stronger walls and lost their pits. These modified cells became the sclerenchyma cells we now know as xylem fibers (see figure 24.8). So the two functions of xylem, conduction and support, became separated into tracheids and fibers.

Were the vessel elements in the xylem of flowering plants also derived from early tracheids? No one knows for sure. In one study, scientists Sherwin Carlquist and Edward Schneider of the Santa Barbara Botanic Garden in California used electron microscopy to examine pits, end walls, and overall shapes of tracheids and vessel elements from many species. They found that some plants have water-conducting cells that seem intermediate between tracheids and vessel elements. They concluded that an "either/or" view of tracheids and vessel elements is too simplistic; the two types of water-conducting cells are closely related, but the question of vessel origin remains open.

The fine details of xylem cells can also help scientists learn more about the selective forces that act on plants. Belgian biologist Steven Jansen and colleagues from the Netherlands and United Kingdom reviewed the scientific literature on xylem vessels in plants from different habitats, from the tundra to the tropics. They tallied the incidence of a specific type of pit, called a vestured pit, in thousands of plant species. A vestured pit has an unusually narrow opening, which may protect against the movement of air bubbles into the xylem. If so, the scientists reasoned, vestured pits should be more common in plants from drought-prone areas. Their study revealed that this was indeed the case: plants native to deserts and tropical seasonal woodlands, both seasonally dry areas, were most likely to have vestured pits (**table 25.1**).

Just as detailed knowledge of animal anatomy can help biologists learn more about evolution in our own kingdom (see section 23.9), the careful study of plant anatomy can reveal many details of plant history. So the next time you admire the beauty of a gleaming hardwood floor, remember what you are really looking at: the remnants of an amazing set of organic pipes whose microscopic parts have a rich evolutionary history.

Carlquist, Sherwin, and Edward L. Schneider. 2002. The tracheid-vessel element transition in angiosperms involves multiple independent features: cladistic consequences. *American Journal of Botany,* vol. 89, pp. 185–195.

Jansen, Steven, Pieter Baas, Peter Gasson, et al. June 8, 2004. Variation in xylem structure from tropics to tundra: Evidence from vestured pits. *Proceedings of the National Academy of Sciences,* vol. 101, no. 23, pp. 8833–8837.

Table 25.1 *Climate Predicts Vestured Pit Frequency*

Climatic Category	Frequency of Plant Species with Vestured Pits (%)
Tropical woodlands, seasonal	51
Deserts	48
Tropical lowlands, rainforest	31
Subtropical warm temperate	31
Tropical (sub)montane	22
Cool temperate	22
Boreal-arctic	3

Question: Besides vestured pits, can you predict other adaptations that might prevent the formation of air bubbles in the xylem of desert plants?

Burning Questions

Where does maple syrup come from?

The source of maple syrup is the xylem of the sugar maple tree (*Acer saccharum*). During the fall, the tree stores carbohydrates in the living cells of its stem and roots. As the temperature warms in early spring, the stockpiled sugars fuel the production of new leaves.

Most sap flow in sugar maples occurs before noon on warm days that follow a freezing night in winter or spring. This flow was once a bit of a puzzle, especially since the cohesion–tension theory that explains most water movement in the xylem does not apply. (During the sap flow period, sugar maples lack leaves, which are required for transpiration.) Instead, the sap flow apparently results from alternating freezing and thawing of the xylem tissues. During the day, respiring cells in the stem produce CO_2. At night, compressed CO_2 bubbles are trapped in ice that forms in the xylem. When the xylem thaws during the day, the gases expand once more, pushing the sap up the tree. Thus, sugar maple sap flows under positive, not negative, pressure.

To harvest the sap, collectors drill a hole and insert a spout through the tree's bark and into the xylem. The xylem sap drips off the end of the spout and into a container (**figure 25.B**). Each tap produces about 40 L (10 gallons) of raw xylem sap each year, which syrup producers boil down to yield about one liter of finished syrup. The distinctive maple syrup flavor develops as heat alters some of the sap's chemical constituents.

Have a Burning Question of your own? Submit it to marielle_hoefnagels@mcgraw-hill.com for possible inclusion in future editions of this book!

FIGURE 25.B Sap Flow. The sugar maple tree (*Acer saccharum*) yields sweet xylem sap, the precursor to maple syrup.

CHAPTER SUMMARY

25.1 Soil and Air Provide Water and Nutrients

A. Plants Require 16 Essential Elements

- Like all organisms, plants require water and **essential nutrients.**
- In all plants, the essential **macronutrients** are C, H, O, P, K, N, S, Ca, and Mg. The **micronutrients** are Cl, Fe, B, Mn, Zn, Cu, and Mo.

B. Soils Have Distinct Layers

- **Soil** consists of rock and mineral particles mixed with living organisms and decaying organic molecules. Soil layers are called horizons.
- **Topsoil** contains **humus,** a major source of water and nutrients for plants. **Erosion** destroys topsoil and hinders plant growth.

C. Leaves and Roots Absorb Essential Elements

- Plants get CO_2 and O_2 from the atmosphere, and they get hydrogen and oxygen from H_2O in soil. The other elements also come from soil.
- Several types of bacteria live in root **nodules** and **fix nitrogen,** converting it into forms that plants can use.

25.2 Water and Dissolved Minerals Are Pulled Up to Leaves

A. Water Vapor Is Lost from Leaves Through Transpiration

- Water and dissolved minerals (**xylem sap**) are pulled up through xylem to replace water lost through **transpiration** in leaves. This is called the **cohesion–tension theory.**

B. Xylem Transport Relies on Cohesion and Adhesion

- Because of water's cohesive properties, water molecules evaporating from leaves or used in metabolism are replaced by those pulled up from below.
- Water adhering to xylem walls draws water upward by capillary action and prevents the formation of air bubbles.

C. Water Enters Plants in Roots

- Water enters roots by osmosis because the solute concentration in the soil is less than that of root cells.
- Root branches, root hairs, and mycorrhizal fungi provide abundant surface area for absorption.
- Water and dissolved minerals move through the root's epidermis and cortex either extracellularly or through cell interiors. The endodermis, with its impermeable **Casparian strip,** controls which substances enter the nearby xylem.

D. The Cuticle and Stomata Help Conserve Water

- The waxy, waterproof cuticle prevents water loss in aerial plant parts. In woody plants, bark also conserves water.
- Plants can open and close pores called stomata in response to water availability.

25.3 Organic Compounds Are Pushed to Nonphotosynthetic Cells

A. Phloem Sap Contains Sugars and Other Organic Compounds

- **Phloem sap** includes sugars, hormones, viruses, and other organic molecules along with water and minerals from xylem.

B. The Pressure Flow Theory Explains Phloem Function

- According to the **pressure flow theory,** phloem sap flows under positive pressure through sieve tubes from a **source** to a **sink.** Sources are photosynthetic or sugar-storing parts; sinks are nonphotosynthetic plant parts such as roots, flowers, and fruits.

25.4 Investigating Life: Pondering the Pipes in the Plumbing of Plants

- Careful study of vessels and tracheids show that the two xylem cell types are closely related.
- Drought apparently selects for constrictions called vestured pits in xylem vessels.

MULTIPLE CHOICE QUESTIONS

1. What is the difference between a macronutrient and a micronutrient?
   a. The size of the atoms
   b. The amount required by the plant
   c. The use by multicellular (macro-organisms) versus microorganisms
   d. The source of the nutrient

2. A carnivorous plant consumes insects as a source of
   a. O_2.
   b. CO_2.
   c. organic compounds.
   d. N.

3. Which soil layer is most important for supporting plant growth?
   a. The A horizon
   b. The B horizon
   c. The C horizon
   d. Both b and c

4. How do nitrogen-fixing bacteria contribute to the nitrogen cycle?
   a. They decompose dead plant material, releasing N_2.
   b. They convert NH_4^+ to N_2 for release into the atmosphere.
   c. They reduce N_2 to NH_4^+ for use by plants.
   d. They release N_2 from root nodules.

5. What is transpiration?
   a. The evaporation of water from leaves
   b. The movement of water through the xylem
   c. The uptake of water through root hairs
   d. All of the above

6. The cohension-tension theory depends on
   a. the presence of the Casparian strip.
   b. hydrogen bond formation between water molecules.
   c. environmental factors such as heat or humidity.
   d. the diameter of the tracheids and vessels.

7. What determines whether a mineral dissolved in the soil becomes part of a plant?
   a. The concentration of the mineral within the soil
   b. The presence of specific transport proteins along the epidermis of the root hair
   c. The presence of specific transport proteins in the endodermis of the root
   d. The osmotic balance between the root and the soil

8. Why do desert plants that live in very hot and dry conditions have stomates?
   a. To allow for transpiration
   b. To allow for gas exchange
   c. To allow for the release of excess heat
   d. Both a and b

9. Where do the simple sugars in phloem sap ultimately come from?
   a. The roots
   b. Storage organs like fruits
   c. Photosynthesis
   d. Starch

10. Which of the following is NOT a sink?
    a. A root
    b. A leaf
    c. A flower
    d. A meristem

TESTING YOUR KNOWLEDGE

1. List the macronutrients and the micronutrients essential for plant growth.
2. How do plants obtain carbon and nitrogen?
3. On a planet with an atmosphere similar to Earth's, which elements would have to be present in the soil for earthly plants to grow?
4. How do the cells of xylem and phloem differ?
5. Trace the path of water and dissolved minerals from soil, into the root's xylem, and up to the leaves.
6. Does xylem transport cost energy? What about phloem transport? Explain your answers.
7. Distinguish between cohesion and adhesion, and describe how each plays a part in xylem transport.
8. Are roots necessary for transpiration to occur? Are leaves?
9. Distinguish between a source and a sink. How can the same plant part act as both a source and a sink?
10. How does xylem flow influence phloem flow?
11. At which points in the transport of water and nutrients in plants does osmosis occur? At which points does active transport occur?
12. Explain this statement in terms of the cohesion–tension and pressure flow theories: "The flow of xylem sap is unidirectional, whereas the flow of phloem sap is bidirectional."
13. Explain two ways that leaf mesophyll cells and root cortical cells interact.

THINKING AS A SCIENTIST

1. How might transgenic technology (see chapter 12) be used to endow plants with the ability to fix nitrogen? In what ways would this change agriculture?
2. Suppose that a scientist exposes a leaf to CO_2 labeled with carbon-14, and the radioactive carbon is incorporated into organic compounds in photosynthetic cells. At various times after exposure, the scientist can determine the location of the radioactive carbon in the plant. In what tissues do you expect to find the radioactive material immediately after exposure to the labeled carbon? What about during transport? When transport is complete, will the radioactive material be in plant parts above the leaf, below the leaf, or both?
3. In the early 1600s, Jean Baptista van Helmont did an experiment to determine how plants acquire new mass as they grow. He weighed and planted a willow shoot into soil that he had also weighed. After 5 years of adding nothing but water to the plant, he found that the soil had lost only a little weight, while the plant had grown from 2 kg to about 76 kg. He therefore concluded, incorrectly, that water was the sole source of the added plant material. What other source did he fail to consider in his experiment?
4. Scientists studying phloem function find that if they use a tiny instrument to damage sieve tubes, the plant begins to plug the wound within a minute. Why is it adaptive for the plant to prevent leakage of phloem sap?
5. "Root pressure theory" proposes that water from roots is pushed up a stem. Although root pressure does occur in some circumstances, it does not account for all xylem flow. What evidence would you look for to determine whether water in xylem moves under tension or pressure?
6. During what time of day is a plant at greatest risk for the formation of air bubbles in the xylem stream?

Visit www.mhhe.com/hoefnagels for practice quizzes, animations, videos, and activities designed to help you master the material in this chapter.

Reproduction and Development of Flowering Plants

Partners. Many flowering plants depend on healthy populations of pollinating animals such as honeybees for reproduction. This beekeeper is tending her hives.

What's the Point?

Imperiled Pollinators

Nearly all terrestrial ecosystems depend ultimately on plants for food and oxygen. Yet animals are also essential to many plant species. Many angiosperms (flowering plants) lure pollinators with a sweet substance called nectar. As insects, birds, and small mammals explore the flowers for food, they unwittingly help the plants reproduce by transferring pollen from one flower to another.

*Bees are particularly proficient pollinators. European honeybees (*Apis mellifera*) pollinate many flowering plant species in the United States. Europeans introduced these bees in Jamestown, Virginia, in 1621. Because these insects can use pollen and nectar from many plant species in a variety of habitats, they rapidly displaced native bees, reaching a population pinnacle of nearly 6 million colonies in 1947.*

*But bee populations worldwide are plummeting. In North America, increased use of pesticides in agriculture began killing bees shortly after World War II and had depleted more than a million colonies by 1972. A study of *Cereus* cactus along the Mexico and Arizona border revealed the effects of herbicides used to clear land to build new homes. Where herbicide was used, 27% of the plants monitored were pollinated, and 5% developed fruit. Without herbicide, 60% to 100% of the cacti were pollinated, and 75% to 100% developed fruit.*

*Moths, mites, and infectious disease threaten bees too. Caterpillars of the greater wax moth (*Galleria mellonella*) tunnel through honeycombs and eat beeswax, pollen, and cocoons. Tracheal mites (*Acarapis woodi*) first appeared in Florida bee colonies in 1984 and rapidly spread throughout the United States, killing up to half of the bees in individual hives. Another type of mite (*Varroa*) began invading hives in Wisconsin in 1987 and likewise spread, eventually destroying 60% of honeybee colonies in the United States. Africanized bees invaded at the end of the 1980s, bringing yet other deadly mites to many honeybee hives. In addition, a bacterial disease called American foul brood kills bee larvae and spreads easily from hive to hive.*

Adding to the problems is a mysterious ailment called Colony Collapse Disorder, in which the bees in a hive die for no known reason. The insects have weakened immune systems and harbor unusually high populations of harmful bacteria and fungi, but research has so far been unable to identify the culprit responsible for the bee deaths.

Bee experts, called apiculturists, predict that the falling bee populations will soon lead to declining crops of many edible fruits and seeds. As we shall see in this chapter, pollination is a vital link in the life cycles of flowering plants.

LEARNING OUTLINE

26.1 Angiosperms Reproduce Asexually and Sexually

Flowering plants (angiosperms) dominate many terrestrial landscapes. From grasslands to deciduous forests, and from garden plots to large-scale agriculture, angiosperms have been extremely successful. Humans have used them as sources of food, shelter, clothing, furniture, medicines, and fuel.

Angiosperms owe their widespread distribution to three adaptations. First, pollen (which also occurs in gymnosperms) enables sperm to fertilize the egg in the absence of free water; in contrast, the sperm cells of mosses and ferns must swim to the egg, restricting these plants to moist habitats. Second, angiosperms and gymnosperms produce seeds, structures that protect the embryo during dormancy and nourish the developing seedling. Third, only angiosperms have flowers, which promote pollination and develop into fruits that help disperse the seeds far from the parent plant.

FIGURE 26.1 **Asexual Reproduction.** Quaking aspen (*Populus tremuloides*) trees that cover more than 100 acres in Utah are clones that remain attached to the parent, producing huge individuals. The picture shows a portion of *Pando* (Latin for "I spread"), a clone that consists of 47,000 tree trunks and weighs more than 6 million kilograms (13.2 million pounds).

The majority of plants reproduce sexually, although some species also reproduce asexually. This section describes asexual reproduction and introduces sexual reproduction, the topic of sections 26.2 and 26.3.

A. Asexual Reproduction Yields Clones

Many plants reproduce asexually, forming new individuals by mitotic cell division. In **asexual reproduction,** a parent organism produces offspring that are genetically identical to it and to each other—they are clones. Asexual reproduction, also called vegetative reproduction, is advantageous when environmental conditions are stable and plants are well adapted to their surroundings, because the clones will also be suited to that environment.

Plants often reproduce asexually by forming new plants from portions of their roots, stems, or leaves (**figure 26.1**). This works because each nonreproductive cell contains a complete set of genetic instructions. For example, buds form on the roots of cherry, pear, apple, and black locust plants, and when they sprout, aerial shoots grow upward. If these shoots, called "suckers," are cut or broken away from the parent plant, they can become new individuals.

Asexual reproduction has many commercial applications. Most houseplants arise from rooted cuttings taken from parent plants. In addition, most fruit and nut trees are produced by grafting a scion (part of a parent tree) to rootstock taken from a different but closely related plant. The grower selects the scion for the quality of its fruit. Usually the rootstock is either disease- and pest-resistant or especially well adapted to dry or salty soil. Grafting the scion to the rootstock gives the grower the advantages of both.

Scientists can use tissue culture to produce huge numbers of identical plants in a laboratory. In tissue culture, a plant biologist grows pieces of plant tissue in a dish with nutrients and hormones. After a few days, the cells lose their specialized characteristics and form a white lump called a callus. The callus grows, its cells dividing, for a few weeks. Then some callus cells develop into tiny identical plantlets with shoots and roots (**figure 26.2**). Callus growth is unique to plants. The human equivalent would be a cultured skin cell multiplying into a blob of unspecialized tissue and then sprouting tiny humans!

B. Sexual Reproduction Generates Variability

Asexually generated plants have predictable characteristics because they are essentially identical to their parents (except for mutations). In contrast, sexual reproduction yields off-

FIGURE 26.2 **Asexual Reproduction in the Laboratory.** When plant tissue is cultured with the correct combination of hormones and nutrients, a callus gives rise to plantlets. When separated from one another, the plantlets grow into genetically identical plants, producing a uniform crop.

spring with a mix of traits derived from two parents, and each offspring is different from its siblings and its parents. Sexual reproduction is adaptive in a changing environment. After all, a gene combination that is successful today might not work at all if selective pressures change in the future. Producing variable offspring improves reproductive success in an uncertain world.

The major groups of multicellular organisms (fungi, animals, and plants) all have the same basic pattern of sexual reproduction: Meiosis produces the cells that begin the haploid generation, and fertilization unites the gametes to begin the diploid generation. But organisms vary in the proportion of the life cycle spent as haploid or diploid cells. For example, nearly all of an animal's cells are diploid; sperm and egg cells are usually the only haploid animal cells.

The basic plant life cycle has multicellular diploid and haploid stages (see figure 20.2). The **sporophyte,** or diploid generation, produces haploid spores through meiosis. A spore divides mitotically to produce a multicellular haploid **gametophyte,** which produces haploid egg cells or sperm (gametes) by mitosis. These gametes fuse to form a diploid zygote. The zygote grows and develops as its cells divide mitotically, eventually becoming a mature sporophyte, and the cycle begins anew. **Figure 26.3** applies this life cycle to the angiosperms.

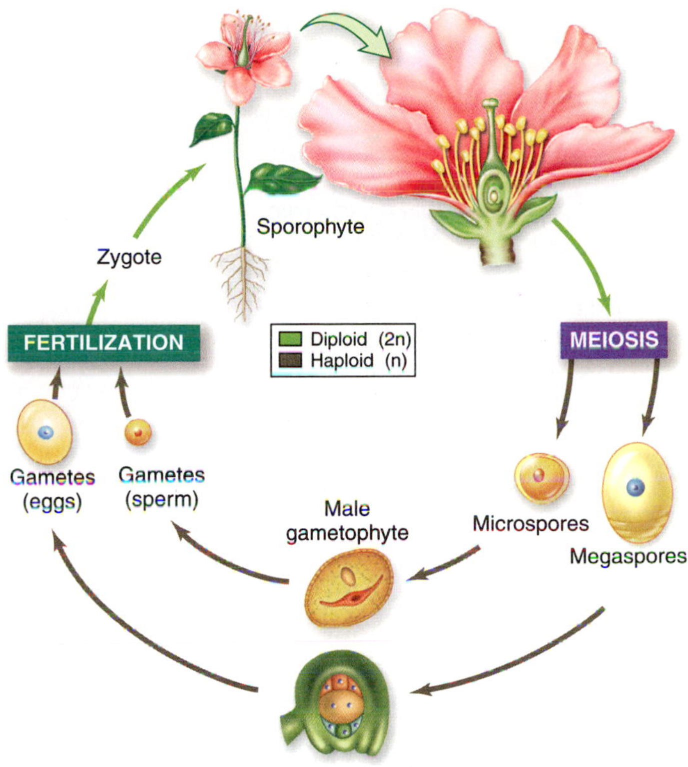

FIGURE 26.3 **Alternation of Generations in Flowering Plants.** The flowering plant life cycle includes an alternation of haploid and diploid generations. Megaspores and microspores divide mitotically to produce female and male gametophytes, which produce gametes.

Summary *In asexual reproduction, identical new individuals form on a plant's roots, stems, or leaves. Many commercially important plants are produced asexually. Sexual reproduction yields genetically variable offspring.*

26.1 MASTERING CONCEPTS

1. What are three adaptations that contribute to the reproductive success of angiosperms?
2. What are some examples of asexual reproduction in plants?
3. When are sexual and asexual reproduction each adaptive?
4. How does a plant life cycle compare with that of an animal?

26.2 Egg and Sperm Unite in Female Flower Parts

When humans reproduce, sexual intercourse brings sperm near an egg cell, and the human embryo develops into a fetus. Childbirth separates woman from baby. How does the sperm get to the egg in angiosperms? How does the angiosperm embryo develop (along with surrounding tissues) into a seed? How do the seeds separate from the "mother" plant? The answers begin with a look at flower structure.

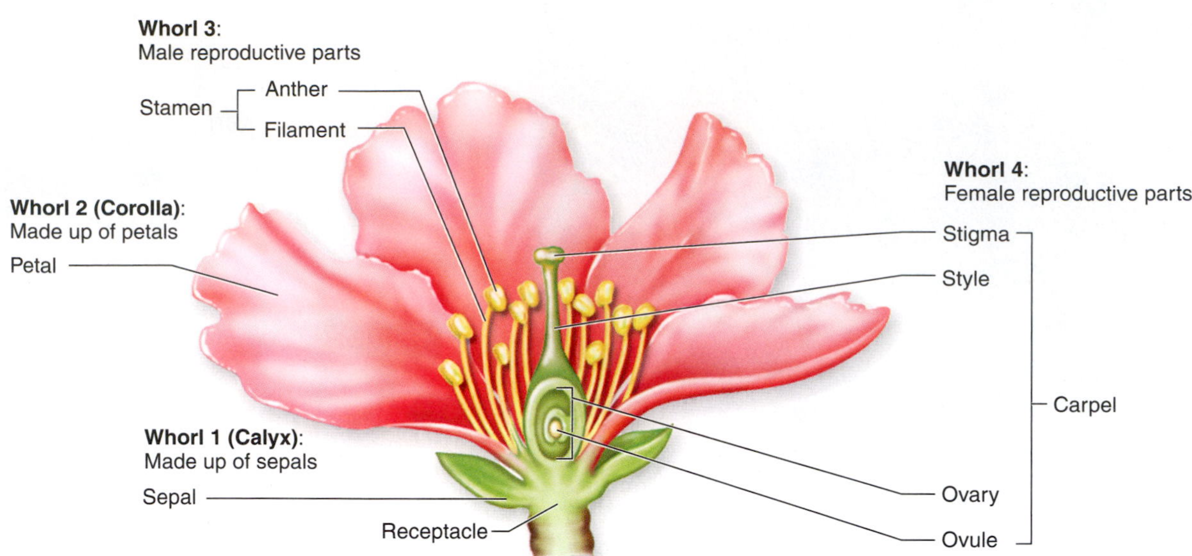

FIGURE 26.4
Parts of a Flower. A flower is made up of four whorls. The outermost whorl (the calyx) is sepals, whorl 2 (the corolla) is petals, whorl 3 is stamens (male parts), and the innermost whorl (whorl 4) consists of one or more carpels (female parts). The flower shown here has only one carpel.

A. Flowers Are Reproductive Organs

In angiosperms, **flowers** are reproductive organs that bring together eggs and sperm, but they also have other functions. You'll learn in the next section that a flower protects its seeds as they develop. Also, certain parts of the flower develop into a **fruit,** which protects and disperses the seeds.

A flower begins to form on the mature sporophyte when a shoot apical meristem turns into a floral meristem. The cue that initiates the transformation may be day length (see section 26.6), temperature, or the stage in the plant's life cycle. Whatever the trigger, a previously unexpressed set of genes switches on in the meristematic cells, which divide to produce all the parts of the flower.

A part of the floral stalk called the **receptacle** is the attachment point for all flower parts (**figure 26.4**). A typical flower has four types of structures, all of which are modified leaves composed mostly of parenchyma cells (see chapter

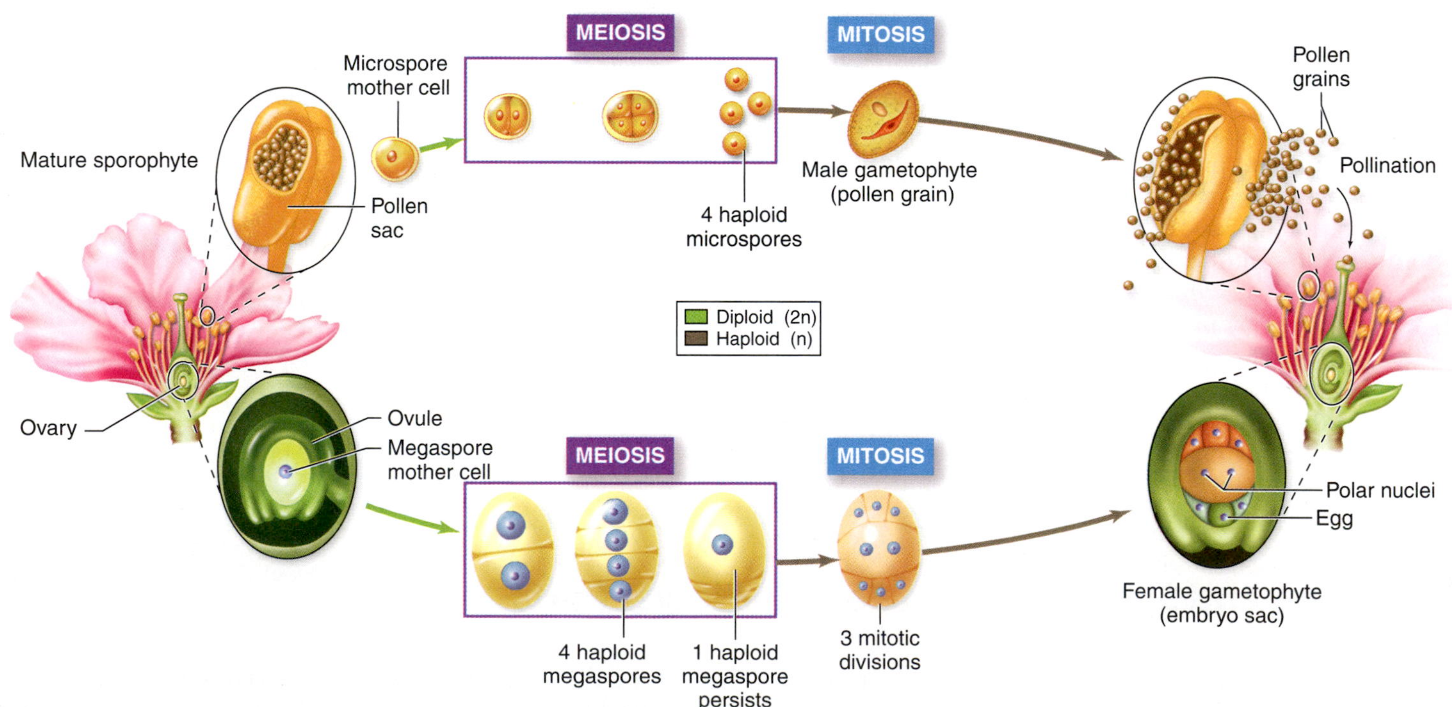

FIGURE 26.5 **Sperm and Egg Formation in a Flowering Plant.** Pollen, which gives rise to sperm nuclei, is produced in pollen sacs on stamens. Egg cells are produced in ovules inside the ovary.

24). The outermost whorl is called the calyx. It consists of **sepals,** which are green, leaflike structures that enclose and protect the inner floral parts. Inside the calyx is the corolla, which is a whorl of **petals.** The calyx and the corolla do not play a direct role in sexual reproduction, although in many flowers, colorful petals attract pollinators.

The two innermost whorls of a flower are essential for sexual reproduction. The whorl within the corolla consists of the **stamens,** which are stalklike filaments that bear pollen-producing bodies called **anthers** at their tips. The whorl at the center of a flower is composed of one or more **carpels,** which are leaflike structures enclosing the egg-bearing **ovules.** The bases of carpels and their enclosed ovules make up the **ovary.** The upper part of each carpel is a stalklike **style** that bears a structure called a **stigma** at its tip. Stigmas receive pollen.

The number of flower parts is one feature that distinguishes monocots from eudicots. Most monocots, such as lilies and tulips, have petals, stamens, and other flower parts in multiples of three (typically three or six). Most eudicots, on the other hand, have flower parts in multiples of four or five. Buttercups and geraniums are eudicots with five prominent petals on each flower.

B. The Pollen Grain and Embryo Sac Are Gametophytes

Male and female gametophytes arise from two types of spores produced by the sporophyte. Inside the anther's four **pollen sacs,** diploid cells divide by meiosis to produce four haploid **microspores** (**figure 26.5**). Each microspore then divides mitotically and produces a two-celled, thick-walled structure called a **pollen grain,** which is the young male gametophyte. One of the haploid cells inside the pollen grain divides by mitosis to form two sperm nuclei, either before or after the pollen sac opens and releases millions of pollen grains.

Meanwhile, meiosis also occurs in the ovary (see figure 26.5). The ovary may contain one or more ovules, inside of which a diploid cell divides by meiosis to produce four haploid **megaspores.** In many species, three of these cells quickly disintegrate, leaving one large megaspore. The megaspore undergoes three mitotic divisions to form the **embryo sac,** which is a mature female gametophyte. The female gametophyte consists of eight haploid nuclei but only seven cells, one of which is the egg. A large, central cell contains two **polar nuclei;** both the egg and polar nuclei participate in fertilization.

The prefix *mega* may suggest huge size, but that is only in comparison with the size of microspores. In flowering plants, the spores, gametophytes, and gametes are all microscopic; the entire haploid gametophyte generation consists of only a few cells.

C. Pollination Brings Pollen to the Stigma

The next step in angiosperm sexual reproduction is **pollination:** the transfer of pollen from an anther to a receptive stigma. Many agents transfer pollen from plant to plant, including animals (**figure 26.6**), wind, and even water.

FIGURE 26.6 Animals as Pollinators. Animal pollinators include insects (see chapter opening figure), (**A**) birds, (**B**) bats, and many others that brush past flowering plants.

Flower color, shape, and odor attract animals. For example, birds are attracted to red flowers. Beetles respond to dull-colored flowers with spicy scents, whereas blue or yellow sweet-smelling blooms attract bees. Bee-pollinated flowers often have markings that are visible only at ultraviolet wavelengths of light, which bees can perceive (**figure 26.7**). Moths and bats pollinate white or yellow, heavily scented

a.

b.

FIGURE 26.7 Ultraviolet Flower Markings. Bees can detect ultraviolet wavelengths of light, so this black-eyed Susan, which appears fully yellow to us, has a dark patch in the center to bees. Humans require special photographic equipment to see these patterns, but bees use them as guides.

flowers, which are easy to locate at night, when these animals are most active.

Many animals benefit from their association with plants: they obtain food in the form of pollen or sugary nectar, they may seek shelter among the petals, or they use the flower for a mating ground (see chapter 6's Investigating Life: Plants' "Alternative" Lifestyles Yield Hot Sex). Some flowers, however, lure pollinators with a false reward. Skunk cabbage and the "carrion" flowers of South African *Stapelia* plants emit foul odors that attract flies. Flies that land on the flowers gain nothing, but the plant benefits if the insect unwittingly carries pollen grains to another flower of the same species.

Some connections between flower and pollinator are so strong that the species directly influence each other's evolution. In **coevolution,** a genetic change in one species selects for subsequent change in the genome of another species. Coevolution is likely when a plant has an exclusive relationship with just one pollinator species. Sometimes, the flower and its animal partner have matching parts. For example, some hummingbirds have long, curved bills that fit precisely into the tubular flowers from which they sip nectar.

About 10% of angiosperms (and most gymnosperms) use wind, not animals, to carry pollen. Wind-pollinated flowers are small, greenish, and odorless; perfume and showy flowers are not necessary for wind to disperse pollen. One advantage of wind pollination is that the plant does not spend energy on nectar or other lures. On the other hand, the wind drops pollen where and when it slackens, whereas an animal delivers pollen directly to another plant. Wind-pollinated plants therefore manufacture abundant pollen. The large quantities of pollen produced by oaks, cottonwoods, ragweed, and grasses provoke allergies in many people. **allergies, p. 740**

D. Double Fertilization Yields Zygote and Endosperm

After a pollen grain lands on a stigma of the correct species, a pollen tube emerges (**figure 26.8**). The pollen grain's two haploid sperm nuclei enter the pollen tube as it grows through the tissue of the style toward the ovary. When the pollen tube reaches an ovule, it discharges its two sperm nuclei into the embryo sac.

Then, in **double fertilization,** these sperm nuclei fertilize the egg and the two polar nuclei. That is, one sperm nucleus fuses with the haploid egg nucleus and forms a diploid zygote, which will develop into the embryo. The second sperm nucleus fuses with the two haploid polar nuclei. The resulting triploid nucleus divides to form a tissue called **endosperm,** which is composed of parenchyma cells that store food for the developing embryo. Familiar endosperms

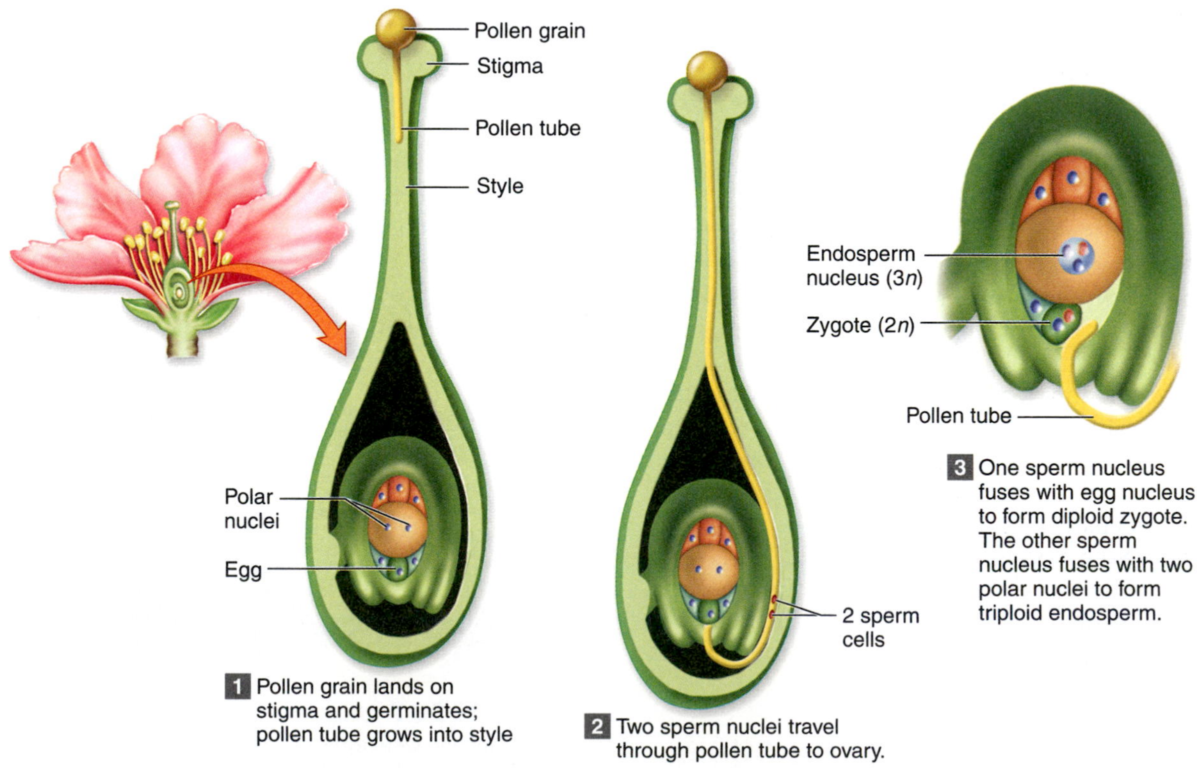
Pollen grain
Stigma
Pollen tube
Style
Polar nuclei
Egg
1 Pollen grain lands on stigma and germinates; pollen tube grows into style
2 sperm cells
2 Two sperm nuclei travel through pollen tube to ovary.
Endosperm nucleus (3n)
Zygote (2n)
Pollen tube
3 One sperm nucleus fuses with egg nucleus to form diploid zygote. The other sperm nucleus fuses with two polar nuclei to form triploid endosperm.

FIGURE 26.8 Double Fertilization. Pollen sticks to a stigma on a flower of the same species. A pollen tube grows toward the ovule and transports two sperm nuclei. One sperm nucleus fertilizes the egg to form a zygote, and the other fertilizes the polar nuclei to yield the endosperm.

are the "milk" and "meat" of a coconut and the starchy parts of wheat, rice, and corn grains.

Summary *Specialized cells in flowers undergo meiosis, yielding the haploid spores that develop into gametophytes. The gametophytes generate haploid gametes by mitosis. Wind, water, or animals carry pollen grains (the male gametophytes) to other flowers, where double fertilization yields both a diploid zygote and a triploid endosperm.*

26.2 MASTERING CONCEPTS

1. What are the parts of a flower, and what are their functions?
2. What are the roles of spores and gametophytes in sperm and egg formation in angiosperms?
3. Why is pollination essential for angiosperm sexual reproduction?
4. Describe the events of double fertilization.
5. How is endosperm important to plants? To humans?

26.3 Seeds Develop Inside Fruits

Flowers, seeds, and fruits are intimately related (**figure 26.9**). This section explains how development of the embryo inside the ovule triggers the changes that result in seed and fruit formation.

A. A Seed Is an Embryo and Its Food Supply Inside a Seed Coat

Immediately after fertilization, the ovule contains an embryo sac with a diploid zygote and a triploid endosperm. Initially, the endosperm cells divide more rapidly than the zygote and thus form a large multicellular mass. This endosperm supplies nutrients to the developing embryo.

The shoot and root apical meristems form at opposite ends of the embryo. The developing embryo also forms **cotyledons,** or seed leaves (**figure 26.10**). Recall that angiosperms that have one cotyledon are monocots and those with two cotyledons are dicots (see figure 24.1). In monocots, the cotyledon transfers stored nutrients from the endosperm to the embryo during germination. In many eudicots,

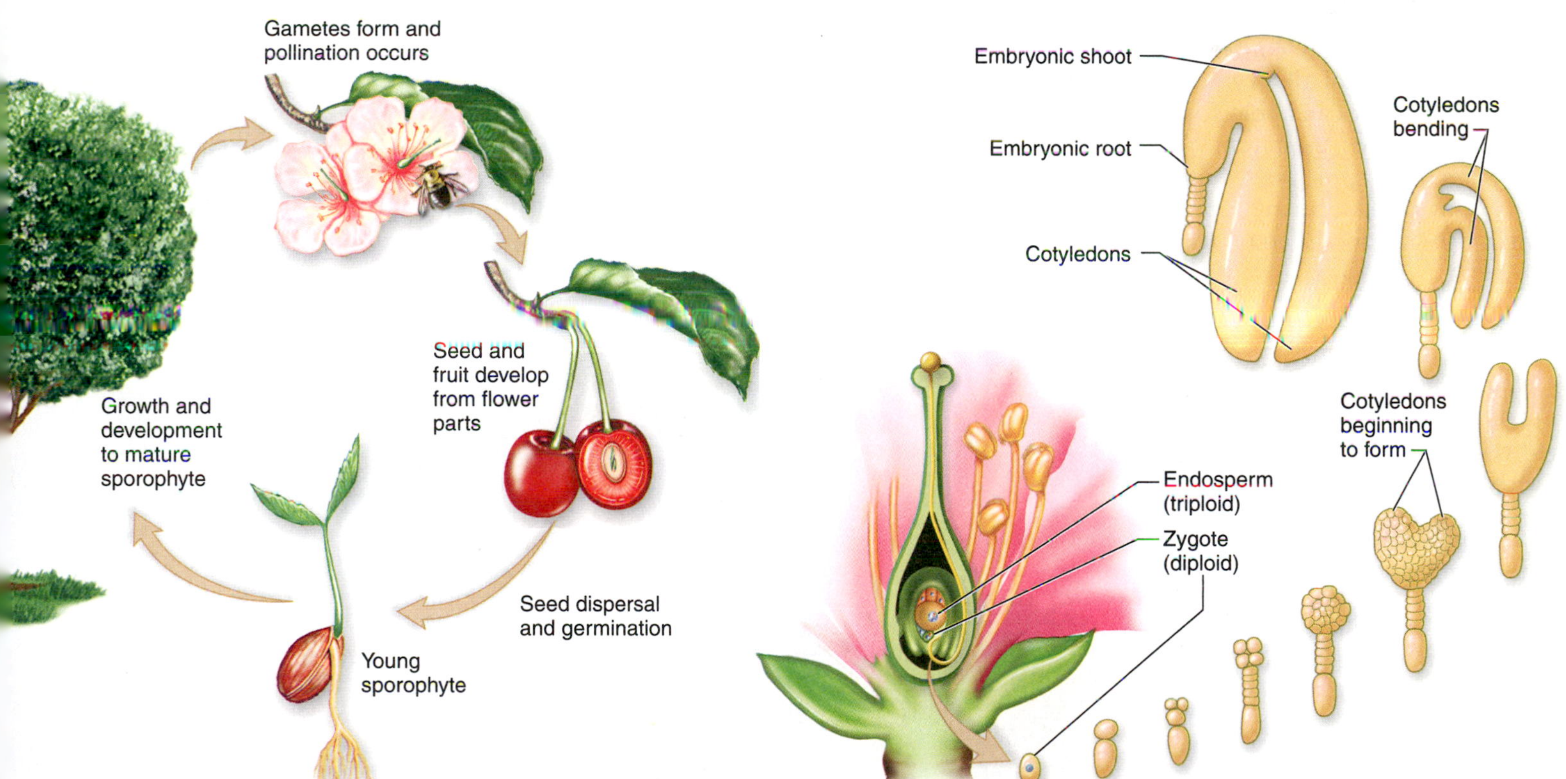

FIGURE 26.9 Fruits Come From Flowers. After fertilization, the ovule develops into a seed, which contains the embryo. The ovary matures into a fruit enclosing the seed. The seed is dispersed, perhaps with the help of a hungry or passing animal, and germinates, developing into the mature sporophyte.

FIGURE 26.10 Embryonic Development. Inside the flower, the zygote divides repeatedly to form the tiny embryonic plant. As the embryo develops, other flower parts become the endosperm, seed coat, and fruit.

the cotyledons completely absorb the endosperm as they develop, becoming thick and fleshy (**figure 26.11**). **meristems, p. 503**

At some point in embryonic development, in response to hormonal signals, cell division and growth stop, and the embryo becomes dormant. The **seed coat,** a tough, sclereid-rich outer layer, protects the dormant plant embryo and its food supply. A **seed** is a plant embryo together with its stored food and seed coat.

Seed dormancy is a crucial adaptation that enables seeds to postpone development when the environment is unfavorable, such as during a drought or frost. Favorable conditions trigger embryo growth to resume when young plants are more likely to survive.

B. The Fruit Develops from the Ovary

A flower begins to change as seeds develop. When a pollen tube begins growing, the stigma produces large amounts of ethylene, a plant hormone discussed later in this chapter. Ethylene triggers unneeded flower parts such as stamens and petals to wither and fall to the ground. Developing seeds also produce another hormone, auxin, that stimulates fruit to form. This chapter's Burning Question: How can a fruit be seedless? explains the paradox of seedless fruits.

In many angiosperms, the ovary grows rapidly to form the fruit, which may contain one or more seeds. In some species, additional plant parts also join in fruit development. The pulp of an apple, for example, derives from a cup-shaped region of the receptacle. **Figure 26.12** shows how the parts of an apple flower give rise to the fruit.

Fruits come in many forms (**table 26.1**), and they owe their textures to different tissue compositions. Soft, wet

Burning Questions

How can a fruit be seedless?

Given the role of seeds and fruits in plant reproduction, seedless fruits might seem a bit of a puzzle. After all, from a plant's point of view, what's the point of producing a fruit without seeds inside? Natural selection clearly would not favor such a trait in the wild!

Seedless fruits form in two main ways. In most seedless plant varieties, including oranges and watermelons, fruit develops even if fertilization has not occurred. Grapes and bananas illustrate another way seedless fruits form: fertilization occurs, but the embryos die during development. Tiny, immature seeds remain inside the fruit, but they do not interfere with eating.

Since seedless fruits lack seeds, how do farmers grow more of them? Most are produced asexually by grafting or taking cuttings. But, surprisingly, seedless watermelons do come from seeds! To make "seedless watermelon seeds," plant breeders first cross a regular diploid watermelon plant with a tetraploid plant (with four sets of chromosomes). Farmers then plant the triploid seeds arising from this union. The new triploid plants produce flowers, which the grower pollinates with diploid pollen. Pollination stimulates fruit production, but, like a mule, the triploid plant is sterile. The "seeds" that do develop inside seedless watermelons are actually empty, soft hulls that are easy to chew and swallow.

Have a Burning Question of your own? Submit it to marielle_hoefnagels@mcgraw-hill.com for possible inclusion in future editions of this book!

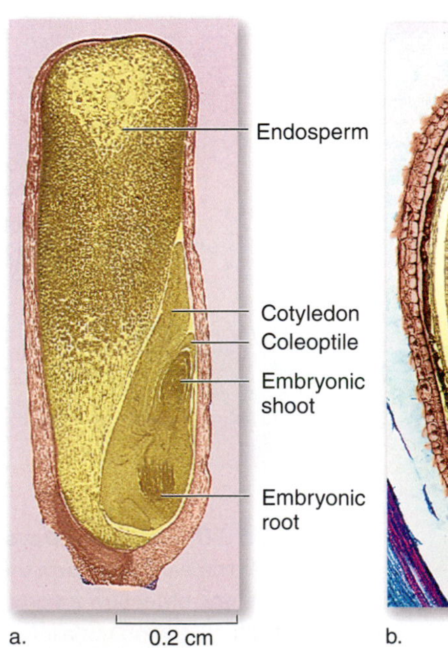

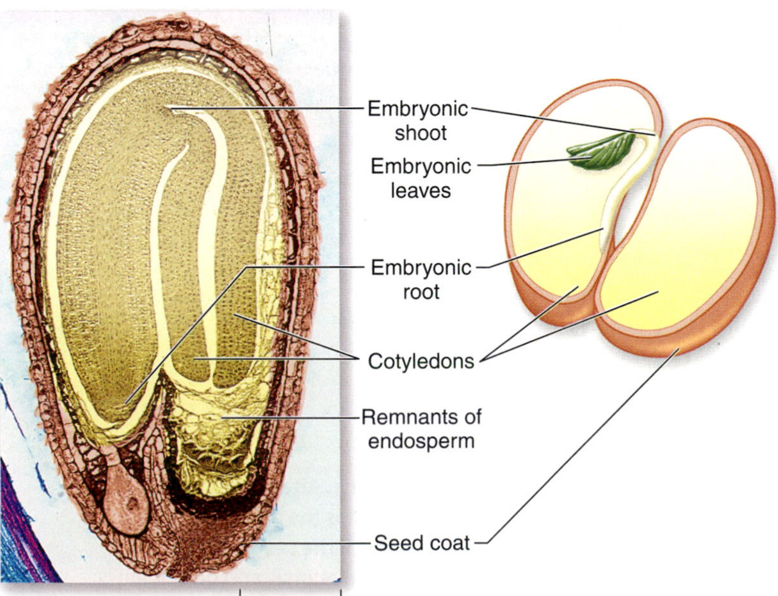

FIGURE 26.11
Mature Seeds. The nutrient-packed endosperm fuels development of the embryo. (**A**) A monocot seed, such as the one inside a grain of corn, retains most of the endosperm until after germination, when the single cotyledon absorbs its nutrients. (**B**) The cotyledons of a eudicot embryo may absorb much of the endosperm as the seed develops. The photo shows an embryo of shepherd's purse (*Capsella bursa-pastoris*); the diagram shows the corresponding structures in a bean seed.

FIGURE 26.12 Fruit Parts Derive from Flower Parts. After a flower is pollinated and the egg is fertilized, the flower begins to develop into a fruit. The fleshy part of an apple develops from an enlargement of the receptacle, which grows along with the ovary wall as the fruit develops. The "core" is derived from the carpel walls, which enclose the seeds.

fruits such as tomatoes and apples have more parenchyma than any other cell type. But the hard, papery apple core, the outer wall of a dry fruit such as a nut, and the gritty particles in pear flesh are all made of sclereids. The more sclereids, the tougher the fruit.

Because flowers always form on shoots, it may seem surprising that at least one species produces fruits underground: peanuts. Its yellow flowers form aboveground. After fertilization, the petals wither, and the young fruit produces a peg that turns downward and buries itself in the soil. Three to five months later, farmers dig up the plants to harvest the mature fruits: fibrous shells each containing one to three peanuts (the seeds).

C. Fruits Protect and Disperse Seeds

Fruits have two main functions. Besides protecting developing seeds from desiccation or destruction, fruits also facilitate seed dispersal by animals, wind, and water.

Many animal species eat nutritious fruits and seeds (**figure 26.13A**). Colored berries attract birds and other animals that carry the ingested seeds to new locations, only to release them in their feces. Squirrels and other animals that hide acorns and other nuts to eat later also disperse seeds when they forget some of their cache locations. Birds and mammals also spread seeds when spiny fruits attach to their feathers or fur (figure 26.13B).

Table 26.1 ***Types of Fruits***

Fruit Type	Characteristics	Example(s)
Simple	Derived from one flower with one carpel	Olive, cherry, peach, plum, coconut, grape, tomato, pepper, eggplant, apple, pear
Aggregate	Derived from one flower with many separate carpels	Blackberry, strawberry, raspberry, magnolia
Multiple	Derived from tightly clustered flowers whose ovaries fuse as fruit develops	Pineapple

a.

b.

c.

FIGURE 26.13 Seed Dispersal. (**A**) This cedar waxwing helps disperse the seeds of winterberry, a type of holly. (**B**) Velcro resembles the burdock fruit, which hitches a ride on a passing animal. (**C**) Dandelion fruits have fluff that enables them to float on a breeze.

The seeds are often deposited far from the parent plant, minimizing competition between parent and offspring.

Wind and, less commonly, water can also distribute seeds. Wind-dispersed fruits, such as those of dandelions and maples, have wings or other structures that catch air currents (figure 26.13C). Coconuts are water-dispersed fruits that travel long distances before colonizing distant islands. The coconut's familiar hard brown "shell" is actually the innermost layer of the fruit wall; the green husk that forms the rest of the fruit wall is removed before transport and distribution to grocery stores.

Summary *A seed is a dormant plant embryo and its food supply, all packaged in a tough seed coat. Fruits, which develop from flower parts, protect seeds and facilitate their dispersal.*

26.3 MASTERING CONCEPTS

1. How does embryo development differ in monocots and eudicots?
2. Which flower parts develop into a fruit?
3. What are some ways that seeds disperse to new habitats?

26.4 Development Begins with Seed Germination

Germination is the resumption of growth and development after a period of seed dormancy is broken. It usually requires water, oxygen, and a favorable temperature. First, the seed absorbs water. The incoming water swells the seed, rupturing the seed coat and exposing the plant embryo to oxygen. Water also may cause the embryo to release hormones that stimulate the production of starch-digesting enzymes. The stored starch in the embryo breaks down to sugars, providing energy for the now-growing embryo.

Growth and development continue after the growing embryo bursts out of the seed coat (**figure 26.14**). Rapidly dividing cells in apical meristems add length to

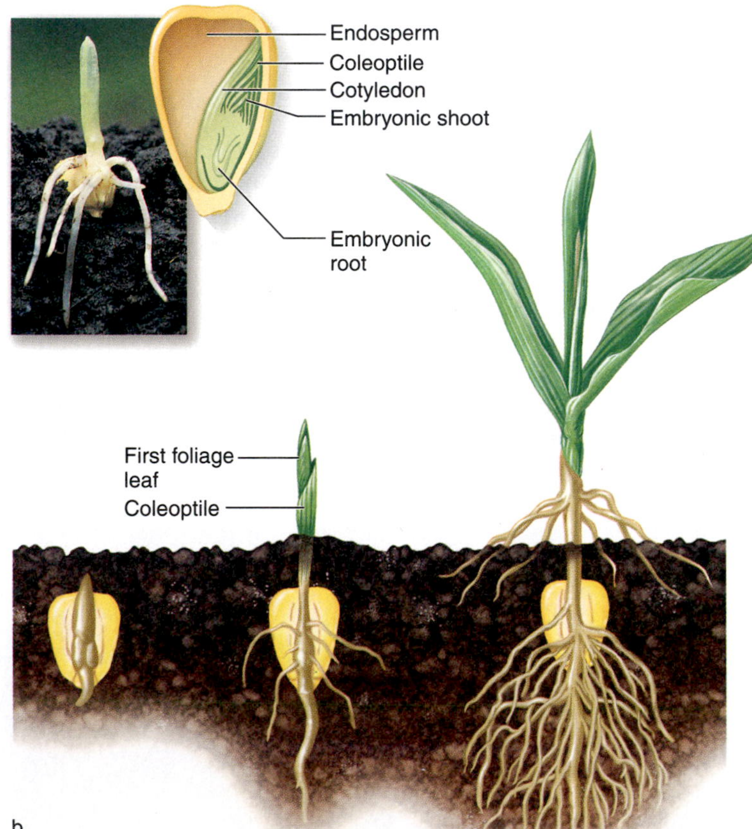

FIGURE 26.14 Seeds Germinate to Yield New Plants. (**A**) In a eudicot like this green bean, the shoot apical meristem forms at the tip of the stemlike region above the cotyledons, and the root apical meristem differentiates near the tip of the embryonic root. (**B**) In a monocot such as corn, a sheathlike structure called the coleoptile covers the shoot's embryonic leaves. Each corn kernel is an individual fruit containing a single seed.

both shoots and roots. In response to gravity, young roots grow downward. The root system anchors the plant in the soil and provides water and minerals required for growth. By the time the seedling has used its embryonic food reserves, the young shoot is already producing its own food by photosynthesis.

If conditions are favorable, the plant keeps growing for weeks, years, decades, or even centuries, depending on the species. When it reaches reproductive maturity, it too will develop flowers, seeds, and fruits, continuing the life cycle.

Summary ***Seed germination marks the resumption of sporophyte growth.***

26.4 MASTERING CONCEPTS

1. Why must seeds absorb water before germinating?
2. How is early seedling growth different for monocots and eudicots?

26.5 Hormones Influence Plant Growth and Development

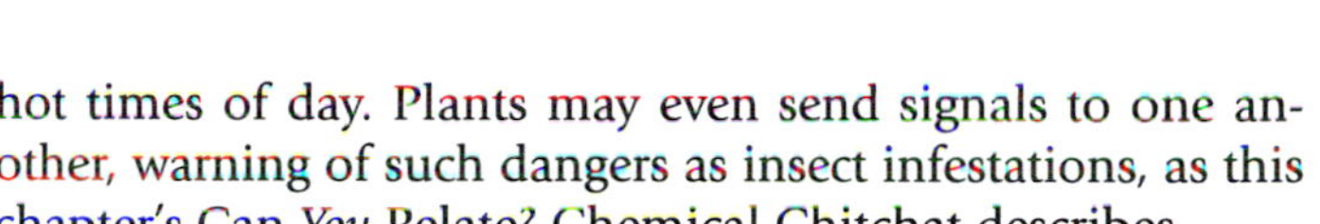

We now turn to some of the ways in which a plant responds to environmental stimuli as it grows and develops, from seed germination through senescence (aging and death). Shoots grow toward light and against gravity; roots grow down. Many plants leaf out in spring, produce flowers and fruits, then return to dormancy in autumn—all in response to seasonal changes. Other responses are immediate, such as when stomata close to reduce transpiration during hot times of day. Plants may even send signals to one another, warning of such dangers as insect infestations, as this chapter's Can You Relate? Chemical Chitchat describes.

Plant responses usually seem much more subtle than those of many animals. Plants cannot hide, bite, or flee; instead they must adjust to the environment. Their responses are partly mediated by chemicals called hormones. Classically,

Can *You* Relate?

Chemical Chitchat

If a small animal tried to take a bite of your flesh, you would probably swat the pest away and yell to warn your friends of the danger. Likewise, ravenous insect larvae can devastate a plant by eating its leaves. In response, tomato plants (and other species) produce a hormone called jasmonic acid that not only activates the plant's own defenses but also warns the neighbors and recruits an army of killer wasps! The plants "shout" by using chemicals instead of sound.

The reaction begins when a caterpillar chews a leaf. The damage stimulates cells in the leaf to secrete a small peptide hormone that travels in the phloem to cells in other parts of the plant, where it stimulates the production of jasmonic acid. This biochemical is a key player in the plant's defenses. It activates genes that direct cells to produce protease inhibitors, molecules that destroy the insects' digestive enzymes. But that is not all. Jasmonic acid also forms a gas that wafts over to nearby plants, stimulating them to manufacture their own protease inhibitors. At the same time, the gas attracts parasitic wasps that deposit their eggs in the caterpillars, which are then destroyed from the inside out (**figure 26.A**).

FIGURE 26.A
Reinforcements Arrive.
Plants can use jasmonic acid to signal neighboring plants. The chemical also helps vanquish hungry caterpillars by attracting parasitic wasps to the scene.

Jennifer Thaler, while a graduate student at the University of California at Davis, investigated this subtle effect. She set up 103 field plots, each containing four to six tomato plants. She sprayed half of the plants with jasmonic acid and half with water. Hungry armyworm caterpillars arrived soon thereafter. Three weeks later, the plants treated with jasmonic acid had twice as many wasps as did the plants sprayed with just water. Was the jasmonic acid somehow attracting the exact species of insect that would attack the caterpillars?

To demonstrate that the plants were attracting the wasps, and not that the armyworms on treated tomato leaves were simply vulnerable because they ingested protease inhibitors, Thaler ran another series of experiments. She kept the armyworms in cups placed under tomato leaves, so that they could not suffer any ill effects of the protease inhibitors. These "sentinel caterpillars" were not in contact with the leaves—but the wasps still reached them. A day after setting the cups out, the caterpillars beneath the jasmonic acid-treated tomato plants had 37% more visits by parasitic wasps than did the caterpillars on the water-treated plants. Thaler's conclusion: The plants under attack produce a chemical that attracts the wasps.

Plant defenses go way beyond waxy coverings and thorns. Jasmonic acid and protease inhibitors join with an arsenal of many other biochemical weapons that help plants survive damage from animals. So if you thought animals were the only organisms with the ability to communicate, think again.

a **hormone** is a biochemical synthesized in small quantities in one part of an organism and transported to another location, where it stimulates or inhibits a response from target cells. This definition must be modified slightly for plants, however, because some plant hormones act close to or at their site of synthesis. Also, a plant can produce the same hormone at multiple sites throughout the body. Unlike animals, plants do not have glands dedicated to hormone production.

Hormones interact with one another and the plant's DNA to regulate many aspects of growth, flower and fruit development, senescence (aging), and responses to environmental change. The interactions and responses are extremely complex, for several reasons. First, each type of hormone exerts a wide variety of effects. Second, the same hormone can either stimulate or inhibit a process, depending on its concentration and the developmental stage of the plant. Third, the functions of plant hormones may overlap: at least three different compounds promote stem elongation, for example.

The "classic five" plant hormones are auxins, cytokinins, gibberellins, ethylene, and abscisic acid (**table 26.2**). A plant must produce auxins and cytokinins if it is to develop at all. Both of these hormones occur in all major organs of all plants at all times, and biologists have never found a mutant plant lacking either one. Other hormones are required for normal development, but plants can complete their life cycles (albeit with altered morphology) without them.

A. Auxins and Cytokinins Are Essential for Plant Growth

Auxins (from the Greek meaning "to increase") are hormones that promote cell elongation in stems and fruits but have the opposite effect in roots. Auxins also control plant responses to light and gravity. The most active auxin is indoleacetic acid (IAA), which is produced in shoot tips, embryos, young leaves, flowers, fruits, pollen, and coleoptiles. These hormones act rapidly, spurring noticeable growth in a grass seedling in minutes.

The first plant hormones described were auxins. In the late 1870s, decades before researchers determined the chemical structures of plant hormones, Charles Darwin and his son Francis learned that a plant-produced "influence" caused plants to grow toward light (**figure 26.15**). This influence was later discovered to be auxin.

Auxins have commercial uses. These hormones stimulate the growth of adventitious roots in cuttings, which is important in the asexual production of plants. A synthetic compound with auxinlike effects, called 2,4-D (2,4-dichlorophenoxyacetic acid),

Table 26.2 *The "Classic Five" Plant Hormones*

Class	Selected Actions	Synthesis Site(s)
Auxins	• Elongate cells in seedlings, shoot tips, leaves, embryos (by triggering H^+ export to cell wall) • Control phototropism, gravitropism, thigmotropism • Promote growth of adventitious roots from stem cuttings • Inhibit growth of lateral buds (apical dominance)	Developing leaves and seeds, shoot tips
Cytokinins	• Stimulate cell division in seeds, roots, young leaves, fruits • Delay leaf senescence • Stimulate growth of lateral buds	Root tips
Gibberellins	• Stimulate cell division and elongation in roots, shoots, young leaves • Break seed dormancy	Young shoot, developing seeds
Ethylene	• Hastens fruit ripening • Stimulates leaf and flower senescence • Stimulates leaf and fruit abscission • Participates in thigmotropism	All parts, especially under stress, aging, or ripening
Abscisic acid	• Inhibits shoot growth and maintains bud dormancy • Induces and maintains seed dormancy • Stimulates protein storage in seeds • Stimulates stomatal closing • Promotes leaf, flower, and fruit abscission (perhaps by stimulating ethylene release)	Mature leaves, plants under stress

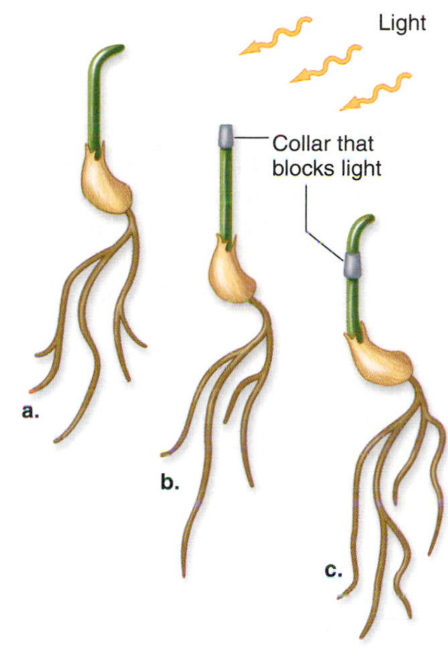

FIGURE 26.15 **Plants Bend Toward Light.** (**A**) Shoots normally bend toward a unidirectional light source. (**B**) When Charles Darwin and his son blocked the coleoptile at the tip of an oat seedling, the shoot no longer bent toward the light. (**C**) Blocking the shoot beneath the tip did not have this effect. Therefore, a substance (later identified as auxin) in the growing tip enables the plant to sense and respond to light.

is used extensively as an herbicide. The plant cannot completely break it down, and it accumulates to lethal levels. For reasons that are not completely understood, 2,4-D kills eudicots ("broadleaf weeds") but not grasses, which are monocots. **weed killers, p. 109**

Cytokinins earned their name because they stimulate cytokinesis, the division of the cell after DNA has replicated and separated. In flowering plants, most cytokinins affect roots and developing organs such as seeds, fruits, and young leaves. Cytokinins also slow leaf senescence, and they can be used to extend the shelf lives of leafy vegetables.

The actions of cytokinins and auxins compete with each other. Cytokinins are more concentrated in the roots, and auxins more concentrated in shoot tips. Cytokinins move upward within the xylem and stimulate lateral bud sprouting. In a counteracting effect called **apical dominance,** the terminal bud of a plant secretes auxins that move downward and suppress the growth of lateral buds. If the shoot tip is cut off, the concentration of auxins in lateral buds decreases, and cytokinins increase. Meristem cells in the buds then begin dividing. This is why gardeners can promote bushier growth by pinching off a plant's tip.

B. Gibberellins, Ethylene, and Abscisic Acid Influence Plant Development in Many Ways

In 1926, Japanese scientists studying "foolish seedling disease" in rice discovered **gibberellins,** another class of plant hormone that causes shoot elongation. A fungus *(Gibberella fujikuroi)* causes affected plants to grow rapidly, becoming so spindly that they fall over and die. The researchers soon discovered that a chemical extract of the fungus produced the same symptoms. In 1934, scientists isolated the active compound and named it gibberellin. We now know of at least 84 naturally occurring gibberellins.

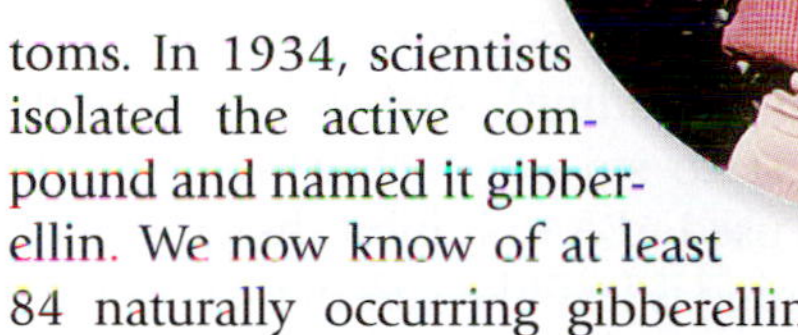

FIGURE 26.16 **Gibberellins Induce Shoot Elongation.** Gibberellins applied to grapes lengthen the stems and increase the size of the fruit. Treated grapes are on the right.

Gibberellins are present in all plant parts, in varying amounts. These hormones have several functions. Gibberellins stimulate shoot elongation in trees, shrubs, and a few other plants. They promote cell division and elongation, and they stimulate seed germination by inducing the production of enzymes that digest starch in the seed. Farmers use them in agriculture to stimulate stem elongation and fruit growth in seedless grapes (**figure 26.16**).

Ethylene is a gaseous hormone that ripens fruit in many species. The expression "one bad apple spoils the bushel" refers to the way ethylene released from one overripe apple can hasten the ripening, and eventual spoiling, of others nearby. Exposure to ethylene also ripens fruit that farmers pick while still green. For example, shipping can damage soft, vine-ripened tomatoes. Farmers therefore pick the fruit while it is still hard and green. Ethylene treatment just before distribution to supermarkets yields ripe-looking (if not good-tasting) tomatoes.

All parts of flowering plants synthesize ethylene, particularly the shoot apical meristem, nodes, flowers, and ripening fruits. Like many hormones, ethylene has several effects. In most species, it causes flowers to fade and wither (**figure 26.17**). When a damaged plant produces the hormone, it

FIGURE 26.17 **Ethylene's Effects on Flowers.** All four of these petunia flowers were treated with ethylene gas (2 parts per million) for 18 hours. The wild-type flowers on the left withered, but the flowers on the right, genetically modified to be insensitive to ethylene, remained fresh. The fresh flowers have mutant ethylene receptor genes.

hastens aging of the affected part so that it sheds before the problem spreads to other regions of the plant. This effect was noticed in Germany in 1864, when ethylene in a mixture of gases used to light streetlamps caused nearby trees to lose their leaves.

A fifth plant hormone, **abscisic acid** (abbreviated ABA), counters the growth-stimulating effects of many other hormones. It inhibits seed germination, opposing the effects of gibberellins. ABA also closes stomata, which helps plants conserve water during drought. This hormone promotes leaf, flower, and fruit abscission (shedding). ABA is produced in higher amounts in response to stresses such as drought or frost. Commercial growers used ABA to inhibit the growth of nursery plants so that shipping is less likely to damage them.

C. Biologists Continue to Discover Additional Plant Hormones

Researchers once thought plant hormones included only the five previously listed groups. Biologists now classify several other groups of molecules as plant hormones as well. Two examples include:

- **Brassinosteroids:** These steroid hormones occur throughout the plant but are most abundant in pollen and immature seeds. Plants lacking brassinosteroids have dwarfed growth forms, indicating that these hormones are essential for stem elongation.
- **Jasmonic acid:** This molecule and its volatile relative, methyl jasmonate, induce defenses against insects and pathogens. As described in this chapter's Can *You* Relate? Chemical Chitchat, a plant that produces jasmonic acid not only defends itself, but it also signals neighboring plants to beef up their own defenses.

Summary *Hormones orchestrate nearly every aspect of a plant's life, from seed germination through death. They interact with one another and the environment to regulate cell division, cell expansion, enzyme activity, aging, defenses, and many other facets of plant growth and development.*

26.5 MASTERING CONCEPTS

1. How does the definition of "hormone" differ for plants and animals?
2. List the major classes of plant hormones and name some of their functions.
3. Give an example of how plant hormones interact.

26.6 Light Influences Germination, Growth, Flowering, and Daily Rhythms

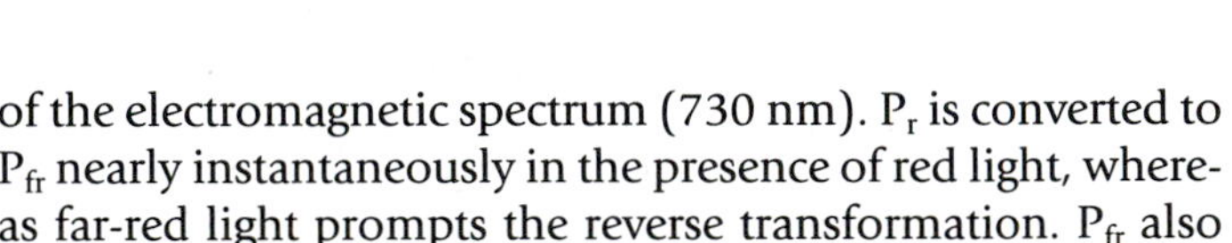

Like many other organisms, plants are exquisitely attuned to light. Their lives depend on it, because light is their sole energy source for photosynthesis (see chapter 5). This section begins by describing how plants sense light, then illustrates some of the other ways that light influences nearly every aspect of plant life, from germination to body form to flowering.

A. Phytochrome Is a Pigment that Detects Light

Angiosperms have several types of **photoreceptors,** molecules that detect the quality and quantity of light. All have the same type of structure: a protein bound to a pigment molecule that is "tuned to" certain wavelengths of light. Light absorption triggers a change in the protein, which then promotes transcription of specific genes. The likely outcome: a change in the plant's appearance or behavior.

One type of photoreceptor, **phytochrome,** is a blue pigment molecule that exists in two interconvertible forms (**figure 26.18**). The P_r form absorbs red wavelengths of light (660 nm). The other form, P_{fr}, absorbs in the far-red portion of the electromagnetic spectrum (730 nm). P_r is converted to P_{fr} nearly instantaneously in the presence of red light, whereas far-red light prompts the reverse transformation. P_{fr} also converts slowly to P_r in the dark.

Red light (660 nm)
Far-red light (730 nm)
P_r
P_{fr}
Responses: seed germination, flowering
Slow spontaneous conversion in darkness

FIGURE 26.18 The Blue Pigment Phytochrome Comes in Two Forms. P_r absorbs red light and is rapidly converted into P_{fr}; the reverse occurs rapidly when P_{fr} absorbs far-red light. P_{fr} is also converted slowly to P_r in the absence of light. P_{fr} has a variety of biological effects.

FIGURE 26.19 Growing Up in the Dark. An overlying log was lifted to reveal these spindly seedlings, which grew in the absence of light. Note the pale, elongated stems and tiny leaves.

B. Seed Germination May Require Light

Although some seeds will germinate in total darkness, others require light. In seeds of lettuce and many weeds, for example, red light stimulates germination, and far-red light inhibits it. The phytochrome system "informs" the seed whether sunlight is available. If seeds are buried too deeply in the soil, P_{fr} is absent (due to lack of sunlight needed to convert P_r to P_{fr}), and germination does not occur. Exposure to red light stimulates the conversion of P_r to P_{fr}. The presence of P_{fr} apparently induces the transcription of genes required for seed germination.

After germination, phytochrome and another photoreceptor, cryptochrome, control early seedling growth. Seedlings grown in the dark have elongated stems, small roots and leaves, a pale color, and a spindly appearance (**figure 26.19**). The pale bean sprouts used in Chinese cooking are grown without light. Once exposed to light, normal growth begins: stem elongation slows, root and leaf development accelerate, and chlorophyll synthesis begins.

C. Phototropism Is Growth Toward Light

A **tropism** is the orientation of a plant part toward or away from a stimulus such as light, gravity, or touch. All tropisms result from differential growth, in which one side of the responding organ grows faster than the other.

Phototropism, or growth toward light, occurs when cells on the shaded side of a stem elongate more than cells on the opposite side. Photoreceptors somehow cause auxins to migrate to the shaded side of the stem. The auxin influx causes proteins in the cell membrane to pump protons (H^+) into the cell wall. The protons, in turn, activate enzymes that separate the cellulose fibers, enabling the wall to expand and elongate against turgor pressure. When this happens to several cells on one side of a stem, the plant bends (**figure 26.20**). **cell walls, p. 71**

a.

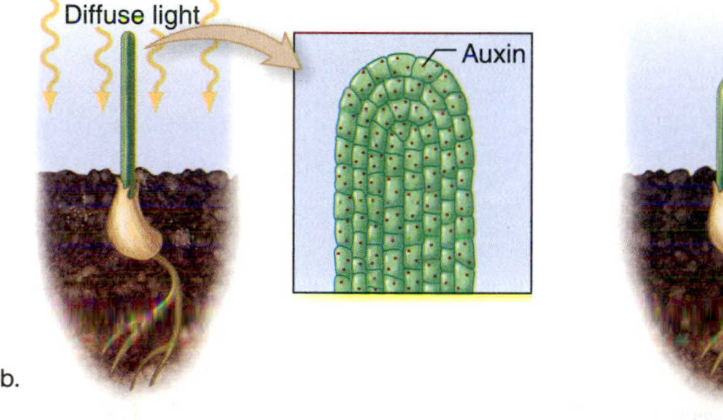
Diffuse light
Auxin
b.

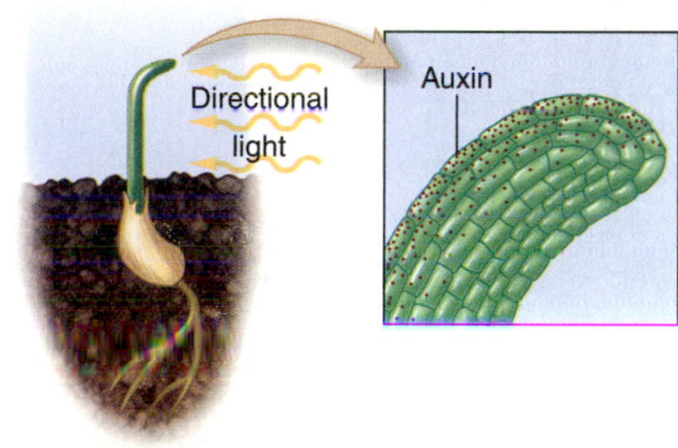
Directional light
Auxin

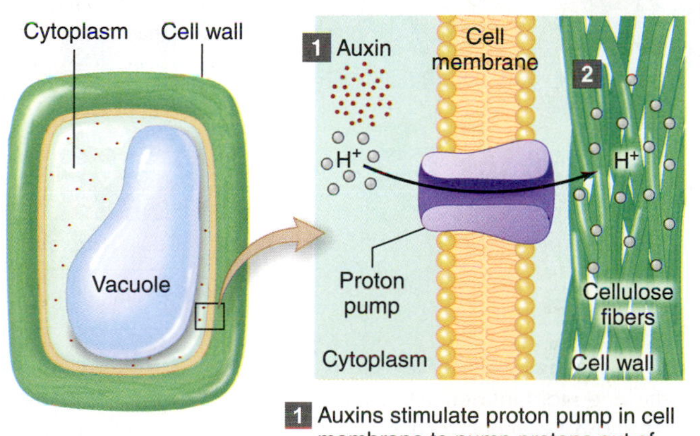
c.
Cytoplasm
Cell wall
Vacuole
1 Auxin
Cell membrane
H^+
2
H^+
Proton pump
Cytoplasm
Cellulose fibers
Cell wall
1 Auxins stimulate proton pump in cell membrane to pump protons out of cytoplasm into the cell wall.
2 High acidity in cell wall loosens bonds between cellulose fibers.

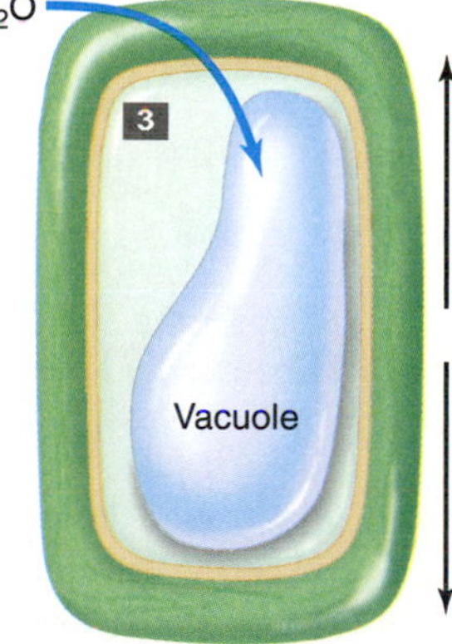
H_2O
3
Vacuole
3 Cell elongates as water moves in by osmosis and turgor pressure stretches the weakened cell wall.

FIGURE 26.20 Phototropism. (**A**) These autumn crocuses (*Colchicum autumnale*) show strong phototropism when the sun is off to the side. (**B**) Phototropism occurs because auxin moves to the shaded side of a shoot. (**C**) This hormonal action bends the plant toward the light because auxins promote acidification of the cell walls and subsequent lengthening of the shaded cells.

The plant commonly sold as "lucky bamboo" often has a curled stem, illustrating the effects of phototropism. Farmers grow the plants for a year or more in greenhouses, exposing only one side to light. Periodically rotating each plant directs the stem's growth into a twist.

D. Flowering May Be a Response to Photoperiod

Plants can also respond to seasonal changes in **photoperiod,** or day length. Many seasonal events, including bud formation, leaf abscission (shedding), and dormancy, are responses to changing photoperiod—in this case, the shortening of days as winter approaches. In the spring, buds resume growth and rapidly transform a barren deciduous forest into a leafy canopy. These seasonal changes illustrate the interactions among environmental signals, hormones and other biochemicals, and the plant's genes.

Some plants flower in response to photoperiod; biologists divide these plants into two main groups. **Long-day plants** flower when light periods are longer than a critical length, usually 9 to 16 hours. These plants typically bloom in the spring or early summer and include lettuce, spinach, beets, clover, corn, and iris. **Short-day plants** flower when light periods are shorter than some critical length, usually in late summer or fall. For example, ragweed plants flower only when exposed to 14 hours or fewer of light per day. Asters, strawberries, poinsettias, potatoes, soybeans, and goldenrods are short-day plants.

Experiments with long- and short-day plants have confirmed that flowering actually requires a specific period of uninterrupted darkness, rather than uninterrupted light (**figure 26.21**). Thus, short-day plants are really long-night plants, because they flower only if their uninterrupted dark period exceeds a critical length. Similarly, long-day plants are really short-night plants.

Phytochrome is the photoreceptor that plants use to measure the lengths of night and day. Sunlight has more red than far-red wavelengths; the resulting rapid production of P_{fr} somehow tells the plant that it is day. As night falls, the P_{fr} form of phytochrome slowly converts back to P_r. A flash of red light during the dark period can therefore inhibit flowering in a short-day plant, but a subsequent flash of far-red light cancels this effect (**figure 26.22**).

Plant hormones, which help control such fundamental functions as cell division and elongation, surely play some part in flowering. Researchers have long sought a flower-inducing hormone ("florigen"), so far without success. The precise interaction between phytochrome and plant hormones thus remains unknown.

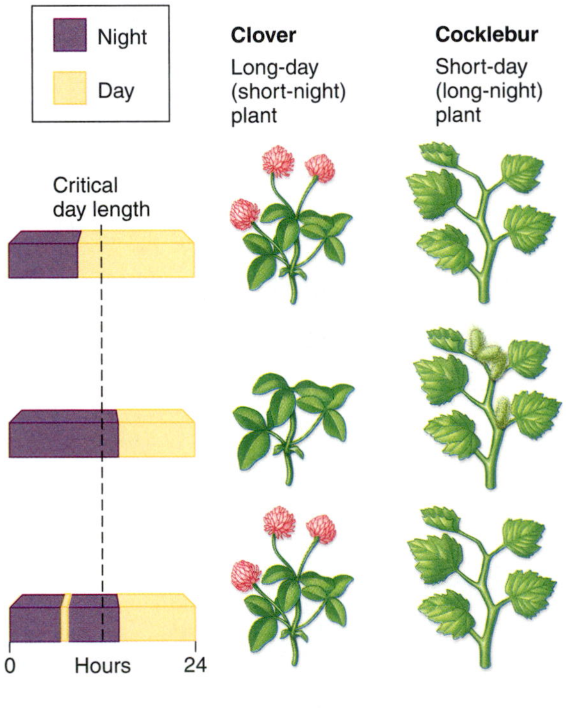

Night
Day
Clover
Long-day (short-night) plant
Cocklebur
Short-day (long-night) plant
Critical day length
0 Hours 24

FIGURE 26.21 Night-Time Light Flashes Inhibit Flowering in Short-Day Plants. Length of day and night influence flowering in long-day plants such as clover and short-day plants such as cocklebur. Long-day plants require a dark period shorter than a critical length. Short-day plants require an uninterrupted dark period longer than a critical period. Interrupting the dark period of a short-day plant inhibits flowering.

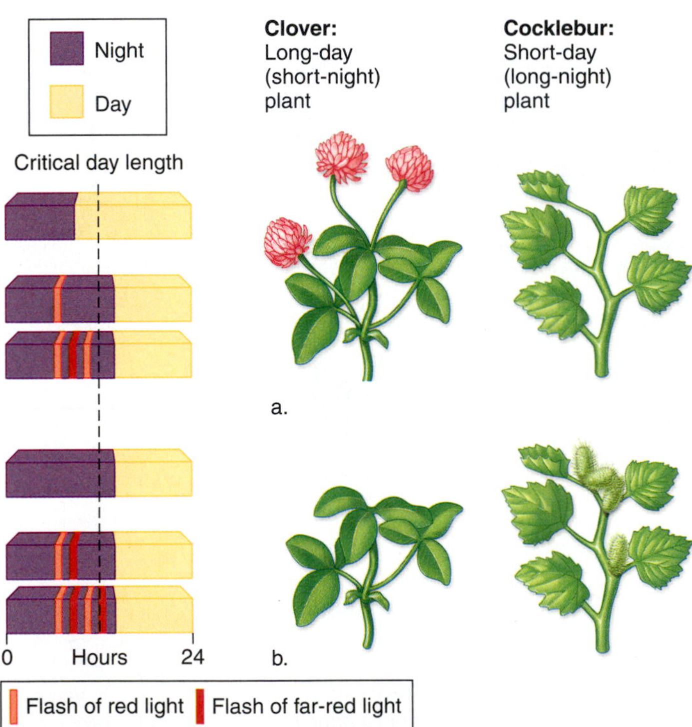

FIGURE 26.22 The Last Flash Matters. (**A**) Interrupting night with a flash of red light shortens continuous darkness. A long-day plant flowers, but a short-day plant does not because the time of uninterrupted darkness is too short. (**B**) A flash of far-red light closely following a flash of red cancels the effect of the red. The last flash matters because it determines the prevalent form of phytochrome, which plants use to detect the length of the night.

FIGURE 26.23 **Sleep Movements.** The prayer plant, *Maranta*, exhibits rhythmic sleep movements. When the sun goes down, the plant's leaves fold inward in a configuration resembling hands folded in prayer.

E. Light Entrains a Plant's Internal Clock

Many processes in plants recur at about the same time each day. In most plants, for example, stomata close at night and reopen in the morning. Some plants, such as the four-o'clock and the evening primrose, open their flowers daily in late afternoon or at nightfall, when their pollinators are most likely to visit. The prayer plant, *Maranta*, is an ornamental houseplant that exhibits nightly "sleep movements." It folds its leaves vertically each night, then moves them to a horizontal position during the day (**figure 26.23**). These examples all illustrate **circadian rhythms,** which are physiological cycles that repeat daily.

Circadian rhythms occur in many protists, fungi, and animals. In all species studied, pigments detect light and transmit signals to a "central oscillator," a set of genes and proteins that keep the clock on track. The details of the central oscillator, well understood in animals, remain unclear in plants.

Circadian rhythms often continue under laboratory conditions of constant light or dark. They are ingrained. Nevertheless, external conditions such as a change in photoperiod can reset ("entrain") the plant's internal clock. But entrainment to a new environment is limited. If a new photoperiod differs too much from a plant's biological clock, the plant reverts to its internal rhythms.

Summary *Photoreceptors absorb light energy, providing cues for seed germination and seedling development. Auxins and photoreceptors interact to mediate growth toward light (phototropism). Flowering is keyed to seasonal changes in day length. Internally controlled circadian clocks and photoperiodic responses together control rhythmic plant movements.*

26.6 MASTERING CONCEPTS

1. How does phytochrome control seed germination and photoperiodism?
2. How does a seedling grown in the dark look different from one exposed to light?
3. What is auxin's role in phototropism?
4. How are long-day and short-day plants different from one another?
5. How do biologists know whether photoperiodism is a response to duration of light or dark?
6. What is a circadian rhythm, and what is the evidence that both genes and the environment influence biological clocks?

26.7 Plants Respond to Gravity and Touch

Besides light, a developing plant also responds to countless other environmental cues. For example, the more CO_2 in the atmosphere, the lower the density of stomata on leaves. Likewise, a plant in soil with abundant nitrate (NO_3^-) produces fewer lateral roots than in nutrient-poor soil. Plants can also sense temperature; many require a prolonged cold spell before producing buds or flowering. A warm period in December, before temperatures have really plummeted, does not stimulate apple and cherry trees' buds to "break," but a similar warm-up in late February induces growth. Two other important environmental cues are gravity and mechanical stimulation.

As a seed germinates, its shoot points upward toward light, and its root grows downward into the soil. **Gravitropism,** the seedling's response to gravity, is another tropism that Charles Darwin and his son Francis investigated. In the late 1800s, scientists already knew that something in the root cap is necessary for gravitropism. Despite more than a century of subsequent research, the role of the root cap in sensing gravity is still not completely understood.

One hypothesis centers on root cap cells that have **statoliths,** starch-containing plastids that function as gravity detectors (**figure 26.24**). Statoliths normally sink to the bottoms of the cells, somehow telling the cells which direction is down. Turning a root causes the statoliths to move, reorienting growth downward. Movement of the statoliths is coupled to calcium ion and auxin movement in a way that bends the root. **chloroplasts, p. 66**

Besides ever-present gravity, a plant also experiences a changing variety of mechanical stimuli, including contact with wind, rain, and animals. Even roots encounter obstacles as they grow. Over time, repeated touching produces shorter, stockier roots and shoots. The mechanism is unknown, but perhaps plants can detect slight changes in cell shape, which somehow induces the expression of genes that regulate cell expansion. Because of the touch response, the cells expand radially instead of lengthening.

The coiling tendrils of twining plants exhibit **thigmotropism,** a directional response to touch (**figure 26.25**). Specialized epidermal cells detect contact with an object, which induces differential growth of the tendril. In only 5 to 10 minutes, the tendril completely encircles the object. Auxin and ethylene apparently control thigmotropism.

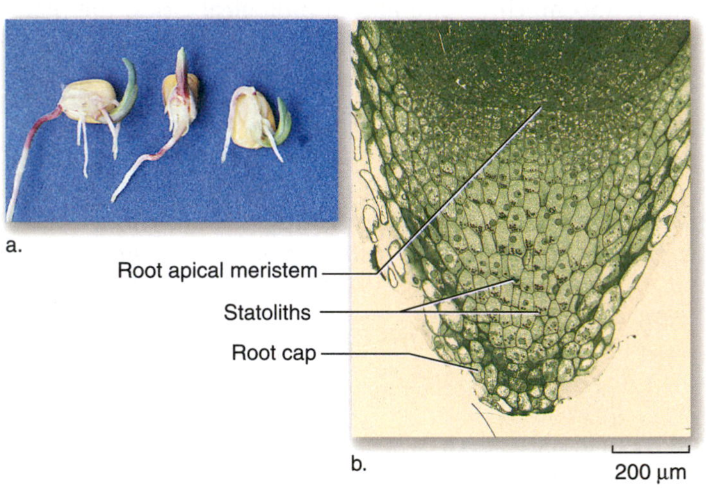

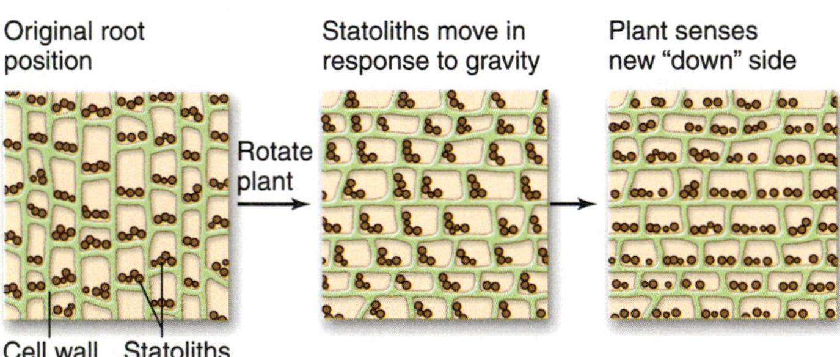

FIGURE 26.24 Gravity Influences Plant Growth. (**A**) Gravitropism causes shoots to grow up and roots to grow down, no matter how the seed was oriented. (**B**) Starch-filled organelles that function as statoliths may be the plant's mechanism for detecting gravity. (**C**) When a root is turned sideways, the statoliths move to occupy the new "down" side. This action somehow signals changes that set the root back on its normal downward course.

FIGURE 26.25 Thigmotropism. A tendril's epidermis is sensitive to touch. This tendril of a passion vine wraps around a blackberry stem.

Summary *A plant can adjust its direction and rate of growth in response to gravity and touch.*

MASTERING CONCEPTS

1. How are statoliths thought to participate in gravitropism, and how is this response adaptive?
2. How does thigmotropism help some plants climb?

26.8 Plant Parts Die or Become Dormant

A normal part of every plant's life is **senescence,** or aging, when metabolism switches from synthesis to breakdown. Annual plants, whose entire lives span just one growing season, senesce and die after they produce seeds. Trees and other perennial plants live for more than a year, sometimes for centuries. But even their tissues senesce, often seasonally.

The most spectacular example of senescence is the changing colors of the leaves of deciduous trees in autumn. As days begin to shorten, enzymes digest proteins, chlorophyll, and other large molecules in the leaves. Yellow, orange, and red carotenoid pigments, previously masked by chlorophyll, then become visible. These newly exposed carotenoids, along with the production of purplish anthocyanin pigments, combine to create the spectacular colors of autumn leaves. **why leaves change color, p. 107**

By the time a leaf falls, it is little more than a collection of cell walls and remnants of nutrient-depleted cytoplasm. The leaf separates from the plant at its **abscission zone,** a specialized layer of cells at the base of its petiole (**figure 26.26**). This cell layer forms early in development, but abscission does not occur until ethylene stimulates the production of digestive enzymes that degrade the cellulose and pectin that bond the cells together.

Why would a deciduous tree spend energy on leaf production each spring, only to discard its investment each autumn? The reason is that green leaves lose moisture to cold, dry winter winds, but the tree's roots cannot take up water from frozen soil. The seasonal loss of leaves removes a large surface area from which water would otherwise be lost. But what about evergreen trees and shrubs? Their leaves also die, but not all at once. Instead, they retain leaves for several years, periodically shedding the oldest ones.

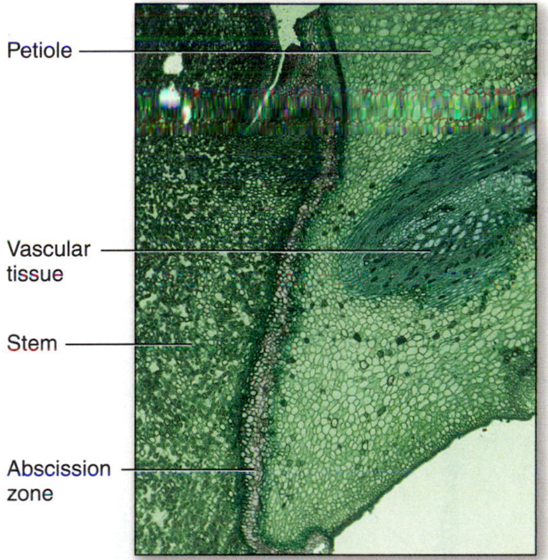

FIGURE 26.26 Leaf Abscission. The abscission zone is a region of separation that forms near the base of a leaf's petiole, minimizing the risk of infection and nutrient loss when the leaf is shed. Abscission layers also form at the bases of flower petals and fruit stalks.

FIGURE 26.27 Dormancy. Plants enter a seasonal state of dormancy, which enables them to survive harsh weather. The protective scales give the buds of this species a light brown color.

As a deciduous tree sheds its leaves before winter, other plant parts often become **dormant,** entering a state of decreased metabolism. Cells synthesize sugars and amino acids, which function as antifreeze that minimizes cold damage. Growth inhibitors accumulate in buds, transforming them into winter buds covered by thick, protective scales (**figure 26.27**).

The longer days and warmer temperatures of spring may break dormancy, although some plants use cues other than the seasons. In many desert plants, for example, rainfall alone releases the plant from dormancy. In contrast, potatoes require a dry period before renewing growth.

Summary *Senescence is the aging and death of some or all of a plant's tissues. Plants can also enter dormancy, a period of extremely low metabolic activity. Environmental conditions trigger release from dormancy.*

26.8 MASTERING CONCEPTS

1. What events occur when the leaves of a deciduous tree senesce and fall?
2. How is dormancy different from senescence?

26.9 INVESTIGATING LIFE
The ABCs of Flower Formation

Have you ever wondered how your head, arms, and legs formed in the right places as you developed from a fertilized egg into an adult? Like all multicellular organisms, humans and other animals have genes that operate early in development and ensure that the right parts form in the right places. These "master switches" are called homeotic genes. Figure 12.13 shows how a mutation in a homeotic gene causes legs to grow in place of antennae in the fruit fly *Drosophila melanogaster*. Similar genes control the development of flowers.

Flower formation is straightforward. A flower is actually a highly compressed stem, and flower parts are modified leaves. The stem's floral meristem produces cells that differentiate into sepals, petals, stamens, or carpels. Mutations in homeotic genes that control this differentiation should result in flowers with parts in the wrong places. To learn more about these genes, California Institute of Technology researcher Elliot Meyerowitz studies a small plant called *Arabidopsis thaliana*. (This plant, which is an important model organism in biological research, is described in chapter 20.)

In the 1980s, Meyerowitz assembled a collection of unusual *Arabidopsis* flowers that lacked one or more parts (**figure 26.28**). He deduced that three groups of genes, which he called *A*, *B*, and *C*, control flower formation. Each group of genes is expressed in certain parts of the flower:

- *A* genes are expressed in sepals and petals
- *B* genes are expressed in petals and stamens
- *C* genes are expressed in stamens and carpels

Sepals therefore express only class *A* genes, petals express *A* and *B*, stamens express *B* and *C*, and carpels express only *C*. Meyerowitz also hypothesized that when an *A* gene is inactive, a *C* gene takes over, and vice versa—when a *C* gene is inactive, an *A* gene takes over.

Meyerowitz tested his hypothesis by inducing mutations that inactivated each class of gene and then noting the effects on flower development. For example, when he mutated class *A* genes, he observed flowers with the pattern carpels–stamens–stamens–carpels; the flowers lacked sepals and petals. Here is how Meyerowitz explained this observation (remember that if an *A* gene is inactivated, a *C* gene takes over):

- The outermost whorl, which should consist of sepals, expresses *C* instead of *A* and therefore develops as carpels.
- The next whorl, normally made up of petals, expresses *C* and *B* instead of *A* and *B* and therefore develops as stamens.
- The two innermost whorls develop as stamens and carpels, as they would in a normal flower; these structures are unaffected by a mutation in an *A* gene.

Class *C* mutations, in contrast, not only yielded flowers consisting entirely of sepals and petals, but also somehow lifted controls on whorl formation. Mutant flowers developed many whorls, in the pattern sepals–petals–petals (again, remember that if a *C* gene is inactivated, an *A* gene takes over):

- The two outermost whorls are unaffected by a mutation in a *C* gene and therefore develop as sepals and petals.
- The third whorl, normally stamens, expresses *B* and *A* instead of *B* and *C* and therefore develops petals.
- The innermost whorl, normally carpels, does not develop at all, for unknown reasons.

Class *B* mutations yielded flowers with the whorl pattern sepals–sepals–carpels–carpels; they lacked petals and stamens. (Using the rules listed at the beginning of this section, can you explain why?)

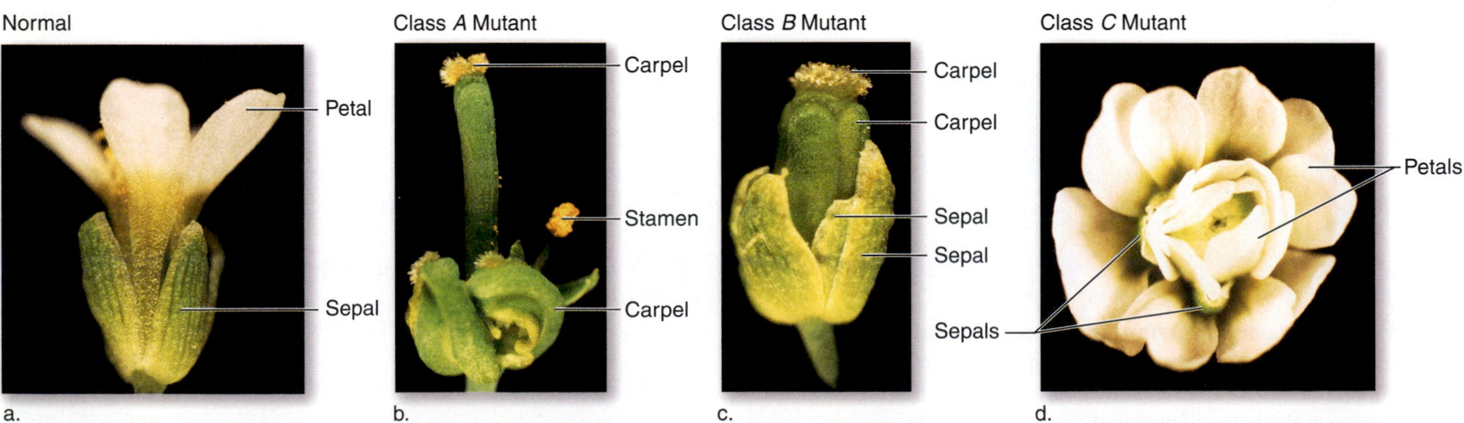

FIGURE 26.28 Homeotic Mutations in *Arabidopsis*. (**A**) Side view of a normal *Arabidopsis thaliana* flower. The stamens and carpel are hidden within the petals. (**B, C, D**) Mutating different classes of genes yields different abnormal phenotypes. **Question**: Why do class *B* mutations yield flowers with the whorl pattern sepals–sepals–carpels–carpels?

Although the DNA sequences in plants and animals are not the same, these two types of organisms do share a fundamental aspect of their development—control by a very few genes. Like homeotic genes of animals, the *A*, *B*, and *C* genes encode transcription factors, which are proteins that activate transcription of other genes. Meyerowitz and his colleagues continue to study how homeotic genes in plants are activated and what other genes these factors might regulate.

Bowman, John L., David R. Smyth, and Elliot M. Meyerowitz. 1991. Genetic interactions among floral homeotic genes of *Arabidopsis*. *Development*, vol. 112, no. 1, pp. 1–20.

Meyerowitz, Elliot M. November 1994. The genetics of flower development. *Scientific American*, vol. 271, pp. 56–65.

CHAPTER SUMMARY

Angiosperms Reproduce Asexually and Sexually

A. Asexual Reproduction Yields Clones

- In **asexual reproduction,** clones develop from the roots, stems, or leaves of a parent plant. It is advantageous in a stable environment where plants are well adapted to their surroundings.
- Grafting is a form of asexual reproduction. Tissue culture techniques help growers produce clones.

B. Sexual Reproduction Generates Variability

- Sexual reproduction produces genetically variable offspring, which increases reproductive success in a changing environment. Plant life cycles include alternation of haploid (**gametophyte**) and diploid (**sporophyte**) generations.

Egg and Sperm Unite in Female Flower Parts

A. Flowers Are Reproductive Organs

- **Flowers** are reproductive structures built of whorls of parts attached to a **receptacle.** The calyx, made of **sepals,** and the corolla, made of **petals,** are accessory parts. Inside the corolla, the **stamens** have pollen-containing **anthers** at their tips. **Carpels** occupy the center of the flower. The **ovary** contains one or more **ovules.** The **stigma** tops the **style,** which extends from the ovary.

B. The Pollen Grain and Embryo Sac Are Gametophytes

- Inside the flower, meiosis produces haploid **megaspores** and **microspores,** which develop into haploid female and male gametophytes, respectively. The gametophytes produce haploid gametes by mitosis.
- In the anther, specialized cells in **pollen sacs** divide meiotically, each yielding four haploid microspores. The microspores divide mitotically to yield haploid cells, one of which gives rise to two identical sperm nuclei. The **pollen grain** is the immature male gametophyte.
- In ovules within the ovary, specialized cells divide meiotically to yield four haploid cells, one of which persists as a haploid megaspore that divides mitotically three times. The resulting female gametophyte, or **embryo sac,** contains seven cells. One is the egg, and another has two **polar nuclei.**

C. Pollination Brings Pollen to the Stigma

- Animals or wind transfer pollen from anthers to a stigma. Flower structures and odors are usually adapted to either animal or wind **pollination.** Animal pollinators and flowers select for changes in one another in **coevolution.**

D. Double Fertilization Yields Zygote and Endosperm

- Once on a stigma, a pollen grain grows a pollen tube, and its two sperm nuclei move through the tube toward the ovary.
- In the embryo sac, one sperm nucleus fertilizes the egg to form the zygote, and the other sperm nucleus fertilizes the polar nuclei to form the **endosperm.** This phenomenon is termed **double fertilization.**

26.3 Seeds Develop Inside Fruits

A. A Seed Is an Embryo and Its Food Supply Inside a Seed Coat

- A **seed** is an embryo, endosperm, and **seed coat.** The endosperm nourishes the developing embryo as cells in apical meristems divide to produce the embryonic shoot and root. As the embryo grows, **cotyledons** develop.
- Seeds enter a dormancy period in which the embryo postpones development.

B. The Fruit Develops from the Ovary

- After fertilization, nonessential floral parts fall off, and hormones influence the ovary and sometimes other plant parts to develop into a **fruit.** The fruit protects the seeds and aids in dispersal.

C. Fruits Protect and Disperse Seeds

- Animals, wind, and water disperse seeds to new habitats, reducing competition between parent plants and their offspring.

26.4 Development Begins with Seed Germination

- Seed **germination** requires oxygen, water, and a favorable temperature. When the embryo bursts from the seed coat, the plant's primary growth begins.

26.5 Hormones Influence Plant Growth and Development

- Plants respond to the environment with changes in growth and movement, mediated by the action of **hormones.** Hormone interactions are complex.

A. Auxins and Cytokinins Are Essential for Plant Growth

- **Auxins** stimulate cell elongation in shoot tips, embryos, young leaves, flowers, fruits, and pollen. Auxins are most concentrated at the main shoot tip, which blocks growth of lateral buds (**apical dominance**).
- **Cytokinins** stimulate mitosis in actively developing plant parts, including lateral buds.

B. Gibberellins, Ethylene, and Abscisic Acid Influence Plant Development in Many Ways

- **Gibberellins** stimulate cell division and elongation and help break seed dormancy.
- **Ethylene** is a gas that speeds ripening, senescence, and leaf abscission.
- **Abscisic acid** counters the growth-inducing effects of other hormones by inducing dormancy and inhibiting shoot growth.

C. Biologists Continue to Discover Additional Plant Hormones

- **Brassinosteroids,** the only known steroid hormones in plants, are required for stem elongation. **Jasmonic acid** stimulates plant defenses.

26.6 Light Influences Germination, Growth, Flowering, and Daily Rhythms

A. Phytochrome Is a Pigment that Detects Light

- **Photoreceptors,** including phytochrome, absorb light energy of specific wavelengths and trigger specific responses.

B. Seed Germination May Require Light

- **Phytochrome** controls seed germination and early seedling growth. Seedlings grown in the dark are typically pale and spindly.

C. Phototropism Is Growth Toward Light

- A **tropism** is a growth response toward or away from an environmental stimulus. Typically, different parts of an organ or structure grow at different rates. In **phototropism,** light sends auxin to the shaded portion of the plant, stimulating growth toward the light.

D. Flowering May Be a Response to Photoperiod

- **Photoperiodism** is the ability of a plant to measure length of day and night; phytochrome senses the ratio of red light to far-red light to determine day and night length.
- Among plants that use photoperiod as a cue to produce flowers, **short-day plants** flower only when the duration of light is less than a critical length. **Long-day plants** require a light period longer than a critical length. Plants probably respond to length of darkness rather than length of daylight.

E. Light Entrains a Plant's Internal Clock

- Internal biological clocks control daily responses, or **circadian rhythms.** Environmental changes can alter, or entrain, these clocks.

26.7 Plants Respond to Gravity and Touch

- **Gravitropism** is growth toward or away from the direction of the force of gravity. The upward growth of shoots and downward growth of roots are gravitropic responses. The positions of starch-rich **statoliths** in cells apparently help plants detect gravity.
- **Thigmotropism** is growth directed toward or away from a mechanical stimulus such as wind or touch.

26.8 Plant Parts Die or Become Dormant

- **Senescence** is an active and passive cessation of growth. Senescent leaves detach at an **abscission zone.**
- Growth becomes **dormant** during cold or dry times and resumes when environmental conditions are more favorable.

26.9 Investigating Life: The ABCs of Flower Formation

- The pattern of homeotic gene expression in a developing flower determines the number and positions of sepals, petals, stamens, and carpels.

MULTIPLE CHOICE QUESTIONS

1. The new gene combinations associated with sexual reproduction in plants are the result of
   a. mitosis. c. pollination/fertilization.
   b. meiosis. d. both b and c.
2. Which of the following is NOT a flower structure associated with reproduction?
   a. Anther c. Sepal
   b. Stigma d. Carpel
3. Where would you find a microgametophyte?
   a. Inside a pollen grain c. Inside the embryo sac
   b. Inside an ovule d. Both b and c
4. What are the products of double fertilization?
   a. Two diploid zygotes
   b. A diploid zygote and a triploid endosperm
   c. A haploid sperm and a diploid zygote
   d. A triploid zygote
5. What is the function of a cotyledon?
   a. To support photosynthesis for the embryo
   b. To protect the dormant embryo
   c. To transfer nutrients from the endosperm
   d. To produce more cells for the growth of the embryo
6. Predict what would happen if a plant produced an excess of auxin.
   a. The plant would be very bushy.
   b. The plant would be very tall.
   c. The plant would have very long roots.
   d. The leaves of the plant would not die.

7. How does auxin cause bending in a plant?
   a. It causes an increase in cell division on one side of the plant.
   b. It suppresses cell division on one side of the plant.
   c. It triggers elongation of cells at the apical meristems of the plant.
   d. It triggers elongation of cells on one side of the plant.
8. Which plant hormone is most associated with ripening fruit?
   a. Abscisic acid
   b. Gibberellins
   c. Ethylene
   d. Brassinosteroids
9. Why does far-red light inhibit seed germination?
   a. It increases the levels of the P_{fr} form of phytochrome.
   b. It increases the levels of the P_r form of phytochrome.
   c. It prevents the synthesis of gibberellin.
   d. Both b and c.
10. What is a statolith?
    a. A starchy plastid
    b. A type of phytochrome
    c. An immature chloroplast
    d. A touch receptor

TESTING YOUR KNOWLEDGE

1. Give an example of asexual reproduction in a plant.
2. How does a flowering plant's life cycle differ from that of other multicellular organisms?
3. State the location and function of each of the following floral parts: sepal, petal, carpel, stamen, stigma, corolla, calyx, style, anther.
4. What happens to the two sperm cells that form inside a pollen tube?
5. Name tissues or cells in an angiosperm that are haploid, diploid, and triploid.
6. In what ways do plants "manipulate" animals to help them carry out their life cycles?
7. What are some examples of environmental cues that plants detect and respond to?
8. List the major plant hormones and describe some of their actions.
9. What is the function of photoreceptors?
10. Describe four effects of light on plant growth and development.
11. Name and describe three tropisms. Which hormone is common to all three?
12. Describe how photoperiodism controls flowering and entrains the circadian clock.
13. How does senescence occur?
14. What are some conditions that can release a plant from dormancy?

THINKING AS A SCIENTIST

1. The most common variety of banana in grocery stores, the "Cavendish" cultivar, comes from trees propagated by cloning. A serious fungal pathogen threatens Cavendish banana trees worldwide. How does cloning contribute to the problem?
2. Chefs consider a food a fruit or a vegetable according to how it is prepared and eaten, and how sweet it tastes. How does this differ from the biological definition of a fruit?
3. A single oak tree may produce thousands of acorns, which squirrels bury or eat. Why does the tree make so many acorns? Why might a plant whose seeds disperse far from the parent have better reproductive success than one whose fruits fall at the base of the parent plant?
4. Why is it adaptive for a plant to flower in response to photoperiod rather than temperature or another environmental cue?
5. Your friend bought a house last year from a previous owner who had planted apple trees in the yard. She is disappointed that her trees didn't bear much fruit in her first year. You discover that she regularly sprayed insecticides to kill mosquitoes in her yard. What might the pesticides have to do with the poor fruit yield?

Visit www.mhhe.com/hoefnagels for practice quizzes, animations, videos, and activities designed to help you master the material in this chapter.

UNIT 6

CHAPTER 27

Animal Tissues and Organ Systems

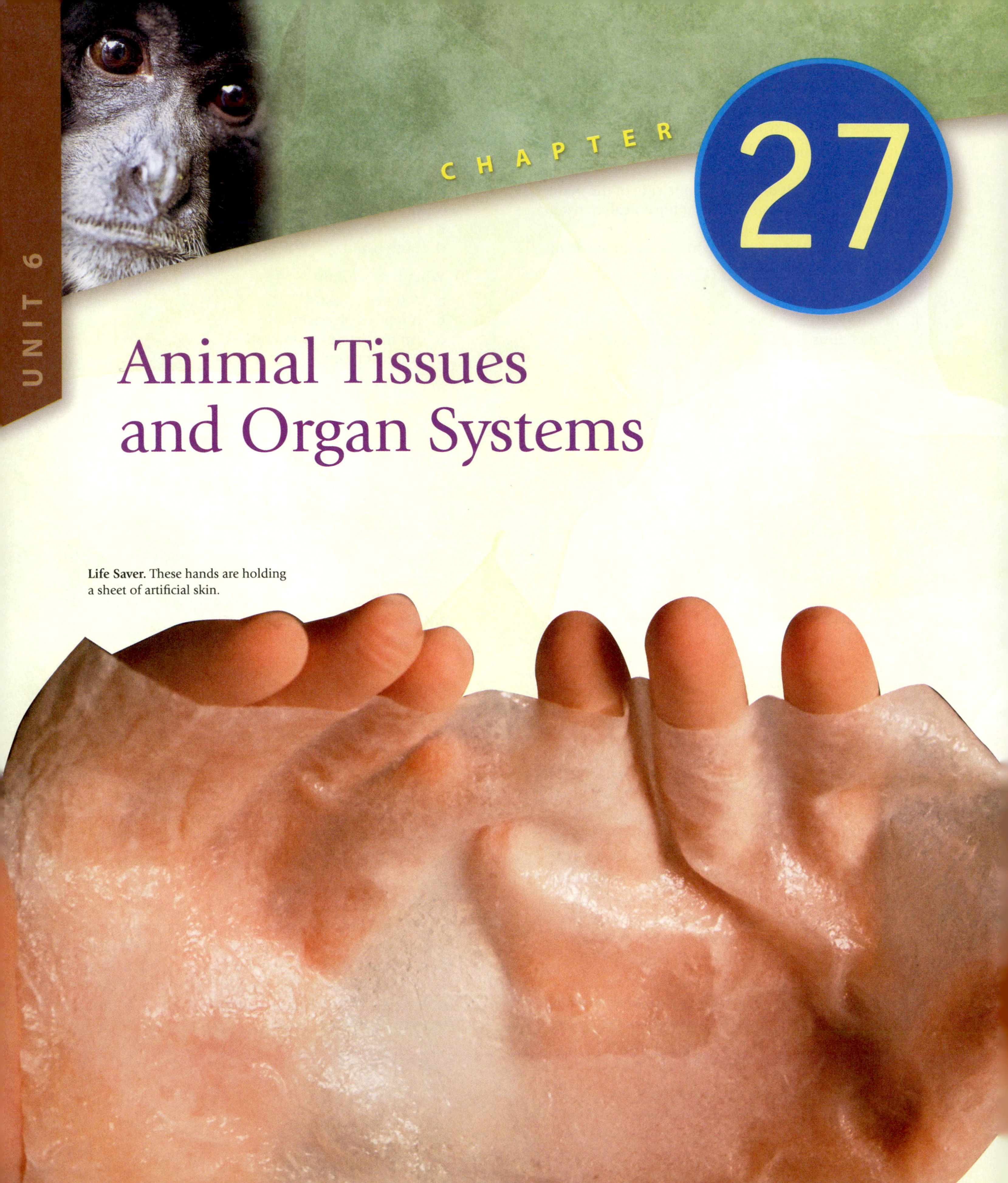

Life Saver. These hands are holding a sheet of artificial skin.

What's the Point?

Artificial Tissues and Organs

Healthy organs are a precious resource. In the United States alone, thousands of people with diseased or damaged kidneys, hearts, lungs, or livers endure an agonizing wait for someone else to die so that they might live. For many, the wait is fruitless. Every day, about 16 people in the United States die awaiting transplants. The shortage of healthy, compatible, transplantable organs creates a tremendous demand for substitutes. Thanks to a discipline called tissue engineering, however, the production of replacement organs may one day become routine.

One approach to tissue engineering is to grow tissues outside the body. Burn victims, for example, often lose large amounts of skin, increasing their risk of dehydration and infection. Physicians can graft skin from another part of the body to the burn site, but that is impossible when burns cover most of the body. Artificial skin is a combination of human cells and artificial scaffolding that helps solve this problem. The skin cells come from donated foreskins removed during circumcision of male babies. In one method of producing artificial skin, cells called fibroblasts separated from the donated tissue are grown on a biodegradable scaffold. After a few weeks, the cells form a layer of artificial skin, and the scaffold has dissolved. Best of all, the artificial tissue is free of bacteria and viruses, can be stored frozen, and does not provoke an immune response.

Scaffolds are already being used to grow three-dimensional organs outside the body. The first step in producing a replacement urinary bladder, for example, is to harvest healthy bladder cells and allow them to divide in culture. Biologists then "sow" the cells onto a biodegradable polymer, a temporary physical structure to which cells cling as they divide. This scaffold disintegrates after surgeons implant the replacement bladder into a patient. A similar technique might one day produce engineered blood vessels, kidney tubules, liver, or cartilage.

Tissue engineering may also include embedded devices that induce the body's cells to regenerate themselves. For example, physicians may someday be able to treat spinal cord injuries by creating porous "bridges" that would secrete signal molecules that promote cell division. The implant's channels would provide the architecture needed to guide and support the new cells.

All of these potentially life-saving advances are possible because of basic research into the biology of cells, tissues, and organs. This chapter introduces these levels of organization in the animal body; subsequent chapters in this unit explore each organ system in detail.

LEARNING OUTLINE

27.1 Specialized Cells Build Animal Bodies

Everywhere we look, form and function are entwined (**figure 27.1**). The broad, flat surface of a plant's leaf maximizes its exposure to light. A neuron's many branches permit the cell-to-cell connections essential to communication in the nervous system. In birds, fluffy down feathers trap pockets of air and conserve warmth.

Anatomy is the study of an organism's structure, that is, the parts that compose it and their location in the body. **Physiology** is a related discipline that describes how those parts work. Unit 5 described the anatomy and physiology of plants; unit 6 turns to the animals.

Most animals have cells organized into **tissues,** which are groups of cells that interact and provide a specific function (**figure 27.2**). Blood is a tissue, as is bone. An **organ,** in turn, consists of two or more interacting tissues. An eye is an organ that consists of light-sensitive cells, blood vessels, muscle, and other types of tissue. (Organ donation is the topic of this chapter's Burning Question: Which types of organs can be transplanted in humans?) Two or more organs may be joined, either physically or functionally, into **organ systems.** The eyes, ears, spinal cord, and brain are just a few components of the human nervous system. This organizational hierarchy is similar to that of plants; cells make up the tissues that build a plant's roots, stem, leaves, and other organs (see chapter 24).

FIGURE 27.1 Form and Function. A penguin is a flightless bird that spends much of its life in water. Its adaptations include a streamlined body, wings that function as flippers, and an insulating layer of air trapped beneath its smooth plumage.

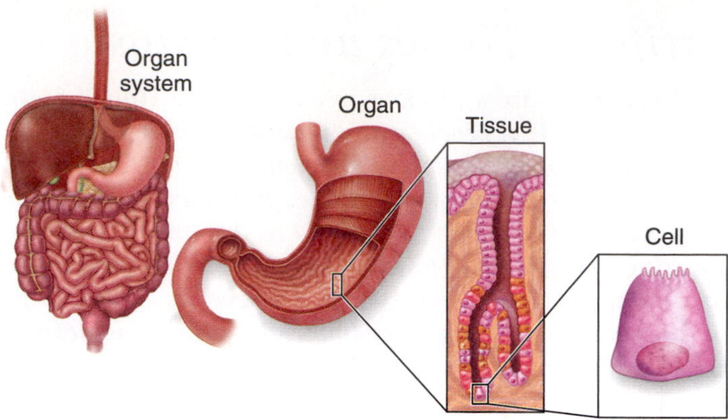

FIGURE 27.2 Organizational Hierarchy. Cells build tissues, which build organs, which interact to form organ systems.

To understand how the human body grows, develops, and comes to consist of trillions of specialized cells, turn the clock back to the beginning of your life. Your father's sperm fertilized your mother's egg. Hours later, that zygote divided and became two cells. Both cells divided again, and the embryo consisted of four cells, and so on. The cells formed a ball, and then the ball hollowed out. Up until this point, cells were unspecialized and appeared to be pretty much alike.

Then changes occurred. A few cells collected on the inner face of the ball and spread out to form sheets. The sheets then folded, eventually forming the three primary germ layers that gave rise to all of your body's tissues and organs (see figure 22.4). Ectoderm, for example, developed into your skin and nervous system. Your digestive tract, liver, and lungs all derive from endoderm. Mesoderm gave rise to muscles, bones, reproductive organs, and several other structures. **gastrulation, p. 451**

This chapter introduces the four basic tissue types that build animal organs, focusing on humans. Chapters 28 through 37 then consider the organ systems one at a time.

Summary *An animal's body is built of differentiated cells that associate as tissues, which begin forming early in embryonic development.*

27.1 MASTERING CONCEPTS

1. What is the difference between anatomy and physiology?
2. What is the relationship among cells, tissues, organs, and organ systems?
3. Trace the body's early development from one fertilized egg to a many-celled organism.

Burning Questions

Which types of organs can be transplanted in humans?

Medical technology can help replace body parts damaged by disease or injury. Transplantable organs include corneas, pancreases, kidneys, skin, livers, lungs, bone marrow, parts of the digestive tract, and hearts.

One of the challenges in transplanting an organ is to prevent the recipient's immune system from rejecting the foreign tissues. Drugs can suppress the immune system, but this increases the risk of infection from bacteria and viruses. Transplant physicians also minimize rejection by matching organ donors with recipients based on cell surface compatibility.

Unfortunately, the demand for transplantable organs far exceeds the supply. The chapter opening essay described tissue engineering as one approach to solving this problem. Transplanting organs from other species into humans is another possible solution (**figure 27.A**). Heart valves from pigs, for example, can replace malfunctioning ones in humans. But cross-species transplants may transmit viruses from the donor animal to the recipient. We do not know what effect pig viruses can have on a human body, and because many viral infections take years to cause symptoms, a new infectious disease in the future could be the trade-off for using nonhuman animal parts today.

Have a Burning Question of your own? Submit it to marielle_hoefnagels@mcgraw-hill.com for possible inclusion in future editions of this book!

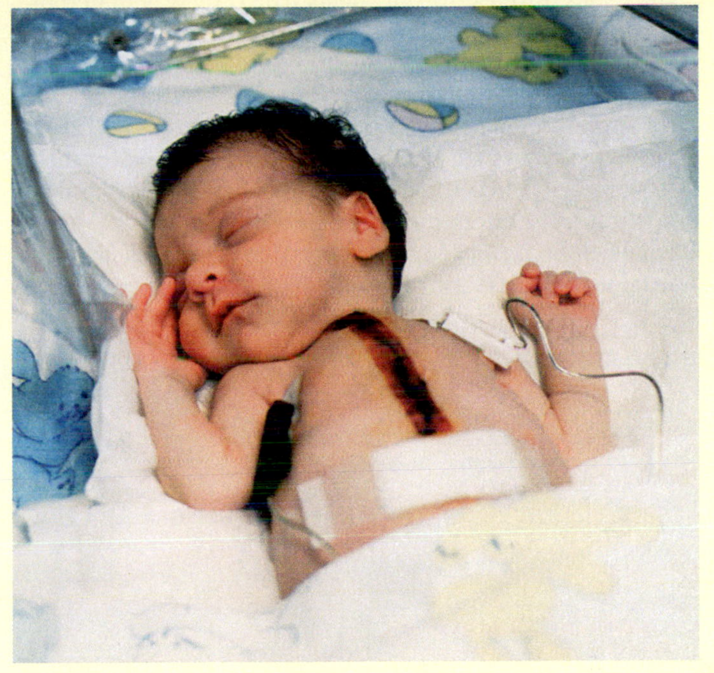

FIGURE 27.A Xenograft. In 1984, a California newborn, Baby Fae, lived 20 days with a baboon heart. She was born with an underdeveloped left half of the heart.

27.2 Animals Consist of Four Tissue Types

All together, the human body (and that of other vertebrates) has at least 260 cell types, forming tissues that fall into four broad categories (**table 27.1**): epithelial, connective, nervous, and muscle. These four tissue types are widely distributed throughout an animal's body.

Table 27.1 *Functions and Origins of Four Tissue Types*

Tissue Type	General Functions	Embryonic Origin
Epithelial	Covers and lines organs	Endoderm, ectoderm, or mesoderm
Connective	Provides support, adhesion, insulation, attachment, and transportation	Mesoderm
Nervous	Forms rapid communication networks among cells	Ectoderm
Muscle	Contracts to power movement	Mesoderm

A. Epithelial Tissue Covers Surfaces

Epithelial tissues coat the body's internal and external surfaces with one or more layers of tightly packed cells. They cover organs and line the inside of hollow organs and body cavities. The diverse functions of epithelial tissues include protection, nutrient absorption along the intestinal tract, and gas diffusion in the lungs. These tissues also form **glands,** organs that secrete substances such as breast milk, sweat, saliva, tears, mucus, hormones, enzymes, and other substances.

Epithelial tissues always have a "free" surface, exposed either to the outside or to a space within the body. On the opposite side, epithelium is anchored to other tissues by a noncellular layer called the basement membrane (although it is not the same as the membrane that surrounds all cells). Connections called tight junctions join cells of many epithelial tissues into leak-proof sheets. The only way materials can pass through these cell layers is by passing through their cell membranes. The tightly knit structure of epithelial tissue is closely tied to its function as a border between the body's tissues and an open space. **animal cell junctions, p. 72**

Epithelial tissues are classified partly by the shapes of their cells: squamous (flattened), cuboidal (cube-shaped), or columnar (tall and thin). The number of cell layers is also important. Simple epithelial tissues consist of a single layer of cells, whereas stratified epithelial tissues are made of multiple cell layers (**figure 27.3**). Pseudostratified epithelium appears stratified because of the staggered arrangement of nuclei in a single layer of columnar cells.

B. Connective Tissues Include Blood, Bone, and Cartilage

Much of a vertebrate's body consists of **connective tissues,** which are the most variable of the body's tissues. They fill spaces, attach epithelium to other tissues, protect and cushion organs, and provide mechanical support. Unlike epithelial tissues, connective tissues never coat any body surface.

The cells of connective tissues are embedded in an **extracellular matrix** consisting of nonliving substances. (Can *You*

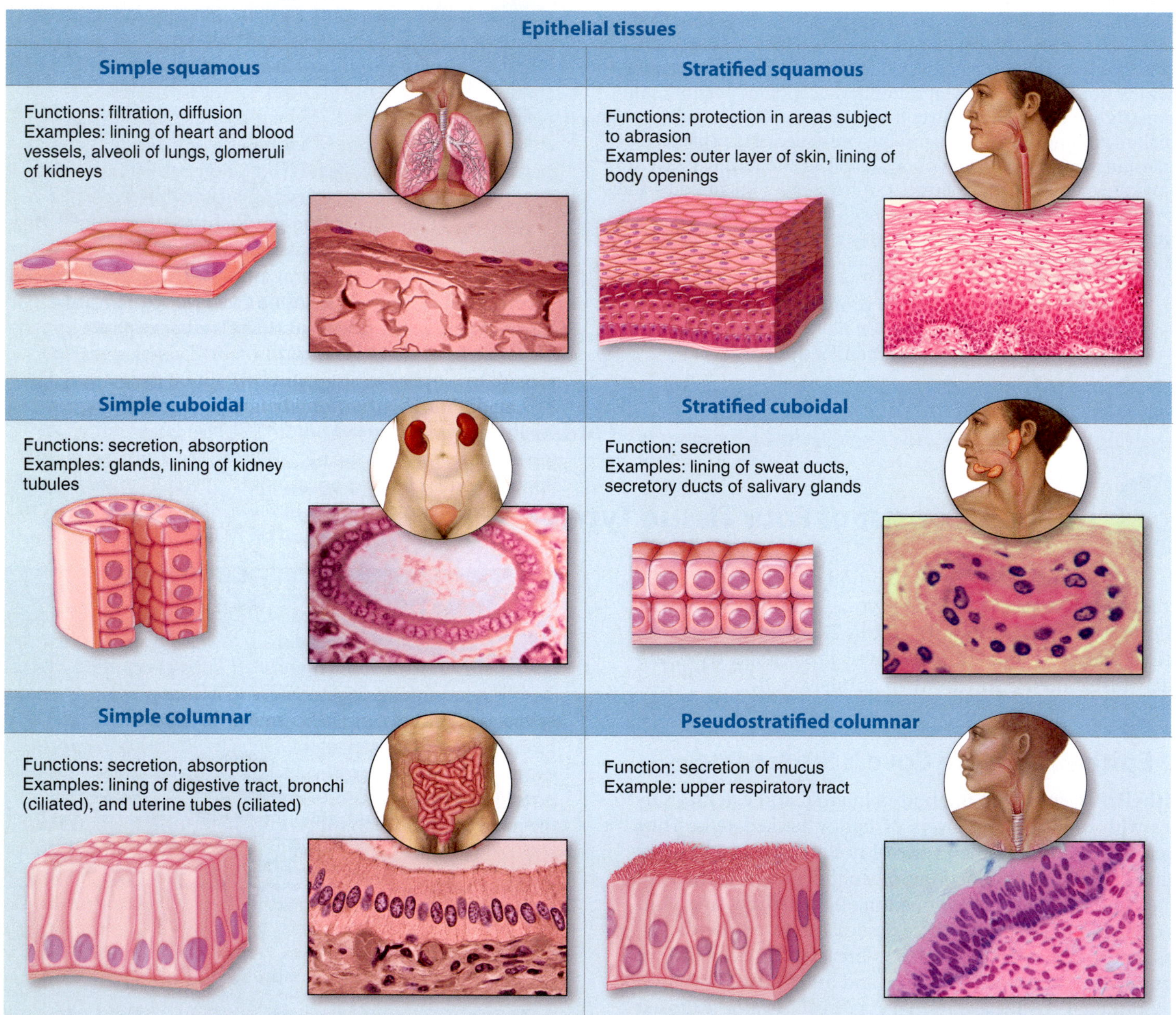

Epithelial tissues
Simple squamous
Functions: filtration, diffusion
Examples: lining of heart and blood vessels, alveoli of lungs, glomeruli of kidneys
Stratified squamous
Functions: protection in areas subject to abrasion
Examples: outer layer of skin, lining of body openings
Simple cuboidal
Functions: secretion, absorption
Examples: glands, lining of kidney tubules
Stratified cuboidal
Function: secretion
Examples: lining of sweat ducts, secretory ducts of salivary glands
Simple columnar
Functions: secretion, absorption
Examples: lining of digestive tract, bronchi (ciliated), and uterine tubes (ciliated)
Pseudostratified columnar
Function: secretion of mucus
Example: upper respiratory tract

FIGURE 27.3 Types of Epithelial Tissue. Epithelial tissues are composed of tightly packed cells in single or multiple layers that rest atop a basement membrane. Epithelial tissues cover body surfaces and line hollow organs and body cavities.

Relate? Cancers, Tissues, and the Extracellular Matrix explains the role of the cell's surroundings in the spread of cancer.) In all connective tissues except blood, cells called fibroblasts manufacture and secrete the collagen and elastin protein fibers in the matrix. Fibroblasts also secrete the extracellular matrix's "ground substance," a mixture of water, proteins, carbohydrates, lipids, and (in the case of bone) minerals. The ground substance that surrounds the connective tissue's cells and fibers may be solid (as in bone), liquid (as in blood), or semisolid (as in cartilage).

Connective tissues, which may contain many other cell types besides fibroblasts, are distinguished based on cell specializations, matrix composition, and the proportion of cells relative to matrix. **Figure 27.4** summarizes the six major types of connective tissues.

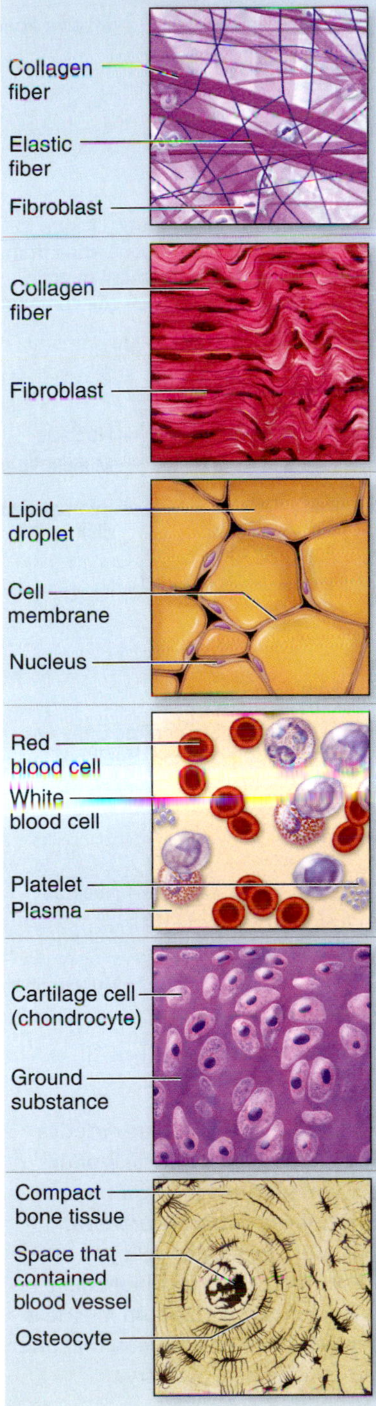

	Type of connective tissue	Cells	Matrix composition	Proportion of cells to matrix	Site
Collagen fiber; Elastic fiber; Fibroblast	**Loose connective tissue**	Fibroblasts	Loose elastin and collagen networks	Low	Under skin
Collagen fiber; Fibroblast	**Dense connective tissue**	Fibroblasts	Dense elastin and collagen networks	Low	Ligaments and tendons
Lipid droplet; Cell membrane; Nucleus	**Adipose tissue**	Fat cells (adipocytes)	Minimal	High	Beneath skin, between muscles, around heart and joints
Red blood cell; White blood cell; Platelet; Plasma	**Blood**	Red and white blood cells, platelets	Plasma	Low	In vessels throughout the body
Cartilage cell (chondrocyte); Ground substance	**Cartilage**	Chondrocytes	Fine fibers of collagen	Low	Ears, joints, bone ends, respiratory passages, embryonic skeleton
Compact bone tissue; Space that contained blood vessel; Osteocyte	**Bone**	Osteoclasts, osteoblasts, osteocytes	Collagen, minerals	Low	Skeleton

FIGURE 27.4 Types of Connective Tissue. Connective tissues are highly diverse in form and function, but they all consist of cells embedded in a nonliving matrix.

C. Nervous Tissue Connects and Integrates the Body

Nervous tissue conveys information within an animal's body. Sensory cells detect stimuli such as the scent of a rose or a prick of its thorn. Other cells then convey that information along nerves to the central nervous system (brain and spinal cord), which helps you interpret what you experience.

The nervous tissue that makes up nerves, the spinal cord, and the brain consists of two cell types (**figure 27.5**). **Neurons** form communication networks that receive, process, and transmit information. The cell may connect to another neuron at a junction called a synapse, or it may stimulate a muscle or gland. **Neuroglia** are cells that support neurons and assist in their functioning. Schwann cells, for example, are neuroglia with lipid-rich cell membranes that form insulating sheaths of the lipid myelin, which speeds nerve impulse conduction along a neuron. Astrocytes are star-shaped neuroglia with many functions in the brain.

Nervous tissue

a. Neuron

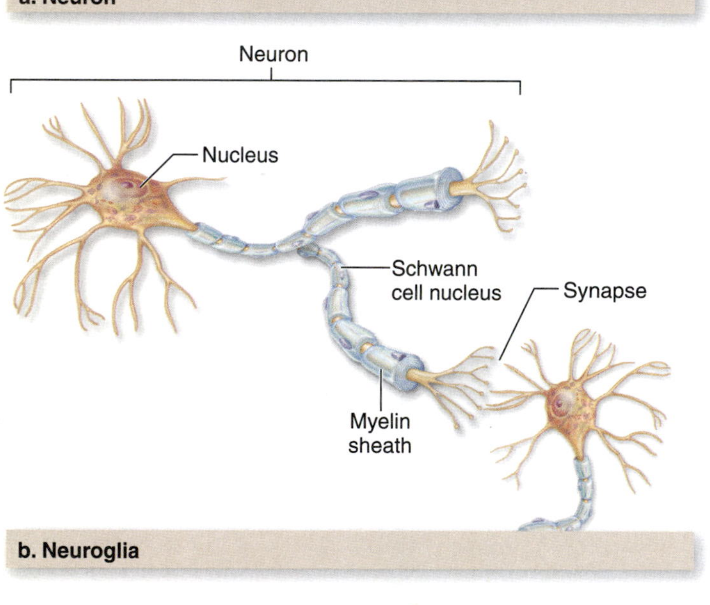

b. Neuroglia

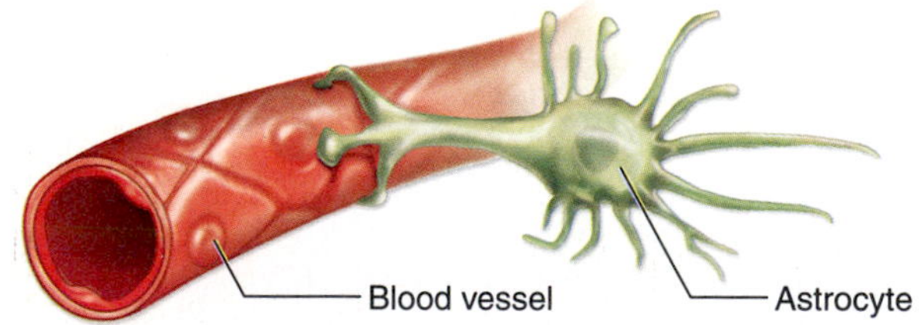

FIGURE 27.5 Nervous Tissue. Neurons and several types of neuroglia make up nervous tissue. (**A**) A neuron's cell body has several extensions that receive messages. The long axon relays impulses to other cells. (**B**) This astrocyte is a type of neuroglia that attaches to blood vessels in the brain. Neuroglia called Schwann cells also make up the myelin sheath coating the axon of the neuron in (A).

a. Skeletal muscle

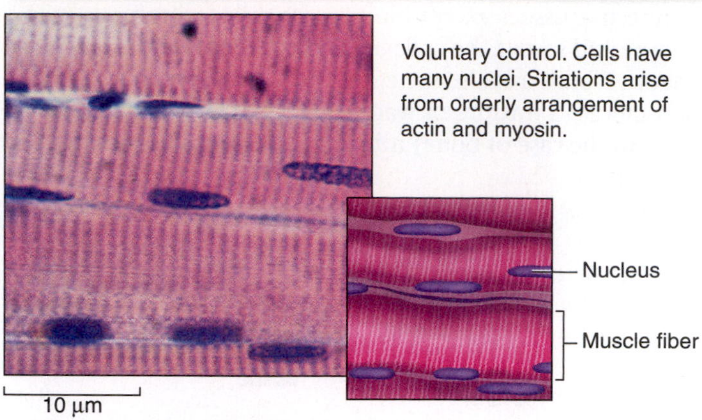

b. Cardiac muscle

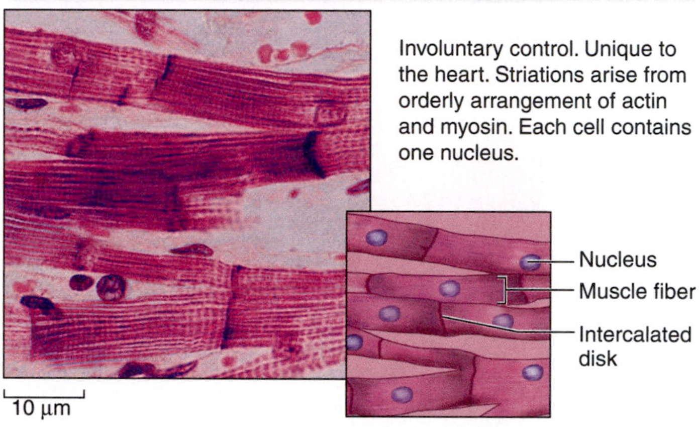

c. Smooth muscle

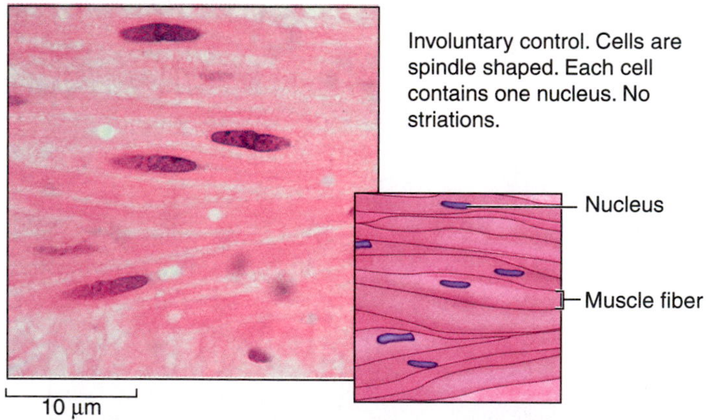

Involuntary control. Cells are spindle shaped. Each cell contains one nucleus. No striations.
Nucleus
Muscle fiber
10 μm

FIGURE 27.6 Types of Muscle Tissue. Muscle tissue includes skeletal, cardiac, and smooth muscle. (**A**) Skeletal muscle is under voluntary control, and the cells have many nuclei. The cells are striated due to the orderly organization of contractile proteins. (**B**) Cardiac muscle is unique to the heart. Its striated cells join at connections called intercalated disks. The cells form a branching network, and their contraction is involuntary. (**C**) Smooth muscle is composed of spindle-shaped cells, each with one nucleus, that often form layers. Contraction of smooth muscle is involuntary.

D. Muscle Tissue Provides Movement

Muscle tissue consists of cells that contract (become shorter) when protein filaments slide past one another. Abundant mitochondria provide the energy for contraction. Muscle cells attach to soft tissue or bone; when the muscle contracts, the body part moves.

Muscle tissue is of three types (**figure 27.6**). **Skeletal muscle** tissue consists of long cells called muscle fibers, each containing many nuclei. Under a microscope, skeletal muscle tissue appears striped, or striated, because many protein filaments align. Most skeletal muscle attaches to bone, and it provides the voluntary movements that an animal can consciously control. **Cardiac muscle** tissue, which is found only in the heart, is also striated, but the cells are shorter than those in skeletal muscle, and their control is involuntary. Cardiac muscle cells are electrically coupled with one another, so they contract simultaneously to produce the heart beat. **Smooth muscle** tissue is not striated, and its contraction is involuntary. This type of muscle pushes food along the intestinal tract, regulates the diameter of blood vessels, and controls the size of the pupil of the eye.

Summary ***Cells associate to form four major tissue types: epithelial tissue, connective tissue, nervous tissue, and muscle tissue.***

27.2 MASTERING CONCEPTS

1. Where in the body does epithelial tissue occur?
2. How are epithelial tissues named?
3. What is the relationship between the cells and extracellular matrix of connective tissue?
4. List and describe six types of connective tissue.
5. What are the two main cell types in nervous tissue?
6. List and describe the three types of muscle tissue.

Can You Relate?

Cancers, Tissues, and the Extracellular Matrix

Cancer is a disease of the cell cycle. As described in section 8.4, when cells escape the normal limits on cell division, they replicate continuously and form a mass called a tumor. Some of the abnormal cells may break away, travel in the blood or lymph, settle in other tissues, and seed new tumors.

Cancers are named for the type of tissue in which they originate. Carcinomas, which represent 90% of human cancers, arise in epithelial tissues. Common examples include breast, lung, prostate, and colon cancers. Leukemias and lymphomas originate in the tissues that give rise to blood and immune system cells; they account for about 8% of human cancers. Sarcomas arise in muscle or connective tissue that originates from mesoderm, but they are rare in humans.

Most normal body cells cannot survive or replicate if they are removed from the extracellular matrix. Ordinarily, cell membrane proteins bind to collagen or other matrix molecules on the outside, and to the cytoskeleton inside the cell. A normal cell fails to divide if not anchored in this way. Somehow, cancer cells escape this "anchorage dependence," breaking away from the extracellular matrix yet retaining the ability to divide. Furthermore, these abnormal cells often secrete enzymes that destroy the fibers of extracellular matrix, clearing the way for the cancer to spread to adjacent tissues.

Tumor cells also have another unusual property. They secrete molecules called growth factors that stimulate epithelial cells in nearby blood vessels to divide and sprout new capillaries into the tumor. Anticancer drugs called angiogenesis inhibitors block these signals, starving the tumor of the nutrients and oxygen it needs to grow.

27.3 Organ Systems Are Interconnected

Organ systems have distinctive functions, but they interact in many ways. The following sections provide a brief overview of human organ systems, organized by their contributions to the body's function (**figure 27.7**).

A. The Nervous and Endocrine Systems Coordinate Communication

The human **nervous system** is a vast, interconnected network of trillions of neurons and neuroglia. Some neurons are sensory receptors that detect stimuli in the environment or inside the body; others relay or interpret the sensory input. Still other neurons carry impulses from the brain or spinal cord to muscles or glands, which contract or secrete products in response. The nervous system specializes in rapid communication.

The **endocrine system** includes glands that secrete hormones, which are communication biochemicals that affect development, reproduction, mental health, metabolism, and many other functions. Most hormones travel within the

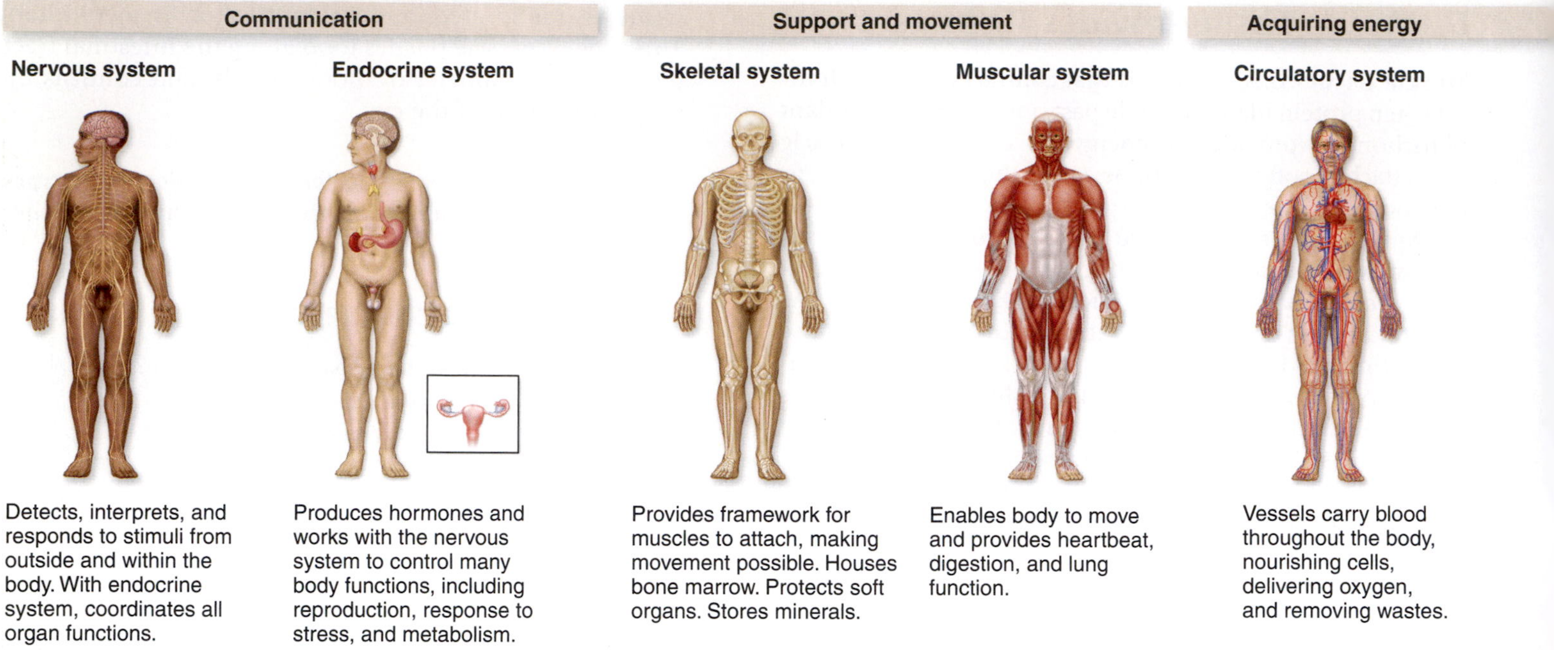

Communication

Nervous system

Detects, interprets, and responds to stimuli from outside and within the body. With endocrine system, coordinates all organ functions.

Endocrine system

Produces hormones and works with the nervous system to control many body functions, including reproduction, response to stress, and metabolism.

Support and movement

Skeletal system

Provides framework for muscles to attach, making movement possible. Houses bone marrow. Protects soft organs. Stores minerals.

Muscular system

Enables body to move and provides heartbeat, digestion, and lung function.

Acquiring energy

Circulatory system

Vessels carry blood throughout the body, nourishing cells, delivering oxygen, and removing wastes.

FIGURE 27.7 **Human Organ Systems.**

circulatory system and stimulate a characteristic response in target organs. For example, the pituitary gland in the brain produces the hormone prolactin, which promotes milk secretion in the breasts of a new mother. Endocrine signals act slowly but last longer than nerve impulses.

B. The Skeletal and Muscular Systems Support and Move the Body

The **skeletal system** consists of bones, ligaments, and cartilage. Bones provide frameworks and protective shields for soft tissues and serve as attachment points for muscles. The marrow within some bones produces the components of blood; bones also store minerals such as calcium.

Individual muscles are the organs that make up the **muscular system.** Muscle contraction provides the forces that move body parts, help maintain posture, keep food moving through the digestive tract, and enable the lungs and heart to do their jobs. The heat released by contracting muscles also helps maintain the body's temperature.

C. The Digestive, Circulatory, and Respiratory Systems Work Together to Acquire Energy

The **digestive system** supplies the nutrients that sustain all cells. Blood, which is part of the **circulatory system,** carries the digested food molecules to all of the body's cells. Blood also transports O_2 gas, which the lungs of the **respiratory system** acquire from the atmosphere. Cells use these raw materials in respiration to generate energy, producing carbon dioxide gas (CO_2) as a waste product. The circulatory system delivers this gas to the lungs to be exhaled. **cellular respiration, p. 121**

D. The Integumentary, Urinary, Immune, and Lymphatic Systems Protect the Body

The body protects itself in several ways. For example, the **integumentary system,** which consists of the skin and its outgrowths, is a barrier between an animal and its environment. The kidneys, which are part of the **urinary system,** remove toxins and wastes from blood, reabsorb useful substances, and maintain the concentrations of many ions in body fluids.

The body also fights infection, injury, and cancer. The **immune system** is a huge army of specialized cells, organs, and transport vessels. This system not only launches an attack against invading viruses, microorganisms, and cancer cells, but it also develops immunity to subsequent infection by the same pathogen.

The **lymphatic system** is a bridge between the immune system and the circulatory system. Lymph originates as fluid that leaks out of blood capillaries and fills the spaces between the body's cells. Lymphatic capillaries absorb the excess fluid and pass it through the lymph nodes, where immune system cells screen out foreign substances. The cleansed fluid then returns to the circulatory system.

Acquiring energy

Respiratory system

Delivers oxygen to blood and removes carbon dioxide. Helps control blood pH.

Digestive system

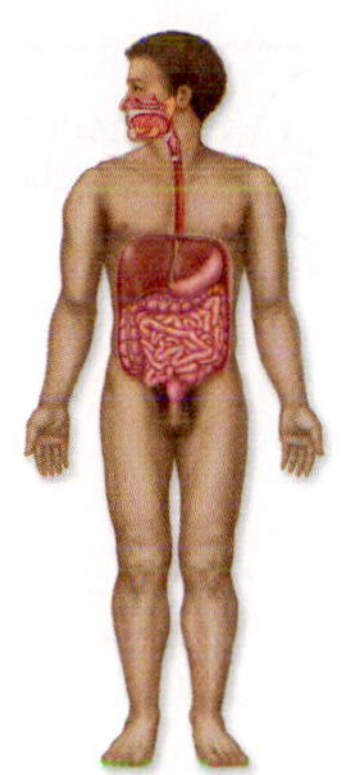

Breaks down nutrients into chemical components that are small enough to enter the circulation. Eliminates undigested food.

Protection

Integumentary system

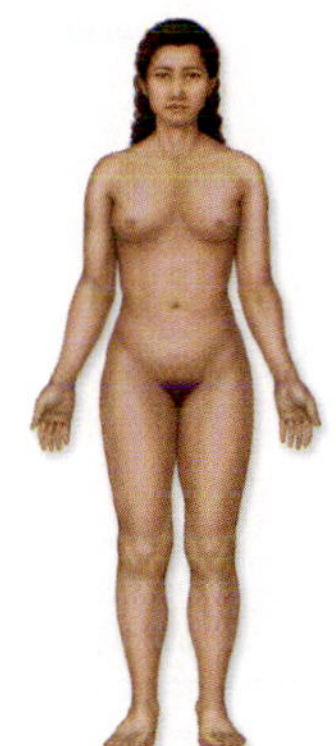

Protects the body, controls temperature, and conserves water.

Urinary system

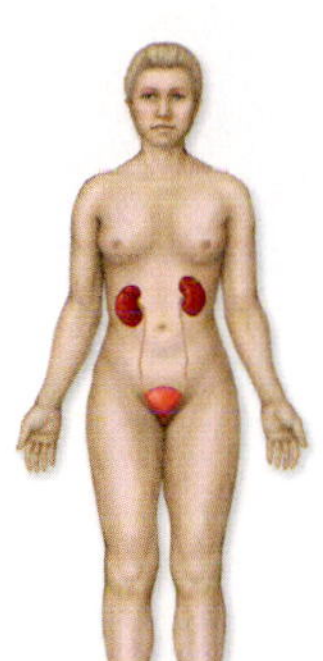

Excretes nitrogenous wastes and maintains volume and composition of body fluids.

Immune and lymphatic systems

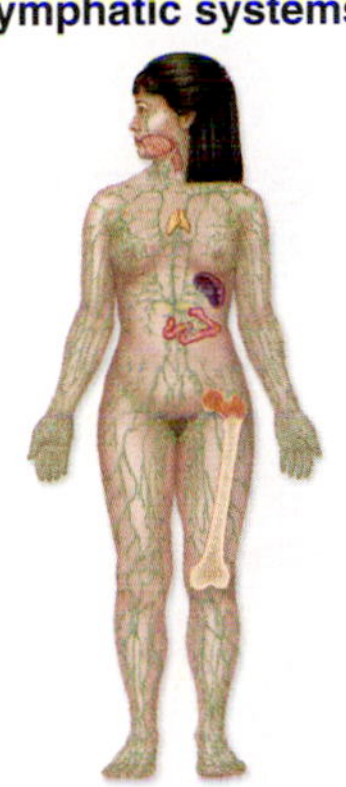

Protects body from infection, injury, and cancer.

Reproduction

Reproductive system

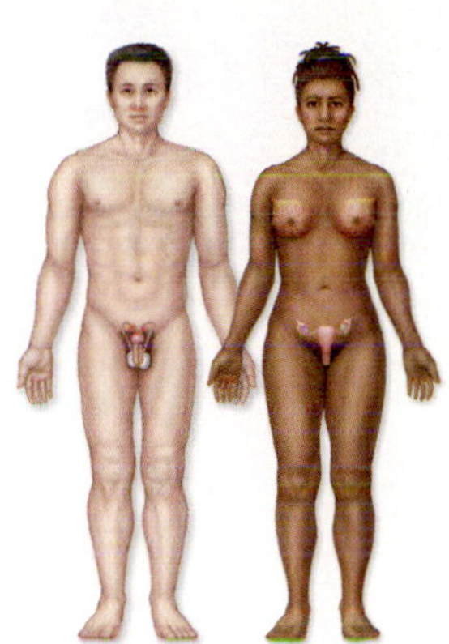

Manufactures gametes and enables the female to carry and give birth to offspring.

E. The Reproductive System Produces the Next Generation

Although the **reproductive system** is not vital to the functioning of an individual, it is essential for producing offspring. The reproductive system consists of organs that produce gametes, glands that enable the gametes to mature and function, and tubules that transport the gametes. The female body also can nurture developing offspring.

The reproductive system illustrates how the organ systems are, in a sense, not separate at all. Consider the uterus, the pear-shaped sac that houses the embryo and fetus. The majority of the uterus is muscle. It also contains nervous tissue, which is why a woman feels cramps when it contracts. Hormones from the endocrine system stimulate these contractions. The entire system is richly supplied with the circulatory system's blood vessels, which also deliver cells and biochemicals of the immune system.

Summary *An animal's organ systems provide communication, support and movement, energy, protection, and the ability to reproduce. These systems interact in many ways.*

27.3 MASTERING CONCEPTS

1. What are some examples of interactions between organ systems?
2. Which organ systems contribute to each of the five general functions of life?

27.4 Organ System Interaction Promotes Homeostasis

So far, this chapter has emphasized the body's cells, tissues, organs, and organ systems. An animal's body, however, consists mostly of water. Some of this moisture makes up the cytoplasmic "soup" that fills every cell. The rest of it forms blood plasma and the **interstitial fluid** that bathes the body's cells. Because interstitial fluid is inside the body but outside the cells, biologists consider it part of the "internal environment." Maintaining the correct concentrations of sugars, salts, hydrogen ions, and dissolved gases in body fluids is vital.

Yet the external environment changes constantly. Temperatures rise and fall; food may be abundant or scarce; water comes and goes. In the midst of this great variability, an animal's body must maintain its internal temperature, its blood

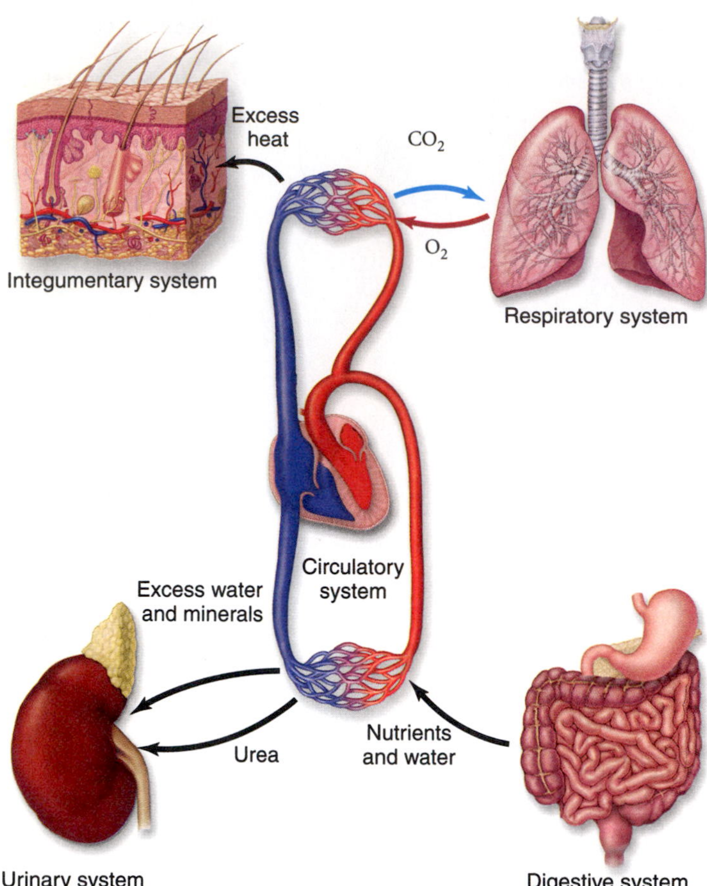

FIGURE 27.8 Examples of Homeostasis in the Human Body. Many organ systems work together to maintain the body's temperature and concentrations of salts, carbon dioxide, oxygen, and urea.

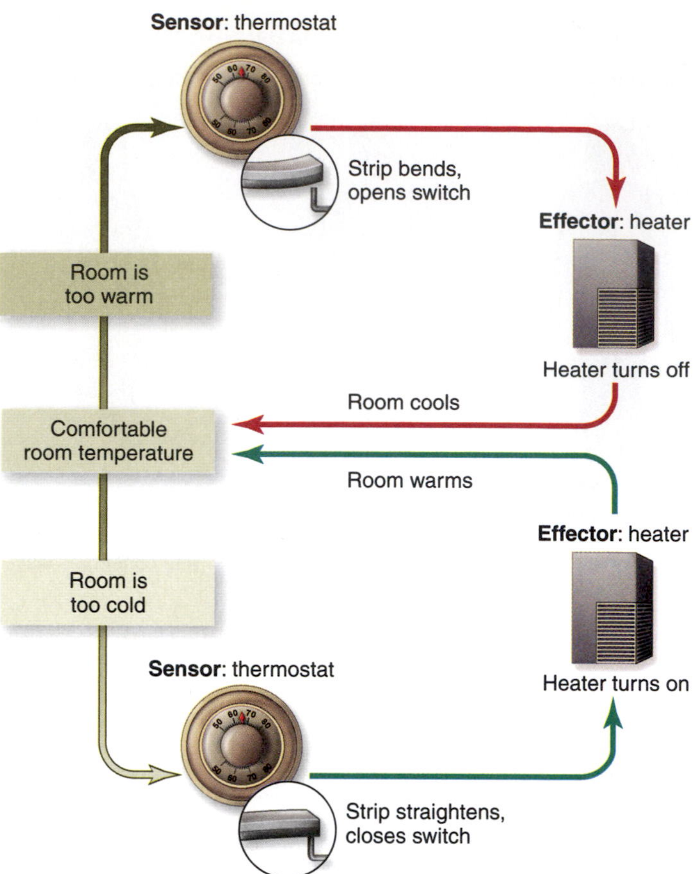

FIGURE 27.9 Negative Feedback Systems Promote Homeostasis. In negative feedback systems, some variable is controlled within limits. Sensors monitor changes in the controlled variable and activate an effector if the value drifts too high or too low. The effector's response counteracts the original change in the controlled variable.

pressure, and the chemical composition of its fluids within certain limits. **Homeostasis** is this state of internal constancy (**figure 27.8**).

If a body system cannot maintain homeostasis, it may stop functioning, and the organism may die. As just one example, consider what happens if the lungs fill with water. The body can no longer acquire O_2 or dispose of CO_2, yet cells continue to respire. Soon, all available O_2 is consumed, and CO_2 accumulates to toxic levels. Cells begin to die, and the person will drown unless rescuers arrive quickly.

The body's interconnected tissues and organs work together to maintain homeostasis, often under the direction of the nervous system. If the concentration of glucose in the blood is too high, the pancreas secretes a hormone called insulin, which stimulates other cells to absorb more sugar. If the concentration of salts is too high, the kidney releases more salt into urine. If oxygen is scarce, the brain stimulates faster breathing. If the body is too cold, muscles shiver and generate heat.

As these examples illustrate, a common way to maintain homeostasis is by **negative feedback,** which is an action that counters an existing condition (**figure 27.9**). In all negative feedback systems, **sensors** monitor changes in some variable. If the value of the variable is too high or too low, the sensor activates one or more effectors. The **effector's** response counteracts the original change.

Figure 27.9 illustrates negative feedback in a familiar situation: maintaining room temperature. When the temperature is low, the thermostat signals the heater to switch on. When the room gets too warm, the heater turns off. In this example, the thermostat acts as the sensor, and the heater is the effector. Likewise, the body uses negative feedback to maintain blood pressure. If blood pressure rises too high, receptors in the walls of blood vessels signal the brain to slow the contraction of the heart. The pressure drops. If blood pressure falls too low, the brain center signals the heart to speed up, sending out more blood.

Only a few biological functions demonstrate **positive feedback,** in which the body reacts to a change by amplifying it. Blood clotting and milk secretion are examples of positive feedback—once started, they perpetuate their activity. Positive feedback therefore does not maintain homeostasis. Ultimately, however, other controls take over and restore equilibrium.

Summary *An animal's body must maintain its temperature and the chemical composition of its fluids within a certain range. Organ systems interact to provide this internal constancy, or homeostasis.*

27.4 MASTERING CONCEPTS

1. Which fluids are part of the body's internal environment?
2. What happens to an organism that fails to maintain homeostasis?
3. Distinguish between negative and positive feedback.

27.5 The Integumentary System Regulates Temperature and Conserves Moisture

The integumentary system, which consists of the skin, hair, nails, and several types of glands, is an organ system that beautifully illustrates the main themes of this chapter. Not only does the integumentary system consist of multiple interacting tissue types, but it also helps the body maintain homeostasis.

Skin is an extensive organ. It is only 1 to 2 mm thick on average, although skin on the soles of the feet and several other places is thicker. Nevertheless, it makes up about 15% by weight of an adult's body. Human skin has two major layers (**figure 27.10**): the epidermis and the dermis. The **epidermis**, or outermost layer, consists mostly of stratified

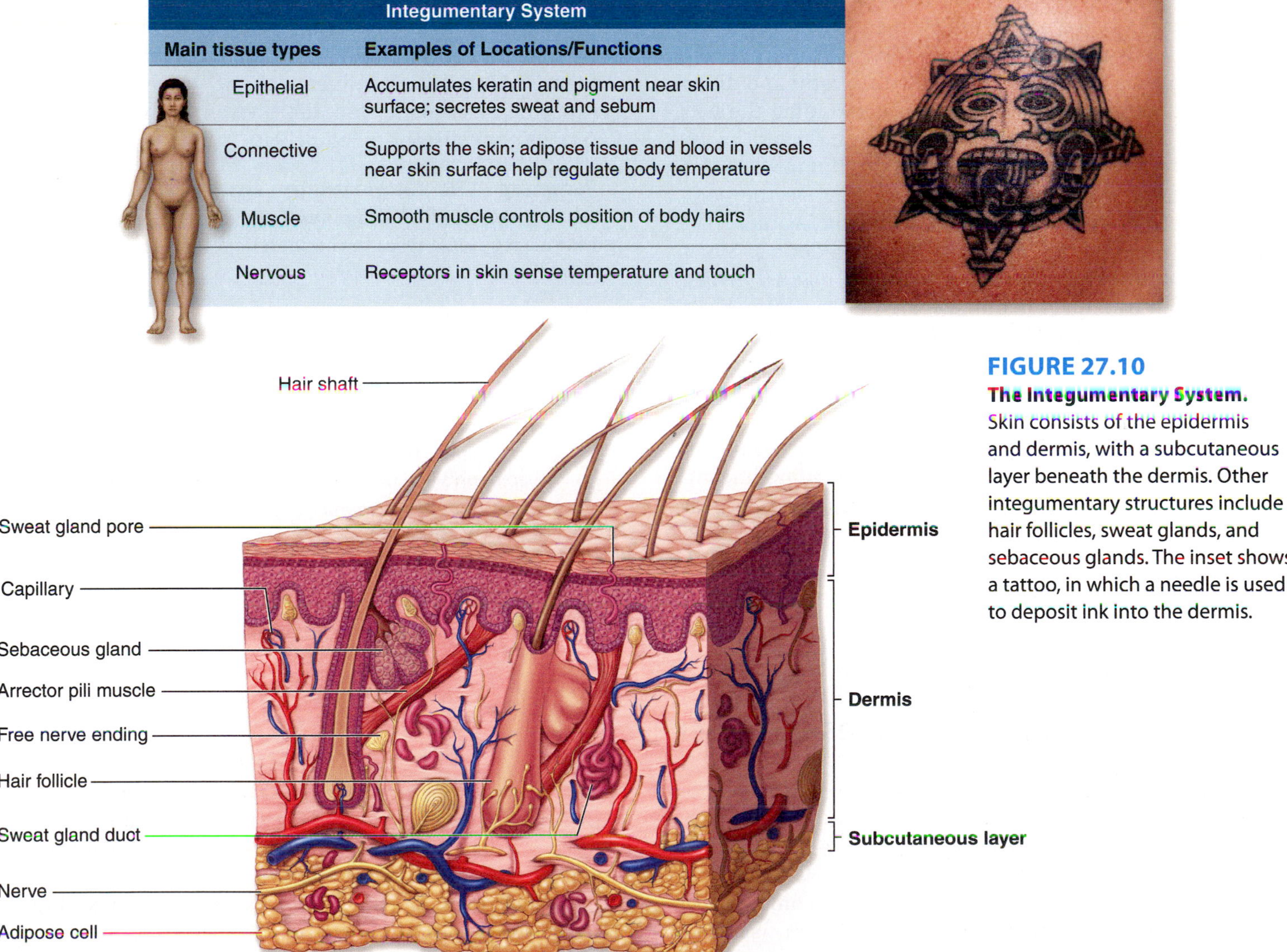

Integumentary System	
Main tissue types	**Examples of Locations/Functions**
Epithelial	Accumulates keratin and pigment near skin surface; secretes sweat and sebum
Connective	Supports the skin; adipose tissue and blood in vessels near skin surface help regulate body temperature
Muscle	Smooth muscle controls position of body hairs
Nervous	Receptors in skin sense temperature and touch

FIGURE 27.10 **The Integumentary System.** Skin consists of the epidermis and dermis, with a subcutaneous layer beneath the dermis. Other integumentary structures include hair follicles, sweat glands, and sebaceous glands. The inset shows a tattoo, in which a needle is used to deposit ink into the dermis.

squamous epithelium. Anchoring (adhering) junctions join the cells together, somewhat like rivets. Below the epidermis is the **dermis,** which is composed of dense, collagen-rich connective tissue. A thin basement membrane binds the two layers together, much as a staple seals two sheets of cardboard together. Beneath the dermis lies a layer of loose connective tissue and adipose tissue that is not technically part of the skin.

Cells within the epidermis have specialized functions. Keratinocytes are abundant cells that produce the protein keratin. (A callus is a thick, scaly accumulation of keratin.) The tough, dry keratin protects skin from disease-causing organisms and abrasion. The skin's color derives from melanocytes, cells that produce melanin. This pigment absorbs ultraviolet radiation that can damage DNA and therefore protects against some types of skin cancer. Other epidermal cells detect touch, and yet others help fight infection if the skin is injured.

Like the epidermis, the dermis has several functions. The collagen fibers of the dermis support the skin. Contracting the muscles attached to fibers in the skin of the face yields a wide array of expressions. Nerve endings in the dermis stimulate glands and convey information from sensory receptors to the brain.

The dermis also produces hair or fur. A hair grows from a group of cells called a hair follicle, which is anchored in the dermis. Cells at the base of a hair follicle divide, pushing daughter cells up. These cells stiffen with keratin and die to form the hair. Smooth muscle tissue surrounding the hair follicle can contract and make the hair "stand on end" when skin senses cold—this is the basis of goose bumps. Fur also camouflages many animals. In humans, hair on the body and face help to visibly differentiate men from women.

Several types of glands originate in the dermis. Sweat glands produce perspiration. Mammary glands, which produce milk in female mammals, are derived from sweat glands. Many animals, including humans, have sebaceous glands, which secrete an oily substance that softens and helps protect the skin and hair.

Injuries can damage the dermis. Suppose a person cuts her finger on broken glass. As the injury heals, the replacement skin contains more collagen fibers than the surrounding undamaged tissue. Neither hair follicles nor sweat glands develop in the damaged area. As a result, the hand retains a scar consisting of new skin that does not look or work exactly like the old. A tattoo is an intentional injury to the dermis. To create a tattoo, a needle deposits ink directly into the dermis. The injury eventually heals, but the ink remains just beneath the boundary between the epidermis and dermis (see figure 27.10).

Skin helps the body maintain homeostasis in several ways; the most obvious example is body temperature regulation. Specialized nerve endings sense the temperature of the skin and convey the information to a region of the brain called the hypothalamus. If the temperature is too high, the hypothalamus stimulates blood vessels in the skin to dilate, releasing heat from deeper tissues to the environment. Perspiration pours out of sweat glands. When the temperature is too cold, the skin's blood vessels constrict, keeping warm blood away from the body's surface. Muscle cells surrounding hair follicles contract; the erect fur provides extra insulation that helps many non-human mammals retain body heat.

The integument also maintains homeostasis in several other ways. For example, skin conserves water. People who suffer extensive burns lose large amounts of skin and may die of dehydration. Burn patients are also especially vulnerable to infection because intact skin is the first defense against disease-causing microorganisms. In addition, the initial step in vitamin D synthesis occurs in the skin. Vitamin D helps regulate the concentrations of calcium and phosphate in the blood, which in turn influences bone structure. Inadequate exposure to sunlight, or a diet poor in vitamin D, causes a serious bone disease called rickets.

Summary ***The integumentary system includes skin and its derivatives, including hair and sweat glands. Multiple interacting tissue types in the integument protect the body and help maintain homeostasis.***

27.5 MASTERING CONCEPTS

1. What are the layers of human skin, and what tissue types are they made of?
2. What are some examples of specialized cells of the epidermis and dermis?
3. How does the integumentary system help the body maintain homeostasis?

27.6 INVESTIGATING LIFE
When a Chair Becomes a Ladder

A small child quickly learns the many uses of a chair. She can sit on the seat and drink a glass of juice at the table, or she can drag the chair to the counter, stand on the seat, and reach the cookie jar on the counter. If the child's brother hangs his jacket on the chair when he comes home from school, the chair acquires yet another function—a makeshift coat rack. As versatile as the chair is, however, it cannot perform all of its functions at the same time. But if the family has more than one chair, the child can remain seated in one while her brother uses another as a ladder.

Many times in evolutionary history, organisms have essentially turned a chair into a ladder—that is, they have created "redundant" structures, then used them in multiple ways. This repeated expansion of function eventually produced the diverse, specialized cells and tissues that we find today. The same "repurposing" also happens on a molecular level. For instance, if a gene's DNA sequence changes, the amino acid sequence of the encoded protein may also change (see chapter 12). The resulting protein might not work at all, might do its old job better, or might even take on an entirely new function.

Ordinarily, such repurposing would mean the loss of the protein's former job. Sometimes, however, errors during meiosis can cause genes to become duplicated. The result is a pair of redundant genes, one of which can retain its old function while the other can mutate and perhaps acquire a new function—a little like having two chairs in a kitchen, one to sit in and one to climb. **errors in meiosis, p. 185**

Gene duplication has been an important force in evolution, and many specialized functions in our own tissues originated as slight variations on a previous theme. Consider two calcium-rich substances in the human body: teeth and breast milk. Teeth consist mostly of calcium-hardened dentin and enamel, whereas milk contains lipids, proteins, sugars, and calcium. The functions of teeth and milk are very different, but making either substance depends on cells that can do the same thing: secrete calcium into their extracellular matrix.

Did mammals acquire the ability to secrete milk by finding a new use for an old protein, one required in tooth production? Not long ago, it would have been difficult to answer this question, but the biotechnology revolution has led to an explosion in scientific knowledge about the history of our genes. Researchers now deposit all publicly available DNA sequences—for any species—into online databases such as GenBank, an ever-expanding resource containing tens of millions of searchable records. Comparing the nucleotide sequences of similar genes for multiple species has yielded invaluable clues to evolution.

Pennsylvania State University researchers Kazuhiko Kawasaki and Kenneth M. Weiss wanted to learn more about the evolution of teeth and milk. They therefore studied genes encoding proteins that regulate extracellular concentrations of calcium and phosphate. Some of these proteins occur in calcium-hardened tooth enamel; others, called caseins, provide calcium and phosphate to an infant in milk. Still other proteins occur in saliva and regulate the deposition of minerals onto the enamel surface.

Kawasaki and Weiss hypothesized that the genes encoding all of these proteins descended from a common ancestor by gene duplication and subsequent mutation. To test their hypothesis, Kawasaki and Weiss retrieved from GenBank the sequence for human chromosome 4, so they could pinpoint the physical locations of the human versions of the genes. They also obtained the gene sequences encoding calcium-phosphate regulating proteins for humans, other mammals (pig and cow), and reptiles (caiman and alligator). The researchers then compared the gene sequences and constructed evolutionary tree diagrams for the proteins.

Kawasaki and Weiss found 12 functional genes encoding calcium-rich proteins in teeth, milk, and saliva clustered on chromosome 4. The researchers then used DNA sequences from multiple species to construct a phylogenetic tree covering hundreds of millions of years of evolutionary history (**figure 27.11**). Their results suggest that the "primordial" gene encoded an enamel matrix protein in a toothed fish called a conodont. This gene was duplicated in other fishes and reptiles, leading to multiple enamel proteins. In mammals, additional gene duplications eventually gave rise to the current diversity of milk, saliva, and tooth proteins.

Kawasaki and Weiss' study helps scientists understand how cells and tissues can use the same basic tool to do different jobs. In this case, the researchers showed that gene duplication has created new genes, and that subsequent changes have led to brand-new adaptations. The study therefore reveals an important pathway of evolutionary change. In addition, teeth and milk are especially interesting because these adaptations have literally changed the history of vertebrate life on Earth. Eventually, the researchers might be able to discover the genetic events that produced the vertebrate skeleton—an important event for us if ever there was one.

Kawasaki, Kazuhiko, and Kenneth M. Weiss. 2003. Mineralized tissue and vertebrate evolution: the secretory calcium-binding phosphoprotein gene cluster. *Proceedings of the National Academy of Sciences*, vol. 100, pages 4060–4065.

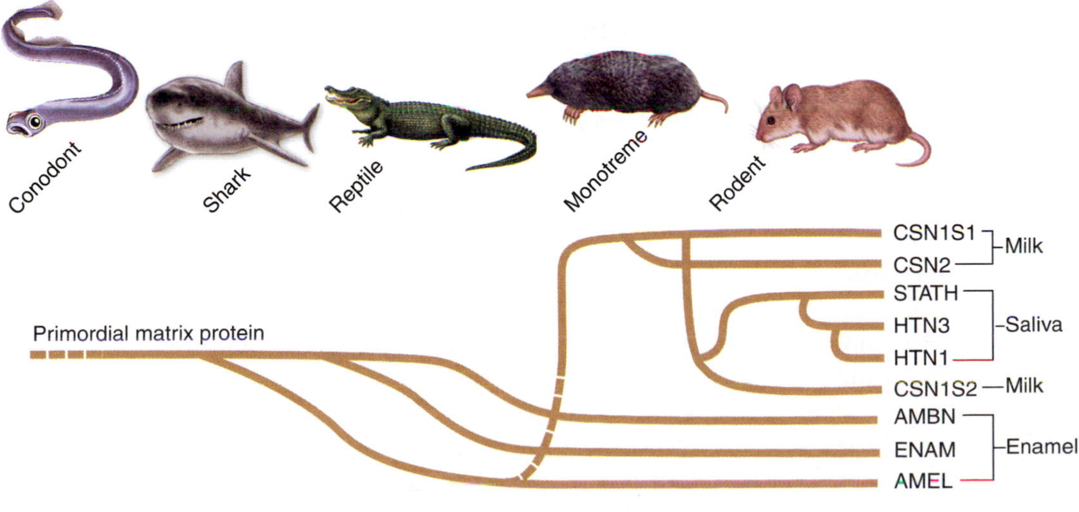

FIGURE 27.11 All in the Gene Family. One ancestral enamel matrix protein diverged over evolutionary time to give rise to milk and saliva proteins in humans. Except for *AMEL*, all of the genes encoding proteins listed on the right side of the diagram are on human chromosome 4. **Question:** Blood clotting results from a cascade of chemical reactions, each requiring the participation of a slightly different protein. How might you test the hypothesis that the genes encoding the blood-clotting proteins arose from multiple gene duplication events?

CHAPTER SUMMARY

27.1 Specialized Cells Build Animal Bodies

- **Anatomy** and **physiology** are interacting studies of the structure and function of organisms.
- Specialized cells express different genes. These cells aggregate and function together to form **tissues.** Tissues build **organs,** and interacting organs form **organ systems**.

27.2 Animals Consist of Four Tissue Types

A. Epithelial Tissue Covers Surfaces

- **Epithelial tissue** lines organs and forms **glands.** This tissue protects, senses, and secretes.
- Epithelium may be simple (one layer) or stratified (more than one layer), and the cells may be squamous (flat), cuboidal (cube-shaped), or columnar (tall and thin).

B. Connective Tissues Include Blood, Bone, and Cartilage

- **Connective tissues** consist of cells within an **extracellular matrix** that may be solid, semisolid, or liquid. Fibroblasts secrete collagen and elastin.
- The six major types of connective tissues are loose connective tissue, dense connective tissue, adipose tissue, blood, cartilage, and bone.

C. Nervous Tissue Connects and Integrates the Body

- **Neurons** and **neuroglia** make up **nervous tissue.**
- A neuron functions in rapid communication; neuroglia support neurons.

D. Muscle Tissue Provides Movement

- **Muscle tissue** provides movement when protein filaments slide past one another.
- Three types of muscle tissue are **skeletal, cardiac,** and **smooth muscle.**

27.3 Organ Systems Are Interconnected

A. The Nervous and Endocrine Systems Coordinate Communication

- The **nervous system** and **endocrine system** coordinate all other organ systems.
- Neurons form networks of cells that communicate rapidly, whereas hormones produced by the endocrine act more slowly.

B. The Skeletal and Muscular Systems Support and Move the Body

- The bones of the **skeletal system** protect and support the body, and they act as a reservoir for calcium and other minerals.
- The **muscular system** enables body parts to move and generates body heat.

C. The Digestive, Circulatory, and Respiratory Systems Work Together to Acquire Energy

- The **digestive system** provides nutrients. The **respiratory system** obtains O_2, and the **circulatory system** delivers it to tissues.
- The body's cells use O_2 to extract energy from food molecules. The circulatory and respiratory systems eliminate the waste CO_2.

D. The Integumentary, Urinary, Immune, and Lymphatic Systems Protect the Body

- The **integumentary system** provides a physical barrier between an animal and its surroundings.
- The **urinary system** removes wastes from the blood and reabsorbs useful substances.
- The **immune system** protects against infection, injury, and cancer.
- The **lymphatic system** connects the circulatory and immune systems, filtering the body's fluids through the lymph nodes.

E. The Reproductive System Produces the Next Generation

- The male and female **reproductive systems** are essential for the production of offspring.

27.4 Organ System Interaction Promotes Homeostasis

- **Homeostasis** is the maintenance of a stable internal environment, including regulation of body temperature and the chemical composition of blood plasma and **interstitial fluid.**
- **Negative feedback** restores the level of a substance or parameter to within a normal range. **Sensors** detect changes in the internal environment and activate **effectors** that counteract the change.
- **Positive feedback** perpetuates an action.

27.5 The Integumentary System Regulates Temperature and Conserves Moisture

- The integument in many animals consists of an **epidermis** over a **dermis,** plus specialized structures such as hairs and sweat glands. A basement membrane joins the epidermis to the dermis, and a layer of connective tissue underlies the dermis.
- In the epidermis, keratinocytes accumulate keratin, and melanocytes provide pigment.
- The integumentary system helps animals regulate their temperature, conserves moisture, and contributes to the production of vitamin D.

27.6 Investigating Life: When a Chair Becomes a Ladder

- Gene duplication is an important force in evolution because it allows cells to acquire new functions.
- Ancient gene duplications have created a set of genes responsible for the formation of teeth and milk, two important adaptations in vertebrates.

MULTIPLE CHOICE QUESTIONS

1. Which of the following represents the correct order of organization of an animal's body?
   a. Cells; organs; organ systems; tissues
   b. Cells; tissues; organ systems; organs
   c. Tissues; cells; organs; organ systems
   d. Cells; tissues; organs; organ systems
2. The cells of the nervous system are most closely related to cells that form your
   a. muscles.
   b. bones.
   c. skin.
   d. digestive tract.
3. The cells of an epithelium must ______ in order to function properly.
   a. form a single layer
   b. attach to one another by tight junctions
   c. secrete substances
   d. transport chloride ions
4. A common property of all types of connective tissue is the
   a. formation of solid or semi-solid arrangements of cells.
   b. production of collagen.
   c. arrangement of cells in an extracellular matrix.
   d. arrangement of cells into multiple layers.
5. Blood is an example of what type of tissue?
   a. Epithelium
   b. Nervous
   c. Connective
   d. Muscle
6. Smooth muscle is *different* from skeletal muscle because
   a. smooth muscle contraction is involuntary.
   b. skeletal muscle is striated.
   c. smooth muscle contains actin and myosin.
   d. both a and b.
7. Which of the following represents an example of an interaction between organ systems?
   a. Nervous; muscular; skeletal
   b. Respiratory; lymphatic; reproductive
   c. Reproductive; skeletal; immune
   d. Integumentary; circulatory; endocrine
8. Which of the following is NOT an organ system associated with protection?
   a. Immune
   b. Lymphatic
   c. Endocrine
   d. Urinary
9. The inner layer of skin is composed of _______ while the outer layer is ________ .
   a. epithelium; connective tissue
   b. dermis; epidermis
   c. epidermis; epithelium
   d. epithelium; epidermis
10. How does the integumentary system influence homeostasis?
    a. By preventing water loss
    b. By sensing external temperatures
    c. By preventing infection
    d. All of the above

TESTING YOUR KNOWLEDGE

1. Identify and describe the four basic tissue types.
2. Distinguish between:
   a. neurons and neuroglia.
   b. negative and positive feedback.
   c. organs and organ systems.
3. What is homeostasis, and how is it important?
4. When a person gets cold, he or she may begin to shiver. If the weather is too hot, the heart rate increases and blood vessels dilate, sending more blood to the skin. How does each scenario illustrate homeostasis?

THINKING AS A SCIENTIST

1. How would you design an experiment to determine whether a new brand of artificial skin is safe for use in humans?
2. Marfan syndrome and osteogenesis imperfecta are two examples of heritable disorders of connective tissue. Why do people with such disorders have many interrelated symptoms?
3. Explain how negative feedback regulates the size of your eye's pupil when the light in a room becomes brighter or dimmer.
4. List three examples of organ system interaction.

Visit www.mhhe.com/hoefnagels for practice quizzes, animations, videos, and activities designed to help you master the material in this chapter.

CHAPTER 28

The Nervous System

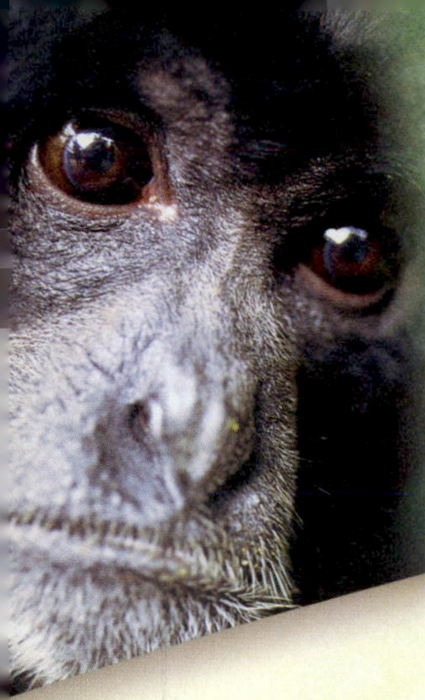

Just Relax. Stanford University sleep researcher Dr. William Dement holds a dog, Tucker, before and after a cataplexy attack.

What's the Point?

The Narcolepsy Gene

A sleeping cat looks inactive, but appearances can be deceiving—a whirlwind of chemical activity is occurring inside the brain of a dozing animal. Although scientists do not fully understand the function of sleep, it is clearly essential to good health. Among other physiological effects, a lack of sleep apparently increases the production of stress hormones and causes high blood pressure. Some studies link sleep deprivation to obesity, heart disease, and other serious illnesses.

Several sleep disorders plague humans. The most familiar is insomnia, the inability to sleep. Causes include everything from worrying to consuming too much caffeine. Another common disorder is sleep apnea, in which a person briefly stops breathing several times a night. In some cases, the brain's "breathing center" briefly quits working and then resumes. Usually, however, an obstructed airway causes sleep apnea.

Of all the sleep disorders, narcolepsy may be the strangest. People with narcolepsy are excessively sleepy during the day, even if they slept enough the night before. Some patients also experience vivid hallucinations or paralysis during or after sleep. The most dramatic symptom, however, is cataplexy, a state in which the person abruptly loses control of his or her muscles, often in response to intense emotion such as joy or surprise. The effects vary widely, from a slight drooping of the head to a total collapse. Cataplexy usually only lasts for a few minutes, but it can be very dangerous. For example, a person who loses muscle control, even for a moment, while driving a car may cause a deadly accident.

Sleep research took a step forward in 1999, when scientists reported the discovery of a defective gene associated with narcolepsy in Doberman pinschers and Labrador retrievers. In healthy dogs, the gene encodes a protein expressed on cell surfaces in a part of the brain called the hypothalamus. The protein is a receptor that receives neurotransmitters, chemical messages from other cells. Narcoleptic dogs have a mutated version of the gene, so the receptor protein is absent or fails to bind to the neurotransmitter. Apparently, affected brain cells do not transmit the "stay awake" message, especially when the narcoleptic dogs become excited.

Is this marriage of genetic and neurological research important to anyone other than dog-fanciers? Yes, because humans have the gene encoding the receptor, too. The discovery of this gene may someday yield not only better narcolepsy treatments but also, on the opposite end of the sleep disorder spectrum, better sleeping pills.

LEARNING OUTLINE

28.1 The Nervous System Forms a Rapid Communication Network

Love, happiness, tranquility, sadness, jealousy, and rage—all of these emotions spring from the cells of the nervous system. So do the ability to understand language, the sensation of warmth, memories of your childhood, and your perception of pain, color, sound, smell, and taste. The muscles that move when the heart beats or when you chew, blink, or breathe all are under the control of the nervous system, as is the unseen motion that propels food along the digestive tract.

The most critical functions of the nervous system are the "behind the scenes" activities that maintain a state of internal constancy, or homeostasis (see chapter 27). The feedback systems that maintain homeostasis require communication between the sensors (the cells that detect internal and external conditions) and the effectors, which make adjustments. The **nervous system** and the **endocrine system** provide this essential communication. A major difference between these two organ systems is the speed with which they act. The nervous system's electrical impulses travel so rapidly that their effects are essentially instantaneous. The moment you decide to pick up a pencil or type a letter, for example, you can carry out your plan. The endocrine system, the subject of chapter 30, acts much more slowly. Endocrine glands secrete chemical messages called hormones that take hours (or longer) to exert their effects.

To understand how the nervous system regulates virtually all other organ systems, imagine a lynx hunting a hare (**figure 28.1**). The cat can hear, see, and smell its prey. The lynx's nervous system interprets this sensory input and decides how to act, and then motor systems coordinate the skeletal muscles that move the lynx into position to catch the hare. Meanwhile, the cat's heart pumps blood, and its lungs inhale and exhale.

The three major roles of the nervous system—sensory input, sensory integration, and motor response—result from interactions between the nervous system's two main divisions:

- The **central nervous system** integrates sensory information and coordinates the body's response. In vertebrates, the central nervous system consists of the **brain** (inside the skull) and the tubular **spinal cord.**
- The **peripheral nervous system** carries information between the central nervous system and the rest of the body. The sensory pathways carry information to the spinal cord and brain; the motor pathways transmit nerve impulses from the central nervous system to muscle or gland cells.

The neurons and neuroglia that make up nervous tissue control mood, appetite, blood pressure, coordination, and perception of pain and pleasure. Neurons enable animals to sense the environment, screen out unimportant stimuli, move, learn, and remember. Yet despite their diverse functions, all neurons communicate in a similar manner. This chapter first explores how neurons function and then considers the human nervous system in more detail.

Sensory input

Sensory integration (brain and spinal cord interpret sensory input)

Motor response (muscles and glands react)

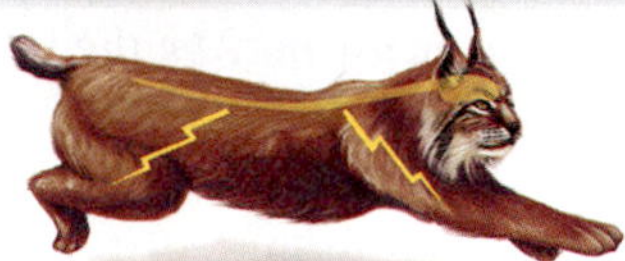

FIGURE 28.1 Three Nervous System Roles. Sensory organs such as eyes, ears, and the nose receive sensory input. The central nervous system integrates the information and sends signals that initiate appropriate motor responses.

Summary *The nervous system is a complex, rapid communication network that enables an organism to maintain homeostasis. The two main subdivisions of the nervous system are the central nervous system and the peripheral nervous system.*

28.1 MASTERING CONCEPTS

1. How is the nervous system's role in maintaining homeostasis different from that of the endocrine system?
2. What are the three main roles of the nervous system?
3. Distinguish between the central and peripheral nervous systems.

28.2 Neurons Are Functional Units of a Nervous System

All components of the nervous system include the same two cell types: interconnected neurons and their associated neuroglial cells (see chapter 27). The **neurons** are the cells that communicate, and the more numerous **neuroglia** play supporting roles.

A. A Typical Neuron Consists of a Cell Body, Dendrites, and an Axon

All neurons have the same basic parts (**figure 28.2**). The rounded **cell body** contains the nucleus, mitochondria that supply ATP, and ribosomes that manufacture proteins. **Dendrites** are short, branched extensions that transmit information toward the cell body. They may number from one to thousands, and each branching dendrite can receive input from many other neurons.

The **axon,** also called the nerve fiber, is typically a single long extension of the cell body. The axon is finely branched at its tip, and each tiny terminal extension connects to another cell at a junction called a synapse. Usually, the axon conducts nerve impulses from the cell body to a muscle, gland, or other neuron. The axon that permits

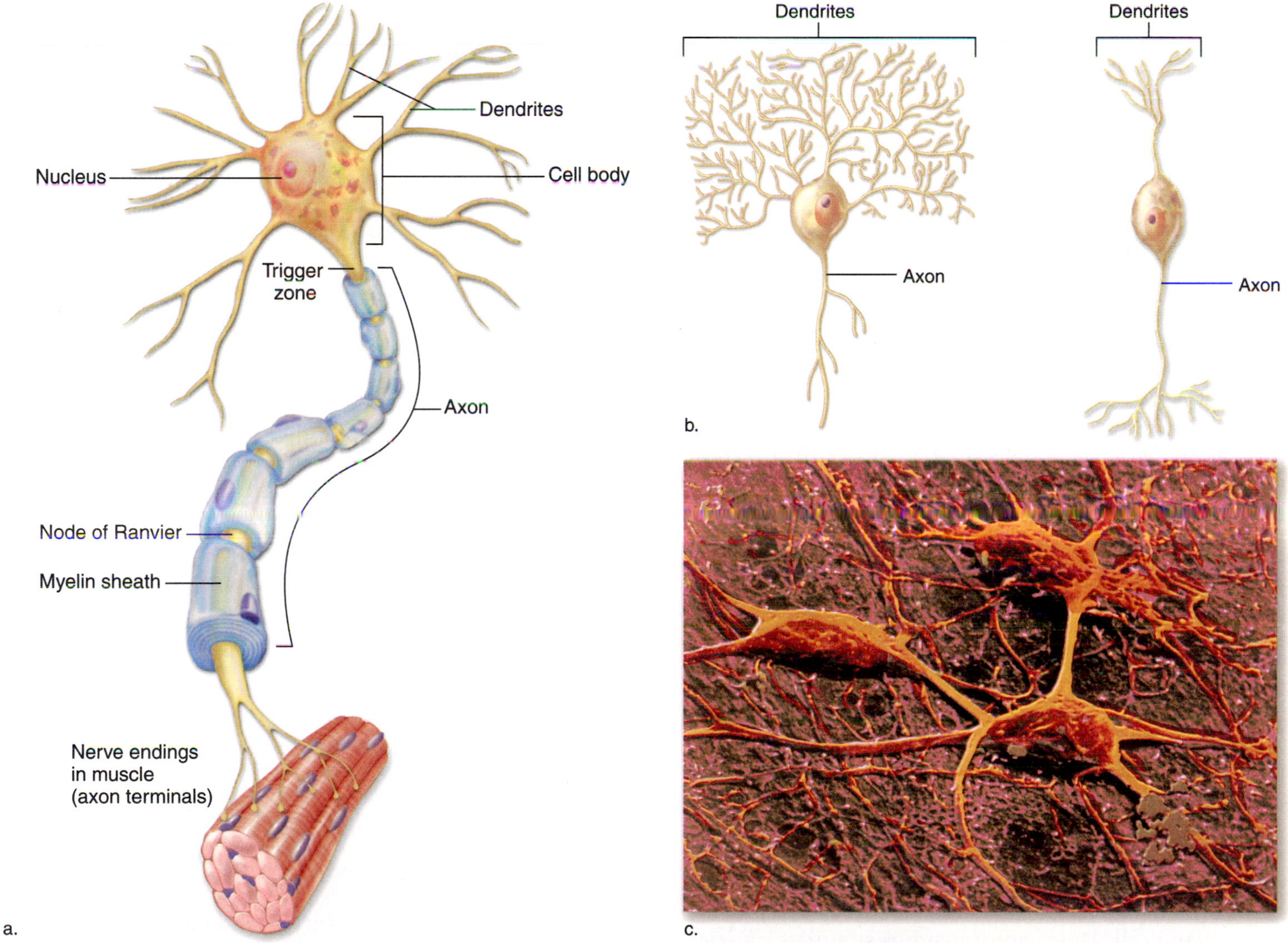

FIGURE 28.2 Parts of a Neuron. Neurons receive, integrate, and conduct messages. (**A**) A neuron consists of a rounded cell body, "receiving" branches called dendrites, and a "sending" branch called an axon. The junction of one neuron and an adjacent neuron is called a synapse. Many axons are encased in fatty myelin sheaths. Unmyelinated regions between adjacent myelin sheath cells are called nodes of Ranvier. (**B**) Neuron shape varies considerably. (**C**) These unmyelinated neurons from the cortex of a human brain are magnified 500 times. Note their entangled axons and dendrites.

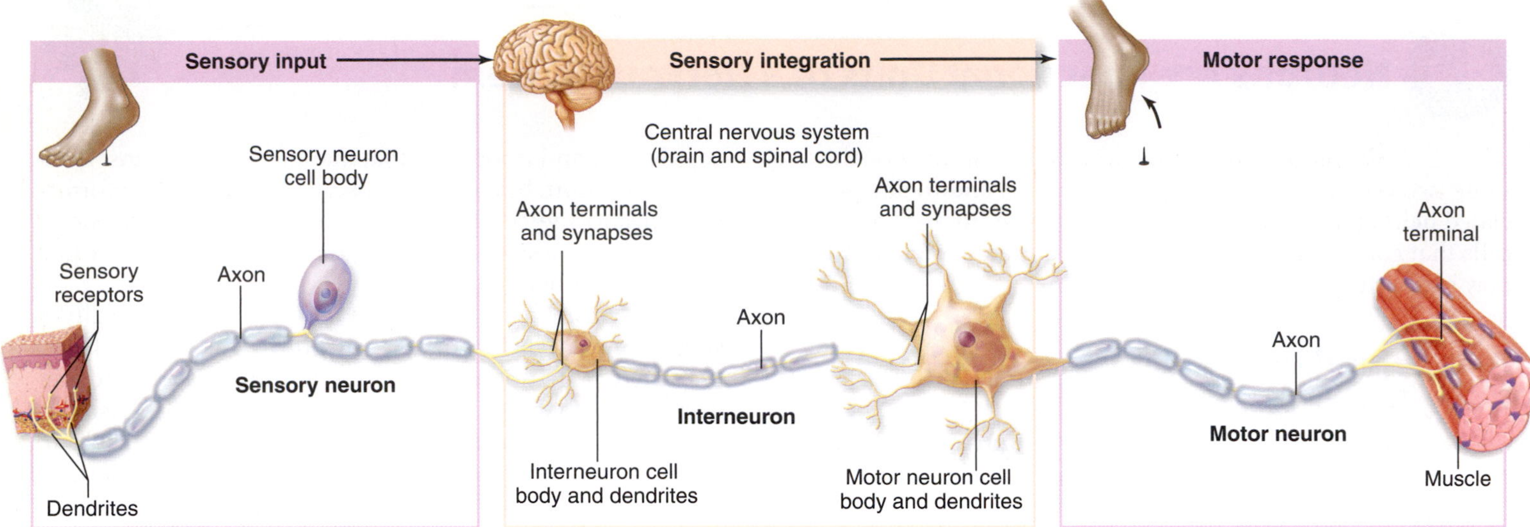

FIGURE 28.3 Categories of Neurons. Sensory neurons transmit information from sensory receptors to the central nervous system. Motor neurons send information from the central nervous system to muscles or glands. Interneurons connect sensory and motor neurons.

you to wiggle a big toe, for example, extends about a meter from the base of the spinal cord to the toe.

In many neurons, fatty material called a **myelin sheath** coats all or part of the axon, speeding nerve impulse conduction (see section 28.4). In the peripheral nervous system, neuroglia called **Schwann cells** form the myelin sheath; **nodes of Ranvier** are short intervals between individual Schwann cells. Other types of neuroglia make up the myelin sheaths in the central nervous system.

To picture the relative sizes of a typical neuron's parts, imagine its cell body is the size of a tennis ball. The axon might then be up to 1.5 km (about a mile) long but only a few centimeters thick. The mass of dendrites extending from the cell body would fill an average-size living room.

B. The Nervous System Includes Three Classes of Neurons

A neuron may have one of three general functions (**figure 28.3**):

- A **sensory neuron** brings information toward the central nervous system. Sensory neurons respond to light, pressure from sound waves, heat, touch, pain, and chemicals detected as odors or taste.
- A **motor neuron** conducts its message from the central nervous system toward an effector: a muscle or gland cell. Thus, motor neurons stimulate muscle cells to contract and glands to secrete. (They are called motor neurons because most lead to muscle cells.) Figure 28.2A shows a motor neuron innervating a skeletal muscle.
- **Interneurons** connect one neuron to another within the central nervous system. Large, complex networks of interneurons receive information from sensory neurons, process this information, and generate the messages that the motor neurons carry to effector organs. About 90% of all neurons are interneurons.

Only interneurons reside entirely within the central nervous system. Sensory and motor neurons are bridges between the central and peripheral nervous systems. The dendrites, cell body, and most of the axon of each sensory neuron lie in the peripheral nervous system, whereas the axon's endings reside in the central nervous system. In contrast, a motor neuron's cell body and dendrites reside in the central nervous system, but its axon extends into the peripheral nervous system.

Summary *Networks of neurons and neuroglia build nervous systems. A typical neuron has a cell body, numerous dendrites, and an axon. Myelin sheaths coat many axons. Three classes of neurons are sensory neurons, motor neurons, and interneurons.*

28.2 MASTERING CONCEPTS

1. What are the roles of neurons and neuroglia in the nervous system?
2. Describe the three parts of a typical neuron.
3. What is the usual direction in which a message moves within a neuron?
4. What are the functions of each of the three classes of neurons?

28.3 A Neuron at Rest Has a Negative Charge

A neuron sends messages by conveying a neural impulse, which is an electrical signal that results from the movement of ions across the cell membrane. (Recall that an ion is an atom or molecule with an electrical charge.) **ions, p. 23**

To understand how ions move in a neural impulse, it helps to be familiar with the **resting potential,** which is the state a neuron is in when it is not conducting a message. At rest, a neuron's membrane is polarized (**figure 28.4A**). That is, the inside carries a slightly negative electrical charge relative to the outside. This separation of charge creates an electrical "potential" that measures around –70 millivolts (mV). (A volt measures the difference in electrical charge between two points.)

A neuron has a resting potential because it maintains an unequal distribution of ions, notably potassium (K^+), across its membrane. One membrane protein that helps maintain this gradient is the **sodium–potassium pump,** which pumps three sodium ions (Na^+) out of the cell for every two K^+ that enter, at a cost of one ATP molecule per cycle. The sodium–potassium pump operates continuously. In a neuron at rest, the concentration of K^+ is therefore much higher inside the cell than outside, while the reverse is true for Na^+. **active transport, p. 93**

The neuron's resting potential reflects a balance between two opposing forces on K^+ (figure 28.4B). On one hand, the sodium–potassium pump concentrates K^+ inside the cell. Channel proteins in the membrane allow some of this K^+ to diffuse back out of the cell along its concentration gradient. On the other hand, charge interactions hold K^+ inside the cell. Positive charges from the Na^+ ions outside the cell repel K^+, while large, negatively charged proteins (and other negative ions) trapped inside the cell attract K^+. When the two opposing forces—the concentration gradient and charge interactions—are equal, the membrane has a net positive charge on the outside and a net negative charge on the inside. This difference in charge is the resting potential.

The term *resting potential* is a bit misleading because the neuron consumes a tremendous amount of energy while "at rest;" the nervous system devotes about three quarters of its total energy budget to maintaining the resting potential of its neurons. The resulting state of readiness allows the neuron to respond more quickly than it could if it had to generate a potential difference across the membrane each time it received a stimulus. This is analogous to holding back the string on a bow to be constantly ready to shoot an arrow.

Summary *The interior of a resting neuron is negatively charged compared with the outside. This polarity results from the unequal distribution of sodium ions, potassium ions, and proteins across the cell membrane.*

28.3 MASTERING CONCEPTS

1. What are the main factors that determine whether an ion will cross a cell's membrane?
2. Describe the distribution of K^+ and Na^+ across the cell membrane in a neuron at rest.
3. What are the mechanisms that establish and maintain the resting potential? Which of these mechanisms uses ATP?
4. In what way is the term *resting potential* misleading?

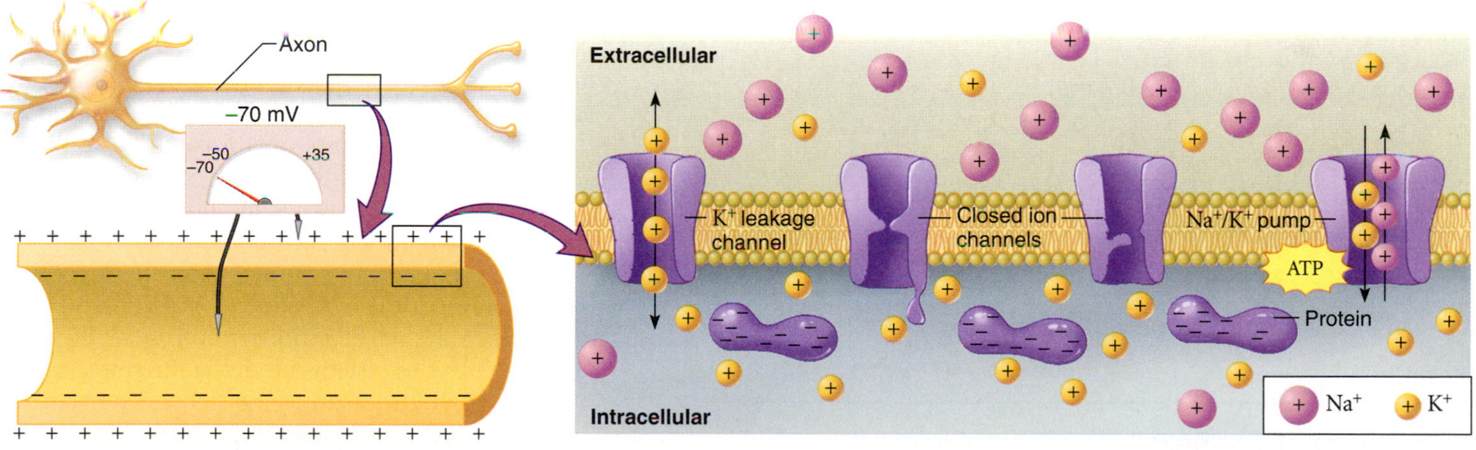

FIGURE 28.4 The Resting Potential. (**A**) The separation of positive (+) and negative (–) charges across the cell membrane produces an electrical potential measuring approximately –70 mV inside the cell relative to the outside. (**B**) At rest, the concentration of Na^+ is greatest outside the cell, and the concentration of K^+ is greatest inside the cell. Large, negatively charged proteins inside the cell contribute to the net negative charge. The Na^+/K^+ pump operates continuously, using ATP to move three Na^+ out of and two K^+ into the cell. The concentration gradient driving K^+ out of the cell is balanced by negative charges attracting K^+ back into the cell.

28.4 Action Potentials Convey Messages

A neuron's resting potential keeps it primed to convey messages at any moment. When a stimulus does arrive, ions begin to move across the neuron's membrane. If the neuron "decides" to convey the message to another cell, action potentials will occur along the neuron's axon. During an **action potential,** Na^+ and K^+ quickly redistribute across a small patch of the axon's membrane, creating a series of electrochemical changes that propagate like a wave along the nerve fiber. A neural impulse is the spread of action potentials along an axon.

A. A Neuron Transmitting an Impulse Undergoes a Wave of Depolarization

A neural impulse begins when a stimulus (a change in pH, a touch, or a signal from another neuron) causes some sodium channel gates in a neuron's membrane to open, usually at the dendrites or cell body. In a local flow of electrical current called a **graded potential,** some Na^+ begins to leak into the cell, causing the interior to become less negative. (It is called "graded" because it weakens with distance from the source of the stimulus and because the amount of depolarization depends on the signal's strength.) If the stimulus is very weak, the graded potential is small, and no action potential will occur. But if the stimulus is strong enough, the depolarization may spread to a "trigger zone" near the cell body (**figure 28.5**). This zone has a higher concentration of sodium channels than the dendrites or cell body, so it can continue the chain of events that leads to an action potential.

FIGURE 28.5 Sodium Channel Concentration. Dendrites and the cell body have a low concentration of sodium channels in comparison to the trigger zone and axon. Only areas with a high concentration of sodium channels can propagate action potentials.

Once enough Na^+ enters to depolarize the trigger zone's membrane to a **threshold potential** of about −50 mV, additional sodium gates open, further increasing permeability to Na^+ and triggering the action potential (**figure 28.6**). Driven by both the electrical gradient and the concentration gradient, Na^+ pours into the cell. As a result, the cell's interior briefly has a positive charge. Near the peak of this action potential, however, membrane permeability changes again. Sodium channel gates close, again preventing Na^+ from entering the cell. But permeability to K^+ suddenly increases as delayed potassium gates open. Repelled by the positively charged Na^+ ions inside the cell, K^+ diffuses out of the cell. The loss of K^+ repolarizes the cell membrane to its resting potential. The sodium–potassium pump, which operates continuously, returns Na^+ to the membrane's exterior. The entire process takes only 1 to 5 milliseconds to complete. (Burning Questions: Why does a scorpion's sting hurt? describes what happens when the ion channels required for the spread of action potentials become disrupted.)

How does the electrical charge spread along the length of the axon? Some of the Na^+ ions that rush into the cell diffuse to adjacent parts of the membrane. These ions, in turn, cause the neighboring part of the axon to reach its threshold potential, triggering a new influx of Na^+ there and carrying the impulse forward. A neural impulse is similar to people "doing the wave" in a football stadium, when successive groups of spectators stand and then quickly sit. The participants do not change their locations, yet the wave appears to travel around the stadium.

A neural impulse usually begins at the trigger zone and moves down the axon toward the axon terminal. It does not spread "backward" because of a refractory period during which the membrane reestablishes its resting potential and cannot generate another action potential. The refractory period lasts only a couple of milliseconds, but by the time it is over for one patch of membrane, the neural impulse has moved on.

Unlike a graded potential, an action potential is an "all-or-none" event; that is, it either proceeds to completion or it doesn't occur at all. Although all action potentials are identical, the nervous system can detect the strength of a stimulus by measuring the frequency of action potentials. Whereas a light touch to the skin might produce 10 impulses per second, a hard hit might generate 100 impulses—along with a more intense sensation. Neurons also distinguish the type of stimulus. For example, we can tell light from sound because light stimulates neurons that transmit impulses to one part of the brain, whereas sound-generated impulses go to another part.

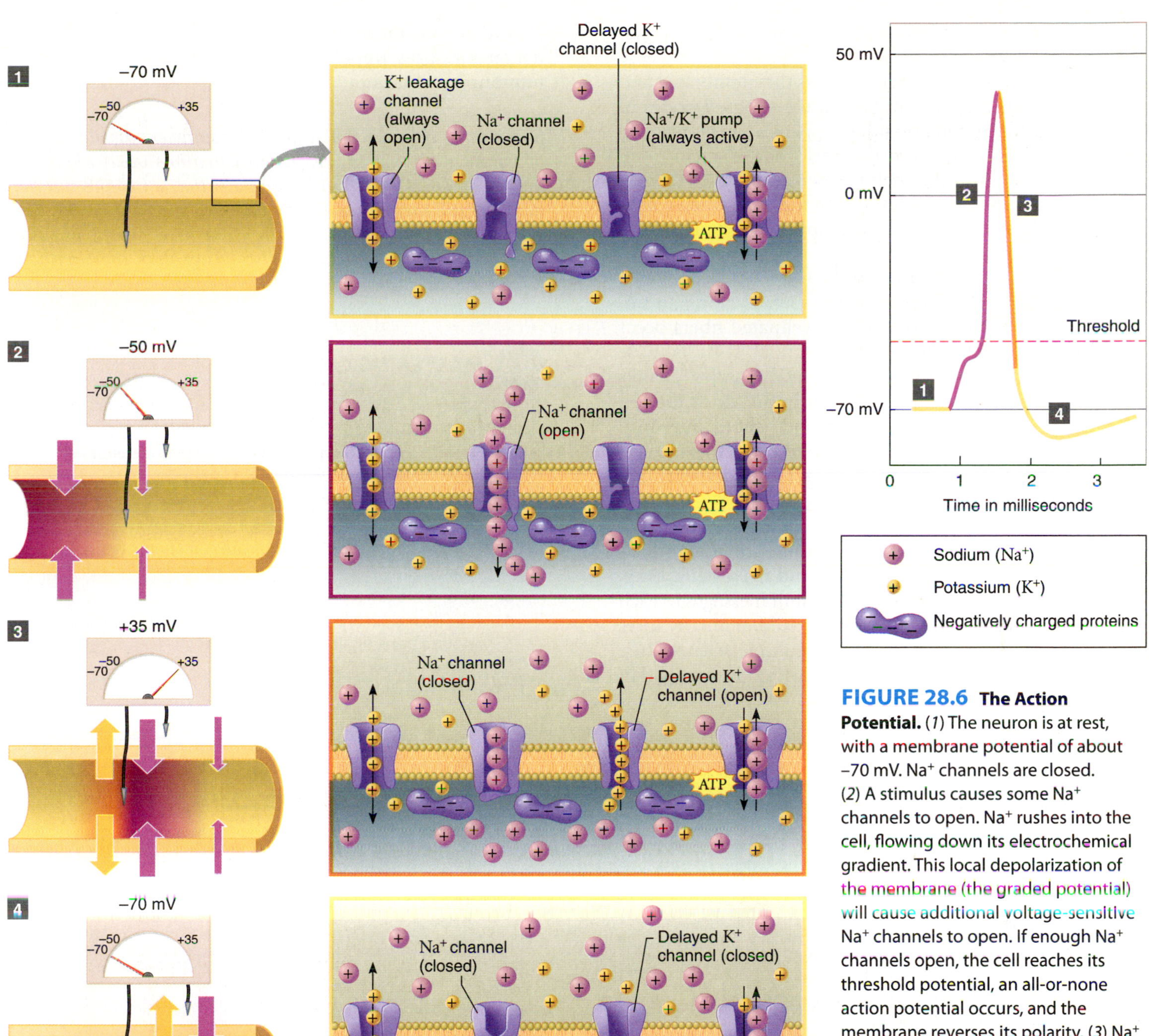

FIGURE 28.6 The Action Potential. (*1*) The neuron is at rest, with a membrane potential of about −70 mV. Na^+ channels are closed. (*2*) A stimulus causes some Na^+ channels to open. Na^+ rushes into the cell, flowing down its electrochemical gradient. This local depolarization of the membrane (the graded potential) will cause additional voltage-sensitive Na^+ channels to open. If enough Na^+ channels open, the cell reaches its threshold potential, an all-or-none action potential occurs, and the membrane reverses its polarity. (*3*) Na^+ channels close after a split second, and K^+ leaves the axon, restoring the resting potential. (*4*) The membrane briefly hyperpolarizes as the resting potential is reestablished.

B. The Myelin Sheath Speeds Impulse Conduction

The greater the diameter of an axon, the faster it conducts an impulse. A squid's "giant axons" are up to 1 mm in diameter. (Much of what biologists know about action potentials comes from studies on these large-diameter nerve fibers.) Axons from vertebrates are a hundredth to a thousandth the diameter of the squid's. Yet even thin vertebrate nerve fibers can conduct impulses very rapidly when they are coated with a myelin sheath. Schwann cells, whose membranes contain enormous amounts of lipid, form myelin sheaths around axons in the peripheral nervous system (**figure 28.7A**).

The myelin prevents ion flow across the membrane, and sodium channels are concentrated at the exposed nodes of Ranvier. When an action potential happens at one node, Na^+ entering the axon diffuses to the next node. The Na^+ then stimulates the sodium channels to open at the second node, triggering an action potential there. In this way, when a neural impulse travels along the axon, it appears to "jump" from node to node (figure 28.7B).

The neural impulse moves up to 100 times faster when it leaps between nodes than when it spreads along an unmyelinated axon. Not surprisingly, myelinated fibers occur in neural pathways where speed is essential, such as those that transmit motor commands to skeletal muscles. Thanks to myelin, a sensory message travels from the toe to the spinal cord in less than 1/100 of a second (about one third the speed of sound). Unmyelinated fibers occur in pathways where speed is less important, such as in the neurons that trigger the secretion of stomach acid.

Illness can result if an axon has too much or not enough myelin. In Tay–Sachs disease, for example, cell membranes accumulate excess lipid and wrap around nerve cells, burying them in fat so that they cannot transmit messages to each other and to muscle cells. An affected child gradually loses the ability to see, hear, and move. In multiple sclerosis, the reverse happens. The Schwann cells coating the axons that lead to certain muscles lack myelin. The resulting blocked neural messages impair vision and movement and cause numbness and tremors.

Summary *If graded potentials cause sufficient depolarization of the neuron's membrane, action potentials occur along the neuron's axon. During an action potential, an axon's interior near the cell membrane briefly becomes positively charged, a condition that spreads down the axon as a neural impulse. Myelinated axons conduct impulses faster than those without a fatty sheath.*

28.4 MASTERING CONCEPTS

1. What is the difference between a graded potential, the threshold potential, and an action potential?
2. How do changing cell membrane ion permeabilities generate and transmit a neural impulse?
3. What prevents action potentials from spreading in both directions along an axon?
4. How do action potentials indicate stimulus intensity and type?
5. What is the function of Schwann cells?
6. How do myelin and the nodes of Ranvier speed neural impulse transmission along an axon?

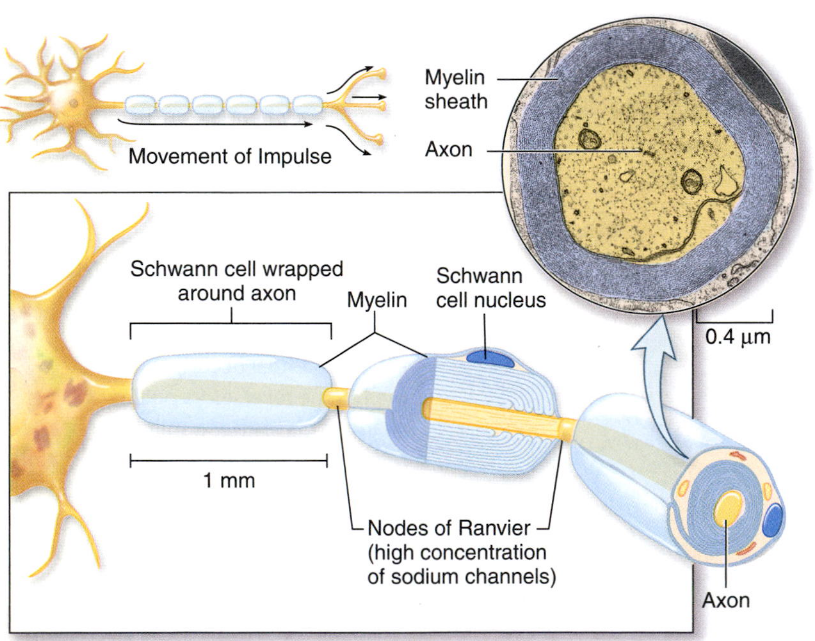

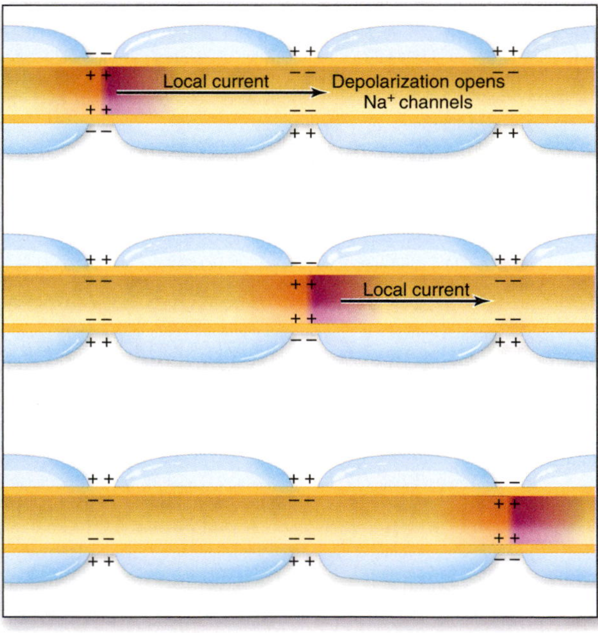

FIGURE 28.7 The Myelin Sheath. (**A**) A myelin sheath forms when a Schwann cell wraps many times around a small segment of an axon, so that several layers of its lipid-rich cell membrane form a whitish coating. The sections of axon between Schwann cells are called nodes of Ranvier. The inset shows a transmission electron micrograph of a myelinated axon in cross section. (**B**) In myelinated axons, the inward diffusion of Na^+ can occur only at the nodes of Ranvier. Action potentials appear to "jump" from one node to the next, which speeds impulse transmission along the axon.

Burning Questions

Why does a scorpion's sting hurt?

The distinctive, venomous stinger on a scorpion's tail is arguably one of the most recognizable arthropod structures. Despite their fearsome reputation, however, only one scorpion species in the United States has venom potent enough to threaten human life. Still, it hurts to be stung by any scorpion. Why is it so painful?

Scorpion venom is a complex cocktail of toxins, enzymes, salts, and many other substances. Proteins called neurotoxins (poisons that act on the nervous system) cause most of the sting's worst effects. Some scorpion neurotoxins cause an axon's sodium channels to become stuck in the "open" position, whereas others block potassium channels. The resulting membrane depolarization triggers a continuous barrage of action potentials, accompanied by simultaneous sensations of pain, heat, cold, and touch. The venom also depolarizes the motor neurons that control the body's muscles, sometimes causing convulsions. In the most severe cases, the victim may die of cardiac or respiratory failure because the body can no longer control the muscles required for heartbeat and breathing.

Have a Burning Question of your own? Submit it to marielle_hoefnagels@mcgraw-hill.com for possible inclusion in future editions of this book!

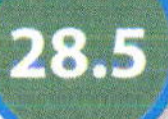

28.5 Neurotransmitters Pass the Message from Cell to Cell

To form a communication network, a neuron must convey an impulse to another cell. Most neurons do not touch each other, so the impulse cannot travel directly from cell to cell. Instead, the impulse causes release of a **neurotransmitter,** a chemical signal that travels from a "sending" cell to a "receiving" cell across a tiny space (**figure 28.8**). A **synapse** is

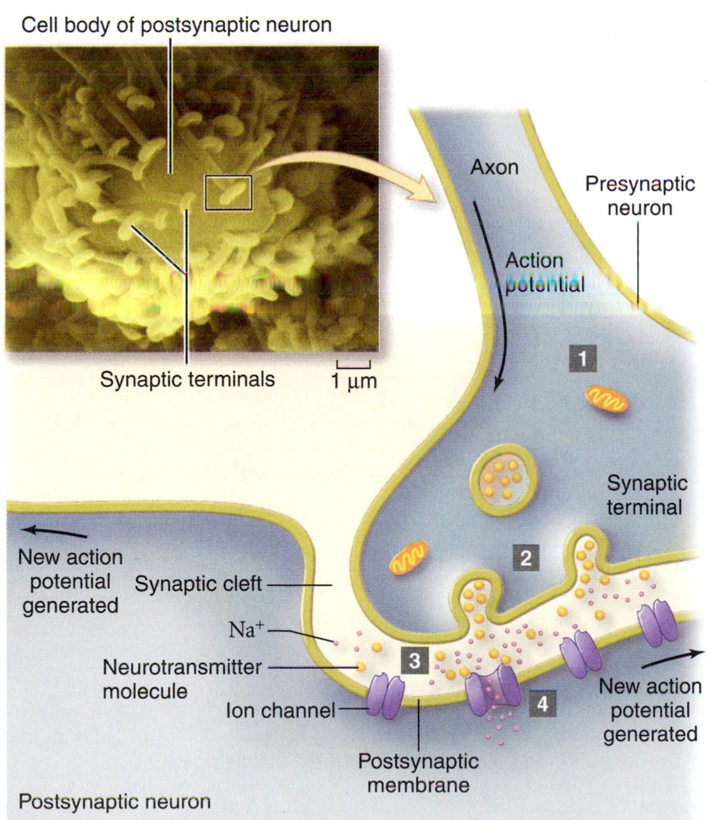

1 Action potential arrives at presynaptic neuron.

2 Vesicle loaded with neurotransmitter fuses with presynaptic cell membrane.

3 Neurotransmitters are released into synaptic cleft.

4 Neurotransmitters bind to receptor proteins in postsynaptic cell membrane. Ion gates open, and neural impulse is initiated.

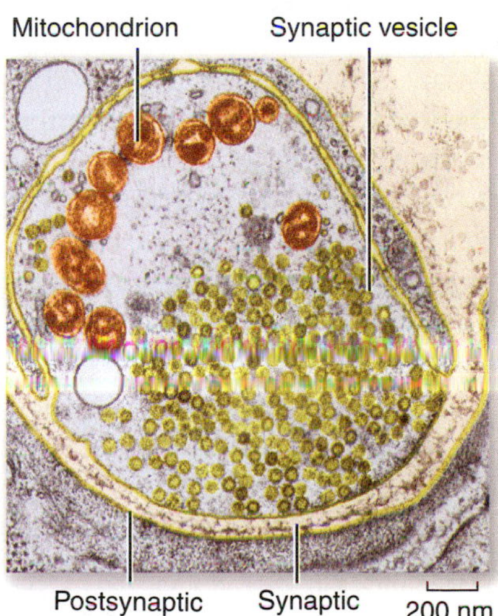

FIGURE 28.8 Transmission Across a Synapse. (**A**) An action potential triggers release of neurotransmitter from an axon of a presynaptic neuron. The neurotransmitter diffuses across the synaptic cleft and binds with receptors in the postsynaptic neuron's cell membrane. Ion channels then open, either increasing or decreasing the likelihood that an action potential will occur in the postsynaptic cell. The inset shows synaptic terminals from many neurons converging on the cell body of a single postsynaptic neuron. (**B**) This transmission electron micrograph shows synaptic vesicles and a synaptic cleft.

the combination of the tip of a sending cell's axon, the membrane of the receiving cell, and the **synaptic cleft,** the space between the two cells.

The **presynaptic cell** is the neuron sending the message. The end of its axon has **synaptic terminals,** tiny knobs that enlarge at the tips. These knobs contain many small sacs, or vesicles, that hold neurotransmitter molecules. Action potentials that reach the membrane of the synaptic terminal trigger the loaded vesicles to dump their neurotransmitter contents into the synaptic cleft by exocytosis. **exocytosis, p. 94**

Neurotransmitter molecules diffuse across the synaptic cleft and attach to protein receptors on the membrane of the **postsynaptic cell**—the neuron, muscle cell, or gland cell receiving the message. Each neurotransmitter fits only into a specific receptor type, as a key fits only one lock. When the neurotransmitter contacts the receptor, the shape of the receptor changes. Ion channels then open in the postsynaptic membrane, either increasing or decreasing the probability that an action potential will occur.

If a neurotransmitter stayed in the synapse indefinitely, its effect on the receiving cell would be continuous. The nervous system would be bombarded with stimuli. Instead, the neurotransmitter can diffuse away from the synaptic cleft, be destroyed by an enzyme, or be taken back into the presynaptic axon soon after its release, an event called reuptake.

So far, we have considered the events that occur at one synapse. But the human brain consists of billions of neurons, each of which has synaptic connections to a thousand other neurons (see figure 28.8A). With so many synapses, how does a neuron "decide" whether to pass a neural impulse to the next cell in a pathway? The cell uses a process called **synaptic integration,** a sort of "voting" system, to evaluate the incoming messages. A neurotransmitter at an excitatory synapse increases the probability that an action potential will occur in the postsynaptic cell; a neurotransmitter at an inhibitory synapse inhibits an action potential. If a neuron receives more excitatory "votes," the postsynaptic cell is stimulated; if inhibitory "votes" predominate, it is not.

Clearly, neurotransmission is very much a matter of balance. This chapter's Can *You* Relate? Drugs and Neurotransmitters illustrates how drugs tamper with neurotransmission, and **table 28.1** lists disorders caused by too much or too little of a neurotransmitter.

Summary *Neurotransmitters are released from terminals on axons in response to an action potential. These chemical messengers diffuse across the synaptic cleft and bind to receptors on the receiving cell membrane. Neurotransmitter binding alters the permeability of the membrane in a way that stimulates or blocks depolarization. A neuron integrates incoming excitatory and inhibitory messages to determine whether to transmit an impulse.*

28.5 MASTERING CONCEPTS

1. What is a synapse?
2. What happens to a neurotransmitter after its release?
3. What is synaptic integration?
4. How do excitatory and inhibitory synapses constitute a "voting" system that controls whether a neuron conducts action potentials?

Table 28.1 *Disorders Associated with Neurotransmitter Imbalances*

Condition	Imbalance of Neurotransmitter in Brain	Symptoms
Alzheimer disease	Deficient acetylcholine (caused by death of acetylcholine-producing cells)	Memory loss, depression, disorientation, dementia, hallucinations, death
Epilepsy	Excess GABA* leads to excess norepinephrine and dopamine	Seizures, loss of consciousness
Huntington disease	Deficient GABA*	Personality changes, loss of coordination, uncontrollable movement, death
Hypersomnia	Excess serotonin	Excessive sleeping
Insomnia	Deficient serotonin	Inability to sleep
Myasthenia gravis	Deficient acetylcholine at neuromuscular junctions	Progressive muscular weakness
Parkinson disease	Deficient dopamine	Tremors of hands, slowed movements, muscle rigidity
Schizophrenia	Deficient GABA* leads to excess dopamine	Inappropriate emotional responses, hallucinations

*GABA, gamma aminobutyric acid.

Can You Relate?

Drugs and Neurotransmitters

Understanding how neurotransmitters work helps explain the action of some mind-altering illicit and pharmaceutical drugs. The following are some examples, organized by the neurotransmitter affected.

Norepinephrine

Amphetamine drugs are chemically similar to norepinephrine; they bind to norepinephrine receptors and trigger the same changes in the postsynaptic membrane. The resulting enhanced norepinephrine activity heightens alertness and mood. Cocaine is chemically related to amphetamine. It produces a short-lived feeling of euphoria, both by blocking reuptake of norepinephrine and by binding to molecules that transport dopamine to postsynaptic cells.

Acetylcholine

Poisonous nerve gases and some insecticides prevent acetylcholine from breaking down in the synaptic cleft. The resulting excess acetylcholine activity overstimulates skeletal muscles, causing them to contract continuously. The twitching legs of a cockroach sprayed with insecticide demonstrate the effects.

Serotonin

A deficiency of norepinephrine or serotonin can cause depression, a debilitating feeling of sadness. Drugs called selective serotonin reuptake inhibitors (SSRIs) block reuptake of serotonin. The neurotransmitter accumulates in the synapse, offsetting a deficit that presumably causes the symptoms (**figure 28.A**). The reverse situation, excess serotonin in the synaptic cleft, causes sleepiness. This is why people become sleepy after eating turkey. The meat contains abundant tryptophan, an amino acid that the body uses to synthesize serotonin.

Endorphins

Humans produce several types of endorphins, molecules that influence mood and perception of pain. Opiate drugs such as morphine, heroin, codeine, and opium are potent painkillers that bind endorphin receptors in the brain. In doing so, they elevate mood and make the pain easier to tolerate.

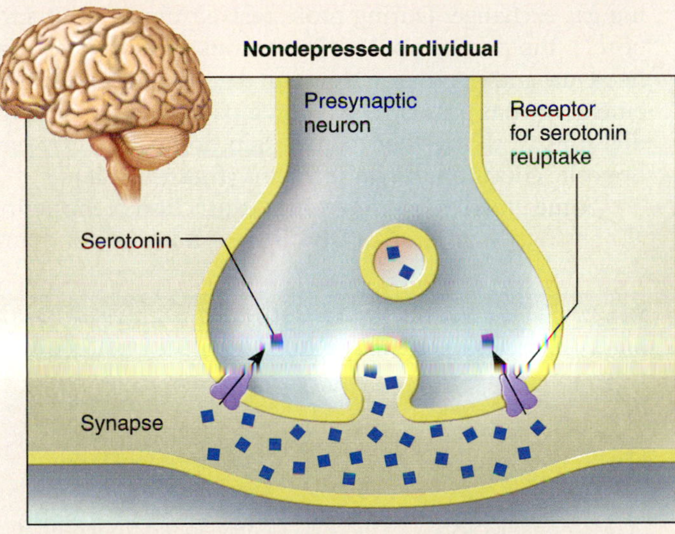

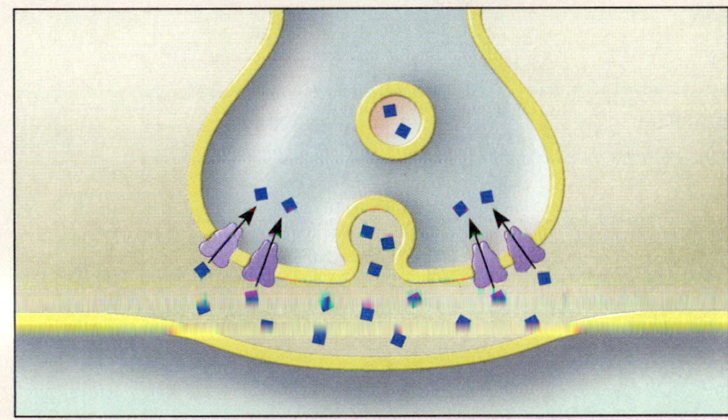

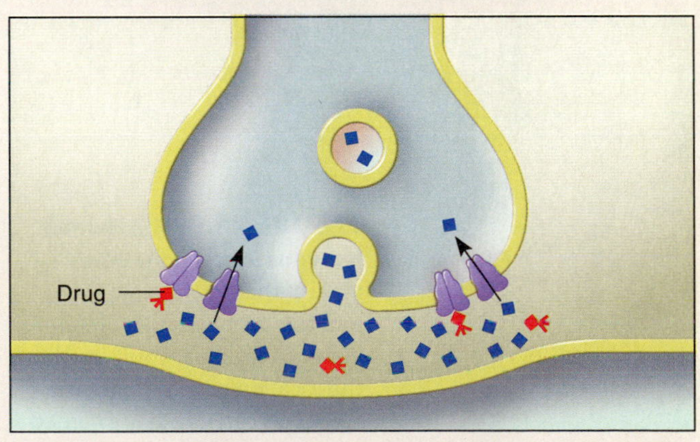

FIGURE 28.A **Anatomy of an Antidepressant.** Selective serotonin reuptake inhibitors (SSRIs) block the reuptake of serotonin, making more of the neurotransmitter available in the synaptic cleft. The precise mechanism of SSRIs is not well understood.

28.6 The Peripheral Nervous System Consists of Nerve Cells Outside the Central Nervous System

The brain and spinal cord interact constantly with the peripheral nervous system—the nerve cells outside the central nervous system (**figure 28.9**). The peripheral nervous system consists mainly of **nerves,** which are bundles of axons encased in connective tissue.

The peripheral nervous system is functionally divided into sensory and motor divisions (**table 28.2** and **figure 28.10**). Sensory pathways carry signals to the central nervous system from sensory receptors in the skin, skeleton, muscles, and other organs. Motor pathways convey information in the opposite direction: from the central nervous system to muscles and glands. In most nerves, the sensory and motor nerve fibers form a single cable.

The motor pathways of the peripheral nervous system are divided into the somatic (voluntary) nervous system and the autonomic (involuntary) nervous system. The **somatic nervous system** carries signals to the voluntary muscles of the body, such as those that enable you to ride a bicycle, shake hands, or talk. The **autonomic nervous system** transmits impulses to smooth muscle, cardiac muscle, and glands, enabling internal organs to function without conscious awareness.

The autonomic nervous system is further subdivided into the sympathetic and parasympathetic nervous systems (see figure 28.10). The **sympathetic nervous system** dominates under stress, preparing the body to face emergencies. It accelerates heart rate and breathing rate; shunts blood away from the digestive system and to the heart, brain, and the skeletal muscles necessary for "fight or flight;" and dilates airways, easing gas exchange. During more relaxed times ("rest and repose"), the **parasympathetic nervous system** is in charge; heart rate and respiration slow and digestion resumes after the emergency has passed. The parasympathetic and sympathetic divisions work together to maintain homeostasis by having opposite effects on the same organs (**figure 28.11**).

Some illnesses interfere with the function of the peripheral nervous system. In Guillain–Barré syndrome, for example,

Nervous System

Main tissue types	Examples of Locations/Functions
Connective	Surrounds nerves
Nervous	Makes up brain, spinal cord, and nerves; functions in sensation, communication, and information storage

FIGURE 28.9 The Human Nervous System. (**A**) Dorsal view of the brain, spinal cord, and spinal nerves. (**B**) Lateral view of the spinal cord showing the location of the cranial and spinal nerves, which are named according to the point at which they leave the spinal cord.

Table 28.2 Functional Divisions of the Peripheral Nervous System

Division	Function(s)
Sensory pathways	Carry signals to central nervous system from sensory receptors
Motor pathways	Carry signals from central nervous system to muscles and glands
Somatic (voluntary) motor neurons	Carry signals from central nervous system to voluntary muscles
Autonomic (involuntary) motor neurons	Carry signals from central nervous system to involuntary muscles and glands
Sympathetic nervous system	Dominates during times of stress
Parasympathetic nervous system	Dominates during times of "rest and repose"

Brain and spinal cord
Sensory pathways
Motor pathways
Sensory neurons registering external stimuli
Sensory neurons registering internal stimuli
Autonomic motor neurons (involuntary)
Sympathetic nervous system "fight or flight"
Parasympathetic nervous system "rest and repose"
Somatic motor neurons (voluntary)
Central nervous system
Peripheral nervous system

FIGURE 28.10 Hierarchical Organization of the Vertebrate Nervous System. The peripheral nervous system is divided into motor and sensory pathways. In the motor pathways, somatic neurons innervate skeletal (voluntary) muscles, and autonomic neurons stimulate glands and muscles under involuntary control. The autonomic neurons are further subdivided into the sympathetic and parasympathetic divisions.

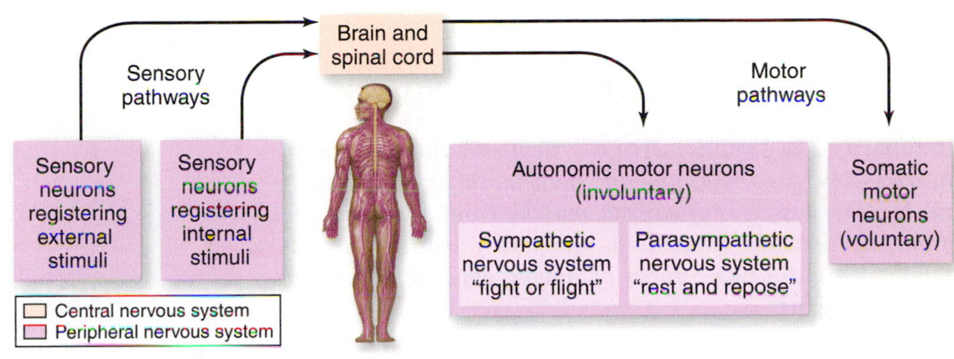

Sympathetic division "fight or flight"
Parasympathetic division "rest and repose"
Dilates pupil
Eyes
Contracts pupils
Inhibits salivation
Stimulates salivation
Salivary glands
Relaxes bronchi
Bronchi
Constricts bronchi
Cranial nerves
Accelerates heartbeat, strengthens contractions
Heart
Slows heartbeat
Stimulates secretions of adrenaline plus noradrenaline
Adrenals
T1
T2
T3
T4
T5
T6
T7
T8
T9
T10
T11
T12
Thoracic nerves
L1
L2
Lumbar nerves
Stimulates glucose release by liver
Liver and gallbladder
Stimulates gallbladder
Inhibits activity
Stimulates activity
Stomach and pancreas
Inhibits activity
Stimulates activity
Inhibits activity
Stimulates activity
S2
S3
S4
Sacral nerves
Colon and small intestine
Relaxes bladder
Contracts bladder
Bladder
Stimulates ejaculation and vaginal contractions
Stimulates erection of sex organs
Genitals

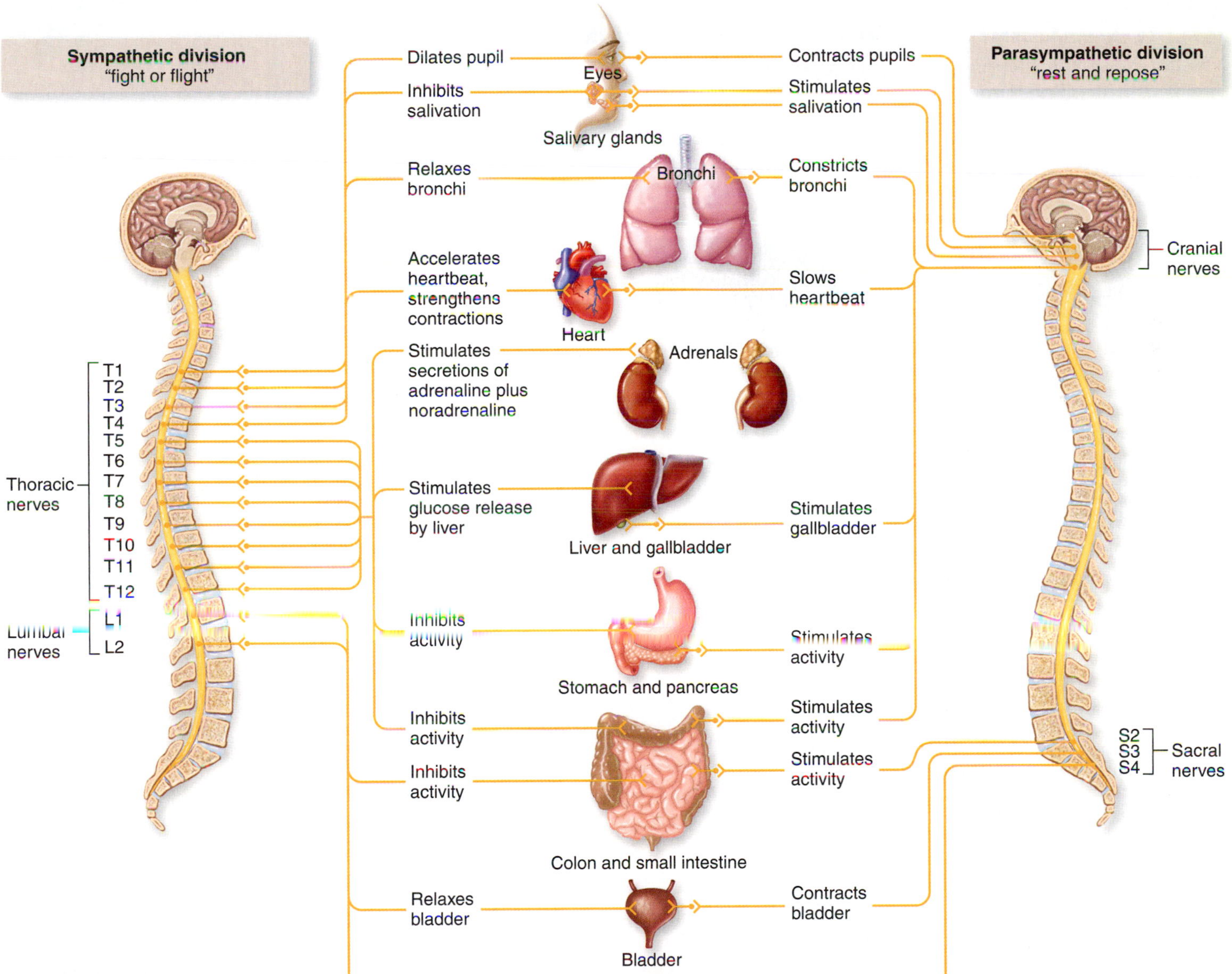

FIGURE 28.11 The Autonomic Nervous System. The parasympathetic and sympathetic divisions of the autonomic nervous system have opposite effects on the same organs.

the immune system attacks and destroys the nerves of the peripheral nervous system. The disease can be life-threatening if it causes paralysis, breathing difficulty, and heart problems.

Summary *The somatic nervous system stimulates voluntary muscles, whereas the autonomic nervous system controls the involuntary responses of smooth muscle, cardiac muscle, and gland secretion. The sympathetic and parasympathetic nervous systems maintain homeostasis by having opposing actions on the same organs.*

28.6 MASTERING CONCEPTS

1. How does the direction of information flow differ between the sensory and motor divisions of the peripheral nervous system?
2. What is the relationship between the sensory, motor, somatic, autonomic, sympathetic, and parasympathetic nervous systems?
3. Which division of the autonomic nervous system dominates during stress, and which division dominates when the situation returns to normal?

28.7 The Central Nervous System Consists of the Spinal Cord and Brain

The nerves of the peripheral nervous system spread across the body, but the brain and spinal cord form the largest part of the nervous system. **Table 28.3** summarizes their parts and functions.

Table 28.3 *Some Parts of the Central Nervous System*

Structure	Function(s)
Spinal Cord	Conducts information to and from the brain; centralizes rapid involuntary responses such as reflexes
Brain	
Brainstem	
Hindbrain	
Medulla oblongata	Regulates essential physiological processes such as blood pressure, heartbeat, and breathing
Pons	Connects brain with spinal cord
Cerebellum	Coordinates muscular movements
Midbrain	Passes sensory information to forebrain
Forebrain	
Thalamus	Processes information and relays it to the cerebrum
Hypothalamus	Homeostatic control of most organs
Cerebrum	
White matter	Transmits information within brain
Gray matter (cerebral cortex)	Sensory, motor, and association areas
Limbic system	Amygdala is important in emotion; hippocampus is important in memory

Two types of nervous tissue occur in the central nervous system (**figure 28.12**). **Gray matter** consists of neuron cell bodies and synapses. The outer surface of the brain, and a few inner structures, are composed of gray matter, as is the central core of the spinal cord. Information processing occurs in the gray matter. **White matter** consists of myelinated axons transmitting information throughout the central nervous system. It lies at the periphery of the spinal cord. In addition, most of the inner structures of the brain consist of white matter.

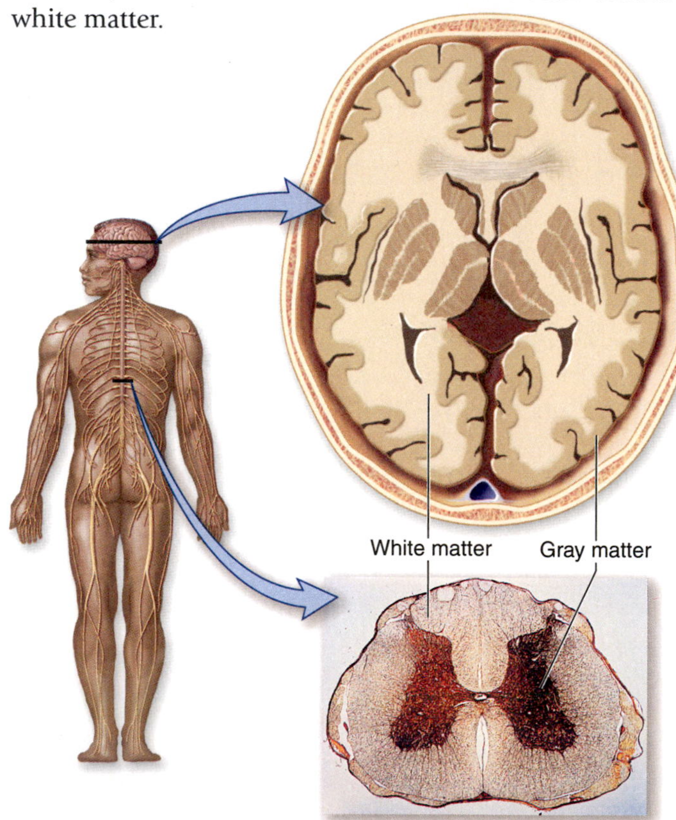

FIGURE 28.12 Gray Matter and White Matter. Gray matter makes up the exterior of the brain and some internal structures. It also makes up the central core of the spinal cord. Myelin-rich white matter is at the periphery of the spinal cord and forms most of the brain's interior.

A. The Spinal Cord Transmits Information Between Body and Brain

In adult humans, the spinal cord extends from the base of the brain to a couple of centimeters below the last rib. The spinal cord is a tube of neural tissue encased in the bony armor of the vertebral column, or backbone (**figure 28.13**). The backbone protects the delicate nervous tissue and provides points of attachment for muscles.

A cross section of the spinal cord shows a central H-shaped core of gray matter surrounded by white matter (see figure 28.12). In the white matter, ascending bundles of axons carry sensory information to the brain, and descending axons carry motor information from the brain to muscles and glands.

Spinal reflex neurons interact at the spinal cord. A reflex is a rapid, involuntary response to a stimulus that may come from within or outside the body. The spinal cord handles reflexes without interacting with the brain (although impulses must be relayed to the brain for awareness to occur).

The two neurons that cause you to pull your hand away from a painful stimulus such as a thorn make up a **reflex arc,** a neural pathway that links a sensory receptor and an effector such as a muscle (**figure 28.14**). The arc begins with a sensory neuron whose dendrites reside in your fingertip. When you touch the sharp thorn, an action potential is generated along the neuron's axon. Inside the spinal cord, the axon synapses with a motor neuron. The motor neuron's axon exits the spinal cord, and its activation stimulates a skeletal muscle cell to contract. When enough muscle fibers contract, you pull your hand away from the thorn. The original sensory neuron also synapses with interneurons. Some of these send action potentials to the brain, perhaps prompting you to yell in pain.

B. The Human Brain Is Divided into Several Regions

The human brain weighs, on average, about 1.4 to 1.6 kg (about 3.5 pounds) and looks and feels like grayish pudding. The brain requires a large and constant energy supply to oversee organ systems and to provide the qualities of "mind"—learning, reasoning, and memory. At any time,

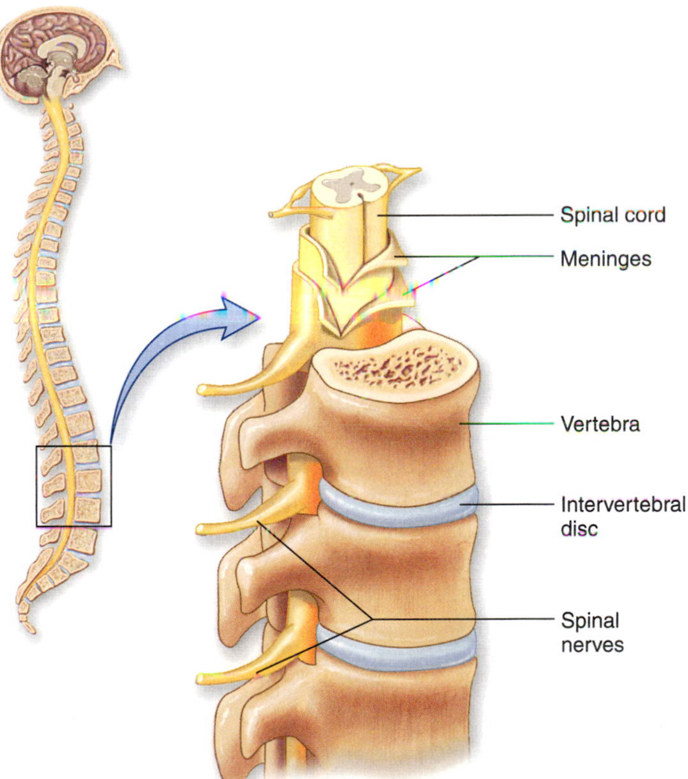

FIGURE 28.13 The Backbone. The spinal cord runs through the tunnel formed within the stacked vertebrae of the vertebral column. The spinal nerves exit laterally between the vertebrae. Membranes called meninges surround the spinal cord.

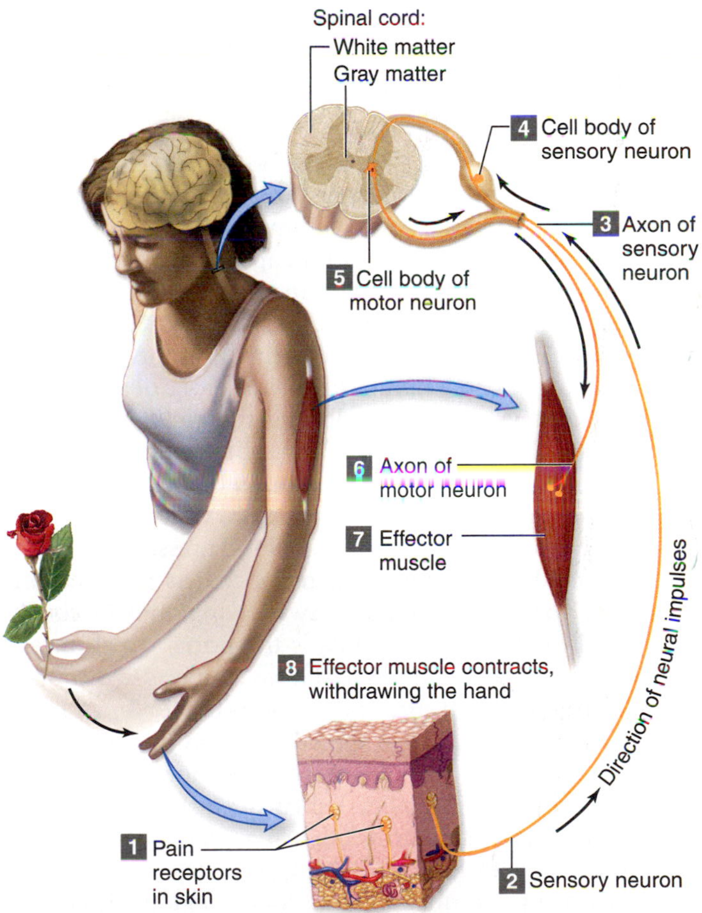

FIGURE 28.14 A Reflex Arc. A reflex arc links a sensory receptor to an effector. A sensory receptor, such as a sensory neuron dendrite that ends in the skin, senses an environmental stimulus and relays an action potential to its cell body in the spinal cord. This neuron sends the information to a motor neuron, which, in turn, stimulates a muscle cell to contract.

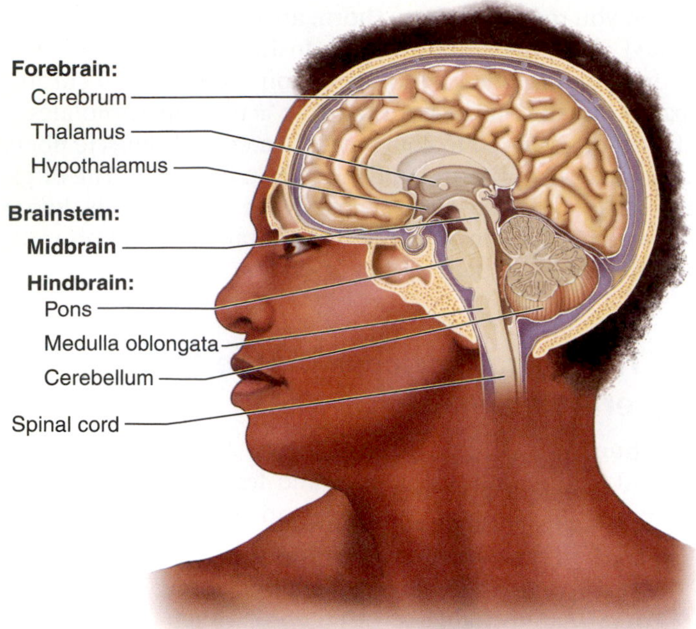

FIGURE 28.15 The Human Brain. A vertical section through the human skull shows the three major areas of the vertebrate brain: the hindbrain, the midbrain, and the forebrain.

Table 28.4 ***Functional Divisions of the Brain***

Division	Function(s)	Brain Region(s) Involved
Limbic	Emotion, motivation	Amygdala, hippocampus, hypothalamus
Sensory	Senses of sight, hearing, olfaction (smell), taste, and touch	Primary visual cortex, primary auditory cortex, olfactory bulbs, gustatory cortex, primary somatosensory cortex
Motor	Voluntary movements	Primary motor cortex
Associative	Learning, creativity	Frontal lobe and parts of the parietal, occipital, and temporal lobes of the cerebrum

brain activity consumes 20% of the body's oxygen and 15% of its blood glucose. Permanent brain damage occurs after just 5 minutes of oxygen deprivation.

Three subdivisions of the brain appear early in development: the hindbrain, the midbrain, and the forebrain. The hindbrain and midbrain together make up the **brainstem,** the stalklike lower portion of the brain. As embryonic development proceeds, the forebrain grows rapidly, obscuring the midbrain and parts of the hindbrain (**figure 28.15**).

The **hindbrain** is the part of the brainstem located toward the back of the skull. The section of hindbrain closest to the spinal cord is the **medulla oblongata,** which regulates essential functions such as breathing, blood pressure, and heart rate. The medulla also contains reflex centers for vomiting, coughing, sneezing, defecating, swallowing, and hiccoughing. The area above the medulla is the **pons,** which means "bridge." White matter in this oval mass connects higher brain centers with the spinal cord and connects the forebrain to the cerebellum. The neurons of the **cerebellum** ("little brain"), the largest part of the hindbrain, refine motor messages and coordinate muscle movements subconsciously.

The **midbrain** is a narrow brainstem region that connects the hindbrain with the forebrain. In humans, the midbrain receives sensory information from touch, sound, visual, and other receptors and passes it to the forebrain.

The **forebrain,** or front of the brain, contains structures important in complex functions such as learning, memory, language, motivation, and emotion (**table 28.4**). The **thalamus** is a mass of gray matter that acts as a relay station for sensory input, processing incoming information and sending it to the appropriate part of the cerebrum. The **hypothalamus,** which lies below the thalamus, occupies less than 1% of the brain volume, but it plays a vital role in maintaining homeostasis. It regulates body temperature, heartbeat, water balance, blood pressure, hunger, thirst, sexual arousal, and feelings of pain, pleasure, anger, and fear. The hypothalamus also regulates hormone secretion from the anterior pituitary gland. Thus, the hypothalamus links the nervous and endocrine systems, the body's two communication systems (see chapter 30).

The other major region of the forebrain is the **cerebrum,** which controls the qualities of what we consider the "mind;" that is, intelligence, learning, perception, and emotion. It consists mostly of white matter—myelinated axons that transmit information within the cerebrum and between the cerebrum and other parts of the brain. In humans, the cerebrum is large (occupying 83% of the brain's volume) and highly developed. It is divided into two **hemispheres** that gather and process information simultaneously. In most people, parts of the left hemisphere are associated with speech, language skills, mathematical ability, and reasoning, whereas the right hemisphere specializes in spatial, intuitive, musical, and artistic abilities. The cerebral hemispheres work together, interconnected by a thick band of nerve fibers called the corpus callosum.

The outer layer of the cerebrum, the **cerebral cortex,** consists of gray matter where neural integration occurs (**figure 28.16**). Sensory areas receive and interpret messages from sense organs about temperature, body movement, pain, touch, taste, smell, sight, and sound. Motor areas send impulses to skeletal muscles. (Damage to the motor regions

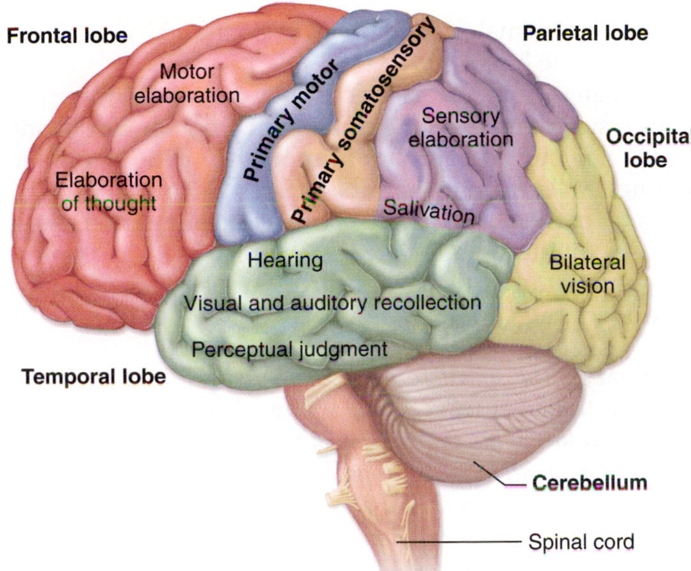

FIGURE 28.16 Cerebral Specialization. The surface of the cerebrum contains four main sections: the frontal, temporal, occipital, and parietal lobes. A band of tissue called the primary somatosensory area receives sensory input from the muscles, joints, bones, and skin. The primary motor area controls voluntary muscles. Association areas are located in the front of the frontal lobe and in parts of the occipital, parietal, and temporal lobes.

of the brain can cause cerebral palsy, a disorder of posture or muscle movement.) Association areas are the seats of learning and creativity.

The **limbic system** is a collection of forebrain structures that surrounds the corpus callosum and thalamus. Among others, it includes the amygdala, which is important in emotion, and the **hippocampus,** which is involved in memory.

C. Damage to the Central Nervous System Can Be Devastating

The central nervous system is well protected. The bones of the skull and vertebral column shield nervous tissue from bumps and blows. **Meninges** are layered membranes that jacket the central nervous system. The **blood–brain barrier,** formed by specialized brain capillaries, helps protect the brain from extreme chemical fluctuations. The tilelike epithelial cells that form these capillaries fit so tightly together that only some chemicals can cross into the **cerebrospinal fluid** that bathes and cushions the brain and spinal cord. This fluid, made by cells that line ventricles in the brain, further insulates the central nervous system from injury. **tight junctions, p. 72**

Accidents can damage the spinal cord and prevent motor impulses from descending from the brain, causing full or partial paralysis (**figure 28.17**). Christopher Reeve, who broke his neck in a horseback riding accident in 1995, was among the most famous spinal cord injury patients. He died in 2004 because of complications from being bedridden, a common problem in paralyzed people.

Accidents and head trauma can also cause brain damage, as can a more subtle killer: stroke. In a stroke, a burst or blocked blood vessel can flood the brain with blood or interrupt the flow of blood to part of the brain. Deprived of oxygen, some brain cells die, often so many that the stroke is fatal. In other cases, the patient may slowly recover as other brain parts learn to compensate for the damaged area.

Infectious diseases can also affect the central nervous system. Several viruses and bacteria can cause meningitis, or inflammation of the meninges. Bacterial meningitis is rare but can be fatal. Prions, the infectious proteins discussed in chapter 17, also damage the brain. Fungal infections of the central nervous system often afflict people with compromised immune systems. **prions, p. 365**

Many serious illnesses of the central nervous system remain a mystery. Parkinson disease is a slowly progressing, degenerative disease in which the death of brain cells causes muscle tremors, weakness, slow movement, and loss of coordination. In most cases, such as that of actor

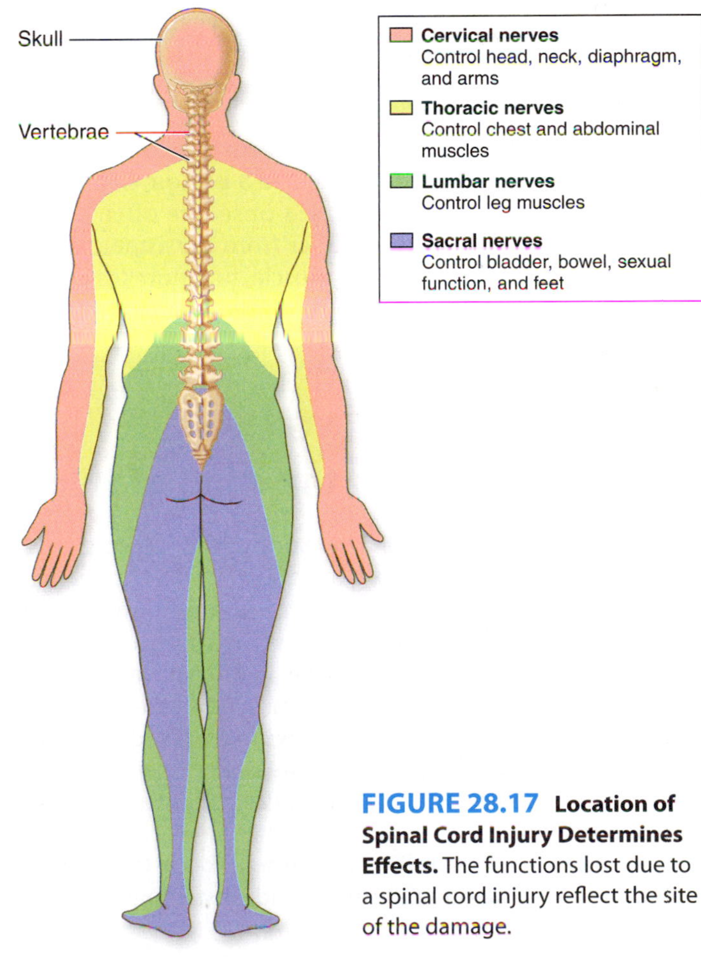

FIGURE 28.17 Location of Spinal Cord Injury Determines Effects. The functions lost due to a spinal cord injury reflect the site of the damage.

FIGURE 28.18 **Testimony.** Boxer Muhammad Ali (left) and actor Michael J. Fox joke around before asking the U.S. Senate to fund research on Parkinson disease.

Michael J. Fox (**figure 28.18**), no one knows what triggers the cell death. The death of brain cells also causes Alzheimer disease. The patient, who is usually elderly, suffers from memory loss, confusion, and personality changes. Autopsy reveals tangled neurons and clusters of degenerating nerve endings; neurotransmitter production is also reduced (see table 28.1). Unlike in Parkinson disease, motor function usually remains unaffected.

Another mysterious nervous system disease is amyotrophic lateral sclerosis, also called ALS or Lou Gehrig disease. The first clues to this disease's onset are often subtle; a person may drop things or suffer from persistent muscle twitches. Eventually, impaired muscle function leaves the patient unable to speak, breathe, swallow, or move, yet the senses and mental function remain unaffected. This devastating disease has no cure.

Whatever its cause, part of the difficulty in reversing nervous system damage is that mature neurons do not divide, so the nervous system cannot simply heal itself by producing new cells as your skin does after a minor cut. Current research on stem cells, growth factors (chemicals that stimulate cell division), and gene therapy may one day improve the outlook for patients with neurological damage or disease. **stem cells, p. 158**

Summary *The spinal cord receives impulses from the rest of the body, conducts reflexes, and communicates with the brain. Structures in the hindbrain control vital functions and connect to other regions, and neurons in the midbrain process and integrate sensory input. In the forebrain, the cerebrum receives sensory input and directs motor responses, and the hypothalamus regulates secretion of some hormones. Brain and spinal cord injuries are usually either deadly or extremely debilitating.*

28.7 MASTERING CONCEPTS

1. What is the relationship between gray matter and white matter in the spinal cord?
2. Describe the functions of the neurons that form a reflex arc.
3. What are the functions of the major structures in the hindbrain, midbrain, and forebrain?
4. Distinguish between the functions of the somatosensory and motor areas of the cerebral cortex.
5. List some structures that protect the central nervous system from damage.
6. What are some examples of diseases that affect the central nervous system?
7. To what extent can the nervous system regenerate?

28.8 INVESTIGATING LIFE

Looking into the Mind of Humankind

Taking an algebra examination, talking with friends, and playing video games are a few experiences that are unique to humans. These and many other complex behaviors are possible because of the evolution of the human brain. Scientists have wondered at the differences among the nervous systems and brains of other animals and humans. Charles Darwin remarked, "Certain actions which we recognize as expressive of certain states of the mind, are the direct result of the constitution of the nervous system."

Scientists have traditionally learned about the nervous system by focusing on anatomical and physiological differences among brains of humans and other animals. We understand much less about the influence of genes on the evolution of the brain and the mind of humankind. Recently, however, researchers have begun to use DNA sequence data to peer back into our evolutionary history.

Chapter 12 explained that a gene encodes the sequence of amino acids in a protein; a mutation is a change in a gene's DNA sequence. Because genetic mutations fuel evolution, mutation rates are a measure of evolutionary change. Some mutations are neutral; that is, they do not result in a change in the amino acid sequence of the encoded protein. Such mutations are called synonymous (the two genes encode identical proteins, so the original gene is essentially a "synonym" of the mutated one). If the mutated gene encodes a different amino acid sequence than the original gene, the mutation is nonsynonymous. Because they may change an organism's phenotype, nonsynonymous mutations are subject to natural selection; synonymous ones are not.

Researchers can use these two types of mutations to measure the evolutionary rate of a gene. They begin by obtaining DNA sequences for two or more organisms that have the gene of interest. They then compute the ratio of nonsynonymous (K_a) to synonymous (K_s) mutations. If the ratio K_a/K_s is low, most of the gene's mutations are neutral, so the evolutionary rate is low. For example, the genes encoding essential "housekeeping" proteins in the cell tend to have low K_a/K_s ratios; most nonsynonymous mutations in those genes are fatal. A high K_a/K_s ratio, in contrast, indicates a high evolutionary rate.

University of Chicago biologist Bruce Lahn and his colleagues hypothesized that genes involved in nervous system biology have evolved more rapidly in primates than nonprimate mammals. To test their hypothesis, they compared the evolutionary rates of nervous system genes between two mammalian groups, rodents (mouse and rat) and primates (macaque monkey and human). The evolutionary time separating mice and rats (about 20 million years) is roughly equal to the divergence time between the macaque and the human (**figure 28.19**).

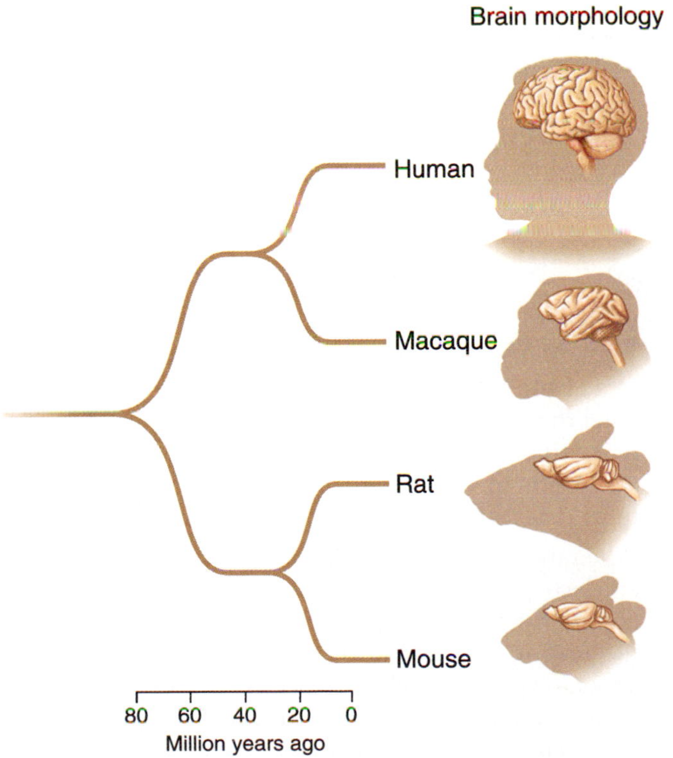

FIGURE 28.19 **Brain Shapes in Rodents and Primates.**

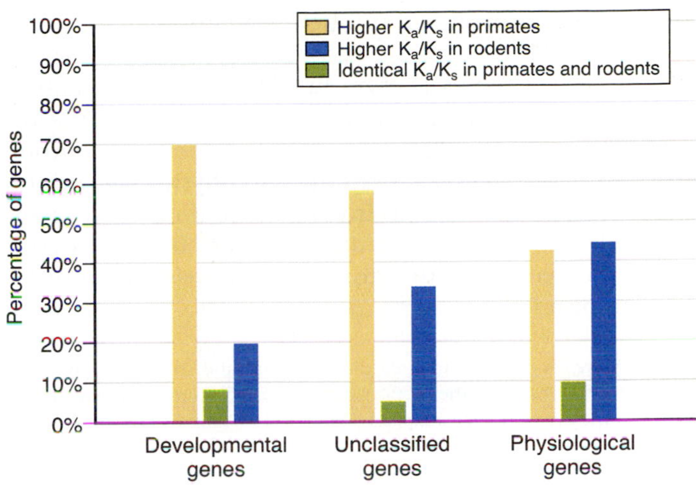

FIGURE 28.20 **Rapid Evolution in Primate Brains.** Of 214 genes studied in primates and rodents, 53 were related to development of the physical structure of the brain, 95 were related to brain physiology, and 66 were unclassified. Genes classified as "developmentally biased" evolved much faster in primates than in rodents. **Question:** Although correlated with the development of language, tool use, and culture in humans, it is impossible to say whether changes in brain structure caused our species to become more intelligent. Why is it so difficult to develop an objective measure of intelligence that applies to multiple species?

Lahn and his colleagues analyzed 214 nervous system-related genes that were common to all four animal species. They found that the K_a/K_s ratio for these genes (and thus the rate of protein evolution) was significantly higher in the primates than in the rodents. To test the importance of the finding, the researchers also measured the K_a/K_s ratio of housekeeping genes in the study animals. The K_a/K_s ratio of the primate housekeeping genes was equal to that of the rodent genes. So the difference in the evolutionary rates was specific to the nervous system genes and not simply a genome-wide phenomenon.

One of the most obvious trends in the evolution of the human brain was a dramatic increase in brain size and complexity—a change that is correlated with language, tool use, and the development of culture. To learn more about how this change occurred, Lahn and his team divided the 214 nervous system genes into three categories, based on function. Developmental genes greatly influence nervous system development, physiological genes regulate physiological processes, and unclassified genes include those that could not be assigned to the other two categories. The K_a/K_s ratios for the three categories supported the hypothesis that the developmental genes seemed to be most active in shaping the human brain (**figure 28.20**).

Finally, the team investigated K_a/K_s ratios among primates including squirrel monkeys, macaques, chimpanzees, and humans. The data reinforced the notion that K_a/K_s values of nervous system genes in primates is elevated in the lineage leading from ancestral primates and that this trend has continued through recent human evolution (**figure 28.21**).

In 1871, Charles Darwin first speculated that natural selection played an essential role in developing the human brain. Nearly 150 years later, researchers combine the traditional fields of anatomy and physiology with the newer tools of genetics to investigate brain evolution and its implications for the history of our species. Future studies will soon add to the story of the development of the human mind.

Dorus, Steve, Eric J. Vallender, Patrick D. Evans, and colleagues, including Bruce T. Lahn. 29 December 2004. Accelerated evolution of nervous system genes in the origin of *Homo sapiens*. *Cell*, vol. 119, pages 1027–1040.

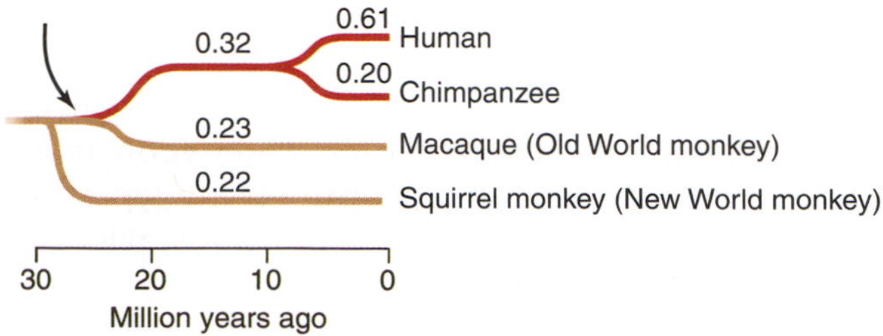

FIGURE 28.21 Humans Have the Fastest Evolving Brains. This phylogenetic tree shows K_a/K_s values for nervous system genes in each lineage along the primate family tree. The farther along the human lineage, the higher the number, and the more rapid the evolutionary rate.

CHAPTER SUMMARY

28.1 The Nervous System Forms a Rapid Communication Network

- The **nervous system** and **endocrine system** work together to coordinate the feedback systems that maintain homeostasis. The nervous system's electrical signals produce much more rapid effects than the endocrine system.
- The nervous system receives sensory information, integrates it, and coordinates a response.
- The **central nervous system** consists of the **brain** and **spinal cord.** The **peripheral nervous system** conveys information between the central nervous system and the rest of the body.

28.2 Neurons Are Functional Units of a Nervous System

A. A Typical Neuron Consists of a Cell Body, Dendrites, and an Axon

- A **neuron** has a **cell body, dendrites** that receive impulses and transmit them toward the cell body, and an **axon** that conducts impulses away from the cell body.
- **Neuroglia** are cells that support neurons. Fatty neuroglia called **Schwann cells** wrap around some axons to form the **myelin sheath.** The gaps between these insulating cells are **nodes of Ranvier.**

B. The Nervous System Includes Three Classes of Neurons

- A **sensory neuron** carries information toward the central nervous system.
- A **motor neuron** carries information away from the central nervous system and stimulates an effector (a muscle or gland).
- An **interneuron** conducts information between two neurons and coordinates responses.

28.3 A Neuron at Rest Has a Negative Charge

- In a neuron at rest, the K^+ concentration is much greater inside the cell than outside, whereas the Na^+ concentration is greater outside than inside. The **sodium–potassium pump** uses ATP to maintain this chemical gradient.
- The concentration gradient combined with negatively charged proteins within the cell give the interior a negative charge, called the **resting potential**.

28.4 Action Potentials Convey Messages

A. A Neuron Transmitting an Impulse Undergoes a Wave of Depolarization

- In a **graded potential,** a stimulus causes some Na^+ to enter the cell, depolarizing the membrane in proportion to the strength of the stimulus. If enough Na^+ comes in, the membrane may further depolarize to its **threshold potential**.
- When the membrane reaches its threshold potential, an electrical change called an **action potential** begins. Membrane channels open and allow still more Na^+ in, creating a positive charge inside. At the peak of depolarization, Na^+ channels close. Repolarization occurs as K^+ leaves the cell, restoring the resting potential. The action potential spreads as Na^+

diffuses laterally inside the axon, triggering additional Na^+ channels to open farther along the nerve fiber.

B. The Myelin Sheath Speeds Impulse Conduction

- Myelination increases the speed of neural impulse transmission. In a myelinated fiber, the neural impulse rapidly "jumps" from one node of Ranvier to the next.

28.5 Neurotransmitters Pass the Message from Cell to Cell

- Neurons communicate at **synapses.** An action potential reaching the end of an axon causes vesicles in the **synaptic terminals** of the **presynaptic neuron** to approach the cell membrane and release **neurotransmitters** into the **synaptic cleft.** These chemicals diffuse across the cleft and bind to receptors on the **postsynaptic neuron.**
- Used neurotransmitter diffuses away, is enzymatically destroyed, or is reabsorbed into the presynaptic cell.
- An excitatory neurotransmitter makes an action potential more probable; an inhibitory neurotransmitter has the opposite effect. **Synaptic integration** sums excitatory and inhibitory messages, providing fine control of neuron activity.

28.6 The Peripheral Nervous System Consists of Nerve Cells Outside the Central Nervous System

- **Nerves** are bundles of axons that convey information in the peripheral nervous system.
- The peripheral nervous system is divided into the sensory and motor pathways, and it includes the cranial and spinal nerves that transmit sensations from sensory receptors and stimulate muscles and glands.
- The motor pathways of the peripheral nervous system consist of the **somatic** (voluntary) division and the **autonomic** (involuntary) division. The autonomic nervous system receives sensory information and conveys impulses to smooth muscle, cardiac muscle, and glands.
- Within the autonomic nervous system, the **sympathetic nervous system** controls physical responses to stressful events, and the **parasympathetic nervous system** dominates during rest.

28.7 The Central Nervous System Consists of the Spinal Cord and Brain

A. The Spinal Cord Transmits Information Between Body and Brain

- **White matter** on the periphery of the spinal cord conducts impulses to and from the brain; the central **gray matter** processes information.
- The spinal cord is a reflex center. A reflex is a quick, automatic, protective response that travels through a **reflex arc.**

B. The Human Brain Is Divided into Several Regions

- The **brainstem** consists of the hindbrain and midbrain.
- The **hindbrain** includes three main subdivisions: the **medulla oblongata,** which controls many vital functions; the **cerebellum,** which coordinates unconscious movements; and the **pons,** which bridges the medulla and higher brain regions and connects the cerebellum to the cerebrum.
- The **midbrain** receives and integrates visual and auditory sensory information.
- The major parts of the **forebrain** are the **cerebrum;** the **thalamus,** a relay station between lower and higher brain regions; the **hypothalamus,** which regulates vital physiological processes and regulates levels of some pituitary hormones; and the pineal and pituitary glands. The **limbic system,** which includes the amygdala and **hippocampus,** also resides in the forebrain.
- The outer layer of the cerebrum is the **cerebral cortex,** where information is processed and integrated. The cerebrum's two **hemispheres** each receive sensory input from and direct motor responses to the opposite side of the body.

C. Damage to the Central Nervous System Can Be Devastating

- Bones of the skull and vertebrae, **cerebrospinal fluid,** the **blood–brain barrier,** and **meninges** protect the central nervous system.
- Trauma, infectious agents, strokes, and degenerative diseases all can damage the nervous system.

28.8 Investigating Life: Looking into the Mind of Humankind

- By comparing hundreds of genes from rodents and primates, scientists have determined that genes related to nervous system development have evolved most rapidly in the lineage leading to humans.

MULTIPLE CHOICE QUESTIONS

1. Some cells of the central nervous system are located in the
   a. spinal cord.
   b. muscles.
   c. glands.
   d. both a and c.
2. What is the function of an axon?
   a. Metabolic support for the neuron
   b. Insulation to speed impulse conduction
   c. Conduction of an impulse
   d. Input of signals to the nerve cell
3. Which class(es) of neuron would you expect to find in the peripheral nervous system?
   a. Interneurons
   b. Sensory neurons
   c. Motor neurons
   d. Both b and c
4. The function of the sodium–potassium pump is to move
   a. sodium in and potassium out of the neuron.
   b. potassium into the neuron and sodium out.
   c. both potassium and sodium into the neuron.
   d. both potassium and sodium out of the neuron.
5. What event triggers an action potential?
   a. Opening of sodium channels
   b. Opening of delayed potassium channels
   c. High concentration of negative ions outside the cell
   d. Activation of the sodium-potassium pump
6. What is the likely effect of a loss of myelin along an axon?
   a. It causes the action potential to speed up because more of the membrane is exposed.
   b. It causes the action potential to slow down.
   c. It speeds up the transport of the sodium and potassium across the membrane.
   d. It increases the size of the action potential.
7. Which of the following examples of synaptic integration is most likely to lead to an action potential?
   a. All excitatory and no inhibitory synaptic inputs
   b. An equal mix of excitatory and inhibitory inputs
   c. A majority of inhibitory inputs with only a few excitatory synaptic inputs
   d. Both a and b
8. Which type of nervous system would be responsible for a rapid heart beat?
   a. Autonomic
   b. Sympathetic
   c. Parasympathetic
   d. Both a and b
9. The part of the human brain involved in coordinating muscle movements is the
   a. cerebrum.
   b. medulla oblongata.
   c. cerebellum.
   d. hypothalamus.
10. Which of the following is NOT among the structures that protects the central nervous system?
    a. Meninges
    b. Vertebrae
    c. Pons
    d. Cerebrospinal fluid

TESTING YOUR KNOWLEDGE

1. How do the nervous and endocrine systems communicate differently?
2. In the illustration, label the following:
   a. dendrites
   b. axons
   c. cell bodies
   d. a Schwann cell
   e. a myelin sheath
   f. a synapse

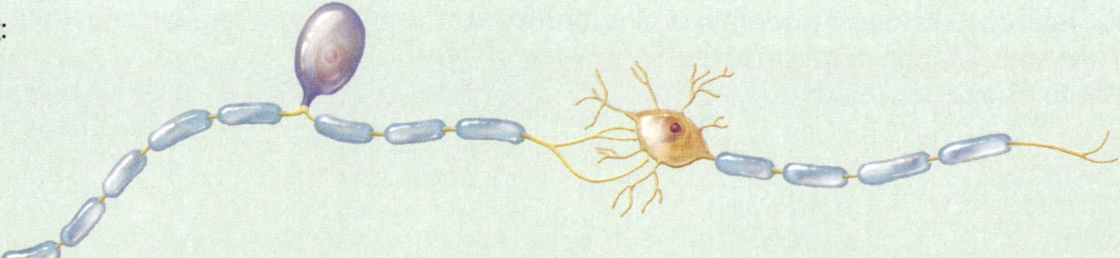

3. How do the functions of sensory neurons, motor neurons, and interneurons differ?
4. Describe the distribution of charges in the membrane of a resting neuron.
5. What causes the wave of depolarization and repolarization constituting an action potential?
6. How does myelin alter conduction of a neural impulse along a nerve fiber?
7. How does a neuron use neurotransmitters to communicate with other cells?
8. How does a neuron use information from other neurons to "decide" whether to send a neural impulse?
9. What are the main subdivisions of the human nervous system?
10. In what part of the brain do the qualities of "mind" lie?
11. In carpal tunnel syndrome, a nerve in the wrist becomes compressed, causing numbness or pain in the forearm and hand. Is this a disease of the peripheral or central nervous system?

THINKING AS A SCIENTIST

1. Cyanide is a poison that disables the sodium–potassium pump. Explain how cyanide prevents nerve transmission and causes death.
2. What would happen to neural impulse transmission in a myelinated axon without nodes of Ranvier?
3. In what ways does an action potential resemble "the wave" in a football stadium? In what ways does a graded potential resemble a cheerleader's attempts to get "the wave" started?
4. What symptom might result from an overdose of an SSRI drug? (Don't try this!)
5. How would you test the hypothesis that a nonhuman animal feels pain or thinks? Which animals would you choose to investigate this question?
6. All adult human brains are about the same size, contain the same major structures, and function in similar ways. How, then, does each of us develop a distinct personality?
7. Albert Einstein's brain was of normal size, but a part of his parietal lobe was about 15% wider than normal. This area controls mathematical reasoning, imagery, and the ability to visualize objects in space. A particular groove in the area appeared much reduced, leading researchers to speculate that this might have allowed more synaptic connections to form than normal. What additional information would help to determine whether Einstein's brain distinctions could have accounted for his genius?
8. Neuroglia outnumber neurons in the nervous system by about 10 to 1. In addition, neuroglia retain the ability to divide, unlike neurons. How do these two observations relate to the fact that most brain cancers begin in glial cells?
9. Dentists apply local anesthetics to deaden the pain associated with filling a cavity. These drugs block sodium channels in neurons surrounding the affected tooth. Major surgery requires general anesthetics that act on the brain, causing the patient to become unconscious and unaware of his or her surroundings. Use what you have learned about the nervous system to explain how local and general anesthetics temporarily eliminate pain.
10. Scientists know little about many common illnesses, including migraines and Alzheimer disease. What ethical considerations make research on these diseases difficult? What are the limitations of using animals as models to study the nervous system?

Visit www.mhhe.com/hoefnagels for practice quizzes, animations, videos, and activities designed to help you master the material in this chapter.

The Senses

Eclectic Senses. A snake's forked tongue detects chemical gradients, enabling the animal to follow a scent trail.

What's the Point?

Different Views on the Same World

Most people know that dogs have extremely sensitive noses and that cats see well in the dark. Here are three lesser-known adaptations that other animals use to sense the world.

Platypus Electroreception

A platypus eats half its body weight daily as it navigates along the river bottoms of Australia, nudging rocks aside with its bill to find crayfish, worms, insect larvae, frogs, and fish. But the bill doesn't sense its targets by touch alone; it can also detect the weak electrical fields that the muscles of their prey emit in water. The sensors in the platypus bill are specialized neurons called electroreceptors, which were unknown until Australian researchers placed hungry platypuses in pools with live and dead batteries. The animals explored only the live batteries, suggesting that they are attracted to electricity. Next, the researchers recorded platypuses' brain waves while stimulating their bills electrically and detected a response. A close look at the bills revealed threadlike ends of nerve fibers deep within the pits that form mucous glands. These were the electroreceptors. The tiniest flick of a shrimp's tail can stimulate the electroreceptors to trigger 20 to 50 action potentials per second!

The Snake's Forked Tongue

A snake uses its forked tongue to sense odors at two points simultaneously, so it can detect whether a signal is changing in strength over a distance. The snake collects odor molecules by flicking its tongue. Inside the mouth, the tongue tips pass through openings in the palate and contact sensory cells in the snout. The receptors transmit action potentials to the brain, which compares the two signals and interprets whether the trail is weakening or strengthening. The snake integrates this information along with the nature of the chemical stimulus to determine whether it is approaching a potential mate or food.

Scorpion Pectines

A scorpion searches for food and mates by tasting the ground surface with a pair of comb-shaped structures called pectines, located just behind the last pair of legs on the animal's "chest." Each pecten's flexible spine carries a row of "teeth" adorned with thousands of tiny pegs. A close look at each peg reveals an elongated pore. As the animal walks, it taps its pectines on the ground. Chemicals left on the ground by other animals enter the pegs' pores and bind to sensory receptors inside, signaling the nervous system that a meal or mate is nearby. Pectines may also detect ground surface texture, which helps males decide where to deposit sperm packets.

The sensory world of the platypus, snake, and scorpion may seem strange, but that is only because human sense organs respond to different signals. This chapter focuses on our own senses.

LEARNING OUTLINE

29.1 Diverse Senses Operate by the Same Principles

The sensory abilities of animals are tremendously diverse. Consider a person and a dog walking through the woods. The human relies mostly on sights and sounds, but the dog detects odors that our own noses miss: fox urine left as a calling card near an oak and the scent revealing a deer's bed in a shrub. If this is an evening stroll, another animal, the little brown bat, "sees" the scene with its ears. Bats emit high-frequency pulses of sound and analyze the resulting echoes to "picture" their surroundings in a way that we cannot.

The senses help animals maintain homeostasis. Recall from chapter 27 that homeostasis relies on feedback loops in which sensors report on internal and external conditions (see figure 27.9). The central nervous system uses that information to coordinate the actions of muscles and glands, which make the necessary adjustments. Many feedback loops operate without our awareness; for example, we can't directly "feel" our blood pH or thyroid hormone concentration. But we are aware of temperature, light, sounds, tastes, odors, and many other stimuli. The organs that produce these sensations are the topic of this chapter (**figure 29.1**).

Sensory System	
Main tissue types	**Examples of Locations/Functions**
Epithelial	Makes up some sensory receptors (e.g., taste cells, rods, and cones)
Connective	Makes up part of nose and outer ear and coverings of brain and nerves
Muscle	Skeletal muscle controls opening of eyes and mouth; smooth muscle controls size of iris
Nervous	Makes up the brain, spinal cord, and nerves; functions in sensation, communication, and information storage

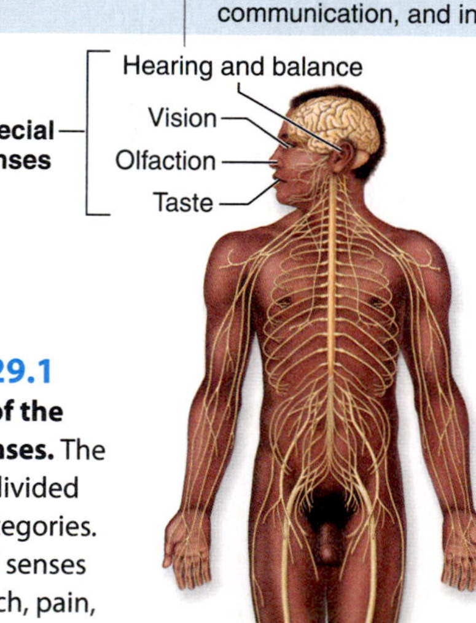

FIGURE 29.1 **Overview of the Human Senses.** The senses are divided into two categories. The general senses include touch, pain, and other senses with receptors located throughout the body. Receptors for the special senses, such as vision and hearing, are limited to the head.

A. Sensory Receptors Convert Stimuli to Action Potentials

All sense organs ultimately derive their information from **sensory receptor** cells that detect stimuli (**table 29.1**). Sensory receptors are portals through which nervous systems experience the world. Some sensory receptors are neurons, such as pain receptors in skin and "smell cells" in the nose. Others, such as the rods and cones of the eye, are specialized epithelial cells that communicate with sensory neurons. Whatever the receptor type, the sensory neurons pass the information to the central nervous system. The brain receives the sensory input, integrates the information, and consults memories to form a perception, or interpretation of the sensation.

How does a sensory receptor **transduce,** or convert, sensory information into an action potential? Generally, the stimulus alters the shape of a protein embedded in a sensory receptor's cell membrane in a way that changes the membrane's permeability to ions. This action triggers a **receptor potential,** which is a graded potential that occurs in a sensory receptor. (Chapter 28 described a graded potential as a local change in the electrical potential of the cell membrane.) The stronger the stimulus, the greater the receptor potential. For example, a loud sound causes greater depolarization than a soft noise, producing a larger receptor potential. If the receptor potential is large enough, an action potential will be generated in the sensory receptor. **graded potential, p. 580**

Figure 29.2 shows a touch receptor in the skin. Very light pressure generates only small receptor potentials (green lines in the figure). The central nervous system would not sense this

Table 29.1 *Types of Sensory Receptors*

Type of Receptor	Stimulus
Chemoreceptor	Airborne or dissolved molecules
Photoreceptor	Light
Mechanoreceptor	Touch, vibration
Thermoreceptor	Heat
Pain receptor (nocireceptor)	Sharp blow, excessive heat
Proprioreceptor	Position of limbs and other body parts
Electroreceptor	Electrical fields

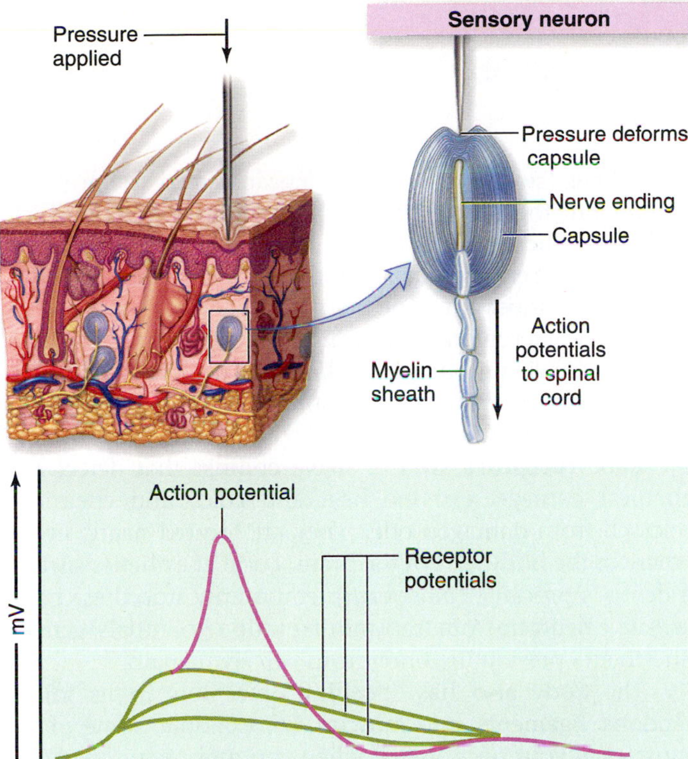

FIGURE 29.2 Receptor Potentials Trigger Action Potentials. Skin contains mechanoreceptors called Pacinian corpuscles that detect touch and pressure. Each corpuscle consists of Schwann cells wrapped around the dendrite of a sensory neuron. When pressure deforms the corpuscle, the neuron undergoes a local flow of electrical current called a receptor potential (green lines). Strong pressure causes stronger receptor potentials, until a threshold is reached that triggers an action potential (red line). The central nervous system detects only those stimuli that provoke action potentials.

touch, because the depolarization does not exceed the touch receptor's threshold potential. A stronger poke causes larger receptor potentials, increasing the likelihood of action potentials (red line in the figure). Only action potentials send information about the touch to the brain. **threshold potential, p. 580**

B. Continuous Stimulation May Cause Sensory Adaptation

You may have noticed that your perceptions of some stimuli can change over time. Your first thought when you roll out of bed may be "I smell coffee." But by the time you stand up, pull your clothes on, and wander to the kitchen, you hardly notice the smell anymore. Likewise, the steaming water in a tub may seem too hot at first, but it soon becomes tolerable, even pleasant. **Sensory adaptation** is a phenomenon in which sensations become less noticeable with prolonged exposure to the stimulus.

Sensory adaptation occurs because the receptors generate fewer action potentials under prolonged stimulation. Without sensory adaptation, we would be distinctly aware of the touch of clothing and every sight and sound. Concentrating on a single stimulus, such as a person speaking, would be difficult. The senses of smell and touch adapt quickly, but other sensory receptors are very slow to adapt, and this is also protective. For example, being very aware of continuous pain helps ensure action to remove its source.

The remainder of the chapter explores some of the senses in more detail. It first describes the general senses, such as touch and pain, whose receptors are located throughout the body. The chapter then turns to the "special" senses, those that are limited to the head: smell, taste, vision, hearing, and balance. **Table 29.2** summarizes the senses.

Table 29.2 *Summary of the Senses*

Sense	Sense Organ	Sensory Receptor
General senses		
Touch	Skin	Mechanoreceptor
Temperature	Skin	Thermoreceptor
Pain	Everywhere except the brain	Pain receptor (nociceptor)
Position of body parts	Joints, tendons, ligaments, muscles	Stretch receptor (proprioceptor)
Special senses		
Smell	Nasal cavity	Chemoreceptor
Taste	Mouth and tongue	Chemoreceptor
Vision	Eyes	Photoreceptor
Hearing	Ears	Mechanoreceptor
Equilibrium	Ears	Mechanoreceptor

Summary *Proteins in the cell membranes of sensory receptors respond to specific stimuli, changing the membrane's permeability and provoking action potentials. The resulting neural impulses travel to the brain, which interprets the input as a perception. Sensory receptors detect the type and strength of a stimulus; they also adapt to prolonged stimulation.*

29.1 MASTERING CONCEPTS

1. What role do the senses play in maintaining homeostasis?
2. What are the major types of sensory receptors?
3. What is a receptor potential?
4. What is sensory adaptation and how is it beneficial?

29.2 The General Senses Detect Touch, Temperature, Pain, and Position

The general senses allow you to feel sensations such as touch, temperature, or pain with any part of your skin. For all of the general senses, the receptors are sensory neurons. Most have their dendrites wrapped (encapsulated) in neuroglia or connective tissue, but some have free (unencapsulated) nerve endings. **integumentary system, p. 569**

Mechanoreceptors are sensory receptors that respond to physical deflection. The sense of touch comes from several types of mechanoreceptors in the skin (**figure 29.3**). Pacinian and Meissner corpuscles each consist of a single encapsulated dendrite. A touch pushes the flexible sides of the corpuscle inward, generating an action potential in the nerve fiber (see figure 29.2). Unencapsulated touch receptors include the dendrites that wrap around each hair follicle and sense when the hair bends.

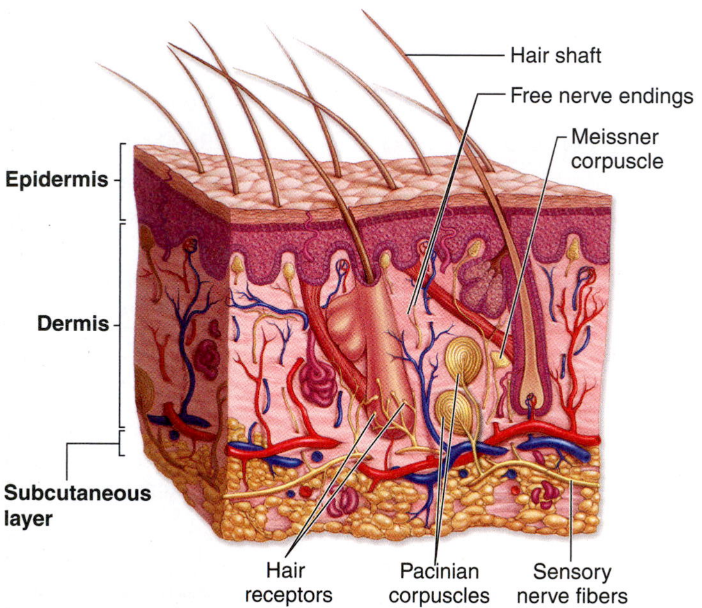

FIGURE 29.3 Skin Senses Many Stimuli. Free nerve endings in skin respond to touch, temperature, and pain. Encapsulated dendrites, including Meissner and Pacinian corpuscles, respond to light touch and vibration.

Free nerve endings that act as **thermoreceptors** in the skin respond to temperature. The brain integrates input from many cold and heat receptors to determine whether a stimulus is cool, hot, or somewhere in between. Thermoreceptors adapt quickly, which is why we soon become comfortable after jumping into a cold swimming pool or easing into a steaming hot tub.

Pain receptors are free nerve endings that detect mechanical damage, extreme heat and cold, and chemicals released from damaged cells. They are located nearly everywhere in the body, except the brain. Local anesthetics such as a dentist's procaine (Novocain) temporarily stop these pain-sensitive neurons from transmitting action potentials; general anesthetics prevent the brain from perceiving pain.

The body also has "position" receptors in its joints, tendons, ligaments, and muscles. For example, some of the muscles in your neck stretch when you move your head. Encapsulated nerve endings wrap around specialized cells in the muscles, and the dendrites of these stretch receptors initiate nerve impulses that tell the brain exactly which way your head is facing. These specialized cells are most abundant in body parts with the finest muscle control, such as the hands.

Summary *Free and encapsulated nerve endings in the skin detect touch, temperature, and pain. Receptors in the muscles and joints report on the position of body parts.*

29.2 MASTERING CONCEPTS

1. Which structures provide the sense of touch, temperature, and pain?
2. How do stretch receptors in muscles help the body detect the position of its parts?

29.3 The Senses of Smell and Taste Detect Chemicals

Chemoreception is probably the most ancient sense. Bacteria and protists use chemical cues to approach food or move away from danger, so the ability to detect external chemicals must have arisen long before animals evolved. **Chemoreceptors** are the sensory cells that detect chemicals in the environment.

Taste and smell depend on chemoreceptors. Both senses require that a stimulus molecule be dissolved in a watery solution, such as saliva or the moist lining of a nasal passage. In addition, the molecule must interact with a receptor on a sensory cell's membrane. Chemoreceptors in taste buds detect very high concentrations of chemicals, so that we can taste an item only at very close range. In contrast, the nose detects low concentrations of volatile odor molecules that travel into the nose with inhaled air. We can therefore smell scents that originate from near or distant objects.

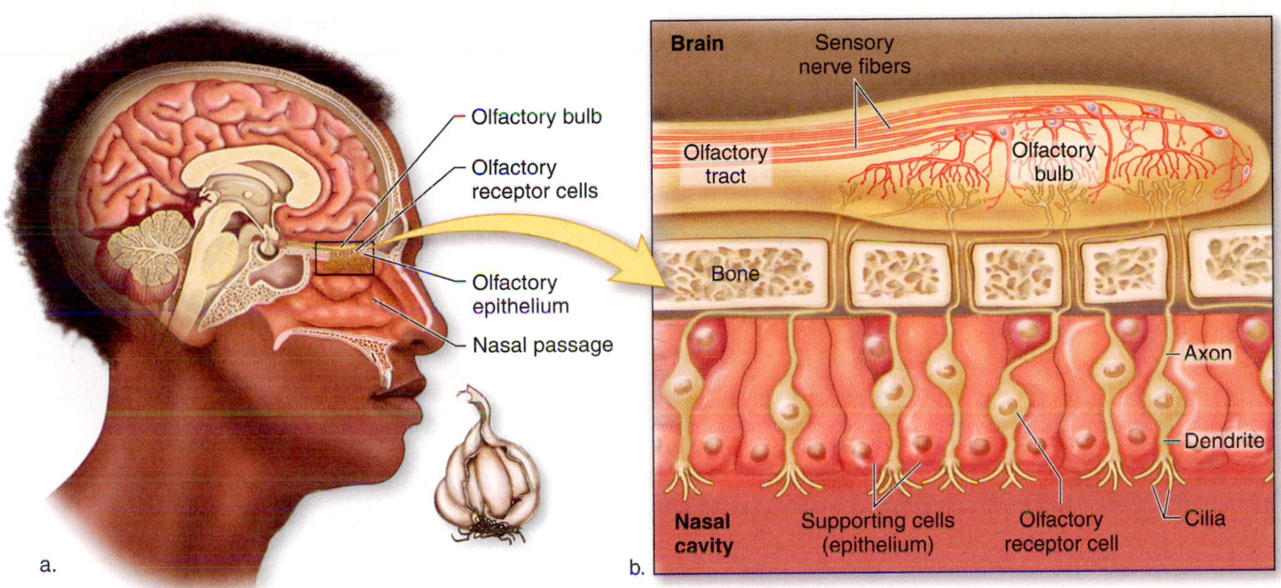

FIGURE 29.4 **The Sense of Smell.** (**A**) Olfaction derives from receptor cells in the nasal cavity. (**B**) An olfactory receptor cell binds an odorant molecule (such as from garlic) and transmits neural impulses to cells in the olfactory bulb. The axons of these neurons pass the information to the brain for interpretation as a particular odor.

A. Chemoreceptors in the Nose Detect Odor Molecules

The **nose** forms the entrance to the nasal cavity inside the head. It functions both in breathing and the sense of smell, or **olfaction.** Specialized olfactory receptor neurons are located in a patch of epithelium high in the nasal cavity (**figure 29.4**). Humans have about 12 million olfactory receptor cells, each with 10 to 20 cilia that increase the surface area for receiving odorant molecules. (In contrast, a bloodhound has 4 billion receptor cells, and its olfactory epithelium has dozens of times more surface area than ours. The dog's sense of smell is therefore much more acute than our own).

To trigger the sense of smell, an odor molecule dissolves in the moist lining of the nasal passage, then binds to proteins on the cilia of olfactory receptor cells. About 500 different types of membrane-bound receptor proteins bind specific odorant molecules. Proteins inside the receptor cell transduce this chemical signal into receptor potentials. Part of each olfactory receptor cell passes through the skull and synapses with neurons in the brain's olfactory bulb. From here, sensory neurons relay the message to the cerebral cortex, which interprets the message as an odor and identifies it. Sensory information from olfactory receptors also travels to the limbic system, the brain center of memory and emotion. This is why a whiff of the perfume Grandma used to wear may bring back a flood of memories. **limbic system, p. 591**

Many insects use chemicals in communication. **Pheromones** are volatile chemical substances that elicit specific responses in other members of the same species. For example, female silk moths (*Bombyx mori*) release pheromones that attract males up to several kilometers away (**figure 29.5**). The male is so sensitive to these odors that a single pheromone molecule is enough to trigger an action potential in a sensory receptor. The role of chemical communication in our own species remains an open question, as this chapter's Burning Question: Do humans have pheromones? describes.

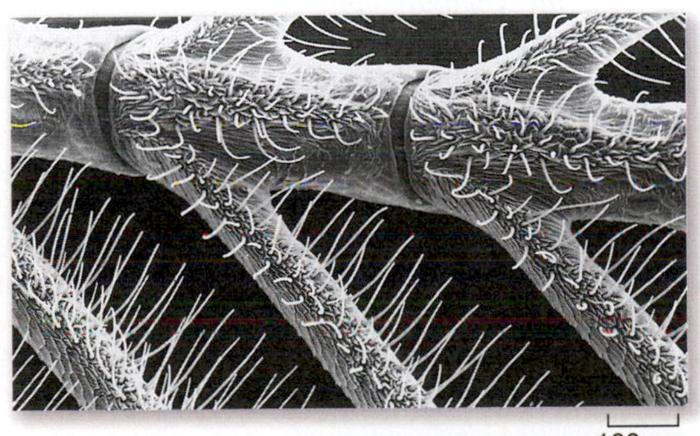

FIGURE 29.5 **Moth Pheromone Chemoreception.** (**A**) The antennae of the male silkworm moth *Bombyx mori* effectively trap wafting chemicals. (**B**) The 50,000 or so sensory hairs detect pheromones that the female emits. Each sensory hair includes a fluid-filled chamber in which the pheromones dissolve.

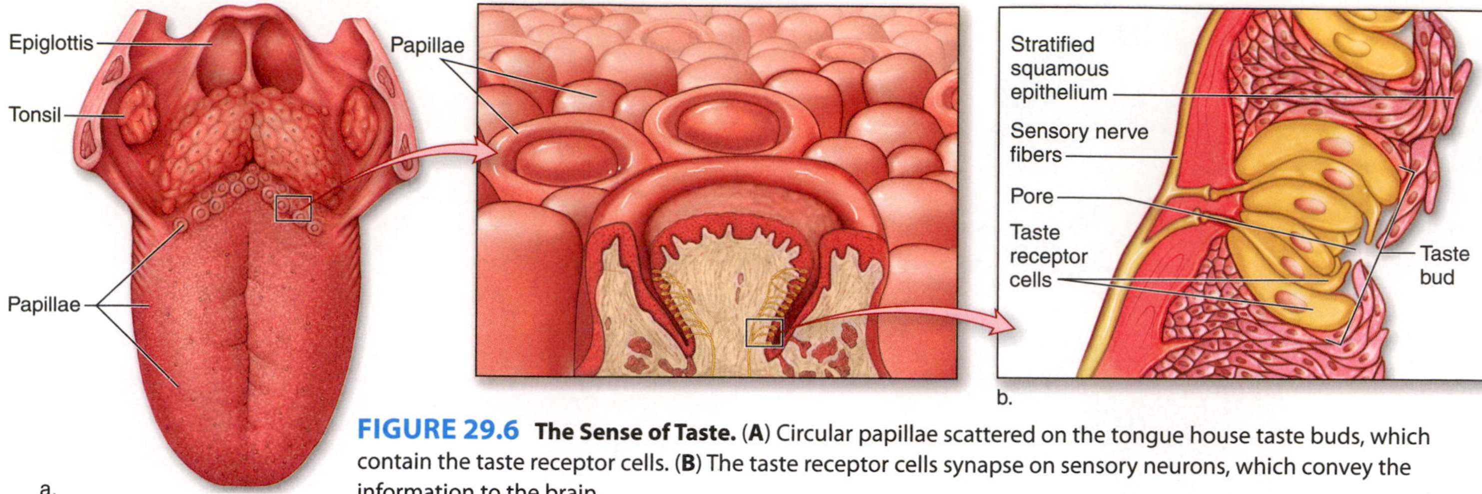

FIGURE 29.6 The Sense of Taste. (**A**) Circular papillae scattered on the tongue house taste buds, which contain the taste receptor cells. (**B**) The taste receptor cells synapse on sensory neurons, which convey the information to the brain.

B. Chemoreceptors in the Mouth Detect Taste

Taste, or **gustation,** is intimately related to smell, as anyone whose sense of taste has been dulled by a stuffy nose can attest. The sense of taste is centered in the **tongue,** a muscular organ arising from the floor of the mouth. Humans have about 10,000 **taste buds,** each containing 50 to 150 chemoreceptor cells that generate action potentials when dissolved food molecules bind to them (**figure 29.6**). The receptors are epithelial cells that synapse at the base of the taste bud onto sensory neurons, which relay information to the medulla of the brain. Taste buds are lightly scattered around the inside of the mouth, but they are concentrated in grooves on the tongue's upper surface.

The human tasting experience includes the four primary sensations of sweet, sour, salty, and bitter. A fifth taste sensation, called umami (from the Japanese word for "delicious"), reflects the ability to sense the amino acid glutamate. It imparts a savory flavor to meat. The food additive monosodium glutamate (MSG) activates the umami sense. Most taste buds sense all four of the "primary" tastes, but each responds most strongly to one or two. Our sense of taste depends on the pattern of activity across all taste neurons. Food's texture, temperature, and aroma also contribute to its flavor.

Summary *Both smell and taste rely on chemoreception. Receptor cells in olfactory epithelia bind odorant molecules and send signals that the brain interprets as odors. Taste receptor cells in the tongue bind food molecules.*

29.3 MASTERING CONCEPTS

1. How does the brain distinguish one odor from another?
2. What are pheromones?
3. How does a taste bud function?
4. What are the five taste sensations that impart a food's flavor?

Burning Questions

Do humans have pheromones?

Advertisements for "human pheromone" colognes appeal to the desire to attract the opposite sex. Dab some on, they say, and watch your love life blossom. But are there really human pheromones?

Pheromones are relatively easy to study in insects, whose behavioral repertoire is limited. But mammals, especially humans, have much more complex behaviors, so it is difficult to find chemicals that elicit predictable responses. Nevertheless, at least some mammals do produce pheromones. A male hamster smeared with vaginal secretions from a female will provoke sexual advances from another male—but only if the responding male has an intact vomeronasal organ, a tiny offshoot of the olfactory system. This structure is apparently the pheromone detector.

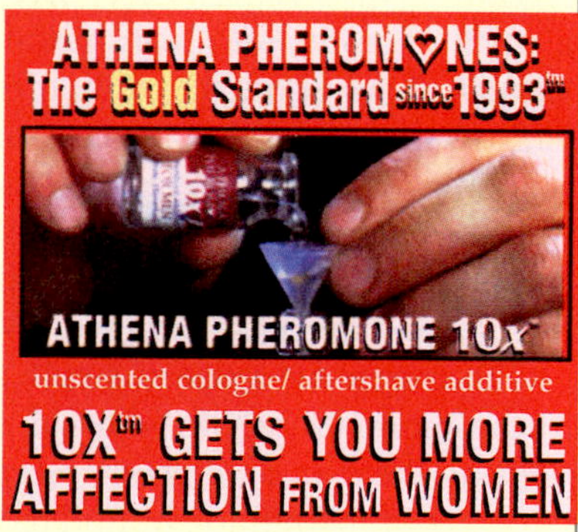

In 1998, the journal *Nature* published a study demonstrating that pheromones from human females influence the menstrual cycles of other women. However, researchers still know little about how humans detect pheromones. We do have a vomeronasal organ, but no one has ever shown that it is functional. Researchers are working to discover the genes that encode the pheromone-binding receptors in a rodent's vomeronasal organ. Comparison with the human genome should provide clues about how human pheromones work and whether the vomeronasal organ plays a role in human life or is just a vestige of our evolutionary history.

Have a Burning Question of your own? Submit it to marielle_hoefnagels@mcgraw-hill.com for possible inclusion in future editions of this book!

29.4 Vision Depends on Light-Sensitive Cells

An **eye** is an organ that produces the sense of sight. Animal eyes contain dense concentrations of **photoreceptors,** which are sensory cells that respond to light. A photoreceptor contains a pigment molecule associated with a membrane. **Rhodopsin** is a common light-sensitive pigment. When rhodopsin absorbs light, its shape changes, altering the charge across the membrane and possibly generating an action potential in a nearby neuron.

A. The Vertebrate Eye Consists of Three Layers

The vertebrate eye has a **lens** that focuses incoming light on the **retina**—a sheet of photoreceptors at the back of the eye. Both structures are housed inside the eyeball, a fluid-filled sphere with a wall consisting of three layers (**figure 29.7**):

- **Sclera:** The **sclera** is the white, outermost layer of the eye. It protects the inner structures. Toward the front of the eye, the sclera is modified into the **cornea,** a transparent curved window that bends incoming light rays and helps to focus them on the retina.
- **Choroid:** The **choroid** lies between the sclera and the retina; blood vessels in the choroid supply nutrients and oxygen to the retina. (Reflection of bright light from these blood vessels produces the "red eye" effect in photographs.) A portion of the choroid thickens into a structure that holds the lens. In front of the lens, the choroid becomes the **iris,** which is the colored part of the eye. The iris regulates the size of the **pupil,** the hole in the middle of the iris. In bright light, the pupil is tiny, shielding the retina from excess stimulation. The pupil grows larger as light becomes dimmer.
- **Retina:** The third and innermost layer, the retina, transduces light energy to action potentials.

Each eyeball also contains fluid that helps bend light rays and focus them on the retina. Behind the lens is the jellylike vitreous humor, which occupies most of the eyeball's volume. The watery aqueous humor lies between the cornea and the lens. This fluid cleanses and nourishes the cornea and the lens and maintains the shape of the eyeball.

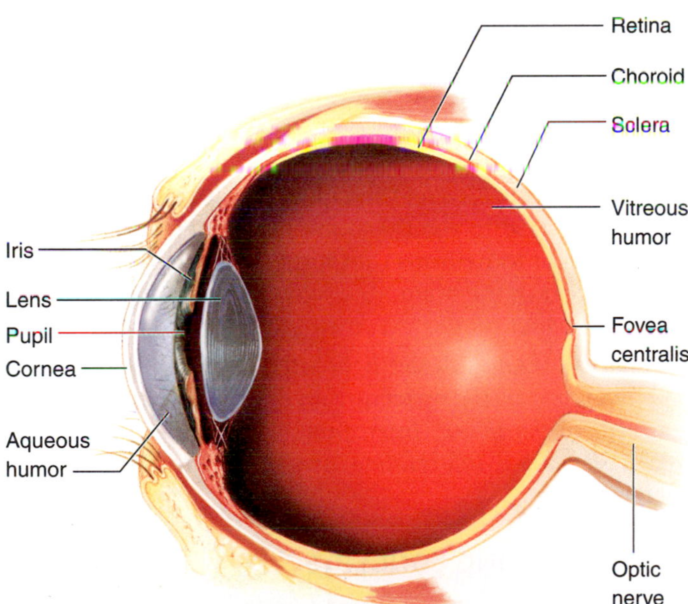

FIGURE 29.7 Anatomy of the Vertebrate Eye. The wall of the eye has three layers: the sclera, the choroid, and the retina. The iris regulates the amount of light entering the eye, and the retina contains the sensory cells that transmit light information to the optic nerve. The liquid inside the eyeball is the vitreous humor.

B. Cells in the Retina Convert Light Energy to Nerve Impulses

Vision begins when light rays pass through the cornea, lens, and humors of the eye and are focused on the retina. (This chapter's Can *You* Relate? Correcting Vision explains how glasses and surgery can improve poor eyesight.) Because of the curved shape of the lens, the image projected onto the retina is actually upside down and backward (**figure 29.8**). The brain processes this information so that we perceive a right-side-up world.

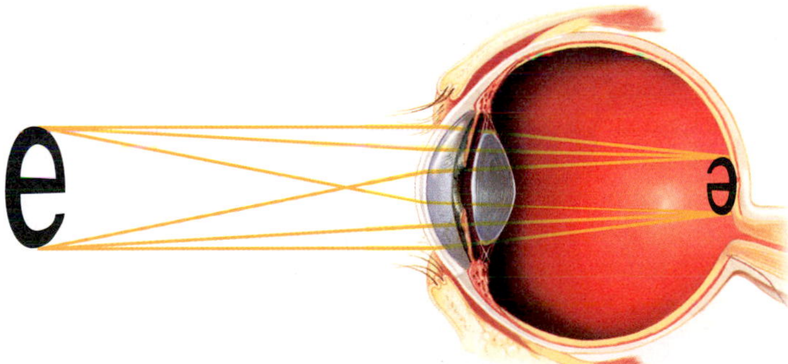

FIGURE 29.8 Focusing the Light. The eye collects, transmits, and focuses light onto photoreceptor cells at the back of the eyeball. The image on the retina is upside down and backward; the brain later processes this information so that the object appears right-side-up.

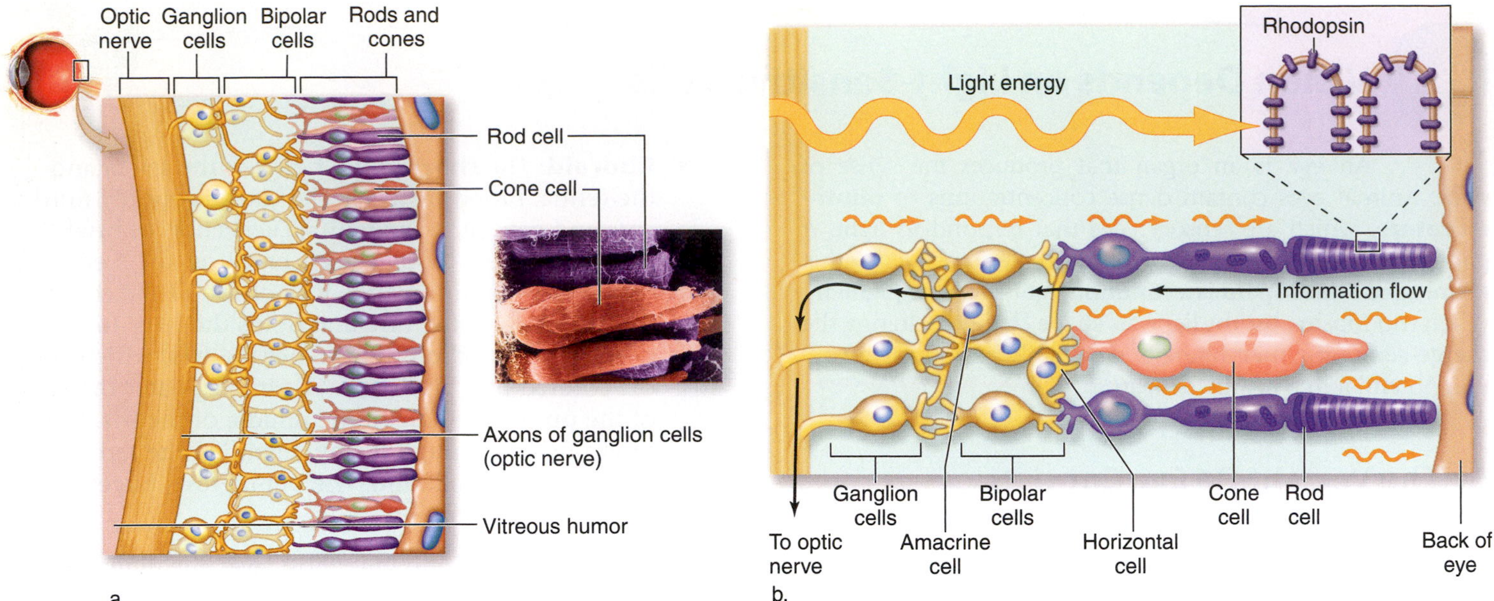

FIGURE 29.9 The Visual Pathway. (**A**) Photons pass through the outer layers of the eye and hit the rods and cones at the back of the eye. These sensory receptors transduce the light energy and transmit the information to bipolar cells, which pass the message to the action potential-generating ganglion cells that form the optic nerve. (**B**) Rod cells and cone cells contain highly folded membranes studded with visual pigments that absorb photon energy and transduce it to an electrochemical message. Light energy alters the conformation of rhodopsin, which may ultimately alter the pattern of action potentials in the optic nerve.

A more detailed look at the retina helps explain how the eye converts light energy to action potentials (**figure 29.9**). Furthest to the rear of the eye are the photoreceptors, epithelial cells called rods and cones. **Rod cells,** which are concentrated around the edges of the retina, provide black-and-white vision in dim light and enable us to see at night. **Cone cells** detect color; they are concentrated toward the center of the retina. The human eye contains about 125 million rod cells and 7 million cone cells. (Section 29.6 explores the evolutionary origin of these cells.)

Both rods and cones consist of many interconnected discs of membranes that are studded with pigment molecules (see figure 29.9). In a rod cell, the pigment is rhodopsin. Cone cells contain rhodopsin-like pigments that absorb light of different wavelengths. Humans have three cone types: "blue" cones absorb shorter wavelengths of light, "green" cones absorb medium wavelengths, and "red" cones absorb long wavelengths. People who lack a cone type entirely, due to a genetic mutation, are color-blind. The most common form of color blindness is a deficiency of pigments sensitive to red and green wavelengths. Because the genes encoding these pigments are on the X chromosome, red–green color blindness is more common in males than females. **X-linked disorders, p. 225**

C. Signals Travel along the Optic Nerve to the Brain

When a rod or cone cell absorbs photon energy, the pigment molecule changes shape and triggers receptor potentials that stimulate the retina's bipolar neurons (see figure 29.9). The bipolar neurons, in turn, transmit the message to the **ganglion cells,** interneurons that make up the innermost layer of the retina. (Two other types of neurons in the retina, called horizontal and amacrine cells, form connections that modify the information sent along the visual pathway.) Ganglion cells are the only neurons in the retina that generate action potentials; all others produce only graded potentials.

The axons of the ganglion cells make up the **optic nerve** that connects the retina to the brain. The point where the optic nerve exits the retina is called the blind spot because it lacks photoreceptors and therefore cannot sense light. Both optic nerves lead to the thalamus. From there, visual information is sent to neurons that form the primary visual cortex at the rear of the brain.

Summary *Animals detect light with specialized cells called photoreceptors. In the vertebrate eye, light stimulates rod and cone cells in the retina. Interconnected neurons in the retina process the visual information and send it to the brain for interpretation.*

29.4 MASTERING CONCEPTS

1. What are the parts of the vertebrate eye?
2. What are the roles of rod cells, cone cells, and light-sensitive pigments in human vision?
3. Trace the pathway of information flow from the retina to the visual cortex of the brain.

Can *You* Relate?

Correcting Vision

The eyeball must have a certain shape for the cornea and lens to focus light rays precisely on the retina. For those of us whose eyeballs are not perfectly formed, corrective lenses (eyeglasses and contact lenses) can alter the path of light (**figure 29.A**). A more recent technology for correcting vision problems is laser eye surgery, which vaporizes tiny parts of the cornea, changing the path of light to the retina.

Sometimes, the cornea becomes clouded or misshapen. Surgeons can replace the defective cornea with one taken from a cadaver. Corneal transplant surgery carries a low risk of immune system rejection because, unlike other transplantable organs, the cornea lacks blood vessels. Another common eye disorder is a cataract, in which the lens of the eye becomes opaque. Cataract surgery is a simple procedure that replaces the clouded lens with a plastic implant.

Even people with perfectly shaped eyeballs and corneas usually need reading glasses after the age of about 40. To focus on a very close object, a muscle inside the eye must curve the lens so that it can bend incoming light rays at sharper angles. As we age, the lens becomes less flexible. It therefore becomes difficult for the muscles in the eye to bend the lens enough to clearly focus on nearby objects or printed words. Laser surgery cannot correct this age-related decline in eyesight.

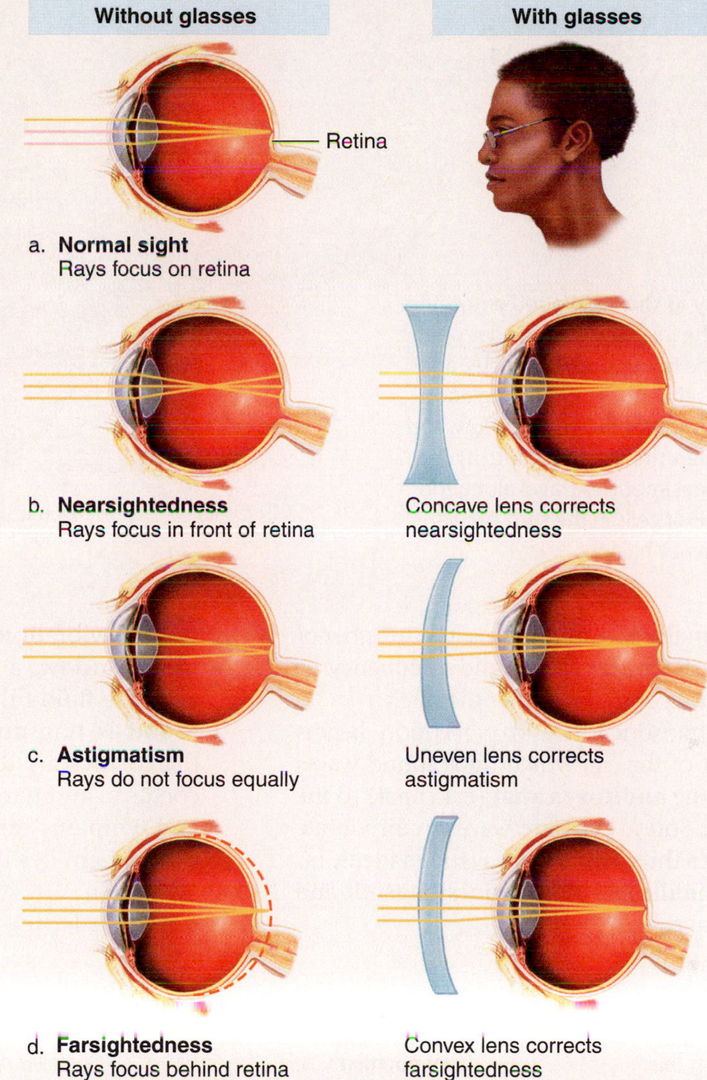

FIGURE 29.A Correcting Vision. (**A**) A normally shaped eyeball focuses light rays on the retina. (**B**) In an elongated eyeball, light rays converge in front of the retina, impairing the ability to see distant objects (nearsightedness). A concave eyeglass lens alters the point of focus to the retina. (**C**) In astigmatism, the lens or cornea is misshapen, and incoming light rays do not focus evenly. An irregularly shaped eyeglass lens can help. (**D**) A short eyeball focuses light beyond the retina, and the person has difficulty seeing close objects (farsightedness). A convex lens corrects this problem.

29.5 The Senses of Hearing and Balance Begin in the Ears

Mechanoreceptors inside the **ear** provide two senses: hearing and balance. In both cases, the sensory receptors are epithelial cells with many hairlike extensions. When the hairs bend, they provoke action potentials in nearby neurons that relay the signals to the brain.

A. Mechanoreceptors in the Inner Ear Detect Sound Waves

The clatter of a train, the sounds of a symphony, a child's wail—what do they have in common? All sounds, regardless of the source, originate when something vibrates and creates

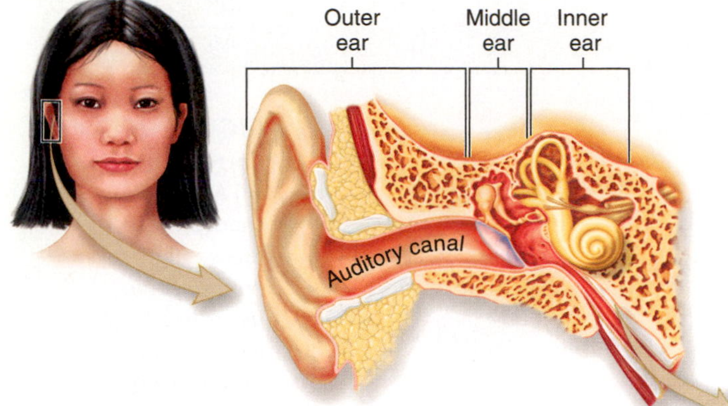

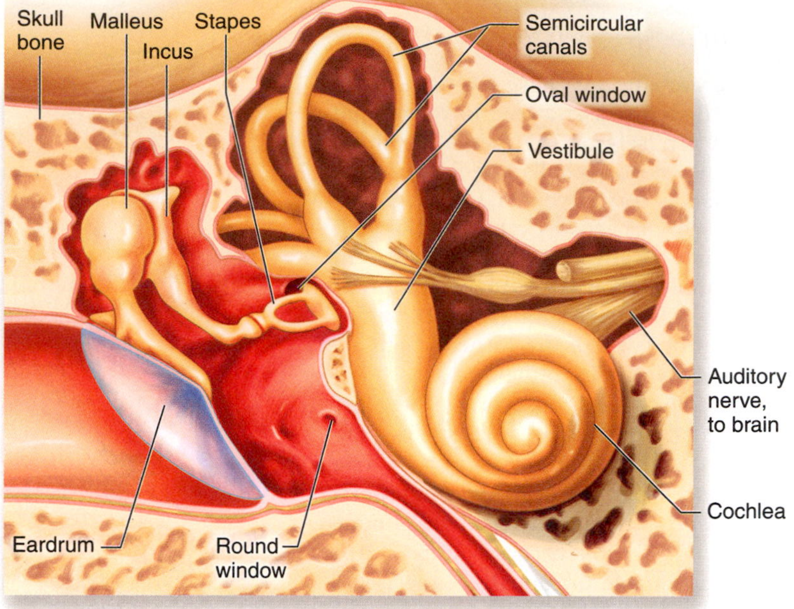

FIGURE 29.10 Anatomy of the Ear. Sound enters the outer ear and, on reaching the middle ear, impinges on the eardrum, which vibrates three bones (malleus, incus, and stapes). The inner ear houses the organs of hearing and balance. The vibrating stapes moves the oval window, and hair cells in the cochlea convert the vibrations into action potentials. The resulting neural impulses travel along the auditory nerve to the brain. Hair cells in the semicircular canals and in the vestibule sense balance.

repeating pressure waves in the surrounding air. The number of waves (cycles) per second determines the sound's frequency, or pitch. The more cycles per second, the higher the pitch.

In humans, sound transduction and perception begins with the fleshy outer part of the ear, which traps sound waves and funnels them down the **auditory canal** (ear canal) to the **eardrum** (**figure 29.10**). Sound pressure waves in air vibrate the eardrum, which moves three small bones (the malleus, incus, and stapes) of the middle ear. These bones transmit and amplify the incoming sound.

The vibrations of the stapes are transmitted through the **oval window,** a membrane that connects the middle ear with the fluid-filled inner ear. At the snail-shaped **cochlea,** sound is transduced into neural impulses. The spirals of the cochlea are three fluid-filled ducts called the vestibular, cochlear, and tympanic canals (**figure 29.11**). The vestibular and tympanic canals form a continuous U-shaped tube. Between them lies the cochlear canal.

When the stapes moves, it pushes on the oval window, transferring the vibration to the fluid in the vestibular

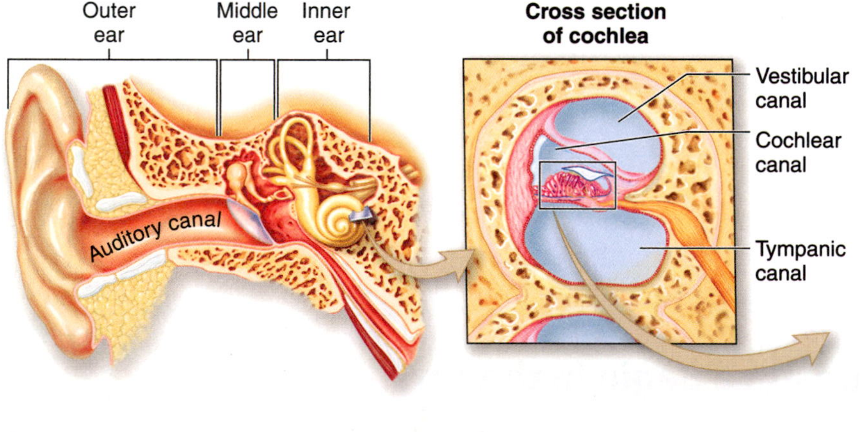

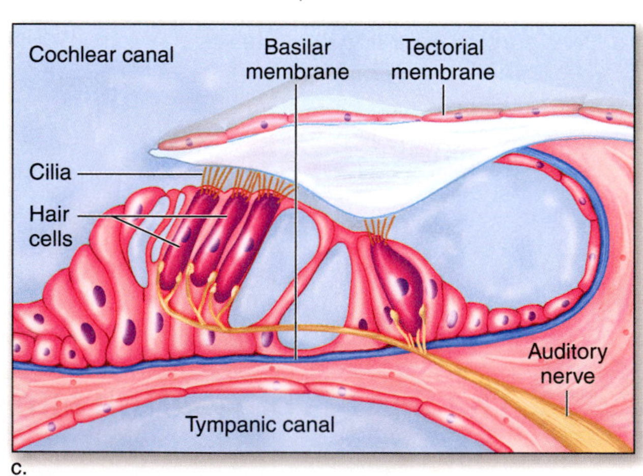

FIGURE 29.11 The Cochlea Transmits Sound to the Auditory Nerve. Hearing is the transduction of vibrations to neural impulses. The cochlea, shown in anatomical context in (**A**) and in cross section in (**B**), is a spiral-shaped structure consisting of three fluid-filled canals. When the stapes moves against the oval window, the fluid in the vestibular canal vibrates. The vibrations press on the hair cells that lie between the basilar and tectorial membranes (**C**). Cilia fringe the tops of the hair cells. When hair cells push against the tectorial membrane, action potentials are triggered in the auditory nerve.

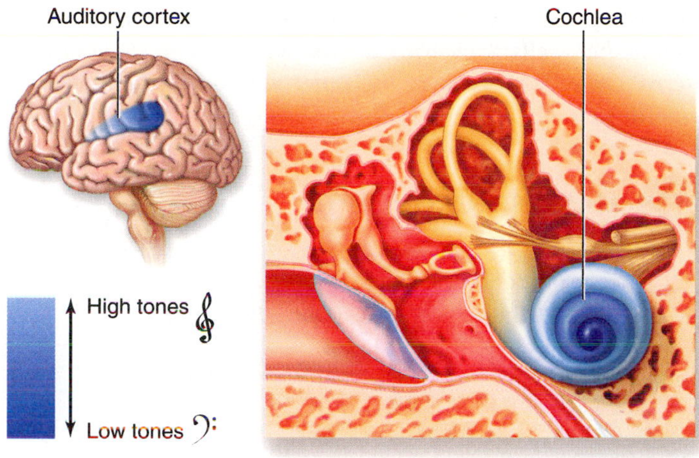

FIGURE 29.12 **Correspondence Between Cochlea and Cortex.** Sounds of different frequencies excite different sensory neurons in the cochlea. These neurons, in turn, send their input to different regions of the auditory cortex.

canal. The pressure of each vibration moves the round window at the other end of the U. The vibration also moves the **basilar membrane,** which forms the lower wall of the cochlear canal. Embedded in the basilar membrane are **hair cells,** specialized epithelial cells that are the cochlea's mechanoreceptors; they initiate the change of mechanical energy to receptor potentials. When a region of the basilar membrane vibrates, the hair cells move relative to the **tectorial membrane,** a sheet of cells resting on the cilia of the hair cells. The hair cells then depolarize and release a neurotransmitter that initiates action potentials in fibers of the **auditory nerve,** which carries the impulses to the brain's auditory cortex.

Each frequency of sound vibrates a different part of the basilar membrane. The high-pitched tinkle of a bell stimulates the narrow region of the basilar membrane at the base of the cochlea, whereas the low-pitched tones of a tugboat whistle stimulate the wide end near the cochlea's tip. The brain interprets the input from different regions as sounds of different pitches (**figure 29.12**). In addition, loud sounds cause the basilar membrane to vibrate more than softer sounds. This increased movement stimulates additional hair cells, each of which triggers action potentials at a higher frequency. The brain interprets the resulting increase in the rate and number of neurons firing as an increase in loudness.

The sense of hearing requires the interaction of many parts of the ear and nervous system. Deafness can occur if any of those components fails to function correctly. Can *You* Relate? Deafness explores the causes and treatments of hearing loss.

B. The Inner Ear Also Provides the Sense of Balance

The **semicircular canals** are three interconnected, fluid-filled loops that tell us when the head is rotating and help us maintain the position of the head in response to sudden movement (**figure 29.13**). The enlarged bases of the semicircular canals are lined with hair cells covered by a caplike structure called a cupula. The semicircular canals are perpendicular to one another, so the fluid that fills a canal may or may not flow back and forth in response to a movement. When the fluid in a canal moves, it bends the cilia on the hair cells, which stimulates action potentials in a nearby cranial nerve. The brain interprets these impulses as body movements.

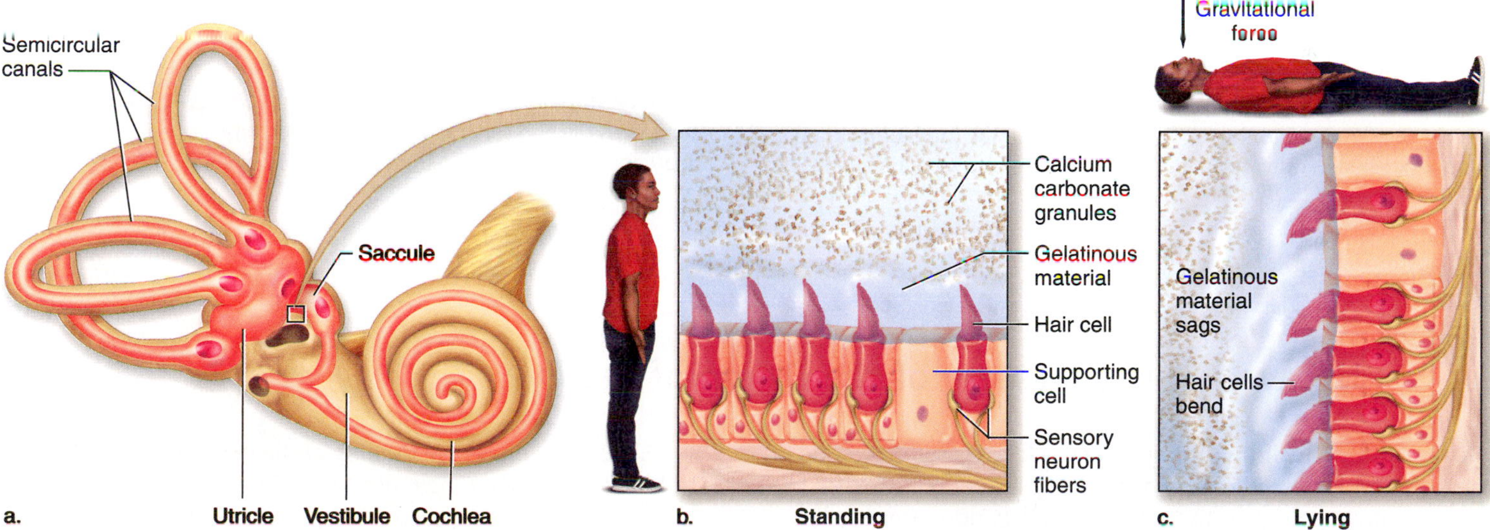

FIGURE 29.13 **Equilibrium and Balance.** (**A**) The senses of equilibrium and balance derive from the semicircular canals and vestibule of the inner ear. (**B**) When the head is upright, calcium carbonate granules balance atop the cilia of hair cells. (**C**) When the head is horizontal, the granules move, bending the cilia. This provokes an action potential.

The inner ear also detects whether the head is moving with respect to gravity (see figure 29.13). The **vestibule** lies between the cochlea and the semicircular canals. It contains two pouches, the utricle and the saccule, both of which are filled with jellylike fluid and lined with ciliated hair cells. Granules of calcium carbonate float on the fluid. When the head accelerates or tilts, the granules move, bending the cilia on the hair cells. The brain interprets this information as a change in velocity or a change in body position. **gravitropism, p. 552**

Motion sickness results from contradictory signals. The inner ear signals the brain that the person is not accelerating. At the same time, the eyes detect passing scenery and signal the brain that the person is moving. The result of these mixed signals is nausea.

Summary ***Structures in the ear transmit sound waves that contact hair cells, triggering action potentials on the auditory nerve. Other structures in the inner ear control balance and equilibrium.***

29.5 MASTERING CONCEPTS

1. What is the role of mechanoreceptors in the senses of hearing and equilibrium?
2. What are the parts of the ear, and how do they transmit sound?
3. How do the semicircular canals and vestibule sense equilibrium and balance?

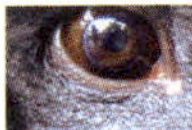

Can *You* Relate?

Deafness

The sense of hearing is so complex that it is not surprising to know that deafness can take multiple forms. In conductive deafness, the middle ear fails to transmit sounds to the inner ear. In nerve deafness, either the auditory nerve does not function, or the brain does not respond to input from the nerve.

What causes hearing loss? Some babies are born deaf because of a genetic mutation, chromosomal abnormality, or prenatal exposure to disease. Other people lose their hearing later because of disease, exposure to loud noise, or injury. Earwax or an ear infection can cause short-term deafness. And nearly everyone suffers some hearing loss later in life as the ear becomes less sensitive to higher frequencies.

Hearing aids can treat some forms of hearing loss. By amplifying sounds, the hearing aid moves the eardrum more than normal, helping the person hear more clearly. If the middle ear cannot transmit sound, however, a conventional hearing aid is useless. Bone-conduction aids solve this problem by transmitting sound waves directly to the bones of the skull. The vibrations stimulate the cochlea directly, bypassing the middle ear.

A cochlear implant may restore some hearing to a person who is profoundly deaf (**figure 29.B**). A surgeon places the device under the skin behind the ear. A microphone in the implant picks up sound, then a sound processor decomposes it into separate frequency components. Electrodes placed directly in the cochlea stimulate the parts of the auditory nerve corresponding to each frequency. By sending signals directly to the nervous system, cochlear implants compensate for nonfunctioning parts of the middle and inner ear.

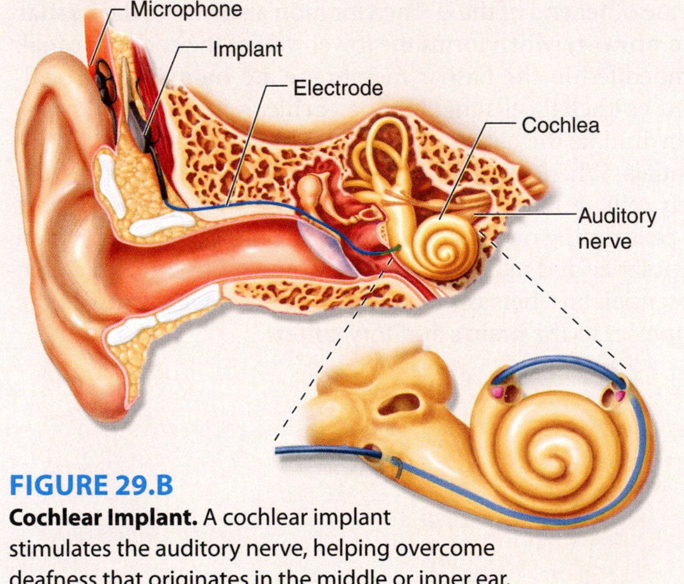

FIGURE 29.B
Cochlear Implant. A cochlear implant stimulates the auditory nerve, helping overcome deafness that originates in the middle or inner ear.

29.6 INVESTIGATING LIFE
Unraveling the Mystery of the Origin of the Eye

In Greek mythology, the Cyclops was a grotesque, one-eyed monster. Of course, the Cyclops never really existed. Modern archaeologists and paleontologists now suggest that the myth originated when ancient Greeks stumbled on the remains of a mastodon fossil. The vertebrae, giant limb bones, and profile of the skull could have resembled a human with a monstrous face and a single eye.

Today, biologists are looking at the origin of the eye itself. In his book *On the Origin of Species*, Charles Darwin calls the notion that the eye formed through natural selection "absurd in the highest possible degree." He follows this statement with great insight, however, when he states, "Yet reason tells me that if numerous gradations from a perfect and complex eye to one very imperfect and simple, each grade being useful to its possessor, can be shown to exist" then this paradox can be overcome. Could a structure as complex as the eye really be the product of evolution?

Animal eyes vary greatly, from the simple "eye cups" of flatworms to the complex, fluid-filled eyes of vertebrates. Despite their diversity, however, all eyes apparently arose from a common ancestor. All organisms with eyes have a gene (*pax6*) that regulates eye formation. All eyes also have light-sensitive pigments (derived from vitamin A) that combine with opsin proteins in the membranes of the photoreceptor cells.

Two major types of photoreceptor cells store opsin pigments in animals. One type, called a rhabdomeric photoreceptor, occurs predominantly in invertebrates. These cells express a unique form of opsin, called r-opsin. Vertebrate rods and cones, as well as cells in the brain's light-sensitive pineal gland, are ciliary photoreceptors that express a different form of opsin (c-opsin). This observation raises a question. If our invertebrate ancestors had rhabdomeric photoreceptors, where did the ciliary photoreceptors in the vertebrate eye come from?

Part of the answer has come from studies of *Platynereis dumerilii*, a small, segmented marine worm commonly called a ragworm (**figure 29.14**). *Platynereis dumerilii* is considered a "living fossil" because it differs little from its 600-million-year-old ancestors. This animal has been the subject of intense molecular and evolutionary research.

The ragworm has two types of eyes at different stages of its life. The adult eyes develop separately from the larval eyes, but the light-sensitive cells in both types of eyes are rhabdomeric, as expected for an invertebrate. Embedded in the worm's brain, however, are cells that resemble ciliary photoreceptor cells. In the worm, these cells do not contribute to vision.

A team of researchers led by Detlev Arendt and Joachim Wittbrodt at the European Molecular Biology Laboratory noticed that the ciliary photoreceptor cells in the brain of *P. dumerilii* strongly resemble rods and cones in the human eye. To test the hypothesis that these cells share the same evolutionary origin, the scientists used a tool called "molecular fingerprinting." In this procedure, researchers analyze the proteins inside living cells and attempt to identify a "fingerprint" (combination of proteins) that is unique to a cell type. If cells in multiple species have matching fingerprints, then they probably have a common evolutionary origin.

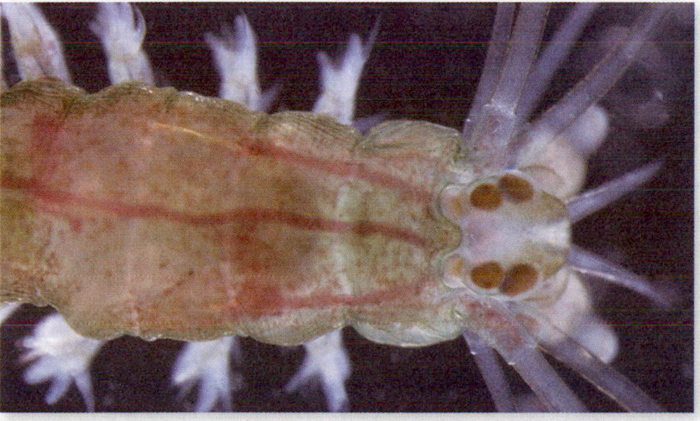

FIGURE 29.14 The marine ragworm, *Platynereis dumerilii*.

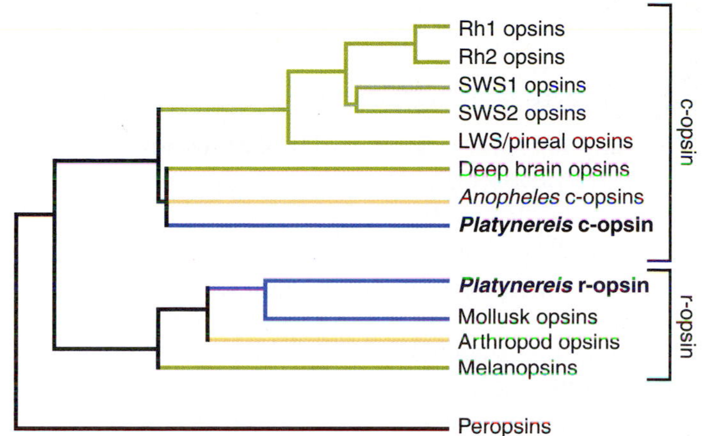

FIGURE 29.15 Two Groups of Opsins. This tree shows that the opsin from *Platynereis* rhabdomeric photoreceptors groups with other invertebrate r-opsins, and that the c-opsin from *Platynereis* ciliary cells groups with vertebrate c-opsins. **Question:** How might you test the researchers' hypothesis that an ancestor of *P. dumerilii* had one photoreceptor and both types of opsins?

The researchers screened the ragworm's cells for opsin proteins. They discovered a new type of opsin that was present only in the ciliary cells of the worm's brain, not in the rhabdomeric photoreceptor cells of the worm's eyes. When they compared the amino acid sequence of the newly discovered protein with known opsins, the researchers found it to be more similar to the c-opsins of vertebrates than the invertebrate r-opsins (**figure 29.15**). The team concluded that the ragworm's ciliary cells are homologous to vertebrate rods and cones. That is, the cell types share a common evolutionary origin.

Arendt and his colleagues propose that early animals had one type of photoreceptor cell containing one type of opsin. Long ago, this ancestral protein diverged into r-opsin and c-opsin, paving the way for the photoreceptor cell to diverge into rhabdomeric and ciliary subtypes. The rhabdomeric cells developed into simple eyes, and the ciliary cells remained in the brains of ancient invertebrates. The vertebrate retina, however, incorporated both types of cells. The rhabdomeric cells became the ganglion cells of the vertebrate retina, and the ciliary cells evolved into rods and cones. Vertebrate ciliary cells therefore took on a new role—vision—that they never had in their invertebrate ancestors.

For all its marvelous complexity, the basic components of the vertebrate eye were already present hundreds of millions of years ago. Our ever-expanding understanding of the evolution of the eye reinforces and extends the revolutionary ideas of Charles Darwin. Like the myth of the Cyclops, many of our old questions about the evolution of the eye have been put to rest.

Arendt, Detlev, Kristin Tessmar-Raible, Heidi Snyman, Adriaan W. Dorresteijn, and Joachim Wittbrodt. 2004. Ciliary photoreceptors with a vertebrate-type opsin in an invertebrate brain. *Science* vol. 306, pages 869–871.

CHAPTER SUMMARY

29.1 Diverse Senses Operate by the Same Principles

A. Sensory Receptors Convert Stimuli to Action Potentials

- **Sensory receptors** are sensory neurons or specialized epithelial cells that detect, **transduce,** and amplify stimuli, allowing an animal to perceive its environment.
- A sensory receptor selectively responds to a single form of energy and transduces it to **receptor potentials,** which change membrane potential in proportion to stimulus strength.

B. Continuous Stimulation May Cause Sensory Adaptation

- In **sensory adaptation,** sensory receptors cease to respond to a constant stimulus.

29.2 The General Senses Detect Touch, Temperature, Pain, and Position

- The skin's **mechanoreceptors,** such as Pacinian corpuscles, Meissner corpuscles, and free nerve endings, respond to mechanical deflection. Free nerve endings also include **thermoreceptors** and **pain receptors.**
- Sensory neurons in muscles help detect the positions of body parts.

29.3 The Senses of Smell and Taste Detect Chemicals

- **Chemoreceptors** sense chemicals dissolved in watery solutions, such as those in the nose and mouth.

A. Chemoreceptors in the Nose Detect Odor Molecules

- **Olfaction** occurs when odorant molecules bind to receptors in the olfactory epithelium of the **nose.** The brain perceives a smell by evaluating the pattern of olfactory receptor cells that bind odorant molecules.
- **Pheromones** are chemicals that many animals use to communicate with others of the same species.

B. Chemoreceptors in the Mouth Detect Taste

- Humans perceive taste (**gustation**) when chemicals stimulate receptors within **taste buds** on the **tongue.**

29.4 Vision Depends on Light-Sensitive Cells

- **Photoreceptors** contain pigments such as **rhodopsin** associated with membranes.

A. The Vertebrate Eye Consists of Three Layers

- The human **eye's** outer layer, the **sclera,** forms the transparent **cornea** in the front of the eyeball.
- The next layer, the **choroid,** supplies nutrients to the retina. At the front of the eye, the choroid holds the muscle that controls the shape of the **lens,** which focuses light on the photoreceptors. The **iris** adjusts the amount of light entering the eye by constricting or dilating the **pupil.**
- The innermost eye layer is the **retina.**

B. Cells in the Retina Convert Light Energy to Nerve Impulses

- The photoreceptors are **rod cells,** which provide black-and-white vision in dim light, and **cone cells,** which provide color vision in brighter light.
- Light stimulation alters the pigments embedded in membranes of rod and cone cells. The resulting change in the charge across the membrane may generate an action potential.
- Photoreceptor cells synapse with bipolar cells that, in turn, synapse with **ganglion cells.**

C. Signals Travel along the Optic Nerve to the Brain

- Axons of ganglion cells leave the retina as the **optic nerve,** which carries the neural messages to the brain for interpretation.

29.5 The Senses of Hearing and Balance Begin in the Ears

- Mechanoreceptors in the **ear** bend in response to sound waves or the motion of fluids.

A. Mechanoreceptors in the Inner Ear Detect Sound Waves

- Sound enters the **auditory canal,** vibrating the **eardrum.** Three bones in the middle ear amplify these vibrations. The movements of these bones are transmitted through the **oval window,** changing the pressure in fluid within the **cochlea.** At the base of the cochlea, vibration of the **basilar membrane** pushes **hair cells** against the **tectorial membrane.** The **auditory nerve** transmits the impulses to the brain.
- The brain perceives the pitch of the sound through the location of the moving hair cells. The frequency of action potentials and the number of stimulated hair cells conveys information about loudness.

B. The Inner Ear Also Provides the Sense of Balance

- In the inner ear, the **semicircular canals** and the **vestibule** sense body position and movement. Fluid movement within these areas stimulates sensory hair cells. The brain interprets this information, providing a sense of equilibrium.

29.6 Investigating Life: Unraveling the Mystery of the Origin of the Eye

- Molecular fingerprinting studies of a marine segmented worm, *Platynereis dumerilii,* show that the rods and cones of the vertebrate eye originated from light-sensitive cells in the invertebrate brain.

MULTIPLE CHOICE QUESTIONS

1. What is a receptor potential?
   a. The potential for a sensory cell to receive a specific type of information
   b. The graded potential generated from the stimulus received by a sensory cell
   c. The change in membrane permeability in a sensory cell
   d. The threshold potential of a receptor
2. What is the relationship between homeostasis and the sensory system?
   a. Sensory information is used to help maintain internal conditions of an organism.
   b. Homeostasis controls which sensory inputs are perceived by an organism.
   c. Information from the sensory system is used to alter behavior of an organism.
   d. Both a and c.
3. A ______ is an encapsulated nerve ending that is sensitive to ______ .
   a. thermoreceptor; pain
   b. mechanoreceptor; stretch
   c. pain receptor; pain
   d. mechanoreceptor; touch
4. Which of the following is the most important limit to a human's perception of smell?
   a. The size of his or her olfactory epithelium
   b. The number and type of receptor proteins
   c. The number of olfactory receptor cells
   d. All of the above are important.
5. Which of the following is NOT a primary sensation associated with taste?
   a. Spicy
   b. Sweet
   c. Bitter
   d. Sour
6. How would your vision be affected if your photoreceptor cells only contained rhodopsin?
   a. There would be no effect—your vision would be normal.
   b. You would only perceive the color red.
   c. You could only see at night.
   d. You would only perceive black and white—no colors.
7. The region of the ear associated with the sense of balance is the
   a. semicircular canals.
   b. cochlear canal.
   c. basilar membrane.
   d. auditory canal.
8. Why is the basilar membrane of the cochlea wide at one end and narrow at the other?
   a. In order to fit within the spiral of the cochlea.
   b. The wide end responds to loud sounds while the narrow end responds to soft sounds.
   c. Because the different widths of membrane vibrate at different frequencies.
   d. Because the number of hair cells vary along its length.

TESTING YOUR KNOWLEDGE

1. Which sensory receptors are neurons, and which are epithelial cells that pass signals to sensory neurons?
2. Distinguish between a sense organ, a sensory receptor, and a sensory neuron.
3. List the types of sensory receptors that detect pain, heat, touch, and the position of the limbs.
4. In what way are the senses of smell and taste similar?
5. How does the nervous system detect different smells?
6. List the structures of the human eye and their functions.
7. How can a pigment associated with a protein transduce an environmental stimulus into a sensation?
8. In what ways do the three major structures of the inner ear (cochlea, semicircular canals, and vestibule) function similarly?

THINKING AS A SCIENTIST

1. People with Hansen disease (formerly called leprosy) suffer nerve damage that leaves them unable to sense pain in their extremities. How might this situation be dangerous?
2. Natural selection is unequal reproductive success based on inherited differences among individuals. Suppose that some male moths lack a functional pheromone receptor protein in their antennae. Would you expect this trait to spread in the moth population? Explain.
3. In a disorder called macular degeneration, photoreceptors at the center of the retina die. How does macular degeneration impair vision? Would you expect the entire field of view to be affected?
4. When you enter a darkened room, the rod cells in your eyes increase their production of rhodopsin, increasing their sensitivity to light over a period of 10 to 30 min. How do you think your eyes adjust when you emerge from a movie theater on a sunny day?

Visit www.mhhe.com/hoefnagels for practice quizzes, animations, videos, and activities designed to help you master the material in this chapter.

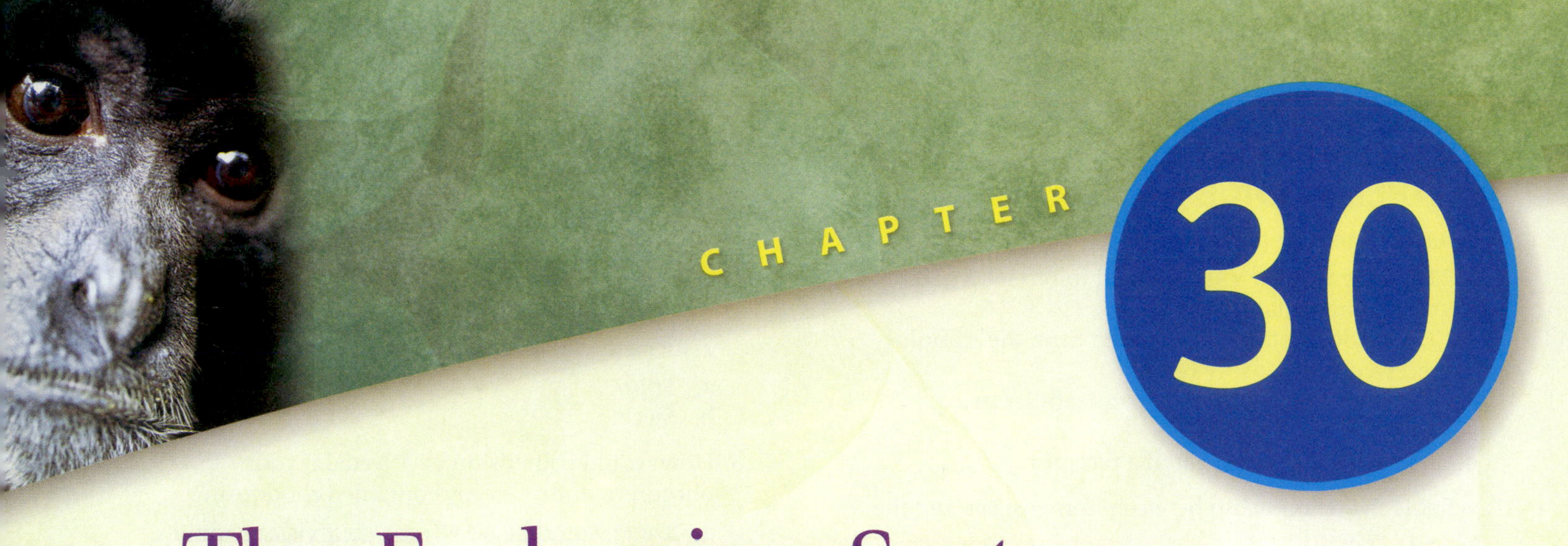

The Endocrine System

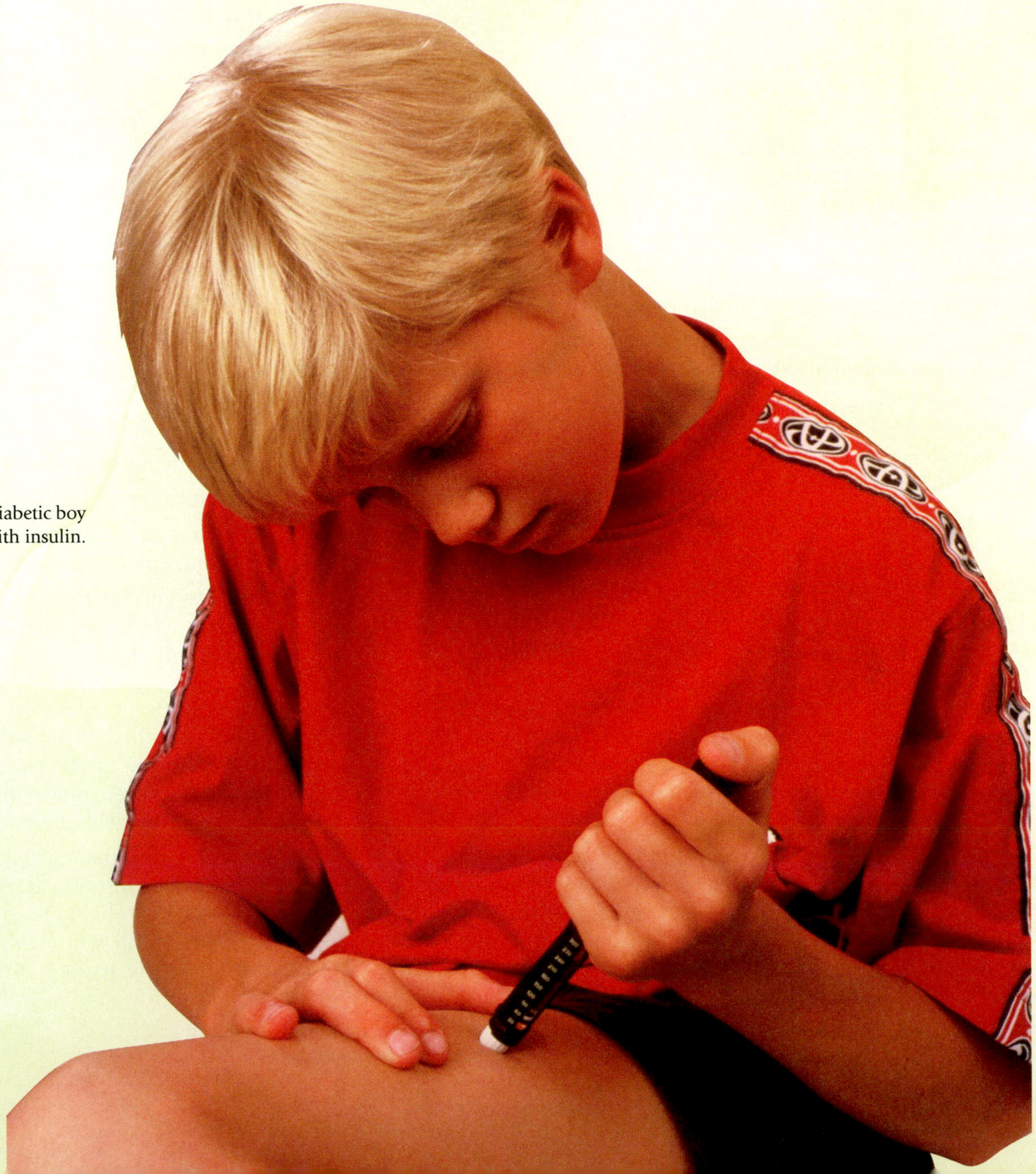

Diabetes Mellitus. This diabetic boy is injecting himself with insulin.

What's the Point?

History of an Illness: Diabetes Mellitus

Physicians have recognized diabetes mellitus for centuries. They knew that the urine of diabetics contained too much sugar, but they watched helplessly as patients withered and died. No one knew what caused the disease, or how to treat it, until early in the twentieth century.

In October 1920, a young surgeon named Frederick Banting was lecturing medical students at the University of Toronto. He described an experiment performed in 1889 in which a dog weakened, lost weight, and died after removal of its pancreas. Banting wondered whether a pancreatic extract might restore such a dog's health. If so, would such an approach help people with the same symptoms? Banting asked his department chairman, J.R.R. Macleod, for help and received a tiny lab, an assistant named Charles Best, and 10 dogs.

In May 1921, Banting and Best removed the pancreas from one dog. In a second dog, they tied off the pancreatic ducts, halting the production of digestive enzymes and causing the pancreas to shrivel. (Banting suspected that digestive enzymes would destroy the other chemicals he sought.) The first dog developed diabetes. Two months later, Banting and Best removed the dried-up pancreas from the second dog, mashed it, and extracted fluid from it. They then injected the fluid into the dog dying from diabetes. An hour later, the diabetic dog was walking about wagging its tail! (The treated dog died the next day, perhaps from infection.) Further experiments showed that dogs require daily doses of a hormone—insulin—to prevent diabetes.

The next step was to give insulin to diabetic people. The treatment saved the lives of two volunteers: a physician-friend of Dr. Banting who was dying of diabetes, and a 14-year-old boy. A large-scale trial on residents of an institution for injured and ill soldiers again achieved astounding success. Soon, people with diabetes flocked to Toronto. By mid-1922, insulin treatment became widely available. Banting and Macleod won the 1923 Nobel Prize in physiology or medicine for their discovery of insulin.

Scientists now know that in type I diabetes, specialized cells in the pancreas fail to produce insulin and release it into the bloodstream. Without this critical hormonal signal, cells throughout the body cannot absorb glucose from blood. Blood sugar reaches dangerously high levels. As the cells weaken for lack of energy, sugar leaves the body in urine.

Insulin is just one of the many hormones that are critically important to health. All of the body's hormones are products of a communication network called the endocrine system—the subject of this chapter.

LEARNING OUTLINE

30.1 The Endocrine System Uses Hormones to Communicate

Animal cells must "talk" to one another to maintain the negative feedback loops critical to homeostasis. The nervous system, described in chapters 28 and 29, is one important communication network in an animal's body; the other is the **endocrine system.** The endocrine system's glands and hormones do not act with the speed of neural impulses, but their chemical messages have something else: staying power.

To illustrate the power of the endocrine system, consider the changes that occur in the body during puberty. Under the influence of hormones produced in the ovaries or testes, the adolescent body transforms: females develop breasts and hips, males acquire a deeper voice and more muscular physique, and new body hair sprouts in both sexes. The same hormones also affect mood, emotions, and feelings of sexual attraction. (Another dramatic hormone-orchestrated event is a larval insect's metamorphosis into its adult form, as described in this chapter's Burning Question: How does a caterpillar "remodel" itself into a butterfly?.) In addition, hormones also regulate growth and development, influence appetite, regulate the concentration of calcium and glucose in blood, and ready the body to confront stress.

A. The Endocrine System Consists of Hormones and Glands

The endocrine system has two main components: hormones and glands. **Hormones** are biochemicals that travel in the bloodstream and alter the metabolism of the **target cells** that respond to the hormone. An **endocrine gland** consists

FIGURE 30.1 Some Human Endocrine Glands. The endocrine system includes several glands that contain specialized hormone-secreting cells. Other cells that secrete hormones are scattered among the other organ systems. The hormones circulate throughout the body in blood vessels, which are not shown in this figure.

Hypothalamus
Produces hormones that stimulate or inhibit the release of hormones from the pituitary gland

Pineal gland
Produces melatonin

Pituitary gland
Produces numerous hormones that affect target tissues directly or stimulate other endocrine glands

Thyroid gland
Releases thyroid hormones, which regulate metabolism

Parathyroid glands
(behind thyroid)
Secrete parathyroid hormone, which helps regulate blood calcium

Adrenal glands
Produce hormones related to sympathetic nervous system and steroids that help regulate body fluids

Pancreas
Releases hormones that regulate blood glucose levels

Ovaries (in female)
Produce estrogen and progesterone, which mediate monthly changes in the uterine lining

Endocrine System	
Main tissue types	**Examples of Locations/Functions**
Epithelial	Makes up the bulk of most glands and secretes many types of hormones
Connective	Blood circulates hormones throughout the body
Nervous	Parts of the brain secrete some hormones and control release of others; some neurons secrete hormones

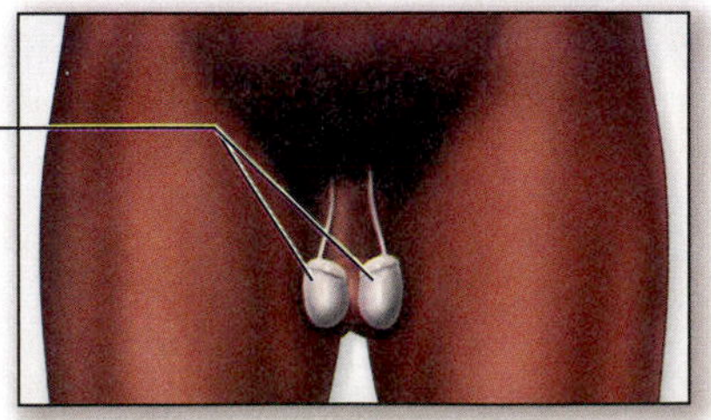

Testes (in male)
Produce testosterone, which promotes sperm maturation and secondary sex characteristics

of groups of hormone-producing cells, connected with a blood supply that carries away the secretions.

The main endocrine organs in vertebrates are the hypothalamus, pituitary gland, pineal gland, thyroid gland, parathyroid glands, adrenal glands, pancreas, and gonads (ovaries and testes) (**figure 30.1**). But not all hormones originate in cells within a gland. The heart, kidneys, liver, stomach, small intestine, and placenta also contain scattered hormone-secreting cells.

Endocrine glands form one of the two major types of animal glands; the other group is exocrine glands (**figure 30.2**). Whereas endocrine cells secrete hormones into the bloodstream, **exocrine glands** release their products into ducts that lead to a body cavity or to the body's exterior. For example, exocrine glands produce saliva, digestive enzymes, sweat, and milk. These products are not hormones, so exocrine glands are not part of the endocrine system.

Some organs have both exocrine and endocrine components. The pancreas, for example, secretes digestive enzymes into ducts that lead to the small intestine (an exocrine function). The same organ also releases the hormone insulin into the bloodstream (an endocrine function). Likewise, the gonads produce gametes (exocrine) and secrete sex hormones (endocrine).

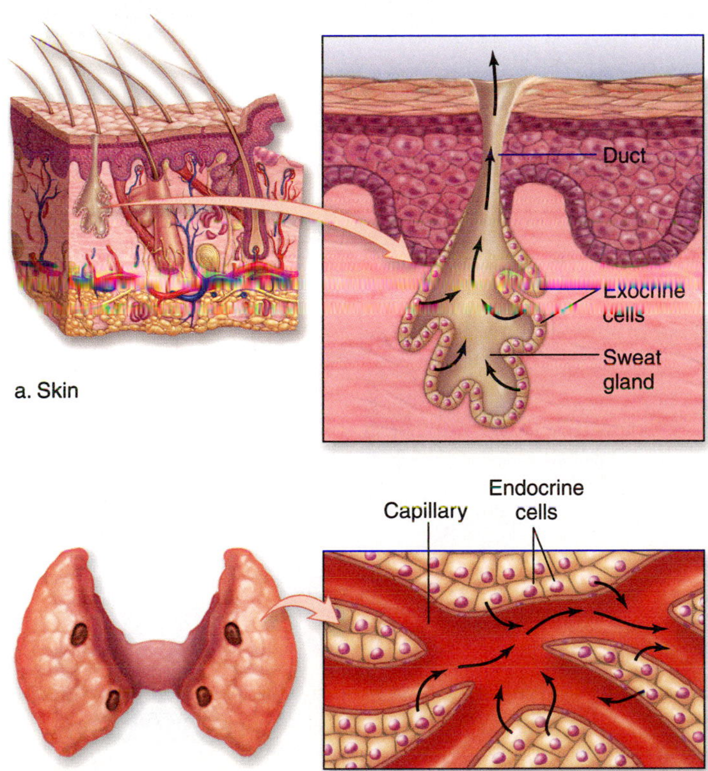

FIGURE 30.2 Exocrine and Endocrine Glands. (**A**) Exocrine glands release their secretions through ducts. (**B**) Endocrine glands release their hormones directly into the circulatory system.

B. The Nervous and Endocrine Systems Work Together

The nervous and endocrine systems are tightly integrated—so much so that some biologists refer to them together as the "neuroendocrine system." The two organ systems share many similarities. For example, some chemicals can act as both neurotransmitters and hormones. Some neurons in the hypothalamus, a brain structure that links the nervous and endocrine systems, release hormones. In addition, neurotransmitters and hormones share some target cells.

But the nervous and endocrine systems also differ in many ways. First, whereas neurons use action potentials and neurotransmitters to send messages, the endocrine system employs hormones. Second, the endocrine system communicates much more slowly than the nervous system. A nervous impulse is virtually instantaneous, and its effects disappear as soon as the stimulus disappears. Hormones take minutes, hours, or even days to exert their effects, which are generally more prolonged. Third, a single neuron influences only a few cells at a time, whereas hormones circulate throughout the body in the blood.

C. Negative Feedback Loops Control Most Hormone Levels

To maintain homeostasis, an animal's body must strictly regulate the levels of hormones in the bloodstream. This tight control often occurs by negative feedback interactions between the hormone (or its effects) and the glands that produce it, ensuring that hormone secretion stops once the concentration is too high (**figure 30.3**). Recall from chapter 27 that in **negative feedback,** a sensor monitors the value of a variable

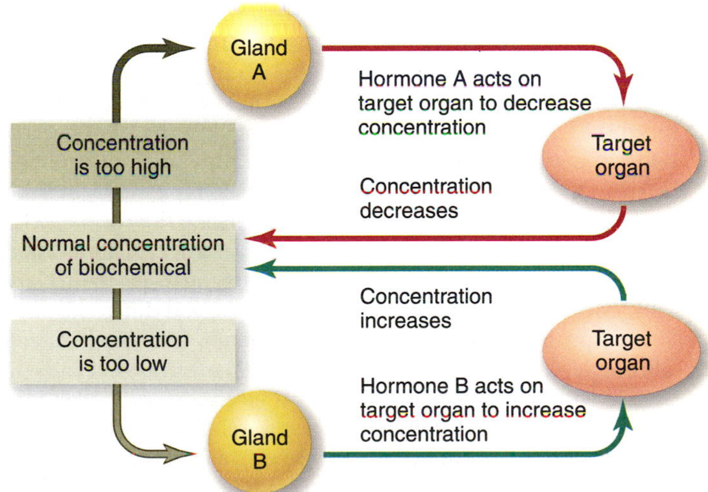

FIGURE 30.3 Feedback Loops Regulate Hormone Action. In this generalized depiction, release of hormone A lowers the levels of an accumulating biochemical, whereas release of hormone B offsets a deficit.

(such as the concentration of the hormone), and effectors counteract the change if the value is too high or too low.

Positive feedback, in which a process reinforces itself, is much less common than negative feedback. For example, at the onset of labor in childbirth, the cervix begins to stretch. Sensory receptors at the cervix relay the message to the hypothalamus, which stimulates release of the hormone oxytocin. Oxytocin stimulates muscles in the uterus to contract, pushing the baby out. This further stretches the cervix, stimulating more oxytocin production, which intensifies the uterine contractions, and so on. When the baby is born, other hormonal changes stop the cycle. **labor and childbirth, p. 763**

D. Some Chemical Signals Are Not Part of the Endocrine System

Animals use many chemical signals that are not part of the endocrine system. Chapter 28 described neurotransmitters, which convey messages across the tiny space between a neuron and another cell. Pheromones are communication molecules that transmit signals to other individuals, rather than within an organism (see section 29.3).

Paracrine secretions are communication molecules that exert their effects locally, stimulating nearby cells rather than traveling in the bloodstream. Prostaglandins, for example, are lipids that are produced throughout the body and that affect many tissues and organs. Some stimulate smooth muscle contraction in blood vessels and airways, adjusting blood flow and oxygen delivery. Others assist in childbirth by causing muscles in the uterus to contract. Still others trigger inflammation and swelling after an injury. Aspirin, ibuprofen, and similar drugs relieve pain by inhibiting prostaglandin production.

Summary *Endocrine systems consist of glands and the hormones they secrete, as well as scattered hormone-secreting cells. The endocrine system interacts with the nervous system. Most hormone levels remain within specific ranges by negative feedback mechanisms.*

30.1 MASTERING CONCEPTS

1. What is the overall function of an endocrine system?
2. How do the nervous and endocrine systems differ?
3. Distinguish between negative and positive feedback loops.
4. How are hormones different from neurotransmitters, paracrine secretions, and pheromones?

Burning Questions

How does a caterpillar "remodel" itself into a butterfly?

The changes that sculpt a butterfly from a caterpillar, a fly from a maggot, or a beetle from a grub are among the most dramatic in the animal kingdom. This metamorphosis from larva to adult is a complex, highly coordinated process orchestrated by changing levels of two hormones (**figure 30.A**).

A caterpillar is a streamlined eating machine that grows at a spectacular rate. Juvenile hormone predominates during this time. To accommodate this rapid growth, the young insect must regularly molt, or shed its exoskeleton. Periodic spikes in levels of molting hormone trigger each molt. Eventually, the cells that produce juvenile hormone shut off, and molting hormone predominates. The insect then secretes a cocoon and becomes a pupa.

The body remodels itself completely inside the cocoon. Many larval cells die, but other cells that were set aside in tiny disklike packets in early larval life obtain energy from the degenerating larval cells. The reawakened cells divide, aggregating and differentiating to form adult body parts such as legs, antennae, and wings. Eventually the pupa case splits, and an adult insect emerges—wet and compacted, with its wings plastered against its abdomen. It takes a few hours to dry out and for fluid to flow into veins in the wings and expand them. Soon the insect is free to take on an entirely new lifestyle. Rather than living to eat, the adult insect lives to mate.

Have a Burning Question of your own? Submit it to marielle_hoefnagels@mcgraw-hill.com for possible inclusion in future editions of this book!

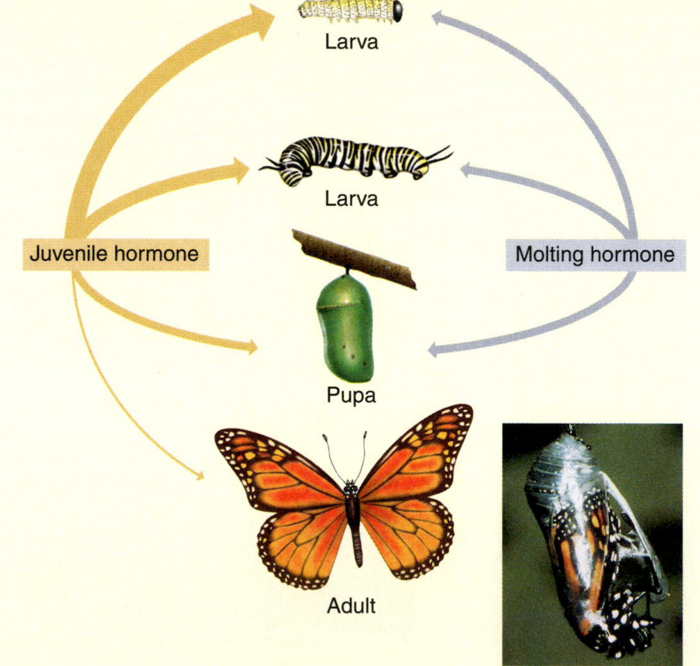

FIGURE 30.A Metamorphosis. The inset shows a monarch butterfly emerging from its chrysalis (pupa). Two hormones control insect metamorphosis. The thickness of the line leading from each hormone denotes the magnitude of its effect at each life stage.

30.2 Hormones Bind to Receptors on or in Target Cells

Just as a key fits a lock, each hormone affects only target cells bearing specific receptor molecules. The term *target cells* is a little misleading, because it implies that hormones somehow travel straight from their source to a limited set of cells. In reality, many hormones circulate throughout the body at the same time. Each hormone's target cells are simply those with the corresponding receptors. In this way, a hormone does not affect every cell it encounters.

A hormone's receptor may be on or in the target cell (**figure 30.4**). Water-soluble hormones, which cannot pass readily through the cell membrane's oily phospholipid bilayer, bind to receptors on cell surfaces. Lipid-soluble hormones, in contrast, move easily through the membrane. Their receptors are inside the target cell, often in the cytoplasm or nucleus. **cell membrane, p. 54**

A. Water-Soluble Hormones Trigger Second Messenger Systems

Peptide hormones, which are chains of a few to several hundred amino acids, are water-soluble. When a peptide

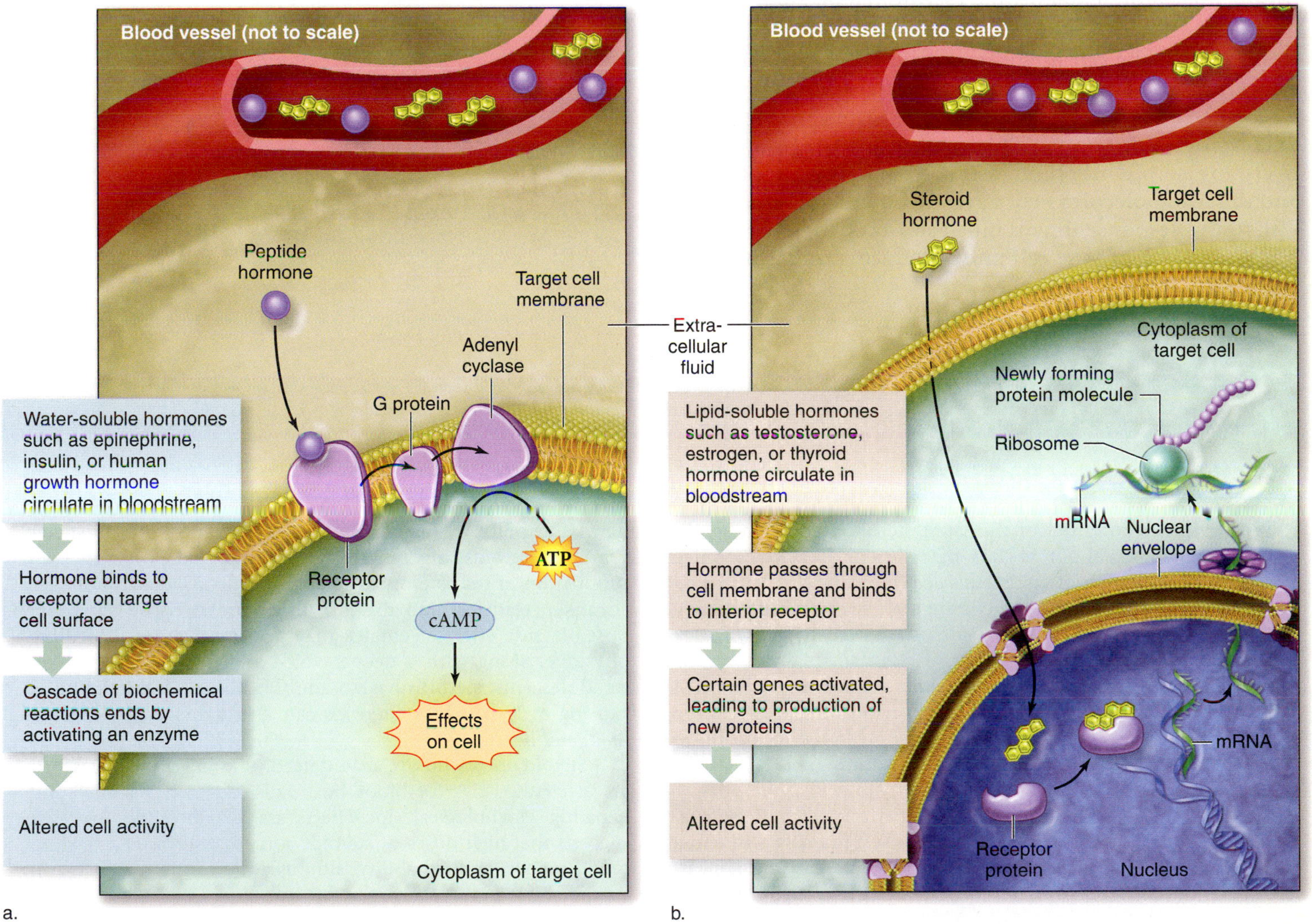

FIGURE 30.4 Target Cells Express Receptor Proteins. Endocrine glands release their chemical signals into the circulatory system. Only target cells express receptors that interact with the signal molecule. (**A**) Peptide hormones are water-soluble and bind to cell-surface receptors. (**B**) Steroid hormones are lipid-soluble. They pass through cell membranes and bind to receptors in the cytoplasm or nucleus. Because steroid hormones alter gene expression, their action is typically slower than that of peptide hormones.

hormone binds to its receptor on a target cell's surface, it usually activates a second messenger inside the cell. This molecule, in turn, activates the enzymes that produce the hormone's effects (see figure 30.4). In general, peptide hormones act rapidly (within minutes of their release) because all of the participating biochemicals are already in place when the hormone binds the receptor.

Some hormones are derived from a single amino acid. Most of these hormones, including epinephrine, norepinephrine, and dopamine, are water-soluble. Like peptide hormones, they bind to receptors on the surface of target cells. **neurotransmitters, p. 583**

B. Lipid-Soluble Hormones Directly Alter Gene Expression

The body synthesizes lipid-soluble **steroid hormones** from cholesterol, which is one reason dietary cholesterol is essential for good health. Unlike peptide hormones, steroids cross the cell membrane unassisted. Once inside the cell, the hormone may enter the nucleus and bind to a receptor associated with DNA, which triggers the production of particular proteins (see figure 30.4B). Alternatively, the steroid may bind to a receptor in the cytoplasm, and the two molecules may travel together to the nucleus. Either way, response time is much slower than for peptide hormones, because the cell must produce proteins before the hormone takes effect. **cholesterol, p. 37; protein synthesis, p. 236**

Testosterone, estrogen, and progesterone are steroid hormones. Testosterone and other anabolic steroids activate genes that direct muscle protein synthesis (see this chapter's Can *You* Relate? Anabolic Steroids Build Muscle, but with Significant Side Effects). Two other lipid-soluble hormones, the thyroid hormones, are not steroids. Rather, they are derived from a single amino acid. They therefore pass freely through the cell membrane and, like steroid hormones, they function by binding to receptors in the nucleus.

Summary *A hormone moves in the bloodstream and binds to receptors on or in target cells. Peptide hormones bind to cell-surface receptors that trigger second messenger cascades. Steroid hormones bind to receptors inside cells, changing the expression of particular genes.*

30.2 MASTERING CONCEPTS

1. How does a hormone affect some cells but not others?
2. What is the role of second messengers in hormone action?
3. Explain the observation that peptide hormones usually act faster than steroid hormones.

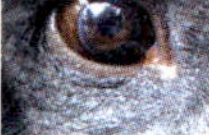

Can *You* Relate?

Anabolic Steroids Build Muscle, but with Significant Side Effects

When news reports tell of the latest steroid scandal in sports, they are talking about a class of synthetic hormones called anabolic-androgenic steroids. *Anabolic* (meaning, "to build") refers to muscle growth; androgens are male sex hormones. Essentially, anabolic steroids (as they are usually called) are synthetic forms of testosterone.

Synthetic steroids have a legitimate place in medicine. A physician might prescribe them for a person who produces too little testosterone, for example, or to someone with an illness that causes muscles to waste away. These drugs are legal only by prescription.

Some athletes, in search of a competitive edge, abuse anabolic steroids as a shortcut to greater muscle mass. It is true that steroid users may improve strength, performance, and appearance in the short term. In the long run, however, the hormones harm the body in many ways. In males, the body mistakes synthetic steroids for the natural hormone and lowers its own production of testosterone, causing infertility later. Impotence, shrunken testicles, and the growth of breast tissue are other possible side effects. Females may acquire a more masculine physique, along with a deeper voice and facial or body hair. In adolescents, steroids hasten adulthood, stunting height and causing early hair loss. Steroids can also damage the kidneys, liver, and heart.

Steroid users who use shared needles to inject the drugs also are at elevated risk for blood-borne diseases such as AIDS and hepatitis. Psychological side effects are also possible. Research suggests that high doses of steroids sometimes cause aggression, mood swings, and irritability, although the connection between steroid abuse and violence is unknown.

30.3 The Hypothalamus and Pituitary Gland Oversee Endocrine Control

Two structures, the hypothalamus and the pituitary gland, regulate all of the glands and hormones of the vertebrate endocrine system (**figure 30.5** and **table 30.1**). The **hypothalamus** is a part of the forebrain, and the **pituitary gland** is a pea-sized gland attached to a stalk extending from the hypothalamus. It is really two glands in one:

Hypothalamus
Neurons from hypothalamus
Anterior pituitary pathway
Neurons in the hypothalamus secrete inhibiting or releasing hormones, which enter the bloodstream and travel to the anterior pituitary.
Posterior pituitary pathway
Neurons in the hypothalamus produce the hormones ADH and oxytocin, which leave the neuron endings in the posterior pituitary. The hormones enter the capillaries and travel to their target organs.
Anterior pituitary gland
Hormones from the hypothalamus act on the cells of the anterior pituitary, inhibiting or stimulating release of anterior pituitary hormones.
Posterior pituitary gland
Hormones secreted by the anterior pituitary enter the bloodstream and travel to their target tissues.
Anterior pituitary
Posterior pituitary
TSH
ACTH
FSH and LH
GH
Prolactin
Endorphins
Oxytocin
ADH
Thyroid
Adrenal cortex
Testes
Ovaries
Mammary glands
Pain receptors in the brain
Contracts cells in mammary glands and uterus
Kidney
Thyroxine and triiodothyronine
Cortisol
Testosterone
Estrogen and progesterone
Most cells in the body

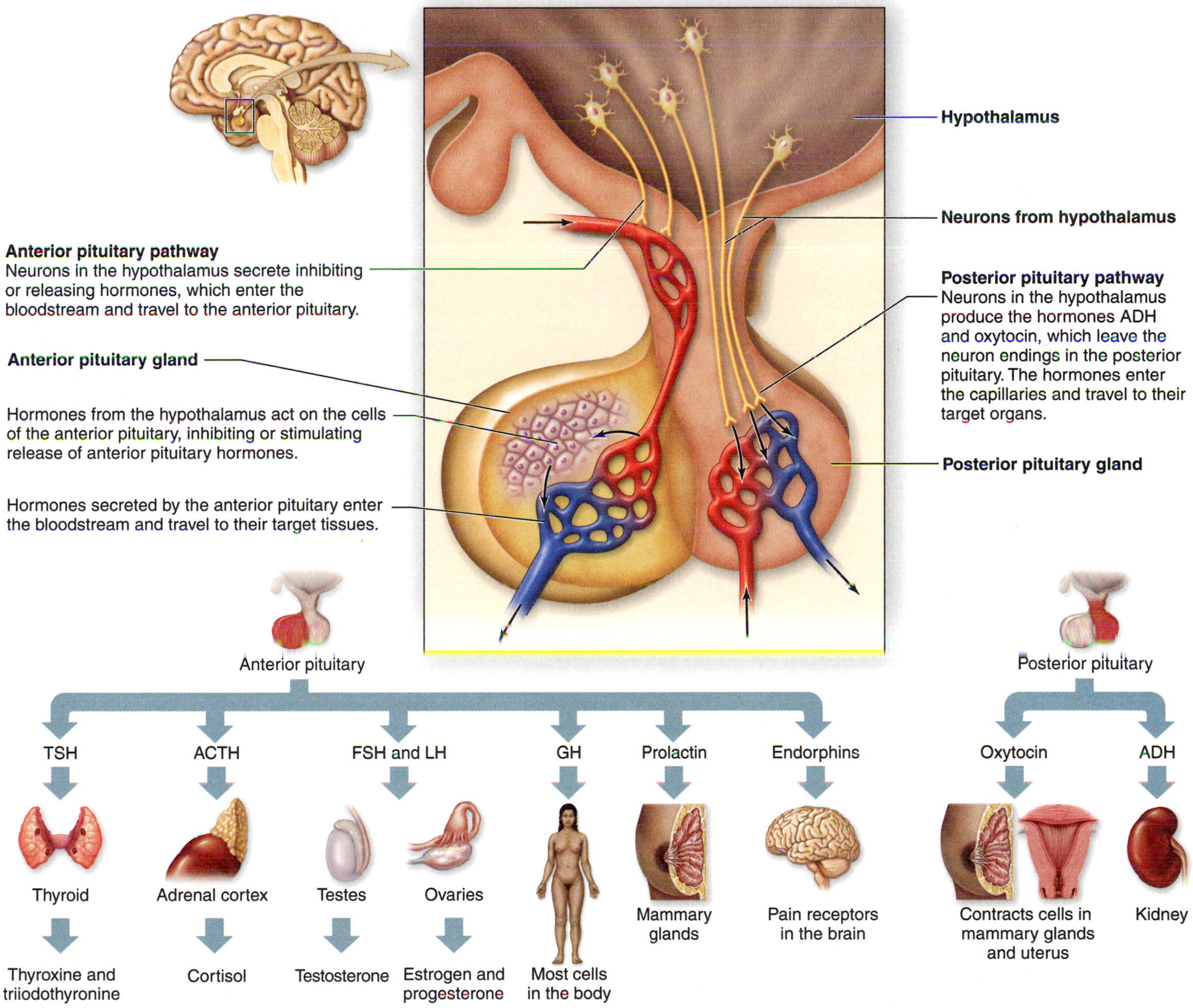

FIGURE 30.5 Master Glands. The hypothalamus and the pituitary gland control the functioning of the endocrine system.

Table 30.1 *Hormones of the Hypothalamus and Pituitary*

Gland	Hormone	Type	Location of Target Cells	Effects
Hypothalamus	Releasing hormones (e.g., thyrotropin-releasing hormone (TRH), gonadotropin-releasing hormone (GnRH))	Peptide	Anterior pituitary	Stimulate release of hormones from anterior pituitary
	Inhibiting hormones (e.g., growth hormone-inhibiting hormone)	Peptide	Anterior pituitary	Inhibit release of hormones from anterior pituitary
Posterior pituitary	Antidiuretic hormone (ADH)	Peptide	Kidneys	Helps maintain composition of body fluids
	Oxytocin	Peptide	Uterus, mammary glands, brain	Stimulates smooth muscle contraction; has role in affection and bonding
Anterior pituitary	Growth hormone (GH)	Protein	Liver	Stimulates production of insulin-like growth factors, which promote tissue growth throughout the body
	Prolactin	Protein	Mammary glands	Stimulates milk secretion
	Follicle-stimulating hormone (FSH)	Protein	Ovaries, testes	In females: stimulates follicle development, oocyte maturation, release of estrogen In males: stimulates sperm production
	Luteinizing hormone (LH)	Protein	Ovaries, testes	In females: stimulates ovulation, progesterone secretion In males: stimulates testosterone secretion
	Thyroid-stimulating hormone (TSH)	Glycoprotein (protein with carbohydrate attached)	Thyroid gland	Stimulates secretion of thyroid hormones
	Adrenocorticotropic hormone (ACTH)	Protein	Adrenal cortex, pancreas	Stimulates secretion of glucocorticoids from adrenal cortex and insulin from pancreas

the **anterior pituitary** (toward the front) and the **posterior pituitary** (toward the back).

The hypothalamus links the nervous and endocrine systems by controlling pituitary secretions. The posterior pituitary does not synthesize hormones, but it does store and release two hormones that the hypothalamus produces. When neural activity in the brain stimulates the cells of the posterior pituitary, they release the hormones. The hypothalamus controls the anterior lobe of the pituitary in a different way—by influencing the production of the hormones it secretes. Hormones from the hypothalamus reach the anterior pituitary through a specialized system of blood vessels.

A. The Posterior Pituitary Stores and Releases Two Hormones

The posterior pituitary stores and releases **antidiuretic hormone (ADH),** also called vasopressin, which the hypothalamus produces. Negative feedback control of ADH secretion helps maintain the chemical balance of body fluids. If the blood contains too little water, receptor cells in the hypothalamus stimulate the posterior pituitary to release more ADH. The hormone signals the kidneys to conserve water, which dilutes the blood. When balance is restored, the receptor cells signal the hypothalamus to slow production of ADH. Chapter 35 explains kidney function in more detail.

Oxytocin is chemically similar to ADH, but it has different target cells. Oxytocin stimulates the contraction of cells in the breasts, causing them to release milk when a baby nurses. It also triggers contraction of muscles in the uterus, which pushes a baby out during labor. Physicians use synthetic oxytocin to induce labor or accelerate contractions in a woman who is giving birth. Oxytocin also plays a role in bonding, affection, and social recognition in both sexes.

B. The Anterior Pituitary Produces and Secretes Six Hormones

The hormones of the hypothalamus that stimulate the anterior pituitary are examples of **tropic hormones,** which are produced in one gland and then influence hormone secretion in another. Some of the tropic hormones that the hypothalamus produces are **releasing hormones** that stimulate hormone secretion. **Inhibiting hormones** from the hypothalamus suppress hormone release from the anterior pituitary.

The anterior pituitary gland produces **growth hormone (GH),** also called somatotropin, which promotes growth and development of all tissues by increasing protein synthesis and cell division rates. Levels of GH peak in the preteen years and, together with rising levels of sex hormones, cause adolescent growth spurts. A severe deficiency of GH during childhood leads to pituitary dwarfism, which produces extremely short stature. At the other extreme, a child with too much GH becomes a pituitary giant. In an adult, GH does not affect overall height because the long bones of the body are no longer growing. However, excess GH can cause acromegaly, a thickening of the bones in the hands and face (**figure 30.6**).

Prolactin is an anterior pituitary hormone that stimulates milk production in a woman's breasts after she gives birth. In males and in women who are not breast-feeding, an inhibiting hormone from the hypothalamus suppresses prolactin synthesis. In nursing mothers, however, a suckling infant triggers nerve impulses that overcome this inhibition.

The other four hormones released by the anterior pituitary are tropic hormones. Two of them are gonadotropic, meaning that they stimulate hormone release from the gonads (ovaries and testes). These two hormones are **follicle-stimulating hormone (FSH)** and **luteinizing hormone (LH)**. Section 30.5 and chapter 37 describe their function in more detail. The other two tropic hormones secreted by the anterior pituitary are thyroid-stimulating hormone and adrenocorticotropic hormone. **Thyroid-stimulating hormone (TSH)** causes the thyroid gland in the neck to release two hormones that control metabolism. **Adrenocorticotropic hormone (ACTH)** stimulates release of the glucocorticoid hormones from the adrenal glands. The glucocorticoids increase the level of glucose in the blood during stress.

The anterior pituitary also produces **endorphins,** which are natural painkillers, in response to prolonged painful stimuli. When endorphins are released into the bloodstream, they bind to receptors on target cells in the brain. Usually, however, endorphins are not detectable in the blood, so their status as hormones is questionable.

Summary *The hypothalamus and the pituitary gland control the endocrine system. The posterior pituitary releases two hormones produced by the hypothalamus. Hormones from the hypothalamus stimulate the anterior pituitary to produce and release six hormones.*

30.3 MASTERING CONCEPTS

1. How do the hypothalamus and the pituitary gland interact?
2. Which pituitary hormones are tropic hormones?
3. What are the functions of ADH and oxytocin, and where are these hormones produced?
4. Which hormones does the anterior pituitary produce, and what are their functions?

a.

b. 16 years

33 years

52 years

FIGURE 30.6 **Growth Hormone Abnormalities.** (**A**) A pituitary giant poses with his father and young brother. **(B)** When GH is overproduced during adulthood, the bones of the hands and face enlarge considerably, as this woman with acromegaly shows.

30.4 Hormones from Many Glands Regulate Metabolism

The thyroid gland, parathyroid glands, adrenal glands, and pancreas secrete hormones that control metabolism (**table 30.2**). Tropic hormones from the anterior pituitary control many, but not all, of their activities (see figure 30.5).

A. The Thyroid Gland Sets the Metabolic Pace

The **thyroid gland** is a two-lobed structure in the neck. The lobes secrete two thyroid hormones, **thyroxine** and **triiodothyronine,** that increase the rate of metabolism in target cells. Under thyroid stimulation, the lungs exchange gases faster, the small intestine absorbs nutrients more readily, and fat levels in cells and in blood plasma decline.

The thyroid hormones illustrate how the hypothalamus and pituitary interact in negative feedback loops. When blood levels of thyroid hormones are low, the hypothalamus secretes thyrotropin-releasing hormone (TRH), which stimulates the anterior pituitary to increase production of thyroid-stimulating hormone (TSH). In response, epithelial cells in the thyroid secrete thyroxine and triiodothyronine (**figure 30.7**).

Both thyroid hormones contain iodine; a deficiency of this essential element causes a goiter, or swollen thyroid gland. Ancient Egyptian doctors treated this disorder with seaweed, not realizing that iodine in the seaweed reversed the condition by enabling the thyroid to produce hormones. Today, iodine-deficient goiter is rare in nations where iodine is added to table salt.

Several other disorders affect the thyroid gland as well. An underactive thyroid gland (hypothyroidism) slows the metabolic rate. Because the body burns fewer calories, weight increases. Heartbeat slows, and blood pressure and body temperature fall. An overactive thyroid, called hyperthyroidism, can cause a short attention span, irritability, hyperactivity, and elevated heart rate, blood pressure, and temperature. Appetite is great, but the high metabolic rate keeps weight off. Graves disease is the most common type of hyperthyroidism. Both former President George H. W. Bush and his wife, Barbara, have this disorder. **autoimmune disorders, p. 738**

Scattered cells throughout the thyroid gland produce a third hormone, **calcitonin,** which decreases blood calcium level under some conditions. Levels of calcitonin greatly increase during pregnancy and milk production. Overall, however, the physiological importance of calcitonin in adult humans seems minimal.

B. The Parathyroid Glands Control Calcium Level

The **parathyroid glands** are four small groups of cells embedded in the back of the thyroid gland. These glands

Table 30.2 *Glands and Hormones that Regulate Metabolism*

Gland	Hormone	Type	Location of Target Cells	Effects
Thyroid gland	Thyroid hormones (thyroxine, triiodothyronine	Amine	All tissues	Increase metabolic rate
	Calcitonin	Peptide	Bone	Increases rate of calcium deposition in bone
Parathyroid glands	Parathyroid hormone (PTH)	Peptide	Bones, digestive organs, kidneys	Releases calcium from bone, increases calcium absorption in digestive tract and kidneys
Adrenal medulla	Epinephrine and norepinephrine	Amine	Blood vessels	Raise blood pressure, constrict blood vessels, slow digestion
Adrenal cortex	Mineralocorticoids	Steroid	Kidneys	Maintain blood volume and electrolyte balance
	Glucocorticoids (e.g., cortisol)	Steroid	All tissues	Increase glucose levels in blood and brain
Pancreas	Insulin	Peptide	All tissues	Increases cellular glucose uptake
	Glucagon	Protein	Liver, adipose tissue	Stimulates breakdown of glycogen to glucose and of fat to fatty acids
Pineal gland	Melatonin	Amine	Other endocrine glands	Regulates effects of light–dark cycle on other glands

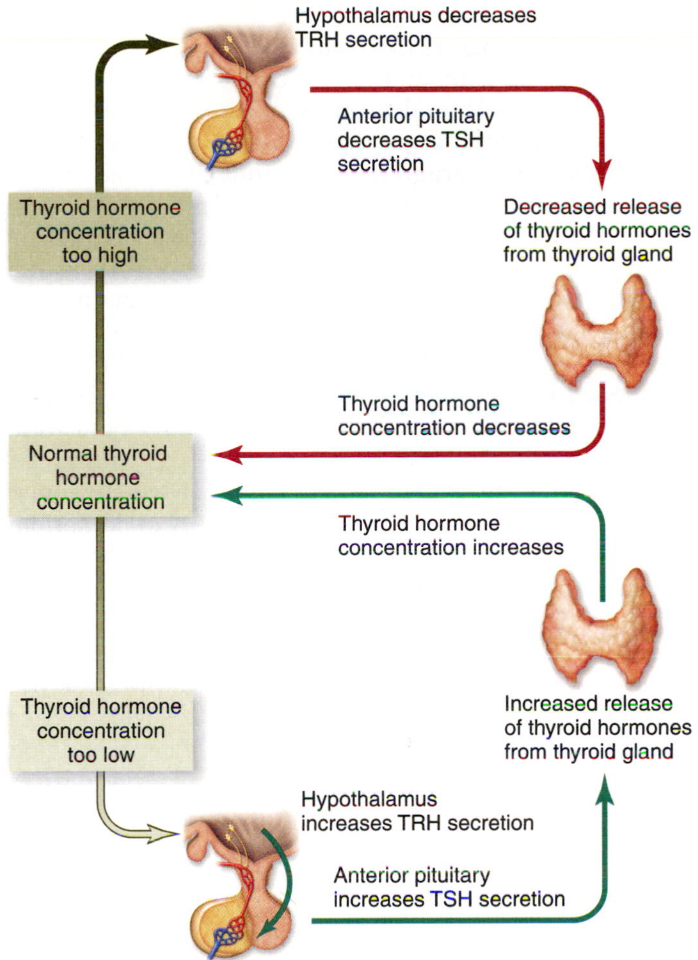

FIGURE 30.7 Control of Thyroid Gland Function. Thyrotropin-releasing hormone (TRH) from the hypothalamus triggers secretion of thyroid-stimulating hormone (TSH) from the anterior pituitary, which causes the thyroid gland to secrete the thyroid hormones thyroxine and triiodothyronine. Blood levels of the two thyroid hormones affect the rate of their synthesis in a negative feedback loop.

secrete **parathyroid hormone (PTH),** which increases calcium levels in blood and tissue fluid by releasing calcium from bones and by enhancing calcium absorption through the digestive tract and kidneys (see figure 31.9). PTH action therefore opposes that of calcitonin.

Calcium is vital to muscle contraction, neural impulse conduction, blood clotting, bone formation, and the activities of many enzymes. Underactivity of the parathyroids can therefore be fatal. Excess PTH can also be harmful if calcium leaves bones faster than it accumulates. This condition, called osteoporosis, is most common in women who have reached menopause (cessation of menstrual periods). The estrogen decrease that accompanies menopause makes bone-forming cells more sensitive to PTH, which depletes bone mass.

C. The Adrenal Glands Coordinate the Body's Stress Responses

The paired **adrenal glands** sit on top of the kidneys ("ad" means near, "renal" means kidney). The **adrenal medulla** is the inner portion of each gland, and the **adrenal cortex** is the outer portion. Each region secretes different hormones, mostly in response to stress.

The adrenal medulla hormones, **epinephrine** (adrenaline) and **norepinephrine** (noradrenaline), help the body respond to short-term emergencies by increasing heart and breathing rates and blood flow (**figure 30.8**A). The sympathetic nervous system stimulates their release from the adrenal medulla. These chemicals also act locally as neurotransmitters. **sympathetic nervous system, p. 586**

Epinephrine can save the lives of people with severe allergic reactions to bee stings or specific foods. Moments after contacting the allergen, a massive immune system reaction causes the airway to constrict. People with known allergies may therefore carry a self-injectable dose of epinephrine with them. The epinephrine temporarily reverses the allergic reaction, allowing the person to survive long enough to seek medical help. **allergies, p. 740**

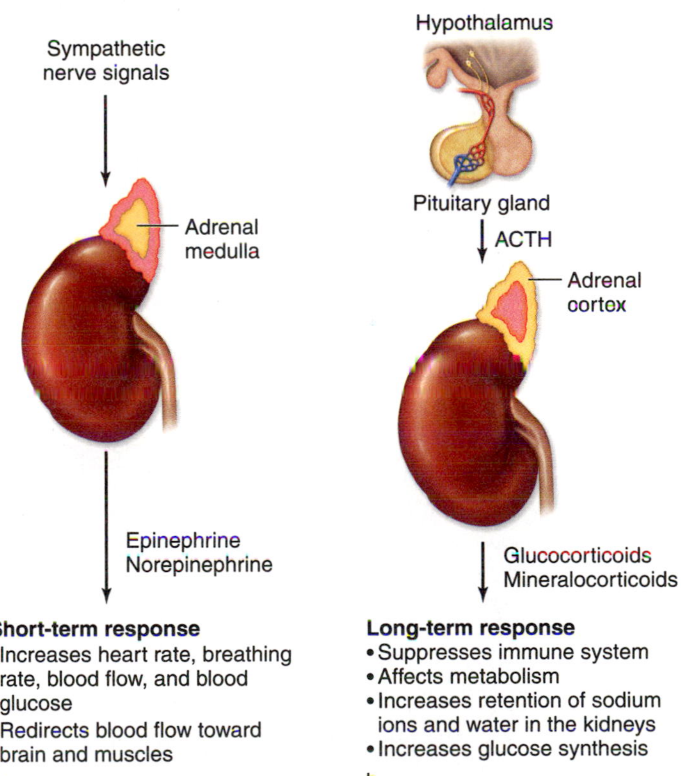

FIGURE 30.8 The Adrenal Glands. (A) The adrenal medulla secretes epinephrine and norepinephrine, which ready the body for an immediate "fight-or-flight" response during emergencies. **(B)** The adrenal cortex secretes mineralocorticoids and glucocorticoids, which enable the body to survive prolonged stress. The adrenal cortex also secretes small amounts of sex hormones (not shown).

The adrenal cortex secretes steroid hormones, including mineralocorticoids and glucocorticoids. During prolonged stress, these hormones mobilize energy reserves while stabilizing blood volume and composition (see figure 30.8B). The **mineralocorticoids** maintain blood volume and salt balance by stimulating the kidneys to return sodium ions and water to the blood while excreting potassium ions. This action is particularly important in compensating for fluid loss from severe bleeding. The **glucocorticoids,** the most important of which is cortisol, are essential in the body's response to prolonged stress. Glucocorticoids stimulate the production of glucose from amino acids, providing energy after the immediate supply of glucose is depleted. Glucocorticoids also indirectly constrict blood vessels, which slows blood loss and prevents tissue inflammation after an injury.

Prednisone, a synthetic glucocorticoid, is an anti-inflammatory drug that can treat arthritis, allergic reactions, and asthma. However, it also suppresses the immune system, leaving a person vulnerable to infection. In addition, with long-term use of the drug, the body may stop producing its own glucocorticoid hormones. Abruptly stopping treatment may therefore be dangerous.

D. The Pancreas Regulates Nutrient Use

The **pancreas** is an elongated gland, about the size of a hand, located between the spleen and the small intestine. Clusters of cells called **pancreatic islets,** also called islets of Langerhans (**figure 30.9**), make up the endocrine portion of the pancreas. The hormones they secrete are polypeptides that regulate the body's use of nutrients.

The islets produce two hormones, glucagon and insulin, that oppose each other in regulating blood glucose levels (**figure 30.10**). After a meal rich in carbohydrates, glucose enters the circulation at the small intestine. The resulting rise in blood sugar triggers beta cells in the pancreas to secrete **insulin,** which stimulates target cells to admit glucose. As glucose leaves the bloodstream to enter cells, insulin secretion slows. If blood sugar dips too low, alpha cells in the pancreas secrete **glucagon,** which stimulates target cells in the liver to release glucose into the bloodstream.

Diabetes mellitus is a disease in which sugar builds up in the blood and in the urine. Fifteen percent of affected individuals have type I diabetes, in which the body does not produce insulin. (Type I diabetes is also called insulin-dependent diabetes because insulin injections can replace the missing hormone.) This disorder usually begins in childhood or early adulthood. Symptoms include thirst, blurred vision, weakness, fatigue, irritability, nausea, and weight loss. Untreated, a severe insulin deficiency can result in a lethal diabetic coma.

A more common form of diabetes is type II, or noninsulin-dependent diabetes, which usually begins in adulthood. The pancreas produces insulin, but the body's cells do not respond to it, a condition called insulin resistance. The incidence of type II diabetes in developed countries (including the United States) is increasing. The disease is strongly associated with obesity, although the cause–effect relationship is unclear. If left untreated, type II diabetes can cause blindness, kidney failure, and poor healing of wounds. Treatments include dietary restrictions, exercise, and drugs that enable the body to use insulin.

The opposite of diabetes is **hypoglycemia,** in which excess insulin production or insufficient carbohydrate intake causes low blood sugar. A person with this condition feels weak, sweaty, anxious, and shaky. A healthy person might temporarily experience hypoglycemia following very strenuous exercise. A diet of frequent, small meals low in carbohydrates and high in protein can prevent insulin surges and help relieve symptoms of hypoglycemia.

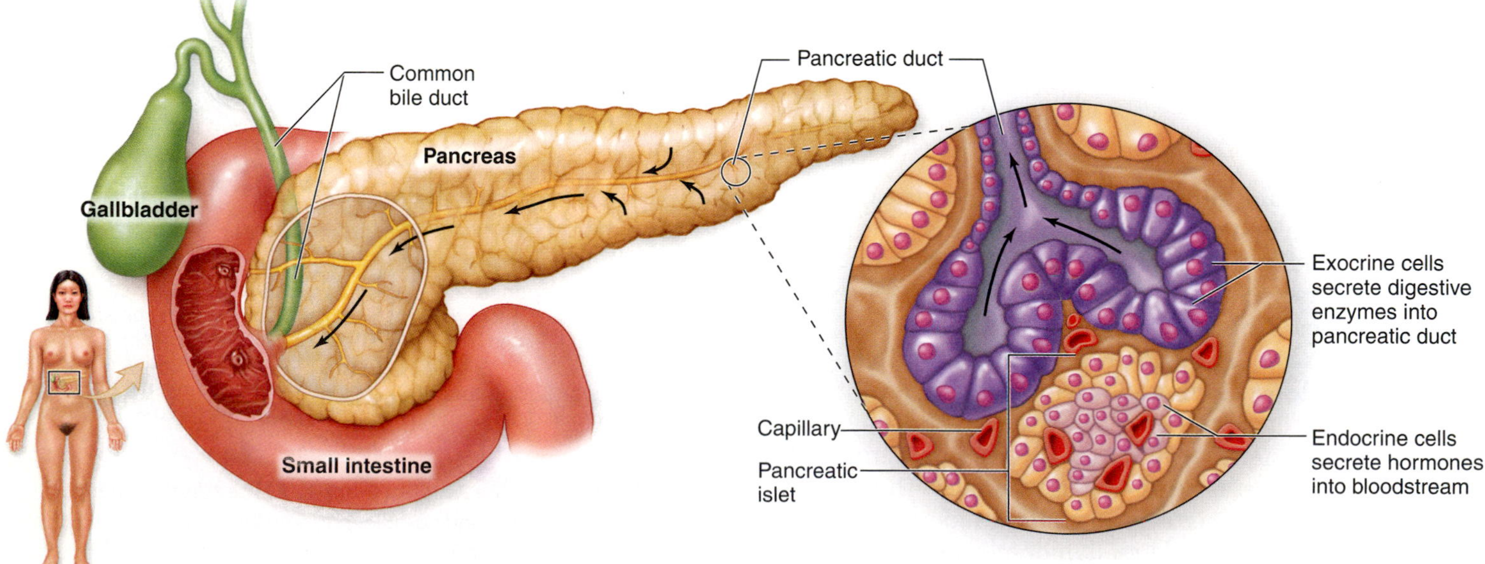

FIGURE 30.9 The Pancreas. The pancreas lies above the small intestine and beneath the stomach. It consists of many lobes and produces digestive enzymes and hormones. The expanded drawing shows the hormone-secreting pancreatic islets next to enzyme-secreting exocrine cells.

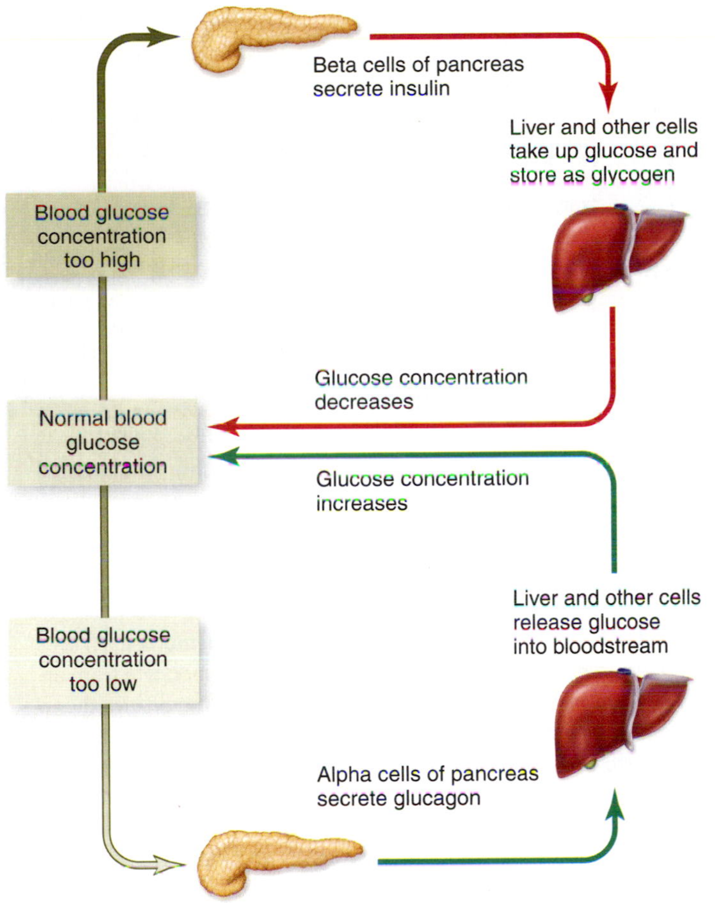

FIGURE 30.10 The Pancreas Regulates Blood Glucose Level. Beta cells secrete insulin, which admits glucose into cells. Alpha cells release glucagon, which increases blood glucose levels.

E. The Pineal Gland Secretes Melatonin

The **pineal gland,** a small brain structure near the hypothalamus, produces the hormone **melatonin.** Exposing the retina of the eye to light inhibits melatonin synthesis in the pineal gland. Darkness, on the other hand, stimulates melatonin production. The amount of melatonin in the bloodstream therefore communicates the amount of light to other cells of the body.

The functions of the human pineal gland are not well understood. Mood swings, particularly a form of depression called seasonal affective disorder (SAD), may be linked to abnormal melatonin secretion. Exposure to additional daylight can elevate mood. Because melatonin levels decrease with age, some people believe that taking extra melatonin might prevent age-related diseases.

Summary *The thyroid gland produces two hormones that speed metabolism. The parathyroid glands regulate blood calcium levels. The adrenal glands secrete hormones important in the body's response to short-term and prolonged stress. Hormones from the pancreas regulate blood sugar, and the pineal gland secretes melatonin.*

30.4 MASTERING CONCEPTS

1. What are the three hormones produced in the thyroid, and what do they control?
2. What is the function of PTH?
3. How do the functions of hormones secreted by the adrenal cortex and adrenal medulla differ?
4. What are the functions of insulin and glucagon?
5. How do darkness and light affect melatonin secretion?

30.5 Hormones from the Ovaries and Testes Control Reproduction

The **gonads** (**ovaries** in females and the **testes** in males) are the glands that manufacture gametes. They also secrete hormones that enable the gametes to mature and that are responsible for development of secondary sexual characteristics (**table 30.3**). The production of reproductive hormones is under negative feedback control by the pituitary

Table 30.3 *Hormones Produced in the Ovaries and Testes*

Gland	Hormone	Type	Location of Target Cells	Effects
Ovaries	Progesterone	Steroid	Uterine lining	Regulates menstrual cycle, maintains secondary sex characteristics in females
	Estrogen	Steroid	Uterine lining	Regulates menstrual cycle, maintains secondary sex characteristics in females
Testes	Testosterone	Steroid	Skin, muscles, sperm-producing cells	Promotes sperm development, maintains secondary sex characteristics in males

gland and the hypothalamus (**figure 30.11**). Chapter 37 details the role of these hormones in reproduction.

In a woman of reproductive age, the levels of several sex hormones cycle approximately every 28 days. Low blood levels of **estrogen** and **progesterone** in the bloodstream prompt the hypothalamus to produce and secrete **gonadotropin-releasing hormone (GnRH)**. GnRH triggers the anterior pituitary to release FSH and LH into the bloodstream. At target cells in the ovary, these two hormones stimulate oocyte maturation and division, growth of follicle cells in the ovary, and ovulation. The growing follicle cells release estrogen and progesterone, which, in turn, exert negative feedback control on both the hypothalamus and pituitary. Estrogen and progesterone also promote development of the female secondary sex characteristics, such as breasts and hips.

In males, reproductive hormone levels do not cycle to the same extent as in females. GnRH from the hypothalamus stimulates the anterior pituitary to release FSH and LH into the bloodstream. At the testes, FSH stimulates the early stages of sperm formation. LH completes sperm production and stimulates cells in the testes to synthesize the male steroid hormone **testosterone.** Testosterone and another hormone, inhibin, prevent overproduction of sperm in a negative feedback loop. Testosterone stimulates development of male structures in the embryo and promotes later development of male secondary sexual characteristics, including facial hair, deepening of the voice, and increased muscle growth.

The adrenal glands also secrete small amounts of sex hormones. The adrenal cortex produces more testosterone than estrogen and progesterone. Adrenal hormones are especially important after menopause, when the production of sex hormones in the ovaries stops.

Summary *Cycling levels of hormones control the female reproductive system. In males, constant hormone levels foster sperm development.*

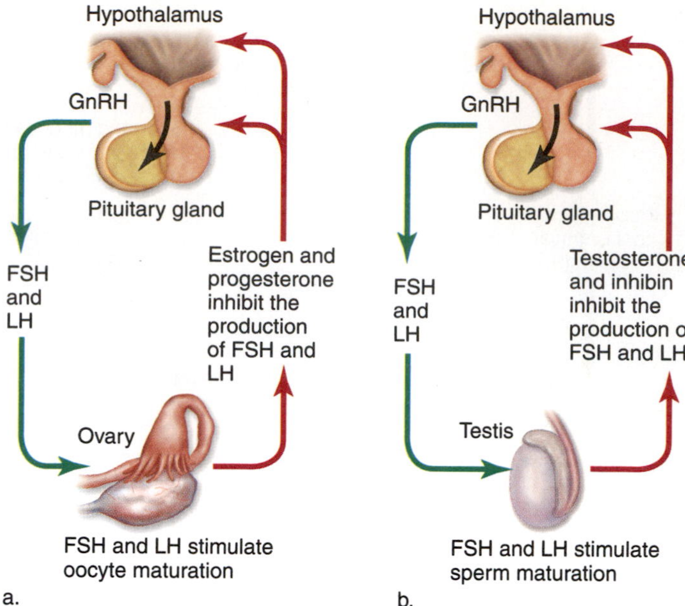

FIGURE 30.11 Control of Reproductive Hormones. (**A**) In females, a monthly surge of LH and FSH expels an oocyte from an ovary. Different hormonal events follow, depending on whether a fertilized ovum implants in the uterine lining. If not, falling levels of progesterone (released from the ovary) initiate the breakdown of the uterine lining. Because progesterone and estrogen inhibit both the hypothalamus and pituitary, lowered levels of these hormones allow LH and FSH levels to increase again, and the cycle begins anew. (**B**) Male sex hormones do not cycle but are still subject to negative feedback controls.

30.5 MASTERING CONCEPTS

1. How do GnRH, LH, and FSH interact to regulate reproduction in males and females?
2. What are the functions of estrogen, progesterone, and testosterone?

30.6 INVESTIGATING LIFE
Something's Fishy in Evolution—the Origin of the Parathyroid Gland

Humans are marvelously inventive at creating new uses for old items. Cut off the handle of an antique silver teaspoon, drill a hole and add a wire, and the old handle becomes an earring. Broken pieces of pottery can find new life in colorful mosaics. Likewise, natural selection can yield new adaptations from existing parts. A bat's wing, for example, contains the same bones as the forelimbs of other mammals (see figure 15.11). It is easy to see that a bat's wings are homologous to our own arms; that is, the structures share a common evolutionary history.

Identifying homologous structures is more difficult when organisms are distantly related. Consider fishes, which look quite different from us. Their heads are fused to their trunks; that is, fishes do not have necks. Their gills extract O_2 from water; our lungs breathe air. Nevertheless, fishes are vertebrates, and the zebrafish *(Danio rerio)* is a model organism that has taught scientists a lot about our own biology. One of the many products of research on zebrafish is insight into the evolutionary origin of the parathyroid glands—findings that are all

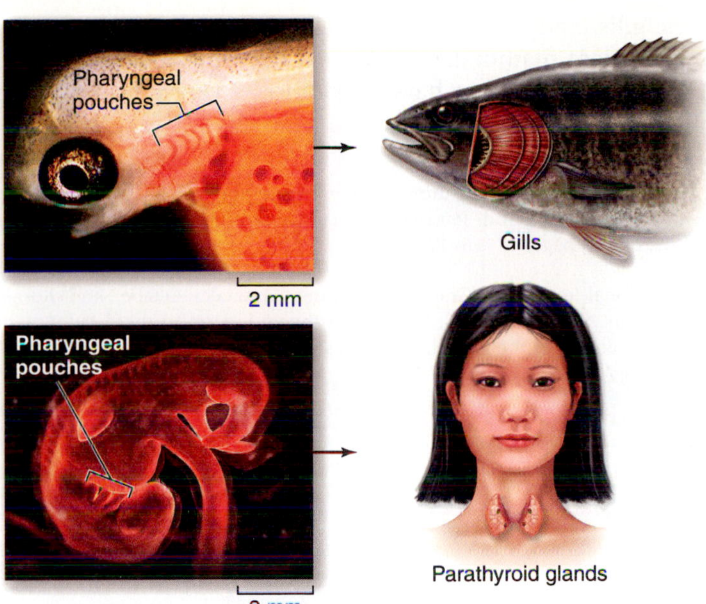

FIGURE 30.12 Pharyngeal Pouches. The same embryonic structures give rise to the gills of fishes and the parathyroid gland of terrestrial vertebrates.

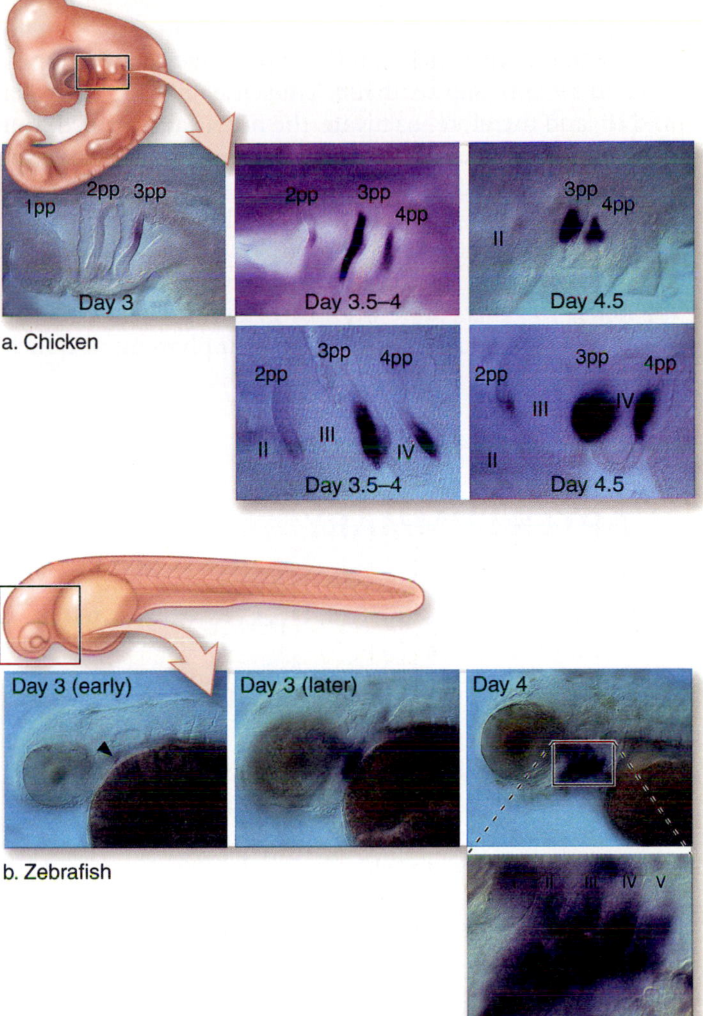

FIGURE 30.13 Different Animals, Same Gene. The dark bluish areas in the (**A**) chicken embryo and (**B**) zebrafish embryo show where the gene *Gcm-2* is expressed. Pharyngeal pouches, indicated by "pp," are separated by arches (indicated by Roman numerals). **Question**: How do you think the embryos would look if the probe indicated expression of a gene required for cellular respiration?

the more surprising because fishes don't even have parathyroid glands!

To understand this research, it helps to know more about the functions of gills. Besides releasing CO_2 and acquiring O_2, a fish's gills also take up calcium dissolved in the water. The emergence of air-breathing vertebrates some 400 MYA (see chapter 23) immediately selected for a new way to regulate calcium levels; this "new way" turned out to be the parathyroid glands. When blood calcium levels are too low, PTH from the parathyroid stimulates the release of calcium from bones and increases its absorption from food.

Where did these new calcium-regulating glands come from? The answer seems to be that the gills of fish transformed into the terrestrial vertebrates' parathyroid glands. Researchers Masataka Okabe and Anthony Graham of Kings College London used multiple lines of evidence to make this argument. First, they pointed to the location of the parathyroid glands in the neck, somewhat similar to the behind-the-head location of a fish's gills. Second, they noted that in fishes, gills arise from embryonic structures called pharyngeal pouches. Parathyroid glands originate from the same structures (**figure 30.12**).

Third, Okabe and Graham checked the embryos of the zebrafish and air-breathing vertebrates for the expression of a gene, called *Gcm-2*, that is known to be involved in the development of the parathyroid gland in mice. (To understand how they did this part of their research, it may be helpful to review sections 12.2 and 12.3.) Gene expression begins when enzymes transcribe DNA; the product of transcription is messenger RNA (mRNA). If researchers know the sequence of mRNA that is transcribed from a gene, they can create complementary nucleic acid "probes" that bind to that mRNA. By treating a developing embryo with a probe and then checking where the probe accumulates, researchers can see precisely which cells express the corresponding gene.

The researchers used this technique to search for *Gcm-2* expression in a developing chick. Early on, cells of the pharyngeal pouches expressed *Gcm-2*. Later, the gene was expressed in the developing parathyroid glands. In zebrafish embryos, *Gcm-2* was expressed in the pharyngeal pouches and in the gill buds that develop from these structures. *Gcm-2* therefore is somehow involved in the development of both the zebrafish's gills and a chicken's parathyroid gland—evidence of a close relationship between the two organs (**figure 30.13**).

Fourth, the researchers checked what would happen to zebrafish development if *Gcm-2* was turned off. Genes do not come with on–off switches, but it is possible stop a gene's expression by applying synthetic "antisense" nucleic acids that bind to, and therefore inactivate, the mRNA transcribed from a gene. When zebrafish embryos were injected with antisense RNA that knocked out the expression of *Gcm-2*, the fish failed to develop normal gill buds. *Gcm-2* is therefore required for the development of gills in fish, just as it is involved in the development of the parathyroid gland in mice.

Taken together, these lines of evidence clearly demonstrate the intimate relationship between pharyngeal pouches, gills, and the parathyroid glands. Just as the forelimbs of ancient mammals gave rise to the wings of the bat, the pharyngeal pouches have proved to be very versatile actors in vertebrate development.

Okabe, Masataka, and Anthony Graham. December 21, 2004. The origin of the parathyroid gland. *Proceedings of the National Academy of Sciences*, vol. 101, no. 51, pages 17716–17719.

Additional reference: Graham, Anthony, Masataka Okabe, and Robyn Quinlan. November 2005. The role of the endoderm in the development and evolution of the pharyngeal arches. *Journal of Anatomy*, vol. 207, no. 5, pages 479–487.

CHAPTER SUMMARY

30.1 The Endocrine System Uses Hormones to Communicate

A. The Endocrine System Consists of Hormones and Glands

- The **endocrine system** includes several **endocrine glands** and scattered cells, plus the **hormones** they secrete into the bloodstream. Hormones interact with **target cells** to exert their effects.
- **Exocrine glands,** which secrete products into ducts or outside the body, are not part of the endocrine system.

B. The Nervous and Endocrine Systems Work Together

- The nervous and endocrine systems interact to maintain homeostasis. The nervous system acts faster and more locally than the endocrine system. Neurons that secrete hormones physically link the two systems.

C. Negative Feedback Loops Control Most Hormone Levels

- **Negative feedback** loops ensure that the levels of a hormone in the bloodstream are not too high or too low. In a **positive feedback** loop, the presence of a hormone increases its production.

D. Some Chemical Signals Are Not Part of the Endocrine System

- Paracrine secretions such as prostaglandins are communication molecules that act near their site of synthesis, and pheromones send signals between individuals.

30.2 Hormones Bind to Receptors on or in Target Cells

- A hormone exerts a physiological effect on **target cells,** which have receptors for it.

A. Water-Soluble Hormones Trigger Second Messenger Systems

- **Peptide hormones** are water-soluble and bind to the surface receptors of target cells, and then a second messenger triggers the hormone's metabolic effect.

B. Lipid-Soluble Hormones Directly Alter Gene Expression

- Most lipid-soluble **steroid hormones** cross target cell membranes, bind to receptors in the cytoplasm or nucleus, and activate genes that direct synthesis of proteins that provide the cell's response.

30.3 The Hypothalamus and Pituitary Gland Oversee Endocrine Control

- Neurons from the **hypothalamus** stimulate or inhibit the release of hormones from the **pituitary gland.** These hormones influence many other processes in the body.

A. The Posterior Pituitary Stores and Releases Two Hormones

- The hypothalamus manufactures two hormones that are stored in and released from the **posterior pituitary: antidiuretic hormone,** which regulates body fluid composition, and **oxytocin,** which contracts the uterus and milk ducts.

B. The Anterior Pituitary Produces and Secretes Six Hormones

- The hypothalamus produces **releasing hormones** and **inhibiting hormones,** which regulate the production and release of hormones from the **anterior pituitary.** Releasing and inhibiting hormones are examples of **tropic hormones**—they affect hormone secretion by other glands.
- **Growth hormone** stimulates cell division, protein synthesis, and growth in cells throughout the body.
- **Prolactin** stimulates milk production.
- **Follicle-stimulating hormone (FSH)** and **luteinizing hormone (LH)** are tropic hormones that stimulate hormone release from the gonads.
- **Thyroid-stimulating hormone (TSH)** prompts the thyroid gland to release hormones.
- **Adrenocorticotropic hormone (ACTH)** stimulates the adrenal cortex to release hormones.
- **Endorphins** are painkillers with target cells in the brain.

30.4 Hormones from Many Glands Regulate Metabolism

A. The Thyroid Gland Sets the Metabolic Pace

- **Thyroxine** and **triiodothyronine** from the **thyroid gland** speed metabolism. The thyroid also releases **calcitonin,** which lowers the level of calcium in the blood.

B. The Parathyroid Glands Control Calcium Level

- The **parathyroid glands** secrete **parathyroid hormone (PTH),** which increases blood calcium level by releasing calcium from bone and increasing its absorption in the gastrointestinal tract and kidneys.

C. The Adrenal Glands Coordinate the Body's Stress Responses

- The inner portion of each **adrenal gland,** the **adrenal medulla,** secretes **epinephrine** and **norepinephrine.** These hormones ready the body to cope with a short-term emergency. The **adrenal cortex** secretes **mineralocorticoids** and **glucocorticoids,** which mobilize energy reserves during stress and maintain blood volume and blood composition.

D. The Pancreas Regulates Nutrient Use

- The **pancreatic islets** of the **pancreas** secrete **insulin,** which stimulates cells to take up glucose. **Glucagon** increases blood glucose levels.
- In **diabetes mellitus,** the pancreas fails to produce insulin, or the body's cells do not respond to insulin (or both). Either way, blood sugar concentrations rise to dangerous levels. **Hypoglycemia** is low blood sugar.

E. The Pineal Gland Secretes Melatonin

- The **pineal gland** may regulate the responses of other glands to light–dark cycles through its hormone, **melatonin.**

30.5 Hormones from the Ovaries and Testes Control Reproduction

- **Gonadotropin-releasing hormone (GnRH)** from the hypothalamus triggers the release of FSH and LH from the anterior pituitary.
- FSH and LH stimulate the **ovaries** to secrete **estrogen** and **progesterone,** hormones that stimulate development of female sexual characteristics and control the menstrual cycle.
- In males, FSH and LH stimulate the **testes** to secrete **testosterone,** which stimulates sperm cell production and the development of secondary sexual characteristics.

30.6 Investigating Life: Something's Fishy in Evolution—the Origin of the Parathyroid Gland

- Multiple lines of evidence, including comparative anatomy and studies of gene expression, indicate that the parathyroid gland of vertebrates is derived from the same structure as the gills of fishes.

MULTIPLE CHOICE QUESTIONS

1. How is a hormone different from a pheromone?
   a. It is produced by a gland.
   b. It functions as a signaling molecule between cells.
   c. It functions as a signaling molecule between individuals.
   d. It influences reproduction and behavior.

2. What is a second messenger?
   a. A hormone produced only in adults
   b. A hormone secreted into a duct
   c. A molecule that initiates a hormone's effects
   d. A molecule participating in a positive feedback loop

3. The effect of a peptide hormone is generally quicker than that of a steroid hormone because
   a. peptide hormones bind on the outside of the cell.
   b. steroid hormones trigger the synthesis of new proteins.
   c. steroid hormones cannot pass through the cell's plasma membrane.
   d. both b and c.
4. A tropic hormone is one that
   a. is produced in one gland and affects hormone secretion in another gland.
   b. is transported a short distance between source and target cells.
   c. triggers an increase in cell division and protein synthesis.
   d. is synthesized in the hypothalamus and stored in the posterior pituitary.
5. Treatment for high blood pressure often involves the use of medication that alters blood volume. Which of the following treatments would decrease a person's blood volume?
   a. Increase in release of thyroid-stimulating hormone
   b. Increase in inhibiting hormones
   c. Activation of prolactin synthesis
   d. Inhibition of antidiuretic hormone
6. Glucocorticoids are secreted by
   a. the thyroid.
   b. the pancreas.
   c. the adrenal glands.
   d. the pineal gland.
7. Would an insulin injection help a person with type II diabetes?
   a. Yes, because their beta cells are not producing insulin.
   b. No, because their target cells do not respond to insulin.
   c. Yes, because the extra insulin will help with glucose uptake.
   d. No, because the extra insulin will trigger excess release of glucagon.
8. Secretion of the hormone melatonin is regulated by
   a. light.
   b. temperature.
   c. stress.
   d. glucose.
9. Birth control pills that contain synthetic progesterone work because progesterone
   a. promotes secondary sexual characteristics.
   b. helps regulate follicle-stimulating hormone.
   c. inhibits gonadotropin-releasing hormone.
   d. both b and c.
10. Which of the following is not a property of a prostaglandin?
    a. Is synthesized from lipid
    b. Can influence blood flow, muscle contraction, and metabolism
    c. Is effective for a short time
    d. Can influence target tissues that are distant from the site of synthesis

TESTING YOUR KNOWLEDGE

1. How does the endocrine system interact with the nervous system? The circulatory system?
2. What prevents a hormone from affecting all body cells equally?
3. How are the endocrine and nervous systems similar, and how are they different?
4. How do hormones regulate their own levels?
5. How do the mechanisms of peptide and steroid hormone function differ?
6. Describe how thyroxine, triiodothyronine, TSH, and TRH interact.
7. How are the pituitary and adrenal glands each really two glands in one?
8. List the hormones released from the posterior pituitary and the anterior pituitary.
9. Which hormone fits each of the following descriptions?
   a. A woman who is breast feeding produces this hormone.
   b. Too little of this hormone causes fatigue.
   c. This hormone increases blood calcium level.
   d. This hormone decreases blood glucose level.
   e. Synthetic steroid drugs mimic the muscle-building effects of this hormone.

THINKING AS A SCIENTIST

1. Imagine you are a researcher who has just found a small glandlike structure in a human. How would you determine whether it is an endocrine gland? If it is, how might you identify the gland's role in the endocrine system?
2. An endocrine disruptor is a molecule that either mimics or blocks the activity of a hormone. Many synthetic chemicals are endocrine disruptors. How might you test the hypothesis that a pesticide such as DDT is an endocrine disruptor? How might you test the hypothesis that endocrine disruptors are responsible for recent declines in sperm counts?
3. A queen honeybee secretes a substance from a gland in her mouthparts that inhibits the development of ovaries in worker bees. Is this substance most likely a hormone, prostaglandin, or pheromone? Cite a reason for your answer.
4. Because steroid hormones are lipid-soluble, they do not travel easily in blood, which is mostly water. To circulate in the blood, they must attach to hydrophilic (water-soluble) transport proteins. What do you think would happen to a person who lacks these transport proteins?
5. Once in the blood, a hormone can be inactivated, degraded, or excreted in urine. A hormone's action also depends on the number of receptors and their affinity for the hormone. How could a genetic mutation affecting a hormone receptor cause illness? How might mutations in different genes cause malfunctions of the same hormones?
6. In an episode of the *X-Files*, a monster sucks out people's pituitary glands. What symptoms might the victims develop?
7. How might insulin-producing stem cells transplanted to the pancreas help people with type I diabetes? Would the same treatment help people with type II diabetes?
8. Many dairy operators inject their cows with bovine growth hormone to stimulate milk production. Cite two reasons that bovine growth hormone might not stimulate growth in people drinking the milk.
9. Some, but not all, breast cancers grow in response to estrogen in blood. If you could examine the proteins in breast cancer cells, what is one difference you would expect between the cancer cells that respond to estrogen and those that do not?
10. Some people advocate treating severely disabled children with hormones that prevent growth, since a small person is much easier to clothe and bathe than a full-grown adult. What factors would you consider in deciding whether to pursue such a treatment for a disabled child?

Visit www.mhhe.com/hoefnagels for practice quizzes, animations, videos, and activities designed to help you master the material in this chapter.

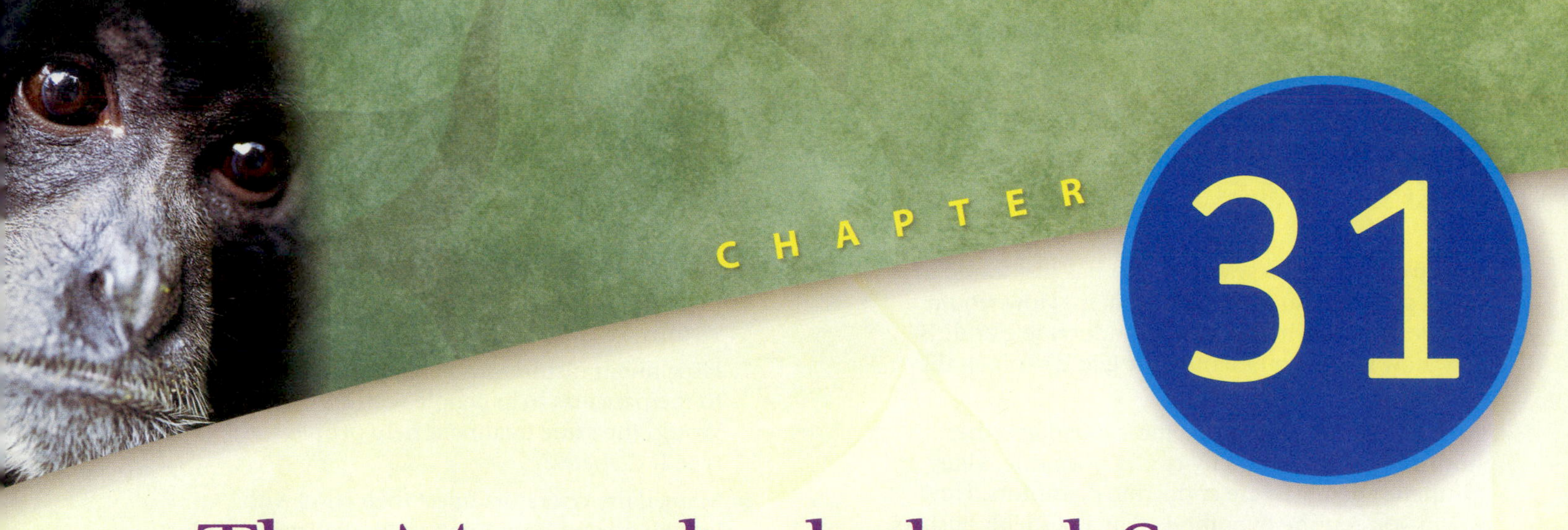

The Musculoskeletal System

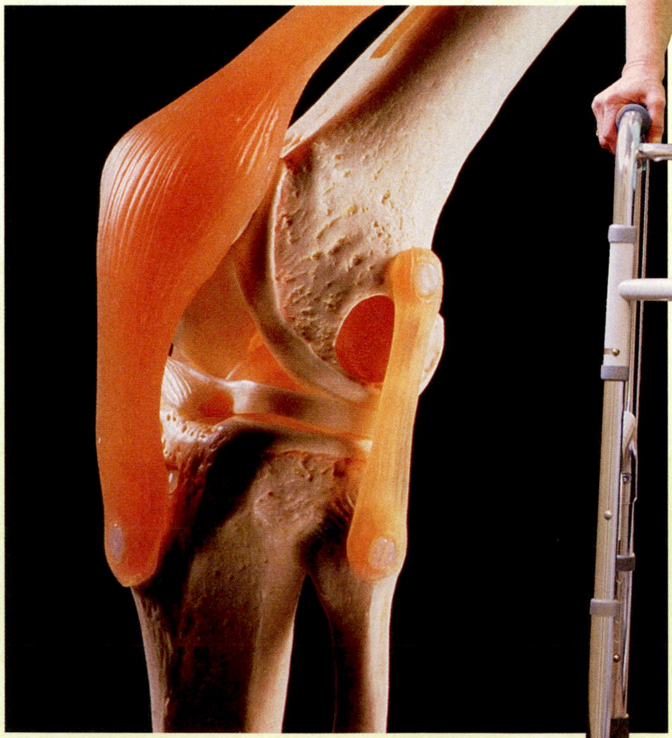

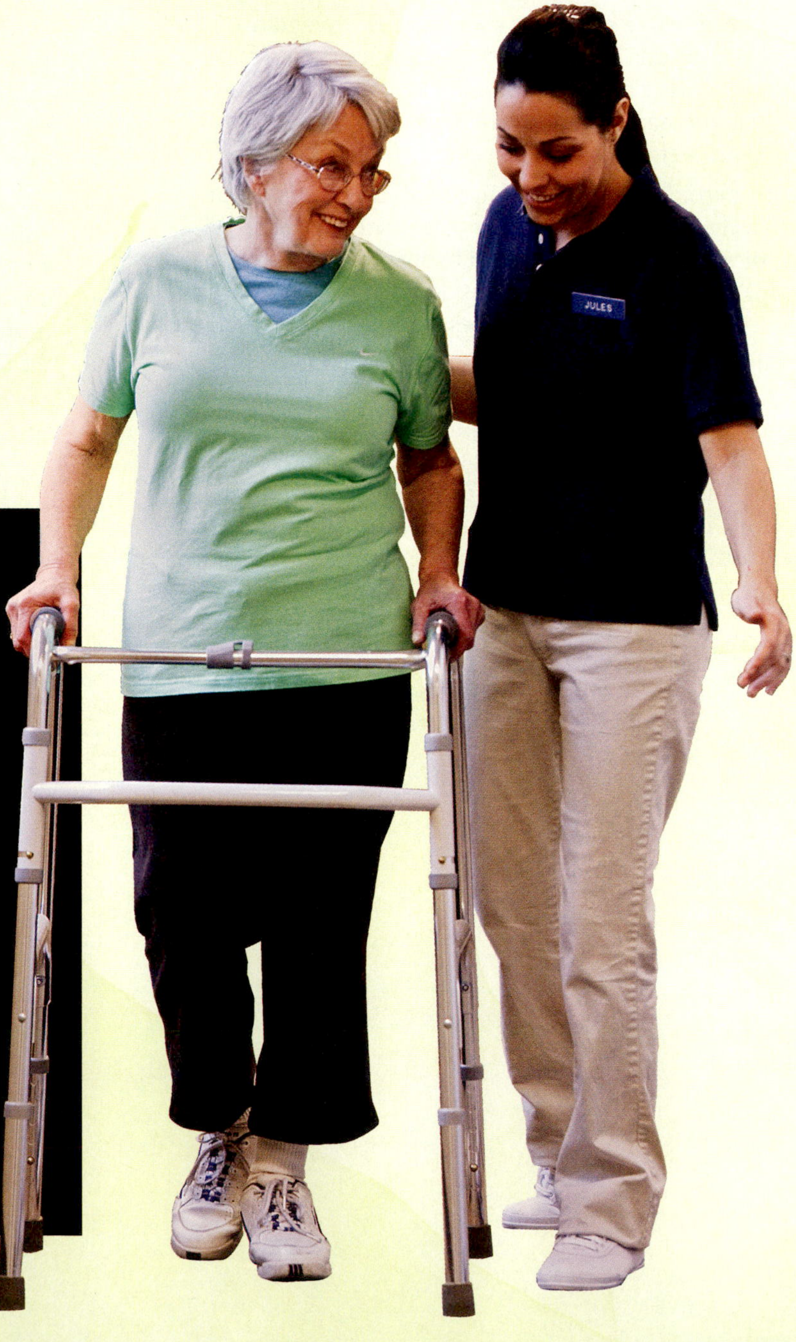

Artificial knee. This woman is recovering from knee replacement surgery.

What's the Point?

Spare Parts Replace Worn Out Joints

An orthopedic surgeon repairs damage to bones and the muscles that move them. Orthopedic literally means "straight child," because when the term was coined in the mid-eighteenth century, orthopedic surgeons focused on correcting birth defects such as a club foot. Since then, the profession has expanded to include many adult surgeries, including hip and knee replacement and the repair of bone fractures and torn tendons and ligaments.

Joint replacements are among the most dramatic of orthopedic surgeries because they substitute synthetic materials for a patient's original bone. Finding the right material is crucial to the success of the operation. The new part must not trigger rejection by the immune system, and it must be strong and flexible enough to take over the function of the original joint. It must also resist wear and tear—after all, many joints are in almost constant motion.

The most commonly replaced joint is the hip, where the thighbone's spherical head swivels in a cup-shaped socket in the pelvis. Osteoarthritis, or inflammation of the joints, commonly causes deformed, swollen hips that make movement stiff and painful as a person ages. In a hip replacement surgery, a surgeon replaces the top part of the thighbone with a ball-shaped substitute and inserts a cup that fits over the ball into the worn-out area of the pelvis. Early hip replacement surgeries in the nineteenth century used ivory in place of the patient's original bones, but surgeons now use ceramic, plastic, and metals such as titanium.

Arthritis also damages the knees, making walking painful. In a knee replacement operation, a surgeon detaches the quadriceps (thigh) muscle from the knee cap. After moving the knee cap to the side, the surgeon replaces the ends of the thighbone and shinbone with metal parts. A piece of plastic between the two metal parts provides cushioning. The surgeon then puts the knee cap back into place, reattaches the quadriceps muscle, and reconstructs the tendons and ligaments that allow the bones to move.

Complete recovery from hip or knee replacement surgery takes several months. Unfortunately, the metal and plastic parts do not last forever. After 15 to 20 years, depending on how active the patient is, the replacement parts may themselves need replacing.

It is easy to take bones and muscles for granted, at least until they stop working correctly. This chapter describes the structure and function of both components of the musculoskeletal system.

LEARNING OUTLINE

31.1 Skeletons Take Many Forms

Ask a child what sets animals apart from other organisms, and he or she will likely answer "movement." This response is technically wrong—many prokaryotes, protists, fungi, and plants have swimming, creeping, or gliding cells. Yet the child's answer correctly implies that animal movements are unmatched in their drama, versatility, and power. The ability to hop, dig, fly, slither, scuttle, or swim comes from two closely allied, interacting organ systems: the **muscular system** and the **skeletal system,** which together function under the direction of the nervous system.

In most animals, contractile cells are organized into muscle tissues, which (together with other tissue types) form organs called **muscles.** Muscles provide motion; the **skeleton** adds a firm supporting structure that muscles pull against. The skeleton also gives an animal's body shape and protects internal organs.

The simplest type of skeleton is a **hydrostatic skeleton** (*hydro-* means water), which consists of fluid constrained within a layer of flexible tissue. Combined with muscle action, a hydrostatic skeleton can provide locomotion. A jellyfish, for example, swims by drawing water into its digestive cavity and then forcibly ejecting it (**figure 31.1**). In earthworms, the fluid-filled coelom is a hydrostatic skeleton. By alternately contracting and relaxing muscles surrounding the coelom, the worm can burrow through soil. **cnidarians, p. 455; coelom, p. 450**

An **exoskeleton** (*exo-* means outside) protects an organism from the outside, much like a suit of armor. Mollusks such as clams and snails produce hard, calcium-containing shells. In lobsters, insects, and other arthropods, the jointed exoskeleton consists of the complex carbohydrate chitin (**figure 31.2**). **carbohydrates, p. 34**

An **endoskeleton** (*endo-* means inner) is an internal support structure. Sea stars and other echinoderms, for example, produce rigid, calcium-rich spines and internal plates. Sharks and rays have endoskeletons made of cartilage (see chapter 23). The endoskeletons of most vertebrates, including humans, are composed primarily of bone. An endoskeleton offers advantages over an exoskeleton. First, it can grow as the animal grows, whereas an arthropod must periodically molt to accommodate its enlarging body. Until the new exoskeleton has hardened, the animal is vulnerable to predators. Second, because an endoskeleton consumes less of an organism's total body mass, it increases mobility.

FIGURE 31.1 Hydrostatic Skeleton. When this jellyfish contracts muscles surrounding its digestive cavity, water shoots out of its body. The animal moves forward.

FIGURE 31.2 Exoskeleton. Insects such as fleas have tough, jointed exoskeletons that consist of chitin.

The skeletal system's capacity to change over evolutionary time is striking. Each type of vertebrate has a distinctive skeleton, yet all are composed of the same types of cells and have similar arrangements (**figure 31.3**). The characteristics of each species' musculoskeletal system reflect common ancestry and the selective forces in its environment. **natural selection, p. 269; homologous structures, p. 316**

Summary *Muscles enable an animal to move, and a skeleton is a supportive framework. The types of skeletons include a hydrostatic skeleton, an exoskeleton, and an endoskeleton. Vertebrates have endoskeletons.*

31.1 MASTERING CONCEPTS

1. How do the skeletal and muscular systems interact?
2. How can a hydrostatic skeleton allow an animal to move?
3. What are two differences between endoskeletons and exoskeletons?
4. How do vertebrate skeletons reveal common ancestry?

FIGURE 31.3 Vertebrate Skeleton Diversity. A bird, a seal, and a fish have different body forms yet their skeletons are composed of similar bones.

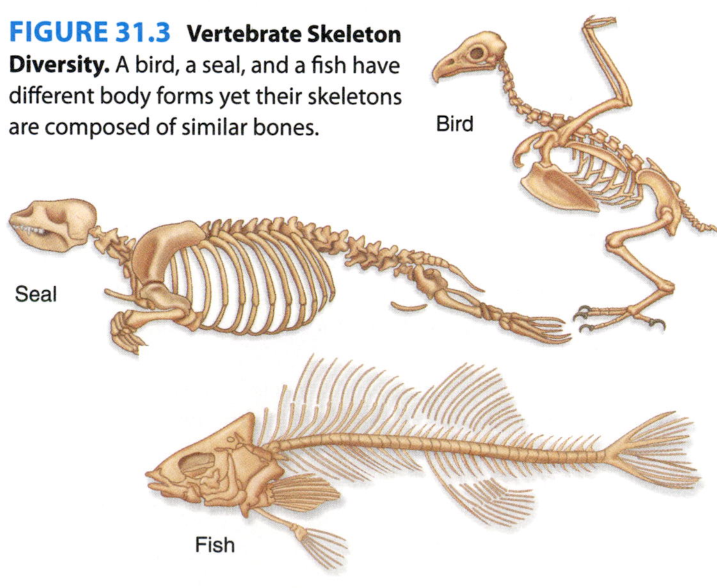

31.2 The Backbone Is Central to the Vertebrate Skeleton

The bones of the vertebrate skeleton are grouped into two parts. The **axial skeleton,** so named because it is located in the longitudinal central axis of the body, consists of the skull, vertebral column, ribs, and sternum (breastbone). The **appendicular skeleton** consists of the appendages (the limbs) and the bones that support them. **Figure 31.4** depicts the human axial and appendicular skeletons.

The axial skeleton shields soft body parts. The skull, which protects the brain and many of the sense organs, consists of hard, dense bones that fit together like puzzle pieces. All of the head bones are attached with immovable joints, except for the lower jaw and the middle ear. These bones move to enable chewing, speech, and hearing. Sinuses are air-filled spaces within the bones of the head that help lighten the skull and create a resonating voice chamber.

FIGURE 31.4 The Human Skeleton. The human skeleton is divided into the axial skeleton and the appendicular skeleton. The axial skeleton in humans includes the skull, vertebral column, and rib cage. The bones that compose and support the limbs constitute the appendicular skeleton.

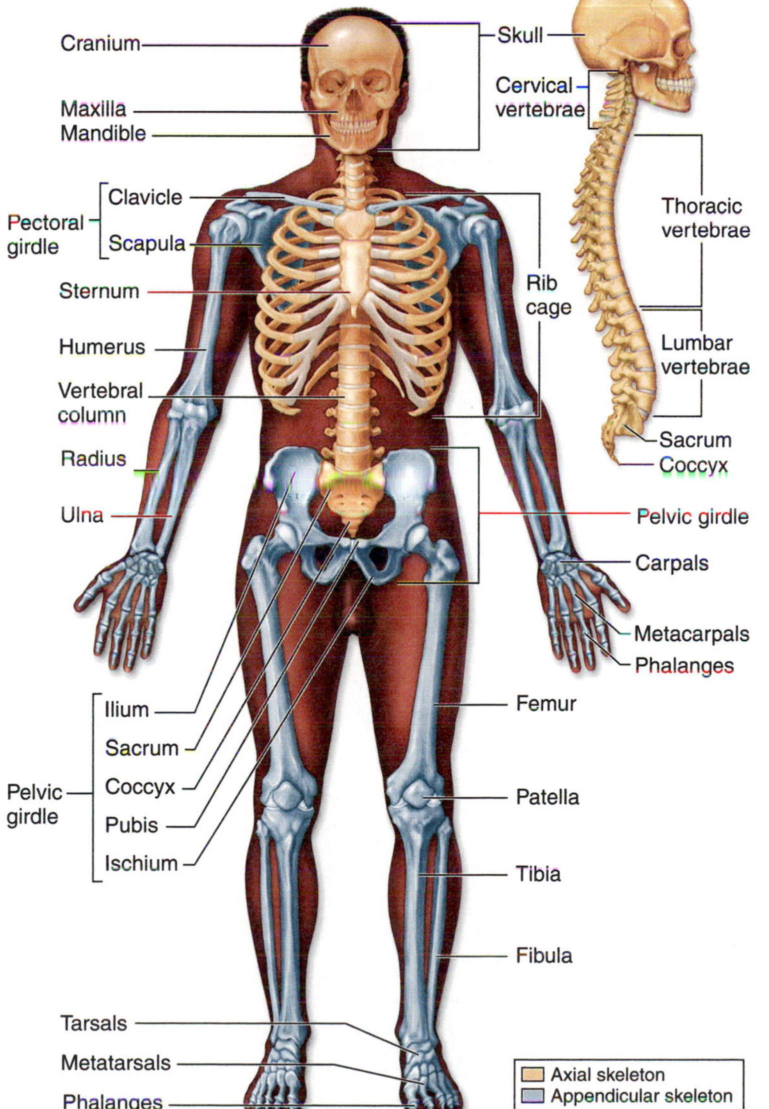

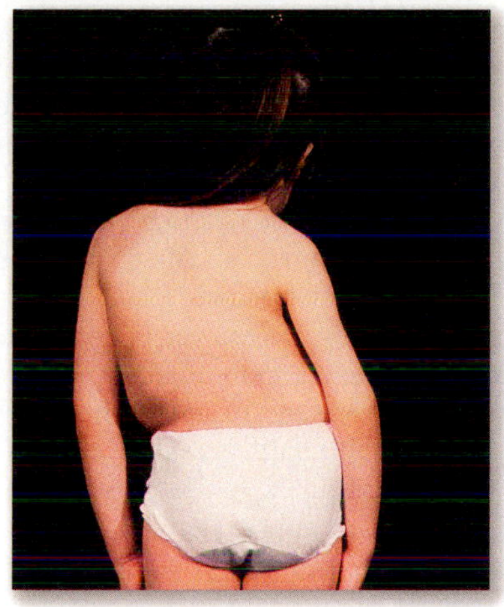

FIGURE 31.5 Scoliosis. This young girl's spine curves to the side. Girls are more often affected with scoliosis than boys.

The **vertebral column** supports and protects the spinal cord (see figure 31.4 inset). A human vertebral column consists of 33 vertebrae, separated by cartilage disks that cushion shocks and enhance flexibility. A "slipped," or herniated, disk occurs when these pads tear or rupture, causing a bulge that presses painfully on a nearby nerve. Scoliosis, in which the vertebral column curves to the side, is also a disorder of the axial skeleton (**figure 31.5**).

Humans have 12 pairs of ribs attached to the vertebral column. The flexibility of the cartilage between the ribs and other bones allows muscles to elevate the ribs, a movement important in breathing. The rib cage protects the heart and lungs.

In the appendicular skeleton, the **pectoral girdle** connects the forelimbs to the axial skeleton; it includes the clavicles (collarbones) and scapulae (shoulder blades). Likewise, the **pelvic**

Skeletal System	
Main tissue types	**Examples of Locations/Functions**
Connective	Makes up bone, cartilage, tendons, ligaments, marrow of vertebrate skeleton
Muscle	Skeletal muscle connects to movable bones, enabling voluntary movements
Nervous	Senses body position and controls muscles

Can *You* Relate?

Bones, Crimes, and Evolution

After death, a person's hard, mineral-rich bones and teeth remain intact long after the soft body parts decay. Skeletal remains therefore make durable, useful clues that help detectives identify murder victims and shed light on evolution.

Males and females have slightly different skeletons. Most obviously, the average male is larger than the average female. In addition, the front of the female pelvis is broader than the male's, and it is larger and has a wider bottom opening. (These differences in the female pelvis are adaptations that accommodate the birth of a baby.) Detectives and anthropologists can use these features to determine the sex of recent murder victims and long-dead human ancestors.

Bones can also reveal events unique to each person's life. Healed breaks may indicate accidents or abuse. (Egypt's King Tut, for example, suffered a severe leg break shortly before he died.) The pattern of thickenings in the bone tells whether a person spent his or her life in hard physical labor. Similarly, anthropologists examine the shapes of fossilized bones to determine whether ancient primates walked upright or on all fours. The shapes of the skulls and teeth in primate fossils have also revealed much about our own evolutionary history. **human evolution, p. 344**

Animal skeletons also tell the larger story of vertebrate evolution. For example, paleontologists can use the shapes and sizes of fossilized bones to determine whether an extinct animal was aquatic or terrestrial; land-dwellers tend to have sturdier skeletons than their relatives in water.

girdle attaches the lower limb bones to the axial skeleton. The two hipbones join the sacrum in the rear and meet each other in front, creating a bowl-like pelvic cavity. (The term *pelvis* is Latin for "basin.") The bony pelvis protects the lower digestive organs, the bladder, and some of the reproductive structures (especially in the female).

This chapter's Can *You* Relate? Bones, Crimes, and Evolution illustrates how skeletal features reveal clues that are useful to people in several different professions.

Summary *Vertebrate endoskeletons are organized into axial and appendicular parts. The pectoral and pelvic girdles connect the appendages to the axial skeleton.*

31.2 MASTERING CONCEPTS

1. What are the two major groups of bones that make up the human skeleton?
2. What are the locations of the pectoral and pelvic girdles?

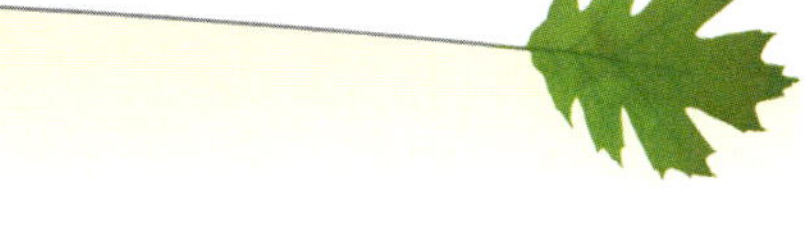

31.3 Bones Provide Support, Protect Internal Organs, and Supply Calcium

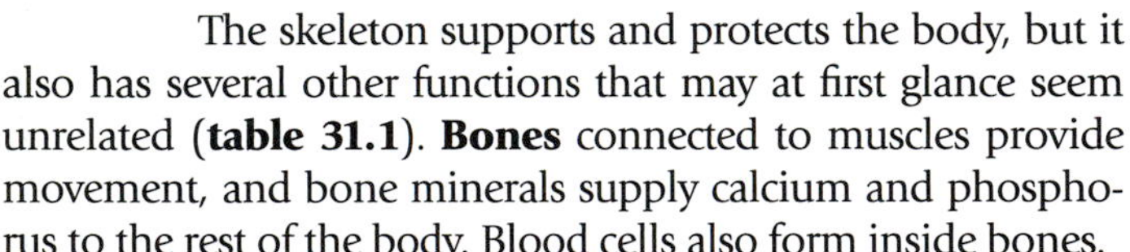

The skeleton supports and protects the body, but it also has several other functions that may at first glance seem unrelated (**table 31.1**). **Bones** connected to muscles provide movement, and bone minerals supply calcium and phosphorus to the rest of the body. Blood cells also form inside bones.

Bones come in multiple shapes. Long bones make up the arms and legs, whereas short bones are in the wrists and ankles. Flat bones include the ribs and skull. Vertebrae are irregularly shaped. No matter what the shape, however, bones are lightweight and strong because they are porous, not solid (**figure 31.6**A). The weight of long bones is further reduced by the **marrow cavity,** a space in the shaft that contains yellow marrow consisting primarily of fat cells.

A. A Bone Consists Mostly of Bone Tissue and Cartilage

Bones are strong, lightweight organs consisting of multiple tissues, including marrow, nerves, and blood vessels. But the majority of the vertebrate skeleton consists of two types of connective tissue: bone and cartilage.

Table 31.1 *Functions of the Vertebrate Endoskeleton*

Function	Explanation
Support	The skeleton is a framework that supports an animal's body against gravity. It largely provides the body's shape.
Movement	The vertebrate skeleton is a system of muscle-operated levers. Typically, the two ends of a skeletal (voluntary) muscle attach to different bones that connect in a structure called a joint. When the muscle contracts, one bone is pulled toward the other.
Protection of internal structures	The backbone surrounds and shields the spinal cord, the skull protects the brain, and ribs protect the heart and lungs.
Production of blood cells	Many bones, such as the long bones of the human arm and leg, contain and protect red marrow, a tissue that produces red blood cells, white blood cells, and platelets.
Mineral storage	The skeleton stores calcium and phosphorus.

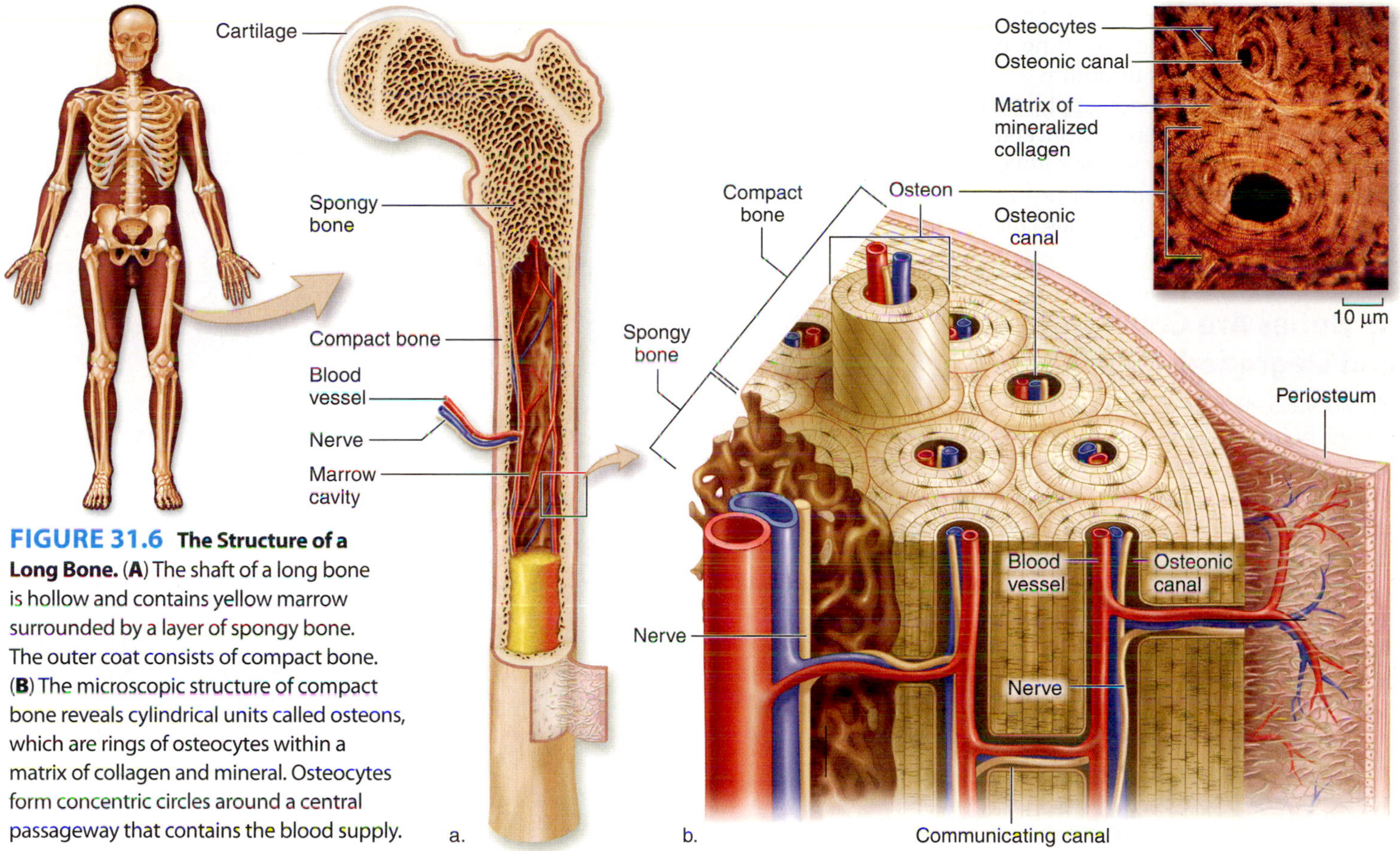

FIGURE 31.6 The Structure of a Long Bone. (**A**) The shaft of a long bone is hollow and contains yellow marrow surrounded by a layer of spongy bone. The outer coat consists of compact bone. (**B**) The microscopic structure of compact bone reveals cylindrical units called osteons, which are rings of osteocytes within a matrix of collagen and mineral. Osteocytes form concentric circles around a central passageway that contains the blood supply.

Bone tissue consists of cells suspended in a matrix of collagen and minerals. Collagen gives the bone flexibility, elasticity, and strength. The hardness and rigidity of bone comes from the minerals, primarily calcium and phosphate, that coat the collagen fibers. Stem cells lining bone passageways give rise to **osteoblasts,** bone-forming cells that secrete matrix. An **osteocyte** is a former osteoblast embedded in the matrix it has produced. Each osteocyte inhabits a small space joined to others by narrow passageways. **Osteoclasts** are cells that degrade the matrix at the bone surface, releasing calcium and phosphorus into the blood.

An **osteon** consists of a set of concentric rings of osteocytes (figure 31.6B). In the center of each osteon is a passageway, the osteonic canal, that contains nervous tissue and a blood supply. Nutrients and oxygen pass from the blood, and from osteocyte to osteocyte, through a "bucket brigade" of up to 15 cells. Still other canals connect the entire labyrinth to the outer surface of the bone and to the marrow cavity.

Compact and spongy bone differ in density. **Compact bone** consists of closely packed osteons. **Spongy bone** is hard, but it has many large spaces between a web of bony struts, which increase the bone's strength. Red marrow, a nursery for blood cells and platelets, fills the spaces within spongy bone. Spongy bone tissue contains few osteons; instead, the osteocytes in spongy bone acquire nutrients and oxygen directly from the nearby bone marrow. Most flat bones consist of spongy bone sandwiched between two layers of compact bone. Long bones consist mostly of compact bone, although their bulbous tips contain spongy bone. In short bones, a thin layer of compact bone covers the spongy bone.

Besides bone tissue, the skeleton also contains abundant cartilage. Recall from chapter 27 that cartilage is a rubbery connective tissue (**figure 31.7**). Cells called chondrocytes secrete the proteins collagen and elastin, which give cartilage

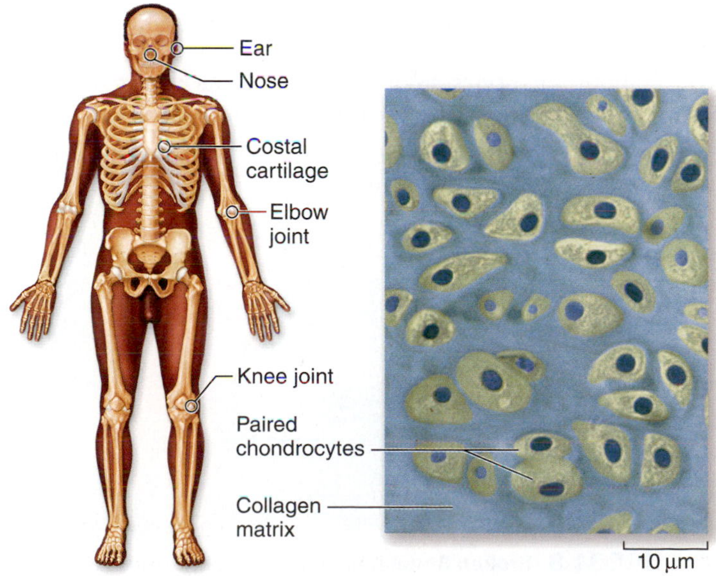

FIGURE 31.7 Tough but Flexible. Cartilage consists of chondrocytes within a matrix of collagen and elastin.

tissue its firmness and flexibility. Strong networks of collagen fibers resist breakage and stretching, even when bearing great weight. Elastin provides flexibility.

Cartilage lacks a blood supply, which means it is slow to heal. But cartilage's protein network entraps a great deal of water. As the body moves, water within cartilage cleanses the tissue and bathes it with dissolved nutrients from nearby blood vessels. The high water content of the matrix also makes cartilage an excellent shock absorber.

B. Bones Are Constantly Built and Degraded

Most bones originate in the embryo as cartilage models shaped like the bones they will become. As the embryo develops, each cartilage model's matrix hardens with calcium salts, cutting off diffusion to the cartilage cells. The cartilage matrix degenerates, and capillaries replace it. Osteoblasts enter and secrete bone matrix, eventually becoming mature osteocytes.

After birth, bone growth becomes concentrated near the ends of the long bones in thin disks of cartilage ("growth plates") that enable bones to elongate as the child grows. Growth continues until the late teens, when bone tissue begins to replace the cartilage plates. By the early twenties, bone growth is complete.

Even after a person stops growing, bone is continually under reconstruction. Bones become thicker and stronger with strenuous exercise such as weight lifting. Less-used bones lose mass as the minerals slowly dissolve. For example, astronauts lose bone density if they are in a prolonged weightless environment because their musculoskeletal systems don't have to work as hard as they do against Earth's gravity. Moreover, broken bones can repair themselves (**figure 31.8**). Osteoblasts near the site of the fracture produce new bone, so that after several weeks the injury is all but healed.

C. Bones Help Regulate Calcium Homeostasis

Throughout life, bones are a reservoir for calcium, a mineral that is vital for muscle contraction, the transmission of neural impulses, blood clotting, the activity of some enzymes, cell adhesion, and cell membrane permeability. The body constantly shuttles calcium between blood and bone. Hormones from the parathyroid and thyroid glands control this exchange in a negative feedback loop (**figure 31.9**). **parathyroid glands, p. 624**

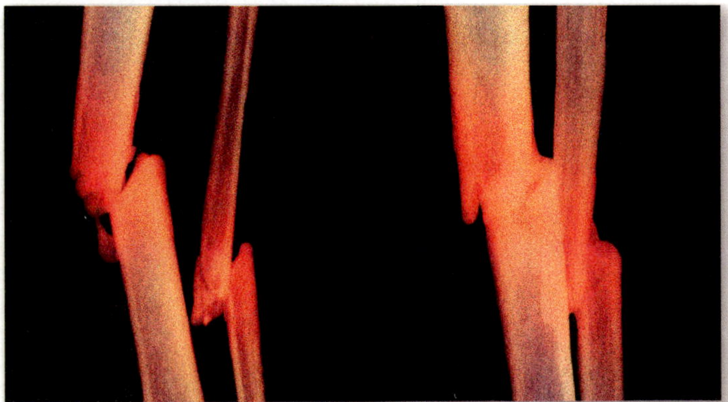

FIGURE 31.8 Broken Bone. A broken bone, or fracture, can result from a forceful impact or from stress on the bone.

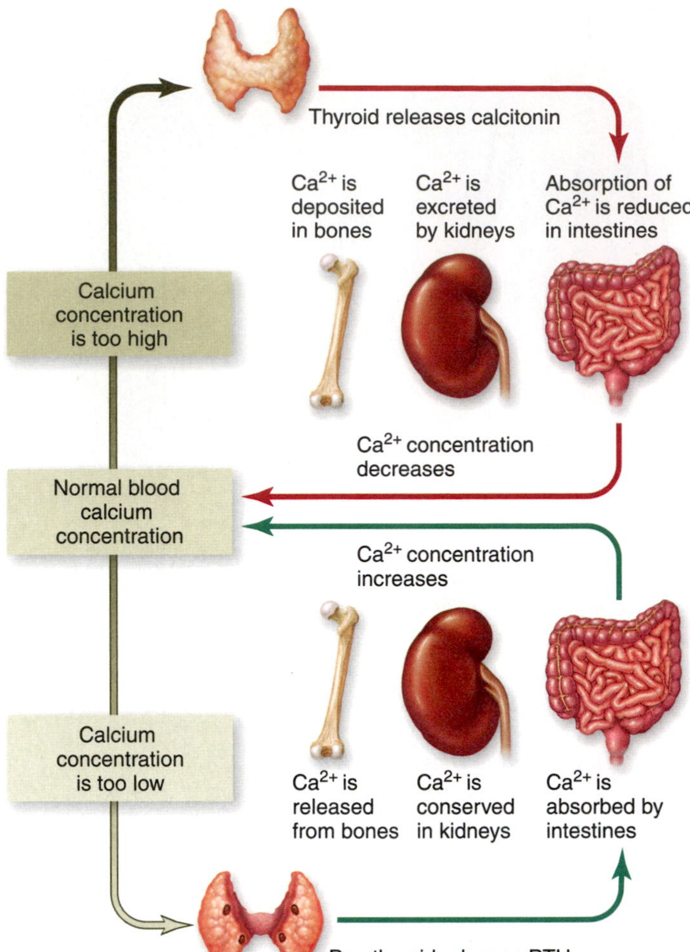

FIGURE 31.9 Regulation of Blood Calcium Level. Bones play a critical role in maintaining calcium homeostasis. When blood contains too much calcium, the thyroid gland releases the hormone calcitonin, which acts on bones and other organs to reduce the calcium concentration. Parathyroid hormone (PTH) acts in reverse, stimulating bone to release calcium (Ca^{2+}) and the kidneys to conserve it. PTH indirectly stimulates the intestine to absorb Ca^{2+}. The resulting increase in blood Ca^{2+} concentration inhibits secretion of PTH.

As a person ages, bones lose more calcium than they add. This leads to **osteoporosis,** a condition in which bones become less dense (**figure 31.10**). Osteoporosis causes shrinking stature, chronic back pain, and frequent fractures. Females are more likely to suffer from osteoporosis than males for several reasons. Women live longer, and the bone mass of the average woman is about 30% less than a man's, so her bones are more easily depleted of calcium. In addition, bone cells have receptors for the steroid hormone estrogen; a decline in the level of estrogen somehow makes a woman more prone

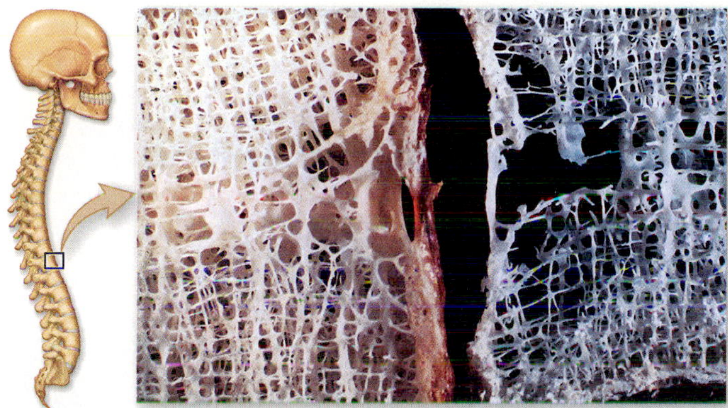

FIGURE 31.10 **Osteoporosis.** In osteoporosis, Ca^{2+} loss weakens bones. The vertebra on the left is normal; osteoporosis has weakened the vertebra on the right.

to osteoporosis. Doctors advise all women to take 1000 to 1500 mg of calcium daily and to exercise regularly. Several drugs can increase bone density or slow its loss.

D. Bone Meets Bone at a Joint

A **joint** is an area where two bones meet. The most common and familiar joints are the freely movable synovial joints, such as those of the fingers, knees, hips, and elbows. A **synovial joint** consists of movable bones joined by a fluid-filled capsule of fibrous connective tissue (**figure 31.11**). A lubricating

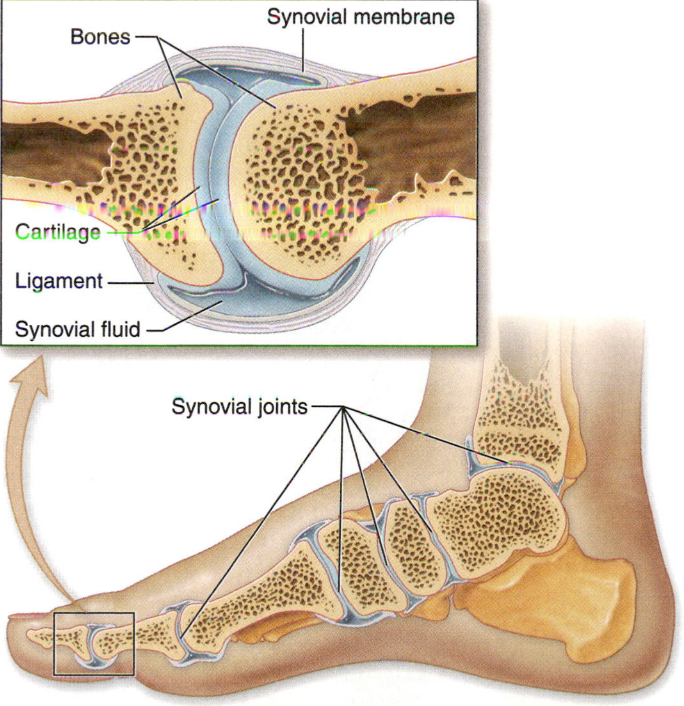

FIGURE 31.11 **The Synovial Joint: Where Bone Meets Bone.** A fluid-filled capsule of fibrous connective tissue surrounds a synovial joint. This illustration shows the synovial joints in the foot and ankle. The enlargement shows bones of the big toe.

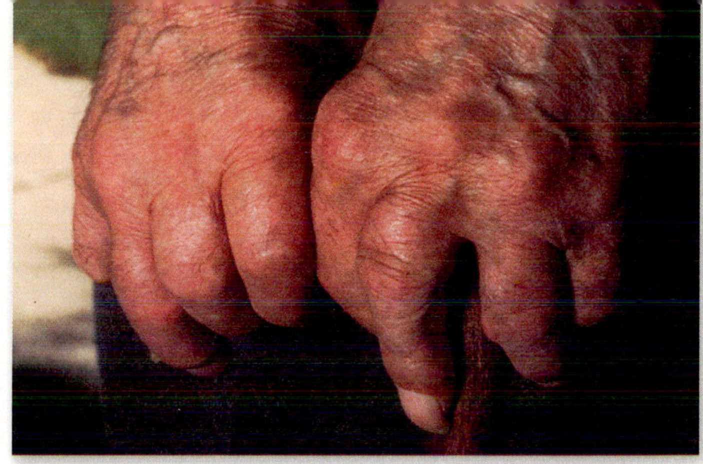

a.

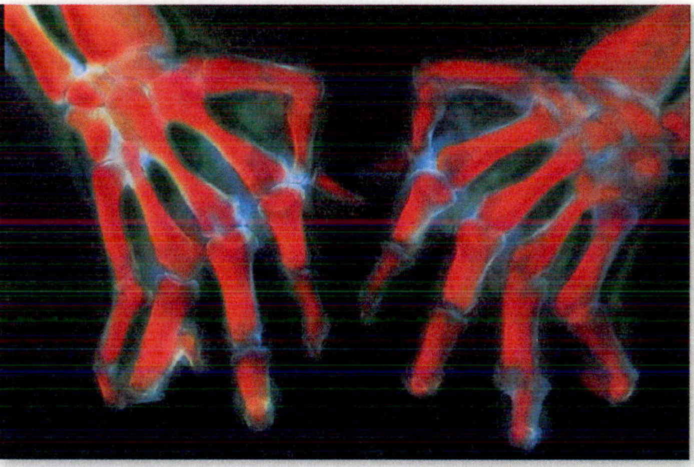

b.

FIGURE 31.12 **Arthritis Inflames Joints.** (**A**) The characteristic gnarled hands of a person with rheumatoid arthritis make many tasks painful or impossible. (**B**) The inflamed joints are visible on this X-ray film.

fluid inside the capsule, combined with slippery cartilage on the bone ends, allows bones to move against each other in a nearly friction-free environment.

Ligaments and tendons help stabilize synovial joints. **Ligaments** are tough bands of connective tissue that attach one bone to another; **tendons** are similar structures that attach bone to muscle. A sprain is a stretched or torn ligament. A common type of sprain, especially in sports such as basketball and volleyball, is a torn anterior cruciate ligament (ACL). The ACL is one of two ligaments that criss-cross at the knee, connecting the thigh bone to the shin bone. Surgical reconstruction of the ACL enables many injured athletes to return to their sports.

Arthritis is a common disorder of joints. A very severe form, rheumatoid arthritis, is an inflammation of the synovial membrane, usually of the small joints of the hands and feet (**figure 31.12**). In the more common osteoarthritis, joint cartilage wears away. As the bone is exposed, small bumps of new bone begin to form. Osteoarthritis usually reveals itself as stiffness and soreness after age 40.

Summary *The vertebrate skeleton is composed mainly of bone and cartilage. Bones are constantly being remodeled in response to injury or stress, and they are crucial in maintaining calcium homeostasis. A synovial joint connects movable bones.*

31.3 MASTERING CONCEPTS

1. Describe the organization and functions of bone tissue and cartilage.
2. What are the functions of each of the three cell types in bone tissue?
3. What are osteons, and in what type of bone tissue are they most abundant?
4. How do osteoclasts and osteoblasts remodel and repair bones throughout life?
5. How do bones participate in calcium homeostasis?
6. What structures form a synovial joint?

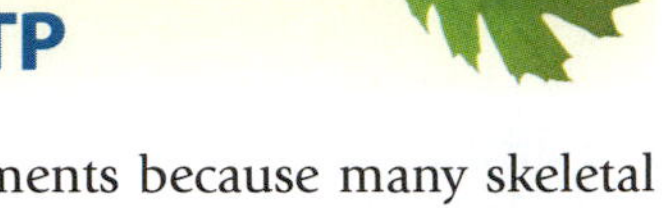

31.4 Muscle Movement Requires Contractile Proteins and ATP

Skeletal muscles generate voluntary movements between pairs of bones. The greater the number of bones connected by movable joints, the finer and more variable the movements an animal can make. The human muscular system is very complex and includes more than 600 skeletal muscles. **Figure 31.13** identifies a few of the major skeletal muscles in a human, and **table 31.2** lists some functions of muscles.

A contracting muscle can pull a bone in only one direction, it cannot push the bone the opposite way. The body can generate back-and-forth movements because many skeletal muscles occur in antagonistic pairs whose members operate in opposite directions (**figure 31.14**). For example, when

Table 31.2 *Functions of Muscles*

Function	Types of Muscle Tissue
Movement (both voluntary and involuntary, including the muscles that control breathing, circulation, digestion, defecation, etc.)	Skeletal, smooth, cardiac
Control body openings (eyelids, pupils, mouth, anus, sphincters that control movement of food through digestive system)	Skeletal, smooth
Maintain posture	Skeletal
Communication (facial expression, speaking, writing, gesturing)	Skeletal
Maintain body temperature	Skeletal

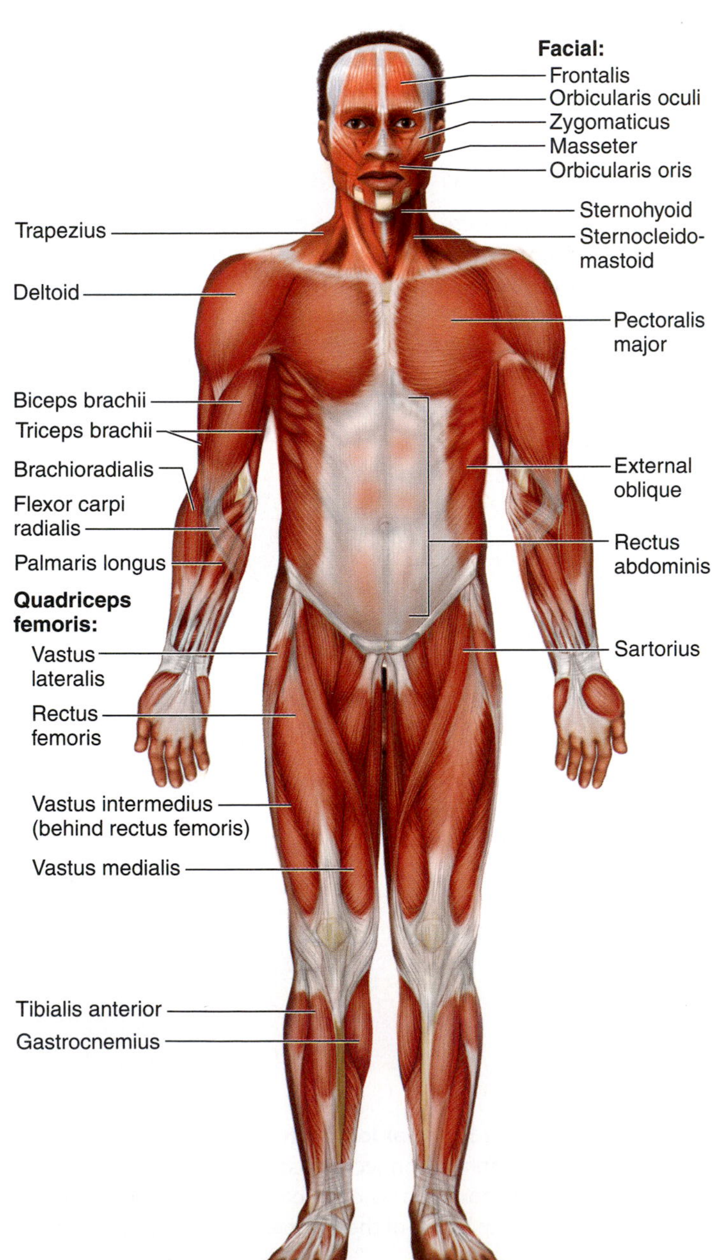

FIGURE 31.13 **The Human Muscular System.** The human body has more than 600 skeletal muscles, a few of which are identified here.

Muscular system	
Main tissue types	**Examples of Locations/Functions**
Connective	Makes up bone, cartilage, marrow of vertebrate skeleton, and tendons that attach muscles to bone; surrounds muscle cells, bundles, and whole muscles
Muscle	Connects to bones and soft tissue, enabling movement of body parts
Nervous	Senses body position and controls muscles

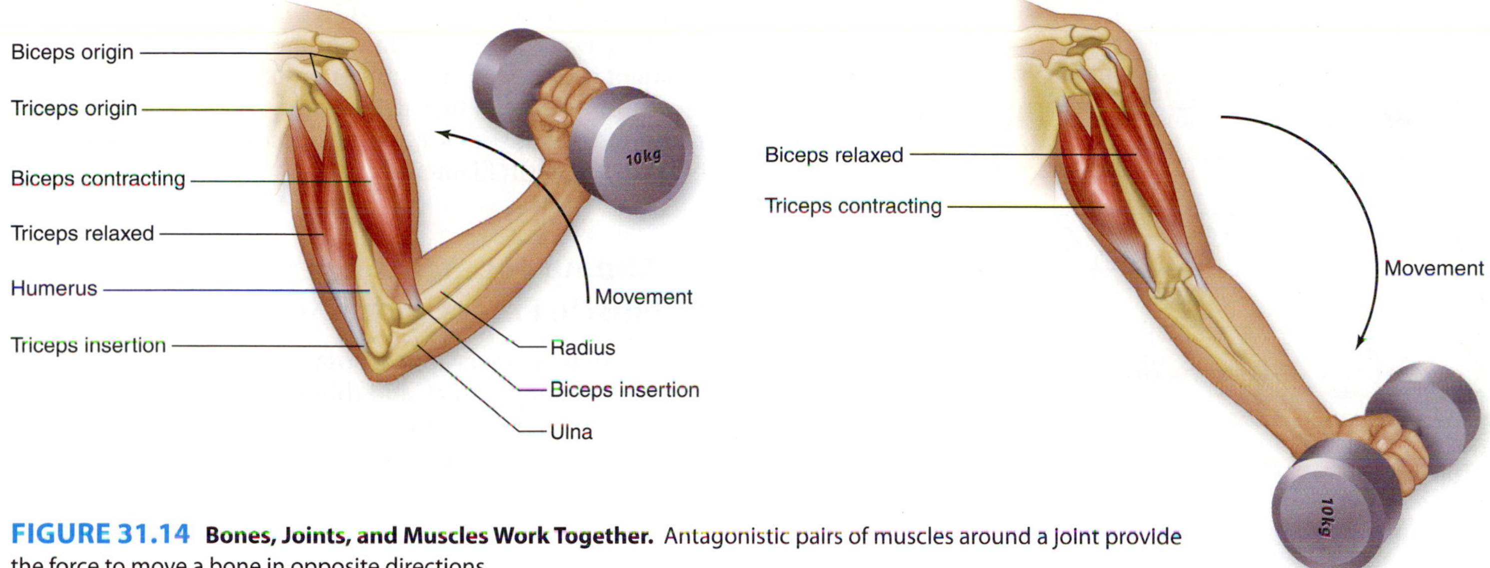

FIGURE 31.14 Bones, Joints, and Muscles Work Together. Antagonistic pairs of muscles around a joint provide the force to move a bone in opposite directions.

a person contracts the biceps (the bulge that appears when you "make a muscle"), the arm bends at the elbow joint. Contraction of the triceps muscle straightens the arm.

A. Actin and Myosin Filaments Fill Muscle Cells

The powerful arm muscles that propel a tennis player's racquet illustrate several levels of organization (**figure 31.15**). Each muscle is an organ consisting of several tissue types. Muscle tissue consists of cells called **muscle fibers,** which are combined into bundles that then form parallel clusters. A connective tissue sheath wraps around each fiber, bundle, and whole muscle. Blood vessels between the bundles of muscle fibers provide nutrients and oxygen, and they remove wastes. Nerves trigger and control muscle contraction. Each muscle lies along the length of a bone, and a tendon attaches each end of the muscle to a different bone.

A skeletal muscle fiber can be more than 30 cm (about 1 foot) long and contains many nuclei. Most of the volume of a muscle fiber consists of hundreds of thousands of cylindrical myofibrils. Each **myofibril,** which runs the length of a muscle cell, is a contractile bundle that consists of two types

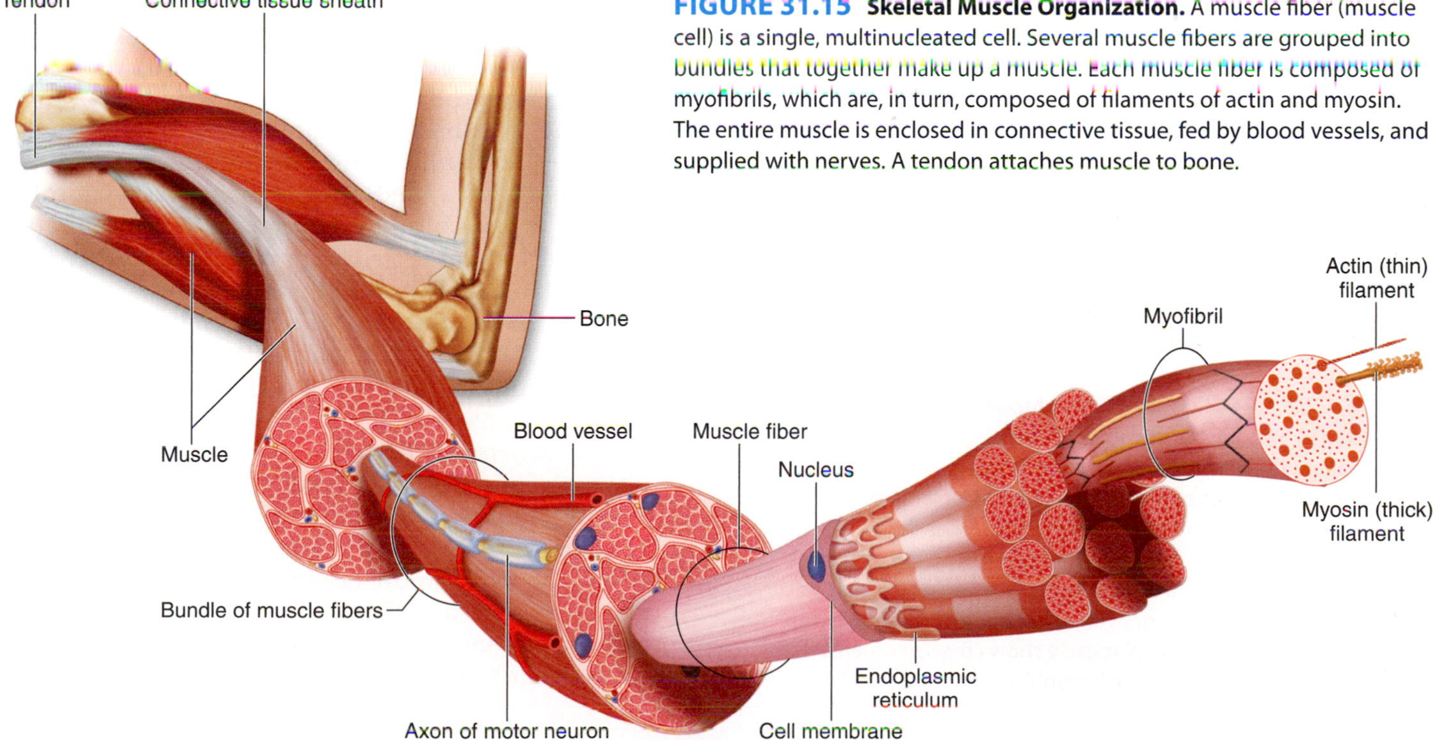

FIGURE 31.15 Skeletal Muscle Organization. A muscle fiber (muscle cell) is a single, multinucleated cell. Several muscle fibers are grouped into bundles that together make up a muscle. Each muscle fiber is composed of myofibrils, which are, in turn, composed of filaments of actin and myosin. The entire muscle is enclosed in connective tissue, fed by blood vessels, and supplied with nerves. A tendon attaches muscle to bone.

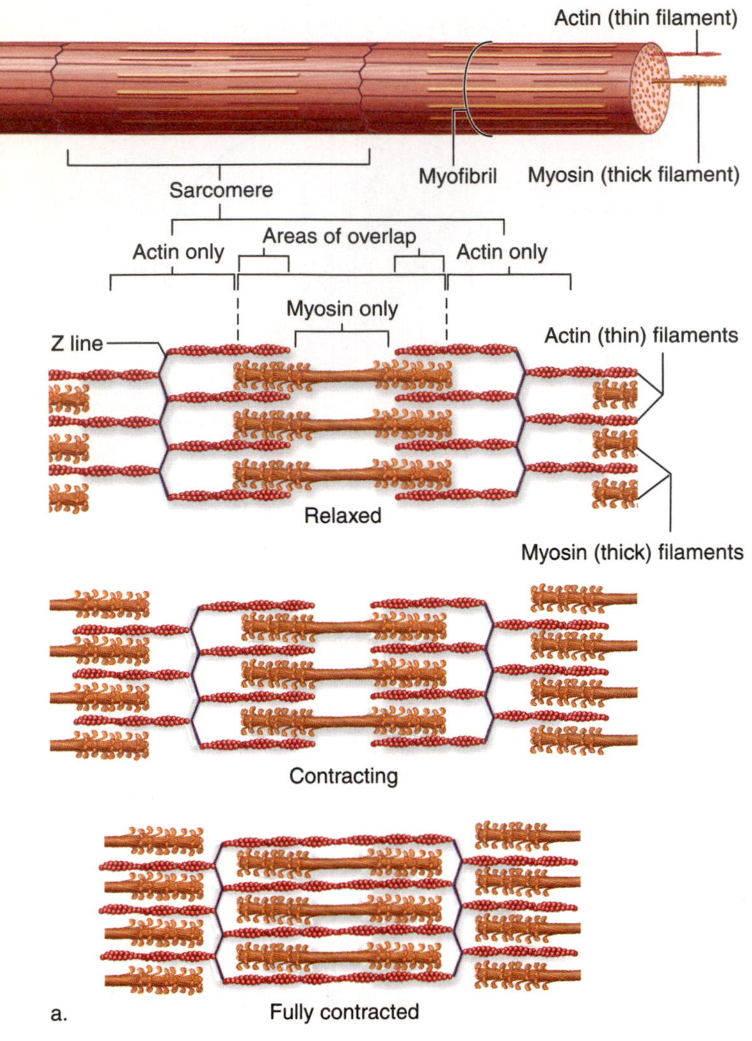

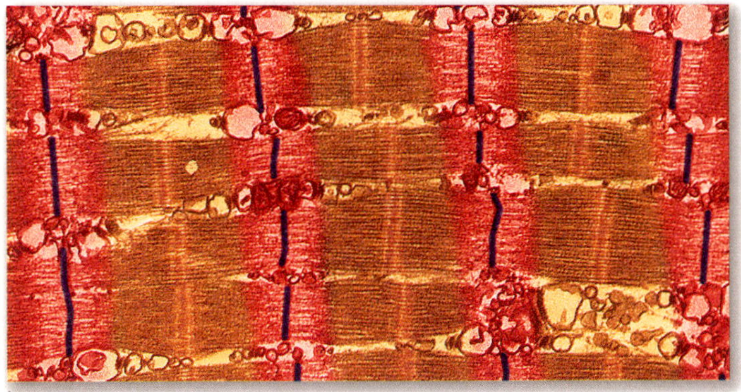
b.

FIGURE 31.16 The Sliding Filament Model of Muscle Contraction. (**A**) Sarcomeres, the units of muscle contraction, occur end to end along the length of a myofibril. Within each sarcomere, six actin thin filaments surround each myosin thick filament. Actin (thin) filaments attach to the membranous boundaries of a sarcomere. Areas where the thick and thin filaments overlap appear dark under a microscope. As a muscle cell contracts, sarcomeres shorten by increasing the area of overlap between actin and myosin. (**B**) This section of striated skeletal muscle shows the sarcomeres that make up the contractile units of each myofibril.

of filaments: thin and thick. A **thin filament** is composed primarily of two entwined strands of the protein **actin;** a **thick filament** is composed of another protein, **myosin.** (You may recall from chapter 3 that actin and myosin also power movements of the cytoskeleton.) Six thin actin filaments surround each thick myosin filament.

B. Sliding Filaments Are the Basis of Muscle Fiber Contraction

Skeletal muscle fibers appear striated because of the orderly arrangement of the thick and thin filaments (**figure 31.16**). A **sarcomere** is one of the many repeated units of a myofibril. **Z lines** define the boundaries of each sarcomere; they are membranes to which thin (actin) filaments attach.

According to the **sliding filament model,** a muscle fiber contracts when the thin filaments slide between the thick ones. This motion shortens the sarcomere; that is, it brings the Z lines closer together without shortening either type of filament. The effect is a little like fitting your fingers together to shorten the distance between your hands.

For thick and thin filaments to move past each other and contract a muscle fiber, actin and myosin must touch. To understand how this occurs, it helps to take a closer look at myosin. Each myosin molecule is shaped like a golf club. Each club-shaped head attaches to its shaft with a hingelike connection that allows the head to pivot. A myosin head forms a **cross bridge** when it swings out and contacts an actin molecule.

ATP provides the energy that powers muscle contraction. First, a myosin head attaches to an actin monomer on a thin filament. The cross bridge bends, which pulls on actin and causes it to slide past myosin in the same way the oar's motion moves a boat. The myosin head then binds a molecule of ATP, which causes the cross bridge to release the actin. (After death, muscles run out of ATP. The cross bridges cannot release from actin, and the muscles remain in a stiff position for several hours. This stiffening is called rigor mortis.) ATP hydrolysis causes the myosin head to swivel back to its original position, ready to contact an actin monomer farther down the thin filament. This sliding repeats about a hundred times per second on each of the hundreds of myosin molecules of a thick filament. Although each individual movement is minuscule, a skeletal muscle contracts quickly and forcefully due to the efforts of many thousands of "rowers." **ATP, p. 85**

Muscle fibers do not continuously contract when ATP is available, because two regulatory proteins control muscle fiber contraction. In a resting muscle, one of the proteins holds the other in a groove between the actin strands, which prevents myosin cross bridges from binding to actin. These proteins block muscle contraction until signals from the nervous system tell them to "let go."

C. Motor Neurons Stimulate Muscle Fiber Contraction

Muscles contract in response to commands from the nervous system. Recall from chapter 28 that **motor neurons** convey signals from the central nervous system to muscles (or glands). A motor neuron releases the neurotransmitter acetylcholine at a **neuromuscular junction,** a synapse between a neuron and a muscle cell (**figure 31.17**). When the neurotransmitter binds to a receptor on the cell surface of the muscle fiber, an electrical wave races along the cell membrane. **synapse, p. 583; action potential, p. 580**

The muscle cell membrane is highly folded into tunnel-like structures extending deep into the cell's interior. These tubules touch the muscle cell's endoplasmic reticulum, which surrounds the myofibrils. When the neural impulse reaches the cell membrane, the endoplasmic reticulum releases calcium ions (Ca^{2+}) into the cytoplasm. The Ca^{2+}, in turn, binds to the regulatory proteins attached to the actin filaments. The proteins move aside, and the muscle is free to contract.

Skeletal muscle fibers also relax on cue. Filament sliding stops when Ca^{2+} is no longer available. Shortly after the release of Ca^{2+}, the ions are actively transported back to the endoplasmic reticulum. The regulatory proteins return to their original position. When actin can no longer interact with myosin, the sarcomere relaxes. **active transport, p. 93**

Muscular dystrophies are inherited diseases caused by absent or abnormal dystrophin, a protein that links the muscle cell membrane to the cytoskeleton. Dystrophin makes up only 0.002% of the total muscle protein, but it is critically important to muscle function. Without its support of the cell membrane, some or all of the body's muscles weaken and fail.

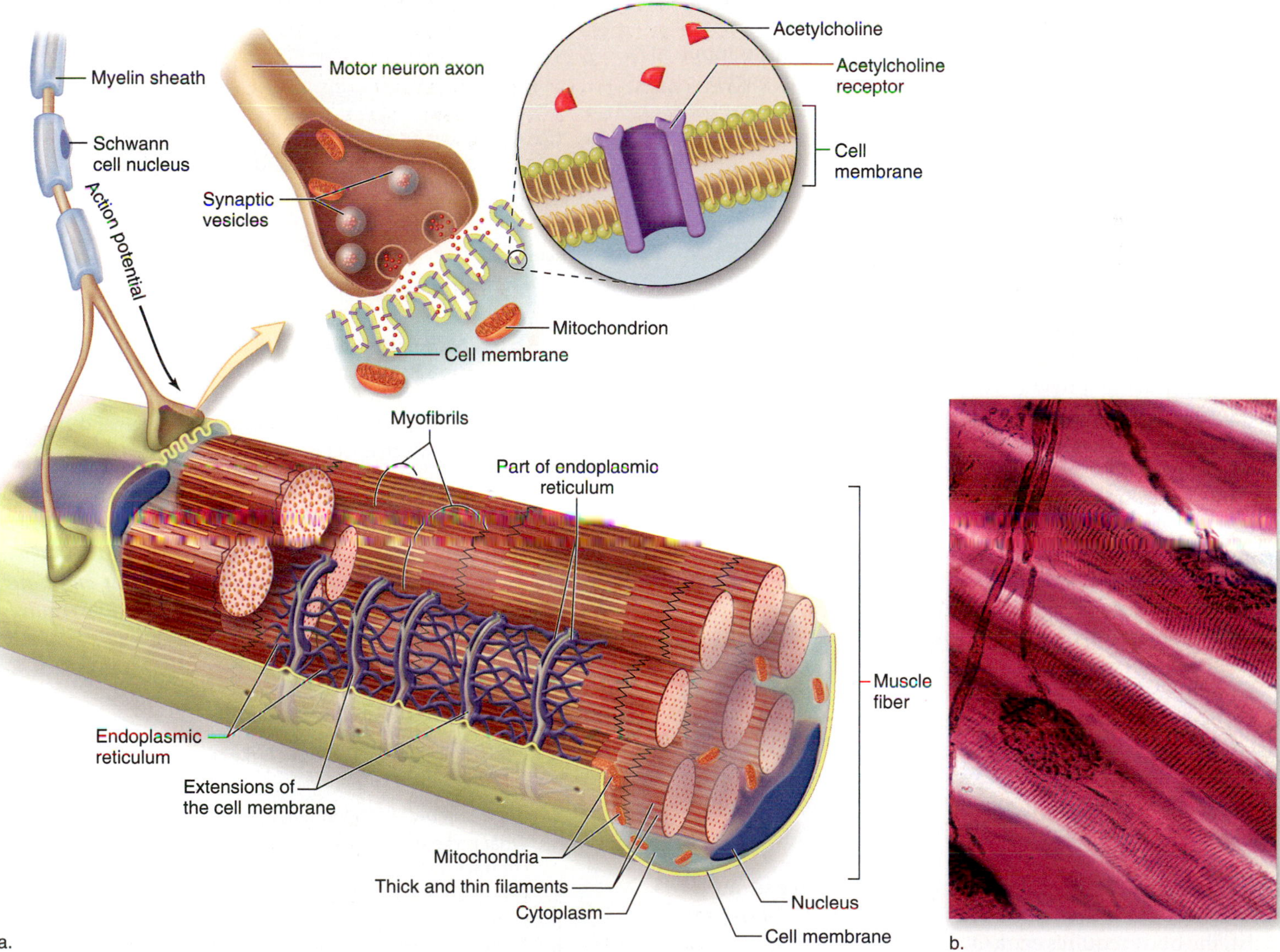

FIGURE 31.17 Neurons Stimulate Muscle Contraction. (**A**) An action potential in a motor neuron's axon releases a neurotransmitter that triggers electrical changes in the muscle cell membrane. Inside the muscle fiber, the highly folded endoplasmic reticulum stores Ca^{2+}. The action potential triggers the release of the Ca^{2+}, leading to contraction. Infoldings of the cell membrane conduct impulses to the interior of the muscle cell. (**B**) This photo shows a neuromuscular junction, a synapse between a neuron and a muscle fiber.

Some infectious diseases can cause muscle paralysis by interfering with the neural signals that stimulate contraction. Botulinum toxin, produced by the bacterium *Clostridium botulinum*, blocks the release of acetylcholine. Ingesting botulinum in tainted foods can cause paralysis, which can be fatal. Injecting tiny amounts of this "botox" into the face, however, temporarily helps reduce the appearance of wrinkles. The virus that causes polio causes paralysis in a different way—it destroys the cell bodies of motor neurons, making them unable to stimulate muscle contraction.

The sliding filament model also applies to the contraction of cardiac and smooth muscle. Cardiac muscle is striated and has visible Z lines. Smooth muscle cells do not appear striated; in these cells, actin and myosin filaments are not arranged into sarcomeres. Both cardiac and smooth muscle are considered involuntary because neither requires stimulation from the central nervous system to contract. Cardiac muscle is "self-excitable;" smooth muscle responds to input from the autonomic nervous system, signals from neighboring cells, or hormones. Chapters 32 and 34 further explain the functions of cardiac and smooth muscle.

D. Muscle Contraction Requires Abundant Energy

Skeletal muscle contraction requires huge amounts of ATP, both to power the return of Ca^{2+} to the endoplasmic reticulum and to break the connection between actin and myosin, allowing a new cross bridge to form. This intense metabolic activity generates abundant heat; skeletal muscle contraction therefore helps maintain body temperature homeostasis.

An animal's body has several ways to generate ATP (**figure 31.18**). A resting muscle cell generates ATP by aerobic respiration (see chapter 6). When muscle activity begins, a molecule called **creatine phosphate** rapidly replenishes ATP by donating a high-energy phosphate to ADP. (This chapter's Burning Question: Is creatine a useful dietary supplement? discusses creatine.) After the supply of creatine phosphate falls, the body must generate ATP from other sources. If exercise is intense, the lungs and blood may not supply enough O_2 to permit aerobic respiration. In that case, the muscle cell switches to fermentation, a less efficient metabolic route that rapidly leads to muscle fatigue. Lactic acid buildup from fermentation pathways causes muscle pain (see figure 6.12).

Heavy breathing for several minutes after intense muscle activity is a sign of oxygen debt. During a period of **oxygen debt,** the body requires extra O_2, both to restore resting levels of ATP and creatine phosphate and to recharge the oxygen-carrying proteins, hemoglobin and myoglobin. (Myoglobin is a red pigment that binds oxygen inside muscle cells.)

hemoglobin, p. 665

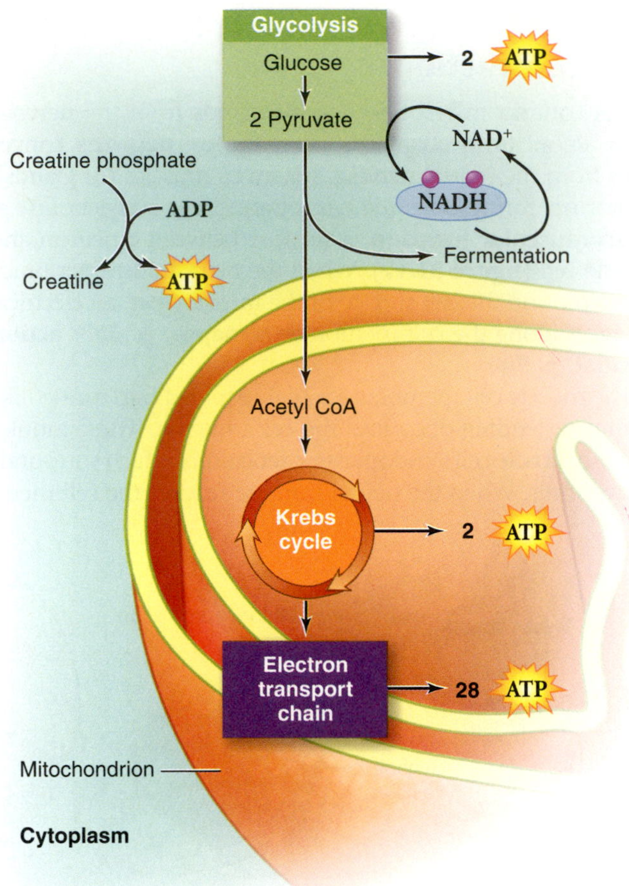

FIGURE 31.18 Energy Sources. Once a muscle cell uses available ATP, it uses its store of creatine phosphate. As long as O_2 remains available, the cell can generate ATP aerobically. If the cardiovascular system cannot keep up with oxygen demand, the cell uses fermentation, which is far less efficient but does not require O_2.

Summary *Tendons attach muscles to bones, forming lever systems that generate movement. Muscle fibers contract when signals from motor neurons stimulate filaments of actin and myosin to slide past each other. Muscle contraction requires abundant ATP.*

31.4 MASTERING CONCEPTS

1. What is an antagonistic pair of muscles?
2. Describe the levels of organization of a vertebrate muscle.
3. Describe how protein interactions move muscles.
4. What is the source of energy that powers muscle contraction?
5. How do motor neurons control skeletal muscle contraction?
6. What is the role of creatine phosphate in muscle contraction?
7. Under what conditions would a muscle cell switch from aerobic respiration to fermentation to generate ATP?

Burning Questions

Is creatine a useful dietary supplement?

You may have seen jars of creatine on nutrition store shelves. Manufacturers market this powder as a muscle-building aid. But does this supplement really work?

Recall that creatine phosphate donates its P to ADP, quickly regenerating ATP soon after muscle activity starts. The idea behind creatine supplementation is that an increase in creatine levels should help skeletal muscle cells generate ATP, providing an energy boost during brief, intense bouts of exercise.

Despite this logic, people differ in their response to creatine supplementation. For some, taking creatine powder increases the amount of creatine phosphate inside skeletal muscle cells. (Vegetarians are most likely to see an effect, because meat and fish are the richest dietary sources of creatine.) But not everyone sees improved athletic performance. Results vary from no effect to small but measurable gains in sprints and other short-term intense exercises.

Long-term creatine supplementation may be harmful. Much of the supplemental creatine ends up in urine, indicating extra stress on the kidneys. The possibility of kidney toxicity requires further study.

Have a Burning Question of your own? Submit it to marielle_hoefnagels@mcgraw-hill.com for possible inclusion in future editions of this book!

31.5 Many Muscle Fibers Combine to Form One Muscle

This chapter has so far described individual muscle fibers, each of which is either contracted or relaxed. But a whole muscle is an organ that may contract a lot, a little, or not at all. How do the activities of individual muscle fibers contribute to the behavior of whole muscles?

A. Neurons Control How Strongly a Muscle Contracts

Electrical stimulation of an isolated muscle produces a **twitch,** a single rapid cycle of contraction and relaxation (**figure 31.19**). Each twitch yields one jerky movement. Yet in the body, the nervous and musculoskeletal systems interact to produce smooth, coordinated movements. Two integrated mechanisms maintain this fine control of whole-muscle movement.

The first mechanism operates at the level of the individual muscle cell. A muscle cell contracts whenever a motor neuron stimulates it, but Ca^{2+} is quickly cleared from the cytoplasm of the muscle fiber after each action potential. For a muscle cell to contract to its full range, repeated action potentials must sustain Ca^{2+} availability. **Tetanus** is the maximal contraction caused by continuous stimulation (not to be confused with the infection caused by *Clostridium tetani*). With a high enough rate of action potentials,

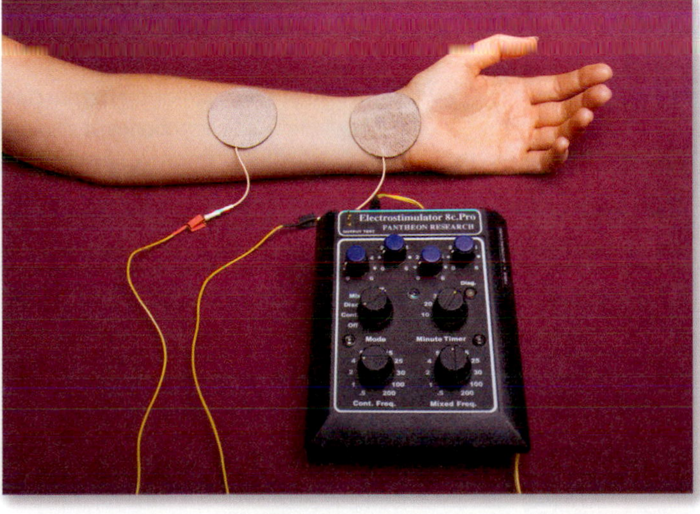

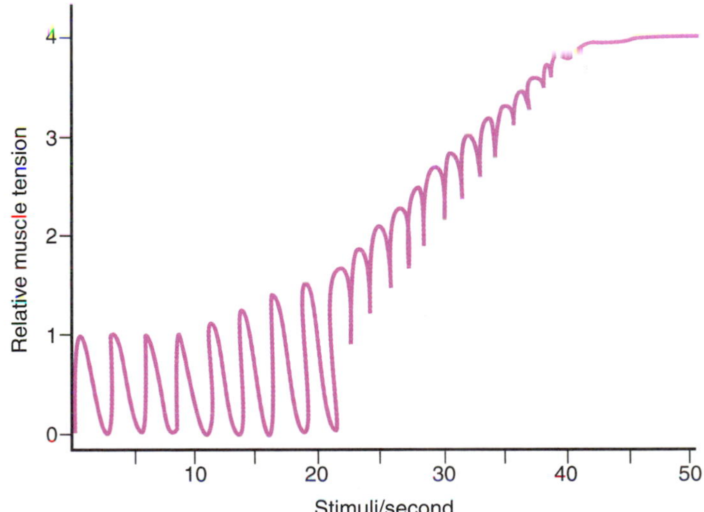

FIGURE 31.19 Muscle Twitch and Tetanus. When a muscle is electrically stimulated at low frequency, it contracts and relaxes following each stimulus. Each such cycle is called a twitch. Under higher frequency stimulation, however, the muscle does not have time to relax between stimuli, so it contracts further. Finally, if the stimulation frequency is high enough, the muscle contracts as much as it can. This sustained maximum contraction is called tetanus.

tetanus ensures smooth, prolonged contractions of individual muscle cells.

The second mechanism that ensures smooth muscle movements involves groups of muscle fibers. Each motor neuron's axon branches extensively at its tip, with each branch leading to a different muscle fiber. A **motor unit** consists of one neuron and all the muscle fibers it contacts. When a neural impulse arrives, all of these fibers contract at the same time. A motor nerve cell that controls only a few muscle fibers produces fine, small-scale responses, such as the eye movements required for reading. A motor unit consisting of hundreds of muscle cells produces large, coarse movements, such as those required for throwing a ball. Within a given muscle, motor units vary in size from tens to hundreds of cells per motor neuron. This is how the same hand can both grip a hammer and pick up a tiny nail.

B. Muscles Contain Fast-Twitch and Slow-Twitch Fibers

Most skeletal muscles contain fibers of two main twitch types, distinguished by how quickly they contract and tire. Slow-twitch (fatigue-resistant) fibers contract slowly because the myosin heads split ATP slowly. They resist fatigue because their plentiful mitochondria are well supplied with O_2 by many molecules of reddish myoglobin. Therefore, these slow-twitch, slow-fatiguing muscles are called red, or dark, fibers. They predominate in muscles that are active for extended periods, such as the flight muscles ("dark meat") of ducks and geese.

Fast-twitch (fatigable) muscle fibers, in contrast, split ATP quickly. These fibers are white because they lack myoglobin and a rich blood supply. They are involved in short bouts of rapid contraction, fueled by anaerobic pathways. The white breast muscle of a chicken, for example, can power barnyard flapping for a short time but cannot support sustained long-distance flight.

The proportion of fast-twitch to slow-twitch fibers affects athletic performance. Most people have about equal numbers of fast- and slow-twitch muscle fibers. Those who have a higher proportion of slow-twitch fibers excel at endurance sports, such as long-distance biking, running, and swimming. Athletes who have a higher proportion of fast-twitch fibers perform best at short, fast events, such as sprinting, weight lifting, and hurling the shot put.

C. Exercise Strengthens Muscles

Regular exercise strengthens the muscular system and enables it to use energy more efficiently (**figure 31.20**). During the few months after a runner begins training, leg muscles noticeably enlarge. This increase in muscle mass comes from the growth of individual skeletal muscle cells rather than from an increase in their number. Exercise-induced muscle growth is even more pronounced in a weight lifter, because the resistance of the weights greatly stresses the muscles. Anabolic steroids boost muscle growth by activating the genes encoding muscle proteins, but with significant side effects. **anabolic steroids, p. 620**

A trained runner's muscle fibers use energy more efficiently than the skeletal muscle cells of an inactive person. The athlete's cells contain more active and more numerous enzymes, and more mitochondria, so his or her muscles can withstand far more exertion before anaerobic metabolism begins. The runner's muscles also receive more blood and store more glycogen than those of an untrained person.

Atrophy is the opposite condition, in which muscles degenerate from lack of use. After just 2 days of inactivity, mitochondrial enzyme activity drops in skeletal muscle cells. After a week without exercise, aerobic respiration efficiency falls by 50%. The number of small blood vessels surrounding muscle fibers declines, lowering the body's ability to deliver O_2 to the muscle. Lactic acid metabolism becomes less efficient, and glycogen reserves fall. After 2 or 3 months of inactivity, the benefits of regular exercise all but disappear.

Summary *A muscle twitch is a sequence of contraction and relaxation. Muscles contract more forcefully when a higher frequency of action potentials arrives from motor neurons. Muscle fibers use ATP at different rates. Exercise strengthens muscles, whereas lack of use causes them to degenerate.*

31.5 MASTERING CONCEPTS

1. How can the same muscle generate both small and large movements?
2. How do slow- and fast-twitch muscle fibers differ?
3. How do muscles grow as a result of increased use?

FIGURE 31.20 **The Benefits of Exercise.** Regular exercise not only strengthens muscles and bones, it also improves flexibility, reduces blood pressure, and elevates mood.

31.6 INVESTIGATING LIFE
Did a Myosin Gene Mutation Make Humans Brainier?

The old admonition not to "bite off more than you can chew" seems to apply especially well to humans. Our chewing muscles are considerably smaller than those of most primates, including chimpanzees and gorillas. We favor soft foods such as bread, cheese, cake, and ice cream, and we would have a hard time chewing the bark, stems, and seeds that some of our primate relatives savor. Fossils show that our chewing muscles are even more delicate than those of the ancestral hominines *Australopithecus* and *Paranthropus*. These differences have spurred researchers to investigate the evolution of the muscles that connect the lower jaw to the other bones of the skull. **fossil hominines, p. 348**

Part of the evidence that has helped scientists understand the evolution of jaw muscles came from an unexpected source. It all began with the Human Genome Project, which has allowed researchers to comb through human DNA in search of particular genes. Hansell H. Stedman and associates from the University of Pennsylvania and The Children's Hospital of Philadelphia hoped to catalog every myosin gene in the human genome, in an effort to better understand diseases such as muscular dystrophy. During their search of chromosome 7, they stumbled on an inactive gene that encoded a nonfunctional myosin protein. **Human Genome Project, p. 251**

This chapter has already explained the role of myosin in muscle contraction. But myosin is not just one protein; it's a family of proteins encoded by at least 40 closely related genes. Myosins participate in cell division, the transport of organelles within a cell, and other cell processes that require movement. Mutations in many of the myosin-encoding genes can lead to loss of muscle function and other serious disorders.

At first, Stedman's team thought the inactive gene they found represented a quirk in the sequences they were searching. They therefore searched the genomes of six, widely dispersed human populations originating in locations ranging from Africa to Iceland. All of the human groups had the mutated gene. The team also compared the gene to a homologous DNA sequence in seven species of nonhuman primates, including gorillas and chimpanzees. The results were clear: the human myosin gene contains a mutation that is not present in nonhuman primates (**figure 31.21**).

Next, the researchers wanted to know which cells in primates express the gene. Recall from chapter 12 that the first step in gene expression is transcription, in which a gene's DNA sequence is transcribed to mRNA. A search for mRNA showed that, in both humans and macaque monkeys, only two muscles express the gene, and both of those muscles are involved in the up-and-down jaw movements required for chewing. Because the protein is nonfunctional in humans but functional in macaques, this finding suggested that an ancient genetic mutation made our chewing muscles small and weak, at least compared with our closest relatives.

That result prompted Stedman's research team to learn more about the chewing apparatus in fossil primates. Most fossils consist of bones, which bear marks indicating where

```
Non-human                  10        20        30        40        50        60        70        80
  Woolly monkey    GAGCAGCTGAACAAGCTGATGACCACCCTCCACAGCACTGTACCCCATTTTGTCCGCTGTATTGTGCCCAATGAGTTTAAGCAGTCAG
  Pigtail macaque  GAGCAGCTGAACAAGCTGATGACCACCCTCCACAGCACTGTACCCCATTTTGTCCGCTGTATTGTGCCCAATGAGTTTAAGCAGTCAG
  Rhesus           GAGCAGCTGAACAAGCTGATGACCACCCTCCACAGCACTGTACCCCATTTTGTCCGCTGTATTGTGCCCAATGAGTTTAAGCAGTCAG
  Orangutan        GAGCAGCTGAACAAGCTGATGACCACCCTCCACAGCACTGTACCCCATTTTGTCCGCTGTATTGTGCCCAATGAGTTTAAGCAGTCAG
  Gorilla          GAGCAGCTGAACAAGCTGATGACCACCCTCCACAGCACTGTACCCCATTTTGTCCGCTGTATTGTGCCCAATGAGTTTAAGCAGTCAG
  Bonobo           GAGCAGCTGAACAAGCTGATGACCACCCTCCACAGCACTGTACCCCATTTTGTCCGCTGTATTGTGCCCAATGAGTTTAAGCAGTCAG
  Chimpanzee       GAGCAGCTGAACAAGCTGATGACCACCCTCCACAGCACTGTACCCCATTTTGTCCGCTGTATTGTGCCCAATGAGTTTAAGCAGTCAG

Human
  Africa (pygmy)   GAGCAGCTGAACAAGCTGATGACCACCCTCCATAGC--CGCACCCCATTTTGTCCGCTGTATTATCCCCAATGAGTTTAAGCAATCGG
  Spain (Basque)   GAGCAGCTGAACAAGCTGATGACCACCCTCCATAGC--CGCACCCCATTTTGTCCGCTGTATTATCCCCAATGAGTTTAAGCAATCGG
  Iceland          GAGCAGCTGAACAAGCTGATGACCACCCTCCATAGC--CGCACCCCATTTTGTCCGCTGTATTATCCCCAATGAGTTTAAGCAATCGG
  Japan            GAGCAGCTGAACAAGCTGATGACCACCCTCCATAGC--CGCACCCCATTTTGTCCGCTGTATTATCCCCAATGAGTTTAAGCAATCGG
  Russia           GAGCAGCTGAACAAGCTGATGACCACCCTCCATAGC--CGCACCCCATTTTGTCCGCTGTATTATCCCCAATGAGTTTAAGCAATCGG
  South America    GAGCAGCTGAACAAGCTGATGACCACCCTCCATAGC--CGCACCCCATTTTGTCCGCTGTATTATCCCCAATGAGTTTAAGCAATCGG
```

FIGURE 31.21 Myosin Mutation. The DNA sequences for a small portion of the myosin gene are shown for nonhuman primates (upper seven sequences) and humans (lower six sequences). Note the two-base deletion just after position 36 in the human gene. This frameshift mutation causes the encoded protein to be truncated (shortened) and nonfunctional. **Question**: Use what you learned in chapter 12 to predict the amino acid sequence corresponding to the two gene sequences. How does the deletion result in a truncated protein?

the muscles were once attached. Until about 2 MYA, primates had large, robust skull bones and chewing muscles. These large complexes occurred in *Australopithecus* and *Paranthropus*, as well as in contemporary primates such as macaques and gorillas (**figure 31.22**). A more delicate chewing apparatus appeared in *Homo erectus/ergaster*, however, between 1.8 and 2.0 MYA.

The researchers used a "molecular clock" technique to estimate that the mutation in the myosin gene occurred approximately 2.4 MYA—before the migration of *Homo* out of Africa. The timing coincides with a significant trend in human evolution: the increasing size of the brain. Stedman and his colleagues argue that the myosin mutation may have changed the muscles in a way that released a constraint on the size of the brain. The increased brain power may have eventually led to the spread of culture, profoundly changing the course of human evolution.

Of course, one mutation in a myosin gene is, by itself, not likely to have set in motion the entire course of human history. Other changes in the skeletal and nervous systems must also have occurred to spur the evolution of the brain. Researchers also continue to debate how the myosin gene mutation became "fixed" in the human population. Did it become more common as humans began eating softer foods, or did humans only begin eating softer foods once the gene mutation occurred?

The story of this myosin gene illustrates how a seemingly routine study can have thoroughly unexpected results; what began as a simple search through the human genome ended up having much wider implications for the study of human evolution. It also points to the connections between different areas of biology. To tell their story, Stedman and his colleagues needed to understand both the muscular and the skeletal systems. In the process, they combined detailed observations of ancient fossils with modern studies of gene mutation and protein function. Their results give us something to chew on as we contemplate the evolutionary history of our own species.

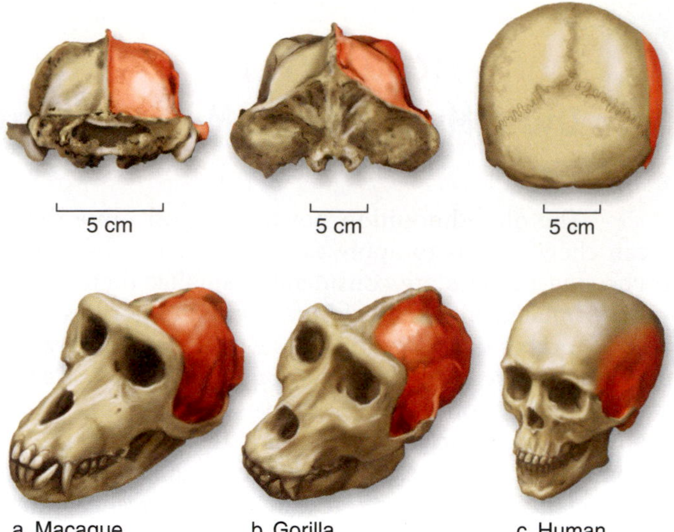

FIGURE 31.22 Primate Comparison. The temporalis muscle is one of the four main chewing muscles. In macaques and gorillas, the temporalis attachment area (*red*) occupies most of the side of the skull. In humans, the temporalis attachment area is much smaller, resulting in a much weaker chewing apparatus.

Stedman, Hansell, Benjamin W. Kozyak, Anthony Nelson, et al. 2004. Myosin gene mutation correlates with anatomical changes in the human lineage. *Nature*, vol. 428, pages 415–418.

CHAPTER SUMMARY

 31.1 Skeletons Take Many Forms

- The **muscular system** and the **skeletal system** together enable an animal to move.
- An animal's **skeleton** supports its body and protects soft tissues. **Muscles** acting on the skeleton enable an animal to move.
- A **hydrostatic skeleton** consists of tissue containing constrained fluid. An **exoskeleton** is on the organism's exterior, and an **endoskeleton** forms inside the body.

31.2 The Backbone Is Central to the Vertebrate Skeleton

- The **axial skeleton** consists of the skull, **vertebral column,** breastbone, and ribs.
- The **appendicular skeleton** includes the limbs and the limb girdles (**pectoral** and **pelvic**) that attach them to the axial skeleton.

 31.3 Bones Provide Support, Protect Internal Organs, and Supply Calcium

- **Bones** are strong and lightweight because they are porous. Long bones have a **marrow cavity** that helps reduce weight.

A. A Bone Consists Mostly of Bone Tissue and Cartilage

- Bone tissue derives its strength from collagen and its hardness from minerals. Bone cells include **osteoblasts, osteoclasts,** and **osteocytes.** In **compact bone,** osteocytes are arranged in concentric rings that form **osteons. Spongy bone** has few osteons but many spaces.
- Cartilage consists of chondrocytes embedded in a matrix of collagen and elastin. It entraps a great deal of water, which makes it an excellent shock absorber.

B. Bones Are Constantly Built and Degraded

- Bone continually degenerates and builds up, based on the opposing activities of osteoclasts and osteoblasts.
- Weight-bearing exercise strengthens bones; conversely, bones weaken with disuse.

C. Bones Help Regulate Calcium Homeostasis

- Hormones control the exchange of calcium between blood and bones, maintaining homeostasis. **Osteoporosis** results when a person's bones lose more calcium than they replace.

D. Bone Meets Bone at a Joint

- **Joints** attach bones to each other. Some joints, such as those holding the skull bones in place, are immovable.
- Freely moving **synovial joints** consist of cartilage and a connective tissue capsule that contains lubricating fluid. **Ligaments** and **tendons** stabilize the joint.

31.4 Muscle Movement Requires Contractile Proteins and ATP

- When a skeletal muscle contracts, a bone moves. Muscles form antagonistic pairs, which enable bones to move in two directions.

A. Actin and Myosin Filaments Fill Muscle Cells

- **Skeletal muscle** fibers are elongated, multinucleate, striated, and voluntary.
- Each skeletal **muscle fiber** is a long, cylindrical cell that contains **myofibrils** composed of two types of protein filaments. The **thick filaments** are **myosin,** and the **thin filaments** are composed primarily of **actin.**

B. Sliding Filaments Are the Basis of Muscle Fiber Contraction

- A myofibril is a chain of contractile units called **sarcomeres. Z lines** define the borders of each sarcomere. The orderly arrangement of thick and thin filaments within a sarcomere gives skeletal muscle tissue its striated appearance.
- According to the **sliding filament model,** muscle contraction occurs when thick and thin filaments move past one another. Once myosin **cross bridges** touch actin, ATP attached to the myosin head splits. The head moves, causing the actin filament to slide past the myosin filament. A new ATP then binds to the myosin head, and the cross bridge to actin breaks. The myosin head returns to its original position, and a new cross bridge forms further along the filament.

C. Motor Neurons Stimulate Muscle Fiber Contraction

- When a **motor neuron** stimulates a muscle fiber, acetylcholine is released at a **neuromuscular junction.** Electrical waves spread along the muscle cell membrane, causing the endoplasmic reticulum to release Ca^{2+}, which binds to regulatory proteins that normally block muscle contraction. As a result, actin can bind to myosin, and the muscle can contract.
- After the electrical impulse, Ca^{2+} is actively pumped back into the endoplasmic reticulum, and the regulatory proteins once again prevent actin and myosin interactions. The muscle relaxes.

D. Muscle Contraction Requires Abundant Energy

- The energy that powers muscle contraction comes first from stored ATP, then from **creatine phosphate** stored in muscle cells, and then from aerobic respiration, and finally from fermentation.
- **Oxygen debt** is a temporary deficiency of O_2 after intense exercise.

31.5 Many Muscle Fibers Combine to Form One Muscle

A. Neurons Control How Strongly a Muscle Contracts

- When a muscle cell is stimulated once, it contracts and relaxes (a **twitch**). At a high rate of stimulation, muscle cells reach a sustained state of maximal contraction, **tetanus.**
- A nerve cell and the muscle fibers it touches form a **motor unit.** The more motor units stimulated, the greater the contraction of the muscle.

B. Muscles Contain Fast-Twitch and Slow-Twitch Fibers

- Slow-twitch muscle fibers use ATP slowly and regenerate it by aerobic respiration. Fast-twitch cells use ATP quickly and use mostly anaerobic pathways to replenish it.
- People vary in their proportion of fast- and slow-twitch muscles.

C. Exercise Strengthens Muscles

- A muscle exercised regularly increases in size because each muscle cell thickens. An unused muscle shrinks (atrophies). Regular exercise causes changes in muscle cells that enable them to use energy more efficiently.

31.6 Investigating Life: Did a Myosin Gene Mutation Make Humans Brainier?

- Researchers have discovered in humans a mutated myosin gene that is expressed only in the muscles required for chewing. Nonhuman primates lack the mutation.
- The evolution of weaker, smaller chewing muscles may have paved the way for increased brain capacity in humans.

MULTIPLE CHOICE QUESTIONS

1. A hydrostatic skeleton
   a. is only found in invertebrates.
   b. is filled with fluid.
   c. changes shape due to muscle contraction.
   d. All of the above.
2. Exoskeletons differ from endoskeletons in
   a. the material that they are composed of.
   b. their ability to change as an organism grows.
   c. their role in forming a framework for muscle attachment.
   d. both a and b.
3. The bones of your hand are part of the ______, whereas your backbone is part of the ______.
   a. axial skeleton; pectoral girdle
   b. appendicular skeleton; axial skeleton
   c. axial skeleton; appendicular skeleton
   d. axial skeleton; vertebral column
4. Which cell type is most likely to cause the bone loss associated with osteporosis?
   a. Osteoblasts
   b. Osteocytes
   c. Osteoclasts
   d. Both a and b
5. Which type of joint connects movable bones?
   a. Fibrous joint
   b. Synovial joint
   c. Cartilaginous joint
   d. Both b and c
6. Muscle tissue is made up of what type of cells?
   a. Sarcomeres
   b. Myofibrils
   c. Myomeres
   d. Muscle fibers
7. How does a muscle change from being contracted to being relaxed?
   a. The myosin heads of the thick filaments work in reverse.
   b. The actin thin filaments stretch.
   c. The sarcomeres of an antagonistic muscle contract.
   d. Both a and b.
8. What is the source of energy for muscle contraction?
   a. ATP
   b. Acetylcholine
   c. Creatine phosphate
   d. Calcium
9. Why do slow-twitch muscle cells have a high concentration of myoglobin?
   a. To provide oxygen to the mitochondria to promote ATP synthesis
   b. To help promote the growth of the muscle cells in response to exertion
   c. To prevent tetanus
   d. To slow down the rate of ATP hydrolysis by the myosin
10. Which of the following is NOT a value of exercise?
    a. Increased bone density
    b. Enhanced muscle cell metabolism
    c. Increase in the number of muscle cells
    d. Increase in the size of the muscle cells

TESTING YOUR KNOWLEDGE

1. Distinguish among a hydrostatic skeleton, an exoskeleton, and an endoskeleton. Give an example of an animal with each.
2. What advantages and disadvantages does a jointed exoskeleton have over a shell? What advantages and disadvantages does an endoskeleton have over a jointed exoskeleton?
3. What are the main components of the axial and appendicular skeletons?

4. What role does cartilage play in the vertebrate skeletal system?
5. What are the structural differences between spongy bone and compact bone?
6. What are the two major components of bone matrix? How do they work together to give bone its characteristics?
7. How do antagonistic muscle pairs move bones? Give an example of such a pair.
8. What is the role of calcium in bones? In muscle contraction?
9. Describe two muscle proteins and their functions.
10. How does the muscular system interact with each of the following organ systems?
    a. the nervous system
    b. the skeletal system
    c. the circulatory system
11. How do the effects of exercise (or lack thereof) illustrate homeostasis in muscles and bones?

THINKING AS A SCIENTIST

1. How can the musculoskeletal systems of diverse vertebrate species consist of the same molecules, cells, and tissues, and have similar organization, yet also be adapted to a species' way of life?
2. What roles does fluid play in hydrostatic skeletons, cartilage, and synovial joints?
3. Would an overactive parathyroid prevent osteoporosis, or contribute to it? Explain your answer.
4. The following table shows recent mens' world-record times for various running events. Graph the distance traveled against the average running speed, in meters per second. How does muscle use of ATP over time explain the graph?

Distance (m)	Time	Average meters per second
100	9.78 s	10.22
200	19.32 s	10.35
400	43.18 s	9.26
800	1 min, 41.11 s	7.91
1500	3 min, 26.00 s	7.28
5000	12 min, 37.37 s	6.60

5. The aerobics instructor chants, "Just concentrate, and feel your muscles expand and contract." Is her statement an accurate description of muscle action? Why or why not?
6. In some European nations, the ratio of fast- to slow-twitch fibers is used as a predictor of athletic success. Athletes give small samples of muscle tissue to test for twitch fiber proportions. How would the ratio probably differ between a champion sprinter and a long-distance runner?
7. A father wants to test his healthy daughter's muscles to determine the percentage of slow-twitch and fast-twitch muscle fibers to decide if she should try out for soccer or cross-country (long-distance) running. Do you think this is a valid reason to test muscle tissue? Why or why not?
8. Explain how each of the following would affect musculoskeletal function:
    a. a cramp (involuntarily contracted muscle)
    b. a pulled muscle (torn muscle fibers)
    c. tendonitis (inflammation of a tendon)
    d. osteogenesis imperfecta (inherited disease that causes a lack of collagen in bones)
9. Meat packers sometimes add carbon monoxide gas (CO) to packaged meat. The gas binds to myoglobin in the meat and keeps it from turning brown when exposed to air. This practice is controversial, because it may mask the color change that is an early sign of spoilage. Who benefits from this practice, and what are the risks? Do you advocate labelling CO-treated meat?

ARIS™ **Visit www.mhhe.com/hoefnagels for practice quizzes, animations, videos, and activities designed to help you master the material in this chapter.**

The Circulatory System

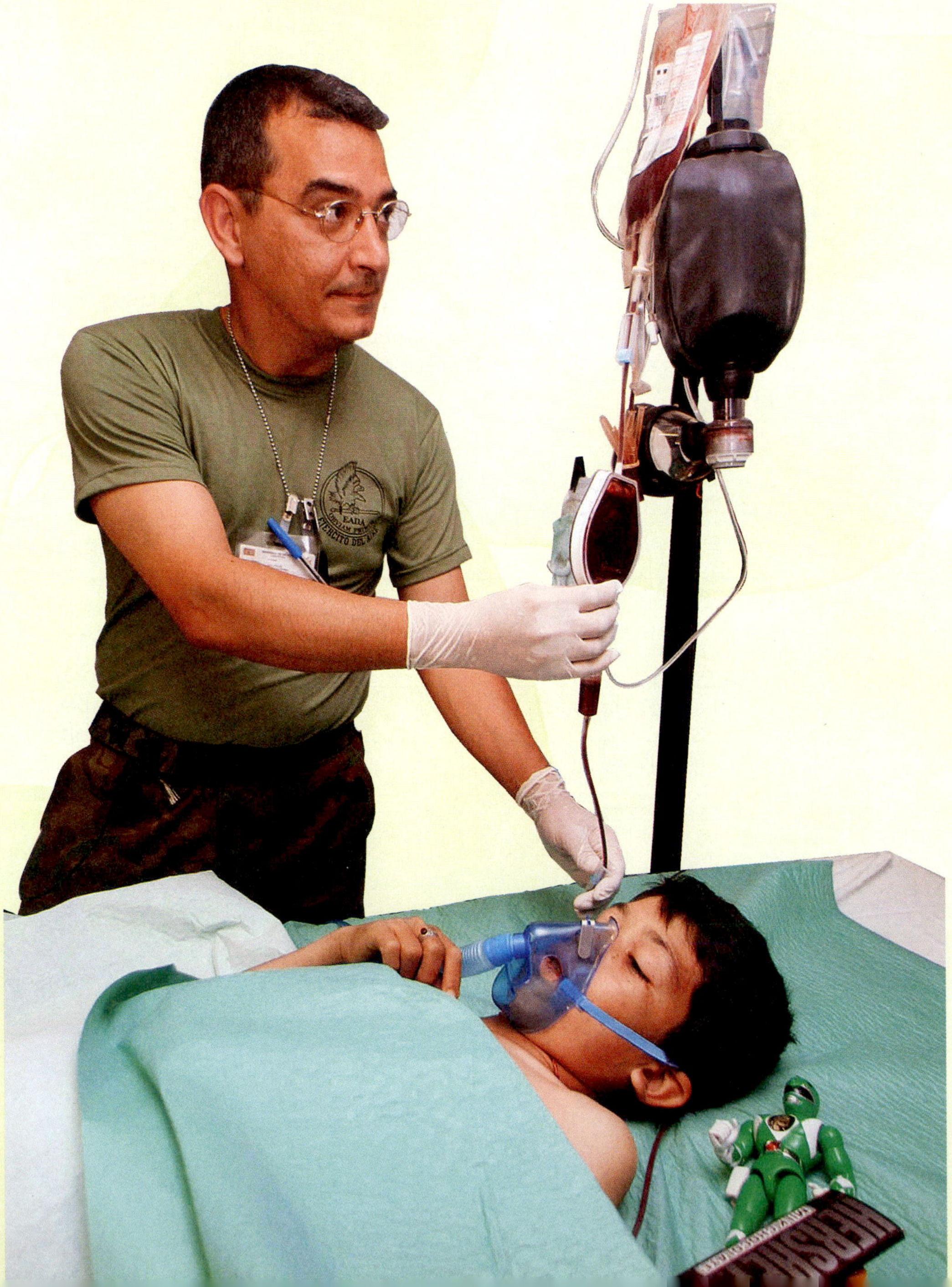

River of Life. A 12-year-old boy with a severe blood disorder receives a transfusion.

What's the Point?

Using Stem Cells to Replace Hearts and Blood

The heart is the circulatory system's only pump; failure to do its job can be fatal. Treatment options for heart failure currently include surgery, drugs, and lifestyle changes, but in the future, physicians may have a new tool: stem cells.

Scientists may one day be able to coax stem cells to divide and produce daughter cells that differentiate into exactly what is needed to heal damaged tissue. Hearts transplanted from women to men provide evidence that this differentiation happens naturally. In one study, at varying times after the transplant, researchers detected cells containing Y chromosomes in the donor hearts from women. This meant that the male recipient's cells had migrated to the donor heart and specialized into connective tissue, cardiac muscle tissue, and epithelium—precisely what was required to accept the new part.

A heart is useless without blood to transport. Blood transfusions have been common since the nineteenth century, but donors may transmit diseases to the recipients of the blood. Artificial blood, or a blood substitute, would eliminate this risk.

Blood substitutes used (unsuccessfully) in the past include wine, milk, urine, plant resins, and even opium. More recent efforts try to copy the oxygen-carrying capacity of the red blood cells. Chemicals called perfluorocarbons, for example, carry dissolved O_2. *In the 1960s, a famous photo showed a mouse swimming in the stuff, literally breathing in the oxygen-rich chemical. In another type of blood substitute, the* O_2*-carrying hemoglobin molecules from red blood cells are isolated and then linked in various ways. A cow hemoglobin preparation saved the life of a young woman whose immune system was attacking her own blood, maintaining her circulation for several days until the illness subsided.*

Future red blood cell replacements may come from stem cells. In one experiment, researchers placed stem cells from human umbilical cord blood in a laboratory dish. They added growth factors that directed development toward forming mature red blood cells. The cells divided and developed until just before the final stage, when they jettison their nuclei. The researchers then infused the cells into mice that lack immune systems, and the animals soon produced mature human red blood cells. Not only did the new cells lack nuclei, but the hemoglobin they produced was of the adult variety, not like the fetal hemoglobin that was in the original cord blood. The stem cells had responded to signals from the host to differentiate.

Stem cell research is a young science, and researchers still have much to learn before the medical applications become widespread. Success will require a basic understanding of how the body works, including the circulatory system—the subject of this chapter.

LEARNING OUTLINE

32.1 Circulatory Systems Deliver Nutrients and Remove Wastes

Watch a crime drama on TV, and it won't be long until a stabbing or gunshot wound leaves someone lying in an alarming pool of bright red blood. Life quickly fades from the unfortunate victim, a vivid reminder of blood's critical importance (**figure 32.1**).

A **circulatory system** transports blood (or a comparable fluid) in a one-directional circuit throughout the body. **Blood** is a liquid connective tissue with many functions, including the delivery of nutrients, water, and oxygen. Without these resources, the body's cells could not generate the ATP required for life. A central pump, the **heart,** forces the fluid throughout the body. Most animals have some type of circulatory system that transports nutrients, O_2, and wastes.

A. Circulatory Systems Are Open or Closed

Circulatory systems are of two basic types: open and closed (**figure 32.2**). In an **open circulatory system,** a heart pumps fluid through short, open-ended vessels that lead to spaces where the fluid bathes the body's cells. Most mollusks and arthropods have open circulatory systems. **mollusks, p. 458; arthropods, p. 465**

In a **closed circulatory system,** blood remains within vessels. **Arteries** conduct blood away from the heart. They branch into **arterioles,** smaller vessels that then diverge into a network of **capillaries,** the body's tiniest blood vessels. Water and dissolved substances diffuse between the capillary and the **interstitial fluid,** the liquid that bathes the body's cells. The interstitial fluid exchanges materials with the tissue cells. Capillaries empty into slightly larger vessels, called **venules,** which unite to form the **veins** that carry blood back to the heart.

Annelids are the simplest animals with a closed circulatory system. Cephalopod mollusks such as squids and octopuses have closed circulatory systems too, as do vertebrates. **annelids, p. 460**

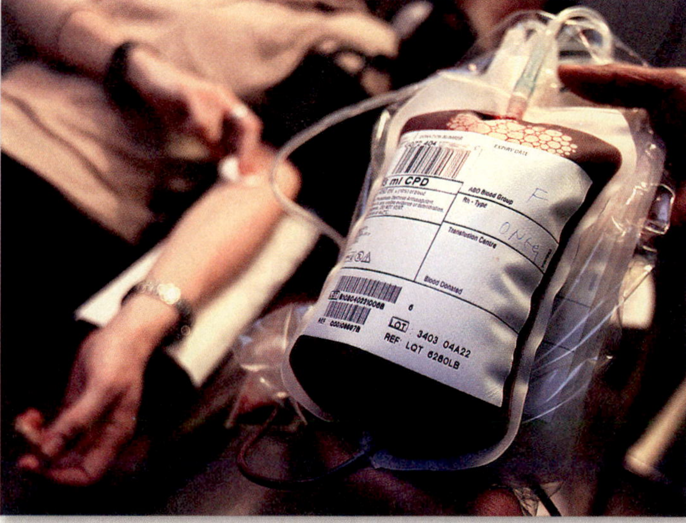

FIGURE 32.1 Vital Fluid. Millions of people in the United States receive life-saving blood transfusions each year as a result of injury, surgery, or cancer treatment.

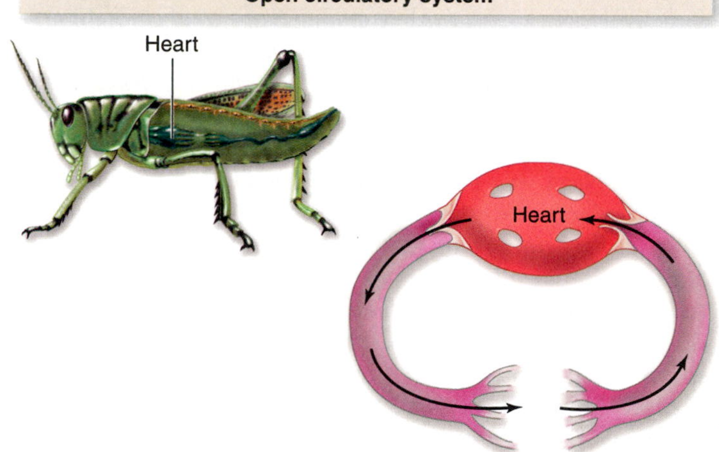

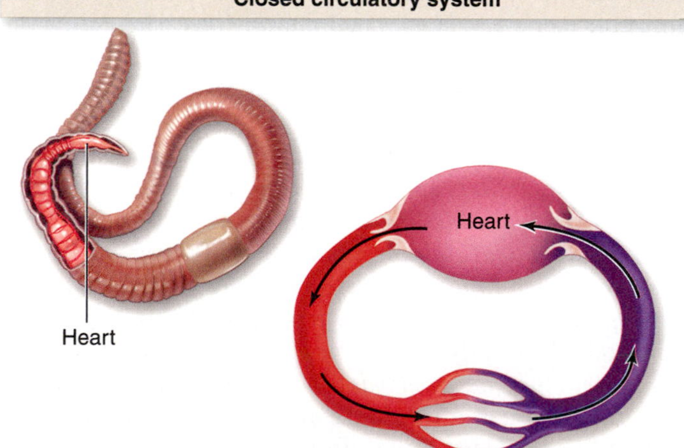

FIGURE 32.2 Open and Closed Circulatory Systems. (**A**) In an open circulatory system, fluid leaves vessels and bathes cells directly. (**B**) In a closed circulatory system, blood flows within vessels.

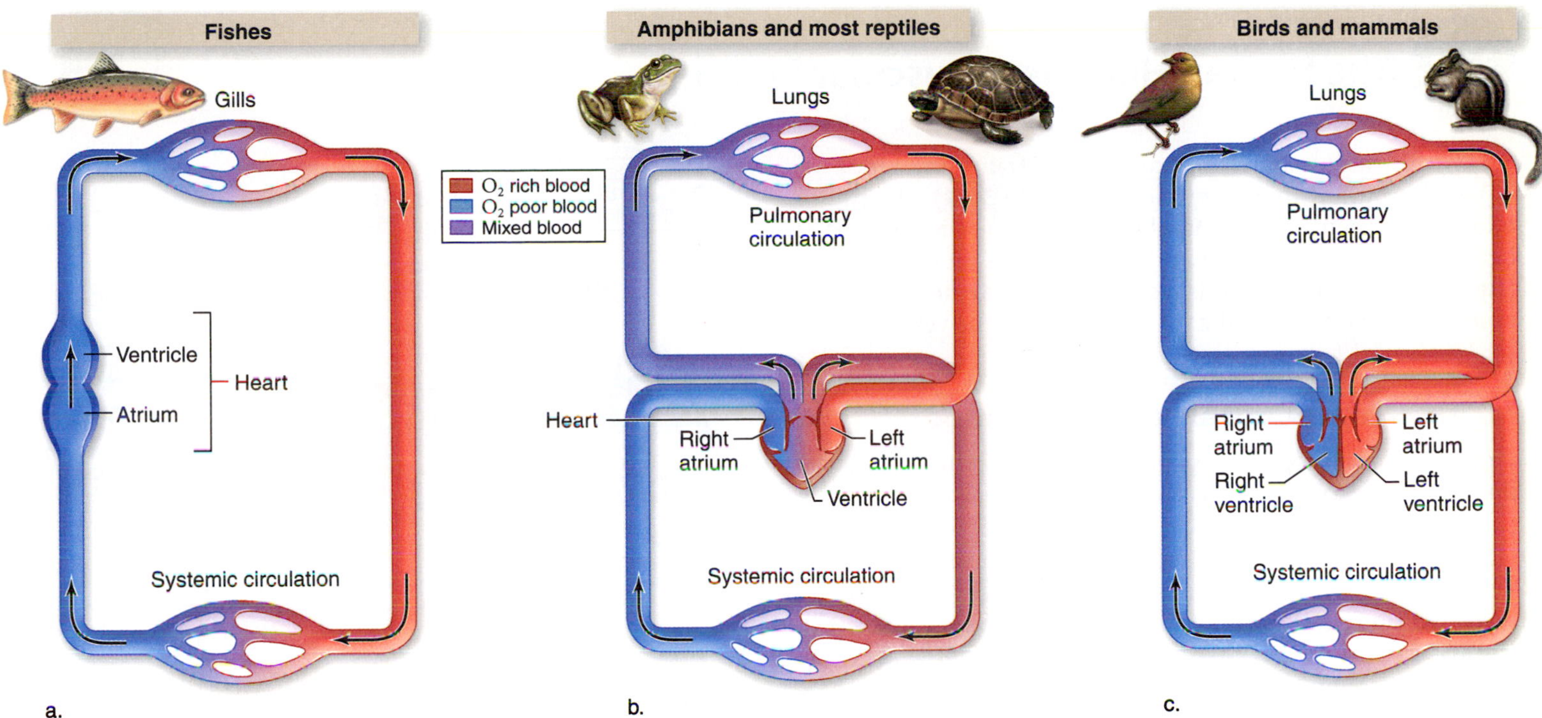

FIGURE 32.3 Vertebrate Circulatory Systems. (**A**) A fish's two-chambered heart allows blood to flow in a single circuit around the body. (**B**) An amphibian's heart has three chambers: two atria and a large ventricle. (**C**) A bird or mammal has a four-chambered heart, supporting two trips through the heart in each circuit. This more powerful pumping system enables birds and mammals to be more active.

B. Natural Selection Has Shaped Vertebrate Circulatory Systems

Among vertebrates, circulatory systems have become increasingly complex over time. A fish's heart has just two chambers (**figure 32.3**A): an **atrium** (where blood enters) and a **ventricle** (from which blood exits). A fish's blood flows through the heart only once in each one-way circuit around the body. The heart pumps oxygen-poor blood through the gills to pick up oxygen, and then to the rest of the body before it returns to the heart.

In amphibians, reptiles, birds, and mammals, blood makes two trips through the heart in each circuit around the body (figure 32.3B, C). In the **pulmonary circulation,** blood circulates to the lungs and back to the heart; in the **systemic circulation,** blood circulates throughout the rest of the body and back to the heart.

The three-chambered heart of an amphibian has two atria and one ventricle. Oxygenated and deoxygenated blood mix in the ventricle, making the amphibian heart relatively inefficient. These animals obtain much of their O_2 through the mouth surface and skin. In most reptiles, the heart also has three chambers, but the ventricle is partially divided. A divided ventricle helps to separate oxygenated from deoxygenated blood. The metabolic rate, blood pressure, and O_2 requirements of reptiles exceed those of amphibians.

Birds and mammals independently evolved four-chambered hearts with two atria and two ventricles. Crocodiles, which share a clade with birds, have similar hearts. The four-chambered heart separates oxygenated blood from deoxygenated blood more effectively than the amphibian or reptilian heart. This arrangement maximizes the amount of oxygen reaching tissues, an adaptation that benefits an endothermic ("warm-blooded") animal with enormous energy needs.

Summary *Circulatory systems pump blood (or similar fluid) throughout the body. Open circulatory systems have hearts and vessels that lead to spaces between cells; closed circulatory systems contain blood in a system of vessels. Vertebrate hearts have increased in complexity over evolutionary time.*

32.1 MASTERING CONCEPTS

1. What are the components of a circulatory system?
2. Which animals have open circulatory systems, and which have closed circulatory systems?
3. Distinguish among arteries, veins, arterioles, venules, and capillaries.
4. Are vertebrate circulatory systems open or closed?
5. Describe the circulatory systems of fishes, amphibians, reptiles, birds, and mammals.

32.2 The Human Heart Is a Muscular Pump

The human circulatory system transports blood, a complex fluid with many functions (**table 32.1**). The two main components of the circulatory system are a fist-sized heart and a continuous network of tubes, the blood vessels (**figure 32.4**). This system is also called the **cardiovascular system** (*cardio-* refers to the heart, *vascular* to the vessels).

A. The Heart Contains Four Chambers

The heart is enclosed in the **pericardium** (meaning "around the heart"), a tough connective tissue sac that consists of two layers with fluid between them. The pericardium protects the heart but allows it to move, even during vigorous beating. The wall of the heart itself is three layers thick. The **myocardium,** or muscular middle layer, is by far the thickest (**figure 32.5**). Contraction of the cardiac muscle that makes up the myocardium provides the force that propels blood. The other two layers of the heart wall consist of thin sheets of epithelium.

The heart has two upper chambers (atria) and two lower chambers (ventricles). The atria are "primer pumps" that send blood to the ventricles, which pump the blood throughout the body. The **atrioventricular valves (AV valves)** are thin flaps of tissue that prevent blood from moving back into the atrium when a ventricle contracts. The other two heart valves, the **semilunar valves,** prevent blood in arteries from reentering the ventricles.

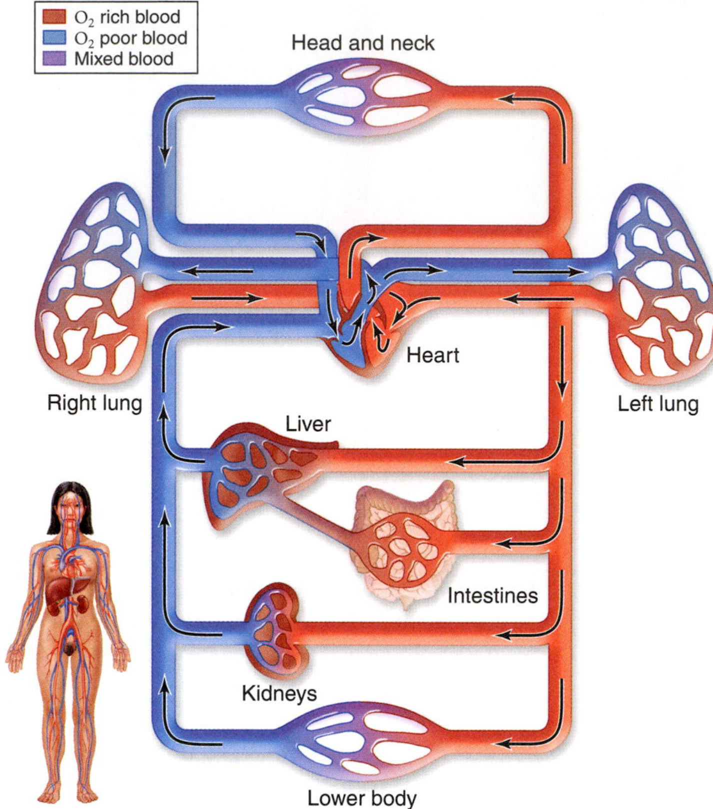

FIGURE 32.4 Blood's Journey in the Circulatory System. Oxygen-depleted blood leaving the right side of the heart goes to the lungs to pick up O_2. The oxygenated blood enters the left side of the heart, from where it is pumped throughout the body.

Table 32.1 *Blood Functions*

Function	Explanation
Gas exchange	Carries O_2 from lungs to tissues; carries CO_2 as bicarbonate ion (HCO_3^-) to the lungs to be exhaled
Nutrient transport	Carries nutrients absorbed by the digestive system throughout the body
Waste transport	Carries urea (a waste product of protein metabolism produced in the liver) to the kidneys, which excrete it in urine
Hormone transport	Carries hormones secreted by endocrine glands throughout body
Creation of interstitial fluid	Interstitial fluid that surrounds cells forms from blood plasma
Maintain homeostasis	Regulates blood pH; regulates cells' water content; generates pressure gradient that keeps plasma in capillaries; absorbs heat and dissipates it at the body's surface
Protection	Blood clots plug damaged vessels; white blood cells destroy foreign particles and participate in inflammatory response, which dilutes toxins at an injury site.

B. Each Atrium Delivers Blood to a Ventricle

Each day, the human heart sends a volume equal to more than 7,000 L of blood through the body, and it contracts more than 2.5 billion times in a lifetime. The heart is sometimes considered two pumps that beat in unison. Both sides of the heart function in the same way, but the blood from each side has a different destination.

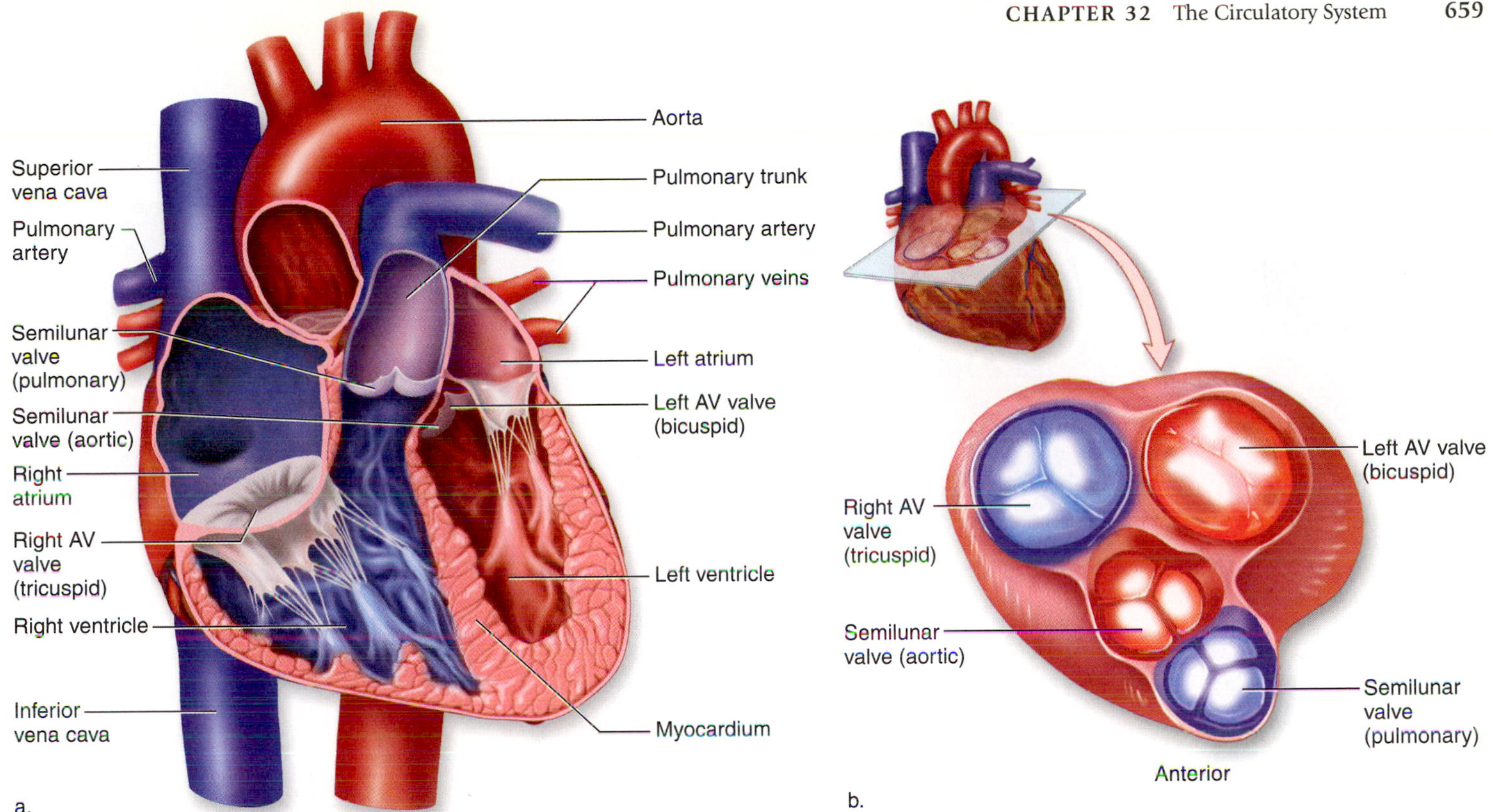

FIGURE 32.5 A Human Heart. (**A**) This illustration depicts the four chambers, the valves, and the major blood vessels of the human heart. (**B**) A cross-sectional view shows the relationship of the four major heart valves, which keep blood flow unidirectional.

The two largest veins in the body, the superior vena cava and the inferior vena cava, deliver blood from the systemic circulation to the right atrium. From there, blood passes into the right ventricle. Next, the blood flows out of the right ventricle and through the **pulmonary arteries** to the lungs, where blood picks up O_2. The **pulmonary veins** carry bright red oxygen-rich blood from the lungs to the left atrium of the heart. The blood then flows from the left atrium into the left ventricle. The massive force of contraction of the left ventricle, the most powerful heart chamber, sends blood into the **aorta,** the largest artery in the body. The blood then circulates throughout the body before returning to the right side of the heart.

How does the heart muscle itself receive its blood supply? Blood does not seep from the heart's chambers directly to the myocardium. Instead, two vessels that branch off of the aorta, the **coronary arteries,** supply blood to the heart muscle (**figure 32.6**). A smaller vein that enters the right atrium returns blood that has been circulating within the walls of the heart. Blockage of a coronary artery can cause a heart attack.

C. Cardiac Muscle Cells Produce the Heartbeat

A **cardiac cycle,** or a single beat of the heart, consists of two events: the contraction and relaxation of the heart muscle. Many **cardiac muscle** cells contract in unison without input from the central nervous system. How do they beat at just the right time?

Cardiac muscle cells branch, forming an almost netlike pattern (see figure 27.6). Where individual cells meet, gap junctions spread synchronized waves of electrical impulses from cell to cell. A heartbeat begins in the **sinoatrial (SA) node,** a region of specialized cardiac muscle cells in the

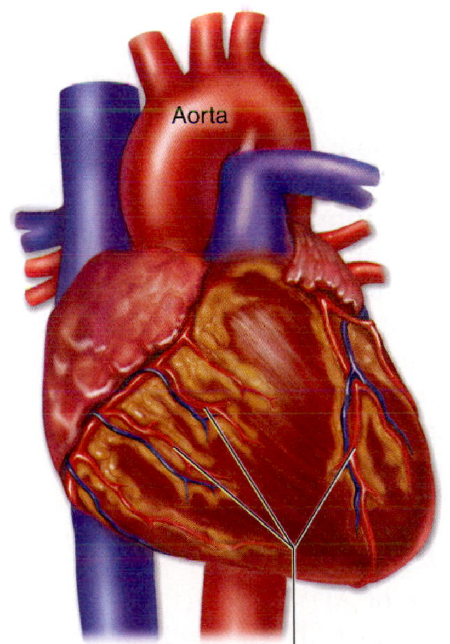

FIGURE 32.6
Coronary Arteries. The blood vessels on the outside of the heart supply the cardiac muscle cells with nutrients and O_2 and remove wastes.

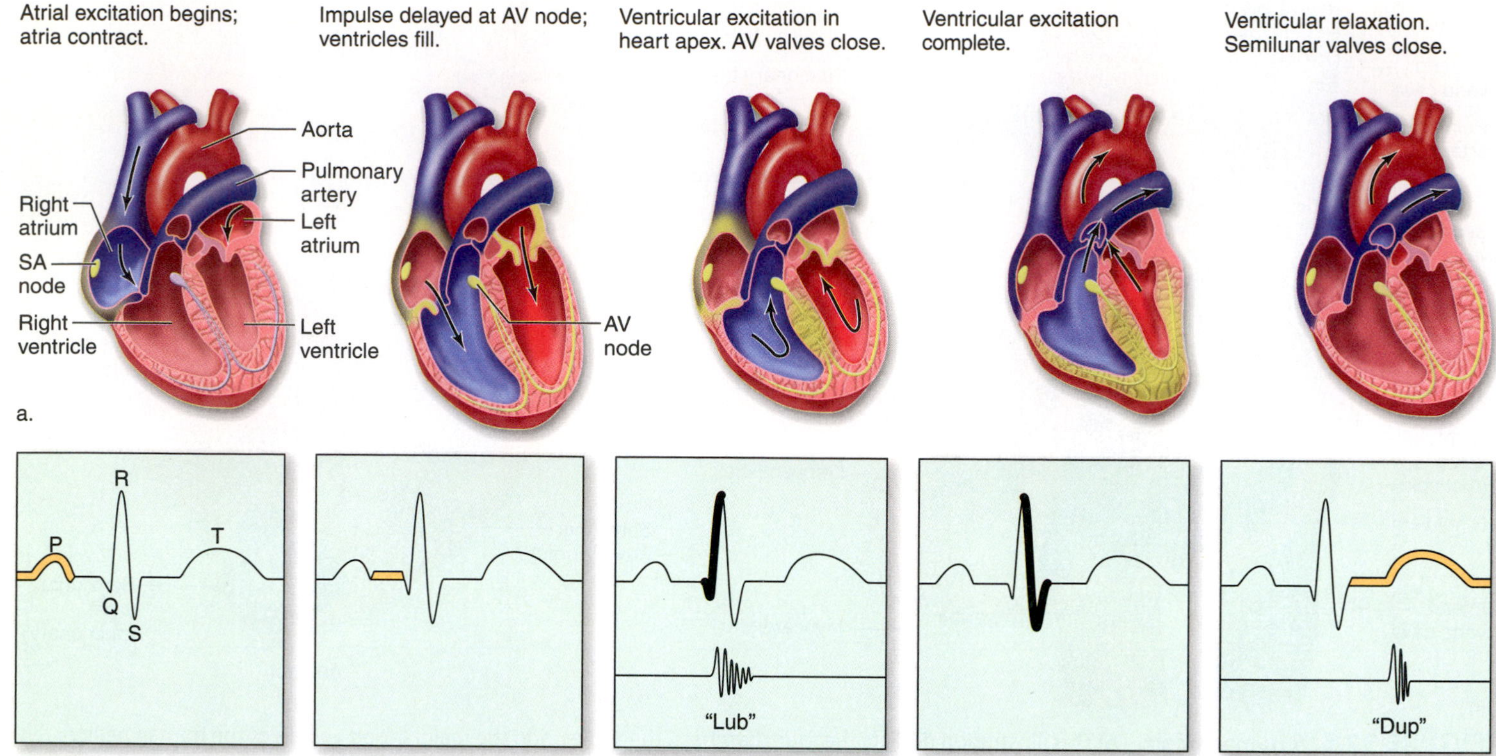

FIGURE 32.7 Heartbeat. (**A**) Electrical changes start in the sinoatrial (SA) node and travel through the arterial wall to the AV node and then to the ventricle walls. (**B**) An electrocardiogram (ECG) is a printout of electrical activities of the heart. The P wave corresponds to the depolarization of the atria, the QRS complex corresponds to the depolarization of the ventricles, and the T wave corresponds to the repolarization of the ventricles. The wave from the repolarization of the atria is obscured by the massive QRS wave of the ventricles. The closing of the atrioventricular (AV) valves during ventricle contraction produces the "lub" sound of the heartbeat; the closing of the semilunar valves during ventricle relaxation causes the "dup" sound.

upper wall of the right atrium (**figure 32.7**A). The SA node is called the **pacemaker** because it sets the tempo of the beat. Once begun, the impulses triggering contraction race across the atrial wall to the **atrioventricular (AV) node** in the wall of the lower right atrium. The AV node gives rise to specialized cardiac muscle cells that rapidly conduct electrical stimulation throughout the ventricle walls. These electrical signals stimulate the cells of the myocardium to contract in unison. **gap junctions, p. 72**

A device called an electrocardiograph records the electrical changes that accompany the contraction of the heart (figure 32.7B), producing a chart called an electrocardiogram (ECG). The "lub-dup" sound of the beating heart comes from the two sets of heart valves closing. A heart murmur is a variation on the normal "lub-dup" sound, and it often reflects abnormally functioning valves.

D. Exercise Strengthens the Heart

Active muscles require lots of ATP to power contraction, and they need O_2 to generate that ATP efficiently. When you exercise, your heart meets the increased demand for O_2 by increasing its **cardiac output,** a measure of the volume of blood that the heart pumps each minute.

Cardiac output can increase in two ways: a faster heartbeat or a greater volume of blood pumped per stroke. Within an exercise session, however, stroke volume is essentially fixed; this is why the most noticeable effect of exercise is elevated heart rate. With regular exercise, however, the stroke volume will increase. An active person can therefore pump the same amount of blood at 50 beats per minute as the sedentary person's heart pumps at 75 beats per minute. (At the peak of his career, cyclist Lance Armstrong had a resting heart rate of about 32 beats per minute.)

Exercise provides several other cardiovascular benefits as well. The number of red blood cells increases in response to regular exercise, and these cells are packed with more hemoglobin, delivering more O_2 to tissues. Exercise can also lower blood pressure and reduce the amount of cholesterol in blood. Moreover, regular activity spurs the development of extra blood vessels within the walls of the heart, which may help prevent a heart attack by providing alternative pathways for blood to flow to the heart muscle.

To achieve the most benefit from exercise, the heart rate must be elevated to 70 to 85% of its "theoretical maximum" for at least half an hour three times a week. One way to calculate your theoretical maximum is to subtract your age from 220 (**table 32.2**). If you are 18 years old, your theoretical maximum is 202 beats per minute. Seventy to 85% of this value is 141 to 172 beats per minute. Tennis, skating, skiing, handball, vigorous dancing, hockey, basketball, biking, or brisk walking can elevate heart rate to this level.

Table 32.2 *Target Heart Rates by Age*

Age	Theoretical Maximum Heart Rate (beats per minute)	Target Heart Rate During Exercise (beats per minute)
20	200	140–170
25	195	137–166
30	190	133–162
40	180	126–153
50	170	119–145
60	160	112–136
70	150	105–128

Summary *The heart's two atria receive blood, and the two ventricles propel it throughout the body. Heart valves keep blood flowing in the correct direction. The pacemaker dictates the heart rate. Regular exercise strengthens the heart.*

32.2 MASTERING CONCEPTS

1. Describe the path of blood through the heart's chambers and valves and through the pulmonary and systemic circulations.
2. How does heartbeat originate and spread?
3. How does exercise affect the circulatory system?
4. Explain why exercise makes the heart beat faster.

32.3 Blood Vessels Form the Circulation Pathway

An adult human body has more than 96,000 km (about 60,000 miles) of blood vessels, and blood journeys through them in a predictable pathway (**figure 32.8**).

A. Blood Flows from Arteries to Capillaries to Veins

In systemic circulation, blood leaves the heart through the aorta. It enters a series of arteries, then arterioles, which lead to the capillaries that exchange materials with the interstitial

Cardiovascular System	
Main tissue types	**Examples of Locations/Functions**
Epithelial	Forms inner lining of heart wall; lines veins and arteries; makes up capillary walls
Connective	Surrounds heart (part of pericardium); forms outer layers of veins and arteries; blood and lymph are connective tissues
Nervous	Regulates heart rate and blood pressure
Muscle	Heart wall is mostly cardiac muscle; smooth muscle forms middle layer of arteries and veins; skeletal muscle propels blood in veins and lymph in lymph vessels

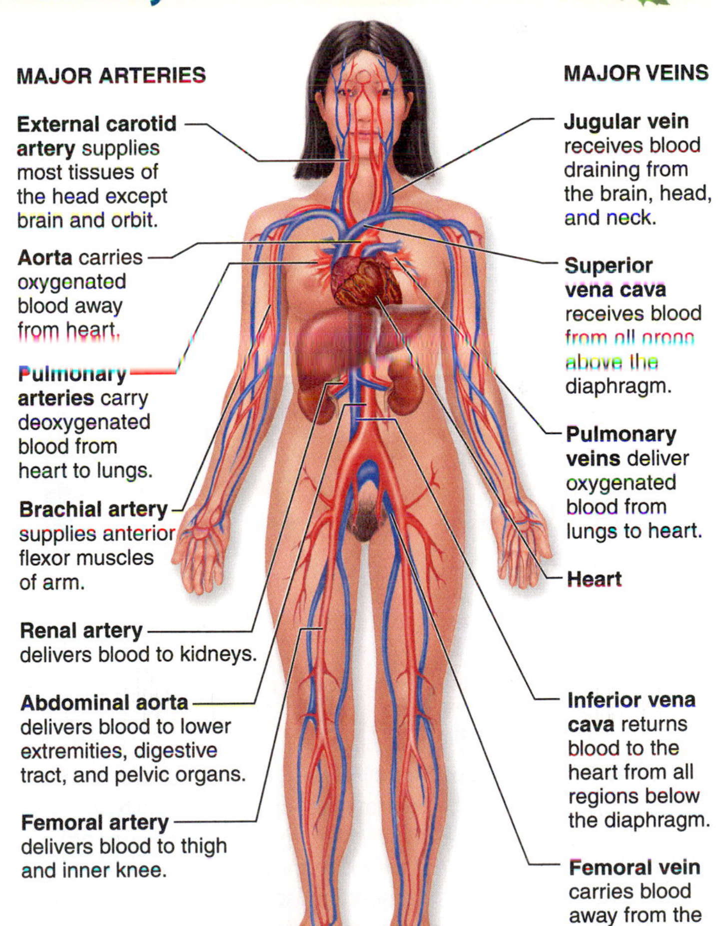
MAJOR ARTERIES
External carotid artery supplies most tissues of the head except brain and orbit.
Aorta carries oxygenated blood away from heart.
Pulmonary arteries carry deoxygenated blood from heart to lungs.
Brachial artery supplies anterior flexor muscles of arm.
Renal artery delivers blood to kidneys.
Abdominal aorta delivers blood to lower extremities, digestive tract, and pelvic organs.
Femoral artery delivers blood to thigh and inner knee.
MAJOR VEINS
Jugular vein receives blood draining from the brain, head, and neck.
Superior vena cava receives blood from all organs above the diaphragm.
Pulmonary veins deliver oxygenated blood from lungs to heart.
Heart
Inferior vena cava returns blood to the heart from all regions below the diaphragm.
Femoral vein carries blood away from the thigh and inner knee.

FIGURE 32.8 **Human Circulatory System.**

fluid in contact with the body's cells. The capillaries converge into venules, then veins, that lead back to the heart.

The walls of arteries and veins consist of multiple layers (**figure 32.9**). The outermost layer is a sheath of connective tissue. The middle layer is made mostly of **smooth muscle** (see figure 27.6). **Endothelium** forms the innermost layer; it is a one-cell-thick lining of simple squamous epithelium. Arteries are more muscular than veins. As arteries branch farther from the heart, they become thinner, and the outermost layer may taper away. In arterioles, the smooth muscle layer enables the diameter of these small vessels to decrease or increase as the body regulates blood pressure.

Capillary beds are networks of capillaries that connect an arteriole and a venule (**figure 32.10**). Capillaries are tiny but very numerous, providing extensive surface area where the exchange of materials occurs. Because the capillary wall consists of a single layer of endothelial cells, nutrients and gases diffuse easily between the capillary and the interstitial fluid (and from there to body cells).

Venules and veins return blood from the capillary beds to the heart. The smooth muscle layer of a vein is much reduced or even absent, and the vessel collapses when it is empty. Some medium and large veins of the lower legs, where blood moves against gravity to return to the heart, have flaps called venous valves that keep blood flowing in one direction. As skeletal muscles in the leg contract, they squeeze veins and propel blood through the open valves in the only direction it can move: toward the heart (**figure 32.11**).

In some locations, circulating blood detours from the usual route of capillary to venule to vein. In a **portal system,** blood passes through a second set of capillaries before it returns to the heart. For example, in the hepatic portal system,

FIGURE 32.9 Types of Blood Vessels. The walls of arteries and veins have three layers: The outermost layer is connective tissue; the middle layer is elastic and muscular; the inner layer is smooth endothelium. Walls of arteries are much thicker than those of veins, which have a greatly reduced smooth muscle layer. The capillary wall consists only of endothelium. Many veins have valves, which keep blood flowing in one direction.

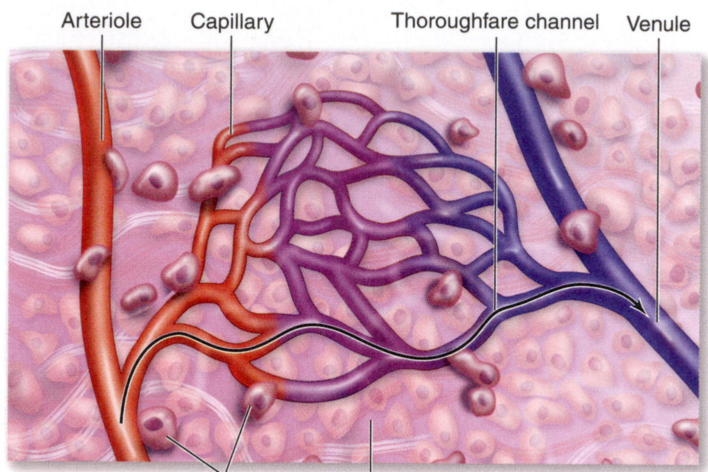

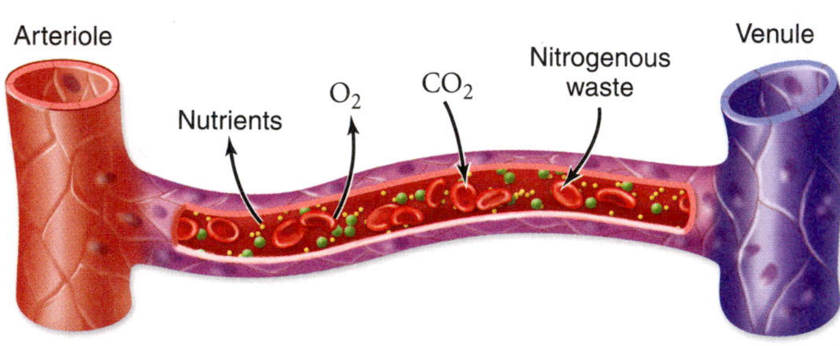

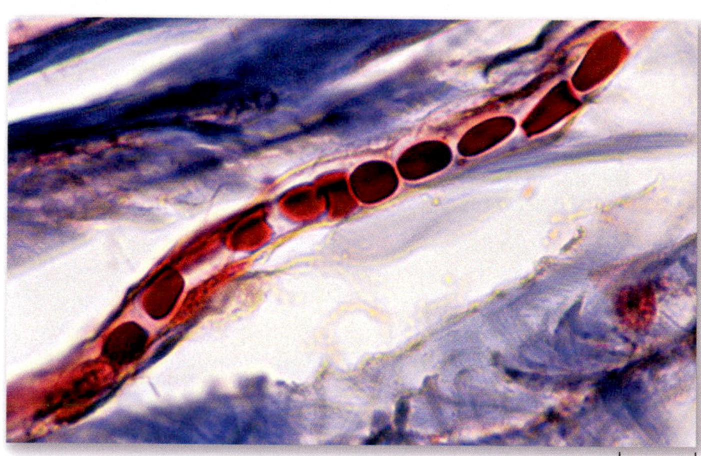

FIGURE 32.10 Capillaries. (**A**) A capillary bed is a network of tiny vessels that lies between an arteriole and a venule. (**B**) At capillaries, O_2 leaves red blood cells and CO_2 enters the circulation. Nutrients are released to the tissues, and nitrogenous wastes are collected from them. (**C**) Capillaries are so small that red blood cells move through them in single file.

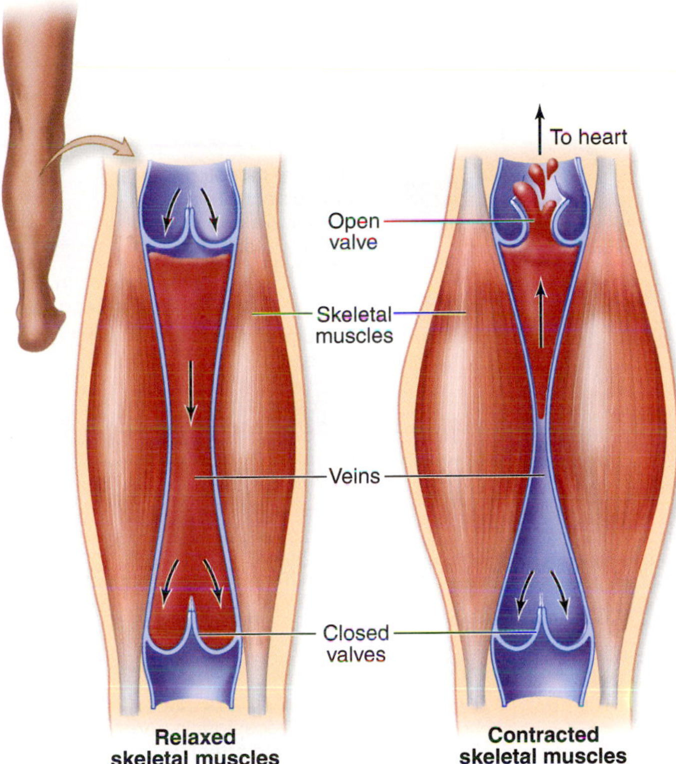

FIGURE 32.11 Valves in Veins Help Return Blood to the Heart. When skeletal muscles relax, valves prevent blood from flowing backward. When skeletal muscles contract, they thicken and squeeze the veins, propelling blood through the open valves. In this illustration, the veins appear much larger than they are relative to real muscles.

veins leaving the intestines diverge into a capillary bed in the liver. (The hepatic portal vein is the blood vessel "bridge" connecting the intestines and liver in figure 32.4.) From there, blood enters veins that return the blood to the heart. Another portal system delivers hormones from the hypothalamus to target cells in the anterior pituitary (see figure 30.5).

B. Blood Pressure and Velocity Differ Among Vessel Types

A routine doctor's office visit always includes a blood pressure reading, a good indication of overall cardiovascular health. **Blood pressure** is the force that blood exerts on artery walls. As the heart drives blood through the vessels, you can feel your blood pressure as a "pulse."

A device called a sphygmomanometer measures blood pressure (**figure 32.12**). The **systolic pressure,** or upper number in a blood pressure reading, reflects the contraction of the ventricles. The **diastolic pressure,** or low point, occurs when the ventricles relax. Blood pressure readings are in units of "millimeters of mercury" (mm Hg). (This is because older devices measured how far blood pressure could push a column of mercury, but these are now considered unsafe.) A typical blood pressure reading for a young adult is 120 mm Hg for the systolic pressure and 80 mm Hg for the diastolic pressure, expressed as "120 over 80" (written 120/80). "Normal" blood pressure, however, varies with age, sex, race, and other factors.

Pressure falls with distance from the heart. It is highest in the arteries, lower in the capillaries, and lower still in the

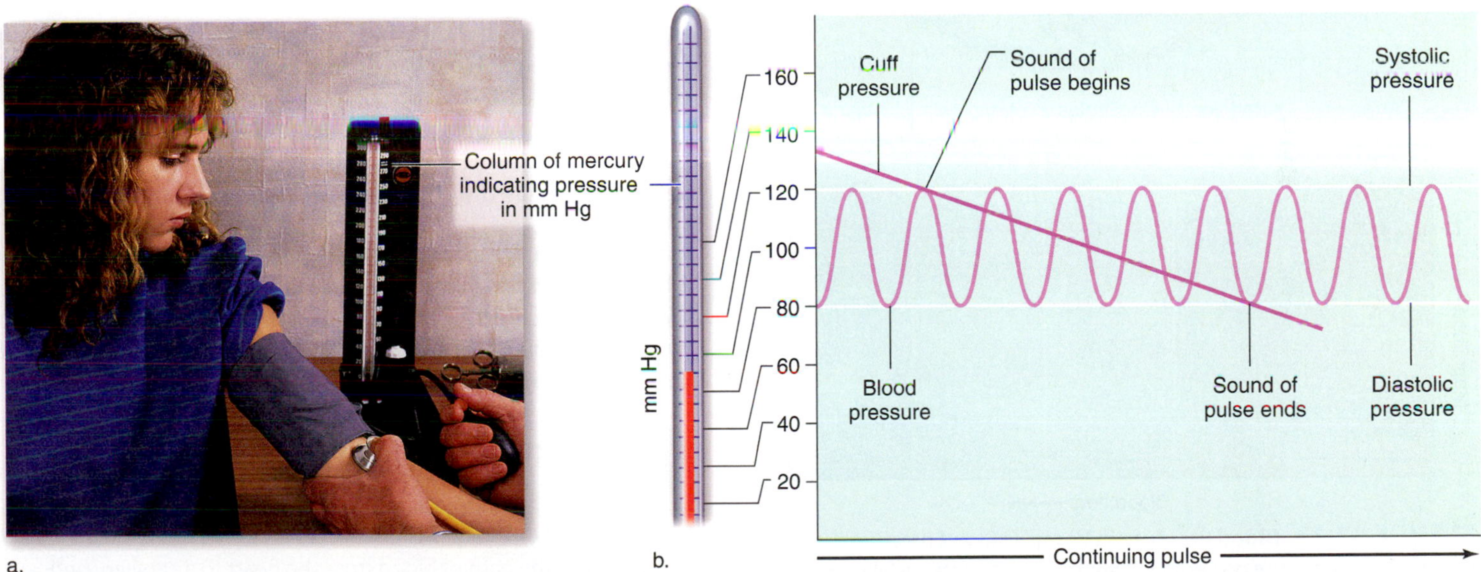

FIGURE 32.12 Blood Pressure. (**A**) A sphygmomanometer, which is an inflatable cuff attached to a pressure gauge, measures blood pressure. The cuff is wrapped around the upper arm and inflated until no pulse is felt in the wrist, which signifies that circulation to the lower arm has been temporarily cut off. A stethoscope placed on the arm just below the cuff detects the sound of returning blood flow when the cuff slowly deflates. (**B**) The listener notes the pressure on the gauge when a thumping is audible. This sound is the blood rushing through the arteries past the deflating cuff as the pressure peaks due to ventricular contraction. The value on the gauge when the sound begins is the systolic blood pressure. The sound fades until it disappears, and the pressure reading at this point is the diastolic blood pressure.

veins (**figure 32.13**). Even though blood pressure is lowest in the veins, blood moves most slowly through the capillaries, which enhances nutrient and waste exchange with the interstitial fluid. Past the capillaries, veins merge and the total cross-sectional area decreases, helping to speed blood flow back to the heart.

Overall, blood pressure reflects both the heart's pumping action and the diameter of the blood vessels. **Vasoconstriction** is the narrowing of blood vessels; the opposite effect, **vasodilation,** is the widening of blood vessels. Blood pressure rises when smooth muscle in blood vessel walls contracts, narrowing their diameter. Altering the diameter of arterioles allows the body to adjust to exercise, when blood supply to muscles increases at the expense of organs not in immediate use, such as those in the digestive tract.

Blood pressure is under constant regulation (**figure 32.14**). Pressure receptors within the walls of major arteries send information about blood pressure to the medulla of the brain. The medulla, via the autonomic nervous system, regulates both heart rate and the diameter of arterioles in a negative feedback loop that maintains homeostasis in the short term. The body can also regulate blood pressure over a longer period by raising or lowering the volume of blood. Chapter 35 describes how the kidneys adjust the blood's volume and salt concentration by controlling the amount of fluid excreted in urine. **autonomic nervous system, p. 586**

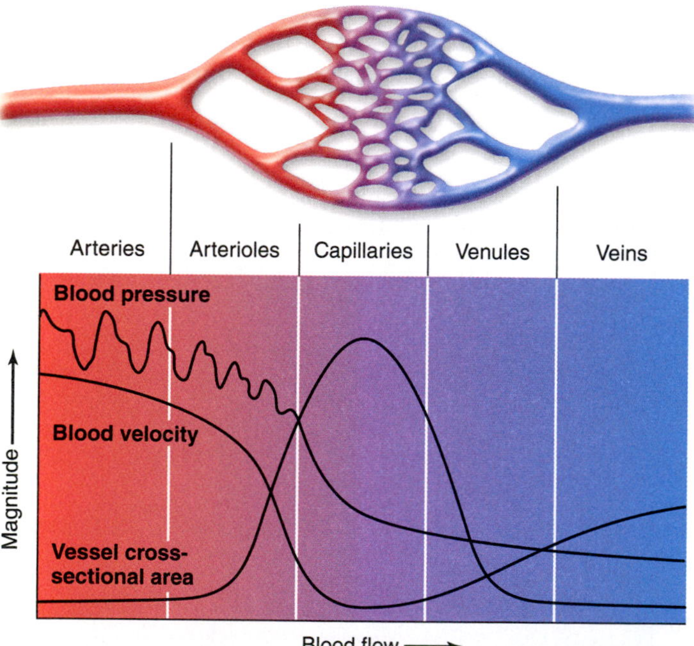

FIGURE 32.13 Blood Pressure in Vessels. Blood pressure drops with distance from the powerful left ventricle. The pressure falls so low in veins that muscle contraction, valves, and gravity must assist circulation. Blood velocity is very high at the aorta, slows to a crawl as it moves through the capillaries, and picks up speed again as it moves through the converging veins. Even though blood pressure in capillaries is greater than in veins, blood velocity is slower because of the expansion in total cross-sectional area across the capillary beds.

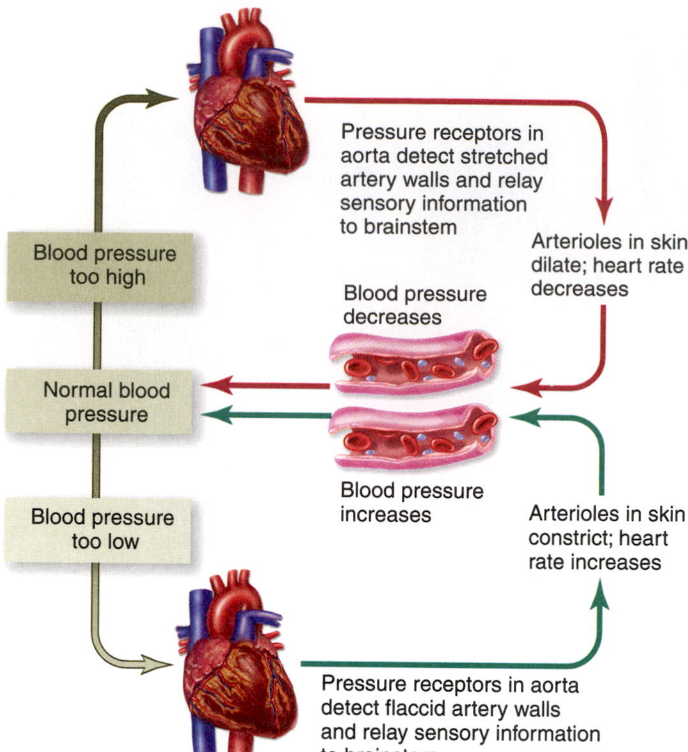

FIGURE 32.14 The Nervous System Maintains Circulatory Homeostasis. This negative feedback loop shows how the nervous system regulates blood pressure. Stretch-sensitive neurons detect pressure in major arteries. If it is too high, the brain decreases the heart rate and causes blood vessels to dilate. If blood pressure is too low, vessels constrict and the heart beats faster.

Blood pressure that is too low or too high can cause health problems. Hypotension, which is blood pressure significantly lower than normal, may cause fainting. Consistently elevated blood pressure, or hypertension, affects 15 to 20% of adults in industrialized nations. High blood pressure may severely damage the circulatory system and other organs.

Summary *Arteries lead from the heart to arterioles to capillaries, which exchange materials with interstitial fluid. Capillaries feed into venules and then veins that lead back to the heart. Blood pressure is highest in the arteries and lowest in the veins, but blood velocity is lowest in the capillaries.*

32.3 MASTERING CONCEPTS

1. How are the structures of arteries, capillaries, and veins similar and different?
2. Trace the path of blood from the heart to a capillary bed in the foot and back to the heart.
3. Across the human circulatory system, how are blood pressure, blood velocity, and vessel diameter related?
4. How is regulation of blood pressure an example of homeostasis?

32.4 Blood Is a Complex Mixture

Like all connective tissues, blood consists of cells within a matrix (**table 32.3** and **figure 32.15**). **Plasma** is the liquid matrix of blood. A cubic millimeter of blood normally contains about 5 million red blood cells, 7000 white blood cells, and 250,000 cell fragments called platelets. Blood cells originate from stem cells in bone marrow (see figure 31.6). This chapter's Burning Question: What is the difference between donating whole blood and donating plasma? explains two ways your blood can help save lives.

Blood plasma, which makes up more than half of the blood's volume, is 90 to 92% water. It contains about 7 to 8% dissolved proteins of more than 70 different types. Antibodies are plasma proteins that participate in the body's immune response; other proteins transport cholesterol. Another plasma protein, fibrinogen, plays a key role in blood clotting. About 1% of plasma consists of dissolved salts, hormones, metabolic wastes, CO_2, and nutrients and vitamins absorbed by the digestive system. The concentrations of these dissolved molecules are low, but they are critical. For example, blood usually contains about 0.1% glucose; if the concentration falls to 0.06%, convulsions begin.

A. Red Blood Cells Transport Oxygen

Mature **red blood cells**, or erythrocytes, are saucer-shaped disks packed with the pigment **hemoglobin,** a protein that carries O_2. Tucked into each hemoglobin molecule are four iron atoms, which can each combine with one O_2 molecule picked up in the lungs.

Red blood cells are by far the most numerous cells in blood (see figure 32.15). Their precursors form within red bone marrow at the rate of 2 to 3 million per second. As they develop, human red blood cells lose their nuclei, ribosomes, and mitochondria. Mature red blood cells that lack nuclei cannot divide or carry out most cellular metabolism. They can, however, generate the ATP they require. During its life of about 120 days, a red blood cell pounds against artery walls and squeezes through tiny capillaries. Eventually, the liver or spleen destroys the cell and recycles most of its components.

Like all cells, red blood cells have molecules embedded in their outer membranes. Some of these molecules play a role in **agglutination,** a clumping reaction caused by antibodies in

Table 32.3 *Components of Blood*

Component	Function
Plasma	Liquid component of blood; exchanges water and many dissolved substances with interstitial fluid
Red blood cells (erythrocytes)	Carry O_2
White blood cells (leukocytes)	Destroy foreign substances, initiate inflammation
Platelets (thrombocytes)	Initiate clotting

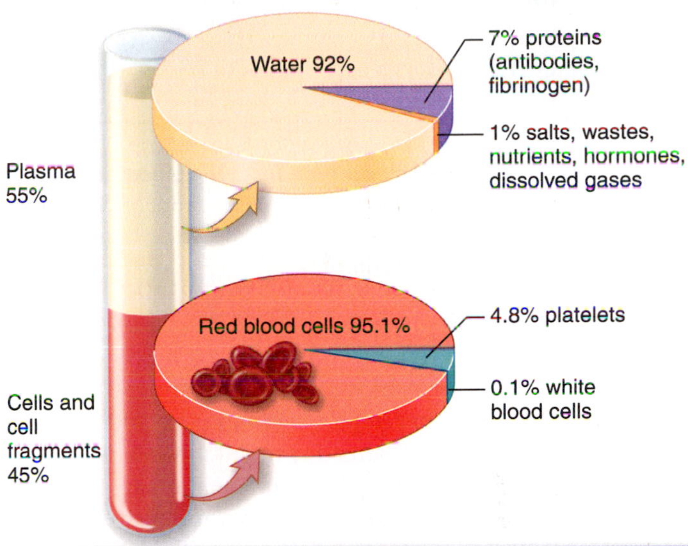

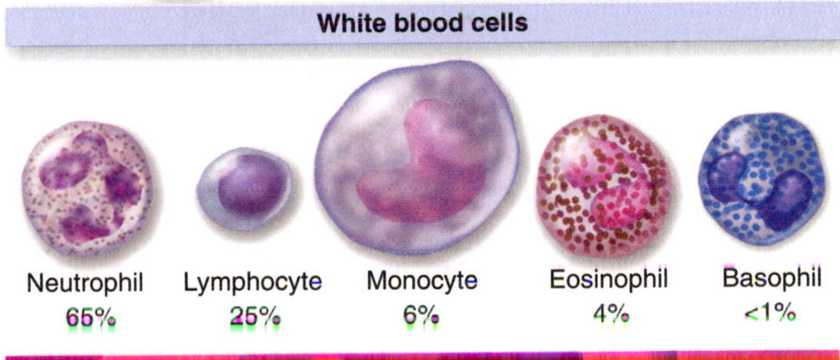

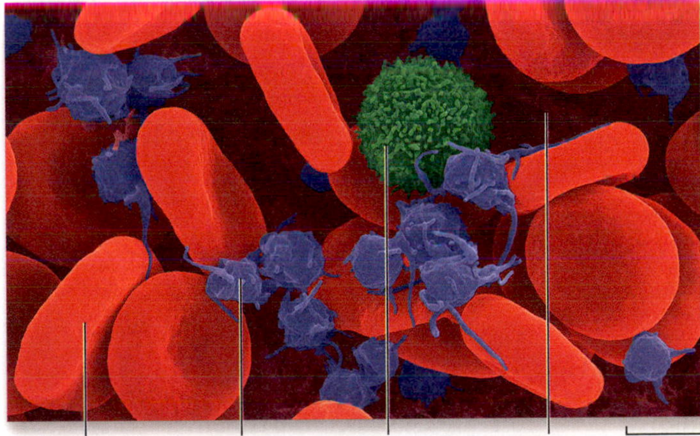

FIGURE 32.15 Blood Composition. Human blood is a complex mixture of platelets, red blood cells, and white blood cells suspended in a liquid plasma of water, proteins, salts, wastes, nutrients, hormones, and dissolved gases. Chapter 36 details the functions of the five types of white blood cells.

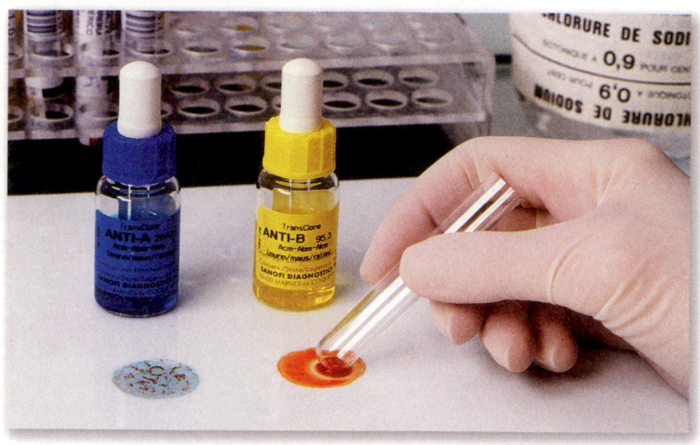
ANTI-A
ANTI-B

FIGURE 32.16 An ABO Blood Type Test. The blood spot on the left has been treated with anti-A antigen, and the spot on the right has been treated with anti-B antigen. The clumping pattern reveals whether the blood type is A, B, AB, or O.

blood plasma (**figure 32.16**). A person's red blood cells may express marker A (type A blood), marker B (type B), both A and B (type AB), or neither (type O). Plasma contains antibodies against any markers not already present in the blood; these antibodies cause cells carrying "foreign" markers to clump together. Type A blood, for example, contains antibodies that will react with blood carrying marker B. People with blood type AB have neither anti-A nor anti-B antibodies, so they can receive a transfusion of any type of blood. Type O blood contains both anti-A and anti-B antibodies, so people with this blood type can only receive blood type O. Figure 10.12 explains the genetic basis of the ABO blood type system.

Rh blood groups also rely on an agglutination reaction. People who express the Rh marker are Rh^+, whereas those who lack it are Rh^-. (Rh typing yields the "+" and "−" designations such as "A positive" or "O negative.") People who lack the Rh marker, however, do not usually have an anti-Rh antibody. Therefore, Rh^- people can receive transfusions from Rh^+ donors, and vice versa. Rh incompatibility can, however, be important in pregnancy (see figure 36.22).

B. White Blood Cells Fight Infection

White blood cells, or leukocytes, are immune system cells. They are larger than red blood cells and retain their nuclei, but they lack hemoglobin. They originate in the red and yellow bone marrow, leave the circulatory system, and wander through tissues. Chapter 36 describes the functions and interactions of the five types of white blood cells. In short, white blood cells secrete signaling molecules that provoke the warmth and swelling of inflammation. They also surround and destroy microbes and produce antibodies that help destroy foreign microbes or particles.

The proportions of white blood cell types may provide a clue to the type of infection or disease present. For example, **leukemias** are cancers in which bone marrow overproduces white blood cells. The abnormal white cells form at the expense of red blood cells, so when the patient's "white cell count" rises, the "red cell count" falls. Thus, leukemia also causes anemia. Having too few white blood cells is also devastating, because the body cannot fight infection without them. HIV destroys T lymphocytes (a type of white blood cell), causing AIDS. Exposure to radiation or toxic chemicals can severely damage bone marrow, killing many white blood cells. Unless the cells are immediately replaced, death occurs rapidly from rampant infection.

C. Blood Clotting Requires Platelets and Plasma Proteins

Platelets, or thrombocytes, are small, colorless cell fragments that initiate blood clotting. A platelet originates as part of a huge cell containing rows of vesicles that divide the cytoplasm into distinct regions, like a sheet of stamps. The vesicles enlarge and join together, "shedding" fragments that become platelets.

In a healthy circulatory system, platelets travel freely within the vessels. Sometimes, however, a wound nicks a blood vessel or a blood vessel's normally smooth inner lining becomes obstructed. Platelets then "catch" on the bumps and shatter, releasing biochemicals that combine with plasma

Endothelial lining
Collagen fiber
Sharp edge
Blood vessel
Platelet
Red blood cell

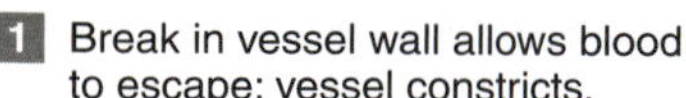
1 Break in vessel wall allows blood to escape; vessel constricts.

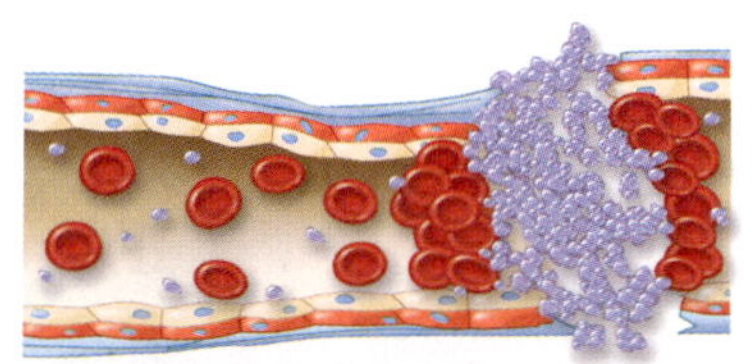
2 Platelets adhere to each other, to end of broken vessel, and to exposed collagen. Platelet plug helps control blood loss.

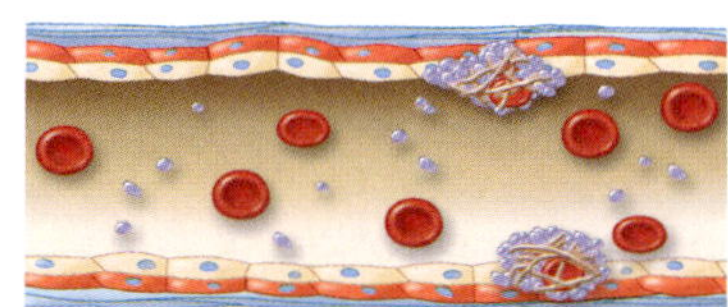
3 Exposure of blood to surrounding tissue activates clotting factors. Trapped red blood cells form a clot.

a.

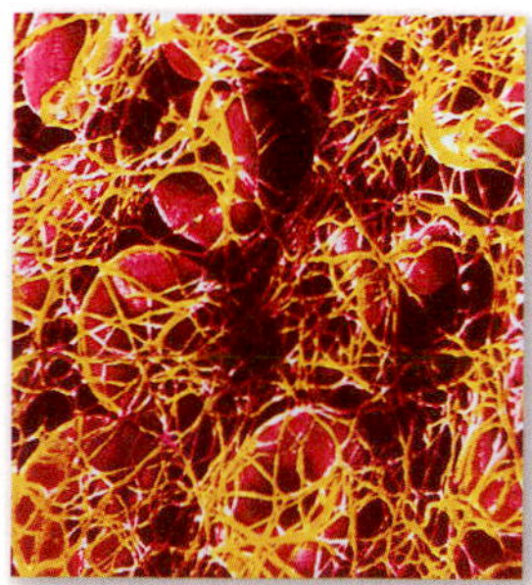

b.

FIGURE 32.17 Blood Clotting. (**A**) Blood clotting results from vessel constriction, platelet aggregation, and a biochemical cascade that produces a meshwork of protein threads. (**B**) A scanning electron micrograph of protein threads in a blood clot.

proteins called clotting factors. The resulting complex series of reactions ends with the formation of a **blood clot**—a plug of solidified blood (**figure 32.17**).

Absent or abnormal clotting factors cause inherited bleeding disorders called hemophilias. Deficiencies of vitamins C or K can also slow clotting and wound healing. Blood that clots too readily is also extremely dangerous. In atherosclerosis, platelets may snag on rough spots in blood vessel linings. The resulting clot may stay in place or travel in the bloodstream to another location; either way, it may cut off circulation and sometimes even cause death.

Clotting disorders are just one example of things that can go wrong with the human circulatory system; this chapter's Can *You* Relate? The Unhealthy Circulatory System considers several others.

Summary ***Blood consists of liquid plasma, red blood cells, white blood cells, and platelets. Red blood cells carry O_2, and white blood cells help fight infection. Platelets aid in clotting.***

MASTERING CONCEPTS

1. What substances are dissolved in blood plasma?
2. What cell type carries hemoglobin, and what is this protein's function?
3. Where do red and white blood cells originate?
4. What is the basis for the ABO and Rh blood groups in humans?
5. How does blood clot?
6. How is blood clotting that is too fast or too slow dangerous?

Can *You* Relate?

The Unhealthy Circulatory System

Anemia

Anemia is a decrease in the oxygen-carrying capacity of blood. Symptoms include fatigue and lack of tolerance to cold. Anemia can result from red blood cells that are too small, contain too little hemoglobin, are manufactured too slowly, or die too quickly. Iron deficiency is the most common cause of anemia; sickle cell disease is a type of inherited anemia (see figure 12.14).

Atherosclerosis

Fatty deposits inside the walls of coronary arteries reduce blood flow to the heart muscle (**figure 32.A**). This "hardening of the arteries" is called atherosclerosis (*athero* is from the Greek word for "paste," and *sclerosis* meaning "hardness"). Atherosclerosis can cause several ailments, including angina pectoris (chest pain), heart attack, arrhythmia, and aneurysm.

Heart Attack

Blocked blood flow in a coronary artery kills part of the myocardium, the heart muscle. This is a heart attack (myocardial infarction), and it may come on suddenly. A common treatment for a blocked coronary artery is a bypass operation. A surgeon creates a bridge around the blockage by sewing pieces of blood vessel taken from the patient's chest or leg onto the blocked artery. The procedure's name comes from the number of arteries repaired. A quadruple bypass operation, for example, bridges four obstructed arteries.

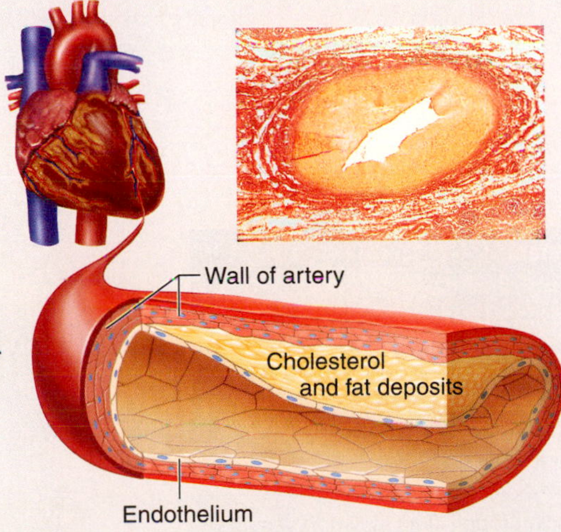

FIGURE 32.A **Atherosclerosis.** Deposits of cholesterol and other fatty materials beneath the inner lining of the coronary arteries cause atherosclerosis, or "hardening of the arteries."

Arrhythmia

An arrhythmia is an abnormal heartbeat. Some arrhythmias originate in the atria, causing transient flutters or racing that lasts only a few seconds. An electronic pacemaker implanted under the skin is a common treatment. In ventricular fibrillation, the ventricles contract wildly, causing sudden cardiac arrest. Death may occur within minutes. First aid for cardiac arrest includes cardiopulmonary resuscitation (CPR); in addition, external defibrillators that shock an erratically beating heart back to normal are available in many public places. People with chronic arrhythmia may have a defibrillator implanted under the skin of the chest.

Aneurysm

Atherosclerosis can so weaken the wall of an artery that a region of the vessel forms a pulsating, enlarging sac called an aneurysm. If it bursts, blood loss may be great. Aneurysms may also result from a congenitally weakened area of an arterial wall, trauma, infection, persistently high blood pressure, or an inherited disorder such as Marfan syndrome.

The Effects of Smoking on Cardiovascular Health

Smoking is the most common preventable cause of death. Cigarette smoke damages the lungs, impairing their ability to deliver O_2 to the heart (and increasing the chance of lung cancer). Nicotine stimulates the secretion of epinephrine and norepinephrine, increasing both heart rate and blood pressure. Nicotine also damages blood vessels and stimulates the formation of blood clots, increasing the risk of stroke. Cigarette smoking is clearly unhealthy, yet many people find it difficult or impossible to quit; chapter 3 described nicotine's addictive qualities.

Burning Questions

What is the difference between donating whole blood and donating plasma?

A person who "gives blood" donates 450 to 500 mL (about a pint) of blood to a nonprofit blood bank. After being screened for disease-causing agents, the blood may go to patients who need transfusions following trauma or surgery. More commonly, however, the blood is separated into its components, such as red blood cells, platelets, or clotting proteins. In this way, a single blood donation can help several different patients.

Plasma donation is another option. In this process, whole blood is removed from a donor's body, then a machine separates out the plasma. The red blood cells and other components return to the donor. The plasma center sells the fluid to pharmaceutical companies, which use it to manufacture treatments for hemophilia, hepatitis, and other diseases. A person can donate plasma more frequently than whole blood, because it takes only about a day or two to replenish the fluid lost in plasma donation. Whole blood takes longer to replace.

In the United States, it is illegal to pay a donor for whole blood. This law promotes a safe blood supply, because donors have no incentive to lie about illnesses that might disqualify them from donating. Plasma donors, however, can receive money. The companies that process the plasma can purify each fraction separately, removing viruses and other harmful components.

Have a Burning Question of your own? Submit it to marielle_hoefnagels@mcgraw-hill.com for possible inclusion in future editions of this book!

32.5 The Lymphatic System Maintains Circulation and Protects Against Infection

The blood circulatory system and the lymphatic system work together. Blood carries a continuous supply of nutrients and O_2 to cells and removes wastes from them. The **lymphatic system** collects fluid that has leaked from blood capillaries, removes bacteria, debris, and cancer cells, and returns the cleansed fluid to the blood (**figure 32.18**). In addition, special lymph vessels in the small intestine absorb dietary fats.

Lymph, the colorless fluid that the lymphatic system transports, originates in **lymph capillaries**—tiny, dead-end vessels that absorb fluid that seeps out of blood capillaries (**figure 32.19**A). The chemical composition of lymph is therefore similar to that of blood plasma, minus the proteins that are too large to leave blood capillaries. The cells of lymph capillaries, however, are not joined as tightly together as those of blood capillaries. Lymph capillaries therefore also admit bacteria, viruses, cancer cells, and other large particles in body tissues. These capillaries then converge into

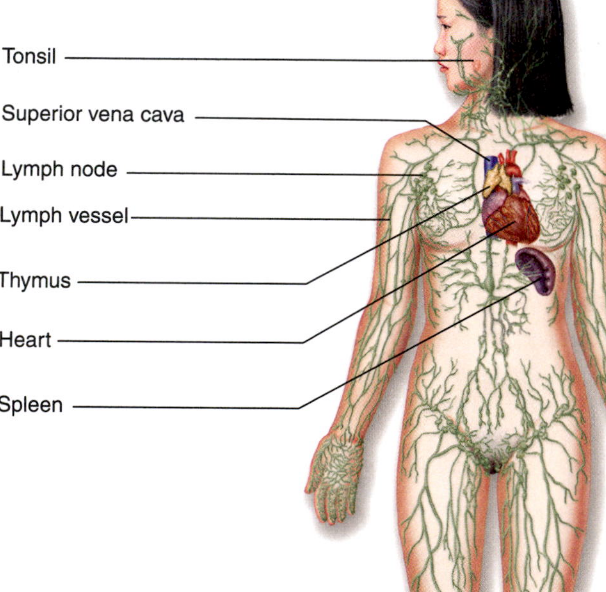

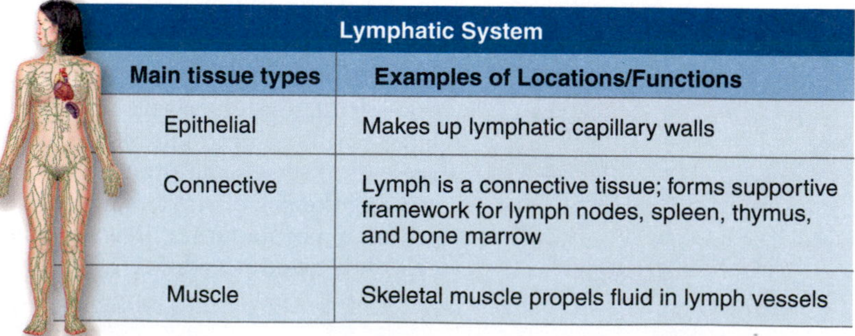

Lymphatic System	
Main tissue types	**Examples of Locations/Functions**
Epithelial	Makes up lymphatic capillary walls
Connective	Lymph is a connective tissue; forms supportive framework for lymph nodes, spleen, thymus, and bone marrow
Muscle	Skeletal muscle propels fluid in lymph vessels

FIGURE 32.18 Human Lymphatic System. This network of vessels and lymphoid organs collects excess fluid that leaks from the blood capillaries, filters it, and returns it to the bloodstream.

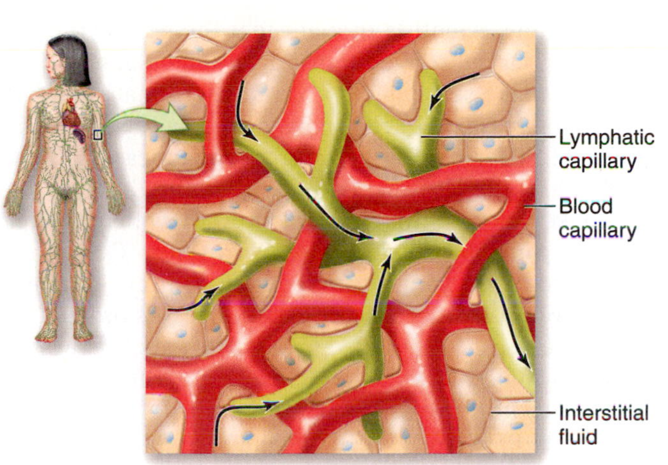

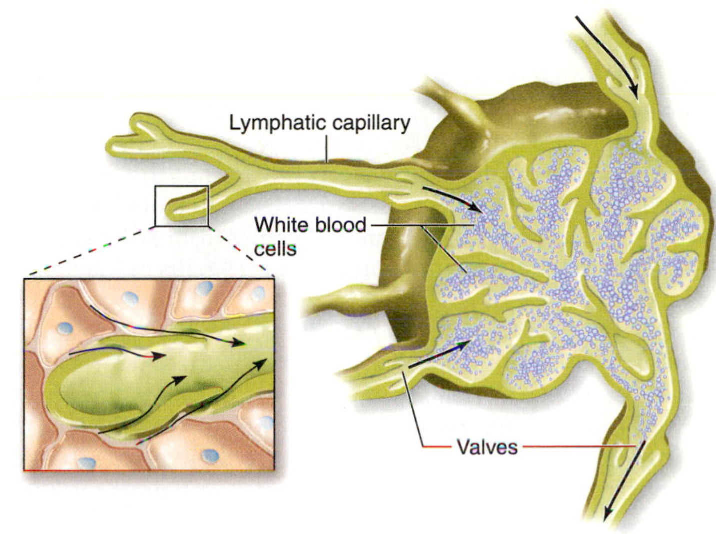

FIGURE 32.19 Lymph Vessels and Lymph Nodes. (**A**) Lymph vessels collect excess fluid that leaks out of capillaries in the tissues of the body. (**B**) Lymph is filtered through infection-fighting white blood cells in lymph nodes on its way back to veins.

larger lymph vessels that eventually empty into veins in the chest, where the fluid returns to the blood.

Along the way, lymph passes through **lymph nodes,** kidney-shaped organs that contain infection-fighting white blood cells (figure 32.19B). Lymph nodes range in size from microscopic to much larger masses, such as those that can be felt as "swollen glands" along the sides of the neck during respiratory infections (see figure 36.4). A lymph node consists of fibrous tissue with many pockets, each filled with millions of white blood cells that filter cellular debris and bacteria from the lymph flow. The lymph nodes also add white blood cells to the lymph, which transports them to the blood. White blood cells can detect and destroy cancer cells. But if cancer originates in or spreads to a lymph node and is not destroyed, it may spread to the rest of the body in the lymph fluid.

White blood cells occur in other organs of the lymphatic system, too (see figure 32.18). The **spleen** destroys dead or damaged red blood cells, filters and stores blood, and releases white blood cells. The **thymus** is an organ in which specialized white blood cells called T cells learn to distinguish body cells from foreign cells.

Lymph flow is sluggish, because the lymphatic system has no pump. Contractions of surrounding skeletal muscles and valves in lymph vessels help move the fluid. The lymphatic system normally returns less than 30 mL (1 ounce) of fluid per minute to the veins, in contrast to the 5 or 6 L (4 or 5 quarts) pumped through the circulatory system in the same amount of time. But if the lymphatic system fails, the excess fluid accumulates in the body's tissues and causes swelling (edema). In an infectious disease called elephantiasis, parasitic nematode worms block the flow of lymph and cause grotesquely swollen tissues. **nematodes, p. 462**

Summary *Lymph capillaries collect excess interstitial fluid and pass it to larger lymph vessels that return the fluid to veins. Lymph nodes filter lymph and release white blood cells.*

32.5 MASTERING CONCEPTS

1. Where does lymph come from?
2. List the components of the lymphatic system and their functions.
3. How do the blood and lymph circulatory systems work together?

32.6 INVESTIGATING LIFE
Living Fossils Reveal the Long History of Blue Blood

A person born into an old, aristocratic family has "blue blood"—an expression that denotes prestige and social status. But did you know that some of the humblest of animals really do have blue blood? Evolutionary biologists are studying the pedigrees of these organisms to learn more about the origin of respiratory pigments.

Scientists have long puzzled over the evolutionary origins of hemoglobin and other proteins that carry O_2 within animal circulatory systems. New evidence comes from an obscure group of animals called the Onychophorans, also known as velvet worms. The worms' plush-sounding name comes from their skin, which sports tiny bumps that give the Onychophorans a velvety appearance. The name of their phylum, Onychophora ("claw-bearers"), comes from another feature of the velvet worm's anatomy: about 15 pairs of lobelike, clawed, stumpy legs.

All 110 living species of Onychophorans are terrestrial animals, yet fossil evidence indicates that the ancestors of the velvet worms were marine. Comparison of fossil marine velvet worms and their contemporary counterparts illustrates why these animals are "living fossils"—their bodies have changed little since the Cambrian period, approximately 500 to 550 MYA (**figure 32.20**). During that time, an enormous variety of animal body plans originated and underwent great change. Today's terrestrial velvet worms are among our closest ties to this important period in the history of animal evolution. **Cambrian explosion, p. 340**

Velvet worms appear to be the closest existing relatives of the arthropods, the largest and most diverse animal phylum. Among the similarities between the two groups is an open circulatory system in which a colorless fluid called hemolymph bathes cells in a watery solution of minerals, organic molecules, and gases (see figure 32.2). An elaborate network of air-filled tubes connects the animal's hemolymph with the atmosphere (see figure 33.2).

The function of hemolymph is somewhat similar to that of our own blood, except that we use iron-containing hemoglobin as an oxygen-transport protein, whereas arthropod hemolymph contains a different protein, hemocyanin. Hemocyanins form a large family of copper-containing proteins in arthropods and mollusks. When bound to O_2, hemocyanins lend a blue tinge to an arthropod's hemolymph.

a.

b.

FIGURE 32.20 Velvet Worms, Past and Present. (**A**) A living velvet worm and its young. (**B**) This fossil velvet worm, called *Aysheaia*, was a soft-bodied, segmented animal that lived hundreds of millions of years ago. Its resemblance to a modern velvet worm is evident.

At one time, scientists thought that the hemolymph of velvet worms lacked oxygen transport proteins of any kind. This would make sense, given that these animals are considered living fossils. But researchers Kristina Kusche, Hilke Ruhberg, and Thorsten Burmester from Johannes Gutenberg University and the University of Hamburg in Germany decided to take a closer look. First, they withdrew hemolymph from *Epiperipatus* sp., a living velvet worm from Costa Rica. They purified hemocyanin from the hemolymph—clear evidence that velvet worms do indeed produce this protein.

Next, the researchers wanted to compare the velvet worm's hemocyanin molecule with that of existing arthropods. They extracted RNA from two adult worms, created DNA complimentary to the RNA, and looked for hemocyanin-specific DNA sequences. Once they found the gene they were seeking, they used the gene's DNA sequence to predict the amino acid sequence of the velvet worm's protein (see chapter 12). The team compared that protein with the sequences of similar proteins in 18 arthropod species and used the differences to construct a phylogenetic tree. The "outgroup," or ancestral protein sequence that formed the base of the tree, was a group of copper-containing proteins called phenoloxidases, which are ancient members of the same protein family as hemocyanins.

The analysis revealed that the velvet worm's hemocyanin protein sequence was most closely related to the phenoloxidases (**figure 32.21**). Apparently, hemocyanins were

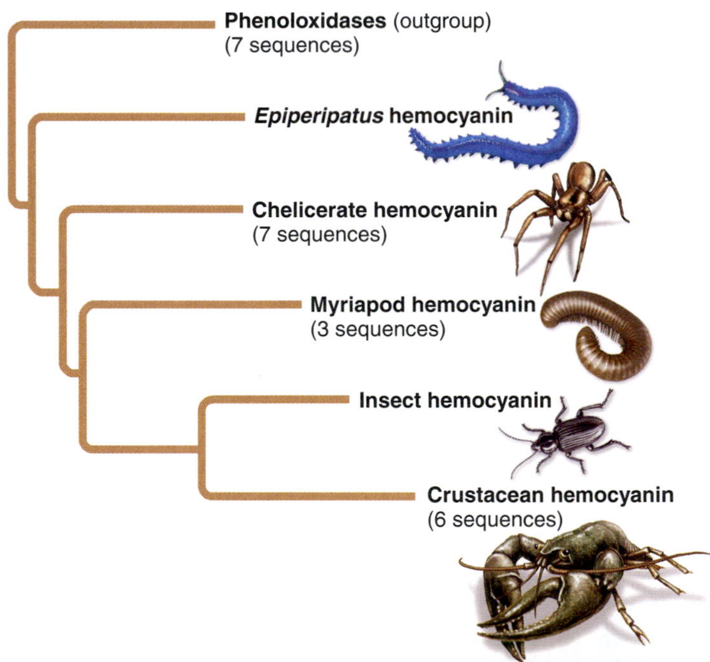

FIGURE 32.21 Hemocyanin Phylogenetic Tree. This tree shows differences among the amino acid sequences of hemocyanins from *Epiperipatus* velvet worms and four groups of arthropods. **Question**: The researchers used this tree to hypothesize that hemocyanins originated from copper-containing enzymes called phenoloxidases. What other types of evidence could you collect to test this hypothesis?

used as respiratory pigments before the split that divided the velvet worms from the chelicerates, an ancient group of arthropods that includes horseshoe crabs and scorpions. Since both of these groups occur in fossils from the Cambrian period, the use of hemocyanin as an O_2 carrier most likely evolved even earlier. Kusche and the other researchers propose that an ancestor common to both arthropods and Onychophorans "recruited" the copper-based phenoloxidase proteins as O_2 carriers, perhaps following gene duplication and subsequent mutations.

Biologists sometimes seem hooked on model organisms such as *C. elegans*, *E. coli*, *Drosophila*, and *Arabidopsis*. These species have made immense contributions to our knowledge of biology and evolution. The velvet worm researchers know, however, that even the most obscure species—those with the very first blue blood—have something to tell us about the early events of animal evolution.

Kusche, Kristina, Hilke Ruhberg, and Thorsten Burmester. August 6, 2002. A hemocyanin from the Onychophora and the emergence of respiratory proteins. *Proceedings of the National Academy of Sciences*, vol. 99, 10545–10548.

CHAPTER SUMMARY

32.1 Circulatory Systems Deliver Nutrients and Remove Wastes

- A **circulatory system** consists of **blood** or a similar fluid, a network of vessels, and a **heart**. The fluid delivers nutrients and O_2 to the **interstitial fluid** surrounding the body's cells, removes metabolic wastes, and transports other substances.

A. Circulatory Systems Are Open or Closed

- In many animals, a heart pumps the fluid to the body cells. In an **open circulatory system,** the fluid bathes tissues directly in open spaces before returning to the heart.
- In a **closed circulatory system,** such as that of vertebrates, the heart pumps the fluid through a system of vessels to cells and back.

B. Natural Selection Has Shaped Vertebrate Circulatory Systems

- Vertebrate circulatory systems increased in complexity as animals became better adapted to dry land. **Pulmonary circulation** delivers oxygen-depleted blood to the lungs, and then **systemic circulation** brings freshly oxygenated blood to the rest of the body.

32.2 The Human Heart Is a Muscular Pump

- The heart is the muscular pump that drives blood through the vessels of the human **cardiovascular system.**

A. The Heart Contains Four Chambers

- A sac of connective tissue, the **pericardium,** surrounds the heart. The **myocardium** is the muscular portion of the heart wall.
- The heart has two **atria** that receive blood and two **ventricles** that propel blood in the body. The heart's **atrioventricular (AV) valves** and **semilunar valves** ensure one-way blood flow.

B. Each Atrium Delivers Blood to a Ventricle

- **Pulmonary arteries** and **pulmonary veins** transport blood between the right side of the heart and the lungs.
- The left side of the heart pumps blood to the rest of the body.
- **Coronary arteries** supply blood to the heart muscle itself.

C. Cardiac Muscle Cells Produce the Heartbeat

- A **cardiac cycle** consists of a single contraction and relaxation of the heart muscle.
- The **pacemaker** or **sinoatrial (SA) node,** a collection of specialized **cardiac muscle** cells in the wall of the right atrium, sets the heart rate. From there, heartbeat spreads to the **atrioventricular (AV) node** and then along specialized fibers through the ventricles.

D. Exercise Strengthens the Heart

- Exercise increases the heart's **cardiac output** and lowers blood pressure.

32.3 Blood Vessels Form the Circulation Pathway

- The walls of **arteries** and **veins** consist of an inner layer of **endothelium,** a middle layer of **smooth muscle,** and an outer layer of connective tissue. Arteries have thicker, more elastic walls than veins.

A. Blood Flows from Arteries to Capillaries to Veins

- Blood leaves the heart through the **aorta** and travels in increasingly narrower arteries and **arterioles** to the **capillary beds,** where nutrient and waste exchange occur.
- Blood typically flows from the capillaries to **venules** and then to veins, and it reenters the heart through the venae cavae. In a **portal system,** however, blood passes through a second set of capillaries before returning to the heart.

B. Blood Pressure and Velocity Differ Among Vessel Types

- The pumping of the heart and constriction of blood vessels produces **blood pressure. Systolic pressure** reflects the force exerted on blood vessel walls when the ventricles contract. The low point, **diastolic pressure,** occurs when the ventricles relax.

- Blood pressure is highest in the arteries and lowest in the veins. Because of high total cross-sectional area, blood velocity is lowest in the capillaries.
- The autonomic nervous system speeds or slows heart rate. **Vasoconstriction** and **vasodilation** adjust blood pressure.

32.4 Blood Is a Complex Mixture

- Human blood is a mixture of cells and cell fragments, proteins, and molecules that are dissolved or suspended in **plasma.** Blood cells originate in bone marrow.

A. Red Blood Cells Transport Oxygen

- **Red blood cells** contain abundant **hemoglobin,** a pigment that binds O_2 molecules. Surface markers on these cells react with antibodies in **agglutination** reactions that reveal a person's blood type.

B. White Blood Cells Fight Infection

- The five types of **white blood cells** provoke inflammation, destroy infectious organisms, and secrete antibodies. **Leukemia** is a type of cancer in which bone marrow produces too many white blood cells.

C. Blood Clotting Requires Platelets and Plasma Proteins

- **Platelets** are cell fragments that collect near a wound. Damaged tissue activates plasma proteins that trigger the formation of a network of fibers, trapping additional platelets and perpetuating **blood clot** formation.

32.5 The Lymphatic System Maintains Circulation and Protects Against Infection

- The **lymphatic system** includes a network of **lymph capillaries** that collect fluid from the body's tissues, purify it, and return it to the blood.
- **Lymph nodes** filter **lymph,** and the **spleen** and **thymus** release specialized white blood cells.

32.6 Investigating Life: Living Fossils Reveal the Long History of Blue Blood

- Researchers investigating the origin of oxygen-carrying hemocyanin proteins have studied velvet worms, terrestrial "living fossils" that are close relatives of arthropods.
- The hemolymph of velvet worms contains hemocyanin, suggesting that these copper-containing proteins originated in a common ancestor of velvet worms and arthropods.

MULTIPLE CHOICE QUESTIONS

1. A property of an open circulatory system includes
   a. the absence of a heart.
   b. movement of fluid into spaces in the tissues of an organism.
   c. the absence of vessels.
   d. All of the above.

2. What is the advantage of a four-chambered heart?
   a. It can support blood flow through a larger organism.
   b. It separates the blood from the pulmonary and systemic circulation.
   c. It enhances the mixing of blood from the pulmonary and systemic circulation.
   d. Both a and c.

3. The function of the pericardium is to
   a. prevent the flow of blood back into an atrium.
   b. contract, causing the blood to move.
   c. allow blood to flow into a ventricle from an atrium.
   d. protect the heart.

4. Which chamber of the heart collects the oxygenated blood from the lungs?
   a. Left atrium
   b. Left ventricle
   c. Right atrium
   d. Right ventricle

5. How would the sinoatrial node of an athlete at rest differ from that of a sedentary person?
   a. It would establish a higher rate of contraction.
   b. It would not be any different.
   c. It would establish a lower rate of contraction.
   d. It would trigger a stronger contraction.

6. Which type of blood vessel uses a thick layer of smooth muscle contractions to control blood flow?
   a. A capillary
   b. A venule
   c. A vein
   d. An arteriole

7. If a person suffers from high blood pressure, then it is possible that
   a. his blood vessels are abnormally dilated.
   b. his blood vessels are abnormally constricted.
   c. his kidneys are removing too much fluid.
   d. his medulla is underactive.

8. Why are people with type O blood considered to be universal donors?
   a. Because type O cells do not carry surface antigens.
   b. Because there are more people with type O than any other blood type.
   c. Because everyone has the O antigen.
   d. Because type O cells lack an Rh group.

9. Which type of blood cell is responsible for clot formation?
   a. Red blood cells
   b. White blood cells
   c. Plasma
   d. Platelets

10. How is lymph fluid related to blood plasma?
    a. They are both protein-rich fluids.
    b. Blood cells are transported in both.
    c. Lymph is plasma that has leaked out of blood vessels.
    d. Only lymph carries white blood cells.

TESTING YOUR KNOWLEDGE

1. Describe the relationship between blood and the interstitial fluid.
2. How do open and closed circulatory systems differ?
3. What is the advantage of separating the pulmonary and systemic circulatory pathways?
4. What causes the heartbeat? Why is the heart sometimes called "two hearts that beat in unison?"
5. Trace the pathway of an O_2 molecule from the lungs to a respiring cell at the tip of your finger.
6. Explain why blood pressure is highest in the arteries and lowest in the veins.
7. What process discussed in the chapter illustrates positive feedback? Negative feedback?
8. Maintaining the proper proportions of cells and platelets in the blood is essential for health. What can happen when the blood contains too few or too many red blood cells? Too few or too many white blood cells? Too few or too many platelets?
9. Why can a person with type A blood not receive a transfusion of type B blood? If a person has type O blood, what blood type(s) can he or she receive in a transfusion? Does it matter if the Rh type is "–" or "+"?
10. Where does lymph originate? What propels it through the lymphatic vessels? How is the lymphatic system connected with the circulatory system? The immune system?
11. What are the roles of cardiac, smooth, and skeletal muscle in the circulatory system?
12. Describe the interactions between the circulatory system and the respiratory, immune, digestive, and endocrine systems.

THINKING AS A SCIENTIST

1. What types of changes in blood vessels would raise blood pressure?
2. HIV colonizes lymph nodes before it appears in the bloodstream. How is the presence of HIV in lymph nodes dangerous?
3. Explain how women who have had cancerous breasts and the associated lymph nodes removed can develop swollen arms.
4. Name three ways that the circulatory system helps maintain homeostasis.
5. Athletes usually are slim and strong, have low blood pressure, do not smoke, and alleviate stress through exercise. How might these characteristics complicate a study designed to assess the effects of exercise on the circulatory system?
6. For reasons that scientists do not fully understand, diabetes (see chapter 30) impairs the circulatory system. Explain how damage to the blood vessels and peripheral nerves of the kidneys, eyes, and extremities cause many of the complications of diabetes, including blindness, kidney failure, and poor wound healing.

ARIS™ **Visit www.mhhe.com/hoefnagels for practice quizzes, animations, videos, and activities designed to help you master the material in this chapter.**

CHAPTER 33

The Respiratory System

Early Arrival. A premature baby is tiny and fragile and requires round-the-clock medical care.

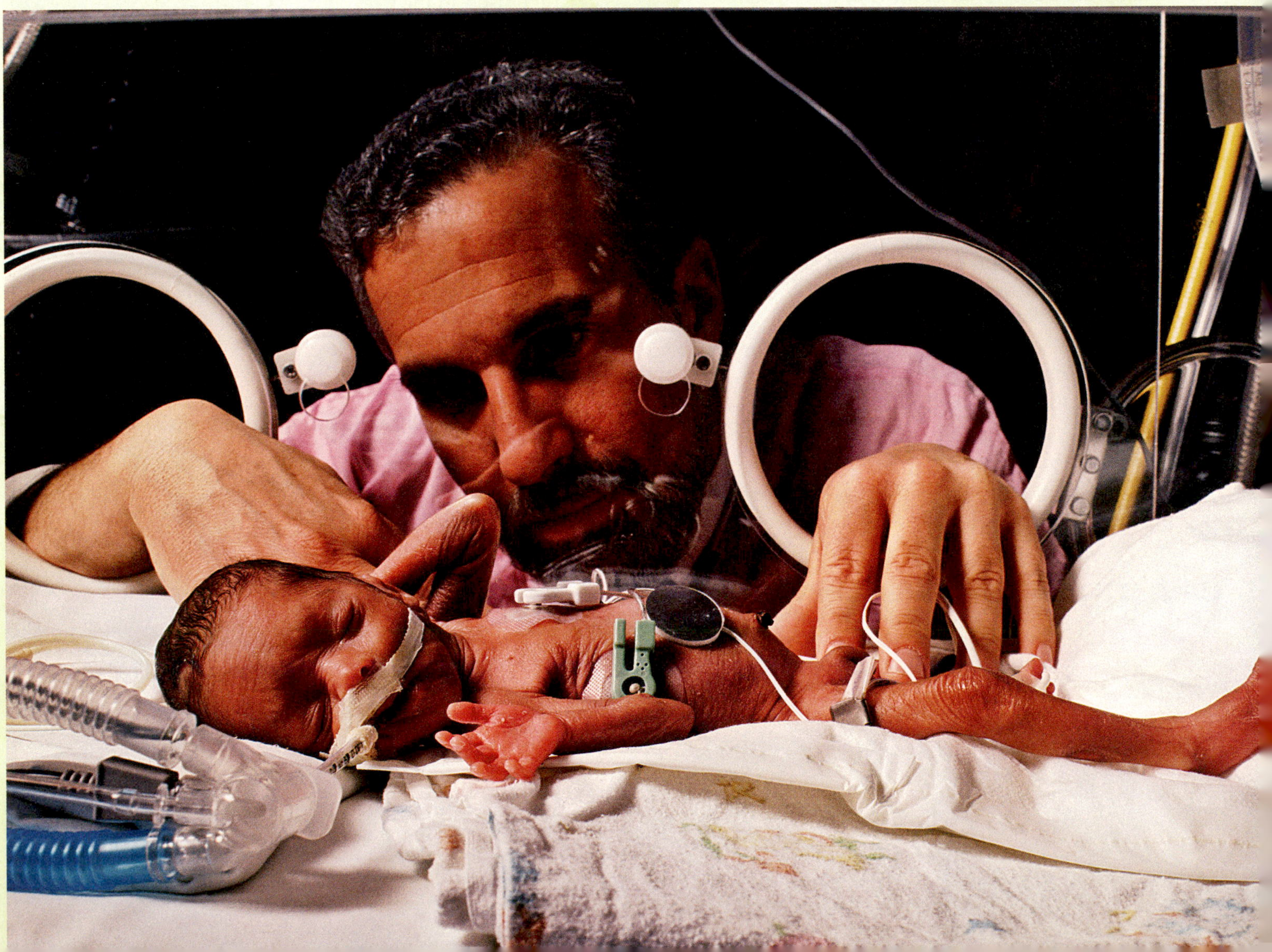

What's the Point?

Premature Babies Have a Tough Time Breathing

A baby often emerges from the birth canal with a bluish color that may be frightening to new parents. The first breath of a newborn requires 15 to 20 times the strength needed for subsequent breaths. The infant must force air into millions of partially inflated air sacs for the first time. As oxygen rapidly diffuses into the bloodstream and reaches the tissues, the baby turns a robust pink and lets out a yowl.

Life is especially difficult for premature babies. Normally, a pregnancy lasts about 9 months, or from 38 to 42 weeks. A baby is considered premature if it is born at 37 weeks or earlier. The shorter the gestation, the more health problems the infant will have. Usually a baby born after about 25 weeks of gestation can survive with intensive care; a fetus younger than that is usually not mature enough to survive.

While in the womb, a fetus gets all the oxygen it needs through the placenta. After birth, however, the baby must acquire all of his or her own oxygen. One problem that premature babies face is that their lungs are not yet ready to breathe air. In fully developed lungs, oxygen gas enters the blood through tiny, moist air sacs called alveoli. The sacs would collapse if not for a chemical called lung surfactant that counters the cohesive force of water molecules. This surfactant is especially important at birth. Once the first breath expands the lungs, the surfactant that has accumulated for the preceding weeks keeps the air sacs open, easing subsequent breaths.

Many premature babies have not yet produced enough surfactant to prevent the alveoli from collapsing after each breath, which causes infantile respiratory distress syndrome. The premature newborn must fight as hard for every breath as a full-term infant does for the first one. For this reason, a pregnant woman who knows she will deliver a premature baby may receive hormones that speed the production of surfactant in the fetus. After birth, synthetic surfactant dripped into the respiratory tract can help these tiny patients breathe.

In addition, a mechanical ventilator or fine breathing tubes may help keep a premature infant's lungs inflated. A monitor constantly measures the amount of oxygen in the blood—too much can be as harmful as too little. Even if the baby can breathe on its own, its underdeveloped brain may not control breathing as it should. Neonatal intensive care wards therefore continuously monitor babies for apnea, the cessation of breathing.

The challenges that premature babies face illustrate the importance of the respiratory system in supplying the oxygen that all of the body's tissues need to function properly. This chapter describes the structure and function of this vital organ system.

LEARNING OUTLINE

33.1 Animals Exchange Gases Across Respiratory Surfaces

FIGURE 33.1 It's Cold Outside. Warm air exhaled from the lungs contains moisture that condenses when it hits cold outside air, making "foggy breath" a sure sign of frigid weather.

Breathing comes so naturally that it is easy to forget why we do it (**figure 33.1**). Cells require ATP to power movement, growth, DNA replication, protein synthesis, and countless other activities that require energy. As described in chapter 6, animal cells generate ATP in **aerobic respiration,** which consumes oxygen gas (O_2) and generates carbon dioxide gas (CO_2) as a waste product. The **respiratory system** works with the circulatory system to acquire and deliver O_2 to cells and to eliminate the CO_2 waste. Without gas exchange, cells die. **ATP, p. 85**

Animals use different areas of the body in gas exchange (**figure 33.2**). These varied **respiratory surfaces** all have two things in common. First, they consist of moist membranes that allow gases to diffuse freely into and out of the body. (Figure 4.12 illustrates diffusion, the spontaneous movement of a substance from a region where it is more concentrated to a region where it is less concentrated). Second, respiratory surfaces must have enough surface area to meet metabolic requirements.

A. In Invertebrate Animals, Gas Exchange May Occur in Several Locations

In the simplest gas exchange system, cells are close enough to the external environment that gases can diffuse across the body surface and reach all cells. For example, cnidarians such as *Hydra* and sea anemones can obtain oxygen by diffusion across their thin, extended body parts. Flatworms are also thin enough for gas exchange and distribution to occur by diffusion. In larger animals, diffusion cannot effectively distribute gases. Some organisms, such as earthworms, exchange gases across the body surface, but they also have a circulatory system that transports gases between cells and the body surface. **cnidarians, p. 455; flatworms, p. 456; annelids, p. 460**

In most terrestrial arthropods, including insects, centipedes, millipedes, and some spiders, tracheae are the main organs of gas exchange. **Tracheae** are extensively branched internal tubules that connect to the atmosphere through openings of the exoskeleton (spiracles) that occur along the animal's segments. Inside the animal, tracheae branch into tiny fluid-filled tubules that extend around individual cells. These tubules bring the outside environment close enough to nearly every cell for gases to diffuse in and out.

B. Gills Exchange Gases in Water

In aquatic organisms, the respiratory surface usually takes the form of **gills,** highly folded structures that exchange gases directly with water. Gills occur in invertebrates, such as mollusks, and in vertebrates such as fishes and amphibians. The

FIGURE 33.2 Diversity in Respiratory Surfaces. An earthworm has the simplest type of respiratory surface: diffusion across a thin layer of cells at the body surface. Many land-dwelling insects exchange gases using an extensive network of tracheae. A fish uses gills, and terrestrial vertebrates use lungs.

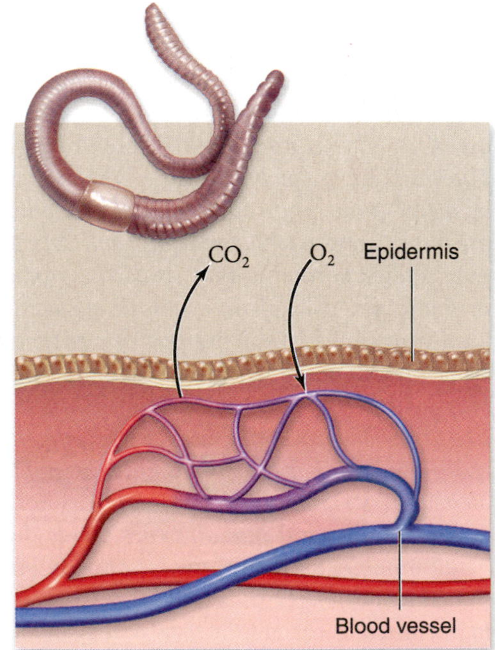

Body surface: Gases diffuse into the cells through the body's surface. Found in many invertebrates and amphibians.

adult fish's complex internal gills create an extensive surface area (**figure 33.3**A,B). A continuous supply of water flows from the mouth over the gills. Within each gill is a dense network of **capillaries,** the tiniest of blood vessels. Oxygen diffuses from the water into the blood, and CO_2 diffuses in the opposite direction. **fishes, p. 484**

Blood flows through the capillaries in the opposite direction that the water flows over the gill membrane (figure 33.3C). This anatomical arrangement is one example of **countercurrent exchange,** in which two adjacent currents flow in opposite directions and exchange materials with each other (chapter 35 describes other examples). Countercurrent exchange maximizes the amount of oxygen extracted from the water. Oxygen diffuses along its concentration gradient from water to blood along the entire length of the capillary bed.

Tadpoles and other amphibian larvae have external gills that close off as lungs begin working after metamorphosis. Some salamanders, however, retain external gills through adulthood. **amphibians, p. 486**

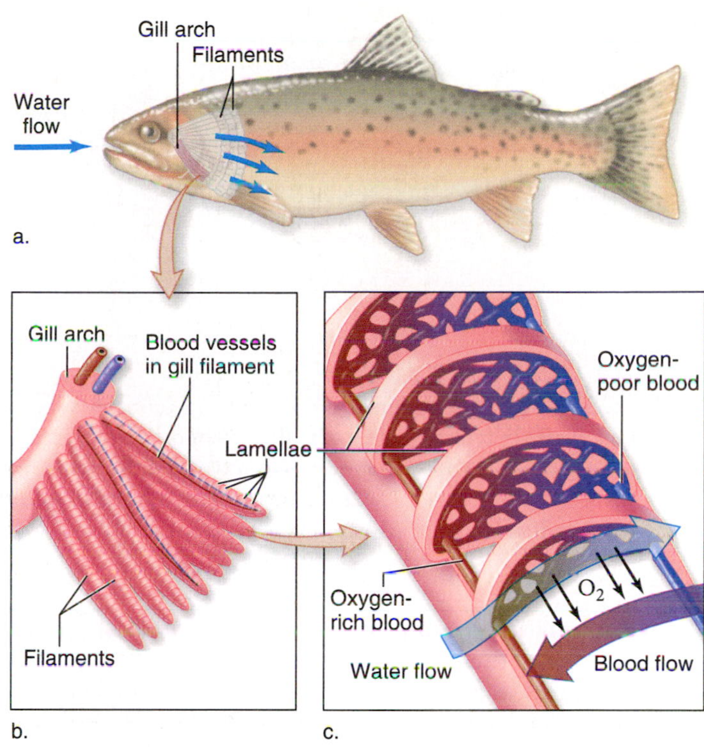

FIGURE 33.3 **Fish Gills.** (**A**) The feathery gills of a fish lie beneath a protective cover. Each side of the head has four gill arches, and each arch consists of many filaments. (**B**) A filament houses blood vessels and is divided into many lamellae. (**C**) The direction of water flow across the lamellae opposes that of blood flow in the capillaries. This countercurrent relationship maximizes the amount of oxygen that the blood receives.

C. Terrestrial Vertebrates Exchange Gases in Lungs

As organisms ventured onto the land hundreds of millions of years ago, gills became obsolete; they would simply dry up without a watery environment. The movement of vertebrates to terrestrial habitats drove the evolution of saclike **lungs** in these air-breathing organisms. The location of lungs inside bodies kept the gas exchange surfaces moist.

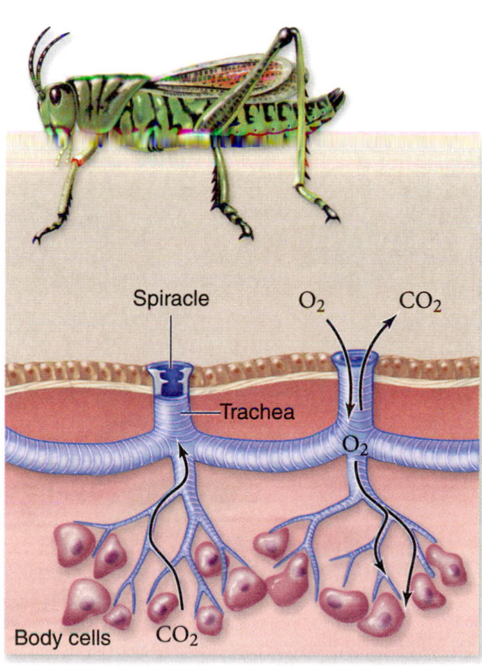

Tracheae: Gases enter tracheae and diffuse directly into the cells; they do not enter capillaries first as in other respiratory systems. Found in arthropods.

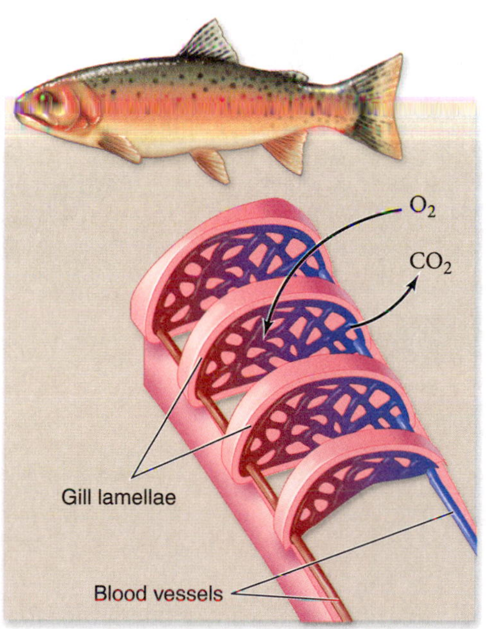

Gills: Found in fishes, amphibian larvae, and many invertebrates.

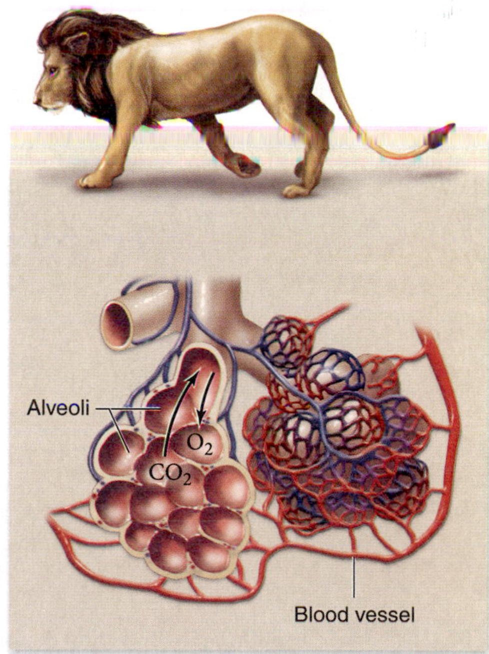

Lungs: Two-way airflow. Found mainly in terrestrial vertebrates.

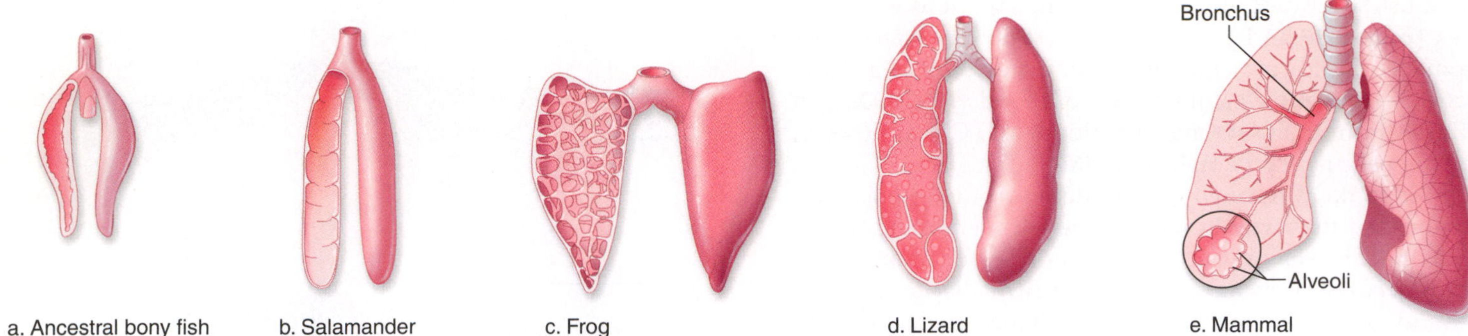

FIGURE 33.4 Lung Evolution. (**A**) The first lungs probably supplemented gills during droughts and were little more than paired sacs with rich capillary networks in their smooth walls. (**B**) As vertebrates moved onto land, lungs became more elaborate, first developing a few subdivisions in salamanders, then (**C**) becoming more extensively subdivided in frogs and toads. (**D**) The reptile lung, illustrated by that of the lizard, is even more clearly compartmentalized. (**E**) The mammalian lung takes maximization of surface area to an extreme, with millions of microscopic air sacs wrapped in capillaries.

A look at lung complexity in modern species gives an idea of how these organs evolved (**figure 33.4**). The simplest lungs, such as those in lungfishes, are little more than air-filled pouches that apparently evolved from swim bladders. Salamander lungs resemble those of lungfishes. The lungs of frogs and toads, however, introduce a few segments, an adaptation that increases surface area. A reptile's lung has much more extensive subdivisions, and the mammalian lung is truly spectacular in its maximization of surface area; millions of tiny air sacs sit within baskets of capillaries. The complex mammalian lung supports the high metabolic rate necessary to maintain body temperature and a very active existence. **reptiles, p. 487; mammals, p. 491**

Lungs introduced a new challenge: moving air in and out to renew the oxygen supply. Frogs force air into their

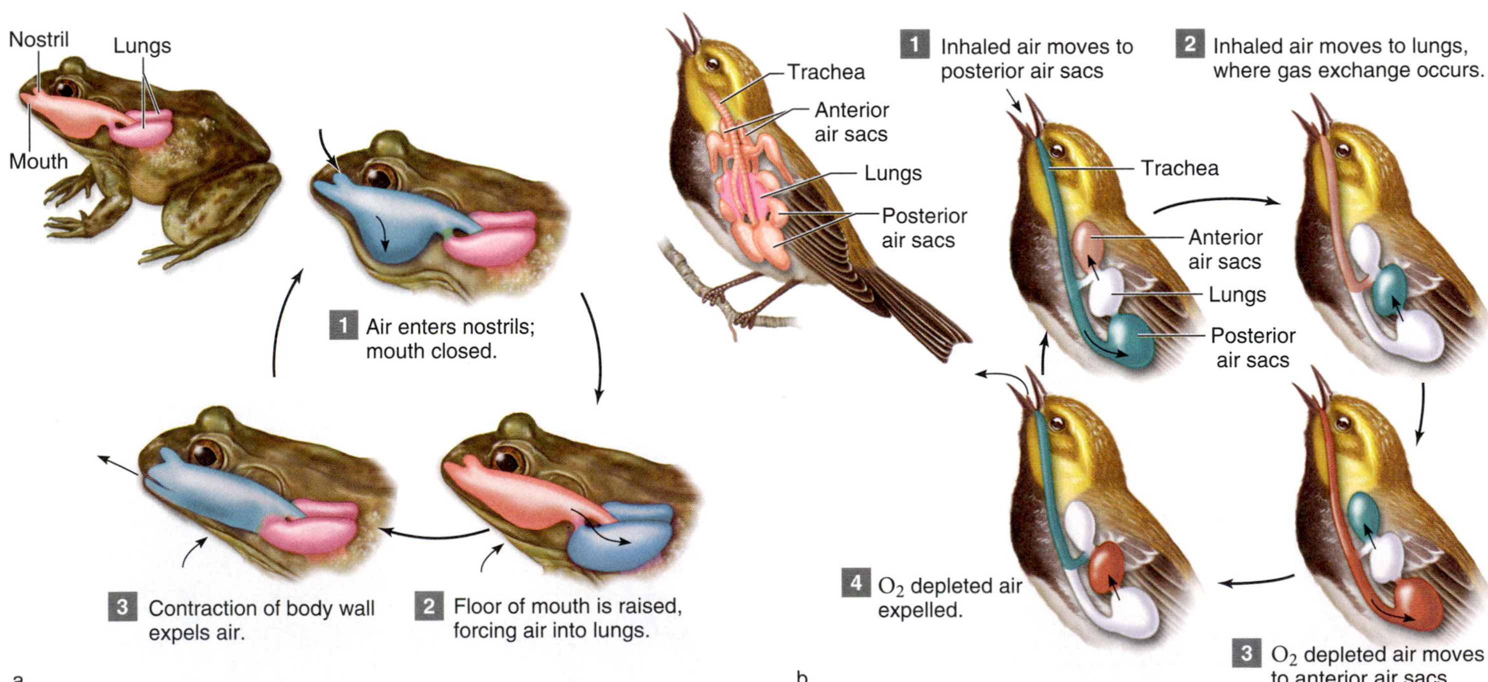

FIGURE 33.5 Breathing in Frogs and Birds. (**A**) A frog draws air into its nostrils with its mouth shut, raises the floor of the mouth, and forces the air into the only place it can go—the lungs. The air is expelled when the body wall contracts and the lungs recoil. A frog obtains additional oxygen through its moist skin and the lining of its mouth. (**B**) In a bird, air moves in one direction through the lungs, and each breath remains inside the animal for two cycles of inhaling and exhaling. Air goes first to the posterior air sacs, then passes through the lungs and enters the anterior air sacs before being released. This figure traces the path of a single gulp of air (*blue*) as it makes its way through the air sacs and lungs.

lungs, and then expel it by contracting muscles in the body wall (**figure 33.5**A). Birds, on the other hand, have an extensive respiratory system through which air flows in one direction (figure 33.5B). Oxygen remains in the respiratory system even during exhalation, providing the continuous oxygen supply required to power flight. In mammals, including humans, a muscular diaphragm expands the chest area, pulling air into the lungs. Section 33.2 describes this topic in more detail. **birds, p. 489**

Summary *Respiratory systems acquire oxygen and eliminate carbon dioxide. The exchange of gases occurs across respiratory surfaces such as the body surface, tracheae, gills, and lungs.*

33.1 MASTERING CONCEPTS

1. What process in animal cells requires oxygen and generates carbon dioxide?
2. What is the relationship between the circulatory and respiratory systems?
3. What size or shape are animals that use the body wall as the main respiratory surface?
4. Describe how tracheae, gills, and lungs participate in gas exchange.
5. Describe how lungs differ among the main groups of terrestrial vertebrates.

33.2 The Human Respiratory System Delivers Air to the Lungs

The human respiratory system is a continuous network of tubules that delivers oxygen to the circulatory system. **Figure 33.6** presents an overview of the system, and **table 33.1** summarizes its functions.

Respiratory System	
Main tissue types	**Examples of Locations/Functions**
Epithelial	Enables diffusion across walls of alveoli and capillaries; secretes mucus along respiratory tract
Connective	Blood (a connective tissue) exchanges gases with lungs; cartilage makes up part of the nose, trachea, bronchi, and larynx
Nervous	Autonomic nervous system controls smooth muscle in bronchi
Muscle	Smooth muscle in lungs regulates airflow to alveoli; skeletal muscle in diaphragm expands lungs

FIGURE 33.6

The Human Respiratory System. Inhaled air passes through the mouth and trachea and then into increasingly narrow tubes until it arrives in the lung's microscopic air sacs, the alveoli, where gas exchange occurs.

Table 33.1 *Functions of the Human Respiratory System*

Function	Explanation
Gas exchange	Lungs exchange oxygen (O_2) and carbon dioxide (CO_2) with blood.
Olfaction (sense of smell)	Breathing moves air across the nose's olfactory epithelium, which detects odors.
Production of sounds, including speech	Movement of air across the vocal cords in the larynx produces sounds.
Maintaining blood pH homeostasis	Breathing more slowly or rapidly adjusts the concentration of CO_2 in blood, which affects blood pH.

A. The Nose, Pharynx, and Larynx Form the Upper Respiratory Tract

The **nose,** which forms the external entrance to the nasal cavity, functions in both breathing and the sense of smell. Stiff hairs at the entrance of each nostril block the inhalation of large particles. Most airborne bacteria and dust particles that pass the hairs catch in sticky mucus secreted by the nose's epithelium. If they are inhaled further, they are trapped by mucus lower in the respiratory tract and swept back out by waving cilia. (A large inhaled particle may trigger a sensory cell in the nose that signals the brain to orchestrate a sneeze, which forcefully ejects the particle.) The nose also adjusts the temperature and humidity of incoming air. **sense of smell, p. 602**

The back of the nose and mouth leads into the **pharynx,** or throat. The pharynx conducts both food and air. A reflex action steers food and fluid toward the digestive system during swallowing.

Just below and in front of the pharynx is the **larynx,** or Adam's apple, a boxlike structure that produces the voice. Stretched over the larynx are the **vocal cords,** two elastic bands of tissue that vibrate as air from the lungs passes over them through a slit-like opening called the **glottis.** These vibrations produce the sounds of speech. The expansion of the Adam's apple and lengthening of the vocal cords causes the deepening voice of a male during puberty.

B. The Lower Respiratory Tract Consists of the Trachea and Lungs

The **trachea,** or windpipe, is the tube just beneath the larynx. C-shaped rings of cartilage hold the trachea open and accommodate the expansion of the esophagus—the tube leading from the mouth to the stomach—during swallowing. You can feel these rings in the lower portion of your neck. Along the inside surface of the trachea, cilia and mucus filter, warm, and moisten incoming air.

The trachea branches into two **bronchi,** one leading to each lung. The bronchi branch repeatedly, each branch decreasing in diameter and wall thickness (**figure 33.7**). **Bronchioles** ("little bronchi") are the finest branches. The bronchioles have no cartilage, but their walls contain smooth muscle. The autonomic nervous system controls contraction of these muscles, adjusting airflow in response to metabolic demands. **autonomic nervous system, p. 586**

Each bronchiole narrows into several alveolar ducts, and each duct opens into a grapelike cluster of alveoli, where gas exchange occurs (**figure 33.8**). Each **alveolus** is a tiny sac with a wall of epithelial tissue that is one cell layer thick. A vast network of capillaries surrounds each cluster of alveoli. Oxygen and CO_2 diffuse through the thin walls of the alveoli and the neighboring capillaries. The interface between the alveoli and the circulatory system's capillaries is the respiratory surface in humans.

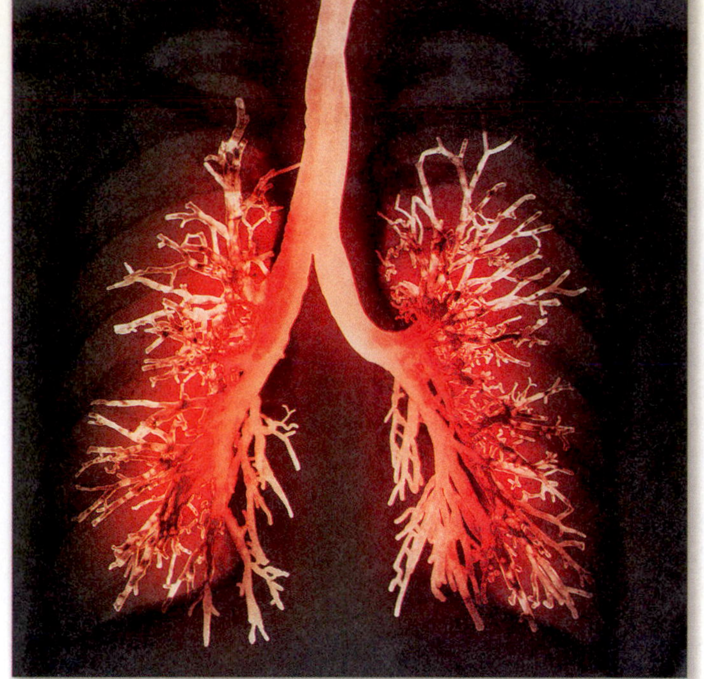

FIGURE 33.7 The Bronchial Tree. Respiratory passages in the lungs form a complex branching pattern.

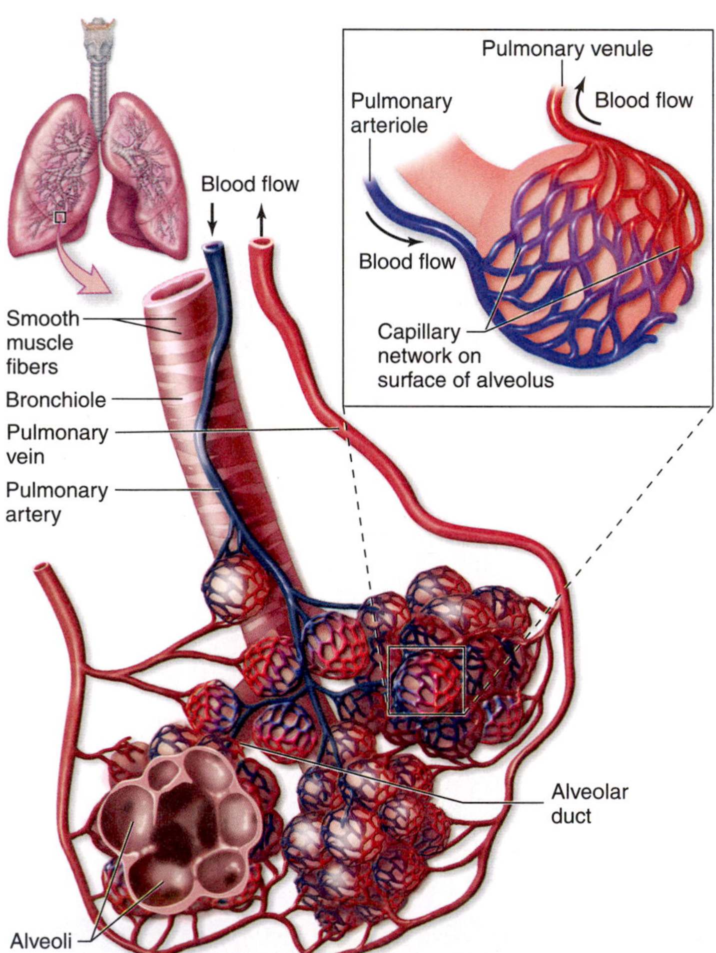

FIGURE 33.8 Alveoli. Most of the lung is composed of some 300 million alveoli, which makes the lung structure similar to that of foam rubber. Gas exchange between the lungs and the bloodstream occurs at the lush capillary network that surrounds each cluster of alveoli. The respiratory surface is enormous—the total surface area of the alveoli in one set of human lungs is about 50 times the area of the skin!

C. Breathing Requires Pressure Changes in the Lungs

Breathing moves air back and forth between the atmosphere and the lungs. Air moves from areas of high pressure to areas of low pressure. Therefore, air moves into the lungs when the air pressure in the lungs is lower than the pressure outside the body. Conversely, air moves out when the pressure in the lungs is greater than the atmospheric pressure.

Pay attention to your own breathing for a moment. Each **respiratory cycle** consists of one inhalation and one exhalation (**figure 33.9**). Each time you **inhale,** air moves into the lungs. How does this happen? The skeletal muscles of the rib cage and diaphragm contract, expanding and elongating the chest cavity. The resulting increase in volume lowers the air pressure within the space between the lungs and the outer wall of the chest. This causes the lungs to expand, lowering pressure in the alveoli. Air rushes in. Inhalation requires energy because muscle contraction uses ATP. **muscle movement, p. 642**

Exhalation moves air out of the lungs. When the muscles of the rib cage and the diaphragm relax, the rib cage falls to its former position, the diaphragm rests up in the chest cavity again, and the elastic tissues of the lung recoil. When the pressure in the lungs exceeds atmospheric pressure, air moves out. At rest, exhalation is passive—that is, it requires only muscle relaxation, not contraction—and therefore does not require ATP.

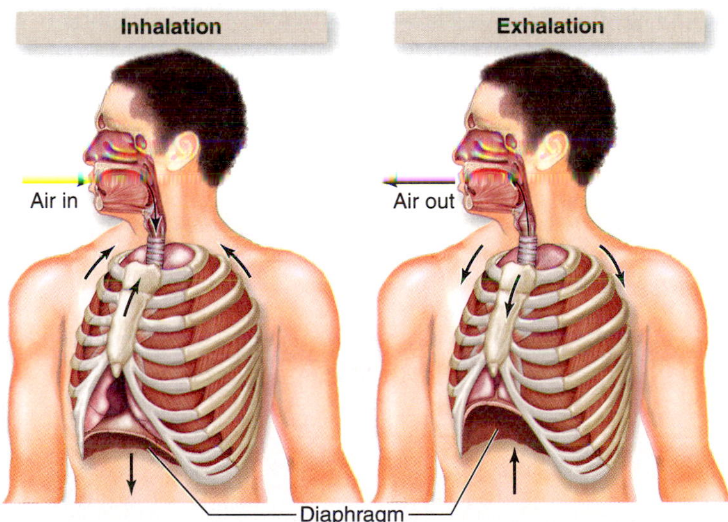
Inhalation
Exhalation
Air in
Air out
Diaphragm
Contraction of muscles in diaphragm and rib cage expands chest cavity, drawing air in
Relaxation of muscles in diaphragm and rib cage allows lungs to recoil, expelling air

FIGURE 33.9 How We Breathe. Inhalation requires contraction of the muscles of the diaphragm and rib cage. The expanding chest cavity has lower air pressure than the atmosphere, so air moves into the lungs. When we exhale, the diaphragm relaxes and the rib cage lowers, reversing the process and pushing air out of the lungs.

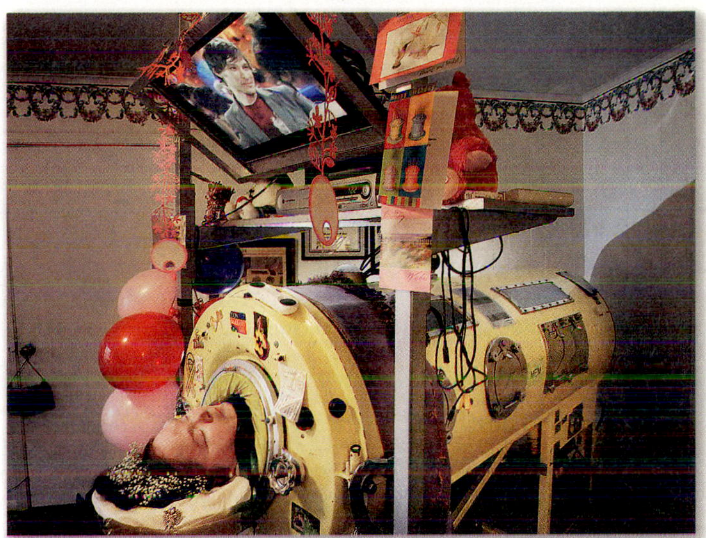

FIGURE 33.10 Breathing Machine. This 60-year-old woman has been in an iron lung for 57 years. She was a small child when she contracted polio, a viral disease that caused her muscles to deteriorate. Because her diaphragm is paralyzed, she cannot inhale on her own.

Old-fashioned "iron lungs" and modern mechanical ventilators replace the role of the diaphragm and rib muscles (**figure 33.10**). An iron lung is an airtight chamber in which the patient lies with his or her head sticking out. Lowering the air pressure inside the machine draws air into the lungs through the nose or mouth; raising the pressure causes exhalation.

This chapter's Can *You* Relate? The Unhealthy Respiratory System describes some breathing disorders.

Summary *Air is cleansed and warmed in the nose, then transported through the pharynx, larynx, and trachea to the lungs. Here, air continues moving through bronchi and bronchioles, ending at the alveoli, where gas exchange occurs. Air moves into and out of lungs because of pressure differences between the lungs and the outside air. Contraction of muscles in the diaphragm and rib cage causes the lungs to inflate.*

33.2 MASTERING CONCEPTS

1. What are the components of the upper respiratory tract? The lower respiratory tract?
2. How does the nose clean inhaled air?
3. Which structures prevent swallowed food from entering the respiratory system?
4. Describe the structures and functions of the trachea, bronchi, bronchioles, and alveoli.
5. What is the relationship between the volume of the chest cavity and the air pressure in the lungs?
6. Describe the events of one respiratory cycle (inhalation and exhalation).

Can *You* Relate?

The Unhealthy Respiratory System

Asthma

During an asthma attack, spasms occur in the smooth muscle lining the lung's bronchi, slowing airflow and causing wheezing. An allergy to pollen, dog or cat dander (skin particles), or tiny mites in house dust trigger most asthma attacks. Inhalant drugs that treat asthma usually relax the bronchial muscles.

Lung Cancer

All cancers, including lung cancer, result from cells dividing out of control. Patients with lung cancer experience chest pain, chronic coughing, and shortness of breath. If the tumor grows large enough, it may obstruct the airway. Pieces of the tumor may also break away and spread to other parts of the body. Eighty-five percent of lung cancer cases occur in smokers (**figure 33.A**). Other risk factors include exposure to asbestos or radioactivity. **cancer, p. 166**

Apnea

Apnea is the cessation of breathing. It can be voluntary for a short time, as when you "hold your breath." Premature infants may stop breathing because the part of the brain that regulates breathing is not fully developed (see the chapter opening essay). Overweight adults are especially susceptible to sleep apnea. The fleshy folds of the throat sag into the airway, causing breathing to stop for a short time. The person reflexively clears the throat, and breathing resumes. Because sleep is interrupted many times a night, people with sleep apnea may be extremely tired during the day.

Pneumonia

Pneumonia is an inflammation of the alveoli, usually resulting from an infection. A variety of bacteria cause pneumonia. Symptoms include green or yellow mucus, fever, chest pain, coughing, and shortness of breath.

Common Cold

The hundreds of viruses that cause the common cold infect cells in the upper respiratory tract (see chapter 17). A day or so after infection, the immune system's efforts to get rid of the invading viruses cause the typical symptoms of a cold: coughs, runny nose, and sneezes. There is no cure. Fortunately, colds are usually not a serious health problem, although clogged sinuses can be vulnerable to bacterial infections.

Tuberculosis

Infection with the bacterium *Mycobacterium tuberculosis* causes tuberculosis (also called TB). Symptoms of advanced cases of TB include painful coughs, bloody sputum, fever, and weight loss. When a patient with an active TB infection coughs or sneezes, he or she expels bacteria-laden droplets that nearby people can inhale. Once in the lungs, the bacteria replicate inside immune system cells in alveoli. Other immune system cells clump around the infected cells, a response that keeps the bacteria from spreading. But these clusters also provide a place for the bacteria to become dormant, producing a latent infection that may later reemerge as full-blown (active) TB. Tuberculosis is especially important today because of the emergence of *Mycobacterium* strains that resist antibiotic treatment.

Emphysema

The term *emphysema,* which comes from the Greek word for "inflate," is an enlargement of the lungs. Long-term exposure to cigarette smoke and other irritants cause the bronchi to swell, trapping air in the lungs. The walls of the alveoli tear, reducing the surface area for gas exchange. The patient experiences shortness of breath, an expanded chest, and hyperventilation. Most patients are men older than 50 years of age.

Cystic Fibrosis

Cystic fibrosis is among the most common inherited diseases. A faulty transport protein in the membranes of epithelial cells causes sticky mucus to accumulate in the lungs. The thick mucus prevents cilia from beating. Bacteria also thrive in the warm, moist mucus, causing persistent infections. (The same faulty transport protein also causes digestive problems, as described in chapter 4.)

The Effects of Smoking on the Respiratory System

Besides increasing the risk of heart attack and stroke (see chapter 32), tobacco use also disrupts the respiratory system. The very first inhalation of cigarette smoke slows the beating of cilia. With time, the cilia become paralyzed, and they eventually vanish. Without cilia to remove mucus, coughing alone must clear particles from the airways. Smoking also causes production of excess mucus, which favors the reproduction of disease-causing microorganisms. Smokers are therefore especially susceptible to respiratory infections.

The smoker's cough leads to chronic bronchitis. As the linings of the bronchioles thicken and become less elastic, they can no longer absorb the pressure changes that accompany coughing. A cough can therefore rupture the delicate walls of alveoli, causing a worsening cough, fatigue, wheezing, and impaired breathing. As these changes progress to emphysema, lung cancer may also develop and spread throughout the body. Tobacco use also increases the risk of cancers of the mouth, larynx, esophagus, and jaw.

Although the addictive properties of nicotine make quitting difficult, it pays to stop smoking. Cilia may reappear, and the thickening of alveolar walls can reverse, although ruptured alveoli are gone forever.

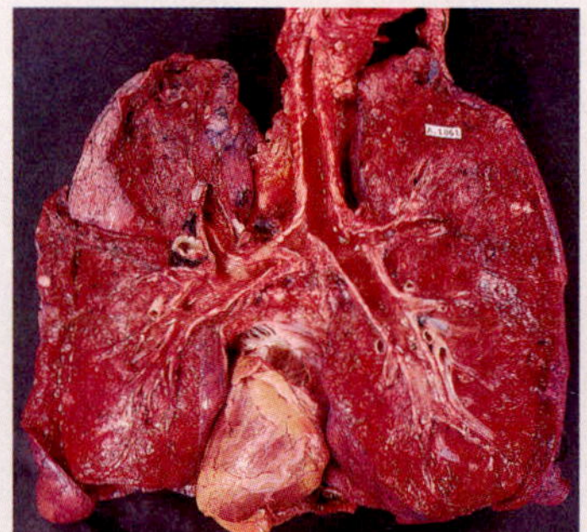

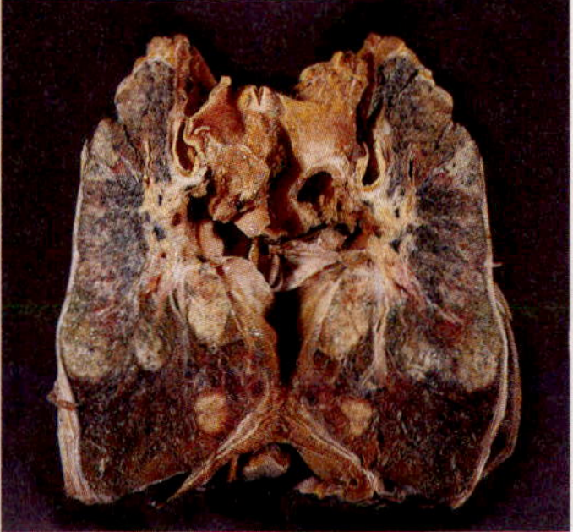

FIGURE 33.A Smoking Causes Cancer. Long-term exposure to the chemicals in cigarette smoke causes lung cancer. Compare the healthy lungs on the left to the cancer-ridden lungs on the right.

33.3 Blood Delivers Oxygen and Removes Carbon Dioxide

The respiratory and circulatory systems are intimately related. The respiratory system is responsible for breathing—the lungs take in O_2 and expel CO_2. The circulatory system, in turn, transports and delivers these gases (see figure 32.4). In the **pulmonary circulation,** oxygen-poor blood travels from the heart to the capillaries surrounding the alveoli, expels CO_2 and picks up O_2, and returns to the heart. In **systemic circulation,** arteries deliver freshly oxygenated blood to the rest of the body's tissues, delivering O_2 and picking up CO_2. Then, the oxygen-depleted but CO_2-rich blood travels back to the heart, and from there to the lungs. As a result, the air we exhale has a different composition than inhaled air (**table 33.2**).

A. Blood Carries Gases in Several Forms

How does blood carry oxygen? **Blood** is a connective tissue consisting of cells suspended in **plasma,** the liquid component of blood (see figure 32.15). **Red blood cells** are packed with **hemoglobin,** an iron-rich pigment that carries O_2. (Only about 1% of O_2 in blood is dissolved in plasma.) Each hemoglobin molecule can carry up to four O_2 molecules. A red blood cell becomes almost completely saturated with O_2 after spending only a second or two in the alveolar capillaries.

Not surprisingly, illness or death results when hemoglobin cannot bind O_2. Carbon monoxide (CO), for example, is a colorless, odorless gas in cigarette smoke and in exhaust from car engines, kerosene heaters, woodstoves, and home furnaces. CO binds to hemoglobin more readily than O_2 does, and it is less likely to leave the hemoglobin molecule. When 30% of the hemoglobin molecules carry CO instead of O_2, a person loses consciousness and may go into a coma or even die.

Blood carries CO_2 in three ways. A small amount is dissolved in plasma. Hemoglobin transports about 23% of the CO_2. (Hemoglobin can carry both O_2 and CO_2 at the same time because the gases attach to different sites on the molecule.) About 70% of CO_2 is transported as bicarbonate ions (HCO_3^-) dissolved in plasma. CO_2 reacts with water to form carbonic acid (H_2CO_3), which dissociates to hydrogen ions (H^+) and HCO_3^-, as the following chemical equations show (the two double arrows indicate that the reactions are reversible):

$$CO_2 + H_2O \longleftrightarrow H_2CO_3$$

$$H_2CO_3 \longleftrightarrow H^+ + HCO_3$$

Carbonic acid and bicarbonate are an important part of blood's ability to buffer pH changes and maintain homeostasis. Ordinarily, the respiratory and excretory systems work together to maintain the pH of blood at about 7.4. Several metabolic processes, however, release H^+ into the blood; an example is the production of lactic acid in strenuously exercising muscles. If blood pH declines, excess H^+ reacts with HCO_3^- and forms carbonic acid, raising the pH. If the pH is too high, H_2CO_3 absorbs the excess OH^- ions, releasing additional H^+ into the blood. **pH, p. 32**

The transfer of gases between the lungs and the body's other tissues relies on simple diffusion (**figure 33.11**). At the lungs, O_2 diffuses from the alveoli into the bloodstream, where the concentration of O_2 is lower. At the same time, CO_2 diffuses from the blood to the air spaces of the lungs. The heart then pumps the freshly oxygenated blood to the rest of the body. O_2 diffuses from the bloodstream to the tissue fluid and then into the body's respiring cells, which have the lowest O_2 concentration. CO_2 diffuses in the opposite direction. **diffusion, p. 90**

B. Blood Gas Levels Help Regulate the Breathing Rate

The regulation of breathing rate helps maintain homeostasis in blood gas concentrations. The most important signal that the brain uses to set the breathing rate is the blood CO_2 level (or, more precisely, the concentration of H^+ formed when CO_2 reacts with H_2O and forms H_2CO_3). Because CO_2 is a byproduct of aerobic respiration, monitoring blood CO_2 levels is a good way to determine how quickly cells use oxygen.

Table 33.2 *Comparison of Inhaled and Exhaled Air*

Gas	Proportion by Volume in Inhaled Air (at sea level) (%)	Proportion by Volume in Exhaled Air (%)
Oxygen (O_2)	21	16.5
Carbon dioxide (CO_2)	0.03	4
Nitrogen (N_2)	78	79.5
Other gases	≈ 1	≈ 1

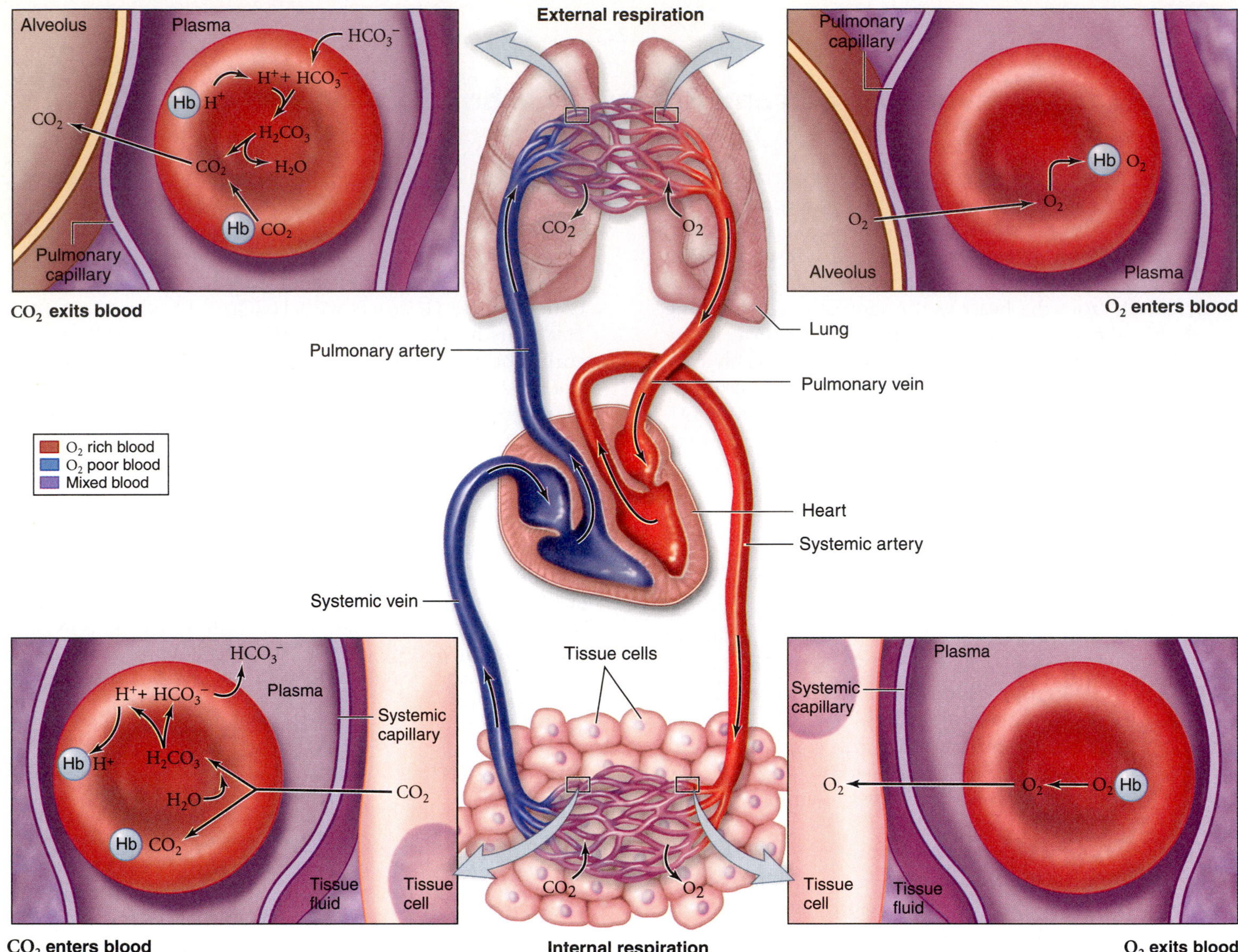

FIGURE 33.11 Gas Exchange at the Lungs and Tissues. Most O_2 in blood is bound to hemoglobin (Hb), whereas CO_2 moves in the blood in three ways: dissolved in plasma, carried in hemoglobin, and as bicarbonate ion (HCO_3^-). In the lungs, CO_2 diffuses out of red blood cells and plasma and into the alveoli of the lungs. O_2 moves in the opposite direction. In the systemic circulation, O_2 diffuses from blood and to tissues, while CO_2 moves into the bloodstream.

Chemoreceptors in the medulla of the brain, and to some extent those in the body's largest arteries, monitor blood CO_2 levels (**figure 33.12**). When the blood CO_2 level increases, the resulting decline in pH stimulates chemoreceptors to send messages to the medulla, which triggers an increase in the breathing rate. Small changes in the CO_2 concentration of the air we breathe can trigger a tremendous increase in breathing rate. For example, an increase to 0.5% CO_2 in the air (from the normal 0.03%) makes us breathe 10 times faster.

Oxygen level is less important than CO_2 level in regulating breathing. In fact, the concentration of O_2 in blood only affects breathing rate if it falls dangerously low. (This chapter's Burning Question: How does the body respond to high elevations? discusses mountaintop breathing.) This does not normally happen, however, because the great amount of O_2 bound to hemoglobin provides a large safety margin. Sudden infant death syndrome, in which a baby dies while asleep, may occur when chemoreceptors fail to detect low oxygen levels in arterial blood.

Summary *Red blood cells deliver O_2 to tissues. Blood carries CO_2 mostly as HCO_3^-, but some CO_2 is dissolved in plasma or bound to hemoglobin. Gases move between tissues, blood, and alveoli by simple diffusion. Chemoreceptors detect blood pH, which reflects CO_2. The brain regulates breathing rate, which helps maintain blood pH homeostasis.*

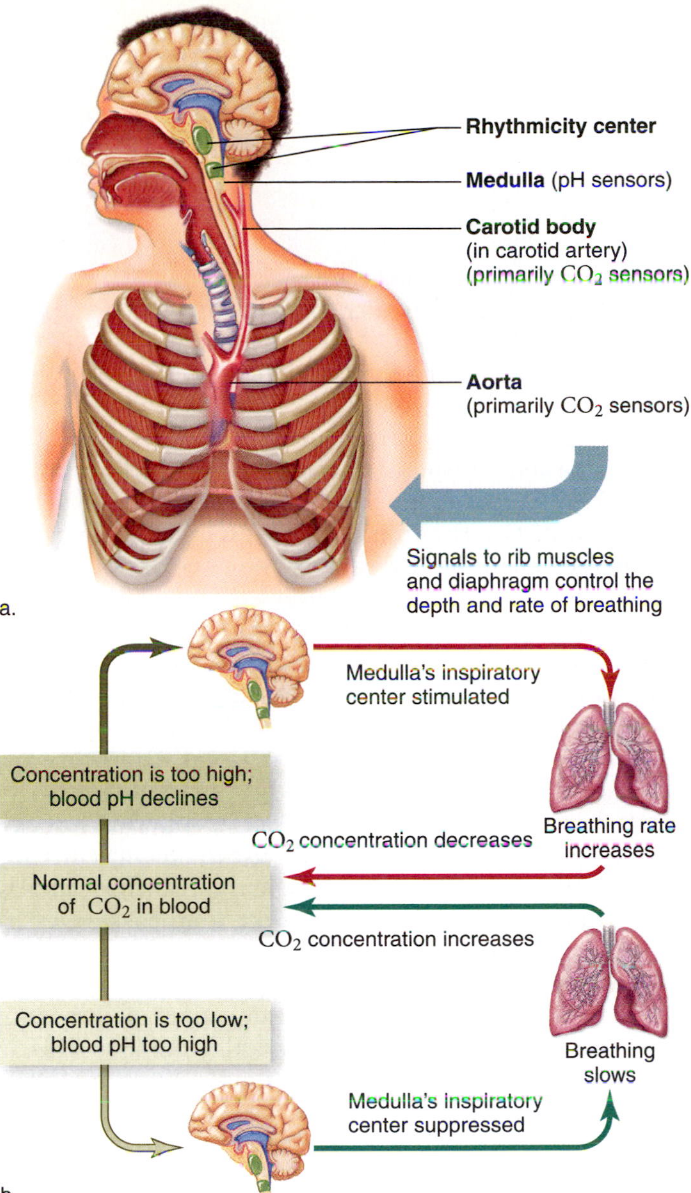

FIGURE 33.12 Breathing Control. (**A**) The brain's medulla has receptors that monitor H^+ concentration in the cerebrospinal fluid. Receptors in the aorta and carotid artery detect CO_2, H^+, and O_2 in the bloodstream and pass the information to the rhythmicity center in the medulla. (**B**) The control of breathing rate is an example of a negative feedback loop.

33.3 MASTERING CONCEPTS

1. What protein in the red blood cell delivers oxygen to the body's tissues?
2. How does carbon monoxide reduce blood's ability to carry oxygen?
3. What is the relationship between carbon dioxide, carbonic acid, and bicarbonate ions?
4. Describe the diffusion gradients for O_2 and CO_2 in the lungs and in the rest of the body.
5. How does the brain detect and respond to blood CO_2 level?

Burning Questions

How does the body respond to high elevations?

The human respiratory system functions best near sea level. At high elevations, air density and oxygen availability fall gradually, so that at about 3000 m (almost 10,000 feet) above sea level, an individual inhales a third less O_2 with each breath. Each year, 100,000 mountain climbers and high-altitude exercisers experience altitude sickness (**table 33.A**). More than 180 climbers have died attempting to reach Mt. Everest's 8850-m (29,028-foot) summit, many from the effects of low oxygen. At high altitudes, the body's effort to get more O_2, by increasing breathing and heart rate, cannot keep pace with declining O_2 concentration. Altitude sickness can be cured by descending slowly at the first appearance of symptoms.

The body also reacts to the low O_2 supply at high altitudes by increasing red blood cell production. When cells in the kidneys do not receive enough O_2, they trigger production of the hormone erythropoietin (also called EPO), which stimulates red blood cell production. Synthetic erythropoietin is a treatment for anemia (a reduction in the O_2-carrying capacity of blood), but some athletes inject themselves with EPO to gain a performance edge. This illicit practice is a form of blood doping, which improves an athlete's stamina but is difficult for sports authorities to detect. It can be dangerous because it thickens the blood and therefore strains the heart.

Have a Burning Question of your own? Submit it to marielle_hoefnagels@mcgraw-hill.com for possible inclusion in future editions of this book!

Table 33.A ***The Effects of High Altitude***

Condition	Altitude	Symptoms
Acute mountain sickness	1,800 m (5,900 feet)	Headache, weakness, nausea, poor sleep, shortness of breath
High-altitude pulmonary edema (fluid accumulation in lungs)	2,700 m (9,000 feet)	Severe shortness of breath, cough, gurgle in chest, stupor, weakness; person can drown in accumulated fluid in lungs
High-altitude cerebral edema (fluid accumulation in brain)	4,000 m (13,000 feet)	Brain swells, causing severe headache, vomiting, altered mental status, loss of coordination, hallucinations, coma, and death

33.4 INVESTIGATING LIFE
Why Do Bugs Hold Their Breath?

A child having a tantrum may scream, "I'll hold my breath until I turn blue!" in an effort to get his way. Luckily, the parents have nothing to fear; even if the child does begin to carry out his plan, his brain's breathing center will eventually override his effort to hold his breath. In insects, the situation is different. An insect can hold its breath much longer than we can, and its parents might actually approve, because evidence suggests that an insect that periodically holds its breath may live a long, healthy life.

An insect's respiratory system consists of air-filled tracheae extending inward from valvelike openings (spiracles) along the abdomen and thorax (see figure 33.2). Gases diffuse significantly more rapidly in air than in water or blood, and calculations show that an insect can easily exchange all the O_2 and CO_2 it needs by keeping its spiracles open all the time. Yet some insects close their spiracles for extended periods; in effect, the animal periodically holds its breath. In one breathing cycle, the spiracles may be open for 15 min, closed for 20 min, and then "flutter" open and closed for over an hour before opening again for 15 min.

Scientists have speculated for years about how this strange pattern benefits an insect. Some thought that discontinuous breathing reduces water loss, much as plants conserve water by closing pores (stomata) in their leaves during hot or windy weather. Yet subsequent studies did not support this hypothesis. For example, grasshoppers close their spiracles at night, when water loss is minimal, but not during the much drier daytime. Another idea was that discontinuous breathing helps underground insects such as ants cope with low O_2 and high CO_2 conditions. But what about aboveground insects that breathe discontinuously?

Biologists Stefan Hetz (from Humboldt University in Berlin) and Timothy Bradley (from the University of California, Irvine) suggested a third explanation. They proposed that discontinuous breathing guards against *too much* O_2. Oxygen, of course, is necessary for aerobic respiration; without it, an animal dies. But O_2 is also toxic. The higher the concentration of O_2, the more harmful free radicals produced in the mitochondria. The more free radicals, the more damage to cells—in fact, studies have shown that insects experimentally exposed to the highest concentrations of O_2 live the shortest time.

Hetz and Bradley thought that discontinuous breathing might help the insect's body maintain a constant, safe level of O_2. To test this idea experimentally, they studied pupae of the Atlas moth (*Attacus atlas*; **figure 33.13**). The pupa is the stage of the life cycle during which the animal transforms from a caterpillar into an adult moth. Hetz and Bradley poked tiny oxygen sensors into spiracles of the Atlas moth pupa and measured the concentration of O_2 inside the tracheae. At the same time, they changed the external concentration of O_2. The results of the experiment supported the "oxygen-guarding" hypothesis: the internal oxygen concentration stayed the same regardless of how much O_2 the insects were exposed to (**figure 33.14**).

The oxygen-guarding hypothesis explains one puzzling aspect of discontinuous breathing—it happens only in resting insects. During periods of high metabolic activity, the O_2 concentration inside the tracheae remains low because cells consume O_2 as quickly as it arrives. An active insect therefore has no reason to close its spiracles. Just as a race car's engine is too powerful for use in everyday driving, however, the insect respiratory system has excess capacity for periods of rest. An inactive insect compensates for its supercharged respiratory system by opening its spiracles only when necessary to prevent toxic buildup of CO_2.

A toddler who refuses to breathe does nothing but annoy his parents. In contrast, a logical conclusion of Hetz and Bradley's work is that discontinuous breathing helps insects live longer by guarding against excess O_2 exposure. If they live longer, they may have more mating opportunities, which may mean greater reproductive success—and an explanation for how natural selection maintains the behavior in insect populations.

Hetz, Stefan K. and Timothy J. Bradley. 2005. Insects breathe discontinuously to avoid oxygen toxicity. *Nature*, vol. 433, pages 516–519.

FIGURE 33.13 **The Atlas Moth, *Attacus atlas*.**

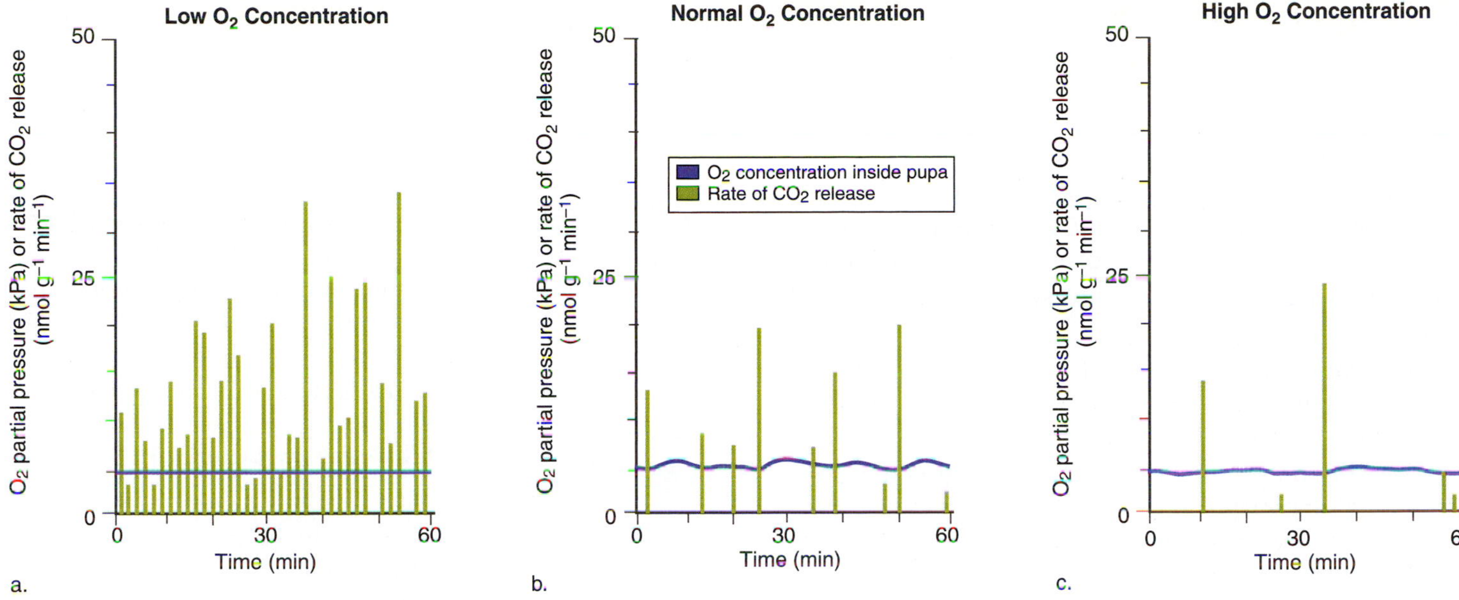

FIGURE 33.14 The Benefit of Discontinuous Breathing. The concentration of O_2 inside of an Atlas moth pupa stays constant, regardless of the external concentration of O_2. **Question:** Cells respire more rapidly at higher temperatures, increasing the demand for O_2. Do you predict that Atlas moth pupae would close their spiracles more frequently or less frequently as the temperature increases? Design an experiment to test your hypothesis.

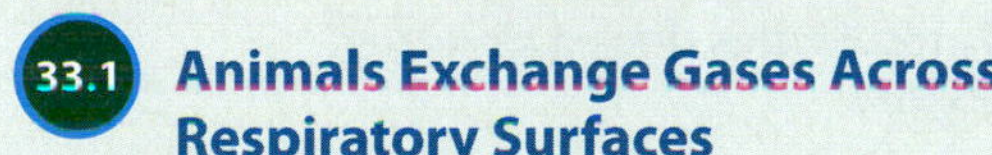

CHAPTER SUMMARY

33.1 Animals Exchange Gases Across Respiratory Surfaces

- Cells use oxygen gas (O_2) in **aerobic respiration** to store nutrient energy in ATP. Carbon dioxide (CO_2) forms as a byproduct of respiration and must be eliminated from the body.
- O_2 and CO_2 are exchanged by diffusion across a moist **respiratory surface.** Body size, metabolic requirements, and habitat have affected the evolution of **respiratory systems.**

A. In Invertebrate Animals, Gas Exchange May Occur in Several Locations

- Simple organisms exchange O_2 and CO_2 directly across the body surface.
- Most terrestrial arthropods bring the environment into contact with almost every cell through a highly branched system of **tracheae.**

B. Gills Exchange Gases in Water

- Many aquatic animals exchange gases across **gill** membranes enclosing networks of **capillaries.** In bony fishes, water flows over the gills in the direction opposite blood flow, an arrangement called **countercurrent exchange.**

C. Terrestrial Vertebrates Exchange Gases in Lungs

- Vertebrate **lungs** create a moist internal surface for two-way gas exchange. Amphibians have the simplest lungs; birds and mammals have the most complex.

33.2 The Human Respiratory System Delivers Air to the Lungs

A. The Nose, Pharynx, and Larynx Form the Upper Respiratory Tract

- The **nose** purifies, warms, and moisturizes inhaled air. The air then flows through the **pharynx** and **larynx.**
- **Vocal cords** stretched over the larynx produce the voice as air passes through the **glottis.**

B. The Lower Respiratory Tract Consists of the Trachea and Lungs

- Cartilage rings hold open the **trachea,** which branches into **bronchi** that deliver air to the lungs. The bronchi branch extensively and form tinier air tubules, **bronchioles,** which end in clusters of thin-walled, saclike **alveoli.**
- Many capillaries surround each alveolus. O_2 diffuses into the blood from the alveolar air, while CO_2 diffuses from the blood into the alveoli.

C. Breathing Requires Pressure Changes in the Lungs

- When the diaphragm and rib cage muscles contract, the chest cavity expands. This reduces air pressure in the lungs, drawing air in. When these muscles relax and the chest cavity shrinks, the pressure in the lungs increases and pushes air out.
- A **respiratory cycle** consists of one **inhalation** and one **exhalation.**

33.3 Blood Delivers Oxygen and Removes Carbon Dioxide

- The **pulmonary circulation** moves **blood** between the heart and lungs, whereas the **systemic circulation** moves blood between the lungs and the rest of the body.

A. Blood Carries Gases in Several Forms

- Almost all oxygen transported to cells is bound to **hemoglobin** in **red blood cells.** Carbon monoxide (CO) poisoning occurs when CO prevents O_2 from binding hemoglobin.
- Some CO_2 in the blood is bound to hemoglobin or dissolved in **plasma.** Most CO_2, however, is transported as bicarbonate ion, generated from carbonic acid that forms when CO_2 reacts with water. An enzyme in red blood cells speeds this reaction.
- In the lungs, where the concentration of O_2 is high and that of CO_2 is low, hemoglobin binds O_2 and releases CO_2 for exhalation. At the body's other tissues, the concentration of O_2 is low, so O_2 diffuses out of the blood and into the respiring cells. CO_2 moves in the opposite direction along its concentration gradient.

B. Blood Gas Levels Help Regulate the Breathing Rate

- Breathing rate adjusts to the body's demands. Chemoreceptors in the medulla and major arteries sense a rise in CO_2 concentration and blood acidity and trigger faster breathing.
- Only critically low O_2 levels change breathing rate.

33.4 Investigating Life: Why Do Bugs Hold Their Breath?

- At rest, some insects breathe discontinuously by alternately opening and closing the openings (spiracles) that connect their tracheae to the atmosphere.
- Experimental evidence suggests that discontinuous breathing helps the insects maintain a safe, low level of O_2 inside their bodies.

MULTIPLE CHOICE QUESTIONS

1. Why do organisms breathe?
   a. To eliminate CO_2
   b. To support aerobic respiration in mitochondria
   c. To support ATP production through glycolysis
   d. Both a and b
2. What is a common feature of tracheae and gills?
   a. Countercurrent exchange
   b. High surface area
   c. Gases diffuse through water
   d. Lungs
3. Which of the following organisms do NOT have lungs?
   a. Earthworms
   b. Amphibians
   c. Lizards
   d. Birds
4. Gas exchange in a human lung occurs at the
   a. bronchioles.
   b. alveoli.
   c. bronchi.
   d. trachea.
5. The voice is produced at the
   a. olfactory epithelium.
   b. bronchioles.
   c. larynx.
   d. trachea.
6. How many O_2 molecules does a saturated hemoglobin carry?
   a. Two c. Four
   b. Three d. Eight
7. What happens to CO in the blood?
   a. It is carried in form of carbonic acid and bicarbonate.
   b. It competes with O_2 for binding to hemoglobin.
   c. It is used instead of O_2 to generate ATP.
   d. Both a and c.
8. Which reaction would most likely occur near the aveoli of the lung?
   a. Formation of CO_2 and water from carbonic acid
   b. Conversion of bicarbonate into carbonic acid
   c. Formation of carbonic acid from CO_2 and water
   d. Conversion of carbonic acid to bicarbonate
9. The release of O_2 from hemoglobin is triggered by
   a. an O_2 concentration gradient.
   b. an acidic environment.
   c. a warmer temperature.
   d. all of the above.
10. What regulates the breathing rate?
    a. Level of O_2
    b. Level of CO_2
    c. Hemoglobin content of blood
    d. Demand for ATP

TESTING YOUR KNOWLEDGE

1. What is the function of breathing?
2. What is the connection between breathing and aerobic cellular respiration?
3. How do an animal's size, activity level, and environment influence the structure and function of its respiratory system?
4. Compare and contrast the structures and functions of the human trachea and an insect's tracheae.
5. Trace the path of an O_2 molecule from a person's nose to a red blood cell at an alveolar capillary.
6. How is air cleaned before it reaches the lungs?
7. Describe inhalation. Which muscles are active in the process?
8. How does the body transport O_2 to cells?
9. How does the respiratory system transport most CO_2? In what other ways is CO_2 transported?
10. How does the brain establish breathing rhythm?

THINKING AS A SCIENTIST

1. A person can choke if a hard candy or other small object obstructs the airway. In drowning, a person's lungs fill with water. Explain how each of these events can cause death.
2. Describe an example of homeostasis mentioned in this chapter.
3. One recommended treatment for anxiety-related hyperventilation (overbreathing) is to breathe into a paper bag for several minutes. After several breaths, how does the composition of air inside the bag compare with that in the atmosphere? How would breathing air from the bag help relieve hyperventilation? How might it be dangerous to breathe into a paper bag for more than a few minutes?
4. A single-celled organism can exchange gases by simple diffusion through its cell membrane, which provides enough surface area for oxygen to diffuse into all parts of the cell. Animals, however, have many cells. Why do animals need respiratory systems to perform a function that one-celled organisms can do alone?
5. It is well below freezing outside, but the dedicated runner bundles up and hits the roads anyway. "You're crazy," shouts a neighbor. "Your lungs will freeze." How is the well-meaning neighbor wrong?
6. How does breathing through the mouth instead of the nose dry out the throat?
7. Why can't you commit suicide by holding your breath?
8. The concentration of oxygen in the atmosphere declines with increasing elevation. Why do you think the times of endurance events at the 1968 Olympics, held in 2200-m (7218-feet) elevation Mexico City, were relatively slow?

Visit www.mhhe.com/hoefnagels for practice quizzes, animations, videos, and activities designed to help you master the material in this chapter.

Digestion and Nutrition

Take a Pill. Many patients with ulcers now take antibiotics plus antacids to combat infection with *Helicobacter pylori* (inset).

What's the Point?

Bacteria Can Cause Gastric Ulcers

A gastric ulcer is a common, painful condition in which the stomach wall becomes irritated and eroded. From 1910 until the mid-1990s, physicians thought ulcers were the direct result of excess stomach acid secretion, which they blamed on stress and eating spicy foods—in short, "hurry, worry, and curry." The finding that drugs that block acid production relieve ulcers (albeit temporarily) seemed to support this hypothesis in the 1970s. The standard treatment for ulcers was a bland diet, stress reduction, acid-blocking drugs, or surgery.

In the 1980s, however, a bacterium called *Helicobacter pylori* was identified in the laboratory of J. Robin Warren at Royal Perth Hospital in Western Australia. Warren had found *H. pylori* in stomach tissue samples from people suffering from gastritis (an inflamed stomach). Were the bacteria attracted to inflamed tissue, or did they cause the inflammation? Warren's assistant, medical resident Barry Marshall, helped choose between the hypotheses. Marshall knew he had a healthy stomach and had never had gastritis or an ulcer. If the bacteria caused the irritation, they would do so in him. So on a hot July day in 1984, Marshall drank a brew containing about a billion of the microbes. (Note that human experiments such as this are potentially dangerous and unethical, regardless of a person's willingness to volunteer!)

Marshall suffered for a few days with gastritis, which cleared up without treatment. A second volunteer, however, was ill for several months. Although bismuth subcitrate (Pepto Bismol) gave temporary relief, the pain returned. Only when he took two antibiotic drugs for a few weeks did the gastritis vanish completely.

Other researchers, using laboratory animals rather than themselves, soon confirmed the causative link between *H. pylori* and gastritis and then a link to ulcers, too. Still, it took until 1994, a full decade after Marshall's self-experiment, for the National Institutes of Health to advise doctors that a 2-week course of two antibiotic drugs plus an antacid should become the standard treatment for gastritis and ulcers in patients with an infection. More recently, researchers have discovered that nonsteroidal antiinflammatory drugs, such as aspirin and ibuprofen, can also cause gastric ulcers. Together, the two factors raise risk significantly; a person who takes these drugs and harbors the bacteria has a 61-times higher chance of developing an ulcer than an individual with neither exposure.

H. pylori inhabits the stomach, part of an extensive internal network of food-processing tubes and compartments. This organ system—the digestive system—is the subject of this chapter.

LEARNING OUTLINE

34.1 Animals Have Diverse Diets and Feeding Strategies

34.2 Animal Digestive Tracts Take Many Forms

34.3 The Human Digestive System Consists of Several Organs

A. Digestion Begins in the Mouth and Esophagus
B. The Stomach Stores, Digests, and Pushes Food
C. The Small Intestine Digests and Absorbs Nutrients
D. The Large Intestine Completes Nutrient and Water Absorption
E. The Pancreas, Liver, and Gallbladder Aid Digestion

34.4 A Healthy Diet Includes Essential Nutrients and the Right Number of Calories

A. A Varied Diet Is Essential to Good Health
B. Body Weight Reflects Food Intake and Activity Level
C. Starvation: Too Few Calories to Meet the Body's Needs
D. Obesity: More Calories than the Body Needs

34.5 Investigating Life: Brawny Vegetarian Lizards Hit the Salad Bar

34.1 Animals Have Diverse Diets and Feeding Strategies

Is it true that "you are what you eat?" In some ways, the answer is yes. After all, the atoms and molecules that make up your body came from food that you ate (or that your mother ate before you were born). But in other ways, the answer is no. A cow may eat grass, yet the animal looks and acts nothing like grass. Clearly, food is not incorporated whole into each animal's body, even though atoms and molecules from food build each animal.

The resolution of this paradox lies in the **digestive system.** A **nutrient** is a substance that an organism uses for metabolism, growth, maintenance, and repair of its tissues. For an animal to use a meal, its body must dismantle large nutrient molecules into smaller components that then enter the bloodstream, which delivers them to the body's cells. The cells use the small molecules from food to obtain energy and raw materials required for growth, repair, and maintenance of the body. In addition, food provides vitamins and minerals that animals cannot manufacture themselves. **ATP, p. 85**

The overall process of obtaining and using food has four major steps. First, **ingestion** is the assimilation of food into the digestive tract (**figure 34.1**). The second stage, **digestion,** is the physical and chemical breakdown of food. Chewing, a form of mechanical digestion, breaks food into small pieces, increasing the surface area exposed to digestive enzymes. In chemical digestion, enzymes add water molecules between the building blocks of large nutrient molecules, splitting them apart. In the third stage, **absorption,** the nutrients enter the cells lining the digestive tract and the bloodstream. Fourth, in **elimination,** the animal's body expels (egests) undigested food. **Feces** are the solid wastes that leave the digestive tract. **hydrolysis, p. 34; enzymes, p. 87**

FIGURE 34.1 Ingestion. Animals, such as this robin, ingest their food. The digestive system breaks down the food, absorbs the nutrients, and eliminates the waste.

a.

b.

FIGURE 34.2 Specialized Feeders. (**A**) The giant panda has a very limited diet, consuming enormous quantities of bamboo. It is an herbivore. (**B**) A baleen whale is a filter feeder. It can swim through a school of krill and filter them from the water column, thanks to a curtain of baleen plates.

Biologists divide animals into four broad categories based on what they eat. **Herbivores,** such as cows and rabbits, eat only plants. Eagles, cats, wolves, and other **carnivores** hunt other animals for food. **Detritivores** consume decomposing organic matter; dung beetles and earthworms illustrate this diet. Finally, some animals are **omnivores,** eating a broad variety of foods including plants and animals. Humans are omnivores, as are raccoons and chickens.

Many animals rely on one or a few kinds of food. Insectivores such as frogs, bats, and spiders eat only insects. A piscivore eats fish, and a frugivore eats fruits. The giant panda (**figure 34.2A**) is a folivore (leaf-eater). Because it eats only bamboo, the giant panda can survive only where that plant thrives. Animals with flexible diets, such as the robin and human, can live in a broader range of habitats.

Humans and most other animals are **bulk feeders,** ingesting large pieces of food. Some, such as snakes and lizards, swallow meals whole, as the opening essay to chapter 6 vividly shows. **Fluid feeders,** in contrast, drink their food: a mosquito takes a "blood meal" after piercing human skin with its mouthparts. Still other animals, called **substrate feeders,** live in their food and eat it from the inside. A female parasitic fly, for example, may lay eggs inside a live grasshopper. When the eggs hatch, the larvae feed on the host's tissues. A **deposit feeder** such as an earthworm is also a type of substrate feeder. These animals strain partially decayed organic matter from soil or other sediments.

Sponges, corals, clams, and mussels are examples of **filter feeders** (also called suspension feeders), which strain particles from water. This strategy is adaptive in animals that do not move about much, because it brings the food to them. Nevertheless, some filter feeders, such as the baleen whale in figure 34.2B, are active swimmers.

Summary *Animals use nutrient molecules in food to acquire energy, raw materials, vitamins, and minerals. The use of food involves four main steps. Animals eat plants, other animals, and decaying organic matter, and they obtain food in various ways.*

34.1 MASTERING CONCEPTS

1. In what way is it true that "you are what you eat?"
2. What four processes does food undergo when an animal eats?
3. Define the terms *herbivore, carnivore, detritivore,* and *omnivore.*
4. List four ways that animals obtain nutrients.

34.2 Animal Digestive Tracts Take Many Forms

Because animals and their food are composed of the same types of chemicals, digestive enzymes could just as easily attack an animal's body as its food. Digestion therefore occurs within specialized compartments.

The simplest organisms, including protists such as *Paramecium,* take in nutrients by phagocytosis and enclose the food in a food vacuole (**figure 34.3**A). A loaded food vacuole fuses with another sac containing digestive enzymes that break down nutrient molecules. Digestion occurs entirely inside the cell (intracellularly), but it remains separate from other functions. Sponges are the only animals that rely solely on intracellular digestion. **phagocytosis, p. 94; sponges, p. 454**

More complex animals use extracellular digestion, producing hydrolytic enzymes in a digestive cavity connected with the outside world (figure 34.3B). The enzymes dismantle large food particles, then cells lining the cavity absorb the products of digestion. In this system, food remains outside the body's cells until it is digested and absorbed. Extracellular digestion eases waste removal because indigestible components of food never enter the body's cells. Instead, the digestive tract simply ejects the waste.

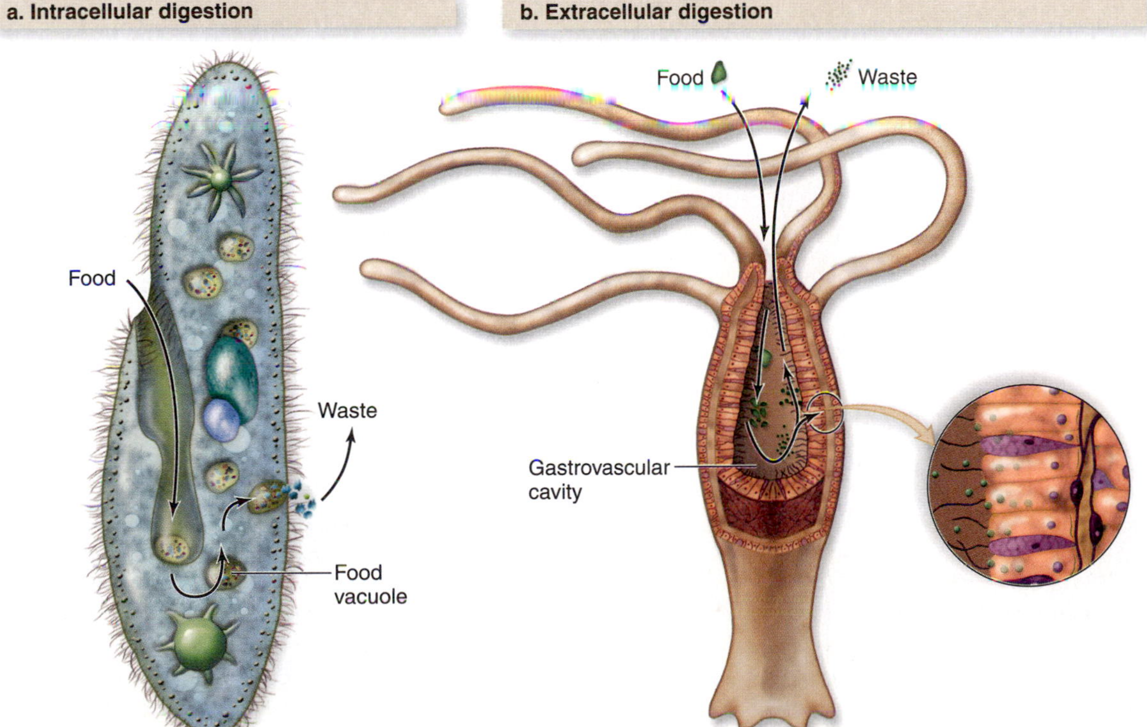

FIGURE 34.3
Intracellular and Extracellular Digestion. (**A**) In intracellular digestion, a vacuole containing food fuses with a sac containing digestive enzymes. Sponges and protists such as *Paramecium* use this as their sole means of digestion. (**B**) Most multicellular animals, including *Hydra*, digest food extracellularly, inside a cavity. The cells lining the cavity absorb the nutrients and continue the breakdown process by intracellular digestion.

The cavity in which extracellular digestion occurs may have one or two openings (**figure 34.4**). An **incomplete digestive tract** has only one opening: a mouth that both ingests food and ejects wastes. The animal must digest food and expel the residue before the next meal can begin. This two-way traffic limits the potential for specialized compartments that might store, digest, or absorb nutrients. Cnidarians such as jellyfish and *Hydra* have incomplete digestive tracts, as do flatworms. In these organisms, the digestive tract is also called a **gastrovascular cavity** because it doubles as a circulatory system that distributes nutrients to the body cells. **cnidarians, p. 455; flatworms, p. 456**

Most familiar animals, including earthworms, mollusks, insects, sea stars, and vertebrates, have complete digestive tracts. A **complete digestive tract** has two openings, and food passes through in one direction. The tubelike digestive cavity is called the **alimentary canal** or **gastrointestinal (GI) tract.** The mouth is the entrance, and the **anus** is the exit. Capillaries near the digestive tract absorb nutrients into the bloodstream for delivery to the body's cells. One advantage of a two-opening digestive system is that regions of the tube can develop specialized areas that break food into smaller particles, digest it, absorb the nutrients, and eliminate

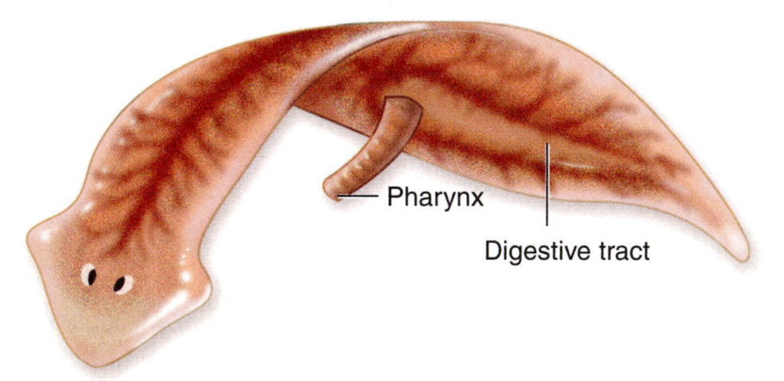

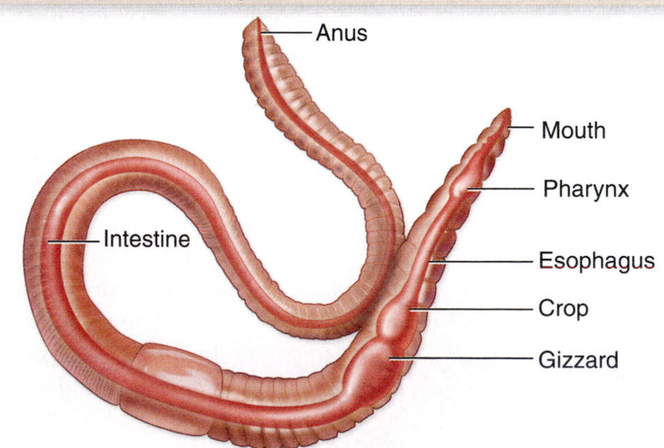

FIGURE 34.4 One-Opening and Two-Opening Digestive Systems. (A) In an animal with a gastrovascular cavity such as this flatworm, food enters and wastes exit through the same opening. **(B)** An earthworm's alimentary canal has two openings: the mouth and anus.

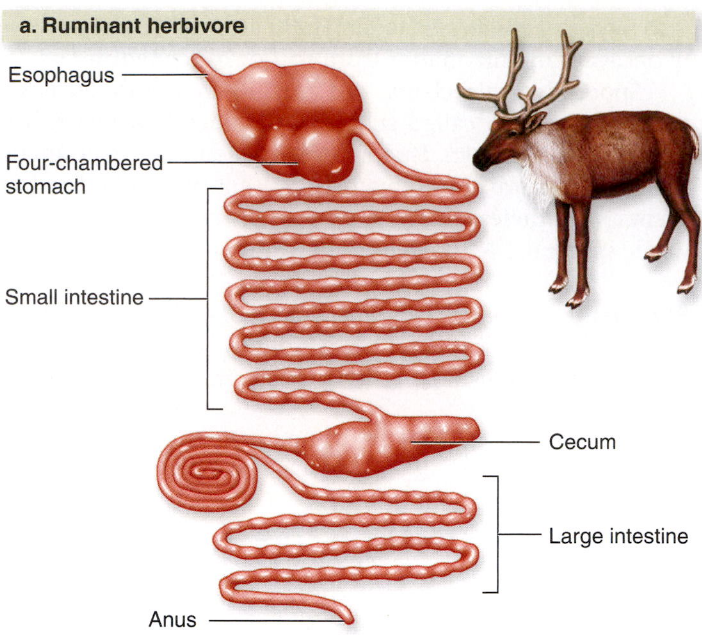

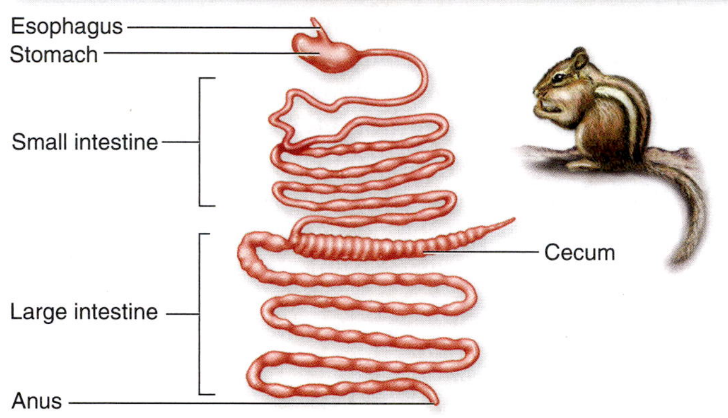

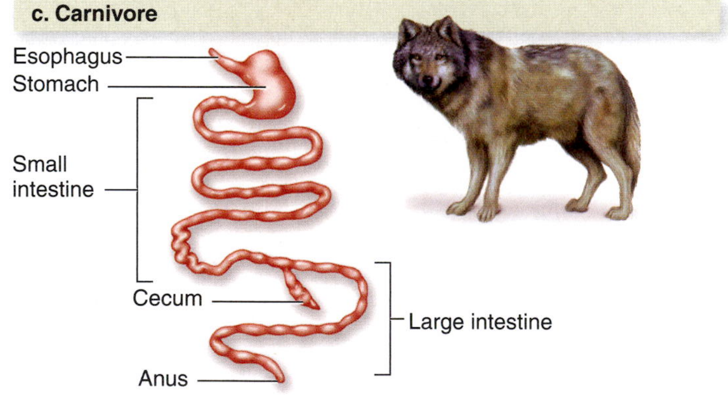

FIGURE 34.5 Digestive System Adaptations. (A) Ruminant herbivores have a characteristic four-part stomach and an extensive gastrointestinal tract. Bacteria break down the cellulose in their fibrous diet. **(B)** The intestines of nonruminant herbivores such as rabbits and rodents also harbor anaerobic microbes that break down cellulose in plants. **(C)** The protein-rich diet of carnivores is easy to digest, so the digestive system is much shorter and has a reduced cecum.

wastes. Digestion is therefore more efficient than in an incomplete digestive tract.

Diet and lifestyle differences select for digestive system adaptations in all animals, including mammals (**figure 34.5**). An herbivore's diet is rich in hard-to-digest cellulose from the cell walls of plants. The long digestive tract allows extra time for digestion. In addition, **ruminants** such as cows, sheep, deer, and goats have a four-chambered stomach. Grass enters the **rumen,** the first section of the stomach. Bacteria break the plant matter down into balls of cud, which the cow regurgitates into its mouth. Chewing the cud breaks the food down further. When the cow swallows again, the food bypasses the rumen, continuing digestion in the other three sections of the stomach.

Farther along the digestive tract, a pouch called the **cecum** forms the entrance to the large intestine. In herbivores, the cecum is large, and it houses bacteria that ferment plant matter. Carnivores, like the wolf, eat mostly protein. Their digestive tracts have short intestines and a small or absent cecum. The cecum is medium-sized in omnivores. **fermentation, p. 129**

Summary *Most animals digest food in extracellular compartments. Gastrovascular cavities have one opening, whereas alimentary canals have two openings. Digestive systems are adapted to each animal's diet.*

34.2 MASTERING CONCEPTS

1. What do digestive enzymes do?
2. Distinguish between intracellular and extracellular digestion.
3. What is a limitation of an incomplete digestive tract?
4. Compare and contrast the digestive systems of a cow, a chipmunk, and a wolf.

34.3 The Human Digestive System Consists of Several Organs

The human digestive system consists of the alimentary canal and accessory structures (**figure 34.6**). Some accessory structures, such as the salivary glands and pancreas, produce digestive enzymes. Two others, the liver and gallbladder, produce and store bile, which assists fat digestion. The teeth and tongue are also accessory organs.

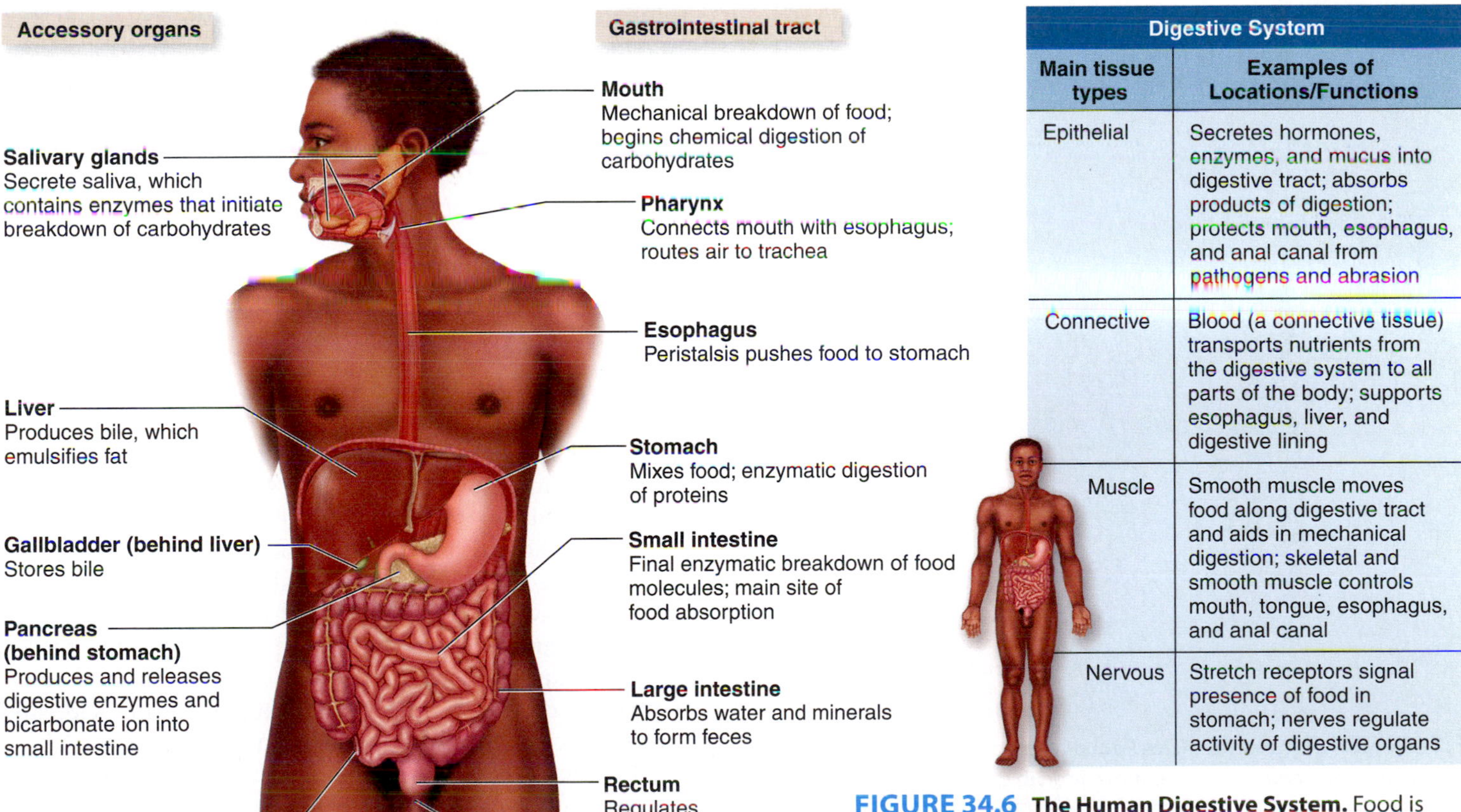
Accessory organs
Salivary glands
Secrete saliva, which contains enzymes that initiate breakdown of carbohydrates
Liver
Produces bile, which emulsifies fat
Gallbladder (behind liver)
Stores bile
Pancreas (behind stomach)
Produces and releases digestive enzymes and bicarbonate ion into small intestine
Appendix
Gastrointestinal tract
Mouth
Mechanical breakdown of food; begins chemical digestion of carbohydrates
Pharynx
Connects mouth with esophagus; routes air to trachea
Esophagus
Peristalsis pushes food to stomach
Stomach
Mixes food; enzymatic digestion of proteins
Small intestine
Final enzymatic breakdown of food molecules; main site of food absorption
Large intestine
Absorbs water and minerals to form feces
Rectum
Regulates elimination of feces
Anus

Digestive System	
Main tissue types	**Examples of Locations/Functions**
Epithelial	Secretes hormones, enzymes, and mucus into digestive tract; absorbs products of digestion; protects mouth, esophagus, and anal canal from pathogens and abrasion
Connective	Blood (a connective tissue) transports nutrients from the digestive system to all parts of the body; supports esophagus, liver, and digestive lining
Muscle	Smooth muscle moves food along digestive tract and aids in mechanical digestion; skeletal and smooth muscle controls mouth, tongue, esophagus, and anal canal
Nervous	Stretch receptors signal presence of food in stomach; nerves regulate activity of digestive organs

FIGURE 34.6 The Human Digestive System. Food is broken down as it moves through the chambers of the digestive tract. Accessory organs break food into small pieces and deliver enzymes and other chemicals that aid digestion.

How does food move within the alimentary canal? Layers of smooth muscle produce rhythmic waves of contraction, called **peristalsis,** that propel food along the entire digestive tract (**figure 34.7**). These contractions also churn food, mixing it with enzymes into a liquid. In addition, **sphincters** are rings of muscle that can open and close. The sphincters at the mouth and anus are under voluntary control, so we can decide when to open our mouths or eliminate feces. Other sphincters in the digestive tract control entry to, and exit from, the stomach. The intestines also contain muscular valves that help control the movement of food.

A. Digestion Begins in the Mouth and Esophagus

The thought, smell, or taste of food triggers salivary glands to secrete saliva into the mouth. Chemical digestion of ingested food begins when salivary amylase, an enzyme in saliva, starts to break down starch into maltose. Meanwhile, the **teeth**—mineral-hardened structures embedded in the jaws—grasp and chew the food (**figure 34.8**). (Chapter 27's Investigating Life: When a Chair Becomes a Ladder describes the evolutionary history of teeth.) Water and mucus in saliva aid the teeth as they tear food into small pieces, increasing the surface area available for chemical digestion. The muscular **tongue** at the floor of the mouth mixes the saliva and food and pushes it to the back of the mouth to be swallowed.

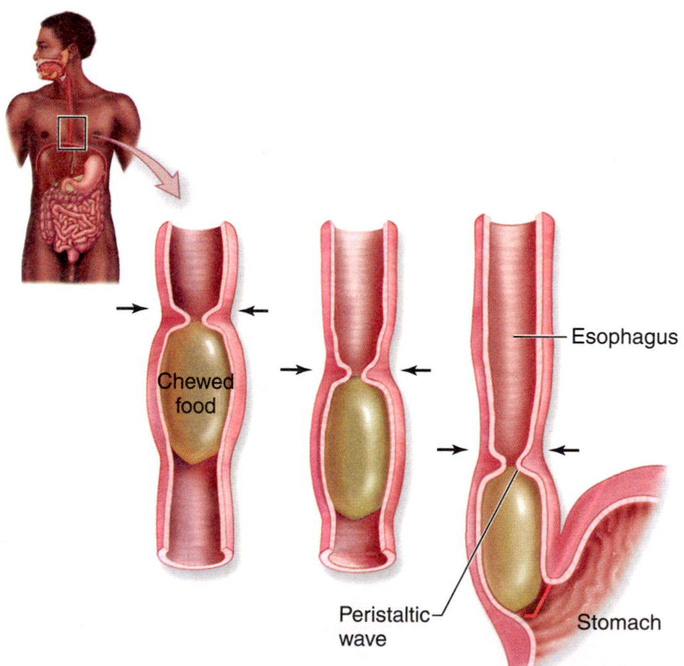

FIGURE 34.7 Muscles Move Food. Layers of smooth muscle coordinate their contractions to move food in one direction through the digestive tract. Circular muscles constrict the tube when they contract; longitudinal muscles shorten it. When food expands the walls of part of the tube, the circular muscles immediately behind it contract, squeezing the mass forward.

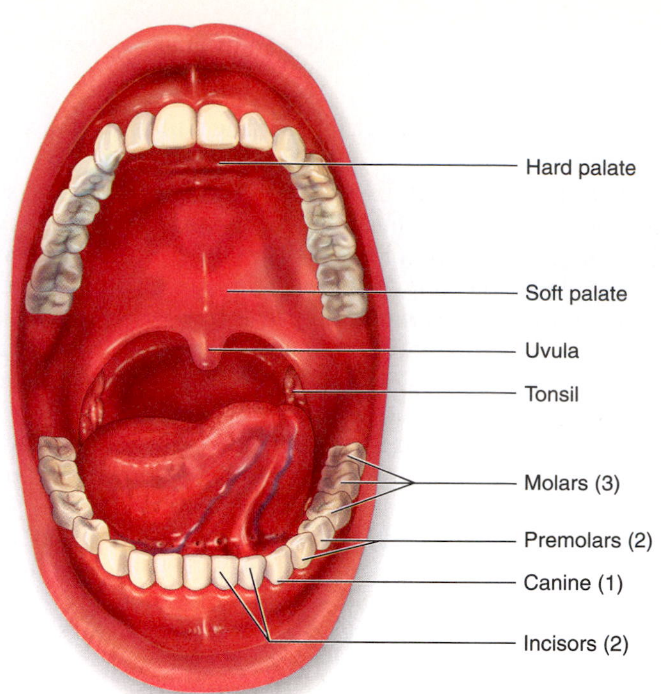

FIGURE 34.8 Human Teeth. Processing food begins with the teeth, which grasp, cut, tear, shred, crush, and grind food into pieces small enough to be swallowed.

The chewed food passes first through the **pharynx,** or throat, which also conducts air to the trachea. During swallowing, the **epiglottis** covers the opening to the trachea so that food enters the digestive tract instead of the lungs. Swallowed food and liquids next pass to the **esophagus,** a muscular tube leading from the pharynx to the stomach. Food does not merely slide down the esophagus under the influence of gravity; instead, contracting muscles push it along in a wave of peristalsis (see figure 34.7).

B. The Stomach Stores, Digests, and Pushes Food

The **stomach** is a J-shaped muscular bag that receives food from the esophagus (**figure 34.9A**). The stomach is about the size of a large sausage when empty, but when very full it can expand to hold as much as 3 or 4 L (about 3 or 4 quarts) of food. Folds in the stomach's lining can unfold like the pleats of an accordion to accommodate a large meal.

Both mechanical and chemical digestion occur in the stomach. Waves of peristalsis push food against the stomach bottom, churning it backward, and breaking it into pieces. Chemical digestion in the stomach requires **gastric juice,** a mixture of water, mucus, salts, hydrochloric acid, and enzymes produced at the stomach lining.

Gastric pits in the epithelium of the stomach lining house glands that secrete gastric juice at a rate of about 2 to 3 L/day (figure 34.9B). Some cells in the gastric pits secrete mucus, whereas others release hydrochloric acid (the pH of gastric juice normally ranges from 1.5 to 2 and may be as

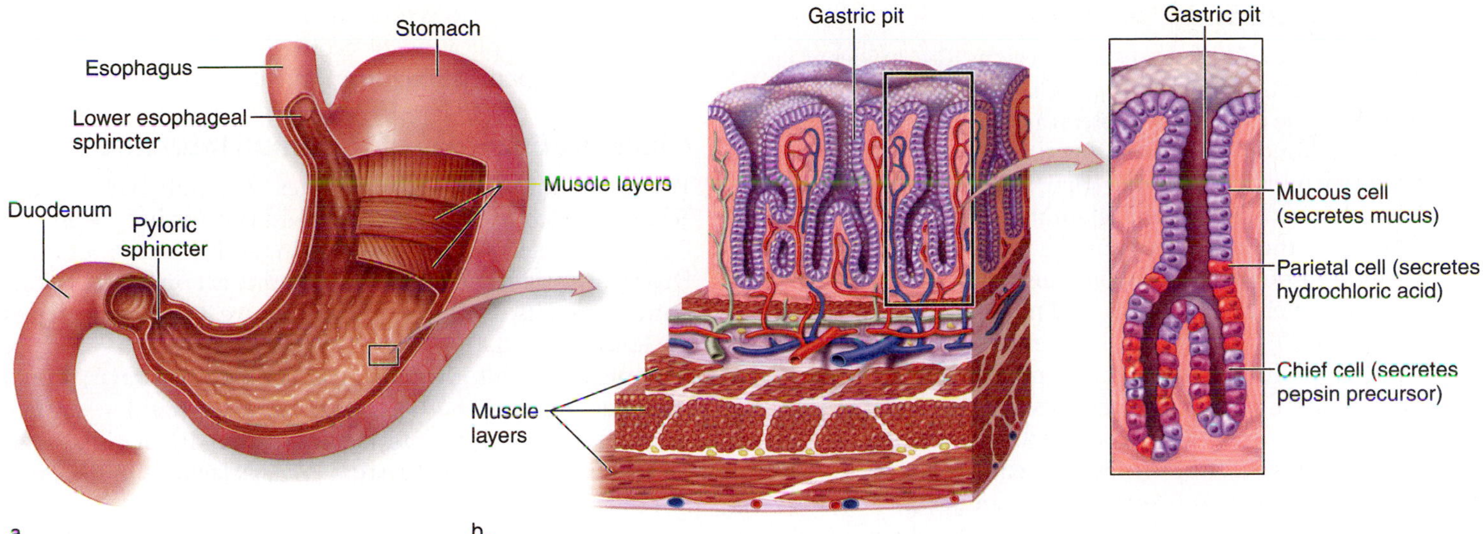

FIGURE 34.9 The Stomach. (**A**) The stomach is a J-shaped bag that stores, mixes, and digests food until it is fluid enough to move on to the small intestine. The stomach's two sphincters control the movement of food. (**B**) The lining of the stomach contains cells that secrete mucus and gastric pits that produce the gastric juice. Layers of muscle in the stomach wall enable the organ to move food and break it into smaller pieces.

low as 0.8). The acidity denatures proteins in food and kills most disease-causing organisms. Gastric juice also contains **pepsin,** the enzyme that begins the digestion of proteins. **epithelial tissue, p. 561; pH, p. 32**

If gastric juice breaks down protein in food, how does the stomach keep from digesting itself? First, the stomach produces little gastric juice until food is present. Second, mucus coats and protects the stomach lining. Tight junctions between cells in the stomach lining also prevent gastric juice from seeping through to the tissues below. The chapter opening essay describes the research that led to a better understanding of gastric ulcers, a common condition in which the stomach's protection fails.

The stomach absorbs very few nutrients, but it can absorb some water and salts (electrolytes), a few drugs (e.g., aspirin), and, like the rest of the digestive tract, alcohol. As a result, we feel alcohol's intoxicating effects quickly.

Chyme is the semifluid mass of food and gastric juice in the stomach. Small amounts of chyme squirt through a sphincter (the pyloric sphincter) that links the stomach and the duodenum, the upper part of the small intestine. A negative feedback loop controls the opening and closing of the pyloric sphincter (**figure 34.10**).

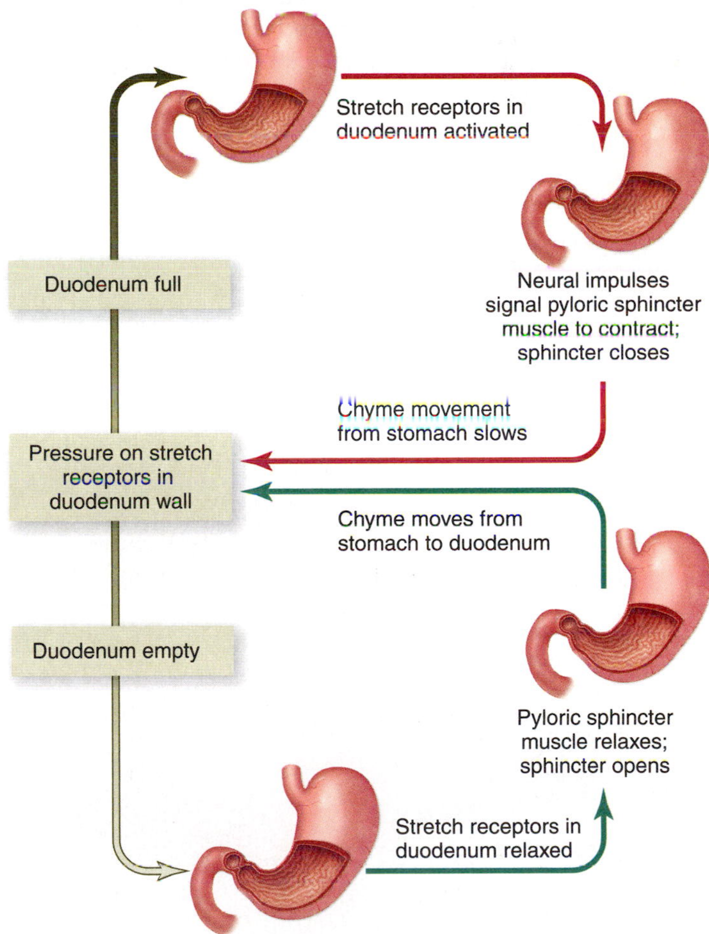

FIGURE 34.10 A Feedback Loop Controls Movement of Chyme to the Small Intestine. Stretch receptors signal the pyloric sphincter to open or close, depending on the quantity of food in the duodenum.

C. The Small Intestine Digests and Absorbs Nutrients

The **small intestine,** a 7-m (approximately 23-foot) tubular organ, completes digestion and absorbs nutrients. (Although its diameter is small compared with the large intestine, the small intestine is the longest organ in the digestive system.)

Anatomy of the Small Intestine

The small intestine contains three main regions. The duodenum makes up the first 25 cm (10 inches). Glands in the wall of the duodenum secrete mucus that protects and lubricates the small intestine. In addition, ducts from the pancreas and liver open into the duodenum, contributing some digestive enzymes and bile. The other two regions, the jejunum and the ileum, form the majority of the small intestine. The 2.5-m long jejunum absorbs most carbohydrates and proteins, and the 3.5-m ileum absorbs most water, fats, vitamins, and minerals.

Close examination of the hills and valleys along the small intestine's lining reveals millions of **villi,** tiny fingerlike projections that absorb nutrients (**figure 34.11**). The epithelial cells on the surface of each villus bristle with **microvilli,** extensions of the plasma membrane. Each villus cell and its 500 microvilli increase the surface area of the small intestine at least 600 times, allowing for the efficient extraction of nutrients from food. The capillaries that snake throughout each villus absorb nutrients and water, then empty into veins that carry the nutrient-laden blood to the liver. Also inside each villus is a lymph capillary that absorbs digested fats.

The small intestine is home to a community of microorganisms. Although "germs" have a bad reputation, these normal inhabitants of the intestinal tract are beneficial. Most notably, they help prevent infection by harmful microorganisms—an interesting application of ecology's competitive exclusion principle. **competitive exclusion, p. 807**

Digestive Enzymes of the Small Intestine

The small intestine absorbs water, minerals, free amino acids, cholesterol, and vitamins without further digestion. But most molecules in food require additional processing. Digestive enzymes in the small intestine act on proteins, carbohydrates, fats, and nucleic acids. **Figure 34.12** summarizes the locations and actions of the main digestive enzymes. For example, the small intestine produces carbohydrases, which break down disaccharides into monosaccharides. People who lack one such enzyme, lactase, cannot digest milk sugar; this chapter's Burning Question: What's lactose intolerance? discusses this condition.

Fats present an interesting challenge to the digestive system. Lipases, the pancreatic enzymes that digest fats, are water soluble, but fats are not. Therefore, lipase can only act at the surface of a fat droplet, where it contacts water. **Bile** is a biochemical that helps solve this problem by dispersing the fat into tiny globules, increasing the surface area exposed to lipase.

Considering that cells of the small intestine are made of the same molecules as food, what prevents the small intestine from digesting itself? First, digestive biochemicals are

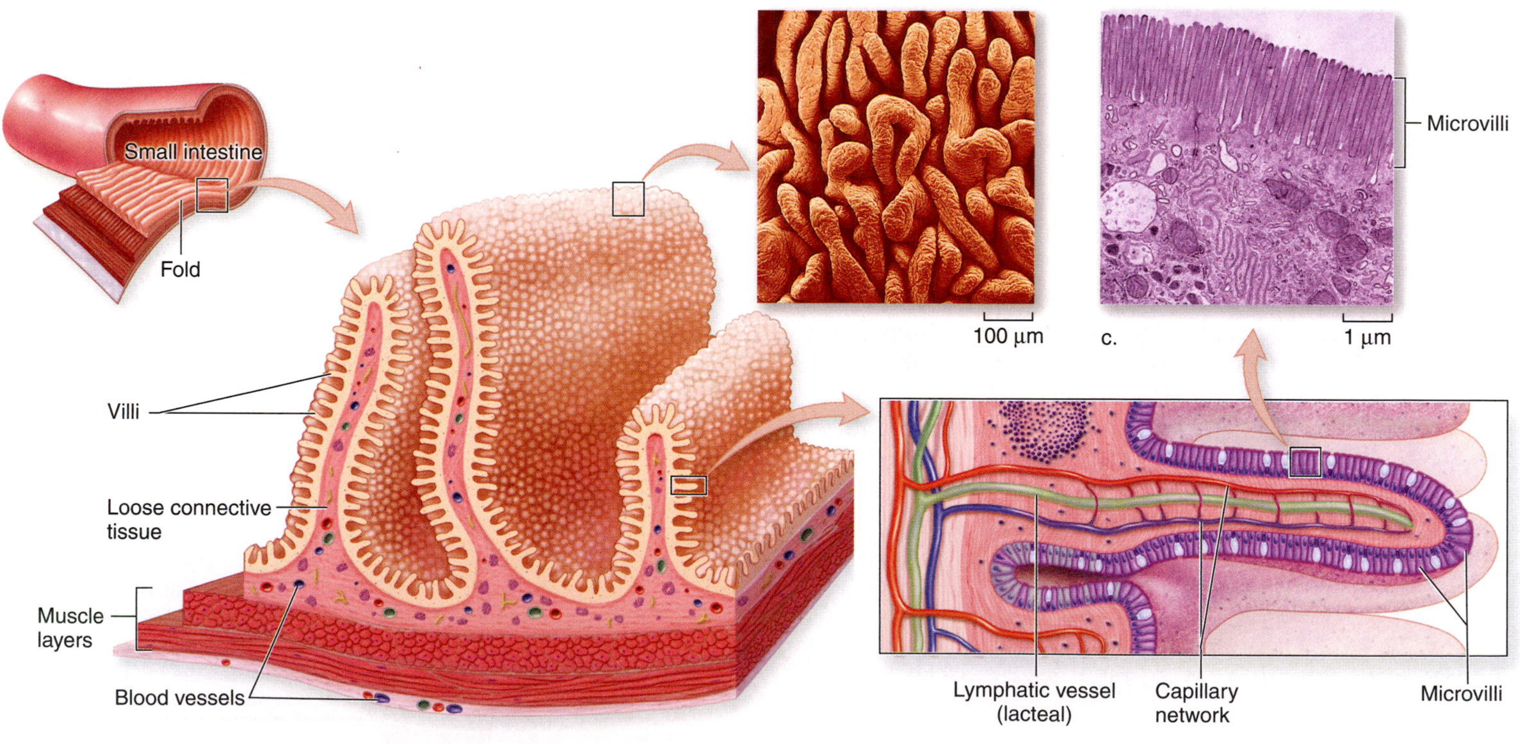

FIGURE 34.11 **The Lining of the Small Intestine.** (**A**) The lining of the small intestine has ridges, each of which is folded into villi. (**B**) Within each villus, blood-filled capillaries absorb digested carbohydrates and proteins, and lymph capillaries absorb digested fats. (**C**) Each villus cell's membrane is folded into microvilli with extensive surface area.

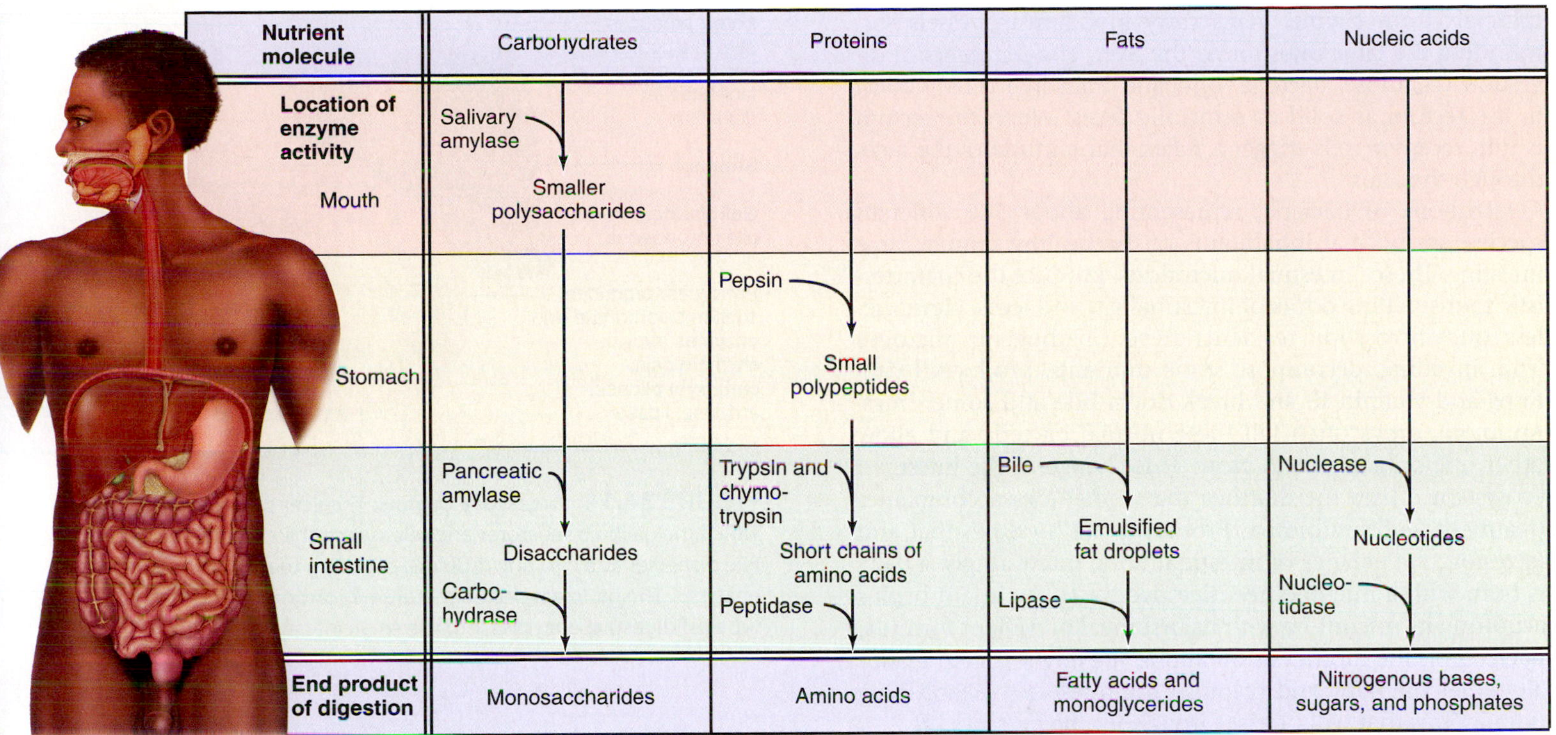

Nutrient molecule
Carbohydrates
Proteins
Fats
Nucleic acids
Location of enzyme activity
Mouth
Salivary amylase
Smaller polysaccharides
Stomach
Pepsin
Small polypeptides
Small intestine
Pancreatic amylase
Disaccharides
Carbo-hydrase
Trypsin and chymo-trypsin
Short chains of amino acids
Peptidase
Bile
Emulsified fat droplets
Lipase
Nuclease
Nucleotides
Nucleo-tidase
End product of digestion
Monosaccharides
Amino acids
Fatty acids and monoglycerides
Nitrogenous bases, sugars, and phosphates

FIGURE 34.12 Overview of Chemical Digestion. Although digestion begins in the mouth and stomach, the small intestine digests and absorbs most molecules.

produced only when food is present. In addition, mucus protects the intestinal wall from digestive juices and neutralizes stomach acid. Nevertheless, many intestinal lining cells die in the caustic soup. Rapid division of the small intestine's epithelial cells compensates for the loss, replacing the lining every 36 hours. Nearly one-quarter of the bulk of feces consists of dead epithelial cells from the small intestine.

D. The Large Intestine Completes Nutrient and Water Absorption

The material remaining in the small intestine after nutrients are absorbed enters the **large intestine,** which extends to the anus (**figure 34.13**). This 1.5-m (5-foot) tube surrounds the small intestine, roughly in the shape of a question mark. The large intestine is much shorter than the small intestine, but its diameter is greater, about 6.5 cm (2.5 inches). Its main functions are to receive the indigestible components of food from the small intestine, absorb water and salts, and eliminate the remainder as feces.

The start of the large intestine is the pouchlike cecum. The appendix is a thin, worm-shaped tube dangling from the cecum. Trapped bacteria or undigested food can cause the appendix to become irritated, inflamed, and infected, producing severe pain. If it bursts, it can spill its contents into the abdominal cavity and spread the infection.

The colon forms the majority of the large intestine. Here, the large intestine absorbs most of the water, electrolytes, and

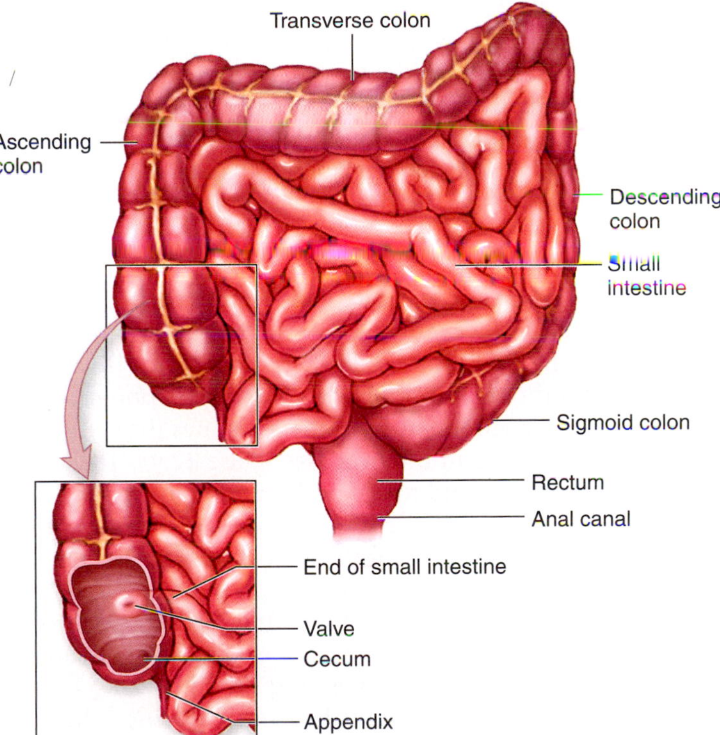

FIGURE 34.13 The Large Intestine. The large intestine absorbs water, salts, and minerals and temporarily stores feces. *Inset:* The appendix attaches to the pouchlike cecum, which receives food from the small intestine.

minerals from chyme. Veins carry blood from vessels surrounding the large intestine to the liver. The remnants of digestion (cellulose, bacteria, bile, and intestinal cells) collect in the rectum as solid or semisolid feces. When the rectum is full, receptor cells trigger a reflex that eliminates the feces through the anus.

Trillions of bacteria, representing about 500 different species, are normal inhabitants of the healthy human large intestine. These "intestinal microflora" produce the characteristic foul-smelling odors of intestinal gas and feces. Nevertheless, our microscopic residents prevent pathogenic microbes from invading, decompose some nutrients, produce B vitamins and vitamin K, and break down bile and some drugs. Antibiotic drugs often kill these normal bacteria and allow other microorganisms to grow. This change in the intestinal ecosystem causes the diarrhea that sometimes accompanies treatment with antibiotics. Probiotics are products that aim to restore the balance of intestinal flora. Interestingly, a baby is born with a microbe-free digestive tract. The infant begins acquiring its microflora with its first meal of milk or formula. Bacteria on the mother's skin and in the environment gradually enter the baby and colonize the intestines, establishing populations that will persist throughout the person's life.

E. The Pancreas, Liver, and Gallbladder Aid Digestion

Several accessory organs are essential to digestion (**figure 34.14**). As you learned in chapter 30, the **pancreas** is an endocrine gland, regulating blood sugar levels by producing the hormones insulin and glucagon. The pancreas's role in digestion, however, is as an exocrine gland. It sends about a liter of pancreatic "juice" to the duodenum each day. This fluid contains many digestive enzymes, including trypsin, chymotrypsin, pancreatic amylase, pancreatic lipase, and nucleases (see figure 34.12). Pancreatic juice also contains sodium bicarbonate that neutralizes the hydrochloric acid from the stomach.

At a weight of about 1.4 kg (3 pounds), the **liver** is the largest solid organ in the body. The hepatic portal vein passes nutrient-laden blood from the intestines through extensive capillary beds in the liver (see figure 32.4). This organ removes bacteria and toxins and gets "first dibs" on the nutrients in the blood before it is pumped to the rest of the body. In a condition called cirrhosis, scar tissue blocks this vital blood flow through the liver.

The liver has more than 200 functions, including detoxifying harmful substances in the blood, storing glycogen and fat-soluble vitamins, and synthesizing blood-clotting proteins. The liver's only *direct* contribution to digestion, however, is the production of greenish yellow bile. The **gallbladder** stores this bile until chyme triggers its release into the small intestine. The cholesterol in bile can crystallize, forming gallstones that partially or completely block the duct to the small intestine. Gallstones are very painful and may require removal of the gallbladder. This chapter's Can *You* Relate? The Unhealthy Digestive System describes some other examples of illnesses affecting the gastrointestinal tract.

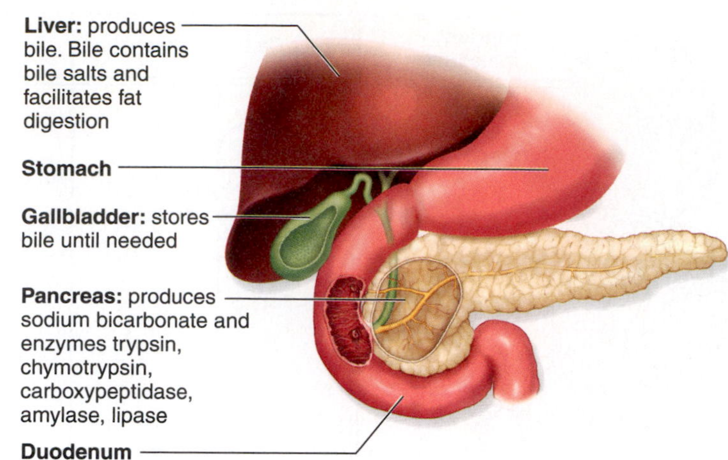

FIGURE 34.14 Accessory Organs. The liver produces bile, which the gallbladder stores and releases to the small intestine. Bile disperses fats into tiny droplets, exposing them to digestive enzymes. The pancreas secretes sodium bicarbonate and several types of digestive enzymes into the small intestine.

Summary *Salivary amylase begins chemical digestion as the teeth mechanically break down food. The esophagus conducts chewed food to the stomach, where it mixes with gastric juice. The small intestine digests most nutrients, which intestinal villi absorb. The large intestine completes digestion and absorbs most of the water. Undigested waste leaves through the rectum and anus. The pancreas, liver, and gallbladder also play essential roles in digestion.*

34.3 MASTERING CONCEPTS

1. Explain the action and importance of peristalsis and sphincters in digestion.
2. Which structures in the mouth and throat participate in digestion?
3. Describe the mechanical and chemical digestion that occurs in the stomach.
4. How does the small intestine maximize surface area?
5. What are the products of digestion of carbohydrates, lipids, proteins, and nucleic acids?
6. Describe the events that occur as food passes through the large intestine.
7. How do secretions of the pancreas, liver, and gallbladder aid digestion?

Burning Questions

What's lactose intolerance?

Many adults have difficulty digesting milk products because of lactose intolerance, which results from the absence of the enzyme lactase in the small intestine. This enzyme breaks down the disaccharide lactose (milk sugar) into the monosaccharides glucose and galactose. When lactase is absent, bacteria in the large intestine ferment undigested lactose, producing abdominal pain, gas, diarrhea, bloating, and cramps. A person with lactose intolerance can avoid these symptoms by consuming fermented dairy products such as yogurt, buttermilk, and cheese instead of fresh dairy products, because bacteria have already broken down the lactose in those foods. Taking lactase tablets can also prevent symptoms of lactose intolerance.

Have a Burning Question of your own? Submit it to marielle_hoefnagels@mcgraw-hill.com for possible inclusion in future editions of this book!

Can *You* Relate?

The Unhealthy Digestive System

The entire length of the human digestive tract is subject to numerous disorders. Some are a nuisance or easily treated, whereas others can be deadly. A few are listed here:

- **Tooth decay:** Bacteria living on the tooth surface secrete acids that eat through the surface of the teeth, causing cavities. The decayed area can extend to the interior of the tooth, eventually killing the tooth's nerve and blood supply.
- **Acid reflux:** Gastric juice, normally confined to the stomach, passes through the esophageal sphincter and burns the esophagus. This painful condition is commonly known as "heartburn."
- **Vomiting:** When a person vomits, the medulla (in the brainstem) coordinates the contraction of diaphragm and abdominal muscles and relaxes the sphincter at the entrance of the stomach, forcing chyme out of the stomach and up the esophagus. Alcohol, bacterial toxins from spoiled foods, and excessive eating can trigger queasiness and vomiting.
- **Hepatitis:** Hepatitis literally means inflammation of the liver. Many conditions can cause hepatitis, including alcohol abuse, some pharmaceutical drugs, and eating poisonous mushrooms. Five viruses (hepatitis A–E) cause most cases of infectious hepatitis; chapter 17 describes these viruses.
- **Gallstones:** The gallbladder stores bile and secretes it into the small intestine. Sometimes, parts of the bile crystallize and accumulate as gallstones, which can obstruct the ducts that distribute bile. Most gallstones are made primarily of cholesterol. An affected person may have one or thousands of gallstones.
- **Appendicitis:** The appendix has no known function, but it can nevertheless cause problems when it becomes inflamed or infected. If it ruptures, it can release bacteria into the abdominal cavity, causing serious infection and sometimes even death when left untreated.
- **Diarrhea:** If the intestines fail to absorb as much water as they should, the feces become loose and watery. The risk of dehydration is high, and diarrhea is a major cause of death in underdeveloped countries. Many food- and waterborne viruses, bacteria, and protists cause diarrhea, as can treatment with some antibiotics. (Chapter 4's Investigating Life: Does Natural Selection Maintain Some Genetic Illnesses? describes the faulty protein implicated in cystic fibrosis, which may protect against some causes of diarrhea.)
- **Constipation:** A constipated person eliminates feces less frequently than three times a week, and the feces are hard, dry, and difficult to eliminate. Constipation has many causes, including an obstructed large intestine, loss of peristalsis, dehydration, starvation, and anxiety.
- **Colon (colorectal) cancer:** Cancerous tumors may arise in the rectum, colon, or appendix. This is among the most common cancer types, and it is a leading cause of death worldwide.
- **Hemorrhoids:** Hemorrhoids are swollen, distended veins that protrude into the rectum or anus. They cause painful defecation and bleeding.

34.4 A Healthy Diet Includes Essential Nutrients and the Right Number of Calories

Healthy eating has two main components. One consideration is the caloric content of food, which must balance a person's activity level. Second, a healthy diet includes all of the essential nutrients. Fortunately, packaged foods have labels that depict both the nutrient content and the number of calories per serving of food (**figure 34.15**). Labels may also indicate other information; for example, chapter 2's Burning Question explains the distinction between "natural" and "organic."

Nutrition Facts
Serving Size 3/4 Cup (27g)

Amount Per Serving	Cereal	With 1/2 Cup Skim Milk
Calories	90	130
Calories from Fat	10	10
	% Daily Value	
Total Fat 1g*	**2%**	**2%**
Saturated Fat 0g	**0%**	**0%**
Trans Fat 0g	**0%**	**0%**
Cholesterol 0mg	**0%**	**0%**
Sodium 190mg	**8%**	**11%**
Potassium 85mg	**2%**	**8%**
Total Carbohydrate 23g	**8%**	**10%**
Dietary Fiber 5g	**20%**	**20%**
Sugars 5g		
Protein 2g		
Vitamin A	0%	4%
Vitamin C	10%	15%
Calcium	0%	15%
Iron	2%	2%

FIGURE 34.15 Nutritional Information. The packaging of processed foods, such as cereal, includes a standard nutritional label that indicates the calorie and nutrient content in each serving.

A. A Varied Diet Is Essential to Good Health

Nutrients fall into two categories. **Macronutrients** are required in large amounts. Water is a macronutrient because all living cells require water as a solvent and as a participant in many reactions. Organisms use three other

Table 34.1 *Minerals in the Human Diet*

Mineral	Food Sources	Functions in the Human Body
Bulk Minerals		
Calcium	Milk products, green leafy vegetables	Bone and tooth structure, blood clotting, hormone release, nerve transmission, muscle contraction
Chlorine	Table salt, meat, fish, eggs, poultry, milk	Digestion in stomach
Magnesium	Green leafy vegetables, beans, fruits, peanuts, whole grains	Muscle contraction, nucleic acid synthesis, enzyme activity
Phosphorus	Meat, fish, eggs, poultry, whole grains	Bone and tooth structure
Potassium	Fruits, potatoes, meat, fish, eggs, poultry, milk	Body fluid balance, nerve transmission, muscle contraction, nucleic acid synthesis
Sodium	Table salt, meat, fish, eggs, poultry, milk	Body fluid balance, nerve transmission, muscle contraction
Sulfur	Meat, fish, eggs, poultry	Hair, skin, and nail structure, blood clotting, energy transfer, detoxification
Trace Minerals		
Chromium	Yeast, pork kidneys	Regulates glucose use
Cobalt	Meat, eggs, dairy products	Part of vitamin B_{12}
Copper	Organ meats, nuts, shellfish, beans	Part of many enzymes, storage and release of iron in red blood cells
Fluorine	Water (in some areas)	Maintains dental health
Iodine	Seafood, iodized salt	Part of thyroid hormone
Iron	Meat, liver, fish, shellfish, egg yolk, peas, beans, dried fruit, whole grains	Transport and use of oxygen (as part of hemoglobin and myoglobin), part of some enzymes
Manganese	Bran, coffee, tea, nuts, peas, beans	Part of some enzymes, bone and tendon structure
Selenium	Meat, milk, grains, onions	Part of some enzymes, heart function
Zinc	Meat, fish, egg yolk, milk, nuts, some whole grains	Part of some enzymes, nucleic acid synthesis

macronutrients—carbohydrates, proteins, and lipids—to build cells and to generate ATP (see figure 6.10). **Micronutrients** are required in very small amounts. Two examples are vitamins (organic molecules that often function as coenzymes) and minerals (inorganic elements). Many processed foods are fortified with micronutrients, so that deficiencies in developed countries are rare. **Tables 34.1** and **34.2** provide details on the sources and functions of vitamins and minerals.

Eating a variety of foods helps meet nutritional requirements. The U.S. government's food pyramid emphasizes whole grains, fresh vegetables, and low-fat dairy products,

Table 34.2 *Vitamins and Health*

Vitamin	Function(s)	Food Sources	Deficiency Symptoms
Water-Soluble Vitamins			
B complex vitamins			
Thiamine (vitamin B_1)	Growth, fertility, digestion, nerve cell function, milk production	Pork, beans, peas, nuts, whole grains	Beriberi (neurological disorder), loss of appetite, swelling, poor growth, heart problems
Riboflavin (vitamin B_2)	Energy use	Liver, leafy vegetables, dairy products, whole grains	Hypersensitivity of eyes to light, lip sores, oily dermatitis
Pantothenic acid*	Growth, cell maintenance, energy use	Liver, eggs, peas, potatoes, peanuts	Headache, fatigue, poor muscle control, nausea, cramps
Niacin	Growth, energy use	Liver, meat, peas, beans, whole grains, fish	Dark rough skin, diarrhea, mouth sores, mental confusion (pellagra)
Pyridoxine (vitamin B_6)*	Protein use	Red meat, liver, corn, potatoes, whole grains, green vegetables	Mouth sores, dizziness, nausea, weight loss, neurological disorders
Folic acid (folate)	Manufacture of red blood cells, metabolism	Liver, navy beans, dark green vegetables	Anemia, neural tube defects
Biotin*	Metabolism	Meat, milk, eggs	Skin disorders, muscle pain, insomnia, depression
Cobalamin (vitamin B_{12})	Manufacture of red blood cells, growth, cell maintenance	Meat, organ meats, fish, shellfish, milk	Pernicious anemia
Ascorbic acid (vitamin C)	Growth, tissue repair, bone and cartilage formation	Citrus fruits, tomatoes, peppers, strawberries, cabbage	Weakness, gum bleeding, weight loss (scurvy)
Fat-Soluble Vitamins			
Retinol (vitamin A)	Night vision, new cell growth	Liver, dairy products, egg yolk, vegetables, fruit	Night blindness, rough dry skin
Cholecalciferol (vitamin D)	Bone formation	Fish liver oil, milk, egg yolk	Skeletal deformation (rickets)
Tocopherol (vitamin E)*	Prevents oxidation of some compounds	Vegetable oil, nuts, beans	Anemia in premature infants
Vitamin K*	Blood clotting	Liver, egg yolk, green vegetables	Bleeding, liver problems

*These vitamin deficiencies are rare in humans, but they have been observed in experimental animals.

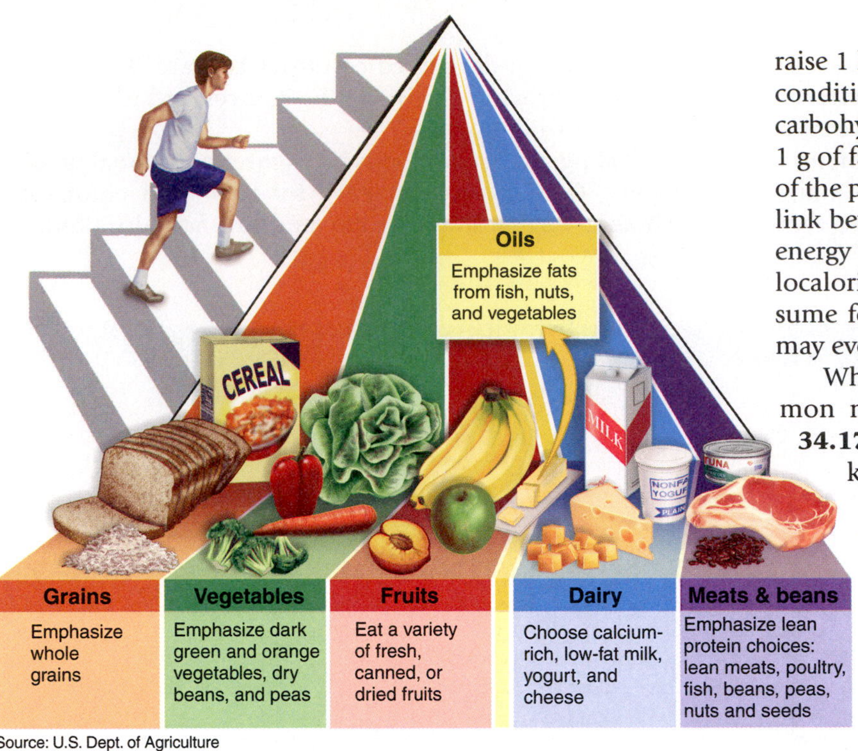

FIGURE 34.16 **One Healthful Diet.** The U.S. Department of Agriculture's food pyramid emphasizes whole grains, fruits, vegetables, and low-fat dairy products, along with exercise.

along with a variety of fruits and limited amounts of meat and fat (**figure 34.16**). The Harvard School of Public Health suggests a somewhat different diet that minimizes dairy products, red meat, and starchy processed grains. Whole grains and vegetable oils, along with abundant vegetables, make up the base of this pyramid.

Even indigestible components of food help maintain good health. Dietary fiber, for example, is composed of cellulose from plant cell walls. Because humans cannot digest cellulose, it contributes only bulk, not nutrients, to food. This increased mass eases movement of the food through the digestive tract. As a result, harmful ingredients in food contact the walls of the intestines for a shorter period. The result is a lower incidence of colorectal cancer among people with abundant fiber in their food. A high-fiber diet also reduces blood cholesterol and helps regulate blood sugar. **cellulose, p. 35**

B. Body Weight Reflects Food Intake and Activity Level

A food's caloric content is determined by burning it in a bomb calorimeter, a chamber immersed in water. Energy released from the burning food raises the water temperature. One **kilocalorie** (one food **Calorie**) is the energy needed to raise 1 kg of water from 14.5°C to 15.5°C under controlled conditions. Bomb calorimetry studies have shown that 1 g of carbohydrate yields 4 kcal, 1 g of protein yields 4 kcal, and 1 g of fat yields 9 kcal. Although the body cannot extract all of the potential energy in food, these values help explain the link between a fatty diet and weight gain; fats supply more energy than most people can use. When we take in more kilocalories than we expend, weight increases; those who consume fewer kilocalories than they expend lose weight and may even starve.

What constitutes a "healthful" weight? The most common measure is the body mass index, or BMI (**figure 34.17**). To calculate BMI, divide a person's weight (in kilograms) by his or her squared height (in meters): BMI = weight/(height)^2. Many health professionals consider a person whose BMI is less than 19 underweight. A BMI between 19 and 25 is healthy; an overweight person has a BMI greater than 25; and a BMI greater than 30 denotes **obesity**. Morbid obesity is defined as a BMI greater than 40. One limitation of BMI is that it cannot account for many of the details that affect health. An extremely muscular person, for example, will have a high BMI because muscle is denser than fat, yet he or she would not be considered overweight.

Another useful measure is the ratio of waist diameter to hip diameter. People whose fat accumulates at the waistline ("apples") are more susceptible to health problems such as insulin resistance than are "pears," who are bigger around the hips.

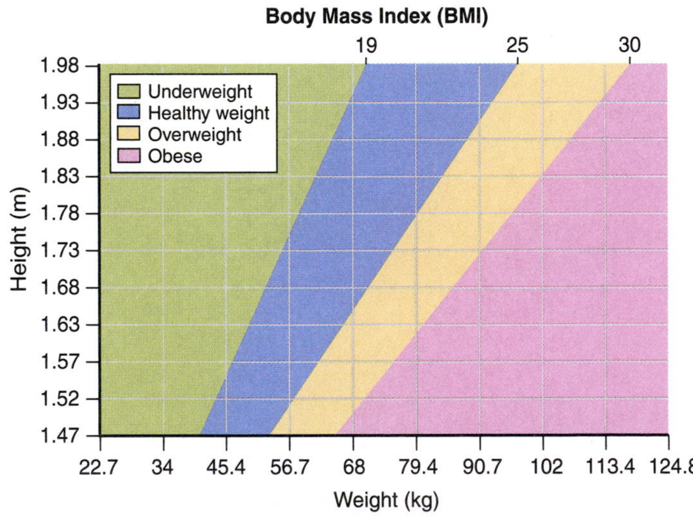

FIGURE 34.17 **Body Mass Index.** This rough measure indicates whether a person's weight is healthful. The formula is BMI = weight/(height)^2 if weight is measured in kilograms and height is measured in meters. Alternatively, you can multiply your weight (in pounds) by 704.5, and divide by the square of your height (in inches).

C. Starvation: Too Few Calories to Meet the Body's Needs

A healthy human can survive for 50 to 70 days without food—much longer than without air or water. In some areas of the world, famine is a constant condition, and millions of people starve to death every year. Hunger strikes, inhumane treatment of prisoners, and eating disorders can also cause starvation (**figure 34.18**).

The starving human body essentially digests itself. After only a day without food, reserves of sugar and glycogen are gone, and the body begins extracting energy from stored fat and then from muscle protein. By the third day, hunger eases as the body uses energy from fat reserves. Gradually, metabolism slows, blood pressure drops, the pulse slows, and chill sets in. Skin becomes dry and hair falls out as the proteins that form these structures are digested. When the body dismantles the immune system's antibody proteins, protection against infection declines. Mouth sores and anemia develop, the heart beats irregularly, and bone begins to degenerate. Near the end, the starving human is blind, deaf, and emaciated.

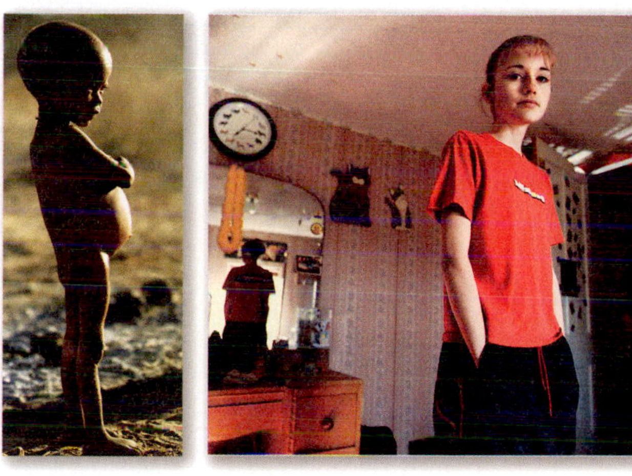

FIGURE 34.18 Two Forms of Starvation. The malnourished girl on the left has thin legs and a distended belly, a sign of a severe protein deficiency. Her country was stricken by famine after repeated droughts and crop failures. The girl on the right suffers from anorexia nervosa, or self-imposed starvation.

Anorexia nervosa, or self-imposed starvation, is refusal to maintain normal body weight. The condition affects about 1 in 250 adolescents, more than 90% of whom are female. The sufferer perceives herself as overweight and eats barely enough to survive, losing as much as 25% of her original body weight. She may further lose weight by vomiting, taking laxatives and diuretics, or exercising intensely. Intravenous feedings, psychotherapy, and nutritional counseling may help, but 15 to 21% of people with anorexia die from the disease. Bulimia is another eating disorder that mainly affects females. Rather than avoiding food, a person with **bulimia** eats large quantities and then intentionally vomits or uses laxatives shortly afterward, a pattern called "binge and purge." A person with bulimia may or may not be underweight.

D. Obesity: More Calories than the Body Needs

Obesity is increasingly common in the United States (**figure 34.19**), and the health consequences can be serious. People who accumulate fat around their waists are susceptible to type II diabetes, high blood pressure, and atherosclerosis. High body weight is also correlated with congestive heart failure, acid reflux disease, urinary incontinence, low back pain, stroke, sleep disorders, and many other health problems. In addition, obese people also face higher risk of cancers of the colon, breast, and uterus.

Excess body weight accumulates when a person consumes more calories than he or she expends. A diet loaded

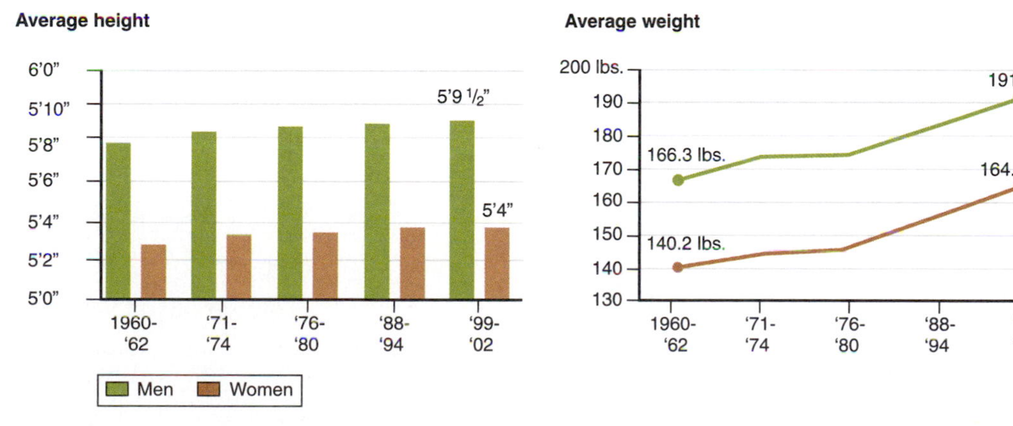

FIGURE 34.19 A Growing Nation. Americans are a little taller and a lot heavier than they were in 1960.

with sugar and fat, coupled with an inactive lifestyle, certainly contributes to the problem. For many people, however, there is more to the story. Because many genes contribute to appetite, digestion, and metabolic rate, the combination of alleles that a person inherits influences his or her risk for obesity (**figure 34.20**).

The concern over expanding waistlines has fueled demand for low-calorie foods such as artificial sweeteners and fats (see chapter 2). Although fad diets remain popular, the most healthful way to lose weight is to exercise and reduce calorie intake while maintaining a balanced diet. For people who have difficulty losing weight in this way, drugs and stomach-reduction surgery offer other options.

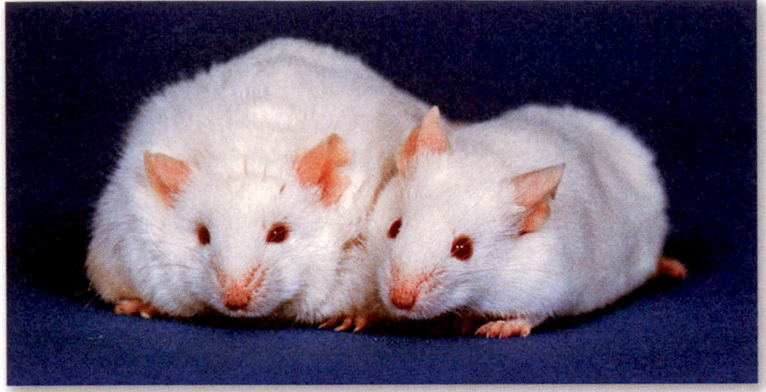

FIGURE 34.20 **Hormone Deficiency.** The obese mouse on the left cannot produce a hormone called leptin. In a normal mouse, leptin secreted by fat cells affects target cells in the hypothalamus, helping the brain regulate appetite and metabolism.

Summary *Macronutrients such as water, carbohydrates, proteins, and lipids are required in large amounts. Vitamins and minerals, which are micronutrients, are required in small amounts. Weight loss occurs when a person spends more energy than he or she consumes in food. If a person eats more Calories than he or she spends, weight gain will result.*

34.4 MASTERING CONCEPTS

1. What information appears on a food nutrition label?
2. What are some examples of essential nutrients in a human diet?
3. How does indigestible fiber contribute to a healthy diet?
4. How do scientists measure the calorie content of food?
5. What is body mass index?
6. Describe the events of starvation.
7. What are some of the causes and effects of obesity?

34.5 INVESTIGATING LIFE
Brawny Vegetarian Lizards Hit the Salad Bar

It is maddening that plants are all around us, but we can't eat most of them. Of course, we can eat lettuce, celery, and carrots, but the cellulose in the plants' cell walls passes through our digestive systems as indigestible fiber. Ironically, cellulose is a polysaccharide composed of energy-rich glucose molecules. Fewer people would starve if only we could do what the herbivores do—derive energy from cellulose-rich grass, tough leaves, or even wood.

Figure 34.5 illustrated digestive specializations in several groups of mammals, including ruminants, nonruminant herbivores, and carnivores. Why do the digestive tracts look so different? Is it because the animals have different diets, or because they have different evolutionary histories? In other words, do lions and antelope have different digestive tracts because lions eat meat and antelope eat grass? Or is it just another one of the many differences between big cats and ruminants? To get around this problem, it is useful to study closely related animals that eat different foods, minimizing the confounding effect of evolutionary history. The lizard family Liolaemidae fits the bill (**figure 34.21**). These reptiles are well suited for a comparison of dietary adaptations because the family includes large numbers of herbivorous, omnivorous, and insectivorous species. **taxonomic hierarchy, p. 9**

Shannon O'Grady and Denise Dearing, from the University of Utah, along with Argentine colleagues Mariana Morando and Luciano Avila, studied diet and digestive system specialization in lizards of the family Liolaemidae. Their study began with the capture of 204 adult lizards representing 22 species in Argentina. After euthanizing the lizards, the researchers measured each animal from the tip of its nose to its anus. (This measurement, called the "snout to vent" length,

FIGURE 34.21 **Lava Lizard.** This lizard (*Microlophus delanonis*), in the family Liolaemidae, is an omnivore that eats everything from insects to cactus flowers.

is common in research on reptiles.) They also measured the stomach, small intestine, and large intestine, and recorded the location and species of microscopic worms in the digestive tract. The researchers also classified the contents of each digestive organ and determined that of the 22 species collected, 7 were insectivores, 8 were omnivores, and 7 were herbivores.

After collecting all of their data, the researchers used statistical tools to determine which (if any) of the measurements were correlated with diet. Three results stood out. First, the overall body size of the herbivores was 33% larger than omnivores, which were (in turn) 18% larger than insectivores. This made sense. Plants take longer to digest than animal protein, so herbivores should have longer digestive tracts. It is not surprising that animals with longer intestinal tracts are larger in other dimensions as well.

Herbivory was also correlated with a long small intestine (**figure 34.22**A). This second finding was a bit of a surprise. The investigators had predicted that the large intestine would be elongated, since that is where the fermentation of cellulose occurs in many herbivores. This result prompted speculation that these lizards eat fruit, which contains sugars and other carbohydrates that are easily digested in the small intestine, rather than leaves and other fibrous plant material. Unfortunately, the researchers could not settle this question because they had not classified the origin of the plant material in the lizards' digestive tracts.

The third striking result was that herbivores were more likely to harbor microscopic worms (nematodes) than their omnivorous or insectivorous counterparts (figure 34.22B). Recall from chapter 22 that nematodes, also called roundworms, occur nearly everywhere in the world, including animal digestive tracts. The lizards' roundworms appear to be beneficial, and they probably eat bacteria, but it is unclear why herbivores house more of them than do omnivores or insectivores. Perhaps their huge populations simply slow the passage of plant material through the intestine, increasing the time available for digestion.

Correlation does not necessarily imply causation, and a study such as this one leaves many unanswered questions.

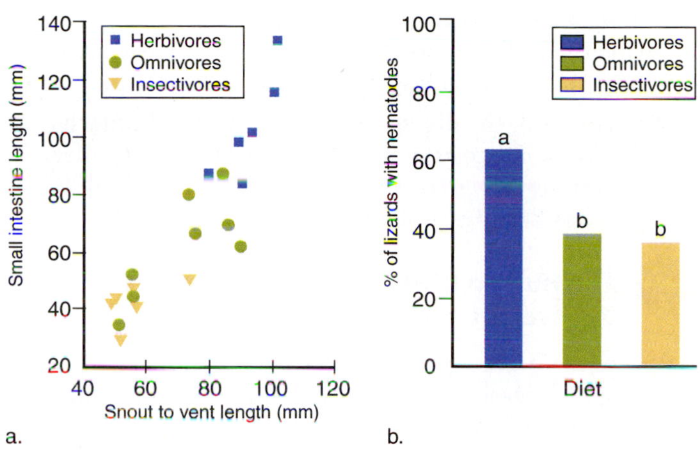

FIGURE 34.22 Diets and Digestive Adaptations in Lizards. (**A**) Herbivores have larger bodies and longer small intestines than either omnivores or insectivores. (**B**) More herbivores harbor nematode worms than do lizards with other diets. **Question:** How could you design an experiment that would determine whether the nematodes in the guts of lizards are beneficial, harmful, or neutral?

Because these animals were captured in the wild, it is impossible to say how food availability and food-gathering activities may have affected the contents of the lizards' digestive tracts. Would the results have been different during another season, or in a different location? A controlled laboratory study could address these questions but would require an enormous number of captive lizards.

The animal kingdom has many examples of herbivores being larger than their omnivorous, insectivorous, and carnivorous counterparts—this could even have been the driving force behind the evolution of the largest dinosaurs. The study of Argentine lizards adds another morsel of scientific knowledge to our understanding of the role of diet in the evolutionary history of animals.

O'Grady, Shannon P., Mariana Morando, Luciano Avila, and M. Denise Dearing. 2005. Correlating diet and digestive tract specialization: Examples from the lizard family Liolaemidae. *Zoology*, vol. 108, pages 201–210.

CHAPTER SUMMARY

34.1 Animals Have Diverse Diets and Feeding Strategies

- The **digestive system** acquires food. **Nutrients** in food supply energy, raw materials for growth and maintenance, and essential vitamins and minerals.
- Food is **ingested, digested,** and **absorbed** into the bloodstream; indigestible wastes are **eliminated** as **feces.**
- **Herbivores** eat plants, **carnivores** eat meat, **detritivores** consume decaying organic matter, and **omnivores** have a varied diet.
- Animals obtain food in diverse ways. They may be **bulk feeders, fluid feeders, substrate feeders,** or **filter feeders.** A **deposit feeder** is a type of substrate feeder.

34.2 Animal Digestive Tracts Take Many Forms

- Protists and sponges have intracellular digestion. Their cells engulf food and digest it in food vacuoles.
- Animals that are more complex have extracellular digestion in a cavity outside cells. An **incomplete digestive tract** (also called a **gastrovascular cavity**) has one opening. A **complete digestive tract,** or **alimentary**

canal (gastrointestinal tract), has two openings. Food enters through the mouth and is digested and absorbed; undigested material leaves through the **anus.**

- The length of the digestive tract, number of stomachs, and size of the **cecum** are adaptations to specific diets. In **ruminants,** bacteria inhabiting the **rumen** help break down hard-to-digest plant matter.

34.3 The Human Digestive System Consists of Several Organs

- Waves of contraction called **peristalsis** move food along the digestive tract. Muscular **sphincters** control movement from one compartment to another.

A. Digestion Begins in the Mouth and Esophagus

- In the mouth, **teeth** break food into small pieces. Salivary glands produce saliva, which moistens food and begins starch digestion.
- With the help of the **tongue,** swallowed food moves past the **pharynx** and through the **esophagus** to the stomach. The **epiglottis** prevents food from entering the trachea.

B. The Stomach Stores, Digests, and Pushes Food

- The **stomach** stores food, mixes it with **gastric juice,** and churns it into liquefied **chyme.** Hydrochloric acid in the gastric juice kills microorganisms and denatures proteins. The protein-splitting enzyme **pepsin** begins protein digestion.

C. The Small Intestine Digests and Absorbs Nutrients

- The **small intestine** is the main site of digestion and nutrient absorption.
- Enzymes break down polypeptides, carbohydrates, nucleic acids, and lipids. **Villi** absorb the products of digestion; **microvilli** on each villus provide tremendous surface area.

D. The Large Intestine Completes Nutrient and Water Absorption

- Material remaining after absorption in the small intestine passes to the **large intestine,** which absorbs water, minerals, and salts, leaving feces. Bacteria digest the remaining nutrients and produce useful vitamins that are then absorbed. Feces exit the body through the anus.

E. The Pancreas, Liver, and Gallbladder Aid Digestion

- The **pancreas** supplies pancreatic amylase, trypsin, chymotrypsin, lipases, and nucleases. The **liver** supplies **bile,** which emulsifies fat, and the **gallbladder** stores bile.

34.4 A Healthy Diet Includes Essential Nutrients and the Right Number of Calories

A. A Varied Diet Is Essential to Good Health

- Metabolism, growth, maintenance, and repair of body tissues require nutrients from food. **Macronutrients** include carbohydrates, proteins, fats, and water, whereas **micronutrients** include vitamins and minerals.

B. Body Weight Reflects Food Intake and Activity Level

- **Kilocalories** measure the energy food provides. Fat has more **Calories** per gram than either carbohydrate or protein.

C. Starvation: Too Few Calories to Meet the Body's Needs

- If a person does not eat enough over a long period, the body uses reserves of fat and protein to fuel essential processes. **Anorexia nervosa** and **bulimia** are eating disorders that may reduce calorie intake to dangerously low levels.

D. Obesity: More Calories than the Body Needs

- A person who eats more Calories than he or she expends will gain weight. **Obesity** is associated with many health problems.

34.5 Investigating Life: Brawny Vegetarian Lizards Hit the Salad Bar

- A study of 22 species of field-collected Argentine lizards has revealed that an herbivorous diet selects for larger body size, a longer small intestine, and an internal community of nematode worms.

MULTIPLE CHOICE QUESTIONS

1. At which stage does an organism's cells gain nutrients?
   a. Ingestion
   b. Digestion
   c. Absorption
   d. Elimination
2. A human is a
   a. filter feeder.
   b. deposit feeder.
   c. fluid feeder.
   d. bulk feeder.
3. A flatworm has a(n) __________ .
   a. incomplete digestive tract.
   b. alimentary canal.
   c. mouth and anus.
   d. Both b and c.
4. Which of the following parts of a human's digestive tract is NOT directly involved in digestion?
   a. The mouth
   b. The small intestine
   c. The large intestine
   d. Both b and c
5. Making the pH of the stomach closer to neutral (pH7) would inhibit the
   a. digestion of proteins.
   b. prevent chyme from moving to the duodenum.
   c. absorption of nutrients.
   d. Both a and c.

6. What is the function of bile?
   a. To break proteins into amino acids
   b. To digest fat molecules
   c. To break fat into small droplets
   d. To digest nucleic acids
7. How do the normal intestinal microflora prevent infection by pathogenic bacteria?
   a. Pathogens cannot find a place to grow.
   b. Pathogens are degraded by digestive enzymes.
   c. Pathogens are decomposed by the existing prokaryotes.
   d. Pathogens are killed by B vitamins.
8. An amino acid that can only be obtained from the diet is an example of
   a. an essential nutrient.
   b. a macronutrient.
   c. a micronutrient.
   d. both a and b.
9. What is the value of dietary fiber?
   a. It is a carbohydrate source.
   b. It helps material move along the digestive tract.
   c. It helps trigger the production of gastric juice.
   d. All of the above.
10. A person's body mass index is calculated based on
    a. caloric intake.
    b. height.
    c. weight.
    d. Both b and c.

TESTING YOUR KNOWLEDGE

1. What are the functions of food in an animal's body?
2. List four ways that animals obtain food.
3. Name an organism that has each of the following:
   a. extracellular digestion
   b. a gastrovascular cavity
   c. an alimentary canal
   d. filter feeding
4. What are the four stages of food use in animals, and where in the human digestive tract does each of these stages occur?
5. Identify the part of the digestive system that includes the following:
   a. duodenum
   b. cecum, appendix, rectum, anus
   c. villi and microvilli
   d. pyloric sphincter
6. How does mechanical breakdown of food speed chemical digestion?
7. How does the structure of the small intestine maximize surface area?
8. What are the digestive products of carbohydrates, proteins, and fats?
9. Trace the movement of food in the digestive tract, from mouth to anus.
10. How do the circulatory and muscular systems interact with the digestive system?
11. What is the difference between a macronutrient and a micronutrient? Give examples of each.
12. What factors determine whether a person will gain, lose, or maintain weight?

THINKING AS A SCIENTIST

1. The protein in a hamburger is mostly myosin, a contractile protein abundant in muscle. Why doesn't a burger-eating human simply use the cow version of myosin in his or her own muscles? What happens instead?
2. How does it benefit an organism to have a digestive system with an extensive surface area?
3. Why can't a person "eat" by finely grinding meat and injecting it directly into the bloodstream?
4. Compare and contrast the digestive systems of a whale and *Paramecium*.
5. Explain the observation that lactose intolerance is more common in adults than infants.
6. Compare and contrast the alveoli of the lungs with the villi of the small intestine.
7. Orlistat is a weight-loss drug that inhibits the activity of lipases in the small intestine. Why would this be more effective than a drug that blocks absorption of proteins or carbohydrates? What might be a side effect of blocking fat absorption?
8. Design an experiment to test the hypothesis that intestinal bacteria are essential to nutrient absorption in mice.
9. Many children believe that a piece of swallowed chewing gum will remain undigested in the body for 7 years. Does this idea make sense?
10. Evolutionary biologists suggest that we inherited from our ancestors both our taste for sugary, fatty foods and our ability to store excess energy as fat. How could these traits have been beneficial to early humans? How have circumstances changed in modern society?
11. Calculate your body mass index, using the formula in the text. How could you change your BMI?

Visit www.mhhe.com/hoefnagels for practice quizzes, animations, videos, and activities designed to help you master the material in this chapter.

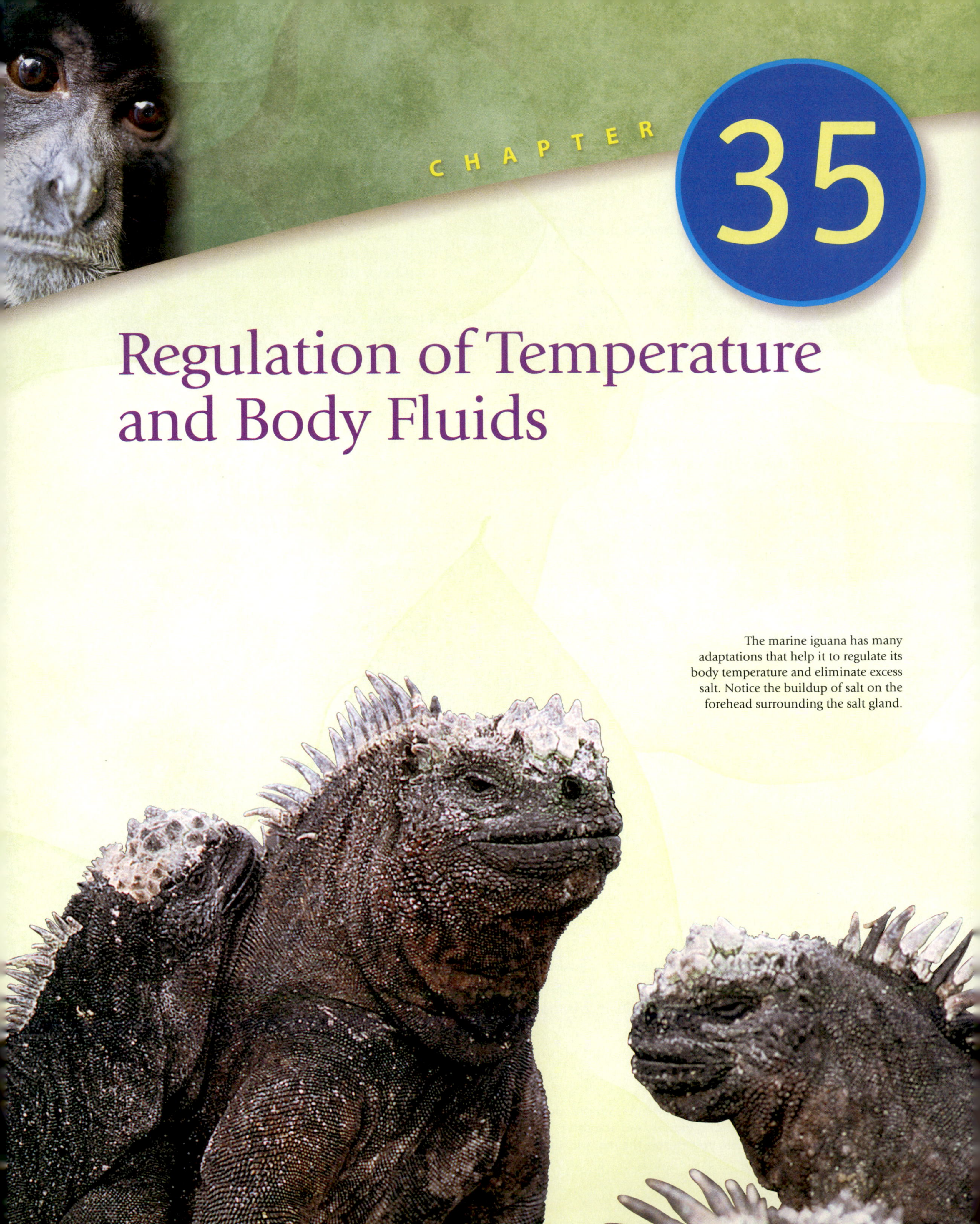

CHAPTER 35

Regulation of Temperature and Body Fluids

The marine iguana has many adaptations that help it to regulate its body temperature and eliminate excess salt. Notice the buildup of salt on the forehead surrounding the salt gland.

What's the Point?

A Day in the Life of a Marine Iguana

Many environments are hostile to cell survival—they are too hot, too cold, too salty, or too dilute. Humans cope with some of these extremes in familiar ways; we sweat when we are hot and shiver when we are cold, for example. We can eat salty pretzels, but only to the extent that our kidneys can dispose of the excess salt. We cannot drink seawater. Other animals, however, have radically different behavioral and physiological adaptations to both temperature and salt.

Consider the marine iguana lizard. These reptiles spend much of their time regulating their body temperatures. Iguanas and other reptiles are sometimes called "cold-blooded" (as opposed to "warm-blooded" birds and mammals). But this term is misleading. An iguana's body temperature is not necessarily low; rather, its temperature fluctuates as the lizard derives its heat from the environment.

Marine iguanas on the Galápagos Islands begin their days basking in the rising sun, draped on boulders and hardened lava, sunning their backs and sides. After about an hour, they turn, raise their bodies, and aim their undersides at the sun.

By midmorning, the air temperature is rising rapidly. Iguanas cannot sweat as humans do; instead, they must escape the blazing sun. They lift their bodies by extending their short legs, which removes their bellies from the hot rocks and allows breezes to fan them. By noon, these push-ups are insufficient to stay cool. The iguanas retreat to the shade of rock crevices.

The animals are hungry by midday. The iguanas dive into the ocean, although it is too cold for them to stay in for more than a few minutes. They eat green algae on the ocean floor or seaweed by the shore, hanging off rocks to reach it. The water is so cold that the lizards' body temperatures rapidly drop. Arteries near their body surfaces constrict, helping conserve heat. After feeding, the iguanas stretch out on the rocks again, warming sufficiently to digest their meal. They continue basking as the day ends, absorbing enough heat to sustain them through the cooler night temperatures until a new day begins.

The iguana's food, which comes from the ocean, is salty. A gland in the iguana's head rids the animal's body of extra salt. Cells lining the interior of the salt gland secrete a fluid that travels in branching tubules, which empty through the nostrils. This fluid carries the salts with it.

Marine iguanas, and every other animal, maintain homeostasis in many ways. This chapter describes how animals regulate their temperatures and control the composition of their body fluids.

LEARNING OUTLINE

35.1 Animals Must Regulate Their Internal Temperature

Animals live nearly everywhere on Earth (**figure 35.1**). The extremely dry, scorching hot home of a camel contrasts sharply with a frigid arctic habitat or with the perpetual humidity of a tropical rain forest. These habitats select for very different ways of regulating body temperature, conserving water, and disposing of wastes. In hot, dry areas, an animal must conserve enough water for cells to function, and, at the same time, produce enough sweat to tolerate the heat. At another extreme, frigid water surrounds an animal living in an arctic freshwater lake. An animal in this habitat must continually pump excess water out of its body and conserve salts, all while surviving the cold.

Most animals live in more moderate environments, but each species has adaptations that enable it to maintain homeostasis. This chapter begins by describing how animals regulate body temperature and the concentrations of salt and water in their tissues. It moves next to the types of nitrogenous wastes that animals produce, then concludes with the structure and function of the human urinary system.

FIGURE 35.1 Different Habitats. Penguins thrive on Antarctic ice, whereas parrots inhabit the warm, humid tropics. Although penguins and parrots are both birds, their adaptations reflect their diets, the temperatures of their habitats, and many other selective forces.

A. An Animal's Body Temperature Reflects Heat Gains and Losses

Whether it lives in the Arctic circle or the Amazonian rainforest, an animal's body temperature must remain within certain limits. Part of the reason is that extreme temperatures alter biological molecules. Excessive heat can alter a protein's three-dimensional shape and disrupt its function. Extreme cold solidifies lipids and therefore inhibits the function of biological membranes. In addition, if the temperature inside a cell deviates from an animal's customary body temperature, enzymes function less efficiently, and vital biochemical reactions slow. **protein folding, p. 39; lipids, p. 35; enzymes, p. 87**

Thermoregulation is the control of body temperature, and it requires the ability to balance heat gained from, and lost to, the environment. An animal can also maintain homeostasis by controlling how much heat it produces. When cells generate ATP in aerobic respiration, they also produce metabolic heat. The more active the animal, the higher its metabolic rate, and the more heat it produces. **aerobic respiration, p. 124**

The main source of an animal's body heat may be external or internal (**figure 35.2**). An **ectotherm** lacks an internal temperature-regulating mechanism. It thermoregulates by moving to areas where it can gain or lose heat, so its temperature varies with external conditions. The vast majority of animals are ectotherms, including all invertebrates, plus fishes, amphibians, and reptiles such as the iguana described

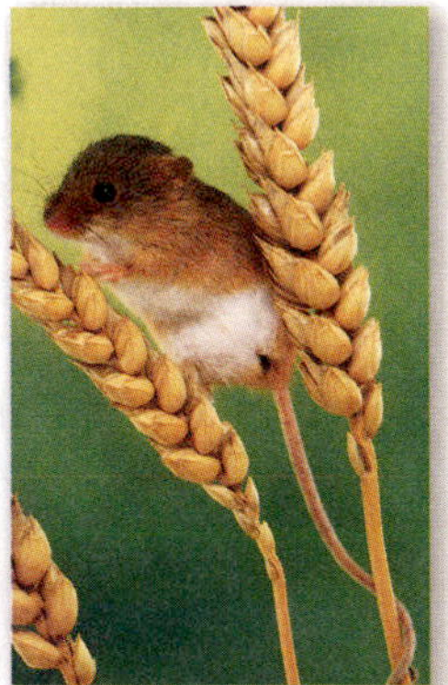

FIGURE 35.2 Ectothermy and Endothermy. An ectotherm such as a snake alters its behavior to manage the gain of heat from, or losses of heat to, the environment. In contrast, an endotherm's metabolism generates most of its body heat. A mouse is an endotherm.

in the chapter opening essay. An **endotherm,** such as a camel, regulates its body temperature internally. Most endotherms maintain a relatively constant body temperature by balancing heat generated in metabolism (especially in the muscles) with heat lost to the environment. Mammals and birds are endotherms; insulation in the form of fat, feathers, or fur help retain their body heat.

Both ectothermy and endothermy have advantages and disadvantages. The ectotherm uses much less energy, and therefore requires less food and O_2, than an endotherm. However, an ectothermic animal must be able to obtain or escape environmental heat. An injured iguana that could not squeeze into a crevice to avoid the broiling noonday sun would cook to death. Ectotherms also become sluggish when the temperature is low, which can make it hard for them to escape from predators. On the other hand, an endotherm maintains its body temperature even in cold weather or in the middle of the night. But this internal constancy comes at a cost. The metabolic rate of an endotherm is generally five times that of an ectotherm of similar size and body temperature. The endotherm therefore requires much more food than an ectotherm.

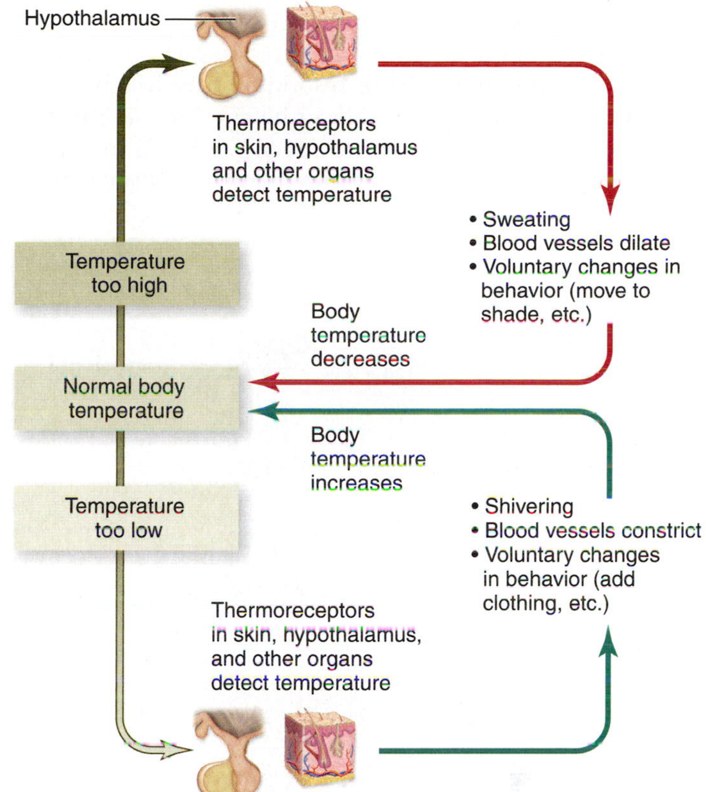

FIGURE 35.3 Thermoregulation in Humans. Thermoreceptors signal the hypothalamus to keep body temperature within a certain range.

B. Several Adaptations Enable an Animal to Adjust Its Temperature

Neurons in the hypothalamus receive temperature information from thermoreceptors in the skin and other organs, and then control many of the responses that maintain homeostasis (**figure 35.3**). These responses may be physiological or behavioral. Sweating is a physiological reaction to heat, but a person who is too hot might also move into the shade—a behavioral response. Thermoregulation therefore depends on the interactions of the musculoskeletal, nervous, endocrine, respiratory, integumentary, and circulatory systems.

One adaptation to extreme cold is a **countercurrent exchange** system, in which two adjacent currents flow in opposite directions and exchange materials or heat with each other (see figure 33.3). Countercurrent exchange can move heat from an artery containing warm blood (from the body's interior) to a vein carrying cold blood from near the body's surface (**figure 35.4**). In this way, the body conserves heat rather than allowing it to escape through the extremities. Countercurrent exchange allows penguins to spend hours in frigid water and enables arctic mammals such as wolves to hunt in extreme cold.

Many birds and mammals, including humans, also make other physiological changes that conserve heat. In cold weather, blood vessels in the extremities constrict, retaining more blood in the warmer core of the body. The animal may also shiver; the contraction of skeletal muscle generates heat.

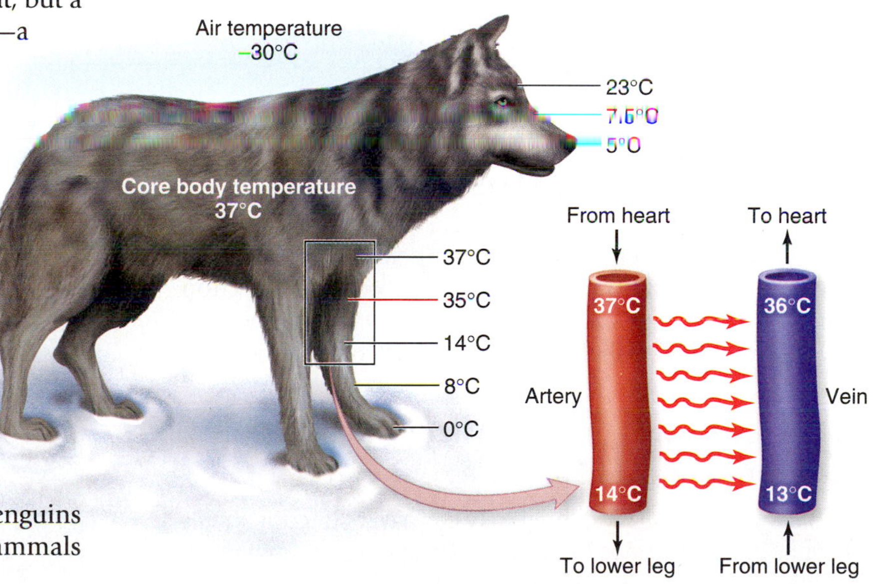

FIGURE 35.4 Countercurrent Heat Exchange. A mammal's extremities chill more easily than interior body parts. In a countercurrent exchange system, arteries carrying warm blood lie near veins carrying cooler blood, thereby conserving heat rather than losing it to the environment.

At the same time, muscles in the skin cause feathers or fur to stand erect, trapping an insulating air layer next to the skin. (In humans, this hair-raising response is useless because we have so little body hair. Nevertheless, "goose bumps" form when the hair muscles contract.) Animals may also huddle together, migrate, or hibernate to conserve heat (**figure 35.5**).

FIGURE 35.5 Huddling Conserves Heat. These snow monkeys (Japanese macaques) are adapted to cold winters. Their thick fur retains body heat, as does their huddling behavior; some snow monkeys soak in warm springs to escape winter weather!

Animals also must maintain homeostasis in warm environments. Evaporative cooling lowers body temperature. Humans sweat. A coyote pants, and an owl flutters loose skin under its throat to move air over moist surfaces in the mouth. In addition, when the environment is warm, blood vessels in the extremities dilate and allow more blood to approach the relatively cool body surface. Tiny veins in the face and scalp also reroute blood cooled near the body's surface toward the brain. This adaptation causes the face of a person who is vigorously exercising to turn red.

Behavioral strategies can also help an animal cool off. Many animals escape the sun's heat by retreating to the shade; some burrow underground and emerge only at night. Humans also consume cold food and drinks, swim, or fan themselves to increase heat loss by convection. We also shed extra layers of clothing.

Summary *An animal's body temperature must remain within a certain range for cells to function. Ectotherms obtain heat from the surroundings; endotherms derive heat from metabolism. Adaptations to temperature extremes are physiological and behavioral.*

35.1 MASTERING CONCEPTS

1. Which animals are ectotherms, and which are endotherms?
2. What are the advantages and disadvantages of ectothermy and endothermy?
3. What are some physiological and behavioral adaptations to extreme cold and extreme heat?

35.2 Animals Must Regulate Water and Ion Balance and Excrete Wastes

Besides regulating its temperature, an animal must also maintain homeostasis in other ways. This section describes why animals regulate water, salts, and wastes. Section 35.3 then discusses how the kidneys produce urine.

A. Osmoregulation Controls Water and Ions in Body Fluids

The balance of solutes and water between a cell's interior and its surroundings is critical to life. Cells must retain water even when conditions are dry, yet too much water is damaging. Sodium, hydrogen, and other ions are vital to life, but not in excess. In most habitats, therefore, organisms must **osmoregulate;** that is, they control the concentration of ions in their body fluids as the environment changes (**figure 35.6**). Osmoregulation requires managing the gain and loss of ions, water, or both.

FIGURE 35.6 Osmoregulation in a Crab. Salt water and fresh water meet in estuaries, and the salinity changes with the tide. When the surrounding water becomes less salty, crabs must pump out excess water that enters the body by osmosis. This crab is shooting dilute urine from antennal glands near its eyes.

A brief review of how water and ions move across membranes will help explain how osmoregulation works. Osmosis is the diffusion of water across a membrane that is permeable to water but not to ions and other charged solutes. When the concentration of solutes in fluid is higher outside the cell than inside, water moves out of the cell. If the cell's solute concentration is higher than the environment's, water moves into the cell (see figure 4.13). Most ions, on the other hand, cross membranes via protein channels (see figure 4.16). Cells often use active transport to move ions against their concentration gradient; water then follows by osmosis.

Land animals use a combination of strategies to obtain and conserve water (**figure 35.7**). Whereas humans ingest most of their water in food and drink, kangaroo rats derive most of their water as a byproduct of respiration (see chapter 6). Animals lose water through evaporation from lungs and skin surfaces, in feces, and in urine. Similarly, electrolytes enter the body in food and drink and leave the body in urine.

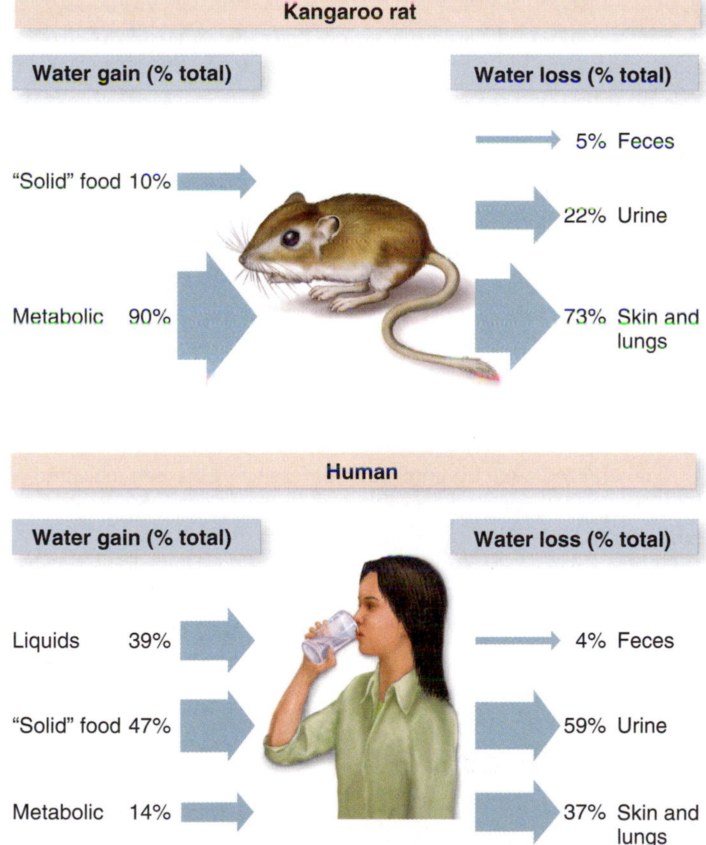

FIGURE 35.7 Water Gain and Loss. The desert-dwelling kangaroo rat must conserve water. It gets most of its water from its metabolism and loses little through feces and urine. In contrast, a human gets most water from food and drink and loses most of it through urine.

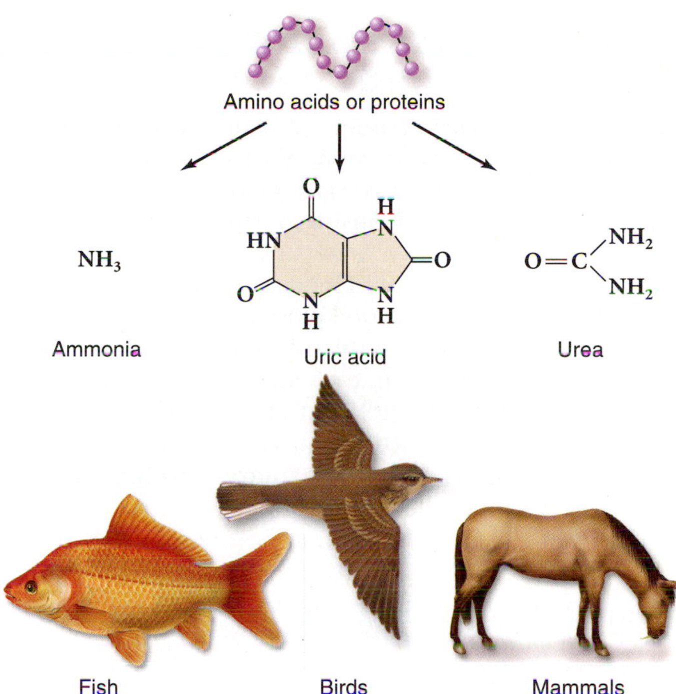

FIGURE 35.8 Three Nitrogenous Wastes. Ammonia is the simplest byproduct of protein breakdown, but it is toxic. Urea and uric acid are more costly to produce but much less toxic.

B. Nitrogenous Wastes Include Ammonia, Urea, and Uric Acid

Animals produce two main categories of waste. One type is feces, which contains undigested food that passes through the digestive tract without ever entering the body's cells (see chapter 34). The other category is waste produced by the body's cells; **excretion** is the elimination of these metabolic wastes. For example, as described in chapter 33, the respiratory system excretes CO_2, a byproduct of aerobic cellular respiration.

The urinary system excretes nitrogen-containing (nitrogenous) wastes, which cells produce mainly during the breakdown of proteins (**figure 35.8**). As a protein is degraded, amino groups ($-NH_2$) are released from amino acids. Each amino group then picks up a hydrogen ion and becomes **ammonia,** NH_3. Excreting ammonia is energetically efficient, but ammonia is very toxic and therefore animals must excrete it in a dilute solution.

Land animals expend energy to change ammonia to less toxic substances that can be stored and excreted in a relatively concentrated form. Mammals, amphibians, sharks, and some bony fishes form **urea,** which the liver produces as cells break down proteins. Insects, reptiles, and birds change most of their nitrogenous wastes to **uric acid** and excrete it in an almost solid form. In birds, uric acid mixes with undigested food to form the familiar "bird dropping."

C. The Urinary System Produces, Stores, and Eliminates Urine

In the human **urinary system,** the paired **kidneys** are the major excretory organs (**figure 35.9**). Each kidney is about the size of an adult fist and weighs about 230 g (about half a pound). These blood-cleansing organs not only excrete urea, but they also conserve water and nutrients, adjust salt concentrations, and regulate blood pH and blood volume (which in turn influences blood pressure).

As the kidneys process blood, a liquid waste called **urine** forms. The urine from each kidney drains into a **ureter,** a muscular tube about 28 cm (11 inches) long. Waves of smooth muscle contraction squeeze the fluid along the two ureters and squirt it into the **urinary bladder,** a saclike muscular organ that collects urine.

The **urethra** is the tube that connects the bladder with the outside of the body. In females, the urethra opens between the clitoris and vagina. In males, the urethra extends the length of the penis. The urethra also carries semen in males (see chapter 37). The term *urogenital tract* reflects the intimate connection between the urinary and reproductive systems.

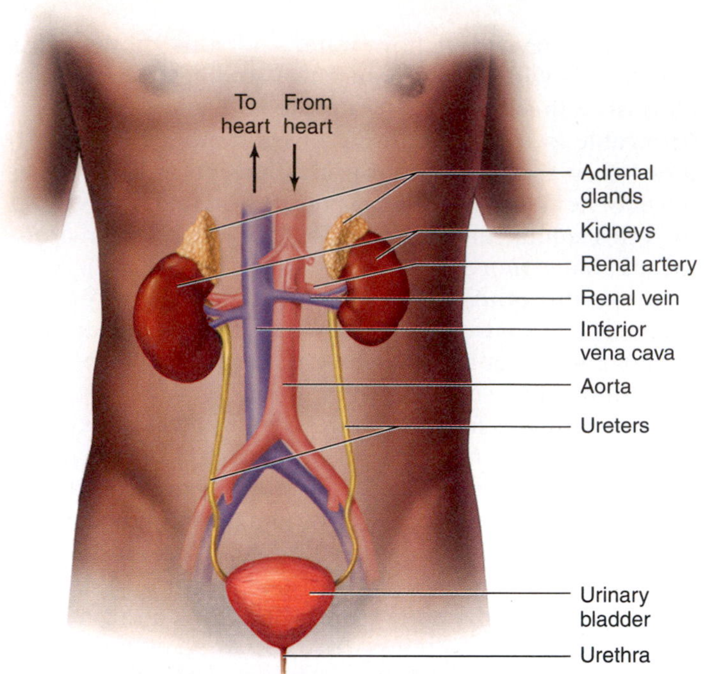

Urinary System	
Main tissue types	**Examples of Locations/Functions**
Epithelial	Enables diffusion across glomerular capsules and kidney tubules; also lines ureters and bladder.
Connective	Blood (which kidneys filter) is a connective tissue.
Muscle	Smooth muscle controls flow of blood to and from nephrons; smooth and skeletal muscle sphincters control urine release.
Nervous	Sensory cells in hypothalamus coordinate negative feedback loops that maintain homeostasis.

FIGURE 35.9 Human Urinary System. The human urinary system includes the kidneys, ureters, urinary bladder, and urethra. This generalized depiction omits the differences between male and female organs. Figures 37.3 and 37.6 show the sex differences in more detail.

Summary ***Animals must maintain the concentrations of ions in their body fluids within a specific range. Animals must also excrete nitrogen-rich wastes resulting from the breakdown of proteins and nucleic acids. In the human urinary system, paired kidneys drain urine into ureters, which empty into the bladder. Urine exits the body through the urethra.***

35.2 MASTERING CONCEPTS

1. Why is osmoregulation important?
2. What are three examples of nitrogenous wastes in animal bodies?
3. What are the functions of the kidneys?
4. What are the organs that make up the human urinary system?

35.3 The Nephron Is the Functional Unit of the Kidney

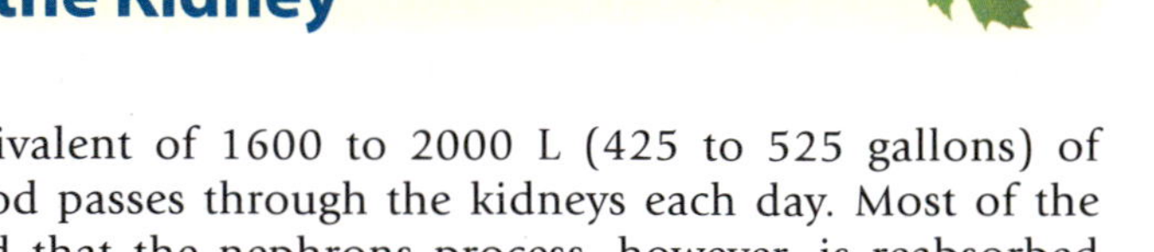

The kidney illustrates a familiar biological organization: extensive surface area packed into a relatively small volume. Each kidney contains 1.3 million microscopic tubular **nephrons**—the functional units of the kidney. The body's entire blood supply courses through the kidney's blood vessels every 5 min. At that rate, the equivalent of 1600 to 2000 L (425 to 525 gallons) of blood passes through the kidneys each day. Most of the fluid that the nephrons process, however, is reabsorbed into the blood and not released in urine. As a result, a person produces only about 1.5 L (about 1.5 quarts) of urine daily.

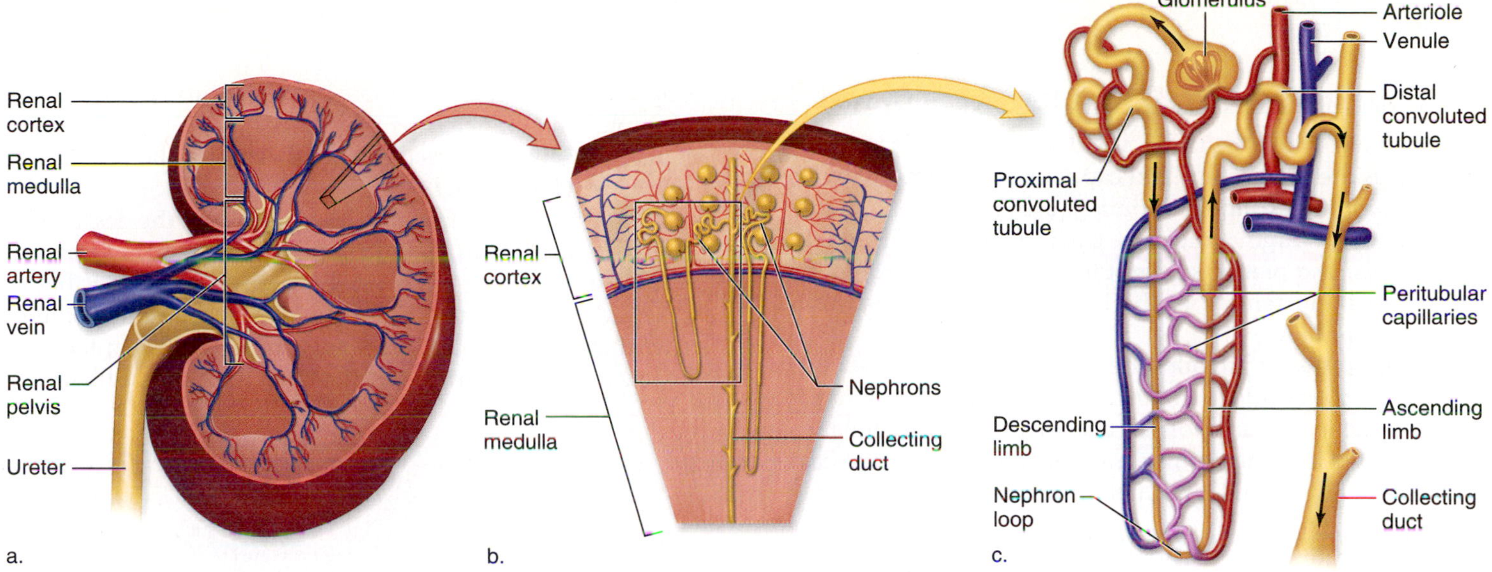

FIGURE 35.10 Anatomy of a Kidney. (**A**) A kidney sliced down the middle reveals an outer area, the renal cortex, and an inner section, the renal medulla. (**B**) The nephrons are aligned; as a result, the same nephron parts are found in the same region of the kidney. (**C**) The nephron is the structural and functional unit of the kidney.

Not surprisingly, nephrons and blood vessels are intimately entwined inside each kidney (**figure 35.10**). A renal artery arising from the aorta enters each kidney (see figure 35.9) and then branches into multiple arteries and arterioles. Each arteriole delivers blood to a **glomerulus,** a tuft of capillaries containing blood to be filtered. These capillaries then converge into another arteriole, which leads to the **peritubular capillaries** that snake around each nephron. These blood vessels reabsorb water and useful ions from the nephron. The capillaries then empty into a venule, which joins the renal vein carrying cleansed blood out of the kidney and (ultimately) to the heart.

Each nephron, when stretched out, is about 12 mm (about half an inch) long and consists of two main parts: a glomerular capsule and a renal tubule. The **glomerular (Bowman's) capsule** receives fluid from the glomerulus. From there, the solution travels along the **renal tubule,** which adjusts the composition of the filtrate. A **collecting duct** receives the resulting urine from the other end of the renal tubule.

The chemical composition of urine reflects three processes (**figure 35.11**):

1. **Filtration**: Urea, nutrients, and water from the blood enter the glomerular capsule, leaving large proteins and blood cells in the blood.
2. **Reabsorption:** Salts, water, and nutrients are recycled back into the blood from the renal tubule into the peritubular capillaries.
3. **Secretion:** Toxic substances, drug residues, hydrogen ions (H^+), and some electrolytes are actively transported out of the peritubular capillaries and into the renal tubule for elimination in urine.

When the kidney's nephrons fail to do their job, nitrogenous wastes and other toxins accumulate in the blood to harmful levels, or a person can lose too much water and become dehydrated. Without treatment, kidney failure can be fatal (see Can *You* Relate? Kidney Failure, Dialysis, and Transplants).

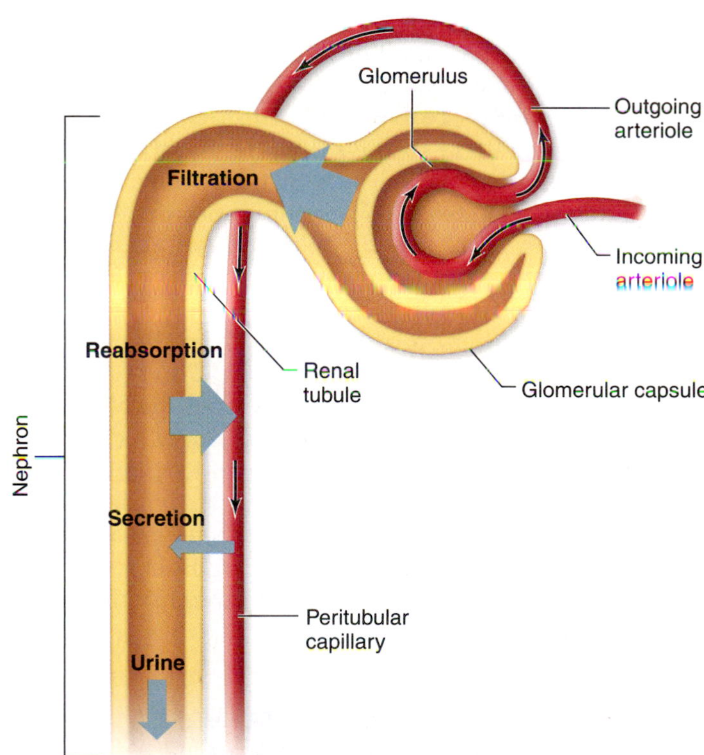

FIGURE 35.11 Overview of Urine Formation in the Nephron. Urine forms from the processes of filtration, reabsorption, and secretion that selectively move substances between the blood and the renal tubules. This is a stylized, schematic view that does not reflect nephron anatomy.

A. The Glomerular Capsule Filters Blood

Like a catcher's mitt catching a ball, the glomerular capsule surrounds the glomerulus and captures water, urea, glucose, salts, amino acids, and creatinine (a byproduct of muscle contraction). Large structures such as plasma proteins, blood cells, and platelets do not fit through the pores in the glomerulus and remain in the bloodstream. **composition of blood, p. 665**

Blood pressure provides the force that drives substances out of the glomerulus. The incoming arteriole has a larger diameter than the arteriole that leaves the glomerulus. As blood moves into the thinner arteriole, the resulting increase in pressure forces fluid and small dissolved molecules across the capillary walls.

The filtrate in the glomerular capsule is the product of the first step in urine manufacture. The 1600 to 2000 L (425 to 525 gallons) of blood per day that pass through the kidneys produces approximately 180 L (45 gallons) of glomerular filtrate.

B. Blood Recovers Water and Other Valuable Substances from the Proximal Convoluted Tubule

The remainder of the renal tubule consists of a winding passageway with three functional regions (see figure 35.10C): the proximal convoluted tubule, the nephron loop (loop of Henle), and the distal convoluted tubule.

The **proximal convoluted tubule,** which leads from the glomerular capsule to the nephron loop, is the most important site of selective reabsorption. All along the tubule, specialized cells transport sodium (Na^+) and other important electrolytes into the interstitial fluid surrounding the tubule. Water follows the ions by osmosis. Overall, almost two thirds of the water and ions in the filtrate is reabsorbed into the blood from the proximal convoluted tubule.

Secretion also occurs in the proximal convoluted tubule. The antibiotic penicillin, for example, leaves the blood and enters the filtrate in the renal tubule by secretion. Patients must take penicillin several times a day to compensate for this loss. Together, the secretion of H^+ and reabsorption of bicarbonate ions (HCO_3^-) help maintain the pH of blood between 7 and 8—any variance from this range is deadly.

C. The Nephron Loop Conserves Additional Water and Exchanges Ions with Blood

Next, the filtrate moves into the hairpin-shaped **nephron loop,** which connects the proximal and distal convoluted tubules. Blood in the peritubular capillaries flows in the opposite direction of the filtrate in the nephron loop, another example of countercurrent exchange. The nephron loop returns positively charged ions and water to the peritubular capillaries (**figure 35.12**).

Fluid in the nephron first moves through the descending limb of the nephron loop. In this region, the solute concentration in the peritubular capillaries is higher than that of the fluid inside the nephron loop. The concentration difference draws water out of the renal tubule, into the interstitial spaces of the renal medulla, and then into the capillaries by

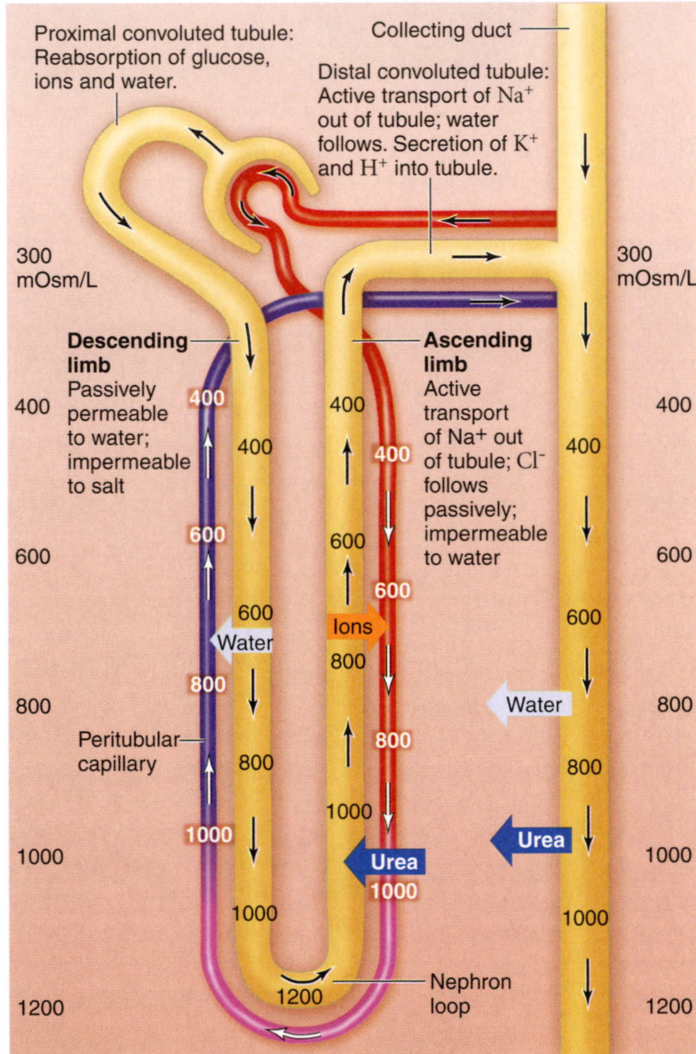

FIGURE 35.12 Nephron Structure and Function. This schematic view shows the direction of ion, water, and urea movement in the nephron loop. Active transport of Na^+ from the ascending limb creates a briny broth around the loop, which causes water to move from the renal tubule into the bloodstream. Some urea diffuses out of the collecting duct and further adds to the high solute concentration around the nephron loop. The countercurrent flow of fluids in the tubules and adjacent blood vessels improves the exchange of substances. ("mOsm" stands for milliosmole, a measure of ion concentration.)

osmosis. The cells of the descending limb are permeable to water, but impermeable to ions and urea. As water leaves the descending limb, the sodium ion (Na^+) concentration inside the tubule rises until it reaches its maximum at the bottom of the loop.

The filtrate next moves around the bend and into the ascending limb, which is impermeable to water but permeable to ions and urea. Near the bend, the wall of the ascending limb is relatively thin. Here Na^+ diffuses from the filtrate along its concentration gradient into the more dilute blood of the peritubular capillaries. At the thick-walled portion of the ascending limb, near the distal convoluted tubule, Na^+ and other ions move out of the filtrate by active transport. As a result, the filtrate is less concentrated than the tissues and blood in the surrounding cortex.

D. The Distal Convoluted Tubule and Collecting Duct Conserve More Water

The **distal convoluted tubule** in the kidney's cortex connects the nephron loop to the collecting duct. Na^+ continues to be reabsorbed into the peritubular capillaries from the distal convoluted tubule by active transport (see figure 35.12). The walls of the distal convoluted tubule are permeable to water. As Na^+ accumulates outside the distal convoluted tubule, water moves by osmosis into the capillaries. At the same time, if excess K^+ is present in the blood, it may be secreted into the distal convoluted tubule and collecting duct. Once the filtrate has passed through the distal convoluted tubule, 97% of the water in the original glomerular filtrate has been reabsorbed, and little salt remains.

The fluid from several renal tubules drains into one collecting duct. The collecting duct descends back into the Na^+-rich medulla, as did the nephron loop. Much of the remaining water in the filtrate therefore diffuses out of the collecting duct. Some urea also diffuses out of the collecting duct and contributes to the high solute concentration surrounding the nephron loop.

After reabsorption and secretion, the filtrate is urine. Urine contains water, urea, a small amount of uric acid, creatinine, and several ions (see Burning Question: What can urine reveal about health and diet?). From the collecting duct, urine accumulates in the renal pelvis before entering the ureter and urinary bladder and moving out of the body through the urethra.

E. Hormones Regulate Kidney Function

The activities of the kidneys are adjusted continuously to maintain homeostasis. For example, when we drink too much water, our kidneys allow more fluid to pass into the urine. If water is scarce, however, our kidneys produce more concentrated urine. The body makes these adjustments by altering the permeability of the renal tubules in a negative feedback loop (**figure 35.13**). When we are dehydrated, osmoreceptor cells in the hypothalamus send impulses to the posterior pituitary gland, which secretes a peptide hormone called **antidiuretic hormone (ADH).** The more ADH in the bloodstream, the more water-permeable the walls of the distal convoluted tubule and collecting duct. The blood reabsorbs more water. Conversely, if blood plasma is too dilute, ADH production stops, and more water remains in the urine. **posterior pituitary, p. 622**

A diuretic is a substance that increases the volume of urine. The ethyl alcohol in alcoholic beverages is a diuretic. Alcohol stimulates urine production partly by reducing ADH secretion, thereby decreasing the permeability of the tubules to water. Because it increases water loss to urine, an

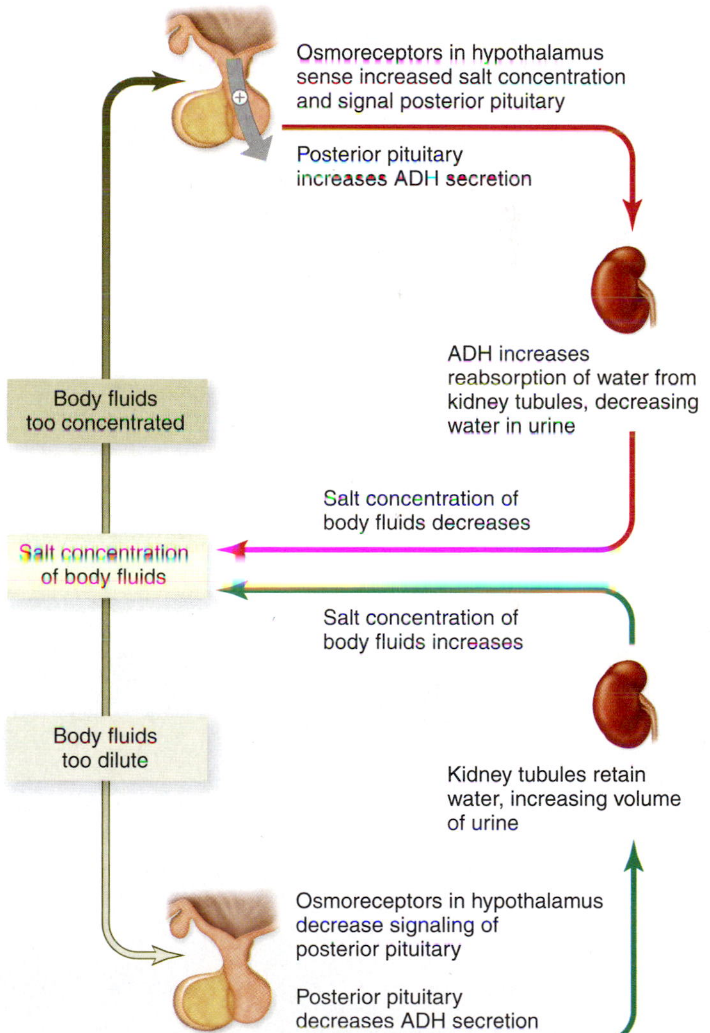

FIGURE 35.13 Negative Feedback Control of Ion Concentration. A feedback loop connecting the hypothalamus, the posterior pituitary, and the blood controls the amount of water reabsorbed from the kidneys into the bloodstream.

alcoholic beverage actually intensifies thirst. Dehydration resulting from drinking too much alcohol causes the discomfort of a hangover.

Hormones also regulate blood pressure. For example, when blood pressure and blood volume dip too low, the adrenal cortex releases the steroid hormone **aldosterone.** This mineralocorticoid stimulates Na^+ movement out of the distal convoluted tubules and into the bloodstream. Water follows, and blood pressure rises. **adrenal glands, p. 625**

Summary *A nephron is a tubule that filters blood and secretes wastes into the forming urine. Blood reabsorbs water and other valuable materials from the nephron. Antidiuretic hormone controls water reabsorption from the distal convoluted tubule, and aldosterone increases salt and water reabsorption in the kidneys.*

35.3 MASTERING CONCEPTS

1. What pathway does blood take as it moves through a kidney?
2. What three processes occur in the kidneys?
3. What occurs in the glomerular capsule and the proximal convoluted tubule?
4. What is the function of the countercurrent exchange system in the nephron loop?
5. What happens in the distal convoluted tubule and collecting duct?
6. Describe the negative feedback loop that keeps the solute concentration of the blood constant, regardless of water intake.
7. How does aldosterone control blood volume and sodium level?

Can *You* Relate?

Kidney Failure, Dialysis, and Transplants

Kidneys can fail for many reasons. The most common causes are diabetes, high blood pressure, and chronic inflammation that causes progressive loss of nephrons. Other causes are polycystic kidney disease (the accumulation of cysts in the kidneys), scarring from childhood kidney infections, and obstructed urine flow.

One common treatment option for kidney patients is dialysis, in which a machine takes over the kidney's function (**figure 35.A**). The dialysis machine pumps blood out of the patient's body and past a semipermeable membrane. Wastes and toxins, along with water, diffuse across the membrane to a waste fluid called "dialysate," but blood cells do not. The cleaned blood then circulates back to the patient's body. The procedure requires hours a day, several times a week.

Dialysis membranes cannot replace all of the kidney's functions. For example, nephrons selectively recycle useful components such as glucose and salts to the blood. The dialysis machine cannot do this, although a technician can adjust the concentrations of these dissolved compounds in the dialysate to promote or inhibit diffusion from the blood.

Transplantation is a second option. Kidneys are among the most commonly transplanted organs, and the success rate is very high. The transplanted kidney may come from a cadaver, or a living person may donate one healthy kidney to a recipient. (The remaining kidney has enough blood cleansing power for the donor to live.) Surgeons connect the new kidney to the recipient's blood supply and attach its ureter to the bladder, usually leaving the old kidneys in place (**figure 35.B**).

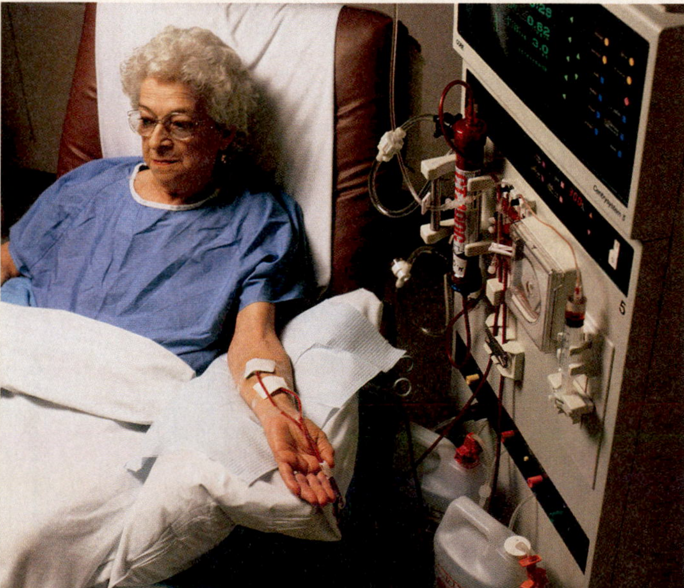

FIGURE 35.A Kidney Dialysis. A dialysis machine can take over some, but not all, of a healthy kidney's functions.

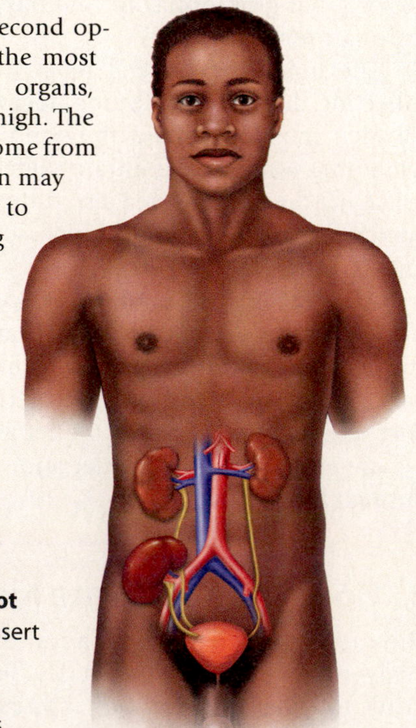

FIGURE 35.B Three Is Not a Crowd. Surgeons usually insert a new kidney in a cavity low in the abdomen, leaving the patient's original two kidneys in place.

Burning Questions

What can urine reveal about health and diet?

Urinalysis is a routine part of many medical examinations. Laboratory tests of the chemical components of urine can reveal many health problems:

- More than a trace of glucose may be a sign of diabetes mellitus, a high-carbohydrate diet, or stress. Stress causes release of excess epinephrine, which stimulates the liver to break down more glycogen into glucose.
- Albumin may be a sign of damaged nephrons. This plasma protein is too large to fit through the pores of intact glomerular capsules.
- Pus and an absence of glucose indicate a urinary tract infection. The pus consists of infection-fighting white blood cells along with bacteria, which consume the glucose.
- Traces of marijuana, cocaine, and other substances may appear in the urine. Athletes and employees of some organizations undergo routine drug testing.

Although urine is usually odorless, some foods impart a distinctive aroma. Asparagus, for example, contains a sulfur-rich molecule called mercaptan. Many people produce an enzyme that breaks down mercaptan, and its breakdown products have a strong odor.

What about urine's color—why is it normally pale yellow? The pigment that lends its color to urine is a byproduct of the liver's breakdown of dead blood cells. Colorless urine usually indicates excessive water intake or the ingestion of diuretics such as coffee or beer. A reddish tinge may suggest anything from bleeding in the urinary tract to beet consumption to mercury poisoning. Either vitamin C or carrot consumption can color urine orange. It is always best to check with a physician when urine changes color.

Have a Burning Question of your own? Submit it to marielle_hoefnagels@McGraw-Hill.com for possible inclusion in future editions of this book!

35.4 INVESTIGATING LIFE

Sniffing Out the Origin of Fur and Feathers

While visiting the zoo and appreciating the beauty of peacocks, parrots, and flamingos, have you wondered about the origin of their beautiful plumage (**figure 35.14**)? Have you ever admired the fur coats of a giraffe, jaguar, or red fox and asked yourself why hair evolved? Evolutionary biologists have been working to answer the same questions.

Fossil and DNA evidence clearly indicate that mammals and birds arose from two different lineages of reptiles. Existing nonavian reptiles are ectothermic, whereas both birds and mammals are endothermic. Because the ancestors of birds were not closely related to mammalian ancestors, researchers surmise that endothermy evolved independently in these two animal lineages.

It makes sense that feathers and fur evolved at the same time as the elevated metabolic rates associated with birds and mammals. After all, these integumentary adaptations provide insulation that helps the animals maintain their body temperature. To test this hypothesis of simultaneous evolution, scientists need to trace animal ancestry back in time. Ideally, they would examine fossils of animals from several points along the mammal and bird lineages, before and after the evolution of endothermy, fur, and feathers. Such an analysis would not prove a direct cause-and-effect relationship between endothermy and insulation, but it would show whether the adaptations arose at the same time.

This task is easier said than done. Both fur and feathers occasionally show up in fossils, but determining the metabolic style of an extinct animal is difficult. How can we know whether a long-extinct animal was an endotherm?

FIGURE 35.14 Pretty Bird. Feathers have many functions, including insulation, flight, and communication. Scientists are working to determine how these structures evolved.

John A. Ruben and his colleagues at Oregon State University believe they have hit on the ideal indicator of endothermy: the inside of the nose. Endothermy requires a high metabolic rate, which in turn means a huge demand for O_2 to fuel respiration. Mammals and birds therefore have high breathing rates relative to their ectothermic counterparts. Along with this greater volume of air moving in and out of the lungs, however, comes a higher potential loss of both heat and water from an animal's body. Endotherms minimize this loss with the help of structures called turbinates that direct the flow of air within the nasal cavity (**figure 35.15**A).

Turbinates are baffles made of bone or cartilage covered with epithelium. These structures warm and humidify inhaled air before it enters the lungs; they also cool and remove water vapor from air as it is exhaled. The nasal cavities of more than 99% of all existing birds and mammals contain turbinates; no known ectotherms have them. Ruben and his colleagues reasoned that turbinates are therefore the next best thing to direct evidence of endothermy in extinct animals. They might be useful for approximating when endothermy first appeared in the ancestors of mammals and birds.

The researchers began by looking for evidence of bony turbinates in numerous fossils of reptilian ancestors to mammals, including therapsids (see figure 16.17). They found them in therapsids that lived during the Permian period, some 250 MYA. The first evidence of insulating fur appears only in fossils of true mammals, which did not arise until millions of years later, during the Triassic period. The researchers therefore concluded that the reptilian ancestors of mammals developed endothermy long before the origin of fur.

But birds posed a problem. The earliest known bird, *Archaeopteryx*, lived some 150 MYA and clearly had feathers. Did this animal have turbinates? It is hard to know, because turbinates in birds are made of cartilage, a soft tissue that does not leave fossils. Nevertheless, Ruben's team suggested an indirect way to detect whether turbinates were present. The researchers predicted that, in general, endotherms should have broader nasal cavities than ectotherms, both to accommodate the presence of turbinates and to allow for a high breathing rate.

To make sure this was the case, Ruben's research team measured the cross-sectional areas of the nasal cavities of 21 living species of birds, mammals, and reptiles. In every case, the cross-sectional area of the nasal cavity was larger for an endotherm than for an ectotherm of equal size (figure 35.15B). The consistent, strong relationship suggested that measuring the nasal cavity in a fossil should be a good way to learn if an extinct animal was an endotherm or an ectotherm.

The researchers used modern imaging technologies to measure nasal cavities inside the fossilized skulls of three dinosaur species that lived about 70 MYA and are closely related to modern birds. The results indicate that the reptilian ancestors to birds probably were ectotherms. Feathers therefore preceded endothermy in birds by tens of millions of years.

Fur and feathers are adaptations that help mammals and birds stay warm, so it is easy to assume they evolved hand-in-hand with endothermy. Ruben's team paired old-fashioned comparative anatomy with modern technology to turn this assumption on its head. Endothermy was evidently present in mammalian ancestors before there was fur, and feathers apparently assisted in flight long before birds became endothermic. Thanks to the efforts of this research team, we now have part of the answer to the puzzle—and it is right in front of our nose.

Ruben, John A., Willem J. Hillenius, Nicholas R. Geist, et al. 1996. The metabolic status of some late Cretaceous dinosaurs. *Science*, vol. 273, pages 1204–1207.

Ruben, John A., Terry D. Jones. 2000. Selective factors associated with the origin of fur and feathers. *American Zoologist*, vol. 40, pages 585–596.

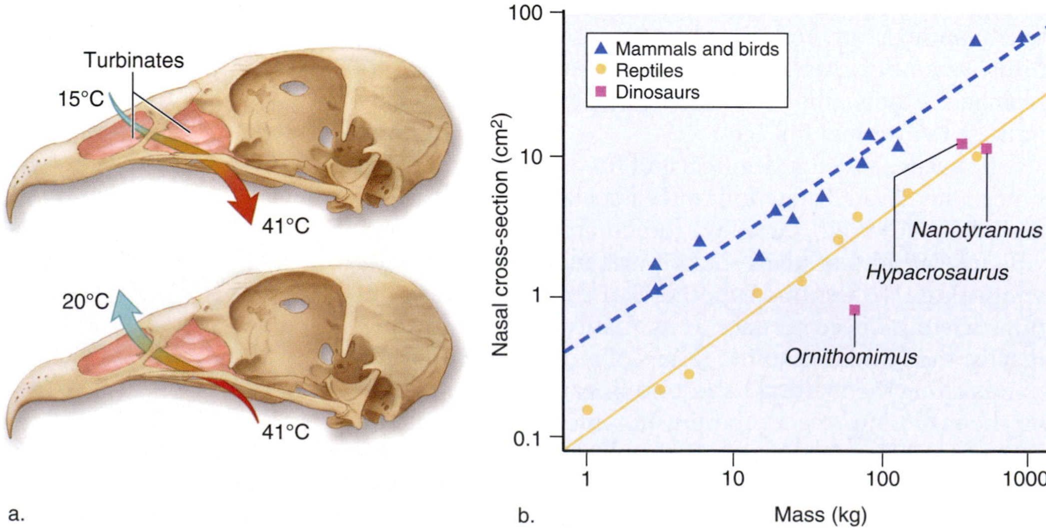

FIGURE 35.15 The Nose Knows. (**A**) Turbinates direct air flow within the nasal cavity of endotherms, warming and humidifying inhaled air. The same structures cool and dehumidify exhaled air, conserving both heat and water. (**B**) The nasal cavities of existing endotherms (ranging in size from herons to African cape buffaloes) have a higher cross-sectional area than those of ectotherms of equal size. Dinosaurs had nasal cavities similar in size to those of existing ectotherms. **Question**: How would it affect Ruben's conclusions if a 100-kg dinosaur were found to have a nasal cavity greater than 10 cm^2?

CHAPTER SUMMARY

35.1 Animals Must Regulate Their Internal Temperature

A. An Animal's Body Temperature Reflects Heat Gains and Losses

- **Ectotherms** regulate body temperature by seeking an environment with the appropriate temperature. **Endotherms** use internal metabolism to generate heat.

B. Several Adaptations Enable an Animal to Adjust Its Temperature

- Animals regulate their body temperatures (**thermoregulate**) with physiological and behavioral adaptations.
- Adaptations to cold include **countercurrent exchange** systems of blood vessels; fat, feathers, and fur; and hibernation.
- Evaporative cooling, circulatory specializations, and behaviors are adaptations to heat.

35.2 Animals Must Regulate Water and Ion Balance and Excrete Wastes

A. Osmoregulation Controls Water and Ions in Body Fluids

- **Osmoregulation** is the control of ion concentrations in body fluids. Terrestrial animals have adaptations that conserve water and regulate ion composition.

B. Nitrogenous Wastes Include Ammonia, Urea, and Uric Acid

- Animals must **excrete** metabolic wastes. Most nitrogenous wastes come from protein breakdown and include **ammonia, urea,** and **uric acid.**
- Ammonia formation requires the least energy but releases the most water. Urea and uric acid require more energy but are less toxic and help conserve water.

C. The Urinary System Produces, Stores, and Eliminates Urine

- The **urinary system** excretes nitrogenous wastes (mostly urea) and regulates water and electrolyte levels.
- The **kidneys,** each of which drains into a **ureter,** produce **urine**. Ureters drain urine into the **urinary bladder** for temporary storage. Urine leaves the body through the **urethra.**

35.3 The Nephron Is the Functional Unit of the Kidney

- The **nephron** is closely associated with blood vessels of the **glomerulus** and **peritubular capillaries.**
- The nephron **filters** blood and **secretes** wastes and other substances into the filtrate, while blood in the peritubular capillaries **reabsorbs** useful components.

A. The Glomerular Capsule Filters Blood

- Some components of blood are forced from the glomerulus into the **glomerular capsule.** The filtrate passes from the glomerular capsule to the **renal tubule,** where its composition is adjusted.

B. Blood Recovers Water and Other Valuable Substances from the Proximal Convoluted Tubule

- The **proximal convoluted tubule** returns glucose, amino acids, water, ions, and other solutes to the blood.
- H^+ and HCO_3^- adjustments at the proximal convoluted tubule help maintain blood pH.

C. The Nephron Loop Conserves Additional Water and Exchanges Ions with Blood

- The filtrate moves into the **nephron loop** that dips into and then out of the medulla of the kidney. The nephron loop helps to concentrate urine. Water leaves the filtrate because an osmotic gradient forms in the fluid around the nephron loop by a countercurrent exchange system.

D. The Distal Convoluted Tubule and Collecting Duct Conserve More Water

- The filtrate then moves to the **distal convoluted tubule** and the **collecting duct,** where more Na^+ and water are reabsorbed into the blood.
- Collecting ducts deliver urine to the renal pelvis, which drains the fluid into the ureter.

E. Hormones Regulate Kidney Function

- **Antidiuretic hormone (ADH)**, secreted by the posterior pituitary gland, regulates the amount of water reabsorbed from the distal convoluted tubule. ADH increases permeability of the distal convoluted tubule and the collecting duct so more water is reabsorbed and urine is more concentrated.
- The adrenal glands release **aldosterone** in response to either low Na^+ concentration in the plasma or low blood pressure. Aldosterone causes additional Na^+ to be reabsorbed into the blood, and water follows by osmosis.

35.4 Investigating Life: Sniffing Out the Origin of Fur and Feathers

- Researchers have tested the hypothesis that fur and feathers evolved at the same time as endothermy. Their findings suggest that whereas endothermy preceded fur in the ancestors of mammals, feathers came first in birds.

MULTIPLE CHOICE QUESTIONS

1. Why do endotherms require more energy than ectotherms?
   a. Because they need more energy to move between hot and cold environments.
   b. Because they use metabolic energy to maintain an internal temperature.
   c. Because they are more active.
   d. Both a and c.

2. Which of the following does NOT contribute to thermoregulation?
   a. Sweating
   b. Shivering
   c. Moving toward shade
   d. All of the above contribute to thermoregulation

3. Countercurrent heat exchange in birds or mammals functions to
   a. warm the blood at the surface of the organism.
   b. cool the blood at the surface of the organism.
   c. equalize the temperature of the blood throughout the organism.
   d. warm the blood as it returns to the core of the organism.

4. The sensors that detect body temperature are in the:
   a. hypothalamus.
   b. nephrons.
   c. fat cells.
   d. musculoskeletal system.

5. What would happen to a deep-sea fish if it were placed in fresh water?
   a. Water would move into the organism.
   b. Ions would move into the organism.
   c. Water would move out of the organism.
   d. There would be no effect.

6. What form of nitrogenous waste would you expect to find produced by an insect?
   a. An amino group
   b. Ammonia
   c. Urea
   d. Uric acid

7. Urine passes through the ______ as it exits the body.
   a. ureter
   b. bladder
   c. urethra
   d. kidney

8. The proximal convoluted tubule is responsible for
   a. filtration.
   b. reabsorption.
   c. secretion.
   d. both b and c.

9. Water moves out of the filtrate of the descending limb of the nephron loop because
   a. there is a high concentration of solutes in the surrounding capillaries.
   b. the cells of the descending limb are permeable to ions.
   c. the difference in the diameter of the arterioles creates a pressure gradient.
   d. there is a high concentration of solutes in the descending limb.

10. A common treatment for high blood pressure is the use of a *diuretic* medication. How would this help control blood pressure?
    a. By increasing water permeability of the distal convoluted tubule and collecting duct
    b. By decreasing water permeability of the distal convoluted tubule and collecting duct
    c. By enhancing the production of aldosterone
    d. By increasing Na^+ reabsorption in the distal convoluted tubule and collecting duct

TESTING YOUR KNOWLEDGE

1. Cite three adaptations that help keep an animal's feet warm.
2. How do humans and iguanas differ in body temperature regulation?
3. What are the three types of nitrogenous wastes?
4. List the organs that make up the human urinary system.
5. Draw a nephron and label the parts. Indicate which regions of the renal tubule participate in filtration, reabsorption, and secretion.
6. How does the kidney reduce the volume of urine to a small fraction of the volume of filtrate that enters the glomerular capsule?
7. How do ADH and aldosterone control kidney function?
8. Shortly after you drink a large glass of water, you will feel the urge to urinate. Explain this observation. Begin by tracing the path of the water, starting at the glomerulus and ending with the arrival of urine in the bladder. As you do so, be sure to consider the role of sodium, osmoreceptors in the hypothalamus, and hormones.

THINKING AS A SCIENTIST

1. Imagine you are adrift at sea. If you drink seawater, you will dehydrate much faster than if you have access to fresh water. Explain.
2. Urinary tract infections frequently accompany sexually transmitted diseases, especially in women. Why?
3. Kidney stones are calcium-rich crystals that form inside the kidney. What symptoms would you expect if the stones lodge in a ureter?
4. Which of the substances in the following table are excreted, and which are reabsorbed into the bloodstream?

Concentrations (mg/100mL)			
Substance	**Plasma**	**Glomerular Filtrate**	**Urine**
Glucose	100	100	0
Urea	26	26	1820
Uric acid	4	4	53
Creatinine	1	1	196

5. Why is protein in the urine a sign of kidney damage? What structures in the kidney are probably affected?
6. How could very low blood pressure impair kidney function?
7. An amphibian's kidneys lack nephron loops. Why can't a human also live without nephron loops?
8. Many pharmaceutical drugs leave the body in urine. As we age, the number of nephrons in the kidneys declines. Do you predict that an older person would need a higher or lower dose of a drug to compensate for the amount lost in urine?
9. In a disease called diabetes insipidus, ADH activity is insufficient. Would a person with this disease produce more or less urine than normal? Explain.
10. Birds and insects frequently collect nectar from plants. Birds are endothermic, and insects are ectothermic. For the same investment of nectar, do you think a plant can support a greater mass of insects or of birds?

Visit www.mhhe.com/hoefnagels for practice quizzes, animations, videos, and activities designed to help you master the material in this chapter.

CHAPTER 36

The Immune System

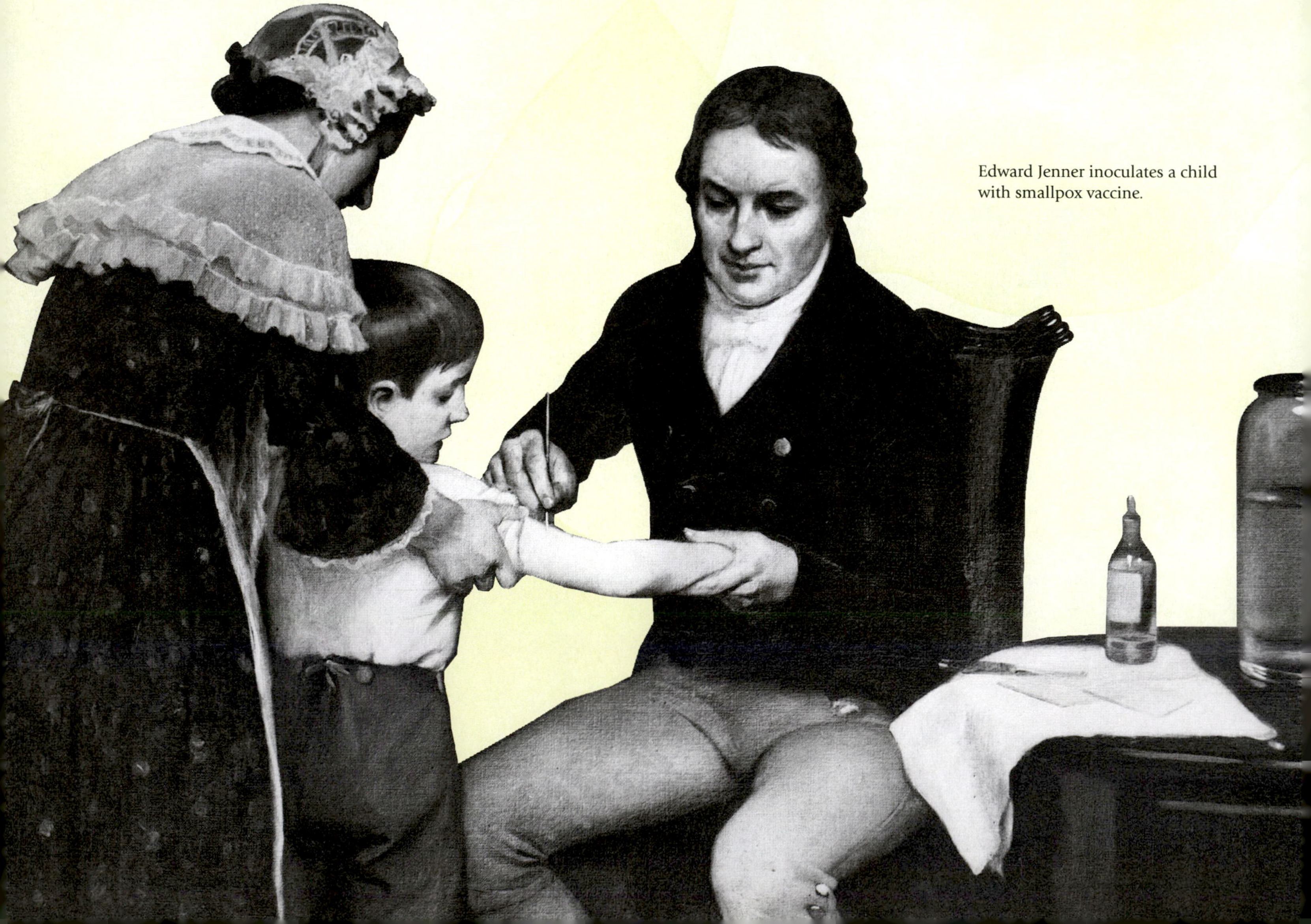

Edward Jenner inoculates a child with smallpox vaccine.

What's the Point?

The Demise of a Deadly Disease

Once called "the speckled monster," smallpox is a viral infection that killed many millions. The disease began with abrupt headache, backache, and fever. Within days, small red marks appeared, mostly on the face and extremities. These lesions would enlarge, rise, form bubbles, and crust over by the seventh day, after which a crop of larger lesions formed. There was no treatment, the fatality rate was about 25%, and about two thirds of survivors were left horribly scarred and sometimes blind.

Long ago, people recognized that smallpox survivors did not contract the infection again. People therefore intentionally applied crusts from affected individuals to scratched skin, often inducing a mild case that protected against recurrence.

In the late eighteenth century, a British country physician, Edward Jenner, improved the technique, using material from the sores of milkmaids who had cowpox, a much milder infection. In a now famous story, Jenner rubbed cowpox lesion crusts from a young milkmaid, Sarah Nelmes, into the scratched arm of 8-year-old James Phipps. Young James developed a fever and a pustule at the site of the treatment, but was otherwise fine. Six weeks later, Jenner scratched a smallpox sore into the boy's skin to show that he was now protected (an experiment that would be unethical by today's standards). Fortunately, James lived to a very old age, unmarked by any sort of pox.

Jenner had invented the first vaccine, a term that comes from the Latin vaca *for "cow." Vaccines work by "teaching" the recipient's immune system to recognize a disease-causing organism without actually causing illness. The smallpox vaccine is made from a live virus called vaccinia, a smallpox relative. Both viruses share surface molecules, called antigens, that the immune system recognizes.*

The vaccine was so effective that a worldwide smallpox eradication campaign began in 1967. The May 1980 issue of the World Health Organization's flagship magazine showed a globe with "smallpox is dead" *emblazoned across it. The declaration heralded a milestone in medicine—the eradication of a disease. Routine smallpox vaccinations ended worldwide.*

In the meantime, scientists have developed vaccines against many other illnesses, including influenza, chickenpox, measles, diphtheria, and cholera. All rely on the amazing ability of the immune system—the subject of this chapter—to recognize and destroy foreign particles.

LEARNING OUTLINE

36.1 Many Cells, Tissues, and Organs Defend the Body
- A. White Blood Cells Play a Major Role in the Immune System
- B. The Lymphatic System Consists of Several Tissues and Organs

36.2 Innate Defenses Are Nonspecific and Act Early
- A. Barriers Form the First Line of Defense
- B. Redness and Swelling Indicate Inflammation
- C. Innate Chemical Defenses Include Complement Proteins and Cytokines
- D. Fever Helps Fight Infection

36.3 Adaptive Immunity Defends Against Specific Pathogens
- A. Genetic Recombination Yields an Endless Variety of Antibodies and Antigen Receptors
- B. Macrophages, T Cells, and B Cells Have Distinct Roles in Adaptive Immunity
- C. T Cells Coordinate Cell-Mediated Immunity
- D. B Cells Direct the Humoral Immune Response
- E. The Immune Response Turns Off Once the Threat Is Gone
- F. The Secondary Immune Response Is Faster than the Primary Response

36.4 Several Disorders Affect the Immune System
- A. Immune Deficiencies Lead to Opportunistic Infections
- B. Autoimmune Disorders Are Devastating and Mysterious
- C. Allergies Misdirect the Immune Response
- D. A Pregnant Woman's Immune System Occasionally Attacks Her Developing Fetus

36.5 Investigating Life: Sea Urchins Lead Researchers from RAGs to Riches

36.1 Many Cells, Tissues, and Organs Defend the Body

Disease-causing organisms are nearly everywhere. Viruses, bacteria, protists, fungi, and worms are in water, food, soil, and air. Unit 4 describes some of these pathogens, which can enter our bodies whenever we eat, drink, breathe, or interact with people and other animals. Yet we are not constantly sick. The cells of the **immune system** enable the body to recognize its own cells and to destroy any cell or other structure perceived as "not self." The vertebrate immune system is a network of cells, defensive chemicals, and fluids that permeate the body (**figure 36.1**).

Immune System	
Main tissue types	**Examples of Locations/Functions**
Epithelial	Thymus, spleen, and tonsils consist partly of epithelial tissue; lines lymphatic and blood vessels; lymphoid tissue lies beneath epithelial tissues of the digestive, respiratory, and urinary tracts, guarding potential points of entry for pathogens
Connective	Lymphocytes, antibodies, and other immune system chemicals circulate in blood and lymph, which are connective tissues; bone marrow is connective tissue that produces lymphocytes; thymus, spleen, and lymph nodes consist partly of connective tissue

FIGURE 36.1 Organs of the Human Immune System. Many specialized organs and cells contribute to the human immune response.

A. White Blood Cells Play a Major Role in the Immune System

Blood is a connective tissue consisting of cells in a fluid matrix called plasma (see figure 32.15). Red blood cells do not directly participate in the immune system, but plasma carries defensive proteins called antibodies.

In addition, white blood cells are essential to immune function (**figure 36.2**). Stem cells in **bone marrow,** the spongy tissue inside bones, give rise to all white blood cells. Basophils, the least abundant type of white blood cell, release chemical signals that trigger inflammation (see section 36.2). Two types of white blood cells, neutrophils and eosinophils, function primarily as **phagocytes,** which are scav-

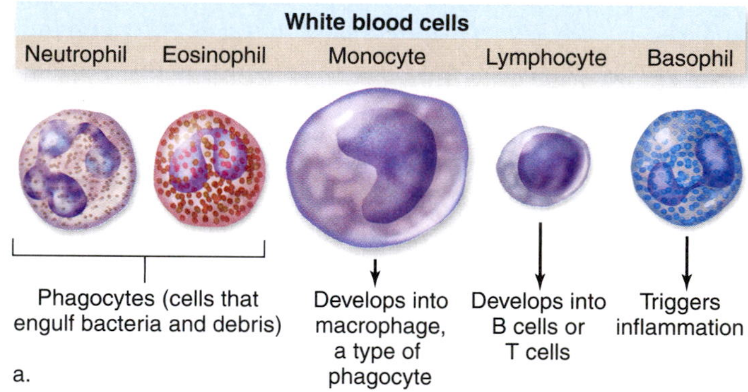

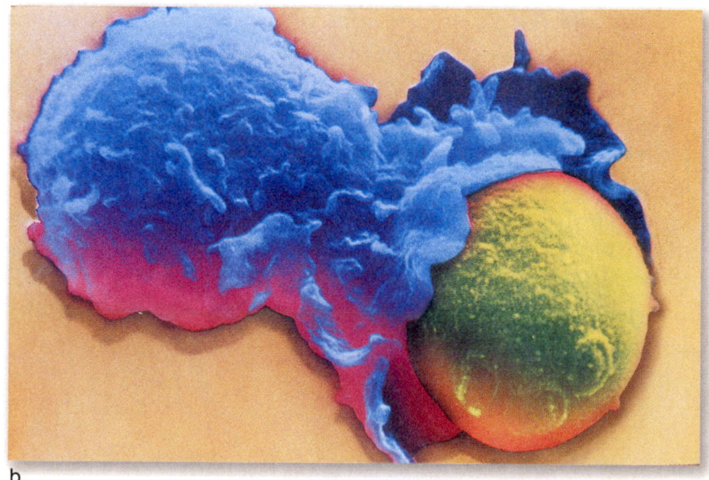

FIGURE 36.2 White Blood Cells. (**A**) Human blood contains five types of white blood cells. (**B**) A phagocyte (blue) engulfs a yeast cell.

enger cells that travel in the bloodstream and engulf bacteria and debris. White blood cells called monocytes give rise to **macrophages,** another type of phagocyte. Many different types of macrophages exist; some wander throughout the body, whereas others remain in just one tissue. **bone marrow, p. 639; phagocytosis, p. 94**

Lymphocytes (B and T cells) are white blood cells that travel not only in blood, but also in lymph and between the tissues of the lymphatic system and the body's connective tissues (**figure 36.3**). **B cells** mature in bone marrow, then migrate to lymphoid tissues and into the blood. **T cells** also originate in bone marrow, but they mature in the **thymus** ("T" is for thymus). From the thymus, T cells migrate throughout the body. Together, B cells and T cells coordinate the body's response to specific pathogens, as described in section 36.3.

B. The Lymphatic System Consists of Several Tissues and Organs

We have already seen that the **lymphatic system** helps regulate the volume of tissue fluid (see chapter 32) and absorbs fat from the small intestine (see chapter 34). A third role is to help defend the body from disease-causing organisms.

Lymph is the colorless fluid that the lymphatic system transports, cleanses, and returns to the bloodstream (see figure 32.19A). Lymph forms from plasma that seeps out of blood capillaries and into the interstitial fluid. Dead-end, thin-walled lymph capillaries absorb the fluid, along with bacteria, viruses, and other large particles. Lymph also carries white blood cells.

Lymphoid organs are collections of lymphocytes embedded in loose connective tissue. The thymus is one example, as is the **spleen,** which is the largest lymphoid organ in the body. Surrounding the spleen's blood vessels are masses of lymphocytes and macrophages, which detect and destroy pathogens and foreign substances circulating in the blood. The body's many **lymph nodes** are lymphoid organs that cleanse lymph (see figure 32.19B). These masses of fibrous tissue are located along the lymph vessels; you can sometimes feel them as "swollen glands" in the neck, armpits, or groin when you have an infection (**figure 36.4**). Inside each lymph node, millions of lymphocytes filter dead cells and pathogens from the lymph. Lymph nodes also release B and T cells to lymph, which carries them to the blood.

Many pathogens enter the body through the mucous membranes lining the digestive, respiratory, urinary, and reproductive tracts. Scattered concentrations of lymphoid tissues guard these linings. Examples include the tonsils (near the throat), appendix, and patches of lymphoid tissue in the small intestine.

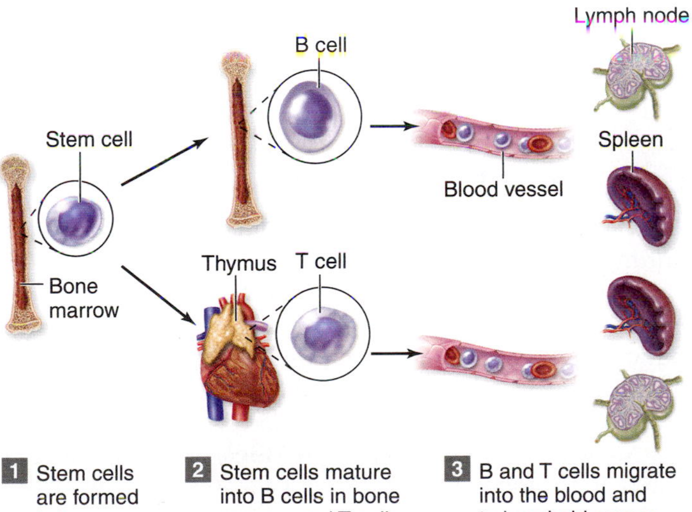
B cell
Lymph node
Stem cell
Spleen
Blood vessel
Bone marrow
Thymus
T cell
1 Stem cells are formed in bone marrow.
2 Stem cells mature into B cells in bone marrow and T cells in the thymus.
3 B and T cells migrate into the blood and to lymphoid organs.

FIGURE 36.3 B Cells and T Cells. Lymphocytes originate in bone marrow. B cells mature in the bone marrow before migrating to lymphoid organs, whereas T cells mature in the thymus.

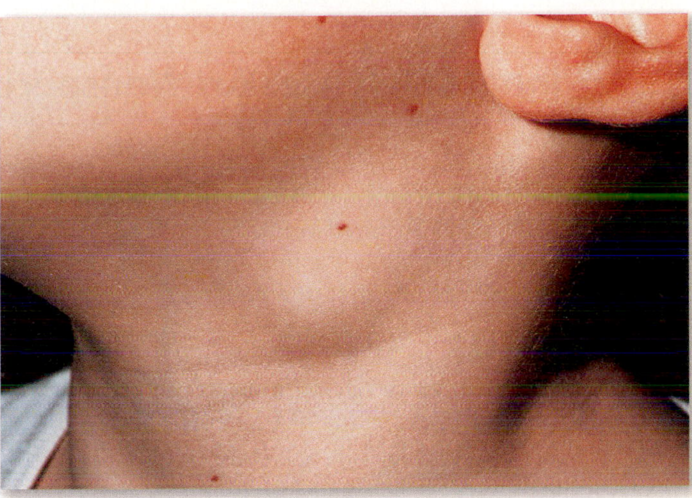

FIGURE 36.4 Swollen Lymph Nodes. Lymph nodes become swollen and tender as immune system cells divide in response to an infection.

Summary *Immune system cells occur throughout the body, either fixed in place or circulating in blood and lymph. Several types of white blood cells participate in the immune response. Some engulf and dismantle invaders and damaged cells; others recognize specific pathogens. Lymphoid tissues and organs produce and house immune cells.*

36.1 MASTERING CONCEPTS

1. What are the five types of white blood cells, and what role does each play in the body's defenses?
2. What is the relationship between lymph and blood?
3. What are the main lymphoid organs, and what are their functions?

36.2 Innate Defenses Are Nonspecific and Act Early

The human immune system has two separate parts that interact in highly coordinated ways: innate defenses and adaptive (once called acquired) immunity. This section describes innate defenses. Adaptive immunity, which is a complex response to a specific part of a pathogen, is the subject of section 36.3.

Innate defenses provide a rapid, broad defense against any infectious agent. *Innate* refers to the fact that these defenses are always present and ready to function. Some elements of the innate response can distinguish invading cells, such as bacteria, from host cells. However, there is no "memory" of this encounter that would assist future responses.

A. Barriers Form the First Line of Defense

Physical barriers block microbes from entering the body. Unpunctured skin is the most extensive and obvious wall (**figure 36.5**). Other physical barriers include mucus that traps inhaled dust particles in the nose; wax in the ears; tears that wash irritants from the eyes and contain an antimicrobial substance called lysozyme; and cilia that sweep bacteria out of the respiratory system. Most microorganisms that reach the stomach die in a vat of acidic secretions. **integumentary system, p. 569; stomach, p. 696**

An often underappreciated component of this first line of defense is the body's normal microflora. Resident microbes on the skin, in the gut, and elsewhere help prevent colonization by pathogens. Studies of laboratory animals reared in special microbe-free environments show that the immune system does not develop correctly without stimulation from our microscopic companions.

FIGURE 36.5 First Line of Defense. Intact skin is a physical barrier that prevents microbes from entering the body.

B. Redness and Swelling Indicate Inflammation

Inflammation is the body's immediate, localized reaction to an injury or any pathogen that breaches the body's barriers. The familiar response to a minor cut illustrates the effects of inflammation: the area immediately surrounding the wound becomes red, warm, swollen, and painful. This nonspecific defense recruits immune components, helps clear debris, and creates an environment hostile to microorganisms around the site of an injury or infection (**figure 36.6**).

Basophils and **mast cells** trigger inflammation. Mast cells do not circulate in the blood. Rather, they settle in tissues, especially those near the skin, digestive tract, and respiratory system. Tissue damage provokes basophils and mast cells to release **histamine,** a biochemical that dilates (widens) blood vessels and causes them to become more permeable. This response, in turn, eases entry of the macrophages and neutrophils that engulf and destroy bacteria and damaged cells. Pus may also accumulate at the injured site; this whitish fluid contains white blood cells, bacteria, and debris from dead cells. Plasma at the wound site brings in antimicrobial substances and dilutes the toxins secreted by bacteria. Increased blood flow warms the area, turning it swollen and red.

In medical terminology, the suffix *-itis* indicates inflammation. For example, dermatitis (a rash) signifies inflamed skin, often resulting from direct contact with an allergen such as poison ivy. Appendicitis is inflammation of the appendix, usually caused by bacterial infection. Aspirin and ibuprofen reduce pain and swelling after an injury by blocking the enzymes required for inflammation to occur.

C. Innate Chemical Defenses Include Complement Proteins and Cytokines

As inflammation begins, cells produce and release many antimicrobial biochemicals. For example, the 25 or so **complement** proteins all help to destroy pathogens in the body. When activated, some trigger a chain reaction that punctures bacterial cell membranes (**figure 36.7**). Others cause mast cells to release histamine, and still others attract phagocytes.

Many immune cells release **cytokines,** messenger proteins that bind to other immune cells and alter their activity. For example, cells infected by viruses release **interferons,** which are cytokines that alert other components of the immune system to the infection. White blood cells release **interleukins,** the largest group of cytokines. Their name comes from their role in communicating (*inter-*) between leukocytes (*-leukins*). One

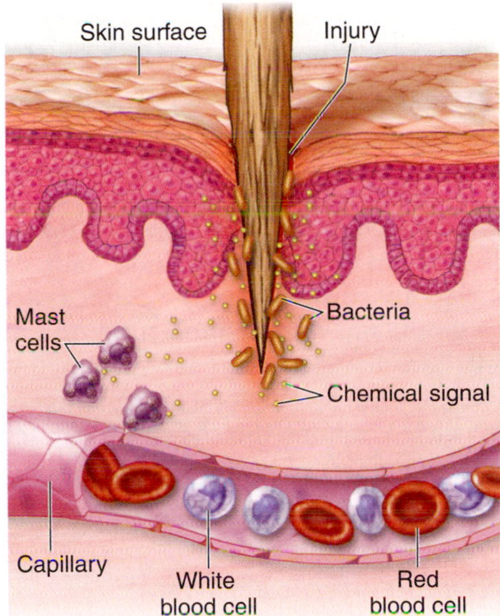

1 After a splinter penetrates the skin, damaged cells trigger mast cells and basophils to release histamine and other chemical signals.

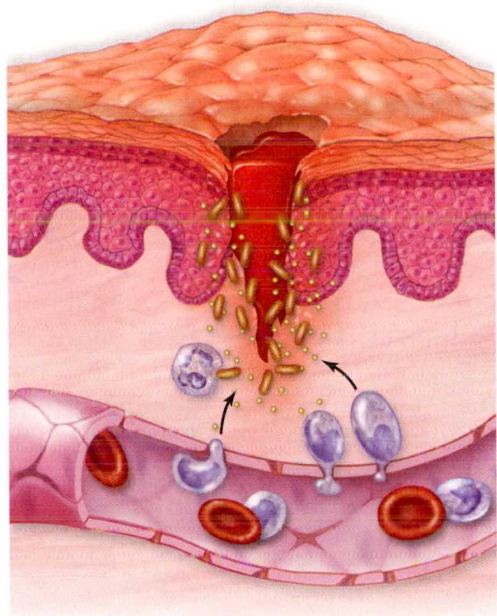

2 Histamine causes blood vessels to dilate and become more permeable, causing swelling. White blood cells move into the damaged area by squeezing between the cells in the capillary walls.

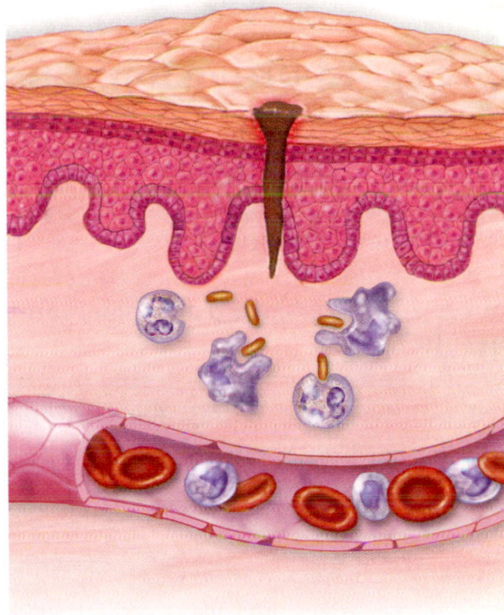

3 White blood cells engulf and destroy bacteria and damaged cells.

FIGURE 36.6 Inflammation. Chemicals released at the site of an injury set into motion the inflammatory response.

type of interleukin, produced by macrophages, activates B and T cells, provokes inflammation, and causes fever.

D. Fever Helps Fight Infection

Cytokines travel throughout the body in the bloodstream. At the hypothalamus, they can trigger a temporary increase in the set-point of the body's thermostat. **Fever,** a rise in the body's temperature, is therefore a common reaction to infection. Although it feels uncomfortable, a mild fever can help fight infection. A higher body temperature kills some bacteria and viruses. Fever also counters microbial growth indirectly, because higher body temperature reduces the iron level in the blood. Bacteria and fungi require more iron as the temperature rises, so a fever stops the replication of these pathogens. Phagocytes also attack more vigorously when the temperature climbs.

Summary *Innate defenses include nonspecific barriers, inflammation, chemical defenses, and fever. Inflammation is a complex general response to infection or injury that delivers phagocytes and antimicrobial substances.*

36.2 MASTERING CONCEPTS

1. Distinguish between innate defenses and adaptive immunity.
2. What are some barriers to infection in the human body?
3. What are the roles of mast cells, histamine, and phagocytes in inflammation?
4. What are some examples of antimicrobial biochemicals?
5. How is fever protective?

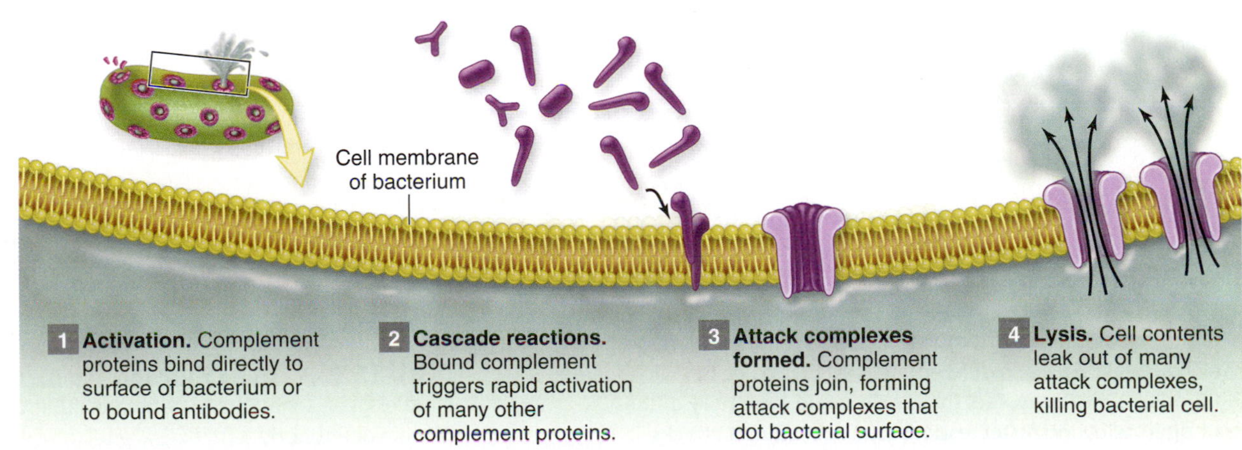

1 **Activation.** Complement proteins bind directly to surface of bacterium or to bound antibodies.

2 **Cascade reactions.** Bound complement triggers rapid activation of many other complement proteins.

3 **Attack complexes formed.** Complement proteins join, forming attack complexes that dot bacterial surface.

4 **Lysis.** Cell contents leak out of many attack complexes, killing bacterial cell.

FIGURE 36.7 Complement Kills Bacteria. Triggered by a bound antibody, complement proteins combine and riddle a bacterium's cell membrane with holes, shattering its physical integrity. The bacterial cell quickly dies.

36.3 Adaptive Immunity Defends Against Specific Pathogens

In **adaptive immunity,** the body's immune cells not only recognize specific pathogens, but they also "remember" previous encounters. Together, innate defenses and adaptive immunity defend the body against a never-ending assault from many kinds of pathogens (**figure 36.8**).

An **antigen** is any molecule that stimulates an immune reaction by B and T cells. The word *antigen* (short for *anti*body-*gen*erating) reflects a crucial part of adaptive immunity: the production of **antibodies,** which are Y-shaped proteins that recognize specific antigens. Most antigens are carbohydrates or proteins. Examples include parts of a bacterial cell wall or a virus, proteins on the surface of a mold spore or pollen grain, and unique molecules on the surface of a cancer cell. Each B and T cell is genetically programmed to recognize and bind to only one target antigen. Since every foreign particle contains dozens of molecules that can act as antigens, many sets of lymphocytes respond to invasion by one pathogen.

Bacteria
Viruses
Innate responses
Physical and chemical barriers
Resident microorganisms
Phagocytosis
Antimicrobial proteins
Inflammatory response
Fever
Adaptive response
Macrophages present antigens
Cellular response
Humoral response
Helper T cells
Cytokines
B cells
Memory B cells
Memory T cells
Plasma cells
Cytotoxic T cells
Antibodies

FIGURE 36.8 Levels of Immune Protection. Bacteria and viruses first encounter barriers that prevent their entry into the body. If they breach these barriers, an array of nonspecific cells and molecules attacks the pathogen. In the adaptive immune response, antigen-presenting cells stimulate T cells to produce cytokines, which activate B cells to differentiate into plasma cells, which secrete antibodies. Once activated, these cells retain a memory of the pathogen, allowing faster responses to subsequent attacks.

A. Genetic Recombination Yields an Endless Variety of Antibodies and Antigen Receptors

How can one person's lymphocytes produce a billion or so different antibody proteins and antigen receptor proteins? Generating these diverse molecules is a little like using the limited number of words in a language to compose an infinite variety of stories.

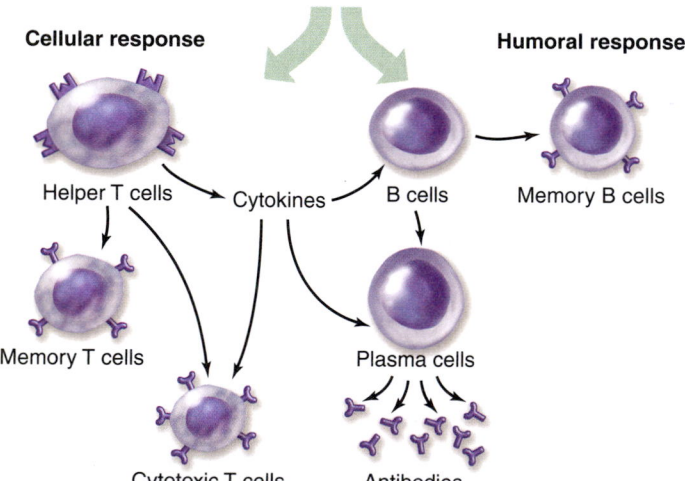

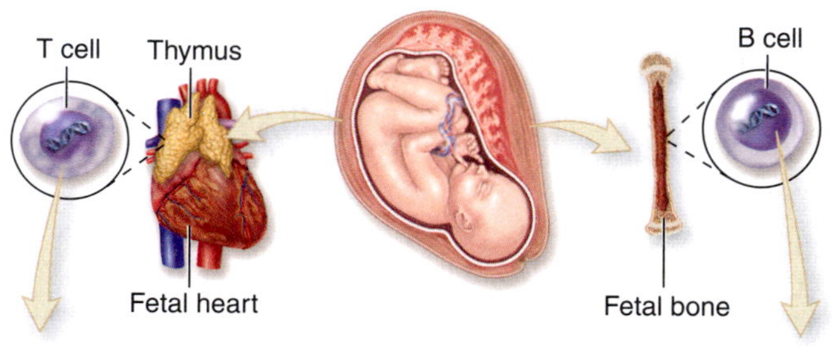
T cell
Thymus
B cell
Fetal heart
Fetal bone

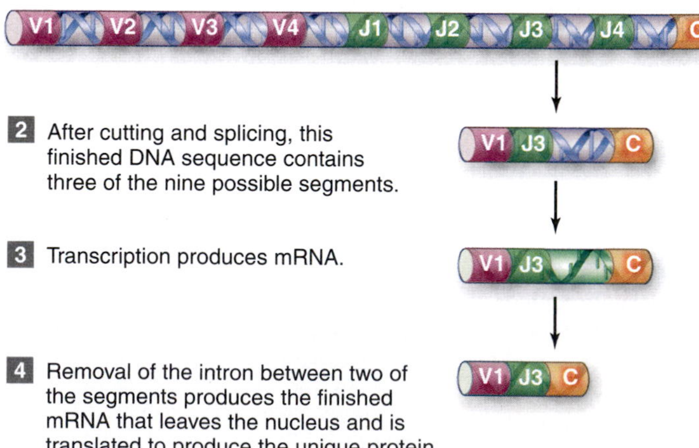
1 DNA encoding antibodies contains several segments that the developing lymphocyte can mix and match.
V1 V2 V3 V4 J1 J2 J3 J4 C
2 After cutting and splicing, this finished DNA sequence contains three of the nine possible segments.
V1 J3 C
3 Transcription produces mRNA.
V1 J3 C
4 Removal of the intron between two of the segments produces the finished mRNA that leaves the nucleus and is translated to produce the unique protein.
V1 J3 C

FIGURE 36.9 Gene Shuffling. A developing lymphocyte shuffles gene segments to generate a unique antigen receptor protein or antibody.

Of the human genome's 25,000 or so genes, fewer than 250 encode proteins that specifically bind to antigens. But these genes contain hundreds of small DNA segments that are rearranged in developing lymphocytes (**figure 36.9**). The result: countless lineages of cells that each produce a unique antigen receptor and antibody. Most lymphocytes will never encounter a pathogen with the corresponding antigen, but a few will. Thanks to genetic recombination, the immune system can respond to even newly emerging pathogens.

Some of these antigen receptors, however, will no doubt correspond to the body's own molecules. In a process called clonal deletion, apoptosis (programmed cell death) weeds out lymphocytes that recognize the body's own cell surfaces and molecules (**figure 36.10**). This process begins before birth. At the same time, the immune system somehow also "learns" not to attack antigens in food. **apoptosis, p. 168**

B. Macrophages, T Cells, and B Cells Have Distinct Roles in Adaptive Immunity

One of the first cell types to respond to infection is the macrophage, which both participates in innate defenses and triggers adaptive immunity. The macrophage engulfs a pathogen, dismantles it, and links each antigen to a self protein on the macrophage surface (**figure 36.11**).

Antigen-presenting macrophages travel in lymph to the lymph nodes, where they encounter collections of T and B cells. **Helper T cells** are "master cells" of the immune system because they initiate and coordinate the adaptive immune response. When an antigen-presenting macrophage meets a helper T cell specific to the antigen it is displaying, the two cells bind. This event simultaneously initiates the

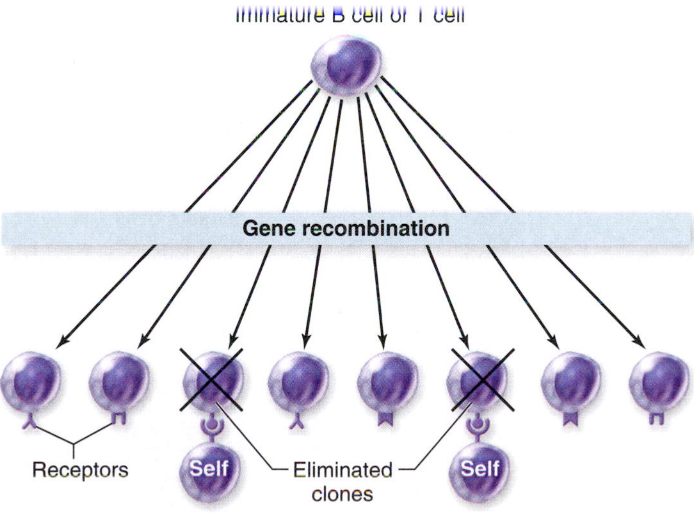
Immature B cell or T cell
Gene recombination
Receptors
Self
Eliminated clones
Self

FIGURE 36.10 Clonal Deletion. As lymphocytes develop, they are tested against proteins and polysaccharides already present on the body's own cell surfaces. Clones that match self antigens are eliminated by programmed cell death (apoptosis).

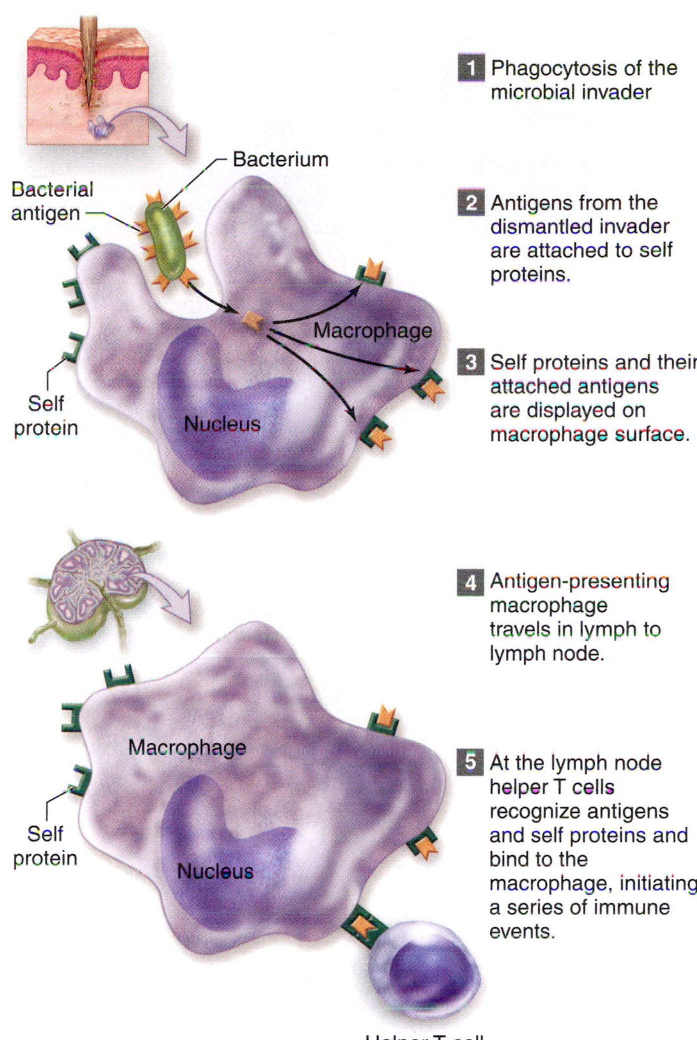
1 Phagocytosis of the microbial invader
Bacterium
Bacterial antigen
2 Antigens from the dismantled invader are attached to self proteins.
Macrophage
3 Self proteins and their attached antigens are displayed on macrophage surface.
Self protein
Nucleus
4 Antigen-presenting macrophage travels in lymph to lymph node.
Macrophage
5 At the lymph node helper T cells recognize antigens and self proteins and bind to the macrophage, initiating a series of immune events.
Self protein
Nucleus
Helper T cell

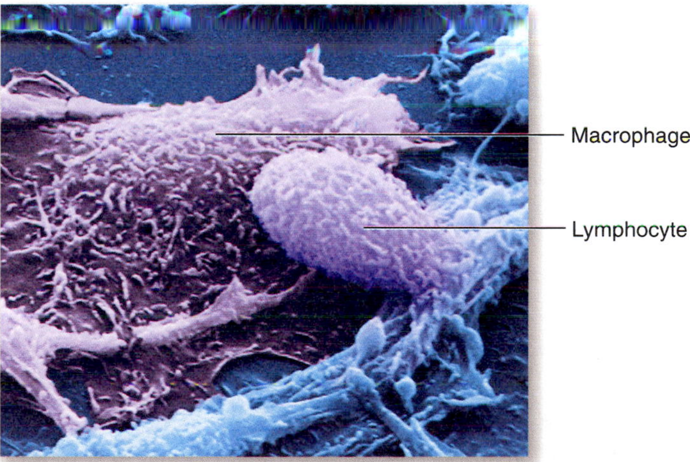
Macrophage
Lymphocyte

FIGURE 36.11 Macrophages Are Antigen-Presenting Cells. A macrophage engulfs a bacterium, then displays foreign antigens on its surface. This event sets into motion many immune reactions. Each bacterial cell has many antigens, all of which the macrophage would display. For simplicity, this figure depicts only one.

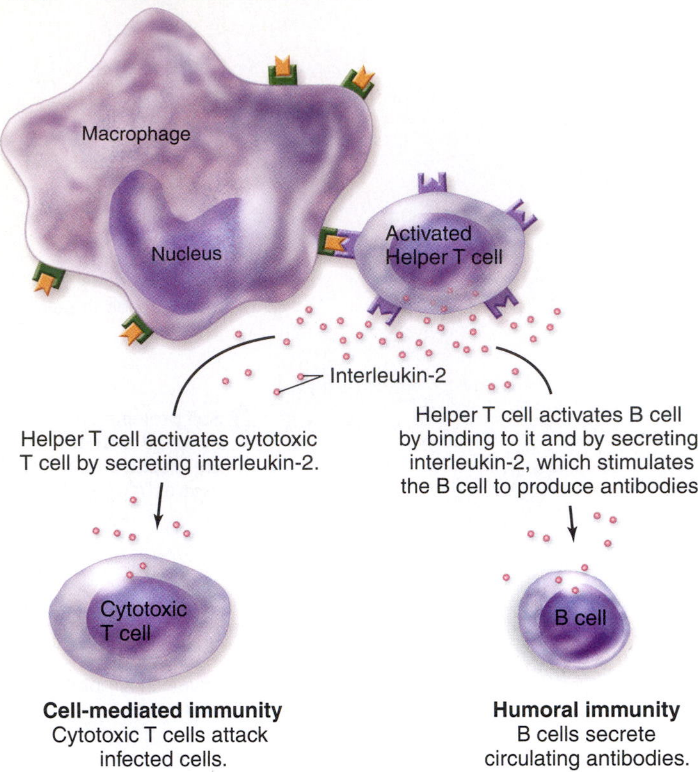
Macrophage
Nucleus
Activated Helper T cell
Interleukin-2
Helper T cell activates cytotoxic T cell by secreting interleukin-2.
Helper T cell activates B cell by binding to it and by secreting interleukin-2, which stimulates the B cell to produce antibodies.
Cytotoxic T cell
B cell
Cell-mediated immunity
Cytotoxic T cells attack infected cells.
Humoral immunity
B cells secrete circulating antibodies.

FIGURE 36.12 Helper T Cells Are the Heart of an Immune Response. An activated helper T cell initiates both cell-mediated and humoral immunity.

two arms of the adaptive immune response: cell-mediated and humoral immunity (**figure 36.12**).

In **cell-mediated immunity,** defensive cells attack and kill invaders by direct cell-to-cell contact. The activated helper T cell secretes cytokines that activate **cytotoxic T cells,** which destroy virus-infected or damaged cells. The cytokines also activate B cells specific to the same antigen and trigger **humoral immunity,** which relies primarily on antibodies. On stimulation by T cells, B cells proliferate explosively and differentiate into plasma cells and memory B cells. **Plasma cells** secrete huge numbers of antibodies, whereas **memory cells** "remember" antigens the immune system has already encountered.

Figure 36.13 shows the interactions between many immune system cells. You may find it helpful to refer to this figure as you read the remainder of this chapter.

C. T Cells Coordinate Cell-Mediated Immunity

Cytotoxic T cells (sometimes called killer T cells) provide cell-mediated immunity. Receptors on the surface of a cytotoxic T cell bind to a specific antigen on a foreign cell. The T cell then releases a biochemical that cuts holes in the cell's membrane, killing the invader. Cytotoxic T cells also recog-

Cell-mediated immunity:
Cytotoxic T cells attack cells directly
T cells
Thymus
Small intestine
Skin
Mature T cells
Cytotoxic T cells
T cells mature in the thymus gland, in the small intestine, and in the skin.
Interleukins
Helper T cells
Helper T cells initiate cell-mediated and humoral immunity.
Bone marrow:
T cells, B cells, and macrophages originate in the bone marrow and migrate into the blood.
B cells
Spleen
Lymph nodes
Mature B cells
Humoral immunity:
Antibodies
Plasma cells
B cells mature in bone marrow and travel to lymphoid tissues, such as the spleen and lymph nodes.
B cells are released from lymphoid tissue and secrete antibodies.
Memory cells
Macrophages
Macrophages engulf bacteria and stimulate helper T cells to proliferate and activate B cells.

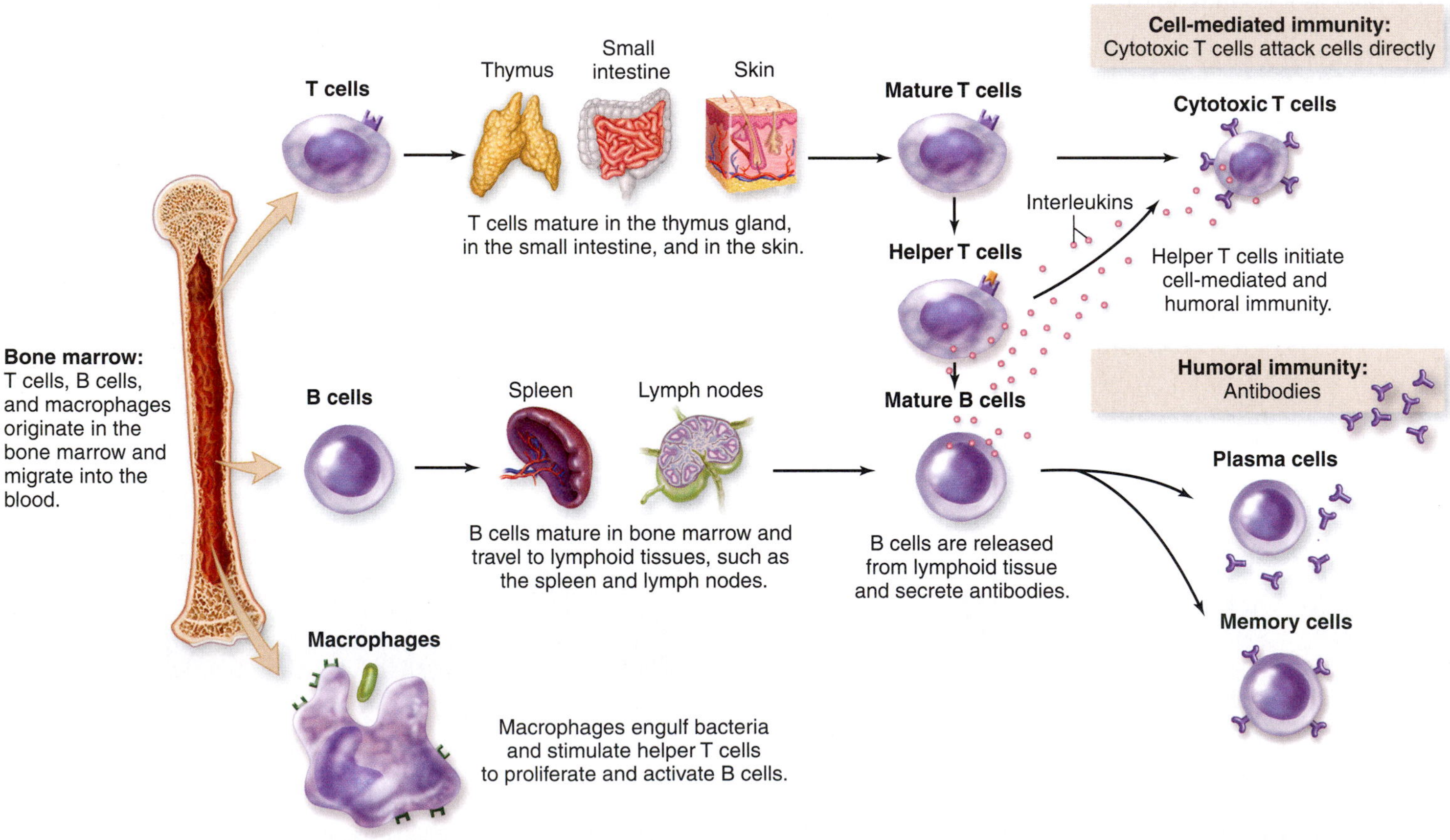

FIGURE 36.13 Immune Cells Are Diverse. T cells, B cells, and macrophages each contribute to the adaptive immune response.

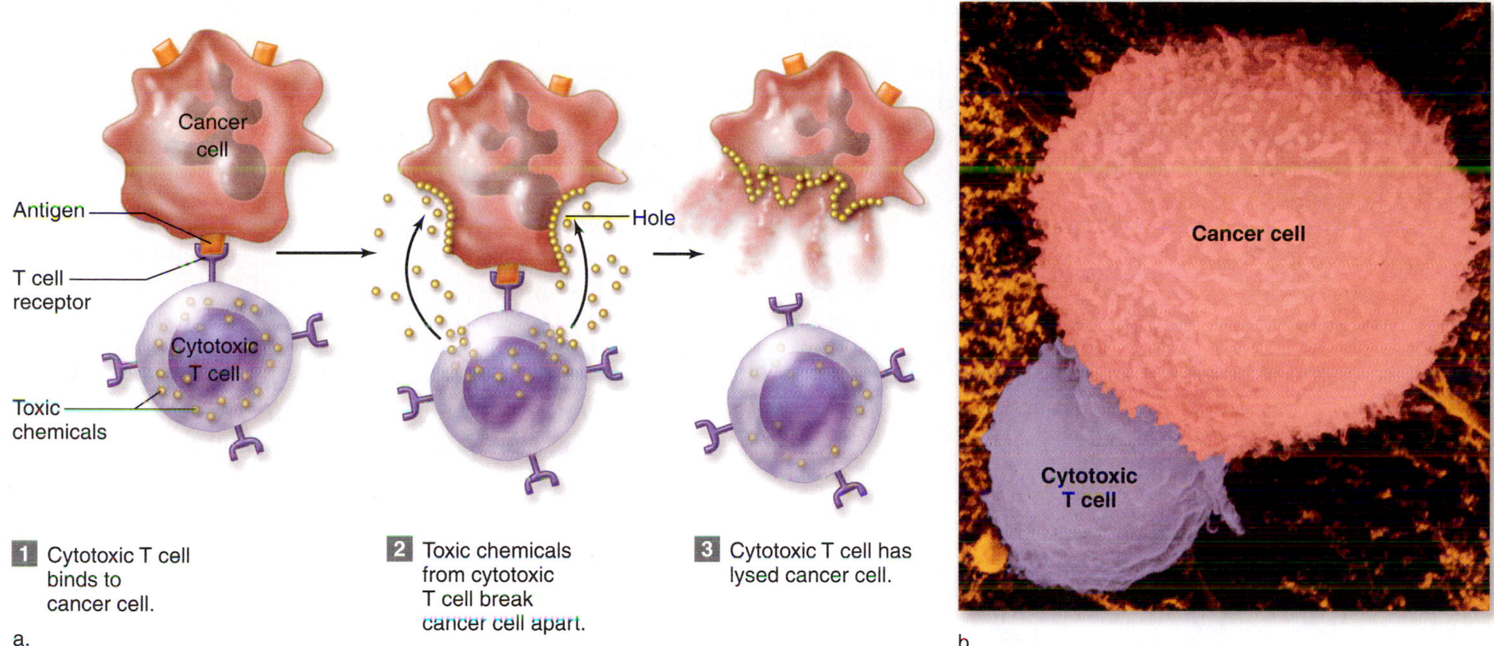

FIGURE 36.14 **Death of a Cancer Cell.** (**A**) An activated cytotoxic T cell binds to a cancer cell and injects a protein that pokes holes in the cancer cell's membrane. As the holes form, the cancer cell dies, leaving behind debris that macrophages clear away. (**B**) A cytotoxic T cell homes in on the surface of the large cancer cell above it.

nize and destroy cells infected with viruses. By destroying the cell before the virus can replicate, the infection is stopped. Cytotoxic T cells also bind to tissues transplanted from other individuals and to abnormal molecules found on the surface of cancer cells, destroying them (**figure 36.14**).

Another type of lymphocyte, the memory T cell, helps the immune system retain a long-term "memory" of a pathogen. If the body encounters the same pathogen again, memory T cells differentiate immediately into cytotoxic T cells.

D. B Cells Direct the Humoral Immune Response

The humoral immune response includes millions of different B cells, each producing a unique antibody. This type of immunity may be passive or active (**table 36.1**). In **passive immunity,** a person receives intact antibodies from another individual. For example, an infant acquires antibodies from its mother in breast milk. Eventually, however, the infant begins to produce antibodies for itself. This is **active immunity,** because the baby's immune system makes its own antibodies after exposure to antigens in the environment.

Antibodies Are Defensive Proteins

Antibodies (also called immunoglobulins) are the main weapons of humoral immunity. These large proteins circulate freely in blood plasma, lymph, and interstitial fluid. The simplest antibody molecule consists of four polypeptide chains that together form a shape like the

Table 36.1 *Ways to Acquire Immunity*

Type	Description	Examples
Passive immunity	Individual acquires antibodies from another individual	• Fetus acquires antibodies from mother via placenta or milk • Dog bite victim receives injections of antibodies to rabies virus
Active immunity	Individual produces antibodies to an antigen	• Having chickenpox confers future immunity to that disease ("natural" active immunity) • Influenza vaccine triggers production of memory cells specific to antigens in the vaccine ("artificial" active immunity)

letter Y (**figure 36.15**). Each antibody has constant and variable regions. The amino acid sequences of **constant regions** are very similar in all antibody molecules. The upper portions of each polypeptide chain, the **variable regions,** differ a great deal among antibodies. The variable regions determine the specific target antigen to which an antibody binds.

The binding of an antibody to an antigen can inactivate a microbe or neutralize its toxins. Antibodies can cause pathogens to clump, making them more apparent to macrophages, which then destroy the pathogens. They can coat viruses, preventing them from contacting target cells. Antibodies also activate complement proteins, which destroy microorganisms.

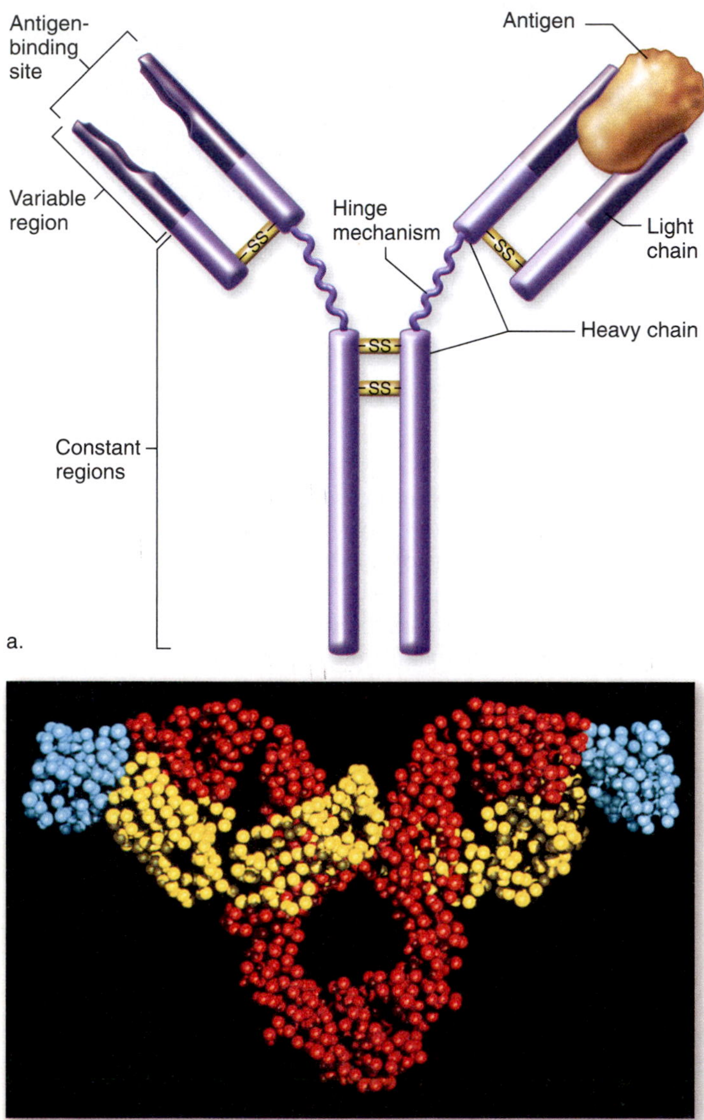

FIGURE 36.15 Antibody Structure. (**A**) The simplest antibody molecule consists of four polypeptide chains. Part of each polypeptide chain has a constant sequence of amino acids, and the remainder of the sequence is variable. The tops of the Y-shaped molecules form antigen-binding sites. (**B**) A three-dimensional view of an antibody molecule.

Clonal Selection Explains the Surge of Identical Antibodies

Each B cell has on its surface a version of the antibody that the cell will produce. Until a B cell encounters the antigen it is genetically programmed to recognize, it remains dormant. But when the antigen is encountered, binding to this surface antibody triggers the B cell to begin to activate. Cytokines from the corresponding helper T cell complete the activation.

In a process called **clonal selection,** a stimulated B cell divides rapidly, generating an army of memory cells and plasma cells that are clones of the original B cell (**figure 36.16**). The plasma cells are efficient protein factories, making thousands of

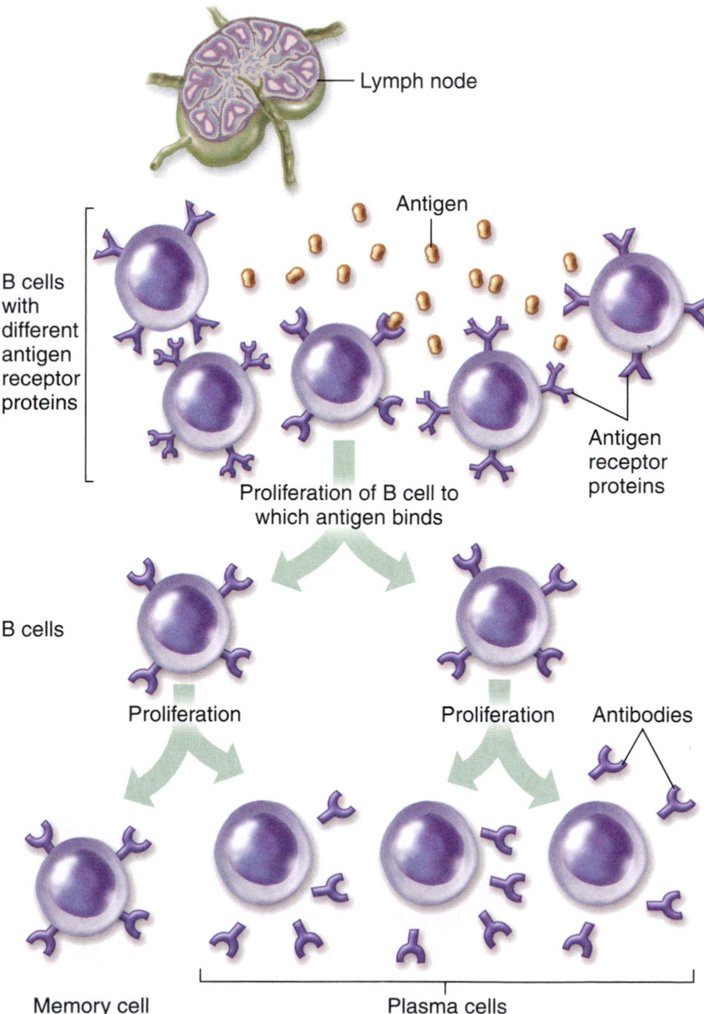

FIGURE 36.16 Clonal Selection. According to clonal selection theory, only the lymphocyte that binds the antigen proliferates; its descendants develop into memory cells or plasma cells.

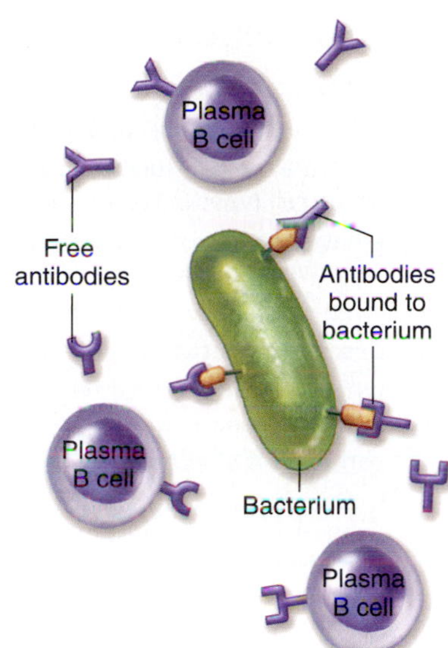

FIGURE 36.17 An Immune Response Recognizes Many Targets. In the humoral immune response, different B cells (plasma cells) produce antibody proteins that recognize and bind to different features of a foreign cell's surface.

antibody molecules each second. Each plasma cell that descends from one activated B cell produces the same antibody, but even the simplest of bacteria have many different antigens (**figure 36.17**). The immune response therefore involves several different antibodies, each made by a specific B cell and its clones.

E. The Immune Response Turns Off Once the Threat Is Gone

Turning off an immune response once an infection has been halted is as important as turning it on because powerful immune biochemicals can also attack the body's healthy tissues. Immunologists continue to learn more about the precise cytokine combinations that signal the immune system to "back down" after a threat is removed.

Figure 36.18 shows a simplified version of the negative feedback that regulates lymphocyte and antibody concentrations in blood. Once the threat is removed, suppressor T cells somehow reduce the number of dividing B and T cells, and the army of plasma cells shrinks until only the memory cells remain.

F. The Secondary Immune Response Is Faster than the Primary Response

The **primary immune response** is the adaptive immune system's first reaction to a nonself antigen, and it takes days or weeks to reach an effective level. During this time, the pathogen can cause severe damage or death. If a person survives, however, the memory B cells and memory T cells leave a lasting impression—that is, immunological memory.

As a result, the **secondary immune response**—the immune system's reaction the next time it detects the same foreign antigen—is much more rapid than the primary response (**figure 36.19**). Memory B cells transform

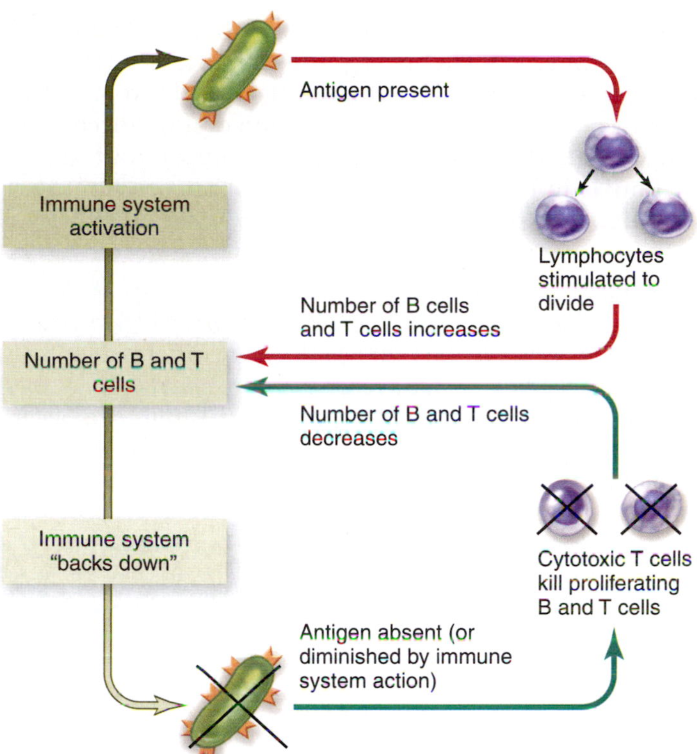

FIGURE 36.18 Homeostasis in the Immune System. This simplified negative feedback loop summarizes how the immune system regulates itself so that immune attack stops when an antigen's concentration declines.

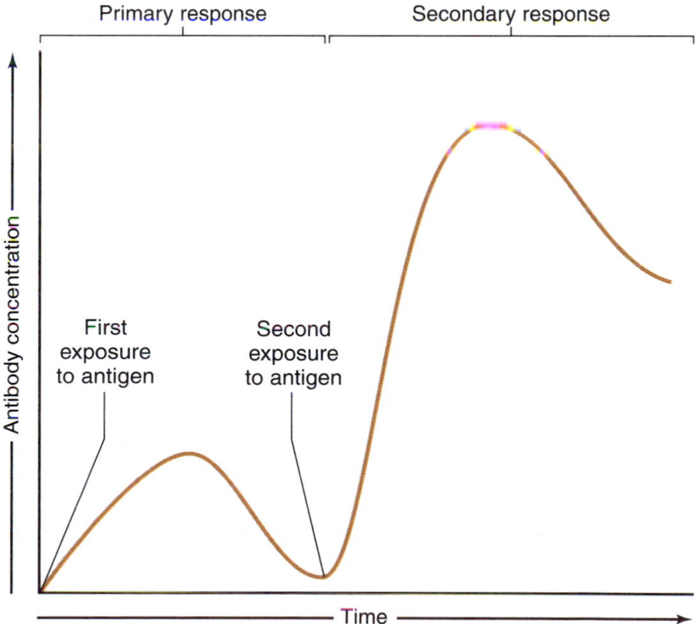

FIGURE 36.19 Primary and Secondary Immune Response. The primary immune response leaves memory cells that stimulate a faster, stronger immune response on subsequent exposure to the foreign antigen.

into rapidly dividing plasma cells. Within hours, billions of antibodies are circulating throughout the host body, destroying the pathogen before it takes hold. Usually, there is no hint that a second infection ever occurred. **Vaccines** create this immunological memory without risking an initial infection. Can *You* Relate? Vaccines Jumpstart Immunity describes how immunization saves lives.

Summary *Antigens trigger adaptive immunity. A helper T cell alerted by an antigen-presenting macrophage activates cytotoxic T cells. B cells also proliferate, differentiate, and secrete antibodies. Genetic recombination enables both types of lymphocytes to respond to many potential pathogens. The primary immune response is slower than the secondary response.*

36.3 MASTERING CONCEPTS

1. How can a relatively small number of genes encode billions of antibodies and antigen receptors?
2. What are the two subdivisions of adaptive immunity, and which cell types participate in each?
3. What is the role of macrophages in innate and adaptive immunity?
4. What do cytotoxic T cells do?
5. Describe the structure and function of an antibody.
6. What happens after a B cell is stimulated?
7. What happens if an immune reaction persists after a pathogen is eliminated?
8. Explain the difference between the primary and secondary immune response.

Can *You* Relate?

Vaccines Jumpstart Immunity

The immune system is remarkably effective at keeping bacteria, viruses, and tumor cells from taking over our bodies. Can we improve on nature?

People have been altering immune system function since the ninth-century Chinese stuffed smallpox crusts up their noses to prevent infection. Today's vaccines are more sophisticated, but the mode of action is the same. Once taken into the body, a vaccine simulates a primary immune response. A subsequent encounter with the real pathogen triggers the rapid secondary immune response.

Vaccines take several different forms (**table 36.A**). Measles and mumps vaccines, for example, are weakened viruses. Others, such as the vaccine against polio, contain killed viruses. Diphtheria and tetanus vaccines incorporate a toxin that bacterial pathogens produce. Still others, such as the hepatitis A and hepatitis B vaccines, incorporate only a part of the pathogen's surface. The "cervical cancer" vaccine, approved by the FDA in 2006, contains proteins identical to those on the human papillomavirus, a pathogen that is strongly correlated with cervical cancer in women.

Immunization has saved countless lives since Jenner developed the smallpox vaccine. Yet vaccines cannot prevent all diseases. Influenza viruses, for example, mutate so rapidly that each vaccine is effective only for one season. Similarly, HIV is changeable enough that a vaccine has so far eluded researchers. And it is impossible to develop one vaccine that will prevent infection by the many different viruses that cause the familiar symptoms of a cold. **influenza, p. 355**

Table 36.A *Types of Vaccines*

Vaccine Formulation	Examples
Live, weakened (attenuated) pathogen	Polio (oral vaccine), measles, mumps, rubella, chickenpox
Inactivated (killed) pathogen	Polio (injectable vaccine), influenza, hepatitis A
Inactivated toxins	Tetanus, diphtheria
Parts of killed pathogens	Cholera, whooping cough
Recombinant vaccines; component vaccines	Lyme disease, hepatitis B, human papillomavirus

36.4 Several Disorders Affect the Immune System

The immune system may fail to respond to disease-causing organisms, or it may turn against the body's own cells. In addition, harmless substances sometimes trigger an immune response.

A. Immune Deficiencies Lead to Opportunistic Infections

A weakened immune system leaves a person vulnerable to **opportunistic** pathogens that do not normally infect people with healthy immune systems. Viruses such as HIV can weaken the immune system, as can some inherited diseases and drugs.

Human immunodeficiency virus (HIV) causes acquired immune deficiency syndrome (AIDS). A person can acquire HIV by sexual contact or by using contaminated needles when injecting drugs. (Routine screening of the blood supply has virtually eliminated blood transfusions as an exposure route.) A mother can also transmit HIV to her baby, either during delivery or in breast milk.

When HIV enters the body, it initially targets helper T cells (see figure 17.7). Viral RNA enters the cell, where the enzyme reverse transcriptase transcribes the RNA to DNA. The freshly synthesized DNA inserts itself into the cell's chromosome. HIV's genes encode the raw materials needed to manufacture new viruses. Some types of infected cells, including helper T cells, die as they assemble and release new viruses. These viral particles infect additional helper T cells, spreading the infection within the body.

For months to a decade or more, however, no AIDS symptoms appear, because the body can produce enough new T cells to compensate for the loss. During this latent period, B cells manufacture antibodies to the virus, making possible rapid tests for HIV exposure (see the Burning Question: How do HIV tests work?). Unfortunately, the antibodies do not stop the production of new viruses.

HIV-positive people track their disease progress with blood tests that measure the number of helper T cells. As helper T cell counts decline, the ability of cytotoxic T cells to destroy infected cells also weakens. Eventually, the immune system fails entirely, and the infections and cancers of AIDS begin. Kaposi sarcoma (**figure 36.20**) and pneumonia caused by *Pneumocystis jiroveci* (formerly called *Pneumocystis carinii*) are two common AIDS-associated illnesses in North America, but AIDS patients are susceptible to many other infectious diseases as well.

AIDS is a consequence of viral infection, but immune deficiency can also be inherited. Each year, a few children are born defenseless against infection due to **severe combined immune deficiency (SCID),** a disorder in which neither T nor B cells function. David Vetter was one such youngster. Born in Texas in 1971, David had no thymus gland and spent the 12 years of his life in a vinyl bubble, awaiting a treatment that never came. Today, children with SCID have several treatment options that were unavailable to David. For example, gene therapy has been used to replace faulty genes in some SCID patients.

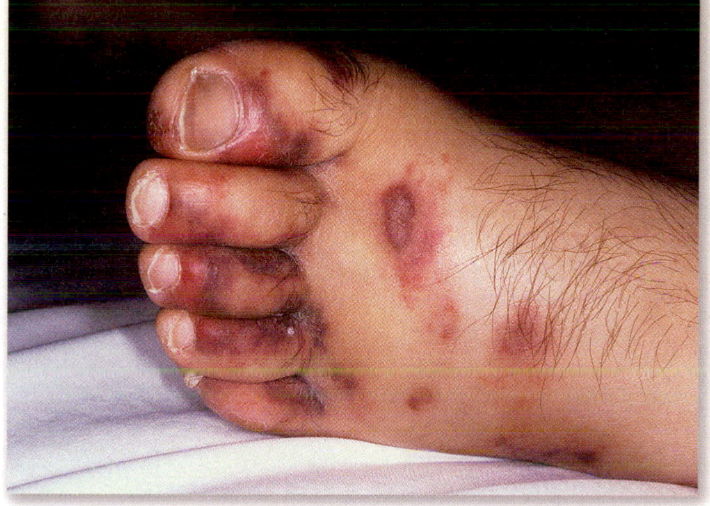

FIGURE 36.20 Kaposi Sarcoma, an AIDS-Related Illness. Kaposi sarcoma is a form of cancer that is prevalent in people with reduced immune system function.

Physicians use drugs to induce immunodeficiency in organ transplant recipients. The body perceives any foreign object, including a donated kidney, heart, or skin graft, as something to be destroyed. To avoid rejection of a donated organ, transplant recipients must take immune-suppressing drugs such as cyclosporine for the rest of their lives. Like other people with immunodeficiencies, these patients are vulnerable to opportunistic infections.

B. Autoimmune Disorders Are Devastating and Mysterious

Ideally, the immune system does not attack the body's own cells—as a person develops, lymphocytes corresponding to molecules already present in the body are eliminated. In **autoimmunity,** however, the immune system attacks the body's "self antigens," damaging vital tissues and organs. **Table 36.2** lists some autoimmune disorders.

Immunologists have much to learn about autoimmunity. For example, the body develops self-tolerance against

Table 36.2 *Autoimmune Disorders*

Disorder	Symptoms	Targets of Antibody Attack
Glomerulonephritis	Lower back pain, kidney damage	Kidney cell antigens that resemble *Streptococcus* antigens
Graves disease	Restlessness, weight loss, irritability, increased heart rate and blood pressure	Thyroid gland
Juvenile (Type I) diabetes	Thirst, hunger, weakness, emaciation	Pancreatic beta cells
Myasthenia gravis	Muscle weakness	Nerve message receptors on skeletal muscle cells
Rheumatic heart disease	Weakness, shortness of breath	Heart valve cell antigens that resemble *Streptococcus* antigens
Rheumatoid arthritis	Joint pain and deformity	Cells lining joints
Scleroderma	Thick, hard, pigmented skin patches	Connective tissue cells
Systemic lupus erythematosus	Red rash on face, prolonged fever, weakness, kidney damage	DNA, neurons, blood cells

some antigens by deleting lymphocytes in the bone marrow and thymus. Other mechanisms of self-tolerance act on mature lymphocytes. And the immune system simply ignores some self antigens for unknown reasons. Which self antigens induce which type of tolerance, and how? The answers will lead to a better understanding of, and perhaps better treatments for, autoimmunity.

C. Allergies Misdirect the Immune Response

In an **allergy,** the immune system is overly sensitive, launching an exaggerated attack on a harmless substance. Common **allergens,** or antigens that trigger an allergy, include foods, dust mites, pollen, fur, and oils in the leaves of plants such as poison ivy. The allergens activate B cells to produce antibodies. A first exposure to the allergen initiates a step called sensitization, in which antibodies bind to mast cells. On subsequent exposure, the allergens bind to the molecules attached to the mast cells, causing them to explosively release histamine and other allergy mediators (**figure 36.21**).

The symptoms of an allergic response depend on the site in the body where mast cells release mediators. Many mast cells are in the skin, respiratory passages, and digestive tract, so allergies often affect these organs. The result: hives, runny nose and eyes, asthma, nausea, vomiting, and diarrhea. Antihistamine drugs relieve these symptoms by preventing the release of histamine or blocking its binding to target cells.

Some individuals react to allergens with a terrifying and potentially life-threatening reaction, called anaphylactic shock, in which mast cells release allergy mediators throughout the body. The person may at first feel an inexplicable apprehension. Then, suddenly, the entire body itches and erupts in hives. Histamine causes blood vessels to dilate. As blood rushes to the skin, not enough of it reaches the brain, and the person becomes dizzy and may lose consciousness. Breathing becomes difficult as the bronchioles in the lungs become constricted. Meanwhile, the face, tongue, and larynx begin to swell. Unless the person receives an injection of epinephrine and sometimes an incision into the trachea to restore breathing, death can come within minutes.

Anaphylactic shock most often results from an allergy to penicillin, insect stings, or foods. Peanut allergy, for

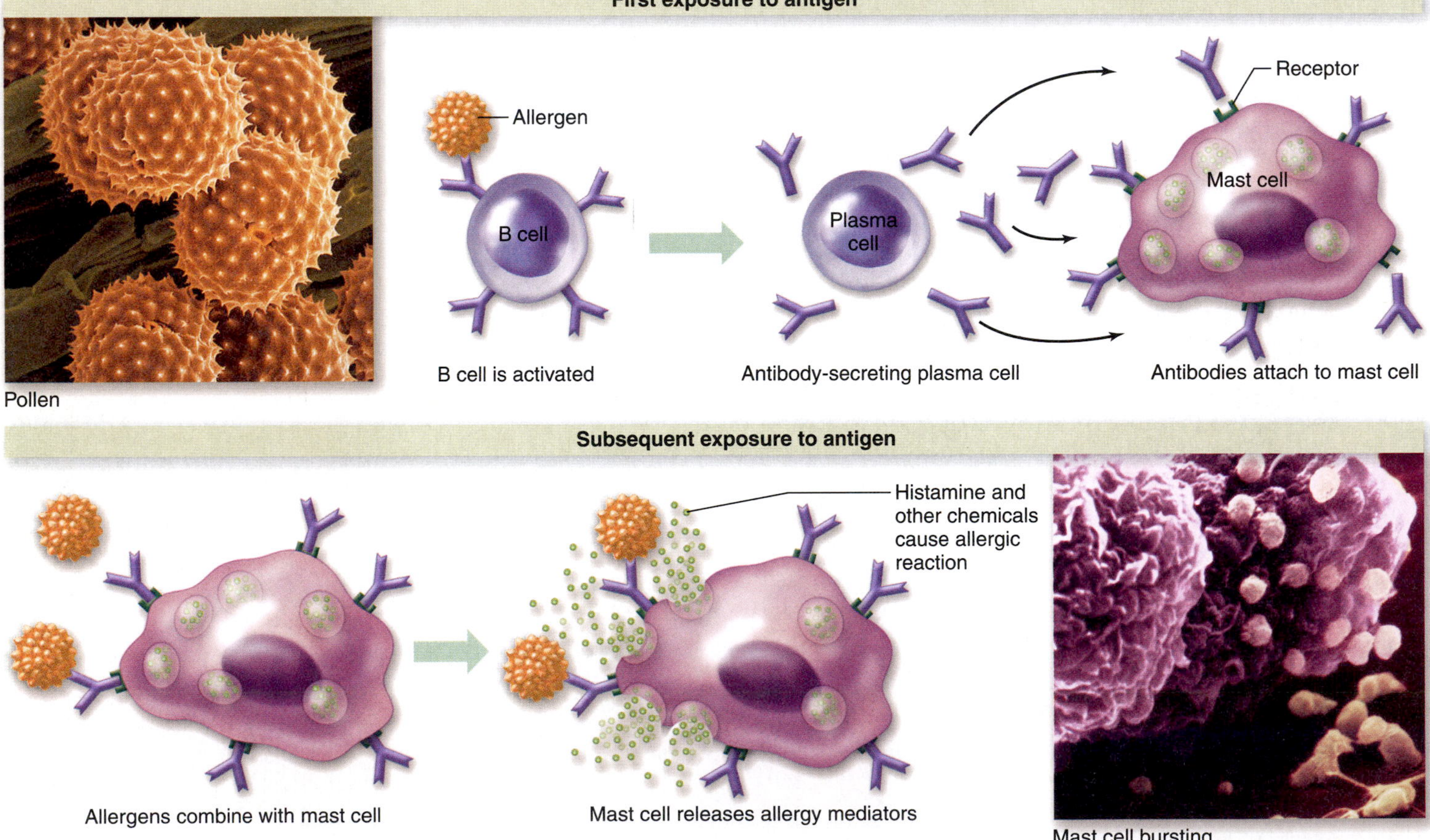

FIGURE 36.21 Allergy. Pollen enters the eyes, nose, and lungs, triggering B cells to differentiate into antibody-secreting plasma cells. The antibodies attach to mast cells. When pollen is encountered again, it binds to antibodies on the mast cells. The mast cells burst, releasing the chemicals that cause itchy eyes and a runny nose.

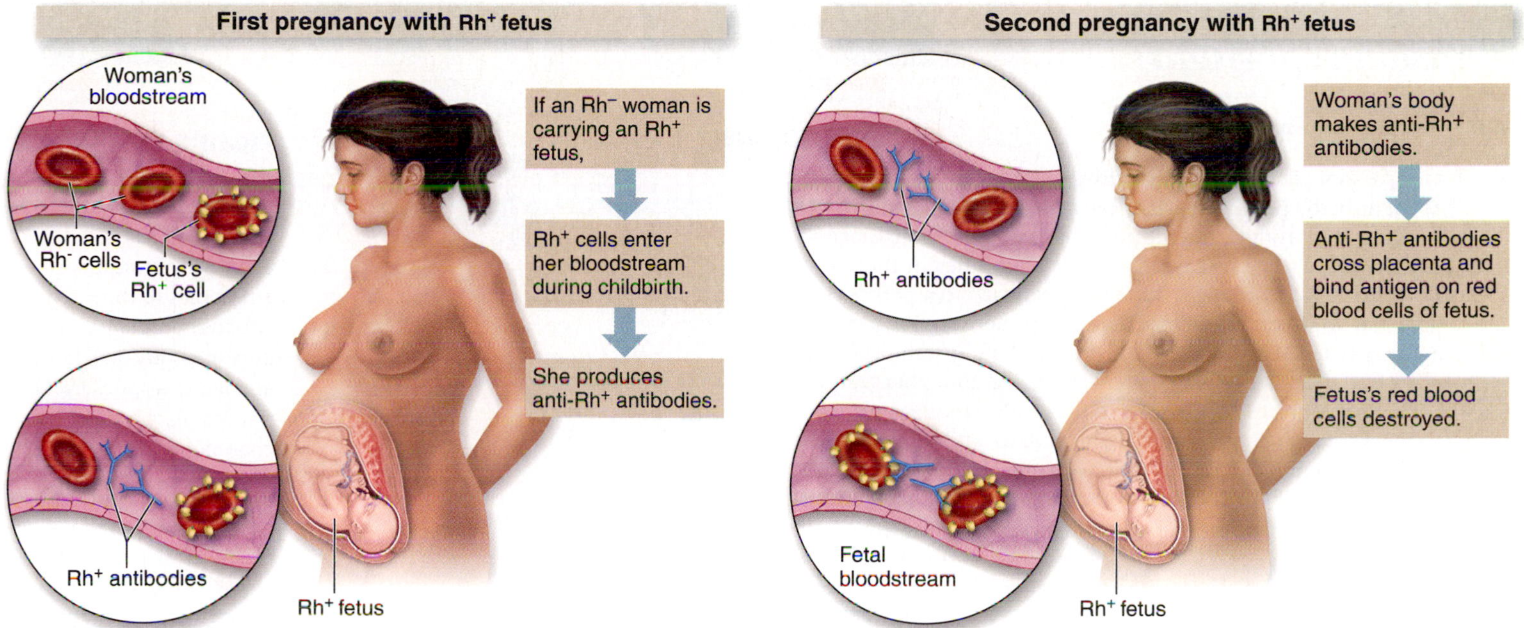

FIGURE 36.22 Rh Incompatibility. Fetal cells entering the pregnant woman's bloodstream can stimulate her immune system to make anti-Rh antibodies, if the fetus is Rh^+ and she is Rh^-. A drug called RhoGAM prevents attacks on subsequent fetuses.

example, affects 6% of the U.S. population and is on the rise. The allergens are seed storage proteins that enter the bloodstream undigested. The fact that the initial allergic reaction to peanuts occurs at an average age of 14 months, typically after eating peanut butter, suggests that sensitization occurs even earlier, during breastfeeding or before birth.

Early exposure to the microorganisms and viruses we share the planet with may be crucial to the development of the immune system. A growing body of evidence has led to the "hygiene hypothesis," which suggests that excessive cleanliness has contributed to recent increases in the prevalence of asthma and some allergies. Apparently, ultraclean surroundings decrease stimulation of the immune system early in life.

D. A Pregnant Woman's Immune System Occasionally Attacks Her Developing Fetus

Since the immune system should reject nonself cells, it may seem surprising that a woman's body does not destroy her fetus. After all, the developing child is not genetically identical to its mother. In general, the female immune response dampens during pregnancy so that it doesn't reject the embryo and fetus. Full immune function returns after the woman gives birth.

One possible source of problems, however, traces to an antigen called the Rhesus (Rh) factor. A person can be Rh-positive (Rh^+) or Rh-negative (Rh^-). If your blood type is positive (such as "A-positive"), the Rh antigen is on the surface of your red blood cells. If your blood is Rh-negative (such as "O negative"), your cells lack the Rh antigen.

The immune system of an Rh^- women can destroy the blood of her Rh^+ fetus (**figure 36.22**). A transfusion of Rh^- blood at birth can save the newborn's life, but this is rarely necessary. Blood tests identify couples who face this potential problem, and a drug is given that inactivates the woman's anti-Rh^+ antibodies.

Summary ***Immunodeficiency, as in AIDS and SCID, leads to infection and cancer. An autoimmune response damages the body's own tissues. The immune system can be misdirected against harmless substances, producing allergies. A pregnant woman's immune system may attack her developing fetus.***

36.4 MASTERING CONCEPTS

1. Which immune system cells does HIV attack, and what is the consequence?
2. How does SCID differ from HIV infection?
3. What events might lead to autoimmunity?
4. Which cells and biochemicals participate in an allergic reaction?
5. How do antihistamines relieve symptoms of allergies?
6. How is the Rh-factor important in determining whether a pregnant woman's immune system attacks her fetus?

Burning Questions

How do HIV tests work?

When a person gets an infectious disease caused by bacteria, one way to identify the pathogen is to grow the cells in pure culture. But viruses cannot reproduce in culture as bacteria do. Some diagnostic tests therefore rely on the presence of antibodies in the saliva, blood, or urine of people who have been exposed to a pathogen.

The oral swab test is one tool that determines whether a person has antibodies to HIV, the virus that causes AIDS. A technician first soaks a cotton swab between a person's cheek and gum, then puts the fluid into a solution containing viral antigens plus another substance that changes color when the viral antigen binds to antibodies from the saliva (**figure 36.A**). The results are available in about 20 minutes. If the test is positive for antibodies, then other tools, including the Western blot test, can confirm the presence of specific HIV proteins.

For most people, 2 to 3 months elapse between HIV infection and the production of detectable antibodies. Although antibody tests for HIV are both accurate and sensitive, people who have recently been exposed to HIV may have "false-negative" test results. To be safe, a person who suspects exposure to HIV should be tested both immediately and a few months later.

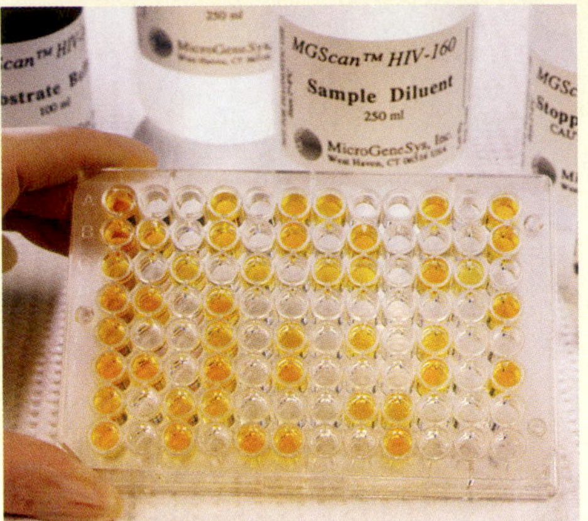

FIGURE 36.A **HIV Test.** A person who has been exposed to HIV will eventually produce antibodies to the virus. When the person's body fluid is applied to a specially treated plate, a color change reveals the presence of anti-HIV antibodies.

Have a Burning Question of your own? Submit it to marielle_hoefnagels@mcgraw-hill.com for possible inclusion in future editions of this book!

36.5 INVESTIGATING LIFE

Sea Urchins Lead Researchers from RAGs to Riches

Think about the objects you encounter during a typical day: water, food, dust, plants, paper, pet hair, mosquitoes, houseflies ... the list is long and varied. In addition, parasitic worms, bacteria, fungal spores, and viruses are virtually everywhere, and some of them can make you very sick. Danger lurks around every corner, and yet, as long as your immune system is working, your body seems to know exactly what to fight off and what to leave alone. Moreover, your defenses can recognize and conquer brand new invaders.

The complexity of the vertebrate immune system is astounding, especially when you consider that less than 1% of human genes encode antibody proteins. Part of this versatility traces to DNA fragments that can be shuffled and reshuffled to encode antibodies that recognize a virtually limitless variety of antigens (see figure 36.9). This gene shuffling, which occurs in all jawed vertebrates, is a hallmark of adaptive immunity. Many researchers have accepted the notion that the components of adaptive immunity arose in a sort of immunological "big bang" in the common ancestor of all fishes some 450 to 500 MYA.

The genome of the purple sea urchin, *Strongylocentrotus purpuratus* (**figure 36.23**), however, has changed many scientists' minds about the history of the vertebrate immune system. Sea urchins are echinoderms, a group of invertebrates that also includes sea stars and sea cucumbers. Echinoderms make interesting subjects for immunity research because they form a sister group to our own phylum, Chordata (see figure 22.2). The purple sea urchin was the first echinoderm to have its DNA sequenced, a project that was completed in 2006. **echinoderms, p. 469**

Immunologist Jonathan Rast of the University of Toronto was one of the leaders of the sea urchin genome project. He and his colleagues decided to search the echinoderm's DNA for evidence of genes required for vertebrate-style adaptive immunity. They focused on genes encoding a pair of enzymes that actually carry out the DNA shuffling. The enzymes are called RAG proteins, for *r*ecombination *a*ctivating *g*ene. In vertebrates, two RAG proteins (called Rag1 and Rag2) must be present for lymphocytes to mature properly. The two proteins interact with each other to recognize sites where a gene is to be cut and spliced.

FIGURE 36.23 **The Purple Sea Urchin, *Strongylocentrotus purpuratus*.**

The genes encoding Rag1 and Rag2 occur only in jawed vertebrates. But when Rast and his team searched the sea urchin genome for DNA sequences similar to these genes, they found a gene that they called *SpRag1L* (for *S. purpuratus Rag1*-like). The researchers used the DNA sequence of *SpRag1L* to predict the amino acid sequence of the encoded protein, a process described in chapter 12. A significant fraction (31%) of the predicted amino acids in the sea urchin's protein were identical to those of the Rag1 protein in the mouse, *Mus musculus*. The researchers subsequently found a DNA sequence similar to the *Rag2* gene in the sea urchin genome and named this second gene *SpRag2L*.

The newly discovered sea urchin genes, *SpRag1L* and *SpRag2L*, are similar in sequence to *Rag1* and *Rag2*, their counterparts in the vertebrate genome. Do the encoded proteins also bind to each other, as do the DNA-shuffling Rag1 and Rag2 proteins in our own immune systems? If so, a sea urchin should produce both proteins at the same time. Rast's team measured the expression of *SpRag1L* and *SpRag2L* genes in sea urchin embryos for 3 days after fertilization and found that both genes were indeed expressed simultaneously (**figure 36.24**). The researchers subsequently discovered that the two sea urchin proteins bind to each other.

The sea urchin gene sequences and protein properties led Rast's team to conclude that *Rag1* and *Rag2* genes in vertebrates share a common ancestor with the *SpRag1L/SpRag2L* gene pair in sea urchins. The function of the newly discovered sea urchin genes remains a mystery—these animals lack the adaptive immunity that characterizes vertebrates, so the proteins probably do not have the same DNA-shuffling function as our own Rag1 and Rag2. Until we understand the answer to that question, we cannot know how the genes took on new functions in the vertebrate immune system. One thing, however, is clear: at least some parts of our complex immune systems were already present in our invertebrate ancestors. Vertebrate adaptive immunity did not appear in one "big bang."

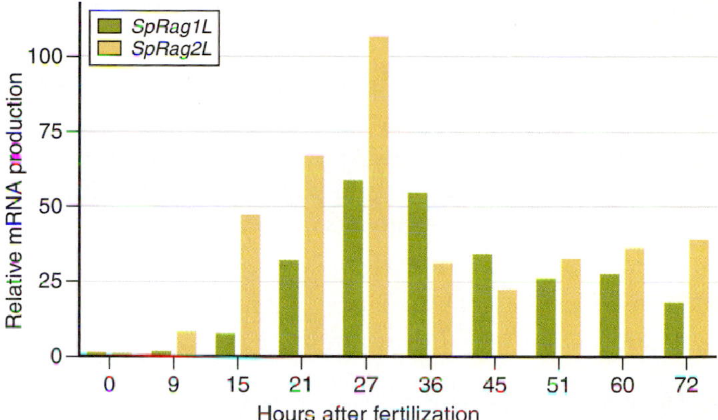

FIGURE 36.24 Gene Expression in the Sea Urchin. The genes *SpRag1L* and *SpRag2L* are expressed at the same time early in the development of the purple sea urchin. **Question:** Suppose that *SpRag1L* or *SpRag2L* were expressed at different times during the development of a sea urchin. Would that finding weaken the researchers' conclusion about the importance of these genes?

Fugmann, Sebastian D., Cynthia Messier, Laura A. Novack, R. Andrew Cameron, and Jonathan P. Rast. 2006. An ancient evolutionary origin of the *Rag1/2* gene locus. *Proceedings of the National Academy of Sciences*, vol. 103, pages 3728–3733.

CHAPTER SUMMARY

36.1 Many Cells, Tissues, and Organs Defend the Body

- The **immune system** protects the body against pathogens and cancer cells.

A. White Blood Cells Play a Major Role in the Immune System

- Five types of white blood cells—neutrophils, eosinophils, monocytes, **lymphocytes,** and basophils—participate in immune responses, as do several types of noncirculating cells.
- Neutrophils, eosinophils, and **macrophages** (which arise from monocytes) are all **phagocytes,** cells that engulf and destroy bacteria and debris.
- **B cells** and **T cells** are lymphocytes that mature in the **bone marrow** and the **thymus,** respectively.

B. The Lymphatic System Consists of Several Tissues and Organs

- The **lymphatic system** plays a crucial role in the immune response. The vessels of the lymphatic system collect and distribute a fluid called **lymph.**
- Besides the thymus, other lymph organs include the **spleen** and **lymph nodes.** Immune cells are also concentrated in the tonsils, appendix, and digestive tract.

36.2 Innate Defenses Are Nonspecific and Act Early

- **Innate defenses** provide broad protection against all pathogens.

A. Barriers Form the First Line of Defense

- Intact skin, mucous membranes, tears, earwax, and cilia block pathogens.

B. Redness and Swelling Indicate Inflammation

- Basophils and **mast cells** trigger **inflammation,** which is an immediate reaction to injury. These cells release **histamine,** a biochemical that causes blood vessels to dilate.
- Redness, warmth, swelling, and pain are associated with inflammation.

C. Innate Chemical Defenses Include Complement Proteins and Cytokines

- **Complement** proteins interact in a cascade that bursts bacterial cells.
- **Cytokines** are antimicrobial molecules that communicate with immune system cells and trigger fever. **Interferons** and **interleukins** are examples of cytokines.

D. Fever Helps Fight Infection

- The elevated body temperature of a mild **fever** helps discourage microbial replication.

36.3 Adaptive Immunity Defends Against Specific Pathogens

- **Adaptive immunity** is directed against specific **antigens,** so that immunity to one pathogen does not confer immunity to others. Only adaptive immunity produces immunological memory, which greatly speeds the reaction after a second encounter with a pathogen.

A. Genetic Recombination Yields an Endless Variety of Antibodies and Antigen Receptors

- Antibodies and antigen receptors are incredibly diverse because DNA segments shuffle during early lymphocyte development.

B. Macrophages, T Cells, and B Cells Have Distinct Roles in Adaptive Immunity

- Macrophages display on their cell surfaces the foreign antigens from pathogens they have engulfed.
- A **helper T cell** binding to the antigen-presenting cell triggers the **cell-mediated** and **humoral** components of the adaptive immune response. Activated helper T cells activate other T cells and B cells.

C. T Cells Coordinate Cell-Mediated Immunity

- **Cytotoxic T cells** release biochemicals that kill bacteria and cells infected with viruses. Memory T cells help provide long-term immunity.

D. B Cells Direct the Humoral Immune Response

- In **passive immunity,** a person receives antibodies from someone else. In **active immunity,** a person makes his or her own antibodies.
- An **antibody** is a Y-shaped protein composed of four polypeptide chains. Each chain has a **constant** amino acid sequence and a **variable** sequence. Antibodies bind antigens and form complexes that attract other immune system components.
- In **clonal selection,** an activated B cell multiplies rapidly, generating an army of identical **plasma cells** that all churn out the same antibody. Some also differentiate into **memory B cells.**

E. The Immune Response Turns Off Once the Threat Is Gone

- Tissue damage can occur if the immune response does not dampen after eliminating a pathogen.

F. The Secondary Immune Response Is Faster than the Primary Response

- The first encounter with an antigen provokes the **primary immune response,** which is relatively slow. Its legacy is memory cells that greatly speed the **secondary immune response** on subsequent exposure to the same antigen.
- A **vaccine** "teaches" the immune system to recognize specific components of a pathogen, bypassing the primary immune response.

36.4 Several Disorders Affect the Immune System

A. Immune Deficiencies Lead to Opportunistic Infections

- **HIV** enters helper T cells and replicates inside them, killing them directly. This leaves the body vulnerable to infection and cancer. **Opportunistic** illnesses make the AIDS sufferer sick.
- **Severe combined immune deficiency (SCID)** is an inherited disease. Immune deficiency can also be induced by drugs that prevent organ transplant rejection.

B. Autoimmune Disorders Are Devastating and Mysterious

- **Autoimmunity** occurs when the immune system produces antibodies that attack the body's own tissues.

C. Allergies Misdirect the Immune Response

- An **allergy** is an immune reaction to a harmless substance. An **allergen** triggers production of antibodies, which bind mast cells. On subsequent exposure, these mast cells release allergy mediators such as histamine.

D. A Pregnant Woman's Immune System Occasionally Attacks Her Developing Fetus

- The immune system of a woman whose cells lack the Rh-factor can reject an Rh-positive fetus. A drug can prevent this problem.

36.5 Investigating Life: Sea Urchins Lead Researchers from RAGs to Riches

- Scientists once thought that the components of vertebrate-style adaptive immunity arose all at once in the common ancestor to jawed vertebrates.
- Researchers have found counterparts of some genes required for adaptive immunity in the purple sea urchin, indicating that vertebrate adaptive immunity did not arise in one "big bang."

MULTIPLE CHOICE QUESTIONS

1. Histamine acts on the ______ , which causes redness and swelling.
   a. white blood cells
   b. cells lining blood vessels
   c. smooth muscle cells
   d. red blood cells
2. How do complement proteins contribute to the innate immune response?
   a. They cause invading bacteria to burst.
   b. They attract phagocytes to the injury.
   c. They cause mast cells to release histamine.
   d. All of the above.
3. An advantage of the innate immune response is its
   a. rapid response to invading pathogens.
   b. ability to "remember" pathogens it has already encountered.
   c. ability to produce antibodies.
   d. Both b and c.
4. Antibody function requires that the shape of the ______ matches the shape of the antigen.
   a. constant region
   b. light chain
   c. variable region
   d. heavy chain
5. What cellular process is responsible for clonal selection?
   a. Meiosis
   b. Mitosis
   c. Homeostasis
   d. Phagocytosis
6. Why is the secondary immune response so much more rapid than the primary response?
   a. Because the specific antibodies are already present.
   b. Because the phagocytes present the antigens more rapidly.
   c. Because memory cells can rapidly convert to plasma cells.
   d. Because protein synthesis occurs more quickly in memory cells.
7. What is the function of a cytotoxic T cell?
   a. It "presents" antigen to a helper T cell.
   b. It secretes proteins that destroy foreign cells.
   c. It secretes cytokines to stimulate antibody production.
   d. It triggers clonal selection of B cells.
8. Which type of immune cell is attacked by HIV?
   a. B memory cells
   b. Helper T cells
   c. Plasma cells
   d. Cytotoxic T cells
9. An allergic reaction is an example of
   a. a secondary immune response.
   b. a primary immune response.
   c. autoimmunity.
   d. passive immunity.

TESTING YOUR KNOWLEDGE

1. What are the components of the lymphatic system?
2. Describe three ways that phagocytes participate in the body's defenses.
3. List four examples of innate defenses.
4. State the functions of antibodies, cytokines, and complement proteins.
5. What is the difference between cell-mediated and humoral immunity?
6. What do a plasma cell and a memory cell descended from the same B cell have in common, and how do they differ?
7. How does each of the following illnesses affect immunity?
   a. SCID
   b. AIDS
   c. hay fever
8. What part do antibodies play in allergic reactions and in autoimmune disorders?
9. How does the immune system interact with the circulatory system?

THINKING AS A SCIENTIST

1. Explain the observation that lymphoid tissues are scattered in the skin, lungs, stomach, and intestines.
2. Dead phagocytes are one component of pus. Why is pus a sure sign of infection?
3. How can inflammation be both helpful and harmful?
4. How do the innate defenses and adaptive immunity cooperate to eradicate an infection?
5. Which do you think is more dangerous, a deficiency of T cells or a deficiency of B cells? Explain your reasoning.
6. How might a drug advertised as a "histamine blocker" relieve allergy symptoms?
7. Influenza viruses mutate rapidly, whereas the chickenpox virus does not. Why are people encouraged to receive vaccinations against influenza every year, whereas immunity to chickenpox lasts for decades?
8. One benefit of sexual reproduction is a genetically variable population. According to the red queen hypothesis (see section 9.8), genetic variability helps a population stay "one step ahead" of pathogen populations. Describe how genetic variability can enhance immunity.

Visit www.mhhe.com/hoefnagels for practice quizzes, animations, videos, and activities designed to help you master the material in this chapter.

CHAPTER 37

Human Reproduction and Development

Tiny Model Organisms. In-depth studies of nematode worms have revealed much about animal development.

What's the Point?

Worms and Flies Teach Lessons About Human Development

The study of animal development is perhaps most exciting in our own species. But investigating human development before birth is difficult for legal, ethical, and practical reasons. To avoid these problems, and because many developmental processes are very similar in closely related species, researchers often work with model organisms to learn more about human development (see chapter 22's Focus on Model Organisms: Caenorhabditis elegans *and* Drosophila melanogaster*).*

The tiny, transparent nematode worm Caenorhabditis elegans *has taught biologists a lot. The worm's life begins as a single fertilized egg cell. In just 3 days of mitotic cell division and programmed cell death, this one cell transforms into the final 959 cells that make up the adult worm. The cells of the early embryo are totipotent, which means they can develop into any cell type. But as new cells accumulate, some commit to follow particular pathways. Gradually, distinct structures begin to appear, specialize, interact, and organize in the embryo. Researchers understand nematode development in detail, describing and mapping the fates of all cells produced from the original fertilized egg cell.*

In C. elegans *and other animals, genes orchestrate development. The role of genes and protein gradients is well studied in the fruit fly* Drosophila melanogaster*. For example, female flies pack a protein called bicoid into one side of each developing egg cell. Bicoid instructs the embryo where to develop a head end. Normally, a high concentration of bicoid protein in the front end of the embryo causes the tissues of the head to differentiate. At the rear of the animal, higher concentrations of another protein, nanos, specify tail formation. When the gene encoding bicoid is mutated in the mother, the embryo develops two rear ends!*

These protein gradients are signals that stimulate cells to produce yet other proteins, which ultimately regulate the formation of a specific structure (see figure 15.16). A cell destined to be part of the adult fly's antenna, for example, has different concentrations of specific proteins than a cell whose future lies in the eye. Sometimes, when genes encoding these proteins mutate, developmental signals go off in the wrong part of the animal, with striking results. The fly in figure 12.13 is an example; it has legs in place of its antennae.

Thanks to C. elegans*,* Drosophila*, and many other research organisms, scientists now understand many details of reproduction and development down to the molecular level. Nevertheless, with today's heated debates over human cloning and stem cell therapies, developmental biology—the subject of this chapter—clearly remains a life science that is very much in the public eye.*

LEARNING OUTLINE

37.1 Animal Development Begins with Reproduction

A monarch butterfly emerges from its chrysalis. A baby bird hatches from an egg. A kitten becomes a full-grown cat. All illustrate growth and development, a property that all multicellular organisms share (**figure 37.1**).

Animal development begins with reproduction, which may be asexual or sexual (see section 9.1). **Asexual reproduction,** which requires only one individual and produces identical clones, is advantageous in environments that do not change much over time. In **sexual reproduction,** two parents contribute the DNA in each offspring. The genetic diversity in a population of sexually reproducing organisms increases reproductive success in a changing environment (see chapter 9's Investigating Life: An Arms Race at a Snail's Pace).

In organisms that reproduce sexually, haploid **gametes** are the sex cells that carry the genetic information from each parent (see figure 9.4). The gametes (sperm cells from males and egg cells from females) are the product of meiosis. In meiosis, a diploid cell containing two sets of chromosomes divides into four haploid cells, each containing just one chromosome set. **Fertilization** unites the gametes and produces the **zygote**, the diploid first cell of the new offspring.

Sperm and egg come together in a variety of ways. In **external fertilization,** males and females release gametes into the same environment, and fertilization occurs outside the body (**figure 37.2**A). Salmon, for example, spawn in streams. Females lay eggs in gravelly nests, and then males shed sperm over them. Other animals with external fertilization include sponges, corals, and amphibians.

FIGURE 37.1 Growing Up. A baby tortoise emerges from its egg. Like all animals, this tortoise has developed from a single cell, and it will continue to grow as it reaches reproductive maturity.

a.

b.

FIGURE 37.2 External and Internal Fertilization. (**A**) A tropical sea urchin releases sperm cells into the water. Meanwhile, females release eggs, and fertilization is external. (**B**) A male black-necked stilt mates with a female. The offspring will develop inside the female's body until she lays three or four hard-shelled eggs in a nest.

In **internal fertilization,** a male uses a copulatory organ such as a penis to deposit sperm inside a female's body, where fertilization occurs (figure 37.2B). Reptiles, birds, and mammals, including humans, use internal fertilization. Depending on the species, the female may then lay hard-shelled, fertilized eggs that provide both nutrition and a protected environment in which young develop. A chicken egg is a familiar example. Alternatively, the female may bear live young, as humans and other mammals do. In a few species, males become pregnant! In seahorses, for example, males have a brood pouch in which females deposit eggs. After fertilization, the young develop in the male's pouch for a few weeks, after which he gives birth.

No matter what the reproductive strategy, development of sexually reproducing animals begins with the zygote, which forms when a sperm cell penetrates an egg cell and the nuclei of the parents merge. From this point, an animal develops into its distinctive form. Cell division, cell death, and growth all maintain body form and function (see figure 8.1). Developmental biologists study how cells specialize and interact to form tissues, organs, and organ systems.

In the pages that follow, we focus on human reproduction and development. We begin by describing the male and female **reproductive systems,** which are similar in many ways. Each system includes primary sex organs: the paired **gonads** (testes or ovaries), which contain the **germ cells** that give rise to gametes. Secondary sex organs include the network of tubes that transport the gametes. In both sexes, hormones and glandular secretions control reproduction. Males and females also have **secondary sex characteristics** that distinguish the sexes but do not participate directly in reproduction. Breast development and menstruation in females, and facial hair and a deep voice in males, are examples of secondary sex characteristics.

After describing the male and female reproductive systems, we then turn to the stages of sperm and egg cell production, the events of fertilization, and the development of a human from conception to birth. The chapter ends with a discussion of birth defects.

Summary ***Sexual reproduction creates offspring by bringing together gametes from two parents. Fertilization may be external or internal. The reproductive systems of human males and females share many similarities.***

37.1 MASTERING CONCEPTS

1. What is the difference between asexual and sexual reproduction?
2. How is internal fertilization different from external fertilization?
3. How are the human male and female reproductive systems similar?
4. What is the relationship among gonads, germ cells, gametes, and the zygote?

37.2 Males Produce Sperm Cells

Although the male and female reproductive systems share some similarities, there are also obvious differences. This section details the features unique to males, beginning with the internal and external anatomy of the reproductive system (**figure 37.3**).

A. Male Reproductive Organs Are Inside and Outside the Body

The paired **testes** (singular: testis) are the male gonads. The testes lie in a sac called the **scrotum;** a **testicle** is one testis plus its surrounding tissues. Their location outside of the abdominal cavity allows the testes to maintain a temperature about 3°C cooler than the rest of the body, which is necessary for sperm to develop properly.

Each testis contains about 200 tightly coiled, 50-cm long **seminiferous tubules,** which produce the sperm cells. Large **sustentacular cells** extend the entire thickness of a seminiferous tubule, surrounding, supporting, and nourishing developing sperm cells. **Interstitial cells** fill the spaces between the seminiferous tubules; they are endocrine cells that secrete male sex hormones.

A maze of small ducts carries developing sperm to the left or right **epididymis,** a tightly coiled tube that receives and stores sperm from one testis. From each epididymis, sperm cells move into a **vas deferens,** a duct that bends behind the bladder and connects with the left or right **ejaculatory duct.** The two ejaculatory ducts empty into the **urethra,** the tube that extends the length of the cylindrical **penis** and carries both urine and semen out of the body.

Semen, the fluid that carries the sperm cells, includes secretions from several accessory glands. The two **seminal vesicles,** one of which opens into each vas deferens, secrete most of the fluid in semen. The secretions include sugar, which supplies energy, plus prostaglandins, which may stimulate contractions in the female reproductive tract that help propel sperm. The walnut-sized **prostate gland** wraps around part of the urethra and contributes a thin, milky, alkaline fluid that activates the sperm to swim. The **bulbourethral glands,** each about the size of a pea, attach to the urethra where it passes through the body wall. These glands secrete alkaline mucus, which coats the urethra before sperm are released.

During sexual arousal, the penis becomes erect, enabling it to penetrate the vagina and deposit semen in the female reproductive tract. (Chapter 2's opening essay describes erectile dysfunction and the drugs that treat the condition.) At the peak of sexual stimulation, a pleasurable sensation called **orgasm** occurs, accompanied by rhythmic muscular contractions that eject the semen through the urethra and out the penis. **Ejaculation** is the discharge of semen from the penis. One human ejaculation typically delivers more than 100 million sperm cells.

Two cancers of the male reproductive system originate in the prostate gland and testes. Prostate cancer is the second

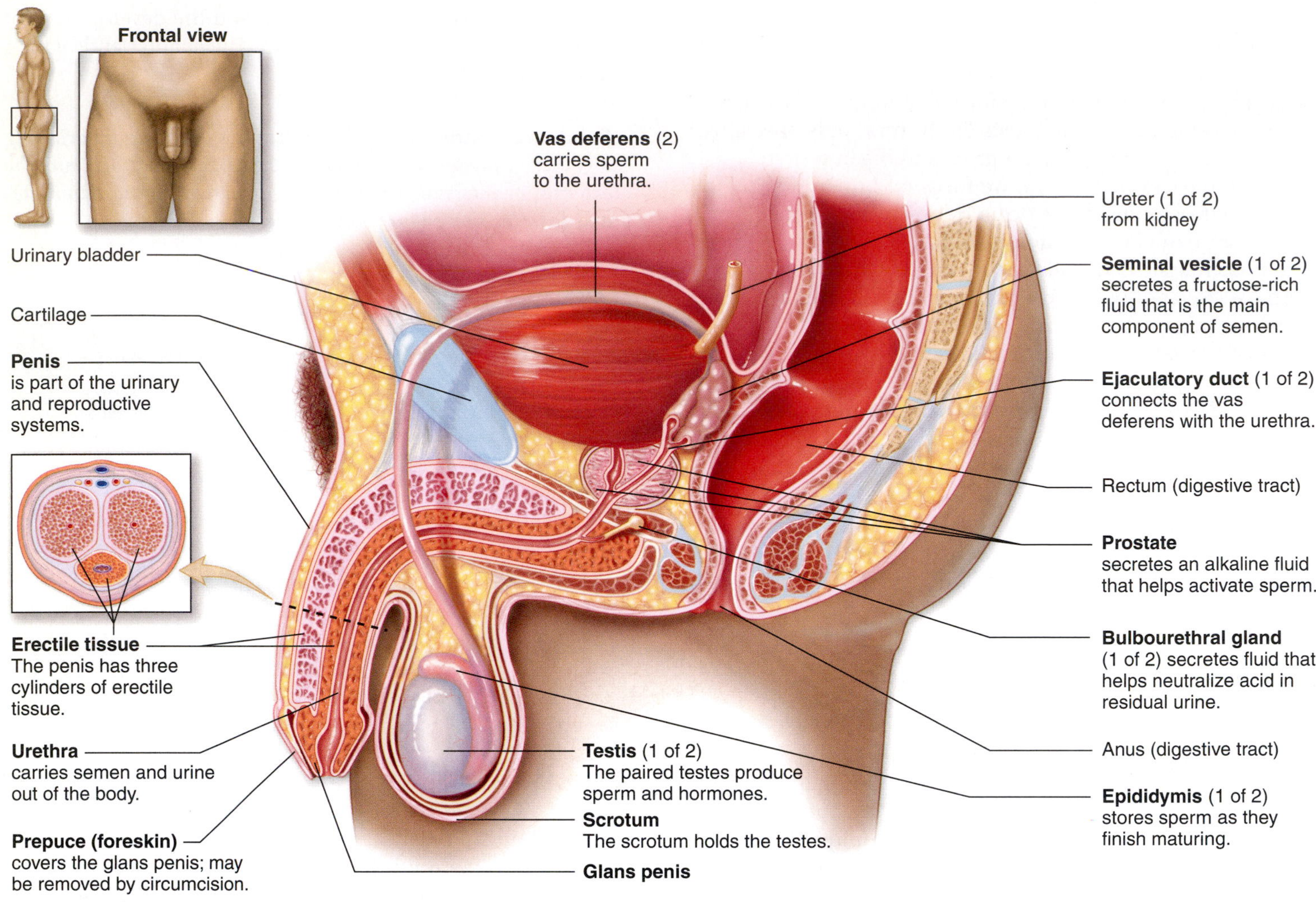

Frontal view
Vas deferens (2) carries sperm to the urethra.
Ureter (1 of 2) from kidney
Urinary bladder
Seminal vesicle (1 of 2) secretes a fructose-rich fluid that is the main component of semen.
Cartilage
Penis is part of the urinary and reproductive systems.
Ejaculatory duct (1 of 2) connects the vas deferens with the urethra.
Rectum (digestive tract)
Prostate secretes an alkaline fluid that helps activate sperm.
Erectile tissue The penis has three cylinders of erectile tissue.
Bulbourethral gland (1 of 2) secretes fluid that helps neutralize acid in residual urine.
Urethra carries semen and urine out of the body.
Testis (1 of 2) The paired testes produce sperm and hormones.
Anus (digestive tract)
Epididymis (1 of 2) stores sperm as they finish maturing.
Scrotum The scrotum holds the testes.
Prepuce (foreskin) covers the glans penis; may be removed by circumcision.
Glans penis

Reproductive System (Male)	
Main tissue types	**Examples of Locations/Functions**
Epithelial	Lines ducts of reproductive tract; produces sperm cells in seminiferous tubules; produces secretions in accessory glands
Connective	Makes up walls of testes; makes up erectile cylinders in penis
Nervous	Penis contains sensory nerve fibers and nerve endings; hypothalamus secretes hormones that affect the anterior pituitary
Muscle	Smooth muscle surrounds ducts of reproductive tract, propelling sperm out of the body

FIGURE 37.3 The Human Male Reproductive System. Sperm cells are manufactured in the paired testes. Sperm mature and are stored in the left or right epididymis and move to a vas deferens. Each vas deferens joins the urethra, through which semen exits the body. The prostate gland, seminal vesicles, and bulbourethral glands add secretions to the semen. During sexual arousal, three cylinders of erectile tissue (*inset*) fill with blood and cause the penis to become erect.

most common type of cancer in men (lung cancer is most common). The resulting prostate enlargement constricts the urethra and may interfere with urination and ejaculation. Prostate cancer usually affects men older than 50. In testicular cancer, which usually strikes men younger than 40, mutated germ cells divide out of control, forming lumps that may be detected in a self-examination. Testicular cancer has a very high cure rate, if detected before it spreads to other parts of the body. **cancer, p. 166**

B. Spermatogenesis Yields Sperm Cells

Spermatogenesis, the production of sperm, is a continuous process that begins when a male reaches puberty and continues throughout life. It occurs in several stages, which correspond to the stages of meiosis depicted in chapter 9. In reading this section, you may want to refer back to figure 9.15, which summarizes how meiosis I and II apply to spermatogenesis.

Sperm production occurs in the seminiferous tubules of the testes (**figure 37.4**). Germ cells called **spermatogonia**

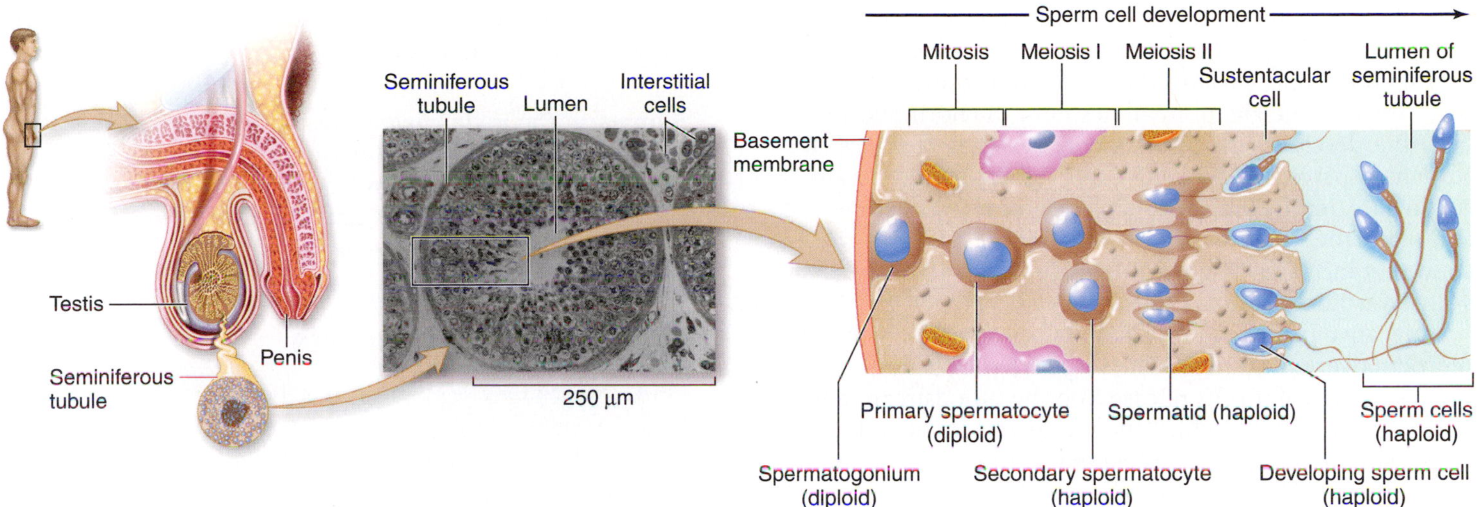

FIGURE 37.4 Spermatogenesis Occurs in the Testes. Diploid spermatogonia divide mitotically in the linings of the seminiferous tubules. Some of the daughter cells then undergo meiosis, producing four haploid secondary spermatocytes that differentiate into spermatids, then mature sperm cells.

reside within a seminiferous tubule, farthest from the lumen (the central cavity). The spermatogonia are diploid, so their nuclei contain 46 chromosomes. When a spermatogonium divides mitotically, one daughter cell remains in the tubule wall and acts as a stem cell, continually giving rise to cells that specialize into sperm. The other cell becomes a diploid **primary spermatocyte** that moves closer to the lumen and accumulates cytoplasm. **mitosis, p. 161**

In the wall of the seminiferous tubule, the primary spermatocyte undergoes the first division of meiosis, yielding two haploid **secondary spermatocytes.** These cells undergo meiosis II, forming four round, haploid cells called **spermatids** that approach the lumen of the seminiferous tubule. Each spermatid contains 23 chromosomes.

As they move into the lumen of the seminiferous tubule, the spermatids complete their differentiation into **spermatozoa,** or mature sperm cells. They separate into individual cells and develop tails (although they will not be able to swim or fertilize an egg until they reach the epididymis). They also lose much of their cytoplasm, acquire a streamlined shape, and package their DNA into a distinct head (**figure 37.5**). Mitochondria just below the head region generate the ATP the sperm needs to move toward an egg cell. The caplike **acrosome** covers the head and releases enzymes that will help the sperm penetrate the egg cell. The entire process, from spermatogonium to spermatozoon, takes 74 days in the human. **mitochondria, p. 67**

C. Hormones Influence Male Reproductive Function

Hormones play a critical role in male reproductive function. In the brain, the hypothalamus secretes **gonadotropin-releasing hormone (GnRH).** This peptide hormone travels in the bloodstream to the anterior pituitary, where it stimulates the release of two other peptide hormones (see figure 30.11): **follicle-stimulating hormone (FSH)** and **luteinizing hormone (LH).** Blood carries FSH and LH throughout the body. **peptide hormones, p. 619**

In the testes, LH signals interstitial cells to release the steroid hormone **testosterone** and other male sex hormones

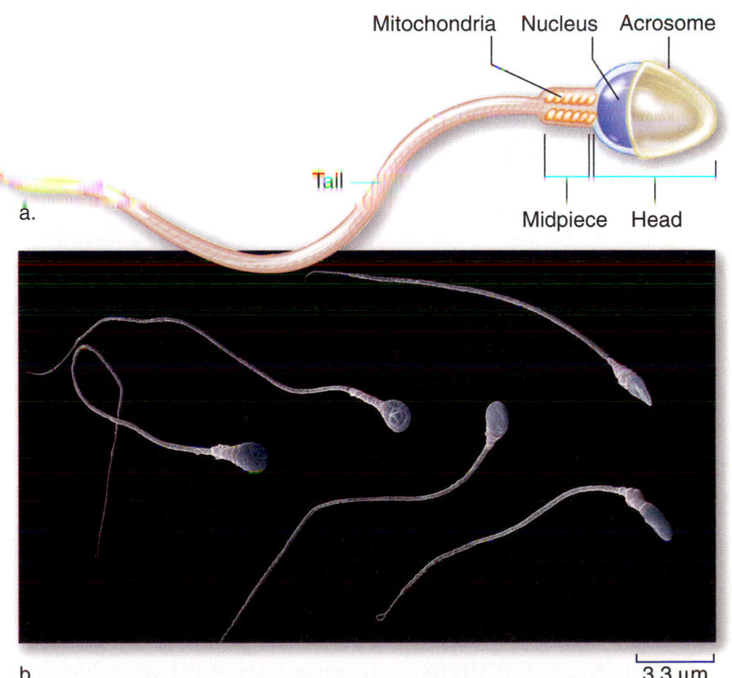

FIGURE 37.5 Human Sperm. (**A**) A sperm has distinct structures that assist in delivering DNA to an egg cell. (**B**) Scanning electron micrograph of human sperm cells.

(androgens). In the presence of FSH, testosterone affects the body in multiple ways. In adults, testosterone stimulates sperm production, sustains the libido, and controls the activity of the prostate gland. In adolescents, the hormone stimulates the development of secondary sex characteristics. The testes and penis begin to enlarge at puberty, and hair grows on the face, in the armpits, and at the groin. Testosterone also stimulates the secretion of growth hormone, causing a growth spurt that increases height and muscle mass and deepens the voice. **steroid hormones, p. 620**

Negative feedback loops maintain homeostasis in male reproductive hormones. When sperm production is too high, sustentacular cells release into the bloodstream a hormone called inhibin, which blocks the release of additional FSH from the pituitary. Lower sperm counts suppress inhibin production, increasing FSH production. One well-known consequence of abusing anabolic steroids is infertility or low sperm counts (see chapter 30's Can *You* Relate? Anabolic Steroids Build Muscle, but with Significant Side Effects). Negative feedback helps explain this phenomenon. The body mistakes the synthetic steroids for testosterone, causing the testes to produce less of the real sex hormone. Without testosterone, sperm do not form.

Summary ***The male gonads are the testes; other sex organs include the penis, ducts that carry sperm, and glands that contribute secretions. Spermatogenesis produces sperm cells in several stages. Hormones control sperm production, libido, and the development of male secondary sex characteristics.***

37.2 MASTERING CONCEPTS

1. What is the role of each part of the male reproductive system?
2. Where in the testes do sperm develop?
3. What are the stages of sperm development and maturation?
4. What are the parts of a mature sperm cell?
5. How do hormones regulate sperm production throughout life?

37.3 Females Produce Egg Cells

Egg cell production in females is somewhat more complicated than is sperm production in males, for at least two reasons. First, in females, meiosis begins before birth, pauses, and resumes at sexual maturity. It is not completed until after a sperm cell fertilizes the egg cell. Second, egg cell production is cyclical, under the control of several interacting hormones whose levels fluctuate monthly during a woman's reproductive years. Keep these differences in mind as you read this section.

A. Female Reproductive Organs Are Inside the Body

The female sex cells develop within the **ovaries,** paired gonads in the abdomen (**figure 37.6**). Ovaries produce both egg cells and sex hormones. They do not contain ducts comparable to the seminiferous tubules of the male's testes. Instead, within each ovary of a newborn female are about a million oocytes, the cells that give rise to mature egg cells. Nourishing **follicle cells** surround each oocyte.

Approximately once a month, beginning at puberty, one ovary releases the single most mature oocyte. Beating cilia sweep the mature oocyte into the fingerlike projections of one of the two **uterine tubes** (also called fallopian tubes). If sperm are present, fertilization occurs in a uterine tube. The tube carries the oocyte or zygote into a muscular saclike organ, the **uterus.** During pregnancy, the fetus develops inside the uterus, also called the womb. Its inner lining, the **endometrium,** has a rich blood supply that is important in both menstruation and pregnancy.

The **cervix** is the necklike narrowing at the lower end of the uterus. The cervix opens into the **vagina,** the tube that leads outside the body. The vagina receives the penis during intercourse, and it is also the birth canal. Like many other areas of the body, the vagina harbors a community of resident microorganisms. These bacteria lower the pH of the vagina, which helps prevent colonization by harmful bacteria and the yeast, *Candida albicans*. Taking antibiotics can disrupt this microbial community and create an opportunity for *Candida* to overgrow, causing a vaginal yeast infection.

Two pairs of fleshy folds protect the vaginal opening on the outside: the labia majora (major lips) and the thinner, underlying flaps of tissue they protect, called the labia minora (minor lips). The **clitoris** is a 2-cm-long (about 1 inch) structure at the upper junction of both pairs of labia. Rubbing the clitoris stimulates females to experience orgasm. Together, the labia, clitoris, and vaginal opening constitute the **vulva,** or external female genitalia.

The female secondary sex characteristics include the wider and shallower shape of the pelvis, the accumulation

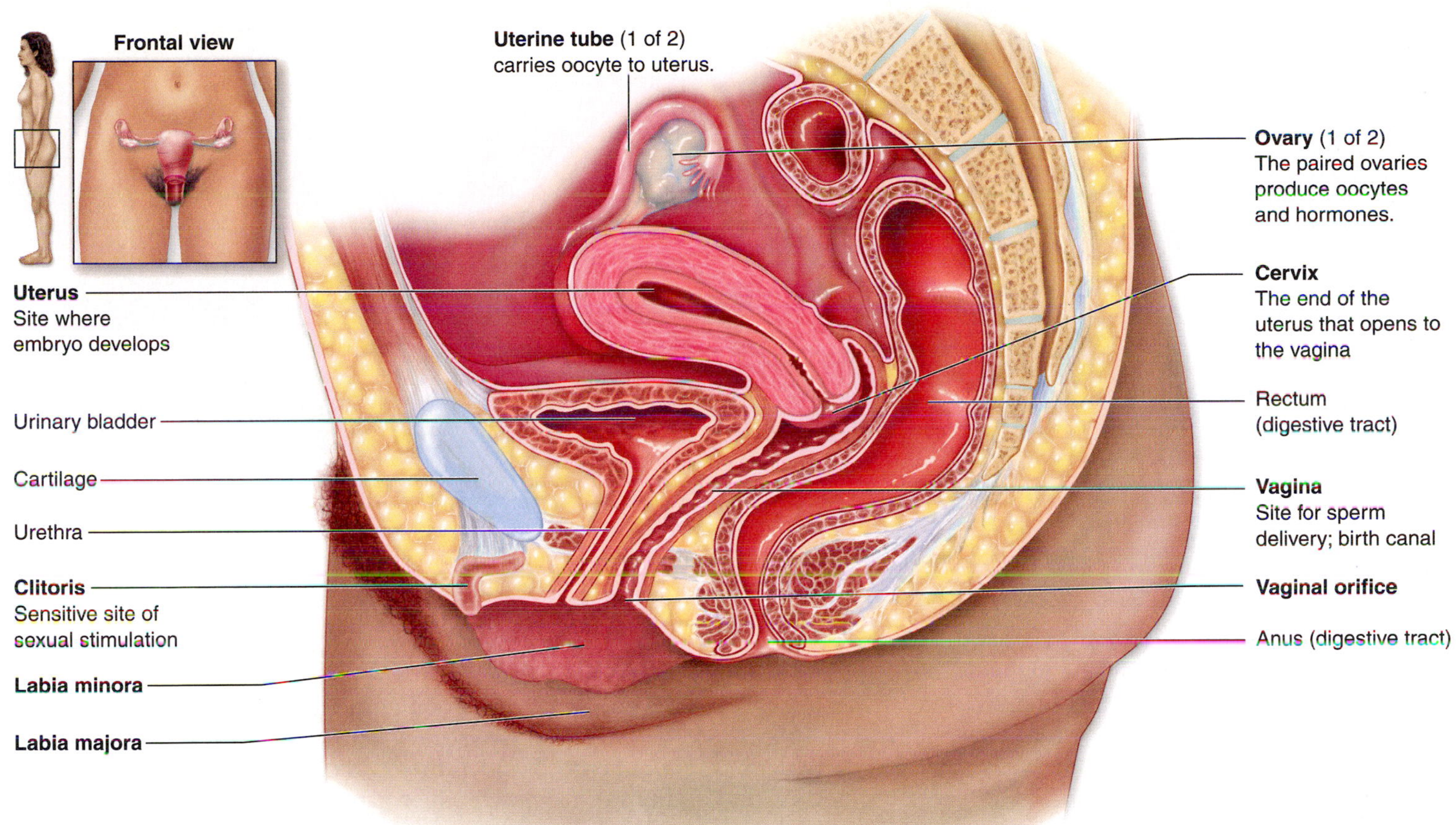

Reproductive System (Female)	
Main tissue types	**Examples of Locations/Functions**
Epithelial	Lines uterus, uterine tubes, and vagina; produces oocytes in ovaries; forms external surface of umbilical cord
Connective	Makes up walls of ovaries, uterus, and vagina
Nervous	Clitoris contains sensory nerve fibers and nerve endings; hypothalamus secretes hormones that affect the anterior pituitary
Muscle	Smooth muscle surrounds uterine tubes, uterus, and vagina

FIGURE 37.6 The Human Female Reproductive System. The paired ovaries contain immature egg cells (oocytes). Once a month, one ovary releases an oocyte, which is drawn into the fingerlike projections of a nearby uterine tube. If a sperm cell fertilizes the oocyte in the uterine tube, the embryo attaches to the wall of the uterus, where it develops over the next 9 months. The vagina is the birth canal. If fertilization does not occur, the oocyte is expelled with the built-up uterine lining as menstrual flow through the cervix and vagina. The external genitalia consist of the labia minora, labia majora, and clitoris.

of fat around the hips, a higher pitched voice than that of the male, and the breasts. Each **breast** contains fatty tissue, collagen, and milk ducts. The nipple delivers milk to the nursing infant. (Chapter 27's Investigating Life: When a Chair Becomes a Ladder describes the evolutionary history of milk in mammals.)

Cancers of the female reproductive system often develop in the breast, cervix, or ovaries. Breast cancer is the most common cancer type in women. The abnormally dividing cells may originate in the milk-forming tissues or in the milk ducts of the breast. Some, but not all, forms of breast cancer have a strong genetic component. A family history is also the leading risk factor for ovarian cancer, which usually originates in the outer lining of the ovary. In contrast, nearly all cases of cervical cancer are associated with the human papillomavirus. Pap tests detect the abnormal cells of cervical cancer. In 2006, the FDA approved for girls and young women a "cervical cancer vaccine" that contains proteins of the human papillomavirus. **vaccine, p. 738**

B. Oogenesis Yields Egg Cells

The making of an egg cell—**oogenesis**—begins with an **oogonium,** a diploid germ cell containing 46 chromosomes. Each oogonium grows, accumulates cytoplasm, replicates its DNA, and divides mitotically, becoming two

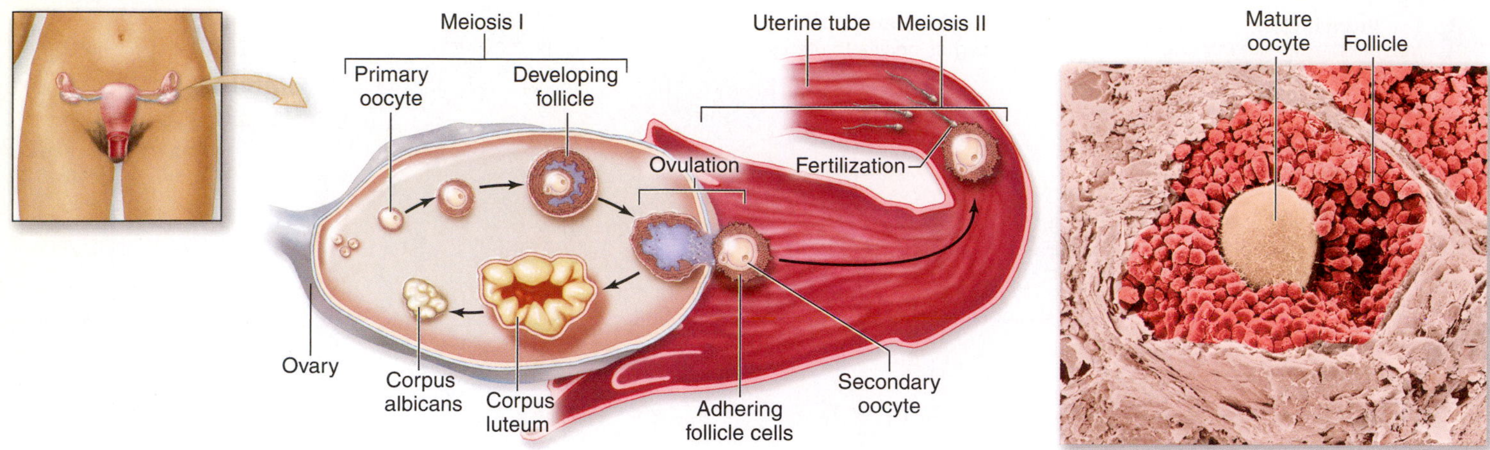

FIGURE 37.7 Oogenesis Occurs in the Ovaries. Oocytes develop within each ovary in protective follicles. In a maturing follicle, a diploid primary oocyte undergoes meiosis I to yield one secondary oocyte and a polar body. Every month between puberty and menopause, the most mature follicle ruptures and the secondary oocyte bursts out, an event called ovulation.

primary oocytes (**figure 37.7**). The ensuing meiotic divisions divide the cytoplasm unequally, so that oogenesis (unlike spermatogenesis) produces cells of different sizes. In meiosis I, the primary oocyte divides into a small haploid **polar body** and a larger haploid **secondary oocyte** (see figure 9.16), each containing 23 chromosomes. In meiosis II, the tiny polar body may divide into two polar bodies of equal size, or it may decompose. The secondary oocyte, however, again divides unequally in meiosis II to produce a small second polar body and the mature haploid egg cell (or ovum), which contains 23 chromosomes and a large amount of cytoplasm.

The egg cell, in receiving most of the cytoplasm, contains all of the biochemicals and organelles that the zygote will use until its own DNA begins to function. A woman's body absorbs polar bodies, which normally play no further role in development. Rarely, however, sperm can fertilize polar bodies, and a mass of tissue that does not resemble an embryo grows until the woman's body rejects it. A fertilized polar body accounts for about 1 in 100 spontaneous abortions. (Can *You* Relate? When a Pregnancy Ends describes other causes of miscarriage and stillbirths.)

Between puberty and menopause (the cessation of menstruation), monthly hormonal cues prompt an ovary to release a secondary oocyte into a uterine tube. If a sperm penetrates the oocyte membrane, meiosis in the oocyte completes, and the two nuclei combine to form the diploid zygote. If the secondary oocyte is not fertilized, it degenerates and leaves the body with the endometrium in the menstrual flow.

Oogenesis is somewhat similar to spermatogenesis (**figure 37.8** and **table 37.1**). Each process starts with a diploid germ cell (spermatogonium or oogonium) that gives rise to the haploid gametes. Also, like a testis containing sperm cells in various stages of development, each ovary houses oocytes in different stages of development. Of course, the two processes also differ. For example, spermatogenesis gives rise to four equal-sized sperm cells, whereas one oogonium yields one functional egg cell and three smaller polar bodies.

Also, the timetable for oogenesis differs greatly from that of spermatogenesis. A male takes about 74 days to produce a sperm cell. In contrast, oogenesis stretches from before birth until after puberty in a woman. Three months after conception, the ovaries of a female fetus contain 2 million or more primary oocytes. From then on, the number of primary oocytes declines. A million are present by birth, and about 400,000 remain by the time of puberty. At birth, the primary oocytes arrest in prophase I. After puberty, one or a few oocytes complete meiosis I each month. These secondary oocytes stop meiosis again, this time at metaphase II. Meiosis is completed only if fertilization occurs.

Table 37.1 *Spermatogenesis and Oogenesis Compared*

Diploid Starting Cell	Product of Mitosis (Diploid)	Products of Meiosis I (Haploid)	Products of Meiosis II (Haploid)
Spermatogonium in seminiferous tubule	Primary spermatocyte	Two secondary spermatocytes	Four equal-sized spermatids
Oogonium in ovary	Primary oocyte	One large secondary oocyte + one small polar body	One large egg cell + three small polar bodies

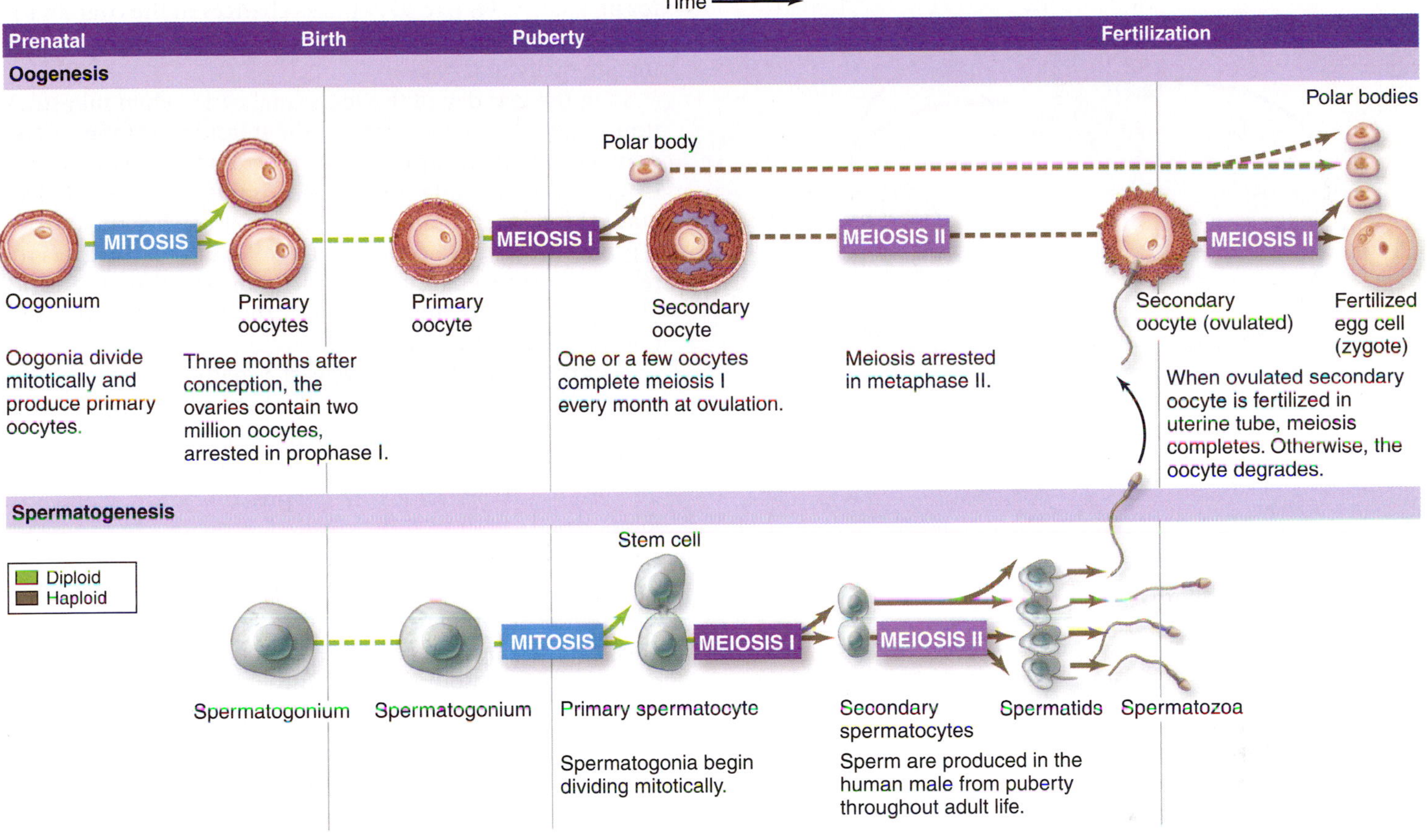

FIGURE 37.8 Oogenesis and Spermatogenesis in the Human Compared. The steps of sperm and oocyte formation are similar, and the products of each process are haploid. But the male and female gametes look different, and they form according to different timetables.

C. Hormones Influence Female Reproductive Function

The male and female reproductive systems rely on many of the same hormones, but in different quantities and on a different schedule (**table 37.2**). Unlike in males, hormone levels in females fluctuate cyclically. The result is two interrelated cycles, called the ovarian cycle and the menstrual cycle. The **ovarian cycle** controls the timing of oocyte maturation in the ovaries, and the **menstrual cycle** prepares the uterus

Table 37.2 *The Roles of GnRH, FSH, and LH in Human Males and Females*

Hormone	Released from:	Acts on:	Function in males	Function in females
GnRH (gonadotropin-releasing hormone)	Hypothalamus	Anterior pituitary	Secreted at constant frequency; stimulates release of LH and FSH	Amount secreted depends on stage of menstrual cycle; stimulates release of LH and FSH
LH (luteinizing hormone)	Anterior pituitary	Testes, ovaries	Stimulates production of testosterone	Stimulates production of androgens and estrogens; LH surge midway through menstrual cycle triggers ovulation; LH stimulates development of corpus luteum, which secretes progesterone
FSH (follicle-stimulating hormone)	Anterior pituitary	Testes, ovaries	Enhances production of androgen-binding protein; required for spermatogenesis	Stimulates follicle maturation

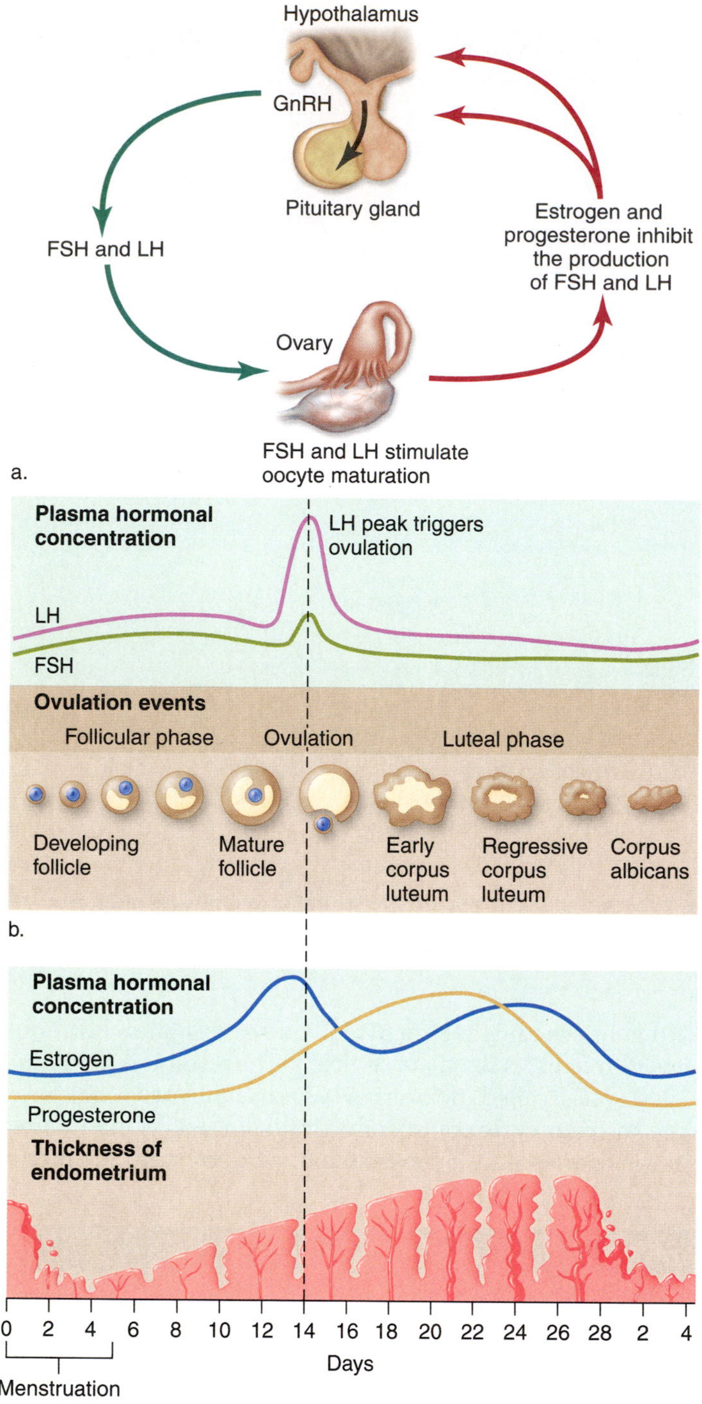

FIGURE 37.9 The Ovarian and Menstrual Cycles. (**A**) The anterior pituitary releases follicle-stimulating hormone (FSH) and luteinizing hormone (LH) when stimulated by gonadotropin-releasing hormone (GnRH) from the hypothalamus. (**B**) In the ovarian cycle, changes in FSH and LH stimulate follicle maturation and ovulation. (**C**) In the menstrual cycle, fluctuating concentrations of estrogen and progesterone prompt changes in thickness of the uterine lining. The curves in the upper part of (B) and (C) indicate changing levels of FSH, LH, estrogen, and progesterone throughout the ovarian and menstrual cycles.

for pregnancy. **Figure 37.9** tracks changes in the ovarian follicle, the uterine lining, and the levels of four hormones during the menstrual cycle.

On the first day of the menstrual cycle, when menstruation begins, low blood levels of the steroid hormones **estrogen** and **progesterone** signal the hypothalamus to secrete GnRH. This prompts the anterior pituitary to release FSH and LH into the circulation. When FSH reaches the ovaries, it promotes the development of a follicle. It also triggers the release of estrogens, progesterone, and small quantities of androgens, setting into motion hormonal interactions that thicken the uterine lining.

Around day 14 of the cycle, the mature follicle secretes enough estrogen to trigger the release of additional LH from the anterior pituitary. This LH surge in the bloodstream triggers **ovulation,** the release of an oocyte from an ovary into the nearby uterine tube. The **corpus luteum,** a gland that forms from ruptured follicle cells left behind in the ovary, secretes progesterone and estrogen (see figure 37.7). Together, progesterone and estrogen promote the thickening of the endometrium, preparing the uterus for possible pregnancy. The same two hormones also feed back to the hypothalamus, inhibiting production of GnRH, LH, and FSH.

The corpus luteum degenerates into the corpus albicans, an inactive scar made of collagen. Levels of progesterone and estrogen decline. The reduced levels of these hormones no longer maintain the endometrium, which then exits the body through the cervix and vagina as menstrual flow. Lowered progesterone and estrogen levels also release their inhibition of LH and FSH in the brain, and the cycle begins anew.

Contraception is the use of devices or practices that work "against conception" (**table 37.3**). Birth control pills contain progesterone, which mimics the hormonal effects of pregnancy. By preventing ovulation, "the pill" prevents conception. Other birth control methods kill sperm, block the meeting of sperm and oocyte, or prevent a developing embryo from implanting in the lining of the uterus.

Summary *Ovaries are the female gonads; other female sex organs include the uterine tubes, uterus, cervix, vagina, and clitoris. Oogenesis begins before birth, halts, and then resumes at puberty. Hormones control the timing of menstruation and egg cell release during a woman's reproductive years.*

37.3 MASTERING CONCEPTS

1. What is the role of each part of the female reproductive system?
2. Where do egg cells develop?
3. What are the stages of egg cell development and maturation?
4. What is the role of polar bodies in the production of an egg cell?
5. How do hormones regulate the ovarian and menstrual cycles?

Table 37.3 *Birth Control Methods*

Method	Mechanism	Advantages	Disadvantages	Likelihood of Success (%)
Barriers and Spermicides				
Condom and spermicide	Worn over penis or inserted into vagina, keeps sperm out of vagina, and kills sperm that escape	Protects against sexually transmitted diseases	Disrupts spontaneity, reduces sensation	95–98
Diaphragm and spermicide	Kills sperm and blocks cervix	Inexpensive	Disrupts spontaneity, must be fitted	83–97
Cervical cap and spermicide	Kills sperm and blocks cervix	Inexpensive, can be kept in for 24 hours	May slip out of place, must be fitted	80–95
Spermicidal foam or jelly	Kills sperm and blocks cervix	Inexpensive	Messy	78–95
Spermicidal suppository	Kills sperm and blocks cervix	Easy to use and carry	Irritates 25% of users	85–97
Hormonal				
Combination birth control pill	Prevents ovulation and implantation, thickens cervical mucus	Does not interrupt spontaneity, lowers cancer risk, lightens menstrual flow	Raises risk of heart disease in some women, causes weight gain and breast tenderness	90–100
Minipill	Blocks implantation, deactivates sperm, thickens cervical mucus	Fewer side effects than other birth control pills	Weight gain	91–100
Depo-Provera	Prevents ovulation, alters uterine lining	Easy to use, lasts 3 months	Menstrual changes, weight gain, injection	99
Norplant	Prevents ovulation, thickens cervical mucus	Easy to use, lasts 5 years	Menstrual changes, doctor must implant	99.8
Behavioral				
Rhythm method	No intercourse during fertile times	No cost	Difficult to do, hard to predict timing	79–87
Withdrawal	Removal of penis from vagina before ejaculation	No cost	Difficult to do; sperm may leak from penis before it is withdrawn, even without ejaculation	75–91
Surgical				
Vasectomy	Cuts vas deferentia, so sperm cells never reach urethra	Permanent, does not interrupt spontaneity	Requires minor surgery, difficult to reverse	99.85
Tubal ligation	Cuts uterine tubes, so oocytes never reach uterus	Permanent, does not interrupt spontaneity	Requires surgery, risk of infection, difficult to reverse	99.6
Other				
Intrauterine device	Prevents implantation of preembryo	Does not interrupt spontaneity	Severe menstrual cramps, risk of infection	95–99

Can You Relate?

When a Pregnancy Ends

Many conditions can cause miscarriage (also called a spontaneous abortion), in which the placenta and fetus spontaneously separate from the wall of the uterus. The woman expels them from her body, ending the pregnancy. This may occur any time before 7 months of development, when a fetus can survive outside the womb. Usually a miscarriage that occurs early in a pregnancy results from a chromosomal abnormality in the fetus. If it occurs during the second trimester, the problem usually lies with the woman. Diabetes, high blood pressure, hormonal abnormalities, infectious diseases, problems with the uterus, or drug use may be to blame, but often the cause of a miscarriage is unclear. **nondisjunction, p. 185**

Just as a miscarriage is the loss of a pregnancy before the fetus is viable, a stillbirth occurs if a baby dies in the uterus after about the twenty-fourth week. The same conditions that contribute to miscarriage are also associated with stillbirths.

An induced abortion is the deliberate termination of a pregnancy. During the first 15 weeks of pregnancy, the most common method of abortion is for a medical professional to use a manual syringe or electric pump to remove the embryo or fetus from the uterus. If the woman terminates the pregnancy between the fifteenth and eighteenth weeks, the cervix must first be dilated before removing the fetus. In the third trimester, a variety of chemical substances can be administered to induce premature delivery, or the fetus can be surgically removed from the uterus.

37.4 The Human Infant Begins Life as a Zygote

So far, this chapter has described gamete production in the human male and female reproductive systems. This section now turns to the events that produce the human baby. As you shall see, fertilization produces the zygote, a single cell that will divide many times as it develops into a preembryo, embryo, and fetus (**table 37.4**). This section describes the events of fertilization, prenatal development, and childbirth.

A. Fertilization Joins Genetic Packages and Initiates Pregnancy

After intercourse, sperm cells swim toward the oocyte. As they travel inside the woman's body, the sperm cells undergo chemical changes that enable them to enter an oocyte. Of the 100 million sperm that begin the journey, only about 200 approach the egg cell in the uterine tube.

Those that make it must penetrate two layers to contact the ovum. A thin, clear layer of proteins and carbohydrates encases the oocyte and is itself surrounded by a layer of follicular cells (**figure 37.10**). On contact with the cells surrounding the oocyte, each sperm's acrosome bursts, spilling enzymes that digest both outer layers.

Fertilization (conception) begins when the outer membranes of one sperm cell and the secondary oocyte touch. At that time, a wave of electricity spreads physical and chemical changes across the oocyte surface, preventing other sperm from fertilizing the same egg cell. As the sperm enters the secondary oocyte, the female cell completes meiosis and becomes a fertilized egg cell, or zygote. The nuclear membrane

Table 37.4 *Stages in Development from Conception to Birth*

Name	Duration	Description
Zygote	About 24 hours after ovulation	Fertilized egg cell
Preembryonic stage		
Morula	3–4 days after fertilization	Solid ball of 16 or more cells
Blastocyst	4 days to 2 weeks after fertilization	Hollow sphere consisting of outer layer (trophoblast) and inner cell mass; implants into endometrium
Gastrula	2 weeks after fertilization	Three germ layers form: endoderm, mesoderm, ectoderm
Embryonic stage	From implantation until about 8 weeks after fertilization	Germ layers differentiate into organ systems
Fetal stage	From end of eighth week until birth (end of first trimester, plus second and third trimesters)	Organ systems become functional

of the ovum disappears, and the two sets of chromosomes mingle. This diploid cell has 23 pairs of chromosomes; one chromosome of each pair comes from each parent. The cell is still within a uterine tube. **homologous chromosomes, p. 179**

Occasionally a woman ovulates two or more egg cells at once, and a different sperm fertilizes each one (see figure 9.9). If all the zygotes complete development, the result is twins, triplets, quadruplets, or even higher order multiple births. Because each zygote results from a separate sperm and egg cell, the siblings will not be genetically identical, and they may be of different sexes. **multiple births, p. 190**

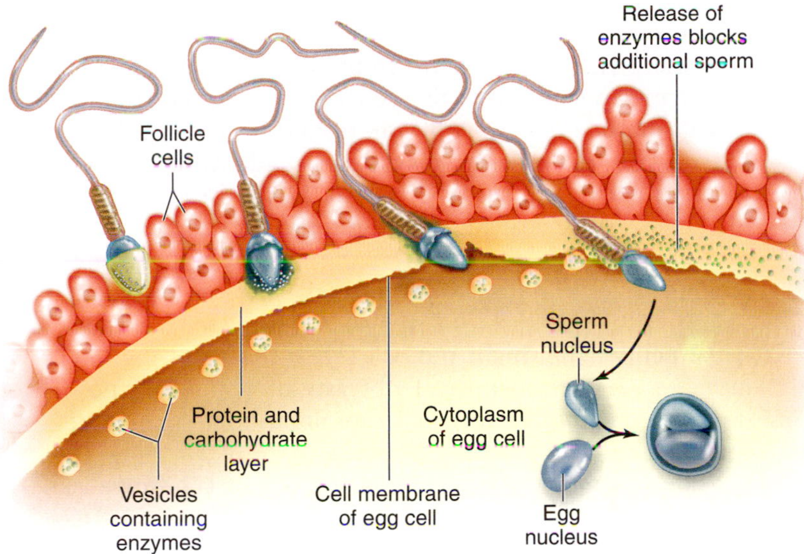

FIGURE 37.10 Fertilization. Before fertilization occurs, the sperm's acrosome bursts, spilling enzymes that help the sperm's nucleus enter the oocyte. A series of chemical reactions helps ensure that only one sperm nucleus enters an oocyte.

B. Preembryonic Events Include Cleavage, Implantation, and Gastrulation

The first 2 weeks of prenatal development have a variety of names, but we will call this period the **preembryonic stage.**

About 3 hours after fertilization, the zygote divides for the first time, beginning a period of rapid mitotic cell division called **cleavage** (**figure 37.11**). The result is a **morula,** a solid ball of 16 or more cells. *Morula* is Latin for "mulberry," which the preembryo resembles. Three days after fertilization, the morula is still within the uterine tube, moving toward the uterus. It is about the same size as the zygote, because each cleavage division produces daughter cells that are about half the size of the parent cell. Soon cell size levels off, and the morula enlarges as cells accumulate. It reaches the uterus 3 to 6 days after fertilization. It then hollows out, its center filling with fluid that seeps in from the uterus. The preembryo is now called a **blastocyst.**

The blastocyst's outer layer, the trophoblast, will eventually form the fetal portion of the placenta. The cells inside the blastocyst form the **inner cell mass,** the cells that will develop into the embryo itself. (The inner cell mass is also the source of embryonic stem cells.) In a process

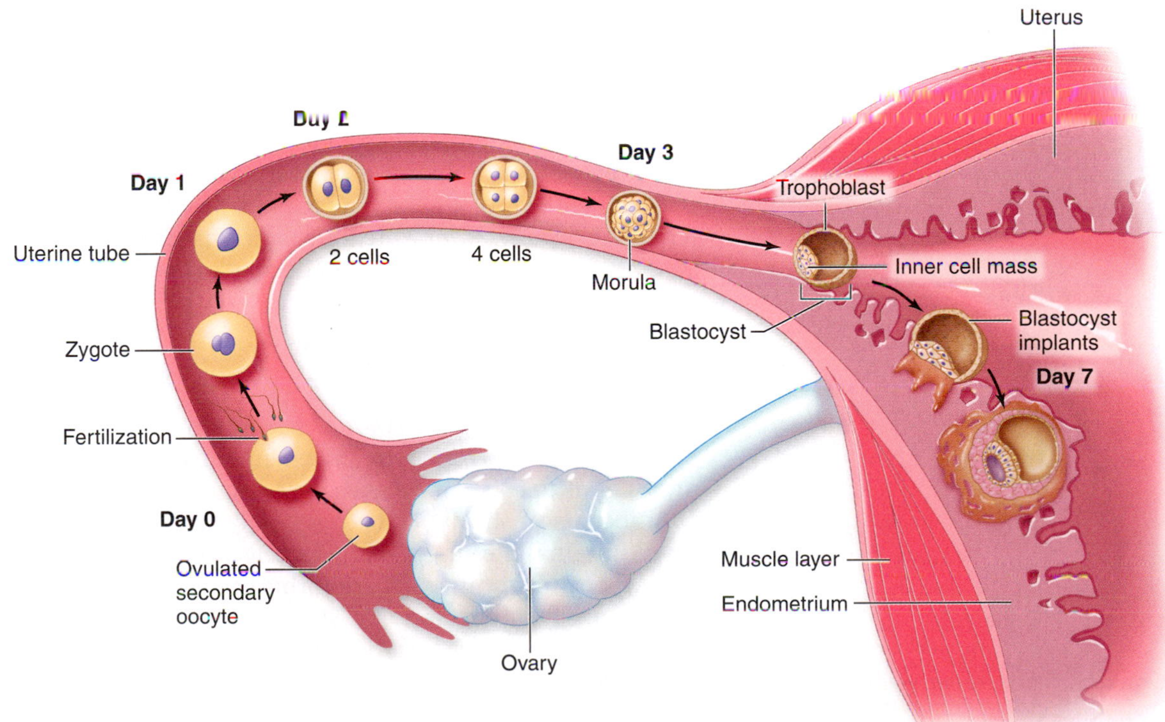

FIGURE 37.11 From Ovulation to Implantation. Fertilization occurs in the uterine tube to produce the zygote. The first mitotic divisions occur while the preembryo moves toward the uterus. By day 7, the preembryo begins to implant in the uterine lining.

called **implantation,** the blastocyst becomes embedded in the lining of the uterus (**figure 37.12**). Implantation occurs within a week of fertilization. The trophoblast not only secretes digestive enzymes that eat through the outer layer of the uterine lining, but it also sends fingerlike projections into the uterine lining. **stem cells, p. 158**

The trophoblast cells now secrete **human chorionic gonadotropin (hCG)**, the hormone that is the basis of pregnancy tests. For a while, hCG keeps the cells of the corpus luteum producing progesterone, which prevents menstruation and further ovulation. In this way, the blastocyst helps to ensure its own survival; if the uterine lining were shed, the blastocyst would leave the woman's body too. The trophoblast cells produce hCG for about 10 weeks.

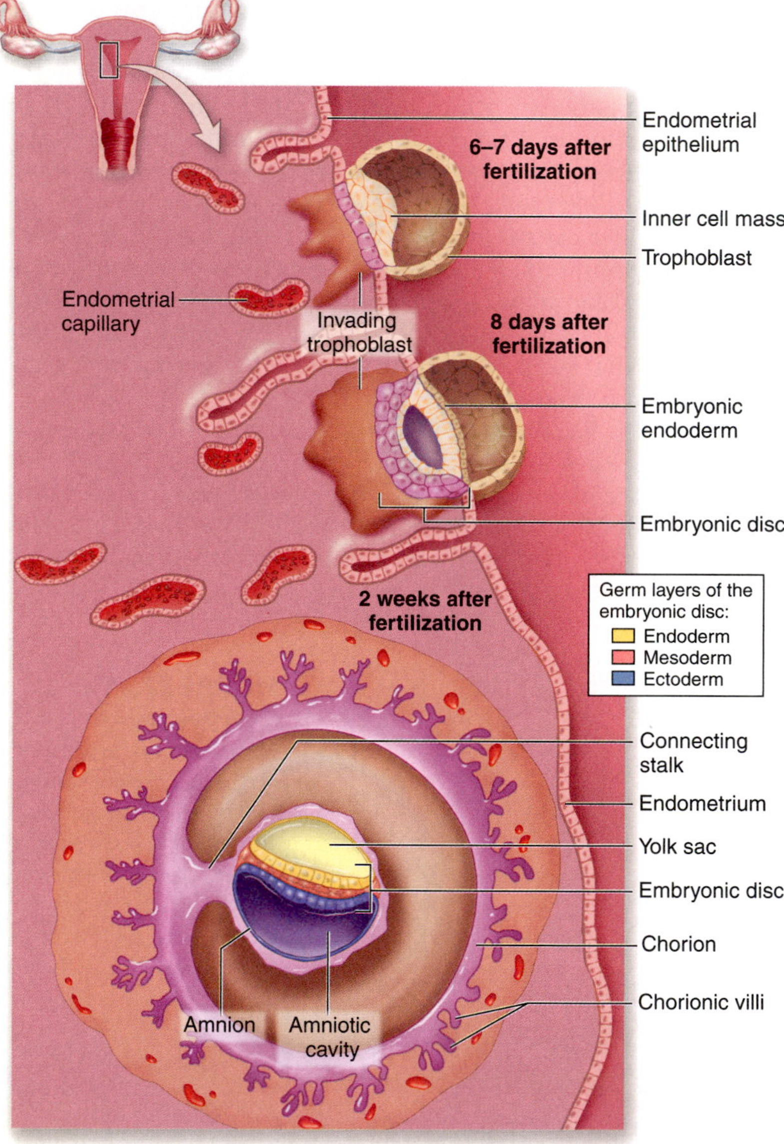

FIGURE 37.12 Implantation and Gastrulation. About a week after fertilization, the inner cell mass settles against the uterine lining. The trophoblast then sends out fingerlike extensions. Meanwhile, the enlarging embryo begins to fold into the three primary germ layers. Implantation is complete two weeks after fertilization.

During the second week of development, the blastocyst completes implantation, and the inner cell mass changes. A space called the amniotic cavity forms within a sac called the amnion, which lies between the inner cell mass and the trophoblast (see figure 37.12). The inner cell mass then flattens and forms the **embryonic disc,** which will develop into the embryo.

One of the layers of the embryonic disc will become ectoderm, and another will become endoderm. Soon, a middle mesoderm layer will form from the ectoderm. These three layers (endoderm, mesoderm, and ectoderm) are called primary germ layers (see figure 22.4). The **gastrula** is this three-layered structure. The formation of the gastrula is called gastrulation, and its start marks the end of the preembryonic stage at about 2 weeks after conception (see figure 37.12). Although the woman has not yet missed her menstrual period, she might notice effects of her shifting hormones, such as swollen, tender breasts and fatigue. By now, her urine contains enough hCG for an at-home pregnancy test to detect.

The preembryo may split during the first 2 weeks of development, forming identical twins (see figure 9.9). Depending on when the split occurs, the twins may or may not share the same amnion and placenta. The later in development that twinning occurs, the more structures the twins will share. If a preembryo splits after day 12 of pregnancy, the twins are unlikely to separate completely, and they will be conjoined.

C. Organs Take Shape During the Embryonic Stage

The **embryonic stage** lasts from the end of the second week until the end of the eighth week. During this stage of development, cells of the three layers continue to divide and differentiate, forming all of the body's organs. Structures that support the embryo also develop during this period.

Four Membranes, the Placenta, and the Umbilical Cord Support the Embryo

Four extraembryonic membranes support, protect, and nourish the embryo. The **yolk sac** manufactures blood cells until about the sixth week, when the liver takes over, and then it starts to shrink. Parts of the yolk sac eventually develop into the intestines and germ cells. By the third week, an outpouching of the yolk sac forms the **allantois,** another extraembryonic membrane. It, too, manufactures blood cells, and it gives rise to blood vessels in the umbilical cord. The **amnion** is the transparent sac that contains the amniotic fluid. This fluid cushions the embryo, maintains a con-

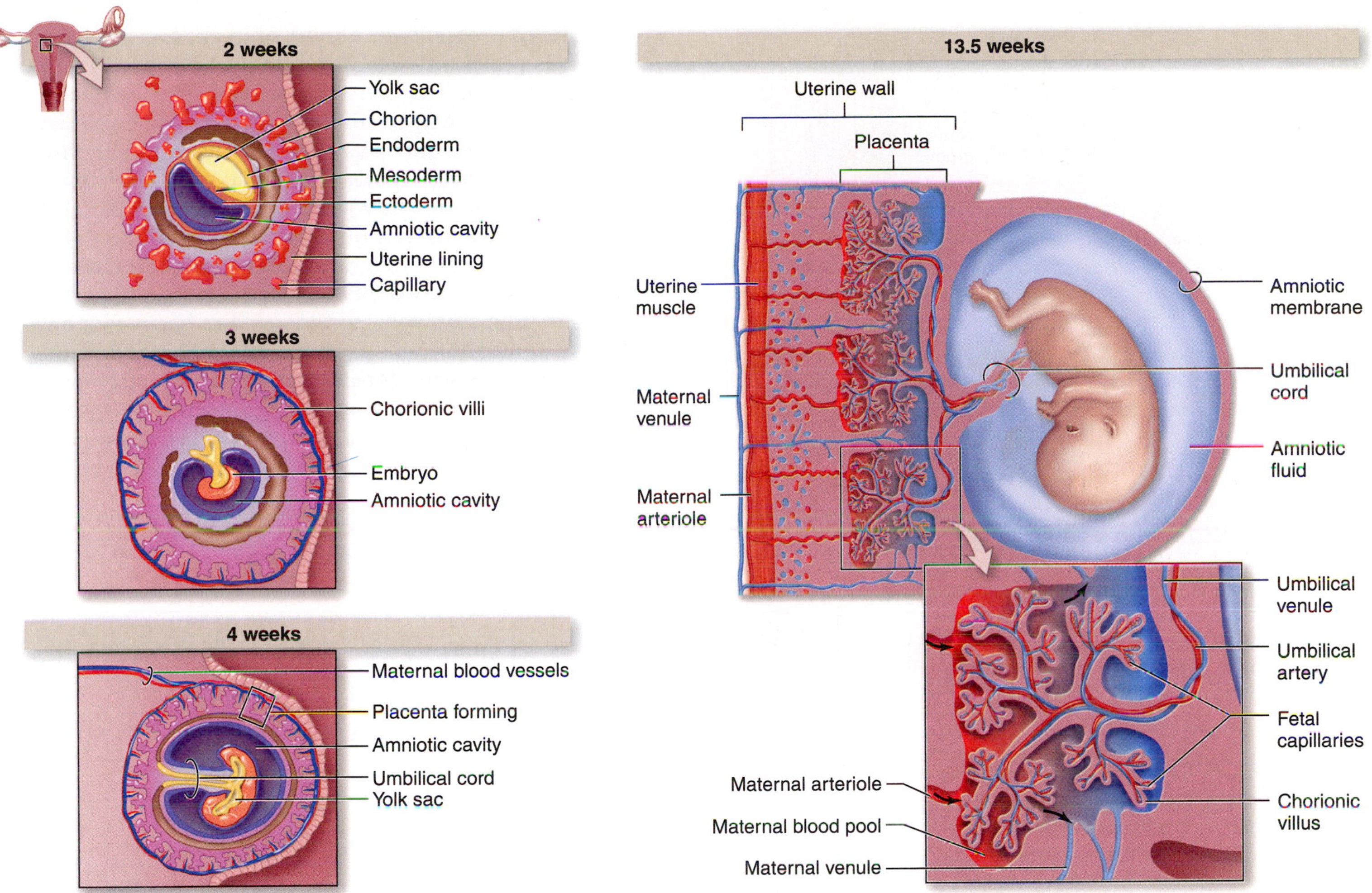

FIGURE 37.13 Development of the Placenta. As the embryo develops, the yolk sac forms blood cells, immune system stem cells, and part of the embryo's digestive system. By week 4, the yolk sac has been incorporated into the umbilical cord. Meanwhile, extensions from the chorion establish the beginning of the placenta. In the placenta, chorionic villi that extend from the embryo contact pools of maternal blood. Nutrients and oxygen pass from the maternal blood to the embryo, while wastes diffuse in the opposite direction.

stant temperature and pressure, and protects the embryo if the woman falls.

The **chorion** is the outermost extraembryonic membrane (**figure 37.13**). By the end of the preembryonic stage, the **chorionic villi,** or fingerlike projections from the chorion, extend farther into the uterine lining. This establishes the beginnings of the **placenta,** a structure that will connect the developing embryo with its mother's uterus. One side of the placenta comes from the embryo; the other side consists of blood from the pregnant woman's circulation. Cells collected from amniotic fluid, or sampled chorionic villi cells, form the basis of many prenatal medical tests.

At 10 weeks (2 weeks into the fetal period) the placenta is completely developed. Nutrients and oxygen diffuse from the woman's circulatory system across the chorionic villi cells to the embryo, and wastes from the embryo's circulation diffuse in the opposite direction. The **umbilical cord** connects the embryo to the placenta. Umbilical cord blood is a rich source of stem cells used to treat an ever-expanding list of disorders.

Organ Formation Begins in the Third Week of Development

As the extraembryonic membranes, placenta, and umbilical cord develop, so does the embryo itself. Beginning during the third week of prenatal development, distinct organs form in the embryo. Over several months, ectoderm cells develop into the nervous system, sense organs, outer skin layers, hair, nails, and skin glands. Splits in the mesoderm form the coelom, which becomes the chest and abdominal cavities. Mesoderm cells develop into bone, muscle, blood, the inner skin layer, and reproductive organs. Endoderm cells form the organs and the linings of the digestive and respiratory systems. **coelom, p. 450**

During the third week of development, a furrow called the **primitive streak** appears along the back of the embryonic disc, forming a longitudinal axis around which other structures organize as they develop (**figure 37.14**). For example, the primitive streak gives rise to the **notochord,** a flexible rodlike structure that forms the basic framework of

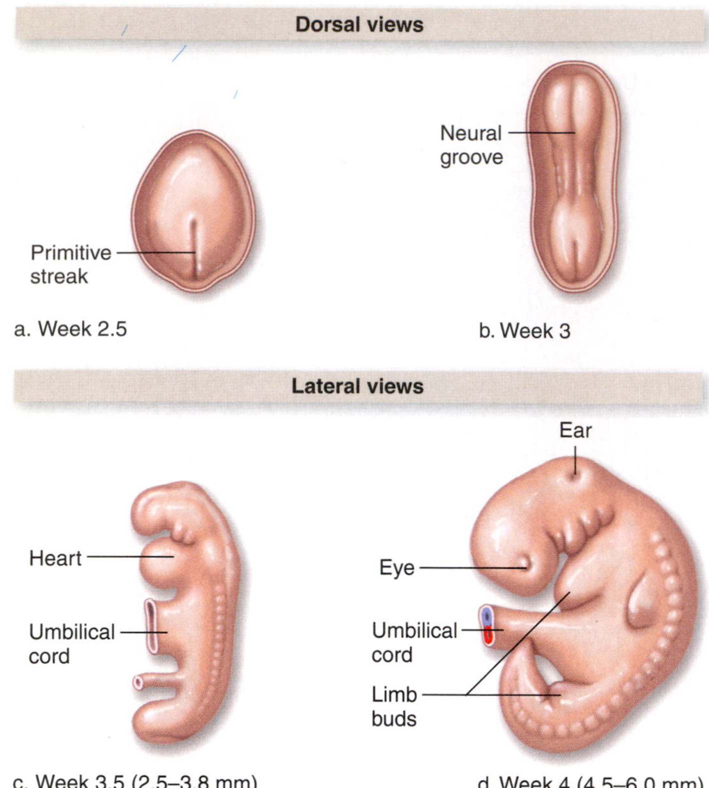

FIGURE 37.14 Early Embryos. It takes about a month for the embryo to take on typical vertebrate features. (**A**) At first, all that can be distinguished is the primitive streak. (**B**) At a signal from the notochord, ectoderm folds into the neural tube, which will gradually form the brain and spinal cord. (**C**) The heart becomes prominent during week 3. (**D**) By week 4, the embryo is beginning to look like a typical vertebrate.

the vertebral column. The notochord also induces overlying ectoderm to differentiate into a hollow **neural tube,** which eventually develops into the brain and spinal cord. Then the central nervous system begins to elaborate. At about the same time, a reddish bulge containing the heart appears. It begins to beat around day 22. **notochord, p. 478; central nervous system, p. 588**

The fourth week of the embryonic period is a time of rapid growth and differentiation. Blood cells begin to form and to fill developing blood vessels. Immature lungs and kidneys appear. If the neural tube does not close normally at about day 28, a neural tube defect such as spina bifida results. (In a neural tube defect, nervous tissue protrudes from an open area of the spine, causing paralysis from the site downward.) Small buds appear that will develop into arms and legs. The 4-week embryo has a distinct head and jaw and early evidence of eyes, ears, and nose. The digestive system appears as a long, hollow tube that will develop into the intestines. A woman carrying this embryo, which is now only about 6 mm (1/4 inch) long, may suspect that she is pregnant because her menstrual period is about 2 weeks late.

By the fifth week, the embryo's head appears disproportionately large. Limbs extending from the body end in platelike structures. Tiny ridges run down the plates, and by week 6, the ridges deepen as certain cells die, molding fingers and toes. The eyes open, but they do not yet have eyelids or irises. Cells in the brain are rapidly differentiating. The embryo is now about 1.3 cm (0.5 inch) from head to buttocks.

All early embryos have unspecialized reproductive structures. At week 7, however, a gene on the Y chromosome, called *SRY* (for "sex-determining region of the Y"), is activated in male embryos. Hormones then begin to stimulate development of male reproductive organs and glands. If there is no *SRY* gene (or if it never switches on), female reproductive structures develop (**figure 37.15**). Overall, many enzymes and membrane proteins participate in sex determination. This chapter's Burning Question: Do human

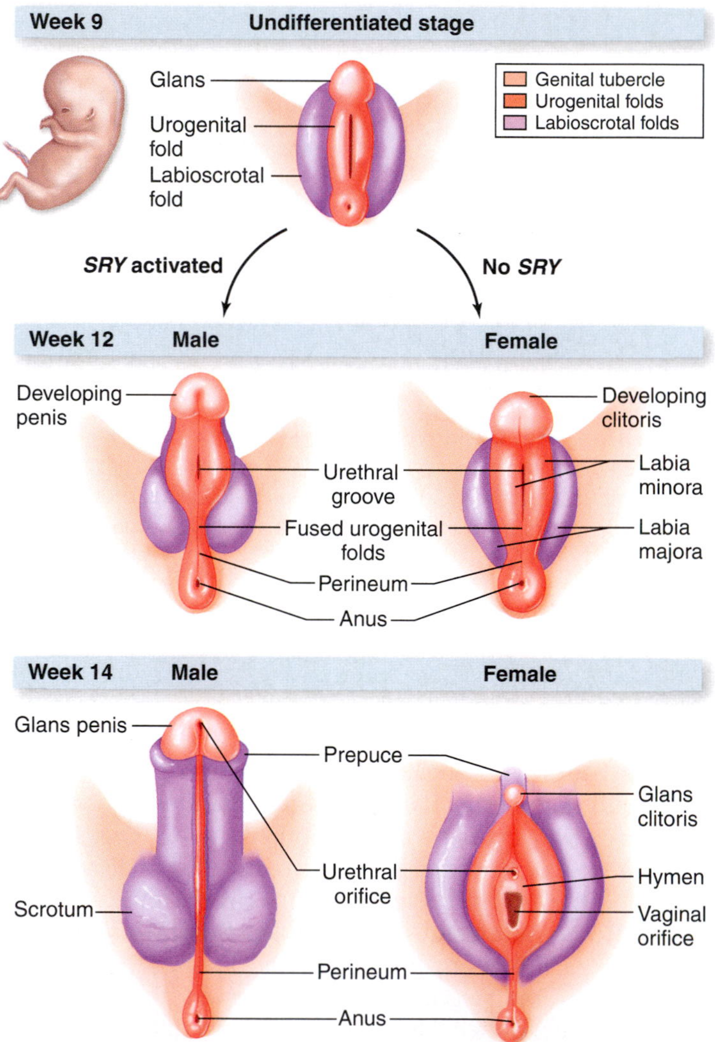

FIGURE 37.15 Development of Male and Female Genitalia. Until about week 9 of development, embryos have undifferentiated genitalia. If the *SRY* gene on the Y chromosome is activated, development continues as a male. Otherwise, female genitalia develop.

hermaphrodites exist? describes how abnormalities in these molecules cause a condition called intersex.

During the seventh and eighth weeks, a cartilage skeleton appears. The placenta is now almost fully formed and functional, secreting estrogen and progesterone that maintain the blood-rich uterine lining. The embryo is about the size and weight of a paper clip. By the end of the eighth week, all organ systems are in place, and the embryonic stage of prenatal development is complete. The offspring is now a fetus.

D. Organ Systems Become Functional in the Fetal Period

The third stage of prenatal development is the **fetal period,** lasting from the beginning of the ninth week through the full 38 weeks of development. The fetal period therefore begins near the end of the first trimester (3-month period) of pregnancy and ends 9 months after fertilization. During this time, the fetus grows considerably (**figure 37.16**). Organs begin to function and coordinate, forming organ systems.

As the first trimester progresses, the body proportions of the fetus begin to appear more like those of a newborn. Bone begins to form and will eventually replace most of the cartilage, which is softer. Soon, as the nerves and muscles coordinate, the fetus will move its arms and legs. Physical differences between the sexes are usually detectable by ultrasound after the twelfth week. From this time, the fetus sucks its thumb, kicks, and makes fists and faces, and baby teeth begin to form in the gums. More than half of all pregnant women experience the nausea and vomiting of morning sickness in their first trimester.

During the second trimester, body proportions become even more like those of a newborn. By the fourth month, the fetus has hair, eyebrows, lashes, nipples, and nails. Bone continues to replace the cartilage skeleton. The fetus's muscle movements become stronger, and the woman may begin to feel a slight fluttering in her abdomen. By the end of the fifth month, the fetus curls into the classic head-to-knees position. As the second trimester ends, the woman feels distinct kicks and jabs and may even detect a fetal hiccup. The fetus is now about 30 cm (12 inches) long.

In the final trimester, fetal brain cells rapidly connect into networks, and organs differentiate further and grow. A layer of fat develops beneath the skin. The digestive and respiratory systems mature last, which is why infants born prematurely often have difficulty digesting milk and breathing. The fetus may move vigorously, causing back pain and frequent urination in the woman as it presses against her bladder. About 266 days (38 weeks) after a single sperm burrowed into an oocyte, a baby is ready to be born.

E. Muscle Contractions in the Uterus Drive Labor and Childbirth

Labor refers to the strenuous work a pregnant woman performs in the hours before giving birth. Labor may begin with an abrupt leaking of amniotic fluid as the fetus presses down and ruptures the sac ("water breaking"). Labor may also begin with a discharge of blood and mucus from the vagina, or a woman may feel mild contractions in her lower abdomen about every 20 minutes.

As labor proceeds, hormones prompt the smooth muscle that makes up the wall of the uterus to contract with increasing frequency and intensity. During the first stage of labor, the cervix dilates (opens) a little more each time the baby presses against it. By the end of the first stage of labor, the cervix stretches open to about 10 cm. The second stage

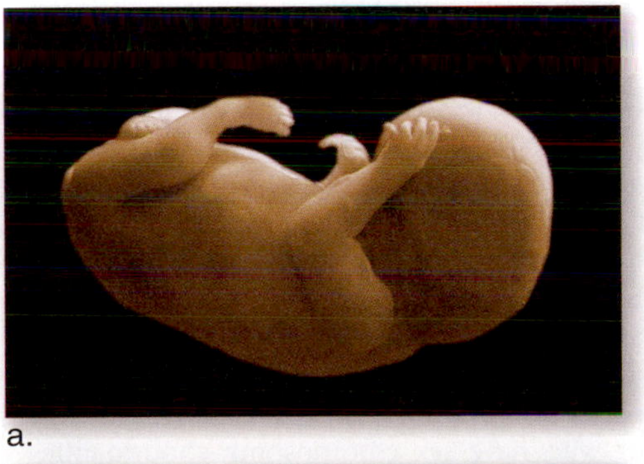
a.

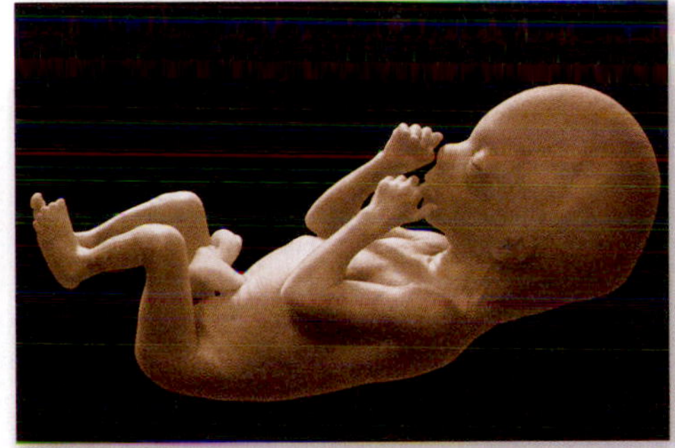
b.

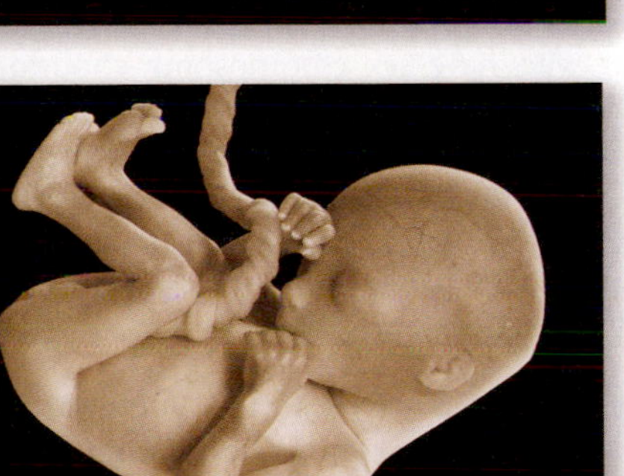
c.

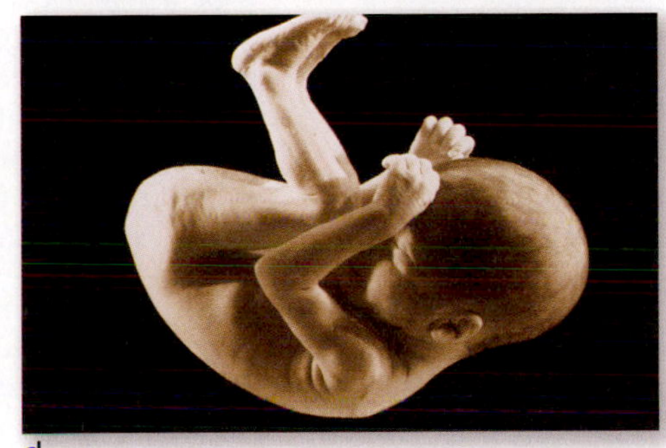
d.

FIGURE 37.16 The Fetal Period. The fetal period lasts from week 9 until birth. These photographs show fetuses at 9 weeks, 14 weeks, 16 weeks, and 7 months gestation.

of labor is delivery (**figure 37.17**), during which the baby descends through the cervix and vagina and is born. In the third and last stage of labor, the uterus expels the placenta. (A baby may also be delivered surgically in a procedure called a cesarean section.)

Sometimes, a pregnancy ends with a premature birth. An infant is premature if it is born before completing 35 weeks (about 8 months) of gestation. A fetus born before 22 weeks of gestation is not viable, but babies born after that time may survive. Premature infants are at risk for many health problems, so they typically spend their first weeks or months in incubators at a neonatal intensive care unit. The incubators control the temperature and protect the infants from infection. **premature babies, p. 675**

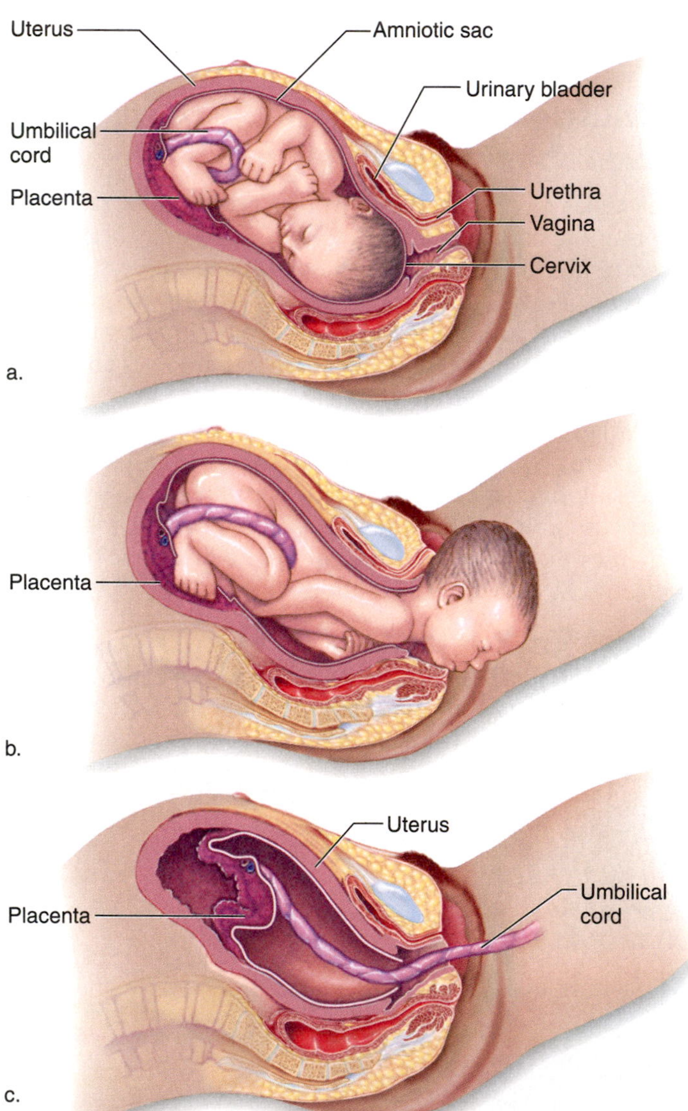

FIGURE 37.17 Labor and Birth. (**A**) About 2 weeks before birth, the fetus "drops" in the woman's pelvis, and the cervix may begin to dilate. (**B**) At the onset of labor, the amniotic sac may break as the baby begins to emerge. (**C**) The baby is pushed out of the birth canal, followed by the placenta and extraembryonic membranes.

Summary *During fertilization, the sperm cell's nucleus enters the egg cell. The zygote divides rapidly, forming a solid ball of cells that hollows out. Three cell layers form and gradually fold into an embryo. The structures that support the embryo develop during the embryonic stage, as do the foundations of all major organ systems. As the second and third trimesters progress, fetal organ systems develop. During childbirth, muscle contractions expel the fetus and placenta from the uterus.*

37.4 MASTERING CONCEPTS

1. What is the role of the acrosome in fertilization?
2. What is the relationship among the zygote, morula, blastocyst, inner cell mass, and gastrula?
3. What is implantation, and when does it occur?
4. Which supportive structures develop during the embryonic period? What are their functions?
5. When do sex differences appear, and what triggers them?
6. What are the events of the second and third trimesters?
7. Which events make up the three stages of labor?

Burning Questions

Do human hermaphrodites exist?

Many people believe that there are just two kinds of people: males and females. But the truth is considerably more interesting. The development of genitalia actually depends on a complex interplay of genes, hormones, and embryonic tissues.

In a child conceived as a boy, genes on the Y chromosome signal the embryonic testes to produce testosterone and other masculinizing hormones (androgens). If cells in the early embryo have a membrane receptor that binds to the androgens, the fetus develops normal male genitalia. Sometimes, however, a child inherits a mutated version of the gene encoding the receptor. The testes produce androgens, but the target tissues cannot bind to the hormones because the receptor is misshapen or absent. This condition is called androgen insensitivity syndrome. The child (conceived as a male) may develop external genitalia that appear female, ambiguous, or male.

Androgen insensitivity syndrome is just one of many so-called *intersex* conditions, in which a person's anatomy does not match the "standard" male or female. Others include 5-alpha reductase deficiency, Klinefelter syndrome, and Turner syndrome. Intersex conditions affect people who are genetically male (XY) or female (XX) or who have sex chromosome abnormalities (such as XO or XXY). **sex chromosome abnormalities, p. 186**

Although it is tempting to confuse the terms *intersex* and *hermaphrodite*, they are not the same. A true hermaphrodite, such as a flatworm or earthworm, is an organism that produces both male and female sex cells. In humans, this is impossible, and there are no human hermaphrodites.

Have a Burning Question of your own? Submit it to marielle_hoefnagels@mcgraw-hill.com for possible inclusion in future editions of this book!

37.5 Birth Defects Have Many Causes

The birth of a live, healthy baby seems against the odds, considering the complexity of human development from conception. Of every 100 egg cells exposed to sperm, 84 are fertilized. Of these, 69 implant in the uterus, 42 survive a week or longer, 37 survive 6 weeks or longer, and only 31 are born alive. Of those that do not survive, about half have chromosomal abnormalities too severe to maintain life.

About 97% of newborns are apparently normal. In the remaining cases, genetic abnormalities, vitamin deficiencies, toxins, or viruses disrupt prenatal development and cause a **birth defect**—any abnormality that causes death or disability in the child.

Some birth defects result from a chromosomal abnormality or a faulty gene that acts during prenatal development. An extra copy of chromosome 21, for example, is the most common cause of Down syndrome. Faulty genes cause Tay-Sachs disease, phenylketonuria, and other metabolic disorders. **trisomy 21, p. 186**

Teratogens are substances that cause birth defects. For example, a drug called thalidomide was once prescribed as a treatment for morning sickness in pregnant women. Unfortunately, it also caused severe limb shortening in the children of women who took thalidomide (**figure 37.18**). People may also encounter teratogens in the workplace. Women who work with textile dyes, lead, some photographic chemicals, semiconductor materials, mercury, and cadmium face increased risk of miscarriage and birth defects in their children.

Several other teratogens include:

- **Alcohol:** A pregnant woman who consumes alcohol risks fetal alcohol syndrome. A child with fetal alcohol syndrome has a small head, misshapen eyes, and a flat face and nose. The child has impaired intellect, ranging from minor learning disabilities to mental retardation. Because everyone metabolizes alcohol slightly differently, physicians advise all pregnant women to avoid alcohol.
- **Cigarettes:** Smoking during pregnancy increases risk of miscarriage, stillbirth, and prematurity. Carbon monoxide crosses the placenta and robs rapidly growing fetal tissues of oxygen. Other chemicals in cigarette smoke prevent nutrients from reaching the fetus. The placentas of women who smoke lack important growth factors, thus lowering birth weight.
- **Excess or insufficient vitamins:** The acne medicine isotretinoin (Accutane), derived from vitamin A, causes miscarriages and defects of the heart, nervous system, and face. Excess vitamin C can also harm a fetus. Vitamin deficiencies can also cause birth defects. Neural tube defects such as spina bifida, for example, are associated with insufficient folic acid in the mother's diet. **vitamins, p. 703**
- **Starvation:** Inadequate nutrition during pregnancy increases the incidence of miscarriage and damages the placenta, causing low birth weight, short stature, tooth decay, delayed sexual development, learning disabilities, and possibly mental retardation. Starvation in the uterus greatly raises the risk of developing obesity, type II diabetes, and other conditions later in life. **starvation, p. 705**
- **Viral infection:** HIV, the virus that causes AIDS, infects 15 to 30% of infants born to HIV-positive women. This risk drops sharply if an infected woman takes antiviral drugs while pregnant. Fetuses infected with HIV are at risk for low birth weight, prematurity, and stillbirth. The virus that causes rubella (German measles) and herpes simplex viruses can also harm a fetus and newborn.

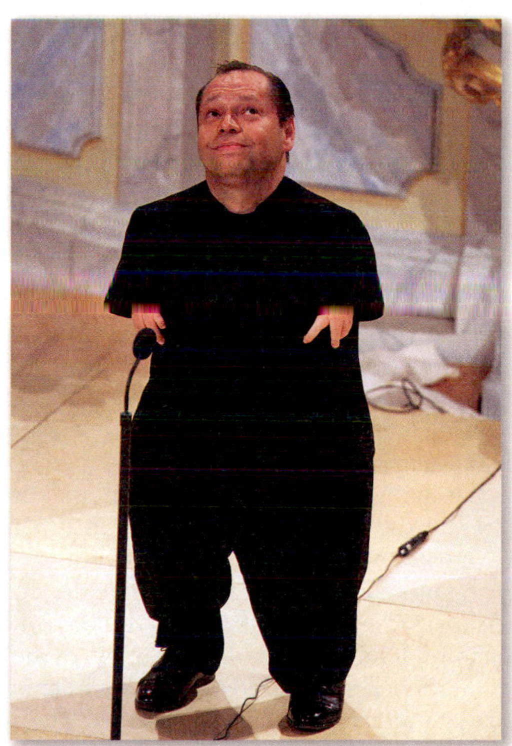

FIGURE 37.18 Teratogenic Effects of Thalidomide. Grammy award-winning baritone Thomas Quasthoff receives the European Culture Award in 2006. Quasthoff's mother took thalidomide to combat morning sickness early in pregnancy. The limb bones in his arms and legs are unusually short, and his hands resemble flippers.

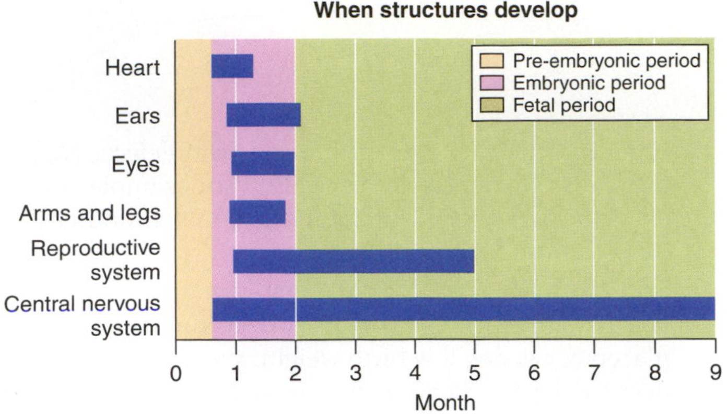

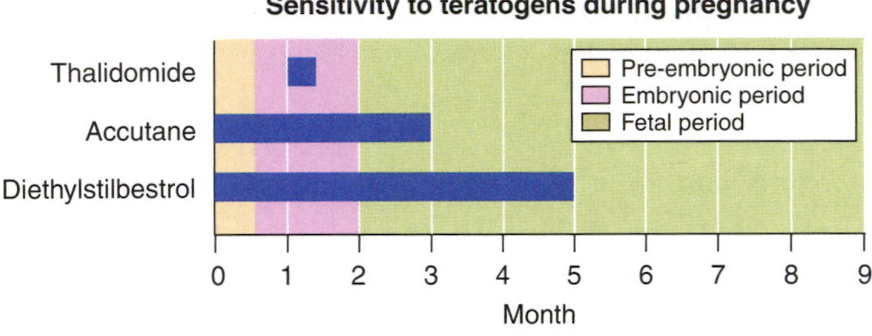

FIGURE 37.19 Sensitive Periods of Development. Body parts differ in their sensitive periods. The type of birth defect resulting from each drug depends on which structures are developing at the time of the exposure.

The "critical period" of a structure is the time during which its development is most susceptible to damage by a genetic abnormality, toxin, or virus (**figure 37.19**). About two thirds of birth defects stem from a disruption during the embryonic period, because developing organs are especially sensitive to damage. The brain is vulnerable to damage throughout prenatal development, as well as during the first 2 years of life. Because of the brain's long critical period, many birth defect syndromes include mental retardation. The continuing sensitivity of the brain after birth explains why toddlers who accidentally ingest lead-based paint suffer impaired learning.

Summary *Chromosomal abnormalities, genetic mutations, dietary deficiencies, or external conditions cause birth defects. Damage to the embryo or fetus may occur early or late in development, and the effects range from mild to severe.*

37.5 MASTERING CONCEPTS

1. What are some examples of genetic conditions that cause birth defects?
2. What is a teratogen? Name some examples.
3. What is the critical period in the development of a structure?

37.6 INVESTIGATING LIFE
Infertility Clues in the Sperm of the Worm

What is something you desperately want but cannot have? Money? A better car? Better health? Less stress? One more chance to talk to a loved one who has passed away?

For millions of couples, the answer is "a baby." About 12% of women of childbearing age in the United States have trouble conceiving a child. Assisted reproductive technologies such as *in vitro* fertilization have helped many couples to conceive, but many people eventually give up. The expense and emotional rollercoaster of infertility treatment simply become overwhelming.

Knowing more about the causes of infertility could lead researchers to new or better ways to help couples conceive. For about one third of infertile couples, the origin of the problem is unclear. Another one in three couples are infertile because of low-quality sperm. Either the man produces too few gametes, or the ones he does produce do not swim correctly and are therefore unable to reach his partner's oocyte.

One way to approach male infertility is to focus on everything that must go right for a man to produce viable sperm cells. The correct packaging of genetic material, for example, is one essential element of sperm quality. A catalog of all DNA-associated proteins in sperm might help researchers identify common problems and eventually lead to new solutions for male infertility.

Biologist Diana Chu of San Francisco State University, along with colleagues at the Scripps Research Institute and the University of California at Berkeley, recognized the importance of DNA packing in male fertility. But the human genome encodes well over 20,000 proteins. Finding just the ones associated with sperm DNA is a bit like searching for the proverbial needle in a haystack. Moreover, studying the step-by-step production of gametes in humans is difficult, because it requires uncomfortable surgeries that expose patients to infection and other risks.

The researchers have therefore turned to a surrogate study animal, the nematode worm *Caenorhabditis elegans*. The differences between humans and the microscopic *C. elegans* seem to outnumber the similarities. But the worm has many useful features. Like us, they are sexually reproducing animals. In addition, researchers understand the worm's reproductive biology in detail, and the *C. elegans* genome has been sequenced. The animal has a short life cycle and is easy to maintain in the lab, making the study of its genes relatively easy and cheap.

Chu and her team set out to develop a list of proteins that fit two criteria: they are essential for sperm production, and they do not occur in the worm's other cell types. A process of elimination helped the researchers achieve their objective. The first step was to isolate DNA from *C. elegans* cells that were in the process of becoming either sperm or egg cells. They then used a complex protein identification technology to identify 1099 DNA-associated proteins in developing sperm and 812 in developing egg cells.

The researchers next narrowed their search to the 502 proteins that were most abundant. After eliminating the 370 proteins that occurred in both sperm and egg cells, the team had a manageable list of 132 that should be unique to sperm production (**figure 37.20**). The researchers used fluorescent microscopy to confirm that some of the proteins were physically associated with DNA in only sperm cells. **Figure 37.21** shows one example.

But not every protein unique to sperm is necessarily essential for fertility. To find out which of the 132 proteins are required for reproduction, Chu and her team used a technique called RNA interference (see figure 12.22). First, they determined the sequences of the 132 genes encoding the target proteins. From that information, they could figure out the sequence of the messenger RNA (mRNA) transcribed from each gene. They then created RNA strands complementary to each of these mRNA molecules. The name *RNA interference* comes from the base-pairing interaction between mRNA and a complementary strand. When the two molecules bind together in a cell, the resulting double-stranded nucleic acid cannot be translated at a ribosome. The introduced RNA strand therefore "interferes" with the production of a protein.

For each of the 132 mRNA molecules, each corresponding to one gene, they injected complementary RNA into separate virgin larval worms. They then observed each worm's reproductive success. Of the 132 genes, 50 were deemed essential to reproduction. Researchers are still working to discover the function of each one.

What does *C. elegans* have to do with our own reproductive problems? The nematode worm is an animal, so it

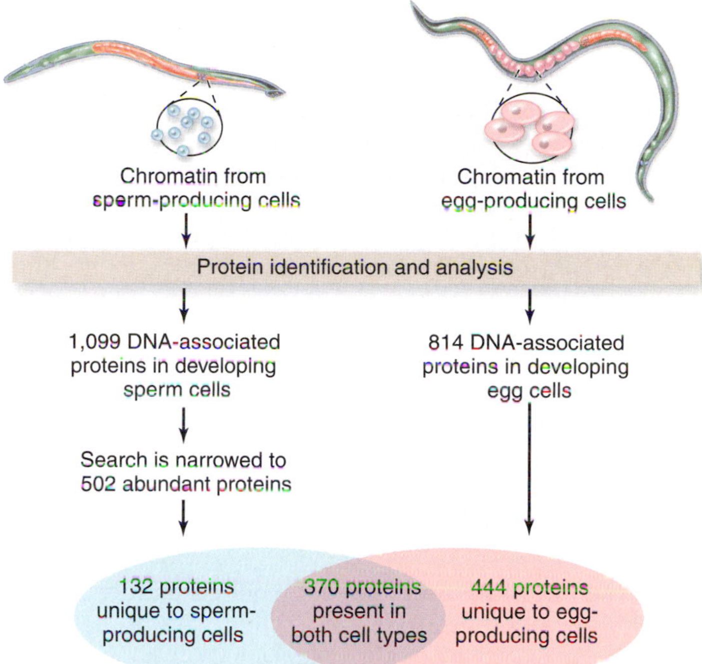

FIGURE 37.20 Protein Hunting Strategy. This diagram shows how researchers narrowed their search for sperm-specific proteins from 1099 candidates to a manageable 132. **Question:** Think of two other specific cell types that might express different proteins. Draw a diagram similar to this one that depicts a possible strategy you could use to find proteins that are unique to one of the two cell types.

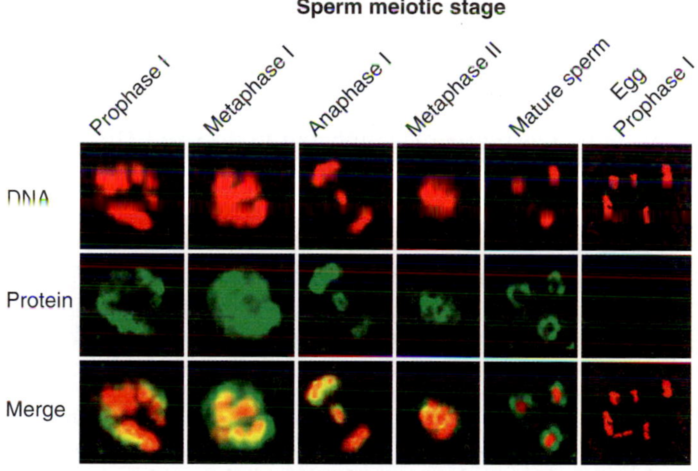

FIGURE 37.21 You Light up My Cells. Antibodies tagged with fluorescent markers were used to check the location of DNA (*red*) and associated proteins (*green*) in sperm and egg cells during multiple stages of meiosis. This example shows a protein that encases the chromosomes during meiosis in sperm-producing cells. The protein does not occur in egg-producing cells, although the cells clearly contain DNA.

shares much of our evolutionary history. Indeed, at least 70 of the 132 worm genes have versions in the human genome, providing an excellent place to start looking for fertility-related genes and proteins in human males. Ironically, the same research could also lead to new types of male contraceptives. Theoretically, a "male birth control pill" could stop the production of sperm, perhaps by preventing DNA-associated proteins from carrying out their functions.

Thanks to Chu and her colleagues, millions of couples who desperately want a child may one day get their wish. Who would have thought that the fulfillment of their hearts' desires might lie—of all places—in the sperm of a worm?

Chu, Diana S., Hongbin Liu, Paola Nix, et al. 7 September 2006. Sperm chromatin proteomics identifies evolutionarily conserved fertility factors. *Nature*, vol. 443, pages 101–105.

CHAPTER SUMMARY

37.1 Animal Development Begins with Reproduction

- **Asexual reproduction** does not require a partner and yields identical offspring.
- In **sexual reproduction,** two haploid **gametes** unite and form a **zygote,** the first cell of a new offspring. **Fertilization** may occur in the environment (**external fertilization**) or inside an animal's body (**internal fertilization**).
- The **reproductive systems** of both males and females include **gonads,** which house the **germ cells** that give rise to gametes. Other sex organs deliver or nurture the gametes. Males and females have different **secondary sex characteristics,** traits that do not directly participate in reproduction.

37.2 Males Produce Sperm Cells

A. Male Reproductive Organs Are Inside and Outside the Body

- Developing sperm originate in **seminiferous tubules** within the paired **testes.** Also inside the testes are **sustentacular cells** that surround the seminiferous tubules and **interstitial cells** that secrete hormones. A **testicle** consists of one testis plus the surrounding **scrotum.**
- Sperm travel through the **epididymis, vas deferens,** and **ejaculatory duct,** and they exit the body with **semen** through the **urethra** (within the **penis**) during **orgasm** and **ejaculation.**
- The **prostate gland, seminal vesicles,** and **bulbourethral glands** add secretions to semen.

B. Spermatogenesis Yields Sperm Cells

- **Spermatogenesis** begins with diploid **spermatogonia,** which divide mitotically to yield a stem cell and a **primary spermatocyte.** The first meiotic division produces two haploid **secondary spermatocytes.** In meiosis II, the secondary spermatocytes divide, yielding four **spermatids.** The spermatids develop into **spermatozoa,** or mature sperm cells.
- A mature sperm cell has a tail and a head covered with a caplike **acrosome.**

C. Hormones Influence Male Reproductive Function

- **Gonadotropin-releasing hormone (GnRH)** from the hypothalamus stimulates the anterior pituitary gland to release **follicle-stimulating hormone (FSH)** and **luteinizing hormone (LH).** In males, these hormones affect the testes, triggering the release of **testosterone** necessary for sperm formation and the development of secondary sex characteristics.

37.3 Females Produce Egg Cells

A. Female Reproductive Organs Are Inside the Body

- Egg cells (and their nourishing **follicle cells**) originate in the **ovaries.** Each month after puberty, one ovary releases an egg cell into a **uterine tube,** which leads to the **uterus.** A blood-rich **endometrium** lines the inside of the uterus. The **cervix** leads to the **vagina,** which connects the uterus with the outside of the body.

- The **vulva,** or external genitalia, consists of the labia, **clitoris,** and vaginal opening. **Breasts** deliver milk to infants.

B. Oogenesis Yields Egg Cells

- In **oogenesis, oogonia** divide mitotically to form two **primary oocytes.** In meiosis I, the cytoplasm of the primary oocyte divides unevenly as it splits into one large, haploid **secondary oocyte** and a much smaller **polar body.** In meiosis II, the secondary oocyte again divides unequally, yielding the large ovum and another small polar body.
- Meiosis in the female begins before birth and completes at fertilization.

C. Hormones Influence Female Reproductive Function

- The ovaries secrete **estrogen** and **progesterone,** hormones that stimulate development of female sexual characteristics. Together with GnRH, FSH, and LH from the hypothalamus and anterior pituitary gland, these hormones control the **ovarian cycle** and the **menstrual cycle.**
- After **ovulation,** the ruptured follicle develops into the **corpus luteum,** which secretes hormones that prepare the body for pregnancy. If pregnancy does not occur, the endometrium is shed in menstrual flow.

37.4 The Human Infant Begins Life as a Zygote

A. Fertilization Joins Genetic Packages and Initiates Pregnancy

- In fertilization, a sperm cell burrows through the two layers surrounding a secondary oocyte. The two united cells constitute the diploid zygote.

B. Preembryonic Events Include Cleavage, Implantation, and Gastrulation

- The **preembryonic** stage lasts from conception until the end of the second week of development.
- After fertilization, **cleavage** divisions produce the **morula.** Between days 3 and 6, the morula arrives at the uterus and hollows to form a **blastocyst.** An outer layer of cells (the trophoblast) and an **inner cell mass** form. **Implantation** occurs between days 6 and 14.
- During the second week, the amniotic cavity forms as the inner cell mass flattens, forming the **embryonic disc.** The **primitive streak** appears. Ectoderm and endoderm form, and then mesoderm appears, establishing the three primary germ layers of the **gastrula.**

C. Organs Take Shape During the Embryonic Stage

- The **embryonic** stage lasts from the second week through the eighth week of development.
- **Chorionic villi** extending from the **chorion** start to develop into the **placenta.** The **yolk sac, allantois,** and **umbilical cord** form as the **amniotic sac** swells with fluid.
- Cells of the chorion secrete **human chorionic gonadotropin (hCG),** which prevents menstruation.
- Organs form throughout the embryonic period. Gradually, structures appear, including the **notochord, neural tube,** arm and leg buds, heart, facial structures, skin specializations, sex organs, and skeleton.

D. Organ Systems Become Functional in the Fetal Period

- Structures continue to elaborate during the **fetal period,** which lasts from the ninth week of development until the baby is ready for birth.

E. Muscle Contractions in the Uterus Drive Labor and Childbirth

- Labor begins as the fetus presses against the cervix. Uterine contractions expel the baby and placenta.

37.5 Birth Defects Have Many Causes

- Genetic abnormalities, dietary deficiency, or exposure to **teratogens** such as chemicals or viruses can cause **birth defects.**
- Each body structure has a different critical period, the time in development during which it is especially vulnerable to teratogens.

37.6 Investigating Life: Infertility Clues in the Sperm of the Worm

- The isolation of proteins associated with the DNA of sperm cells in the nematode worm *C. elegans* may lead to new contraceptives or treatments for male infertility.

MULTIPLE CHOICE QUESTIONS

1. What is a zygote?
   a. The male gamete
   b. The product of fertilization
   c. The female germ cell
   d. The female gamete

2. What is the relationship between a primary spermatocyte and a spermatogonium?
   a. They are genetically identical.
   b. The spermatocyte is haploid and the spermatogonium is diploid.
   c. The spermatocyte has a flagellum.
   d. Both b and c.

3. Why is a fertilized polar body unable to produce a fetus?
   a. Because it has too few chromosomes.
   b. Because it has too many chromosomes.
   c. Because it lacks cytoplasm and organelles.
   d. Because it carries excess RNA.

4. The acrosome reaction allows a sperm cell to penetrate the
   a. proteins and carbohydrates encasing the oocyte.
   b. follicular cells surrounding the oocyte.
   c. plasma membrane of the secondary oocyte.
   d. Both a and b.

5. Why do the cells produced during cleavage become progressively smaller?
   a. Cleavage is so rapid there is less time for DNA synthesis.
   b. The main source of cytoplasm and organelles is the fertilized egg.
   c. Mitosis is shorter.
   d. Cell walls prevent the cells from growing.

6. A human embryo forms from the
   a. chorion.
   b. placenta.
   c. inner cell mass.
   d. trophoblast.

7. What is the function of the umbilical cord?
   a. To transport nutrients to the fetal digestive system.
   b. To transfer nutrients and oxygen from maternal blood to fetal blood.
   c. To carry the blood of the mother to the developing fetus.
   d. To transfer nutrients from the yolk to the developing fetus.

8. The ______ first establishes the head-to-tail axis of a developing embryo.
   a. primitive streak
   b. notochord
   c. neural tube
   d. spinal cord

9. A chemical that can cause a birth defect is known as a
   a. toxin.
   b. trisomy.
   c. mutagen.
   d. teratogen.

10. The hormones that determine the development of male or female reproductive structures first become active during
    a. fertilization.
    b. the preembryonic stage.
    c. the embryonic stage.
    d. the fetal period.

TESTING YOUR KNOWLEDGE

1. What are the advantages and disadvantages of asexual and sexual reproduction? Of internal and external fertilization?
2. How are the human male and female reproductive tracts similar, and how are they different? How are the structures of the testis and ovary similar and different?
3. Another name for LH in the male is interstitial cell stimulating hormone (ICSH). What does this hormone stimulate the interstitial cells to do?
4. What happens if two sperm fertilize the same egg cell? If two sperm fertilize two egg cells?
5. How does menstruation stop when a woman becomes pregnant?
6. How are the timetables different for oogenesis and spermatogenesis in humans?
7. Is each of the following cell types haploid or diploid?
   a. an oogonium
   b. a primary spermatocyte
   c. a spermatid
   d. a secondary oocyte
   e. a polar body derived from a primary oocyte
8. What events must take place for fertilization to occur?
9. Arrange these structures from youngest to oldest: morula, gastrula, zygote, fetus, blastocyst.
10. What are the events of childbirth?
11. List some of the causes of birth defects.

THINKING AS A SCIENTIST

1. How do protein gradients cause cells to differentiate?
2. How do the structures of the male and female human gametes aid them in performing their functions?
3. This chapter used the term *villus* in describing part of the chorion; the same term appeared in chapter 34's description of the small intestine. How are the chorionic and intestinal villi similar and different in structure and function?
4. What technology would be necessary to enable a fetus born in the fourth month to survive in a laboratory setting?
5. What kinds of studies and information would be necessary to determine whether exposure to a potential teratogen during a war can cause birth defects a year later? How would such an analysis differ if it were a man or a woman who was exposed?
6. One risk factor for developing breast cancer or ovarian cancer is a family history of either disease. Researchers have identified some alleles associated with inherited breast and ovarian cancers, making genetic tests possible. If you (or a female close to you) had a family history of either cancer, what considerations would help you decide whether to have your DNA tested for these alleles?

Visit www.mhhe.com/hoefnagels for practice quizzes, animations, videos, and activities designed to help you master the material in this chapter.

UNIT 7

CHAPTER 38

Animal Behavior

A Stotting Thomson's Gazelle

What's the Point?

Risky Business: Stotting

It is a hot morning in the African bush, and a herd of Thomson's gazelles grazes on shoots of new grass. The gazelles are alert as they feed. Their ears swivel and their tails twitch away flies. Every now and then one pauses, looks up, and scans the horizon. Death crouches in the tall grasses in the form of a hungry cheetah. The wind shifts, and one gazelle jerks up its head. Then it does something that looks awfully odd. Instead of running away, the gazelle leaps into the air. Its nose points to the ground as it bounces up and down on stiff legs. In a flash it races away, but the odd bouncing, called stotting, seems to be contagious. As they race away, other gazelles also stott, sometimes in midstride. The cheetah charges, but it is a half-hearted attempt, and the spotted cat stops short and watches the herd disappear.

Mobbing is another risky animal behavior, but you can see it closer to home. Terns (a type of gull) will mob any animal that intrudes into their nesting colony, from cats to foxes to humans. When a peregrine falcon sails over a colony of terns, for example, the birds fly up from their nests and the mob joins the attack. Screaming, they dive-bomb the falcon. This truly is risky business. The peregrine is such an agile flier that it is perfectly capable of rolling over on its back in midair and striking at a mobber or grabbing it in lethal talons.

Other birds also mob predators. If a crow spies a great horned owl or red-tailed hawk, it will send a loud alarm call. The sound draws other crows and often smaller birds that gather where the predator is perched. Sometimes a huge mixed flock is recruited. Giving their different alarm calls, the birds settle on tree branches near the predator. Some fly straight at it, veering away only at the last second. Eventually the predator flies away, out of the nesting territories of the bird. They soon lose interest and cease harassing the predator.

Stotting and mobbing are spectacular behaviors, but if you watch any animal long enough, something interesting is bound to happen. It might spin a web, curl into a ball, start barking, whistle, jump into the air, yawn, or swish its tail. Even the most mundane behavior can prompt interesting questions: Why does the animal do that? How does it do that? Was it born knowing how to perform the behavior, or did it learn by imitating other animals?

In pondering such questions, the mind may wander to some of the strange things we do, like kissing, playing basketball, or holding hands. How did those behaviors ever start, and why do they persist? The answers to many such questions remain elusive, but this chapter explores some of what we do know about a wide spectrum of animal behaviors, both simple and complex.

LEARNING OUTLINE

38.1 Animal Behaviors Have Proximate and Ultimate Causes

In earlier times, the rhythms of the natural world were integrated into people's lives, and behaviors of animals were familiar. After all, a hunter's knowledge of an animal's habits might mean the difference between survival and starvation, and dangerous animals might lurk around the bend of any trail. Nowadays, although many people own pets (see Can *You* Relate? Puppy Love), few have firsthand experience with wild animals.

As popular knowledge of animal behavior has moved to the periphery, the science of animal behavior has developed and blossomed into several related disciplines, including behavioral ecology and evolutionary psychology. In the 1930s, the naturalists Niko Tinbergen, Karl von Frisch, and Konrad Lorenz introduced the term **ethology** for the scientific study of animal behavior, especially in the natural environment. An ethologist must understand *how* animals behave, gathered by firsthand observations that do not interfere with the animals (**figure 38.1**). The emphasis in ethological studies, however, is on experiments. Karl von Frisch, for example, used experiments to help decode the meaning of the honeybees' famous "waggle dance."

The study of the causes of animal behavior falls into two general subdivisions. Ethologists may study the **proximate** causes of a behavior; that is, they may ask "how" questions. How do genes and hormones contribute to feelings of attachment to a mate or a baby? How do the nervous and muscular systems translate instinct into action? How does a male songbird learn the songs that attract the females of the species? Researchers who ask proximate questions seek to understand the underlying mechanisms that produce a behavior.

Ethologists studying the **ultimate** causes of an animal's behavior focus on evolutionary explanations—the "why" questions. Why does natural selection favor the production of huge numbers of offspring in some species, but just a few offspring in others? Why do parents risk their lives to keep their offspring safe? The emphasis of these studies is on the evolution of animal behaviors and on their survival or reproductive value.

This chapter begins by examining the range of animal behaviors. Then it delves into genetic influences on animal behavior and puts some familiar behaviors into an evolutionary context.

Summary ***Early people were familiar with common animal behaviors. Modern studies of animal behavior try to determine how and why animals do what they do.***

38.1 MASTERING CONCEPTS

1. What is ethology?
2. Distinguish between the proximate and ultimate causes of an animal's behavior.

FIGURE 38.1 Field Notebook. An ethologist records observations of animal behavior in a journal such as this.

Can *You* Relate?

Puppy Love

If you own—and love—a pet, you may wonder how the animal's experience of your relationship compares with your own. The cat may jump in your lap, curl up, and begin to purr. The dog may greet you exuberantly when you come home. Do these behaviors mean that your pet loves you?

However tempting it is to project love onto our animal companions, it is unscientific to do so. Scientists have no way to measure objectively the love that humans feel, let alone the emotions that other animals may have. Your cat may seem to enjoy her time in your lap, but is she simply seeking a soft, warm, comfortable place to rest? Your dog may seem happy to see you, but perhaps he is simply behaving in ways that he has learned will earn attention or treats.

This question of animal emotions illustrates the sometimes frustrating reality that science cannot answer every question that interests us. We may never have the technology to know for sure what humans feel (although it is interesting to contemplate what life would be like if we could). Our inability to peer into the animal mind adds mystery to our relationships with our nonhuman companions. Perhaps this mystery helps make pet ownership one of the world's most popular hobbies.

38.2 Animal Behaviors Combine Innate and Learned Components

Without a doubt, animals do fascinating and sometimes puzzling things. Consider a praying mantis that you might spot on a bush (**figure 38.2**A). The green, brown, or gray of the insect's body blends almost perfectly into the leafy background. The mantis holds its first pair of legs folded below its head. It looks as though it is praying, but it is actually waiting to snag its prey.

The mantis makes a sudden lunge and impales a beetle on the needle-sharp spines of its forelegs (figure 38.2B). Like a tiny robot, the mantis's head swivels down to its prey. Its mouthparts work swiftly and delicately as they slice into the beetle's neck. It is not long before the beetle stops struggling. Fascinated by this grim little spectacle, you step closer. In a wink, the mantis transforms. Black eyes glare at you from the bush. It takes a moment before you realize that the threatening eyes are merely a pattern on the mantis's wings, which now fan out above its body. You make another movement and the mantis plays another card: it flings its forelegs wide open, threatening a spiny embrace (figure 38.2C). What is the mantis doing? And why?

Before further discussing the mantis's behavior, it would be helpful to introduce the major kinds of animal behaviors. Biologists traditionally classify animal behaviors based on the way they are acquired and how changeable they are.

A. Innate Behaviors Do Not Require Experience

An **innate behavior** is inborn and does not require learning or experience to be performed correctly. Innate behaviors occur in a regular sequence once stimulated, and they are always performed in the same way. Most are related to feeding, communicating, navigating through the environment, responding to predators, capturing prey, or reproducing.

Innate behaviors include reflexes, taxes, and fixed action patterns. A **reflex** is an instantaneous, automatic response to a stimulus. For example, you blink if something suddenly comes near your eye, you jerk your hand away from a hot pot, and your pupil constricts in bright light. A **taxis** (plural: taxes) is a movement toward or away from a stimulus. Given the choice, earthworms crawl toward moist soil, cockroaches seek darkness, and moths are drawn to light.

A **fixed action pattern** is a sequence of innate behaviors that is performed to completion when triggered by a **releaser**—an environmental cue that elicits the behavior. The behaviors of the praying mantis as it caught and ate its prey are fixed action patterns. Tiny, newly hatched mantids display the same prey-catching lunge and prey-handling behavior of fully grown adults. The mantis's startle display is also a fixed action pattern.

a.

b.

c.

FIGURE 38.2 Praying Mantis. (**A**) The position of the insect's front legs gives it the appearance of praying. (**B**) A mantis consumes its prey. (**C**) The defensive display of a praying mantis.

FIGURE 38.3 **Begging Baby Birds.** As a parent approaches, these young birds display a characteristic behavior. They fling their mouths open and screech, begging the parent for food.

A releaser may be a visual, auditory, or chemical stimulus. For example, when a parent bird appears at the nest, the nestlings begin the fixed action pattern of begging for food (**figure 38.3**). Their insistent screeching, the bright markings inside their mouths and throats, and their quivering movements are releasers that stimulate the parent bird to begin its own fixed action pattern. The parent stuffs food into a begging mouth and then flies off to gather more food.

Humans have several innate behaviors. As in all mammals, a touch on a baby's cheek triggers the suckling response. The nuzzling behavior that allows a human infant to locate the nipple of its mother's breast is innate. Laughing, sneezing, yawning, and crying are innate behaviors, as is jerking away from an uncomfortable sensation by reflex. Like many other animals, human babies also have an innate response that causes them to back away from a visual cliff (**figure 38.4**).

FIGURE 38.4 **Backing Away.** Despite the mother's encouragement, most babies will not cross the glass spanning a "visual cliff."

B. Learning Requires Experience

Learned behaviors result from an animal's experiences, and they are more changeable than innate behaviors. For instance, think of how you learned to ride a bicycle. If you were like most beginners, at first you were hopeless at balancing your body and coordinating the actions of your arms and legs. But through the combination of trying, making mistakes, changing your behavior to avoid mistakes, and the satisfaction of mastering the new skill, you learned to ride your bike.

This example illustrates a form of **trial-and-error learning,** in which behaviors that are rewarded (in this case, riding your bike without falling) become more frequent than those that are "punished." Dogs and cats learn by trial and error to beg for table scraps at mealtime, as long as that behavior is reinforced with treats. In your own experience, you have probably learned by trial and error which foods you like best and which ones to avoid.

Imprinting is a kind of rapid learning that occurs only during a restricted time, early in the life of an animal. The animal retains the memory throughout life. In the 1930s, Lorenz found that during a critical period of their lives, goslings would imprint on the first moving object that they encountered. In one experiment, goslings raised in an incubator accepted Lorenz as "mother" and followed him wherever he went (**figure 38.5**A). Animals that have not been properly imprinted, however, may have trouble recognizing their own species when it comes time to mate. For this reason, biologists who raise endangered species in captivity use hand puppets resembling parent birds to feed the young (figure 38.5B).

Habituation is a behavior in which an animal learns *not* to respond to a stimulus. Have you ever noticed how you tune out the familiar sounds, sights, and smells of your own home? A behaviorist would say that you are habituated to your surroundings. A city apartment dweller hardly hears the noise of traffic in the streets, but he may not be able to sleep in the country—the sounds of nocturnal insects will keep him awake. Habituation allows animals to ignore normal stimuli while remaining alert to the out-of-the-ordinary.

C. Innate Behaviors and Learning Are Interrelated

Biologists once placed innate and learned behaviors at opposite ends of a spectrum. Genes were thought to control innate behaviors, whereas learned behaviors supposedly stemmed from experience with the environment. We now know that the reality is considerably more complex. Innate behaviors do not occur without some interaction with the environment, and learning has a strong genetic component.

Innate behaviors and learning are therefore inseparable. Genes control the structure of sensory receptors, the development of muscle cells, and the production of hor-

a.

b.

FIGURE 38.5 Imprinting. (**A**) Konrad Lorenz and his flock of greylag geese. (**B**) A hand puppet modeled on an adult condor is used to feed an endangered condor chick.

mones, pheromones, and neurotransmitters. These molecules form the foundation of *all* behaviors (**figure 38.6**). Genes also direct the overall development of the nervous system, which determines how much any animal can learn. But the expression of those genes is not fixed. The environment provides the external stimuli that trigger each behavior, and it shapes the expression of an individual's genes (see figure 10.15).

The important point is that animal behavior does not really follow a strict "innate or learned" dichotomy. Every animal begins life with a range of possible behaviors, some of which have more genetic involvement than others. As time goes by, the animal's experiences help determine how it actually behaves.

Summary *The spectrum of animal behaviors traditionally stretches from innate behaviors to learned behaviors. Reflexes, taxes, and fixed action patterns are innate behaviors. Learned behaviors include trial-and-error learning, imprinting, and habituation. Innate and learned behaviors are intimately related.*

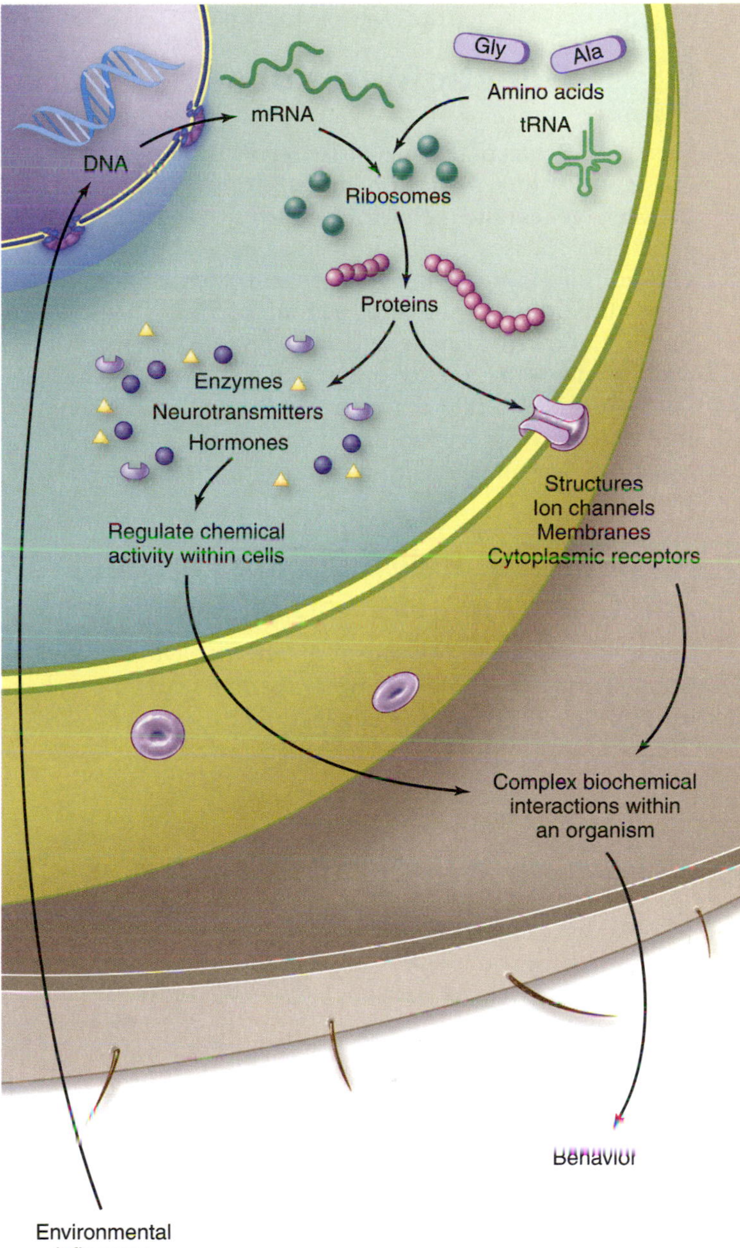

FIGURE 38.6 Genes Influence All Behaviors. Most behaviors require the interaction of multiple proteins, encoded by multiple genes. External stimuli also influence gene expression.

38.2 MASTERING CONCEPTS

1. How did biologists originally define innate and learned behaviors?
2. What is a fixed action pattern, and what is its relationship to a releaser?
3. Define and give examples of reflexes, taxes, trial-and-error learning, imprinting, and habituation.
4. Why is it impossible to classify any behavior as innate or learned?

38.3 Genetic Mutations Can Alter Some Behaviors

Behavioral geneticists delve into the proximate causes of animal behavior, investigating the relationship between genes and behaviors. Much work of behavioral geneticists focuses on the fruit fly, *Drosophila melanogaster*. One common way to find genes in fruit flies is to induce genetic mutations by irradiating a group of flies. The irradiated flies are allowed to mate and lay eggs, and then researchers carefully watch for abnormal behaviors in the offspring. Such studies indicate that genes influence many complex fruit fly behaviors, including foraging (searching for food), courtship, learning, and even memory. **fruit fly, p. 464**

A. Genes Strongly Influence Foraging and Courtship Behaviors of Fruit Flies

Foraging was one of the first fruit fly behaviors traced to genes. Larvae and adults have two food-searching behavior patterns: they are either rovers or sitters (**figure 38.7**). When reared on plates of agar that had two discrete patches of yeast, rover larvae crawled equally within and between the yeast patches, while sitter larvae concentrated on feeding in one patch. Although there is some overlap in the two traits, populations usually average 70% rovers and 30% sitters.

These behavioral differences arise from variation in a gene named *foraging* (*for*). The "rover" allele (for^R) is dominant to the "sitter" allele (for^s). When an environment is crowded, a selective advantage goes to rovers who can move to a different, less-congested patch of yeast. With less crowding, sitters have the selective advantage over rovers, perhaps because they don't waste time and energy wandering around in search of other food patches. **dominant and recessive alleles, p. 196**

A gene named *fruitless* (*fru*, for short) is responsible for most aspects of a male fruit fly's sexual behavior. *Drosophila melanogaster* has six courtship behaviors that occur in sequence as a fixed action pattern (**figure 38.8**). Males with mutated *fru* genes court one another, moving in conga lines several flies long. The gene is active in only about 500 of the 100,000 nerve cells in the brain and nerve cord of the male fly. Research is ongoing, but it seems likely that male sexual behavior occurs after information from sense organs reaches the fruit fly's central nervous system. Neurons that express the protein encoded by *fru* process this sensory information. Sexual behavior may result as these proteins act on muscles and nerves.

Rovers

Sitters

Yeast

Agar

FIGURE 38.7 Foraging Behavior of *Drosophila melanogaster*: Rovers versus Sitters. Rovers tend to move between and within patches of yeast, whereas sitters are more sedentary.

B. Some Human Behaviors Have a Strong Genetic Component

What part do genes play in human behaviors? This question is very difficult to answer, in part because many of our behaviors involve a complex interplay between our surroundings, sensory receptors, central nervous system, and the cells that actually carry out the behavior. In addition, it is obviously unethical to manipulate human genes intentionally as we do in fruit flies.

FIGURE 38.8 Courtship Behavior of *Drosophila melanogaster*. (**A**) Sequence of normal courtship behaviors. (**B**) A line of courting males with mutated *fru* genes.

Even so, you probably have heard or read stories about a newly identified gene "for" some human behavior (see chapter 12's Burning Question: Is there a gay gene?). This language incorrectly implies that genes are responsible for behavior on a one-to-one basis. As chapter 12 describes in detail, a gene is a stretch of nucleotide bases that encodes a protein. That is all a gene does. The protein encoded by a gene may produce the behavior, but only when the organism has all the other biochemical parts of the behavioral puzzle in place. Most often, many genes work together with the environment to influence a behavior.

One line of indirect evidence for the influence of genes on human behavior comes from twin studies (**figure 38.9**). Unlike identical twins, fraternal twins are no more genetically similar than any other siblings in a family. If identical twins are more alike for some trait than are fraternal twins, this is evidence that the characteristic has a genetic component. To rule out the influence of a shared environment, researchers seek identical twins who have been reared apart. **twins, p. 183**

Researchers have used twin studies to examine the heritability of everything from alcoholism to reading ability. Studies on schizophrenia and bipolar disorder, for example, show that genes can predispose people to mental illness, but so can exposure to drugs or stress. From this and other studies of twins reared apart, it is reasonable to conclude that nature (genetics) and nurture (external influences) both influence human behavior. However, there are problems with these studies. Whether reared apart or together, twins once shared the environment of their mother's womb. In addition, shared mitochondrial genes may influence the behaviors of the twins. (See chapter 10's Can *You* Relate? The Roots of Addiction).

FIGURE 38.9 Double the Fun. These identical twin sisters enjoy reading books in the library.

Summary *Behavioral geneticists have demonstrated that much of the behavior of fruit flies is genetically controlled. Research is beginning to reveal the genetic components of behaviors of other animals, including humans.*

38.3 MASTERING CONCEPTS

1. Describe the connection between genes and animal behavior. What is the evidence for that connection?
2. What is the difference between a fruit fly that is a rover and one that is a sitter?
3. Explain the connection between the *fru* gene and courtship behavior in *Drosophila*.
4. What is the evidence that many human behaviors have a genetic component?

38.4 Animals May Live Alone or in Groups

Animals vary widely in their interactions with others. Mountain lions live alone, searching out others only when it is time to mate, whereas giraffes and schooling fish live their entire lives in large groups. This section describes two mammals at opposite ends of this spectrum: the solitary bobcat (*Lynx rufus*) and the highly social wolf (*Canis lupus*).

A. Bobcats Are Solitary Animals

The bobcat (**figure 38.10**) is a predator that lives in most of the continental United States. Bobcats, especially males, spend most of their lives alone. A male and a female come together in the fall or winter to mate. For a few days, the pair will travel, hunt, and eat together. They part company soon after mating, with the male leaving the female's home range, presumably to find another mate. The female gives birth to two or three kittens about 70 days later. She nurses

FIGURE 38.10 The Bobcat, *Lynx rufus*.

them for about 2 months and then begins to wean them and teach them to hunt. When her kittens are between 8 and 11 months old, the mother mates again. Soon she forces the juveniles from her first litter out of her home range. They must wander in search of a suitable habitat where they can set up their own home ranges.

Like many animals, bobcats use **territorial behaviors** to mark and defend a home range against other animals. They deposit urine or feces on top of a pile of dirt and dry leaves, which allows breezes to carry the bobcat's scent. Bobcats also mark trees with vertical scratches tinged with scent from the sweat glands in the bobcat's paws. Other bobcats investigate these territorial markings and evaluate the pheromones in the urine, feces, and claw marks. **Pheromones** are chemicals that are released into the environment and that influence the behavior of other organisms, usually of the same species. For instance, pheromones from a female bobcat will broadcast the news that she is is ready to mate.

Why do adult bobcats go to such lengths to stay away from one another? The answer seems to be related to prey availability. Bobcats usually feed on small mammals, reptiles, and birds. An average female bobcat and three kittens will eat about 3800 rats, 700 rabbits, and 3200 mice in a year. She must compete with other predators such as foxes, weasels, snakes, and coyotes that hunt the same small animals. By excluding other bobcats from her home range, the female bobcat ensures that she and her kittens will have enough to eat. If she did not force juvenile bobcats to leave her home range, they would compete for food and might even kill and eat her next litter of kittens. Solitary behavior therefore improves the bobcat's reproductive success.

FIGURE 38.11 **The Wolf, *Canis lupus*.** A pack of timber wolves feeds on a white-tailed deer.

B. Wolves Are Social Animals

Wolves are among the world's most social mammals, living together in packs that are usually composed of closely related animals. How can these carnivores live together without depleting their ranges of prey? For one thing, wolves generally take large herbivores such as elk, moose, and caribou (**figure 38.11**). Although a lone wolf can bring down an animal of this size, it is easier and safer for a pack to tackle such large prey. After the kill, members of the pack eat according to a strict social order. The alpha male and alpha female, the pack's dominant animals, feed first. Then less dominant members of the pack take their share.

The same social order also governs reproduction within the wolf pack. Only the alpha animals mate and produce young. Other pack members do not reproduce, but help care for the young of the alpha pair, whose genes they also share. The social order of a wolf pack becomes established when wolves are pups, and it is reinforced throughout life by social interactions that include aggressive and submissive behaviors. Tail wagging, ear positions, displays of teeth, and sounds such as whines, snarls, growls, and barks reinforce an individual's position within the pack.

Like bobcats, wolves are territorial animals, but the home ranges they defend belong to the pack, not an individual. A wolf pack's territory may span hundreds of square kilometers. Howls and scents mark the boundaries of the territory, and the pack chases out or kills any "outsider" wolf that wanders in.

C. Kin Selection May Explain Many Altruistic Behaviors

Selfish behaviors foster the survival of the individual at the expense of others. For example, when a wolf pack feeds, each member selfishly fights for a bit of the kill. Some of the pack members then return to the den, where hungry wolf pups come out and greet them. Some of the adults regurgitate food for the pups. This is altruism, a self-sacrificing behavior.

Altruistic behavior is common among humans (**figure 38.12A**). For instance, humans donate blood and organs to their children, spouses, or relatives. Some parents postpone their own schooling and many go hungry so that their children may have food. In humans and other animals, most altruism is expressed toward relatives. Yet altruism also extends to larger social groups and even to communities (see the Burning Question: Why do people engage in rituals that are risky, expensive, painful, and dangerous?).

Altruism is a bit of a puzzle, because it reduces one animal's fitness for the sake of another. One explanation is **kin selection,** an evolutionary mechanism that sacrifices an individual's genes for the sake of the genes it shares with related animals. A bird called a killdeer, for example, will risk her life to perform a "broken wing display" that draws a predator away from her nest or young (figure 38.12B). Kin selection may also explain why many burrowing animals

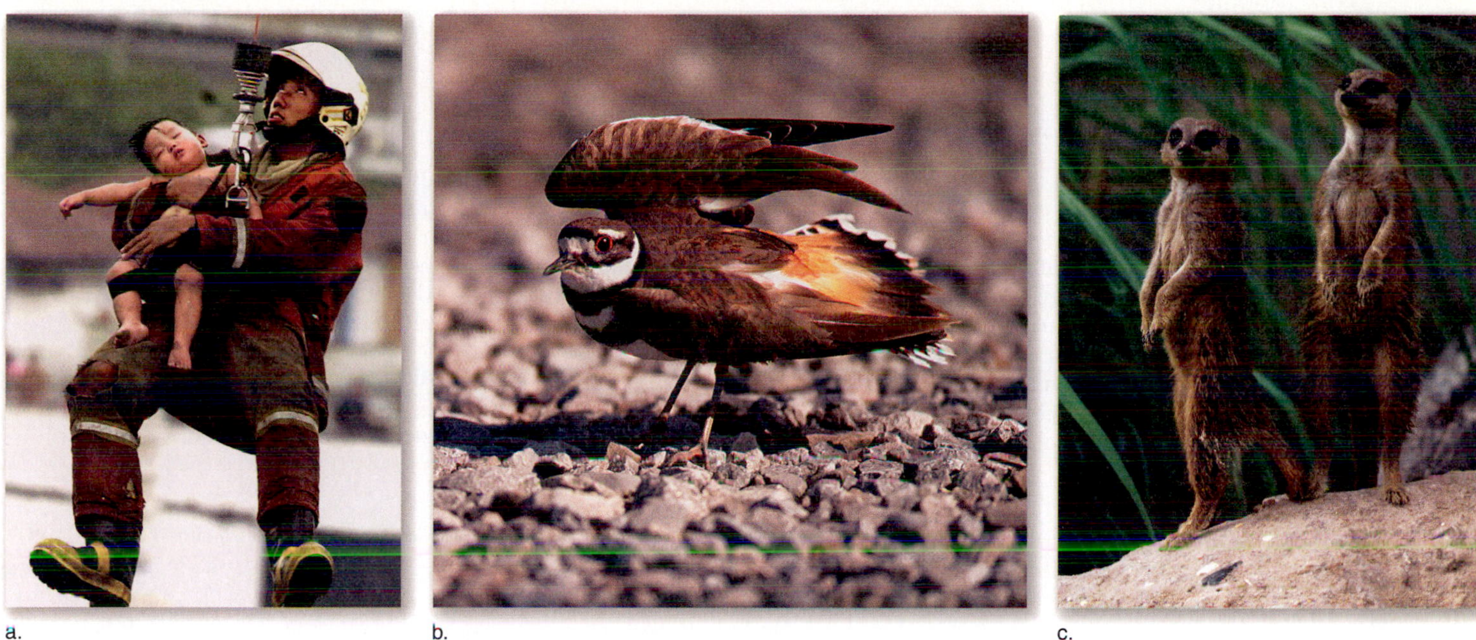

FIGURE 38.12 **Altruistic Behaviors.** (**A**) Human altruism. (**B**) A killdeer's broken wing display distracts predators. (**C**) Meerkat sentries.

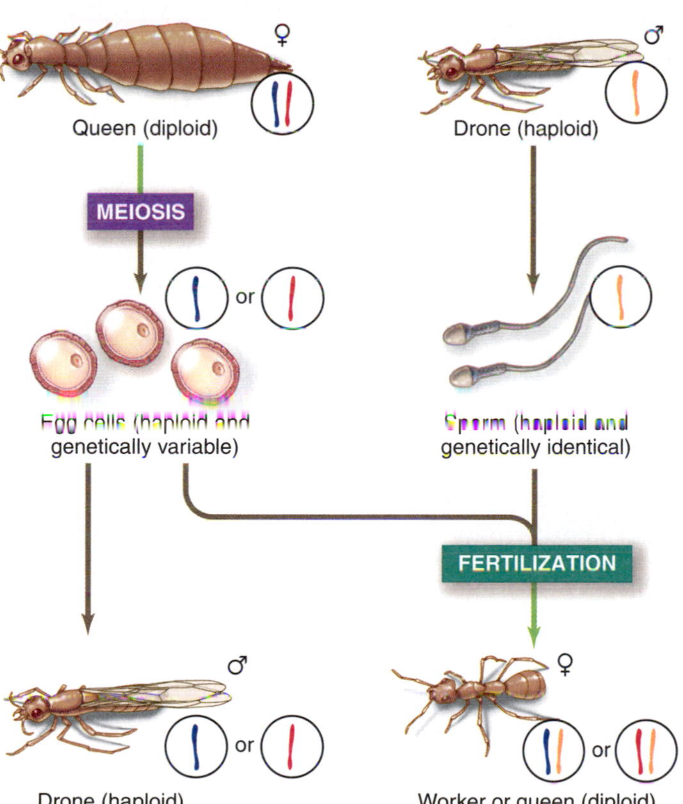

FIGURE 38.13 **Sex Determination in Ants, Bees, and Wasps.** In social insects, the female queens are diploid, whereas the males (drones) are haploid. The queen produces genetically variable, haploid eggs. If the eggs remain unfertilized, a male develops; if they are fertilized, a female develops. Because the drone's sperm are genetically identical, the sisters share more of their genes than do siblings in most species.

take turns acting as lookouts that keep watch while others feed, warning of the approach of predators (figure 38.12C). When the sentry barks, the entire community dashes for safe underground burrows. By barking, the sentry may draw attention to itself and away from the other animals, many of which may be its close relatives.

Likewise, the social behavior of ants, bees, and wasps is rooted in the way that sex is determined in these insects (**figure 38.13**). Male ants, bees, and wasps develop from unfertilized eggs; they are therefore haploid. Females develop from fertilized eggs, so they are diploid. Because each male bee produces genetically identical sperm cells, full sister bees share an average of 75% of their genes. A female worker bee is therefore more closely related to her sisters than she is to her brothers, her mother, or even her own offspring. It is to her advantage to care for her sisters instead of her own offspring!

Summary *Solitary and social behaviors allow animals to thrive in different ecological situations. Altruistic and selfish behaviors emphasize the survival of others or of self. Animals are most likely to exhibit altruistic behaviors toward their close relatives.*

38.4 MASTERING CONCEPTS

1. Why do animals display territorial behaviors?
2. Compare the social behavior in a wolfpack to the solitary behavior of the bobcat.
3. List the disadvantages and advantages of selfish behavior and compare them with the disadvantages and advantages of altruism.

Burning Questions

Why do people engage in rituals that are risky, expensive, painful, and dangerous?

Fervent believers of every religion risk their lives to go on pilgrimages, endure long and uncomfortable ceremonies, inflict excruciatingly painful tortures on themselves, contribute wealth, wear odd or uncomfortable clothing, fast for long periods, or abstain from certain foods—all in the name of faith. How can we reconcile these religious practices with the idea of natural selection? Can there be a selective advantage to behaviors that threaten an individual's survival?

Interestingly, there is a connection to stotting, in which an antelope communicates to a predator its health and ability to escape. Behaviorists call this an "honest signal," one that is too costly to fake. Similarly, religious rituals can also be viewed as behaviors that communicate your honest intentions to other members of your group. This tends to promote trust and social cooperation among group members, who may need to rely on one another in difficult times. Studies are ongoing, but behavioral ecologists suggest that these rituals may foster greater survival of group members.

Have a Burning Question of your own? Submit it to marielle_hoefnagels@mcgraw-hill.com for possible inclusion in future editions of this book!

38.5 INVESTIGATING LIFE
Addicted to Affection

The sexual behavior of animals fascinates many people, perhaps because of the insight it can lend into human relationships. In our own species, sexual attraction and feelings of love are often intertwined. Love is difficult to study in humans (and impossible to study in other animals), but scientists can examine patterns of sexual behavior in many species other than our own.

Some animals are faithful to their sexual partners for life, whereas others are much more promiscuous. One of the rarest and least understood social behaviors is monogamy. To qualify as monogamous, an animal must mate for life, live with its mate, help with care of the young, and defend the young, nest, and mate from intruders. Only a tiny fraction (about 3%) of mammal species are monogamous.

Two closely related species of snub-nosed rodents have given scientists the opportunity to investigate the biological basis of monogamy. Whereas prairie voles (*Microtus ochrogaster*) are monogamous and highly social, montane voles (*Microtus montanus*) are promiscuous and solitary (**figure 38.14**). What accounts for the difference in lifestyle? A chemical messenger called antidiuretic hormone (abbreviated ADH and also called vasopressin) apparently plays a role. In the 1990s, researchers learned that if male prairie voles were given ADH, they rapidly formed pair bonds, even with females they had not mated with. When ADH's effects were blocked, bonds did not form.

One logical explanation for the difference in sexual behavior between the two species is that the monogamous prairie voles have naturally higher levels of ADH than do the promiscuous montane voles. Yet this is not the case; both vole species have similar levels of the pair-bonding hormone. The location of receptors for ADH in the brain, however, does differ between the two species. In monogamous prairie voles, ADH receptors occur in the same brain region where addictive drugs act, and males seem to derive feelings of reward from being with their mates and young. In contrast, ADH receptors in promiscuous montane voles are located in brain regions associated with aggression.

FIGURE 38.14 Prairie Vole. These small rodents live in underground colonies. They are highly social and typically monogamous, so biologists study them to learn more about the genetic basis of sexual fidelity.

A team of researchers, led by Lauren Pitkow and Larry Young from Emory University in Atlanta, wondered what would happen to prairie voles with extra ADH receptors in the brain areas associated with pleasure and rewards. Would the animals form social attachments even more readily? To find out, the researchers inserted the gene encoding the ADH receptor into a virus. They injected the modified virus into the reward areas of the brains of prairie voles, effectively increasing the number of ADH receptors. Two groups of control voles also received injections. One control group received a different gene (one not related to ADH) into the reward area of the brain. Voles in the second control group received the ADH receptor gene, but in a different brain area.

Afterward, each male spent 17 hours in a cage with an adult female that was not sexually receptive. The voles then underwent "partner-preference" tests. The researchers placed each male into a choice chamber, where he was free to spend as much time as he wanted with his female "partner" or with a different female who was a stranger to him. Male prairie voles do not ordinarily pair-bond with females after less than 24 hours together, unless they have mated. Nevertheless, the voles with the extra ADH receptors in the reward region of the brain spent much more time in contact with their partners than did either group of control voles (**figure 38.15**). The extra ADH receptors evidently made the male prairie voles especially likely to form pair bonds.

Of course, the evolution of complex mating behaviors required more than just one change in the brain chemistry of the vole. Nevertheless, this study is interesting because it explicitly links genes, brain chemistry, and social behavior. This research also raises the startling possibility that a genetically modified virus can transmit genes that increase social attachment, at least in prairie voles. Could such a "love bug" infect humans too? Researchers already know that the location of ADH receptors in the brain may play a role in human social behaviors; examples include sexual fidelity and autism, a disorder in which individuals have difficulty forming social attachments. Researchers now are investigating primate ADH receptors to learn more about the biochemistry of attachment in our closest relatives.

Pitkow, Lauren, and five co-authors (including Larry Young). 2001. Facilitation of affiliation and pair-bond formation by vasopressin receptor gene transfer into the ventral forebrain of a monogamous vole. *Journal of Neuroscience*, vol. 21, no. 18, pages 7392–7396.

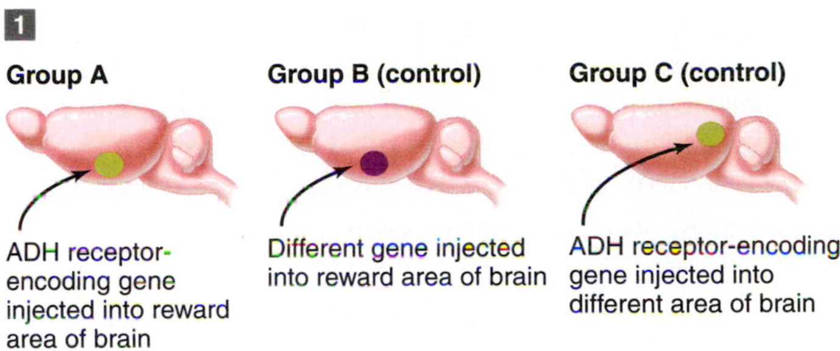

2 Each male spends 17 hours in cage with non-sexually receptive female

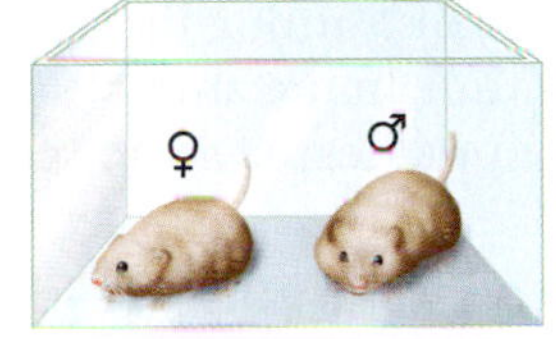

3 In a 3-hour partner-preference test, the male can choose to spend time with his "partner" or a stranger

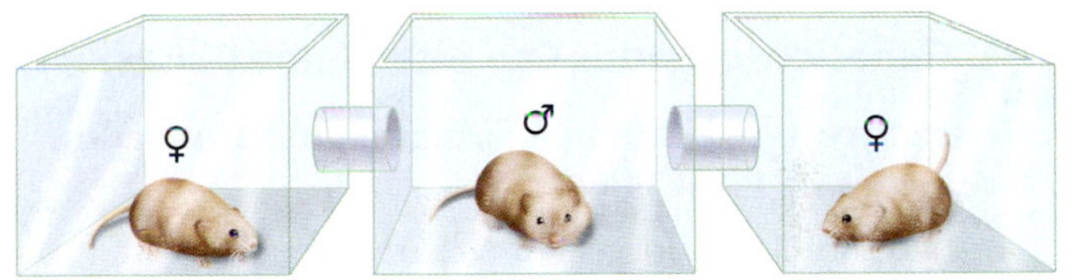

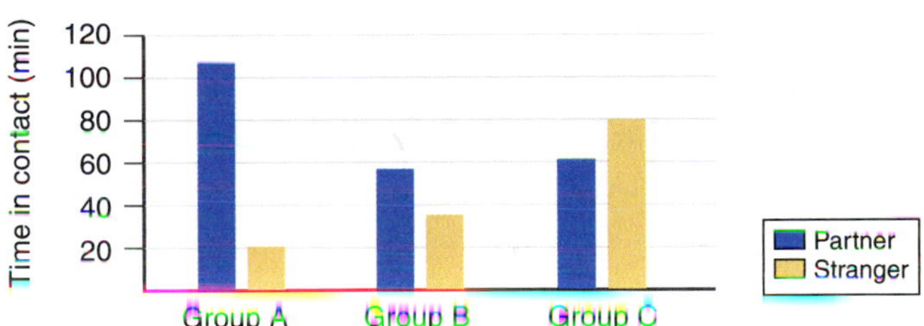

FIGURE 38.15 More Receptors, More Bonding. Prairie voles injected with ADH receptor genes in the reward center of the brain formed pair bonds much more easily than either group of control voles. **Question:** Could the researchers have drawn the same conclusions if they had omitted one of the control groups? Why or why not?

CHAPTER SUMMARY

38.1 Animal Behaviors Have Proximate and Ultimate Causes

- The science of animal behavior has its roots in observations by people whose everyday environment includes wild animals.
- **Ethology** emphasizes experiments into the causes of animal behavior in natural environments. The focus may be on the **proximate** causes, such as the hormones that contribute to a behavior. Questions about **ultimate** causes reflect a behavior's evolution or adaptive significance.

38.2 Animal Behaviors Combine Innate and Learned Components

A. Innate Behaviors Do Not Require Experience

- **Innate behaviors** are inborn and performed perfectly the first time they occur.

- Examples of innate behaviors include **reflexes,** involuntary responses that do not vary between individuals of a species; **taxes,** automatic movements toward or away from an environmental stimulus; and **fixed action patterns,** sequences of behaviors that are triggered by a **releaser.**

B. Learning Requires Experience

- **Learned behaviors** result from an animal's experiences. They are changeable and open to modification by subsequent experiences.
- In **trial-and-error-learning,** an animal alters its behavior to achieve rewards or avoid punishment. **Imprinting** is a type of learning that occurs only during a restricted time in the animal's life. In **habituation,** an animal learns *not* to respond to a stimulus.

C. Innate Behaviors and Learning Are Interrelated

- All learned behaviors have a genetic component, and all innate behaviors are triggered by interactions with the environment.

38.3 Genetic Mutations Can Alter Some Behaviors

A. Genes Strongly Influence Foraging and Courtship Behaviors of Fruit Flies

- Fruit flies are either rovers or sitters, reflecting how much they move once they have found a food source. A gene, *for*, contributes to this foraging behavior.
- Mutations in the *fru* gene cause male fruit flies to court other males.

B. Some Human Behaviors Have a Strong Genetic Component

- Evidence for genetic control of human behaviors comes from studies of identical twins who have been reared apart.
- The common image of a gene "for" a human behavior gives the erroneous impression that genes are responsible for behavior on a one-to-one basis. A protein encoded by a gene will produce a behavior only if all the other necessary biochemicals, produced by the interactions of the products of other genes, are also present.

38.4 Animals May Live Alone or in Groups

A. Bobcats Are Solitary Animals

- **Territorial behaviors** ensure that individuals do not intrude on one another's hunting territories.
- Many animals, including bobcats, produce chemical signals called **pheromones** that influence the behavior of other individuals of the same species.

B. Wolves Are Social Animals

- Social groups generally consist of related individuals. The social hierarchy typically includes an alpha pair that feeds first and does most of the mating.
- In social animals, sharing resources promotes the success of their shared genes.

C. Kin Selection May Explain Many Altruistic Behaviors

- Selfish behaviors promote the survival of an individual, whereas altruistic behaviors sacrifice the interests of an individual for its children, mate, relatives, or even total strangers.
- Behavioral ecologists propose that **kin selection** explains why animals are most likely to sacrifice themselves for close relatives.

38.5 Investigating Life: Addicted to Affection

- Monogamy is unusual among animals, but the prairie vole forms a monogamous bond with its mate. Researchers have traced pair-bonding behavior to a receptor that binds antidiuretic hormone in the pleasure-seeking area of the vole's brain.
- Increasing the number of ADH receptors made male voles more likely to bond with a female partner.

MULTIPLE CHOICE QUESTIONS

1. Which of the following studies is dealing with an ultimate cause of a behavior?
   a. A study of the ability of a squid to alter its appearance to avoid predation
   b. A study of the aerodynamics involved in a hummingbird's ability to hover
   c. A study of the social behavior of prairie dogs
   d. A study of the migration behavior of monarch butterflies
2. A *fixed action pattern* is an example of
   a. a learned behavior.
   b. an innate behavior.
   c. a reflex.
   d. a taxis.
3. What is the environmental trigger of an innate behavior?
   a. A gene
   b. A taxis
   c. A releaser
   d. A reflex
4. Although the trains that rumble through your town once bothered you, now you barely even hear them. This is an example of
   a. habituation.
   b. trial-and-error learning.
   c. a reflex.
   d. imprinting.
5. What is the link between a gene and a behavior?
   a. External stimuli
   b. Mutations
   c. Proteins
   d. Muscles

6. The appearance of a behavior in an organism could be the result of
   a. interactions between multiple genes.
   b. environmental factors.
   c. chromosomal abnormalities.
   d. all of the above.
7. How do animals use pheromones?
   a. To attract a mate
   b. To establish a territory
   c. To repel predators
   d. Both a and b
8. What is the advantage of kin selection?
   a. It allows the individual to survive and reproduce.
   b. It ensures the survival of the species.
   c. It increases the chance that an individual's genes will be maintained in a population.
   d. It increases reproduction between related individuals.
9. In what way is parental investment an example of altruistic behavior?
   a. Time spent rearing offspring represents a sacrifice by an individual.
   b. Time spent rearing offspring increases the genetic fitness of the individual.
   c. It benefits the community.
   d. It teaches the offspring the best way to rear their own children.
10. Why is it difficult to classify a human behavior as being either innate or learned?
   a. Because it is impossible to fully know the history of an individual
   b. Because behavior is a blend of genetics and experience
   c. Because not all genes affecting behavior have been identified
   d. All of the above

TESTING YOUR KNOWLEDGE

1. Differentiate between the proximate and ultimate causes of a particular behavior.
2. Compare and contrast innate and learned behaviors.
3. How are taxes different from reflexes? How are they similar?
4. Give an example of a fixed action pattern.
5. How is trial-and-error learning different from imprinting?
6. Explain how biologists identify genes controlling behavior in a fruit fly. Why is it difficult to identify similar genes in humans?
7. What advantages are there to social behavior? To solitary behavior?
8. What is kin selection and how can it explain some types of altruistic behavior?

THINKING AS A SCIENTIST

1. People tend to see what they expect to see. How can ethologists prevent themselves from falling into this trap when observing an animal's behavior?
2. Once accustomed to captivity, praying mantids tend to stop making the startle response when their caregivers come near. Given what you now know about animal behavior, construct a hypothesis to explain this change. How would you test your hypothesis?
3. Are humans solitary or social animals? Explain your answer.
4. At home, you are usually a sound sleeper, but when you travel, you toss and turn and get little sleep. Give a behavioral explanation for this observation.
5. How is it advantageous for a species to have a fixed action pattern that controls mating behavior?
6. European cuckoos and American cowbirds are "brood parasites" that lay their eggs in the nests of other birds. The chicks are generally raised by "adoptive parents," whose own chicks receive less nourishment and often are killed by the larger intruders. What kinds of behaviors does natural selection favor in the cuckoo and cowbird chicks? In the adoptive parents?
7. Who is the most altruistic person that you know of? How does altruism benefit that person? How does it benefit a philanthropist to donate millions or billions of dollars to charity?
8. How is it incorrect to refer to a gene "for" alcoholism?
9. Suppose an American couple moves to Australia and raises all of their children in their new country. The children all speak English with an Australian accent. What does this mean about the influence of environment on a person's accent? Does this mean that speech does not have a genetic component?

Visit www.mhhe.com/hoefnagels for practice quizzes, animations, videos, and activities designed to help you master the material in this chapter.

CHAPTER 39

Population Ecology

The brown tree snake,
Boiga irregularis.

What's the Point?

A Population Out of Control: Snakes Decimate an Island's Biodiversity

When a new predator colonizes an island, populations of native species can plummet. This fate befell the island of Guam, thanks to the brown tree snake, Boiga irregularis. *The brown tree snake likely arrived on Guam during World War II, hitching a ride on cargo shipped to the island from its native East Asia. Once there, the population grew explosively. Because this predator was new to the island, nature had not selected for defensive adaptations that might have enabled more birds and reptiles to escape. The snakes feasted. More than a dozen species of native and introduced birds and small reptiles have disappeared.*

It wasn't until the 1960s that ecologists began to notice that something was wrong on Guam. Birds had vanished, first in the south, then farther north. At first, biologists thought a pesticide or disease was at fault. Further study, however, revealed several pieces of evidence implicating the snake:

- *The snake is extremely common, and it stalks and kills a variety of prey. It can also eat hamburgers, dog food, and garbage.*
- *Areas where snakes were common matched areas where birds disappeared.*
- *Birds and lizards were not disappearing on nearby snake-free islands.*
- *No known insecticide or disease was killing the animals.*

Once the snake hypothesis was accepted and researchers began surveying the forests of Guam, they found an enormous snake population. Whereas most snakes live in populations of 1 to 10 animals per hectare (2.47 acres), on Guam, 100 brown tree snakes occupied a hectare! The birds and lizards of Guam were hopelessly outnumbered.

Many efforts to control the snake population on Guam focus on increasing the death rate. The most common approach is to capture and kill the snakes; researchers are developing pheromone-based attractants to make the traps more alluring. Another tactic is leaving dead rats laced with acetaminophen, a commonly used painkiller that destroys the snakes' livers. Since a population's growth rate reflects both its birth and death rates, however, researchers are also investigating ways to either prevent reproduction or keep juveniles from reaching reproductive age. To keep brown tree snakes from leaving Guam and causing disasters elsewhere, specially trained Jack Russell terriers inspect outgoing cargo for snakes.

The story of the brown tree snake illustrates the importance of studying how populations grow, stabilize, and decline. This chapter introduces the factors that affect populations, including that of our own species.

LEARNING OUTLINE

39.1 A Population Consists of Individuals of One Species

The brown tree snakes on Guam, described in the chapter opening essay, represent a **population:** a group of organisms of one species occupying a geographic location at the same time. The potential for interbreeding defines a population's size. Members of the same population are much more likely to breed with one another than with a member of another population (**figure 39.1**). A population may include all individuals over the entire range of a species, or only those that live in one part of the range.

A population's **habitat** is simply the physical location where its members live. The ocean, desert, or rain forest are typical examples, but an organism's habitat might even be another organism. Your body, for example, is home to billions of microbes. Because many different species typically inhabit an area, populations interact. A **community** includes all the organisms in a given area. The size of a population is often affected by interactions with predators, disease-causing organisms, and other members of the community, among other factors. **normal microflora, p. 381**

Such interactions are part of the science of **ecology,** the study of the relationships among organisms and the environment. Because the environment can change, the populations within a community can also change. This chapter examines the factors that influence populations over time, and chapter 40 looks at how populations interact.

FIGURE 39.1 Pretty Population. Individuals that belong to the same species and can potentially interbreed belong to the same population.

A. Density and Dispersion Patterns Are Static Measures of a Population

The study of population ecology is relevant to everything from disease prediction to land management. Typical questions population ecologists might ask include: "Which weather conditions favor the spread of rodents that transmit human diseases?" "How large a patch of old-growth forest does a breeding pair of spotted owls require?" "How many deer should hunters cull to keep the herd healthy?" "How many humans can Earth support?" To answer such questions, ecologists begin by describing the population.

Population density is the number of individuals of a species per unit area or unit volume of habitat. For some species, simply observing the organisms can provide a measure of population density. For others, more sophisticated techniques are required, as the Can *You* Relate? Keeping Track of Animal Populations describes.

Two populations can have the same density, but the individuals may be distributed throughout the habitat in different ways. **Population dispersion** patterns describe how individuals are scattered through the habitat space (**figure 39.2**). Organisms may occur in a random pattern if the environment is relatively uniform (for example, the deep shade under a dense forest canopy) and individuals neither strongly attract nor repel one another. More uniform spacing may occur if individuals repel one another, as in the case of strongly territorial animals. Most often, however, a population's distribution is somewhat clumped, with the highest densities occurring where the habitat is most favorable.

Population density and distribution measurements provide static "snapshots" of a population at one time. The next section explains the factors that determine whether a population will grow, stay the same, or shrink.

B. Population Size May Increase, Decrease, or Remain Stable

Population dynamics is the study of the factors that influence changes in a population's size. Regionally, migration can increase or decrease a population; **immigration** is the movement of individuals into a population, and **emigration** is migration out. For example, 200 or so years of immigration has tremendously increased the human population in the United States. Most Americans are either immigrants or descended from immigrants, and today migration accounts for over 30% of population growth in this country.

FIGURE 39.2 Three Patterns of Population Dispersion. (**A**) Random spacing, as illustrated by these lichens, is rare because it requires an environment in which resources are very evenly distributed and the organisms neither attract nor repel one another. (**B**) These penguins defend the space immediately surrounding their nests, leading to a uniform dispersion pattern. (**C**) An uneven distribution of resources may yield regions of high density, as in the human population of the United States.

Ultimately, however, births and deaths are the most obvious ways to add to and subtract from a population. If a population adds more individuals than are subtracted, the population grows. If the opposite happens, the population shrinks (and may become extinct). The population size remains unchanged if additions exactly balance subtractions.

Births Add Individuals to a Population

A population's **birth rate** is the number of new individuals produced per unit time. The number of offspring an individual produces over its lifetime depends on the number of times it reproduces and the number of offspring per reproductive episode. Age at first reproduction is also important. All other things being equal, the earlier reproduction begins, the faster the population will grow.

A population's **age structure**, or distribution of age classes, determines whether a population is growing, stable, or declining (**figure 39.3**). A population with a large fraction of prereproductive individuals will grow. As these individuals enter their reproductive years and produce offspring, the prereproductive age classes swell further, building a foundation that ensures future growth. This pattern is common in less developed countries, where death rates have declined faster than have birth rates, ensuring that more people reach reproductive age. Conversely, a population that consists mainly of older individuals will be stable or may even decline. This is happening in Spain and many other developed countries. A similar situation can doom a population of endangered plants if, for example, habitat destruction makes it impossible for seedlings to establish themselves. With few individuals of reproductive age, the population may go extinct.

Survivorship Curves Show the Probability of Dying at a Given Age

A population's **death rate** is the number of deaths per unit time. Each individual in a population will eventually die; the only question is when. In some species, the probability

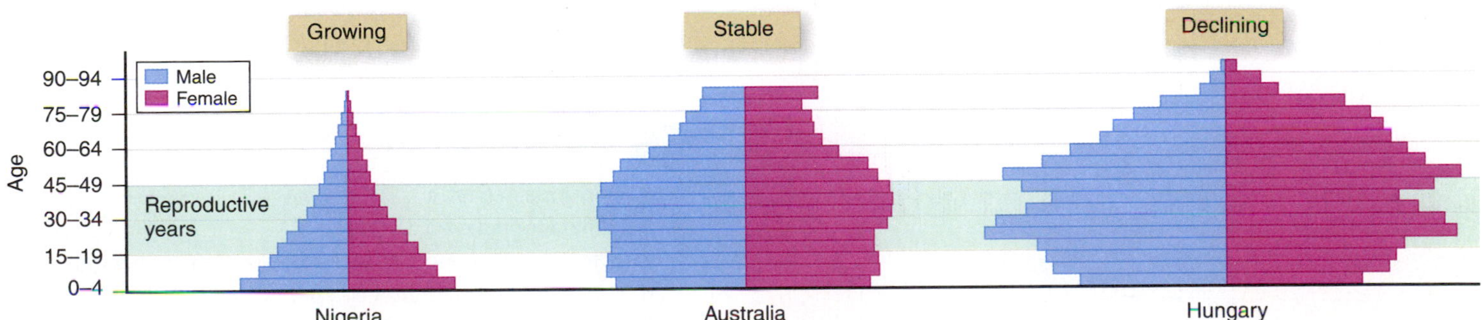

FIGURE 39.3 Population Age Structures Predict the Future. In age structure diagrams, the width of each bar is proportional to the percent of individuals in that age class. Populations are likely to grow if they have a high proportion of individuals in prereproductive age classes, assuming that the individuals are healthy and fertile. A stable population has roughly equal numbers of people of each age group. A declining population has most of its members in reproductive or postreproductive age classes. (Data from U. S. Census Bureau)

of dying before reaching reproductive age is much higher than it is in others.

A **life table** is a chart that shows the probability of surviving to any given age (**table 39.1**). The values in a life table account for predation, disease, food availability, and other factors that prevent an individual from reaching its theoretical life span. (Life insurance companies use life tables to compute premiums for clients of different ages.)

A **survivorship curve** is a graph of the proportion of surviving individuals at any given age. The survivorship curves of many species follow one of three general patterns (**figure 39.4**). Type I species, such as humans and elephants, invest a great deal of energy and time into each offspring. The mortality rate is highest as individuals approach the maximum life span. Type II species, including many birds and mammals, have an equal probability of dying at any age. Type III species, such as most marine fishes and invertebrates, most insects, and many plants, may produce many offspring but invest little in each one. Most offspring of type III species therefore die at a very young age. Section 39.3 looks more closely at the evolutionary trade-offs that these survivorship curves reflect.

Of course, these generalized examples do not describe all populations. Many species have survivorship curves that fall between two patterns. The scorpion from table 39.1, for example, has a survivorship curve that has features of both type I and type II curves.

Table 39.2 summarizes some of the factors affecting a population's growth rate.

Table 39.1 ***Age-Specific Life Table for the Desert Sand Scorpion,* Paruroctonus mesaensis**

Age (Months)	Proportion Surviving to Age
0	1.000
1	0.849
2	0.688
4	0.612
6	0.589
8	0.567
10	0.542
12	0.500
16	0.451
20	0.419
24	0.326
36	0.044
48	0.007
60	0.001
72	0.000

Source: Data from G. A. Polis, R. D. Farley. 1980. Population Biology of a Desert Scorpion: Survivorship, Microhabitat, and the Evolution of Life History Strategy. *Ecology*, vol. 61, p. 620.

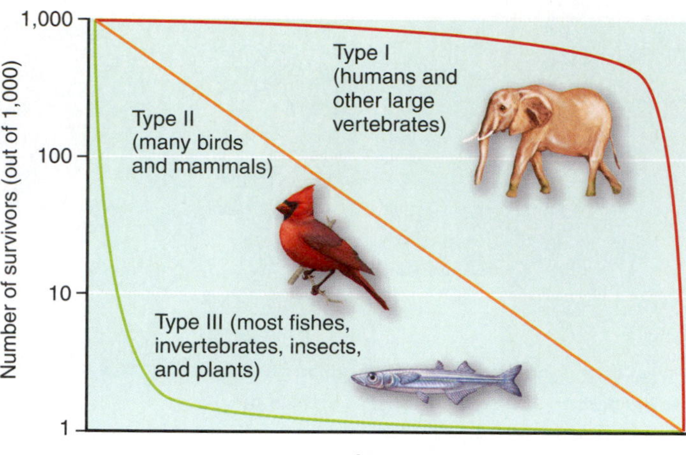

FIGURE 39.4 Three Survivorship Curves. This graph depicts the number of survivors out of a cohort of 1000 individuals as age increases. The scale of the *y*-axis is logarithmic, which means that straight-line portions of the curves reflect a constant survivorship rate.

Table 39.2 ***Factors Affecting Population Growth***

Additions to a Population	Losses from a Population
Births	Deaths
Number of reproductive episodes per lifetime	
Number of offspring per reproductive episode	
Age at first reproduction	
Population age structure (proportion at reproductive age)	
Immigration	Emigration

Summary *Ecologists study interactions among organisms and the living and nonliving parts of the environment. Populations, which are composed of members of a single species, make up communities of interacting organisms. The size of a population reflects birth and death rates, along with migration. Since an organism's chances of reproducing and dying vary with its age, the growth rate of a population also reflects its age structure.*

39.1 MASTERING CONCEPTS

1. What is a habitat?
2. What is population density?
3. Distinguish among the types of dispersion in a habitat.
4. Under what conditions will a population grow?
5. What factors determine the birth rate in a population?
6. What are the three patterns of survivorship curves?
7. How does a population's age structure predict its future growth?

Can You Relate?

Keeping Track of Animal Populations

Collecting long-term data on wildlife populations can help scientists measure the effects of environmental catastrophes such as fires, volcanic eruptions, or oil spills. Hunting, trapping, and fishing regulations are also based on population estimates, as are decisions on where to build (or not to build) houses, dams, bridges, and pipelines. But news reports of booming deer populations or plummeting amphibian populations rarely report on how wildlife biologists arrive at their estimates.

One way to estimate the size of a population is to count the number of droppings within a defined area. The more excrement piles, the higher the population. Alternatively, many wildlife surveys count migrating herds of mammals and flocks of waterfowl from the ground or from the air. Researchers even sometimes set up animal-triggered cameras. Rotting meat or fresh sardines are draped in a tree, along with a heat detector attached to a camera. When a bear saunters over to collect the treat, the detector picks up the bear's body heat, and the camera snaps its picture. Tags are used to trace bear identities.

The mark-recapture technique is widely used to estimate animal populations. Suppose researchers want to know how many squirrels inhabit a park. They might place nest boxes in trees and then collect them when squirrels wander in. After recording statistics such as weight, sex, age, and health status, the researchers could also tattoo the squirrels before releasing them. The proportion of tattooed squirrels that they subsequently catch can help the biologists estimate the population size.

For example, suppose researchers mark and release 50 squirrels (M = number marked = 50). Subsequently, they capture 25 squirrels, of which five bear the tattoo (f = fraction of marked animals in "recapture" sample = 5/25 = 0.20). The number of marked squirrels (M) divided by this fraction (f) estimates the total squirrel population size. For the area sampled, then, the estimated total squirrel population is $M/f = 50/0.20 = 250$.

The potential for sampling bias is one weakness of the mark–recapture technique. Baited traps may repeatedly lure some individuals in search of a free meal. On the other hand, if an animal finds the capture or marking technique unpleasant, it may learn to avoid the traps.

39.2 Population Growth May Be Exponential or Logistic

Any population will grow if the number of individuals added exceeds the number removed. The **per capita rate of increase,** r, is the difference between the birth rate and the death rate. If r is negative, the population shrinks; a positive r means the population is growing. For example, a population in which there are 35 births and 10 deaths per 1000 individuals is growing at a rate (r) of 25 people per 1000, or 2.5%. (The growth rate may also reflect migration into or out of the population.)

A. Growth Is Exponential When Resources Are Unlimited

The number of individuals added during any time interval depends on the initial size of the population, as expressed by this equation:

$$G = rN$$

G is the number of individuals added per unit time, r is the per capita rate of increase, and N is the number of individuals at the start of the time interval. This pattern describes **exponential growth,** in which the number of new individuals is proportional to the size of the population. A **J-shaped curve** emerges when exponential growth is plotted over time (**figure 39.5**). Growth resulting from repeated doubling (1, 2, 4, 8, 16, 32 . . .), such as in bacteria, is exponential. Species introduced to an area where they are

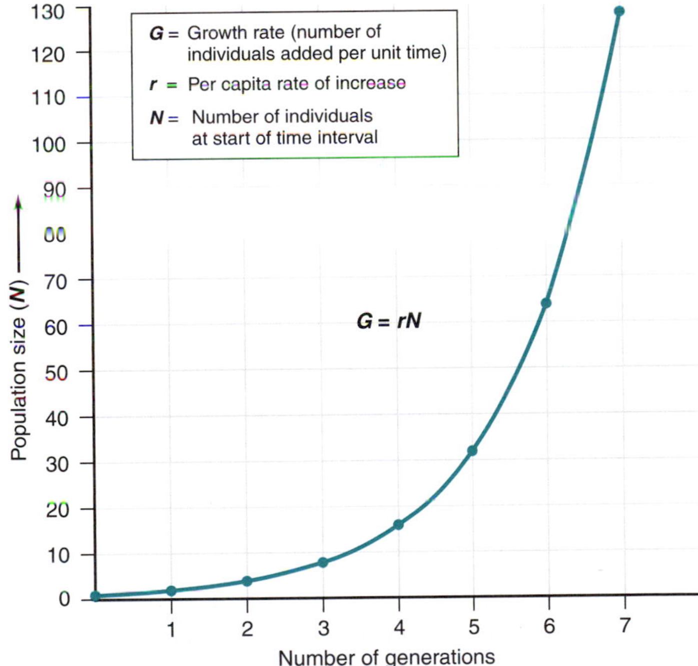

FIGURE 39.5 Exponential Population Growth. During each time interval, the population's growth rate, G, is the product of the per capita rate of increase, r, and the number of individuals in the population at the beginning of the interval, N. With each successive interval, G increases, even though r remains constant.

not native may also proliferate exponentially for a time, since they often have no natural population controls.

Cancer illustrates one feature of exponential growth. Cancer cells are not organisms and therefore do not technically meet the definition of a population. Nevertheless, their reproduction inside the body can be exponential; one cell divides into two, then four, then eight, and so on. Surgery, chemotherapy, and radiation treatment may reduce the number of cancerous cells, but the tumor may come back rapidly as cells that survived the treatment continue to divide. **cancer, p. 166**

B. Population Growth Eventually Slows

Exponential growth may continue for a short time, but it cannot continue indefinitely because some resource is eventually depleted. **Environmental resistance** is the combination of external factors that keep a population from reaching its maximum growth rate. Anything that reduces reproduction and immigration, or that increases mortality and emigration, contributes to environmental resistance.

Every habitat has a **carrying capacity**, the maximum number of individuals that the habitat can support indefinitely. As the population approaches the habitat's carrying capacity, growth slows, then stops. The **logistic (or S-shaped) growth curve** depicts the leveling off of a population in response to environmental resistance (**figure 39.6**). The number of new individuals added after each time interval follows this simple equation:

$$G = rN(K - N)/K$$

G equals the number of individuals added per unit time; r is the per capita rate of increase; N is the number of individuals in the population at a given time; and K is the carrying capacity. In the right side of the equation, the term rN is the growth rate that would occur if resources were not limiting. The term $(K - N)/K$ factors in the increasing environmental resistance as the population approaches the carrying capacity. When N is very small relative to K, this expression yields a numerical value near 1, so the population growth rate is high. When N is near K, the value of the expression drops to near 0, and the growth rate is low.

In reality, the carrying capacity of a habitat is not fixed. A drought that lasts for a decade may be followed by a year of exceptionally heavy rainfall and a sudden increase in the availability of food. A catastrophic flood can drastically reduce the carrying capacity as habitat is destroyed. In addition, populations differ in how they respond as they approach the carrying capacity. The population may grow exponentially and fluctuate wildly about the carrying capacity. In other cases, population growth may slow and stabilize as the carrying capacity nears.

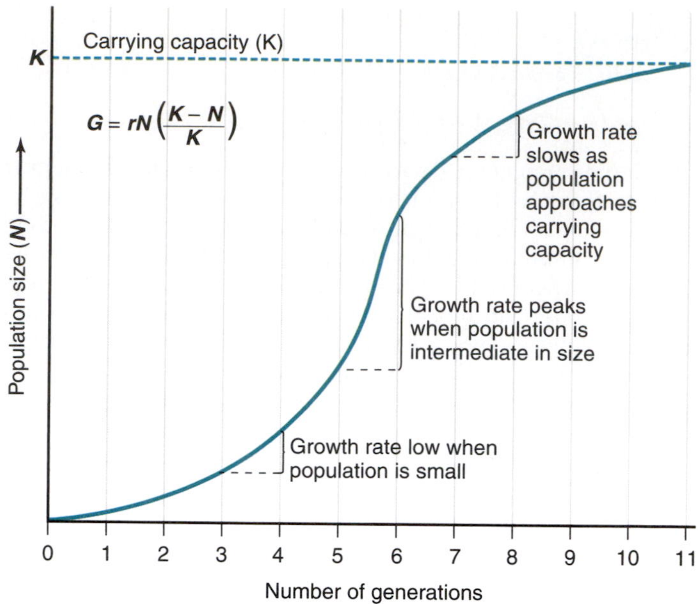

FIGURE 39.6 Logistic Growth. Although exponential growth may occur for some time, growth will slow when resources become limiting. According to the logistic growth model, the growth rate slows to zero as the population approaches the habitat's carrying capacity (*K*).

C. Many Conditions Limit Population Size

A combination of factors regulates the size of most populations. Consider the population of songbirds in your town. Some lose their lives to cold weather or food shortages, whereas others succumb to infectious disease or the jaws of a free-roaming cat. These and other limits on the growth of the bird population fall into two general categories: density-dependent and density-independent.

Density-dependent factors are conditions whose growth-limiting effects increase as a population grows. Infectious disease is one example. Many viruses, bacteria, fungi, and protists spread by direct contact between infected individuals and new hosts. The higher the host population density, the more opportunity for these disease-causing organisms to find new hosts. Likewise, a higher population density leads to a higher probability of death by predation. Competition for space, nutrients, sunlight, and food is also density-dependent. When many individuals share limited resources, few may be able to reproduce, and population growth slows or even crashes (**figure 39.7**). Since multiple species often share a habitat, competition may also limit population growth. **interspecific competition, p. 806**

If competition for prey limits predator populations, how can the brown tree snake described in the chapter opening essay continue to thrive on Guam? Perhaps small reptiles such as geckos and skinks provide food for the snakes in the short term. But ecologists predict that snake reproduction should eventually slow as food becomes less abundant.

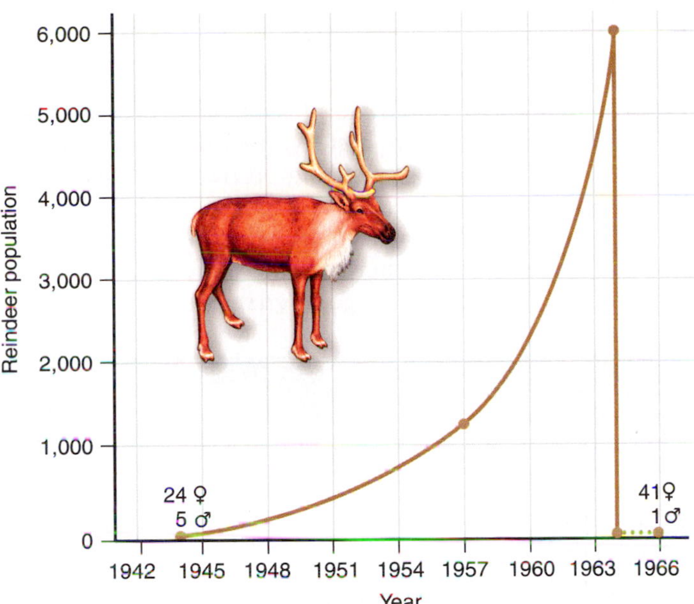

FIGURE 39.7 Population Crash. In 1944, 29 reindeer were introduced to an optimal habitat on St. Matthew Island in the Bering Sea. Free of large predators, the population grew exponentially through the summer of 1963, but food was becoming scarce. During the winter of 1963–1964, severe weather and a depleted food supply led to mass starvation.
Source: David R. Klein, April 1968, *Journal of Wildlife Management*, vol. 32, no. 2, p. 352.

FIGURE 39.8 Density-Independent Limit. The unstoppable flow of lava from a volcanic eruption destroys everything in its path, regardless of the population.

Density-independent factors exert effects that are unrelated to population density (**figure 39.8**). Natural disasters such as earthquakes, floods, volcanic eruptions, and severe weather conditions are typical density-independent factors, as are oil spills and other industrial accidents.

Summary *The difference between the birth rate and death rate determines a population's growth rate. With unlimited resources, population growth is exponential, but environmental resistance prevents unlimited population growth. Growth slows as the population size approaches the habitat's carrying capacity. The factors that limit population growth may or may not depend on population density.*

39.2 MASTERING CONCEPTS

1. What is the per capita rate of increase?
2. What conditions support exponential population growth?
3. How does logistic growth differ from exponential growth?
4. What is environmental resistance, and what is its relationship to the carrying capacity?
5. How does a habitat's carrying capacity fluctuate over time?
6. Distinguish between density-dependent and density-independent factors that limit population size, and give three examples of each.

39.3 Natural Selection Influences Reproductive Strategies

Ecologists often talk of a species' **life history**, which includes all events that influence reproduction. Life history traits include such features as whether an organism reproduces sexually or asexually, its life span and age at maturity, when and how often it reproduces, and the number and size of its offspring.

A. *r*- and *K*-Selected Species Have Different Life Histories

Species vary widely in their life history characteristics. One female moth may lay hundreds of fertilized eggs before she dies, and a plant such as pigweed may shed 100,000 seeds in just one growing season. Humans, elephants, and century plants, on the other hand, may produce only a few offspring throughout their long lives.

These examples illustrate two extremes in a continuum of life history types. At one extreme are ***r*-selected species**, in which individuals tend to be short-lived, reproduce at an early age, and have many offspring that receive little care (**figure 39.9**A). Each offspring has a very low probability of surviving to reproduce, a pattern typical of species with type III survivorship curves (see figure 39.4). Weeds, crop pests such as insects, and many other invertebrates live in *r*-selected populations. Populations can skyrocket when conditions are favorable, thanks to a high reproductive rate. Density-independent factors such as frost or drought tend to limit the growth of these species.

At the other extreme are ***K*-selected species**, in which individuals tend to be long-lived, to be late-maturing, and to produce a small number of offspring that receive extended parental care (figure 39.9B). High parental investment in each offspring means that most live long enough to reproduce. Density-dependent factors such as competition keep *K*-selected populations close to the carrying capacity. Many birds and large mammals (organisms with survivorship curves that approximate type I or type II) live in *K*-selected populations.

The concept of *r*- and *K*-selection dates to the 1960s, and ecologists have since found that strict adherence to *r*- or *K*-selection is the exception, not the rule. For example, even populations of large animals with *K*-selected life histories fluctuate greatly in response to changes in their environments. Nevertheless, these two strategies illustrate the important point that natural selection shapes a species' life history characteristics. Each species' life history must balance the competing demands of reproduction and survival.

B. Guppies Illustrate the Importance of Natural Selection

Experiments show that natural selection directly influences the details of a life history, including age at maturity and the number of offspring per reproductive event. For example, ecologists have studied Caribbean fish called guppies (*Poecilia reticulata*), which live in rivers and streams on the South American mainland and on the islands of Trinidad and Tobago (**figure 39.10**). In streams

FIGURE 39.10 **The Guppy, *Poecilia reticulata*.** Guppies are brightly colored fish that are popular subjects for studies of population ecology. A female guppy typically bears from 30 to 60 live young per reproductive episode.

FIGURE 39.9 ***r*- and *K*-Selected Species.** Plants that produce many small offspring and invest little in each are *r*-selected. Some plants, however, produce large fruits that each represent a relatively large investment of energy. Coconut trees are an example.

with high predation intensity, predators called cichlids eat adult guppies. In a second set of streams, with moderate predation intensity, the predators are omnivorous fish called *Rivulus*, 10% of whose diet consists of juvenile guppies. In a third set of streams, the guppies do not face significant predation.

The researchers hypothesized that predation would be a selective force influencing the guppies' life histories. For example, they predicted that guppies from streams with cichlids would reach reproductive maturity at a younger age and devote more energy to reproduction than guppies from other streams. To test their hypothesis, they collected guppies from 16 sites in Trinidad: 4 with low predation intensity, 5 with moderate predation intensity, and 7 with high predation intensity. They dissected each female fish and weighed any offspring she carried inside her, and they measured the mature males and females from each stream to estimate the age at maturity.

Guppies from the cichlid stream matured earlier, had more and smaller offspring, and reproduced more frequently than guppies in the other streams. Other studies confirmed that these differences were genetic and therefore were subject to natural selection. The investigators also used field and laboratory experiments to confirm that the predators (not some other condition in the streams) were responsible for the life history differences. For example, they transplanted guppies from streams with cichlids to streams with *Rivulus*. Since the guppies were moving from a habitat with high adult mortality to a habitat with high juvenile mortality, the researcher predicted that the guppies' life histories should evolve. That is exactly what happened. The transplanted guppies began taking longer to reach reproductive maturity, they had fewer broods, and each brood had fewer, larger offspring.

Summary *Each population's life history reflects its reproductive characteristics and the trade-offs between "quantity" and "quality" of offspring. Two types of reproductive strategies are* **r-selection** *and* **K-selection**. *Natural selection shapes life histories.*

39.3 MASTERING CONCEPTS

1. Distinguish between *K*-selected and *r*-selected species.
2. How have studies with guppies shown that natural selection shapes life histories?

39.4 The Human Population Continues to Grow

The principles of population ecology apply to all species, including our own. By the middle of 2007, the world's human population exceeded 6.6 billion. The average population density is about 43 people per square kilometer of land area, but the distribution is far from random. Two countries—China and India—alone account for one third of all humans. For the most part, the highest population densities worldwide occur along the coastlines and in the valleys of major rivers.

For much of human history, the human population has been growing exponentially (**figure 39.11**). As section 39.2 explains, exponential growth cannot continue over the long term. This rule applies to humans as well, and indeed, our population's growth rate has slowed in recent years.

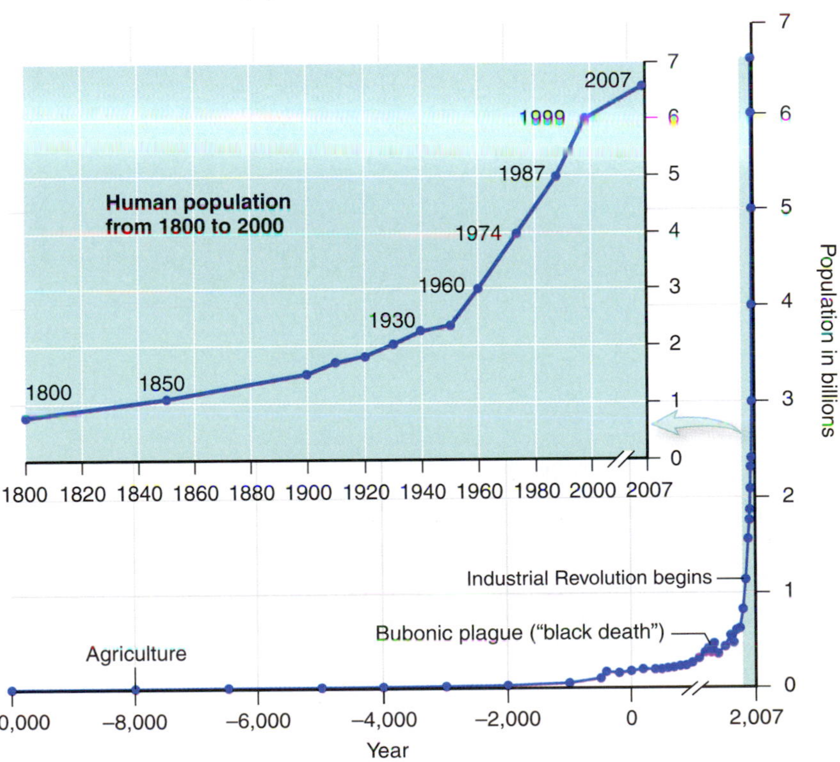

FIGURE 39.11 The Human Population Continues to Grow. Human population growth is on the rise, as indicated by the J-shaped curve. The most rapid population growth has occurred in the past 200 years (*inset*). (Data from U.S. Census Bureau.)

A. Population Dynamics Reflect the Demographic Transition

Overall, the human population growth rate is about 1.2% per year and declining, but the rate is not the same everywhere (**figure 39.12**). Fertility, mortality, and age structure differ worldwide and produce a mosaic of regional human subpopulations: some growing, some stable, and some declining.

The more and less developed regions of the world are in different phases of the **demographic transition**, during which birth and death rates shift from high to low (**figure 39.13**). In the first stage of the demographic transition, population growth is minimal because both birth and death rates are high. Then, during the second stage, improved living conditions and disease control lower the death rate, but birth rates remain high. This transitional period therefore sees the rapid population growth typical of the world's less developed countries. During the third stage of the demographic transition, birth rates fall; the difference between birth and death rates is once again small. The world's more developed countries have entered this stage, and a few even have declining populations because death rates exceed birth rates.

Birth rates are projected to continue to decline in less developed countries, and demographers project that zero population growth may happen during the twenty-second century, but no one is certain. Nor do we know how many people will inhabit Earth when that occurs (see Burning Question: What will happen to the human population?).

Factors Affecting Birth Rates

How do birth rates decline? Family planning programs that widely distribute information and services are relatively inexpensive and have immediate results. Social and economic factors play an important role as well. For example, delayed childbearing slows population growth. In addition, as family income increases, fertility drops. Women with more education also typically have smaller families than less educated women. Educated women are most likely to learn about and use family planning services, have more opportunities outside the home, and may delay marriage and childbearing until after they enter the work force.

Sometimes, government policies directly affect family planning options. The governments of France, Canada, and Japan offer financial incentives to encourage citizens to have children. At the opposite extreme, China controlled runaway population growth with drastic measures, rewarding one-child families and revoking the first child's benefits if a second child is born.

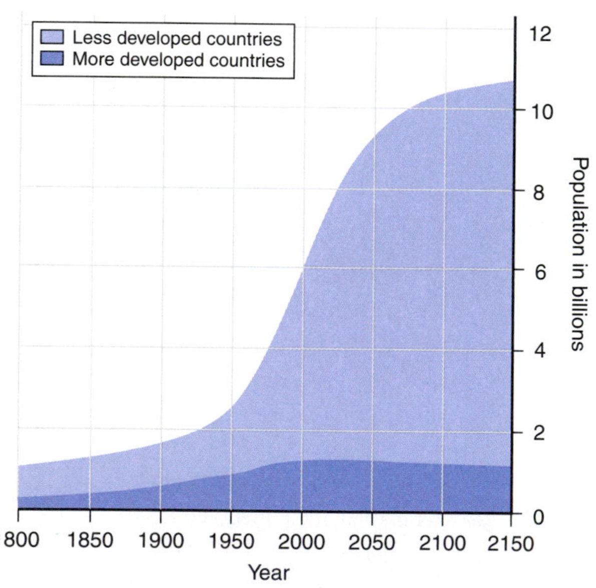

FIGURE 39.12 Projected Human Population Growth Through 2150. Future population growth will be concentrated in less developed countries.

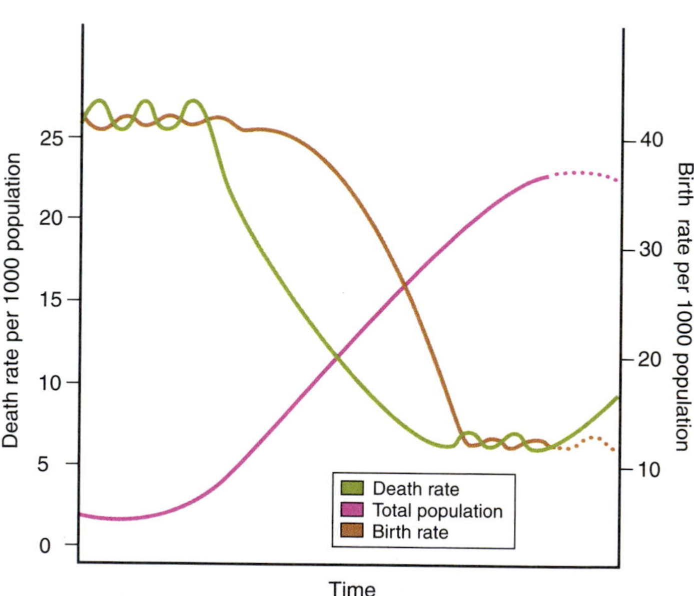

FIGURE 39.13 The Demographic Transition. During the demographic transition, birth and death rates shift from high to low. Eventually, the rates are approximately equal, and the population stabilizes. Population growth is rapid during the second stage, when birth rates exceed death rates.

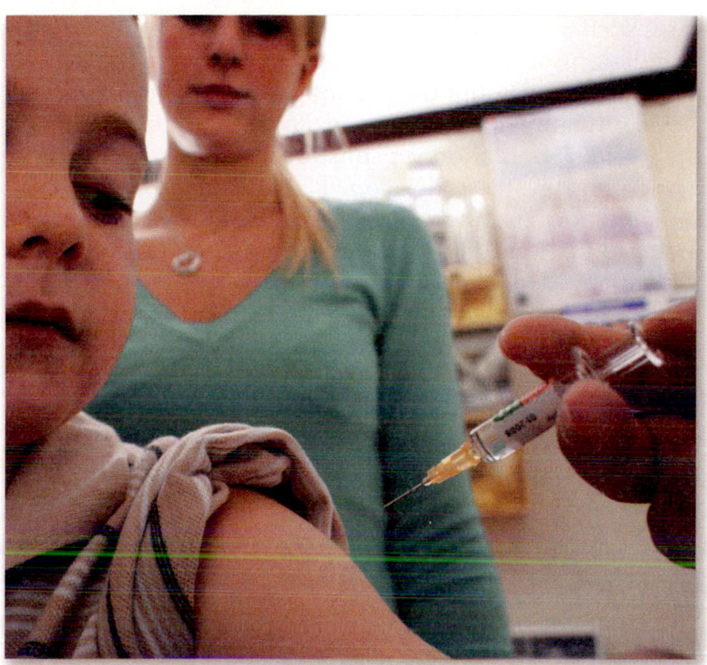

FIGURE 39.14 Lifesaver. Vaccines have saved countless lives since Edward Jenner invented the smallpox vaccine in the eighteenth century. Disease prevention is one factor that has allowed the human population to continue to grow.

Factors Affecting Mortality

Average life expectancy has increased steadily throughout history, from 22 during the Roman Empire to 33 in the Middle Ages. Today, average life expectancy is about 75 in the most developed countries and about 50 in the least developed countries. Improved public health and medical technology account for most of the historical increase (**figure 39.14**). In the future, life expectancy in developed countries will continue to increase as mortality from cancer and heart disease declines.

The spread of AIDS, however, threatens to erase progress in increasing life expectancy. In Sub-Saharan Africa, for example, the AIDS epidemic has significantly affected mortality rates and life expectancy has actually begun to decline. **Table 39.3** shows the effect of AIDS on regional human populations in 2006. The relatively low number of deaths in North America and Western Europe reflect ready access to life-saving anti-HIV medications. **anti-HIV drugs, p. 363**

B. Population Growth Affects Environmental Quality

Worldwide, increasing numbers of people will mean greater pressure on land, water, and air as people demand more resources and generate more waste. The increasing need for food, clean water, and energy all will continue to take their toll.

Food production has kept pace with population growth as farmers have taken advantage of improved crop varieties to increase yields. But to produce more food, people often expand their farms into forest lands. Deforestation is a significant threat to biodiversity, and as these marginal farmlands are abandoned, they may not return to their previous condition. Improper cultivation promotes soil erosion, chemical fertilizers pollute water, and pesticides kill nonpest species. Poor irrigation practices increase soil salinity, forcing farmers to abandon some lands.

Table 39.3 *The AIDS Situation in 2006*

Region	Number of Adults and Children Living with HIV in 2006 (Estimated)	Number of Adult and Child Deaths due to AIDS in 2006 (Estimated)
Sub-Saharan Africa	24,700,000	2,100,000
South and Southeast Asia	7,800,000	590,000
Eastern Europe and Central Asia	1,700,000	84,000
Latin America	1,700,000	65,000
North America	1,400,000	18,000
East Asia	750,000	43,000
Western and Central Europe	740,000	12,000
North Africa and Middle East	460,000	36,000
Caribbean	250,000	19,000
Australia and New Zealand	81,000	4,000

Source: UN AIDS Epidemic Update, December 2006

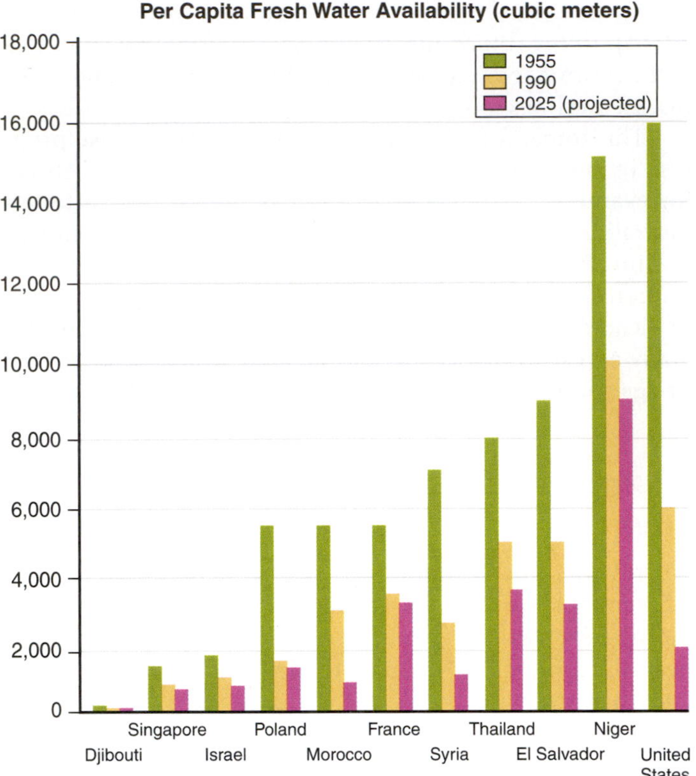

FIGURE 39.15 Water Availability. Access to fresh water is declining worldwide as the population continues to expand.

The availability of fresh water has also declined as people have demanded more water for agriculture, industry, and household use (**figure 39.15**). In many less developed countries, less than half the population has access to safe water for drinking and cooking. As a result, waterborne diseases such as cholera periodically surge through the dense populations of India, Africa, and other locations.

Energy consumption is another significant issue. The world's wealthiest countries make up less than 20% of the world's population yet consume more than half of the world's energy. Less developed countries, however, will probably increase their share of the world's energy use as their populations grow and their economies become more industrialized. Since the vast majority of the energy comes from burning oil, coal, and natural gas, the result will be increased air pollution, acid rain, and carbon dioxide buildup in the atmosphere.

Chapter 42 further explores deforestation, species extinctions, increased fuel consumption, and other environmental problems related to the expanding human population.

Summary *The world's human population is highest near coastlines and rivers. The growth rate of the population is positive but decreasing, reflecting declines in both birth and mortality rates. The highest growth rates occur in less developed countries. Human population biology is complex because it reflects not only ecological principles but also diverse cultural, social, and economic influences. As the human population grows, environmental damage accumulates.*

39.4 MASTERING CONCEPTS

1. Which parts of the world have the highest and lowest rates of human population growth?
2. How does the demographic transition reflect changes in mortality and birth rates?
3. What factors affect birth and mortality rates worldwide?
4. What are some of the environmental consequences of human population growth?

Burning Questions

What will happen to the human population?

No one knows exactly how growth rates will change in the future, so no one can tell exactly when the human population will level off, or what size it will be at that time. In 2006, the United Nations issued three projections for the world's population, assuming high, medium, and low growth rates. The highest projection says the Earth's population will be around 10.8 billion (and still growing) in 2050. The medium estimate shows the population leveling off at around 9 billion in 2050. The low estimate predicts that the population will peak at about 7.8 billion in 2040, then it will decline to just over 7.7 billion by 2050.

These projections are only as good as their assumptions. Will fertility in developing countries continue to decline? If so, by how much, and how fast will it happen? Will fertility decline by the same amount in all countries? Will more developed countries be willing and able to provide family planning services even as the rural populations of less developed countries continue to grow? We will need the answers to these questions before we can know the future of Earth's human population.

Have a Burning Question of your own? Submit it to marielle_hoefnagels@mcgraw-hill.com for possible inclusion in future editions of this book!

39.5 INVESTIGATING LIFE
Let Your Love Light Shine

In sexually reproducing organisms, population growth—the production of new individuals—requires that sperm find their way to egg cells. Many species conduct elaborate games of seduction, in which the spoils of mating go to those who play their cards just right. But why does the game exist in the first place? What prompts a ram to butt heads with a rival, a frog to croak, or a peacock to flourish its magnificent tail? The answer is that these actions draw the attention of choosy mates (usually females) who often decide which males will get to pass their genes to the next generation.

In some species, the signals of the mating game are easy to interpret. The strongest ram gets the ewe by driving out rivals with jaw-jarring furor. His display of brute strength suggests that his offspring also will be able to fend for themselves. In other species, however, winning strategies are more difficult to sort out. Why does an ornamental set of tail feathers turn the drab female peahen's eye instead of, say, a large beak? And what is it about a male frog's croak that is so irresistible to the female of the species? Couldn't she just as easily make a mating decision based on the color of his belly or the length of his toes?

The mystery of sexual selection has long intrigued evolutionary biologists. Chapter 13 described sexual selection as a version of natural selection that results from variation in the ability to get mates. Not surprisingly, individuals with the most mating opportunities usually leave the most offspring. In sexual selection, the female's choice is critical. Her contribution to the next generation's gene pool—that is, her fitness—depends in part on a mate that offers the best provisions for the next generation. Clearly, there is strong selective pressure on females to choose mates based on honest signals. Yet every male conveys many possible signals, including his size, weight, strength, color, calls, and movements. Sorting out which signals are the most important in the female's choice is a challenge.

Tufts University biologists Christopher Cratsley and Sara Lewis sought to unravel the mystery of the sexual dialogue. They studied flashes of light from the firefly *Photinus ignitus* (**figure 39.16**). Though the light charms us on a warm summer's evening, the real function of the bioluminescent flashes is to deliver precise mating information between the sexes.

The consummate act in firefly mating is the transfer of a protein-rich sperm packet, called a spermatophore, from the male to the female's reproductive tract. Because fireflies (which are actually beetles) do not feed as adults, the spermatophore is a nuptial gift complete with genetic information and nutritious amino acids for the developing eggs.

Cratsley and Lewis figured that a fitness advantage went to females that choose males with larger spermatophores. The larger the nuptial gift, the more protein would be available for the offspring, and the better their chance of survival. But how would a female know which male packed the largest spermatophore? The researchers hypothesized that male fireflies might code this information in their flashes.

To test their hypothesis, the researchers watched 36 male fireflies copulate with females in the laboratory. They dissected the females immediately after spermatophore transfer and removed each packet of sperm and protein. The spermatophore weights were analyzed against male flash duration, body mass, lantern width, and lantern area. Of these, flash duration carried the most reliable

FIGURE 39.16 Firefly. A male firefly lets his love light shine.

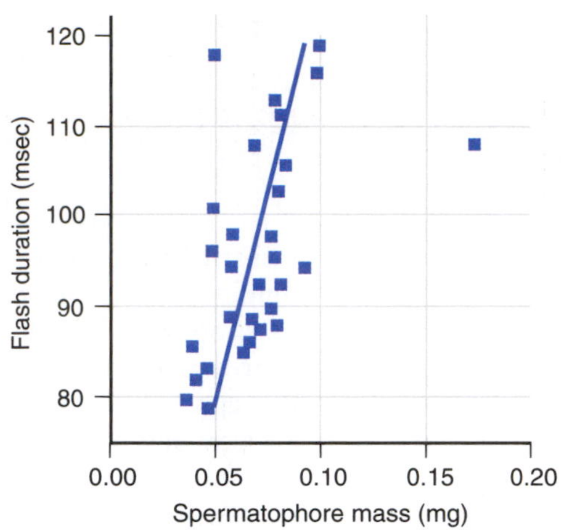

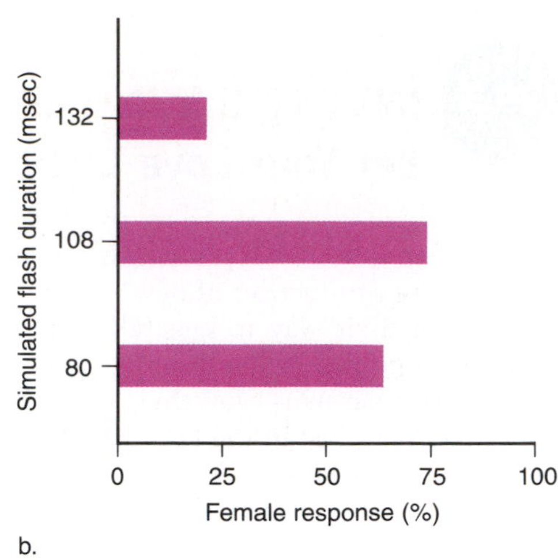

FIGURE 39.17 The Firefly Flash: An Honest Signal. (**A**) Flash duration is positively correlated with the mass of a male firefly's spermatophore, which contains both sperm and nutrition for the future offspring. (**B**) Females responded the most to flashes that were long—but not too long. **Question**: In the short term, what would be the consequence of a male firefly with a small spermatophore that nevertheless "lies" by using a long flash? In the long term, what would happen to the reliability of the flash duration signal if many males "cheated" in this way?

information: the longer the flash, the heavier the spermatophore (**figure 39.17A**).

Next, the team let female fireflies tell them what they liked about the males. Would they prefer the long flashes that signal a large nuptial gift? One way to answer this question would have been to let the females choose between males with different flash durations. However, because males differ in other ways as well, it would have been difficult to conclude that the female preference was for flash duration and not some other variable that the researchers never thought to measure.

The researchers therefore created artificial males. In a series of clever experiments, they flashed a yellow light-emitting diode (LED), and the females responded to the artificial males by giving a return flash if they liked what they saw. The researchers tested females against a range of flash durations: from very fast, 55-ms flashes, to ones much longer than those the males of the species can produce (130 ms). They found that females prefer longer flashes, but within the limit of what the males normally produce (figure 39.17B). Why the upper limit? Apparently, very long flashes resemble those of predatory fireflies. Furthermore, wasting time and energy responding to flashes from the wrong species would reduce the female's reproductive success.

Cratsley and Lewis' experiments support the notion that firefly flashes are much more than the sweet nothings of a romantic tryst. Males advertise the size of their spermatophore gifts, and females respond when the size and flash duration are right. In some ways, these advertisements are similar to commercial messages on TV. Manufacturers of everything from deodorant to dishwashing detergents clamor for your attention, each claiming that their products are best. If you are unsatisfied with your purchase, the worst that can happen is that you buy another brand on your next trip to the store. In the mating game, however, the consequence of a "wrong" decision may make or break an individual's contribution to the next generation.

Cratsley, Christopher K., Sara M. Lewis. 2003. Female preferences for male courtship flashes in *Photinus ignitus* fireflies. *Behavioral Ecology*, vol. 14, pages 135–140.

CHAPTER SUMMARY

A Population Consists of Individuals of One Species

- A **population** is a group of organisms of the same species sharing a geographic region at the same time. A **habitat** is the location where an individual lives.
- **Ecology** considers interrelationships between organisms and their environments. It includes the interactions of individuals within populations and their interactions with individuals of other species in **communities**.

A. Density and Dispersion Patterns Are Static Measures of a Population

- **Population density** is a measure of the number of individuals per unit area, and **population dispersion** describes how the individuals are distributed within the habitat.

B. Population Size May Increase, Decrease, or Remain Stable

- The factors that determine a population's size over time are part of the study of **population dynamics**.
- A population grows when more individuals are added through birth or **immigration** than leave due to death or **emigration**.
- **Birth rates** reflect the number of reproductive episodes, the number of offspring per reproductive episode, and the age at which individuals begin to reproduce. A population's **age structure** predicts whether the population will grow, remain stable, or shrink.
- The **death rate** reflects the number of deaths per unit time.
- A **life table** reflects the chance of dying at any given age. **Survivorship curves** fall into three patterns, reflecting the balance between number of offspring and the amount of parental investment in each.

39.2 Population Growth May Be Exponential or Logistic

A. Growth Is Exponential When Resources Are Unlimited

- The difference between the birth rate and the death rate is *r*, the **per capita rate of increase**. Population growth that is proportional to the size of the population is **exponential** and produces a **J-shaped curve.**

B. Population Growth Eventually Slows

- In response to **environmental resistance**, the population may stabilize at its **carrying capacity**, which is the number of individuals an environment can indefinitely support.
- Plotting population size over time produces a **logistic growth curve**, also called an **S-shaped curve**.

C. Many Conditions Limit Population Size

- **Density-dependent factors** such as infectious disease, predation, and competition have the greatest effect on large populations.
- **Density-independent** factors such as natural disasters kill the same fraction of the population regardless of the population's density.

Natural Selection Influences Reproductive Strategies

A. *r*- and *K*-Selected Species Have Different Life Histories

- Different **life histories** reflect a trade-off between the number of offspring and parental investment in each.
- ***K*-selected species** invest heavily in rearing relatively few young.
- In contrast, ***r*-selected species** produce many offspring but expend little energy on each.

B. Guppies Illustrate the Importance of Natural Selection

- Predation experiments on guppies show that natural selection influences a population's life history characteristics.

39.4 The Human Population Continues to Grow

A. Population Dynamics Reflect the Demographic Transition

- Population growth is uneven across the world, reflecting different stages in the **demographic transition**.
- In less-developed countries, birth rates are high, but death rates are low, producing rapid population growth. As economic development increases, birth rates decline, and population growth slows.
- Education, access to contraceptives, and government policies affect birth rates worldwide.
- Although average life expectancy has increased steadily for centuries, AIDS is reducing life expectancy in some parts of the world.

B. Population Growth Affects Environmental Quality

- Sustained human population growth will continue to strain supplies of natural resources such as farmland, clean water, and fossil fuels.

39.5 Investigating Life: Let Your Love Light Shine

- Males use many signals to display their qualities to females. Understanding the relationship between fitness, a suitor's signal, and the female's detection system is the key to understanding sexual selection.
- Female fireflies prefer males that generate long flashes of light, a signal that is positively correlated with the size of his spermatophore. The larger the spermatophore, the more nutrition available to the firefly pair's offspring.

MULTIPLE CHOICE QUESTIONS

1. What is the definition of a population?
   a. A group of multiple species coexisting within a defined location
   b. A group of animals that exist within a specific habitat
   c. A group of organisms of a single species within a specific location
   d. A group of genetically identical organisms
2. How does population dispersion affect population density?
   a. It affects the pattern of density across the habitat.
   b. It causes the number of individuals to decrease as dispersion increases.
   c. Individuals become more dispersed as their numbers increase.
   d. There is no relationship between these two concepts.
3. What process, other than birth, can affect population density?
   a. Immigration
   b. Emigration
   c. Death
   d. All of the above
4. Use Table 39.1 to determine the age that half of the scorpion cohort would be predicted to reach.
   a. 1 month
   b. 6 months
   c. 12 months
   d. 16 months
5. A population that has a consistent survivorship rate is a ______ species.
   a. Type I
   b. Type II
   c. Type III
   d. Type IV
6. In a population with exponential growth, how many individuals will there be after four generations if the starting population is 20?
   a. 20
   b. 40
   c. 320
   d. 640
7. What determines the carrying capacity of a habitat?
   a. The resources available within the environment
   b. The doubling time of a species
   c. The birth rate for a population
   d. Both b and c
8. Which of the following is NOT an example of a density-dependent factor?
   a. Predation
   b. Weather
   c. Disease
   d. Competition for food
9. The life histories of guppy populations with high predation intensity on adults shifted toward
   a. constant survivorship.
   b. *r*-selection.
   c. *K*-selection.
   d. a mix of *r*- and *K*-selection.
10. A decline in growth rates can be attributed to a large percentage of a population in its
    a. post-reproductive years.
    b. reproductive years.
    c. pre-reproductive years.
    d. Both b and c.

TESTING YOUR KNOWLEDGE

1. State two descriptive measures of population structure.
2. Construct a survivorship curve for the desert sand scorpion (see table 39.1), using a logarithmic scale for the *y*-axis.
3. How does a population's age structure help predict the population's future?
4. Define the following terms: per capita rate of increase, environmental resistance, carrying capacity.
5. What are the differences between the conditions that result in exponential versus logistic population growth?
6. The rat (*Rattus norvegicus*) has a per capita rate of increase (r) of 5.48/year. Assuming exponential growth and an initial population of 50 rats, what will be the population size after 1 year? The bacterium *E. coli* has an r of 2.5/h. Assuming exponential growth and an initial colony size of 10 cells, how many cells will be in the colony after 1 week of growth? List three factors that prevent each species from achieving this reproductive potential.
7. In 2006, the birth rate in the United States was 14.14 births per 1000 population, and the death rate

was 8.26 deaths per 1000. Immigration added 3.18 migrants per 1000 population. Calculate the overall growth rate of the U.S. population in 2006.

8. How do density-independent and density-dependent factors affect population growth?
9. What keeps predator populations from eliminating prey populations?
10. Distinguish between *K*-selection and *r*-selection, and give an example of an organism with each type of life history.
11. How can the global human population continue to increase for several more decades even though the rate of increase is declining?

THINKING AS A SCIENTIST

1. Biologists often use rectangular sampling plots called quadrats in studies of plant population ecology. Suppose you want to determine which type of habitat is most favorable for the growth of a particular weed species. How could you use a quadrat of known area to measure the weed's population density in different habitats? Since plants are rarely distributed uniformly throughout a habitat, how might you ensure random sampling?
2. Describe a method for estimating the number of elk in an area. What clues might help you understand whether the elk were below or above their habitat's carrying capacity?
3. Decades of overfishing led to the collapse of the cod fishing industry off the coast of North America in the 1980s and 1990s. Given the effect of birth rates, death rates, and age structure on a population, propose an explanation for the decline of the cod population. What sorts of policies might prevent the cod from extinction?
4. In Pakistan, 43% of the population is younger than age 15, and 4% is older than 65. In the United Kingdom, 19% of the population is younger than 15, and 15% is older than 65. Which population will increase faster in the future? Explain.
5. Cite three recent environmental upheavals that may have had a density-independent influence on wildlife populations.
6. Because of historical, cultural preferences for boys, an unintended result of China's "one child" policy has been a skewing of the sex ratio toward males. In China, about 120 boys are born for every 100 girls. What effects might this have on future population growth in China? Can you foresee any social consequences of having so many females "missing" from the population?
7. The fur of the rock pocket mouse (*Chaetodipus intermedius*), a small mammal that lives in the southwestern United States, comes in two colors. Light variants typically live on sandy-colored rocks, and the dark mice live on black basalt. Propose a hypothesis that explains this observation, then design one or more experiments to test your hypothesis.

Visit www.mhhe.com/hoefnagels for practice quizzes, animations, videos, and activities designed to help you master the material in this chapter.

CHAPTER 40

Communities and Ecosystems

Sea otters are keystone species in the Pacific Northwest.

What's the Point?

Otter Deaths Topple a Food Web in the Pacific Northwest

Populations interact. Individuals of different species live in or on one another, eat one another, and compete with one another. Some species are called keystone species because they are so important to their communities that their removal can dismantle a food web. The keystone species in the vast underwater kelp forests is the sea otter, Enhydra lutris, *which eats sea urchins that would otherwise devour the kelp.*

Otters are disappearing in ocean and coastline ecosystems near the Aleutian Islands in the Pacific Northwest. Marine ecologist James Estes and his coworkers first noticed a decline in the sea otter population in the 1990s. Populations go down when fertility decreases, mortality increases, or individuals migrate (see chapter 39). However, by tracking otters with radio collars, they knew that birth rates were not changing, and they detected no redistribution of otters from one island to another. They therefore concluded that an increase in mortality was responsible. Further study with radio-tagged otters ruled out disease, pollution, or reproductive problems as the cause of the apparent otter depletion. One clue was that no dead otters were washing up on beaches. Could a new type of predator have been responsible?

The next clue came from an observant researcher in the group, who saw an orca (killer whale) kill an otter, unusual because the two species had lived together for decades. Then other researchers noticed similar attacks. Whales were killing the otters. Adak Island had two lagoons with otters: one where killer whales could enter, and one where they could not. Where orcas had access, the otter population plunged by two thirds, but where they didn't, only 12% of the otter population perished.

By considering these clues and taking into account other members of the ocean and coastline ecosystems, the researchers developed a hypothesis to explain the shifting food web. The events leading to the otters' demise began with a decrease in populations of oily, nutrient-rich fishes such as herring and ocean perch. When Steller sea lions and harbor seals ate a less nutrient-packed fish called pollock instead of their usual oilier fish fare, they began to die from malnutrition. The orcas that normally eat sea lions and seals sought another food source, and they turned to the coastal ecosystem and its abundant otters. As a result, by 1997 the population of otters declined by a factor of 10 over an 800-km (500-mile) expanse of coastline.

Interacting populations sustain every ecosystem on Earth, and the ocean is no exception. The killer whales and their prey form just one strand in its complex web of interactions. This chapter describes the ties that bind species together, on land and in the sea.

LEARNING OUTLINE

40.1 Communities Are the Living Parts of Ecosystems

Just as a human community consists of many people that interact in different ways, an ecological **community** is a group of interacting populations that inhabit the same region. A downed Douglas fir, for example, houses a bustling **biotic,** or living, community of mosses, beetles, parasitic wasps, wood decay fungi, and bacteria (**figure 40.1**). Of course, no community lives in isolation of its physical and chemical surroundings. An **ecosystem** includes all the organisms plus the **abiotic,** or nonliving, environment within a defined area.

The first half of this chapter describes community-level interactions between individual populations. Sections 40.3 and 40.4 consider ecosystem-level interactions.

A. Populations Interact in Many Ways to Form Communities

Communities usually consist of enormous numbers of species. Some are easily visible, whereas others are microscopic. One on one, their interactions may seem simple—an orca eats an otter, or a wasp kills a caterpillar. But an attempt to map all interactions within a community quickly becomes complicated. Individuals of different species compete for limited resources, live in or on one another, eat one another, and try to avoid being eaten. These community-level interactions are such important forces in natural selection that they shape many of the adaptations that characterize each species.

Some connections between species are so strong that the species directly influence one another's evolution. In **coevolution,** a genetic change in one species selects for subsequent change in the genome of another species. Of course, all interacting species in one community have the potential to influence one another, and they are all "evolving together." These genetic changes are considered coevolution only if scientists can demonstrate that adaptations specifically result from the interactions among the species.

FIGURE 40.1 **A Rotten Home.** Rotting wood houses a changing community of insects, worms, fungi, and microorganisms.

FIGURE 40.2 **Coevolution of Ants and Plants.** Ants and acacias have adaptations that suggest the two species have coevolved.

Some flowering plants, for example, have coevolved with insects. A plant may rely exclusively on one insect species for pollination, and the insect eats nothing but nectar from that plant (see chapter 26). Another example is the intimate relationship between ants (*Pseudomyrmex* species) and acacia trees (**figure 40.2**). Each species has adaptations not found among related species that lack the relationship. For example, the acacia produces swollen, hollow thorns in which the ants live, along with nectar and nutritious nodules that the ants consume. These adaptations have, in turn, selected for ant behaviors that defend "their" tree from other herbivores.

Competition Prevents Species from Occupying the Same Niche

Each species in a community has a characteristic home and way of life. Recall from chapter 39 that a **habitat** is the physical place where members of a population typically live, such as a rain forest canopy or the bottom of a river. The habitat is a subset of the **niche,** which is the total of all the resources a species exploits for its survival, growth, and reproduction. The niche also includes the salinity, temperature, light, water availability, and other abiotic conditions where the species lives.

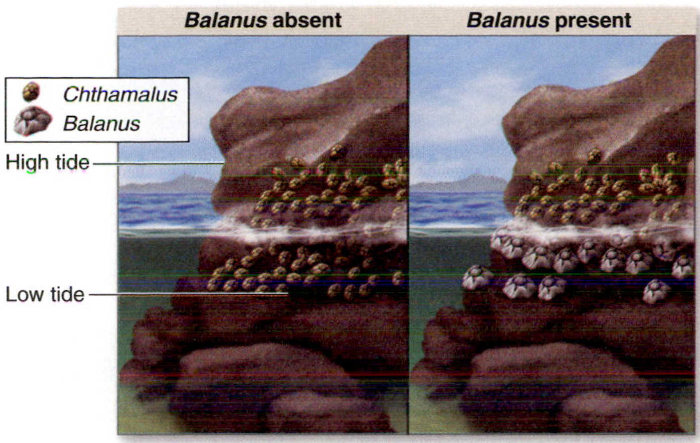

FIGURE 40.3 Competition Restricts a Barnacle's Niche. When the barnacle *Balanus* is absent, *Chthamalus* adults occupy the entire intertidal zone. *Balanus* grows faster than *Chthamalus*, however, and when *Balanus* is present, competition for space limits *Chthamalus* to the upper intertidal zone.

Competition occurs when two or more species vie for the same limited resource, such as food, shelter, nutrients, water, or light. If two or more species with similar niches live in the same area, the competition between them may restrict each species to only some of the resources available. Two species of barnacles, for example, live in the intertidal zone along Scotland's shoreline (**figure 40.3**). By itself, either type of barnacle can survive throughout the intertidal zone. When both species are present, however, the faster-growing species crowds out its competitor in the lower, moister region of the intertidal zone. The slower growing species, however, better tolerates dehydration while the tide is out. It can therefore more efficiently use the resources of the upper region of the intertidal zone.

The barnacle example illustrates the **competitive exclusion principle,** which says that two species cannot coexist indefinitely in the same niche. The two species will compete for the limited resources that they both require, such as food, nesting sites, or soil nutrients. The species that acquires more of the resources will eventually replace the other, less successful species. (Competitive exclusion explains how the microbes that normally occupy the human intestinal tract provide such good protection against harmful invaders; see chapter 34.)

The displacement of native species by introduced species occasionally disrupts whole communities. Zebra mussels (*Dreissena polymorpha*), for example, are native to the Caspian Sea in Asia. They were accidentally introduced to the Great Lakes in the 1980s and have since spread to many waterways in the United States and Canada. The tiny filter feeders have crowded out native mussel species by competitive exclusion, and they have changed plant communities by greatly increasing water clarity. Consequently, the community of fishes has also changed in waterways that zebra mussels inhabit.

Note that the competitive exclusion principle does not rule out the possibility of niche overlap. Species with similar needs can (and do) coexist in communities. In **resource partitioning,** multiple species use the same resource in a slightly different way or at a different time (**figure 40.4**). For example, five species of warblers, all small insect eaters, coexist in the same trees in New England forests. Birds of each species feed in different ways and on different parts of the tree.

Species Live in and on One Another

In a **symbiosis** (literally, "living together"), one species lives in or on another. (Chapter 16 describes the ancient

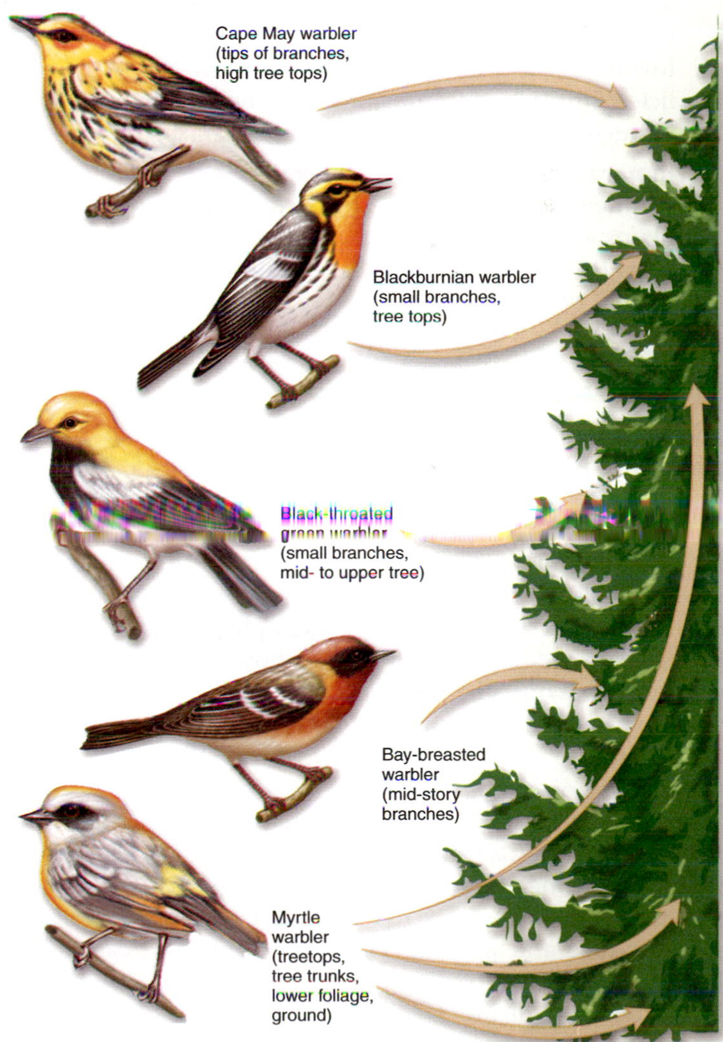

FIGURE 40.4 Resource Partitioning. These five species of warblers live together in conifer forests, but each has a different primary feeding zone.

Table 40.1 *Types of Symbiosis*

Type	Definition	Examples
Mutualism	Both partners benefit	Algae in coral animals; mycorrhizal fungi in plant roots
Commensalism	One partner benefits with no effect on the other	Moss plants on tree bark
Parasitism	One partner benefits to the detriment of the other	Tick on a deer; tapeworm in a human (see chapters 18, 19, 21, and 22 for many examples of disease-causing organisms)

"endosymbiosis" that is thought to account for the origin of eukaryotic cells.) The relationship between symbiotic species may take many forms (**table 40.1**):

- **Mutualistic** relationships are beneficial to both partners, such as when a cow derives simple sugars from the cellulose-digesting microbes in its rumen. The ant-acacia alliance is another example.
- **Commensalism** is a type of symbiosis in which one species benefits, but the other is not significantly affected. Most humans, for example, never notice the tiny mites that live, eat, and breed in their hair follicles. Similarly, trees are apparently neither helped nor harmed by mosses that grow on their trunks.
- In **parasitism,** one species benefits at the expense of another. The parasites may be bacteria, protists, fungi, plants, or animals. Mistletoe, for example, is a parasitic plant that taps into the water- and nutrient-conducting "pipes" of a host plant. **vascular tissue, p. 508**

Predators Eat Other Organisms

Species also interact when one, a **predator,** eats another, the **prey.** Predation exerts strong selective pressure on prey species, which often have adaptations that help them avoid being eaten (**figure 40.5**). Camouflage and warning coloration are two examples. An interesting variation on the theme of warning coloration is mimicry, in which different species develop similar appearances. For example, a harmless species of fly may deter predators with yellow and black stripes similar to that of a stinging bee.

Prey species also commonly have weapons and structural defenses. Some plant species produce thorns, distasteful or poisonous chemicals, or milky sap that deters herbivores. Many animals have hard shells, pinchers, or other defensive adaptations (some of which are also useful in capturing their own prey). Likewise, only predators that can defeat prey defenses will live long enough to reproduce and provide for their young. The caterpillars of monarch butterflies, for example, tolerate the noxious chemicals in milkweed plants. Tigers and other big cats have markings that hide their shape against their surroundings, which helps them sneak up on their prey. Hunting in groups helps many predators capture prey efficiently (**figure 40.6**).

a.

b.

c.

FIGURE 40.5 Prey Defenses. (**A**) Camouflage helps prey species hide from predators. This insect from Madagascar resembles the leaves in its habitat. (**B**) Warning coloration advertises a poisonous species' defenses. (**C**) These animals look like ants, but each has eight legs, not six. They are ant-mimicking jumping spiders, with none of the weaponry of the ants they resemble.

FIGURE 40.6 Predators Sometimes Cooperate. By working together, a pack of lions can bring down prey animals that are much larger than the lions themselves.

B. A Keystone Species Has a Pivotal Role in the Community

Sometimes, many species in a community depend on one type of organism, such as the otter described in the chapter opening essay. A **keystone species** makes up a small portion of the community by weight, yet exerts a disproportionate influence on community diversity. (The producer species that support all communities are not considered keystone species because they make up the bulk of a community.)

Many keystone species are versatile predators, such as sea stars that prey on diverse tide pool invertebrates, or birds that eat herbivorous insects. But not all keystone species are animals. For example, root-associated fungi help coniferous trees acquire nutrients from soil, and they produce underground fruiting bodies that small rodents eat. Owls and other predators hunt these small mammals. The fungi are keystone species because neither the trees nor the animals can live without them. **mycorrhizal fungi, p. 440**

Summary *Populations that share a community exert selective forces on one another. Competition may lead a species to exploit a subset of the resources in its habitat. Other interactions among species include symbiosis and predation. Keystone species have especially important roles in communities.*

40.1 MASTERING CONCEPTS

1. Distinguish among communities, ecosystems, and populations.
2. Name some of the abiotic and biotic components of your environment.
3. Define *coevolution,* and describe an example.
4. Distinguish between habitat and niche.
5. What is the competitive exclusion principle?
6. List three examples of symbiotic relationships between species.
7. Describe some adaptations of predator and prey species.
8. How are keystone species important in communities?

40.2 Communities Change Over Time

Communities may appear deceptively stable, because we usually only observe them over a relatively short period. **Succession** is a gradual change in a community's species composition. Succession occurs as competing organisms, especially plants, respond to and modify the physical environment. When pine trees invade a site, for example, they shade out lower growing plants while attracting species that grow or feed on pines. Eventually, succession may lead to a **climax community,** which is a community that remains fairly constant.

A. In Primary Succession, a New Community Arises

Ecologists define two major types of succession: primary and secondary. **Primary succession** occurs in an area where no community previously existed. When a volcano erupts, for example, lava may obliterate existing life, a little like suddenly replacing an intricate painting with a blank canvas. Road cuts and glaciers that scour the landscape also expose virtually lifeless areas on which new communities eventually arise.

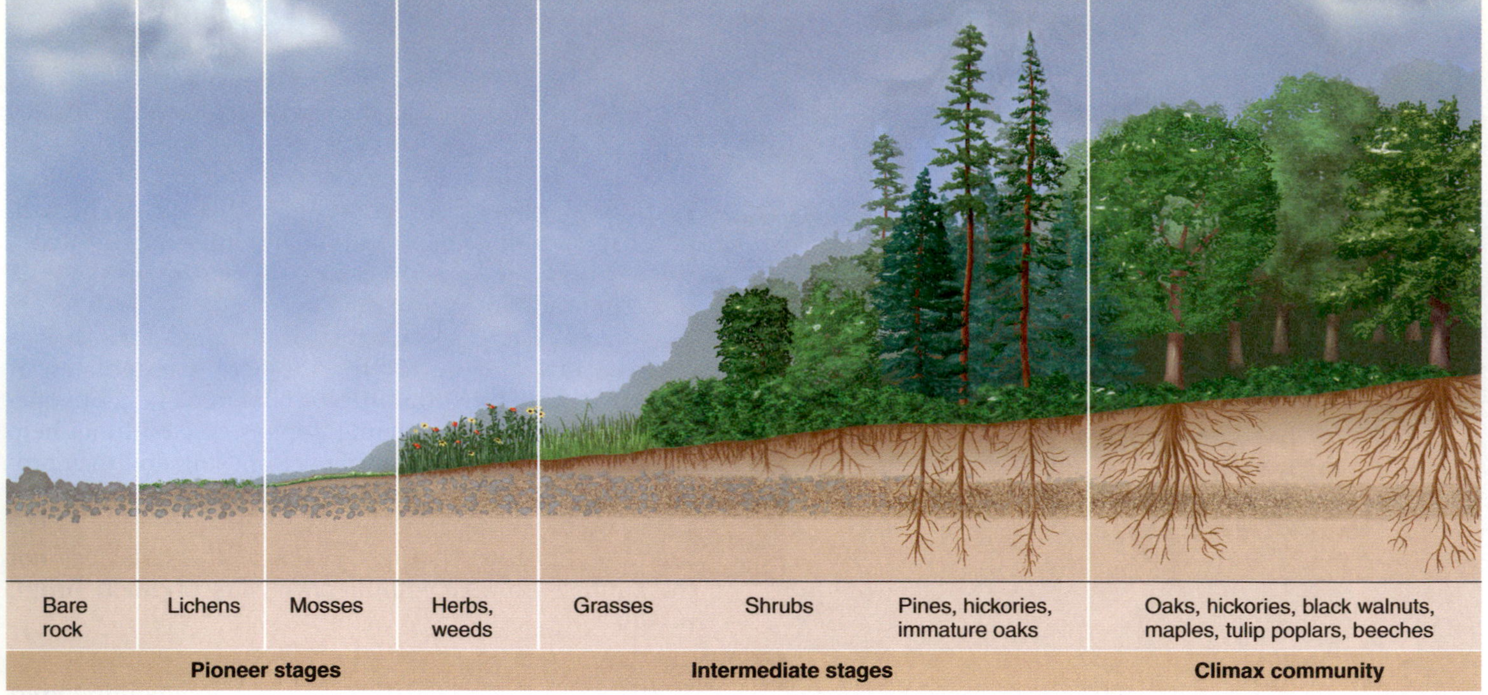

FIGURE 40.7 Primary Succession. It takes centuries for a climax community to develop on a patch of bare rock. The community shown here includes species typical of New England.

A patch of bare rock also provides a clear view of primary succession (**figure 40.7**). **Pioneer species** are first to colonize the area. These hardy organisms, such as lichens and mosses, can grow on smooth rock. As lichens produce organic acids that erode the rock, sand and dust accumulate in the crevices. Decomposing lichens add organic material, eventually forming a thin covering of soil. Microorganisms, worms, and insects colonize the forming soil, contributing organic matter as they live and die. Then rooted plants such as herbs and grasses invade. Soil continues to form, and larger plants, such as shrubs, appear. Larger animals move into the area. Next come aspens and conifers, such as jack pines or black spruces. Finally, hundreds of years after lichens first colonized bare rock, the soil becomes rich enough to support other deciduous trees, and a climax community of an oak–hickory forest may develop. **lichens, p. 441**

B. Secondary Succession Replaces a Disturbed Community

Secondary succession occurs where a community is disturbed but not destroyed, with some soil and life remaining. Because the area isn't completely devastated, secondary succession occurs faster than primary succession. Fires, hurricanes, and agriculture commonly trigger secondary succession (**figure 40.8**).

An abandoned farm in the eastern United States illustrates secondary succession. This "old field" succession begins when the original deciduous forest is cut down for farmland. As long as crops are cultivated, succession stops. When the land is no longer farmed, fast-growing pioneer species, such as black mustard, wild carrot, and dandelion, move in, followed by slower growing, taller goldenrod and perennial grasses. In a few years, trees such as pin cherries and aspens arrive; pine and oak even-

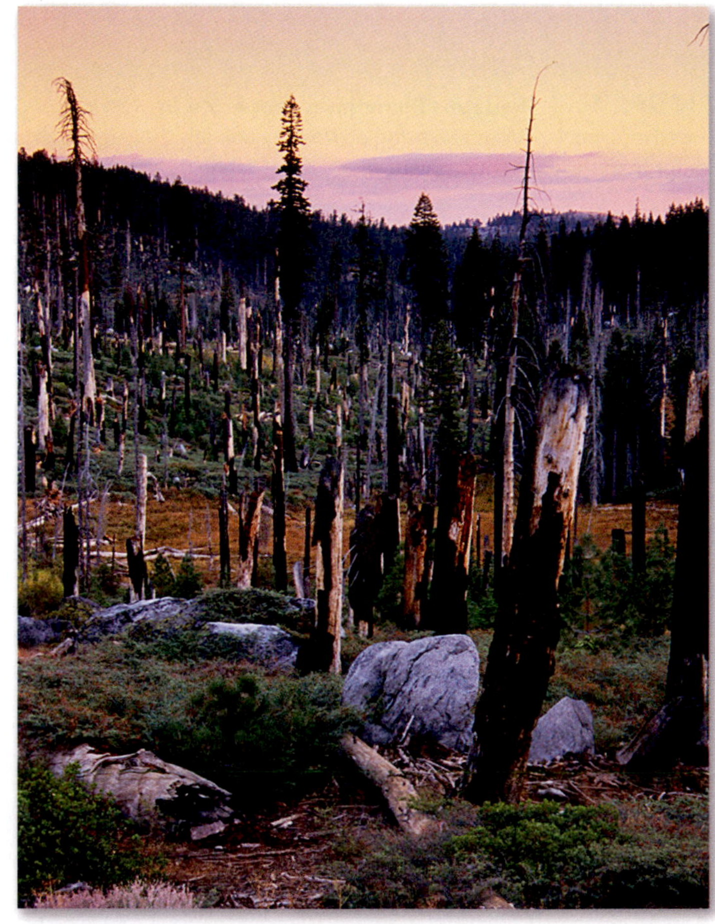

FIGURE 40.8 Secondary Succession. A forest fire can devastate an existing community. But it does not take long for seedlings to sprout and take advantage of the minerals in the tree's ashes. Soon, fresh green foliage will obscure the burned stumps.

tually replace them. A century or so later, the climax community of beech and maple may again be well developed.

C. Succession Can Be Complex

Ecologists recognize a common set of processes in primary and secondary succession. First, pioneer species colonize a bare or disturbed site. Recall from chapter 39 that the hardy pioneer species are usually *r*-selected, with rapid reproduction and efficient dispersal. Interestingly, these early colonists often alter the physical conditions in ways that enable other species to become established. These new arrivals continue to change the environment. Some early colonists do not survive the new challenges, further altering the community. The species that appear in the climax community are usually long-lived, late-maturing, *K*-selected species that are strong competitors in a stable environment.

Few communities ever reach true climax conditions. Pockets of local disturbance, such as the area affected when a large tree blows over, create a patchy distribution of successional stages across a landscape. Major disturbances such as fire, disease, and severe storms can influence successional patterns for centuries. Usually, the rate of change slows late in the process, but succession never ceases. In the Pacific Northwest, for example, old-growth forests are 500 to 1000 years old, yet they are still changing in their structure and composition.

Summary ***In primary succession, a new community forms; in secondary succession, the types of resident species change after a disturbance. In both types of succession, pioneer species pave the way for changes in the community and the physical environment.***

40.2 MASTERING CONCEPTS

1. What is succession?
2. What processes and events contribute to succession?
3. How common are climax communities?

40.3 Ecosystems Require Continuous Energy Input

All ecosystems share two properties. First, all ecosystems rely on a continuous supply of energy from some outside source. Second, nutrient cycles continuously recycle the atoms that make up every object in an ecosystem. This section and the next describe these two properties and their consequences for ecosystem function. This chapter's Burning Question: Could human life be supported in space or on Mars? considers the possibility of building artificial ecosystems.

A. Energy Passes Through Food Chains

Unlike chemicals, which cycle and recycle between organisms and their environment, energy flows through ecosystems in one direction only. The ultimate source of energy that fuels most ecosystems on Earth is sunlight. Plants, algae, and some microorganisms use photosynthesis to trap solar energy in the bonds of organic chemicals such as glucose. But a few ecosystems do not rely on sunlight. Some bacteria and archaea can extract energy from inorganic chemicals such as hydrogen (H_2). As the opening essay to chapter 41 describes, hydrothermal vents house countless microbes that form the bases of complex communities that never see the sun. **photosynthesis, p. 102**

A **food chain** is a series of organisms that successively eat one another. An organism's **trophic level** is its position in the food chain, relative to the ecosystem's energy source (**figure 40.9**). The first trophic level in any food chain is a primary producer. A **primary producer,** or **autotroph**

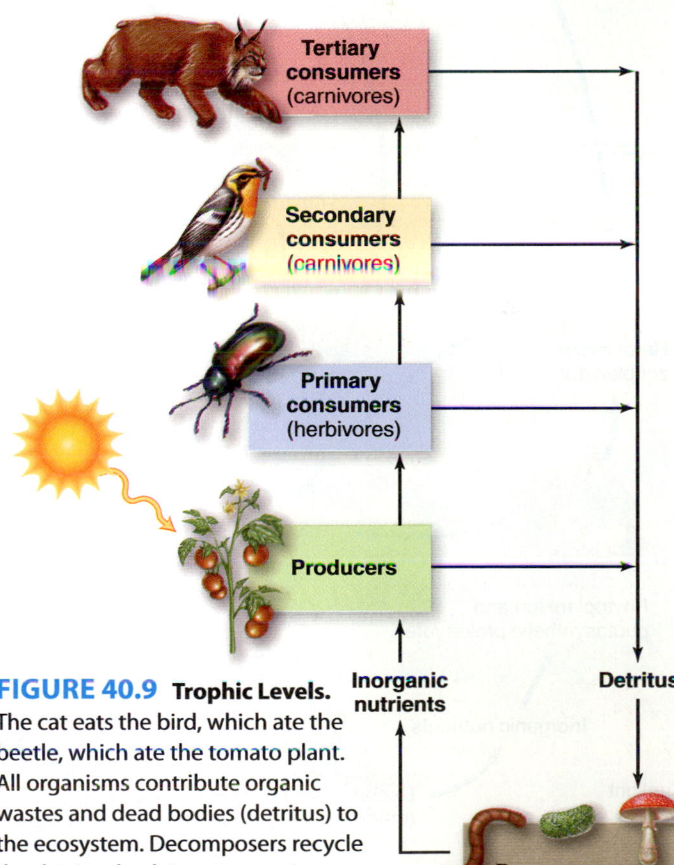

FIGURE 40.9 Trophic Levels. The cat eats the bird, which ate the beetle, which ate the tomato plant. All organisms contribute organic wastes and dead bodies (detritus) to the ecosystem. Decomposers recycle the detritus back into inorganic nutrients that producers can use.

("self-feeder"), is any organism that can use energy, CO_2, H_2O, and other inorganic substances to produce all the organic material it requires. Directly or indirectly, primary producers provide energy for all other organisms. **organic molecules, p. 33**

Consumers, or **heterotrophs** ("other eaters") obtain energy from producers or other consumers. Primary consumers are organisms such as herbivores (plant-eaters) that eat producers. They form the second trophic level. Secondary consumers are carnivores (meat-eaters) that eat primary consumers, and tertiary consumers eat secondary consumers. **Decomposers,** such as many fungi, bacteria, insects, and worms, break down **detritus** (dead organisms and feces).

Like autotrophs, decomposers are critical to ecosystem function. Whereas autotrophs absorb inorganic nutrients and produce organic molecules, decomposers do the opposite—they return those organic molecules to their inorganic form. (Can *You* Relate? What Happens After You Flush? describes how we employ decomposers in community wastewater treatment facilities.) Without decomposers, dead bodies and organic wastes would tie up all useful nutrients, and ecosystems would grind to a halt.

Of course, feeding relationships in an ecosystem are more complex than a simple food chain might suggest. A **food web** is a network of interconnected food chains, such as the Antarctic web in **figure 40.10**. Careful examination of

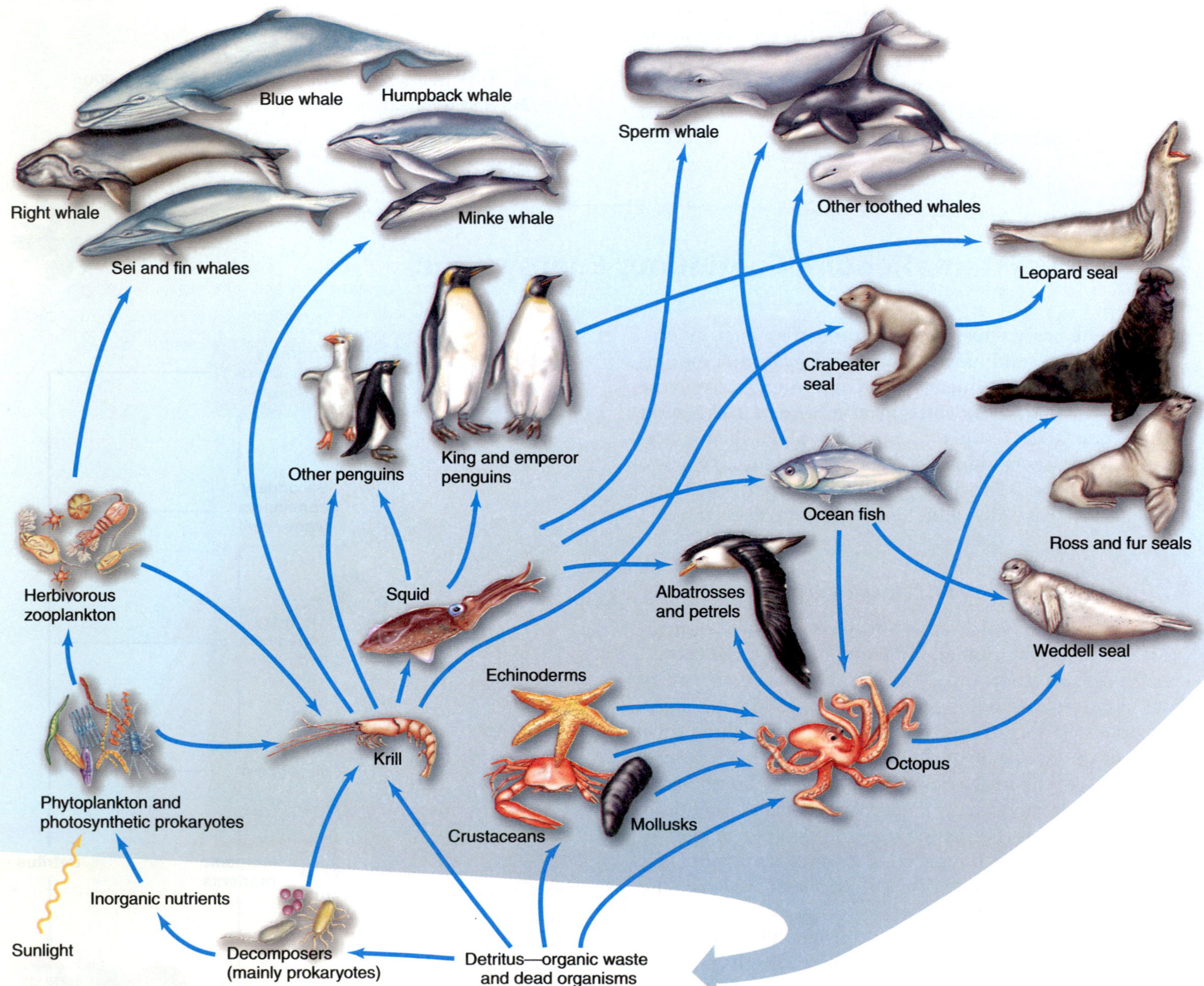

FIGURE 40.10 The Antarctic Web of Life. As in most food webs, the multiple interactions among Antarctic residents form a very complex network. Note that producers, consumers, and scavengers are all present, and that decomposers return inorganic nutrients that producers require.

a food web diagram reveals that even the fiercest top predator relies on other organisms, many of them microscopic. As the chapter opening essay describes, food webs can unravel if a keystone species vanishes.

B. Every Trophic Level "Wastes" Energy

The total amount of energy that is trapped, or "fixed," by all autotrophs in an ecosystem is called gross primary production. Autotrophs use much of this energy to generate ATP for their own needs. As they do so, they lose heat energy (the second law of thermodynamics, described in section 4.1, explains this loss). The remaining energy in the producer level is called **net primary productivity;** it is the amount of energy available for consumers to eat. Primary productivity varies widely across the globe. Wetlands and tropical rain forests have among the highest rates of net primary productivity per square meter; deserts have the lowest.

Consumers are no better than producers at conserving energy; they cannot digest every bit of food they eat, and they lose energy to heat. Because of these inefficiencies, only a small fraction of the potential energy stored in the bonds of organic molecules at one trophic level fuels growth and reproduction of organisms at the next trophic level. As an overall average, about one-tenth of the energy at one trophic level is available to the next highest rank in the food chain.

The "10% rule" provides a convenient estimate, but it ignores the fact that food quality varies widely. The transfer efficiency from one trophic level to the next actually ranges from about 2% to 30%. Primary consumers that eat hard-to-digest plants convert only a small percentage of the energy available to them into animal tissue, whereas meat is easy to digest. In addition, ectothermic animals such as invertebrates and reptiles use energy much more efficiently than do endothermic mammals and birds. A trophic level consisting of lizards therefore consumes much less energy than a trophic level consisting of an equal weight of birds. **endotherms and ectotherms, p. 712**

Eventually, as organic molecules pass from trophic level to trophic level, all the stored energy is lost as heat. The heat energy leaves the ecosystem forever, because no organism can use heat as its energy source. Thus, energy flows through an ecosystem in one direction: from source (usually the sun), through organisms, to heat. For the ecosystem to persist, it must have a continual supply of energy. If the energy source goes away, so does the ecosystem.

C. Ecological Pyramids Describe Ecosystem Characteristics

A **pyramid of energy** represents each trophic level as a block whose size is directly proportional to the energy stored in new tissues per unit time (**figure 40.11**A). Because every organism loses energy to heat, the energy pyramid explains why food chains rarely extend beyond four trophic levels. An organism in a still higher trophic level would have to expend tremendous effort just to find the small amount of food available, and that small amount would not be enough to make all that effort pay off. The energy pyramid also helps explain why the world's largest animals, such as sauropod dinosaurs and elephants, have been herbivores. Only the producer level contains sufficient energy to fuel the growth of an enormous consumer.

The loss of energy at each trophic level suggests a way to maximize the benefit we get from crops we grow for food. The most energy available in an ecosystem is at the producer

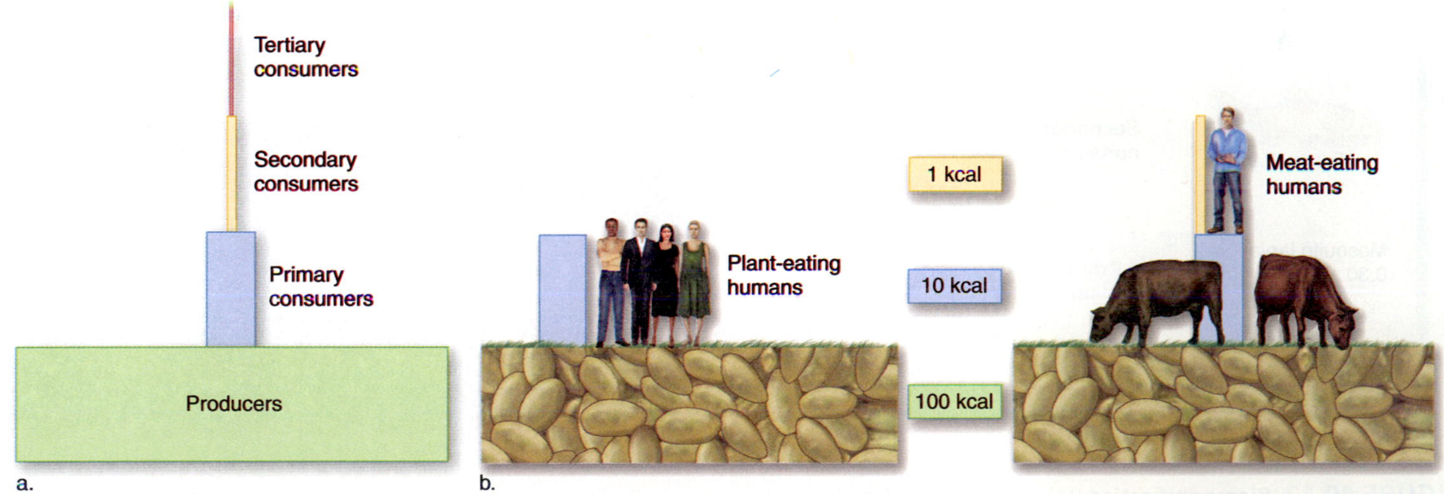

FIGURE 40.11 Pyramid of Energy. **(A)** A small percentage of the energy stored at one trophic level per unit time is transferred and stored in new growth and reproduction at the next level. Decomposers are omitted in this figure. **(B)** Humans who derive energy by eating meat are getting only a small fraction of the energy originally present in grain. This example assumes an average of 10% of the energy in any trophic level is available to the next. In reality, the value varies from about 2 to 30%, depending on the food source and the consumer.

level. The lower we eat on the food chain, the more people we can feed. A person can do this by getting protein from beans, grains, and nuts instead of meat (figure 40.11B).

D. Harmful Chemicals May Accumulate in the Highest Trophic Levels

The shape of the energy pyramid has another consequence for ecosystems. In **biomagnification,** a chemical becomes more and more concentrated in organisms at successively higher trophic levels. Biomagnification happens for pollutants and other chemicals that share two characteristics. First, they dissolve in fat. (Animals eliminate water-soluble chemicals in their urine.) Second, chemicals that biomagnify are not readily degraded. A highly degradable chemical would not persist long enough in the environment to ascend food chains.

DDT is a fat-soluble, persistent chemical that illustrates biomagnification (**figure 40.12A**). This insecticide was once widely used to kill pests such as mosquitoes and body lice. Researchers soon found that it also harmed other organisms indirectly. Imagine DDT being sprayed over a waterway for mosquito control. Its concentration in the water is low, but once it enters an organism's body, it stays there for a long time. As one organism eats another, all of the DDT stored in the prey ends up in the predator. Each predator eats many prey, so the DDT concentrates in the predator's tissues. In organisms at the fourth level of the food chain, DDT concentrations may be 2000 times greater than in organisms at the base of the food web.

The United States banned use of DDT in 1971, after much evidence showed it to become biomagnified, cause cancer, and disrupt reproduction in birds of prey such as bald eagles and peregrine falcons (figure 40.12B). Nevertheless, DDT is still used in other countries, especially to control the mosquitoes that transmit malaria.

Summary *Food chains and food webs rely on primary producers that pass energy to successive levels of consumers. Pyramid diagrams represent energy flow in an ecosystem. Some chemicals become concentrated as they ascend food webs.*

40.3 MASTERING CONCEPTS

1. Identify the trophic levels of a food chain.
2. What sources of energy sustain ecosystems?
3. What is the role of decomposers in an ecosystem?
4. How efficient is energy transfer between trophic levels in food webs?
5. Draw an energy pyramid for an ecosystem with three levels of consumers.
6. Explain how biomagnification disproportionately affects organisms at the top of a food chain.

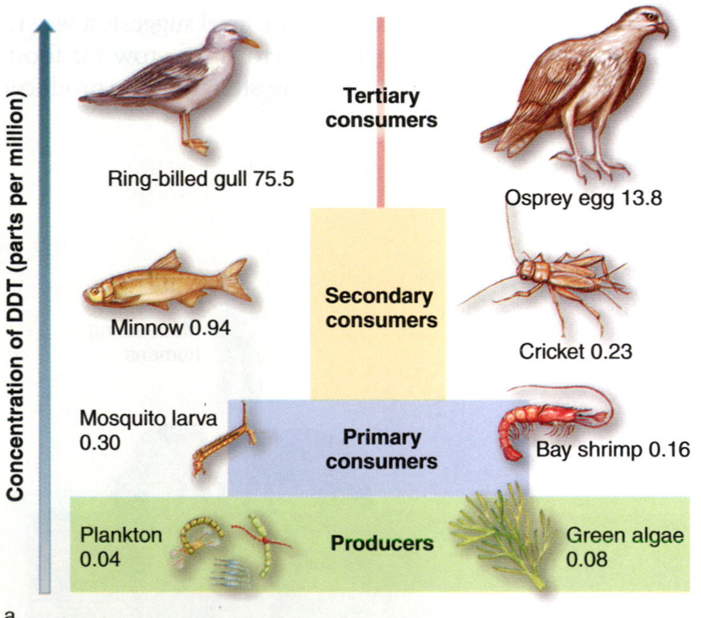

FIGURE 40.12 Biomagnification. (**A**) The concentration of DDT in organisms' bodies increases up the food web. Values represent concentrations of DDT per unit of tissue, measured in parts per million (ppm). (**B**) The lifeless body of this about-to-hatch peregrine falcon is testament to the high concentration of DDT at the top of the food web, a position the animal occupies as a predatory bird. The pesticide caused birds to produce weak eggshells, which broke when adult birds attempted to incubate the eggs. Few young birds hatched, and populations plummeted until a ban on DDT took effect.

Can You Relate?

What Happens After You Flush?

Most people probably give little thought to what happens to whatever they flush down the toilet. But the engineers and technicians who run community sewage treatment facilities think about it all the time.

Decades ago, many towns simply released untreated sewage into rivers and oceans. Because untreated sewage harbors disease-causing organisms, this practice posed an obvious threat to human health. It was also harmful for another reason: the raw sewage killed fish and other aquatic animals. The problem is not that human diseases spread to aquatic organisms. Rather, the fish kills occurred as aquatic microbes decayed the feces and other organic wastes in the sewage. As the microbes respired, they used all the available O_2 dissolved in the water, in effect strangling the fish.

Federal law now mandates that communities treat wastewater before releasing it. Treatment plants harness the power of microorganisms to remove organic matter from sewage before it enters waterways. In trickling filters, for example, sewage-eating bacteria are given "dream homes"—all the organic matter they can eat, along with plenty of moisture and O_2 (see figure 18.12). As they degrade the sewage, they release CO_2 into the atmosphere.

And if you've ever wondered why communities prohibit dumping used motor oil or organic solvents down the drain, the answer again relates to the microbes. These substances can poison the bacteria that degrade the sewage, making water treatment impossible.

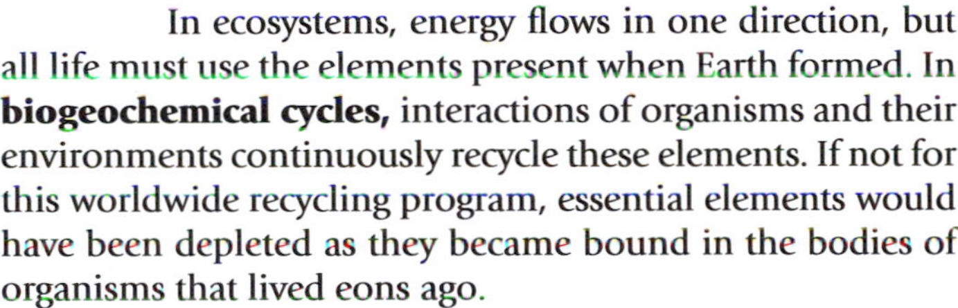

40.4 Chemicals Cycle Within Ecosystems

In ecosystems, energy flows in one direction, but all life must use the elements present when Earth formed. In **biogeochemical cycles,** interactions of organisms and their environments continuously recycle these elements. If not for this worldwide recycling program, essential elements would have been depleted as they became bound in the bodies of organisms that lived eons ago.

Whatever the element, all biogeochemical cycles have steps in common. Generally, autotrophs such as plants take up an inorganic form of the element from the environment, then incorporate it into organic molecules. If an animal eats the plant, the element may become part of animal tissue. If another animal eats this animal, the element may be incorporated into the second animal's body. Eventually, decomposers release the elements from wastes or dead tissues back into the environment in inorganic form.

A. Water Circulates Between the Land and the Atmosphere

Water covers much of Earth's surface, primarily as oceans but also as lakes, rivers, streams, ponds, swamps, snow, and ice (**figure 40.13**). Water also occurs below the land surface as groundwater.

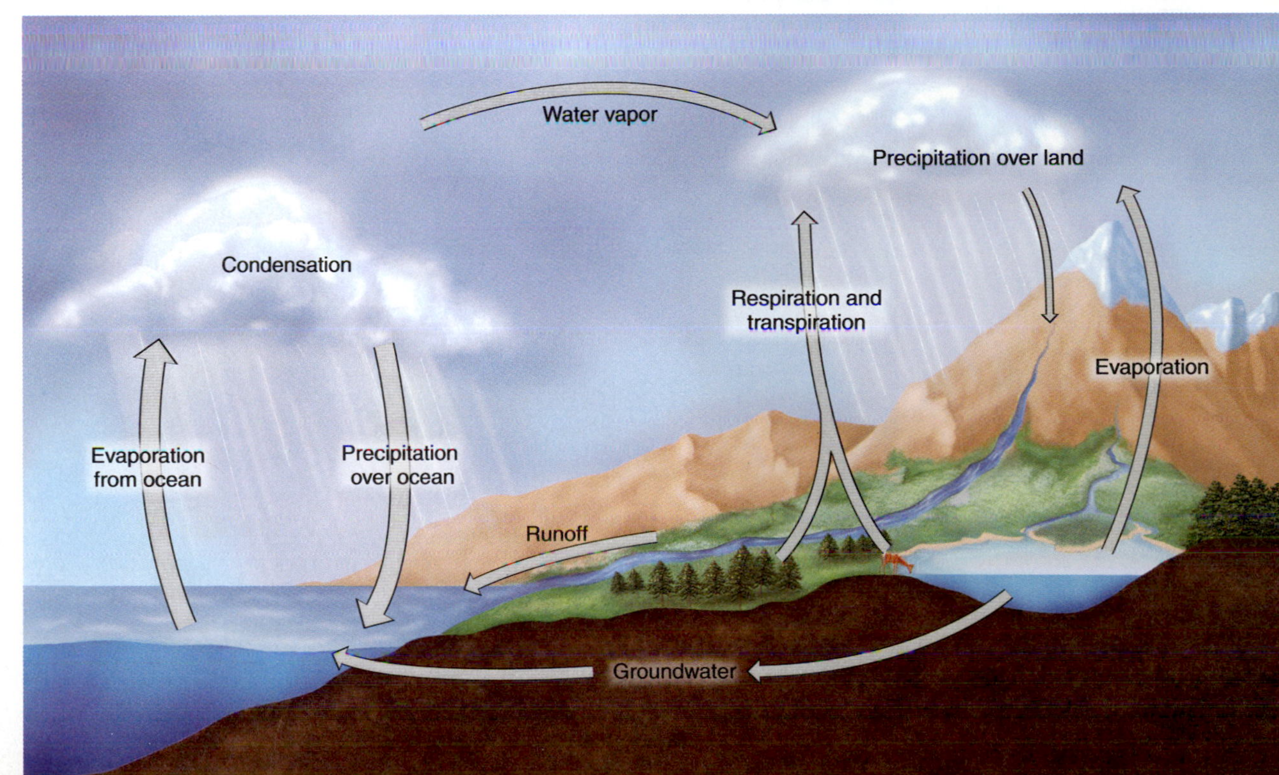

FIGURE 40.13
The Water Cycle
Water falls to Earth as precipitation. Organisms use some water, and the remainder evaporates, runs off into streams, or enters the ground. Animals return water to the environment by respiring and excreting, and plants do so by transpiring.

The sun's heat evaporates water from land and water surfaces. Water vapor may rise on warm air currents, cooling and forming clouds. If air currents carry this moisture higher or over cold water, more cooling occurs, and the vapor condenses into water droplets that fall as rain, snow, hail, fog, sleet, or freezing rain. Some of this precipitation falls on land, where it may run along the surface. Streams unite into rivers that lead back to the ocean, where the sun's energy again heats the surface, evaporating the water and continuing the cycle.

Rain and snowmelt may also soak into the ground, restoring soil moisture and groundwater. This underground water supplies wells and feeds springs that many species use. Spring water may evaporate or flow into streams, linking groundwater to the overall water cycle.

How do organisms participate in the water cycle? Plant roots absorb water from soil and release much of it from their leaves in transpiration. The lush plant life of the tropical rainforests draws huge amounts of water from soil and releases it to the atmosphere. Animals drink water and consume it with their food, returning it to the environment through evaporation and urination. Of course, human life also depends on fresh water; figure 39.15 shows that the availability of fresh water is dwindling as the human population grows. **transpiration, p. 525**

B. The Carbon Cycle Is Strongly Linked to Earth's Climate

In the carbon cycle, autotrophs use atmospheric CO_2 to synthesize organic compounds that they incorporate into their tissues (**figure 40.14**). Cellular respiration releases carbon back to the atmosphere as CO_2. Dead organisms and excrement return organic carbon to soil or water. Invertebrates, bacteria, and fungi complete the breakdown of these organic compounds to release simple carbon compounds into the soil, air, and water. **cellular respiration, p. 120**

Some geological deposits contain carbon from past life. Limestone, for example, consists mostly of exoskeletons and shells of ancient sea inhabitants. Fossil fuels, such as coal and oil, form from the remains of long-dead organisms. When these fuels burn, carbon returns to the atmosphere as CO_2. Decades of accumulation of CO_2 and

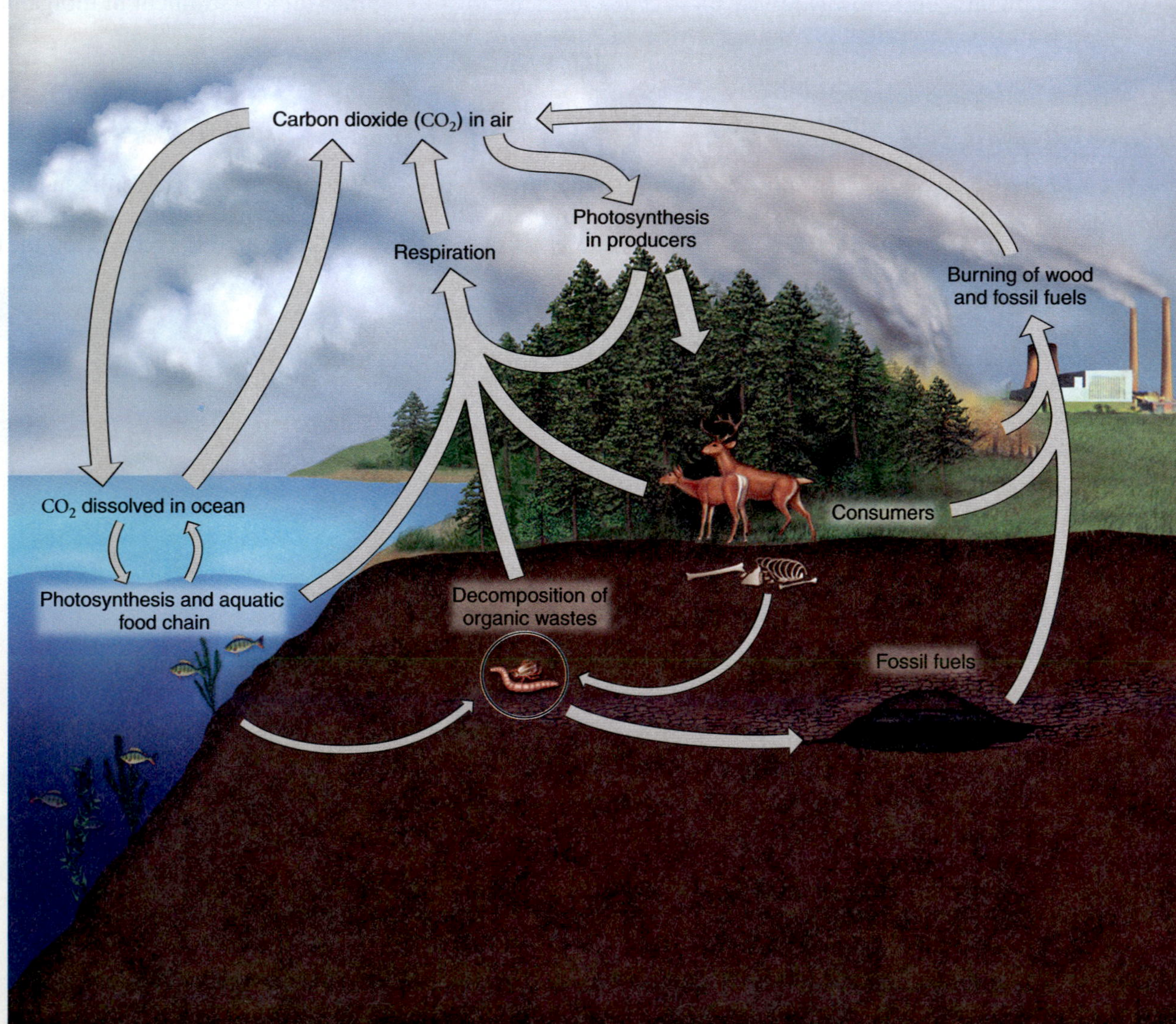

FIGURE 40.14
The Carbon Cycle. Carbon dioxide in the air and water enters ecosystems through photosynthesis and then passes along food chains. Respiration and combustion return carbon to the abiotic environment. Carbon can be retained in geological formations and fossil fuels for long periods.

other so-called greenhouse gases in the atmosphere is likely responsible for Earth's gradually warming climate. **global climate change, p. 849**

C. The Nitrogen Cycle Relies on Bacteria

Nitrogen is an essential component of proteins, nucleic acids, and other biochemicals in living cells. Although the atmosphere is about 78% nitrogen gas (N_2), most organisms cannot use this form of nitrogen. The nitrogen cycle therefore depends on free-living or symbiotic **nitrogen-fixing** bacteria that convert N_2 into ammonia (present in soil as ammonium ion, NH_4^+), a form that plant roots can take up (**figure 40.15**). For example, *Rhizobium* bacteria live in nodules on the roots of legume plants such as beans, peas, and clover (see figure 18.10). Many farmers alternate nonlegume crops, such as corn, with legumes to enrich the soil with biologically fixed nitrogen.

In a process called **nitrification,** bacteria convert ammonia to nitrites (NO_2^-) and eventually to nitrates (NO_3^-), which is another form that plants can use. (Some nitrate is also produced when lightning fixes atmospheric nitrogen.) Once plants incorporate the nitrogen into proteins and nucleic acids, herbivores acquire it by eating the plants, and so on up the food chain. Decomposers release some ammonia when they decay the dead bodies and wastes.

Yet another group of bacteria complete the cycle. In **denitrification,** bacteria return nitrogen to the atmosphere as they convert nitrites and nitrates to N_2.

D. The Phosphorus Cycle Begins with the Weathering of Rocks

Phosphorus occurs in nucleic acids, ATP, and membrane phospholipids; in vertebrates, it is also a major component of bones and teeth. Unlike the other biogeochemical cycles, the inorganic portion of the phosphorus cycle occurs mostly in sediments and rocks, rather than the atmosphere (**figure 40.16**). As they erode, these geological sources gradually release phosphate ions (PO_4^{-3}) to ecosystems. Autotrophs take up some of this phosphorus, but much of it ultimately returns to the oceans and other bodies of water, where it becomes part of sediments. Geological

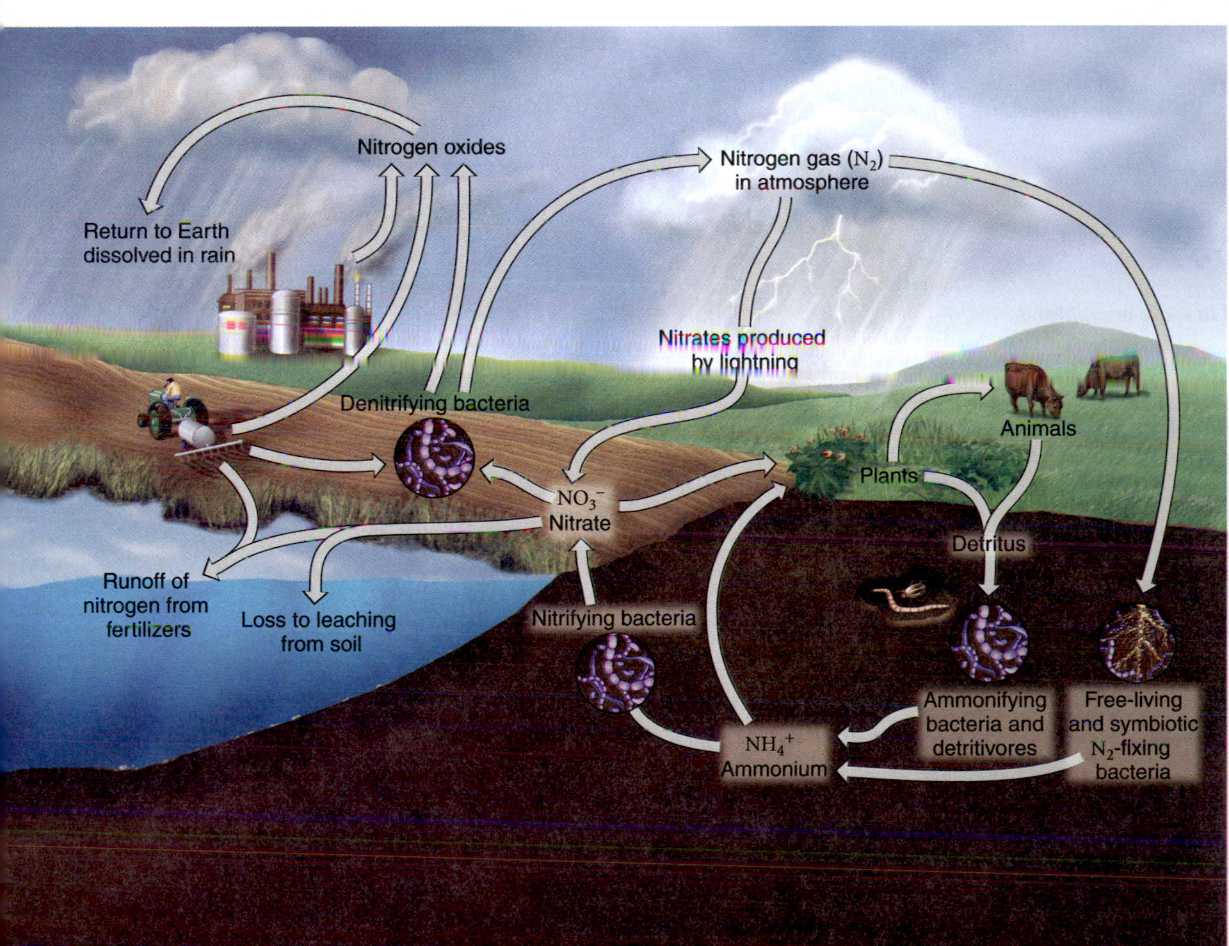

FIGURE 40.15
The Nitrogen Cycle. Nitrogen-fixing bacteria convert atmospheric nitrogen gas (N_2) to organic forms and ammonium ions, which plants can use. Plants use the nitrogen to synthesize amino acids, nucleic acids, and chlorophyll. They pass these biochemicals along food chains. Nitrogen returns to the abiotic environment in urine and feces and by decomposition of dead organic matter. Specific groups of bacteria convert ammonium ions to nitrate (another form plants can use), and nitrate to nitrogen gas, completing the cycle.

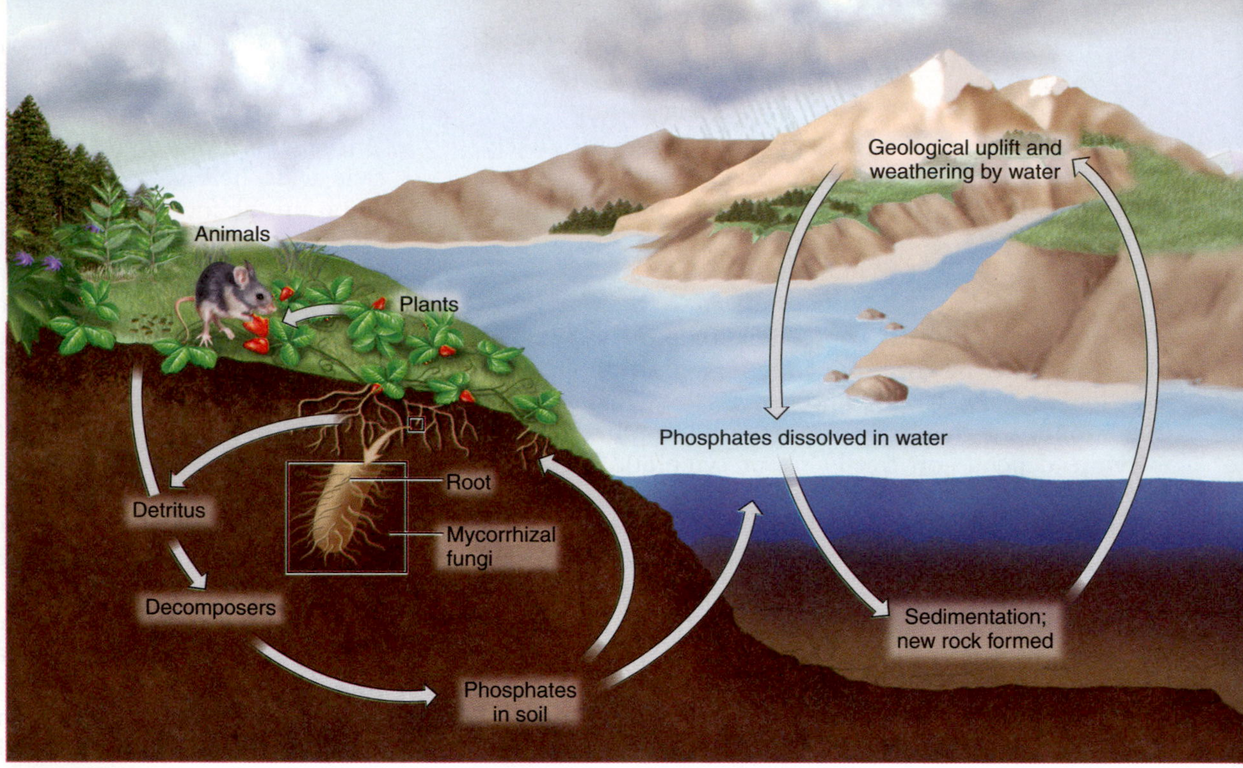

FIGURE 40.16
The Phosphorus Cycle. Phosphorus comes from rock, which slowly erodes. Plant roots take up soluble phosphorus (often with the help of mycorrhizal fungi) and pass it up food chains. Decomposers return phosphorus to the abiotic environment. Mining and the distribution of the element in fertilizer (not shown) has increased phosphorus availability to both terrestrial and aquatic organisms.

uplift eventually returns underwater sedimentary rock to the land. Decomposers return phosphorus from living organisms to soil and water. **eutrophication, p. 834**

Summary *Biogeochemical cycles describe how elements move between organisms and the environment. Elements ascend food webs and return to the environment when an organism decomposes.*

40.4 MASTERING CONCEPTS

1. What are the basic steps of a biogeochemical cycle?
2. Describe the steps of the water, carbon, nitrogen, and phosphorus cycles.
3. Describe how a terrestrial ecosystem can interact with a faraway aquatic ecosystem.

Burning Questions

Could human life be supported in space or on Mars?

A human mission to Mars would take as long as 10 months, just to get there. During that time, the space capsule would have to become a fully self-contained ecosystem. It would have to have a constant energy source, probably the sun, to fuel the growth of primary producers, probably plants.

FIGURE 40.A **Biosphere II.**

Without a way to make "pit stops" to acquire new resources, the capsule's occupants would have to be fanatical about recycling all essential elements. Water would have to be purified and reused, over and over. Feces would likely be composted into a form that on-board plants could use. Luckily, the byproduct of photosynthesis is O_2. But imagine the catastrophic effect if an unwanted stowaway, such as a plant pathogen, killed all the on-board plants!

The Biosphere II project in Oracle, Arizona, was an attempt to create an artificial ecosystem that space travelers could eventually replicate on Mars or on the moon (**figure 40.A**). Eight crew members lived in Biosphere II for 2 years beginning in September, 1991. They recycled all water and wastes to grow crops, which supplied about 80% of their food (the rest came from an initial 3-month supply grown in Biosphere II before the crew's residence began). Still, the system did not remain self-contained. Soil bacteria produced more CO_2 than anticipated, so Biosphere II's operators had to pump in O_2 from outside. The Biosphere II story illustrates just how hard it will be to design a new human habitat on another planet.

Have a Burning Question of your own? Submit it to marielle_hoefnagels@mcgraw-hill.com for possible inclusion in future editions of this book!

40.5 INVESTIGATING LIFE

Two Kingdoms and a Virus Team Up to Beat the Heat

In his 1733 poem entitled *On Poetry, a Rhapsody,* Jonathan Swift wrote:

So, naturalists observe, a flea
Has smaller fleas that on him prey;
And these have smaller still to bite 'em;
And so proceed ad infinitum.

Swift's observation, though not literally true, showed amazing insight into the relationships between the merely small and the truly microscopic. We now know that symbiotic arrangements, where one species lives on or in another, can cross all phylogenetic lines. Symbioses exist between animals and plants (ants and acacias), plants and fungi (mycorrhizae), protists and animals (protozoa in termite guts), and bacteria or protists and fungi (lichens). Mutualism is a particularly interesting form of symbiosis, in which both species benefit from the deal. It is not that each species is advocating the success of the other's genome. Rather, if an alliance gives both parties a selective advantage, then "*vive la difference*!"

Researchers at the Noble Foundation in Ardmore, Oklahoma, have discovered a new three-way symbiosis and documented its survival benefit to the partners. Biologist Luis Márquez knew of a type of panic grass (*Dichanthelium lanuginosum*) that endured the scorching geothermal soils of Yellowstone National Park. Previous studies had shown that the grass benefited from a fungus growing in its roots. The fungus (*Curvularia protuberate*) is an endophyte, an organism that lives between a plant's cells without causing disease. By itself, neither plant nor fungus could grow at temperatures above 38°C (100°F). The grass–fungus relationship somehow allows the plant to survive in soil temperatures up to 65°C (about 150°F). As long as the grass survives, so does the fungus, but each is doomed without the other.

By itself, a plant–endophyte relationship is not unusual; section 21.7 described endophytes that help cacao plants fight off other fungi. But Márquez and his team dug a bit deeper. They knew that many fungi are themselves infected with viruses. For example, a mycovirus (a virus that infects fungi) helps shape the biology of the fungus responsible for a disease of chestnut trees. The team wondered whether the fungus within the grass might also have a partnership with a virus.

The first thing to do was to see if a mycovirus was present. Mycoviruses usually have genomes made of double-stranded RNA, which is distinctive from the fungus's own double-stranded DNA. The researchers extracted nucleic acids from the fungal cells and found telltale fragments of double-stranded RNA within the fungal tissue.

Next, the team wanted to see if the presence of the mycovirus within the fungal endophytes helped plants survive in hot soils. They found a colony of the fungus that contained very low amounts of viral RNA. By repeatedly drying, freezing, and thawing this fungus in the laboratory, they "cured" the fungus of its virus. The researchers then set up an experiment with three treatment groups: grass with the wild-type (virus-infected) fungus, grass without any fungus at all, and grass inoculated with virus-free fungus. They submitted all three groups of plants to a tough regimen of 65°C soil temperatures for 10 hours a day, followed by 14 hours at 37°C. After 14 days, they checked to see how the plants were doing. The results were clear: only the plants with the virally infected fungus survived the ordeal (**figure 40.17**).

The results seemed promising, but the team had to be certain it was the mycovirus that did the trick. After all, some unseen difference between the two fungi (something other than the virus) may have accounted for the heat tolerance. They therefore inserted a special genetic marker in one of their virus-free fungus samples and grew this marked fungus

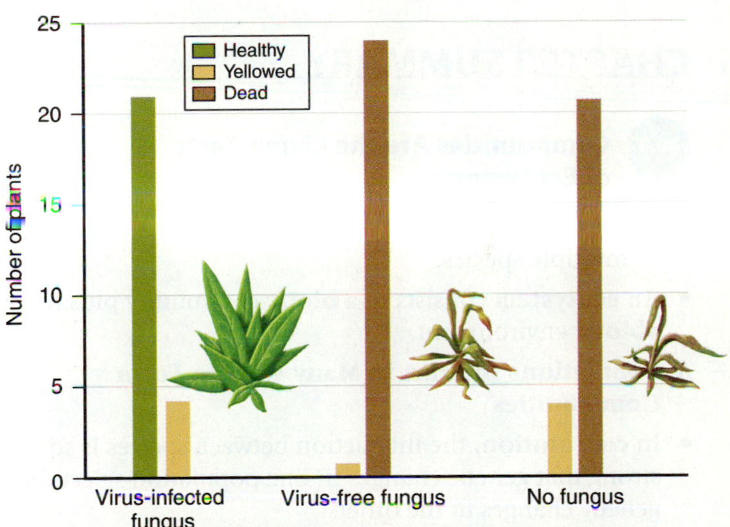

FIGURE 40.17 The Virus Matters. Plants that were inoculated with the normal virus-infected fungus survived the heat. Not so for plants infected with virus-free fungi or for nonsymbiotic plants without fungal endophytes. **Question:** The fungus–virus partnership helped these young grass plants survive at very high temperatures. What would be the benefits of inoculating all of our food crop plants with the fungus–virus team? What else would you need to know before recommending that strategy? Can you think of any possible drawbacks?

on a plate next to a fungus that contained the mycovirus. They let the two fungi grow across each other and took samples from the entwined hyphae. They moved these samples to new plates, and in one of the 35 subcultures they found that the marked fungus had picked up the virus. They inoculated grass with this newly infected fungus, planted the plants, and turned up the heat. Sure enough, these plants survived too (**figure 40.18**).

Finally, the scientists decided to see if the mycovirus–fungus partnership could work its magic in a very distant relative of grass—a tomato plant. They repeated the treatments and found that the plant–fungus–virus trio survived the heat, but the others did not.

Nearly 300 years after Jonathan Swift penned his whimsical poem, biologists are coming to realize that symbiotic relationships are the rule rather than the exception. Over hundreds of millions of years of shared evolutionary history, the struggle for existence has generated some unusual working relationships. The grass–fungus–virus partnership apparently allows all of the partners to colonize habitats that were previously unavailable. No one knows how the mycovirus-infected endophytes help their hosts beat the heat. Whatever the mechanism, however, it appears to be deeply "rooted" in plant–fungus–virus evolutionary history.

Luis M. Márquez, Regina S. Redman, Russell J. Rodriguez, and Marilyn J. Roossinck. 2007. A virus in a fungus in a plant: three-way symbiosis required for thermal tolerance. *Science*, vol. 315, pages 513–515.

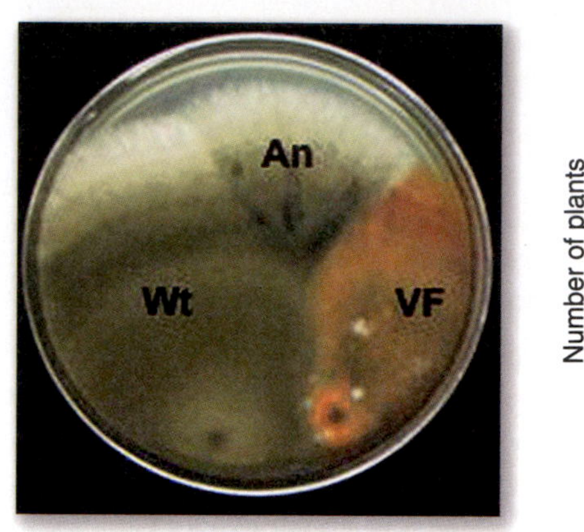

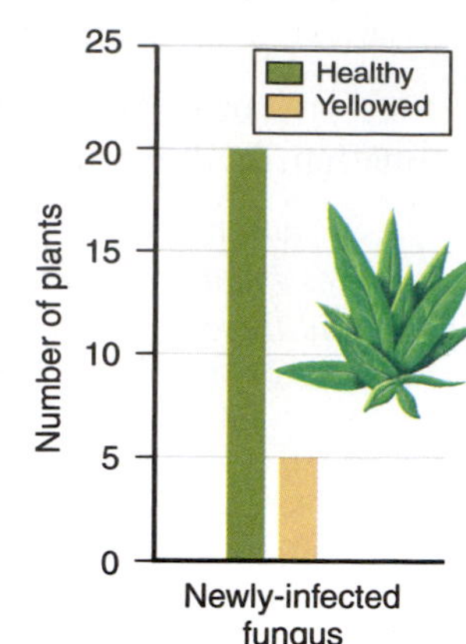

FIGURE 40.18 Reinfecting a "Cured" Fungus. (**A**) The researchers grew virus-infected (Wt, or wild-type) and virus-free (VF) fungi on agar in a petri dish. Wt and VF fungi are growing together in the zone marked "An," where previously virus-free fungi reacquired the mycovirus. (**B**) When the virus was reintroduced into fungi that were previously virus-free, the plants were heat-tolerant once more.

CHAPTER SUMMARY

40.1 Communities Are the Living Parts of Ecosystems

- **Communities** are composed of coexisting populations of multiple species.
- An **ecosystem** consists of a **biotic** community plus its **abiotic** environment.

A. Populations Interact in Many Ways to Form Communities

- In **coevolution,** the interaction between species is so strong that genetic changes in one population select for genetic changes in the other.
- Each species has characteristic conditions where it lives **(habitat)** and resources necessary for its life activities **(niche).** Populations that share a habitat often **compete** for limited resources.
- According to the **competitive exclusion principle,** two species cannot indefinitely occupy the exact same niche. In **resource partitioning,** slight differences in niche allow different species to share surroundings.
- **Symbiotic** relationships include **mutualism, commensalism,** and **parasitism.**
- **Predator** and **prey** species have camouflage, warning coloration, mimicry, and other adaptations that help them capture food or avoid being captured.

B. A Keystone Species Has a Pivotal Role in the Community

- **Keystone species** disproportionately affect community composition.

40.2 Communities Change Over Time

- As species interact with one another and their physical habitats, they change the community, a process called **succession.** Succession may lead toward a stable **climax community.**

A. In Primary Succession, a New Community Arises

- **Primary succession** occurs in a previously unoccupied area, beginning with **pioneer species** that allow soil to develop, paving the way for additional organisms to thrive.

B. Secondary Succession Replaces a Disturbed Community

- **Secondary succession** is more rapid than primary succession because soil does not have to build anew.

C. Succession Can Be Complex

- Complete stability is rare in a community. Pockets of local disturbance mean that most communities are a patchwork of successional stages.

40.3 Ecosystems Require Continuous Energy Input

A. Energy Passes Through Food Chains

- A **food chain** begins when **primary producers (autotrophs)** harness energy from the sun or inorganic chemicals, forming the first **trophic level.**
- **Consumers (heterotrophs)** make up the next trophic levels. Primary consumers (herbivores) eat the primary producers. A secondary consumer may eat the primary consumer, and a tertiary consumer may eat the secondary consumer. **Decomposers** break down **detritus** (nonliving organic material) into inorganic nutrients.
- Interconnected food chains form **food webs.**

B. Every Trophic Level "Wastes" Energy

- The total amount of energy converted to chemical energy is gross primary production. The energy remaining after producers' metabolism is **net primary production.**
- Food chains rarely extend beyond four trophic levels because only a small percentage of the energy in one trophic level transfers to the next level.

C. Ecological Pyramids Describe Ecosystem Characteristics

- A **pyramid of energy** is a diagram that depicts the amount of energy at each trophic level in a food chain.

D. Harmful Chemicals May Accumulate in the Highest Trophic Levels

- **Biomagnification** concentrates stable, fat-soluble chemicals in the highest trophic levels because the chemical passes to the next consumer rather than being metabolized or excreted.

40.4 Chemicals Cycle Within Ecosystems

- **Biogeochemical cycles** are geological and chemical processes that recycle chemicals essential to life.

A. Water Circulates Between the Land and the Atmosphere

- Water cycles from the atmosphere as precipitation over land or water, then into organisms that release water in transpiration, evaporation, or urination.

B. The Carbon Cycle Is Strongly Linked to Earth's Climate

- Autotrophs use atmospheric carbon in CO_2 to manufacture carbohydrates. Cellular respiration and burning fossil fuels release CO_2. Decomposers release carbon from once-living material.

C. The Nitrogen Cycle Relies on Bacteria

- **Nitrogen-fixing** bacteria convert atmospheric nitrogen to ammonia, which plants can incorporate into their tissues. Decomposers convert the nitrogen in dead organisms back to ammonia. **Nitrifying** bacteria convert the ammonia to nitrites and nitrates. **Denitrifying** bacteria convert nitrates and nitrites to nitrogen gas.

D. The Phosphorus Cycle Begins with the Weathering of Rocks

- As rain falls over land, rocks release phosphorus as useable phosphates. Decomposers return phosphorus to the soil.

40.5 Investigating Life: Two Kingdoms and a Virus Team Up to Beat the Heat

- Researchers have discovered a three-way symbiosis between a grass plant, a fungus, and a virus that enables plants to survive extremely high temperatures.

MULTIPLE CHOICE QUESTIONS

1. What is the definition of a community?
   a. A group of organisms of a single species within a specific location
   b. A group of multiple species coexisting within a defined location
   c. A group of animals that exist within a specific habitat
   d. The abiotic (physical and chemical) qualities of a defined location

2. Picocyanobacteria are photosynthetic bacteria that live in the ocean. Two species, isolated from the same habitat, have been found to use different wavelengths of light in photosynthesis. This is an example of
   a. competitive exclusion.
   b. commensalism.
   c. resource partitioning.
   d. realized niche.

3. Why is the example of cellulose-digesting microorganisms in a cow's gut an example of mutualistic symbiosis?
   a. Because the cow needs the microorganisms to break down the grass it eats
   b. Because the microorganisms have a source of food provided by the cow
   c. Because the microorganisms don't use all of the sugar released from cellulose
   d. All of the above
4. Monarch butterflies contain a substance that causes predators to vomit. Viceroy butterflies look like monarchs but lack the substance that sickens predators. This is an example of
   a. warning coloration.
   b. camouflage.
   c. mimicry.
   d. structural defense.
5. Primary succession would be found
   a. on a newly formed volcanic island.
   b. following a massive fire.
   c. in an area that has been graded for new construction.
   d. both a and c.
6. What two properties do all ecosystems share?
   a. The organisms and the abiotic environment
   b. A biome and an energy source
   c. Biogeochemical cycles and an energy source
   d. A biome and a biogeochemical cycle
7. Primary producers are at the first trophic level because they are
   a. autotrophs that do not rely on any other organism for food.
   b. the least complex organisms in an ecosystem.
   c. the most abundant life form.
   d. the first organisms to appear within an ecosystem.
8. A moth that eats plants is a(n) ______ while the bat that eats the moth is a ______.
   a. heterotroph; decomposer
   b. secondary consumer; decomposer
   c. primary consumer; secondary consumer
   d. autotroph; secondary consumer
9. The loss of energy between each trophic level accounts for the
   a. size of each trophic level.
   b. biomass at each trophic level.
   c. observation that large organisms are typically herbivores.
   d. All of the above.
10. Which biogeochemical cycle does not rely on decomposers?
    a. Carbon cycle
    b. Nitrogen cycle
    c. Phosphorus cycle
    d. Water cycle

TESTING YOUR KNOWLEDGE

1. How does a community differ from an ecosystem?
2. How does a habitat differ from a niche? Describe your habitat and niche.
3. How can two species share the same habitat?
4. Describe and give examples of three types of symbiotic relationships.
5. List examples of adaptations that enable an organism to compete with other species, live inside another species, find food, and avoid predation. How does each example contribute to the organism's reproductive fitness?
6. What is a keystone species?
7. Distinguish between primary and secondary succession.
8. Use the second law of thermodynamics and a pyramid of energy to explain why most food chains have four or fewer levels.
9. Some forms of mercury biomagnify in aquatic food chains. High concentrations of mercury in fish are prompting health advisories that warn pregnant women not to eat tuna and some other fish species. From this information, do you think tuna is an herbivore or a carnivore? Explain your reasoning. What properties do you predict that mercury shares with DDT and other pollutants that biomagnify as they ascend food chains?
10. How do organisms return water, carbon, nitrogen, and phosphorus to the abiotic environment?

THINKING AS A SCIENTIST

1. In one type of mimicry, a harmless species such as a jumping spider physically resembles a noxious species such as an aggressive type of ant. Explain why this type of mimicry can exist only if the spiders are less abundant than the ants.
2. How is natural selection apparent in ecological succession?
3. Why are true climax communities rare?
4. How can an organism be both a producer and a consumer?
5. Identify a consumer in an ecosystem not mentioned in the text.
6. Krill are tiny, shrimplike plankton that are abundant in Antarctic waters. Some people have suggested that we "farm" the krill in Antarctic waters and use it to feed people who are starving elsewhere. Predict the effects of krill farming on the Antarctic ecosystem.
7. After fires destroyed much of Yellowstone National Park in 1988, forest managers suggested humans could help the areas recover by feeding deer, bringing in plants, and planting trees. What are some advantages and disadvantages of intervening in recovery from a disaster?
8. In a long-term study of a tropical rain forest in Puerto Rico, researchers studied the aftermath of hurricanes Hugo and Georges. Tree debris supported the growth of detritivores. Beetles and flies ate the detritivores. Propose a food chain and trace how the booming insect populations might eventually benefit top predators such as birds and snakes in the forest. What do you predict happened to fruit- and nectar-eating animals immediately after the storm?
9. Suppose your friend says "I hate germs! I wish we could kill all the bacteria in the world!" What would happen to your friend if she didn't harbor bacteria in her own body? What would happen to nutrient cycles without bacteria?

Visit www.mhhe.com/hoefnagels for practice quizzes, animations, videos, and activities designed to help you master the material in this chapter.

Biomes

The submersible *Alvin* completes a research mission in the Atlantic Ocean.

What's the Point?

Life in Deep-Sea Hydrothermal Vents

In 1977, the submersible Alvin *took a group of geologists to the sea bottom near the Galápagos Islands, directly above cracks in Earth's crust called deep-sea hydrothermal vents. In these areas of intense pressure and temperature extremes where Earth's crust is born, the researchers were astounded to see abundant life. Tubelike worms waved, anemones clung, crabs crawled, and shrimp grazed (see figure 4.3). Most of these animals were unknown to marine biologists.*

Since that time, many more vent ecosystems have been discovered along the ridge of undersea volcanoes that dot the floor of the Pacific Ocean. In the early 1980s, researchers discovered chimneylike structures extending from some hydrothermal vents. Black, intensely hot, mineral-laden water shot through the chimneys, which formed as molten minerals from Earth's mantle crystallized in the cold water.

The environment at a hydrothermal vent could hardly be more different from our own familiar surroundings. The pressure is 300 times greater than at Earth's surface, and the water temperature can exceed 400°C. Animals cannot survive in these hottest spots, but tube worms can live at 100°C. Moreover, the pH can be as low as 2.8, about as acidic as orange juice. There is no oxygen, and the hydrogen sulfide that spews from the Earth would be toxic to most organisms.

Too far beneath the surface to use sunlight, the producers in these vibrant communities are bacteria and archaea that tap energy from the chemicals in and of Earth, predominantly hydrogen sulfide. Thick mats of these microorganisms cover the rock surfaces. Consumers include worms and giant clams that harbor symbiotic prokaryotes inside their tissues. Some of these animals have no digestive tracts; instead, they absorb nutrients directly from their live-in microbes. Other organisms prey on the tube worms, and crabs are scavengers and decomposers.

Deep-sea hydrothermal vents may hold clues to conditions at the time when life began. The discovery of these ecosystems has also helped to dispel the long-held idea that life exists only on the thin layer between bedrock and Earth's atmosphere. Nevertheless, the life on and near Earth's surface is by far more familiar. This chapter describes the forests, grasslands, waters, and other ecosystems that are critically important to our own survival.

LEARNING OUTLINE

41.1 The Physical Environment Determines Where Life Exists
- A. Abiotic Factors Influence All Life
- B. Earth Has Diverse Climates

41.2 Terrestrial Biomes Range from the Lush Tropics to the Frozen Tundra
- A. Towering Trees Dominate the World's Forests
- B. Grasslands Occur in Tropical and Temperate Regions
- C. Whether Hot or Cold, All Deserts Are Dry
- D. Arctic and Alpine Tundras Occupy High Latitudes and High Elevations
- E. Fire- and Drought-Adapted Plants Dominate Mediterranean Shrublands (Chaparral)

41.3 Freshwater Biomes Include Lakes, Ponds, and Streams
- A. Lakes and Ponds Contain Standing Water
- B. Streams Carry Running Water

41.4 Oceans Make Up Earth's Largest Ecosystem
- A. Land Meets Sea at the Coast
- B. The Open Ocean Remains Mysterious

41.5 Investigating Life: Some Like It Hot

41.1 The Physical Environment Determines Where Life Exists

Life abounds almost everywhere on Earth, even in places once thought to be much too harsh to support it (see chapter 18's Burning Question: Are there areas on Earth where no life exists?). Scientists have discovered life in arctic ice, salt flats, hot springs, hydrothermal vents, and miles below Earth's surface in mines. All of these areas are part of the **biosphere,** the portion of Earth where life exists. Water, air, sediments, and organisms travel freely from one part of the biosphere to another. The biosphere is therefore one huge **ecosystem,** an interconnected group of organisms and their physical environment.

Ecologists divide the biosphere into distinctive regions that broadly correspond to the major climatic regions on Earth. **Biomes** are the major types of ecosystems. Forests, deserts, and grasslands are examples of terrestrial biomes. Lakes, streams, and oceans are water-based ecosystems. Whether terrestrial or aquatic, each type of ecosystem supports a characteristic community. How do these distinctive assemblages of species develop?

A. Abiotic Factors Influence All Life

Gardeners know that every species is adapted to a limited set of conditions; no "superorganism" can exist everywhere. A plant adapted to bright sunlight will not thrive on the shady side of the house, nor will a cactus survive in a pond. Fish need water to breathe, and earthworms cannot live on a sandy beach. Along with community interactions (see chapter 40), the **abiotic,** or nonliving, features of an organism's habitat shape the adaptations that contribute to its survival and reproductive success (**figure 41.1**).

Many abiotic factors determine the limits of each species' distribution. Some of the major ones include resources such as sunlight, moisture, and nutrients. Sunlight is the ultimate energy source for most ecosystems (the exceptions are ecosystems that are based on chemical energy, such as deep-sea hydrothermal vent ecosystems). All life requires water—the more the better, if the lush greenery of the tropical rain forest is any indication. Along with sunlight and water, nutrients are often limiting factors that determine how much productivity an ecosystem can sustain.

Other abiotic factors include temperature and salinity. All organisms are adapted to a limited temperature range; trees, for example, cannot live at high elevations where the temperature is low (see the Burning Question: Why is there a "tree line" above which trees won't grow?). Many organisms are adapted to seawater or salty soils, but others are not. The problem of life in a salty habitat lies mostly in the difficulty of extracting water from a saline solution. This is why human castaways adrift on the ocean can die of thirst, even while surrounded by water. Likewise, saline soils prevent agriculture in many parts of the world, because most crop plants are not salt-tolerant.

Fire is also an abiotic condition. Although flames may seem destructive, fire is essential in many ecosystems. In grasslands, for example, periodic fires kill trees that might otherwise take over. In coniferous forests, many adult trees die in fires, but their cones open only after

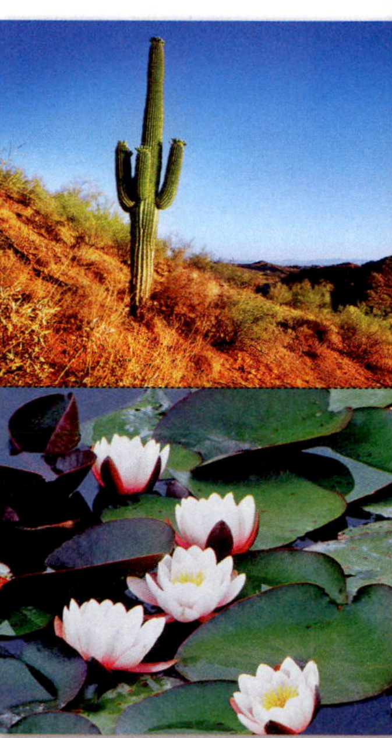

FIGURE 41.1
Habitat Variety. The giant kelp bed that is home to a damselfish has an entirely different set of abiotic conditions than the dry, rocky habitat of a young manzanita tree. A saguaro cactus has a stem that expands like an accordion, helping the plant to store water during rare rainstorms. In contrast, the broad, flat leaf of a lily pad soaks up sunshine in a habitat that is always wet.

prolonged exposure to heat. The seeds germinate after the fire, and the young trees thrive with little competition for sunlight or nutrients.

B. Earth Has Diverse Climates

Sunlight, temperature, and moisture are the major determinants of climate. From the year-round warmth and moisture of the tropics to the perpetually chilly poles, Earth has a wide variety of climates. These differences reflect the fact that the planet is a sphere that rotates at an angle as it orbits the sun (**figure 41.2**). Solar energy is most intense at the equator, where the sun is directly overhead. Because of Earth's curvature, the sun's rays hit other parts of the surface at oblique angles. The average temperature falls with distance from the equator because the same amount of solar energy is distributed over a larger area.

Precipitation also is distributed unevenly across the globe (**figure 41.3**). When intense sunlight heats the air over the equator, the air rises and expands. As the air rises, it cools. Because cool air cannot hold as much moisture as warm air, the excess water vapor condenses, forming the thick cloud cover that pours near-constant rain over the tropics.

Air from the equatorial region also travels north and south. As air cools at higher latitudes and elevations, its density increases, and it sinks back down to Earth at about 30° North and South latitude. Here it absorbs moisture from the land, creating the vast deserts of Asia, Africa, the Americas, and Australia. Some of the air continues toward the poles, rising and cooling at about 60° North and South latitude, bringing the rains that support temperate (midlatitude) forests in these areas. The air rises, and some again continues toward the poles, where precipitation is quite low, and some returns to the equator. Near the equator, the air is heated again, and the cycle begins anew.

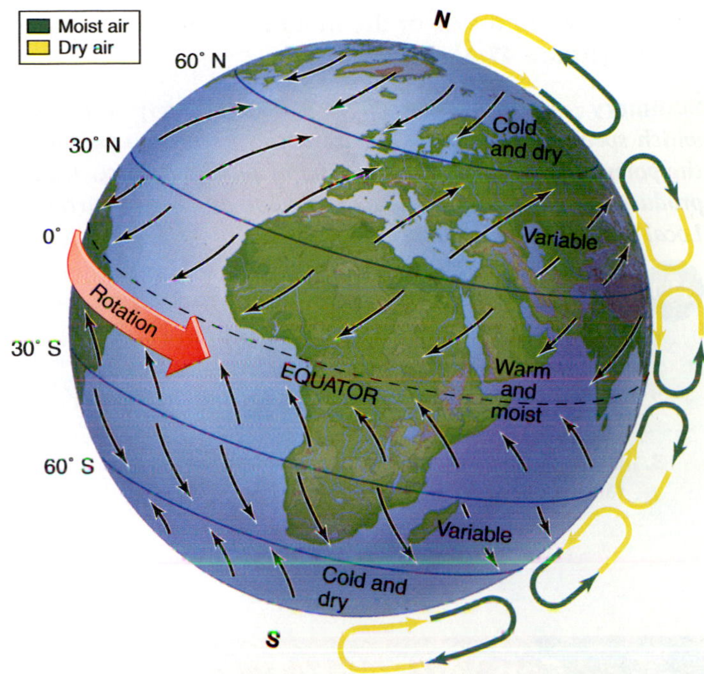

FIGURE 41.3 Patterns of Air Circulation and Moisture. Clouds form over the tropics as equatorial air cools and releases moisture. Air heated over the equator rises, cools, and drops its moisture. The air falls at about 30° North and South latitude, absorbs moisture from the land, then some spreads farther north and south. Six patterns of air circulation (convection cells) form. The winds deflect to the right in the northern hemisphere and to the left in the southern hemisphere because Earth rotates on its axis.

A cycle of heating and cooling, rising and falling air is called a convection cell. The planet has three such convection cells above the equator and three below (see figure 41.3). Earth's major winds correspond to these convection cells. Together, these winds power the major ocean currents, which also contribute to climatic patterns. Changes in these winds can trigger widespread ecological upheaval, as this chapter's Can *You* Relate? El Niño Years describes.

Local features can also influence climate. For example, coasts often have milder climates than their latitude alone might suggest. The reason is that ocean water stores a great deal of heat energy from the sun, producing cooling winds during the summer and releasing stored heat during the cooler months. Mountains also affect climate, in two ways. First, the top of a mountain is generally cooler than its base. This is why many residents of Phoenix, Arizona escape the broiling summer heat by migrating to nearby mountains. Second, mountains often block wind and moisture-laden clouds on their upwind side. The resulting **rain shadow**

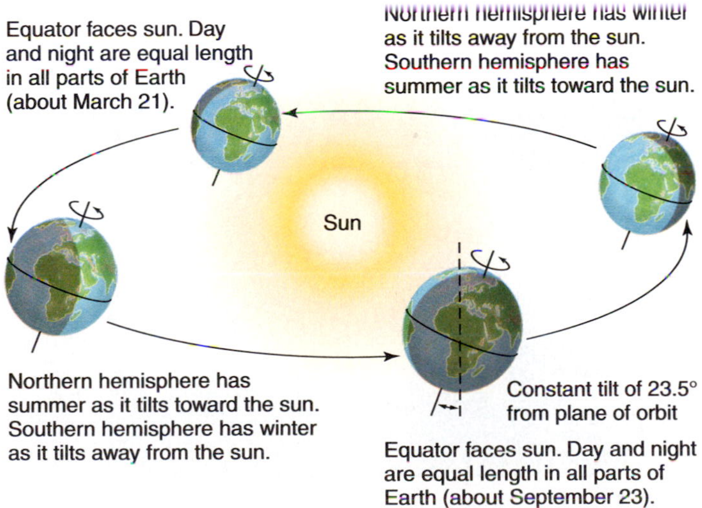

FIGURE 41.2 Earth's Seasons. The tilt of Earth's axis produces distinct seasons in the northern and southern hemispheres as Earth travels around the sun.

on the downwind side of the mountain has a much drier climate (**figure 41.4**).

Summary *Abiotic (nonliving) factors strongly influence which species live where. Climate varies from the equator to the poles because of uneven heating of Earth's surface, which produces temperature gradients and wind and ocean currents. Local features also influence climate.*

41.1 MASTERING CONCEPTS

1. What are the major physical and chemical factors that influence the distribution of species in the biosphere?
2. Moving outward from the equator, what are the major climatic regions of the world?
3. How do prevailing winds, ocean currents, and mountain ranges affect climate?

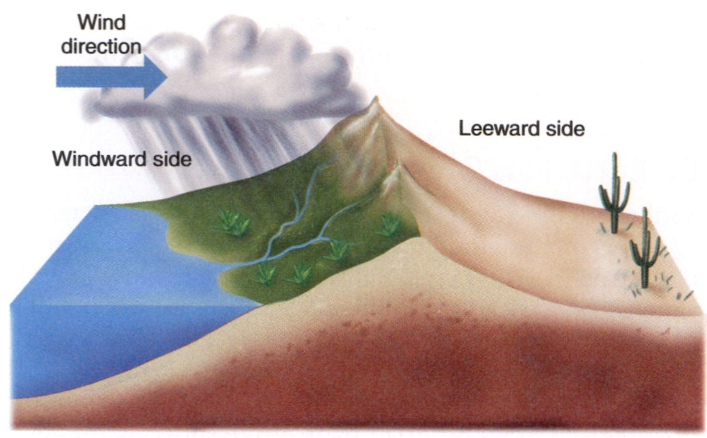

FIGURE 41.4 Rain Shadow. Precipitation falls on the windward side of the mountain, leaving the leeward side with a dry climate.

Can *You* Relate?

El Niño Years

An El Niño is a periodic global climate shift in which a slack in the trade winds sends unusually warm water from the western Pacific eastward (**figure 41.A**). The resulting changes in wind patterns, rainfall, and temperature drastically alter food webs. Along the coast, scallop and shrimp populations explode, yet many types of fishes become scarce. With the fishes gone, the fishing industry collapses, and birds, seals, and sea lions starve.

El Niño is Spanish for "the child," and refers to the Yuletide beginnings of the phenomenon. The event usually occurs every 2 to 7 years, and typically lasts a year or two. An El Niño is often followed by a reversal of conditions called a La Niña.

Although El Niño results from events in the western Pacific, the effects are widespread. Shifting wind directions can cause powerful storms and raise sea surface temperatures along a wide belt in the Pacific around the equator. Severe or unusual weather may occur worldwide: monsoons in the central Pacific, typhoons in Hawaii, torrential rains in South America, blizzards in the Rockies, warm winters in the Northeast, flooding in southern California, and severe droughts in southern Africa.

The effects on ecosystems are severe. Organisms adapted to tropical and subtropical waters move toward the poles, disrupting shore ecosystems. For example, as warm water moves east, it prevents cold, nutrient-rich water below from rising to the surface ("upwelling"). Huge numbers of anchovies and sardines that live in the colder waters move down, then the birds and seals that normally scoop these smaller fishes from the water can no longer reach them. Many animals starve, grimly illustrating the connection between nutrients, producers, and consumers.

Normal Conditions December 1993

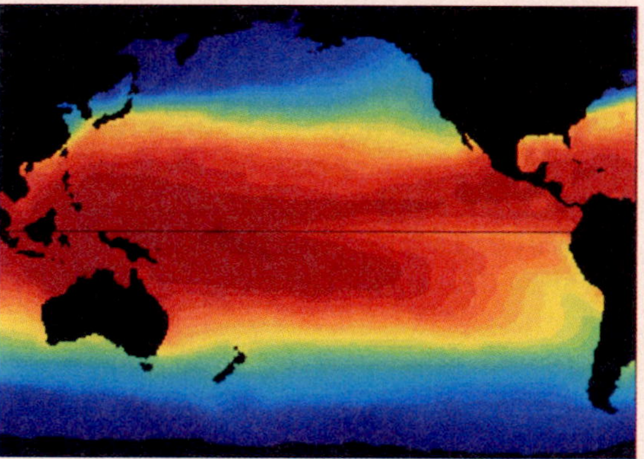

El Niño Conditions December 1997

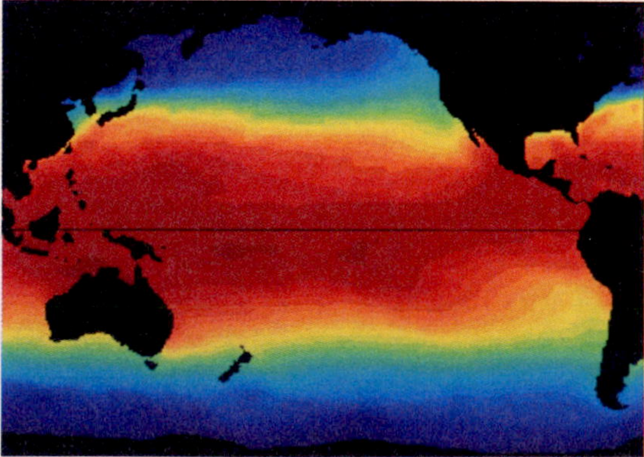

FIGURE 41.A El Niño. An El Niño brings warm water from west to east. In these maps, red areas represent warm water.

Burning Questions

Why is there a "tree line" above which trees won't grow?

The tree line, or timberline, is an edge beyond which trees will not grow. The Arctic and Antarctic tree lines are the farthest points north and south that trees can grow; the alpine tree line is the highest elevation at which they can grow. In each case, the tree line usually defines the point at which the environment simply becomes too cold to support trees.

Most species at the tree line are evergreen conifers such as pine, spruce, larch, and fir. Their needles have a waxy coating and arrangement of stomata that minimizes water loss in the thin, dry air. Eventually, however, the increasingly chilly temperatures and biting winds get the best of even these hardy trees. At the timberline, they shorten to low, stunted bushes, and beyond the tree line, it is simply too cold year-round for seeds to germinate. **stomata, p. 508**

Wind, salt, and a dry climate also produce tree lines. For example, along coasts, a tree line can result from high winds and salt spray that make life impossible for trees. Beyond the desert tree line, rainfall is insufficient to support trees.

Have a Burning Question of your own? Submit it to marielle_hoefnagels@mcgraw-hill.com for possible inclusion in future editions of this book!

41.2 Terrestrial Biomes Range from the Lush Tropics to the Frozen Tundra

Temperature and moisture influence plant growth so much that Earth's climatic zones give rise to huge bands of characteristic types of vegetation. The overall pattern of vegetation influences which microorganisms and animals can live in a biome. In turn, animals also contribute to the structure of the biome.

Soils form the framework of terrestrial biomes, for they directly support plant life. Although it may seem like "just dirt," soil is actually a complex mixture of rock fragments, organic matter, and microbes (see figure 25.3). Climate influences soil development in many ways. Heavy rain may leach soluble materials from surface layers and deposit them in deeper layers, or it may remove them entirely from the soil. In addition, in a warm, moist climate, rapid decomposition may leave little humus (organic material) in the soil. In cold, damp areas, undecomposed peat may accumulate in the soil.

The following sections consider some of the world's major terrestrial biomes, as shown in **figure 41.5.** The map shows the original range of each biome. It is important to remember, however, that humans have drastically altered many natural biomes, replacing them with farmland, suburban housing,

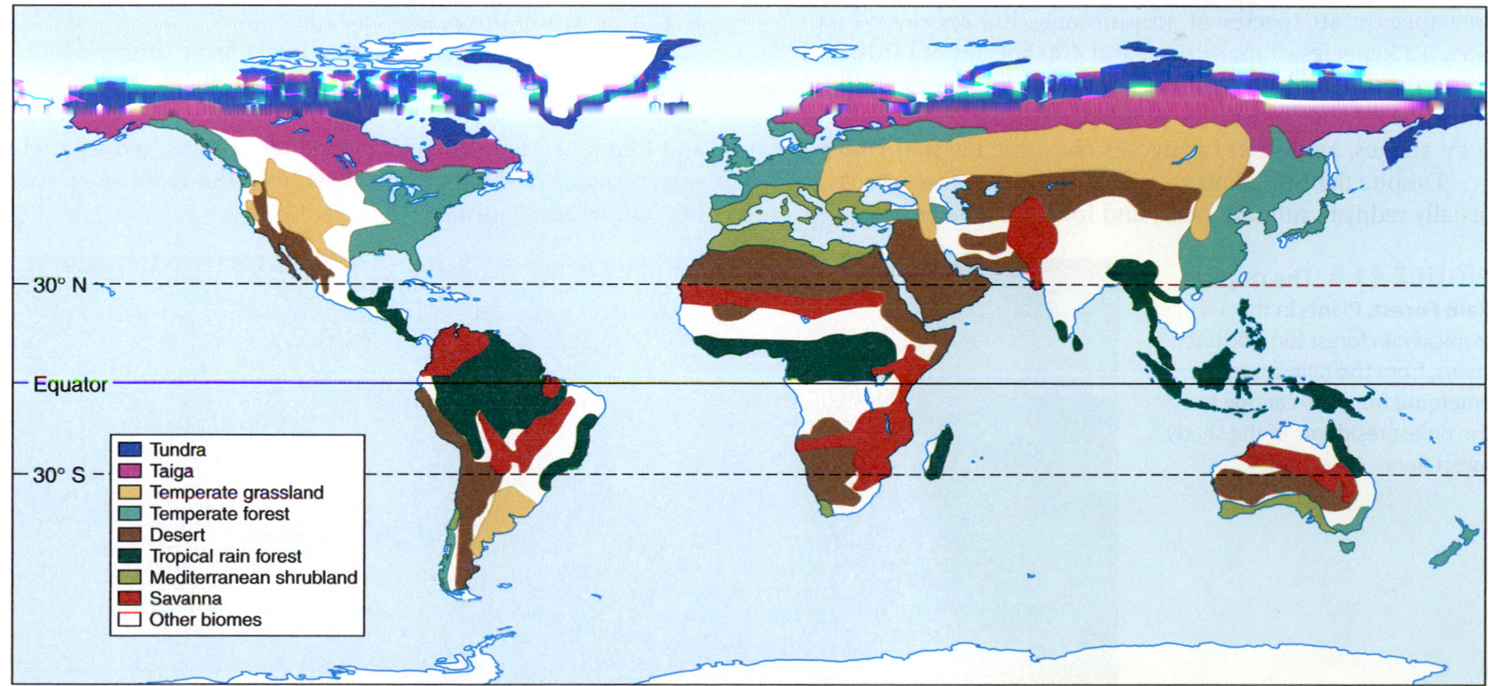

FIGURE 41.5 Earth's Major Terrestrial Biomes. This map shows the large terrestrial areas that have characteristic plant communities.

and cities. In addition, biomes under permanent snow and ice, such as the North and South Poles and the highest mountaintops, are not included.

A. Towering Trees Dominate the World's Forests

Forests supply many of the resources that we use every day: lumber, paper, furniture, and foods such as wild mushrooms and nuts. The trees in the forest also provide wildlife habitat and protect soil from erosion. Moreover, like all plants, trees absorb CO_2 from the atmosphere and use it in photosynthesis, a process that also releases O_2. The wood of a living tree is an especially important long term "carbon sink" that helps offset CO_2 released when humans burn fossil fuels. **global climate change, p. 849**

Tropical Rain Forests

The **tropical rain forests** encircling the equator are home to a stunning diversity of life. An intricate network of relationships connects the species together into a highly complex community. Their position near the equator in Africa, Southeast Asia, and Central and South America ensures that the climate is almost constantly warm and moist.

The warm, wet climate means favorable conditions for plant growth year-round (**figure 41.6**). Broadleaf evergreen trees form the forest canopy. Plants beneath the canopy include vines and epiphytes (small plants that grow on the branches, bark, or leaves of another plant), with tree saplings and countless smaller species growing in the deep shade of the forest floor. Animal life is similarly diverse. A 10 km^2 area of tropical rain forest is likely to house, among its 750 tree species, 60 species of amphibians, 100 species of reptiles, 125 species of mammals, and 400 species of birds, and thousands or even millions of insect species. The herbivores eat leaves, fruit, and other plant parts. Large cats, birds of prey, snakes, and other carnivores consume the herbivores.

Despite the lush plant growth, tropical rain forest soils are usually reddish, nutrient-poor, and low in organic matter. The heat and humidity speed decomposition, but the mycorrhizal roots of giant trees absorb the nutrients efficiently. Animals recycle nutrients too. Leaf-cutting ants, for example, farm gardens of fungi on decomposing leaf fragments. **fungi interact with other organisms, p. 440**

Worldwide, people are logging and burning tropical rain forests to make room for crops and domesticated animals. As described in chapter 42, tropical rain forest destruction threatens indigenous people and global water and carbon cycles. As plants near extinction, we lose potential medicines and the ancestors of many of our most important domesticated plants. These older varieties are a valuable resource to plant breeders, who are always searching for disease-resistance genes to breed into modern crop plants.

Temperate Deciduous and Coniferous Forests

Temperate forests are dominated by either **deciduous** trees, which shed their foliage seasonally, or evergreen **coniferous** trees, which lose only a few leaves at a time (**figure 41.7**). These forests once covered parts of Asia, Western Europe, North America, South America, Australia, and New Zealand, but logging, agriculture, and urbanization have decimated most of the world's native temperate forests. **deforestation, p. 845**

Deciduous trees predominate where winters are relatively mild and rainfall is approximately constant throughout the year. Usually one or two tree species predominate, as in oak–hickory or beech–maple forests. Shrubs grow beneath the towering trees. Below them, small flowering plants grow in early spring, when light penetrates the leafless tree canopy. Herbivores include seed- and nut-eating mice, whitetail deer, ruffed grouse, and gray squirrels. Red foxes and snakes are common carnivores. Raccoons eat a wide variety of foods, including insect grubs, acorns, frogs, and bird eggs.

Summer drought and severe winters favor conifers. Most trees in the temperate coniferous forest are evergreens such as spruce, pine, fir, and hemlock. Understory shrubs include alder and hazelnut. Herbivores include whitetail deer, red squirrel, spruce grouse, and moose. Bobcats hunt the herbivores, and black bears are omnivores.

FIGURE 41.6 The Tropical Rain Forest. Plants in the tropical rain forest form distinct layers, from the tallest trees emerging from the canopy to the tiniest residents of the shady forest floor.

a. b.

FIGURE 41.7 Temperate Forests. (**A**) Trees that lose their leaves each fall dominate temperate deciduous forests. (**B**) Temperate coniferous forests have evergreen trees such as these western hemlocks in Washington State.

Taiga (Boreal Forest)

North of the temperate zone in the northern hemisphere lies the cold, snowy **taiga** (**figure 41.8**). This biome is also called the boreal forest or the northern coniferous forest. The long, harsh winter can last more than 6 months, so the growing season is short. Moisture can be scarce in winter, when water may remain frozen for months.

Soils in the taiga are cold, damp, acidic, and nutrient-poor. The low temperature and acidic pH slow decomposition, and nutrients tend to stay in the leaf litter above the soil, rather than entering the topsoil. Spruce, fir, pine, and tamarack (larch) are the dominant trees. Mycorrhizal fungi help the plants maximize nutrient uptake from the leaf litter. Seed- or leaf-eating herbivores include the woodland caribou, porcupines, red squirrels, chipmunks, snowshoe hares, and moose. Carnivores include lynx, gray wolves, and wolverines. Migratory birds visit during the short growing season.

Caribou, reindeer, berries, and fungi have provided food for millennia, but expanded logging, hunting, and trapping are rapidly depleting the boreal forests.

FIGURE 41.8 Taiga. Woodland caribou forage for food in the Canadian boreal forest.

B. Grasslands Occur in Tropical and Temperate Regions

Like forests, grasslands provide critical wildlife habitat. Vast seas of grasses sustain huge herds of large, grazing animals such as bison and zebras. But the rich soils that support the grasslands also make these prime areas for agriculture. Grasslands are therefore endangered around the world.

Tropical Savannas

Tropical **savannas** are grasslands with scattered trees or shrubs and bands of woody vegetation along stream courses (**figure 41.9**). The weather is warm year-round, with distinct wet and dry seasons. Perennial grasses dominate the savanna along with scattered patches of drought- and fire-resistant trees and shrubs such as palms, acacias, and baobab trees. The Australian savanna is home to many birds and kangaroos, whereas the grassland in Africa features herds of zebra, giraffes, wildebeests, gazelles, and elephants. Lions, cheetahs, wild dogs, and hyenas prey on the herbivores, and vultures and other scavengers eat the leftovers. In many tropical savannas, termites are major detritivores, and their huge nest mounds dot the landscape. It was from the African savanna that our own ancestors evolved, emerged, and spread. **human evolution, p. 344**

FIGURE 41.9 Tropical Savanna. The savanna is home to grasses, scattered trees, and large herbivores such as these elephants.

Widespread overgrazing by domesticated animals threatens to turn large areas of tropical savanna to desert. Hunting and poaching also endanger some populations of large animals. **desertification, p. 845**

Temperate Grasslands

The **temperate grasslands** are also known as the prairies of North America, the steppes of Russia, and the pampas of Argentina (**figure 41.10**). The word *grassland* calls to mind endless rolling hills covered with grasses and wandering herds of grazing bison. Temperate grasslands have few if any trees, partly because annual rainfall is often not sufficient to support them. Grazing and fire also suppress tree growth. The growing regions of trees (i.e., the tips of branches) are easily destroyed by fire or herbivores. In contrast, the perennial buds of grasses lie protected below the soil surface.

Grasses dominate this biome, although sunflowers, thistles, and other flowering plants are also common. Bison, elk, and pronghorn antelope were originally the large, grazing herbivores of North America, but hunting has caused their numbers to dwindle. Other herbivores include prairie chickens, insects such as grasshoppers, and burrowing rodents such as prairie dogs and mice. Coyotes, snakes, and birds of prey feed on these smaller animals.

The deep, black prairie soils of the North American grasslands are famously fertile and rich in organic matter. Because these soils are ideal for cultivation, only small patches remain of the vast grasslands that once occupied three continents. Farmland has replaced prairie, with wheat and corn taking the place of diverse grasses. Most prairie remnants are too small to sustain the herds of large herbivores that once roamed the plains.

FIGURE 41.10 Temperate Grassland. Bison once dominated the North American prairie, a temperate grassland.

C. Whether Hot or Cold, All Deserts Are Dry

All **deserts** are dry, receiving less than 20 cm (8 inches) of rainfall per year. They ring the globe at 30° North and South latitude, and some occur in the rain shadows of tall mountains. Sparse desert life means soils are low in organic matter.

Deserts have a reputation for being hot, but the temperature can vary dramatically. In a hot desert such as the Sonoran, which spans parts of Arizona and Mexico, the days can be scorchingly hot because few clouds filter the sun's strongest rays. In China's cold Gobi desert, in contrast, the average annual temperature is below freezing.

Parts of the Sahara are so dry that they are nearly devoid of life. Other desert habitats support only a few species, and still others are species-rich (**figure 41.11**). Desert plants often have long taproots, quick life cycles that exploit the brief rainy periods, fleshy stems or leaves that store water, and spines that guard against thirsty herbivores. Most desert animals burrow or seek shelter during the day and become active when the sun goes down.

Humans are drawn to the warm climates of the desert southwest of the United States, then change the ecosystem as they divert huge amounts of water from faraway rivers

FIGURE 41.11 Desert. Saguaro cacti and many other plants that have water-conserving adaptations live in the Sonoran Desert in the southwestern United States.

to support their lifestyles. Even so, deserts are among the few biomes that are expanding worldwide, as unsustainable agriculture eats away at forests and savannas.

D. Arctic and Alpine Tundras Occupy High Latitudes and High Elevations

The **tundra** takes two forms. A band of Arctic tundra runs across the northern parts of Asia, Europe, and North America. Winter is bitterly cold and dark, but temperatures venture above freezing for a few months each year. Alpine tundra occupies high mountaintops between the tree line and areas of permanent snow and ice cover. Summer sunlight is intense, but snow covers the tundra in winter.

Because cold temperatures slow decomposition, soils are rich in organic matter. Below the top layer of soil is a zone called **permafrost,** where the ground remains frozen year-round. Permafrost limits rooting depth, which prevents the establishment of large plants (**figure 41.12**). The shallow tundra soil supports reindeer lichens, mosses, dwarf shrubs, and low-growing perennial plants such as sedges, grasses, and broad-leafed herbs. Animal inhabitants of the arctic tundra include caribou, musk oxen, reindeer, lemmings, hares, snowy owls, foxes, and wolverines. Polar bears sometimes visit coastal areas of the tundra to den. In the summer, migratory birds raise their young and feed on the insects that flourish in the tundra and its ponds.

The shallow soil, short growing season, and slow decomposition make the tundra a very fragile environment that recovers slowly from disturbance. In recent decades, oil drilling in the arctic tundra has significantly increased the human presence.

FIGURE 41.12 Arctic Tundra. In the tundra, permafrost limits plant life to shallow-rooted shrubs. Lichens are also abundant.

FIGURE 41.13 Mediterranean Shrubland. Dry summers, wet winters, and fire shape the chaparral biome.

E. Fire- and Drought-Adapted Plants Dominate Mediterranean Shrublands (Chaparral)

Despite the name, **Mediterranean shrublands** (also called **chaparral**) occur not only around the Mediterranean, but also in other small areas along the west coasts of North and South America, Australia, and South Africa. Summers are hot and dry; winters are mild and moist. As the name suggests, the dominant vegetation is shrubby plants (**figure 41.13**), including poison oak, manzanita, and scrub oak. The plants have thick bark and small, leathery, evergreen leaves with thick cuticles and hairs that slow moisture loss during the dry summers. Herbivores include jackrabbits, mule deer, and birds and rodents that forage for seeds under the shrub canopy; some of their predators include coyotes, foxes, snakes, and hawks.

Mediterranean shrublands are especially susceptible to fires because the vegetation dries out during the summer. The sandy soils retain little water. The fire-adapted plants resprout from underground parts or produce seeds that germinate only after the heat of a fire. People once burned chaparral to make room for grazing livestock. Since then, large human communities have moved onto the shrublands, which nevertheless remain at great risk for fire. Burning the plant cover makes the soil susceptible to erosion, raising the risk of mudslides and further property damage.

Summary *The planet is a patchwork of large geographic areas characterized by specific climates, soils, and communities of plants and animals. Fire and grazing help maintain some biomes.*

41.2 MASTERING CONCEPTS

1. How do climate and soil composition determine the characteristics of terrestrial biomes?
2. Describe the climate, soils, and inhabitants of each of the major terrestrial biomes.

41.3 Freshwater Biomes Include Lakes, Ponds, and Streams

Although terrestrial biomes are most familiar to us, aquatic ecosystems occupy much more space. Water covers about 71% of Earth's surface. Most is in the ocean (see section 41.4); only about 1% of all water has a low enough salt content to be considered "fresh." Glaciers and the great polar ice sheets of Greenland and Antarctica tie up about 77% of this fresh water. Most of the rest is in underground aquifers. Lakes, ponds, and streams on Earth's surface contain only about 0.3% of the freshwater supply. Yet this tiny sliver of the global water "pie" is vital to humans, for both drinking water and irrigation. Most other terrestrial species rely on surface fresh water as well.

A. Lakes and Ponds Contain Standing Water

Standing water includes lakes and ponds, which differ from one another mainly in size and depth. A lake is generally larger and deeper than a pond.

Zones of a Lake or Pond

Light penetrates the regions of a lake to differing degrees, creating zones with characteristic groups of organisms (**figure 41.14**). The **littoral zone** is the shallow shoreline region where enough light reaches the bottom for photosynthesis to occur. Productivity is high, thanks to rooted plants such as cattails and floating plants such as water lilies. The **limnetic zone** is the layer of open water where light penetrates. Most photosynthesis in a lake occurs here, courtesy of **phytoplankton:** microscopic, free-floating, photosynthetic organisms such as cyanobacteria and green algae. Together, the littoral and limnetic zones make up the lake's photic zone. In contrast, the **profundal zone** is the deep region of water where light does not penetrate. Scavengers and decomposers, both here and in the **benthic zone** (the sediment at the lake bottom), rely on a gentle rain of organic material from above.

Littoral zone:
The richest area of a lake. Water is shallow enough for rooted plants, algae, and cyanobacteria, which support diverse animal life.
Limnetic zone:
The layer of open water through which light can penetrate, supporting plankton and fishes
Profundal zone:
The deepest region of open water, where light does not penetrate. Organisms here are mostly scavengers and decomposers.
Benthic zone:
Sediments at bottom, with scavengers and decomposers

FIGURE 41.14 Zones of a Lake. The littoral, photic, profundal, and benthic zones of a lake each house different types of organisms.

Lakes and Ponds Change Over Time

Lakes age. Younger lakes are often deep, steep-sided, and low in nutrients. These lakes are termed **oligotrophic,** which means they are low in productivity. They are clear and sparkling blue, because phytoplankton aren't abundant enough to cloud the water. As a lake ages, however, nutrients accumulate from decaying organisms and sediment (**figure 41.15**). These lakes are termed **eutrophic,** which means they are nutrient-rich and high in productivity. The rich algal growth turns the water green and murky. In time, the lake becomes a bog or marsh. These wetlands often host spectacularly diverse assemblages of plants and animals that rely on the interface between land and water. Eventually, however, the wetland fills in completely and becomes dry land.

Nutrient-rich urban wastewater and farm runoff carrying phosphate-rich fertilizers can speed this process. In extreme cases, the nutrients promote the excessive growth of algae, which sink to the lake bottom after they die. As the dead algae decompose, deep waters are rapidly depleted of oxygen; fish kills and unpleasant odors often follow.

B. Streams Carry Running Water

Streams include brooks, creeks, and rivers that carry water and sediment from all portions of the land toward the ocean (or an interior basin such as the Great Salt Lake). Along the way, streams provide moisture and habitat to aquatic and terrestrial organisms.

Rivers are the largest streams. They change as they flow toward the ocean (**figure 41.16**). At the headwaters, the water is relatively clear, and the stream channel is narrow. Where the current is swift, turbulence mixes air with water, so the water is rich in oxygen. As the river flows toward the ocean, it continues to pick up sediment and nutrients from the channel. The river widens as small streams draining additional land areas contribute more water. As the land flattens, the current slows. The river is now murky, restricting

a.

b.

FIGURE 41.15 Nutrients Make a Difference. (**A**) Algae are accumulating in the shallow, nutrient-rich waters of this pond. (**B**) Oregon's Crater Lake is oligotrophic. It formed about 6850 years ago, and few nutrients have made their way into the crystal-clear water.

photosynthesis to the banks and water surface. As a result, the oxygen content is low relative to the river upstream.

Rivers and streams depend heavily on the land for water and nutrients. Dead leaves and other organic material that fall into a river add to the nutrients. Rivers also return nutrients to the land. Many rivers flood each year, swelling with meltwater and spring runoff and spreading nutrient-rich silt onto their floodplains. When a river approaches the ocean, its current diminishes, which deposits fine, rich soil that forms new delta lands at the mouth of the river.

Summary *Lakes are divided into several zones, each with a different characteristic community. Lakes and ponds change over time as nutrient concentrations increase. Stream life is adapted to changing water velocities, nutrient and oxygen concentrations, and drought and flooding conditions.*

41.3 MASTERING CONCEPTS

1. Describe the types of organisms that live in each zone of a lake or pond.
2. What is the difference between an oligotrophic and a eutrophic lake?
3. Describe the ways a river changes from its headwaters to its mouth.

a.

b.

c.

FIGURE 41.16 A River Changes Along Its Course. (**A**) A narrow, swift stream in the mountains becomes a slow-moving river as it accumulates water and sediments and approaches the ocean. (**B**) Many small mountain streams contribute water to (**C**) the vast Mississippi River.

41.4 Oceans Make Up Earth's Largest Ecosystem

The oceans, covering 70% of Earth's surface and running 11.2 km (7 miles) deep in places, form the world's largest biome. Most photosynthesis on Earth occurs in the vast oceans, contributing enormous amounts of oxygen to the atmosphere. Moreover, oceans absorb so much of the sun's energy that they help stabilize Earth's climate.

A. Land Meets Sea at the Coast

The world's coasts are a vital source of food, transportation, and recreation. By the year 2010, 65% of the world's people will reside within 16 km (10 miles) of an ocean. As the human population expands, concerns about coastal ecosystems will likely increase. Several types of ecosystems border shorelines.

Estuaries

An **estuary** is an area where the fresh water of a river meets the salty ocean (**figure 41.17A**). The salinity is constantly changing. When the tide is out, the water may not be much saltier than water in the river. The returning tide, however, may make the water nearly as salty as the sea. Organisms able to withstand these extremes receive nutrients from both the river and the tides. An estuary therefore houses a very productive ecosystem. Moreover, almost half of an estuary's photosynthetic products go out with the tide and nourish coastal communities.

Estuaries are also nurseries for many sea animals. More than half the commercially important fish and shellfish species spend some part of their life cycle in an estuary. Migratory waterfowl feed and nest here as well. Human activities can threaten these important ecosystems. **vanishing estuaries, p. 846**

Intertidal Zone

Along coastlines lie the sandy, muddy, or rocky areas of the **intertidal zone,** the area between the high tide and low tide marks. This region of constant change is alternately exposed and covered with water as the tide rises and falls.

Species-rich mangrove forests occupy intertidal areas in the tropics (figure 41.17B). At the other extreme, a sandy beach features long strips of bare sand that make for beautiful vistas and all-day exposure to the sun. Constantly shifting sands mean that few producers can take root on the beach, but ocean water delivers a constant supply of dead organisms and other nutrients. Crabs and shorebirds feed on the organic debris and then burrow into the sand or fly away to escape the pounding waves.

Some intertidal zones are rocky, not sandy (figure 41.17C). Here, organisms often attach to rocks, preventing wave action from carrying them away. Holdfasts attach large marine algae (seaweeds) to rocks. Threads and suction fasten mussels to rocks. Sea anemones, sea urchins, sea stars, and snails live in pools of water that form between rocks.

Coral Reefs

Colorful and highly productive coral reef ecosystems border some tropical coastlines. **Coral reefs** are vast underwater structures of calcium carbonate whose nooks and crannies provide habitats for a huge variety of organisms (figure 41.17D). The Great Barrier Reef of Australia, for example, is composed of some 400 species of coral and supports more than 1500 species of fishes, 400 of sponges, and 4000 of mollusks. Other residents include algae, snails, sea stars, sea urchins, octopuses, and countless types of microorganisms. Food is abundant because the sun penetrates the shallow

a.

b.

FIGURE 41.17
Coastal Ecosystems. (**A**) Salt and fresh water meet at estuaries. (**B**) Mangrove trees provide habitat for many other organisms. (**C**) The rocky intertidal zone is frequently pounded by waves. (**D**) Coral reefs house a spectacular diversity of species.

water, allowing photosynthesis to occur, and constant wave action brings in additional nutrients.

Individual coral animals, called polyps, build the reefs and house symbiotic algae that are essential for the coral's, and the ecosystem's, survival. The living coral is but a thin layer atop the remains of ancestors. A coral reef, then, is at the same time an immense graveyard and a thriving ecosystem. It is rich in biodiversity, yet fragile. Chapter 42 considers threats to coral reefs.

B. The Open Ocean Remains Mysterious

The oceans cover most of Earth's surface, but we know less about biodiversity there than we do about biodiversity in a single tree in a tropical rain forest. Not only is this ecosystem vast, but it also houses populations that are sometimes small, usually very dispersed, and nearly always difficult for us to observe. Biologists have explored only 5% of the ocean floor and 1% of the huge volume of water above.

The most productive ocean environments arise in zones of **upwelling,** where cold, nutrient-rich lower layers of water move upward. The influx of nutrients causes phytoplankton to "bloom," and with this widening of the food web base, many ocean populations grow. Upwelling generally occurs on the western side of continents, such as along the coasts of southern California, South America, parts of Africa, and the Antarctic. About every seven years, a climatic event called El Niño shifts these ocean currents, greatly disrupting coastal ecosystems (see this chapter's Can *You* Relate? El Niño Years).

Marine biologists divide the ocean into zones (**figure 41.18**). The intertidal zone, discussed in the previous section, is the shoreline. The **neritic zone** is the area from the coast to the edge of the continental shelf. Beyond the continental shelf is the **oceanic zone;** it is, in turn, subdivided according to depth. The **pelagic zone** consists of all of the water above the sea floor. The upper layer of the pelagic zone, the photic zone, is the only area where photosynthesis can occur. At the very top of the pelagic zone, the sea surface is rich in phytoplankton and the small animals that feed on them (see figure 41.18).

Below the photic zone, light is too dim for photosynthesis. Nevertheless, the bodies and wastes of top-dwelling organisms provide a continual rain of nutrients to the species below, including great numbers and varieties of fishes, mollusks, echinoderms, crustaceans, and organisms yet to be discovered. The unusual communities that occupy hydrothermal vents add biodiversity to the benthic zone (see the chapter opening essay).

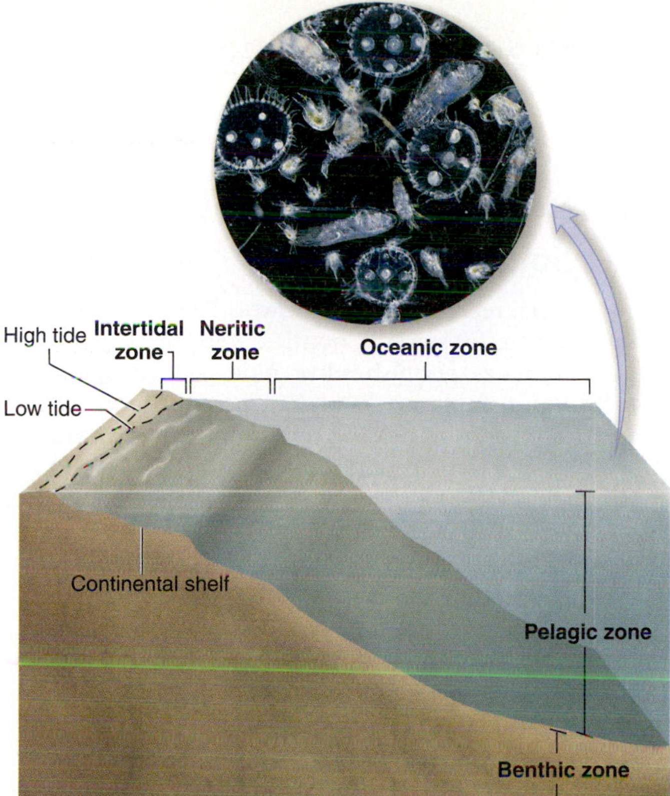

FIGURE 41.18 Zones of the Ocean. The intertidal, neritic, pelagic, and benthic zones of the ocean each harbor unique types of organisms. The inset shows the sea surface microlayer, which brims with phytoplankton, zooplankton, tiny animals, eggs, larvae, and nutrients.

c.

d.

Summary *Coastal areas, where land meets sea, can be as productive as an estuary or as barren as a sandy beach. The oceans, which form the world's largest biome, are divided into zones, each with a characteristic set of organisms.*

41.4 MASTERING CONCEPTS

1. Describe some of the adaptations that characterize organisms living in estuaries, intertidal zones, and coral reefs.
2. How is upwelling important to coastal ecosystems?
3. List and define the major zones of the ocean.

41.5 INVESTIGATING LIFE
Some Like It Hot

Coral reefs are among the world's most fascinating and important ecosystems. Countless species of microbes, algae, invertebrates, and fishes live, hide, feed, and breed in the reef's innumerable cracks and crevices. Coral reefs also protect coastlines from erosion and attract the commercial fishing boats and tourists that support entire economies.

Although coral looks like a brightly colored plant, it is in reality a type of animal called a cnidarian. These animals secrete stony calcium carbonate skeletons that protect their soft bodies from injury and predation. Coral animals feed in two ways. First, like its sea anemone relatives, a coral animal has stinging tentacles that can grab small animals and stuff them into its saclike digestive cavity. Second, the animal's tissues harbor huge populations of symbiotic single-celled algae (see the opening essay for chapter 5). As long as light is abundant, the golden brown algae carry out photosynthesis, providing their hosts with organic carbon in exchange for shelter and a stable home. **cnidarians, p. 455; symbiosis, p. 807**

The relationship between coral animals and their photosynthetic residents is surprisingly vulnerable to disruption. For example, the coral animals will expel the symbionts when the temperature of the seawater increases by just a few degrees. If that occurs, the coral reef "bleaches;" that is, the animals lose their vividly colored algal partners and then die, leaving behind only the white calcium carbonate skeleton (**figure 41.19**).

The prospect of global warming therefore raises the possibility of widespread coral bleaching and the eventual collapse of the reef ecosystem. As chapter 42 describes in more detail, human activities are causing the concentration of CO_2 in the atmosphere to increase. The additional CO_2, a so-called greenhouse gas, is causing the average surface temperature on Earth to slowly increase. As sea surface temperatures rise, coral bleaching may become a more widespread problem.

FIGURE 41.19 Coral Bleaching. This colony of staghorn coral has a large bleached area at the center. The photosynthetic algae that normally inhabit the tissues of a healthy coral have died, leaving the white calcium carbonate skeletons of the coral animals.

Or will it? In 1993, a group of scientists proposed the "adaptive bleaching hypothesis," which suggests that bleaching allows coral animals to expel algae that are poorly suited for warm water and replace them with other, better adapted varieties. This idea makes sense. After all, coral animals appeared in the fossil record more than 500 million years ago; somehow, they must have adjusted to many periods of heating and cooling.

One way to test the adaptive bleaching hypothesis is to study reefs during and after El Niño events, when the sea surface temperature spikes in some coastal areas. Biologist Andrew Baker of the Wildlife Conservation Society, working with colleagues from Columbia University and the University of Miami in Florida, used the 1997–1998 El Niño event to do just that. They knew from previous research that the genus *Symbiodinium* is the predominant genus of symbiotic algae living in the tissues of coral animals. Studies of ribosomal DNA have revealed at least eight major genetically different varieties, or genotypes, of *Symbiodinium*.

Baker and his team surveyed the genotypes of *Symbiodinium* in hundreds of coral samples collected from five reefs along the coasts of Kenya, Mauritius, Saudi Arabia, and Panama. The researchers compared reefs that had bleached during the El Niño event (in the Persian Gulf and along the coast of Kenya) with reefs that had not bleached (in the Red Sea and off the island of Mauritius). They also took samples on a Panamanian reef before, during, and after the El Niño event. **Figures 41.20**A and B show that recent bleaching was associated with the presence of one algal genotype (type D) and a shift away from another (type C).

Does this result mean that *Symbiodinium* group D functions better in warm water than does group C? One way to find out is to measure photosynthetic rates at different temperatures in controlled conditions. Rob Rowan of the University of Guam. He collected corals in genus *Pocillopora* and used genetic tests to identify colonies that harbored *Symbiodinium* types C or D. He kept the corals alive in cups in his laboratory. After allowing the animals to become acclimated to 28.3°C water for 14 days, Rowan divided the corals randomly into two groups. Controls remained at 28.3°C for the duration of the experiment. The water temperature of the treatment group climbed to 31.3°C for 4.5 h every day for 5 days. For 5 days after that, the temperature spiked to 32°C for 4.5 h.

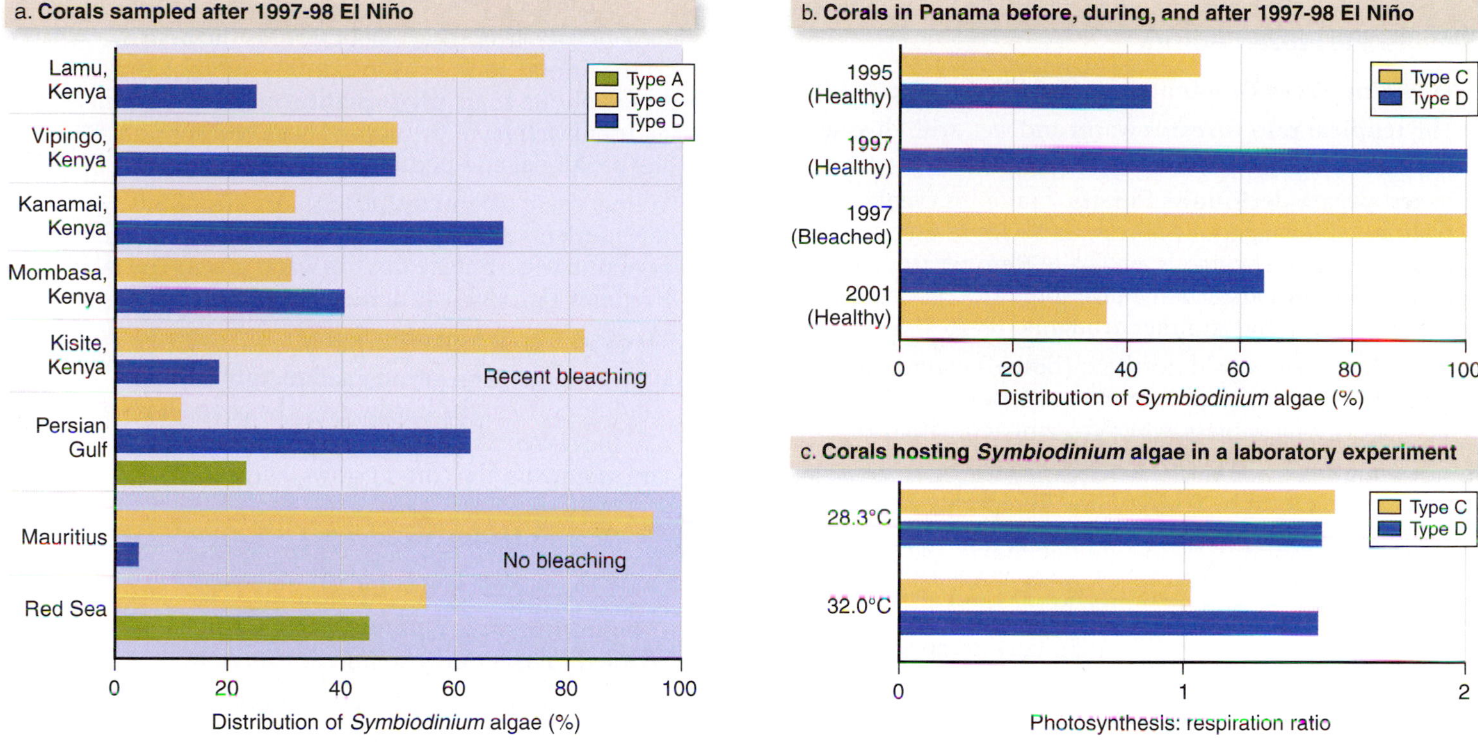

FIGURE 41.20 Adjustments to Warm Water. (**A, B**) A survey of coral reefs suggests that warm waters prompt coral animals to shift from *Symbiodinium* type C to type D. (**C**) Laboratory experiments show that *Symbiodinium* type D tolerates warm water temperatures better than type C. **Question:** What are the benefits and limitations of field experiments such as those described in parts (A) and (B), and of a lab experiment such as the one in part (C)?

Rowan monitored O_2 concentrations to estimate the rates of photosynthesis and respiration at each temperature. A healthy coral produces more O_2 in photosynthesis than it consumes in respiration, so the photosynthesis-to-respiration ratio should exceed 1. That is exactly what he found at 28.3°C, regardless of the alga's genotype (figure 41.20C). But when Rowan cranked up the heat, corals colonized with *Symbiodinium* type C had a photosynthesis-to-respiration ratio of about 1. Those harboring type D were unaffected. Rowan concluded that type D is "a high-temperature specialist."

Together, these two studies illustrate the importance of genetic diversity within a species. If all of the symbionts were identical, corals might be unable to adjust to temperature shifts in their habitat. Instead, changing conditions select for different genotypes of *Symbiodinium* algae. This adaptability is good news for the coral reefs, and for other organisms, including humans, that depend on them. But a big question remains: will the corals be able to adjust to the current warming period, which is apparently occurring more rapidly than previous climate shifts? Only time will tell.

Baker, Andrew C., Craig J. Starger, Tim R. McClanahan, and Peter W. Glynn. 2004. Corals' adaptive response to climate change. *Nature*, vol. 430, page 741.

Rowan, Rob. 2004. Thermal adaptation in reef coral symbionts. *Nature*, vol. 430, p. 742.

CHAPTER SUMMARY

41.1 The Physical Environment Determines Where Life Exists

- The **biosphere** is subdivided into **biomes,** which are major types of **ecosystems** that occupy large geographic areas. Each biome has a characteristic group of species.

A. Abiotic Factors Influence All Life

- The major **abiotic** (nonliving) factors that limit a species' distribution include sunlight, temperature, moisture, salinity, and fire.

B. Earth Has Diverse Climates

- Temperature and rainfall define Earth's major climatic regions. Uneven heating due to the angle of solar rays hitting the planet's curved surface generates wind and moisture patterns.
- Local features such as a mountain range can influence climate by producing a **rain shadow.**

41.2 Terrestrial Biomes Range from the Lush Tropics to the Frozen Tundra

A. Towering Trees Dominate the World's Forests

- The **tropical rain forest** is warm and wet, with diverse life. Nutrients cycle rapidly, leaving soils relatively poor.
- In **temperate deciduous forests,** rainfall is evenly distributed throughout the year, and winters are relatively mild. Their soils are fertile. **Temperate coniferous forests** usually have somewhat poorer soils, colder winters, and summer droughts.
- The **taiga** is a very cold northern (boreal) coniferous forest. Conifers are well adapted to conserve moisture during the long winters, when freezing temperatures mean liquid water is scarce.

B. Grasslands Occur in Tropical and Temperate Regions

- Tropical **savannas** have alternating dry and wet seasons and are dominated by grasses, with sparse shrubs and woody vegetation and migrating herds of herbivores.
- Fire, grazing, and seasonal drought keep **temperate grasslands** free of trees.

C. Whether Hot or Cold, All Deserts Are Dry

- **Deserts** are dry; plants of that biome have adaptations that help them obtain and store water. Animals minimize water loss by becoming active at night.

D. Arctic and Alpine Tundras Occupy High Latitudes and High Elevations

- **Tundra** can occur in the Arctic region or high on mountaintops in more temperate zones.
- Arctic tundra has very cold and long winters. A layer of frozen soil called **permafrost** lies beneath the surface and prevents the growth of trees.

E. Fire- and Drought-Adapted Plants Dominate Mediterranean Shrublands (Chaparral)

- **Mediterranean shrublands** such as California's **chaparral** have dry summers and moist, mild winters. Many of the plants are fire-adapted.

41.3 Freshwater Biomes Include Lakes, Ponds, and Streams

A. Lakes and Ponds Contain Standing Water

- The **littoral zone** of a lake is the shallow area where light reaches the bottom; the **limnetic zone** is the lit upper layer of open water; the **profundal zone** is the dark deeper layer. The lake bottom is the **benthic zone.** In the littoral zone, most producers are rooted plants. In the photic zone, **phytoplankton** predominate. Nutrients fall from the upper layers and support life in the profundal and benthic zones.
- Young, deep, **oligotrophic** lakes are clear blue, with few nutrients to support algae. Nutrients gradually accumulate, and algae tint the water green. The lake becomes a productive, or **eutrophic,** lake.

B. Streams Carry Running Water

- In rivers, organisms are adapted to local current conditions. Near the headwaters, the channel is narrow, and the current is swift. As the river accumulates water and sediments, the current slows, and the channel widens.

41.4 Oceans Make Up Earth's Largest Ecosystem

A. Land Meets Sea at the Coast

- In **estuaries,** rivers empty into oceans. Life here is adapted to fluctuating salinity.
- Residents of the rocky **intertidal zone** are adapted to stay in place as the tide ebbs and flows.
- **Coral reefs** support many thousands of species in and around 400 or so types of coral.

B. The Open Ocean Remains Mysterious

- The region of ocean near the shore is the **neritic zone.** Open water is the **oceanic zone** and includes the benthic zone (the bottom), and the **pelagic zone** (open water above the ocean floor). The most productive areas are in the neritic zones where **upwelling** occurs.

41.5 Investigating Life: Some Like It Hot

- Coral animals maintain populations of symbiotic algae, often of genus *Symbiodinium*, in their tissues. High water temperature causes coral bleaching as the corals expel their symbionts.
- Researchers have found that some genotypes of *Symbiodinium* tolerate high temperatures better than others, which may help coral reefs survive the current period of climate change.

MULTIPLE CHOICE QUESTIONS

1. Which of the following is a biome?
   a. A rotting log
   b. A freshwater lake
   c. A forest
   d. Both b and c

2. Which abiotic factors likely affect the type of organisms found in Utah's Great Salt Lake?
   a. Sunlight
   b. Salinity
   c. Temperature
   d. All of the above

3. A biome composed of trees that seasonally shed their leaves is a
   a. chaparral.
   b. temperate forest.
   c. tropical dry forest.
   d. taiga.
4. What is a distinguishing feature of the tundra?
   a. Permafrost
   b. Infertile soil
   c. Grasses
   d. Boreal forest
5. What two abiotic factors influence the fertility of the soil of a biome?
   a. Temperature and rainfall
   b. Fire and sunlight
   c. Sunlight and salinity
   d. Both a and c
6. Why is use of tropical rainforests for human agriculture impractical?
   a. Because it reduces biodiversity
   b. Because the soil is nutrient poor and erodes quickly
   c. Because the climate is warm and wet
   d. Because only a limited number of crop plants will grow
7. An organism that lives in lake sediment would be found in the __________ zone.
   a. limnetic
   b. profundal
   c. benthic
   d. littoral
8. Why can't the photosynthetic phytoplankton of a eutrophic lake maintain deep-water oxygen levels?
   a. Because oxygen production by photosynthesis only occurs where there is light
   b. Because high productivity means lots of material for decomposers that use up oxygen
   c. Because water temperature affects photosynthesis
   d. Both a and b
9. How is an estuary different from a mangrove swamp?
   a. The salinity of the water changes with the tides
   b. The presence of plants that are salt tolerant
   c. The presence of a benthic zone
   d. All of the above
10. The deepest waters of the ocean would be found in
    a. the oceanic zone.
    b. the neritic zone.
    c. the pelagic zone.
    d. both a and c.

TESTING YOUR KNOWLEDGE

1. How does the fact that Earth is a sphere tilted on its axis influence the distribution of life?
2. Explain why the climate on the west side of Oregon's Cascade Mountains is much wetter than on the east side of the mountain range.
3. List adaptations that characterize organisms in each of the following biomes: tropical rain forest, savanna, temperate grassland, tundra, desert, taiga, the rocky intertidal zone, the bottom of a lake.
4. How can the tropical rain forest support diverse and abundant life with such poor soil?
5. What is permafrost?
6. Describe how photosynthetic activity differs in the zones of a lake or ocean.
7. Describe the physical and chemical differences between the water in a mountain stream and the water near the mouth of the Mississippi River.
8. What are the sources of energy in the various zones of the ocean?

THINKING AS A SCIENTIST

1. Some scientists are currently attempting to catalog all of the world's biodiversity. What are some of the technical problems they may encounter?
2. Suppose you are exploring the chaparral ecosystem in California. You encounter a shrub species that you think may be fire-adapted, and you wonder whether the plant can reproduce in the absence of fire. Design an experiment that would help answer your question.
3. Polar bears live on the ice cap near the North Pole. Their numbers are dwindling, apparently because of both pollution and global warming. Ice on Canada's Hudson Bay, for example, is melting about 3 weeks earlier in the year than it was some 30 years ago. List some specific ways that this change in habitat might affect polar bear populations.
4. Poultry farmers apply large amounts of nutrient-rich animal waste onto the land, where it runs off into nearby lakes and streams. What effect might this nutrient input have on the aquatic ecosystems? Lawmakers in some states have debated whether animal waste, a natural substance, should legally qualify as a hazardous waste. Do you think it should?
5. Researchers and citizens in Prairie City, Iowa, are reconstructing the prairie by collecting seeds from remnants of native grasslands and reintroducing animals. Which other biomes discussed in the chapter might be possible to reconstruct, and which not?

Visit www.mhhe.com/hoefnagels for practice quizzes, animations, videos, and activities designed to help you master the material in this chapter.

CHAPTER 42

Preserving Biodiversity

Herons, crayfish, and countless other species make up the biodiversity of the Everglades.

What's the Point?

The Endangered Everglades

Although the world has no shortage of degraded ecosystems, the Florida Everglades region is remarkable for the scope of its destruction. Before 1900, much of Florida was a continuous waterway. This vast area included estuaries, saw grass plains, mangrove swamps and other wetlands, and tropical hardwood forests. The Everglades was home to marsh grasses, cypress trees, egrets, eagles, herons, panthers, and alligators.

Damage to the Everglades began in the twentieth century. From 1903 until 1917, the "Everglades reclamation project" built four large canals that drained the area to prevent flooding. In 1930, following severe hurricanes in 1926 and 1928, levees were built around Lake Okeechobee. Farmers, reassured of their safety from floods, doubled their sugarcane crops.

In 1947 and 1948, hurricanes dumped 274 cm (108 inches) of rain, nearly twice the normal rainfall. Many farms and homes were submerged. Congress decided to control the flooding with a new water supply system. Engineers built a system of canals, levees, and pumps that covered nearly 1600 km (1000 miles) to direct water to agricultural and urban areas. Meanwhile, the human population swelled from about 750,000 to more than 4 million today.

Then came disaster—nature changed. In the early 1960s, a drought started that would last until 1975. Nearly 300,000 hectares of land burned. The great effort to prevent flooding now backfired, causing a terrible water shortage. The situation worsened with a drought in 1981 and torrential rains in 1982–1983. In 1986, a fifth of Lake Okeechobee fell under an algal bloom, the result of nitrogen and phosphorus from fertilizer and dairy operations washing into the water. Native saw grass could not compete with species that could tolerate the nutrient pollution. The influx of fresh water robbed mangroves of salt water, which led to declines in populations of sea grasses and various animals. Wading bird populations plummeted to 10% of their levels a century earlier. Dozens of species became endangered.

Efforts have been underway to undo the damage to the Florida ecosystem. In 1984, flooding was restored to a small area of the historic Kissimmee River floodplain. This action drew waterfowl, aquatic invertebrates, and fishes as native vegetation returned. A more ambitious plan, approved in 2000, will remove canals and levees; it will also restore natural channels and floodplain wetlands to the central third of the Kissimmee River. With a price tag estimated at \$7.8 billion, the restoration plan is "the world's largest ecosystem restoration effort." Monitoring its progress will require a small army of conservation biologists. This chapter describes some of the many challenges these scientists face worldwide.

LEARNING OUTLINE

42.1 Earth's Biodiversity Is Dwindling

Humans simply cannot live without a rich array of other species, both obvious and unseen (**figure 42.1**). We use other organisms for food, shelter, energy, clothing, and drugs. Microbes carry out indispensable tasks, from digesting food in our intestines, to decaying organic matter, to fixing nitrogen, to producing oxygen (O_2). Plants and microbes together purify the air, soil, and water. Wetland species reduce the severity of floods. Insects pollinate our crops. The remains of species that lived millions of years ago provide the fossil fuels that sustain our economies now.

FIGURE 42.1 We Need Other Species. This child is standing in a rice field with a water buffalo. Humans have cultivated rice for thousands of years, using the grain as a source of protein and starch, and feeding the rice straw to livestock. The water buffalo is not only a work animal but also a source of dairy products. The animal's dung fertilizes the rice fields.

Clearly, our existence as a species depends on **biodiversity**—the variety of life on Earth. Yet many biologists are concerned that Earth is in the midst of a biodiversity crisis (**table 42.1, figure 42.2**). The current extinction rate of vertebrates is some 100 to 1000 times the "background" species extinction rate, which estimates how quickly species disappeared before human intervention (see figure 14.12). About one third of amphibian species are currently threatened (likely to become endangered in the future). About 20% of mammals and at least 10% of birds are also threatened.

Conservation biologists study the preservation of biodiversity. These scientists try to determine why species disappear, and they develop strategies for maintaining diversity. This final chapter focuses first on the main causes of the loss of biodiversity: habitat destruction and degradation, the introduction of nonnative species, and overexploitation. The chapter ends on a hopeful note, with some ways people can help counteract the biodiversity crisis.

A. Human Activities Destroy Habitats

Habitat destruction is the primary cause of diminishing biodiversity. Humans have altered nearly 50% of the land, replacing grasslands with wheat fields, wetlands with suburbs, and forests with rangeland. The link to biodiversity is obvious: destroying habitat makes it difficult or impossible for its occupants to survive and reproduce.

Table 42.1 *A Few of the World's Endangered Species*

Species	Former Range	Threat
California condor *(Gymnogyps californianus)*	Western United States	Habitat loss, shooting, lead poisoning, toxic substances in environment
Green pitcher plant *(Sarracenia oreophila)*	Alabama, Tennessee, Georgia	Habitat loss, overharvest for commercial trade
Red-cockaded woodpecker *(Picoides borealis)*	Eastern United States from Florida to New Jersey and Maryland, inland to Texas, Oklahoma, Missouri, Kentucky, and Tennessee	Habitat loss
American burying beetle *(Nicrophorus americanus)*	Arkansas, Kansas, Oklahoma, Nebraska, South Dakota	Competition for food (carrion)
Spix's macaw (bird) *(Cyanopsitta spixii)*	Arid savanna scrubland in northeastern Brazil	Excessive trapping
Sumatran rhino *(Dicerorhinus sumatrensis)*	Indonesia, Malaysia	Poachers
Vaquita (cetacean) *(Phocoena sinus)*	North end of Gulf of California	Caught in fishing nets

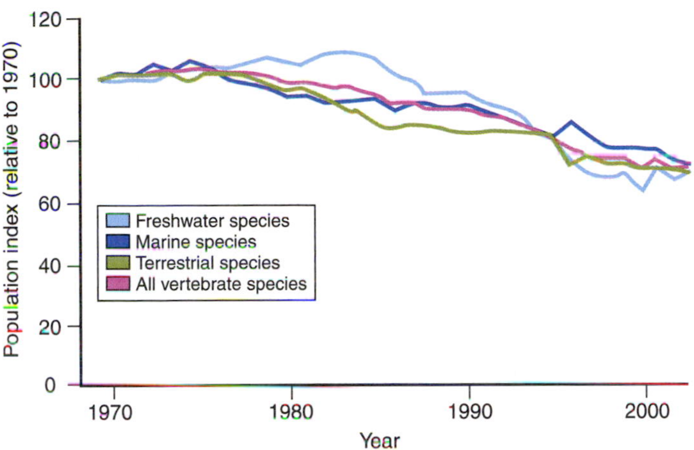

FIGURE 42.2 Biodiversity Crisis. The World Wildlife Fund, an environmental advocacy group, has used scientific reports to monitor 1300 species of freshwater, marine, and terrestrial vertebrate species. The population index measures overall population trends relative to a baseline year of 1970. Overall, biodiversity has declined by about 30% since 1970.

a.

b.

FIGURE 42.3 Tropical Deforestation. (**A**) A new road into the rain forest brings new possibilities for human settlement and agriculture. Poor soils mean the crops will not last long. (**B**) This cattle pasture in western Brazil was once a lush tropical rain forest.

Shrinking Forests, Expanding Deserts

Deforestation is the removal of all tree cover from a forested area. Forests harbor a tremendous diversity of plant, animal, and microbial life, which may become extinct as the forests vanish. Deforestation also promotes soil erosion, which reduces soil fertility and contributes to water pollution. Transpiration by forest plants contributes water vapor to the atmosphere, affecting global climates. Burning these forests not only removes an important component of the global water cycle, but also releases stored carbon into the atmosphere, contributing to the greenhouse effect (see section 42.2). **forests, p. 830**

Satellite imaging data indicate that nearly half of the world's moist tropical forests have already been cleared, mostly to make room for subsistence agriculture (**figure 42.3**). Ironically, the same soils that support the lush tropical rain forest produce poor crop yields. The warm temperatures promote rapid decomposition of organic matter, and heavy rains deplete soil nutrients. Once native plants give way to crops or grazing animals, the nutrient-poor soils harden into a cementlike crust. Food webs topple, threatening many native people as well as other inhabitants.

The disappearance of the native North American temperate forest has paralleled settlement by Europeans. Since the early eighteenth century, people have cleared the land from east to west to create farmland, obtain fuel, and make room for railroads and towns. Today, although vast areas of managed pine forests and plantations occupy the region, less than 1% of the original temperate forest survives. Likewise, only a small percentage of the original "old growth" Douglas-fir forest remains in the Pacific Northwest.

Whereas forests are shrinking worldwide, deserts are expanding into surrounding areas in a process called **desertification**. Desert is displacing savanna in Africa, where the Sahara is the driest and largest desert in the world. Along the Sahara's southern edge and in countries farther south, a combination of drought and overgrazing have destroyed grasses and compacted the soil. New plants cannot establish themselves, so patches of desert enlarge and join. As the desert spreads, farmers cut down more forest to create more grazing land. This land is destined to become desert too, if these practices continue. **savanna, p. 831; desert, p. 832**

Changing Rivers, Depleted Lakes, and Vanishing Estuaries

Freshwater habitats are also vulnerable to destruction. Damming for flood control or power generation, for example, completely alters river ecosystems. Deep reservoirs replace waterfalls, rapids, and wetlands, where many species breed. Areas that were seasonally flooded become dry. Water temperature, oxygen content, and nutrient levels all change, triggering shifts in species diversity and food webs both above and below the dam. Dams also disrupt the migration of fish and other aquatic animals. The number of large (over 15 m high) dams worldwide has increased from 5000 in 1950 to more than 45,000 today (**figure 42.4**).

Changing a river's path also destroys habitat. Along the banks of the Mississippi River, levees built to prevent flooding alter the pattern of sediment deposition. Channelization increases the water's flow rate, eroding sediments and choking out downstream communities. Nutrients that once spread over the floodplain during periodic floods are now confined to the river channel, which carries them to the Gulf of Mexico. There, the nutrients feed algae, causing their populations to bloom. Red and brown tides, the result of such blooms, may kill fishes, manatees, and other sea life. **red tide, p. 391**

Another striking example of aquatic habitat destruction is the death of central Asia's Aral Sea, once the world's fourth largest lake. Since the 1920s, when its water was first used to irrigate cotton, water has been removed much faster than it has been replenished by rainfall in this naturally dry area. Today, the Aral Sea is but three small remnants of what it once was (**figure 42.5**). The remaining water has grown so salty from fertilizer and pesticides that all 24 native fish species have become locally extinct. Salt deposits, stranded boats, and toxic dust storms carrying fertilizer and pesticide residues paint an eerie landscape. Efforts are underway to correct the damage. Smaller volumes of water are being diverted from the streams that feed the lake, and local officials hope that the water table will rise and the Aral Sea might begin to refill with rainwater.

Coastlines are also suffering from habitat destruction. Many fishes and invertebrates spend part of their lives in estuaries, and diverse algae and flowering plants support the food web. Yet humans have drained and filled estuaries for urbanization, housing, tourism, dredging, mining, and agriculture. These activities affect life in the oceans too. The loss of coastal habitats can threaten populations of commercially important species of marine animals such as bluefin tuna, grouper, and cod. **estuaries, p. 836**

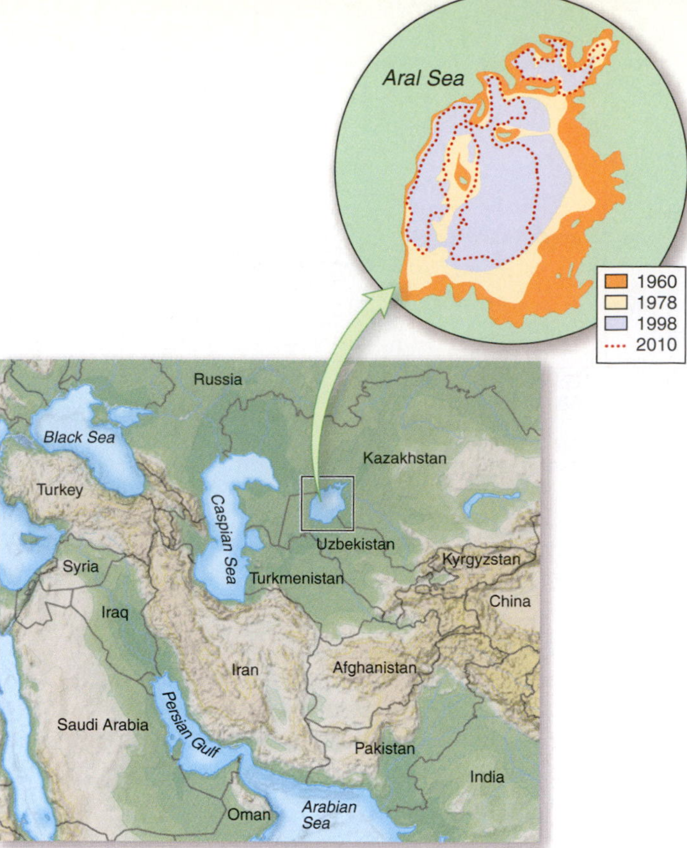

FIGURE 42.5 Loss of an Inland Sea. Due to a short-sighted irrigation plan, the once-huge Aral Sea is disappearing.
Reprinted with permission from Richard Stone, Coming to grips with the Aral Sea's grim legacy, *Science,* April 2, 1999, vol. 284. *Source:* Philip Micklin, Western Michigan University, Copyright © 1999 American Association for the Advancement of Science.

FIGURE 42.4 Big Dam. The Ataturk Dam on the Euphrates River provides power to Turkey. Dams help control flooding and provide irrigation water, but they also cause the loss of streamside habitat and disrupt the migration of fishes and other animals.

B. Pollution Degrades Habitats

Pollution is any chemical, physical, or biological change in the environment that harms living organisms. Pollution degrades the quality of air, water, and land, threatening biodiversity worldwide.

Air Pollution

Smog, a contraction of *sm*oke and f*og,* is a type of air pollution that forms a visible haze in the lower atmosphere (**figure 42.6**). Industrial smog occurs in urban and industrial regions where power plants, factories, and households burn coal and oil. The resulting smoke and sulfur dioxide (SO_2) may form a dark haze. Photochemical smog forms when nitrogen oxides and emissions from vehicle tailpipes

FIGURE 42.6 Smog. Air pollution continues to plague many cities. This is Santiago, Chile.

undergo chemical reactions in the presence of light, producing ozone (O_3) and other harmful chemicals that injure plants and cause severe respiratory problems in humans. Warm, sunny areas with heavy automobile traffic have the most photochemical smog, but winds may carry it to sparsely populated areas.

Air also carries suspended **particulates,** tiny bits of matter that float in the air. Examples include road dust, volcanic ash, soot from partially burned fossil fuels, mold spores, pollen, and acidic particles. The damage they cause extends beyond the occasional need to dust off bookshelves and window sills. Most harmful are particles that are 2.5 μm in diameter or smaller. Not only do they become trapped deep within the lungs, but the heavy metals and toxic organic compounds in these particles makes them especially likely to trigger inflammation, shortness of breath, asthma, or even cancer.

Other forms of air pollution are less visible but perhaps more harmful than smog. One example is **acid deposition:** acidic rain, snow, fog, dew, or dry particles. Because the atmosphere contains carbon dioxide (CO_2) and water, all rainfall includes some carbonic acid (H_2CO_3) and is therefore slightly acidic, with a pH around 5.6. Burning fossil fuels, however, release sulfur and nitrogen oxides (SO_2 and NO_2) into the atmosphere, where they join water and form sulfuric acid and nitric acid. These acids return to the Earth as acid deposition. **pH scale, p. 32**

Coal-burning power plants release the most acid-generating emissions. In the United States, winds carry acids hundreds of miles east and northeast of the power plants in the Midwest. As a result, the average rainfall in the eastern United States has a pH of about 4.4. Acid deposition also affects the Pacific Northwest, the Rockies, Canada, Europe, East Asia, and the former Soviet Union.

Most lakes have a pH between 6 and 8; acid deposition may lower it to 5 or less. The acid can leach toxic metals such as aluminum or mercury from soils and sediments, causing fish eggs to die or yield deformed offspring. Lake-clogging algae replace aquatic flowering plants. Organisms that feed on the doomed species must seek alternative food sources or starve, which disrupts or topples food webs. Eventually, lake life dwindles to a few species that can tolerate increasingly acidic conditions. Acid deposition alters forests too (**figure 42.7**). As soil pH drops, aluminum ions released from soil enter roots and stunt tree growth. The trees become less able to resist infection or to survive harsh weather. As a result, acid deposition is thinning high-elevation forests throughout Europe and on the U.S. coast from New England to South Carolina.

Chemicals that destroy ozone can also be extremely harmful. Ozone is an atmospheric molecule with two faces. In photochemical smog at Earth's surface, ozone is a harmful pollutant. In the upper atmosphere, however, this gas protects life on Earth from harmful ultraviolet radiation. The **ultraviolet (UV)** portion of the electromagnetic spectrum includes wavelengths shorter than 400 nm. UV-B radiation (wavelength 290 to 320 nm), along with radiation of even shorter wavelengths, can damage biological molecules such as DNA. The stratospheric **ozone layer,** about 11 to 50 km (7 to 30 miles) above Earth's surface, forms when UV radiation from the sun reacts with oxygen gas (O_2). The ozone layer thus absorbs much of the harmful radiation that would otherwise strike Earth and harm organisms.

In recent decades, the ozone layer has thinned over parts of Asia, Europe, North America, Australia, and New Zealand

FIGURE 42.7 Acid Deposition. Acid rain has severely damaged this fir forest in the Czech Republic.

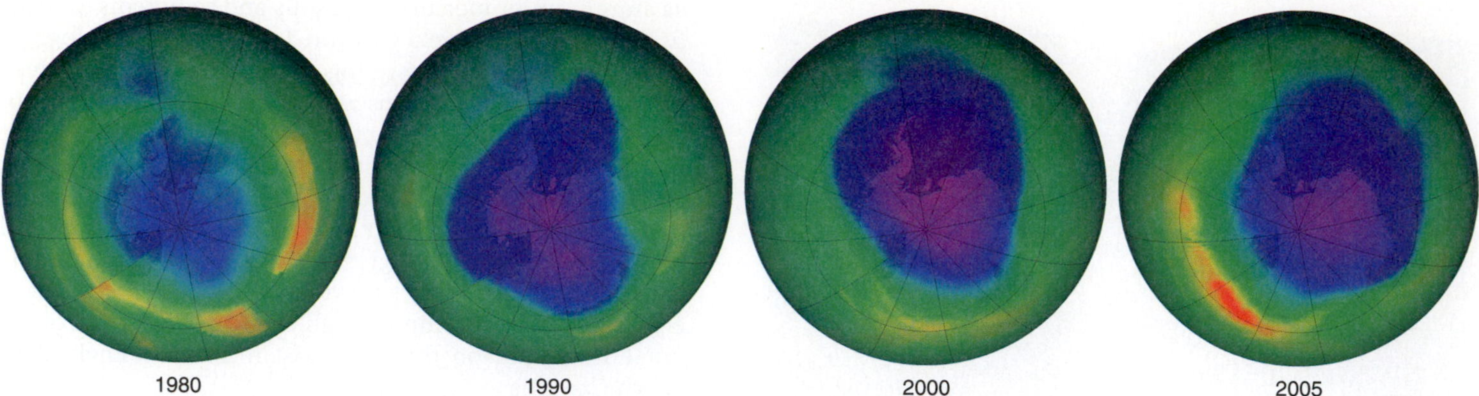

FIGURE 42.8 **The Ozone Hole and Ultraviolet Radiation.** Satellite images taken over time reveal the ozone hole over Antarctica. In these images, purple and blue represent thinned areas of the ozone layer.
Source: NASA; Data from Richard McKenzie, et al., "Increased Summertime UV Radiation in New Zealand in Response to Ozone Loss" in *Science*, vol. 285, September 10, 1999.

and formed a "hole" over Antarctica (**figure 42.8**). The main culprits are chlorine, fluorine, and bromine gases, some of which enter the atmosphere as chlorofluorocarbon (CFC) compounds. These compounds were once used in refrigerants such as Freon, as propellants in aerosol cans, and to produce foamed plastics. They can persist for decades in the upper atmosphere, catalyzing chemical reactions that break down ozone. An international treaty signed in 1987, the Montreal Protocol, banned the use of CFCs.

As the ozone layer thins, UV radiation is increasing at Earth's surface. Exposure to UV-B can cause skin cancer or cataracts. Ozone depletion may also indirectly contribute to species extinctions. UV-B can harm phytoplankton, which form the base of the food web in many aquatic ecosystems. Increasing UV-B radiation may also be one of many factors causing the global decline of amphibian populations.

Water Pollution

A diverse array of pollutants affects rivers, lakes, and groundwater (**table 42.2**). For example, raw sewage can be a major pollutant. In addition to carrying disease-causing organisms, sewage also contains organic matter and nutrients such as nitrogen and phosphorus. When released into waterways, organic matter fuels the growth of bacteria, whose respiration depletes the water of oxygen. Fish and other organisms die. At the same time, in a process called **eutrophication,** nitrogen and phosphorus fertilize algae in the water (see section 41.3). The resulting algal blooms are unsightly, and when the algae die, the microbes that decompose their dead bodies deplete the water of oxygen. Eutrophication also occurs after fertilizer or animal waste runs off soil and into water. **what happens after you flush?, p. 815**

Other pollutants are toxic chemicals, such as heavy metals or cyanide from mining operations or petroleum from oil spills. **Persistent organic pollutants** are carbon-containing molecules such as polychlorinated biphenyls (PCBs), polycyclic aromatic hydrocarbons (PAHs), DDT, and other pesticides that contaminate ecosystems over long periods. Some persistent organic pollutants may cause cancer and disturb reproduction in some species.

Table 42.2 *Examples of Chemical Water Pollutants*

Organic	Inorganic
Sewage	Chloride ions
Detergents	Heavy metals (mercury, lead, chromium, zinc, nickel, copper, cadmium)
Pesticides	Nitrogen from fertilizer
Wood-bleaching agents	Phosphorus from fertilizer and sewage
Hydrocarbons from petroleum	Cyanide
Humic acids	Selenium
Polychlorinated biphenyls (PCBs)	
Polycyclic aromatic hydrocarbons (PAHs)	

Organisms living in polluted areas are often exposed to several toxins. In the aftermath of Hurricane Katrina in 2005, floodwaters contained raw sewage, garbage, crude oil, gasoline, lead and other heavy metals, pesticides, and countless other pollutants spilled from damaged refineries and chemical plants (**figure 42.9**). The toxic soup threatened not only human health, but also the microbes, producers, and animals in the aquatic food chain. Moreover, when the floodwaters receded, they left behind their poisonous residues in soil.

FIGURE 42.9 Deadly Waters. These automobiles were submerged in the aftermath of Hurricane Katrina, which devastated the city of New Orleans in 2005. The floodwaters contained toxic chemicals that will continue to threaten ecosystems for some time.

Katrina's deadly legacy may extend far into the future, well beyond the lives and homes lost in the hurricane itself.

Some pollutants seem deceptively harmless. Sediments, for example, reduce photosynthesis by blocking light penetration into water. Even heat can be a pollutant. Hot water discharged from power plants reduces the ability of a river to carry dissolved oxygen, affecting fishes and other aquatic organisms.

Toxic chemicals, nutrients, and trash also pollute the open ocean. Fishing nets and plastic entrap and kill millions of seabirds and marine mammals. Some birds build their nests with plastic and feed their nestlings plastic bits. Sea turtles mistake floating blobs of plastic for their natural food, jellyfishes, and swallow them. The plastic lodges in their intestines and kills them.

Summary *Human activities are causing biodiversity to decline worldwide. Urbanization, agriculture, deforestation, and desertification destroy habitats on land. Dams, channelization, water depletion, and the loss of estuaries harm aquatic ecosystems. Atmospheric pollution includes smog, acid deposition, particulates, and ozone-depleting chemicals. Toxic chemicals, nutrients, and trash threaten freshwater and marine ecosystems.*

42.1 MASTERING CONCEPTS

1. What is the value of diversity to humans and to ecosystems as a whole?
2. What events have led to deforestation and desertification worldwide?
3. How do dams and channelization alter river ecosystems?
4. Why is damage to estuaries especially devastating?
5. What are major sources of industrial smog, photochemical smog, and acid deposition?
6. What effects do smog, acid deposition, particulates, and the thinning ozone layer have on life?
7. How do toxic chemicals, nutrients, sediments, and heat affect aquatic ecosystems?

42.2 Global Climate Change Alters and Shifts Habitats

The news is full of stories about global climate change and the greenhouse effect. In the past, scientists debated whether human activities could actually change something as complex as Earth's overall climate. Now, the scientific consensus is clear: we can and do.

A. Greenhouse Gases Warm Earth's Surface

Carbon dioxide is a colorless, odorless gas present in the atmosphere at a concentration of 380 parts per million. Although it is a minor atmospheric constituent, CO_2 is one of several gases that contribute to the **greenhouse effect,** an increase in surface temperature caused by heat-trapping gases in Earth's atmosphere (**figure 42.10**). Sunlight passes through the atmosphere and reaches Earth's surface, where some of the energy is absorbed and reradiated as heat. Carbon dioxide and other so-called greenhouse gases absorb this radiation and reradiate some of it back toward Earth's

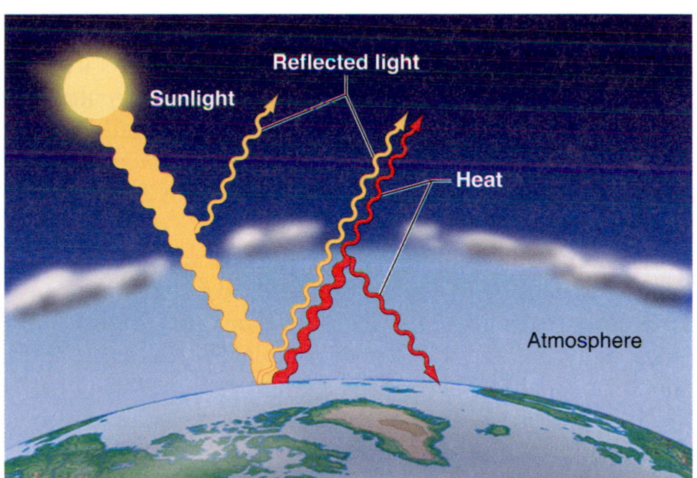

FIGURE 42.10 The Greenhouse Effect. Solar radiation heats Earth's surface. Some of this energy is reradiated to the atmosphere as heat, some of which is trapped near the surface by CO_2 and other greenhouse gases.

surface. Greenhouse gases therefore block heat escape, much as do the glass panes of a greenhouse.

Other greenhouse gases include methane, nitrous oxide, and CFCs. These gases trap heat much more efficiently than does CO_2, but because they are less abundant, they contribute only half as much to global warming (see this chapter's Burning Question: What does the ozone hole have to do with global climate change?).

In a sense, the greenhouse effect supports life, because Earth's average temperature would be much lower without its blanket of greenhouse gases. But CO_2 has been steadily accumulating in the atmosphere since monitoring began in the 1950s. The increase in atmospheric concentration of CO_2 is largely caused by burning fossil fuels such as coal, oil, and gas. This, plus tropical deforestation and other combustion activities, releases some 6 million metric tons of CO_2 into the atmosphere each year. Photosynthesis temporarily removes some of this carbon from the atmosphere, but overall, more CO_2 is added than is removed.

B. Global Climate Change May Have Severe Consequences

This accumulation of CO_2 was accompanied by an increase in average global temperatures in the twentieth century (**figure 42.11**). According to the 2007 report of the Intergovernmental Panel on Climate Change, the primary cause is "very likely" to be the rising concentration of CO_2 and other greenhouse gases in the atmosphere. Computer models predict that these trends will continue. Depending on the future concentration of CO_2 in the atmosphere, the IPCC estimates that the global average surface temperature could increase by anywhere from 1.8 to 4.0°C by the end of the twenty-first century, accompanied by a sea level rise of 18 to 59 cm (7 to 23 inches).

Because Earth's average temperature is rising, this phenomenon is often called "global warming." In reality, however, some areas will become warmer, and others will become cooler. **Global climate change** is therefore a more accurate term for past and future changes in Earth's weather patterns.

An increase of 1.8 to 4.0°C may seem too small to make much difference. Yet the 2.5°C warming of the twentieth century has been associated with the shrinking of Arctic sea ice and alpine glaciers (**figure 42.12**), a decrease in the amount of permafrost, more intense and longer droughts, fewer cold snaps, and increased incidence of both heat waves and heavy precipitation events. Changes in temperature or moisture can alter populations, kill some organisms outright, stress others, or cause migrations.

Evidence is mounting that species' ranges are shifting. For example, the ranges of at least 34 species of butterflies are moving northward. At the southern ends of the ranges, where temperatures are rising, some species have become locally extinct. In addition, scientists are tracking events known to occur at the same time each year. In the United Kingdom, butterflies are emerging and amphibians are mating a few

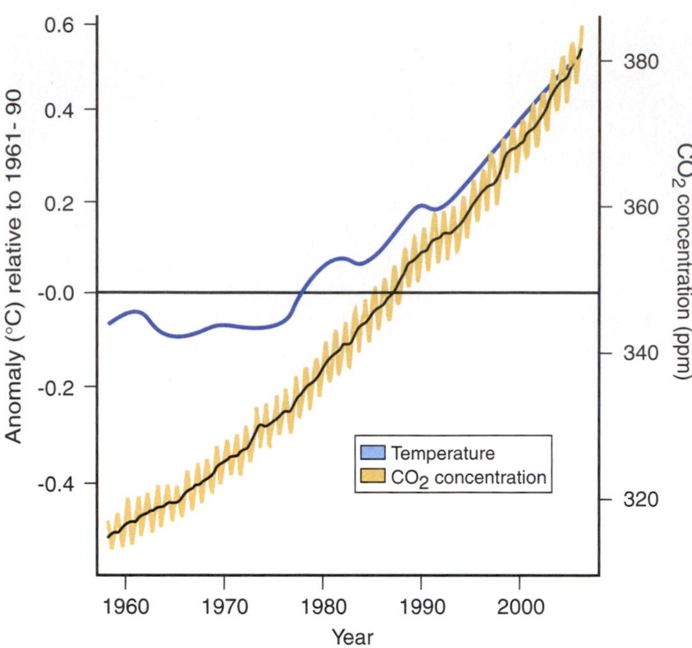

FIGURE 42.11 Carbon Dioxide Level and Global Average Temperature Are Correlated. Atmospheric CO_2 concentrations continue to rise annually, as measured in an observatory at Mauna Loa, Hawaii. As CO_2 accumulates, the average global temperature may also be increasing.
Source: NOAA

a.

b.

FIGURE 42.12 Shrinking Glaciers. (**A**) Switzerland's Steigletscher glacier in the summer of 1994. The position of the glacier's terminus reflects the balance between the melting rate and the amount of snowfall. (**B**) By the summer of 2006, the glacier had retreated substantially.

days earlier than usual; in North America, many plants are flowering and birds are migrating earlier. Climate change has also shifted the migration routes of birds whose excrement passes influenza viruses, infecting seals and whales. High sea temperatures could also lead to an increase in coral bleaching and threaten the world's reefs (see section 41.5).

Continued climate change will affect not only wild organisms but also agriculture and public health. Growing seasons in temperate areas are lengthening, and the U.S. South may become too dry to sustain many traditional crops. Water shortages worldwide may affect more than a billion people. Infectious disease patterns may also shift. Tropical diseases such as malaria, African sleeping sickness, dengue fever, and river blindness may move into northern areas.

Summary *Carbon dioxide and other atmospheric greenhouse gases trap heat at Earth's surface. Increasing CO_2 concentration is contributing to global climate change.*

42.2 MASTERING CONCEPTS

1. Which human activities contribute to the accumulation of CO_2 in Earth's atmosphere?
2. Describe how Earth's climate changed during the twentieth century.
3. How might continued warming trends alter migration patterns and species distributions?

Burning Questions

What does the ozone hole have to do with global climate change?

Many people confuse two of the main environmental issues affecting the atmosphere: global climate change and the increasing ozone hole. These two problems are largely separate, but they do share two common threads. First, the chlorofluorocarbon gases that deplete the ozone layer are also greenhouse gases, contributing to a warmer atmosphere. Second, the greenhouse effect may cause the hole in the ozone layer to grow. A thick heat-trapping "blanket" of greenhouse gases in the troposphere (the lowest part of the atmosphere) means less heat reaches the stratosphere, where the ozone layer is. A cooler stratosphere, in turn, extends the time that stratospheric clouds blanket the polar regions in winter. These clouds of ice and nitric acid speed the chemical reactions that consume stratospheric ozone.

Have a Burning Question of your own? Submit it to marielle_hoefnagels@mcgraw-hill.com for possible inclusion in future editions of this book!

42.3 Exotic Invaders and Overexploitation Devastate Many Species

Besides habitat destruction, the two other most important threats to biodiversity are invasive species and overexploitation.

A. Invasive Species Displace Native Organisms

An introduced species (also called a nonnative, alien, or exotic species) is one that humans bring to an area where it did not previously occur. When we move from one location to another, we often bring along our pets, crops, livestock, and ornamental plants. We also unintentionally introduce microorganisms, parasites, and stowaways such as rodents and insects on ships, cars, and planes.

This transport may seem harmless at first, and many introduced species never do cause problems. For example, the house sparrow, *Passer domesticus*, was introduced to the United States from Europe in the 1850s. Although it has spread throughout the North American continent, it has not caused obvious ecological problems. In addition, at least 5000 nonnative plant species live in U.S. ecosystems, introduced from agriculture and urbanization. Most have apparently done no harm.

If a nonnative species becomes invasive, however, it can cause immense destruction. To be considered **invasive,** an introduced species must begin breeding in its new location and spread widely from the original point of introduction. In addition, according to some definitions, the species must harm the environment, human health, or the economy. Of every 100 species introduced, only 1 persists to take over a niche. Nonetheless, more than 85 plants, animals, and microorganisms are considered invasive in the United States alone, with many more worldwide.

Some invasive species are staggeringly destructive. The opening essay to chapter 39 (A Population Out of Control: Snakes Decimate an Island's Biodiversity) describes how brown tree snakes have devastated Guam's wildlife, and section 40.1 refers to the invasion of zebra mussels in the Great Lakes. Other examples of invasive species include:

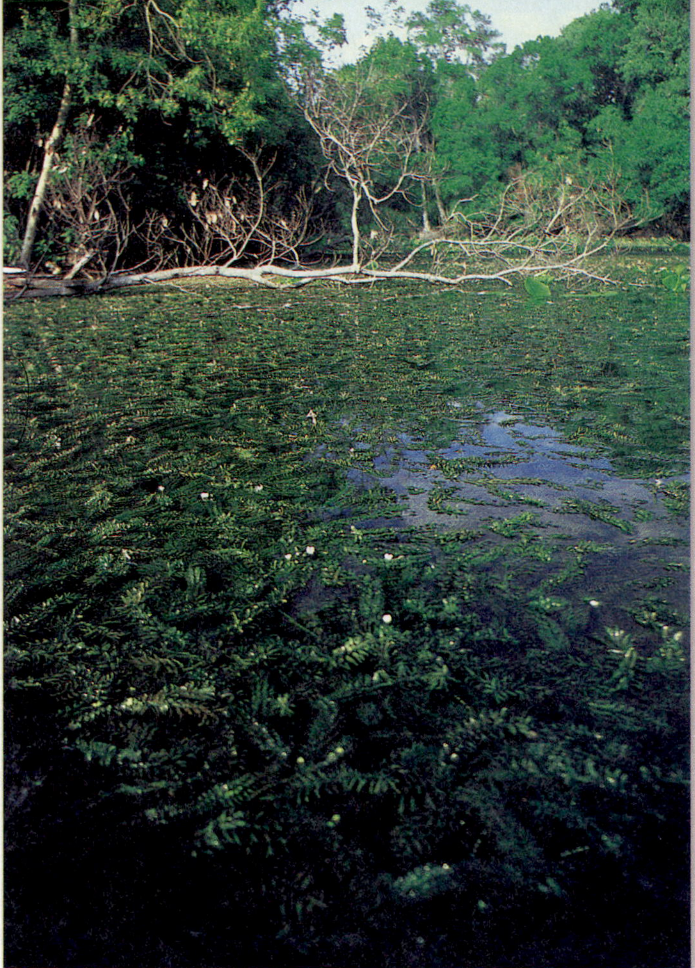

FIGURE 42.13 **Invasive Plant.** *Hydrilla verticillata* clogs this waterway in central Florida. This species came to the United States from southeast Asia as an aquarium plant and quickly became a nuisance, after residents put the imported plant into local waters.

- The marine toad *(Bufo marinus)* is a voracious omnivore that competes with and preys on native amphibians in Florida.
- Hydrilla *(Hydrilla verticillata)* is an aquatic plant that alters nutrient cycles, affects aquatic animals, and reduces recreational use of lakes and rivers (**figure 42.13**).
- Purple loosestrife *(Lythrum salicaria)* is a wetland plant that displaces native plants, threatening the turtles and other animals that would otherwise feed on them.
- Hungry caterpillars of the gypsy moth *(Lymantria dispar)* strip the foliage from hundreds of tree species in North America.
- Fungi have all but eradicated American chestnut and American elm trees.

When an invasion does occur, the harm may be ecological and economic. A nonnative species can not only change the composition of a community but can also carry diseases that spread to native species. The economic costs include everything from the purchase of herbicides that kill invasive weeds, to the loss of grain to hungry rodents, to the decrease of tax revenue when invasive aquatic plants interfere with boating and recreation.

B. Overexploitation Can Drive Species to Extinction

The third most common cause of extinction is **overexploitation:** harvesting a species faster than it can reproduce (**figure 42.14**). Chapter 23's Can You Relate? Wild Caught Pets describes the effect of the pet trade on native populations. Many of the most famous examples of species extinctions result from overhunting of terrestrial animals. The dodo, for example, was a flightless bird that once lived on the Indian Ocean island of Mauritius. In the late seventeenth century, humans began hunting the dodo for food while introducing other species to the island. The dodo soon went extinct. In the United States, the passenger pigeon and Carolina parakeet shared a similar fate in the 1900s, victims of overhunting and habitat destruction.

The best illustration of widespread overexploitation is the recent collapse of the ocean fisheries (see section 13.5). Since the 1950s, some 90% of the world's large, predatory ocean fishes (including tuna, flounder, halibut, swordfish, and cod) have disappeared. Superefficient fishing boats harvest the adults faster than the fishes can reproduce. Moreover, fishing pressure has shifted to other species as predatory fishes have vanished. Fishing equipment scrapes and scours the sea floor, destroying habitat for many other species. In addition, many marine mammals, seabirds, sea turtles, and nontarget fish species are killed accidentally as they are caught up in the nets set for the target species (**figure 42.15**). Even farmed seafood may contribute to the problem, because ocean fishes are used to feed farmed shrimp and salmon, depleting marine food webs.

FIGURE 42.14 **Unsustainable Harvest.** Customs officials in Lhasa, the capital of Tibet, inspect confiscated tiger skins. Trade in tiger skins is illegal, yet poaching remains one of the main threats to tiger populations in Asia.

FIGURE 42.15 "By-catch." Nontarget species sometimes get caught in fishing nets. Fishermen call these nontarget species "by-catch" and often discard them, dead or alive.

Summary *Nonnative species eat or compete with many types of native organisms. Overexploitation means members of a population are harvested faster than they can reproduce*

42.3 MASTERING CONCEPTS

1. Which human activities introduce nonnative species to new areas?
2. What are some characteristics of invasive species, and how do they disrupt ecosystems?
3. List examples of species declines caused by overexploitation.

42.4 Some Biodiversity May Be Recoverable

As the human population continues to grow (see figure 39.12), pressure on natural resources will only increase. One key to reversing environmental decline will therefore be to slow the growth of the human population. In addition, although some species are gone forever, humans may have the power to undo some of our past mistakes (see this chapter's Can *You* Relate? Environmental Legislation). For example, thanks to the Endangered Species Act of 1973, some species that faced extinction, such as the bald eagle, are recovering (**figure 42.16**).

One important conservation tool is to set aside natural areas to protect them from destruction, invasive species, and overexploitation. Preserving critical habitat is a good conservation tool because it saves not just one endangered species, but also the species that share its habitat. For example, the red-cockaded woodpecker *(Picoides borealis)* is native to the southeastern United States, where it builds its nest in a cavity that it excavates high on the trunk of a mature pine tree (**figure 42.17**). Although most of its preferred habitat has been logged, private property owners and the government have cooperated to save some of the remaining habitat. Other animals that use the nesting cavities, including birds, mammals, reptiles, amphibians, and insects, also benefit from the woodpecker recovery plan.

FIGURE 42.17 At Home in the Pines. The endangered red-cockaded woodpecker builds its nest in a hole in the trunk of a longleaf pine. The tree produces resin that helps protect the nest from tree-climbing snakes.

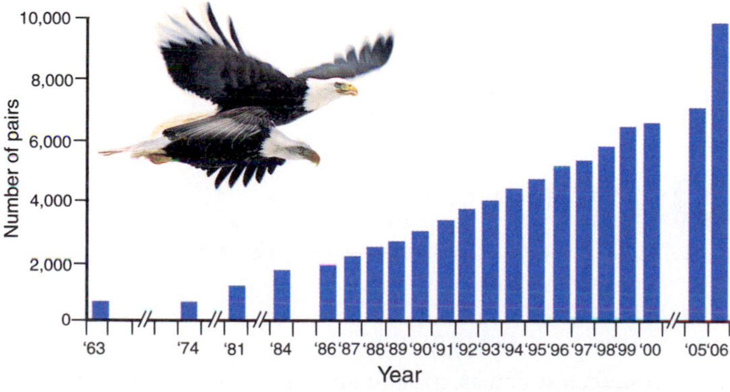

FIGURE 42.16 Good News for Bald Eagles. After nearing extinction in the 1960s, bald eagles have made a steady recovery.

Reversing habitat destruction is another important conservation tool. Major restoration projects such as the Everglades plan described in the chapter opening essay show that recovery, although costly and difficult, may yet reverse some of the trends described in this chapter. In addition, environmental legislation forbids some of the practices that degrade habitats. We can also help species bypass degraded habitats by supplying wildlife corridors through housing developments or building "fish ladders" over dams.

Sometimes, biologists can breed a species in captivity and return it to its former habitat. The California condor, red wolf, and black-footed ferret are notable examples. But this solution does not work for species whose habitat is gone (submerged after dam construction, for example) or still besieged by the same pressures that threatened the species in the first place. Native birds will not recover on Guam as long as the brown tree snake roams the island.

Yet another conservation tool is to manage harvests. Northern and southern right whales, for example, were nearly hunted to extinction for their blubber in the 1800s. They remain endangered, but it is now illegal to kill them. The catastrophic decline of Atlantic cod in the past few decades prompted the closure of some fisheries off Newfoundland's coast, along with strict quotas for the overall catch. Whether conservation efforts are on time to save the cod fishery remains to be seen. On a much smaller scale, it is illegal to collect endangered carnivorous plant species such as Venus flytraps and pitcher plants in the wild.

Introduced predators are the greatest threat to some species. For example, rats, weasels, and other predators brought by European settlers endanger the great spotted kiwi *(Apteryx haastii)*, a flightless bird native to New Zealand. A program of predator control in a nature preserve has made life much easier for the endangered kiwis.

All conservation efforts require a scientific approach. To get a true measure of Earth's biodiversity, taxonomists must continue to catalog all organisms, not just vertebrates and plants (**figure 42.18**). Evolutionary biologists must continue to analyze the relationships among all species. Preserving biodiversity also requires an understanding of which species need help, whether current conservation efforts are working, and the consequences to ecosystems as species disappear.

But not every important question has a scientific answer. Are the only species worth saving the photogenic ones, like giant pandas? Or do we also commit to saving the worms, algae, bacteria, and fungi so essential to global ecology? How much money should we spend on conservation? Should developed countries help poor nations with their efforts? How do we balance the need for conservation with the need for economic growth? Which of the tangle of threads that tie all life together should we sacrifice to other interests?

Life has had many millions of years to adapt, diversify, and occupy nearly every part of the planet's surface. It would be very difficult to halt life on Earth completely, short of a global catastrophe such as a meteor collision or a nuclear holocaust. Just the presence of life, however, does not guarantee that the surviving species will have the diversity that humans value. It is safest to try to protect the remaining resources for the future, while maintaining a reasonable standard of living for all people. Scientists and politicians, as well as ordinary citizens, share this heavy burden. Part of the solution lies within you and how you choose to live (see the Burning Question: What can an ordinary person do to help the environment?). Do whatever you can to preserve the diversity of life, for in diversity lies resiliency and the future of life on Earth.

FIGURE 42.18 Biologist at Work. This entomologist is placing bait to check the numbers of phorid flies at a research site in Texas. The small, inconspicuous flies may hold the key to controlling fire ant populations in the southern United States.

Summary ***Counteracting the loss of biodiversity requires well-planned conservation strategies. We can all be part of maintaining the health of the planet.***

42.4 MASTERING CONCEPTS

1. What is the relationship between population growth and conservation biology?
2. List and describe five tools that conservation biologists use to preserve biodiversity.
3. How can scientists, governments, and ordinary citizens work together for conservation?

Can *You* Relate?

Environmental Legislation

Congress has passed several laws to combat some of the worst environmental problems in the United States. Below are a few major pieces of environmental legislation:

- The **Endangered Species Act** of 1973 requires that the U.S. Secretary of the Interior identify threatened and endangered species and provide for their conservation. Since the act was implemented, more than 500 species of vertebrate and invertebrate animals and more than 700 species of plants and lichens have been classified as threatened or endangered. Only a few dozen species have been removed from the list because they have either recovered or become extinct, or because new information revealed that their populations are larger than had been thought.
- The **Clean Air Act,** passed in 1970 and amended several times since, sets minimum air quality standards for many types of air pollutants. Since 1970, emissions of nitrogen and sulfur oxides, lead, carbon monoxide, particulates, and other pollutants have declined, leading to significant improvements in regional air quality. In the United Kingdom, similar measures have decreased acid precipitation by about half over the past 15 years.
- Among other provisions, the **Clean Water Act** of 1972 required nearly every city to build and maintain a sewage treatment plant, drastically reducing discharge of raw sewage into rivers and lakes. The 1987 **Water Quality Act** followed up on the Clean Water Act, regulating water pollution from industry, agricultural runoff, sewage overflows during storms, and runoff from city streets. Many of the nation's surface waters have recovered from past unregulated discharge of phosphorus, other nutrients, and toxic chemicals.

42.5 INVESTIGATING LIFE

The Case of the Missing Frogs: Is Climate the Culprit?

With so many different strains on biodiversity, how can we ever know what drives a species to extinction? It's one thing to document that a species that once occupied a habitat has vanished; it's another thing to explain how it happened. Not knowing what has gone wrong in the past makes it difficult to prevent future extinctions.

The decline of amphibians worldwide has been especially difficult to unravel. Amphibians are frogs, salamanders, and caecilians. All of these animals require fresh water to reproduce. Most breathe through their thin skin, which must therefore remain moist. According to some estimates, about one third of the world's amphibians, representing thousands of species, have declined. Hundreds of species are critically endangered, and hundreds more have already vanished. Researchers have pointed at pollution, habitat loss, and overhunting as possible causes. But what are we to make of widespread extinctions in pristine habitats and in species that humans do not hunt for food or medicine? **amphibians, p. 486**

A multinational research team, led by J. Alan Pounds of the Monteverde Cloud Forest Preserve and Tropical Science Center in Costa Rica, may have found part of the answer. They studied the extinction of harlequin frogs in the genus *Atelopus* (**figure 42.19**). These animals are a good choice for this type of research because their populations are relatively easy to monitor. Harlequin frogs are brightly colored and active during the day, so they are much easier to observe than amphibians that either blend in with their surroundings or are active at night.

Pounds and his team examined a database that cataloged the last reported sighting of more than 100 species

FIGURE 42.19 Golden Frog. A golden frog, *Atelopus zeteki*, sits on a mushroom in the rain forest.

of harlequin frogs in mountainous regions of Central and South America. Many had not been observed since the 1980s or 1990s and were apparently extinct. The pace of the extinctions was puzzling, considering that these species have thrived in the same habitat for millions of years. What caused them all to vanish in such a short time?

One clue came from looking closely at where the extinctions were occurring (**figure 42.20**A). None of the species that had vanished were from lowland areas (less than 200 m elevation). The most vulnerable species occupied the middle elevations, between 1000 and 2400 m.

Another clue was the timing of the extinctions, which coincided with a time of increasing temperatures in the tropics. The researchers decided to take a closer look at weather patterns in one study area, Costa Rica's Monteverde Cloud Forest Preserve. This area is perpetually shrouded in fog, which forms when moist air blows up the mountainside. As the air rises, it cools, causing moisture in the air to condense into clouds. At Monteverde, the cloud forest extends from an elevation of about 1500 m to about 1850 m, which closely overlaps the habitats of the amphibians that were most likely to go extinct.

Pounds and his team used weather station records at Monteverde to learn that days had gotten cooler, and nights warmer, between 1977 and 1997 (figure 42.20B). This made sense. Air temperatures and sea surface temperatures were climbing throughout the tropics at that time, causing more evaporation and therefore more clouds. The researchers surmised that the thicker cloud cover blocked the sun during the day, but trapped heat at night.

How did these changes affect the frogs? According to Pounds and his team, the new conditions made the amphibians more susceptible to a skin disease caused by a type of fungus called a chytrid (see figure 21.7). This organism, whose scientific name is *Batrachochytrium dendrobatidis*, has an optimal growth temperature that matches the cooler days and warmer nights at Monteverde. Furthermore, the chytrid dies if the temperature climbs above 30°C. The additional cloud cover may have meant fewer sunny spots where the frogs could raise their skin temperatures high enough to kill the fungi. Without this defense, perhaps the chytrids finally gained the advantage over their amphibian hosts, driving many to extinction.

Overall, the researchers concluded that higher temperatures in the tropics created conditions that favored the spread of skin disease, producing a wave of midelevation amphibian extinctions. According to Pounds, "The disease was the bullet killing the frogs, but climate was pulling the trigger." Nevertheless, one critical link in this chain of events is missing: no one knows whether the frogs that vanished in recent decades actually died from the chytrid. Without evidence of a direct cause-and-effect relationship, the meaning of this study remains open to debate.

This study illustrates the difficulty that scientists will face in assessing the biological effects of global climate change. Clearly, Earth's average temperature is rising, local weather patterns are shifting, and habitats are changing. Will it ever be possible to "connect the dots," linking global climate change directly to species extinctions? Scientists around the globe continue to study these issues, knowing that the first step toward preserving biodiversity is to understand what affects it.

Pounds, J. Alan, Martín R. Bustamante, Luis A. Coloma, et al. 2006. Widespread amphibian extinctions from epidemic disease driven by global warming. *Nature*, vol. 439, 161–167.

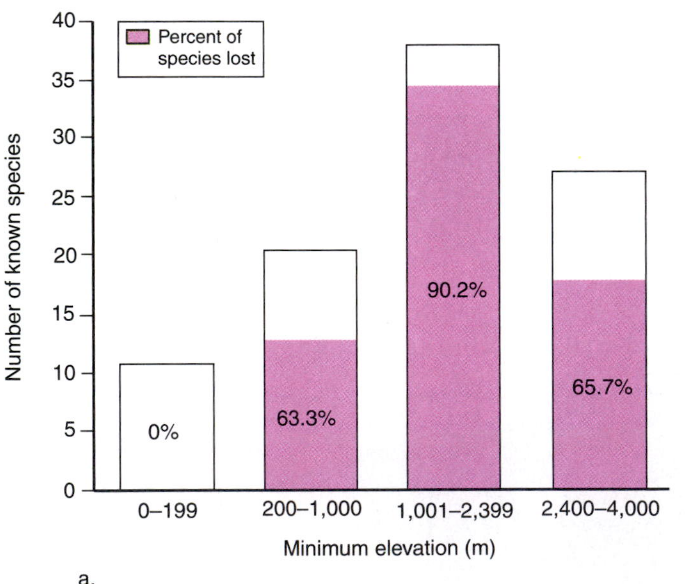

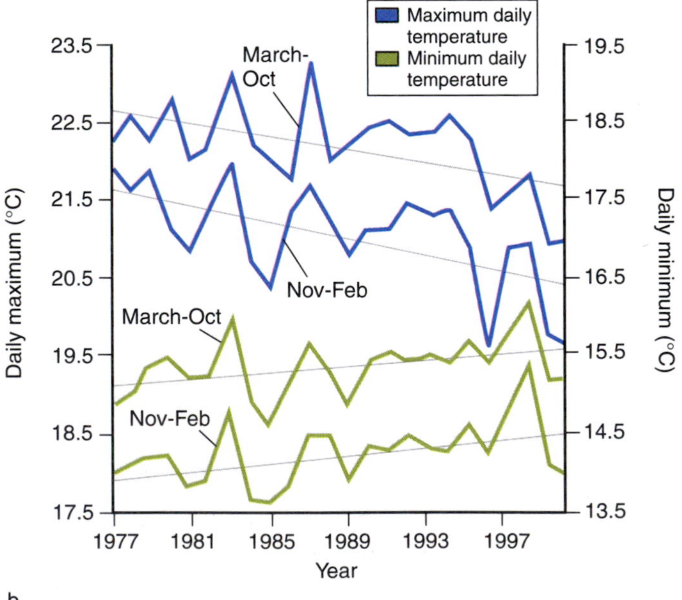

FIGURE 42.20 Elevation Matters. (**A**) In Central and South America, the harlequin frogs (genus *Atelopus*) at middle elevations are at greatest risk of extinction. (**B**) Days are getting cooler, and nights are getting warmer, at the Monteverde Cloud Forest Preserve in Costa Rica. **Question:** Other than climate change, what is an alternative hypothesis that explains that extinction pattern of harlequin frogs? How would you test this hypothesis?

Burning Questions

What can an ordinary person do to help the environment?

Even ordinary citizens can join forces to clean and preserve the environment. The number of small actions that together make a big difference is endless, but here are a few ideas:

- Choose foods and products that reflect sustainable practices. Buying organic, for example, reduces the use of pesticides in agriculture.
- Encourage manufacturers to change packaging to reduce threats to wildlife.
- Conserve energy—replace conventional light bulbs with compact fluorescent bulbs, recycle, carpool, drive a smaller car, ride a bicycle, or turn down the thermostat.
- Pay attention to what you discard and pour down the drain.
- Encourage your local governments to set aside land for parks.
- If you garden, use native plant species that attract wildlife.
- Donate time or money to groups that save critical habitat.

Have a Burning Question of your own? Submit it to marielle_hoefnagels@mcgraw-hill.com for possible inclusion in future editions of this book!

CHAPTER SUMMARY

42.1 Earth's Biodiversity Is Dwindling

- **Biodiversity** means the variety of life on Earth.
- Human activities (habitat destruction, introduced species, and overexploitation) threaten an unknown number of species. **Conservation biologists** study and attempt to preserve biodiversity.

A. Human Activities Destroy Habitats

- Agriculture, logging, and urbanization contribute to **deforestation** in tropical and temperate regions.
- Drought and agriculture promote **desertification,** the expansion of desert into surrounding areas.
- Dams and levees alter the species that live in and near rivers.
- Overuse of water is shrinking freshwater supplies, including the Aral Sea.
- Preserving estuaries is important because they are breeding grounds for many species.

B. Pollution Degrades Habitats

- **Pollution** is any change in the environment that harms living organisms.
- Air pollutants include heavy metals, **particulates,** and emissions from fossil fuel combustion in automobiles and industries. Some of these pollutants react in light to form photochemical **smog.**
- **Acid deposition** forms when nitrogen and sulfur oxides react with water in the upper atmosphere to form nitric and sulfuric acids. These acids return to Earth as dry particles or in precipitation.
- Particulates are bits of matter suspended in the air. When inhaled, the tiniest particles can cause lung problems.
- Use of chlorofluorocarbon compounds has thinned the stratospheric **ozone layer,** which protects life from **ultraviolet radiation.**
- Excessive nutrient levels cause **eutrophication.** Sediments and heat also pollute aquatic ecosystems.
- Water and sediments can be contaminated by a mixture of toxic substances such as **persistent organic pollutants,** heavy metals, spilled oil, and plastics.

42.2 Global Climate Change Alters and Shifts Habitats

A. Greenhouse Gases Warm Earth's Surface

- The **greenhouse effect** results from CO_2 and other gases trapping heat near Earth's surface. Agriculture, burning fossil fuels, and destruction of tropical rain forests generate greenhouse gases that are accumulating in the atmosphere.

B. Global Climate Change May Have Severe Consequences

- As Earth's average temperature increases, polar ice is melting and sea level is rising.
- Shifting vegetation patterns and species ranges are responses to **global climate change.**

42.3 Exotic Invaders and Overexploitation Devastate Many Species

A. Invasive Species Displace Native Organisms

- **Invasive** species reduce biodiversity by eating or outcompeting native organisms.

B. Overexploitation Can Drive Species to Extinction

- **Overexploitation** means individuals are harvested faster than they can reproduce. Global fisheries in particular are in trouble.

42.4 Some Biodiversity May Be Recoverable

- Protected reserves, habitat restoration, captive breeding, harvest management, and predator exclusion are important tools for conservation biology.
- Every human can choose actions that preserve or deplete biodiversity.

42.5 Investigating Life: The Case of the Missing Frogs: Is Climate the Culprit?

- Researchers studying the disappearance of amphibians in Central and South America have suggested a direct link to climate change.
- Increasing cloud cover has altered temperatures in a way that favors the spread of a fungus that infects frogs.

MULTIPLE CHOICE QUESTIONS

1. Which of the following is not one of the main causes of today's biodiversity crisis?
   a. Natural disasters, such as earthquakes
   b. Habitat destruction and degradation
   c. Overexploitation
   d. Introduction of nonnative species
2. Periodic flooding is beneficial to an ecosystem because it
   a. redistributes sediment.
   b. deposits nutrients.
   c. creates habitat for migratory birds.
   d. both a and b.
3. Which human activity has had the greatest impact on biodiversity in tropical forests?
   a. Housing
   b. Transportation
   c. Agriculture
   d. Tourism
4. The burning of fossil fuels produces
   a. smog.
   b. acid deposition.
   c. particulates.
   d. all of the above.
5. How does destruction of the ozone layer affect life on Earth?
   a. It alters global temperatures.
   b. It leads to DNA damage.
   c. It changes the spectrum of light reaching the surface of the Earth.
   d. It reduces the amount of O_2 in the atmosphere.
6. Based on the data in figure 42.8, what can you conclude about the effectiveness of the 1987 Montreal Protocol banning the use of CFCs?
   a. It has helped to stabilize the ozone layer.
   b. It has accelerated the destruction of the ozone layer.
   c. It has helped to restore the ozone layer.
   d. It has caused the ozone hole to disappear.
7. What is the *greenhouse effect*?
   a. The filtering of specific wavelengths of light by Earth's atmosphere
   b. The increase in global plant growth due to enhanced photosynthesis
   c. The trapping of heat by certain gases in the atmosphere
   d. The reduction in the amount of CO_2 in Earth's atmosphere.
8. Why is deforestation associated with global climate change?
   a. Because burning forests adds to the CO_2 levels in the atmosphere
   b. Because loss of trees reduces the amount of photosynthesis occurring on the planet
   c. Because loss of forest frees land for agricultural production
   d. Both a and b
9. An *invasive* species is one that
   a. is introduced into a new habitat.
   b. causes the destruction of an existing species.
   c. establishes a breeding population in a new habitat.
   d. both a and c.
10. What limits a captive breeding program for the restoration of endangered species?
    a. It relies on a limited gene pool.
    b. It does not preserve the habitat of an organism.
    c. Not all animals will breed in captivity.
    d. All of the above.

TESTING YOUR KNOWLEDGE

1. Which human activities promote habitat destruction?
2. What are some examples of air and water pollutants that degrade habitats?
3. How can too much of a nutrient alter an aquatic ecosystem?
4. In what ways is the greenhouse effect both beneficial and detrimental?
5. Cite biological evidence of global climate change.
6. Why are invasive species harmful?

7. Name three ways you can alter your lifestyle in a way that promotes conservation practices.
8. Give an example of an environmental problem that can immediately reduce biodiversity, and one that has a delayed effect.
9. What is being done to restore or prevent further destruction of the Everglades and the Aral Sea?

THINKING AS A SCIENTIST

1. In an article in *Nature* magazine, Sean Nee writes: "Earth's real biodiversity is invisible, whether we like it or not." What does that statement mean?
2. Several polar bears were discovered that have reproductive organs of both sexes. These bears also have high concentrations of PCBs in their blood, but researchers do not know whether exposure to PCBs is related to the disturbed sexual development. Given that many heavily polluted ecosystems are tainted with several types of pollutants, how could you link exposure to one chemical to a specific biological effect?
3. How does the combustion of fossil fuels influence such different phenomena as acid deposition and global climate change?
4. Particulate air pollution damages animal respiratory tracts. How might dust-covered leaves harm a plant?
5. Select a biome from chapter 41, and list three ways that an earlier spring and later fall resulting from global climate change might affect biodiversity.
6. In the Aral Sea and in the Everglades, humans diverted natural water flow. Compare and contrast the effects of these actions in these two very different ecosystems.
7. One approach to combat species invasion is to kill the invaders. In Hawaii, officials shoot feral cats, goats, and pigs. In Australia, the government added chlorine and copper to a bay to kill zebra mussels, killing everything living in the water. Do you think that this is an effective approach? Suggest an alternative.
8. In the southeastern United States, several species of freshwater mussels are extinct or threatened because of habitat destruction. In the past, they were also harvested for the button trade. How would a population ecologist (see chapter 39) approach the problem of species recovery for these animals?
9. Your friend reads an article about scientists that are collecting DNA from endangered species. She says, "I don't know why we spend so much money to save these species when we can just sequence their DNA!" Given what you know about the importance of biodiversity, how would you respond?

Visit www.mhhe.com/hoefnagels for practice quizzes, animations, videos, and activities designed to help you master the material in this chapter.

Appendix A

Answers to Multiple Choice Questions

CHAPTER 1

1. d
2. b
3. c
4. a
5. d
6. c
7. c
8. d
9. c
10. a

CHAPTER 2

1. d
2. b
3. c
4. c
5. a
6. b
7. d
8. a
9. d
10. d

CHAPTER 3

1. b
2. c
3. b
4. d
5. a
6. c
7. a
8. c
9. a
10. b

CHAPTER 4

1. b
2. b
3. a
4. c
5. a
6. c
7. a
8. d
9. b
10. b

CHAPTER 5

1. a
2. b
3. a
4. b
5. c
6. d
7. b
8. c
9. b
10. d

CHAPTER 6

1. d
2. c
3. b
4. d
5. c
6. d
7. d
8. a
9. c
10. c

CHAPTER 7

1. c
2. a
3. b
4. c
5. d
6. d
7. b
8. b
9. b
10. c

CHAPTER 8

1. d
2. c
3. b
4. d
5. c

6. a
7. b
8. c
9. c
10. a

CHAPTER 9

1. c
2. d
3. a
4. b
5. d
6. a
7. d
8. c
9. b
10. a

CHAPTER 10

1. b
2. d
3. b
4. c
5. b
6. b
7. a
8. d
9. d
10. c

CHAPTER 11

1. d
2. b
3. d
4. a
5. c
6. c
7. b
8. d
9. b
10. c

CHAPTER 12

1. b
2. d
3. c
4. d
5. a
6. c
7. d
8. d
9. b
10. a

CHAPTER 13

1. b
2. c
3. c
4. d
5. d
6. c
7. b
8. c
9. a
10. a

CHAPTER 14

1. a
2. d
3. b
4. c
5. b
6. c
7. d
8. b
9. d
10. b

CHAPTER 15

1. d
2. c
3. d
4. a
5. b
6. b
7. d
8. a
9. b
10. c

CHAPTER 16

1. c
2. d
3. a
4. c
5. d
6. d
7. a

CHAPTER 17

1. a
2. d
3. a
4. a
5. b
6. a
7. c
8. d
9. c
10. b

CHAPTER 18

1. c
2. b
3. a
4. b
5. c
6. b
7. d
8. b
9. a
10. d

CHAPTER 19

1. a
2. d
3. b
4. d
5. c
6. c
7. c
8. d
9. b
10. d

CHAPTER 20

1. d
2. a
3. b
4. b
5. d
6. b
7. c
8. d
9. c
10. c

CHAPTER 21

1. b
2. b
3. c
4. c
5. b
6. a
7. c
8. c
9. b
10. a

CHAPTER 22

1. c
2. d
3. b
4. c
5. b
6. b
7. d
8. a
9. c
10. d

CHAPTER 23

1. d
2. d
3. a
4. b
5. c
6. d
7. c
8. d
9. a
10. c

CHAPTER 24

1. c
2. b
3. a
4. b
5. d
6. b
7. d
8. a
9. a
10. c

CHAPTER 25

1. b
2. d
3. a
4. c
5. a
6. b
7. c
8. d
9. c
10. b

CHAPTER 26

1. d
2. c
3. a
4. b
5. c
6. b
7. d
8. c
9. b
10. a

CHAPTER 27

1. d
2. c
3. b
4. c
5. c
6. d
7. a
8. c
9. b
10. d

CHAPTER 28

1. a
2. c
3. d
4. b
5. a
6. b
7. a
8. d
9. c
10. c

CHAPTER 29

1. b
2. d
3. d
4. d
5. a
6. d
7. a
8. c

CHAPTER 30

1. b
2. c
3. b
4. a
5. d
6. c
7. b
8. a
9. d
10. d

CHAPTER 31

1. d
2. d
3. b

4. c
5. b
6. d
7. c
8. a
9. a
10. c

CHAPTER 32

1. b
2. b
3. d
4. a
5. c
6. d
7. b
8. a
9. d
10. c

CHAPTER 33

1. d
2. b
3. a
4. b
5. c
6. c
7. b
8. a
9. a
10. b

CHAPTER 34

1. c
2. d
3. a
4. c
5. a
6. c
7. a
8. d
9. b
10. d

CHAPTER 35

1. b
2. d
3. d
4. a
5. a
6. d
7. c
8. d
9. a
10. b

CHAPTER 36

1. b
2. d
3. a
4. c
5. b
6. c
7. b
8. b
9. a

CHAPTER 37

1. b
2. a
3. c
4. d
5. b
6. c
7. b
8. a
9. d
10. c

CHAPTER 38

1. a
2. b
3. c
4. a
5. c
6. d
7. d
8. c
9. a
10. d

CHAPTER 39

1. c
2. a
3. d
4. c
5. b
6. c
7. a
8. b
9. b
10. a

CHAPTER 40

1. b
2. c
3. d
4. c
5. d
6. c
7. a
8. c
9. d
10. d

CHAPTER 41

1. d
2. d
3. b
4. a
5. a
6. b
7. c
8. d
9. a
10. d

CHAPTER 42

1. a
2. d
3. c
4. d
5. b
6. a
7. c
8. d
9. d
10. d

Appendix B

Metric Units and Conversions

METRIC PREFIXES

Symbol	Prefix	Increase or Decrease	
G	giga	One billion	1,000,000,000
M	mega	One million	1,000,000
k	kilo	One thousand	1,000
h	hecto	One hundred	100
da	deka (or deca)	Ten	10
d	deci	One-tenth	0.1
c	centi	One-hundredth	0.01
m	milli	One-thousandth	0.001
m	micro	One-millionth	0.000001
n	nano	One-billionth	0.000000001

METRIC UNITS AND CONVERSIONS

	Metric Unit	Metric to English Conversion	English to Metric Conversion
Length	1 meter (m)	1 km = 0.62 mile 1 m = 1.09 yards = 39.37 inches 1 cm = 0.394 inch 1 mm = 0.039 inch	1 mile = 1.609 km 1 yard = 0.914 m 1 foot = 0.305 m = 30.5 cm 1 inch = 2.54 cm
Mass	1 gram (g) 1 metric ton (t) = 1,000,000 g = 1,000 kg	1 t = 1.102 tons (U.S.) 1 kg = 2.205 pounds 1 g = 0.0353 ounce	1 ton (U.S.) = 0.907 t 1 pound = 0.4536 kg 1 ounce = 28.35 g
Volume (liquids)	1 liter (l)	1 l = 1.06 quarts 1 ml = 0.034 fluid ounce	1 gallon = 3.79 l 1 quart = 0.95 l 1 pint = 0.47 l 1 fluid ounce = 29.57 ml
Temperature	Degrees Celsius (°C)	°C = (°F – 32)/1.8	°F = (°C × 1.8) + 32
Energy and Power	1 joule (J)	1 J = 0.239 calorie 1 kJ = 0.239 kilocalorie ("food calorie")	1 calorie = 4.186 J 1 kilocalorie ("food calorie") = 4186 J
Time	1 second (sec)		

Appendix C
Amino Acid Structures

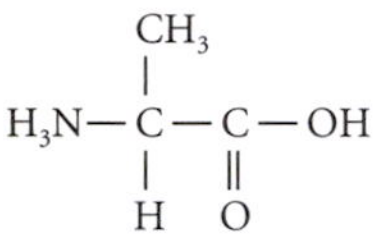

Alanine
(Ala; A)

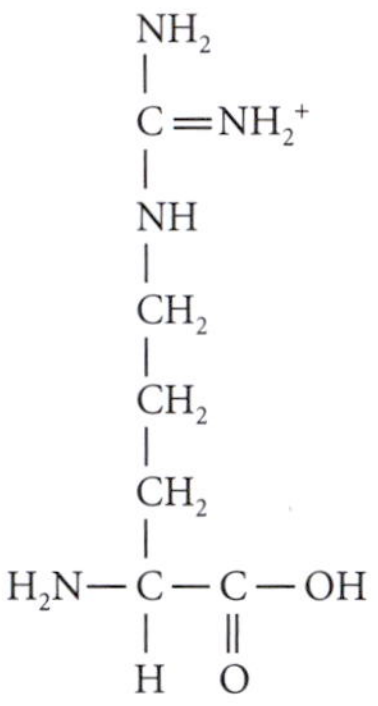

Arginine
(Arg; R)

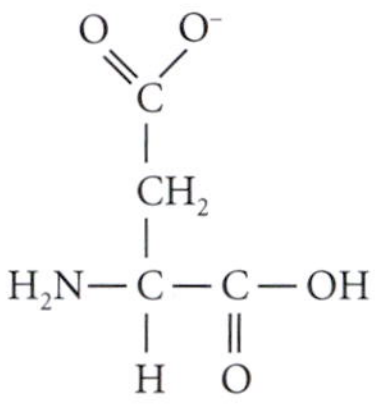

Aspartic
acid (Asp; D)

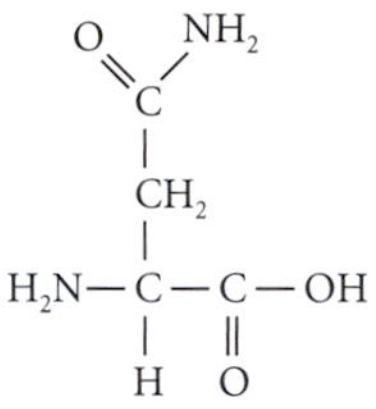

Asparagine
(Asn; N)

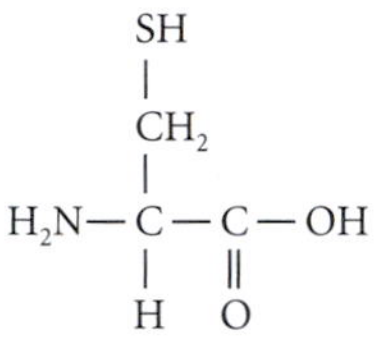

Cysteine
(Cys; C)

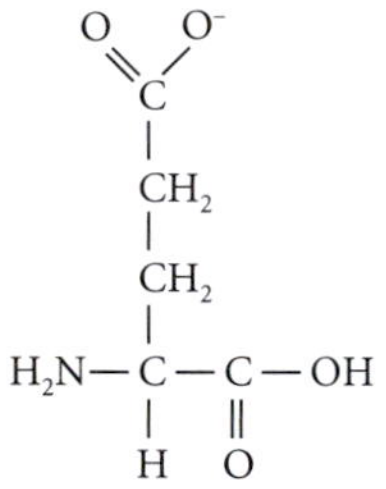

Glutamic
acid (Glu; E)

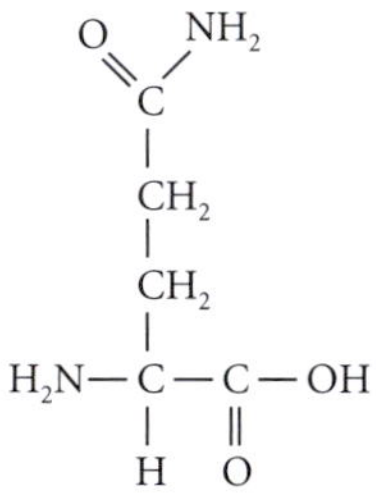

Glutamine
(Gln; Q)

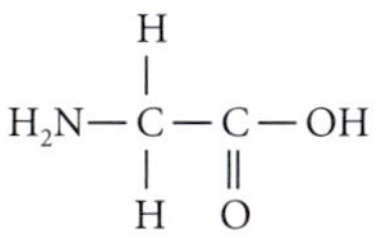

Glycine
(Gly; G)

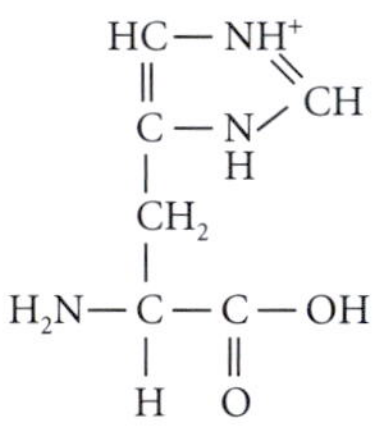

Histidine
(His; H)

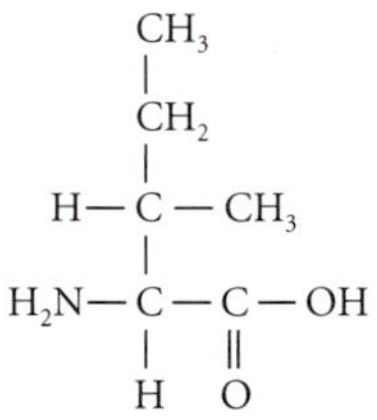

Isoleucine
(Ile; I)

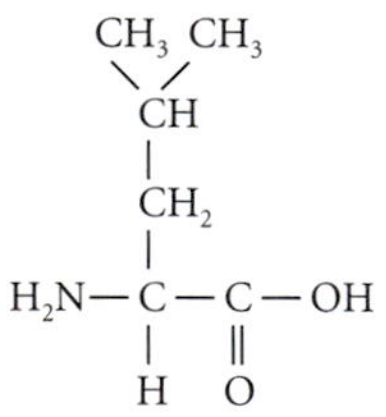

Leucine
(Leu; L)

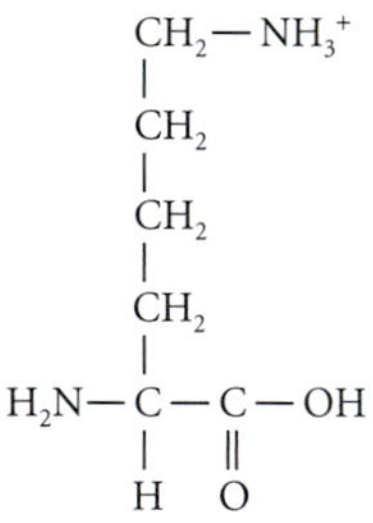

Lysine
(Lys; K)

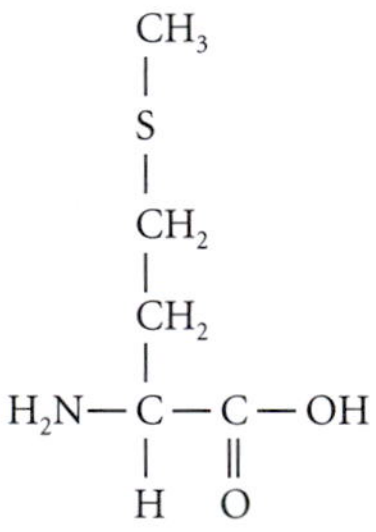

Methionine
(Met; M)

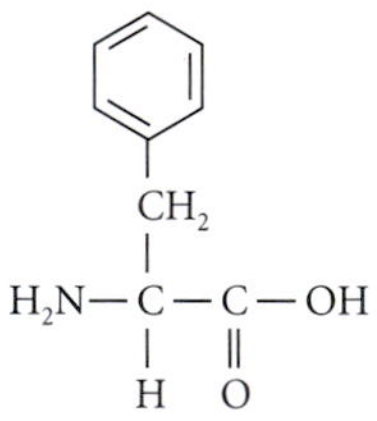

Phenylalanine
(Phe; F)

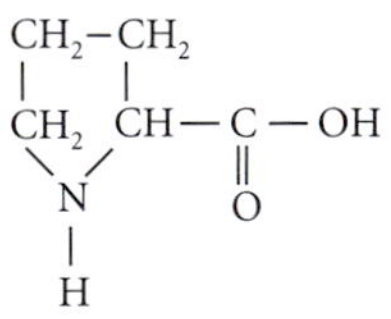

Proline
(Pro; P)

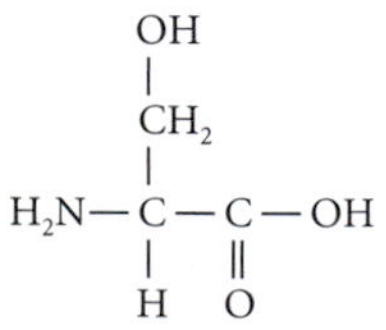

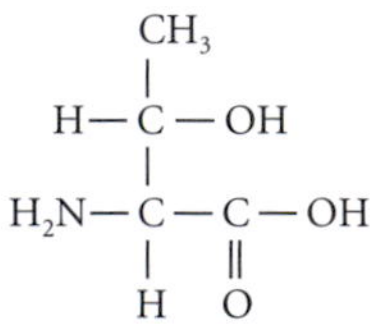

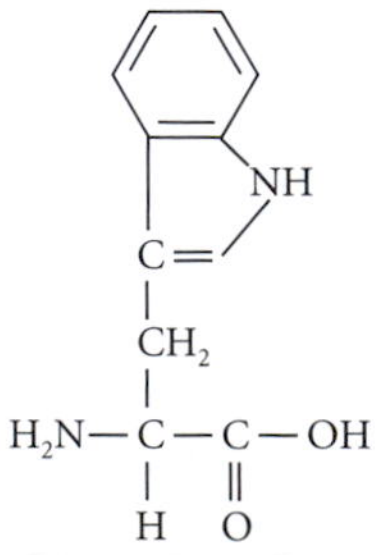

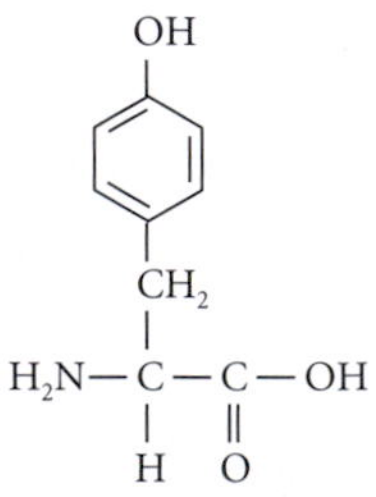

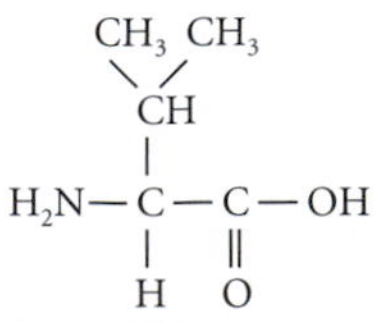

Appendix D

Answers to genetics problems

CHAPTER 10

1. No calves will inherit citrullinemia; each has a 50% chance of being a carrier
2. a. 100%
   b. 1/8 chance that a child will have the same genotype as either parent
3. Type A: 50%; type B: 50%; type AB: 0%; type O: 0%
4. *Ll Mm Pdpd*
5. 50% chance for each color
6. a. Eileen, Gary, and Marvin
   b. Suzanne, Michael, and Jackie
   c. Eileen
   d. 50% (given that Marvin's genotype must be *Ll hh*)

CHAPTER 11

1. 50% for sons; 50% for daughters
2. Most likely X-linked recessive; the doctor is wrong because only the mother need be a carrier to pass the trait to a son
3. a. hh ee
   b. linked
   c. In the brown-haired, brown-eyed F_1 individuals, one chromosome bears alleles *h* and *e*, and the other bears alleles *H* and *E* (because each individual had a blue-eyed, blond parent). A crossover between these two genes, when paired with the homolog from the blond-haired, blue-eyed F_1 parent would yield the unusual trait combinations of blond hair/brown eyes and brown hair/blue eyes.
4. a. X chromosome inactivation
   b. 50%
5. 50% if Lydia is a carrier; 0% if she is not
6. BPES syndrome: autosomal dominant; lysinuric protein intolerance: autosomal recessive
7. a. 50%
   b.

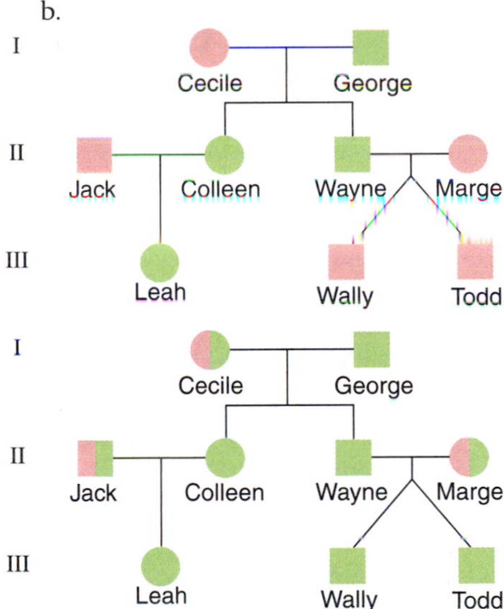

8. a. autosomal recessive; boys and girls are equally affected, and the parents are unaffected.
   b. 100%
9. 50% chance the woman inherits the allele; if she did, her son has a 50% chance of having hemophilia

CHAPTER 12

1. 26,927
2. 125 (not counting synonymous codons)
3. one
4. 3,777 DNA bases (plus promoter sequence)
5. a. smaller
   b. smaller
   c. larger
   d. larger
   e. same
6. a. start of a gene
   b. tRNA
   c. pseudogene
   d. mRNA
7. exon shuffling, pseudogenes, DNA sequences where exons and introns are parts of different genes
8. a. patient D
   b. 357
9. a. UGG to CGG
   b. GGU to GUU
   c. UAU to CAU
10. a. the enzyme is truncated (too short)
    b. CGA to CAA; CGG to CAG
    c. 3
    d. patient A

Glossary

A

abiotic: nonliving

abscisic acid: plant hormone that inhibits many other hormones

abscission zone: specialized layer of cells from which leaf petiole detaches from plant

absolute dating: determining the age of a fossil in years

absorption: the process of taking in and incorporating nutrients or energy

accessory pigment: photosynthetic pigment other than chlorophyll *a* that extends the range of light wavelengths useful in photosynthesis

accommodation: changing the shape of the lens to bring an object into focus

acetyl CoA: molecule that enters the Krebs cycle in cellular respiration; product of partial oxidation of pyruvate

acid: a molecule that releases hydrogen ions into a solution

acid deposition: low pH precipitation or particles that form when air pollutants react with water in the upper atmosphere

acrosome: protrusion covering the head of a sperm cell, containing digestive enzymes that enable the sperm to penetrate layers around the oocyte

actin: protein that forms thin filaments in muscle cells; also part of cytoskeleton

action potential: all-or-none electrochemical change across the cell membrane of a neuron; the basis of a nerve impulse

active immunity: immunity generated by an organism's production of antibodies

active site: the part of an enzyme to which substrates bind

active transport: movement of a solute across a membrane against its concentration gradient, using a carrier protein and energy from ATP

adaptation: inherited trait that permits an organism to survive and reproduce

adaptive immunity: defense system in vertebrates that recognizes and remembers specific antigens

adaptive radiation: divergence of multiple new species from a single ancestral type in a relatively short time

adenosine triphosphate (ATP): a molecule whose high-energy phosphate bonds power many biological processes

adhesion: the tendency of water to hydrogen bond to other compounds

adipose tissue: type of connective tissue consisting of cells laden with lipid

adrenal cortex: outer portion of an adrenal gland; secretes mineralocorticoids, glucocorticoids, and small amounts of sex hormones

adrenal gland: one of two endocrine glands atop the kidneys; produces catecholamines, mineralocorticoids, glucocorticoids, and sex hormones

adrenal medulla: inner portion of an adrenal gland; secretes epinephrine and norepinephrine

adrenocorticotropic hormone (ACTH): hormone produced in the anterior pituitary; stimulates the adrenal cortex to secrete glucocorticoid hormones

adventitious root: roots arising from stem or leaf

aerobic respiration: complete oxidation of glucose to CO_2 in the presence of O_2, producing ATP

age structure: distribution of age classes in a population

agglutination: clumping reaction

alcoholic fermentation: metabolic pathway in which NADH from glycolysis reduces pyruvate, producing ethanol and CO_2

aldosterone: mineralocorticoid hormone produced in the adrenal cortex; increases sodium ion reabsorption in the renal tubule

alga: aquatic, photosynthetic protist

alimentary canal: two-opening digestive tract; also called gastrointestinal (GI) tract

alkaline: having a pH greater than 7

allantois: extraembryonic membrane that forms as an outpouching of the yolk sac

allele: an alternative form of a gene

allergen: antigen that triggers an allergic reaction

allergy: exaggerated immune response to a harmless substance

allopatric speciation: formation of new species after a physical barrier separates a population into groups that cannot interbreed

alternate leaf arrangement: pattern in which each node has only one leaf

alternation of generations: the sexual life cycle of plants and many green algae, which alternates between a diploid sporophyte stage and a haploid gametophyte stage

alveolate: protist with flattened sacs, or alveoli, just beneath the cell membrane

alveolus (pl. alveoli): microscopic air sac in the mammalian lungs, where gas exchange occurs

amino acid: an organic molecule consisting of a central carbon atom bonded to a hydrogen atom, an amino group, a carboxyl group, and an R group

amino group: a nitrogen atom single-bonded to two hydrogen atoms

ammonia: nitrogenous waste (NH_3) generated by deamination of amino acids

amnion: extraembryonic membrane that contains amniotic fluid

amniote: vertebrate in which protective membranes surround the embryo (amnion, chorion, and allantois); reptiles, birds, and mammals

amniote egg: bird, reptile, or monotremes egg containing fluid and nutrients within membranes

amphibian: tetrapod vertebrate that can live on land, but requires water to reproduce

anaerobic respiration: cellular respiration using an electron acceptor other than O_2

analogous: similar in function but not in structure because of convergent evolution, not common ancestry

anaphase I: anaphase of meiosis I, when spindle fibers pull homologous chromosomes toward opposite poles of the cell

anaphase II: anaphase of meiosis II, when centromeres split and spindle fibers pull sister chromatids toward opposite poles of the cell

anatomy: the study of an organism's structure

ancestral character: characteristic already present in the ancestor of a group being studied

anchoring (or adhering) junction: connection between two adjacent animal cells that anchors intermediate filaments in a single spot on the cell membrane

anemia: reduction in the oxygen-carrying capacity of blood

angiosperm: a seed plant that produces flowers and fruits; includes monocots and dicots

annelid: segmented worm; phylum Annelida

anorexia nervosa: eating disorder characterized by refusal to maintain normal body weight

antenna pigment: photosynthetic pigment that passes photon energy to the reaction center of a photosystem

anterior pituitary: the front part of the pituitary gland; produces and secretes FSH, LH, ACTH, GH, TSH, and prolactin

anther: pollen-producing bodies at tip of stamen

antheridium: multicellular sperm-producing structure in bryophytes and seedless vascular plants

antibody: protein that B cells secrete that recognizes and binds to an antigen

anticodon: a three-base portion of a tRNA molecule; the anticodon is complementary to one codon

antidiuretic hormone (ADH): hormone released from the posterior pituitary; acts on kidneys to maintain the composition of body fluids

antigen: molecule that elicits an immune reaction by B and T cells

antigen binding site: portion of an antibody's variable region that binds to an antigen

anus: exit from a complete digestive tract

aorta: the largest artery leaving the heart

apical complex: group of microtubules and organelles at one end of an apicomplexan's cell

apical dominance: intact terminal bud of a plant suppresses growth of lateral buds

apical meristem: meristem at tip of root or shoot

Apicomplexan: non-motile protist with cell containing an apical complex; obligate animal parasite

apoptosis: programmed cell death that is a normal part of development

appendicular skeleton: the limb bones and the bones that support them in the vertebrate skeleton

appendix: thin sac extending from the cecum in the human digestive system

arachnid: type of chelicerate arthropod; spiders, ticks, mites, and scorpions

archegonium: multicellular egg-producing structure in bryophytes and seedless vascular plants

arteriole: small artery

artery: vessel that carries blood away from the heart

arthropod: segmented animal with an exoskeleton and jointed appendages; phylum Arthropoda

artificial selection: selective breeding strategies in which a human chooses which organisms breed based on one or a few desired traits

ascending limb: region of the nephron loop that leads to the distal convoluted tubule

Ascomycete: fungus that produces sexual spores in a sac called an ascus

ascospore: haploid spore produced in an ascus

ascus (pl. asci): in ascomycetes, a saclike structure in which ascospores are produced

asexual reproduction: any form of reproduction that does not require the fusion of gametes

atom: a particle composed of protons, neutrons, and electrons, that cannot be further broken down by chemical means

atomic mass: the average mass of all isotopes of an element

atomic number: the number of protons in an atom's nucleus

ATP synthase: enzyme complex that admits protons through a membrane, where they trigger phosphorylation of ADP to ATP

atrioventricular (AV) node: specialized cardiac muscle cells that help control heartbeat

atrioventricular (AV) valve: flap of heart tissue that prevents the flow of blood from a ventricle to an atrium

atrium: upper heart chamber

auditory canal: ear canal; funnels sounds from the pinna to the tympanic membrane

auditory nerve: nerve fibers that connect the cochlea in the inner ear to the brain

autoimmunity: immune reaction to the body's own cells

autonomic nervous system: in the peripheral nervous system, motor pathways that lead to smooth muscle, cardiac muscle, and glands

autosomal dominant: inheritance pattern of a dominant allele on an autosome. The phenotype affects males and females equally and does not skip generations

autosomal recessive: inheritance pattern of a recessive allele on an autosome. The phenotype affects males and females equally and can skip generations

autosome: a nonsex chromosome

autotroph: organism that produces organic molecules by acquiring carbon from inorganic sources; primary producer

auxin: plant hormone that promotes cell elongation in stems and fruits

axial skeleton: the central axis of a vertebrate skeleton; consists of the skull, vertebral column, ribs, and sternum

axillary bud: undeveloped shoot in the angle between stem and petiole

axon (nerve fiber): extension of a neuron that transmits messages away from the cell body and toward another cell

B

B cell: type of lymphocyte

bacillus (plural bacilli): rod-shaped prokaryote

background extinction rate: steady, gradual loss of species through natural competition or loss of genetic diversity

bacteriophage (phage): a virus that infects bacteria

balanced polymorphism: condition in which multiple alleles persist indefinitely in a population

bark: tissues outside the vascular cambium

base: a molecule that either releases hydroxide ions into a solution or removes hydrogen ions from it

Basidiomycete: fungus that produces sexual spores on a basidium

basidiospore: haploid spore produced on a basidium

basidium (pl. basidia): in basidiomycetes, a club-shaped structure that produces basidiospores

basilar membrane: the membrane in the cochlea of the inner ear; vibrates in response to sound

basophil: type of white blood cell

benthic zone: sediment at the bottom of an ocean or lake

bilateral symmetry: body form in which only one plane divides the animal into mirror image halves

bile: digestive biochemical that emulsifies fats

binary fission: type of asexual reproduction in which a prokaryotic cell divides into two identical cells

biodiversity: the variety of life on Earth

biogeochemical cycle: geological and biological processes that recycle elements vital to life

biogeography: the study of the distribution patterns of species across the planet

biological species: a population, or group of populations, whose members can interbreed and produce fertile offspring

biomagnification: increasing concentrations of a chemical in higher trophic levels

biomass: total dry weight of organisms in a given area at a given time

biome: one of several major types of ecosystems

biosphere: part of Earth where life can exist

biotic potential: maximum reproductive capacity of a population under ideal conditions

biotic: living

bipedalism: type of locomotion in which an animal walks upright on two legs

bird: tetrapod vertebrate with feathers, wings, and an amniote egg

birth defect: abnormality of structure or metabolism that causes death or disability in a newborn

birth rate: the number of new individuals produced per unit time

bivalve: type of mollusk

blade: flattened part of a leaf

blastocyst: preembryonic stage consisting of a fluid-filled ball of cells

blastomere: preembryonic cell produced in cleavage divisions

blastula: stage of early animal embryonic development; a sphere of cells surrounding a fluid-filled cavity

blood: connective tissue consisting of cells suspended in a liquid matrix; delivers nutrients to cells and removes wastes

blood clot: plug of solidified blood

blood pressure: force that blood exerts against artery walls

blood-brain barrier: close-knit cells that form capillaries in the brain, limiting the substances that can enter

bolus: mass of chewed food

bone: connective tissue consisting of osteoblasts, osteocytes, and osteoclasts, embedded in a mineralized matrix; also, in the vertebrate skeleton, an organ consisting of bone tissue, cartilage, and other tissues

bone marrow: spongy tissue inside bones

bony fish: jawed fish with a skeleton made of bone; ray-finned fishes, lungfishes, and lobe-finned fishes

book lung: site of gas exchange in spiders and scorpions; contains pagelike membranes folded into the abdomen and connected with the air through a small opening

boreal forest: taiga

braced framework: skeleton built of solid structural components strong enough to resist collapsing

brachiation: type of locomotion in which an animal swings from one arm to the other while the body dangles below

brain: a distinct concentration of nervous tissue, often encased within a skull, at the anterior end of an animal

brainstem: continuation of the spinal cord into the vertebrate hindbrain; controls vital functions

breast: milk-producing organ in female mammals

bronchiole: small branched airway that connects bronchi to alveoli

bronchus (pl. bronchi): one of the large tubes that branch from the trachea as it reaches the lungs

brown alga: multicellular photosynthetic aquatic protist with swimming spores and containing brownish accessory pigments

bryophyte: plant that lacks vascular tissue; includes liverworts, hornworts, and mosses

buffer system: weak acid/base pair that maintains body fluid pH

bulbourethral gland: small gland near the male urethra that secretes mucus

bulimia: eating disorder in which a person eats large quantities and then intentionally vomits or uses laxatives shortly afterward

bulk element: an element that an organism requires in large amounts

bulk feeder: animal that ingests large pieces of food

bundle-sheath cell: thick-walled plant cell surrounding veins; site of Calvin cycle in C_4 plants

bursa (pl. bursae): small pocket that contains lubricating fluid in a synovial joint

C

C_3 plant: plant that uses only the Calvin cycle to fix CO_2

C_4 pathway: in photosynthesis, a carbon fixation pathway in which CO_2 combines with a three-carbon molecule to form a four-carbon compound

caecilian: type of amphibian

calcitonin: thyroid hormone that decreases blood calcium levels

Calorie: one kilocalorie

calorie: the energy required to raise the temperature of 1 gram of water by 1°C under standard conditions

Calvin cycle: in photosynthesis, metabolic pathway in which CO_2 is fixed and incorporated into glucose

CAM plant: plant that reduces photorespiration by fixing carbon at night for use in the Calvin cycle during the day

camouflage: adaptation that helps an organism blend in with its surroundings

cancer: class of diseases characterized by uncontrolled division of cells that invade or spread to other tissues

capacitation: chemical changes that enable a sperm cell to fertilize an egg cell

capillary: tiny vessel that connects an arteriole with a venule

capillary bed: network of capillaries

capsid: protein coat of a virus

capsule: distinct, gelatinous glycocalyx

carbohydrate: compound containing carbon, hydrogen, and oxygen in a ratio 1:2:1

carbon fixation: the initial incorporation of carbon from CO_2 into an organic compound

carbon reactions: the reactions of photosynthesis that use ATP and NADPH to synthesize glucose from carbon dioxide

carboxyl group: a carbon atom double-bonded to an oxygen and single-bonded to a hydroxyl group

cardiac cycle: sequence of contraction and relaxation that makes up the heartbeat

cardiac muscle tissue: involuntary muscle tissue composed of branched, striated, single-nucleated contractile cells

cardiac output: the volume of blood that the heart pumps each minute

cardiovascular system: circulatory system

carinate: type of bird

carnivore: animal that eats animals

carpel: leaflike structure enclose an angiosperm's ovule

carrying capacity: maximum number of individuals that the habitat can support indefinitely

cartilage: type of connective tissue consisting of cells surrounded by a rubbery matrix

cartilaginous fish: jawed fish with a skeleton made of cartilage; sharks, skates, and rays

Casparian strip: waxy barrier in root endodermis

caspase: apoptosis-specific enzyme that destroys a cell's proteins

catastrophism: a theory of geological change that stated that a series of brief upheavals such as floods, volcanic eruptions, and earthquakes are responsible for most geological formations

cavitation: formation of air bubbles in xylem

cecum: the entrance to the large intestine

cell: smallest unit of life that can function independently

cell body: enlarged portion of a neuron that contains most of the organelles

cell cycle: sequence of events that occur in an actively dividing cell

cell membrane: the boundary of cells, consisting of proteins embedded in a phospholipid bilayer

cell theory: the ideas that all living matter consists of cells, cells are the structural and functional units of life, and all cells come from preexisting cells

cell wall: a rigid boundary surrounding cells of many prokaryotes, protists, plants, and fungi

cell-mediated immunity: branch of adaptive immune system in which defensive cells kill invaders by direct cell-cell contact

cellular slime mold: protist in which feeding stage consists of individual cells that come together as a multicellular "slug" when food runs out

centipede: type of mandibulate arthropod

central nervous system (CNS): brain and the spinal cord

centromere: small section of a chromosome where sister chromatids attach to each other

centrosome: part of the cell that organizes the mitotic spindle

cephalization: development of sensory structures and a brain at the head end of an animal

cephalopod: type of mollusk

cerebellum: area of the hindbrain that coordinates muscular responses

cerebral cortex: outer layer of the cerebrum

cerebrospinal fluid: fluid that bathes and cushions the central nervous system

cerebrum: region of the forebrain that controls intelligence, learning, perception, and emotion

cervix: lower, narrow part of the uterus

chaparral: Mediterranean shrubland

charophyte: type of green algae thought to be most closely related to terrestrial plants

chelicerate: arthropod with clawlike mouthparts (chelicerae); horseshoe crabs and arachnids

chemical bond: attractive force that holds atoms together

chemical equilibrium: condition in which a chemical reaction proceeds in both directions at the same rate

chemical reaction: interaction in which bonds break and new bonds form

chemiosmotic phosphorylation: reactions that produce ATP using ATP synthase and the potential energy of a proton gradient

chemoreceptor: sensory receptor that responds to chemicals

chemotroph: organism that derives energy by oxidizing inorganic or organic chemicals

chiton: type of mollusk

chlorophyll *a*: green pigment that plants, algae, and cyanobacteria use to harness the energy in sunlight

chloroplast: organelle housing the reactions of photosynthesis in eukaryotes

choanoflagellate: unicellular or colonial flagellated protist that may be the ancestor of sponges

cholecystokinin (CCK): hormone that suppresses appetite

chordate: animal that at some time during its development has a notochord, hollow nerve cord, pharyngeal pouches or slits, and a postanal tail; phylum Chordata

chorion: outermost extraembryonic membrane that develops into the placenta

chorionic villus: fingerlike projection extending from the chorion to the uterine lining

choroid: middle layer of the eyeball, between the sclera and the retina

chromatid: a continuous strand of DNA comprising one-half of a replicated chromosome

chromatin: DNA and its associated proteins in the nucleus of a eukaryotic cell

chromosome: a continuous molecule of DNA wrapped around protein in the nucleus of a eukaryotic cell; also, the genetic material of a prokaryotic cell

chyme: semifluid mass of food and gastric juice that moves from the stomach to the small intestine

Chytridiomycete (chytrid): microscopic fungus that produces motile zoospores

ciliate: protist with cilia-covered cell surface

cilium (pl. cilia): one or many short, movable protein projections from a cell

circadian rhythm: physiological cycle that repeats daily

circulatory system: organ system that distributes blood (or a comparable fluid) throughout the body

clade: monophyletic group of organisms consisting of a common ancestor and all of its descendants

cladistics: phylogenetic system that defines groups by shared derived characters

cladogram: treelike diagram built using shared derived characteristics

class: taxonomic category between phylum and order

cleavage: period of rapid cell division following fertilization

cleavage furrow: in animals, the initial indentation between two daughter cells in mitosis

climax community: community that persists indefinitely if left undisturbed

clitoris: small, highly sensitive female sexual organ

cloaca: an opening common to the digestive, reproductive, and excretory systems in birds and many other animals

clonal selection: rapid division of a stimulated B cell, generating memory B cells and plasma cells that are clones of the original B cell

cloning vector: a structure that carries DNA from cell to cell

closed circulatory system: circulatory system that confines blood to vessels

club moss: a type of seedless vascular plant

Cnidaria: phylum with animals that have radial symmetry, two germ layers, a jellylike interior, and nematocysts

cnidocyte: cell in cnidarians that can fire a toxic barb for predation or defense

coccus (plural cocci): spherical prokaryote

cochlea: spiral-shaped part of the inner ear, where vibrations are translated into nerve impulses

codominance: mode of inheritance in which two alleles are fully expressed in the heterozygote

codon: a triplet of mRNA bases that specifies a particular amino acid

coelom: fluid-filled animal body cavity that forms completely within mesoderm

coevolution: genetic change in one species selects for subsequent change in another species

cofactor: inorganic or organic substance required for activity of an enzyme

cohesion: the attraction of water molecules to each other

cohesion-tension theory: theory that explains how water moves under tension in xylem

cohort: group of individuals that begin life at the same time

collagen: a type of flexible connective tissue protein

collecting duct: tubule in the kidney into which nephrons drain urine

collenchyma: elongated plant cells with thick, elastic cell walls

colon: longest portion of the large intestine, where water is absorbed

columnar: tall and thin

commensalism: type of symbiosis in which one member benefits without affecting the other member

community: group of interacting populations that inhabit the same region

compact bone: solid, hard bone tissue consisting of tightly packed osteons

companion cell: parenchyma cells adjacent to sieve tube elements

competition: struggle between organisms for the same limited resource

competitive exclusion principle: the idea that two or more species cannot indefinitely occupy the same niche

competitive inhibition: an inhibitor binds to an enzyme's active site, competing with the enzyme's normal substrate

complement: group of proteins that help destroy pathogens

complementary: in DNA and RNA, the precise pairing of purines (A and G) to pyrimidines (C, T, and U) such that each strand defines the sequence of the other

complete digestive tract: digestive tract through which food passes in one direction from mouth to anus

compound: a molecule including different elements

compound eye: eye that consists of multiple ommatidia

compound leaf: leaf that is divided into leaflets

concentration gradient: difference in solute concentrations between two adjacent regions

cone: a pollen- or ovule-bearing structure in many gymnosperms

cone cell: photoreceptor cell in the retina that detects colors

conifer: a type of gymnosperm

conjugation: a form of gene transfer in prokaryotes

connective tissue: tissue type consisting of widely spaced cells in a matrix; includes loose and fibrous connective tissues, cartilage, bone, fat, and blood

conservation biology: study of the preservation of biodiversity

constant region: amino acid sequence that is the same for all antibodies

consumer (heterotroph): organism that uses organic sources of energy and carbon

contact inhibition: property of most noncancerous eukaryotic cells; inhibits cell division when cells contact one another

control: untreated group used as a basis for comparison with a treated group in an experiment

convergent evolution: the evolution of similar adaptations in organisms that do not share the same evolutionary lineage

coral reef: underwater deposit of calcium carbonate formed by colonies of coral animals

cork cambium: lateral meristem that produces cork cells and parenchyma in woody plant

cornea: in the eye, a modified portion of the sclera that forms a transparent curved window that admits light

corona radiata: layer of cells surrounding the zona pellucida of an egg cell

coronary artery: artery that provides blood to the heart muscle

corpus luteum: gland formed from a ruptured ovarian follicle that has recently released an oocyte

correlation: relationship between events that occur together but may not be causally related

cortex: ground tissue between epidermis and vascular tissue in roots and stems

cotyledon: seed leaf in angiosperms

countercurrent exchange: arrangement in which two adjacent currents flow in opposite directions and exchange materials or heat

coupled reactions: two simultaneous chemical reactions, one of which provides the energy that drives the other

covalent bond: type of chemical bond in which two atoms share electrons

cranial nerve: peripheral nerve that exits the vertebrate central nervous system from the brain

craniate: animal with a cranium

cranium: part of the skull that encloses the brain

creatine phosphate: molecule stored in muscle fibers that can donate its high-energy phosphate to ADP to regenerate ATP

crista (pl. cristae): fold of the inner mitochondrial membrane along which many of the reactions of cellular respiration occur

crocodilian: type of reptile; crocodiles, alligators, caimans, gavials

crop: food storage compartment in birds and some other animals

cross bridge: contact between a myosin molecule's head and an actin molecule in a myofibril

crossing over: exchange of genetic material between homologous chromosomes during prophase I of meiosis

Crustacea: types of mandibulate arthropods

cuboidal: cube-shaped

culture: the knowledge, beliefs, and behaviors that humans transmit from generation to generation

cuticle: waterproof layer covering the aerial epidermis of a plant

cycad: a type of gymnosperm

cyclic adenosine monophosphate (cAMP): a second messenger formed from ATP

cyst: thick-walled, dormant form of a protist

cytokine: messenger protein synthesized in immune cells that influences the activity of other immune cells

cytokinesis: distribution of cytoplasm into daughter cells in cell division

cytokinin: plant hormone that stimulates cell division

cytoplasm: the watery soup of salts, organic molecules, and other substances inside the cell. In eukaryotic cells, it consists of all materials, including organelles, between the nuclear membrane and the cell membrane

cytoskeleton: framework of protein rods and tubules in eukaryotic cells

cytotoxic T cell: lymphocyte that kills invading cells by binding them and releasing chemicals

D

day-neutral plant: plant that does not rely on photoperiod to stimulate flowering

death rate: the number of deaths per unit time

decomposer: organism that consumes wastes and dead organic matter, returning inorganic nutrients to the ecosystem

deforestation: removal of tree cover from a previously forested area

dehydration synthesis: formation of a covalent bond between two molecules by loss of water

deletion: loss of one or more genes from a chromosome

demographic transition: phase of economic development during which birth and death rates shift from high to low

demography: statistical study of populations

denaturation: modification of a protein's shape so that its function is destroyed

dendrite: thin neuron branch that receives neural messages and transmits information to the cell body

dendritic cell: type of macrophage

denitrification: conversion of nitrites and nitrates to N_2

dense connective tissue: type of connective tissue with dense collagen tracts

density-dependent factor: population-liming condition whose effects increase when populations are large

density-independent factor: population-limiting condition that acts irrespective of population size

dentin: bonelike substance beneath a tooth's enamel

deoxyribonucleic acid (DNA): double-stranded nucleic acid composed of nucleotides

dependent variable: response that may be under the influence of an independent variable

deposit feeder: type of substrate feeder that strains partially decayed organic matter from sediment

derived character: characteristic not found in the ancestor of a group being studied

dermal tissue: tissue covering a plant's surface

dermis: layer of connective tissue that lies beneath the epidermis

descending limb: region of the nephron loop that extends from the proximal convoluted tubule

descent with modification: Darwin's term for evolution, describing gradual change from an ancestral type of organism

desert: type of terrestrial biome; very low precipitation

desertification: encroachment of desert into surrounding areas

determinate growth: growth that halts at maturity

detritivore: animal that eats decomposing organic matter

detritus: feces and dead organic matter

Deuteromycete: fungus that lacks a sexual phase

deuterostome: clade of bilaterally symmetrical animals in which the first opening in the gastrula develops into the anus

diabetes mellitus: disease resulting from an inability to produce or use insulin

diaphragm: sheet of muscle separating the thoracic and abdominal cavities

diastole: relaxation of the heart muscle

diastolic pressure: lower number in a blood pressure reading; reflects relaxation of the ventricles

diatom: photosynthetic aquatic protist with two-part silica wall

dicotyledon (dicot): an angiosperm with seeds that have two cotyledons

diffusion: movement of a substance from a region where it is highly concentrated to an area where it is less concentrated

digestion: the physical and chemical breakdown of food

digestive system: organ system in which food is broken down into nutrient molecules that are absorbed into capillaries

dihybrid cross: mating between two individuals that are heterozygous for two genes

dikaryotic: containing two genetically distinct haploid nuclei

dinoflagellate: unicellular aquatic protist with two flagella of unequal length and cellulose plates

dioecious: describes species in which different individuals produce male and female gametes.

diploid: containing two different copies of each chromosome, one from each parent; also called 2n

direct development: gradual development of a juvenile animal into an adult, without an intervening larval stage

directional selection: form of natural selection in which one extreme phenotype is fittest, and the environment selects against the others

disaccharide: a simple sugar that consists of two bonded monosaccharides

distal convoluted tubule: region of the renal tubule that connects the nephron loop to the collecting duct

DNA (deoxyribonucleic acid): protein-encoding informational molecule passed from generation to generation.

DNA microarray: collection of short DNA fragments of known sequence placed in defined spots on a small square of glass

DNA polymerase: enzyme that adds new DNA nucleotides and corrects mismatched base pairs in DNA replication

DNA profiling: biotechnology tool that uses only DNA to detect genetic differences between individuals

domain: highest (most inclusive) taxonomic category

dominant: describes an allele that is expressed whenever it is present

dormancy: state of decreased metabolism

dorsal hollow nerve cord: hollow nerve cord that forms dorsal to the notochord; one of the four characteristics of chordates

double-blind: type of experiment in which neither participants nor researchers know which subjects received a placebo and which received the treatment being evaluated

double fertilization: in angiosperms, one sperm cells fertilizes the egg and another fertilizes the polar nuclei.

duodenum: first section of the small intestine

E

ear: sense organ of hearing and equilibrium

echinoderm: unsegmented deuterostome with a five-part body plan, radial symmetry in adults, and spiny outer covering; phylum Echinodermata

ecology: study of relationships among organisms and the environment

ecosystem: a community and its nonliving environment

ectoderm: outermost germ layer in an animal embryo

ectotherm: animal that lacks an internal mechanism that keeps its temperature within a narrow range; invertebrates, fishes, amphibians, and reptiles

effector: in homeostatic responses, a structure that counteracts a change in a variable

ejaculation: discharge of semen through the penis

elastin: a type of stretchy connective tissue protein

electromagnetic spectrum: range of naturally occurring radiation

electron transport chain: membrane-bound molecular complex that shuttles electrons to slowly extract their energy

electron: a negatively charged subatomic particle with negligible mass that orbits the atom's nucleus

electronegativity: an atom's tendency to attract electrons

element: a pure substance consisting of atoms containing a characteristic number of protons

elimination: the expulsion of waste from the body

embryo sac: mature female gametophyte in angiosperms

embryonic disc: during the second week of development, a flattened, two-layered mass of cells that develop into the embryo

embryonic stage: stage of human development lasting from the end of the second week until the end of the eighth week of gestation

emergent property: quality that results from interactions of a system's components

emigration: movement of individuals out of a population

enamel: hard substance covering a tooth

endergonic reaction: a chemical reaction that requires a net input of energy

endocrine gland: concentration of hormone-producing cells in an animal

endocrine system: organ system consisting of glands and cells that secrete hormones

endocytosis: form of transport in which the cell membrane engulfs extracellular material

endoderm: innermost germ layer in an animal embryo

endodermis: the innermost cell layer of root cortex

endometrium: inner uterine lining

endoplasmic reticulum: interconnected membranous tubules and sacs that wind from the nuclear envelope to the cell membrane, along which proteins are synthesized (in rough ER) and lipids synthesized (in smooth ER)

endorphin: pain-killing protein produced in the anterior pituitary

endoskeleton: braced framework skeleton on the inside of an animal

endosperm: triploid tissue that stores food for embryo in an angiosperm seed

endospore: dormant, thick-walled structure that enables some bacteria to survive harsh conditions

endosymbiont theory: the idea that mitochondria and chloroplasts originated as free-living bacteria engulfed by other prokaryotic cells

endothelium: layer of epithelial tissue that lines blood vessels

endotherm: animal that maintain its body temperature by using heat generated from its own metabolism; birds and mammals

energy: the ability to do work

energy of activation: energy required for a chemical reaction to begin

energy shell: group of electron orbitals that share the same energy level

entropy: randomness or disorder

envelope: the membrane layer surrounding the protein coat of some viruses

environmental resistance: combination of factors that limit population growth

enzyme: a protein that catalyzes a specific type of chemical reaction without being consumed

eosinophil: type of phagocyte

epidermis: the outermost layer of skin, consisting of epithelial tissue

epididymis: tube that receives and stores sperm from one testis

epiglottis: cartilage that covers the glottis, routing food to the digestive tract and air to the respiratory tract

epinephrine (adrenaline): hormone secreted by the adrenal medulla; raises blood pressure, constricts blood vessels, and slows digestion; also can act as a neurotransmitter

epistasis: one gene masks another gene's expression

epithelial tissue (epithelium): tissue type consisting of tightly packed cells that form linings, coverings, and glands

erosion: wearing away of soil by water and wind

esophagus: muscular tube that leads from the pharynx to the stomach

essential nutrient: chemicals vital for metabolism, growth, and reproduction that must come from food because an organism cannot synthesize it

estrogen: steroid hormone produced in ovaries of female vertebrates; helps regulate reproductive cycles

estuary: area where fresh water in a river meets salty water of an ocean

ethology: the study of animal behavior, especially in the natural environment

ethylene: volatile plant hormone that ripens fruit

euglenoid: unicellular flagellated protist with elongated cell

eukaryote: organism composed of cells containing a nucleus and other membrane-bounded organelles

eukaryotic cell: a complex cell containing membrane-bounded organelles; organism in domain Eukarya

eumetazoan: animal with true tissues and in which the embryo forms a gastrula stage

Eustachian tube (auditory tube): tube that connects the middle ear with air passageways at the back of the nose; allows for the adjustment of pressure on the inside of the tympanic membrane

eutrophic: describes nutrient-rich water

eutrophication: addition of nutrients to a body of water

evaporation: the conversion of a liquid to a vapor (gas)

evolution: change in allele frequencies in a population over time

excretion: elimination of these metabolic wastes

exergonic reaction: an energy-releasing chemical reaction

exhalation: movement of air out of the lungs

exocrine gland: structure that secretes substances into a duct

exocytosis: form of transport in which vesicles containing cell secretions fuse with the cell membrane

exon: portion of an mRNA that is translated after introns are removed

exoskeleton: braced framework skeleton on the outside of an animal

expanding repeat mutation: type of mutation in which the number of copies of a three- or four- nucleotide sequence increases over several generations

experiment: a test of a hypothesis under controlled conditions

exponential growth: the number of new individuals is proportional to the size of the population

external fertilization: release of gametes by males and females into the same environment

extinction: disappearance of a species

extracellular matrix: nonliving complex of substances that surrounds cells of connective tissue

eye: organ that detects light and produces the sense of sight

F

F_1 (first filial) generation: the offspring of the P generation in a genetic cross

F_2 (second filial) generation: the offspring of the F_1 generation in a genetic cross

facilitated diffusion: form of passive transport in which a solute moves down its concentration gradient with the aid of a transport protein

facultative anaerobe: organism that can live with or without O_2

fall turnover: seasonal mixing of the upper and lower layers of a lake

family: taxonomic category between order and genus

fatty acid: long-chain hydrocarbon terminating with a carboxyl group

feather: epidermal outgrowth composed of keratin that distinguishes birds

feces: solid waste that leaves the digestive tract

fermentation: metabolic pathway in the cytoplasm in which NADH from glycolysis reduces pyruvate

fertilization: the union of two gametes

fetal period: stage of human development lasting from the beginning of the ninth week of gestation through birth

fever: rise in the body's temperature

fiber: elongated sclerenchyma cell

fibrous root system: branching root system arising from stem

filter feeder: animal that strains food particles from water

filtration: removal of water and solutes from the blood, as occurs at the glomerulus

first law of thermodynamics: energy cannot be created or destroyed, just converted from one form to another

fish: vertebrate animal with fins and external gills

fitness: an organism's contribution to the next generation's gene pool

fixed action pattern: type of innate behavior; stereotyped sequence of events that is performed to completion once initiated by a releaser

flagellum (pl. flagella): a long whiplike appendage a cell uses for motility; in eukaryotes, it is composed of microtubules

flatworm: unsegmented worm lacking a coelom; phylum Platyhelminthes

flower: the reproductive structure in angiosperms that produces pollen and eggs

fluid feeder: animal that drinks its food

fluid mosaic: two-dimensional fluid of phospholipids and proteins that form biological membranes

follicle cell: nourishing cell surrounding an oocyte

follicle stimulating hormone (FSH): hormone produced in the anterior pituitary; controls follicle development, oocyte maturation, and the release of estrogen in females and stimulates sperm production in males

food chain: series of organisms that successively eat each other

food web: network of interconnecting food chains

foot: ventral muscular structure that provides movement in mollusks

foramen magnum: the large hole in the skull where the spinal cord leaves the brain

foraminiferan: protist with a calcium carbonate shell that produces pseudopods

forebrain: front part of the vertebrate brain

fossil: any evidence of an organism from more than 10,000 years ago

founder effect: genetic drift that occurs when a small, nonrepresentative group of individuals leaves their ancestral population and begins a new settlement

fovea centralis: indentation in the retina opposite the lens; it has only cones and provides visual acuity

frameshift mutation: type of mutation in which nucleotides are added or deleted by any number other than a multiple of three, altering the reading frame

free-living flatworm:

frog: type of amphibian

frond: leaf of a true fern

fruit: seed-containing structure in angiosperms

fruiting body: multicellular spore-bearing organ of a fungus

fundamental niche: all the resources that a species could possibly use

G

G_0 phase: resting phase of the cell cycle in which the cell continues to function but does not divide

G_1 phase: gap stage of interphase in which the cell grows and carries out its basic functions

G_2 phase: gap stage of interphase in which the cell synthesizes and stores membrane components and spindle proteins

gallbladder: organ that stores bile from the liver and releases it into the small intestine

gamete: a sex cell; sperm or egg

gametophyte: haploid, gamete-producing stage of the plant life cycle

ganglion cell: interneuron in the retina that generates action potentials; ganglion cell axons make up the optic nerve

ganglion: cluster of neuron cell bodies

gap junction: connection between two adjacent animal cells that allows cytoplasm to flow between them

gastric juice: mixture of water, mucus, salts, hydrochloric acid, and enzymes produced at the stomach lining

gastrin: peptide hormone that stimulates secretion of additional gastric juice

gastrointestinal (GI) tract: two-opening digestive tract; also called alimentary canal

gastropod: type of mollusk

gastrovascular cavity: digestive chamber with a single opening

gastrula: stage of early animal embryonic development during which three tissue layers form

gemma (pl. gemmae): asexual reproductive structure in bryophytes

gene: sequence of DNA that codes for a specific protein or RNA molecule

gene flow: the movement of alleles between populations

gene pool: all of the genes and their alleles in a population

gene therapy: treatment that replaces a faulty gene in a somatic cell with a functioning version of the gene

genetic code: correspondence between specific nucleotide sequences and amino acids

genetic drift: change in allele frequencies that occurs purely by chance

genetic mutation: change in an organism's DNA sequence

genome: all the genetic material in an organism

genotype: genetic makeup of an individual

genus: taxonomic category between family and species

geological time scale: a division of Earth's history into eons, eras, periods, and epochs defined by major geological or biological events

germ cell: cell that gives rise to gametes in an animal

germination: resumption of growth after seed dormancy is broken

germline mutation: a mutation in a cell that gives rise to gametes

ghrelin: hormone that stimulates appetite

gibberellin: plant hormone that promotes shoot elongation

gill: respiratory surface that exchanges gases in water

ginkgo: a type of gymnosperm

gizzard: muscular part of the stomach in birds and other animals that mechanically digests food with the aid of sand and small pebbles

gland: organ that secretes substances into the bloodstream, into a cavity, or outside the body

global climate change: long term changes in Earth's weather patterns

glomerular (Bowman's) capsule: cup-shaped end of a renal tubule; surrounds the glomerulus

glomerulus: ball of capillaries containing blood to be filtered at a nephron Glossary manuscript—Chapter 1

glottis: slit like opening between the vocal cords

glucagon: pancreatic hormone that raises blood sugar level by stimulating liver cells to break down glycogen into glucose

glucocorticoid: hormone secreted by the adrenal cortex; enables the body to survive prolonged stress

glycerol: a three-carbon alcohol that forms the backbone of triglycerides and phospholipids

glycocalyx: sticky layer composed of proteins and/or polysaccharides that surrounds some prokaryotic cell walls; slime layer or capsule

glycolysis: a metabolic pathway occurring in the cytoplasm of all cells; one molecule of glucose splits into two molecules of pyruvate

gnetophyte: a type of gymnosperm

goiter: swelling of the thyroid gland

golden alga: photosynthetic aquatic protist with two flagella and carotenoid accessory pigments

Golgi apparatus: a system of flat, stacked, membrane-bounded sacs that packages cell products for export

gonad: gland that manufactures hormones and gametes in animals; ovary or testis

gonadotropin-releasing hormone (GnRH): hormone produced in the hypothalamus; stimulates the anterior pituitary to release follicle-stimulating hormone (FSH) and luteinizing hormone (LH)

graded potential: a local flow of electrical current that occurs when Na^+ leaks into a neuron, weakening with distance from the source of the stimulus

gradualism: theory that proposes that evolutionary change occurs gradually, in a series of small steps

Gram stain: technique for classifying bacteria into two main groups based on cell wall structure

granum (pl. grana): a stack of flattened thylakoid discs in a chloroplast

gravitropism: directional growth response to gravity

gray matter: nervous tissue in the central nervous system; consists of neuron cell bodies, unmyelinated fibers, interneurons, and neuroglial cells

green alga: photosynthetic protist that has pigments, starch, and cell walls similar to those of land plants

greenhouse effect: increase in surface temperature caused by carbon dioxide and other atmospheric gases

ground tissue: plant tissue derived from ground meristem, comprised mostly of parenchyma

growth factor: a protein that binds a specific receptor type on certain cells, starting a chain reaction of chemical messages that signals the cell to divide

growth hormone (GH): hormone produced in the anterior pituitary; promotes growth and development of tissues by increasing protein synthesis and cell division rates

guard cells: pair of cells flanking a stoma

gustation: the sense of taste

gymnosperm: a plant with seeds that are not enclosed in a fruit; includes conifers, *Ginkgo*, gnetophytes, and cycads

H

habitat: physical place where an organism lives

habituation: type of learning in which an animal learns not to respond to irrelevant stimuli

hagfish: jawless animal with a cranium but not vertebrae

hair cell: mechanoreceptor that initiates sound transduction in the cochlea

half-life: the time it takes for half the atoms in a sample of a radioactive substance to decay

haploid: containing one copy of each chromosome; also called 1n

Hardy–Weinberg equilibrium: situation in which allele frequencies do not change from one generation to the next

heart: muscular organ that pumps blood (or a comparable fluid) throughout the body

heartwood: dark-colored, nonfunctioning secondary xylem in woody plant

heavy chain: large polypeptide of an antibody subunit

helper T cell: lymphocyte that coordinates activities of other immune system cells

hemoglobin: pigment that carries oxygen in red blood cells

herbaceous plant: plant with green, soft stem at maturity

herbivore: animal that eats plants

heritability: proportion of a trait attributable to heredity

hermaphrodite: individual that produces both sperm and eggs

heterotroph: organism that obtains carbon by eating another organism; consumer

heterozygote advantage: condition in which a heterozygote has greater fitness than homozygotes, maintaining balanced polymorphism in a population

heterozygous: possessing two different alleles for a particular gene

hibernation: state of low metabolic rate and low body temperature for extended periods during winter

hindbrain: lower, posterior portion of the vertebrate brain

histamine: biochemical that dilates blood vessels; involved in inflammation and allergies

homeostasis: the ability of an organism to maintain constant body temperature, fluid balance, and chemistry

homeotic: describes any gene that, when mutated, leads to organisms with structures in the wrong places

hominid: type of hominoid that includes the "great apes" (orangutans, gorillas, chimpanzees, and humans)

hominine: type of hominid; gorilla, chimpanzee, or human

hominoid: type of primate; any ape, including humans

homologous pair: chromosomes that look alike and have the same sequence of genes

homologous: similar in structure or position because of common ancestry

homozygous: possessing two identical alleles for a particular gene

horizontal gene transfer: transfer of genetic information from one cell to another cell that is not its descendant

hormone: biochemical synthesized in small quantities in one place and transported to another

hornwort: a type of bryophyte

horseshoe crab: type of chelicerate arthropod

horsetail: a type of seedless vascular plant

human chorionic gonadotropin (hCG): hormone secreted by an embryo; prevents menstruation

human immunodeficiency virus (HIV): virus that causes acquired immune deficiency syndrome (AIDS)

humoral immunity: branch of adaptive immune system in which B cells secrete antibodies in response to a foreign antigen

humus: chemically complex jellylike organic substance in soil

hydrogen bond: weak chemical bond between oppositely charged portions of molecules

hydrolysis: splitting a molecule by adding water

hydrophilic: attracted to water

hydrophobic: repelled by water

hydrostatic skeleton: skeleton consisting of flexible tissue surrounding a constrained fluid

hypertonic: describes a solution in which the solute concentration is greater than on the other side of a semipermeable membrane

hypha (pl. hyphae): a fungal filament; the basic structural unit of a multicellular fungus

hypoglycemia: low blood sugar caused by excess insulin or insufficient carbohydrate intake

hypothalamus: small forebrain structure beneath the thalamus that controls homeostasis and links the nervous and endocrine systems

hypothesis: a testable, tentative explanation based on prior knowledge

hypotonic: describes a solution in which the solute concentration is less than on the other side of a semipermeable membrane

I

immigration: movement of individuals into a population

immune system: organ system consisting of cells that defends the body against infections, cancer, and foreign cells

impact theory: idea that mass extinctions were caused by impacts of extraterrestrial origin

implantation: embedding of the blastocyst into the uterine lining

imprinting: type of learning that usually occurs during a critical period early in life and occurs without obvious reinforcement

incomplete digestive tract: digestive tract with one opening that takes in food and ejects wastes

incomplete dominance: mode of inheritance in which a heterozygote's phenotype is intermediate between the phenotypes of the two homozygotes

independent variable: hypothesized influence on a dependent variable

indeterminate growth: growth that persists indefinitely

indirect development: development of a juvenile animal into an adult while passing through intervening larval stages

inferior vena cava: lower branch of the largest vein that leads to the heart

inflammation: immediate, localized reaction to an injury or any pathogen that breaches the body's barriers

ingestion: the taking in of food into the digestive tract

inhalation: movement of air into the lungs

inheritance of acquired characteristics: the idea that an organism can inherit the traits that its parent acquired during its lifetime

inhibiting hormone: hormone produced by the hypothalamus; inhibits hormone production in the anterior pituitary

innate behavior: an instinctive behavior that develops independently of experience

innate defense: cell or substance that provides generalized protection against all infectious agents

inner cell mass: cells in the blastocyst that develop into the embryo

insect: type of mandibulate arthropod

insulin: pancreatic hormone that lowers blood sugar level by stimulating body cells to take up glucose from the blood

integumentary system: organ system consisting of skin and its outgrowths

intercalary meristem: meristem between the nodes of a mature stem

interferon: type of cytokine

interleukin: type of cytokine

intermediate filament: component of the cytoskeleton intermediate in size between a microtubule and a microfilament

intermembrane compartment: the space between a mitochondrion's two membranes

internal fertilization: use of a copulatory organ to deposit sperm inside a female's body

interneuron: neuron that connects one neuron to another in the central nervous system

internode: stem areas between the points of leaf attachment

interphase: stage preceding mitosis or meiosis, when the cell carries out its functions, replicates its DNA, and synthesizes proteins, lipids, and carbohydrates

interstitial cell: testosterone-secreting endocrine cell in a testis

interstitial fluid: liquid that bathes cells in a vertebrate's body

intertidal zone: region along a coastline between the high and low tide marks

intrinsic rate of increase (r_{max}): value of r (the per capita rate of increase) when the environment does not restrict reproduction and survival

intron: portion of an mRNA molecule that is removed before translation

invasive species: introduced species that establishes a breeding population in a new location and spreads widely from the original point of introduction

inversion: a portion of a chromosome that flips and reinserts itself

invertebrate: animal without a backbone

ion: an atom or group of atoms that has lost or gained electrons, giving it an electrical charge

ionic bond: attraction between oppositely charged ions

iris: colored part of the eye; regulates the size of the pupil

isotonic: condition in which a solute concentration is the same on both sides of a semipermeable membrane

isotope: any of the forms of an element, each having a different number of neutrons in the nucleus

J

jaws: bones that frame the entrance to the mouth

joint: area where two bones meet

J-shaped curve: plot of exponential growth over time

K

karyotype: a size-ordered chart of the chromosomes in a cell

kelp: the largest type of brown algae

keystone species: species whose effect on community structure is disproportionate to its biomass

kidney: excretory organ in vertebrates

kilocalorie (kcal): energy required to raise 1 kilogram of water 1 °C; one food Calorie.

kin selection: the sacrifice of one individual's genes for the sake of the genes it shares with related animals

kinetic energy: energy being used to do work; energy of motion

kinetochore:

kinetoplast: DNA-containing structure that forms part of the mitochondrion in some flagellated protists

kingdom: taxonomic category between domain and phylum

knuckle-walking: type of locomotion in which an animal with tree-dwelling ancestors runs rapidly on the ground on all fours

Koch's postulates: rules used to verify that an organism causes a particular disease

Krebs cycle: stage in cellular respiration that completely oxidizes the products of glycolysis

***K*-selected species:** species consisting of long-lived, late-maturing individuals that have few offspring, with each receiving heavy parental investment

L

***lac* operon:** in *E. coli*, three lactose-degrading genes plus the promoter and operator that control their transcription

lacteal: lymph capillary that absorbs fat in the small intestine

lactic acid fermentation: metabolic pathway in which NADH from glycolysis reduces pyruvate, producing lactic acid

lamprey: type of jawless fish

lancelet: type of invertebrate chordate

large intestine: part of the digestive tract that connects the small intestine to the anus

larva: immature stage of animal development; does not resemble the adult of the species

larynx: the "voice box" and a conduit for air

latent: describes an infection in which viral genetic material is integrated into the host cell's chromosome and not causing symptoms

lateral line system: network of canals that extends along the sides of fishes and houses receptor organs that detect vibrations

lateral meristem: meristem whose daughter cells thicken a root or stem

law of independent assortment: Mendel's law stating that during gamete formation, the segregation of the alleles for one gene does not influence the segregation of the alleles for another gene

law of segregation: Mendel's law stating that the two alleles of each gene are packaged into separate gametes

leaf hair: trichome that slows movement of air over the leaf surface

learned behavior: behavior that results from an animal's experiences and can be modified over time

leech: type of annelid

lens: structure in the eye through which light passes and is focused

lenticel: spongy areas that permits gas exchange between living bark tissue and atmosphere

leptin: hormone that suppresses appetite and signals fat storage to the brain

leukemia: cancer in which bone marrow overproduces white blood cells

lichen: association of a fungus and a green alga or cyanobacterium

life expectancy: prediction of how long an individual will live, based on current age and mortality rates

life history: the events that influence reproduction for a species

life span: the longest a member of a species can live

life table: chart that shows the probability of surviving to any given age

ligament: band of fibrous connective tissue that connects bone to bone across a joint

ligase: enzyme that catalyzes formation of covalent bonds in the DNA sugar-phosphate backbone

light reactions: photosynthetic reactions that harvest light energy and store it in molecules of ATP or NADPH

lignin: tough, complex molecule that strengthens the walls of some plant cells

limbic system: collection of forebrain structures (including the amygdala and hippocampus) involved in emotion and memory

limnetic zone: layer of open water in a lake or pond where light penetrates

linkage group: group of genes that tend to be inherited together because they are on the same chromosome

linkage map: diagram of gene order and spacing on a chromosome, based on crossover frequencies

linked genes: genes on the same chromosome

lipid: hydrophobic organic molecule consisting mostly of carbon and hydrogen

littoral zone: shallow region along the shore of a lake or pond where sufficient light reaches to the bottom for photosynthesis

liver: organ that produces bile, detoxifies blood, stores glycogen and fat-soluble vitamins, synthesizes blood proteins, and monitors blood glucose level

liverwort: a type of bryophyte

lizard: type of reptile

lobe-finned fish: type of bony fish; coelacanth

logistic growth curve: a graph that depicts the leveling-off of a population in response to environmental resistance

long-day plant: plant that flowers when light periods are longer than a critical length

loose connective tissue: type of connective tissue with widely spaced fibroblasts and a few fat cells

lung surfactant: mixture of phospholipids and proteins that reduce the cohesive forces between water molecules, keeping alveoli open

lung: sac-like structure where gas exchange occurs in air-breathing vertebrates

lungfish: type of bony fish

luteinizing hormone (LH): hormone produced in the anterior pituitary; promotes ovulation and progesterone secretion in females and stimulates testosterone secretion in males

lymph: fluid in lymph capillaries and lymph vessels

lymph capillary: dead-end vessel that collects lymph

lymph node: structure in the lymphatic system that contains white blood cells and fights infection

lymphatic system: organ system consisting of lymphoid organs and lymph vessels that transport lymph

lymphocyte: type of white blood cell; T or B cell

lysogenic conversion: a change in a bacterium's phenotype that is caused by a prophage (latent viral infection)

lysogenic infection: type of viral infection in which the genetic material of a virus "hides" in a host cell's chromosome until conditions are right for replication

lysosome: organelle in a eukaryotic cell that buds from the Golgi apparatus and enzymatically dismantles molecules, bacteria, and worn-out cell parts

lytic infection: type of viral infection in which a virus enters a cell, replicates, and causes the host cell to burst (lyse) as it releases the new viruses

M

macroevolution: large-scale evolutionary changes, such as the appearance of new species, genera and higher taxonomic levels

macronutrient: essential nutrient required in large amounts

macrophage: type of phagocyte

major histocompatibility complex (MHC): cluster of genes that encode cell surface proteins

mammal: tetrapod amniote vertebrate with hair and milk

mammary gland: milk-producing gland in mammals

mandibulate: arthropod with jawlike mouthparts (mandibles); crustaceans, insects, centipedes, and millipedes

mangrove swamp: tropical wetland dominated by salt-tolerant trees

mantle: dorsal fold of tissue that secretes a shell in most mollusks

marrow cavity: space in a bone shaft that contains yellow marrow

marsupial: pouched mammal

marsupium: pouch in which the immature young of marsupial mammals nurse and develop

mass extinction: the disappearance of many species over relatively short expanses of time

mass number: the total number of protons and neutrons in an atom's nucleus

mast cell: immune system cell that releases histamine when stimulated

matrix: the inner compartment of a mitochondrion

matter: substance that takes up space and is made of atoms

mechanoreceptor: sensory receptor sensitive to mechanical energy

Mediterranean shrubland: type of terrestrial biome; rainy winters and dry summers (also called chaparral)

medulla oblongata: part of the brainstem nearest the spinal cord; regulates breathing, heartbeat, blood pressure, and reflexes

medusa: free-swimming form of a cnidarian

megagametophyte: in seed plants, the female gametophyte

megaspore: in seed plants, spore that gives rise to female gametophyte

meiosis: division of genetic material that halves the chromosome number and yields genetically variable gametes

melatonin: hormone produced in the pineal gland; plays a role in sensing light and dark cycles

memory: the ability to retain and recall information

memory cell: lymphocyte that responds quickly when an antigen is encountered again

meninges: membranes that cover and protect the central nervous system

menopause: cessation of a woman's menstrual periods

menstrual cycle: hormonal cycle that prepares the uterus for pregnancy

meristem: localized regions of active cell division in a plant

mesoderm: embryonic germ layer that forms between ectoderm and endoderm in an animal embryo

mesophyll: photosynthetic ground tissue in leaves

messenger RNA (mRNA): a molecule of RNA that encodes a protein

metabolic pathway: a series of connected reactions in a cell

metabolism: the biochemical reactions of a cell

metamorphosis: developmental process in which an animal changes drastically in body form between the juvenile and the adult

metaphase I: metaphase of meiosis I, when homologous chromosome pairs align down the center of a cell

metaphase II: metaphase of meiosis II, when replicated chromosomes align down the center of a cell

metastasis: spreading of cancer

microevolution: relatively short-term changes in allele frequencies within a population or species

microfilament: component of the cytoskeleton; made of the protein actin

microgametophyte: in seed plants, the male gametophyte

micronutrient: essential nutrient required in small amounts

microspore: in seed plants, spore that gives rise to male gametophyte

microtubule: component of the cytoskeleton; made of subunits of tubulin protein

microvillus (pl. microvilli): extension of the plasma membrane of a villus

midbrain: part of the brain between the forebrain and hindbrain; important in vision and hearing

millipede: type of mandibulate arthropod

mimicry: similar appearances of different species

mineralocorticoid: hormone secreted by the adrenal cortex; helps maintain blood volume and electrolyte balance

mitochondrion: organelle that houses the reactions of cellular respiration in eukaryotes

mitosis: division of genetic material that yields two genetically identical cells

mitotic spindle: a structure of microtubules that aligns and separates chromosomes in mitosis

molecular clock: application of the rate at which DNA mutates to estimate when two types of organisms diverged from a shared ancestor

molecule: two or more atoms joined by chemical bonds

mollusk: unsegmented animal with a soft body, mantle, muscular foot, and visceral mass; phylum Mollusca

monocotyledon (monocot): an angiosperm with seeds that have one cotelydon

monoecious: describes individual that produces male and female gametes.

monohybrid cross: mating between two individuals that are heterozygous for the same gene

monomer: a single link in a polymeric molecule

monophyletic: describes a group of organisms consisting of a common ancestor and all of its descendants

monosaccharide: a sugar that is one five- or six-carbon unit

monotreme: egg-laying mammal

morula: preembryonic stage consisting of a solid ball of cells

moss: a type of bryophyte

motor neuron: neuron that transmits a message from the central nervous system toward a muscle or gland

motor unit: a neuron and all the muscle fibers it contacts

muscle: organ that powers movements in animals by contracting; consists of muscle tissue and other tissue types

muscle fiber: muscle cell

muscle spindle: group of small, modified muscle cells innervated by stretch-sensitive neurons; monitors muscle tension

muscle tissue: tissue type consisting of contractile cells that provide motion

muscular system: organ system consisting of muscles whose contractions form the basis of movement

mutagen: any external agent that causes a mutation

mutant: a phenotype or allele that is not the most common for a certain gene in a population or that has been altered from the "typical" (wild type) condition

mutation: a change in a DNA sequence

mutualism: type of symbiosis that benefits both partners

mycelium: assemblage of hyphae that forms an individual fungus

mycorrhiza: association of a fungus and the roots of a plant

mycorrhizal fungi:

myelin sheath: fatty material that insulates some nerve fibers in vertebrates, speeding nerve impulse transmission

myocardium: thick, muscular middle layer of the heart wall

myofibril: cylindrical subunit of a muscle fiber, consisting of actin and myosin

myosin: protein that forms thick filaments in muscle cells

N

natural selection: differential reproduction of organisms whose genetic traits better adapt them to a particular environment

negative feedback (feedback inhibition): pathway in which the product of a reaction inhibits the enzyme that controls its formation

nephron: functional unit of the kidney

nephron loop: hairpin-shaped region of the renal tubule that connects the proximal and distal convoluted tubules

neritic zone: region of an ocean from the coast to the edge of the continental shelf

nerve: bundle of nerve fibers (axons) bound together in a sheath of connective tissue

nerve net: diffuse network of neurons

nervous system: organ system that transmits information via a network of neurons and neuroglia

nervous tissue: tissue type whose cells (neurons and neuroglia) form a communication network

net primary production: energy available to consumers in a food chain, after cellular respiration and heat loss by producers

neural tube: embryonic precursor of the central nervous system

neuroendocrine cell: neuron that secretes hormones

neuroglia: cell type that makes up nervous tissue and that supports, nourishes, or assists neurons

neuromodulator: peptide that alters a neuron's response to a neurotransmitter or blocks the release of a neurotransmitter

neuromuscular junction: chemical synapse of a neuron onto a muscle cell

neuron: cell type that makes up nervous tissue; consists of a cell body, an axon, and numerous dendrites

neurotransmitter: chemical passed from a neuron to receptors on another neuron or on a muscle or gland cell

neutral: neither acidic nor basic. Also, not electrically charged

neutron: a particle in an atom's nucleus that is electrically neutral and has one mass unit

neutrophil: type of phagocyte

niche: all resources a species uses for survival, growth, and reproduction

nitrification: conversion of ammonia into nitrites and nitrates

nitrogen fixation: bacterial conversion of N_2 to forms plants can use

nitrogenous base: a nitrogen-containing compound that forms part of a nucleotide

node: point at which leaves attach to a stem

node of Ranvier: short region of exposed axon between Schwann cells on neurons of the vertebrate peripheral nervous system

nodule: root growth housing nitrogen-fixing bacteria

noncompetitive inhibition: an inhibitor binds to an enzyme at a site other than the active site, altering the enzyme's shape

nondisjunction: failure of chromosomes to separate at anaphase I or anaphase II of meiosis

nonpolar covalent bond: a covalent bond in which atoms share electrons equally

norepinephrine: hormone secreted by the adrenal medulla; raises blood pressure, constricts blood vessels, and slows digestion; also can act as a neurotransmitter

nose: organ that forms the entrance to the nasal cavity inside the head; functions in breathing and olfaction

notochord: in embryonic development, a flexible rod that forms the framework of the vertebral column and induces formation of the neural tube; one of the four characteristics of chordates

nuclear envelope: a two-layered structure bounding a cell's nucleus

nuclear pore: a hole in the nuclear envelope

nucleic acid: a long polymer of nucleotides; DNA or RNA

nucleoid: the part of a prokaryotic cell where the DNA is located

nucleolus: a structure within the nucleus where components of ribosomes are assembled

nucleosome: the basic unit of chromatin; consists of DNA wrapped around eight histone proteins

nucleotide: building block of a nucleic acid, consisting of a phosphate group, a nitrogenous base, and a five-carbon sugar

nucleus: membrane-bounded sac that contains DNA in a eukaryotic cell

nutrient: any substance that an organism uses for metabolism, growth, maintenance, and repair of its tissues

O

obesity: unhealthy amount of body fat

obligate aerobe: organism that requires O_2 for generating ATP

obligate anaerobe: organism that cannot live in the presence of O_2

oceanic zone: open sea beyond the continental shelf

olfaction: the sense of smell

oligochaete: type of annelid

oligosaccharide: intermediate-length carbohydrate consisting of 3 to 100 monosaccharides

oligotrophic: describes water containing few nutrients

ommatidium: one unit of a compound eye

omnivore: animal that eats many types of food, including plants and animals

oncogene: gene that normally controls cell division but when overexpressed leads to cancer

oogenesis: the production of egg cells

oogonium: in oogenesis, diploid germ cell that divides mitotically to yield two primary oocytes

open circulatory system: circulatory system in which blood circulates freely through the body cavity

operator: in an operon, the DNA sequence between the promoter and the protein-encoding regions

operon: group of related bacterial genes plus a promoter and operator that control the transcription of the entire group at once.

opportunistic: describes a pathogen that cannot cause disease in a healthy individual

opposite leaf arrangement: pattern in which each node has two leaves

optic nerve: nerve fibers that connect the retina to the brain; consist of ganglion cell axons

orbital: volume of space where a particular electron is most of the time

order: taxonomic category between class and family

organ: two or more tissues that interact and function as an integrated unit

organ system: two or more physically or functionally linked organs

organelle: compartment of a eukaryotic cell that performs a specialized function

organic molecule: compound containing both carbon and hydrogen

organogenesis: development of organs in an embryo

orgasm: pleasurable sensation, accompanied by involuntary muscle contractions, associated with sexual activity

osmoconformer: organism whose ion concentrations match those of its surroundings

osmoregulator: organism that actively controls its ion concentrations in a changing environment

osmosis: simple diffusion of water through a semipermeable membrane

osteoblast: bone cell that secretes matrix

osteoclast: bone cell that degrades matrix

osteocyte: mature bone cell (former osteoblast) surrounded by matrix

osteon: one set of concentric rings of osteocytes

osteoporosis: condition in which bones become less dense

ostracoderm: extinct jawless fish

oval window: membrane between the middle ear and the inner ear

ovarian cycle: hormonal cycle that controls the timing of oocyte maturation in the ovaries

ovary: female gonad

ovary: in angiosperms, base of carpel plus enclosed ovules

overexploitation: harvesting a species faster than it can reproduce

ovulation: release of an oocyte from an ovary

ovule: egg-bearing structure that develops into seed in gymnosperms and angiosperms

oxidation: the loss of one or more electrons by a participant in a chemical reaction

oxidation-reduction (redox) reaction: chemical reaction in which one reactant is oxidized and another is reduced

oxygen debt: after vigorous exercise, a period in which the body requires extra oxygen to restore ATP and creatine phosphate to muscle and to recharge hemoglobin and myoglobin

oxytocin: hormone released from the posterior pituitary; stimulates muscle contraction in the mammary glands and the uterus and plays a role in affection and bonding

ozone layer: atmospheric zone rich in ozone gas (O_3), which absorbs the sun's ultraviolet radiation

P

P (parental) generation: the first, true-breeding generation in a genetic cross

pain receptor: sensory receptor that detects mechanical damage, temperature extremes, or chemicals released from damaged cells

paleoanthropologist: scientist who studies fossil humans

paleontology: the study of fossil remains or other clues to past life

pancreas: gland between the spleen and the small intestine; an endocrine part produces hormones (somatostatin, insulin, and glucagon) and an exocrine part produces digestive enzymes

pancreatic islet (islet of Langerhans): cluster of cells in the pancreas that secretes somatostatin, insulin, and glucagon

parapatric evolution: formation of new species when part of a population enters a habitat bordering the parent species' range, and the two groups become reproductively isolated

paraphyletic: describes a group of organisms that contains a common ancestor and some, but not all, of its descendants

parasitism: type of symbiosis in which one member benefits at the expense of the other

parasitoid: parasitic insect that kills its host from within

parasympathetic nervous system: part of the autonomic nervous system; controls vital functions such as respiration and heart rate and opposes the sympathetic nervous system

parathyroid gland: one of four small groups of cells behind the thyroid gland; secretes parathyroid hormone

parathyroid hormone (PTH): hormone produced in the parathyroid gland; maintains blood calcium levels

parenchyma: unspecialized plant cells making up majority of ground tissue

parental chromosome: chromosome containing genetic information from only one parent

parental investment: the time and resources a parent spends on producing and raising its offspring

parsimonious: in cladistics, describes an evolutionary tree that requires the fewest steps to construct

particulate: small piece of matter suspended in air

passive immunity: immunity generated when an organism receives antibodies from another organism

passive transport: movement of a solute across a membrane without the direct expenditure of energy

pectoral girdle: bones that connect the forelimbs to the axial skeleton; the collarbones and shoulder blades

Pedigree: chart showing relationships of relatives and which ones have a particular trait

peer review: evaluation of scientific results by experts before publication in a journal

pelagic zone: all water above the ocean floor

pellicle: a protective layer that supports the cell membrane of many protists

pelvic girdle: bones that connect the hind limbs to the axial skeleton

penis: male organ of copulation and urination

pepsin: enzyme that begins the digestion of proteins in the stomach

peptide bond: a covalent bond between adjacent amino acids; results from dehydration synthesis

peptide hormone: water-soluble, amino acid–based hormone that cannot freely diffuse through a cell membrane

per capita rate of increase (*r*): difference between the birth rate and the death rate in a population

perception: interpretation of a sensation

pericardium: connective tissue sac that houses the heart

pericycle: outermost layer of root vascular cylinder; produces branch roots

periderm: dermal tissue covering woody plant part

periodic table: chart that lists elements according to their properties

peripheral nervous system: neurons that transmit information to and from the central nervous system

peristalsis: waves of muscle contraction that propel food along the digestive tract

peritubular capillaries: blood vessels that surround a renal tubule

permafrost: permanently frozen ground in tundra

peroxisome: membrane-bounded sac that houses enzymes that break down fatty acids and dispose of toxic chemicals

persistent organic pollutant: carbon-based chemical pollutants that remain in ecosystems for long periods

petal: flower part interior to sepals

petiole: stalk that supports a leaf blade

pH scale: a measurement of how acidic or basic a solution is

phagocyte: white blood cell that engulfs and digests foreign material and cell debris

phagocytosis: form of endocytosis in which the cell engulfs a large particle

pharyngeal pouch (or slit): openings in the pharynx of a chordate embryo; one of the four characteristics of chordates

pharynx: tube just behind the oral and nasal cavities; the throat

phenotype: observable expression of a genotype

pheromone: biochemical an organism releases that elicits a response in another member of the species

phloem: vascular tissue that transports sugars and other dissolved substances in plants

phloem sap: solution of water, minerals, sucrose, and other biochemicals in phloem

phospholipid: molecule consisting of two fatty acids and a phosphate; hydrophobic at one end and hydrophilic at the other end

phospholipid bilayer: double layer of phospholipids that forms in water, with hydrophilic parts facing outwards and hydrophobic tails forming the interior

phosphorylation: the addition of a phosphate to a molecule

photon: a packet of light or other electromagnetic radiation

photoperiod: day length

photoreceptor: molecule that perceives quality and quantity of light

photorespiration: a metabolic pathway in which rubisco reacts with O_2 instead of CO_2, counteracting photosynthesis

photosynthesis: biochemical reactions that enable plants to harness sunlight energy to manufacture organic molecules

photosystem: cluster of pigment molecules and proteins in a chloroplast's thylakoid membrane

phototroph: organism that derives energy from sunlight

phototropism: directional growth response to unidirectional light

phylogenetics: field of study that attempts to explain the evolutionary relationships among species

phylogeny: depiction of evolutionary relationships among species, based on descent from shared ancestors

phylum: taxonomic category between kingdom and class

physiology: the study of the functions of organisms and their parts

phytochrome: a type of photoreceptor in plants

phytoplankton: microscopic photosynthetic organisms that drift in water

phytoremediation: use of plants to remove dangerous chemicals from the environment

pilus: short projection made of protein on a prokaryotic cell

pineal gland: small gland in the brain that secretes melatonin

pioneer species: the first species to colonize an area devoid of life

pith: ground tissue inside a ring of vascular bundles in roots and stems

pituitary gland: pea-sized endocrine gland attached to the hypothalamus

placebo: inert substance used as an experimental control

placenta: structure that connects the developing fetus to the maternal circulation in placental mammals

placental mammal: mammal in which the developing fetus is nourished by a placenta

placoderm: extinct lineage of giant fishes with jaws

plasma: watery, protein-rich fluid that forms the matrix of blood

plasma cell: B cell that secretes large quantities of one antibody type

plasmid: small circle of double-stranded DNA separate from a cell's chromosome

plasmodesma (pl. plasmodesmata): connection between plant cells that allows cytoplasm to flow between them

plasmodial slime mold: protist in which feeding stage consists of a multinucleated plasmodium

plasmodium: mass of thousands of nuclei enclosed by a single cell membrane

plate tectonics: theory that Earth's surface consists of several plates that move in response to forces acting deep within the planet

platelet: cell fragment that orchestrates clotting in blood

pleiotropic: describes a genotype with multiple expressions

point mutation: type of mutation in which one DNA base substitutes for another

polar body: small cell produced in female meiosis

polar covalent bond: a covalent bond in which electrons are attracted more to one atom's nucleus than to the other

polar nucleus: one of two nuclei fertilized to yield endosperm

pollen grain: immature male gametophyte in seed plants (gymnosperms and angiosperms)

pollen sac: pollen-producing cavity in anther

pollen tube: a structure, formed upon germination of a pollen grain, that grows through the ovule and carries sperm to the egg

pollination: transfer of pollen from anther to stigma

pollution: physical, chemical or biological change in the environment that adversely affects organisms

polychaete: type of annelid

polygenic: caused by more than one gene

polymer: a long molecule composed of similar subunits (monomers)

polymerase chain reaction (PCR): biotechnology tool that rapidly produces millions of copies of a DNA sequence of interest

polyp: sessile form of a cnidarian

polypeptide: a long polymer of amino acids

polyphyletic: describes a group of organisms that excludes the most recent common ancestor of all members of the group

polyploid: having extra chromosome sets

polysaccharide: carbohydrate consisting of hundreds of monosaccharides

pons: oval mass in the brainstem where white matter connects the medulla to higher brain structures and gray matter helps control respiration

population: members of the same species occupying a region

population bottleneck: genetic drift that occurs when many members of a population die, causing the loss of alleles that were present in the larger ancestral population

population density: number of individuals of a species per unit area or volume of habitat

population dispersion: pattern in which individuals are scattered throughout a habitat

population dynamics: study of the factors that influence changes in a population's size

portal system: capillary bed connected by two sets of venules

positive feedback: a process that reinforces an existing condition

postanal tail: muscular tail that extends past the anus; one of the four characteristics of chordates

posterior pituitary: the back part of the pituitary gland; secretes ADH and oxytocin produced in the hypothalamus

postsynaptic cell: neuron, muscle cell, or gland cell that receives a message at a synapse

postzygotic reproductive isolation: separation of species due to nonviability or infertility of a hybrid embryo or offspring

potential energy: stored energy available to do work

prebiotic simulation: experiment that attempts to recreate the conditions on early Earth that gave rise to the first cell

predator: organism that kills another for food

preembryonic stage: first two weeks of human development

pressure flow theory: theory that explains how phloem sap moves from source to sink

presynaptic cell: neuron that releases neurotransmitters into a synaptic cleft

prey: organism killed by another for food

prezygotic reproductive isolation: separation of species due to factors that prevent the formation of a zygote

primary growth: growth from primary meristems

primary (1°) structure: the amino acid sequence of a protein

primary immune response: immune system's response to its first encounter with a foreign antigen

primary meristem: meristem arising from apical meristem; protoderm, procambium, and ground meristems

primary oocyte: in oogenesis, a diploid cell that undergoes the first meiotic division and yields a haploid polar body and a haploid secondary oocyte

primary producer: species forming the base of a food web; autotroph

primary spermatocyte: in spermatogenesis, a diploid cell that undergoes the first meiotic division and yields two haploid secondary spermatocytes

primary succession: appearance of organisms in an area previously devoid of life

primate: mammal with opposable thumbs, eyes in front of the skull, relatively large brain, and flat nails instead of claws; includes prosimians, simians, and hominoids

primitive streak: furrow that appears along the back of the embryonic disc in the third week of human development

principle of superposition: the idea that lower rock layers are older than those above them

prion: infectious protein particle

producer (autotroph): organism that uses inorganic sources of energy and carbon

product: the result of a chemical reaction

product rule: the chance of two independent events occurring equals the product of the chances of either event occurring

profundal zone: deep region of a lake or pond where light does not penetrate

progenote: collection of nucleic acid, protein, and lipids that was the forerunner to cells

progesterone: steroid hormone produced in ovaries of female vertebrates; helps regulate reproductive cycles

prokaryote: a cell that lacks a nucleus and other membrane-bounded organelles; bacteria and archaea

prokaryotic cell: a cell that lacks nuclei and other organelles; organisms in domains Bacteria and Archaea

prolactin: hormone produced in the anterior pituitary; stimulates milk production

prometaphase: stage of mitosis just before metaphase, when the nuclear membrane breaks up and spindle fibers attach to kinetochores

promoter: a control sequence at the start of a gene; attracts RNA polymerase and (in eukaryotes) transcription factors

prophage: DNA of a lysogenic bacteriophage that is inserted into a host cell's chromosome

prophase I: prophase of meiosis I, when chromosomes condense and become visible, and crossing over occurs

prophase II: prophase of meiosis II, when chromosomes condense and become visible

prosimian: type of primate; a lemur, aye-aye, loris, tarsier, or bush baby

prostaglandin: lipid released locally and transiently at the site of a damaged cell

prostate gland: structure that produces a milky, alkaline fluid that activates sperm

protein: a polypeptide folded into its functional three-dimensional shape

proteome: all of the proteins that an organism expresses

Protista: kingdom that includes mostly unicellular, eukaryotic organisms that are not plants, fungi, or animals

proton: a particle in an atom's nucleus carrying a positive charge and having one mass unit

protostome: clade of bilaterally symmetrical animals in which the first opening in the gastrula develops into the mouth

protozoan: unicellular protist that shares some characteristics of animals, including heterotrophy and (usually) motility

proximal convoluted tubule: region of the renal tubule between the glomerular capsule and the nephron loop

proximate: describing the mechanistic causes of behavior, including the interactions among genes, neurons, and hormones

pseudocoelom: fluid-filled animal body cavity lined by endoderm and mesoderm

pseudogene: a DNA sequence that is very similar to that of a gene; a pseudogene is transcribed but its mRNA is not translated

pseudostratified epithelial tissue: epithelium consisting of one layer of cells whose nuclei are staggered, giving the appearance of multiple layers

pulmonary artery: artery that leads from the right ventricle to the lungs

pulmonary circulation: blood circulation between the heart and lungs

pulmonary vein: vein that leads from the lungs to the left atrium

pulp: the soft inner part of a tooth, consisting of connective tissue, blood vessels, and nerves

punctuated equilibrium: theory that life's history has had long periods of stasis interrupted by bursts of rapid evolutionary change

Punnett square: diagram that uses the genotypes of the parents to reveal the possible results of a genetic cross

pupil: opening in the iris that admits light into the eye

pus: a whitish fluid that contains white blood cells, bacteria, and debris from dead cells

pyramid of biomass: diagram depicting dry weight of organisms at each trophic level at a given time

pyramid of energy: diagram depicting energy stored at each trophic level at a given time

pyruvate: the product of glycolysis

Q

quaternary (4°) structure: the shape arising from interactions between multiple polypeptide subunits of the same protein

R

R group: an amino acid side chain

radial symmetry: body form in which any plane passing through the body from the mouth to the opposite end divides the body into mirror images

radioactive isotope: atom that emits particles or rays as its nucleus disintegrates

radiolarian: protist with a silica shell that produces pseudopods

radiometric dating: a type of absolute dating that uses known rates of radioactive decay to date fossils

radula: a chitinous, tonguelike strap that mollusks use to eat

rain shadow: downwind side of a mountain, with a drier climate than the upwind side

ratite: type of bird

ray: parenchyma cells extending from center of woody stem or root

ray-finned fish: type of bony fish

reabsorption: renal tubule's return of useful substances to the blood

reactant: a starting material in a chemical reaction

reaction center: a molecule of chlorophyll *a* (and associated proteins) that participates in the light reactions of photosynthesis

realized niche: the resources that a species actually uses, considering competition and other limitations

receptacle: attachment point for flower parts

receptor potential: localized change in membrane potential (a graded potential) in a sensory receptor

recessive: describes an allele whose expression is masked if a dominant allele is present

recombinant chromosome: chromosome containing genetic information from both parents

recombinant DNA: genetic material spliced together from multiple sources

rectum: portion of the large intestine that collects feces

red alga: multicellular, photosynthetic, marine protist with red or blue accessory pigments

red blood cell: disc-shaped blood cell that contains hemoglobin and lacks a nucleus

reduction: the gain of one or more electrons by a participant in a chemical reaction

reflex: type of innate behavior; an instantaneous, automatic response to a stimulus

reflex arc: neural pathway that links a sensory receptor and an effector

relative dating: placing a fossil into a sequence of events without assigning it a specific age

releaser: stimulus that triggers a fixed action pattern

releasing hormone: hormone produced by the hypothalamus; stimulates hormone production in the anterior pituitary

renal cortex: outer portion of a kidney

renal medulla: middle part of a kidney

renal pelvis: inner part of a kidney

renal tubule: portion of a nephron that adjusts the composition of filtrate

renin: protein that stimulates release of aldosterone

repressor: in an operon, a protein that binds to the operator and prevents transcription

reproductive system: organ system that produces and transports gametes and may nurture developing offspring

reptile: tetrapod vertebrate with an amniote egg and a dry scaly body covering

reservoir: an organism that harbors a virus without developing disease but acts as a source of the virus to infect other species

residual volume: air that remains in the lungs and inflates the alveoli after a maximal exhalation

resource partitioning: use of the same resource in different ways or at different times by multiple species

respiratory surfaces:

respiratory system: organ system that acquires oxygen gas and releases carbon dioxide

resting potential: electrical potential inside of a neuron not conducting a nerve impulse

restriction enzyme: enzyme that cuts double-stranded DNA at a specific base sequence

reticular formation: diffuse network of nerve tracts that extends through the brainstem and into the thalamus; screens sensory input to the cerebrum

retina: sheet of photoreceptors at the back of the vertebrate eye

reverse transcriptase: enzyme that uses RNA as a template to construct a DNA molecule

rhizoid: rootlike extension on gametophytes of some nonvascular plants

rhizopod: protist without a shell that moves when its cytoplasm flows into projections called pseudopods; also called an amoeba

rhodopsin: pigment that transduces light into an electrochemical signal in photoreceptors

ribosomal RNA (rRNA): a molecule of RNA that, along with proteins, forms a ribosome

ribosome: A structure built of RNA and protein where mRNA anchors during protein synthesis

ribulose bisphosphate (RuBP): the five-carbon intermediate of the carbon reactions of photosynthesis

RNA polymerase: enzyme that uses a DNA template to produce a molecule of RNA

RNA world: the idea that the first independently replicating life form was RNA

rod cell: photoreceptor in the retina that provides black-and-white vision

root: belowground part of most plants

root cap: cells that protect root apical meristem from abrasion

root hair: epidermal outgrowth that increases root surface area

rough endoplasmic reticulum: ribosome-studded portion of the ER where secreted proteins are synthesized

roundworm: unsegmented worm with a pseudocoelom; phylum Nematoda

***r*-selected species:** species consisting of short-lived, early-maturing individuals that have many offspring, with each receiving little parental investment

rubisco: enzyme that adds CO_2 to ribulose bisphosphate in the carbon reactions of photosynthesis

rumen: the first chamber of a ruminant's stomach

ruminant: herbivore with a four-chambered stomach

S

S phase: the synthesis phase of interphase, when DNA replicates

saccule: one of two pouches in the vestibule of the inner ear

salamander: type of amphibian

salivary amylase: enzyme in saliva that begins chemical digestion of starch

sample size: number of subjects in each experimental group

sapwood: light-colored, functioning secondary xylem in woody plant

sarcomere: one of many repeated units in a myofibril of a muscle cell

saturated fatty acid: a fatty acid with single bonds between all carbon atoms

savanna: type of terrestrial biome; grassland with scattered trees

Schwann cell: type of neuroglia that forms a myelin sheath around some axons

scientific method: a systematic approach to understanding the natural world; involves asking questions, formulating and testing hypotheses, collecting data, drawing conclusions, and putting results into perspective with existing knowledge

sclera: the outermost layer of the eye; the white of the eye

sclereid: relatively short sclerenchyma cell

sclerenchyma: rigid plant cells that support mature plant parts

scrotum: the sac containing the testes

second law of thermodynamics: every reaction loses some energy to the surroundings as heat; entropy always increases

secondary (2°) structure: a "substructure" within a protein, such as an alpha helix, resulting from hydrogen bonds between parts of the peptide backbone

secondary growth: increase in girth from cell division in lateral meristem

secondary immune response: immune system's response to subsequent encounters with a foreign antigen

secondary oocyte: in oogenesis, a haploid cell that undergoes the second meiotic division and yields a polar body a haploid egg cell

secondary sex characteristic: trait that distinguishes the sexes but does not participate directly in reproduction

secondary spermatocyte: in spermatogenesis, a haploid cell that undergoes the second meiotic division and yields two haploid spermatids

secondary succession: change in a community's species composition following a disturbance

secretin: hormone that triggers the release of bicarbonate from the pancreas, stimulates the liver to secrete bile, and inhibits the release of gastrin

secretion: addition of substances to the fluid in a renal tubule

seed: in gymnosperms and angiosperms, a plant embryo packaged with a food supply inside a tough outer coat

seed coat: protective outer layer of seed

seedless vascular plant: plant with specialized vascular tissues but that does not produce seeds; includes true ferns, club mosses, whisk ferns, and lycopods

segmentation: division of an animal body into repeated subunits

selective permeability: admitting some substances but not others

semen: fluid that carries sperm cells out of the body

semicircular canal: one of three perpendicular fluid-filled structures in the inner ear; provides information on the position of the head

semilunar valve: flap of heart tissue that prevents the flow of blood from an artery to a ventricle

seminal vesicle: structure that contributes most fluid, along with fructose and prostaglandins, to semen

seminiferous tubule: tubule within a testis where sperm form and mature

senescence: aging

sensation: awareness of a stimulus

sensor: in homeostatic responses, a structure that monitors changes in a variable

sensory adaptation: lessening of sensation with prolonged exposure to a stimulus

sensory neuron: neuron that transmits information from a stimulated body part to the central nervous system

sensory receptor: cells that detects and passes stimulus information to a sensory neuron

sepal: flower part that encloses inner whorls

severe combined immune deficiency (SCID): immune system disorder in which neither T nor B cells function

sex chromosome: a chromosome that carries genes that determine sex

sex pilus: a bridge of cytoplasm transfers DNA from one cell to another in conjugation

sex-linked: describes genes or traits on the X or Y chromosome

sexual reproduction: the combination of genetic material from two individuals to create a third individual

sexual selection: type of natural selection resulting from variation in the ability to obtain mates

shoot: aboveground part of a plant

short-day plant: plant that flowers when light periods are shorter than a critical length

sieve cell: long, tapering conducting cell in phloem

sieve plate: porous area at end of sieve tube element

sieve tube element: conducting cell that comprises sieve tube in phloem

signal transduction: the biochemical transmission of a message from outside the cell to inside

simian: type of primate; a monkey

simple diffusion: form of passive transport in which a solute moves down its concentration gradient without the use of a transport protein

simple epithelial tissue: epithelium consisting of one layer of cells

simple leaf: leaf with undivided blade.

sink: plant part that does not photosynthesize

sinoatrial (SA) node: specialized cardiac muscle cells that set the pace of the heartbeat; the pacemaker

skeletal muscle: muscle tissue consisting of long, unbranched, striated, multinucleated cells under voluntary control; also, an organ composed of bundles of skeletal muscle cells and other tissue types that generates voluntary movements between pairs of bones

skeletal muscle tissue: voluntary muscle tissue composed of unbranched, multinucleated cells

skeletal system: organ system consisting of bones and ligaments that support body structures and attach to muscles

skeleton: structure that supports an animal's body

sleep: a state of in which voluntary muscle movement and responsiveness to sensory input are reduced

sliding filament model: sliding of actin and myosin past each other to shorten a muscle cell

slime layer: diffuse, irregular glycocalyx

small intestine: part of the digestive tract that connects the stomach with the large intestine; site of most chemical digestion and absorption

smog: type of air pollution that forms a visible haze in the lower atmosphere

smooth muscle tissue: involuntary muscle tissue consisting of nonstriated, spindle-shaped cells

snake: type of reptile

sodium-potassium pump: protein that uses energy released from splitting ATP to transport Na^+ out of cells and K^+ into cells

soil: rock and mineral particles mixed with organic matter

solute: a chemical that dissolves in another, forming a solution

solution: a homogenous mixture of a solute dissolved in a solvent

solvent: a chemical in which others dissolve, forming a solution

somatic cell: body cell that does not give rise to gametes

somatic mutation: mutation in any cell other than one that gives rise to gametes

somatic nervous system: in the peripheral nervous system, motor pathways that carry signals to skeletal (voluntary) muscles

somatostatin: pancreatic hormone; inhibits the production of insulin and glucagon

sorus (pl. sori): collection of sporangia on the underside of a fern frond

source: plant part that produces or releases sugar

speciation: formation of new species

species: a distinct type of organism

species evenness: measure of biodiversity; considers the relative sizes of the populations that make up a community

species richness: measure of biodiversity; number of species in a community

spermatogenesis: the production of sperm

spermatogonium: in spermatogenesis, diploid germ cell that divides mitotically to yield a stem cell and a primary spermatocyte

spermatozoon (pl. spermatozoa): mature sperm cell

sphincter: muscular ring that can open and close

spinal cord: tube of nervous tissue that extends through the vertebral column

spinal nerve: peripheral nerve that exits the vertebrate central nervous system from the spinal cord

spirillum (plural spirilla): spiral-shaped prokaryote

spleen: abdominal organ that produces and stores lymphocytes and destroys worn-out red blood cells

sponge: simple, asymmetrical animal lacking true tissues and gastrulation; phylum Porifera

spongy bone: bone tissue with large spaces between a web of bony struts

sporangium: structure in which spores are produced

spore: reproductive cell of a fungus

sporophyte: diploid, spore-producing stage of the plant life cycle

spring turnover: seasonal mixing of the upper and lower layers of a lake

squamous: flat

S-shaped curve: logistic growth curve

stabilizing selection: form of natural selection in which extreme phenotypes are less fit than the optimal intermediate phenotype

stamen: male flower part

standardized variable: any factor held constant for all subjects in an experiment

statistically significant: unlikely to be attributed to chance

statolith: starch-containing plastid in root cap cell that functions as a gravity detector

stem cell: undifferentiated cell that divides to give rise to additional stem cells and cells that specialize

steroid hormone: a lipid-soluble hormone that can freely diffuse through a cell membrane and bind to receptor inside the cell

sterol: lipid consisting of four interconnected carbon rings

stigma: in angiosperms, pollen-receiving tip of stigma

stipules: paired outgrowths that flank the base of a leaf's petiole

stoma (pl. stomata): pore in a plant's epidermis through which gases are exchanged between the plant and the atmosphere

stomach: J-shaped compartment in the digestive tract that receives food from the esophagus

stramenopile: protist with two flagella, one smooth and one covered with tubular "hairs" including brown algae, diatoms, and water molds. 418

stratified epithelial tissue: epithelium consisting of multiple cell layers

stroma: the fluid inner region of the chloroplast

style: in angiosperms, the stalklike upper part of a carpel

substrate feeder: animal that lives in its food and eats it from the inside

substrate-level phosphorylation: ATP formation from transferring a phosphate group from a high-energy donor molecule to ADP

succession: change in the species composition of a community over time

superior vena cava: upper branch of the largest vein that leads to the heart

suppressor T cell: lymphocyte that dampens the immune response

survivorship curve: graph of the proportion of individuals that survive to a particular age

sustainability: use of resources while not depleting them for the future

sustentacular cell: cell within a seminiferous tubule that supports developing sperm cells

swim bladder: organ that allows bony fishes to adjust their buoyancy

symbiosis: one species living in or on another

sympathetic nervous system: part of the autonomic nervous system; mobilizes the body to respond quickly to environmental stimuli and opposes the parasympathetic nervous system

sympatric speciation: formation of a new species within the boundaries of a parent species

synapse: junction between two neurons or between a neuron and a muscle or gland cell

synapsid: vertebrate with a single opening behind each eye orbit; mammals and their immediate ancestors

synaptic cleft: space into which neurotransmitters are released between two cells at a synapse

synaptic integration: a neuron's overall response to many incoming neural messages

synaptic terminal: enlarged tip of an axon; contains synaptic vesicles

synovial joint: joint between two freely movable bones connected by a fluid-filled capsule of fibrous connective tissue

systematics: field of study that includes taxonomy and phylogenetics

systemic circulation: blood circulation between the heart and the rest of the body, except the lungs

systole: contraction of the ventricles of the heart

systolic pressure: upper number in a blood pressure reading; reflects contraction of the ventricles

T

T cell: type of lymphocyte

taiga: type of terrestrial biome; the northern coniferous forest (also called boreal forest)

taproot: large central root that persists throughout the life of the plant

target cell: cell that a hormone binds to and directly affects

taste bud: cluster of cells that detect chemicals in food

taxis: directed movement toward or away from a stimulus

taxonomy: classification of organisms

tectorial membrane: membrane in contact with hair cells of the cochlea

tectum: thickened area of gray matter at the roof of the midbrain

telomerase: enzyme that extends telomeres, enabling cells to divide continuously

telomere: the tip of a eukaryotic chromosome

telophase I: telophase of meiosis I, when homologs arrive at opposite poles

telophase II: telophase of meiosis II, when chromosomes arrive at opposite poles and nuclear envelopes form

temperate coniferous forest: type of terrestrial biome; coniferous trees dominate

temperate deciduous forest: type of terrestrial biome; deciduous trees dominate

temperate grassland: type of terrestrial biome; grazing, fire, and drought restrict tree growth

template strand: the strand in a DNA double helix that is transcribed

tendon: band of fibrous connective tissue that attaches a muscle to a bone

teratogen: substance that causes birth defects

terminator: sequence in DNA that signals where the gene's coding region ends

territorial behavior: behavior that marks and defends an animal's home range

tertiary (3°) structure: the overall shape of a protein, resulting mostly from interactions between amino acid R groups and water

test cross: breeding an individual of unknown genotype to a homozygous recessive individual to reveal the unknown genotype

test: hard shell, such as those secreted by foraminiferans

testicle: a testis plus its surrounding tissues

testis: male gonad

testosterone: steroid hormone that regulates sperm production and development of male characteristics

tetanus: maximal muscle contraction caused by continual stimulation

tetrapod: vertebrate with four limbs

thalamus: tight grouping of nerve cell bodies beneath the cerebrum in the forebrain; relays sensory input to the cerebrum

theory: well-supported scientific explanation

thermocline: layer of water in which temperature changes quickly

thermogenesis: production of metabolic heat

thermoreceptor: sensory receptor that responds to temperature

thermoregulation: control of an animal's body temperature

thick filament: in muscle cells, filament composed of myosin

thigmotropism: directional growth response to touch

thin filament: in muscle cells, filament composed of actin

threshold potential: potential to which a neuron's membrane must be depolarized to trigger an action potential

thylakoid: disclike structure that makes up the inner membrane of a chloroplast

thylakoid space: the inner compartment of the thylakoid

thymus: lymphoid organ in the upper chest where T cells learn to distinguish foreign from self antigens

thyroid gland: gland in the neck that secretes two thyroid hormones (thyroxine and triiodothyronine) and calcitonin

thyroid-stimulating hormone (TSH): hormone produced in the anterior pituitary; stimulates the thyroid gland to release two types of hormones

thyroxine: one of two thyroid hormones that increases the rate of cellular metabolism

tidal volume: volume of air inhaled or exhaled during a normal breath

tight junction: connection between two adjacent animal cells that prevents fluid from flowing past the cells

tissue: group of cells that interact and provide a specific function

tongue: muscular structure on the floor of the mouth

tonsil: collection of lymphatic tissue in the throat

tooth: mineral-hardened structure embedded in the jaw

topsoil: uppermost soil layer

total fertility rate (TFR): average number of children per woman of child-bearing age

trace element: an element that an organism requires in small amounts

trachea (pl. tracheae): in vertebrates, the respiratory tube just beneath the larynx; the "windpipe." In invertebrates, a branched tubule that brings air in close contact with cells, facilitating gas exchange

tracheid: long, narrow conducting cell in xylem

transcription: production of RNA using DNA as a template

transcription factor: in a eukaryotic cell, a protein that binds a gene's promoter and regulates transcription

transduction: conversion of energy from one form to another

transfer RNA (tRNA): a molecule of RNA that binds an amino acid at one site and an mRNA codon at its anticodon site

transformation: type of horizontal gene transfer in which an organism takes up naked DNA without cell-to-cell contact

transgenic: containing DNA from multiple species

translation: assembly of an amino acid chain according to the sequence of nucleotides in mRNA

translocation: exchange of genetic material between nonhomologous chromosomes

transmissible spongiform encephalopathy: a disease caused prions that riddle the brain with holes

transpiration: evaporation of water from a leaf

transposable element (transposon): DNA sequence that can move within a genome

trial-and-error learning: type of learned behavior in which an animal repeats behaviors that bring success

trichome: outgrowth of plant epidermis

triglyceride: lipid consisting of one glycerol bonded to three fatty acids

triiodothyronine: one of two thyroid hormones that increases the rate of cellular metabolism

trilobite: extinct type of arthropod

trophic level: an organism's position along a food chain

tropic hormone: hormone that one gland produces to influence another gland's hormone secretion

tropical dry forest: type of terrestrial biome; yearly dry and rainy seasons

tropical rain forest: type of terrestrial biomes; year-round high temperatures and precipitation

tropism: orientation toward or away from a stimulus

true fern: a type of seedless vascular plant

trypanosome: flagellated protist that causes human diseases transmitted by biting flies

tuatara: type of reptile

tumor: abnormal mass of tissue resulting from cells dividing out of control

tumor suppressor gene: gene that normally prevents cell division but when inactivated or suppressed causes cancer

tundra: type of terrestrial biome; low temperature and short growing season

tunicate: type of invertebrate chordate

turgor pressure: the force of water pressing against the cell wall

turtle: type of reptile

twitch: contraction and relaxation of a muscle cell following a single stimulation

tympanic membrane: eardrum; transmits sound from air to the middle ear

U

ultimate: describing the evolutionary origin or adaptive advantage of an animal's behavior

ultraviolet (UV) radiation: portion of the electromagnetic spectrum with wavelengths shorter than 400 nm

umbilical cord: ropelike structure that connects an embryo or fetus with the placenta

uniformitarianism: the idea that modern geological processes of erosion and sedimentation have also occurred in the past, producing changes in Earth over time

unsaturated fatty acid: a fatty acid with at least one double bond between carbon atoms

upwelling: upward movement of cold, nutrient-rich lower layers of a body of water

urea: nitrogenous waste ($(NH_2)_2CO$) derived from ammonia

ureter: muscular tube that transports urine from the kidney to the bladder

urethra: tube that transports urine (and semen in males) out of the body

uric acid: nitrogenous waste ($C_5H_4N_4O_3$) derived from ammonia

urinary bladder: muscular sac where urine collects

urinary system: organ system that filters blood and helps maintain concentrations of body fluids

urine: liquid waste produced by kidneys

uterine tube (fallopian tube): tube leading from near an ovary to the uterus

uterus: muscular, saclike organ where embryo and fetus develop

utricle: one of two pouches in the vestibule of the inner ear

V

vaccine: substance that initiates a primary immune response so that when an infectious agent is encountered, the secondary immune response can rapidly deactivate it

vacuole: membrane-bounded storage sac in a cell, especially the large central vacuole in a plant cell

vagina: conduit from the uterus to the outside of the body

valence shell: outermost occupied energy shell of an atom

variable: any changeable element in an experiment

variable region: amino acid sequence that is different for every antibody

vas deferens: tube that transports sperm from an epididymis to an ejaculatory duct

vascular bundle: collection of xylem, phloem, parenchyma, and sclerenchyma in plants

vascular cambium: lateral meristem that produces secondary xylem and phloem

vascular tissue: conducting tissue for water, minerals, and sucrose in plants

vasoconstriction: decrease in the diameter of a blood vessel

vasodilation: increase in the diameter of a blood vessel

vegetative plant parts: nonreproductive parts (roots, stems, and leaves)

vein: vascular bundles inside leaf; vessel that returns blood to the heart

ventricle: heart chamber beneath an atrium

venule: small vein

vertebra: one unit of the vertebral column composed of bone or cartilage that supports and protects the spinal cord

vertebral column: bone or cartilage that supports and protects the spinal cord

vertebrate: animal with a backbone

vesicle: a membrane-bounded sac that transports materials in a cell

vessel element: short, wide conducting cell in xylem

vestibule: structure in the inner ear between the cochlear canal and the semicircular canals; provides information on the position of the head and changes in velocity

vestigial: having no apparent function in one organism, but homologous to a functional structure in another species

villus (pl. villi): tiny projection on the inner lining of the small intestine

viroid: infectious RNA molecule

virus: infectious agent that consists of genetic information enclosed in a protein coat

visceral mass: part of a mollusk that contains the digestive and reproductive systems

vital capacity: maximal volume of air that can be forced out of the lungs during one breath

vocal cord: elastic tissue band that covers the larnyx and vibrates as air passes, producing sound

vulva: external female genitalia

W

warning coloration: bright or distinctive display that advertises an organism's defenses

water mold: filamentous, heterotrophic protist; also called an oomycetes

water vascular system: system of canals in echinoderms; provides locomotion and osmotic balance

wavelength: the distance a photon moves during a complete vibration

wax: lipid consisting of fatty acids connected to alcohol or other molecules

whisk fern: a type of seedless vascular plant

white blood cell: immune system cell that lacks hemoglobin and circulates in blood

white matter: nervous tissue in the central nervous system; consists of myelinated axon

whorled leaf arrangement: pattern in which each node has three or more leaves

wild type: the most common phenotype, genotype, or allele for a gene

woody plant: plant with stems and roots made of wood and bark

X

X inactivation: turning off all but one X chromosome in each cell of a mammal (usually female) early in development

X-linked: describes genes or traits on the X chromosome

xylem: vascular tissue that transports water and dissolved minerals in plants

xylem sap: solution of water and dissolved minerals in xylem

Y

yeast: unicellular fungus

yolk sac: extraembryonic membrane that forms beneath the embryonic disc and manufactures blood cells

Z

Z lines: membranes that define the boundaries of a sarcomere

zona pellucida: thin, clear layer of proteins and sugars that surrounds an egg cell

zooplankton: microscopic heterotrophs that drift in water

zoospore: flagellated spore produced by chytrids

Zygomycete: fungus that produces zygospores

zygospore: diploid resting spore produced by fusion of haploid cells in zygomycetes

zygote: the fused egg and sperm that develops into a diploid individual

Credits

Preface and Table of Contents

Front matter leaf graphics: © PhotoLink (RF)/Getty Images; **iii both:** Courtesy of Robert H. Taylor/University of Oklahoma; **iv & Unit 5:** © PhotoLink/Photodisc (RF)/Getty Images; **vi left & Unit 6:** © Comstock (RF)/Punchstock; **vi right & Unit 2:** © Eye of Science/Photo Researchers, Inc.; **vii & Unit 1:** © Photodisc Collection (RF)/Getty Images; **ix burning questions flames:** © Stockbyte (RF)/PunchStock; **Unit 3:** © Creatas (RF)/PunchStock; **Unit 4:** © Comstock Images (RF)/PictureQuest; **Unit 7:** © Royalty-Free/Corbis; **end-matter texture graphics:** © MaryBeth Theilhelm (RF)/Getty Images;

Chapter 1

Opener: © Science VU/DOE/Visuals Unlimited; **1.1:** © SMC Images/The Image Bank/Getty Images; **1.2 top:** © Royalty-Free/Corbis; **1.2 second:** © Manoj Shah/The Image Bank/Getty Images; **1.2 third:** © Todd Gustafson/Danita Delimont; **1.2 bottom:** © Gregory G. Dimijian, M.D./Photo Researchers, Inc.; **1.5a:** © Dennis Kunkel/Phototake; **1.5b:** © Runk/Schoenberger/Grant Heilman; **1.5c:** © Corbis Animals in Action CD; **1.6 both:** © Michael and Patricia Fogden/Animals Animals - Earth Scenes; **1.7a left:** © Dennis Kunkel Microscopy, Inc.; **1.7a right:** © Ron Occalea/The Medical File/Peter Arnold, Inc.; **1.9 left:** © Kwangshin Kim/Photo Researchers, Inc.; **1.9 second:** © James King-Holmes/Science Photo Library; **1.9 third:** © Wim van Egmond/Visuals Unlimited; **1.9 fourth:** © Dwight Kuhn; **1.9 fifth:** © David Dennis/Animals Animals - Earth Scenes; **1.9 right:** © Colombini Medeiros, Fabio/Animals Animals - Earth Scenes; **1.10:** © Comstock (RF)/Alamy; **1.12:** © Geoff McIlleron: Firefly Images/photographersdirect.com; **p. 15 left:** © Time Inc./Time Life Pictures/Getty Images; **p. 15 middle:** © Ted Thai/Time Magazine/Time & Life Pictures/Getty Images; **p. 15 right:** © Andrew Unangst/Time Magazine/Time & Life Pictures/Getty Images; **1.13:** Courtesy Chris Adami & Charles Ofria; **TA1.1:** © Bettmann/Corbis

Chapter 2

Opener: © Roy Morsch/Corbis; **2.6a:** Courtesy Diane R. Nelson; **2.6b:** © F. Schussler/PhotoLink (RF)/Getty Images; **2.6c:** Courtesy Diane R. Nelson; **2.10:** © Herman Eisenbeiss/Photo Researchers, Inc.; **2.11:** © Lester V. Bergman/Corbis; **2.16 top:** © Dr. Dennis Kunkel/Visuals Unlimited/Getty Images; **2.16 middle:** © Gary Gaugler/Visuals Unlimited; **2.16 bottom:** © Marshall Sklar/Science Photo Libray; **2.19:** © Stephen J. Krasemann/DRK Photo; **2.22a:** © Topham/The Image Works; **2.22b:** © Gerald Lacz/Peter Arnold, Inc.; **2.22c:** © John Cancelosi/Peter Arnold, Inc.

Chapter 3

Opener: © AP Photo/The Columbian, Janet L. Mathews; **3.1a:** © Kathy Talaro/Visuals Unlimited; **3.1b:** © Dr. Jeremy Byrgess/Science Photo Library; **3.3a bottom:** © Comstock (RF)/Alamy; **3.3a top:** © Michael Abbey/Visuals Unlimited; **3.3b bottom:** © Inga Spence/Visuals Unlimited; **3.3b top:** © Dr. Dennis Kunkel/Visuals Unlimited; **3.3c bottom:** © Inga Spence/Visuals Unlimited; **3.3c top:** © Microworks Color/Phototake; **3.3d bottom:** © Inga Spence/Visuals Unlimited; **3.3d top:** © Steve Gschmeissner/Science Photo Library; **3.3e bottom:** © Veeco Instruments Inc.; **3.3e top:** Courtesy of Michael Schmid, Institut F. Allgemeine Physik, TU Wien; **3.5b:** © Dr. Don W. Fawcett/Visuals Unlimited; **3.10b:** © CNRI/Photo Researchers, Inc.; **3.11a:** © Mediscan/Corbis; **3.11b:** © Eye of Science/Photo Researchers, Inc.; **3.11c:** © Science VU/Visuals Unlimited; **3.12:** © Dr. T.J. Beveridge/Visuals Unlimited; **3.a:** © Dr. Philippa Uwins, Whister Research Pty/Science Photo Library; **3.13:** © Dr. Gopal Murti/Visuals Unlimited; **3.14:** © Dr. George Chapman/Visuals Unlimited; **3.15:** © Tim Flach/Getty Images; **3.16c:** © David M. Phillips/The Population council/Science Source/Photo Researchers, Inc.; **3.17:** © R. Bolender - D. Fawcett/Visuals Unlimited; **3.18 & 3.19:** © Prof. P. Motta & T. Naguro/Science Photo Library/Photo Researchers, Inc.; **3.20a:** © D. Friend - D. Fawcett/Visuals Unlimited; **3.20b:** From S.E. Frederick and E.H. Newcomb, Journal of Cell Biology, 4:343-53, 1969; **3.21:** © Biophoto Assoc./Photo Researchers, Inc.; **3.22:** © Bill Longcore/Photo Researchers, Inc.; **3.23a:** © Innerspace Imaging/Photo Researchers, Inc.; **3.23b:** © Collection CNRI/Phototake; **3.23c:** © Edwin A. Reschke/Peter Arnold, Inc.; **3.23d:** © Robert Down Photography/photographersdirect.com; **3.24:** © Visuals Unlimited/Corbis; **3.25 left:** © Thomas Deerinck/Visuals Unlimited; **3.25 middle:** © Dr. Dennis Kunkel/Visuals Unlimited; **3.25 right:** © AgResearch, Natural Sciences Image Library; **3.26a top:** © Dr. David M. Phillips/Visuals Unlimited; **3.26a bottom:** © David M. Phillips/Visuals Unlimited; **3.26b left:** © D.W. Fawcett/Photo Researchers, Inc.; **3.26b right:** © Dr. Tony Brain/Photo Researchers, Inc.; **3.27 left:** © BioPhoto Associates/Photo Researchers; **3.27 right:** © C. Gerald Van Dyke/Visuals Unlimited; **3.28 top:** © David M. Phillips/Visuals Unlimited; **3.28 middle:** © D.W. Fawcett/Photo Researchers, Inc.; **3.28 bottom:** © D. Albertini - D. Fawcett/Visuals Unlimited; **3.29a & b:** Courtesy Prof. Jeffery Errington

Chapter 4

Opener: © Hulton-Deutsch/Corbis; **4.1b:** © Mitchell Layton/Getty Images; **4.1c:** © Stephen Dalton/NHPA; **4.3:** © Ralph White/Corbis; **4.4 left:** © Mario Tama/Getty Images; **4.4 right:** © Justin Sullivan/Getty Images; **4.6 top:** © Blair Seitz/Photo Researchers, Inc.; **4.6 bottom:** © Gamma Liason/Getty Source; **4.A:** © Darwin Dale/Photo Researchers, Inc.; **4.14a-c:** © Dr. David M. Phillips/Visuals Unlimited; **4.15a & 4.16b:** © Nigel Cattlin/Photo Researchers, Inc.; **4.18:** © Biology Media/Photo Researchers, Inc.; **4.19 all:** The Company of Biologist, Ltd.

Chapter 5

Opener: © Reinhard Dirscherl/Peter Arnold, Inc.; **5.1:** © Steve Raymer/NGS Image Collection; **5.4:** Courtesy of Dr. Eldon Newcomb, University of Wisconsin, Madison; **p. 107 top:** © Royalty-Free/Corbis; **5.8:** Courtesy Professor So Iwata; **5.13a:** © Tony Sweet/Digital Vision (RF)/Getty Images; **5.13b left:** © Kay Chernush/The Image Bank/Getty Images; **5.13b right:** © Biophot; **5.14:** © Colin Weston/Planet Earth Pictures; **5.15:** Courtesy Julian M. Hibberd

Chapter 6

Opener: © Gunter Ziesler/Bruce Coleman, Inc.; **6.1:** © Thomas Deerinck, NCMIR/Science Photo Library; **6.3:** © SPL/Custom Medical Stock Photos; **6.8:** Courtesy of Peter Hinkle; **6.10:** © Digital Vision (RF)/Getty Images; **6.12a:** © Adam Woolfitt/Corbis; **6.12b:** © Scimat/Science Photo Library; **6.14 both:** © Marc Gibernau, CNRS, Toulouse, France

Chapter 7

Opener: © Stock Montage; **7.1a:** © Dr. Gopal Murti/Science Photo Library/ Photo Researchers, Inc.; **7.1b left:** © Universal Pictures/Photofest; **7.1b right:** © Louie Psihoyos/Science Faction/Getty Images; **7.4b:** © Oliver Meckes/MPI - Tubingen/Photo Researchers, Inc.; **7.6a-b:** © Science Source/Photo Researchers, Inc.; **7.6c:** © Bettmann/Corbis; **7.7a:** © SMC Images/The Image Bank/Getty Images; **7.13:** © David Tietz/Editorial Image, LLC; **7.15a:** © Charles Thatcher/ Stone/Getty Images; **7.15b:** © Josh Westrich/zefa/Corbis; **7.16:** © Sam Ogden/ Photo Researchers, Inc.; **7.17b:** © Courtesy of Cellmark Diagnostics, Germantown, Maryland; **7.18:** Courtesy of David Gilichinsky

Chapter 8

Opener: © AP Photo/Ben Margot; **8.3b:** © Dr. A.T. Sumner, "Mammalian chromosomes/© Springer Verlag; **8.4:** © CNRI/Photo Researchers, Inc.; **8.b:** © AP/Wide World Photos; **8.6:** © Dr. Conly L. Rieder and Dr. Alexey Khodjakov/Visuals Unlimited; **8.7 top all:** © Ed Reschke; **8.7 bottom all:** © Dr Alexey Khodjakov/Photo Researchers, Inc.; **8.8a:** © Dr. Mark W. Kirschner Dept. of Cell Biology Harvard Medical School; **8.8b:** © R. Calentine/Visuals Unlimited; **8.10a:** From L. Chong "A Human Telomeric Protein", Science, 270: 1663-1667, © American Association for the Advancement of Science, Photo courtesy Dr. Titia DeLange; **8.11:** © Eye of Science/Photo Researchers, Inc.; **8.13a:** © Stockbyte (RF); **8.13b:** © GK Hart/Vikki Hart/Digital Vision (RF)/Getty Images

Chapter 9

Opener: © Breck P. Kent/Animals Animals - Earth Scenes; **9.1a all:** © Carolina Biological Supply Company/Photo Take; **9.1b:** © Jane Burton/Getty Images; **9.2:** © David Scharf/Science Faction/Getty Images; **9.5:** © James Cavallini/Photo Researchers, Inc.; **9.6 all:** © Ed Reschke/Peter Arnold, Inc.; **9.9 top:** © Stuart Westmorland/Science Faction/Getty Images; **9.9 bottom:** © Image Source Black (RF)/Getty Images; **9.12a both:** © CNRI/Photo Researchers, Inc.; **9.12b:** © George Doyle/Stockbyte (RF)/Getty Images; **9.13b:** © Addenbrookes Hospital/Photo Researchers, Inc.; **9.14:** © Francis Leroy, Biocosmos/Photo Researchers, Inc.; **9.16:** © Dr. Y. Nikas/Phototake; **9.17 left:** © 2007 Warren Rosenberg, Fundamental Photographs, NYC; **9.17 right:** © Dr. Keith Wheeler/Science Photo Library; **9.18a:** © Gabe Harp; **9.18b:** © Peter E. Smith, Natural Sciences Image Library

Chapter 10

Opener left: © James King-Holmes/Photo Researchers, Inc.; **Opener right:** © Burke Triolo Productions (RF)/Getty Images; **10.a:** © David Tietz/Editorial Image, LLC; **10.7:** © Pat Pendarvis; **10.13:** © North Wind Picture Archives; **10.14:** © Carolyn A. McKeone/Photo Researchers, Inc.; **10.15a-b:** © Peter Morenus/University of Connecticut; **10.16:** © Sarah Leen/National Geographic Image Collection; **10.17:** © Nigel Cattlin/Visuals Unlimited

Chapter 11

Opener: © Nancy Ney/Digital Vision (RF)/Getty Images; **11.1:** © Clouds Hill Imaging Ltd./Corbis; **11.2 both:** © CNRI/Photo Researchers, Inc.; **11.3:** © Science Vu/Visuals Unlimited; **11.7a:** © Lester Bergman; **11.7b:** © Joe McDonald/ Visuals Unlimited; **11.a:** © Image Source Pink (RF)/Getty Images; **11.8:** © Biophoto Assoc./Photo Researchers, Inc.; **11.9:** © Minden Pictures/Getty Images; **11.11 both:** © Andrew Syred/Photo Researchers, Inc.; **11.12:** © Hulton Archive/ Getty Images; **11.14a:** © Horst Schaefer/Peter Arnold, Inc.; **11.14b:** © William E. Ferguson; **11.14:** © StockFood Creative/Getty Images

Chapter 12

Opener: © Gregory Heisler/Time Magazine/Time & Life Pictures/Getty Images; **TA12.1 middle:** Courtesy of RNA Center, Sinsheimer Labs, UCSC; **TA12.1 bottom:** © Tom Pantages/Phototake; **12.6c:** Courtesy of RNA Center, Sinsheimer Labs, UCSC; **12.7b:** © Tom Pantages/Phototake; **12.9b:** © Kiseleva and Donald Fawcett/Visuals Unlimited; **12.13a:** © Andrew Syred/Science Photo Library; **12.13b:** © Science VU/Dr. F. R. Turner/Visuals Unlimited; **12.14a:** © Micro Discovery/Corbis; **12.14b:** © Dr. Gopal Murti/Science Photo Library; **12.17a:** © Erich Schlegel/Dallas Morning News/Corbis; **12.17b:** © Pallava Bagla/Corbis; **12.17c:** © Scott Olson/Getty Images; **12.18:** © C Squared Studios (RF)/Getty Images; **12.19:** © Edward Kinsman/Photo Researchers, Inc.; **12.23 left, second, fourth:** © Creatas (RF)/PictureQuest; **12.23 third:** © Paul Souders/Corbis; **12.23 fifth:** © A & M SHAH/Animals Animals - Earth Scenes; **12.23 right:** © Naturfoto Honal/Corbis

Chapter 13

Opener: © Bettmann/Corbis; **13.1 left:** © Science Source/Photo Researchers; **13.1 second & third:** © The Print Collector/Imagestate; **13.1 fourth:** © George Bernard/Photo Researchers, Inc.; **13.1 fifth:** © Bettmann/Corbis; **13.1 sixth:** © Corbis; **13.1 right:** © Richard Milner/Handout/epa/Corbis; **13.2:** © Jeff Greenberg/Peter Arnold, Inc.; **13.3:** © David Zurick (RF); **13.4 both:** © Tom McHugh/Photo Researchers, Inc.; **13a top:** © Brand X Pictures (RF)/PunchStock; **13a bottom:** © D. Cavagnaro/Visuals Unlimited; **13.6a:** © Garrett W. Ellwood/NBAE via Getty Images; **13.6b:** © Perennou Nuridsany/Photo Researchers, Inc.; **13.7a left:** © Dennis Kunkel Microscopy, Inc.; **13.7a right:** © Ron Occalea/The Medical File/Peter Arnold, Inc.; **13.8 both:** © Peter Grant/PrInceton University; **13.9:** © John Cancalosi/Peter Arnold, Inc.; **13.10a:** © Comstock (RF)/JupiterImages; **13.10b:** © Royalty-Free/Corbis; **13.15a:** © KONRAD WOTHE/Minden Pictures; **13.15b:** © Robert and Linda Mitchell; **13.15c:** © Tom McHugh/Photo Researchers, Inc.; **13.15d:** Dennis Sheridan © David Liebman PinkGuppy; **13.16:** Courtesy of Dr. Victor A. McKusick/Johns Hopkins Hospital; **13.18:** © Lawson Wood/Corbis

Chapter 14

Opener: © Chris Johns/National Geographic/Getty Images; **14.1a left:** S. Lowry/University Ulster/© Tony Stone Images; **14.1a middle:** Corbis #CB030324; **14.1a right:** © Robert Fried/Stock Boston; **14.1b left:** © Corbis AAWI029 Animals & Wildlife; **14.1b middle:** Corbis #CB009677; **14.1b right:** PhotoDisc #WL003983; **14.2:** David Liittschwager & Susan Middeton/© 1973 Reprinted with permission of Discover Magazine; **14.3a top left:** © GRAHAM HATHERLEY/naturepl.com; **14.3a top right:** © D. Robert Franz/Bruce Coleman, Inc.; **14.3a top middle:** © Ray Coleman/Photo Researchers, Inc.; **14.3a middle:** © Bob Gossington/Bruce Coleman, Inc.; **14.3a bottom middle:** © Phil Banko/ Corbis; **14.3a bottom left:** © MARK CHAPPELL/Animals Animals - Earth Scenes; **14.3a bottom right:** © AD VAN ROOSENDAAL/FOTO NATURA/Minden Pictures; **14.3b top left:** © S.E.Arndt/Peter Arnold, Inc.; **14.3b top right:** © David Dalton/Bruce Coleman, Inc.; **14.3b bottom:** © GERARD LACZ/Animals Animals - Earth Scenes; **14.5 left:** © ROBYN BECK/AFP/Getty Images; **14.5 right:** © Barry Mansell/npl/Minden Pictures; **14a:** Discover Magazine; **14.10 top-fifth:** © Kevin deQueiroz; **14.10 bottom:** © Jonathan B. Losos; **14.13:** © Francois Gohier/Photo Researchers, Inc.; **14.15:** Courtesy of Bill Dahl, Botanical Society of America; **14.16:** Courtesy of Dr. Pamela S. Soltis

Chapter 15

Opener: © O. Louis Mazzatenta/NGS Image Collection; **15.2 far left:** Reprinted with permission from SCIENCE elect. Vol. 282 No. 5394 pages 1601-1772 November 27 © 1998; **15.2 top left:** © PhotoLink/Getty Images; **15.2 top middle left:** © University of the Witwatersrand/epa/Corbis; **15.2 top middle right:** © Dr. Hendrik Poinar/Max-Planck Institute; **15.2 top right:** © Siede Preis/Getty Images; **15.2 bottom left:** © Phil Degginger/Carnegie Museum/Alamy; **15.2 bottom middle:** © Biophoto Associates/Photo Researchers, Inc.; **15.2 bottom right:** © Francois Gohier/Photo Researchers, Inc.; **15.3:** © Jean-Claude Carton/Bruce Coleman, Inc.; **15.4a:** © William E. Ferguson; **15.4b:** © Dr. John D. Cunningham/Visuals Unlimited; **15.4c:** © Robert Gossington/Bruce Coleman, Inc.; **15.4d:** © John Reader/Photo Researchers, Inc.; **15.4e:** © Louie Psihoyos/ Science Faction/Getty Images; **15.5a left, b, c:** © Dr. Luis M. Chiappe; **15.5a right:** © Tim Boyle/Getty Images; **15.6c:** © Staffan Widstrand/naturepl.com; **15.9 left & right:** From James H. Asher and Thomas B. Friedman, Journal of Medical Genetics 27:618-626, 1990; **15.9 middle:** © Vickie Jackson; **15.12a:** © E.R. Degginger/Animals Animals - Earth Scenes; **15.12b:** © Science Vu/Visuals Unlimited; **15.13 left:** © Dr. Richard Kessel/Visuals Unlimited; **15.13 middle:** © Dr. John Cunningham/Visuals Unlimited; **15.13 right:** © 3D4Medical.com/Getty Images; **15.16a:** Courtesy of Christine Nusslein-Volhard; **15.16b-d:** Courtesy of Jim Langeland, Steve Paddock, Sean Carroll/Howard Hughes Medical Institute, University of Wisconsin; **15.A:** © Tom McHugh/Photo Researchers, Inc.; **15.20 left:** © FRANS LANTING/Minden Pictures; **15.20 right:** © Gerald & Buff Corsi/Visuals Unlimited

Chapter 16

Opener: © JOE TUCCIARONE/Science Photo Library; **16.4:** © David McCarthy/Photo Researchers, Inc.; **16.8:** © Dr. Malcolm Walter; **16.10b:** © M. Abbey/Visuals Unlimited; **16.10c:** © Charles O'Kelly, Bigelow Laboratory for Ocean Sciences; **16.10d:** © W. Bergdorfer/Visuals Unlimited; **16.11:** © Dr. N.J. Butterfield; **16.12a:** © J.G. Gehling; **16.12b:** © J.G. Gehling from R.Raff, The Shape of Life, The University of Chicago Press; **16.13:** © A.J. Copley/Visuals Unlimited; **16.14:** Photo by W.A. Taylor/permission NATURE 9/4/97 vol 389 p. 35 fig 2d; **16.15c:** © De Agostini Picture Library/Getty Images; **16.16a:** © Field Museum of Natural History, Chicago (Geo 75400C); **16.16b:** © Kevin Schafer/Corbis; **16.20 left:** Skulls Unlimited International, Inc © David Liebman Pink Guppy; **16.20 right:** © Manfred Kage/Peter Arnold, Inc.; **16.21 top left:** Skulls Unlimited International, Inc © David Liebman Pink Guppy; **16.21 top second:** © Pascal Goetgheluck/Photo Researchers, Inc.; **16.21 top third:** Skulls Unlimited International, Inc © David Liebman Pink Guppy; **16.21 top right:** © Ralph Hutchings/Visuals Unlimited; **16.21 bottom first -third:** Skulls Unlimited International, Inc © David Liebman Pink Guppy; **16.21 bottom right:** © The McGraw-Hill Companies, Inc./Photo by Christine Eckel; **16.24:** © John Reader/Photo Researchers, Inc.; **16.26:** © Creatas (RF)/PunchStock

Chapter 17

Opener: © Culver Pictures, Inc.; **17.1a:** © Dr. O. Bradfute/Peter Arnold, Inc.; **17.1b:** © Oliver Mexkes/MPI - Tubingen/Photo Researchers, Inc.; **17.1c:** © E. O.S./Gelderblom/Photo Researchers, Inc.; **17.1d:** © Peter Arnold, Inc.; **17.1e:** © Hans Gelderblom/Visuals Unlimited; **17.2 left:** © Steve Maslowski/Visuals Unlimited/Getty Images; **17.2 middle:** © HEIDI & HANS-JURGEN KOCH/Minden Pictures; **17.2 right:** © Thomas Kitchin & Victoria Hurst/Getty Images; **17.3:** © Syracuse Newspapers/Stephen D. Cannerelli/The Image Works Image; **17.6:** © E.C.S. Chan/Visuals Unlimited; **17.8:** © Runk/Schoenberger/Grant Heilman; **17.11 left:** © Nigel Dickinson/Peter Arnold, Inc.; **17.11 right:** © Ralph Eagle Jr./Photo Researchers, Inc.

Chapter 18

Opener bottom: © BSIP/Phototake; **Opener top:** © Dennis Kunkel/Phototake; **18.3:** © G. Murti/SPL/Photo Researchers, Inc.; **18.4:** © Jack Bostrack/Visuals Unlimited; **18.5a:** © George Musil/Visuals Unlimited; **18.5b:** © CNRI/Photo Researchers, Inc.; **18.5c:** © Dr. Fred Hossler/Visuals Unlimited; **18.6:** Courtesy of American Society of Microbiology; **18.7a:** © David M. Phillips/Visuals Unlimited; **18.7b:** © SciMAT/Photo Researchers, Inc.; **18.7c:** © Thomas Tottleben/Tottleben Scientific Company; **18.9:** Courtesy of E.L. Wollman; **18.10a:** © Sylvan Wittwer/Visuals Unlimited; **18.10b:** © C.P. Vance/Visuals Unlimited; **18.12a:** © Joe Munroe/Photo Researchers, Inc.; **18.12b:** © David Wrobel/Visuals Unlimited; **18.12c:** © Jonathan A. Meyers/Photo Researchers, Inc.

Chapter 19

Opener: © J. Burkholder; **19.2a:** © Serguei Karpov; **19.3a:** © Dr. David M. Phillips/Visuals Unlimited; **19.3c:** © Bill Bachman/Photo Researchers, Inc.; **19.A:** © Michael Marten/Photo Researchers, Inc.; **19.4:** © Wim van Egmond/Visuals Unlimited/Getty Images; **19.5:** © Carolina Biological Supply Company/Phototake; **19.6:** © Daniel W. Gatshall/Visuals Unlimited; **19.7:** © GEORGETTE DOUWMA/Science Photo Library; **19.8:** © Andrew J. Martinez/Photo researchers, Inc.; **19.9 top:** © Tom Stack/Tom Stack & Associates; **19.9 second left:** © Ed Reschke; **19.9 second right:** © Wim van Egmond/Visuals Unlimited/Getty Images; **19.9 third left:** © Wim van Egmond/Visuals Unlimited; **19.9 third right:** © Darlyne A. Murawski/National Geographic/Getty Images; **19.9 bottom:** © Linda Sims/Visuals Unlimited; **19.10:** © Beatty/Visuals Unlimited; **19.11 both:** © Carolina Biological Supply/Phototake; **19.12a:** © M.S. Fuller; **19.12b:** © W.E. Fry, Plant Pathology, Cornell University; **19.12c:** © Darlyne A. Murawski/National Geographic/Getty Images; **19.13:** © Tony Stone Images/Robert Brons/BPS; **19.14a:** © Ric Ergenbright/Corbis; **19.14a inset:** © Manfred Kage/Peter Arnold, Inc.; **19.14b:** © Dr. Richard Kessel & Dr. Gene Shih/Visuals Unlimited; **19.15:** © Eric V. Grave/Photo Researchers, Inc.; **19.16:** © Dr. Dennis Kunkel/Visuals Unlimited; **19.17a-b:** From M. Schaechter, G. Medoff, and D. Schlessinger (Eds.) Mechanism of Microbial Disease, 1989 Williams and Wilkins; **19.18a:** © Dr. Charles Stratton/Visuals Unlimited; **19.18b:** © Dr. Dennis Kunkel/Visuals Unlimited/Getty Images; **19.19:** © M. Abbey/Visuals Unlimited; **19.B:** © MedicalRF.com/Visuals Unlimited; **19.20:** Courtesy of E.L. Wollman; **19.22:** © Ric Ergenbright/Corbis; **19.23 all:** © Edward C. Theriot

Chapter 20

Opener: © G. R. "Dick" Roberts/Natural Sciences Image Library; **20.1:** © Royalty-Free/Corbis; **20.6a:** © Steve Kaufman/Peter Arnold, Inc.; **20.6b:** © Edward S. Ross; **20.6c:** © William E. Ferguson; **20.7:** © John D. Cunningham/Visuals Unlimited; **20.8:** © Ed Reschke; **20.10a:** © Rod Planck/Tom Stack & Associates; **20.10b:** © Kjell B. Sandved/Butterfly Alphabet; **20.10c:** © Bud Lehnhausen/Photo Researchers, Inc.; **20.10d:** © Ed Reschke; **20.11:** © W. Ormerod/Visuals Unlimited, Inc.; **20.12 top:** © David M. Dennis/Tom Stack & Associates; **20.12 bottom:** © Stan Elems/Visuals Unlimited; **20.14a bottom:** © Ed Reschke/Peter Arnold, Inc.; **20.14a top:** © Jack Dykinga/naturepl.com; **20.14b top:**© Walter H. Hodge/Peter Arnold, Inc.; **20.14b bottom:** © Pat Pendarvis; **20.15a top:** © Richard Shiell/Animals Animals - Earth Scenes; **20.15a bottom:** © G. R. "Dick" Roberts/Natural Sciences Image Library; **20.15b top:** © Gerald & Buff Corsi/Visuals Unlimited; **20.15b bottom:** © Edward S. Ross; **20.16 left:** © Gerald & Buff Corsi/Visuals Unlimited; **20.16 right:** © Jeff Foott/Discovery Channel Images/Getty Images; **20A:** © C Squared Studios (RF)/Getty Images; **20.18a:** © Richard Weiss/Peter Arnold, Inc.; **20.18b:** © Dwight Kuhn; **20.18c:** © Pat Pendarvis; **20.18d:** © Hans Reinhard/OKAPIA/Photo Researchers, Inc.; **20.20 all:** Courtesy of Toby Bradshaw and Doug Schemske

Chapter 21

Opener: © Ed Young/Corbis; **21.1a:** © Nigel Cattlin/Visuals Unlimited; **21.1a inset:** © Biodisc/Visuals Unlimited; **21.1b:** © Thomas J. Volk, University of Wisconsin-La Crosse (www.wisc.edu/botany/fungi/volkmyco.html); **21.2:** © Scimat/Photo Researchers, Inc.; **21.4:** © Dr. Dennis Kunkel/Visuals Unlimited; **21.6a:** © J. Robert Waaland/Biological Photo Service; **21.6b:** Supplied by MS Fuller, from Reichle & Fuller, 1967; **21.7:** © Lee Berger/CSIRO Australian Animal Health Lab, P.O. Bag 24, Geelong VIC, 3220, Australia; **21.7 inset:** © John M. Parker; **21.8a:** © Dwight Kuhn; **21.8b:** © Carolina Biological/Phototake; **21.9 left:** © Runk/Schoenberger/Grant Heilman; **21.9 right:** © Ed Reschke; **21.10:** Courtesy of N. Allin and G.L. Barron, University of Guelph; **21.11a:** © DEA/G. COZZI/De Agostini Picture Library/Getty Images; **21.11b:** © Doug Sherman/Geofile; **21.A:** Prof. Dr. Peter Ruoff, University of Stavanger; **21.12 left:** © Dr. James W. Richardson/Visuals Unlimited; **21.12 right:** © Dr. Matthew Springer, University of California, San Francisco; **21.13a:** © Bill Keogh/Visuals Unlimited; **21.13b:** © Hans Reinhard/Bruce Coleman, Inc.; **21.13c-d:** Courtesy of G. L. Barron, The University of Guelph; **21B:** © E. Chan/Visuals Unlimited; **21.14:** © Stanley Flegler/Visuals Unlumited; **21C:** © Hold Studios, Ltd/Animals Animals - Earth Scenes; **21.15a:** John Dennis/Canadian Forest Service; **21.15b:** © R.L. Peterson/Biological Photo Service; **21.15c:** Courtesy of CSIRO; **21.16:** © Martin Harvey/Corbis; **21.17a2:** © V. Ahmadijian/Visuals Unlimited; **21.17b:** © William H. Mullins/Photo Researchers, Inc.; **21.18:** © Robert van der Hilst/Corbis; **21.19:** © Dr. Elizabeth Arnold

Chapter 22

Opener: © Jodi Hilton/Getty Images; **22.1b:** © Creatas (RF)/PunchStock; **22.4 top:** © Science Source/Photo Researchers, Inc.; **22.4 bottom:** © Science Source/Photo Researchers, Inc.; **22.7:** © Rob & Ann Simpson/Visuals Unlimited; **22A:** © Andrew Syred/Photo Researchers, Inc.; **22.9:** © Getty Images (RF); **22.11a:** © T. E. Adams/Visuals Unlimited; **22.11b:** © Ihoko Saito (RF)/Toshiyuki Tajima/Getty Images; **22.11c:** © Leslie Newman & Andrew Flowers/Photo Researchers, Inc.; **22.14a:** © Carolina Biological Supply/Phototake; **22.14b:** © Volker Steger/Photo Researchers, Inc.; **22.14c:** © Dr. Richard Kessel & Dr. Gene Shih/Visuals Unlimited; **22.17a:** © Franklin J. Viola; **22.17b:** © Kjell B. Sandved/Photo Researchers, Inc.; **22.17c:** © Digital Vision Ltd. (RF); **22.17d:** © Fred Bavendam/Peter Arnold, Inc.; **22.19a:** © E.R. Degginger/Bruce Coleman, Inc.; **22.19b:** © Edward Kinsman/Photo Researchers, Inc.; **22.19c:** © Ken Lucas/Visuals Unlimited; **22.21:** © Arthur Siegelman/Visuals Unlimited; **22.22 left:** © Sinclair Stammers/Photo Researchers, Inc.; **22.22 right:** © R. Umesh Chandran, TDR, WHO/Photo Researchers, Inc.; **22.26:** © Francois Gohier/Photo Researchers, Inc.; **22.28a:** © Nature's Images/Photo Researchers, Inc.; **22.28b:** © Luis C. Marigo/Peter Arnold, Inc.; **22.28c:** © Philip Brownell/Oregon State University; **22.29a:** © (Zigmond) Carmella Leszczynski/Animals Animals -

Earth Scenes; **22.29b:** © G/C Merker/Visuals Unlimited; **22.29c:** © Dr. James L. Castner; **22.29d:** Davies & Starr/© Tony Stone Images; **22.29e:** © Susan Beatty/Animals Animals - Earth Scenes; **22.30a:** © E.R. Degginger/Animals Animals - Earth Scenes; **22.30b:** © Nancy Sefton/Photo Researchers, Inc.; **22.30c:** © Andrew Martinez/Photo Researchers, Inc.; **22.30d:** © Norbert Wu/Peter Arnold, Inc.; **22.31b:** © David Fleetham/Visuals Unlimited; **22.32a:** © PETER PARKS/OSF/Animals Animals - Earth Scenes; **22.32b:** © D. P. Wilson/FLPA/Minden Pictures/Getty Images; **22.33:** © Georgette Douwma/Photo Researchers, Inc.

Chapter 23

Opener: © Mark Moffett/Minden Pictures/Getty Images; **23.5:** © Creatas (RF)/PunchStock; **23.6 top left:** © JEFFREY L. ROTMAN/Photonica/Getty Images; **23.6 top middle & right:** © Siede Preis (RF)/Getty Images; **23.6 bottom left:** © Heidi & Hans-Jurgen Koch/Minden Pictures; **23.6 bottom right:** © Royalty-Free/Corbis; **23.7:** © Nancy Sefton/Photo Researchers, Inc.; **23.8:** © Norbert Wu/Minden Pictures/Getty Images; **23.9 left:** © Brandon D. Cole/Corbis; **23.9 right:** © Rob Wood - Wood Ronsaville, Harlin, Inc.; **23.10:** © Russ Kinne/Photo Researchers, Inc.; **23.11a:** © Hal Beral/Visuals Unlimited; **23.11b:** © W. Gregory Brown/Animals Animals - Earth Scenes; **23.12a:** © E.R. Degginger; **23.12b:** © Tom McHugh/Photo Researchers, Inc.; **23.12c:** © Peter Scoones/Planet Earth Pictures; **23.13a:** © Creatas (RF)/PunchStock; **23.13b:** © Suzanne L. Collins & Joseph T. Collins/Photo Researchers, Inc.; **23.13c:**© E.D. Brodie Jr.; **23.14a:** © Ed Reschke/Peter Arnold, Inc.; **23.14b:** © Creatas (RF)/PunchStock; **23.14c-d:** © Joe McDonald/Animals Animals - Earth Scenes; **23.16b:** © Panoramic Images/Getty Images; **23.16a:** © Royalty-Free/Corbis; **23.15b:** © Meckes/Ottawa/Photo Researchers, Inc.; **23.17a:** © Fritz Prenzel/Animals Animals - Earth Scenes; **23.17b:** © Martin Harvey/Gallo Images/Corbis; **23.17c:** © Ansel Horn/Phototake; **23.17d:** © Tim Davis/Photo Researchers, Inc.; **23.17e:** © S. Dalton/Animals Animals - Earth Scenes; **23.18a:** © Ted Daeschler/VIREO/Academy of Natural Sciences; **23.19a:** © Dr. Hussam Zaher

Chapter 24

Opener: © Gustavo Gilabert/Corbis SABA; **24.5a:** © Franz Krenn/Photo Researchers, Inc.; **24.5b:** © Dwight Kuhn; **24.5c:** © G.C. Kelly/Photo Researchers, Inc.; **24.5d:** © Kenneth W. Fink/Photo Researchers, Inc.; **24.6a:** © William E. Ferguson; **24.6b:** © Richard Carlton/Visuals Unlimited; **24.6c:** © Jeffrey Lepore/Photo Researchers, Inc.; **24.7:** © Jack M. Bostrack/Visuals Unlimited; **24.8a:** © Dr. Ken Wagner/Visuals Unlimited; **24.8b:** © Biophoto Associates/Photo Researchers, Inc.; **24.8c:** © Biodisc/Visuals Unlimited; **24.8d:** © BioPhot; **24.9:** © Dr Jeremy Burgess/Photo Researchers, Inc.; **24.10:** © George Wilder/Visuals Unlimited; **24.12a:** © BioPhot; **24.12b:** © Dwight Kuhn; **24.13a left:** © Cabisco/Visuals Unlimited; **24.13a right:** © Dwight Kuhn; **24.13b left:** © Ed Reschke/Peter Arnold, Inc.; **24.13b right:** © Science Vu/Visuals Unlimited; **24.14:** © M. I. Walker/Photo Researchers, Inc.; **24.15a:** © Dwight Kuhn; **24.15b:** © Dwight Kuhn; **24.16:** © John D. Cunningham/Visuals Unlimited; **24.17:** © Runk/Schoenberger/Grant Heilman; **24.18a:** © John D. Cunningham/Visuals Unlimited; **24.18b both:** © Stan Elems/Visuals Unlimited; **24.20b:** © Manfred Kage/Peter Arnold, Inc.; **24.20c:** © A.J. Copley/Visuals Unlimited; **24.21a:** © Dr. Hans Kerp; **24.21b:** © Hans Steur, The Netherlands

Chapter 25

Opener: © Dan Suzio/Photo Researchers, Inc.; **25.2 both:** © 1991 Regents University of California Statewide IPM Project; **25.3:** © Deborah A. Kopp/Visuals Unlimited; **25.A:** © David Tietz/Editorial Image, LLC; **25.8 both:** © Ray Simon/Photo Researchers, Inc.; **25.9 left:** © E.R. Degginger; **25.B:** © Frederick Mckinney/Photographer's Choice RR/Getty Images

Chapter 26

Opener left: © Stephen Dalton/Photo Researchers, Inc.; **Opener right:** © LIU JIN/AFP/Getty Images; **26.1:** © Dr. Jeffery B. Milton/University of Colorado; **26.2:** © Inga Spence/Visuals Unlimited; **26.6a:** © Robert A. Tyrrell; **26.6b:** © Merlin D. Tuttle/Bat Conservation International/Photo Researchers, Inc.; **26.7 both:** © Leonard Lessin/Photo Researchers, Inc.; **26.11a:** © John D. Cunningham/Visuals Unlimited; **26.11b:** © Biodisc.com; **26.12a-e:** © Brent Seabrook; **tb26.1 top:** © Ingram Publishing (RF)/Alamy; **tb26.1 middle:** © Royalty-Free/Corbis; **tb26.1 bottom:** © Ingram Publishing (RF)/Alamy; **26.13a:** © Rod Planck/Photo Researchers, Inc.; **26.13b:** © W.H. Hodge/Peter Arnold, Inc.; **26.13c:** © Adam Hart-Davis/SPL/Photo Researchers, Inc.; **26.14a:** © Ed Reschke; **26.14b:** © Dwight Kuhn; **26.A:** Courtesy University of California Statewide Integrated Pest Management Project, Photo by Jack Kelly Clark; **26.16:** © Sylvan H. Wittwer; **26.17:** © David G. Clark; **26.19:** © John D. Cunningham/Visuals Unlimited; **26.20a:** © Leonard Lessin/Photo Researchers, Inc.; **26.23 both:** © Tom McHugh/Photo Researchers, Inc.; **26.24a:** © C. Calentine/Visuals Unlimited; **26.24b:** © BioPhot; **26.25:** © William E. Ferguson; **26.26:** © Ed Reschke; **26.27:** © L. West/Photo Researchers, Inc.; **26.28 all:** © Davis Littschwager, Susan Middleton/Discover Magazine

Chapter 27

Opener: © Dan McCoy - Rainbow/Science Faction/Getty Images; **27.1:** © DLILLC (RF)/Corbis; **27.A:** © Bettmann/Corbis; **27.3 top left:** © Ed Reschke/Peter Arnold, Inc.; **27.3 top right:** © Biophoto Associates/Photo Researchers, Inc.; **27.3 middle left:** © Dr. David M. Phillips/Visuals Unlimited; **27.3 middle right:** © Dr. Richard Kessel/Visuals Unlimited; **27.3 bottom left:** © Ed Reschke/Peter Arnold, Inc.; **27.3 bottom right:** © Gladden Willis, M.D./Visuals Unlimited; **27.6a:** © Ed Reschke; **27.6b:** © Manfred Kage/Peter Arnold, Inc.; **27.6c:** © Ed Reschke; **27.10:** © Lee Davenport/McGraw-Hill

Chapter 28

Opener: © Louie Psihoyos/Science Faction; **28.2c:** © Secchi-League/Roussel - UCLAF/CNRI/SPL/Photo Researchers, Inc.; **28.7:** © Fawcett/Coggeshall/Photo Researchers, Inc.; **28.8a:** © E.R. Lewis, Y.Y. Zaevi, T.E. Evenhart/University of California Berkely; **28.8b:** © Dan Fawcett/Photo Researchers, Inc.; **28.12:** © Manfred Kage/Peter Arnold, Inc.; **28.18:** © AP Photo/Kenneth Lambert

Chapter 29

Opener: © Art Wolfe/Photo Researchers, Inc.; **29.5a:** © Hans Pfletschinger/Peter Arnold, Inc.; **29.5b:** © R.A. Steinbrecht; **p. 604:** © David Tietz/Editorial Image, LLC; **29.9:** © Frank S. Werblin; **29.14:** © EMBL/EIROforum

Chapter 30

Opener: © Saturn Stills/Photo Researchers, Inc.; **30A:** © Dan Kline/Visuals Unlimited; **30.6a:** © AP Photo; **30.6b all:** Clinical Pathological Conference on Acromegaly, Diabetes, Hypemetabolism, Protien Use & Heart Failure, American Journal of Medicine 20: 133 (1956); **30.12 top:** Coutesy of Dr Glen Sweeney, Cardiff Uinversity, UK; **30.12 bottom:** © 3D4Medical.com/Getty Images; **30.13a&b all:** From Masataka and Graham 2004 PNAS 101:17716-17719."The origin of the parathyroid gland".Masataka Okabe and Anthony Graham, used by permission

Chapter 31

Opener left: Princess Margaret Rose Orthopaedic Hospital/Science Photo Library; **Opener right:** © Andersen Ross/Digital Vision (RF)/Getty Images; **31.1:** © Corbis Royalty Free; **31.2:** © David Scharf/Peter Arnold, Inc.; **31.5:** © Southern Illinois University/Photo Researchers, Inc.; **31.6:** © Ed Reschke; **31.7 right:** © Chuck Brown/Photo Researchers; **31.8:** © BIOPHOTO ASSOCIATES/Science Photo Library; **31.10:** © Michael Klein/Peter Arnold, Inc.; **31.12a:** © Custom Medical Stock Photos; **31.12b:** © B.S.I.P./Custom Medical Stock Photos; **31.16b:** © Biology Media/Photo Researchers, Inc.; **31.17b:** © Victor B. Eichler; **31.19:** © David Tietz/Editorial Image, LLC; **31.20:** © Claudio Peri/epa/Corbis

Chapter 32

Opener: © AP Photo/Greg Baker; **32.1:** © TOBY MELVILLE/Reuters/Corbis; **32.10c:** © Ed Reschke/Peter Arnold, Inc.; **32.12a:** © Sheila Terry/SPL/Photo Researchers, Inc.; **32.15:** © Dr. Dennis Kunkel/Visuals Unlimited; **32.16:** © Jean Claude Revy - ISM/Phototake, Inc.; **32.17b:** © SPL/Photo Researchers, Inc.; **32.A:** © J&L Weber/Peter Arnold, Inc.; **32.20a:** © Dennis Paulson/Visuals Unlimited; **32.20b:** © Smithsonian Institution

Chapter 33

Opener: © Louie Psihoyos/Corbis; **33.1:** © STAN HONDA/AFP/Getty Images; **33.7:** © Innerspace Imaging/Photo Researchers, Inc.; **33.10:** © AP Photo/The Tennessean, John Partipilo; **33.A both:** © USCS; **33.13:** Courtesy of Dr. Stefan K. Hetz

Chapter 34

Opener left: © NIBSC/Photo Researchers, Inc.; **Opener right:** © Darama/Corbis; **34.1:** © Dwight Kuhn; **34.2a:** © Tom McHugh/Photo Researchers, Inc; **34.2b:** © Jeff Foott/Discovery Channel Images/Getty Images; **34.11a:** © David Scharf/Peter Arnold, Inc.; **34.11c:** Courtesy of David H. Alpers, M.D.; **p. 701:** © Burke/Triolo Productions (RF)/Getty Images; **34.15:** © David Tietz/Editorial Image, LLC; **34.18 left:** © AP Photo/Brennan Linsley; **34.18 right:** © AP Photo/The Grand Island Independent, Barrett Stinson; **34.20:** © Science VU/Jackson/Visuals Unlimited; **34.21:** © Chris Mattison Photography

Chapter 35

Opener: © Fred Bavendam/Peter Arnold, Inc.; **35.1 bottom:** © IT Stock (RF)/PunchStock; **35.1 top:** © Johnny Johnson/The Image Bank/Getty Images; **35.2 left:** © IT Stock (RF)/PunchStock; **35.2 right:** © JOHN A.L. COOKE/Animals Animals - Earth Scenes; **35.5:** © Akira Kaede/Digital Vision (RF)/Getty Images; **35.6:** © Roger Eriksson; **35.A:** © Y.BEAULIEU, PUBLIPHOTO DIFFUSION/Science Photo Library; **35.14:** © Pat Pendarvis

Chapter 36

Opener: © Bettmann/Corbis; **36.2b:** © Biology Media/Photo Researchers, Inc.; **36.4:** © Dr. P. Marazzi/Photo Researchers, Inc.; **36.5:** © Glowimages (RF)/Getty Images; **36.11:** © Manfred Kage/Peter Arnold, Inc.; **36.14b:** © DR. ANDREJS LIEPINS/Science Photo Library; **36.15b:** © Len Lessin/Peter Arnold, Inc.; **36.20:** © Dr. Ken Greer/Visuals Unlimited; **36.21 top:** © David Scharf/Peter Arnold, Inc.; **36.21 bottom:** © Phil Harrington/Peter Arnold, Inc.; **36.A:** © Hank Morgan/Photo Researchers, Inc.; **36.23:** © David Wrobel/Visuals Unlimited

Chapter 37

Opener: © KEITH WELLER/US DEPARTMENT OF AGRICULTURE/Science Photo Library; **37.1:** © Daniel Heuclin/Photo Researchers, Inc.; **37.2a:** © Andrew J. Martinez/Photo Researchers, Inc.; **37.2b:** © Tim Fitzharris/Minden Pictures/Getty Images; **37.4:** Larry Johnson, Dept. of Veterinary Anatomy and Public Health; **37.5b:** © Eye of Science/Photo Researchers, Inc.; **37.7:** © Prof. P.M. Motta, G. Macchiarelli, S.A. Nottola/Photo Researchers, Inc.; **37.16 all:** © Ralph Hutchings/Visuals Unlimited; **37.18:** © AP Photo/Peer Grimm, Pool; **37.21:** Courtesy of Diana S. Chu, Ph.D, from Nature 443, 101-105 (7 Sept 2006)

Chapter 38

Opener: © GERALD HINDE/ABPL/Animals Animals - Earth Scenes; **38.1 left:** © Dan Tenaglia, Missouriplants.com; **38.1 right:** © Biosphoto/BIOS; Bios - Auteurs (droits geres); Huguet Pierre;/Peter Arnold, Inc.; **38.2a:** © Stuart Westmorland/Stone/Getty Images; **38.2b:** © Stockbyte (RF)/Getty Images; **38.2c:** © Konrad Wothe/Minden Pictures/Getty Images; **38.3:** © S. JANSSEN/FOTO NATURA/Minden Pictures; **38.4:** © Elizabeth Crews/The Image Works; **38.5a:** © Nina Leen//Time Life Pictures/Getty Images; **38.5b:** © Corbis; **38.8b:** Courtesy of Professor Daisuke Yamamoto, Tohoku University/JST-ERATO project; **38.9:** © Tanya Constantine/Blend Images (RF)/Getty Images; **38.10:** © W. Perry Conway/Corbis; **38.11:** © JIM BRANDENBURG/Minden Pictures; **38.12a:** © Daily News/epa/Corbis; **38.12b:** © Adam Jones/Photo Researchers, Inc.; **38.12c:** © JIM TUTEN/Animals Animals - Earth Scenes; **38.14:** © Gary Meszaros/Visuals Unlimited

Chapter 39

Opener: © AUSTIN J. STEVENS/Animals Animals - Earth Scenes; **39.1:** © TORU YAMANAKA/AFP/Getty Images; **39.2a:** © Wayne P. Armstrong; **39.2b:** © John A. Novak/Animals Animals - Earth Scenes; **39.2c:** © NOAA/NGDC DMSP Digital Archive/Science Source/Photo Researchers, Inc.; **39.8:** © Stringer/AFP/Getty Images; **39.9 left:** © Inga Spence/Visuals Unlimited/Getty Images; **39.9 right:** © Dorling Kindersley/Getty Images; **39.10:** Reinhard/ARCO/naturepl.com; **39.14:** © Peter Cade/Iconica/Getty Images; **39.16:** © Phil Degginger/Alamy

Chapter 40

Opener: © Johnny Johnson/Photographer's Choice/Getty Images; **40.1:** © Gregory K. Scott/Photo Researchers, Inc.; **40.1 inset:** © Stephen Dalton/Photo Researchers, Inc.; **40.2:** © Mark Moffett/Minden Pictures/Getty Images; **40.5a:** © Kevin Schafer/The Image Bank/Getty Images; **40.5b:** © James L. Castner; **40.5c:** © Simon D. Pollard/Photo Researchers, Inc.; **40.6:** © Beverly Joubert/National Geographic/Getty Images; **40.8:** © Jerry Dodrill/Aurora/Getty Images; **40.12b:** © R.T. Smith/Ardea; **40.A:** © Joseph Sohm/Visions of America/Corbis; **40.18:** Courtesy Marilyn J. Roossinck. Luis M. Márquez, Regina S. Redman, Russell J. Rodriguez, and Marilyn J. Roossinck. 2007. A virus in a fungus in a plant: three-way symbiosis required for thermal tolerance. Science 26 January 2007: Vol. 315. no. 5811, pp. 513 - 515. DOI: 10.1126/science.1136237

Chapter 41

Opener: © Perry Thorsvik/National Geographic Image Collection; **41.1 left:** © David Hall/Photo Researchers, Inc; **41.1 middle:** © Zandria Muench Beraldo/Corbis; **41.1 top right:** © Kristy-Anne Glubish/Design Pics (RF)/Corbis; **41.1 bottom right:** © Terry W. Eggers/Corbis; **41.A:** © Science VU/NOAA/Visuals Unlimited; **41.6:** © Frans Lanting/Corbis; **41.7a:** © Larry Lefever/Grant Heilman; **41.7b:** © E.R. Degginger; **41.8:** © Thomas Kitchin/Photo Researchers, Inc.; **41.9:** © E.R. Degginger; **41.10:** © PhotoDisc #WL002106; **41.11:** © Ed Reschke; **41.12:** © Michael DeYoung/Corbis; **41.13:** © Gerry Ellis/Minden Pictures/Getty Images; **41.15a:** © Jerome Wexler/Photo Researchers, Inc.; **41.15b:** © ML Sinibaldi/Corbis; **41.16b:** © Byron Jorjorian/Photo Researchers, Inc.; **41.16c:** © Joseph Sohm/Visions of America/Corbis; **41.17a:** © Joe/Arrington/Visuals Unlimited; **41.17b:** © Mike Parry/Minden Pictures/Getty Images; **41.17c:** © D. Brown/Animals Animals - Earth Scenes; **41.17d:** © Chris Newbert/Minden Pictures/Getty Images; **41.18:** © D. P. Wilson/FLPA/Minden Pictures/Getty Images; **41.19:** © Fred Bavendam/Minden Pictures/Getty Images

Chapter 42

Opener: © Pat Pendarvis; **42.1:** © Keren Su/Stone/Getty Images; **42.3a:** © Wayne Lawler, Ecoscene/Corbis; **42.3b:** © Michael Nichols/Magnum; **42.4:** © Ed Kashi/Corbis; **42.6:** © Reuters/Corbis; **42.7:** © Oliver Strewe/Stone/Getty Images; **42.8a-d:** NASA; **42.9:** © AP Photo/Robert Galbraith, pool; **42.12a-b:** © Dr. Juerg Alean/SPL/Photo Researchers, Inc.; **42.13:** © M.T. O'Keefe/Bruce Coleman, Inc.; **42.14:** © China Photos/Getty Images News; **42.15:** © Paul A. Souders/Corbis; **42.17:** © Barry Mansell/npl/Minden Pictures; **42.18:** © Francesco Tomasinelli/Photo Researchers, Inc.; **42.19:** © Alex Kerstitch/Visuals Unlimited/Getty Images;

Index

Note: Page numbers followed by an f indicate figures; numbers followed by a t indicate tables.

Periodic Table of Elements

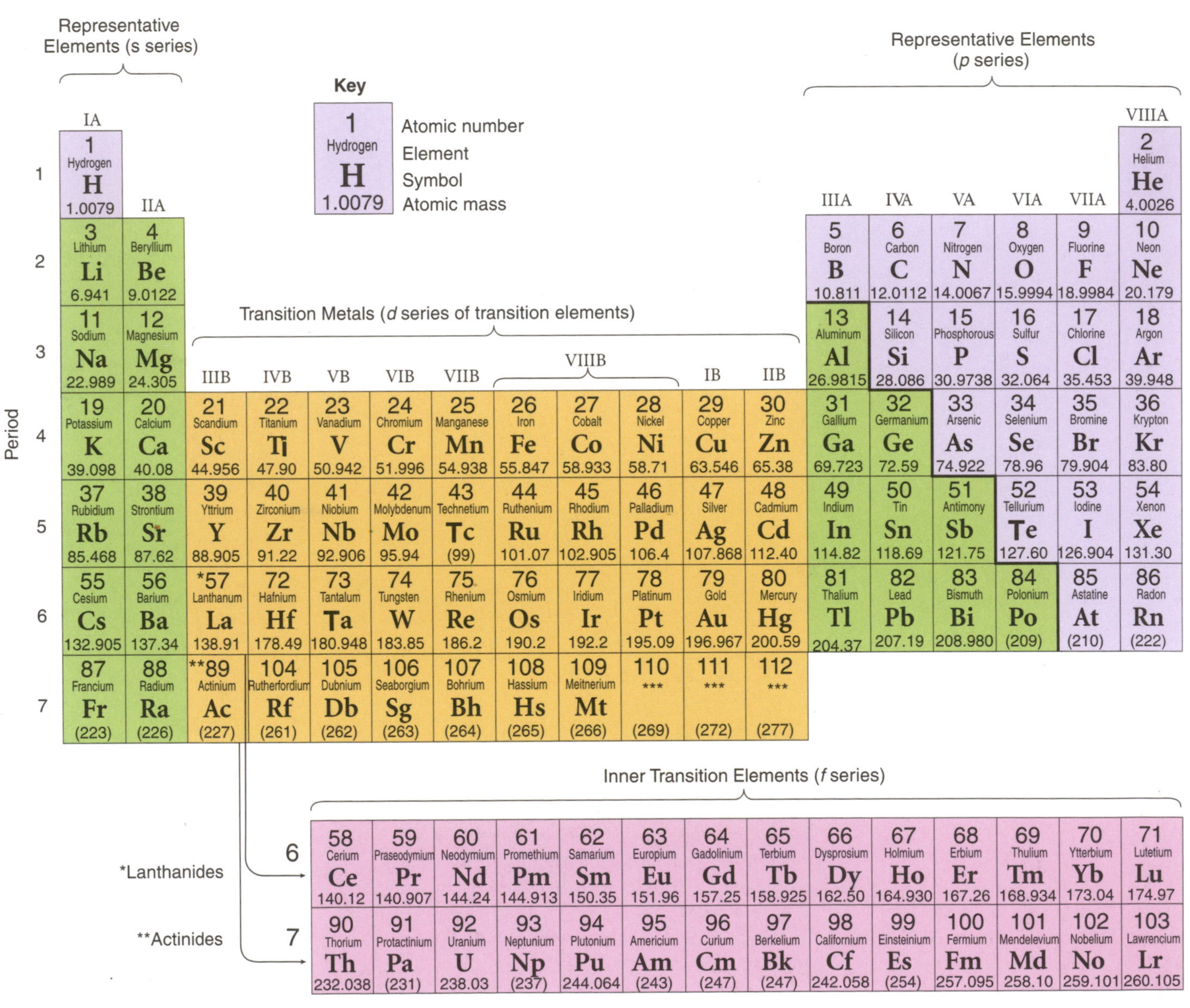

Representative Elements (s series)

Representative Elements (p series)

Transition Metals (d series of transition elements)

Inner Transition Elements (f series)

Key

1 Hydrogen H 1.0079	
1	Atomic number
Hydrogen	Element
H	Symbol
1.0079	Atomic mass

Period	IA	IIA	IIIB	IVB	VB	VIB	VIIB	VIIIB	VIIIB	VIIIB	IB	IIB	IIIA	IVA	VA	VIA	VIIA	VIIIA
1	1 Hydrogen H 1.0079																	2 Helium He 4.0026
2	3 Lithium Li 6.941	4 Beryllium Be 9.0122											5 Boron B 10.811	6 Carbon C 12.0112	7 Nitrogen N 14.0067	8 Oxygen O 15.9994	9 Fluorine F 18.9984	10 Neon Ne 20.179
3	11 Sodium Na 22.989	12 Magnesium Mg 24.305											13 Aluminum Al 26.9815	14 Silicon Si 28.086	15 Phosphorous P 30.9738	16 Sulfur S 32.064	17 Chlorine Cl 35.453	18 Argon Ar 39.948
4	19 Potassium K 39.098	20 Calcium Ca 40.08	21 Scandium Sc 44.956	22 Titanium Ti 47.90	23 Vanadium V 50.942	24 Chromium Cr 51.996	25 Manganese Mn 54.938	26 Iron Fe 55.847	27 Cobalt Co 58.933	28 Nickel Ni 58.71	29 Copper Cu 63.546	30 Zinc Zn 65.38	31 Gallium Ga 69.723	32 Germanium Ge 72.59	33 Arsenic As 74.922	34 Selenium Se 78.96	35 Bromine Br 79.904	36 Krypton Kr 83.80
5	37 Rubidium Rb 85.468	38 Strontium Sr 87.62	39 Yttrium Y 88.905	40 Zirconium Zr 91.22	41 Niobium Nb 92.906	42 Molybdenum Mo 95.94	43 Technetium Tc (99)	44 Ruthenium Ru 101.07	45 Rhodium Rh 102.905	46 Palladium Pd 106.4	47 Silver Ag 107.868	48 Cadmium Cd 112.40	49 Indium In 114.82	50 Tin Sn 118.69	51 Antimony Sb 121.75	52 Tellurium Te 127.60	53 Iodine I 126.904	54 Xenon Xe 131.30
6	55 Cesium Cs 132.905	56 Barium Ba 137.34	*57 Lanthanum La 138.91	72 Hafnium Hf 178.49	73 Tantalum Ta 180.948	74 Tungsten W 183.85	75 Rhenium Re 186.2	76 Osmium Os 190.2	77 Iridium Ir 192.2	78 Platinum Pt 195.09	79 Gold Au 196.967	80 Mercury Hg 200.59	81 Thallium Tl 204.37	82 Lead Pb 207.19	83 Bismuth Bi 208.980	84 Polonium Po (209)	85 Astatine At (210)	86 Radon Rn (222)
7	87 Francium Fr (223)	88 Radium Ra (226)	**89 Actinium Ac (227)	104 Rutherfordium Rf (261)	105 Dubnium Db (262)	106 Seaborgium Sg (263)	107 Bohrium Bh (264)	108 Hassium Hs (265)	109 Meitnerium Mt (266)	110 *** (269)	111 *** (272)	112 *** (277)						

	Period														
*Lanthanides	6	58 Cerium Ce 140.12	59 Praseodymium Pr 140.907	60 Neodymium Nd 144.24	61 Promethium Pm 144.913	62 Samarium Sm 150.35	63 Europium Eu 151.96	64 Gadolinium Gd 157.25	65 Terbium Tb 158.925	66 Dysprosium Dy 162.50	67 Holmium Ho 164.930	68 Erbium Er 167.26	69 Thulium Tm 168.934	70 Ytterbium Yb 173.04	71 Lutetium Lu 174.97
**Actinides	7	90 Thorium Th 232.038	91 Protactinium Pa (231)	92 Uranium U 238.03	93 Neptunium Np (237)	94 Plutonium Pu 244.064	95 Americium Am (243)	96 Curium Cm (247)	97 Berkelium Bk (247)	98 Californium Cf 242.058	99 Einsteinium Es (254)	100 Fermium Fm 257.095	101 Mendelevium Md 258.10	102 Nobelium No 259.101	103 Lawrencium Lr 260.105

*** These elements have not yet been named.